2016 IEEE Applied Power Electronics Conference and Exposition (APEC 2016)

Long Beach, California, USA
20-24 March 2016

Pages 2973-3726

IEEE Catalog Number: CFP16APE-POD
ISBN: 978-1-4673-9551-9

5/5

Copyright © 2016 by the Institute of Electrical and Electronic Engineers, Inc
All Rights Reserved

Copyright and Reprint Permissions: Abstracting is permitted with credit to the source. Libraries are permitted to photocopy beyond the limit of U.S. copyright law for private use of patrons those articles in this volume that carry a code at the bottom of the first page, provided the per-copy fee indicated in the code is paid through Copyright Clearance Center, 222 Rosewood Drive, Danvers, MA 01923.

For other copying, reprint or republication permission, write to IEEE Copyrights Manager, IEEE Service Center, 445 Hoes Lane, Piscataway, NJ 08854. All rights reserved.

******This publication is a representation of what appears in the IEEE Digital Libraries. Some format issues inherent in the e-media version may also appear in this print version.***

IEEE Catalog Number: CFP16APE-POD
ISBN (Print-On-Demand): 978-1-4673-9551-9
ISBN (Online): 978-1-4673-9550-2
ISSN: 1048-2334

Additional Copies of This Publication Are Available From:

Curran Associates, Inc
57 Morehouse Lane
Red Hook, NY 12571 USA
Phone: (845) 758-0400
Fax: (845) 758-2633
E-mail: curran@proceedings.com
Web: www.proceedings.com

TECHNICAL PAPERS

Session T01: Three-Phase AC-DC Converters
Location: 101A
March 22, 2016 8:30 - 12:00
Session Chairs: Gerry Moschopoulos, *Western University, Canada*
Patrick Wheeler, *University of Nottingham*

Hardware Implementation and Characterization of SiC-Based Hybrid Three-Phase Rectifier Employing Third Harmonic Injection 1
M. Makoschitz, *Technische Universität Wien, Austria*
M. Hartmann, *Schneider Electric SE, Austria*
H. Ertl, *Technische Universität Wien, Austria*

Voltage Oriented Control of the Three-Level Vienna Rectifier using Vector Control Method 9
Jeevan Adhikari, *National University of Singapore, Singapore*
Prasanna IV, *National University of Singapore, Singapore*
S.K. Panda, *National University of Singapore, Singapore*

Compensation of Neutral Point Deviation in 3-Level NPC Converter Under Unbalanced Grid Conditions 17
Kyungsub Jung, *Chungbuk National University, Korea, South*
Yongsug Suh, *Chungbuk National University, Korea, South*

High Power Factor Modular Polyphase AC/DC Converters with Galvanic Isolation based on Resistor Emulators 25
Javier Sebastián, *Universidad de Oviedo, Spain*
Ignacio Castro, *Universidad de Oviedo, Spain*
Diego G. Lamar, *Universidad de Oviedo, Spain*
Aitor Vázquez, *Universidad de Oviedo, Spain*
Kevin Martín, *Universidad de Oviedo, Spain*

Reduced Duty-Cycle Loss and Output Inductor Current Ripple in a ZVS Switched Three-Phase Isolated PWM Rectifier 33
Jahangir Afsharian, *Ryerson University, Canada*
Dewei David Xu, *Ryerson University, Canada*
Tao Zhao, *Ryerson University, Canada*
Bing Gong, *Murata Power Solution, Canada*
Zhihua Yang, *Murata Power Solution, Canada*

Analysis, Design, and Evaluation of Three-Phase Three-Wire Isolated AC-DC Converter Implemented with Three Single-Phase Converter Modules 38
Laszlo Huber, *Delta Products Corporation, United States*
Misha Kumar, *Delta Products Corporation, United States*
Milan M. Jovanović, *Delta Products Corporation, United States*
Dinggang Ping, *Delta Electronics Shanghai Co., Ltd., China*
Gang Liu, *Delta Electronics Shanghai Co., Ltd., China*

Startup Procedure for Three-Phase Three-Wire Isolated AC-DC Converter Implemented with Three Single-Phase Converter Modules 46
Misha Kumar, *Delta Products Corporation, United States*
Laszlo Huber, *Delta Products Corporation, United States*
Milan M. Jovanović, *Delta Products Corporation, United States*
Dinggang Ping, *Delta Electronics Shanghai Co., Ltd., China*
Gang Liu, *Delta Electronics Shanghai Co., Ltd., China*

Control of a Single-Stage Three-Phase Boost Power Factor Correction Rectifier 54
Ayan Mallik, *University of Maryland, United States*
Bryan Faulkner, *Virginia Polytechnic Institute and State University, United States*
Alireza Khaligh, *University of Maryland, United States*

A Bidirectional Single-Stage Three-Phase Rectifier with High-Frequency Isolation and Power Factor Correction 60
Bruno Ricardo de Almeida, *Universidade Federal do Ceará, Brazil*
Demercil de Souza Oliveira Jr., *Universidade Federal do Ceará, Brazil*
Paulo P. Praça, *Universidade Federal do Ceará, Brazil*

Session T02: High Frequency and Fast-Response DC-DC Converters
Location: 104A
March 22, 2016 8:30 - 12:00
Session Chairs: Olivier Trescases, *University of Toronto*
Jeff Nilles, *Texas Instruments*

A 5 MHz, 12 V, 10 A, Monolithically Integrated Two-Phase Series Capacitor Buck Converter 66
Pradeep S. Shenoy, *Texas Instruments Inc., United States*
Orlando Lazaro, *Texas Instruments Inc., United States*
Ramanathan Ramani, *Texas Instruments Inc., United States*
Mike Amaro, *Texas Instruments Inc., United States*
Wlodek Wiktor, *Texas Instruments Inc., United States*
Joseph Khayat, *Texas Instruments Inc., United States*
Brian Lynch, *Texas Instruments Inc., United States*

A 10-MHz Isolated Class-Φ_2 Synchronous Resonant DC-DC Converter 73
Yuan Zhou, *Nanjing University of Aeronautics and Astronautics, China*
Zhiliang Zhang, *Nanjing University of Aeronautics and Astronautics, China*
Xue-Wen Zou, *Nanjing University of Aeronautics and Astronautics, China*
Zhou Dong, *Nanjing University of Aeronautics and Astronautics, China*
Xiaoyong Ren, *Nanjing University of Aeronautics and Astronautics, China*

865 MHz Switching-Speed Step-Down DC-DC Power Converter for Envelope Tracking 79
Vivek Mehrotra, *Teledyne Scientific Company, United States*
Andrea Arias, *Teledyne Scientific Company, United States*
Joshua Bergman, *Teledyne Scientific Company, United States*
Charles Neft, *Teledyne Scientific Company, United States*
Miguel Urteaga, *Teledyne Scientific Company, United States*
Berinder Brar, *Teledyne Scientific Company, United States*

Current Parking Regulator for Zero Droop/Overshoot Load Transient Response 86
Sudhir S. Kudva, *Nvidia Corporation, United States*
William J. Dally, *Nvidia Corporation, United States*
Thomas H. Greer III, *Nvidia Corporation, United States*
C. Thomas Gray, *Nvidia Corporation, United States*

A 5MHz, 24V-to-1.2V, AO^{2}T Current Mode Buck Converter with One-Cycle Transient Response and Sensorless Current Detection for Medical Meters 94
Xugang Ke, *University of Texas at Dallas, United States*
Joseph Sankman, *Texas Instruments Inc., United States*
Dongsheng Ma, *University of Texas at Dallas, United States*

Capacitively-Aided Switching Technique for High-Frequency Isolated Bus Converters 98
Seungbum Lim, *Massachusetts Institute of Technology, United States*
Alex J. Hanson, *Massachusetts Institute of Technology, United States*
Juan A. Santiago-González, *Massachusetts Institute of Technology, United States*
David J. Perreault, *Massachusetts Institute of Technology, United States*

A 10 MHz, 48-to-5V Synchronous Converter with Dead Time Enabled 125 ps Resolution Zero-Voltage Switching 106
Alexander Barner, *Robert Bosch GmbH, Germany*
Jürgen Wittmann, *Hochschule Reutlingen, Germany*
Thoralf Rosahl, *Robert Bosch GmbH, Germany*
Bernhard Wicht, *Hochschule Reutlingen, Germany*

Plug-and-Play Electronic Capacitor for VRM Applications 111
Or Kirshenboim, *Ben-Gurion University of the Negev, Israel*
Alon Cervera, *Ben-Gurion University of the Negev, Israel*
Bar Halivni, *Ben-Gurion University of the Negev, Israel*
Eli Abramov, *Ben-Gurion University of the Negev, Israel*
Mor Mordechai Peretz, *Ben-Gurion University of the Negev, Israel*

Adaptive Voltage Positioning (AVP) Design of Multi-Phase Constant On-Time I^2 Control for Voltage Regulators with Ramp Compensations 118
Kuang-Yao Cheng, *Texas Instruments Inc., United States*
Yipeng Su, *Texas Instruments Inc., United States*

Session T03: Microgrids and Hybrid Systems
Location: 104B
March 22, 2016 8:30 - 12:00
Session Chairs: Yunwei Li, *University of Alberta*
Joesep Guerrero, *Aalborg University*

Reactive Power Support Capabilities of Nonsynchronous Interconnection Systems in Microgrid Applications 125
Yong-Duk Lee, *University of Connecticut, United States*
Sung-Yeul Park, *University of Connecticut, United States*

Zero Standby Power High Efficiency Hot Plugging Outlet for 380VDC Power Delivery System 132
Kai Tan, *North Carolina State University, United States*
Chang Peng, *North Carolina State University, United States*
Pengkun Liu, *North Carolina State University, United States*
Xiaoqing Song, *North Carolina State University, United States*
Alex Q. Huang, *North Carolina State University, United States*

Design of Control System for Smooth Mode Transfer in Smart Microgrid Application 138
Mingzhi Gao, *Zhejiang University, China*
Canhui Zhang, *Zhejiang University, China*
Maohang Qiu, *Zhejiang University, China*
Min Chen, *Zhejiang University, China*
Aron Levy, *Technology Dynamics Inc., United States*

Resonance Propagation Modeling and Analysis of AC Filters in a Large-Scale Microgrid ... 143
Yusi Liu, *University of Arkansas, United States*
Chris Farnell, *University of Arkansas, United States*
H. Alan Mantooth, *University of Arkansas, United States*
Juan Carlos Balda, *University of Arkansas, United States*
Roy A. McCann, *University of Arkansas, United States*
Cheng Deng, *University of Arkansas, United States*

A New Bidirectional DC-DC Converter for Fuel Cell, Solar Cell and Battery Systems 150
Ankur Patel, *Vicor Corporation, United States*

A Multiport Isolated DC-DC Converter 156
Yan-Kim Tran, *École Polytechnique Fédérale de Lausanne, Switzerland*
Drazen Dujic, *École Polytechnique Fédérale de Lausanne, Switzerland*

A Seamless Transfer Control Method with High Load Sharing Performance for Modular ESS ... 163
Jung-Hoon Ahn, *Sungkyunkwan University, Korea, South*
Won-Yong Sung, *Sungkyunkwan University, Korea, South*
Chang-Yeol Oh, *Sungkyunkwan University, Korea, South*
Byoung-Kuk Lee, *Sungkyunkwan University, Korea, South*
Yun-Sung Kim, *Dongahelecomm Corporation, Korea, South*

A Plug-and-Play Ripple Mitigation Approach for DC-Links in Hybrid Systems 169
Sinan Li, *University of Hong Kong, Hong Kong*
Albert T.L. Lee, *University of Hong Kong, Hong Kong*
Siew-Chong Tan, *University of Hong Kong, Hong Kong*
S.Y. Ron Hui, *University of Hong Kong, Hong Kong*

Active Control of Low Frequency Common Mode Voltage to Connect AC Utility and 380 V DC Grid 177
Fang Chen, *Virginia Polytechnic Institute and State University, United States*
Rolando Burgos, *Virginia Polytechnic Institute and State University, United States*
Dushan Boroyevich, *Virginia Polytechnic Institute and State University, United States*
Xuning Zhang, *Virginia Polytechnic Institute and State University, United States*

Session T04: Control Strategies for Inverters and Motor Drives
Location: 103C
March 22, 2016 8:30 - 12:00
Session Chairs: Bilal Akin, *Univeristy of Texas, Dallas*
Babak Nahid-Mobarakeh, *University of Lorraine*

A Three-Level Space Vector Modulation Scheme for Paralleled Two Converters to Reduce Zero-Sequence Circulating Current and Common Mode Voltage 185
Zhongyi Quan, *University of Alberta, Canada*
Yunwei Li, *University of Alberta, Canada*

Nonlinearity Analysis and Linear Modulation Method for Two Level Voltage Source Inverter with Low Switching to Operating Frequency Ratio 193
Yongjae Lee, *Seoul National University, Korea, South*
Jung-Ik Ha, *Seoul National University, Korea, South*

Synchronization Strategies in Cascaded H-Bridge Multi Level Inverters for Carrier based Sinusoidal PWM Techniques 199
Saroj Kumar Sahoo, *Indian Institute of Technology Kharagpur, India*
Tanmoy Bhattacharya, *Indian Institute of Technology Kharagpur, India*

Design and Implementation of a Sinusoidal Flux Controller for Core Loss Measurements 207
Burak Tekgun, *University of Akron, United States*
Ali R. Boynuegri, *University of Akron, United States*
Md Asif Mahmood Chowdhury, *University of Akron, United States*
Yilmaz Sozer, *University of Akron, United States*

Implementation of Deadbeat-Direct Torque and Flux Control for Synchronous Reluctance Machines to Minimize Loss Each Switching Period 215
Michael Saur, *Universität der Bundeswehr München, Germany*
Francisco Ramos, *Universität der Bundeswehr München, Germany*
Aday Perez, *Universität der Bundeswehr München, Germany*
Dieter Gerling, *Universität der Bundeswehr München, Germany*
Robert D. Lorenz, *University of Wisconsin at Madison, United States*

Addressing the Unbalance Loading Issue in Multi-Drive Systems with a DC-Link Modulation Scheme for Harmonic Reduction 221
Yongheng Yang, *Aalborg University, Denmark*
Pooya Davari, *Aalborg University, Denmark*
Firuz Zare, *Danfoss Power Electronics A/S, Denmark*
Frede Blaabjerg, *Aalborg University, Denmark*

Input Current Interharmonics in Adjustable Speed Drives Caused by Fixed-Frequency Modulation Techniques 229
Hamid Soltani, *Aalborg University, Denmark*
Pooya Davari, *Aalborg University, Denmark*
Poh Chiang Loh, *Aalborg University, Denmark*
Frede Blaabjerg, *Aalborg University, Denmark*
Firuz Zare, *Danfoss Power Electronics A/S, Denmark*

Low-Frequency Voltage Ripples in the Flying Capacitors of the Nested Neutral-Point-Clamped Converter 236
Amer M.Y.M. Ghias, *University of Sharjah, U.A.E.*
Josep Pou, *University of New South Wales, Australia*
Salvador Ceballos, *TECNALIA, Spain*
Vassilios G. Agelidis, *University of New South Wales, Australia*

DC Bus Capacitor Discharge of Permanent Magnet Synchronous Machine Drive Systems for Hybrid Electric Vehicles 241
Ziwei Ke, *Oregon State University, United States*
Julia Zhang, *Oregon State University, United States*
Michael W. Degner, *Ford Motor Company, United States*

Session T05: Si Devices and Power Module Packaging
Location: 101B
March 22, 2016 8:30 - 12:00
Session Chairs: Iulian Nistor, *Corporate Research, ABB Inc.*
Brian Rowden,

C_{OSS} Hysteresis in Advanced Superjunction MOSFETs 247
J.B. Fedison, *Enphase Energy, Inc., United States*
M.J. Harrison, *Enphase Energy, Inc., United States*

Compact Electrothermal Models for Unbalanced Parallel Conducting Si-IGBTs 253
Roozbeh Bonyadi, *University of Warwick, United Kingdom*
Olayiwola Alatise, *University of Warwick, United Kingdom*
Ji Hu, *University of Warwick, United Kingdom*
Zarina Davletzhanova, *University of Warwick, United Kingdom*
Yeganeh Bonyadi, *University of Warwick, United Kingdom*
Jose Ortiz-Gonzalez, *University of Warwick, United Kingdom*
Li Ran, *University of Warwick, United Kingdom*
Philip Mawby, *University of Warwick, United Kingdom*

General 3D Lumped Thermal Model with Various Boundary Conditions for High Power IGBT Modules 261
Amir Sajjad Bahman, *Aalborg University, Denmark*
Ke Ma, *Aalborg University, Denmark*
Frede Blaabjerg, *Aalborg University, Denmark*

Improved 6.5kV FREEMD-Pair based on SiC JFET and Si IGBT 269
Xiaoqing Song, *North Carolina State University, United States*
Alex Q. Huang, *North Carolina State University, United States*
Chang Peng, *North Carolina State University, United States*
Liqi Zhang, *North Carolina State University, United States*

On the Comparative Assessment of 1.7 kV, 300 a Full SiC-MOSFET and Si-IGBT Power Modules 276
Muhammad Nawaz, *ABB Corporate Research, Sweden*
Kalle Ilves, *ABB Corporate Research, Sweden*

Suppression of Reverse Recovery Ringing 3.3kV/450A Si/SiC Hybrid in Low Internal Inductance Package Next High Power Density Dual; nHPD^2 283
Katsuaki Saito, *Hitachi Europe Ltd., United Kingdom*
Daisuke Kawase, *Hitachi Power Semiconductor, Ltd., Japan*
Masamitsu Inaba, *Hitachi Power Semiconductor, Ltd., Japan*
Keiichi Yamamoto, *Hitachi Power Semiconductor, Ltd., Japan*
Katsunori Azuma, *Hitachi Power Semiconductor, Ltd., Japan*
Seiichi Hayakawa, *Hitachi Power Semiconductor, Ltd., Japan*

New Layout Concepts in MW-Scale IGBT Modules for Higher Robustness during Normal and Abnormal Operations 288
Paula Diaz Reigosa, *Aalborg University, Denmark*
Francesco Iannuzzo, *Aalborg University, Denmark*
Stig Munk-Nielsen, *Aalborg University, Denmark*
Frede Blaabjerg, *Aalborg University, Denmark*

Design, Package, and Hardware Verification of a High Voltage Current Switch 295
Ankan De, *North Carolina State University, United States*
Adam Morgan, *North Carolina State University, United States*
Vishnu Mahadeva Iyer, *North Carolina State University, United States*
Haotao Ke, *North Carolina State University, United States*
Xin Zhao, *North Carolina State University, United States*
Kasunaidu Vechalapu, *North Carolina State University, United States*
Subhashish Bhattacharya, *North Carolina State University, United States*
Douglas C. Hopkins, *North Carolina State University, United States*

Investigation of Short Circuit in a IGBT Power Module with Three-Level Neutral Point Clamped Type 2 (NPC2, T-NPC, Mixed Voltage) Topology 303
Kevin Lenz, *Danfoss Silicon Power, Germany*
Vladan Jerinic, *Danfoss Silicon Power, Germany*
Reiner Hinken, *Danfoss Silicon Power, Germany*

Session T06: DC-DC Converter Control
Location: 102AB
March 22, 2016 8:30 - 12:00
Session Chairs: Sombuddha Chakraborty, *Texas Instruments*
Rafael Pena Alzola, *University of British Columbia*

Closed-Loop Design and Time-Optimal Control for a Series-Capacitor Buck Converter 308
Timur Vekslender, *Ben-Gurion University of the Negev, Israel*
Ofer Ezra, *Ben-Gurion University of the Negev, Israel*
Yevgeny Bezdenezhnykh, *Ben-Gurion University of the Negev, Israel*
Mor Mordechai Peretz, *Ben-Gurion University of the Negev, Israel*

Unified Constant On/Off-Time Hybrid Compensation for Fast Recovery in Digitally Current-Mode Controlled Point-of-Load Converters 315
K. Hariharan, *Indian Institute of Technology Kharagpur, India*
Santanu Kapat, *Indian Institute of Technology Kharagpur, India*
Siddhartha Mukhopadhyay, *Indian Institute of Technology Kharagpur, India*

Digital Implementation of Adaptive Synchronous Rectifier (SR) Driving Scheme for LLC Resonant Converters 322
Chao Fei, *Virginia Polytechnic Institute and State University, United States*
Fred C. Lee, *Virginia Polytechnic Institute and State University, United States*
Qiang Li, *Virginia Polytechnic Institute and State University, United States*

Digital Synchronous Rectification Controller for LLC Resonant Converters 329
Maryam S. Amouzandeh, *University of Toronto, Canada*
Behzad Mahdavikhah, *University of Toronto, Canada*
Aleksandar Prodić, *University of Toronto, Canada*
Brent McDonald, *Texas Instruments Inc., United States*

A Novel Adaptive Synchronous Rectification Method for Digitally Controlled LLC Converters ... 334
Fan Wang, *Texas Instruments Inc., United States*
Brent A. McDonald, *Texas Instruments Inc., United States*
Jeff Langham, *Texas Instruments Inc., United States*
Bo Fan, *Texas Instruments Inc., China*

Influence of the ADC Zero Bin on the Performance of an Integrated DC-DC Converter 339
S. Vesti, *Infineon Technologies Austria AG, Austria*
M. Agostinelli, *Infineon Technologies Austria AG, Austria*
H. Koltsov, *Infineon Technologies Austria AG, Austria*
S. Marsili, *Infineon Technologies Austria AG, Austria*

Improved Current-Mode Control with Single-Cycle Load Transient 343
Virginia Li, *Virginia Polytechnic Institute and State University, United States*
Pei-Hsin Liu, *Virginia Polytechnic Institute and State University, United States*
Qiang Li, *Virginia Polytechnic Institute and State University, United States*
Fred C. Lee, *Virginia Polytechnic Institute and State University, United States*

A Mixed-Signal Ripple-Based Controller for a 16 V, 10 MHz Integrated Buck Converter 350
Sergii Tkachov, *Infineon Technologies Austria AG, Austria*
Matteo Agostinelli, *Infineon Technologies Austria AG, Austria*

New Control Concept for Soft-Switching Flyback Converters with Very High Switching Frequency 355
A.M. Connaughton, *Technische Universität Graz, Austria*
K. Krischan, *Technische Universität Graz, Austria*
K.K. Leong, *Infineon Technologies AG, Austria*
A. Muetze, *Technische Universität Graz, Austria*

Session T07: Solar Energy Systems

Location: 104C

March 22, 2016 8:30 - 12:00

Session Chairs: Babak Fahimi, *UT- Dallas*
Morgan Kiani, *Texas Christian University*

Analysis, Modeling and Control of an Interleaved Isolated Boost Series Resonant Converter for Microinverter Applications 362
Luciano A. Garcia-Rodriguez, *University of Arkansas, United States*
Cheng Deng, *University of Arkansas, United States*
Juan Carlos Balda, *University of Arkansas, United States*
Andrés Escobar-Mejía, *Universidad Tecnologica de Pereira, Colombia*

Benchmarking of Constant Power Generation Strategies for Single-Phase Grid-Connected Photovoltaic Systems 370
Ariya Sangwongwanich, *Aalborg University, Denmark*
Yongheng Yang, *Aalborg University, Denmark*
Frede Blaabjerg, *Aalborg University, Denmark*
Huai Wang, *Aalborg University, Denmark*

Advanced Slip Mode Frequency Shift Islanding Detection Method for Single Phase Grid Connected PV Inverters 378
Bahador Mohammadpour, *Queen's University, Canada*
Majid Pahlevani, *Queen's University, Canada*
Sajjad Makhdoomi Kaviri, *Queen's University, Canada*
Praveen Jain, *Queen's University, Canada*

Direct MPPT Control of PWM Converters for Extreme Transient PV Applications 386
Ignacio Galiano Zurbriggen, *University of British Columbia, Canada*
Francisco Paz, *University of British Columbia, Canada*
Martin Ordonez, *University of British Columbia, Canada*

Feeding Partial Power into Line Capacitors for Low Cost and Efficient MPPT of Photovoltaic Strings 392
Ali Elrayyah, *Qatar Environment and Energy Research Institute, Qatar*
Mohammed Badawey, *University of Akron, United States*
Yilmaz Sozer, *University of Akron, United States*

Single Phase Cascaded H5 Inverter with Leakage Current Elimination for Transformerless Photovoltaic System 398
Xiaoqiang Guo, *Yanshan University, China*
Xiaoyu Jia, *Yanshan University, China*
Zhigang Lu, *Yanshan University, China*
Josep M. Guerrero, *Aalborg University, Denmark*

Optimal Low Switching Frequency Pulse Width Modulation of Current-Fed Three-Level Inverter for Solar Integration 402
Gnana Sambandam Kulothungan, *National University of Singapore, Singapore*
Akshay K. Rathore, *National University of Singapore, Singapore*
Amarendra Edpuganti, *National University of Singapore, Singapore*
Dipti Srinivasan, *National University of Singapore, Singapore*

Low Leakage Current Single-Phase PV Inverters with Universal Neutral-Point-Clamping Method 410
Liwei Zhou, *Shandong University, China*
Feng Gao, *Shandong University, China*

Modular Subpanel Photovoltaic Converter System: Analysis and Control 417
Yuan Li, *Sichuan University / Northeastern University, China*
Yue Zheng, *Northeastern University, United States*
Su Sheng, *Northeastern University, United States*
Brad Scandrett, *PowerFilm, Inc., United States*
Brad Lehman, *Northeastern University, United States*

Session T08: Advanced Converter for Power Systems used in Transportation
Location: 103AB
March 22, 2016 8:30 - 12:00
Session Chairs: Omer Onar, *Oak Ridge National Laboratory*
Khurram Afridi, *University of Colorado, Boulder*

Integrated DC-DC Converter Design for Electric Vehicle Powertrains 424
Saeed Anwar, *University of Tennessee, United States*
Weimin Zhang, *University of Tennessee, United States*
Fred Wang, *University of Tennessee, United States*
Daniel J. Costinett, *University of Tennessee, United States*

A 1 MHz Bi-Directional Soft-Switching DC-DC Converter with Planar Coupled Inductor for Dual Voltage Automotive Systems 432
Chenhao Nan, *Arizona State University, United States*
Raja Ayyanar, *Arizona State University, United States*

A Bridgeless Totem-Pole Interleaved PFC Converter for Plug-In Electric Vehicles 440
Yichao Tang, *University of Maryland, United States*
Weisheng Ding, *University of Maryland, United States*
Alireza Khaligh, *University of Maryland, United States*

Stability Analysis of Hybrid AC/DC Power Systems for More Electric Aircraft 446
Mehdi Karbalaye Zadeh, *Norwegian University of Science and Technology, Norway*
Roghayeh Gavagsaz-Ghoachani, *Université de Lorraine, France*
Babak Nahid-Mobarakeh, *Université de Lorraine, France*
Serge Pierfederici, *Université de Lorraine, France*
Marta Molinas, *Norwegian University of Science and Technology, Norway*

On the Concept of the Multi-Source Inverter 453
Lea Dorn-Gomba, *McMaster University, Canada*
Pierre Magne, *McMaster University, Canada*
Clement Barthelmebs, *McMaster University, Canada*
Ali Emadi, *McMaster University, Canada*

Time-Domain Analysis of a Wide-DC-Range Series Resonant Dual-Active-Bridge Bidirectional Converter with a New Passive Auxilliary Circuit 460
Alireza Safaee, *Queen's University, Canada*
Praveen Jain, *Queen's University, Canada*
Alireza Bakhshai, *Queen's University, Canada*

A New High Capacity Compact Power Modules for High Power EV/HEV Inverters 468
Seiichiro Inokuchi, *Mitsubishi Electric Corporation, Japan*
Shoji Saito, *Mitsubishi Electric Corporation, Japan*
Arata Izuka, *Mitsubishi Electric Corporation, Japan*
Yuki Hata, *Mitsubishi Electric Corporation, Japan*
Shinji Hatae, *Mitsubishi Electric Corporation, Japan*
Toshiya Nakano, *Powerex, Inc., United States*
Eric R. Motto, *Powerex, Inc., United States*

Modular Pet, Two-Phase Air-Cooled Converter Cell Design and Performance Evaluation with 1.7kV IGBTs for MV Applications 472
Frederick Kieferndorf, *ABB Switzerland Ltd, Switzerland*
Uwe Drofenik, *ABB Switzerland Ltd, Switzerland*
Francesco Agostini, *ABB Switzerland Ltd, Switzerland*
Francisco Canales, *ABB Switzerland Ltd, Switzerland*

A Phase Shift Full Bridge based Reconfigurable PEV Onboard Charger with Extended ZVS Range and Zero Duty Cycle Loss 480
Haoyu Wang, *ShanghaiTech University, China*

Session T09: Gate Drives, Failure Analysis, and Protection

Location: 102C

March 22, 2016 8:30 - 12:00

Session Chairs: Zhiliang Zhang, *Nanjing University of Aeronautics and Astronautics*
Indumini Ranmuthu, *Texas Instruments*

Series Arc Fault Detection Method based on Statistical Analysis for DC Microgrids 487
Gab-Su Seo, *Seoul National University, Korea, South*
Jung-Ik Ha, *Seoul National University, Korea, South*
Bo-Hyung Cho, *Seoul National University, Korea, South*
Kyu-Chan Lee, *Smart Power Supply Co., Ltd., Korea, South*

Arc Welding Inverter with Embedded Digital Active EMI Controller 493
Junpeng Ji, *Xi'an Jiaotong University, China*
Wenjie Chen, *Xi'an Jiaotong University, China*
Xu Yang, *Xi'an Jiaotong University, China*

A Thermo-Sensitive Electrical Parameter with Maximum dI_C/dt during Turn-Off for High Power Trench/Field-Stop IGBT Modules 499
Yuxiang Chen, *Zhejiang University, China*
Haoze Luo, *Zhejiang University, China*
Wuhua Li, *Zhejiang University, China*
Xiangning He, *Zhejiang University, China*
Jun Ma, *Shanghai Electric, China*
Guodong Chen, *Shanghai Electric, China*
Ye Tian, *Shanghai Electric, China*
Enxing Yang, *Shanghai Electric, China*

A Software Frequency Response Analysis Method to Monitor Degradation of Power MOSFETs in Basic Single-Switch Converters 505
Serkan Dusmez, *University of Texas at Dallas, United States*
Manish Bhardwaj, *Texas Instruments Inc., United States*
Lei Sun, *University of Texas at Dallas, United States*
Bilal Akin, *University of Texas at Dallas, United States*

A New Capacitance Estimation Method of Supercapacitor Bank using a Bank Impedance and Current Injection 511
Junwon Lee, *Chungnam National University, Korea, South*
Hyunsik Jo, *Chungnam National University, Korea, South*
Hanju Cha, *Chungnam National University, Korea, South*

Gate Driver Design for 1.7kV SiC MOSFET Module with Rogowski Current Sensor for Shortcircuit Protection 516
Jun Wang, *Virginia Polytechnic Institute and State University, United States*
Zhiyu Shen, *Virginia Polytechnic Institute and State University, United States*
Christina Dimarino, *Virginia Polytechnic Institute and State University, United States*
Rolando Burgos, *Virginia Polytechnic Institute and State University, United States*
Dushan Boroyevich, *Virginia Polytechnic Institute and State University, United States*

2 MHz High-Density Integrated Power Supply for Gate Driver in High-Temperature Applications 524
Remi Perrin, *Université Claude Bernard Lyon 1, France*
Bruno Allard, *Université Claude Bernard Lyon 1, France*
Cyril Buttay, *Université Claude Bernard Lyon 1, France*
Nicolas Quentin, *Université Claude Bernard Lyon 1, France*
Wenli Zhang, *Virginia Polytechnic Institute and State University, United States*
Rolando Burgos, *Virginia Polytechnic Institute and State University, United States*
Dushan Boroyevich, *Virginia Polytechnic Institute and State University, United States*
Philippe Preciat, *Labinal Power Systems, France*
Donatien Martineau, *Labinal Power Systems, France*

Design Consideration of Gate Driver Circuits and PCB Parasitic Parameters of Paralleled E-Mode GaN HEMTs in Zero-Voltage-Switching Applications 529
Juncheng Lu, *Kettering University, United States*
Hua Bai, *Kettering University, United States*
Alan Brown, *Hella Corporate Center USA Inc., United States*
Matt McAmmond, *Hella Corporate Center USA Inc., United States*
Di Chen, *GaN Systems Inc., Canada*
Julian Styles, *GaN Systems Inc., Canada*

A Gate Driver of SiC MOSFET for Suppressing the Negative Voltage Spikes in a Bridge Circuit 536
Qi Zhou, *Shandong University, China*
Feng Gao, *Shandong University, China*

Session T10: Control of AC-DC Converters

Location: 102AB
March 23, 2016 8:30 - 10:10
Session Chairs: Tsorng-Juu Liang, *National Cheng-Kung University (Taiwan)*
Laszlo Balogh, *Fairchild Semiconductor*

Interleaved Boost based AC/DC Bidirectional Converter with Four Quadrant Power Control based on One-Cycle Controller (OCC) 544
Snehal Bagawade, *Queen's University, Canada*
Praveen Jain, *Queen's University, Canada*

A New Control Scheme to Improve Load Transient Response of Single Phase PWM Rectifier with Auxiliary Current Injection Circuit 552
Naga Brahmendra Yadav Gorla, *National University of Singapore, Singapore*
Sandeep Kolluri, *National University of Singapore, Singapore*
Pritam Das, *National University of Singapore, Singapore*
Sanjib Kumar Panda, *National University of Singapore, Singapore*

Active Capacitor with Ripple-Based Duty Cycle Modulation for AC-DC Applications 558
Ching-Chieh Yang, *National Taiwan University, Taiwan*
Yang-Lin Chen, *National Taiwan University, Taiwan*
Yaow-Ming Chen, *National Taiwan University, Taiwan*

Novel Approach to Current-Mode Control in DCM/CCM Boundary Boost PFC 564
Giovanni Gritti, *STMicroelectronics, Italy*
Claudio Adragna, *STMicroelectronics, Italy*

Reducing the Switching Frequency Variation Range for CRM Buck PFC Converter by Variable On-Time Control 572
Xiaoping Wang, *Nanjing University of Science and Technology, China*
Kai Yao, *Nanjing University of Science and Technology, China*
Junfang Zhang, *Nanjing University of Science and Technology, China*

Session T11: GaN-Based DC-DC Converters
Location: 104A
March 23, 2016 8:30 - 10:10
Session Chairs: Alexis Kwasinski, *University of Pittsburgh*
Regan Zane, *Utah State*

High Efficiency 20-400 MHz PWM Converters using Air-Core Inductors and Monolithic Power Stages in a Normally-Off GaN Process 580
Alihossein Sepahvand, *University of Colorado at Boulder, United States*
Yuanzhe Zhang, *University of Colorado at Boulder, United States*
Dragan Maksimović, *University of Colorado at Boulder, United States*

Thermal Evaluation of Chip-Scale Packaged Gallium Nitride Transistors 587
David Reusch, *Efficient Power Conversion Corporation, United States*
Johan Strydom, *Efficient Power Conversion Corporation, United States*
Alex Lidow, *Efficient Power Conversion Corporation, United States*

Over 300kHz GaN Device based Resonant Bidirectional DCDC Converter with Integrated Magnetics 595
Gang Liu, *Fudan University, China*
Dan Li, *Fudan University, China*
Yungtaek Jang, *Delta Products Corporation, United States*
Jianqiu Zhang, *Fudan University, China*

Effective Control & Software Techniques for High Efficiency GaN FET based Flexible Electrical Power System for Cube-Satellites 601
Ashish Shrivastav, *North Carolina State University, United States*
Shikhar Singh, *IBM, United States*
Anirudh Mahajan, *North Carolina State University, United States*
Subhashish Bhattacharya, *North Carolina State University, United States*

A 98.8% Efficient Bidirectional Full-Bridge Isolated DC-DC GaN Converter 609
Rakesh Ramachandran, *University of Southern Denmark, Denmark*
Morten Nymand, *University of Southern Denmark, Denmark*

Session T12: Electric Machines
Location: 101A
March 23, 2016 8:30 - 10:10
Session Chairs: Bilal Akin, *Univeristy of Texas, Dallas*
Bulent Sarlioglu, *University of Wisconsin - Madison*

Comparison of Lateral- and Cylindrical-Stator Electrical Machines for High-Speed Direct-Drive Applications in Confined Spaces 615
Arda Tüysüz, *ETH Zürich, Switzerland*
Johann W. Kolar, *ETH Zürich, Switzerland*

Novel Contactless Axial-Flux Permanent-Magnet Electromechanical Energy Harvester 623
Michael Flankl, *ETH Zürich, Switzerland*
Arda Tüysüz, *ETH Zürich, Switzerland*
Ivan Subotic, *Liverpool John Moores University, United Kingdom*
Johann W. Kolar, *ETH Zürich, Switzerland*

Design of Rare-Earth Free Five-Phase Outer-Rotor IPM Motor Drive for Electric Bicycle 631
Md. Zakirul Islam, *University of Akron, United States*
Seungdeog Choi, *University of Akron, United States*

Transverse Flux Machines with Rotary Transformer Concept for Wide Speed Operations without using Permanent Magnet Material 638
Iftekhar Hasan, *University of Akron, United States*
Md Wasi Uddin, *University of Akron, United States*
Yilmaz Sozer, *University of Akron, United States*

Field Oriented Modeling and Control of Six Phase, Open-Delta Winding, Interior Permanent Magnet Synchronous Machines Considering Current Unbalance and Zero Sequence Currents 643
Murat Senol, *RWTH Aachen University, Germany*
Michael Schubert, *RWTH Aachen University, Germany*
Georges Engelmann, *RWTH Aachen University, Germany*
Rik W. De Doncker, *RWTH Aachen University, Germany*
Thorben Grosse, *RWTH Aachen University, Germany*
Kay Hameyer, *RWTH Aachen University, Germany*

Session T13: Advances in Magnetics

Location: 101B
March 23, 2016 8:30 - 10:10
Session Chairs: Matthew Wilkowski, *Enpirion*
Charles Sullivan, *Dartmouth*

Passive Integration using FMLF Technique for Integrated Boost Resonant Converters 651
Cheng Deng, *University of Arkansas, United States*
Luciano Andres Garcia Rodriguez, *University of Arkansas, United States*
Juan Zou, *Xiangtan University, China*
Juan Carlos Balda, *University of Arkansas, United States*

Magnetic Characterization Technique and Materials Comparison for Very High Frequency IVR 657
Dongbin Hou, *Virginia Polytechnic Institute and State University, United States*
Fred C. Lee, *Virginia Polytechnic Institute and State University, United States*
Qiang Li, *Virginia Polytechnic Institute and State University, United States*

Large-Signal Power Circuit Characterization of On-Silicon Coupled Inductors for High Frequency Integrated Voltage Regulation 663
S. Kulkarni, *Tyndall National Institute, Ireland*
Z. Pavlovic, *Tyndall National Institute, Ireland*
S. Kubendran, *Tyndall National Institute, Ireland*
C. Carretero, *Universidad de Zaragoza, Spain*
N. Wang, *Tyndall National Institute, Ireland*
C. O'Mathuna, *Tyndall National Institute / University College Cork, Ireland*

Point-of-Load Inductor with High Swinging and Low Loss at Light Load 668
Ting Ge, *Virginia Polytechnic Institute and State University, United States*
Khai Ngo, *Virginia Polytechnic Institute and State University, United States*
Jim Moss, *Texas Instruments Inc., United States*

Iron Loss Evaluation of Three-Phase Inductor for Three-Phase PWM Inverter 676
Hiroaki Matsumori, *Tokyo Metropolitan University, Japan*
Toshihisa Shimizu, *Tokyo Metropolitan University, Japan*
Koushi Takano, *Iwatsu Test Instrument Corporation, Japan*
Ishii Hitoshi, *Iwatsu Test Instrument Corporation, Japan*

Session T14: System Design and Layout for Improved Performance

Location: 102C

March 23, 2016 8:30 - 10:10

Session Chairs: Jeff Nilles, *Texas Instruments*
Ernie Parker, *Crane Aerospace & Electronics*

CMOS Gate Drive IC with Embedded Cross Talk Suppression Circuitry for SiC Devices 684
Jeffery Dix, *University of Tennessee, United States*
Zheyu Zhang, *University of Tennessee, United States*
Benjamin J. Blalock, *University of Tennessee, United States*

Optimal Design of a Voltage Regulator based Resonant Switched-Capacitor Converter IC 692
Eli Abramov, *Ben-Gurion University of the Negev, Israel*
Alon Cervera, *Ben-Gurion University of the Negev, Israel*
Mor Mordechai Peretz, *Ben-Gurion University of the Negev, Israel*

Novel Highly Integrated Current Measurement Method for Drive Inverters 700
N. Langmaack, *Technische Universität Braunschweig, Germany*
G. Tareilus, *Technische Universität Braunschweig, Germany*
M. Henke, *Technische Universität Braunschweig, Germany*

A Novel DBC Layout for Current Imbalance Mitigation in SiC MOSFET Multichip Power Modules 704
Helong Li, *Aalborg University, Denmark*
Stig Munk-Nielsen, *Aalborg University, Denmark*
Szymon Bęczkowski, *Aalborg University, Denmark*
Xiongfei Wang, *Aalborg University, Denmark*

A Double-End Sourced Multi-Chip Improved Wire-Bonded SiC MOSFET Power Module Design 709
Miao Wang, *Ohio State University, United States*
Fang Luo, *Ohio State University, United States*
Longya Xu, *Ohio State University, United States*

Session T15: Modeling of AC Energy Converters and Systems
Location: 104B
March 23, 2016 8:30 - 10:10
Session Chairs: Jaber Abu Qahouq, *The University of Alabama*
Xiongfei Wang, *Aalborg University*

Comparing Extended Kalman Filter and Particle Filter for Estimating Field and Damper Bar Currents in Brushless Wound Field Synchronous Generator for Stator Winding Fault Detection and Diagnosis 715
Sivakumar Nadarajan, *National University of Singapore, Singapore*
S.K. Panda, *National University of Singapore, Singapore*
Bicky Bhangu, *Rolls-Royce Singapore Pte. Ltd., Singapore*
Amit Kumar Gupta, *Rolls-Royce Singapore Pte. Ltd., Singapore*

Analytical Determination of Conduction Power Losses for Active Neutral-Point-Clamped Multilevel Converter 720
Vahid Dargahi, *Clemson University, United States*
Arash Khoshkbar Sadigh, *Extron Electronics, United States*
Keith Corzine, *Clemson University, United States*

Multifrequency Small-Signal Model of Voltage Source Converters Connected to a Weak Grid for Stability Analysis 728
Xing Li, *Huazhong University of Science and Technology, China*
Hua Lin, *Huazhong University of Science and Technology, China*

A New Approach to Control the Modified LinVerter for High Frequency Applications 733
Peyman Farhang, *University of Southern Denmark, Denmark*
Stefan Mátéfi-Tempfli, *University of Southern Denmark, Denmark*

Small-Signal Terminal Characteristics Modeling of Three-Phase Boost Rectifier with Variable Fundamental Frequency 739
Zeng Liu, *Xi'an Jiaotong University, China*
Jinjun Liu, *Xi'an Jiaotong University, China*
Dushan Boroyevich, *Virginia Polytechnic Institute and State University, United States*

Session T16: Manufacturing, Test, and Reliability
Location: 103C
March 23, 2016 8:30 - 10:10
Session Chairs: Jim Marinos, *Payton Group*
Brian Narveson, *Narveson Innovative Consulting*

Reliability Analysis of a High-Efficiency SiC Three-Phase Inverter for Motor Drive Applications 746
Juan Colmenares, *KTH Royal Institute of Technology, Sweden*
Diane-Perle Sadik, *KTH Royal Institute of Technology, Sweden*
Patrik Hilber, *KTH Royal Institute of Technology, Sweden*
Hans-Peter Nee, *KTH Royal Institute of Technology, Sweden*

RCP Evaluation of Electrolytic Capacitor Degradation for SMPS Failure Prediction 754
Hiroshi Nakao, *Fujitsu Laboratories Ltd., Japan*
Yu Yonezawa, *Fujitsu Laboratories Ltd., Japan*
Yoshiyasu Nakashima, *Fujitsu Laboratories Ltd., Japan*
Fujio Kurokawa, *Nagasaki University, Japan*

Modular Test System Architecture for Device, Circuit and System Level Reliability Testing 759
Roland Sleik, *Kompetenzzentrum Automobil- und Industrieelektronik GmbH, Austria*
Michael Glavanovics, *Kompetenzzentrum Automobil- und Industrieelektronik GmbH, Austria*
Sascha Einspieler, *Kompetenzzentrum Automobil- und Industrieelektronik GmbH, Austria*
Annette Muetze, *Technische Universität Graz, Austria*
Klaus Krischan, *Technische Universität Graz, Austria*

EMI Noise Cancelation by Optimizing Transformer Design without Need for the Traditional Y-Capacitor 766
Yongjiang Bai, *Xi'an Jiaotong University, China*
Wenjie Chen, *Xi'an Jiaotong University, China*
Ruirui He, *Xi'an Jiaotong University, China*
Dan Zhang, *Silergy Corp., China*
Xu Yang, *Xi'an Jiaotong University, China*

Manufacturing, Assembly and Production Qualifications of High Density, High Reliability POL DC-DC Converters 772
Fariborz Musavi, *CUI Inc., United States*

Session T17: Soft-Switching Converters in Renewable Energy Systems

Location: 104C

March 23, 2016 8:30 - 10:10

Session Chairs: Khurram Afridi, *University of Colorado at Boulder*
Katherine Kim, *Ulsan NIST*

Power Flow Control and ZVS Analysis of Three Limb High Frequency Transformer based Three-Port DAB 778
Ritwik Chattopadhyay, *North Carolina State University, United States*
Subhashish Bhattacharya, *North Carolina State University, United States*

A Novel Multi-Input Converter using Soft-Switched Single-Switch Input Modules with Integrated Power Factor Correction Capability for Hybrid Renewable Energy Systems 786
Sanjida Moury, *York University, Canada*
John Lam, *York University, Canada*
Vineet Srivastava, *Cistel Technology Inc., Canada*
Ron Church, *Cistel Technology Inc., Canada*

Analysis and Design of Impulse Commutated ZCS Three-Phase Current-Fed Push-Pull DC/DC Converter 794
Radha Sree Krishna Moorthy, *National University of Singapore, Singapore*
Akshay Kumar Rathore, *National University of Singapore, Singapore*

ZCS Resonant Converter based Parallel Balancing of Serially Connected Batteries String 802
Ilya Zeltser, *Rafael Advanced Defense Systems Ltd., Israel*
Or Kirshenboim, *Ben-Gurion University of the Negev, Israel*
Nadav Dahan, *Ben-Gurion University of the Negev, Israel*
Mor Mordechai Peretz, *Ben-Gurion University of the Negev, Israel*

A Novel Topology of High Voltage and High Power Bidirectional ZCS DC-DC Converter based on Serial Capacitors 810
Lejia Sun, *Xi'an Jiaotong University, China*
Fang Zhuo, *Xi'an Jiaotong University, China*
Feng Wang, *Xi'an Jiaotong University, China*
Tianhua Zhu, *Xi'an Jiaotong University, China*

Session T18: Solid State Lighting
Location: 103AB
March 23, 2016 8:30 - 10:10
Session Chairs: Jim Spangler, *Spangler Prototype Inc*
Nan Chen, *ABB*

Control Scheme for TRIAC Dimming High PF Single-Stage LED Driver with Adaptive Bleeder Circuit and Non-Linear Current Reference 816
Weizhong Ma, *Hangzhou Dianzi University, China*
Xiaogao Xie, *Hangzhou Dianzi University, China*
Yang Han, *Hangzhou Dianzi University, China*
Hao Deng, *Hangzhou Dianzi University, China*

Three Phase Converter with Galvanic Isolation based on Loss-Free Resistors for HB-LED Lighting Applications 822
Ignacio Castro, *Universidad de Oviedo, Spain*
Diego G. Lamar, *Universidad de Oviedo, Spain*
Manuel Arias, *Universidad de Oviedo, Spain*
Javier Sebastián, *Universidad de Oviedo, Spain*
Marta M. Hernando, *Universidad de Oviedo, Spain*

A ZV-ZCS Electrolytic Capacitor-Less AC/DC Isolated LED Driver with Continous Energy Regulation 830
John Lam, *York University, Canada*
Nader A. El-Taweel, *York University, Canada*

High Efficiency and Power Density GaN-Based LED Driver 838
Eric Faraci, *Texas Instruments Inc., United States*
Michael Seeman, *Texas Instruments Inc., United States*
Bin Gu, *Texas Instruments Inc., United States*
Yogesh Ramadass, *Texas Instruments Inc., United States*
Paul Brohlin, *Texas Instruments Inc., United States*

A Novel LED Drive System based on Matrix Rectifier 843
Baoping Shi, *Nanjing University of Aeronautics and Astronautics, China*
Bo Zhou, *Nanjing University of Aeronautics and Astronautics, China*
Jiadan Wei, *Nanjing University of Aeronautics and Astronautics, China*
Xianhui Qin, *Nanjing University of Aeronautics and Astronautics, China*
Yuanyu Yang, *Nanjing University of Aeronautics and Astronautics, China*
Bing Liu, *Nanjing University of Aeronautics and Astronautics, China*

Session T19: Resonant and Soft Switching DC-DC Converters
Location: 101A
March 23, 2016 14:00 - 17:30
Session Chairs: Mahshid Amirabadi, *Northeastern University*
Ray Orr, *Solantro*

LLC Synchronous Rectification using Coordinate Modulation 848
Mehdi Mohammadi, *University of British Columbia, Canada*
Navid Shafiei, *University of British Columbia, Canada*
Martin Ordonez, *University of British Columbia, Canada*

Low Parasitics Planar Transformer for LLC Resonant Battery Chargers 854
Mohammad Ali Saket, *University of British Columbia, Canada*
Navid Shafiei, *University of British Columbia, Canada*
Martin Ordonez, *University of British Columbia, Canada*
Marian Craciun, *Delta-Q Technologies Corporation, Canada*
Chris Botting, *Delta-Q Technologies Corporation, Canada*

New Symmetrical Bidirectional L3C Resonant DC-DC Converter with Wide Voltage Range 859
Minjae Kim, *Seoul National University of Science and Technology, Korea, South*
Shinyoung Noh, *Seoul National University of Science and Technology, Korea, South*
Sewan Choi, *Seoul National University of Science and Technology, Korea, South*

Influence of the Junction Capacitance of the Secondary Rectifier Diodes on Output Characteristics in Multi-Resonant Converters 864
Stefan Ditze, *Fraunhofer Institute for Integrated Systems and Device Technology, Germany*
Thomas Heckel, *Fraunhofer Institute for Integrated Systems and Device Technology, Germany*
Martin März, *Fraunhofer Institute for Integrated Systems and Device Technology, Germany*

A Triple Active Bridge DC-DC Converter Capable of Achieving Full-Range ZVS 872
Ling Jiang, *University of Tennessee, United States*
Daniel Costinett, *University of Tennessee, United States*

A Novel High Gain Step-Up Resonant DC-DC Converter for Automotive Application 880
Fei Shang, *Illinois Institute of Technology, United States*
Mahesh Krishnamurthy, *Illinois Institute of Technology, United States*
Alexander Isurin, *Vanner Inc., United States*

Series Injection Enabled Full ZVS Light Load Operation of a 15kV SiC IGBT based Dual Active Half Bridge Converter 886
Awneesh Tripathi, *North Carolina State University, United States*
Sachin Madhusoodhanan, *North Carolina State University, United States*
Krishna Mainali, *North Carolina State University, United States*
Kasunaidu Vechalapu, *North Carolina State University, United States*
Subhashish Bhattacharya, *North Carolina State University, United States*

Soft Switching for Half Bridge Current Doubler for High Voltage Point of Load Converter in Data Center Power Supplies 893
Yutian Cui, *University of Tennessee, United States*
Weimin Zhang, *University of Tennessee, United States*
Leon M. Tolbert, *University of Tennessee, United States*
Daniel J. Costinett, *University of Tennessee, United States*
Fred Wang, *University of Tennessee, United States*
Benjamin J. Blalock, *University of Tennessee, United States*

An Algorithm to Analyze Circulating Current for Multi-Phase Resonant Converter 899
Hongliang Wang, *Queen's University, Canada*
Yang Chen, *Queen's University, Canada*
Zhiyuan Hu, *Queen's University, Canada*
Laili Wang, *Queen's University, Canada*
Tianshu Liu, *Queen's University, Canada*
Wenbo Liu, *Queen's University, Canada*
Yan-Fei Liu, *Queen's University, Canada*
Jahangir Afsharian, *Murata Power Solutions, Canada*
Zhihua Yang, *Murata Power Solutions, Canada*

Session T20: Control Applications and Modulation Schemes
Location: 102C
March 23, 2016 14:00 - 17:30
Session Chairs: Masoud Karimi Ghartemani, *Mississippi state University*
Paul Bauer, *University of Lorraine*

A Simple Active Damping Method for Active Power Filters 907
Huawei Yuan, *Tsinghua University, China*
Xinjian Jiang, *Tsinghua University, China*

Simultaneous Voltage and Current Compensation of the 3-Phase Electric Spring with Decomposed Voltage Control 913
Shuo Yan, *University of Hong Kong, Hong Kong*
Tianbo Yang, *University of Hong Kong, Hong Kong*
C.K. Lee, *University of Hong Kong, Hong Kong*
Siew-Chong Tan, *University of Hong Kong, Hong Kong*
S.Y. Ron Hui, *University of Hong Kong / Imperial College London, Hong Kong*

Self-Synchronization Operation of Global Synchronous Pulsewidth Modulation with Communication Fault Tolerant and Simplified Calculation Capabilities 921
Tao Xu, *Shandong University, China*
Feng Gao, *Shandong University, China*

Design Considerations and Predictive Direct Current Control of Active Regenerative Rectifiers for Harmonic and Current Ripple Reduction 928
Alberto Berzoy, *Florida International University, United States*
A.A.S. Mohamed, *Florida International University, United States*
Osama Mohammed, *Florida International University, United States*

A Robust Controller for Medium Voltage AC Collection Grid for Large Scale Photovoltaic Plants based on Medium Frequency Transformers 936
Bahaa Hafez, *Texas A&M University, United States*
Prasad Enjeti, *Texas A&M University, United States*
Shehab Ahmed, *Texas A&M University at Qatar, Qatar*

Optimal Low Switching Frequency Pulse Width Modulation of Current-Fed Five-Level Inverter for Solar Integration 943
Gnana Sambandam Kulothungan, *National University of Singapore, Singapore*
Akshay K. Rathore, *National University of Singapore, Singapore*
Amarendra Edpuganti, *National University of Singapore, Singapore*
Dipti Srinivasan, *National University of Singapore, Singapore*

Design and Implementation of D-Σ Digital Controlled Multi-function Inverter to Achieve APF, Active Power Injection and Rectification 951
T.-F. Wu, *National Tsing Hua University, Taiwan*
H.-C. Hsieh, *National Chung Cheng University, Taiwan*
L.-C. Lin, *National Tsing Hua University, Taiwan*
C.-H. Chang, *National Tsing Hua University, Taiwan*

Operation and Analysis of an Improved Transformerless Unified Power Flow Controller 959
Yang Liu, *Michigan State University, United States*
Shuitao Yang, *Michigan State University / Ford Motor Company, United States*
Fang Zheng Peng, *Michigan State University, United States*

Design Consideration of Converter based Transmission Line Emulation 966
Bo Liu, *University of Tennessee, United States*
Shuoting Zhang, *University of Tennessee, United States*
Sheng Zheng, *University of Tennessee, United States*
Yiwei Ma, *University of Tennessee, United States*
Fred Wang, *University of Tennessee, United States*
Leon M. Tolbert, *University of Tennessee, United States*

Session T21: Advances in Wide BandGap Devices

Location: 104A
March 23, 2016 14:00 - 17:30
Session Chairs: Doug Hopkins, *North Carolina State University*
Alex Huang, *North Carolina State University*

Short-Circuit Characterization of 10 kV 10A 4H-SiC MOSFET 974
Emanuel-Petre Eni, *Aalborg University, Denmark*
Szymon Bęczkowski, *Aalborg University, Denmark*
Stig Munk-Nielsen, *Aalborg University, Denmark*
Tamas Kerekes, *Aalborg University, Denmark*
Remus Teodorescu, *Aalborg University, Denmark*

Record-Low 10mΩ SiC MOSFETs in TO-247, Rated at 900V 979
Vipindas Pala, *Wolfspeed, A Cree Company, United States*
Gangyao Wang, *Wolfspeed, A Cree Company, United States*
Brett Hull, *Wolfspeed, A Cree Company, United States*
Scott Allen, *Wolfspeed, A Cree Company, United States*
Jeffrey Casady, *Wolfspeed, A Cree Company, United States*
John Palmour, *Wolfspeed, A Cree Company, United States*

Performance Evaluation of Multiple Si and SiC Solid State Devices for Circuit Breaker Application in 380VDC Delivery System 983
Kai Tan, *North Carolina State University, United States*
Pengkun Liu, *North Carolina State University, United States*
Xijun Ni, *North Carolina State University, United States*
Chang Peng, *North Carolina State University, United States*
Xiaoqing Song, *North Carolina State University, United States*
Alex Q. Huang, *North Carolina State University, United States*

Evaluation of High Voltage Cascode GaN HEMTs in Parallel Operation 990
He Li, *Ohio State University, United States*
Xuan Zhang, *Ohio State University, United States*
Lucheng Wen, *Ohio State University, United States*
John Alex Brothers, *Ohio State University, United States*
Chengcheng Yao, *Ohio State University, United States*
Ke Zhu, *Ohio State University, United States*
Jin Wang, *Ohio State University, United States*
Liming Liu, *ABB Inc., United States*
Jing Xu, *ABB Inc., United States*
Joonas Puukko, *ABB Inc., United States*

A New Driving Concept for Normally-On GaN Switches in Cascode Configuration 996
Bernhard Zojer, *Infineon Technologies Austria AG, Austria*

Avoiding Divergent Oscillation of Cascode GaN Device Under High Current Turn-Off Condition 1002
Weijing Du, *Virginia Polytechnic Institute and State University, United States*
Xiucheng Huang, *Virginia Polytechnic Institute and State University, United States*
Fred C. Lee, *Virginia Polytechnic Institute and State University, United States*
Qiang Li, *Virginia Polytechnic Institute and State University, United States*
Wenli Zhang, *Virginia Polytechnic Institute and State University, United States*

Temperature-Dependent Turn-On Loss Analysis for GaN HFETs 1010
Edward A. Jones, *University of Tennessee, United States*
Fred Wang, *University of Tennessee, United States*
Daniel Costinett, *University of Tennessee, United States*
Zheyu Zhang, *University of Tennessee, United States*
Ben Guo, *United Technologies Research Center, United States*

Analysis of Parasitic Elements of SiC Power Modules with Special Emphasis on Reliability Issues 1018
Diane-Perle Sadik, *KTH Royal Institute of Technology, Sweden*
Juan Colmenares, *KTH Royal Institute of Technology, Sweden*
Hans-Peter Nee, *KTH Royal Institute of Technology, Sweden*
Konstantin Kostov, *Acreo Swedish ICT AB, Sweden*
Florian Giezendanner, *Alstom Power Sweden AB, Sweden*
Per Ranstad, *Alstom Power Sweden AB, Sweden*

Static and Dynamic Characterization of GaN HEMT with Low Inductance Vertical Phase Leg Design for High Frequency High Power Applications 1024
Nidhi Haryani, *Virginia Polytechnic Institute and State University, United States*
Xuning Zhang, *Virginia Polytechnic Institute and State University, United States*
Rolando Burgos, *Virginia Polytechnic Institute and State University, United States*
Dushan Boroyevich, *Virginia Polytechnic Institute and State University, United States*

Session T22: Motor Drive Design and Inverter Topologies
Location: 101B
March 23, 2016 14:00 - 17:30
Session Chairs: Yingying Kuai, *Caterpillar Inc.*
Jin Wang, *The Ohio State University*

A Family of Single-Phase Current Source Converters with Double Outputs 1032
Louelson A. Costa, *Universidade Federal de Campina Grande, Brazil*
Maurício B.R. Corrêa, *Universidade Federal de Campina Grande, Brazil*
Montiê A. Vitorino, *Universidade Federal de Campina Grande, Brazil*
Gutemberg G. Dos Santos, *Universidade Federal de Campina Grande, Brazil*
Darlan A. Fernandes, *Universidade Federal da Paraíba, Brazil*

Multiple-Output Boost Resonant Inverter for High Efficiency and Cost-Effective Induction Heating Applications 1040
Hector Sarnago, *Universidad de Zaragoza, Spain*
Oscar Lucia, *Universidad de Zaragoza, Spain*
José M. Burdío, *Universidad de Zaragoza, Spain*

Development of 2-kW Interleaved DC-Capacitor-Less Single-Phase Inverter System 1045
Runruo Chen, *Michigan State University, United States*
Hulong Zeng, *Michigan State University, United States*
Deepak Gunasekaran, *Michigan State University, United States*
Yunting Liu, *Michigan State University, United States*
Fang Z. Peng, *Michigan State University, United States*

Single Stage Transformer Isolated High Frequency AC Link based Open End Drive 1051
Srikant Gandikota, *University of Minnesota, United States*
Ned Mohan, *University of Minnesota, United States*

A Quasi-Z-Source Integrated Multi-Port Power Converter with Reduced Capacitance for Switched Reluctance Motor Drives 1057
Fan Yi, *University of Texas at Dallas, United States*
Wen Cai, *University of Texas at Dallas, United States*

A Fault-Tolerant Topology of T-Type NPC Inverter with Increased Thermal Overload Capability 1065
Jiangbiao He, *Marquette University, United States*
Nathan Weise, *Marquette University, United States*
Lixiang Wei, *Rockwell Automation, United States*
Nabeel A.O. Demerdash, *Marquette University, United States*

A Novel Analysis and Design Method of Phase Lead Filters in Repetitive Controllers for Pulse-Width Modulated Inverters 1071
Shunfeng Yang, *Nanyang Technological University, Singapore*
Peng Wang, *Nanyang Technological University, Singapore*
Yi Tang, *Nanyang Technological University, Singapore*
Michael Zagrodnik, *Rolls-Royce Singapore Pte. Ltd., Singapore*
Xiaolei Hu, *Nanyang Technological University, Singapore*
King Jet Tseng, *Nanyang Technological University, Singapore*

Research on the Filter of Load Side Converter in BDFG based Ship Shaft Power Generation System 1078
Meilin Wang, *Huazhong University of Science and Technology, China*
Hua Lin, *Huazhong University of Science and Technology, China*
Hongbin Yang, *Huazhong University of Science and Technology, China*
Xingwei Wang, *Huazhong University of Science and Technology, China*

Investigation of Common Mode Current Related DC-Bus Overvoltage in Multiple Converter Systems 1084
Jiangbiao He, *Rockwell Automation, United States*
Zoran Vrankovic, *Rockwell Automation, United States*
Patrick E. Ozimek, *Rockwell Automation, United States*
Craig Winterhalter, *Rockwell Automation, United States*

Session T23: Modeling of Magnetic Circuits and Systems

Location: 102AB
March 23, 2016 14:00 - 17:30
Session Chairs: Ed Herbert,
Jin Ye, *San Francisco State University*

High Frequency AC Inductor Analysis and Design for Dual Active Bridge (DAB) Converters .. 1090
Zhe Zhang, *Technical University of Denmark, Denmark*
Michael A.E. Andersen, *Technical University of Denmark, Denmark*

A Comprehensive Assessment of PM Motor Topology Impact on Magnet Defect Fault Signatures 1096
Mohsen Zafarani, *University of Texas at Dallas, United States*
Taner Goktas, *University of Texas at Dallas, United States*
Bilal Akin, *University of Texas at Dallas, United States*

High Frequency Modeling for Transformer Common Mode Noise Coupling Path based on Multiconductor Transmission Line Theory 1102
Peipei Meng, *Wuhan University of Technology, China*
Xiangming Zhang, *Naval University of Engineering, China*

Leakage Flux Modelling of Multi-Winding Transformer using Permeance Magnetic Circuit ... 1108
Min Luo, *École Polytechnique Fédérale de Lausanne, Switzerland*
Drazen Dujic, *École Polytechnique Fédérale de Lausanne, Switzerland*
Jost Allmeling, *Plexim GmbH, Switzerland*

Modeling Magnetic Devices using SPICE: Application to Variable Inductors 1115
J. Marcos Alonso, *Universidad de Oviedo, Spain*
Gilberto Martínez, *Continental Automotive R&D, Mexico*
Marina Perdigão, *Universidade de Coimbra, Portugal*
Marcelo Cosetin, *Universidade Federal de Santa Maria, Brazil*
Ricardo N. do Prado, *Universidade Federal de Santa Maria, Brazil*

Investigation of a Thermal Model for a Permanent Magnet Assisted Synchronous Reluctance Motor 1123
Joseph Herbert, *University of Akron, United States*
A.K.M. Arafat, *University of Akron, United States*
Guo-Xiang Wang, *University of Akron, United States*
Seungdeog Choi, *University of Akron, United States*

Design Procedure for Multi-Phase External Rotor Permanent Magnet Assisted Synchronous Reluctance Machines 1131
Sai Sudheer Reddy Bonthu, *University of Akron, United States*
Seungdeog Choi, *University of Akron, United States*

Applicability and Limitations of an M2Spice-Assisted "Planar-Magnetics-in-the-Circuit" Simulation Approach 1138
Samantha J. Gunter, *Massachusetts Institute of Technology, United States*
Minjie Chen, *Massachusetts Institute of Technology, United States*
Stephanie A. Pavlick, *Massachusetts Institute of Technology, United States*
Rose A. Abramson, *Massachusetts Institute of Technology, United States*
Khurram K. Afridi, *University of Colorado at Boulder, United States*
David J. Perreault, *Massachusetts Institute of Technology, United States*

Session T24: Inverter/Converter Control

Location: 103C
March 23, 2016 14:00 - 17:30
Session Chairs: Siavash Pakdelian, *UMass Lowell*
Behrooz Mirafzal, *Kansas State University*

Solution of Input Double-Line Frequency Ripple Rejection for High-Efficiency High-Power Density String Inverter in Photovoltaic Application 1148
Xiaonan Zhao, *Virginia Polytechnic Institute and State University, United States*
Lanhua Zhang, *Virginia Polytechnic Institute and State University, United States*
Rachael Born, *Virginia Polytechnic Institute and State University, United States*
Jih-Sheng Lai, *Virginia Polytechnic Institute and State University, United States*

Fractional-Order Phase Lead Compensation for Multi-Rate Repetitive Control on Three-Phase PWM DC/AC Inverter 1155
Zhichao Liu, *University of South Carolina, United States*
Bin Zhang, *University of South Carolina, United States*
Keliang Zhou, *University of Glasgow, United Kingdom*

A Robust Modified Model Predictive Control (MMPC) based on Lyapunov Function for Three-Phase Active-Front-End (AFE) Rectifier 1163
M. Parvez, *University of Malaya, Malaysia*
S. Mekhilef, *University of Malaya, Malaysia*
Nadia M.L. Tan, *Universiti Tenega Nasional, Malaysia*
Hirofumi Akagi, *Tokyo Institute of Technology, Japan*

Adaptive Reference Model Predictive Control for Power Electronics 1169
Yun Yang, *University of Hong Kong, Hong Kong*
Siew-Chong Tan, *University of Hong Kong, Hong Kong*
Shu-Yuen Ron Hui, *Imperial College London, United Kingdom*

Power Switch Lifetime Extension Strategies for Three-Phase Converters 1176
Serkan Dusmez, *University of Texas at Dallas, United States*
Enes Ugur, *University of Texas at Dallas, United States*
Bilal Akin, *University of Texas at Dallas, United States*

Current Controller Modeling for an Interleaved Boost with Voltage Multiplier Cells for PV Applications 1183
Alessandro Pevere, *Katholieke Universiteit Leuven, Belgium*
Urmimala Chatterjee, *Katholieke Universiteit Leuven, Belgium*
Johan Driesen, *Katholieke Universiteit Leuven, Belgium*

New Active Capacitor Voltage Balancing Method for Five-Level Stacked Multicell Converter 1191
Arash Khoshkbar Sadigh, *Extron Electronics, United States*
Vahid Dargahi, *Clemson University, United States*
Keith Corzine, *Clemson University, United States*

Gate Signal Jitter Elimination and Noise Shaping Modulation for High-SNR Class-D Power Amplifiers 1198
M. Mauerer, *ETH Zürich, Switzerland*
A. Tüysüz, *ETH Zürich, Switzerland*
J.W. Kolar, *ETH Zürich, Switzerland*

Analysis and Compensation of Inverter Nonlinearity for Three-Level T-Type Inverters 1206
Hyeon-Sik Kim, *Seoul National University, Korea, South*
Yong-Cheol Kwon, *Seoul National University, Korea, South*
Seung-Jun Chee, *Seoul National University, Korea, South*
Seung-Ki Sul, *Seoul National University, Korea, South*

Session T25: Topics in Renewable Energy Systems I
Location: 104B
March 23, 2016 14:00 - 17:30
Session Chairs: Fei Gao, *University of Technology of Belfort-Montbéliard*
Kent Wanner, *John Deere*

Front-End Isolated Quasi-Z-Source DC-DC Converter Modules in Series for Photovoltaic High-Voltage DC Applications 1214
Yushan Liu, *Texas A&M University at Qatar, Qatar*
Haitham Abu-Rub, *Texas A&M University at Qatar, Qatar*
Baoming Ge, *Texas A&M University, United States*

Analysis of Non Detection Zone for Multiple Distributed PCS based on Equivalent Single PCS using Reactive Power Approach 1220
Byeong-Heon Kim, *Seoul National University, Korea, South*
Seung-Ki Sul, *Seoul National University, Korea, South*

Optimal Power Scheduling for a Grid-Connected Hybrid PV-Wind-Battery Microgrid System .. 1227
Adriana Luna, *Aalborg University, Denmark*
Nelson Diaz, *Aalborg University, Denmark*
Mehdi Savaghebi, *Aalborg University, Denmark*
Juan C. Vásquez, *Aalborg University, Denmark*
Josep M. Guerrero, *Aalborg University, Denmark*
Kai Sun, *Tsinghua University, China*
Guoliang Chen, *Shanghai Solar Energy & Technology Co., Ltd., China*
Libing Sun, *Shanghai Solar Energy & Technology Co., Ltd., China*

High Efficiency Power Converter for a Doubly-Fed SOEC/SOFC System 1235
Kevin Tomas-Manez, *Technical University of Denmark, Denmark*
Alexander Anthon, *Technical University of Denmark, Denmark*
Zhe Zhang, *Technical University of Denmark, Denmark*

A Hierarchical Active Balancing Architecture for Li-Ion Batteries 1243
Han-Dong Gui, *Nanjing University of Aeronautics and Astronautics, China*
Zhiliang Zhang, *Nanjing University of Aeronautics and Astronautics, China*
Dong-Jie Gu, *Nanjing University of Aeronautics and Astronautics, China*
Yang Yang, *Nanjing University of Aeronautics and Astronautics, China*
Zhouyu Lu, *Nanjing University of Aeronautics and Astronautics, China*
Yan-Fei Liu, *Queen's University, Canada*

A Series-DG based Autonomous Islanding Microgrid 1249
Beihua Liang, *Tianjin University, China*
Yun Wei Li, *University of Alberta, Canada*
Jinwei He, *Tianjin University, China*
Chengshan Wang, *Tianjin University, China*

An Enhanced Droop Control Scheme for Resilient Active Power Sharing in Paralleled Two-Stage PV Inverter Systems 1253
Hongpeng Liu, *Harbin Institute of Technology, China*
Yongheng Yang, *Aalborg University, Denmark*
Xiongfei Wang, *Aalborg University, Denmark*
Poh Chiang Loh, *Aalborg University, Denmark*
Frede Blaabjerg, *Aalborg University, Denmark*
Wei Wang, *Harbin Institute of Technology, China*
Dianguo Xu, *Harbin Institute of Technology, China*

Voltage Closed-Loop Virtual Synchronous Generator Control of Full Converter Wind Turbine for Grid-Connected and Stand-Alone Operation 1261
Yiwei Ma, *University of Tennessee, United States*
Liu Yang, *University of Tennessee, United States*
Fred Wang, *University of Tennessee, United States*
Leon M. Tolbert, *University of Tennessee, United States*

DC Voltage Ripple Quantification for a Flywheel-Battery based Hybrid Energy Storage System 1267
Christopher R. Lashway, *Florida International University, United States*
Ahmed T. Elsayed, *Florida International University, United States*
Osama A. Mohammed, *Florida International University, United States*

Session T26: Electric Vehicle Charging Systems
Location: 104C
March 23, 2016 14:00 - 17:30
Session Chairs: Jim Spangler, *Spangler Prototype Inc*
Hadi Malek, *Ford*

Adaptive Loss Reduction Charging Strategy Considering Variation of Internal Impedance of Lithium-Ion Polymer Batteries in Electric Vehicle Charging Systems 1273
Nari Kim, *Sungkyunkwan University, Korea, South*
Jung-Hoon Ahn, *Sungkyunkwan University, Korea, South*
Dong-Hee Kim, *Sungkyunkwan University, Korea, South*
Byoung-Kuk Lee, *Sungkyunkwan University, Korea, South*

A Pulse Width Modulated LLC Type Resonant Topology Adpated to Wide Output Voltage Range 1280
Haoyu Wang, *ShanghaiTech University, China*

A Series Resonant Circuit for Voltage Equalization of Series Connected Energy Storage Devices 1286
Yanqi Yu, *University of British Columbia, Canada*
Raed Saasaa, *University of British Columbia, Canada*
Wilson Eberle, *University of British Columbia, Canada*

Implementation of 3.3-kW GaN-Based DC-DC Converter for EV On-Board Charger with Series-Resonant Converter that Employs Combination of Variable-Frequency and Delay-Time Control 1292
Yungtaek Jang, *Delta Products Corporation, United States*
Milan M. Jovanović, *Delta Products Corporation, United States*
Juan M. Ruiz, *Delta Products Corporation, United States*
Misha Kumar, *Delta Products Corporation, United States*
Gang Liu, *Delta Electronics Shanghai Co., Ltd., China*

Dual Active Bridge-Based Full-Integrated Active Filter Auxiliary Power Module for Electrified Vehicle Applications with Single-Phase Onboard Chargers 1300
Ruoyu Hou, *McMaster University, Canada*
Ali Emadi, *McMaster University, Canada*

All-SiC Inductively Coupled Charger with Integrated Plug-In and Boost Functionalities for PEV Applications 1307
M. Chinthavali, *Oak Ridge National Laboratory, United States*
O.C. Onar, *Oak Ridge National Laboratory, United States*
S.L. Campbell, *Oak Ridge National Laboratory, United States*
L.M. Tolbert, *Oak Ridge National Laboratory, United States*

Switching Condition and Loss Modeling of GaN-Based Dual Active Bridge Converter for PHEV Charger 1315
Lingxiao Xue, *Virginia Polytechnic Institute and State University, United States*
Dushan Boroyevich, *Virginia Polytechnic Institute and State University, United States*
Paolo Mattavelli, *Università degli Studi di Padova, Italy*

Analysis of Cascaded Multi-Output-Port Converter for Wireless Plug-In Hybrid/On-Board EV Chargers 1323
Erdem Asa, *Hevo Power Inc. / New York University, United States*
Kerim Colak, *Istanbul Ulasim A.S., Turkey*
Dariusz Czarkowski, *New York University, United States*

Comparative Analysis of High Step-Down Ratio Isolated DC/DC Topologies in PEV Applications 1329
Zhiqing Li, *ShanghaiTech University, China*
Haoyu Wang, *ShanghaiTech University, China*

Session T27: Utility Interface and Inverter Applications
Location: 103AB
March 23, 2016 14:00 - 17:30
Session Chairs: Akshay Kumar Rathore, *Concordia University*
Yichao Tang, *Texas Instruments*

DC to Single-Phase AC Voltage Source Inverter with Power Decoupling Circuit based on Flying Capacitor Topology for PV System 1336
Hiroki Watanabe, *Nagaoka University of Technology, Japan*
Keisuke Kusaka, *Nagaoka University of Technology, Japan*
Keita Furukawa, *Nagaoka University of Technology, Japan*
Koji Orikawa, *Nagaoka University of Technology, Japan*
Jun-Ichi Itoh, *Nagaoka University of Technology, Japan*

GaN FET and Hybrid Modulation based Differential-Mode Inverter 1344
Sudip K. Mazumder, *NextWatt LLC, United States*
Ankit Gupta, *University of Illinois at Chicago, United States*
Shirish Raizada, *University of Illinois at Chicago, United States*
Harshit Soni, *University of Illinois at Chicago, United States*
Nikhil Kumar, *University of Illinois at Chicago, United States*
Paromita Mazumder, *NextWatt LLC, United States*
Parijat Bhattachaarjee, *NextWatt LLC, United States*

Thermal and Electrical Co-Design of a Modular High-Density Single-Phase Inverter using Wide-Bandgap Devices 1350
Steven Chung, *University of Toronto, Canada*
Miad Nasr, *University of Toronto, Canada*
David Guirguis, *University of Toronto, Canada*
Masafumi Otsuka, *University of Toronto, Canada*
Shahab Poshtkouhi, *University of Toronto, Canada*
David K.W. Li, *University of Toronto, Canada*
Vishal Palaniappan, *University of Toronto, Canada*
David Romero, *University of Toronto, Canada*
Cristina Amon, *University of Toronto, Canada*
Ray Orr, *Solantro Semiconductor, Canada*
Olivier Trescases, *University of Toronto, Canada*

Reactive Power Compensation with Improvement of Current Waveform Quality for Single-Phase Buck-Type Dynamic Capacitor 1358
Xinwen Chen, *Huazhong University of Science and Technology, China*
Ke Dai, *Huazhong University of Science and Technology, China*
Chen Xu, *Huazhong University of Science and Technology, China*
Ziwei Dai, *Huazhong University of Science and Technology, China*
Li Peng, *Huazhong University of Science and Technology, China*

Circulating Current Reduction for a D-Σ Digital Controlled Transformerless UPS 1364
T.-F. Wu, *National Tsing Hua University, Taiwan*
T.-H. Shiu, *National Tsing Hua University, Taiwan*
P.-H. Lin, *National Tsing Hua University, Taiwan*
L.-C. Lin, *National Tsing Hua University, Taiwan*
J.-W. Huang, *Industrial Technology Research Institute, Taiwan*

A Multi-Function Three-Level Dynamic Voltage Corrector with Wide Correction Range and Short Circuit Fault Isolation 1371
Jiankun Cao, *Nanjing University of Aeronautics and Astronautics, China*
Pengling Ding, *Nanjing University of Aeronautics and Astronautics, China*
Haichun Liu, *Nanjing University of Aeronautics and Astronautics, China*
Shaojun Xie, *Nanjing University of Aeronautics and Astronautics, China*

Effects and Analysis of Minimum Pulse Width Limitation on Adaptive DC Voltage Control of Grid Converters 1376
Bo Sun, *Aalborg University, Denmark*
Ionut Trintis, *Aalborg University, Denmark*
Stig Munk-Nielsen, *Aalborg University, Denmark*
Josep M. Guerrero, *Aalborg University, Denmark*

Improved Three-Phase Micro-Inverter using Dynamic Dead Time Optimization and Phase-Skipping Control Techniques 1381
S. Milad Tayebi, *University of Central Florida, United States*
Xianmin Mu, *University of Central Florida, United States*
Issa Batarseh, *University of Central Florida, United States*

Correcting Current Imbalances in Three-Phase Four-Wire Distribution Systems 1387
Vinson Jones, *University of Arkansas, United States*
Juan Carlos Balda, *University of Arkansas, United States*

Session T28: Isolated DC-DC Converters
Location: 104A
March 24, 2016 8:30 - 11:20
Session Chairs: Dragan Maksimovic, *UC Boulder*
Zhong Ye, *Texas Instruments*

New Design Methdology for Megahertz-Frequency Resonant DC-DC Converters using Impedance Control Network Architecture 1392
Yushi Liu, *University of Colorado at Boulder, United States*
Ashish Kumar, *University of Colorado at Boulder, United States*
Jie Lu, *University of Colorado at Boulder, United States*
Dragan Maksimovic, *University of Colorado at Boulder, United States*
Khurram K. Afridi, *University of Colorado at Boulder, United States*

Dual Voltage Regulations of Single Switch Flyback Converter using Variable Switching Frequency 1398
Jin-Woong Kim, *Seoul National University, Korea, South*
Jung-Ik Ha, *Seoul National University, Korea, South*

On-Chip PLL-Based Methods for Synchronizing Active Switches Across the Isolation Boundary in DC-DC Converters 1403
Shahab Poshtkouhi, *University of Toronto, Canada*
Miad Fard, *University of Toronto, Canada*
Olivier Trescases, *University of Toronto, Canada*

An Isolated Soft-Switching Buck-Boost Converter Utilizing Two Transformers and Embedded Bidirectional Switches on Secondary-Side for Wide Voltage Applications 1410
Tingting Liu, *Nanjing University of Aeronautics and Astronautics, China*
Hongfei Wu, *Nanjing University of Aeronautics and Astronautics, China*
Yan Xing, *Nanjing University of Aeronautics and Astronautics, China*
Kai Sun, *Tsinghua University, China*

Effect of Transformer Design on Operation of Fundamental Duty Modulation for Dual-Active-Bridge Converter 1416
Wooin Choi, *Seoul National University, Korea, South*
Moonhyun Lee, *Seoul National University, Korea, South*
Bo-Hyung Cho, *Seoul National University, Korea, South*

A High Step-Up Bidirectional Isolated Dual-Active-Bridge Converter with Three-Level Voltage-Doubler Rectifier for Energy Storage Applications 1424
Xiaohai Zhan, *Nanjing University of Aeronautics and Astronautics, China*
Hongfei Wu, *Nanjing University of Aeronautics and Astronautics, China*
Yan Xing, *Nanjing University of Aeronautics and Astronautics, China*
Hongjuan Ge, *Nanjing University of Aeronautics and Astronautics, China*
Xi Xiao, *Tsinghua University, China*

Digitized Self-Oscillating Loop for Piezoelectric Transformer-Based Power Converters 1430
Marzieh Ekhtiari, *Technical University of Denmark, Denmark*
Thomas Andersen, *Technical University of Denmark, Denmark*
Zhe Zhang, *Technical University of Denmark, Denmark*
Michael A.E. Andersen, *Technical University of Denmark, Denmark*

Session T29: Multilevel Converters
Location: 101A
March 24, 2016 8:30 - 11:20
Session Chairs: Maryam Saeedifard, *Georgia Tech*
Julia Zhang, *Oregon State University*

An Isolated Topology for Reactive Power Compensation with a Modularized Dynamic-Current Building-Block 1437
Hao Chen, *Georgia Institute of Technology, United States*
Anish Prasai, *Varentec, Inc., United States*
Deepak Divan, *Georgia Institute of Technology, United States*

Design and Control of a Compact MMC Submodule Structure with Reduced Capacitor Size using the Stacked Switched Capacitor Architecture 1443
Yuan Tang, *University of Warwick, United Kingdom*
Minjie Chen, *Massachusetts Institute of Technology, United States*
Li Ran, *University of Warwick, United Kingdom*

Fundamental Frequency Sorting Strategy for Capacitor Voltage Balance of Modular Multilevel Converters with Phase Disposition PWM 1450
Kun Wang, *Zhejiang University, China*
Yan Deng, *Zhejiang University, China*
Wenyu Li, *Zhejiang University, China*
Hao Peng, *Zhejiang University, China*
Guipeng Chen, *Zhejiang University, China*
Xiangning He, *Zhejiang University, China*

Active Voltage Balancing Control for 10kV Three-Level Converter using Series-Connected HV-IGBTs 1456
Shiqi Ji, *Tsinghua University, China*
Ting Lu, *Tsinghua University, China*
Zhengming Zhao, *Tsinghua University, China*
Hualong Yu, *Tsinghua University, China*
Fred Wang, *University of Tennessee, United States*

Average-Value Model of Modular Multilevel Converters Considering Capacitor Voltage 1462
Heya Yang, *Zhejiang University, China*
Yuxiang Chen, *Zhejiang University, China*
Wuhua Li, *Zhejiang University, China*
Xiangning He, *Zhejiang University, China*
Wei Sun, *China Electric Power Research Institute, China*
Yongning Chi, *China Electric Power Research Institute, China*
Yan Li, *China Electric Power Research Institute, China*

New Submodule Circuits for Modular Multilevel Current Source Converters with DC Fault Ride through Capability 1468
Xinyu Yu, *Tsinghua University, China*
Yingdong Wei, *Tsinghua University, China*
Qirong Jiang, *Tsinghua University, China*

Voltage and Power Balance Control Strategy for Three-Phase Modular Cascaded Solid Stated Transformer 1475
Zhiyu Zhang, *Zhejiang University, China*
Hengyang Zhao, *Zhejiang University, China*
Shihang Fu, *Zhejiang University, China*
Jianjiang Shi, *Zhejiang University, China*
Xiangning He, *Zhejiang University, China*

Session T30: Multilevel and Matrix Converters for Motor Drives
Location: 102C
March 24, 2016 8:30 - 11:20
Session Chairs: SeonHwan Hwang, *Kyungnam University, Korea*
Xiaohu Liu, *GE*

New Flying-Capacitor-Based Multilevel Converter with Optimized Number of Switches and Capacitors Controlled with a New Logic-Form-Equation based Active Voltage Balancing Technique 1481
Vahid Dargahi, *Clemson University, United States*
Arash Khoshkbar Sadigh, *Extron Electronics, United States*
Keith Corzine, *Clemson University, United States*

New Low-Cost Five-Level Active Neutral-Point Clamped Converter 1489
Hongliang Wang, *Queen's University, Canada*
Lei Kou, *Queen's University, Canada*
Yan-Fei Liu, *Queen's University, Canada*
Paresh C. Sen, *Queen's University, Canada*
Sucheng Liu, *Anhui University of Technology, China*

Medium Voltage (≥ 2.3 kV) High Frequency Three-Phase Two-Level Converter Design and Demonstration using 10 kV SiC MOSFETs for High Speed Motor Drive Applications .. 1497
Sachin Madhusoodhanan, *North Carolina State University, United States*
Krishna Mainali, *North Carolina State University, United States*
Awneesh Tripathi, *North Carolina State University, United States*
Kasunaidu Vechalapu, *North Carolina State University, United States*
Subhashish Bhattacharya, *North Carolina State University, United States*

Novel Three Phase Multi-Level Inverter Topology with Symmetrical DC-Voltage Sources . 1505
Ahmed Salem, *Aswan University, Egypt*
Emad M. Ahmed, *Aswan University, Egypt*
Mahrous Ahmed, *Aswan University, Egypt*
Mohamed Orabi, *Aswan University, Egypt*

A 2 kW, Single-Phase, 7-Level, GaN Inverter with an Active Energy Buffer Achieving 216 W/in^3 Power Density and 97.6% Peak Efficiency 1512
Yutian Lei, *University of Illinois at Urbana-Champaign, United States*
Christopher Barth, *University of Illinois at Urbana-Champaign, United States*
Shibin Qin, *University of Illinois at Urbana-Champaign, United States*
Wen-Chuen Liu, *University of Illinois at Urbana-Champaign, United States*
Intae Moon, *University of Illinois at Urbana-Champaign, United States*
Andrew Stillwell, *University of Illinois at Urbana-Champaign, United States*
Derek Chou, *University of Illinois at Urbana-Champaign, United States*
Thomas Foulkes, *University of Illinois at Urbana-Champaign, United States*
Zichao Ye, *University of Illinois at Urbana-Champaign, United States*
Zitao Liao, *University of Illinois at Urbana-Champaign, United States*
Robert C.N. Pilawa-Podgurski, *University of Illinois at Urbana-Champaign, United States*

Indirect Matrix Converter based Open-End Winding AC Drives with Zero Common-Mode Voltage 1520
Saurabh Tewari, *MTS Systems Corporation, United States*
Ranjan K. Gupta, *First Solar, Inc., United States*
Apurva Somani, *Dynapower Company LLC, United States*
Ned Mohan, *University of Minnesota, United States*

Precharging Strategy for Soft Startup Process of Modular Multilevel Converters based on Various SM Circuits 1528
Jiangchao Qin, *Arizona State University, United States*
Suman Debnath, *Oak Ridge National Laboratory, United States*
Maryam Saeedifard, *Georgia Institute of Technology, United States*

Session T31: System Design Techniques for Reduced EMI

Location: 101B
March 24, 2016 8:30 - 11:20
Session Chairs: John Vigars, *Allegro Microsystems*
Doug Hopkins, *North Carolina State University*

Conducted EMI Analysis and Filter Design for MHz Active Clamp Flyback Front-End Converter 1534
Xiucheng Huang, *Virginia Polytechnic Institute and State University, United States*
Junjie Feng, *Virginia Polytechnic Institute and State University, United States*
Fred C. Lee, *Virginia Polytechnic Institute and State University, United States*
Qiang Li, *Virginia Polytechnic Institute and State University, United States*
Yuchen Yang, *Virginia Polytechnic Institute and State University, United States*

EMC Investigation of a Very High Frequency Self-Oscillating Resonant Power Converter 1541
Jeppe A. Pedersen, *Technical University of Denmark, Denmark*
Arnold Knott, *Technical University of Denmark, Denmark*
Michael A.E. Andersen, *Technical University of Denmark, Denmark*

Numerical Optimization of Passive Line Filter Components for Suppression of Electromagnetic Interference (EMI) 1547
Carsten Henkenius, *Universität Paderborn, Germany*
Norbert Fröhleke, *Universität Paderborn, Germany*
Joachim Böcker, *Universität Paderborn, Germany*
Heiko Figge, *Delta Energy Systems GmbH, Germany*

Electromagnetic Noise Coupling and Mitigation for Fast Response On-Die Temperature Sensing in High Power Modules 1554
Chengcheng Yao, *Ohio State University, United States*
Pengzhi Yang, *Ohio State University, United States*
Mingzhi Leng, *Ohio State University, United States*
He Li, *Ohio State University, United States*
Lixing Fu, *Ohio State University, United States*
Jin Wang, *Ohio State University, United States*
Ke Zou, *Ford Motor Company, United States*
Chingchi Chen, *Ford Motor Company, United States*

Ultra-Low Inductance Vertical Phase Leg Design with EMI Noise Propagation Control for Enhancement Mode GaN Transistors 1561
Xuning Zhang, *Virginia Polytechnic Institute and State University, United States*
Zhiyu Shen, *Virginia Polytechnic Institute and State University, United States*
Nidhi Haryani, *Virginia Polytechnic Institute and State University, United States*
Dushan Boroyevich, *Virginia Polytechnic Institute and State University, United States*
Rolando Burgos, *Virginia Polytechnic Institute and State University, United States*

Decoupling of Interaction between WBG Converter and Motor Load for Switching Performance Improvement 1569
Zheyu Zhang, *University of Tennessee, United States*
Fred Wang, *University of Tennessee, United States*
Leon M. Tolbert, *University of Tennessee, United States*
Benjamin J. Blalock, *University of Tennessee, United States*
Daniel J. Costinett, *University of Tennessee, United States*

Control and Characterization of Electromagnetic Emissions in Wide Band Gap based Converter Modules for Ungrounded Grid-Forming Applications 1577
Robert Cuzner, *University of Wisconsin at Milwaukee, United States*
Rasoul Hosseini, *University of Wisconsin at Milwaukee, United States*
Andrew Lemmon, *University of Alabama, United States*
James Gafford, *Mississippi State University, United States*
Michael Mazzola, *Mississippi State University, United States*

Session T32: Modeling of DC Energy Converters and Systems
Location: 102AB
March 24, 2016 8:30 - 11:20
Session Chairs: Santanu Kapat, *IIT Kharagpur*
Sombuddha Chakraborty, *Texas Instruments*

A Practical Switching Time Model for Synchronous Buck Converters 1585
Yuan Rao, *Texas Instruments Inc., United States*
Surinder P. Singh, *Texas Instruments Inc., United States*
Taisuke Kazama, *Texas Instruments Inc., United States*

Off-Line Identification of Digitally Controlled Power Converters using an Analog Frequency Response Analyzer 1591
Marco Meola, *Zentrum Mikroelektronik Dresden AG, Germany*
Anthony Kelly, *Altera Corporation, Ireland*

Extended Wide-Load Range Model for Multi-Level DC-DC Converters and a Practical Dual-Mode Digital Controller 1597
Nenad Vukadinović, *University of Toronto, Canada*
Aleksandar Prodić, *University of Toronto, Canada*
Brett A. Miwa, *Maxim Integrated, United States*
Cory B. Arnold, *Maxim Integrated, United States*
Michael W. Baker, *Maxim Integrated, United States*

Burst Mode Control and Switched-Capacitor Converters Losses 1603
Michael Evzelman, *Utah State University, United States*
Regan Zane, *Utah State University, United States*

Equivalent Circuit Modeling of LLC Resonant Converter 1608
Shuilin Tian, *Virginia Polytechnic Institute and State University, United States*
Fred C. Lee, *Virginia Polytechnic Institute and State University, United States*
Qiang Li, *Virginia Polytechnic Institute and State University, United States*

Small Signal Modeling of the Hysteretic Modulator with a Current Ripple Synthesizer 1616
Yi Huang, *Intersil Corporation, United States*
Chun Cheung, *Intersil Corporation, United States*

A Black-Box Modeling Approach for DC Nanogrids 1624
A. Francés, *Universidad Politécnica de Madrid, Spain*
R. Asensi, *Universidad Politécnica de Madrid, Spain*
O. García, *Universidad Politécnica de Madrid, Spain*
R. Prieto, *Universidad Politécnica de Madrid, Spain*
J. Uceda, *Universidad Politécnica de Madrid, Spain*

Session T33: Gate Drive Techniques
Location: 103C
March 24, 2016 8:30 - 11:20
Session Chairs: Christopher Bridge, *SIMPLIS Technologies*
Martin Ordonez, *University of British Columbia*

Design and Evaluation of Isolated Gate Driver Power Supply for Medium Voltage Converter Applications 1632
Krishna Mainali, *North Carolina State University, United States*
Sachin Madhusoodhanan, *North Carolina State University, United States*
Awneesh Tripathi, *North Carolina State University, United States*
Kasunaidu Vechalapu, *North Carolina State University, United States*
Ankan De, *North Carolina State University, United States*
Subhashish Bhattacharya, *North Carolina State University, United States*

General-Purpose Clocked Gate Driver (CGD) IC with Programmable 63-Level Drivability to Reduce IC Overshoot and Switching Loss of Various Power Transistors 1640
Koutarou Miyazaki, *University of Tokyo, Japan*
Seiya Abe, *Kyushu Institute of Technology, Japan*
Masanori Tsukuda, *Kyushu Institute of Technology, Japan*
Ichiro Omura, *Kyushu Institute of Technology, Japan*
Keiji Wada, *Tokyo Metropolitan University, Japan*
Makoto Takamiya, *University of Tokyo, Japan*
Takayasu Sakurai, *University of Tokyo, Japan*

An Integrated SiC CMOS Gate Driver 1646
Matthew Barlow, *University of Arkansas, United States*
Shamim Ahmed, *University of Arkansas, United States*
H. Alan Mantooth, *University of Arkansas, United States*
A. Matt Francis, *Ozark Integrated Circuits, Inc., United States*

Digital Active Gate Drives using Sequential Optimization 1650
Daniel J. Rogers, *University of Oxford, United Kingdom*
Boris Murmann, *Stanford University, United States*

One Adaptive Turn-Off Method for PFC Converter with Voltage Spike Limitation 1657
Qunfang Wu, *Nanjing University of Aeronautics and Astronautics, China*
Qin Wang, *Nanjing University of Aeronautics and Astronautics, China*
Lan Xiao, *Nanjing University of Aeronautics and Astronautics, China*
Jialin Xu, *Nanjing University of Aeronautics and Astronautics, China*
Hongxu Li, *Nanjing University of Aeronautics and Astronautics, China*

A Digital Implementation for PWM Phase-Frequency Synchronization in SMPS Systems 1663
Luca Bizjak, *Infineon Technologies Austria AG, Austria*
Emanuele Bodano, *Infineon Technologies Austria AG, Austria*
Ante Gotovac, *Infineon Technologies Austria AG, Austria*
Sergii Tkachov, *Infineon Technologies Austria AG, Austria*

A High Accuracy and High Bandwidth Current Sense Circuit for Digitally Controlled DC-DC Buck Converters 1670
David Stack, *Altera Corporation, Ireland*
Anthony Kelly, *Altera Corporation, Ireland*
Thomas Conway, *University of Limerick, Ireland*

Session T34: Energy Storage Systems
Location: 104B
March 24, 2016 8:30 - 11:20
Session Chairs: Wei Qiao, *University of Nebraska Lincoln*
Yilmaz Sozer, *University of Akron*

Modular Multilevel Dual Active Bridge DC-DC Converter with ZVS and Fast DC Fault Recovery for Battery Energy Storage Systems 1675
Yuxiang Shi, *Florida State University, United States*
Rui Li, *Florida State University, United States*
Hui Li, *Florida State University, United States*

An Analytical Framework to Design a Dynamic Frequency Control Scheme for Microgrids using Energy Storage 1682
Ajit A. Renjit, *Ohio State University, United States*
Feng Guo, *NEC Laboratories America, Inc., United States*
Ratnesh Sharma, *NEC Laboratories America, Inc., United States*

Comparative Evaluation of $LiFePO_4$ Cell SOC Estimation Performance with ECM Structure and Noise Model/Data Rejection in the EKF for Transportation Application 1690
Hyun-jun Lee, *Soongsil University, Korea, South*
Joung-hu Park, *Soongsil University, Korea, South*
Jonghoon Kim, *Chosun University, Korea, South*

A Power Sharing Scheme for Series Connected Offshore Wind Turbines in a Medium Voltage DC Collection Grid 1695
Michael T. Daniel, *Texas A&M University, United States*
Prasad N. Enjeti, *Texas A&M University, United States*

Fault Ride-Through Performance Evaluation of an Interleaved Grid-Connected Converter Employing Low Switching Frequency 1702
Lorand Bede, *Aalborg University, Denmark*
Ghanshyamsinh Gohil, *Aalborg University, Denmark*
Mihai Ciobotaru, *University of New South Wales, Australia*
Tamas Kerekes, *Aalborg University, Denmark*
Remus Teodorescu, *Aalborg University, Denmark*
Vassilios G. Agelidis, *University of New South Wales, Australia*

Analysis of Two Charging Modes of Battery Energy Storage System for a Stand-Alone Microgrid 1708
Jongmin Jo, *Chungnam National University, Korea, South*
Hanju Cha, *Chungnam National University, Korea, South*

Proposition and Experimental Verification of a Bi-Directional Isolated DC/DC Converter for Battery Charger-Discharger of Electric Vehicle 1713
Ryota Kondo, *Mitsubishi Electric Corporation, Japan*
Yusuke Higaki, *Mitsubishi Electric Corporation, Japan*
Masaki Yamada, *Mitsubishi Electric Corporation, Japan*

Session T35: Topics on Inductive and Capacitive Wireless Power Transfer
Location: 104C
March 24, 2016 8:30 - 11:20
Session Chairs: Chris Mi, *San Diego State University*
Omer Onar, *Oak Ridge National Laboratory*

A CLLC-Compensated High Power and Large Air-Gap Capacitive Power Transfer System for Electric Vehicle Charging Applications 1721
Fei Lu, *University of Michigan at Ann Arbor, United States*
Hua Zhang, *Northeastern Polytechnical University, China*
Heath Hofmann, *University of Michigan at Ann Arbor, United States*
Chris Mi, *San Diego State University, United States*

A Large Air-Gap Capacitive Power Transfer System with a 4-Plate Capacitive Coupler Structure for Electric Vehicle Charging Applications 1726
Hua Zhang, *Northwestern Polytechnical University, China*
Fei Lu, *University of Michigan at Ann Arbor, United States*
Heath Hofmann, *University of Michigan at Ann Arbor, United States*
Weiguo Liu, *Northwestern Polytechnical University, China*
Chris Mi, *San Diego State University, United States*

Dynamic Wireless Power Transfer System for Electric Vehicles to Simplify Ground Facilities – Power Control and Efficiency Maximization on the Secondary Side – 1731
Katsuhiro Hata, *University of Tokyo, Japan*
Takehiro Imura, *University of Tokyo, Japan*
Yoichi Hori, *University of Tokyo, Japan*

Uniform-Gain Frequency Tracking of Wireless EV Charging for Improving Alignment Flexibility 1737
Yabiao Gao, *University of Georgia, United States*
Antonio Ginart, *University of Georgia / Sonnenbatterie GmbH, United States*
Kathleen Blair Farley, *Southern Company Services, Inc., United States*
Zion Tsz Ho Tse, *University of Georgia, United States*

Design and Optimization of a Multi-Coil System for Inductive Charging with Small Air Gap .. 1741
Christopher Joffe, *Fraunhofer Institute for Integrated Systems and Device Technology, Germany*
Andreas Roßkopf, *Fraunhofer Institute for Integrated Systems and Device Technology, Germany*
Stefan Ehrlich, *Fraunhofer Institute for Integrated Systems and Device Technology, Germany*
Christian Dobmeier, *Fraunhofer Institute for Integrated Systems and Device Technology, Germany*
Martin März, *Fraunhofer Institute for Integrated Systems and Device Technology, Germany*

Core Design for Better Misalignment Tolerance and Higher Range of Wireless Charging for HEV 1748
Mostak Mohammad, *University of Akron, United States*
Sangshin Kwak, *Chung-ang University, Korea, South*
Seungdeog Choi, *University of Akron, United States*

A 25 kW Industrial Prototype Wireless Electric Vehicle Charger 1756
Mariusz Bojarski, *Hevo Power Inc., United States*
Erdem Asa, *Hevo Power Inc. / New York University, United States*
Kerim Colak, *Istanbul Ulasim A.S., Turkey*
Dariusz Czarkowski, *New York University, United States*

Session T36: Wireless Power Transfer
Location: 103AB
March 24, 2016 8:30 - 11:20
Session Chairs: Sriram Jala Reddy, *Ford Motors*
Michael Masquelier, *WAVE*

Full-Bridge Series Resonant Multi-Inverter Featuring New 900-V SiC Devices for Improved Induction Heating Appliances 1762
Mario Pérez-Tarragona, *Universidad de Zaragoza, Spain*
Héctor Sarnago, *Universidad de Zaragoza, Spain*
Óscar Lucía, *Universidad de Zaragoza, Spain*
José M. Burdío, *Universidad de Zaragoza, Spain*

A Novel Phase Control of Single Switch Active Rectifier for Inductive Power Transfer Applications 1767
Kerim Colak, *Istanbul Ulasim A.S., Turkey*
Erdem Asa, *Hevo Power Inc. / New York University, United States*
Dariusz Czarkowski, *New York University, United States*

Optimal Shaped Dipole-Coil Design and Experimental Verification of Inductive Power Transfer System for Home Applications 1773
Duy T. Nguyen, *Korea Advanced Institute of Science and Technology, Korea, South*
Eun S. Lee, *Korea Advanced Institute of Science and Technology, Korea, South*
Byeung G. Choi, *Korea Advanced Institute of Science and Technology, Korea, South*
Chun T. Rim, *Korea Advanced Institute of Science and Technology, Korea, South*

A Novel Time-Sharing Current-Fed ZCS High Frequency Inverter-Applied Resonant DC-DC Converter for Inductive Power Transfer 1780
Kyohei Konishi, *Kobe University, Japan*
Tomokazu Mishima, *Kobe University, Japan*
Mutsuo Nakaoka, *University of Malaya, Malaysia*

Optimization of Coils for Magnetically Coupled Resonant Wireless Power Transfer System based on Maximum Output Power 1788
Dan Jiang, *Nanjing University of Aeronautics and Astronautics, China*
Yong Yang, *Nanjing University of Aeronautics and Astronautics, China*
Fuxin Liu, *Nanjing University of Aeronautics and Astronautics, China*
Xinbo Ruan, *Nanjing University of Aeronautics and Astronautics, China*
Xuling Chen, *Nanjing University of Aeronautics and Astronautics, China*

Online Regulation of Receiver-Side Power and Estimation of Mutual Inductance in Wireless Inductive Link based on Transmitter-Side Electrical Information 1795
Jeff Po Wa Chow, *City University of Hong Kong, Hong Kong*
Henry Shu-Hung Chung, *City University of Hong Kong, Hong Kong*
Chun Sing Cheng, *City University of Hong Kong, Hong Kong*

Dynamic Period Switching of PRS-PWM with Run-Length Limiting Technique for Spurious and Ripple Reduction in Fast Response Wireless Power Transmission 1802
Takahiro Moroto, *Keio University, Japan*
Toru Kawajiri, *Keio University, Japan*
Hiroki Ishikuro, *Keio University, Japan*

Session T37: Single-Phase AC-DC Converters
Location: 102AB
March 24, 2016 14:00 - 17:30
Session Chairs: Dusty Becker, *Emerson Network Power*
Pritam Das, *National University of Singapore*

A Flyback AC/DC Converter using Power Semiconductor Filter for Input Power Factor Correction 1807
Chung-Pui Tung, *City University of Hong Kong, Hong Kong*
Henry Shu-Hung Chung, *City University of Hong Kong, Hong Kong*

Reducing the Variation Range of the Switching Frequency for CRM Boost PFC Converter by Injecting 3rd Harmonic into the Input Current 1815
Yi Wang, *Nanjing University of Science and Technology, China*
Kai Yao, *Nanjing University of Science and Technology, China*

A Sustained Increase of Input Current Distortion in Active Input Current Shapers to Eliminate Electrolytic Capacitor for Designing AC to DC HB-LED Drivers for Retrofit Lamps Applications 1823
D.G. Lamar, *Universidad de Oviedo, Spain*
M. Arias, *Universidad de Oviedo, Spain*
A. Rodriguez, *Universidad de Oviedo, Spain*
J. Sebastian, *Universidad de Oviedo, Spain*
A. Fernandez, *European Space Agency, Netherlands*
J.A. Villarejo, *Universidad de Cartagena, Spain*

Reduced Current Stress Bridgeless Cuk PFC Converter with New Voltage Multiplier Circuit ... 1831
Yi-Hung Liao, *National Penghu University of Science and Technology, Taiwan*

Implementation of Multi-Level Bridgeless PFC Rectifiers for Mid-Power Single Phase Applications 1835
Trong Tue Vu, *Eisergy Ltd., Ireland*
George Young, *Eisergy Ltd., Ireland*

US Mains Stacked Very High Frequency Self-Oscillating Resonant Power Converter with Unified Rectifier 1842
Jeppe A. Pedersen, *Technical University of Denmark, Denmark*
Mickey P. Madsen, *Technical University of Denmark, Denmark*
Jakob D. Mønster, *Technical University of Denmark, Denmark*
Thomas Andersen, *Technical University of Denmark, Denmark*
Arnold Knott, *Technical University of Denmark, Denmark*
Michael A.E. Andersen, *Technical University of Denmark, Denmark*

Digital-Based Interleaving Control for GaN-Based MHz CRM Totem-Pole PFC 1847
Zhengyang Liu, *Virginia Polytechnic Institute and State University, United States*
Zhengrong Huang, *Virginia Polytechnic Institute and State University, United States*
Fred C. Lee, *Virginia Polytechnic Institute and State University, United States*
Qiang Li, *Virginia Polytechnic Institute and State University, United States*

A Novel AC-to-DC Adaptor with Ultra-High Power Density and Efficiency 1853
Yan-Cun Li, *Virginia Polytechnic Institute and State University, United States*
Fred C. Lee, *Virginia Polytechnic Institute and State University, United States*
Qiang Li, *Virginia Polytechnic Institute and State University, United States*
Xiucheng Huang, *Virginia Polytechnic Institute and State University, United States*
Zhengyang Liu, *Virginia Polytechnic Institute and State University, United States*

A Single-Stage Single-Phase Isolated AC-DC Converter based on LLC Resonant Unit and T-Type Three-Level Unit for Battery Charging Applications 1861
Yikai Gao, *University of Texas at Dallas, United States*
Wen Cai, *University of Texas at Dallas, United States*
Fan Yi, *University of Texas at Dallas, United States*

Session T38: Non-Isolated DC-DC Converters
Location: 101A
March 24, 2016 14:00 - 17:30
Session Chairs: Pradeep Shenoy, *Texas Instruments*
Juan Rivas-Davila, *Stanford*

DC-DC Power Converter Controller for SOC Balancing of Paralleled Battery System 1868
Jaber A. Abu Qahouq, *University of Alabama, United States*
Lin Zhang, *University of Alabama, United States*
Yuan Cao, *University of Alabama, United States*
Bharat Balasubramanian, *University of Alabama, United States*

Ultra-Step-Up DC-DC Converter with Integrated Autotransformer and Coupled Inductor ... 1872
Yam P. Siwakoti, *Aalborg University, Denmark*
Frede Blaabjerg, *Aalborg University, Denmark*
Poh Chiang Loh, *Aalborg University, Denmark*

Optimal Dynamic Phase Add/Drop Mechanism in Multiphase DC-DC Buck Converters 1878
Anandha Ruban T T, *Texas Instruments India Pvt. Ltd., India*
Preetam Tadeparthy, *Texas Instruments India Pvt. Ltd., India*
Sankaran Aniruddhan, *Indian Institute of Technology Madras, India*
Vikram Gakhar, *Texas Instruments India Pvt. Ltd., India*
Muthusubramanian Venkateswaran, *Texas Instruments India Pvt. Ltd., India*

A Universal Self-Calibrating Dynamic Voltage and Frequency Scaling (DVFS) Scheme with Thermal Compensation for Energy Savings in FPGAs 1882
Shuze Zhao, *University of Toronto, Canada*
Ibrahim Ahmed, *University of Toronto, Canada*
Carl Lamoureux, *University of Toronto, Canada*
Ashraf Lotfi, *Altera Corporation, United States*
Vaughn Betz, *University of Toronto, Canada*
Olivier Trescases, *University of Toronto, Canada*

Morphing Switched-Capacitor Step-Down DC-DC Converters with Variable Conversion Ratio 1888
Song Xiong, *University of Hong Kong, Hong Kong*
Ying Huang, *University of Hong Kong, Hong Kong*
Siew-Chong Tan, *University of Hong Kong, Hong Kong*
Shu-Yuen Ron Hui, *University of Hong Kong, Hong Kong*

Compact Modular Switched-Capacitor DC/DC Converters with Exponential Voltage Gain 1894
Ying Huang, *University of Hong Kong, Hong Kong*
Song Xiong, *University of Hong Kong, Hong Kong*
Siew-Chong Tan, *University of Hong Kong, Hong Kong*
Shu-Yuen Ron Hui, *University of Hong Kong, Hong Kong*

Study and Implementation of a High Step-Up Voltage DC-DC Converter using Coupled-Inductor and Cascode Techniques 1900
Tsorng-Juu Liang, *National Cheng Kung University, Taiwan*
Yung-Ting Huang, *National Cheng Kung University, Taiwan*
Jian-Hsing Lee, *National Cheng Kung University, Taiwan*
Lo Pang-Yen Ting, *National Cheng Kung University, Taiwan*

20 mV Input, 4.2 V Output Boost Converter with Methodology of Maximum Output Power for Thermoelectric Energy Harvesting 1907
Taichi Ogawa, *Toshiba Corporation, Japan*
Takeshi Ueno, *Toshiba Corporation, Japan*
Takayuki Miyazaki, *Toshiba Corporation, Japan*
Tetsuro Itakura, *Toshiba Corporation, Japan*

Clarification of Relationship between Current Ripple and Power Density in Bidirectional DC-DC Converter 1911
Hoai Nam Le, *Nagaoka University of Technology, Japan*
Koji Orikawa, *Nagaoka University of Technology, Japan*
Jun-Ichi Itoh, *Nagaoka University of Technology, Japan*

Session T39: Inverter Applications and Technologies

Location: 101B

March 24, 2016 14:00 - 17:30

Session Chairs: Ali Khajehoddin, *University of Alberta*
Wen Cai, *University of Texas, Dallas*

Grid-Voltage Feedforward based Control for Grid-Connected LCL-Filtered Inverter with High Robustness and Low Grid Current Distortion in Weak Grid 1919
Jinming Xu, *Nanjing University of Aeronautics and Astronautics, China*
Qiang Qian, *Nanjing University of Aeronautics and Astronautics, China*
Shaojun Xie, *Nanjing University of Aeronautics and Astronautics, China*
Binfeng Zhang, *Nanjing University of Aeronautics and Astronautics, China*

Evaluation of PV Frequency-Watt Function for Fast Frequency Reserves 1926
J. Neely, *Sandia National Laboratories, United States*
J. Johnson, *Sandia National Laboratories, United States*
J. Delhotal, *Sandia National Laboratories, United States*
S. Gonzalez, *Sandia National Laboratories, United States*
M. Lave, *Sandia National Laboratories, United States*

A Systematic Design Method and Verification for a Zero-Ripple Interface for PV/Battery-to-Grid Applications 1934
Suvankar Biswas, *University of Minnesota, United States*
Ned Mohan, *University of Minnesota, United States*
William Robbins, *University of Minnesota, United States*

Grid-Voltage-Feedforward Active Damping for Grid-Connected Inverter with LCL Filter 1941
Minghui Lu, *Aalborg University, Denmark*
Xiongfei Wang, *Aalborg University, Denmark*
Frede Blaabjerg, *Aalborg University, Denmark*
S.M. Muyeen, *Petroleum Institute, U.A.E.*
Ahmed Al-Durra, *Petroleum Institute, U.A.E.*
Siyu Leng, *Petroleum Institute, U.A.E.*

A High Power Density Single-Phase Inverter using Stacked Switched Capacitor Energy Buffer 1947
Colin McHugh, *University of Colorado at Boulder, United States*
Sreyam Sinha, *University of Colorado at Boulder, United States*
Jeffrey Meyer, *University of Colorado at Boulder, United States*
Saad Pervaiz, *University of Colorado at Boulder, United States*
Jie Lu, *University of Colorado at Boulder, United States*
Fan Zhang, *University of Colorado at Boulder, United States*
Hua Chen, *University of Colorado at Boulder, United States*
Hyeokjin Kim, *University of Colorado at Boulder, United States*
Usama Anwar, *University of Colorado at Boulder, United States*
Ashish Kumar, *University of Colorado at Boulder, United States*
Alihossein Sepahvand, *University of Colorado at Boulder, United States*
Scott Jensen, *University of Colorado at Boulder, United States*
Beomseok Choi, *University of Colorado at Boulder, United States*
Daniel Seltzer, *University of Colorado at Boulder, United States*
Robert Erickson, *University of Colorado at Boulder, United States*
Dragan Maksimovic, *University of Colorado at Boulder, United States*
Khurram K. Afridi, *University of Colorado at Boulder, United States*

A Novel Single-Stage Dual-Active Bridge based Isolated DC-AC Converter 1954
Shiladri Chakraborty, *Indian Institute of Technology Kharagpur, India*
Souvik Chattopadhyay, *Indian Institute of Technology Kharagpur, India*

Ultra-Low Ripple Inverters for Distributed Generation Applications 1962
Ang Shen, *Missouri University of Science and Technology, United States*
Pourya Shamsi, *Missouri University of Science and Technology, United States*
Mehdi Ferdowsi, *Missouri University of Science and Technology, United States*

A 15 kV SiC MOSFET Gate Drive with Power Over Fiber based Isolated Power Supply and Comprehensive Protection Functions 1967
Xuan Zhang, *Ohio State University, United States*
He Li, *Ohio State University, United States*
John A. Brothers, *Ohio State University, United States*
Jin Wang, *Ohio State University, United States*
Lixing Fu, *Texas Instruments Inc., United States*
Mico Perales, *MH GoPower Co., Ltd., Taiwan*
John Wu, *MH GoPower Co., Ltd., Taiwan*

A 15-kV Class Intelligent Universal Transformer for Utility Applications 1974
Jih-Sheng Lai, *Virginia Polytechnic Institute and State University, United States*
Wei-Han Lai, *Enertronics, Inc., United States*
Seung-Ryul Moon, *Virginia Polytechnic Institute and State University, United States*
Lanhua Zhang, *Virginia Polytechnic Institute and State University, United States*
Arindam Maitra, *Electric Power Research Institute, United States*

Session T40: Modeling, Modulation and Control of Motor Drive
Location: 102C
March 24, 2016 14:00 - 17:30
Session Chairs: Jin Wang, *The Ohio State University*
River-TinHo Li, *ABB*

Modulation Technique for Common Mode Voltage Reduction in a Matrix Converter Drive Operating with High Voltage Transfer Ratio 1982
Varsha Padhee, *Rockwell Automation, United States*
Ashish Kumar Sahoo, *University of Minnesota, United States*
Ned Mohan, *University of Minnesota, United States*

Soft-Switched Discontinuous Pulse-Width Pulse-Density Modulation Scheme 1989
Arash Rahnamaee, *University of Illinois at Chicago, United States*
Alireza Mojab, *University of Illinois at Chicago, United States*
Hossein Riazmontazer, *University of Illinois at Chicago, United States*
Sudip K. Mazumder, *University of Illinois at Chicago, United States*
Milos Zefran, *University of Illinois at Chicago, United States*

A Novel Flux Estimator based on SOGI with FLL for Induction Machine Drives 1995
Rende Zhao, *China University of Petroleum, China*
Zhen Xin, *Aalborg University, Denmark*
Poh Chiang Loh, *Aalborg University, Denmark*
Frede Blaabjerg, *Aalborg University, Denmark*

Performance Characterization of Random Pulse Width Modulation Algorithms in Industrial and Commercial Adjustable Speed Drives 2003
Kevin Lee, *Eaton Corporation, United States*
Guangtong Shen, *Purdue University, United States*
Wenxi Yao, *Zhejiang University, China*
Zhengyu Lu, *Zhejiang University, China*

Stability Analysis and Controller Synthesis for Digital Single-Loop Voltage-Controlled Inverters 2011
Xiongfei Wang, *Aalborg University, Denmark*
Poh Chiang Loh, *Aalborg University, Denmark*
Frede Blaabjerg, *Aalborg University, Denmark*

High Efficiency, Hybrid Selective Harmonic Elimination Phase-Shift PWM Technique for Cascaded H-Bridge Inverters to Improve Dynamic Response and Operate in Complete Normal Modulation Indices 2019
Amirhossein Moeini, *University of Florida, United States*
Zhao Hui, *University of Florida, United States*
Shuo Wang, *University of Florida, United States*

Implementation and Experimental Validation of Efficiency Improvement in PMSM Drives through Switching Frequency Reduction 2027
Parag Kshirsagar, *United Technologies Research Center, United States*
Krishnan Ramu, *Virginia Polytechnic Institute and State University, United States*

Sensorless Speed Control of Symmetrical Triple-Star Nine-Phase Interior Permanent Magnet Machines 2035
Olorunfemi Ojo, *Tennessee Technological University, United States*
Medhi Ramezani, *Tennessee Technological University, United States*

Mitigation of Common-Mode Noise in Wide Band Gap Device based Motor Drives 2043
Sneha Narasimhan, *Rockwell Automation, United States*
Saurabh Tewari, *MTS Systems Corporation, United States*
Eric Severson, *University of Minnesota, United States*
Rohit Baranwal, *University of Minnesota, United States*
Ned Mohan, *University of Minnesota, United States*

Session T41: Gate Drivers and Integrated Packaging
Location: 103C
March 24, 2016 14:00 - 17:30
Session Chairs: Qiang Li, *Virginia Tech*
Jean-Luc Schanen, *Ecole Nationale Supérieure de l'Energie*

A High-Efficient Driving Isolated Drive-by-Microwave Half-Bridge Gate Driver for a GaN Inverter 2051
Shuichi Nagai, *Panasonic Corporation, Japan*
Yasufumi Kawai, *Panasonic Corporation, Japan*
Osamu Tabata, *Panasonic Corporation, Japan*
Songbaek Choe, *Panasonic Corporation, Japan*
Noboru Negoro, *Panasonic Corporation, Japan*
Tesuzo Ueda, *Panasonic Corporation, Japan*

Sensing Gallium Nitride HEMT Junction Temperature using Gate Drive Output Transient Properties 2055
He Niu, *University of Wisconsin at Madison, United States*
Robert D. Lorenz, *University of Wisconsin at Madison, United States*

Design and Application of a 1200V Ultra-Fast Integrated Silicon Carbide MOSFET Module ... 2063
Suxuan Guo, *North Carolina State University, United States*
Liqi Zhang, *North Carolina State University, United States*
Yang Lei, *North Carolina State University, United States*
Xuan Li, *North Carolina State University, United States*
Wensong Yu, *North Carolina State University, United States*
Alex Q. Huang, *North Carolina State University, United States*

Active Gate Charge Control Strategy for Series-Connected IGBTs 2071
Fan Zhang, *Xi'an Jiaotong University, China*
Xu Yang, *Xi'an Jiaotong University, China*
Yu Ren, *Xi'an Jiaotong University, China*
Ying Chen, *Xi'an Jiaotong University, China*
Ruifeng Gou, *Xi'an XD Power Systems Co., LTD, China*

A MV Intelligent Gate Driver for 15kV SiC IGBT and 10kV SiC MOSFET 2076
Awneesh Tripathi, *North Carolina State University, United States*
Krishna Mainali, *North Carolina State University, United States*
Sachin Madhusoodhanan, *North Carolina State University, United States*
Akshat Yadav, *North Carolina State University, United States*
Kasunaidu Vechalapu, *North Carolina State University, United States*
Subhashish Bhattacharya, *North Carolina State University, United States*

Linear Temperature Sensors in High-Voltage GaN-HEMT Power Devices 2083
Richard Reiner, *Fraunhofer Institute for Applied Solid State Physics, Germany*
Patrick Waltereit, *Fraunhofer Institute for Applied Solid State Physics, Germany*
Beatrix Weiss, *Fraunhofer Institute for Applied Solid State Physics, Germany*
Matthias Wespel, *Fraunhofer Institute for Applied Solid State Physics, Germany*
Dirk Meder, *Fraunhofer Institute for Applied Solid State Physics, Germany*
Michael Mikulla, *Fraunhofer Institute for Applied Solid State Physics, Germany*
Rüdiger Quay, *Fraunhofer Institute for Applied Solid State Physics, Germany*
Oliver Ambacher, *Fraunhofer Institute for Applied Solid State Physics, Germany*

An Innovative Power Module with Power-System-in-Inductor Structure 2087
Laili Wang, *Sumida Corporation, Canada*
Doug Malcolm, *Sumida Corporation, Canada*
Yan-Fei Liu, *Queen's University, Canada*

Thermal Analysis of a Magnetic Packaged Power Module 2095
Laili Wang, *Sumida Corporation, Canada*
Doug Malcolm, *Sumida Corporation, Canada*
Wenbo Liu, *Queen's University, Canada*
Yan-Fei Liu, *Queen's University, Canada*

Analysis of a Low-Inductance Packaging Layout for Full-SiC Power Module Embedding Split Damping 2102
Yu Ren, *Xi'an Jiaotong University, China*
Xu Yang, *Xi'an Jiaotong University, China*
Fan Zhang, *Xi'an Jiaotong University, China*
Linlin Tan, *Xi'an Jiaotong University, China*
Xiangjun Zeng, *Xi'an Jiaotong University, China*

Session T42: Component Modeling

Location: 103AB

March 24, 2016 14:00 - 17:30

Session Chairs: Sheldon Williamson, *University of Ontario Institute of Technology*
Abhijit Pathak, *Infineon/IR*

Comprehensive Parametric Analyses of Thermally Aged Power MOSFETs for Failure Precursor Identification and Lifetime Estimation based on Gate Threshold Voltage 2108
Serkan Dusmez, *University of Texas at Dallas, United States*
Bilal Akin, *University of Texas at Dallas, United States*

Modeling and Design Guidelines of High Density Power Inductor for Battery Power Unit 2114
Zhigang Dang, *University of Alabama, United States*
Jaber A. Abu Qahouq, *University of Alabama, United States*

Degradation of Low Voltage Metal Oxide Varistors in Power Supplies 2122
Dawood Talebi Khanmiri, *Northeastern University, United States*
Roy Ball, *Mersen USA, United States*
Jerry Mosesian, *Mersen USA, United States*
Brad Lehman, *Northeastern University, United States*

Characterization and Modeling of SiC MOSFET Body Diode 2127
Kang Peng, *University of South Carolina, United States*
Soheila Eskandari, *University of South Carolina, United States*
Enrico Santi, *University of South Carolina, United States*

A Simple Behavioral Electro-Thermal Model of GaN FETs for SPICE Circuit Simulation 2136
Liyao Wu, *Georgia Institute of Technology, United States*
Maryam Saeedifard, *Georgia Institute of Technology, United States*

Decomposition and Electro-Physical Model Creation of the CREE 1200V, 50A 3-Ph SiC Module 2141
Adam J. Morgan, *North Carolina State University, United States*
Yang Xu, *North Carolina State University, United States*
Douglas C. Hopkins, *North Carolina State University, United States*
Iqbal Husain, *North Carolina State University, United States*
Wensong Yu, *North Carolina State University, United States*

A Three-Legged MATLAB/Simulink Transformer Model using a Fictitious Delta Winding 2147
Thomas A. Nondahl, *Rockwell Automation, United States*
Jingbo Liu, *Rockwell Automation, United States*
Peter B. Schmidt, *Rockwell Automation, United States*

A Lifetime Prediction Method for LEDs Considering Mission Profiles 2154
Xiaohui Qu, *Southeast University, China*
Huai Wang, *Aalborg University, Denmark*
Xiaoqing Zhan, *City University of Hong Kong, Hong Kong*
Frede Blaabjerg, *Aalborg University, Denmark*
Henry Shu-Hung Chung, *City University of Hong Kong, Hong Kong*

Enhanced Li-Ion Battery Modeling using Recursive Parameters Correction 2161
Jae-Gu Kim, *Sungkyunkwan University, Korea, South*
Jung-Hoon Ahn, *Sungkyunkwan University, Korea, South*
Byoung-Kuk Lee, *Sungkyunkwan University, Korea, South*

Session T43: Grid and Utility Interface

Location: 104A
March 24, 2016 14:00 - 17:30
Session Chairs: Manish Bhardwaj, *Texas Instruments*
Nan Chen, *ABB*

Robust Sensorless Control of Grid Connected Converters with LCL Line Filters using Frequency Adaptive Observers as AC Voltage Estimators 2167
Vlatko Miskovic, *Danfoss Drives, United States*
Vladimir Blasko, *United Technologies Research Center, United States*
Thomas Jahns, *University of Wisconsin at Madison, United States*
Robert Lorenz, *University of Wisconsin at Madison, United States*
Haojiong Zhang, *Danfoss Drives, United States*

Active Stabilization of Direct Matrix Converter Input Side Filter through Grid Current Control 2175
Martin Leubner, *Technische Universität Dresden, Germany*
Nico Remus, *Technische Universität Dresden, Germany*
Marc Stübig, *Technische Universität Dresden, Germany*
Wilfried Hofmann, *Technische Universität Dresden, Germany*

Impedance-Based Stability Analysis of Single-Phase Inverter Connected to Weak Grid with Voltage Feed-Forward Control 2182
Jiangfeng Wang, *Nanjing University of Aeronautics and Astronautics, China*
Jianhui Yao, *Nanjing University of Aeronautics and Astronautics, China*
Haibing Hu, *Nanjing University of Aeronautics and Astronautics, China*
Yan Xing, *Nanjing University of Aeronautics and Astronautics, China*
Xiaobin He, *Shanghai Institute of Space Power-Sources, China*
Kai Sun, *Tsinghua University, China*

New Configuration of Dynamic Voltage Restorer for Medium Voltage Application 2187
Arash Khoshkbar Sadigh, *Extron Electronics, United States*
Vahid Dargahi, *Clemson University, United States*
Keith Corzine, *Clemson University, United States*

Studies on the Clustered Voltage Balancing Mechanism for Cascaded H-Bridge STATCOM ... 2194
Daorong Lu, *Nanjing University of Aeronautics and Astronautics, China*
Haibing Hu, *Nanjing University of Aeronautics and Astronautics, China*
Yan Xing, *Nanjing University of Aeronautics and Astronautics, China*
Xiaobin He, *Shanghai Institute of Space Power-Sources, China*
Kai Sun, *Tsinghua University, China*
Jianhui Yao, *Nanjing University of Aeronautics and Astronautics, China*

Design of a Fast Response Time Single-Phase PLL with DC Offset Rejection Capability ... 2200
Abhijit Kulkarni, *Indian Institute of Science, India*
Vinod John, *Indian Institute of Science, India*

Four New Applications of Second-Order Generalized Integrator Quadrature Signal Generator 2207
Zhen Xin, *Aalborg University, Denmark*
Rende Zhao, *China University of Petroleum, China*
Xiongfei Wang, *Aalborg University, Denmark*
Poh Chiang Loh, *Aalborg University, Denmark*
Frede Blaabjerg, *Aalborg University, Denmark*

Three-Phase Multiple Harmonic Sequence Detection based on Generalized Delayed Signal Superposition 2215
Yong Lu, *Xi'an Jiaotong University, China*
Guochun Xiao, *Xi'an Jiaotong University, China*
Xiongfei Wang, *Aalborg University, Denmark*
Frede Blaabjerg, *Aalborg University, Denmark*

Hybrid Modelling and Control of Single-Phase Grid-Connected NPC Inverters 2223
Xingda Yan, *University of Southampton, United Kingdom*
Zhan Shu, *University of Southampton, United Kingdom*
Suleiman M. Sharkh, *University of Southampton, United Kingdom*

Session T44: Topics in Renewable Energy Systems II
Location: 104B
March 24, 2016 14:00 - 17:30
Session Chairs: Akshay Kumar Rathore, *Concordia University*
Yichao Tang, *Texas Instruments*

Stability Criterion and Controller Parameter Design of Radial-Line Renewable Systems with Multiple Inverters 2229
Wenchao Cao, *University of Tennessee, United States*
Xuan Zhang, *University of Tennessee, United States*
Yiwei Ma, *University of Tennessee, United States*
Fred Wang, *University of Tennessee, United States*

Stability Analysis and Improvement of Solid State Transformer (SST)-Paralleled Inverters System using Negative Impedance Feedback Control 2237
Qing Ye, *Florida State University, United States*
Hui Li, *Florida State University, United States*

Compensator-Less Structures for Droop Control of Single Phase Inverters in a Flexible Microgrid 2245
Onkar Vitthal Kulkarni, *Indian Institute of Technology Bombay, India*
Suryanarayana Doolla, *Indian Institute of Technology Bombay, India*
B.G. Fernandes, *Indian Institute of Technology Bombay, India*

Comparative Evaluation of the Loss and Thermal Performance of Advanced Three Level Inverter Topologies 2252
Alexander Anthon, *Technical University of Denmark, Denmark*
Zhe Zhang, *Technical University of Denmark, Denmark*
Michael A.E. Andersen, *Technical University of Denmark, Denmark*
Grahame Holmes, *RMIT University, Australia*
Brendan McGrath, *RMIT University, Australia*
Carlos Teixeira, *RMIT University, Australia*

Dual Buck Inverter with Series Connected Diodes and Single Inductor 2259
Liwei Zhou, *Shandong University, China*
Feng Gao, *Shandong University, China*

Magnetic Integration of the Harmonic Filter Inductor for Dual-Converter Fed Open-End Transformer Topology 2264
Ghanshyamsinh Gohil, *Aalborg University, Denmark*
Lorand Bede, *Aalborg University, Denmark*
Remus Teodorescu, *Aalborg University, Denmark*
Tamas Kerekes, *Aalborg University, Denmark*
Frede Blaabjerg, *Aalborg University, Denmark*

Mechanism Analysis and Mitigation of Instability in Grid-Connected Voltage Source Inverter with LCL Filters based on Terminal Impedance 2272
Teng Liu, *Xi'an Jiaotong University, China*
Zeng Liu, *Xi'an Jiaotong University, China*
Jinjun Liu, *Xi'an Jiaotong University, China*
Qingyun Dou, *Xi'an Jiaotong University, China*

Seven-Switch Five-Level Active Neutral-Point Clamped Converter and Optimal Modulation Strategy 2278
Hongliang Wang, *Queen's University, Canada*
Lei Kou, *Queen's University, Canada*
Yan-Fei Liu, *Queen's University, Canada*
Paresh C. Sen, *Queen's University, Canada*
Sucheng Liu, *Anhui University of Technology, China*

A Simple Variable Step Size Method for Maximum Power Point Tracking using Commercial Current Mode Control DC-DC Regulators 2286
Su Sheng, *Northeastern University, United States*
Brad Lehman, *Northeastern University, United States*

Session T45: Envelope Tracking and Resonant Conversion

Location: 104C
March 24, 2016 14:00 - 17:30
Session Chairs: Brian Zahnstecher, *PowerRox*
Davide Giacomini, *Infineon*

Envelope Tracking GaN Power Supply for 4G Cell Phone Base Stations 2292
Yuanzhe Zhang, *University of Colorado at Boulder, United States*
Johan Strydom, *Efficient Power Conversion Corporation, United States*
Michael de Rooij, *Efficient Power Conversion Corporation, United States*
Dragan Maksimović, *University of Colorado at Boulder, United States*

Envelope Tracking Power Supply for Volume-Sensitive Low-Power Applications based on a Resonant Switched-Capacitor Converter 2298
Alon Cervera, *Ben-Gurion University of the Negev, Israel*
Mor Mordechai Peretz, *Ben-Gurion University of the Negev, Israel*

A Passive-Impedance-Matching Concept for Multi-Phase Resonant Converter 2304
Hongliang Wang, *Queen's University, Canada*
Yang Chen, *Queen's University, Canada*
Yan-Fei Liu, *Queen's University, Canada*

LLC Converter with Auxiliary Switch for Hold Up Mode Operation 2312
Yang Chen, *Queen's University, Canada*
Hongliang Wang, *Queen's University, Canada*
Yan-Fei Liu, *Queen's University, Canada*
Jahangir Afsharian, *Murata Power Solutions, Canada*
Zhihua Yang, *Queen's University, Canada*

A Common Capacitor Multi-Phase LLC Resonant Converter 2320
Hongliang Wang, *Queen's University, Canada*
Yang Chen, *Queen's University, Canada*
Zhiyuan Hu, *Queen's University, Canada*
Laili Wang, *Queen's University, Canada*
Yajie Qiu, *Queen's University, Canada*
Wenbo Liu, *Queen's University, Canada*
Yan-Fei Liu, *Queen's University, Canada*
Jahangir Afsharian, *Murata Power Solutions, Canada*
Zhihua Yang, *Murata Power Solutions, Canada*

LLC Resonant Converter Design for Bendable Power Converter 2328
Kwun Yuan Godwin Ho, *University of Hong Kong, Hong Kong*
M.H. Bryan Pong, *University of Hong Kong, Hong Kong*
Shu-Yuen Ron Hui, *University of Hong Kong, Hong Kong*

Design Consideration of MHz Active Clamp Flyback Converter with GaN Devices for Low Power Adapter Application 2334
Xiucheng Huang, *Virginia Polytechnic Institute and State University, United States*
Junjie Feng, *Virginia Polytechnic Institute and State University, United States*
Weijing Du, *Virginia Polytechnic Institute and State University, United States*
Fred C. Lee, *Virginia Polytechnic Institute and State University, United States*
Qiang Li, *Virginia Polytechnic Institute and State University, United States*

A New Capacitor Voltage Balancing Control for Hybrid Modular Multilevel Converter with Cascaded Full Bridge 2342
Mahendra B. Ghat, *Indian Institute of Technology Bombay, India*
Anshuman Shukla, *Indian Institute of Technology Bombay, India*
Richa Mishra, *Indian Institute of Technology Bombay, India*

Sensorless Scheduling of the Modular Multilevel Series-Parallel Converter: Enabling a Flexible, Efficient, Modular Battery 2349
Stefan M. Goetz, *Duke University, United States*
Zhongxi Li, *Duke University, United States*
Angel V. Peterchev, *Duke University, United States*
Xinyu Liang, *North Carolina State University, United States*
Chengduo Zhang, *North Carolina State University, United States*
Srdjan M. Lukic, *North Carolina State University, United States*

Session D01: AC-DC Converters
Location: Poster Area
March 24, 2016 11:30 - 14:00
Session Chairs: Nathan Weise, *Marquette*
Daniel Costinett, *University of Tennessee-Knoxville*

An Input Current Calculation Switching Driver for High Power-Factor and Phase-Cut Dimmer Compatibility 2355
Hyunchul Eum, *Fairchild Semiconductor International, Inc., Korea, South*
Youngjong Kim, *Fairchild Semiconductor International, Inc., Korea, South*
Kuohsien Huang, *Fairchild Semiconductor International, Inc., Taiwan*

High Frequency Range Conducted Common-Mode Noise Suppression in SMPS 2360
Jinping Zhou, *Delta Electronics Shanghai Co., Ltd., China*
Yicong Xie, *Delta Electronics Shanghai Co., Ltd., China*
Min Zhou, *Delta Electronics Shanghai Co., Ltd., China*

Improved Medium Voltage AC-DC Rectifier based on 10kV SiC MOSFET for Solid State Transformer (SST) Application 2365
Qianlai Zhu, *North Carolina State University, United States*
Li Wang, *North Carolina State University, United States*
Liqi Zhang, *North Carolina State University, United States*
Wensong Yu, *North Carolina State University, United States*
Alex Q. Huang, *North Carolina State University, United States*

Suppression of Circulating Current in Parallel Operation of Three-Level Converters 2370
Young-Kwang Son, *Seoul National University, Korea, South*
Seung-Jun Chee, *Seoul National University, Korea, South*
Younggi Lee, *Seoul National University, Korea, South*
Seung-Ki Sul, *Seoul National University, Korea, South*
Changjin Lim, *LG Electronics, Korea, South*
Sungjae Huh, *LG Electronics, Korea, South*
Jaeyoon Oh, *LG Electronics, Korea, South*

Hybrid Bridgeless DCM SEPIC Rectifier Integrated with a Modified Switched Capacitor Cell ... 2376
Paulo Junior Silva Costa, *Universidade Federal de Santa Catarina, Brazil*
Telles Brunelli Lazzarin, *Universidade Federal de Santa Catarina, Brazil*
Carlos Henrique Illa Font, *Universidade Tecnológica Federal do Paraná, Brazil*

LCL Filter Design for Three-Phase Two-Level Power Factor Correction using Line Impedance Stabilization Network 2382
Alireza Kouchaki, *University of Southern Denmark, Denmark*
Morten Nymand, *University of Southern Denmark, Denmark*

Sensorless Current Rebuilding Strategy in a Single Phase Bridgeless PFC 2389
Felipe López, *Universidad de Cantabria, Spain*
Paula Lamo, *Universidad de Cantabria, Spain*
Alberto Pigazo, *Universidad de Cantabria, Spain*
F.J. Azcondo, *Universidad de Cantabria, Spain*

A Compact Electrolytic-Free Two-Stage Universal Input Offline LED Driver 2395
Saad Pervaiz, *University of Colorado at Boulder, United States*
Ashish Kumar, *University of Colorado at Boulder, United States*
Khurram K. Afridi, *University of Colorado at Boulder, United States*

Session D02: DC-DC Converters I

Location: Poster Area
March 24, 2016 11:30 - 14:00
Session Chairs: Charles Sullivan, *Dartmouth*
Mahshid Amirabadi, *Northeastern University*

Design Methodology for a High Insulation Voltage Power Transmission Function for IGBT Gate Driver 2401
Sokchea Am, *Grenoble Institute of Technology, France*
Pierre Lefranc, *Grenoble Institute of Technology, France*
David Frey, *Grenoble Institute of Technology, France*
Mahmoud Ibrahim, *Grenoble Institute of Technology, France*

Optimized Design of GaN Switching Capacitor based Envelope Tracking Power Supply for Satellite Applications 2409
Qian Jin, *Nanjing University of Aeronautics and Astronautics, China*
M. Vasić, *Universidad Politécnica de Madrid, Spain*
O. Garcia, *Universidad Politécnica de Madrid, Spain*
P. Alou, *Universidad Politécnica de Madrid, Spain*
J.A. Oliver, *Universidad Politécnica de Madrid, Spain*
J.A. Cobos, *Universidad Politécnica de Madrid, Spain*

An Isolated High Step-Up Converter with Continuous Input Current and LC Snubber 2415
K.I. Hwu, *National Taipei University of Technology, Taiwan*
W.Z. Jiang, *National Taipei University of Technology, Taiwan*
Y.T. Yau, *National Taipei University of Technology, Taiwan*

Output-Inductor-Less Full-Bridge Converter with SiC-MOSFETs for Low Noise and ZVS Operation 2422
Kazuhide Domoto, *Nagasaki University, Japan*
Yoichi Ishizuka, *Nagasaki University, Japan*
Seiya Abe, *Kyushu Institute of Technology, Japan*
Tamotsu Ninomiya, *Green Electronics Research Institute, Kitakyushu, Japan*

Reduction Technique of Leakage Flux Effects on GaN-HEMTs in 5 MHz / 100 W Isolated DC-DC Converters 2430
Akinori Hariya, *Nagasaki University, Japan*
Tomoya Koga, *Nagasaki University, Japan*
Ken Matsuura, *TDK Corporation, Japan*
Hiroshige Yanagi, *TDK-Lambda Corporation, Japan*
Satoshi Tomioka, *TDK-Lambda Corporation, Japan*
Yoichi Ishizuka, *Nagasaki University, Japan*
Tamotsu Ninomiya, *City of Kitakyushu, Japan*

A High-Voltage Level Shifter with Sub-Nano-Second Propagation Delay for Switching Power Converters 2437
Ahmed Abdelmoaty, *Ohio State University, United States*
Mohammad Al-Shyoukh, *TSMC Inc., United States*
Ayman Fayed, *Ohio State University, United States*

Dual-Output, Three-Level GaN-Based DC-DC Converter for Battery Charger Applications 2441
Ren Ren, *Nanjing University of Aeronautics and Astronautics, China*
Bo Liu, *University of Tennessee, United States*
Edward A. Jones, *University of Tennessee, United States*
Fred Wang, *University of Tennessee, United States*
Zheyu Zhang, *University of Tennessee, United States*
Daniel Costinett, *University of Tennessee, United States*

Quadruple Active Bridge DC-DC Converter as the Basic Cell of a Modular Smart Transformer 2449
Levy F. Costa, *Christian-Albrechts-Universität zu Kiel, Germany*
Giampaolo Buticchi, *Christian-Albrechts-Universität zu Kiel, Germany*
Marco Liserre, *Christian-Albrechts-Universität zu Kiel, Germany*

Analytical Model of a Phase-Shift Controlled Three-Level Zero-Voltage Switching Converter 2457
Cas Bakker, *Prodrive Technologies, Netherlands*
Bas Vermulst, *Technische Universiteit Eindhoven, Netherlands*
Anton Driessen, *Prodrive Technologies, Netherlands*

High Efficiency Design for ISOP Converter System with Dual Active Bridge DC-DC Converter 2465
Masaki Sato, *Nagasaki University, Japan*
Kazuhide Domoto, *Nagasaki University, Japan*
Yoichi Ishizuka, *Nagasaki University, Japan*
Masahiro Yamaguchi, *Tohoku University, Japan*
Shinya Manabe, *RICOH Electronic Devices Co., Ltd., Japan*
Hiizu Okubo, *RICOH Electronic Devices Co., Ltd., Japan*
Atsushi Itagaki, *Ryowa Electronics Co., Ltd., Japan*

Wide Input Range Power Converters using a Variable Turns Ratio Transformer 2473
Ziwei Ouyang, *Technical University of Denmark, Denmark*
Michael A.E. Andersen, *Technical University of Denmark, Denmark*

Design Approaches for Fast Supercapacitor Chargers for Applications like SCATMA, SRUPS 2479
Nicoloy Gurusinghe, *University of Waikato, New Zealand*
Nihal Kularatna, *University of Waikato, New Zealand*
W. Howell Round, *University of Waikato, New Zealand*
D. Alistair Steyn-Ross, *University of Waikato, New Zealand*

Stack Multiphase Asymmetrical Half-Bridge Topology Offering Advance Performance and Efficiency 2485
Trong Tue Vu, *Eisergy Ltd., Ireland*
George Young, *Eisergy Ltd., Ireland*

Session D03: DC-DC Converters II
Location: Poster Area
March 24, 2016 11:30 - 14:00
Session Chairs: Jason Stauth, *Dartmouth*
Yan-Fei Liu, *Queens*

Design of a Novel APWM Half-Bridge DC-DC Resonant Converter with Load-Independent Soft-Switching and Reduced Circulating Current 2491
Kawsar Ali, *National University of Singapore, Singapore*
Sandeep Kolluri, *National University of Singapore, Singapore*
Naga Brahmendra Yadav Gorla, *National University of Singapore, Singapore*
Pritam Das, *National University of Singapore, Singapore*
Sanjib Kumar Panda, *National University of Singapore, Singapore*

A Low-Volume Hybrid Step-Down DC-DC Converter based on the Dual use of Flying Capacitor 2497
S.M. Ahsanuzzaman, *University of Toronto, Canada*
Yingxian Ma, *University of Toronto, Canada*
Abrar Ahmed Pathan, *University of Toronto, Canada*
Aleksandar Prodić, *University of Toronto, Canada*

Fractional Pulse Skipping in Digitally Controlled DC-DC Converters for Improved Light-Load Efficiency and Power Spectrum 2504
Bipin Chandra Mandi, *Indian Institute of Technology Kharagpur, India*
Santanu Kapat, *Indian Institute of Technology Kharagpur, India*
Amit Patra, *Indian Institute of Technology Kharagpur, India*

A New Compact and High Efficiency Resonant Converter 2511
Sheng-Yang Yu, *Texas Instruments Inc., United States*

A 10-MHz eGaN FETs based Isolated Class-Φ_2 DCX 2518
Xuewen Zou, *Nanjing University of Aeronautics and Astronautics, China*
Zhiliang Zhang, *Nanjing University of Aeronautics and Astronautics, China*
Zhou Dong, *Nanjing University of Aeronautics and Astronautics, China*
Yuan Zhou, *Nanjing University of Aeronautics and Astronautics, China*
Xiaoyong Ren, *Nanjing University of Aeronautics and Astronautics, China*
Qianhong Chen, *Nanjing University of Aeronautics and Astronautics, China*

Multi-Level Capacitor Clamped DC-DC Multiplier/Divider with Variable and Fractional Voltage Gain – An (n/m)X DC-DC Converter 2525
Deepak Gunasekaran, *Michigan State University, United States*
Liang Qin, *Wuhan University, China*
Ujjwal Karki, *Michigan State University, United States*
Yuan Li, *Sichuan University, China*
Fang Z. Peng, *Michigan State University, United States*

Multi-Mode Quasi-Z-Source Series Resonant DC/DC Converter for Wide Input Voltage Range Applications 2533
Dmitri Vinnikov, *Ubik Solutions LLC, Estonia*
Andrii Chub, *Tallinn University of Technology, Estonia*
Indrek Roasto, *Ubik Solutions LLC, Estonia*
Liisa Liivik, *Tallinn University of Technology, Estonia*

Hybrid Serial-Output Converter for Integrated LED Lighting Applications 2540
T. McRae, *University of Toronto, Canada*
A. Prodić, *University of Toronto, Canada*
G. Lisi, *Texas Instruments Inc., United States*
W. McIntrye, *Texas Instruments Inc., United States*
A. Aguilar, *Texas Instruments Inc., United States*

Analysis and Modeling of a Modular ISOP Full Bridge based Converter with Input Filter ... 2545
P. Zumel, *Universidad Carlos III de Madrid, Spain*
E. Oña, *Universidad Carlos III de Madrid, Spain*
C. Fernandez, *Universidad Carlos III de Madrid, Spain*
M. Sanz, *Universidad Carlos III de Madrid, Spain*
A. Lazaro, *Universidad Carlos III de Madrid, Spain*
A. Barrado, *Universidad Carlos III de Madrid, Spain*
A. Vazquez, *Universidad de Oviedo, Spain*
D.G. Lamar, *Universidad de Oviedo, Spain*

Wide-Input High Power Density Flexible Converter Topology for DC-DC Applications 2553
Parth Jain, *University of Toronto, Canada*
Aleksandar Prodić, *University of Toronto, Canada*
Alexander Gerfer, *Würth Elektronik eiSos GmbH & Co. KG, Germany*

High Efficiency LLC Converter Design for Universal Battery Chargers 2561
Navid Shafiei, *University of British Columbia, Canada*
Ali Arefifar, *University of British Columbia, Canada*
Mohammad Ali Saket, *University of British Columbia, Canada*
Martin Ordonez, *University of British Columbia, Canada*

A New High Power Density Modular Multilevel DC-DC Converter with Localized Voltage Balancing Control for Arbitrary Number of Levels 2567
Ahmed Morsy, *Texas A&M University, United States*
Yong Zhou, *Texas A&M University, United States*
Prasad Enjeti, *Texas A&M University, United States*

Design and Control of a Fault Tolerant Soft Switching DC-DC Converter for High Power High Voltage Applications 2573
Tao Li, *Rensselaer Polytechnic Institute, United States*
Leila Parsa, *Rensselaer Polytechnic Institute, United States*

Accurate Parametric Steady State Analysis and Design Tool for DC-DC Power Converters 2579
Mohammad Daryaei, *University of Alberta, Canada*
Mohammad Ebrahimi, *University of Alberta, Canada*
S. Ali Khajehoddin, *University of Alberta, Canada*

Analysis of Multi-Output Half-Wave Semi-Synchronous Rectifier with a Uniform Magnetic Field Transmitter 2587
Erdem Asa, *Hevo Power Inc. / New York University, United States*
Kerim Colak, *Istanbul Ulasim A.S., Turkey*
Dariusz Czarkowski, *New York University, United States*

High Gain QZS DC/DC Converter with Coupled Inductor 2592
Rafael V. Silva, *Universidade Federal do Ceará, Brazil*
Antônio A.A. Freitas, *Universidade Federal do Ceará, Brazil*
Marcus R. Castro, *Universidade Federal do Ceará, Brazil*
Fernando L.M. Antunes, *Universidade Federal Rural do Semi-Árido, Brazil*
Edilson M. Sá Jr., *Universidade Federal do Ceará, Brazil*

Session D04: Utility Interface
Location: Poster Area
March 24, 2016 11:30 - 14:00
Session Chairs: Ali Khajehoddin, *University of Alberta*
Julia Zhang, *Oregon State University*

A Power Decoupling Method with Small Capacitance Requirement based on Single-Phase Quasi-Z-Source Inverter for DC Microgrid Applications 2599
Dingyi He, *University of Texas at Dallas, United States*
Wen Cai, *University of Texas at Dallas, United States*
Fan Yi, *University of Texas at Dallas, United States*

Operation Analysis of High Efficiency Grid Connected Bi-Directional Power Conversion System for Various Storage Battery Systems with Bi-Directional Switch Circuit Topology 2607
Go Yamada, *Panasonic Corporation, Japan*
Takaaki Norisada, *Panasonic Corporation, Japan*
Fumito Kusama, *Panasonic Corporation, Japan*
Keiji Akamatsu, *Panasonic Corporation, Japan*
Masakazu Michihira, *Kobe City College of Technology, Japan*

Fault Tolerant Control of MMC with Redundant Sub-Modules based on Carrier Phase Shift Modulation 2613
Kai Li, *Tsinghua University, China*
Zhengming Zhao, *Tsinghua University, China*
Liqiang Yuan, *Tsinghua University, China*
Sizhao Lu, *Tsinghua University, China*
Bing Pan, *State Grid Smart Grid Research Institute, China*
Zhengang Lu, *State Grid Smart Grid Research Institute, China*

A New Topology of Multilevel VSC Converter for Hybrid HVDC Transmission System 2620
Jae-Jung Jung, *Seoul National University, Korea, South*
Shenghui Cui, *RWTH Aachen University, Germany*
Seung-Ki Sul, *Seoul National University, Korea, South*

Performance of Solid State Transformers Under Imbalanced Loads in Distribution Systems 2629
Tao Yang, *University College Dublin, Ireland*
Ronan Meere, *University College Dublin, Ireland*
Cathal O'Loughlin, *University College Dublin, Ireland*
Terence O'Donnell, *University College Dublin, Ireland*

Steady-State Analysis of Modular Multilevel Converter (MMC) Under Unbalanced Grid Conditions 2637
Xiaojie Shi, *University of Tennessee, United States*
Yalong Li, *University of Tennessee, United States*
Zhiqiang Wang, *University of Tennessee, United States*
Bo Liu, *University of Tennessee, United States*
Leon M. Tolbert, *University of Tennessee, United States*
Fred Wang, *University of Tennessee, United States*

Design and Control of a Compensated Submodule Testing Scheme for Modular Multilevel Converter 2645
Yuan Tang, *University of Warwick, United Kingdom*
Li Ran, *University of Warwick, United Kingdom*
Olayiwola Alatise, *University of Warwick, United Kingdom*
Philip Mawby, *University of Warwick, United Kingdom*

A Voltage Independent Islanding Detection Method and Low Voltage Ride through of a Two-Stage PV Inverter 2652
Partha Pratim Das, *Indian Institute of Technology Kharagpur, India*
Souvik Chattopadhyay, *Indian Institute of Technology Kharagpur, India*
Shiladri Chakraborty, *Indian Institute of Technology Kharagpur, India*

Low Cost and High Efficiency Topology for Flexible Integration of Multi-PV and Batteries in Resonant-Based Converters 2660
Ali Elrayyah, *Qatar Environment and Energy Research Institute, Qatar*

Real-Time Integrated Model of a Micro-Grid with Distributed Clean Energy Generators and their Power Electronics 2666
Weiqiang Chen, *University of Connecticut, United States*
Ali M. Bazzi, *University of Connecticut, United States*
James Hare, *University of Connecticut, United States*
Shalabh Gupta, *University of Connecticut, United States*

Minimization of Inter-Module Leakage Current in Cascaded H-Bridge Multilevel Inverters for Grid Connected Solar PV Applications 2673
V.V.S. Pradeep Kumar, *Indian Institute of Technology Bombay, India*
B.G. Fernandes, *Indian Institute of Technology Bombay, India*

Effect of Grid Inductance on Grid Current Quality of Parallel Grid-Connected Inverter System with Output LCL Filter and Closed-Loop Control 2679
Wooyoung Choi, *University of Wisconsin at Madison, United States*
Woongkul Lee, *University of Wisconsin at Madison, United States*
Bulent Sarlioglu, *University of Wisconsin at Madison, United States*

Small Signal Modeling and Control of a Grid Tied Converter without a Syncronization Unit 2687
Subhajyoti Mukherjee, *Missouri University of Science and Technology, United States*
Pourya Shamsi, *Missouri University of Science and Technology, United States*
Mehdi Ferdowsi, *Missouri University of Science and Technology, United States*

Bridgeless SEPIC PFC Converter for Low Total Harmonic Distortion and High Power Factor ... 2693
Yasemin Onal, *Bilecik Seyh Edebali University, Turkey*
Yilmaz Sozer, *University of Akron, United States*

Effectiveness of Pareto-Front Analysis Applied to the Design of a Single-Phase PFC Rectifier 2700
Mahmoud Ibrahim, *Eaton Corporation, France*
Luc Gonnet, *Eaton Corporation, France*
Pierre Lefranc, *Grenoble Institute of Technology, France*
David Frey, *Grenoble Institute of Technology, France*
Jean-Paul Ferrieux, *Grenoble Institute of Technology, France*
Sokchea Am, *Grenoble Institute of Technology, France*

State Space Analysis and Duty Cycle Control of a Switched Reactance based Center-Point-Clamped Reactive Power Compensator 2706
Pankaj Kumar Bhowmik, *University of North Carolina at Charlotte, United States*
Somasundaram Essakiappan, *University of North Carolina at Charlotte, United States*
Madhav Manjrekar, *University of North Carolina at Charlotte, United States*

A SiC-Based Power Converter Module for Medium-Voltage Fast Charger for Plug-In Electric Vehicles 2714
Srdjan Srdic, *North Carolina State University, United States*
Chi Zhang, *North Carolina State University, United States*
Xinyu Liang, *North Carolina State University, United States*
Wensong Yu, *North Carolina State University, United States*
Srdjan Lukic, *North Carolina State University, United States*

Shunt Active Power Filter based on Cascaded Transformers Coupled with Three-Phase Bridge Converters 2720
Gregory A. de Almeida Carlos, *Universidade Federal de Campina Grande, Brazil*
Cursino B. Jacobina, *Universidade Federal de Campina Grande, Brazil*
João Paulo R. Méllo, *Universidade Federal de Campina Grande, Brazil*
Euzeli C. dos Santos Jr., *Indiana University - Purdue University, United States*

Independent DC Link Voltage Control of Cascaded Multilevel PV Inverter 2727
Qingyun Huang, *North Carolina State University, United States*
Wensong Yu, *North Carolina State University, United States*
Alex Q. Huang, *North Carolina State University, United States*

New Active Damping Method for LCL Filter Resonance based on Two Feedback System 2735
Mahmoud A. Gaafar, *Kyushu University, Japan*
Gamal M. Dousoky, *Minia University, Egypt*
Masahito Shoyama, *Kyushu University, Japan*

Static Synchronous Generator Model for Investigating Dynamic Behaviors and Stability Issues of Grid-Tied Inverters 2742
Liansong Xiong, *Xi'an Jiaotong University, China*
Xiaokang Liu, *Xi'an Jiaotong University, China*
Feng Wang, *Xi'an Jiaotong University, China*
Fang Zhuo, *Xi'an Jiaotong University, China*

Session D05: Motor Drives and Inverters: Modeling and Control I

Location: Poster Area
March 24, 2016 11:30 - 14:00
Session Chairs: Liming Liu, *ABB Inc.*
Thomas Gietzold, *United Technologies Aerospace Systems*

Initial Orientation and Sensorless Starting Strategy of Wound-Rotor Synchronous Starter/Generator 2748
Jichang Peng, *Northwestern Polytechnical University, China*
Weiguo Liu, *Northwestern Polytechnical University, China*
Jinhao Meng, *Northwestern Polytechnical University, China*
Tao Meng, *Northwestern Polytechnical University, China*
Guangzhao Luo, *Northwestern Polytechnical University, China*

A Novel Method for Polarity Detection of Non-Salient PMSMs in Initial Position Estimation 2754
Bing Liu, *Nanjing University of Aeronautics and Astronautics, China*
Bo Zhou, *Nanjing University of Aeronautics and Astronautics, China*
Jiadan Wei, *Nanjing University of Aeronautics and Astronautics, China*
Long Wang, *Nanjing University of Aeronautics and Astronautics, China*
Tianheng Ni, *Nanjing University of Aeronautics and Astronautics, China*

A Speed Adaptive Sensorless Flux Observer for the Induction Motor Drive using Sylvester Criterion Design 2759
Mihai Comanescu, *Penn State Altoona, United States*

Discontinuous PWM for Low Switching Losses in Indirect Matrix Converter Drives 2764
Yeongsu Bak, *Ajou University, Korea, South*
Kyo-Beum Lee, *Ajou University, Korea, South*

Model Predictive Control for Extended Kalman Filter based Speed Sensorless Induction Motor Drives 2770
Jie Li, *Xi'an University of Technology, China*
Li-Heng Zhang, *Xi'an University of Technology, China*
Ying Niu, *Xi'an University of Technology, China*
Hai-Peng Ren, *Xi'an University of Technology, China*

Research on Excitation Control Methods for the Two-Phase Brushless Exciter of Wound-Rotor Synchronous Starter/Generators in the Starting Mode 2776
Ningfei Jiao, *Northwestern Polytechnical University, China*
Weiguo Liu, *Northwestern Polytechnical University, China*
Tao Meng, *Northwestern Polytechnical University, China*
Jichang Peng, *Northwestern Polytechnical University, China*
Shuai Mao, *Northwestern Polytechnical University, China*

A High Performance Speed Regulator Design for AC Machines 2782
Adil Khurram, *American University of Sharjah, U.A.E.*
Habibur Rehman, *American University of Sharjah, U.A.E.*
Shayok Mukhopadhyay, *American University of Sharjah, U.A.E.*

Zero-Sequence Current Suppression for Open-End Winding Induction Motor Drive with Resonant Controller 2788
Hajime Kubo, *Meidensha Corporation, Japan*
Yasuhiro Yamamoto, *Meidensha Corporation, Japan*
Takeshi Kondo, *Meidensha Corporation, Japan*
Kaushik Rajashekara, *University of Texas at Dallas, United States*
Bohang Zhu, *University of Texas at Dallas, United States*

Optimized Control of High-Performance Servo-Motor Drives in the Field-Weakening Region 2794
Jack Bermingham, *Moog Ireland Ltd, Ireland*
Gerard O'Donovan, *Moog Ireland Ltd, Ireland*
Ray Walsh, *Moog Ireland Ltd, Ireland*
Michael Egan, *University College Cork, Ireland*
Gordon Lightbody, *University College Cork, Ireland*
John G. Hayes, *University College Cork, Ireland*

Motor Current Reference Generation for Reducing Motor Currents in Drive Systems with Single-Phase Diode Rectifier and Small DC-Link Capacitor 2801
Young-Ho Chae, *Seoul National University, Korea, South*
Jung-Ik Ha, *Seoul National University, Korea, South*

A Simple Double Mapping based SVPWM Method for Balancing DC-Link Capacitor Voltages of Five-Level Diode-Clamped Converters 2806
Aparna Saha, *University of Akron, United States*
Ali Elrayyah, *Qatar Environment and Energy Research Institute, Qatar*
Yilmaz Sozer, *University of Akron, United States*

Session D06: Motor Drives and Inverters: Modeling and Control II
Location: Poster Area
March 24, 2016 11:30 - 14:00
Session Chairs: Bulent Sarlioglu, *University of Wisconsin - Madison*
Yichao Tang, *Texas Instruments*

Capacitor-Clamped Inverter based Transient Suppression Method for Azimuth Thruster Drives 2813
Shantha Gamini Jayasinghe, *Australian Maritime College, University of Tasmania, Australia*
Viknash Shagar, *Australian Maritime College, University of Tasmania, Australia*
Hossein Enshaei, *Australian Maritime College, University of Tasmania, Australia*
Danyal Mohammadi, *Boise State University, United States*
Mahinda Vilathgamuwa, *Queensland University of Technology, Australia*

Active Common-Mode Voltage Reduction in a Fault-Tolerant Three-Phase Inverter 2821
Danyal Mohammadi, *Boise State University, United States*
Said Ahmed-Zaid, *Boise State University, United States*

Power Cycling Lifetime Improvement of Three-Level NPC Inverters with an Improved DPWM Method 2826
Jiangbiao He, *Marquette University, United States*
Lixiang Wei, *Rockwell Automation, United States*
Nabeel A.O. Demerdash, *Marquette University, United States*

Synchronous Optimal Pulsewidth Modulation Digital Implementation Concept for Multilevel Converters 2833
Jackson Lago, *Universidade Federal de Santa Catarina, Brazil*
Marcelo Lobo Heldwein, *Universidade Federal de Santa Catarina, Brazil*

Analytical Determination of Conduction Losses for Modified Flying Capacitor Multicell Converters 2840
Vahid Dargahi, *Clemson University, United States*
Arash Khoshkbar Sadigh, *Extron Electronics, United States*
Keith Corzine, *Clemson University, United States*

Comparison of Electrical Losses in an Inverter-Fed Five-Phase and Three-Phase Permanent Magnet Assisted Synchronous Reluctance Motor 2847
Akm Arafat, *University of Akron, United States*
Seungdeog Choi, *University of Akron, United States*

A Hybrid Adaptive Observer for the Speed and Flux Estimation of Induction Motors 2855
Mihai Comanescu, *Penn State Altoona, United States*

Determination of CM Choke Parameters for SiC MOSFET Motor Drive based on Simple Measurements and Frequency Domain Modeling 2861
Di Han, *University of Wisconsin at Madison, United States*
Casey Morris, *University of Wisconsin at Madison, United States*
Woongkul Lee, *University of Wisconsin at Madison, United States*
Bulent Sarlioglu, *University of Wisconsin at Madison, United States*

An Improved Model Predictive Current Control of Permanent Magnet Synchronous Motor Drives 2868
Yongchang Zhang, *North China University of Technology, China*
Sugu Gao, *North China University of Technology, China*
Wei Xu, *Huazhong University of Science and Technology, China*

Analysis of Magnet Defect Faults in Permanent Magnet Synchronous Motors through Fluxgate Sensors 2875
Taner Goktas, *University of Texas at Dallas, United States*
Kun Wang Lee, *University of Texas at Dallas, United States*
Mohsen Zafarani, *University of Texas at Dallas, United States*
Bilal Akin, *University of Texas at Dallas, United States*

Session D07: Motor Drives and Inverters: Topologies
Location: Poster Area
March 24, 2016 11:30 - 14:00
Session Chairs: Amirnaser Yazdani, *Ryerson University*
Babak Nahid-Mobarakeh, *University of Lorraine*

Performance Comparison of Transfer Switch Topologies in Switched-Doubly-Fed Machine Drives 2881
Arijit Banerjee, *Massachusetts Institute of Technology, United States*
Steven B. Leeb, *Massachusetts Institute of Technology, United States*
James L. Kirtley, *Massachusetts Institute of Technology, United States*

Multilevel Converter Topologies for High-Power High-Speed Switched Reluctance Motor: Performance Comparison 2889
Devendra Patil, *University of Texas at Dallas, United States*
Shiliang Wang, *University of Texas at Dallas, United States*
Lei Gu, *University of Texas at Dallas, United States*

Bidirectional Magnetically Coupled T-Source Inverter for Extra Low Voltage Application 2897
Thomas Baier, *Friedrich-Alexander-Universität Erlangen-Nürnberg, Germany*
Bernhard Piepenbreier, *Friedrich-Alexander-Universität Erlangen-Nürnberg, Germany*

Active Virtual Ground: Single Phase Grid-Connected Voltage Source Inverter Topology ... 2905
River Tin-Ho Li, *ABB China Ltd., China*
Carl Ngai-Man Ho, *University of Manitoba, Canada*

Design and Evaluation of 30kVA Inverter using SiC MOSFET for 180°C Ambient Temperature Operation 2912
Feng Qi, *Ohio State University, United States*
Miao Wang, *Ohio State University, United States*
Longya Xu, *Ohio State University, United States*
Bo Zhao, *State Grid Corporation of China, China*
Zhe Zhou, *State Grid Corporation of China, China*
Xizhou Ren, *State Grid Corporation of China, China*

A DC to Three-Phase Boost-Buck Inverter with Stored Energy Modulation and a Tiny DC Link Capacitor 2919
Mahima Gupta, *University of Wisconsin at Madison, United States*
Giri Venkataramanan, *University of Wisconsin at Madison, United States*

Drive Circuits for Ultra-Fast and Reliable Actuation of Thomson Coil Actuators used in Hybrid AC and DC Circuit Breakers 2927
Chang Peng, *North Carolina State University, United States*
Alex Huang, *North Carolina State University, United States*
Iqbal Husain, *North Carolina State University, United States*
Bruno Lequesne, *E-Motors Consulting, LLC, United States*
Roger Briggs, *Energy Efficiency Research, LLC, United States*

Improved Transformerless Dual Buck Inverters with Buffer Inductors 2935
Liwei Zhou, *Shandong University, China*
Feng Gao, *Shandong University, China*

A 99% Efficiency SiC Three-Phase Inverter using Synchronous Rectification 2942
Shan Yin, *Nanyang Technological University, Singapore*
K.J. Tseng, *Nanyang Technological University, Singapore*
C.F. Tong, *Nanyang Technological University, Singapore*
Rejeki Simanjorang, *Rolls-Royce Singapore Pte. Ltd., Singapore*
C.J. Gajanayake, *Rolls-Royce Singapore Pte. Ltd., Singapore*
Amit K. Gupta, *Rolls-Royce Singapore Pte. Ltd., Singapore*

Comparison and Evaluation of Common Mode EMI Filter Topologies for GaN-Based Motor Drive Systems 2950
Casey T. Morris, *University of Wisconsin at Madison, United States*
Di Han, *University of Wisconsin at Madison, United States*
Bulent Sarlioglu, *University of Wisconsin at Madison, United States*

Analysis of Thermal Cycling Stress on Semiconductor Devices of the Modular Multilevel Converter for Drive Applications 2957
Xiangyu Han, *Georgia Institute of Technology, United States*
Qichen Yang, *Georgia Institute of Technology, United States*
Liyao Wu, *Georgia Institute of Technology, United States*
Maryam Saeedifard, *Georgia Institute of Technology, United States*

Fault Tolerant Topologies of Five-Level Active Neutral-Point-Clamped Converters 2963
Jun Li, *ABB Inc., United States*

Session D08: Advanced Components and Devices

Location: Poster Area
March 24, 2016 11:30 - 14:00
Session Chairs: Abhijit Pathak, *Infineon/IR*
Doug Hopkins, *North Carolina State University*

Dynamic Characterization of the Input and Reverse Transfer Capacitances in Power MOSFETs under High Current Conduction 2969
Cristino Salcines, *Universität Stuttgart, Germany*
Ingmar Kallfass, *Universität Stuttgart, Germany*
Hisao Kakitani, *Keysight Technologies International, Japan*
Atsushi Mikata, *Keysight Technologies International, Japan*

Medium Voltage Power Switch based on SiC JFETs 2973
Xueqing Li, *United Silicon Carbide, Inc., United States*
Hao Zhang, *United Silicon Carbide, Inc., United States*
Peter Alexandrov, *United Silicon Carbide, Inc., United States*
Anup Bhalla, *United Silicon Carbide, Inc., United States*

Numerical Model and Experimental Study on Comparison of Semiconductor Pulsed Power Devices 2981
Lin Liang, *Huazhong University of Science and Technology, China*
Changdong Chen, *Huazhong University of Science and Technology, China*
Fang Luo, *Ohio State University, United States*

A Normalization Procedure of DC-Side Stray Inductance for High-Speed Switching Circuit 2986
Masato Ando, *Tokyo Metropolitan University, Japan*
Keiji Wada, *Tokyo Metropolitan University, Japan*

Thermal Network Parameter Identification of IGBT Module based on the Cooling Curve of Junction Temperature 2992
Xiong Du, *Chongqing University, China*
Tengfei Li, *Chongqing University, China*
Jun Zhang, *Chongqing University, China*
Heng-Ming Tai, *University of Tulsa, United States*
Pengju Sun, *Chongqing University, China*
Luowei Zhou, *Chongqing University, China*

Design and Evaluation of High Current PCB Embedded Inductor for High Frequency Inverters 2998
Mehrdad Biglarbegian, *University of North Carolina at Charlotte, United States*
Neel Shah, *University of North Carolina at Charlotte, United States*
Iman Mazhari, *University of North Carolina at Charlotte, United States*
Johan Enslin, *University of North Carolina at Charlotte, United States*
Babak Parkhideh, *University of North Carolina at Charlotte, United States*

Prognosis of Wire Bond Lift-Off Fault of an IGBT based on Multisensory Approach 3004
Moinul Shahidul Haque, *University of Akron, United States*
Jeihoon Baek, *Korean Rail Research Institute, Korea, South*
Joseph Herbert, *University of Akron, United States*
Seungdeog Choi, *University of Akron, United States*

Electrical Parasitics and Thermal Modeling for Optimized Layout Design of High Power SiC Modules 3012
Amir Sajjad Bahman, *Aalborg University, Denmark*
Frede Blaabjerg, *Aalborg University, Denmark*
Atanu Dutta, *University of Arkansas, United States*
Alan Mantooth, *University of Arkansas, United States*

Calculation of Losses in PCB Windings for Multi-Coil Contactless Charging Systems 3020
J. Serrano, *Universidad de Zaragoza, Spain*
J. Acero, *Universidad de Zaragoza, Spain*
I. Lope, *BSH Home Appliances Group, Spain*
C. Carretero, *Universidad de Zaragoza, Spain*
J.M. Burdío, *Universidad de Zaragoza, Spain*
R. Alonso, *Universidad de Zaragoza, Spain*

Design of Efficient Loads for Domestic Induction Heating Applications by Means of Non-Magnetic Thin Metallic Layers 3026
Jesús Acero, *Universidad de Zaragoza, Spain*
Claudio Carretero, *Universidad de Zaragoza, Spain*
Rafael Alonso, *Universidad de Zaragoza, Spain*
José Miguel Burdío, *Universidad de Zaragoza, Spain*

A New Evaluation Circuit with a Low-Voltage Inverter Intended for Capacitors used in a High-Power Three-Phase Inverter 3032
Kazunori Hasegawa, *Kyushu Institute of Technology, Japan*
Ichiro Omura, *Kyushu Institute of Technology, Japan*
Shin-Ichi Nishizawa, *Kyushu Institute of Technology / National Institute of Advanced Industrial Science and Technology, Japan*

Energy Absorption Capability of Low Voltage Metal Oxide Varistors in AC and Impulse Currents 3038
Dawood Talebi Khanmiri, *Northeastern University, United States*
Roy Ball, *Mersen USA, United States*
Craig McKenzie, *Mersen USA, United States*
Brad Lehman, *Northeastern University, United States*

Optimization and Experimental Validation of Medium-Frequency High Power Transformers in Solid-State Transformer Applications 3043
M.A. Bahmani, *Chalmers University of Technology, Sweden*
T. Thiringer, *Chalmers University of Technology, Sweden*
M. Kharezy, *SP Technical Research Institute of Sweden, Sweden*

Evaluation of Core Loss in Magnetic Materials Employed in Utility Grid AC Filters 3051
Remus Beres, *Aalborg University, Denmark*
Xiongfei Wang, *Aalborg University, Denmark*
Frede Blaabjerg, *Aalborg University, Denmark*
Claus Leth Bak, *Aalborg University, Denmark*
Hiroaki Matsumori, *Tokyo Metropolitan University, Japan*
Toshihisa Shimizu, *Tokyo Metropolitan University, Japan*

A Novel Gate Assisted Circuit to Reduce Switching Loss and Eliminate Shoot-Through in SiC Half Bridge Configuration 3058
Shan Yin, *Nanyang Technological University, Singapore*
K.J. Tseng, *Nanyang Technological University, Singapore*
C.F. Tong, *Nanyang Technological University, Singapore*
Rejeki Simanjorang, *Rolls-Royce Singapore Pte. Ltd., Singapore*
C.J. Gajanayake, *Rolls-Royce Singapore Pte. Ltd., Singapore*
Amit K. Gupta, *Rolls-Royce Singapore Pte. Ltd., Singapore*

Session D09: System Design Considerations for Power Electronics
Location: Poster Area
March 24, 2016 11:30 - 14:00
Session Chairs: John Vigars, *Allegro Microsystems*
Ernie Parker, *Crane Aerospace & Electronics*

Methods to Enhance the Thermal Performance of a 3D Power Package 3065
Jonathan Noquil, *Texas Instruments Inc., United States*
Ozzie Lopez, *Texas Instruments Inc., United States*
Tianyi Luo, *Lehigh University, United States*

Highly Reliable and Cost Effective Thick Film Substrates for Power LEDs 3069
Paul Gundel, *Heraeus Deutschland GmbH & Co. KG, Germany*
Ryan Persons, *Heraeus Deutschland GmbH & Co. KG, Germany*
Melanie Bawohl, *Heraeus Deutschland GmbH & Co. KG, Germany*
Mark Challingsworth, *Heraeus Deutschland GmbH & Co. KG, Germany*
Christoph Czwickla, *Heraeus Deutschland GmbH & Co. KG, Germany*
Virginia Garcia, *Heraeus Deutschland GmbH & Co. KG, Germany*
Christina Modes, *Heraeus Deutschland GmbH & Co. KG, Germany*
Ilias Nikolaidis, *Heraeus Deutschland GmbH & Co. KG, Germany*
Jessica Reitz, *Heraeus Deutschland GmbH & Co. KG, Germany*
Caitlin Shahbazi, *Heraeus Deutschland GmbH & Co. KG, Germany*
Torsten Nowak, *Fraunhofer-Institut für Zuverlässigkeit und Mikrointegration, Germany*

Design and Evaluation of SiC-Based High Power Density Inverter, 70kW/Liter, 50kW/kg ... 3075
Koji Yamaguchi, *IHI Corporation, Japan*

An Improved Automatic Layout Method for Planar Power Module 3080
Puqi Ning, *Chinese Academy of Sciences, China*
Xuhui Wen, *Chinese Academy of Sciences, China*
Yaohua Li, *Chinese Academy of Sciences, China*
Xiongxuan Ge, *Chinese Academy of Sciences, China*

Practical Implementation Schemes of Motor Speed Measurement by Magnetic Encoder on Electric Power Steering Applications 3086
Jae-Hyun Lee, *Hyundai Mobis, Korea, South*

Low-Cost Input Impedance Estimator of DC-to-DC Converters for Designing the Control Loop in Cascaded Converters 3090
M. Sanz, *Universidad Carlos III de Madrid, Spain*
A. Lázaro, *Universidad Carlos III de Madrid, Spain*
M. Bermejo, *Universidad Carlos III de Madrid, Spain*
D. López del Moral, *Universidad Carlos III de Madrid, Spain*
P. Zumel, *Universidad Carlos III de Madrid, Spain*
C. Fernández, *Universidad Carlos III de Madrid, Spain*
A. Barrado, *Universidad Carlos III de Madrid, Spain*

On-Chip High Performance Magnetics for Point-of-Load High-Frequency DC-DC Converters 3097
Dragan Dinulovic, *Würth Elektronik eiSos GmbH & Co. KG, Germany*
Mahmoud Shousha, *Würth Elektronik eiSos GmbH & Co. KG, Germany*
Martin Haug, *Würth Elektronik eiSos GmbH & Co. KG, Germany*
Alexander Gerfer, *Würth Elektronik eiSos GmbH & Co. KG, Germany*
Mike Wens, *MinDCet NV, Belgium*
Jef Thone, *MinDCet NV, Belgium*

Effects of Auxiliary Source Connections in Multichip Power Module 3101
Helong Li, *Aalborg University, Denmark*
Stig Munk-Nielsen, *Aalborg University, Denmark*
Szymon Bęczkowski, *Aalborg University, Denmark*
Xiongfei Wang, *Aalborg University, Denmark*
Emanuel-Petre Eni, *Aalborg University, Denmark*

Session D10: Modeling and Simulation

Location: Poster Area
March 24, 2016 11:30 - 14:00
Session Chairs: Marco Meola, *ZMD AG*
Mehdi Ferdowsi, *Missouri University of Science & Technology*

Modelling Technique Utilizing Modified Sigmoid Functions for Describing Power Transistor Device Capacitances Applied on GaN HEMT and Silicon MOSFET 3107
H.L. Yeo, *Nanyang Technological University, Singapore*
K.J. Tseng, *Nanyang Technological University, Singapore*

Design and Precise Modeling of a Novel Digital Active EMI Filter 3115
Junpeng Ji, *Xi'an Jiaotong University, China*
Wenjie Chen, *Xi'an Jiaotong University, China*
Xu Yang, *Xi'an Jiaotong University, China*

Development of a Hybrid Emulation Platform based on RTDS and Reconfigurable Power Converter-Based Testbed 3121
Shuoting Zhang, *University of Tennessee, United States*
Yiwei Ma, *University of Tennessee, United States*
Liu Yang, *University of Tennessee, United States*
Fred Wang, *University of Tennessee, United States*
Leon M. Tolbert, *University of Tennessee, United States*

Online Temperature Estimation for Phase Change Composite – 18650 Lithium Ion Cells based Battery Pack 3128
Mohamad Salameh, *Illinois Institute of Technology, United States*
Ben Schweitzer, *AllCell Technologies, United States*
Peter Sveum, *AllCell Technologies, United States*
Said Al-Hallaj, *AllCell Technologies, United States*
Mahesh Krishnamurthy, *Illinois Institute of Technology, United States*

Modeling and Fault Diagnosis of Inter-Turn Short Circuit for Five-Phase PMSM based on Particle Swarm Optimization 3134
Jianwei Yang, *Northwestern Polytechnical University, China*
Manfeng Dou, *Northwestern Polytechnical University, China*
Zhiyong Dai, *Northwestern Polytechnical University, China*
Dongdong Zhao, *Northwestern Polytechnical University, China*
Zhen Zhang, *Northwestern Polytechnical University, China*

Comprehensive Modeling, Testing, and Experimental Validation of Ultracapacitor Open Circuit Voltage Characteristics 3140
Amandeep Singh, *University of Ontario Institute of Technology, Canada*
Najath Abdul Azeez, *University of Ontario Institute of Technology, Canada*
Sheldon S. Williamson, *University of Ontario Institute of Technology, Canada*

Novel SPICE Model for Common Mode Choke Including Complex Permeability 3146
Katsuya Nomura, *Toyota Central R&D Labs., Inc., Japan*
Naoto Kikuchi, *Toyota Central R&D Labs., Inc., Japan*
Yoshitoshi Watanabe, *Toyota Central R&D Labs., Inc., Japan*
Shuntaro Inoue, *Toyota Central R&D Labs., Inc., Japan*
Yoshiyuki Hattori, *Toyota Central R&D Labs., Inc., Japan*

Session D11: Control I
Location: Poster Area
March 24, 2016 11:30 - 14:00
Session Chairs: Bilal Akin, *Univeristy of Texas, Dallas*
Brian Zahnstecher, *PowerRox LLC*

Analysis and Design of Capacitive Power Transmission System Employing Out-of-Band Wireless Feedback Link 3153
Sung-Jin Choi, *University of Ulsan, Korea, South*
Hee-Su Choi, *University of Ulsan, Korea, South*

Introducing Fourier-Based Modeling and Control of Active-Bridge Converters 3158
B.J.D. Vermulst, *Technische Universiteit Eindhoven, Netherlands*
J.L. Duarte, *Technische Universiteit Eindhoven, Netherlands*
C.G.E. Wijnands, *Technische Universiteit Eindhoven, Netherlands*
E.A. Lomonova, *Technische Universiteit Eindhoven, Netherlands*

A Stability Analysis and Efficiency Improvement of Synchronverter 3165
Prasanna Piya, *Mississippi State University, United States*
Masoud Karimi-Ghartemani, *Mississippi State University, United States*

Compensation of Switching Dead-Time Effects in Voltage-Fed PWM Inverters using FPGA-Based Current Oversampling 3172
Bastian Weber, *Leibniz Universität Hannover, Germany*
Tobias Brandt, *Leibniz Universität Hannover, Germany*
Axel Mertens, *Leibniz Universität Hannover, Germany*

Control Strategy of High Power Converters with Synchronous Generator Characteristics for PMSG-Based Wind Power Application 3180
Yuzhi Zhang, *University of Arkansas, United States*
Haoyan Liu, *University of Arkansas, United States*
H. Alan Mantooth, *University of Arkansas, United States*

Phase Compensation, ZVS Operation of Wireless Power Transfer System based on SOGI-PLL 3185
Pingan Tan, *Xiangtan University, China*
Haibing He, *Xiangtan University, China*
Xieping Gao, *Xiangtan University, China*

A Novel Low-Cost Online State of Charge Estimation Method for Reconfigurable Battery Pack 3189
Ni Lin, *University of Nebraska at Lincoln, United States*
Song Ci, *University of Nebraska at Lincoln, United States*
Dalei Wu, *University of Tennessee at Chattanooga, United States*

Effect of Decoupling Terms on the Performance of PR Current Controllers Implemented in Stationary Reference Frame 3193
Sizhan Zhou, *Xi'an Jiaotong University, China*
Jinjun Liu, *Xi'an Jiaotong University, China*

Fuzzy Predictive DTC of Induction Machines with Reduced Torque Ripple and High Performance Operation 3200
Alberto Berzoy, *Florida International University, United States*
Osama Mohammed, *Florida International University, United States*
Johnny Rengifo, *Universidad Simon Bolivar, Venezuela*

Session D12: Control II
Location: Poster Area
March 24, 2016 11:30 - 14:00
Session Chairs: Martin Ordonez, *University of British Columbia*
Jiangbiao He, *GE Global Research*

Fixed-Frequency Generalized Peak Current Control (GPCC) for Inverters 3207
Mohammad Ebrahimi, *University of Alberta, Canada*
S. Ali Khajehoddin, *University of Alberta, Canada*

Improved Control Strategy of 1 MHz LLC Converter for High Frequency Resolution 3213
Hwa-Pyeong Park, *Ulsan National Institute of Science and Technology, Korea, South*
Jee-Hoon Jung, *Ulsan National Institute of Science and Technology, Korea, South*

Bumpless Control for Reduced THD in Power Factor Correction Circuits 3219
Joel Steenis, *Microchip Technology, United States*
Alex Dumais, *Microchip Technology, United States*

Mixed-Signal Hysteretic Internal Model Control of Buck Converters for Ultra-Fast Envelope Tracking 3224
V. Inder Kumar, *Indian Institute of Technology Kharagpur, India*
Santanu Kapat, *Indian Institute of Technology Kharagpur, India*

A Continuous Actor-Critic Maximum Power Point Tracker Applied to Low Power Wind Turbine Systems 3231
J.L. Wattes, *Universidade Federal do Ceará, Brazil*
A.J.S. Dias Jr., *Universidade Federal do Ceará, Brazil*
A.P.S. Braga, *Universidade Federal do Ceará, Brazil*
P.P. Praça, *Universidade Federal do Ceará, Brazil*
A.U. Barbosa, *Universidade Federal do Ceará, Brazil*
D.S. de Souza Oliveira Jr., *Universidade Federal do Ceará, Brazil*

Multi-Band Mixed-Signal Hysteresis Current Control for EMI Reduction in Switch-Mode Power Supplies 3237
Arindam Mandal, *Indian Institute of Technology Kharagpur, India*
V. Inder Kumar, *Indian Institute of Technology Kharagpur, India*
Santanu Kapat, *Indian Institute of Technology Kharagpur, India*

A Parabolic Current Control based Digital Current Control Strategy for High Switching Frequency Voltage Source Inverters 3243
Lanhua Zhang, *Virginia Polytechnic Institute and State University, United States*
Rachael Born, *Virginia Polytechnic Institute and State University, United States*
Xiaonan Zhao, *Virginia Polytechnic Institute and State University, United States*
Jih-Sheng Jason Lai, *Virginia Polytechnic Institute and State University, United States*
Hongbo Ma, *Southwest Jiaotong University, China*

Finite Control Set Model Predictive Control of Dual-Output Four-Leg Indirect Matrix Converter Under Unbalanced Load and Supply Conditions 3248
Ozan Gulbudak, *University of South Carolina, United States*
Enrico Santi, *University of South Carolina, United States*

A Silicon Carbide Integrated Circuit Implementing Nonlinear-Carrier Control for Boost Converter Applications 3255
Richard Kyle Harris, *University of Tennessee, United States*
Benjamin M. McCue, *University of Tennessee, United States*
Benjamin D. Roehrs, *University of Tennessee, United States*
Charles Roberts II, *University of Tennessee, United States*
Benjamin J. Blalock, *University of Tennessee, United States*
Daniel J. Costinett, *University of Tennessee, United States*
Kouros Sariri, *Frequency Management International, United States*
George Megyei, *Frequency Management International, United States*
Cheng-Po Chen, *GE Global Research, United States*
Avinash Kashyap, *GE Global Research, United States*
Reza Ghandi, *GE Global Research, United States*

A New Current Mode Constant on Time Control with Ultrafast Load Transient Response 3259
Syed Bari, *Virginia Polytechnic Institute and State University, United States*
Qiang Li, *Virginia Polytechnic Institute and State University, United States*
Fred C. Lee, *Virginia Polytechnic Institute and State University, United States*

A Web-Based Tool for Compensation Design of Power Converters using Hybrid Optimization 3266
Srikanth Pam, *Texas Instruments Inc., India*
Yudhister Satija, *Texas Instruments Inc., India*
Pradeep Chawda, *Texas Instruments Inc., United States*
Makram Mansour, *Texas Instruments Inc., United States*
Robert Hanrahan, *Texas Instruments Inc., United States*
Jeff Perry, *Texas Instruments Inc., United States*

Second Order Sliding Mode Controlled Point of Load Power Supply 3273
Prasanta K. Achanta, *University of Colorado at Boulder, United States*
David C. Jones, *University of Colorado at Boulder, United States*
Dragan Maksimovic, *University of Colorado at Boulder, United States*
Serhii M. Zhak, *Linear Technology Corporation, United States*
Brett Miwa, *Maxim Integrated, United States*
Cory Arnold, *Maxim Integrated, United States*

Vibration and Torque Ripple Reduction of Switched Reluctance Motors through Current Profile Optimization 3279
Cong Ma, *University of Nebraska at Lincoln, United States*
Liyan Qu, *University of Nebraska at Lincoln, United States*
Rakesh Mitra, *Nexteer Automotive, United States*
Prerit Pramod, *Nexteer Automotive, United States*
Rakib Islam, *Nexteer Automotive, United States*

Modified Predictive Current Control of Neutral-Point Clamped Converter with Reduced Switching Frequency 3286
Dinto Mathew, *Indian Institute of Technology Bombay, India*
Anshuman Shukla, *Indian Institute of Technology Bombay, India*
Santanu Bandyopadhyay, *Indian Institute of Technology Bombay, India*

Implicit Finite Control Set Model Predictive Current Control for Modular Multilevel Converter based on IPA-SQP Algorithm 3291
Hamed Nademi, *ABB AS, Norway*
Lars Einar Norum, *Norwegian University of Science and Technology, Norway*

Resolution Requirements to Avoid Limit Cycling in LLC Resonant Converter 3297
Shadi Dashmiz, *University of Toronto, Canada*
Behzad Mahdavikhah, *University of Toronto, Canada*
Aleksandar Prodić, *University of Toronto, Canada*
Brent McDonald, *Texas Instruments Inc., United States*

Session D13: Renewable Energy Systems I

Location: Poster Area
March 24, 2016 11:30 - 14:00
Session Chairs: Akshay Kumar Rathore, *Concordia University*
Xiaoqiang Guo, *Yanshan University, China*

Reduction of Storage Capacity in DC Microgrids using PV-Embedded Series DC Electric Springs 3302
Ming-Hao Wang, *University of Hong Kong, Hong Kong*
Siew-Chong Tan, *University of Hong Kong, Hong Kong*
Shu-Yuen Ron Hui, *University of Hong Kong, Hong Kong*

A Vector Control Strategy of Grid-Connected Brushless Doubly Fed Induction Generator based on the Vector Control of Doubly Fed Induction Generator 3310
Sheng Hu, *Wuhan University of Technology, China*
Guorong Zhu, *Wuhan University of Technology, China*

An Energy Router based on Multi-Winding High-Frequency Transformer 3317
Xianzhuo Liu, *Tsinghua University, China*
Zedong Zheng, *Tsinghua University, China*
Kui Wang, *Tsinghua University, China*
Yongdong Li, *Tsinghua University, China*

Noise Suppression of the DWT-Based MRA on Mother Wavelet and Decomposition Level Optimization for a Robust Adaptive SOC Estimator in Multi-Cell Battery String 3322
Jonghoon Kim, *Chosun University, Korea, South*
Chang Yoon Chun, *Seoul National University, Korea, South*
Woonki Na, *California State University, Fresno, United States*

A Feedforward Control based Power Decoupling Scheme for Voltage-Controlled Grid-Tied Inverters 3328
Baojin Liu, *Xi'an Jiaotong University, China*
Zeng Liu, *Xi'an Jiaotong University, China*
Jinjun Liu, *Xi'an Jiaotong University, China*
Teng Wu, *Xi'an Jiaotong University, China*
Shike Wang, *Xi'an Jiaotong University, China*

Light Load Efficiency Improvement of Solar Farms Three-Phase Two-Stage Module Integrated Converter 3333
Ahmadreza Amirahmadi, *University of Central Florida, United States*
Utsav Somani, *University of Central Florida, United States*
Mahmood Alharbi, *University of Central Florida, United States*
Charlie Jourdan, *University of Central Florida, United States*
Issa Batarseh, *University of Central Florida, United States*

Switching System Stability Analysis of DC Microgrids with DBS Control 3338
Na Zhi, *Xi'an University of Technology, China*
Hui Zhang, *Xi'an University of Technology, China*
Xi Xiao, *Tsinghua University, China*

A Grid-Connected WECS with Power Limiting Control 3346
Jéssica Santos Guimarães, *Universidade Federal do Ceará, Brazil*
Demercil de Souza Oliveira Jr., *Universidade Federal do Ceará, Brazil*
Juliano de Oliveira Pacheco, *Universidade Federal do Ceará, Brazil*
Paulo P. Peixoto, *Universidade Federal do Ceará, Brazil*

Overshoot Control of the Electromagnetic Torque during Fault Recovery for an SCIG with a STATCOM 3353
Zahra Mahmoodzadeh, *Washington State University, United States*
Mehrdad Yazdanian, *Washington State University, United States*
Hooman Ghaffarzadeh, *Washington State University, United States*
Ali Mehrizi-Sani, *Washington State University, United States*

A Self-Adaptive Power Balance Control Strategy for PV Inverters in Islanded Microgrids 3358
Zhenxiong Wang, *Xi'an Jiaotong University, China*
Hao Yi, *Xi'an Jiaotong University, China*
Fang Zhuo, *Xi'an Jiaotong University, China*
Zhigang Zhang, *Xi'an Jiaotong University, China*

High Performance ZVT with Bus Clamping Modulation Technique for Single Phase Full Bridge Inverters 3364
Yinglai Xia, *Arizona State University, United States*
Raja Ayyanar, *Arizona State University, United States*

Small AC Signal Droop based Secondary Control for Microgrids 3370
Teng Wu, *Xi'an Jiaotong University, China*
Zeng Liu, *Xi'an Jiaotong University, China*
Jinjun Liu, *Xi'an Jiaotong University, China*
Baojin Liu, *Xi'an Jiaotong University, China*
Shike Wang, *Xi'an Jiaotong University, China*

Mode Transition Control Strategy for Multiple Inverter based Distributed Generators Operating in Grid-Connected and Stand-Alone Mode 3376
Onkar Vitthal Kulkarni, *Indian Institute of Technology Bombay, India*
Suryanarayana Doolla, *Indian Institute of Technology Bombay, India*
B.G. Fernandes, *Indian Institute of Technology Bombay, India*

An Autonomous Power Management Strategy based on DC Bus Signaling for Solid-State Transformer Interfaced PMSG Wind Energy Conversion System 3383
Rui Gao, *North Carolina State University, United States*
Iqbal Husain, *North Carolina State University, United States*
Alex Q. Huang, *North Carolina State University, United States*

An Isolated Buck-Boost Type High-Frequency Link Photovoltaic Microinverter 3389
Shiladri Chakraborty, *Indian Institute of Technology Kharagpur, India*
Souvik Chattopadhyay, *Indian Institute of Technology Kharagpur, India*

Energy Management and Stabilization of a Hybrid DC Microgrid for Transportation Applications 3397
Mehdi Karbalaye Zadeh, *Norwegian University of Science and Technology, Norway*
Louis-Marie Saublet, *Université de Lorraine, France*
Roghayeh Gavagsaz-Ghoachani, *Université de Lorraine, France*
Babak Nahid-Mobarakeh, *Université de Lorraine, France*
Serge Pierfederici, *Université de Lorraine, France*
Marta Molinas, *Norwegian University of Science and Technology, Norway*

A Low-Cost Solar Micro-Inverter with Soft-Switching Capability Utilizing Circulating Current 3403
Xiaohu Liu, *GE Global Research, United States*
Mohammed Agamy, *GE Global Research, United States*
Dong Dong, *GE Global Research, United States*
Maja Harfman-Todorovic, *GE Global Research, United States*
Luis Garces, *GE Global Research, United States*

Session D14: Renewable Energy Systems II
Location: Poster Area
March 24, 2016 11:30 - 14:00
Session Chairs: Haoyu Wang, *Shanghai Tech University*
Robert Pilawa-Podgurski, *University of Illinois at Urbana-Champaign*

Design and Stability Analysis for an Autonomous DC Microgrid with Constant Power Load ... 3409
Qianwen Xu, *Nanyang Technological University, Singapore*
Xiaolei Hu, *Nanyang Technological University, Singapore*
Peng Wang, *Nanyang Technological University, Singapore*
Jianfang Xiao, *Nanyang Technological University, Singapore*
Leonardy Setyawan, *Nanyang Technological University, Singapore*
Changyun Wen, *Nanyang Technological University, Singapore*
Lee Meng Yeong, *Rolls-Royce Singapore Pte. Ltd., Singapore*

MPC-SVM Method for Vienna Rectifier with PMSG used in Wind Turbine Systems 3416
June-Seok Lee, *Korea Railroad Research Institute, Korea, South*
Yeongsu Bak, *Ajou University, Korea, South*
Kyo-Beum Lee, *Ajou University, Korea, South*
Frede Blaabjerg, *Aalborg University, Denmark*

An Equivalent Circuit Model for State of Energy Estimation of Lithium-Ion Battery 3422
Kaiyuan Li, *Nanyang Technological University, Singapore*
King Jet Tseng, *Nanyang Technological University, Singapore*

Distributed Optimal Control of Reactive Power and Voltage in Islanded Microgrids 3431
Yanbo Wang, *Aalborg University, Denmark*
Xiongfei Wang, *Aalborg University, Denmark*
Zhe Chen, *Aalborg University, Denmark*
Frede Blaabjerg, *Aalborg University, Denmark*

New Start-Up Scheme for HF Transformer Link Photovoltaic Inverter 3439
Abhijit Kulkarni, *Indian Institute of Science, India*
Vinod John, *Indian Institute of Science, India*

Analysis and Improvement of Harmonic Quasi Resonant Control for LCL-Filtered Grid-Connected Inverters in Weak Grid 3446
Qiang Qian, *Nanjing University of Aeronautics and Astronautics, China*
Jinming Xu, *Nanjing University of Aeronautics and Astronautics, China*
Shaojun Xie, *Nanjing University of Aeronautics and Astronautics, China*
Lin Ji, *Nanjing University of Aeronautics and Astronautics, China*

Model Predictive Control Method to Reduce Common-Mode Voltage and Balance the Neutral-Point Voltage in Three-Level T-Type Inverter 3453
Xiangyang Xing, *Shandong University, China*
Alian Chen, *Shandong University, China*
Zicheng Zhang, *Shandong University, China*
Jie Chen, *Shandong University, China*
Chenghui Zhang, *Shandong University, China*

Convergence Analysis of Distributed Control for Operation Cost Minimization of Droop Controlled DC Microgrid based on Multiagent 3459
Chendan Li, *Aalborg University, Denmark*
Juan C. Vásquez, *Aalborg University, Denmark*
Josep M. Guerrero, *Aalborg University, Denmark*

A Novel Model Predictive Control Algorithm to Suppress the Zero-Sequence Circulating Currents for Parallel Three-Phase Voltage Source Inverters 3465
Zicheng Zhang, *Shandong University, China*
Alian Chen, *Shandong University, China*
Xiangyang Xing, *Shandong University, China*
Chenghui Zhang, *Shandong University, China*

Design of Dynamic Voltage Restorer and Active Power Filter for Wind Power Systems Subject to Unbalanced and Harmonic Distorted Grid 3471
Woei-Luen Chen, *Chang Gung University, Taiwan*
Meng-Jie Wang, *Chang Gung University, Taiwan*

Dynamic Variable Coupling Analysis and Modeling of Proton Exchange Membrane Fuel Cells for Water and Thermal Management 3476
Daming Zhou, *Université de Technologie de Belfort-Montbéliard, France*
Elena Breaz, *Université de Technologie de Belfort-Montbéliard, France*
Alexandre Ravey, *Université de Technologie de Belfort-Montbéliard, France*
Fei Gao, *Université de Technologie de Belfort-Montbéliard, France*
Abdellatif Miraoui, *Université de Technologie de Belfort-Montbéliard, France*
Ke Zhang, *Northwestern Polytechnical University, China*

Voltage and Frequency Control of Electric Spring based Smart Loads 3481
Yun Yang, *University of Hong Kong, Hong Kong*
Siew-Chong Tan, *University of Hong Kong, Hong Kong*
Shu-Yuen Ron Hui, *University of Hong Kong, Hong Kong*

Second Harmonic Current Compensator with Improved One-Cycle-Control 3488
Li Zhang, *Nanjing University of Aeronautics and Astronautics, China*
Xinbo Ruan, *Nanjing University of Aeronautics and Astronautics, China*
Xiaoyong Ren, *Nanjing University of Aeronautics and Astronautics, China*

Frequency Adaptive Control of a Smart Transformer-Fed Distribution Grid 3493
Zhi-Xiang Zou, *Christian-Albrechts-Universität zu Kiel, Germany*
Giovanni De Carne, *Christian-Albrechts-Universität zu Kiel, Germany*
Giampaolo Buticchi, *Christian-Albrechts-Universität zu Kiel, Germany*
Marco Liserre, *Christian-Albrechts-Universität zu Kiel, Germany*

A Synchronization Scheme for Single-Phase Grid-Tied Inverters under Harmonic Distortion and Grid Disturbances 3500
Lenos Hadjidemetriou, *University of Cyprus, Cyprus*
Elias Kyriakides, *University of Cyprus, Cyprus*
Yongheng Yang, *Aalborg University, Denmark*
Frede Blaabjerg, *Aalborg University, Denmark*

Series-Parallel Connection of Low-Voltage Sources for Integration of Galvanically Isolated Energy Storage Systems 3508
Ramy Georgious, *Universidad de Oviedo, Spain*
Jorge Garcia, *Universidad de Oviedo, Spain*
Angel Navarro, *Universidad de Oviedo, Spain*
Sarah Saeed, *Universidad de Oviedo, Spain*
Pablo Garcia, *Universidad de Oviedo, Spain*

Saturation Controller-Based Direct Power Control for Doubly-Fed Induction Generator 3514
Chun Wei, *University of Nebraska at Lincoln, United States*
Zhe Zhang, *Nexteer Automotive, United States*
Wei Qiao, *University of Nebraska at Lincoln, United States*
Liyan Qu, *University of Nebraska at Lincoln, United States*

Inductance-Simulating Control for DFIG-Based Wind Turbine to Ride-Through Grid Faults 3521
Donghai Zhu, *Huazhong University of Science and Technology, China*
Xudong Zou, *Huazhong University of Science and Technology, China*
Yong Kang, *Huazhong University of Science and Technology, China*
Lu Deng, *Wuhan NARI Limited Company of State Grid Electric Power Research Institute, China*
Qingjun Huang, *State Key Laboratory of Disaster Prevention & Reduction for Power Grid Transmission and Distribution Equipment, China*

Session D15: Transportation Power Electronics
Location: Poster Area
March 24, 2016 11:30 - 14:00
Session Chairs: Ted Bohn, *Argonne National Labs*
Khurram Afridi, *University of Colorado, Boulder*

Misalignment Effect on Efficiency of Wireless Power Transfer for Electric Vehicles 3526
Yabiao Gao, *University of Georgia, United States*
Antonio Ginart, *University of Georgia / Sonnenbatterie GmbH, United States*
Kathleen Blair Farley, *Southern Company Services, Inc., United States*
Zion Tsz Ho Tse, *University of Georgia, United States*

Genetic Algorithm Design of a 3D Printed Heat Sink 3529
Tong Wu, *University of Tennessee, United States*
Burak Ozpineci, *Oak Ridge National Laboratory, United States*
Curtis Ayers, *Oak Ridge National Laboratory, United States*

Evaluation of Power Flow Control for an All-Electric Warship Power System with Pulsed Load Applications 3537
J. Neely, *Sandia National Laboratories, United States*
L. Rashkin, *Sandia National Laboratories, United States*
M. Cook, *Sandia National Laboratories, United States*
D. Wilson, *Sandia National Laboratories, United States*
S. Glover, *Sandia National Laboratories, United States*

Reduced Active Switch AC to DC Rectifier with High Frequency Isolation for Electric Vehicle Chargers 3545
José Juan Sandoval, *Texas A&M University, United States*
Taeyong Kang, *Texas A&M University, United States*
Prasad Enjeti, *Texas A&M University, United States*

A Wide Bandgap Device based Multilevel Switched-Capacitor Converter 3553
Diogo Cesar Santos de Moura, *North Dakota State University, United States*
Boris Curuvija, *North Dakota State University, United States*
Dong Cao, *North Dakota State University, United States*

Session D16: Power Topologies, Distribution, and Control
Location: Poster Area
March 24, 2016 11:30 - 14:00
Session Chairs: Tiefu Zhao, *Eaton*
Xiaonan Lu, *Argonne National Laboratory*

Novel Circulating Current Suppression Strategy for MMC based on Quasi-PR Controller 3560
Shengbao Geng, *Shanghai Jiao Tong University, China*
Yiliang Gan, *Shanghai Jiao Tong University, China*
Yungui Li, *Shanghai Jiao Tong University, China*
Lijun Hang, *Shanghai Jiao Tong University, China*
Guojie Li, *Shanghai Jiao Tong University, China*

Assymmetric Duty-Cycle Phase-Shift Modulation for Power Management in Double Half-Bridge Inverter with Partly Coupled Inductive Loads 3566
C. Carretero, *Universidad de Zaragoza, Spain*
H. Sarnago, *Universidad de Zaragoza, Spain*
O. Lucia, *Universidad de Zaragoza, Spain*
J. Acero, *Universidad de Zaragoza, Spain*
J.M. Burdío, *Universidad de Zaragoza, Spain*

Control Implementation for a Wide Voltage Range High Efficiency Power Supply Utilizing Low Voltage MOSFETs 3570
Werner Konrad, *Technische Universität Graz, Austria*
Gerald Deboy, *Infineon Technologies AG, Austria*
Annette Muetze, *Technische Universität Graz, Austria*

A Single-Phase Dual Frequency Inverter based on Multi-Frequency Selective Harmonic Elimination 3577
Chongwen Zhao, *University of Tennessee, United States*
Daniel Costinett, *University of Tennessee, United States*
Brad Trento, *University of Tennessee, United States*
Daniel Friedrichs, *Medtronic, United States*

Grid Connected DC Distribution Network Deploying High Power Density Rectifier for DC Voltage Stabilization 3585
Danillo B. Rodrigues, *Universidade Federal do Triângulo Mineiro, Brazil*
Paulo R. Silva, *Universidade Federal de Uberlândia, Brazil*
Gustavo B. Lima, *Universidade Federal do Triângulo Mineiro, Brazil*
Ernane A.A. Coelho, *Universidade Federal de Uberlândia, Brazil*
Luiz C.G. Freitas, *Universidade Federal de Uberlândia, Brazil*

Even-Harmonic Repetitive Control for Circulating Current Suppression in Modular Multilevel Converters 3591
Shunfeng Yang, *Nanyang Technological University, Singapore*
Peng Wang, *Nanyang Technological University, Singapore*
Yi Tang, *Nanyang Technological University, Singapore*
Michael Zagrodnik, *Rolls-Royce Singapore Pte. Ltd., Singapore*
Xiaolei Hu, *Nanyang Technological University, Singapore*
King Jet Tseng, *Nanyang Technological University, Singapore*

A New DSC-PLL using Recursive Discrete Fourier Transform for Robustness to Frequency Variation 3598
Jaedo Lee, *Korea Institute of Nuclear Safety, Korea, South*
Hanju Cha, *Chungnam National University, Korea, South*

A Four-Quadrant Modulation Technique for Cascaded Multilevel Inverters to Extend Solution Range for Selective Harmonic Elimination/Compensation 3603
Hui Zhao, *University of Florida, United States*
Shuo Wang, *University of Florida, United States*

Online Battery Impedance Spectrum Measurement Method 3611
Jaber A. Abu Qahouq, *University of Alabama, United States*

Analysis and Control of a Reduced Switch Converter for Active Magnetic Bearings 3616
Dong Jiang, *Huazhong University of Science and Technology, China*
Parag Kshirsagar, *United Technologies Research Center, United States*

A Novel Balanced Winding Topology to Mitigate EMI without the Need for a Y-Capacitor 3623
Yongjiang Bai, *Xi'an Jiaotong University, China*
Xu Yang, *Xi'an Jiaotong University, China*
Xinlei Li, *Silergy Corp., China*
Dan Zhang, *Silergy Corp., China*
Wenjie Chen, *Xi'an Jiaotong University, China*

Topology and Control Strategy for Accelerated Lifetime Test Setup of DC-Link Capacitor of Wind Turbine Converter 3629
Youngjong Ko, *Christian-Albrechts-Universität zu Kiel, Germany*
Holger Jedtberg, *Christian-Albrechts-Universität zu Kiel, Germany*
Giampaolo Buticchi, *Christian-Albrechts-Universität zu Kiel, Germany*
Marco Liserre, *Christian-Albrechts-Universität zu Kiel, Germany*

Voltage Droop Compensation based on Resonant Circuit for Generalized High Voltage Solid-State Marx Modulator 3637
Hiren Canacsinh, *Instituto Superior de Engenharia de Lisboa, Portugal*
Luís M. Redondo, *Instituto Superior de Engenharia de Lisboa, Portugal*
J. Fernando Silva, *Instituto Superior Técnico, Portugal*
Beatriz Borges, *Instituto Superior Técnico, Portugal*

Four H-Bridge based Shunt Active Power Filter for Three-Phase Four Wire System 3641
Edgard L.L. Fabricio, *Universidade Federal da Paraíba, Brazil*
Cursino B. Jacobina, *Universidade Federal de Campina Grande, Brazil*
Gregory A.A. Carlos, *Universidade Federal de Campina Grande, Brazil*
Maurício B.R. Correa, *Universidade Federal de Campina Grande, Brazil*

High-Frequency AC Distributed Power Delivery System 3648
Mengqi Wang, *University of Michigan at Dearborn, United States*
Qingyun Huang, *North Carolina State University, United States*
Wensong Yu, *North Carolina State University, United States*
Alex Q. Huang, *North Carolina State University, United States*

Effect of the Capacitance Distribution on the Output Impedance of the Half-Wave Cockcroft-Walton Voltage Multiplier 3655
Liran Katzir, *Tel Aviv University, Israel*
Doron Shmilovitz, *Tel Aviv University, Israel*

Session D17: Emerging and Renewable Power
Location: Poster Area
March 24, 2016 11:30 - 14:00
Session Chairs: Katherine Kim, *Ulsan NIST*
Dimitri Torregrossa, *EPFL*

A Cost Effective High Performance LED Driver Powered by Electronic Ballasts 3659
Jianwen Shao, *STMicroelectronics, United States*
Thomas Stamm, *STMicroelectronics, United States*

Model Predictive Control of Z-Source Four-Leg Inverter for Standalone Photovoltaic System with Unbalanced Load 3663
Sertac Bayhan, *Gazi University, Turkey*
Mohamed Trabelsi, *Texas A&M University at Qatar, Qatar*
Haitham Abu-Rub, *Texas A&M University at Qatar, Qatar*

Efficiency Optimization of an Integrated Wireless Power Transfer System by a Genetic Algorithm 3669
Rosario Pagano, *Integrated Device Technology Inc., United States*
Siamak Abedinpour, *Integrated Device Technology Inc., United States*
Angelo Raciti, *Università degli Studi di Catania, Italy*
Salvatore Musumeci, *Università degli Studi di Catania, Italy*

Loss Analysis of a High Efficiency GaN and Si Device Mixed Isolated Bidirectional DC-DC Converter 3677
Fei Xue, *North Carolina State University, United States*
Ruiyang Yu, *North Carolina State University, United States*
Alex Q. Huang, *North Carolina State University, United States*

Dynamic Efficiency Tracking Controller for Reconfigurable Four-Coil Wireless Power Transfer System 3684
Yuan Cao, *University of Alabama, United States*
Zhigang Dang, *University of Alabama, United States*
Jaber A. Abu Qahouq, *University of Alabama, United States*
Evan Phillips, *University of Alabama, United States*

Wireless Power and Data Transfer System for Smart Bridge Sensors 3690
Yujin Jang, *Korea Advanced Institute of Science and Technology, Korea, South*
Jung Kyu Han, *Korea Advanced Institute of Science and Technology, Korea, South*
Shin Young Cho, *Korea Advanced Institute of Science and Technology, Korea, South*
Gun-Woo Moon, *Korea Advanced Institute of Science and Technology, Korea, South*
Ji-Min Kim, *Korea Advanced Institute of Science and Technology, Korea, South*
Hoon Sohn, *Korea Advanced Institute of Science and Technology, Korea, South*

Inrush Transient Current Analysis and Suppression of Photovoltaic Grid-Connected Inverters during Voltage Sag 3697
Zhongyu Li, *China University of Petroleum, China*
Rende Zhao, *China University of Petroleum, China*
Zhen Xin, *Aalborg University, Denmark*
Josep M. Guerrero, *Aalborg University, Denmark*
Mehdi Savaghebi, *Aalborg University, Denmark*
Peide Li, *Shandong Jinan Power Equipment Factory Co., LTD, China*

A Highly Reliable Single-Stage Converter for Electric Vehicle Applications 3704
S.A.Kh. Mozaffari Niapour, *Northeastern University, United States*
Mahshid Amirabadi, *Northeastern University, United States*

Simple and Efficient Low Power Photovoltaic Emulator for Evaluation of Power Conditioning Systems 3712
Jesus Gonzalez-Llorente, *Universidad Sergio Arboleda, Colombia*
Andres Rambal-Vecino, *Universidad Sergio Arboleda, Colombia*
Luciano A. Garcia-Rodriguez, *University of Arkansas, United States*
Juan C. Balda, *University of Arkansas, United States*
Eduardo I. Ortiz-Rivera, *University of Puerto Rico at Mayaguez, Puerto Rico*

Data Transmission Method without Additional Circuits in Bidirectional Wireless Power Transfer System 3717
Yeongrack Son, *Seoul National University, Korea, South*
Jung-Ik Ha, *Seoul National University, Korea, South*

Improved Impedance Source Inverter for Hybrid/Electric Vehicle Application with Continuous Conduction Operation 3722
Thilak Senanayake, *University of Tsukuba, Japan*
Ryuji Iijima, *University of Tsukuba, Japan*
Takanori Isobe, *University of Tsukuba, Japan*
Hiroshi Tadano, *University of Tsukuba, Japan*

Medium Voltage Power Switch Based on SiC JFETs

Xueqing Li, Hao Zhang, Peter Alexandrov, and Anup Bhalla
United Silicon Carbide, Inc., 7 Deer Park Drive, Suite E, Monmouth Junction, NJ 08852, USA
{lli, szhang, palexandrov, abhalla}@unitedsic.com

***Abstract*—This paper presents a high performance cost-effective medium voltage power switch based on the series-connection of low voltage SiC normally-on JFETs using an innovative approach. A prototype module of the proposed switch is experimentally demonstrated having an on-resistance of 127 milliohm and a blocking voltage above 4,500V at room temperature. The prototype module is capable of very low switching losses. Under a 3kV-54A inductive load, the turn-on rise time is 73ns and the turn-off fall time is 37ns; the turn-on energy loss is 14.8mJ and the turn-off energy loss is 1.5mJ. The prototype module is constructed with reliable, proven and commercially available SiC and Si parts, ensuring the high reliability on the component level. This approach exploits all of the advantages of SiC material in a cost-effective way, allowing quick widespread adoption of SiC technology in next generation high-frequency power conversion and motor control applications.**

Keywords—Silicon Carbide; Normally-on JFET; Series Connection; Medium Voltage; Power Switch

I. INTRODUCTION

The unique electrical and physical properties of silicon carbide (SiC) material provide the SiC-based power devices with the capability to operate at higher voltages, higher temperatures, and higher switching frequencies compared to their Si counterparts. These characteristics promise SiC-based power conversion systems with high efficiency and high power density. All these advantages of SiC technology have been experimentally demonstrated over the last decade. However, the widespread adoption of SiC power switches is still not happening due to the high cost of SiC devices [1]. At present, the SiC unipolar power switches including MOSFET and normally-on JFETs are commercially available at voltage rating up to 1200V from several manufactures, such as Cree, Rohm, Infineon and USCi, and higher voltage devices are still under development. Higher-voltage blocking capability can be achieved by series-connecting multiple lower voltage devices, which also improves switching performance and reduces the cost because a single high voltage switch usually has a much worse performance and is more expensive than low-voltage devices. Series-connection of low voltage devices has been widely used in the high voltage power conversion applications, such as utility-scale photovoltaic power inverters, railway traction, and high voltage direct current transmission systems, where 1.2kV or 1.7kV Si IGBTs are connected in series to form medium-voltage power switches [2, 3]. Recently, the series-connections of SiC MOSFETs [4] and normally-on JFETs [5, 6] have been reported. Normally-on JFETs are more suitable for series-connection because they can be connected in series with a simple cascoded circuit configuration that requires a simple voltage-balancing circuit and no power supply, from either self-powered circuits or externally powered circuits, for driving the JFET gates. This paper presents a medium-voltage power switch constructed by series-connecting five 1.2kV SiC normally-on JFETs according to the circuit topology proposed in [5, 7]. Detailed design information and experimental results are reported.

II. SWITCH CIRCUIT TOPOLOGY

Fig. 1 shows the circuit schematic of the proposed power switch including five SiC normally-on JFETs J1-5 and a low-voltage Si MOSFET connected in series with a cascaded circuit topology. The low-voltage MOSFET serves as the control unit to trigger the turn-on and turn-off processes of the whole switch. Thus, the conventional IGBT/MOSFET gate drivers can be directly used to drive the proposed power switch. In addition, the threshold voltage of the power switch is determined by the threshold voltage of the MOSFET. High threshold voltage of 4V to 5V can be easily realized if high noise immunity is required.

The diodes D2-D5 are avalanche diodes used to clamp the gate potentials of the JFETs J2-5, respectively, which, in turn, clamp the blocking voltages of the JFETs J1-4 during off-state, respectively. The maximum blocking voltage of JFETs J1-4 is clamped to a value equal to the sum of the turn-off source-gate voltage of JFETs J1-4 and the avalanche voltage of the avalanche diodes D2-5. The blocking voltage of the last JFET J5 is not clamped. Therefore, the JFET J5 can go into avalanche breakdown. The resistors R1-5 are load resistors used to provide the bias currents for the diodes D2-5 during off-state so that the diodes D2-5 can operate stably during

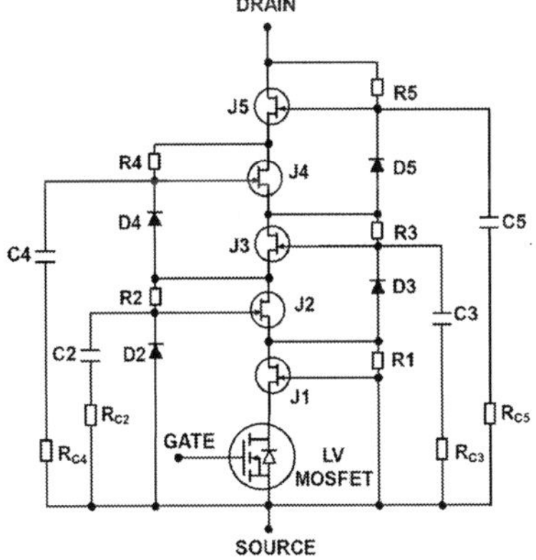

Fig. 1. Circuit schematic of the proposed medium voltage power switch formed by five SiC normally-on JFETs connected in series.

avalanche.

The capacitors C2-5 are used to provide a current path during switching transients for charging and discharging the reverse transfer (Crss) capacitances of JFETs J2-5, respectively, since the capacitance of the diodes D2-5 are usually very small. By optimizing the capacitance values of C2-5, the turn-on and turn-off processes of the JFETs can be easily synchronized and the dynamic blocking voltages among JFETs during switching transients can be balanced very well. The resistors RC2-5 are used to damp the possible oscillations caused by these voltage balancing capacitors C2-5.

The unique connection arrangement of the avalanche diodes D2-5, load resistors R2-5, and voltage-balancing capacitors C2-5, as shown in Fig.1, provide the following advantages. First, the bias currents of the diodes D2-5 are determined only by the load resistors R2-R5 and the blocking voltages of the JFETs J2-5, respectively, not by the JFET parameters, such as leakage current and threshold voltage. Second, the total additional leakage current of the power switch caused by biasing the avalanche diodes D2-5 is at most twice the bias current of one avalanche diode and is not dependent on the number of JFETs in series. This approach requires no pre-selection of the JFETs and is suitable for mass production. The detailed descriptions of the operation principle and the advantages of this circuit can be found in [5, 7].

III. Experimental Setup

The ultimate goal of this work is to build a medium-voltage power switch module with commercial available 1200V SiC normally-on JFETs. Therefore, all passive components and diodes used in the experiments are surface mounted, and SiC JFETs and Si MOSFET are in die form. In order to optimize the circuit components and evaluate the static and dynamic performance, the proposed power switch is built on a polyimide PCB with soft gold finish suitable for die attach and aluminum wire bonding. The optimized PCB layout can be directly transferred to DBC board for later module construction. During the test, the PCB is immersed in fluorinert to avoid arcing. Once the component values are fixed, the assembly can be encapsulated.

A. SiC JFETs

At present, there are two types of SiC normally-on JFETs available on the market. One has a horizontal-channel [8] and the other has a trenched-and-implanted vertical channel [9]. The JFETs with horizontal-channel has a large drain-source capacitance C_{DS} ($C_{DS}= C_{oss} - C_{rss}$). It is seen from Fig.1 that the drain-source capacitances of the JFETs J1-5 and the low-voltage MOSFETs are connected in series, which may cause an issue of voltage unbalance among JFETs and MOSFETs during off-state when the voltage across the switch is rapidly changing. The JFETs with vertical-channel have no such issue because they have a negligible drain-source capacitance C_{DS}. Thus, the JFETs with vertical-channel are recommended for constructing the proposed power switch.

In this work, the commercially available 1200V-45mΩ SiC normally-on JFETs (UJN1205Z) are used to build the proposed power switch. These JFETs have a trenched-and-implanted vertical channel and are available in die form. Five JFET dies are connected in series to provide medium-voltage blocking capability.

B. Low-Voltge Si MOSFET

Low-voltage Si MOSFET is the control-unit of the whole switch and forms a conventional cascode configuration with the JFET J1 as shown in Fig.1. The blocking voltage of the MOSFET should be higher than the turn-off source-gate voltage of the JFET J1, which is typically from 15V to 20V. A critical aspect for the MSOFET in the cascode configuration is that, during turn-off transients, the MOSFET may be driven into avalanche breakdown for a very short period. Therefore, the repetitive avalanche capability is required for the low-voltage MOSFET in order to ensure high reliability.

The MOSFET used in this work is a custom designed 20V-5mΩ Si MOSFET that has a built-in voltage clamp in order to make it operate safely in avalanche mode. Fig. 2 shows the general concept of how this is accomplished by creating a low breakdown region between cells to carry the entire breakdown current. The MOSFET has been qualified by the burn-in test when biased into avalanche for 1000 hours at 150°C junction temperature. Negligible parameter shifts are observed after the burn-in test [10]

C. Voltage-Clamping Diodes

The voltage-clamping diodes D2-D5 are used to clamp the gate potential of the JFETs J2-J5 and further limit the blocking voltages of the JFETs J1-J4 at off-state. Since the avalanche voltage of the diodes in general increases with the temperature, the selection of the diodes should be based on the diode maximum junction temperature in the actual applications. In a practical module design, the diodes are placed on the same DBC board with JFETs, meaning the maximum junction temperature of the diodes will be approximately the same as the module maximum case temperature. In this work, the module maximum case temperature is assumed to be 125°C and the blocking voltages of the JFETs J1-4 are limited to about 80% (960V) of the rated voltage of 1200V. According to the above design considerations, the commercially available avalanche diodes AU2PJ-M3/86A are selected. This diode has a stable avalanche voltage of about 860V and 980V at 25°C and 125°C, respectively, as shown in Fig.3, very close to the design target.

D. Passive Components

The load resistors R1-5 should have a high-voltage pulse-withstanding capability. The value of the resistance is determined by the bias current of the avalanche diodes at the designed maximum junction temperature, which is 125°C in this work. It is seen from Fig.3 that, at T_J = 125°C, the avalanche breakdown of the diode AU2PJ-M3/86A starts at 980V with an avalanche current of about 50μA. A bias current of 100μA is sufficient to maintain a stable avalanche operation. The voltage across the load resistors is approximately the same as the diode avalanche voltage. Thus a resistance value of 10MΩ is selected for the load resistors R1-5.

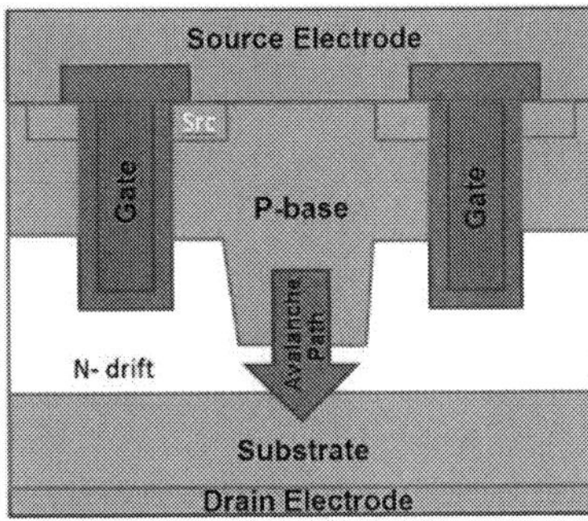

Fig. 2. Custom-designed low-voltage silicon MOSFET equipped with an avalanche clamp to ensure avalanche robustness.

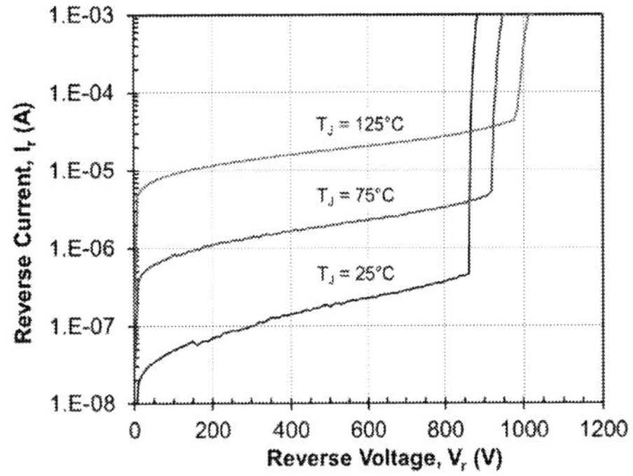

Fig. 3. Measured reverse characteristics of the avalanche diode AU2PJ-M3/86A at the junction temperature up to 125°C.

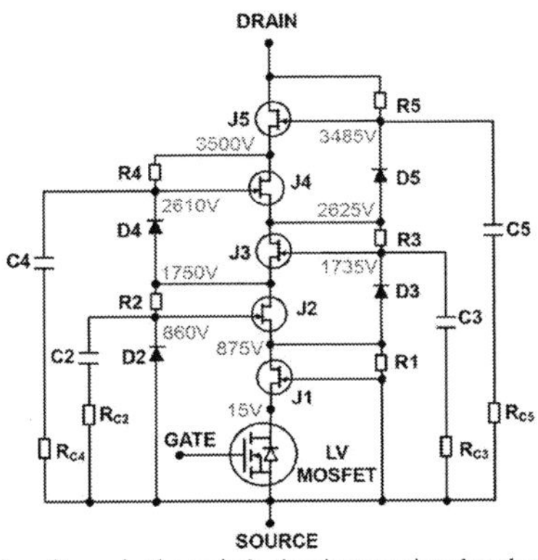

Fig. 4. Node voltages in the switch circuit assuming that the switch is in fully off-state, the JFET turn-off source-gate voltage is 15V and the diode avalanche voltage is 860V.

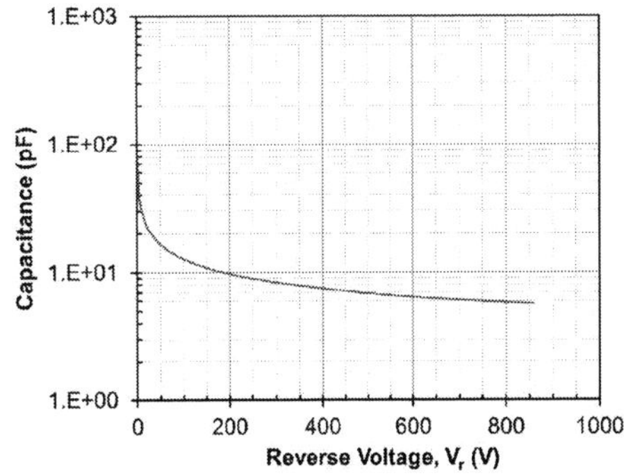

Fig. 5. Measured junction capacitance of the avalanche diode AU2PJ-M3/86A at 1MHz.

The capacitors C2-5 provide a current path for charging and discharging Crss of the JFETs J2-5. The capacitance values of the capacitors C2-5 need to be optimized to achieve the synchronization of the turn-on and turn-off transients and the voltage balance among the JFETs. The first-order approximate capacitance values of C2-5 can be estimated by assuming, at off-state, the capacitive charge at the gate node of the JFETs J2-5, respectively, is balanced as follows: for the gate node of each of the JFETs J2-5, the total charge stored in the voltage-balancing capacitor and the charge stored in the avalanche diode is equal to the total charge stored in the reverse-transfer capacitance Crss and gate-source capacitance Cgs of the JFET. For example, for the gate node of the JFET J3, the total charge stored in the capacitor C3 and the diode D3 is equal to the total charge stored in the Crss and Cgs of the JFET J3.

Fig.4 shows the node voltages in the switch circuit schematic to describe the voltages supported by the capacitors C2-5, the diodes D2-5 and the capacitors Crss and Cgs of the JFETs J2-5 assuming that the switch is in fully off-state, the JFET turn-off source-gate voltage is 15V and the diode avalanche voltage is 860V. It is seen that, for all of the JFETs J2-5, the voltage across the Crss is 890V and the voltage across the Cgs is 15V. Based on the capacitances presented in the datasheet of UJN1205Z, the total charge stored in the Crss and Cgs is calculated to be about 104.7nC. All of the diodes D2-5 block the same voltage of 860V. Fig.5 shows the measured junction capacitance of the avalanche diode AU2PJ-M3/86A. By integrating this C-V curve, the charge stored in the diode junction capacitance at 860V is found to be about 7.6nC. Thus, all of the voltage balancing capacitors C2-5 should have the same charge, which is equal to 97.1nC (104.7nC-7.6nC).

It is seen from Fig.4 that the voltage across the capacitors C2-5 is 860V, 1735V, 2610V, and 3485V, respectively, so the capacitance of the capacitors C2-5 is calculated to be 113pF (97.1nC/860V), 56pF (97.1nC/1735V), 37pF (97.1nC/2610V), and 28pF (97.1nC/3485V), respectively. These are the initial approximate values used for the capacitors C2-5 to start the experiments and may need adjusting according to experimental results. In this work, the excellent dynamic voltage balancing and switching transient synchronization have been achieved when C2 = 100pF, C3 = 68pF, C4 = 50pF, and C5 = 34pF.

The resistors R_{C2-5}, which are connected in series the capacitors C2-5, respectively, are used to damp the oscillations. The resistors R_{C2-5} can also be used to tune the di/dt rate and dv/dt rate of the switch.

IV. Experimental Results

A. Descriptions of the Power Switch

The switch under test includes five 1200V-45mΩ SiC normally-on JFETs (UJN1205Z) in die form and one 20V-5mΩ Si MOSFET in die form. The bias resistors R1-5 are all 10MΩ. The diodes D2-5 are avalanche diodes (AU2PJ-

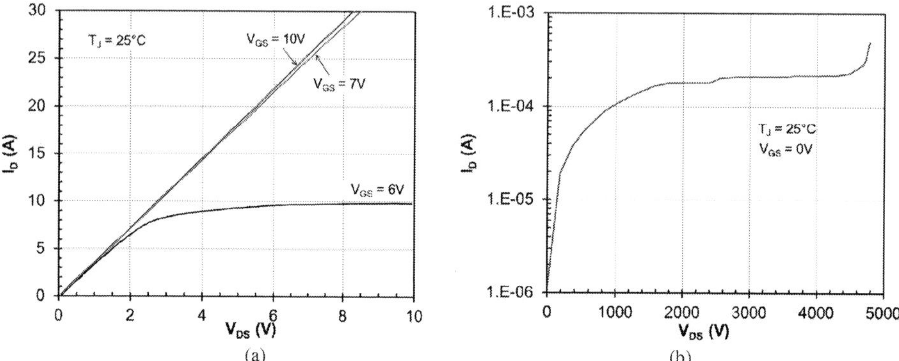

Fig. 6. Measured on-state characteristics (a) and blocking characteristics (b) of the power switch at room temperature.

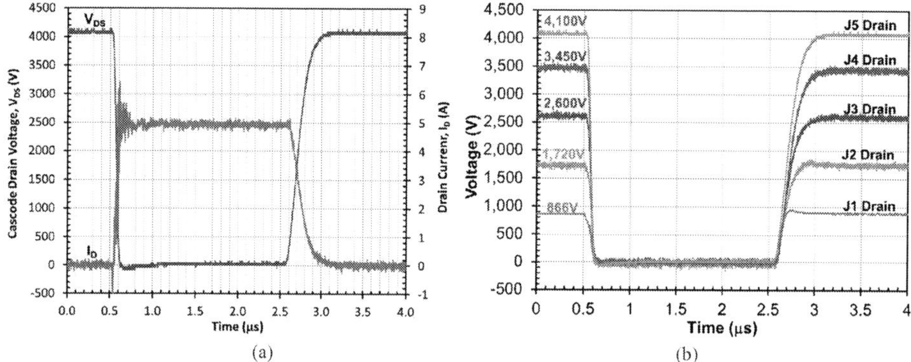

Fig. 7. Measured switching waveforms of the power switch (a) and measured drain voltage waveforms of the JFETs J1-5 during turn-on and turn-off transients (b) under a resistive load of 800Ω at room temperature.

M3/86A) with a stable avalanche voltage of about 860V. The capacitors C2, C3, C4 and C5 are high voltage ceramic capacitors with capacitance values of 100pF, 68pF, 50pF and 34pF, respectively.

B. DC Characteristics

Fig.6 shows the measured on-state and blocking characteristics at room temperature. It is seen that the switch is already fully on when V_{GS} =10V, meaning the switch can be driven with a standard IGBT/MOSFET gate driver. The on-resistance is 275mΩ at V_{GS} = 10V, which is about the sum of the on-resistances of the five JFETs and MOSFET. The blocking voltage is 4,780V at V_{GS} = 0V. Each of the JFETs J1-4 is designed to block about 860V, which is determined by the avalanche voltage of the corresponding diodes D2-5. The blocking voltage of the top JFET J5 is not clamped. So the JFET J5 needs to support any voltage above 3,440V (4 x 860V) and may go into avalanche. It is also seen that the leakage current increases with the drain voltage at first, and then is level off when the drain voltage is greater than about 1800V because, at V_{DS} = 1800V, the bias currents have been established for all of the avalanche diodes D2-5 and no more bias current is needed after 1800V. It is seen from Fig. 4 that the voltage across the load resistors R2 is 890V, so the bias current for the diodes D2 and D4 is 890V/10MΩ =89μA. Similarly, the bias current for the diodes D3 and D5 is also 890V/10MΩ =89μA. Thus, the total bias current is 178μA, which is very close to the switch leakage current of 180μA at 1800V. The slightly increase in the leakage current when V_{DS} > 1800V is because the JFET leakage currents increase with the drain voltage.

C. Dynamic Voltage Sharing Characteristics

Resistive load switching tests with a load resistor of 800Ω are performed to evaluate the dynamic voltage sharing characteristics of the JFETs during switching transients. Fig. 7 shows the measured switching waveforms of the switch and the measured drain voltage waveforms of the JFETs J1-5 during turn-on and turn-off transients. An excellent dynamic voltage sharing among the JFETs J1-5 has been achieved with very well synchronized turn-on and turn-off transients of the JFETs J1-5.

D. Power Switching Performance

The switching performance of the device is investigated under a 3,000V-29A inductive load condition. Three 1.2kV-60mΩ SiC cascode devices (UJC1206K) are connected in series to serve as the freewheeling diode. The output voltage of the gate driver is from -12V to +12V. The gate resistor R_G is 27Ω. A ferrite bead is connected to the gate of the device to reduce the oscillations on the gate signal. Fig. 8 presents the measured switching waveforms of the device at room temperature, showing a very fast switching speed. The turn-on rise time t_r is 41ns measured on drain voltage waveform from 90% to 10% of the peak drain voltage. The turn-off fall time t_f

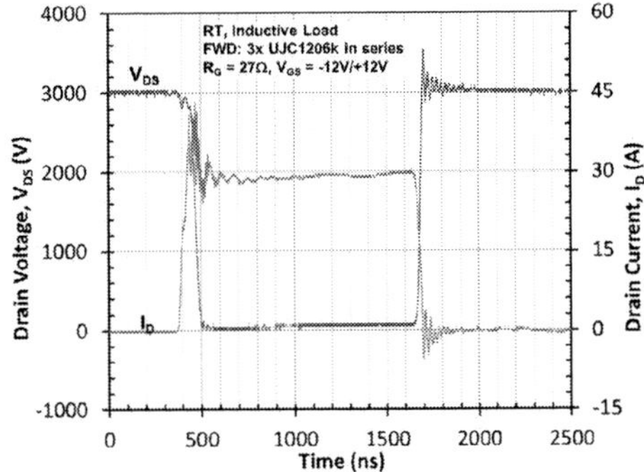

Fig. 8. Measured switching waveforms of the power switch under a 3kV-29A inductive load condition at room temperature.

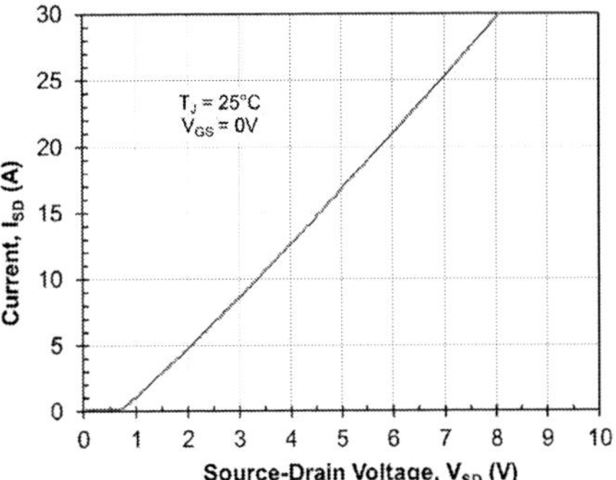

Fig. 9. Measured reverse conduction characteristics of the power switch at V_{GS} = 0V and room temperature.

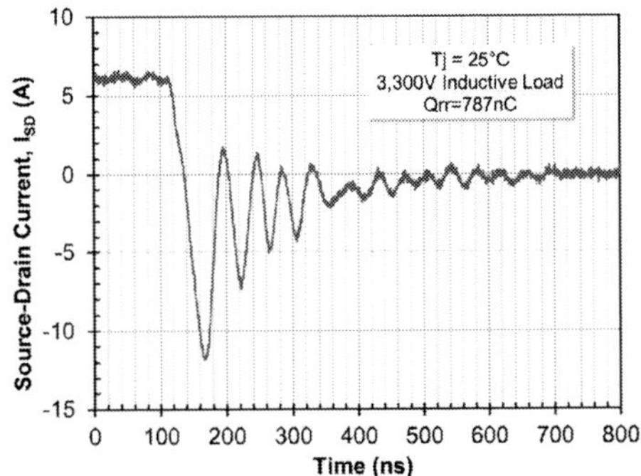

Fig. 10. Measured reverse recovery waveforms of the power switch under a 3.3kV-6A inductive load condition at room temperature.

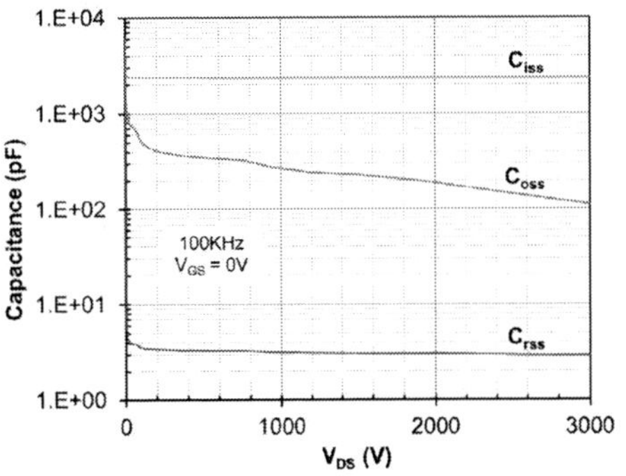

Fig. 11. Measured capacitance of the power switch at V_{GS} = 0V and frequency = 100kHz.

is 24ns measured on drain voltage waveform from 10% to 90% of the peak drain voltage. The turn-on energy loss E_{on} is 6.92mJ and turn-off energy loss E_{off} is 0.6mJ. The oscillations during turn-on and turn-off transients are caused by the large parasitic inductances in the testing circuit.

E. Reverse Conduction Characteristics

The proposed power switch can conduct a reverse current from SOURCE to DRAIN as described in [5, 7], which eliminates the requirement of an additional freewheeling diode and reduces the cost and size significantly. Fig.9 shows the measured reverse conduction (diode mode) characteristics at V_{GS} = 0V and room temperature. The knee voltage is as low as 0.7V, much smaller than the knee voltage of the body diode of SiC MOSFET that is typically 3V.

Fig. 10 shows the measured reverse recovery waveform of the power switch under 3300V-6A inductive load condition. The reverse recovery charge Qrr is measured to be about 787nC obtained by the integral of the measured reverse recovery current waveform. To determine whether this Qrr is storage charge or capacitive charge, the capacitances of the device are measured and presented in Fig. 11. Fig. 12 shows the device capacitive charge Qc as a function of the drain voltage V_{DS} obtained by integrating the curve of Coss vs. V_{DS} in Fig.11. By extrapolating the Qc vs. V_{DS} curve in Fig. 12 to 3,300V, the device capacitive charge Qc is estimated to be about 816nC at 3300V, which is basically equal to the measured Qrr of 787nC when considering the measurement and calculation error. This means the power switch has basically no storage charge in the diode conduction mode and its reverse recovery characteristics is very similar to that of SiC Schottky diodes.

V. MEDIUM-VOLTAGE POWER MODULE

A medium-voltage power module has been designed and fabricated by scaling up the PCB-version of the power switch discussed above. The module includes five JFETs and one 20V-5mΩ Si MOSFET connected in series. Each JFET contains two 1200V-45mΩ SiC JFETs connected in parallel to scale up the current rating of the power module. The module uses the same bias resistors R1-5 of 10MΩ and the same avalanche diodes D2-5 (AU2PJ-M3/86A) as the PCB-version power switch. The capacitance values of the voltage-balancing capacitors C2-5 are doubled to 200pF, 136pF, 100pF and 68pF in order to maintain the dynamic voltage balance and synchronization of the switching speeds among the JFETs during switching transients. The module has a custom-designed package as illustrated in Fig.13.

A. Module DC Characteristics

Fig.14 shows the measured DC characteristics of the power module at room temperature. The module is fully on at V_{GS} = 10V, meaning the standard IGBT/MOSFET gate drivers can be used to drive the module. The on-resistance is 127mΩ at V_{GS} = 10V. The blocking voltage is measured up to 4,600V at V_{GS} = 0V with a leakage current of 256μA at 4000V. Even though the

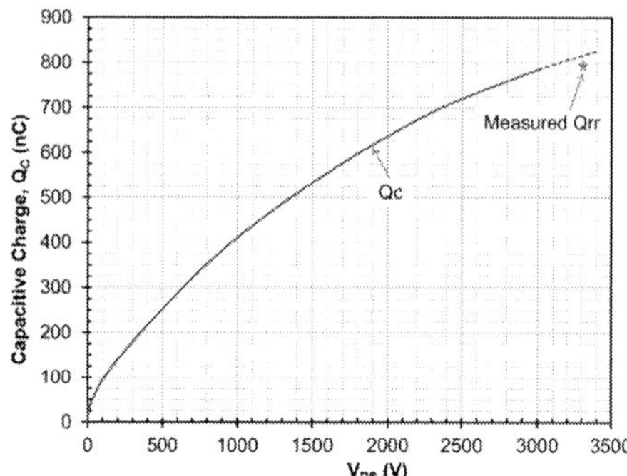

Fig. 12. Capacitive charge Qc of the power switch as a function of the drain voltage V_{DS} obtained by integrating the curve of Coss vs. V_{DS} in Fig.11.

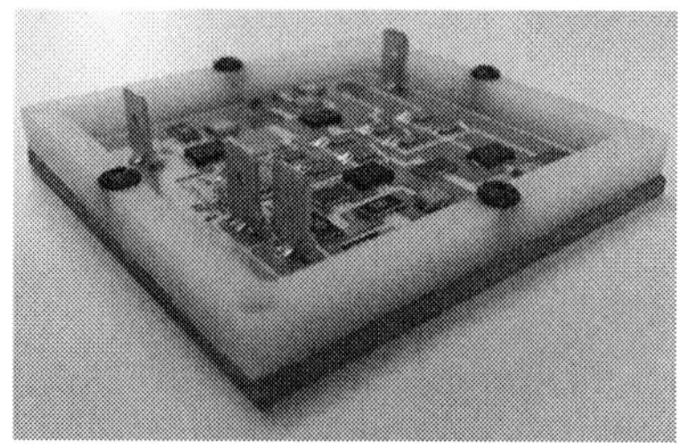
Fig. 13. Photograph of the proposed SiC JFET-based power switch module.

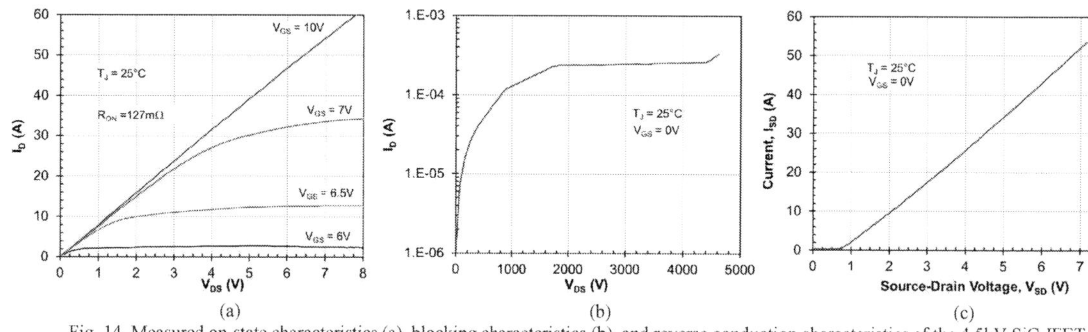

Fig. 14. Measured on-state characteristics (a), blocking characteristics (b), and reverse conduction characteristics of the 4.5kV SiC JFET based power module at room temperature.

avalanche breakdown voltage of the module is only about 4,600V, the module can be safely used for 3,000V class switching applications because the internal JFETs J1-4 are clamped, meaning the voltage-derating have been applied to these internal JFETs J1-4 already. Fig.14c shows the reverse conduction characteristics of the module with a low knee voltage of 0.7V.

B. Module Switching Characteristics

The switching performance of the power module is investigated under a 3,000V-54A inductive load condition. Three 6.5kV-15A SiC JBS diodes are connected in parallel to serve as the freewheeling diode. The output voltage of the gate driver is from -12V to +12V. The gate resistor R_G is 22Ω. A ferrite bead is connected to the gate of the module to reduce the oscillations on the gate signal. Fig. 15 presents the measured switching waveforms of the module at room temperature. The turn-on rise time t_r is 73ns measured on drain voltage waveform from 90% to 10% of the peak drain voltage. The turn-off fall time t_f is 37ns measured on drain voltage waveform from 10% to 90% of the peak drain voltage. The turn-on energy loss E_{on} is 14.8mJ and the turn-off energy loss E_{on} is 1.5mJ.

Fig. 16 shows the measured reverse recovery waveform of the power module under 3000V-22A inductive load condition. The reverse recovery charge Qrr is measured to be about 1410nC obtained by the integral of the measured reverse recovery current waveform, which is about twice the reverse recovery charge of the PCB-version power switch as expected.

VI. COMPARISON WITH 6.5KV SIC NORMALLY-OFF JFETS AND SI IGBTS

At present, the reported medium-voltage power switches suitable for 3000V-class power switching applications include 6.5kV SiC normally-off JFETs and 6.5kV Si IGBTs. The 6.5kV-100mΩ SiC normally-off JFET half-bridge modules have been reported in [11], which are based on the trenched-and-implanted vertical channel JFET structure [9, 12]. One of the disadvantages of the trenched-and-implanted vertical channel structure is that the device will have a large miller capacitance (Crss), which has significant effect on the gate drive and switching performance. Fig.17 shows the capacitances of the proposed 127mΩ medium-voltage power switch module and the capacitances of the 6.5kV-100mΩ SiC normally-off JFET module. It is seen that, compared to the 6.5kV SiC normally-off JFET module, the proposed power module has much lower miller capacitance Crss and much lower input capacitance Ciss. Because of the smaller capacitances, the proposed power module can be switched much faster as shown in Fig. 18, which compares the switching performance of the proposed module with the 6.5kV SiC normally-off JFET module under 3,000V-54A inductive load condition. A summary of the switching performance of these two modules is presented in Table 1. The switching performance of the latest generation 6.5kV-85A Si IGBT module QID6508001 [13] are also included in Table 1 for reference. The switching data of the IGBT module are extracted from the manufacturer's datasheet [13] and are taken at a bus voltage of 3.6kV and a junction temperature of 125°C.

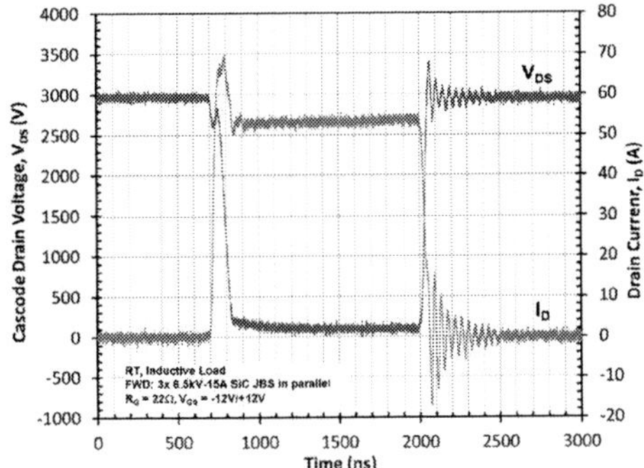

Fig. 15. Measured switching waveforms of the power switch module under a 3kV-54A inductive load condition at room temperature.

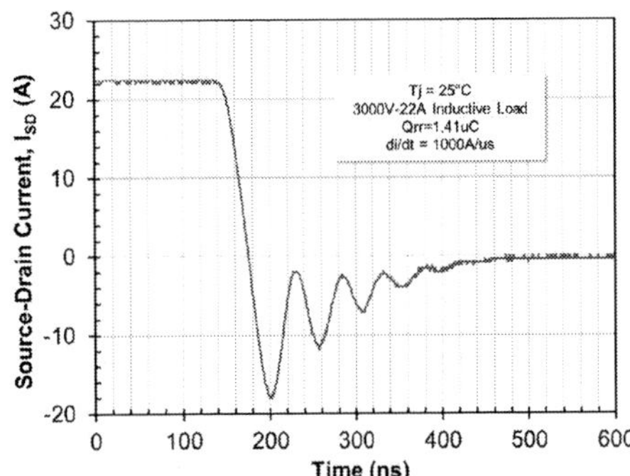

Fig. 16. Measured reverse recovery waveforms of the power switch module under a 3.0kV-22A inductive load condition at room temperature.

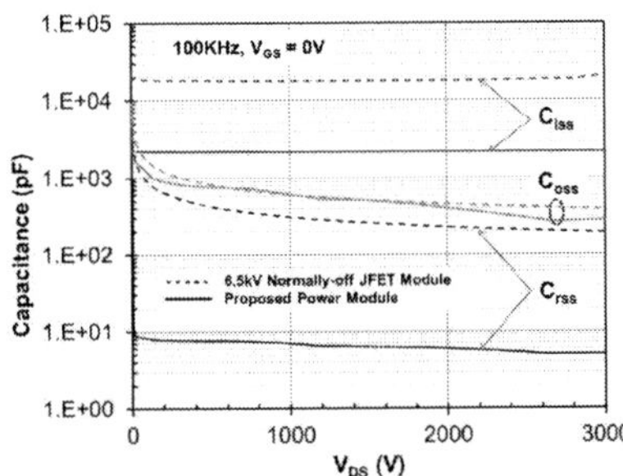

Fig. 17. Measured capacitances of the proposed power module and the 6.5kV-100mΩ SiC normally-off JFET module.

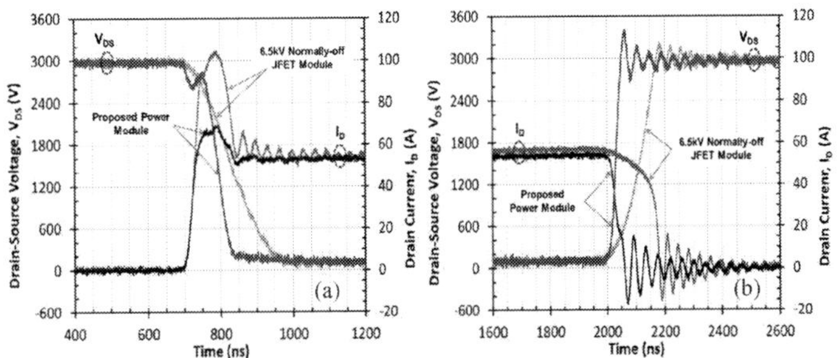

Fig. 18. Measured turn-on (a) and turn-off (b) waveforms of the proposed power module and the 6.5kV-100mΩ SiC normally-off JFET module under a 3.0kV-54A inductive load condition at room temperature.

However, both the proposed power module and the 6.5kV SiC normally-off JFET module are unipolar devices, their switching performance in general will not change much with the junction temperature. So, a first-order comparison can still be made with the data in Table 1.

TABLE I. COMPARISONS OF THE PROPOSED POWER MODULE, 6.5KV-100MΩ SIC NORMALLY-OFF JFET MODULE [11], AND 6.5KV-80A SI IGBT MODULE [13]

	Test Conditions	*Proposed Module*	*6.5kV SiC N-Off JFET Module*	*6.5kV-85A IGBT Module*
Turn-on Rise Time t_r	T_J =25°C 3kV-54A Inductive Load IGBT: T_J =125°C 3.6kV-54A	73ns	185ns	-
Turn-off Fall Time t_f		37ns	130ns	-
Turn-on Energy Loss E_{on}		14.8mJ	28.6mJ	323mJ
Turn-off Energy Loss E_{off}		1.5mJ	9.2mJ	375mJ
Reverse Capacitive/ Recovery Charge Qc/Qrr	T_J = 25°C 3kV-22A IGBT: T_J =125°C 3.6kV-85A	1.41μC	-	100μC

It is seen from Table 1 that the proposed power module has much lower switching energy losses than the other two modules. The turn-on energy loss of the proposed module is almost 2 times lower than the 6.5kV SiC normally-off JFET module, and 20 times lower than the 6.5kV IGBT module. The turn-off energy loss of the proposed module is 6 times lower than the 6.5kV SiC normally-off JFET module, and 250 times lower than the 6.5kV IGBT module. The reverse capacitive charge of the proposed module is negligible compared to reverse recovery charge of the 6.5kV IGBT module.

VII. CONCLUSIONS

A medium-voltage power switch formed by series-connecting five 1.2kV-45mΩ SiC normally-on JFETs is designed and experimentally demonstrated. The experimental results show that the excellent dynamic voltage sharing and switching transient synchronization can be achieved among JFETs. The power switch exhibits very fast switching speed. Under 3kV-29A inductive load condition, the turn-on rise time t_r is 41ns and the turn-off fall time t_f is 24ns. The power switch can conduct a reverse current from source to drain during off-state with a low knee voltage of only 0.7V, indicating the power switch can serve as an anti-parallel diode during off-state. The reverse recovery characteristics of the power switch is capacitive in nature, very similar to that of SiC Schottky diodes.

A medium-voltage power module is constructed by scaling up the proposed power switch. The module includes five JFETs and one 20V-5mΩ Si MOSFET connected in series. Each JFET contains two 1200V-45mΩ SiC JFETs connected in parallel to scale up the current rating of the power module. The module has an on-resistance of 127mΩ and can block above 4,500V at room temperature. The power module has a very fast switching speed and very low switching losses. Under 3kV-54A inductive load condition, the turn-on rise time t_r is 73ns and the turn-off fall time t_f is 37ns; the turn-on energy loss E_{on} is 14.8mJ and the turn-off energy loss E_{off} is 1.5mJ which are much lower than those of the 6.5kV-100mΩ SiC normally-off JFET module and the latest generation 6.5kV-85A Si IGBT module.

With these excellent switching performances, the proposed medium voltage power switch will be a strong candidate for the high-frequency and high-power density medium-voltage power conversion applications.

REFERENCES

[1] T. D. Heidel, D. Henshall, and P. Gradzki, "Strategies for wide bandgap, inexpensive transistors for controlling high-efficiency systems," CS MANTECH Conference, May 18th - 21st, 2015, Scottsdale, Arizona, USA, pp.7-10.

[2] T.V. Nguyen, P.O. Jeannm, E. Vagnon, D. Frey, and J.C. Crebier, "Series connection of IGBT," in Applied Power Electronics Conference and Exposition (APEC), 2010 Twenty-Fifth Annual IEEE , pp.2238 - 2244.

[3] D. Vinnikov "High-voltage switch realization possibilities for the 3.0kV dc fed voltage converters", 5th Int. Conf. Elect. Electron. Eng. (ELECO), 2007.

[4] Q. Xiao, Y. Yan, X. Wu, N. Ren and K. Sheng, "A 10kV/200A SiC MOSFET module with series-parallel hybrid connection of 1200V/50A Dies," Proceedings of the 27th ISPSD, May 10-14, 2015, Hong Kong, pp. 349-352.

[5] Xueqing Li, A. Bhalla, P. Alexandrov, J. Hostetler, and L. Fursin, "Series-connection of SiC normally-on JFETs," Proceedings of the 27th ISPSD, May 10-14, 2015, Hong Kong, pp.221-224.

[6] J. Biela, D. Aggeler, D. Bortis, J.W. Kolar, "Balancing circuit for a 5kV/50ns pulsed power switch based on SiC-JFET Super Cascode," IEEE Pulsed Power Conference, 2009, pp.635 - 640.

[7] U.S. patent No. 9190993.

[8] W. Bergner, F. Bjoerk, D. Domes, and G. Deboy, "Infineon's 1200V SiC JFET – The new way of efficient and reliable high voltages switching," Infineon Technologies, Tech. Rep., 2012.

[9] Xueqing Li, A. Bhalla, P. Alexandrov, and L. Fursin, "Study of SiC Vertical JFET Behavior during Unclamped Inductive Switching," APEC 2014 conference, March 16-20, 2014, Fort Worth, Texas, USA, pp.2588-2592. Also see the datasheet of UJN1205K at http://www.unitedsic.com.

[10] Anup Bhalla, Xueqing Li, and John Bendel, "Switching Behavior of USCi's SiC Cascodes" Magazine of Bodo's Power Systems, June 2015, pp.22-26.

[11] John Hostetler, Xueqing Li, Peter Alexandrov, Xing Huang, Anup Bhalla, "6.5kV enhancement mode SiC JFET based power module," The 3rd IEEE Workshop on Wide Bandgap Power Devices and Applications (WiPDA), November 2-4, 2015, Session B2L-C.

[12] J. L. Hostetler, P. Alexandrov, X. Li, L. Fursin, A. Bhalla, "6.5 kV SiC normally-off JFETs – technology status," Proceedings of The 2nd IEEE Workshop on Wide Bandgap Power Devices and Applications (WiPDA), October 13-15, 2014, pp.143-146.

[13] See the datasheet of QID6508001 at http://www.pwrx.com.

Numerical Model and Experimental Study on Comparison of Semiconductor Pulsed Power Devices

Lin Liang, Changdong Chen
School of Optical and Electronic Information
Huazhong University of Science and Technology
Wuhan, China
lianglin@hust.edu.cn

Fang Luo
Center for High Performance Power Electronics
Ohio State University
Columbus, USA

***Abstract*—The characteristics applied in pulsed power area for the power semiconductor devices of RSD(reversely switched dynistor), thyristor and IGBT are compared synthetically in this paper. By establishing the two-dimensional electro-thermal coupling models for the three types of devices, the turn-on performances of each one at low and high current density are discussed, separately. The results show that the special type device RSD always acquires the lowest turn-on voltage and power dissipation. This advantage is more obvious with the current density increase. The experimental results have approved the trend.**

Keywords—RSD, thyristor, IGBT, pulse power, electro-thermal coupling models

I. Introduction

Modern pulsed power technology has been widely applied in the military areas, such as inertial confinement fusion, electromagnetic gun, etc, and civil areas, such as environmental protection, new material development, etc[1]. Due to the advantage of small volume and high reliability, semiconductor switches have been more and more applied in pulsed power area[2]. RSD(reversely switched dynistor) is a special type of pulse power device proposed by Prof. I. V. Grekhov from Russia[3], which has low dissipation and high di/dt capability[4]. Other common power semiconductor devices such as thyristor and IGBT, etc, could also be used in pulsed power area[5-6], with special working condition.

The two-dimensional numerical models for RSD, thyristor and IGBT are established in this paper, separately. The electro-thermal coupling simulations are carried out at low and high current density. The influences of the triggering conditions on the turn-on process have been discussed for each device. The turn-on performances for the three kinds of devices are compared synthetically for the first time. RSD has acquired the lowest turn-on dissipation on the same condition, which has also been approved by the experiments.

II. Operation Principle and Models

A. Operation Principle

This work was supported by the National Natural Science Foundation of China (51377069).

Figure 1 shows the cell structure of RSD. As a two-electrode device, it turns on based on the principle of the controllable plasma layer commutation and has low dissipation. From the anode, it consists of many alternating P^+ and N^+ small units, so it is the parallel structure of P^+NPN^+ thyristor cell and N^+NPN^+ transistor cell. A short reverse current from cathode to anode makes the plasma layer P_1 form near the central collector junction J_2. In the forward discharge process, the electric field pointing from anode to cathode makes the electrons of P_1 layer move to N base and holes move to P base, which leads to the minority carrier injections from the corresponding emitter regions. Then the conductance modulation is realized and the device turns on.

As common power semiconductor devices, the operation principles of the thyristor and IGBT are not stated in detail here. Put simply, thyristor is current triggered and pure bipolar device, the advantage of which is low forward voltage drop and high power capacity, while IGBT is voltage triggered and unipolar-bipolar hybrid device, the advantage of which is balance between power and speed, and simpler triggering circuit, etc. It is special operation status for them to apply in the pulsed power circuits.

B. Circuit Models

Figure 2 shows the simulated circuits of pulse discharge for the three devices. Thereinto, RSD is triggered by reverse pre-charge current, thyristor is triggered by gate current and IGBT is triggered by gate voltage. The main discharge circuits

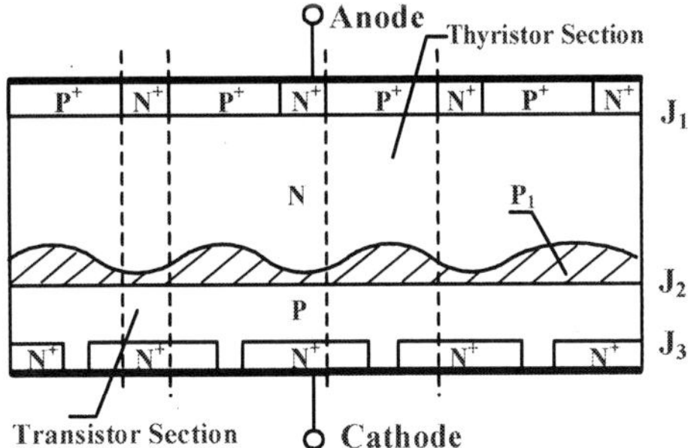

Fig. 1. RSD structure.

keep the exact same conditions.

Figure 2(a) shows the RSD-based discharge schematic with resonant triggering. When the triggering switch S is closed, the triggering capacitor C_C begins to discharge. When its lower plate gets positive charge, the resonant current provides pre-charge for RSD. The C_1 is the main capacitor. The MS is the saturable core choke. The R_m is the resistive load and the L_m is the series inductance in the main loop.

Figure 2(b) and (c) show the simulated schematics of thyristor and IGBT, respectively. The main loops keep the same conditions as that of the RSD. The triggering circuits are simulated by the series of the voltage source U_i and resistor R_i, providing gate current for the thyristor, or gate voltage for the IGBT, separately.

C. Device Models

The two-dimensional electro-thermal coupling models for the three types of devices are established in TCAD tools. Figure 3 shows the doping profile designs and the blocking characteristics simulations. All designs are for the blocking voltage of 1kV with optimized structures.

The structural parameters(the width and doping concentration) in the drift region are as follows: (1) RSD, 2.1E14 cm^{-3}, 95μm; (2) Thyristor, 1.2E14 cm^{-3}, 115μm; (3) IGBT, 1.8E14 cm^{-3}, 94μm. The simulated blocking voltages are 1040V, 1016V and 1004V, respectively. The electric field distribution curves shown in Fig. 3 are acquired when 1000V is applied to the devices. The electric field for each device almost expands to the boundary of the drift region, thus the conducting resistor is reduced as much as possible with the required blocking voltage.

Except for the Poisson equations and the carrier continuity equations, the lattice temperature equations are also solved. The physical models mainly contain the mobility model, the generation-recombination rate model and heavy doping model, etc, and the electro-thermal effect is considered.

III. Simulation Results and Discussions

Based on the device and circuit models established in Section II, the two working conditions of the low current density(~1.4kA/cm^2) and the high current density(~22kA/cm^2) of the pulse turn-on for the three types of devices are simulated. Figure 4 to Figure 6 are for low current density. Figure 7 to Figure 9 are for high current density. From Figure 4 and Figure 7, there exists a critical turn-on condition for the three devices. The turn-on voltage reaches a saturated minimum value with the critical condition satisfied. Figure 5 and Figure 8 show the electro-thermal coupling simulated results for the three devices at low and high current density, respectively, with the critical triggering condition, which contain the transient distributions for voltage, current density, power and maximum lattice temperature. RSD always gets the lowest voltage drop and this advantage is more obvious at high current density. Figure 6 and Figure 9 show the switching energy loss.

A. Low Current Density

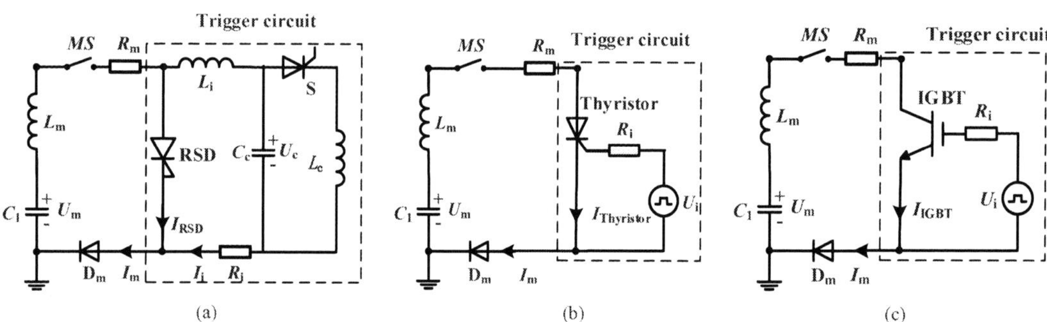

Fig. 2. Simulated circuit of pulse discharge. (a) RSD-based. (b) Thyristor-based. (c) IGBT-based.

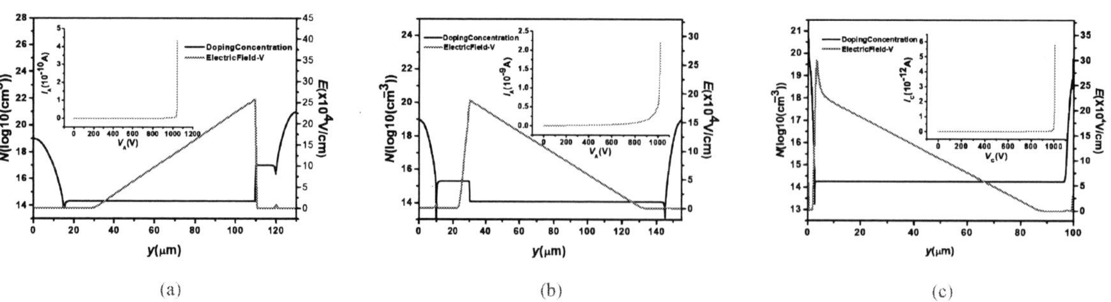

Fig. 3. Doping profile and blocking characteristics of devices. (a) RSD with y=0 at anode. (b) Thyristor with y=0 at cathode. (c) IGBT with y=0 at emitter.

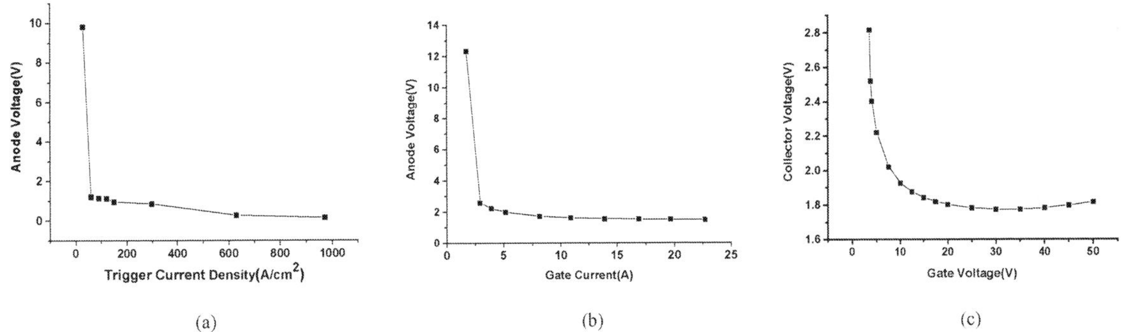

Fig. 4. Turn-on voltage on the devices with different triggering conditions at low current density. (a) RSD. (b) Thyristor. (c) IGBT.

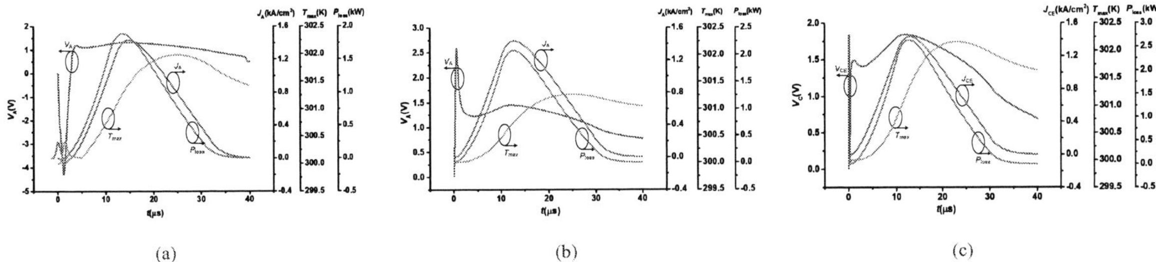

Fig. 5. Simulated results for voltage, current density, power and maximum lattice temperature versus time in pulse turn-on process at low current density. (a) RSD. (b) Thyristor. (c) IGBT.

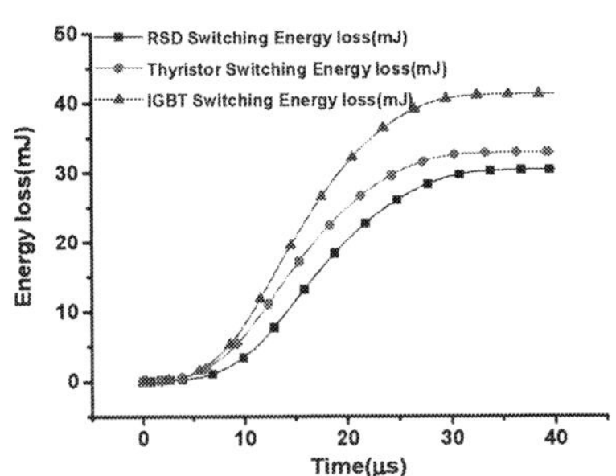

Fig. 6. Switching energy loss of the three devices at low current density.

The C_1 in Fig. 2 is 10μF and the original voltage is 1000V. The L_m is 1.5μH and the R_m is 0.6Ω. The chip area is 1cm^2 and the acquired peak current is about 1.4kA. The turn-on voltage of the RSD is reduced as the triggering current is increased. The critical value of the triggering current density is 58.96A/cm^2. The decrease of the turn-on voltage becomes slow after that point. The triggering characteristics of the thyristor are similar to that of the RSD. Its critical gate triggering current is 2.91A. For the IGBT, it could be well triggered with the gate voltage of 15V. When the gate voltage is above 35V, the turn-on voltage increases with the increase of the gate voltage instead. Figure 5 shows the turn-on waveforms corresponding to the critical triggering conditions for the three types of devices. For the RSD, thyristor and IGBT, the turn-on dissipation energy is 304.70mJ, 329.70mJ and 412.90mJ,

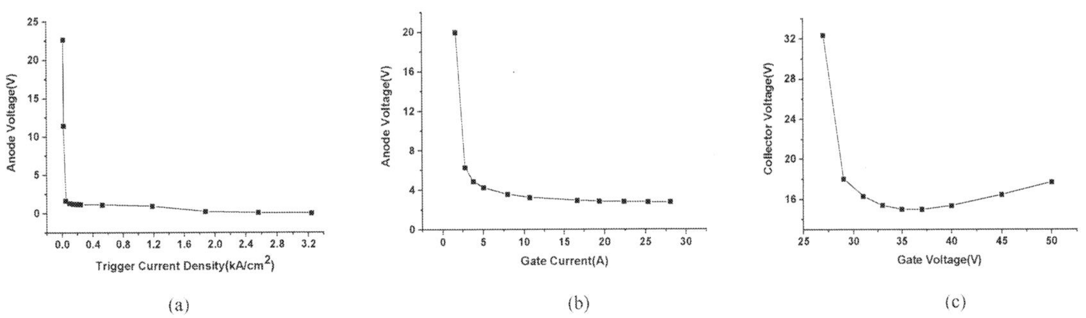

Fig. 7. Turn-on voltage on the devices with different triggering conditions at high current density. (a) RSD. (b) Thyristor. (c) IGBT.

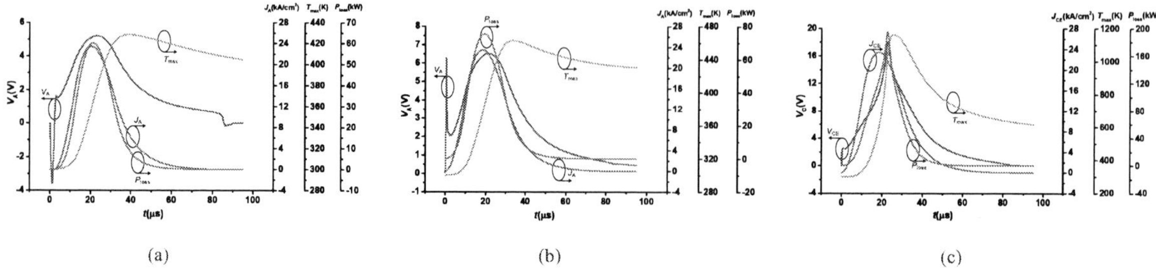

Fig. 8. Simulated results for voltage, current density, power and maximum lattice temperature versus time in pulse turn-on process at high current density. (a) RSD. (b) Thyristor. (c) IGBT.

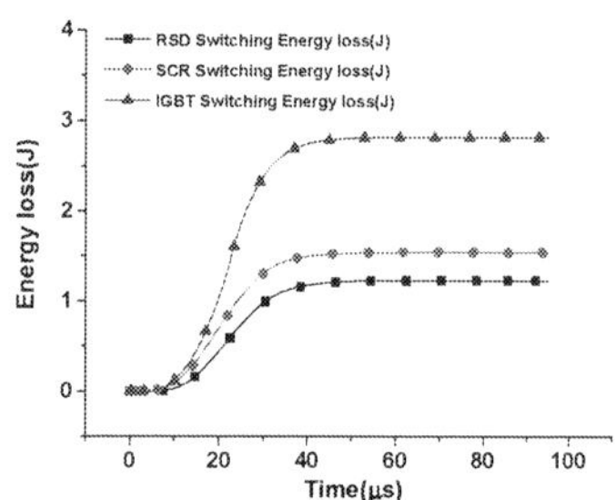

Fig. 9. Switching energy loss of the three devices at high current density.

respectively. Generally speaking, the lattice temperature rise is little and the difference is small. All the three devices could work normally on this condition.

B. High Current Density

The C_1 in Fig. 2 is 10μF and the original voltage is 1000V. The L_m is 0.5μH and the R_m is 0.05 Ω. The chip area is 0.5cm^2 and the acquired peak current is about 22kA. Similar to the low current density cases, the critical triggering conditions still exist. The critical value of the triggering current density is 46.56A/cm^2 for the RSD. The critical gate triggering current is 3.82A for the thyristor. The critical gate voltage is 27V for the IGBT and the turning point is at 37V.

On the critical triggering conditions, the temperature rise in the RSD caused by the high current density is 128K. The lattice temperature of the RSD is 428K. The lattice temperature of the thyristor is 463K, which is 34K higher than that of the RSD. It is because that the RSD could turn on uniformly over the whole chip area due to its special operation principle based on the controllable plasma layer. The power dissipation is small and the temperature rise is low. Anyway, according to the intrinsic temperature for the silicon material 550K, both of them are still in the normal range. But the simulation results show that the peak power dissipation is 195kW and the maximum lattice temperature is 1161K for the IGBT. Actually it has failed already, showing that it is not fit for the current commutation of such high current density.

IV. Experimental Verifications

According to the simulated circuits in Figure 2, a practical pulse discharge platform is established as shown in Figure 10. Similarly, the main circuit conditions keep the same and each one gets good triggering. The voltage on the device is measured by the high-voltage differential probe P5200A from Tektronix. The main capacitor C_1 is 10μF and the original voltage is 1000V. The load resistor R_L is 0.2 Ω. The rated voltage and current for the three devices are: (1) RSD, 2500V/200A; (2) Thyristor, 1200V/200A; (3) IGBT, 1200V/600A. Though the rated values are not the same, in fact the RSD is provided with hasher conditions. The experimental current and voltage waveforms are shown in Figure 11. Though due to the measurement error caused by contact voltage drop, etc, the measured value is higher than the simulated value, the trend of the maximum turn-on voltage has been verified.

V. Conclusions

The two-dimensional electro-thermal coupling models for RSD, thyristor and IGBT are established in this paper. Their

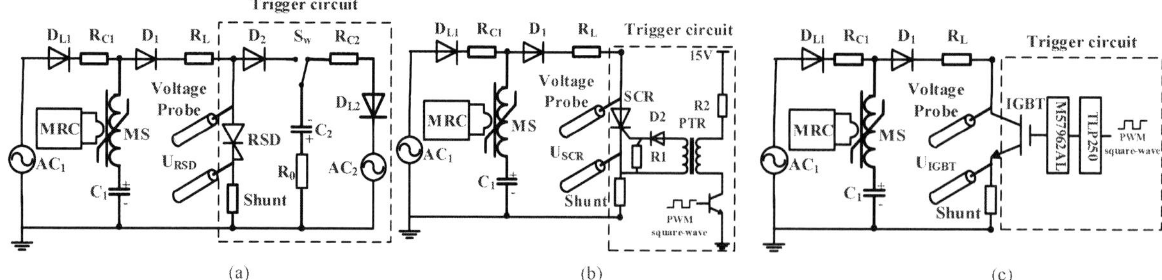

Fig. 10. Experimental circuit of pulse discharge. (a) RSD-based. (b) Thyristor-based. (c) IGBT-based.

978-1-4673-9551-9/16 $31.00 © 2016 IEEE

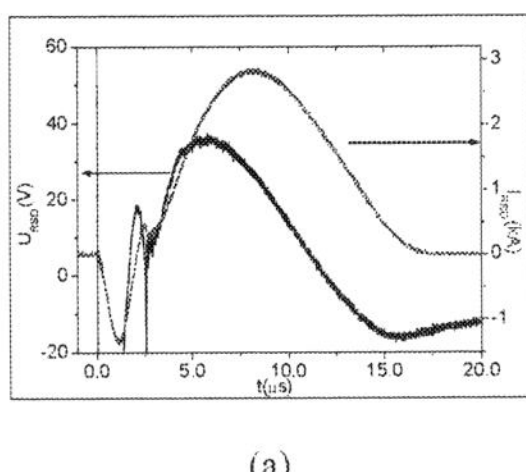
(a)

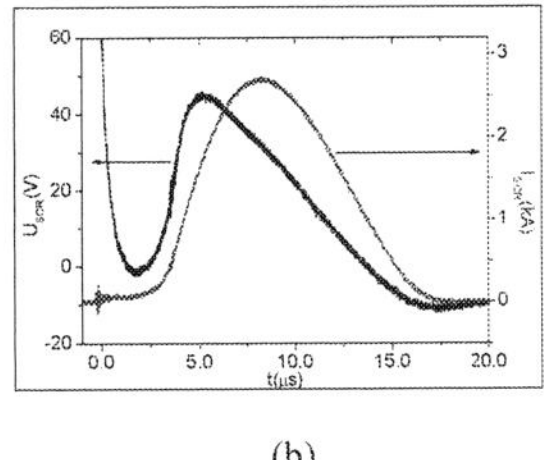
(b)

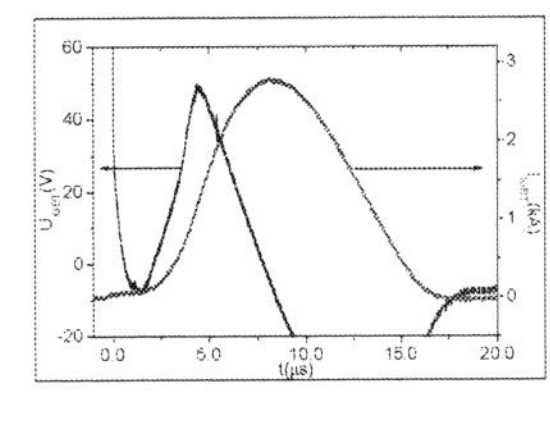
(c)

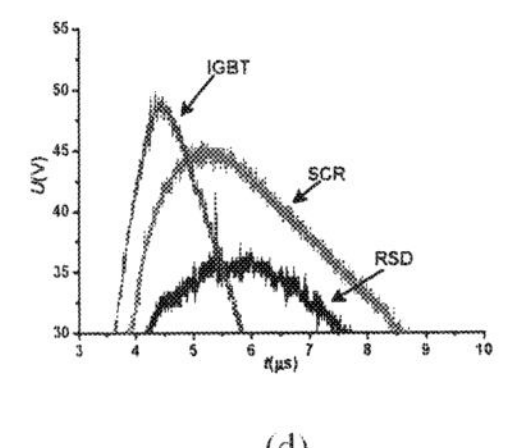

(d)

Fig. 11. Experimental turn-on voltage and current waveforms. (a) RSD. (b) Thyristor. (c) IGBT. (d) Zoom-in comparison for the maximum part of voltage.

turn-on characteristics applied in pulse power are compared systematically and quantitatively. The results show that the gap is not large at low current density, while RSD reflects obvious advantage of low dissipation at high current density. In addition, the critical triggering conditions exist for each device. The conclusion provides some basis for users on choosing switches. It has been verified by experiment in some extent.

REFERENCES

[1] M. V. Fazio, H. C. Kirbie, "Ultracompact pulsed power," *Proceedings of the IEEE, Pulsed Power: Technology and Applications*, vol. 92, no. 7, pp. 1197-1203, 2004.

[2] W. Jiang, K. Yatsui, K. Takayama, et al, "Compact solid-State switched pulsed power and its applications," *Proceedings of the IEEE, Pulsed Power: Technology and Applications*, vol. 92, no. 7, pp. 1180-1195, 2004.

[3] I. V. Grekhov, A. V. Nalivkin, "Theory of quasi-diode operation of reversely switched dynistors," *Solid-State Electron.*, vol. 31, no. 10, pp. 1483–1491, Oct. 1988.

[4] L. Liang, Q. Wei, Y. Yu, "Two-Dimensional Numerical Model and Turn-On Performance Simulation of Reversely Switched Dynistor," *IEEE Trans. Power Electron.*, vol. 29, no. 1, pp. 522-528, Jan. 2014.

[5] C. Strowitzki, M. Dahlke "Parametric Measurements of switching Losses of IGBT's in Pulsed Power Applications," *Power Modulator and High Voltage Conference (IPMHVC), 2012 IEEE International,* June, 2012, San Diego, pp. 324-327.

[6] D. Wang, K. Liu, J. Qiu, "Investigation on the performance of thyristors for pulsed power applications", *IEEE Trans. Dielectrics and Electrical Insulation.*, vol. 19, no. 2, pp. 182-185 , April 2012.

A Normalization Procedure of DC-side Stray Inductance for High-speed Switching Circuit

Masato Ando and Keiji Wada
Department of Electrical Engineering
Tokyo Metropolitan University
1-1, Minami-Osawa, Hachioji, Tokyo, 192-0397, Japan
Email: kj-wada@tmu.ac.jp

***Abstract*—Recently, high-speed switching circuits using SiC and GaN power devices have been developed for realizing higher efficiency. Stray inductance caused by the wiring structure between a DC capacitor and power devices is one of the most critical parameters for these high-speed switching circuits. In this paper a DC-side stray inductance design procedure for a high-speed switching circuit is presented based on a normalization procedure. The stray inductance is presented not as the absolute value [H] but as the percent value [%] based on the power rating of the converter circuit. By applying the proposed method, the stray inductance can be designed for a circuit depending on the switching time and the voltage and current ratings. To verify the normalization method, experimental results are shown using an all-SiC module at voltage and current ratings of 500 V and 100 A, respectively.**

I. Introduction

Wide-bandgap power devices (SiC and GaN) have been developed to achieve high-speed switching. Because stray inductance affects over-voltages under turn-off operations and increases switching losses of the power devices, it is important to analyze and design the DC-side stray inductance of the converter circuit used in these power devices.

Fig. 1 shows the relationship between the rated impedance of power electronics circuits and percentage of over-voltages [1]~[5]. The circuit impedance of the DC-side circuits is defined as $V_{\mathrm{DC}}/I_{\mathrm{D}}$, where V_{DC} and I_{D} are the voltage and current ratings, respectively. It is clear that the percentage of the over-voltage for the circuit in wide-bandgap power devices is much lower compared to that of Si devices. This means that the stray inductance has a significant effect on high-speed switching circuits. Table I lists the detail circuit parameters of Fig. 1. From this table, it is clear that the over-voltage values vary with the voltage and current ratings of the circuit.

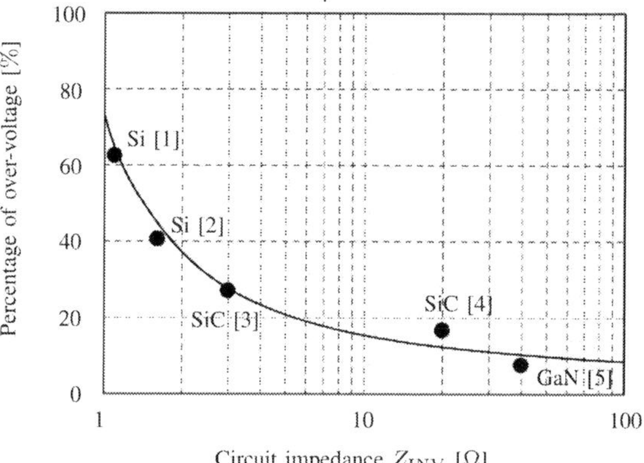

Fig. 1. Relationship between rated impedance of the power electronics circuits and percentage of over-voltage for various power devices and applications.

The relationship between the stray inductance and the switching characteristics through experiments and simulations has been demonstrated in many papers [6]~[12]. However, how to design the stray inductance of the DC-side circuit has not been sufficiently shown from these results. Generally, a DC capacitor and power devices are selected by the circuit designer from the parts list depending on both applications and power ratings. In this case, equivalent series inductances (ESL) of those components are found from their datasheets. Therefore, the circuit designer has to calculate the upper limit of the stray inductance of the wire connected between the DC capacitor and the power devices. In general, the stray inductance depends on the wiring structure from the DC capacitor to the power devices; hence, wiring structures that reduce the stray inductance have been proposed [7]-[8]. In addition, the authors have proposed an inductance map for designing the stray inductance of a bus bar without using a three-dimensional finite element method (FEM) [9]. In these papers case studies of the relationship between the absolute value of the stray inductance and the wiring structure have been discussed. Therefore, it is unclear whether the upper limit of the stray inductance and the acceptable over-voltage value depend on the voltage and current ratings of the circuit. As a result, it is necessary to discuss a generalized design procedure for the stray inductance depending on the voltage and current ratings of the circuit under high-speed switching operation.

This paper shows a stray inductance design method based on a normalization procedure by considering the voltage and current-ratings and the switching time of the power devices. This paper treats the relationship among the stray inductance of the DC-side wire and ESL of the DC capacitor and the power devices, where the stray inductance means the sum of those inductances. Normalization implies that the impedance of the stray inductance is shown as a percentage of the rated impedance of the power converter circuit. As a result, the stray inductance has to be shown not as the absolute value

978-1-4673-9551-9/16 $31.00 © 2016 IEEE

TABLE I
EXAMPLES OF STRAY INDUCTANCE FOR VARIOUS APPLICATIONS

Items	UPS [1]	PV-PCS [2]	Traction [3]	Industry [4]	Industry [5]
Voltage and current rating of the circuit	440 V, 400 A	565 V, 350 A	600 V, 200 A	600 V, 30 A	400 V, 10 A
Power devices	Si-IGBT	Si-IGBT	SiC-MOSFET	SiC-MOSFET	GaN-HEMT
Percentage of over-voltage ($\frac{v_{\rm L}(t)}{V_{\rm DC}} \times 100$)	62.5%	40.5%	27%	16.7%	7.5%

[H] but as the percent value [%], which takes the switching time into consideration. Analytical and experimental results can be obtained using a SiC-MOSFET and a SiC-SBD, rated at 500 V and 100 A, buck chopper circuit, and it is clear that the over-voltage value depends on both the turn-off current and the stray inductance. The objective of this normalization procedure is to clarify the relationship between the voltage and current ratings of the circuit and the upper limit of the stray inductance.

II. ANALYSIS METHOD USING A MOSFET DEVICE MODEL

A. Circuit configuration and method for analyzing over-voltage

Fig. 2 shows the circuit configuration of the buck chopper circuit used for analysis and experiment. $L_{\rm bus}$ indicates the stray inductance of the DC-side wire, $L_{\rm d}$ is the ESL of the DC capacitor and power devices, and $L_{\rm s}$ indicates the source-side inductance of the MOSFET. $R_{\rm g}$ is the external gate resistance of the MOSFET, and $I_{\rm D}$, the output current through the inductor L is the turn-off current.

In this circuit, $L_{\rm bus}$ and $L_{\rm d}$ affect the amplitude of the over-voltage across the power devices under the turn-off operation of the MOSFET. Fig. 3 shows a typical turn-off waveform of the MOSFET including the occurrence of over-voltage. The over-voltage $V_{\rm p}$ can be calculated by adding the DC voltage $V_{\rm DC}$ and the EMF $v_{\rm L}(t)$ (electromotive force) of the stray inductance, as follows:

$$V_{\rm p} = V_{\rm DC} + v_{\rm L}(t) = V_{\rm DC} + (L_{\rm d} + L_{\rm bus}) \times \frac{{\rm d}i_{\rm d}(t)}{{\rm d}t}. \quad (1)$$

According to Fig. 3, the switching speed ${\rm d}i_{\rm d}(t)/{\rm d}t$, which causes the over-voltage, can be replaced by the turn-off current $I_{\rm D}$ and the turn-off time $\delta_{\rm f}$:

$$\frac{{\rm d}i_{\rm d}(t)}{{\rm d}t} = \frac{I_{\rm D}}{\delta_{\rm f}} \quad (2)$$

Here, the drain current $i_{\rm d}(t)$ can be determined by the following equation using the equivalent device model of the MOSFET [13]:

$$i_{\rm d}(t) = g_{\rm m}\{v_{\rm gs}(t) - V_{\rm t}\}, \quad (3)$$

where $g_{\rm m}$ is the transconductance of the MOSFET, $v_{\rm gs}(t)$ is the gate-to-source voltage of the MOSFET, and $V_{\rm t}$ is the threshold voltage of the MOSFET. From (1) and (3), the $v_{\rm L}(t)$ can be calculated by using the equation

$$v_{\rm L}(t) = (L_{\rm d} + L_{\rm bus}) g_{\rm m} \frac{{\rm d}v_{\rm gs}(t)}{{\rm d}t}. \quad (4)$$

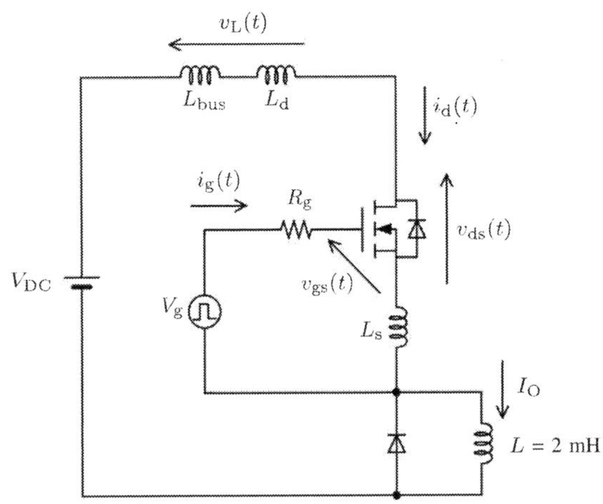

Fig. 2. Circuit configuration of the buck chopper circuit.

During the turn-off operation, $v_{\rm gs}(t)$ is represented by the following equation:

$$A\frac{{\rm d}^2 v_{\rm gs}(t)}{{\rm d}t^2} + B\frac{{\rm d}v_{\rm gs}(t)}{{\rm d}t} + v_{\rm gs}(t) = 0, \quad (5)$$

$$\text{where} \quad A = R_{\rm g} C_{\rm gd} g_{\rm m}(L_{\rm d} + L_{\rm bus} + L_{\rm s}), \quad B = R_{\rm g}(C_{\rm gd} + C_{\rm gs}) + g_{\rm m} L_{\rm s}. \quad (6)$$

The second-order differential equation (5) is solved when the following equation is satisfied:

$$B^2 - 4A \geq 0. \quad (7)$$

Then, the switching speed ${\rm d}i_{\rm d}(t)/{\rm d}t$ is given as

$$\frac{{\rm d}i_{\rm d}(t)}{{\rm d}t} = (g_{\rm m} V_{\rm th} + I_{\rm o})\sqrt{\frac{4}{B^2 - 4A}} \exp\left(-\frac{B}{2A}t\right) \times \cosh\left(\frac{\sqrt{B^2 - 4A}}{2A}t\right). \quad (8)$$

B. Procedure for normalizing stray inductance

The percent impedance $\%Z$ of the stray inductance, which is proposed in this paper, is defined by the following equation:

$$\%Z = \frac{Z_{\rm L}}{Z_{\rm INV}} \times 100\ [\%], \quad (9)$$

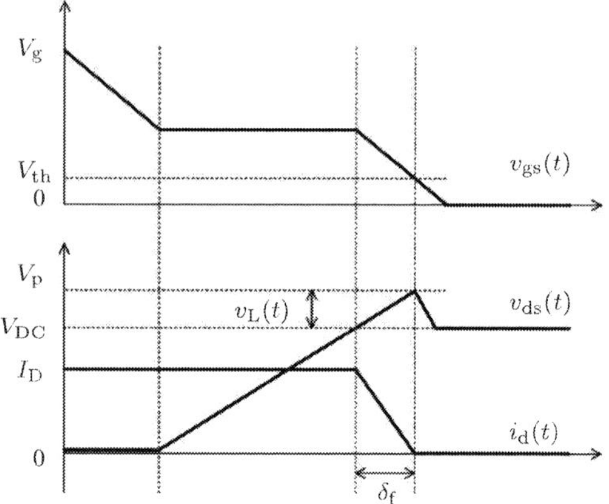

Fig. 3. Typical turn-off waveforms of the MOSFET including the occurrence of over-voltage.

TABLE II
ANALYTICAL PARAMETERS OF THE SiC-MOSFET

Parameter	Value
Gate-Drain Capacitance C_{gd}	13 pF
Gate-Source Capacitance C_{gs}	1902 pF
Trans-conductance g_m	7.3 S
Gate Threshold Voltage V_t	2.5 V
Internal Gate Resistance R_g	5 Ω

where Z_L is the impedance of the stray inductance. The impedance Z_L [Ω] of the stray inductance can be defined in henrys per second ([H]/[s]) as follows:

$$Z_L = \frac{(L_d + L_{bus})}{\delta_f}. \ [\Omega] \tag{10}$$

The rated circuit impedance Z_{INV} [Ω] is also defined as the voltage per current ([V]/[A]):

$$Z_{INV} = \frac{V_{DC}}{I_D}. \ [\Omega] \tag{11}$$

From the above equation, the percent impedance $\%Z$ is defined by the following equation:

$$\%Z = \frac{Z_L}{Z_{INV}} \times 100 = \frac{(L_d + L_{bus})/\delta_f}{V_{DC}/I_D} \times 100 \ [\%]. \tag{12}$$

III. Analytical Results and Experimental Verification Using a SiC-MOSFET and a SiC-SBD

A. Analytical and experimental results for the over-voltage

In this analysis and experiment, a SiC-MOSFET (CMF20120D, Cree) and a SiC-SBD (C2D10120, Cree) rated at 1,200 V were used. Table II lists the analytical parameters of the SiC-MOSFET. These device parameters are taken from the datasheets. This experimental circuit has no snubber circuit around the power devices and no external gate resistance is connected to this discrete SiC-MOSFET for achieving high-speed switching operation. This is because to clarify the over-voltage occurrence remarkably. However, the SiC-MOSFET is connected to a 5 Ω gate resistance inside the package. In this experiment, the DC capacitor and power devices are connected through a laminated bus bar, whose thickness and width are set to 0.5 and 50 mm, respectively. Two types of bus bar with different lengths and distances are used to vary the stray inductance of the bus bar.

Fig. 4 shows analytical and experimental results for the relationship between $v_L(t)$ and the turn-off current I_D. Here, $v_L(t)$ is described as the percentage of the DC voltage, which is 500 V. The solid line shows the analysis results with the source-side inductance L_s of 2.5 nH considered, and the dashed line presents the results in the case of no source-side inductance calculated using (4) and (5). The source-side inductance is the sum of the inductances of the bonding wire and external lead wire in the package (TO-247) of the power devices, and it is on the order of several nanohenrys [14].

It is clear that the over-voltage with the source-side inductance considered is lower than that without the source-side inductance. In addition, even if the stray inductance $L_d + L_{bus}$ is set to a constant value, the over-voltage increases with a larger turn-off current of the MOSFET. The experimental results are indicated by solid circles •. From the analytical and experimental results, it is apparent that analyzing the over-voltage accurately requires consideration of the source-side inductance. In the case of using the same MOSFET and gate drive circuit, the upper limits of the stray inductance are different for different turn-off currents; that is, the stray inductance depends on the voltage and current ratings of the circuit using the same power devices and a gate drive circuit.

Fig. 5 shows the relationship between the over-voltage and the stray inductance. It is clear that the over-voltage increases with increasing stray inductance. In addition, it is clear that the stray inductance has to be designed appropriately to prevent the breakdown of the power devices by the over-voltage.

Figs. 6 and 7 show the experimental waveforms of the drain-to-source voltage $v_{ds}(t)$ and drain the current $i_d(t)$ of the MOSFET when the SiC-MOSFET is turned off. For different turn-off current I_D of 10 A and 60 A, the over-voltage V_p increases from 11.2% (556 V) to 55.0% (775 V) under a stray inductance of constant value: $L_d + L_{bus}$ = 134 nH. This is because the switching speed $di_d(t)$/dt becomes higher (going from 0.66 kA/μs to 2.05 kA/μs,) and so the over-voltage V_p increases from (1).

B. Normalization of stray inductance

The percent impedance $\%Z$ of the stray inductance can be converted using (1) and (12):

$$\%Z = \frac{(L_d + L_{bus})/t_f}{V_{DC}/I_D} \times 100 = \frac{v_L(t)}{V_{DC}} \times 100 \ [\%]. \tag{13}$$

From this equation, it is observed that the percentage of the EMF $v_L(t)$ for DC voltage ($v_L(t)/V_{DC} \times 100$) under the turn-off operation has the same meaning as the percent impedance $\%Z$ of the stray inductance.

Fig. 8 shows the relationship between the circuit impedance Z_{INV} [Ω] and the percent impedance $\%Z$ [%] of the stray inductance. The solid line shows the analysis results with the

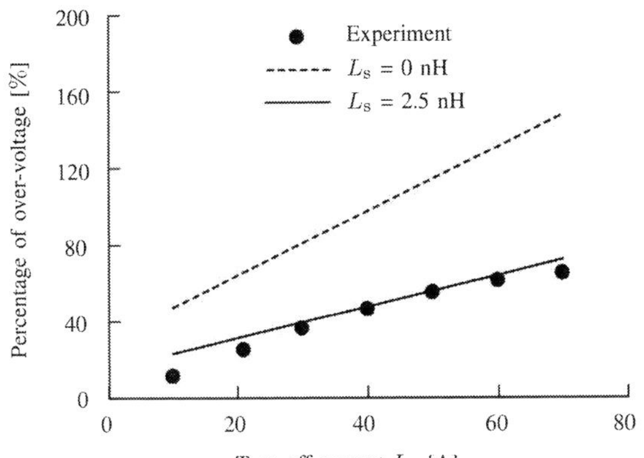

Fig. 4. Relationship between the turn-off current and $v_L(t)$, when $L_d + L_{bus}$ is set to 134 nH.

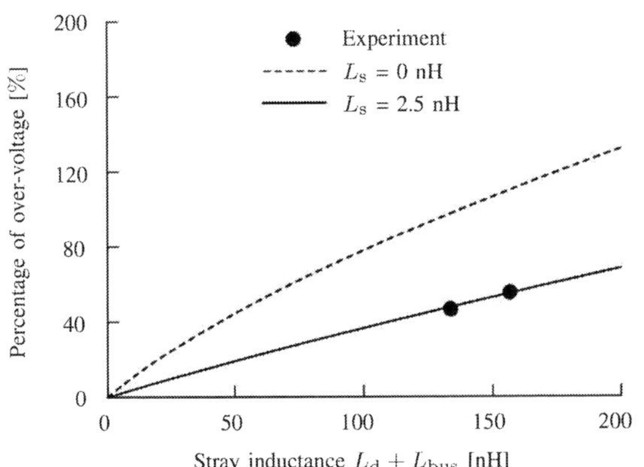

Fig. 5. Relationship between the stray inductance and $v_L(t)$, when the turn-off current I_D is set to 40 A.

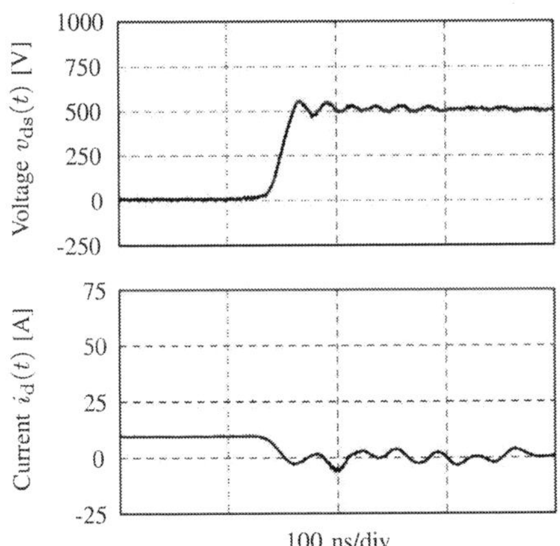

Fig. 6. Turn-off waveforms of the SiC-MOSFET when $L_d + L_{bus}$ = 134 nH and I_D = 10 A. The upper side represents the drain-to-source voltage $v_{ds}(t)$, and the lower side represents the drain current $i_d(t)$.

stray inductance $L_d + L_{bus}$ at a constant value of 134 nH. It is clear that the percent impedance $\%Z$ of the stray inductance becomes greater with lower circuit impedance Z_{INV}. Since

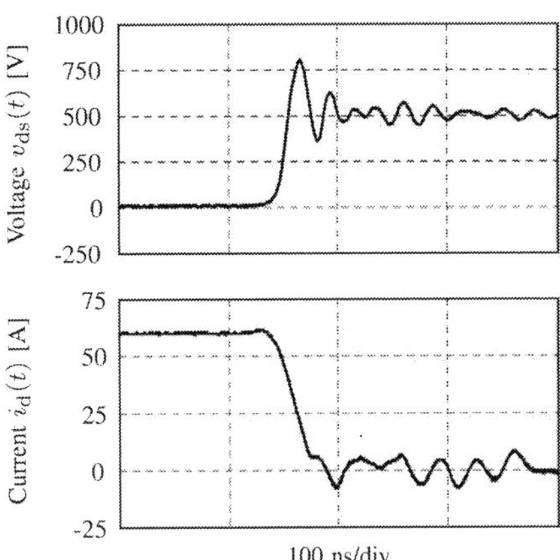

Fig. 7. Turn-off waveforms of the SiC-MOSFET when $L_d + L_{bus}$ = 134 nH and I_D = 60 A. The upper side represents the drain-to-source voltage $v_{ds}(t)$, and the lower side represents the drain current $i_d(t)$.

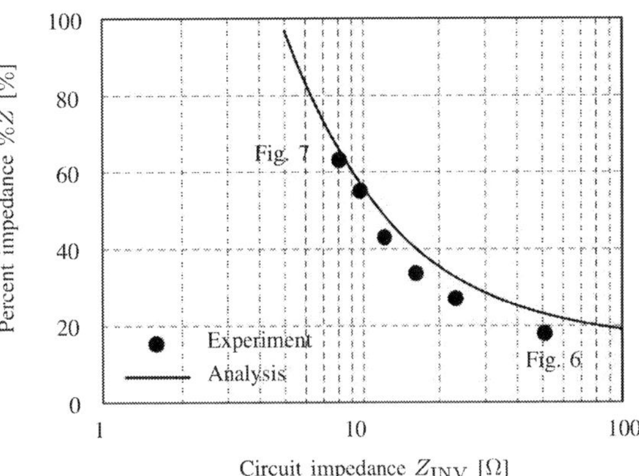

Fig. 8. Normalization of stray inductance with circuit impedance considered.

the percent impedance $\%Z$ of the stray inductance and the percentage of the EMF $v_L(t)$ have the same value, higher over-voltage occurs with lower impedance. Therefore, the stray inductance $L_d + L_{bus}$ has to be designed for each power electronics circuit. These experimental results are indicated by the solid circles • from Figs. 6 and 7. The experimental results correspond to the analytical results, thus verifying the effectiveness of the proposed method.

IV. Design of Stray Inductance Using the Normalization Procedure for an All-SiC Circuit

A. Target value of stray inductance

This section shows the design method for a stray inductance for 500-V and 113-A high-speed switching circuit using the normalization procedure.

To keep a voltage-rating under abnormal conditions for the power electronics circuit, the percent impedance $\%Z$ of the stray inductance, which equals percentage of the over-voltage, has to be $<50\%$. This is because, once a fault operation such as a short circuit occurs, the maximum turn-off current may reach double compared to that in normal circuit operations. In this case, the switching time δ_f is designed to be 40 ns for stable switching and the percent impedance $\%Z$ of the stray inductance is set to 35%. The following equation is given from (12):

$$L_d + L_{bus} = \frac{35\% \times 40 \text{ ns} \times (500 \text{ V}/113 \text{ A})}{100} = 62 \text{ nH}. \quad (14)$$

Fig. 9 shows the upper limit of $L_d + L_{bus}$ for each current rating. L_d separates the ESL of the DC capacitor and power module, and the stray inductance value is taken from the datasheet. To achieve a higher current rating, the DC capacitor is connected in parallel, which lowers the ESL and increases the current.

In this circuit, the all-SiC module (CAS100H12AM1, Cree) and DC capacitor (PEH200YX4470M, KEMET) are used. The stray inductance values of the all-SiC module and DC capacitor taken from the datasheets are 20 and 17 nH, respectively. The stray inductance of an additional wire for inserting the current sensor (CWT3, PEM) to measure the current waveform $i_d(t)$ is 22 nH. Therefore, the stray inductance of the bus bar has to be designed to be 3 nH confirm to (14).

Fig. 10 shows the simplified laminated bus bar structure. The laminated bus bar consists of N-side and P-side bus bars, which have a cross section of thickness $t \times$ width w and a length ℓ. An insulator is inserted maintain keep a small distance d between the two bus bars. Then, the stray inductance of the bus bar can be designed by specifying the length ℓ and distance d. To design the stray inductance of the bus bar, an inductance map is applied [7].

Fig. 11 shows the inductance map for designing the stray inductance of the bus bar. The thickness t and width d of the bus bars are set to 35 μm and 42 mm, respectively. In this experiment, a glass epoxy printed board is used, so that the minimum distance d is set to 1.6 mm. The horizontal axis indicates the length ℓ and the vertical axis indicates the distance d. The solid line shows the equivalent stray inductance of the bus bar of 3 nH. Therefore, it is clear that to design the stray inductance of the bus bar of 3 nH, the bus bar length has to be set to 75 mm.

B. Experimental verification

Fig. 12 shows the voltage and current waveforms during turn-off operation of the all-SiC module. The DC voltage V_{DC} and the turn-off current I_D are set to 500 V and 113 A, respectively. To adjust the switching speed $di_d(t)$/dt, a 15 Ω external resistance is connected to the all-SiC module as the gate resistor. From the experimental waveform, a $di_{d(max)}(t)$/dt value of 3.1 kA/μs. Therefore, from (2) and (12) the switching time δ_f and percent impedance $\%Z$ of the stray inductance can be calculated as 36.5 ns and 38.4%, respectively. Since the target of percent impedance $\%Z$ of the stray inductance is 35%, the experimental result almost corresponds to the designed value. In addition, the EMF $v_L(t)$ occurs at 35% (175 V) for 500 V DC voltage during the turn-off operation. This means that the percentage of the over-voltage is 35%, which

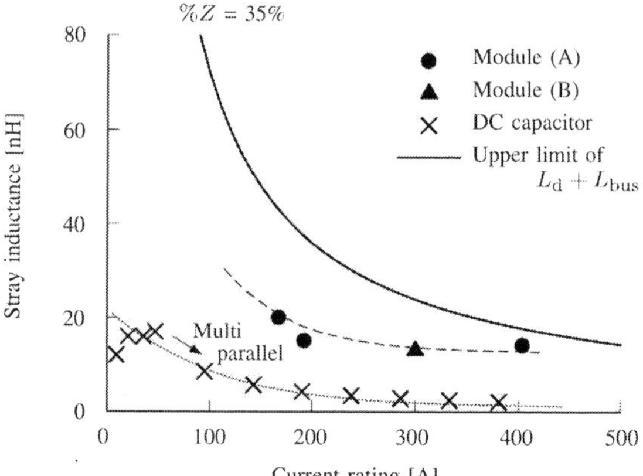

Fig. 9. Upper limits of the stray inductance for each current rating of the circuit, when $\%Z$=35%.

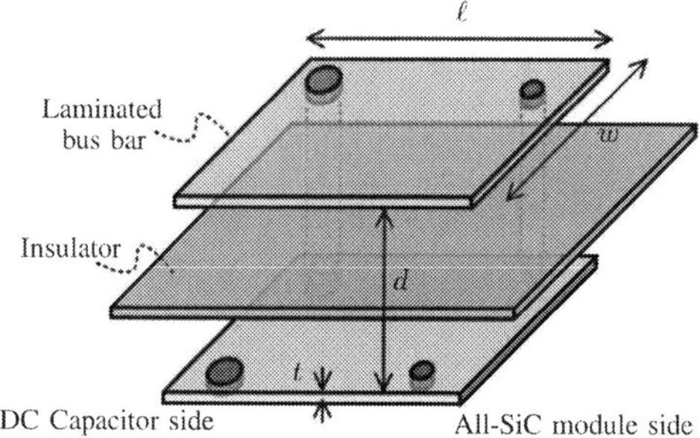

Fig. 10. Laminated bus bar structure connected from the DC capacitor to the power module.

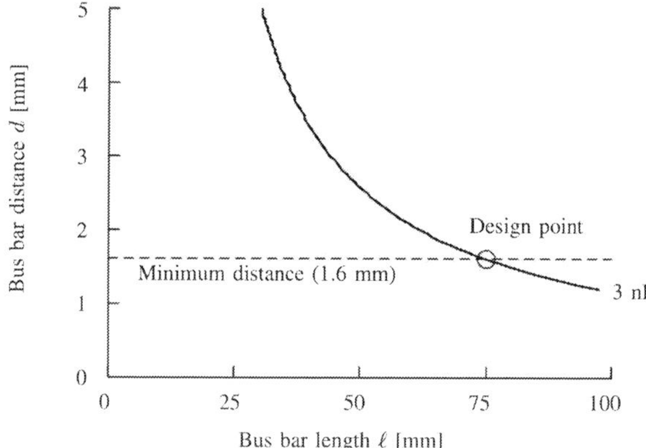

Fig. 11. Inductance map for designing stray inductance of the bus bar structure.

equals the percent impedance $\%Z$ of the stray inductance. As a result, it is clear that the bus bar structure with over-voltage considered for the power electronics circuit can be designed using the proposed percent impedance $\%Z$ of the stray inductance.

As mentioned before, an additional wire for measuring the current waveform is inserted, and it has an inductance of 22 nH. This wire is not necessary for actual manufacture, so in practice the percent impedance $\%Z$ that can be realized is 23%

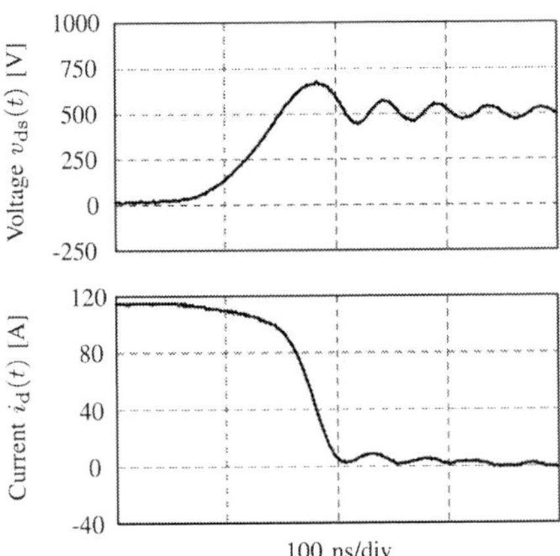

Fig. 12. Turn-off waveforms of the all-SiC module under 500 V and 113 A. The upper side represents the drain-to-source voltage $v_{ds}(t)$, and the lower side represents the drain current $i_d(t)$.

(=35% × 40 nH/62 nH). This means that the stray inductance design has a greater margin for the actual circuit structure.

V. Conclusions

This paper presented a normalization procedure for the DC-side stray inductance based on the voltage and current rating of the power electronics circuit. It is clear that the normalization procedure enables to design the stray inductance by considering the circuit ratings and the switching time. To verify the proposed method, the experimental results under 500 V and 100 A using an all-SiC power module were shown. Such a stray inductance design procedure based on normalization can be applied in future to converter circuits.

References

[1] A. Abrishamifar, R. Lourakzadegan, R. Esmaili, and M. Arefian, "Design and Construction of a Bus Bar Spike Reduction in an Industrial Inverter," *IEEE Power Electronics & Drive Systems & Technologies Conf. (PEDSTC)*, 2010, pp. 13-17.

[2] L. Yuan, H. Yu, X. Wang, and Z. Zhao, "Design, Simulation and Analysis of the Low Stray Inductance Bus Bar for Voltage Source Inverters," *IEEE International Conf. on Electrical Machines and Systems (ICEMS)*, 2011, pp. 1-5.

[3] J. Fabre, P. Ladoux, and M. Piton, "Characterization and Implementation of Dual-SiC MOSFET Modules for Future Use in Traction Converters," *IEEE Trans. on Power Electronics*, vol. 30, no. 8, pp. 4079-4090, August 2015.

[4] P. Nayak, V. M. Krishna, K. Vasudevakrishna, and K. Hatua, "Study on the Effects of Parasitic Inductances and Device Capacitances on 1200 V, 35 A SiC MOSFET Based Voltage Source Inverter Design," *IEEE International Conf. on Power Electronics, Drives and Energy Systems (PEDES)*, 2014, pp. 1-6.

[5] K. Shirabe, M. M. Swamy, J-K. Kang, M. Hisatsune, Y. Wu, D. Kebort and J. Honea, "Efficiency Comparison Between Si-IGBT-Based Drive and GaN-Based Drive," *IEEE Trans. on Industry Applications*, vol. 50, no. 1, pp. 566-572, January/February 2014.

[6] H. Li, S. M-Nielsen, X. Wang, R. Maheshwari, S. Beczkowski, C. Uhrenfeldt, and W. Franke : "Influences of Device and Circuit Mismatches on Paralleling Silicon Carbide MOSFETs," *IEEE Trans. on Power Electronics*, 2015.

[7] K. Wada, and M. ando : "Limitation of DC-side stray inductance by considering over-voltage and short-circuit current", *IEEE European Conference on Power Electronics (EPE)*, 2013.

[8] M. C. Caponet, F. Profumo, R. W. De Doncker, and A. Tenconi : "Low Stray Inductance Bus Bar Design and Construction for Good EMC Performance in Power Electronic Circuits", *IEEE Trans. on Power Electronics*, vol. 17, no. 2, pp. 225-231, 2002.

[9] K. Wada, A. Hino, and M. Ando : "High-Speed Analysis of Bus Bar Inductance for a Laminated Structure," *IEEJ Journal of Industry Applications*, vol. 2, no. 4, pp. 189-194, 2013.

[10] Z. Chen, D. Boroyevich, and R. Burgos : "Experimental Parametric Study of the Parasitic Inductance Influence on MOSFET Switching Behavior", *IEEE International Power Electronics Conference (IPEC)*, pp. 164-169, 2010.

[11] Y. Xiao, H. Shah, T. P. Chow, and R. J. Gutmann : "Analytical Modeling and Experimental Evaluation of Interconnect Parasitic Inductance on MOSFET Switching Characteristic", *IEEE Applied Power Electronics Conference and Exposition (APEC)*, vol. 1, pp. 516-521, 2004.

[12] M. Ando, and K. Wada : "Design of DC-side stray inductance for high speed switching inverter based on normalization procedure", *IEEE Applied Power Electronics Conference and Exposition (APEC)*, pp. 2432-2437, 2015.

[13] Y. Ren, M. Xu, J. Zhou, and F. C. Lee : "Analytical Loss Model of Power MOSFET", *IEEE Trans. on Power Electronics*, vol. 21, no. 2, pp. 310-319, 2006.

[14] S.Hashino, and T. Shimizu : "Separation measurement of parasitic impedance on a power electronics circuit board using TDR", *IEEE Energy Conversion Congress and Exposition (ECCE)*, pp. 2700-2705, 2010.

Thermal Network Parameter Identification of IGBT Module Based on the Cooling Curve of Junction Temperature

Xiong Du[1], Tengfei Li[1], Jun Zhang[1], Heng-Ming Tai[2], Pengju Sun[1] and Luowei Zhou[1]
[1] State Key Laboratory of Power Transmission Equipment & System Security and New Technology, Chongqing University
[2]Department of Electrical and Computer Engineering at the University of Tulsa
Email: duxiong@cqu.edu.cn

***Abstract*—The thermal network parameters of IGBT module are wildly used for reliability evaluation, lifetime calculation, health monitoring, and predictive maintenance. Commonly used thermal network parameter identification methods include the finite element method (FEM) and the curve fitting of thermal impedance (Zth) method. The former method calculates directly using the IGBT geometric structure and material properties, but it is difficult to consider the aging procedure of IGBT. The curve fitting method mainly depends on constant loss constraints, which is difficult to guarantee in practical application. In this paper, we propose an IGBT thermal network parameter identification method using the measured junction temperature cooling curve in the procedure of converter shutdown. The relationship between the time constants of junction temperature curve and the thermal network parameters is investigated. We show that the RC parameters can be identified in two different cooling conditions. The obtained RC parameters are in good agreement with the results by JESD51-14 method. The proposed method is simple and easy to implement. In particular, it does not need to consider the dissipation loss of IGBT before shutdown, and is suitable for RC thermal parameter quasi-online measurement.**

Keywords—IGBT module; thermal network; parameter identification; junction temperature; cooling curve

I. Introduction

Insulated gate bipolar transistor (IGBT) power modules have been widely used in high-voltage and high-power applications such as electric vehicles(EVs) [1, 2], ships [3], aircraft [4], wind turbines [5], smart grids [6], and industrial drives [7]. This is mainly due to their superior performance in terms of power density, switching frequency, energy efficiency, and cost effectiveness. On the other hand, considering the increasing role of power electronic converters in critical functions, particularly in human transport applications with significant safety requirements, features such as reliability, lifetime, health monitoring, and predictive maintenance become increasingly important [8, 9].

All the features mentioned above are closely related to the thermal network parameter of IGBT module. Characterization of the thermal performance helps to determine the production consistency and provide the basis for enhancement of thermal quality of manufacturing [10]. Knowledge of instantaneous junction temperature is essential for effective reliability evaluation and thermal management of power converters. However, measuring the junction temperature on-line is not an easy task. In general, the junction temperature is determined by RC parameter and power loss [11]. Scheuermann and Hecht [12] suggested that a 20% increase of the thermal resistance indicates the solder fatigue failure. Therefore the thermal network parameter measurement is critical in revealing aspects of the internal module structure and as a criterion of aging.

Each RC parameter of thermal network in power devices can be derived by the corresponding thermal response characteristics [13]. Theoretically, 3-deimental heat transfer reveals the process of thermal conduction. Existed study shows that one-dimensional (1-D) multi-layered structures could simulate the thermal conduction very well [11]. So, the one-dimensional (1-D) thermal network structure is employed in this paper. The cross-section of an IGBT power module with heat sink is shown in Fig. 1 [14]. During the operation of IGBT, the heat is assumed to be generated at the top surface of a die and spreads down through intern layers to the bottom of the baseplate, which is cooled by a heat sink or a cold plate.

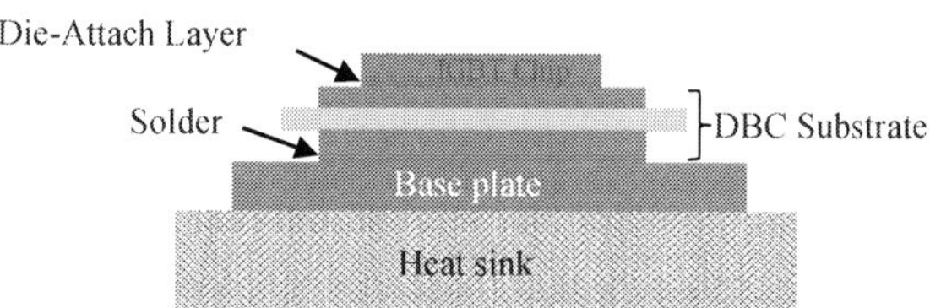

Fig. 1. Cross section of a power module with heat sink.

The thermal behaviour of IGBT module can be described using different equivalent circuit models. The Foster- and Cauer-based networks are the widely used thermal networks [15]. In the Foster network, the individual RC elements are in series, which can be easily extracted from a measured cooling curving of the module, while the Cauer network closely related to the real process of thermal conduction [16]. This paper chooses the Cauer-network for analytical study.

Theoretically, each layer of one type of material needs one order RC network to model its thermal characteristics [13,17], while the 7-layer structure of IGBT module will need a 7-order RC network to model it, this 7-order RC network will be

This work was supported in part by the National Natural Science Foundation project of CQ CSTC under Grant CSTC2012JJJQ90004, and in part by the National Natural Science Foundation of China under Grant 51577020.

complex for practical application. Actually, some different but adjacent layers within typical IGBT module have same thermal characteristics [17, 18]. Existed studies show that the thermal conduction process in an IGBT module can be equivalent to a 3-layer structure [19, 20], thus we can simplify the thermal network of Fig. 1 to a cascade of 3-order of copper layer, insulator layer and aluminum layer, as depicted in Fig. 2. The cold plate is modeled with a 1st order RC network [21]. The ambient temperature is denoted as T_a.

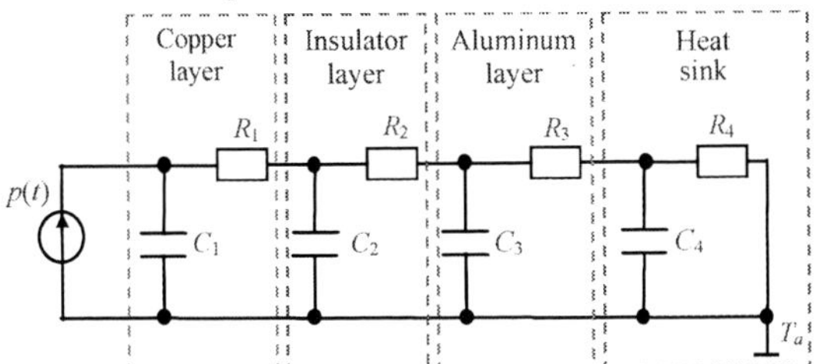

Fig. 2. Cauer model of an IGBT module and heat sink.

It is known that for obtaining the thermal parameter online or quasi-online is challenging. Values of thermal parameters can be obtained from IGBT manufacturer's datasheet [22, 23], or from the calculation in multi-physical environment such as COMSOL and ANSYS [18, 19]. These methods usually could not reflect the aging process. And the equivalent parameters after aging are not exactly equal to the original ones [24]. JEDEC established a standards to measure the semiconductor thermal resistance according to the transient cooling curve of junction temperature in 2010 [25]. Yin and van Wyk [26] obtained the total thermal impedance and its components from the numerical analysis of the cool-down temperature curve. Ref [27] described a PRBS method to separate the thermal parameter. Ref [28] presented a method to monitor the solder fatigue inside a module by identifying the increased internal thermal resistance due to solder fatigue, taking into account the masking effect of the variable operating point. Thermal resistance analysis by the induced transient (TRAIT) was proposed in [13], which relies on the different characteristics of material.

The thermal parameter estimation methods in [13], [25]-[28] directly rely on a complete loss model. Accuracy of the above mentioned methods are constraint by the loss and the aging effect is not considered. And modeling and measurement of loss in AC converter are difficult [29]. The experimental results shown that a large number of power cycling could result an obvious change in thermal parameter, while in a short time range, the thermal parameter will nearly be invariant [24]. According to this property, this paper proposes a new method for identifying the thermal network parameters from the junction temperature curves measured in two different cooling conditions at two converter shutdown situations. Compared with the reported methods, the proposed method enables the measurement of thermal network parameters online or quasi-online, and does not need to measure or model the IGBT loss.

II. Problem Description

A typical three-phase converter is depicted in Fig. 3, where u_a, u_b, u_c are the phase voltages of AC side and U_d represents the DC bus voltage. S_a, S_b, S_c are the half-bridge module. This topology is widely used in wind power, photovoltaic and other 3-phase DC-AC or AC-DC applications.

When the converter is on working, each switch S_{ap}-S_{cn} would switch the load current with their corresponding switching signals. Hence the junction temperature of IGBT fluctuates with the changing of load current as shown with the red line in Fig. 4. It is difficult to measure the junction temperature on line [11]. While when the converter is shutdown, the junction temperature will decrease as shown in Fig 4 with blue line after time t_o due to no power processed by the converter, and it is possible to measure the junction temperature in the shutdown procedure. With the auxiliary circuit as shown in Fig. 3 in the red-dashed rectangle area, the junction temperature can be measured.

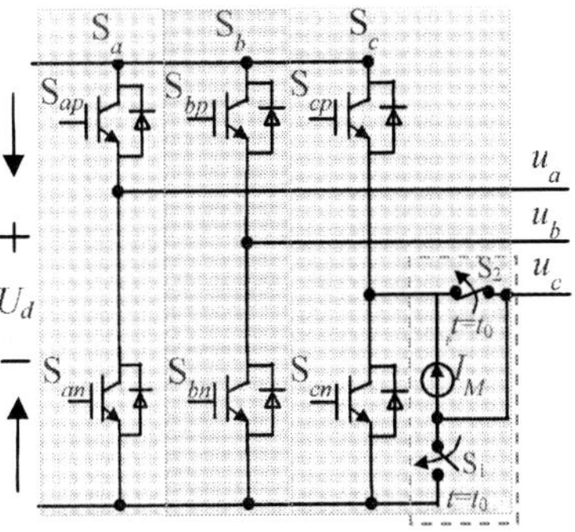

Fig. 3. A typical three-phase converter with the junction temperature measurement circuit.

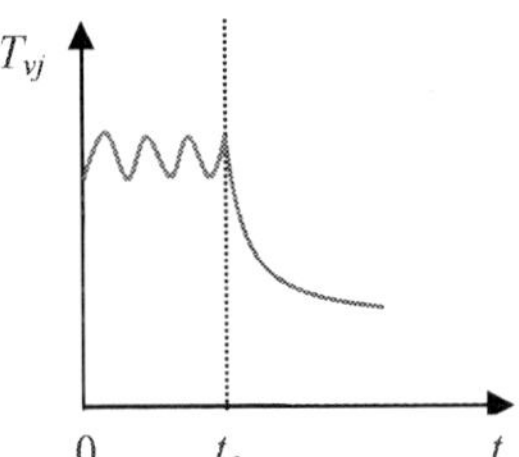

Fig. 4. T_{vj} of S_{cn} under converter shutdown process.

When the operating condition exceeds the limitation of power converter, the converter would cut off. Taking the wind power converter as an example, the power converter is cut off when the wind speed exceeds the operation range. After shutdown, the temperature sensitive parameter (TSP), v_{cesat}, the saturation voltage of IGBT is employed here to measure the junction temperature [25]. At time t_o, the converter is shutdown, then the power processed by the converter is changed to zero instantaneously, and all the switches S_{ap} to S_{cn} turn off. At the same time, S_1 is switched on and S_2 is switched off to impose the measurement I_M (e.g., 100mA) flowing through IGBT S_{cn}. In the shutdown procedure, the voltage v_{cesat} across S_{cn} is recorded, and v_{cesat} can be converted to junction temperature $T_j(t)$ of IGBT S_{cn} in the shutdown procedure with the K-FACTOR, which specifies the relationship between $T_j(t)$ and V_{cesat}.

Thus, we can measure the $T_j(t)$ online or quasi-online in the procedure of converter shutdown. According to the recorded information of $T_j(t)$ in the process of cooling, the time constants corresponding to the thermal model in Fig. 2 can be identified through mathematical manipulation. If we can set up the relationship between the time constants and the thermal R, C parameters, then it is possible to identify the thermal parameter. In the following, the relationship between the time constants and the RC parameters will be investigated.

III. Relationship Between Tj(t) and RC Parameters

This section describes the relationship between the time constants of junction temperature curve and thermal RC parameters according to the electrical-thermal analogy.

The Cauer thermal equivalent circuit model during a converter's cut down situation and in the cooling process of IGBT module is shown in Fig. 5.

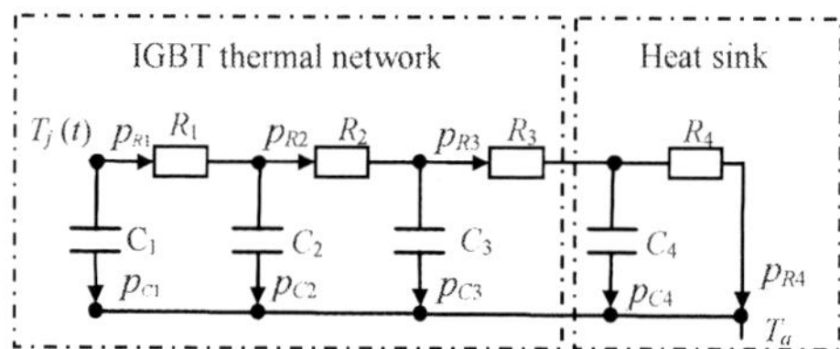

Fig. 5. Cauer model of thermal equivalent circuit for IGBT and heat sink during cut off ($t > 0$).

Due to the electrical-thermal analogy, the response of junction temperature $T_j(t)$ can be considered as a zero-input response. For $t > 0$, the response of $T_{vj}(t)$ can be described by

$$\frac{d^4T_{vj}(t)}{dt^4} + k_3\frac{d^3T_{vj}(t)}{dt^3} + k_2\frac{d^2T_{vj}(t)}{dt^2} + k_1\frac{dT_{vj}(t)}{dt} + k_0T_{vj}(t) = 0 \quad (1)$$

where $T_{vj}(t) = T_j(t) - T_a$. Coefficients $k_0 \cdots k_3$ are functions of the RC parameters R_1-R_4 and C_1-C_4 and can be expressed as

$$k_0 = \frac{1}{C_1C_2C_3C_4R_1R_2R_3R_4} \quad (2)$$

$$k_1 = \frac{C_1 + C_2 + C_3 + C_4}{C_1C_2C_3C_4R_1R_2R_3} + \frac{C_1R_1 + (C_1 + C_2)R_2 + (C_1 + C_2 + C_3)R_3}{C_1C_2C_3C_4R_1R_2R_3R_4} \quad (3)$$

$$\begin{aligned} k_2 = {} & \frac{C_1C_2R_1 + C_3(C_1R_1 + (C_1 + C_2)R_2) + C_4(C_1R_1 + (C_1 + C_2)R_2}{C_1C_2C_3C_4R_1R_2R_3} \\ & + \frac{(C_1 + C_2 + C_3)R_3) + (C_1 + C_2)R_2 + (C_1 + C_2 + C_3)R_3)}{C_1C_2C_3C_4R_1R_2R_3} \\ & + \frac{C_1C_2R_1R_2 + (C_1C_2R_1 + C_3(C_1R_1 + (C_1 + C_2)R_2))R_3}{C_1C_2C_3C_4C_1R_2R_3R_4} \end{aligned} \quad (4)$$

$$\begin{aligned} k_3 = {} & \frac{C_1C_2C_3R_1R_2 + C_4(C_1C_2R_1R_2 + (C_1C_2R_1 + C_3R_3C_1R_1))}{C_1C_2C_3C_4R_1R_2R_3} \\ & + \frac{C_4C_3R_3(C_1 + C_2)R_2}{C_1C_2C_3C_4R_1R_2R_3} + \frac{C_1C_2C_3R_1R_2R_3}{C_1C_2C_3C_4R_1R_2R_3R_4} \end{aligned} \quad (5)$$

For the Cauer network, the general solution of Eq. (1) can be expressed as [11]

$$T_{vj}(t) = \alpha_1 e^{-\frac{t}{\tau_1}} + \alpha_2 e^{-\frac{t}{\tau_2}} + \alpha_3 e^{-\frac{t}{\tau_3}} + \alpha_4 e^{-\frac{t}{\tau_4}} \quad (6)$$

where α_i and time constants τ_i are functions of RC parameters. Moreover, coefficients α_i are also related to the initial voltage of each capacitor at the time zero. The relationship between τ_i and RC parameters can be expressed as

$$\begin{cases} k_0 = \frac{1}{\tau_1}\cdot\frac{1}{\tau_2}\cdot\frac{1}{\tau_3}\cdot\frac{1}{\tau_4} \\ k_1 = \frac{1}{\tau_1}\cdot\frac{1}{\tau_2}\cdot\frac{1}{\tau_3} + \frac{1}{\tau_1}\cdot\frac{1}{\tau_2}\cdot\frac{1}{\tau_4} + \frac{1}{\tau_1}\cdot\frac{1}{\tau_3}\cdot\frac{1}{\tau_4} + \frac{1}{\tau_2}\cdot\frac{1}{\tau_3}\cdot\frac{1}{\tau_4} \\ k_2 = \frac{1}{\tau_1}\cdot\frac{1}{\tau_2} + \frac{1}{\tau_1}\cdot\frac{1}{\tau_3} + \frac{1}{\tau_1}\cdot\frac{1}{\tau_4} + \frac{1}{\tau_2}\cdot\frac{1}{\tau_3} + \frac{1}{\tau_2}\cdot\frac{1}{\tau_4} + \frac{1}{\tau_3}\cdot\frac{1}{\tau_4} \\ k_3 = \frac{1}{\tau_1} + \frac{1}{\tau_2} + \frac{1}{\tau_3} + \frac{1}{\tau_4} \end{cases} \quad (7)$$

Equations (2) – (5) and (7) show the relationship between the coefficients of $k_0, \cdots, k_3$ and RC parameters.

Waveforms of the junction temperature $T_{vj}(t)$ can be measured for curving fitting to obtain the time constant τ_i. It follows from (7) that there are eight variables, R_1 to R_4 and C_1 to C_4, but only four equations are available. So the RC parameter identification could not be realized just from one junction temperature cooling curve.

IV. Additional Dissipation Condition

In this section, we describe the means to create an additional dissipation condition so as to determine the values of RC parameters. Close examination of Fig. 2 shows that R_4 and C_4 are the equivalent thermal parameters of the heat sink, which is measurable and can be changed in the converter's cut down procedure [16]. For example, in one cut down procedure, the water velocity or fan speed of the heat sink is set to be a specific value and the corresponding thermal network parameter of the heat sink is R_4 and C_4, the waveform of $T_{vj}(t)$ can be measured to get Eq. (6). While in another cut down procedure, the water velocity or fan speed can be changed to another different value and the corresponding thermal network parameter changes to R'_4 and C'_4.

Under this changed cooling condition, a new state equation can be rebuilt as Eq. (1), except that R_4 and C_4 are changed to R'_4 and C'_4. Following the procedure described in Section III, we can obtain a new set of equations relating τ'_1- τ'_4, R_1-R_3, R'_4, C_1-C_3, and C'_4 parameters. The response of $T_{vj}(t)$ with the new values can be described as

$$\frac{d^4T_{vj}(t)}{dt^4} + k'_3\frac{d^3T_{vj}(t)}{dt^3} + k'_2\frac{d^2T_{vj}(t)}{dt^2} + k'_1\frac{dT_{vj}(t)}{dt} + k'_0T_{vj}(t) = 0 \quad (8)$$

where coefficients k'_0 – k'_3 are

$$k'_0 = \frac{1}{C_1C_2C_3C'_4R_1R_2R_3R'_4} \quad (9)$$

$$k'_1 = \frac{C_1 + C_2 + C_3 + C'_4}{C_1C_2C_3C'_4R_1R_2R_3} + \frac{C_1R_1 + (C_1 + C_2)R_2 + (C_1 + C_2 + C_3)R_3}{C_1C_2C_3C'_4R_1R_2R_3R'_4} \quad (10)$$

$$k_2' = \frac{C_1C_2R_1 + C_3(C_1R_1 + (C_1 + C_2)R_2) + C_4'(C_1R_1 + (C_1 + C_2)R_2}{C_1C_2C_3C_4'R_1R_2R_3} + \frac{(C_1 + C_2 + C_3)R_3) + (C_1 + C_2)R_2 + (C_1 + C_2 + C_3)R_3)}{C_1C_2C_3C_4'R_1R_2R_3} + \frac{C_1C_2R_1R_2 + (C_1C_2R_1 + C_3(C_1R_1 + (C_1 + C_2)R_2))R_3}{C_1C_2C_3C_4'C_1R_2R_3R_4'} \quad (11)$$

$$k_3' = \frac{C_1C_2C_3R_1R_2 + C_4'(C_1C_2R_1R_2 + (C_1C_2R_1 + C_3R_3C_1R_1))}{C_1C_2C_3C_4'R_1R_2R_3} + \frac{C_4'C_3R_3(C_1 + C_2)R_2}{C_1C_2C_3C_4'R_1R_2R_3} + \frac{C_1C_2C_3R_1R_2R_3}{C_1C_2C_3C_4'R_1R_2R_3R_4'} \quad (12)$$

Similarly, the relationship between the new time constants τ'_1 - τ'_4 and RC parameters can be expressed as

$$\begin{cases} k_0' = \frac{1}{\tau_1'}\cdot\frac{1}{\tau_2'}\cdot\frac{1}{\tau_3'}\cdot\frac{1}{\tau_4'} \\ k_1' = \frac{1}{\tau_1'}\cdot\frac{1}{\tau_2'}\cdot\frac{1}{\tau_3'} + \frac{1}{\tau_1'}\cdot\frac{1}{\tau_2'}\cdot\frac{1}{\tau_4'} + \frac{1}{\tau_1'}\cdot\frac{1}{\tau_3'}\cdot\frac{1}{\tau_4'} + \frac{1}{\tau_2'}\cdot\frac{1}{\tau_3'}\cdot\frac{1}{\tau_4'} \\ k_2' = \frac{1}{\tau_1'}\cdot\frac{1}{\tau_2'} + \frac{1}{\tau_1'}\cdot\frac{1}{\tau_3'} + \frac{1}{\tau_1'}\cdot\frac{1}{\tau_4'} + \frac{1}{\tau_2'}\cdot\frac{1}{\tau_3'} + \frac{1}{\tau_2'}\cdot\frac{1}{\tau_4'} + \frac{1}{\tau_3'}\cdot\frac{1}{\tau_4'} \\ k_3' = \frac{1}{\tau_1'} + \frac{1}{\tau_2'} + \frac{1}{\tau_3'} + \frac{1}{\tau_4'} \end{cases} \quad (13)$$

Combination of Eq. (8) and (13) from two different cooling conditions, we have eight equations and eight variables, R_1-R_3, R'_4, C_1-C_3, C'_4 since R_4 and C_4 are treated as measurable in this study. So the RC parameter identification can be achieved with two measured junction temperature cooling curves under two cut down procedures with two different cooling conditions.

V. CASE STUDY

This section investigates the steps that how to obtain the thermal network parameters according to the time constants of junction temperature curve under two dissipation conditions. An experiment platform is built to certify the method proposed in this paper.

The IGBT module under test is Infineon FF50R12RT4 module. The test circuit for S_{cn} in the half-bridge module, S_c, is shown in Fig.6. A 50A heating current is substituted for the conducting current in IGBT, which results in the rise of junction temperature. When S_{cn} is heating, S_1 and S_3 is on while S_2, S_4 is off. When S_{cn} is heated up, the heating current is removed through switching on S_4 and switching off S_1. At the same time, S_2 is switched on and S_3 is switched off to impose a 100mA measurement current flowing through S_{cn}. A driving voltage is applied to the gate of IGBT during the test.

The IGBT module under test is mounted on a water cold plate. A thin layer of thermal grease is inserted between the IGBT module and the cold plate to change the thermal parameter of heat sink from R_4 and C_4 to R'_4 and C'_4 in this study. The first measurement is performed without any thermal interface material between IGBT Module and cold-plate. For the second measurement a thin layer of thermal grease is applied at the interface. The increased interface resistance in the second measurement due to the microscopic surface roughness between package and cold-plate ensures a clear difference between R_4, C_4 and R'_4, C'_4.

In order for convenient measurement, a cyclic heating and cooling is triggered in the test. In one shut down procedure, the recorded saturation voltage v_{cesat} across S_{cn} is shown in Fig. 7.

The linear relationship between V_{cesat} and T_j of the module employed in this test is

$$V_{ce} = 0.6376 - 0.00212T_j \quad (14)$$

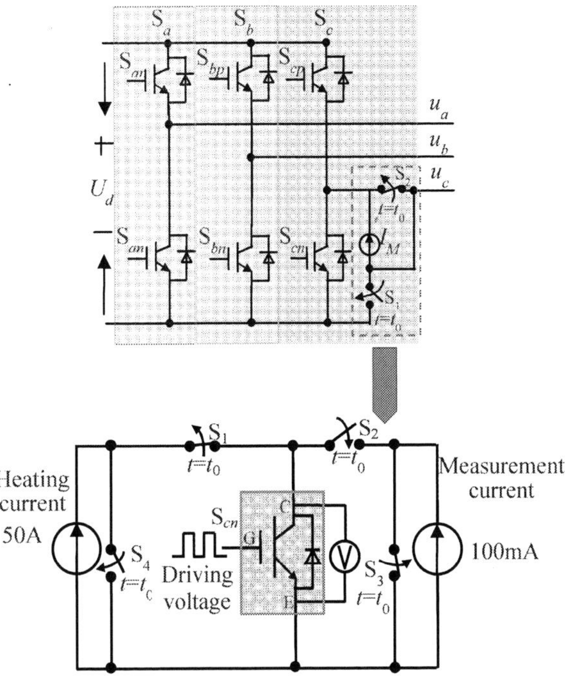

Fig.6. Experiment schematic.

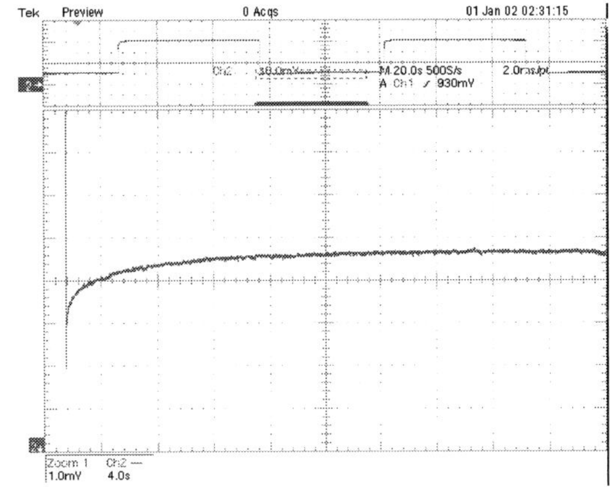

Fig.7. Recorded v_{cesat} when cooling.

Based on Eq. (14), the recorded v_{cesat} curve is converted to junction temperature curve $T_j(t)$, which is shown in Fig. 8.

In order to reconstruct the junction temperature T_{j0} at t=0, an "offset-correction" is performed in this paper. For short times there exists an almost linear relationship between temperature change $\Delta T_j(t)$ and the square-root of time $\sqrt{t}$ that can be used to extrapolate the $T_j(t)$. Fig. 9 is an example of the offset-correction in the situation with thermal grease. In this figure, the red line is the downtrend of $T_j(t)$ theoretically, which can replace the measure signal at the beginning.

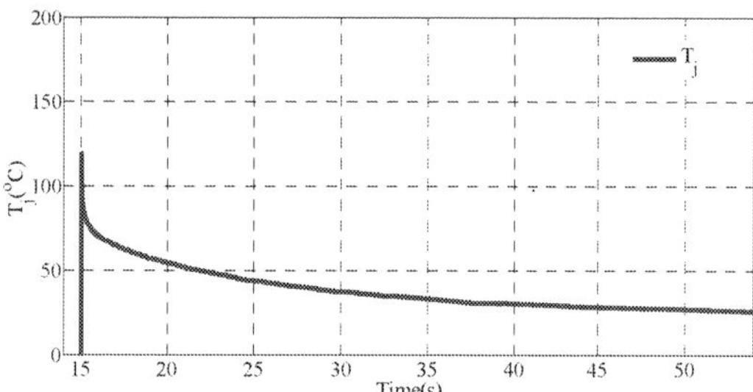

Fig.8. Converted $T_j(t)$ from v_{cesat}.

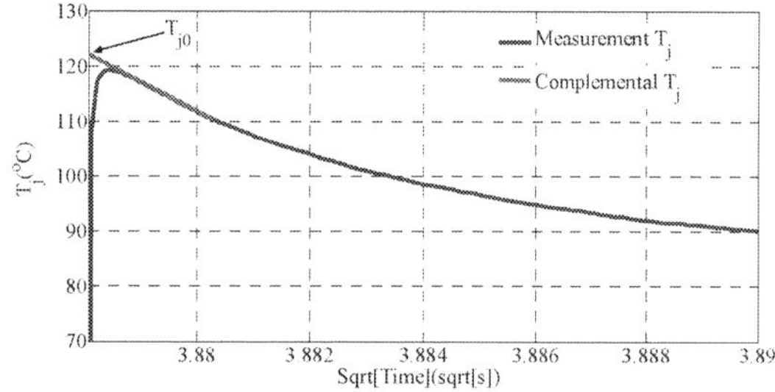

Fig. 9. Determination of initial $T_j(t)$

MATLAB program with nonlinear least-squares data fitting is used to realize the curve fitting. The expression form of fitted curve is

$$T_{vj}(t) = \alpha_1 e^{-\frac{t}{\tau_1}} + \alpha_2 e^{-\frac{t}{\tau_2}} + \alpha_3 e^{-\frac{t}{\tau_3}} + \alpha_4 e^{-\frac{t}{\tau_4}} \quad (15)$$

According to the two measured junction temperature curves, the obtained time constants under two cooling conditions with and without grease is shown in Table. I.

TAB. I THE RESULT OF FITTING AT DIFFERENT COOLING CONDITIONS

heat dissipation condition	constant time τ			
	1	2	3	4
without grease(τ)	1.472	9.8232	0.04065	0.002805
with grease(τ')	7.7189	18.003	0.04065	0.002805

From Tab. I, it follows that the smaller constant time remain unchanged. The reason is that water cold plate's constant time is not at the same level with IGBT, so changing heat dissipation condition doesn't affect the smaller constant time.

The RC parameter of the heatsink without grease is measured as R_4= 0.2872K/W, and C_4= 20.90788J/K.

According to equations with two different cooling conditions and time constants in Tab. I, and the measured R_4 and C_4. Substituting the constant time and R_4, C_4 to Eq. (7) and (13), we can get equation sets (16),

$$\begin{cases} \dfrac{1}{C_1C_2C_3C_4R_1R_2R_3R_4} = 606.5209 \\ \quad\cdots\cdots \\ \dfrac{C_1C_2C_3R_1R_2 + C_4(C_1C_2R_1R_2 + (C_1C_2R_1 + C_3R_3C_1R_1))}{C_1C_2C_3C_4R_1R_2R_3} + \\ + \dfrac{C_4C_3R_3(C_1+C_2)R_2}{C_1C_2C_3C_4R_1R_2R_3} + \dfrac{C_1C_2C_3R_1R_2R_3}{C_1C_2C_3C_4R_1R_2R_3R_4} = 381.8876 \\ \dfrac{1}{C_1C_2C_3C_4R_1R_2R_3R_3} = 631.1118 \\ \quad\cdots\cdots \\ \dfrac{C_1C_2C_3R_1R_2 + C_4(C_1C_2R_1R_2 + (C_1C_2R_1 + C_3R_3C_1R_1))}{C_1C_2C_3C_4'R_1R_2R_3} + \\ + \dfrac{C_4C_3R_3(C_1+C_2)R_2}{C_1C_2C_3C_4'R_1R_2R_3} + \dfrac{C_1C_2C_3R_1R_2R_3}{C_1C_2C_3C_4'R_1R_2R_3R_4'} = 381.29 \end{cases} \quad (16)$$

Eq. (16) is solved through Mathematica program, the 8 parameters solved from Eq. (16) is listed in Tab. II.

TAB. II RESULTS OF IDENTIFICATION

Parameters	Proposed method	JESD51-14 method	Error
R_1/(K/W)	0.075	0.08	-6.28%
R_2/(K/W)	0.212	0.2	6%
R_3/(K/W)	0.24	0.235	2.13%
Total R	0.527	0.515	2.33%
C_1/ (J/K)	0.1	0.095	5.26%
C_2/ (J/K)	0.073	0.07	4.29%
C_3/ (J/K)	9.85	9.835	0.153%

For accuracy comparison, we employ the JESD51-14 method as a benchmark. The result through JESD51-14 is also shown in Tab. II. It is observed from Tab. II that the errors obtained by the proposed method and the JESD51-14 is about ±6%. The total thermal resistance by the proposed method is 2.33% more than that by JESD51-14. Therefore, it suffices to say that the result obtained by the information of junction temperature can identify the RC thermal network quasi-online with considerable accuracy.

VI. CONCLUSION

This paper proposes a method to identify the thermal network RC quasi-online. The key issue is to get the relationship between the time constants of junction temperature curve in cooling procedure and the thermal network parameters. Then the RC parameters can be identified by solving equations derived at two different cooling conditions. The proposed method does not need to consider the complex loss model of power device. This makes it easy to implement and extend the range of applications, especially in the quasi-online condition monitoring of IGBT modules. Furthermore, this method can be applied to real converter system with the measurement of junction temperature under converter's cut off situation.

References

[1] B. Ji, V. Pickert, W. Cao, and B. Zahawi, "In situ diagnostics and prognostics of wire bonding faults in IGBT modules for electric vehicle drives," *IEEE Trans. Power Electron.*, vol. 28, no. 12, pp. 5568–5577, Dec. 2013.

[2] Z. Xu, M. Li, F. Wang, and Z. Liang, "Investigation of Si IGBT operation at 200°C for traction applications," *IEEE Trans. Power Electron.*, vol. 28, no. 5, pp. 2604–2615, Sep. 2013.

[3] B. Zahedi and L. E. Norum, "Modeling and simulation of all-electric ships with low-voltage DC hybrid power systems," *IEEE Trans. Power Electron.*, vol. 28, no. 10, pp. 4525–4537, Oct. 2013

[4] W. Cao, B. C. Mecrow, G. J. Atkinson, J. W. Bennett, and D. J. Atkinson, "Overview of electric motor technologies used for more electric aircraft (MEA)," *IEEE Trans. Ind. Electron.*, vol. 59, no. 9, pp. 3523–3531, Sep. 2012.

[5] O. S. Senturk, L. Helle, S. Munk, Nielsen, P. Rodriguez, and R. Teodorescu, "Power capability investigation based on electrothermal models of presspack IGBT three-level NPC and ANPC VSCs for multimegawatt wind turbines," *IEEE Trans. Power Electron.*, vol. 27, no. 7, pp. 3195–3206, Jul. 2012.

[6] M. Yilmaz and P. T. Krein, "Review of the impact of vehicle-to-grid technologies on distribution systems and utility interfaces," *IEEE Trans. Power Electron.*, vol. 28, no. 12, pp. 5673–5689, Dec. 2013.

[7] G. Xun and J. A. Ferreira, "Comparison and reduction of conducted EMI inSiC JFET and Si IGBT-based motor drives," *IEEE Trans. Power Electron.*, vol. 29, no. 4, pp. 1757–1767, Apr. 2014.

[8] V. Smet, F. Forest., J. J. Huselstein, F. Richardeau, Z. Khatir, S. Lefebvre, and M. Berkani, "Ageing and failure modes of IGBT modules in high-temperature power cycling", *IEEE Trans. Ind. Electron.*, vol.58, no.10, pp. 4931–4941, Oct. 2011.

[9] H. Lu, C. Bailey, and C. Yin, "Design for reliability of power electronics modules," *Microelectron. Reliab.*, vol. 49, no. 9–11, pp. 1250–1255, 2009.

[10] Sofia, J. W. "Fundamental of thermal resistance measurement". Appl. Notes Anal.Tech., Wakefield, MA, 1995

[11] M. A. Eleffendi and C. M. Johnson, "Application of Kalman filter to estimate junction temperature in IGBT power modules", *IEEE Trans. Power Electron.* vol. 31, no. 2, pp. 1576-1587, 2016.

[12] U. Scheuermann and U. Hecht, "Power cycling lifetime of advanced power modules for different temperature swings," in *Proc. of the PCIM Nuremberg*, 2002, pp. 59–64.

[13] P. E. Bagnoli, C. Casarosa, M. Ciampi, and E. Dallago, "Thermal resistance analysis by induced transient (TRAIT) method for power electronic devices thermal characterization—part I: fundamentals and theory", *IEEE Trans. Power Electron.*, vol.13, no.6, pp. 1208–1215, Nov.1998.

[14] X. Cao, T. Wang, Khai D. T. Ngo, and G. Q. Lu, "Characterization of lead-free solder and sintered nano-silver die-attach layers using thermal impedance," *IEEE Trans. Compon. Packag. Manuf.Technol*, vol.1, no.4, pp.495-501,Apr. 2011

[15] Thermal equivalent circuit models, June. 2008, http://www.infineon.com/

[16] Y. C. Gerstenmaier, W. Kiffe, and G. Wachutka, "Combination of thermal subsystems modeled by rapid circuit transformation", in *Proc. of the 13th THERMINIC*, 2007, pp. 115-120

[17] D. Schweitzer, H. Pape, and L. Chen, "Transient measurement of the junction-to-case thermal resistance using structure functions: chances and limits," in *Proc. of the 28th SEMI-THERM*, 2008, pp. 191-197

[18] T. K. Gachovska, B. Tian, J. L. Hudgins, W. Qiao, and J. F. Donlon,"A real-time thermal model for monitoring of power semiconductor devices," *IEEE Trans. Ind. Appl.*, vol. 51, no. 4, pp. 3361-3367, Jul/Aug. 2015.

[19] Y. Luo, Y. Kajita, T. Hatakeyama, S. Nakagawa, and M. Ishizuka, "Thermal transient test based thermal structure function analysis of IGBT package", in *Proc. of the ICEP*, 2014, pp. 596-599

[20] A. V. Varnai, S. Gao, Z. Sarkany, J. Kim, S. Choi, G. Farkas, A. Poppe, and M. Rencz, "Issues in junction-to-case thermal characterization of power packages with large surface area", in *Proc. of the 30th SEMI-THERM*, 2010, pp. 158-164

[21] A. A. Merrikh and A. J. McNamara, "Parametric evaluation of foster RC-network for predicting transient evolution of natural convection and radiation around a flat plate," in *Proc. of the ITherm*, 2014, pp. 1011-1018.

[22] K. Ma, M. Liserre, F. Blaabjerg, and T. Kerekes, "Thermal loading and lifetime estimation for power device considering mission profiles in wind power converter", *IEEE Trans. Power Electron.*, vol. 30, no. 2, pp. 590–602, Feb. 2015.

[23] L. Wei, R. J. Kerkman, R. A. Lukaszewski, H. H Lu, and Z. H. Yuan," Analysis of IGBT power cycling capabilities used in doubly fed induction generator wind power system" *IEEE Trans. Ind. Appl.*, vol. 47, no. 4, pp.1794–1801, July-Aug. 2011.

[24] Z. Sarkany, A. V. Varnai, S. Laky, and M. Rencz, "Thermal transient analysis of semiconductor device degradation in power cycling reliability tests with variable control strategies," in *Proc. of the 34th SEMI-THERM*, 2014, pp. 236-241

[25] Electronic Industries Association, "Transient dual interface test method for the measurement of the thermal resistance junction-to-case of semiconductor devices with heat flow through a single path", EIA/JEDEC Standard, JESD51-14, 2010 [www.jedec.org].

[26] J. Yin and J. D. van Wyk, "Comparison of transient thermal parameters for different die connecting approaches," *IEEE Trans. Ind. Appl.*, vol. 42, no. 6, pp. 1403-1411, Nov. 2006.

[27] J. N Davidson, D. A. Stone and M. P. Foster, "Real-time prediction of power electronic device temperatures using PRBS-generated frequency-domain thermal cross coupling characteristics", *IEEE Trans. Power Electron.* vol. 30, no. 6, pp. 2950-2961, 2015.

[28] Xiang, L. Ran, "Monitoring solder fatigue in a power module using case-above-ambient temperature rise," *IEEE Trans. Ind. Appl.*, vol. 47, no. 6, pp. 2578-2591, Nov.-Dec. 2011.

[29] T. Lei, M. Barnes, A. C. Smith, "Thermal cycling evaluation for DFIG wind turbine power converter based on joint modelling", in *Proc. ECCE*, 2013, pp. 3845-3851

Design and Evaluation of High Current PCB Embedded Inductor for High Frequency Inverters

Mehrdad Biglarbegian, Neel Shah, Iman Mazhari, Johan Enslin and Babak Parkhideh
Electrical and Computer Engineering Department
Energy Production and Infrastructure Center (EPIC)
University of North Carolina at Charlotte
Charlotte, USA
emails: {mbiglarb, nshah49, imazhari, jenslin,bparkhideh}@uncc.edu

***Abstract*— This paper describes the design and evaluation of high current toroidal Printed Circuit Board (PCB) embedded inductor applicable for high frequency DC/AC converters (lower than 10MHz). The equivalent circuit model of the inductor is presented and the filtering effectiveness is verified by testing three different geometries with its own characteristics and function through a GaN converter. Besides, the temperature rise analysis is provided for all the prototypes by using finite element methods, and the obtained results are also verified by experiments. The reason behind asymmetric trend of temperature rise for the toroidal PCB embedded inductor is also discussed through both simulation and experiment, and the capability of using a heat sink to lower the temperature increment for different geometries is explained.**

***Keywords*— air-core PCB embedded inductor; temperature rise analysis; high frequency; GaN**

I. Introduction

The demand for ever-increasing switching frequency using wide band gap devices, higher power density, keep driving the development of passive integration technologies in the modern power electronics. With increasing the frequency to the MHz range, magnetic core loss increases drastically [1]. So, air core PCB embedded inductors become a viable solution, as the sizes are shrinking, and the core losses are avoided [2], [3]. It will also lead to a cost reduction, as smaller passive components are less expensive. In general, the advantages of passive components integration will be size reduction, better thermal conduction, high parameter repeatability, high reliability, and cost reduction [3], [4].

Among several structures for the design of PCB embedded inductors described in the previous literature [3], [5], toroidal inductors are chosen in this paper due to their internal encapsulation of the magnetic field avoiding EMI problems. Also, compared to other common structures for a PCB embedded inductor, toroid has some drawbacks addressed in this paper such as relatively lower quality factor and inductance per area, less flexibility, and high resistance caused by thin petals, and the poor fill factor of the vias used in both inner and outer edges [3], [6].

Previous researches have been conducted combining RF circuits and power electronics to design switch mode power supplies (SMPSs) operable at very high switching frequencies (VHF) in the range of (30-300MHz), but this range of frequency is more suitable for resonant DC/DC converters [7].

Contrarily, this paper conducts a comprehensive analysis, and design investigations for three different high efficient toroidal PCB embedded inductors suitable for smoothing the AC current of high frequency hard switched inverters. Moreover, temperature Rise analysis including asymmetric distribution, proximity and skin effect, total harmonic distortion (THD), and thermal resistance are investigated and formulation for the built designs are presented. In section II, the equivalent circuit model of a toroidal PCB embedded inductor is expressed and its temperature rise profile is formulated. Section III shows all the three implemented designs, and their main characteristics such as filtering capability in THD reduction, AC resistance, and temperature profile variations. In section IV, temperature rise analysis including asymmetric distribution, proximity and skin effect, and effective factors are discussed and how it can be affected by using a heat sink will be explained.

II. Temperature Rise Formulation for the PCB Embeededd Inductor

A. Equivalent Circuit Model of Toroidal PCB Embedded Inductor

For the full bridge inverters, it has been assumed that current and voltage ripples set to be 1%. Designing an inductor with the above specifications, not only limit the ripple amplitude, it also helps to effectively filter our high frequency harmonics [3]. Following equation shows the resonant frequency of parallel LC circuit:

$$f_{cutoff} = \frac{1}{2\pi\sqrt{L_{\min}C_{\min}}} = \frac{1}{\sqrt{\frac{4\pi^2\left(V_{in}-V_{out}\right)D_{\max}}{\Delta I\times f_{sw}}\times\frac{\Delta ID_{\max}}{f_{sw}\Delta V}}} \qquad (1)$$

Typically, in order to be far enough from the high frequency harmonics and decrease the output noise at 10MHz switching frequency, the cutoff frequency is set to be lower than 1/10th of the switching frequency. In (1), D_{max} shows the maximum duty cycle , f_{sw} is the switching frequency of the inverter and f_{cutoff}

978-1-4673-9551-9/16 $31.00 © 2016 IEEE

is the cutoff frequency of the AC filter, respectively. Therefore, the inductance can be calculated in (2) after simplifying (1):

$$L = \frac{(V_{in} - V_{out}) D_{max}^{2}}{\Delta V \times f_{sw}^{2} \times C} \tag{2}$$

By choosing a low ceramic capacitor (<1μF) to have a low equivalent series resistance (ESR), the 50-300nH range of the inductance will be obtained for DC/AC converter applications in the specified range of switching frequency ($f_{sw} < 10MHz$). The geometric parameters of a typical air-core PCB embedded inductor is shown in Fig.1.

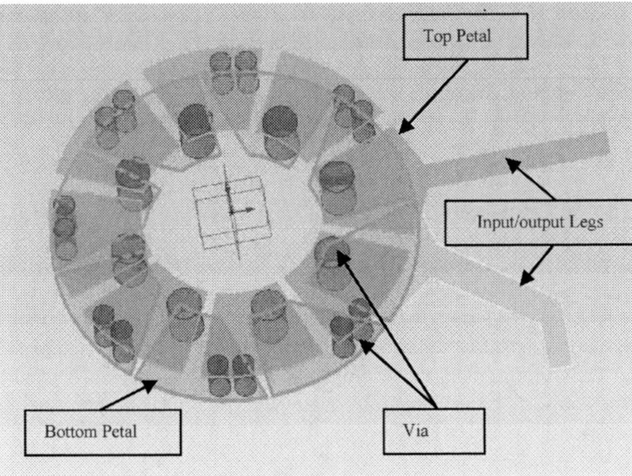

Fig. 1. Geometric parameters of an air-core PCB embedded inductor (yellow: bottom petals, cyan: top petals)

Fig. 2a. depicts the equivalent electrical circuit model of the proposed embedded inductor. Due to the encapsulation of magnetic flux inside the toroid, the path of current in each petal can be represented in a LR circuit configuration. The behavior of this impedance over a wide range of frequencies has been studied through the spectrum analyzer shown in the Fig. 2b. As can be observed, in very high frequency (~280MHz), the inductor has a resonance, which mainly caused by the parasitic capacitance of adjacent petals and vias.

The occurrence of the resonant at this range of frequency shows that the parasitic capacitance would be in the range of (~pF). As the proposed inductor is normally operating at much lower than this resonant frequency, the effect of the capacitance in the equivalent impedance can be neglected for further analysis.

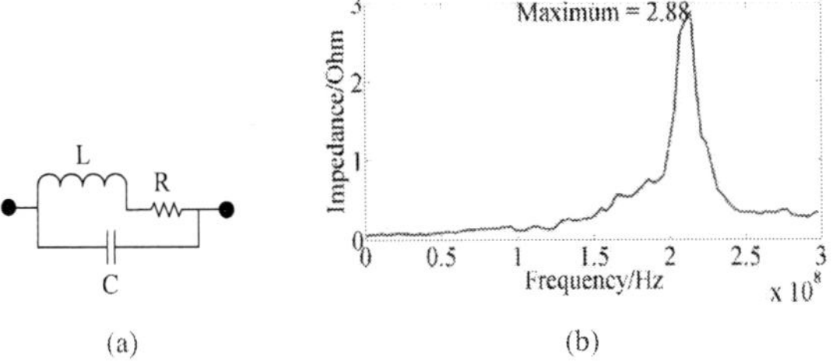

Fig. 2. Equivalent circuit model and the impedance spectrum vs. frequency

Therefore, inductance and AC resistance of the toroidal PCB embedded inductor can be calculated as follow:

Based on the previous studies [6], [8], in a toroidal air-core PCB embedded inductor, the inductance can be calculated in (3):

$$L = \frac{N^2 h \mu_0}{2\pi} \ln\left(\frac{d_o}{d_i}\right) + \frac{d_o + d_i}{4} \mu_0 \left[\ln(8\frac{d_o + d_i}{d_o - d_i}) - 2\right] \tag{3}$$

In (3), N is the number of petals, h is the height of PCB/inductor and μ_0 is the permeability of the air. Furthermore, outer and inner diameters of the inductor are shown by d_i and d_o, respectively. As shown in [3], [9], the inductance can be simplified as below:

$$L \propto \mu_0 k N^2 h \tag{4}$$

where, k at the defined outer/inner geometries is a constant value.

As shown in [6], [10], the resistance can be expressed in (5). As shown in (5), the resistance consists of two main parts: resistance of petals and resistance of vias.

$$R = N\left(2 \times R_{petal} + \frac{R_{inner\_via}}{N_{inner}} + \frac{R_{outer\_via}}{N_{outer}}\right) \tag{5}$$

The resistance of each petal is proportional to the copper resistivity, number of turns, and outer diameter of the inductor, and it is inversely related to the skin depth, petals' clearance and inner diameter of vias. On the other side, the vias resistance will be increased directly by copper resistivity and inversely with the via diameter, and the skin depth, which is a measure of the penetration of a plane electromagnetic wave into the copper. By considering the impact of skin and proximity effects on the total resistance of the vias, the current tends to pass through the edges of the conductor (copper) as the current density is largest near the surface. Also, proximity effect due to the relatively bigger size of each petal respect to their clearance can be ignored. By simplifying the equations and applying the skin effect relation with switching frequency, the AC resistance can be expressed in (6) as shown also[3], [9]:

$$R_{ac} \propto k_1 N^2 \sqrt{f} \tag{6}$$

where, k_1 is constant factor which depends on the geometry of the inductor.

B. Temperature Rise Formulation of Toroidal PCB Embedded Inductor

As shown in [11], the temperature rise in convection mode at the steady-state conditions can be defined as (7):

$$P \times R_{th} = \Delta T \tag{7}$$

In this formula, P is the total dissipative power of the inductor, and R_{th} is the total thermal resistivity between the board and the environment.

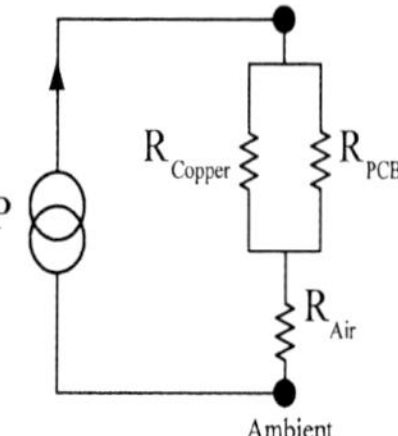

Fig. 3. Equivalent thermal circuit model without heatsink

In order to quantitatively demonstrate and compare the temperature rise variations of the inductor, the temperature rise can be normalized to the specific reference. The temperature rise of the PCB embedded inductor is normalized per quality factor, which can be obtained regardless of the current conditions:

$$\Delta T_{P.U} = {}^{P\times R_{th}}\!/_{Q} \tag{8}$$

Therefore, $\Delta T_{P.U}$ is the temperature rise per the quality factor. After simplifying the equations (4)-(8):

$$\Delta T_{P.U} = k_3 \frac{I^2 Z}{x^3\sqrt{f}} = k_4 \frac{I^2(\sqrt{f}+k_5)}{x^3} \tag{9}$$

In (9), x defines the ratio of the surface area to the effective length and describes the inductor geometry for heat transferring, I is the rms current, k_3 ,k_4, are the constant values depending on inductor geometry and k_5 depicts AC resistance variation of the inductor. In the PCB embedded inductor, x is defined as a degree of freedom of a design which would be area divided by the perimeter of the inductor with considering the outer diameter. Larger x corresponding to bigger surface can bring benefits for better heat transferring capability; however, it exacerbates achieving higher power density/efficiency [3].

III. Implemented PCB Embedded Inductor Designs

By considering the initial constraints and imposing the equations from (3)-(9), different designs of PCB embedded inductors are simulated through using finite element methods in JMAG. The finite element simulations verify the predicted results for temperature distributions, current density, AC resistance, and inductance of the PCB embedded inductor. Then, after evaluation of various parameters of toroid inductors in the specified range of frequency ($f_{sw} < 10MHz$), three patterns are selected for prototyping as shown in Fig. 4. In all implemented patterns, allocating a minimum surface area for high power density, and keeping the temperature rise lower than 45°C at the rated current are considered as the main design objectives, and the pros and cons of each design are investigated.

In the first model with 26-turn (L: 240nH/ R(DC): 160mΩ/ R(AC): 200mΩ-5MHz), the inductor is designed such that the highest inductance to filter out the high frequency ripples is provided. Based on (3), the higher number of turns makes via's diameter smaller and also increases the petals resistance drastically.

Fig. 4. Prototyped PCB inductor designs with 26-turn (60 mm outer diameter), 13-turn (60 mm outer diameter), and 8-turn (30mm outer diameter)

By considering (9), the proposed inductor has been tested with relatively lower continuous current (5A) to keep the temperature rise limited in the predefined range. The second prototype with 13-turn (L: 150nH/ R(DC): 32mΩ/ R(AC): 50mΩ-5MHz), is proposed such that by decreasing the number of turns and increasing the height of the inductor, the inductance value is high enough to filter out the high frequency ripples and decrease the DC conduction losses simultaneously. It can be shown that because of its relatively high resistance, the temperature rise for continuous currents bigger than 10A will be increased to 50°C. Finally the third prototype with 8-turn (L: 50nH/ R(DC): 2.4mΩ/ R(AC): 8mΩ-5MHz) is designed and implemented in 4-oz copper to decrease the resistance and increase the continuous current capacity of inverter to enhance power density. The 50nH has been tested with GaN converter (EPC9033) in a Half-Bridge mode with 2.5MHz switching frequency, as shown in Fig. 5:

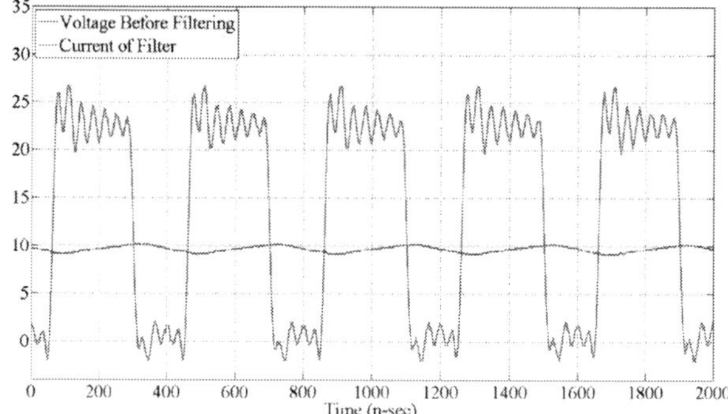

Fig. 5. Experimental Results: Voltage/Current of Inductor design-3 with Half Bridge GaN converter

The total harmonic distortion (THD) of the inductors has been calculated for all the three samples to evaluate how effective they can filter out high frequency switching harmonics. Table I summarizes the THD calculated for all three prototypes and their frequency spectrums have been shown in Fig. 6.

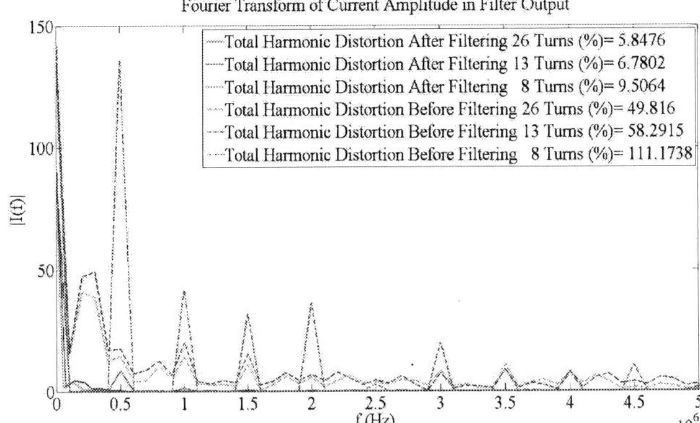

Fig. 6. Experimental Results: THD waveforms of the current for all three designs before (dash lines) and after (solid lines) filtering.

TABLE I. THD CALCULATION OF THREE PROTOTYPES

THD	Design-1	Design-2	Design-3
Before Filtering	49.81%	58.29%	111.17%
After Filtering	5.84%	6.78%	9.51%

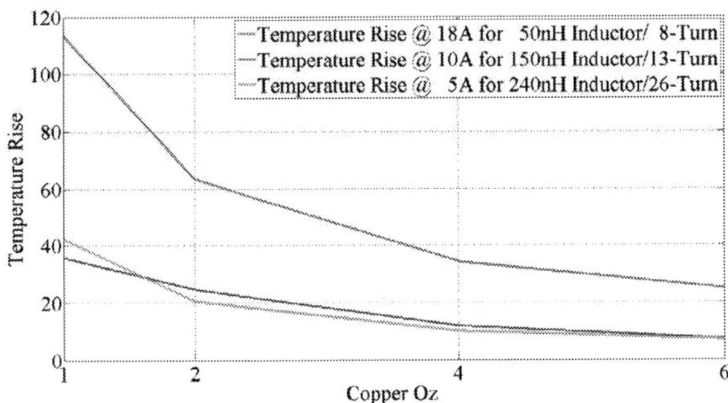

Fig. 7. Simulation Results: Temperature rise variations of three prototypes respect to copper thickness at rated current at 5MHz Switching Frequency

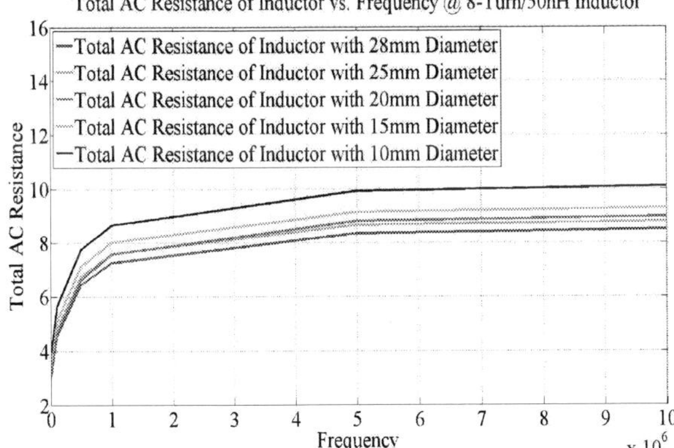

Fig. 8. Simulation Results: AC resistance of PCB embedded inductor vs. frequency by variable vias' diameter in 8-turn (50nH) inductor at 18A current.

In order to evaluate the impact of the resistance on the temperature rise, two different analyses are investigated from different aspects: copper thickness, and via diameters. The temperature rise variations of the petals in each design have been simulated in JMAG as shown in Fig. 7. The continuous current of each inductor is selected such that the maximum temperature rise cannot exceed 45°C.

For the second analysis, by changing the diameter of via, the trend of total AC resistance respects to the frequency is shown for the 8-turn inductor in Fig. 8. As it can be seen, with increasing the diameter of via, the total AC resistance of inductor will be notably reduced. The slope of resistance variations in high frequencies drops significantly due the considerable reduction in skin depth variations.

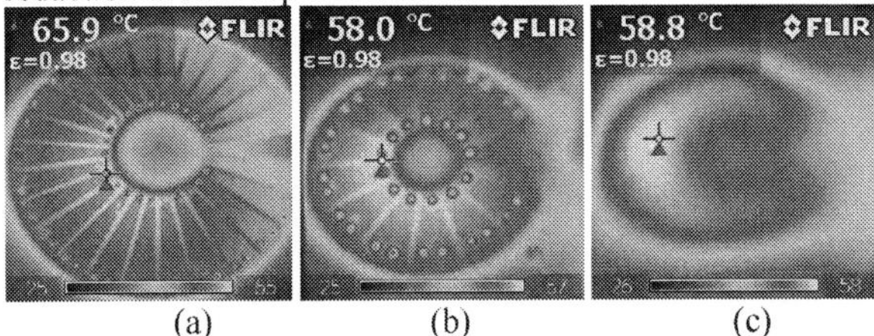

(a) (b) (c)

Fig. 9. Experimental Results: Temperature rise variations of PCB embedded inductors in three different prototypes. a) design-1:26-turn b) design-2:13-turn

Fig. 10. Simulation Results: Asymmetric temperature distribution for different

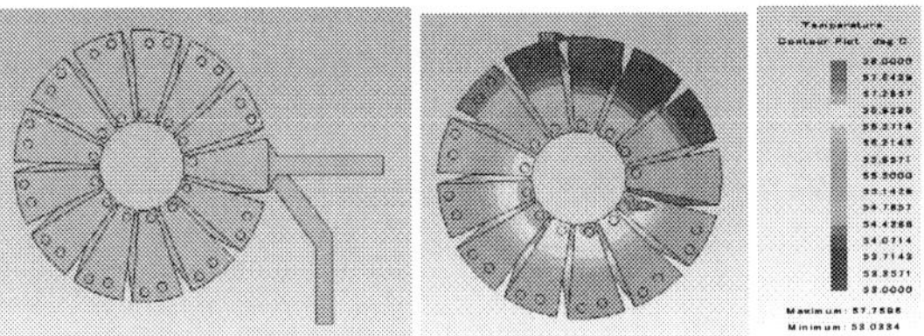

(a) Simulation Reuslts: Assymetric temperature distribuition in outer petals is more than 4°C

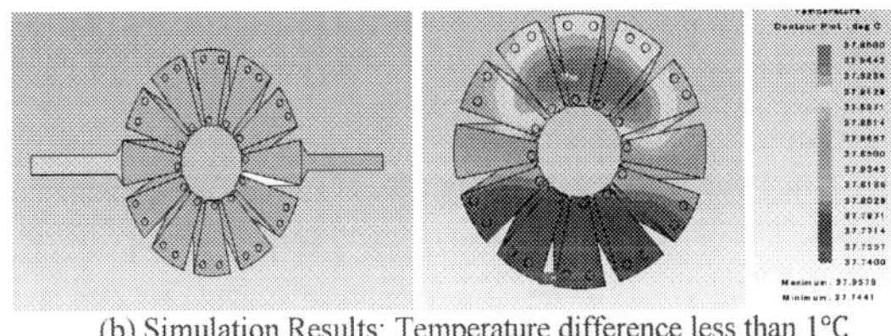

(b) Simulation Results: Temperature difference less than 1°C

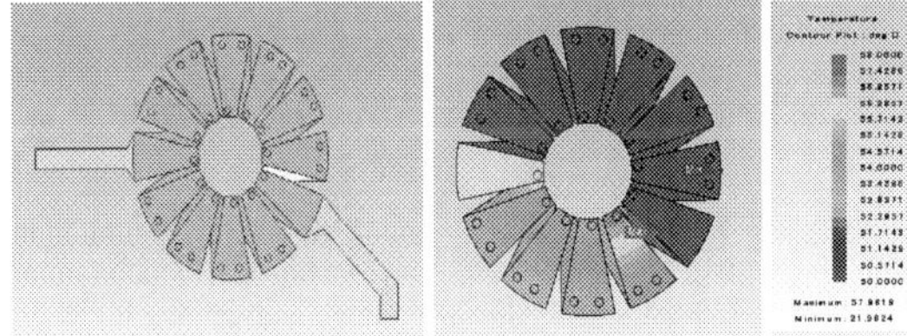

(c) Simulation Results: Temperature difference is around 3°C

input/output configurations in design-2: 13-turn inductor at the nominal current (10A current)

IV. Temperature Rise Analysis and Results

A. Asymmetrical Temperature Rise Evaluations

After analyzing three fabricated PCB embedded inductors, they are tested continuously under the nominal currents (5A for 26-turns, 10A for 13-turn, and 18A for 8-turn) for 20 minutes to reach the steady state conditions to extract the temperature rise profile as shown in Fig. 9. The asymmetric temperature rise distribution among the petals is observed for all three porotypes in both simulation and hardware results as described in [3].

To prove the impact of location of input/output legs, different configurations, for the 13-turn inductor are proposed in Fig. 10. As expected, similar results are also observed after simulating both design-1: 26-turn and design-3: 8-turn not presented in this paper for reasons of space. Both simulations, and experimental results verify that the temperature reaches the highest point where the output leg is bent, and the current path is forced to be distracted. So, the convention of the temperature increment in adjacent regions is observed in all provided results. It should also be noted that the spectrum range of the temperature variations is not the same for all the three implemented designs. The highest attainable temperature is expectedly related to the inductor with 26-turn, which has higher conduction losses through the petals resulting in higher temperature rise.

Regardless of the electrical characteristics, this asymmetric temperature profile relates to different geometries. To prove this fact, all three designs have been tested under hard switching frequency with Fluorinert as a non-electrical conductive liquid. The FC-40 flourinert is elected because of the advantage of having a wide range of boiling temperature, and very high volume resistivity. These capabilities provide a symmetric temperature distribution, isotherms behavior, under multiple scenarios such as fully and partially immersed in the fluorinert. Due to the constant coefficient dispatch of the liquid, the heat is smoothly dissipated along the petals, and no assymetric thermal distribution is observed.

B. Temperature Rise Analysis of PCB Embedded Inductor with Heat Sink

Practically, using a heat sink for cooling down the converter switches due to their fast operation when the dissipative losses exceed more than 1W is inevitable. This means heat sink adds a new dimension in the converter design, and decreases the overall power density. Therefore, by designing a customized heat sink, there is a chance to make inductor smaller such that the AC resistance of the inductor is decreased, which results higher efficiency, and enhancing the power density. Considering only conduction and natural convection transferring the heat through the heat sink and air respectively, the equivalent thermal circuit model can be described as shown in Fig. 11.

As the surface area of the heat sink cannot exceed the total area of the inductor, and assuming its height is 25mm, the total thermal resistance can be calculated as below [11]:

$$R_{th\_heatsink} = \frac{1}{h(A_{heatsink} + \eta N_{fin} A_{fin})} \quad (10)$$

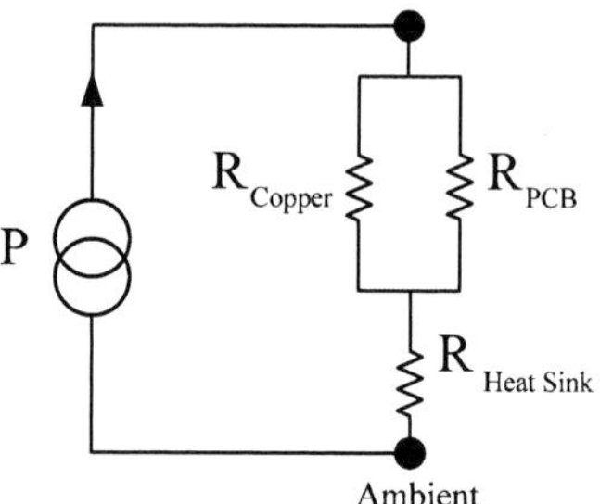

Fig. 11. Equivalent Thermal circuit model without heatsink

where, h is convective heat transfer coefficient, $A_{heatsink}$ is the base surface area, A_{fin} is the surface area of the fins and N_{fin} is the total number of the heat sink (Fig. 12). So, the total thermal resistivity of the PCB embedded inductor based on Fig. 11 will be calculated as:

$$R_{th\_total} = R_{th\_heatsink} + (R_{th\_PCB} \| R_{th\_copper}) \quad (11)$$

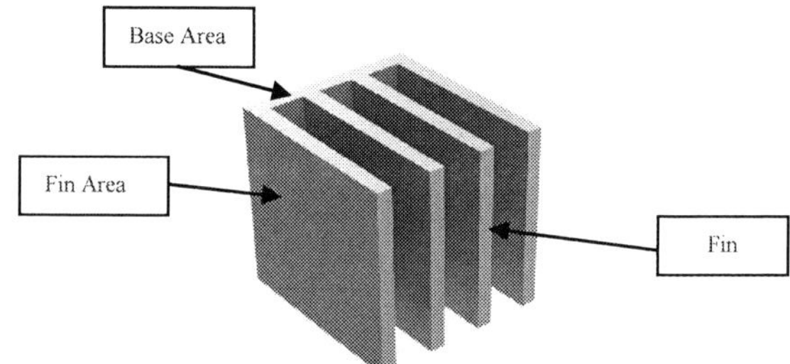

Fig. 12. Heat sink geometries selected for the PCB embedded inductor

$$R_{th\_PCB} = \frac{1}{h_{PCB} A_{inductor}} \quad (12)$$

Now, by considering the designed values for the heat sink from table-II and imposing the equations (10)-(12), there is a possibility to shrink the inductor diameter from 60mm to 39mm for the design-1 and 2 (26-turn and 13-turn respectively) and from 30mm to 25mm for the design-3: 8-turn with the constant temperature rise (45°C). This also means that the total resistance of the PCB embedded inductor will be decreased by 20% for the design-1 and design-2 and 10% of the design-3, which results in higher efficiency. The analysis shows that because of the smaller size of the heat sink in design-3 and relatively smaller x in (9), the rate of size reduction in design-3 is much lower than design-1 and design-2.

TABLE II. PARAMETERS FOR THE DESIGN OF HEAT SINK

Parameters	Design1	Design 2	Design 3	Units
Number of Fins	5	5	3	-
Heat Sink Height	25	25	25	mm
Fins' clearance	5	5	5	mm
Fins' thickness	2	2	2	mm
Convective Coefficient of Heat Sink	5000	5000	5000	W/m^2K
Convective Coefficient of Air	20	20	20	W/m^2K

V. CONCLUSION

In summary, the aim of this paper is to formulate the temperature rise variations of toroidal PCB embedded inductor and evaluates the design considerations for high switching frequency inverters. Equivalent circuit model of these inductors at the range of lower than 10MHz switching frequency is presented and three different prototypes are also fabricated. These designs have been tested under 2.5MHz switching frequency with GaN type half bridge converter. The hardware results show that all the prototypes can effectively filter high switching frequency signals with THD lower than 10%. The outcome of the analysis shows: the design-1 with the highest filtering effect has the lowest THD, but its AC resistance is very high, which is not suitable for efficient converter design. The design-2 has acceptable THD, but there is a room for improvement of power density and efficiency. In the last design, the THD is slightly scarified to improve the power density and AC resistance of the inductor. Furthermore, the effective parameters to impact AC resistance of PCB embedded inductors as well as asymmetrical temperature variations of the prototypes were simulated and explained. Eventually, an approach to improve power density and efficiency through size reduction of PCB embedded inductor by adding a heat sink to the converter is also investigated.

ACKNOWLEDGMENT

This work is supported by Energy Production and Infrastructure Center (EPIC), University of North Carolina at Charlotte. The authors would like to thank the JMAG Corporation for using FEA software, EPC Corporation, and 3M for their great support. The authors also would like to appreciate Ivan Chan from EPC due to his tremendous help and guidance during this work.

REFERENCES

[1] Y. Han, G. Cheung, A. Li, C. R. Sullivan, D. J. Perreault, and S. Member, "Evaluation of Magnetic Materials for Very High Frequency Power Applications," vol. 27, no. 1, pp. 425–435, 2012.

[2] J. Qiu and C. R. Sullivan, "Design and fabrication of VHF tapped power inductors using nanogranular magnetic films," IEEE Trans. Power Electron., vol. 27, no. 12, pp. 4965–4975, 2012.

[3] M. Biglarbegian, N. Shah, I. Mazhari, and B. Parkhideh, "Design Considerations for High Power Density / Efficient PCB Embedded Inductor," in 3rd IEEE workshop on Wide Band Gap Devices and Applications, Nov. 2015.

[4] S. Mao and Y. Zhang, "Design and characterization of planar integrated passive component for power converters," Proc. 2011 14th Eur. Conf. Power Electron. Appl., pp. 1–6, 2011.

[5] S. Orlandi, B. Allongue, G. Blanchot, S. Buso, F. Faccio, C. Fuentes, M. Kayal, S. Michelis, and G. Spiazzi, "Optimization of shielded PCB air-core toroids for high efficiency dc-dc converters," 2009 IEEE Energy Convers. Congr. Expo., pp. 2073–2080, Sep. 2009.

[6] M. Madsen, A. Knott, M. a. E. Andersen, and A. P. Mynster, "Printed circuit board embedded inductors for very high frequency Switch-Mode Power Supplies," 2013 IEEE ECCE Asia Downunder, pp. 1071–1078, Jun. 2013.

[7] H. Schneider, T. Andersen, J. D. Mønster, M. P. Madsen, A. Knott, and M. A. E. Andersen, "Investigation of a Hybrid Winding Concept for Toroidal Inductors using 3D Finite Element Modeling," pp. 1–4, 2013.

[8] C. R. Sullivan, W. Li, S. Prabhakaran, and S. Lu, "Design and Fabrication of Low-Loss Toroidal Air-Core Inductors," 2007 IEEE Power Electron. Spec. Conf., pp. 1754–1759, 2007.

[9] D. J. Perreault, J. H. J. Hu, J. M. Rivas, Y. H. Y. Han, O. Leitermann, R. C. N. Pilawa-Podgurski, a. Sagneri, and C. R. Sullivan, "Opportunities and Challenges in Very High Frequency Power Conversion," 2009 Twenty-Fourth Annu. IEEE Appl. Power Electron. Conf. Expo., 2009.

[10] P. Kamby, A. Knott, and M. a E. Andersen, "Printed circuit board integrated toroidal radio frequency inductors," IECON Proc. (Industrial Electron. Conf., pp. 680–684, 2012.

[11] T. L. Bergman, A. S. Lavine, F. P. Incropera, and P. Dewitt, Heat and Mass Transfer. 2011.

978-1-4673-9551-9/16 $31.00 © 2016 IEEE

Prognosis of Wire Bond Lift-Off Fault of an IGBT Based on Multisensory Approach

Moinul Shahidul Haque
The University of Akron
Email: msh100@zips.uakron.edu

Jeihoon Baek
Korean Rail Research Institute
Email: baek@krri.re.kr

Joseph Herbert
The University of Akron
Email: jph64@zips.uakron.edu

Seungdeog Choi
The University of Akron
Email: schoi@ uakron.edu

***Abstract* — The safe and continuous operation of the power switches of the inverter units is vital in assisting the functioning of the inverter units. This paper proposes an algorithm for the condition monitoring and prognosis of wire-bond fault in an IGBT- extensively used power switch in inverter systems. The proposed algorithm employs both on-state collector emitter voltage and collector current in the estimation of IGBT junction temperature. These two signals complement each other as thermal sensitive electrical parameters and results in better estimation of junction temperature. This algorithm has integrated a stochastic method and Coffin Manson model to introduce criticality matrix which is used to determine the necessity for IGBT replacement. Both simulation result and experiment data verified the proposed. The proposed algorithm has the scope to be further modified for the condition monitoring of a more complex system in the future.**

Keywords—IGBT, WireBond Lift-off, TSEPs, Stochastic Approach, Multi Sensor Approach

I. INTRODUCTION

HYBRID electric vehicle (HEV), high speed train systems are environmental friendly and potentially low cost alternative to traditional transport system. It therefore becomes imperative that these systems need to be more efficient and reliable, to further accentuate their appeal for more frequent usage [1]. Inverter systems employed in these traction systems play a crucial role in providing an adequate and reliable power supply. Semiconductor switches (IGBT, MOSFET) are one of the vulnerable components in an inverter system. IGBTs are considered more often than MOSFETs as power switches because of their lower losses at higher switching frequencies. IGBTs are the most frequently used devices (42%) among power semiconductor devices followed by MOSFETs (27%), thyristors (14%), PiN diodes (10%) [2].

Wire bond IGBTs, used extensively due to their cost advantage over press pack IGBT, experience wire bond lift off due to temperature swing at the point of contact between Aluminum wire and silicon chip [3]. This temperature swing causes different thermal expansion (CTE) of materials at junction which creates mechanical stress at the joint and thus, results in accelerated degradation of the IGBT at its junction. This makes the monitoring of IGBT junction temperature (T_j) highly pertinent in ensuring the reliability of both the IGBT and converter system.

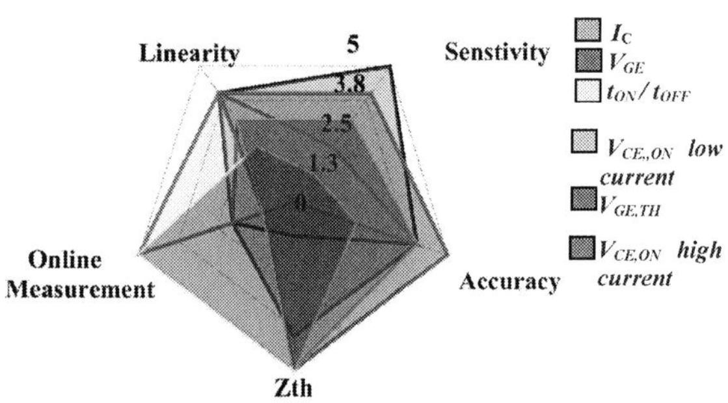

Fig. 1. Spider web charts depicting the characteristics of TSEPs- Collector current (I_c), Gate Emitter Voltage (V_{GE}), Switching Time(t_{ON}/t_{OFF}), V_{CE} at low current, Gate Emitter Threshold Voltage($V_{GE,TH}$) and V_{CE} at high current.

Junction temperature Measurement techniques of IGBT can be classified into three groups- optical, physical contact and electrical. Optical methods such as luminescence, Raman Effect, refraction index, reflectance, or laser deflection are formulated. Besides, methods based on the variation of infrared (IR) radiation with temperature also developed to measure junction temperature utilizing temperature dependent optical properties of semiconductor devices [4]. The major drawback of these methods is that they require modifying the power module. On the other hand, the physical contact methods only give the local temperature. However, electrical methods employ the thermal sensitive electrical parameters (TSEPs) for temperature measurement. The advantage of electrical method is that it provides a wide temperature range and a fast response time compared to the other two methods[4]. Moreover, low cost and non-invasive T_j estimation technique employ TSEPs such as Collector Emitter On-Voltage ($V_{CE,ON}$) at low and high currents, gate emitter threshold voltage ($V_{GE,th}$), collector current(I_C) and switching time (t_{on}, t_{off}). A comparison of the performance of the different types of TSEPs used for IGBT junction temperature measurement is shown in Fig. 1 [3].

The $V_{CE,ON}$ at low current has high accuracy, linearity and sensitivity in T_j estimation but no online measurement capability. This makes it the least favorable TSEP for T_j online estimation. Similarly, $V_{GE,th}$ also has low online capability [5]. In addition this TSEP needs additional circuit for its measurement. Switching time has a lower sensitivity, of the order 2 ns/^{0}C [6] but better Linearity and online measurement capability but is still highly prone to noise. However, I_C has high linearity, sensitivity, and accuracy but poor online measurement capability while $V_{CE,ON}$ at high current has positive temperature coefficient, high online measurement capability but its accuracy, sensitivity and linearity is low [7,8]. $V_{CE,ON}$ is the most popular TSEP for T_j estimation but this estimate is prone to noise and less accurate. It is therefore proposed that $V_{CE,ON}$ could be combined with I_C and the combined estimation will have high sensitivity, accuracy and online measurement capability. Moreover, the junction temperature is result of change in collector current. Thus considering both of these signals provides better estimation of T_j. This estimated T_j over a cycle is then collected and analyzed stochastically to determine if the junction temperature has reached the critical T_j. In conjunction with the above stochastic theory a junction temperature based lifetime estimation model is employed and a matrix called criticality matrix is introduced to categorize the health of the IGBT and the necessity for unit replacement.

The subsequent sections of the paper will cover the coverage of the proposed algorithm including an explanation of the stochastic approach and the lifetime model. Finally, the simulation and experimental setups and results in each case will be addressed and discussed in detail to further supplement the algorithm.

In the light of the advantage of the proactive approach, the technique considered in this paper is a prognostic one to accentuate the reliability of the system. Additionally, to further supplement this proposition a multi-sensor method has been considered as it gives a superior result compared to a single sensor approach.

II. Estimation of Junction Temperature from TSEPs

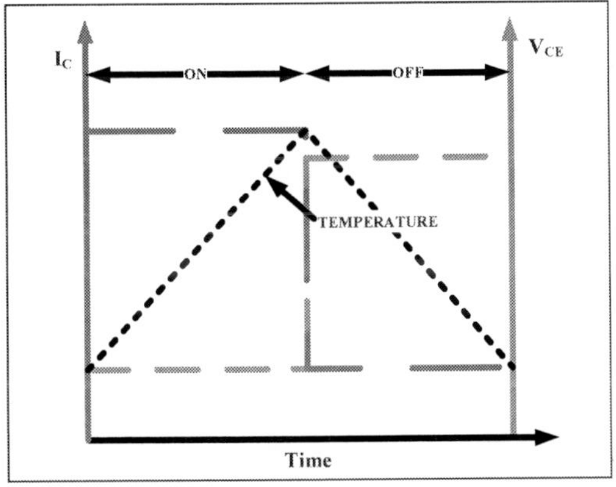

Fig. 2. V_{CE}, I_C and Temperature with time of Ideal IGBT.

From the Fig. 2, the variation of the voltage, V_{CE} and the current I_C, during one switching cycle of an ideal IGBT is shown along with the variation of the temperature.

However, in reality there are finite amount of switching and conduction losses in an IGBT. It is assume that both $V_{CE,ON}$ and I_C are in the linear region. The collector to emitter on voltage equation in this region is as follows: [9-10]

$$V_{CE,ON} = V(p+n-) + V_B + V_{MOS} + V_R \quad (1)$$

$$V_{CE,ON} = V(p+n-) + V_B + V_{MOS} + I_C R \quad (2)$$

where R is the resistance of bond wire of IGBT during the on time, $V_{CE,ON}$ is the on-state Collector-Emitter Voltage and I_C is the Collector Current of an IGBT, $V_{(p+n-)}$ is the junction drop; V_B is the voltage drop across the lightly doped storage region, V_{MOS} is the voltage drop over the space charge region and V_R is the ohmic voltage drop due to the bond wire.

Therefore, the degradation of the wire-bond is exhibited by the increase in resistance of wire-bond. So any increment of $V_{CE,ON}$ is caused by increase of this resistance. This change in resistance results in the change in junction temperature.

The temperature dependency of R can be expressed as-

$$R_t = R_{ref}\,[1 + \alpha(T_{j,t} - T_{j,ref})] \quad (3)$$

where $T_{j,t}$ is the temperature at any time t and $T_{j,ref}$ is the reference temperature, R_t is the on-state resistance at temperature T, R_{ref} is the on-state resistance at temperature $T_{j,ref}$ and α is the temperature coefficient at a particular collector current.

Equation (3) can be rearranged as follows:

$$T_{j,t} = T_{j,ref} + \frac{R_t - R_{ref}}{\alpha R_{ref}} \quad (4)$$

At any time t of IGBT operation, from $V_{CE,ON}$ and I_C, R_t can be calculated and from this R_t, the junction temperature at that instant can be calculated using equation (4).

A. Probability Distribution Functions (PDFs)

Probability Distribution (Density) Function (PDF) of a random variable defines the probability distribution of that variable. The nature of probability distribution of a PDF depends on the underlying distribution of the random variable. To fit the junction temperature distribution Gamma, Normal and Weibull are considered. These distributions consider continuous data which resembles temperature distribution of IGBT. From the best fitted distribution, mean temperature and change is temperature per cycle will be estimated and moreover, it will be observed if the IGBT is entering into critical temperature region.

Gamma Distribution: If any random variable x has the Gamma variable shape parameter k and scale parameter b then the PDF of x can be defined as [11]-

$$f_X(x) = \frac{1}{\Gamma(k)\,b^k} x^{k-1} e^{\frac{-x}{b}}\,; x \in (0,\infty) \quad (5)$$

where $\Gamma(k)$ is the gamma function evaluated at k.

Weibull Distribution: For any random variable, x that has a Weibull distribution with shape parameter, k and scale parameter, b then the PDF can be defined as follows: [12]

Table I. Different Distribution with Their Mean and Variance [11].

NAME OF THE DISTRIBUTION	MEAN	VARIANCE
Normal Distribution	μ	σ^2
Weibull Distribution	$b\Gamma(1+\frac{1}{k})$	$b^2\left[\Gamma\left(1+\frac{2}{k}\right)-\left(\Gamma\left(1+\frac{1}{k}\right)\right)^2\right]$
Gamma Distribution	bk	bk^2

$$f(x;\lambda,b)=\begin{cases}\frac{k}{b}\left(\frac{x}{b}\right)^{k\text{-}1}e^{-(x/b)^k} & x\geq 0,\\ 0 & x<0,\end{cases} \tag{6}$$

Normal Distribution: Any random variable, x that has a normal distribution with location parameter μ and scale parameter σ has a PDF that can be defined as [13]-

$$f_x(x)=\frac{1}{\sigma}\phi(\frac{x-\mu}{\sigma})=\frac{1}{\sigma\sqrt{2\Pi}}e^{\frac{-(x-\mu)^2}{2\sigma^2}} \quad \text{where } x\in\Re \tag{7}$$

The mean and the variance of a distribution can be calculated using the parameters of the distributions. The mean and the variance of these distributions are given in Table I.

Standard deviation (σ) of a distribution is the square root of the variance of that distribution. Two standard-deviations around the mean of the best fitted distribution will be considered for this algorithm. When the critical junction temperature reaches within two standard deviation of the mean it can be said this temperature is more evident at junction. To find the best fitted distribution for a given data set, the parameters with minimum negative log likelihood will be chosen for a distribution.

B. Maximum Likelihood Estimate (MLE)

The likelihood function of any parameter of a distribution function is the probability of that parameter for the given points or data set. For x, any random variable with the distribution function f depending on parameter θ, the likelihood function of θ can be written as [14]

$$L(\theta\mid x)=\Pi_{i=1}^{n}f_\theta(x) \tag{8}$$

The maximum likelihood estimation determined the distribution for which there exists a maximum likelihood of a parameter. Usually, parameters are found out by setting the derivative of equation (8) to zero. But to ensure maximum likelihood, the second derivative of the function should be negative.

From the optimization point of view, rather than finding a maximum, finding a minimum is more convenient. Taking the logarithm of a function makes computation easier and brings consistency in numerical value. Moreover, the logarithm of a function reaches maximum at the same point where function itself reaches maximum. Negative Log Likelihood (NLL) minimizes at the same point where log likelihood becomes maximum. So the following steps are followed to find the parameter:

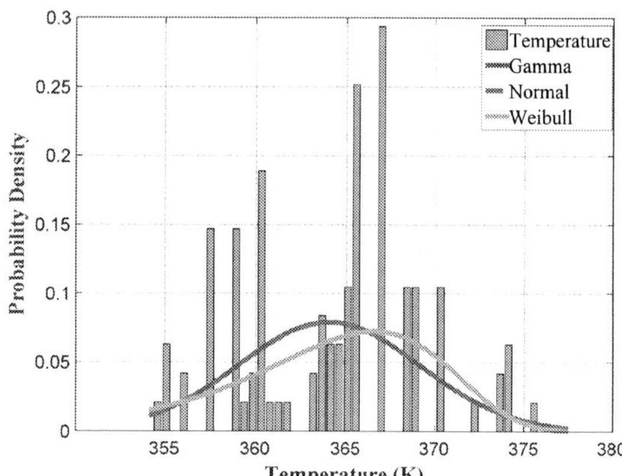

Fig. 3. Distribution fitting of a given data set.

Table II. Negative Log Likelihood and Parameters for Different Distrbituions.

Distribution Name	Negative Log Likelihood	Parameter 1	Parameter 2
Gamma	3.2634e+03	Shape (4.7575)	Scale (3.1219)
Weibull	3.2790e+03	Shape (2.3532)	Scale (16.7966)
Normal	3.3248e+03	Location (14.8525)	Scale (6.7286)

1) Negative log of a likelihood function is taken.

$$l(\theta\mid x)=-\log[L(\theta\mid x)]=\Pi_{i=1}^{n}-\log[f_\theta(x)] \tag{9}$$

2) Derivative of this NLL is taken and set to zero to find out the parameter.

3) To ensure the minimum NLL point (maximum likelihood), check if $\left(d^2l(\theta\mid x)\right)/dx^2>0$

For a given data set, the parameters of different continuous distribution functions could be determined using the MLE principle. The distribution which has minimum NLL is the best fitted distribution function.

For a random data set, the best fits of different distributions using MATLAB® are shown in Fig. 3 and the corresponding NLL of Weibull, Normal and Gamma distributions are shown in Table II.

C. Life Time Model:

Various residual life time models are employed to calculate the number of cycles to failure for a system. These models can be used for the early detection of degradation and thus a detection of a fault. In this paper, the damaged is induced in the IGBT operating it at higher load current(I_C). The effect of I_C on IGBT is observed through change in junction temperature (ΔT_j) [15]. Among the life time estimation models, Coffin Manson Model (CMM) provides an account of damage each temperature cycle does to the power switch.

The CMM is an analytical lifetime model that assumes the number of cycles to failure depends only on the temperature swing of the junction temperature (ΔT_j) and the mean temperature (T_m). The model can be expressed as shown below [16].

$$N_f = A\Delta T_j^{-n} e^{\frac{E_a}{kT_m}} \quad (10)$$

where N_f is number of cycles to failure; E_a is the activation energy; K is the Boltzman's constant; A and n are constants.

III. Proposed Algorithm

The flow chart for the proposed algorithm is illustrated in the Fig. 4. In this flow chart, both the deterministic and stochastic approach has been integrated for condition monitoring and prognosis purpose. The processed data of both approaches is fed as input to a 'criticality matrix' where based on fulfilling conditions a corresponding action is triggered. This section includes an overview of the proposed algorithm from acquisition of TSEPs from the system to the final triggering of corresponding condition based action.

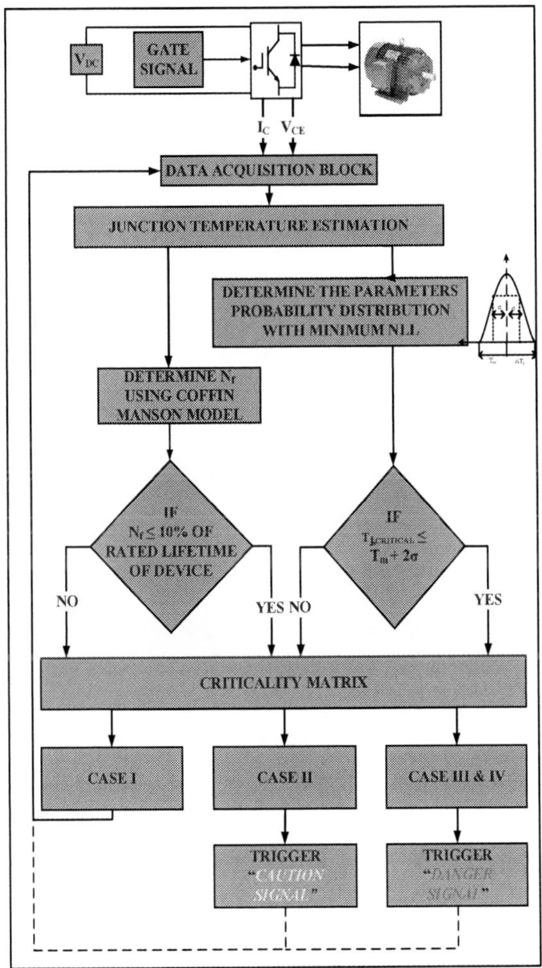

Fig. 4. Algorithm for fault diagnosis using multisensory approach.

During the operation of an IGBT, the $V_{CE,ON}$ and I_C will be measured and the corresponding T_j will be estimated using equation (4). The precision of $V_{CE,ON}$ and I_C will depend on the sensitivity of the sensors and the number of data points to be stored will depend on the robustness of the system.

The distribution of this data will be approximated by Normal, Weibull or Gamma distribution. Next, the best fitted distribution will be filtered out using minimum NLL method. The parameters for this distribution will be utilized to calculate the mean and standard deviation of that distribution. The range of temperature within two standard deviations will be determined from these parameters. Manufacturers provide safe region of operation for T_j in the data sheet from which $T_{j,cr}$ can be determined. For discrete type of IGBT this temperature region is 298K to 448K. 448K has been considered as the critical temperature for this algorithm [17]. If $T_{j,cr}$ falls within the range of two standard deviations of the best fitted distribution of T_j determined from the NLL method, the system will be notified that the IGBT is operating at a higher temperature.

Simultaneously, the average (T_m), maximum ($T_{j,max}$) and minimum junction temperature ($T_{j,min}$) will be determined during the on-time of each power cycle from T_j. Temperature swing per cycle can be calculated using

$$\Delta T_j = T_{j,\max} - T_{j,\min} \quad (11)$$

This will be used along with T_m to estimate the number of cycles to failure using Coffin Manson Model equation (10).

It is evident from equation (10) that N_f has an inverse relationship with variation in junction temperature (ΔT_j) and that N_f decreases with the increases of ΔT_j. This means, that an IGBT despite being in the safe operating region may still fail due to a large ΔT_j or it may continue operation despite being in the high temperature region because the ΔT_j is small. This leads to the establishment of the Criticality Matrix as shown in Table III.

Table III. Criticality Matrix.

	N_f> 10% of lifetime	N_f< 10% of lifetime
$T_{j,cr}$ > (T_m+2σ)	Case I: safe	Case III: Immediate replacement
$T_{j,cr}$ < (T_m+2σ)	Case II: Replacement necessary	Case IV: Immediate replacement

Based on the occurrence any of the above configuration of above conditions, a corresponding case of criticality of the unit will be determined based on which a subsequent action would have to be taken

The Case I corresponds to the safe area of operation where the critical temperature is greater than two standard deviations from the mean temperature and the number of cycles to failure are above the 10% threshold of the total life time of the system.

Case II corresponds to a cautionary area of operation as in this region, even though the number of cycles to failure may be well above the 10% threshold of the system lifetime, the junction temperature is outside the safe operating region.

Case III and Case IV both trigger immediate replacement signals. In the case III, even though the IGBT is within the safe area of operating temperature, the "Immediate Replacement" trigger will be initiated due to the higher temperature swing. Similarly, in the case IV, the same trigger will be initiated as in this case apart from the high temperature swing, unit will be operating outside the safe temperature area.

An IGBT at a particular temperature swing may not fail but it may fail with the similar swing while operating at higher temperature. So this algorithm integrate stochastic model with deterministic model to take both mean temperature and temperature swing into consideration.

IV. SIMULATION

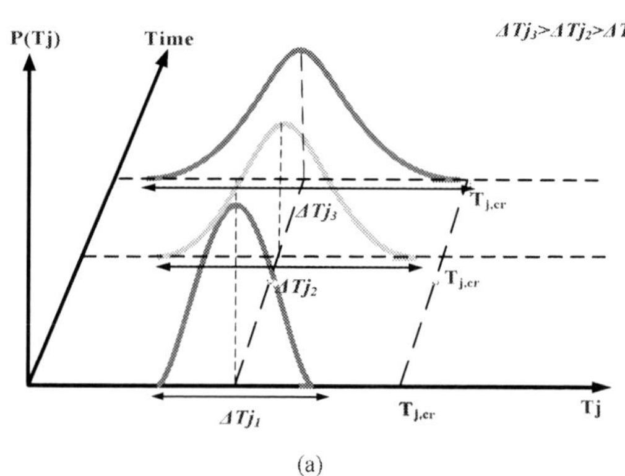

(a)

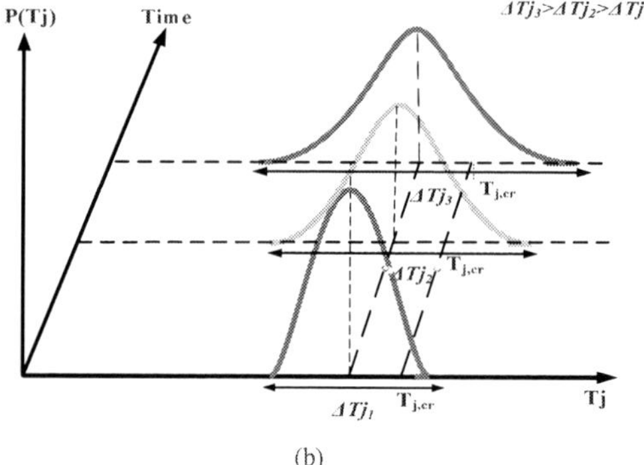

(b)

Fig. 5. Distribution curve (a) when critical temperature lies outside two standard deviation and (b) when critical temperature is inside standard deviation.

The proposed algorithm was simulated using MATLAB® and both of these cases are illustrated in Fig. 5(a) and 5(b) respectively. In order to be able to verify the algorithm, a random signal was generated to simulate the temperature variation over the on state of a cycle of the IGBT operation. The signal being generated could be set for a particular mean value and for a corresponding variation. Given this, the temperature signal being considered over a cycle was chosen for two cases of mean temperature – for a safe operating temperature region (403K) and one outside the safe operating temperature region (443K).

Table IV. Results of Simultion.

Mean (K)	Junction Temperature Swing, ΔT_j (K)	Standard Deviation (K)	Best Fitted Distribution	Number of cycles to failure, N_f	Area of operation
403.19	27.65	6.91	Normal	2.61X10^{13}	safe
402.87	70.46	17.02	Normal	1.94X10^{13}	safe
443.63	28.23	6.62	Normal	1.12X10^{12}	cautionary
443.34	70.13	17.76	Gamma	8.38X10^{11}	unsafe

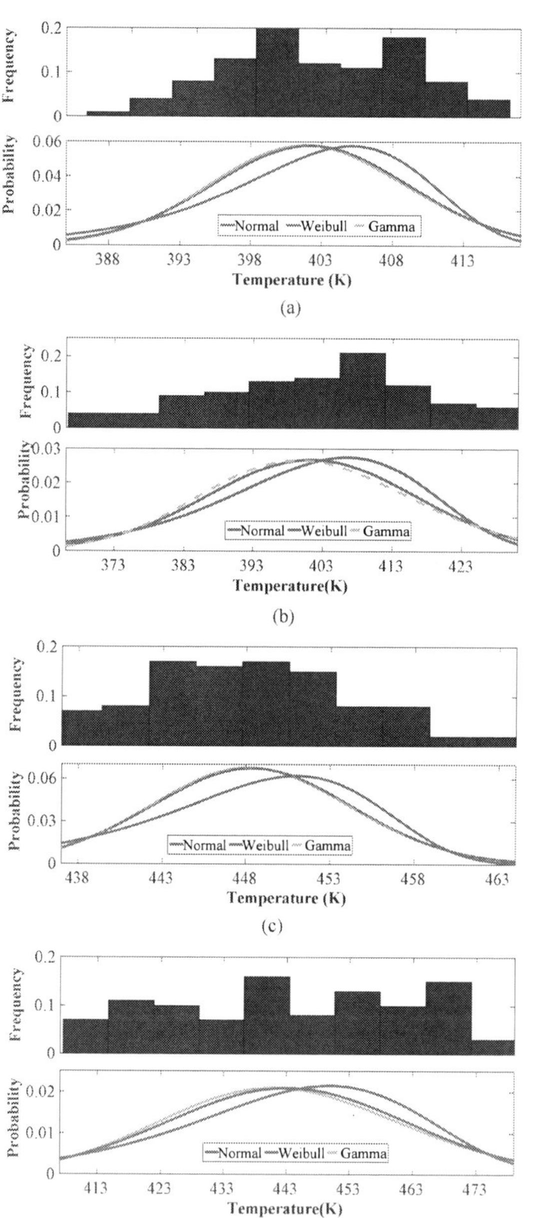

Fig. 6. Results of Simulation for (a) Tm = 403 K with ΔTj = 28K (b) Tm = Tm = 403 K with ΔTj = 70K (c) Tm = 443K with ΔTj = 28K (d) Tm = 443K with ΔTj = 70K.

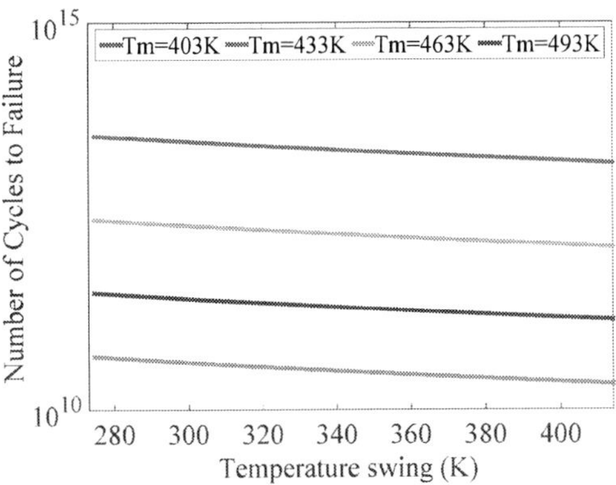

Fig. 7. Simulation results for the variation of temperature swing for a fixed value of mean temperature and the corresponding variation of the number of cycles to failure.

In each case, the temperature swing was varied for 401K and 343K to illustrate the impact on the number of cycles to failure by the temperature swing for the same value of mean temperature. This provides us a total of four test cases, the values and results for which are mentioned in table IV and The various PDFs obtained in each case are as shown in the Fig. 6.

In each case the best fitted distribution and the number of cycles to failure were determined for the temperature considered. The evident impact on the number of cycles to failure with an increase in the temperature swing should be noted.

Another set of simulations were conducted using the temperature signal where for a fixed mean temperature, the temperature swing was varied and the number of cycles to failure were calculated and plotted against the temperature swing. This was repeated for a sequence of mean temperatures and the consolidated plots obtained for the varying mean temperatures is shown in the Fig. 7.

Once again, for a fixed value of mean temperature, the clear impact of the increase in the temperature swing on the reduction of the number of cycles to failure of the IGBT unit is evident.

V. EXPERIMENTAL RESULTS AND INFERENCES

An experimental setup was arranged for a single IGBT from which the TSEPs – $V_{CE,ON}$ and I_C were measured to obtain junction temperature, T_j. The obtained data were used to determine the Probability Density Functions of Tj. Using the MLE method, the best fitted PDF curve was established and highlighted in the obtained data. The following section is a description of the setup used and a synopsis of the collected and analyzed data.

The setup implemented consisted of a single IGBT unit connected in series to a resistor bank and supplied via a variable DC voltage source. The IGBT was provided a gate signal from a Digital Signal Processor (DSP) to control the switching time period. A block diagram of the experimental circuit setup is shown in the Fig. 8.

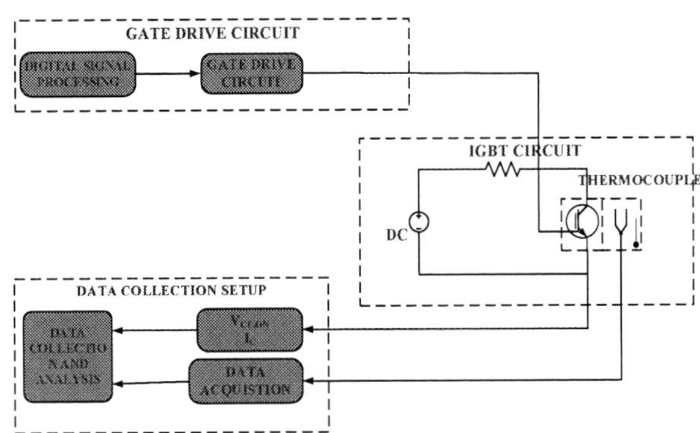

Fig. 8: Block diagram of experimental setup.

Fig. 9. Experimental setup.

The IGBT implemented was the International Rectifier® IRG4BC20KDPBF and its specifications are illustrated in the [17]. The experimental setup is shown in the Fig. 9:

In order to check the effect of I_C and ageing on the IGBT, two sets of experiments were conducted – one in which the I_C was varied and the other was an ageing test where the I_C was constant but the test was conducted until a failure occurred. In both cases, the values of T_j were estimated and number of cycles to failure N_f was estimated as well. Simultaneously, the temperature of the IGBT was measured using a Type K thermocouple and the value of N_f estimated from this measured data as well. Additionally in each case, the temperature variation over one cycle was determined and the best fitting PDF was determined using MLE from which the presence of T_{cr} within two standard deviations from mean was checked and noted.

In the first case the value of I_C was increased in steps from 12A to 22A to encourage an accelerated heating of the IGBT over its rated value of 16A. The output PDF waveforms in each case of increased I_C is shown in Fig. 10 and the output results in each case for the determination of N_f is shown in Table V. In addition to the PDFs, a histogram of the normalized frequency of occurrence has been plotted as well. The best fitted curve would have a greater fitting based on this histogram.

Table V. Experimental Data Results for Different I_c Values.

I_c (P-P) (A)	ΔTj measured (K)	ΔTj estimated (K)	N_f measured	N_f estimated	Variation in N_f
12	8.19	10.3	9.29E+17	8.57E+17	8%
13	16.7	20.1	1.22E+17	1.12E+17	8%
14	17.3	26.6	9.21E+15	9.70E+15	-5%
15	30.5	27	2.08E+15	2.13E+15	-2%
16	30.1	32.3	7.74E+14	8.59E+14	-11%
17	19	23.8	2.88E+14	3.21E+14	-11%
18	23	28.2	6.17E+13	5.45E+13	12%
19.8	40	42.1	8.13E+12	6.02E+12	26%
22	34	38.7	1.31E+12	1.19E+12	9%

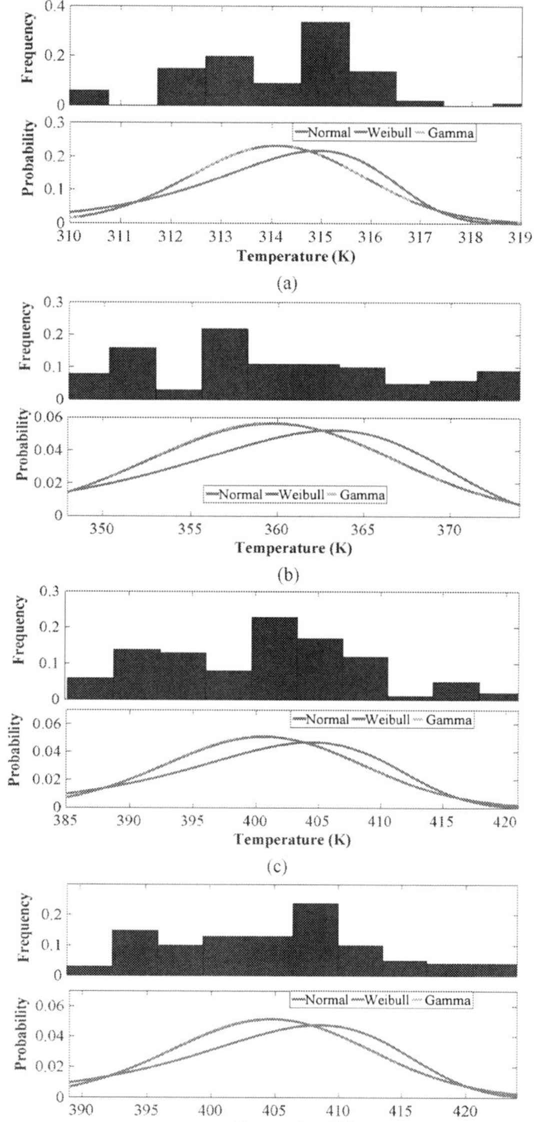

Fig. 10. Experimental results for (a) I_c = 12A (b) I_c = 15A (c) I_c 18A(d) I_c = 19.8A.

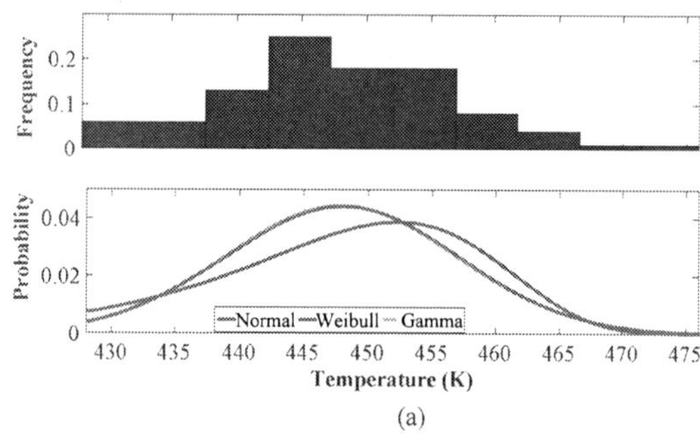

(a)

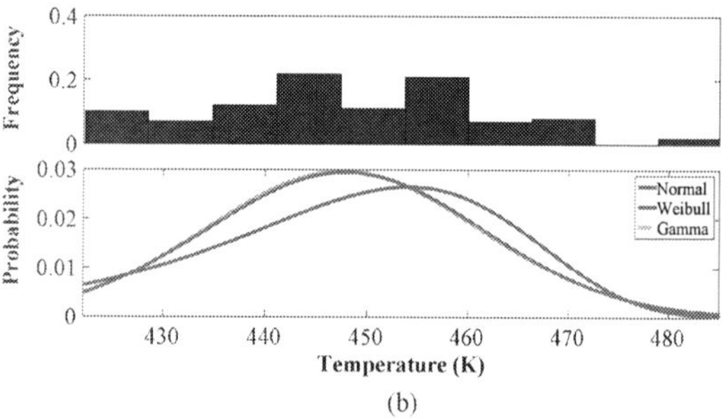

(b)

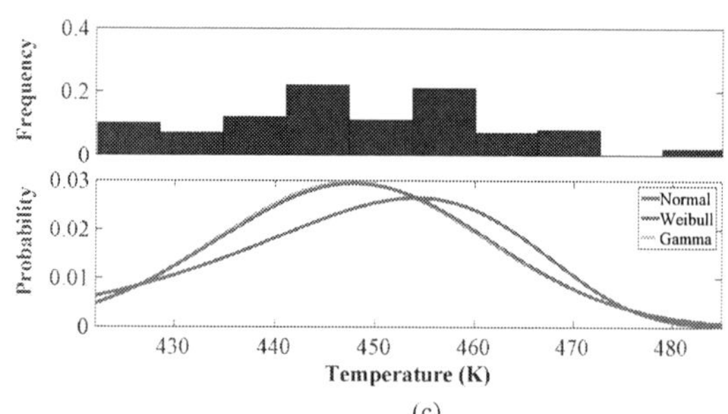

(c)

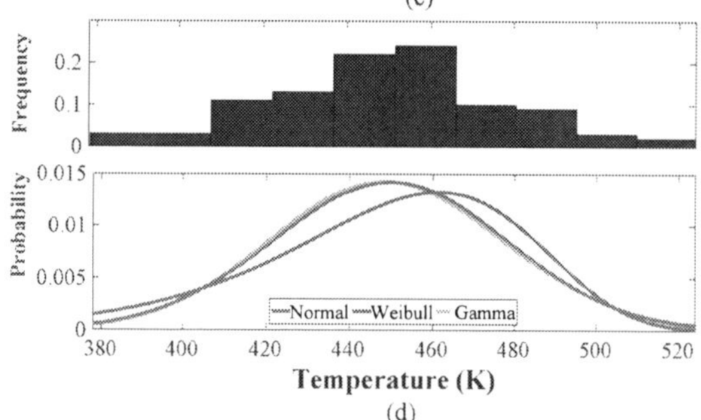

(d)

Fig. 11. Aging test of IGBT at 22 A(P-P). (a) t=24 hr; (b) t=36 hr; (c) t=42 hr; (d) t=48 hr.

The second set of experiments conducted for ageing were conducted for I_C = 22A p-p. The results obtained are shown fig. 11. The N_f calculated from temperature sensor data and estimated temperature data shows variation within the range of 12% from each other. It is observed from table VI, ΔT_j increases with time because of ageing and subsequently the number of cycles to failure decreases with increasing ΔT_j.

Table VI. Experimental Results for Ageing Test.

Time (hr)	ΔTj measured (K)	ΔTj Estimated (K)	N_f measured	N_f Estimated	Variation in N_f	RUL (in year)
2	32	35	1.01E+12	9.30E+11	8%	2.9904
12	38.8	42.9	9.69E+11	8.70E+11	10%	2.7974
24	50.2	52.7	9.03E+11	8.70E+11	4%	2.7974
36	77.13	75.1	7.70E+11	7.90E+11	-3%	2.5402
42	84	80.4	7.40E+11	6.80E+11	8%	2.1864
48	98.1	101.4	6.90E+11	6.10E+11	12%	1.9614

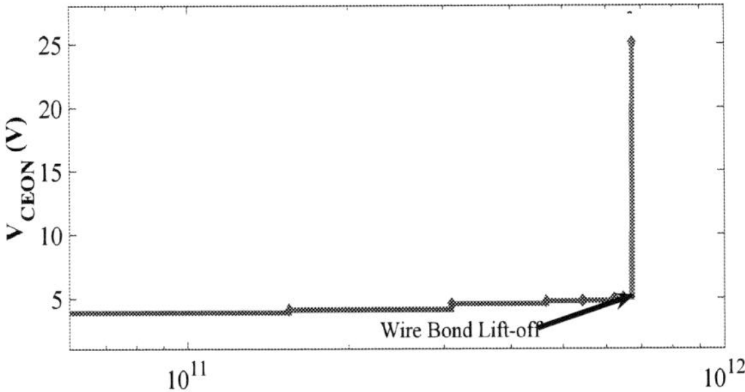

Fig. 12. Variation of on state collector emitter voltage ($V_{CE,ON}$)over time in the case of an IGBT run for a long duration above rated value.

On running the IGBT for a long duration, the variation of the on state coltlector emitter voltage $V_{CE,ON}$ with time was recorded for a case where the IGBT was run long enough to wire-bond failure. A plot of the same is shown in Fig. 12. The collector emitter voltage increases slowly over time, eventually increasing drastically at the instant of failure occurrence.

VI. Conclusion

In this paper, an algorithm based on multi sensor approach for prognosis of individual IGBT switch is proposed. This algorithm integrated stochastic and deterministic methods and introduced a matrix for prognosis of the wire-bond lift off of an IGBT. Maximum likelihood estimate has been used to bet fit the distribution of junction temperature. From the best fitted distribution the mean temperature and temperature swing have been calculated for number of cycles to failure calculation. In experiment, it has been shown that the health of IGBT depend not only the mean temperature but also on the temperature swing. But for the same temperature swing, high mean temperature has more detrimental effect on IGBT. This algorithm can be extended for the prognosis of multiple IGBT switches in a converter for online condition monitoring.

Acknowledgement

This publication was partially supported by Korean Rail Research Institute (KRRI). The statements made herein are solely the responsibility of the authors.

References

[1] Ciappa, M.; Castellazzi, A., "Reliability of High-Power IGBT Modules for Traction Applications," *45th Annual. IEEE International Reliability Physics Symposium, 2007*, pp.480-485, April 2007

[2] H. Oh, B. Han, P. Mcluskey, C. Han, and B. Youn," Physics-of-Failure, Condition Monitoring, and Prognostics of Insulated Gate Bipolar Transistor Modules: A Review," *IEEE Transactions on Power Electronics*, vol. 30, no. 5, pp. 2413-2426, May 2015.

[3] Bo Tian, Wei Qiao, Ze Wang, T. Gachovska, J.L. and Hudgins, "Monitoring IGBT's health condition via junction temperature variations," *Applied Power Electronics Conference and Exposition (APEC), 2014*, pp. 2550-2555, March 2014.

[4] Y. Avenas, L. Dupont, and Z. Khatir, "Temperature Measurement of Power Semiconductor Devices by Thermo-Sensitive Electrical Parameters — A Review," *IEEE Transactions on Power Electronics*, vol. 27, no. 6, pp. 3081-3091, June 2012.

[5] C. Rodriguez , G. Amaratunga,"Long-lifetime Power Inverter for Photovoltaic AC Modules", *IEEE Transactions on Industrial Electronics*, vol. 55, Issue. 7, pp. 2593-2601, June 2008.

[6] Z. Xu; F. Xu; F. Wang, "Junction Temperature Measurement of IGBTs Using Short-Circuit Current as a Temperature-Sensitive Electrical Parameter for Converter Prototype Evaluation," *IEEE Transactions on Industrial Electronics*, vol.62, no.6, pp.3419-3429, June 2015

[7] U. Choi, F. Blaabjerg, K. Lee, "Study and Handling Methods of Power IGBT Module Failures in Power Electronic Converter Systems", *IEEE transactions in Power Electronics*, vol. 30, No. 5, pp. 2517-2532, May 2015.

[8] Patil, N., Das, D., Goebel, K., Pecht, M., "Identification of Failure Precursor Parameters for Insulated Gate Bipolar Transistors (IGBTs)," *International Conference on Prognostics and Health Management (PHM), 2008* , pp.1-5, Oct. 2008

[9] A. babel, A. Muetze, R. Seebacher, K. Krischan and E. Sttangas,"Condition Monitoring and Failure Prognosis of IGBT Inverters based on On-Line Characterization" *Energy Conversion Congress and Exposition (ECCE) 2014*, pp. 3059-3066.

[10] F. Protiwa., O. Apeldoorn., N. Groos., "New IGBT Model for PSpice",*5th European Conference on Power Electronics*, vol. 2, pp. 226-231, 1993.

[11] Papoulis, "A. Probability, Random Variables, and Stochastic Processes", 2nd ed. New York: McGraw-Hill, pp. 100-101, 1984.

[12] Spiegel, M. R." Theory and Problems of Probability and Statistics.", New York: McGraw-Hill, p. 119, 1992.

[13] Feller, W. "An Introduction to Probability Theory and Its Applications", Vol. 2, 3rd ed. New York: Wiley, p. 45, 1971.

[14] Harris, J. W., Stocker, H., "Handbook of Mathematics and Computational Science. "New York: Springer-Verlag, 1998.

[15] C. Busca, R. Teodorescu, F. Blaabjerg, S. Munk-Nielsen, L. Helle, T. Abeyasekera, and P. Rodriguez, "An overview of the reliability prediction related aspects of high power IGBTs in wind power applications," *Microelectronics Reliability*, vol. 51, pp. 1903–1907, 2011.

[16] V. Smet, F. Forest, J.-J. Huselstein, A. Rashed, and F. Richardeau, "Evaluation of V_{CE} Monitoring as a Real-Time Method To Estimate Aging of Bond Wire-IGBT Modules Stressed By Power Cycling ," *IEEE Trans. Industrial Electronics*, vol. 60, no. 7, pp. 2760–2770, Jul. 2013

[17] International Rectifier, IRG4BC20KDPBF datasheet, available: http:// www.irf.com/product-info/datasheets/data/irg4bc20kdpbf.pdf [Accessed 5 October 2015].

Electrical Parasitics and Thermal Modeling for Optimized Layout Design of High Power SiC Modules

Amir Sajjad Bahman, Frede Blaabjerg
Center of Reliable Power Electronics (CORPE)
Department of Energy Technology
Aalborg University, 9220 Aalborg, Denmark
asb@et.aau.dk, fbl@et.aau.dk

Atanu Dutta, Alan Mantooth
Department of Electrical Engineering
University of Arkansas, Fayetteville, AR, USA
adutta@uark.edu, mantooth@uark.edu

***Abstract*—The reliability of power modules is closely depended on their electrical and thermal behavior in operation. As power modules are built to operate more integrated and faster, the electrical parasitic and thermal stress issues become more critical. This paper investigates simplified thermal and parasitic inductance models of SiC power modules. These models can replace the models by Finite Element Methods (FEM) to predict temperatures and electrical parasitics of power modules with much faster speed and acceptable errors and will be used for study of real operation of power modules. As a case study, the presented models are verified by a conventional and an optimized power module layout. The optimized layout is designed based on the reduction of stray inductance and temperature in a P-cell and N-cell half-bridge module. The presented models are verified by FEM simulations and also experiment.**

Keywords—Electrical parasitic modeling, thermal modeling, SiC power module, optimization, reliability.

I. INTRODUCTION

In recent years power electronic devices have found wide applications in renewable energy systems, electric drives and automation. Since industries demand for higher power densities and higher efficiencies, power electronic systems need to be designed more integrated and operate at higher switching frequencies [1]. However, SiC power modules as fast growing devices, which can operate in higher frequencies in more dense packages, may be exposed to more reliability problems. Integration of semiconductor chips increases the temperatures due to thermal coupling of the chips that may lead to bond-wire lift-off or solder cracking [2], [3]. Moreover, poor layout design such as distancing the semiconductor chips in the power module increases the parasitic inductances that lead to voltage spikes and ringing with the parasitic capacitances of the device [4]. Therefore, a poor layout design of the power module may lead to severe thermal stress, severe turn-off overvoltage of the chips and electromagnetic interference problems that seriously affect the reliable operation of the power module [5].

In conventional power modules, stray inductances are mainly introduced by power module terminals, copper trace on the upper layer of Direct Copper Bonding (DCB) and chip bond-wires. A lot of efforts have been made to optimize the layout of power module with minimized parasitic inductances. For power module electrodes, laminated busbar structure has been adopted, which will considerably reduce the stray inductance of electrodes [6]. The common approach to reduce the copper trace stray inductance is to widen the trace width and shorten the trace length. Moreover, the typical method to reduce bond-wires stray inductances are shortening the wire length and increase the number and diameter of wires [7]. However, thermal management of power module becomes an important issue as the power module package becomes more compact and the heat spread capability decreases. Therefore, a trade-off should be made in thermal and electrical design of the power module layout [8].

Currently, the optimized layouts for power module are achieved by multidisciplinary design tools which integrate several commercial software tools that in turn demand for high computational time and effort [9]. Some power module design and evaluation tools are based on 3D modeling and Finite Element Analysis (FEA) to extract the electrical parasitics and temperatures of the power module [10]. These tools are usually time-consuming if re-design process is targeted to achieve minimized parasitics and temperature of layout. On the other hand, many analytical and lumped models have been introduced to estimate the temperatures and electrical parasitic parameters in the power modules [11]-[14]. However, there is a lack of effort for multidisciplinary modeling of power modules that can evaluate the optimized layout and electro-thermal parameters in a transient model, which can be used in reliability studies. Some software tools have been developed, which provide simple parasitics and thermal models and are also used for automatic layout design. For example in [8] PowerSynth, a power module layout design tool, is introduced where simplified electrical parasitics models based on micro-strip transmission line structure and thermal models based on a thermal spreading angle approach have been applied to evaluate the layout of the power module. Using a multi-objective optimization such as genetic algorithm, the best solution for minimized switching loop inductance and chip temperatures depending on the application is achieved. Then, by using simplified thermal and parasitics models, a fast evolution of the designed layouts can be implemented for the redesign process.

978-1-4673-9551-9/16 $31.00 © 2016 IEEE

This paper investigates a simplified geometry-dependent commutation loop inductance index to evaluate a commercial SiC half-bridge power module and the optimized layout. The optimized layout is designed based on P-cell and N-cell layout structure, which is explained in [7] and is optimized by the Non-dominated Sorting Genetic Algorithm II (NSGA-II) procedure [15]. A simplified index to evaluate the stray inductance of both layouts is used and compared for the main switching loops. Then by applying a 3D lumped thermal network that includes the thermal cross coupling effects, the thermal stress on the chips will be identified for steady-state and also dynamic operation. The optimized layout will be validated by FEA as well as experiments.

II. Stray Inductance Evaluation Index

As previously described in [7], P-cell and N-cell structure for phase-leg in inverters has some benefits rather than traditional phase-leg anti-parallel diode structure. Fig. 1 shows the schematics of the two configurations. Under inductive load operation as shown in Fig. 1 (a), the current commutates between S1 and D2when the current direction is from positive DC bus to negative DC bus or between S2 and D1 when the current direction is from negative DC bus to positive DC bus. Traditionally, the anti-parallel diodes are either packaged within the transistors or placed closely in parallel. However, as it is shown in Fig. 1 (b), it is more reasonable to use a P-cell N-cell structure by pairing the upper anti-parallel diodes with lower transistors and lower anti-parallel diodes with higher transistors. So, the load current flows into the N-cell and flows out through the P-cell. The stray inductances exist all over the phase-leg current commutation path as shown in Fig. 1. The closer the devices in each current commutation path are paired together, the lower the stray inductance between the devices. This lower stray inductance reduces the inductive voltage spikes seen across the transistor, while undergoing a turn-off commutation and current overshoot during turn-on.

Usually it may be time-consuming and costly to use FEA software tools for calculation of stray inductances in the complex design of power module layouts. In the design stage of a power module, the designer may try several layouts and calculate the stray inductances by FEA tool, which is too time-consuming. So a simple stray inductance model is needed for fast evaluation of layouts. For simplification, all commutation paths on the substrate are considered as homogenous conductors with similar cross-section area, which is very small compared to the length of conductors. In the power module, the critical commutation paths are open loops, which makes the calculation of their stray inductances more complicated compared to closed loops. Using the principles of partial inductances, the stray inductance in the conductors is calculated by adding a closing path to the open loop [16]. The self-inductance of a single-turn rectangular loop is calculated by

$$L=\frac{\mu_0}{\pi}\left[-2(w+h)+2\sqrt{h^2+w^2}+h\ln\left(\frac{w}{h+\sqrt{h^2+w^2}}\right)+w\ln\left(\frac{h}{w+\sqrt{h^2+w^2}}\right)+h\ln\left(\frac{h}{d}\right)+w\ln\left(\frac{w}{d}\right)\right]\times10^{-3} \quad (1)$$

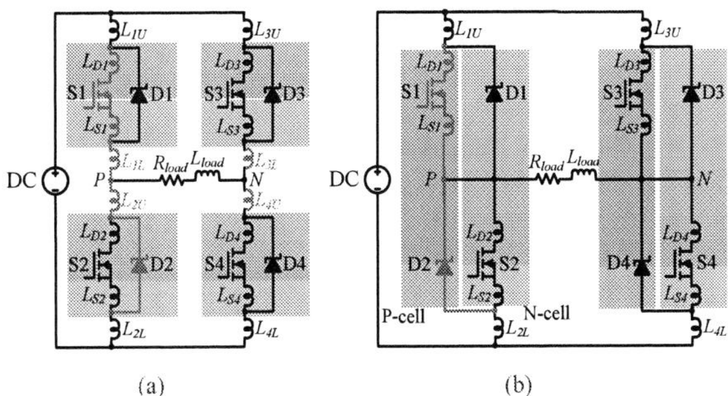

Fig. 1. Circuit schematic showing package parasitics. (a) Conventional inverter, (b) P-cell N-cell phase-leg inverter.

where w and h are the width and height of the rectangular loop in mm, d is the diameter of the loop cross section in mm ($d<<w,h$), μ_0 is the vacuum permeability, and L is the loop inductance in Henry.

The partial self-inductance of a single straight conductor is given by [10]

$$L=\frac{\mu_0 l}{2\pi}\left(\ln\frac{2l}{d}-1\right)\times10^{-3} \quad (2)$$

where l is the length of the conductor in mm, and d is the diameter of the conductor cross section ($d<<l$) both in mm. From eq. (2), and assuming $d<<w,h$, the self-inductance of a rectangular loop is approximated by

$$\begin{aligned}L&\approx\frac{\mu_0}{\pi}(w+h)\cdot\ln(w\cdot h)\times10^{-3}\\&=\frac{\mu_0}{2\pi}\cdot C\cdot\ln(S)\cdot10^{-3}\end{aligned} \quad (3)$$

where C=2·(w+h) is the perimeter of the loop in mm, and S=w·h is the area of the loop in mm^2. So the self-inductance of the rectangular loop is linearly proportional to $C\cdot\ln(S)$. As derived from eq. (2), and assuming $d<<$l, the self-inductance of a single straight conductor is approximated by

$$L\approx\frac{\mu_0}{2\pi}\cdot l\cdot\left[\ln(2l)-1\right]\cdot10^{-3} \quad (4)$$

In order to simplify the model for calculation of the partial inductance of the commutation loop, the shortest closing path, which is a straight line between two ends of the open loop, is chosen [17]. In the presented model, the mutual inductance associated with the path is neglected and only the partial self-inductance is taken into account. Consequently, by subtracting the partial self-impedance from the inductance of the closed commutation loop, the inductance of the open loop between the intended points is achieved. Here, by using eq. (3) and eq. (5), an index is derived to evaluate the inductance of the commutation loop.

$$\alpha=C\cdot\ln(S)-l\cdot\left[\ln(2l)-1\right] \quad (5)$$

In a commutation loop, a large value of the index α indicates large stray inductance. So, in the substrate layout design for the power module, the purpose is to minimize the α values of the commutation loops. However, due to the assumptions and approximations adopted, the index α is not used for an accurate stray inductance calculation and it is used as a scale for comparison between the stray inductances of different layout designs.

III. 3D Thermal Modeling

In this paper, both transient and steady-state thermal behaviors are focused for thermal evaluation of the power module. The method for modeling the transient operation is based on 3D thermal network [18]. The 3D thermal network includes 9 temperature monitoring points in the junction layer, die attach layer and substrate attach layer. For simplification, a simplified model including the junction and case layers is shown in Fig. 2. The network extraction method is based on step response analysis, i.e. applying volumetric step power loss to the semiconductor chips in the FEA software and identification of the temperature responses on the monitoring points. Then the transient thermal impedance between the aforementioned layers are identified by

$$Z^{self}_{th(a-b)}(t) = \frac{T_a(t) - T_b(t)}{P_{self}} \tag{6}$$

where $Z^{self}_{th(a-b)}(t)$ is the transient self-thermal impedance between adjacent layer 'a' and 'b', T_a and T_b are temperatures in the two aforementioned layers and P_{self} is the power loss generated in the chip. Moreover, the thermal coupling effect is considered in the 3D thermal network. The transient coupling-thermal impedance is given by

$$Z^{coupl}_{th(a-b)}(t) = \frac{T_a(t) - T_b(t)}{P_{coupl}} \tag{7}$$

where $Z^{coupl}_{th(a-b)}(t)$ is the transient coupling-thermal impedance between adjacent layer 'a' and 'b' and P_{coupl} is the power loss generated in the neighbor chips. As using FEA simulation is time-consuming for transient long-term studies, circuit simulator is used for temperature calculation. In order to use the 3D thermal network in the circuit simulator, the derived transient thermal impedance curves by eq. (6) and eq. (7) are curve-fitted mathematically using a sum of exponential functions in order to obtain an equivalent lumped RC thermal network in the Foster configuration

$$Z^{self}_{th}(t) = \sum_{i=1}^{n} R^{self}_{th_i}.(1 - e^{-t/R^{self}_{th_i}.C^{self}_{th_i}}) \tag{8}$$

where R^{self}_{thi} is the equivalent thermal resistance and C^{self}_{thi} is the equivalent thermal capacitance when the target chip is powered. Similarly the equivalent mathematical model for the transient coupling-thermal impedance is given by

$$Z^{coupl}_{th}(t) = \sum_{i=1}^{n} R^{coupl}_{th_i}.(1 - e^{-t/R^{coupl}_{th_i}.C^{coupl}_{th_i}}) \tag{9}$$

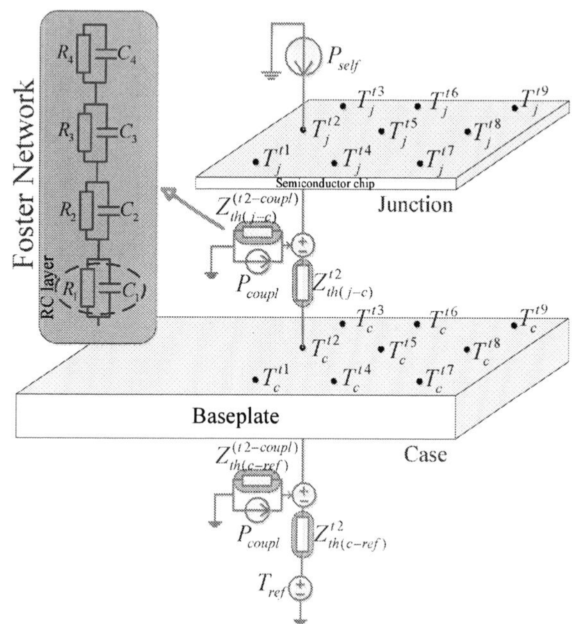

Fig. 2. 3D thermal network structure from chip (junction) to reference (cooling system).

where R^{coupl}_{thi} is the equivalent thermal resistance and C^{coupl}_{thi} is the equivalent thermal capacitance, when the neighbor chips are powered. The equivalent Foster network is highlighted in Fig. 2. In this work, the equivalent thermal impedance curves are fitted by four layers of RC elements for better accuracy. It should be mentioned that all RC elements are flexible by the variation of temperature, power loss and cooling system. By shortening the distance between the chips, thermal coupling effects and thus temperature become higher. So, the thermal coupling impedances are used to evaluate the distance of the chips to each other.

IV. Substrate Layout Design Based on P-cell N-cell Concept

The main optimization objectives for design of the power module layout include minimization of 5 main criteria: loop inductance in the current path from the positive DC terminal to the negative DC terminal, loop inductance from the load terminal to the positive DC terminal, loop inductance from the load terminal to the negative DC terminals, steady-state maximum device temperature and temperature variation of the device. By minimizing the total stray inductances from the positive DC terminal to the negative DC terminal, the inductance of both current commutation paths from load terminal to positive and negative terminals are simultaneously minimized. Therefore, it merges the related optimization criterion to one. Reducing the steady-state maximum temperature and dynamic temperature variation are also helpful in terms of increasing the reliability and thermal stability of the power module. Generally, the lower the operating temperature of the devices, the lower thermal expansion and Coefficient of

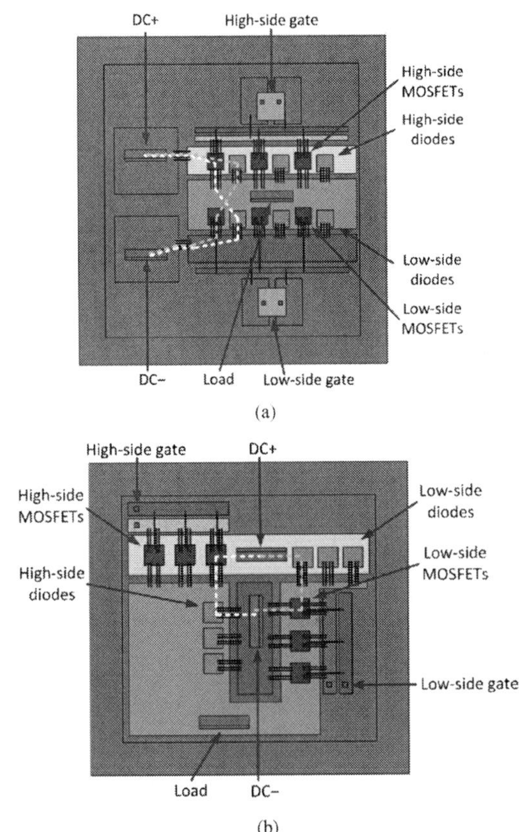

Fig. 3. Substrate layout designs for a 1200V/100A module. (a) conventional module, (b) P-cell N-cell based module

Thermal Expansion (CTE) mismatch between the device and sub-layers.

In this section, two different substrate layout designs for a 1200V/100A (at 25 °C) SiC half-bridge inverter module with separate anti-parallel diodes are studied: one conventional layout and one adopting P-cell N-cell layout. Adding anti-parallel diodes separately to the MOSFET chip allows for reduction of stray inductance in the package using the P-cell N-cell technique. Conventionally the anti-parallel diode is either embedded within the MOSFET packages or placed closely with them. In the P-cell N-cell technique, the stray inductance of the main current commutation loop in a module is reduced. This technique is performed by pairing the upper MOSFETs with the lower anti-parallel diodes and vice-versa. The closer these devices are paired together, the lower the stray inductance in the current commutation loop. This lower inductance reduces the voltage spikes across the MOSFET during a turn-off event and current overshoot during the turn-on event.

A. Electrical Parasitics Evaluation

The power module under study includes a total of 6 MOSFETs and 6 anti-parallel diodes in the design (3 pairs for the high side of the phase-leg and 3 pairs for the low side of the phase-leg). As the module is a half-bridge topology, only a single phase-leg inverter is implemented. The MOSFETs used in the design are CREE CPM2-1200-0080B and the diodes are CPW4-1200S020B. The power MOSFET chips and Schottky diode chips used for the half-bridge modules have the ratings of 1200V/36A and 1200V/20A respectively. The baseplate and substrate dimensions are constrained to a fixed size. ANSYS Q3D is used to initially design the layout of the P-cell N-cell based layout. Then, by using the NSGA-II optimization procedure used in the power module design tool, PowerSynth, the optimization of the layout in terms of device placement, trace widths/lengths are performed. In the optimization procedure, the position of the chips, size and dimension of traces are changed in order to get the best solution. Fig. 3 shows two layouts designed based on the conventional device pairing and the optimized P-cell N-cell based device pairing. The critical commutation paths of two designed layouts are highlighted with dashed lines of different colors on the layout. To form the commutation paths, soldering positions of the terminals and chips on the substrate have been denoted. The current commutation loops including the closing paths, the geometrical dimensions and the related inductance index α are given in Table I. Both the designed layouts have been imported in the ANSYS Q3D Extractor to calculate the stray inductances in the switching loops. As a high critical frequency of turn-off current for the present SiC MOSFET, the simulations have been implemented by using a 50 MHz AC excitation [17].

TABLE I. COMMUTATION LOOPS AND STRAY INDUCTANCE OF CONVENTIOANL MODULE AND P-CELL N-CELL BASED MODULE.

Module Layout	Conventional Layout	P-cell N-cell layout
DC+ to DC- commutation loop	DC+ DC-	DC+ DC-
DC+ to DC- loop dimensions	S=267.88 mm^2 C=66.8 mm l=18.48 mm	S=53.23mm^2 C=29.2 mm l=7.3 mm
Stray inductance index (α)	α =325.21	α=103.77
Extracted stray inductances (L)	12.8 nH	9 nH

It can be seen in Table I that the positive correlation between the index α and the stray inductances is valid. Moreover, the DC+ terminal to DC- terminal stray inductance values confirm that the lowest overall loop inductance is the P-cell N-cell layout. In the P-cell N-cell layout, the actual commutation devices (S1/D2 and S2/D1) are placed closer to

Fig. 4. Optimized SiC power module layout based on P-cell N-cell techniques.

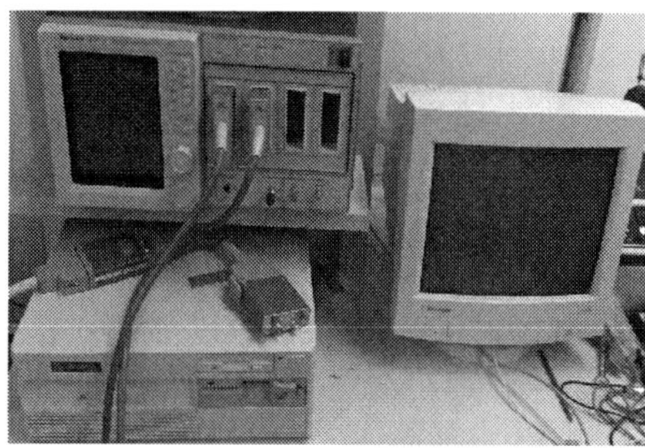

Fig. 5. Experimental measurement setup; Oscilloscope 11801B connected to SD-24 sampling head, terminated with 50Ω impedance and PC (loaded with IPA310 software) interfaced with oscilloscope.

each other. So the P-cell N-cell layout configuration significantly reduces the loop inductance rather than the conventional layout. Moreover, to confirm for the other current loops, the stray inductance values for DC+ terminal loop to the load terminal and DC- terminal to the load terminal are listed.

For verification, a sample module designed by the studied layout has been fabricated that is shown in Fig. 4. Tests of the switching characteristics of the MOSFETs have been done upon the sample modules in a Time Domain Reflectometer (TDR) test bed. The experimental setup is shown in Fig. 5. Parasitics are extracted for the power modules using the TDR method in the form of differential inductance waveforms, from which the absolute parasitic values are computed using IPA310 software. The TDR method involves propagation of a high frequency step signal (incident signal) through the Region of Interest (ROI) in Device under Test (DUT) and capturing the reflected signal. The captured incident, reflected, shorted waveforms are monitored and the corresponding resultant impedance and parasitic inductance waveforms. The frequency in the level of 50 MHz is used for the measurement. High frequency is critical for the small inductance to get measurable impedance. So when a current is applied to the stray inductance, a decent amount of voltage is provided and the accuracy of the measurement is ensured [7]. Moreover, since the wires with alligator probe introduce inductances that are comparable to the parasitic inductance in the module, a pin probe is adopted in the measurement. The parasitics with this

TABLE II. COMMUTATION LOOPS AND MEASURED STRAY INDUCTANCE OF CONVENTIONAL MODULE AND P-CELL N-CELL BASED MODULE.

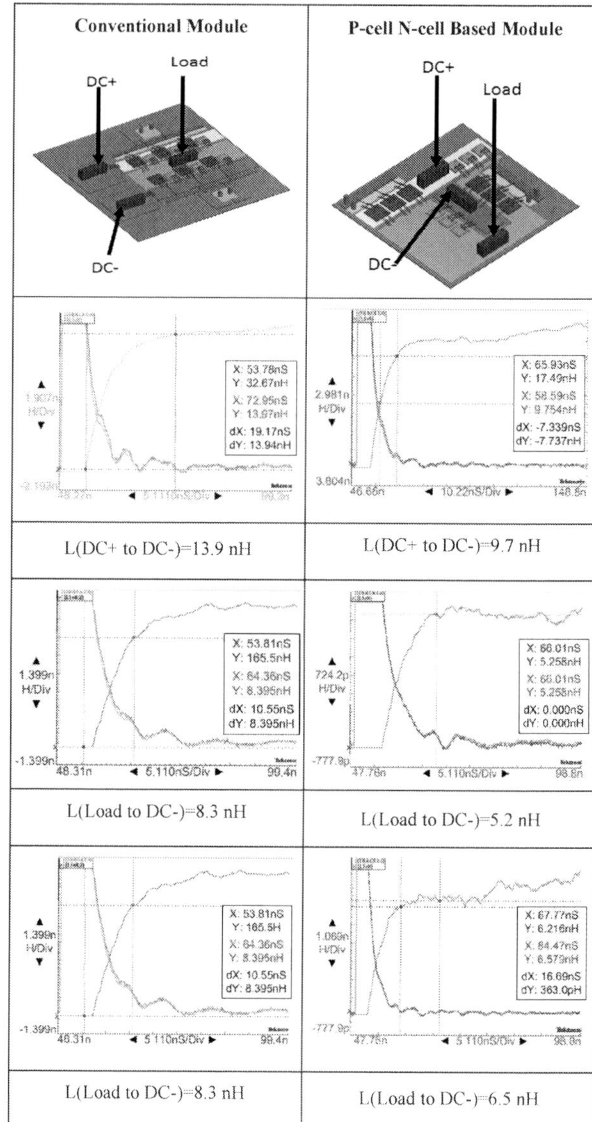

Conventional Module	P-cell N-cell Based Module
L(DC+ to DC-)=13.9 nH	L(DC+ to DC-)=9.7 nH
L(Load to DC-)=8.3 nH	L(Load to DC-)=5.2 nH
L(Load to DC-)=8.3 nH	L(Load to DC-)=6.5 nH

probe are small and are compensated through the calibration process. The TDR responses of the traces of three switching conditions (DC+ to DC-, load to DC+, and load to DC-) in the two layouts are shown in Table II. It can be seen from the test results that in the new layout design, the stray inductances are reduced significantly compared to the conventional layout.

B. Steady-state thermal evaluation

For thermal evaluation, the module is assumed to operate at an ambient temperature of 25 °C and at a switching frequency of 20 kHz. The bottom face of the baseplate is mounted on a heat transfer coefficient (h) of 100 W/m^2·K. The equivalent heat transfer coefficient is a measure, which stands for the amount of heat, which is transferred by convection between a

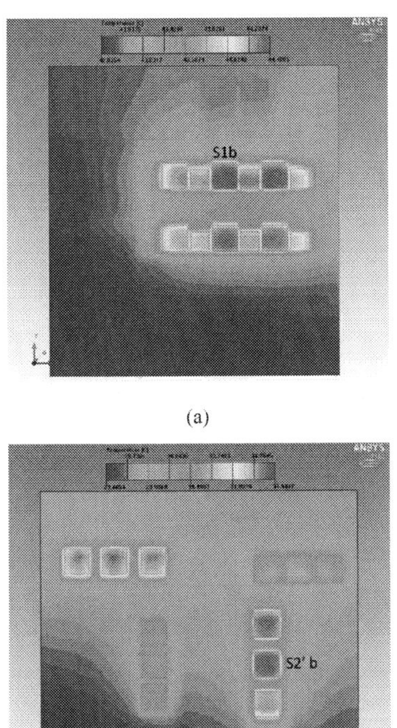

(a)

(b)

Fig. 6. Steady-state temperature profile on power modules. (a) conventional layout, (b) P-cell N-cell based layout.

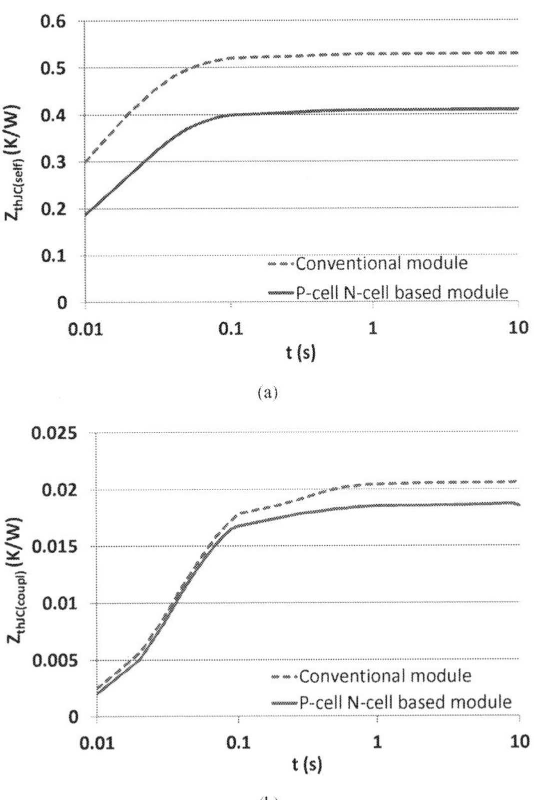

(a)

(b)

Fig. 7. Extracted transient thermal impedance curves for conventional module (point S1b in Fig. 6(a)) and P-cell N-cell based module (point S2'b in Fig. 6(b)). (a) self-thermal impedance, (b) Coupling-thermal impedance.

solid and a fluid [19]. So, to simplify the real cooling system, its equivalent h is used. Each MOSFET is estimated to dissipate 4 Watts of heat flow and each diode 2 Watts of heat flow. Both the designed layouts are imported in the ANSYS Icepak to calculate the temperatures in static condition. All materials used in the models are selected to be temperature dependent. The selected materials are: SiC for the MOSFET chips and diode chips, SAC305 for the die-attach and substrate-attach, Al_2O_3 for the DCB and Copper for the baseplate and traces. The steady-state temperatures are shown in Fig. 7. It is shown that the maximum temperature on the conventional layout is 44.4 °C, but the maximum temperature on the P-cell N-cell based layout is 31.5 °C that shows about 13°C temperature reduction. Moreover, it is seen that in the conventional layout, all chips are concentrated in the middle of the substrate that leads to an increase of temperature due to thermal coupling effects. However, in the P-cell N-cell layout, MOSFETs and diodes are dispersed and placed in positions further to each other. This enables to reduce the thermal coupling effects among the chips and to use larger area of the substrate for heat spreading. Using the P-cell N-cell layout will reduce the cooling requirements of the module, while maintaining the power density of the module.

C. Transient thermal evaluation

As explained in section III, for transient thermal evaluation, a 3D thermal network is used. In order to extract the thermal

TABALE III. Thermal RC Elements for the Hottest Chip in the Conventional (Point S1b) and P-cell N-cell (Point S2'b) Modules

Layout	Conventional module		P-cell N-cell module	
RC elements in Foster network	Self-thermal impedance (j-c)	Coupling-thermal impedance (j-c)	Self-thermal impedance (j-c)	Coupling-thermal impedance (j-c)
R1 (Ω)	0.1182	0.02049	0.2125	0.01856
C1 (F)	0.3059	2.373	0.07604	2.34
R2 (Ω)	0.07548	0	0.09747	0
C2 (F)	0.1693	0	0.4379	0
R3(Ω)	0.1645	0	0.04612	0
C3 (F)	0.007741	0	0.08218	0
R4 (Ω)	0.1697	0	0.05328	0
C4 (f)	0.03413	0	0.3016	0
Total R (Ω)	0.52788	0.02049	0.40937	0.01856

network elements, a step response analysis is implemented and the junction to case transient thermal impedance curves are extracted for the hottest chips in two layouts, which are *S1b* in the conventional module and *S2'b* in the P-cell N-cell based layout (see Fig. 6). The same conditions as the steady-state FEA simulation are applied to models. To extract the transient self-thermal impedance curves, 4 W is applied to the chips *S1b*

TABLE IV. PARAMETERS OF THE PV CONVERTER USING THE DESIGNED MODULE

Rated power P_o	40 kW
Rated load current I_{load}	80 A
DC bus voltage V_{dc}	400 V
Switching frequency f_{sw}	20 kHz
Fundamental frequency f_o	50 Hz
Grid-side power factor	1.0
Filter inductor L_1	0.35 mH
Power module	SiC MOSFET/diode 1200V/100A

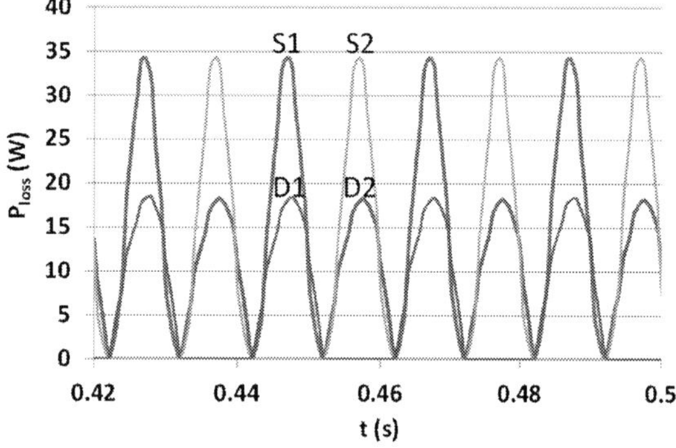

Fig. 8. Average power losses of the MOSFET chips (S1 and S2) and diode chips (D1 and D2) in the phase-leg module. For a PV-converter at 40 kW production.

and *S2'b* and the other chips are not powered. To extract the transient coupling-thermal impedance curves, *S1b* and *S2'b* are not powered, 4 W is applied to all other MOSFETs and 2 W is applied to all other diodes. By a step response analysis, the transient thermal impedance curves are extracted for a monitoring point at the center of the chip and it is shown in Fig. 7. As it is seen, by using the new layout design, it reduces the transient thermal impedance curves. In the new layout, the substrate area is used efficiently to place chips. In the new design, larger Copper traces are used that increase the heat spreading on the surface of the substrate. Moreover, since the chips are placed further, the thermal coupling effect is reduced. Depending on the different operating conditions of the power module (grid-side or converter-side devices) high-side or low-side devices experience different power losses. So, by a decrease of the thermal coupling between the chips, the thermal stress on the devices will be less and the device thermal behavior will be more independent from other devices.

In order to use 3D thermal network in a circuit simulator for fast transient thermal evaluation, RC elements in the Foster networks are extracted for both self-thermal impedances and coupling-thermal impedances (see Fig. 2). Curve-fitting of transient thermal impedance is implemented by using *cftool* in MATLAB. The extracted RC elements are shown in Table IV.

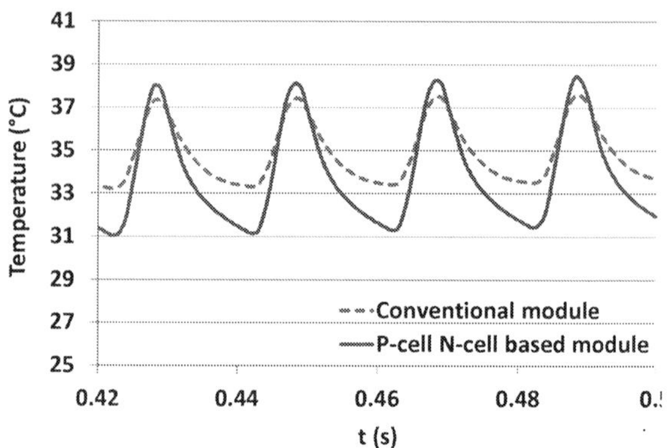

Fig. 9. Dynamic temperature profiles in the center of hottest chips in the conventional module (point S1b in Fig. 6(a)) and P-cell N-cell based module (point S2'b in Fig. 6(b)).

D. *Thermal dynamic evaluation in converter application*

For the transient thermal evaluation, the designed power modules are used as a phase-leg in a 40 kW grid-connected three-phase PV inverter. The parameters of the inverter are listed in Table IV. The average power losses of devices are calculated by using the datasheet of the power MOSFETs and Schottky diodes. For more accurate power loss calculations, temperature-dependent output characteristics and switching losses of the devices are imported as look-up tables in PLECS. Then by the method explained in [20], the power losses of devices are calculated. The calculated loss profiles for each high side and each low side devices are presented in Fig. 8. Although the power loss of the devices also depends on stray inductances of the main circuit and module layout design, the preliminary power loss results are helpful for the thermal evaluation of the module layout.

Using the proposed 3D thermal model, dynamic thermal behavior of the devices can be studied very fast. For this reason, the extracted RC networks are used in a circuit simulator; S1 power loss is applied as P_{self} and a summation of S2, D1 and D2 are applied as P_{coupl}. T_{ref} is defined as 25°C and the thermal network for case to reference is derived based on heat transfer coefficient of 100 W/m^2·K. Extraction of the cooling system thermal impedance curve ($Z_{th(c\text{-}ref)}$) follows the same instructions of module thermal impedance. The dynamic temperature profiles for the hottest chips in both layouts are shown in Fig. 9. This confirms both the minimum and average temperature in the device have been reduced in the P-cell N-cell module.

V. CONCLUSIONS

In this paper, simple thermal and electrical models to evaluate SiC power modules have been studied. The thermal model is based on 3D Foster network and can estimate the transient temperature of the devices accurately in a circuit simulator. Moreover, a simplified index has been used for evaluation of the stray inductances in the power module. To verify the presented models, an optimized SiC power module based on P-cell N-cell configuration has been compared with conventional module. The optimized layout shows better performance compared to the conventional module by shortening the switching loops and distancing the chips verified by both parasitic inductance model and transient thermal model. Measured stray inductances of the conventional and optimized layouts confirm the validity of the stray inductance model. The presented methods can be used for fast pre-design evaluation of multi-chip power modules and also reliability studies in real operating conditions.

REFERENCES

[1] F. C. Lee, J. D. van Wyk, D. Boroyevich, G. Q. Lu, Z. Liang, P. Barbosa, "Technology Trends Toward a System-in-a-module in Power Electronics," *IEEE Circuits Syst. Mag.*, vol.2, no.4, pp.4-22, 2002.

[2] Y. Li Yang, P. A. Agyakwa, C. M. Johnson, "Physics-of-Failure Lifetime Prediction Models for Wire Bond Interconnects in Power Electronic Modules," *IEEE Trans. Dev. Mat. Rel.*, vol.13, no.1, pp.9-17, March 2013.

[3] H. Wang, M. Liserre and F. Blaabjerg, "Toward reliable power electronics: challenges, design tools, and opportunities," *IEEE Ind. Electron. Mag.*, vol. 7, no. 2, pp.17-26, Jun. 2013.

[4] W. Teulings, J. L. Schanen, and J. Roudet, "MOSFET switching behavior under influence of PCB stray inductance," *in Proc. IEEE Ind. Appl. Conf.*, 1996, pp. 1449–1453.

[5] P. Ning, F. Wang, K.D.T. Ngo, "Automatic layout design for power module," *IEEE Trans. Power Electron.*, vol.28, no.1, pp.481-487, Jan. 2013.

[6] D. Medaule, Y. Arita, Y. Yu, "Latest Technology Improvements of Mitsubishi IGBT Modules," *1996 IEE Colloquium on New Developments in Power Semiconductor Devices*, pp. 5/1 - 5/5.

[7] L. Shengnan, l. M. Tolbert, W. Fei, Z. P. Fang, "Stray Inductance Reduction of Commutation Loop in the P-cell and N-cell-Based IGBT Phase Leg Module," *IEEE Trans. Power Electron.,* vol.29, no.7, pp.3616-3624, July 2014.

[8] B.W. Shook, A. Nizam, Z. Gong; A.M. Francis, H.A. Mantooth, "Multi-objective layout optimization for Multi-Chip Power Modules considering electrical parasitics and thermal performance," in *Proc. IEEE 14th Workshop on Control and Modeling for Power Electronics (COMPEL), 2013, pp.1-4.*

[9] N. Hingora, X. Liu, Y. Feng, B. McPherson, A. Mantooth, "Power-CAD: A novel methodology for design, analysis and optimization of Power Electronic Module layouts," in *Proc. IEEE Energy Conversion Congress and Exposition (ECCE)*, 2010, pp.2692-2699.

[10] S. Forster, A. Lindemann, "Combined optimisation of thermal behaviour and electrical parasitics in Power Semiconductor components," *in Proc. 13th European Conference on Power Electronics and Applications (EPE), 2009, pp.1-10.*

[11] H. Wang, A. M. Khambadkone, and X. Yu, "Dynamic electro-thermal modeling in Power Electronics Building Block (PEBB) applications," *in Proc. Energy Conversion Congress and Exposition (ECCE), 2010 IEEE*, pp. 2993–3000.

[12] Y. C. Gerstenmaier, W. Kiffe, and G. Wachutka, "Combination of thermal subsystems modeled by rapid circuit transformation," *in Proc. Of Thermal Investigation of ICs and Systems (THERMINIC) 2007*, pp. 115–120.

[13] M. Rencz and V. Szekely, "Dynamic thermal multiport modeling of IC packages," *IEEE Trans. on Comp. and Packag. Technolog.*, vol. 24, no. 4, pp. 596–604, Dec. 2001.

[14] S. H. Hall and H. L. Heck, *Advanced Signal Integrity for High-Speed Digital Designs*, Hoboken, New Jersey.: John Wiley & Sons, Inc., 2009.

[15] K. Deb, A. Pratap, S. Agarwal, and T. Meyarivan, "A fast and elitist multiobjective genetic algorithm: NSGAII," *IEEE Trans. Evol. Comput.*, vol. 6, no. 2, pp. 182 – 197, Apr. 2002.

[16] M.K. Mills, "Self inductance formulas for multi-turn rectangular loops used with vehicle detectors," in *Proc. 33rd IEEE Vehicular Technology Conference*, 1983, pp.65-73.

[17] N. Zhu, M. Chen, D. Xu, "A Simple Method to Evaluate Substrate Layout for Power Modules," in *Proc. 8th International Conference on Integrated Power Systems (CIPS), 2014, pp.1-6.*

[18] A. S. Bahman, K. M, F. Blaabjerg, "A novel 3D thermal impedance model for high power modules considering multi-layer thermal coupling and different heating/cooling conditions," *in Proc. Applied Power Electronics Conference and Exposition (APEC)* 2015, pp.1209-1215.

[19] J. H. Lienhard IV, and J. H. Lienhard V, *A Heat Transfer Textbook*, 3rd ed. Cambridge, MA: Phlogiston Press, 2006.

[20] K. Ma, A. S. Bahman, S. Beczkowski and F. Blaabjerg, "Complete Loss and Thermal Model of Power Semiconductors Including Device Rating Information," *IEEE Trans. Power Electron.*, vol. 30, no. 5, pp. 2556-2569, May 2015.

Calculation of Losses in PCB Windings for Multi-Coil Contactless Charging Systems

J. Serrano(1), J. Acero(1), I. Lope(2), C. Carretero(3), J.M. Burdío(1), R. Alonso(3).

(1)Department of Electronic Engineering and Communications. Universidad de Zaragoza.
Maria de Luna, 1. 50018 Zaragoza. Spain.
(2)B/S/H/ Home Appliances Group.
Avda. de la Industria, 49. 50016 Zaragoza. Spain.
(3)Department of Applied Physics. Universidad de Zaragoza.
Pedro Cerbuna, 12. 50009 Zaragoza, Spain.
E-mail: jserrano@unizar.es

***Abstract*—This work presents a method for calculating losses in PCB windings used in multi-coil contactless charging systems. This includes an electromagnetic model which combines analytical developments with FEM simulations in order to calculate the frequency-dependent resistance in a rectangular cross-sectional conductor. The obtained resistance per unit length is then applied to a PCB winding composed of multi-stranded turns in which strands rotate their positions building a Litz structure. The proposed model considers a generic case of a multi-coil assembly in which one of the inductors is supplying power to the load. As other coils may be positioned in the immediate vicinity of an alternating magnetic field source, these coils will produce a loss contribution due to proximity effect. The model has been experimentally validated and the results present consistence when compared with the calculations.**

***Keywords*—Losses Modeling, Contactless Charging, PCB Implementation, Inductive Charging, Multi-Coil Assemblies, Planar Winding, Home Appliances, Induction Heating.**

I. Introduction

Contactless power transfer applications are gaining importance among market trends due to their versatility and user convenience [1–3]. However, some aspects can be significantly improved. Low adaptability is achieved by single-coil contactless chargers, as misalignment of the coupled coils leads to high flux leakage and consequently, low efficiency. For this reason, multi-coil power transference devices were proposed [4]. These charging systems are arranged by means of several primary coils in a way that the most aligned coil is supplied while the rest remain in open circuit. Such a coil layout implies that coils in open circuit are suspended in the alternating magnetic field generated by the supplied coil.

Due to the elevated manufacturing cost of multi-stranded windings, and the trend to increase the number of implemented coils, cost-saving alternatives are being evaluated. Among them, PCB implemented windings stand out due to their low cost, standardized manufacturing process and repeatability.

Fig. 1. Concept of multi-coil contactless charger.

Previous studies [5] proposed a method for calculating losses in rectangular cross-section conductors, which was applied to PCB windings with consistent results [6]. However, this method considers 2D-axisymmetric systems - i.e. isolated coils and concentric coils-, which do not fit the case of multi-coil assemblies with partially overlapped coils. In this work, a general 3D case, with several non-concentric coils will be studied. This analysis will be valid for different kinds of energy transfer systems such as charging devices (Fig. 1.) and induction heating appliances.

In this study, 2-layer ultra-flat PCB windings [7, 8] will be used as they present fitting properties for the application: reduced distance from the copper in the primary coil to the secondary coil and compact design, appropriate for consumer electronics devices.

In order to address this problem, in Section II, the resistance of a single isolated rectangular cross-sectional conductor will be calculated, including conduction resistance and proximity resistance caused by an alternating magnetic field with 3 non-null components. In Section III, the developed model will be applied to a winding in order to calculate the resistance of multi-stranded PCB-coils with Litz structure [9, 10] and

978-1-4673-9551-9/16 $31.00 © 2016 IEEE

the case of partially overlapped coils will be considered and modeled. In Section IV, the model will be experimentally validated by measuring the resistance of a coil under the presence of a neighboring coil at different positions of space. Finally, Section V gathers the main conclusions of the investigation and sets the next steps in terms of winding optimization for multi-coil assemblies.

II. Resistance of a Rectangular Cross-Sectional Conductor

Losses in a single rectangular conductor can be divided into two groups depending on their origin [11, 12]. On one hand, conduction losses are caused by an applied voltage source and include dc losses and skin effect. On the other hand, proximity losses are those caused by induced currents generated by an external alternating magnetic field. Previous studies proposed semi-analytical expressions [5] for the conduction resistance per unit length, $R_{\text{cond } u.l.}$, of a rectangular wire (1).

$$R_{\text{cond } u.l.} = \frac{1}{\sigma wh}\Phi_{\text{cond}}\left(\frac{wh}{\delta^2}, \frac{w}{h}\right) \tag{1}$$

Where σ is the material conductivity, w and h are the conductor width and height, δ is the skin depth and Φ_{cond} is a function that includes the geometric and frequency dependences of the resistance. This function was obtained by means of FEM simulations and tabulated [5].

In the case of proximity losses, the incident magnetic field needs to be decomposed into its three spatial components. Each of them was proven to generate decoupled losses [13], a fact which allows calculating separately the losses produced by each component and adding them arithmetically. The losses generated by the transversal field components $\hat{x}$ and $\hat{z}$ were also modeled [5] and expressed in terms of resistance per unit length, $R_{\text{prox } u.l. x}$, $R_{\text{prox } u.l. z}$.

$$R_{\text{prox } u.l.\ x} = \frac{4\pi}{\sigma}\Phi_{\text{prox},t}\left(\frac{wh}{\delta^2}, \frac{w}{h}\right)\bar{H}_{0,x}^2 \tag{2}$$

$$R_{\text{prox } u.l.\ z} = \frac{4\pi}{\sigma}\Phi_{\text{prox},t}\left(\frac{wh}{\delta^2}, \frac{h}{w}\right)\bar{H}_{0,z}^2 \tag{3}$$

Where $\bar{H}_{0,x}$ and $\bar{H}_{0,z}$ are the transversal magnetic fields per Ampere in the direction $\hat{x}$ and $\hat{z}$ respectively and $\Phi_{\text{prox},t}$ includes the geometric and frequency dependence. Due to the symmetry of the problem, the same function can be used for both components by switching the role of w and h. This function was also obtained by FEM simulations [5]. Note that these two components are sufficient for calculating losses in coils or systems with axial symmetry, where there is no longitudinal component of the field ($\bar{H}_{0,y} = 0$). However, when two coils are partially overlapped, one of the coils is suspended in an alternating magnetic field with three non-null components, and as a consequence, a longitudinal component of the field will strike on the conductor ($\bar{H}_{0,y} \neq 0$).

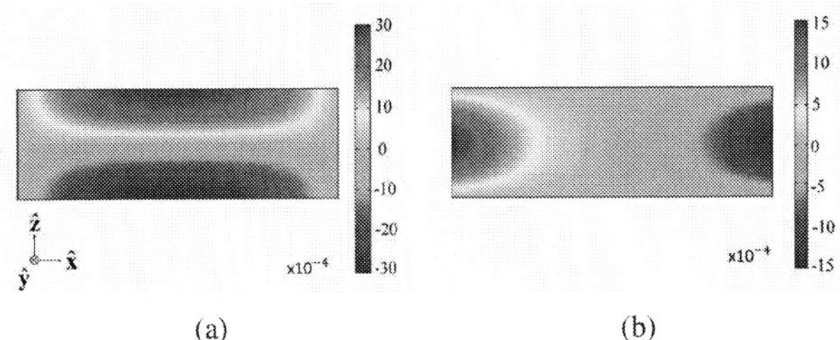

(a) (b)

Fig. 2. Induced current density $\vec{\mathbf{J}}$ when applying an alternating magnetic field (10 kHz) normal to the conductor cross section in (A/m^2). (a) Component J_x. (b) Component J_z.

The longitudinal component incident on the conductor induces currents in azimuthal direction (circulating within the section). These currents produce power losses that can be modeled by a resistance. In order to calculate these losses, in this work, an analogous expression is proposed:

$$P_{\text{prox } u.l.\ y} = \frac{2\pi}{\sigma}\Phi_{\text{prox},l}\left(\frac{wh}{\delta^2}, \frac{w}{h}\right)\bar{H}_{0,y}^2 \tag{4}$$

Considering that $P_{\text{prox } u.l.\ y} = \frac{1}{2}R_{\text{prox } u.l.\ y}\,|I|^2$, this power can also be expressed in terms of resistance:

$$R_{\text{prox } u.l.\ y} = \frac{4\pi}{\sigma}\Phi_{\text{prox},l}\left(\frac{wh}{\delta^2}, \frac{w}{h}\right)\bar{H}_{0,y}^2 \tag{5}$$

In this case, the function including frequency and geometric dependences cannot be the same, as the induced current flows within the section and not along the conductor.

Analogously to the case of transversal fields, the function containing the geometric and frequency dependences of the proximity resistance for the case of a longitudinal field, $\Phi_{\text{prox},l}$, has been calculated and tabulated based on FEM simulations. The simulation imposes a unitary magnetic field in $\hat{\mathbf{y}}$, i.e. normal to the conductor cross section. By calculating the induced current, J, the power losses per unit length $P_{\text{prox,y,u.l.}}$, can be obtained integrating in the section, S_c.

$$P_{\text{prox},y,u.l.} = \frac{1}{2\sigma}\int_{S_c} \vec{\mathbf{J}}\cdot\vec{\mathbf{J}}^*\,dS \tag{6}$$

An example of the induced currents is depicted in Fig. 2.

The resulting values of $\Phi_{\text{prox},l}$ are shown in Fig. 3. as a function of the conductor aspect ratio w/h and the parameter wh/δ^2 which introduces, apart from geometric dependences, the frequency and material influence. On the other hand, Fig. 4. represents the frequency dependence of the losses for

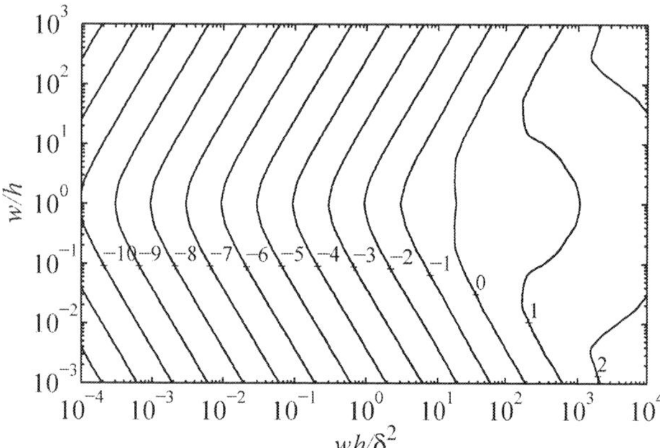

Fig. 3. $\log_{10}(\Phi_{\mathrm{prox},l})$ values for the parameters $\frac{wh}{\delta^2}$ and $\frac{w}{h}$.

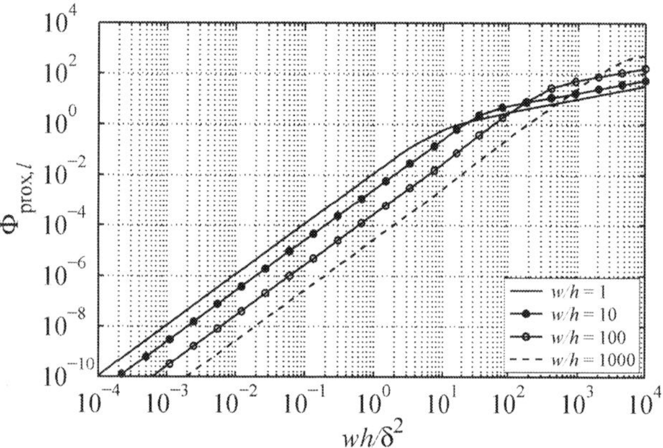

Fig. 4. $\Phi_{\mathrm{prox},l}$ values for different conductor aspect ratios.

several fixed aspect ratios.

At this point, the resistances per unit length for a single isolated conductor can be applied to a winding suspended in an alternating magnetic field generated by a non-concentric coil.

III. Losses in a Multi-Coil System

Consider the case where several coils are arbitrarily placed in space. One of them (coil i) is powered in such a way that an alternating current flows through it. Another coil (coil j), in this case in open circuit, is in close vicinity (Fig. 5.). In this situation, the power dissipation in the windings caused by the current flowing through coil i, $P_{w,ii}$, can be decomposed in: dc and skin losses in coil i, $P_{\mathrm{cond},ii}$, proximity losses in coil i, $P_{\mathrm{prox},ii}$, and proximity losses in coil j, $P_{\mathrm{prox},ij}$.

$$P_{w,ii} = P_{\mathrm{cond},ii} + P_{\mathrm{prox},ii} + P_{\mathrm{prox},ij} \tag{7}$$

As seen in the previous section, these power losses can by modeled by their corresponding resistances:

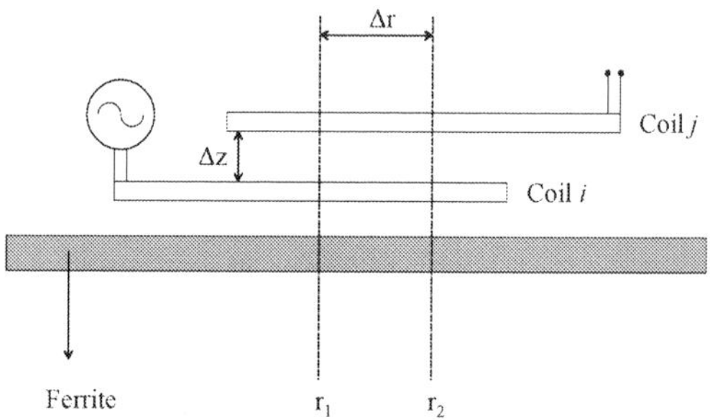

Fig. 5. Schematic of the analyzed system.

$$R_{w,ii} = R_{\mathrm{cond},ii} + R_{\mathrm{prox},ii} + R_{\mathrm{prox},ij} \tag{8}$$

Where $R_{w,ii}$ denotes the resistance of coil i in presence of coil j, $R_{\mathrm{cond},ii} + R_{\mathrm{prox},ii}$ represents the resistance of coil i in the absence of coil j, and $R_{\mathrm{prox},ij}$ represents the resistive contribution caused by the presence of coil j.

In case of coils with several multi-stranded turns, each resistive contribution can be calculated for an average single-stranded turn. This virtual single wire turn, has a length equal to the average length of all turns, and is suspended in a magnetic field equal to the average magnetic field incident on all turns. The conduction and proximity resistance of the average single-wire turn is denoted as $R_{\mathrm{cond},11}$ and $R_{\mathrm{prox},11}$. The average turn's length, l_{eq}, is calculated as the coil volume, V_{coil}, divided by the coil seccion, S_{coil}. For the calculation of the proximity resistance, the incident magnetic field corresponds to the averaged field in the coil volume for each coil and field component ($k = r, a, \varphi$). Therefore, the terms of (8), without considering the number of turns and strands of the coil, are obtained as follows:

$$R_{\mathrm{cond},ii,11} = \frac{l_{eq}}{\sigma wh}\Phi_{\mathrm{cond}}\left(\frac{wh}{\delta^2}, \frac{w}{h}\right) \tag{9}$$

$$\begin{aligned} R_{\mathrm{prox},ii,k,11} = {} & \tfrac{4\pi}{\sigma}\Phi_{\mathrm{prox,k}}\left(\tfrac{wh}{\delta^2}, \tfrac{w}{h}\right) \\ & \tfrac{1}{S_{\mathrm{coil}}}\iiint \overline{\mathrm{H}_k}\cdot\overline{\mathrm{H}_k}^{*} r dr_i\, da_i\, d\varphi_i \end{aligned} \tag{10}$$

$$\begin{aligned} R_{\mathrm{prox},ij,k,11} = {} & \tfrac{4\pi}{\sigma}\Phi_{\mathrm{prox,k}}\left(\tfrac{wh}{\delta^2}, \tfrac{w}{h}\right) \\ & \tfrac{1}{S_{\mathrm{coil}}}\iiint \overline{\mathrm{H}_k}\cdot\overline{\mathrm{H}_k}^{*} r dr_j\, da_j\, d\varphi_j \end{aligned} \tag{11}$$

Being $\{\hat{r}_i, \hat{a}_i, \hat{\varphi}_i\}$ and $\{\hat{r}_j, \hat{a}_j, \hat{\varphi}_j\}$ the local cylindrical coordinates for each coil and $\Phi_{\mathrm{prox,k}}$ can be either $\Phi_{\mathrm{prox,t}}$ when k is a transversal component of the field with respect to the conductor, or $\Phi_{\mathrm{prox,l}}$ when k is the longitudinal component of the field with respect to the conductor.

As the proximity resistance depends on the squared magnetic field, for the sake of simplicity,

TABLE I. Magnetic field distribution and average magnetic field in two neighboring coils expressed in cylindrical local coordinates for each coil.

k	$\overline{\mathrm{H}_k} \cdot \overline{\mathrm{H}_k}^*$	$\left\langle \overline{\mathrm{H}_k}^2 \right\rangle$

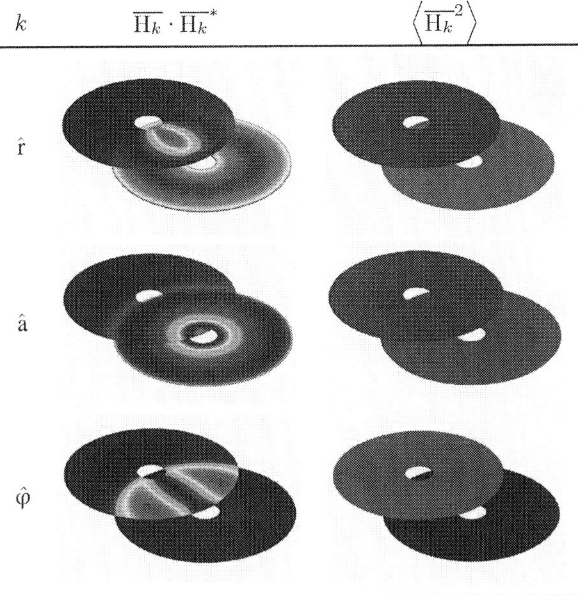

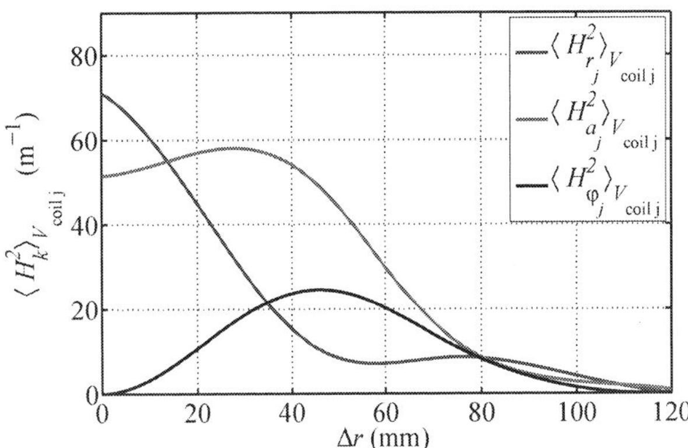

Fig. 6. Average squared magnetic field in the coil in open circuit for each component r̂, â, φ̂. Both, supplied and open-circuit coil, have a 60 mm external radius and 10 mm internal radius.

the average squared magnetic field will be denoted as $\left\langle \overline{\mathrm{H}}^2 \right\rangle = \iiint_{V_{\text{coil}}} \overline{\mathrm{H}} \cdot \overline{\mathrm{H}}^* r dr\, da\, d\varphi$ and its concept is shown in Table I.

The magnetic field has been calculated by means of FEM simulations by modeling the coil by a ring-type ideal current density. Assuming ideal litz structure [9, 14], the same current flows through every strand, and therefore, the current density in the ideal ring-type coil can be considered constant in its section.

It is important to note that the average squared magnetic field in coil j depends on its position in space, and consequently the proximity losses too. The average squared magnetic field in coil j is depicted in Fig. 6. as a function of misalignment, Δr.

Moreover, considering that a winding is a series connection of n turns made of n_s parallel strands, the number of turns and strands, take effect on the winding conduction resistance as [15]:

$$R_{ii,\text{cond}} = \frac{n_i}{n_{s,i}} \cdot R_{ii,\text{cond},11} \tag{12}$$

On the other hand, proximity losses are additive both with the number of strands and with the number of turns, which introduces a $n \cdot n_s$ factor. Moreover, the magnetic field generated by every turn is also additive with the number of turns. Taking into account that the magnetic field has a quadratic influence on the resistance (11), we obtain: $|\bar{H}_0|^2 = |n\bar{H}_{0,1}|^2 = n^2|\bar{H}_{0,1}|^2$, and therefore [15]:

$$R_{ii,\text{prox,k}} = n_i^3 \cdot n_{s,i} \cdot R_{ii,\text{prox},k,11} \tag{13}$$

$$R_{ij,\text{prox,k}} = n_i^2 \cdot n_j \cdot n_{s,j} \cdot R_{ij,\text{prox},k,11} \tag{14}$$

The general case for multi-coil assemblies in which one primary coil (i) is supplied in presence of N open circuit coils (j) can be expressed as follows.

$$R_{w,ii} = \frac{n_i}{n_{s,i}} R_{ii,\text{cond},11} + n_i^3 n_{s,i} R_{ii,\text{prox},11} + \sum_{j=1}^{N} n_i^2 n_j n_{s,j} \left(R_{ij,\text{prox},11,r_j} + R_{ij,\text{prox},11,a_j} + R_{ij,\text{prox},11,\varphi_j} \right) \tag{15}$$

IV. Experimental Validation

The model has been validated on the laboratory large-scale prototype shown in Fig. 7. The tested coils have a $r_{\text{ext}} = 60$ mm external radius, $r_{\text{int}} = 10$ mm internal radius, 9 turns and 13 strands. The strand thickness is h = 140 μm and the strand width is w = 440 μm. The strands rotate their positions by means of vias building a PCB litz structure and they switch position 96 times along the winding achieving a quasi-ideal litz structure [14].

The coils were measured by means of an Agilent E4980A Precision LCR Meter.

For example, the resistance of one supplied coil ($i = 1$) under presence of another coil ($j = 1$) placed at $\Delta z = 0.5$ mm and $\Delta r = 60$ mm is shown in Fig. 8. where blue represents the resistance of coil i when it is isolated, $R_{\text{cond},ii}|_{i=1} + R_{\text{prox},ii}|_{i=1}$, black represents the total resistance measured at coil i terminals when Coil j is placed in close vicinity, $R_{w,ii}|_{i=1}$, and the red line is the contribution of coil j to the resistance of coil i, $R_{\text{prox},ij}|_{i=1,j=1}$, (not directly measurable, but easily calculated as shown in

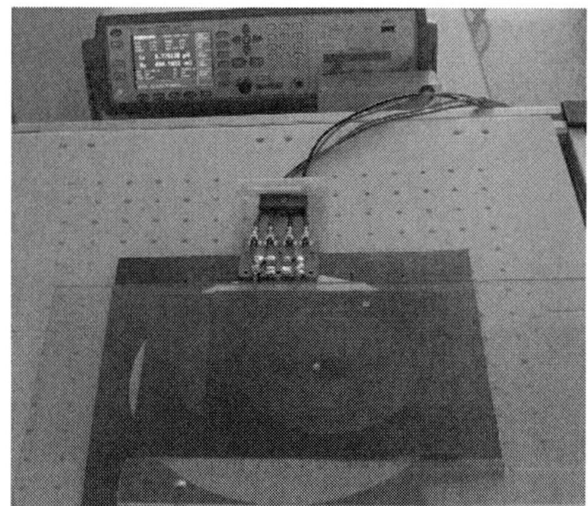

Fig. 7. Small signal resistance measurement.

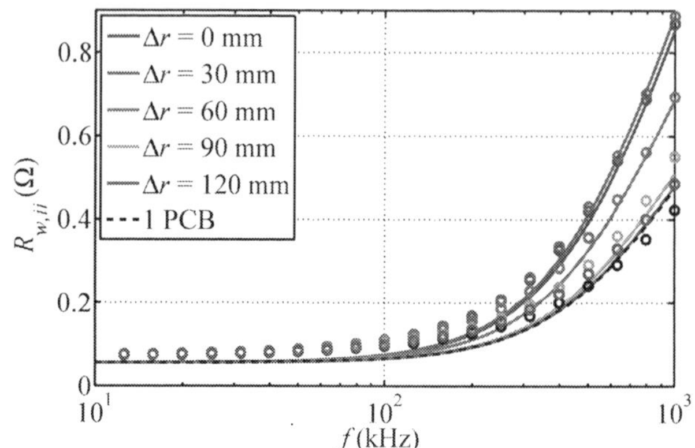

Fig. 9. Resistance as a function of frequency. Calculated R in solid line. Measured R in circles.

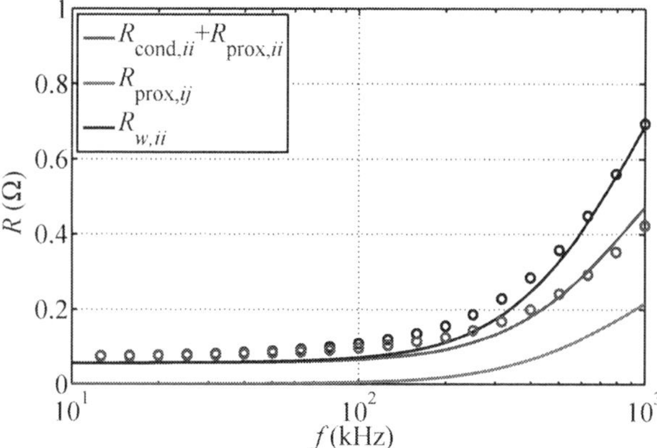

Fig. 8. Resistance for $\Delta z = 0.5$ mm and $\Delta r = 60$ mm. Calculated in solid line. Measured in circles.

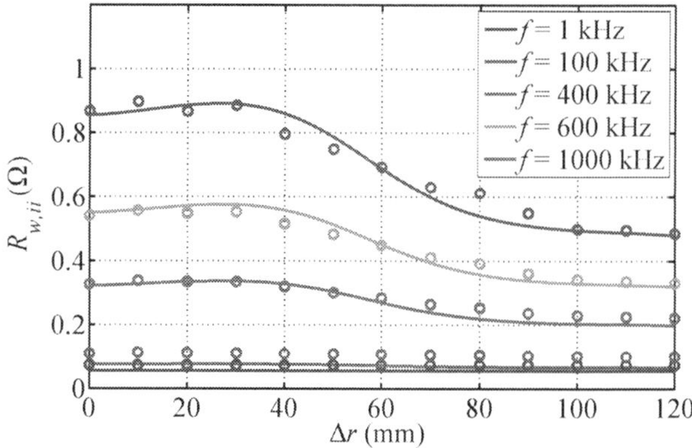

Fig. 10. Resistance as a function of displacement. Calculated R in solid line. Measured R in circles.

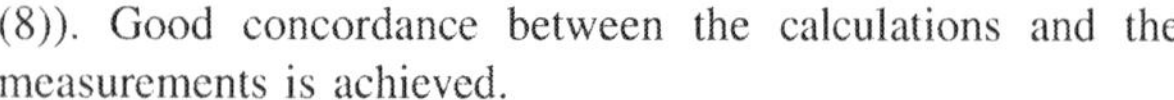

(8)). Good concordance between the calculations and the measurements is achieved.

By sweeping the radial displacement, the loss resistance can be calculated for a wide range of frequencies and any overlapping position. Fig. 9. shows the resistance as a function of frequency, with the isolated PCB winding as a reference in black dashed line. On the other hand Fig. 10. depicts resistance as a function of misalignment.

As expected (see Fig. 6.), the resistance significantly decreases as the inner diameter of the supplied coil is being uncovered i.e. $r_{\text{ext}} - r_{\text{int}} \leq \Delta_r \leq r_{\text{ext}} + r_{\text{int}}$. Given that, misalignments over $r_{\text{ext}} + r_{\text{int}}$ are recommended in order to reduce losses, specially when working at high frequencies where proximity losses are of major importance.

V. Conclusions and Further Work

In this study a model for calculating losses in PCB windings for multi-coil contactless chargers is presented. Firstly, an existing model for single conductors was extended in order to include proximity losses caused by a longitudinal field. This allows the analysis of situations with overlapped coils. In the next place, the model was applied to a general case with one supplied coil and a number of neighbor coils in open circuit suspended in an alternating magnetic field with three non-null components. The total losses have been calculated by adding the losses produced in every coil. The model has been experimentally verified throwing consistent results.

Once the frequency-dependent resistance of every coil is modeled, the basis for the optimization of multi-coil assemblies are set: by selecting the optimal parameters with influence on the loss resistance (strand width, strand thickness, number of turns, number of strands, coil misalignment, coil vertical separation...) the losses can be effectively minimized, improving the device efficiency and reducing the operating temperature. Moreover, the same analysis can be applied to induction heating appliances in which efficency becomes a key issue due to the high power supplied.

VI. Acknowledgements

This work was partly supported by the Spanish MINECO under Project TEC2013-42937-R, Project CSD2009-00046, and Project RTC-2014-1847-6, by the DGA-FSE, by the University of Zaragoza under Project JIUZ-2014-TEC-08, and by the BSH Home Appliances Group.

References

[1] C.-G. Kim, D.-H. Seo, J.-S. You, J.-H. Park, and B.-H. Cho, "Design of a contactless battery charger for cellular phone," in *Applied Power Electronics Conference and Exposition, 2000. APEC 2000. Fifteenth Annual IEEE*, vol. 2, pp. 769–773 vol.2, 2000.

[2] B. Choi, J. Nho, H. Cha, T. Ahn, and B. Choi, "Design and implementation of low-profile contactless battery charger using planar printed circuit board windings as energy transfer device," *Industrial Electronics, IEEE Transactions on*, vol. 51, pp. 140–147, Feb 2004.

[3] C.-H. Hu, C.-M. Chen, Y.-S. Shiao, T.-J. Chan, and L.-R. Chen, "Development of a universal contactless charger for handheld devices," in *Industrial Electronics, 2008. ISIE 2008. IEEE International Symposium on*, pp. 99–104, June 2008.

[4] J. Achterberg, E. Lomonova, and J. de Boeij, "Coil array structures compared for contactless battery charging platform," *Magnetics, IEEE Transactions on*, vol. 44, pp. 617–622, May 2008.

[5] I. Lope, C. Carretero, J. Acero, R. Alonso, and J. Burdío, "Ac power losses model for planar windings with rectangular cross-sectional conductors," *Power Electronics, IEEE Transactions on*, vol. 29, pp. 23–28, Jan 2014.

[6] I. Lope, J. Acero, J. Burdio, C. Carretero, and R. Alonso, "Design and implementation of pcb inductors with litz-wire structure for conventional-size large-signal domestic induction heating applications," *Industry Applications, IEEE Transactions on*, vol. 51, pp. 2434–2442, May 2015.

[7] M. Ludwig, M. Duffy, T. O'Donnell, P. McCloskey, and S. OMathuna, "Design study for ultraflat PCB-integrated inductors for low-power conversion applications," *Magnetics, IEEE Transactions on*, vol. 39, pp. 3193–3195, Sept 2003.

[8] J. Serrano, I. Lope, J. Acero, J. Burdio, C. Carretero, and R. Alonso, "Design and optimization of small inductors on extra-thin pcb for flexible cooking surfaces," in *Applied Power Electronics Conference and Exposition (APEC), 2015 IEEE*, pp. 177–182, March 2015.

[9] I. Lope, C. Carretero, J. Acero, R. Alonso, and J. Burdío, "Design and implementation of PCB inductors with litz-wire structure for conventional-size large-signal domestic induction heating applications," in *Applied Power Electronics Conference and Exposition (APEC), 2014 Twenty-Ninth Annual IEEE*, pp. 732–737, March 2014.

[10] S. Wang, M. de Rooij, W. Odendaal, J. van Wyk, and D. Boroyevich, "Reduction of high-frequency conduction losses using a planar litz structure," *Power Electronics, IEEE Transactions on*, vol. 20, pp. 261–267, March 2005.

[11] J. Ferreira, "Analytical computation of ac resistance of round and rectangular litz wire windings," *Electric Power Applications, IEE Proceedings B*, vol. 139, pp. 21–25, Jan 1992.

[12] J. Acero, R. Alonso, J. Burdío, L. Barragán, and D. Puyal, "Frequency-dependent resistance in litz-wire planar windings for domestic induction heating appliances," *Power Electronics, IEEE Transactions on*, vol. 21, pp. 856–866, July 2006.

[13] C. Carretero, J. Acero, and R. Alonso., "TM-TE decomposition of power losses in multi-stranded litz-wires used in electronic devices," *Progress In Electromagnetics Research*, vol. 123, pp. 83–103, December 2012.

[14] I. Lope, J. Acero, J. Serrano, C. Carretero, R. Alonso, and J. Burdio, "Minimization of vias in pcb implementations of planar coils with litz-wire structure," in *Applied Power Electronics Conference and Exposition (APEC), 2015 IEEE*, pp. 2512–2517, March 2015.

[15] I. Lope, C. Carretero, J. Acero, R. Alonso, and J. Burdío, "Frequency-dependent resistance of planar coils in printed circuit board with litz structure," *Magnetics, IEEE Transactions on*, vol. 50, pp. 1–9, Dec 2014.

Design of efficient loads for domestic induction heating applications by means of non-magnetic thin metallic layers

Jesús Acero[1], Claudio Carretero[2], Rafael Alonso[2], and José Miguel Burdío[1]
[1] Department of Electronic Engineering and Communications. Universidad de Zaragoza.
María de Luna 1. 50018 Zaragoza, Spain

[2] Department of Applied Physics. Universidad de Zaragoza.
Pedro Cerbuna 12. 50009 Zaragoza, Spain

***Abstract*—The inductive performance of non-magnetic thin metallic layers is investigated with the objective of designing convenient loads for domestic induction heating applications. Thin layers of copper and aluminum with thicknesses ranging from one hundred of nanometers to tens of micrometers are fabricated using the phase vapor deposition (PVD) technique. Performance of layers of different thicknesses and a ferromagnetic conventional material are compared by using the inductive resistance and the inductive efficiency as figures of merit.**

Keywords*— Eddy currents, energy efficiency home appliances, induction heating, impedance matching.

I. Introduction

Nowadays, the induction heating phenomenon is being applied to the culinary sphere because this technology provides advantages as safety, cleanliness and high efficiency. Part of these advantages derives from the fact that the heat is directly generated in the materials and, therefore, the usual high-temperature heat sources are avoided. Other advantages also provided by these applications derive from the used high-performance power electronics converters. Among these advantages we can list high rated power, power controllability, and appropriate temperature control.

In the literature many works addressing different aspects of the electronics [1-4] and the inductor system of induction cookers [5-8] have been reported. However, the number of works devoted to the optimal design of the pots is comparatively small. Utensils for domestic induction heating appliances are mainly made of ferromagnetic steel because this material naturally provides good performance. Especially, high induction efficiency is achieved at the usual operation frequency range, from 20 to 100 kHz [5]. Nevertheless, pots made of non-magnetic metals (as aluminum, copper or stainless steel) present high quality factor Q, low equivalent impedance [9] and also present very low efficiency. Therefore, the electronics integrated in the cooker could be damaged when non-magnetic pots are heated. Nowadays, cookers automatically detect and avoid heating these materials. This is the reason of the well-known disadvantage of induction cookers, namely the inability of heating non-magnetic metallic pots.

Apart from the magnetic hysteresis, the other main difference between magnetic and non-magnetic metals is the skin depth of the fields at the interest frequency range. At 100 kHz, the skin depth of a typical F114 ferromagnetic steel (σ_{Fe}=2x10^6 (Ω·m)$^{-1}$, $\mu_{r,Fe}$=1000), is δ=35.5 µm, whereas for a conventional aluminum (σ_{Al} =1x10^7 (Ω·m)$^{-1}$, $\mu_{r,Al}$=1) is δ=503 µm. Considering these values, the equivalent sheet resistance of the F114 steel and the aluminum are R_s=14 (mΩ/□) and R_s=0.2 (mΩ/□) (mΩ per square), respectively. Physically, in the F114 steel current is constrained to flow through a thin layer, which implies high sheet resistance. However, for the same reasons non-magnetic loads have low sheet resistance and low power dissipation.

There are two ways to adjust the sheet resistance: on the one hand by selecting materials with appropriate conductivity and, on the other hand, by building layers with specific thickness. Considering the first one, it has been reported in some works the suitability of non-metals with an intermediate electric conductivity, as the graphite, for induction heating purposes [10]. However, in the nature the set of materials with proper conductivities is small and therefore this factor should be combined with the thickness of the material [10].

Considering the second possibility, in this work is explored the feasibility of using thin layers of non-magnetic good conductors in order to achieve convenient sheet resistances for the application of the domestic induction heating. This is equivalent to force the flow of the current through a layer of a good conductor with small cross section. For this purposes, different layers with different thicknesses and materials were manufactured and tested. Layers were made of copper and aluminum and they were manufactured by means of the phase vapor deposition (PVD) technique, which allows achieving high precision of thickness. In order to model the inductive properties of the different layers, an analytical model of a simple induction heating consisting of a planar coil and a layer of a material was also derived. The model provides a resistance which is equivalent to the induced power in the layer and also provides the induction efficiency. Both

978-1-4673-9551-9/16 $31.00 © 2016 IEEE

parameters are used to compare the inductive properties of different layers.

II. Characterization of the Planar Induction Heating System

A. Parameters of Study

The performance of a planar induction heating system can be characterized by means of two parameters, the induced load resistance R_{ind} [11] and the induction efficiency η_{ind} [5]. The first parameter is a resistance which represents the power dissipated in a piece by means of the induction phenomenon. The second parameter corresponds to the part of the power delivered into the piece with respect to the electrical power supplied into the coil. This parameters is defined as follows [5]:

$$\eta_{ind} = \frac{R_{ind}}{R_{ind} + R_{w}} \quad (1)$$

where R_w is the resistance of the windings of coil. The resistance R_{ind} depends on the frequency, the properties of the load, the geometry of the system and the number of turns of the coil. In its turn, R_w also depends on the frequency, the properties of the load, the geometry of the system, the number of turns of the coil and the kind of cable used for windings. The resistance R_{ind} can be used to compare the inductive performance of a load for a fixed coil. Efficiency η_{ind} is mainly used to design a specific induction system including the windings.

Both R_{ind} and η_{ind} can be obtained by means of either Finite Element Analysis (FEA) tools or analytical approaches. However, the simulation of conductive layers of nanometers of thickness (combined with total system sizing of centimeters) could be problematic with FEA tools because fine meshing is required. Therefore, in this work we propose an analytical approach, which is presented in the next section.

B. Model of an Ideal Induction System

The induction system under study is shown in Fig. 1. This system consist of an ideal filamentary current of radius a which carries a current of amplitude I_ϕ at angular frequency ω. Three layers numbered as 0, 1 and 2 are placed above the filamentary current. Layers 0 and 1 are air and the layer 1 is a material of properties σ_l and μ_{rl}, which is placed at a distance z_l above the current. According to the geometry shown in Fig. 1, the thickness of the layer 1 is:

$$t_l = z_2 - z_1 \quad (2)$$

If the radial dimension of the layer 1 is higher than the diameter of the current, an analytical solution of the electromagnetic fields of this system can be found. This analytical solution can be also applied to obtain the induced impedance Z_{ind} of this system on the basis of several works previously published [10, 12, 13]. As it can be seen in these references, the induced impedance consists of real and imaginary parts. Considering the ideal model of the coil, the real part corresponds to the induced resistance R_{ind}, whereas the imaginary part is the inductance of the system:

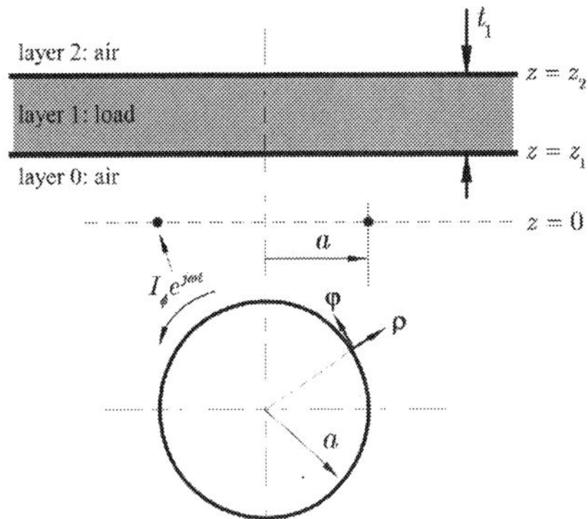

Fig. 1. Ideal induction system used to study the resistive contribution of a layer of specific thickness.

$$Z_{ind} = R_{ind} + j\omega L_{ind} \quad (3)$$

where L_{ind} corresponds to the stored magnetic energy of the system. This value is required when designing the resonant converter which fed an inductor system.

Impedance Z_{ind} for a general system consisting of several stacked layers (or multilayer arrangement) is obtained in some of the mentioned references. This general solution is particularized for the case of Fig. 1. Therefore, according to the mentioned references, the induced impedance is:

$$Z_{ind} = j\omega\mu_0\pi\int_0^\infty\left(1+\phi_{l\,eq}e^{-2\beta z_1}\right)\left[a\cdot J_1\left(a\beta\right)\right]^2 d\beta \quad (4)$$

where $\phi_{l\,eq}$ depends on the geometry and the properties of the layer. In this integral equation the integration variable β corresponds to the argument of the Fourier-Bessel transform [14].

The following additional magnitudes are necessary to calculate $\phi_{l\,eq}$:

$$\eta_i\left(\beta\right) = \sqrt{\beta^2 + j\omega\mu_i\sigma_i} \quad 0 \le i \le 2 \quad (5)$$

$$\phi_i\left(\beta\right) = \frac{\beta\mu_{ri} - \eta_i}{\beta\mu_{ri} + \eta_i} \quad 0 \le i \le 2 \quad (6)$$

$$\lambda_i = \frac{\eta_i}{\mu_i} \quad 0 \le i \le 2 \quad (7)$$

A matrix relating the properties of two adjacent layers can be also defined:

$$\mathbf{R}_{i,i+1} = \frac{1}{2}\begin{bmatrix} \left(1+\frac{\lambda_{i+1}}{\lambda_i}\right)e^{-\eta_i t_i} & \left(1-\frac{\lambda_{i+1}}{\lambda_i}\right)e^{-\eta_i t_i} \\ \left(1-\frac{\lambda_{i+1}}{\lambda_i}\right)e^{\eta_i t_i} & \left(1+\frac{\lambda_{i+1}}{\lambda_i}\right)e^{\eta_i t_i} \end{bmatrix} \quad 0 \le i \le 1 \quad (8)$$

where t_i is the thickness of the i-th layer.

Moreover, an additional matrix is defined as the product of matrices $\mathbf{R}_{0,1}$ and $\mathbf{R}_{1,2}$:

$$\mathbf{T}^{l\,\mathrm{eq}}(\beta)=\begin{bmatrix}\mathbf{T}_{11}^{l\,\mathrm{eq}}(\beta) & \mathbf{T}_{12}^{l\,\mathrm{eq}}(\beta)\\ \mathbf{T}_{21}^{l\,\mathrm{eq}}(\beta) & \mathbf{T}_{22}^{l\,\mathrm{eq}}(\beta)\end{bmatrix}=\mathbf{R}_{0,1}\cdot\mathbf{R}_{1,2} \quad (9)$$

Therefore the parameter $\phi_{l\,\mathrm{eq}}$ is defined as:

$$\phi_{l\,\mathrm{eq}}(\beta)=\frac{\mathbf{T}_{12}^{l\,\mathrm{eq}}(\beta)}{\mathbf{T}_{22}^{l\,\mathrm{eq}}(\beta)} \quad (10)$$

Operating, this parameter has the following expression:

$$\phi_{l\,\mathrm{eq}}(\beta)=\frac{\left(1-e^{-2\eta_1 t_1}\right)}{\left(1-\phi_1^2 e^{-2\eta_1 t_1}\right)}\phi_1 \quad (11)$$

where η_1, ϕ_1 are defined in (5), (6), respectively.

Usual induction heating systems consists of n turns rather than only one turn. However, as it was reported in previous works [10], the equation (4) can be adapted to the case of n turns by replacing the term $[a\cdot J_1(a\beta)]^2$ of (4) for this factor:

$$G(a_i,\beta)=\left[\sum_{i=1}^{n}a_iJ_1(\beta a_i)\right]^2= \sum_{i=1}^{n}a_i^2J_1^2(\beta a_i)+2\sum_{i=1}^{n}\sum_{k>i}^{n}a_ia_kJ_1(\beta a_i)J_1(\beta a_k) \quad (12)$$

where a_i is the radius of the ith turn. Considering this factor and the expression of $\phi_{l\,\mathrm{eq}}$, the equation (4) can be rewritten as follows:

$$Z_{ind}=j\omega\mu_0\pi\int_0^{\infty}\left[1+\frac{\left(1-e^{-2\eta_1 t_1}\right)\phi_1}{\left(1-\phi_1^2 e^{-2\eta_1 t_1}\right)}e^{-2\beta z_1}\right]\left[\sum_{i=1}^{n}a_iJ_1(\beta a_i)\right]^2 d\beta \quad (13)$$

Usually, commercial induction appliances include some ferrite bars beneath the windings. In general, ferrite bars increase the value of R_{ind} independently of the load properties and, therefore, the extracted conclusions (relatives to R_{ind}) are not essentially modified. However, depending on the kind of cable, ferrite bars can also affect the induction efficiency. Considering that this work is focused on the use of thin non-magnetic layers rather than the design of the windings, ferrite bars are not considered.

C. Simulation Results

In this section some simulation results are presented. The analytical model has been implemented in MATLAB because it provides appropriate functions and features. In this simulation, a system with the structure of layers shown in Fig. 1 is considered and the value of R_{ind} as a function of the layer thickness and different materials is calculated. For this purpose, the layer thickness t_1 is swept from 10 nm to 1 mm. The study is carried out at a frequency f = 100 kHz and a coil of n=23 turns with a_1= 10 mm and a_{23} = 47.5 mm was also considered. The distance between the coil and the layer was z_1=2.5 mm. In simulations, the tested materials were conventional copper (σ_{Cu}=5.8x10^7 ($\Omega\cdot$m)$^{-1}$, $\mu_{\mathrm{r,Cu}}$=1), aluminum (σ_{Al}=3x10^7 ($\Omega\cdot$m)$^{-1}$, $\mu_{\mathrm{r,Al}}$=1) and ferromagnetic steel (σ_{Fe}=2x10^6 ($\Omega\cdot$m)$^{-1}$, $\mu_{\mathrm{r,Fe}}$=1000).

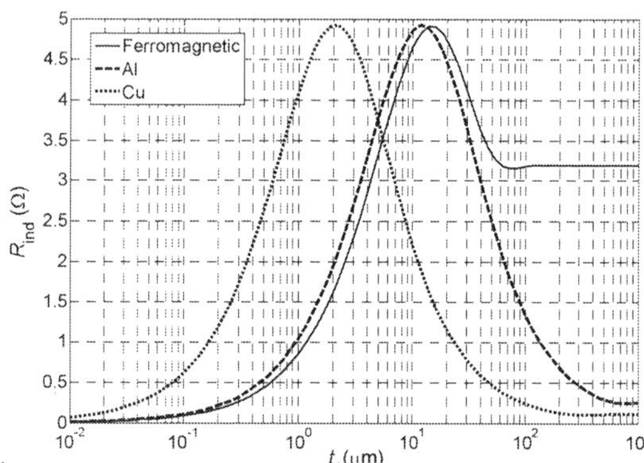

Fig. 2. Induction resistace of a layer of variable thickness of different materials at 100 kHz.

The calculated values of R_{ind} for the different materials are shown in Fig. 2. Regarding the non-magnetic specimens, R_{ind} is very small for both thin and thick layers. Thin layers correspond to very high sheet resistance. As a consequence, currents are hardly induced and therefore small power is induced. On the opposite, the case of thick layers corresponds to the bulk material and for that case the power dissipation is also very small because the sheet resistance is very low. However, as it can be observed in this figure, at a certain thickness the value of R_{ind} is maximized. This thickness is different for each material and frequency. Considering copper, at 100 kHz this thickness is about 2 μm, whereas it is about 15 μm for aluminum.

It is also interesting the case of the ferromagnetic layer. For thin layers, the R_{ind} is similar to the other metals due to the high sheet resistance. However, R_{ind} of thick layers (i.e. R_{ind} of the bulk material) is high, which corresponds to a good material for domestic induction heating purposes. However, it is worth to note that the resistance at the optimum thickness is higher that the resistance of the bulk material, which means that the bulk material is not the best arrangement to maximize the power dissipation even for ferromagnetic materials.

It is also remarkable that the maximum resistance achievable is very similar for the three materials but at different thicknesses. This fact could be extended to other materials and it could mean that for good conductors, if the parameters of an induction heating system are fixed (i.e. geometry, number of turns and frequency) the maximum R_{ind} is independent of the material of the load. This fact is pending of rigorous demonstration and could be a noticeable result.

In order to check these results, some samples were built and checked. The characteristics of the built samples are explained in the next section.

TABLE I
CHARACTERISTICS OF SAMPLES

material	thicknesses (μm)
copper	0.1, 0.2, 0.5, 1, 2, 5
aluminum	0.3, 0.4, 0.8, 2, 2.8, 14

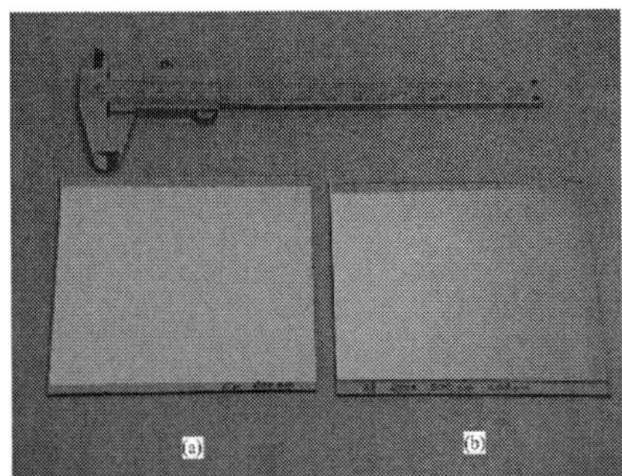

Fig. 3. Image of two samples. (a) copper, (b) aluminum.

III. DESCRIPTION OF THE SET OF SAMPLES

In order to verify the simulated results, several samples of different thicknesses were constructed with aluminum and copper. The thickness of layers varies from 100 nm to 14 μm. Layers were constructed using the phase vapor deposition (PVD) technique, which mainly consists of sputtering particles from a bulk material into a target (i.e. into a substrate) by means of an electron beam under vacuum atmosphere [15]. The PVD technique enables thin films to be sputtered with controlled thickness and texture, and allows the electrical conductivity of bulk materials to be practically preserved. This technique is used in a wide variety of applications as optics, decorative architecture and magnetic recording among others. In this case, the substrate is a piece of vitroceramic glass of dimensions 120x120x2.5 mm. Considering the external diameter of the above mentioned coil, the samples can be assumed as infinite in the radial dimension and the analytical model can be applied. The list of samples and their thickness is shown in Table I. Images of copper and aluminum samples are shown in Fig. 3.

As it has been mentioned, the PVD technique allows manufacturing layer with practically the same conductivity of the original materials. Although it was experimentally proved that the conductivity of the samples was similar to the precursors copper and aluminum, in order to perform a better comparison, the actual conductivity of the different samples was inductively measured. Results are shown in Fig. 4 for the copper samples and in Fig. 5 for the aluminum samples. In each figure the conductivity corresponding to the bulk material has been also added. Moreover it has been also included a piecewise fit in order to see the tendency of the conductivity with respect to the thickness.

As it can be observed, the conductivity of the thinnest copper layer is lesser than the half part of the conductivity of the bulk copper whereas the conductivity of the thinnest aluminum layer is about the fifth part of the conductivity of the bulk material. In both cases, the tendency of the conductivity with respect to the deposited thickness is very similar.

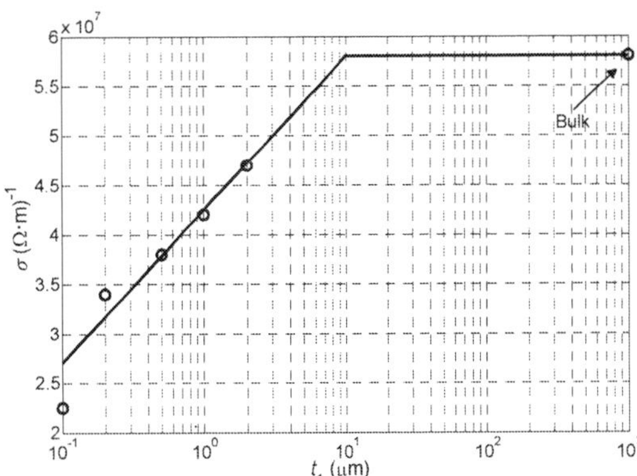

Fig. 4. Conductivity of the copper samples with respect to the thickness of the layer. A piecewise fit is also shown.

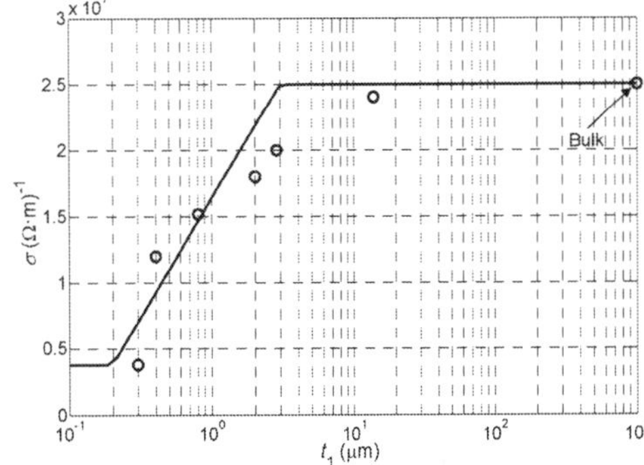

Fig. 5. Conductivity of the aluminum samples with respect to the thickness of the layer. A piecewise fit is also shown.

IV. EXPERIMENTAL RESULTS

A testing spiral coil of n=23 turns with the same external and internal radii of the coil used in the simulations of Section II was built to measure the samples. A small round wire of diameter ϕ_o=0.2 mm was used for the windings in order to avoid frequency-dependent resistance contributions caused by the skin and proximity effects. The resistance of this coil, R_w, is appreciable and therefore it wouldn't be appropriate for heating purposes due to its low efficiency. Considering that this resistance is accounted for in measurements, i.e., the actual measured resistance comprises both R_w and R_{ind}, it follows that the measured resistance is in fact the equivalent resistance of the system, R_{eq}, defined as:

$$R_{eq} = R_{ind} + R_w \tag{14}$$

Considering that the ac resistance of the coil is small, or in other words the resistance of the coil is constant for the measured frequency range, the resistance of the coil in air was added to the calculated R_{ind} and the resultant resistance was compared with the measured R_{eq}.

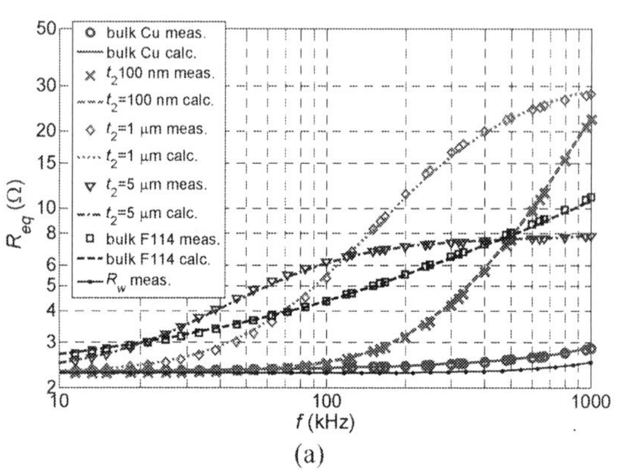

(a)

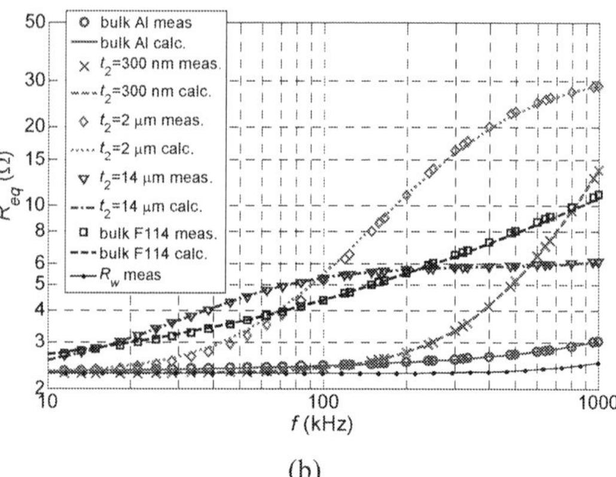

(b)

Fig. 6. Comparison between calculated and measured equivalent resistances for the manufactured samples, bulk materials and F114 disk. (a) Copper samples. (b) Aluminum samples.

The frequency-dependent equivalent resistance of the coil was measured with the Keysight E4980A LCR meter, therefore measurements are under small signal conditions and thermal effects are not considered. The frequency of measurements ranged from 1 kHz to 1 MHz.

In addition to the thin-layer samples, a disk of F114 ferromagnetic steel was also used to compare the manufactured samples with a conventional load used in real applications.

Fig. 6 shows a comparison between calculated and measured equivalent resistances for some samples and the F114 disk. The resistance of the coil in air is also shown. In general it is proved that equivalent resistances higher than those obtained with the bulk F114 can be obtained. Obviously, it is also proved that the resistance of thin layers of metallic materials is higher than the resistance obtained with the corresponding bulk materials. The results also show that at the interest frequency range the thickness of the layer can be selected in order to optimize the inductive performance. Thinner samples present higher resistances at the higher frequencies whereas at the interest frequencies range (from 20 to 100 kHz) the thickness of choice would be in the range of few micrometers. It is also noted the interest in going at higher frequencies. As it can be observed, the agreement between calculations and measurements is in general good.

From the point of view of a real application, it is also interesting to compare the induction efficiency η_{ind} achievable by using this samples with respect to a conventional load. For this purposes, Fig. 7 shows a comparison of the theoretical η_{ind} that would be achieved with a 23-turn commercial coil of external diameter of 21 cm with a conventional ferromagnetic load and two thin-layer samples. One of them is the copper layer of 5 μm of thickness and the other is the aluminum layer of 14 μm. The coil is made of a litz wire of 37 strands of diameter 0.3 mm. A ferrite plane beneath the coil is also considered in the system. Calculations show that, at the interest frequency range, efficiencies higher than the obtained with conventional loads could be theoretically achieved.

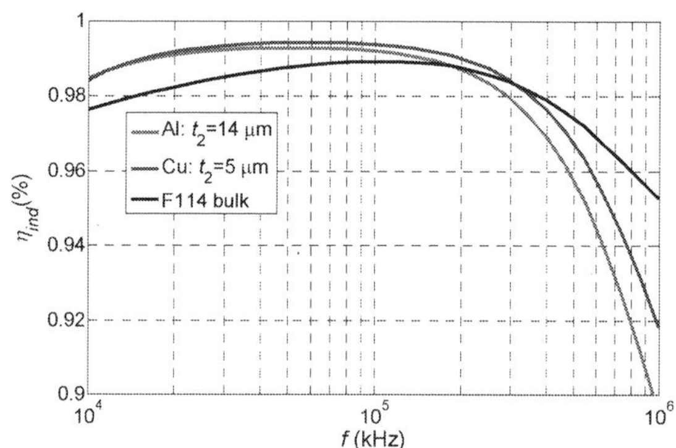

Fig. 7. Comparison of calculated efficiencies for different loads.

V. Conclusion

The suitability of non-magnetic metallic thin layers as loads for domestic induction heating applications is investigated in this paper, with promising results. The inductive performance is characterized by the equivalent resistance of a loaded spiral coil and the induction efficiency. These parameters are obtained by means of an analytical model. The equivalent resistance is also measured and it is found that calculated and measured results shows good agreement. The results are applied to estimate the hypothetical induction efficiency of different conventional-size thin-layer loads. Calculations show that values higher than the obtained with the conventional ferromagnetic loads could be achieved. The results also show that an optimization of the thickness of the layer could be carried out to achieve the highest efficiency at a specific frequency range.

Acknowledgment

This work was partly supported by the Spanish MINECO under Project TEC2013-42937-R, Project CSD2009-00046, and Project RTC-2014-1847-6, by the DGA-FSE, by the

University of Zaragoza under Project JIUZ-2014-TEC-08, and by the BSH Home Appliances Group.

REFERENCES

[1] H. Sarnago, O. Lucia, A. Mediano, and J. M. Burdio, "Efficient and cost-effective ZCS direct ac-ac resonant converter for induction heating," *IEEE Transactions on Industrial Electronics*, vol. 61, pp. 2546-2555, May 2014.

[2] H. Sarnago, O. Lucia, A. Mediano, and J. M. Burdio, "Direct AC–AC resonant boost converter for efficient domestic induction heating applications," *IEEE Transactions on Power Electronics*, vol. 29, pp. 1128-1139, Mar. 2014.

[3] O. Jimenez, O. Lucia, I. Urriza, L. A. Barragan, and D. Navarro, "Design and evaluation of a low-cost high-performance sigma-delta ADC for embedded control systems in induction heating appliances," *IEEE Transactions on Industrial Electronics*, vol. 61, pp. 2601-2611, May 2014.

[4] O. Jimenez, O. Lucia, I. Urriza, L. A. Barragan, and D. Navarro, "Power measurement for resonant power converters applied to induction heating applications," *IEEE Transactions on Power Electronics*, vol. 29, pp. 6779-6788, 2014.

[5] J. Acero, C. Carretero, R. Alonso, and J. M. Burdio, "Quantitative evaluation of induction efficiency in domestic induction heating applications," *IEEE Transactions on Magnetics*, vol. 49, pp. 1382-1389, Apr. 2013.

[6] J. Acero, C. Carretero, I. Lope, R. Alonso, and J. M. Burdio, "FEA-based model of elliptic coils of rectangular cross section," *IEEE Transactions on Magnetics*, vol. 50, pp. 1-7, 2014.

[7] C. Carretero, O. Lucia, J. Acero, and J. M. Burdio, "Computational modeling of two partly coupled coils supplied by a double half-bridge resonant inverter for induction heating appliances," *IEEE Transactions on Industrial Electronics*, vol. 60, pp. 3092-3105, Aug. 2013.

[8] I. Lope, J. Acero, J. M. Burdio, C. Carretero, and R. Alonso, "Design and implementation of PCB inductors with litz-wire structure for conventional-size large-signal domestic induction heating applications," *IEEE Transactions on Industry Applications*, vol. 51, pp. 2434-2442, Jul. 2015.

[9] H. N. Pham, H. Fujita, K. Ozaki, and N. Uchida, "Dynamic analysis and control for resonant currents in a zone-control induction heating system," *IEEE Transactions on Power Electronics*, vol. 28, pp. 1297-1307, Mar. 2013.

[10] J. Acero, I. Lope, J. M. Burdio, C. Carretero, and R. Alonso, "Performance evaluation of graphite thin slabs for induction heating domestic applications," *IEEE Transactions on Industry Applications*, vol. 51, pp. 2398-2404, May-Jun. 2015.

[11] J. Acero, R. Alonso, J. M. Burdio, L. A. Barragan, and D. Puyal, "Analytical equivalent impedance for a planar circular induction heating system," *IEEE Transactions on Magnetics*, vol. 42, pp. 84-86, Jan. 2006.

[12] W. G. Hurley and J. G. Kassakian, "Induction heating of circular ferromagnetic plates," *IEEE Transactions on Magnetics*, vol. 15, pp. 1174-1181, Jul. 1979.

[13] C. Carretero, R. Alonso, J. Acero, and J. M. Burdio, "Coupling impedance between planar coils inside a layered media," *Progress In Electromagnetics Research, PIER*, vol. 112, pp. 381-396, Jan. 2011.

[14] M. Abramowitz and I. A. Stegun, Handbook of Mathematical Functions: With Formulas, Graphs, and Mathematical Tables. Washington, D.C.: U.S. Dept. of Commerce, 1970.

[15] H. K. Pulker, Coatings on Glass. Amsterdam, The Netherlands: Elsevier Science B.V., 1999.

A New Evaluation Circuit with a Low-Voltage Inverter Intended for Capacitors Used in a High-Power Three-Phase Inverter

Kazunori Hasegawa
Department of Biological Functions and Engineering
Kyushu Institute of Technology
Kitakyushu, Fukuoka, Japan

Ichiro Omura
Department of Electrical Engineering and Electronics
Kyushu Institute of Technology
Kitakyushu, Fukuoka, Japan

Shin-ichi Nishizawa
National Institute of Advanced Industrial Science and Technology
Tsukuba, Ibaraki, Japan
Department of Biological Functions and Engineering
Kyushu Institute of Technology
Kitakyushu, Fukuoka, Japan

***Abstract*— DC-link capacitors in power electronic converters are a major constraint on improvement of power density as well as reliability. Evaluation of the dc-link capacitors in terms of power loss, ageing, and failure rate will play an important role in design stages of the next-generation power converters. This paper proposes a new evaluation circuit for dc-link capacitors used in a high-power three-phase inverter, which is intended for testing power loss, failure rate, ageing, and so on. The evaluation circuit produces a practical ripple current waveform and a dc bias voltage into a capacitor under test, in which the ripple current is equivalent to that generated by the three-phase inverter on the dc link. The evaluation circuit employs a full-scale current-rating and downscaled voltage-rating inverter for producing the ripple current, so that the power rating of the evaluation circuit is much smaller than that of a full-scale current rating and full-scale voltage rating inverter.**

Keywords—DC-link capacitors, high-power density, reliability, three-phase inverters.

I. Introduction

Power density of power electronic converters are continuously becoming higher and higher as their market size is getting larger and larger, which is accompanied by smaller power loss, lower volume, and lower weight in the converter. The market size growth also requires improvement of reliability not only in power semiconductor devices but also in passive components [1]. Hence, the next-generation power converters will be designed with managing both power density and reliability.

DC-link capacitors in power electronic converters are a major constraint on improvement of power density [2]. They tend to include a design margin of size or capacitance due to power loss. Thus, the minimum design margin of the capacitors is desirable, which should be considered in design stages of the converters. Furthermore, a lifetime of the capacitors is usually shorter than that of semiconductor devices or magnetic devices, which would degrade reliability of the power converters. Evaluation of the capacitors in terms of power loss, ageing, and failure rate will play an important role in design stages of the next-generation power converters [3-12]. However, characteristics of the capacitors are usually evaluated by a single sinusoidal current such as 120 Hz, 1 kHz, and so on [7, 8, 12-14]. There are some kinds of "ripple current tester" instruments that provide a sinusoidal ripple current as well as a dc-bias voltage into the capacitor [14]. Actual current flowing out of the converter into the capacitor

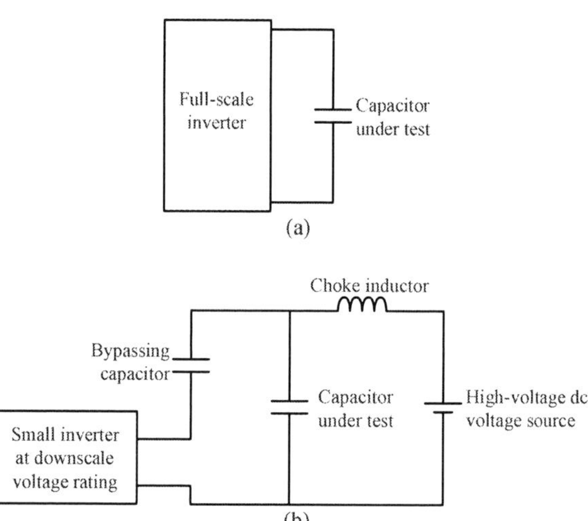

Fig. 1 Basic concepts of an evaluation circuit for a high-power capacitor. (a) Using full-scale inverter. (b) Using full-scale current-rating and downscale voltage-rating inverter.

978-1-4673-9551-9/16 $31.00 © 2016 IEEE

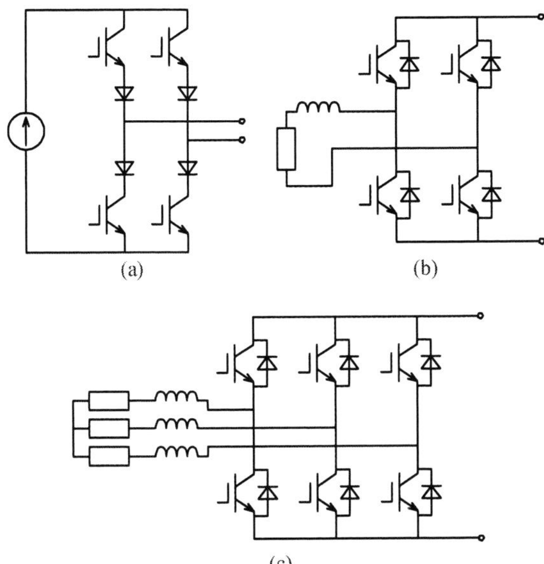

Fig. 2 Possible configurations of the small inverter. (a) Single-phase CSI. (b) Single-phase VSI. (c)Three-phase VSI

contains multiple frequency components [15], so that characteristics of the capacitors cannot be exactly estimated. Although the so-called fast Fourier transform (FFT) can extract the multiple frequency components from the actual current, a power loss of the capacitor cannot be estimated using the multiple frequency components because power loss in general has a nonlinear characteristic. In addition, the dc bias voltage across the capacitors affects power loss and ageing [1, 5, 11, 12]. Thus, existing power converters often employ more capacitors than necessary.

It is important to develop an evaluation circuit for component testing of capacitors, which will be utilized in design stages or in tests before shipment of the converters. Note that the system should behave as an existing inverter in terms of the dc bias voltage and ripple current waveform of the capacitor.

This paper proposes a new evaluation circuit with a low-voltage inverter intended for dc-link capacitors used in a high-power three-phase inverter, which presents a practical ripple current waveform that is equivalent to that of the existing inverter. The circuit will be utilized for evaluating the capacitor by electric or thermal measurement such as the followings:

1. Electrical measurement of ESR and capacitance [6, 9].
2. Power loss measurement by electrical or calorimetric measurement [10, 16].
3. Accelerated aging [12].

Although this paper does not pay attention to measuring characteristics of the capacitor under test, it just focuses on circuit configurations and design of the evaluation circuit. In addition, this paper discusses power rating of the evaluation circuit.

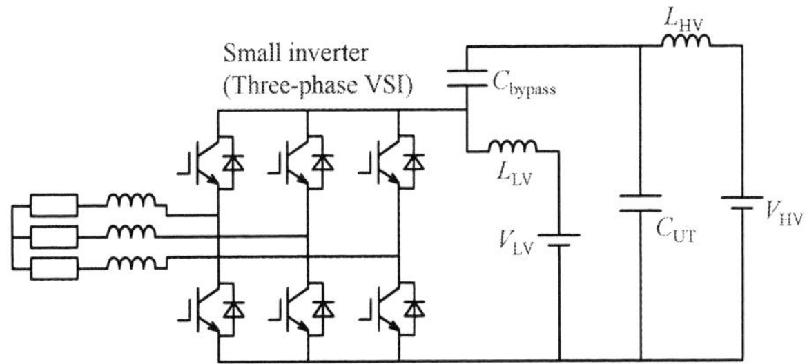

Fig. 3 Proposed evaluation circuit using a three-phase VSI.

II. Basic Concept

A. Evaluation circuit for capacitors

The most effective way to evaluate dc-link capacitors is measuring their characteristics with an existing converter in operation. For example, references [4, 6, 9] discuss real-time monitoring for capacitor condition using equivalent series resistance (ESR) and capacitance. Therefore, a basic idea of the evaluation circuit would utilize an existing power-rating inverter. Fig. 1 (a) shows the basic idea of the evaluation circuit, in which a full-scale current-rating and full-scale voltage-rating inverter is connected to a capacitor under test, CUT. The inverter provides a practical ripple current and dc bias voltage for the capacitor.

There are some special circuits that evaluate the capacitors [8, 12]. Reference [8] presents a simple circuit to evaluate an electrolytic capacitor, which consists of a combination of a dc-voltage supply providing a dc bias voltage, and a line-frequency transformer injecting a sinusoidal ripple current. The circuit is useful for estimating the capacitance and equivalent series resistance (ESR). Reference [12] presents a test circuit for accelerated aging of metalized film capacitors, which consists of a combination of a resonant inverter producing a sinusoidal ripple current and a dc voltage supply providing a dc bias voltage. Although both the two combinations contribute to reducing the overall power rating of the evaluation circuit, the ripple current waveform is different from the practical one generated by the inverter.

Fig. 1 (b) shows the basic concept of the proposed evaluation circuit that employs a small inverter, the capacitor under test, a bypassing capacitor, a choke inductor, and a high-voltage dc supply. The concept is similar to the circuits proposed in [8] and [12] in terms of the combination of a ripple current source and a dc voltage supply, whereas it presents a practical ripple current waveform, i.e., the same current waveform as that generated by the inverter. Current rating of the small inverter is full-scale, while voltage rating of that is downscale. The high-voltage dc supply keeps the capacitor voltage a desired dc bias voltage. The bypassing capacitor is used for circulating the ripple current generated by the inverter through the capacitor under test. The choke inductor is used for blocking the ripple current, through which only dc current flows. Hence, the proposed circuit operates as a full-scale voltage-rating and full-scale current-rating inverter from the standpoint of the dc bias voltage and ripple current. Thus,

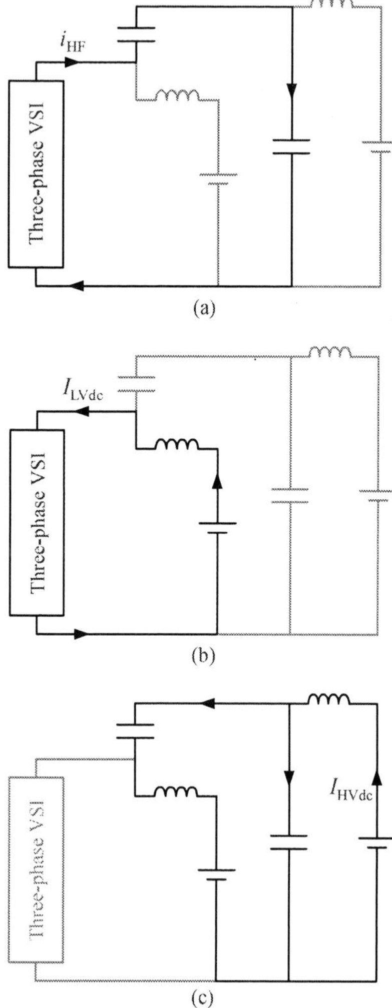

Fig. 4 Current paths of the proposed circuit. (a) AC current flowing out of the VSI. (b) DC current flowing out of the low-voltage dc supply. (c) DC current flowing out of the high-voltage dc supply.

power-rating of the small inverter is much smaller than that of the full-scale inverter.

B. Possible configurations of the small inverter

Candidates for the small inverter can be classified into the followings:

1. Single-phase current-source inverter (CSI) (Fig. 2 (a))
2. Single-phase voltage-source inverter (VSI) (Fig. 2 (b))
3. Three-phase voltage-source inverter (VSI) (Fig. 2 (c))

The evaluation circuit using the single-phase CSI behaves as a full-scale three-phase or single-phase inverter if it can provide the same ripple current waveform as the current generated by the full-scale single-phase or three-phase inverter, respectively. However, not only pulse width but also amplitude should be modulated to synthesize the ripple current waveform. In practice, therefore, quite complex control would be required for the CSI.

The single-phase VSI can be used for an evaluation circuit for the full-scale single-phase inverter. DC-link terminal of the VSI is connected to the bypassing capacitor and capacitor under test, so that the voltage across the bypassing capacitor is slightly lower than that across the capacitor under test by the dc-link voltage. Since instantaneous power in a single-phase circuit fluctuates at double the fundamental frequency, a ripple amplitude of the dc-link voltage in the single-phase VSI tends to be large. It will be a constraint on the voltage rating of the VSI because the dc-link voltage should be larger than the ripple amplitude.

The three-phase VSI is a candidate for an evaluation circuit for the full-scale three-phase inverter. DC-link terminal of the three-phase VSI is connected to the bypassing capacitor and capacitor under test like the single-phase VSI. On the other hand, a ripple amplitude of the dc-link voltage in the three-phase VSI is much lower than that in the single-phase VSI because the instantaneous power of the three-phase inverter is constant in a steady state [17].

This paper deals with the three-phase VSI as the small inverter because of the following reasons:

1. The three-phase inverter is widely used.
2. Lower dc-link voltage allows smaller power rating.
3. General-purpose inverters are available as the small inverter.
4. No special control is required to the ripple current

Note that the three-phase VSI has only to control its output voltage, i.e., open-loop PWM control is applicable.

III. Proposed Circuit Configuration

A. Circuit Configuration

Fig. 3 shows the proposed evaluation circuit consisting of the three-phase VSI, a low-voltage dc supply V_{LV} with a choke inductor L_{LV}, the high-voltage dc supply V_{HV} with a choke inductor L_{HV}, the bypassing capacitor C_{bypass}, and the capacitor under test, C_{UT}. The low-voltage dc supply is used for driving the three-phase VSI. The high-voltage dc supply provides a dc bias voltage to C_{UT}.

Fig. 4 shows current paths flowing out of the two dc voltage supplies and the three-phase VSI. The high-frequency ripple current generated by the VSI, i_{HF} circulates through C_{bypass} and C_{UT} because it does not flow into the two choke inductors L_{LV} and L_{HV} as shown in Fig. 4(a). The low-voltage dc supply V_{LV} provides a dc current I_{LVdc} to the VSI through L_{LV} as shown in Fig. 4(b). The high-voltage dc supply charges C_{UT} and C_{bypass} to their operating voltages, and then supplies a small amount of leakage dc current flowing into the capacitors, I_{HVdc} as shown in Fig. 4(c). Hence, the power rating of the high-voltage dc supply is quite small.

B. Power rating of the small inverter

Since the current rating of the small inverter is the same as that of the full-scale inverter, the relation between the power

Table I Rating and circuit parameters of the proposed circuit used in experiment.

Power rating of the system	P	6400 VA
Power rating of the inverter	P_{inv}	640 VA
AC current rating	I_O	5 A
AC voltage rating	V_O	73 V
Low-voltage dc source	V_{LV}	120 V
High-voltage dc source	V_{HV}	1200 V
Load inductor	L_O	27 mH (100%)
Load resistor	R_O	80 mΩ (1%)
Switching frequency	f_{SW}	6.1 kHz
Output frequency	f_O	50 Hz
High-voltage choke inductor	L_{HV}	10 mH
Low-voltage choke inductor	L_{LV}	10 mH
Capacitor under test	C_{UT}	320 µF
Unit capacitance constant of the capacitor under test [22]	H	36 ms
Bypassing capacitor	C_{bypass}	320 µF
Damping resistor	R_{damp}	3.3 Ω

() is based on 640 VA, 73 V, 5 A, and 50 Hz

Table II Rating and circuit parameters of the full-scale inverter used in simulation.

Power rating	P	6400 VA
AC current rating	I_O	5 A
AC voltage rating	V_O	730 V
DC link voltage	V_{dc}	1.2 kV
Switching frequency	f_{SW}	6.1 kHz
Output frequency	f_O	50 Hz
Load inductor	L_O	270 mH (100%)
Load resistor	R_O	780 mΩ (1%)

() is based on 6400 VA, 730 V, 5A, and 50 Hz

rating of the small inverter, P_{small} and that of the full-scale inverter, P_{FS} is given by

$$\frac{P_{small}}{P_{FS}} = \frac{V_{DClink-S}}{V_{DClink-FS}} = \frac{V_{LV}}{V_{HV}}, \tag{1}$$

where $V_{DClink-FS}$ and $V_{DClink-S}$ are dc-link voltages of the full-scale inverter and the small inverter, respectively, and P_{FS} is the power rating of the full-scale inverter. As the dc-link voltage of the small inverter, $V_{DClink-S}$ contains ripple voltages of C_{UT} and C_{bypass}, the dc mean of $V_{DClink-S}$ should be designed to be more than the ripple voltages. The low-voltage dc supply should provide the dc mean voltage according to the ripple voltages. Since the instantaneous power of the three-phase inverter is constant, dc-link voltage of the three-phase inverter contains only switching ripple component. Thus, one can pay attention to the switching frequency, the current rating, and capacitance of C_{UT}. In practice, however, attention should also be paid to imbalance of the three-phase load of the inverter because it would cause ripple voltage on the dc link. This paper introduces a condition that the dc-link voltage of the small inverter is 1/20 to 1/10 of that of the full-scale inverter.

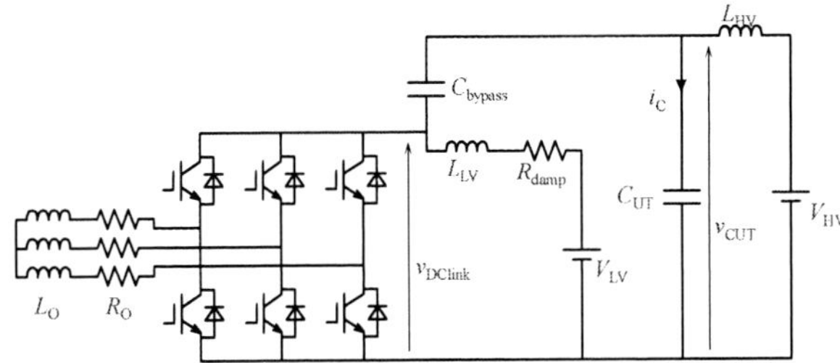

Fig. 5 Experimental circuit configurations of the proposed circuit.

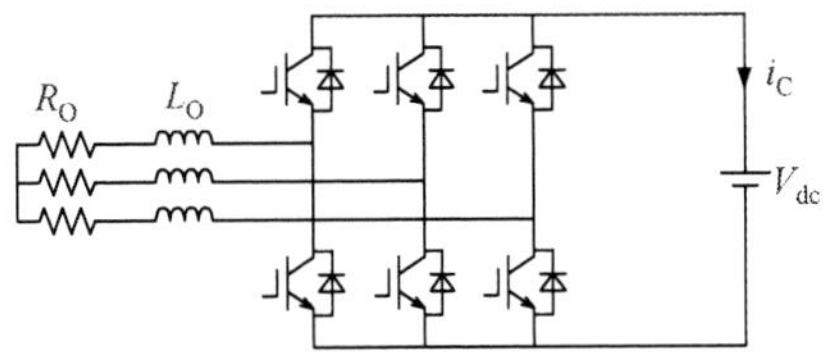

Fig. 6 Full-scale three-phase inverter used for simulation.

C. Design of Choke Inductors

Reactances of the two choke inductors should be much larger than that of capacitors C_{UT} and C_{bypass}.

$$\omega L_{choke} \gg \frac{1}{\omega C_{dc}}, \tag{2}$$

where $\omega = 2\pi f_{sw}$, f_{sw} is the switching frequency of the inverter, L_{choke} indicates a choke inductor of L_{HV} or L_{LV}, and C_{dc} stands for C_{UT} or C_{bypass}. The current rating of L_{choke} is determined by the leakage dc current of C_{dc}. Since volume of inductor is in proportion to 3/4th power to the maximum stored energy of $^1/_2 LI^2$ [18], the volumes of the choke inductors are quite small.

D. Example of practical use

Low-voltage (200 or 400 V) general-purpose inverters are available to the small inverter, so that medium-voltage capacitors are suitable for the proposed circuit. The current rating of the general-purpose inverters are up to 1000 A [19, 20]. Thus, the proposed circuit using the general-purpose inverter can evaluate capacitors used in a high-power inverter with power rating up to 1-10 MVA.

IV. EXPERIMENT AND SIMULATION

This section presents experimental results of the proposed circuit with comparing to the full-scale three-phase inverter by simulation, where a software package of the "PLECS" is carried out [21].

A. Circuit configurations

Fig. 5 illustrates experimental circuit configuration of the proposed evaluation circuit. Sinusoidal pulse-width modulation (SPWM) is applied to the inverter. Table I summarizes ratings and circuit parameters of the experiment. The damping resistor R_{damp} prevents an oscillation caused by the capacitors and the choke inductors. The power

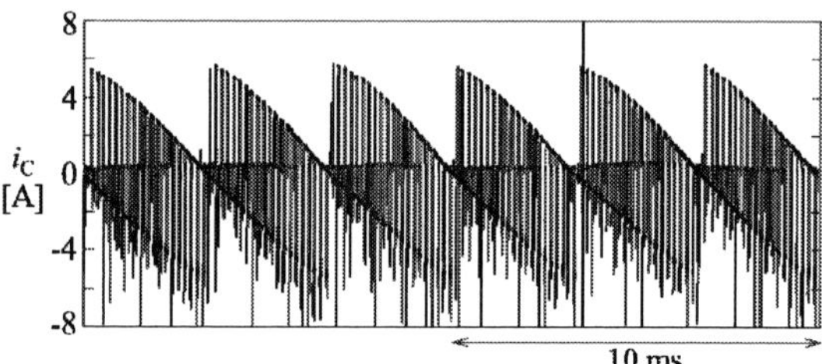

(a)Proposed circuit by experiment.

(b)Full-scale inverter by simulation.

Fig. 7 Current waveforms flowing into the capacitor under test.

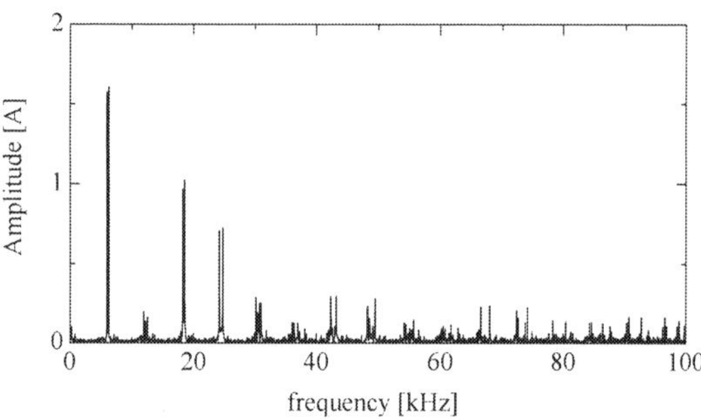

(a)Proposed circuit by experiment.

(b)Full-scale inverter by simulation.

Fig. 8 FFT results of Current waveforms flowing into the capacitor under test.

consumption of the damping resistor is only 0.1% of the supplied power from the high-voltage dc supply. Note that the proposed circuit can also supply a ripple current in case the load power factor is changed although the low-voltage dc supply has to provide an amount of active power to the small inverter.

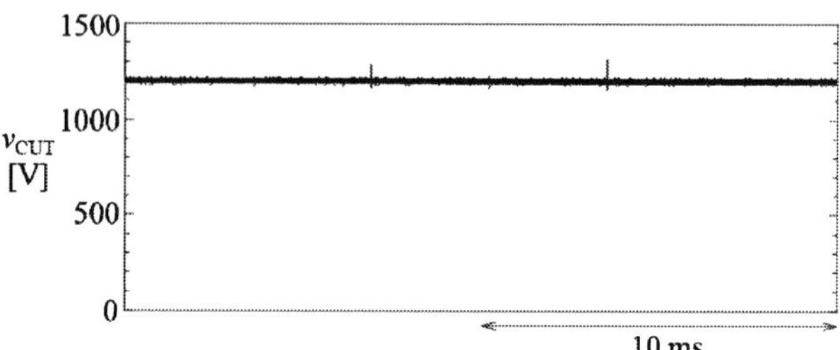

Fig. 9 Experimental waveform of the voltage across C_{UT}, v_{CUT}.

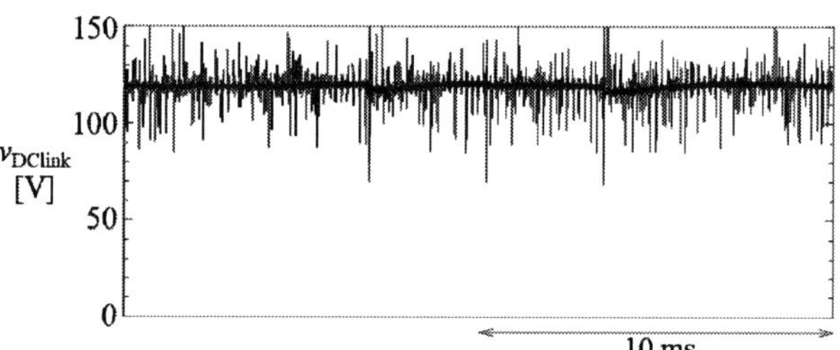

Fig. 10 Experimental waveform of the voltage across C_{UT}, v_{CUT}.

Fig. 6 shows the full-scale three-phase inverter for simulation, where a dc voltage source is connected instead of the capacitor under test on the dc link. Table II summarizes the ratings and circuit parameters of the full-scale inverter.

B. Results

Fig. 7 shows waveforms of the capacitor current i_C. The waveforms of the proposed circuit almost agree with that of the full-scale inverter. In addition, fig. 8 illustrates FFT results of the capacitor current, in which the FFT result of the proposed circuit almost agree with that of the full-scale inverter.

Fig. 8 shows the voltage across the capacitor under test, v_{CUT} in the proposed circuit. They stay at 1.2 kV that is the dc bias voltage applied by the high-voltage dc supply. Fig. 9 shows the dc-link voltages of the small inverter in the proposed circuit, v_{dclink}, which looks a constant dc current except for switching ripple components.

CONCLUSION

This paper has proposed a new evaluation circuit for dc-link capacitors used in a high-power three-phase inverter. The proposed circuit is characterized by combining a small three-phase inverter and a small voltage supply, where an existing low-voltage inverter would be applicable to the small inverter.

Experimental results obtained from a 1200-V 6.4-kVA prototype verifies that the evaluation circuit produces a practical ripple current waveform and a dc bias voltage into a capacitor under test, in which the ripple current is almost equivalent to that generated by the full-scale inverter. Moreover, the results confirms that the power rating of the evaluation circuit can be less than 1/10 of that of the full-scale inverter.

REFERENCES

[1] H. Wang and F. Blaabjerg, "Reliability of capacitors for dc-ink applications in power electronic converters—an overview," *IEEE Trans. Ind. Appl.* vol. 50, no. 5, pp. 3569-3578, 2014.

[2] J. W. Kolar, U. Drofenik, J. Biela, M. Heldwein, H. Ertl, T. Friedli, and S. Round, "PWM converter power density barriers," *IEE Japan Trans. Ind. Appl.*, vol. 128, no. 4, pp.468-480, 2008.

[3] K. Harada, A. Katsuki, and M. Fujiwara, "Use of ESR for deterioration diagnosis of electrolytic capacitor," *IEEE Trans. Power Electron.*, vol. 8, no. 4, pp. 355-361, Oct. 1993.

[4] P. Venet, F. Perisse, M. H. El-Husseini, and G. Rojat, "Realization of a smart electrolytic capacitor circuit," *IEEE Ind. Appl. Mag.*, vol. 8, no. 1,pp. 16–20, Jan./Feb. 2002.

[5] O. Ondel, E. Boutleux, and P. Venet, "A decision system for electrolytic capacitors diagnosis," *in Proc. of IEEE Power Electronics Specialist Conference (PESC)*, pp. 4360-4364, 2004.

[6] E. C. Aeloiza, J. H. Kim, P. Ruminot, and P. N. Enjeti, "A Real Time Method to Estimate Electrolytic Capacitor Condition in PWM Adjustable Speed Drives and Uninterruptible Power Supplies," *IEEE Power Electronics Specialists Conference (PESC)*, pp. 2867-2872, 2005.

[7] A. M. R. Amaral, and A. J. M. Cardoso, "A Simple Offline Technique for Evaluating the Condition of Aluminum–Electrolytic–Capacitors," *IEEE Trans. Ind. Electron.*, vol. 56, no. 8, pp. 3230-3237, Aug. 2009.

[8] A. M. R. Amaral, and A. J. M. Cardoso, "Estimating aluminum electrolytic capacitors condition using a low frequency transformer together with a dc power supply," *in Proc. of IEEE ISIE*, pp. 815-820, 2010.

[9] K. Abdennadher, P. Venet, G. Rojat, J. M. Retif, and C. Rosset, "A Real-Time Predictive-Maintenance System of Aluminum Electrolytic Capacitors Used in Uninterrupted Power Supplies," *IEEE Trans. Ind. Appl.* vol. 46, no. 4, pp. 1644-1652, Jul./Aug., 2010.

[10] J. M. Miller, C. W. Ayers, L. E. Seiber, and D. B. Smith, "Calorimeter evaluation of inverter grade metalized film capacitor ESR," *in Proc. of IEEE ECCE*, pp. 2157-2163, 2012.

[11] M. Makdessi, A. Sari, and P. Venet, "Metallized polymer film capacitors ageing law based on capacitance degradation," *ELSEVIER Microelectronics Reliability*, vol. 54, pp. 1823-1827, 2014.

[12] M. Makdessi, A. Sari, P. Venet, P. Bevilacqua, and C. Joubert, "Accelerated Ageing of Metallized Film Capacitors Under High Ripple Currents Combined With a DC Voltage," *IEEE Trans. Power Electron.*, vol. 30, no. 5, pp. 2435-2444, May 2015.

[13] NICHICON CORPORATION. "General Description of Aluminum Electrolytic Capacitors," [Online]. available: http://www.nichicon.co.jp/english/products/pdf/aluminum.pdf

[14] *RIPPLE CURRENT TESTER MODEL 11800/11801/11810*, Chroma ATE Inc. 2014. [Online]. available: http://www.chromaate.com/File/DownLoad/42014

[15] B. P. McGrath and D. G. Holmes, "A general analytical method for calculating inverter DC-link current harmonics," *IEEE Trans. Ind. Appl.* vol. 45, no. 5, Sep./Oct. 2009

[16] D. Christen, U. Badstuebner, J. Biela, and J.W. Kolar, "Calorimetric power loss measurement for highly efficient converters," *in Conf. Rec. of International Power Electronics Conference (IPEC)*, pp. 1438-1445, 2010.

[17] H. Akagi, Y. Kanazawa, and A. Nabae, "Instantaneous reactive power compensators comprising switching devices without energy storage components," *IEEE Trans. Ind. Appl.*, vol. IA-20, no. 3, pp. 625-630, 1984.

[18] W. T. McLyman, *Transformer and inductor design handbook*. New York: Marcel Dekker, 1988.

[19] *High Performance Vector Control Inverter FRENIC-VG Series*, Fuji Electric Co., Ltd., Tokyo, Japan, 2014. [Online]. available: https://felib.fujielectric.co.jp/download/index.htm?site=global&lang=en

[20] *F700 Series Brochure*, Mitsubishi Electric Automation, Inc. 2011. [Online]. available: https://us.mitsubishielectric.com/fa/en/support/technical-support/knowledge-base/getdocument/?docid=3E26SJWH3ZZR-38-1348

[21] [Online]. available: http://www.plexim.com/plecs

[22] H. Fujita, S. Tominaga, and H. Akagi, "Analysis and design of a dc voltage-controlled static var compensator using quad-series voltage-source inverters," *IEEE Trans. Ind. Appl.*, vol. 32, no. 4, pp. 970--977, 1996.

978-1-4673-9551-9/16 $31.00 © 2016 IEEE

Energy Absorption Capability of Low Voltage Metal Oxide Varistors in AC and Impulse Currents

Dawood Talebi Khanmiri [1] *, Roy Ball [2], Craig Mckenzie [2], Brad Lehman [1]

[1] Electrical & Computer Engineering Department
Northeastern University
Boston, MA, USA
Talebikhanmiri.d@husky.neu.edu

[2] Mersen USA Newburyport LLC
Newburyport, MA, USA

***Abstract*— Metal Oxide Varistors (MOVs) are widely used surge protective components in electronics, and their reliability is important in reliability of the whole system. In this paper, Energy Absorption Capability (EAC) of low voltage MOVs is characterized and measured by experimental tests in a UL-certified lab. It is demonstrated that EAC curve for the low voltage MOV is different than that of high voltage station class MOV. Failure modes of MOVs are identified and presented. Furthermore, causes and consequences of different behavior are discussed.**

Keywords— Surge Protective Device (SPD); Energy Absorption Capability; Metal Oxide Varistor (MOV); Lightning Protection

I. Introduction

External surges and overvoltage transients, resulting from lightning, utility switching, and fault clearing, appear on the utility entrance of the load. However, 70% to 80% [1] of the surges are actually caused by switching surges that are generated internally from load switching, such as air conditioner or electric motor turn-on. These surges and overvoltages are responsible for many failures of power electronic power supplies, particularly in home appliances, consumer electronics, automotive electronics, and industrial control and automation systems. Protection of these sensitive electronic circuits is critical to investors and consumers.

The continued trend of miniaturization and use of lower voltages in electronics have made electronic circuits and devices more susceptible to overvoltage transients and surges [2-5]. Smaller components with lower energy capacity, closer traces and dense circuits have triggered attention to overvoltage transients that might not be seen before. Furthermore, widespread inclusion of electronics in industrial control and automation has made their protection more important. For example, use of electronics in automotive industry has encountered problems of relatively unstable supply voltage and interference from the vehicle's ignition system [6]. Surge protection becomes indispensable, in particular, in order to prevent voltage disturbances in the control circuit to cause costly consequences, such as lost productivity, increased downtime, expensive repairs, and unsatisfied customers.

Metal Oxide Varistors (MOVs) are highly nonlinear components that are used to protect the sensitive circuits against surges and transient overvoltages. An MOV is placed in parallel with the load and diverts the surge energy away by providing a low impedance path to the ground for the surge current and maintains the voltage within acceptable limits. Recently, new applications of MOVs have also emerged, such as energy absorbing element in solid state breakers [7]. Capability of handling the surge energy has a primary role in selecting the MOV and protection design.

MOVs fail in four different ways [8-15]: thermal runaway, puncture, cracking, and flashover. Thermal runaway occurs when the generated heat by an increased leakage current rises beyond the heat dissipation capability of the MOV. This excess of heat would increase the MOV temperature and consequently, its leakage current. Increased leakage current would generate more heat and would rise the temperature again. This positive feedback could lead to rapid increase in the MOV's temperature and a thermal runaway. Puncture and cracking occur because of imperfect microstructure of the MOV. Non-uniformity of grain sizes and breakdown voltages of potential barriers at grain boundaries create heterogeneous current densities within the MOV. This current concentration into a few narrow paths heats up those areas more than adjacent areas and increases the conductivity of those hot spots and paths, leading to even a higher current density and increased locally generated heat. This heat creates a temperature gradient that causes thermal stresses between hot spots and neighbor portions, causing MOV block to crack, or increases the temperature of the hot spots up to the melting point of the ceramic, leading to a puncture failure. Usually, puncture occurs in low density currents, where there is enough time for the hot spots to form. On the other hand, cracking failure generally occurs when the current density is high and stresses become more than stability strength of the ceramic and crack the MOV before reaching to the melting point. Flashover is current concentration on the edge of the MOV, instead of a current flowing through it. In low-voltage varistors, usually an insulating coating covers the whole body of the MOV and prevents a flashover. Any failure of the coating due to high

*The author gratefully acknowledges the support through grant by Mersen USA.

Fig. 1. Samples in AC EAC test after the test

temperatures during discharge activities, humidity, pollution, or material defects can lead to the flashover failure in the MOV. It is shown in literature that flashover can occur in current well below the maximum currents that MOVs can handle.

Energy Absorption Capability (EAC) is the energy that an MOV can absorb before it fails. Previous studies [8, 16-17] on station-class high voltage MOVs have shown EAC of an MOV is dependent on its current density. Based on their results, EAC increases with increasing current density. For current densities around 3 A/cm^2 EAC ranges from 600 to 900 J/cm^3. For high density impulse currents the average EACs increase to 1700 J/cm^3 for some MOVs [17]. Tests in various current densities are carried out with different current waveforms, due to test equipment's limitations. The tests require several samples for each data point, because the behaviors of MOVs are slightly different from each other due to random positions of grains in their microstructure, even if they are acquired from one production batch.

Although MOV manufacturers might have EAC information from internal tests, these curves are not included in datasheets. Instead, a nominal energy is declared for the MOV, which usually is a conservatively calculated value and for only one current density. The nominal energy determines the energy that the MOV can absorb without changing its characteristics, and is used in the design. However, it does not give a proper measure to compare MOVs and does not provide any insight to their EAC, or possible changes due to degradation. This might lead to improper application, unrealistic concerns, and cost due to early replacements.

In this paper, we measure the EAC for low voltage MOVs in different current densities. The tests are conducted in a UL-certified lab. Results of low density and high density current tests are used to develop an EAC curve for the low voltage MOV. It is shown in this paper that low voltage MOVs have different EAC characteristics than station class MOVs. Probable causes of the different behavior is proposed and discussed in this research. Also, failure modes of the low voltage MOVs are identified and presented. The main contributions of this paper can be summarized as below:

- Energy Absorption Capability (EAC) tests for typical 150-volt MOVs in low density AC and high density impulse currents are conducted in a UL-certified lab and the results are reported. Normally, the outcomes of these tests are not public.
- It is shown that for studied low voltage MOV, EAC is decreasing when the current density increases. This is different than previously reported EAC for station-class MOVs.
- Probable causes of different behavior are discussed.
- Failure modes of low voltage MOVs are identified and presented. Puncture, cracking, and coating peeling are the main failure modes in AC, high impulse surge, and lower impulse surge currents, respectively.

II. Test Procedure and Results

A. AC Energy Absorption Capability

Energy Absorption Capability test in low current densities is performed in AC 60 Hz current. All samples are new MOVs. Relation between previous discharge activity and EAC of an MOV is investigated and presented in another paper [18]. The supply voltage of 240 volts is applied to 20 specimens of 150-volt MOV until their destruction. The short circuit current of the source is adjusted to be 5 kA, which is a standard Short Circuit Current Rating (SCCR) based on UL1449 edition 4 [19]. The energy is calculated by recording voltage and current waveforms until the moment of failure. The MOVs were in an enclosure during the test in order to avoid the heat dissipation to the environment. The results are summarized in Table I.

Comparison of the AC EAC results of 150-volt MOV with results from [17] shows that the EAC of low voltage 150-volt MOV is slightly higher than EACs of station-class MOVs. The average EAC per unit volume for 150-volt MOV is 954 J/cm^3 when the average test current density is 3.2 A/cm^2. High AC EAC of the low voltage MOV shows its high energy handling capability in temporary over voltage incidents.

When an overvoltage is applied to the MOV, it conducts excessive currents according to its current-voltage characteristics. However, the real behavior of the MOV is more complicated. The largest peak of the AC current always occurs in the first cycle. Then, the amplitude of the current decreases. During the test, temperature of the MOV increases until a failure occurs and current rapidly increases to the SCCR level. In the MOVs with longer Time-to-Failure (TtF) after the

TABLE I. AC Energy Absorption Capability for studied MOV

	Average	Max	Min	
AC EAC	**2624**	3141	**1901**	**J**
AC EAC / volume	**954**	1142	**691**	**J/cm^3**
RMS current	**35.8**	76.8	6.6	**A**
RMS current / area	**3.2**	6.8	0.6	**A/cm^2**

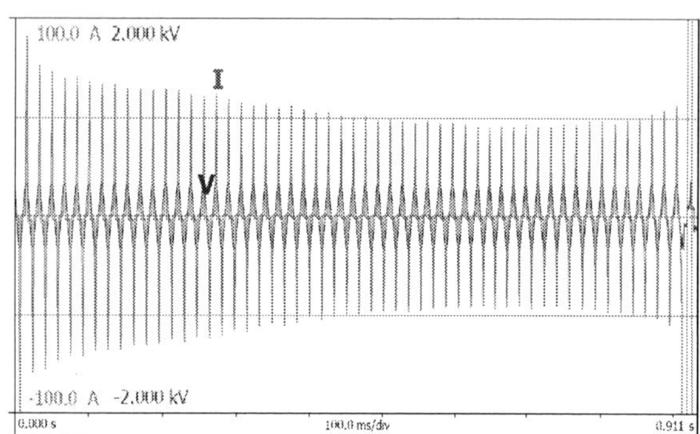

Fig. 2. AC EAC, 240 V, SCCR 5 kA

current peak reaches a minimum, it starts to increase again before the failure. A sample current and voltage waveforms are seen in the Fig. 2.

B. Impulse Energy Absorption Capability

For Impulse EAC test, the varistors are taken out of their case for the test and thermal disconnection mechanism is disabled. The surge generator is capable of delivering up to 135 kA, 8/20 μs current impulse for the above-mentioned low voltage MOV. The test starts with higher current amplitudes and decreases it in every step. If three MOVs tolerate the surges in a step, the test stops and the average of the energies of the three corresponding impulse current surges in that step is assumed as the EAC. If any of three MOV fails in a step, the energy level is decreased for the next step. Only new MOVs are used for the test and each MOV is subjected to only one surge.

The failure criteria are defined as below. All criteria should be met for a pass outcome.

1. No mechanical damage is seen.

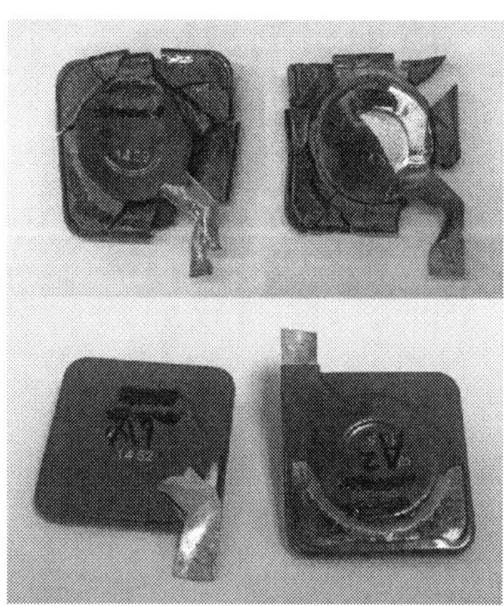

Fig. 3. Samples in AC EAC test after the test

2. Short circuit does not occur during or after the test.

3. Thermal runaway does not occur after the test, when Maximum Continuous Operating Voltage (MCOV) is applied to the MOV.

Small fractures alongside the strap terminals are not considered as a failure. Also, after the surge, specimens are allowed to cool down to room temperature (30 minutes) before the MCOV is applied. Based on the tests, the maximum surge current that the low voltage 150-volt MOV tolerates is 80 kA. The average energy absorbed in these three surges is 1578 Joules or 574 J/cm3, which is approximately one third of the values reported by references [8, 16-17] for station class MOVs. Table II summarizes the test results for nine MOVs used in this test.

Fig. 3 shows the test specimens after the test. From the test specimens (except A0 and A00 that exploded), only A1 showed increased leakage current after the test, which can lead to thermal runaway. Fig. 4 shows the applied current surges and the clamping voltage waveforms for the samples A7 and A0. In sample A0, it is shown that after almost 20 μs from the beginning of the surge, the MOV fails and an arc occurs, which causes fluctuations in the voltage.

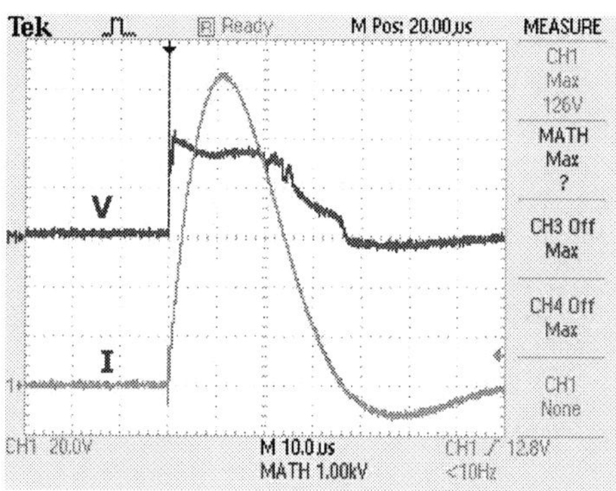

(a)

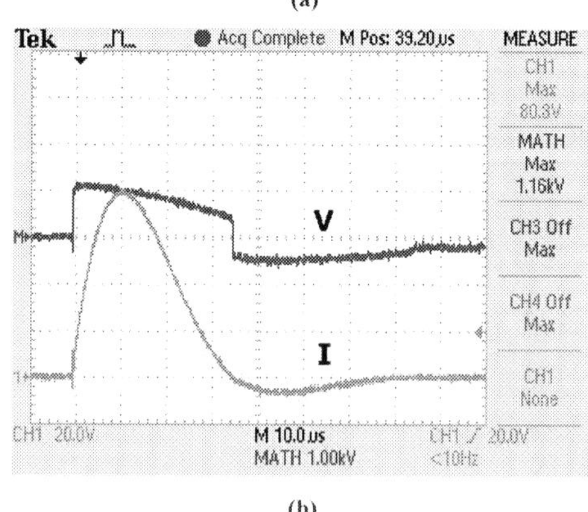

(b)

Fig. 4. Impulse current surges (8/20 μs) on samples (a) A0, 126 kA and (b) A7, 80.8 kA

TABLE II. IMPULSE EAC TEST RESULTS

Sample #	Surge Current [kA]	Energy [J]	Result
A00	133	4924	Failed
A0	126	4133	Failed
A1	100	2493	Failed
A2	90.4	1904	Passed
A3	90.4	-	Failed
A4	84	1800	Failed
A5	80.8	1577	Passed
A5 .	80.8	1607	Passed
A7	80.8	1549	Passed
EAC [J]		1578	
EAC [J/cm³]		574	

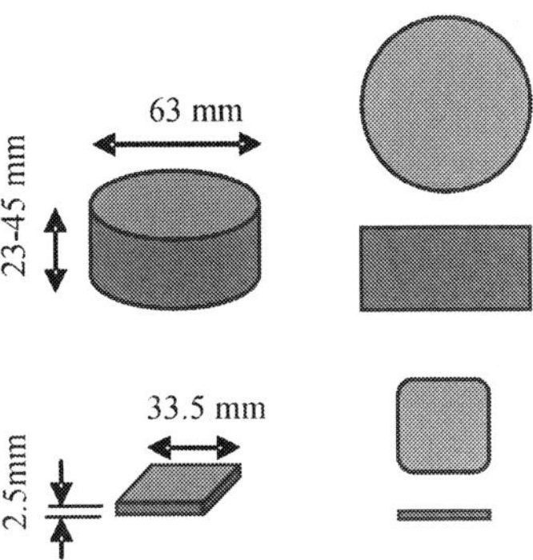

Fig. 5. Sizes of the station class varistors [10-11] (up) and studied low voltage MOVs (down)

III. DISCUSSION ON RESULTS AND FAILURE MODES

Analysis of the failure modes of the test samples provides more insight into the test results. All of failed samples are examined to determine the failure modes after the test.

The main failure modes in impulse tests are cracking and coating peeling. In impulses higher than 120 kA MOVs crack into several pieces. Usually, cracking occurs outside of the electrode area and a piece under the electrode does not crack. It seems that the heat generated in the electrode might be a contributing factor to this failure mode. In lower currents, the ceramic body does not crack and the main failure mode is coating peeling. Epoxy coating is peeled off mostly in one side, where a round electrode lies between coating and the ceramic body. Again, excessive heat, generated in the electrode is the main culprit. Also, the footprint of surface flashover is seen in some cases.

In low density AC current tests, the main failure modes are puncture and cracking. In samples with lower current densities, puncture is the only failure mode. When the test current is low, which is the case in MOVs with high varistor voltage (V_{1mA}, varistor voltage at 1 mA DC), TtF is long enough for current concentration and formation of a puncture failure. Puncture is seen in different areas (under, outside and also on the edge of the electrode) of the ceramic in different samples. In samples with higher AC currents, cracking is seen beside puncture. The cracking failure here is different than the cracking caused by impulse current and in most cases, splits the whole ceramic body into two or three pieces.

Non-uniformity of the microstructure of the MOV explains the failure modes. In high impulse surges, the current is concentrated into a few paths due to non-uniform microstructure, creating hot spots and paths. Thus, high temperature gradient between grains leads to ceramic cracking. However, in lower currents, hot spots reach the melting temperature of the ceramic, before it cracks, leaving a hole through the ceramic body. In surge currents that do not crack the MOV, the rapid heating of the electrode and its expansion may cause the epoxy coating to crack and leave the MOV vulnerable to surface flashovers.

IV. ENERGY ABSORPTION CAPABILITY CURVE

Energy Absorption Capability curve for an MOV shows the energy needed for its destruction or failure in different current densities. Previous studies on station class MOVs showed EAC increases with increasing current density. While the EAC tests in this paper (3.2 A/cm² and 7200 A/cm²) shows the EAC is decreasing. EAC curve is shown in Fig. 6.

Impulse Energy Absorption Capability (IEAC) of the tested low voltage MOVs compared to those of station class MOVs reported in [8, 17] is low. While station class varistors show IEACs of 1500 J/cm³ and higher, IEACs of our test samples are less than 600 J/cm³.

The shape and dimensions of an MOV can be effective factors in handling thermal stresses and formation of hot spots and paths. As it is seen in Fig. 5, the low voltage MOVs have larger area/volume ratios. Furthermore, unlike the station class MOVs of [17], the electrodes in the low voltage MOVs do not cover the entire ceramic surface. This contributes to uneven heating of the MOV surface and can create thermal stresses

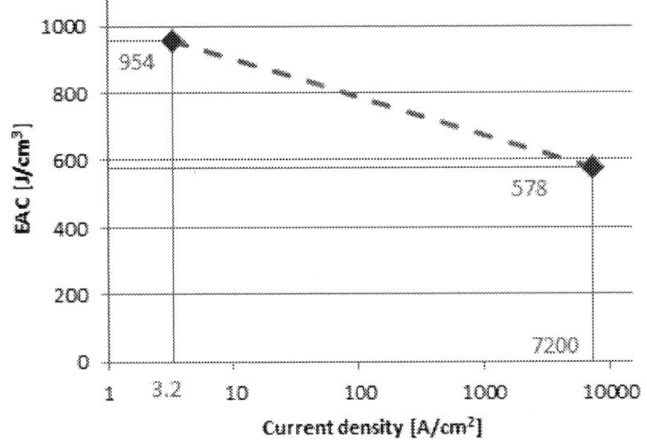

Fig. 6. Energy Absorption Capability curve for studied MOV

between portions of the MOV that is under the electrodes and the outer portions.

Electrodes, or the connection of metal terminals to the ceramic body, in the MOVs can be a source of failure. The impulse currents of high density can generate a large amount of heat and melt the material. Also, microscopic intrusion of the metal into ceramic can contribute to the puncture failure. Furthermore, lack of coordination in thermal expansion coefficients of ceramic body and metal termination plate can lead to ceramic body's cracking [20-21] and peeling off the coating.

Another important factor in low IEAC of a low voltage MOV is the coating. Impulse current surge happens in less than 100 μs and the heat generation in the MOV is almost instantaneous. Since the heating process is adiabatic during an impulse surge event, the epoxy coating is of no importance in temperature rise. However, the epoxy coating, as a susceptible part of the MOV, is damaged in all failed cases, and in that manner, contributed to the lower IEAC.

V. Conclusion

With growing trend of miniaturization, closer and thinner traces, smaller components, and use of lower voltages in electronics, increased susceptibility to overvoltage transients and surges has become a problem. Considering proliferation of electronic circuits in devices, automotive industry, communication systems, industry automation and critical safety and control systems, the need for more reliable surge protection devices are undeniable.

In this paper, the Energy Absorption Capability (EAC) for a low voltage 150-volt MOV is measured by experimental tests in high current densities by applying 8/20 μs impulse current surges and in low density currents by 60Hz AC current in a UL-certified lab. The results are used to develop the EAC curve for the low voltage MOV. The results show that the studied MOV has EAC of 954 J/cm^3 in AC currents, while the EAC decreases to 574 J/cm^3 in high density impulse currents. This is completely different than published EAC curves for station class MOVs in the literature which show an increasing tendency with increasing current density. Failure modes of samples are identified and discussed. Puncture, cracking, and coating peeling are the main failure modes in AC, high impulse surge, and lower impulse surge currents, respectively.

References

[1] Surge Protection Reference Guide, Emerson Network Power, 2011

[2] J. C. Gomez, M. M. Morcos, "Latin-American Experience about Appliance Damage Caused by System Perturbations," IEEE/PES Transmission & Distribution Conference and Exposition: Latin America, 2006.

[3] Chai Yajing, et al., "Test on Lightning Characteristics of Electronic Equipment's Power Supply," IEEE 2007 International Symposium on Microwave, Antenna, Propagation, and EMC Technologies for Wireless Communications, 2007.

[4] I. N. Gondim, et al., "Electronic Equipment Dielectric and Thermal Withstand Capability Curves for Refunding Analysis Purposes," 11th International Conference on Electrical Power Quality and Utilization (EPQU), 2011, pp. 271–350.

[5] Toshihisa Shimizu, et al., "A Motor Surge Voltage Suppression Method With Surge Energy Regeneration," *IEEE TRANSACTIONS ON POWER ELECTRONICS*, VOL. 27, NO. 7, JULY 2012.

[6] B. van Beneden, "Varistors: Ideal Solution to Surge Protection," Power Electronics Technology, May 2003, pp. 26-30.

[7] J. Magnusson, et al., "Separation of the Energy Absorption and Overvoltage Protection in Solid-State Breakers by the Use of Parallel Varistors," *IEEE TRANSACTIONS ON POWER ELECTRONICS*, VOL. 29, NO. 6, JUNE 2014.

[8] M. Bartkowiak, et al., "Failure Modes and Energy Absorption Capability of ZnO Varistors," *IEEE Transactions on Power Delivery*, Vol. 14, No. 1, January 1999

[9] Jinliang He, et al, "Discussions on Nonuniformity of Energy Absorption Capabilities of ZnO Varistors," *IEEE Transactions on Power Delivery*, VOL. 22, NO. 3, JULY 2007

[10] Eda, et al., "Destruction mechanism of ZnO varistors due to high currents," *J. Appl. Phys.* 56 (10),15 November 1984

[11] Mat Darveniza and Tapan K Saha, "SURFACE FLASHOVERS ON METALOXIDE VARISTOR BLOCKS," 1998 IEEE International Conference on Conduction and Breakdown in Solid Dielectrics, 1998, Sweden.

[12] Tsukamoto, et al., "Surge Withstand Capability of Metal Oxide Varistors for 10/350 us Waveform," 2011 International Symposium on Lightning Protection (XI SIPDA), Fortaleza, Brazil, October 3-7, 2011.

[13] Vojta, et al., "Microstructural origin of current localization and "puncture" failure in varistor ceramics," J. Appl. Phys. 81 (2), 15 January 1997.

[14] Arindam Paul, et al., "Evaluation and failure analysis of high-resistance collar coating of varistor," 2012 IEEE 10th International Conference on the Properties and Applications of Dielectric Materials July 24-28, Bangalore, India, 2012.

[15] Steven Boggs, et al., "Increased Energy Absorption in ZnO Arrester Elements Through Control of Electrode Edge Margin," IEEE TRANSACTIONS ON POWER DELIVERY, VOL. 15, NO. 2, APRIL 2000.

[16] K. G. Ringler, et al., "THE ENERGY ABSORPTION CAPABILITY AND TIME-TO-FAILURE OF VARISTORS USED IN STATION-CLASS METAL-OXIDE SURGE ARRESTERS," *IEEE Transactions on Power Delivery*, Vol. 12, No. 1, January 1997

[17] M.N. Tuczek, V. Hinrichsen, "Recent Experimental Findings on the Single and Multi-Impulse Energy Handling Capability of Metal–Oxide Varistors for Use in High-Voltage Surge Arresters," *IEEE Transactions on Power Delivery*, Vol. 20, No. 5, 2014.

[18] D. Talebi Khanmiri, R. Ball, J. Mosesian, B. Lehman, "Degradation of Low Voltage Metal Oxide Varistors in Power Supplies," 2016 IEEE Applied Power Electronics Conference and Exposition (APEC), in press.

[19] Standard for Surge Protective Devices, UL 1449 standard, 4th edition, 2014.

[20] Zhang Nanfa, et al., "Long Duration Impulse Withstand Capability of SPD," 2010 Asia-Pacific International Symposium on Electromagnetic Compatibility, April 12 -16, 2010, Beijing, China

[21] Steven Boggs, et al., "Increased Energy Absorption in ZnO Arrester Elements through Control of Electrode Edge Margin," *IEEE Transactions on Power Delivery*, VOL. 15, NO. 2, APRIL 2000.

Optimization and Experimental Validation of Medium-Frequency High Power Transformers in Solid-State Transformer Applications

M. A. Bahmani and T. Thiringer
Chalmers University of Technology
Gothenburg, Sweden
Email: bahmani@chalmers.se
torbjorn.thiringer@chalmers.se

M. Kharezy
SP Technical Research Institute of Sweden
Borås, Sweden
Email: Mohammad.Kharezy@sp.se

Abstract—**High power isolated DC-DC converters are likely to provide solutions for many technical challenges associated with power density, efficiency and reliability in potential applications such as offshore wind farms, inter-connection of DC grids, MVDC in data centers and in future solid state transformer applications. The high power medium frequency transformer (HPMFT) is one of the key elements of such a converter to realize the voltage adaption, isolation requirements, as well as high power density. This paper describes a design and optimization methodology taking into account the loss calculation, isolation requirements and thermal management. Incorporating this design methodology, an optimization process with a wide range of parameter variations is applied on a 50 kW, 1 / 3 kV, 5 kHz transformer to find the highest power density while the efficiency, isolation, thermal and leakage inductance requirements are all met. The optimized transformers are then manufactured and will be presented in this paper.**

I. Introduction

Nowadays, power electronic converters are considered as one of the enabling technologies that can address many technical challenges in future power grids from the generation phase to the transmission and consequently distribution at different voltage levels [1]. In contrast to the medium power converters (5 to 100 kW) which have been essentially investigated in automotive and traction applications, the megawatt and medium voltage range isolated converters with several kilohertz isolation stage are still mainly in an expansive research phase [2].

Under the scope of grid applications, one of the most cited terminologies for these kinds of high power converters is called solid-state transformer (SST) which is in fact an AC-AC or DC-DC high power converters whereby the voltage adaptation and high-frequency isolation, to reduce the weight and volume, are achieved. Medium-frequency or high-frequency power transformer (MFPT/HFPT) are numerously cited as the key element of SSTs [3] which can potentially replace the conventional LF transformers. The main requirements of SSTs, i.e., high power density, lower specific losses, voltage adaptation and isolation requirements are entirely or to a great extent fulfilled through a careful design of medium frequency transformers.

However, taking high power, high voltage and high frequency effects into account, there are several challenges to be addressed. These challenges are basically related to the extra losses as a result of eddy current in the magnetic core, excess losses in the windings due to enhanced skin and proximity effects [4] and parasitic elements, i.e., leakage inductance and winding capacitances, causing excess switching losses in the power semiconductors which are usually the dominant power losses at higher frequencies [5]. These extra losses together with the reduced size of the transformer lead to higher loss densities requiring a proper thermal management scheme in order to dissipate the higher power losses from a smaller component.

Utilizing a set of more accurate loss evaluation methods, this paper presents a design methodology addressing the applicability of rectangular litz conductors, the high isolation requirements of offshore based transformers as well as the required inductance of the dual active bridge (DAB) topology. In the proposed design, the aforementioned inductance is integrated within the medium frequency transformer as its leakage inductance as shown in Fig. 1. Utilizing this design methodology, an optimization process with a wide range of parameter variations is performed to get the highest power density while the efficiency, isolation, thermal and leakage inductance requirements are all met [6]. Using the proposed design and optimization approach, two optimized 50 kW, 1 / 3 kV, 5 kHz medium frequency transformers have been designed and manufactured. This paper describes different steps in order to design, optimize and validate the mentioned prototype.

II. DC-DC Converter Topology

Recently, there has been growing interest in utilizing dual Active bridge (DAB) converters in high power applications. The equivalent circuit of a DAB converter is shown in Fig. 1(a) in which two square wave voltage waveforms on the two sides of the transformer has been shifted by controlling the input and output bridges, applying full voltage over the inductance, L_σ, which is used to shape the current as a power transfer element [7]. The steady state transformer voltage and current waveforms of a DAB converter with simple phase shift modulation are shown in Fig. 1(b). It is worth to point out that in order to have soft switching at turn on, the anti-parallel diode of each switch should start conducting prior to the turn on moment. In order to achieve zero voltage switching (ZVS)

978-1-4673-9551-9/16 $31.00 © 2016 IEEE

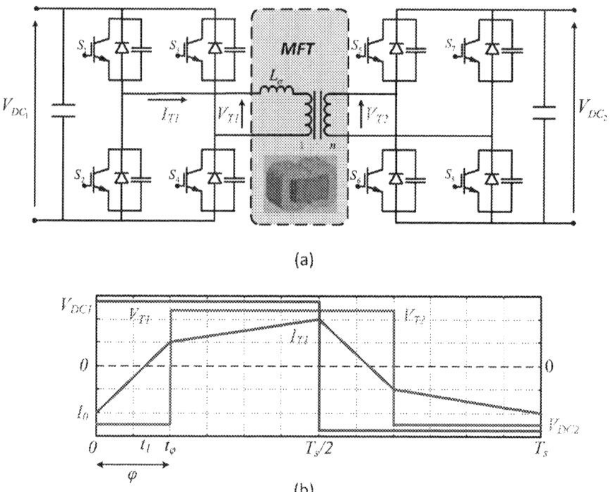

Fig. 1: (a) Dual active bridge circuit. (b) Steady state medium frequency transformer voltage and current waveforms

at turn on, the phase shift between the bridges, φ, should be higher than a certain value resulting in a minimum value of the series inductance as presented in (1). This inductance is preferably integrated as the leakage inductance of the medium frequency transformer, Fig. 1(a), in order to reduce the number of components, hence achieving higher power densities.

$$L_{\sigma(min)} = \frac{V_{DC_1} V_{DC_2} \varphi_{min}(\pi - \varphi_{min})}{2 P_{out} \pi^2 f_s n} \tag{1}$$

where φ_{min} which is associated with the worst case condition, in which the maximum output voltage deviation occurs, can be calculated as.

$$\begin{aligned} \varphi_{min} &= \frac{\pi(d-1)}{2d} \quad for \quad d > 1 \\ \varphi_{min} &= \frac{\pi(1-d)}{2} \quad for \quad d < 1 \end{aligned} \tag{2}$$

and d, the DC conversion ratio for the worst case condition, is defined as

$$d = \frac{V_{DC2}(worst\ condition)}{n V_{DC1}}. \tag{3}$$

It should be noted that in normal conditions, d equals 1. The waveforms illustrated in Fig. 1(b) will later, in this paper, be used as the transformer excitations and consequently they will affect the transformer losses, i.e. copper and core losses.

III. Optimization Procedure

Fig. 2 shows the proposed optimization flowchart used for designing a medium frequency high power transformer which is supposed to meet the mentioned requirements. Different parts of this design and optimization process are explained in details below.

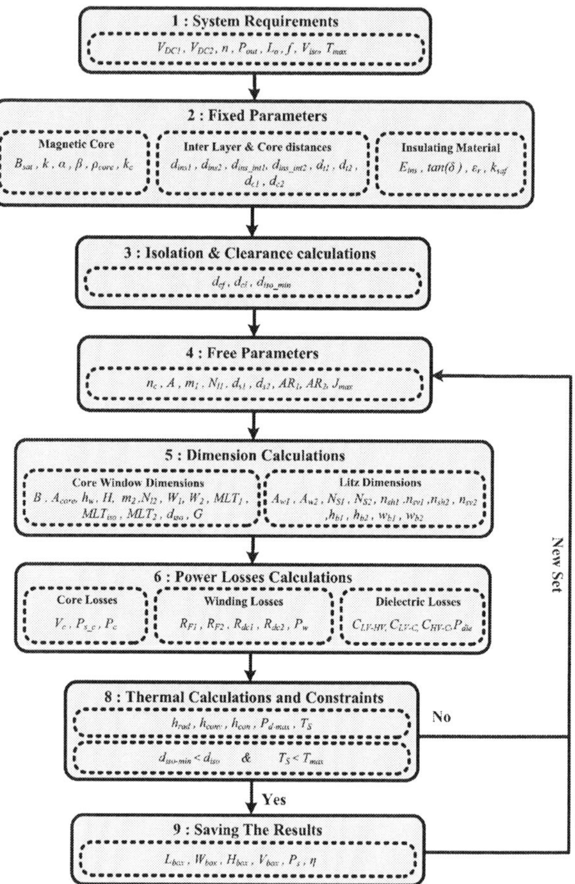

Fig. 2: Design Algorithm.

A. System Requirements and Considerations

The first step is to introduce the converter level requirements, i.e, output power, P_{out}, DC voltage levels, V_{DC1} and V_{DC2}, operating frequency, f, transformer turn ratio, n, required leakage inductance, L_σ, and isolation requirements, V_{iso}, as well as the maximum operating temperature T_{max}. The transformer is exposed to the voltage and current waveforms shown in Fig. 1(b) as the typical waveforms of a DAB converter. However, the proposed design methodology can easily be applied on other types of modulations with duty cycles, D_1 and D_2, being below 0.5. It is worth to point out that because of the power rating limitations of modern semiconductor devices, a modular concept comprising of building blocks enabling parallel connection on the low voltage side and series connection on the high voltage side, will most likely be considered [8], particularly in high power and high voltage applications.

B. Fixed Parameters

Prior to the iterative optimization part, the magnetic core material, insulation material, the windings type and inter-layer distances need to be set.

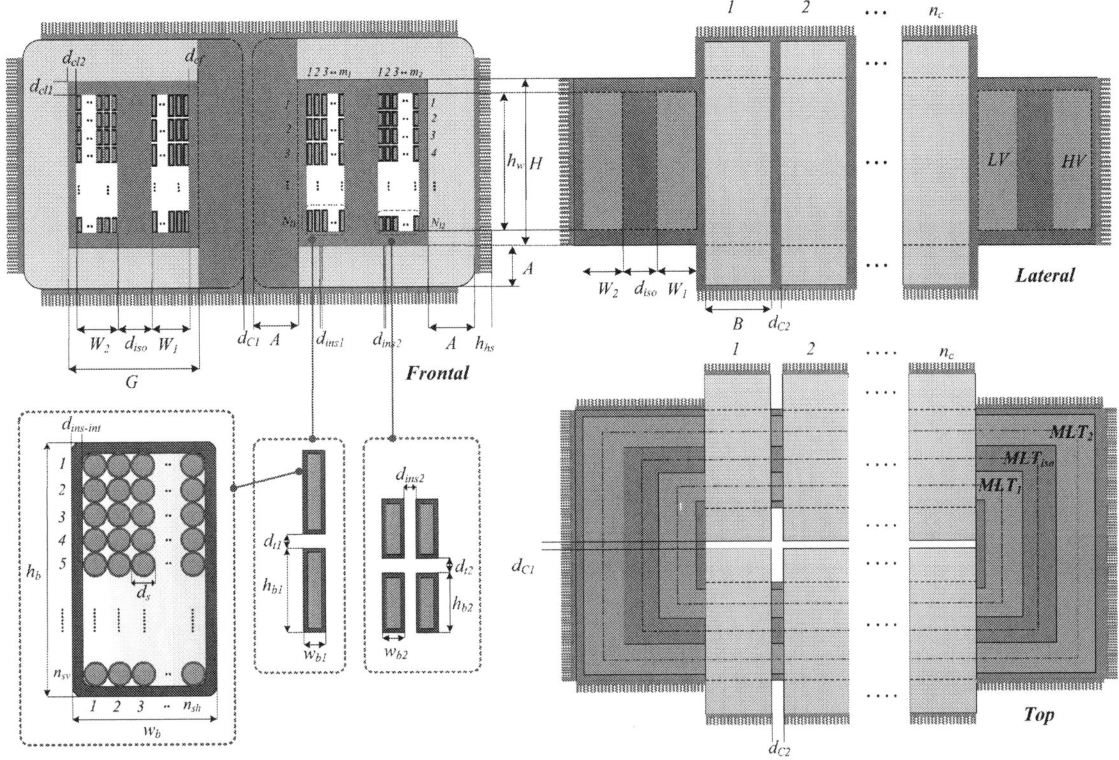

Fig. 3: Design sketch.

C. Free Parameters

For the purpose of this study, nine free parameters which are swept over a wide range are designated as the number of the magnetic core stacks, n_c, the frontal side of the core cross section, A, the number of layers and the number of litz bundle turns per layer in the primary windings portion, m_1 and N_{l1}, the diameter of the strands in the primary and secondary litz wires, d_{s1} and d_{s2}, the aspect ratios of the primary and secondary litz bundles which is the ratio between the respective total height and width, AR_1 and AR_2, and finally the maximum allowed RMS value of the current density through a conductor, J_{max}. The selection of free parameters highly depends on the optimization targets and restrictions, as well as the considered core and winding topologies. For instance, in case of foil conductors, one can consider the thickness of the foil conductors as the complementary free parameter.

All possible combinations of these free parameters will be applied to the optimization flowchart shown in Fig. 2, resulting in distinct transformer geometries corresponding to its set of free variables. Utilizing the modified and developed expressions, the core, windings and dielectric losses are then being evaluated for each set of free parameters [9], [10]. The efficiency, power density and temperature rise of each transformer are then extracted and compared in order to obtain the optimum combination that meets the requirements.

D. Geometry Determination

Having the fixed and free parameters, the transformer geometry can be defined as follows. First, the required magnetic core cross section is defined as

$$A_c = \frac{V_{DC1}}{\frac{2\sqrt{2D-\frac{8}{3}R}}{D-R} k_c m_1 N_{l1} B_m f}. \tag{4}$$

Considering the geometry structure shown in Fig. 3 and considering the number of core stacks, n_c, the lateral side length of the magnetic core can be calculated as

$$B = \frac{A_c}{2n_c A} \tag{5}$$

where A is the frontal side length of each core stack. With a specific set of strand diameters and the allowed current density given by the free parameters, the number of strands at the primary and secondary litz bundles can be respectively calculated as

$$N_{s1} = \left\lceil \frac{4I_{T1}}{\pi d_{s1}^2 J_{max}} \right\rceil \quad , \quad N_{s2} = \left\lceil \frac{4I_{T1}}{n\pi d_{s2}^2 J_{max}} \right\rceil \tag{6}$$

where I_{T1}, the transformer primary RMS current, is the function of the applied phase shift, φ, which is in this case the minimum allowed value ensuring the soft switching and yet having a reasonable amount of reactive power circulation. On the basis of the waveforms presented in Fig. 1(b), I_{T1} can be calculated from

$$I_{T1}\,(rms) = Z\sqrt{\frac{4t_1^2 t_\varphi + T_s t_1^2 - 4t_\varphi^2 t_1 - T_s t_1 t_\varphi + T_s t_\varphi^2}{3T_s}} \tag{7}$$

where Z, t_1 and t_φ are respectively calculated by

$$Z = \frac{nV_{DC1} + V_{DC2}}{nL_\sigma} \tag{8}$$

$$t_1 = \frac{\pi + 2\varphi d - \pi d}{4\pi f\,(1+d)} \quad , \quad t_\varphi = \frac{\varphi}{2\pi f}. \tag{9}$$

After determining the number of strands in each litz bundle, the number of strands in each internal row of the litz bundle, n_{sh} as shown in Fig. 3 , can be calculated by (10) as

$$n_{sh} = \left\lfloor \frac{k + \sqrt{k^2 + N_s d_s^2.AR}}{d_s.AR} \right\rfloor \tag{10}$$

where k is

$$k = d_{ins-int}(1 - AR). \tag{11}$$

It should be pointed out that (10) and (11) are used for both the primary and secondary windings, n_{sh1} and n_{sh2}, with their corresponding free parameters. This internal arrangement of the litz bundle will be later used to calculate the winding losses. Accordingly, the number of strands in each internal column within a litz bundle of primary and secondary windings, with corresponding free parameters, can be defined by

$$n_{sv1} = \left\lceil \frac{N_{s1}}{n_{sh1}} \right\rceil \quad , \quad n_{sv2} = \left\lceil \frac{N_{s2}}{n_{sh2}} \right\rceil. \tag{12}$$

As a result, the winding height, h_w, the core window height, H, and the primary windings build are respectively calculated as

$$h_w = (N_{l1}+1)\,(n_{sv1}d_{s1} + 2d_{ins-int1}) + N_{l1}d_{t1} \tag{13}$$

$$H = h_w + 2d_{cl} \tag{14}$$

$$W_1 = m_1(n_{sh1}d_{s1} + 2d_{ins-int1}) + (m_1 - 1)d_{ins1} \tag{15}$$

where d_{t1}, d_{ins1}, $d_{ins-int1}$ and other parameters were already illustrated in Fig. 3. Hence, the mean length turn of the primary windings, MLT_1, which will be later used for the windings loss calculation is obtained from

$$MLT_1 = 2\,(2A + d_{c1} + 4d_{cf} + n_c B + (n_c - 1)\,d_{c2} + 2W_1). \tag{16}$$

Likewise, the number of turns per layer and the number of layers at the secondary litz winding are respectively calculated by

$$N_{l2} = \left\lfloor \frac{h_w - (n_{sv2}d_s + 2d_{ins-int2})}{(n_{sv2}d_s + 2d_{ins-int2}) + d_{t2}} \right\rfloor \tag{17}$$

$$m_2 = \left\lceil \frac{nm_1 N_{l1}}{N_{l2}} \right\rceil. \tag{18}$$

Accordingly, the secondary winding build is calculated from

$$W_2 = m_2\,(n_{sh2}d_{s2} + 2d_{ins-int2}) + (m_2 - 1)\,d_{ins2}. \tag{19}$$

1) Isolation Distance, d_{iso} : In order to be able to calculate the rest of the geometrical dimensions, i.e, the core window width, the total length and width of the transformer box and consequently the total volume of the box, one first needs to calculate the required isolation distance, d_{iso}, fulfilling the minimum isolation requirements, V_{iso}, as well as providing the desired leakage inductance, L_σ, ensuring the ZVS operation of the DC-DC converter.

For this purpose, the analytical expression presented in [10], accounting for the high frequency effect on the leakage inductance value, is rearranged as

$$d_{iso} = \frac{-8k_1 - k_2 m_1 + \sqrt{(k_2 m_1 + 8k_1)^2 - 16k_3 m_1}}{8m_1} \tag{20}$$

where k_1 to k_3 can be calculated according to the expressions presented in [6] for the foil and round conductors. However, in case of the rectangular litz wire used in this article, those coefficients should be adapted as follows

$$\begin{aligned} k_1 = {} & \frac{m_1(m_2-1)(2m_2-1)}{6m_2}d_{ins2} \\ & + \frac{m_1'}{m_2'}\frac{sin(\frac{2\Delta_2}{\alpha\delta})4\alpha\delta^2({m_2'}^2-1) + 4d_{s2}'(2{m_2'}^2+1)}{24sin^2(\frac{2\Delta_2}{\alpha\delta}))} \\ & + \frac{m_1'}{m_2'}\frac{-\alpha\delta^2 sin(\frac{4\Delta_2}{\alpha\delta})(2{m_2'}^2+1) + 8d_{s2}'(1-{m_2'}^2)cos(\frac{2\Delta_2}{\alpha\delta})}{24sin^2(\frac{2\Delta_2}{\alpha\delta}))} \end{aligned} \tag{21}$$

where the number of layers, strand diameters and the penetration ratios are adapted for the rectangular litz conductor as

$$m_1' = n_{sh1}m_1 \quad , \quad m_2' = n_{sh2}m_2 \tag{22}$$

$$d_{s1}' = d_{s1}\frac{\sqrt{\pi}}{2} \quad , \quad d_{s2}' = d_{s2}\frac{\sqrt{\pi}}{2} \tag{23}$$

$$\Delta_1 = \frac{d_{s1}}{2\delta}\sqrt{\frac{\pi N_{l1}n_{sv1}d_{s1}\sqrt{\pi}}{2H}} \; , \; \Delta_2 = \frac{d_{s2}}{2\delta}\sqrt{\frac{\pi N_{l2}n_{sv2}d_{s2}\sqrt{\pi}}{2H}} \tag{24}$$

where δ is the skin depth at any particular frequency and α is defined as $\frac{1+j}{\delta}$. Furthermore, k_2 can be calculated as

$$k_2 = MLT_1 + 4W_1 \tag{25}$$

and k_3 are calculated as

$$k_3 = (MLT_1 + 4(W_1 + W_2))k_1 - k_4 \tag{26}$$

where k_4 with the rectangular litz conductor's parameters can be calculated as

$$\begin{aligned} k_4 = & \frac{h_w}{\mu m_1 N_{l1}^2} L_{\sigma 1} \\ & - MLT_1 \frac{(m_1 - 1)(2m_1 - 1)}{6} d_{ins1} \\ & - MLT_1 \frac{sin(\frac{2\Delta_1}{\alpha\delta}) 4\alpha\delta^2 ({m'_1}^2 - 1) + 4d'_{s1}(2{m'_1}^2 + 1)}{24 sin^2(\frac{2\Delta_1}{\alpha\delta}))} \\ & + MLT_1 \frac{\alpha\delta^2 sin(\frac{4\Delta_1}{\alpha\delta})(2{m'_1}^2 + 1) - 8d'_{s1}(1 - {m'_1}^2) cos(\frac{2\Delta_1}{\alpha\delta})}{24 sin^2(\frac{2\Delta_1}{\alpha\delta}))} \end{aligned} \tag{27}$$

Thus, the expression in (20) gives the required isolation distance providing the desired leakage inductance as one of the design specifications. It should be noted that the obtained value of d_{iso} must be sufficient enough to withstand the isolation voltage level, V_{iso}. Otherwise, the design is not acceptable and the next transformer geometry construction with a new set of the free parameters will be initiated.

Having V_{iso} determined, one can uniquely draw the transformer sketch, shown in Fig. 3, with all the geometrical details. Therefore, it is possible to utilize the loss evaluation methods, in order to calculate the power losses of each transformer corresponding to each set of free parameters.

E. Losses Calculations

The magnetic components volume reduction at higher frequencies is at the expense of enhanced core losses as well as increased winding losses due to intensified skin and proximity effects. Although the main focus is to mitigate this loss enhancement by utilizing the proper magnetic material, winding type as well as the right thermal management strategy, it is also important to more accurately evaluate the aforementioned power losses in order to properly implement a thermal management scheme. For this purpose, various analytical and empirical loss evaluation methods have been studied and modified in some cases. Finally the most suitable ones have been selected to be used in this paper.

1) Core Losses: In order to evaluate the core losses, the modified expression of the improved generalized Steinmetz equation (IGSE) adapted for the square-wave excitation with arbitrary duty cycles and rise times presented in [6] has been used as

$$P_{core} = \left(2D - \frac{4\alpha}{\alpha + 1} R\right) \frac{2^\beta}{(D - R)^\alpha} k.k_i f^\alpha B_m^\beta V_c \rho \tag{28}$$

where D and R are equal to 0.5 and 0 respectively for the standard phase shift DAB converter, ρ is the magnetic core density, α and β are the Steinmetz core loss coefficients. Moreover, the coefficient, k_i, which is a function of α and β can be calculated as

$$k_i = \frac{1}{(2\pi)^{\alpha - 1} \int_0^{2\pi} |cos(\theta)|^\alpha \, 2^{\beta - \alpha} d\theta} \tag{29}$$

Accordingly, the magnetic core volume, V_c, can be determined as follows

$$V_c = 4n_c AB\,(H + 2A) + 4n_c ABG \tag{30}$$

in which H is the window height previously calculated in (14) and G is the core window width calculated from

$$G = d_{cf} + W_1 + d_{iso} + W_2 + d_{cl} \tag{31}$$

2) Windings Losses : The winding losses can be calculated using the harmonic contents of the excitation current, thus

$$P_{w1} = \sum_{h=1}^{n} R_{DC1}.RF_{1_h}.I_{1_h}^2 \;\;,\;\; P_{w2} = \sum_{h=1}^{n} R_{DC2}.RF_{2_h}.I_{2_h}^2 \tag{32}$$

where R_{DC1} and R_{DC2} are the DC resistance of the primary and secondary windings portion, RF_{1_h} and RF_{2_h} are respectively the AC resistance factor of the primary and secondary windings portion at the hth harmonic. In addition, I_{1_h} and I_{2_h}, are respectively the rms values of the primary and secondary currents through the transformer windings for the hth harmonic. The DC resistance of the primary and secondary windings portion associated with rectangular litz winding arrangement shown in Fig. 3 are calculated from

$$R_{DC1} = \frac{4m_1 N_{l1} MLT_1}{\sigma\pi d_{s1}^2 N_{s1}} \quad , \quad R_{DC2} = \frac{4m_2 N_{l2} MLT_2}{\sigma\pi d_{s2}^2 N_{s2}} \tag{33}$$

where σ is the conductivity of the conductors, here copper, MLT_1 is previously defined in (16) and MLT_2 is the mean length turn of the secondary winding which is calculated, as can be seen in top view of Fig. 3, as

$$MLT_2 = MLT_1 + 4W_1 + 4W_2 + 8d_{iso}. \tag{34}$$

As shown in Fig. 1(b), the transformer voltages and currents are not sinusoidal. Therefore, AC resistance factors, RF_{1_h} and RF_{2_h}, should be, separately for each harmonic content, calculated from the expression in [11] which, in this paper, is adapted to account for the rectangular litz configuration shown in Fig. 3.

$$\begin{aligned} RF\,(\Delta_h) = & \Delta_h \left(\frac{sinh(2\Delta_h) + sin(2\Delta_h)}{cosh(2\Delta_h) - cos(2\Delta_h)}\right) \\ & + \frac{2\Delta_h (m^2 n_{sh}^2 - 1)}{3} \left(\frac{sinh(\Delta_h) - sin(\Delta_h)}{cosh(\Delta_h) + cos(\Delta_h)}\right) \end{aligned} \tag{35}$$

where Δ_h is the penetration ratio of the hth harmonic defined in (24) in which, it is clear that Δ_h increases at higher harmonic numbers, because it is inversely proportional to the skin depth.

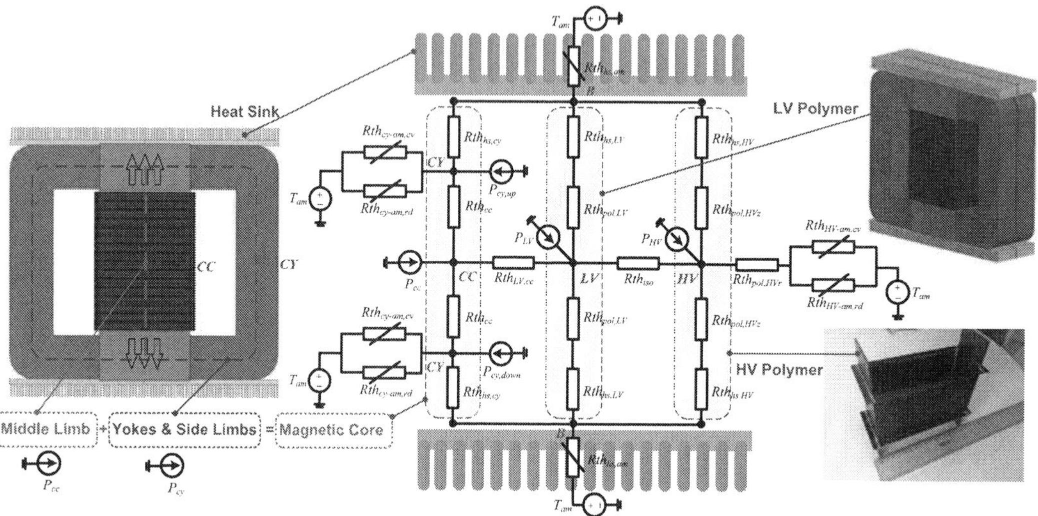

Fig. 4: Equivalent thermal network of the proposed transformer topology with six temperature nodes, i.e, (a) HV winding temperature, T_{HV} (b) LV winding temperature, T_{LV} (c) Center limb core temperature, T_{cc} (d) Yokes and side limbs temperature, T_{cy} (e) heat sink base temperature, T_b and (f) Ambient temperature, T_{am} .

Finally, the rms value of the hth harmonic of the transformer currents, to be used in (32), are derived as

$$I_{1_h} = \frac{4V_{DC1}\sqrt{1+d^2-2dcos(h\varphi)}}{2\sqrt{2}\pi^2 fh^2 L_\sigma} \quad , \quad I_{2_h} = \frac{I_{1_h}}{n} \tag{36}$$

F. Thermal Analysis

The combination of two thermal management methods, shown in Fig. 4, are implemented in the current design approach. First of all, as in planar transformers, heat sinks are considered to be placed on the core surfaces in order to increase the effective surface area resulting in thermal resistance reduction and consequently higher power dissipation capability of the design. The main drawback of this method, besides increasing the total volume of the transformer box, is that due to the poor thermal contact of the core and the primary windings, the heat around the middle limbs, usually the hottest spot of the transformer, can not easily be conducted to the heat sink. For this reason, a complementary thermal management method has been incorporated placing a thermally conductive material between the primary windings and the core in order to directly conduct the heat to the top and bottom part of the transformer where heat sinks are assembled.

The equivalent thermal network of the proposed transformer is depicted in Fig. 4 in which six different temperature nodes connected by the thermal resistances have been considered. These temperature nodes are the HV winding temperature, T_{HV}, LV winding temperature, T_{LV}, center limb core temperature, T_{cc}, yokes and side limbs temperature, T_{cy}, heat sink base temperature, T_b and finally the ambient temperature, T_{am}. In shell type transformers the middle limb is surrounded by the windings. Hence its surfaces are not exposed to the air and its corresponding heat should be first conducted to the surface areas which are exposed to the open air and then be dissipated by means of convection and radiation. Therefore, the magnetic core is divided into two areas, the middle limb with the temperature, T_{cc}, and the rest of the core consisting of side limbs and the yokes with the temperature, T_{cy}. The main assumption here is that each of the aforementioned sections are assumed to have a uniform temperature distribution, e.g., all copper areas inside the LV winding portion is assumed to have a one temperature, T_{LV}.

It should be mentioned that the heat exchange inside the transformer is mainly taking place due to the heat conduction mechanism, whereas the outer parts, the heat sinks and the open surfaces, transfer heat to the surrounding fluid, air, by means of convection and radiation mechanisms. Therefore, the thermal resistances illustrated in Fig. 4 are composed of conduction thermal resistances, as well as the convection and radiation ones. These thermal resistances are here calculated with respect to the transformer design sketch shown in Fig. 3.

Identifying the thermal network of the proposed design and calculating its corresponding thermal resistances, one can generally define the matrix representation as

$$AT + BU = 0 \Rightarrow T = -A^{-1}BU \tag{37}$$

where the matrices , T, U and B can be respectively represented as

$$T = \begin{bmatrix} T_{HV} \\ T_{LV} \\ T_{cc} \\ T_{cy} \\ T_B \end{bmatrix}_{5\times 1} \quad , \quad U = \begin{bmatrix} P_{HV} \\ P_{LV} \\ P_{cc} \\ P_{cy} \\ P_B \\ T_{am} \end{bmatrix}_{6\times 1} \tag{38}$$

TABLE I: Design specification of the prototypes

Output Power, P_{out}	50 kW
Low voltage side, V_{LV}	1 kV
High voltage side, V_{HV}	3 kV
Leakage inductance, L_σ	38 and 29 μH
Transformer turn ratio, n	3
Isolation level, V_{iso}	6 kV
Switching frequency, f	5 kHz

$$B = \begin{bmatrix} 1 & 0 & 0 & 0 & 0 & \frac{1}{R_{th_{pol,HV_r}} + R_{th_{HV,am}}} \\ 0 & 1 & 0 & 0 & 0 & 0 \\ 0 & 0 & 1 & 0 & 0 & 0 \\ 0 & 0 & 0 & 1 & 0 & \frac{2}{R_{th_{cy,am}}} \\ 0 & 0 & 0 & 0 & 0 & \frac{2}{R_{th_{hs,am}}} \end{bmatrix}_{5\times 6} \tag{39}$$

and A can be presented by

$$A = \begin{bmatrix} R_1 & \frac{1}{R_{th_{iso}}} & 0 & 0 & \frac{2}{R_{th_{HV_z}}} \\ \frac{1}{R_{th_{iso}}} & R_2 & \frac{1}{R_{th_{LV,cc}}} & 0 & \frac{2}{R_{th_{LV_z}}} \\ 0 & \frac{1}{R_{th_{LV,cc}}} & R_3 & \frac{2}{R_{th_{cc}}} & 0 \\ 0 & 0 & \frac{2}{R_{th_{cc}}} & R_4 & \frac{2}{R_{th_{hs,cy}}} \\ \frac{2}{R_{th_{HV_z}}} & \frac{2}{R_{th_{LV_z}}} & 0 & \frac{2}{R_{th_{hs,cy}}} & R_5 \end{bmatrix} \tag{40}$$

Unlike the conduction heat transfer coefficient, the convection and radiation heat transfer coefficients are strongly temperature dependent. Therefore, the matrix equation in (37) should be solved iteratively to achieve the steady state temperature of each node. The iterations start with the initial temperature as the ambient temperature for all the nodes and continue with the updated thermal resistances in each iteration until it yields the steady state value.

IV. Down-Scaled Prototype Design and Optimization

To validate the proposed design approach, two down-scaled 50 kW, 1 / 3 kV, 5 kHz medium frequency transformer prototypes with two different core material, i.e., nanocrystalline material, Vitroperm500F, and the ferrite, N87, have been designed and manufactured. The specification of the prototypes are tabulated in Table. I. The transformer prototypes are designed to operate with the voltage and current waveforms shown in Fig. 1(b) which are the typical waveforms of a DAB converter, although, the proposed design methodology can easily be applied on other types of modulations with duty cycles, D_1 and D_2, being below 0.5. The isolation level, 6 kV, is considered to account for two times the voltage of the high voltage side. The desired values of the leakage inductances are on the basis of maximum 4% and 3% output voltage deviation for the prototypes with N87 and Vitroperm500F, respectively.

A. Design and Optimization Results

The proposed design methodology explained in this article has been applied on the prototype specification shown in Table. I. For this purpose, each free parameter, shown in the forth step in Fig. 2, is swept over a wide range resulting in more than 400000 different combination of free parameters. Each unique combination of free parameters corresponds to a unique transformer geometry based on the design approach explained in this article. The great majority of the considered cases are discarded since they do not fulfill either the thermal requirements or the isolation distances required to obtain the desired leakage inductance. All the dimensional and electrical data of the remaining passed transformers are then saved, step 9 in Fig. 2, and will be used to form the Pareto-front optimization curve and consequently to choose the optimum point.

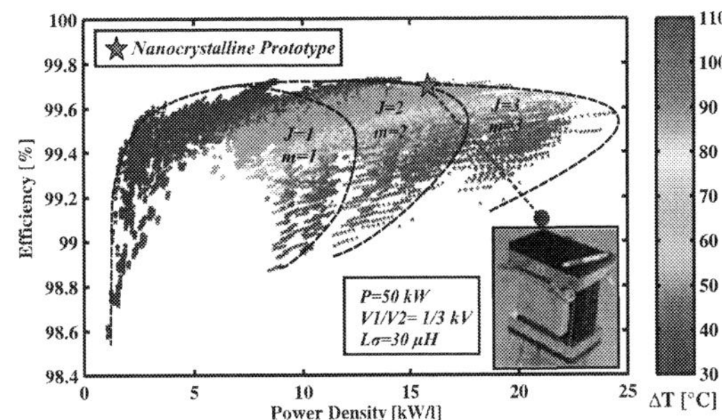

Fig. 5: Efficiency versus Power density of feasible nanocrystalline-based transformers according to the proposed design methodology.

The optimization outcome showing the efficiencies and power densities of the nanocrystalline based transformers corresponding to the accepted sets of free parameters illustrated in Fig. 5 in which each colored star represents a distinct transformer which meet all the design requirements and its color indicates the corresponding temperature rise of that particular transformer. The effect of the maximum allowed current density, J_{max}, and the number of primary layers, m_1, as two of the optimization free parameters are illustrated in Fig. 5. It can be seen that the achieved power densities increases by allowing higher current densities and number of winding layers; however, this is on the expense of higher loss densities and consequently higher temperature increase as one of the design constraints. The approximate borders between different J_{max} and m_1 are drawn in Fig. 5.

The nanocrystalline-based transformer design is highlighted in Fig. 5 with the large red star. According to the design approach, it is expected to reach a power density of 15.1 kW/l and the efficiency of 99.67%. These quantities will be evaluated on the manufactured prototype in the next parts of this article. The reason that the highest achievable power density is not selected for manufacturing the prototype, is that the author had limited selection of magnetic core dimensions, particularly for the nanocrysstalline prototype, as well as the limited dimensions of rectangular litz conductors which are available on the market. Similarly, the optimation outcome and the selected prtotype design of the Ferrite based transformers are shown in Fig. 6 in which the selected design, highlighted with a large red star, is expected to exhibit 99.54% efficiency while the power density is about 11.5 kW/l.

As stated before, two different types of magnetic material, i.e., Vitroperm500F and N87, are used to design and manufacture the prototype transformers. Both noncrystalline and Ferrite material are suitable for high frequency applications and it can be of interest to investigate their performance on two transformer prototypes with the same specifications. As can be seen in Fig. 5 and Fig. 6, Vitroperm500F demonstrates higher efficiencies for the same power density which can be explained by the lower specific losses of Vitroperm500F compared to N87. Furthermore, Fig. 5 and Fig. 6 show that Vitroperm500F based designs at 5 kHz can achieve substantially higher power densities of up to about 23 kW/l with maximum 70$^\circ C$ temperature increase compared to the one of about 13 kW/l for N87. This also can be explained by the relatively higher saturation level of Vitroperm500F compared to the one with N87 ferrite core. It is worth to mention that the high isolation requirements of the case study

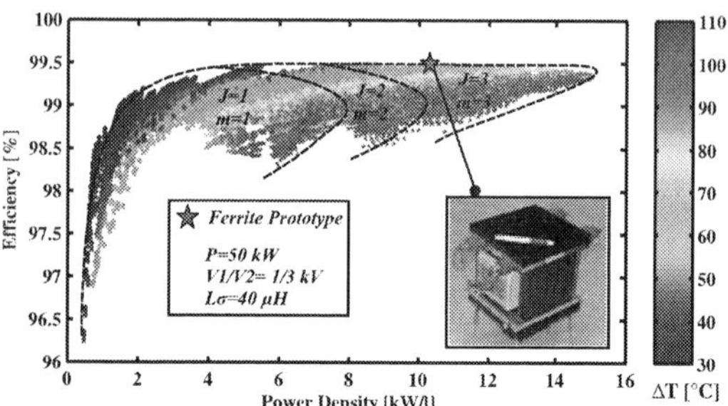

Fig. 6: Efficiency versus Power density of feasible ferrite-based transformers according to the proposed design methodology.

Fig. 7: Built prototypes

transformer, 6 kV isolation for a 3/6 kV transformer, together with the desired leakage inductance as two of the design inputs, play an important role in determining the power density of the final optimum design. The finally built nanocrystalline and ferrite-based transformer prototypes are presented in Fig. 7 in which the relatively smaller volume of the nanocrystalline-based prototype is clear.

B. Experimental Verification

The prototypes are exposed to extensive open and short circuit measurements for core and winding loss evaluations, respectively. The total expected losses of the nanocrystalline-based transformer is 166.1 W with the expected efficiency of 99.67%. The measurement results indicate a power loss of 176.1 W with the corresponding efficiency of 99.66%. The ferrite-based transformer is expected to have the efficiency of 99.52% with 241.5 W power losses. The experimental results showed 99.48% efficiency with the total power losses of 259.8 W. In order to measure the core losses, the actual waveform of the transformer has been applied on the LV side of the prototype under the test while the HV side was open. The voltage waveform, as well as the

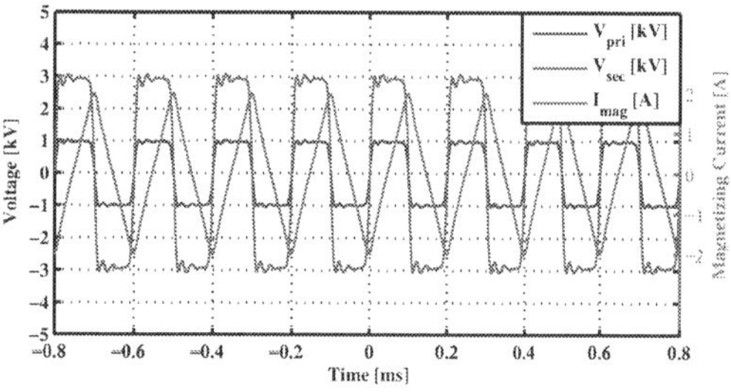

Fig. 8: Open circuit voltages and current waveforms used for the core loss measurements at 5 kHz.

corresponding magnetizing current is shown in Fig. 8. The measured leakage inductance of the nanocrystalline-based transformer is 29.9 μH which is only 3% higher than the intended design value, i.e, 29 μH. In case of the ferrite-based transformer, the design target of the leakage inductance was 38 μH while the measurement shows 40 μH which is a 5% deviation from the initial target.

V. Conclusions

This paper presented a design and optimization method for medium-frequency transformers in SST applications. Tuned value of the leakage inductance, high isolation requirements, applicability of the square-type litz conductors and nanocrystalline magnetic materials are different design aspects addressed in the proposed approach. Utilizing the proposed design method, two down-scaled prototype transformers have been design and manufactured. The nanocristalline-based prototype reached an efficiency of 99.66% which is almost the same as the theoretically predicted one. Also their leakage inductance were rather accurately predicted with the theoretical designs compared to the measured values.

Acknowledgment

Thanks to the Swedish Energy Agency for the financial support of this research. The authors would also like to thank Dr. Lars Kvarnsjö from VAC for providing us with the Vitroperm500F cores.

References

[1] K. Vechalapu, A. Kadavelugu, and S. Bhattacharya, "High voltage dual active bridge with series connected high voltage silicon carbide (sic) devices," in *Energy Conversion Congress and Exposition (ECCE), 2014 IEEE*, Sept 2014, pp. 2057–2064.

[2] M. Bahmani, T. Thiringer, A. Rabiei, and T. Abdulahovic, "Comparative study of a multi-mw high power density dc transformer with an optimized high frequency magnetics in all-dc offshore wind farm," *Power Delivery, IEEE Transactions on*, vol. PP, no. 99, pp. 1–1, 2015.

[3] I. Villar, L. Mir, I. Etxeberria-Otadui, J. Colmenero, X. Agirre, and T. Nieva, "Optimal design and experimental validation of a medium-frequency 400kva power transformer for railway traction applications," in *Energy Conversion Congress and Exposition (ECCE), 2012 IEEE*, Sept 2012, pp. 684–690.

[4] M. A. Bahmani, T. Thiringer, and H. Ortega, "An accurate pseudoempirical model of winding loss calculation in hf foil and round conductors in switchmode magnetics," *Power Electronics, IEEE Transactions on*, vol. 29, no. 8, pp. 4231–4246, Aug 2014.

[5] R. T. Naayagi, A. Forsyth, and R. Shuttleworth, "High-power bidirectional dc-dc converter for aerospace applications," *Power Electronics, IEEE Transactions on*, vol. 27, no. 11, pp. 4366–4379, 2012.

[6] M. A. Bahmani, "Design and optimization of hf transformers for high power dc-dc applications," Licentiate Thesis, Chalmers University of Technology, Gothenburg, Sweden, April 2014.

[7] R. De Doncker, D. Divan, and M. Kheraluwala, "A three-phase soft-switched high-power-density dc/dc converter for high-power applications," *Industry Applications, IEEE Transactions on*, vol. 27, no. 1, pp. 63–73, 1991.

[8] C. Meyer, "Key components for future offshore dc grids," Ph.D. dissertation, RWTH Aachen University, Germany, 2007.

[9] M. A. Bahmani, E. Agheb, T. Thiringer, H. K. Hoidalen, and Y. Serdyuk, "Core loss behavior in high frequency high power transformers—i: Effect of core topology," *Journal of Renewable and Sustainable Energy*, vol. 4, no. 3, p. 033112, 2012.

[10] M. A. Bahmani and T. Thiringer, "Accurate evaluation of leakage inductance in high frequency transformers using an improved frequency-dependent expression," *Power Electronics, IEEE Transactions on*, vol. PP, no. 99, pp. 1–1, 2014.

[11] W.-J. Gu and R. Liu, "A study of volume and weight vs. frequency for high-frequency transformers," in *Power Electronics Specialists Conference, 1993. PESC '93 Record., 24th Annual IEEE*, Jun 1993, pp. 1123–1129.

Evaluation of Core Loss in Magnetic Materials Employed in Utility Grid AC Filters

Remus Beres, Xiongfei Wang, Frede Blaabjerg, Claus Leth Bak
Department of Energy Technology
Aalborg University
Aalborg, Denmark
rnb@et.aau.dk, xwa@et.aau.dk, fbl@et.aau.dk, clb@et.aau.dk

Hiroaki Matsumori, Toshihisa Shimizu
Dept. of Electrical Engineering
Tokyo Metropolitan University
Tokyo, Japan
hrak.mtmr@gmx.com, shimizut@tmu.ac.jp

***Abstract*—Inductive components play an important role in filtering the switching harmonics related to the pulse width modulation in voltage source converters. Particularly, the filter reactor on the converter side of the filter is subjected to rectangular excitation which may lead to significant losses in the core, depending on the magnetic material of choice and current ripple specifications. Additionally, shunt or series reactors that exists in *LCL* or *trap* filters and which are subjected to sinusoidal excitations have different specifications and requirements. Therefore, the core losses of different magnetic materials adopted in utility grid ac filters have been investigated and measured for both sinusoidal and rectangular excitation, with and without dc bias condition. The core loss information can ensure cost-effective passive filter designs and may avoid trial-error design procedures of the passive components parameters.**

***Keywords*—core loss; harmonic filters; passive components; PWM inverters.**

I. INTRODUCTION

Inductive components are a critical component in today's power electronics based power systems [1]. To achieve lower size and cost, high-order passive filters [2] are preferred in general, to limit the switching harmonics from Pulse Width Modulation (PWM). Passive filters solutions for voltage-source converters (VSC) may include the *LCL* [3] or *trap* filters [4], [5]. To reduce the size and cost of the passive filters, the total stored energy in the inductive components is minimized [6]–[9]. However, lower inductances in the filter, results in higher current ripple which adds loss to the overall system, decreasing the efficiency [10].

Particularly, the filter reactor on the converter side of the filter is subjected to high frequency rectangular excitation which may lead to significant losses in the core [11], [12], depending on the magnetic material of choice [13] and current ripple specifications [10]. Additionally, shunt or series reactors that exists in *LCL* or *trap* filters and which are subjected to sinusoidal excitation have different specifications and requirements compared to the reactor on the converter side of the filter. Then, the difficulties to describe the associated frequency-dependent core losses, especially under the non-sinusoidal excitation [11], [14] and dc bias condition [11], leads to rather conservative design approach of passive filters. In general, there is a gap between the parameter selection of the passive filters which are well documented in literature [3], [6], [10] and the physical design of passive components which can be used for filter optimization [12]. Therefore, in most situations, trial and error based passive filter design methods cannot be avoided.

In the case of single-phase inductors, recent works [11], [15] established improved calculation methods of the core losses under rectangular excitation waveform and dc bias with good accuracies. For three-phase inductors, a core loss calculation and a measurement method was developed in [16]. However, since these models are mainly based on a loss map which stores the actual measurements of the core loss as a function of the dc pre-magnetization, flux density ripple and operating frequency, it is more difficult to be used and extended to other materials without additional measurements. Additionally, the core loss under sinusoidal excitation given in datasheets with Steinmetz coefficients are typically not accurate for low operating frequencies (lower than 20 kHz), which is the case of utility grid ac filters. Since the core losses under rectangular excitation with and without dc bias are not yet available in datasheets, it became more difficult to design PWM reactors with reasonable knowledge about power loss. Therefore, the core loss under sinusoidal and rectangular excitation with and without dc bias condition is measured and reported for several core materials which can be used for the physical design of the passive filters. The core losses are reported as a function of the magnetic field strength which can be used directly to relate the loss information with the current ripple in the reactors. The core loss information can ensure cost-effective passive filter designs and may avoid trial-error design procedures of the passive components parameters.

The operating principle of passive filters used in ac applications is described in the next section. The size and loss characteristics of typical magnetic cores and their applications to ac filter design are briefly summarized afterwards. The actual measurements of the core losses for the investigated magnetic materials conclude this paper.

This work was supported by European Research Council (ERC) under the European Union's Seventh Framework Program (FP/2007-2013) /ERC Grant Agreement n. [321149-Harmony].

978-1-4673-9551-9/16 $31.00 © 2016 IEEE

II. Description of AC Filters

Typical ac filters employed in utility grid applications are illustrated in Fig. 1 where L_1 is the reactor on the VSC output (PWM reactor); L_2 is the reactor on the grid side (series reactor); L_f and C_f are the filter shunt reactor and capacitor, respectively. The circuit in Fig. 1 assumes ideal passive components. The parasitic resistances or the detailed model of the passive components can be lumped into the complex impedances Z_1, Z_2 and Z_3 with Z_g being the impedance of the utility grid. Note that all the impedances are time and frequency-dependent and vary with the operating point of the converter and utility grid.

The high-order filter illustrated in Fig. 1 and delimited by the VSC output and the Point of Common Coupling (PCC) can be used to analyze the three types of reactors, independently. The input (i_1) and output current (i_2) of the filter can be written as function of the admittance system as:

$$\begin{bmatrix} i_1 \\ i_2 \end{bmatrix} = \begin{bmatrix} Y_{11} \\ Y_{21} \end{bmatrix} v_1 + \begin{bmatrix} Y_{12} \\ Y_{22} \end{bmatrix} v_2 \tag{1}$$

where v_1 and is v_2 are the output converter voltage and PCC voltage, respectively (phase to ground voltages); Y_{11} and Y_{22} are the primary and secondary short-circuit admittances, respectively; Y_{12} and Y_{21} are the transfer short-circuit admittances. The admittance system in (1) applies to current controlled VSCs. For filter design, the high frequency content in the PCC voltage (v_2) can be assumed very low therefore it can be neglected from (1). As result, the high frequency harmonic content is given mainly form the PWM harmonics found in the converter output voltage (v_1) and which should be minimized by the filter.

The amplitude of the fundamental current (I_{1h}) of the VSC can be calculated as:

$$I_{1h} = \frac{2}{3}\frac{S_{VSC}}{V_{1h}}\bigg|_{h=1} \tag{2}$$

where S_{VSC} is the rated apparent power of the VSC and V_{1h} is the amplitude of the converter voltage fundamental component. I_{1h} can be used to derive the inductance and current ratings needed for the physical design of the three types of reactors.

A. PWM reactor

PWM filters are mainly represented by a reactor subjected to rectangular excitation (which depends on the PWM method) of frequencies up to tens of kHz. The time varying switched excitation waveform of the PWM reactor can be written in the general form as [17]:

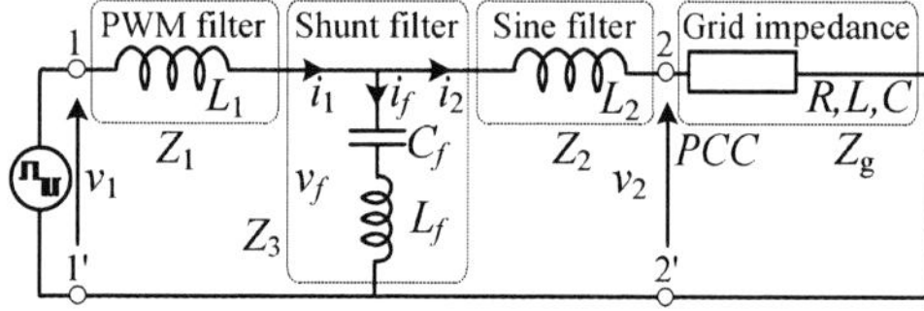

Figure 1: One-phase schematic of typical AC filters employed in utility grids.

$$F(t) = \frac{A_{00}}{2} + \sum_{n=1}^{\infty}\{A_{0n}\cos(n\omega_0 t) + B_{0n}\sin(n\omega_0 t)\}$$
$$+\sum_{m=1}^{\infty}\{A_{m0}\cos(m\omega_c t) + B_{m0}\sin(m\omega_c t)\} \tag{3}$$
$$+\sum_{m=1}^{\infty}\sum_{\substack{n=-\infty \\ n\neq 0}}^{\infty}\{A_{mn}\cos[(m\omega_c + n\omega_0)t] + B_{mn}\sin[(m\omega_c + n\omega_0)t]\}$$

where A_{00} is the dc offset of the time varying signal, the first summation term represents the fundamental and baseband harmonics, the second summation term represents the carrier harmonics while the third summation term represents the carrier sideband harmonics. For a given PWM method, the magnitude of $[m\omega_c+n\omega_0]$ harmonic voltage components can be found evaluating the double integral Fourier form [17]:

$$C_{mn} = A_{mn} + jB_{mn} = \frac{1}{2\pi^2}\int_{-\pi}^{\pi}\int_{-\pi}^{\pi} F(x,y)e^{j(mx+ny)}dxdy \tag{4}$$

where $F(x,y)$ is the switched waveform for one fundamental period, ω_c is the carrier frequency and $x=\omega_c t$, ω_0 is the fundamental frequency and $y=\omega_0 t$, m is the carrier index (ω_c/ω_0); n is the baseband index. The analytical solutions of the Fourier coefficients for the most known PWM methods for single and three phase inverters are given in [18].

Once the harmonic spectrum of v_1 is known, the inductance and current rating (to avoid saturation) of the PWM reactor can be found. The inductance of the PWM reactor is limited by the current ripple requirement. The inductance L_1 at the rated current is directly proportional with the dc-link voltage (V_{dc}) and inversely proportional with the maximum current ripple (Δi_{1pk}) and switching frequency (f_{sw}):

$$L_1 \approx \frac{V_{dc}}{r\Delta i_{1pk} f_{sw}} \tag{5}$$

For a two-level three-phase VSC which uses the conventional Space Vector Modulation (SVM) (assuming a modulation index of 0.9), the ripple factor $r = 24$ [2] while for unipolar modulation in the case of a single-phase VSC, $r = 8$. However, to evaluate the time varying current ripple, it is required to analyze (1) around the switching frequencies and multiples. Above the resonance frequency of the system (ω_{res}), the individual current harmonics in the PWM reactor that contributes to the current ripple can be found as:

$$I_{1h} = \frac{V_{1h}}{h\omega_0 L_1}\bigg|_{h>h_{res},\ h_{res}=\frac{\omega_{res}}{\omega_0}} \tag{6}$$

Then, the time varying current ripple waveform results as:

$$i_{1r}(t) = \sum_{h>h_{res}}^{n} I_{1h}\cos(h\omega_0 t) \tag{7}$$

978-1-4673-9551-9/16 $31.00 © 2016 IEEE

The current rating to design the PWM reactor in order to avoid the core saturation can be found as:

$$I_{L1pk} = \max\left(i_{11}(t)+i_{1r}(t)\right) = \frac{A_c B_{\max} N}{L_1} \quad (8)$$

where $i_{11}(t)=I_{1h}\cos(\omega_0 t)$; the maximum operating flux density of the reactor is chosen lower than the saturation flux density of the magnetic core ($B_{max} < B_{sat}$); A_c is the cross-section area of the core and N is the number of turns. The key trade-off in the PWM reactor design is the optimum selection (to minimize loss) of the ripple current for a given magnetic material. The current ripple creates minor hysteresis loops related to the switching frequency modulation and are the major part of the core loss [16]. The core loss of each minor loop corresponding to each switching interval can be characterized and measured with a dc buck-chopper circuit and a B-H analyzer [11]. A smaller part of the core loss in a PWM reactor is related to the fundamental output frequency of the VSC, which can be measured with low frequency sinusoidal excitation.

B. Shunt Reactor

Shunt filters can be used for power factor correction, voltage support or harmonic compensation. The common operating frequencies of ac reactors in a harmonic shunt filter can be up to 3 kHz [19]. Shunt filtering can be also used to cancel out the ripple current from VSCs, especially in trap filter configurations [20]. For trap filters, the operating frequency of the reactors can reach tens of kHz [21]. The core loss of a shunt filter reactor can be measured with a high frequency (equal to the tuned frequency) and small amplitude sinusoidal voltage. A fixed inductance value is desirable in shunt reactors, in order to avoid de-tuning of the filter during operation. The shunt reactor inductance can be derived as:

$$L_f = \frac{1}{\omega_t^2 C_f} \quad (9)$$

The current rating of the shunt reactor must consider the fundamental current given by the impedance of the shunt capacitor and the high frequency ripple current given in (7) as:

$$I_{Lfpk} = \max\left(i_{Cf1}(t)+i_{1r}(t)\right) = \frac{A_c B_{\max} N}{L_f} \quad (10)$$

where $i_{Cf1}(t)= \omega_0 C_f V_1\cos(\omega_0 t)$ is the fundamental current in the shunt reactor.

C. Series Reactor

Sine wave filters or series reactors used to smoothen out the grid current are driven by sinusoidal voltages at the fundamental grid frequency. Since full current (rated current of the VSC) have to be handled by the reactors, high energy storage capability and low cost is preferable for this type of filter.

III. Application of Magnetic Materials to AC Filter Design

A. Magnetic Materials Properties

The key properties of common magnetic materials are illustrated in Table I. The magnetic cores can be divided in two categories: with distributed gap (powder cores) and with physical gap (cut cores). The B-H curves of the cheapest materials from the two categories are illustrated in Fig. 2.

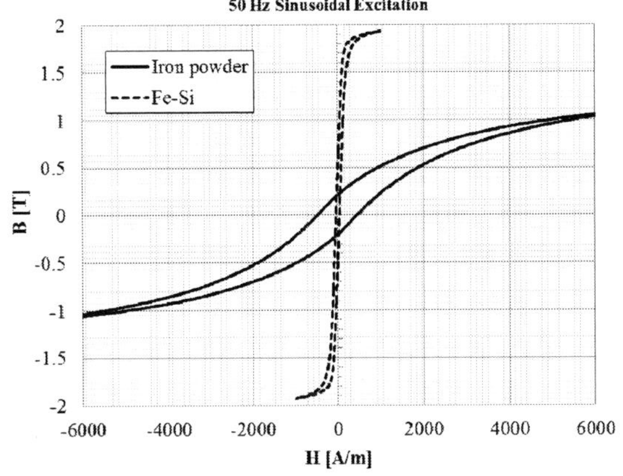

Figure 2: B-H dependence of laminated Fe-Si and Fe powder.

Table I: Comparison of different magnetic core materials [22]

Materials		Permeability (μ)	B_{sat} (T)	Core Loss	DC Bias	Relative Cost	Temp. Stability	Curie Temp. (°C)
Powder	MPP (Ni-Fe-Mo)	14-200	0.7	Lower	Better	High	Best	450
	High Flux (Ni-Fe)	26-160	1.5	Low	Best	Medium	Better	500
	Sendust (Fe-Si-Al)	26-125	1	Low	Good	Low	Good	500
	Mega Flux (Fe-Si)	26-90	1.6	Medium	Best	Low	Better	700
	Iron (Fe)	10-100	1	High	Poor	Lowest	Poor	770
Strip (Gapped)	Silicon Steel (Fe-Si)	Up to 10000	1.8	High	Best	Lowest	Good	740
	Amorphous (Fe-Si-Bo)		1.5	Low	Better	Medium	Good	400
(Gapped)	Ferrite (Mn-Zn)		0.45	Lowest	Poor	Lowest	Poor	100~300

The cut cores have linear B-H dependence for $B<0.95B_{sat}$ (constant permeability) and exhibits a hard saturation while the powder cores have a non-linear B-H dependence (decreased permeability with increased excitation) and exhibits soft saturation. Due to the distributed air gap, powder cores can reach a maximum magnetic field strength (H_m) in the range of 10-80 kA/m. On the other hand, the ungapped laminated steel or ferrites class can reach about 1 kA/m. By adding air gap, the H range can be increased, leading to higher storage capability.

The permeability (μ) dependence (or inductance decrease) as function of the dc bias magnetic field (H_0) are illustrated in Fig. 3 for magnetic core samples of equal volume (except the ferrite which is 1.5 times larger than others). The inductance factor (A_L) or the permeance of the core samples is defined as:

$$A_{L@0A} = \frac{L_0}{N^2} = \frac{\mu A_c}{l_e} = \frac{1}{\Re} \quad (11)$$

$$N = \frac{H_m l_e}{I} \quad (12)$$

where L_0 is the inductance of the core at zero current, A_c is the core cross-section area, l_e is the effective core magnetic path length and N is the number of turns, which can be written as in (12) as a function of H_m, l_e and the inductor current I.

B. Selection of Magnetic Material Based on Energy Storage and DC Bias Charactersitics

For designing reactors, it is reasonable to assume a 70% of the initial permeability for powder cores, in order to "fully" exploit the stored energy capability. For gapped cores it is considered to be around 95-100% of the initial permeability in order to avoid saturation. Assuming a 1 mm diameter round wire for the inductor windings (with a current density of 5A/mm^2) and considering a fixed H_{max} to avoid saturation, the number of turns is readily available from (12). Since the inductance factor is fixed for a given material, the inductance value is given only by the number of turns, assuming fixed current and H_{max}. By using the inductance factor from Fig. 3 and using (11), the inductance value can be found. In Table II, L, N and the energy storage capability (LI^2) of the core samples are shown assuming 95% of the initial permeability at rated current for cut cores and 70% for the powder cores (including 95% and 50% for the Mega Flux core).

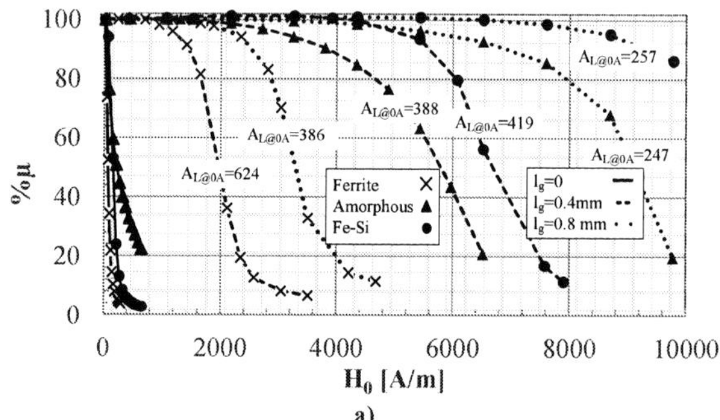

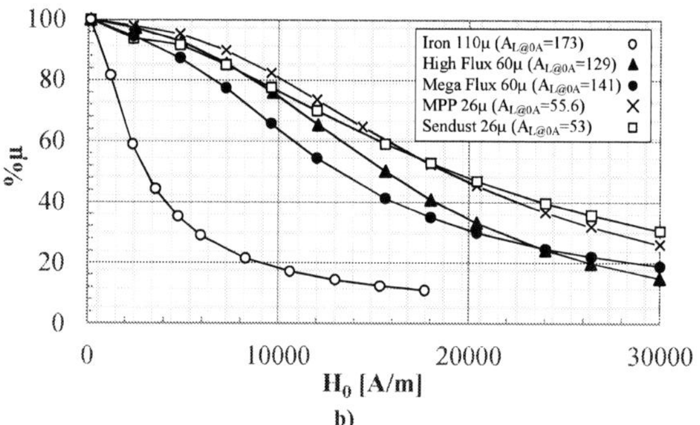

Figure 3: Relative permeability vs. H$_0$ curves, measured with a Wayne Kerr Precision Magnetic Analyzer 3260B at 500 Hz and 10 V drive voltage for: a) cut cores with physical gap (data not available in datasheets); b) powder materials with distributed gap (within ±10 % deviation from the datasheet values).

Eq. (11)-(12) are used only to evaluate the relative size of the inductors. The window utilization factor (K_u) has been assumed 0.45. However, as can be seen from Table II, the number of turns that lead to K_u=0.45 is exceeded for some reactors design. To avoid excessive temperature rise in reactors, the number of turns should be decreased or a larger core should be adopted. From the considered magnetic cores, the silicon steel and amorphous offers the best cost/energy storage trade-offs. For the powder cores, it is possible to increase the energy storage capability by adopting lower initial permeability which in turn will increase the size of the winding and associated cost/loss.

Table II: Design example of sample inductor with different materials

Samples	Ferrite (E55 N87)		Amorphous (AMCC10)		Fe-Si (SC10)		Iron (SK-36M)	High Flux (CH571)	Mega Flux (CK571)			MPP (CM571)	Sendust (CS571)
	l_g=0.4 mm	l_g=0.8 mm	l_g=0.4 mm	l_g=0.8 mm	l_g=0.4 mm	l_g=0.8 mm	70 μ%	70 μ%	95 μ%	70 μ%	50 μ%	70 μ%	70 μ%
H_m (kA/m)	1.2	2.3	3	6	5	8.7	1.8	11	3	9	13	13	12
l_e (mm)	124		138		138		142.4	125	125			125	125
L (mH)	0.6	1.95	4.06	10.3	12.2	22	0.5	11	1.23	8.13	12	6.67	5.41
I (A)	4	4	4	4	4	4	4	4	4	4	4	4	4
N	32	73	105	210	175	300	65	350	96	287	414	414	382
LI^2 (mHA^2)	10	30	60	160	190	340	8	170	19	125	186	100	83
N (K_u=0.45)	185		220		220		400	300	300			300	300

IV. Evaluation of Core Loss

The core loss per volume (P_{cv}) for sinusoidal and rectangular excitation is measured with an IWATSU SY-8219 B-H analyzer. The core loss under dc-bias is measured with an IWATSU SY-8232 B-H analyzer and a dc-chopper. The magnetic core samples are shown in Fig. 4.

Figure 4: Magnetic core samples

A. Core Loss Under Sinusiodal Excitation

The core losses measured for 50 Hz sinusoidal excitation are shown in Fig. 5 as function of the applied magnetic field strength. The core loss for cut cores does not change significantly by adding air gap (if reported as function of magnetic field density). For a given reactor data, (12) can be used to associate the core loss to the rated current. The loss information shown in Fig. 5 can be used to design series or shunt reactors. For shunt reactors, the core loss under higher frequency sinusoidal excitation is needed.

B. Core Loss Under PWM Excitation and DC Bias

The core losses under rectangular excitation without dc bias ($H_0 = 0$) are illustrated in Fig. 6 for cut cores, while for powder cores are shown in Fig. 7. The dc-bias influence on the core loss is illustrated in Fig. 8 for powder cores also as a function of the magnetic field ripple H_m which can be used as design parameter for PWM reactors. Hence, for a given H_m, the instantaneous core loss is readily available regardless of the inductance value (L is already included in A_L in Fig. 3).

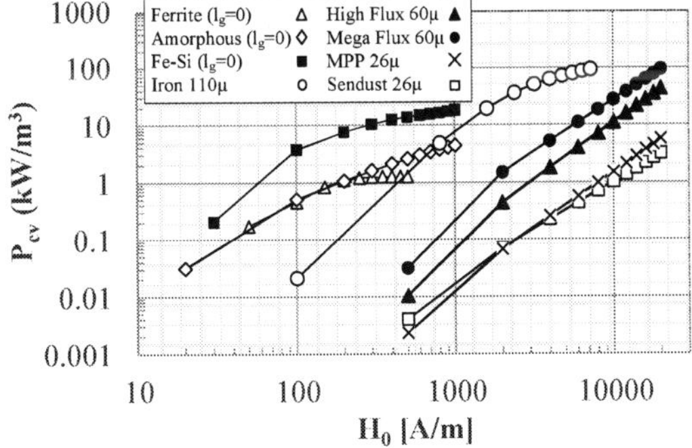

Figure 5: Core loss measured at 50 Hz under sinusoidal excitation (data extrapolated for $H_0 > 7000$ A/m).

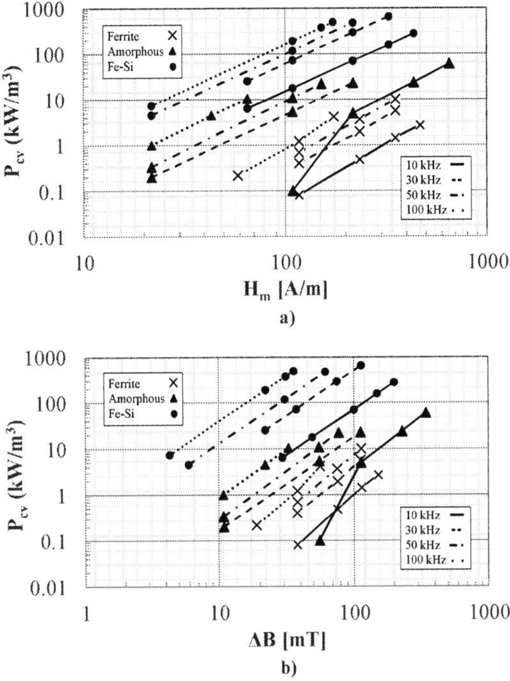

Figure 6: Core loss under rectangular excitation ($H_0 = 0$) for cut cores ($l_g = 0.4$ mm) as function of: a) maximum field ripple, H_m; b) flux density ripple, ΔB.

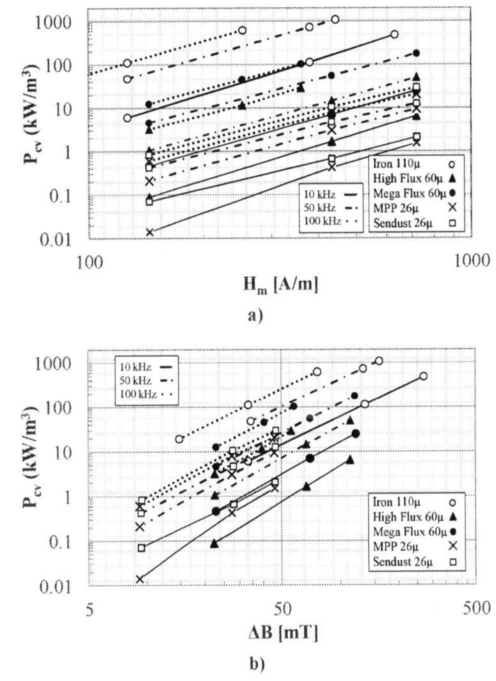

Figure 7: Core loss under rectangular excitation ($H_0 = 0$) for powder cores as function of: a) maximum field ripple, H_m; b) flux density ripple, ΔB.

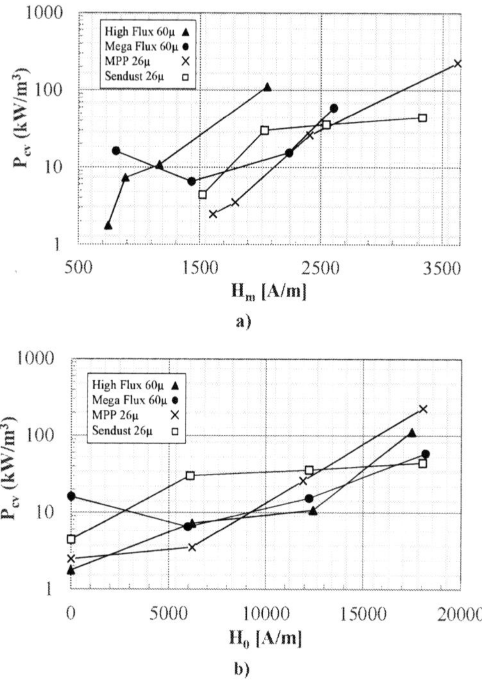

Figure 8: Core loss under rectangular excitation and dc bias condition measured at 10 kHz and ΔB=0.09T as function of: a) maximum field ripple, H_m; b) dc-premagnetization field, H_0.

C. Core Loss in PWM Converters

In Figure 9, the instantaneous B-H curves for a PWM reactor with a Mega Flux core (designed for 70% of the initial permeability or around 30% inductance drop at rated current) are shown for a single phase and three-phase VSC configuration, respectively. In both situations, the maximum current ripple is around 9 % (defined as half the peak to peak ripple). For a modulation index of 0.95, the maximum frequency of the minor loop loss is 20 f_{sw} (where f_{sw} is the switching frequency of the VSC) while the minimum frequency is around 1 f_{sw}. Both occur when the bias level is highest. For no dc bias condition, the minor loop loss frequency is 2 f_{sw}. For sine filters, the low frequency component of the magnetic field (illustrated as H_0 in Fig. 9 by applying a low-pass filter to the instantaneous H) can be used as design parameter. Fig. 9 can be used to associate the core loss information to the actual operating waveforms of the filter as presented in [16]. For this particular PWM reactor design example for single-phase VSC, maximum ΔB_{max}=0.2 T and occurs at no bias, while the minimum ΔB_{min}=0.06 T. Comparing the loss information for sinusoidal excitation and rectangular excitation it can be seen how the core losses specific to one minor loop can easily exceed the loss of the fundamental loop given by the sinusoidal excitation. Therefore, the loss information is more crucial for the PWM reactor design, rather than series or shunt reactors.

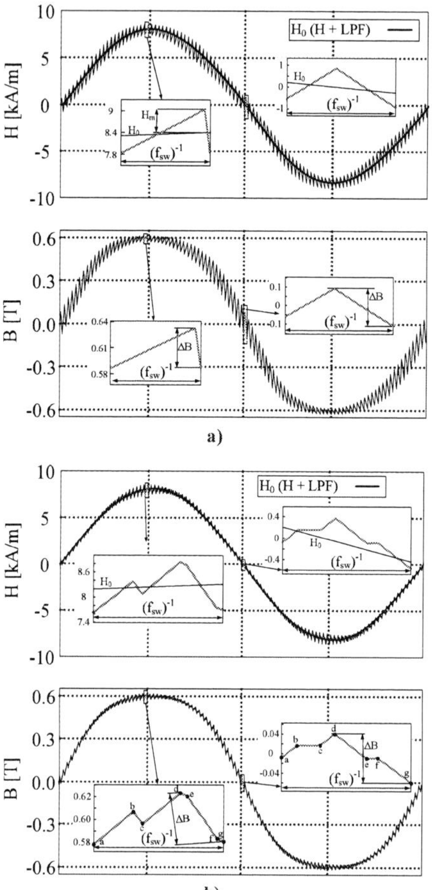

Figure 9: One fundamental period of instantaneous B-H curves for Mega Flux core: a) bipolar modulation (single phase full bridge 5 kHz VSC); b) sinusoidal PWM (three phase - 2 level, 5 kHz VSC).

D. Discussion on Magnetic Core Selection

For the PWM reactor design is required to choose the ripple current in such a way to minimize the core loss. From Fig. 6 and Fig. 7 it can be seen that Ferrite provides the lowest core loss while the iron powder and laminated steel are the worst candidate in terms of core losses. Under dc bias, measurement from Fig. 8 reveals that MPP is the most suited to be used. The best trade-off between size and core loss is given by the amorphous cut core. The Sendust material can be adopted for high ripple condition since with increasing the magnetic field strength ripple, the loss remain relatively constant.

A fixed inductance value is desirable in shunt reactors, therefore cut cores are favored for this application. For series reactors, high energy storage capability and low cost imply the use of the laminated steel.

V. Conclusions

In this paper, the core losses of several magnetic core materials have been evaluated. From the investigated core materials, the laminated steel provides the best energy storage capability. However, it provides around 10 times higher core loss than the equivalent Mega Flux powder under rectangular excitation. Hence, the laminated steel is the best candidate for sine wave filters applications. The powder materials are good candidates for PWM filters, especially at high ripple current. The Amorphous core can be seen as the material with the best trade-off between size and loss. The Sendust core can ensure low loss under very high ripple condition. Ferrites do provide lower core loss, but the energy storage capability is the lowest. In addition, the permeability and loss of ferrites are highly dependent on temperature (the measurements reported in this paper are made at room temperature). For shunt filter applications, cut cores are preferred in order to keep a fixed inductance value as function of the dc bias.

References

[1] R. Teichmann, M. Malinowski, and S. Bernet, "Evaluation of Three-Level Rectifiers for Low-Voltage Utility Applications," *IEEE Trans. Ind. Electron.*, vol. 52, no. 2, pp. 471–481, Apr. 2005.

[2] R. Beres, X. Wang, F. Blaabjerg, M. Liserre, and C. Bak, "Optimal Design of High-Order Passive-Damped Filters for Grid-Connected Applications," *IEEE Trans. Power Electron.*, vol. 53, no. 4, pp. 1–1, 2015.

[3] M. Liserre, F. Blaabjerg, and S. Hansen, "Design and Control of an LCL-Filter-Based Three-Phase Active Rectifier," *IEEE Trans. Ind. Appl.*, vol. 41, no. 5, pp. 1281–1291, Sep. 2005.

[4] J. M. Bloemink and T. C. Green, "Reducing Passive Filter Sizes with Tuned Traps for Distribution Level Power Electronics," in *Proc. 14th European Conference on Power Electronics and Applications (EPE 2011), vol., no., pp.1-9, Aug. 30 2011-Sept. 1 2011.*

[5] W. Wu, Y. He, and F. Blaabjerg, "An LLCL Power Filter for Single-Phase Grid-Tied Inverter," *IEEE Trans. Power Electron.*, vol. 27, no. 2, pp. 782–789, Feb. 2012.

[6] K. Jalili and S. Bernet, "Design of LCL Filters of Active-Front-End Two-Level Voltage-Source Converters," *IEEE Trans. Ind. Electron.*, vol. 56, no. 5, pp. 1674–1689, May 2009.

[7] G. Zeng, T. W. Rasmussen, and R. Teodorescu, "A novel optimized LCL-filter designing method for grid connected converter," in *Proc. International Symposium on Power Electronics for Distributed Generation Systems*, 2010, vol. 2, pp. 802–805.

[8] J. Verveckken, J. F. Silva, and J. Driesen, "Optimal analytic LCL filter design for grid-connected voltage-source converter," in *Proc. of Eurocon 2013*, 2013, no. July, pp. 823–830.

[9] L. Shen, G. Asher, P. Wheeler, and S. Bozhko, "Optimal LCL filter design for 3-phase Space Vector PWM rectifiers on variable frequency aircraft power system," *2013 15th Eur. Conf. Power Electron. Appl. EPE 2013*, 2013.

[10] Y. Jiao and F. C. Lee, "LCL Filter Design and Inductor Current Ripple Analysis for a Three-Level NPC Grid Interface Converter," *IEEE Trans. Power Electron.*, vol. 30, no. 9, pp. 4659–4668, Sep. 2015.

[11] T. Shimizu and S. Iyasu, "A practical iron loss calculation for AC filter inductors used in PWM inverters," *IEEE Trans. Ind. Electron.*, vol. 56, no. 7, pp. 2600–2609, 2009.

[12] J. Muhlethaler, M. Schweizer, R. Blattmann, J. W. Kolar, and A. Ecklebe, "Optimal Design of LCL Harmonic Filters for Three-Phase PFC Rectifiers," *IEEE Trans. Power Electron.*, vol. 28, no. 7, pp. 3114–3125, Jul. 2013.

[13] M. S. Rylko, K. J. Hartnett, J. G. Hayes, and M. G. Egan, "Magnetic Material Selection for High Power High Frequency Inductors in DC-DC Converters," in *Proc. IEEE Applied Power Electronics Conference and Exposition*, 2009, pp. 2043–2049.

[14] K. Venkatachalam, C. R. Sullivan, T. Abdallah, and H. Tacca, "Accurate prediction of ferrite core loss with nonsinusoidal waveforms using only Steinmetz parameters," *2002 IEEE Work. Comput. Power Electron. 2002. Proceedings.*, no. June, 2002.

[15] J. Muhlethaler, J. Biela, J. W. Kolar, A. Ecklebe, M. Jonas, and A. Ecklebe, "Core Losses Under the DC Bias Condition Based on Steinmetz Parameters," *IEEE Trans. Power Electron.*, vol. 27, no. 2, pp. 953–963, 2012.

[16] T. Shimizu, H. Matsumori, K. Takano, and H. Ishii, "Evaluation of Iron Loss of AC Filter Inductor used in Three-Phase PWM inverters Based on an Iron Loss Analyzer (ILA)," *IEEE Trans. Power Electron.*, vol. 8993, no. c, pp. 1–1, 2015.

[17] D. G. Holmes, "A general analytical method for determining the theoreticalharmonic components of carrier based PWM strategies," *Conf. Rec. 1998 IEEE Ind. Appl. Conf. Thirty-Third IAS Annu. Meet. (Cat. No.98CH36242)*, vol. 2, no. 2, 1998.

[18] D. G. Holmes and T. A. Lipo, *Pulse Width Modulation for Power Converters: Principles and Practice*. John Wiley & Sons, 2003.

[19] J.C. Das, *Power System Harmonics and Passive Filter Designs*. Piscataway, NJ 08854: Wiley-IEEE Press, 2015.

[20] Y. Patel, D. Pixler, and A. Nasiri, "Analysis and design of TRAP and LCL filters for active switching converters," *2010 IEEE Int. Symp. Ind. Electron.*, pp. 638–643, Jul. 2010.

[21] J. Xu, J. Yang, J. Ye, Z. Zhang, and A. Shen, "An LTCL Filter for Three-Phase Grid-Connected Converters," *IEEE Trans. Power Electron.*, vol. 29, no. 8, pp. 4322–4338, Aug. 2014.

[22] CSC catalog, "Magnetic Powder Cores.", ver. 13.

A Novel Gate Assisted Circuit to Reduce Switching Loss and Eliminate Shoot-through in SiC Half Bridge Configuration

Shan Yin, K. J. Tseng, C. F. Tong
Rolls-Royce@NTU Corporate Lab, School of Electrical and Electronic Engineering
Nanyang Technological University
Singapore

Rejeki Simanjorang, C. J. Gajanayake, Amit K. Gupta
Advanced Technology Centre
Rolls-Royce Singapore Pte. Ltd.
Singapore

***Abstract*—The trend towards SiC-based high power density converter requires to drive SiC MOSFET with high-speed switching. However, it tends to aggravate *dv/dt* effect due to the impact of parasitic parameters, resulting in shoot-through and high device stress in the half bridge configuration. In this work, a novel gate assisted circuit is proposed to eliminate shoot-through issue without compromise on switching speed. The mathematical model for the gate assisted circuit is developed. Then the switching test experiment is set up to validate its performance with commercial SiC half bridge module, which shows a high *dv/dt* up to 20 and 31 V/ns at turn-on and turn-off respectively and a 34% reduction of switching loss without any shoot-through issue.**

Keywords—Gate assisted circuit; half bridge; high-speed switching; shoot-through; SiC MOSFET.

I. INTRODUCTION

Silicon carbide (SiC) is regarded as the next-generation wide bandgap material for the power devices used in high power density and high temperature applications like electric vehicle (EV) and more electric aircraft (MEA) [1, 2]. The SiC power converter needs to switch at high speed to allow high operating frequency, and thereby reducing the size of passive components for higher power density. However, the increased *dv/dt* tends to aggravate the switching noise due to the effects of parasitic elements. Considering the lower threshold voltage of SiC MOSFET (2.5 V typically), it is more susceptible to the switching noise [3].

In half-bridge configuration, the high *dv/dt* during turn-on transition of one transistor will affect the complimentary transistor via the Miller capacitance C_{GD} and common source inductance L_{CSI} [4], leading to a momentary arm shoot-through (typically 10~100 ns). It results in an increasing switching loss or even device destruction. The shoot-through issue can be mitigated by the following two strategies.

From the aspect of device design, the Miller capacitance can be reduced by using smaller polysilicon gate width [5]. From the aspect of circuit design, there are four commonly used solutions: 1) negative bias, 2) auxiliary gate-source capacitor, 3) turn-off diode, 4) active Miller clamp [6-11]. The negative bias is costly but most effective, since the gate bias is directly shifted by a negative value. However, the maximum negative gate voltage of SiC MOSFET is only -5 V due to the limitation of gate oxide capability [3]. Hence, the negative bias cannot be lower than -5 V and the outcome is therefore limited. Using auxiliary capacitor in parallel with gate-source terminals of transistor increases the effective input capacitance so as to suppress the gate voltage spike. However, it slows down the switching speed and increases the switching loss. For both turn-off diode and active Miller clamp methods, a separate low impedance path is to be provided for the *Cdv/dt* current. Then a large enough R_G can be used to limit *dv/dt* at turn-on. For the gate driver IC with high output impedance, the effect of turn-off diode is limited. The active Miller clamp has been proposed to eliminate the need of negative power supply in shoot-through mitigation. However, due to the internal gate resistance already inside the package or module of SiC power MOSFET, the active Miller clamp is not applicable for this application. Hence, improved active Miller clamp circuits were proposed.

A gate assisted circuit (GAC) was proposed for driving of SiC JFET in a three-phase inverter [8]. It utilized a PNP transistor and an auxiliary capacitor. And this circuit is active only when the main transistor is off to avoid any influence on the normal turn-on process. However, as the reverse breakdown voltage of base-emitter junction of PNP transistor is normally below 10 V, it cannot be directly used in SiC MOSFET, whose recommended gate-source voltage reaches up to 20 V. Replacement of PNP with NMOS makes this kind of circuit practical for use in SiC MOSFET [10], since the maximum gate-source voltage of NMOS can be as high as 25 V. However, gate drivers with complementary outputs are required, since the logic of auxiliary NMOS is inverse of that of the main transistor.

In our study, a novel GAC to eliminate the shoot-through issue by replacing NMOS used in [10] with PMOS is proposed. The mathematical models for shoot-through in half bridge configuration with conventional gate driver and the proposed GAC are developed. Furthermore, a switching test is experimentally set up to validate the performance of GAC.

978-1-4673-9551-9/16 $31.00 © 2016 IEEE

II. Shoot-Through in Half Bridge Configuration

As Fig. 1(a) shows, during turn-on transition of low side (LS) transistor M_L, a voltage ramp occurs across the drain-gate terminals of high side (HS) transistor M_H. Then a *Cdv/dt* current flows via C_{GD} and gate resistor, resulting in a voltage drop across gate-source terminals of M_H. In addition, it is further aggravated by L_{CSI}. Parasitic turn-on of M_H may happen if it exceeds the threshold voltage. Similarly, during the turn-off transition of M_L shown in Fig. 1(b), a *Cdv/dt* current is induced by C_{GD}, but in an opposite direction. Hence a negative voltage drop across gate-source terminals of M_H is observed, which brings no adverse side-effect to the switching transition but tends to stress the gate oxide.

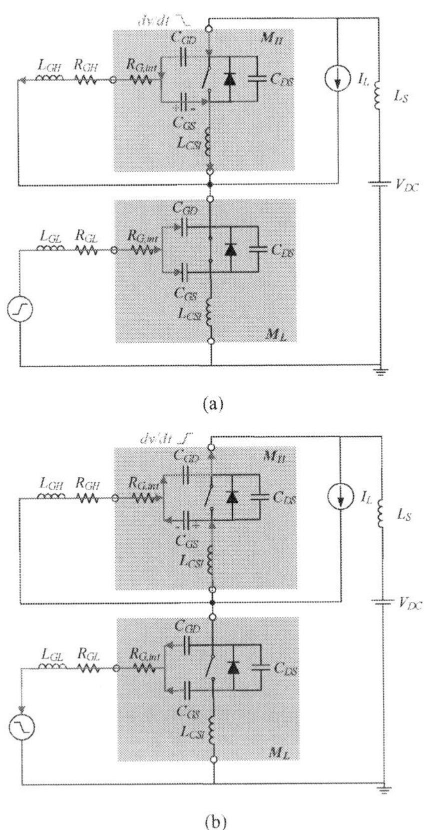

Fig. 1. (a) Turn-on and (b) turn-off transitions of LS transistor.

A mathematical model for shoot-through in half bridge configuration is developed firstly. The simplified equivalent impedance network without consideration of stray inductance is shown by Fig. 2(a). The drain-source voltage is simplified as a ramp function $v_{DS}(t) = kt$, and hence Laplace transform is given by $v_{DS}(s) = k/s^2$,where k is the dv/dt ratio. According to the voltage-divider relationship, $v_{GS}(s)$ is given by

$$v_{GS}(s) = \frac{\frac{kC_{rss}}{C_{iss}}}{s\left(s + \frac{1}{R_G C_{iss}}\right)} \tag{1}$$

where R_G is a sum of internal and external gate resistances, and all the capacitances are assumed to be constant. Applying inverse Laplace transform to Eq.(1), the time-domain relationship is given by

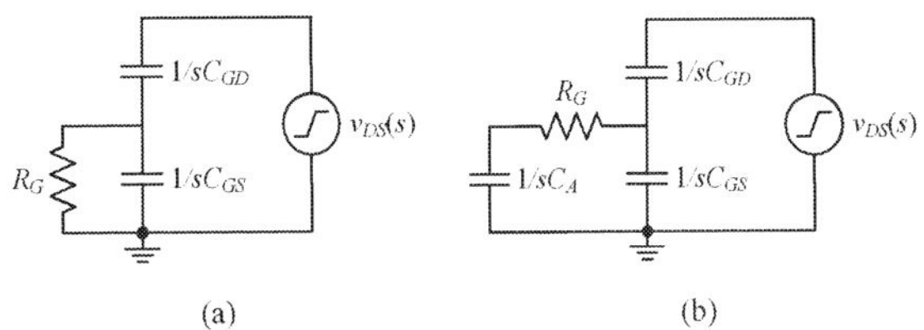

Fig. 2. Equivalent impedance network: (a) conventional gate driver, (b) novel gate driver with gate assisted circuit.

$$v_{GS}(t) = kR_G C_{rss}\left[1 - \exp\left(-\frac{t}{R_G C_{iss}}\right)\right] \tag{2}$$

The peak gate-source voltage $V_{GS,pk}$ is obtained at the end of drain voltage ramp

$$V_{GS,pk} = \pm kR_G C_{rss}\left[1 - \exp\left(-\frac{V_{DC}}{kR_G C_{iss}}\right)\right] \tag{3}$$

During turn-on transition, $V_{GS,pk} > 0$, while during turn-off transition, $V_{GS,pk} < 0$. For the ideal situation, R_G becomes 0, and shoot-through is non-existing. For the worst situation, R_G becomes infinity and $V_{GS,pk}$ can be approximated by

$$\lim_{R_G \to \infty} V_{GS,pk} = \pm \frac{C_{rss}}{C_{iss}} V_{DC} \tag{4}$$

Based on the discussions above, shoot-through issue can be mitigated by smaller gate resistance, smaller C_{rss}/C_{iss} ratio or negative bias. Compared with Si IGBT, SiC MOSFET has a smaller C_{rss}/C_{iss} ratio. For example, the C_{rss}/C_{iss} ratio of Cree 1.2 kV, 30 A SiC MOSFET at 25 V is 0.02 [12], while the ratios of two commercial IGBTs [13, 14] with the same rating are 0.051 and 0.038 respectively. Hence, gate driver design becomes the major challenge to eliminate shoot-through.

III. Proposed Gate Assisted Circuit to Eliminate Shoot-Through

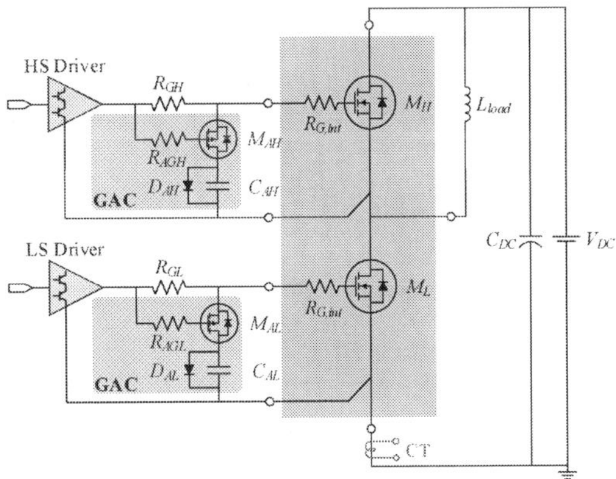

Fig. 3. Switching testing circuit for SiC half bridge module with proposed gate assisted circuit.

A novel GAC to mitigate shoot-through in SiC half bridge module is proposed, which solve the issues in [8, 10]. As Fig. 3 shows, it consists of an auxiliary PMOS transistor, an auxiliary capacitor, a gate resistor and a diode. The gate resistor damps the gate signal oscillation of auxiliary transistor. The Schottky diode paralleling with the auxiliary capacitor is optional, which clamps the capacitor voltage.

Due to the use of PMOS transistor, the auxiliary transistors (M_{AH} and M_{AL}) can share the same gate driving voltage with the main transistors, which simplifies the gate driver circuit significantly. The 30 V, 8.8 A, 20 mΩ P-channel trench MOSFET FDS4435BZ from Fairchild Semiconductor is used as the auxiliary transistor. The peak value of V_{GS} is ±25 V, which has enough margin for application in SiC MOSFET. In addition, inside the SO-8 package, a pair of Zener diodes is internally parallel with the PMOS die to prevent breakdown of gate oxide. During hardware implement, the GAC should be put as close as possible to the main transistor to minimize the loop inductance.

The equivalent impedance network with the proposed GAC is shown by Fig. 2(b). Similar to earlier discussion, according to the voltage-divider relationship, $v_{GS}(s)$ can be derived by

$$v_{GS}(s)=\frac{\frac{kC_{rss}}{C_{iss}}\left(s+\frac{1}{R_G C_A}\right)}{s^2\left(s+\frac{C_{iss}+C_A}{R_G C_{iss} C_A}\right)} \quad (5)$$

Applying inverse Laplace transform, the time-domain relationship is given by

$$v_{GS}(t)=\frac{kC_{rss}}{C_{iss}+C_A}t+\frac{kR_G C_{rss} C_A^2}{\left(C_{iss}+C_A\right)^2}\left[1-\exp\left(-\frac{C_{iss}+C_A}{R_G C_{iss} C_A}t\right)\right] \quad (6)$$

The peak gate-source voltage is also obtained at the end of drain voltage ramp. After considering the initial voltage V_{EE} on C_a, it is given by

$$\begin{aligned}V_{GS,pk}&=V_{EE}\pm\frac{C_{rss}}{C_{iss}+C_A}V_{DC}\\&\pm\frac{kRC_{rss}C_A^2}{\left(C_{iss}+C_A\right)^2}\left[1-\exp\left(-\frac{C_{iss}+C_A}{kRC_{iss}C_A}V_{DC}\right)\right]\end{aligned} \quad (7)$$

where the +/- symbol is for turn-on/off transition.

The operating principle of GAC is discussed by Fig. 4. It is assumed only M_L is active, which means M_H is always off and M_{AH} is always on. For the complementary situation when M_H is active, it behaves in a similar way. The stray inductance of PCB layout and packaging is neglected. The switching time of auxiliary transistors are neglected, as C_{iss} of auxiliary transistor is much smaller than that of main transistor. The switching waveforms of the gate-source voltage of main transistor, auxiliary capacitor voltage and current are shown by Fig. 5.

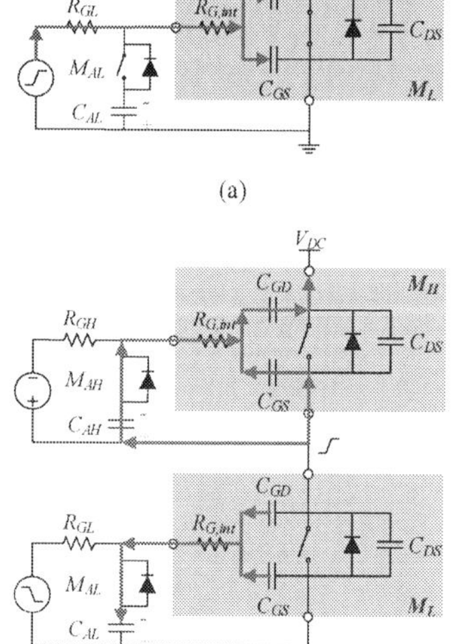

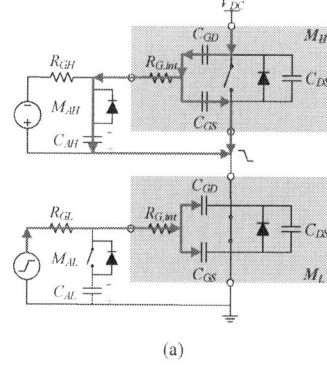

(a)

(c)

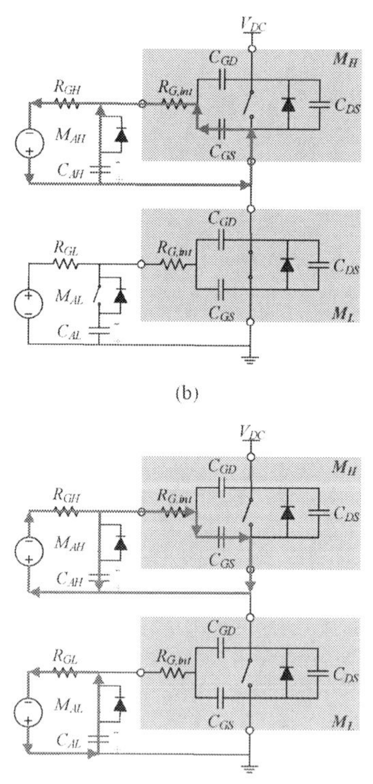

(b)

(d)

Fig. 4. Operating principle of GAC, (a) subinterval $t_0 \sim t_1$, (b) subinterval $t_1 \sim t_2$, (c) subinterval $t_2 \sim t_3$, (d) subinterval $t_3 \sim t_4$.

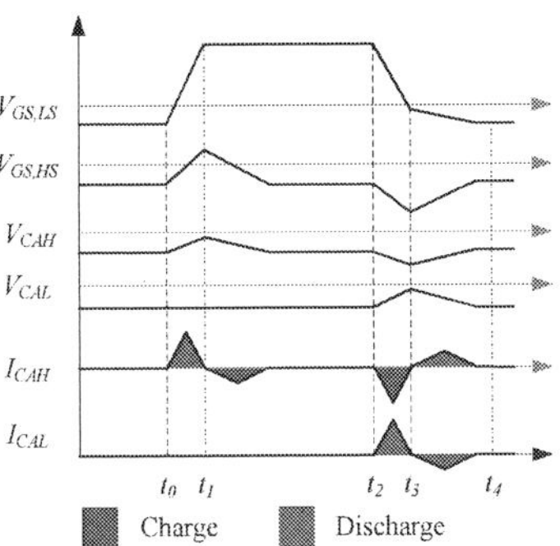

Fig. 5. Switching waveforms of gate-source voltage, auxiliary capacitor voltage and current.

TABLE 1. PARAMETERS FOR EVALUATION OF GATE-SOURCE VOLTAGE SPIKE.

Parameters	Value
R_G	1.8 Ω
$R_{DS,on}$	13 mΩ
C_{iss}	6.3 nF
C_{rss}	600 pF at 10 V, 37 pF at 750 V
dv/dt	750/50 V/ns
V_{EE}	-5 V
V_{DC}	10 V, 750V

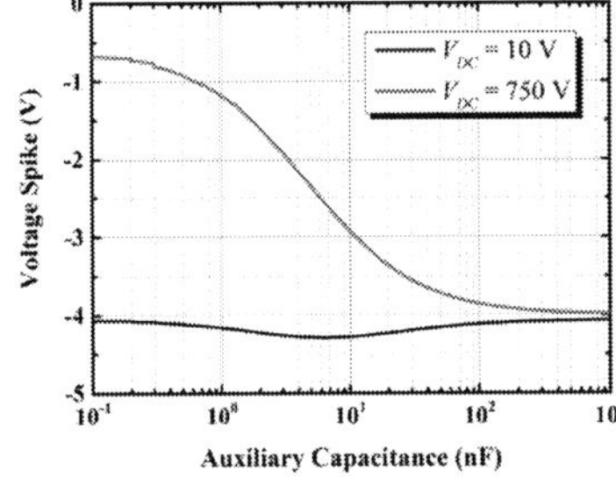

Fig. 6. Impact of auxiliary capacitance on gate-source peak voltage.

Subinterval $t_0 \sim t_1$: M_L is turned on and M_{AL} is off, which has no influence on the normal turn-on process of M_L. The common terminal between M_H and M_L is now pulled down to ground. Then a sudden voltage ramp occurs across the drain-source terminals of M_H. M_{AH} provides a low impedance path for the Cdv/dt current induced by Miller capacitance. Since the initial voltage on C_{AH} (V_{CAH} in Fig. 5) is negative, C_{AH} is discharged and the potential increases. Then $V_{GS,HS}$ is a sum of voltage drops across $R_{G,int}$, M_{AH} on-resistance $R_{DS,on}$ and C_{AH}. The small positive voltage drop across M_{AH} and the negative potential on C_{AH} help to mitigate shoot-through. The auxiliary Schottky diode is optional for the extreme situation when the potential on C_{AH} rises beyond zero, hence C_{AH} is clamped to avoid a floating positive potential. This period lasts until $V_{GS,LS}$ is fully pulled up to V_{CC} and M_L is fully turned on.

As Eq.(7) illustrates, the voltage spike of $V_{GS,HS}$ can be mitigated by a large auxiliary capacitance. To reach the optimized design, the relationship between the voltage spike and auxiliary capacitance needs to be derived. Due to the nonlinear property of Miller capacitance, two extreme situations are assumed with high bias/low capacitance and low bias/high capacitance to simplify the analytical model. The parameters are list in Table 1. From the computational results shown by Fig. 6, $V_{GS,pk}$ is far below the threshold voltage of SiC MOSFET with the proposed GAC. For high bias assumption, the voltage spike can be significantly reduced when the auxiliary capacitance is large enough. And the decreasing tendency will become insignificant when the auxiliary capacitance is above 100 nF. For low bias assumption, there will be an optimized value for the auxiliary capacitance, which is close to the input capacitance. And this voltage spike is much lower than the assumption of high bias.

Subinterval $t_1 \sim t_2$: M_L is in on and M_{AL} is still off. The negative power supply starts to charge C_{AH} via R_{GH} and channel of M_{AH} until it is completely charged to V_{EE}. Meanwhile, C_{GS} of M_H is also charged via R_{GH} and $R_{G,int}$. This period lasts until M_L is turned off.

Subinterval $t_2 \sim t_3$: M_L is turned off while M_{AL} is on. A low impedance path is provided for the discharging current of M_L and the turn-off speed is improved. C_{AL} is discharged and the potential increases (V_{CAL} in Fig. 5). Meanwhile, a Cdv/dt current is induced by Miller capacitance similar to subinterval $t_0 \sim t_1$, but in an opposite direction. Hence, C_{AH} is charged and the potential further decreases. To estimate the value of C_{AL}, an ideal assumption is made that all the positive charge on C_{iss} of M_L is neutralized by the negative charge on C_{AL}. Hence C_{AL} should hold enough negative charge to make sure the potential always below zero during this period. According to the charge balance equation

$$C_{AL}V_{EE} + C_{iss}V_{CC} < 0 \quad (8)$$

Considering V_{EE} = -5 V and V_{CC} = 20 V, C_{AL} should be at least 4 times of C_{iss}. C_{AH} is same as C_{AL} due to the circuit symmetry. Considering the effect of C_{AH} on the voltage spike, a 100 nF, 50 V ceramic capacitor is selected as the auxiliary capacitor, which has enough capability for shoot-through mitigation as well as turn-off enhancement.

During this period, M_{AL} mainly acts as a speed-enhancement device. Considering the threshold voltage of PMOS, it becomes less effective as $V_{GS,LS}$ gradually approaches V_{EE}. And the discharging current starts to flow through R_{GL}. This transition only occurs after $V_{GS,LS}$ is already below zero, hence the turn-off speed is unaffected actually. This period lasts until $V_{GS,LS}$ is fully pulled down to V_{EE} and M_L is fully turned off.

Subinterval $t_3 \sim t_4$: both M_H and M_L are off, which is the dead zone for half-bridge configuration. Meanwhile, both M_{AH} and M_{AL} are on. As the potential on C_{AH} is below V_{EE} after the previous subinterval, it is discharged via R_{GH} and M_{AH}. And C_{AL} is charged by the negative power supply via R_{GL} and M_{AL}.

IV. EXPERIMENTAL SETUP

A double pulse testing (DPT) experiment is set up to test the performance of gate driver with GAC, as Fig. 7 shows. And the schematic of DPT is shown by Fig. 3. The 1.2 kV, 120 A SiC half bridge module CAS120M12BM2 from Wolfspeed is investigated in present work [15]. The control signal is generated by the TI DSP TMS320F28335. A single-layer winding, air core inductor with low equivalent parallel capacitance is used as the current load. The other parts of DPT experimental will be discussed as follow.

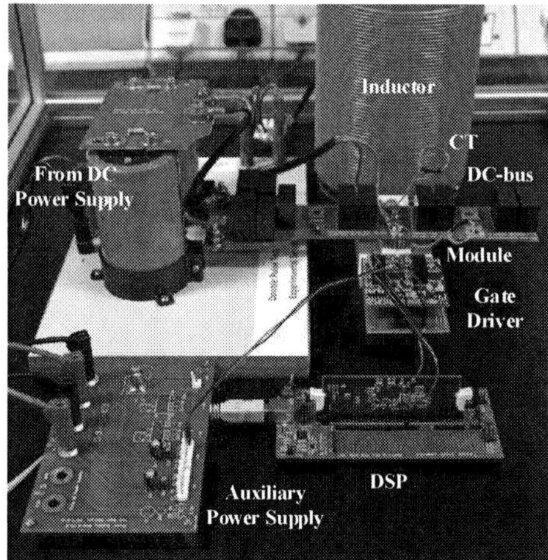

Fig. 7. Experimental setup.

A. Gate Driver

The schematic of isolated gate driver with GAC is shown by Fig. 8(a). The isolated single gate driver IC 1ED020I12-B2 from Infineon is used, which provides a galvanic isolation up to 1.2 kV based on coreless transformer (CLT) technology. Since the driver IC 1ED020I12-B2 can only provide a rail-to-rail output current of 2 A, a totem-pole driver with the co-packaged NPN and PNP transistors is used as the following amplifier stage. It provides a peak source/sink current of 10 A with a maximum power supply of 40 V. And the power supply is provided by the isolated DC/DC converter module.

The high *dv/dt* ratio resulted from the high-speed switching of SiC power converter will induce perturbation to logic circuits via Miller capacitor and coupling capacitor (coupling effect between signal side and power side). Hence, care must be taken in the isolation barrier between the two sides in PCB layout [16]. In addition, the GAC should be put as close as possible to the gate-source terminals to minimize the loop inductance, as Fig. 8(b) shows. The multi-layer PCB of half bridge gate drivers are designed to exactly fit the module and mounted onto it by solder, as Fig. 9 shows.

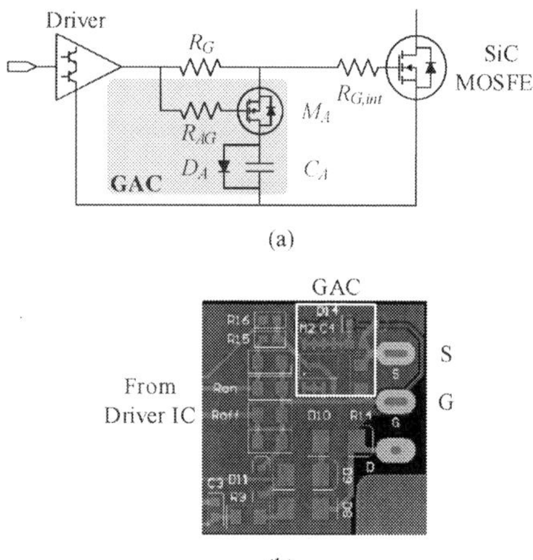

Fig. 8. Schematic (a) and layout (b) of gate assisted circuit.

Fig. 9. Prototype of gate driver.

B. DC-bus

The laminated structure is adopted for DC-bus to reduce the stray inductance contributed by bus layout [17, 18]. The positive- and negative-planes are overlapped with each other, with an isolation layer between the two planes. Hence, the two planes are strongly coupled to give a large mutual inductance. The DC-link capacitor consists of 2 film capacitors (10 μF, 500 V) in series to clamp the neutral point, and another 9 parallel film capacitors (4 μF, 900 V).

C. Current Measuring Technology

To measure the current, usually a bulky current probe has to be inserted into the circuit loop, which will increase the parasitic inductance significantly. In this work, a scale-down current transformer (CT) is made [19]. The primary side of CT only comprises a single turn, while the secondary side comprises 10 turns of isolated wire wound on a ferrite toroid. The CT is installed on the negative terminal of module by screw. A spacer also needs to be installed on the positive terminal. Finally, the output terminal of secondary side of CT is connected with Tektronix current probe TCP0030 (30A, 120MHz) for current measurement.

V. Results and Discussion

A switching test up to 750 V and 100 A is conducted, as Fig. 10 shows. The conventional gate driver with $R_{on} = R_{off} = 10\ \Omega$ is used as a comparison. The turn-off speed of M_L is significantly improved with GAC. As Fig. 11 shows, the turn-off loss of proposed gate driver with GAC almost keeps constant with load current increasing. By comparison, the turn-off loss of conventional gate driver increases significantly. For the normal turn-on process, it is totally unaffected, since M_{AL} is off during this period. And it allows a 34% reduction of total switching loss at 100 A, as Table 2 illustrates.

The *dv/dt* and *di/dt* ratios for both conventional and proposed gate drivers are summarized in Table 3. The increasing *di/dt* with GAC introduces an increasing voltage overshoot (115 V) compared with the conventional (90 V). And the voltage overshoot will be reduced significantly to 75 V if removing the CT. Hence, the proposed GAC permits the maximum turn-off speed without the voltage overshoot issue.

The waveforms of LS and HS gate voltage are measured to evaluate the shoot-through possibility. The CT is removed since the current measurement is unnecessary in this step. And the complementary gate signals are applied to LS and HS transistors.

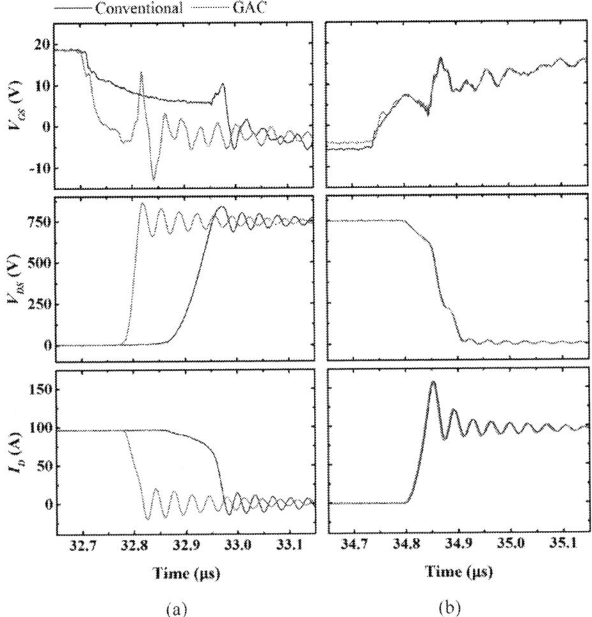

Fig. 10. Switching characterizations of conventional gate driver and proposed gate driver with gate assisted circuit: (a) turn-off, (b) turn-on.

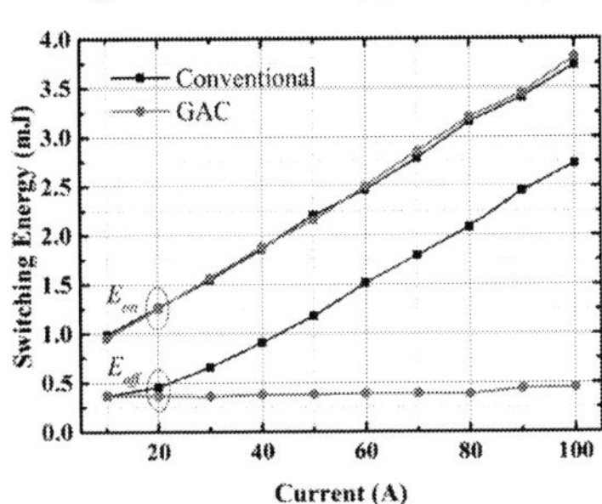

Fig. 11. Switching loss versus load current at 750 V.

TABLE 2. SWITCHING LOSS AT 750 V AND 100 A.

	Unit	Conventional	GAC
E_{off}	mJ	2.73	0.45
E_{on}		3.73	3.81
E_{tot}		6.46	4.26

TABLE 3. DV/DT AND DI/DT AT 750 V AND 100 A.

	Unit	Conventional		GAC	
		Turn-on	Turn-off	Turn-on	Turn-off
dv/dt	V/ns	20	11	20	31
di/dt	A/ns	4.2	3.6	4.2	4.2

As Fig. 12(a) shows, during the turn-off transition of LS transistor, the gate voltage across LS transistor shows more serious ringing due to the extremely fast turn-off speed. However, it brings no side-effect of shoot-through due to the dead time. As Fig. 12(b) shows, during the turn-on transition of LS transistor, an obvious gate voltage spike across HS transistor is observed for the conventional gate driver, which already exceeds the threshold voltage. By comparison, the gate voltage spike with proposed GAC is far below threshold voltage. Hence, the shoot-through issue can be completely eliminated.

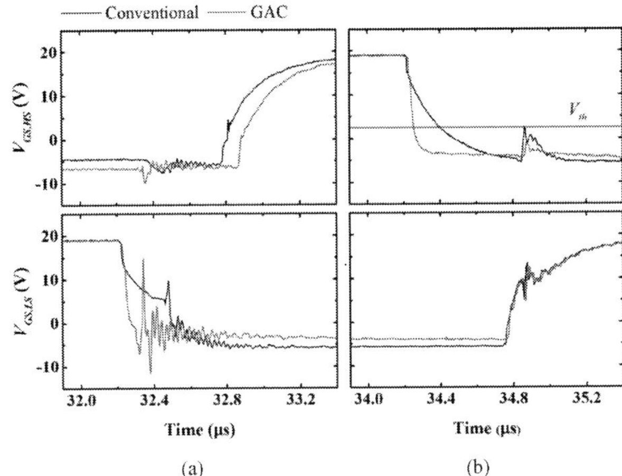

Fig. 12. Effect of gate assisted circuit on shoot-through mitigation: (a) turn-off, (b) turn-on.

VI. CONCLUSION

The shoot-through issue in the classical half bridge configuration has to be completely addressed before applying SiC power converter to the high-speed switching. In this work, a novel gate assisted circuit is proposed for shoot-through elimination. In addition, it also achieves fast turn-off speed and reduced switching loss. This has been experimentally demonstrated in our study with a switching test up to 750 V and 100 A. The proposed gate assisted circuit can drive the SiC MOSFET module with the maximum switching speed, but without such issues such as shoot-through and voltage overshoot.

ACKNOWLEDGEMENT

This work was conducted within the Rolls-Royce@NTU Corporate Lab with support from the National Research Foundation (NRF) Singapore under the Corp Lab@University Scheme.

REFERENCES

[1] T. Funaki, J. C. Balda, J. Junghans, A. S. Kashyap, H. A. Mantooth, F. Barlow, *et al.*, "Power conversion with SiC devices at extremely high ambient temperatures," *IEEE Trans. Power Electron.*, vol. 22, pp. 1321-1329, 2007.

[2] Z. Chen, Y. Yao, D. Boroyevich, K. D. Ngo, P. Mattavelli, and K. Rajashekara, "A 1200-V, 60-A SiC MOSFET multichip phase-leg module for high-temperature, high-frequency applications," *IEEE Trans. Power Electron.*, vol. 29, pp. 2307-2320, 2014.

[3] B. Callanan, "Application considerations for silicon carbide MOSFETs," *Cree Inc. Application Note*, 2011.

[4] A. Volke and M. Hornkamp, *IGBT modules, technologies, driver and application*: Infineon Technologies AG, Munich, 2012.

[5] R. Francis, P. Wood, and A. Alderman, ""Positive only" gate drive IGBTs created by Cres minimization," in *Proc. Power Convers. Intelligent Motion*, 2001.

[6] "Mitigation methods for parasitic turn-on effect due to Miller capacitor," *Avago Technologies Application Note*, 2010.

[7] B. Callanan, "SiC MOSFET isolated gate driver," *Cree Inc. Application Note*, 2012.

[8] Y. Zushi, S. Sato, K. Matsui, Y. Murakami, and S. Tanimoto, "A novel gate assist circuit for quick and stable driving of SiC-JFETs in a 3-phase inverter," in *Proc. IEEE Appl. Power Electron. Conf. Expo.*, 2012, pp. 1734-1739.

[9] T. Funaki, "A study on self turn-on phenomenon in fast switching operation of high voltage power MOSFET," in *Proc. IEEE CPMT Symp. Japan*, 2013, pp. 1-4.

[10] Z. Zhang, F. Wang, L. M. Tolbert, and B. J. Blalock, "Active gate driver for crosstalk suppression of SiC devices in a phase-leg configuration," *IEEE Trans. Power Electron.*, vol. 29, pp. 1986-1997, 2014.

[11] S. Yin, K. J. Tseng, C. F. Tong, R. Simanjorang, C. J. Gajanayake, A. Nawawi, *et al.*, "Gate driver optimization to mitigate shoot-through in high-speed switching SiC half bridge module," in *Proc. IEEE Power Electron. Drv. Syst.*, 2015, pp. 484-491.

[12] SiC MOSFET C2M0080120D, Wolfspeed Power [Online]. Available: http://www.wolfspeed.com/Power/Products/MOSFETs/TO247/C2M0080120D

[13] Trench gate and field-stop IGBT IKW15N120H3, Infineon Technologies [Online]. Available: http://www.infineon.com/cms/en/product/igbts/IKW15N120H3

[14] Trench gate and field-stop IGBT STGW15M120DF3, STMicroelectronics [Online]. Available: http://www.st.com/web/catalog/sense_power/FM100/CL826/SC953/PF260491

[15] SiC half bridge module CAS120M12BM2, Wolfspeed Power [Online]. Available: http://www.wolfspeed.com/Power/Products/SiC-Power-Modules/SiC-Modules/CAS120M12BM2

[16] M. Z. Choo, "Keep hybrid powertrain drives noise free by rejecting dvdt noise with isolated-gate drivers," *Avago Technologies Application Note*, 2011.

[17] M. C. Caponet, F. Profumo, R. W. De Doncker, and A. Tenconi, "Low stray inductance bus bar design and construction for good EMC performance in power electronic circuits," *IEEE Trans. Power Electron.*, vol. 17, pp. 225-231, 2002.

[18] K. Wada, M. Ando, and A. Hino, "Design of DC-side wiring structure for high-speed switching operation using SiC power devices," in *Proc. IEEE Appl. Power Electron. Conf. Expo.*, 2013, pp. 584-590.

[19] B. Callanan, "SiC MOSFET double pulse fixture," *Cree Inc. Application Note*, 2011.

978-1-4673-9551-9/16 $31.00 © 2016 IEEE

Methods to Enhance the Thermal Performance of a 3D Power Package

Jonathan Noquil, Ozzie Lopez and *Tianyi Luo
Texas Instruments in Lehigh Valley
*Lehigh University
Bethlehem, Pennsylvania, U.S.A

Abstract— **Integration of power semiconductor devices poses many challenges, i.e. in a synchronous buck converter package where stacking multiple dies concentrates heat dissipation within a given area, the effects of which could result in elevated junction temperatures and degradation in its performance and long term reliability. To counter these challenges one needs to have a semiconductor solution with extremely good efficiency and packaging solutions that can conduct heat extremely well in multiple directions.**

Since TI NexFET™ addresses the silicon efficiency and performance; we focused our work on enhancing the thermal performance of a 3D power package. By implementing new design modifications in the clips and reducing the mold cap thickness, we were able to improve the overall thermal performance of stacked synchronous buck converters. There are many methods to improve the thermal performance of the 3D power package and in this paper we will present three methods evaluated for achieving that.

I. Introduction

Integration of devices in a single package is one way to improve efficiency in power electronics. Among the popular families are in the DC-DC power supply circuits particularly in the synchronous buck converter applications. Common approaches to improve efficiency in DC-DC synchronous buck converters include reducing interconnect parasitics, reducing conduction losses in the MOSFETs through lowering the on state resistance of the silicon and lowering the switching losses by reducing Gate charges.

One way to eliminate interconnects parasitic resistance and inductance is via an innovative power stage packaging approach presented (see Figure 1). The innovative power stage package integrates MOSFETs vertically with a collocated driver. This reduces the package size footprint by more than half compared to the conventional side by side approach and both reduces and eliminates critical resistances and inductances that detract from efficiency. With the side by side integration (MOSFETs and IC) as the number of die to be assembled increases, the package footprint and size increases which means it will consume much more real estate when mounted on a pc board.

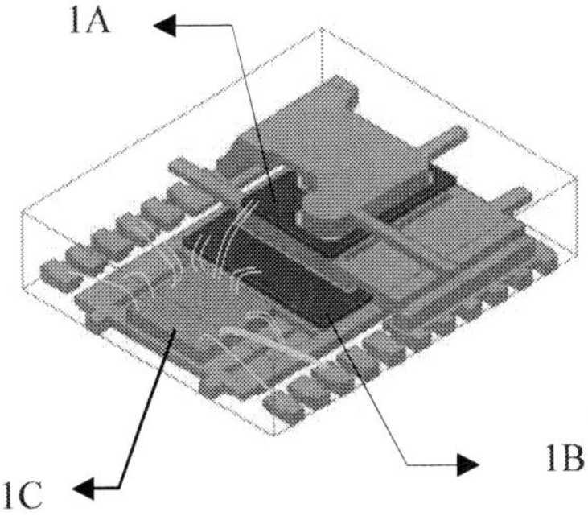

Figure 1: Power Stage Integrated Package (5 x6 mm)

II. Power Stage Integration Approach

Texas Instruments' NexFET Power Stack technology is integrated with the driver IC to form a 3 die combination called power stage. This achieves a small footprint and the lowest parasitics possible with a stacking topology that is used in the NexFET Power Block package design [2]. A patented source down silicon technology allows high-side die (Fig. 1A) to be stacked on top of the low-side transistor (Fig 1B) to implement a synchronous buck converter topology in a very simple and cost-effective manner. An integrated driver IC chip (Fig. 1C) is attached in close proximity to the stacked NexFETs eliminating lengthy electrical paths.

In order to realize a cost effective and reliable process, a simultaneous reflow and a gang placement technique of the assemblies are being utilized here. First, a low cost screen printing solder process to cater the die attachment of the driver IC and the bottom FET (low-side transistor). Next, the

978-1-4673-9551-9/16 $31.00 © 2016 IEEE

paste is then being dispensed on top of the bottom FET for the gang clip placement (multiple clips in one batch placement). A high side die is attached on top of the low-side clip and a high-side gang clip is then attached on the high-side die. The whole assembly then undergoes a simultaneous reflow which makes the process shorter, reliable, and cost effective with high manufacturing throughput.

To achieve better package efficiency, thick copper clips for the stack NexFET MOSFETs have been used. This enables high current connections which substantially reduces the package resistance and inductances typically associated with wire-bonded approaches. In addition, the implementation of thin silicon dies substantially reduces conduction losses by reducing the contribution of the device's substrate to on state resistance.

Finally, the stack die topology virtually eliminates the parasitic inductance and resistance between high- and low-side MOSFETS. In the stack configuration the drain of the low-side FET is connected directly to the source of the high side FET (control FET) or it is referred as the switching node of the package.

III. THERMAL AND RELIABILITY CHALLENGES

A. *Enhancing the Thermal Performance*

Power Stage NexFETs power stack technology eliminates switching and conduction losses due to its low inductive and resistive parasitics resulting in less power dissipation or heat generation [3]. TI's NexFET technology allows the source terminal to be at the bottom rather than on top of the MOSFET device which routes the heat dissipation directly onto the paddle and to the power ground node of the PCB board (see Figure 2) unlike conventional MOSFET technologies, where the source terminal of the device is always on the top

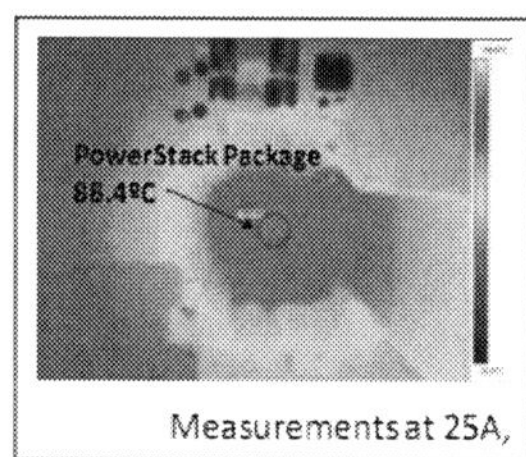

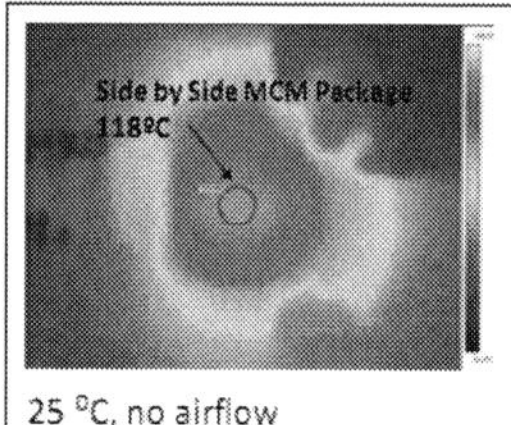

Measurements at 25A, 25 ºC, no airflow

Fig.2: Thermal measurements comparing Power Stack and conventional packaging with Power Stack cooler in temperature.

As power densities are increased further and the packaging complexity increases, thermal conduction from the device junctions to top of the package become critical aspects in enabling proper top side heat sinking. To enhance the thermal performance of the power stage 3D power package, three designs and assemblies were evaluated in order to compare the most feasible solution to meet this customer critical requirement.

A.1) Direct Heat Slug Attach

One approach evaluated on TI NexFET 3D power stage packages to improve its thermal performance was to attach a heat slug to the top of the overall stack structure of the package via soldering or welding as illustrated in Figures 3A and 3B.

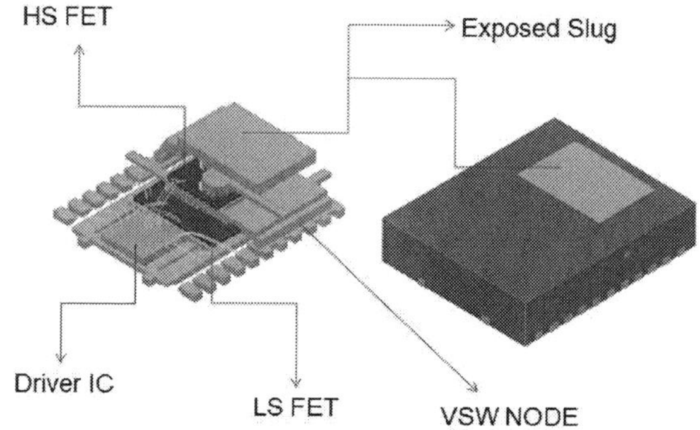

Fig.3a: Heat Slug Attach by Solder Dispense and Reflow

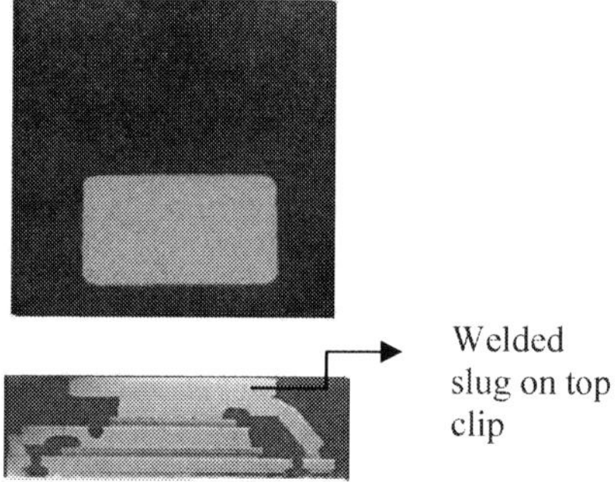

Fig. 3b: A Clip with Slug Welded on Top

As shown in figure3a, the heat slug is being attached on the top most clip of the package. This is done by a gang slug placement after a solder paste is being dispensed on top of the clip, and the whole structure is then reflowed at the same reflow temperature. Figure 3b is a clip with a pre-welded heat slug from the supplier which eliminates a solder dispense step. Dual film assisted molding exposes both top and bottom surface of the stack enabling a direct flow of heat on both sides. From the data measured, the overall thermal performance from junction to top of the 3D power package had been greatly improved by as much 92 % percent as compared to the standard 3D power stage package (Fig.1).

A.2) Expanding Clip Area and Reduction of Mold Cap Thickness

In the presence of an unexposed top of stack, simulation data indicated that mold compound and mold cap thickness can greatly impact the overall thermal performance of the package. By thinning the molding compound between the top of the clip and increasing the clip surface area it was possible to reduce an undesirable low K-thermal path of the package thereby improving the thermal performance. Figure 4 is an illustration of the structure of the thermally enhance 3D power package with expanded clip and reduced mold cap thickness. The top clip area has been expanded for better heat spreading and the mold cap thickness has been reduced to shorten the path of heat flow to the top of the package. In comparison from the standard 3D power package in figure 1, the mold cap thickness has been reduced from 0.44 mm to 0.140 mm.

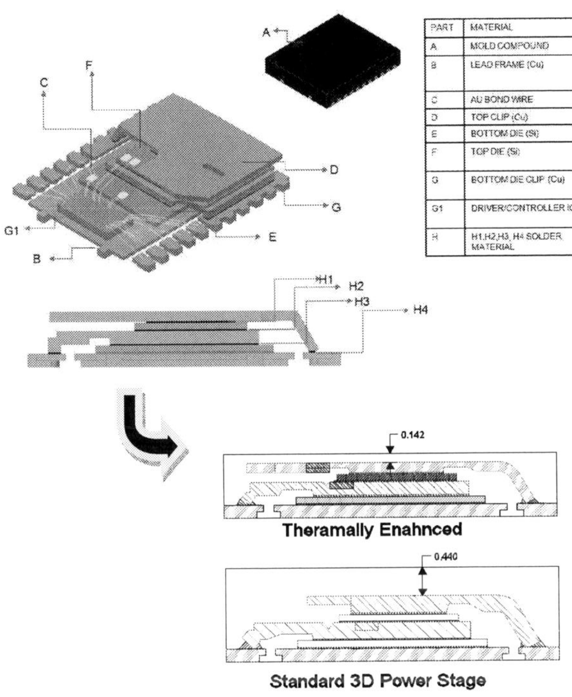

PART	MATERIAL
A	MOLD COMPOUND
B	LEAD FRAME (Cu)
C	AU BOND WIRE
D	TOP CLIP (Cu)
E	BOTTOM DIE (Si)
F	TOP DIE (Si)
G	BOTTOM DIE CLIP (Cu)
G1	DRIVER/CONTROLLER IC
H	H1,H2,H3, H4 SOLDER MATERIAL

Fig. 4: Thermally Enhanced 3D Power Package. Top Clip (D) was expanded to increase clip surface area for better heat dissipation.

Actual samples were assembled to validate the simulation results and measured using the ANALYSIS TECH thermal resistance measurement instrument as illustrated in Graph1.

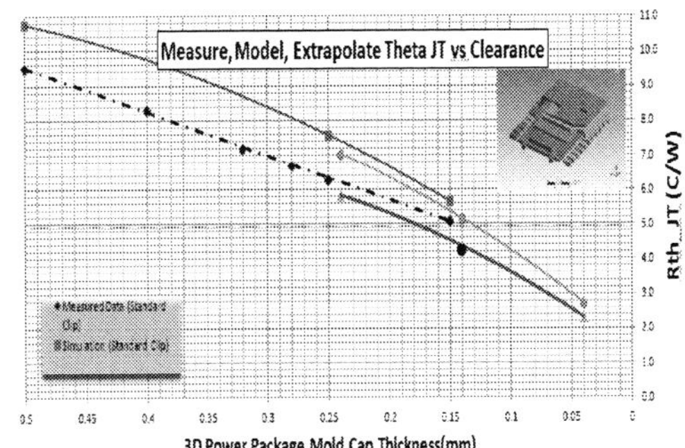

Graph 1.Mold Cap Thickness Vs. Rth_JT (C/W) Measurements.

Comparing the 3D power package described in Fig. 1 and the thermally enhanced 3D power package in Fig. 4, the Rth_JT was improved by about 60% which is a huge reduction in the thermal resistance.

Another benefit of the new design (thermally enhanced power package) was the ease of assembly since there were no new process steps added to the standard 3D power package.

A.3) Dual Side -Top and Bottom Exposed

Exposure of the clip without attaching any heat slug is another way to enhance the thermal performance of the 3D power package. After the package has been molded or encapsulated, a mechanical polish is applied to thin down the mold compound until the clip is exposed (Fig. 5)

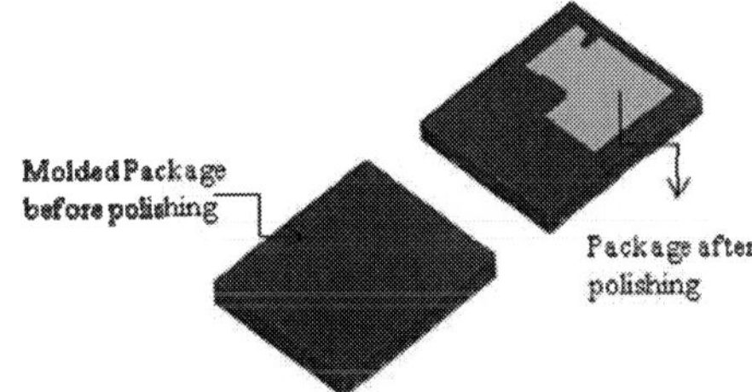

Fig. 5a: Exposed Top Clip by Mechanical Polishing

Theta JT= 5C/W

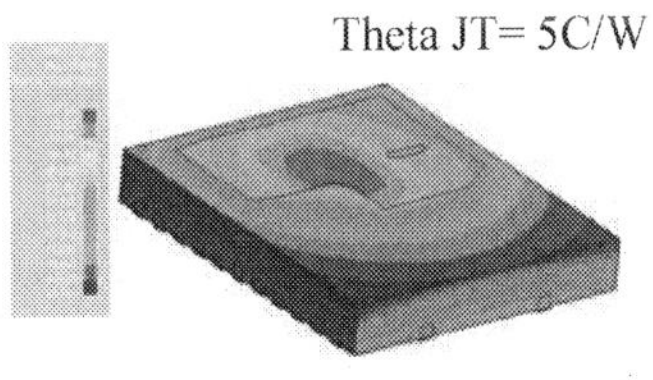

Fig. 5b: Model showing the thermal resistance of the package with exposed clip

Model data showed about 92% improvement of the thermal performance as compared to standard 3D power package from figure 1.

B. *Reliability Performance*

The thermally enhanced 3D Power Package reliability was evaluated through simulations and actual reliability testing. Simulations had predicted that the highest stress solder joint area can achieve the reliability requirements, i.e. thermal cycling at minus 40 to 125 deg C for more than 2000 cycles. The table below summarizes the simulation results. Solder part 1 is the life of solder joint for IC die attach. Solder part 2 to 5 are the solder joints of stack die interfaces beginning from bottom die attach up to top clip attach (Illustrated in Fig. 4).

Table 1: Solder joint life prediction for power stage.

Solder Part	Plastic Work density	Nf-QFN	Nf-Schubert
1	0.1321	2186	2720
2	8.11E-02	4397	4473
3	8.43E-02	4162	4301
4	7.95E-02	4530	4569
5	9.35E-02	3584	3868

Actual measurements demonstrated that the thermally enhanced 3D power package was very reliable and able to support an excess of 1000 thermal cycles without any degradation. This package was also able to successfully pass without degradation additional reliability tests such as IOL, HTGB, HTRB and ACLV with pre-conditioning level 2.

IV. Conclusions

The new improved designs of the 3D power package presented in this study concluded the following

1. The designs demonstrated excellent thermal performance (Rth_JT) on all three designs with excellent results in the direct heat slug attach (pre-welded clip and exposed clip). Theta JT is at 1.0 C/W which is an improvement of about 92%.
2. Thermally enhanced design with expanded clip area and thinner mold cap thickness demonstrated a huge reduction in thermal resistance as compared to standard 3D power package. It was verified that this design meets customer performance requirement.
3. Other than exposing the heat slug, there are two ways to improve thermal performance of the 3D power package without adding a new step of the assembly process. These are via reduction of the mold cap thickness and increasing the surface area of the clips.
4. Finally, the reliability performance of the design demonstrated beyond qualification requirements. With the actual reliability testing completed and passed, we can conclude that the thermally enhanced 3D power package is robust enough to meet customer's critical requirements.

References

[1] "High Frequency Power Module Having High Efficiency and High Current," Patern Application, U.S.PTO, November 2010.

[2] "Innovative 3D Integration of Power MOSFETs for Synchronous Buck Converters," J.Herbsommer, J.Noquil, O.Lopez and D. Jaurregui, Texas Instruments, APEC 2011.

[3] "3D Packaging Advancements Drive Performance, Power and Denssity in Power Devices," M. Romig and O. Lopez. Texas Instruments, white paper, July 2011: http://www.ti.com/lit/an/slit126/slit126.pdf

[4] "Next Generation of Power MOSFETs," Jacek Korec and Shuming Xu, Power Electronics Europe, May 2009: http://www.power-mag.com/pdf/issuearchive/29.pdf.

[5] "NexFET: A New Power Device," S. Xu, J. Korec, C. Kocon, D. Jauregui, Texas Instruments, International Electron Devices meeting, December 2009.

[6] "History of FET Technology and the move to NexFET," Jacek Korec and Chris Bull, Bodo's Power Systems, May 2009

[7] http://www.bodospower.com/pe/restricted/downloads/bp_2009_05.pdf.

[8] "Next Generation of Power MOSFETs," Jacek Korec and Shuming Xu, Texas Instruments, PCIM proceedings, 2009.

Highly Reliable and Cost Effective Thick Film Substrates for Power LEDs

Paul Gundel, Ryan Persons, Melanie Bawohl, Mark Challingsworth, Christoph Czwickla, Virginia Garcia, Christina Modes, Ilias Nikolaidis, Jessica Reitz, Caitlin Shahbazi
Heraeus Electronics – Innovation Thick Film
Heraeus Deutschland GmbH & Co. KG
Hanau, Germany
Paul.Gundel@heraeus.com

Torsten Nowak
Fraunhofer-Institut für Zuverlässigkeit und Mikrointegration IZM
Berlin, Germany
Torsten.Nowak@izm.fraunhofer.de

***Abstract*— This paper presents Thick Printed Copper (TPC) as substrate technology for High Brightness LEDs, which features a strongly improved reliability combined with a significant cost advantage over incumbent technologies for High Brightness LED substrates. The advantages of TPC over Direct Bonded Copper (DBC) and Direct Plated Copper (DPC) substrates will be demonstrated by thermal shock test results and a demonstration of the versatile design options. The good thermal performance of TPC is demonstrated by comparative Finite Element Modelling of TPC and DBC substrates. A detailed review is given discussing the different copper to ceramic bonding mechanisms, which prevent void formation and are responsible for the excellent thermo-mechanical reliability.**

Keywords—LED, packaging, substrates, thermal performance

I. Introduction

Thick-film technology for electric circuits is always used, when a circuit is subject to harsh environmental conditions, such as aggressive chemicals, high and low temperatures, high humidity, mechanical stress and strong temperature variations. Due to the high reliability and resistance to adverse conditions combined with excellent thermal and electrical properties thick-film circuits are well established in a wide range of industrial applications covering fuel level sensors, solar cells, headlights for cars and motor controllers.

Each thick-film circuit manufacturing consists of at least three steps:

1. Printing a thick-film paste on a suitable substrate. The paste can act as conductor, resistor or insulator. Common substrate materials are ceramics, metals and glasses. Typical pastes consist of metal or insulator powder, glass or metal oxides as adhesion promoter and an organic, to make the powders printable.
2. Drying the paste to release solvents.
3. Firing the substrate at temperatures between ~ 500 °C and 950 °C for bonding and powder sintering.

In the thick-film industry the use of screen printing is most widespread, but pad printing, dispensing, co-extrusion and jet printing are also used for certain applications.

Thick Print Copper (TPC) substrates are a highly reliable and cost effective alternative to Direct Bonded Copper (DBC) and Direct Plated Copper (DPC) substrates for High Brightness LEDs in level 1 as well as in level 2 packaging or for Chip on Board configurations. An exemplary TPC substrate for LED applications is outlined in Fig. 1. The metallization including copper via fills on TPC substrates is applied by screen printing of copper pastes on a suitable ceramic. TPC substrates can be built on alumina, aluminium nitride (AlN), Zirconia Toughened Alumina (ZTA) and on beryllia. After printing, the paste is dried at 125 °C on air for 10 minutes.

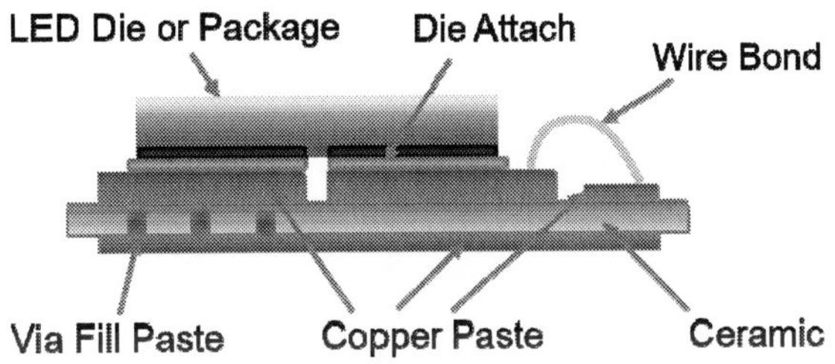

Fig. 1. Sketch of a LED application of TPC (flip chip in this case). TPC can be used as level 1 and level 2 packaging substrate, offering highest thermal conductivity and reliability. The back-side of the substrate can be mounted on a heat sink. Different copper thicknesses can be easily applied on the same substrate.

Before firing, additional copper layers can be printed on the same or on the opposite side of the ceramic substrate. Firing is done under nitrogen atmosphere at temperatures between 850 °C and 930 °C. By subsequent printing, drying and firing, copper layers with a thickness of up to 300 μm can be built. By choosing different screens for printing, different thicknesses can be formed on one substrate without any etching, laser or plating processes, giving engineers a freedom of circuit design and saving significant cost. For circuit assembly and die attach all common assembly technologies known on DBC and DPC

978-1-4673-9551-9/16 $31.00 © 2016 IEEE

can be used, including soldering, gluing, wire and ribbon bonding and silver sintering. TPC for automotive High Brightness LEDs is already proven in mass production [1] as highly reliable and cost effective technology, which can also be applied in power electronics. Technical details of TPC will be discussed in section 2.

Thermal simulation results of High Brightness IC's flip chip mounted on DBC and TPC substrates are discussed in the last section. The results include full thermal FEM simulations with different copper thicknesses, cooling modes and ceramic types (AlN and alumina). These simulations demonstrate, that the lower thermal conductivity of TPC results in negligible temperature differences below 0.6 K. This is in good agreement with [2], where the influence of the copper thickness on DBC substrates for LED substrates was simulated in detail.

II. Thick Print Copper

In this section the main physical properties of TPC substrates relevant for LEDs are discussed covering bondability, fine line capability, warpage variation, bonding mechanism, reliability under thermal cycling. And thermal performance based on Finite Element Modelling (FEM) of a LED Flip Chip Array will be given.

A. Thick Print Copper – Bondability

The bondability of TPC substrates is crucial for many LED applications. The excellent bondability of TPC is demonstrated by shear force test results of 300 µm thick Al H11 wires and 300 µm thick CuCorAl wires in comparison to DBC (Fig. 2). The bonding parameters are the same for TPC and DBC substrates.

For 300 µm Al H11 wires the shear forces on DBC and TPC are very similar, while with CuCorAl wires even higher shear forces are achieved on TPC.

To prove the long term stability of bonding wires on TPC the pull forces of 300 µ Al H11 wires is measured before and after 500 thermal shock cycles (-40 °C; +150 °C): Before the cycling the minimum pull force is 6.1 N, the maximum pull force 7.2 N and the average pull force 6.7 N. After 500 thermal shock cycles, these values are effectively unaltered with a minimum pull force of 5.2 N, a maximum of 7.7 N and an average of 6.5 N.

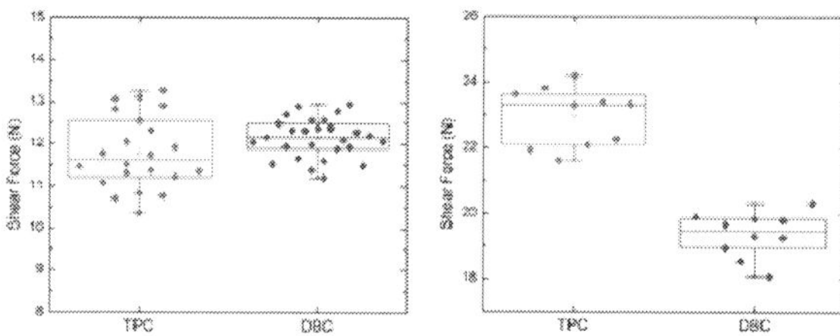

Fig. 2. Shear forces on TPC and DBC of 300 µm H11 (left side) and 300 µm CuCorAl (right side) bonding wires. While for Al H11 wires the shear forces are effectively the same, for CuCorAl wires the shear force on TPC is even higher than on DBC.

B. Thick Print Copper – Fine Lines

For LED applications a high spatial resolution of the metallization layer is required. The high spatial resolution of TPC is demonstrated in Fig. 3. In this case line and pitch dimensions of 75 µm are achieved. If required, thick-film pastes are able to provide fine-line resolutions of below 50 µm for example in solar cell mass production. The combination of fine structures with very thick structures on one substrate gives engineers additional options for circuit layout. Even higher spatial resolution can be achieved by laser structuring of TPC.

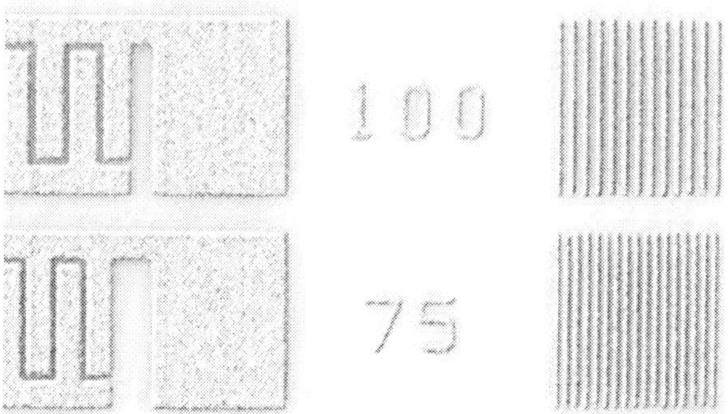

Fig. 3. Microscope image of a TPC test pattern on alumina. In the bottom row line and space dimension of 75 µm are realized.

C. Thick Print Copper – Warpage Variation

For the stability of LED systems, a low warpage variation on temperature changes is desirable, since additional warpage puts mechanical stress on solder/sinter/adhesive joints and bonding wires. In [3] the warpage variation upon heating of TPC was compared to DBC and DPC. The results of this publication are replicated in Fig 4, showing, that while for DPC and DBC the warpage varies more than 350 µm when heated from room temperature to 260 °C, the variation for TPC is below 30 µm for the same substrate layout. This means a significantly reduced stress on joints on the substrate for TPC in comparison to DPC and DBC. The low warpage of TPC can be ascribed to the remaining pores within the copper layers, which act as buffer for the mismatch in coefficients of thermal expansion. In addition, TPC substrates are completely symmetrical, since front and back-side are fired in one firing step, whereas for DBC each side is bonded in separate steps leading to hysteresis effects.

978-1-4673-9551-9/16 $31.00 © 2016 IEEE

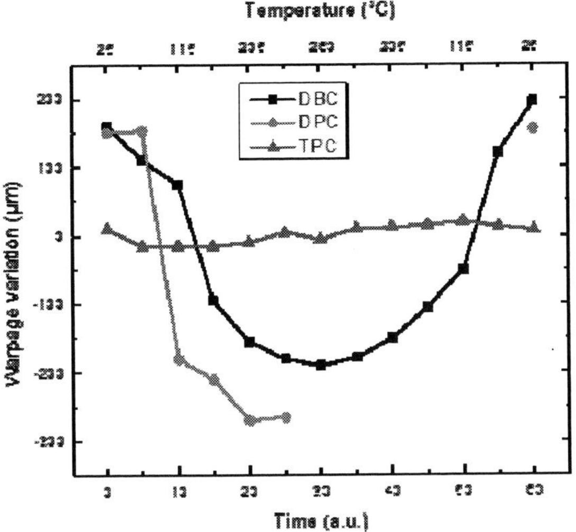

Fig. 4. Warpage variation around a median for DBC, DPC and TPC following [3]. For DBC and DPC the warpage changes by more than 350 µm from 25 °C to 260 °C. TPC provides a significantly lower warpage variation of less than 30 µm for the same temperature range and layout.

D. Thick Print Copper - Bonding Mechanism

The bonding mechanism of copper on alumina for DBC by eutectic formation at 1065 °C is well known [4]. Here we analyze the bonding interface of TPC on alumina and ZTA by means of SEM analysis. The SEM images in Fig. 5 show the wide interface layer for TPC, which extends 20 µm or more into the ceramic giving no room for voids.

To clarify the composition of the interface area, EDX analysis is performed for the yellow marked areas in Fig. 5. The results of this analysis is displayed in Table 1.

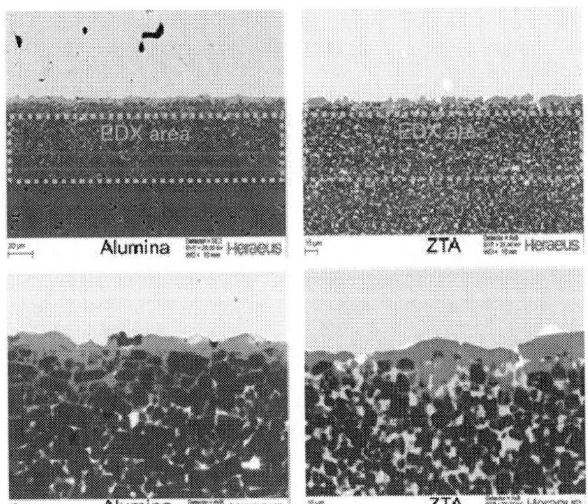

Fig. 5. Cross section SEM images of TPC substrates on alumina (left side) and ZTA (right side). The bright areas on top represent the copper layer, the speckled areas on the bottom represent the ceramic. The two images on top are the overview images, below details are shown. The bright dots within the alumina, represent the adhesion promoter and copper from the paste that soaks into the ceramic. In ZTA the contrast is less clear, but EDX reveals the same mechanism as for alumina.

TABLE I. EDX RESULTS OF THE YELLOW MARKED AREAS WITHIN THE CERAMICS IN FIG. 6. THE ELEMENT ANALYSIS PROVES, THAT COPPER AND ADHESION PROMOTER FROM THE CU PASTE SOAK INTO THE CERAMICS FORMING A STRONG BOND BETWEEN COPPER LAYER AND CERAMIC.

TABLE II. MATERIAL EDX	Alumina	ZTA
Ceramic	94 %	96 %
Adhesion promoter	5 %	3 %
Copper	1 %	1%

E. Thick Print Copper - Reliability

To prove how minimized warpage variation (section 2.3) and wide interface area (section 2.4) contribute to a strongly improved reliability of TPC, DBC and TPC substrates are subjected to thermal shock tests. For this, the same test-layout is prepared on alumina for DBC and on AlN, ZTA and alumina for TPC with a copper thickness of 280 µm. During the shock cycling the samples are held at -40 °C for 15 minutes, then the parts are transferred within 15 s to a chamber at +150 °C, where they are held for another 15 minutes before being transferred back to -40 °C. Each 100 cycles the substrates are optically checked for delamination between copper and ceramic.

It is well known, that for DBC conchoidal fracture occurs after around 100 cycles [5]. Our measurements confirm this phenomenon after 200 cycles. In Fig. 6 a DBC substrate on alumina after 200 thermal shock cycles is compared to a TPC substrate on AlN after 2500 cycles.

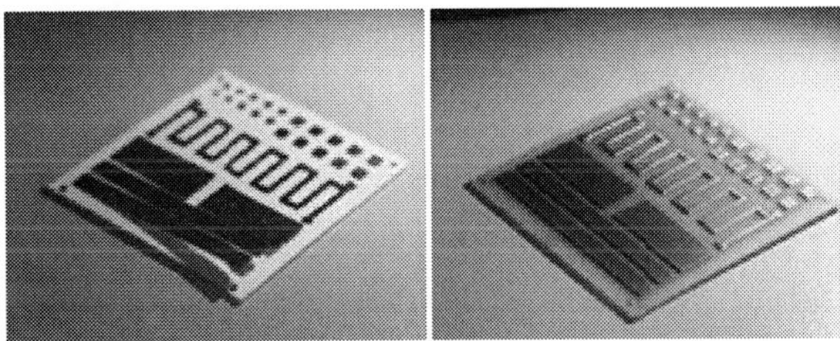

Fig. 6. A fractured DBC substrates on alumina after 200 thermal shock cycles (left side) and an intact TPC on AlN after 2500 thermal shock cycles. DBC on AlN is known to delaminate even before 100 cycles [5].

For a comprehensive overview, the measured numbers of cycles before delamination are summarized in the graph in Fig. 7. This clearly demonstrates the strong reliability advantage of TPC over DBC. In [6] it was demonstrated, that TPC is even more stable than AMB substrates on silicon nitride.

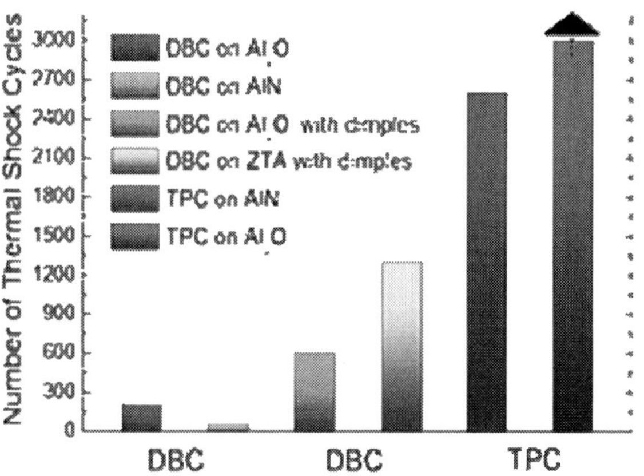

Fig. 7. Comparison of reliability for DBC and TPC substrates on alumina, aluminium nitride and. DBC alumina and AlN substrates fail before 200 thermal shock cycles, DBC substrates on ZTA with dimples fail before 1500 cycles, while TPC is still stable after 2600 cycles on AlN and at least 3000 cycles on alumina for the same layout and cycling parameters. The cycling of the TPC substrates on alumina is still being continued, since no failures have occurred so far.

F. Thick Print Copper – Thermal Performance

As can be seen in Fig. 5 the copper layer of sintered TPC substrates contains some residual pores. In combination with the non-conductive adhesion promoter embedded in the copper layer this leads to a reduced electrical and thermal conductivity. The electrical conductivity equals 70 % of the conductivity of copper [3] and the thermal conductivity was measured by the laser flash method [7] to be 290 W/m·K at 25 °C compared to copper on DBC substrates of 385 W/m·K [e.g. 8].

To evaluate the impact of the reduced conductivity on the overall thermal performance of TPC substrates in comparison to DBC substrates thermal simulations for a LED package were executed at Fraunhofer IZM, which uses ANSYS Workbench v15.0 for Finite Element Modelling (FEM).

1) Simulation Model

The LED used in the model are Luxeon's 1 mm² flip chip LED package L0F2-B4501 [9]. This LED has a 1 W power dissipation at standard conditions. The LED is attached to the substrate with a lead-free SnAg2.5Cu0.5In2 solder, which provides a thermal conductivity of 60 W/m·K [10] and a thickness of 35 µm. The substrate is covered by copper of the same thickness on both sides. A cross section of the model with a typical temperature distribution is depicted in Fig. 8.

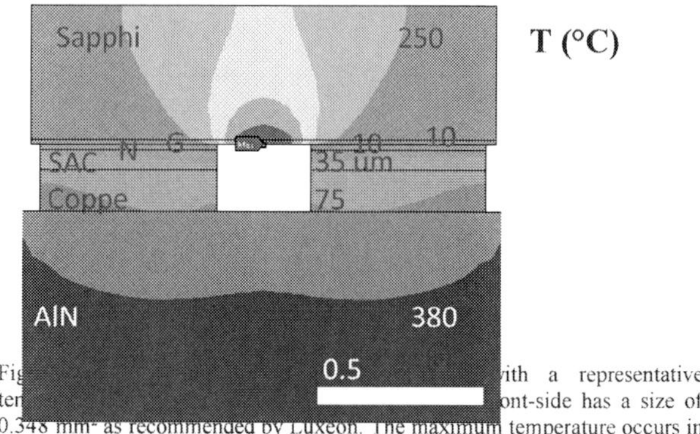

Fi… …ith a representative te… …ont-side has a size of 0.348 mm² as recommended by Luxeon. The maximum temperature occurs in the middle of the active layer (red arrow).

2) Simulation Variable

In order to assess the thermal performance of TPC under a wide range of conditions the following input parameters are varied: ceramic type, copper layer thickness and cooling mode. This is done for TPC and DBC in comparison. The used parameters of these variables are summarized in Tab. 1. The thermal conductivity of AlN and alumina are extracted from [12].

TABLE III. VARIABLE INPUT PARAMETERS FOR THE THERMAL FEM SIMULATIONS. ALL 18 POSSIBLE COMBINATIONS OF THE VARIABLES ARE SIMULATED FOR DBC AS WELL AS FOR TPC.

Variable	Ceramic	Copper thickness	Cooling mode at substrate back-side
Value 1	Alumina	75 µm	Fixed at T_{const} = 20 °C
Value 2	AlN	130 µm	Est. Air cooling, h_c = 150 W/m²·K
Value 3		250 µm	Est. Fluid cooling, h_c = 1500 W/m²·K

The first "fixed" cooling mode represents the ideal case with a fixed temperature of 20 °C on the back-side, the second "air" mode represents a typical value with air cooling and the third "fluid" value represents a fluid cooled back-side. The copper thicknesses are typical values for thick print copper substrates as they are already used in industrial mass production. The thickness of the alumina and of the AlN ceramics is set at 0.38 mm.

3) Results on Alumina

A very critical parameter for LED operation is the maximum temperature within the active layer. In Fig. 9 this maximum temperature on alumina substrates is shown for the three different cooling modes. The temperature increases with increasing copper thickness. On alumina the temperature difference is between 0.2 K for forced cooling and 1 K for air cooling lower on DBC than on TPC. For DBC thin copper thicknesses of less than 150 µm are difficult to achieve and expensive in contrast to TPC, where lower copper thicknesses are cheaper and can be easily realized with thick film printing. Therefore, it is important to note, that TPC with 75 µm thickness performs better or equal to DBC at a realistic DBC copper thickness of 250 µm.

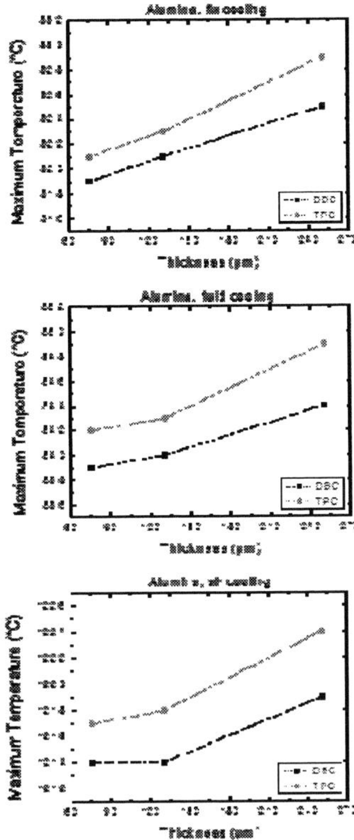

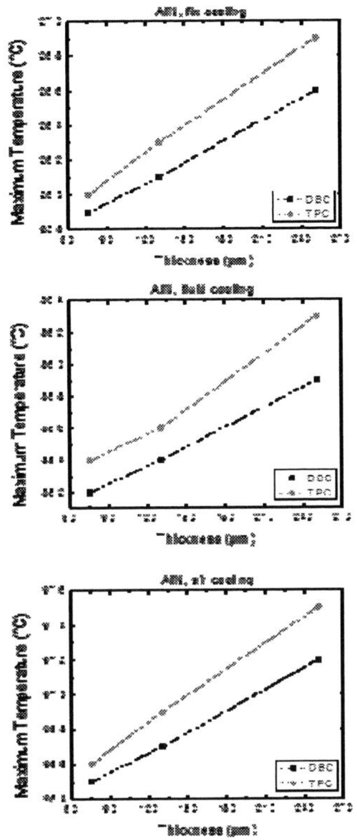

Fig. 9. Maximum temperature within the LED on alumina substrates for the three simulated cooling modes: fixed on the left side, fluid cooling in the middle and air cooling on the right side. The dashed lines are only guides to the eye. The maximum temperature increases with increasing copper thickness. Due to the lower thermal conductivity of TPC the temperature is always slightly higher on TPC substrates. The simulated differences are within 1 K.

Fig. 10. Maximum temperature within the LED on aluminum nitride substrates for the three simulated cooling modes: fixed on the left side, fluid cooling in the middle and air cooling on the right side. The dashed lines are only guides to the eye.

4) Results on Aluminum Nitride

In Fig. 10 the simulated maximum temperature in the LED on AlN substrates is displayed. While the temperature on AlN is lower than on alumina due to the superior thermal conductivity of AlN, the temperature trends are the same on both ceramics. For practical purposes, it is important, that also on AlN TPC at 75 µm thickness performs better than DBC with a realistic DBC copper thickness of 250 µm.

III. Summary

This paper demonstrates the advantages of copper thick-film substrates for High Brightness LED applications. The focus is laid on the reliability aspects of the substrates and the thermal properties. The high reliability of TPC substrates on alumina and AlN under thermal shock cycling is demonstrated, showing stability of TPC up to at least 2500 thermal shock cycles from -40 °C to +150 °C. This excellent reliability is attributed to the wide interface region between copper and ceramic and the low warpage variation upon temperature changes. The low warpage variation also reduces the stress to any joint on the substrate in comparison to DPC or DBC substrates, which can further increase the lifetime of LED packages.

While the residual porosity in the sintered copper layer contributes to the high reliability of TPC, it also reduces the thermal and electrical conductivity by about 30 % compared to the copper layers of DBC substrates. The effect of the reduced conductivity is investigated based on a Finite Element Model for a high brightness LED substrate. The simulation shows for air cooling conditions with a heat transfer coefficient value of 150 W/m²·K on AlN the highest temperature is within 0.1 K for DBC and TPC for the best performing copper thickness (75 µm). On alumina the difference is 0.2 K for 75 µm thick copper. For fixed cooling conditions with a back-side temperature of 25 °C or fluid cooling with 1500 $W/m^2 \cdot K$ at the back-side these differences become even smaller. The smaller difference on AlN than on alumina is caused by the good thermal spreading in AlN, which has a more than six times higher thermal conductivity compared to alumina.

While thin copper layers are impossible or expensive to achieve with DBC technology, thin copper layers down to 20 μm can be easily achieved with TPC and are even cheaper than thick layers due to the additive nature of TPC. Therefore, it is crucial to note, that the thermal performance of TPC at 75 μm copper thickness is better than that of DBC at for realistic DBC copper thicknesses of more than 150 μm. This can be explained by the fact, that the thinner layers that are possible with TPC compensate the higher thermal conductivity, when the thermal resistance of the full system is taken into account. This proves that the thermal performance of TPC is well suited for high brightness LEDs and power electronics applications.

In combination with an attractive cost structure, thick-film copper substrates are highly competitive with DBC and DPC for High Brightness LED applications.

[1] Press release: http://www.pressebox.de/pressemitteilung/heraeus-holding-gmbh/Heraeus-auf-der-PCIM-Materialien-und-Materialsysteme-im-Fokus/boxid/738952 (2015).

[2] Min Seok Ha, "Thermal analysis of high power LED arrays", Master thesis Georgia Institute of Technologyy (2009).

[3] V. Wei, M. Huang, R. Lai, R. Persons, "A comparison study for metalized ceramic substrate technologies: for high power module applications", In Proc. of IMPACT 2014 (2014).

[4] C.I. Helgesson, "Ceramic to Metal Bonding", Bost. Tech. Publ., Boston, Massachusetts (1969).

[5] D. C. Hopkins, "Direct Bonded Copper", Ph.D. thesis (2003).

[6] D. Hamilton, "High temperature reliability of power module substrates", In Proc. of PCIM Europe 2015 (2015).

[7] W.J. Parker, R.J. Jenkins, C.P. Butler, G.L. Abbott; "Flash method of determining thermal Diffusivity, heat capacity, and thermal conductivity", J. Appl. Physics.32 (1961).

[8] P. Gundel, M. Bawohl, M. Challingsworth, et al., "*Thick Printed Copper as Highly Reliable Substrate Technology for Power Electronics*", In Proceedings of PCIM Europe (2015).

[9] H. Cotal, J. Frost, "Heat transfer modelling of concentrator multijunction solar cell assemblies using finite difference techniques", In Proceedings of 35th IEEE PVSC (2010).

[10] Datasheet „LUXEON FlipChip Royal Blue", http://www.lumileds.com/uploads/436/DS116-pdf (2015).

[11] J. Bilek, J. Atkinson, W. Wakeham, "Thermal conductivity of molten lead free solders", In Proceedings of EMPS (2004).

[12] CINDAS LLC material database.

Design and Evaluation of SiC-Based High Power Density Inverter, 70kW/liter, 50kW/kg

Koji Yamaguchi
Power Electronics G., Electrical System Dept.
IHI Corporation
Yokohama, Japan
Koji_yamaguchi@ihi.co.jp

***Abstract*—This paper introduces the development and experimental performance of SiC-Based high power density inverter. The Power density of the developed inverter is about 70kW/litter in volumetric, 50kW/kg in gravimetric. The inverter is forced air cooled 2-level voltage source inverter. In order to achieve higher power density than conventional inverters, we need to reduce losses of inverters or improve cooling systems of inverters. First, we developed small light-weight SiC-MOSFET power module which has good thermal conductivity. We also developed gate driver to reduce switching losses and switching delay time. The prototype high power density inverter is developed with the developed power module and the proposed gate driver. The volume of the prototype inverter is about 0.5 liter and the weight is about 660g. Experimental results confirmed that the prototype inverter is able to operate up to 35kW, which means the power density of the prototype inverter is about 70kW/liter and 50kW/kg. The efficiency of the prototype inverter is more than 98%.**

Keywords—SiC; MOSFET; gate drive; high power density; inverter

I. INTRODUCTION (*Heading 1*)

High power density has become one of the key topics of the next generation power electronics applications. Research and development for higher power density seem to be accelerated by emerging devices such as Silicon Carbide (SiC) and Gallium Nitride (GaN). It is predicted that the power density of power converters will approach 30kW/liter around 2015 in [1]. Power density of converters was investigated well in [2].

Making switching frequency higher is one promising choice to reduce volume of passive components. 2.5MHz switching is investigated to achieve 30kW/liter in [3]. New cooling concept for magnetic devices is also proposed for high power density in [4]. The purpose of our research is to achieve higher power density than 30kW/liter or other latest works.

The next generation aircrafts such as More Electric Aircraft (MEA) and All Electric Aircraft (AEA) strongly need high power density converters [5] - [10]. 3.03kW/liter and 3.59kW/kg three phase ac-dc-ac converter is developed in [5]. High power density converters are getting more important for vehicles as well [11] [12]. Highly integrated double side cooling power module is developed for the next generation inverters for vehicles in [11]. Other double side cooling structures are proposed in [13] – [15]. Double side cooling methods are effective for integrated modules, but it might be too complicated to implement easily. Single side cooling structure was chosen for the prototype inverter in this paper to avoid the inverter from being too complicated for the first step. The prototype inverter has heat sink which is direct soldered to the base plate of the power module without thermal interface material such as silicon paste.

The prototype SiC power module removed Schottky Diodes (SBDs) to be smaller, while some commercialized power modules have anti-parallel SBDs. Some positive effects of removing SBDs are introduced in [16] such as high speed and low noise switching capability.

Active gate drive methods are proposed to improve switching characteristics for Silicon IGBTs in [17]. Active gate drivers for SiC hybrid modules are also proposed in [18] [19], but active gate drivers for full SiC MOSFETs have not been proposed before. The author proposed gate driver for SiC MOSFETs to reduce switching losses and delay time in [20]. The proposed gate driver in [20] is applied for the high power density inverter in this paper to reduce losses and dead time as much as possible.

The proposed inverter uses synchronous rectification to reduce conductive losses of body diode of SiC MOSFETs while freewheeling. The first section introduces the prototype power module, and then the second section shows some experimental performance of the prototype high power density inverter. The last section provides the conclusions and future works of this research.

II. PROTOTYPE SIC POWER MODULES

A. Structure of the modules

Fig. 1 shows cross section structure of the developed power module. The cupper heat sink is directly soldered to the cupper base plate of substrate. Conventional structures have thermal interface material such as silicon grease instead of solder. The proposed direct soldered structure significantly improves thermal conductivity of power modules. Improved thermal conductivity allows power modules to handle bigger power.

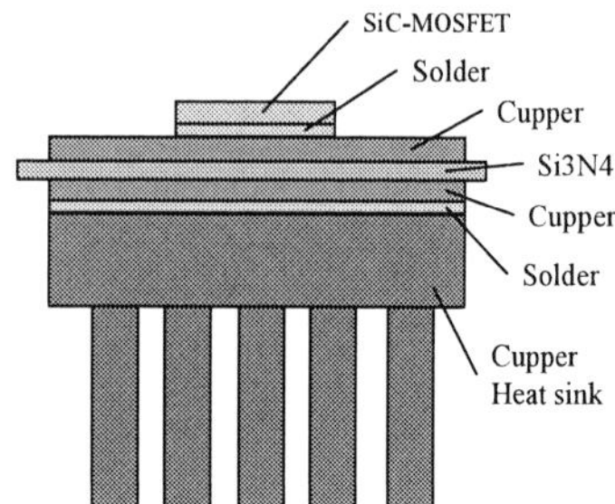

Fig. 1. The proposed structure of the power modules.

The thermal resistance from junction of SiC-MOSFET to heat sink *Rth(j-f)* is about 1.56K/W when silicon grease G-765 (Shinetsu silicon) is used. *Rth(j-f)* of the proposed structure is about 0.58K/W, so the proposed structure reduce about 60% of thermal resistance between junction of SiC- MOSFET and heat sink.

Fig. 2 shows the prototype of the developed power module. The power module has 8 paralleled 1200V SiC-MOSFETs and ceramic capacitors between dc input terminals to reduce inductance of commutation loop. Fig. 3 shows the void proportion between cupper heat sink and cupper base plate. The void proportion is about 2.7%.

Fig. 4 shows switching waveforms of turn-off transient at 400V, 100A, room temperature about 25 degree C. Fig. 5 shows turn-off waveforms of 1200V All SiC power module from ROHM (BSM120D12P2C005) at same conditions with Fig. 4.

The voltage overshoot of the proposed power module is just about 50V and almost no oscillation, so there is still some possibility to make switching faster and switching losses smaller.

B. Removing SBDs

The proposed power modules removed SBDs. Some positive effects of removing SBDs from SiC power modules are introduced in [16]. Fig. 6 shows positive effect of removing SBDs for turn-off transients. Fig. 6 (a) shows turn-off waveforms of SiC-MOSFET without SBDs. Fig. 6 (b) shows turn-off waveforms of SiC-MOSFET with SBDs.

Removing SBDs can reduce switching ringing and switching delay time, because of power modules without SBDs have smaller junction capacitance and bigger damping factor. Apparently, removing SBDs make it possible to reduce size and cost of SiC power modules.

III. High Power Density Inverters

This section shows the prototype high power density inverter of this research with experimental results.

A. Prototype inverter

Fig. 7 shows the prototype of high power density inverter. The size and weight of the prototype inverter are about 0.5 liter and 660g respectively.

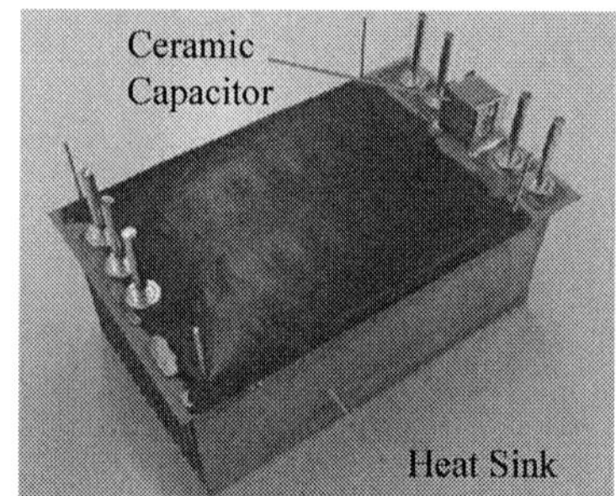

53mm x 32mm x31mm (52cc, 95g)

Fig. 2. The prototype proposed power module.

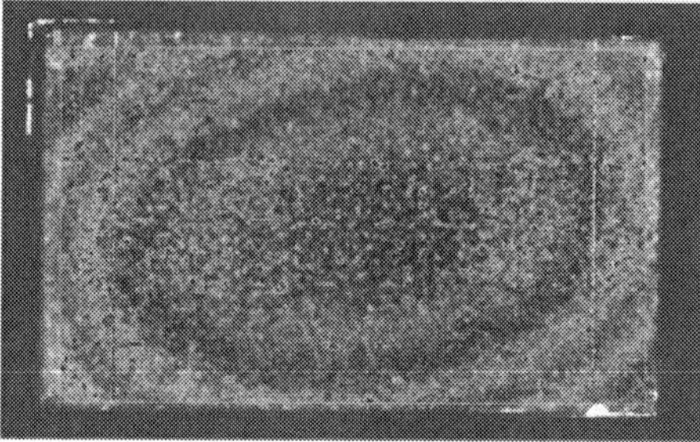

Ultrasonic inspection (Void proportion 2.7%)

Fig. 3. Result of ultrasonic inspection to evaluate void propostion.

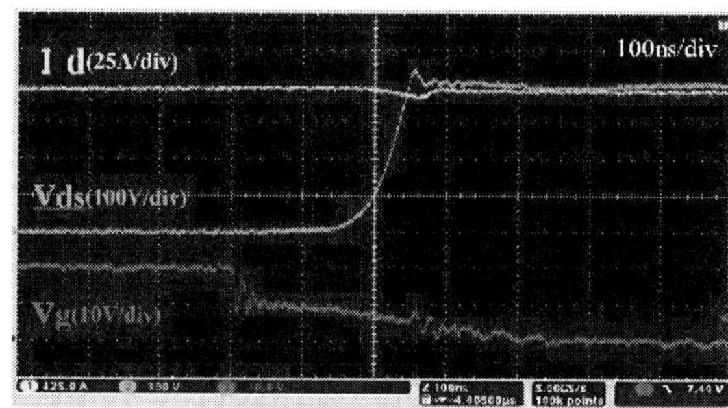

Fig. 4. Turn-off waveforms of the prototype proposed power module.

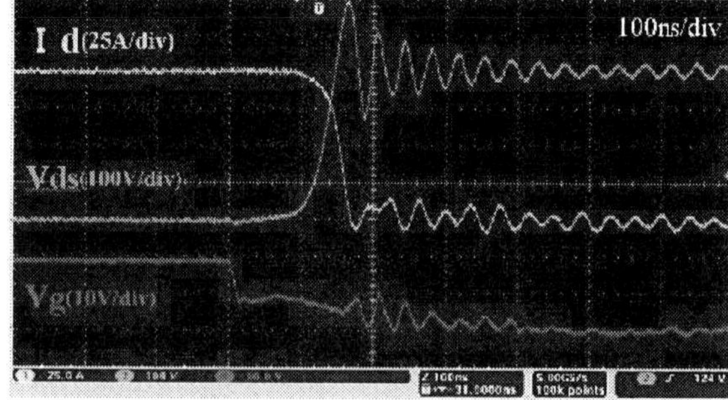

Fig. 5. Turn-off waveforms of ROHM SiC power module (BSM120D12P2C005).

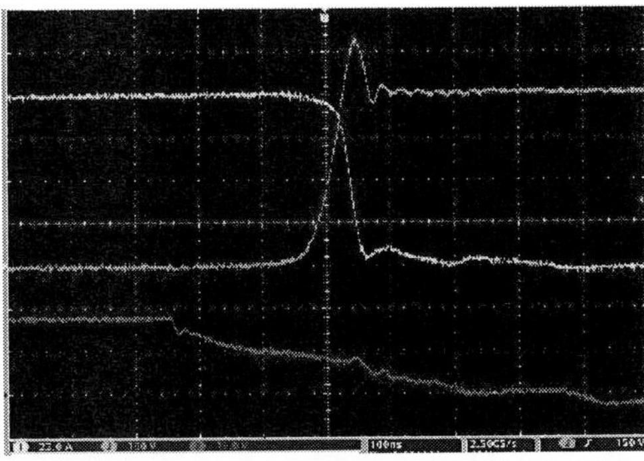

(a) SiC-MOSFET without SBDs , Vds=400V, Id=100A

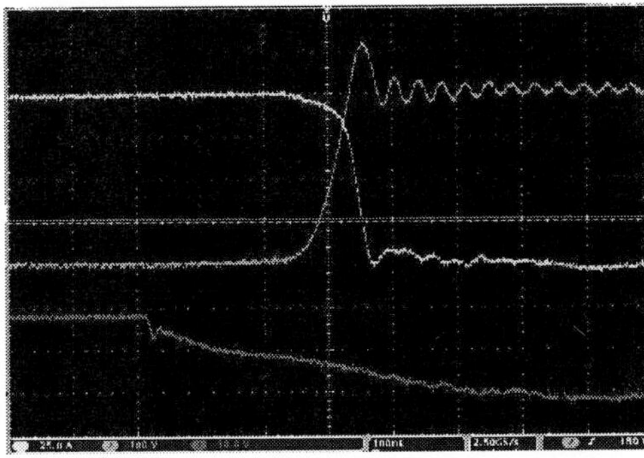

(b) SiC-MOSFET with SBDs , Vds=400V, Id=100A

Fig. 6. Affect of removing SBDs against turn-off waveforms.

The prototype inverter consists of three prototype SiC power modules, gate driver board, dc-link capacitors, control circuit board, and cooling fans.

B. *Experimental results*

Fig. 8 shows experimental results to evaluate maximum continuous operation. Fig. 8 (a) shows power meter screen shot, Fig. 8 (b) shows three phase output currents of inverter and output voltage waveforms.

The experiment conditions are followings, dc input voltage is 400V, ambient temperature is room temperature about 15 degree C, switching frequency is 20kHz, output ac current frequency is 300Hz, the load is resistive inductive load which is 1.567 ohm and 100uH.

The maximum output power is more than 35kW, and efficiency of the inverter is more than 98.6 %, it is measured after heat sink temperature has saturated enough. The experimental results confirmed that the power density of the prototype inverter is about 70kW/liter and 50kW/kg. The prototype inverter achieved much higher power density than 30kW/liter as predicted in [1] and other previous works and challenges.

Fig. 9 shows experimental results to evaluate the inverter efficiency. The prototype inverter can operate efficiently more than 98% at very wide range.

(a) Photo of the prototype inverter

(b) Photo of the inside structure of the prototype inverter

Fig. 7. Photos of the prototype inverter.

U_{rms1}	292.74	V	CH B	0.000k	
U_{rms2}	292.75	V	CH A	- 0.023	
U_{rms3}	292.77	V	P_1	35.68k	W
U_{rms4}	397.63	V	P_{123}	35.18k	W
I_{rms1}	83.99	A	P_n	- 0.00k	W
I_{rms2}	83.21	A	η_1	98.61	%
I_{rms123}	83.41	A	η_2	0.00	%
I_{rms4}	90.35	A	η_3	0.00	%

(a) Maximum output of the prototype inverter.

(b) Output current and voltage waveforms of the prototype inverter

Fig. 8. Maximum output of the prototype inverter.

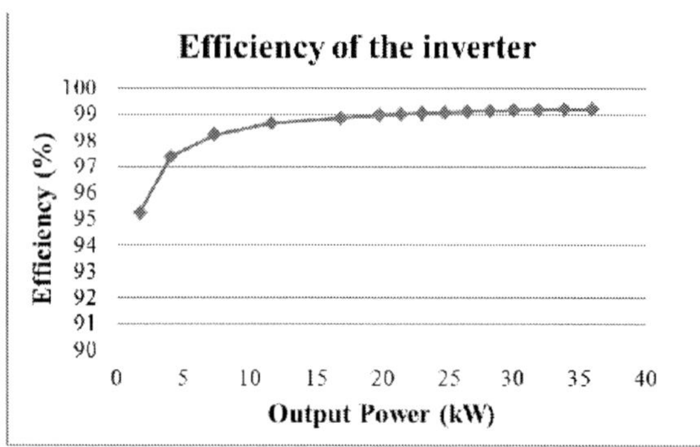

Fig. 9. Efficienny of the prototype inverter.

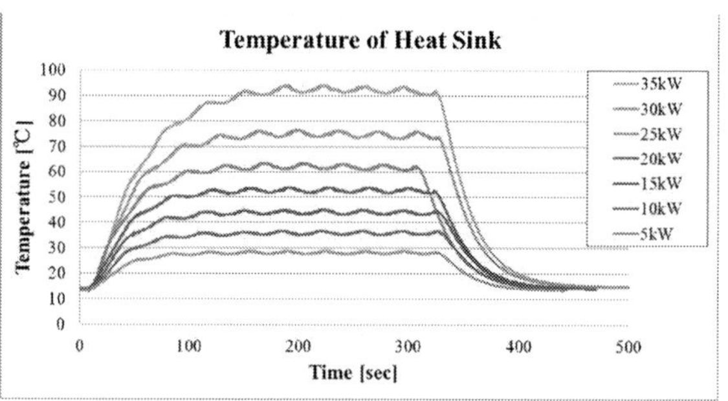

Fig. 10. Temperature of heat sink at various outputs.

Fig. 10 shows measured data of temperature of heat sink. According to the heat sink temperature data, estimated junction temperature of SiC MOSFETs is less than 125 degree C which is much lower than rated junction temperature.

Fig. 11 shows experimental results of the inverter efficiency with various switching frequency. The inverter is still able to work more than 95% at 50 kHz.

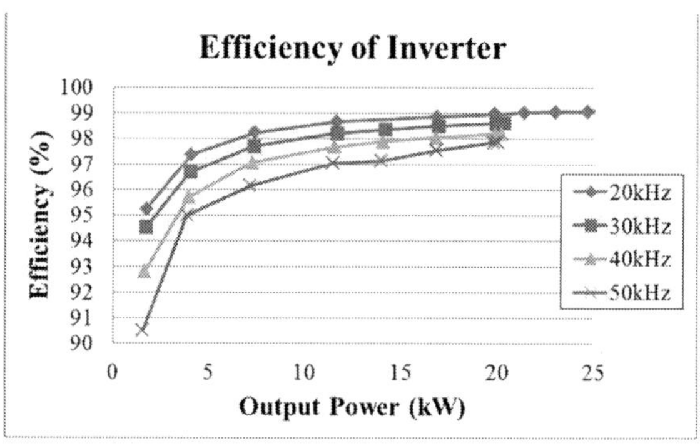

Fig. 11. Efficiency ot the inverter with various switching frequencies.

Fig. 12 shows experimental results of the inverter efficiency with various dead times. Since the prototype inverter uses synchronous rectification, shorter dead time reduce conductive loss during freewheeling. Since the prototype power module has removed SBDs, shortening dead time works more effectively than module with SBDs. The fast switching capability of the proposed gate driver can reduce dead time and conductive losses.

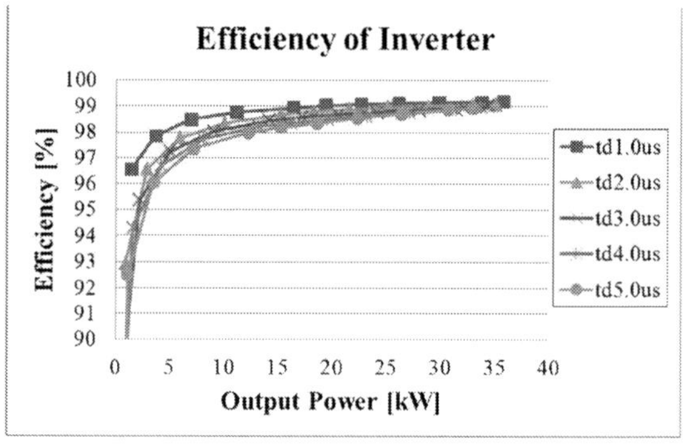

Fig. 12. Efficiency of the inverter with various dead time.

IV. Conclusions and Future Works

This paper presented SiC-based high power density inverter with experimental results, achieving 70kW/liter and 50kW/kg, which is more than two times higher than predicted value in [1] and much higher than previous works.

The proposed SiC power module with direct soldered heat sink is able to handle high power in small volume and weight. The proposed gate driver in [20] can reduce switching losses and dead time.

The inverter efficiency is investigated various conditions. The challenge to achieve higher power density is still on going, and evaluation test of EMC and research and development of filters are need to be done for the next step.

REFERENCES

[1] H. Ohashi, I. Omura, S. Mutsumoto, Y. Sato, H. Tadano, and I. Ishii, "Power electronics innovation with next generation advanced power devices," IEICE Transactions on Communications, vol. E87-B, no. 12, pp. 3422-3429, Dec. 2004.

[2] J. W. Kolar, U. Drofenik, J. Biela, M. Heldwein, H. Ertl, T. Friedli, and S. Round, "PWM converter power density barriers," in Proc. Power Conversion Conference (PCC), Nagoya, pp. 9-29, Apr. 2007.

[3] S. D. Round, P. Karutz, M. L. Heldwein, and J. W. Kolar, "Towards a 30kW/liter Three-Phase uniti power factor rectifier," in Proc. Power Conversion Conference (PCC), Nagoya, pp. 1251-1259, Apr. 2007.

[4] J. Biela, and J. W. Kolar, "Cooling concepts for high power density magnetic. devices," in Proc. Power Conversion Conference (PCC), Nagoya, pp. 1-8, Apr. 2007.

[5] R. Lai, F. Wang, P. Ning, D. Zhang, D. Jiang, R. Burgos, D. Boroyevich, K. J. Karimi, and V. D. Immanuel, "A High-Power-Density converter," IEEE Industrial Electronics Magazine, vol. 4, pp. 4-12, Dec. 2010.

[6] X. Roboam, B. Sareni, and A. D. Andrade, "More electricity in the air," IEEE Industrial Electronics Magazine, vol. 6, pp. 6-17, Dec. 2012.

[7] N. Morioka, M. Takeuchi, and H. Oyori, "Moving to an All-Electric aircraft system," in IHI Engineering Review, vol. 47, no. 1, pp. 33-39, 2014.

[8] A. Nawai, C. F. Tong, Y. Liu, A. Sakanova, S. Yin, Y. Liu, K. Men, K. Y. See, K. J. Tseng, R. Simanjorang, C. J. Gajanayake, and A. K. Guputa, "Design of high power density converter for aircraft applications," in Proc. ESARS 2015, Archen, March. 2015.

[9] A. K. Singh, P. Das, and S. K. Panda, "A high power density three phase AC-DC converter for more electric aircraft (MEA) ," in Proc. 9th International Conference on Power Electronics (ICPE 2015 ECCE-Asia), Seoul, June. 2015.

[10] J. Bourdon, P. Asfaux, and A. M. Etayo, "Review of power electronics opportunities to integrate in the more electrical aircraft," in Proc. ESARS 2015, Archen, March. 2015.

[11] K. Nakatsu, and R. Saito, "The Next-Generation high power density inverter technology for vehicle," in Proc. The 2014 International Power Electronics Conference (IPEC 2014 ECCE-Asia), pp. 1925-1928, Hiroshima, May. 2014.

[12] S. Kwak, T. Chung, G. Kang, and J. Jung, "The design of an integrated power control unit for hybrid/electric vehicle," in Proc. 9th International Conference on Power Electronics (ICPE 2015 ECCE-Asia), Seoul, June. 2015.

[13] A. K. Solomon, R. Skuriat, A. Castellazzi, and P. Wheeler, "Modular integration of a matrix converter," in Proc. The 2014 International Power Electronics Conference (IPEC 2014 ECCE-Asia), pp. 2920-2925, Hiroshima, May. 2014.

[14] A. K. Solomon, A. Castellazzi, N. Delmonte, and P. Cova, "Modular power converter integration based on non-conventional power switch asembly and interconnects," in Proc. 9th International Conference on Power Electronics (ICPE 2015 ECCE-Asia), Seoul, June. 2015.

[15] O. Goualard, N. Videau, T. Meynard, T. B. Doan, T. Labey, V. Bley, and E. Sarraute, "Integrated dc-dc based-GaN converter with screen printed capacitors allowing double side cooling," in Proc. 40th Annual Conference of the IEEE (IECON 2014), pp. 1549-1555, Dallas, Oct. 2014.

[16] K. Yamaguchi, and K. Katsura, "Research on optimization design of high efficient low noise SiC-MOSFET inverters," in Proc. 9th International Conference on Power Electronics (ICPE 2015 ECCE-Asia), Seoul, June. 2015.

[17] N. Idir, R. Bausiere, and J. J. Franchaud, "Active gate voltage control of turn-on di/dt and turn-off dv/dt in insulated gate transistors," IEEE Transactions on Power Electronics, vol. 21, no. 4, pp. 849-855, Jul. 2006.

[18] K. Onda, A. Konno, and J. Sakano, "New concept high-voltage IGBT gate driver with self-adjasting active gate control function for SiC-SBD hybrid module," in Proc. 25th International Symposium on Power Semiconductor Devices and ICs (ISPSD), pp. 343-346, Kanazawa, May. 2013.

[19] N. Zhu, D. Xu, X. Zhang, M. Cheng, S. Igarashi, and T. Fujihara, "Switching noise suppression for hybrid IGBT modules," in Proc. 9th International Conference on Power Electronics (ICPE 2015 ECCE-Asia), Seoul, June. 2015.

[20] K. Yamaguchi, Y. Sasaki, and T. Imakubo, "Low loss and low noise gate driver for SiC-MOSFET with gate boost circuit," in Proc. 40th Annual Conference of the IEEE (IECON 2014), pp. 1594-1598, Dallas, Oct. 2014.

978-1-4673-9551-9/16 $31.00 © 2016 IEEE

An Improved Automatic Layout Method for Planar Power Module

Puqi Ning, Xuhui Wen, Yaohua Li, Xiongxuan Ge
Laboratory of Power Electronics and Power Conversion
Institute of Electrical Engineering Chinese Academy of Sciences
Beijing, China

***Abstract*—The layout of power modules is one of the key points in power module design. In this paper, along with design examples, an improved automatic layout method for planar power module is presented. Some practical considerations and implementations are introduced in the optimization procedure.**

Keywords—Power module; Layout

I. INTRODUCTION

As shown in [1, 2], the parasitic parameters of a power module may have detrimental influences on switching losses and dynamic behaviors. This phenomenon is exaggerated when the power density increases, because a higher switching rating leads to faster variations. In order to ensure high efficiency and high performance of a high-density power module, efforts are required for power module layout designs[3].

Recently, in order to further reduce parasitic parameters of wirebond modules, many advanced power semiconductor packaging were invented[4] (shown in Fig.1). In these packages, top sides of dies are soldered or sintered directly to DBC, copper strip or terminals. However, literatures rarely demonstrate any method to optimize the layout of these planar packages.

As the next generation power devices, silicon carbide (SiC) chips have many advantages over conventional silicon devices. However, limited by manufacture capability, the rated current for each die is commonly limited up to 50 A[5]. Many dies need to be paralleled for renewable energy, automotive application and other medium power applications (at least over 300 A requirement). It means that a phase-leg module needs to contain at least 12 SiC switches and 12 SiC diodes. It is very difficult to layout a power module only with manual design procedure and experiences. .

During the conventional manual design iterations, devices and power terminals, and other basic components in layout design, first are geometrically placed by considering the electrical connections. The second step is the routing process, in which the dies and pads are connected with copper traces and wire bonds. Then the parasitic parameters in the designed layout can be obtained from finite element analysis simulation tools or theoretical equation-based calculations [3–6]. According to a survey of the relevant literature, there is still no figure of merit that can be used to find the best layout design for minimum power loss. The fitness of a layout can only be determined based on overall evaluation of the parasitic parameters.

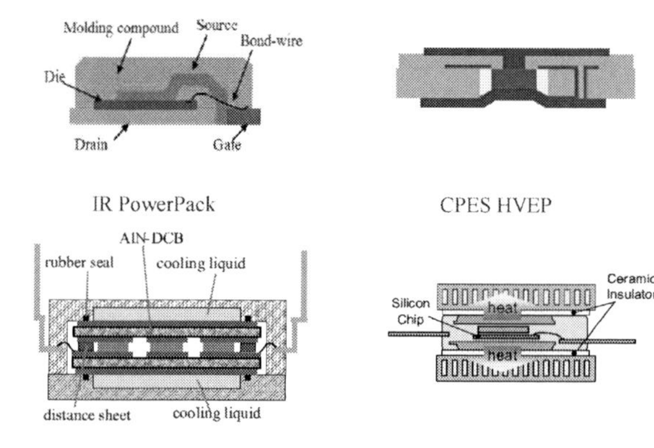

Fig. 1. Planar package to reduce parasitics.

There are many manual layout design procedures presented in the literatures [3-7]. The main issues for these methods are the design speed and the limited choice of candidates. In [8] and [9], along with design examples, automatic design methods by using genetic algorithms were presented. With these automatic methods, layout design spaces can be better searched and optimized.

However, for planar package layout, the design procedure becomes very slow when the number of dies is over 10. The main issue is the complex layout interpretation steps and the inductance calculation steps.

In this paper, the previous design procedure is improved and becomes faster and more universal. Furthermore, By adding analytical equations which can describe electrical, thermal and thermo-mechanical behaviors of the packaging

This research is support from the National Natural Science Foundation of China (No. 51507166) and the National High Technology Research and Development Program of China (No. 2015AA034501).

materials, the optimized power module can better consider multi-physics factors and thus be suitable for operations.

II. Power Module Layout Interpretations

In the layout iterations, the representations of devices are the most important key to solve a layout problem. In [8], the sequence pair (SP) method, which is particularly suitable for stochastic algorithms such as GA, has been proposed as the representation of a solution for this problem. However, the SP method becomes slower and much more complicated when the number of components increases over 10. In [9], typical elements were listed and used in the DNA defination and the optimization speed was increased. The DNA interpretation in [9] is still not simple enough for SiC planar module with a large number of dies.

In order to interpret a power module layout into a DNA string, typical elements in planar modules are sorted and numbered in Fig.2.

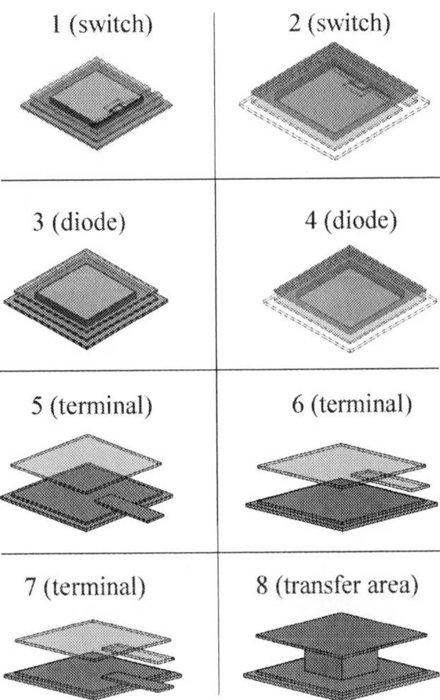

Fig. 2. Basic elements for planar package.

Since the shapes of devices and terminal are rectangle or square, a matrix representing a grid layout region can be used to interpret device positions. Their positions can be then converted to one dimension vector and expressed as integral numbers. the length of position DNA is $i+j+k$, including i power terminals, j switches and k diodes.

The orientations of elements are also critical to parasitic parameter reduction and the balance between paralleled devices. It is because the pads on dies may be asymmetrically distributed while the power terminal connections are restricted by mechanical limits or manufacture limits. The basic element orientations for planar package are depicted in Fig.3.

When considering edge limits for a planar package, the orientations of power terminals are always faces out of the edges. Thus, the orientations in x-y plane can be assigned only to power switches with 2 digit binary. The facing up and facing down elements can be combined with different z axis orientation (1 digit binary). The transfer area can be formed in the copper growing step when required (in Section III). With these simplification, the length of orientation DNA (a binary string) is reduced to $i+3j+k$, including i power terminals, j switches and k diodes.

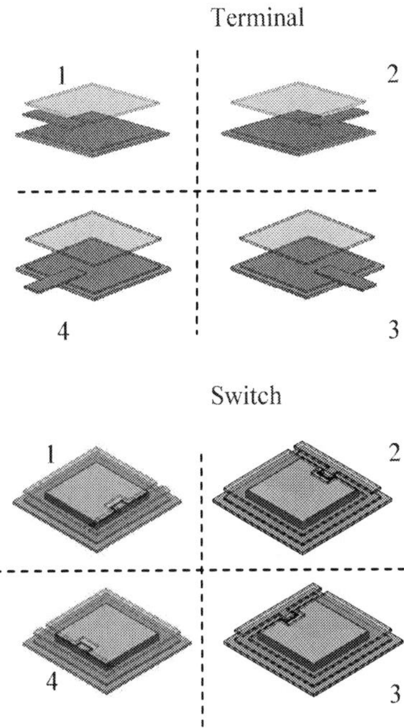

Fig. 3. Basic orientations in planar module.

As an example, the phase-leg module presented in Fig.5 can be represented with 9 elements, including 5 power terminals, 2 power switches and 2 power diodes. The sequenced elements is shown in the left side of Fig.5. In the design, the positions of these elements are limited to a 4×4 grid layout plane. The position DNA is labeled from 0 to 15 and defined as [9,1,10,0,11,4,7,5,6] in this design.

XX...X | XX...X | XX...X

i terminals *j* switches *k* diodes

Position DNA length: $i+j+k$

XX...X | XXX XXX...XXX | XX...X

i terminals *j* switches *k* diodes

Orientation DNA length: $i+3j+k$

Fig. 4. DNA string for positions and orientations.

In this commonly used phase-leg module, there are 5 types of terminals, and they are, positive pin, neutral pin, negative pin, up gate pin and bottom gate pin. If in an orientation DNA, terminal part is defined as [0,1,0,1,0], it actually means [face up, face down, face up, face down, face up]. For device part, the first IGBT DNA is [100], the second IGBT DNA is [001], the first diode DNA is 0, and the second diode DNA is 1. The full interpreted layout is depicted in the right side of Fig.5.

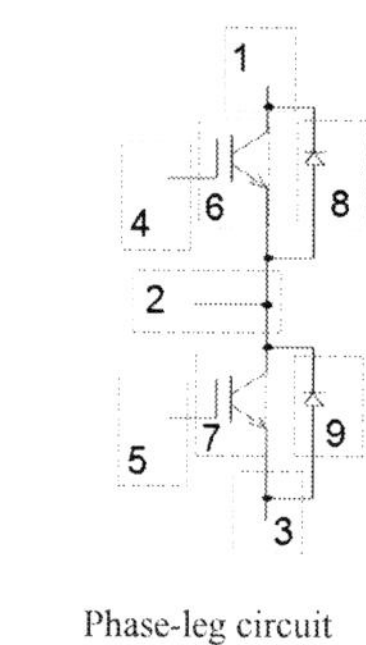

Phase-leg circuit

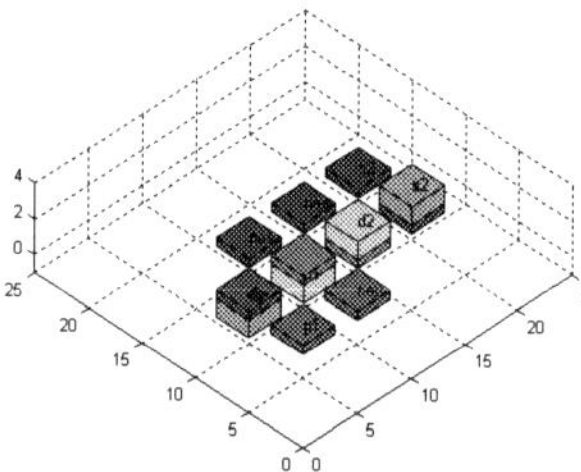

Optimized layout in MATLAB

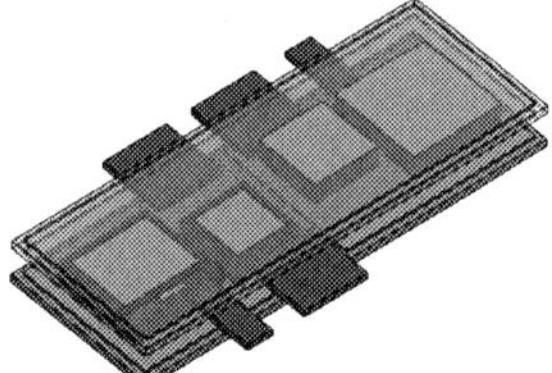

Interpreted power module

DNA: position[9,1,10,0,11,4,7,5,6] ,orientation [0,1,0,1,0,1,0,0,0,0,1,0,1]

Fig.5. Planar package layout representation example.

III. LAYOUT DESIGN PROCEDURE

The genetic algorithm was used in the optimization iterations. Genetic algorithms (GAs) are based on natural genetics and are a particular class of evolutionary algorithms that use techniques inspired by the mechanisms of evolutionary biology such as inheritance, mutation, selection, and crossover [8]. The GA approach is categorized as a global search heuristic, and as such, GA is a search technique used in computing to find exact or approximate solutions to optimization and search problems. The GA is a stochastic search method that deals with the individuals composed of candidate solutions (populations), each of which is generally encoded in a problem-independent representation. During the genetic process, new candidate solutions are evaluated by their fitness and are reproduced using the genetic operators such as crossover and mutation.

As shown in Fig.6, the proposed layout design includes two loops both utilizing genetic algorithm (GA). The outer loop is the element placement loop, in which the components are placed in grid layout plane by following the electrical netlist. The inner loop arranges the orientations of elements by following the packaging limits and rules. At the end of each inner loop, the best candidate with the largest fitness will be transferred to the outer loop. After several iterations, the best layout design can be found to meet certain criterion.

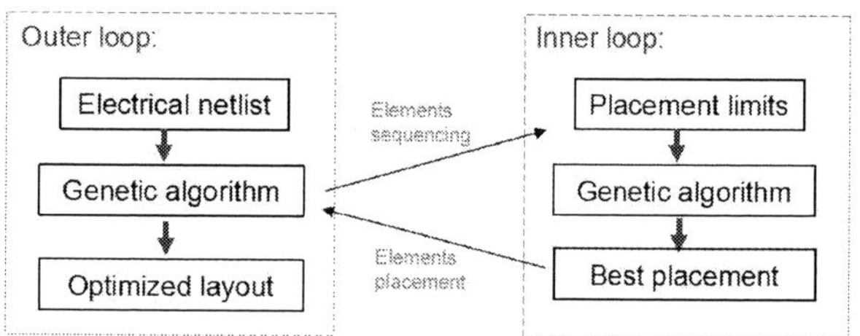

Fig. 6. Two loop genetic algorithm.

In the outer loop, all the elements are placed in a grid layout plane with a certain matrix size. The matrix size should be larger than the total element number. For the first generation, element sequences are generated randomly without overlapping. Then, the DNA strings will evolve with crossover operation and mutation operation to form a new generation. The algorithm terminates when a maximum number of generations is reached or when the optimization criteria is fit.

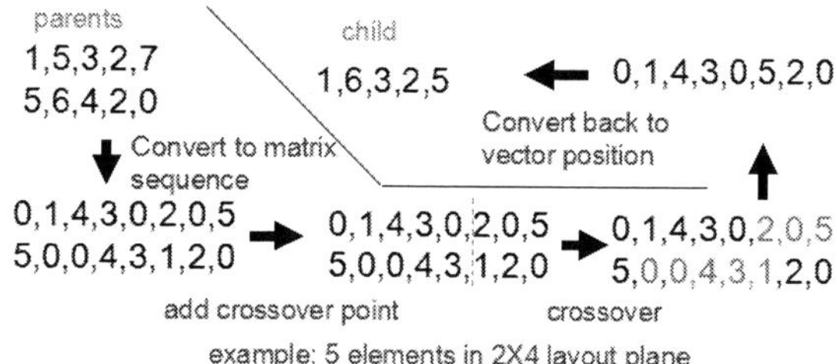

Fig. 7. Crossover operations for outer loop.

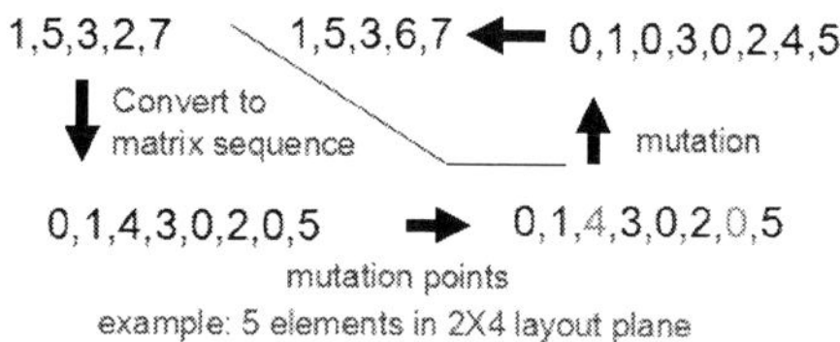

Fig. 8. Mutation operations for outer loop.

Special operations are required and designed for the crossover and mutation. The position DNA is first converted to their matrix sequences. A crossover points is then appointed. The digits before the crossover point of the first parent are preserved in the same sequence. The sequence of rest digits after the crossover point will copy their sequence in the second parents (one example is shown in Fig.7). Finally the matrix sequence of the child is converted back to a position DNA.

Fig.8 shows a mutation operation example, which is similar to crossover operation. The selected two elements can be either a power terminal, a switch, a diodes, or a space.

In the inner loop, the element orientations are stochastically selected. Then evolution is conducted with the most common binary crossover and binary mutation operations. In this procedure, gaps exist between elements and copper areas grow to final layout (as shown in Fig. 9).

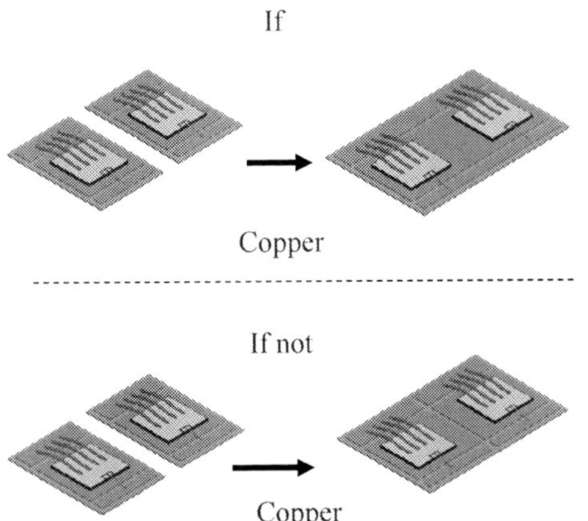

Fig.9. DBC substrate copper growing.

Boundary element method based parasitic parameter calculation is used for fitness evaluation in [6]. It is the main cause for low speed. To increase the optimization speed, the current conducting lengths are used to represent parasitic resistance and the surrounded areas are used to represent the inductance of commutation path. These value can be simply calculated with the positions of the elements in a very fast speed.

In this paper, the calculations were further simplified. The resistance was calculated by multiply the length between two 3-D point and a certain coefficient. the length of two 3-D point A($x1$, $y1$, $z3$) and B($x2$, $y2$, $z2$) can be obtained as (1).

$$\sqrt{(x_1 - x_2)^2 + (y_1 - y_2)^2 + (z_1 - z_2)^2} \quad (1).$$

In the inductance calculation part, the target 3-D layout pattern was first divided into n triangles. These triangles include the center point and two corner points of the 3-D layout pattern. The commutation-path inductance was obtained by multiply the sum up the individual area of each triangle and a certain coefficient.

If the corner point of a 3-D triangle is A($x1$, $y1$, $z3$), B($x2$, $y2$, $z2$),C($x3$, $y3$, $z3$), the area of the triangle s can be calculated from (2).

$$\frac{1}{2}\sqrt{\begin{vmatrix} y_2 - y_1 & z_2 - z_1 \\ y_3 - y_1 & z_3 - z_1 \end{vmatrix}^2 + \begin{vmatrix} z_2 - z_1 & x_2 - x_1 \\ z_3 - z_1 & x_3 - x_1 \end{vmatrix}^2 + \begin{vmatrix} x_2 - x_1 & y_2 - y_1 \\ x_3 - x_1 & y_3 - y_1 \end{vmatrix}^2} \quad (2)$$

.

The best routing results will be passed to outer loop in the end of inner loop iterations. In this paper, the fitness evaluation also considers the parasitic resistance, parasitic inductance and the footprint, as shown in (3).

$$F = \frac{A_{inductance} \, L_{resistance}}{A_{footprint}}$$

(3).

With the best fitness, the layout can provide low parasitics, as well as small footprint. This fitness definition can be further extended to evaluate thermal and mechanical performance by adding proper parameters.

For example, the conventional compact modeling for heat conduction in solids was done by applying finite-difference approximation to the governing equations of heat transfer. Assuming the material has isotropic thermal conductivity, the equation is expressed in (4). C_v is the volumetric specific heat of the material, T is the control cell temperature, and q is the volumetric heat generation.

$$C_v \frac{dT}{dt} + (-k\nabla^2 T) = \dot{q} \quad (4)$$

$$C_v \frac{dT}{dt} + \nabla \cdot (-k\nabla T) + C_v u \nabla T = \dot{q} \quad (5)$$

By combining and energy conservation equation of heat transfer in a control volume of air flow and Stroke's theorem equation, (5) can be derived as a compact convection cell model [7]. In (5), u is the velocity of outflow of the fluid at the surface of the control volume and it can represent the net outflow of heat from the control volume due to the convection.

In [8], a method for using a time domain circuit code to solve partial differential equations was described. The partial differential equation is finite differenced in space and written in state variable form. The spatial derivatives can be finite differenced on a uniform spatial mesh $x_1,\ldots, x_{j-1}, x_j, x_{j+1},\ldots, x_k$, where $\Delta x = x_{j+1} - x_j$. The 1_{st}-order and 2_{nd}-order differential items can be substituted with equation (4) and (5). Equation (6) and (7) shows the discretization result. With these equations, the thermal management part can be simplified as Fig.10, and can be used in the layout optimization.

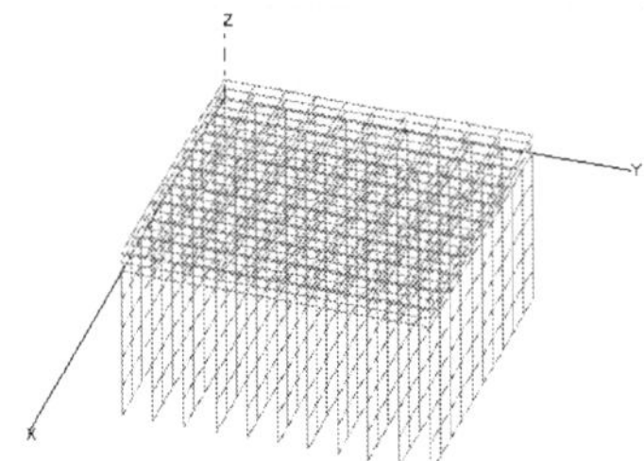

Figure 10. The discretized cell model for solid.

978-1-4673-9551-9/16 $31.00 © 2016 IEEE

$$C_v \Delta x \Delta y \Delta z \frac{T}{t} - k\frac{T_{i+1,j,k}-2T_{i,j,k}+T_{i-1,j,k}}{\Delta x^2} - k\frac{T_{i,j+1,k}-2T_{i,j,k}+T_{i,j-1,k}}{\Delta x^2} - k\frac{T_{i,j,k+1}-2T_{i,j,k}+T_{i,j,k-1}}{\Delta x^2} = \dot{q}\,\Delta x \Delta y \Delta z \quad (6)$$

$$C_v \Delta x \Delta y \Delta z \frac{T}{t} - k\frac{T_{i+1,j,k}-2T_{i,j,k}+T_{i-1,j,k}}{\Delta x^2} - k\frac{T_{i,j+1,k}-2T_{i,j,k}+T_{i,j-1,k}}{\Delta x^2} - k\frac{T_{i,j,k+1}-2T_{i,j,k}+T_{i,j,k-1}}{\Delta x^2} + C_v u_y \Delta A_y (T_{S2} - T_{S1}) = \dot{q}\,\Delta x \Delta y \Delta z \quad (7)$$

IV. EXAMPLES OF AUTOMATIC LAYOUT DESIGN

In order to better explain the proposed method, one design example is demonstrated in Fig. 11. The example focuses on a three-phase wirebond module, which has 12 power switches, 6 diodes and 11 terminals. The project also requires a easy connection of all the phase terminal and then can be convert to a one phase module. The phase terminals need to be in one side of module, and the positive and negative terminal should be in another side.

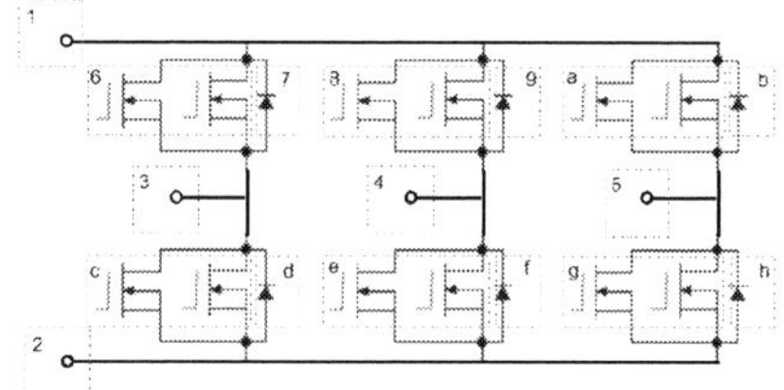

Fig.11. Automatic layout design example.

The design iteration terminates after 45 outer loops and each with 100 inner loops. Two of the initial designs are shown in Fig.12. The fitness is 932 and 824 respectively.

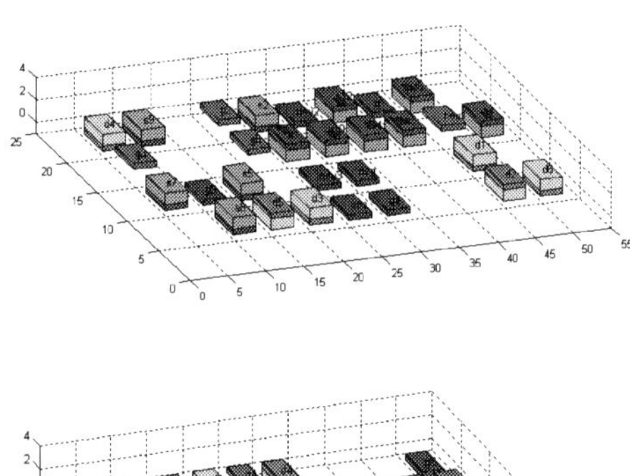

Fig.12 Initial layout designs

The solution with largest fitness (573) is shown in Fig. 13. The interpreted planar power module is in Fig.14.

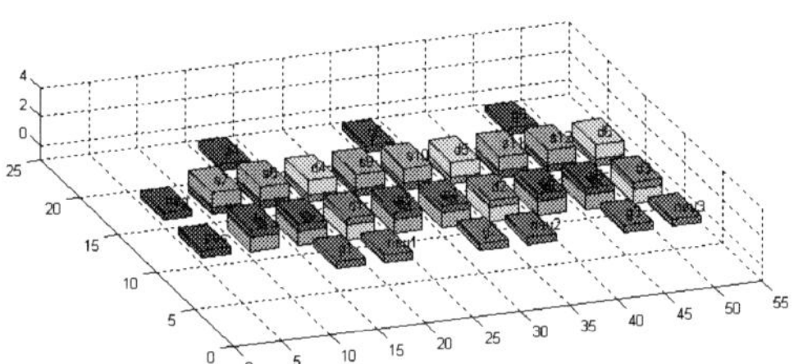

Fig.13. Optimized layout example

Literature [9] presents a wirebond layout optimization (shown in Fig.15). Compared with its design result, the size of planar power module is reduced to 47%, the average parasitic resistance reduced to 27%, and the average parasitic inductance reduced to 63%.

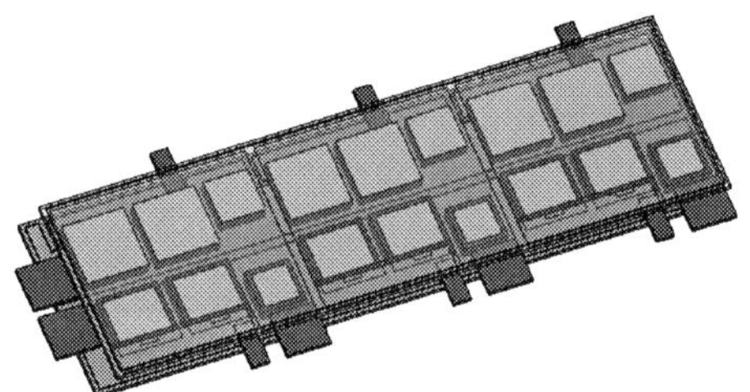

Fig.14 Interpreted planer power module

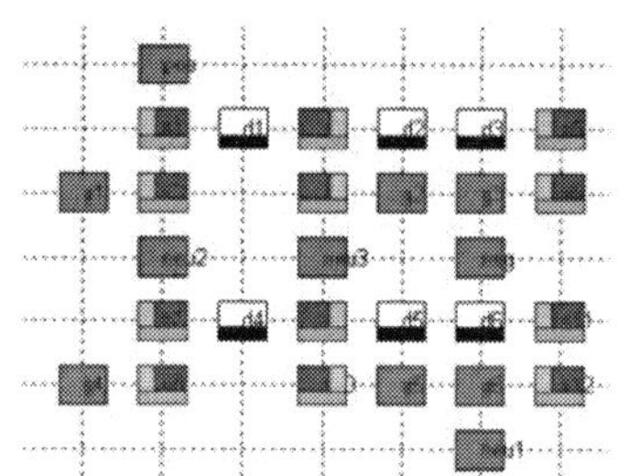

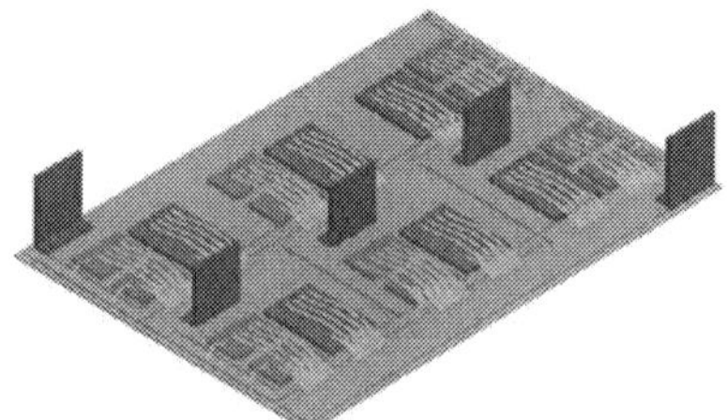

Fig.15. Compared with wirebond module layout.

This design result was also verified by FEM software Ansoft Q3D. The performance was agreed well when compared with the FEM calculation result. With the presented improvement, it is 15 times faster than previous method in [8] and 1.8 times faster than method in [9].

Fig.15. Sputtered IGBT dies for planar module

As shown in Fig. 15, IGBT dies were sputtered for planar module fabrication. A double-pulse setup was built and will be used to evaluate the parasitic parameters of the fabricated module. With the comparison of design and experimental data, this optimization method will be further improved in the near future.

V. SUMMARY AND CONCLUSION

This paper presents an improved automatic layout design method for planar power modules. Some practical considerations are introduced and implemented in the optimization. Along with design examples, the performance of proposed method is demonstrated. It could be a major step to improve the overall performance of future power module.

REFERENCES

[1] Q. Chen, X. Yang, J.Liu, Z. Wang, "Practical Design Considerations for IPEM-based PFC Converter Employing CoolMOS and SiC Diode," in *Proc. IEEE APEC 2006*, pp.6-12.

[2] H. A. Mustain, A. B. Lostetter, W. D. Brown, "Evaluation of gold and aluminum wire bond performance for high temperature (500 /spl deg/C) silicon carbide (SiC) power modules,"in *Proc. Electronic Components and Technology Conference, 2005*, pp, 1623–1628, Vol. 2.

[3] J. D. van Wyk, F. C. Lee, Z. X. Liang, R. G. Chen, S. Wang, B. Lu, "Integrating Active, Passive and EMI-Filter Functions in Power Electronics Systems: A Case Study of Some Technologies,"in *IEEE Trans. on Power electronics*, vol. 20, no. 3, pp. 523–536, May. 2005.

[4] Z. Liang, B. Lu, J. van Wyk, and F. C. Lee, "Integrated CoolMOS FET/SiC-Diode Module for High Performance Power Switching," in *IEEE Trans. on Power Electron.*, vol. 20, no. 3, pp.679-686, MAY 2005.

[5] C. Martin, J. Guichon, and J. Schanen, "Gate Circuit Layout Optimization of Power Module Regarding Transient Current Imbalance," in *IEEE Trans. on Power Electron.*, vol. 21, no. 5, pp. 1176–1184, Sep. 2006.

[6] N. Hingora, X. Liu, B. McPherson, Y. Feng, and H. A. Mantooth, "Power-CAD: A Novel Methodology for Design, Analysis And Optimization of Power Electronic Module Layouts," in *Proc. IEEE ECCE 2010*, pp. 2692-2699.

[7] Jonah Zhou Chen, Ying Feng Pang, Dushan Boroyevich, Elaine P. Scott, Karen A. Thole, "Electrical and Thermal Layout Design Considerations for Integrated Power Electronics Modules," in *Proc. IEEE IAS 2002*, pp. 242–246, vol.1.

[8] P. Ning, F. Wang, and K. D. T. Ngo, "Automatic layout design for power module," in *IEEE Trans. on Power Electronics*, vol 28, issue 1, pp. 481-487, Jan. 2013.

[9] P.Ning, and Xuhui Wen, "A Fast Universal Power Module Layout Method", in *Proc. IEEE ECCE2015*, pp.4132-4237.

Practical Implementation Schemes of Motor Speed Measurement by Magnetic Encoder on Electric Power Steering Applications

Jae-Hyun Lee

R&D Center, HYUNDAI MOBIS, Gyeonggi-do 17-2, Korea

Email: leejaehyun@mobis.co.kr

***Abstract*- In this study, a new perspective yet practical method is proposed for measuring speed of an Electric Power Steering (EPS) drive motor by using a magnetic encoder. The reason that the EPS system measures the speed of a motor is not only to check the speed of the motor, but also to perform control logic of the motor based on the measured speed. Thus, it is important to measure the speed of a motor at each predetermined cycle without changing the speed of the measurement cycle. In this paper, feasible implementation schemes are recommended to improve the performance of a motor speed measurement method based on the M/T method without using additional devices. After the implementation schemes are discussed in detail, their effectiveness will be indicated with further experimental results.**

INTRODUCTION

Power steering of a vehicle is a steering apparatus based on power which assists a driver to operate a steering wheel. Such power steering generally uses hydraulic pressure. Recently, however, the use of an EPS system using the force of a motor has been increasing. This is because the EPS system does not require an oil change and has a lighter weight and occupies a smaller area compared to the existing hydraulic power steering systems. Unlike the existing hydraulic power steering systems, the EPS system generates a torque through current control of a motor by an electronic control unit (ECU) which includes various control logics for controlling the motor.

Such control logics can be divided into three logical parts: logic for realizing a steering feel desired by a driver, logic for improving the stability of the vehicle, and logic for improving the stability of the system. The control unit of the EPS system performs the respective logics based on various parameters; for instance, there is a speed of a vehicle, a torque signal, and a steering angle signal. Among these parameters, a steering angle and the speed of a steering angle are necessary for the driver's delicate steering feel. This can be calculated by pre-processing a signal measured through a steering angle sensor installed on a column assembly.

However, because commonly used steering angle sensor has a low resolution, the steering angle sensor has difficulties in acquiring an exquisite steering feel. Accordingly, the steering angle sensor generally calculates a column angle speed by converting an angular speed of the motor. Therefore, it is important to precisely measure the speed of the drive motor in order to control the EPS system with more accuracy.

In general, the speed of the motor is measured through a rotary encoder. The rotary encoder includes absolute and incremental encoders. Absolute encoder outputs the absolute position of a shaft while the incremental encoder outputs information on a motion of the shaft. Thus, when measuring the speed of a motor, the incremental encoder is mainly used. As illustrated in Figure 1, the methods for measuring the speed of a motor using such incremental encoder may be roughly divided into three methods: the M method, the T method, and the M/T method, as illustrated in Figure 1.

The M method indicates a method for calculating the speed of a motor by counting the number of pulses outputted from an encoder during a predetermined sampling time. The M method can be simply implemented, and has an unchangeable speed measurement cycle. However, since a speed error may occur depending on whether the sampling time is synchronized with an encoder pulse, the M method has relatively low precision.

Secondly, the T method enables to calculate the speed of a motor by measuring the time between the output pulses of the encoder. The T method can accurately measure the speed of a motor at a low-speed region. Yet, the T method requires a high-frequency clock pulse to precisely measure speed at a high-speed region. In this case, as the number of clock pulses to be counted at a low-speed region significantly increases, the production cost will inevitably increase. Also, according to the speed of a motor at an ultra-low speed region, the speed measurement cycle may change.

Finally, the M/T method indicates a method for calculating the speed of a motor by counting the number of pulses obtained from the encoder during a predetermined sampling time similar to the M method. However, when the sampling time and an encoder pulse are not synchronized with each other, the M/T method is able to remove the error as it can measure the time at which the next pulse is collected. Compared to the previous two methods, the M/T method can measure speed with greater accuracy.

Despite these advantages, because the M/T method is very complex and difficult to implement, it increases the production cost conclusively. It may also cause to change the speed measurement cycle when the next pulse output is delayed from the sampling time at an ultra-low speed region.

978-1-4673-9551-9/16 $31.00 © 2016 IEEE

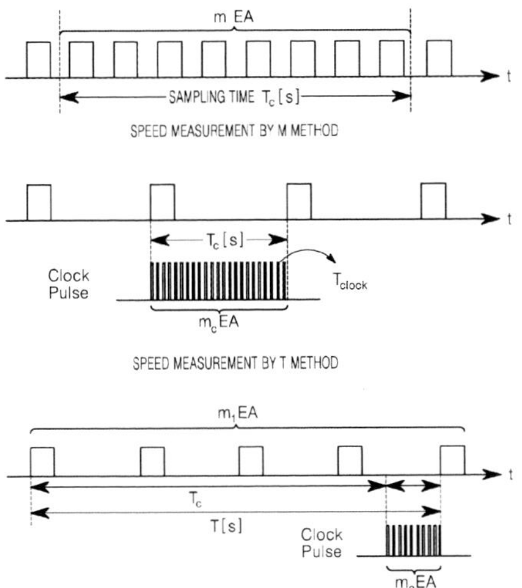

Fig.1. Speed measurement By M, T and M/T method

In particular, the characteristic of the EPS drive motor at this region becomes an essential performance evaluation factor. For instance, when a vehicle is heading straight, it will be difficult for a driver to perform a steering operation for a longer period of time. Therefore, a counter overflow must not occur.

Although the T method or the M/T method can measure speed more accurately than the M method, the T method or the M/T method is not necessarily superior to the M method when measuring the speed of the EPS drive motor. In general, mass-produced EPS systems measure the speed of a motor using the M method. Yet, we still need to confront the problem that it is difficult to precisely measure the speed of the motor when the M method is used.

Consequently, the M/T method is recommended for practical and robust implementation on the EPS Applications.

Design of The Practical M/T Method

A. Configuration of M/T method for EPS Applications

Figure 2. shows the configuration of hardware for the proposed M/T method.

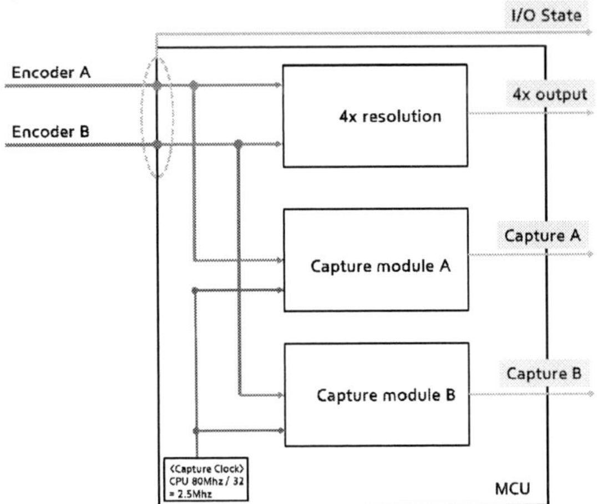

Fig.2. Configuration of hardware for the proposed M/T method

Proposed method doesn't need additional hardware and function for implementation like latch circuit.

B. Implementation of proposed method

Figure 3. shows a flow chart for implementation for the EPS applications.

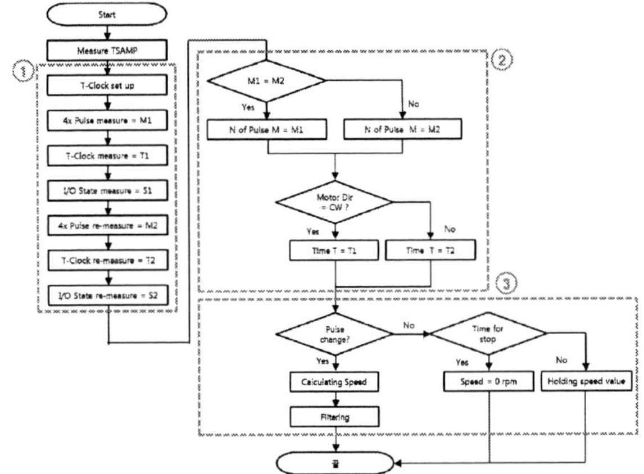

Fig.3. Flow chart of implementation for proposed method

Embodiments of the present invention are directed to an apparatus and method for measuring speed of an EPS drive motor, capable of precisely measuring speed while having a constant speed measurement cycle, without an additional component such as a latch.

In one embodiment, a method for measuring speed of an EPS drive motor may include: receiving, by a controller, A and B pulses having a phase difference of 90 degrees from an encoder during a first reference time, and measuring information of the pulses; receiving A and B pulses having a phase difference of 90 degrees from the encoder again during the first reference time, and remeasuring information of the pulses; selecting any one of the measured pulse information and the remeasured pulse information as data for calculating the speed of the motor, based on the measured pulse information and the remeasured pulse information; and calculating the speed of the motor, based on the selected data.

The pulse information may include a pulse number obtained by multiplying the number of the A and B pulses by four, information on the cycle of the A pulse, information on the cycle of the B pulse, and a pulse state. In the selecting of any one of the measured pulse information and the remeasured pulse information as the data for calculating the speed of the motor, the controller may select the measured pulse information as the data for calculating the speed of the motor when the multiplied-by-four pulse number contained in the measured pulse information is equal to the multiplied-by-four pulse number contained in the remeasured pulse information, and select the remeasured pulse information as the data for calculating the speed of the motor when the multiplied-by-four pulse number contained in the measured pulse information is different from the multiplied-by-four pulse number contained in the remeasured pulse information.

The calculating of the speed of the motor may include: estimating, by the controller, a half-cycle time of any one of the A and B pulses, based on the selected data; when the multiplied-by-four pulse number contained in the selected data is equal to or more than a reference number, calculating the speed of the motor based on the multiplied-by-four pulse number contained in the selected data and the estimated time; and when the multiplied-by-four pulse number contained in the selected data is less than the reference number, determining the speed of the motor based on the continuance time during which the pulse number is less than the reference number.

The determining of the speed of the motor may include: setting, by the controller, the speed of the motor to 0 when the continuance time exceeds a second reference time; and maintaining the speed of the motor when the continuance time does not exceed the second reference time.

The calculating of the speed of the motor may include: estimating, by the controller, a half-cycle time of any one of the A and B pulses based on the selected data, when the multiplied-by-four pulse number contained in the selected data is equal to or more than a reference number; calculating the speed of the motor based on the multiplied-by-four pulse number contained in the selected data and the estimated time; setting the speed of the motor to 0, when the multiplied-by-four pulse number contained in the selected data is less than the reference number and the continuance time during which the pulse number is less than the reference number exceeds the second reference time; and maintaining the speed of the motor, when the multiplied-by-four pulse number contained in the selected data is less than the reference number and the continuance time during which the pulse number is less than the reference number does not exceed the second reference time.

The pulse state may be divided into when the A pulse is high and the B pulse is low (S1), when the A pulse is low and the B pulse is high (S2), when both of the A and B pulses are high (S3), and when both of the A and B pulses are low (S4), and the estimating of the half-cycle time may include: determining, by the controller, the rotation direction of the motor based on a pulse state contained in the selected data; estimating the half-cycle time of the A pulse, when the determined rotation direction of the motor is forward and the last value of the pulse state contained in the selected data is any one of S1 and S2; estimating the half-cycle time of the B pulse, when the rotation direction is forward and the last value is any one of S3 and S4; estimating the half-cycle time of the B pulse, when the rotation direction is backward and the last value is any one of S1 and S2; and estimating the half-cycle time of the A pulse, when the rotation direction is backward and the last value is any one of S3 and S4.

In the calculating of the speed of the motor, the controller may calculate the RPM of the motor through the following equation:

$$\frac{15\times pulse\ number}{PPR\times(first\ reference\ time+\frac{half\ cycle\ time}{2})} \quad (1)$$

where PPR represents the number of output pulses per revolution of the encoder.

In another embodiment, a method for measuring speed of an EPS drive motor may include: receiving, by a controller, A and B pulses having a phase difference of 90 degrees from an encoder during a first reference time, and measuring a multiplied-by-four pulse number and a multiplied-by-four pulse cycle; when the measured pulse number is equal to or more than a reference number, calculating the speed of the motor based on the measured pulse number and the measured pulse cycle; and when the measured pulse number is less than the reference number, determining the speed of the motor based on a continuance time during which the pulse number is less than the reference number.

The determining of the speed of the motor may include: setting, by the controller, the speed of the motor to 0, when the continuance time exceeds a second reference time; and maintaining the speed of the motor, when the continuance time does not exceed the second reference time.

In the calculating of the speed of the motor, the controller calculates the RPM of the motor through the following equation:

$$\frac{15\times pulse\ number}{PPR\times(first\ reference\ time+pulse\ cycle)} \quad (2)$$

where PPR represents the number of output pulses per revolution of the encoder.

In another embodiment, an apparatus for measuring speed of an EPS drive motor may include: an encoder configured to output A and B pulses having a phase difference of 90 degrees, as the motor is rotated; and a controller configured to measure information of A and B pulses received from the encoder during a first reference time, then remeasure information of A and B pulses received from the encoder again during the first reference time, select any one of the measured pulse information and the remeasured pulse information as data for calculating the speed of the motor, based on the measured pulse information and the remeasured pulse information, and calculate the speed of the motor based on the selected data.

The pulse information may include a pulse number obtained by multiplying the number of the A and B pulses by four, information on the cycle of the A pulse, information on the cycle of the B pulse, and a pulse state. When selecting the data for calculating the speed of the motor, the controller may select the measured pulse information as the data for calculating the speed of the motor in case where the multiplied-by-four pulse number contained in the measured pulse information is equal to the multiplied-by-four pulse number contained in the remeasured pulse information, and select the remeasured pulse information as the data for calculating the speed of the motor in case where the multiplied-by-four pulse number contained in the measured pulse information is different from the multiplied-by-four pulse number contained in the remeasured pulse information.

EXPERIMENT AND RESULTS

The proposed method was experimentally validated. Figure 4 shows the experimental results. Figure 4 is a diagram comparatively illustrating a speed measurement result of the method for measuring speed of an EPS drive motor in accordance with the embodiment of the proposed method and a speed measurement result of a conventional M method.

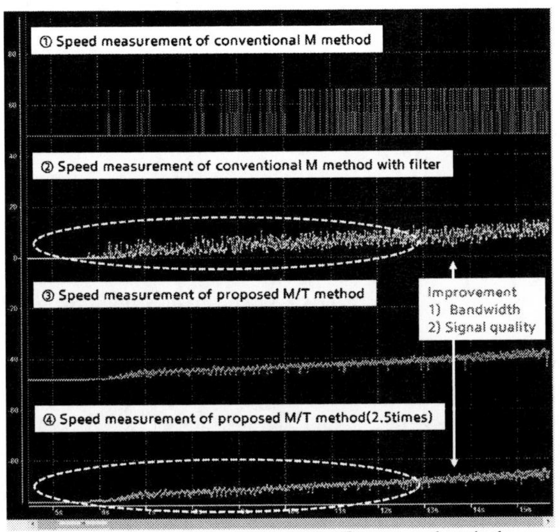

Fig.4. Flow chart of implementation for proposed method

CONCLUSION

In this paper, M/T method for EPS applications was proposed. A method for measuring speed of an EPS drive motor may include: receiving, by a controller, A and B pulses having a phase difference of 90 degrees from an encoder during a first reference time, and measuring information of the pulses; receiving A and B pulses having a phase difference of 90 degrees from the encoder again during the first reference time, and remeasuring information of the pulses; selecting any one of the measured pulse information and the remeasured pulse information as data for calculating the speed of the motor, based on the measured pulse information and the remeasured pulse information; and calculating the speed of the motor, based on the selected data.Lab. based testing and actual production tests validated the proposed method.

ACKNOWLEDGMENT

The present application claims priority to Korean application number 10-2014-0122089, filed on September 15, 2014, which is incorporated by reference in its entirety.

REFERENCES

[1] T. Ohmae, T. Matsuda, K. Kamiyama, M. Tachikawa, in: A microprocessor-controlled high-accuracy wide-range speed regulator for motor drives, IEEE Trans. Ind. Electron., vol. IE-29, no.3, p. 207-221.

[2] Lilit Kovudhikulrungsri,Takafumi Koseki. in:Precise speed estimation from a low-resolution encoder by dual-sampling-rate observer, IEEE/ASME Transactions on Mechatronics, vol.11, no. 6,p.661-670.

[3] Ndubuisi Ekekwe, Ralph Etienne-Cummings, Peter Kazanzides, in: A wide speed range and high precision position and velocity measurements chip with serial peripheral interface, Integration, theVLSI Journal, vol. 41, no. 2, p.297-305.

[4] Richard Bonert, in: Digital Tachometer with Fast Dynamic Response Implemented by a Microprocessor, IEEE Transactions on Industry Applications, vol.IA-19, no.6, p.1052-1056.

[5] Shih-Chin Yang, Robert D. Lorenz, in:Surface Permanent-Magnet Machine Self-Sensing at Zero and Low Speeds Using Improved Observer for Position, Velocity, and Disturbance Torque Estimation, Transactions on Industry Applications, vol.48, no.1,p.151-160

[6] Richard Bonert, in: Design of a high performance digital tachometer with a microcontroller, IEEE Transactions on Instrumentation and Measurement. vol.38, no.6, p. 1104–1108.

[7] N.Hagiwara, Y.Suzuki, H.Murase, in: A method of improving the resolution and accuracy of rotary encoders using a code compensation technique, IEEE Transactions on Instrumentation andMeasurement. vol.41, no.1, pp.98-101.

[8] E.Galvan, A.Torralba, L.G.Franquelo, in: A simple digital tachometer with high precision in a wide speed range, Industrial Electronics, Control and Instrumentation, vol.2, p. 920-923.

[9] E.Galvan,A.Torralba,L.G.Franquelo,in: ASIC implementation of a digital tachometer with high precision in a wide speed range, IEEE Transactions on Industrial Electronics,vol.43, no. 6, p. 655 - 660.

[10] P. Bhatti, B. Hannaford, in: Single-chip velocity measurement system for incremental optical encoders, IEEE Transactions on Control System Technology, vol.5, no.6.

[11] S. K. Sul, *Control of Electric Machine Drive System*, WILEY, 2011.

Low-Cost Input Impedance Estimator of DC-to-Dc Converters for Designing the Control Loop in Cascaded Converters

M. Sanz, A. Lázaro, M. Bermejo, D. López del Moral, P. Zumel, C. Fernández, A. Barrado
Carlos III University of Madrid
Electronic Technology Department
Power Electronics Systems Group
Avda. Universidad, 30; 28911, Leganés, Madrid, SPAIN
Tel.: 34-1-6246024; FAX: 34-1-6249430
E-mail: marina.sanz@uc3m.es

Abstract— **Current trends in power distribution systems are focused on distributed architectures instead of centralized approach. Distributed architectures are more complex since a network of power electronic converters is required to distribute power with appropriate performance. One of the main issues deals with the control loop design of the converters when they are loaded by other converters downstream. Existing design approaches of the converter control loop for system-level stability are typically based on the use of small-signal impedance-based models of the different converters. As a consequence the input impedance of the converters that act as load is required to design the feedback compensation. However, these systems are usually comprised of subsystems made up by a variety of manufacturers, which do not provide the data required to parameterize a model due to reasons of confidentiality. Hence, many power system designers should perform complex measurements and identification techniques to obtain behavioral models of the converters to perform proper control loop design. The aim of this paper is to provide an input impedance estimator of the Dc-to-Dc converters avoiding complex power system measurements and analytical identification techniques. The proposed estimator has been validated with several commercial converters provided by different manufacturers.**

I. INTRODUCTION

In recent years, traditional power distribution systems, based on a centralized architecture, have been progressively replaced by distributed power systems (Fig. 1.a). Distributed architectures comprise a multi-converter network, where the loads are supplied by a number of small power converters which are distributed throughout the system, typically trying to perform the power processing close to the load [1-3]. Distributed power systems were primarily developed for telecommunications and computer applications. Recently, distributed power systems are being adapted in emerging technologies for transportation (Fig. 1.b) such as the More-Electric Aircraft, All-Electric-Ships and Electric/Hybrid-Electric Vehicles [4-6]. These systems include multiple energy sources, energy storage units and numerous passive and active loads.

Then, multiple power electronics converters are required to distribute power throughout the system with the required performance.

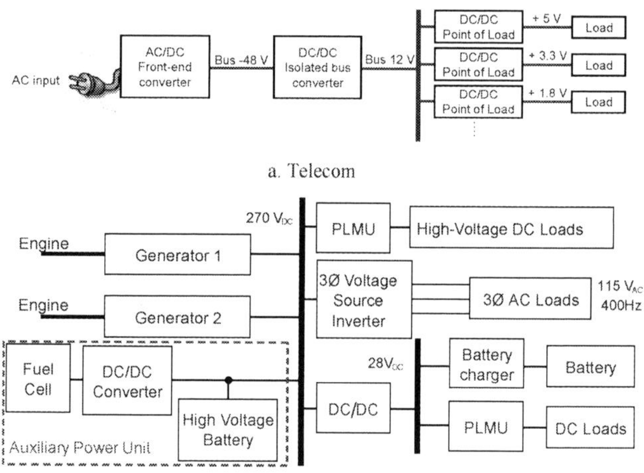

Fig. 1. Distributed power system architectures for Telecom (a) and More Electric Aircraft (b)

Compared to centralized architectures, distributed ones offer different advantages such as the use of standardized converters (commercial off-the-shelf, COTS), the easy addition of redundancy, the capability of on-line replacement (hot-swapping) of damaged converters. However, new challenges regarding integration of a power distribution system made up by multiple converters must be addressed, since the active nature of power electronic converters results

978-1-4673-9551-9/16 $31.00 © 2016 IEEE

in complex dynamic behavior when they are interconnected to each other. Particularly, one important issue is the design of the control loop of the converters when they are loaded by other converters downstream.

The use of analytical models [7] is not possible since power distribution systems are typically comprised of converters made up by a variety of manufacturers, which do not provide the data required to parameterize a model due to reasons of confidentiality.

Hence, power system engineers should be able to design power converter regulators for system stability without having the model of the different power converters. Existing approaches [8], [9] propose to obtain behavioral models of the power converters that act as loads in order to perform the suitable integration of the power distribution system. However, these power converter models require performing additional complex measurements and, hence they need expensive instrumentation. Moreover, the models are not easy to obtain since it requires complex identification methods.

In this paper, an input impedance estimator of Dc-to-Dc converters is proposed. The aim is to provide the required converter dynamic to design the control loop in distributed power systems avoiding complex power system measurements and analytical identification techniques.

II. Stability Design Issues for Dc Power Distributed Systems

First, the main issues regarding stability of Dc power distribution systems are reviewed in a comprehensive way. For a better understanding, two cascaded buck converters with the specifications detailed in Table I has been considered (Fig. 2).

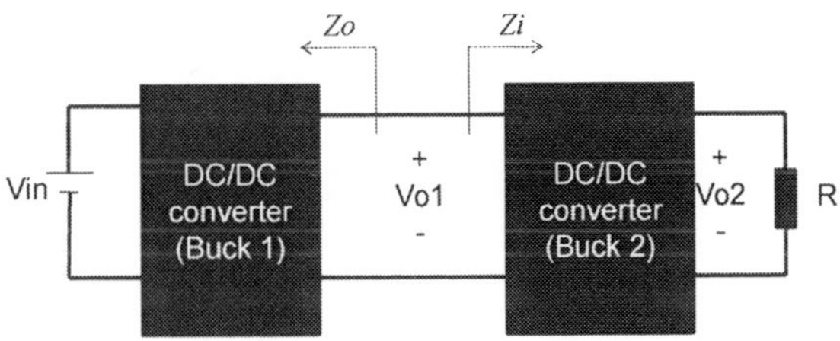

a. Simplified power system architecture

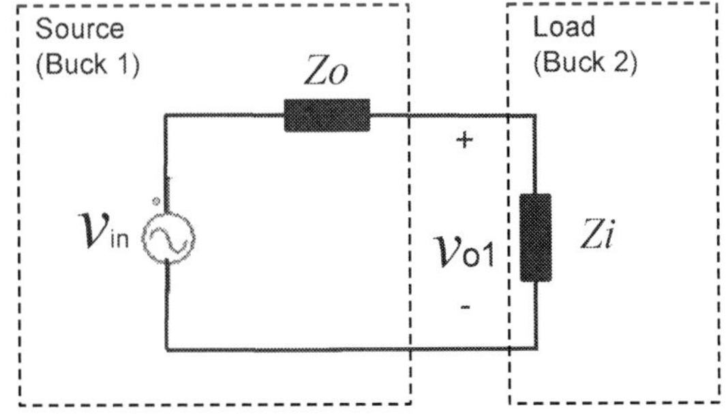

b. Small-signal impedance-based model

Fig.2. Analysis of power converters interaction by means of small-signal impedance-based converter models

Table I. Power Converter Specifications

Magnitude	Buck 1(Source)	Buck 2 (Load)
Input voltage	12 V	5 V
Output Voltage	5 V	1.5 V
Switching frequency	10 kHz	100 kHz
Power	10 W	10 W
Input capacitance	100 µF/1 mΩ	100 µF/1 mΩ
Output capacitance	200 µF/10 mΩ	20 µF/10 mΩ
Inductance	500 µH/1 mΩ	50 µH/1 mΩ
Bandwidth	1.5 kHz	10 kHz

This simplified system only comprises a bus converter and a point of load converter (POL) of the system shown in Fig 2.a. A voltage-mode control of the output voltage has been considered in both converters since it is a typical control in this kind of systems. The system-level stability can be well explained by making use of small-signal impedance-based models as represented in Fig. 2.b [10].

By denoting Z_0(the output impedance of the bus converter (source converter), and Z_i the input impedance of the POL converter (load converter), the small-signal transfer function of the interconnected system is given by (1). Hence, the ratio between Z_o and Z_i, denoted T(s) in (2), can be seen as the minor loop gain of the interconnected system.

$$\frac{v_{o1}}{v_{in}} = \frac{Z_i}{Z_i + Z_o} = \frac{1}{1 + \frac{Z_o}{Z_i}} \quad (1)$$

$$T = \frac{Z_o}{Z_i} \quad (2)$$

Therefore, the Nyquist criterion can be readily applied to T(s) in order to determine system-level stability and so instability is predicted as the point (-1+0j) is encircled by T. According to the Middelbrook criterion [10], this situation can be detected if the magnitude of Z_o is higher than the magnitude of Z_i.

Taking into account the previous stability considerations, the control loop of the source converter of the power system example (Buck 1) has been designed to obtain stable performance in stand-alone conditions by considering as converter load a resistor (R) that consumes the same power (P) than the load converter which value is given by (3).

$$R = \frac{V_{o1}^2}{P} \quad (3)$$

In addition, the control loop has been designed to achieve a magnitude of its output impedance lower than the input impedance magnitude of the load converter (Buck 2) as it is shown in Fig. 3. Therefore, a 1.5 kHz bandwidth of the source converter is required to design the control for power system stability. As seen in Fig. 4, the considered power system is stable since the simulated transient response meets the specifications.

As a consequence, in order to design properly the control loop of the source converter, the input impedance of the load converter is required.

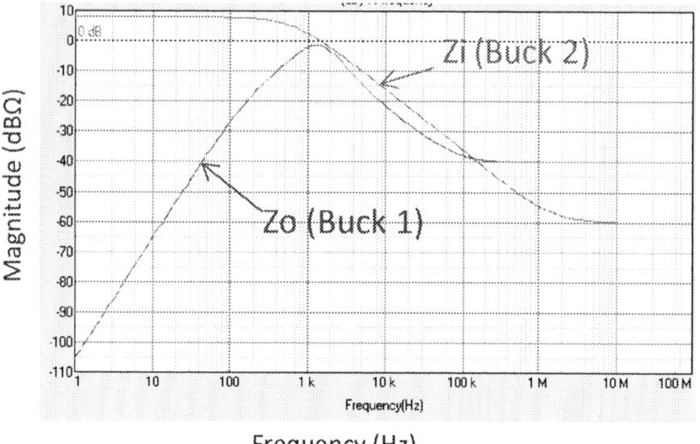

Fig. 3. Magnitude of the input impedance of the load converter (Z_i) and output impedance of the source converter (Z_o)

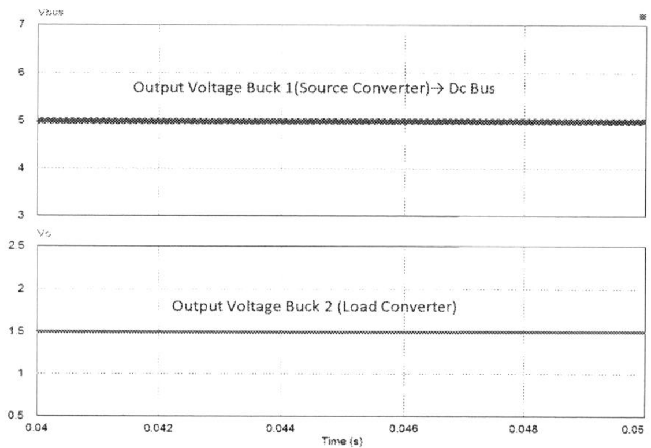

Fig. 4. Transient response of the power system

III. Proposed Input Impedance Estimator

Taking into account the system stability analysis, the aim of this paper is to be able to estimate the small-signal input impedance of the load converter required for the design of the control loop. Hence, the behavior of the Dc-to-Dc converter that acts as load has been analyzed.

For a better understanding, the load converter described in Table I has been used. It is important to highlight that the behavior of the converter regarding its input port does not depend on the considered power converter topology.

In [11], the analytical expression (4) for the input impedance of a buck converter operating in continuous conduction mode and with Voltage Mode Control (Fig. 5) is obtained.

$$Z_i = \frac{V_{o1}^2}{V_{o2}^2} \cdot \frac{CLRs^2 + Ls + R + V_{o1}RG_{mod}K_vR_v}{CRs^2 - V_{o1}G_{mod}K_vR_v + 1} \quad (4)$$

where K_v is the gain of the sensor, R_v is the transfer function of the regulator and G_{mod} is the transfer function of the modulator. This control loop has been calculated with SmartCtrl® and the obtained transfer functions are given by (5), (6) and (7), respectively.

$$K_v = \frac{2}{3} \quad (5)$$

$$R_v = \frac{5.605s^2 + 3.991 \cdot 10^5 s + 7.104 \cdot 10^9}{2 \cdot 10^{-6}s^3 + 4.436 \cdot 10^{-1}s^2 + 2.459 \cdot 10^4 s} \quad (6)$$

$$G_{mod} = 0.3 \quad (7)$$

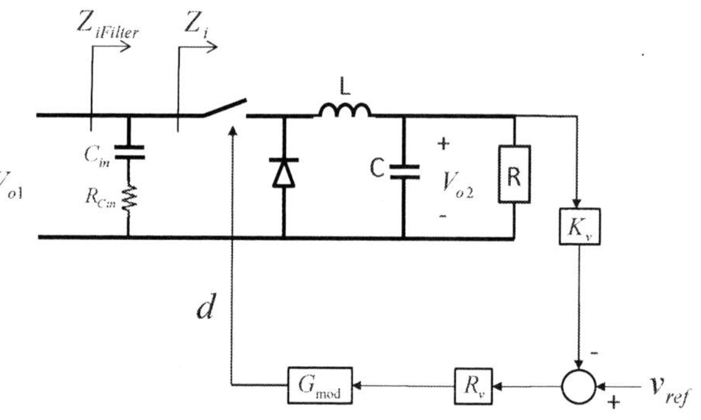

Fig. 5. Block diagram of Buck converter with Voltage Mode Control

The analytical expression of the input impedance of the converter (4) has been compared with the simulated input impedance of the load converter (Buck 2) using PSIM®. As seen in Fig. 6, the simulated and analytical results are similar at frequencies lower the switching frequency of the converter (100 kHz).

From Fig. 6, it can be concluded that due to the action of the control, at frequencies below the crossover frequency of the control loop (10 kHz), the converter behaves as a negative resistor (-180 degrees and constant magnitude) instead of a positive resistor. This small signal behavior corresponds to the constant power load behavior of the Dc-to-Dc converter at large signal [7] given by (8).

$$R_{cpl} = \frac{-V_{in}^2}{P_{in}} \quad (8)$$

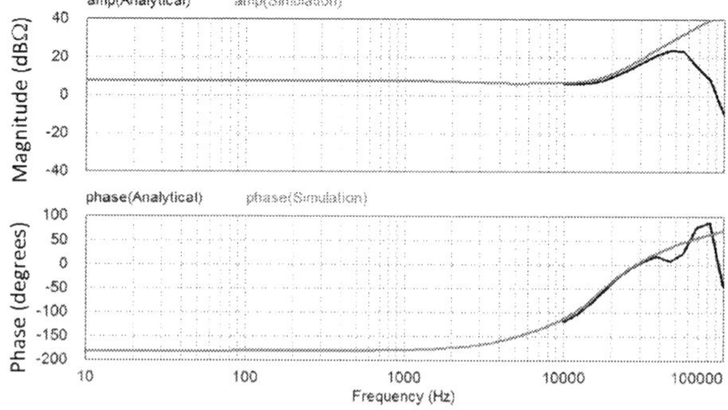

Fig. 6. Analytical versus simulated frequency response of the input impedance of the load converter (Buck 2)

Dc-to-Dc converters include an input filter to reduce the current ripple that typically consists on a capacitor as will be shown in section IV. If a capacitor is being used at the input of the circuit as a filter, the impedance seen from the input of the converter is the result of the parallel of both impedances, giving as result the analytical expression shown in (9).

$$Z_{iFilter} = \frac{Z_i \cdot (1 + R_{Cin} \cdot C_{in} \cdot s)}{1 + (Z_i + R_{Cin}) \cdot C_{in} \cdot s} \qquad (9)$$

In [12], a very simple small-signal model of the converter (Fig. 7) that consists on a negative resistor (R_{cpl}) connected in parallel with the input capacitor (C_{in}) has been proposed to obtain the input impedance of the load converter and so, to design the control loop of the source converter in order to achieve the stability of the power system as has been explained in Section II.

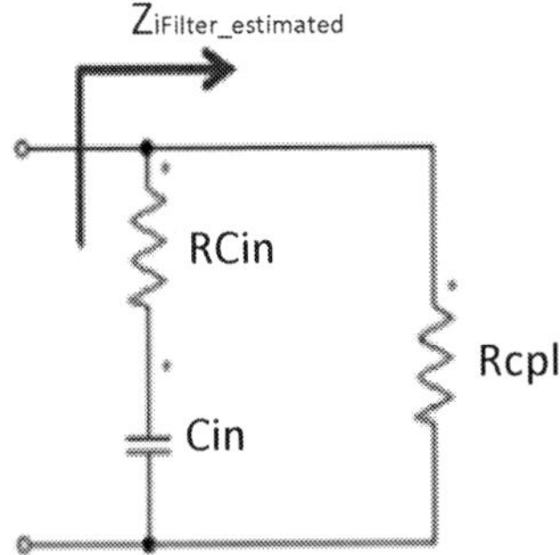

Fig. 7. Small-signal model of the load converter

Using this simple model the input impedance of the converter can be estimated using (10).

$$Z_{iFilter\_estimated} = \frac{R_{cpl} \cdot (1 + R_{Cin} \cdot C_{in} \cdot s)}{1 + (R_{cpl} + R_{Cin}) \cdot C_{in} \cdot s} \qquad (10)$$

where R_{Cin} is the equivalent series resistance of the input capacitor C_{in}.

In this paper, a more detailed analysis and validation of the simplified model is developed.

First, considering a voltage mode control strategy but different values of the input capacitor, a comparison of the simplified model with the exact value of the input impedance of load converter (Buck 2) obtained by simulation has been carried out. As seen in Fig. 7, as the input capacitor decreases (100 μF, 56 μF and 22 μF) the pole frequency that is produced by the input capacitor increases (636 Hz, 1.14 kHz and 2.89 kHz). From Fig. 8 it can be concluded that the frequency response of the input impedance of the converter Buck 2 is dominated by the input capacitor while the frequency pole is a half decade (2.89 kHz) lower than the crossover frequency of the control loop (10 kHz).

Next, Fig. 9 and Fig. 10 show the input impedance of the load converter (buck 2) not only for a voltage-mode control but also for other typical modes of control such as peak current and average current control designed with the same crossover frequency. As seen in Fig. 10, the behavior of the capacitor, which value is 100 μF, introduces a dominant frequency behavior (pole equal to 636 Hz) in comparison with the crossover frequency of the control loop (10 kHz).

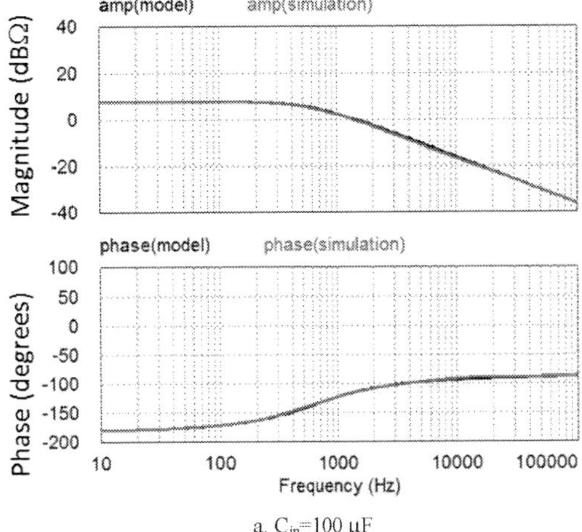

a. C_{in}=100 μF

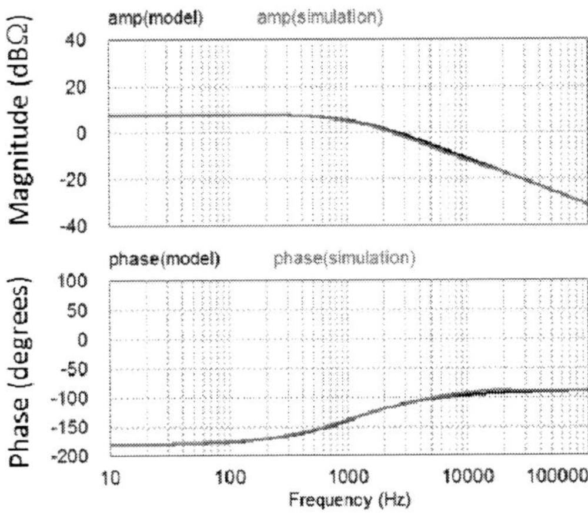

b. C_{in}=56 μF

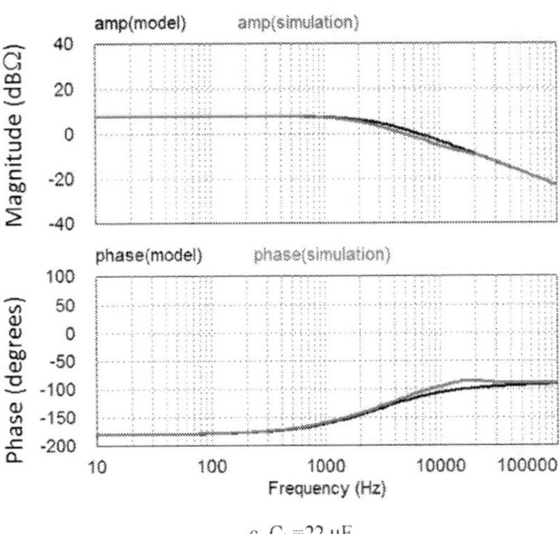

c. C_{in}=22 μF

Fig. 8. Comparison of simulated input impedance and the simplified model for Buck 2 converter with voltage mode control and different input capacitor values: (a) Cin=100 μF, (b) Cin=56 μF, (c) Cin=22 μF.

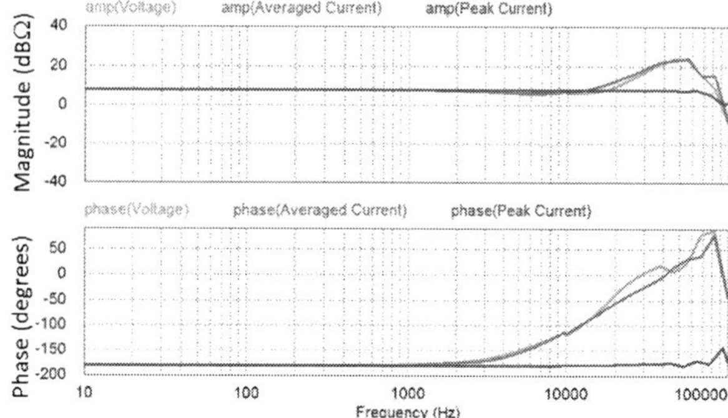

Fig. 9. Small-signal input impedance of load converter (Buck 2) for different control modes without input capacitor

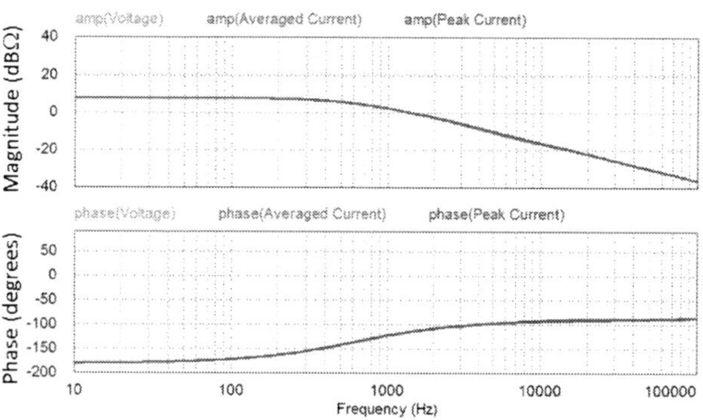

Fig. 10. Small-signal input impedance of load converter (Buck 2) for different control modes with input capacitor (C_{in}=100 μF)

The proposed approach provides two important advantages. On one hand, the analytical calculation of the impedance is very simple in comparison with existing analytical approaches. On the second hand, the most important advantage provided by this model in comparison with existing behavioural models is the easy estimation of the required parameters.

Regarding the value of R_{cpl}, the manufacturer datasheet usually provides the efficiency (η) at different output power levels (P_{out}) and different input voltage values (V_{in}). Thus, using (8), its value can be easily derived using (11).

$$R_{cpl} = \frac{\left(-V_{in}^2\right)\cdot\eta}{P_{out}} \tag{11}$$

However, the manufacturer datasheet only gives information about the type of input filter but not its value. Hence, an additional measurement should be performed.

Notice that the value of R_{Cin} is not a significant parameter regarding the design of the control loop of the source converter, since its effect appears at very high frequencies (more than 1 MHz in this case). Hence, only the value of C_{in} should be estimated. This parameter can be easily obtained in many commercial converters using a very simple and low cost measurement consisting on connecting a multimeter at the input of the converter.

IV. Experimental Validation

One of the main important aspects included in this paper in comparison with previous results is the experimental validation of the proposed simple estimator for the input impedance described in section III.

Two commercial POL Dc-to-Dc converters have been considered for the experimental measurements: C&D TECHNOLOGIES NDTS0503C [13] and MURATA OKX T/10&T/16-D12 Series [14]. The first one is an isolated converter with the following specifications: input voltage (V_{in}) equal to 5V, output voltage (V_{out}) equal to 3.3V and maximum output power (P_{out}) equal to 3W. The second converter is non-isolated with the following specifications: V_{in} equal to 12V, V_{out} equal to 3.3V and maximum P_{out} equal to 80W.

From the datasheet of the manufacturer, the value of R_{cpl} can be estimated using the efficiency data. Regarding the input capacitor, a very simple measurement of the capacitance with a multimeter has been performed to obtain the input capacitor value (Fig. 11). For validating purposes, two different power levels have been considered in both converters. The main parameters of the proposed simple model are summarized in Table II.

Table II. Estimated Parameters of the Proposed Model

Converter	V_{in}	P_{out}	η	R_{cpl}	C_{in}
NDTS0503C	5 V	1.7 W	0.75	-11.03 Ω	5.02 μF
NDTS0503C	5 V	2.8 W	0.70	-6.25 Ω	5.02 μF
OKX T/10&T/16-D12	12 V	24 W	0.875	-19.09 Ω	32.5 μF
OKX T/10&T/16-D12	12 V	72 W	0.93	-6.76 Ω	32.5 μF

In the case of the first converter, additional measurements of the efficiency have been performed since the manufacturer datasheet does not provide the complete information for different power levels. This measurement is very easy and can be performed using only multimeters.

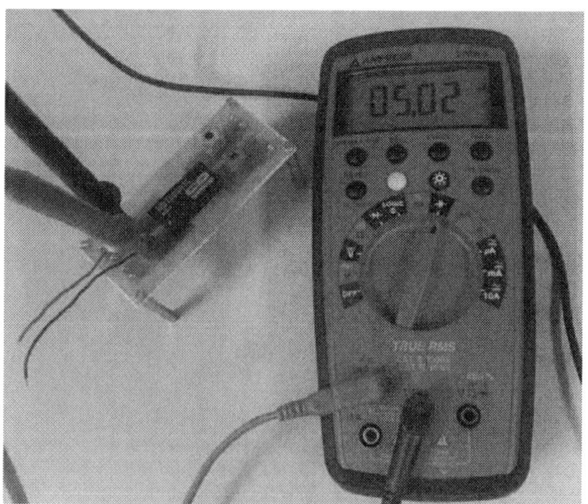

Fig. 11. Measurement of the input capacitor of the converter

In order to validate the proposed model of the converter, the input impedance of each converter has been measured using a frequency response analyzer (VENABLE model 3235) using the experimental setup shown in Fig. 12. A current at the input of the converter has been injected at different frequencies and the input impedance of the converter (Z_i) has been determined by measuring the input voltage (v_i) and current (i_i) of the converter.

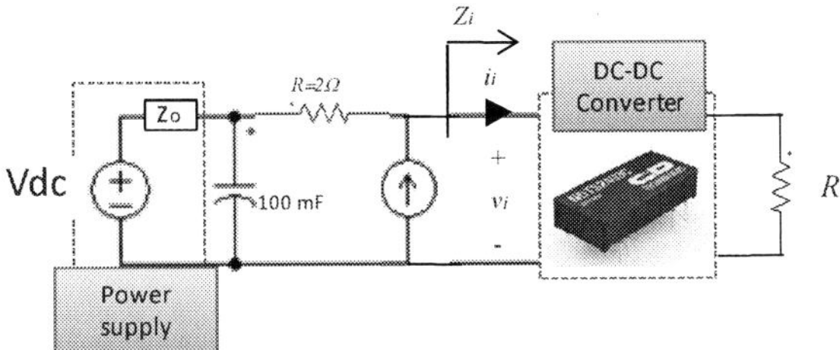

Fig. 12. Measurement of the input impedance of the converter

Fig. 13 and Fig. 14 show the comparison between the experimental input impedance with the estimated value considering the proposed converter model of Fig. 7. As seen, the maximum differences in the input impedance magnitude are lower than 2 dBΩ in the first converter and lower than 4 dBΩ in the second converter.

Hence, it can be concluded that the proposed input impedance estimator provided enough accuracy to design the control loop of the source converter in power distribution systems avoiding complex measurements and using low cost instrumentation.

V. CONCLUSIONS

In this paper, an input impedance estimator of Dc-to-Dc converters is proposed. The aim is to provide power system designers the required converter dynamic for designing the control loop of the converters when they are loaded by other converters downstream in case the power distribution system is comprised of commercial converters, thus there are no data to parameterize a model of the converter.

The proposed estimator has been validated with measurements of several commercial converters provided by different manufacturers. The accuracy provided by the estimator in comparison with measurement results is good enough for the design of the feedback compensation. The main important advantage of the proposed approach in comparison with existing approaches is that a very simple measurement of the input capacitor of the converter with a multimeter is required. Hence, complex setup and expensive instrumentation are avoided in comparison with existing approaches based on behavioral converter models.

ACKNOWLEDGMENTS

This work has been partially supported by the Ministry of Economy and Competitiveness and FEDER funds, through the research project "Storage and Energy Management for Hybrid Electric Vehicles based on Fuel Cell, Battery and Supercapacitors" - ELECTRICAR-AG- (DPI2014-53685-C2-1-R).

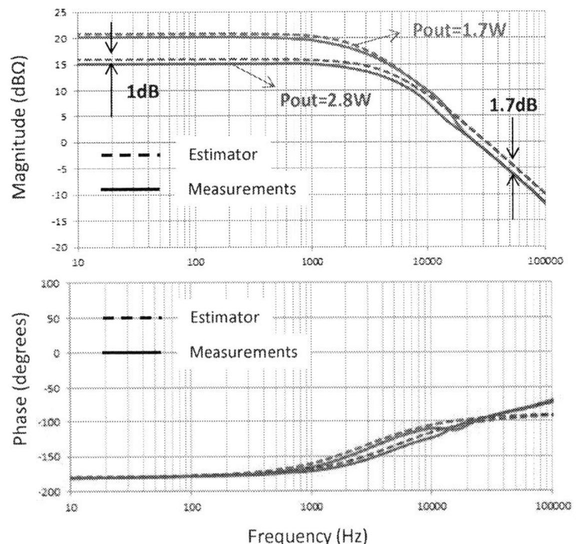

Fig. 13. Input impedance of the converter versus frequency. Model and measurement results for NDTS0503C

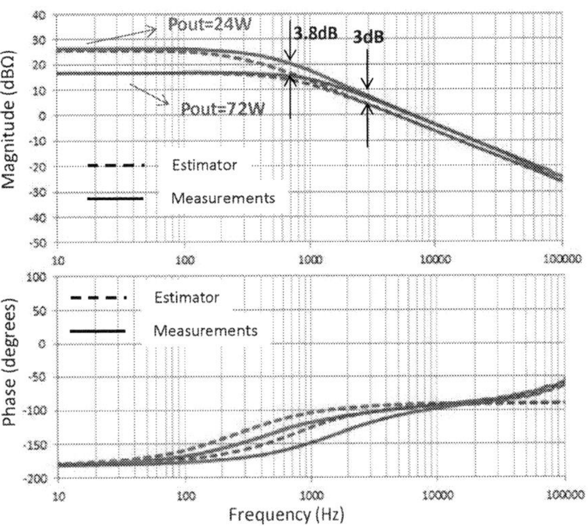

Fig. 14. Input impedance of the converter versus frequency. Model and measurement results for OKX T/10&T/16-D12

REFERENCES

[1] R. Miftakhutdinov, "Power distribution architecture for tele- and data communication system based on new generation intermediate bus converter," proc. INTELEC, 2008, pp. 1-8.

[2] M. Barry, "Design issues in regulated and unregulated intermediate bus converters", Applied Power Electronics Conference and Exposition (APEC), 2004, pp. 1389-1394.

[3] L. Brush, "Distributed power architecture demand characteristics", Applied Power Electronics Conference and Exposition (APEC), 2004, pp.342 2004.

[4] A. Emadi, Sheldon S. Williamson, and Alireza Khaligh, "Power electronics intensive solutions for advanced electric, hybrid electric, and fuel cell vehicular power systems," IEEE Trans. Power Electron., vol. 21, no. 3, pp. 567-577, May 2006.

[5] J. A. Rosero , J. A. Ortega , E. Aldabas and L. Romeral "Moving towards a more electric aircraft", IEEE Aerosp. Electron. Syst. Mag., vol. 22, pp.3 2007.

[6] D. Boroyevich, I. Cvetkovic, D. Dong, "Intergrid: A Future Electronic Energy Network?", in proc. of CPES Annual Power Electronics Conference, 2012.

[7] Reza Ahmadi and Mehdi Ferdowsi "Controller Design Method for a Cascaded Converter System Comprised of Two DC-DC Converters Considering the Effects of Mutual Interactions" IEEE Applied Power Electronics Conference and Exposition (APEC), 2012, 1838 – 1844.

[8] Arnedo, L. ; Boroyevich, D. ; Burgos, R. ; Wang, F., "Black-Box Terminal Characterization Models for the Analysis and Simulation of Distributed Power Electronic Systems", IEEE Power Electronics Specialists Conference, 2007, pp. 1968-1973, 2007

[9] Valdivia, V.; Barrado, A.; Lázaro, A.; Fernández, C.; Zumel, P. "Black-box modeling of DC-DC converters based on transient response analysis and parametric identification methods", Twenty-Fifth Annual IEEE Applied Power Electronics Conference and Exposition (APEC), 2010, pp. 1131 – 1138.

[10] Antonino Riccobono, Enrico Santi "Comprehensive Review of Stability Criteria for DC Power Distribution Systems", IEEE Transactions on Industry Applications, Vol. 50, No. 5, September/October 2014

[11] Reza Ahmadi, Darren Paschedag, Mehdi Ferdowsi . "Analyzing Stability Issues in a Cascaded Converter System Comprised of Two Voltage-Mode Controlled DC-DC Converters", Twenty-Sixth Annual IEEE Applied Power Electronics Conference and Exposition (APEC), 2011, pp. 1769 - 1775.

[12] Sanz, M.; Lazaro, A.; Fernandez, C.; Zumel, P.; Lopez del Moral, D.; Quesada, I.; Barrado, A., "Practicing design method of regulators for cascaded converters", IEEE Workshop on Control and Modeling for Power Electronics (COMPEL), 2014, pp. 1-5.

[13] http://www.datasheetarchive.com/NDTS0503C-datasheet.html

[14] http://power.murata.com/data/power/okx-t10-t16-d12.pdf

On-Chip High Performance Magnetics for Point-of-Load High-Frequency DC-DC Converters

Dragan Dinulovic, Mahmoud Shousha, Martin Haug, and Alexander Gerfer
Würth Eletronik eiSos GmbH &Co KG
EMC & Inductive Solutions
Parkring 29, 85748 Garching, Germany
E-mail: dragan.dinulovic@we-online.de, mahmoud.shousha@we-online.de, martin.haug@we-online.de, alexander.gerfer@we-online.de

Mike Wens and Jef Thone
MinDCet NV
Researchpark Haasrode
Esperantolaan 4
3001 Leuven, Belgium
E-mail: mike@mindcet.com, jef@mindcet.com

***Abstract*—This paper presents the design, fabrication, and characterization of on silicon integrated micro-transformers for high frequency power applications. The microtransformer device is used and tested in DC-DC converter application at high switching frequency. This device has stable L vs. f characteristic up to 50 MHz. The design is improved regarding to the electrical resistance and current capability. The microtransformer shows an inductivity of about 60 nH, resistance of 350 mΩ and can be applied for current up to 1.5 A.**

Keywords — Microtransformer, Magnetisc on Silicon, DC-DC converter, high swetching frequency, power transformer, power conversion, switched-mode power supply

I. Introduction

An increasing trend in power electronics is to miniaturize passive components, like inductors or transformers, to allow for higher power density and increasing mobility. One of the possible techniques to miniaturize passive components is to increase the switching frequency of an electronic power circuit. Recently, first commercial DC-DC converters with switching frequency of 20 MHz are available in the market. A permanent increase of switching frequency sets new requirements for inductive components regarding to size, inductance, and current capability.

At high switching frequencies, inductors and transformers can have a smaller inductivity and hence smaller size, since inductivity reduction allows for inductor profile height reduction. Therefore, thin-film fabricated microinductors and microtransformers are ideal components to be used in high switching frequency applications. Both, microinductors and microtransformers are in many research works fabricated and tested at higher switching frequencies up to 50 MHz [1-7]. Based on these first results, the implementation of micro devices in power electronic applications is expected to be successful.

II. Design and Fabrication of Microtransformer

First fabricated microtransformer device was shown before [8]. This design is improved in terms of increasing current capability and efficiency. The design is simulated and optimized using Finite Element Method (FEM) with the help of the software tool Ansys Maxwell®.

The microtransformer consists of a closed Co-Fe magnetic core and four coils. Two coils are on the primary and two coils are on the secondary transformer side. The size of four coils was restricted by the full size (length of 2.5 mm) of a microtransformer device. An EIA standard size 1008 is chosen as the full size of a microtransformer. Therefore, whole microtransformer device features a footprint of 2.5 mm x 2 mm.

The main goal of simulation is to develop a design with as small as possible electrical resistance and at the same time the design should have maximal electrical inductance. As an optimal design, a system with electrical inductance of about 60 nH and resistance of about 300 mΩ is chosen. Each coil consists of 5 turns. The cross-section of a coil turn is 60 μm (width) x 20 μm (height).

As a magnetic material for the transformer core, the alloy Co-Fe was chosen based on its very good magnetic properties. Co-Fe alloy features saturation flux density Bs up to 2.3 T. Co-Fe magnetic core has been deposited by Co-Fe electroplating. Coils are fabricated using Cu electroplating. Coils and magnetic core are embedded in polyimide [9, 10]. The fabrication procedure of the microtransformer is shown in the Figure 1.

The fabrication starts with fabrication of bottom coil layer by Cu electroplating. The bottom coil Cu layer is 20 μm thick (A). Next step is pattering of first insulation layer (B). As insulation and embedding material a photosensitive polyimide (PI) is used. In the next step, the magnetic core is deposited by using CoFe-electroplating. The magnetic core has a thickness of 10 μm (C). Next, the vias are fabricated using cooper

978-1-4673-9551-9/16 $31.00 © 2016 IEEE

electroplating (D). After that, the vias and the magnetic core are embedded in polyimide (E). The flat surface is achieved by applying of a CMP step. Repeating the same steps as explained above for bottom coil layer, the top coil layer is patterned and electroplated (F). The completed microtransformer device is in the next step embedded in polyimide (G). This third embedding step allows a protection of the device against external environmental influences.

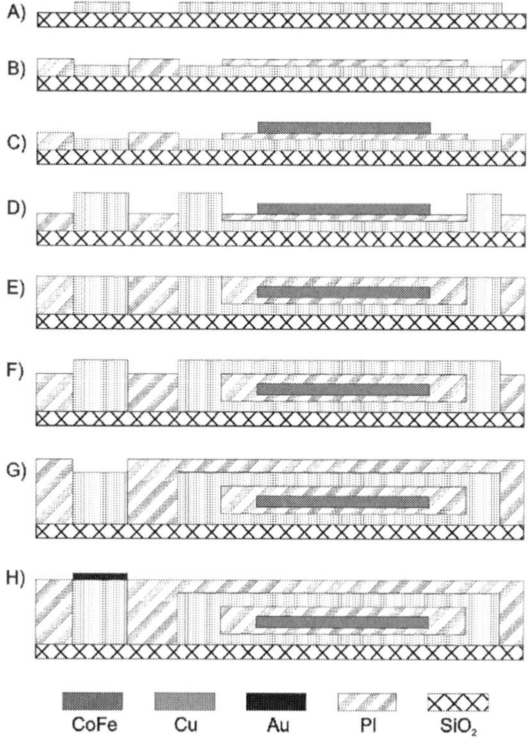

Fig.1: Fabrication procedure for the microtransformer device

Last fabrication step is the fabrication of contact pads by applying Ni and Au electroplating (H). Finally, the completed wafer is separated into chips by dicing. Completed device is shown in the Figure2.

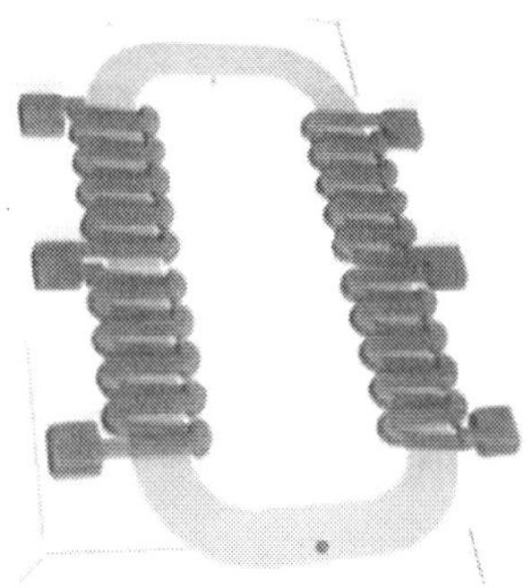

Fig.2: X-ray micrograph of completer microtransformer device

III. Electrical Ppoperties of Microtransformer

Diced micro device chips are prepared for testing. Chips are assembled into open cavity QFN package. The QFN housing 4 mm x 4 mm with 12 leads is used. For our tests only 6 leads were connected. Measurements are carried out by applying Agilent Impedance Analyzer E4991A. The device is measured with a signal oscillating level of 5 mA at 1 MHz.

Figure 3 shows the characteristic of inductance versus frequency. The whole system shows an inductance of more than 50 nH with very good performance up to frequencies up to 100 MHz.

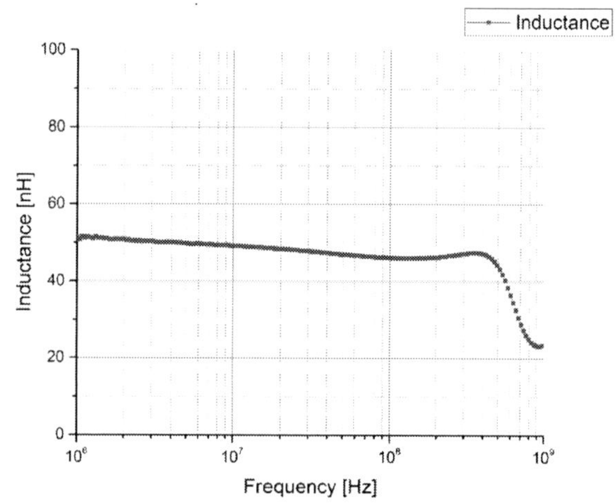

Fig.3: Fabrication procedure for the microtransformer device

Also, other electrical properties of micro device are defined. The electrical resistance of whole system (four coils in series) Rdc is about 350 mΩ. The Q-factor shows maximal value of 18 at frequency of 70 MHz. Figure 4 shows Rdc and Q-factor against frequency of the micro device.

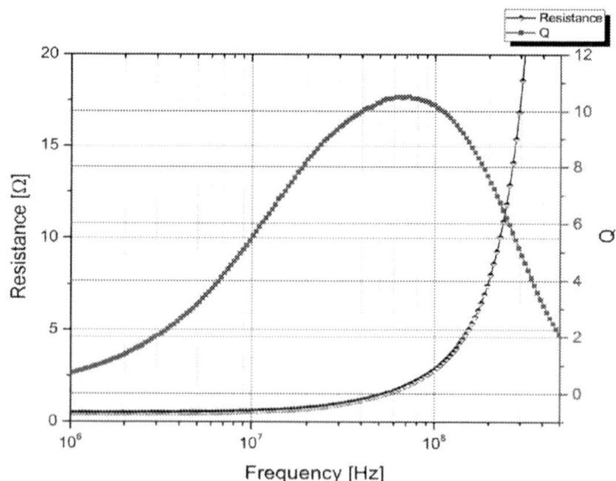

Fig.4: Rdc and Q-factor vs.f for four coils of microtransformer.

Figure 5 shows characteristics of rated current I_R and saturation current I_{SAT} of microtransformer devices. These measurements are done for primary transformer side (two coils in series). The rated current for primary transformer side is 550mA. That means, at current of 550 mA the micro device heats up to 40°C degrees above the ambient temperature.

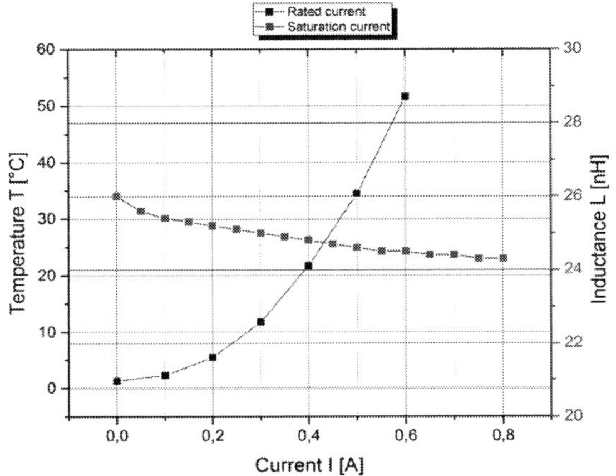

Fig.5: Rated current I_R and saturation current I_{SAT} characteristics of primary side of microtransformer device.

The saturation characteristic of microtransformer is very soft. The inductance decreasing with increasing of the DC-Bias current is therefore small. At DC-Bias current of 800 mA drops the inductance only 2 nH what is about 8 percent. The inductance drop of 30 percent will be achieved at DC-Bias current of about 1.5 A.

IV. Experimental Results in DC-DC Converter at High Switching Frequency

The samples of microtransformer were prepared for testing in HF buck DC-DC converter. For testing, the switcher MDCD073 a dual 7V PN half bridge with pre-drivers in 180 nm SOI technology is developed (Fig. 6).

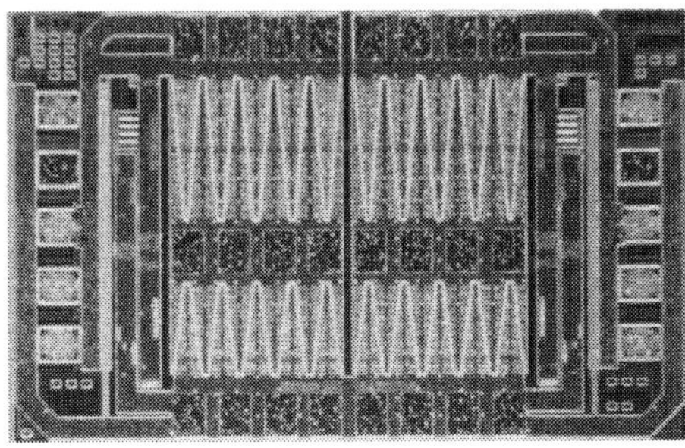

Fig.6: Micrograph of ASIC MDCD073

Figure 7 shows a demo board with assembled microtransformer prepared for efficiency measurements. First measurements show the efficiency of about 65% by switching frequency of 27 MHz. The measurements are done for input voltage of 3.3 V, where the output voltage was 1.8 V. Figure 8 shows the characteristic the efficiency vs. output current.

Fig.7: Prepared demo board for efficiency measurement

The core loss of microtransformer is measured and shows the value of about 100 mW.

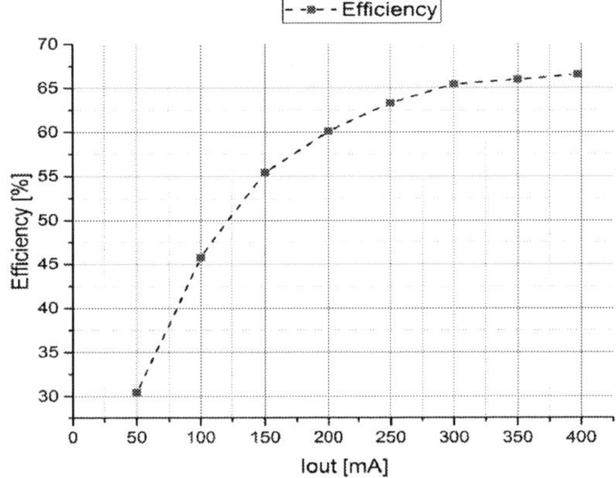

Fig.8: Efficiency measurement for input voltage 3.3 V and output voltage 1.8 V

The measurement at heavy load of 720mW shows that under these loading conditions the inductor conduction losses dominates the overall power losses as expected. The measurement at light load of 7.2mW shows that under these loading conditions the switching losses of both high and low side switches dominate the overall power losses of the converter.

V. Packaging of Device and Assembly

The aim of our development is the integration of micro transformer device together with the controller ASIC MDCD073 and input and output capacitors into one package.

Thanks to the very low profile of the micro transformer the DC-DC buck converter can be entirely embedded into FR4 based laminate. The micro device maximal package height is 1 mm and maximal lateral dimensions are 3 mm x 4 mm. The work on the packaging is still in progress.

The DC-DC converter will be suitable for input voltage between 2.7 V and 5.5 V and for output current up to 0.5 A.

VI. Conclusion and Future Work

New design of microtransformer with the aim to improve the performance of the device is presented in this paper. The design is modified in order to reduce the electrical resistance and to keep inductivity as high as possible. The cross-section of coils is increased and the number of turns is reduced. At the same time, the thickness of magnetic core is increased to 10 µm. The microtransformer device was fabricated using thin-film technology. The magnetic core is fabricated using Co-Fe material by electroplating. New improved device shows very stable inductance characteristic up to frequencies higher than 100 MHz and device is suitable for applications at very high switching frequencies. The maximal inductance of device is about 60 nH (all four coils connected in series). Whole device with all four coils in series show the resistance of only 350 mΩ. The DC-Bias characteristic of mictrotransformer is improved and the saturation current is higher than 1500 mA.

In further work, the microtransformer device will be assembled together with ASIC and capacitors into one package as a module.

Acknowledgment

The authors would like to thank the impt (Institute for Micro Production Technology, Leibniz Universität Hannover) for realization of microtransformer samples and Würth Elektronik CBT for helping on packaging of devices.

References

[1] C. Feeney, N. Wang, C. O'Mathuna, and M. Duffy, "A 20-MHz 1.8-W DC–DC Converter With Parallel Microinductors and Improved Light-Load Efficiency", IEEE Transaction on Power Electronics, Vol. 30, No. 2, pp. 771 – 779, 2015.

[2] X. Xing, N. X. Sun, and B. Chen, "High-Bandwidth Low-Insertion Loss Solenoid Transformers Using FeCoB Multilayers", IEEE Transaction on Power Electronics, Vol. 28, No. 9, pp. 4395 – 4401, 2013.

[3] C. Feeney, M. Duffy, N. Wang; S. Kulkarni, and C. O'Mathuna, "Analysis of coupled microinductors for power-supplyon-chip applications", IEEE Energy Conversion Congress and Exposition (ECCE), pp. 1679 – 1684, 2014.

[4] N. Wang, R. Miftakhutdinov, S. Kulkarni, and C. O'Mathuna, "High Efficiency on Si Integrated Micro-transformers for Isolated Power Conversion Applications", IEEE Transaction on Power Electronics, Vol. 30, No. 10, pp. 5746 – 5754, 2015.

[5] T. Liakopoulos, A. Panda, M. Wilkowski, and A. Lofti, 3rd International Workshop on Power Supply on Chip (PwrSoC 2012), San Francisco, USA, 2012.

[6] C. R. Sullivan, D. V. Harburg, J. Qiu, G. Levey, and D. Yao, IEEE Transaction on Power Electronics, Vol. 28, No. 9, 4342 – 4353, 2013.

[7] D. S. Gardner, G. Schrom, F. Paillet, B. Jamieson, T. Karnik, and S. Borkar, IEEE Transaction on Magnetics, Vol. 45, No. 10, 4760 – 4766, 2009.

[8] D. Dinulovic, M. Kaiser, A. Gerfer, O. Opitz, M. C. Wurz, L. Rissing, MEMS Fabricated Microtransformer with Closed Fe-Co Magnetic Core for High Frequency Power Applications, Journal of Applied Physics 115, 17A317, 2014.

[9] C. Ruffert, J. Chen, L. Rissing, European Congress and Exhibition on Advanced Materials and Processes (Euromat 2011), Montpellier, France, 2011.

[10] C. Ruffert, H. H. Gatzen, Microsystem Technologies, 14, pp. 9 – 11, 2008.

Effects of auxiliary source connections in multichip power module

Helong Li, Stig Munk-Nielsen, Szymon Bęczkowski, Xiongfei Wang, Emanuel-Petre Eni
Department of Energy Technology
Aalborg University
Aalborg, Denmark
Email:{hel, smn, sbe, xwa, epe}@et.aau.dk

Abstract-**Auxiliary source bond wires and connections are widely used to in the power module with paralleled MOSFETs or IGBTs. This paper investigates the working mechanism and the effects of the auxiliary source connections in multichip power modules. It reveals that the auxiliary source connections cannot totally decouple the power loop and the gate loop like how the Kelvin source connection does, because they are still in the loop of power source current. Three effects of the auxiliary source connection are investigated and analyzed: common source stray inductance reduction, transient drain-source current imbalance mitigation and influence on the steady state bond wire current distribution. Simulation and experimental results validate the working mechanism analysis and the effects of the auxiliary source connections.**

I. Introduction

The auxiliary source connection for a single die is called Kelvin source connection. Infineon released the Cool MOS with the Kelvin source connection. The latest Wolfspeed 900V SiC MOSFET in D2PAK package also has the Kelvin source connection, as shown in Fig. 1 [1]. The Kelvin source connection decouples the gate-source current loop and the drain-source current loop. Therefore, for the single die, the Kelvin source connection can avoid the common source stray inductance effect [2], and thus improve the switching speed and reduce the switching losses [1].

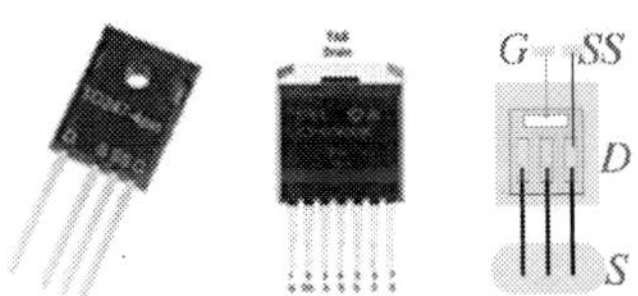

Figure 1: Kelvin source connection for a single device

In multichip power modules with paralleled IGBTs or MOSFETs, there are examples of both power modules without auxiliary source connections [2] and with auxiliary source connections [3]. For the power module customers, the power module with paralleled dies is a black box and usually considered as a single device. For the power module manufacturer, the research emphasis of making multichip SiC MOSFET power module is focused on the reliability of the packaging technology [3–5] and the total stray inductance of the whole package [6, 7]. The circuit mismatch and current distribution in the multichip power module has been investigated [8, 9]. However, there is little discussion about the effects of the auxiliary source connections in the multichip power module. Usually, the auxiliary source connections for the paralleled dies are considered to have the same or similar effect with the Kelvin source connection.

This paper investigates the working mechanism and the effects of the auxiliary source connections. In section II, The description of auxiliary source connections are presented. The difference between Kelvin source connection and auxiliary source connections is discussed with mathematic analysis and circuit modeling. It is found that the auxiliary source connections are still part of the power source current loop. In section III, the effects of the auxiliary source connections are investigated: reduction of common source stray inductance, mitigation of transient drain-source current imbalance and influence on steady state bond wire current distribution. In section IV, the experimental results validate the working mechanism analysis and the effects of the auxiliary source connections.

II. Kelvin Source and Auxiliary Source Connection

A. *Auxiliary source connection*

A single MOSFET with and without the Kelvin source connection is shown in a double pulse test circuit in Fig. 2.

It is very obvious that with the Kelvin source connection, the gate-source current loop and the drain-source current loop are totally decoupled. The gate-source loop inductance L_{gs} will not see the drain-source current.

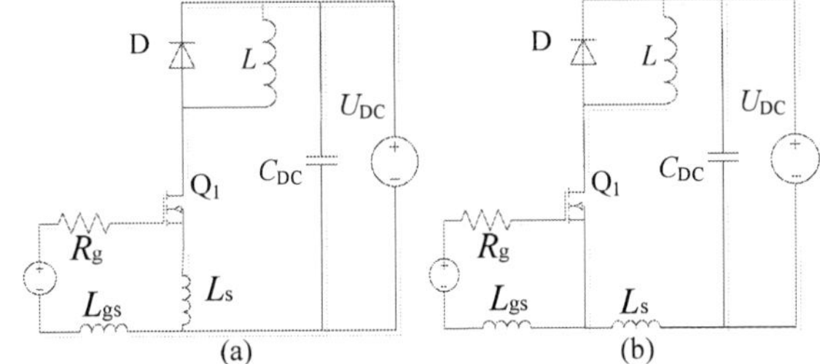

Figure 2: A single MOSFET in a double pulse test circuit (a) without Kelvin source connection (b) with Kelvin source connections

Therefore, the Kelvin source connection can avoid the effect of common source stray inductance L_s, thus improving the switching speed and reducing the switching losses [1].

In case of multichip power modules with paralleled MOSFETs or IGBTs, the auxiliary source connections are used. Fig. 3 shows the schematic of two paralleled MOSFETs with and without auxiliary source connections.

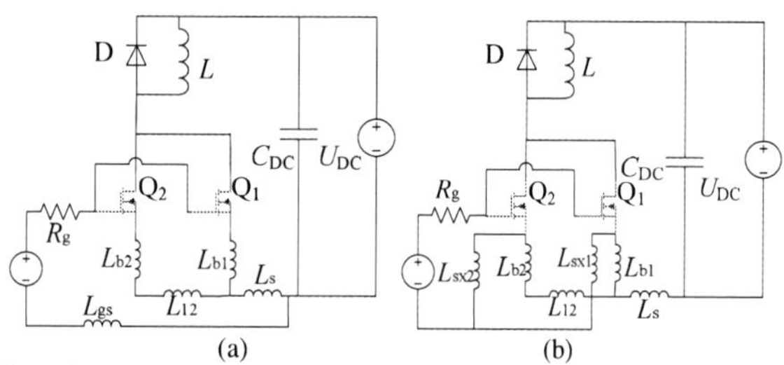

Figure 3: Two paralleled MOSFETs (a) without and (b) with auxiliary source connections

Comparing Fig. 3(b) to Fig. 2(b), it can be seen that the auxiliary source connections L_{sx2} and L_{sx1} form a current loop for the power source current of MOSFETs Q_2 and Q_1. This means that the auxiliary source connections for the paralleled devices cannot totally decouple the gate-source current loop and the drain-source current loop.

The magnitude of the currents going through L_{sx2} and L_{sx1} depends on the impedance relationship of the path L_{sx2}-L_{sx1}-L_{b1} and the path of L_{b2}-L_{12}. It should be mentioned that the impedance relationship ($Z=R+j\omega L$) includes both the resistance and the inductance of the current paths.

B. Auxiliary source connections in multichip power modules

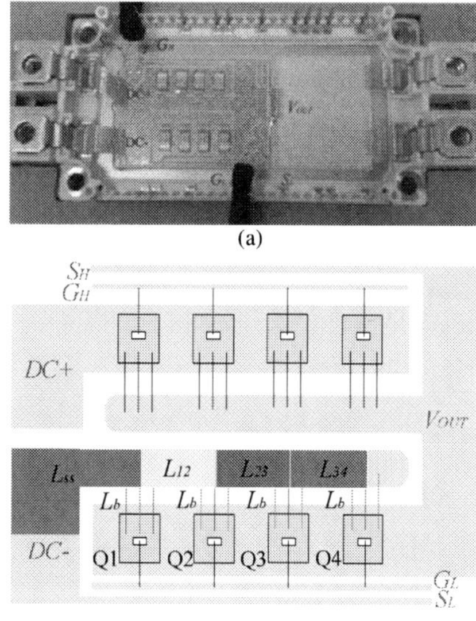

Figure 4: (a) Multichip power module and (b) its DBC layout

A SiC MOSFET multichip half bridge power module is shown in Fig. 4(a). The DBC layout of this power module is shown in Fig. 4(b).

The power module has four dies in parallel and it does not have auxiliary source connections. The DBC layout with added auxiliary source connections is shown in Fig. 5.

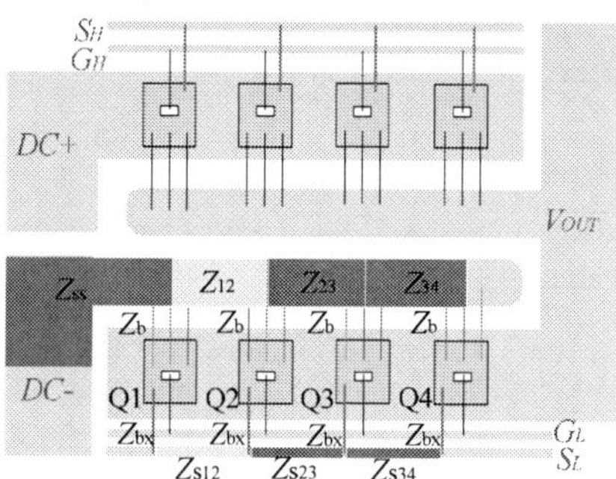

Figure 5: DBC layout modeling (a) without auxiliary source connections (b) with auxiliary source connection

The circuit modeling of the DBC layout with and without auxiliary source connections is shown in Fig. 6. Z_b is the impedance of power source bond wires. Z_{12}, Z_{23}, Z_{34} and Z_{ss} are the impedance of the DBC traces, as stated in Fig. 5. Z_{s12}, Z_{s23}, Z_{s34} are the impedance of the auxiliary DBC traces. Z_{bx} is the impedance of the auxiliary source bond wires.

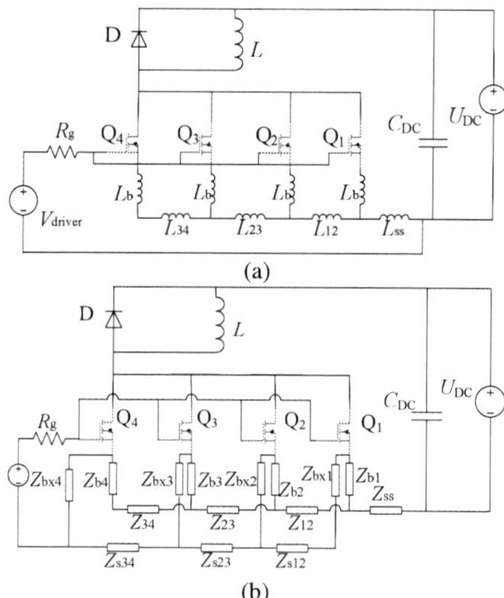

Figure 6: Modeling of DBC layout in a double pulse test circuit (a)without and (b) with auxiliary source connections

Like in the case of paralleling two MOSFETs, the auxiliary source connections for the multichip power modules are also in the power source current loops. For instance, the MOSFET drain current I_{D4} can go to the DC- terminal via Z_{b4}-Z_{34}-Z_{23}-Z_{12}-L_{ss}, which is the expected path. But it can also go through Z_{bx4}-Z_{s34}-Z_{s23}-Z_{s12}-Z_{bx1}-Z_{b1}-Z_{ss}, even though this current loop has much higher impedance than the expected one.

III. Effects of Auxiliary Source Connections

A. Common source stray inductance reduction

Although the auxiliary source connections are in the power source current loop, they can still reduce the common source stray inductance in the power module. For instance, with auxiliary source connections in Fig. 5, the inductance of the DBC traces Z_{ss} is not part of common source stray inductance. Therefore, one of the auxiliary source connection effects is the reduction of the common source stray inductance for the whole power module. Consequently, it can improve the switching speed of the power module and reduce the power losses.

B. Transient current imbalance mitigation effect

The paralleled dies in this DBC layout have mismatched value of common source stray inductance and current coupling effect, which can lead to large transient current imbalance [8]. In the saturation region, the small signal model of the MOSFET drain current and the gate-source voltage is as in (1). And the gate-source voltage is as in (2).

$$i_D = g_{fs}(V_{GS} - V_{th}) \tag{1}$$

$$V_{GS} = V_{driver} - i_G R_G - L_s \frac{di_s}{dt} = V_{driver} - i_G R_G - \Delta V_{LS} \tag{2}$$

The paralleled dies usually have a common gate driver. Compared to drain-source current, the gate-source current is much smaller. The difference in the gate voltage potentials of the paralleled MOSFETs is very small. Therefore, the difference in gate-source voltages between paralleled MOSFETs is the voltage potential difference between the source pads of paralleled MOSFETs, i.e. V_{S1S2}. The transient current imbalance between two paralleled die can be described as in (3).

$$i_{D1} - i_{D2} = g_{fs}(V_{GS1} - V_{GS2}) = g_{fs} V_{s2s1} \tag{3}$$

If the auxiliary connection is considered as a parallel path to the power source path, the auxiliary source connection reduces the total impedance for the power source current path.

If it is assumed that the impedance relationship of the auxiliary source connections and the power source traces is $Z_{bx}=3L_b$ and $Z_{s12}=Z_{s23}=Z_{s34}=3Z_{12}=3Z_{23}=3Z_{34}$, the transient current simulation results of the paralleled four dies with this DBC layout are shown in Fig. 7.The current switching speed d*i*/d*t* in Fig. 7(b) is faster than that in Fig. 7(a). It validates that the auxiliary source connections reduce the total common source stray inductance, the effect of which is similar to that of the Kevin-Source connection. The d*i*/d*t* in Fig. 7(c) is similar to that in Fig. 7(a) by increasing the gate resistance, but the current imbalance in Fig. 7(c) is reduced than that in Fig. 7(a). This validates the analysis of the current imbalance mitigation effect of the auxiliary source connections.

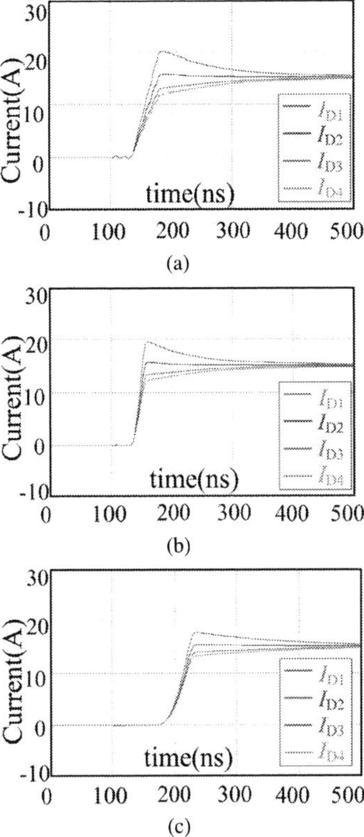

Figure 7: Simulation results of the transient drain current distribution (a) without auxiliary source connections (R_g=10Ω) (b) with auxiliary source connections (R_g=10Ω) (c) with auxiliary source connections (R_g=23Ω)

C. Steady state current distribution

If the impedance relationship of the power trace and the auxiliary source connections is still $Z_{bx}=3L_b$ and $Z_{s12}=Z_{s23}=Z_{s34}=3Z_{12}=3Z_{23}=3Z_{34}$, the steady state current distributions of the power source bond wires Z_{b1}-Z_{b4} and the auxiliary source bond wires can be calculated as in (4). I_D is the steady state drain current of the MOSFET.

$$\begin{cases} I_{Zbx1} = \frac{31}{84} I_D,\ I_{Zbx2} = \frac{1}{84} I_D,\ I_{Zbx3} = \frac{13}{84} I_D,\ I_{Zbx4} = \frac{17}{84} I_D \\ I_{Zb1} = \frac{115}{84} I_D,\ I_{Zb2} = \frac{83}{84} I_D,\ I_{Zb3} = \frac{71}{84} I_D,\ I_{Zb4} = \frac{67}{84} I_D \end{cases} \tag{4}$$

Z_{b1} has the largest current stress while Z_{b4} has the smallest current stress. This proves that not only are the auxiliary source bond wires are carrying the gate current, but they are stressed by the current flowing through the drains. The simulation results of the current distribution in the power source bond wires and the auxiliary source bond wires are shown in Fig. 8.

The impedance of the DBC traces and bond wires in the power module may be different from the case $Z_{bx}=3Z_b$. However, the current distribution should be similar with the theoretical analysis. Based on the mathematic analysis and simulation, the auxiliary source effects can be summarized as below.

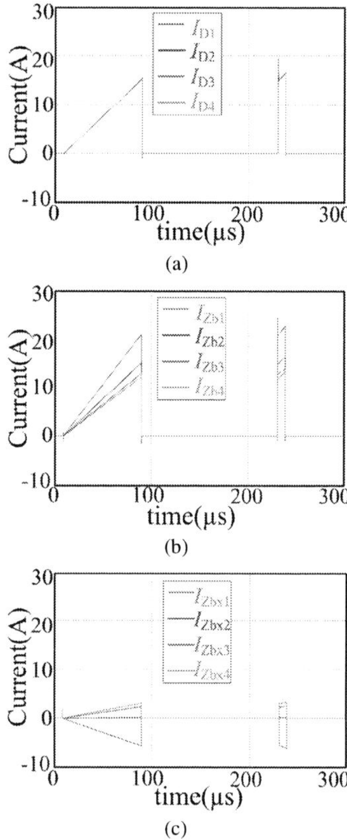

Figure 8: Simulation results of the current distributions (a) four MOSFETs drain currents (b) currents in the power source bond wires (c) currents in the auxiliary source bond wires.

First, the auxiliary source connections can mitigate the transient current imbalance among the paralleled dies. The mitigation mechanism is due to the power currents also going through the auxiliary source connections. Consequently, the current going through the power traces and bond wires are reduced. With the same switching time, the d*i*/dt is reduced. Therefore, the source voltage potential differences are also reduced, which will mitigate the transient current imbalance. On the other hand, the auxiliary source connections are included in the power current loop and will cause the current stresses on the auxiliary source connections and current imbalance in the power source bond wires. There is a trade-off between these two effects.

IV. EXPERIMENTAL STUDY

The experimental study of the current distributions in the multichip power module is performed with PCB circuits, which is designed to have the similar layout of the DBC, as shown in Fig. 9.

A. *Common source stray inductance reduction*

The experimental results of the total current in the four paralleled MOSFETs are shown in Fig. 10. Fig. 10(a)

Figure 9: Experimental study with PCB circuit

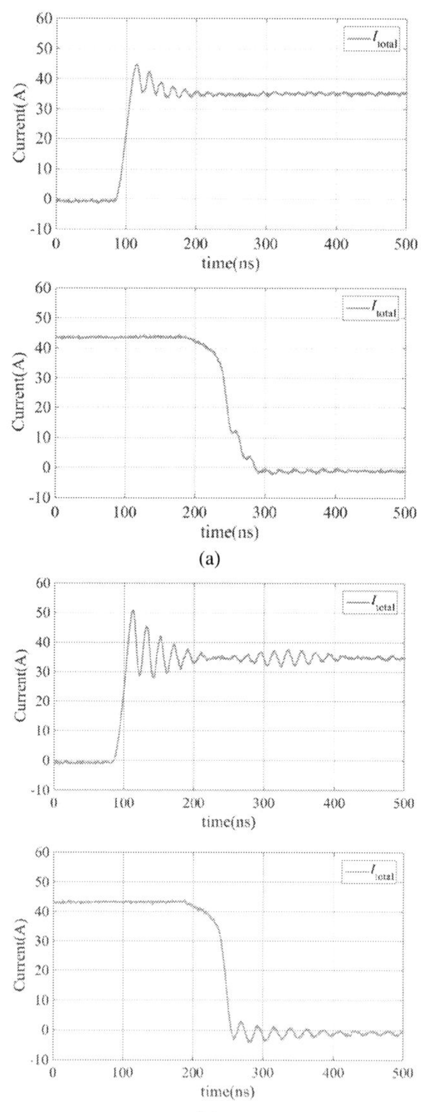

Figure 10: Experimental results of the total SiC MOSFETs current distribution (a) without the auxiliary source connections (b) with the auxiliary source connections

shows the total current without the auxiliary source connections while Fig. 10(b) shows the total current with auxiliary source connections. The total current switching speed with auxiliary source connections is faster than that without the auxiliary source connections. It validates that the auxiliary source connections reduces the total common source stray inductance and improve the 'power module' switching speed.

B. Transient current imbalance mitigation

The transient current distribution among the paralleled MOSFETs is shown in Fig. 11(a) without auxiliary source connections and in Fig. 11(b) with auxiliary source connections. It can be seen that the auxiliary source connections mitigate the transient current imbalance. But the mitigation effect is limited, as the imbalance in the transient drain currents can still be observed.

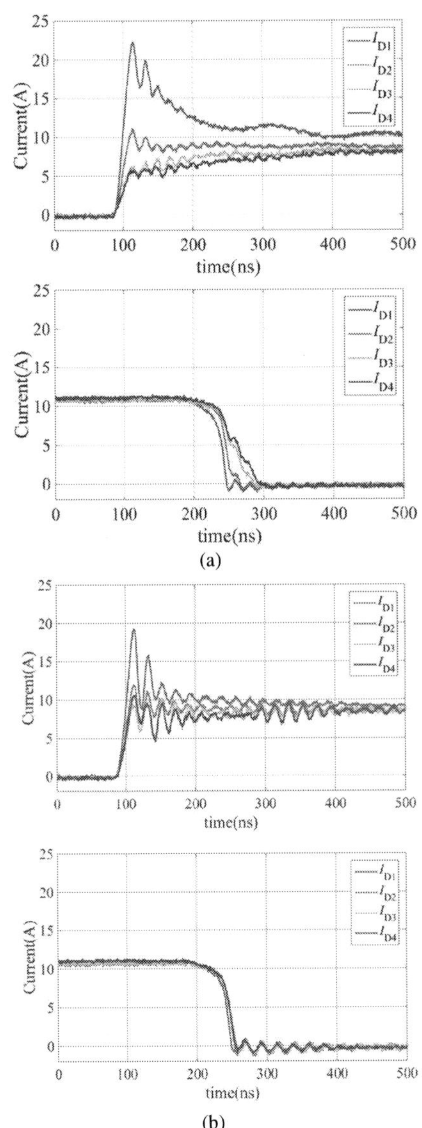

Figure 11: Transient MOSFETs drain current distribution (a) without auxiliary source connections (b) with auxiliary source connections

C. Steady state current distribution

The steady state current distribution is done with a higher current level in order to see the currents in the auxiliary source connections. The MOSFETs drain currents and the currents in the auxiliary source connections are shown in Fig. 12.

In this experimental study, the ratio of the drain currents flowing through the auxiliary source connections is larger than expected to be observed in a power module. This is because the auxiliary source connections were done with wires which have impedance comparable to that of the PCB traces of the drain connection. Thus the relationship between the impedance of the auxiliary source connections and that of the drain connection is much smaller compared to the one in a power module. Normally in a power module, the impedances of the auxiliary source connections are a few orders of magnitude bigger than the ones of the drain connection, thus the current flowing through the auxiliary source connection should be much smaller when compared to the one flowing through the drain connection. Although the ratio of the currents flowing in the auxiliary source connection should be smaller, there should still be currents present as in the case in Fig. 12(b).

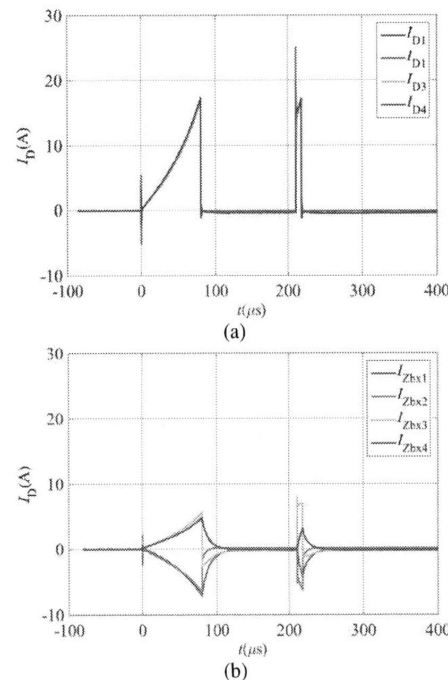

Figure 12: Steady state current distribution of (a) MOSFETs drain currents (b) Auxiliary source connections currents

V. Conclusions

This paper investigates the working mechanism of the auxiliary source connections in multichip power modules with paralleled MOSFETs or IGBTs. The auxiliary source connections can reduce the common source stray inductance for the power module. They cannot totally decouple the gate-source loop and the drain-source loop because they are still in the loop of drain current. Moreover, the auxiliary source connections are shown to be able to mitigate the transient MOSFETs current imbalance. This leads to steady state current imbalance in the bond wires of the power source traces due to currents flowing through the auxiliary source connections. The simulation and experimental results validate the analysis and present the current distribution in the DBC layout, during both transient and steady state period. One possible solution to suppress the auxiliary source current is inserting source resistors in the auxiliary source connections. But with the suppressed current in the auxiliary source connections, the mitigation effect for the transient current imbalance will also be reduced.

REFERENCES

[1] Stueckler F. and Vecino E., "Cool MOS C7 650V switch in a kelvin source configureation," Infineon Application note, 2013.

[2] Jan Leong, "Infineon IGBT/Modules," 2012.

[3] Z. Chen, Y. Yao, D. Boroyevich, K. D. T. Ngo, P. Mattavelli and K. Rajashekara, "A 1200-V, 60-A SiC MOSFET Multichip Phase-Leg Module for High-Temperature, High-Frequency Applications," *IEEE Trans. Power Electron.*, vol. 29, pp. 2307-2320, 2014.

[4] J. D. Scofield, J. N. Merrett, J. Richmond, A. Agarwal and S. Leslie, "Performance and reliability characteristics of 1200 V, 100 A, 200 C half-bridge SiC MOSFET-JBS diode power modules," in *International Conference on High Temperature Electronics, International Microelectronics & PacNaging Society, Albuquerque, NM,* 2010, .

[5] F. Xu, D. Jiang, J. Wang, F. Wang, L. M. Tolbert, T. J. Han, J. Nagashima and S. J. Kim, "High temperature packaging of 50 kW three-phase SiC power module," in *Proc. IEEE ECCE Asia,* 2011, pp. 2427-2433.

[6] S. Li, L. M. Tolbert, F. Wang and F. Z. Peng, "Stray Inductance Reduction of Commutation Loop in the P-cell and N-cell-Based IGBT Phase Leg Module," *IEEE Trans. Power Electron.,* vol. 29, pp. 3616-3624, 2014.

[7] A. Müsing and J. W. Kolar, "Ultra-Low-Inductance Power Module for Fast Switching Semicon-ductors," in *Proc. PCIM Europe*, 2013.

[8] H. Li, S. Beczkowski, M. Stig, Ramkrishan Maheshwari and Toke Franke, "Circuit mismatch and current coupling effect influence on paralleling SiC MOSFETs in multichip power modules," in *Proc. PCIM Europe,* 2015.

[9] H. Li, S. Munk-Nielsen, X. Wang, R. Maheshwari, S. Beczkowski, C. Uhrenfeldt and W. Franke, "Influences of Device and Circuit Mismatches on Paralleling Silicon Carbide MOSFETs," *IEEE Trans. Power Electron.*, vol. PP, pp. 1-1, 2015.

Modelling Technique utilizing Modified Sigmoid Functions for Describing Power Transistor Device Capacitances Applied on GaN HEMT and Silicon MOSFET

H. L. Yeo
School of Electrical and Electronic Engineering
Nanyang Technological University
Singapore
howeli.yeo@sg.bosch.com

K. J. Tseng
School of Electrical and Electronic Engineering
Nanyang Technological University
Singapore
k.j.tseng@pmail.ntu.edu.sg

***Abstract*— In power transistor models, it is very important that device capacitances are modelled accurately so that switching losses, EMI filter requirements and gate timing requirements of the converter can be accurately determined. In this paper, a modelling technique utilizing modified sigmoid functions to describe the device capacitances is applied on a GaN HEMT and a silicon MOSFET to develop their corresponding SPICE models. Comparison of switching energies of the transistors under slow-switching conditions suggest that the technique is suitable for modelling both types of transistors. The developed model for the GaN HEMT is also shown to simulate faster than the manufacturer's model under fast switching conditions.**

Keywords—GaN HEMT; Silicon MOSFET; Modelling and Simulation; Sigmoid Functions

I. Introduction

Gallium nitride high electron mobility transistors (GaN HEMTs) have been cited by various authors to improve converter efficiencies and power densities [1, 2, 3, 4, 5]. Device models of GaN HEMTs are useful for designing converters as they can help calculate the expected losses in the device and subsequently allow the designer to determine cooling requirements. To date, there have been numerous modelling attempts made by various authors [6, 7, 8, 9, 10]. Physics-based models for the GaN HEMT [11, 12, 13, 14] accurately depict device behavior but are too complicated for circuit simulators like SPICE and require intimate knowledge of device parameters like doping and geometry which may not necessarily be readily available. Behavioral models [15, 16, 17, 18] are typically easier to use and develop but accuracy is also typically lower. A few exceptions to this observation exist but typically come with other drawbacks like requiring intimate knowledge of the parasitics of the test setup [19] or significantly more development effort like the development of a new circuit simulator [20].

Accurately modelled device capacitances are critical to the accuracy of power transistor models. In previous literature involving silicon MOSFETs, several approaches were used to address this issue including the piecewise-linear approach [21, 22], lumped charge approach [23, 24] and even use of other device models [25] for exclusively modelling the capacitance. None of these models utilized sigmoid functions. Although sigmoid functions have been employed by a commercial entity to model GaN devices, these models are more complex and involve several fitting parameters.

In this research, a simple and faster model utilizing modified sigmoid functions is proposed for modelling device capacitances in silicon MOSFETs and GaN HEMTs. The paper is organized into the following sections. In section II, the model and associated equations are described. In section III, the parameter extraction procedure is covered. In section IV, a comparison between the simulation and experimental results is made. In the final section, the conclusion is presented.

II. Model Description

A. Device capacitances

The device model is shown in Fig. 1 and consists of three behavioral current sources to simulate the device's i-v characteristics and also three capacitances to simulate the behavior of the charges within the device. C_{gs} is extracted from

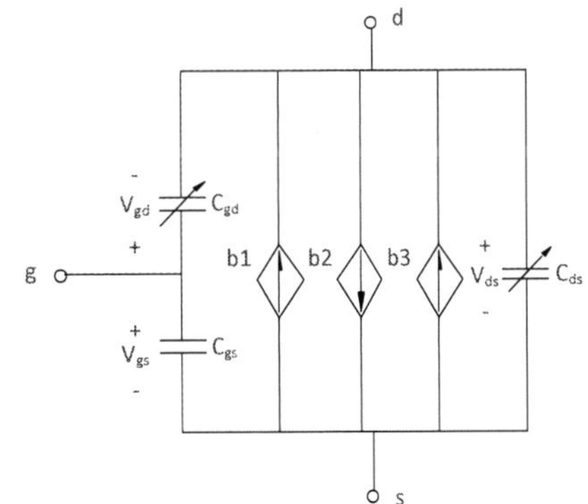

Fig. 1. Proposed power transistor model

gate charge plots and modelled as a constant capacitance whereas C_{gd} and C_{ds} are modelled using sigmoid functions. C_{gd} and C_{ds} are described using equations (1) to (3).

$$C_{gd}(V_{gd}) = \frac{a_1}{1+e^{(-b_1(V_{gd}+c_1))}} + d_1 \tag{1}$$

$$C_{oss}(V_{sd}) = \frac{a_2}{1+e^{(-b_2(V_{sd}+c_2))}} + d_2 \tag{2}$$

$$C_{ds}(V_{sd}) = C_{oss}(V_{sd}) - C_{gd}(V_{sd}) \tag{3}$$

Fitting parameters a_1, b_1, c_1, d_1, a_2, b_2, c_2 and d_2 can be used to adjust the shape of the functions. In order to match the behavior of the C_{gd} capacitance in the model with the miller capacitance of the actual device, graphs of miller charge, Q_{gd}, plotted against V_{gd} must be obtained experimentally and fitted based on the integral of (1). This integral is shown in (4).

$$Q_{gd}(V_{gd}) = \frac{a_1}{b_1} ln\left(1+e^{b_1(x+c_1)}\right) + d_1x + k_1 \tag{4}$$

k1 is a constant offset resulting from the integration that does not influence the behavior of the capacitances in the model. It can be adjusted to fit the simulated and experimental $Q_{gd} - V_{gd}$ curves. C_{ds} is determined by first fitting (2) to datasheet values of C_{oss} using (3).

B. Forward and reverse i-v characteristics

Behavioral current sources b1and b2 are used to model the forward and reverse i-v characteristics when the device is in the on-state. When the device is in the off-state, b3 is used to simulate the body diode of the device. b1's and b2's behavior are described by the typical MOSFET channel equation as listed in (5). b3 is modelled using a form similar to the typical Shockley equation as shown in (6).

$$I_{ds} = \begin{cases} 0, V_{gs} < V_{th} \\ K_n\left(V_{gs}-V_{th}-\frac{V_{ds}}{2}\right) V_{ds}, \ V_{gs} \geq V_{th} \ and \ V_{ds} < (V_{gs}-V_{th}) \\ K_n(V_{gs}-V_{th})^2, \ V_{gs} \geq V_{th} \ and \ V_{ds} \geq (V_{gs}-V_{th}) \end{cases} \tag{5}$$

$$I_{sd} = \begin{cases} 0, V_{sd} \leq 0 \\ I_{sb}\left(e^{k_{tb}V_{sd}}-1\right), V_{sd} > 0 \end{cases} \tag{6}$$

The forward i-v characteristics of the transistor is measured and used to determine K_n and V_{th}. Curves from the data sheet are used to determine I_{sb} and k_{tb}.

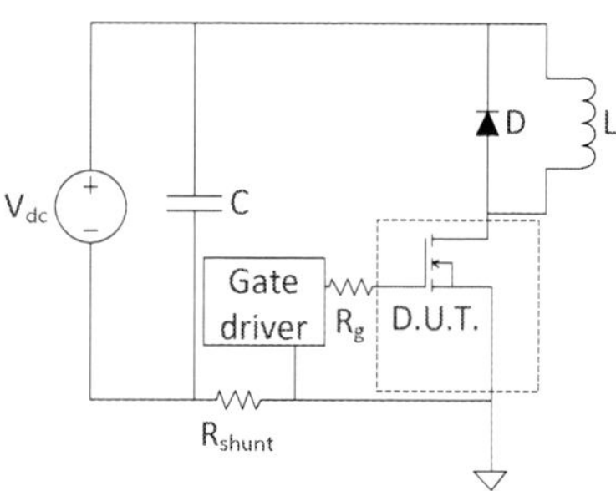

Fig. 2. Double pulse tester schematic

III. Parameter Extraction Procedure

In this section, the equipment used for characterizing the power transistor is described in addition to the procedure used for extracting the fitting parameters for the device capacitances and i-v characteristics.

A. Equipment

A double-pulse tester was used to extract the switching waveforms and i-v characteristics. The typical schematic of a double pulse tester is illustrated in Fig. 2. The actual hardware is shown in Fig. 3 and Fig. 4. The double pulse tester is highly modular and can be used in conjunction with other modules for testing purposes. The device under test (D.U.T.) can be exchanged if needed.

B. Device capacitances

The EPC2001 GaN HEMT and the IPP60R099C6 silicon MOSFET have been chosen as the devices for this study. In order to parameterize the device capacitances, the miller charge must be found. Gate current is first measured by measuring voltage across the gate resistance and scaling down the voltage waveform based on ohm's law. Data from the oscilloscope is then transferred onto a personal computer (PC) and then processed by a MATLAB script.

In order to ensure that the observed waveforms are caused mainly by the semiconductor and not affected by parasitic inductances and capacitances, the gate currents have been limited in order to limit switching speed. In the case of the silicon MOSFET, this was done using a large gate resistance. In the case of the GaN HEMT, the gate current was limited using a 0.47mA constant current diode which is part of a typical gate charge circuit. Waveforms for both tests were

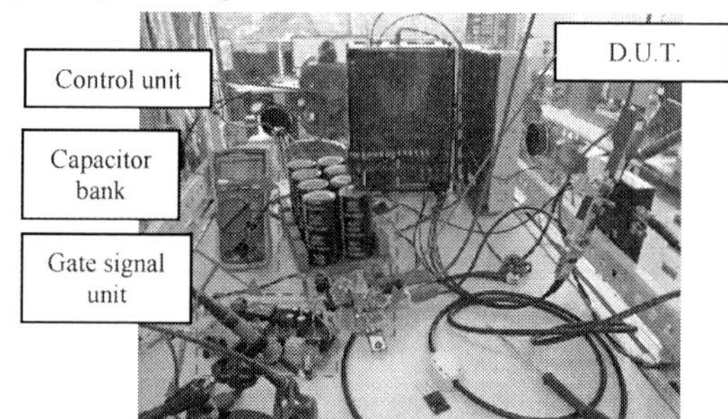

Fig. 3. Various parts of double pulse tester

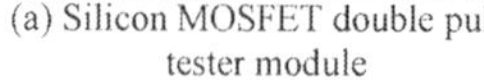

(a) Silicon MOSFET double pulse tester module

(b) GaN HEMT double pulse tester module

Fig. 4. Silicon MOSFET and GaN HEMT test modules

logged and are presented in Fig. 6. MATLAB scripts extract the gate charge values and drain-to-source voltage values corresponding to the miller plateau. Fitting is then done in MATLAB. The curves for the fitting of the miller charge and output capacitance are presented in Fig. 7.

C. Forward and reverse i-v characteristics

As GaN HEMTs are relatively new to the field of power electronics, their i-v characteristics have yet to be fully understood. Due to this reason, and the fact that the datasheet of the EPC2001 GaN HEMT only presents forward i-v characteristics for a very limited range of drain-to-source voltage, it was decided that the forward i-v characteristics had to be extracted through experiment. The results were then fitted using (5). The i-v characteristics were extracted with the double pulse tester configured as described in Fig. 5. A fixed gate voltage is applied to the device under test. T1 is then turned on and current is allowed to ramp up through the inductor and the device under test. An x-y plot is made using the points from the ramp-up period. The results of the fitting for both the manufacturer's model from EPC and the proposed model are seen in Fig. 8.

For the silicon MOSFET i-v characteristics and GaN HEMT body diode i-v characteristics, datasheet values were used for the fitting.

D. Consolidated fitting parameters

The consolidated fitting parameters for both models are given in TABLE I.

IV. Experimental Verification

Simulation results from the proposed models were verified against experimental results under two sets of conditions – slow switching and fast switching. The results from the proposed models were also benchmarked against those given by the manufacturers' models under the same test conditions.

A. Slow switching

Switching energy is influenced by both the semiconductor behavior as well as parasitic elements in the circuit. As estimates of the parasitic elements may not be accurate, it is important to first verify the switching behavior of the device under conditions where the effect of parasitic elements are negligible before proceeding to account for this effect. Under slow switching conditions, the rate of change of currents and voltages are slow. Hence, switching behavior exhibited is purely due to the semiconductor itself. In order to ensure that the switching speed is slow enough, a large gate resistance of 1kΩ was utilized for the double pulse tests of the GaN HEMT and silicon MOSFET. In the case of the GaN HEMT, the gate signal is 5V when it is high and 0V when it is low. In the case of the silicon MOSFET, the gate voltage is 15V when it is high and 0V when it is low. Turn-on and turn-off energies for under specific conditions of applied V_{ds} and I_{ds} as well as the corresponding waveforms for the one set of test conditions for both the silicon MOSFET and GaN HEMT were recorded and compared with those obtained using both the manufacturer's models and the proposed models respectively as shown in Fig. 9 and Fig. 10. A large air core inductor of 2.5mH was used for both experiments in order to clamp the current during switching. The freewheeling diode in the case of the GaN HEMT is the body diode of another GaN HEMT whilst the one in the case of the silicon MOSFET is a IDD04S60C silicon carbide schottky diode.

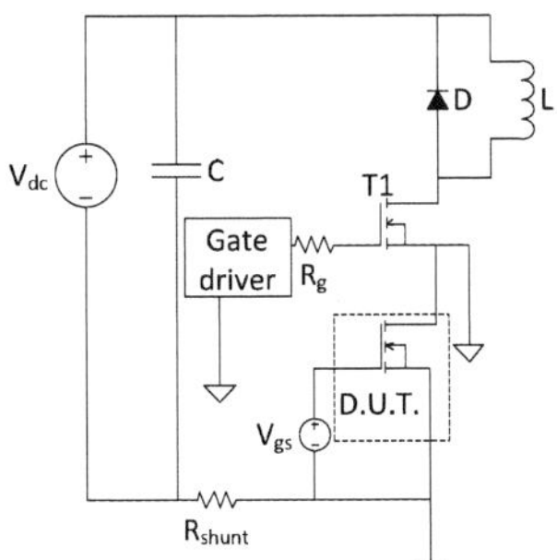

Fig. 5. Double pulse tester configuration for extracting i-v characteristic

The models were simulated as part of a double pulse tester circuit in SPICE. In order to increase the speed of the simulation, the inductor of the SPICE simulation was replaced with a constant current source and the parasitic inductances and capacitances that are typically represented in the circuit have been removed as it is expected that since switching speed is slow, their effect will be negligible. The upper freewheeling diode both cases of the GaN HEMT and silicon MOSFET is assumed to be ideal. This assumption is valid as the reverse recovery charge possessed by the upper freewheeling diode in both cases is small and hence, additional turn-on energy incurred as a result of reverse recovery is negligibly small. Infineon's model of the silicon MOSFET as well as Efficient Power Conversion Corporation's model of the GaN HEMT were used as the benchmarking models.

TABLE I. Consolidated Fitting Parameters

Fitting parameters		
Parameter	***Silicon MOSFET***	***GaN HEMT***
a_1/F	9.054×10^{-9}	0.72×10^{-9}
b_1/V^{-1}	0.2	0.3
c_1/V	0.5	0
d_1/F	9.038×10^{-12}	0.015×10^{-9}
a_2/F	62.92×10^{-9}	1×10^{-9}
b_2/V^{-1}	0.1	0.3
c_2/V	30	10
d_2/F	0.7×10^{-9}	0.4×10^{-9}
C_{gs}/nF	2.48	0.9
K_n/AV^{-2}	5.794	29.75
V_{th}/V	2.344	0.44
I_{sb}/A	1.635×10^{-10}	0.2
k_{tb}/V^{-1}	30.969	2

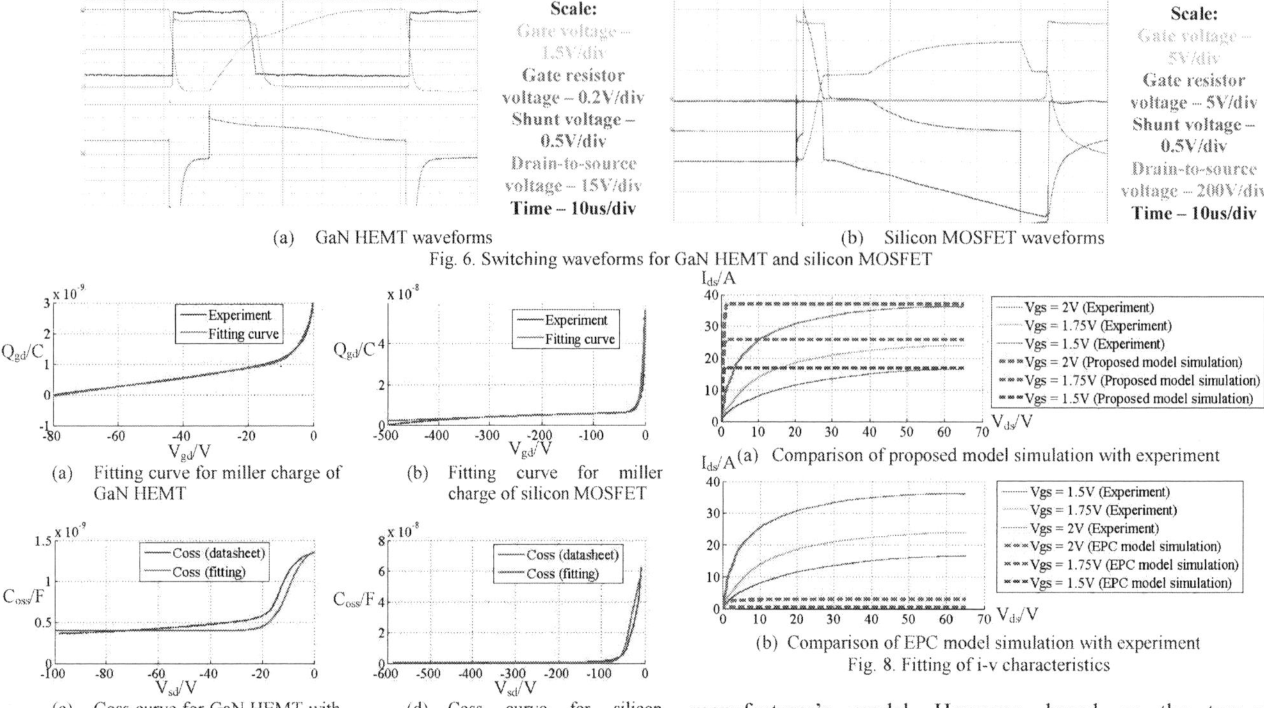

(a) GaN HEMT waveforms (b) Silicon MOSFET waveforms

Fig. 6. Switching waveforms for GaN HEMT and silicon MOSFET

(a) Fitting curve for miller charge of GaN HEMT (b) Fitting curve for miller charge of silicon MOSFET

(c) Coss curve for GaN HEMT with fitting curve (d) Coss curve for silicon MOSFET with fitting curve

Fig. 7. Fitting curves for miller charge and output capacitance of GaN HEMT and silicon MOSFET

(a) Comparison of proposed model simulation with experiment

(b) Comparison of EPC model simulation with experiment

Fig. 8. Fitting of i-v characteristics

In the case of the GaN HEMT as seen by Fig. 9, both the manufacturer's model as well as the proposed model appears to predict the turn-on energy of the GaN HEMT quite accurately. However, both models deviate significantly in terms of predicting the turn-off energies. This deviation is largely due to the fact that the i-v characteristics of both models are not well fitted with experimental values. The i-v characteristics influence the switching trajectory of the device models and as a consequence, the waveforms predicted by both models are expected to deviate as shown in Fig. 10. It can be seen based on the waveforms of Fig. 10 that the proposed model predicts a significantly longer switching time than the experiment. This indicates that the proposed model is not suitable for predicting dead-time.

In the case of the silicon MOSFET, it can be seen from Fig. 9 that the manufacturer's model is quite accurate at predicting the turn-on and turn-off energies at 500V of blocking voltage but not as accurate at the lower values of device blocking voltage like 300V and 400V. This is an expected result as the device is expected to be used at voltages closer to 600Vdc and hence the model should be better optimized for voltages closer to 600Vdc. Based on Fig. 9, the proposed model exhibits a higher accuracy in terms of estimation of switching energies at the lower voltage values, hence, it is more suitable for applications where those voltage levels are employed such as a 400Vdc bus for an automotive dc/dc converter. Based on Fig. 10, the proposed model predicts the shape of turn-on waveforms to a higher degree of accuracy compared with the manufacturer's model. However, based on the turn-off waveforms, it can be seen that the simulated turn-off transition of the proposed model is significantly delayed compared to the experiment. This could lead to overestimation of dead-time.

B. Fast switching

In practical applications, switching speed is typically increased to reduce the switching losses of the transistor. Under these conditions, the parasitic inductances and capacitances of the circuit must be taken into account. Accurately finding the parasitic inductances and capacitances is a difficult task. Typically, authors approximate these parasitic elements by observing the oscillation frequency of the ringing [26] or by assuming the parasitic values as fitting parameters [20]. In this research, it is assumed that the parasitic capacitances of the circuit board are negligibly small. Parasitic inductances are approximated based on datasheet values as well as waveforms observed during experiment. Particular attention is given to the commutation inductance of the setup. Equations (7), (8) and (9) are used to approximate the commutation inductance, L_{loop}, based on oscillation frequency, f_{osc}, observed during turn-on and the voltage overshoot, ΔV_{over}, observed during turn-off.

$$L_{loop1} = \frac{1}{(2\pi f_{osc})^2 C_{oss}} \quad (7)$$

$$L_{loop2} = \Delta V_{over} \div \frac{\Delta i}{t_r} \quad (8)$$

$$L_{loop} \approx \frac{L_{loop1} + L_{loop2}}{2} \quad (9)$$

C_{oss}, Δi and t_r, refer to the output capacitance, change in current and current rise time corresponding to the respective switching conditions.

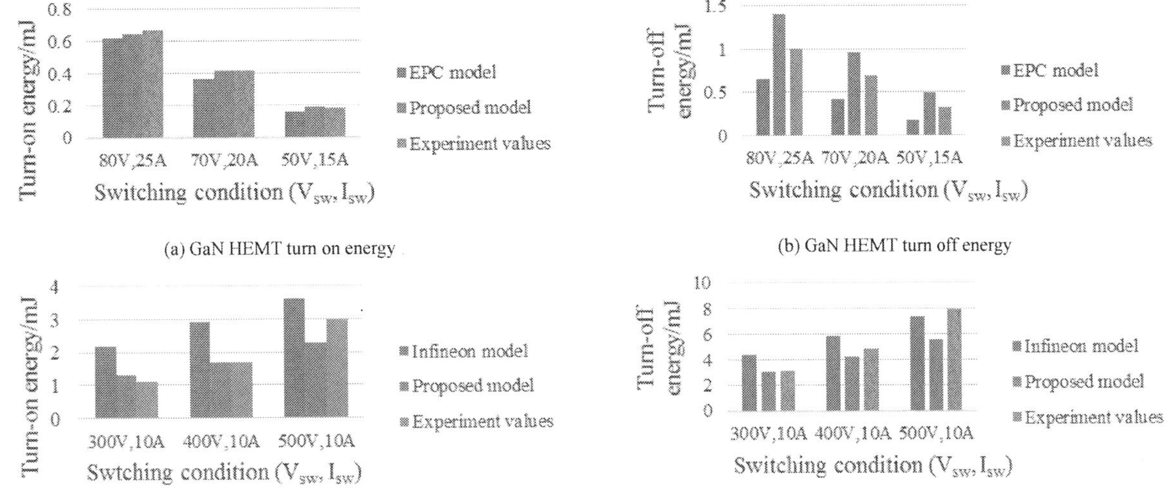

(a) GaN HEMT turn on energy

(b) GaN HEMT turn off energy

(c) Silicon MOSFET turn on energy

(d) Silicon MOSFET turn off energy

Fig. 9. Comparison of turn on and turn off energies under slow switching conditions

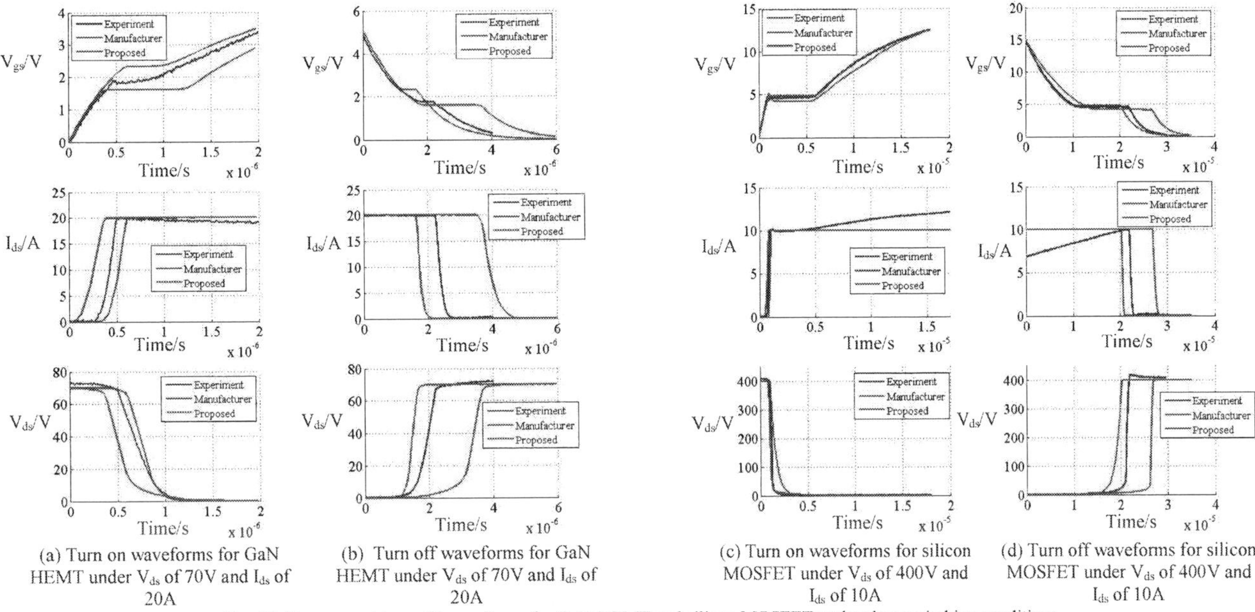

(a) Turn on waveforms for GaN HEMT under V_{ds} of 70V and I_{ds} of 20A

(b) Turn off waveforms for GaN HEMT under V_{ds} of 70V and I_{ds} of 20A

(c) Turn on waveforms for silicon MOSFET under V_{ds} of 400V and I_{ds} of 10A

(d) Turn off waveforms for silicon MOSFET under V_{ds} of 400V and I_{ds} of 10A

Fig. 10. Turn on and turn off waveforms for GaN HEMT and silicon MOSFET under slow switching conditions

Under fast switching conditions, it is important to design the printed circuit board (PCB) to minimize the parasitic inductance so that the device is not overstressed during switching and the switching energy measurement is as close to that in a practical application as possible. A module was designed for the GaN HEMT as seen in order to minimize the stray inductances associated with the commutation cell as much as possible. This is particularly important in the case of the GaN HEMT as a higher rate of change of current with the commutation loop. However, it should be noted that since the current shunt contributes a large portion of the commutation loop inductance, the difference in between the commutation inductance for the case of the silicon MOSFET which utilizes a TO-220 package and the GaN HEMT which utilizes a land grid array (LGA) package is minimal. It has been estimated that the module used for the GaN HEMT possesses a commutation inductance of 9.1 nH whilst the module used for the silicon MOSFET possesses a commutation inductance of 14 nH.

Due to delays in the probes, deskewing must be done so that the observed waveforms conform to theory. At present, only the turn-on behavior of the GaN HEMT under fast switching conditions is examined in detail. The turn-off behavior of the GaN HEMT as well as fast-switching behavior

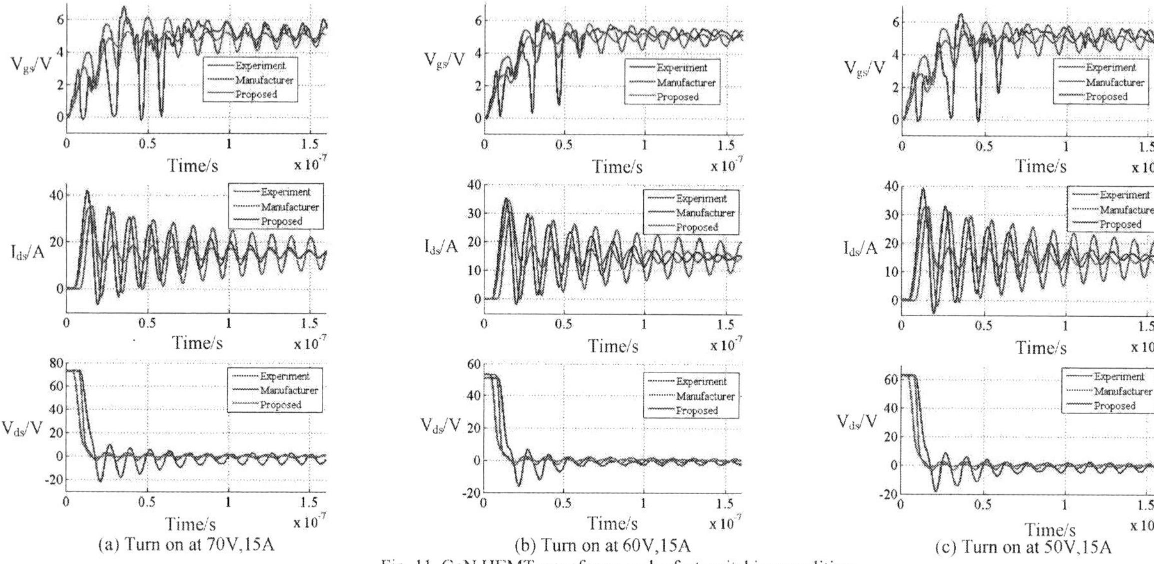

(a) Turn on at 70V,15A (b) Turn on at 60V,15A (c) Turn on at 50V,15A

Fig. 11. GaN HEMT waveforms under fast switching condition

Fig. 12. Simulated and experimental turn-on energies of GaN HEMT under fast switching conditions

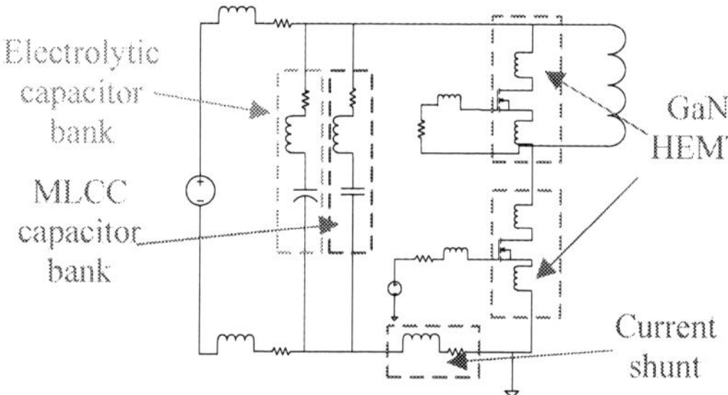

Fig. 13. Simulated double pulse tester

of the silicon MOSFET will be investigated in future work. Deskewing is applied to the I_{ds} waveform of the double pulse test such that the current rises once the gate voltage rises to the threshold voltage of the GaN HEMT. This value is taken as 1.4V as seen in the datasheet. A gate resistance of 5.1 ohms was used for tests involving the GaN HEMT.

Experimental results were compared against simulated results obtained from the circuit of Fig. 13. In Fig. 11, the waveforms of the comparison are seen. It can be seen based on the current waveforms that the manufacturer's model describes the device behavior more accurately under these conditions. It can also be seen that both models predict a similar amount of current overshoot as well as a similar rate of change of current during device turn-on. The switching energies were also documented in Fig. 12. As seen by the charts, neither models accurately predict the switching energies under fast switching conditions. The simulation times of the models were also recorded for the corresponding switching conditions as seen in Table II. The proposed model is seen to be significantly faster in terms of simulation time.

V. Conclusion

In this paper, a new device model which makes use of modified sigmoid functions to model the device capacitances of a GaN HEMT and a silicon MOSFET was proposed. Experimental verification was done under slow switching conditions for both the GaN HEMT and silicon MOSFET to verify the accuracy of the model under those conditions. Experimental verification was also done under fast turn-on conditions in the case of the GaN HEMT and it was found that

Table II. Simulation Times of Models

	Simulation times/s	
Condition	***Manufacturer***	***Proposed***
50V, 15A	97.760	1.026
60V, 15A	103.395	1.157
70V, 15A	96.06	0.994

```
*Generic model for silicon MOSFET and GaN HEMT
.subckt generic gate drain source

***GaN HEMT parameters - comment out silicon MOSFET parameters and uncomment the line below to use***
.params Vth={0.44} Kn={29.75} ktb={2} Isb={0.2}
+a1={0.72E-9} b1={0.3} c1={0} d1={15E-12}
+a2={1E-9} b2={0.3} c2={10} d2={0.4e-9}
+cgs={0.9E-9}

***Silicon MOSFET parameters - comment out GaN HEMT parameters and uncomment the line below to use***
*.params Vth={2.334} Kn={5.794} ktb={30.969} Isb={1.635e-10}
*+a1={9.054E-9} b1={0.2} c1={0.5} d1={9.038E-12}
*+a2={6.292E-8} b2={0.1} c2={30} d2={0.7e-9}
*+cgs={2.48E-9}

***Behavioral source***
B1 drain source I=GDS(V(gate,source),V(drain,source))
.func GDS(Vgs,Vds) = IF
+((Vgs<={Vth})|(Vds<0), 0,
+(IF(Vds<(Vgs-{Vth}), {Kn}*(Vgs-{Vth}-Vds/2)*Vds ,
+{Kn}/2*(square(Vgs-{Vth})))))

***anti-parallel behavioural source
B2 source drain I=GDS(V(gate,source),V(source,drain))
.func GDS(Vgs,Vds) = IF
+((Vgs<={Vth})|(Vds<0), 0,
+(IF(Vds<(Vgs-{Vth}), {Kn}*(Vgs-{Vth}-Vds/2)*Vds ,
+{Kn}/2*(square(Vgs-{Vth})))))

***body diode***
B3 source drain I=IF((V(gate,source)<={Vth})&(V(source,drain)>0),{Isb}*(exp({ktb}*V(source,drain))-1),0)

*** device capacitances ***
Cgs gate source {cgs}
Cgd1 gate drain Q = ((({a1})/({b1}))*ln(1+exp({b1}*(V(gate,drain)+{c1}))))
Cgd2 gate drain {d1}
Cds1 source drain Q = ((({a2})/({b2}))*ln(1+exp({b2}*(V(source,drain)+{c2})))) - ((({a1})/({b1}))*ln(1+exp({b1}*(V(source,drain)+{c1}))))
Cds2 source drain {d2 - d1}
.ends
```

Fig. 14. SPICE code of model

under those conditions, neither the proposed model nor the manufacturer's model were able to accurately obtain the turn-on energies. However, it was found that under these conditions, the proposed model was significantly faster than the manufacturer's model. The SPICE code for the model is presented in Fig. 14.

Acknowledgement

The authors would like to thank Robert Bosch Southeast Asia Private Limited for supplying the equipment and support necessary for this research.

References

[1] F. Lee and Q. Li, "High-Frequency Integrated Point-of-Load Converters: Overview," *IEEE transactions on Power Electronics*, vol. 28, no. 9, pp. 4127-4136, September 2013.

[2] A. Tüysüz, R. Bosshard and J. Kolar, "Performance Comparison of a GaN GIT and a Si IGBT for High-Speed Drive Applications," in *The 2014 international Power Electronics Conference*, Hiroshima, 2014.

[3] Y. Hayashi, "High Power-Density Versatile DC-DC Converter for Environmentally Friendly Data Centre," in *ECCE Europe*, Novi Sad, Serbia, 2012.

[4] K. Shirabe, M. Swarmy, J.-K. Kang, M. Hisatsune, Y. Wu, D. Kebort and J. Honca, "Efficiency Comparison betweeen Si-IGBT based Drive and GaN based Drive," *IEEE Transactions on Industry Applications*, vol. 50, no. 1, pp. 566-572, 2014.

[5] A. Lidow and J. Strydom, *Efficient DC-DC Brick Converters, white paper: WP006*, Efficient Power Conversion Corporation, 2012.

[6] X. Huang, Q. Li, Z. Liu and F. Lee, "Analytical Loss Model of High Voltage GaN HEMT in Cascode Configuration," *IEEE Transactions of Power Electronics 2014*, vol. 29, no. 5, pp. 2208-2218, May 2014.

[7] D. Hou, G. Bilbro and R. Trew, "A Compact Physical AlGaN/GaN HFET Model," *IEEE Transactions on Electron Devices*, vol. 60, no. 2, pp. 639-645, February 2013.

[8] C. Charbonniaud, A. Xiong, S. Deller and T. Gasseling, "A Non Linear Electrothermal Model of AlGaN/GaN HEMT for Switch Applications," in *Compound Semiconductor Integrated Circuit Symposium*, 2012.

[9] J. Waldron and T. Chow, "Physics-Based Analytical Model for High-Voltage Bidirectional GaN Transistors using Lateral GaN HEMT," in *25th International Symposium on Power Semiconductor Devices & ICs*, Kanazawa, 2012.

[10] S. Islam and A. Anwar, "SPICE Model of AlGaN/GaN HEMTs and Simulation of VCO and Power Amplifier," *International journal of high speed electronics and systems*, vol. 14, no. 3, pp. 853-859, March 2004.

[11] Rashmi, A. Kranti, S. Haldar and R. Gupta, "An accurate charge control model for spontaneous and piezoelectric polarization dependent two-dimensional electron gas sheet charge density of lattice-mismatched AlGaN/GaN HEMTs," *Solid State Electronics*, vol. 46, no. 5, pp. 621-630, May 2002.

[12] M. Esposto, A. Chini and S. Rajan, "Analytical Model for Power Switching GaN-Based HEMT Design," *IEEE Transactions on Electron Devices*, vol. 58, no. 5, pp. 1456-1461, 2011.

[13] M. Li and Y. Wang, "2-D Analytical Model for Current–Voltage Characteristics and Transconductance of AlGaN/GaN MODFETs," *IEEE Transactions on Electron Devices*, vol. 55, no. 1, pp. 261-267, 2008.

[14] K. Shreepad and R. Rao, "A Simple Yet Comprehensive Unified Physical Model of the Donor Layer Electrons in Delta-Doped and Uniformly Doped HEMT's," *IEEE Transactions on Microwave Theory and Techniques,* vol. 52, no. 1, pp. 2-9, 2004.
[15] M. Okamoto, G. Toyoda, E. Hiraki, T. Tanaka, T. Hashizume and T. Kaichi, "Loss evaluation of an AC-AC direct converter with a new GaN HEMT SPICE model," in *ECCE'11*, Phoenix, AZ, 2011.
[16] R. Beach, B. A. and R. Strittmatter, *Circuit Simulation Using EPC Device Models, Application note: AN005,* Efficient Power Conversion Corporation, 2011.
[17] O. Jardel, G. Callet, C. Charbonniaud, J. Jacquet, N. Sarazin, E. Morvan, R. Aubry, M.-A. Di Forte Poisson, J.-P. Teyssier, S. Piotrowicz and R. Qu´er´e, "A new nonlinear HEMT model for AlGaN/GaN switch applications," in *4th European Microwave Integrated Circuits Conference*, 2009.
[18] J. Lee and K. Webb, "A temperature-dependent nonlinear analytic model AlGaN-GaN HEMTs on SiC," *IEEE Transactions on Microwave Theory and Technology,* vol. 52, no. 1, pp. 2-9, January 2004.
[19] Z. Liu, X. Huang, F. Lee and Q. Li, "Package Parasitic Inductance Extraction and Simulation Model Development for the High-Voltage Cascode GaN HEMT," *IEEE Transactions on Power Electronics,* vol. 29, no. 4, pp. 1977-1985, April 2014.
[20] A. Nakajima, K. Takao and H. Ohashi, "GaN Power Transistor Modeling for High-Speed Converter Circuit Design," *IEEE Transactions on Electron Devices,* vol. 60, no. 2, pp. 646-652, February 2013.
[21] J. Hancock, "A MOSFET Simulation Model for Use with Microcomputer SPICE Circuit Analysis," in *Power Conversion International Proceedings*, 1987.
[22] H. Yee and P. Lauritzen, "SPICE Models for Power MOSFETs; An Update," in *1988 APEC Conference Proceedings*, 1987.
[23] L. Aubard, G. Verneau, J. Crebier, C. Schaeffer and Y. Avenas, "Power MOSFET Switching Waveforms: An Empirical Model Based on a Physical Analysis of Charge Locations," in *PESC*, 2002.
[24] I. Budihardjo and P. Lauritzen, "The Lumped-Charge Power MOSFET Model, Including Parameter Extraction," *IEEE Transactions on Power Electronics,* vol. 10, no. 3, pp. 379-387, May 1995.
[25] R. Scott, G. Franz and J. Johnson, "An Accurate Model for Power DMOSFET's Including Interelectrode Capacitances," *IEEE Transactions on Power Electronics,* vol. 6, no. 2, pp. 192-198, 1991.
[26] D. Reusch and J. Strydom, "Understanding the Effect of PCB Layout on Circuit Performance in a High-Frequency Gallium-Nitride-Based Point of Load Converter," *IEEE Transactions on Power Electronics,* vol. 29, no. 4, pp. 2008-2015, April 2014.

Design and Precise Modeling of a Novel Digital Active EMI Filter

Junpeng Ji*[†], Wenjie Chen* and Xu Yang*
Email: jijunpeng@xaut.edu.cn, {cwj, yangxu}@mail.xjtu.edu.cn
*School of Electrical Engineering, Xi'an Jiaotong University, Xi'an, Shaanxi, China
[†]Department of Electrical Engineering, Xi'an University of Technology, Xi'an, Shaanxi, China

***Abstract*—Digital Active EMI Filtering (DAEF) technique is a new direction of conductive EMI suppression technologies for switched-mode power converters. However, the conventional DAEF topology only suppresses EMI to ground on the AC (or DC) ports. It doesn't definitely aim to Common Mode (CM) or Differential Mode (DM) EMI. In this paper, a novel DAEF topology is proposed. It can flexibly suppress CM and DM EMI, and also effectively adjust EMI suppression ability as well as prevent interconversion between CM and DM EMI. On the other hand, because of the parasitic parameters of circuit and the time delay of controlling system, the traditional model of DAEF cannot describe the system behavior accurately. This paper proposes a precise model of DAEF system. The model not only considers the passive device parasitic parameters of sense circuit, injection circuit and decoupling circuit, but also considers the time delay of the digital control system. This model can accurately describe filter performance of DAEF. Finally, simulation and experimental results show that the proposed DAEF effectively suppress CM EMI. Moreover, the precise model can predict the system filter performance accurately.**

***Keywords*—EMI filter; Digital control; common mode EMI; different mode EMI**

I. INTRODUCTION

As power density of switched-mode power converters increased year by year and the switching frequency constantly improved, the distance of components becomes closer; therefore the problem of EMI has grown more serious. This generates a higher challenge for EMI filtering technologies. In order to suit the change of new situation, the traditional passive EMI filters [1-4], active EMI filter [5,6] and hybrid active EMI filter [7-9] are making progress.

On the other hand, with the improvement of digital processing technique, such as high speed and high precision FPGA, Analog-Digital Converter (ADC) and Digital-Analog Converter (DAC), the dream of digital active EMI filtering (DAEF) technique is ready to bring into reality.

Since there is no component in series connect with the power circuit, DAEF solves the problem in filter size and power consumption of traditional EMI filter fundamentally. Moreover, DAEF can selectively suppress the CM and DM EMI in accordance with the actual situation of CM and DM EMI in the switched-mode power converters. It can also effectively prevent interconversion between CM and DM interference. Meanwhile, DAEF is an effective method to suppress CM interference without increasing leakage current of converter devices.

The concept of DAEF was first proposed in 2012 by the Canadian academic Hamza, and its effectiveness had been verified in the photovoltaic grid-connected inverter [10]. In the following year, DAEF technology was applied in DC-DC converters of electric vehicle and other switched-mode power converters [11-13]. However, current DAEF only suppresses EMI to ground in the AC (or DC) ports. It doesn't definitely aim to CM or DM interference. This may waste the ability of EMI suppression and cause interconversion problem between CM and DM interference. Besides, current DAEF model doesn't consider the parasitic parameters of circuit and time delay of controlling system. Therefore, it cannot predict the system performance accurately.

On considering the above mentioned problems, a novel DAEF topology is proposed. The goal of this paper is try to find an effective digital active filtering method to eliminate the EMI in high power density converters. The advantage of the proposed circuit lies in that it can neatly suppress CM and DM interference. Meanwhile, it can also adjust EMI suppression ability effectively and prevent interconversion between CM and DM interference. To achieve it, a precise mathematic model of DAEF is built in this paper. It considers the parasitic phenomenon of sense circuit, injection circuit as well as decoupling circuit. What is more, time delay produced by ADC, DAC and FPGA controller are also considered in detail. By means of that, the system behavior may be described with more accuracy.

II. DESIGN OF DAEF SYSTEM

A. Proposed DAEF Topology

The general scheme of proposed DAEF topology is shown in Fig. 1. DAEF system includes six parts such as EMI signal sense circuit, injection circuit, ADC sampling circuit, controller, DAC output and decoupling circuit between injected point and sensed point.

EMI sense circuit is a RC high-pass filter, which consists of resistor and capacitor. The decoupling circuit can reduce the coupling between signal sense point and signal injection point. ADC and DAC need high sample rate to improve system rapidity. EMI injection circuit is a RC low-pass filter. DAEF controller can be embedded in the inherent controller of power converters. It can save total cost of power converters. A high

This work was supported by the National Natural Science Foundation of China (51277145), Science and technology planning of Beilin District of Xi'an in 2015 (GX1508), and Featured research program of Xi'an University of technology (2014TS010).

978-1-4673-9551-9/16 $31.00 © 2016 IEEE

impedance RF inductor is located between the injection point and sense point as the decoupling circuit, which reduced the coupling between the two points to improve the filtering performance. DAEF controller usually use FPGA as control unit, which program with hardware description language and have parallel processing ability, so it can improve system rapidity, too.

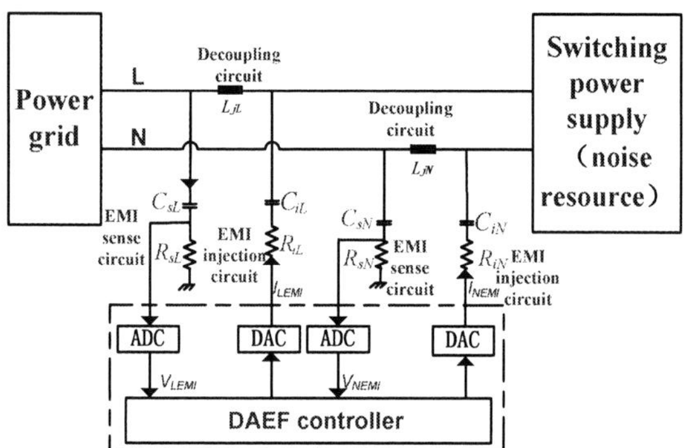

Fig. 1. General scheme of proposed DAEF

B. Scheme and Principles of Proposed DAEF

The control system schematic diagram of proposed DAEF is shown in Fig. 2. EMI signal sense circuit (high-pass filter) picks up conducted EMI signal V_{LEMI} and V_{NEMI} to ground on the L line or N line. The sensed EMI signal voltage is sampled by high-speed ADC to controller. According to equation 1 and equation 2, EMI controller can calculate the CM EMI voltage $V_{Lc}(V_{Nc})$ and DM EMI voltage V_D. By controller $G_C(s)$ compensating, DAEF control system outputs EMI current signal I_{LEMI}, which will cancel the sum of CM EMI on L line and DM EMI by injection circuit (low pass filter). As well as DAEF control system outputs EMI current signal I_{NEMI} to cancel the CM EMI on N line. I_{LEMI} and I_{NEMI} signal is converted back to analog signal for injection back to L line or N line. DAC output is EMI cancel signal of 0~20mA which can suppress 120dBμV EMI signal. The capacitor C_{iL} and C_{iN} of injection can prevent overloading to DAC from the power converter.

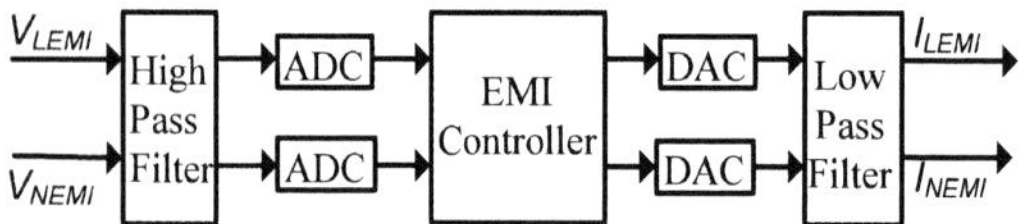

Fig. 2. Control system schematic diagram of DAEF

$$V_{Lc}(V_{Nc}) = \frac{V_{LEMI} + V_{NEMI}}{2} \tag{1}$$

$$V_D = \frac{V_{LEMI} - V_{NEMI}}{2} \tag{2}$$

III. Precise Modeling and Analysis of DAEF System

The closed-loop block diagram of DAEF control precise model is shown in Fig. 3.

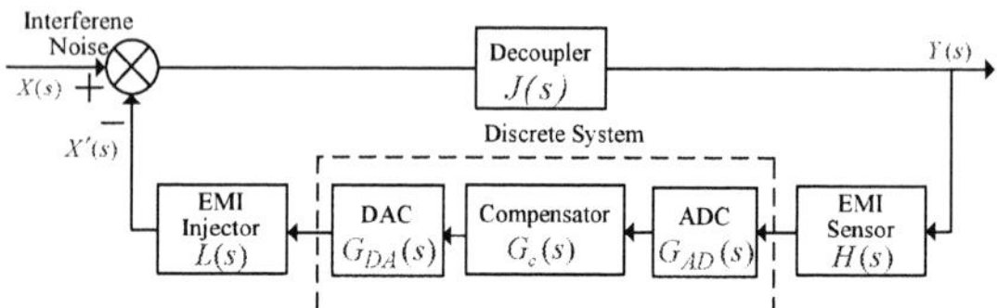

Fig. 3. Closed-loop system diagram of DAEF

In Fig. 3, $H(s)$ is the transfer function of the EMI sense circuit, $G_{AD}(s)$ is the transfer function of ADC. $G_C(s)$ is the transfer function of digital controller, $G_{DA}(s)$ is the transfer function of DAC, $L(s)$ is the transfer function of injection circuit, $J(s)$ is the transfer function of decoupling circuit. X(s) is EMI noise signal before filter, Y(s) is EMI noise signal after filter, X'(s) is the injected EMI noise after digital control. In theory, X(s) is equal in magnitude to X'(s), and their phase is opposite. It can achieve full nullification of the EMI noise which is generated by power converter.

A. Precise Modeling of Sense, Injection and Decoupling Portion

The parasitic parameters of passive components in the sense and injection circuit affect the high frequency filter performance, so the high frequency model needs to be built. The high frequency equivalent model of sense circuit is shown in Fig. 4. Where R_{Cs} and L_{Cs} are series equivalent resistor and inductor of the capacitor C_S. L_{Rs} and C_{Rs} are series equivalent inductor and parallel equivalent capacitor of resistor R_S.

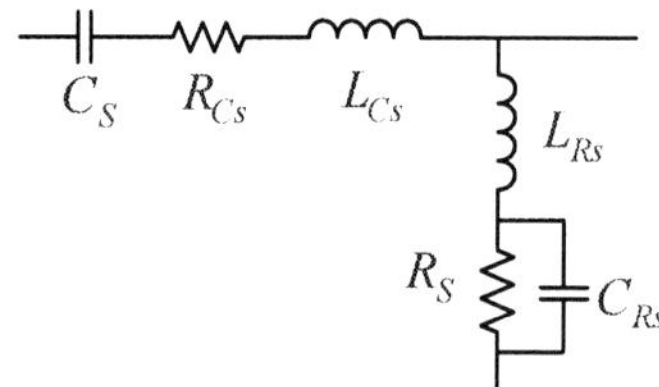

Fig. 4. High frequency equivalent model of sense circuit

The transfer function $H(s)$ of sense circuit with parasitic parameters can be expressed as

$$H(s) = \frac{L_{Rs}C_{Rs}s^3 + \frac{L_{Rs}}{R_S}s^2 + s}{(L_{Cs}C_{Rs} + L_{Rs}C_{Rs})s^3 + (R_{Cs}C_{Rs} + \frac{L_{Cs} + L_{Rs}}{R_S})s^2 + (\frac{C_{Rs}}{C_S} + \frac{R_{Cs}}{R_S} + 1)s + \omega_1} \tag{3}$$

where $\omega_1 = 2\pi f_1 = \frac{1}{R_S C_S}$ is the corner frequency of high-pass filter. According to IEC 62041:2010, f_1 should be under 150kHz to pick EMI signal above 150kHz.

The high frequency equivalent model of injection circuit is shown in Fig. 5. R_{Ci} and L_{Ci} are the series equivalent resistor and inductor of the capacitor C_i. L_{Ri} and C_{Ri} are the series equivalent inductor and parallel equivalent capacitor of resistor R_i.

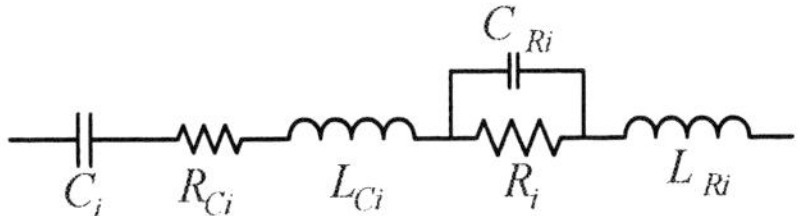

Fig. 5. High frequency equivalent model of injection circuit

The transfer function $L(s)$ of injection circuit with parasitic parameters can be expressed as

$$L(s)=\frac{L_{Ci}C_{Ri}s^3+(R_{Ci}C_{Ri}+\frac{L_{Ci}}{R_i})s^2+(\frac{C_{Ri}}{C_i}+\frac{R_{Ci}}{R_i})s+\omega_2}{(L_{Ri}C_{Ri}+L_{Ci}C_{Ri})s^3+(R_{Ci}C_{Ri}+\frac{L_{Ci}}{R_i}+\frac{L_{Ri}}{R_i})s^2+(\frac{C_{Ri}}{C_i}+\frac{R_{Ci}}{R_i}+1)s+\omega_2} \quad (4)$$

where $\omega_2=2\pi f_2=\frac{1}{R_iC_i}$ is the corner frequency of low-pass filter. According to IEC 62041: 2010, f_2 should be above 30MHz to inject necessary EMI signal below 30MHz.

The high frequency equivalent model of decoupling circuit is shown in Fig. 6. R_F and C_F respectively are the series equivalent resistor and parallel equivalent capacitor of the decoupling inductor L_F, which is implemented by a single-turn RF inductor.

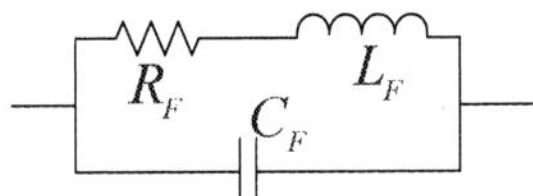

Fig. 6. Precise equivalent model of decoupling circuit

The transfer function $J(s)$ of decoupling circuit with parasitic parameters can be expressed as

$$J(s)=\frac{R_F+L_Fs}{C_FL_Fs^2+R_FC_Fs+1} \quad (5)$$

B. Precise Modeling of Digital Processing Portion

Digital processing portion includes 3 segments as ADC, controller and DAC. There is a delay time in ADC processes, the time t_{AD-d} is a nonlinear function about input voltage, it can be described as

$$t_{AD-d}=\frac{nbC_LV_{in}}{(V_{in}-V_{th})^{\alpha}} \quad (6)$$

where b, n and α are constant, their value varies depending on different ADC chip. C_L is the load capacitor, V_{th} is threshold voltage, V_{in} is input voltage. If input voltage don't change, and is a fixed value in one period, it can be replaced by its average value $t_{AD-d-avg}$. $t_{AD-d-avg}$ can be expressed as

$$t_{AD-d-avg}=\frac{k_{eff}}{V_{in}} \quad (7)$$

where k_{eff} is constant. So, the transfer function $G_{AD}(s)$ of ADC when considering average delay time $t_{AD-d-avg}$ can be expressed as

$$G_{AD}(s)=\frac{1}{T_s}(1+e^{-st_{AD-d-avg}})=\frac{1}{T_s}(1+e^{-s\frac{k_{eff}}{V_{in}}}) \quad (8)$$

Generally, DAC is a zero-order-hold. When considering its delay time and nonlinear potion, the transfer function $G_{DA}(s)$ of DAC can be expressed as

$$G_{DA}(s)=\frac{D_{zoh}(s)B(s)}{s} \quad (9)$$

where $D_{zoh}(s)$ is the transfer function of zero-order holder, $B(s)$ represent the delay time and nonlinear potion, and can be expressed as

$$B(s)=\frac{e^{-st_{DA-d}}}{1+\tau s} \quad (10)$$

The compensator can be designed based on the system indicator requirement. Because this paper focused on study the precise modeling and new DAEF topology, selected the easiest inverted proportion as the proportional part of compensator. Furthermore, the operation time τ_{con} of hardware language cannot be ignored. So, the transfer function $G_C(s)$ of controller with delay time can be expressed as

$$G_C(s)=-Ke^{-s\tau_{con}} \quad (11)$$

So, the transfer function $G_{DAEF}(s)$ of DAEF control precise model can be obtained as

$$G_{DAEF}(s)=\frac{J(s)}{1+J(s)L(s)G_C(s)G_{AD}(s)G_{DA}(s)H(s)} \quad (12)$$

C. Design of ADC Sampling Precision

The DAEF is a closed loop system, the sampling accuracy of sensed EMI signal affected directly EMI suppression performance. If selected the sampling accuracy randomly, the ADC resolution may be greater than the error range of the output EMI signal allowed. When the output signal changes exceeded the error range, the digital signal may have no change, and cannot complete the feedback control. The minimum bit of ADC sampling can be written as

$$n=\mathrm{INT}\left[\log_2\frac{V_{\max}}{\Delta V_o\times H}\right] \quad (13)$$

where $V_{\max}$ is the maximum of ADC, which allowed by input voltage; ΔV_o is the error range of the output signal allowed; H is the output gain of the EMI signal sensed.

In order to ensure the reliable filtering performance, selected the thousandth of the lowest limits of IEC 62041:2010 46dBμV as ΔV_o.

$$\Delta V_o = 0.001 \times 46\text{dB}\mu\text{V} = 0.2\mu\text{V} \tag{14}$$

$H = V_{ref} / V_o = 2/46\text{dB}\mu\text{V} = 10000$, V_{max} is 3V, The minimum bit of ADC sampling can be calculated by

$$n = \text{INT}\left[\log_2 \frac{V_{max}}{\Delta V_o \times H}\right] = \text{INT}\left[\log_2 \frac{3}{2\times 10^{-7} \times 10000}\right] = 11 \tag{15}$$

D. Analysis of DAEF System

Frequency response of traditional DAEF model is the black solid curve showed in Fig. 7. Frequency response of proposed precise DAEF model is the red dotted curve showed in Fig 7.

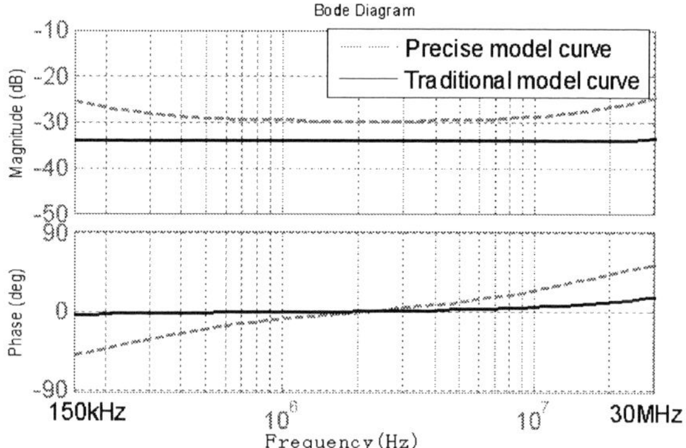

Fig. 7. Frequency response of DAEF system

When considering parasitic phenomenon and delay time, the filter ability decreases to -30dB from -35dB of traditional model in the middle frequency range. In the low and high frequency, the filter ability of precise model decreases to -25dB.

IV. DESIGN OF DECOUPLING CIRCUIT

In order not to effect the advantage in decreasing size, decoupling circuit don't use the inductor which connects in power in series. In this paper, the single-turn inductor is selected.

A. Design of the Single-Turn Inductor Core

In order to suppress EMI interference, this paper chooses ferrite as magnetic core of single-turn inductor. Ferrite cores include Ni-Zn ferrite and Mn-Zn ferrite. Ni-Zn ferrite is used to suppress the electromagnetic interference in high frequency section. Therefore, in this paper, Ni-Zn ferrite is used to be the core material of single-turn inductor. Its operating frequency range is 100kHz-140MHz.

There is a direct relationship between the content of ZnO, NiO and the operating frequency range of Ni-Zn ferrite for Ni-Zn ferrite. The relationship between the formula proportion of Ni-Zn ferrite and relative permeability μ_r , the cut-off frequency is reported in Table I.

Tab. I. The relationship between Ni-Zn ferrite formula and cut-off frequency

Fe_2O_3:NiO:ZnO	μ_r	Cut-off frequency /MHz
50.3:17.5:33.2	320	10
50.2:24.9:24.9	150	30
50.8:31.7:16.5	85	75
51.6:39.0:9.4	44	140

The upper limit frequency of test standard of conductive EMI is 30MHz. So choose the ratio of the Ni-Zn ferrite is: Fe2O3: NiO: ZnO= 50.2:24.9:24.9. The relative permeability of Ni-Zn ferrite is 150.

B. Design and Implementation of Single-Turn Inductor

1) Determination of Impedance Range of Single-Turn Inductor

For design of decoupling inductor, it is first step to determine the impedance range of inductor, the next step is to determine the size of decoupling inductor according impedance range.

Equivalent impedance diagram of the DAEF system is shown in Fig. 8.

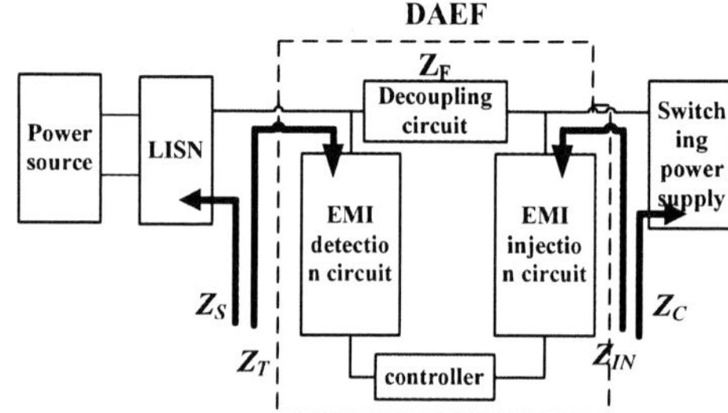

Fig. 8. Equivalent impedance diagram of the DAEF system

In order to prevent the interaction on both sides of the impedance in EMI filter, according to the principle of impedance matching, Z_F should satisfy Equation 16 and Equation 17.

$$Z_F \gg Z_C \parallel Z_{IN} = \frac{Z_C Z_{IN}}{Z_C + Z_{IN}} \tag{16}$$

$$Z_S \gg Z_F \parallel Z_T = \frac{Z_F Z_T}{Z_F + Z_T} \tag{17}$$

By Equation 16 and Equation 17, Z_F impedance range expression of decoupling circuit can be got.

$$\frac{Z_C Z_{IN}}{Z_C + Z_{IN}} = Z_F = \frac{Z_S Z_T}{Z_T - Z_S} \tag{18}$$

Among them, Z_F is the impedance of decoupling circuit, Z_C is the impedance of switching power supply, Z_S is the impedance of LISN, $Z_S = 50\Omega$. Z_T is the impedance of sense circuit and Z_{in} is the impedance of injection circuit. As shown in Fig. 8. The C_S and C_{inj2} can be regarded as short circuit on the high frequency range. So, $Z_T = R_S$, $Z_{in} = R_{inj2}$.

In this paper, the load of the DAEF system is a switching power supply. Z_C is its impedance, which can be derived from two current probe testing method, the impedance values curve is shown in Fig. 9.

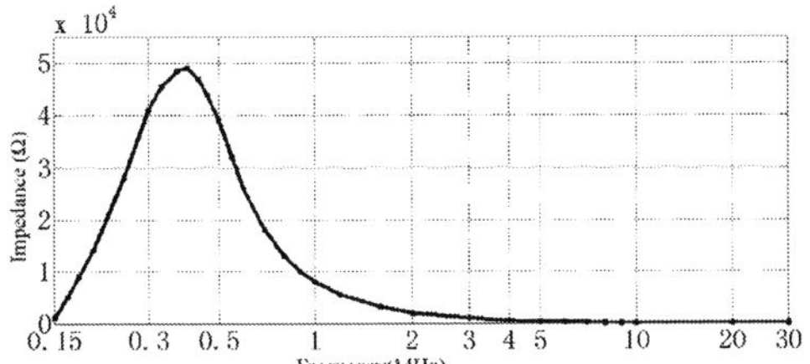

Fig. 9. Impedance tested curve of test switching power supply

The impedance range of decoupling circuit Z_F can be obtained by Equation 18 and impedance range Z_C. As shown in Fig. 10, the impedance range of Z_F changes with frequency, the impedance curve of Z_F should be in between curve 1 and curve 2 in Fig. 10.

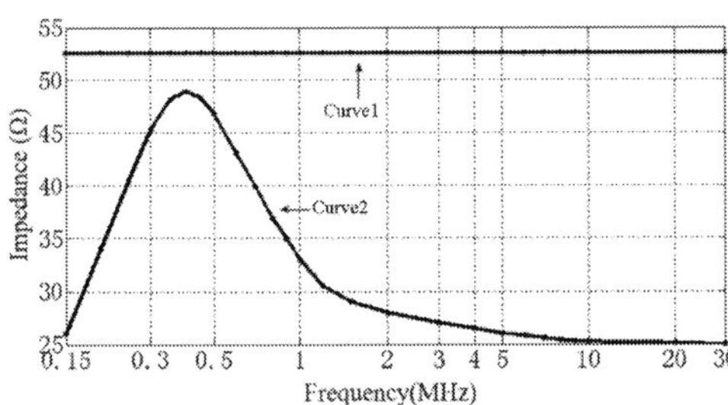

Fig. 10. The range of impedance Z_F of decouple circuit

2) Determination of Single-Turn Inductor Size

The relationship between single-turn inductor and inside diameter, outside diameter, length as:

$$L=\frac{\mu A_C}{h} \tag{19}$$

where h is the length of inductor, A_c is the sectional area of the inductor, $A_c = h \times \left(\frac{D}{2} - \frac{d}{2}\right)$. D is the outside diameter of the inductor, d is the inside diameter of the inductor, μ is the permeability of core, $\mu = \mu_r \times \mu_0 = 150 \times 4\pi \times 10^{-7} = 1.88 \times 10^{-4}$.

According to the standard about Dimensions of uncoated ring cores of magnetic oxides [14], the ratio of outside diameter and inside diameter is: $D/d = 1.67$; The ratio of length and inside diameter ratio is: $h/d = 0.67$. So the expression about inductor value L and inside diameter d is shown in

$$L=0.335\mu d \tag{20}$$

According to Fig.10, we determined that the impedance of single-turn inductor is $Z_F = 50\Omega$. Because $|Z_F| = 2\pi fL$, if f takes 15MHz of center frequency, the calculate inductor value L is obtained, $L=0.54\mu H$.

According to Equation 20, $D/d = 1.67$, $h/d = 0.67$, this can be calculated that dimensions for single-turn inductor is: $D = 14.31\text{mm}$, $h = 5.74\text{mm}$, $d = 8.57\text{mm}$.

V. Experimental Results

To verify the filter performance of DAEF, the DAEF has been tested on a 150W switching power supply, whose parameters are reported in Table II.

Tab. II Parameters of the tested switched-mode power supply

Input Voltage	AC:220V
Output Voltage	DC:48V
Output Current	DC:3A
Filter Capacitor	470μF
Switch Frequency	65kHz

The experimental parameters of the DAEF system based on FPGA are reported in Table III. The DAC and ADC converters are high speed, high resolution types.

Tab. III Parameters of the DAEF system

circuit	Code name	Parameters
Sampling circuit	ADC	14bit、250MSPS
	DAC	14bit、250MSPS
Sense circuit	R_{sL}/C_{sL}	1kΩ/0.1uF
	R_{sN}/C_{sN}	1kΩ/0.1uF
Rejection circuit	R_{iL}/C_{iL}	30Ω/1nF
	R_{iN}/C_{iN}	30Ω/1nF
Decoupling circuit	L_{jL}/L_{jN}	0.54uH

The complete system test setup is depicted in Fig. 11, a switch-mode power supply without internal passive EMI filter is tested as a EMI noise source.

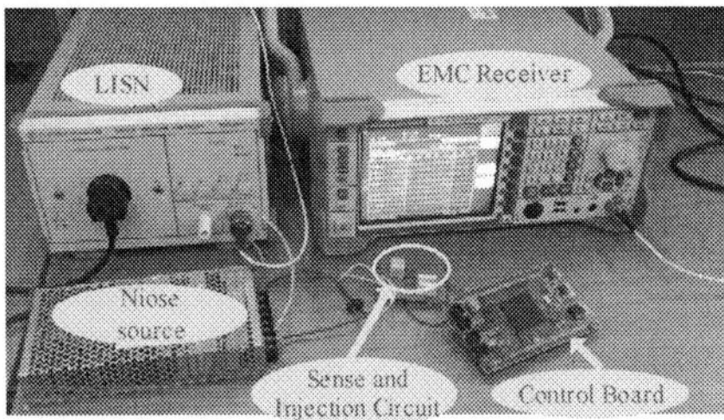

Fig. 11. Test setup of conducted EMI with DAEF

The spectrum of CM EMI emission of test power supply without EMI filter is shown in Fig. 12. The spectrum of CM EMI emission with proposed DAEF is shown in Fig. 13.

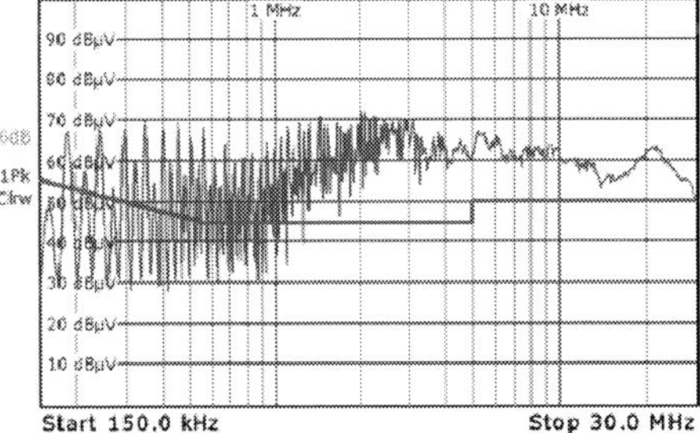

Fig. 12. Spectrum of CM EMI without EMI filter

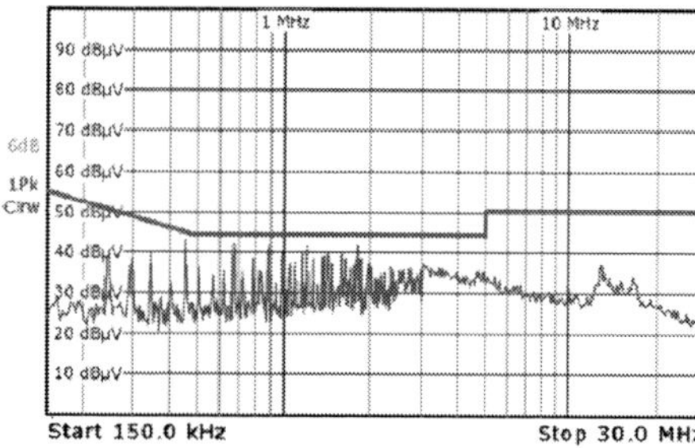

Fig. 13. Spectrum of CM EMI with proposed DAEF

From Fig. 12 and Fig. 13, it can be seen that the spectrum of CM conductive EMI is not only obviously lower than red standard limits curve, but also is lower than curve without EMI filter.

VI. Conclusion

This paper proposed a novel DAEF topology and method, which can suppress CM and DM conductive EMI of switched-mode power converters flexibly and selectively. The method calculates the EMI signals to ground acquired from L and N lines, separately. It obtains actual CM and DM noise of converters. Then cancel CM and DM EMI effectively and separately. Compared to the traditional DAEF, the proposed method may adjust the suppression capacity of CM and DM EMI effectively, and prevent interconversion between CM and DM EMI. Besides, this paper established a precise mathematical model of the proposed DAEF, which consider the parasitic phenomenon of sense circuit, injection circuit and decoupling circuit. Delay time produced by ADC, DAC and FPGA controller is also considered in detailed. Compared with traditional model, this model can describe the system behavior accurately and reflect the filter performance of DAEF.

References

[1] Wang Shishan, Gong Min, Yu Zhenyang. "Theoretical investigation of mutual transformation between differential mode and common mode noise and its implementation in the design of planar EMI filter," Transactions of China Electrotechnical Society, vol. 29, no. 2, pp. 239-246, Feb. 2014.

[2] R. Mrad, G. Pillonnet, F. Morel, et al. "Predicting the impact of magnetic components used for EMI suppression on the base-band of a power amplifier," IEEE Transactions on Power Electronics, vol. 30, no. 8, pp. 4199-4208, Mar. 2015.

[3] M. H. Hedayati, A. B. Acharya, V. John. "Common-mode filter design for PWM rectifier-based motor drives," IEEE Transactions on Power Electronics, vol. 28, no. 11, pp. 5364-5371, May. 2013.

[4] I. F. Kovacevic, T. Friedli, A. M. Musing, et al. "3-D electromagnetic modeling of parasitics and mutual coupling in EMI filters," IEEE Transactions on Power Electronics, vol. 29, no. 1, pp. 135-149, Jan. 2014.

[5] D. Shin, S. Kim, G. Jeong, et al. "Analysis and design guide of active EMI filter in a compact package for reduction of common-mode conducted emissions," IEEE Transactions on Electromagnetic Compatibility, vol. PP, no. 99, pp. 1-12, Feb. 2015.

[6] Xinli Chang, Yuehong Yang, Wenjie Chen, et al. "Implementation of a novel active common-mode filter used in DC-DC converters," in General Assembly and Scientific Symposium, 2014, pp. 1-4.

[7] Shuo Wang, Y. Y. Maillet, Fei Wang, et al. "Investigation of hybrid EMI filters for common-mode EMI suppression in a motor drive system," IEEE Transactions on Power Electronics, vol. 25, no. 4, pp. 1034-1045, Apr. 2010.

[8] Wenjie Chen, Xu Yang, Zhaoan Wang. "A novel hybrid common-mode EMI filter with active impedance multiplication," IEEE Transactions on Industrial Electronics, vol. 58, no. 5, pp. 1826-1834, May. 2011.

[9] Wenjie Chen, Xu Yang, Jing Xue and Fred Wang. "A novel filter topology with active motor CM impedance regulator in PWM ASD system", IEEE Transactions on Industrial Electronics, vol. 61, no. 12, pp.6938-3946, Dec. 2014.

[10] D.Hamza, MeiQiu, P.K.Jain. "Implementation of an EMI active filter in grid-tied PV micro-inverter controller and stability verification," in 2012-38th Annual Conference on IEEE Industrial Electronics Society, 2012, pp. 477-482.

[11] D. Hamza, Mei Qiu. "Digital active EMI control technique for switch mode power converters," IEEE Transactions on Electromagnitic Compatibility, vol. 55, no. 1, pp. 81-88, Feb. 2013.

[12] D. Hamza, MeiQiu, P. K. Jain. "Application and stability analysis of a novel digital active EMI filter used in a grid-tied PV micro-inverter module," IEEE Transactions on Power Electronics, vol. 28, no. 6, pp. 2867-2874, Jun. 2013.

[13] D. Hamza, M. Pahlevaninezhad, P. K. Jain. "Implementation of a novel digital active EMI technique in a DSP-based DC–DC digital controller used in electric vehicle(EV)," IEEE Transactions on Power Electronics, vol. 28, no. 7, pp. 3126-3137, Jul. 2013.

[14] GB/T 28868-2012 Dimensions of uncoated ring cores of magnetic oxides[S]. Implemented on 2012.11.5.

Development of a Hybrid Emulation Platform Based on RTDS and Reconfigurable Power Converter-based Testbed

Shuoting Zhang, Yiwei Ma, Liu Yang, Fred Wang, Leon M. Tolbert
Department of Electrical Engineering and Computer Science, the University of Tennessee
Centre for Ultra-Wide-Area Resilient Electric Energy Transmission Networks (CURENT)
Knoxville, Tennessee, USA
zshuotin@vols.utk.edu

***Abstract*— A real-time hardware testbed (HTB) has been constructed to emulate the power system by modular regenerative converters. This allows system realistic testing and demonstration with the true measurement, communication, and control. However, the size of the system that can be emulated by the HTB is limited, and certain phenomena are not easy or not needed to be modeled in the HTB. A hybrid emulation platform, which combines real time digital simulator (RTDS) and HTB, is developed in this paper to complement the advantages of RTDS and HTB. A power electronics converter is designed to act as the power interface between the RTDS and the HTB, and an integrated interface with two complementary algorithms is implemented to realize the hybrid emulation stably under different system conditions. At the same time, the closed loop control method under dq0 axis is implemented to realize faster response characteristics, and a time delay correction algorithm is integrated into the Park transformation. Experiment results demonstrate the performance and effectiveness of the hybrid emulation compared with the pure HTB emulation and digital simulation.**

I. Introduction

A Hardware Testbed (HTB) platform has been established to emulate power system with interconnected converters by programming them to behave like various electrical elements. This allows system realistic operation and control testing and demonstration without the need of the full scale power equipment but still contains the measurement, communication, and control [1]. However, the size of the system that can be emulated by the HTB is limited by the hardware, and certain phenomena are not easy or not needed to be modeled in the HTB. Hybrid emulation, which emulates the subsystems of a large system in RTDS and HTB separately, complements the advantages of RTDS and HTB. Similar to the Power Hardware-in-the-Loop (PHIL) emulation, an interface, including digital interface and power interface, is implemented to make the two separate emulation subsystems perform as a whole system [2].

Reference [3] chooses an interface algorithm to realize the hybrid emulation of Opal-RT and a real house electric distribution system. Reference [4] set up a hybrid emulation platform based on RTDS to test a motor drive system. Reference [5] combines a real time digital simulator with a microgrid testbed to study the influence between microgrid and utility grid and between different microgrids. For these platforms, the hybrid emulation is stable when the impedance of the digital side is smaller than the hardware side, otherwise there might be a stability problem. Besides, the detailed control algorithm of the interface and the verification method of the hybrid emulation are not mentioned.

All hybrid emulation platforms are designed for specific applications, and there are tradeoff aspects among the stability, bandwidth and accuracy [6-8]. Particularly, as a new type of platform, HTB is full of switching harmonics and the time delay can be large because of the required filters within the interface. A time delay correction method is proposed to improve the accuracy of the two types of hybrid emulation.

This paper presents a hybrid emulation platform based on RTDS and HTB, and presents a closed loop control with a time delay compensation method. The structure of the paper is organized as follows: Part II introduces the basic hardware and software implementation of the hybrid emulation system. Part III gives the models and control strategies used in the interface to keep the stability and accuracy of the system emulation. Part IV presents the experimental results.

II. Basic Implementation of RTDS and HTB Hybrid Emulation System

A. Hardware Implementations

The RTDS and HTB hybrid emulation has a paralleled converter structure as shown in Fig. 1. The converters share the same DC link, which is supported by a DC power supply to keep a constant DC voltage. Thus the AC link connection of the converters is able to be considered as an emulated power system. Although each converter is controlled to export or absorb active power, the total power consumption of the test-bed is only the power loss of the converters because of the power circulation through the DC link. Controlled by a DSP board TMS320F28335, each converter in the HTB serves as

978-1-4673-9551-9/16 $31.00 © 2016 IEEE

an emulating unit. It is programmed to have the same steady-state and dynamic response as the emulated object [9, 10].

As the most critical linkage between RTDS and HTB, the interface is composed of a power interface in HTB side and a digital interface in RTDS side. The power interface, which is the same converter in HTB, is designed to exchange signals with RTDS and controlled as a voltage or current source by tracking the references from RTDS. The digital interface, which is connected to the subsystem in RTDS, is composed of controlled current and voltage source models, whose references come from the power interface.

The ratings of the emulated system are too large to be directly emulated in the HTB, thus the scaled down parameters are implemented in the HTB to keep the same performance with the original system by maintaining the per unit values of the emulated system. Unlike the HTB, the subsystem in RTDS can be a wide range of power levels, so the scale ratio can be totally different from the HTB or simply keeps the original parameters.

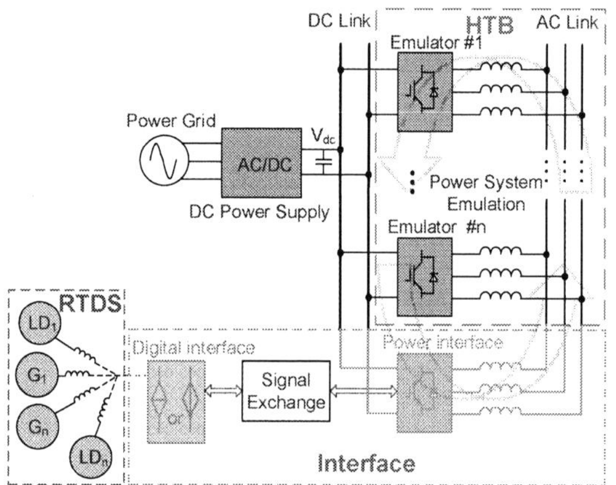

Fig. 1. RTDS and HTB hybrid emulation configuration.

B. Converter Control

Within the HTB, each converter is controlled to have the same steady-state and dynamic characteristics as the emulated object by tracking the voltage or current reference according to the terminal current or voltage, as shown in Fig. 2(a). The power interface converter is controlled to act as the subsystem emulated in RTDS by tracking the voltage or current reference which comes from the interface terminal of the RTDS subsystem, as shown in Fig. 2(b).

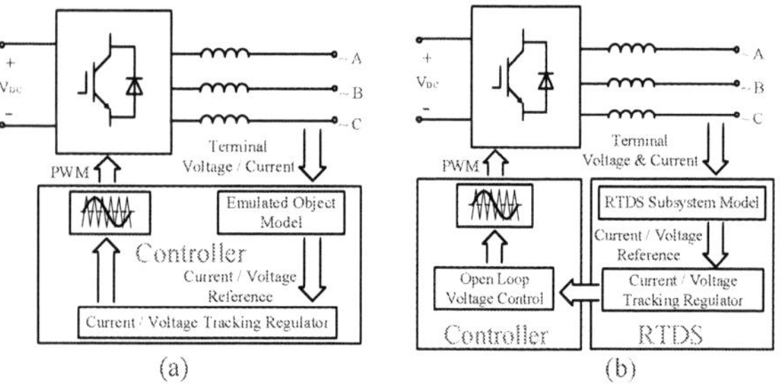

Fig. 2. Basic converter control in RTDS and HTB hybrid emulation (a) HTB emulator converter control (b) Power interface converter control.

III. Interface of Hybrid Emulation

A. Overview

As stated above, the interface is the most critical section among the hybrid emulation system. The hybrid emulation interface consists of a digital interface and a power interface, as shown in Fig. 3. Built in RTDS, the digital interface consists of both controlled current and voltage sources as the main circuit model and the control. The power interface consists of both power electronics converter as the physical parts and the control strategies.

According to [2], when the impedance of the digital subsystem is smaller than the physical subsystem, the controlled voltage source should be applied on the physical side to keep the stability; otherwise, the controlled current source should be implemented. Based on that, two sets of control strategies are integrated to realize either voltage or current tracking. This flexibility enables the choice of different interface algorithms to ensure the stability of the hybrid emulation under different system conditions.

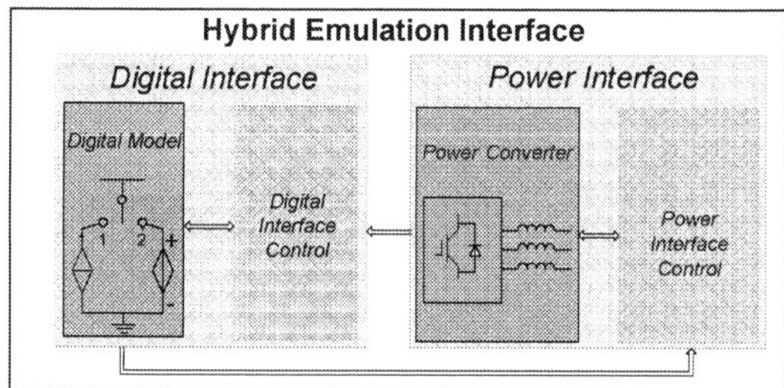

Fig. 3. Block diagram of the hybrid emulation interface.

B. Hybrid Emulation Interface Algorithm

The basic theory of hybrid emulation is the substitution theorem in circuit theory. Within an electrical network, if a subsystem is replaced by a voltage source whose voltage at any instant of time is equal to the voltage at the terminal of the subsystem in the previous network, then the condition in the rest of the network will keep the same. Or alternately, if a subsystem is replaced by a current source whose current at any instant of time is equal to the current at the terminal of the subsystem in the previous network, then the condition in the rest of the network will keep the same.

Based on this idea, the Ideal Transformer Model (ITM) algorithm is implemented to realize the hybrid emulation of a real time digital simulator and hardware platform. In this paper, according to the two different types of controlled source in the hardware platform side, the ITM is divided into two different kinds of algorithms, which are voltage source ITM, as shown in Fig. 4, and current source ITM, as shown in Fig. 5 [11].

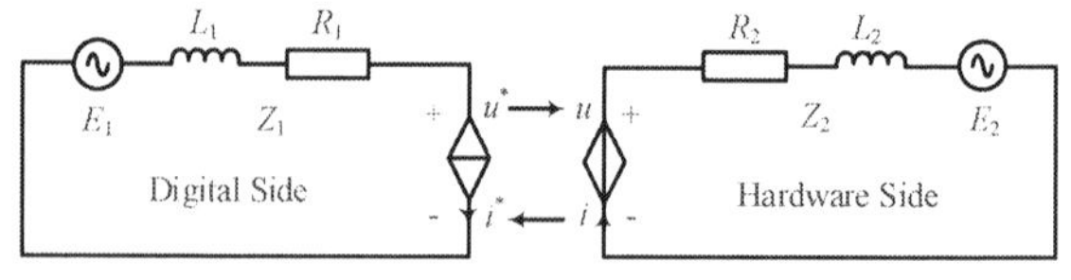

Fig. 4. Voltage source ITM.

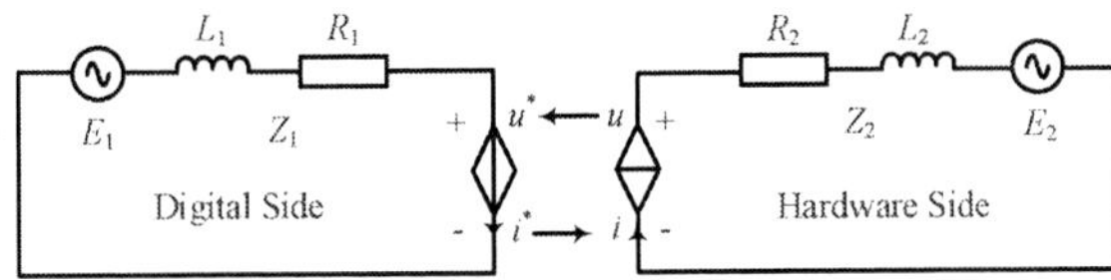

Fig. 5. Current source ITM.

The voltage and current relationships can be derived from Fig. 4:

$$E_1 = R_1 i^* + L_1 \frac{di^*}{dt} + u^* \tag{1}$$

$$E_2 = -R_2 i - L_2 \frac{di}{dt} + u \tag{2}$$

Consider the interface time delay to be Δt, and put Δt in the voltage, that is:

$$i^* = i \tag{3}$$

$$u = u^*(t - \Delta t) \tag{4}$$

Then

$$E_2 = -R_2 i^* - L_2 \frac{di^*}{dt} + u^*(t - \Delta t) \tag{5}$$

Assuming the digital simulation time step to be h, then for the step k

$$E_1(k) = R_1 i^*(k) + \frac{2L_1}{h} i^*(k) + u^*(k) \tag{6}$$

$$E_2(k) = -R_2 i^*(k) - \frac{2L_2}{h} i^*(k) + u^*(k - \frac{\Delta t}{h}) \tag{7}$$

Round $\Delta t / h$ to be Δn, then

$$E_2(k) = -R_2 i^*(k) - \frac{2L_2}{h} i^*(k) + u^*(k - \Delta n) \tag{8}$$

$$\begin{aligned} E_1(k) &= R_1 i^*(k) + \frac{2L_1}{h} i^*(k) + E_2 \\ &+ R_2 i^*(k + \Delta n) + \frac{2L_2}{h} i^*(k + \Delta n) \end{aligned} \tag{9}$$

That is

$$i^*(k + \Delta n) = -\frac{R_1 + \frac{2L_1}{h}}{R_2 + \frac{2L_2}{h}} i^*(k) + \frac{E_1(k) - E_2(k + \Delta n)}{R_2 + \frac{2L_2}{h}} \tag{10}$$

The condition of stability is

$$\frac{R_1 + \frac{2L_1}{h}}{R_2 + \frac{2L_2}{h}} < 1 \tag{11}$$

$$R_1 + \frac{2L_1}{h} < R_2 + \frac{2L_2}{h} \tag{12}$$

With the same assumption for current source ITM

$$\begin{aligned} E_1(k) &= R_1 i^*(k) + \frac{2L_1}{h} i^*(k) + E_2(k - \Delta n) \\ &+ R_2 i^*(k - \Delta n) + \frac{2L_2}{h} i^*(k - \Delta n) \end{aligned} \tag{13}$$

Then

$$i^*(k) = -\frac{R_2 + \frac{2L_2}{h}}{R_1 + \frac{2L_1}{h}} i^*(k - \Delta n) + \frac{E_1(k) - E_2(k - \Delta n)}{R_1 + \frac{2L_1}{h}} \tag{14}$$

The condition of stability is

$$\frac{R_2 + \frac{2L_2}{h}}{R_1 + \frac{2L_1}{h}} < 1 \tag{15}$$

$$R_1 + \frac{2L_1}{h} > R_2 + \frac{2L_2}{h} \tag{16}$$

From the analysis above, the voltage source ITM algorithm is stable if the equivalent impedance of the digital side is smaller than the hardware side; on the contrary, the current source ITM is stable if the equivalent impedance of digital side is larger than the hardware side. Since there is a stability problem with the ITM algorithm, some other interface algorithms are developed to improve the stability.

Damping Impedance Model (DIM) algorithm is shown in Fig. 6, the resistance between the digital and hardware side is used repeatedly in order to improve the stability [12]. The accuracy decreases because of using the resistance repeatedly and there is a resistance needed in the hardware side. Partial Circuit Duplication (PCD) algorithm is the special case of DIM when $R_d = 0$.

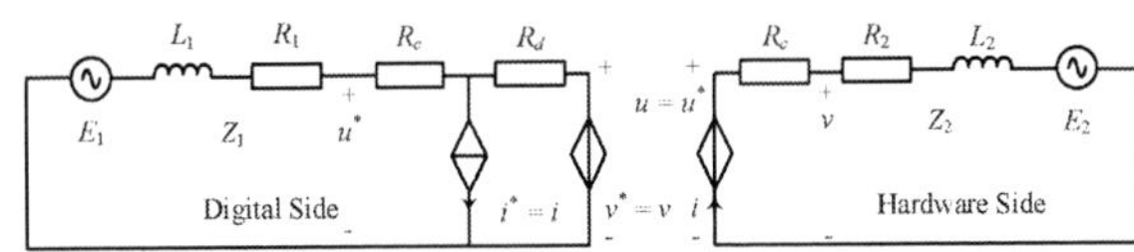

Fig. 6. DIM algorithm of power HIL simulation.

Transmission Line Model (TLM), as shown in Fig. 7, uses the decoupling characteristic of the Bergeron equivalent circuit. The interface stability is improved because of the natural time delay of the transmission line. However, an actual resistor exists in the hardware side which introduces heat and influences the flexibility of the application.

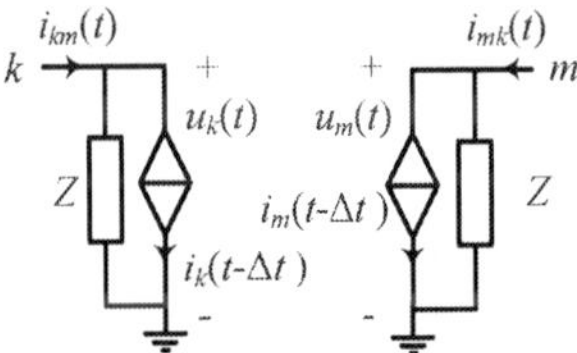

Fig. 7. TLM algorithm of power HIL simulation.

Compared with the purely voltage source or current source ITM algorithm, the algorithms mentioned above are able to improve the interface stability, but there are some other limitations and disadvantages. In general, there are tradeoff aspects among the stability and accuracy within different kinds of interface algorithms.

Since the stability conditions of voltage source ITM and current source ITM are complementary, an interface algorithm that combines these two methods is proposed to realize the hybrid emulation under different situations. When the equivalent impedance in the digital side is smaller than the hardware side, the voltage source ITM is used; when the equivalent impedance in digital side is larger than the hardware side, the current source ITM is used.

C. Accuracy Improvement Methods for Hybrid Emulation

The interface algorithm stability analyzed above only considered the interface time delay based on the simple circuit case, however, the situation is much more complex for the real hybrid emulation application. For example, the transfer function of power interface converter should be considered into the whole system transfer function, and some noise should be added into the system as disturbance.

From the analysis in [7], we can see that the methods to improve the stability are around adding filter into the interface system. However, the filter will influence the accuracy of the results and limit the bandwidth.

In paper [2], as shown in Fig. 8, a method was proposed to increase the stability and accuracy of the hybrid emulation. The RMS values and angle of digital side voltage and the output voltage of the interface converter are compared and adjusted with PI controller, the frequency is set to be fixed 50Hz. Since the RMS value is used, which involves filtering, the high frequency ripples do not penetrate the reference voltage of the interface converter, and there is no phase shift with the fundamental voltage. However, this method will fix the system frequency to be 50Hz and it is not the same with the real power system. At the same time, some harmonics are needed to be considered during system transient process.

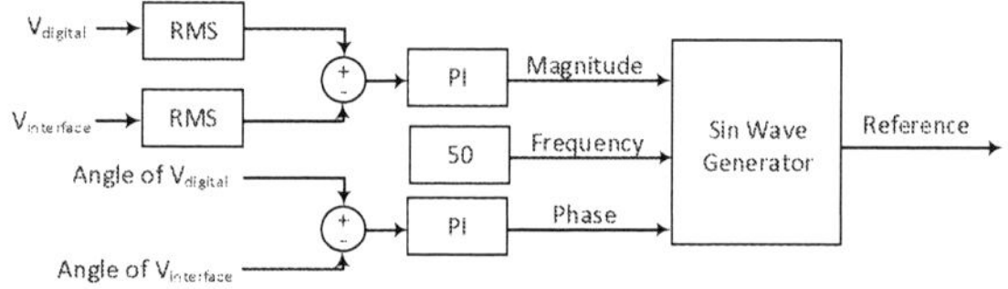

Fig. 8. Voltage RMS and Angle correction with fixed frequency.

In paper [6], as shown in Fig. 9, a method of time delay compensation based on FFT is proposed, and several harmonic components are considered, then the accuracy of the interface can be improved. However, the calculation amount will increase greatly with an increase of the concerned harmonics.

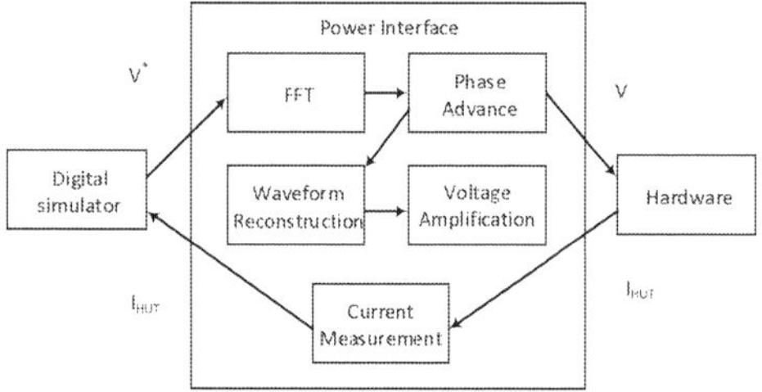

Fig. 9. Time delay compensation with FFT.

Based on the analysis of the time compensation methods above, as shown in Fig. 10, a time delay compensation and amplitude correction method is proposed. In this method, the three phase voltage and current are transferred to be dq0 components based on the frequency measured in real time, and the time delay is corrected by adding a phase angle difference in the Park Transformation. At the same time, a closed-loop controller based on the dq0 value is implemented to realize the voltage and current tracking. Since the Park Transformation does not cut any frequency components, the harmonics are kept.

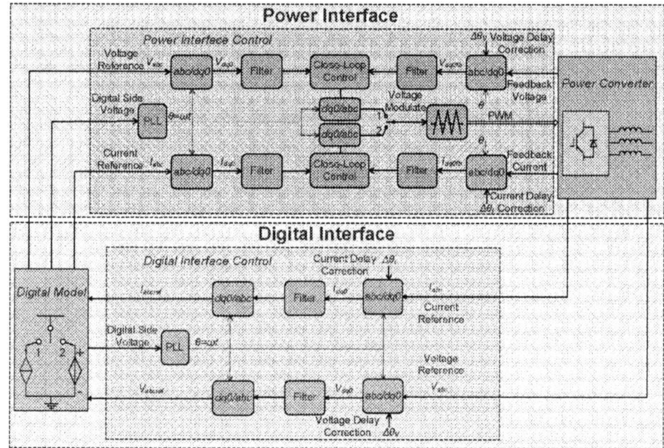

Fig. 10. Proposed method of voltage amplitude and angle correction.

IV. EXPERIMENTAL DEMONSTRATION

Although the hybrid emulation of RTDS and HTB is designed to realize the emulation of large scale systems, a typical two area system is selected as the experimental case in order to simplify the verification of the hybrid emulation, and actually it is easy to extend the subsystem scale of RTDS. The experimental setup with converters rated 600 V and 30 kW is shown in Fig. 11, and the scaled down two area system with the given parameters of the generators and loads is shown in Fig. 12. The pure HTB emulation and digital simulation are implemented as the comparison references to verify the hybrid emulation.

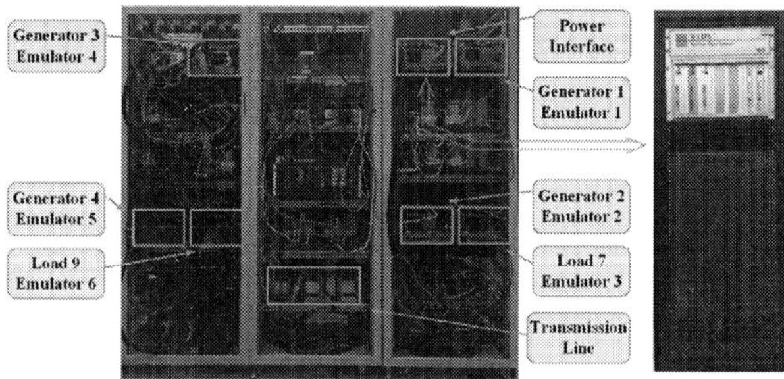

Fig. 11. Hybrid emulation converter cluster.

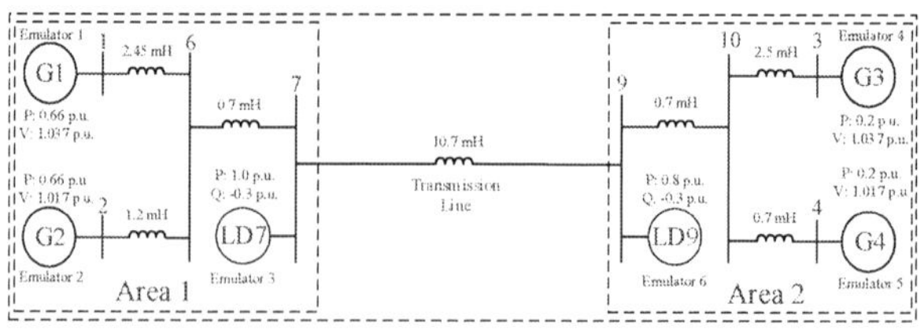

Fig. 12. Scaled down two area system.

In order to explain the way of keeping the hybrid emulation stable under different system conditions, two complementary system division situations are selected to implement the hybrid emulation. As shown in Fig. 13(a), Area 1 is emulated in RTDS and position 1 of the switch is selected to implement the appropriate interface algorithm. For another situation, both Area 1 and the transmission line between two areas are emulated in RTDS and position 2 of the switch is selected, as shown in Fig. 13(b).

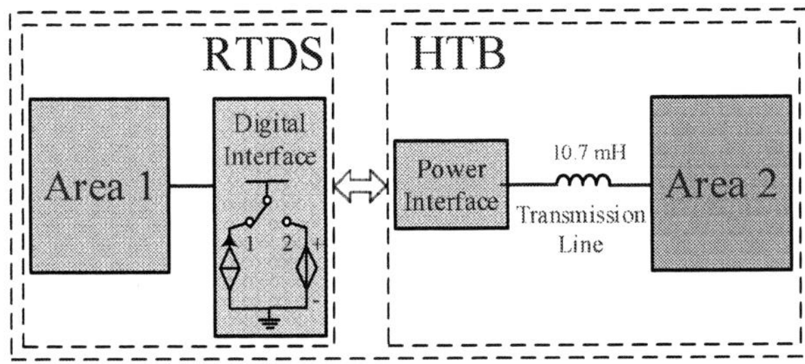

(a) Transmission line in HTB

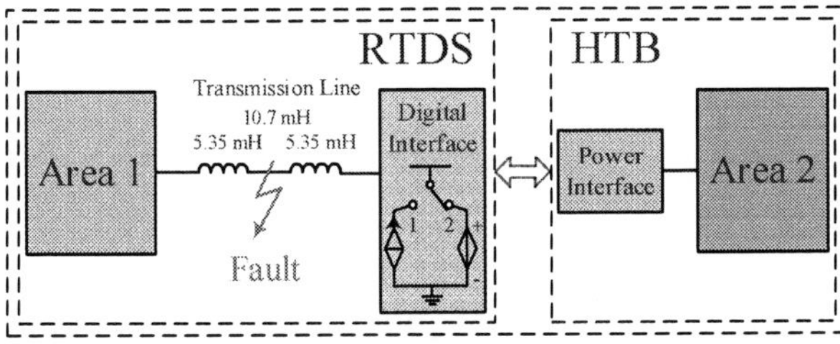

(b) Transmission line in RTDS

Fig. 13. Hybrid emulation of two area system.

A. Case I: Load Step Change Scenario

Fig. 14 and Fig. 15 show the waveform comparison between the hybrid emulation and the pure HTB emulation when there is a step change with load 9 from 0.8 p.u. to 0.4 p.u.. The waveforms indicate that the transient process of the hybrid emulation is the same with the pure HTB emulation.

As shown in TABLE I, the maximum difference of the frequency is 0.03 Hz and the maximum difference of the voltage, active power and reactive power is smaller than 0.02 p.u. during the load step transient. The comparison indicates that the hybrid emulation can match with the pure HTB emulation well, which verifies the validity of hybrid emulation during the load step transient.

TABLE I. DIFFERENCE BETWEEN HYBRID AND HTB EMULATIONS

Element	Maximum difference			
	Δf/Hz	ΔV/p.u.	ΔP/p.u.	ΔQ/p.u.
G3	0.03	0.01	0.015	0.015
LD9	N/A	0.02	0.01	0.003

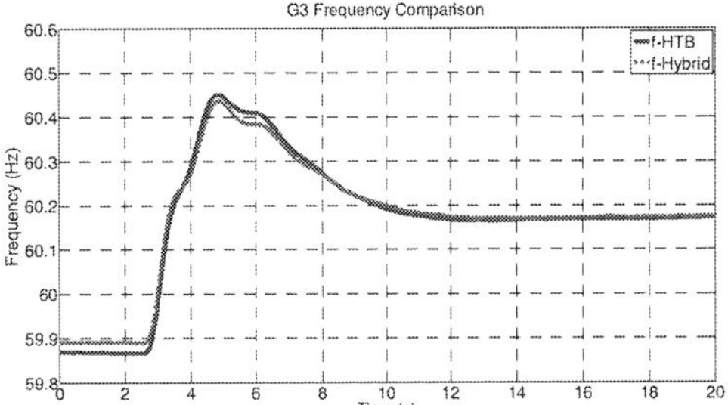

(a) G3 frequency comparison

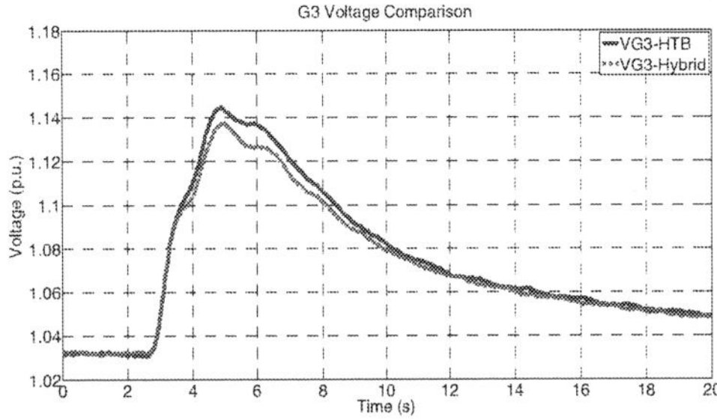

(b) G3 voltage comparison

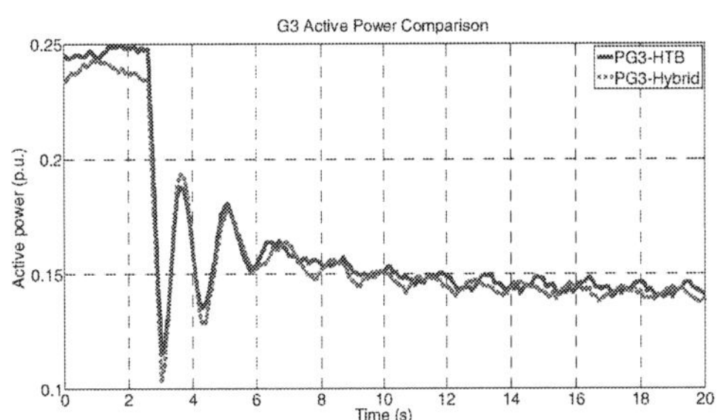

(c) G3 active power comparison

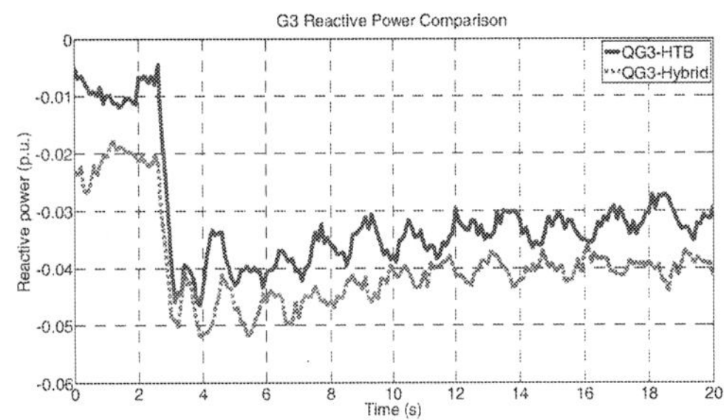

(d) G3 reactive power comparison

Fig. 14. G3 performance comparison between the hybrid emulation and the pure HTB emulation during load step.

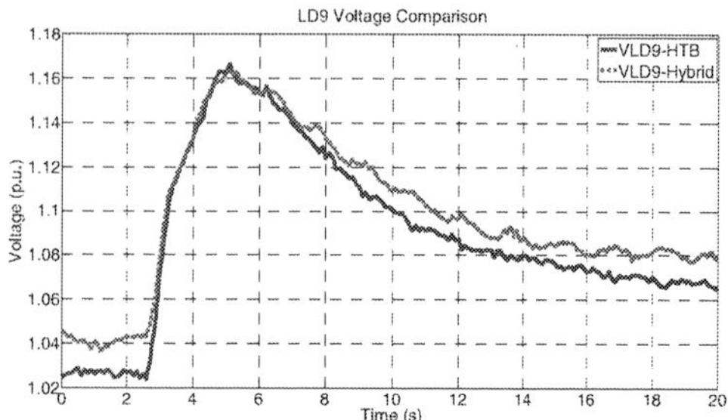

(a) LD9 voltage comparison

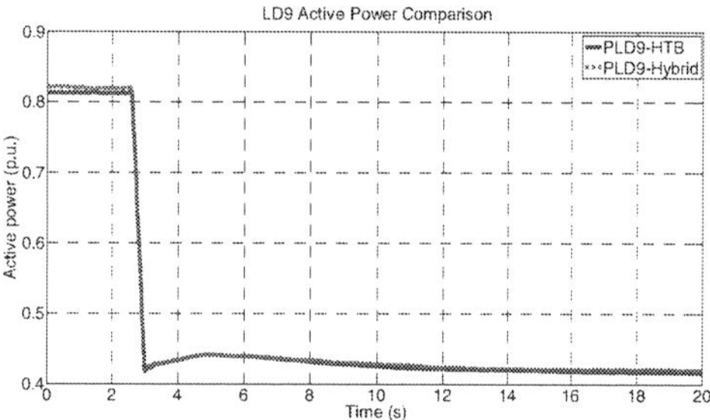

(b) LD9 active power comparison

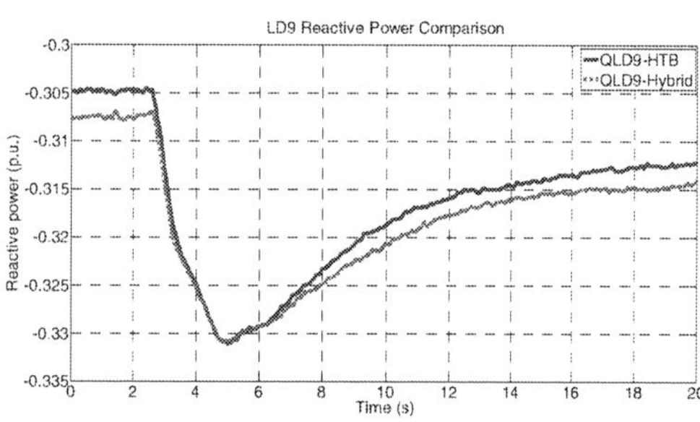

(c) LD9 reactive power comparison

Fig. 15. LD9 performance comparison between the hybrid emulation and the pure HTB emulation during load step.

B. *Case II: Transmission Line Fault Scenario*

Fig. 16 and Fig. 17 show the waveform comparison between the hybrid emulation and the pure digital simulation when a three phase fault happens at the middle point of the transmission line between two areas. The waveforms indicate that the transient process of the hybrid emulation is the same with the pure digital simulation.

As shown in TABLE II, the maximum difference of the frequency is 0.04 Hz and the maximum difference of the voltage, active power and reactive power is smaller than 0.05 p.u. during the fault transient. The difference of fault transient case is larger than the difference of load step change case, but the maximum difference point appears only at the beginning of the fault. The comparison indicates that the hybrid emulation can match with the pure digital simulation well, which verifies the validity of hybrid emulation during fault transients.

TABLE II. DIFFERENCE BETWEEN THE HYBRID EMULATION AND DIGITAL SIMULATION

Element	Maximum difference			
	Δf / Hz	ΔV / p.u.	ΔP / p.u.	ΔQ / p.u.
G3	0.04	0.03	0.05	0.05
LD9	N/A	0.02	0.02	0.015

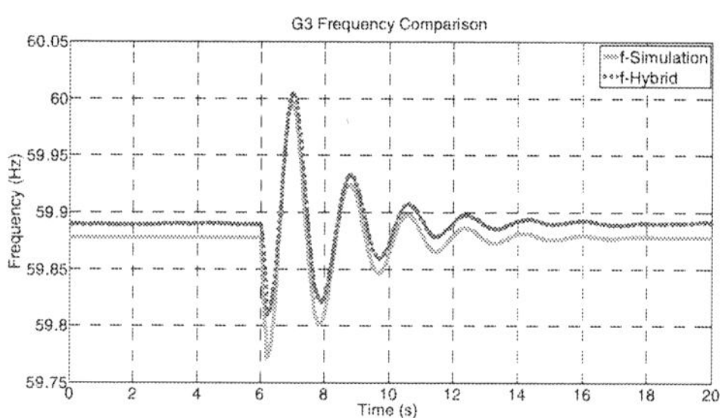

(a) G3 frequency comparison

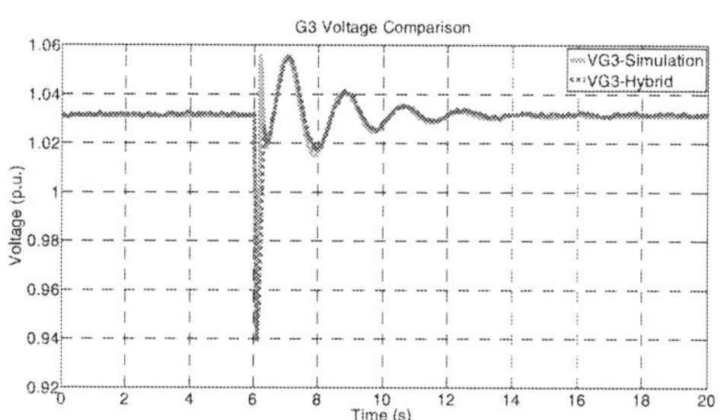

(b) G3 voltage comparison

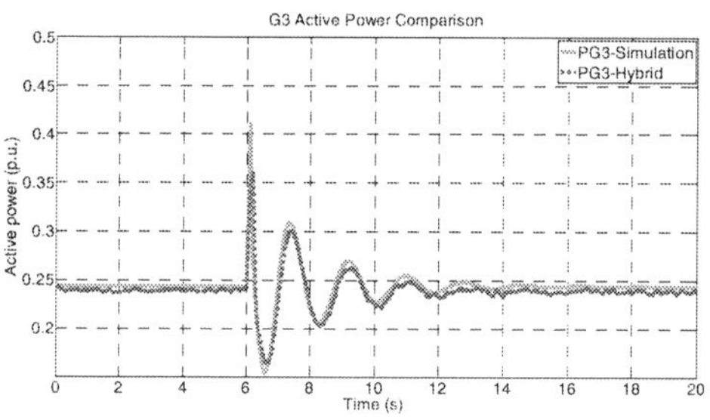

(c) G3 active power comparison

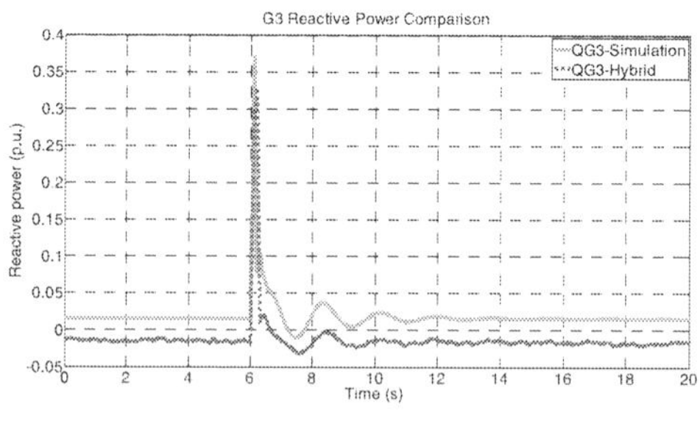

(d) G3 reactive power comparison

Fig. 16. G3 performance comparison between the hybrid emulation and the pure digital simulation during transmission line fault.

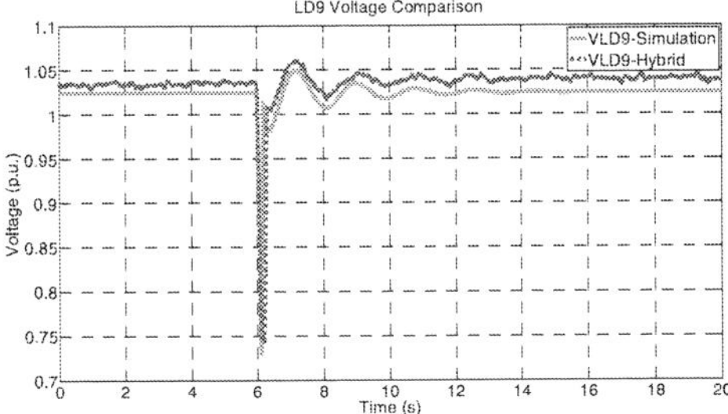

(a) LD9 voltage comparison

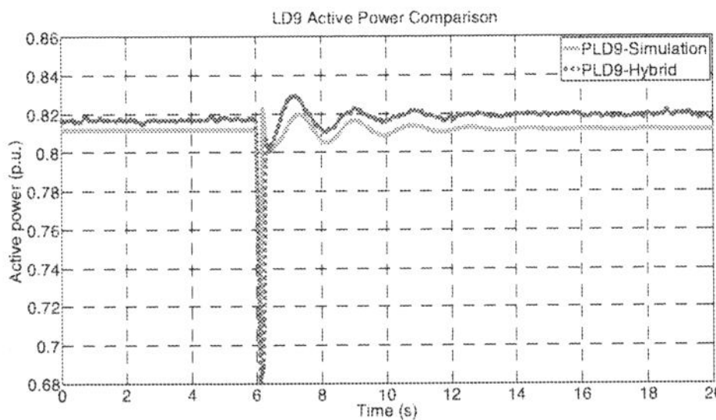

(b) LD9 active power comparison

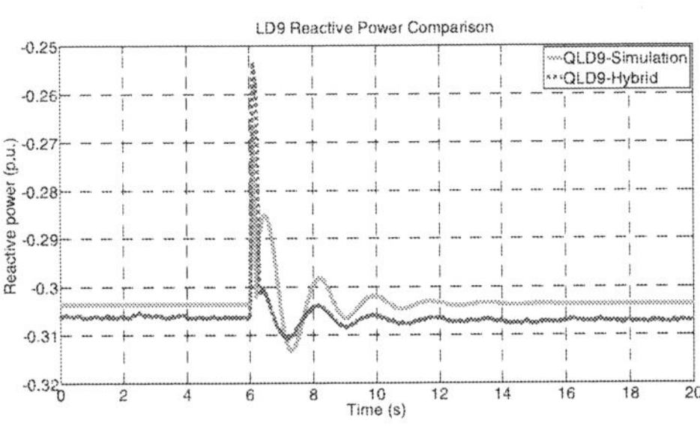

(c) LD9 reactive power comparison

Fig. 17. LD9 performance comparison between the hybrid emulation and the pure digital simulation during transmission line fault.

V. CONCLUSIONS

A hybrid emulation platform based on RTDS and HTB is developed with a multi-converter cluster. By choosing from the two complementary interface algorithms, the system can be stable under different situations. At the same time, phase delay correction and closed loop control algorithm are implemented to enhance the accuracy of the hybrid emulation. Compared with pure HTB emulation and digital simulation, the experimental result indicates the validity of the hybrid emulations.

In the future, different system scenarios will be tested in the HTB environment, and the subsystem in RTDS will be extended to be a larger system.

ACKNOWLEDGMENT

This work was supported primarily by the Engineering Research Center Program of the National Science Foundation and the Department of Energy under NSF Award Number EEC-1041877 and the CURENT Industry Partnership Program.

REFERENCES

[1] J. Wang, L. Yang, Y. Ma, X. Shi, X. Zhang, L. Hang, K. Lin, L. M. Tolbert, F. Wang, and K. Tomsovic, "Regenerative power converters representation of grid control and actuation emulator," in *Proc. 2012 IEEE Energy Conversion Congress and Exposition (ECCE)*, pp. 2460–2465.

[2] M. Dargahi, A. Ghosh, G. Ledwich, and F. Zare, "Studies in power hardware in the loop (PHIL) simulation using real-time digital simulator (RTDS)," in *Proc. 2012 IEEE Int. Conf. on Power Electronics, Drives and Energy Systems (PEDES)*, pp. 1-6.

[3] A. Yamane, W. Li, J. Belanger, T. Ise, I. Iyoda, T. Aizono, and C. Dufour, "A smart distribution grid laboratory," in *IECON 2011 - 37th Annual Conference on IEEE Industrial Electronics Society*, pp. 3708–3712.

[4] W. Ren, M. Steurer, and L. Qi, "Evaluating dynamic performance of modern electric drives via power-hardware-in-the-loop simulation," in *Proc. 2008 IEEE Int. Symp. Ind. Electron.* , pp. 2201–2206.

[5] F. Guo, L. Herrera, M. Alsolami, H. Li, P. Xu, X. Lu, A. Lang, Z. Long, and J. Wang, "Design and development of a reconfigurable hybrid microgrid test bed," in *Proc. 2013 IEEE Energy Conversion Congress and Exposition (ECCE)*, pp. 1350–1356.

[6] E. Guillo-Sansano, A. J. Roscoe, C. E. Jones, and G. M. Burt, "A new control method for the power interface in power hardware-in-the-loop simulation to compensate for the time delay," *2014 49th International Universities Power Engineering Conference (UPEC)*, pp.1–5.

[7] I. D. Yoo, and A. M. Gole, "Compensating for interface equipment limitations to improve simulation accuracy of real-time power hardware in loop simulation," *IEEE Transactions on Power Delivery*, vol. 27, pp. 1284–1291, 2012.

[8] W. Ren, M. Steurer, and T. L. Baldwin, "An effective method for evaluating the accuracy of power hardware-in-the-loop simulations," *IEEE Transaction on Industry Application*, vol. 45, no. 4, pp. 1484–1490, 2009.

[9] L. Yang, X. Zhang, Y. Ma, J. Wang, L. Hang, K. Lin, L. M. Tolbert, F. Wang, and K. Tomsovic, "Hardware implementation and control design of generator emulator in multi-converter system," *2013 IEEE Applied Power Electronics Conference and Exposition (APEC)*, pp. 2316–2323.

[10] J. Wang, Y. Ma, L. Yang, L. M. Tolbert, and F. Wang, "Power converter-based three-phase induction motor load emulator," in Proc. *2013 IEEE Applied Power Electronics Conference and Exposition (APEC)*, pp. 3270–3274.

[11] M. Dargahi, A. Ghosh, P. Davari, and G. Ledwich, "Controlling current and voltage type interfaces in power-hardware-in-the-loop simulations," *IET Power Electronics*, vol. 7, no. 10, pp. 2618–2627, 2014.

[12] B.-I. Craciun, T. Kerekes, D. Sera, R. Teodorescu, R. Brandl, T. Degner, D. Geibel, and H. Hernandez, "Grid integration of PV power based on PHIL testing using different interface algorithms," in *IECON 2013 - 39th Annual Conference on IEEE Industrial Electronics Society*, pp. 5380–5385.

Online Temperature Estimation for Phase Change Composite – 18650 Lithium Ion Cells Based Battery Pack

Mohamad Salameh[1], Ben Schweitzer[2], Peter Sveum[2], Said Al-Hallaj[2], Mahesh Krishnamurthy[1]

[1]Illinois Institute of Technology
3301 S. Dearborn Street
Chicago, Illinois 60616
EML: msalame1@hawk.iit.edu

[2]AllCell Technologies
2321 W 41st St
Chicago, Illinois 60609
EML: PSveum@allcelltech.com

***Abstract*— This paper proposes the design of an online temperature estimation technique for a Li-ion battery pack that utilizes phase change composite (PCC™) for thermal management. The phase change composite allows heat absorption and distribution, enabling lightweight and compact packs with extended cycle-life and safety. A coupled electro-thermal model has been developed for the cylindrical 18650 Li-ion cells, from which the cell heat generation is calculated. The electrical equivalent circuit comprises three RC pairs, where their values are functions of temperature and state of charge. An analytical thermal model is developed for the battery pack, considering the phase change composite and cells, which allows online temperature estimation all over the battery pack.**

***Keywords*—Lithium ion battery, phase change material, temperature estimation, thermal model, cell model.**

I. Introduction

Lithium-ion batteries continue to be the main current commercial battery in use due to their relatively high performance, energy density and specific energy density among batteries [1]. For safe and efficient operation, these batteries require thermal management, as higher temperatures promote the electro-chemical activity and reduce the cell over-potentials, at the same time, higher temperatures can reduce the cell life by promoting side reactions [2-6]. Although, lower temperatures can reduce the rate of the side reactions, they can result in sluggish electrochemical performance [2]. Phase Change Material Composite (PCC™) can be a part of a robust, passive safe thermal management solution, which can simultaneously solve both the performance degradation and safety concerns by preventing overheating and elimination of hotspots within the pack[2], [6]. The proposed thermal management solution will allow the online estimation of temperatures on the pack, based on electro-thermal model of the 18650 cell, without computationally intense finite element analysis.

II. Background

A. Cell Modeling and Battery Packs Thermal Modeling

Several thermal models of lithium ion cells and battery systems have been developed with different approaches and techniques, the choice of the modeling approach is usually based on the objective, accuracy and confidence of predictions, computational resources and applications.

At the cell level, electrochemical models offer high accuracy, but are very complex, which reduce its utilization on system level [7-9]. The electrochemical models can be simplified into reduced order models to allow its application onboard on control application [10]. However models of electrical equivalent circuit using Linear Parameter-Varying (LPV) implemented in [11, 12], can guarantee a good level of accuracy on system level applications, especially that the battery system behavior is dynamic and depends on different variables.

The approaches for the thermal modeling of battery packs are not only related to the pack geometry but also to the thermal management systems and the application. Computationally intensive finite element method (FEM) has been used in [13-15], while Linear Parameter-Varying (LPV) and Partial Differential Equation (PDE) approaches were developed in [16-18].

In this paper a coupled electro-thermal model is developed for an 18650 Li ion cell, in order to estimate the heat generation of the cell, which will be later integrated within an analytical thermal model of a battery pack that utilizes PCC™ as a passive thermal management solution. This solution will allow onboard estimation of temperatures on the pack.

B. Phase Change Composite (PCC™)

The Phase Change Material Composite is a patented wax-graphite material solution, which has been successfully used as a standalone thermal management system for a variety of Li ion batteries. The material is composed of a phase change component that is typically wax or a hydrated salt, and a highly compressible graphite matrix as host. Phase change material can be used to encase lithium-ion cells [2].

When in use, the PCC material in solid phase absorbs the heat generated by the Li ion cells and undergoes phase change into liquid state, thus limiting the temperature rise of the battery. Upon cessation of use, the PCC material can transfer heat to the environment, return through phase change to original state, and be ready for use again. At the same time, the graphite matrix provides high thermal conductivity, which allows even

978-1-4673-9551-9/16 $31.00 © 2016 IEEE

heat distribution resulting in the reduction of hotspots within the battery pack. If one cell goes into thermal runaway, PCC absorbs and wicks away heat to prevent a domino effect within the pack.

III. Electro-Thermal Model

Understanding the electrical and thermal behavior of the Li-ion cells is critical in designing a robust battery that meets the load requirements. Heat generated by the cell must be known to size both Phase Composite material amount and additional cooling/heating components. The voltage under load must be known to ensure adequate power available from the battery across the intended operating range. Both the heat generation and electrical behavior are strongly coupled with two-way interactions. Therefore, a model that accounts for both phenomena and couples them appropriately is necessary to generate accurate simulation results.

A. Electrical Model

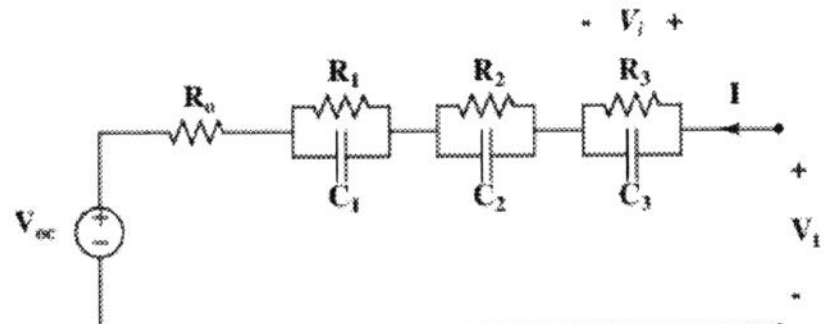

Fig. 1. Equivalent Electrical Circuit

An 18650 cylindrical cell is modeled as an equivalent electric circuit that utilizes the three resistor-capacitor (RC) pairs to model the charge transfer resistance, double layer capacitance, the electrolyte/salt polarization under load and solid state Li ions diffusion in the active material particles respectively. The resistance R_o is equivalent to the internal electrical resistance of the cell current collector [11, 19]. This model is expected to capture the long (500s-3000s) relaxation behavior of the cell, which will enhance the accuracy of the estimation of the cell's terminal voltage and heat generation. As shown in fig. (1), the electrical equivalent circuit comprises from three RC pairs, the values of the resistances and capacitances vary with temperature, state of charge (SoC) and current direction. Based on [11], the three RC pairs can achieve a compromise between the model accuracy and complexity of the identification process. The open circuit voltage of the cell is a function of the state of charge, SoC, which is calculated using Coulombs counting in terms of current into the positive terminal of the cell,

$$SoC(t) = SoC(0) + \int_0^t I.dt \tag{1}$$

Considering the RC pairs, the over-potentials across each RC pair within the cell can be expressed as in (2):

$$\dot{V}_i = -\frac{1}{R_i C_i} V_i + \frac{1}{C_i} I \tag{2}$$

B. Heat Generation

The sources of heat generation in the cell can be estimated by considering two major sources of heat generation, the reversible entropic heat and irreversible ohmic heat and sources such as the heat generated from side reactions. It must also include heat of mixing, which is neglected due to its relatively minor contribution, since side reactions are mostly aging reactions and are slow enough for their heat generation to be neglected [19]. The total heat generation, $\dot{Q}_t$ of the cell can be expressed as shown in (3), the first term corresponds to the ohmic heat generation and the second term presents the reversible heat generation in terms of cell temperature T:

$$\dot{Q}_t = \dot{Q}_{ohmic} + \dot{Q}_{entropic} = I(V_t - V_{oc}) + IT\frac{dV_{oc}}{dt} \tag{3}$$

Where $\frac{dV_{oc}}{dt}$ is the entropic heat coefficient and it is a function of the state of charge.

C. Cell Thermal Model

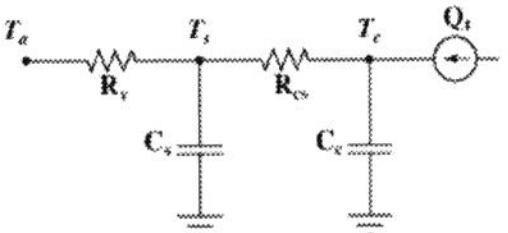

Fig. 2. Cell Thermal Model

The lumped thermal model of the 18650 cell accounts for the cell heat generation $\dot{Q}_t$ and the ambient temperature T_a. The heat transfer between the cell core and surface can be modeled by the cell conduction resistance R_{cs}. The convective heat transfer between the cell surface and the surroundings is modeled with a convection resistance R_v. The rate of change of the core and the surface temperatures is represented in the thermal model with the heat capacity of the cell core C_c and the heat capacity of the cell casing C_s respectively [11]. The equations governing heat transfer for the cell thermal model shown in fig. (2) can be expressed as shown in (4, 5)

$$C_c \frac{dT_c}{dt} = \frac{T_s - T_c}{R_{cs}} + \dot{Q}_t \tag{4}$$

$$C_s \frac{dT_s}{dt} = \frac{T_a - T_s}{R_v} + \frac{T_c - T_s}{R_{cs}} \tag{5}$$

The average of the estimated core and surface temperature that accounts for heat generation and the ambient temperature are fed back to the electrical model as shown in fig.(3), where the values of the electric circuit parameters will be updated based on the new values of temperature and the state of charge.

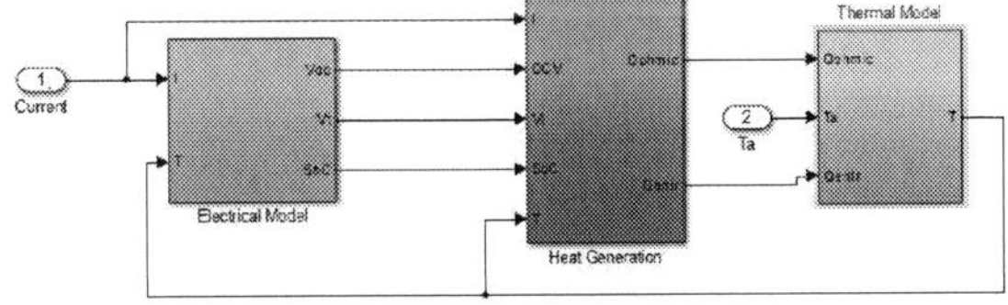

Fig. 3. Coupled Electro-thermal Model

IV. Electrothermal Parameter Identification

A. Electrical Model

A pulse-relaxation test (PRT) was used to identify parameters of the equivalent circuit. The test was performed for a 18650 cylindrical (Li-ion) cell 2.2 Ah, placed in a thermal test chamber MicroClimate Cincinnati Sub-Zero, the charge/discharge cycles were done using Arbin Instruments as shown fig.(4). The test was performed to obtain the values of the seven parameters of the equivalent electric circuit at different temperatures (10°, 25°, 35°and 45° C), different SoC's (varies from 8% - 95%) and at different current directions, i.e. charging and discharging. The test aimed to identify the parameters during discharging, by applying a discharging current pulse of 1C (2.2A) on a fully charged cell for five minutes followed by a relaxation period of 3 hours. This step was repeated until it reached the minimum safe voltage of 2.75V. Similarly, the parameters were identified during charging, by applying a charging current pulse of 1C (2.2A) on a fully discharged cell for five minutes followed by a relaxation period of 3 hours. This step was repeated until the maximum safe voltage 4.20 V is reached.

Fig. 4. Experimental Setup

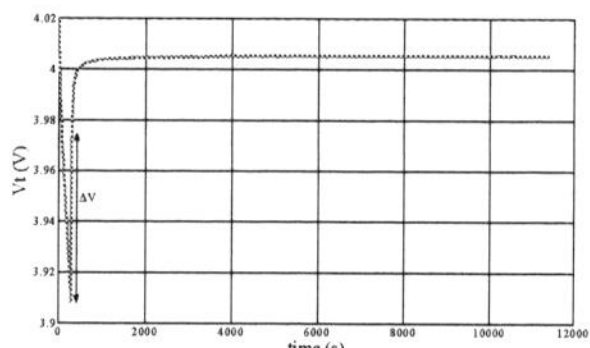

Fig. 5. Terminal Voltage during Pulse-Relaxation

Fig. (5) shows the response of the cell terminal voltage during one cycle of pulse-relaxation. Considering the equivalent circuit and (2) the terminal voltage during relaxation $V_{t,relax}$ can be described as shown in (6-8), assuming that the voltages of the capacitors are zero at the end /before the end of relaxation. The values of the RC pairs were fitted by minimizing the sum of the squared errors (SSE) shown in (9), which corresponds to the difference between the model estimated voltage and the measured experimental data. The curve fitting was done using least square regression, with appropriate choice of the order of three time constants τ_i, where $\tau_i = R_i C_i$, (100, 10, 1000 seconds respectively). Also, constraints and initial values of the identified parameters were set for the solver in order to allow proper parameter identification.

$$V_{t,relax}(t) = V_{oc} + IR_o + V_{RC}(t) \tag{6}$$

$$V_{RC}(t) = V_{1,0}(1-e^{-\frac{t}{R_1C_1}}) + V_{2,0}(1-e^{-\frac{t}{R_2C_2}}) + V_{3,0}(1-e^{-\frac{t}{R_3C_3}}) \tag{7}$$

Where

$$V_{i,0} = IR_i(1-e^{-\frac{300}{R_iC_i}}) \tag{8}$$

$$SSE = \sum (V_{\text{model}} - V_{measured})^2 \tag{9}$$

The value of R_o can be calculated from the instantaneous voltage change ΔV, which occurs when the pulse ends.

$$R_o = \left| \frac{\Delta V}{I} \right| \tag{10}$$

The open circuit voltage V_{oc} can be estimated from the relation between V_{oc} $vs.SoC$, which was obtained by measuring the open circuit voltage at different SoC's, using a small discharging current (C/50). The variation of the open circuit voltage with temperature was measured at different values of SoC to obtain the entropic heat coefficient. Identified parameters and their variation with temperature and SoC are shown in fig. (6).

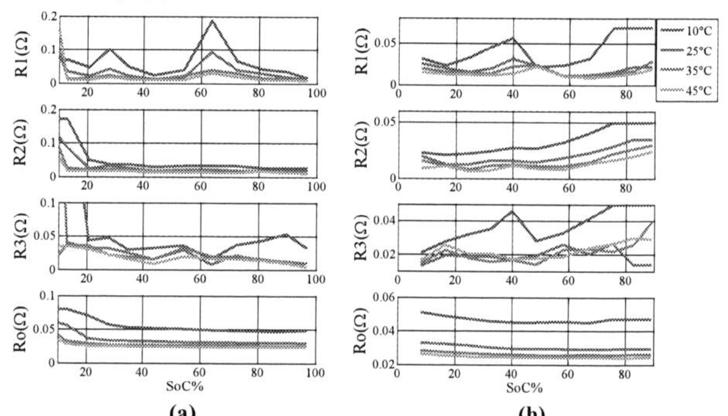

Fig. 6. Electrical Model Identified Parameters, (a) Discharging, (b) Charging

B. Cell Thermal Model

For the measurement of heat capacity C_s and convective thermal resistance R_a, an aluminum cylinder with the same dimension as the cell was used to parameterize. The cylinder was heated using a heater wire wounded around the cylinder. Using the known specific heat capacity of aluminum and the measured temperature of the cylinder during cooling, the value of R_a was calculated by least square regression. Afterwards, the test was re-run with a cell instead of the aluminum cylinder, using the value of R_a obtained from previous test, the value of C_s was identified using least square fitting for the cooling behavior. The rest of the parameters were estimated for the 18650 cylindrical Li-ion cell based on [20]. Table I, shows a summary of the identified parameters.

TABLE I. CELL THERMAL MODEL PARAMETERS

C_c (J/K)	C_s (J/K)	R_{cs} (K/W)	R_a (K/W)
44.3	2.3	1.3	32

V. THERMAL MODELING OF THE BATTERY PACK

A. Thermal Model Description

One of the novel contributions of this paper is the development of the battery pack thermal model. As shown in fig. (7), the battery pack consists of staggered cells placed in phase change composite. This geometry was chosen due to the repeatability of this pattern regardless of the battery pack size, as well, it considers the convective cooling for the cells on the edges of the pack. The block was divided into seven sections, each section is composed from a cell enclosed by PCC of 1mm thickness, and the temperature variations within the same section are neglected. The thermal model of the pack constitutes of seven states $T_1 - T_7$ where the heat exchange between adjacent areas is modeled with a conduction resistance R (Section-Section) (K/W), for the areas where convective cooling is considered, the heat transfer is modeled with a resistance R_a (Section- Ambiance) (K/W). The rate of change of the temperature is modeled with C measured in (J/K), the lumped heat capacity of the section. The equivalent circuit model for one section is shown in fig. (8)

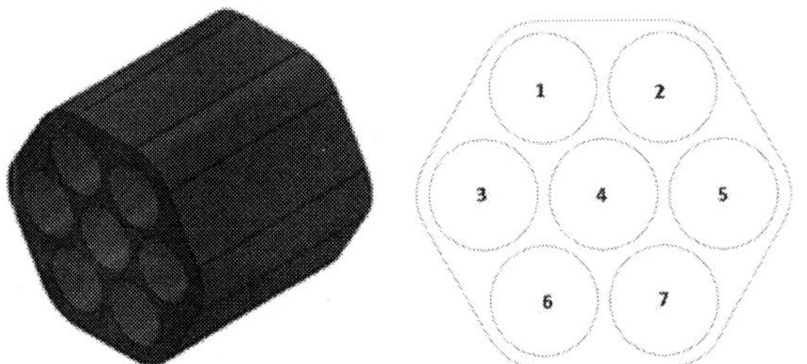

Fig. 7. PCC block and Sections

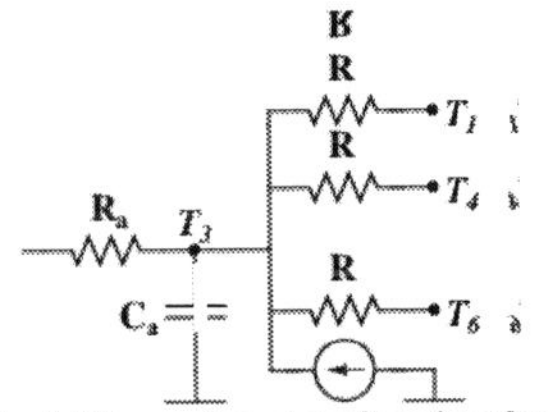

Fig. 8. Thermal model of section three

The differential equations governing the heat transfer within the block are shown in (11).

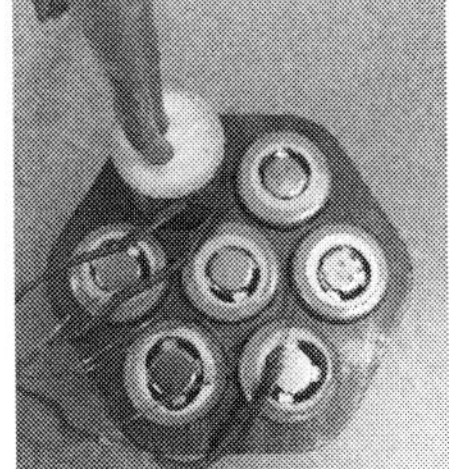
Fig. 9. Battery pack experimental setup

$$
\begin{aligned}
C\dot{T}_1 &= \frac{T_2-T_1}{R}+\frac{T_3-T_1}{R}+\frac{T_4-T_1}{R}+\frac{T_a-T_1}{R_a}+Q_1 \\
C\dot{T}_2 &= \frac{T_1-T_2}{R}+\frac{T_4-T_2}{R}+\frac{T_5-T_2}{R}+\frac{T_a-T_2}{R_a}+Q_2 \\
C\dot{T}_3 &= \frac{T_1-T_3}{R}+\frac{T_4-T_3}{R}+\frac{T_6-T_3}{R}+\frac{T_a-T_3}{R_a}++Q_3 \\
C\dot{T}_4 &= \frac{T_1-T_4}{R}+\frac{T_2-T_4}{R}+\frac{T_3-T_4}{R}+\frac{T_5-T_4}{R}+\frac{T_6-T_4}{R}+\frac{T_7-T_4}{R}+Q_4 \\
C\dot{T}_5 &= \frac{T_2-T_5}{R}+\frac{T_4-T_5}{R}+\frac{T_7-T_5}{R}+\frac{T_a-T_5}{R_a}+Q_5 \\
C\dot{T}_6 &= \frac{T_3-T_6}{R}+\frac{T_4-T_6}{R}+\frac{T_7-T_6}{R}+\frac{T_a-T_6}{R_a}+Q_6 \\
C\dot{T}_7 &= \frac{T_4-T_7}{R}+\frac{T_5-T_7}{R}+\frac{T_6-T_7}{R}+\frac{T_a-T_7}{R_a}+Q_7
\end{aligned}
\tag{11}
$$

Where Q is the cell heat generation (W).

B. Parameters Identifications of Thermal Model

In order to identify the parameters of the equivalent circuit that represents the thermal model of the pack, a heater was used to model the cell heat generation, the cartridge heater was placed on one of the side sections (Section 1), in order to avoid a symmetry condition as shown in fig. (9). The heater was used to supply a measurable amount of heat generation. The waveform of the heat generation input signal used was pulses followed by relaxations. Thermocouples were placed on the surface of each cell (six cells) and the heater, to measure the temperature of each section. The heat generation Q was calculated from the electrical power supplied to the heater, the heater was excited from a DC power supply with measurable voltage and current. Therefore, it was assumed that the electrical power was fully converted to thermal power. The ambient temperature was measured using a thermocouple placed at a reasonable distance from the pack in the thermal chamber. Afterwards, the temperature measurements of the seven sections was recorded and plugged into the MATLAB-based parameters identification code along with the input signals of heat generation and ambient temperature as shown in fig. (10). The identification code combines a Simulink model of the differential equations governing the system behavior and a MATLAB code that performs non-linear least-square optimization for the identification process, in order to find the best estimates of the equivalent circuit parameters.

Fig. 10. Parameters identification procedure

There is no control over the temperature of the thermal chamber, as the chamber is used to create a stable thermal environment that stabilizes the airflow. The parameters are assumed to be constant within the region of sensible heat. The wax used in the PCC has a constant thermal properties at a range below 35°C. Since a heater is used in the place of one cell (Section 1), this section will have a different heat capacity (C_h), and a different conduction resistance with the neighboring sections (2,3,4), which is denoted as (R_1). In conclusion, this will increase the number of the parameters need to be identified to five. A single heating pulse of a constant value of (5W) was supplied to the battery pack through the heater for (14 minutes), followed by a relaxation for (2 hours). The temperature responses shown in fig. (11) were measured and plugged to the parameters identification code with the inputs. The identified parameters are listed in Table II.

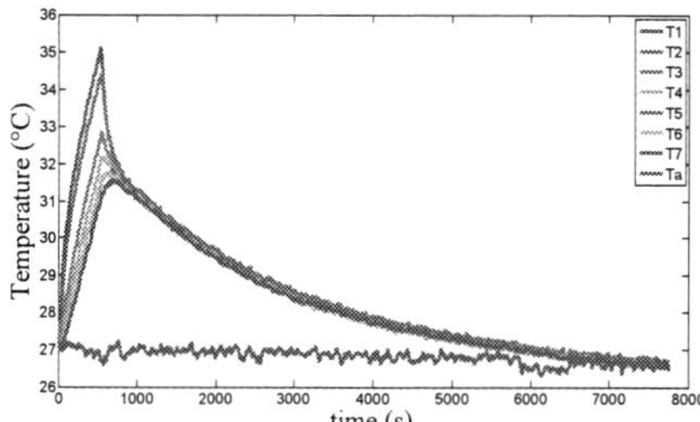

Fig. 11. Temperature responses in the seven sections

TABLE II. IDENTIFIED PARAMETERS OF THE BATTERY THERMAL MODEL

R (K/W)	C (J/K)	R_a (K/W)
3.77	74.1	39.7

VI. EXPERIMENTAL VALIDATION

A. Cell Electro-thermal Model

For the model validation, the Dynamic Stress Test (DST) was performed to compare the cell voltage estimated by the model with the measured experimental cell voltage. The DST current profile, shown in fig. (12), provides different current magnitudes, current directions and duty cycles. To account for different peak C-rates and different temperature of operation, the test was performed with peak discharging current of 2C and 3C at four different temperatures. As shown in fig. (13), the terminal voltage estimated by the model at (45°C, 2C peak) is close to the experimentally measured values, with an average absolute error of (0.29 %). As shown in Table III, the average error percentages varied between (0.29% - 1.01%). The error percentages were calculated for each measurement, by expressing the error as a fraction of the measured experimental value. As shown in (12):

$$\text{error }\% = \frac{\left|V_{\text{model}} - V_{\text{exp}}\right|}{V_{\text{exp}}} \times 100\% \tag{12}$$

On the other hand, the model accuracy deteriorates at low SoC (below 8%). This is due to the inability of the model to capture the behavior at this region of low SoC, which is considered as a limitation of the three RC pairs electrical model for the cell.

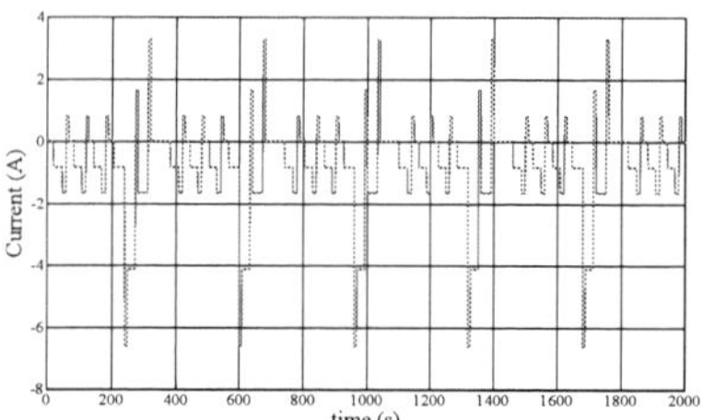

Fig. 12. DST Current Profile

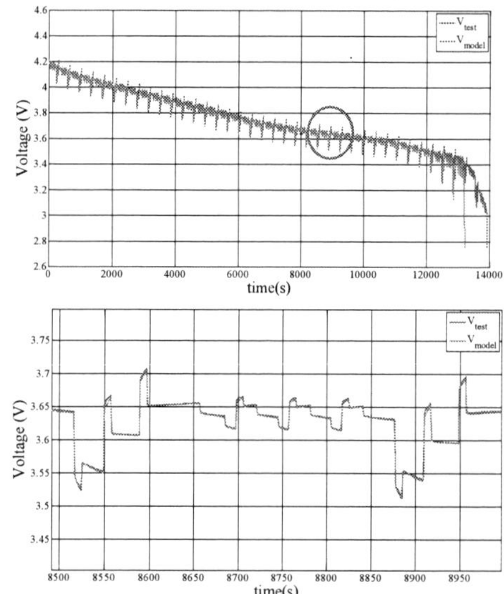

Fig. 13. Measured and model-estimated terminal voltage

TABLE III. CELL TERMINAL VOLTAGE ERROR % SUMMARY

T (°C)	10		25		35		45	
C-Rate	2	3	2	3	2	3	2	3
Average Error %	0.79	0.63	0.49	0.48	0.52	1.01	0.29	0.30

B. Battery Pack Thermal Model

The thermal model was validated by comparing the temperatures predicted by the model with the measured ones. A constant heating (1W) was applied at section 1 followed by a relaxation as shown in fig. (14). The temperatures predicted by the model were close to the measured ones, with a small error value, the resulting average error for the seven temperatures are summarized in table (IV). The average error was calculated using equation (13), where n is the number of measurements.

$$e_T = \frac{\sum\left|T_{\text{model}} - T_{Exp}\right|}{n} \tag{13}$$

TABLE IV. TEMPERATURE ESTIMATION AVERAGE ERROR SUMMARY

	T1	T2	T3	T4	T5	T6	T7
e_T (°C)	0.03	0.10	0.07	0.07	0.07	0.12	0.07

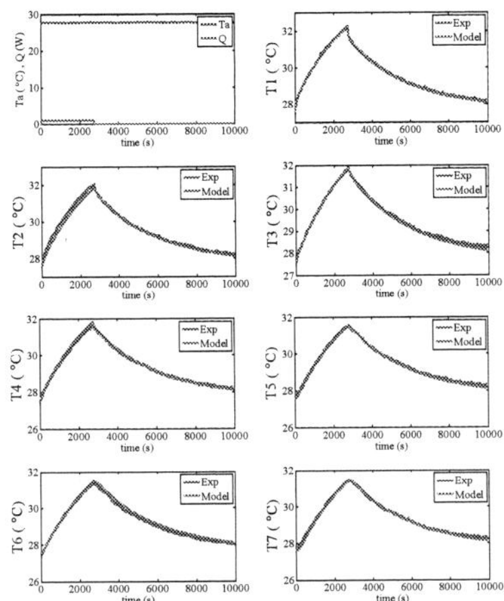

Fig. 14. Comparison between experimental and model estimated temperatures

VII. Conclusion

In this paper, an electro-thermal model for the cylindrical 18650 Li-ion cell has been developed, the electrical part of the model is presented with an equivalent circuit composed from three RC pairs, the circuit parameters were identified using Pulse-Relaxation test for different temperatures, SoC's and current directions. The heat generation of the cell has been calculated based on the ohmic and entropic heat generation. The coupling of the electrical and thermal model allows one to update the values of the electric circuit parameters based on the predicted temperature of the cell thermal model cell during operation. The cell model was validated using dynamic stress test, which demonstrated that the coupling of the three RC-pairs electrical model and the thermal model resulted in accurate prediction of the cell terminal voltage.

An analytical thermal model was developed for the PCC battery pack, a certain pack geometry was considered, which ensures scalability of the model for larger battery packs. The model takes into account the cell heat generation, ambient temperature, and heat transfer within the pack and to the surroundings. The parameters of the thermal model have been identified and validated with a single pulse relaxation test, the developed thermal model allowed accurate estimation of temperatures on the battery pack.

Both models demonstrate promising potential for future development and application, as they offer a trade-off between accuracy and complexity. For future work, the battery thermal model can be expanded to obtain a temperature mapping within the pack over a wider range of temperatures, and also to implement it on the battery management system to demonstrate the expected onboard capability.

References

[1] C. Sen and N. Kar, "Battery pack modeling for the analysis of battery management system of a hybrid electric vehicle", IEEE Vehicle Power and Propulsion Conference, 2009.

[2] B. Schweitzer, S. Wilke, S. Khateeb, S. Al-Hallaj, "Experimental validation of a 0-D numerical model for phase change thermal management systems in lithium-ion batteries", Journal of Power Sources, Vol. 287, August 2015.

[3] S. Bhide, "Novel Predictive Electric Li-Ion Battery Model Incorporating Thermal and Rate Factor Effects", IEEE Transactions on Vehicular Technology, Vol. 60, March 2011.

[4] A. Onda, H. Kameyama,T. Hanamotoand K. Ito, "Experimental Study on Heat Generation Behavior of Small Lithium-Ion Secondary Batteries" J. Electrochem.Soc. 2003, 150.

[5] R.B. Wright, J.P. Christophersen, C.G. Motloch, J.R. Belt, C.D. Ho, V.S. Battaglia, J.A. Barnes, T.Q. Duong, R.A. Sutula, "Power fade and capacity fade resulting from cycle-life testing of Advanced Technology Development Program lithium-ion batteries" J. PowerSources 119–121, 2003.

[6] S. Khateeb, S. Amiruddin, M. Farid, J.R.Selman, S. Al-Hallaj, "Thermal management of Li-ion battery with phase change material for electric scooters: experimental validation, Journal of Power Sources", Volume 142, Issues 1–2, 24 March 2005.

[7] Cheng Zhang, Kang Li, Sean Mcloone, Zhile Yang, " Battery Modeling Methods for Electric Vehicles – A Review", 2014 European Control Conference (ECC), 2014. Strasbourg, France

[8] M. Doyle, T. F. Fuller, and J. Newman, "Modeling of galvanostatic charge and discharge of the lithium/polymer/insertion cell," Journal of the Electrochemical Society, vol. 140, no. 6, 1993.

[9] T. F. Fuller, M. Doyle, and J. Newman, "Simulation and optimization of the dual lithium ion insertion cell," Journal of the Electrochemical Society, vol. 141, no. 1, pp. 1–10, 1994.

[10] D. Di Domenico, G. Fiengo, and A. Stefanopoulou, "Lithium-ion battery state of charge estimation with a kalman filter based on a electrochemical model," in Control Applications, CCA 2008.

[11] H. Perez, J. Siegel, X. Lin; A. Stefanopoulou; Y. Ding; M. Castanier, "Parameterization and validation of integrated electro-thermal cylindrical LFP battery model" , ASME 2012 5th Annual Dynamic Systems and Control Conference Joint with the JSME 2012 11th Motion and Vibration Conference, DSCC 2012-MOVIC 2012.

[12] L.Gao, S.Liu, R.Dougal, "Dynamic Lithium-Ion Battery Model for System Simulation", IEEE Transactions On Components And Packaging Technologies, Vol. 25, No. 3, September 2002.

[13] C. Mi, B. Li, D. Buck, and N. Ota, "Advanced Electro-Thermal Modeling of Lithium-ion Battery System for Hybrid Electric Vehicle Applications", Vehicle Power and Propulsion Conference, 2007, IEEE.

[14] X. Hu, S. Lin, S. Stanton, and W. Lian, "A Foster Network Thermal Model for HEV/EV Battery Modeling", IEEE Transactions On Industry Applications, Vol. 47, No. 4, July/August 2011.

[15] A. Pruteanu, B.V. Florean, G.M. Moraru, R.C. Ciobanu, "Development of a thermal simulation and testing model for a superior lithium-ion-polymer battery", Optim. Electr. Electron. Equip. Optim. IEEE - 2012.

[16] A. Smyshlyaev, M. Krstic, N. Chaturvedi, J. Ahmed, A. Kojic, "PDE model for thermal dynamics of a large Li-ion battery pack" , ACC, IEEE.

[17] X. Hu, S. Asgari, S. Lin, S. Stanton, W. Lian, "A linear parameter-varying modelfor HEV/EV battery thermal modeling", ECCE, IEEE, 2012 .

[18] X. Lin, H. Fu, H. Perez, J. Siegel, A. Stefanopoulou, Y. Ding, M. Castanier, "Parameterization and Observability Analysis of Scalable Battery Clusters for Onboard Thermal Management" Oil & Gas Science and Technology ,2012 IFPEN Energies nouvelles.

[19] N. Damay, C. Forgez, M. Bichat, G. Friedrich, "Thermal modeling of large prismatic LiFePO4/graphite battery. Coupled thermal and heat generation models for characterization and simulation", Journal of Power Sources, Volume 283, 2015.

[20] X. Lin, H. Perez, J. Siegel, A. Stefanopoulou, Y. Li, R. Anderson, Y.Ding and M. Castanier, "Online Parameterization of Lumped Thermal Dynamics in Cylindrical Lithium Ion Batteries for Core Temperature Estimation and Health Monitoring", IEEE Transactions on Control Systems Technology, 2013, Volume 21, Issue 5.

Modeling and Fault Diagnosis of Inter-turn Short Circuit for Five-phase PMSM based on Particle Swarm Optimization

Jianwei Yang, Manfeng Dou, Zhiyong Dai, Dongdong Zhao, Zhen Zhang
School of Automation
Northwestern Polytechnical University
Xi'an 710072, China
yangjianwei100@hotmail.com

***Abstract*—Multiphase PMSMs are widely used in fault-tolerant control applications. And one of the important fault-tolerant control problems is fault diagnosis. In most existing references, the fault diagnosis always focuses on three-phase PMSM. In this paper, a novel model and a fault diagnosis method of inter-turn short circuit (ITSC) fault for five-phase PMSM based on particle swarm optimization (PSO) are presented. There are two contributions in this paper: 1) By analyzing the physical parameters of the PMSM, such as resistance, inductance and flux, the mathematic model of ITSC fault for five-phase PMSM is established. 2) A parameter estimation method based on PSO algorithm, which introduces a fitness function related to the ITSC ratio and reformulates the fault parameters identification problem as the extreme seeking problem, is proposed to track the actual ITSC ratio and detect the fault level for five-phase PMSM. Simulation results provide preliminary verification of the proposed model and fault diagnosis method.**

***Keywords*—modeling; fault diagnosis; ITSC; five-phase; PMSM; PSO**

I. INTRODUCTION

The multiphase PMSMs have been gained more attention in high-power and high-reliability applications such as electric vehicles, aerospace, machine tools, robotics and wind power generation owing to high torque-to-current ratio, large power-to-weight ratio, high efficiency, high power factor, high fault and tolerance robustness [1]-[4]. Compared with the traditional three-phase PMSM, since the phase number is increased, the fault tolerance of the multiphase PMSM is improved, thus the reliability of the multiphase PMSM is higher. Based on these advantages, the multiphase PMSMs are applied in fault-tolerant control systems widely [5][6].

In many fault-tolerant control systems, the fault detection and diagnosis of electrical motor is the precondition for fault-tolerant control. In PMSM, there are three kinds of common faults which include magnetic faults, mechanical faults and electrical faults [7]. In electrical faults, the short circuit faults take account in 21% of the faults in electrical motors. And the commonest short circuit fault in PMSM is the stator winding inter-turn short circuit (ITSC) fault [8]-[10]. It always occurs due to insulation failures but develop into more serious faults very quickly [11]. So the research of effective fault detection and diagnosis methods for the multiphase PMSM with stator winding ITSC fault is meaningful.

To successfully detect and diagnosis the severity of the stator winding ITSC fault for multiphase PMSM, a mathematical model which is able to describe both the healthy and the ITSC fault state is needed firstly. However, the current fault model research of stator winding ITSC fault for multi-phase PMSM is limited [12]-[14]. The current fault model research of stator winding ITSC fault for PMSM is focused on three-phase PMSM and the references of stator winding ITSC fault model for multi-phase PMSM are less [15]. In [16] and [17], W. Liu discussed the basic electrical parameters of normal winding and shorted winding for three-phase PMSM, and established a three-phase PMSM model of ITSC fault. L. Romeral focused on the harmonic of the three-phase PMSM in the condition of stator winding ITSC fault, and divided the motor fault model into the malfunction portion and normal portion [18]. Unfortunately, these methods for ITSC fault modeling are all about three-phase PMSM, and relatively complex. If the fault model of five-phase PMSM with stator winding ITSC fault were modeled as the way before, the fault model would be very complex, the subsequent calculation for the fault detection and diagnosis based on parameter optimization would increase greatly and the efficiency of parameter optimization would be reduced. Therefore, it is important to build a relatively simple stator winding ITSC fault model of five-phase PMSM for fault detection and diagnosis.

After establishing the ITSC fault model, to detect the fault severity of the fault PMSM, the fault severity parameters should to be identified. However, due to the parameters' complex distribution in the fault model, the parameter identification is extremely difficult for nonlinear identification [19]. So the current fault diagnosis methods for the multiphase PMSM with ITSC fault based on parameter optimization is less and the existing method of ITSC fault diagnosis for PMSM are still focused on three-phase PMSM [20]. In [21], K. H. Kim diagnoses the ITSC circuit fault by monitoring the harmonic currents for three-phase PMSM. J. Quiroga proposed a method for ITSC fault diagnosis of three-phase PMSM based on the fuzzy logic decoupling analysis of negative sequence current [22]. J. Urresty analyzed the zero-sequence voltage component (ZSVC) of the rotor three-phase voltage to diagnose the ITSC

978-1-4673-9551-9/16 $31.00 © 2016 IEEE

fault for three-phase PMSM [23]. Thus, it is essential to search an effective fault parameter optimization method for five-phase PMSM fault diagnosis.

In this paper, a novel model and a fault diagnosis method based on the PSO algorithm are proposed for five-phase PMSM with stator winding ITSC fault. It first established a relative simple mathematical model of the five-phase PMSM under ITSC fault state by analyzing the physical parameters of the motor, such as resistance, inductance and flux. Furthermore, a parameter optimization method based on PSO algorithm for five-phase PMSM is proposed to track the actual ITSC ratio. The method introduces a fitness function related to the ITSC ratio and reformulates the fault parameters identification problem as the extreme seeking problem to detect the ITSC fault level for five-phase PMSM. Simulation results validate both the correction of the established models and the effectiveness of the fault diagnosis method.

II. ITSC Model Analysis

A. Five-phase PMSM Healthy Model

To establish the five-phase PMSM healthy model, without loss of generality, the assumptions are as follows. Firstly, the magnetic circuit is linear. It means that the magnetic circuit is not saturation. Secondly, the five-phase PMSM is non-salient pole structure. Thirdly, the rotor magneto motive force (MMF) is sinusoidal and the slot effect is neglected. Fourthly, the stator winding current is sinusoidal, symmetrical and without harmonics. The air gap MMF is sinusoidal. Lastly, the eddy currents and the hysteresis losses are negligible. By the assumptions above, the five-phase PMSM model is shown as:

$$\boldsymbol{V}_s = \boldsymbol{R}_s \boldsymbol{I}_s + p\boldsymbol{\psi}_s \tag{1}$$

$$\boldsymbol{\psi}_s = \boldsymbol{L}_s \boldsymbol{I}_s + \boldsymbol{\psi}_m \tag{2}$$

$$T_e = \frac{\partial W}{\partial \theta} = P\left[\frac{1}{2}\boldsymbol{I}_s^T \frac{\partial \boldsymbol{L}_s}{\partial \theta}\boldsymbol{I}_s + \boldsymbol{I}_s \frac{\partial \boldsymbol{\psi}_m}{\partial \theta}\right] \tag{3}$$

Where, (1) is the voltage balance equation, (2) is the flux equation and (3) is the torque equation. And the stator phase voltage vector $\boldsymbol{V}_s=[V_a\ V_b\ V_c\ V_d\ V_e]^T$; the stator phase current vector $\boldsymbol{I}_s=[i_a\ i_b\ i_c\ i_d\ i_e]^T$; the stator winding resistance $\boldsymbol{R}_s=R_s\times\boldsymbol{I}_{5\times5}$, the stator flux vector $\boldsymbol{\Psi}_s=[\Psi_a\ \Psi_b\ \Psi_c\ \Psi_d\ \Psi_e]^T$; the rotor flux vector $\boldsymbol{\Psi}_m=\Psi_m[\cos\theta\ \cos(\theta-\alpha)\ \cos(\theta-2\alpha)\ \cos(\theta-3\alpha)\ \cos(\theta-4\alpha)]^T$ (θ is the rotor electrical angle and α=72°); $\boldsymbol{L}_s$ is the stator inductance matrix; $p=d/dt$ is the differential operator. P is the number of pole pairs; and T_e is the electromagnetic torque.

Compared to traditional PMSM, for the reason of adding two phase windings, the stator inductance matrix $\boldsymbol{L}_s$ of five-phase PMSM is relatively complex and can be expressed as:

$$\boldsymbol{L}_s = \begin{bmatrix} L_m & M_{ab} & M_{ac} & M_{ad} & M_{ae} \\ M_{ab} & L_m & M_{bc} & M_{bd} & M_{be} \\ M_{ac} & M_{bc} & L_m & M_{cd} & M_{ce} \\ M_{ad} & M_{bd} & M_{cd} & L_m & M_{de} \\ M_{ae} & M_{be} & M_{ce} & M_{de} & L_m \end{bmatrix} \tag{4}$$

Where L_m is the self-inductance of phase winding A (B, C, D, E); and $M_{ab(c,d,e)}$ is the mutual-inductance between phase winding A and B (C, D, E). Actually, the mutual-inductances can be represented by: $M_{ab}=L_m\cos\alpha$, $M_{ac}=L_m\cos2\alpha$, $M_{ad}=L_m\cos3\alpha$, $M_{ae}=L_m\cos4\alpha$.

B. Five-phase PMSM ITSC Fault Model

Without loss of generality, assume that the stator winding of phase A occurs ITSC fault and the other phases are still in healthy state. The schematic of five-phase PMSM with ITSC is shown in Fig.1. It is noticed that a short circuit loop current i_f, which generates braking torque and affects the motor performance seriously, is produced in stator winding of phase A. Besides, the effective turns of the stator winding of phase A is reduced, and the value of other parameters, such as the stator winding resistance, the self-inductance, the mutual-inductance and the flux are all changed at the same time.

In the five-phase PMSM ITSC fault model, one of the most important parameters is the ITSC ratio μ, which is defined as the ratio of the shorted turns number to the total turns number. When the ITSC fault occurs in the stator winding of phase A, the equivalent circuit of five-phase PMSM with ITSC fault is shown in Fig.2, depending on the physical relationship of the stator windings. Where $R_{a(b,c,d,e)\_af}$ is the stator winding resistance of phase A (B, C, D, E); R_{sa_af} is the short circuit winding resistance of phase A; $L_{a(b,c,d,e)\_af}$ is the self-inductance of phase A (B, C, D, E); L_{sa_af} is the self-inductance of the short circuit winding of phase A; L_m is the self-inductance of other phase winding under healthy state. Actually, the resistances and the inductances can be represented by $R_{a\_af}=(1-\mu)R_s$, $R_{b\_af}=R_{c\_af}=R_{d\_af}=R_{e\_af}=R_s$, $R_{sa\_af}=\mu R_s$, $L_{a\_af}=(1-\mu)L_m$, $L_{sa\_af}=\mu^2L_m$, $L_{b\_af}=L_{c\_af}=L_{d\_af}=L_{e\_af}=L_m$. Besides, there are three types of mutual-inductances in the fault model. The first is the mutual-inductance $M_{ab(c,d,e)\_af}$ between the remaining healthy winding of phase A and other healthy windings of phase B, C, D and E, where $M_{ab(c,d,e)\_af}=(1-\mu)M_{ab(c,d,e)}$. The second is the mutual-inductance M_{asa_af} between the remaining healthy winding of phase A and the short circuit winding of phase A, where $M_{asa\_af}=(1-\mu)\mu L_m$. The third is the mutual-inductance $M_{sab(c,d,e)\_af}$ between the short circuit winding of phase A and other healthy windings of phase B, C, D and E, where $M_{sab(c,d,e)\_af}=\mu M_{ab(c,d,e)}$. Because the stator windings of phase B, C, D and E are still healthy, the mutual-inductances between these phases remain the values in healthy state. Moreover, when the stator winding of phase A occurs ITSC fault, the flux linkage can be expressed

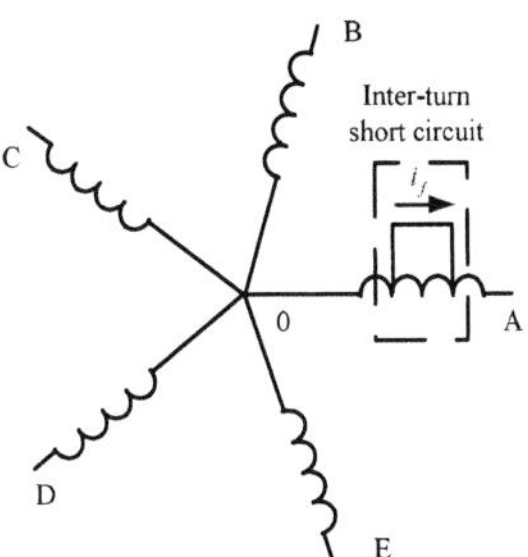

Fig.1 The schematic of five-phase PMSM with ITSC fault

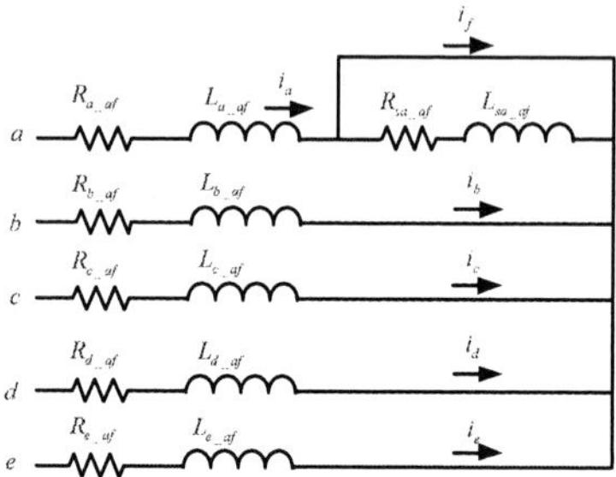

Fig.2 The equivalent circuit of five-phase PMSM with ITSC fault

as: $\Psi_{a\_af}=(1-\mu)\Psi_m\cos\theta$, $\Psi_{b\_af}=\Psi_m\cos(\theta-\alpha)$, $\Psi_{c\_af}=\Psi_m\cos(\theta-2\alpha)$, $\Psi_{d\_af}=\Psi_m\cos(\theta-3\alpha)$, $\Psi_{e\_af}=\Psi_m\cos(\theta-4\alpha)$, $\Psi_{sa\_af}=\mu\Psi_m\cos\theta$. Based on the equivalent circuit and the analysis above, the voltage balance equation of five-phase PMSM with ITSC fault can be derived as:

$$
\begin{cases}
V_a = R_{a\_af} i_a + L_{a\_af}\dfrac{di_a}{dt} + M_{ab\_af}\dfrac{di_b}{dt} + M_{ac\_af}\dfrac{di_c}{dt} + \\
\quad M_{ad\_af}\dfrac{di_d}{dt} + M_{ae\_af}\dfrac{di_e}{dt} + M_{asa\_af}\dfrac{d(i_a-i_f)}{dt} + \dfrac{d\psi_{a\_af}}{dt} \\
V_b = R_{b\_af} i_b + L_{b\_af}\dfrac{di_b}{dt} + M_{ab\_af}\dfrac{di_a}{dt} + M_{bc\_af}\dfrac{di_c}{dt} + \\
\quad M_{bd\_af}\dfrac{di_d}{dt} + M_{be\_af}\dfrac{di_e}{dt} + M_{bsa\_af}\dfrac{d(i_a-i_f)}{dt} + \dfrac{d\psi_{b\_af}}{dt} \\
V_c = R_{c\_af} i_c + L_{c\_af}\dfrac{di_c}{dt} + M_{ac\_af}\dfrac{di_a}{dt} + M_{bc\_af}\dfrac{di_b}{dt} + \\
\quad M_{cd\_af}\dfrac{di_d}{dt} + M_{ce\_af}\dfrac{di_e}{dt} + M_{csa\_af}\dfrac{d(i_a-i_f)}{dt} + \dfrac{d\psi_{c\_af}}{dt} \\
V_d = R_{d\_af} i_d + L_{d\_af}\dfrac{di_d}{dt} + M_{ad\_af}\dfrac{di_a}{dt} + M_{bd\_af}\dfrac{di_b}{dt} + \\
\quad M_{cd\_af}\dfrac{di_c}{dt} + M_{de\_af}\dfrac{di_e}{dt} + M_{dsa\_af}\dfrac{d(i_a-i_f)}{dt} + \dfrac{d\psi_{d\_af}}{dt} \\
V_e = R_{e\_af} i_e + L_{e\_af}\dfrac{di_e}{dt} + M_{ae\_af}\dfrac{di_a}{dt} + M_{be\_af}\dfrac{di_b}{dt} + \\
\quad M_{ce\_af}\dfrac{di_c}{dt} + M_{de\_af}\dfrac{di_d}{dt} + M_{esa\_af}\dfrac{d(i_a-i_f)}{dt} + \dfrac{d\psi_{e\_af}}{dt} \\
0 = R_{sa\_af}(i_a-i_f) + L_{sa\_af}\dfrac{d(i_a-i_f)}{dt} + M_{asa\_af}\dfrac{di_a}{dt} + M_{sab\_af}\dfrac{di_b}{dt} + \\
\quad M_{sac\_af}\dfrac{di_c}{dt} + M_{sad\_af}\dfrac{di_d}{dt} + M_{sae\_af}\dfrac{di_e}{dt} + \dfrac{d\psi_{sa\_af}}{dt}
\end{cases} \tag{5}
$$

Equation (5) can be rewritten as:

$$
\boldsymbol{V}_{af} = \boldsymbol{R}_{af}\boldsymbol{I}_{af} + \boldsymbol{L}_{af}\frac{d\boldsymbol{I}_{af}}{dt} + \frac{d\boldsymbol{\psi}_{af}}{dt} \tag{6}
$$

where

$$
\boldsymbol{V}_{af} = \begin{bmatrix} V_a & V_b & V_c & V_d & V_e & 0 \end{bmatrix}^T \tag{7}
$$

$$
\boldsymbol{I}_{af} = \begin{bmatrix} i_a & i_b & i_c & i_d & i_e & i_f \end{bmatrix}^T \tag{8}
$$

$$
\boldsymbol{R}_{af} = \begin{bmatrix}
R_s & 0 & 0 & 0 & 0 & -\mu R_s \\
0 & R_s & 0 & 0 & 0 & 0 \\
0 & 0 & R_s & 0 & 0 & 0 \\
0 & 0 & 0 & R_s & 0 & 0 \\
0 & 0 & 0 & 0 & R_s & 0 \\
\mu R_s & 0 & 0 & 0 & 0 & -\mu R_s
\end{bmatrix} \tag{9}
$$

$$
\boldsymbol{L}_{af} = \begin{bmatrix}
L_m & M_{ab} & M_{ac} & M_{ad} & M_{ae} & -\mu L_m \\
M_{ab} & L_m & M_{ab} & M_{ac} & M_{ad} & -\mu M_{ab} \\
M_{ac} & M_{ab} & L_m & M_{ab} & M_{ac} & -\mu M_{ac} \\
M_{ad} & M_{ac} & M_{ab} & L_m & M_{ab} & -\mu M_{ad} \\
M_{ae} & M_{ad} & M_{ac} & M_{ab} & L_m & -\mu M_{ae} \\
\mu L_m & \mu M_{ab} & \mu M_{ac} & \mu M_{ad} & \mu M_{ae} & -\mu^2 L_m
\end{bmatrix} \tag{10}
$$

$$
\begin{aligned}
\psi_{af} = \psi_m[&\cos\theta \ \cos(\theta-\alpha) \ \cos(\theta-2\alpha) \\
&\cos(\theta-3\alpha) \ \cos(\theta-4\alpha) \ \mu\cos\theta]^T
\end{aligned} \tag{11}
$$

The electromagnetic torque equation of five-phase PMSM with ITSC fault is:

$$
\begin{aligned}
T_e &= \frac{\partial W}{\partial\theta} = p[\frac{1}{2}\boldsymbol{I}_{af}^T\frac{\partial\boldsymbol{L}_{af}}{\partial\theta}\boldsymbol{I}_{af} + \boldsymbol{I}_{af}\frac{\partial\psi_{af}}{\partial\theta}] \\
&= p[\boldsymbol{I}_{af}\frac{\partial\psi_{af}}{\partial\theta}]
\end{aligned} \tag{12}
$$

And the mechanical motion equation of five-phase PMSM with ITSC fault is still as follow:

$$
T_e - T_L = Jd\omega/dt + B\omega \tag{13}
$$

Where, T_e is the electromagnetic torque; T_L is the load torque; J is the rotational inertia; B is the viscous friction coefficient; ω is the mechanical angular velocity.

III. ITSC Fault Diagnosis

A. PSO Algorithm

Particle Swarm Optimization (PSO) is an intelligence evolutionary algorithm proposed by Dr. Kennedy and Dr. Eberhart in 1995 [24]. The PSO algorithm is started with a random initial position and velocity for each particle in the population [25]. To find the optimal solution, the velocity vector, which each particle flies over the search space with, is dynamically adjusted according to the historical behavior of each particle itself and the neighbors surrounding it [26]. In this way, the particles will get the new position vector, and have the better tendency to fly towards the optimal solution in the searching areas [27]. The extreme searching process is governed by (14) and (15).

$$
\boldsymbol{v}_i(k+1) = w\boldsymbol{v}_i(k) + c_1r_1[\boldsymbol{p}_i - \boldsymbol{x}_i(k)] + c_2r_2[\boldsymbol{p}_g - \boldsymbol{x}_i(k)] \tag{14}
$$

$$
\boldsymbol{x}_i(k+1) = \boldsymbol{x}_i(k) + \boldsymbol{v}_i(k+1) \tag{15}
$$

Where, $\boldsymbol{x}_i$ is the position vector of the ith particle; $\boldsymbol{v}_i$ is the velocity vector for particle $\boldsymbol{x}_i$, $\boldsymbol{p}_i$ is the best previous position of $\boldsymbol{x}_i$; $\boldsymbol{p}_g$ is the best particle among the entire population; c_1 and c_2 are positive acceleration coefficients; $\boldsymbol{w}_i$ is the inertia weight factor; r_1 and r_2 are random functions in the range of [0,1]. Velocity changes in (14) comprise three parts [25]. The first part represents the inertia of the previous velocity and is called as the momentum part, the second part is useful to personal thinking of the particle and is named as the "cognitive" part, and the third part represents the cooperation among particles and is therefore named as the social part [26]. The current position of each particle is then evaluated according to (15), which produces a better position in the solution space [27].

B. *Fault Parameter Identification*

To detect and diagnose the fault severity of the fault five-phase PMSM, the ITSC ratio μ needs to be identified based on the five-phase PMSM ITSC model. Because the parameter in the ITSC fault model is complex distribution, it is extremely difficult to do parameter nonlinear identification. To deal with the problem, firstly, reformulate the fault parameters identification problem as the extreme seeking problem, and then solved it by PSO algorithm. Because the PSO algorithm is relatively simple and the computational efficiency is relatively high, it has been widely used in extreme optimization problem. The PSO algorithm has been applied in many areas for parameter identification. And the PSO algorithm's advantage is that it can be applied without the limit of model structures, like many traditional parameter identification algorithms. As long as the models perform differently with different parameters, the PSO algorithm can achieve the unknown parameter identification in the models.

The principle of fault parameter (the ITSC ratio μ) identification for five-phase PMSM with ITSC fault is as Fig.3 shown. First of all, give an arbitrary constant $\hat{\mu}(0)\in(0,1)$ as the initial estimation value of the ITSC ratio for the five-phase PMSM fault model. At the same time, sample the drive voltage and the phase currents from the actual five-phase fault PMSM. And next, a reference model of the five-phase PMSM with ITSC fault is built as (5), and the drive voltage of the reference model is the same as the actual PMSM. After then, in order to detect and diagnose the fault severity of the five-phase PMSM with ITSC fault, or to identify the fault parameter (the ITSC ratio μ) by PSO algorithm, a quadratic fitness function is utilized as

$$f(\hat{\mu})=\sum_{n=1}^{5}(i_n-i_n^*)^2 \tag{16}$$

where i is the phase current of the actual PMSM, and i^* is the phase current of reference model. Actually, the fitness function is a function which related to the ITSC ratio μ and its estimation $\hat{\mu}$. It can be proved that the fitness function is equal to zero if and only the actual ITSC ratio μ is equal to its estimation $\hat{\mu}$. The last step is using the PSO algorithm to achieve the extreme seeking of ITSC ratio estimation value $\hat{\mu}$ such that the fitness function reaches the minimum value.

IV. Simulation Verification

In order to verify the correctness of the proposed five-phase PMSM model with ITSC fault and the PSO algorithm in the process of the fault parameter identification, the simulation was done by Matlab/Simulink. In the simulation, the proposed model of five-phase PMSM with ITSC fault is simulated with different ITSC ratio. And to compare with the healthy state, the five-phase PMSM healthy model is also simulated. The effect of PSO fault diagnosis method for five-phase PMSM with ITSC fault is also demonstrated in the simulation. The parameters of five-phase PMSM in the simulation are shown in Table 1.

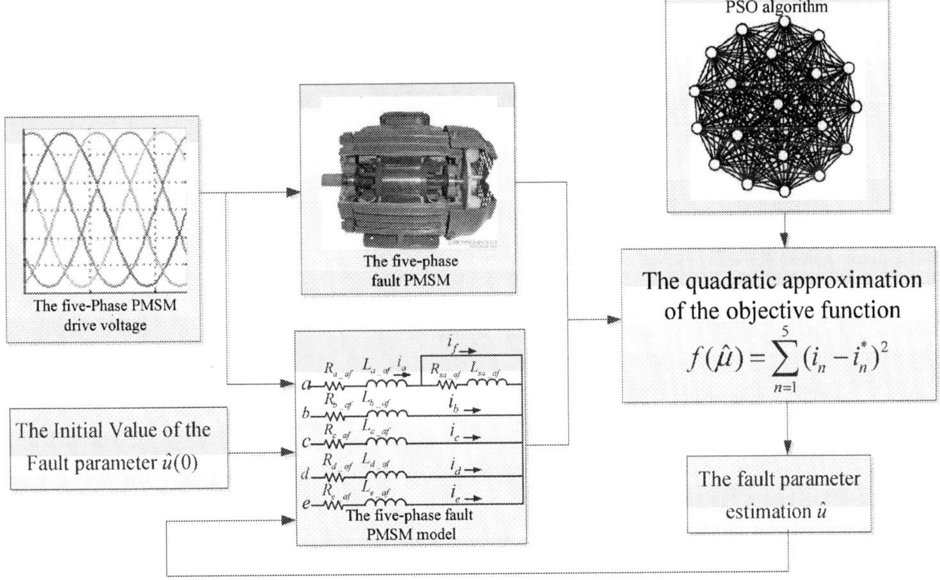

Fig.3. ITSC fault parameter identification for five-phase PMSM based on PSO algorithm

TABLE I. PARAMETERS OF FIVE-PHASE PMSM USED FOR SIMULATION

Parameter	Symbol	Value
Stator winding resistance	R_s	17.4Ω
Self-inductance of phase winding	L_m	4.5e-2H
Flux amplitude	Ψ_m	0.1827Wb
Moment of inertia	J	0.8e-6kgm2
Friction constant	B	1e-3Nms
Number of pole pairs	P	4
Load torque	T_L	0.4Nm
Input sinusoidal voltage amplitude	V	28V
Input voltage frequency	f	20Hz

The Fig.4 shows the phase current waveforms of the five-phase PMSM in healthy state. The Fig.5 to Fig.7 are the phase current waveforms of the five-phase PMSM in ITSC fault state, and the ITSC ratio is 20%, 40%, 50% respectively. It is noticed that the current of phase A (the yellow line) in fault state from Fig.5 to Fig.7 is significantly larger than the healthy ones in Fig.4. And with the increase of the fault parameter μ (the ITSC ratio), the current of phase A in fault state also increases. And the peak current value of the fault phase A is increased from 1.2A to 2A. It is also seen clearly in Fig.7 that the current waveforms of other phases are also affected by the fault phase A when the ITSC ratio increases to 50%. The simulation results from Fig.4 to Fig.7 verified of the correctness of the proposed model for five-phase PMSM under ITSC fault. The Fig.8 is the fitness function variations with the generation numbers under 20% ITSC fault of phase A. It can be seen that within 10 generations and four particles in each generation, the PSO algorithm finishes the optimization process and the fitness function is approached zero with sufficiently small error. The estimation value of fault parameter $\hat{\mu}$ changing with the generation number is shown in Fig.9. As can be seen from the curve, the error of final fault parameter estimation result under 20% ITSC fault in phase A is less than ±0.02%. The simulation results from Fig8 and Fig.9 demonstrated the feasibility of the PSO fault diagnosis method for five-phase PMSM with ITSC fault.

V. CONCLUSION

In this paper, a novel model and an ITSC fault diagnosis method based on the PSO algorithm are proposed for five-phase PMSM. It first established a relative simple mathematical model of the five-phase PMSM with ITSC fault state by analyzing the physical parameters of the motor, such as resistance, inductance and flux. Furthermore, a parameter optimization method based on PSO algorithm is proposed for five-phase PMSM with ITSC fault to track the actual ITSC ratio. The method introduces a fitness function related to the ITSC ratio and reformulates the fault parameters identification problem as the extreme seeking problem to detect the ITSC fault level for five-phase PMSM. Simulation results preliminarily validate both the correction of the established models and the effectiveness of the fault diagnosis method.

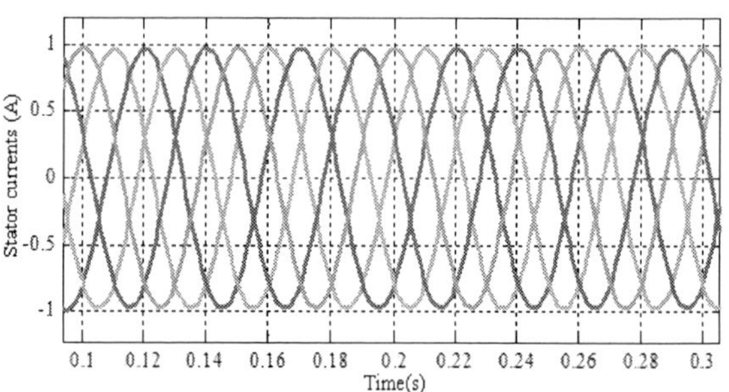

Fig.4. Five-phase stator current waveforms under healthy condition

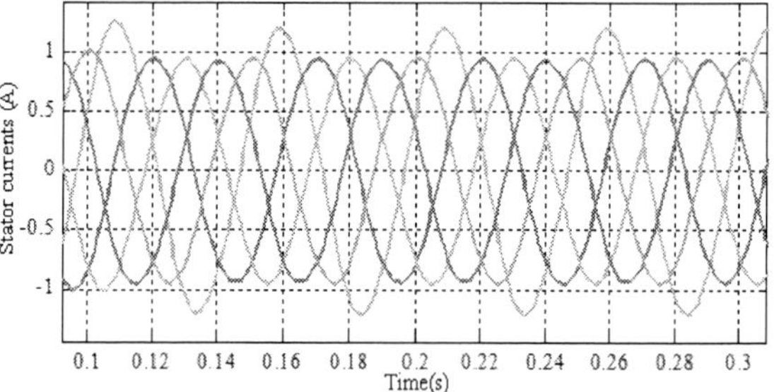

Fig.5. Five-phase stator current waveforms under 20% ITSC fault in phase A

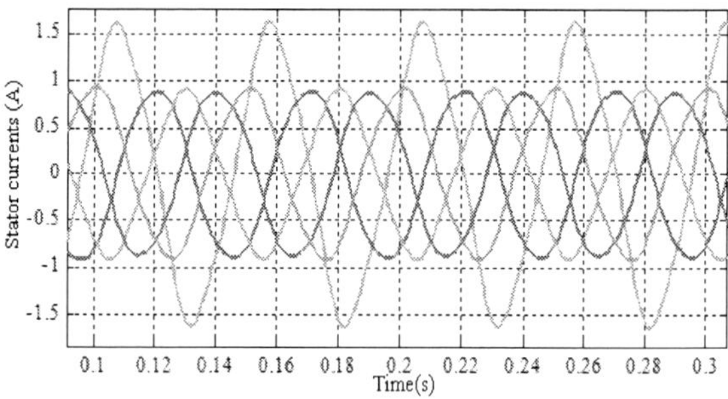

Fig.6. Five-phase stator current waveforms under 40% ITSC fault in phase A

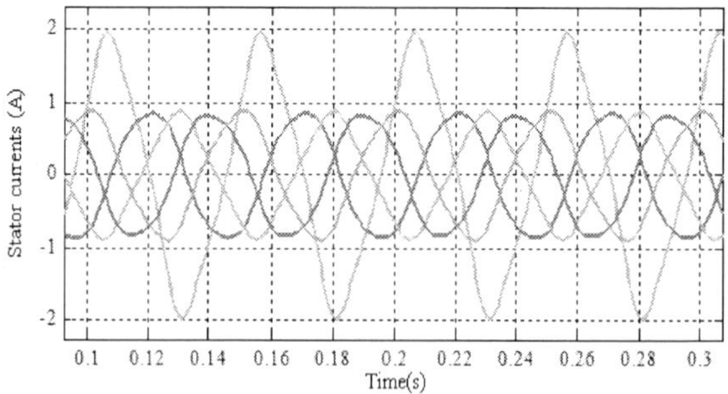

Fig.7. Five-phase stator current waveforms under 50% ITSC fault in phase A

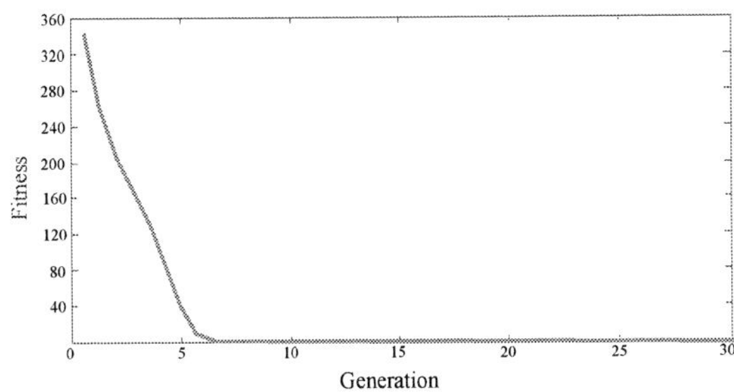

Fig.8. Fitness function curve in optimization process under 20% ITSC fault in phase A

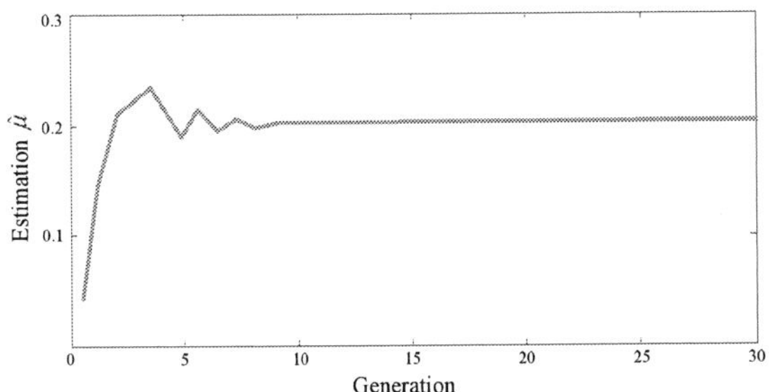

Fig.9 Fault parameter estimation result under 20% ITSC fault in phase A

REFERENCES

[1] K. D. Hoang, Y. Ren, Z. Q. Zhu and M. Foster, "Modified switching-table strategy for reduction of current harmonics in direct torque controlled dual-three-phase permanent magnet synchronous machine drives," *IET Elect. Power Appl.*, vol. 9, no. 1, pp. 10-19, Jan. 2015.

[2] R. Islam, M. Islam, J. Tersigni, and T. Sebastia, "Inter winding short circuit faults in permanent magnet synchronous motors used for high performance applications," in *IEEE Energy Conversion Congress and Exposition (ECCE)*, 2012, pp. 1291-1298.

[3] G. Vinson, M. Combacau, T. Prado, and P. Ribot, "Permanent magnets synchronous machines faults detection and identification," in *IECON 38th Annual Conference on IEEE Industrial Electronics Society*, 2012, pp. 3925-3930.

[4] F. Scuiller, E. Semail, J.-F. Charpentier, and P. Letellier, "Multicriteria based design approach of multiphase permanent magnet low speed synchronous machines," *IET Electr. Power Appl.*, vol. 3, no. 2, pp. 102-110, Mar. 2009.

[5] H. Saavedra, J. R. Riba, and L. Romeral, "Inter-turn fault detection in five-phase pmsms. Effects of the fault severity," in *9th IEEE International Symposium on Diagnostics for Electric Machines, Power Electronics and Drives (SDEMPED)*, 2013, pp. 520-526.

[6] N. Leboeuf, T. Boileau, B. Nahid-Mobarakeh, N. Takorabet, F. Meibody-Tabar, and G. Clerc, "Effects of imperfect manufacturing process on electromagnetic performance and online interturn fault detection in pmsms," *IEEE Trans. Ind. Electron.*, vol. 62, no. 6, pp. 3388–13398, Jun. 2015.

[7] Z. Wang, J. Yang, H. Ye, and W. Zhou, "A review of permanent magnet synchronous motor fault diagnosis," in *IEEE Conference and Expo. Transportation Electrification Asia-Pacific (ITEC Asia-Pacific)*, 2014, pp. 1-5.

[8] Y. Liang, "Diagnosis of inter-turn short-circuit stator winding fault in PMSM based on stator current and noise," in *IEEE 2014 International Conference on Industrial Technology*, 2014, pp. 138-142.

[9] C. Zhang, F. Wang and Z. Wang, "Analysis of stator winding inter-turn short circuit fault of PMSM for electric vehicle based on finite element simulation," in *IEEE 2014 Conference and Expo Transportation Electrification Asia-Pacific*, 2014, pp. 1-6.

[10] T. Boileau, N. Leboeuf, B. Nahid-Mobarakeh, and F. Meibody-Tabar, "Stator winding fault detection using control voltages demodulation," in *IEEE 2012 Transportation Electrification Conference and Expo*, 2012, pp. 1-6.

[11] A. Gandhi, T. Corrigan, and L. Parsa, "Recent advances in modeling and online detection of stator interturn faults in electrical motors," *IEEE Trans. Ind. Electron.*, vol. 58, no. 5, pp. 1564–1575, May. 2011.

[12] M. A. Shamsi Nejad and M. Taghipour, "Inter-turn stator winding fault diagnosis and determination of fault percent in PMSM," in Proc. *IEEE Appl. Power Electron. Conf. Expo.*, 2011, pp. 128–131.

[13] N. H. Obeid, T. Boileau, and B. Nahid-Mobarakeh, "Modeling and diagnostic of incipient inter-turn faults for a three phase permanent magnet synchronous motor," in *IEEE Industry Applications Society Annual Meeting*, 2014, pp. 1-8.

[14] B. G. Gu, J. H. Choi, and I. S. Jung, "A dynamic modeling and a fault detection scheme of a PMSM under an inter turn short," in *IEEE Vehicle Power and Propulsion Conference*, 2012, pp. 1074-1080.

[15] M. Eftekhari, M. Moallem, S. Sadri, and A. Shojaei, "Review of induction motor testing and monitoring methods for inter-turn stator winding faults," in *21st Iranian Conference Electrical Engineering (ICEE)*, 2013, pp. 1–6.

[16] W. Liu, L. Liu, I. Y. Chung, D. A. Cartes, and W. Zhang, "Modeling and detecting the stator winding fault of permanent magnet synchronous motors," *Simulation Modelling Practice and Theory*, vol. 27, pp. 1–16, 2012.

[17] L. Liu, W. Liu, and D. A. Cartes, "Particle swarm optimization based parameter identification applied to permanent magnetic synchronous machine," *Eng. Appl. Artif. Intell.*, vol. 21, no. 7, 2008, pp. 1092–1100.

[18] L. Romeral, J. C. Urresty, J.-R. R. Ruiz, and A. G. Espinosa, "Modeling of surface-mounted permanent magnet synchronous motors with stator winding interturn faults," *IEEE Trans. Ind. Electron.*, vol. 58, no. 5, pp. 1576–1585, May 2011.

[19] B. Park, R. Kim, and D. Hyun, "Fault diagnosis using recursive least squares algorithm for permanent magnet synchronous motor drives," in *Proc. 8th Int. Conf. Power Electron.*, 2011, pp. 2506–2510.

[20] D. Yao, and H. Toliyat, "A review of condition monitoring and fault diagnosis for permanent magnet machines," in *IEEE Power & Energy Society General Meeting*, 2012, pp. 1-4.

[21] K. H. Kim, "Simple online fault detecting scheme for short-circuited turn in a PMSM through current harmonic monitoring," *IEEE Trans. Ind. Electron.*, vol. 58, no. 6, pp. 2565–2568, Jun. 2011.

[22] J. Quiroga, L. Liu, and D. A. Cartes, "Fuzzy logic based fault detection of PMSM stator winding short under load fluctuation using negative sequence analysis," in *IEEE Amer. Control Conf.*, 2008, pp. 11-13.

[23] J. Urresty, J. Riba, and L. Romeral, "Diagnosis of interturn faults in PMSMs operating under non stationary conditions by applying order tracking filtering," *IEEE Trans. Power Electron.*, vol. 28, no. 1, pp. 507-515, Jan. 2013.

[24] J. Kennedy, and R. Eberhart, "Particle swarm optimization," in *Proc. IEEE International Conference on Neural Networks*, 1995, pp. 1942-1948.

[25] F. Grouz, L. Sbita, and M. Boussak, "Modelling for non-salient PMSM with multi-phase inter-turn short circuit," in *10th IEEE International Multi-Conference on Systems, Signals & Devices (SSD)*, 2013, pp. 1-6.

[26] F. Grouz, L. Sbita and M. Boussak, "Particle swarm optimization based fault diagnosis for non-salient PMSM with multi-phase inter-turn short circuit," in *2nd International Conference on Communications, Computing and Control Applications*, 2012, pp. 1-6.

[27] Y. del Valle, G. K. Venayagamoorthy, S. Mohagheghi, J. Hernandez, and R. G. Harley, "Particle swarm optimization: basic concepts, variants and applications in power systems," *IEEE Trans. Evol. Computat.*, vol. 12, no. 2, 2008, pp. 171–195.

Comprehensive Modeling, Testing, and Experimental Validation of Ultracapacitor Open Circuit Voltage Characteristics

Amandeep Singh, *Student Member, IEEE*, Najath Abdul Azeez, *Member, IEEE*, and Sheldon S. Williamson, *Senior Member, IEEE*

Smart Transportation Electrification and Energy Research (STEER) Group
Advanced Storage Systems and Electric Transportation (ASSET) Laboratory
UOIT-Automotive Center of Excellence (UOIT-ACE)
Department of Electrical, Computer, and Software Engineering
Faculty of Engineering and Applied Science
University of Ontario-Institute of Technology
ACE-2025, 2000 Simcoe Street North
Oshawa, ON L1H 7K4, Canada
Tel: +1 (905) 721-8668, ext. 5744
Fax: +1 (905) 721-3178
EML: sheldon.williamson@uoit.ca
URL: http://www.engineering.uoit.ca/; http://ace.uoit.ca/

***Abstract*—Ultracapacitors (UCs) are the significant component of the energy storage systems (ESS) in the electric traction applications. In the field of electric transportation the UCs need to be modeled dynamically in order to capture real-time characteristics. The model should provide precise estimation of UC responses to various charge/discharge conditions and also predict the behavior of UCs under varied load conditions. In order to incorporate the model for testing purposes in various power electronics/drive simulation toolboxes, it is very important for the model to be accurate. This paper aims at the experimental validation of an accurate parameter identification procedure for an UC model using random charging/discharging current pulses of different magnitudes. The UC model can estimate the output voltage without the extraction of the model's branch parameters. The critical parameters have been estimated dynamically from characteristic equations. The estimation technique for an equivalent *RC* branch model is simulated and the output voltage response, controlled between V_{min} (0V) and V_{rated} (2.7V) of a single UC cell, has been presented. Finally, the simulation results obtained from the identification technique are validated against the actual charging results obtained from applying charging pulses to a 1200F UC.**

***Keywords*—Circuit modeling, energy storage, power electronics, electric transportation, supercapacitors.**

I. Introduction

The power peaking devices, such as ultracapacitors (UCs) will be extensively used by the electric vehicles (EVs) of the future as well as the mass electric transportation industry (electric buses, trains, and trams) [1]-[6]. At present Li-ion battery is the most common EV energy storage device. Unfortunately, batteries suffer from critical drawbacks, such as short shelf/cycle life, low power density, and long charging times. Thus, electric mass transit vehicles do not make efficient use of batteries. Hence, there is a need to replace batteries with UCs as the storage medium in such high-power cyclic applications, which need short bursts of power (not energy). UCs can also be used as a power peaking device, in parallel with batteries. UC cells with capacitance values as high as 3400F are now available commercially (Maxwell K2® UCs) [7].

Related literature proves the feasibility of hybrid EV energy storage systems (ESS), using combined battery/UC energy/power storage systems. An on-board hybrid ESS consisting of a battery pack, accompanied by the UC, is used to store energy recovered from kinetic energy, and to provide traction power to the motor, including peak power demands (or peak braking versus steady braking, during regenerative braking events). Thus, the overall ESS achieves larger charge/discharge cycles, while at the same time, battery life can be theoretically doubled. Recent studies show that it is advantageous to use an UC in place of a battery, especially when roundtrip efficiency is concerned [8]. Table 1 shows the advantages of UCs over state of the art batteries in terms of characteristics.

Table I. Comparison of characteristics of Li-ion batteries and electrochemical capacitors [9].

Characteristic	Lithium Ion Battery (24kWh 360V)	Electrochemical Capacitor (650F)
Charge time	5 hours @ level 1 charging (16A)	50 seconds @ 50A charging current
Cycle life	<5000	>500,000
Specific Energy (Wh/kg)	140	5
Specific Power (kW/kg)	2.5	Up to 10
Cycle efficiency (%)	From 50% to 90%	From 75% to 95%
Cost/Wh	$1 to 2/Wh	$10 to 20/Wh
Cost/kW	$75 to 150/kW	$25 to 50/kW

In order to study the behavior of an UC for EV traction power applications, it is critical to build an accurate and

978-1-4673-9551-9/16 $31.00 © 2016 IEEE

dynamic model, which can correctly estimate cell voltage variations. The equivalent *RC*-branch cell models are popular and can reasonably estimate output voltage response of a UC up to an adequate time period [10].

II. Ultracapacitor Structure

UCs store energy in the form of electric field (electrostatically), by essentially polarizing the electrolytic solution [11], [12]. A separator is placed between the activated carbon electrodes, which also acts as an insulator. The electrode material employed in the UCs has a porous structure, Activated carbon having all the required properties is widely used as the electrode material [13].

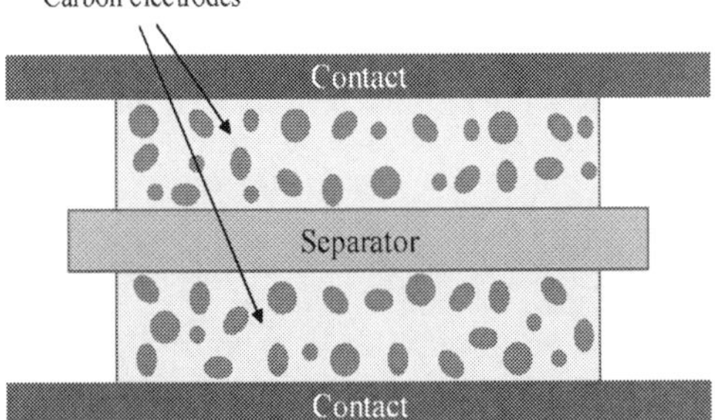

Fig. 1. Simplified structure of a UC.

The simplified structure of a UC is shown in Fig. 1. Activated carbon, which has a high surface area, is used, because it also has near ideal charge/discharge characteristics [14]. As a result, without having to increase the volume of the UC, the contact area of the electrodes, and consequently, the capacitance of the UC, is also increased. The transfer of ions takes place even if there is no physical contact between the electrodes [15], [16].

It is eminent that the capacitance of a capacitor directly depends on the surface area of the interface between electrode and electrolyte. Hence, if the surface area is increased, a large capacitance value can be obtained. Activated carbon, as a porous electrode material, has a specific surface area of a few thousands m^2/gram. The porous structure of the activated carbon electrode has pores of varying sizes: macropores, mesopores, micropores, and sub-micropores [17].

III. UC Model for EV Traction

It is proven in related literature that current profiles for UCs employed in EVs are composed of recurrent charge and discharge cycles [18]-[20]. An *RC* equivalent UC model, with two *RC* branches connected in parallel, for EV traction is shown in Fig. 2. This model is able to capture the dynamic behavior of several UCs for sufficient amount of time, with different capacities and voltage ratings. The choice of two branches is based on the fact that the estimation from two branches gives a correct approximation for a sufficient amount of time.

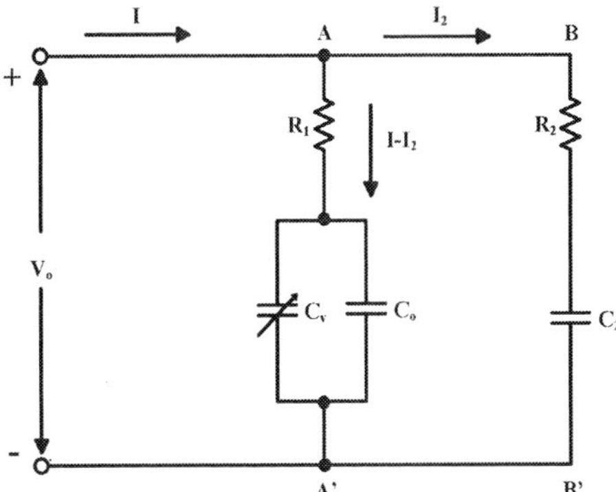

Fig. 2. *RC* equivalent model for EV applications.

The first branch which consists of the resistor R_1 and the parallel combination of C_v and C_o i.e. (C_1) is the fast branch, where C_v is the voltage dependent capacitance. This branch is responsible for the immediate behavior of the UC (charging/discharging) over a duration of seconds. This models the physical behavior of the interaction of the ions with the macropores and mesopores of the electrode surface. The other branch, with resistor R_2 and capacitor C_2, is the slow branch, and models the redistribution of the ions to the micropores and sub-micropores, at the end of charging or discharging. Thus, the *RC* equivalent model can be related to the UC physically.

IV. Characteristics and Mathematical Model

The characteristic equations governing the above *RC* equivalent circuit were introduced by Zubieta and Bonert [21]. Open-circuit voltage (OCV) obtained by employing constant-current charging on the UC is used to obtain these equations. During constant-current charging, a current pulse of short duration is applied to the UC, and the OCV is observed. As long as the current pulse is there, the fast (first) branch gets charged first. Thereafter, the charge gets redistributed, which is modeled by the slow (second) branch. The equations for various time-dependent regions of operation for two branches are hence obtained.

A. Parameters of First Branch

As the voltage across the capacitor is zero at time $t = 0$, the voltage across the UC is mainly due to the voltage drop across the resistance of the first (fast) branch, R_1. Hence, the value of first branch's resistance R_1 is:

$$R_1 = \frac{\Delta V_{R1}}{I} \tag{1}$$

Initially, the capacitor voltage of the first branch is almost zero; hence, the constant part of capacitance can be calculated by assuming the voltage-dependent part of the capacitance is zero. Thus, C_o can be computed as:

$$C_o = I\frac{\Delta t_o}{\Delta v_o} \tag{2}$$

978-1-4673-9551-9/16 $31.00 © 2016 IEEE

The voltage dependent part of the capacitance C_v can be easily computed based on the stored charge relation ($Q = C^*V$) as:

$$C_v = \frac{2}{V_o^2}(Q - C_o V_o) \quad (3)$$

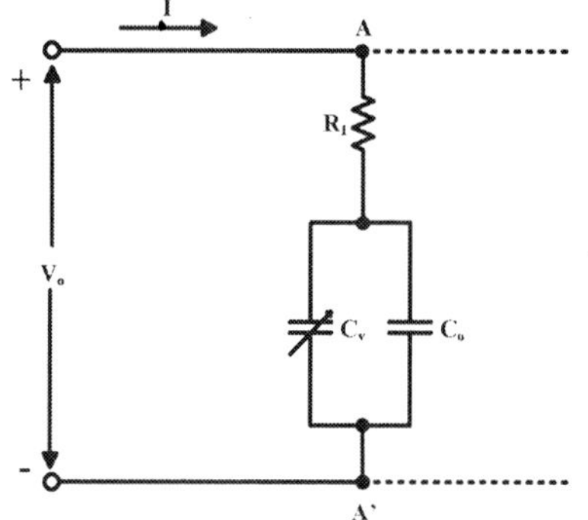

Fig. 3. Equivalent circuit of first branch during charging.

The equivalent circuit of the first branch is given in Fig. 3. This branch of the model is in function during the time of charging i.e. when the ions interact with the macropores and mesopores.

B. Parameters of Second Branch

Estimation of the parameters of the second (slow) branch is carried out after the current pulse diminishes. The resistance of the second branch is calculated using the following equation:

$$R_2 = \frac{(V_o - \frac{\Delta v}{2})\Delta t}{C_1 \Delta v} \quad (4)$$

Finally, the capacitance of the slow branch, C_2, can be calculated from the charge balance equation shown as:

$$C_2 = \frac{Q}{V_o} - (C_o + \frac{1}{2}C_v V_o) \quad (5)$$

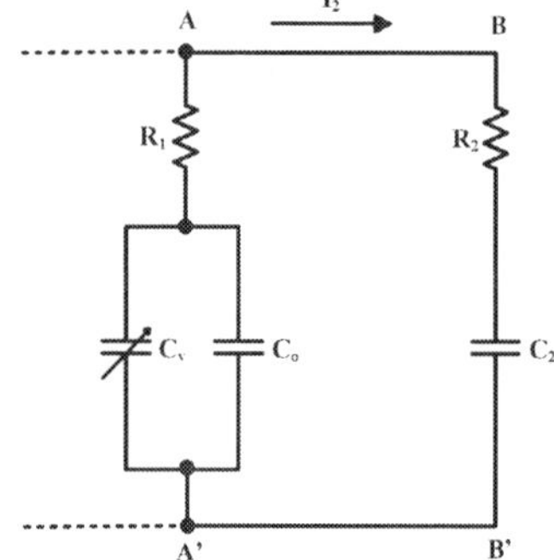

Fig. 4. Equivalent circuit of second branch during charging.

The equivalent RC circuit of the model, when the charging is stopped is shown in Fig. 4. At this stage the charge is transferred from the first branch to second branch, i.e. during distribution of charge in the micro and sub-micropores takes place.

V. Proposed Method to Dynamically Identify the Characteristic Equations Parameters

The characteristic equations' parameters of the UC need to be estimated dynamically in order to estimate the OCV of the UC. The model of UC requires the exact values of Δv_o, Δt_o, Δv, Δt, and ΔV_{R1}, which vary with capacitance and with the current pulse applied. Also, the values of C_o and C_v are different for different UCs and depend on Δv and Δt. Hence, if an accurate estimation has to be made, the values of Δv, Δt, ΔV_{R1}, C_o, C_v, and R_2 need to be identified in real time. The values of the parameters are to be calculated solely with the knowledge of the OCV and characteristic parameters (which are determined by the characteristic equations) through dynamic parameter identification.

Firstly, it can be seen from the characteristic equations of C_o and C_v that these parameters depend on OCV. This forms an inter-dependent system. Hence, in order to compute the OCV and have the model running, the initial values of these parameters must be estimated. Secondly, it is not possible to obtain the values of Δv_o, Δt_o, Δv, and Δt, used in the equations of C_o and R_2 (2) and (4), directly from the output plot, especially if OCV plot is not known. These issues can be addressed, if a close enough initial guess is made possible for the parameters' values, so that the OCV can use the characteristic equations later, to estimate an exact profile. The initial guess includes the estimation of the branch parameters C_o and R_2.

A. Estimation of C_o

The estimation of C_o was done by running simulation tests on UCs for different ratings and values of C_o. It was observed that the value of C_o varies almost linearly with the value of capacitance of the UC. For instance, a 470F UC has a C_o of 275F, while a 1500F capacitor has a C_o of 900F. Hence, a linear relationship can be easily formulated between UC capacity and C_o. This helps determine the initial value of C_o and thereafter, C_o adjusts itself and provides the appropriate output. For example, for a UC with capacitance ranging from 250-500F, the linear equation for C_o will be: C_o = 57.23% of the UC's capacitance value, while for capacitance in the range of 500-750F, the value will be: C_o = 58.57% of the UC's capacitance value.

B. Estimation of R_2

For R_2, the estimation is performed after charging is stopped or the charging current pulse is diminished. The problem with the equation of R_2 is that it has $\Delta v/\Delta t$ in the denominator, which increases instantly after charging is stopped, making estimation of R_2 difficult. R_2 can be estimated by devising an empirical equation at different charging times and observing the decrease in voltage due to the charge redistribution for a fixed time, after charging stops

(40s, for example). This empirical equation will be for a UC with capacitance in the medium range (450-600F). Once the empirical equation for a particular capacitor is obtained, a multiplication factor can be formulated, in order to obtain the value of R_2 for different capacitances. This multiplication factor is obtained by comparing the values of different extracted values of R_2 for the same charging time. For instance, empirical equation for R_2 for 470F capacitor, at different charging times, can be expressed as:

$$R_2 = mf(1.3212 + 0.0003t_{on}^2 + 6.1247e-9t_{on}^4 + 3.9019e-15t_{on}^6 - 8.3281e-12t_{on}^5 - 1.9779e-6t_{on}^3) \quad (6)$$

In (6), mf is the formulated multiplication factor, whose equation for a UC with the capacitance ranging from 50-3000F is given as:

$$mf = 16.4903 + 9.3816e-13C^6 - 0.02841C \quad (7)$$

In (7), C is the capacitance of the UC. Multiplying this multiplication factor with the value obtained from the empirical equation of R_2 (as a base value), the exact value of R_2 can be determined. Hence, accurate values of R_2, for UCs between 50F and 3000F, can be estimated.

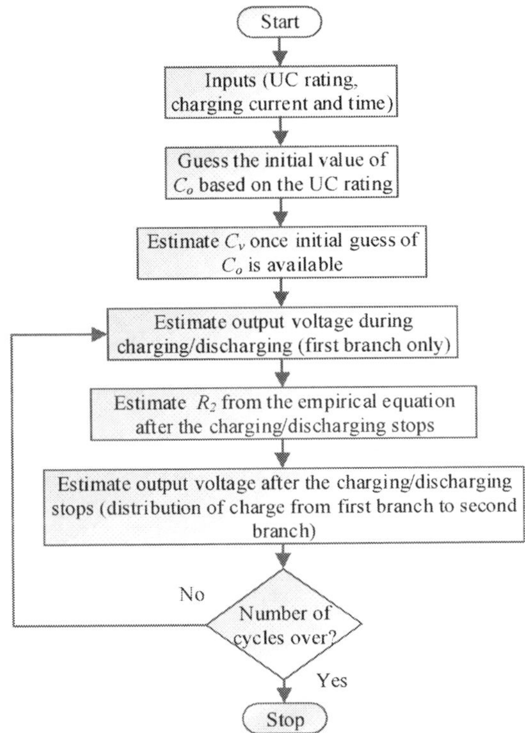

Fig. 5. Flowchart representation of the dynamic estimation method.

The complete dynamic identification/estimation technique in the form of a flowchart is shown in Fig. 5. The steps listed are used to obtain the characteristic equations' parameters.

VI. Equations of the UC Model

The governing equations of the RC equivalent circuit of the UC model shown in Fig. 2 need to be derived in order to use them to run the model and obtain the output voltage V_o. In the model, since the branches are in parallel, the output voltage is equal to the voltage of either of the two branches.

$$V_o = V_{AA'} = V_{BB'} \quad (8)$$

Consider output voltage to be equal to that of the first branch:

$$V_o = V_{AA'} \quad (9)$$

And from Fig. 3:

$$V_{AA'} = V_{R1} + V_{C1} \quad (10)$$

Here, V_{R1} is calculated as:

$$V_{R1} = (I - I_2)R_1 \quad (11)$$

The value of R_1 includes the dependence of ESR on temperature. The effects of temperature on equivalent series resistance is modeled by an empirical relation. The resistance increases at lower temperatures, specifically below 10°C [21].

The voltage across the differential capacitor of the first branch is given as:

$$V_{C1} = V_C' + \frac{1}{C_1}\int I_1 dt \quad (12)$$

The current of the first branch can be written as: $I_1 = I - I_2$. Also, consider V_C' as the initial capacitor voltage. Finally, the output voltage is:

$$V_o = V_{R1} + V_{C1} = (I - I_2)R_1 + V_C' + \frac{1}{C_1}\int I_1 dt \quad (13)$$

Since the voltage across both branches is the same, equal to the output voltage, the value of I_2 can be obtained easily. The value of I_2 can be found by substituting the voltage of the second branch as the output voltage.

$$V_o = V_{R2} + V_{C2} \quad (14)$$

$$V_o = I_2 R_2 + V_{C2} \quad (15)$$

Hence, from the above equation, I_2 can be calculated as:

$$I_2 = \frac{V_o - V_{C2}}{R_2} \quad (16)$$

In the above equation, V_{C2} can be computed as:

$$V_{C2} = V_C' + \frac{1}{C_2}\int I_2 dt \quad (17)$$

$$I_1 = (I - I_2) \quad (18)$$

Finally, current of the first branch, I_1, can be calculated, in order to compute the main voltage, in (13). The second branch's current I_2 has a significant contribution in modeling the charge redistribution when the charging/discharging current pulse is diminished, as the second branch is responsible for the charge redistribution of the ions.

VII. Experimental Validation

The characteristic equations were simulated, in order to obtain the desired OCV for random charging/discharging current pulses. The simulation results for both of the branches responsible for behavior of UC during the charging or discharging and the redistribution of the charge after the charging or discharging pulse is over are given below in Fig 6. Fig. 6 (a) depicts the charging and discharging current pulses of 35A, for a duration of 50s and 15s respectively, and Fig. 6 (b) depicts the OCV plot for the aforementioned charging and discharging currents for a 650F UC.

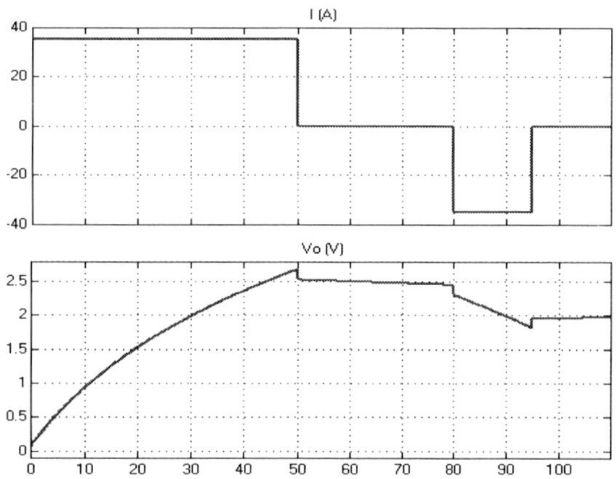

Fig. 6. Current profile and voltage response for a 650F UC (a) charging and discharging current pulses of 35A for 50s and 15s respectively; (b) output voltage response to the current pulses.

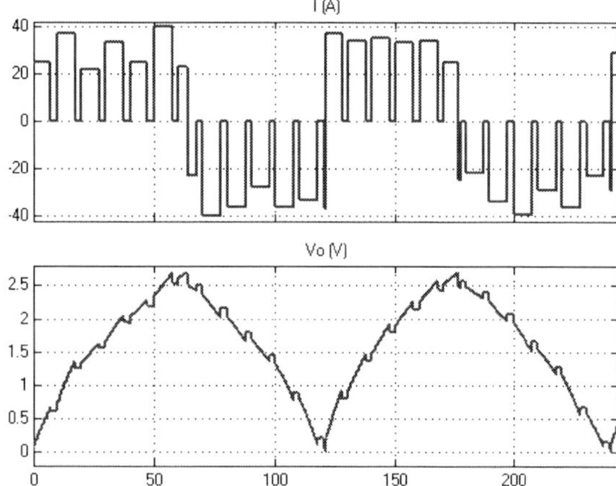

Fig. 7. Current profile and voltage response for a 470F UC (a) random current profile used for testing; (b) output voltage response to the random current profile.

The dynamic UC model was tested for random current pulses between 20A to 40A and the output voltage was controlled and kept between the 0V (V_{min}) and 2.7V (V_{max} *or* V_{rated}). The model was simulated with random charging/discharging current pulses to emulate the extreme drive cycles. The charging/discharging current pulses were applied for 7.5s followed by a rest period of 2.5s, when no current is there. The simulation results for the random tests are shown in Fig. 7. Fig. 7 (a) shows the random charging and discharging current profile (20A to 40A and -20A to -40A) used for a 470F UC, and Fig. 7 (b) shows the OCV plot for the current profile shown in Fig. 7 (a). The results obtained from the parameter estimation technique are validated against the actual charging results in Fig. 8 and 9.

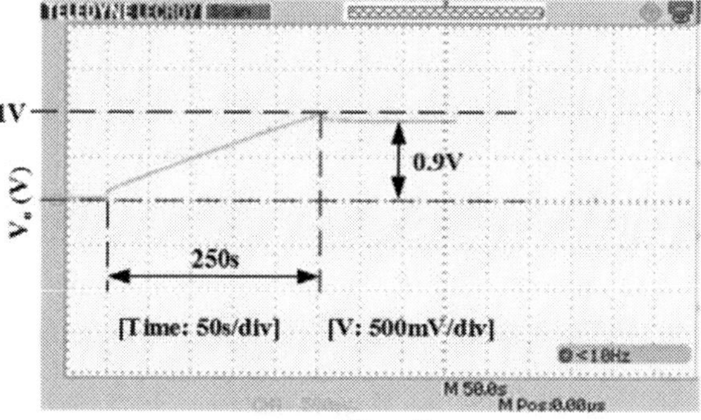

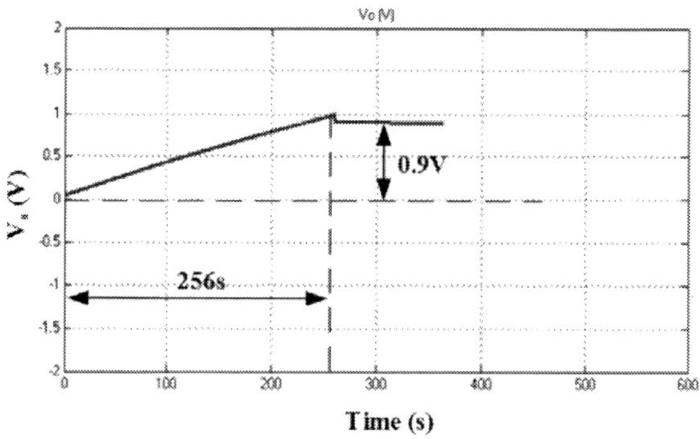

Fig. 8. Output voltage profile for 3.5A charging current and 1V limit (a) actual charging response; (b) simulation result.

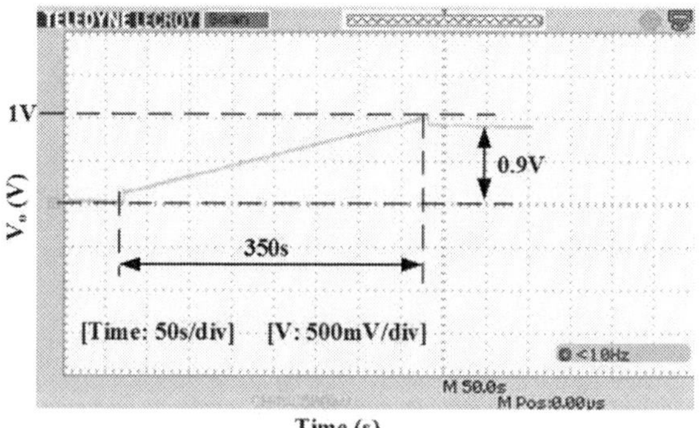

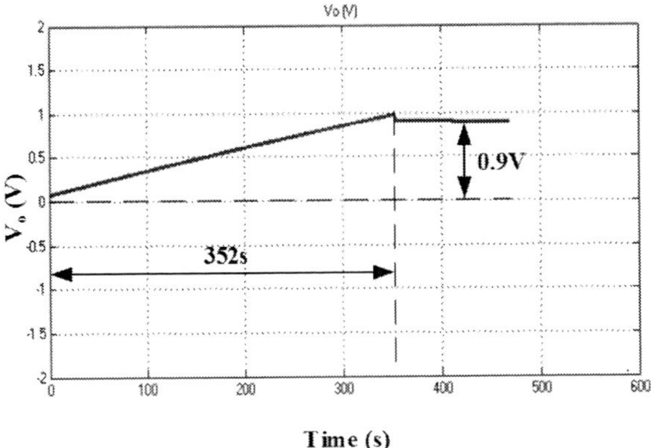

Fig. 9. . Output voltage profile for 2.5A charging current and 1V limit (a) actual charging response; (b) simulation result.

It can be seen from Fig. 8 and 9 that the simulation results obtained from the proposed dynamic parameter identification technique closely matches with the experimental charging response of the UC. The experimental tests were performed at 3.5A and 2.5A of charging currents and a voltage limit of 1V, which was kept to keep the UC voltage low for safety reasons.

VIII. Conclusions

In this paper, a dynamic parameter estimation method, for *RC* equivalent model of ultracapacitor (UC), is presented and validated against actual charging current profiles (with output voltage controlled within the limits i.e. 1V) for open-circuit voltage (OCV) assessment. The proposed estimation technique uses characteristic equations to obtain the OCV response of UCs of varied ratings, without having to perform constant-current tests on and extracting actual branch parameters: R_1, C_1, R_2, and C_2. The estimation method when used along with the UC *RC* equivalent model, accurately estimates the behavior of the UC during a single charge and discharge cycle. It is also capable of precisely estimating the OCV for random charging/discharging current pulses, solely based on the knowledge of the characteristic branch parameters obtained from the dynamic parameter identification technique and the output voltage (V_o).

References

[1] A. Singh, N. A. Azeez and S. S. Williamson, "Dynamic modeling and characterization of ultracapacitors for electric transportation," *IEEE International Symp. On Industrial Electronics*, Buzios, Brazil, June 2015, pp. 275-280.

[2] A. Burke, "Ultracapacitors: why, how, and where is the technology," *Journal of Power Sources*, vol. 91, no. 1, pp. 37-50, Nov. 2000.

[3] S. M. Lukic, J. Cao, R. C. Bansal, F. Rodriguez, and A. Emadi, "Energy storage systems for automotive applications," *IEEE Trans. on Industrial Electronics*, vol. 55, no. 6, pp. 2258–2267, Jun. 2008.

[4] J.-S. Lai, S. Levy, and M. F. Rose, "High energy density double-layer capacitors for energy storage applications," *IEEE Aerospace and Electronic Systems Magazine*, vol. 7, no. 4, pp. 14–19, Apr. 1992.

[5] M. Ortuzar, J. Moreno, and J. Dixon, "Ultracapacitor-based auxiliary energy system for an electric vehicle: Implementation and evaluation," *IEEE Trans. on Industrial Electronics*, vol. 54, no. 4, pp. 2147–2156, Aug. 2007.

[6] S. Lemofouet and A. Rufer, "A hybrid energy storage system based on compressed air and supercapacitors with maximum efficiency point tracking (MEPT)," *IEEE Trans. Industrial Electronics*, vol. 53, no. 4, pp. 1105–1115, Aug. 2006.

[7] Maxwell Technologies®, K2® Ultracapacitors, *Data Sheet*.

[8] H. Gualous, D. Bouquain, A. Berthon, and J. M. Kauffmann, "Experimental study of supercapacitor serial resistance and capacitance variations with temperature," *Journal of Power Sources*, vol. 123, no. 1, pp. 86-93, Sept. 2003.

[9] J. R. Miller and A. F. Burke, "Electrochemical capacitors: Challenges and opportunities for real-world applications," *The Electrochemical Society Interface*, vol. 17, no. 1, pp. 53-57, Spring 2008.

[10] L. Shi and M. L. Crow, "Comparison of ultracapacitor electric circuit models," in *Proc. IEEE Power and Energy Society General Meeting*, Pittsburgh, PA, July 2008, pp. 1-6.

[11] E. G. Bakhoum and M. H. M. Cheng, "Tunable ultracapacitor," *IEEE Trans. on Industrial Electronics*, vol. 60, no. 12, pp. 5613-5619, Dec. 2013.

[12] M. Uno and K. Tanaka, "Accelerated charge-discharge cycling test and cycle life prediction model for supercapacitors in alternative battery applications," *IEEE Trans. on Industrial Electronics*, vol. 59, no. 12, pp. 4704-4712, Dec. 2012.

[13] A. Shukla, A. Banerjee, M. Ravikumar, and A. Jalajakshi, "Electrochemical capacitors: Technical challenges and prognosis for future markets," *Electrochim. Acta*, vol. 84, no. 1, pp. 165–173, Dec. 2012.

[14] D. Torregrossa, M. Bahramipanah, E. Namor, and R. Cherkaoui, "Improvement of dynamic modeling of supercapacitor by residual charge effect estimation," *IEEE Trans. on Industrial Electronics*, vol. 61, no. 3, pp. 1345-1354, March 2014.

[15] R. L. Spyker, and R. M. Nelms, "Optimization of double-layer capacitor Arrays," *IEEE Trans. on Industry Applications*, vol. 36, pp. 194-198, Jan. 2000.

[16] X. del Toro Garcia, P. Roncero-Sanchez, A. Parreno, and V. Feliu, "Ultracapacitor-based storage: Modelling, power conversion and energy considerations," in *Proc. IEEE Innternational Symp. on Industrial Electronics*, Bari, Italy, July 2010, pp. 2493-2498.

[17] A. Hammar, P. Venet, R. Lallemand, G. Coquery, and G. Rojat, "Study of accelerated aging of supercapacitors for transport applications," *IEEE Trans. on Industrial Eelctronics*, vol. 57, no. 12, pp. 3972-3979, Dec. 2010.

[18] Z. Amjadi and S. S. Williamson, "Power-electronics-based solutions for plug-in hybrid electric vehicle energy storage and management systems," *IEEE Trans. on Industrial Electronics*, vol. 57, no. 2, pp. 608-616, Feb. 2010.

[19] Z. Amjadi and S. S. Williamson, "A novel control technique for a switched-capacitor-converter-based hybrid electric vehicle energy storage system," *IEEE Trans. on Industrial Electronics*, vol. 57, no. 3, pp. 926-934, March 2010.

[20] A. Eddahech, O. Briat, M. Ayadi, and J. Vinassa, "Ultracapacitor performance determination using dynamic model parameter identification," in *Proc. IEEE Innternational Symp. on Industrial Electronics*, Taipei, Taiwan, May 2013, pp. 1-5.

[21] L. Zubieta and R. Bonert, "Characterization of double-layer capacitors for power electronics applications," *IEEE Trans. on Industry Applications*, vol. 36, no. 1, pp.199-205, Jan. 2000.

Novel SPICE Model for Common Mode Choke Including Complex Permeability

Katsuya Nomura, Naoto Kikuchi, Yoshitoshi Watanabe, Shuntaro Inoue, Yoshiyuki Hattori
Toyota Central R&D Labs., INC.
41-1 Yokomichi, Nagakute, Aichi 480-1192
Email: k-nomura@mosk.tytlabs.co.jp

Abstract—A highly accurate filter simulation requires a high precision SPICE model for the common mode choke. Conventional SPICE models for chokes consist of parallel RLC circuits and show a substantial difference between the measured and simulated impedance characteristics due to the lack of frequency-dependent complex permeability considerations. Here, we propose a novel SPICE model for a common mode choke that includes the frequency-dependent complex permeability. This model uses a transfer function in the s-domain to describe frequency-dependent elements. Moreover, two model parameter estimation methods are proposed; a measurement based method and a datasheet based method. High accuracy is achieved with the measurement based model by extracting the model parameter using the measured impedance characteristics of the actual coil. In the datasheet based method, the model parameter is estimated from the datasheet and the shape of the coil without the actual coil. Finally, we show that not only the impedance characteristics of a choke can be expressed, but also the insertion loss of a T-type filter can be predicted much more accurately than presently possible using a conventional model. This SPICE model is the first to express the frequency-dependent inductance and resistance of the choke with high accuracy; therefore, it is expected to be useful for filter design.

Keywords—Common mode choke, SPICE model, complex permeability, filter design

I. INTRODUCTION

The electromagnetic noise generated from power electronics converters causes interference with other equipment; therefore, an electromagnetic interference(EMI) filter is often inserted into the power line to suppress the noise.

Efficient filter design requires the filter insertion loss to be calculated with high accuracy using a circuit simulator such as SPICE. A highly accurate filter simulation thus requires a high precision SPICE model for the common mode choke.

Common mode chokes are one of the main components of EMI filters. Common mode chokes consist of two windings and a magnetic toroidal core, as shown in Fig. 1. This coil behaves as an inductor for the common mode current because the two fluxes created by two line currents enhance each other; however, this coil does not behave as an inductor for differential mode current because the two fluxes cancel each other out.

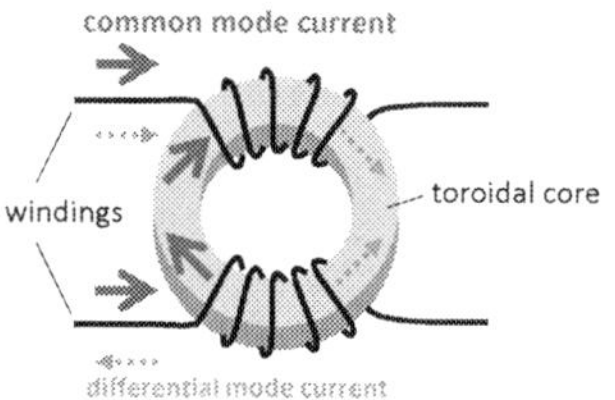

Fig. 1. Schematic of the common mode choke

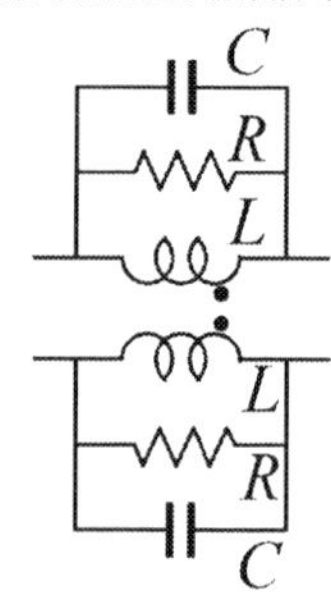

Fig. 2. Conventional equivalent circuit for the common mode choke.

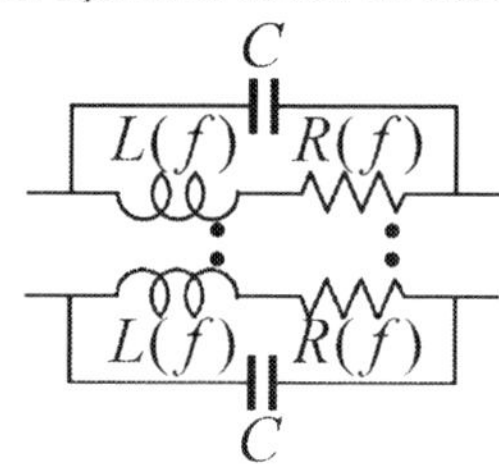

Fig. 3. Equivalent circuit for the common mode choke including complex permeability.

The conventional equivalent circuit with RLC parallel circuits shown in Fig. 2 is widely used as a SPICE model for the common mode choke [1-4], where L is the magnetizing inductance, R is the core loss, and C is the winding capacitance. However, the impedance characteristics simulated with this model are very different from the measured characteristics, primarily because the frequency dependence of the complex permeability is not considered. Inductance L, and resistance R, have frequency dependence because they are proportional to the real μ' and imaginary μ'' parts of the frequency-dependent complex permeability, respectively. The conventional model expresses the frequency-dependent inductance and resistance with constant values; therefore, this model causes significant differences in the impedance characteristics.

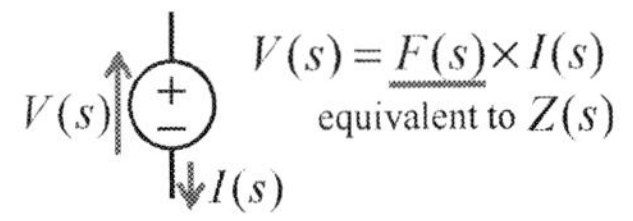

Fig. 4. Frequency-dependent impedance expression based on an arbitrary voltage source.

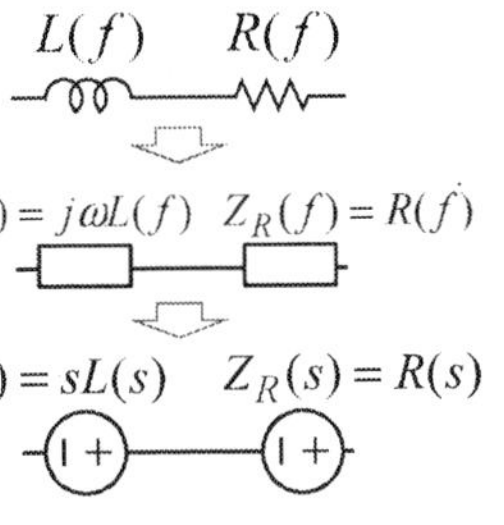

Fig. 5. Expression for frequency-dependent inductance and resistance based on an arbitrary voltage source.

The equivalent circuit shown in Fig. 3 includes complex permeability [5-8]. The circuit has the frequency-dependent inductance and resistance in series; however, a SPICE model that expresses the frequency dependence for these elements has not yet been reported.

Here, we propose a novel SPICE model for the common mode choke that includes the frequency-dependent complex permeability. This model uses a transfer function in the $s(j\omega)$-domain to describe frequency-dependent elements. Moreover, two model parameter estimation methods are proposed; a measurement based method and a datasheet based method. The prediction accuracy of the impedance and filter characteristics for these two methods are verified by comparing the measured and simulated results.

II. PROPOSED SPICE MODEL

To construct a SPICE model that includes frequency-dependent elements, we propose the use of a frequency-dependent function in the SPICE model and curve fitting of the frequency-dependent elements.

Details for the use of a frequency-dependent function in the SPICE model are as follows. Figure 4 shows that the use of an arbitrary voltage source in the SPICE model enables the frequency-dependent impedance $Z(s)$ to be expressed as a transfer function $F(s)$ in the $s(j\omega)$- domain by defining the voltage $V(s)$, as:

$$V(s) = F(s) \times I(s), \tag{1}$$

where $I(s)$ is the current of the voltage source. As a result, the frequency-dependent inductance $L(f)$ and resistance $R(f)$ can be expressed using an arbitrary voltage source, as shown in Fig. 5.

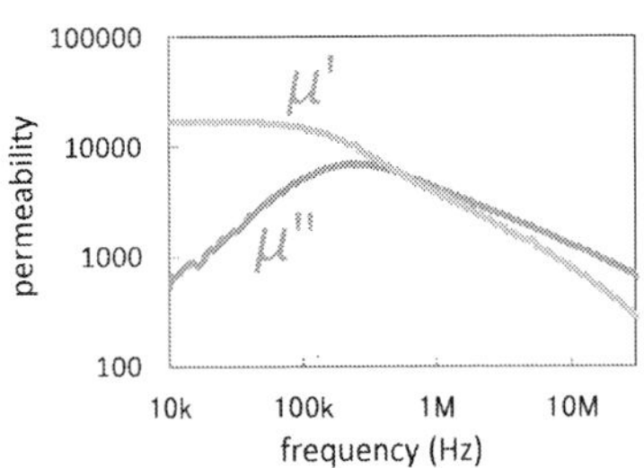

Fig. 6. Frequency-dependent complex permeability for FT-3KL.

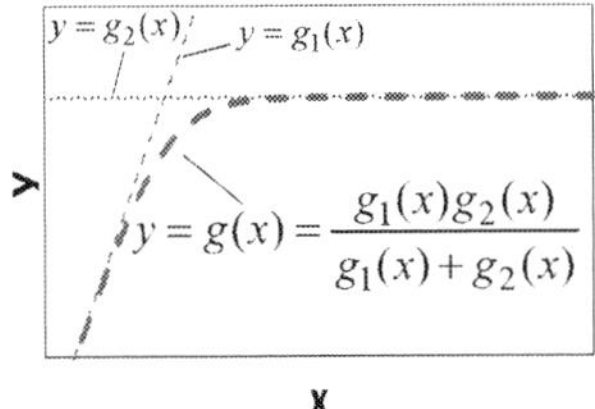

Fig. 7. Curve fitting with two straight lines.

Details for curve fitting of the frequency-dependent elements are as follows. $L(f)$ and $R(f)$ are proportional to the real μ' and imaginary μ'' parts of the complex permeability, respectively. Figure 6 shows the real part μ' and the imaginary part μ'' of the FT-3KL complex permeability [9], which is a representative magnetic material used for common mode chokes. Both curves in Fig. 6 have different slopes in the low and high frequency regions. Therefore, we propose a fitting method using two straight lines on a log-log plot, as shown in Fig. 7. With this method, not only can μ' and μ'' be fitted, but also $L(f)$ and $R(f)$. The equation for the $y = g(x)$ curve with two straight lines, $y = g_1(x)$ and $y = g_2(x)$, is:

$$y = g(x) = \frac{10^{k_2 \log_{10} x + l_2}}{1 + 10^{(k_2 - k_1)\log_{10} x + (l_2 - l_1)}}. \tag{2}$$

Here, the slopes k_1 and k_2, and the y-intercepts l_1 and l_2 are model parameters.

Based on these methods, a SPICE model can be constructed for the common mode choke that includes the frequency-dependent complex permeability. The proposed SPICE model is shown in Fig. 8. There are two kinds of models, model A and model B. In model A, the frequency-dependent inductance $L(f)$ and resistance $R(f)$ are fitted directly, while in model B, $L(f)$ and $R(f)$ are expressed using the fitted permeabilities, μ' and μ'', and the proportional coefficients, K_L and K_R.

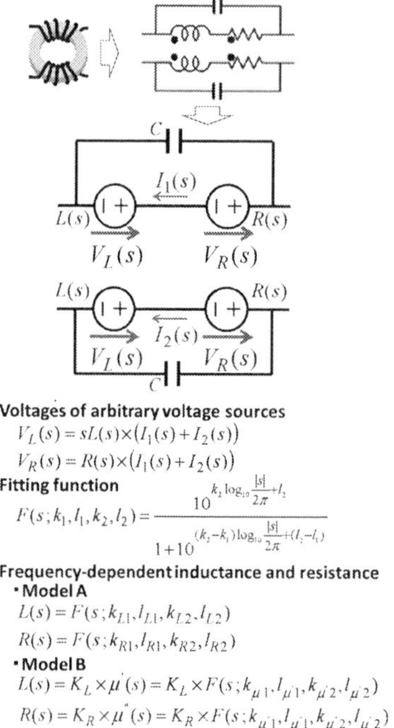

Fig. 8. Proposed common mode choke SPICE model.

III. MODEL PARAMETER ESTIMATION

In this section, we propose measurement based and datasheet based parameter estimation methods for the model shown in Fig. 8.

The measurement based method can construct a model with high accuracy through extraction of the model parameter from the measured impedance characteristics of an actual coil. This method is used to fit the inductance and resistance. The datasheet based method can estimate the model parameter without an actual coil and is used to fit the permeabilities, μ' and μ''.

To explain the model parameter estimation procedure and validate these two methods, a simulation model is constructed for a common mode choke using a toroidal core F1AH0972 [10] that consists of the FT-3KL with 9 windings.

A. Measurement based method

The procedure for model parameter estimation consists of the following four steps.

1) Measurement of the impedance characteristics: The two windings of the common mode choke are shorted and the common mode impedance is then measured with an impedance analyzer (4294A, Agilent Technology Inc.), as shown in Fig. 9. The measured impedance characteristics are shown in Fig. 10.
2) Extraction of winding capacitance: Figure 10 shows the winding capacitance is dominant over 7 MHz because the absolute value of the impedance decreases with an increase in the frequency. In this frequency region, the winding capacitance C_{cmn} can be extracted using $|Z_C| = 1/(2\pi f C_{cmn})$.
3) Capacitance subtraction and derivation of the inductance and resistance: The capacitance C_{cmn} is subtracted from the total coil admittance, Y:

$$\begin{aligned} Z_{RL} &= \frac{1}{Y_{RL}} = \frac{1}{Y - Y_C} \\ R &= \mathrm{Re}(Z_{RL}) \\ X &= \mathrm{Im}(Z_{RL}), \end{aligned} \tag{3}$$

where Z_{RL} and Y_{RL} are the impedance and admittance of the magnetizing inductance L and core loss R, respectively, and Y_C is the admittance of the winding capacitance. Figure 11 shows the absolute value of impedance $|Z|$, reactance X, and resistance R, before and after capacitance subtraction. The reactance dip due to LC oscillation can thus be removed after subtraction of the capacitance.
The inductance L is calculated next using the derived reactance X, from $L = X/(2\pi f)$. The calculated inductance is shown in Fig. 12 as a solid line.

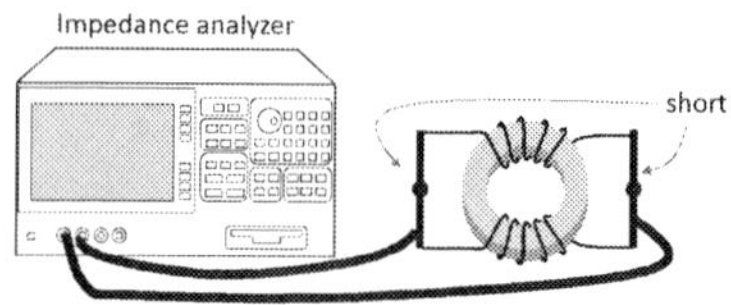

Fig. 9. System for measurement of the common mode impedance of a common mode choke.

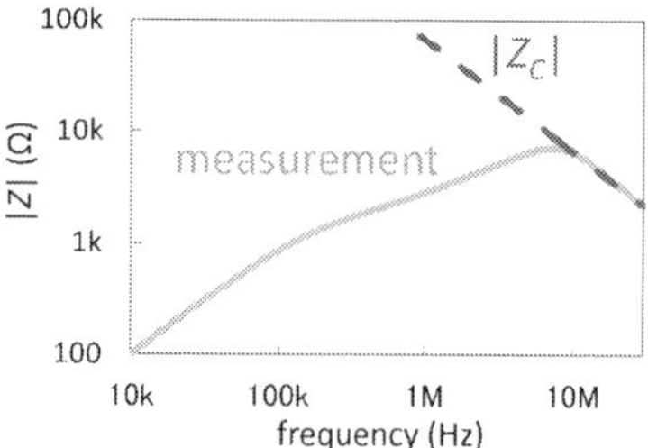

Fig. 10. Impedance and extraction of winding capacitance.

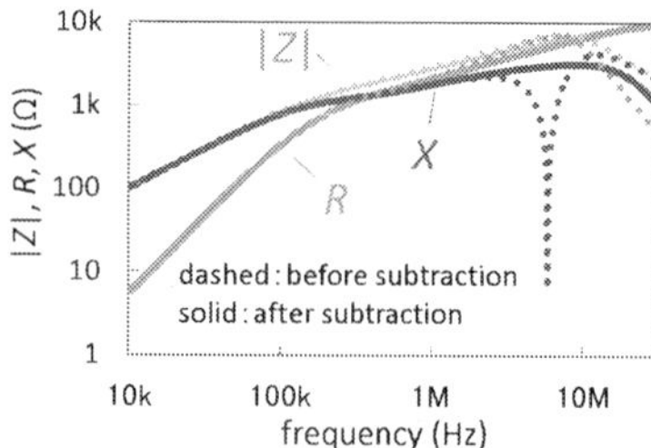

Fig. 11. $|Z|$, R, X before and after subtraction of the capacitance.

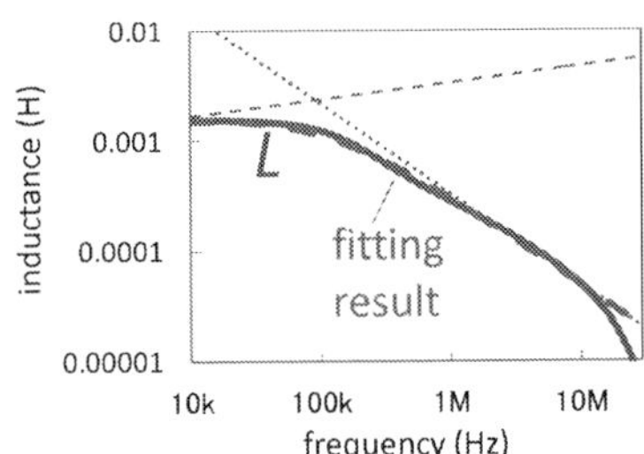

Fig. 12. Fitting result for inductance.

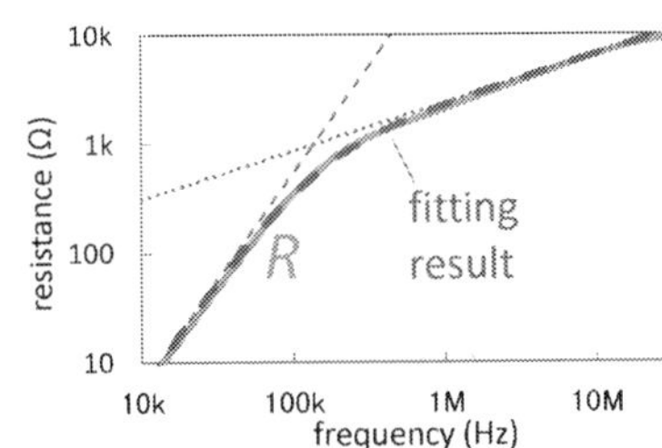

Fig. 13. Fitting result for resistance.

4) Curve fitting of inductance and resistance:
The inductance and resistance can be fitted using the method with two lines explained in the section II. Figures 12 and 13 show the fitting results for the inductance and resistance, respectively, where both characteristics are fitted with high accuracy.

Figure 14 shows the proposed measurement based model constructed in LTSpice [11]. B_L1 and B_L2 are frequency dependent inductors, and B_R1 and B_R2 are frequency dependent resistances. The function "twoline" defines the fitting function (2) using the ".func" sentence. The model parameters are k_1, k_2, k_3, k_4, l_1, l_2, l_3, l_4, which are obtained through fitting of the inductance and resistance, and the winding capacitance, C_{cmn}. The capacitance C for each winding is a half of C_{cmn} because the capacitance of the common mode impedance is equal to the capacitances of two windings.

Figure 15 shows the measured and simulated impedance characteristics. The result obtained using the conventional model based on RLC parallel circuits is also shown in Fig. 15. In the conventional model, L is extracted by the low frequency characteristics, C is extracted by the high frequency characteristics, and R is extracted by the maximum absolute value of the impedance, $|Z|$. The results with the proposed model have better accuracy than those obtained with the conventional model, as shown in Fig. 15.

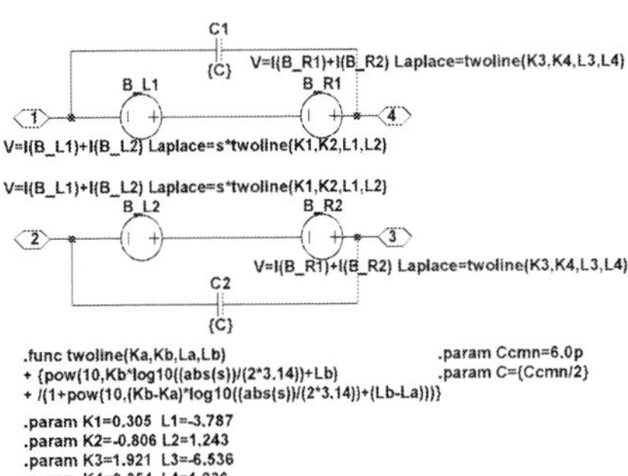

Fig. 14. Measurement based model constructed in LTSpice.

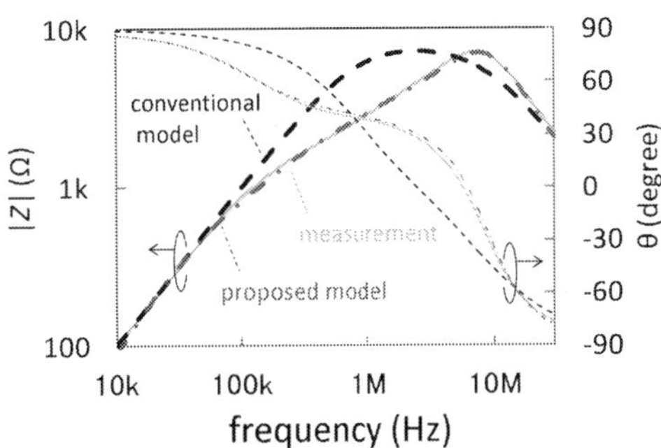

Fig. 15. Measured and fitted impedance characteristics.

B. Datasheet based method

The procedure for estimation of the model parameter consists of the following four steps.

1) Curve fitting of permeability:
The real $\mu^{'}$ and imaginary $\mu^{''}$ parts of the complex permeability are typically on a datasheet, such as in [9]. $\mu^{'}$ and $\mu^{''}$ can then be fitted using the method of two lines, as explained in section II. Figure 16 and 17 show the $\mu^{'}$ and $\mu^{''}$ fitting results for FT-3KL, respectively, where it is shown that both characteristics are fitted with high accuracy.
2) Calculation of core shape coefficient:
The core shape coefficient K_{core} is calculated using the AL value. The core shape coefficient is a proportional coefficient between μ and L that is determined by the core shape. The AL value is typically listed in the datasheet [10] to help a coil designer to calculate the inductance L according to $AL = L/N^2$, where N is the number of turns. The relationship between the core shape coefficient K_{core} and the AL value is expressed as:

$$K_{core} = \frac{AL}{\mu}. \tag{4}$$

The AL value for the F1AH0972 toroidal core used in this work is 16 $\mu H/N^2$ at 100 kHz and μ for FT-3KL is 15800 at 100 kHz; therefore, the core shape coefficient is 0.000805 from (4).

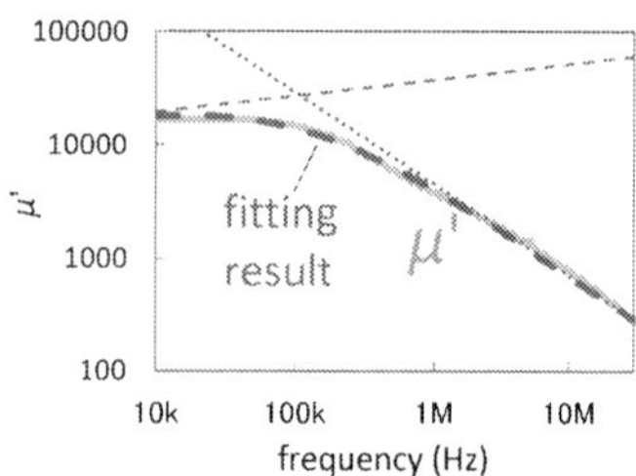

Fig. 16. Fitting result for $\mu^{'}$.

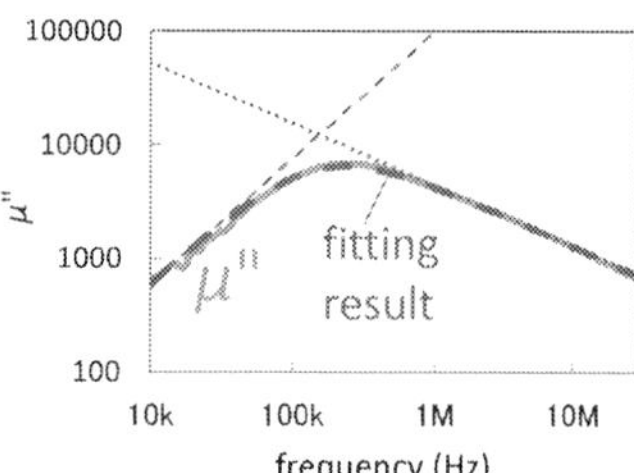

Fig. 17. Fitting result for $\mu^{''}$.

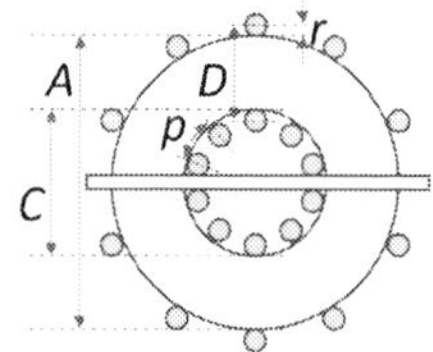

Fig. 18. Cross-section of a common mode choke.

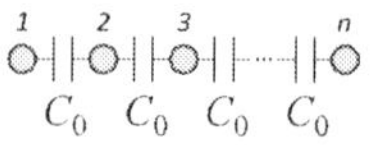

Fig. 19. Turn to turn capacitance.

3) Estimation of winding capacitance:
There are several equations to estimate the winding capacitance, such as for a solenoid coil [5], [12], and for a toroidal coil [8]. The equation in [8] requires a complex calculation due to the multiple integration. In this paper, the following simple equation given in [12] was used to estimate the turn-to-turn winding capacitance:

$$C_0 = \frac{\pi^2 D \epsilon_0}{\ln\left(p/2r + \sqrt{(p/2r)^2 - 1}\right)}, \tag{5}$$

where, D is the diameter of the winding, p is the pitch of the windings, r is the radius of the winding, as shown in Fig. 18. D and p can be calculated using the outer diameter A, and the inner diameter C, by $D = (A - C)/2 + 2r$ and $p = \pi C/N$, where N is the number of turns. For the present case, the coil has a 1 mm diameter winding with 9 turns. From the datasheet, A and C are 27.2 and 13.0 mm, respectively. The capacitance C_0 is thus calculated to be 0.32 pF. Figure 19 shows the winding capacitance C, calculated using the number of turns N, by $C = C_0/(N-1)$ because the turn-to-turn capacitance is connected in series. In this case, N is 9, so that C is 0.040 pF.

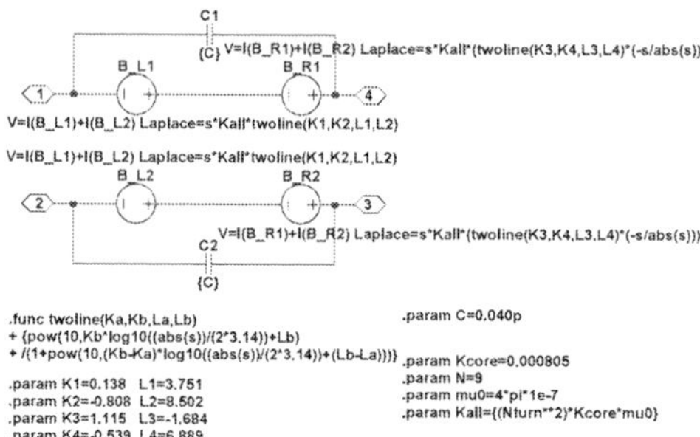

Fig. 20. Datasheet based model constructed in LTSpice.

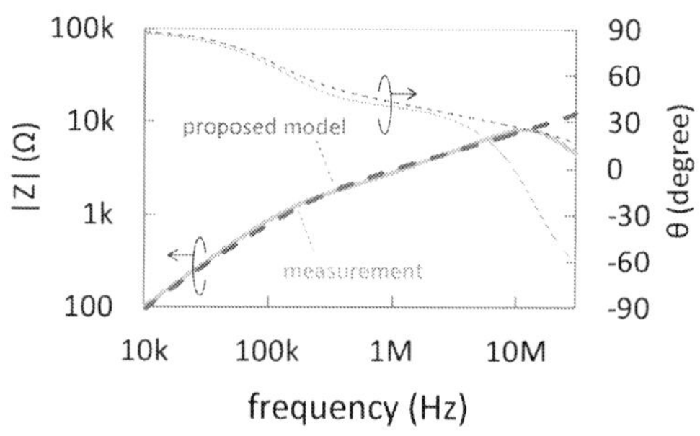

Fig. 21. Measured and predicted impedance characteristics.

4) Impedance calculation:
The impedance is calculated using the obtained permeabilities, $\mu^{'}$ and $\mu^{''}$, the core shape coefficient K_{core}, and the winding capacitance C with the following equation:

$$Z = \frac{Z_L Z_C}{Z_L + Z_C}, \tag{6}$$

where

$$Z_C = \frac{1}{j\omega 2C}, Z_L = j\omega L, \quad L = N^2 K_{core}\mu, \mu = \mu^{'} - j\mu^{''}. \tag{7}$$

The calculation of Z_C requires a coefficient of 2 because the capacitance of common mode impedance is equal to the capacitances of two windings.

Figure 20 shows the proposed datasheet based model constructed in LTSpice. The model parameters are k_1, k_2, k_3, k_4, l_1, l_2, l_3, l_4, which are obtained through curve fitting of $\mu^{'}$ and $\mu^{''}$, the winding capacitance C, the core shape coefficient K_{core}, and the number of turns, N.

Figure 21 shows the measured and simulated impedance characteristics, which are matched closely in terms of both the absolute value and the phase up to the megahertz region. However, the effect of capacitance cannot be seen in the simulated result, although the absolute value of the measured impedance decreases due to the capacitance above 10 MHz. This is because the estimation of the winding capacitance is too small compared to the actual capacitance value. Thus, a more accurate estimation of the winding capacitance will be a goal of our future work.

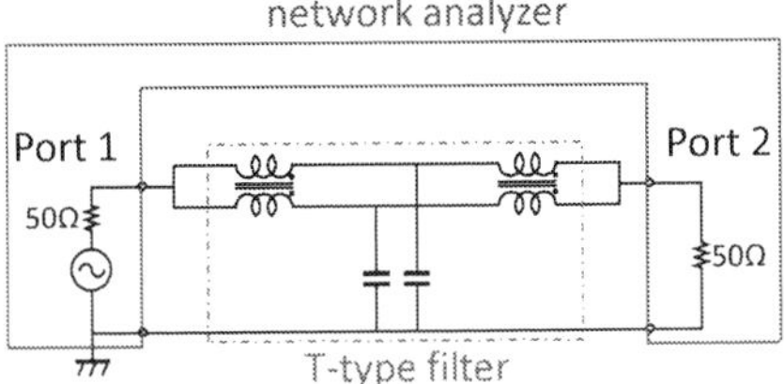

Fig. 22. Schematic diagram of the filter circuit and the measurement system.

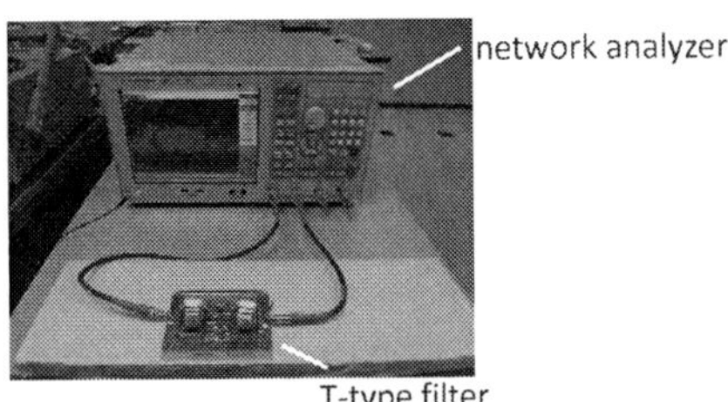

Fig. 23. Photograph of the filter circuit and the measurement system.

IV. Prediction of Filter Characteristics Using Proposed Model

In this section, we compare the measured and simulated insertion losses for a filter using the proposed common mode choke coil. The T-type filter with two common mode chokes shown in Fig. 22 is evaluated. The characteristics of the two chokes are the same as in section III , and a 1 nF film capacitor is used for the y-type capacitor.

Fig. 23 shows a photograph of the filter circuit and the measurement system. The insertion loss for the filter was measured using a network analyzer (E5071C, Agilent Technology Inc.) with a two-port measurement. The average count was set to 999 times, which is the limit of the instrument, and the sweep frequency was set from 100 kHz, which is the lower limit, to 30 MHz.

The proposed and conventional models were used as choke models for the simulation. As the proposed model, both measurement based and datasheet based models were employed. The filter insertion loss IL is calculated using the following equation:

$$IL = 20\log_{10}\frac{V_{out1}}{V_{out2}}, \tag{8}$$

where V_{out1} and V_{out2} are the output voltage without and with a filter, respectively [13]. LTSpice was used for simulation, and IL was obtained from AC analysis.

Figure 24 shows the measured and simulated results for the filter insertion loss. Although the conventional model results are in agreement with the measured results within an error of 12 dB, the measurement based model is accurate to within 3 dB, which demonstrates that the proposed model provides an accurate simulation of the filter insertion loss. Moreover, the datasheet based model is accurate to within 3 dB up to 4 MHz, above which the results are more significantly different due to the problem of winding capacitance estimation, which will be addressed in future work.

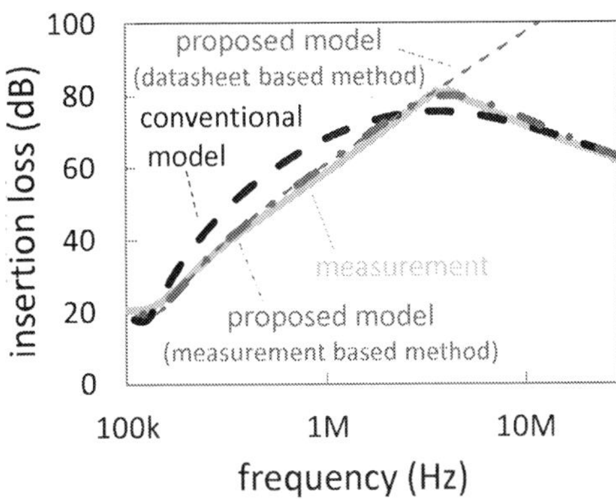

Fig. 24. Measured and simulated filter insertion losses.

V. Conclusion

In this paper, a novel SPICE model was proposed for the common mode choke that includes the frequency-dependent complex permeability. This model used a transfer function in the $s(j\omega)$-domain to describe the frequency-dependent elements.

Moreover, two model parameter estimation methods were proposed: a measurement based method and a datasheet based method. The measurement based model was constructed with high accuracy by extraction of the model parameter using the measured impedance characteristics of the actual coil. The datasheet based method was used to estimate the model parameter from the datasheet and the shape of the coil without using the actual coil.

Finally, it was demonstrated that not only the impedance characteristics of a choke can be expressed, but also the insertion loss of a T-type filter can be predicted much more accurately than possible using a conventional model.

This SPICE model is the first to express the frequency-dependent inductance and resistance of the choke with high accuracy and is thus expected to be useful for filter design.

References

[1] L. Dehong and J. Xanguo, "High frequency model of common mode inductor for emi analysis based on measurements," in Proc. IEEE Int. Symp. Electromagn. Compat., pp. 462-465, 2002.

[2] J.-L. Kotny, X. Margueron, and N. Idir, "High-frequency model of the coupled inductors used in EMI filters," IEEE Trans. Power Electron., vol. 27, no. 6, pp. 2805-2812, 2012.

[3] I. Stevanovic, S. Skibin, M. Masti, and M. Laitinen, "Behavioral modeling of chokes for emi simulations in power electronics," IEEE Trans. Power Electron., vol. 28, no. 2, pp. 695-705, 2013.

[4] W. Tan, C. Cuellar, X. Margueron, and N. Idir, "A high frequency equivalent circuit and parameter extraction procedure for common mode choke in the emi filter," IEEE Trans. Power Electron., vol. 28, no. 3, pp. 1157-1166, 2013.

[5] A. Massarini and M.K. Kazimierczuk, "Self-capacitance of inductors," IEEE Trans. Power Electron., vol. 12, no. 4, pp. 671-676, 1997.

978-1-4673-9551-9/16 $31.00 © 2016 IEEE

[6] M.K. Kazimierczuk, G. Sancineto, G. Grandi, U. Reggiani, and A. Massarini, "High-frequency small-signal model of ferrite core inductors," IEEE Trans. Magn., vol. 35, no. 5, pp. 4185-4191, 1999.

[7] M.L. Heldwein, L. Dalessandro, and J.W. Kolar, "The three-phase common-mode inductor: Modeling and design issues," IEEE Trans. Ind. Electron, vol. 58, no. 8, pp. 3264-3274, 2011.

[8] M. Kovacic, Z. Hanic, S. Stipetic, S. Krishnamurthy, and D. Zarko, "Analytical wideband model of a common-mode choke," IEEE Trans. Power Electron., vol. 27, no. 7, pp. 3173-3185, 2012.

[9] FT-3KL, Hitachi Metals, Ltd., 2014.

[10] FINEMET, EMC Components, Hitachi Metals, Ltd., 2007.

[11] LTspice IV, Linear Technology, 2013.

[12] G. Grandi, M.K. Kazimierczuk, A. Massarini, and U.Reggiani, "Stray capacitances of single-layer solenoid air-core inductors," IEEE Trans. Ind. Appl., vol. 35, no. 5, pp. 1162-1168, 1999.

[13] L. Tihanyi, Electromagnetic Compatibility in Power Electronics, IEEE Press, 1995.

Analysis and Design of Capacitive Power Transmission System Employing Out-of-band Wireless Feedback Link

Sung-Jin Choi and Hee-Su Choi
School of Electrical Engineering
University of Ulsan
Ulsan, 44610, Korea
E-mail: sjchoi@ulsan.ac.kr

***Abstract*— Capacitively-coupled wireless power transmission utilizes electric field applied between physically separated metal plates. Especially in full wireless power system which has no battery inside the target object, how to provide the regulated output voltage through the resonant LC link are challenging task. In this paper, analysis and control design of such system are investigated. Transfer function of the overall system including feedback delay latency is analyzed and control loop is designed in the frequency domain. The theoretical results are verified by PSIM simulation.**

I. Introduction

Capacitive wireless power transmission (C-WPT) utilizes the electric field applied between the physically separated metal barriers. The link capacitance formed between both sides of the barrier provides the ac current path through the barrier, but it should be cancel out by series resonant inductor to improve the efficiency and thus the overall circuit with transmitter and receiver forms resonant converter [1]. However, most of the works on the system has been about the circuit topology and there are few works on its control issue. The reason is because they mainly consider battery charging application. When the battery is present in the receiver circuit, the control doesn't need to be as tight and fast as most power supplies are.

Considering all aspects of the emerging wireless power technology, wireless power application can be divided into two groups. First one is hybrid wireless power system which admits intermittent power transmission because the receiving object already has sufficient battery back-ups therein. Applications such as charging stations for cell phones, tablets, vacuum cleaners, or robot cleaners fall into this group. In these products, battery itself has an inherently large time constant caused by chemical reactions and does not need fast control. It needs just power flow control of charging sequence between the constant current/constant voltage (CC/CV) modes whose time constants are more than a few tens of minutes [2,3].

On the contrary, the other application which needs continuous power transmission through the wireless energy link can be categorized as full wireless power system. It has no battery inside the target object. Therefore, regulating the output voltage in the receiver are challenging task to be investigated. Moreover, because the transmitter and receiver are physically separated, the control loop should be implemented by the wireless feedback link.

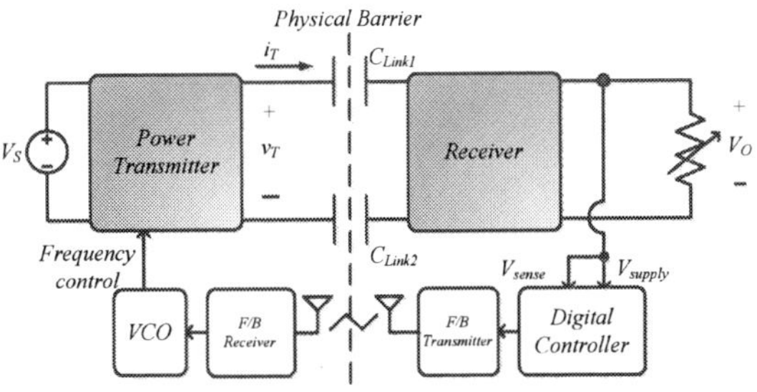

Fig. 1. Full wireless power system based on C-WPT

In this paper, in order to obtain the regulated output voltage in the absence of battery power in the target object, the guideline for the control loop implementation for the C-WPT system has been studied.

II. System Description

Figure 1 shows the block diagram of the target system. It consists of transmitter, receiver, capacitive energy link, feedback (F/B) controller, wireless signal link, gate drive circuitry. Being a resonant converter, the system adopts frequency control where the switching frequency is adjusted by voltage-controlled oscillator (VCO)[4]. The dotted line in the center of the figure depicts the physical barrier between the transmitter and the receiver.

In this paper, the half bridge structure with double matching transformers [5] in Fig. 2 has been adopted. In this configuration, transmitter contains two power MOSFETs turning on alternatively with 50% duty cycle to provide the ac voltage waveforms. The series inductor, L_r, compensates the reactance of the effective link capacitance, C_e which is formed by the series connection of C_{Link1} and C_{Link2}, and the input

978-1-4673-9551-9/16 $31.00 © 2016 IEEE

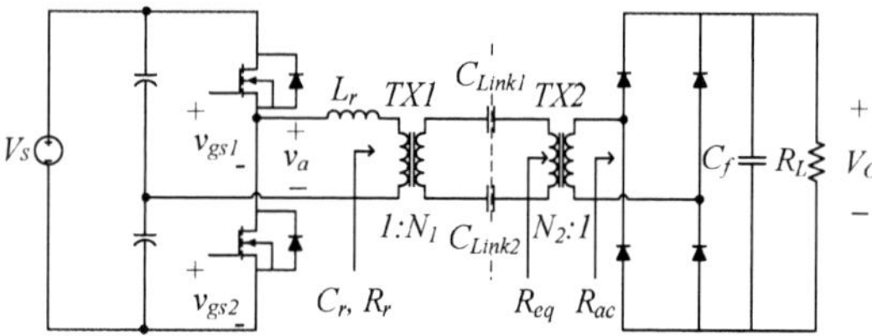

Fig. 2. Power stage topology

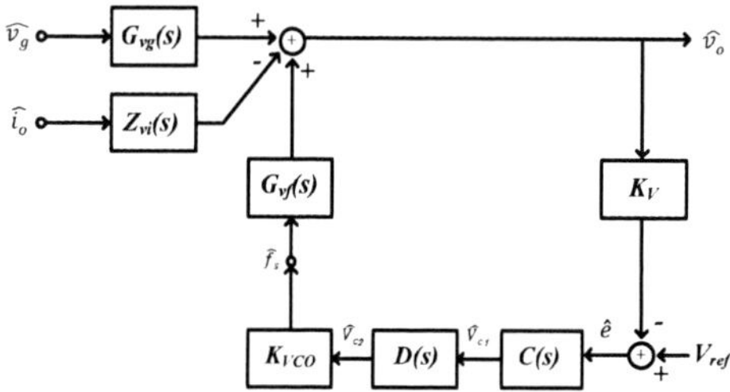

Fig. 3. Control block diagram

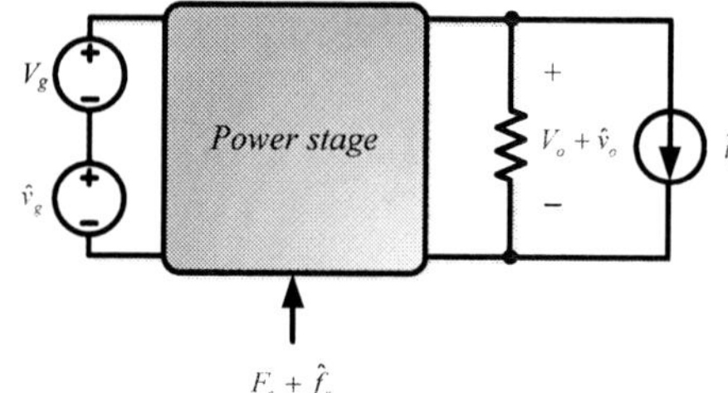

Fig. 4. General method to model resonant converters

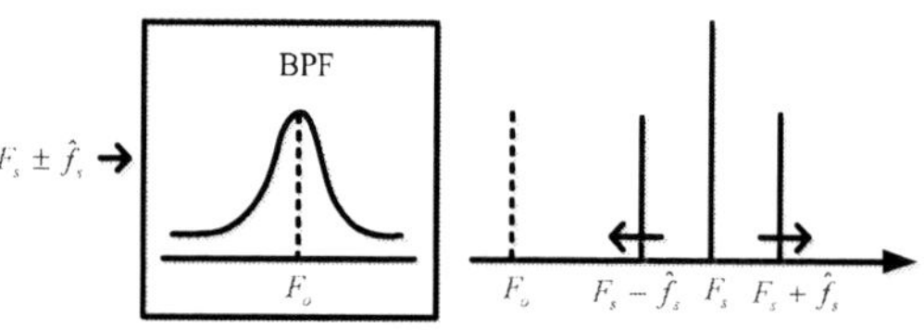

Fig. 5. Frequency modulation and beat frequency

transformer, TX_1 provides the impedance matching to provide zero voltage turn-on features to the main switches. The output matching transformer, TX_2 is adopted to increase the effective output load impedance seen by LC resonant tank. Meanwhile, the resonant waveforms are restored to dc voltage on the load resistance, R_L, by the full-wave rectifier with the output filter capacitor, C_f, in the receiver circuit. Therefore, the overall power stage constitutes a series resonant converter.

Because there is no additional post-regulator or active rectifier in the receiver unit, once the feedback controller generates a control voltage from the error signal between the reference voltage and the sensed voltage, it returns back to the transmitter side by the F/B transmitter-receiver pair. Though it has been shown that feedback signal can be fed back to transmitter side by in band communication where the signal is modulated by the main power waveforms such as in load modulation technique [6], extraction of the modulated signal may not be reliable in the wide operating frequency variation. For these reasons, this scheme is not further investigated, and this paper only studies out-of-band wireless feedback which uses frequency band different from that used in the main power transmission.

III. Analysis and Modeling

The control block diagram of the overall system is shown in Fig. 3. In the figure, $G_{vf}(s)$ is the control to output transfer function, $Z_{vi}(s)$ is output current to voltage transfer function, and $G_{vg}(s)$ is the input voltage to output transfer function of the power stage. K_V is the feedback sensing gain, K_{VCO} is the gain in the voltage controlled oscillator, and C(s) is the compensator block to be designed here. D(s) models the delay latency of the wireless feedback link. The loop gain of the negative feedback system is

$$T(s) = K_{VCO} \cdot K_V \cdot G_{vf}(s) \cdot D(s) \cdot C(s) \tag{1}$$

In order to design the control loop, the individual blocks are analyzed in this section.

A. Forward path : power stage

The modeling of the power stage is performed by injecting small signal perturbation to the power stage as shown in Fig. 4. Especially, the small-signal model of series resonant converters has been investigated either theoretically [7] or heuristically [8]. However, the former is too complex to use and the latter only deals with control to output transfer function. Therefore, in this paper, after the swept sine analysis in PSIM [9] is applied to extract the power stage transfer functions, the salient pole-zero locations are identified by curve fitting algorithms.

Salient pole and zero locations are justified heuristically by the high Q approximation as follows. By time constant analysis, we can expect presence of low frequency pole caused by low filter action of the output capacitor, C_f. It has been also shown in [8] that the LC resonant tank acts as a high pass filter and generates additional two more poles into the system which is given by the beat frequency between the bias switching frequency, F_S, and the resonant frequency, F_o. This kind of frequency modulation phenomena is explained in Fig. 5. For example, when the switching frequency is perturbed with a modulation frequency, there will be a side band in the response of the driving frequency and resonance phenomenon occurs in that beat frequency. If F_s is well above F_o, the system can be approximated as a first order system in the control bandwidth of concern. Such a dominant pole assumption fails if F_s is very close to F_o. In this case, control to output transfer function $G_{vf}(s)$ will be given by the third order system given by

$$G_{vf}(s) = \frac{G_{vf,o}}{\Delta(s)} \tag{2}$$

where $G_{vf,o}$ is the dc gain and $\Delta(s)$ is the system characteristic polynomial as in in the following.

$$\Delta(s) = \left(1 + \frac{s}{\omega_{pf}}\right)\left(1 + \frac{s}{Q\omega_{pn}} + \frac{s^2}{\omega_{pn}^2}\right) \tag{3}$$

As is mentioned above, the low frequency single pole is caused by the output filter, and the 2nd order pole occurs at the beat frequency. Thus it is reasonable to think those pole locations are

$$\omega_{pf} \approx \frac{1}{C_f(R_L \parallel G_{vi,o})} \tag{4}$$

$$\omega_{pn} = \omega_s - \omega_o \tag{5}$$

In the above calculation, $G_{vi,o}$ is dc gain of the current-to-output transfer function.

B. Return path : control signal

In the frequency controlled system, VCO gate pulses and determines the switching frequency. Small signal transfer gain of VCO can be modeled by the following formula.

$$K_{VCO} = \left.\frac{df_s}{dv_{c2}}\right|_{f_s=F_s} \approx \frac{F_{s,max} - F_{s,min}}{V_{c2,max} - V_{c2,min}} \tag{6}$$

where $F_{s,max}$ and $F_{s,min}$ is the maximum and minimum of the possible frequency span and $V_{c2,max}$ and $V_{c2,min}$ is the corresponding input voltage of VCO used in the system.

After a control voltage is generated in the feedback controller from the error signal between the reference voltage and the sensed voltage, it returns back to the transmitter side by the wireless feedback link. Assume the communication is stable and reliable, the latency introduced in the wireless link is modeled by a time delay function

$$D(s) = e^{-s\tau} \approx \frac{1 - s\tau/2 + (s\tau)^2/12}{1 + s\tau/2 + (s\tau)^2/12} \tag{7}$$

where τ is the average delay time of the communication channel and herein is used the second order Padé approximation [10]. It approximates the delay effect as an all-pass-filter which introduces two poles and two right-half-plane (RHP) zeros. Because of the additional strong phase drop it causes, the delay time will mainly determine the control loop bandwidth.

Assume the switching frequency is designed well above the resonant frequency and the load quality factor is sufficiently high, the overall loop gain can be approximated by a single pole system when the delay time is relatively small. On the contrary, if the feedback wireless link suffers severe latency, it can be approximated by double pole and double right-half-plane zero system.

In the former case, a type-2 or PI compensator is enough and its compensator zero is placed to cancel out the power stage pole. Then the loop crossover is placed well below both $1/\pi\tau$ and f_{pn} to prevent additional phase lag and avoid conditional stability situation which may deteriorate the phase margin in the loop gain. If the control crossover frequency, f_C is selected, k_p and k_i of the PI compensator are determined by

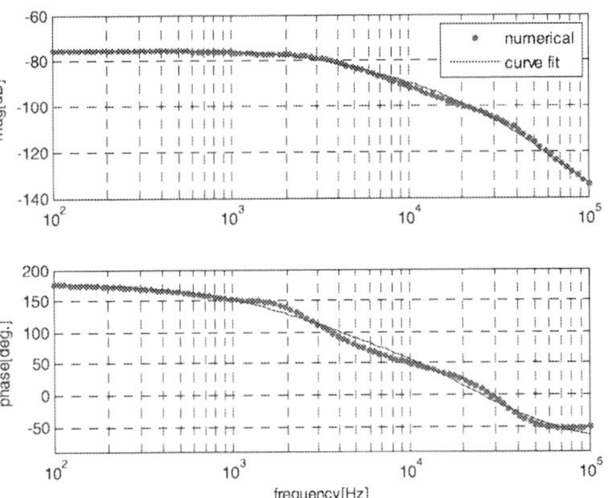

Fig. 6. Control to output transfer function, $G_{vf}(s)$

TABLE II. TRANSFER FUNCTION PARAMETERS

Parameter	$G_{vf}(s)$			
	$G_{vf,o}$	ω_{pf} (krad/s)	ω_{pn} (krad/s)	Q
Value	1.75×10^{-4}	$2\pi\cdot2.24$	$2\pi\cdot22.7$	0.519

TABLE I. SYSTEM PARAMETERS

Parameter	V_s	V_o	I_o	F_s	F_o	Q_o	C_e	L_r	N_1	N_2	C_r	K_{vco}	K_v
Value	19	10	0.5	293	280	1.92	272	21.8	7.5	4.3	10	95.7	1
Unit	V	V	A	kHz	kHz	Ω/Ω	pF	uH			uF	kHz/V	V/V

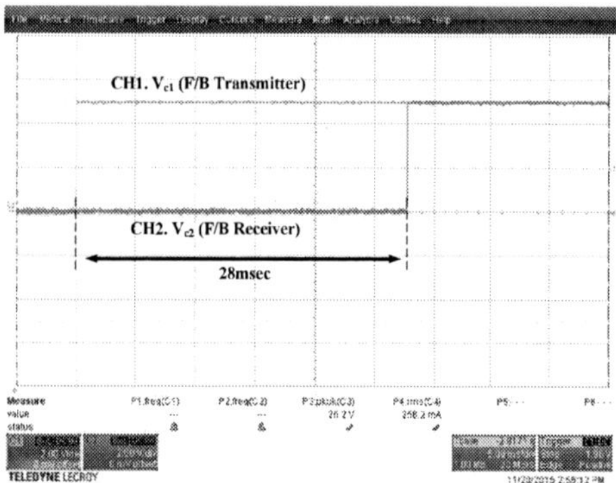

Fig. 7. Measurements of delay in the Bluetooth feedback link (Ch1 : V_{c1} (F/B Transmitter), Ch2 : V_{c2} (F/B Receiver))

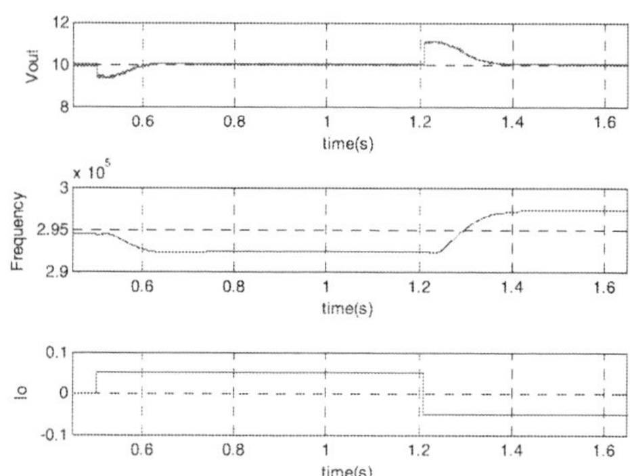

Fig. 9. Output voltage responses to load step changes: simulation

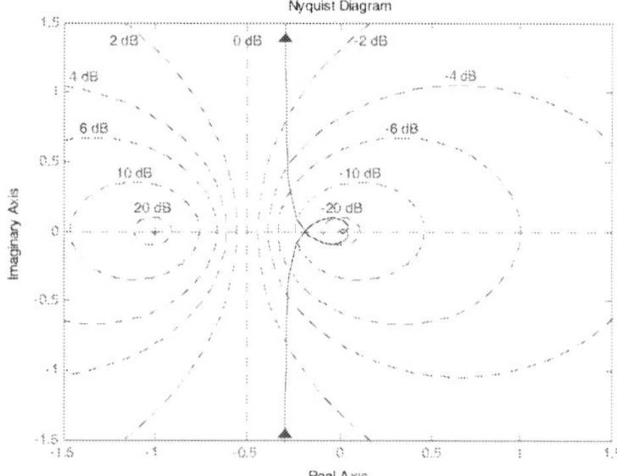

Fig. 8 Nyquist diagram of T(s)

$$k_i = \frac{2\pi f_C}{K_{VCO} G_{vf0} K_V}, \quad k_p = k_i (R \,||\, G_{vi,0}) C_f \tag{8}$$

However, the latter case needs a type-3 controller placing two compensator zeroes to cancel out both the power stage pole and the delay pole.

IV. Design and Implementation

The power stage has been designed to transmit 5W power through a pair of square plates with dimension of 10cm x 10cm separated by 0.15mm air gap and supplies the output voltage of 10V up to 0.5A load current. Table I shows the parameters used in the system construction.

Using the circuit parameter shown in Table I, PSIM schematic has been constructed and transfer functions of the power stage have been extracted by ac sweep analysis. After simulation has been done, pole-zero locations are identified by curve fitting algorithm [11] and are summarized in Table II. The theoretical study on the power stage transfer function given by Eqs. (2)-(5) predicts that the low frequency pole and the high frequency natural poles exit in 2.24kHz and 22.7kHz, respectively. Fig. 6 compares two $G_{vf}(s)$'s obtained by the swept sine analysis and the theoretical predictions, where both results show good agreement.

Meanwhile, the wireless feedback link has been implemented in Serial Port Profile (SPP) of Bluetooth 2.1 protocol compatible with IEEE 802.15.1[12]. The serial communication in 115,200 bps introduces about 87usec delay. However, because of an additional delay caused by internal polling mechanism, its minimum bound is reported as large as 12~13ms [13]. This delay effect can be observed in the hardware results shown in Fig. 7 and hence τ=35msec is used in this paper. Incorporating the double pole and double RHP zero at 16Hz introduced by the wireless link latency, designed compensator provides the f_C=1.33Hz with phase margin of 73.1 degree and gain margin of 14.6dB at the full load condition.

Fig. 8 shows the Nyquist plot of the loop gain, T(s), including the controller that has been designed. Fig. 9 shows the step response when the load current is slightly increased by 10% and then decreased by 10% from the 0.5A full load condition, and the system shows good performance in the output regulation.

V. Conclusions

This paper has analyzed the C-WPT system which involves resonant power link and out-of-band feedback link to regulate the output voltage even without battery in the target object. The practical delay effects are modeled and included in the compensator design. The performance has been verified by PSIM simulation.

Acknowledgement

This research was supported by Basic Science Research Program through the National Research Foundation of Korea (NRF) funded by the Ministry of Education (grant number: NRF-2014R1A1A2059772).

References

[1] C. Liu, A. P. Hu, and N. K. C. Nair, "Modelling and analysis of a capacitively coupled contactless power transfer system," *IET Power Electronics*, vol. 4, no. 7, pp. 808-815, 2011.

[2] T. Diekhans and R. W. De Donker, "A Dual-Side Controlled Inductive Power Transfer System Optimized for Large Coupling Factor Variations and Partial Load, *IEEE Transactions on Power Electronics*, Vol. 30, No. 11, pp. 6320-6328, Nov., 2015.

[3] C.-G. Kim, D.-H. Seo, J.-S. You, J.-H. Park, and Bo H. Cho, "Design of a Contactless Battery Charger for Cellular Phone," *IEEE Transactions on Industrial Electronics*, vol. 48, no. 6, pp. 1238-1247, 2001.

[4] R. Rosshard, U. Badsübner, J. W. Kolar, and I. Stevanovic, "Comparative evaluation of control methods for inductive power transfer," *Proceedings of ICRERA*, 2012, pp.1-6.

[5] S.-J. Choi and H.-S. Choi, "Capacitive Wireless Power Transfer System with Dobule Matching Transformers for Reduced Stress and Extended ZVS Range," *IEEE INTELEC 2015*, Oct., pp.791-796.

[6] K. Finkenzeller, *RFID Handbook: Fundamentals and Applications in Contactless Smart Cards and Identification*, 2_{nd} ed. Wiley & Sons, 2003.

[7] A. F. Witulski and R. W. Erickson, "Small Signal Equivalent Circuit Modelling of the Series Resonant Converter," *IEEE PESC 1987 Record*, pp. 693-704.

[8] V. Vorperian, "Approximate Small-Signal Analysis of the Series and the Parallel Resonant Convertes," *IEEE Transactions on Power Electronics*, vol. 4, no. 1, Jan. 1989.

[9] POWERSIM, *PSIM User's Guide*, ver. 9, May, 2010.

[10] J. R. Partington, "Some Frequency-domain Approaches to the Model Reduction of Delay Systems," *Annual Reviews in Control*, vol. 28. pp. 65-73, 2004.

[11] Levi, E. C. "Complex-Curve Fitting," *IRE Trans. on Automatic Control*, vol. AC-4, pp. 37-44, 1959.

[12] Bluetooth SIG , *Bluetooth 2.1 Specification*, https://www.bluetooth.org/

[13] M.J. Moron, R. Luque, E. Casilari and A. Diaz-Estrella, "Minimum Delay Bound in Bluetooth Transmissions with Serial Port Profile," *Electronics Letters*, vol. 44, pp.1099-1100, Aug., 2008.

Introducing Fourier-based modeling and control of Active-Bridge Converters

B.J.D. Vermulst, J.L. Duarte, C.G.E. Wijnands, E.A. Lomonova
Electromechanics and Power electronics group
Eindhoven University of Technology
Eindhoven, the Netherlands

Abstract—**Modeling of power converters is usually done using cycle-average or piecewise-linear methods. Piecewise-linear methods operate by selecting the model corresponding to the current switching state. When the amount of possible states increase, the number of models increases as well. For model-predictive control (MPC), this results in an increasing computational effort, which limits the applicability of MPC to relatively simple models. This paper presents a Fourier-based modeling method to significantly reduce power-converter model complexity. Furthermore, a "voltage-balance" control algorithm is proposed using this method, which is applied to Active-Bridge type converters. The resulting closed-form algebraic solution yields a reduced circulating current and has considerably lower computational effort compared to piecewise-linear models (256 models vs. 1 model). An arbitrary amount of Active-Bridges can be used, which enables the use of scalable Active-Bridge converters. Results show a substantial reduction of the converter currents when the proposed method is compared to phase-shift control, which also significantly improves the EU efficiency, from 88.3% to 99.0% when applied to a QAB converter.**

CONVENTIONS

$x(t)$	instantaneous value of x at time t
$\langle x(t)\rangle$	average value of x over all t
X	RMS or dc value of $x(t)$ in mains frequency time domain
X'	RMS or dc value of $x'(t)$ in switching frequency time domain
$\hat{X}$	peak value of $x(t)$
$\underline{X}$	complex scalar
$\mathbf{x}$, $\mathbf{X}$	single-column arrays
$\mathbf{x}^*$	complex conjugate of $\mathbf{x}$
$\mathbf{x}^T$	transpose of $\mathbf{x}$
$\mathbf{1}^i$	single-column array with i elements, all 1
$\Re(\underline{X})$	real part of $\underline{X}$
$\Im(\underline{X})$	imaginary part of $\underline{X}$
$\sum \mathbf{x}$	sum of all elements in $\mathbf{x}$
j	imaginary unit

where x can be substituted by an arbitrary symbol.

I. INTRODUCTION

ACTIVE-Bridge topologies are an increasingly popular choice for converters in high-performance applications. For example, the Dual Active-Bridge (DAB) topology is used in high-power dc-dc converters, and the Quad Active-Bridge (QAB, Fig. 2) topology is used for both dc-dc and ac-dc conversion [1].

Active-Bridge (AB) converters have numerous advantages, with features such as soft-switching, isolation, bi-directional power flow and current controlled operation with a straight-forward phase-shift control algorithm [1], [2]. However, the circulating current through the converter increases when the voltage ratio between its ABs is increased. To alleviate this issue, more advanced control algorithms have been developed to extend the soft-switching range and reduce these currents [3]–[6]. These algorithms are generally based on a multi-mode (piecewise-linear) evaluation of the waveforms for all operating conditions, where model-predictive control or look-up tables are implemented to select suitable switching states. This improves the converter efficiency by a large amount, at the cost of having a relatively complex and computational intensive controller.

When the amount of ABs increases, for instance with a Quad Active-Bridge (QAB) with its 256 switching states, the multi-mode methods become complex to design as well as to implement, because the number of operating modes increases with a factor 4 for each additional port in such converters. Therefore, these control methods are not suitable for converters with a large number of ABs. The phase-shift algorithm [1], on the other hand, scales linearly and stays comprehensible even for a large number of ABs. Using this method, however, the converter will suffer from large circulating currents and a limited soft-switching operating region.

The Fourier-based modeling method proposed in this paper, uses a Fourier analysis of the waveforms that are produced in power converters to find an analytical converter model. The resulting model is used to derive a system decoupler and control algorithm which minimizes undesired properties, such as circulating currents, while maintaining the requested output power(s). The proposed method is applied to a generic model for AB-converters and a comparison is made between phase-shift control and the proposed method.

A. Outline

The paper is structured as follows. First, in Section II, the generic Fourier-based modeling method of the AB-converter is described. This model is used in Section III, where the optimization method is described and the closed-form solutions are deduced. The results of this optimization are applied to a QAB-converter in Section IV, and conclusions are written in Section V.

978-1-4673-9551-9/16 $31.00 © 2016 IEEE

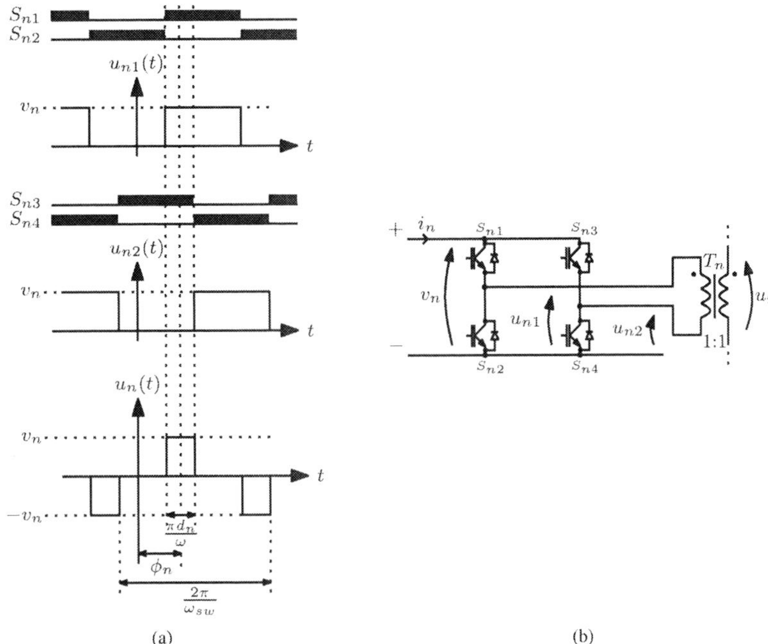

Fig. 1. Generic representation of AB n shown in (b), which has the waveforms shown in (a) with leg voltages u'_{n1} and u'_{n2}, duty cycle d_n, phase shift ϕ_n, bus voltage v'_n, and switching frequency ω_{sw}. The gate signals are indicated by black bars.

II. Fourier-based modeling

To obtain a converter model based on Fourier-analysis, the following steps are required. First, all switched-node voltages must be represented by waveforms. These voltage waveforms can have, for instance, an amplitude, a duty cycle, a period time and a phase shift, as shown in Fig. 1. Then, the waveforms are transformed to Fourier-series, which can be used to derive the currents through inductors and the voltages of capacitors in the circuit. Finally, the output currents, powers and voltages can be found by calculating their Fourier-coefficients, resulting in a closed-form converter model.

This paper uses a QAB ac-dc converter topology to demonstrate the Fourier-based modeling method. This topology is depicted in Fig. 2. All transformers are assumed to have 1:1 winding ratio. In view of the quantities in Fig. 1, the following arrays of variable values can be defined when the number of ABs equals i:

$$\mathbf{u}'(t) = \begin{pmatrix} u'_1(t) \\ u'_2(t) \\ \vdots \\ u'_i(t) \end{pmatrix}, \ \mathbf{v}' = \begin{pmatrix} v'_1 \\ v'_2 \\ \vdots \\ v'_i \end{pmatrix}, \ \mathbf{d} = \begin{pmatrix} d_1 \\ d_2 \\ \vdots \\ d_i \end{pmatrix}, \ \mathbf{\Phi} = \begin{pmatrix} \phi_1 \\ \phi_2 \\ \vdots \\ \phi_i \end{pmatrix} \tag{1}$$

where $\mathbf{u}'(t)$ contains the set of block-shaped voltages $u'_n(t)$ at the secondary side of the high-frequency transformers. Moreover, $\mathbf{v}'$ is the set of AB bus voltages in the switching-frequency time domain. It is assumed that the switching frequency is much higher than the frequencies present in v_n such that the may can be assumed that $v_n(t) = v_n(t+\Delta t)$, with Δt sufficiently small, such that the amplitudes v_n of the waveforms generated by the full bridges can be considered constant in Δt. Variables that use this assumption are indicated by the $'$-symbol. The set of phase-shifts between the AB switching legs is represented by $\mathbf{\Phi}$, and $\mathbf{d}$ is the set of resulting duty-cycles for $u'_n(t)$.

The voltage generated by the ABs of an AB-converter can be schematically represented by a network of voltage sources, as shown in Fig. 3. For example, $n \in [R, S, T, dc]$ for the QAB converter. Based on the definitions (1), a compact array notation for the Fourier-series representation of the block-shaped voltages $u'_n(t)$ is used, as follows:

$$\mathbf{u}'(t) = \frac{2\mathbf{v}'}{\pi} \sum_{k=1}^{\infty} \frac{\sin((\omega_{sw}t + \mathbf{\Phi} + 0.5\pi\mathbf{d})(2k-1))}{2k-1} - \frac{\sin((\omega_{sw}t + \mathbf{\Phi} - 0.5\pi\mathbf{d})(2k-1))}{2k-1}. \tag{2}$$

Furthermore, the function $\mathbf{u}'(t)$ can be re-arranged into

$$\mathbf{u}'(t) = \sum_{k=1}^{\infty} \mathbf{Ua}'(k) \cos((2k-1)\omega_{sw}t) + \mathbf{Ub}'(k) \sin((2k-1)\omega_{sw}t) \tag{3}$$

with

$$\mathbf{Ua}'(k) = \mathbf{U}'(k)\Big(\sin((2k-1)(\mathbf{\Phi} + 0.5\pi\mathbf{d})) - \sin((2k-1)(\mathbf{\Phi} - 0.5\pi\mathbf{d}))\Big), \tag{4}$$

$$\mathbf{Ub}'(k) = \mathbf{U}'(k)\Big(\cos((2k-1)(\mathbf{\Phi} + 0.5\pi\mathbf{d})) - \cos((2k-1)(\mathbf{\Phi} - 0.5\pi\mathbf{d}))\Big) \tag{5}$$

978-1-4673-9551-9/16 $31.00 © 2016 IEEE

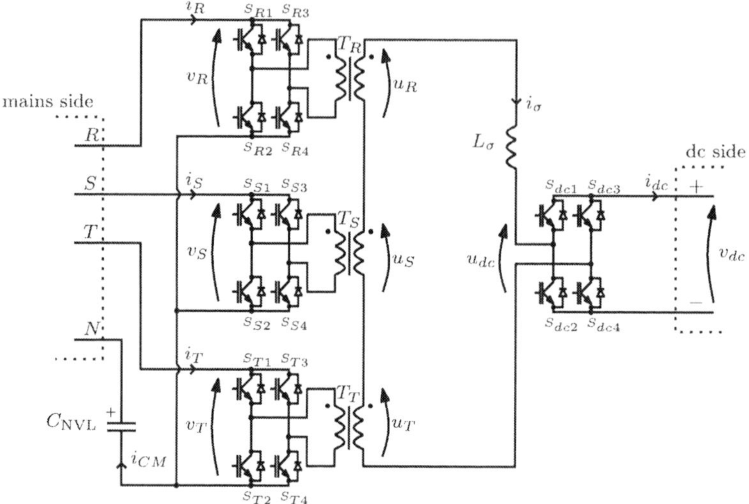

Fig. 2. Example Active-Bridge topology: the Quad Active-Bridge (QAB) for ac-dc conversion [1].

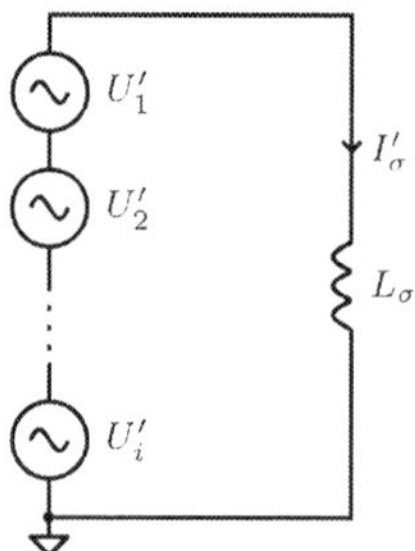

Fig. 3. Generic model of Active-Bridge type converters.

and $\mathbf{U}'(k) = \frac{2\mathbf{v}'}{\pi(2k-1)}$. The current contribution through L_σ of source voltage $v'_n(t)$ is equal to the integral of the sum of the inductor voltage divided by the reactance of the inductor, being

$$\mathbf{i}'_\sigma(t) = \frac{\int \mathbf{u}'(t)\,dt}{L_\sigma} = \frac{\sum_{k=1}^{\infty} -\mathbf{Ub}'(k)\cos((2k-1)\omega_{sw}t)}{(2k-1)\omega_{sw}L_\sigma} + \frac{\sum_{k=1}^{\infty}\mathbf{Ua}'(k)\sin((2k-1)\omega_{sw}t)}{(2k-1)\omega_{sw}L_\sigma} \quad (6)$$

Hence, the total inductor current is equal to the superposition of all currents, and therefore found to be

$$i'_\sigma(t) = \sum_{k=1}^{\infty} I'_{\sigma a}(k)\cos((2k-1)\omega_{sw}t) + I'_{\sigma b}(k)\sin((2k-1)\omega_{sw}t) \quad (7)$$

with the Fourier-coefficients

$$I'_{\sigma a}(k) = \frac{-\sum_{n=1}^{i}\mathbf{Ub}'_n(k)}{(2k-1)\omega_{sw}L_\sigma},\quad I'_{\sigma b}(k) = \frac{\sum_{n=1}^{i}\mathbf{Ua}'_n(k)}{(2k-1)\omega_{sw}L_\sigma}. \quad (8)$$

The inductor RMS current can be found by

$$I'_\sigma = \sqrt{0.5\sum_{k=1}^{\infty}\left((I'_{\sigma a}(k))^2 + \left(I'_{\sigma b}(k)\right)^2\right)}. \quad (9)$$

The active power delivered by each source for harmonic k can then be found by

$$\mathbf{P}'(k) = 0.5 I'_{\sigma a}(k)\,\mathbf{Ua}'(k) + 0.5 I'_{\sigma b}(k)\,\mathbf{Ub}'(k), \quad (10)$$

and the total power equals

$$\mathbf{P}' = \sum_{k=1}^{\infty}\mathbf{P}'(k). \quad (11)$$

The AB input current can be found by

$$i'_n = \frac{P'_n}{v'_n}. \quad (12)$$

A. Truncation

The above equations describe the infinite series. When truncating, the equations become

$$I'_\sigma = \sqrt{0.5\sum_{k=1}^{k_{\max}}\left((I'_{\sigma a}(k))^2 + \left(I'_{\sigma b}(k)\right)^2\right)}. \quad (13)$$

$$\mathbf{P}' = \sum_{k=1}^{k_{\max}} 0.5 I'_{\sigma a}(k)\,\mathbf{Ua}'(k) + 0.5 I'_{\sigma b}(k)\,\mathbf{Ub}'(k). \quad (14)$$

To obtain a relatively simple expression that can be used for optimization purposes, use only the first harmonic ($k_{\max} = 1$). By doing this, the coefficients become

$$\mathbf{Ua}' = \mathbf{U}'\left(\sin(\boldsymbol{\phi} + 0.5\mathbf{d}\pi) - \sin(\boldsymbol{\phi} - 0.5\mathbf{d}\pi)\right) = 2\mathbf{U}'\cos(\boldsymbol{\phi})\sin(0.5\mathbf{d}\pi) \quad (15)$$

$$\mathbf{Ub}' = \mathbf{U}'\left(\cos(\boldsymbol{\phi} + 0.5\mathbf{d}\pi) - \cos(\boldsymbol{\phi} - 0.5\mathbf{d}\pi)\right) = -2\mathbf{U}'\sin(\boldsymbol{\phi})\sin(0.5\mathbf{d}\pi). \quad (16)$$

with $\mathbf{U}' = 2\mathbf{v}'/\pi$. Now, by defining

$$\mathbf{X} = \sin(0.5\pi\mathbf{d}) \quad (17)$$

the power of source n when using only the first harmonic is found by

$$\begin{aligned} P'_n &= 0.5 I'_{\sigma a}(1)\,\mathrm{Ua}'_n(1) + 0.5 I'_{\sigma b}(1)\,\mathrm{Ub}'_n(1) \\ &= \frac{8}{\omega_{sw} L_\sigma \pi^2} v'_n X_n (v'_1 X_1 \sin(\phi_1 - \phi_n) + ... + v'_i X_i \sin(\phi_i - \phi_n)) \end{aligned} \tag{18}$$

and the current through L_σ is found by

$$I'_\sigma = \sqrt{0.5 \left(\frac{4 v'_n X_n}{\pi \omega_{sw} L_\sigma}\right)^2 \left(\left(\sum_{n=1}^{i} \sin(\phi_n)\right)^2 + \left(\sum_{n=1}^{i} \cos(\phi_n)\right)^2\right)}. \tag{19}$$

III. Voltage-balance control

With phase-shift control the efficiency at certain operating points is compromised [1], because the resistive losses will become a dominant factor due to the large circulating currents. In this section the voltage-balance control (VBC) method is provided which gives control over the current through L_σ while maintaining the same AB output current i_n. Via Newton-optimization a control algorithm is established that reduces the circulating current to a minimum.

A. Optimization

The RMS value of the current through the leakage inductance L_σ determines the resistive losses in the converter. To minimize the RMS current through the inductor, the following optimization should be carried out:

$$\text{minimize} \quad I'_\sigma \tag{20}$$

$$\text{with} \quad \begin{pmatrix} P'_1 \\ P'_2 \\ \vdots \\ P'_i \end{pmatrix} = \begin{pmatrix} P'_{1,\text{set}} \\ P'_{2,\text{set}} \\ \vdots \\ P'_{i,\text{set}} \end{pmatrix}, \tag{21}$$

with P'_n the active power of AB n, resulting from the first-harmonic approximation ($k_{\max} = 1$). $P'_{n,\text{set}}$ is the set-point of the active power of AB n. Here, the optimization algorithm selects the duty cycles d_n and phase shifts ϕ_n to have the desired output power with the minimum amount of (RMS) current through the converter.

Because (18) is non-invertible due to the trigonometric functions present, first a linearization of this function is required. These trigonometric functions can be linearized by assuming $\sin(x) = x$, which is valid when it holds that $|x \bmod 2\pi| \ll \pi/2$. Because in general it holds that the phase shifts will always be in the range $(-\pi/2 < \phi_n - \phi_m < \pi/2)\ \forall\ n, m$ where $\{n, m \in \mathbb{N} | n \leq i, m \leq i\}$, the linearization is accurate. This results in

$$\begin{aligned} P'_n &= \frac{8}{\omega_{sw} L_\sigma \pi^2} v'_n X_n (v'_1 X_1 \sin(\phi_1 - \phi_n) + ... + v'_i X_i \sin(\phi_i - \phi_n)) \\ P'_n &\approx \frac{8}{\omega_{sw} L_\sigma \pi^2} v'_n X_n \sum_{m=1}^{i} v'_m X_m (\phi_m - \phi_n). \end{aligned} \tag{22}$$

This linearization is then used to calculate the phase shifts, as follows:

$$\phi_n = \frac{\omega_{sw} L \pi^2}{8\left(v'_1 X_1 + v'_2 X_2 + ... + v'_i X_i\right)} \left(\frac{P_1}{v'_1 X_1} - \frac{P_n}{v'_n X_n}\right). \tag{23}$$

1) Cost function: Minimizing I'_σ in (19) yields the same $\boldsymbol{\phi}$ and $\mathbf{d}$ as minimizing

$$2\left(\frac{\pi \omega_{sw} L_\sigma I'_\sigma}{4}\right)^2, \tag{24}$$

because the parameters ω_{sw} and L_σ are both kept constant in the optimization procedure. Moreover, it is assumed that the first harmonic is dominant in causing circulating currents. The cost function W is therefore defined as follows.

$$\begin{aligned} W &= 2\left(\frac{\pi \omega_{sw} L_\sigma I'_\sigma}{4}\right)^2 \\ &= \left(\sum_{n=1}^{i} v'_n X_n \sin(\phi_n)\right)^2 + \left(\sum_{n=1}^{i} v'_n X_n \cos(\phi_n)\right)^2 \end{aligned} \tag{25}$$

2) Constraints: While the minimization of W results in the lowest circulating currents, it does not guarantee the AB power P'_n equals the set-point $P'_{n,\text{set}}$. This is, however, a constraint which needs to be met. Therefore

$$\sum_{n=1}^{i} \left(P'_n - P'_{n,\text{set}}\right)^2 = 0, \tag{26}$$

to ensure all outputs have the desired power. Equation (26) is valid when $|\phi_m - \phi_n| \ll \pi/2$, according to the linearization used for (22). It can be shown that a strong approximation of the criterion in (26) can be accomplished when the sum of all squared phase-shift cross-differences $(\phi_m - \phi_n)$ equals $2\alpha^2$, as follows

$$\left\| \boldsymbol{\phi} \times \left(\mathbf{1}^i\right)^T - \left(\boldsymbol{\phi} \times \left(\mathbf{1}^i\right)^T\right)^T \right\|_2 = 2\alpha^2, \tag{27}$$

where α is a variable introduced to fix the maximum cross-difference between the phase shifts, to provide a trade-off between model accuracy (linearization error) and the circulating current. The variable should be in the linear range of the model, equal to $0 < \alpha < \frac{\pi}{2}$. Moreover, $\mathbf{1}^i$ is a single-column array with i elements all equal to 1, and $\|\mathbf{x}\|_2 = \sum_{m=1}^{i} \sum_{n=1}^{i} x_{m,n}^2$ with $\mathbf{x}$ an $i \times i$ array. Note that, when (27) is met, the desired output powers are guaranteed to be within the linearization errors of (23), and the error becomes smaller when α is smaller.

3) Solution for two ABs: Using the method of Lagrange multipliers [7], the constraint (27) is taken into account together with the cost function (25) as follows:

$$\begin{aligned} &\text{minimize} \quad H = W + \lambda(\Gamma - 2\alpha^2) \\ &\text{where} \quad \Gamma = \left\| \boldsymbol{\phi} \times \left(\mathbf{1}^i\right)^T - \left(\boldsymbol{\phi} \times \left(\mathbf{1}^i\right)^T\right)^T \right\|_2. \end{aligned} \tag{28}$$

Now, the optimum is found when for each parameter (X_n, ϕ_n) the derivatives are equal to zero. In order to reduce the complexity of the optimization problem, a two-AB model is used ($i = 2$). In subsection III-B, the results are converted to fit a model with i ABs.

978-1-4673-9551-9/16 $31.00 © 2016 IEEE

First, the cost function (25) can be simplified for $i = 2$ by using the approximation

$$W = \left(\sum_{n=1}^{i} v'_n X_n \sin(\phi_n)\right)^2 + \left(\sum_{n=1}^{i} v'_n X_n \cos(\phi_n)\right)^2 \tag{29}$$

$$= (v'_1 X_1)^2 + (v'_2 X_2)^2 + 2(v'_1 X_1)(v'_2 X_2)\cos(\phi_1 - \phi_2) \tag{30}$$

$$\approx (v'_1 X_1)^2 + (v'_2 X_2)^2, \tag{31}$$

which is valid because $\cos(\phi_1 - \phi_2) \to 0$ as α is selected to be large to optimize efficiency, as described in the next paragraph.

Newton's method is used to find the optimum for $i = 2$, where each term of the Jacobian is set to zero. The solution is obtained as follows:

$$\text{minimize} \quad H = (v'_1 X_1)^2 + (v'_2 X_2)^2 + \lambda\left((\phi_1 - \phi_2)^2 - \alpha^2\right) \tag{32}$$

After substitution of ϕ_n with (23) yields the following optimization assignment:

$$\begin{aligned}\text{minimize} \quad H = &(v'_1 X_1)^2 + (v'_2 X_2)^2 \\ &+ \lambda\left(\left(\frac{\omega_{sw} L_\sigma \pi^2}{8(v'_1 X_1 + v'_2 X_2)}\right)^2 \left(\frac{P'_1}{v'_1 X_1} - \frac{P'_2}{v'_2 X_2}\right)^2 - \alpha^2\right).\end{aligned} \tag{33}$$

By power conservation it holds that $P_2 = -P_1$, such that the Jacobian of H is equal to

$$\mathbf{J} = \begin{pmatrix} 2\left(v'_1 X_1\right) - \lambda\left(\frac{\omega_{sw} L_\sigma \pi^2}{8}\right)^2 \left(\frac{2\left(\frac{P'_1}{v'_1 X_1} + \frac{P'_1}{v'_2 X_2}\right)^2}{(v'_1 X_1 + v'_2 X_2)^3} + \frac{2P'_1\left(\frac{P'_1}{v'_1 X_1} + \frac{P'_1}{v'_2 X_2}\right)}{(v'_1 X_1)^2 (v'_1 X_1 + v'_2 X_2)^2}\right) \\ 2\left(v'_2 X_2\right) - \lambda\left(\frac{\omega_{sw} L_\sigma \pi^2}{8}\right)^2 \left(\frac{2\left(\frac{P'_1}{v'_1 X_1} + \frac{P'_1}{v'_2 X_2}\right)^2}{(v'_1 X_1 + v'_2 X_2)^3} + \frac{2P'_1\left(\frac{P'_1}{v'_1 X_1} + \frac{P'_1}{v'_2 X_2}\right)}{(v'_2 X_2)^2 (v'_1 X_1 + v'_2 X_2)^2}\right) \\ \left(\frac{\omega_{sw} L_\sigma \pi^2}{8}\right)^2 \frac{\left(\frac{P'_1}{v'_1 X_1} + \frac{P'_1}{v'_2 X_2}\right)^2}{(v'_1 X_1 + v'_2 X_2)^2} - \alpha^2 \end{pmatrix} \tag{34}$$

with

$$\mathbf{J} = \begin{pmatrix} \frac{\partial H}{\partial (v'_1 X_1)} \\ \frac{\partial H}{\partial (v'_2 X_2)} \\ \frac{\partial H}{\partial \lambda} \end{pmatrix}. \tag{35}$$

Solving $\mathbf{J} = \mathbf{0}$ for $v'_1 X_1$ and $v'_2 X_2$ results in

$$\mathbf{v'X} = \frac{\sqrt{|\mathbf{P'}|\pi^2 \omega_{sw} L_\sigma}}{\sqrt{8\alpha}} \tag{36}$$

$$\mathbf{X} = \frac{\sqrt{|\mathbf{P'}|\pi^2 \omega_{sw} L_\sigma}}{\sqrt{8\alpha}\mathbf{v'}} \tag{37}$$

Note that the multiplication in (36) and the division in (37) are both element-wise. Then, continue by calculating the elements of $\mathbf{d}$ and $\boldsymbol{\phi}$ using (17) and (18), resulting in

$$d_n = \frac{2}{\pi} \arcsin(X_n) \tag{38}$$

$$\phi_n = \frac{\omega_{sw} L_\sigma \pi^2}{8} \frac{1}{v'_1 X_1 + v'_2 X_2} \left(\frac{P'_1}{v'_1 X_1} - \frac{P'_n}{v'_n X_n}\right). \tag{39}$$

Substitution with $\alpha = \pi/2$ and assuming $P'_1 = -P'_2 \geq 0$ yields

$$\phi_1 \propto P'_1 - P'_1 = 0 \rightarrow \phi_1 = 0 \tag{40}$$

$$\begin{aligned}\phi_2 &= \frac{\omega_{sw} L_\sigma \pi^2}{8} \frac{1}{v'_1 X_1 + v'_2 X_2} \left(\frac{P'_1}{v'_1 X_1} - \frac{P'_2}{v'_2 X_2}\right) \\ &= \frac{\pi^2 \omega_{sw} L_\sigma}{8} \frac{\sqrt{8\alpha}}{\sqrt{\pi^2 \omega_{sw} L_\sigma}\left(\sqrt{|P'_1|} + \sqrt{|P'_2|}\right)} \\ &\qquad \frac{\sqrt{8\alpha}}{\sqrt{\pi^2 \omega_{sw} L_\sigma}} \left(\frac{P'_1}{\sqrt{|P'_1|}} - \frac{P'_2}{\sqrt{|P'_2|}}\right) \\ &= \frac{\pi^2 \omega_{sw} L_\sigma}{8} \frac{\sqrt{8\alpha}}{\sqrt{\pi^2 \omega_{sw} L_\sigma}\, 2\sqrt{P'_1}} \frac{\sqrt{8\alpha}\, 2\sqrt{P'_1}}{\sqrt{\pi^2 \omega_{sw} L_\sigma}} \\ &= \alpha = \pi/2,\end{aligned} \tag{41}$$

which meets the constraint $(\phi_1 - \phi_2)^2 = \alpha^2$. This shows that the proposed solution utilizes a fixed phase shift and a varying first-harmonic magnitude, opposed to the varying phase shift and a static fist-harmonic magnitude utilized by the phase-shift control algorithm. Furthermore, $v'_1 X_1 = v'_2 X_2$ because their power is equal, and the maximum output power is achieved when $\max(\mathbf{X}) = 1$.

B. Extension to i ABs: voltage-balance method

To extend the solution from the previous section to i ABs, the following method is used.When assuming ideal efficiency, the sum of the powers of ABs delivering power (P'_{pos}) is equal to the sum of the powers of ABs receiving power (P'_{neg}), as follows:

$$P'_{\text{pos,set}} = \sum_{n=1}^{i} \left\{P'_{\text{n,set}} : P'_{\text{n,set}} \geq 0\right\} \tag{42}$$

$$P'_{\text{neg,set}} = \sum_{n=1}^{i} \left\{P'_{\text{n,set}} : P'_{\text{n,set}} < 0\right\} \tag{43}$$

$$P'_{\text{pos,set}} = P'_{\text{neg,set}}. \tag{44}$$

This is similar to an $i = 2$ situation where all positive power is processed by one source $U'_{\text{pos}} = v'_{\text{pos}} X_{\text{pos}}$, as well as the negative power is processed by another source $U'_{\text{neg}} = v'_{\text{neg}} X_{\text{neg}}$, as shown in Fig. 4(a). The following therefore assumes a two-ABs problem. From (36), it can be seen that for constant power P_{pos}, the magnitude $v'_{\text{pos}} X_{\text{pos}}$ is also constant, with its value equal to

$$v'_{\text{pos}} X_{\text{pos}} = \frac{\sqrt{P_{\text{pos}} \pi^2 \omega_{sw} L}}{\sqrt{8\alpha}} \tag{45}$$

with P_{pos} the sum of the powers of the positive sources, and $v'_{\text{pos}} X_{\text{pos}}$ the sum of the phasor lengths of all positive sources. According to (36) it holds that because $P_{\text{pos}} = P_{\text{neg}}$ the phasor lengths are equal, being

$$v'_{\text{pos}} X_{\text{pos}} = v'_{\text{neg}} X_{\text{neg}}. \tag{46}$$

Therefore, the proposed method is named the "voltage-balance method". It is shown that the phase shift is a constant $\pi/2$ by

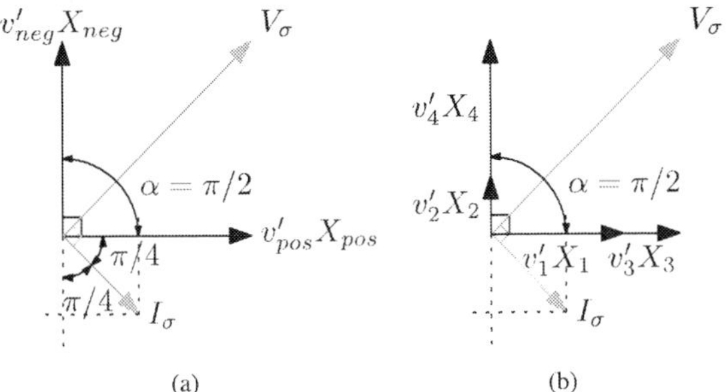

Fig. 4. (a) Phasor distribution of optimal solution, and (b) example for dividing powers across sources according to the voltage-balance method.

TABLE I
SIMULATION PARAMETERS

Parameter	Value	Unit
Switching frequency	45	kHz
DC voltage	425	V
AC RMS voltage	400	V
C_{NVL} voltage	375	V
C_{NVL}	100	μF
L_σ	90	μH

(41). Therefore, the phase shift of the positive and negative sources equals respectively

$$\phi_{\text{pos}} = -\alpha/2 \tag{47}$$

$$\phi_{\text{neg}} = \alpha/2, \tag{48}$$

with $\alpha = \pi/2$ for minimum circulating current. To distribute the correct powers across all i ABs in the converter, the total phasor lengths $v'_{\text{pos}}X_{\text{pos}}$ and $v'_{\text{neg}}X_{\text{neg}}$ are distributed according to the required power per AB as follows

$$v'_nX_n = v'_{\text{pos}}X_{\text{pos}}\frac{|P_n|}{P_{\text{pos}}} \tag{49}$$

This is displayed graphically in Fig. 4(b), where it can be seen that the total positive and negative vectors have equal length, and the phase shift is maintained at a constant value.

IV. RESULTS

This section shows the results obtained by phase-shift control compared to the algebraic solutions described in Section III. The performance of the proposed voltage-balance control algorithm has been verified for a Quad Active-Bridge (QAB) ac-dc converter. Simulations have been carried out using MATLAB Simulink with the Plexim PLECS blockset. The simulation parameters are written in Table I.

Fig. 5 shows the wave form of the QAB converter using phase-shift control. The load is set to 20%, meaning phase shifts are relatively small, while the volt-seconds across the inductor L_σ are large, resulting in a large triangular current. On the other hand, Fig. 6 shows the same converter at the same operating point, however operated using the proposed voltage-balance method. As can be seen, the volt-seconds across the inductor and the peak current are reduced greatly. This is confirmed by Fig. 7, which shows the proposed Fourier-based control method has much lower inductor current at low powers.

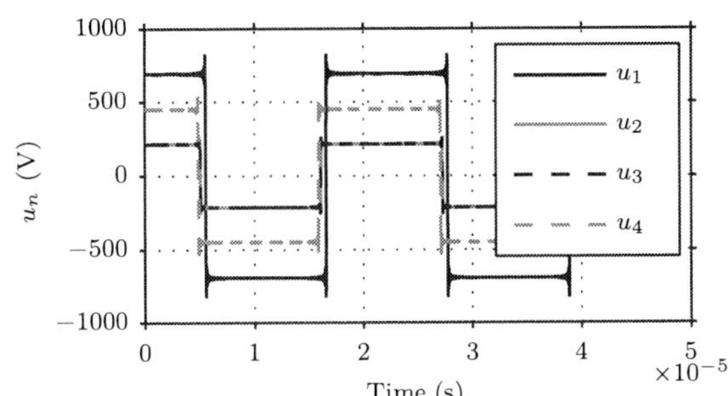

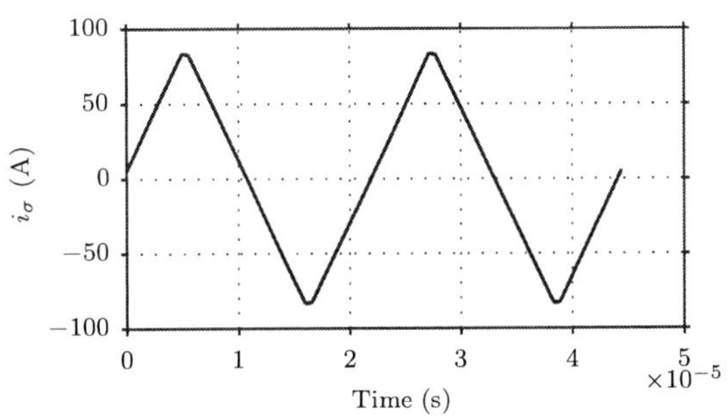

Fig. 5. Wave forms using phase-shift control in QAB-converter. Output power is 2kW (20% load).

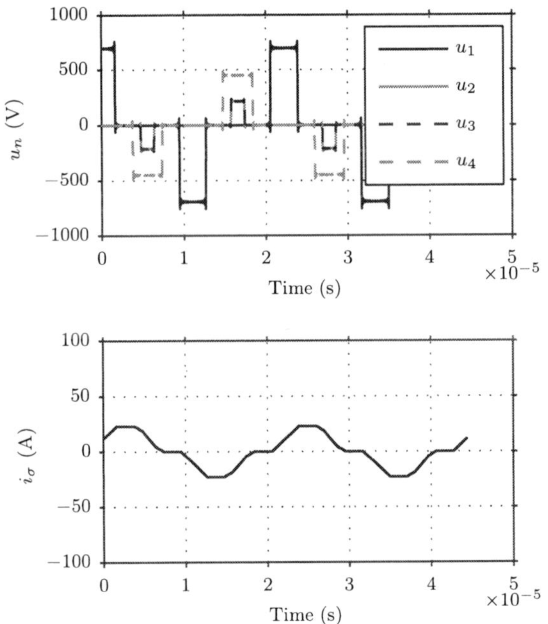

Fig. 6. Wave forms using proposed Fourier-based voltage-balance method in a QAB-converter. Output power is 2kW (20% load).

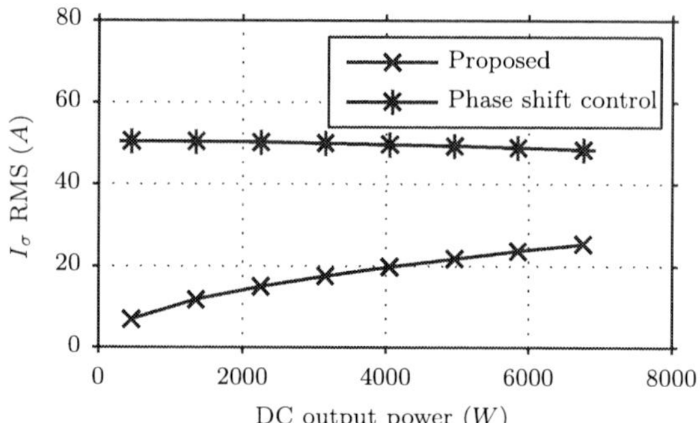

Fig. 7. Comparison of RMS current through L_σ in QAB converter with standard phase-shift control and the proposed Fourier-based control scheme.

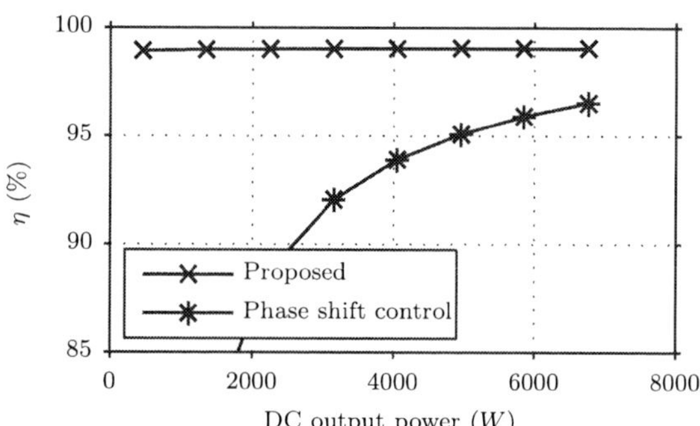

Fig. 8. Comparison of converter efficiency in QAB converter with standard phase-shift control and the proposed Fourier-based control scheme, assuming $100m\Omega$ resistance.

This cuts down the losses in the converter substantially. Fig. 8 shows a comparison in efficiency which assumes 100 mΩ of resistance in the converter. Whereas the phase-shift control algorithm has a remarkably low efficiency at output powers lower than 5 kW, the proposed control algorithm achieves an efficiency of approximately 99% across the whole power range. The EU efficiency standard was used to benchmark the converter, which is defined as

$$\eta_{EU} = 0.03\eta_{@0.05Pmax} + 0.06\eta_{@0.1Pmax} + 0.13\eta_{@0.2Pmax} + 0.1\eta_{@0.3Pmax} + 0.48\eta_{@0.5Pmax} + 0.2\eta_{@Pmax}. \quad (50)$$

When using the phase-shift control algorithm, an EU efficiency of $\eta_{EU} \approx 88.3\%$ was obtained. On the other hand, with the Fourier-based control algorithm, the efficiency improved considerably, to $\eta_{EU} \approx 99.0\%$.

V. Conclusion

In this paper, a novel generic modeling approach of Active-Bridge converters is proposed. The method is applicable to any Active-Bridge converter (DAB, QAB, ...), and models the switched-node waveforms in Fourier-series in order to find an analytical model. Furthermore, the results are used find an optimized converter parameter decoupler and control method using Lagrange operators and Newton's optimization algorithm.

The proposed method is applied to an Active-Bridge type converter, namely the Quad Active-Bridge (QAB) to minimize circulating currents in the converter while maintaining output power performance. The results show a significant improvement of the EU efficiency in the QAB converter, which increases from 88.3% to 99.0% when comparing the phase-shift control algorithm with the proposed voltage-balance control algorithm. The RMS current through converter drastically reduces, especially at reduced output powers.

References

[1] B. Vermulst, J. Duarte, C. Wijnands, and E. Lomonova, "Single-stage three-phase ac to dc conversion with isolation and bi-directional power flow," in *Industrial Electronics Society, IECON 2014 - 40th Annual Conference of the IEEE*, Oct 2014, pp. 4378–4383.

[2] R. De Doncker, D. Divan, and M. Kheraluwala, "A three-phase soft-switched high-power-density dc/dc converter for high-power applications," *Industry Applications, IEEE Transactions on*, vol. 27, no. 1, pp. 63–73, 1991.

[3] J. Everts, F. Krismer, J. Van den Keybus, J. Driesen, and J. Kolar, "Optimal zvs modulation of single-phase single-stage bidirectional dab ac-dc converters," *Power Electronics, IEEE Transactions on*, vol. 29, no. 8, pp. 3954–3970, Aug 2014.

[4] N. Baars, H. Huisman, J. Duarte, and J. Verschoor, "A 80 kw isolated dc-dc converter for railway applications," in *Power Electronics and Applications (EPE'14-ECCE Europe), 2014 16th European Conference on*, Aug 2014, pp. 1–10.

[5] H. Tao, J. Duarte, and M. Hendrix, "Multiport converters for hybrid power sources," in *Power Electronics Specialists Conference, 2008. PESC 2008. IEEE*, June 2008, pp. 3412–3418.

[6] F. Krismer and J. Kolar, "Efficiency-optimized high-current dual active bridge converter for automotive applications," *Industrial Electronics, IEEE Transactions on*, vol. 59, no. 7, pp. 2745–2760, July 2012.

[7] B. Ragg, "Geometric explanation of the method of lagrange multipliers," *IEEE Transactions on Education*, vol. 12, no. 2, pp. 141–142, 1969.

A Stability Analysis and Efficiency Improvement of Synchronverter

Prasanna Piya and Masoud Karimi-Ghartemani
Department of Electrical and Computer Engineering
Mississippi State University, MS, USA
E-mail: pp483@msstate.edu, karimi@ece.msstate.edu

***Abstract*—Synchronverter is an inverter equipped with proper control algorithms to mimic the operation of a synchronous generator. In this paper, a stability analysis of the synchronverter is performed and it is shown that the synchronverter's stability margins are directly related to the presence of an external passive resistance. This resistance compromises its efficiency as much as five percent in a typical low-power setting. A modified synchronverter structure is proposed where the passive resistor is removed from the power circuit and its impact is emulated in the control structure. This modification avoids degradation of the efficiency without compromising the stability. The analytical results are verified by digital and real-time simulations.**

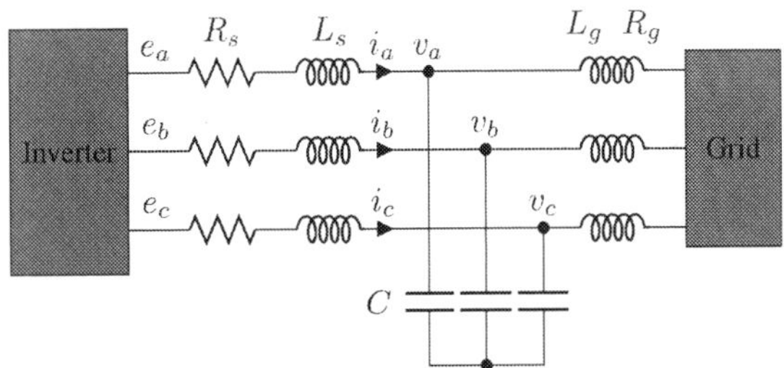

Fig. 1. Three-phase inverter connected to the grid

I. Introduction

Renewable and distributed energy sources such as photovoltaic and wind energy systems are generally interfaced to utility grid using power electronic inverters [1], [2]. These inverters should transfer the power from the energy sources to the grid; maintain the synchronization with the grid and do not compromise its stability and resilience; and respect the power quality codes and standards [3]–[5]. With recent developments in the concept of microgrid for the achievement of smart grid goals and integration of distributed and renewable energy systems at high penetration level, the use of the inverter has significantly increased [1], [6]–[8]. Within this context, and along with the aforementioned requirements, the efficiency of the inverter is also very important. Various control methods are used to control the inverters in microgrid application. Droop characteristics combined with the voltage and current control loops [5], [9]–[11], methods based on the virtual synchronous generator [12], [13], Synchronverter [14], [15], the universal controller based on droop method and optimization [16], and other heuristic methods have been proposed in the literature.

Synchronverter, first proposed in [14], is a control method used for an inverter that is based on mimicking the principles of operation of a conventional synchronous generator (SG). The basic idea was to operate the inverter in a similar way as to an SG and to benefit from the advantages associated with such systems. The physical quantities of an SG (such as its inertia) are emulated within the Synchronverter and can be adjusted. Several subsequent works on the synchronverter-based control algorithm are reported in [15], [17]–[20]. In [17], synchronverter algorithm is used to mimic the operation of a synchronous motor for operation and control of the three phase PWM rectifiers. The paper [18] improves [17] by adding the ability to track grid frequency and phase, and eliminating the need of a phase-locked loop (PLL). Synchronverter-based operation of a STATCOM mimicking synchronous condenser is presented in [19]. The improved synchronverter in [20] models and adds more blocks from an actual SG's control system such as the exciter and governor control units. The synchronverter model of [15] improves the original one of [14] by including a self-synchronization scheme to remove the need for a PLL and thus the complexity of the controller is reduced.

In all reported works on synchronverter, there is a resistor R_s in series with the output filter of the inverter as shown in Fig. 1. No study is reported on the impacts of this resistor on the stability and efficiency of the synchronverter. In this paper, such a study is performed. Specifically, the paper presents the following contributions. 1) It is analytically shown that the synchronverter's stable operation is dependent on the existence of R_s. The minimum value of R_s to maintain the stability is derived analytically and compared with the simulation results. 2) The losses corresponding with the presence of R_s are calculated and it is shown that this resistance causes an efficiency loss of about 5% in a typical low-power inverter. 3) A strategy is proposed based on emulating the resistor R_s within the synchronverter structure. 4) The analytical results are validated using digital and real-time simulations.

II. Review of Synchronverter

Figure 1 shows the power circuit of a three-phase inverter and Fig. 2 shows the block diagram of the synchronverter [14]. It has been derived based on the mathematical model of the three-phase round rotor synchronous machine [14], [15]. The equations are summarized as

$$\ddot{\theta} = \frac{1}{J}(T_m - T_e - D_p\dot{\theta}) \tag{1}$$

$$T_e = \frac{P}{\omega} = M_f i_f \langle i, \widetilde{\sin}\theta \rangle \tag{2}$$

978-1-4673-9551-9/16 $31.00 © 2016 IEEE

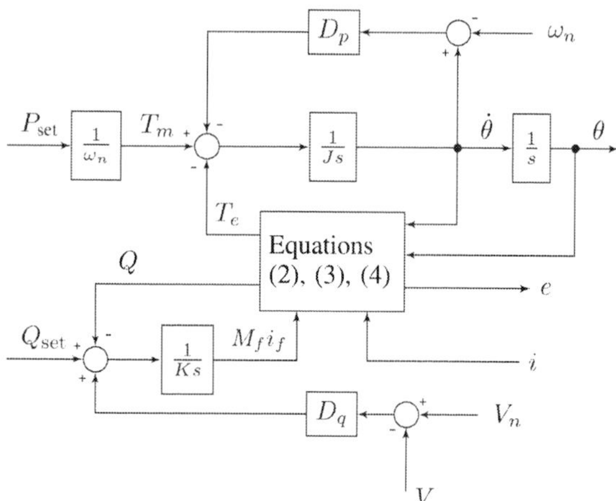

Fig. 2. Block diagram of the synchronverter

$$e = \omega M_f i_f \widetilde{\sin}\theta \tag{3}$$

$$Q = -\omega M_f i_f \langle i, \widetilde{\cos}\theta \rangle \tag{4}$$

where T_m and T_e model the mechanical and electrical torques, respectively, e is the three-phase inverter's generated voltage and i is its current (see Fig. 1), P and Q are the real and reactive powers at the e terminals, θ is the voltage angle, J models the moment of inertia, and $M_f i_f$ models the field excitation. The variable $\omega = \dot{\theta}$ represents the internal frequency of the inverter voltage, $\langle \cdot\,, \cdot \rangle$ is a product defined as $\langle x, y \rangle = x_a y_a + x_b y_b + x_c y_c$, and

$$\widetilde{\sin}\theta = \begin{bmatrix} \sin\theta \\ \sin(\theta - \frac{2\pi}{3}) \\ \sin(\theta + \frac{2\pi}{3}) \end{bmatrix}, \quad \widetilde{\cos}\theta = \begin{bmatrix} \cos\theta \\ \cos(\theta - \frac{2\pi}{3}) \\ \cos(\theta + \frac{2\pi}{3}) \end{bmatrix}. \tag{5}$$

Equations (3) and (5) indicate that the magnitude of the inverter voltage is $E = \omega M_f i_f$. In the synchronverter block diagram of Fig. 2, the real power set-point is defined as $P_{set} = \omega_n T_m$ and a term $D_p \omega_n$ is also included (where ω_n is the nominal frequency). The reactive power set-point is linked to the derivative of the voltage magnitude as shown by the lower portion of the diagram. Moreover, a droop term $D_q(V_n - V)$ is included in it where V and V_n are the magnitudes of the measured output voltage and its nominal value, respectively. In steady-state, assuming the asymptotic stability of the system, the following relationships will hold.

$$V = V_n + \frac{1}{D_q}(Q_{set} - Q) \tag{6}$$

$$\omega = \omega_n + \frac{1}{D_p \omega_n}(P_{set} - P) \tag{7}$$

Equation (7) is an approximation of the accurate relationship of $\omega = \omega_n + \frac{1}{D_p}(\frac{P_{set}}{\omega_n} - \frac{P}{\omega})$. Equation (6) describes a drooping characteristic of the conventional output voltage magnitude versus the reactive power while (7) relates the internal inverter frequency to the real power through a droop characteristics.

III. Stability Analysis of Synchronverter

The three phase inverter connected to the grid is shown in Fig. 1. The system can be described by the equation

$$R i_{abc} + L_s \frac{d i_{abc}}{dt} = e_{abc} - v_{abc} \tag{8}$$

where e_{abc} is the three-phase inverter's generated voltage, i_{abc} is its current, v_{abc} is the three phase grid voltage, and $R = R_s$. Equation (8) can be represented in dq synchronous reference frame by using $abc - \alpha\beta - dq$ transformation. The $abc - \alpha\beta$ transformation is defined as

$$v_{\alpha\beta} = v_\alpha + j v_\beta = \frac{2}{3}(v_a + v_b e^{-j2\pi/3} + v_c e^{j2\pi/3}) \tag{9}$$

where $v_{\alpha\beta}$ is the representation of the voltage in the stationary reference frame and $\alpha\beta - dq$ transformation is defined as

$$v_{dq} = v_d + j v_q = e^{-j(\theta_g - \frac{\pi}{2})} v_{\alpha\beta} \tag{10}$$

where v_{dq} is the representation of voltage in dq synchronous reference frame and θ_g is the grid voltage phase angle. Then, (8) can be presented in dq synchronous reference as

$$\frac{d i_d}{dt} = -\frac{R}{L} i_d + \omega_g i_q + \frac{1}{L} e_d - \frac{1}{L} v_{gd} \tag{11}$$

$$\frac{d i_q}{dt} = -\omega_g i_d - \frac{R}{L} i_q + \frac{1}{L} e_q \tag{12}$$

where e_d and e_q are inverter's generated voltage; i_d and i_q are its currents in dq reference frame; and v_{gd} and ω_g are the grid voltage magnitude and frequency.

In the following, we consider a simplified case where the inverter is operating in grid-connected situation and $P_{set} = Q_{set} = 0$. It will be shown that the synchronverter cannot maintain the stability if R_s does not exist or if it is too small.

Consider the inverter's internal frequency variable $\dot{\theta} = \omega$ and define $\theta = \omega_g t + \delta$ where δ will be the inverter's voltage angle with reference to the grid voltage angle. Then,

$$\dot{\delta} = \omega - \omega_g. \tag{13}$$

From Fig. 2, $\dot{\omega}$ and $M_f i_f$ can be presented as

$$\dot{\omega} = \frac{1}{J}\left(\frac{P_{set}}{\omega_n} - \frac{P}{\omega} - D_p \omega + D_p \omega_g\right) \tag{14}$$

$$M_f i_f = \frac{1}{K} \int \{Q_{set} - Q + D_q(V_n - V)\} \tag{15}$$

where P and Q are the real and reactive powers defined in dq reference frame as $P = \frac{3}{2}(e_d i_d + e_q i_q)$ and $Q = \frac{3}{2}(e_q i_d - e_d i_q)$. The term $D_q(V_n - V)$ can be ignored in grid-connected operation. So the equations (14) and (15) can be modified as

$$\dot{\omega} = -\frac{3}{2J\omega}(e_d i_d + e_q i_q) - \frac{D_p}{J}\omega + \frac{D_p}{J}\omega_g \tag{16}$$

$$\frac{d}{dt}(M_f i_f) = -\frac{3}{2K}(e_q i_d - e_d i_q). \tag{17}$$

From equation (3), the magnitude of the inverter voltage e is defined as $E = \omega M_f i_f$. The synchronverter dynamics indicates that $M_f i_f$ is a fast-changing signal while ω is a slow signal. Neglecting the term $\dot{\omega}(M_f i_f)$ (the validity of this is confirmed by close consistency between the analytical and

978-1-4673-9551-9/16 $31.00 © 2016 IEEE

simulation results reported shortly) and substituting the value for $\frac{d}{dt}(M_f i_f)$ from equation (17) results in

$$\dot{E} = -\frac{3\omega}{2K}(e_q i_d - e_d i_q) \tag{18}$$

The inverter voltage in dq reference frame is defined as $e_d = E\cos\delta$ and $e_q = E\sin\delta$ where $E = \sqrt{e_d^2 + e_q^2}$ and $\delta = \tan^{-1}\left(\frac{e_q}{e_d}\right)$. Therefore, the dynamics of e_d and e_q can be derived as

$$\begin{aligned}\dot{e_d} &= \dot{E}\cos\delta - \dot{\delta}E\sin\delta \\ &= -\frac{3\omega}{2K}(e_q i_d - e_d i_q)\cos\delta - (\omega-\omega_g)\sqrt{e_d^2+e_q^2}\sin\delta \end{aligned} \tag{19}$$

$$\begin{aligned}\dot{e_q} &= \dot{E}\sin\delta + \dot{\delta}E\cos\delta \\ &= -\frac{3\omega}{2K}(e_q i_d - e_d i_q)\sin\delta + (\omega-\omega_g)\sqrt{e_d^2+e_q^2}\cos\delta. \end{aligned} \tag{20}$$

Equations (11), (12), (16), (19), and (20) present the synchronverter-based control method which are reordered and summarized as

$$\dot{i_d} = -\frac{R}{L}i_d + \omega_g i_q + \frac{1}{L}e_d - \frac{1}{L}v_{gd} \tag{21}$$

$$\dot{i_q} = -\omega_g i_d - \frac{R}{L}i_q + \frac{1}{L}e_q \tag{22}$$

$$\dot{e_d} = -\frac{3\omega}{2K}(e_q i_d - e_d i_q)\sin\delta + (\omega-\omega_g)\sqrt{e_d^2+e_q^2}\cos\delta \tag{23}$$

$$\dot{e_q} = -\frac{3\omega}{2K}(e_q i_d - e_d i_q)\sin\delta + (\omega-\omega_g)\sqrt{e_d^2+e_q^2}\cos\delta \tag{24}$$

$$\dot{\omega} = -\frac{3}{2J\omega}(e_d i_d + e_q i_q) - \frac{D_p}{J}\omega + \frac{D_p}{J}\omega_g \tag{25}$$

where $\delta = \tan^{-1}\left(\frac{e_q}{e_d}\right)$.

Equations (21)-(25) are nonlinear differential equations. In order to perform a linear stability analysis, we notice that the equilibrium state of those equations for grid-connected operation at $P_{set} = Q_{set} = 0$ is

$$(i_d^*, i_q^*, e_d^*, e_q^*, \omega^*) = (0, 0, v_{gd}, 0, \omega_g).$$

Linearization of (21)-(25) about the equilibrium point results in the following linear system

$$\dot{X} = \begin{bmatrix} -\frac{R}{L} & \omega_g & \frac{1}{L} & 0 & 0 \\ -\omega_g & -\frac{R}{L} & 0 & \frac{1}{L} & 0 \\ 0 & \frac{3\omega_g v_{gd}}{2K} & 0 & 0 & 0 \\ 0 & 0 & 0 & 0 & v_{gd} \\ -\frac{3v_{gd}}{2J\omega} & 0 & 0 & 0 & -\frac{D_p}{J} \end{bmatrix} X \tag{26}$$

where $X = [i_d, i_q, e_d, e_q, \omega]^T$. The parameters L, R, v_{gd}, and w_g in (26) are constant for a given inverter and grid. And the parameters D_p, J, and K are the control parameters. The control parameters are selected based on the performance requirements. The droop equations (6) and (7) are used to determine the values for parameters D_q and D_p respectively. Consider a change of voltage by ΔV from the nominal value that causes the full change in the reactive power. Then,

$$D_q = \frac{S}{\Delta V}. \tag{27}$$

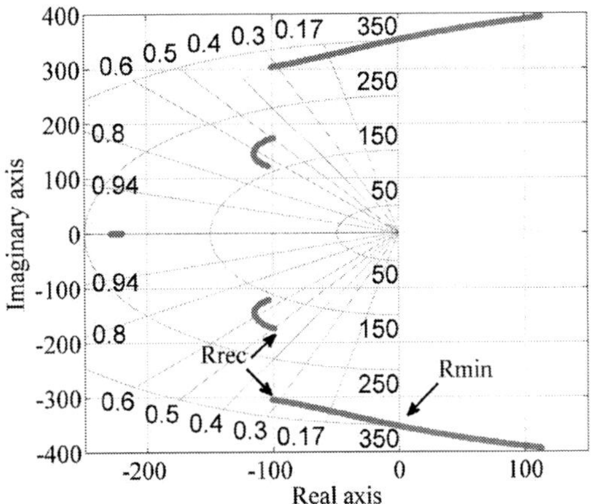

Fig. 3. Eigenvalues of system (26) for varying R

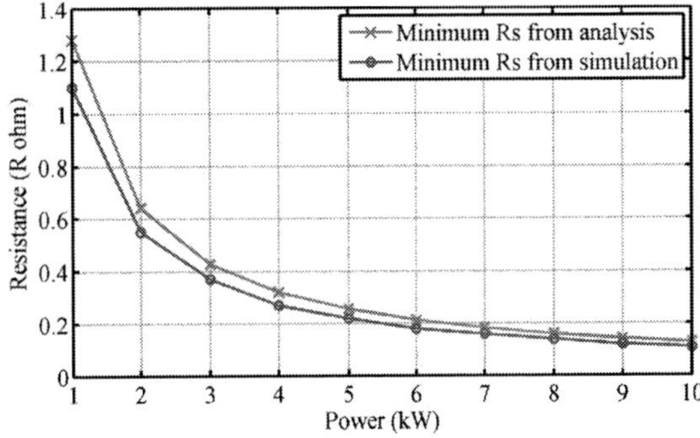

Fig. 4. Minimum value of R_s to stabilize the inverter

Similarly, considering a level of change in frequency of Δf from the nominal frequency corresponding to the full change in real power, D_p can be calculated as

$$D_p = \frac{S}{2\pi\ \Delta f\ w_n}. \tag{28}$$

Here, S is the power rating of inverter and w_n is the nominal frequency. The time constant of the frequency is defined as $\tau_f = \frac{J}{D_p}$ and the time constant of the voltage as $\tau_v = \frac{K}{w_n D_q}$ [14]. By choosing the values for these time constants the parameters J and K can be determined as

$$J = \tau_f D_p \tag{29}$$

$$K = \tau_v w_n D_q. \tag{30}$$

The eigenvalues of system (26) is plotted in Fig. 3 for the values of $V_{rms} = 120$ V, $w_g = 2\pi 60$ rad/s, $v_{gd} = \sqrt{2}V_{rms}$ V, $\tau_f = 0.005$ s, and $\tau_v = 0.005$ s, $\Delta V = 0.1v_{gd}$, $\Delta f = 2$ Hz, and R varies from zero to a positive value. We have performed the study for different values of the inverter rating from $S =$ 1 kVA to $S = 10$ kVA and the inductance is reduced according to $L = 10/S$ H. For $R = 0$, two eigenvalues of the system (26) are in the right half plane. When R is increased, these eigenvalues move towards the left and cross the imaginary axis at

$$R_{\min} = \frac{1.28}{S(\text{kVA})}.$$

The graph of $R_{\min}$ for different values of inverter power rating is also measured through simulations. The analytical and simulation results are very close as indicated in Fig. 4.

IV. On Selection of R_s

The value of $R_{\min}$ given above is the minimum value of R required for the stability of the system. For an acceptable performance of the system, the value of R should be increased. If we set the criterion as having all four (dominant) poles to have equal real parts, then we get a value of $R = \frac{2.15}{S(\text{kVA})}$. For this value, the two high frequency poles give damping of approximately 0.3 and the two low frequency poles give the damping of approximately 0.5. Moreover, all four poles have an equal real part of about -100 as shown in Fig. 3 which signifies a desirable transient response time-constant of about 10 ms. This

$$R_{\text{rec1}} = \frac{2.15}{S(\text{kVA})}$$

is the recommended value of R for the operation of the synchronverter using this criterion.

In another approach, we can select R_s that offers the maximum stability margins (robustness) for the closed-loop system. Consider the synchronverter as shown in Fig. 2. Define P_{set} as an input and the real power P as the output while $Q_{\text{set}} = 0$. Then the stability margins for the system is calculated for different values of R_s increasing from $R_{\min} = \frac{1.28}{S(\text{kVA})}$ to some positive value. Figure 5 shows the phase margin (PM) and gain margin (GM) of the system for varying R_s. When R_s is increased from $R_{\min}$, the GM of the system increases (from 1 to over 2) while the PM decreases (from 52 to about 30 degrees). Consider a cost function

$$J = PM + \alpha GM$$

where α is a constant that brings the GM to a range comparable to the PM (in degrees). The maximum value of this cost function gives the optimal value for R_s. Figure 6 shows the cost function for different values of R_s, when α is changed from 20 to 30. Figure 7 shows the optimal value of R_s for α changing from 20 to 30 obtained from the maximum points of the curves on Fig. 6. The optimal value of R_s can be expressed as $R_{\text{opt}} = \frac{1.9}{S(\text{kVA})}$ for $\alpha = 20$ and $R_{\text{opt}} = \frac{2.1}{S(\text{kVA})}$ for $\alpha = 30$. This analysis gives another recommended value for R_s when $\alpha = 25$:

$$R_{\text{rec2}} = R_{\text{opt}} = \frac{2.01}{S(\text{kVA})}.$$

For this value of resistance, the system has gain margin of about 1.9 and phase margin of about 45^0. With the open-loop bandwidth of $131\ rad/s$ the system can tolerate the time delay of about $6\ ms$.

V. Power Loss Associated with R_s

Assume that an inverter with power rating S is operating at its maximum power situation. Then, $S = 3VI$ where V and I are the rms values of the grid phase voltage and current.

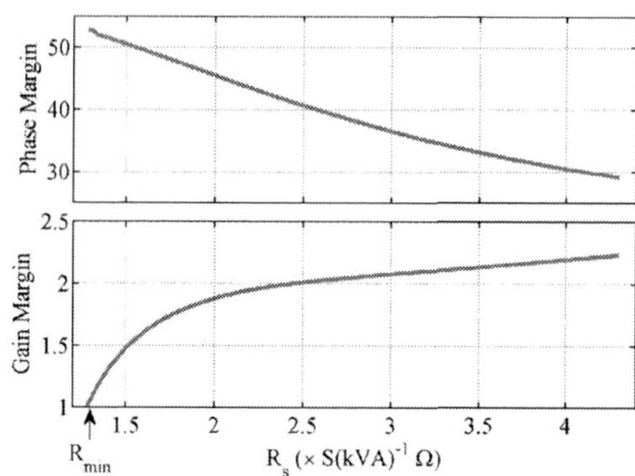

Fig. 5. Phase margin and Gain margin of the system described in section IV

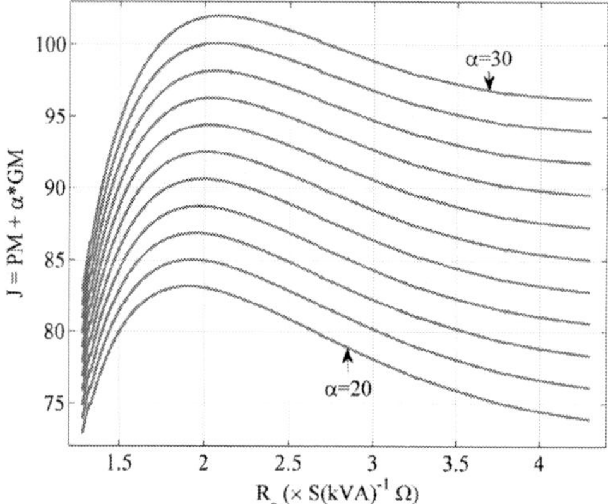

Fig. 6. Cost function for different values of α

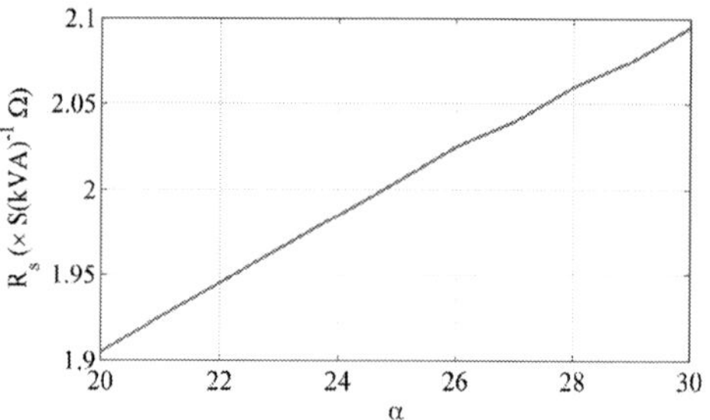

Fig. 7. Optimal value of R_s of different values of α

The power loss in the resistor R is equal to $P_{\text{loss}} = 3RI^2$. Therefore, the percentage power loss is

$$\frac{P_{\text{loss}}}{S} = \frac{3RI^2}{3VI} = \frac{RI}{V} = \frac{RS}{3V^2}.$$

The minimum power loss is obtained for $R = R_{\text{opt}}$ that is

$$\frac{P_{\text{loss}}}{S} = \frac{R_{\text{opt}}S}{3V^2} = \frac{2010}{3V^2} = 4.65\%$$

for the 120 V grid voltage. This is the minimum power loss associated with the resistor R_s.

The commercial inverters, nowadays, have efficiencies greater than 96% [21]. The inverter losses include power

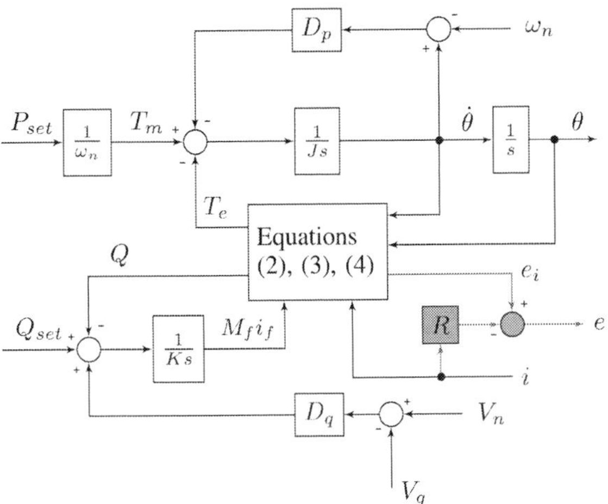

Fig. 8. Improved synchronverter with virtual resistance.

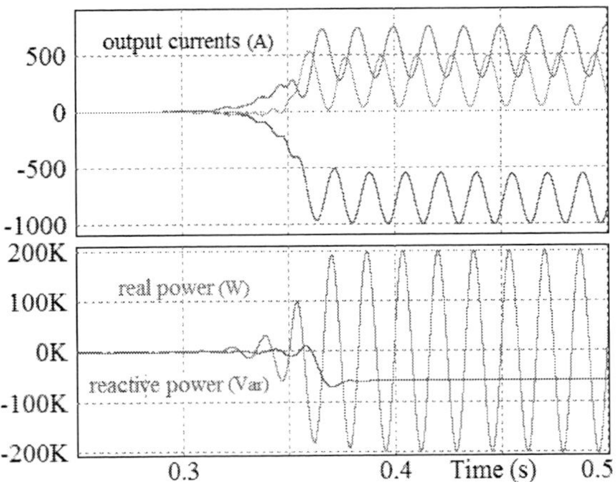

Fig. 9. Unstable performance of synchronverter with $R_s = 0$.

electronic switching losses, magnetic core losses, and the ohmic losses (in switches, inductors and cables). So this value of 4.96% loss caused by the resistor R_s is a significant loss.

VI. Virtual R to Prevent Losses

The power loss due to presence of R_s can be eliminated if we include the effect of resistance R_s in the control part and remove R_s from the physical circuit. Consider the power circuit of Fig (1) without physical resistor R_s and define the virtual inverter voltage

$$e_i(t) = e(t) + Ri(t) \tag{31}$$

where $e(t)$ is inverter voltage, $i(t)$ is its measured current and R is a virtual resistance. This emulates the same actual resistance. Figure 8 shows the synchronverter control circuit with the virtual resistance R. Since the effect of R_s is emulated only in the control loop, there is no real power loss associated with it and the same stability properties are maintained.

VII. Simulation Results

Simulations are performed in PSIM software [22]. A 5 KVA inverter with a dc bus voltage of $500\ V$ and switching frequency of 10 kHz is considered for the study. The LC output filter is used with $L_s = 2\ mH$, $C = 5\ \mu F$. The resistor is calculated according to the minimum value of $R_s = \frac{2010}{S} = 0.4\ \Omega$. The grid is a three-phase 208 V (line-to-line), 60 Hz. The controller is implemented in digital at the sampling frequency of 10 kHz. The grid is $V_{\rm rms} = 120$ V, $w_n = 2\pi 60$ rad/s. Other synchronverter parameters are set at $\tau_f = 0.005$ s, $\tau_v = 0.005$ s, $\Delta V = 0.1\sqrt{2}V_{\rm rms}$, and $\Delta f = 2$ Hz, $D_p = \frac{S}{2\pi\ \Delta f\ w_n} = 1.055$, $D_q = \frac{S}{\Delta V} = 294.6$, $J = 0.0053$, and $K = 555.4$.

Figure 9 shows the simulation result of the synchronverter when $R_s = 0$ which shows the system is unstable. Figure 10 shows the simulation results of the synchronverter when $R_s = 0.4\ \Omega$, the recommended value of R_s for 5 kVA inverter. The synchronverter is connected to the grid at $t = 0.28\ s$ and a real power jump of $P = 4000\ W$ at $t = 0.3\ s$ and reactive power jump of $Q = 3000$ Var at $t = 0.45\ s$ are considered. Figure 10(a) shows the stable response of the synchronverter when R_s is an actual passive resistor in the power circuit. Figure 10(b) shows the simulation result of the synchronverter when $R_s = 0.4\ \Omega$ is removed from the power circuit and is replaced with virtual resistance of the same value inside the control loop hence removing about 5% power loss. The simulations show identical results in both cases.

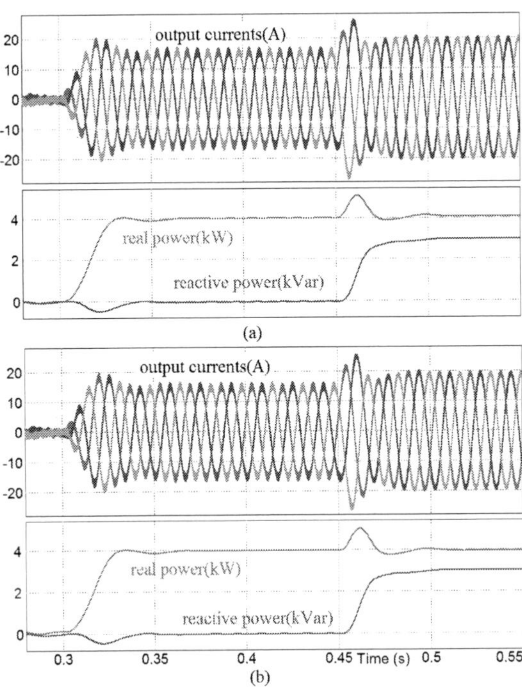

Fig. 10. Stable performance of synchronverter with: (a) Actual resistance $R_s = 0.4\ \Omega$ (b) Virtual resistance $= 0.4\ \Omega$

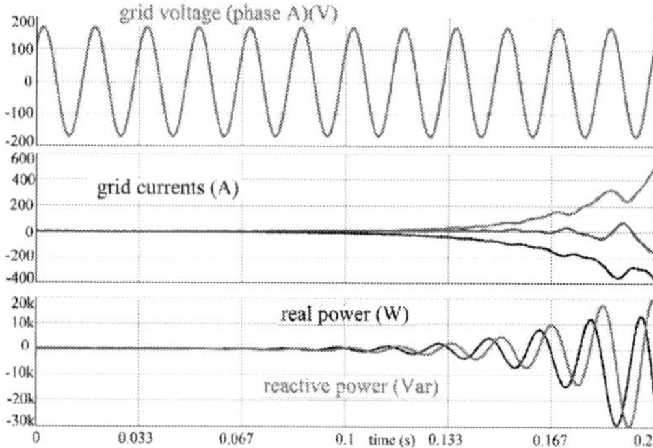

Fig. 11. Real-time simulation of the synchronverter with $R_s = 0$.

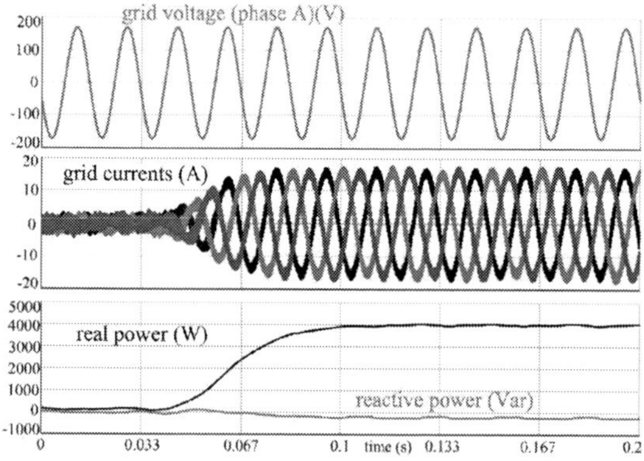

Fig. 12. Real-time simulation results when the real power command increases from 0 to 4 kW (Reactive power set-point = 0, Virtual resistance = 0.4 Ω)

VIII. Real-Time Results

The inverter in section VII is modelled in the RSCAD software for the real time operation in Real Time Digital Simulator (RTDS) [23]. The switching frequency used is 3.3 kHz due to the RTDS limitations and accordingly, the inductance value is increased to 5 mH. The control circuit is also modelled in RSCAD and implemented in RTDS. The control circuit operates at a time step of 50 μs.

Figure 11 shows the real-time simulation results of the synchronverter when $R_s = 0$ which confirms its instability. Figure 12 shows the real-time simulation results of the synchronverter when the virtual resistor $R = 0.4\ \Omega$ is used where the real power jump of $P = 4000\ W$ at $t = 0.033\ s$ is considered. In fig. 13 the reactive power jump of $Q = 3000\ Var$ is considered ($P = 4000\ W$, $R = 0.4\ \Omega$). Both cases confirm stable operation of the synchronverter.

The transient responses of the real-time simulation obtained from RTDS comply with the results obtained from the PSIM simulations. Figure 14 shows another sample of PSIM simulations now with the switching frequency of 3.3 kHz and the inductance value of 5 mH conforming with the real-time simulation conditions.

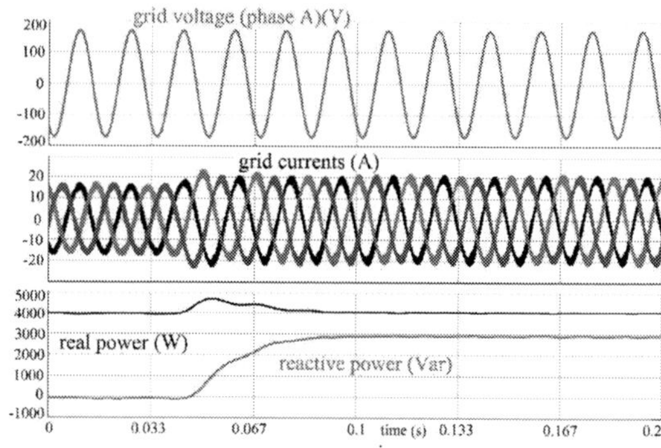

Fig. 13. Real-time simulation results when reactive power command increases from 0 to 3 kVar (Real power set-point = 4 kW, Virtual resistance = 0.4 Ω)

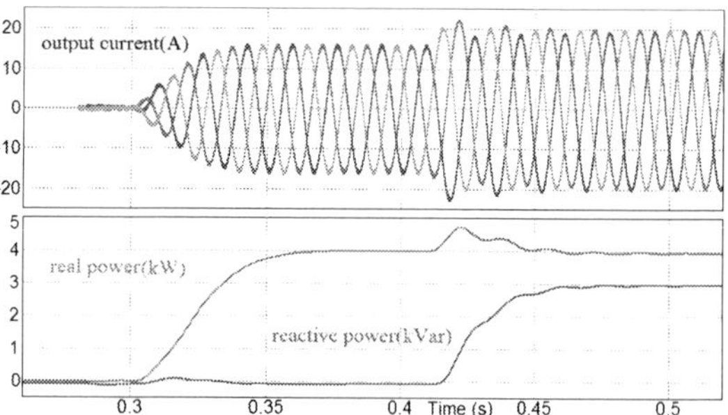

Fig. 14. PSIM simulation with $L_s = 5\ mH$ ($P = 4\ kW$ at $t = 0.3\ s, Q = 3\ kVar$ at $t = 0.41\ s$).

IX. Conclusion

The synchronverter-based control strategy is considered and the impacts of its output interface resistor on the stability and efficiency are studied analytically and using digital real-time simulations. It is shown that the synchronverter becomes unstable as this resistor is removed. A reasonable value for this resistance to yield good performance causes an efficiency degradation as much as about five percents in a typical application. A modified synchronverter structure is proposed to improve the efficiency of the synchronverter by removing the resistor from the power circuit and replacing it by a virtual resistor in the control structure.

References

[1] F. Blaabjerg, Z. Chen, and S. Kjaer, "Power electronics as efficient interface in dispersed power generation systems," *Power Electronics, IEEE Transactions on*, vol. 19, no. 5, pp. 1184–1194, Sept 2004.

[2] F. Blaabjerg, R. Teodorescu, M. Liserre, and A. Timbus, "Overview of control and grid synchronization for distributed power generation systems," *Industrial Electronics, IEEE Transactions on*, vol. 53, no. 5, pp. 1398–1409, Oct 2006.

[3] "IEEE P1547.4/D12 draft guide for design, operation, and integration of distributed resource island systems with electric power systems," *IEEE Standards Coordinating Committee 21 on Fuel Cells, Photovoltaics, Dispersed Generation and Energy Storage*, pp. 1–50, 2011.

[4] "UL1741 standard for inverters, converters, controllers and interconnection system equipment for use with distributed energy resources," *Underwritters Laboratories Inc. Standard for Safety*, pp. 1–144, 2010.

[5] J. Rocabert, A. Luna, F. Blaabjerg, and P. Rodriguez, "Control of power converters in ac microgrids," *Power Electronics, IEEE Transactions on*, vol. 27, no. 11, pp. 4734–4749, 2012.

[6] R. H. Lasseter and P. Paigi, "Microgrid: a conceptual solution," in *Power Electronics Specialists Conference, 2004. PESC 04. 2004 IEEE 35th Annual*, vol. 6. IEEE, 2004, pp. 4285–4290.

[7] B. Kroposki, C. Pink, R. DeBlasio, H. Thomas, M. Simo es, and P. K. Sen, "Benefits of power electronic interfaces for distributed energy systems," *Energy Conversion, IEEE Transactions on*, vol. 25, no. 3, pp. 901–908, 2010.

[8] F. Blaabjerg, F. Iov, T. Terekes, R. Teodorescu, and K. Ma, "Power electronics-key technology for renewable energy systems," in *Power Electronics, Drive Systems and Technologies Conference (PEDSTC), 2011 2nd*. IEEE, 2011, pp. 445–466.

[9] K. De Brabandere, B. Bolsens, J. Van den Keybus, A. Woyte, J. Driesen, and R. Belmans, "A voltage and frequency droop control method for parallel inverters," *Power Electronics, IEEE Transactions on*, vol. 22, no. 4, pp. 1107–1115, July 2007.

[10] J. Guerrero, M. Chandorkar, T. Lee, and P. Loh, "Advanced control architectures for intelligent microgrids- part i: Decentralized and hierarchical control," *Industrial Electronics, IEEE Transactions on*, vol. 60, no. 4, pp. 1254–1262, April 2013.

[11] J. Guerrero, J. Matas, L. G. de Vicuna, M. Castilla, and J. Miret, "Decentralized control for parallel operation of distributed generation inverters using resistive output impedance," *Industrial Electronics, IEEE Transactions on*, vol. 54, no. 2, pp. 994–1004, April 2007.

[12] T. Shintai, Y. Miura, and T. Ise, "Oscillation damping of a distributed generator using a virtual synchronous generator," *Power Delivery, IEEE Transactions on*, vol. 29, no. 2, pp. 668–676, 2014.

[13] J. Alipoor, Y. Miura, and T. Ise, "Distributed generation grid integration using virtual synchronous generator with adoptive virtual inertia," in *Energy Conversion Congress and Exposition (ECCE), 2013 IEEE*. IEEE, 2013, pp. 4546–4552.

[14] Q.-C. Zhong and G. Weiss, "Synchronverters: Inverters that mimic synchronous generators," *Industrial Electronics, IEEE Transactions on*, vol. 58, no. 4, pp. 1259–1267, April 2011.

[15] Q.-C. Zhong, P.-L. Nguyen, Z. Ma, and W. Sheng, "Self-synchronized synchronverters: Inverters without a dedicated synchronization unit," *Power Electronics, IEEE Transactions on*, vol. 29, no. 2, pp. 617–630, Feb 2014.

[16] M. Karimi-Ghartemani, "Universal integrated synchronization and control for single-phase dc/ac converters," *Power Electronics, IEEE Transactions on*, vol. 30, no. 3, pp. 1544–1557, March 2015.

[17] Z. Ma, Q.-C. Zhong, and J. Yan, "Synchronverter-based control strategies for three-phase pwm rectifiers," in *Industrial Electronics and Applications (ICIEA), 2012 7th IEEE Conference on*, July 2012, pp. 225–230.

[18] Q.-C. Zhong, Z. Ma, and P.-L. Nguyen, "Pwm-controlled rectifiers without the need of an extra synchronisation unit," in *IECON 2012 - 38th Annual Conference on IEEE Industrial Electronics Society*, Oct 2012, pp. 691–695.

[19] P.-L. Nguyen, Q.-C. Zhong, F. Blaabjerg, and J. Guerrero, "Synchronverter-based operation of STATCOM to mimic synchronous condensers," in *Industrial Electronics and Applications (ICIEA), 2012 7th IEEE Conference on*, July 2012, pp. 942–947.

[20] C.-H. Zhang, Q.-C. Zhong, J.-S. Meng, X. Chen, Q. Huang, S.-H. Chen, and Z.-P. Lv, "An improved synchronverter model and its dynamic behaviour comparison with synchronous generator," in *Renewable Power Generation Conference (RPG 2013), 2nd IET*, Sept 2013, pp. 1–4.

[21] SMA Inverters. [Online]. Available: http://www.sma-america.com/

[22] PSIM, "Software by powersim technologies," *Professional Version*.

[23] R. Kuffel, J. Giesbrecht, T. Maguire, R. Wierckx, and P. McLaren, "RTDS-a fully digital power system simulator operating in real time," in *WESCANEX 95. Communications, Power, and Computing. Conference Proceedings., IEEE*, vol. 2. IEEE, 1995, pp. 300–305.

Compensation of Switching Dead-Time Effects in Voltage-Fed PWM Inverters Using FPGA-Based Current Oversampling

Bastian Weber, Tobias Brandt and Axel Mertens
Institute for Drive Systems and Power Electronics
Leibniz Universität Hannover, Germany
Email: bastian.weber@ial.uni-hannover.de

Abstract—**This paper presents a novel approach to compensate the voltage error due to switching dead times in voltage-fed inverters using a field-programmable gate array (FPGA) and current oversampling. The FPGA controls an analog-to-digital (AD) converter which operates at a high conversion rate and processes numerous current samples taken at equidistant time steps within each pulse width modulation (PWM) period. The FPGA also generates the inverter's PWM modulation. As phase currents at an inductive load can be approximated by piecewise linear time functions during each switching state of the inverter, the FPGA interpolates the current slopes using a linear least mean squares regression. Based on this regression, a predictor structure in the FPGA predicts the phase current right at the beginning of the switching dead time. The predictor structure can be computed almost instantaneously due to parallel data processing. Based on the estimated phase current, the FPGA uses a lookup table, depending on current and DC link voltage, to correct an upcoming switching instant in order to compensate for the dead-time effect. Depending on current polarity and magnitude, the lookup table stores the respective compensation times for the upcoming switching instants. The lookup table is the result of an offline commissioning process.**

Keywords—Dead-time effect, inverter nonlinearity, FPGA-based control, current oversampling.

I. Introduction

The output voltage deviation during switching dead-time is one of the major sources of output voltage distortion in PWM-operated inverters, [1], [2] . This paper is focused on a three-phase inverter feeding a permanent magnet synchronous machine (PMSM). Due to finite turn-on and turn-off times of power semiconductors and in order to avoid short circuit of the DC link voltage, a dead-time is introduced between turn-off and turn-on of the high and low side switches in each phase leg, [3], [4]. This results in a voltage deviation between the ideal PWM waveform and the actual output waveform. This voltage deviation is primarily dependent on the polarity and magnitude of the phase current in the respective phase leg during the corresponding dead-time. With regard to drive systems, it is known [4] that the voltage deviation causes serious problems such as increased torque ripple at low speeds and light load operation. For this reason the dead-time compensation presented in this paper focusses on the lower speed region, when the inverter is realizing a low modulation index. Concerning a sensorless operated drive system, it is known that the dead-time effect limits the bandwidth of the sensorless observer and causes oscillations of the estimated position [5]. In standard drive systems, the current is sampled once during a PWM half period. This method is also referred to as "single sampling", [6]. Common approaches for compensation rely on the polarity of the sampled current by means of single sampling, [1], [7]. Due to measurement noise, this may cause a false compensation in case of small currents. In [8] a DSP based compensation is introduced using additional phase current measurements. These measurements are taken at time instants specified by the pulse width modulation. The identified voltage error caused by dead-time is then compensated within the next PWM period. In [9], a method is introduced which operates the analog-to-digital converter (A/D converter) at a higher conversion frequency and processes numerous current samples taken at equidistant time steps during a PWM period. The data is processed with an FPGA and is referred to as oversampling. By means of a least mean squares algorithm for the current samples, the machine currents are interpolated, taking into account the switching state of the inverter.

In this paper, a novel approach using FPGA-based current oversampling and linear regression is presented. The improvement is that the approach uses linear regression and a predictor structure computed in the FPGA to predict the value of the current directly before a switching instant. Due to the influence of the parasitic capacitances, as illustrated in [4], [5], [10], the dead-time effect is very sensitive to the magnitude of the phase current when the current is close to zero. The novel current oversampling approach also realizes a high signal-to-noise ratio in view of current measurement noise. This improves the dead-time compensation, particularly when the phase current is close to zero. The FPGA also performs the PWM modulation. Based on the estimated phase current, the FPGA uses an on-chip stored lookup table, depending on current and DC link voltage, to correct an upcoming switching instant in order to compensate for the dead-time effect. The lookup table is the result of an offline commissioning process.

Section II illustrates the inverter nonlinearity effects due to switching dead-time and section III concentrates on the dead-time compensation algorithm. Afterwards section IV describes the auto commissioning and section V shows the experimental results.

II. Inverter Nonlinearity Effects Due to Switching Dead Time

Figure 1 illustrates the 2-level voltage source inverter topology feeding a PMSM. T_1 to T_6 represent the power

978-1-4673-9551-9/16 $31.00 © 2016 IEEE

MOSFETs with the integrated body diodes D_1 to D_6. Also shown are the parasitic capacitances of the MOSFETs C_1 to C_6. The inverter is supplied with a DC link voltage V_{dc}. The inverter's line voltages are v_A, v_B, v_C. The machine in star connection is represented by its resistance R, inductance L and back EMF v_{EMF}. The machine's phase currents are i_A, i_B, i_C and v'_A, v'_B, v'_C are the phase voltages. The zero voltage component is given by v_0.

When the parasitic capacitances are neglected, the line voltage

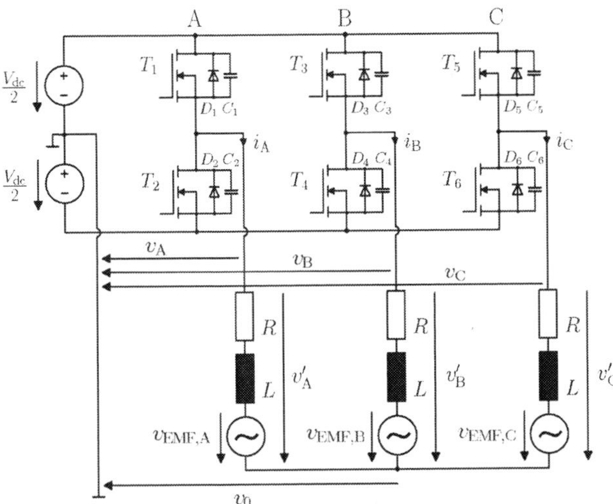

Fig. 1: Voltage source inverter topology

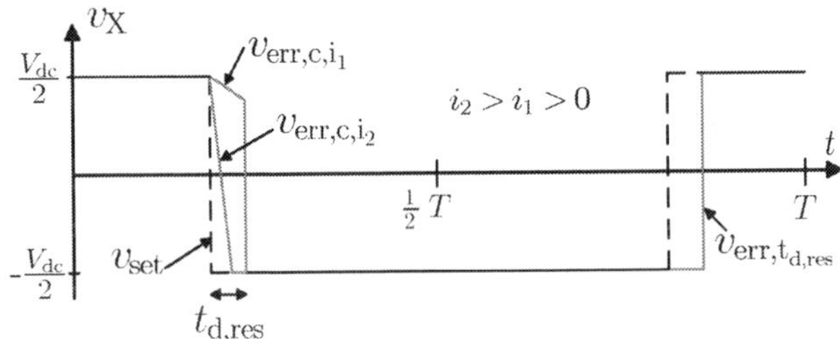

Fig. 2: Inverter output voltage, positive phase current

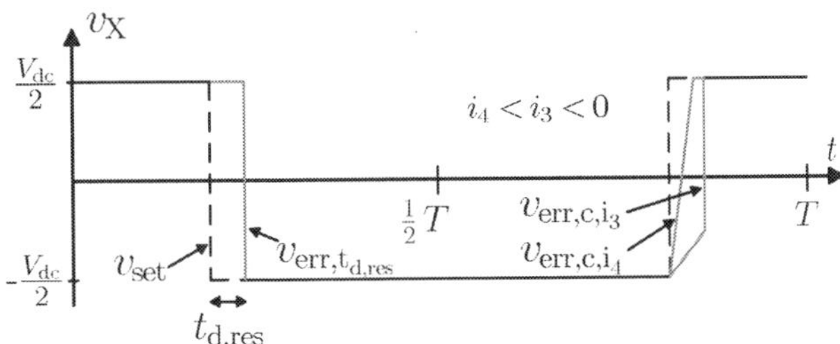

Fig. 3: Inverter output voltage, negative phase current

distortion resulting from the dead time effect for a single phase can be described according to [5] by eq. (1)

$$\begin{aligned} v_{\text{err},t_{\text{d,res}}} &= -\frac{t_{\text{d,res}}}{T} \cdot V_{\text{dc}} \cdot \text{sign}(i) \\ t_{\text{d,res}} &= t_{\text{d}} + t_{\text{d,on}} - t_{\text{d,off}} \end{aligned} \quad (1)$$

where t_d is the implemented dead time of the PWM modulation and $t_{d,on}$ and $t_{d,off}$ are the turn-on and the turn-off delay times of the switching device. These are combined to the resulting dead time $t_{d,res}$. The index X denotes a machine phase. The PWM switching period is given by T. The average voltage error over one switching period is $v_{err,t_{d,res}}$ and i stands for the phase current at the time of switching. The phase current is assumed to be constant during dead time.

The average line voltage error over one switching period that results from the parasitic capacitances can be described according to [5] by eq. (2)

$$v_{\text{err,c}} = \left(\text{sign}(i)\, V_{\text{dc}} - \frac{t_{\text{d,res}}}{2 \cdot (C_1 + C_2)} i\right) \cdot \left(\frac{t_{\text{d,res}}}{T}\right) \quad \text{for} \quad |i| < i_{V_{\text{td}}/2} \quad (2)$$

$$v_{\text{err,c}} = \left(\frac{V_{\text{dc}}^2}{2} \frac{C_1 + C_2}{t_{\text{d}}}\right) \cdot \left(\frac{t_{\text{d,res}}}{T}\right) \cdot \frac{1}{i} \quad \text{for} \quad |i| > i_{V_{\text{td}}/2}$$

$$\text{with} \quad i_{V_{\text{td}}/2} = V_{\text{dc}} \cdot \left(\frac{C_1 + C_2}{t_{\text{d}}}\right)$$

where $i_{V_{td}/2}$ as the current indicates, if the output voltage reaches $\frac{V_{dc}}{2}$ or $-\frac{V_{dc}}{2}$ respectively at the end of the dead time. This results in an average voltage error that is half as big as the dead time error $\left(|v_{err,t_{d,res}}|\right)$. The parasitic capacitances C_1 to C_6 of the switching devices are considered to be constant in this analysis. As illustrated in [5], this assumption is reasonable, because the capacitance changes only moderately, once a voltage level above a few volts is reached. A much higher capacitance exists at small voltages. When the voltage is low, the capacitance is almost discharged, and therefore the resulting voltage error can be neglected. The average line voltage error over one switching period for a single phase is then calculated as eq. (3)

$$v_{\text{err,res}} = v_{\text{err},t_{\text{d,res}}} + v_{\text{err,c}}. \quad (3)$$

As also indicated in [5], Figure 2 and 3 show the resulting line voltage with deviation depending on the phase current. This is compared to the desired reference voltage v_{set}.

III. Dead-time Compensation

At first, the general principle of the dead-time compensation is illustrated. Afterwards, the algorithm will be explained in detail. Figure 5 shows the hardware setup consisting of DSP and FPGA together with the inverter, the current sensing ADC and PMSM. The FPGA generates the PWM modulation for the inverter. The FPGA operates the current sensing A/D converter at a high conversion frequency and performs a linear regression of the current samples taken during the zero voltage vectors of the inverter, see Fig. 6. Directly before (about 700ns)

the next upcoming active voltage vector, the FPGA computes a predictor to estimate the phase currents in the upcoming instants of switching for all phases. This prediction is done in about 100 nanoseconds. Based on the predicted currents, the FPGA processes within nanoseconds data from an on-chip lookup table. The lookup table holds the needed compensation times for the upcoming switching instants dependent on the predicted currents in order to compensate for the dead-time effect. Based on the data from the lookup table, the FPGA directly adjusts the upcoming instants of switching in order to compensate for the dead-time effect. Because of the fast feedback of the current information due to oversampling, and the FPGA's ability to almost instantaneously determine and apply a compensation time to the upcoming instants of switching, the dead-time effect can be directly compensated.
FPGA-based current oversampling and linear regression has been published in [9], where the current oversampling is used to improve the sensorless control of PMSMs at low speed.
As indicated in [9], the linear regression of the current samples requires the computation of four average values $\bar{i}$, $\bar{t}$, $\overline{t^2}$ and $\overline{i \cdot t}$. The current slope is then given by eq. (4)

$$\xi = \frac{\overline{i \cdot t} - \bar{i} \cdot \bar{t}}{\overline{t^2} - \bar{t}^2} \quad . \tag{4}$$

As the current sampling is operated with a fixed sampling frequency, the average values $\bar{t}$ and $\overline{t^2}$ are simple functions of time and can be directly calculated for the respective time span. Based on parallel data processing in the FPGA, the four average values are computed simultaneously for each machine phase. To realize the division in eq. (4) for calculating a current slope, a division unit is installed in the FPGA. By using a time multiplexing state machine, this division unit is subsequently applied to each machine phase. This way the FPGA calculates the current slopes during the zero voltage vectors.
The FPGA uses a simplified model of the PMSM for a single

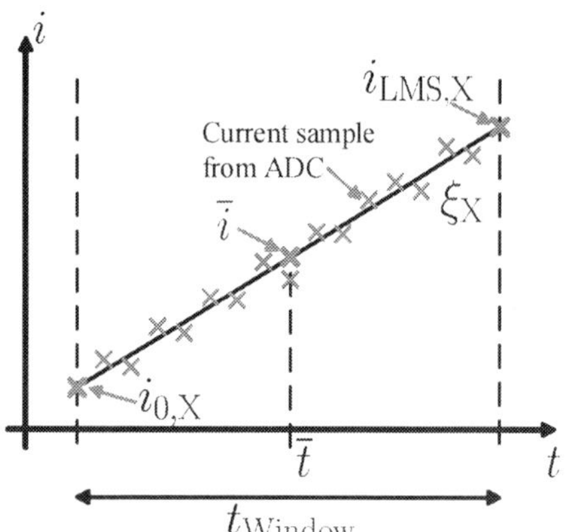

Fig. 4: Parameter of the regression window

machine phase, as in eq. (5).

$$v_X = R\, i_X + L \frac{d}{dt} i_X + u_{\text{EMF,X}} \quad . \tag{5}$$

We define $\tilde{A}_x$ as the current slope of phase X during the zero voltage state. The FPGA identifies $\tilde{A}_x$ as in eq. (6)

$$\frac{d}{dt} i_X = \tilde{A}_X = -\frac{1}{L} \left(R\, i_X + u_{\text{EMF,X}} \right) \quad . \tag{6}$$

Because of parallel data processing in the FPGA, this is performed simultaneously for every machine phase. It should be noted that the current oversampling, as is also illustrated in [6], leads to a high mitigation of current measurement noise. This way, exp. $\tilde{A}_x$ can be identified with a very high signal-to-noise ratio.

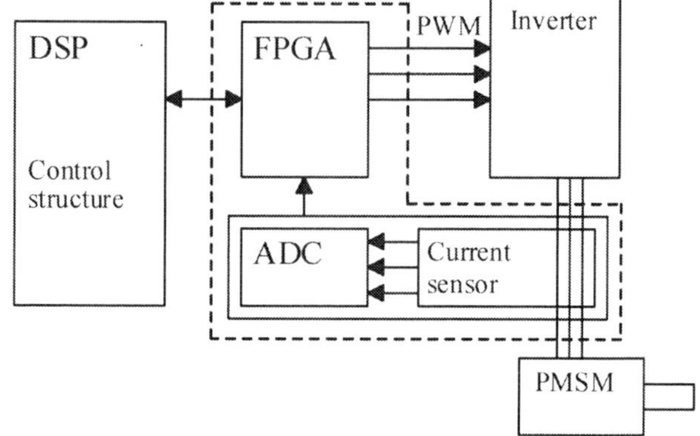

Fig. 5: Hardware setup

At the end of each regression window, the FPGA also computes a "constructed" current sample for each machine phase which is directly related to the end of the regression window, as illustrated in Fig. 4. These samples are also evaluated with the help of the linear regression algorithm, as in eq. (7)

$$\begin{aligned} i_{0,X} &= \bar{i}_X - \tilde{A}_X \bar{t} \\ i_{\text{LMS,X}} &= i_{0,X} + t_{\text{Window}} \tilde{A}_X \quad . \end{aligned} \tag{7}$$

In eq. (7), $i_{0,x}$ describes the current at the beginning of the regression window. The time t_{Window} stands for the time span of the regression window and $i_{\text{LMS,X}}$ describes a "constructed" sample. In Fig. 6, these samples are referred to as "LMS-constructed" samples. Due to FPGA-based current oversampling, these samples have a very high signal-to-noise ratio in view of current measurement noise, and they are computed in parallel for each machine phase.

With the beginning of a new PWM period, the duty cycles are updated. Directly afterwards, the FPGA identifies the half bridge that is the first ①, the second ② and the third ③ to be switched in each half PWM period. This is illustrated for PWM period II in Fig. 6. In PWM period II, half bridge C is the first to be switched, half bridge B is the second and half bridge A the third.

Current prediction

For the dead-time compensation the following assumptions are made. The DC link voltage is constant. The machine has the same resistance in all 3 phases. The machine's EMF is varying slowly compared to the PWM switching frequency.

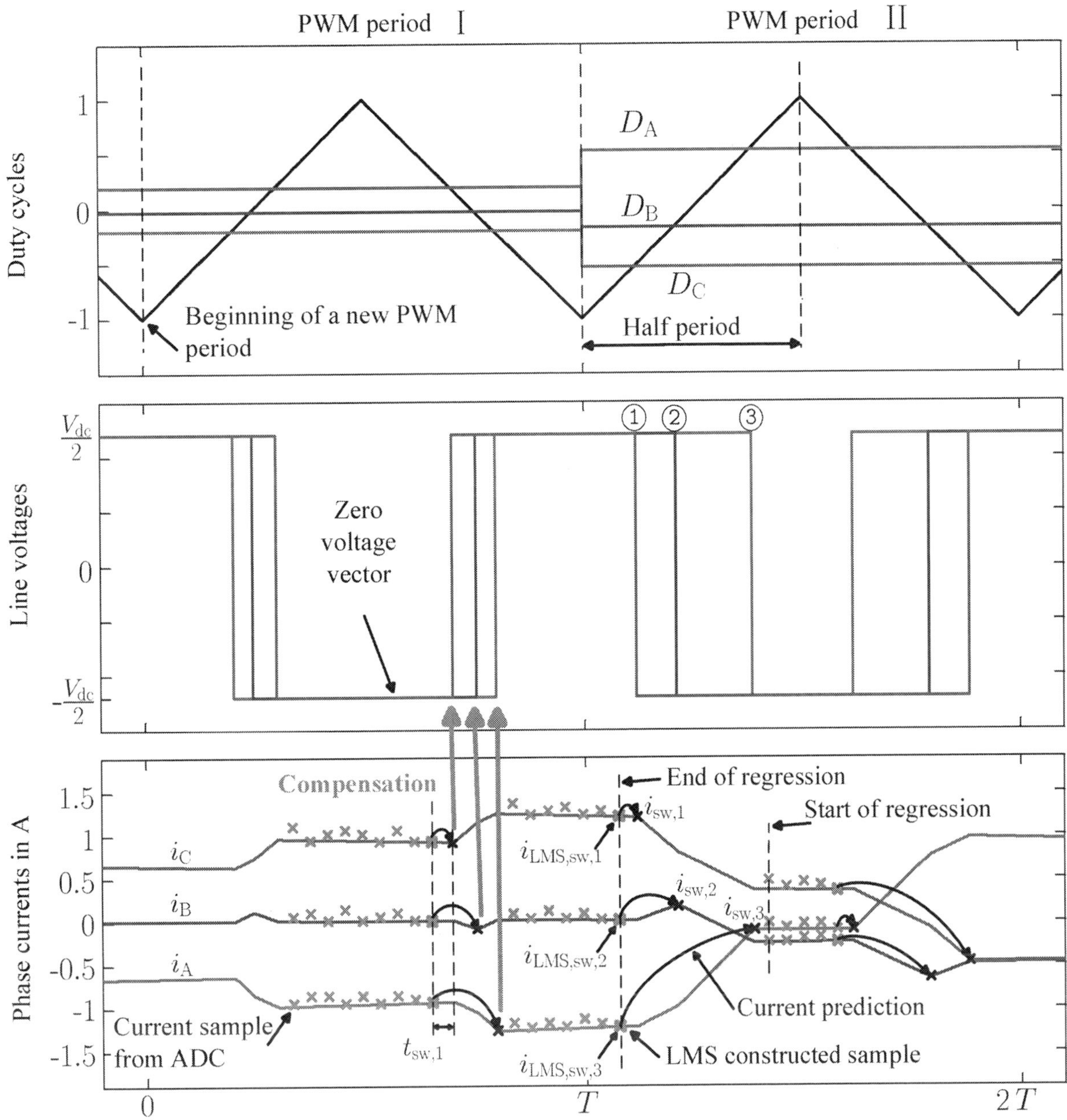

Fig. 6: Linear regression during zero voltage states and current prediction

The time $t_{sw,1}$ before the time span of a zero voltage vector ends, the FPGA performs the current prediction. In Fig. 6, the current $i_{sw,1}$ for half bridge C is the first that has to be predicted. This is done by eq. (8)

$$i_{sw,1} = i_{LMS,sw,1} + \xi_{sw,1}\, t_{sw,1} \quad . \tag{8}$$

In relation to Fig. 6 and considering the first half period of PWM period II, $\xi_{sw,1}$ would be the slope of current i_c during the recent zero voltage vector. The prediction for current $i_{sw,2}$ and $i_{sw,3}$ is based on the approach, as in eq. (9)

$$\begin{aligned} \frac{d}{dt} i_x &= \frac{1}{L} u_x - \frac{1}{L} \left(R\, i_x + u_{EMF,x} \right) \\ \frac{d}{dt} i_x &= \frac{1}{L} u_x + \widetilde{A}_x \quad . \end{aligned} \tag{9}$$

According to eq. (6), the component $\widetilde{A}_x$ depends on the machine's resistance, inductance and EMF. This component is always identified for every machine phase during the recent zero voltage states. Therefore, it is exactly known, and the

current prediction, as in eq. (9), becomes independent from the resistance and the EMF of the machine. This improves the stability and accuracy of the entire dead-time compensation algorithm.
Using a discretization approach, $i_{sw,2}$ can be predicted by eq. (10)

$$i_{sw,2} = i_{LMS,sw,2} + t_{act,1} \frac{1}{L} v_{sw,2} + (t_{act,1} + t_{sw,1}) \widetilde{A}_{sw,2} \quad (10)$$

where $t_{act,1}$ is the duration of the first active voltage vector in each PWM half period. The voltage $v_{sw,2}$ is constant, as it is related to the second half bridge to be switched in each half PWM period, and therefore the prediction can be computed, as in eq. (11)

$$i_{sw,2} = i_{LMS,sw,2} + t_{act,1} \frac{1}{L} \frac{1}{3} V_{dc} + (t_{act,1} + t_{sw,1}) \widetilde{A}_{sw,2} \quad . \quad (11)$$

The prediction of $i_{sw,2}$ is precise, because the voltage distortion related to the switching of the first ① half bridge is compensated, and therefore the first active voltage vector is accurate. As the voltage distortion related to the second ② half bridge is also compensated, the current $i_{sw,3}$ can be predicted, as in eq. (12)

$$i_{sw,3} = i_{LMS,sw,3} + \frac{V_{dc}}{3L} (t_{act,1} + 2\, t_{act,2}) + (t_{act,1} + t_{act,2} + t_{sw,1}) \widetilde{A}_{sw,3} \quad (12)$$

The entire prediction algorithm is computed in parallel in the FPGA and is performed during the time span $t_{sw,1}$. Based on these predictions, the duty cycles are directly adjusted to compensate for the voltage distortion for the upcoming switching instants, as in eq. (13).

$$\begin{aligned} D_A &= D_{A,set} + D_{Corr}(i_{A,sw}) \\ D_B &= D_{B,set} + D_{Corr}(i_{B,sw}) \\ D_C &= D_{C,set} + D_{Corr}(i_{C,sw}) \quad . \end{aligned} \quad (13)$$

The correction duty cycle $D_{Corr}(i_x)$ is stored in a lookup table which is the result of an auto-commissioning process, as described in section IV. In the first half period of PWM period II in Fig. 6, $i_{A,sw}$ equals $i_{sw,3}$, $i_{B,sw}$ equals $i_{sw,2}$ and $i_{C,sw}$ equals $i_{sw,1}$. This correlation is adjusted for each half PWM period. The dead-time compensation compensates every half period of a PWM period separately. This way, it is possible to compensate the dead-time effect, even if the current varies with a high frequency. This is especially advantageous for the sensorless control of PMSMs at low speeds, which is based on a high frequency voltage excitation.

IV. Auto-Commissioning

The lookup table is the result of an offline auto-commissioning process. It is based on the inverter and the connected machine. The machine's rotor is locked in a fixed position. Further hardware is not necessary. The FPGA generates a reference voltage varying with a low frequency to reach stationary operating points of the phase currents. For auto-commissioning generated by the FPGA, a particular characteristic of the PWM modulation is that it is only switching two of the three half bridges. In the following derivations, it is assumed that half bridge C is not switched.
With this modulation, the relation of the inverter's line voltages to the phase voltages is given by eq. (14)

$$\begin{pmatrix} v'_A \\ v'_B \end{pmatrix} = \frac{1}{2} \begin{pmatrix} 1 & -1 \\ -1 & 1 \end{pmatrix} \begin{pmatrix} v_A \\ v_B \end{pmatrix} \quad . \quad (14)$$

The reference line voltages are given by eq. (15)

$$\begin{aligned} v_A &= \hat{V} \cos(\omega t) \\ v_B &= -\hat{V} \cos(\omega t) \quad . \end{aligned} \quad (15)$$

Because half bridge C is not switched, the current i_C stays zero, which is also the initial current value. Generally the inverter's voltage distortion is identified by evaluating the machine's phase current response. This is illustrated below, see eq. (16).

$$\begin{aligned} v'_A - v'_B &= \frac{1}{2} v_A - \frac{1}{2} v_B + \frac{1}{2} v_A - \frac{1}{2} v_B \\ v'_A - v'_B &= v_A - v_B \quad . \end{aligned} \quad (16)$$

As the machine's rotor is locked in a fixed position, the phase voltages v'_A and v'_B can be substituted, as in eq. (17)

$$R\, i_A + L \frac{d}{dt} i_A - R\, i_B - L \frac{d}{dt} i_B = v_A - v_B \quad . \quad (17)$$

As the current i_C is zero, it can be found that, see eq. (18)

$$i_A = -i_B \quad . \quad (18)$$

Because of the low frequency reference voltage, the currents also vary with a low frequency, and therefore eq. (17) can be simplified, as in eq. (19)

$$2\, R\, i_A = \overline{v_A} - \overline{v_B} \quad . \quad (19)$$

Considering the line voltage references $v_{X,set}$ and the line voltage distortion $v_{X,err}$ resulting from the dead time effect leads to eq. (20)

$$2\, R\, i_A = v_{A,set} + v_{A,err} - v_{B,set} - v_{B,err} \quad . \quad (20)$$

Considering eq. (18) and eq. (1), (2), (3) and assuming that all half bridges have the same switching behaviour leads to eq. (21).

$$v_{A,err} = -v_{B,err} \quad . \quad (21)$$

This results in eq. (22)

$$\begin{aligned} R\, i_A &= \frac{1}{2} (v_{A,set} - v_{B,set}) + v_{A,err} \\ R\, i_A &= v'_{A,set} + v_{A,err} \quad . \end{aligned} \quad (22)$$

As the phase current i_A is measured, and the reference phase voltage $v'_{A,set}$ is known, it is possible to identify the line voltage distortion $v_{A,err}$. The identification is still based on the resistance parameter, and therefore an identification strategy for the resistance will be illustrated. Figure 7 shows the resulting phase current, if a reference voltage with a very low frequency, as in eq. (15), is applied to the machine. Eq. (23) considers the two operating points P_1 and P_2:

$$\begin{aligned} R \cdot i_{A,P1} &= v'_{A,P1,set} + v_{A,err} \\ R \cdot i_{A,P2} &= v'_{A,P2,set} + v_{A,err} \quad . \end{aligned} \quad (23)$$

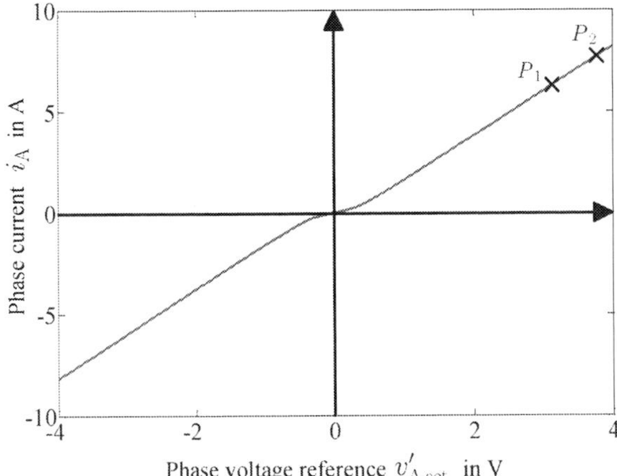

Fig. 7: Identification of the machine's resistance, measurement

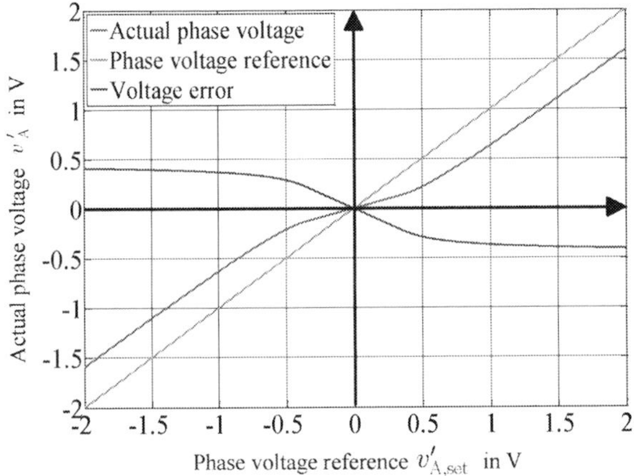

Fig. 8: Identification of the voltage error due to the dead time effect, measurement

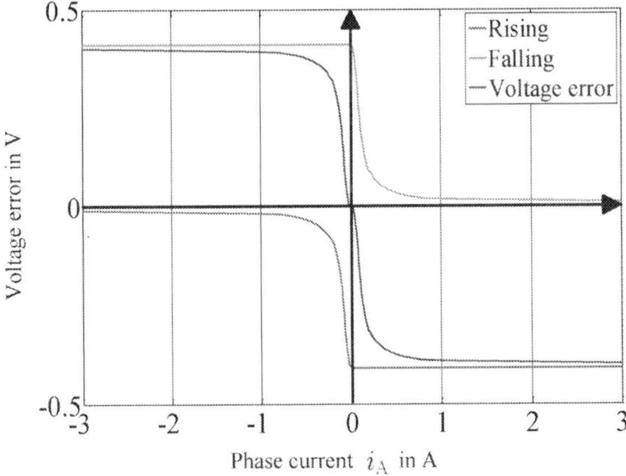

Fig. 9: Separation of the voltage error, measurement

As the voltage distortion of the dead-time effect can be assumed to be constant for a higher phase current [5], the resistance can be calculated, as in eq. (24)

$$R = \frac{v'_{A,P2,set} - v'_{A,P1,set}}{i_{A,P2} - i_{A,P1}} \quad . \tag{24}$$

The location of the points P_1 and P_2 is set manually. An algorithm to find their location automatically will be subject to current work. Because of current oversampling and a resulting high signal-to-noise ratio in view of current measurement noise, the resistance can be identified very precisely. Having the resistance identified, the resulting actual phase voltage can be calculated by eq. (25) and is shown in Fig. 8

$$v'_{A,act} = R \cdot i_A \quad . \tag{25}$$

The resulting line voltage distortion is then given by eq. (26)

$$v_{A,err}(i) = v'_{A,act}(i) - v'_{A,set} \quad . \tag{26}$$

As also illustrated in Fig. 2 and Fig. 3, the influence of the parasitic capacitances is reduced for higher phase currents. Based on the identified resistance and the operating points P_1, P_2, the line voltage distortion, which is nearly constant for a higher phase current, can be identified as in eq. (27). It is mainly determined by the influence of the diodes according to eq. (1).

$$|v_{A,err,t_{d,res}}| = \frac{1}{2}|R\,(i_{A,P1} + i_{A,P2}) - (v'_{A,P1,set} + v'_{A,P2,set})\,| \tag{27}$$

Depending on the polarity of the phase current, the voltage error as shown in Fig. 9 is separated into a component that is related to the rising respectively to the falling edge of the line voltage. This way, the voltage error is separated into components related to each half PWM period. Considering a positive phase current in Fig. 2, the voltage distortion at the falling edge of the line voltage results from the influence of the parasitic capacitances and is dependent on the sign and magnitude of the current. The voltage distortion in Fig. 2 at the rising edge of the line voltage is only dependent on the sign of the current. The component resulting from the parasitic capacitances can be calculated by eq. (28).

$$v_{A,err,C}(i) = v_{A,err}(i) - |v_{A,err,t_{d,res}}| \cdot \mathrm{sgn}(i) \quad . \tag{28}$$

The voltage error components are then used to compute the respective compensation duty cycles, as in eq. (29)

$$\begin{aligned} t_{Corr,X}(i) &= \frac{v_{err,X}(i)}{V_{dc}} T \\ D_{Corr,X}(i) &= 4\frac{t_{Corr,X}(i)}{T} - 1 \quad . \end{aligned} \tag{29}$$

The data for auto-commissioning is taken by the FPGA and is transferred to the DSP unit. The DSP computes the identification algorithm and transfers the duty cycles for dead-time compensation back to the FPGA, where it is stored in a random access memory (RAM). An auto-commissioning process is always performed for a fixed DC link voltage. This way, a lookup table is also referred to the respective DC link voltage. Figure 7 to Fig. 9 are measurements from the auto-commissioning, which are processed by the DSP unit. The DC link voltage is 46V, the PWM switching frequency is 10kHz and the implemented dead-time is $1\mu s$.

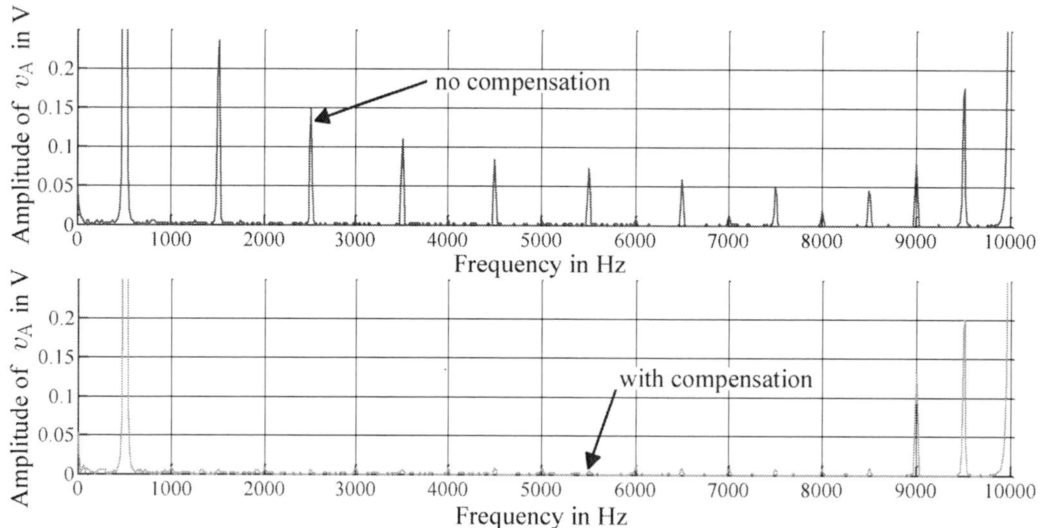

Fig. 10: Measured voltage spectrum for signal injection (500Hz, 3V)

V. Experimental results

The hardware setup uses the Protolar FPGA.Comm realtime extension for the dSpace DS1103 DSP-based rapid prototyping system, as shown in Fig. 14. The FPGA.Comm module carries a Spartan 6 XC6SLX45 FPGA device. The FPGA operates the analog devices AD9220 A/D converter with a resolution of 12 bits at a conversion frequency of 1 mega samples (1MSPS). The PWM switching frequency is 10kHz, and the MOSFET-based inverter's DC link voltage is 46V. The test setup for the electrical machine is shown in Fig. 15. The machine has a phase resistance of 0.5Ω, a d-axis inductance of 0.23mH (idle) and a q-axis inductance of 0.22mH (idle). The machine's torque constant is $0.074\frac{\text{Nm}}{\text{A}}$ and the maximum current is 25A. Figure 11 shows the phase current with and without compensation. The machine is at standstill, and a sinusoidal voltage vector in α-axis with a low frequency is applied. Figure 10 shows the voltage spectrum with and without dead time compensation with a reference voltage as rotating HF injection with 500Hz frequency and 3V amplitude being applied to the machine at standstill. The amplitude of the fundamental component of the voltage spectrum is 2.6V without compensation, and it is 2.9V with the compensation, as described in this paper. For the measurements related to Fig. 12 and Fig. 13 the machine is at standstill and a rotating voltage vector with 2.7V amplitude and 125Hz frequency is applied to the machine. The scaling of the current sensors is $0.1\frac{\text{V}}{\text{A}}$. The measurements show a high mitigation of the dead-time effect. For the simplified model of the machine, the FPGA uses the averaged inductance

$$L_{\text{av}} = \frac{1}{2}\left(L_{\text{d}} + L_{\text{q}}\right) = 0.225\,\text{mH} \quad .$$

When considering a machine with a higher inductance anisotropy this can have a negative influence of the accuracy of the current prediction and can mitigate the accuracy of the dead-time compensation.

The FPGA design consists of the communication interface to the DSP, the PWM modulation, the ADC driver and the described dead-time compensation. The entire design uses 7216 slice registers, 32 DSP48A1 blocks, the number of RAMB8BWERs is 3, the number of RAMB16BWERs is 1. All three half bridges use the same lookup table which has a size of 256 byte. Currently there is still a seperate RAMB8BWER implemented for each half bridge, but all carry the same lookup table. The lookup table has a higher information density in the region of small currents to reduce the needed memory size.

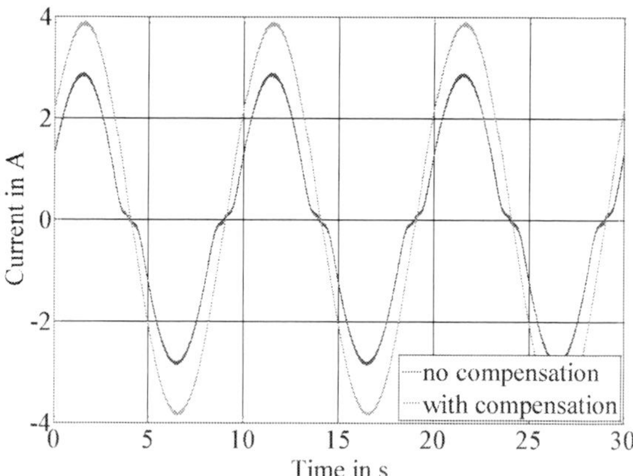

Fig. 11: Phase current with and without compensation, machine at standstill, sinusoidal excitation in α-axis

VI. Conclusions

A novel FPGA-based approach with current oversampling to compensate the inverter's dead-time effect during operation has been presented. The FPGA performs a linear regression of the machine's phase currents during the zero voltage states

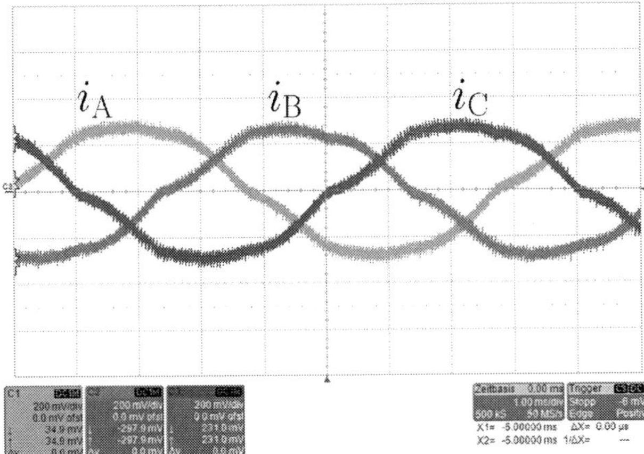

Fig. 12: Phase currents without compensation, machine at standstill

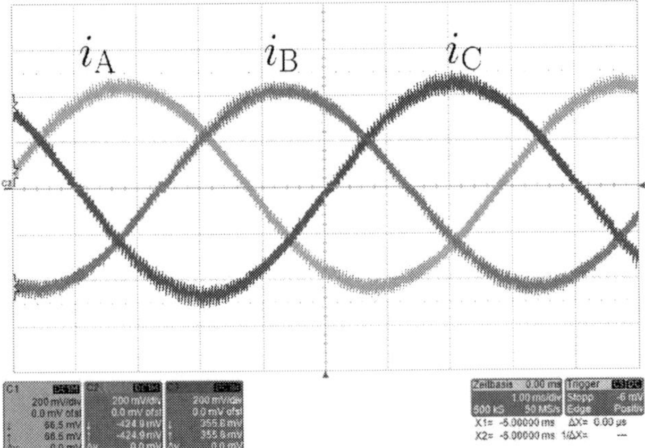

Fig. 13: Phase currents with compensation, machine at standstill

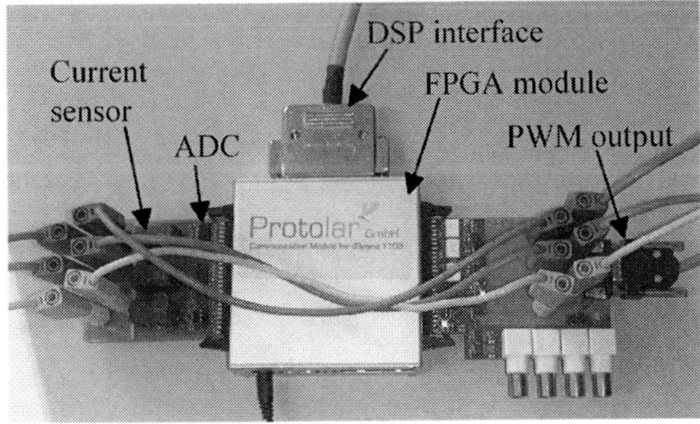

Fig. 14: FPGA unit

of the inverter. Based on an on-chip predictor, the FPGA predicts the currents at the upcoming moments of switching.

Fig. 15: Test setup

Using an on-chip lookup table and the predicted currents, the FPGA adjusts the upcoming instants of switching, in order to compensate the dead-time effect. The lookup table is the result of an offline commissioning process. The online compensation allows an improved quality of the current waveforms at low frequency operation of a PMSM, thus reducing torque ripple. It also allows the injection of a more favourable signal quality for sensorless control of PMSMs in the very low speed region. Since the estimated position is strongly dependent on the injected voltages, this will improve the performance of sensorless PMSM control at low speeds.

REFERENCES

[1] R. Sepe and J. H. Lang, "Inverter nonlinearities and discrete-time vector current control," *Industry Applications, IEEE Transactions on*, vol. 30, no. 1, pp. 62–70, Jan 1994.

[2] J.-W. Choi and S.-K. Sul, "Inverter output voltage synthesis using novel dead time compensation," *Power Electronics, IEEE Transactions on*, vol. 11, no. 2, pp. 221–227, Mar 1996.

[3] U. Buyukkeles and A. Ersak, "Analysis and compensation of nonlinear effects in an inverter with power mosfets," in *Power Electronics and Motion Control Conference (EPE/PEMC), 2012 15th International*, Sept 2012, pp. DS2c.3–1–DS2c.3–6.

[4] K. Yamamoto, K. Shinohara, and H. Ohga, "Effect of parasitic capacitance of power device on output voltage deviation during switching dead-time in voltage-fed pwm inverter," in *Power Conversion Conference - Nagaoka 1997., Proceedings of the*, vol. 2, Aug 1997, pp. 777–782 vol.2.

[5] K. Wiedmann, F. Wallrapp, and A. Mertens, "Analysis of inverter nonlinearity effects on sensorless control for permanent magnet machine drives based on high-frequency signal injection," in *Power Electronics and Applications, 2009. EPE '09. 13th European Conference on*, Sept 2009, pp. 1–10.

[6] B. Weber, K. Wiedmann, and A. Mertens, "Increased signal-to-noise ratio of sensorless control using current oversampling," in *Power Electronics and ECCE Asia (ICPE-ECCE Asia), 2015 9th International Conference on*, June 2015, pp. 1129–1134.

[7] J. Schellekens, R. Bierbooms, and J. Duarte, "Dead-time compensation for pwm amplifiers using simple feed-forward techniques," in *Electrical Machines (ICEM), 2010 XIX International Conference on*, Sept 2010, pp. 1–6.

[8] A. Lewicki, "Dead-time effect compensation based on additional phase current measurements," *Industrial Electronics, IEEE Transactions on*, vol. 62, no. 7, pp. 4078–4085, July 2015.

[9] P. Landsmann, J. Jung, M. Kramkowski, P. Stolze, D. Paulus, and R. Kennel, "Lowering injection amplitude in sensorless control by means of current oversampling," in *Sensorless Control for Electrical Drives (SLED), 2012 IEEE Symposium on*, Sept 2012, pp. 1–6.

[10] T. Mannen and H. Fujita, "Dead-time compensation method based on current ripple estimation," *Power Electronics, IEEE Transactions on*, vol. 30, no. 7, pp. 4016–4024, July 2015.

Control Strategy of High Power Converters with Synchronous Generator Characteristics for PMSG-based Wind Power Application

Yuzhi Zhang, Haoyan Liu, H. Alan Mantooth, Fellow, IEEE
NSF I/UCRC on GRid-connected Advanced Power Electronic Systems (GRAPES)
University of Arkansas, Fayetteville, AR 72701, U.S.A.
mantooth@uark.edu, yxz068@email.uark.edu

Abstract— **A virtual synchronous generator (VSG) control for high power permanent magnetic synchronous generators (PMSG) in wind power generation is investigated in this paper. A power factor correction (PFC) rectifier for each phase of the PMSG is designed to transform variable voltage and frequency of wind power into constant voltage DC power. The PFC boost inductor is replaced by a PMSG armature inductance to reduce the size and cost of the overall system. A three-phase inverter is used as a power interface to feed the generated wind power to the grid or load. By designing an excitation controller and a prime motor controller, the PMSG-based wind power interface can mimic a self-stabilization characteristic of synchronous generator which has rotor inertia, thus helping to improve the stability of wind power generation. Experimental results of the VSG inertia performance are illustrated, and the simulation results of traditional PQ and proposed VSG control are given and compared to validate the proposed converter topology and control algorithm.**

*Keywords— **virtual synchronous generator; permanent magnetic synchronous generator; power factor correction; Direct-Quadrature Frame; PQ control***

I. Introduction

Today, more and more renewable energy such as fuel cell power, wind power, and solar power generation are connected to the grid. Compared with conventional power generation, renewable energy generation usually needs power converters as an interface to connect with the grid or load. Therefore, renewable energy generation has different steady states and dynamic characteristics from conventional power stations because of their own inherent characteristics, such as random and intermittent behavior. This poses potential risks to the secure and stable operation of the grid system and causes difficulty in the centralized power system management and control that has been reported in [1], [2] and [3].

Permanent magnetic synchronous generators (PMSGs) are prevalently used in wind power generation. Compared with the Doubly-Fed Induction Generator (DFIG), the PMSG has higher power density, better controllability, and a simple drive train structure. It does not need an excitation circuit and has much higher efficiency and reliability. There are two common rectification topologies widely used in PMSGs [4], [5]. One is a PMSG through a pulse-width modulation (PWM) rectifier to create a DC voltage for inverters. The structure of this topology is more complex, and needs protection to prevent excitation failure during faults. The other is a PMSG with an uncontrolled rectifier and boost converter to create a DC voltage for inverters. This topology needs an additional boost inductor for the boost converter and has a low power factor due to the uncontrolled rectifier. In order to overcome these drawbacks, a scheme of a Power Factor Correction (PFC) converter is applied in the paper to achieve a steady DC output voltage and unit power factor. It doesn't need an additional inductor, and thus saves on the cost and size of the system [6]. In the conventional method, the active and reactive power decoupling control or PQ control is widely used as a control strategy for inverters [7], [8]. Recently, virtual synchronous generators (VSGs) are under investigation for solar and DFIG-based wind power generation in micro-grids. However, most applications need additional energy storage equipment and rarely focus on the VSG application in high power PMSG [9], [10]. In this paper, a VSG controller is designed for PMSG-based high wind power generation to improve system stability performance such as self-regulation during power demand changes. The power converter combined with the designed VSG control algorithm makes PMSG-based wind power generation more reliable, efficient, and of higher quality to the grid or loads.

The outline of the paper is as follows. Section II describes the circuit diagram of PMSG-based wind power generation. Section III presents the control strategy design for the PFC and VSG control for inverters. Section IV shows the experimental results of the virtual inertia and the simulation results of conventional PQ and proposed VSG control for PMSGs. The conclusion and future work are given in the last section.

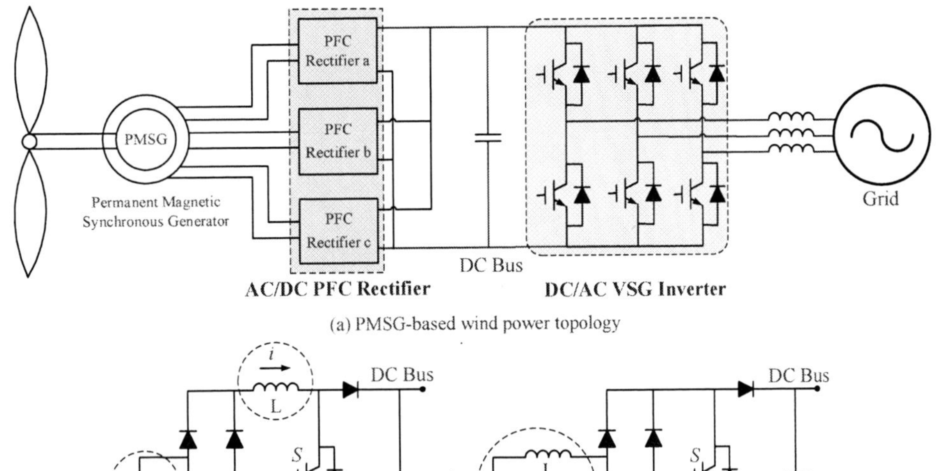

(a) PMSG-based wind power topology

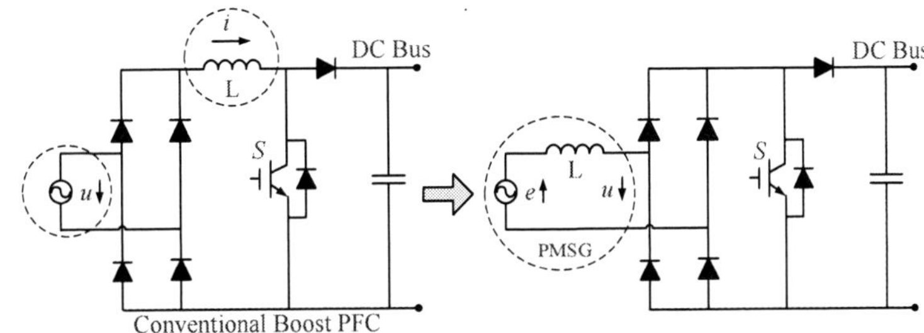

(b) PFC converter

Fig. 1. The scheme of proposed system.

II. THE CIRCUIT DIAGRAM OF THE PMSG-BASED WIND POWER GENERATION

As the proposed topology shown in Fig. 1, through the PFC rectifier which is diode rectifier combined with boost converter, the PMSG generates a steady DC bus and power factor is improved. The boost inductors are utilizing generator isolated coils. Therefore, the system size and cost can be further reduced. Three-phase inverter generates an AC voltage to connect with the grid or directly to the load. The three-phase inverter can be modular multi-level converter such as Neutral Point Clamped (NPC), Flying Capacitor (FC) or Series Connected H-Bridge (SCHB) converter. The detailed inverter control design for the proposed topology in Fig. 1 is presented in Section III.

III. PROPOSED CONTROL STRATEGY

A. Control Scheme of PFC rectifier

The PFC control scheme is shown in Fig. 2. PMSG with three generator coils are chosen as an example. There are two control loops. The voltage error between the DC reference value and DC bus voltage passes through PI control, and is then multiplied by the back electromotive force (EMF) phase of the generator to generate the reference current value. A sensorless control strategy is applied to get phase information of the winding back EMF estimation [11].

B. VSG control for inverter

The proposed VSG control scheme is based on the mathematical equations of a synchronous generator. The fifth-order Park transform equations of a synchronous generator are represented in Eqs. (1) and (2).

$$\begin{bmatrix} u_d \\ u_q \\ u_e \\ u_D \\ u_Q \end{bmatrix} = p\begin{bmatrix} \psi_d \\ \psi_q \\ \psi_e \\ \psi_D \\ \psi_Q \end{bmatrix} + \begin{bmatrix} R_d & 0 & 0 & 0 & 0 \\ 0 & R_q & 0 & 0 & 0 \\ 0 & 0 & R_e & 0 & 0 \\ 0 & 0 & 0 & R_D & 0 \\ 0 & 0 & 0 & 0 & R_Q \end{bmatrix}\begin{bmatrix} -i_d \\ -i_q \\ i_e \\ i_D \\ i_Q \end{bmatrix} + \begin{bmatrix} -\omega\psi_q \\ \omega\psi_d \\ 0 \\ 0 \\ 0 \end{bmatrix} \quad (1)$$

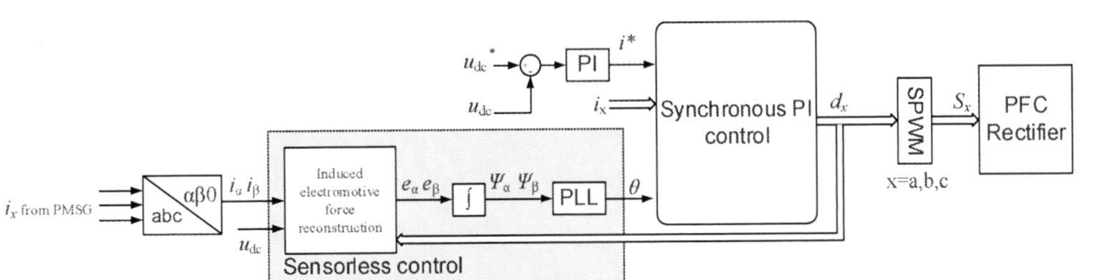

Fig. 2. The PFC control scheme for the PMSG.

$$\begin{bmatrix} \psi_d \\ \psi_q \\ \psi_e \\ \psi_D \\ \psi_Q \end{bmatrix} = \begin{bmatrix} L_d & 0 & M_{de} & M_{dD} & 0 \\ 0 & L_q & 0 & 0 & M_{qQ} \\ M_{ed} & 0 & L_e & M_{eD} & 0 \\ M_{Dd} & 0 & M_{De} & L_D & 0 \\ 0 & M_{Qq} & 0 & 0 & L_Q \end{bmatrix} \begin{bmatrix} -i_d \\ -i_q \\ i_e \\ i_D \\ i_Q \end{bmatrix} \tag{2}$$

where u is the winding voltage, ψ is the flux, ω is the synchronous angular velocity, p is number of pole-pairs. d, q, e, D, and Q are the d-axis winding, q-axis winding, excitation winding, d-axis damper winding, and q-axis damper winding, respectively.

The mechanical characteristic equation of the synchronous generator can be represented by Eq. (3).

$$\begin{cases} T_m - T_e - D\Delta\omega = \dfrac{P_m}{\omega} - \dfrac{P_e}{\omega} - D\Delta\omega = J\dfrac{d\Delta\omega}{dt} \\ \dfrac{d\theta}{dt} = \omega \end{cases} \tag{3}$$

where T_m and T_e are the mechanical and electromagnetic torque; P_M and P_E are mechanical power and electromagnetic power; D is a constant damping coefficient; $\Delta\omega=\omega-\omega_n$, where $\Delta\omega$ is the electrical angular speed difference, ω and ω_n are the actual electrical angular speed and rated electric angular velocity, respectively.

The overall control block diagram of the VSG controller is shown in Fig. 3. It includes primary motion control and excitation control. The output active power P and reactive power Q can be represented by Eqs. (4) and (5), respectively. E and U are the inner potential and the bus voltage amplitude of the VSG, respectively. Φ is the power angle between E and U. In order to analyze the dynamic stability of VSG more accurately, this paper establishes a small-signal model of the VSG controller, then analyzes the stability of the system. Combining Eqs. (1), (2) and (3), the VSG small-signal model is given in Eq. (6). The coefficients of A, B, C, D, E, F, G are given in Eqs. (7), (8), (9), (10), (11), (12) and (13), respectively. The system stability with load power and inertia increases are shown in Figs. 4 (a) and (b), respectively. It shows a load power that is too high or an improper inertia value may lead to an unstable system. The small-signal equations can help to design the controller parameters as well.

$$P = \frac{U}{X_{eq}} E\sin\phi \tag{4}$$

$$Q = \frac{U}{X_{eq}}(E\cos\phi - U) \tag{5}$$

$$As^6\hat{\phi} + Bs^5\hat{\phi} + Cs^4\hat{\phi} + Ds^3\hat{\phi} + Es^2\hat{\phi} + Fs\hat{\phi} + G\hat{\phi} = 0 \tag{6}$$

$$A = J\frac{\omega_n}{n_q} \tag{7}$$

$$B = \frac{m_p}{2\pi n_q} + 2J\frac{\omega_n\omega_c}{Qn_q} \tag{8}$$

$$C = \frac{m_p\omega_c}{Q\pi n_q} + \frac{J\omega_n}{n_q}\left(\frac{\omega_c^2}{Q^2} + 2\omega_c^2\right) + \frac{JU}{X_{eq}}\cos(\phi)\omega_c^2\omega_n \tag{9}$$

$$D = \frac{m_p(\omega_p^2/Q^2 + 2\omega_c^2)}{2\pi n_q} + \frac{2\omega_n J\omega_c^3}{n_q Q} + \frac{U\omega_c^2 m_p\cos\phi}{2\pi X_{eq}} + \frac{JU\cos\phi\omega_c^2\omega_n}{X_{eq}Q} \tag{10}$$

$$E = \frac{m_p\omega_c^3}{n_q\pi Q} + \frac{J\omega_n\omega_c^4}{n_q} + \frac{m_p U\omega_c^2\cos\phi}{2X_{eq}Q\pi} + \frac{JU\omega_n\omega_c^2\cos\phi}{X_{eq}} + \frac{240U\omega_c^2\cos\phi}{X_{eq}n_q} \tag{11}$$

$$F = \frac{m_p\omega_c^4}{2\pi n_q} + \frac{240UQ\cos\phi\omega_c^3}{n_q X_{eq}} + \frac{m_p U\omega_c^4\cos\phi}{2\pi X_{eq}} \tag{12}$$

$$G = \frac{240U^2\omega_c^4}{X_{eq}^{\ 2}} + \frac{240UQ\cos\phi\omega_c^3}{X_{eq}n_q} \tag{13}$$

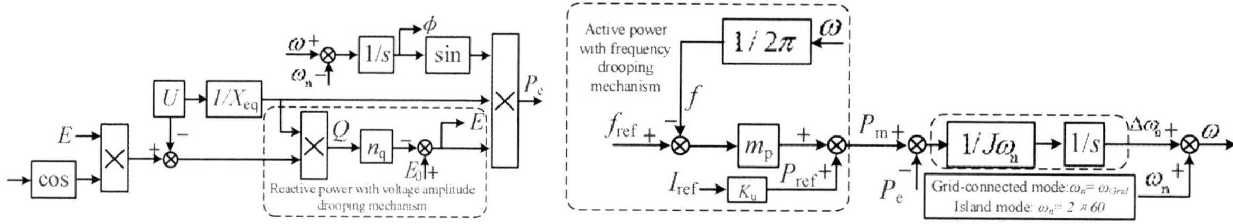

Fig. 3. Diagram of VSG controller for PMSG.

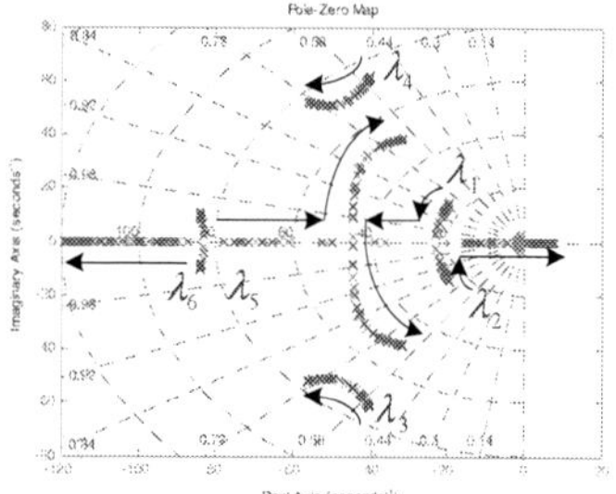

(a) Eigenvalue migration with power angle increased

(b) Eigenvalue migration with inertia increased

Fig. 4. Simulation results of small-signal model.

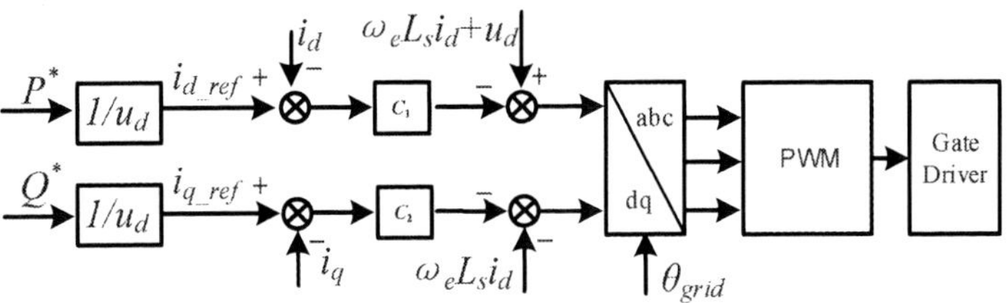

Fig. 5. Conventional PQ control for PMSG-based wind power generation.

C. Conventional PQ control

To compare with the proposed VSG control for PMSG, the conventional PQ control is given in Fig. 5 [8], [12]. The decouple control can be determined by Eqs. (14) and (15).

$$L_s \, {di_d}/{dt} = u_d - S_d - Ri_d + \omega_e L_s i_q \quad (14)$$

$$L_s \, {di_q}/{dt} = u_q - S_q - Ri_q - \omega_e L_s i_d \quad (15)$$

where i_d, i_q, u_d and u_q are the grid currents and voltages in the d, q axis. L_s and R are the inductance and resistance from the grid-side. ω_e is the electrical speed. S_d, S_q are the inverter output voltages.

IV. SIMULATION AND EXPERIMENTAL VERIFICATION

In order to verify the proposed VSG control for the PMSG-based wind power generation, both experimental testing and simulations are conducted. The VSG rotor inertia experimental results and overall system simulation results are shown in Fig. 6 and Fig. 7, respectively. Compared with Fig. 6(a), Fig. 6(b) shows that the dynamic response of the system changes faster due to the smaller inertia J. This phenomenon is the same as the actual synchronous generator. Therefore, the dynamic response performance of the system can be adjusted by changing the virtual inertia parameter in the controller to get a better dynamic response of the system.

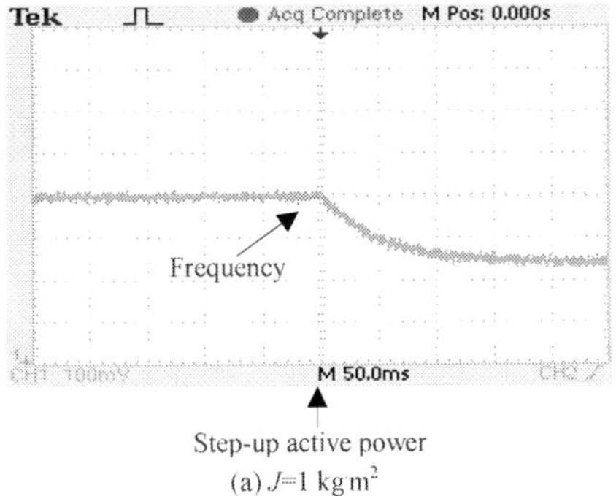

(a) J=1 kg m^2

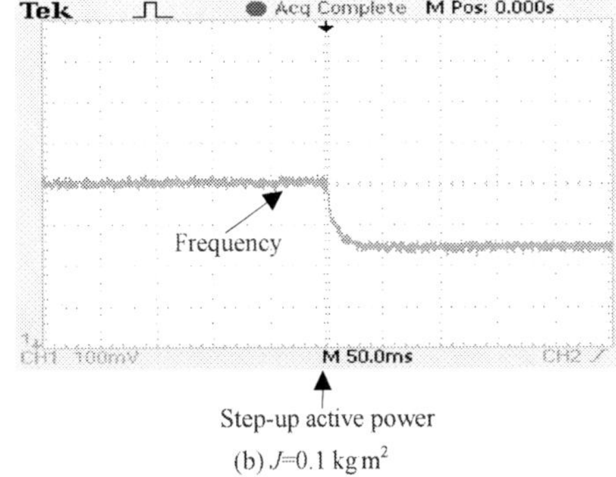

(b) J=0.1 kg m^2

Fig. 6. Frequency responses of different rotary inertia.

Fig. 7(a) and (b) show the simulation results of step-up active power. With the proposed VSG control in PMSG, the output active power achieves a reference value 0.275 MW within 1.5 ms. However, PQ control requires a longer settling time of 5 ms and has higher coupled reactive power Q during the transition time as shown in Fig. 7(b). This is detrimental for inverter voltage stability. Similarly, during the step-up reactive power, Fig. 7(c) and (d) show the VSG control has a shorter response time, and the coupled active power P is much smaller than the PQ control.

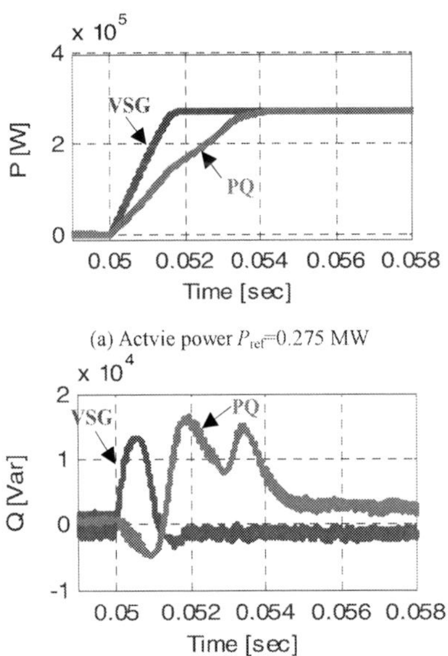

(a) Actvie power P_{ref}=0.275 MW

(b) Coupled reactive power Q

978-1-4673-9551-9/16 $31.00 © 2016 IEEE

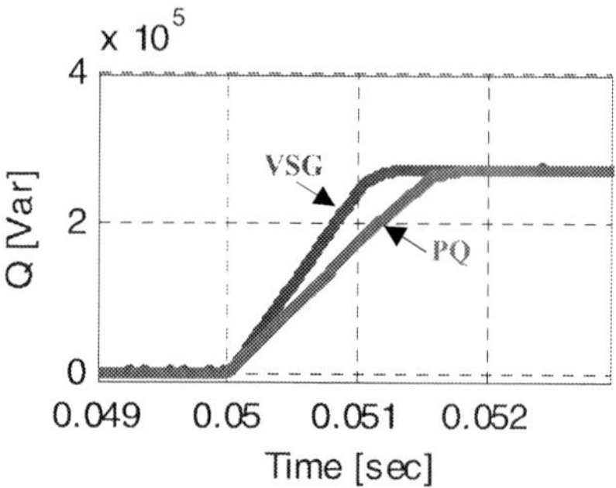

(c) Reactive power Q_{ref}=0.275 MVar

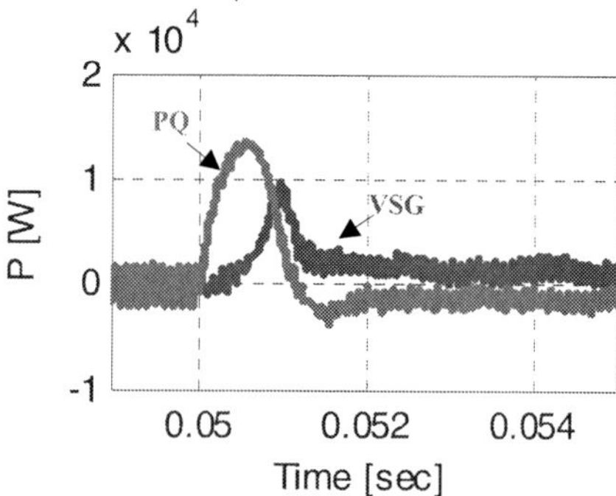

(d) Coupled active power P

Fig. 7. Simulation results comparison between conventional PQ and proposed VSG control.

V. SUMMARY AND FUTURE WORK

This paper presents virtual synchronous generator control for the PMSG-based wind power generation. The excitation and prime motor controllers are designed. The small-signal modeling of synchronous generator is built to optimize the controller parameters. Compared with the actual synchronous generator, the VSG parameters are much easier to design. Compared with the conventional PQ control, the VSG control has better transient dynamics and power decoupled performance in a PMSG-based wind system. Both simulation and experimental results have verified that the proposed topology and its optimized control strategy work as predicted. A more detailed description of the control algorithm and test results will be provided in future work.

ACKNOWLEDGMENTS

Special appreciation goes to the National Science Foundation Industry/University Cooperative Research Center on GRid-connected Advanced Power Electronic Systems (GRAPES) members, the sponsors of this work.

REFERENCES

[1] Jian Xu; Siyang Liao; Yuanzhang Sun; Xi-Yuan Ma; Wenzhong Gao; Xiaomin Li; Junhe Gu; Jianxun Dong; Mi Zhou, "An Isolated Industrial Power System Driven by Wind-Coal Power for Aluminum Productions: A Case Study of Frequency Control," *Power Systems, IEEE Transactions on*, vol.30, no.1, pp.471,483, Jan. 2015.

[2] Kai Tan; Huang, A.Q.; Martin, A., "Development of solid state arc-free socket for DC distribution system," in *Applied Power Electronics Conference and Exposition (APEC), 2014 Twenty-Ninth Annual IEEE*, vol., no., pp.2300-2305, 16-20 March 2014.

[3] Ozkan, M.B.; Karagoz, P., "A Novel Wind Power Forecast Model: Statistical Hybrid Wind Power Forecast Technique (SHWIP)," *Industrial Informatics, IEEE Transactions on*, vol.11, no.2, pp.375,387, April 2015.

[4] Shih-Yu Yang; Yuan-Kang Wu; Huei-Jeng Lin, "The direct torque control of the PMSG based windturbine with two level voltage source converter," *Automatic Control Conference (CACS), 2014 CACS International*, vol., no., pp.39,44, 26-28 Nov. 2014.

[5] Shao Zhang; King-Jet Tseng; Vilathgamuwa, D.M.; Trong Duy Nguyen; Xiao-Yu Wang, "Design of a Robust Grid Interface System for PMSG-Based Wind Turbine Generators," *Industrial Electronics, IEEE Transactions on*, vol.58, no.1, pp.316,328, Jan. 2011.

[6] Lifei Liu; Yingtao Ma; Jianyun Chai; Xudong Sun, "Cascaded single phase PFC rectifier pair for permanent magnet wind generation system," *Electrical Machines and Systems, 2008. ICEMS 2008. International Conference on*, vol., no., pp.2516,2520, 17-20 Oct. 2008.

[7] Lee, C-.T.; Chu, C-.C.; Cheng, P-.T., "A New Droop Control Method for the Autonomous Operation of Distributed Energy Resource Interface Converters," *Power Electronics, IEEE Transactions on*, vol.28, no.4, pp.1980,1993, April 2013.

[8] Adhikari, S.; Fangxing Li, "Coordinated V-f and P-Q Control of Solar Photovoltaic Generators With MPPT and Battery Storage in Microgrids," *Smart Grid, IEEE Transactions on*, vol.5, no.3, pp.1270,1281, May 2014.

[9] Wang, S.; Hu, J.; Yuan, X., "Virtual Synchronous Control for Grid-Connected DFIG-Based Wind Turbines," *Emerging and Selected Topics in Power Electronics, IEEE Journal of*, vol.PP, no.99, pp.1,1.

[10] Albu, M.; Visscher, K.; Creanga, D.; Nechifor, A.; Golovanov, N., "Storage selection for DG applications containing virtual synchronous generators," *PowerTech, 2009 IEEE Bucharest*, vol., no., pp.1,6, June 28 2009-July 2 2009.

[11] Yingtao Ma; Xudong Sun; Jianyun Chai, "Three-phase PFC rectifier with sensorless control for PMSG wind generation system," *Electrical Machines and Systems (ICEMS), 2011 International Conference on*, vol., no., pp.1,4, 20-23 Aug. 2011.

[12] Kai Tan; Qiongxuan Ge; Zhenggang Yin; Congwei Liu; Yaohua Li, "The optimal control strategy for rectifier side of low switching frequency back-to-back converter," in *Applied Power Electronics Conference and Exposition (APEC), 2010 Twenty-Fifth Annual IEEE*, vol., no., pp.1419-1423, 21-25 Feb. 2010.

Phase Compensation, ZVS Operation of Wireless Power Transfer System Based on SOGI-PLL

Pingan Tan, *member, IEEE* and Haibing He
College of Information Engineering
Xiangtan University , Xiangtan, Hunan, 411105, China
E-mail: tanpingan@126.com , hehaibing8561081@126.com

Xieping Gao
MOE Key Laboratory of Intelligent Computing &
Information Processing, College of Information Engineering
Xiangtan University , Xiangtan, Hunan, 411105, China
E-mail: xpgao@xtu.edu.cn

***Abstract*—Wireless power transfer (WPT) technology is now recognized as an efficient means of transferring power because having numerous advantages over conventional wired power transfer system. The phase delays, for example induced in the sampling process, the algorithm implementation process, the signal transduction process, etc. are widely found during the implementation of WPT system, which significantly degrade the system performance. Moreover, it would be extremely necessary to implement the Soft-Switching of the converter and the necessary dead time imposed by the drivers should be compatible with the resonant current phase lag control. This paper proposes a Direct Phase Control (DPC) approach, based on Second-Order Generalized Integrator Phase-Locked Loop (SOGI-PLL), to provide accurate phase compensation and stable ZVS operation in the WPT. The DPC determines the phase difference $\Delta\theta$ of WPT, which include the phase difference θ_l between the output voltage and current of the converter and the phase delay θ_l'' derived from the sampling process. The SOGI-PLL provides the phase of system for adjusting the output voltage frequency of the converter dynamically. Experimental results convincingly demonstrate that with the proposed method the phase delay can be compensated accurately and the ZVS operation can be achieved simultaneously.**

Keywords—WPT; phase Compensation; ZVS operation; DPC; SOGI-PLL

I. INTRODUCTION

WPT technology is a kind of promising technique in our daily life, it provides convenience in such fields as portable electronics, implantable medical instruments and sensor networks [1]-[4]. In the WPT, the current sensing circuit is included and then some algorithms are brought in. As a result, phase delays expressed as θ_l''in Fig.1-(b) are widely found in the WPT, which significantly lead to the performance degradation and even failure of WPT [5], [6]. For the high-frequency application in WPT system, it is essential to assure operation in Soft-Switching mode. What's more, dead time in a voltage source bridge converter is required to prevent shoot-through current.

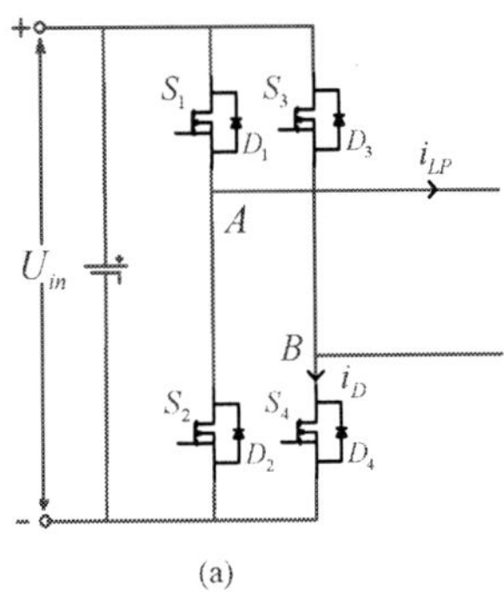

(a)

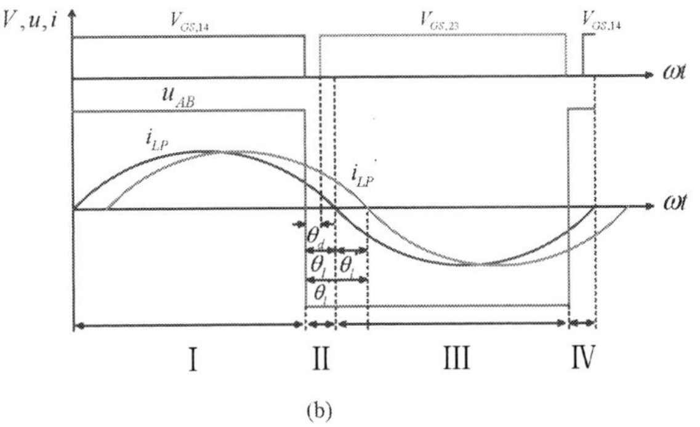

(b)

Fig.1. (a) Converter of WPT; (b) Phase compensation operation.

This paper proposes a DPC approach with SOGI-PLL to provide accurate phase compensation and stable ZVS operation in the WPT. The DPC determines the phase difference $\Delta\theta$ of WPT, which include the phase difference θ_l between the output voltage and current of the converter and the phase delay θ_l'' derived from the sampling process. The SOGI-PLL provides the phase of system for adjusting the output voltage frequency of the converter dynamically. The phase difference of the WPT, the dead angle imposed by the drivers and the compensation phase-angle could be regulated precisely. Moreover, the necessary dead time imposed by the drivers could be compatible with the resonant current phase lag control, which is conducive to the ZVS operation [5], [6]. With the proposed method the phase delay can be compensated accurately, the condition of ZVS operation can be achieved simultaneously.

II. PHASE COMPENSATION, ZVS OPERATION in WPT

The converter of the WPT is presented in Fig.1-(a), and the phase compensation operation is presented in Fig.1-(b). As presented in Fig.1-(b) that the phase delay θ_l'' derived from the sampling process is detected in WPT where i_{LP} is the actual output current of the converter and i_{LP}' is the sampling values of the actual output current. The phase lag θ_l represents the phase difference between the output voltage and current of the converter, which refers to the difference between the zero-crossing point of the current i_{LP} and the falling edge of the voltage V_{AB}. The phase lag θ_l' represents the phase difference between the output voltage and the sampling values of the current, which refers to the difference between the zero-crossing point of the current i_{LP}' and the falling edge of the voltage V_{AB}. Lastly, the phase lag θ_d represents the dead angle imposed by the drivers. In order to assure the ZVS operation, what is the most important is that the resonant current phase lag angle θ_l should be greater than the dead

The Project Supported by National Natural Science Foundation of China(NSFC) (51207134)

angle θ_d [7], [8]. The relationship between the four components can be expressed as

$$\begin{cases} \theta_l' = \theta_l + \theta_l'' \\ \theta_l > \theta_d \end{cases} \quad (1)$$

In this paper, a DPC approach with SOGI-PLL to provide accurate phase compensation and stable ZVS operation in WPT is proposed as shown in Fig.2. By sensing the primary side current, the output phase angle θ' of SOGI-PLL acts as the control signal for the PWM driver. The phase difference between the output signal and input signal of the SOGI-PLL could be regulated by parameter of $\Delta\theta^*$ accurately, which is called as Direct Phase Control (DPC) as shaded area show in Fig.2, and it is of great significance to the phase compensation and ZVS operation.

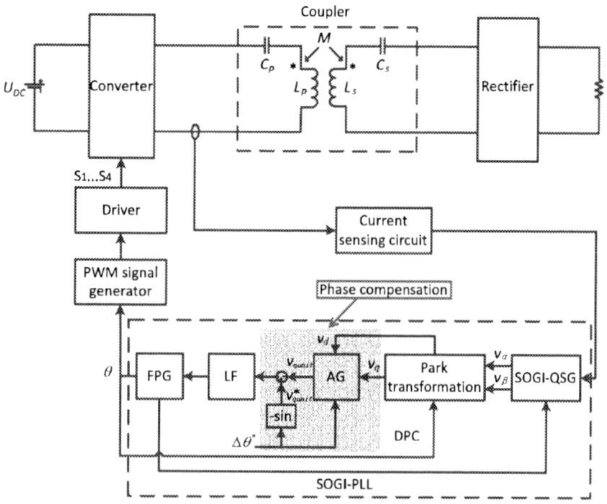

Fig.2. Schematic diagram of the WPT phase compensation.

The SOGI-PLL is the core of the whole control system, and it is composed of five parts as follows, the Second-Order Generalized Integrator Quadrature Signal Generator (SOGI-QSG), the Park transformation, the DPC, the Low-pass Filter (LF), the Frequency/Phase-Angle Generator (FPG). The proposed method has two feedback loops, with which the FPG provides phase and central frequency for the Park transformation and the SOGI-QSG respectively [9], [10].

The SOGI-QSG is a kind of Adaptive Filter (AF) as shown in Fig.3 and it possesses band-pass filtering property[9], [10]. The bandwidth of the SOGI-QSG merely relies on the gain k rather than the central frequency ω'. As a result, it is suitable for those occasions with frequency variation.

According to Fig.3, the transfer functions of the SOGI-QSG can be expressed as (2),(3).

$$D(s) = \frac{v'}{v}(s) = \frac{k\omega' s}{s^2 + k\omega' s + \omega'^2} \quad (2)$$

$$Q(s) = \frac{qv'}{v}(s) = \frac{k\omega'^2}{s^2 + k\omega' s + \omega'^2} \quad (3)$$

where ω' represents the central frequency of SOGI and k is the gain of SOGI-QSG.

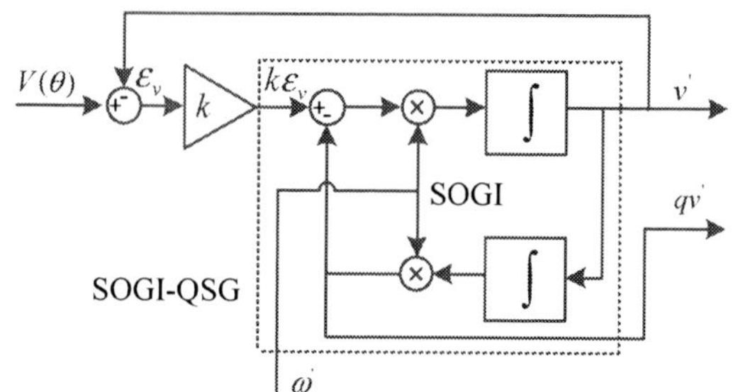

Fig.3. SOGI-QSG.

A traditional filter could merely deal with those signals that lie in the fixed frequency range. What's worse, the parameters of this kind of filter are static and the values have been assigned during the design progress of the filter. However, an AF could adapt its parameters automatically according to the optimization algorithm. In addition, during the design process of an AF the information about the signal to be filtered is not needed at all [11], [12]. The SOGI-QSG could deal with those signals that lie in any frequency range. Moreover, it could be used in those occasions with distortion and disturbance.

With the help of the DPC, the actual phase difference $\Delta\theta$ between the output phase θ' and the input phase θ of SOGI-PLL can be expressed as (4).

$$\Delta\theta = \theta' - \theta = tan^{-1}\left[-\frac{v_{qunit}}{cos(\Delta\theta^*)}\right] \quad (4)$$

Where v_{qunit} represents the output of AG as presented in Fig.2.

At the moment when the SOGI-PLL operates steadily, as presented in Fig.2, the condition of $v_{qunit} = -\sin(\Delta\theta^*)$ will be satisfied. With equation (2), then $\Delta\theta = \Delta\theta^*$. Thus, it's possible to regulate the actual phase difference $\Delta\theta$ by setting the parameter of $\Delta\theta^*$. Fig.4 displays the block diagram of the WPT phase compensation.

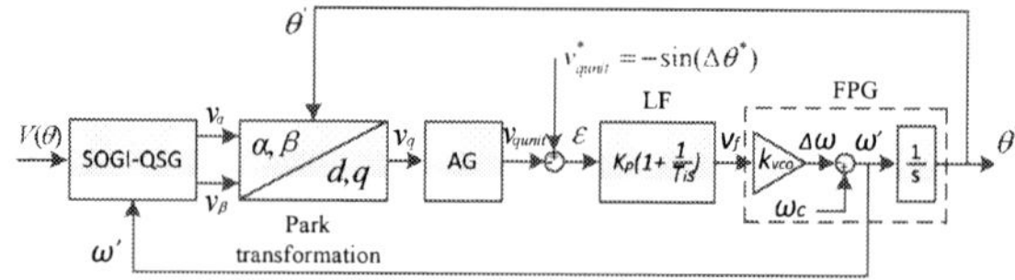

Fig.4. Block diagram of the WPT phase compensation.

What's more, According to the cosine value of θ' the PWM driver signals V_{GS} are obtained as presented in Fig.5. Besides, the dead angle θ_d between the two sets of gate signals $V_{GS,14}$ and $V_{GS,23}$ could be regulated precisely. Where d is a constant whose value is closed to zero nearly and the dead angle θ_d is expressed as:

$$\theta_d = 2\,sin^{-1}\,d \quad (5)$$

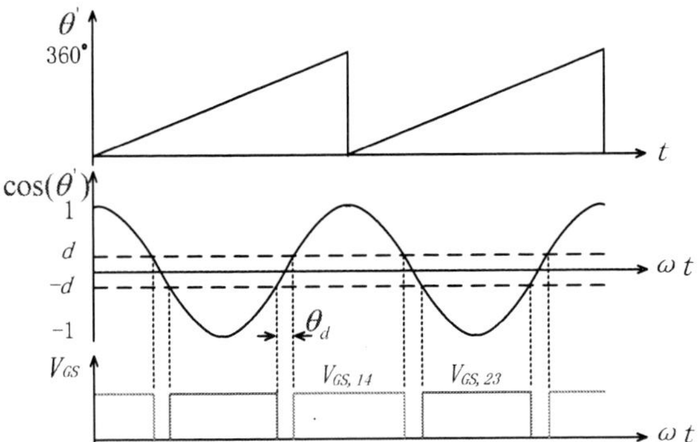

Fig.5. Regulation of the PWM driver signals.

Once the dead angle θ_d and the phase delay-angle θ_l'' have been assigned what should be done is that setting the parameter of $\Delta\theta^*$ that is greater than $\theta_d + \theta_l''$. On the other hand, the minimum input power factor ρ_{min} whose value nearly close to one is introduced. Therefore, we could set the parameter of $\Delta\theta^*$ according to (6).

$$\theta_d + \theta_l'' < \Delta\theta^* < cos^{-1}(\rho_{min}) + \theta_l'' \tag{6}$$

Thus, the phase delay has been compensated successfully. Moreover, the dead time imposed by the drivers is compatible with the resonant current phase lag control. In this case, the ZVS operation can be achieved simultaneously.

As described in the ZVS operation sequence in Fig.6. there are four kinds of operation mode. All of the operation mode represents the operation of the four stages respectively as presented in Fig.1-(b). Mode I: the current i_{LP} is positive and it follows through switch S_1, S_4; Mode II: the current i_{LP} is positive and it follows through the freewheel diode D_2, D_3, at this moment it is proper to turn on switch S_2, S_3, leading to the ZVS; Mode III: the current i_{LP} is negative and it follows through switch S_2, S_3; Mode IV: the current i_{LP} is negative and it follows through the freewheel diode D_1, D_4, at this moment it is proper to turn on switch S_1, S_4, leading to the ZVS.

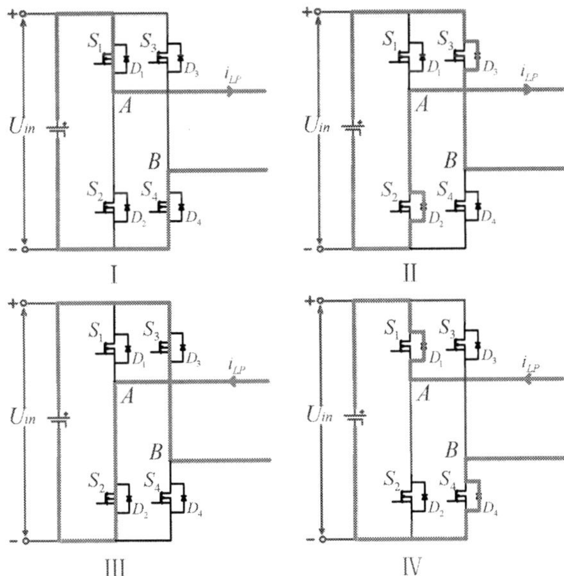

Fig.6. ZVS operation sequence.

III. PERFORMANCES and RESULTS ANALYSIS

In order to verify the validity of the proposed technique further, the hardware implementation has been done and the experimental results are presented in this paper. Fig.7 displays the experimental prototype of the WPT system with phase compensation as well as ZVS operation. The prototype is composed of the following parts, the DC power, the converter, the coupler, the rectifier, the load, the current sensing circuit (a current transducer LA 25-NP, a current conditioning circuit), the DSP controller, the driver. An efficient TMS320F28335 DSP is applied in the experimental prototype, whose clock rate is 150MHz. All of the algorithm is realized with software and there needn't additional hardware implementation.

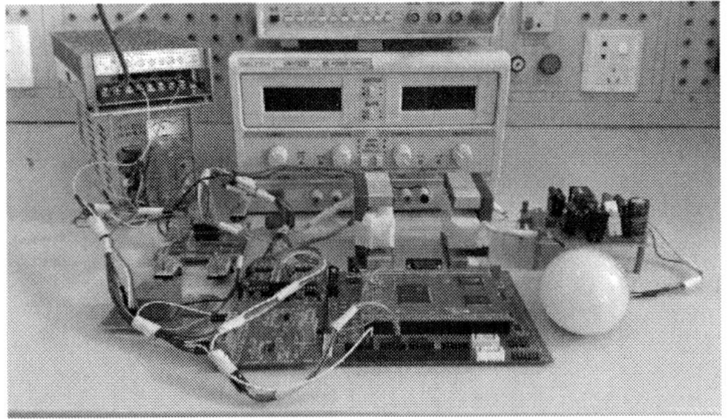

Fig.7. Experimental prototype of the WPT system with phase compensation and ZVS operation.

Fig.8 displays the experimental results of WPT, in which Fig.8-(a) shows the results without phase compensation, ZVS operation and Fig.8-(b) shows the results with phase compensation, ZVS operation. The phase delay θ_l'' is detected in the sampling process. Just as presented in Fig.8-(a) that the driver signals synchronize with the sampling values i_{LP}' rather than the actual output current i_{LP}. On the contrary, with the implementation of the phase compensation as presented in Fig.8-(b), a compensation phase-angle θ_l' is introduced, which includes the dead angle θ_d, the phase lag θ_l and the phase delay θ_l''. As a result, the driver signals synchronize with the output current i_{LP} approximately. What's more, the ZVS operation condition $\theta_l > \theta_d$ is satisfied. According to Fig.8 it's not difficult to conclude that the phase delay has been compensated successfully and the ZVS operation can be achieved simultaneously.

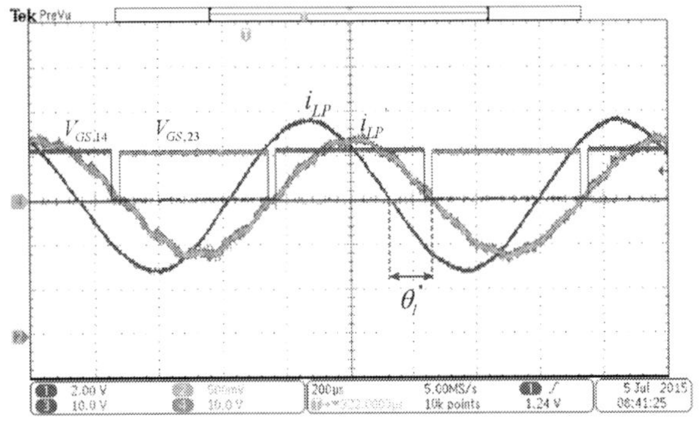

(a)

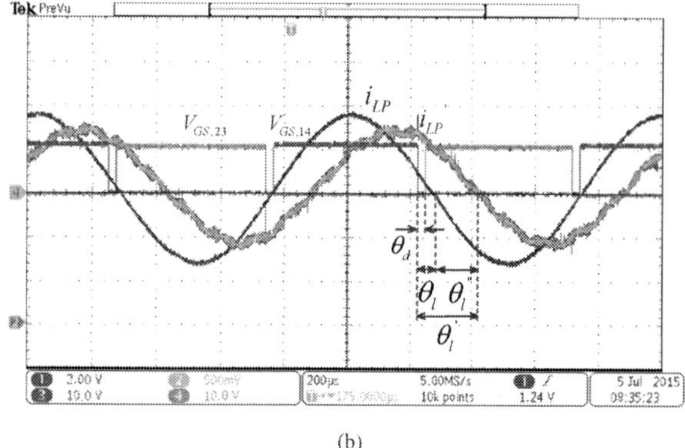

(b)

Fig.8. Experimental results. (a) without phase compensation, ZVS operation; (b) with phase compensation, ZVS operation.

Fig.9 displays the experimental waveforms of u_{AB} and i_{LP} where u_{AB} and i_{LP} represents the output voltage and actual output current of the converter respectively. The switching frequency has been improved compared with the former experiments. As shown in Fig.9 that the phase lag θ_l which represents the phase difference between the output voltage and current of the converter is presented. Compared with Fig.1-(b), it's obvious that the phase compensation as well as the ZVS operation are achieved.

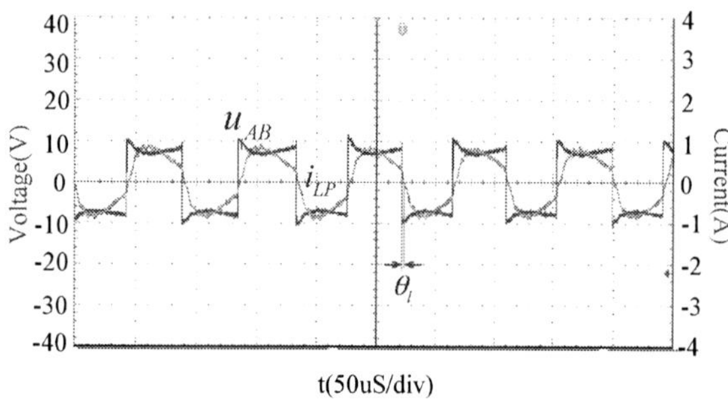

Fig.9. Experimental waveforms of V_{AB} and i_{LP}.

IV. CONCLUSIONS

In this paper, a DPC approach with SOGI-PLL to provide accurate phase Compensation and stable ZVS operation in WPT is proposed. Compared with previous works the contributions of this paper are as follows. The SOGI-PLL could accurately regulate the phase difference of WPT, which include the phase difference between the output voltage and current of the converter and the phase delay derived from the sampling process. The phase difference of the WPT, the dead angle imposed by the drivers and the compensation phase-angle could be regulated precisely. Furthermore, the phase compensation is realized with software and there needn't additional hardware implementation, which is convenient for implementation. With the proposed method, the phase delay can be compensated successfully and accurately, the ZVS operation can be achieved simultaneously.

REFERENCES

[1] M. Solja, E. H. Rafif and A. Karalis, "Coupled-mode Theory for General Free-space Resonant Scattering of Waves," Physical Review, Vol. 75, No.5, pp. 1-5, 2007.

[2] A. Karalis, J. D. Joannopoulos, and M. Soljai, "Efficient Wireless Non-radiative Mid-range Energy Transfer," Annals of Physics, Vol.3, No. 23, pp34-48, 2008.

[3] Jaegue Shin, Seungyong Shin, Yangsu Kim, Seungyoung Ahn, Seokhwan Lee, Guho Jung, Seong-Jeub Jeon, Dong-Ho Cho, "Design and Implementation of Shaped Magnetic-Resonance-Based Wireless Power Transfer System for Roadway-Powered Moving Electric Vehicles," Industrial Electronics, IEEE Transactions on, vol.61, no.3, pp.1179-1192, March 2014.

[4] Yiming Zhang, "Frequency Splitting Analysis of Two-Coil Resonant Wireless Power Transfer," IEEE Antennas and Wireless Propagation Letters,pp.400-402,2014.

[5] Yunhu Yang; Keliang Zhou; Ming Cheng; Bin Zhang, "Phase Compensation Multiresonant Control of CVCF PWM Converters," Power Electronics, IEEE Transactions on, vol.28, no.8, pp.3923,3930, Aug. 2013.

[6] Bin Zhang; Danwei Wang; Keliang Zhou; Yigang Wang, "Linear Phase Lead Compensation Repetitive Control of a CVCF PWM Inverter," Industrial Electronics, IEEE Transactions on, vol.55, no.4, pp.1595,1602, April 2008.

[7] Lopez, V.M.; Navarro-Crespin, A.; Schnell, R.; Branas, C.; Azcondo, F.J.; Zane, R., "Current Phase Surveillance in Resonant Converters for Electric Discharge Applications to Assure Operation in Zero-Voltage-Switching Mode," IEEE Transactions on Power Electronics, vol.27, no.6, pp.2925-2935, June 2012.

[8] Yan Yin; Zane, R., "Digital phase control for resonant inverters," Power Electronics Letters, IEEE, vol.2, no.2, pp.51-53, June 2004.

[9] Rodriguez. P., Teodorescu. R., Candela. I., Timbus. A.V., Liserre. M. and Blaabjerg. F., "New Positive-Sequence Voltage Detector for Grid Synchronization of Power Converters under Faulty Grid Conditions," IEEE Power Electronics Specialists Conference (PESC'06), pp.1-7, June. 2006.

[10] Ciobotaru. M., Teodorescu. R. and Blaabjerg. F., "A New Single-Phase PLL Structure Based on Second Order Generalized Integrator," IEEE Power Electronics Specialists Conference (PESC), pp.1-6, June.2006.

[11] A M Salamah, S J Finney, B W Williams, "Three-phase Phase-lock Loop for Distorted Utilties," IET Electric Power Applications, pp. 937-945, Nov.2007.

[12] Haykin. S.S., Aaptive Filter Theory. Upper Saddle River,NJ:Prentice Hall,2002.

A Novel Low-Cost Online State of Charge Estimation Method for Reconfigurable Battery Pack

Ni Lin, Song Ci*
Department of Computer and Electronics Engineering
University of Nebraska-Lincoln
NE 68182, USA
nlin@huskers.unl.edu, sci@unl.edu

Dalei Wu
Department of Computer Science and Engineering
University of Tennessee at Chattanooga
TN 37403, USA
dalei-wu@utc.edu

***Abstract*— High precision state of charge (SOC) estimation in battery packs is an integral and important part of battery management and maintenance in various applications ranging from smart grid battery energy storage systems to Electric and Hybrid Electric Vehicles (EV/HEV). In practice, measuring voltage of battery cells in a noisy environment will lead to erroneous SOC estimation, thus compromising the management and performance of the entire battery system, especially when sensors with large drifting rates are used to lower the costs of the battery management system. Therefore, in this study, based on our previous work on adaptive reconfigurable battery networks which can carry out online Open Circuit Voltage (OCV) measurements in microseconds by using a low cost voltage sensor, we propose a low cost online SOC estimation method by exploring the one-to-one mapping relationship between OCV and SOC. Experimental data collected from the real testbed indicate the correctness and robustness of the proposed method. Furthermore, the proposed method is applicable to different types of battery cells in terms of capacity and electrochemistry.**

Keywords—state of charge estimation, open circuit voltage.

I. Introduction

Rechargeable battery packs have been more and more pervasively used as a power source and as energy storage for residential, industrial and commercial devices and systems such as EV/HEVs, renewable energy systems, communication base stations, etc. Traditionally, multiple single cells are fixed into series, parallel or hybrid topologies to meet power requirements [1]. However, this fixed design leads to several major deficiencies, such as low reliability, low fault tolerance capability and cell-to-cell unbalance issues. One promising solution to these problems is to develop reconfigurable battery packs where each cell can be controlled and operated independently [2].

A reconfigurable battery pack consists of a group of control units that are designed for turning on/off switches around each battery cell so that the battery cell topology can be reconfigured online. An advanced battery management and control system (ABMCS) is developed to ensure safety and to extend pack lifespan by monitoring and controlling each battery cell in the pack to ensure that it works in the proper state according to the measured data of the battery pack.

Among these parameters, SOC is the most important and the most difficult parameter to be estimated on the fly. The major traditional methods of SOC estimation include Coulomb counting, resistance measurement, and open circuit voltage (OCV) measurement, along with the emerging methods including the neural network method [3], the extended Kalman filter [4], and so on. Although a number of methods have been proposed and validated in the lab environment, it is difficult to be applied in practice. For example, in commercial battery packs, cheaper sensors are usually adopted to reduce the cost rather than the powerful equipment in the lab. As a result, the Coulomb counting method with such low-cost sensors suffers current measurement errors and sensor accuracy drifts, which will lead to error accumulation on SOC estimation. In addition, for some applications like EVs, where onboard computational power is limited and real-time control on battery pack is required, SOC estimation methods such as neural network are not suitable due to its intensive computing and lengthy learning process.

In this work, we propose and validate a novel low-cost online SOC estimation method for practical uses in conjunction with reconfigurable battery packs, based on OCV-SOC mapping and Kalman filter. With only one voltage sensor per cell is required, the proposed method works well with noisy voltage measurements. The experimental data are collected from an Arbin battery testbed. The comparison between test data and the estimated SOC values indicates the accuracy and robustness of the proposed method. This paper is organized as follows. In Section II, system architecture is given. The OCV-SOC mapping is introduced in Section III. Section IV talks about how Kalman filter is applied in SOC estimation with noisy voltage measurement, followed by experimental result in Section V. And Section VI concludes the paper.

II. System Architecture

In literature, there are two types of management architectures that are commonly used in battery systems, namely flat and modular. In flat architectures, one single control module is used to be in charge of all components of a battery pack, which makes it relatively easy for an implementation with low-cost hardware elements. However, an apparent drawback of this type of battery pack architecture is that it does support scalability, since the implementation complexity and control

*Corresponding author: Song Ci

978-1-4673-9551-9/16 $31.00 © 2016 IEEE

latency grow exponentially as the scale of the circuit increases. The fault tolerance ability, energy efficiency and reliability are questionable as well. Modular architecture, on the other hand, is a distributed scheme, in which an individual control module is only in charge of one subset of elements. Much higher monitoring efficiency and energy efficiency can thus be achieved by applying this distributed scheme. However, the cost of the entire system increases quickly when the scale grows.

In our previous work sponsored by National Science Foundation [2], an adaptive reconfigurable battery network (ARBN) was proposed and studied. An ARBN is an integrated smart energy storage device that can serve as power source or backup for commercial, residential, and military applications which usually involves numbers of battery packs. The modular architecture is adopted for large scale battery network design, where the system is divided into several smaller subsystems that can be built and deployed independently. In this work, we adopt a modular battery network as shown in Fig. 1, where a group of battery cells are interconnected through switching circuits forming a large-scale battery pack. The battery pack is built together with the ABMCS and external communication interfaces. One voltage sensor is deployed at each battery cell to measure real-time voltage value. Sensors as such of other types are also adopted in the pack, such as current sensors and temperature sensors, to monitor and ensure battery cells are in a proper working state. Based on the measured data and the current load profile, the battery network provides functions including protection, communication, switching circuit controlling, and topology optimization and adaptation.

III. OCV-SOC Mapping

An accurate SOC estimation should be essential to a battery network to enable the optimal power and energy management as well as to protect battery cells from potential hazards such as the thermal run-away issues [5]. There have been a number of ways of estimating SOC in literature [6][7]. Currently, the Coulomb counting method has been used by most applications. However, due to the dependency of the current measurement accuracy and the ignorance of battery nonlinear effects, the accuracy of this method is questionable. In fact, among all of SOC estimation methods, the OCV-SOC mapping is the simplest and most accurate method [8] and has been applied to recalibrate the SOC estimated by the Coulomb counting method. The OCV of a voltage source, in this case a battery, is an important parameter of the existing potential power, since it varies over the whole range of SOC. However, most research frameworks cannot be fulfilled in traditional multi-cell battery packs, where fixed cell topologies are applied, preventing the OCV measurement of battery cell during the entire operational process. In contrast, in reconfigurable battery packs, it is possible that battery cells be disconnected from the cell topology periodly, which enables the OCV measurement on the fly.

In practice, there are two critical issues on OCV-SOC mapping, namely modification and compensation. The relationship between OCV-SOC is obtained by measuring the open circuit voltage at each SOC level. However, the relationship cannot be exactly the same for each battery cell because of manufacture imperfection, aging process, and any other factors from both inside and outside. Modification is a process dealing with this issue so that each battery cell in the pack will follow the same modified OCV-SOC relationship. Compensation may solve the issue by adding a compensation voltage when individual cells are disconnected from the pack but cannot get enough recovery time to dervie the true OCV. A detailed review of compensation and modification is beyond the scope of this paper, please refer to [9] for more details.

IV. Kalman Filter Based SOC Estimation

Within the mathematical tools that are used for stochastic estimation with measurement noises or noisy sensors, one of the most frequently used methods is the Kalman filter. The Kalman filter is essentially a set of mathematical equations that adopt a prediction-correction strategy to find an optimal estimator by minimizing estimated error covariance. Ever since its introduction, the Kalman filter has been one of the most important research and application topics. In the area of SOC estimation, numerous frameworks have been proposed and validated by using Kalman filters or extended Kalman filters. In this work, we also adopt a Kalman filter as our mathematic tool for estimating SOC with low-cost noisy sensing.

To estimate SOC based on the OCV-SOC mapping, the OCV needs to be estimated first. In this work, a Kalman filter is introduced to extract an OCV value in a noisy environment [10][11]. To use a Kalman filter, known control inputs, the system dynamic model, and multiple sequential measurements are required so that the prediction-correction strategy can be performed. In our work, we apply a Kalman filter in a rather simple manner. We measure the noisy OCV multiple times within a short time period when the battery cell is disconnected from discharging, and these values are multiple sequential measurements, which can be considered as a set of scalar

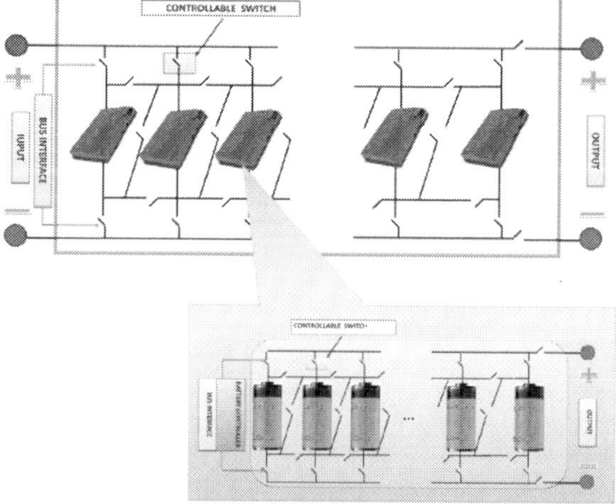

Fig 1: Battery network

random constants. Since a Kalman filter is used to estimate a constant value, the system model is also simply compared with an existing extended Kalman filter in battery state estimations; as a result, a much lower computational complexity can be achieved.

In practice, measurement noises introduced by sensors are uniformly distributed with a maximum error of 2% of the actual values. The noise eliminating process is governed by the linear differential equation with a measurement z. Since the OCV is a constant and does not change from step to step, the state matrix of the filter A=1. The absence of control input makes u=0. The measurement is the voltage measured directly, so the measurement matrix H=1. Then the system function and the measurement function can be expressed as:

$$\xi_K = A\,\xi_{K-1} + B\,\upsilon_K + \omega_K = \xi_{K-1} + \omega_K \quad (1)$$

$$Z_K = H\,\xi_K + \varpi_K = \xi_K + \varpi_K \quad (2)$$

For the estimation of a constant, it is reasonable to set a very small process variance. In our case, we set Q=0.0001. As a result, the time update equations and measurement update equations can be given as equations (3) and (4), respectively:

$$\xi_K = \xi_{K-1}, \qquad P_K^- = P_{K-1} + \Theta \quad (3)$$

$$K_K = P_K^- (P_K^- + P)^{-1} \quad (4)$$

$$\xi_K = \xi_K^- + K_K(\zeta_K - \xi_K^-),\ P_K = (1 - K_K)\,P_K^- \quad (5)$$

V. Experimental Results

The proposed method has been validated by using an Arbin BT-2000 battery testbed and Heter 26650 (LifePO4) lithium-ion battery cells, whose full capacity, nominal voltage and cutoff voltage are 3200 mAh, 3.2V, and 2.5V, respectively. Battery cells are first charged through the well-known method of constant current, constant voltage (CCCV). All the parameters of the battery are calculated by applying a standard least-square estimator, and the OCV-SOC curves are obtained through a pulse charging-discharging process [12]. Then battery cells are discharged with a rest period during discharging. Voltage values are measured and recorded during the rest period. The experimental data are collected by using the Arbin battery testbed of BT2000, and the Esepc BTZ-133 is used to control temperature conditions as shown in Fig. 2. The simulation results of the proposed Kalman filter based algorithm are derived through Matlab.

Fig. 3 shows both the measured OCV curve from the Arbin testbed and the results after processing noisy measurements by the Kalman filter during the entire process. It can be observed that the OCV obtained by the Kalman filter is very close to the tested OCV. The SOC estimation after OCV-SOC mapping is illustrated in Fig. 4. It shows that the estimation error in the middle segment of the curve tends to be larger because the OCV curve is flat in that segment, and a smaller estimation error on voltage measurement will lead to a relatively larger error in SOC estimation. We can also observe that the root mean squared error (RMSE) for the entire process is smaller than 1%, and the maximum absolute error is 3.2%. The comparison between the measured SOC and the estimated results indicates the accuracy of the proposed method.

Fig. 2: Arbin BT 2000 and Esepc BTZ-133

Since the proposed algorithm is designed for multiple applications including EVs and portable devices, where computing hardware capability is quite limited, the computational complexity becomes a critical performance metric. In this work, we also investigate how accuracy varies with the number of iterations of the Kalman filter. It is obvious that a larger number of iterations will lead to higher accuracy at a cost of computation overhead in terms of time and energy. The results of complexity analysis are shown in Table 1.

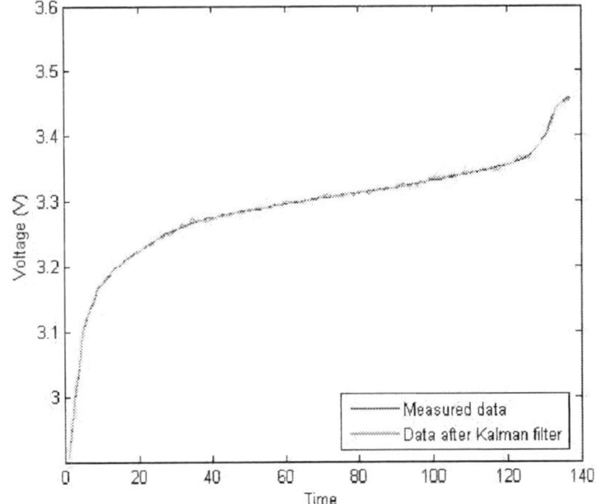

Fig. 3: Measured voltage and estimated voltage.

978-1-4673-9551-9/16 $31.00 © 2016 IEEE

TABLE I. NUMBER OF ITERATIONS VS ERROR RATE

Iterations	50	30	25	20	15
Test 1	0.66%	0.97%	0.94%	0.98%	1.14%
Test 2	0.78%	0.87%	1.01%	1.03%	1.00%
Test 3	0.65%	0.89%	0.89%	1.05%	1.40%
Test 4	0.61%	0.86%	1.02%	1.08%	1.24%
Test 5	0.60%	0.99%	1.06%	1.06%	1.24%
Test 6	0.68%	0.97%	1.05%	1.12%	1.25%
Test 7	0.74%	0.97%	0.99%	1.10%	1.25%
Test 8	0.67%	0.96%	0.91%	1.07%	1.23%
Test 9	0.73%	0.83%	1.01%	1.17%	1.31%
Test 10	0.71%	0.83%	0.97%	1.06%	1.16%
Average	0.683%	0.914%	0.985%	1.072%	1.222%

The average estimation error rate is one of the most important performance metrics that describe the ability of a given battery system. In this test, we set the iteration number as 50, 30, 25, 20 and 15, and each test is repeated 10 times. As can be observed in the table, adopting 50 iterations will lead to a quite low average estimation error rate, which is smaller than 1% and is good enough for most applications. If 20 iterations are used, 60% computational complexity will be reduced compared with applying 50 iterations, and the result is still good enough around 1% average error. In practical use, the choice of iteration number depends on hardware ability, requirements on real-time performance and overhead, as well as other factors.

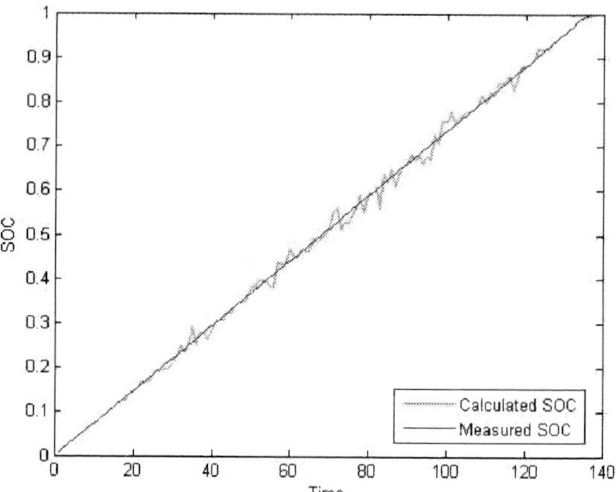

Fig. 4: Measured SOC and estimated SOC.

VI. CONCLUSION AND FUTURE WORK

In this work, we have proposed and validated a novel low-cost approach for online SOC estimation in reconfigurable battery packs. Both experimental and analytical results are collected and presented, and the results indicate the accuracy of the proposed model to estimate battery SOC. Compared with traditional methods, such as Coulomb counting methods, our method is more accurate and robust with higher tolerance of sensor measurement errors and estimation error accumulation. The computational complexity is low and scalable according to hardware cability and estimation accuracy requirements. In addition, there is no need for training or learning process, thus the proposed method is much easier to be implemented than emerging methods such as the neural network method. Future work may focus on dealing with the aging effect of a battery pack. This problem may be solved by adding a learning process and modeling the battery life as a Markov process.

REFERENCES

[1] R. Rao, S. Vrudhula, and N. Chang, "Battery optimization vs energy optimization: Which to choose and when," in Proc. 2005 IEEE/ACM International Conference on Computer-Aided Design, pp. 439-445.

[2] S. Ci, J. Zhang, H. Sharif, and M. Alahmad, "A novel design of adaptive reconfigurable multicell battery for power-aware embedded networked sensing system," in Proc. 2007 IEEE Global Telecommunication Conference, Nov. 26-30, 2007, pp. 1043-1047.

[3] M. Charkhgard, and M. Farrokhi, "State-of-charge estimation for lithium-ion batteries using neural networks and EKF," IEEE Trans. Ind. Electron., vol. 57, no. 12, Dec. 2010.

[4] Liye Wang, Lifang Wang, C. Liao; "Research on Improved EKF Algorithm Applied on Estimate EV Battery SOC", Power and Energy Engineering Conference APPEEC, 2010.

[5] "Boeing 787 Dreamliner battery problems," http://en.wikipedia.org/wiki/Boeing 787 Dreamliner battery problems.

[6] J. Chiasson and B. Vairamohan, "Estimating the state of charge of a battery," IEEE Transactions on Control Systems Technology, pp. 465–470, 2004.

[7] M. Mcintyre, T. Burg, D. Dawson, and B. Xian, "Adaptive State of Charge (SOC) Estimator for a Battery", Proceedings of the 2006 American Control Conference, Minneapolis, Minnesota, USA, June 14-16, 2006.

[8] Bingjun Xiao, Yiyu Shi and Lei He, "A Universal State-of-Charge Algorithm for Batteries", IEEE/CAM Design Automation Conference, June, 2010.

[9] M. Petzl and M. A. Danzer, "Advancements in OCV Measurement and Analysis for Lithium-Ion Batteries," IEEE Transactions on Energy Conversion, vol. 28, no. 3, pp. 675–681, Sep. 2013.

[10] G. L. Plett, "Extended Kalman filtering for battery management systems of LiPB-based HEV battery packs—Part 3. State and parameter estimation," J. Power Sources, vol. 134, no. 2, pp. 277–292, Aug. 2004.

[11] J. Lee, O. Nam, and B. H. Cho, "Li-ion battery SOC estimation method based on the reduced order extended Kalman filtering," J. Power Sources, vol. 174, no. 1, pp. 9–15, Nov. 2007.

[12] Suleiman Abu-Sharkh, Dennis Doerffel, "Rapid test and non-linear model characterization of solid-state lithium-ion batteries," Journal of Power Sources 130 (2004) 266-274.

Effect of Decoupling Terms on the Performance of PR Current Controllers Implemented in Stationary Reference Frame

Sizhan Zhou, Jinjun Liu
School of Electrical Engineering and State Key Lab of Electric Insulation and Power Equipment
Xi'an Jiaotong University
Xi'an China
zhousizhan@gmail.com

***Abstract*—The PI-based current control scheme in *dq*-frame and PR-based one in *αβ*-frame are the two most popular current control schemes. It is commonly known that there exists axes cross-coupling in *dq*-frame, while in *αβ*-frame it doesn't. And decoupling terms are usually performed to decouple *dq*-axes cross-coupling. However, PR control scheme presents different performance, when compared with *dq*-frame PI control scheme containing decoupling terms. In this paper, the meaning of "cross-coupling" was extended to *αβ*-frame. And the decoupling terms can also be implemented for *αβ*-frame control schemes. They can provide benefit to PR current control performance, such as improving the transient performance with less transient phase error and less frequency variation sensitivity, especially for low switching frequency applications with limited current control bandwidth. Finally, simulation and experimental results were shown to verify the analysis.**

Keywords*—cross-coupling; decoupling control; PR controller

I. INTRODUCTION

Grid-connected voltage source converters (VSCs) take a crucial role for a wide range of applications such as renewable/distributed power generation system, energy storage system and power quality conditioners [1-4]. Through controlling the power delivered into or absorbed from the utility, miscellaneous functions can be realized by employing various control strategies. These control strategies usually consist of multi-loop control with the current control as the inner loop. The inner current controller usually regulates the current injected into the grid, following the reference given by the outer loop. The performance of the current control determines the overall system performance, so special care have been devoted to its design in order to achieve fast and accurate current regulation.

Several current control techniques, such as hysteresis control, predictive control, and direct power control, have been proposed to achieve very fast response and negligible steady-state error. Among various current control techniques, linear current controllers, PI in *dq*-frame and PR in *αβ*-frame, are the two most well known controllers [1-3]. This paper takes the fundamental frequency current control of grid-connected voltage source converters (GC-VSCs) as an example. Fig.1 shows the topology of GC-VSCs with the basic current control scheme implemented in *dq*-frame or *αβ*-frame.

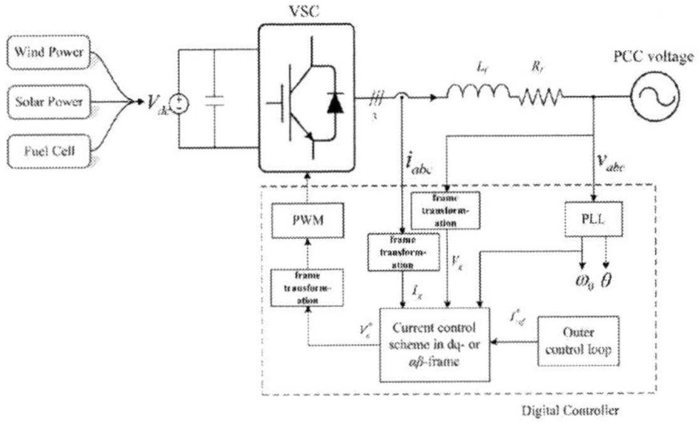

Fig.1 The topology of grid-connected voltage source converter and basic current control scheme.

The *dq*-frame PI control scheme was the first scheme for three-phase current regulation. When controlled in *dq*-frame, the *dq*-axis current components are directly related with the active and reactive power. It is a common knowledge that there exists cross-coupling between *dq*-axes, which will worsen the current control performance. The inductor current state feedback (ICSF) is the most general decoupling technique, employed to deal with this axes cross-coupling [5-9].

PR controllers implemented in *αβ*-frame were proposed to fast track sinusoidal reference with zero steady-state error around 2000 [10, 11]. Since then, PR controllers have gained a large popularity in the following decade. It has been well researched regarding to their parameter design [9,12-14], performance evaluation [9, 14], discretization implementation. The *αβ*-frame PR control scheme shows some advantages over *dq*-frame PI control scheme. 1) PR controllers are implemented in *αβ*-frame, avoiding reference frame rotating transformation. This saves computation burden and makes them much suitable for multi-harmonics current control applications. 2) It is commonly regarded that there is no axes cross-coupling between *αβ*-axes. The decoupling terms are not necessarily required, unlike *dq*-frame PI control scheme. The relationship between *dq*-frame PI controllers and *αβ*-frame PR

This work was supported by the National Natural Science Foundation of China under Grant 51437007, and the Power Electronics Science and Education Development Program of Delta Environmental & Educational Foundation under Grant DREM2014002.

controllers was also researched and deduced without considering decoupling terms [15, 16].

However, dq-frame PI control scheme with decoupling terms differ from $\alpha\beta$-frame PR control scheme in their dynamic behavior. In [17], dq-frame PI control scheme with ICSF decoupling terms was precisely transformed to $\alpha\beta$-frame, deriving a new $\alpha\beta$-frame control scheme (named as PRX2 controller). The PRX2 controller contains a PR controller and two additional cross-coupling branches, one of which is $j\omega_0 L_f$. It directly corresponds to the ICSF decoupling terms in dq-frame PI control scheme. However, the effect of decoupling terms on the performance $\alpha\beta$-frame PR controllers was not fully revealed and discussed in detail.

In this paper, the two current control schemes (PI-based and PR-based) of GC-VSCs and their performance were analyzed and compared based on double frequency side bode diagrams of their complex-coefficient close-loop system transfer functions. When tracking positive-sequence current references, PR control scheme shows almost the same dynamic response waveforms with dq-frame PI control scheme, but worse than dq-frame decoupled PI control scheme. It is pointed out that even though $\alpha\beta$-axes are not cross-coupled with each other as dq-axes do, the cross-coupling phenomenon still exists, presenting as the cross-coupling between the amplitude and phase angle of the AC signals. In dq-frame, the dq-axes cross-coupling is shown as d- or q-axis current change will also disturb the other axis current component. Correspondingly, in $\alpha\beta$-frame, it is presented as there is large transient phase error when tracking the change of the reference amplitude.

The concept of decoupling control can be extended to $\alpha\beta$-frame control scheme. The decoupling terms can benefit the performance of PR controllers with less transient phase error and also improve its frequency sensitivity. However, the conventional decoupling terms based on ICSF will improve control performance for positive-sequence current control, while it will also degrade system dynamic response to negative-sequence current control. In this paper, decoupling terms based on reference current feed-forward are proposed and implemented to reserve the decoupling effect for both positive- and negative-sequence current control. Finally, the effect of decoupling terms on PR-based current control scheme was verified with simulation and experimental results.

II. MODELING OF GRID-CONNECTED VOLTAGE SOURCE CONVERTERS

The topology of the grid-connected VSC is shown in Fig.1. In this paper, we mainly focus on the current control loop, and the DC-link is considered as a DC source. The dynamic of the grid current can be modeled by (1) in $\alpha\beta$-frame as

$$\begin{bmatrix} v_\alpha \\ v_\beta \end{bmatrix} = \begin{bmatrix} sL_f + R_f & 0 \\ 0 & sL_f + R_f \end{bmatrix} \begin{bmatrix} i_\alpha \\ i_\beta \end{bmatrix} + \begin{bmatrix} v_{g\alpha} \\ v_{g\beta} \end{bmatrix} \tag{1}$$

where L_f and R_f are the inductance and resistance of the grid interfacing filter. It will be more convenient to express the model with complex space vector, $i_{\alpha\beta}=i_\alpha+ji_\beta$ and $v_{\alpha\beta}=v_\alpha+jv_\beta$. Based on the concept of complex coefficient transfer function [5, 6, 18], the transfer function $\boldsymbol{P}_{\alpha\beta}(s)$ from the converter terminal voltage to the grid current can be expressed as

$$\boldsymbol{P}_{\alpha\beta}(s) = \frac{i_{\alpha\beta}(s)}{v_{\alpha\beta}(s)} = \frac{1}{sL_f + R_f} \tag{2}$$

Through coordinate transformation, the model of VSCs in dq-frame can be derived as [7]

$$\begin{bmatrix} v_d \\ v_q \end{bmatrix} = \begin{bmatrix} sL_f + R_f & -\omega_0 L_f \\ \omega_0 L_f & sL_f + R_f \end{bmatrix} \begin{bmatrix} i_d \\ i_q \end{bmatrix} + \begin{bmatrix} v_{gd} \\ v_{gq} \end{bmatrix} \tag{3}$$

$$\boldsymbol{P}_{dq}(s) = \frac{i_{dq}(s)}{v_{dq}(s)} = \frac{1}{sL_f + R_f + j\omega_0 L_f} \tag{4}$$

where ω_0 is the grid frequency. The $j\omega_0 L_f$ in (4) represents there exists axes cross-coupling in the plant $\boldsymbol{P}_{dq}(s)$. Fig.2 shows the bode plot of plant model (2) and (4). The plant frequency response in dq-frame is shifted left by f_0 ($f_0=\omega_0/2\pi$), compared with that in $\alpha\beta$-frame. However, they both shows almost 90° phase lagging at their operating frequency (0Hz for dq-frame and +50Hz for $\alpha\beta$-frame). This lagging phase results in the cross-coupling issue in dq-frame. And it will also have the same influence on $\alpha\beta$-frame control schemes.

III. PERFORMANCE COMPARISON BETWEEN DQ-FRAME PI CONTROLLER VERSUS AB-FRAME PR CONTROLLER

Fig.3a and Fig.3b show dq-frame PI-based and $\alpha\beta$-frame PR-based current control schemes, respectively. ICSF decoupling terms are also included in Fig.3a. The control parameter are tuned according to literatures[12-14] with circuit parameters shown in the experiment results part. The digital control delay is considered as $G_d(s)$. And $G_d(s)=e^{-sT_d}$ in $\alpha\beta$-frame, or equivalently $e^{-(s+j\omega_0)T_d}$ in dq-frame. A delay compensation term $e^{j\omega_0 T_d}$ can be employed to cancel the phase leg introduced by the time delay at dc in dq-frame, and the resulting delay coincides with the original one in $\alpha\beta$-frame. This delay compensation term $e^{j\omega_0 T_d}$ provide some benefit to the control performance, especially for control scheme with ICSF [19].

From the control block diagrams, the closed-loop transfer function for each control scheme can be deduced. For dq-

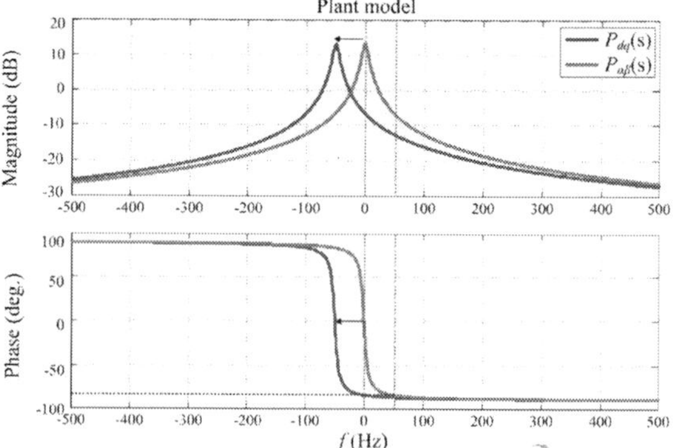

Fig. 2 Plant model in dq-frame and $\alpha\beta$-frame

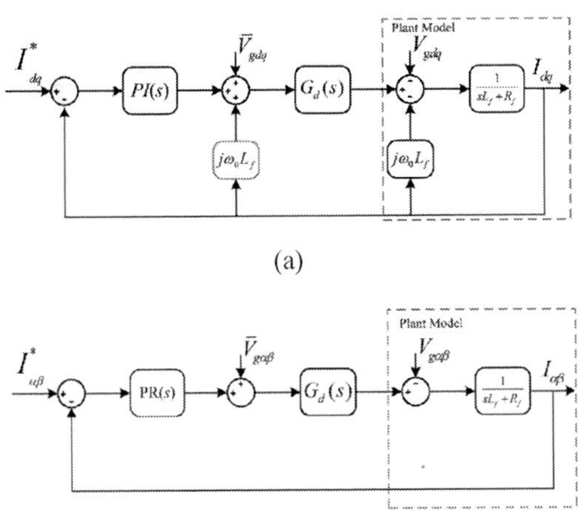

Fig.3 Block diagram of PI current control (a) and PR current control (b)

frame PI control, two closed-loop transfer function can be got with or without the decoupling terms, denoted as $H_{PI}(s)(=I_{dq}(s)/I^*_{dq}(s))$ and $H_{PI\_with\_ICSF}(s)$, respectively. And for $\alpha\beta$-frame PR control scheme, it is denoted as $H_{PR}(s)$ $(=I_{\alpha\beta}(s)/I^*_{\alpha\beta}(s))$. Their bode diagrams are shown in Fig.4.

Two key points can be noticed from Fig.4. Firstly, the bode diagram of $H_{PR}(s)$ around its operating frequency +50Hz is very similar to that of $H_{PI}(s)$ around 0Hz (note that the bode diagram of $H_{PI}(s)$ is intentionally right shifted by 50Hz). However, there is a difference between $H_{PR}(s)$ and $H_{PI\_with\_ICSF}(s)$, especially around +50Hz. $H_{PR}(s)$ shows anomalous peaks, which results in PR controller frequency sensitivity, while $H_{PI\_with\_ICSF}(s)$ presents a relatively much smooth frequency response with given bandwidth. An ideal system response for positive-sequence current control is also shown in Fig.4 with red dashed line, which will provide the best performance for positive-sequence current control. It is possible to get this kind of system response with improved decouple control [20]. Through comparing their system transfer functions, it can be concluded that PR controller will present the same dynamic performance with dq-frame un-decoupled PI controller. Secondly, PR controllers also provide control capability to negative-sequence current, which

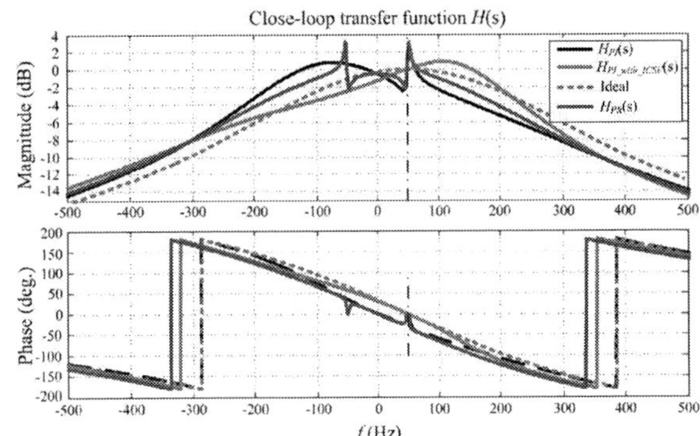

Fig. 4 Bode diagrams of current control scheme closed-loop transfer function

corresponds to -50Hz in the bode diagram.

The performance comparison between dq-frame and $\alpha\beta$-frame control scheme was also verified through simulation. Fig.5 shows the current step responses of each control scheme. The reference was step to 20A at t=0.1s. The current control dynamic of PR controller and un-decoupled dq-frame PI controller are almost the same. And although there is no axes cross-coupling between $\alpha\beta$-axes, the active and reactive power are still coupled with each other. Even though the current magnitude increases to the reference fastly, PR controllers show relatively large transient phase error. And it takes several cycles to eliminate. This transient phase error induces the coupling between active and reactive power. Compared with PR controllers, dq-frame decoupled PI controller with ICSF shows better performance. The dq-axes are almost decoupled with each other, so the same for active and reactive power. In the transient state, the three phase currents almost keep the same phase angle with the references, and their magnitude fastly increases to the reference magnitude. Through above analysis and comparison, the decoupling terms in dq-frame PI control scheme do provide improvement to the control performance. It is possible to extend the decoupling terms to conventional $\alpha\beta$-frame PR control scheme and provide benefit to the current control performance.

IV. DECOUPLING TERMS FOR PR CURRENT CONTROL SCHME

Transform the dq-frame decoupled PI control scheme with ICSF completely into $\alpha\beta$-frame, resulting in a $\alpha\beta$-frame PR

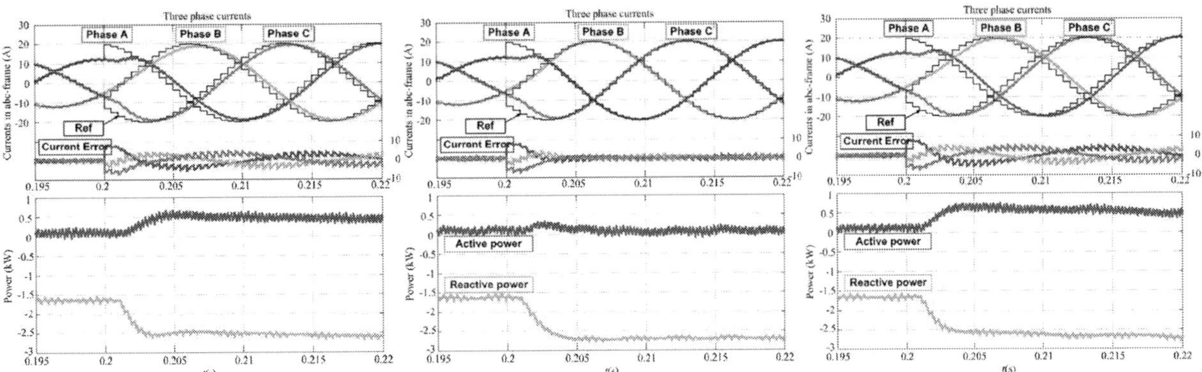

Fig.5 Current tracking performance of (a) PI in dq-frame; (b) PI with ICSF in dq-frame; (3) PR in $\alpha\beta$-frame

control scheme with an additional $j\omega_0 L_f$ decoupling terms, as is shown in Fig.6a [17]. The closed-loop system transfer function can be got and denoted as $H_{PR\ with\ ICSF}(s)$. Its bode diagram is shown in Fig.7 (blue line). It can be observed that the decoupling terms improve the system frequency response around positive frequency side, less stiff slope at +50Hz and much more symmetrical with respect to +50Hz. And the more symmetrical the system transfer function around its operating point is, the better the system dynamic performance is [21]. However, the system response at the negative frequency side is deteriorated. The slope at -50Hz becomes much more stiff, which means more sensitive to frequency variation.

In order to keep the benefit of the decoupling terms and avoid the performance worsening for the negative-sequence current control, decoupling terms based on reference current feed-forward were proposed here [20], as is shown in Fig.6b. The feed-forward filter is expressed as

$$F_P(s) = \frac{\omega_c}{s - j\omega_0 + \omega_c e^{-sT_d + j\omega_0 T_d}} \quad (5)$$

which is a complex-coefficient filter and ω_c is the designed current loop bandwidth. This complex-coefficient filter can be discretized and implemented with a digital filter. The red line in Fig.7 shows the bode diagram of the corresponding system transfer function $H_{PR\ with\ RCFF}$. The system frequency response around the positive working frequency point is further improved, but negative-sequence current control capability is still worsen. However, it is possible to feed forward both the

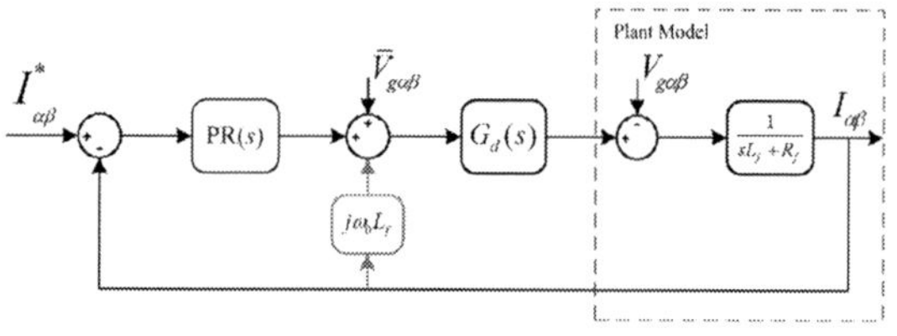

(a) PR with ICSF

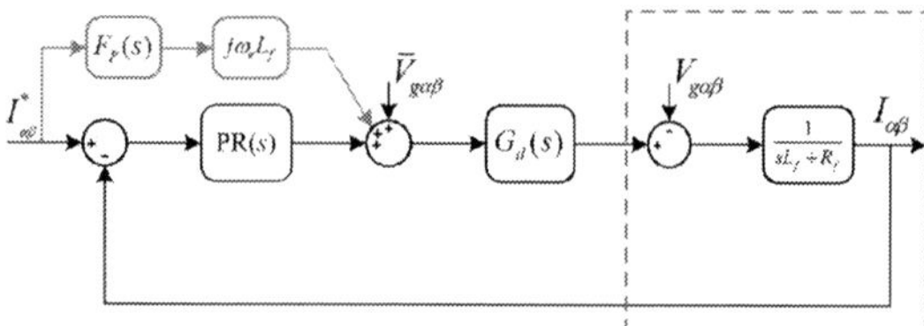

(b) PR with PS-RCFF

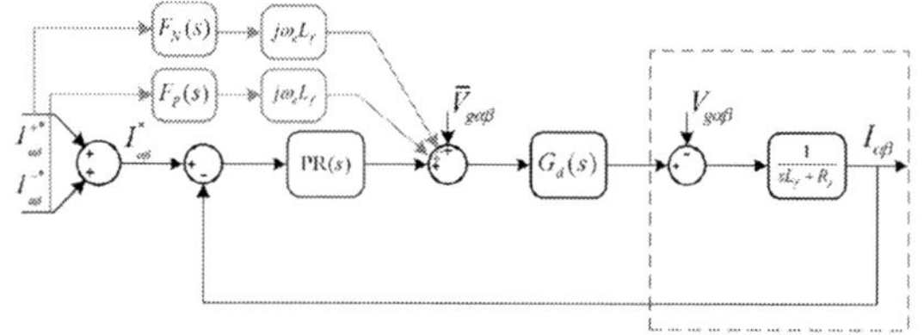

(c) PR with double RCFF

Fig.6 PR current control scheme with ICSF or RCFF decoupling terms

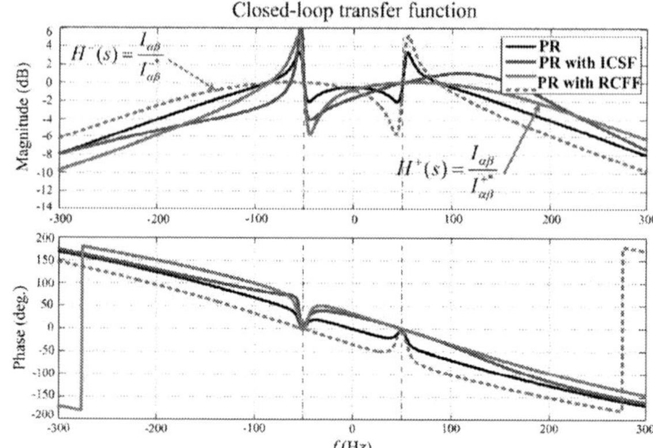

Fig.7 Bode diagrams of closed-loop system transfer functions for PR control scheme with ICSF or RCFF

positive-sequence and negative sequence current references with opposite signs, as is shown in Fig.6c. In many cases, the positive-sequence and negative-sequence current reference are calculated and given separately [22-25], making separate reference feed forward possible. Note that the feed-forward sign is opposite for positive- and negative-sequence. And the negative-sequence feed-forward filter is also different with (5) and expressed as

$$F_N(s) = \frac{\omega_c}{s + j\omega_0 + \omega_c e^{-sT_d - j\omega_0 T_d}} \quad (6)$$

According to Fig.6c, the closed-loop transfer functions for positive- and negative-sequence current are different and should be calculated separately. Their bode diagrams are also shown in Fig.7(red line). As the decoupling terms are applied separately for positive- and negative-sequence current, both the positive- and negative-sequence current tracking performance can be improved.

V. EXPERIMENT RESULTS

A prototype system with its parameters listed in Table I was experimented to verify the decoupling terms effect on the performance of PR control scheme. As high current control bandwidth will significantly reduce the difference between decoupled and un-decoupled system. Therefore, The switching/sampling frequencies are 2.0kHz, and the current loop bandwidth is designed as 120Hz with parameter tuned according to literatures [12, 13]. The simulation and experimental circuit parameters are listed as follows:1) L_f =6.18mH and R_f=0.21Ω; 2) The line-to-line grid voltage is 110V(rms); 3) DC-link voltage is 200V ; 4) The switching/sampling frequencies are 2.0kHz.

A. Positive-sequence current tracking performance

Current control performance with PR control scheme employing decoupling terms or not are tested and compared. Fig.8 shows the step current tracking performance for positive-sequence current control. Comparing the experimental results in Fig.8, it can be noticed that the decoupling terms do benefit the current tracking transient performance, showing less transient phase error. The proposed

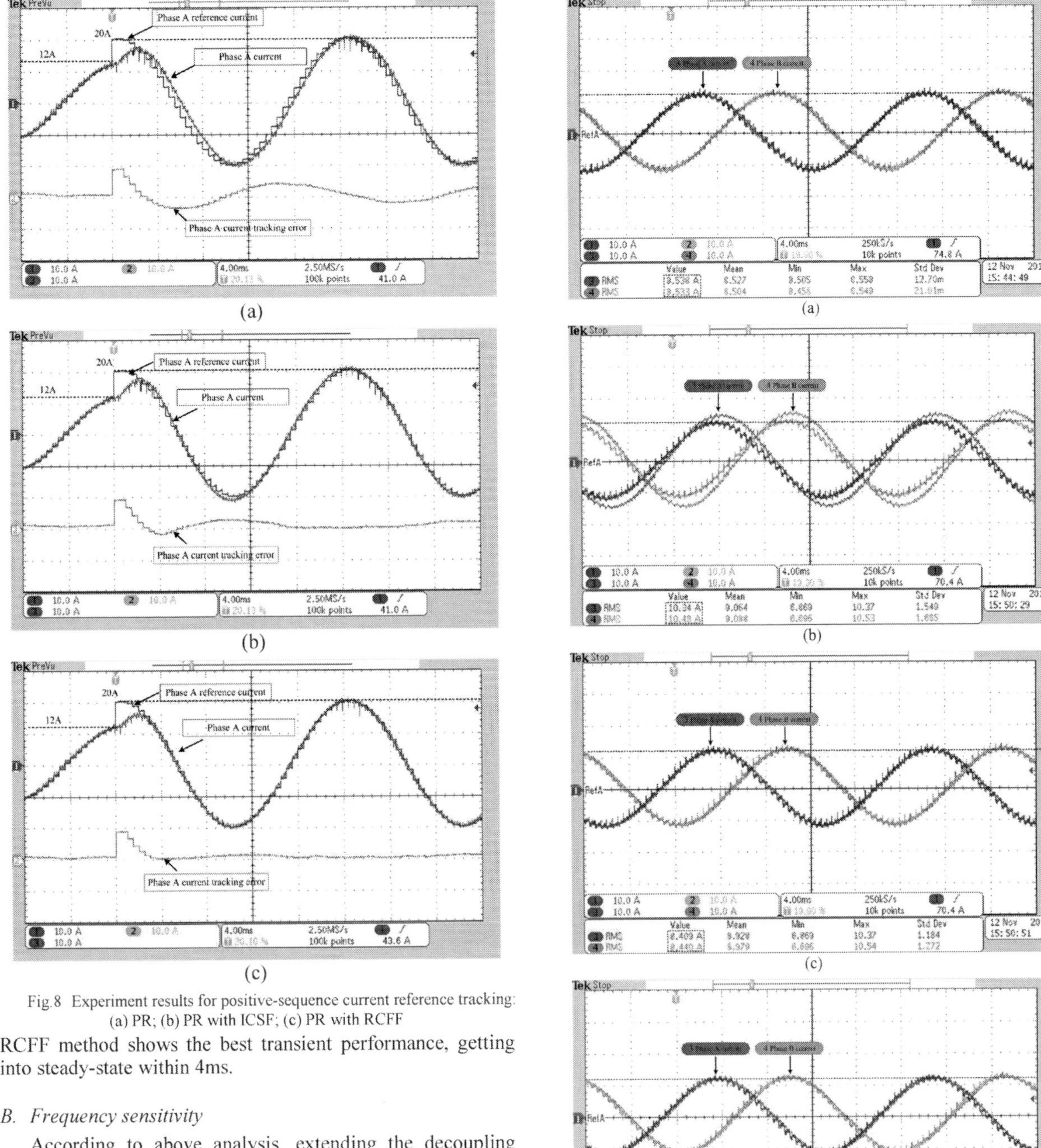

Fig.8 Experiment results for positive-sequence current reference tracking: (a) PR; (b) PR with ICSF; (c) PR with RCFF

RCFF method shows the best transient performance, getting into steady-state within 4ms.

B. Frequency sensitivity

According to above analysis, extending the decoupling terms into $\alpha\beta$-frame PR control scheme not only benefit the transient performance, but also improve the frequency sensitivity of PR controllers to grid frequency variation. When the PR controller resonant frequency coincides with the grid frequency, the three phase currents tracks their references within 1% accuracy (Fig.9a). However, if the grid frequency varies from its rated value, for example 52Hz, the PR controller can't precisely control the current. During experiment test, the grid frequency variation is intentionally set to +2Hz to make the comparison remarkable. For PR

(a)

(b)

(c)

(d)

Fig.8 Experiment results for positive-sequence current reference tracking under grid frequency variation: (a) normal case, PR, f_g=50Hz;(b) PR under frequency variation f_g=52Hz (b) PR with ICSF and f_g=52Hz; (c) PR with RCFF and f_g=52Hz

controller without its resonant frequency dynamically tuned, its performance is severely degraded (Fig.9b). When the decoupling terms are employed based on ICSF or RCFF, the PR controller shows small sensitivity to frequency variation. As is shown in Fig.9c and 9d, there is only a small change in current magnitude and phase.

C. Unbalanced current tracking performance

Under unbalanced grid voltage condition, the GC-VSCs may have to regulate unbalanced three-phase currents. The unbalanced current reference is calculated according to application control objectives [20-23]. Since PR controller have both positive- and negative-sequence current control capability, it is widely employed to regulate the unbalanced current. Due to the lagging phase in the plant, PR controller also shows large transient phase error when tracking negative-sequence current (Fig.10a). With ICSF decoupling terms, positive-sequence current tracking dynamic performance is improved, but the negative-sequence current control capability is degraded, large transient phase error (Fig.10b) and more sensitivity to frequency variation. If the proposed RCFF decoupling terms are implemented for positive-sequence current reference, its control capability to negative-sequence is not affected (Fig.10c). If double RCFF decoupling terms are employed for both sequences, the transient performance regulating unbalanced current is further improved (Fig.10d).

VI. CONCLUSION

In this paper, the effect of decoupling terms on the performance of $\alpha\beta$-frame PR control scheme was analyzed and verified. They can provide benefit to the performance of PR current control scheme, such as improving transient performance with less transient phase error and less frequency sensitivity. Decoupling terms based on reference current feed-forward were further proposed to keep the decoupling effect without worsening dual-sequence current control capability. Experiment results was shown to verify the analysis and the performance of the proposed RCFF decoupling terms.

REFERENCES

[1] F. Blaabjerg, R. Teodorescu, M. Liserre and A.V. Timbus, "Overview of control and grid synchronization for distributed power generation systems," *IEEE Trans Ind. Electron.*, vol. 53, no. 5, pp. 1398-1409, Oct. 2006.

[2] A. Timbus, M. Liserre, R. Teodorescu, P. Rodriguez, and F. Blaabjerg, "Evaluation of current controllers for distributed power generation systems," *IEEE Trans. Power Electron.*, vol. 24, no. 3, pp. 654–664, Mar.2009.

[3] R. Teodorescu, M. Liserre and P. Rodriguez, Grid converters for photovoltaic and wind power systems, IEEE-Wiley, 2011.

[4] B.Wu, S. Li, Y. Liu and K. Ma Smedley, "A New Hybrid Boosting Converter for Renewable Energy Applications," *IEEE Trans. Power Electron.*, vol. 31, no. 2, pp. 1203–1215, Feb. 2016.

[5] Briz, Fernando, Michael W. Degner, and Robert D. Lorenz. "Analysis and design of current regulators using complex vectors." *IEEE Trans. Ind. Appl.*, vol.36, no.3, pp 817-825, May/June 2000.

[6] J. Holtz, J. Quan, J. Pontt, J. Rodriguez, P. Newman and H. Miranda, "Design of fast and robust current regulators for high-power drives based on complex state variables," *IEEE Trans. Ind. Appl.*, vol.40, no.5, pp.1388-1397, Sep/Oct 2004.

[7] B. Bahrani, S. Kenzelmann, and A. Rufer, "Multivariable-PI-based dq

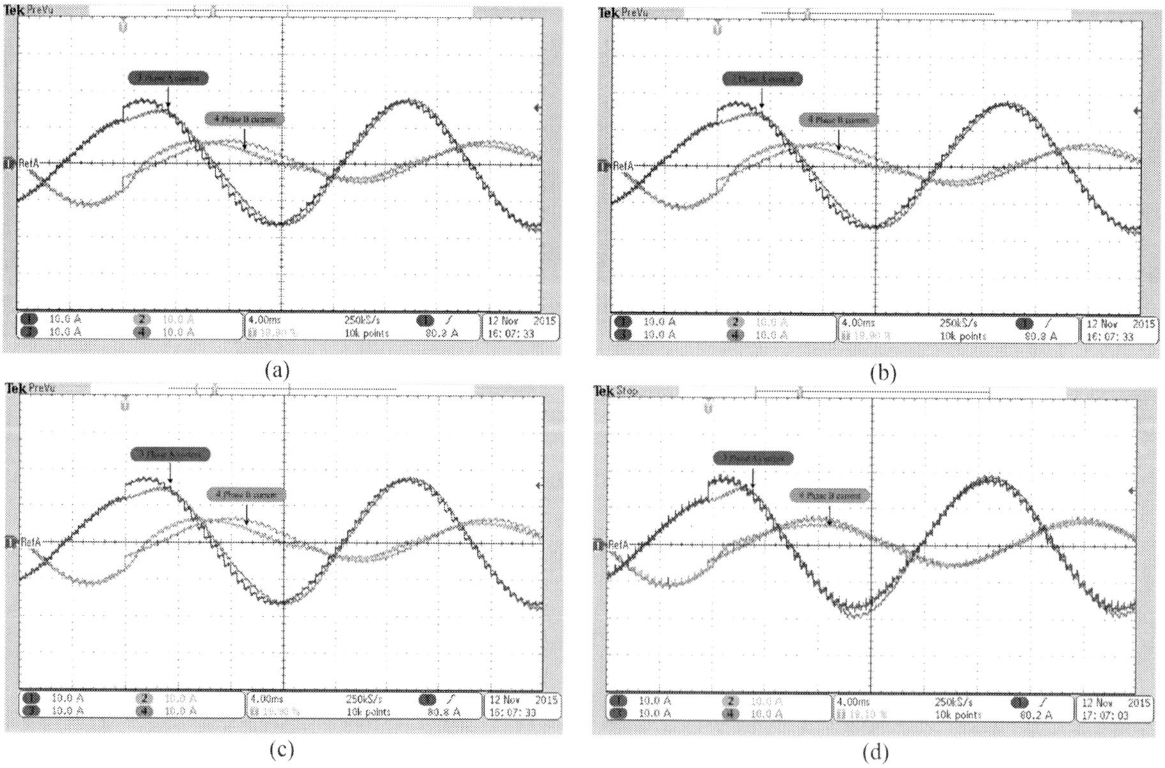

(a)

(b)

(c)

(d)

Fig.10 Experiment results for unbalanced current tracking: (a) PR; (b) PR with ICSF; (c) PR with PS-RCFF;(d) PR with double RCFF

current control of voltage source converters with superior axis decoupling capability," *IEEE Trans Ind. Electron.*, vol. 58, no. 7, pp. 3016–3026,Jul. 2011.

[8] A.G. Yepes, A. Vidal, J. Malvar, O. Lopes and J. Doval-Gandoy, "Tuning method aimed at optimized settling time and overshoot for synchronous proportional-integral current control in electric machines," *IEEE Trans. Power Electron.*, vol. 29, no. 6, pp. 3041-3054, June. 2014.

[9] Yepes, A., Vidal, A., Lopez, O., Doval-Gandoy, "Evaluation of techniques for cross-coupling decoupling between orthogonal axes in double synchronous reference frame current control", *IEEE Trans Ind. Electron.*, vol.61, no.7, pp.3527-3531, July 2014.

[10] D.N. Zmood and D.G. Holms, "Stationary frame current regulation of PWM inverters with zero steady state," in *Proc. IEEE PESC'99*, vol. 2, pp. 1185-1190, 1999.

[11] D. N. Zmood and D. G. Holmes, "Stationary frame current regulation of PWM inverters with zero steady-state error," *IEEE Trans. Power Electron.*, vol. 18, no. 3, pp. 814–822, May 2003.

[12] D.G. Holms, T.A. Lipo, B.P. McGrath and W.Y. Kong, "Optimized design of stationary frame three phase ac current regulators," *IEEE Trans. Power Electron.*, vol. 24, no. 11, pp. 2417-2426, Nov. 2009.

[13] A.G. Yepes, A. Vidal, J. Malvar, O. Lopes and J. Doval-Gandoy, "Tuning method aimed at optimized settling time and overshoot for synchronous proportional-integral current control in electric machines," IEEE Transactions on Power Electronics,, vol. 29, no. 6, pp. 3041-3054, June. 2014.

[14] A. Vidal, F.D. Freijedo, A.G. Yepes, P. Fernandez-Conmesama, J. Malvar et.al, "Assessment and optimization of the transient response of proportional-resonant current controllers for distributed power generation system," *IEEE Trans Ind. Electron.*, vol.60, no.4, pp. 1367-1383, April 2013.

[15] D. N. Zmood, D. G. Holmes, and G. H. Bode, "Frequency-domain analysis of three-phase linear current regulators," *IEEE Trans. Ind. Appl.*, vol. 37, no. 2, pp. 601–610, Mar./Apr. 2001.

[16] X. Yuan, W. Merk, H. Stemmler, and J. Allmeling, "Stationary-frame generalized integrators for current control of active power filters with zero steady-state error for current harmonics of concern under unbalanced and distorted operating conditions," *IEEE Trans. Ind. Appl.*, vol. 38, no. 2, pp. 523–532, Mar./Apr. 2002.

[17] J. G. Hwang, P. W. Lehn, and M. Winkelnkemper, "A generalized class of stationary frame-current controllers for grid-connected AC-DC converters," *IEEE Trans. Power Del.*, vol. 25, no. 4, pp. 2742-2751, Oct. 2010.

[18] L. Harnefors, "Modeling of three-phase dynamic systems using complex transfer functions and transfer matrices", *IEEE Trans Ind. Electron.*, vol.54, no.4,pp.2239-2248, Aug. 2007.

[19] J. Shen, S. Schroder, H. Stagge, R. W. De Doncker, "Precise modeling and analysis of dq-frame current controller for high power conveters with low pulse ratio," In *IEEE Energy Conversion Congress and Exposition, ECCE 2012*, Raleigh, NC, USA, Sept. 15-20, 2012.

[20] S. Zhou, J. Liu, and Y. Zhang, "A Decoupling Method based on Reference Current Feedforward for DQ-Frame PI Current Control of Grid-Connected Voltage Source Converters," in *Proc. IEEE IFEEC 2015,* Taipei, Taiwan, Nov. 1-4, 2015.

[21] S. Zhou, J. Liu, L. Zhou and H. She, "Cross-coupling and decoupling techniques for the current control of grid-connected voltage source converters," in *Proc. IEEE APEC 2015, pp.2821-2827 ,2015.*

[22] H.-S. Song and K. Nam, "Dual current control scheme for PWM converter under unbalanced input voltage conditions," *IEEE Trans. Ind. Electron.*,vol. 46, no. 5, pp. 953–959, Oct. 1999.

[23] Jiabing Hu, Yikang He, Lie Xu and Barry W.Williams. "Improved control of DFIG systems during network unbalance using PI–R current regulators." *IEEE Tran. Indus Electron.*, vol.56, no.2, pp.439-451, Feb.2009.

[24] J.George Hwang, Peter W.Lehn. "A single-input space vector for control of AC–DC converters under generalized unbalanced operating conditions." *IEEE Trans. Power Electro.*,vol.25, no.8, pp. 2068-2081, August 2010.

[25] M. Castilla, J. Miret, A. Camacho, L. Gareia de Vicuna and J. Matas, "Modeling and design of voltage support control schemes for three-phase inverters operating under unbalanced grid conditions," IEEE Trans. Power Electron., vol.29, no.11, pp.6139-6150, Sep, 2014.

Fuzzy Predictive DTC of Induction Machines with reduced Torque Ripple and High Performance Operation

Alberto Berzoy, IEEE Student and Osama Mohammed, IEEE Fellow.
Florida International University
Energy System Research Laboratory
Miami, US
mohammed@fiu.edu

Johnny Rengifo
Simón Bolívar University
Department of Energy Conversion and Delivery
Caracas, Venezuela
jwrengifo@usb.ve

Abstract—**This paper presents an enhanced strategy for Direct Torque Control (DTC) combining artificial intelligent (AI) and predictive algorithms. The advantages of both methodologies are merged to solve the main problems of closed loop controlled induction machines (IM) and, in particular the drawbacks of the classical DTC. Predictive DTC (P-DTC) methods solve the problems of the high torque ripple and poor performance at both starting condition and low mechanical speed operation. However these strategies depend on the IM parameter's knowledge. A new approach of fuzzy logic control (FLC) with dynamic rules based on the laws of predictive DTC is proposed to reduce the parameter dependency and improve the performance of the P-DPC. The predictive rule's main idea is to compute the angle difference in between the lines of constant torque and constant stator flux magnitude expressed as a function of the ($\alpha\beta$) inverter voltage components. For verification purposes, simulations of the DTC, P-DTC and proposed Fuzzy Predictive DTC (FP-DTC) were conducted and compared. Experimental results for the three controllers confirm the expected performance of the proposed algorithm.**

Keywords—Direct Torque Control, Fuzzy Logic, Predictive Control, Induction Machines, PWM.

I. INTRODUCTION

IM are important for industrial applications due to its robustness, ruggedness, high efficiency, lower cost and maintenance. The IM control history starts when the IM were connected directly to the mains or to a scalar controllers. The most common scalar control strategy was the Voltage-Frequency (VF) control. It was a simple control technique however it suffers from limitations at low speed and poor torque response [1]. The appearance of vector control techniques partially solves the IM control problems. One of the most famous was Field Oriented Control (FOC) proposed in [2]–[4]. This methodology provides control of the rotor flux and torque independently, acting over the stator current phase and quadrature components. Typically Proportional-Integral controllers are used in the flux and torque control loops. The FOC design is simple, but its performance depends on the knowledge of the IM model and the load variations [5]. Also, rotor flux observers are difficult to design when the parameters of the IM vary with the frequency and temperature. Solutions to these drawbacks are presented in [6], [7].

Other more recent control vector techniques are the Direct-Self Control (DSC) proposed in [8] and Direct Torque Control (DTC) presented in [9]. The former is better suited for high power applications where the semiconductor devices work at low switching frequency producing high distorted current. The latter is more widely used and best suited for low and medium power applications. DTC produces fast torque response while keeping the IM stator flux and torque decoupled as possible. The main drawbacks of this scheme are the large torque ripple, its high start-up current, variable switching frequency, and poor performance under overload condition and low speed operation. Plethora of research has been done to enhance DTC. Modifications are presented in [10], [11] for improving the starting condition and the low speed operation performance. References [12]–[19] present solutions to these problems with predictive techniques but they are parameters dependent. In [20] it is proposed a variation of DTC to reduce the torque ripple and enhance the response of the control under overload condition with the price of higher parameter sensitivity. In [21]–[24] Fuzzy controllers are proposed to reduce torque ripple and limit the stator current.

In this paper, we propose a DTC that integrates the advantages of FLC and predictive control. The FLC helps in the reduction of the torque ripple and predictive technique aims in the improvement of the performance at both starting condition and low speed operation. All this, with a significant reduction of the parameters dependency due to the FLC properties. In this new approach, the FLC rules are based on a P-DTC where main idea is to observe the non-perpendicular quadrants formed by lines of constant torque and constant stator flux magnitude drawn in the inverter voltage plane. A complete comparative study is performed among the classical DTC, P-DTC and the proposed Fuzzy-Predictive DTC (FP-DTC) showing the advantages and disadvantages of every control under diverse torque profiles and different load conditions.

978-1-4673-9551-9/16 $31.00 © 2016 IEEE

II. PREDICTIVE DTC

The objective of P-DTC is to directly control the electromagnetic torque and the amplitude of the stator flux linkage. For this purpose it is convenient to express the IM state-space dynamic equations in terms of the stator current space vector $\mathbf{i_s}$ and stator flux-linkage vector $\boldsymbol{\lambda_s}$, as given in (1).

$$\begin{aligned} p\mathbf{i_s} &= \frac{v_s}{\hat{L}_s} - \left(\frac{L_s}{L_r}\frac{R_r}{\hat{L}_s} + \frac{R_s}{\hat{L}_s} - jP\omega_m\right)\mathbf{i_s} + \left(\frac{R_r}{L_r} - jP\omega_m\right)\frac{\boldsymbol{\lambda_s}}{\hat{L}_s} \\ p\,\boldsymbol{\lambda_s} &= \mathbf{v_s} - R_s\mathbf{i_s} \\ p\omega_m &= \frac{1}{J}(T_e - T_m) \end{aligned} \quad (1)$$

where R_r and R_s are the resistances, and L_r and L_s are the inductances of the rotor and stator respectively; $\hat{L}_s = L_s - \frac{L_{sr}^2}{L_r}$. L_{sr} is the mutual inductance, ω_m is the mechanical speed, $T_e = NP\,\boldsymbol{\lambda_s} \times \mathbf{i_s}$ and T_m are the electrical and mechanical load torque respectively; NP is the number pole pairs, J is the inertial of the IM and the stator space vectors $\mathbf{i_s}$, $\mathbf{v_s}$ and $\boldsymbol{\lambda_s}$ are written in complex vector notation referred to stationary $(\alpha\beta)$ reference frame, thus $\mathbf{f_s} = f_{\alpha s} + jf_{\beta s}$; $\mathbf{f_s} = \{\mathbf{i_s}, \mathbf{v_s}, \boldsymbol{\lambda_s}\}$; p is the derivative operator $\frac{d}{dt}$. The transformation from the primitive coordinates (abc) to the stationary $(\alpha\beta)$ reference frame is Hermitian or power conservative.

Based on predictive control theory [25] and for the particular case of the IM it is possible to anticipate the error of the output, $\boldsymbol{y} = [T_e, |\boldsymbol{\lambda_s}|]$, in the next cycle of control (t_{k+1}), as the output have a relative degree $\{1,1\}$ which is defined everywhere except at $|\boldsymbol{\lambda_r}| = \mathbf{0}$ [26], [27]. This prediction can be computed based on the time derivatives of the two outputs along the dynamics. Thus, it is obtained:

$$\begin{aligned} pT_e &= p(NP\boldsymbol{\lambda_s} \times \mathbf{i_s}) \\ p|\boldsymbol{\lambda_s}| &= \frac{2}{|\boldsymbol{\lambda_s}|}\boldsymbol{\lambda_s} \times j(\mathbf{v_s} - R_s\mathbf{i_s}) \end{aligned} \quad (2)$$

$$\begin{aligned} pT_e &= \left[\frac{(\mathbf{v_s} - jNP\;\omega_m\boldsymbol{\lambda_s})}{1/NP} \times \left(\mathbf{i_s} - \frac{\boldsymbol{\lambda_s}}{\hat{L}_s}\right) - \left(\frac{L_sR_r + L_rR_s}{L_r\hat{L}_s}\right)T_e\right] \\ p|\boldsymbol{\lambda_s}| &= \frac{2}{|\boldsymbol{\lambda_s}|}\boldsymbol{\lambda_s} \times j(\mathbf{v_s} - R_s\mathbf{i_s}) \end{aligned} \quad (3)$$

where $\mathbf{a} \times \mathbf{b} = a_{\;\alpha}b_{\;\beta} - b_{\;\alpha}a_{\;\beta}$ is the cross product operation between to vectors.

For the case of the P-DTC, the selection of the best inverter voltage vector (*input space* $\mathbf{u} = \mathbf{v_s}$) is done by the comparison of the values with a cost function (ψ) defined in (4), for the 7 inverter switching states (voltages control action).

$$\psi^i = k_1(\{pT_eT_s\}^i - \Delta T_e)^2 + k_2(\{p|\boldsymbol{\lambda_s}|T_s\}^i - \Delta|\boldsymbol{\lambda_s}|)^2 \quad (4)$$

where $(\Delta T_e, \Delta|\boldsymbol{\lambda_s}|)$ are the actual output errors, $(\{pT_eT_s\}^i$, $\{p|\boldsymbol{\lambda_s}|T_s\}^i)$ are the predicted errors or incrementals, the superscript $i = \{1, \ldots, 7\}$ indicates the inverter voltages control action $v_{si} = \{v_{s1}, \ldots, v_{s7}\}$ as shown in Fig. 1, e.g. for $i = 1$, $v_{s1} = [v_{sa}\;v_{sb}\;v_{sc}] = [2V_{dc}/3\;0\;0] = v_{s\alpha} + jv_{s\beta} = \sqrt{2/3}(2V_{dc}/3)$; k_1 and k_2 are weighting constants. The minimum value among the 7 cost functions $[\psi^1, \ldots, \psi^7]$ is the action of control at the instant t_{k+1}.

Setting both derivatives in (3) equal to zero ($pT_e = 0$ and $p|\boldsymbol{\lambda_s}| = 0$), two voltage lines can be expressed as function of the inverter voltages or *input space* $(v_{\alpha s}, v_{\beta s})$ as in (5) and thus plotted in its $(\alpha\beta)$ voltage plane as shown in Fig. 1 and 2.

$$\begin{aligned} v_{s\beta_\lambda} &= R_si_{s\beta} - \frac{\lambda_{s\alpha}}{\lambda_{s\beta}}(v_{s\alpha_\lambda} - R_si_{s\alpha}) \\ v_{s\beta_T} &= \frac{v_{s\alpha_T}(\hat{L}_si_{s\beta} - \lambda_{s\beta}) - \hat{L}_sA}{(\hat{L}_si_{s\alpha} - \lambda_{s\alpha})} \\ A &= \frac{\left(\frac{L_s}{L_r}R_r + R_s\right)T_e}{NP\hat{L}_s} - NP\omega_m\boldsymbol{\lambda_s} \times j\mathbf{i_s} + \frac{NP\omega_m|\boldsymbol{\lambda_s}|^2}{\hat{L}_s} \end{aligned} \quad (5)$$

These two lines correspond to the lines of constant torque variation $(v_{s\beta_T})$ and constant magnitude of the stator flux vector variation $(v_{s\beta_\lambda})$. Each line divide the space of the inverter voltage plane or *input space* plane in two subspaces where the inverter voltage action perpendicular to these lines will create a change in the output variable. If the inverter voltage action is over (under) the line then the tendency is to increment (decrement) this variable for both cases respectively. Generally these two lines are not perpendicular and do not intersect in the origin of the inverter voltage plane, thus four irregular quadrants are created in every cycle of control (t_k). This irregularity generates an unbalance number of input voltage vectors for each of the 4 subspaces created with the two lines. In classical DTC it is assume that these lines are perpendicular and thus even number of input voltage vectors in each quadrant for the four cases of increasing or decreasing the output errors $\Delta T_e = T_{ref} - T_e$ and $\Delta|\boldsymbol{\lambda_s}| = \lambda_{ref} - |\boldsymbol{\lambda_s}|$as in (6). The angle in between this two lines is defined as ϕ (Fig. 1).

$$\begin{aligned} &\Delta T_e > 0 \text{ and } \Delta|\boldsymbol{\lambda_s}| > 0 \\ &\Delta T_e > 0 \text{ and } \Delta|\boldsymbol{\lambda_s}| > 0 \\ &\Delta T_e > 0 \text{ and } \Delta|\boldsymbol{\lambda_s}| > 0 \\ &\Delta T_e > 0 \text{ and } \Delta|\boldsymbol{\lambda_s}| > 0 \end{aligned} \quad (6)$$

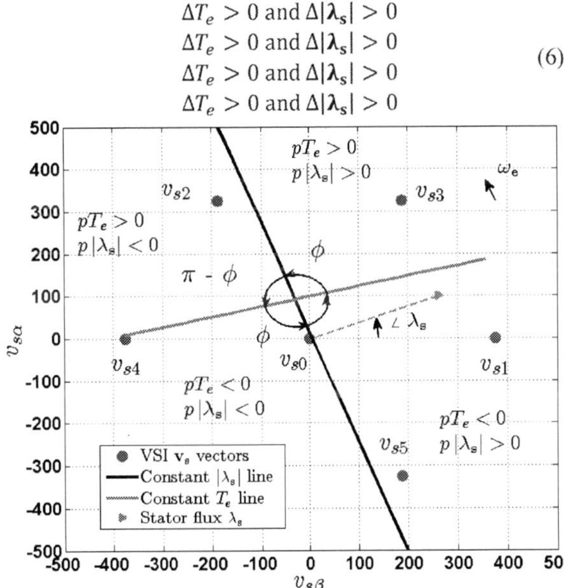

Fig. 1. Torque and stator flux constants lines at the VSI voltage plane

III. FUZZY PREDICTIVE DTC

An efficient solution for mitigating the issues of the classical DTC algorithm is to apply FLC as in [21]. The FL-

978-1-4673-9551-9/16 $31.00 © 2016 IEEE

based DTC provides an adaptive variation of the duty-cycle based on the switching table proposed by Noguchi [9], however the rules are fixed making it to drag the poor performance at both starting and low speed operation.

In the proposed DTC the contribution appears in the integration of FLC with a predictive strategy which allows not only adaptive variation but dynamic rules. For simplicity, the FLC system is designed by choosing minimum number of linear membership functions (MF) for each input and using Takagi-Sugeno (T-S) Fuzzy Inference System (FIS) technique for calculating the control decision. A block diagram for the complete systems is indicated in Fig. 2. In Fig 3 a detailed block diagram of the FIS is depicted. The FIS is a zero-order T-S that has two inputs (ΔT_e and $\Delta|\boldsymbol{\lambda}_s|$) and one output (vector action of control $\mathbf{v}_{\mathbf{Fuzzy}} = ze^{j\theta}$). The universe of discourse of each input is described by three linear MF (N, Z and P). The fuzzy rules which describes the relation between the fuzzy inputs and output are chosen by the angle ϕ (Fig. 1) between the two constant lines of torque $pT_e = 0$ and magnitude of stator flux $p|\boldsymbol{\lambda}_s| = 0$ that form the irregular quadrants as shown in Fig. 1.

Finally the matrix rules is the one presented in Fig. 3. Each fuzzy rule has two parameters: magnitude (z) and angle (θ); e.g. assume first case of (5) $\Delta T_e > 0$ and $\Delta|\boldsymbol{\lambda}_s| > 0$, thus MF_T=P and MF_λ=P. The matrix rules indicate $z = L$ and $\theta = \phi/2$, so the maximum change in both torque and stator flux magnitude is achieve when $\theta = \phi/2$. If the two lines of constant torque and flux are orthogonal and centered at the VSI center plane, the rule would be $\theta = \pi/2$ and the control will be similar to [21]. In this FLC, the rules are dynamic and varying at every moment as the angle ϕ, which varies with the position and the operation condition of the machine.

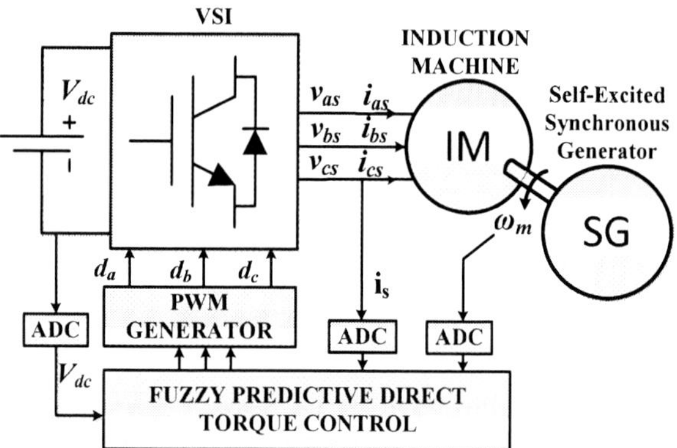

Fig. 2. (a) Block diagram of the complete system

In order to produce the duty cycles for the modulation technique, the FLC output (z and θ) can be implemented in two ways. Using (7) for mounting the fuzzy output over the flux vector or (8) that is the vector summation of Fig. 4.

$$v_{\alpha s} = z\cos(\theta + \angle\boldsymbol{\lambda}_s)$$
$$v_{\beta s} = z\sin(\theta + \angle\boldsymbol{\lambda}_s) \qquad (7)$$

$$\mathbf{v}_s = \mathbf{v}_{\mathbf{Center}} + \mathbf{v}_{\mathbf{Fuzzy}} \qquad (8)$$

where $\mathbf{v}_{\mathbf{Center}}$ can be obtained by making equal $v_{s\beta_\lambda} = v_{s\beta_T}$ in (5). So:

$$v_{Center\beta} = \frac{R_s\left(i_\beta - \frac{\lambda_{\beta s}}{\hat{L}_s}\right)\boldsymbol{\lambda}_s \times j\mathbf{i}_s - A\lambda_{\beta s}}{\boldsymbol{\lambda}_s \times j\mathbf{i}_s - n_p\omega_m\frac{|\boldsymbol{\lambda}_s|^2}{\hat{L}_s}} \qquad (9)$$
$$v_{Center\alpha} = R_s i_{\alpha s} - \frac{\lambda_{\beta s}}{\lambda_{\alpha s}}\left(v_{Center\beta} - R_s i_{\beta s}\right)$$

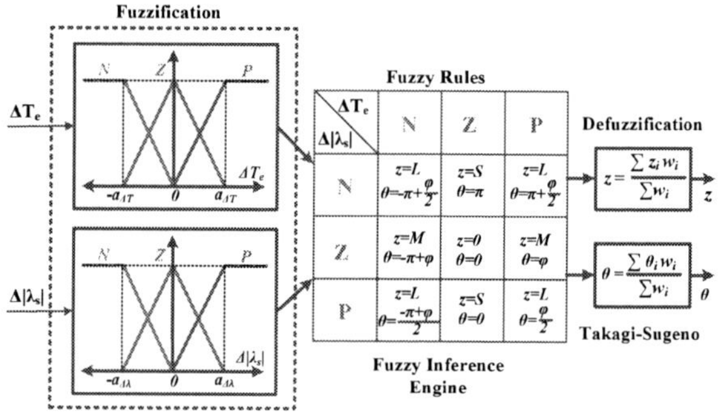

Fig. 3. Block diagram of the proposed fuzzy predictive DTC

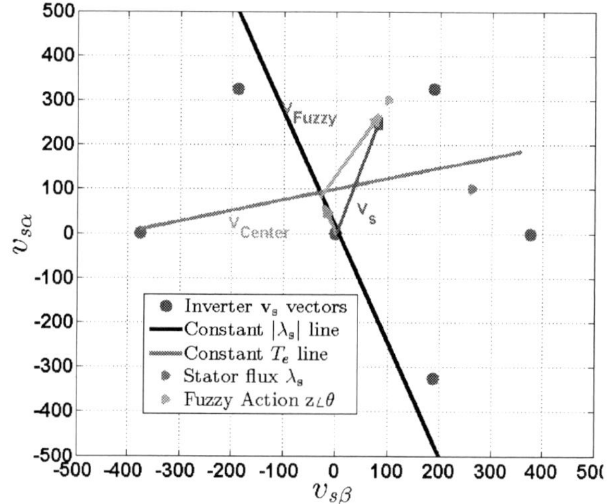

Fig. 4. Vector summation of Fuzzy out and correction

The added computation, in both cases, is just the calculation of two lines (constant torque and flux) and their tangents. The advantage of using a FLC with predictive rules is that the predictive corrects the angle of the control action compared with the table of DTC which sometimes is erratic and it is the reason for such a big torque ripple and poor performance. Also, FLC decrease the parameter dependency of the predictive control which is its main disadvantage. Based on the inputs and the rules table, the output (duty) is calculated by the T-S formula which is the simplest among other FIS.

IV. Simulation and Experimental Results

In order to compare the performance of the proposed FP-DTC a simulation and experimental test comparison among classical DTC, P-DTC and proposed DTC is carried out. The parameters of the induction machine used in this study are found by characterizing the induction motor with a starting of the machine under a step of AC voltage, the procedure followed for the parameter estimation is proposed in [28], [29]. The parameter estimation method adjusts the instantaneous input impedance motor during a start-up. For all the simulations and experimental tests, the machine is belt coupled with a self-excited synchronous generator. The parameter estimation was perform for the no load and loaded case to calculate the inertia of the load. Finally the model parameters are presented in Table I and these values are used in the simulations which are conducted in Simulink MatLab.

The simulation and experimental implementation were performed under the same conditions of DC link capacitor voltage $V_{dc} = 300\ \text{V}$, switching frequency $F_S = 10\ \text{kHz}$, current limitation $I_{limit} = 30/\sqrt{2}\ \text{A}_{\text{rms}}$. For purposes of accuracy and fairness comparison of the simulation and experimental results, the simulation time was $10\ \mu s$ but the control algorithm operates at $1\ \text{ms}$. The PWM in both simulation and experimental is at 10 kHz and centered.

For purposes of control performance verification, a torque profile similar than the torque profile found in electric vehicles applications is implemented in the simulation and experimental tests for the 3 controllers. The torque profile starts with a step at 8 Nm which is the rated torque from the name plate of the machine, then after 2 seconds, a down step of torque reach the 4 Nm. Two sec. later a ramp increases the profile until the nominal torque again keeping it for 2 more sec. and a down ramp ends the profile in 5 Nm (Fig. 5(a) for simulation and 8(a) for experimental). The flux reference in all the simulations and experiments is $\lambda_{s\_ref} = 0.33$ Wb.

TABLE I. NAMEPLATE RATING

Param.	Value	Param.	Value	Rating	Value	Rating	Value
L_{ls}	3.08 mH	R_s	1.506 Ω	P	1.5 kW	PF	0.83
L_{lr}	3.46 mH	R_r	0.6172 Ω	V	208 V	Poles Pairs	2
L_m	119.22 mH	J_m	0.13 kgm^2	I	5.9 A	n_r	1750 *rpm*

A. Simulation Results

In the Fig. 5(a) the torque reference is presented and the electromagnetic torque produced by the IM controlled by DTC, P-DTC and FP-DTC. In this figure it can be noticed that the dynamics of the three algorithms are very fast (well known for DTC) however, the ripple is not the same for the three techniques. In the Fig. 5(b) a zoom of the electromagnetic torque for the three strategies is presented and it is clear that the classical DTC suffer for two problems: steady state error and high torque ripple. In the other hand, the P-DTC corrects the steady state error and reduce the torque ripple however the FP-DTC is even better. The high torque ripple in classical is produced by the selection of the non-optimum voltage vectors as is explained in [14] while FP-DTC chooses the control vector using the constant torque and flux plane allowing to reduce the torque ripple. The locus of the stator flux vector for the DTC, P-DTC and FP-DTC are presented in Fig. 6(a) – 6(c). The three locus present very similar response, however the locus of FP-DTC has less ripple. In the Fig. 6(d) a zoom for the phase A current for the three controllers is shown. It can be notice that the currents are sinusoidal and the current ripple of DTC is higher that P-DTC and FP-DTC, as well as the current ripple of P-DTC is higher that FP-DTC.

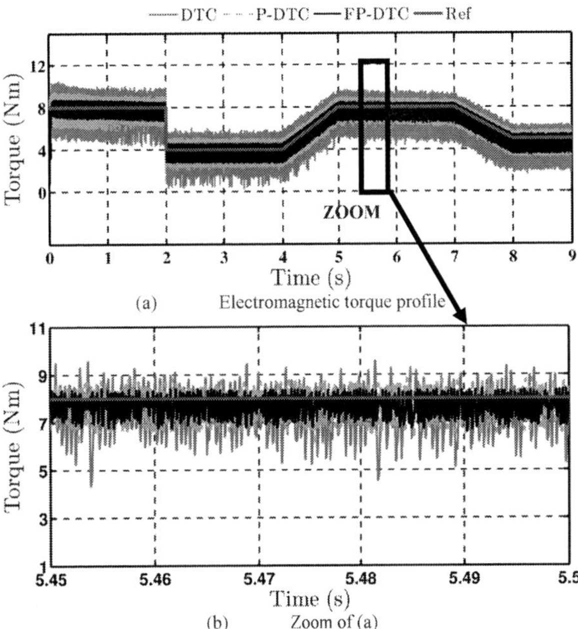

Fig. 5. Simulation results for comparison of the Electromagnetic Torque among DTC, P-DTC and FP-DTC

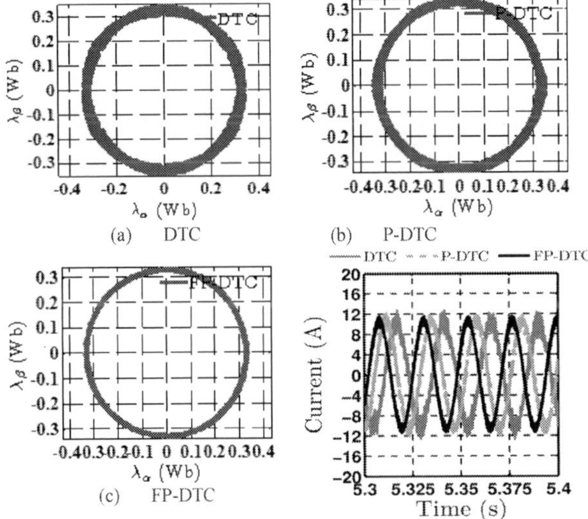

Fig. 6. Simulation results of the stator flux locus and stator current comparison among the strategies.

B. Experimental Results

In order to compare the performance of the proposed FP-DTC an experimental comparison was accomplished. Fig 7 presents the experimental test rig which is composed by a PC, dspace 1104, DC power supply for the control, booster for increase the pulses (0 V off 5 V on) from the dspace to the IGBT drivers (-10 off 15 V on), sensor board (LEM voltage and current transducers), oscilloscope Tecktronics, autotransformer, motor- generator set (motor WEG W21, generator Yanan SLG-164B) and voltage source inverter (VSI) which consist of Semikron IGBTs (1200 V – 50 A), drivers and 2 capacitors of in series of 1200 μF and 525 V.

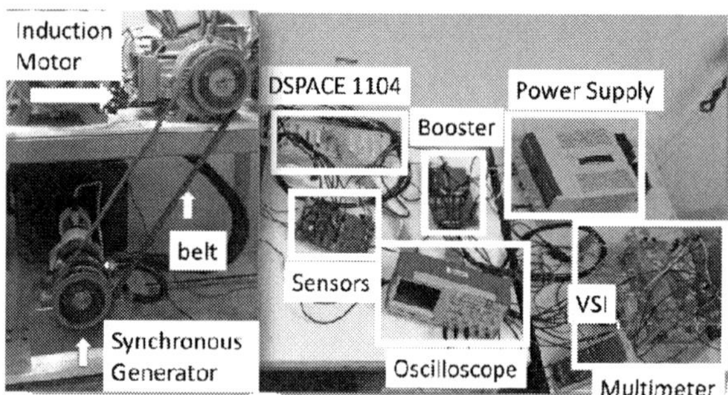

Fig. 7 Experimental Setup

The experimental results for the torque profile response can be observed in Fig. 8a. Fig. 8b presents a zoom of the response for a period of 0.5 s. The experimental results are highly correlated with the simulation results presented in Fig. 5a. As it is expected all the controllers have very fast dynamic response however DTC and P-DTC have more torque ripple and steady state error than FP-DTC.

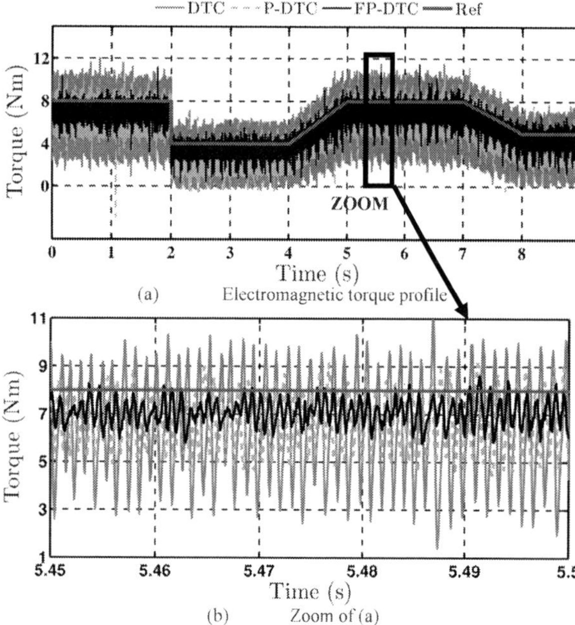

(a) Electromagnetic torque profile

(b) Zoom of (a)

Fig. 8. Experimental comparison of the Electromagnetic Torque among DTC, P-DTC and FP-DTC

In the Fig. 9, the locus of the stator flux vector for the DTC, P-DTC and FP-DTC are presented. It can be noticed, that the 3 locus present similar performance as expected from the simulation, however the FP-DTC demonstrates less ripple. In Fig. 9d, the DC link voltage is presented as it is not an ideal source like in the simulation. The effect on these variations affects the torque response of the experimental resulting in higher torque ripple than simulations. In Fig. 10a, a zoom for the phase current A for the three controllers is shown. Comparing the amplitude of these currents with the simulation it can be observed the same $\frac{12}{\sqrt{2}}A_{rms}$. The current in the FP-DTC present the less ripple content as the simulation indicates. Fig. 10b a frequency spectrum of the currents is conducted corroborating that the FP-DTC has the less harmonic content among the three controllers. Finally, in Fig. 11 shows the mechanical speed for the three control strategies, the control performance obtained is at low speed region. Thus, the torque ripple reduction improves one of the classical DTC problems.

For final comparison among the three controllers and its correlation with the simulation results, a comparative results of the torque ripple, stator flux ripple and current ripple is presented in Table II.

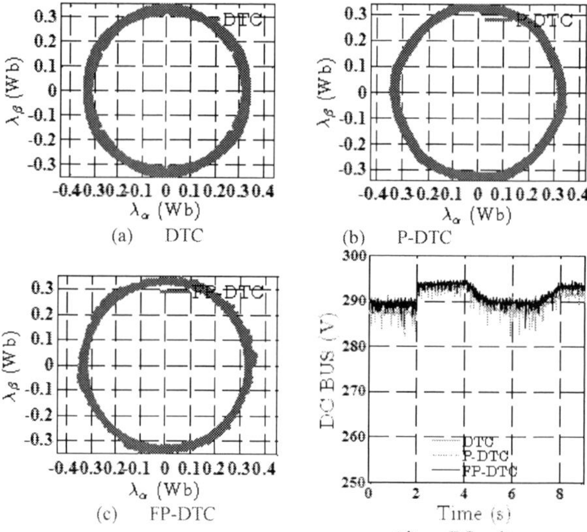

(a) DTC (b) P-DTC (c) FP-DTC (d) DC voltage

Fig. 9. Experimental comparison of the stator flux locus and DC bus voltages among DTC, P-DTC and FP-DTC

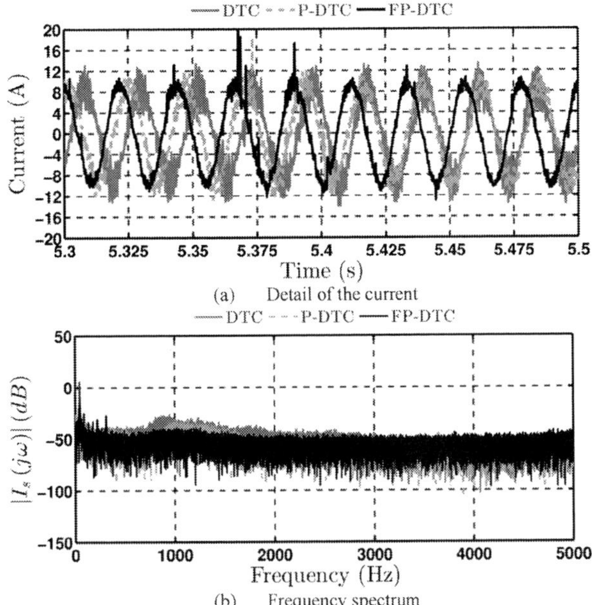

(a) Detail of the current

(b) Frequency spectrum

Fig. 10. Stator current comparison

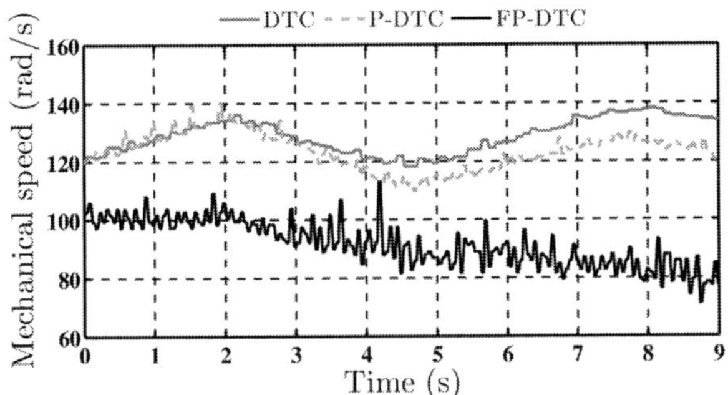

Fig. 11. Experimental mechanical speed

TABLE II. SIMULATION AND EXPERIMENTAL COMPARISON

	Simulation			Experimental		
	DTC	P-DTC	FP-DTC	DTC	P-DTC	FP-DTC
Torque Ripple (Nm)	4.5	2	1	6	2.5	1
Flux Ripple (Wb)	0.05	0.035	0.02	0.05	0.05	0.04
Current Ripple (A)	2	1.5	1	4	2	1

V. CONCLUSIONS

A novel and simple control strategy is presented by based on the integration of a FLC algorithm with minimum and linear MF with a predictive methodology. The advantages of both techniques are exploited to improve the drawbacks and performance of the close loop controllers like classical DTC. Simulations and experimental results show an improvement in the control performance of the torque including ripple and steady state error reduction and a satisfactory performance at the low speed region. The proposed control was tested under 50% variation of the IM parameters obtaining similar results and corroborating its low dependency.

REFERENCES

[1] C. A. Martins and A. S. Carvalho, "Technological trends in induction motor electrical drives," in *Power Tech Proceedings, 2001 IEEE Porto*, 2001, vol. 2, p. 7 pp. vol.2–.

[2] F. Blaschke, "A new method for the structural decoupling of AC induction machines," *Conf Rec IFAC*, pp. 1–15, Oct. 1971.

[3] R. Gabriel, W. Leonhard, and C. J. Nordby, "Field-Oriented Control of a Standard AC Motor Using Microprocessors," *IEEE Trans. Ind. Appl.*, vol. IA-16, no. 2, pp. 186–192, Mar. 1980.

[4] P. Vas and M. Alakula, "Field-oriented control of saturated induction machines," *IEEE Trans. Energy Convers.*, vol. 5, no. 1, pp. 218–224, Mar. 1990.

[5] Y. Miloud and A. Draou, "Fuzzy logic speed control of an indirect field-oriented induction machine drive," in *The 27th Annual Conference of the IEEE Industrial Electronics Society, 2001. IECON '01*, 2001, vol. 3, pp. 2111–2116 vol.3.

[6] W. L. Chan, A. T. P. So, and L. L. Lai, "Evolutionary programming based machine parameters estimation for field oriented control," in *Electric Machines and Drives, 1999. International Conference IEMD '99*, 1999, pp. 534–536.

[7] J.-H. Kim, J.-W. Choi, and S.-K. Sul, "Novel rotor flux observer using observer characteristic function in complex vector space for field oriented induction motor drives," in *Sixteenth Annual IEEE Applied Power Electronics Conference and Exposition, 2001. APEC 2001*, 2001, vol. 1, pp. 615–621 vol.1.

[8] M. Depenbrock, "Direct self-control (DSC) of inverter-fed induction machine," *IEEE Trans. Power Electron.*, vol. 3, no. 4, pp. 420–429, Oct. 1988.

[9] I. Takahashi and T. Noguchi, "A New Quick-Response and High-Efficiency Control Strategy of an Induction Motor," *IEEE Trans. Ind. Appl.*, vol. IA-22, no. 5, pp. 820–827, Sep. 1986.

[10] T. Noguchi, M. Yamamoto, S. Kondo, and I. Takahashi, "High frequency switching operation of PWM inverter for direct torque control of induction motor," in , *Conference Record of the 1997 IEEE Industry Applications Conference, 1997. Thirty-Second IAS Annual Meeting, IAS '97*, 1997, vol. 1, pp. 775–780 vol.1.

[11] M. P. Kazmierkowski and A. B. Kasprowicz, "Improved direct torque and flux vector control of PWM inverter-fed induction motor drives," *IEEE Trans. Ind. Electron.*, vol. 42, no. 4, pp. 344–350, Aug. 1995.

[12] J.-K. Kang and S.-K. Sul, "Analysis and prediction of inverter switching frequency in direct torque control of induction machine based on hysteresis bands and machine parameters," *IEEE Trans. Ind. Electron.*, vol. 48, no. 3, pp. 545–553, Jun. 2001.

[13] T. G. Habetler, F. Profumo, M. Pastorelli, and L. M. Tolbert, "Direct torque control of induction machines using space vector modulation," *IEEE Trans. Ind. Appl.*, vol. 28, no. 5, pp. 1045–1053, Sep. 1992.

[14] G. Papafotiou, J. Kley, K. G. Papadopoulos, P. Bohren, and M. Morari, "Model Predictive Direct Torque Control #x2014;Part II: Implementation and Experimental Evaluation," *IEEE Trans. Ind. Electron.*, vol. 56, no. 6, pp. 1906–1915, Jun. 2009.

[15] R. Vargas, J. Rodriguez, U. Ammann, and P. W. Wheeler, "Predictive Current Control of an Induction Machine Fed by a Matrix Converter With Reactive Power Control," *IEEE Trans. Ind. Electron.*, vol. 55, no. 12, pp. 4362–4371, Dec. 2008.

[16] J. Beerten, J. Verveckken, and J. Driesen, "Predictive Direct Torque Control for Flux and Torque Ripple Reduction," *IEEE Trans. Ind. Electron.*, vol. 57, no. 1, pp. 404–412, Jan. 2010.

[17] F. Wang, S. Li, X. Mei, W. Xie, J. Rodriguez, and R. M. Kennel, "Model-Based Predictive Direct Control Strategies for Electrical Drives: An Experimental Evaluation of PTC and PCC Methods," *IEEE Trans. Ind. Inform.*, vol. 11, no. 3, pp. 671–681, Jun. 2015.

[18] J. Rengifo, J. Aller, A. Berzoy, and J. Restrepo, "Predictive DTC algorithm for induction machines using Sliding Horizon Prediction," in *2014 IEEE 5th Latin American Symposium on Circuits and Systems (LASCAS)*, 2014, pp. 1–4.

[19] E. A. José Sayritupac, J. Rengifo, E. Albanez, J. M. Aller, and J. Restrepo, "Predictive Control Strategy for DFIG Wind Turbines with Maximum Power Point Tracking Using Multilevel Converters," *2nd Workshop Power Electron. Power Qual. Appl.*, pp. 1–6, Jun. 2015.

[20] D. Casadei, F. Profumo, G. Serra, and A. Tani, "FOC and DTC: two viable schemes for induction motors torque control," *IEEE Trans. Power Electron.*, vol. 17, no. 5, pp. 779–787, Sep. 2002.

[21] J. C. Viola, J. A. Restrepo, V. M. Guzman, and M. I. Gimenez, "Direct Torque Control of Induction Motors Using a Fuzzy Inference System for Reduced Ripple Torque and Current Limitation," in *Power Electronics and Motion Control Conference, 2006. EPE-PEMC 2006. 12th International*, 2006, pp. 1161–1166.

[22] M. Ortega, J. Restrepo, J. Viola, M. I. Gimenez, and V. Guzman, "Direct torque control of induction motors using fuzzy logic with current limitation," in *31st Annual Conference of IEEE Industrial Electronics Society, 2005. IECON 2005*, 2005, p. 6 pp.–.

[23] J. A. Restrepo, J. C. Viola, M. Ortega, V. M. Guzman, and M. I. Gimenez, "A fuzzy-PWM direct torque control of induction machines for current limitation and reduced torque ripple," in *2005 European Conference on Power Electronics and Applications*, 2005, p. 8 pp.–P.8.

[24] S. Gdaim, A. Mtibaa, and M. F. Mimouni, "Design and Experimental Implementation of DTC of an Induction Machine Based on Fuzzy Logic Control on FPGA," *IEEE Trans. Fuzzy Syst.*, vol. 23, no. 3, pp. 644–655, Jun. 2015.

[25] Q. Lu, Y. Sun, and S. Mei, *Nonlinear Control Systems and Power System Dynamics*. Springer Science & Business Media, 2001.

[26] R. Ortega, N. Barabanov, G. Escobar, and E. Valderrama, "Direct torque control of induction motors: stability analysis and performance improvement," *IEEE Trans. Autom. Control*, vol. 46, no. 8, pp. 1209–1222, Aug. 2001.

[27] G. Escobar, A. M. Stankovic, E. Galvan, J. M. Carrasco, and R. Ortega, "A family of switching control strategies for the reduction of torque ripple in DTC," *IEEE Trans. Control Syst. Technol.*, vol. 11, no. 6, pp. 933–939, Nov. 2003.

[28] J. W. R. Santana, J. B. Suñe, J. M. A. Castro, A. A. B. Montilla, and J. A. R. Zambrano, "Parameter estimation method for induction machines using instantaneous voltage and current measurements," *Rev. Fac. Ing.*, vol. 0, no. 75, pp. 57–66, May 2015.

[29] J. Rengifo, J. M. Aller, A. Bueno, J. Viola, and J. Restrepo, "Parameter Estimation Method for Induction Machines Using the Instantaneous Impedance During a Dynamic Start-Up," in *Andean Region International Conference (ANDESCON), 2012 VI*, 2012, pp. 11–14.

Fixed-Frequency Generalized Peak Current Control (GPCC) for Inverters

Mohammad Ebrahimi* and S. Ali Khajehoddin†
Department of Electrical & Computer Engineering
University of Alberta
Edmonton, AB, Canada
*m.ebrahimi@ualberta.ca, †khajeddin@ualberta.ca

***Abstract*—A fast and robust fixed switching frequency peak current controller for dc-ac converters is presented. The method is specifically elaborated for single-phase grid-connected distributed generation (DG) applications. This method is called generalized peak current control (GPCC) as it can mimic any known pulse width modulation (PWM) strategy. It is shown that additional control objectives can be achieved by adaptive bands of the GPCC, which are proposed to provide active damping for inverters with LCL output filters. The proposed approach features all the advantages of peak current controllers such as simplicity, fast transient, and optimum dynamic response; with the superiority of fixed switching frequency and harmonic free output. Feasibility and performance of the controller is shown by simulations and experimental results.**

***Index Terms*—Current Control; dc-ac Converters; Generalized Peak Current Control (GPCC); Switching Scheme; Single-Phase Grid-Connected Inverter**

I. Introduction

Peak current controller is a promising candidate for the control of distributed generation (DG) inverters in renewable energy systems and Microgrid applications. Fast transient response, zero tracking error, and inherent maximum current limiting are among the advantages of these controllers [1]. Also peak current controllers perfectly mitigate the injection of dc current in DG systems' inverters. Moreover, they can generate pure sinusoidal currents in highly distorted grid voltage conditions, with no need for additional control actions.

Despite all these advantages, switching frequency of the peak current controllers varies over a fundamental cycle and is susceptible to the operating point and system parameters [1], [2]. This leads to undesired harmonic spectrum of the inverters' output and difficulties in design of the ac filter. In addition, using the existing peak current control or hysteresis control methods, the modulation is not the same as any known pulse width modulation (PWM) scheme. Therefore, harmonic spectrum and the converters' performance are not predictable. Moreover, extensive research has been done on PWM switching scheme and several approaches are proposed to achieve different characteristics for the PWM controlled inverters. Since the link between these PWM strategies and existing peak current controllers is not thoroughly investigated, benefits of these PWM schemes cannot be easily added to the peak current controlled inverters.

Several methods are proposed to address some of the aforementioned limitations and problems. Some methods have tried to synchronize the switching action of peak current controller to an external clock signal. In one approach, bands of the controller are determined based on output of a phase locked loop (PLL), which tries to lock the inverters' switching action to a reference clock [3]. This method employs some filtering actions in the calculation of the bands, which are reported to initiate stability problems. Moreover, the PLL lose its synchronization in large output transients. This method is improved in [4] by replacing the PLL with a predictive band calculation method. In this approach, the external clock is used to calculate inverter bands for next clock cycle, based on slope of the output current in previous cycle. Although they achieve a nearly constant switching frequency, all these methods are complicated and their dynamic performance is degraded, compared to the traditional peak current controllers.

In another approach [5], current bands are adjusted to provide a fixed switching frequency for the inverter. Several research studies have applied the method proposed in [5] to the grid connected inverter applications [6]–[8]. Although these methods achieve a nearly constant switching frequency, their application in DG inverters and the related issues such as resonance damping of the LCL filter are not studied. Moreover, none of them claim to generate any known PWM scheme and as a result, design of the system for electromagnetic interference (EMI), electromagnetic compatibility (EMC), or loss calculation are not possible, and the performance for special cases such as controlling leakage current in photovoltaic (PV) system [9] is not predictable or controllable. None of these studies have addressed a comprehensive solution that can be extended to topologies other than the H-bridge converter, and interfacing filters other than a simple inductor.

A simple but effective approach is presented in this paper to provide fixed switching frequency for the peak current control of an inverter. The proposed method achieves the following objectives: i) in this approach the link between peak current control of inverter and existing PWM schemes is elaborated, which features the advantages of both methods in the resulting controller, ii) the proposed method is proposed for single-phase DG inverter applications and is applied to both L and LCL output filter configurations. Additional control objectives such as active damping of the output filter is also investigated, iii) the approach is extendable to different inverter topologies

978-1-4673-9551-9/16 $31.00 © 2016 IEEE

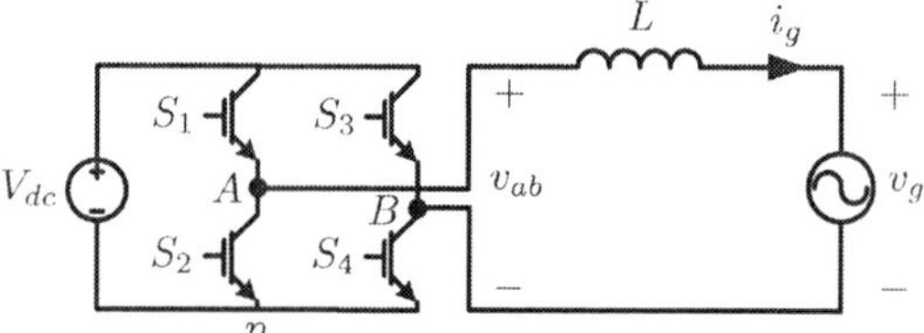

Fig. 1. Single-phase inverter interfaced with the grid through an L filter.

such as multilevel converters and three-phase inverters.

The proposed approach is inspired by the basic operation principle of PWM technique. PWM controlled inverters present a clean output harmonic spectrum, which is a result of their fixed switching frequency. The proposed idea is called generalized peak current control (GPCC) as it will mimic the switching behaviour of any PWM technique, while controlling peak value of the converters' current. In other words, bands of the peak current controller are determined such that its switching characteristic becomes exactly similar to a selected PWM scheme. As a result, all advantages of the peak current controller are inherited along with clean predictable harmonic spectrum, constant switching frequency, and additional desired output characteristics of the original PWM scheme.

The proposed method is a general technique that can be applied to different types of PWMs and different types of converters, resulting in a family of peak current controllers maintaining their original PWM characteristics. In this paper, it is shown how the proposed approach is applied to unipolar PWM. This method is elaborated for a single-phase DG system and it is shown that it can be used with both L and LCL types output filters. Moreover for the LCL case, it is shown that using new adaptive bands, robust active damping is provided for the inverter with no additional sensors. Feasibility of the proposed method and performance of the proposed damping approach is shown by simulations and experimental results.

II. The Proposed GPCC Method Mimicking Unipolar PWM

A. Derivation of the method for the L Filter Case

Circuit diagram of a single-phase inverter interfaced to the grid through an L filter is shown in Fig. 1. In the average sense, this inverter amplifies the modulation index m by gain V_{dc} to generate the output voltage v_{ab}, i.e. $v_{ab} = mV_{dc}$. The inverter is controlled by unipolar PWM. Accordingly, leg A of the inverter is controlled with m, and leg B is controlled with $-m$ signal. To exactly mimic a given PWM scheme, first the current envelop for an inverter using that PWM signal is derived. Then the derived formulas are used to control the current in a peak current control scheme. As a result, the switching patterns will be exactly similar to the original PWM scheme.

Key waveforms of the inverter in Fig. 1 for the case $v_{ab} \geq 0$ (or equally $m \geq 0$) is shown in Fig. 2. This figure shows the switching signals for the two upper switches S_1 and S_3. The duty cycle for switch S_1 and S_3 is called d_1 and d_3,

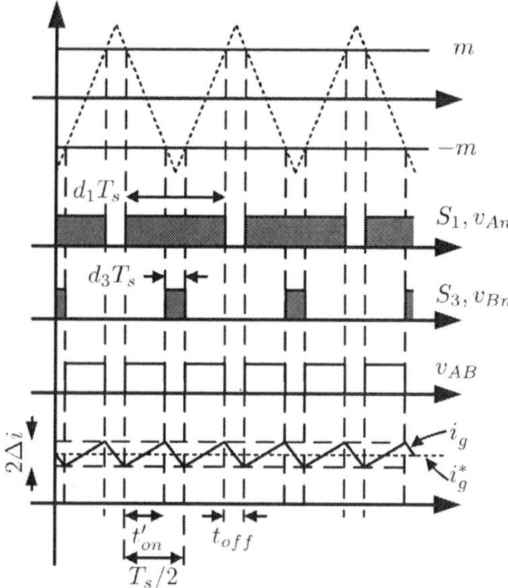

Fig. 2. Operation principle of the inverter controlled with unipolar PWM scheme for $m \geq 0$.

respectively. As per Fig. 2, it can be observed that d_1 and d_3 are related to the modulation index m as $d_1 = (1+m)/2$ and $d_3 = (1-m)/2$. Assuming a sinusoidal grid voltage as $v_g = \widehat{V}_g \sin(\omega t)$ and a grid current such as $i_g^* = \widehat{I}^* \sin(\omega t)$, average voltage at the inverters' output will be

$$v_{ab} = \widehat{V}_{ab} \sin(\omega t + \delta),$$
$$\widehat{V}_{ab} = \sqrt{\left(\omega L \widehat{I}^*\right)^2 + \left(\widehat{V}_g\right)^2}, \quad \delta = \tan^{-1}\left(\frac{\omega L \widehat{I}^*}{\widehat{V}_g}\right).$$

Therefore t_{off} and t'_{on} in Fig. 2 can be calculated as

$$\begin{aligned} t_{off} &= (1-d_1)T_s = \left(1 - \frac{\widehat{V}_{ab}}{V_{dc}} \sin(\omega t + \delta)\right)\frac{T_s}{2}, \\ t'_{on} &= (1-2d_2)\frac{T_s}{2} = \left(\frac{\widehat{V}_{ab}}{V_{dc}} \sin(\omega t + \delta)\right)\frac{T_s}{2}. \end{aligned} \tag{1}$$

Note that this calculations can be extended to the case where the grid current is not in phase with the grid voltage. Now regarding Fig. 2 and assuming constant v_g and i_g^* during each switching period, Δi can be obtained from $V_{dc} - v_g = L(2\Delta i)/(t'_{on})$ as

$$\Delta i = \frac{T_s}{4L}\frac{\widehat{V}_{ab}}{V_{dc}} \sin(\omega t + \delta)\left(V_{dc} - \widehat{V}_g \sin(\omega t)\right) \quad v_{ab} \geq 0. \tag{2}$$

The same procedure gives the current band for $v_{ab} < 0$ as (3).

$$\Delta i = \frac{T_s}{4L}\frac{\widehat{V}_{ab}}{V_{dc}} \sin(\omega t + \delta)\left(V_{dc} + \widehat{V}_g \sin(\omega t)\right) \quad v_{ab} < 0 \tag{3}$$

Equations (2) and (3) calculate the peak value of the grid current in the inverter scheme of Fig. 1, which uses a unipolar PWM. Therefore, if the same equations are used to generate a

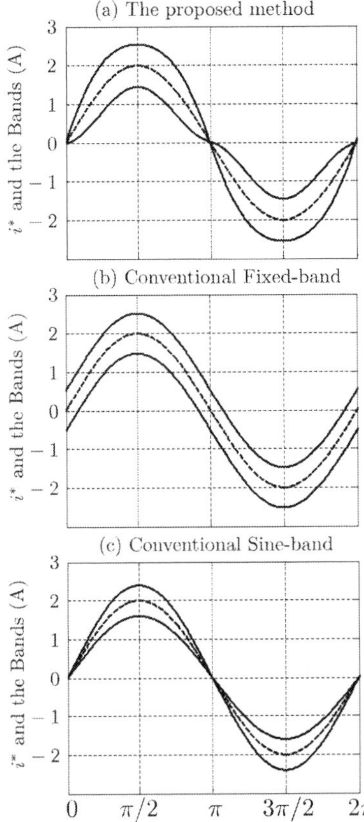

Fig. 3. Comparison of the current bands for the proposed GPCC with the conventional methods.

variable current band, a constant-frequency switching-pattern will be obtained, which mimics the switching action of the unipolar PWM. This is contrary to the conventional peak current controller where the bands are either fixed or sinusoidal [10], generating a variable switching frequency. The proposed and the conventional bands are compared in Fig. 3. As it can be observed, the lower and upper bands in the proposed technique have a double frequency term. This clarifies the essence of employing non-pure-sinusoidal bands to avoid the variable switching frequency. Also it is worth mentioning that maximum amount of Δi obtained from (2) or (3) significantly helps for the optimum magnetic design of the inductor.

The current bands of the proposed technique are sketched for different values of the filter inductance in Fig. 4, where each color shows the upper and lower bands for each inductor value. This figure clearly shows the effect of the inductor value and the double frequency term on the current bands.

Considering the current band formulas (in (2) and (3)) along with $\widehat{V}_{ab}$ and δ definitions, sensitivity of Δi toward different variables can be assessed. Having V_{dc} and v_g as the feedback, Δi just depends on the inductor value, L. Assume that current bands are calculated for a given filter inductance and the inverter is then controlled based on these calculated bands. In this case, if the inverter faces some mismatch in the design

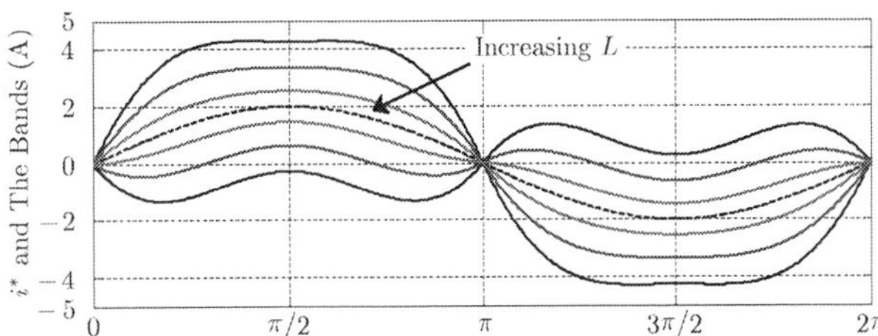

Fig. 4. Upper and lower current bands for different filter inductance values in the proposed GPCC.

value of L compared to its actual value, then the switching frequency will be slightly different from the expected value. Rearranging (2) and (3) shows that switching frequency of the inverter is inversely proportional to the amount of actual L, that is

$$f_s = \frac{1}{T_s} \propto \frac{1}{L}.$$

Therefore, if L slightly deviates from its designed value, output harmonic spectrum of the inverter remains similar to the original PWM scheme, with a slightly different switching frequency.

B. Extension of The GPCC for LCL Type Filters

Circuit diagram of the single-phase DG inverter with an LCL filter is depicted in Fig. 5(a). Although the LCL filter features an effective filtering action at a smaller size, it will introduce an undesired resonant oscillation to the output current. The proposed approach in this section adaptively adjusts the bands in the peak current controller, to actively damp the undesired resonance of the LCL filter.

1) Control Structure and Feedback Selection: PWM inverters control the average value of output grid current indirectly, by controlling output voltage of the inverter. Therefore, as shown in Fig. 5(b) they can be modeled as a voltage source, connected to the grid through an LCL filter. On the contrary, peak current controllers directly control peak value of the inductor current. Therefore, if the inductor L_i current (i_i) is selected as the feedback variable, inverter can be modelled as a current source connected to the rest of circuit as shown in Fig. 5(c). As i_i is tightly controlled, its dynamic does not contribute in the undesired resonance of the output, and only L_g and C will resonate with each other. If capacitor voltage (v_c) is selected as the voltage feedback, equations derived in (2) and (3) for the simple L filter case will be still valid by substitution of v_g by v_c. As discussed later in this section, this configuration provides the opportunity for active damping of the resonance with no need for additional sensors.

2) Resonant Damping Strategy: Figure 6 shows the proposed damping scheme. As it can be seen, branch of i_d is added, which affects the upper and lower bands of controller to provide active damping for the system. i_d is generated using $H(s)$ as

$$H(s) = \frac{I_d(s)}{V_c(s)} = \frac{k\omega_c^2 s}{s^2 + 2\zeta\omega_c s + \omega_c^2} \quad (4)$$

978-1-4673-9551-9/16 $31.00 © 2016 IEEE

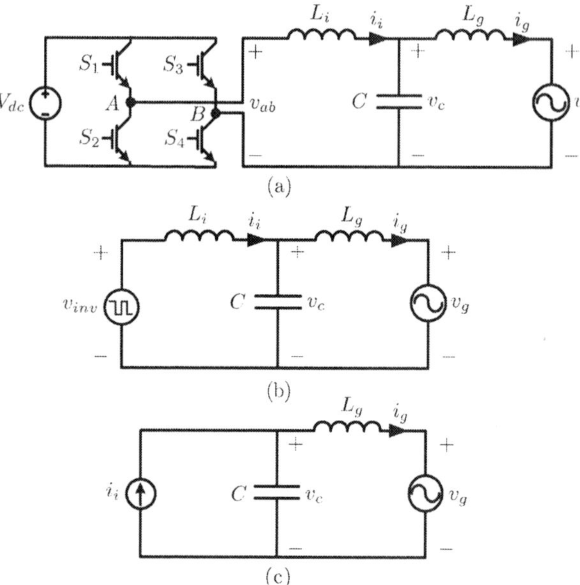

Fig. 5. Single-phase inverter connected to the grid through LCL filter. (a) Circuit diagram (b) PWM controlled inverter circuit diagram (c) Simplified model for a peak current controlled inverter.

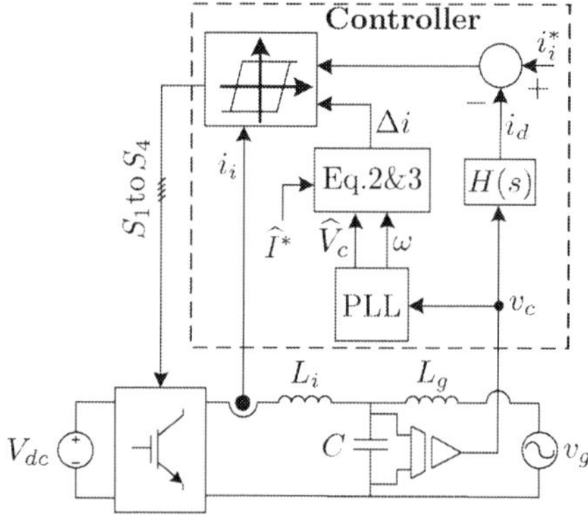

Fig. 6. Block diagram of the proposed controller along with the cost effective active resonant damping technique.

where ω_c and ζ are cut-off frequency and damping ratio of the $H(s)$ low pass filtering action. Using simplified model of Fig. 5(c) and considering the proposed active damping branch for this model (as per Fig.6), equivalent circuit can be sketched as shown in Fig.7. Using this circuit, transfer functions of $G_i(s) = \frac{I_g(s)}{I_i^*(s)}$ and $G_y(s) = \frac{I_g(s)}{V_g(s)}$ can be easily obtained, which are then used to design the parameter k in (4).

Bode diagrams of $G_i(s)$ and $G_y(s)$, for $\omega_c = 2\pi \times 5000$rad/s $\zeta = 0.707$ are shown in Fig. 8. As it can be observed in this figure, when the active damping branch is disabled in the controller ($k = 0$ graphs), both $G_i(s)$ and $G_y(s)$ have resonant peaks. By activating the proposed active damping technique, it is shown in Fig. 8 that different k values change $G_i(s)$ and $G_y(s)$ characteristics. As it can be observed, $k = 10C$ does

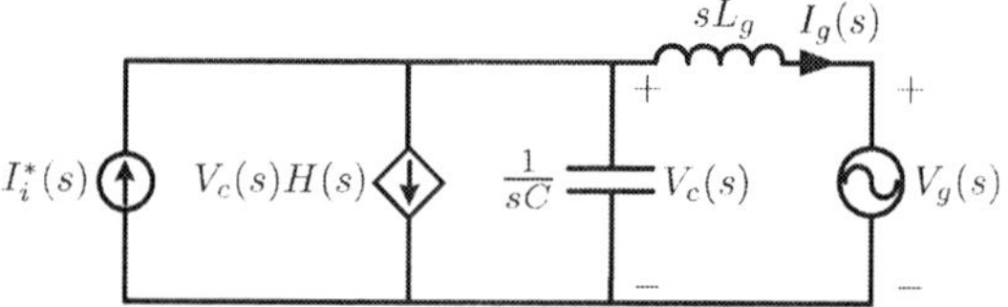

Fig. 7. Equivalent circuit combining proposed active damping technique with simplified model of power circuit.

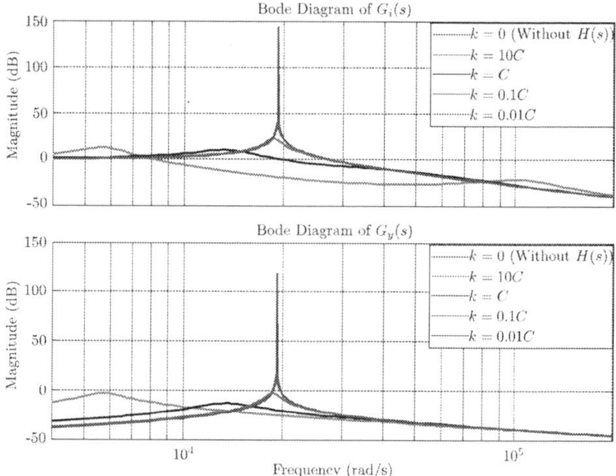

Fig. 8. Bode diagram of $G_i(s)$ transfer function for different k values ($\omega_c = 10^4\pi$ rad/s and $\zeta = 0.707$).

not effectively damp the output resonance, while $k = 0.01C$ totally affects the resonant peak. It can be observed in this figure that $k = C$ gives an effective damping for resonance of the system and hence it is selected as the $H(s)$ gain.

III. Performance Analysis

Performance of the proposed GPCC is evaluated by means of simulations and experimental implementation. System parameters are given in TABLE I.

The LCL filter is simulated with the proposed GPCC, while the damping strategy is disabled. Results are shown in Fig. 9(a) where the resonant current can be observed on the grid current. In another test, the proposed damping technique is added to the system and results are shown in Fig. 9(b). As it can be observed, current bands of i_i are adaptively modified. The bottom plot in Fig. 9(b) shows the effectiveness of proposed method, as the resonant on the grid current (i_g) is actively damped.

Dynamic performance of the GPCC for LCL filter case with the active damping technique is also evaluated and results are presented in Fig. 10. Peak current reference value in this test jumps from 1A to 3A at $t = 0.1$s. As it can be observed, the proposed controller presents a very fast transient response.

The GPCC is also experimentally validated and results are shown in Fig. 11(a). In this test, inverter is injecting 85W to the grid. Output power is set to this low value to better show

TABLE I
SYSTEM PARAMETERS

Dc link voltage	V_{dc}	200	V
Grid RMS voltage	V_g	120	V
Grid angular frequency	ω	$2\pi \times 60$	rad/s
Switching period	T_s	10^{-4}	s
L filter case	L	2	mH
LCL filter case	L_i	2	mH
	L_g	0.5	mH
	C	2	μF
$H(s)$ parameters	ω_c	$2\pi \times 5000$	rad/s
	k	2×10^{-6}	
	ζ	0.707	

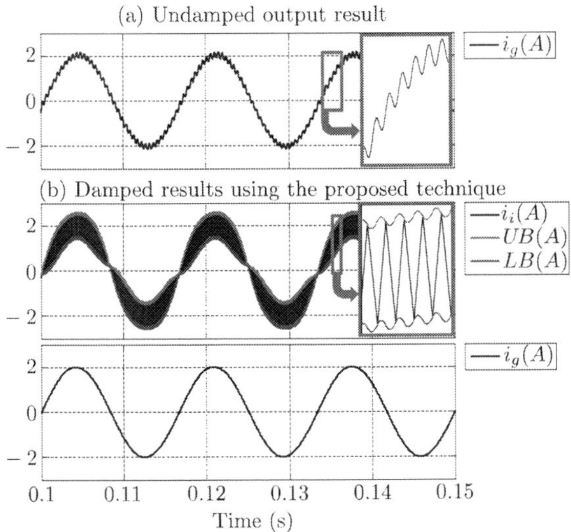

Fig. 9. (a) Grid current for the LCL filter case when the active damping branch is disabled, (b) Current of inverter and grid with the proposed active damping.

the effect of double frequency bands on the output current. A zoomed snapshot of this figure is shown in Fig. 11(b), where the 10kHz switching frequency as a result of the proposed GPCC can be observed. Output voltage of inverter along with its FFT spectrum is shown in Fig. 12. In this test, the system is facing 20% mismatch for the filter inductance. As it can be seen, inverters' output still presents a clean harmonic spectrum similar to that of a PWM controlled inverter, with a slight deviation from the designed value of f_s.

IV. CONCLUSION

A fixed switching frequency Generalized Peak Current Control (GPCC) method for inverters is proposed . While controlling the peak value of the inverters' current, the proposed approach can mimic any known PWM strategy. As a result, the GPCC features all the advantages of peak current controllers, along with a fixed switching frequency and the clean output harmonic spectrum inheriting from the original

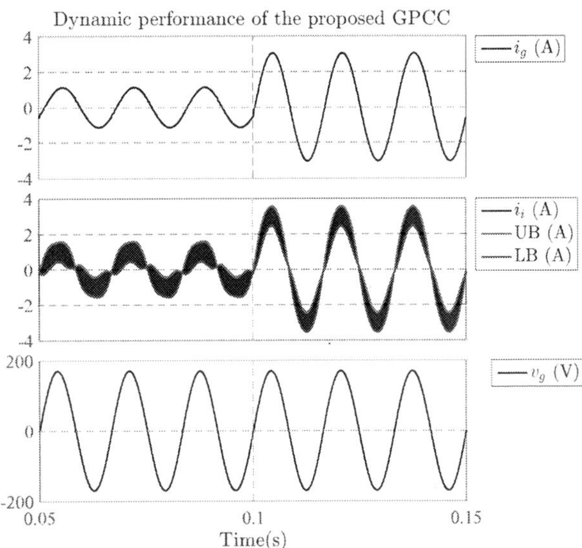

Fig. 10. Dynamic performance evaluation of proposed GPCC. Peak current reference jumps from 1A to 3A at $t = 0.1$s.

PWM scheme. It is shown that the proposed technique is able to obtain additional control objectives by its adaptive bands. As an example, the GPCC is applied to a unipolar PWM scheme and the controller is elaborated for both L- and LCL-type output filters. Demonstrating the advantages of resulting controller, a simple active damping strategy based on adaptive bands of the controller is proposed. Simulations and experimental results are presented to validate the method.

REFERENCES

[1] R. Gupta, "Generalized frequency domain formulation of the switching frequency for hysteresis current controlled vsi used for load compensation," *Power Electronics, IEEE Transactions on*, vol. 27, no. 5, pp. 2526–2535, May 2012.

[2] F. Blaabjerg, R. Teodorescu, M. Liserre, and A. Timbus, "Overview of control and grid synchronization for distributed power generation systems," *Industrial Electronics, IEEE Transactions on*, vol. 53, no. 5, pp. 1398–1409, Oct 2006.

[3] L. Malesani and P. Tenti, "A novel hysteresis control method for current-controlled voltage-source pwm inverters with constant modulation frequency," *Industry Applications, IEEE Transactions on*, vol. 26, no. 1, pp. 88–92, 1990.

[4] L. Malesani, L. Rossetto, and A. Zuccato, "Digital adaptive hysteresis current control with clocked commutations and wide operating range," *Industry Applications, IEEE Transactions on*, vol. 32, no. 2, pp. 316–325, 1996.

[5] B. Bose, "An adaptive hysteresis-band current control technique of a voltage-fed pwm inverter for machine drive system," *Industrial Electronics, IEEE Transactions on*, vol. 37, no. 5, pp. 402–408, Oct 1990.

[6] M. Kale and E. Ozdemir, "An adaptive hysteresis band current controller for shunt active power filter," *Electric Power Systems Research*, vol. 73, no. 2, pp. 113 – 119, 2005.

[7] C.-S. Lam, M.-C. Wong, and Y.-D. Han, "Hysteresis current control of hybrid active power filters," *Power Electronics, IET*, vol. 5, no. 7, pp. 1175–1187, August 2012.

[8] M. Elsaharty, M. Hamad, and H. Ashour, "Digital hysteresis current control for grid-connected converters with lcl filter," in *IECON 2011 - 37th Annual Conference on IEEE Industrial Electronics Society*, Nov 2011, pp. 4685–4690.

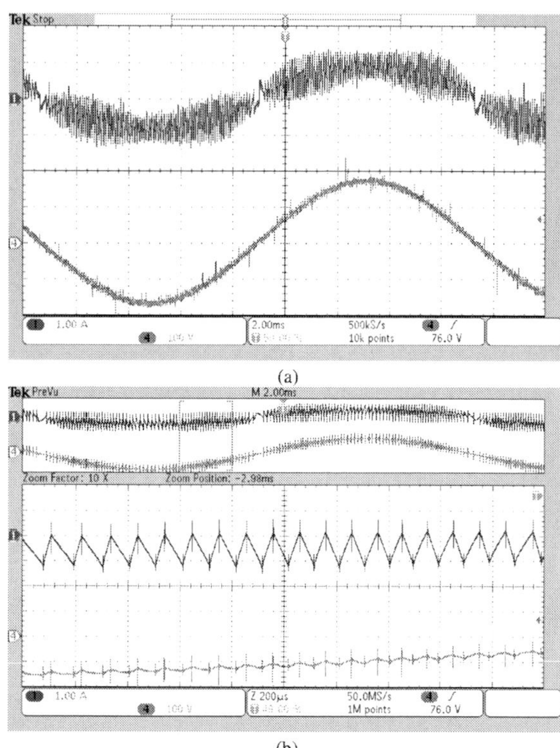

(a)

(b)

Fig. 11. Experimental results. Ch.1: Inverter current (1A/div) and Ch.4: Grid voltage (100 V/div). (a) Inverter is injecting 85W to the grid (2ms/div). (b) Zoomed snapshot of current and voltage waveforms in this test (200μs/div).

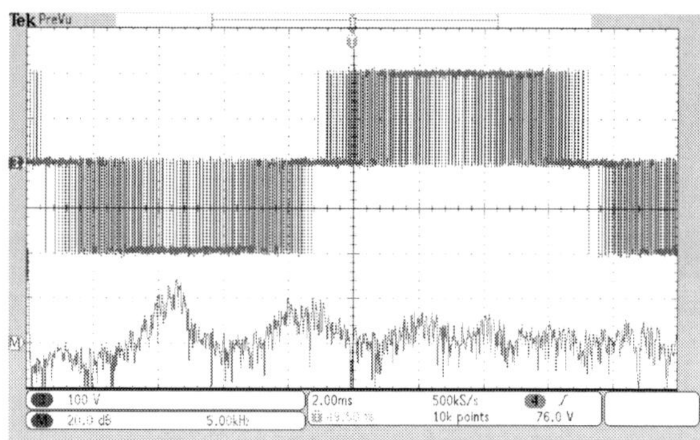

Fig. 12. Experimental results for inductor mismatch test. Ch.3: Inverters' output voltage (100 V/div, 2ms/div) and Ch.M: its FFT spectrum (20dB/div, 5kHz/div).

[9] T. Kerekes, R. Teodorescu, and U. Borup, "Transformerless photovoltaic inverters connected to the grid," in *Applied Power Electronics Conference, APEC 2007 - Twenty Second Annual IEEE*, Feb 2007, pp. 1733–1737.

[10] M. Azizur Rahman, T. Radwan, A. Osheiba, and A. Lashine, "Analysis of current controllers for voltage-source inverter," *Industrial Electronics, IEEE Transactions on*, vol. 44, no. 4, pp. 477–485, Aug 1997.

Improved Control Strategy of 1 MHz LLC Converter for High Frequency Resolution

Hwa-Pyeong Park and Jee-Hoon Jung
Ulsan National Institute of Science and Technology (UNIST)
School of Electrical and Computer Engineering, Ulsan, Korea
Email : darkrla6@unist.ac.kr, jhjung@unist.ac.kr

Abstract—**High switching frequency is one of the effective methods to improve power density for a LLC resonant converter. However, conventional controllers such as a digital signal processor (DSP) and an analog controller have a performance of the limited frequency resolution which introduces high primary and secondary side current variations and poor output voltage regulation at high switching frequency. In this paper, a hybrid control method combining the pulse frequency modulation (PFM) and the pulse width modulation (PWM) is proposed to overcome the limited frequency resolution performance. The proposed hybrid control method is analyzed with its control flow chart and operational principles. The improved voltage regulation performance is compared with the conventional PFM control and verified by simulation and experimental results using a 240 W prototype converter operating at 1 MHz switching frequency.**

Keywords—Pulse width modulation, Pulse frequency modulation, Hybrid control, LLC resonant converter, High frequency operation

I. Introduction

The high switching frequency converter has been researched according to the requirements of industrial fields to achieve high functionality with high power density for various applications. The LLC resonant converter is widely used to implement the high switching frequency operation, because it has high power conversion efficiency and simple structure with low cost. Since the switching performance of state-of-art power switch is improved, the high switching frequency operation is accelerated to become one of industrial trends to achieve high power density [1], [2]. However, in aspect of controller for the power converter, general-purposed digital micro-controller (TI TMS320F28335) which provide a 150 MIPS performance shows not enough frequency resolution for the pulse frequency modulation (PFM) in several mega-hertz switching operations. The limited frequency resolution induces the high variation of the input-output voltage gain which makes the high output voltage ripple in the LLC resonant converter. Moreover, it makes high fluctuation of the primary and secondary currents, which induces the high conduction loss and high current stress of passive components.

One of conventional solutions is high performance field-programmable gate array (FPGA) which has around 8 times higher PWM resolution compared with DSP. However, high cost is required for this method, since the high performance FPGA chip is expensive compared with the conventional DSP. Moreover, it makes hard to obtain cost-effectiveness in the low power-rated SMPS market. Therefore, a PWM and PFM hybrid control method is proposed for the limited performance of the conventional DSP to implement the several mega-hertz switching frequency in the LLC resonant converter.

The PWM and PFM hybrid control strategy has already applied to improve power conversion efficiency at the light load condition [3], [4]. Even though this control method is invented for the steady state operation, the control algorithms only focuses on the light load condition. In another previous research, the PWM is also adopted for the soft start duration to suppress the inrush current with small input-output voltage gain [5]. This soft start method is widely used in the industrial field using analog controller (Champion CM6901). However, this control strategy is only working at the transient operation of the soft start. All the previously mentioned control strategies are proposed to improve not the frequency resolution in the PFM but power conversion efficiency at the light load condition and small inrush current for the soft start operation. Therefore, the conventional PWM and PFM control methods cannot overcome the limited frequency resolution at high switching frequency operation of the LLC resonant converter.

In this paper, the PWM and PFM hybrid control method is proposed to overcome the control performance degradation caused by the limited frequency resolution in the LLC resonant converter. The proposed control algorithm concentrates on the steady state operation for entire load range. The hybrid steady state control method is proposed to improve the control performance for precisely regulating the output voltage and small primary and secondary side peak current at the high switching frequency operation. The proposed control algorithm will be analyzed using control flow charts and operational principles of the proposed converter. The performance of the output voltage regulation using the hybrid control method will be verified by simulation and experimental results compared with the results of the conventional PFM control method. Moreover, operational waveforms and power conversion efficiency will also be presented to show high performance and stable operation in the proposed converter. All analytical details will be verified with a 240 W prototype converter operating at 1 MHz switching frequency.

978-1-4673-9551-9/16 $31.00 © 2016 IEEE

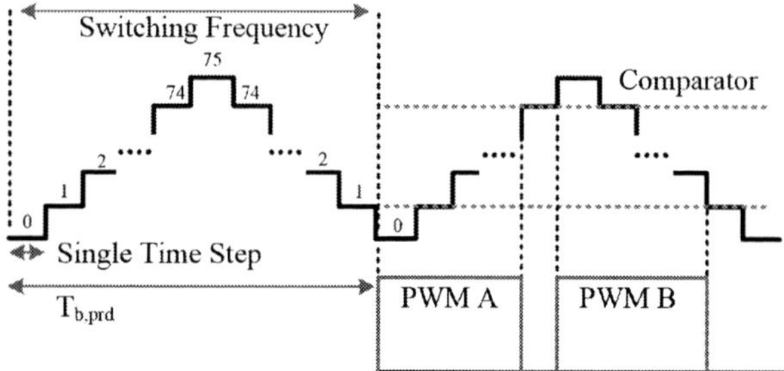

Fig. 1 PWM and PFM generations of the general-purposed DSP

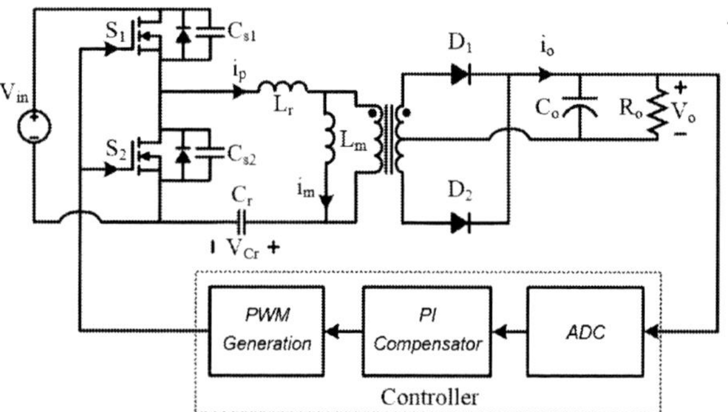

Fig. 2 Schematic of LLC resonant converter and its controller

II. Frequency Resolution Limitation of DSP under High Switching Frequency Operation

The target DSP has 150 MHz clock speed, which sets the time-based period, $T_{b,prd}$. It consists of time steps to generate triangular waveforms for implementing the PFM gate signals. Fig. 1 shows that the total number of the time steps indicates the switching frequency of the PFM signals. The $T_{b,prd}$ of the target DSP can be calculated as follows [6]:

$$T_{b,prd} = \frac{T_s}{2 \cdot T_{tb,clk}} \tag{1}$$

where $T_{b,prd}$ is the total time steps of the DSP, T_s is the switching period of the power converter, and $T_{tb,clk}$ is the system clock period (1/150 MHz) of the DSP. From (1), the calculated $T_{b,prd}$ can represents the variation of the switching frequency according to a single bit change in the DSP control variables. The switching frequency variation by single time step change, Δf_s, can be calculated as follows:

$$\Delta f_s = \frac{1}{2 \cdot T_{tb,clk}} \cdot \left(\frac{1}{T_{b,prd}} - \frac{1}{T_{b,prd}'} \right) \tag{2}$$

where $T_{b,prd}'$ is the modulated $T_{b,prd}$ by the feedback control. The $T_{b,prd}$ of the DSP is inversely proportional to the switching frequency. In addition, it is assumed that the $T_{b,prd}$ is varying single time step by the feedback control. From (1) and (2), the switching frequency variation increases in a square times as increase of switching frequency. At the 1 MHz switching frequency, $T_{b,prd}$ is 75 and its frequency variation step is 13.158 kHz for a single time step change of the DSP. At 100 kHz switching frequency, $T_{b,prd}$ is 750, and its frequency variation step is 133.16 Hz. Therefore, the PFM frequency resolution of the DSP at the 1 MHz switching frequency is 100 times lower than the frequency resolution at the 100 kHz switching frequency.

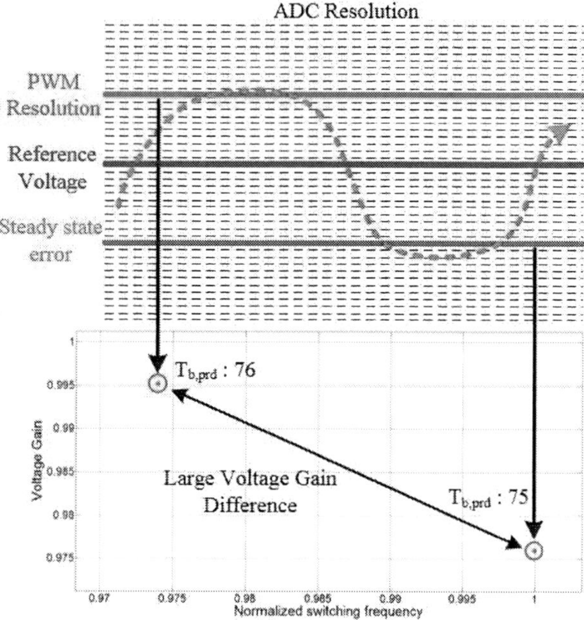

Fig. 3 Single time step change of the voltage gain caused by the limited PWM resolution

In aspect of feedback control loop, the controller is configured with Analog-to-digital converter (ADC), steady state error compensator, and PWM generator. The LLC resonant converter including the controller is described as shown in Fig. 2. The regulated output voltage precision is determined by the ADC resolution and PWM resolution. The ADC resolution can be expressed as follows [7]:

$$\Delta V_{adc} = \frac{V_{out}}{2^{N_{adc}}} \tag{3}$$

where N_{adc} is the bit of ADC, and V_{out} is the output voltage. It shows a quantized output voltage variation in DSP as one bit change of ADC. The PWM resolution can be also expressed as follows:

$$\Delta V_{pwm} = N \cdot V_{in} \cdot Gain(T_{b,prd}) \tag{4}$$

where N is the primary to secondary turn ratio, V_{in} is the input voltage, and $Gain(T_{b,prd})$ is the input-output voltage gain variation according to the $T_{b,prd}$. It shows the quantized output voltage variation in DSP as one $T_{b,prd}$ change of PWM signals.

The calculated ΔV_{adc} and ΔV_{pwm} are around 4.88 mV and 0.5 V at the resonant frequency, respectively. The PWM regulation is 100 times lower resolution compared with ADC. It causes the high steady state error to regulate the output voltage. Therefore, the power converter has frequent switching frequency variations compared with conventional switching frequency condition, even though it is working at full load condition. The poor output voltage regulation is shown in Fig. 3, which shows a single time step change of the voltage gain caused by the limited PFM resolution. Since the low frequency resolution induces high input-output voltage gain variation, the frequent switching frequency variation makes high output voltage ripple and high primary and secondary side currents ripple.

The conventional PFM method regulates the output voltage only using frequency changes according to feedback control. However, the PFM control method is not effective to regulate output voltage as increase of switching frequency caused by high switching frequency variation. The proposed PWM and PFM hybrid method uses both PFM and PWM methods with an intelligent control algorithm to regulate output voltage. With the combined modulation method, the proposed control method can obtain the precisely regulated output voltage under poor PFM frequency resolution.

III. PWM AND PFM HYBRID CONTROL ALGORITHM

To overcome the frequency resolution issue in the general-purposed DSP, the proposed hybrid control method contains not only the PFM control method but also PWM control method with the intelligent selection algorithm. It compensates the limited frequency resolution using the change in the duty cycle. Under the PFM control, the voltage gain of the LLC resonant converter is only affected by the variation of the switching frequency, as follows [8]:

$$\|H_r(f_n)\| = \left| \frac{L_n f_n^2}{(L_n+1)f_n^2 - 1 + j[(f_n^2-1)f_n Q_s L_n]} \right| \quad (5)$$

where $H_r(f_n)$ is the voltage gain according to f_n, L_n is the inductance ratio between the magnetizing inductance and the leakage inductance, f_n is the normalized switching frequency, and Q_s is the quality factor of the power converter. From (5), significant gain oscillation can be expected due to the poor PFM frequency resolution of the general-purposed DSP as shown in Fig. 3.

Using the proposed PWM and PFM hybrid control algorithm, the input-output voltage gain can derived with theoretical method as follows [9]:

$$\frac{V_o}{V_{in}} = \|H_r(f_n)\| \sin \pi \left(\frac{t_d}{T_s} \right) \quad (6)$$

where t_d is the time duration of the duty cycle and T_s is the switching period of PWM. Fig. 4 (a) shows the voltage gain curve of the LLC resonant converter with respect to the variation of the duty cycle at the resonant frequency. In (6), the output voltage can be regulated using both the changes in f_n and t_d, which means that there are two degrees of freedom in the output voltage control. Fig. 4 (b) shows the 3D gain curve of the proposed PWM and PFM hybrid control algorithm according to the switching frequency and the duty cycle. Using two control variables of the pulse width and the switching frequency at the same time, the voltage gain can be precisely controlled in the LLC resonant converter even with the poor PFM frequency resolution.

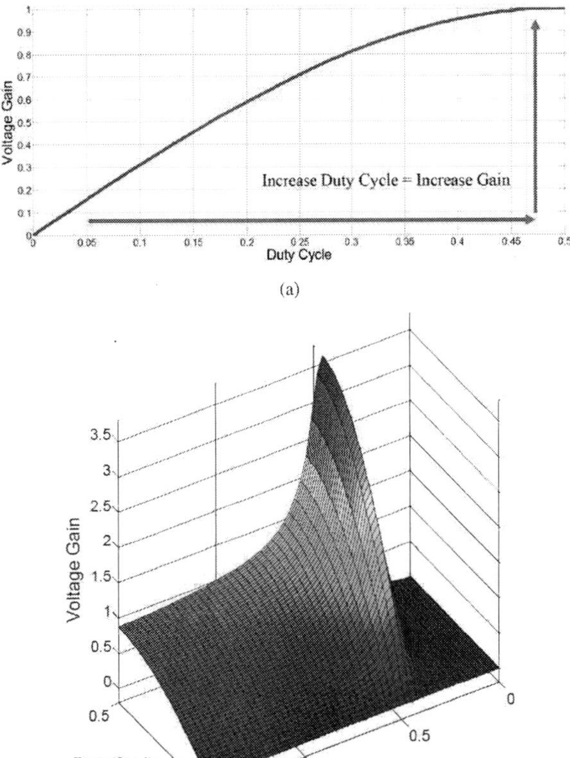

Fig. 4 Voltage gain of the LLC resonant converter: (a) Gain curve according to the duty cycle at the resonant frequency, (b) 3D gain curve according to the normalized switching frequency and the duty cycle.

A. Design Considerations of Hybrid Control

Using (6), the output voltage can be precisely regulated with pulse width control compared with conventional PFM control method. However, this hybrid control method induces the design complexity with duty ratio variation. This additional variable should be reflected to the magnetizing inductance design for zero voltage switching (ZVS) condition. The conventional magnetizing inductance design method can be derived as follows [10]:

$$L_m \leq \frac{t_{dt}(T_s - t_{dt})}{16 \cdot C_{eq}} \quad (7)$$

where t_{dt} is the dead time duration, C_{eq} is the equivalent capacitance of MOSFET, and L_m is the magnetizing inductance of transformer.

Using (7), the proper value of magnetizing inductance can be obtained for ZVS condition according to the duty ratio variation, and switching frequency variation. The calculated magnetizing inductance is shown as Fig. 5. It shows that the power converter operating at the highest switching frequency and the largest duty ratio condition has the smallest magnetizing inductance value for ZVS condition. Therefore,

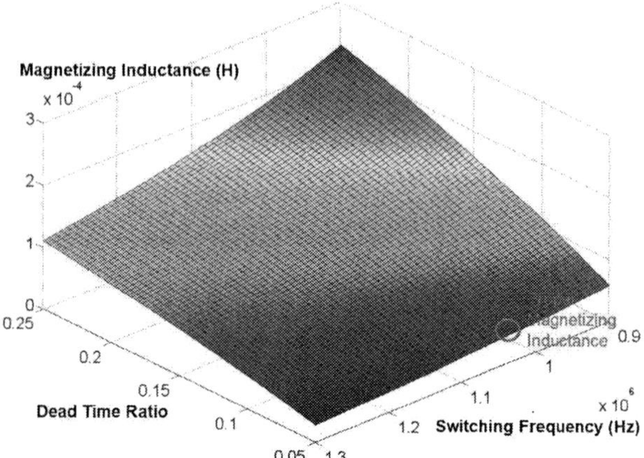

Fig. 5 Calculated magnetizing inductance satisfied with the ZVS condition according to the switching frequency and the duty ratio

the proper magnetizing inductance design criteria should consider the light load condition with the largest duty ratio.

In a practical design manner, the minimum dead time is 10 percent of the switching period to protect arm short between two MOSFETs at the same leg. The magnetizing inductance is designed at the worst condition of 15 percent duty ratio and 1.05 MHz switching frequency to achieve ZVS condition. To obtain enough control performance, the improved voltage control resolution can be derived using the designed duty ratio and the PFM gain as follows:

$$\Delta V_{pwm} = N \cdot V_{in} \cdot Gain(T_{b,prd}) \cdot \sin\pi\left(\frac{t_d}{T_s}\right) \qquad (7)$$

From (7), the output voltage can be precisely regulated with improved PWM resolution which is determined with variation range of duty ratio. Even though the PWM resolution is improved, the ADC resolution is still higher than the PWM resolution. However, the variation of output voltage is smaller than the conventional PFM control method.

B. Operational Principles of Hybrid Control

In Section II, The performance degradation of the output voltage regulation using the general purpose DSP was discussed. In addition, the poor PFM resolution induces high current stress on the primary and secondary sides. The proposed PWM and PFM hybrid control method has two control variables: the pulse width and the pulse frequency modulation to overcome the limited frequency resolution. To implement the PWM in the PFM control method, the priority between two control variables should be determined for the converter's stable operation.

The priority is determined with the control borderline which separates control variables. Fig. 6 shows the operational principle of the proposed hybrid control algorithm. There is single line as a steady state error reference. In the steady state operation, the accumulated steady state error should be converged to the reference line with desired output voltage. Moreover, there is a one set of borderlines to separate the control variables. The outer side of borderlines indicates the

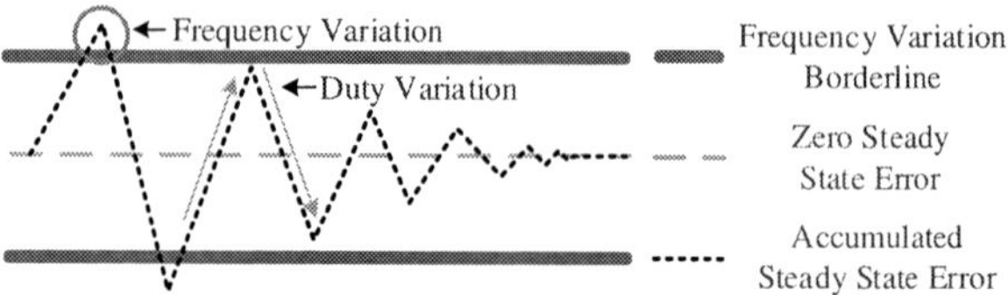

Fig. 6 Operational principle of the proposed hybrid control algorithm

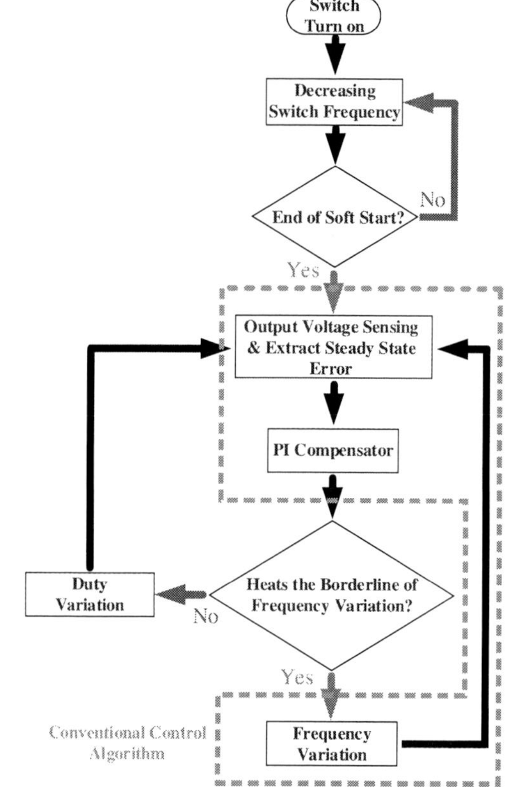

Fig. 7 Flow chart of the proposed PWM and PFM hybrid control algorithm

frequency control region. The inner side of borderline indicates the duty control region. It shows that the priority of duty control is higher than the frequency control method.

Fig. 7 shows the flow chart of the conventional PFM control algorithm, and proposed hybrid control algorithm. The conventional method has only a switching frequency control loop. The steady state error of the output voltage is compensated by a PI controller in the control loop. Using this controller, the compensated steady state error changes the switching frequency to regulate the output voltage. However, the proposed hybrid control method shows that the pulse width and the pulse frequency are selected as control variables by the intelligent selection algorithm. This selection algorithm uses the accumulated steady state error as a criterion to decide the proper control variables for the output voltage regulation.

There are two potential operating conditions to determine control variables. First, if the accumulated steady state error is zero or located in between the duty cycle variation borderlines, the duty cycle changes to regulate the output voltage. Second, if the accumulated steady state error is over the switching frequency variation borderline, the switching frequency changes a single step of $T_{b,prd}$ to regulate the output voltage. In addition, the duty cycle is adjusted to the initial duty ratio. After a single step change of $T_{b,prd}$, the duty control starts to regulate output voltage from the initial duty ratio.

IV. Simulation and Experimental Results

Fig. 8 (a) shows the simulation waveforms of the conventional PFM control algorithm. Fig. 8 (b) show the simulation waveforms of the proposed hybrid control algorithm. In the simulation waveforms, the variation of PFM ($T_{b,prd}$) shows the switching frequency change in the LLC resonant converter. In Fig. 8 (a), there are periodic output voltage oscillation induced by the limited PFM frequency resolution of the conventional control algorithm because it uses only the PFM method with the variation of $T_{b,prd}$. As a result, the primary and secondary side current also shows oscillation. In Fig. 8 (b), however, the output voltage changes is smaller because it is well regulated by the duty cycle control rather than the switching frequency. In addition, there is smaller current oscillation in the primary and secondary sides compared with PFM control.

Fig. 9 shows the operational waveforms using the PFM control method at the full load conditions. It has high output voltage ripple and high primary and secondary side currents according the switching frequency variation. Fig. 10 shows the operational waveforms using proposed PWM and PFM hybrid control method at the full load condition. In this experiment, the duty range of the proposed hybrid control is limited between 60% and 88% of the entire duty cycle because this range allows the controller to compensate the gain variation that occurs when there is a single step change in the switching frequency. Compared with Fig. 10, the proposed hybrid control algorithm shows smaller primary and secondary current ripple (around 91% and 90%, respectively), and smaller output voltage ripple (around 68%). All the simulation and experimental results are summarized in Table I, where $I_{p,p}$ is the primary side peak current, $I_{s,p}$ is the secondary side peak current, and $V_{o,r}$ is the output voltage ripple.

Fig. 11 shows the power conversion efficiency according to load variations at the 1 MHz switching frequency. The proposed PWM and PFM hybrid control method does not affect the power conversion efficiency compared to the conventional PFM control method since it improves only the control performance of the output voltage without changing the operating point of the converter.

V. Conclusion

A PWM and PFM hybrid control method is proposed to obtain sufficient gain reduction during soft start and to overcome the limited frequency resolution of the general-purposed DSP in the LLC resonant converter application. The proposed hybrid control method is focused on the steady state control algorithm to regulate the output voltage. The proposed hybrid algorithm shows not only enhanced output voltage regulation but also lower current spikes in the primary and

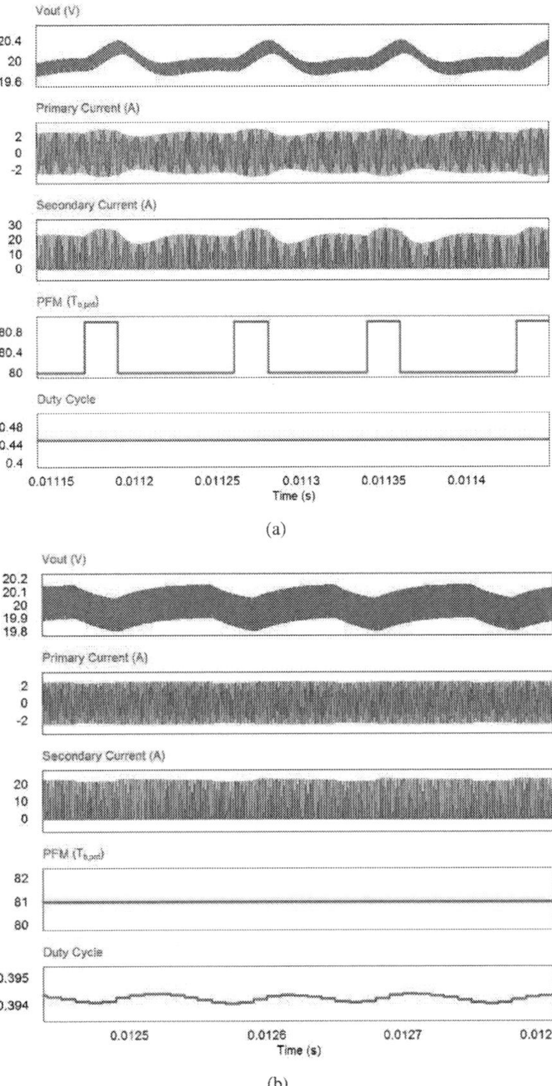

Fig. 8. Simulation waveforms of output voltage regulation performance: (a) Conventional PFM algorithm, (b) Proposed PWM and PFM hybrid algorithm

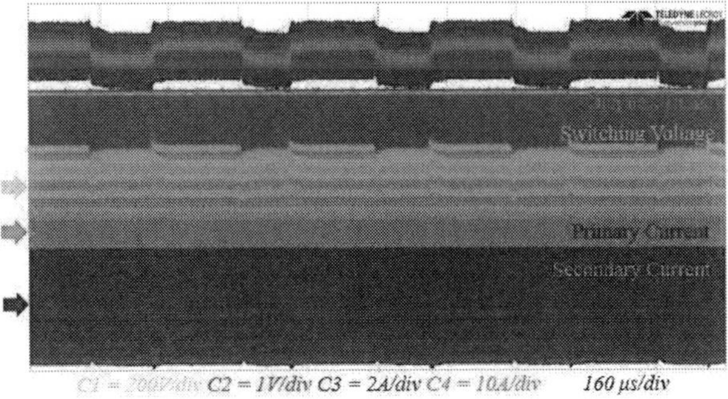

Fig. 9. Experimental waveforms of the output voltage regulation performance with the conventional PFM control

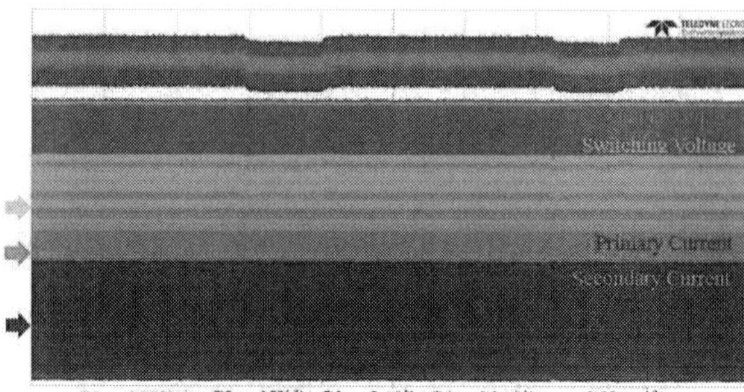

Fig. 10. Experimental waveforms of the output voltage regulation performance with the proposed hybrid control

TABLE I. SIMULATION AND EXPERIMENTAL PERFORMANCE

	Conventional Method		Hybrid Method	
	Simulation	*Experiment*	*Simulation*	*Experiment*
$I_{p,p}$	2.88 A	3.10 A	2.29 A	2.85 A
$I_{s,p}$	28.18 A	21.3 A	22.40 A	19.3 A
$V_{o,r}$	0.70 V_{pp}	1.67 V_{pp}	0.35 V_{pp}	1.14 V_{pp}

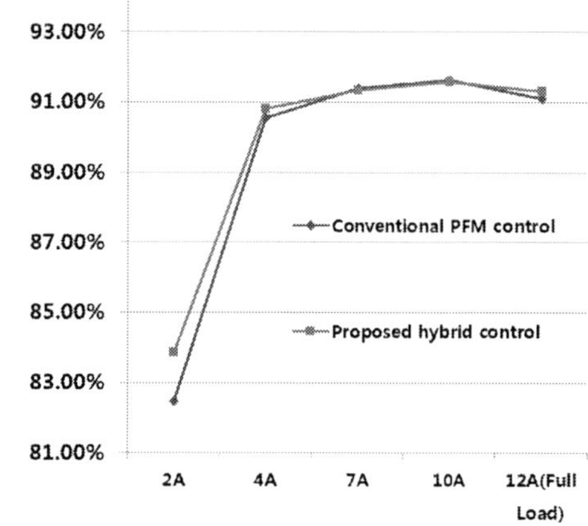

Fig. 11. Power conversion efficiency

secondary sides. Using the 240 W prototype LLC resonant converter operating at the 1 MHz switching frequency, the proposed control method implementation resulted in reduced peak current by 91 % and 90 % in the primary and secondary side, respectively, and reduced output voltage ripple by 68 %.

ACKNOWLEDGMENT

This research was supported by the Basic Science Research Program through the National Research Foundation of Korea (NRF) funded by the Ministry of Science, ICT & Future Planning (NRF-2013R1A1A1009632).

REFERENCES

[1] H. Xiucheng, L. Zhengyang, L. Qiang, and F.C. Lee, " Evaluation and application of 600 V GaN HEMT in cascode structure," *IEEE Trans. Power Electron.*, vol. 29, no. 5, pp. 2453-2461, May 2014

[2] M.D. Seeman, S.R. Bahl, D.I. Anderson, and G.A. Shah, "Advantage of GaN in a high-voltage resonant LLC converter," in *IEEE Applied Power Electronics Conference and Exposition (APEC)* 2014, Texas, USA, vol. 1, 2014, pp. 476-483.

[3] C. Dick, P. Deck, and A. Schmidt, " Optimized Buck – Mode Modulation Strategy and Control of a LLC – Type Resonant Converter in a Solar Application," in *IEEE Power Electronics and Applications (EPE)*, 2014, pp. 1-9.

[4] J. Yamamoto, T. Zaitsu, S. Abe, and T. Ninomiya, "PFM and PWM Hybrid contolled LLC converter," in *IEEE Power Electronics Conference (ECCE-Asia)*, 2014, pp. 177-182.

[5] H.-P Park and J.-H Jung, " Design considerations of 1 MHz LLC resonant converter with GaN E-HEMT," in *Power Electronics and Applications (EPE 2015)*, 2015, pp. 1-10.

[6] Texas Instruments, "TMS320F28335 /F28334 /F28332 /F28235 /F28234 /F28232 Digital Signal Controllers (Rev. M)," Texas Instruments, Tech. Rep, SPRS439M, August 2012.

[7] A.V. Peterchev and S.R. Sanders, " Quantization resolution and limit cycling in digitally controlled PWM converters," IEEE *Trans. Power Electronics*, vol. 18, no. 1, pp. 301-308, Jan 2003.

[8] J.-H. Jung, J.-M. Choi, and J.-G. Kwon, "Design Methodology for Transformers Including Integrated and Center-Tapped Structure for LLC Resonant Converters," *Journal of Power Electronics*, Vol. 9, no. 1, pp. 215-223, Mar. 2009.

[9] J.-H. Jung, H.-S. Kim, M.-H. Ryu, and J.-W. Baek, "Design Methodology of Bidirectional CLLC Resonant Converter for High-Frequency Isolation of DC Distribution Systems," *IEEE Trans. Power Electronics*, Vol. 28, no. 4, pp. 1741-1755, Apr. 2013.

[10] J.-H Jung and J.-G Kwon, "Theoretical analysis and optimal design of LLC resonant converter," in *Power Electronics and Applications (EPE 2007)*, pp.1-10.

Bumpless Control for Reduced THD in Power Factor Correction Circuits

Joel Steenis and Alex Dumais

Abstract—**It is well known that power factor correction (PFC) circuits suffer from two fundamentally different operating modes over a given AC input cycle. These two modes, continuous conduction mode (CCM) and discontinuous conduction mode (DCM), have very different frequency-response characteristics that can make control design for PFC circuits challenging. The problem is exacerbated by attempts to improve efficiency by dynamically adjusting the PWM switching frequency based on the load. Adjusting the PWM frequency based on the load limits controller bandwidth and restricts dynamic performance. Prior work has made use of multiple controllers, however, they have not addressed the discontinuity (bump) that exists when switching between controllers. In this paper, bumpless controllers will be synthesized for a 750 watt, semi-bridgeless PFC for the CCM-DCM operating modes.**

Index Terms—**Bumpless Control, Switched Control, CCM-DCM, Power Factor Correction, Efficiency, Distortion**

I. Introduction

POWER factor correction circuits should conserve energy and have a high power factor, as a matter of course. In the commercial marketplace, certifying a product with an energy-saving program differentiates that product and offers a competative advantage. Two examples of energy-savings certifications are the 80 PLUS program that requires 90% efficiency at a 0.2 per unit (p.u.) load, and the ENERGY STAR version 5.0 standard that requires a power factor of at least 0.65 at a 0.1 p.u. load. Researchers have devised control strategies that may be used to meet these requirements, and typically use multiple controllers [1], [3], [4], [5]. The salient features of these approaches is to partition the PFC operation into discrete regions based on load, CCM-DCM operation, or other criteria; synthesize a controller for each region; and switch between the controllers based on the operating region. Switching between controllers will result in a discontinuous signal that can degrade performance, as illustrated in Figure 1 and Figure 2. This discontinuity may be addressed using bumpless controllers [6].

In the remainder of this paper, bumpless controllers will be introduced, the DCM-CCM models for the inner current loop will be used to illustrate the significant differences between the plant in the two operating modes, two bumpless controllers will be synthesized for the two operating modes, the bumpless controllers will be tested in simulation, and the bumpless controllers will be implemented in hardware using a 750 watt semi-bridgeless PFC.

Joel Steenis and Alex Dumais are engineers at Microchip Technology, 2355 West Chandler Blvd., Chandler, AZ, 85224 USA e-mail: Joel.Steenis@Microchip.com, Alex.Dumais@Microchip.com

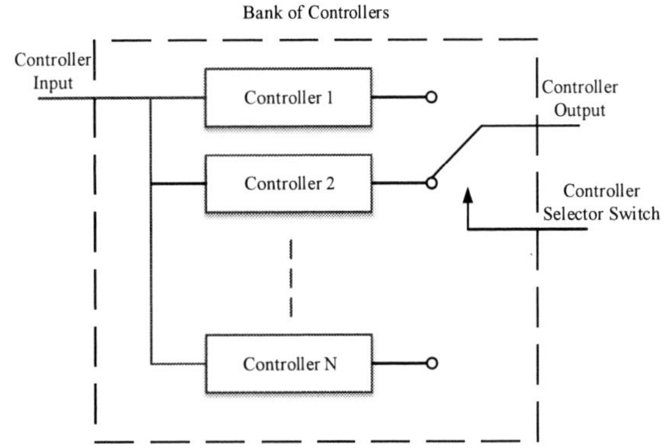

Fig. 1. Bank of selectable controllers. It is assumed that each controller is optimized for a given operating range and selected when operating in that range. The output may have large discontinuities, as shown in Figure 2.

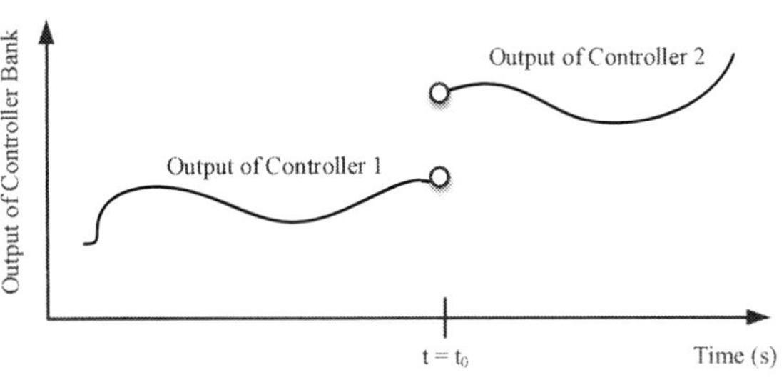

Fig. 2. Output of switched controller bank, as shown in Figure 1. This figure illustrates the hypothetical discontinuity when switching between two controllers at $t = t_0$.

II. Background

A. Semi-Bridgeless PFC Modeling

The bumpless controllers will be designed for a semi-bridgeless topology. The semi-bridgeless PFC topology was chosen for its high efficiency and is shown in Figure 3.

The PFC has two loops, a current loop and a voltage loop, as shown in Figure 4.

The plant for the current loop may be derived using state space averaging, or Vorperian's model, for the PWM switch [7], [8], [9]. The response may be expressed as (1) for DCM [7] and (2) for CCM [10].

978-1-4673-9551-9/16 $31.00 © 2016 IEEE

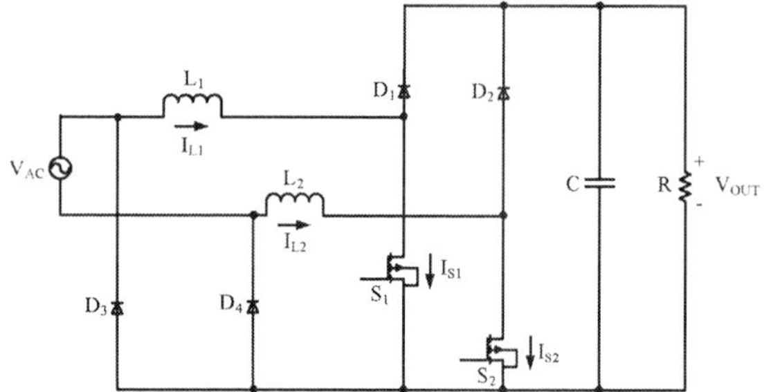

Fig. 3. Schematic of a semi-bridgless PFC. The bumpless controllers will be used with this topology.

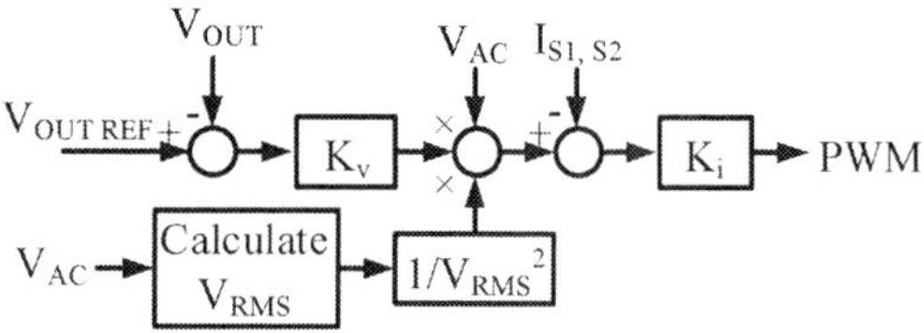

Fig. 4. Control architecture for the semi-bridgeless PFC. The variable "$V_{OUT\ REF}$" is the desired output voltage. This is also referred to as the setpoint for the system.

$$G_{i\ DCM}(s) = \frac{\hat{i}_L(s)}{\hat{d}(s)} = \frac{\frac{4MV_{IN}}{RLC}\left(s\frac{RC}{2}+1\right)}{\left(s+\frac{2M-1}{(M-1)RC}\right)\left(s+\frac{2(M-1)}{D1T_s}\right)} \quad (1)$$

$$G_{i\ CCM}(s) = \frac{\hat{i}_L(s)}{\hat{d}(s)} = \frac{2V_{OUT}}{RD'^2}\frac{1+\frac{sRC}{2}}{1+s\frac{L}{RD'^2}+s^2\frac{LC}{D'^2}} \quad (2)$$

Where: M is the voltage conversion ratio, L is the inductance of either boost converter, T_s is the switching period, D1 is the duty cycle of the converter in DCM ($D1 = T_{ON}/T_s$), and D' is the compliment of the duty cycle ($D' = 1 - D$).

The DCM-CCM boundary occurs when the inductor current I_{lb} satisfies the equality $I_{lb} = \frac{T_s V_{OUT}}{2L} D(1-D)$. Currents less than this will clearly be in the DCM region and the plant model (1) must be used. It is also important to note that the converter will repetatively alternate between the CCM and DCM modes since the input voltage is sinusoidal. A qualitative graph is shown in Figure 5.

The plant for the voltage loop may be derived based on a constant power assumption [10] and is expressed as (3). The voltage control loop is used to control the DC output, and typically has a bandwidth in the 10 Hz to 20 Hz range. It can be seen from the plant model (3) that a simple controller may be used. While a proportional controller may provide satisfactory performance, a second-order controller may be used to filter the second harmonic of the AC line, thereby reducing THD. Because of the simple nature of the voltage loop, bumpless controllers will only be used in the current loop.

$$\frac{\hat{v}_{out}}{\hat{v}_c}(s) = \frac{1}{sCV_{OUT}} \quad (3)$$

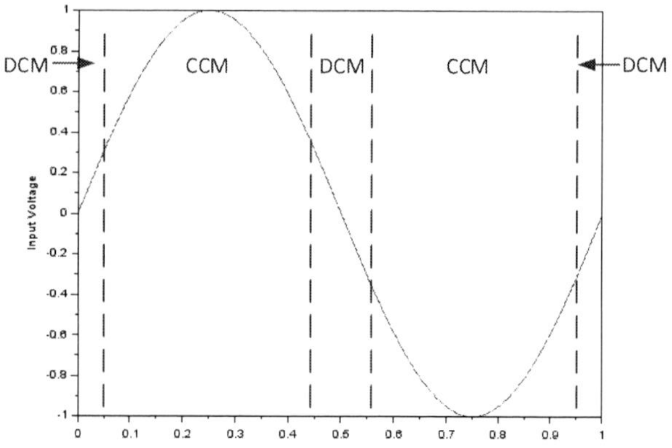

Fig. 5. Qualitative figure illustrating regions where a semi-bridgeless PFC operates in CCM and DCM. The graph ordinate and abscissa refer to the magnitude of the PFC input voltage and appropriate time scale, respectively.

B. Bumpless Control

As illustrated in Figure 6, the fundamental concept of bumpless control is to add an additional feedback path to a controller so that when the switch "S" is closed there is no influence on the controller "i". When the switch is open and another controller is selected, the output of the active controller acts as a setpoint for the inactive controller (or controllers) to track.

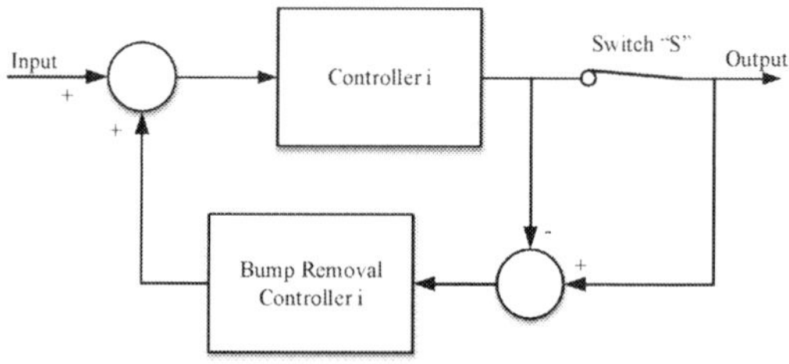

Fig. 6. This figure illustrates one bumpless controller (controller "i") in a bank of switched controllers. Inclusion of the additional feedback path forces the output of controller "i" to follow the output of the active controller when switch "S" is open, and to have no effect when the switch is closed.

It is important to note that bumpless control refers to a unique configuration of multiple controllers that allows the smooth transition between those controllers. The individual controllers in this configuration may be synthesized by whatever means the designer is familiar.

Consider "Controller i" in Figure 7. From this figure it becomes apparent that when synthesizing the "Bump Removal Controller" the "Controller i" is treated as the plant, the input to the active controller is a disturbance, and that the output of the active controller serves as a setpoint for the inactive controllers. The "Bump Removal Controller i" may then be synthesized with characteristics appropriate to track the output of the active controller [2]. The characteristics of the controller output signal vary from system-to-system, however, it is generally advisable for the closed loop poles

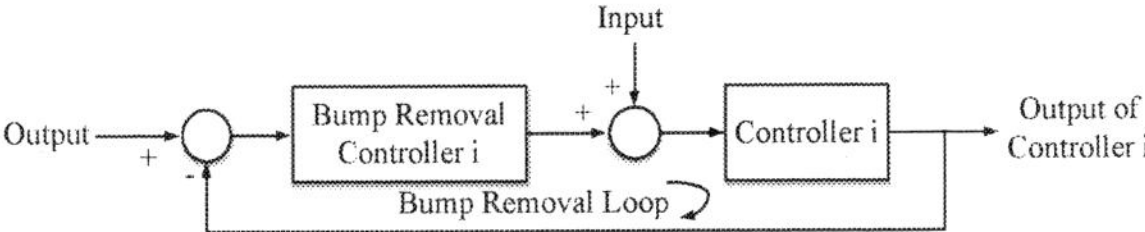

Fig. 7. This figure illustrates one bumpless controller in a bank of switched controllers, when controller "i" is inactive. This figure is the same as Figure 6, but does not have the selector switch "S", and is redrawn to accentuate the "bump removal loop". The figure also accentuates the fact that the output of the active controller, labeled "Output", serves as a reference for the inactive controller. The "Bump Removal Controller i" must be designed to track the signal "Output".

to be approximately five times higher in frequency than the open-loop poles.

III. Practical Example

Parameters for the semi-bridgless PFC are given in Table I. The DCM-CCM boundary occurs for instantaneous voltages of approximately 35 V. Bode magnitude plots corresponding to the plant in CCM and DCM are shown in Figure 8. From this figure, it can be seen that the system has very different characteristics under each operating mode. The controllers may be synthesized using any methodology, as stated in the previous section. The controllers for this design were synthesized for each operating mode using pole zero placement, and are given in (4) and (5).

TABLE I
Parameter values for the semi-bridgeless PFC.

Parameter	Value
L	460 μH
C_{OUT}	1.06 mF
f_s	100 kHz
V_{IN}	110 V_{RMS}
$V_{OUT\ REF}$	400 V
R	1422 Ω

$$K_{i\ DCM}(s) = \frac{8435(s+1)}{s(s/0.1+1)} \tag{4}$$

$$K_{i\ CCM}(s) = \frac{150(s+1)(s/100+1)}{s(s/0.1+1)} \tag{5}$$

The bump-removal controllers are chosen to be proportional gains, as this greatly simplifies implementation, and are given as (6) and (7) for the DCM and CCM controllers, respectively.

$$K_{DCM\ \text{Bump Removal}} = 500 \tag{6}$$

$$K_{CCM\ \text{Bump Removal}} = 100 \tag{7}$$

The PFC was simulated with a single controller and two bumpless controllers using Simulink with the SimPowerSystems toolbox. The time domain responses for the PFC at full load and a single controller, two controllers without bump supression, and two bumpless controllers are shown in Figure 10. The simulations show that the PFC with a single controller has a THD of 17.7%, the PFC with two

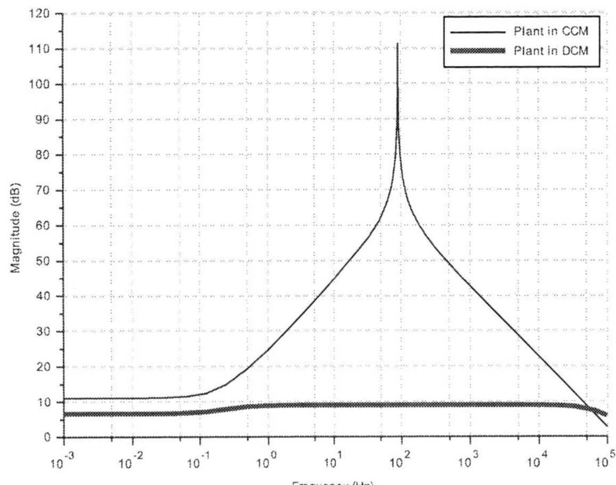

Fig. 8. Bode magnitude plots for the semi-bridgeless PFC in DCM and in CCM. Note the pronounced difference in frequency response for the two operating modes. Also note that the response at crossover is not shown because it is beyond half the switching frequency.

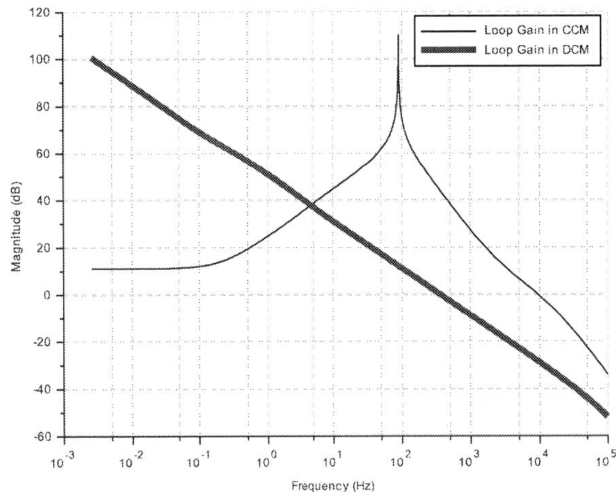

Fig. 9. Bode plots for the semi-bridgeless PFC loop gain in DCM and in CCM using controllers (4) and (5). A lower loop gain in DCM is used to provide some degree of robustness if the DCM controller is inadvertently selected while the converter is operating in CCM.

controllers without bump suppression has a THD of 16.5%, and the PFC with two bumpless controllers has a THD of 8.9%. These values may be regarded as larger than expected or seen in hardware. They do, however, illustrate a qualitative relationship that is consistent in both simulation and hardware.

IV. Hardware Results

A 750 watt, semi-bridgeless PFC was designed to meet the CSCI titanium efficiency specification, has a maximum efficiency of 96%, and fits into the 1U standard form factor. The PFC uses the Microchip dsPIC33EP64GS502 microcontroller, and is shown in Figure 11. The current control loops for this

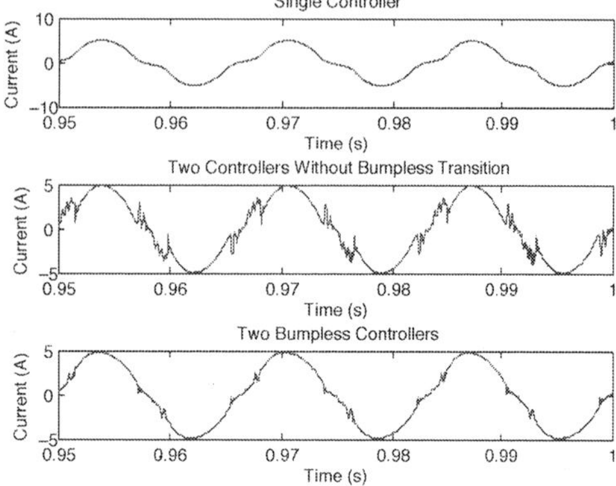

Fig. 10. Time-domain response of PFC when using a single controller, two controllers without any attempt at removing the transition bump, and two bumpless controllers. One can see that the signal is distorted, particularly due to the change in shape when the PFC is operating in DCM. The shape is a closer approximation to a sinewave, when using two controllers, but there are significant transient spikes when switching between controllers. It can be seen that there is still a small discontinuity when switching between the two bumpless controllers, as indicated by the small transient spikes. Note however, that there is a significant improvement compared to two controllers with no attempt to reduce the transition bump, and that the controller design will be refined for the final paper. Note, this figure is for the PFC operating at $110V_{RMS}$ and full load. This figure was included as it clearly illustrates the influence of the controller types.

hardware operate at 100 kHz[1]. At this control-loop rate, the DCM and bumpless control adds roughly 30 cycles, which is 4.3% more MIPS, compared to a single CCM controller.

Fig. 11. Hardware implementation of 750 watt semi-bridgless PFC. The hardware was designed to meet the CSCI titanium efficiency specification (efficiency > 96 %) and fits into the 1U standard form factor.

The PFC THD is tabulated in Table II. From this table, one can make several observations. First is that there is one datapoint for the single DCM controller. This is because the system became unstable at heavier loads. This is to be expected, since the DCM controller has much higher gain compared to the CCM controller. Second, also note that there is no "bumpy" controller (CCM-DCM controller without bump removal). This is because the system with this controller was unstable, regardless of the load. This is likely due to perturbations from the transition bump. Also, see that the bumpless controller reduces THD by approximately 2% at light loads, when compared to a single controller synthesized for CCM operation. While the THD improvement is less than expected, it is still notable. The THD with bumpless control is probably higher than expected due to aberrations in the CCM-DCM detection scheme and un-modeled dynamics. The aberrations in the CCM-DCM detection scheme are likely due to sampling at discrete time instants and system noise.

TABLE II
COMPARISON OF THD FOR BUMPLESS AND NON-BUMPLESS CONTROLLERS AS A FUNCTION OF LOAD AND AN INPUT VOLTAGE OF $110V_{RMS}$. ONLY ONE DATA POINT IS AVAILABLE FOR THE DCM CONTROLLER SINCE THE PFC BECOMES UNSTABLE FOR HIGHER LOADS.

Controller	Percent Load	THD (%)
CCM Controller	10	12.2
-	15	14.2
-	20	8.1
-	100	2.2
DCM Controller	10	10.4
Bumpless Controller	10	10.8
-	15	12.2
-	20	6.4
-	100	2.2

A figure showing the transient response of the PFC with bumpless controller is shown in Figure 12. From this figure, it can be seen that the CCM-DCM mode detection scheme does transition, thereby causing the transitions between the two controllers. Also, be seen that the mode-detection scheme generally occurs near the desired points, but needs to be improved. It can also be seen that the inductor-current tracks the input voltage with relatively low distortion.

Bumpless control provides greater improvements than those highlighted here, when used under other operating conditions. For example, under extremely light load conditions, such as 5% load, the THD of the PFC using bumpless control is 10.5% compared to 14.8% when using a single CCM controller. Similarly, under high line conditions with 230V input, the THD reduction is on the order of 9%. Regardless of the operating condition, the distortion can probably be reduced further for the stated reasons and will be the topic of future research.

V. Conclusion and Future Work

Researchers have devised control strategies for reducing THD that typically use multiple controllers [1], [3], [4], [5]. Switching between controllers will result in a discontinuous control signal that can degrade THD. In this paper, it was shown that bumpless control can yield significant improvements without significant tuning of the controllers or the controllers used to remove the transition bump. Using low-order controllers, the PFC THD was reduced by 9% in simulation and 1.7% in hardware using a 750 watt semi-bridgeless PFC. The THD with bumpless control is higher than expected, and may be attributed to aberrations in the CCM-DCM detection scheme and un-modeled dynamics in the system. The aberrations in the CCM-DCM detection

[1]One should note that in the simulation, the CCM and DCM controller outputs are in terms of duty cyle and vary between 0 and 1. The controller outputs in the hardware are in terms of duty cycle counts and vary from 0 to 2400. Due to this difference between the two implementations, the bump removal gains become much smaller in the hardware. The bump removal gains are 1.2 in the hardware. Similarly, the controllers had to be modified to account for differences in the implementation.

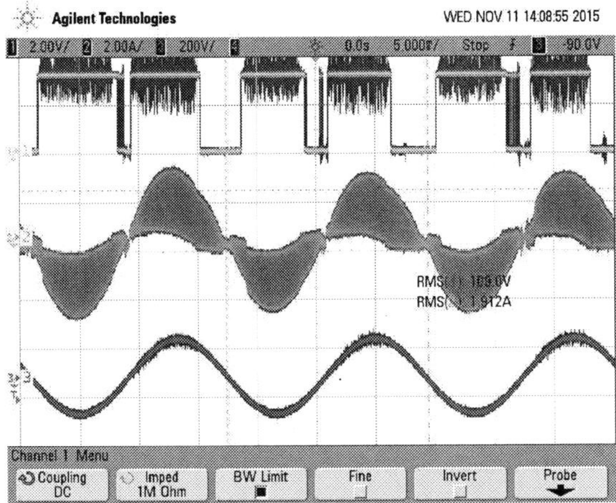

Fig. 12. Time-domain response of PFC with bumpless controller with $110V_{RMS}$ input and 15% load. The trace at the top indicates which controller is active, CCM when the signal is high, and DCM when the signal is low (2V/div). The trace in the center is the inductor current (2A/div). The trace at the bottom is the input voltage (200V/div). The scope timebase is 5 ms/div.

scheme are likely due to sampling at discrete time instants and system noise. The bumpless controller still shows notable improvement when compared to a single controller, and will be the topic of future research.

References

[1] Younghoon Cho and Jih-Sheng Lai. Digital plug-in repetitive concontrol for single-phase bridgeless pfc converters. *IEEE Transactions on Power Electronics*, 2013.

[2] Richard C. Dorf and Robert H. Bishop. *Modern Control Systems*. Prentice-Hall, 2010.

[3] Shu Fan Lim and A.M. Khambadkone. A multimode digital control scheme for boost pfc with higher efficiency and power factor at light load. In *Applied Power Electronics Conference and Exposition (APEC) 2012*, 2012.

[4] Barry Mather and Dragan Maksimovic. A simple digital power factor correction rectifier controller. *IEEE Transactions on Power Electronics*, 2011.

[5] Aleksander Prodic, Dragan Maksimovic, and Robert Erickson. Deadzone digital ccontroller for improved dynamic response of low harmonic rectifiers. *IEEE Transactions on Power Electronics*, 2006.

[6] Joel Steenis, Kostas Tsakalis, and Raja Ayyanar. Bumpless control for lpv modeled inverters in a microgrid. *IEEE Transactions on Power Electronics*, 2014.

[7] Jian Sun, Daniel Mitchell, Matthew Greuel, Philip Krein, and Richard Bass. Averaged modeling of pwm converters operating in discontinuous conduction mode. *IEEE Transactions on Power Electronics*, 2001.

[8] Vatche Vorperian. Simplified analysis of pwm converters using model of pwm switch continuous conduction mode. *IEEE Transactions on Aerospace and Electronic Systems*, 1990.

[9] Vatche Vorperian. Simplified analysis of pwm converters using model of pwm switch ii. *IEEE Transactions on Aerospace and Electronic Systems*, 1990.

[10] Manjing Xie. Digital control for power factor correction. Master's thesis, Virginia Polytechnic Institute and State University, 2003.

Joel Steenis has undergraduate degrees in electrical engineering and mathematics from the University of Arizona, Tucson, AZ, USA, a masters degree in electrical engineering from the University of Southern California, Los Angeles, CA, USA, and the Ph.D. degree in electrical engineering from Arizona State University, Tempe, AZ, USA. He worked as an Applications and Circuit Design Engineer at National Semiconductor and is currently an Applications Engineer at Microchip Technology.

Alex Dumais has an undergraduate degree in electrical engineering from Arizona State University, Tempe, AZ, USA. He has held a variety of positions at Microchip Technology where he is currently manager of the switch-mode power supply (SMPS) group.

Mixed-Signal Hysteretic Internal Model Control of Buck Converters for Ultra-Fast Envelope Tracking

V Inder Kumar, *Student Member, IEEE* and Santanu Kapat, *Member, IEEE*
Embedded Power Management Lab
Department of Electrical Engineering
Indian Institute of Technology Kharagpur, West Bengal, India-721302
Email: indervedula@gmail.com and santanu.kapat@ieee.org

***Abstract*—Envelope tracking (ET) applications demand high tracking bandwidth and high efficiency using a DC–DC converter. Design based on small-signal models using fixed frequency digital pulse width modulator (DPWM) often results in limited closed-loop bandwidth. This paper proposes a mixed-signal hysteresis current control (MSHCC) in a DC–DC buck converter with an analog current-loop and a digital voltage controller $G_c(z)$. This achieves robust stability and parameter insensitive current ripple by sampling the output voltage at the rising edge of high side gate signal. A fixed-gain power amplifier driven by a buck converter can be assumed to be a constant resistive load. Thus the load current information is directly obtained from the reference command, which further improves the tracking performance. The real-time tuning of $G_c(z)$ using the proposed MSHCC achieves fast recovery with an inherent current limiting, and the use of the internal model control (IMC) further minimizes the tracking error. A fixed frequency operation can be achieved through a real-time band adaptation. The proposed controller is implemented using an FPGA device.**

I. Introduction

The last few decades have witnessed a significant advancement in portable electronic devices. Power amplifiers [1] are integral parts of these devices, and they consume considerable portion of the total power. Efficiency of a power amplifier can be improved by dynamically varying its power supply depending on the power demand [2]– [6]. This technique is commonly referred to as Envelope Tracking (ET). Fig. 1 shows the block diagram of ET power supply, where A_v is the gain of RFPA, and T_d is the time-delay.

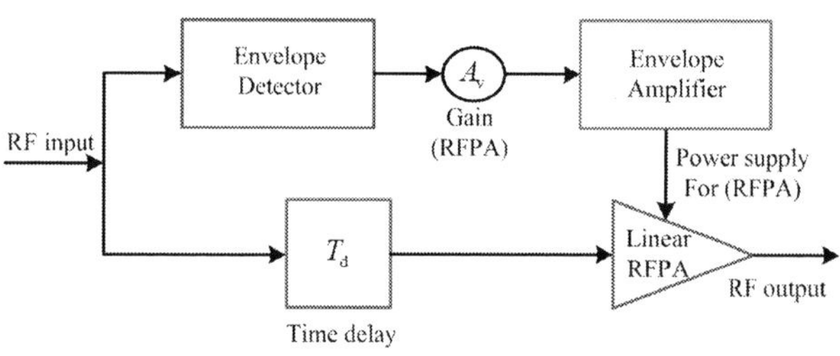

Fig. 1. A block diagram of ET power supply for a linear RFPA

The recent advancement in digital power management facilitates high performance control implementation for envelope tracking (ET) buck converters [7]. Design rule of thumb using linear PWM control considers the switching frequency of the buck converter to be at least 10 times the desired tracking bandwidth [8], [9], which requires a high switching frequency for a fast changing envelope of an RFPA load. Thus, it increases switching and driver losses, thereby degrading the efficiency. However for higher bandwidth envelope signals, combination of DC-DC converters and linear assisted stages may be alternative options [10]– [13]. Multiphase [14], [15], multi-input [16], multilevel [17] converter topologies are all used to enhance tracking bandwidth and efficiency. However, major drawbacks are increased size, cost and component count. Thus it is essential to investigate high performance control to evaluate critical performance limits using a conventional synchronous buck converter. Fixed frequency current-mode DPWM controller [18]– [20] facilitates high performance controller implementation, but suffers from sub-harmonic instability using a higher controller gain. This would lead to EMI problems and increase current and voltage ripples, thereby increasing conduction losses. Hysteresis current control (HCC) offers (i) inherent current loop stability (ii) fast transient response (iii) good accuracy [21]– [25]; however, they suffer from parameter sensitive current ripple and varying switching frequency. A mixed-signal HCC technique is useful as it considers the conventional approach of using inductor current in the analog domain [26]– [28].

This paper proposes mixed-signal hysteretic current control in a synchronous buck converter for ET power supply, with an analog current-loop and a digital voltage controller $G_c(z)$. The real time tuning of $G_c(z)$ achieves fast recovery with inherent current limiting, and the use of IMC [29] alongwith the reference feed-forward further improves the tracking performance. It achieves robust stability and parameter insensitive current ripple by sampling the output voltage at the rising edge of high side gate signal. Further, fixed frequency operation can be achieved through real time band adaptation. This paper analytically shows the improvement in closed-loop bandwidth with the proposed scheme over conventional DPWM current mode control. The proposed scheme can further be extended to state of the art ET power supplies using GaN devices [14], [30], for high bandwidth and efficiency. A buck converter prototype is made and experimental results are obtained using an FPGA device.

978-1-4673-9551-9/16 $31.00 © 2016 IEEE

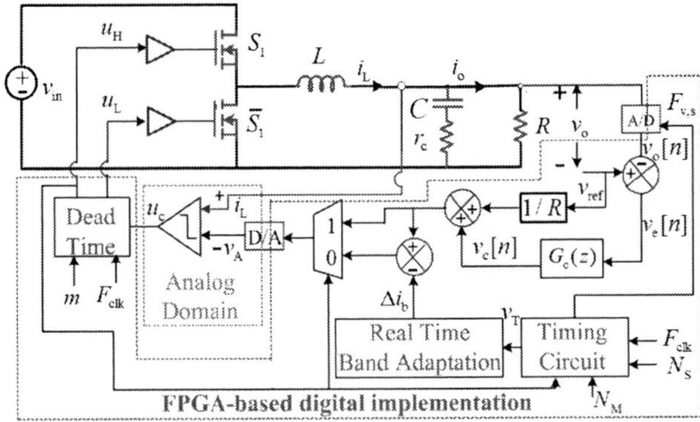

Fig. 2. Proposed MSHCC in a synchronous buck converter

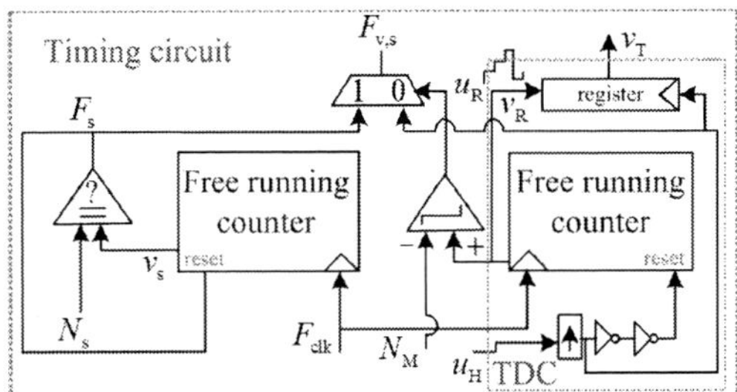

Fig. 3. Schematic of the timing circuit

II. THE PROPOSED MIXED-SIGNAL HCC TECHNIQUE

A. Mixed-signal hysteresis current control

This paper proposes MSHCC in a synchronous buck converter, for high bandwidth and efficient ET power supply. The schematic diagram is shown in Fig. 2. This primarily consists of three blocks: (a) Timing circuit (shown in Fig. 3) which generates the sampling clock $F_{v,s}$ for the voltage-loop, and the digital coded switching period v_T using a time-to-digital converter (TDC). The output voltage is sampled using an A/D converter with a rate $F_{v,s}$ which is also used for the computation of the digital voltage controller $G_c(z)$. The controller output $v_c[n]$ is added along with the reference feed-forward to generate the peak current reference. The valley current reference is generated from the peak reference by simply subtracting the hysteresis band Δi_b. Unlike conventional analog HCC, only one comparator is used, thus one D/A converter is sufficient to generate current references using a MUX which uses u_H as the select line (shown in Fig. 2). The fast changing inductor current is compared directly with output of the D/A converter in the analog domain. Also, variety of power efficient current sensing methods are available in the literature [31] as well as in various commercial products; (b) Real time band adaptation generates the current hysteresis band Δi_b for a fixed switching frequency operation during steady state; (c) Dead time circuit generates the latched gate signals using the comparator output u_c. This dead time circuit attempts to avoid any shoot through current, in which the dead time can be adjusted depending upon the timing parameters of the MOSFETs and the driver.

B. Synchronous/Asynchronous sampling clock

The 'timing circuit in Fig. 3 uses two counters which increment at the rising edge of the controller clock F_{clk} with the time period t_{clk}. The first counter generates v_T using the rising edge of the high side gate signal u_H, while the other counter generates a synchronous fixed frequency clock F_s with the time period $T = (N_s + 1)t_{clk}$. A MUX is used to generate the sampling clock $F_{v,s}$ for the A/D converter (ADC) which accepts synchronous and asynchronous clocks as the inputs (in Fig. 3). As long as the counter value v_r of the TDC is lower than the upper limit N_M, the ADC is sampled asynchronously at the rising edge of u_H; otherwise, a synchronous clock F_s is used. Asynchronous sampling can cause a complete collapse, especially during large signal recovery, and hence should not be used throughout. This is primarily because, error voltage cannot be updated for few cycles when u_H does not change its state. Thus synchronous sampling clock F_s is used during large signal recovery, and also the controller computation will occur at the same rate. This allows the proposed control scheme to successfully operate during large signal transients, and an asynchronous sampling of the output voltage v_o[n] is retained under steady state conditions. Rest of the paper deals with the analysis and design of the proposed control scheme for ET power supply.

III. ANALYSIS OF THE PROPOSED CONTROL SCHEME

A. Stability analysis of MSHCC

Let the inductor current and output voltage at the rising edge of high side gate signal be i_n and $v_o[n]$, and at the next rising edge be i_{n+1} and $v_o[n+1]$. For analytical study, a simple proportional controller is considered in the outer voltage-loop under the assumption of an ESR dominated output voltage ripple. The rising and falling slopes of output voltage are $m_1 r_c$ and $m_2 r_c$, where m_1 and m_2 are the rising and falling slopes of the inductor current and r_c is the ESR of the output capacitor. From Fig. 4, the valley currents at two successive rising edges i_n and i_{n+1} can be obtained as

$$i_n = v_c[n-1] - \Delta i_b \,;\; i_{n+1} = v_c[n] - \Delta i_b \tag{1}$$

where the controller output v_c[n]= $k_p(v_{ref} - v_o[\mathrm{n}])$, and Δi_b is the hysteresis band. It is clear from (1) that the current-loop is inherently stable, provided the voltage loop is stable. Thus it is important to investigate the stability of voltage loop. The final voltage $v_o[n+1]$ is given as

$$v_o[n+1] = v_o[n] + m_1 r_c t_{on} - m_2 r_c (T_h - t_{on}) \tag{2}$$

where,

$$t_{on} = \frac{v_c[n] - i_n}{m_1} = \frac{\Delta v_c[n]}{m_1} + \frac{\Delta i_b}{m_1} \tag{3}$$

$$T_h = \frac{\Delta v_c[n]}{m_1} + \left(\frac{m_1 + m_2}{m_1 m_2}\right)\Delta i_b \tag{4}$$

$$\Delta v_c[n] = v_c[n] - v_c[n-1] = -k_p \Delta v_o[n] \tag{5}$$

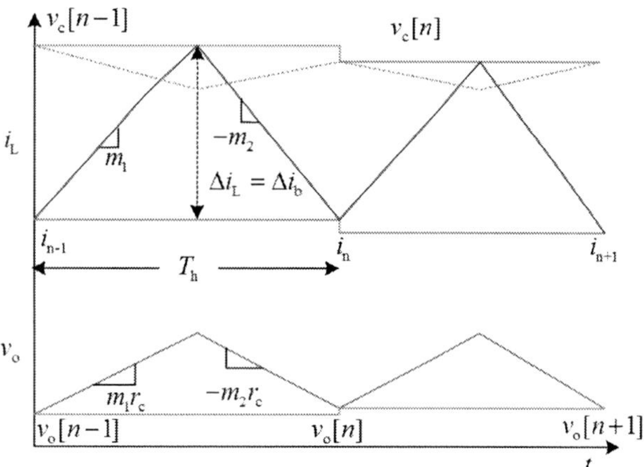

Fig. 4. Control waveforms of the proposed scheme

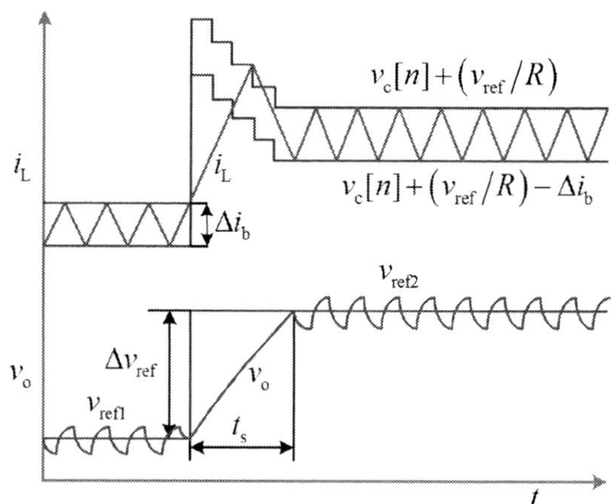

Fig. 5. Large signal recovery under MSHCC

From the above expressions, we get

$$\Delta v_o[n+1] = -k_p r_c \Delta v_o[n] \tag{6}$$

where,

$$\Delta v_o[n+1] = v_o[n+1] - v_o[n]; \ \ \Delta v_o[n] = v_o[n] - v_o[n-1] \tag{7}$$

Thus the voltage-loop is stable for $k_p r_c < 1$, which implies the stability boundary of the closed-loop converter with an ESR dependent upper bound on the proportional gain. Also, the impact of the integral gain is found to be insignificant. Thus, the proposed scheme will remain stable over a wide range of the input voltage. The above derived stability condition is valid under steady state operation; however, the proportional gain k_p during transients can be increased for fast recovery. During steady state lower gain is retained to meet the required stability criteria.

B. Transient performance

1) Closed loop bandwidth under MSHCC: Consider a step-reference transient as shown in Fig. 5, where v_{ref1} and v_{ref2} are the reference voltages before and after the transient respectively, and Δv_{ref} is the step size. The outer voltage loop controller is tuned to set the desired current reference $v_c[n]$, and further the load current information is added using (v_{ref}/R) for a given R to achieve fast transient recovery with settling time t_s. Thus, the peak and the valley current references generated during large signal recovery are $v_c[n]$+(v_{ref}/R) and $v_c[n]$+(v_{ref}/R)-Δi_b respectively. Using capacitor charge balance, the settling time t_s can be obtained as [32]

$$t_s = \sqrt{\frac{2LC\Delta v_{ref} v_{in}}{(v_{in} - v_{ref1})v_{ref1}}} \tag{8}$$

where v_{in} is the input voltage, and L, C are the power circuit parameters in Fig. 2. Under optimal recovery as shown in Fig. 5, the settling time t_s and the rise time t_r are equivalent, and hence the closed loop bandwidth f_{BW} can be approximated as

$$f_{BW} = \frac{0.35}{t_s} = 0.35\sqrt{\frac{(v_{in} - v_{ref1})v_{ref1}}{2LC\Delta v_{ref} v_{in}}} \tag{9}$$

It is seen from (9), that the closed loop bandwidth f_{BW} is primarily governed by the power circuit parameters (L and C), input voltage v_{in} and reference step-size Δv_{ref}. Any bandlimited signal with the maximum frequency component less than f_{BW} can be accurately tracked. For a sinusoidal signal, v_{ref1} is the DC offset voltage and Δv_{ref} is half of its amplitude.

2) Closed loop bandwidth under DPWM current mode control: The closed loop bandwidth under fixed-frequency DPWM current mode control indicates the maximum frequency component in a band-limited signal that can be tracked without significant distortion in magnitude and phase. Design based on small signal modeling often results in limited closed-loop bandwidth because duty cycle saturation violates small signal assumptions. Design rule of thumb considers the crossover frequency f_c to be $(1/10)^{th}$ of the switching frequency f_{sw}. Thus, the closed loop bandwidth $f_{BW,DPWM}$ is approximately given as

$$f_{BW,DPWM} \cong 0.16 f_{sw} \tag{10}$$

Consequently, values of the power circuit parameters (L and C) depends on the switching frequency f_{sw}. From (10), it is observed that higher closed loop bandwidth requires higher switching frequencies; however, it increases switching and driver losses, thereby degrading the efficiency.

C. Efficiency of the proposed scheme

The efficiency of a converter is generally considered at steady state conditions. However, frequent transients can considerably affect the overall efficiency. This is particularly important when the reference step-size increases. Thus, it is important to minimize the number of switching events during large signal recovery. The voltage loop controller in the

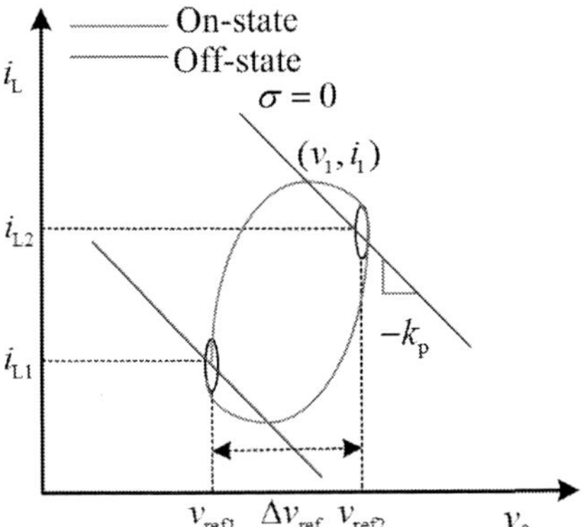

Fig. 6. Phase-plane trajectory depicting optimal transient recovery

proposed scheme is tuned to set the desired current reference so as to reduce the number of switching events under transient, and consequently to reduce the switching and driver losses. Multiple switching events using DPWM current mode control can reduce the overall efficiency. Analysis justifies that the proposed scheme achieves higher efficiency and has better figure of merit (product of losses and RMS tracking error), and it further increases under higher reference frequencies.

IV. Design of the Proposed Scheme

A. Controller design procedure

Using a simple proportional voltage controller with a gain k_p, the overall controller in Fig. 2 can be framed into a switching surface as given below

$$\sigma = k_p (v_{ref} - v_o) + (i_{ref} - i_L) = 0 \tag{11}$$

where v_{ref} and i_{ref} are reference voltage and reference current respectively. The current error $(i_{ref} - i_L)$ can be used as a derivative action which replaces a direct voltage derivative and reduces high frequency noise injection. The state space model of a synchronous buck converter becomes

$$\begin{bmatrix} \dot{x}_1 \\ \dot{x}_2 \end{bmatrix} = \begin{bmatrix} 0 & \frac{-1}{L} \\ \frac{1}{C} & \frac{-1}{RC} \end{bmatrix} \begin{bmatrix} x_1 \\ x_2 \end{bmatrix} + \begin{bmatrix} \frac{u}{L} & 0 \\ 0 & 0 \end{bmatrix} \begin{bmatrix} v_{in} \\ i_o \end{bmatrix} \tag{12}$$

where $x_1 = i_L$, x_2 is the voltage across the capacitor, v_C, R is the load resistance, and i_o is the load current. If the control signal u is set to 'logic 1', the high-side MOSFET is turned on; otherwise, the switch is turned off. The state space solutions of (12) represents the on- and off-state trajectories of the converter. The on- and off-state trajectories are shown in Fig. 6, and the intersection point (v_1, i_1) can be obtained by solving the respective state trajectories. Thus, the optimal value of k_p for fast recovery can be obtained by substituting the intersection point (v_1, i_1) in (11) as

$$k_p = \frac{i_1 - i_{ref}}{v_{ref} - v_1} \tag{13}$$

B. Effect of reference feed-forward

Assuming a fixed gain power amplifier as a constant resistive load [33], the load current feed-forward i_{ref} can be added simply by using v_{ref}/R, where v_{ref} is the reference voltage and R is the load resistance. This results in fast transient response for a dynamically varying reference envelope signal. Also, it significantly improves the output impedance and achieves near load invariant regulation even in the absence of an integral gain.

C. Compensator design with IMC

Internal mode control (IMC) [29] is useful to minimize the tracking error by considering the internal model of the reference envelope into the compensator design. For example, a step reference command requires a discrete-time compensator to be

$$G_c(z) = P(z)/z - 1 \tag{14}$$

Similarly, for a sinusoidal reference signal, the compensator becomes

$$G_c(z) = \frac{P(z)}{(z-1)(z^2 - 2z\cos(wT_s + 1)} \tag{15}$$

where, (z-1) and $(z^2 - 2z\cos(wT_s + 1)$ are the denominator polynomials of the z-transform of the reference signal. $P(z)$ is the numerator polynomial of the compensator. IMC attempts to minimize the tracking error, thereby improving accuracy and linearity of the RFPA. This paper proposes the addition of IMC to the ET power supply, and further it can be extended to existing ET power supplies in the literature.

D. Real time band adaptation for frequency regulation

The error voltage is sampled at the rising edge of gate signal as shown in Fig. 4; thus the peak current $v_c[n]$ remains constant throughout the switching period, and the inductor current ripple $\Delta i_L = \Delta i_b$ at steady-state. For a given period T_1, the current ripple Δi_1 is

$$\Delta i_1 = \left(\frac{(v_{in} - v_o) v_o}{L v_{in}} \right) T_1 \tag{16}$$

If v_{T1} is the digital code corresponding to a time period T_1 and band Δi_1, then the desired time period T_d can be simply achieved by setting a desired band Δi_d as

$$\Delta i_d = \left(\frac{v_{T1}}{v_{Th}} \right) \Delta i_1 \tag{17}$$

Thus, the frequency regulation at steady-state can be achieved by adjusting Δi_d in real-time using the digital coded switching period using the time-to-digital converter.

E. Real time reconfigurability

The proposed control scheme can be reconfigured in real time to achieve fast envelope tracking. During step-up transient, the control scheme can be reconfigured to a valley current mode control, by adaptively varying Δi_b to generate the peak current reference $v_c[n]$+Δi_b. Similarly peak current

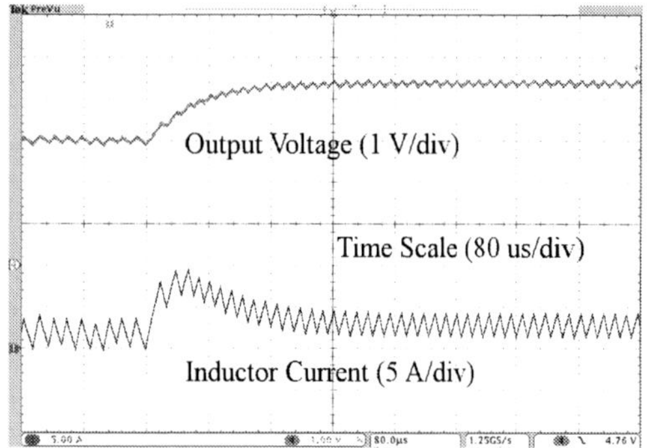

Fig. 7. Reference transient in a CCM buck converter from 3.6V to 5.2V at v_{in}=8V and R=1Ω using MSHCC without IMC and reference feed-forward

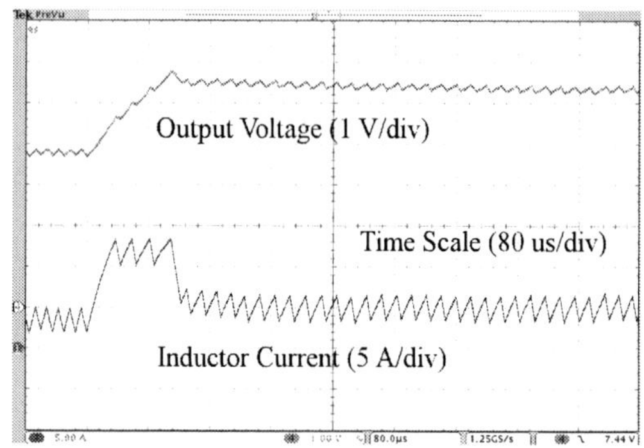

Fig. 9. Reference transient from 3.6V to 5.2V at v_{in}=8V and R=1Ω using MSHCC with IMC, reference feed-forward and current-limit

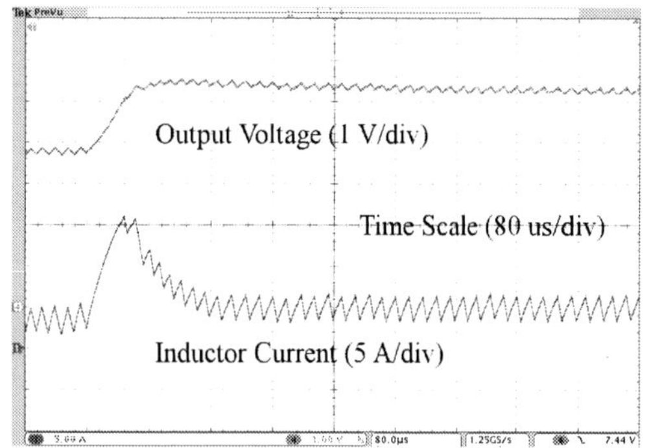

Fig. 8. Reference transient in a CCM buck converter from 3.6V to 5.2V at v_{in}=8V and R=1Ω using MSHCC with IMC and reference feed-forward

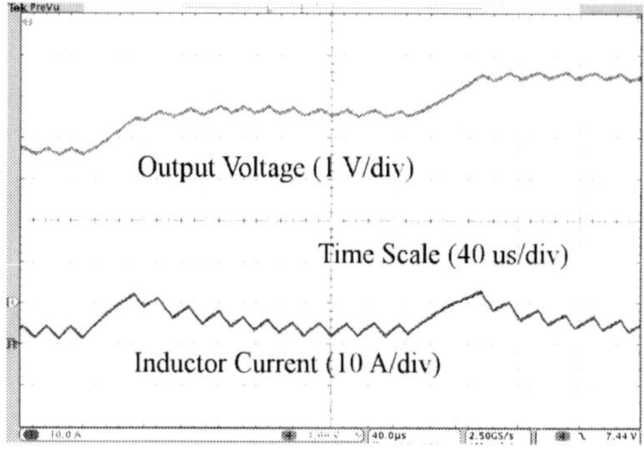

Fig. 10. Reference transient from 3.6V to 4.5V to 5.2V at v_{in}=8V and R=1Ω using MSHCC with IMC and reference feed-forward

mode control can be used during step-down transient. The proposed control scheme can also be extended to average current mode control with peak and valley current references as $v_c[n]$+$(\Delta i_b/2)$ and $v_c[n]$-$(\Delta i_b/2)$ respectively. Hence, it provides complete flexibility to set the desired current reference, and to achieve higher bandwidth.

V. Hardware Implementation

A buck converter prototype has been made, and the proposed control scheme is implemented using an FPGA device. For the prototype signal conditioning circuits, a differential 10-bit pipeline A/D converter (AD-9215) is considered for sampling the output voltage. The ADC uses a differential amplifier (AD-8138) to convert the single ended output voltage into differential form for the reduction of common mode noise. The output of the digital compensator $v_c[n]$ is added with (v_{ref}/R). The output is then converted into an equivalent analog voltage v_A by using a 12-bit D/A converter (AD-9762) followed by a differential amplifier (AD-8130). A high speed comparator (TLV-3501) is used to compare v_A with the sensed inductor current i_L. The comparator output is then latched, and the dead-time circuit generates the respective gate signals. A $10\,\mathrm{m}\Omega$ current sense resistor is used to sense i_L followed by a current sense amplifier (ADM-4073). Hardware results are demonstrated for two separate test conditions to illustrate the effect of power circuit parameters (L and C) on tracking bandwidth.

A. Case study 1: $L = 10\ \mu H$ and $C = 470\ \mu F$

The nominal parameters are: $v_{in} = 8-12$V, $v_{ref} = 1-7.5$V, the capacitor ESR $r_C \cong 35\ \mathrm{m}\Omega$, and the inductor DC resistance $r_L \cong 2.2\ \mathrm{m}\Omega$. The time period of the FPGA controller clock t_{clk} =10ns. Fig. 7 shows the reference transient from 3.6 V to 5.2 V at v_{in}=8 V and R=1 Ω under the proposed scheme. Without considering IMC and the reference feed-forward, Fig. 7 shows a steady state error of 0.7 V with the peak current of 9 A. The use of IMC and reference feed-forward significantly improves the transient response as shown in Fig. 8 with 80 μs settling time and negligible steady state error; however, this results in 15 A peak current. The use of a current limiter at 13 A, Fig. 9 shows the settling time is increased to 160 μs. Moreover, stable behavior is achieved for all the cases. A further reduction in peak current will increase the response time and the proposed MSHCC enables one to customize the current limiting based on device ratings. The proposed scheme is equally applicable for frequent transients as shown in Fig. 10, from 3.6 V to 4.5 V to 5.2 V. Figs. 11(a), 11(b), and 11(c) show a sinusoidal envelope tracking with the frequency of 10 kHz of 1 V peak to peak amplitude and 7 V DC offset at v_{in}=8 V and R=1 Ω.

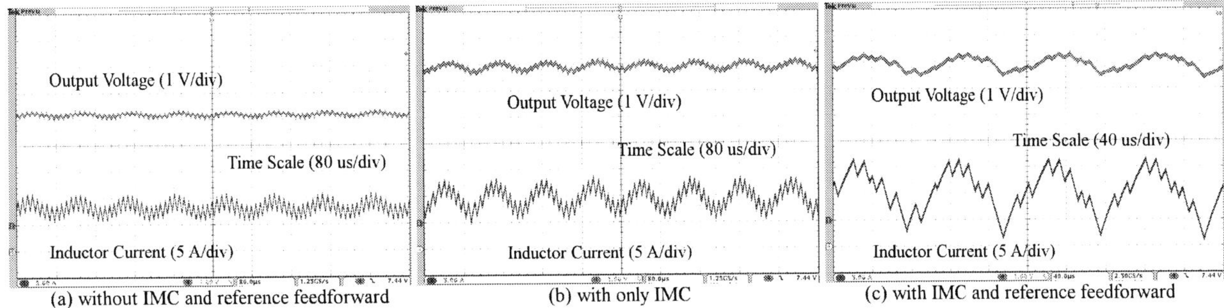

Fig. 11. Envelope tracking of a 10 kHz signal in a synchronous buck converter with peak to peak amplitude of 1V and DC offset of 7V at v_{in}=8V and R=1Ω using MSHCC

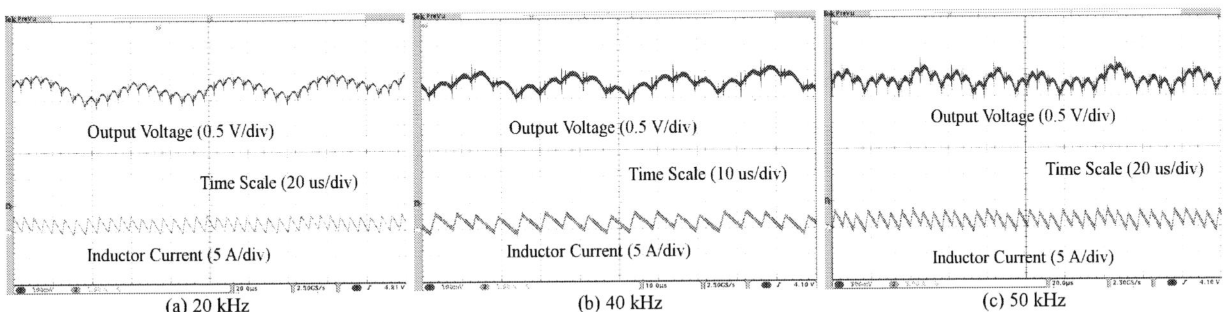

Fig. 12. Envelope tracking in a synchronous buck converter with peak to peak amplitude of 0.5V and DC offset of 2.25V at v_{in}=12V and R=1Ω using MSHCC with IMC and reference-feed-forward

Without IMC and feed-forward, Fig. 11(a) shows the output using the proposed control is attenuated to 0.2 V peak-to-peak with 5 V DC offset. The use of IMC in Fig. 11(b) further improves the tracking performance with 0.5 V peak-to-peak and 6.8 V DC offset. However, an addition of the feed-forward term significantly improves the performance as shown in Fig. 11(c) with 0.9 V peak to peak amplitude and 6.9 V DC offset.

B. Case study 2: $L = 3.3$ μH and $C = 20$ μF

For Fig. 12, $L = 3.3$ μH, $C = 20$ μF, the capacitor ESR $r_C \cong 5$ mΩ, and the inductor DC resistance $r_L \cong 1.8$ mΩ. Figs. 12(a), 12(b), and 12(c) show envelope tracking using the proposed scheme with different frequencies of 20 kHz, 40 kHz, and 50 kHz respectively, with both IMC and reference feed-forward included in the design.

Above two case studies consider two different sets of power circuit parameters. For both the cases, the proposed control attempts to achieve the critical performance without sacrificing the tracking accuracy by using IMC. A further improvement would be possible by redesigning the power circuit, particularly by reducing the size of the output capacitor as given in (9). Thus the proposed digital control scheme can be extended to the state-of-the-art RFPA applications which consider high switching frequencies. This would be helpful to trade-off between the tracking performance, accuracy, and efficiency with reduced hardware requirements.

VI. Conclusion

A mixed signal HCC was proposed in an ET buck converter. This considers a digital voltage controller with reference feed-forward and IMC, which improves the transient response and minimizes the tracking error. Robust stability and parameter insensitive current ripple were achieved using asynchronous sampling at steady-state. The MSHCC achieves fast recovery through tuning the voltage controller in real time, and the band adaptation achieves a fixed frequency at steady-state. The proposed control scheme is shown to have higher bandwidth compared to DPWM current mode control, and further can be extended to other ET power supplies available in the literature. The proposed scheme was realized using FPGA device and was tested using a buck converter.

Acknowledgment

The authors would like to thank Amit K. Singha for his help in the experimentation.

References

[1] S. C. Cripps, "RF Power Amplifiers for Wireless Communication," Norwood, MA: Artech house, 1999.

[2] G. Hanington, P. Chen, P. M. Ashbeck, and L. E. Larsen, "High Efficiency Power Amplifier using Dynamic Power Supply Voltage for CDMA Applications," *IEEE Trans. Microwave Theory Tech.*, vol. 47, no. 8, pp. 1471–1476, Aug. 1999.

[3] J. Staudinger, B. Gilsdorf, D. Newman, G. Norris, G. Sadowniczak, R. Sherman, and T. Quach, "High Efficiency CDMA Power Amplifier using Dynamic Envelope Tracking Technique," *in Proc. IEEE MTT-S Int. Microwave Symp. Dig.*, pp. 873–876, 2000.

[4] D. R. Anderson and W. H. Cantrell, "High-Efficiency High-Level Modulator for use in Dynamic Envelope Tracking CDMA RF Power Amplifiers," *in Proc. IEEE MTT-S Int. Microw. Symp. Dig.*, vol. 3, pp. 1509–1512, May 2001.

[5] A. P. Chandrakasan, S. Sheng, and R. W. Brodersen, "Low-Power CMOS Digital Design," *IEEE J. Solid-State Circuits*, vol. 27, no. 4, pp. 473–474, Apr. 1992.

[6] F. H. Raab, "Envelope Elimination and Restoration System Requirements," *in Proc. RF Technol. Expo.*, pp. 499–512, Feb. 1988.

[7] O. Garcia, M. Vasic, P. Alou, J. A. Oliver, and J. A. Cobos, "An Overview of Fast DC-DC Converters for Envelope Amplifier in RF Transmitters," *IEEE Trans. Power Electron.*, vol. 28, no. 10, pp. 4712–4722, Oct. 2013.

[8] P. Midya, K. Haddad, L. Connell, S. Bergstedt, and B. Roeckner, "Tracking Power Converter for Supply Modulation of RF Power Amplifiers," *in Proc. IEEE PESC*, pp. 1540–1545, 2001.

[9] H. Xi, Q. Jin, X. Ruan, and X. Xiong, "Full Feedforward of the Output Voltage to Improve Efficiency for Envelope Tracking Power Supply using Switch Linear Hybrid Configuration," *IEEE Trans. Power Electron.*, vol. 28, no. 1, pp. 451–456, Jan. 2013.

[10] P. F. Miaja, M. Rodriguez, A. Rodriguez, and J. Sebastian, "A linear Assisted DC/DC Converter for Envelope Tracking and Envelope Elimination and Restoration Applications," *IEEE Trans. Power Electron.*, vol. 27, no. 7, pp. 3302–3309, Jul. 2012.

[11] M. Vasic, O. Garcia, J. A. Oliver, P. Alou, and J. A. Cobos, "Theoretical Efficiency Limits of a Serial and Parallel Linear-Assisted Switching Converter as an Envelope Amplifier," *IEEE Trans. Power Electron.*, vol. 29, no. 2, pp. 719–728, Feb. 2014.

[12] V. Yousefzadeh, E. Alarcon, and D. Maksimovic, "Band Separation and Efficiency Optimization in Linear-Assisted Switching Power Amplifiers," in *IEEE PESC*, pp, 1–7, Jun. 2006.

[13] F. Wang, D. Kimball, J. Popp, A. Yang, D. Lie, P. Asbeck, and L. Larson, "An Improved Power-Added Efficiency 19-dbm Hybrid Envelope Elimination and Restoration Power Amplifier for 802.11g Wlan Applications," *IEEE Transactions on Microwave Theory and Techniques*, vol. 54, pp. 4086–4099, Dec. 2006.

[14] P. F. Miaza, A. Rodriguez, and J. Sebastian, "Buck Derived Converters based on Gallium Nitride Devices for Envelope Tracking Applications," *IEEE Trans. Power Electron.*, vol. 30, no. 4, pp. 2084–2095, Apr. 2015.

[15] A. Soto, J. A. Oliver, J. A. Cobos, J. Cezon, and F. Arevalo, "Power Supply for a Radio Transmitter with Modulated Supply Voltage," *in Proc. IEEE APEC*, vol. 1, pp. 392–398, Feb. 2004.

[16] M. Rodriguez, P. F. Miaja, A. Rodriguez, and J. Sebastian, "Multilevel Converter for Envelope Tracking in RF Power Amplifiers," *in Proc. IEEE ECCE*, 2009.

[17] M. Vasic, O. Garcia, J. A. Oliver, P. Alou, D. Diaz, and J. A. Cobos, "Multilevel Power Supply for High Efficiency RF Power Amplifiers," *IEEE Trans. Power Electron.*, vol. 25, no. 4, pp. 1078–1089, Apr. 2010.

[18] B. Patella, A. Prodic, A. Zirger, and D. Maksimovic, "High-Frequency Digital PWM Controller IC for DC-DC Converters," *IEEE Trans. Power Electron.*, vol.18, no.1, pp. 438–446, Jan. 2003.

[19] J. Chen, A. Prodic, R. W. Erickson, and D. Maksimovic, "Predictive Digital Current Programmed Control," *IEEE Trans. Power Electron.*, vol. 18, no. 1, pp. 411–419, Jan. 2003.

[20] S. Chattopadhyay and S. Das, "A Digital Current-Mode Control Technique for DC-DC Converters," *IEEE Trans. Power Electron.*, vol. 21, no. 6, pp. 1718–1726, Nov. 2006.

[21] W. R. Redl and S. Jian, "Ripple-Based Control of Switching Regulators-An Overview," *IEEE Trans. Power Electron.*, vol. 24, no. 12, pp.2669–2680, Dec. 2009.

[22] J. C. Tsai, C. L. Chen, Y. H. Lee, H. Y. Yang, M. S. Hsu, and K. H. Chen, "Modified Hysteretic Current Control (MHCC) for Improving Transient Response of Boost Converter," *IEEE Trans. Cir. Syst. I*, vol.58, no.8, pp. 1967–1979, Aug. 2011.

[23] S. C. Huerta, P. Alou, O. Garcia, J. A. Oliver, R. Prieto, and J. Cobos, "Hysteretic Mixed-Signal Controller for High-Frequency DC-DC Converters Operating at Constant Switching Frequency," *IEEE Trans. Power Electron.*, vol.27, no.6, pp.2690–2696, Jun. 2012.

[24] Y. Chen and Y. Kang, "The Variable-Bandwidth Hysteresis-Modulation Sliding-Mode Control for the PWM-PFM Converters," *IEEE Trans. Power Electron.*, vol.26, no.10, pp. 2727–2734, Oct. 2011.

[25] M. C. W. Hoyerby and M. A. E. Anderson, "Ultrafast Tracking Power Supply with Fourth Order Output Filter and Fixed Frequency Hysteretic Control," *IEEE Trans. Power Electron.*, vol.23, no.5, pp.2387–2398, Sept. 2008.

[26] S. Saggini, M. Ghioni, and A. Geraci, "An Innovative Digital Control Architecture for Low-Voltage High-Current DC-DC converters with Tight Load Regulation," *IEEE Trans. Power Electron.*, vol. 19, no. 1, pp. 210–218, Jan. 2004.

[27] O. Trescases, A. Prodic, and W. T. Ng, "Digitally Controlled Current-Mode DC-DC Converter IC," *IEEE Trans. Cir. Syst.-I*, vol. 58, no. 1, pp. 219–231, Jan. 2011.

[28] M. Hallworth and S. Shirsavar, "Microcontroller Based Peak Current Mode Control Using Digital Slope Compensation," *IEEE Trans. Power Electron.*, vol. 27, no. 7, pp. 3340–3351, Jul. 2012.

[29] K. Tarakanath, S. Patwardhan, and V. Agarwal, "Internal Model Control of DC-DC Boost Converter Exhibiting Non Minimum Phase Behaviour," *in Proc. IEEE PEDES*, pp. 1–7, Dec. 2014.

[30] Y. Zhang, M. Rodriguez, and D. Maksimovic, "100 MHz, 20 V, 90% Efficient Synchronous Buck Converter with Integrated Gate Driver," *in Proc. IEEE ECCE*, pp. 3664–3671, 2014.

[31] H. P. Forghani-Zadeh and G. A. Rincon-Mora, "Current-Sensing Techniques for DC-DC Converters", *in Proc. Midwest Symp. Cir. Syst. (MWSCAS)*, vol. 2, pp. 577–580, 2002.

[32] V. I. Kumar and S. Kapat, "Geometric Control Breaks Tracking Performance Limits Using Linear Control in a Buck Converter," *in Proc. IEEE COMPEL*, pp. 1–8, Jun. 2014.

[33] V. Yousefzadeh, E. Alarcon, and D. Maksimovic, "Three-Level Buck Converter for Envelope Tracking Applications," *IEEE Trans. Power Electron.*, vol. 21, no. 2, pp. 549–552, Mar. 2006.

A Continuous Actor-Critic Maximum Power Point Tracker Applied to Low Power Wind Turbine Systems

J. L. Wattes, A. J. S. Dias Júnior, A. P. S. Braga, P.P. Praça, A. U. Barbosa, D. S. Oliveira Júnior
Group of Processing of Energy and Control - GPEC
Department of Electrical Engineering, Federal University of Ceara - UFC
Fortaleza, Brazil
E-mail: jorgewatte@alu.ufc.br

***Abstract*—This paper aims to present an application of a continuous states and actions actor-critic algorithm based on artificial neural networks function approximations, capable of tracking the maximum power point of a wind turbine system. The application does not use any mechanical sensor for wind. The proposed system is composed of a three phasic permanent magnet synchronous generator of 1 kW, a diode rectifier, a Boost converter and a resistive load. The algorithm consists of two Multi-layer Perceptron Neural Networks (NN) witch approximate the state value function and the DC bus voltage reference, the power signal feedback adjusts NN parameters through backpropagation algorithm. To validate the proposed algorithm, simulations using average models of the system are made on Simulink/Matlab software with different wind speed profiles. A wind turbine emulator test bench using a DC motor to emulate the torque made by the wind on the generator rotor provides experimental results Over more dynamic wind profiles tested, the proposed MPPT algorithm response overcomes the classic P&O.**

Keywords—Wind Energy Converter, MPPT; Artificial Intelligence, Reinforcement Learning

I. Introduction

Electric Generation Systems based on wind turbine are a growing slice from the global generation chart [1]. Due to its environmental friendly characteristic, many support programs were developed and the number of wind farms keep increasing. However, Wind Energy Converter Systems (WECS) have the characteristic of be highly dependent on the wind speed. Even when the wind has a level in which the kinetic energy is high enough to generate electrical energy, the aerodynamic of the turbine can cause a drop of generation.

This effect can be better understood over the generic turbine power coefficient (C_p) curve [2] given in (1).

$$C_p(\lambda,\beta) = 0.5(116\lambda' - 0.4\beta - 5).e^{-21\lambda'} + 0.01\lambda$$
$$\lambda' = \frac{1}{\lambda + 0.08\beta} - \frac{0.035}{\beta^3 + 1} \quad (1)$$

where β is the blade pit angle, in degrees, and λ is a mathematical value that relates angular rotation speed, in hertz, (ω_m), blade swiped area radius, in meter, and wind speed, in meter per second (v_w), as shown in (2).

$$\lambda = \frac{radius.\omega_m}{v_w} \quad (2)$$

To clarify the behavior of $C_p(\lambda, \beta)$ function presented in (1), the Fig. 1, presents graphics with examples of the function for four different fixed values of β.

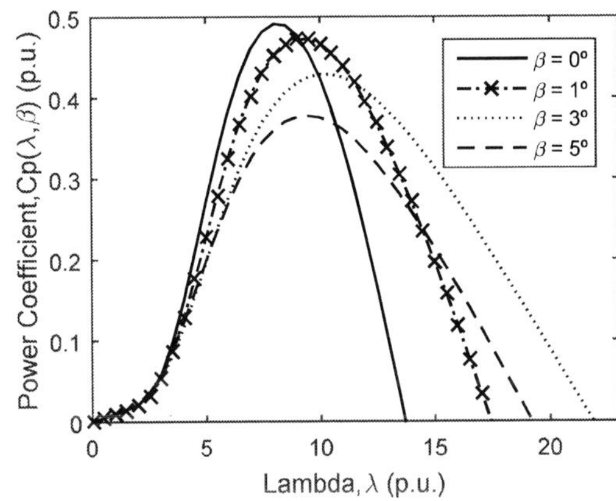

Fig. 1. Generic power coefficient graphic

As seen in the examples of Fig. 1, for each β graphic there is only one value of λ that maximizes C_p. Once Power Coefficient represents the ratio between the absorbed kinetic energy and the air mass total kinetic energy, the λ that maximizes C_p, also maximize the extracted energy.

Assuming a fixed β, the turbine should always operate at a fixed λ. However, as seen in (2), λ is a relation between ω_m and v_w which means that for each wind speed, there is an optimum rotation speed that extracts the maximum energy of the wind. Once v_w can have high variation in short periods, it is a challenge to define and operate with the optimal ω_m.

978-1-4673-9551-9/16 $31.00 © 2016 IEEE

A. Maximum Power Point Trackers

Maximum Power Point Trackers (MPPT) algorithms seek the most efficient operation point continuously [3]. For a wind turbine, it means to seek for a ω_m that for the instantaneous v_w would result in the optimum λ.

There are three basic types of MPPT for wind turbines: Wind Speed Measurement (WSM), Power Signal Feedback (PSF) and Perturb & Observe (P&O)[3]–[5]. The WSM method uses wind speed sensors, which simplify seek for the optimum ω_m, once know the optimal λ. However, this method increases the final cost, as a safer system uses more than one sensor, and all of them require regular maintenances. PSF systems are based in recording, for several values of v_w, the maximum generated power and the related optimal ω_m. During operation, ω_m is set to track the curve of collected data. The main drawback of this method is the recording of data, which must be made through aerodynamic simulation or tests.

The P&O MPPT, has an advantage over WSM and PSF it doesn't need to measure v_w, neither to have a preview knowledge of the turbine C_p curve. P&O MPPT consists of (i) apply a variation in ω_m, (ii) wait the stabilization and (iii) observe the behavior of the generated power. If it grows, the next variation keeps the same direction; otherwise, it will lead to the opposite direction. This is one of the most used MPPT techniques due to its simplicity, however it also have some difficulties - the size of the step can be hard to define. If a small step size is set, a great number of steps might be needed before reaching the maximum power point, once the wind changes in short time, it can be very problematic. Otherwise, if the step size is set too big, the maximum power point can be never reached. In addition, the step size is inversely linked with the oscillation around the maximum power point. Fig. 2. Illustrate the problematic P&O step size definition.

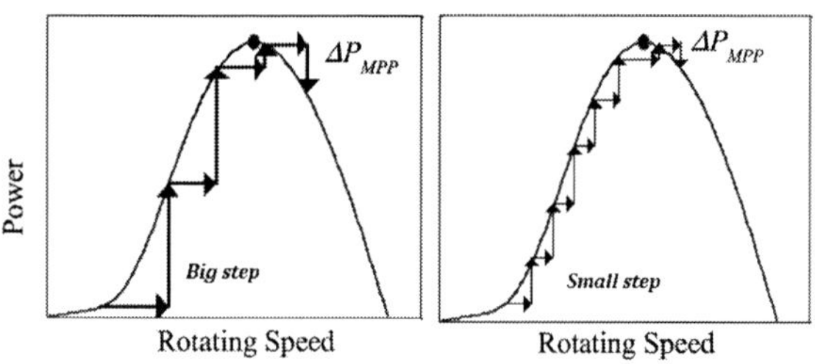

Fig. 2. P&O step size duality between oscillation and number of steps

As noticed, P&O algorithms do not find the exactly maximum power point they just oscillate around it. In addition, it does not have any memory of the already visited operation points. Some work had been done with intelligent algorithms, to improving those flaws.

Chun Wei presents on [6], [7] a MPPT based on reinforcement learning (RL) which uses extensive lookup tables [6], or a set of 3 Multilayer Perceptron (MLP) Neural Network (NN) as the evaluation functions for each possible action [7]. In PSF approach, Hui Li presents in [8] an offline trained NN that estimates the optimum rotational speed based on inputs of power and rotation speed with posterior adjust of draft.

This paper proposes an intelligent wind turbine MPPT based on reinforcement learning [9]. The MPPT applies the Continuous Actor-Critic Learning Automaton (CACLA) algorithm proposed in [10].

II. Power Electronic Converter

In the field of small power wind turbines, it is common to have three-phasic Permanent Magnet Synchronous Generators (PMSG) without use gearbox. In those turbines, the rotation can be controlled through the electrical torque component, which is directly related with the current drained from the generator. To control this electrical variable, a power electronic converter is needed.

Similar as what is done in [11], a three-phasic Graetz bridge rectifier connects the variable frequency output of the turbine to a DC bus. To control the drained current, a Boost converter with input current control is done. Fig. 3. Presents the power electronic circuit schematic.

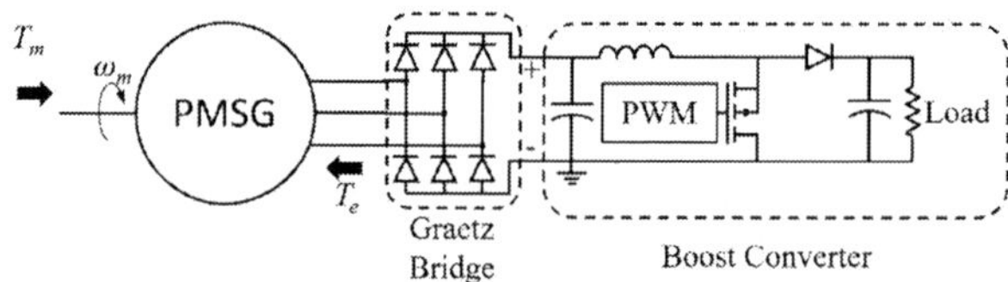

Fig. 3. Power electronic circuit diagram

The current control is done through a PI loop control and its reference comes from ω_m a PI loop control. The CACLA-MPPT algorithm gives the ω_m reference value. Fig. 4. Presents the diagram of the controlled system.

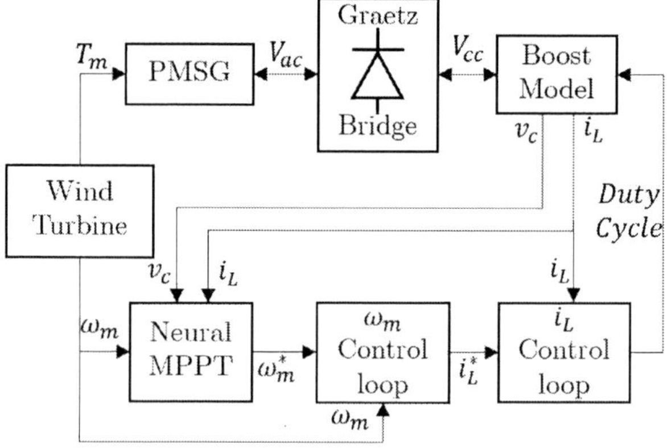

Fig. 4. Schematic of the proposed controlled WECS

The controllers where projected as discrete, the current controller sample rate was defined as 20 kHz, while the rotational speed controller sample rate was set to 2 kHz. Due to the limit of ω_m fluctuates with the wind speed variation, the MPPT algorithm verifies the ω_m controller saturation. If the control error module is bigger than 0.1, the algorithm consider it as saturated.

III. Continuous Actor-Critic

Reinforcement Learning (RL) is a very used technique for model free control algorithms [9]. RL algorithms replicate the way human beings learn tasks even without have been taught, as example the act of walk.

The RL agent observes the state of operation (s_t), takes a partially random action (a_t), observes the new state of operation (s_{t+1}), and define, through a reward (r_t), how good the taken action was. Based on those observations, the RL algorithms estimates a state value function ($V(s_t)$) for each state, or an action value function ($Q(s_t,a_t)$) for each state/action pair. Those functions estimates the maximum accumulated future rewards.

The $V(s_t)$ describe how good is to be in s_t based on the state value of the possible next states ($V(s_{t+1})$) and the reward (r_t) obtained from the transition; the same is valid for $Q(s_t,a_t)$, taking in count the possible future action-state pairs. When a state transition model is available, the $V(s_t)$ function is calculated iteratively as the algorithm wander through the states, and its value is updated according to (3).

$$V_t(s_t) = V_{t-1}(s_t) + \alpha\delta_t \quad (3)$$

where α is the learning rate and δ_t is the Temporal Difference error (TD error) given by (4).

$$\delta_t = r_t + \gamma V_t(s_{t+1}) - V_{t-1}(s_t) \quad (4)$$

where γ is a discount factor applied to forward states values. This parameter is necessary in systems without end episodes to limit the value function. When a state transition model is not available, the $Q(s_t,a_t)$ function is calculated similarly to (3) and (4), but considering the action-state pairs.

For discrete states, the convergence of the algorithm with such equations is proofed in [12]. A problem faced by this algorithm is that for each possible state it is necessary to store the value, which can be unfeasible in problems with large or continuous number of states.

A solution to this problem is to use function approximation technique such as Neural Networks [13] as done in [7]. To deal with the similar problem found in continuous actions the CACLA algorithm was proposed in [10].

The CACLA algorithm is composed of two MLP: the first one approximates the $V(s_t)$ function and it's named Critic, the second MLP approximates the $Q(s_t,a_t)$ function that gives the best action to be taken when in s_t and it is named Actor.

The critic MLP is responsible to establish a dynamic reward for the actor. The adjust of the critic NN weights is done through backpropagation of the TD error as shown in (5).

$$w_{t+1}^{critic} = w_t^{critic} + \alpha.\delta_t.\frac{\partial V_t(s_t)}{\partial w_t^{critc}} \quad (5)$$

The actor normally defines the action value function, however, CACLA algorithm uses the actor to obtain the action with the highest value function, instead of the value function itself. Due to the imperative necessity of explore better policies, a random factor is applied to the action decision process. This factor can be established through ε-greed or a Gaussian exploration.

After every taken action the actor NN weights are adjusted as showed in (6).

$$\begin{aligned} & if(\delta_t > 0): \\ & w_{t+1}^{actor} = w_t^{actor} + \alpha.(a_t^{taken} - a_t^{actor})\delta_t.\frac{\partial a_t^{actor}}{\partial w_t^{actor}} \end{aligned} \quad (6)$$

In which a_{taken} is the taken action and a_{actor} is the action with the highest action value, defined by the actor NN output. The "if" clause, indicates that the Actor will only be adjusted towards the a_{taken}, when the Critic observe a better state value.

In CACLA algorithm, the update is done when a better state value is achieved, since a worse state wouldn't mean that an action in the opposite direction would be better [10]. However, the proposed application has characteristics that allow it to consider that if a worse evaluation was found, the Actor can be adjusted towards the opposite way of the taken action.

IV. The Proposed Maximum Power Point Tracker

As way to validate the application, theoretical wind turbine parameters are set: 350 W of maximum power; 0.69 m blades length, $C_p(\lambda, \beta)$ function with peak at 0.35 C_p at λ of 8 for a nil β.

To apply the CACLA algorithm to a MPPT strategy, some parameters must be defined. The state of operation of the turbine is defined as $S=\{s|s=(P_e, \omega_m)\}$, where P_e is the electrical power output, and ω_m is the turbine rotational speed, estimated through the measured electrical signals. The action is defined as $A=\{a|a=\omega_{m\_ref}\}$, in which ω_{m_ref} stands for the rotational speed reference.

The random parcel of the action is set through ε-greed exploration policy. Which mean a probability ε to take the best action and (1- ε), to take a random action. This random action selection is uniformly distributed in a range of ±σ. The generic code for the ε-greed is presented in Fig. 5.

```
if ε>random(0→1)
    a_taken=a_act
else
    a_taken=a_act · (1+random(-σ→+σ))
```

Fig. 5. Generic code of ε-greed exploration policy

$$\begin{cases} if(|P_e(t) - P_e(t-1)| < dP_{e\_min}) : r = 0 \\ elseif(P_e(t) - P_e(t-1) > dP_{e\_min}) : r = 0.5 \\ elseif(P_e(t) - P_e(t-1) < -dP_{e\_min}) : r = -1 \end{cases} \quad (7)$$

The reward is given as in (7), where $dP_{e\ min}$ indicates the minimum variation of P_e that can be considered a good or bad variation. The asymmetrical rewards given to an increase (0.5) and decrease (-1) on P_e, aims to avoid oscillation of increment and decrement of ω_m reference, favoring a steady behave of it.

The dP_{e_min} value is defined as 2W, based in rated power PMSG.

Other parameters to be defined are the actor learning rate (α) equals to 0.01, the critic learning rate (β) defined as 0.95, the critic discount factor (γ) established as 0.90. The random parcel of the ε-greed exploration policy of maximum value (σ) as 0.025, and the greed coefficient (ε) is fixed at 0.3.

The period of iterations is related to the machine inertia and was set to 1 second, define through simulated data. The flowchart of the algorithm is shown in Fig. 6.

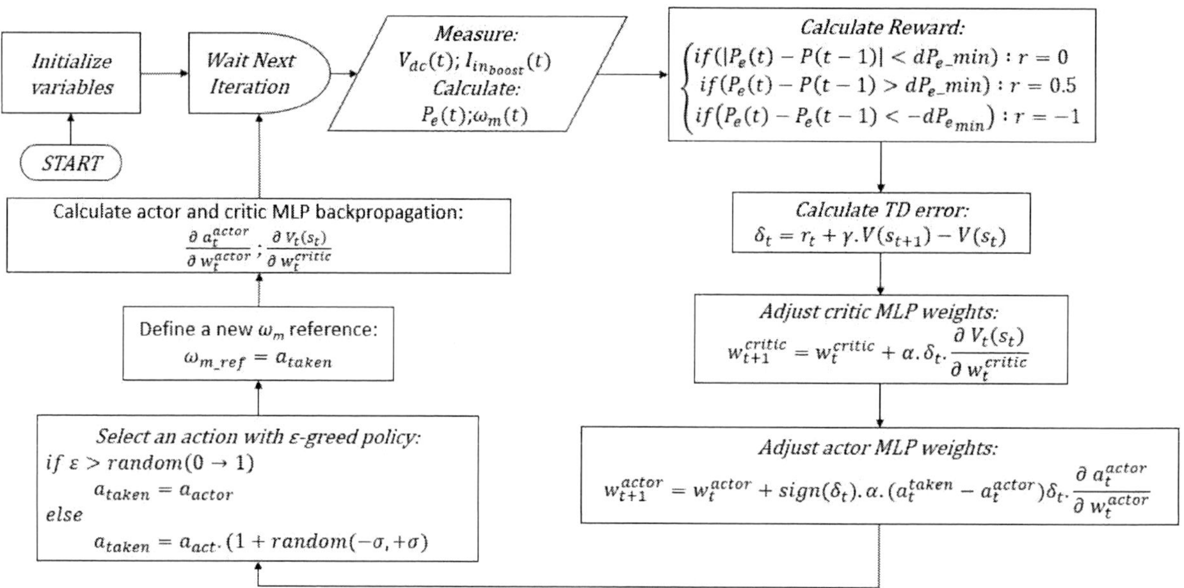

Fig. 6. CACLA-MPPT algorithm flowchart

V. SIMULATIONS RESULTS

The simulation of the preview defined wind turbine and MPPT are done on Matlab/Simulink environment. Due to the time constant of the current controller, the rotation speed controller, and the CACLA-MPPT, validation were done independently aiming to reduce the time consume.

The current and rotation speed PI controller are validated with reference steps of 25% the max values of current and rotation speed. The results are shown in Fig. 7.

The validation of the proposed algorithm is made for difference types of wind profiles. Fig. 8. presents the P_e and ω_m responses for the proposed algorithm over a fixed value of wind. It can be seen that after 500 seconds the algorithm converge. In this case the weights were initialized as zeros. After convergence it presents deviations due the ε-greed exploration policy.

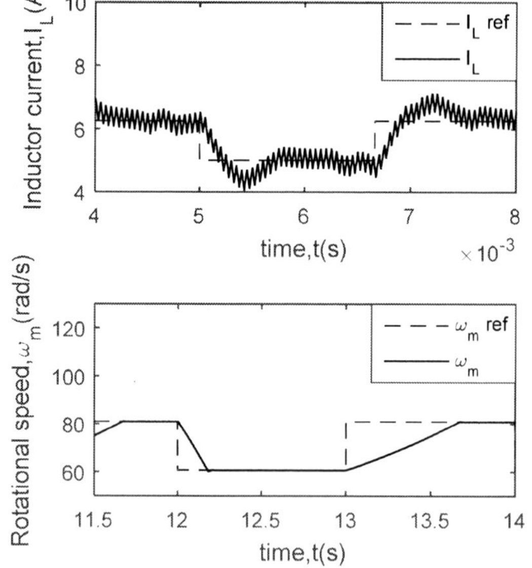

Fig. 7. Current and rotation speed PI controller's validation

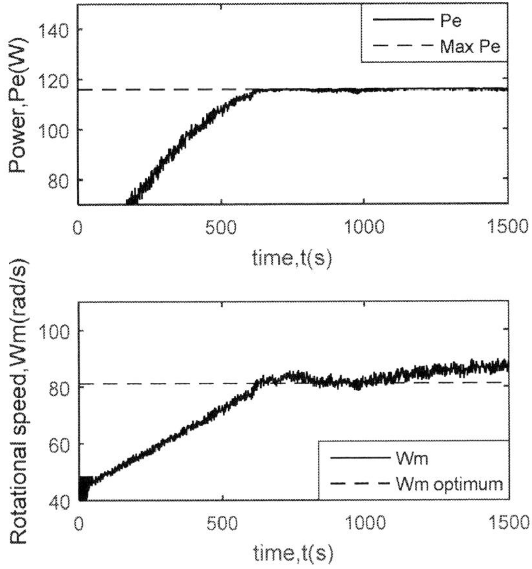

Fig. 8. Electrical power and rotational speed response in a fixed wind speed

Seeking to observe the algorithm response to high variance wind speed profile, a squared shaped wave with steps of 6.5 and 7.5 m/s, with a 120 seconds period is used as wind profile. A comparison is made with a fixed step P&O MPPT algorithm. Fig. 9. presents those results. It can be seen that CACLA algorithm takes some time learning in the beginning, but after some time it began to stablish an operation point that, with lower oscillation, would generate equivalent energy, given that wind profile.

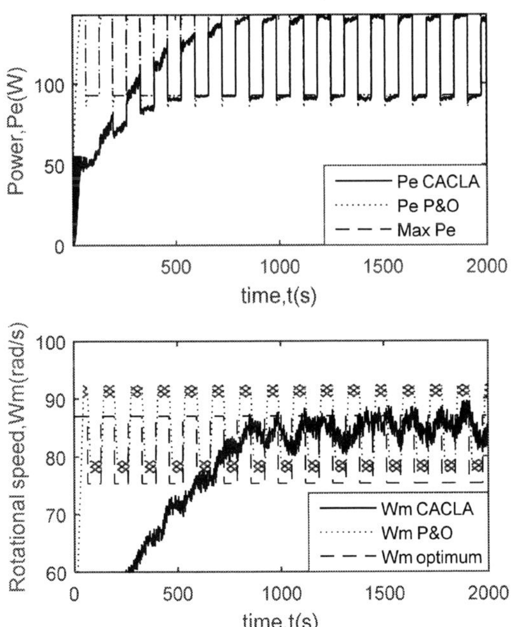

Fig. 9. Electrical power and rotational speed response in a squared shape wind speed profile

A generic multi sinusoidal wave is also used as wind profile for validation. This wind profile is shown in Fig. 10. Fig. 11. and Fig. 12 shows values of P_e and ω_m for the CACLA-MPPT and a fixed step P&O MPPT algorithm.

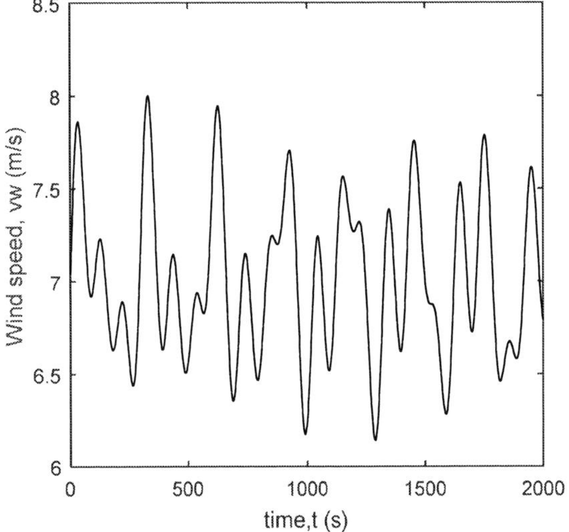

Fig. 10. Multi sinusoidal wind profile

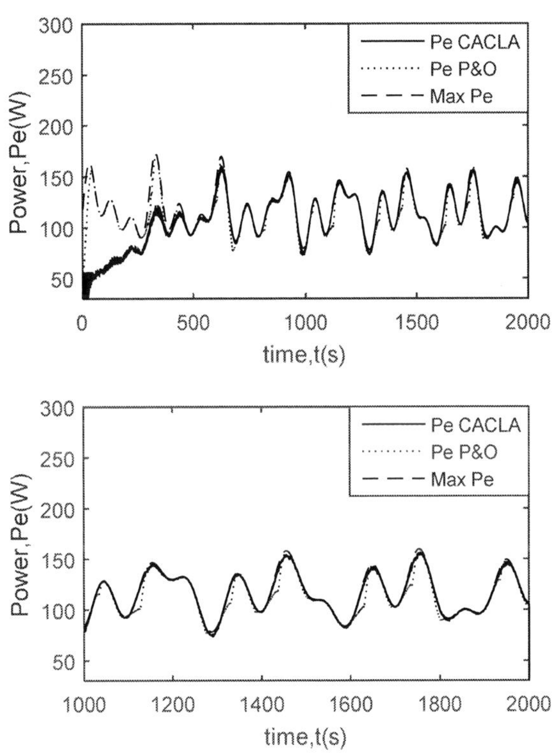

Fig. 11. Electrical power response to dynamic wind profile

As can be seen, in Fig. 11. after convergence, the CACLA algorithm has a better response over wind variation. Fig. 12.

shows that ω_m tends to stablish in an average point, reducing the oscillation, but maintaining the power generation.

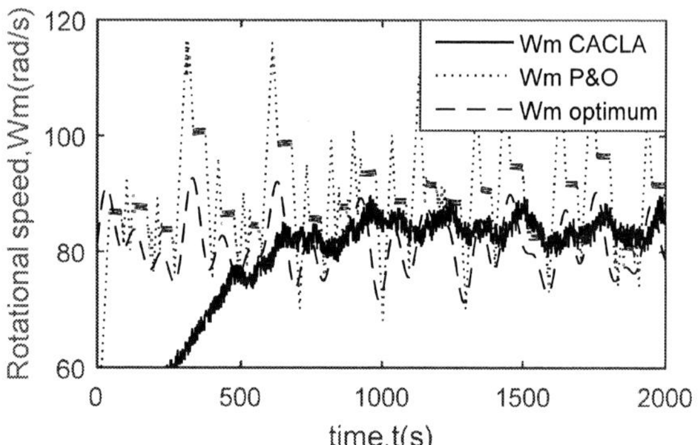

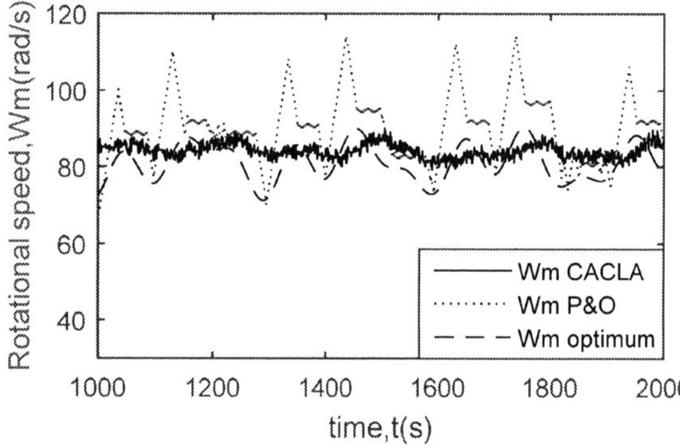

Fig. 12. Rotational speed response to dynamic wind profile.

CONCLUSIONS

As it could be seen, the proposed algorithm is able to reach the MPP for a fixed wind speed, although it take some time in earlier learning stages. As mentioned, the ω_m oscillations around the optimum operation point is due the exploration policy. However, in long term applications, it can be decreased after a wide range of operation points had been accessed.

The CACLA-MPPT algorithm response to wind steps gets improved with time, although it tends to track an average point of operation that would result in the maximum extracted energy with lesser ω_m variation.

As proposed by [10] the algorithm indeed reduced computational cost compared with the look-up table methods, especially on the proposed application where the action discretization would require a large memory storage.

Over more dynamic wind profiles, the proposed algorithm response overcomes the classic P&O. While the wind speed increasing misleads the P&O, the proposed algorithm was succeeding on tracking the MPP.

References

[1] The World Wind Energy Association, "2014 Report Half-year," *2014 half-year Rep.*, vol. 1, p. 8, 2014.

[2] S. Heier, *Grid Integration of Wind Energy: Onshore and Offshore Conversion Systems, 3rd Edition*, 3rd Editio. Wiley, 2014.

[3] A. M. Bhandare, P. J. Bandekar, and S. S. Mane, "Wind energy maximum power extraction algorithms: A review," *2013 Int. Conf. Energy Effic. Technol. Sustain.*, no. 1, pp. 495–500, 2013.

[4] Z. M. Dalala, Z. U. Zahid, and J.-S. Lai, "New Overall Control Strategy for Small-Scale WECS in MPPT and Stall Regions With Mode Transfer Control," *IEEE Trans. Energy Convers.*, vol. 28, no. 4, pp. 1082–1092, 2013.

[5] Q. Wang and L. Chang, "An intelligent maximum power extraction algorithm for inverter-based variable speed wind turbine systems," *Power Electron. IEEE Trans.*, vol. 19, no. 5, pp. 1242–1249, 2004.

[6] C. Wei, Z. Zhang, W. Qiao, and L. Qu, "Reinforcement-Learning-Based Intelligent Maximum Power Point Tracking Control for Wind Energy Conversion Systems," *IEEE Trans. Ind. Electron.*, vol. 62, no. 10, pp. 6360–6370, 2015.

[7] C. Wei, Z. Zhang, W. Qiao, and L. Qu, "Intelligent Maximum Power Extraction Control for Wind Energy Conversion Systems Based on Online Q-learning with Function Approximation," pp. 4911–4916, 2014.

[8] H. Li, K. L. Shi, and P. G. McLaren, "Neural-network-based sensorless maximum wind energy capture with compensated power coefficient," *IEEE Trans. Ind. Appl.*, vol. 41, no. 6, pp. 1548–1556, 2005.

[9] R. S. Sutton, A. G. Barto, and A. B. Book, "Reinforcement Learning□: An Introduction," 1998.

[10] H. Van Hasselt and M. a. Wiering, "Reinforcement learning in continuous action spaces," *Proc. 2007 IEEE Symp. Approx. Dyn. Program. Reinf. Learn. ADPRL 2007*, no. Adprl, pp. 272–279, 2007.

[11] H. M. O. Filho, D. S. O. Jr, R. P. T. Bascopé, C. E. a Silva, and G. J. Almeida, "on the Study of Wind Energy Conversion System Applied To Battery Charching Using Multiblade Turbines," *Control*, pp. 964–971, 2009.

[12] R. S. Sutton, "Learning to predict by the methods of temporal differences," *Mach. Learn.*, vol. 3, no. 1, pp. 9–44, 1988.

[13] S. Haykin, *Neural networks and learning machines*. 2009.

Multi-Band Mixed-Signal Hysteresis Current Control for EMI Reduction in Switch-Mode Power Supplies

Arindam Mandal, *Student Member, IEEE*, V. Inder Kumar, *Student Member, IEEE*, and Santanu Kapat, *Member, IEEE*
Embedded Power Management Lab
Department of Electrical Engineering
Indian Institute of Technology Kharagpur, West Bengal, India-721302
Email: arindam.ee.iitkgp@gmail.com, indervedula@gmail.com, and santanu.kapat@ieee.org

***Abstract*—Switched mode power supply (SMPS) is used in majority of portable devices and plays the central role for energy-efficient power management. Recent developments in digital VLSI allow high performance control implementation with ultra-fast recovery and high power density. However, electromagnetic interference (EMI) is a major concern in SMPS, which interferes with nearby devices and input power lines. This paper proposes a multi-band mixed-signal hysteresis current control (MBMSHCC) technique in a DC-DC converter. This considers the fast-changing current-loop in the analog domain, whereas a digital voltage controller $G_c(z)$ is used to achieve fast recovery through real-time configuration. This achieves robust stability and parameter insensitive current ripple by sampling the output voltage at the rising edge of the high side gate signal. Closed-loop stability is shown to be insensitive to the input voltage, but it is dependent on the capacitor ESR. Thus a stable behaviour is achieved over a wide input voltage range, even under periodic modulation of the hysteresis band. This periodic multi-band operation significantly reduces the spectral peaks with insignificant impact on the ripple parameters and the transient performance. A synchronous buck converter prototype is made, and the proposed scheme is implemented using an FPGA device.**

I. Introduction

Switch mode power supplies suffer from conducted and radiated EMI issues because of fast changing voltage and current in them [1]. This causes interference with nearby devices as well as input power lines. Therefore EMI reduction is a very important issue in DC-DC power converters. Several techniques are used for EMI reduction, such as passive EMI filters, and spectral spreading techniques employing randomized switching frequency modulation [2] or chaos based modulation scheme [3]. However, it is difficult to correctly predict ripple parameters and so as for the frequency. Periodic frequency modulation in [4] can achieve spectral spreading with predictable ripple parameters. A periodic bi-frequency scheme was introduced in [5], where time periods are generated using variable frequency approach. However, closed loop stability as well as dynamic performance have not been reported. A reconfigurable bi-frequency digital PWM technique is presented in [6], [7] in which custom spread spectrum is achieved with predictable ripple parameters and without significant impact on dynamic performance. However, it considers voltage-mode control, and there are scopes to further develop spectral spreading techniques under current-mode control. Hysteresis current control technique [8]– [14] is characterized by its inherent current loop stability, fast transient response, and good accuracy. However, they suffer from parameter sensitive current ripple and varying switching frequency. Its analog implementation requires considerable hardware resources to overcome parameter sensitivity [13], [14]. On the other hand digital implementation may exhibit multiple limit cycle oscillation [15], and it also requires multiple A/D converters [8]. A mixed-signal hysteresis current control is useful as it considers the conventional approach of using inductor current in the analog domain [16]– [18].

A randomized hysteresis current control in [19], [20] attenuates spectral peaks; however, ripple parameters become unpredictable. Thus it is important to investigate a mixed-signal HCC scheme which would ensure (i) reduced switching harmonics, (ii) predictable current ripple, (iii) parameter insensitivity, (iv) robust stability, and (v) reduced hardware resources. This paper proposes a multi-band mixed-signal hysteresis current control scheme with an analog current loop and digital voltage controller $G_c(z)$. Two discrete hysteresis current bands are used to generate two discrete time periods in order to achieve harmonic reduction.

II. Proposed Multi-band Mixed-signal Hysteresis Current Control Technique

A. *Working principle of the proposed MBMSHCC scheme*

The schematic of the proposed multi-band mixed-signal hysteresis current control technique is shown in Fig. 1. The 'timing circuit' block generates the clock F_{vs} to sample the output voltage v_o, which is also used for the computation of digital voltage controller $G_c(z)$. The digital controller $G_c(z)$ takes the error voltage $v_e[n]$ as the input to generate the peak current reference $i_{ref}[n]$. The 'real-time band adaptation' block generates the variable hysteresis band Δi_{Hr} (where $r \in [1, 2]$) for variable frequency operation which is essential for custom

978-1-4673-9551-9/16 $31.00 © 2016 IEEE

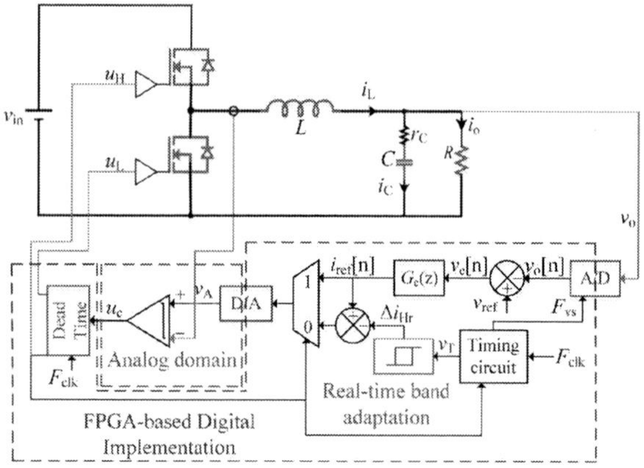

Fig. 1. Proposed multi-band mixed-signal hysteresis current control architecture

harmonic reduction, and the valley current reference is generated as $i_{ref}[n] - \Delta i_{Hr}$. The peak and valley current reference are fed to a multiplexer whose select line is the high side gate signal u_H. Thus one D/A converter is sufficient to generate the analog current reference and only one comparator is required in the proposed HCC scheme. The output of the D/A conveter v_A is compared with the sensed inductor current in the analog domain. The 'dead-time' circuit generates latched gate signals using the comparator output u_c. This dead time circuit attempts to avoid any shoot through current.

B. Generation of synchronous/asynchronous sampling clock

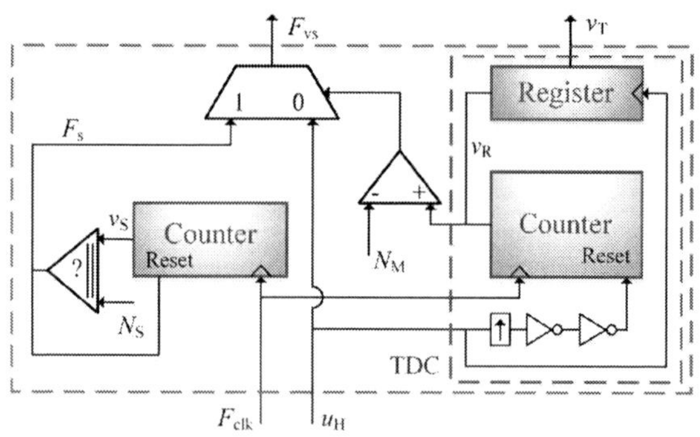

Fig. 2. Schematic of the timing circuit

The 'timing circuit' block shown in Fig. 2 consists of two free running counters which are incremented at the rising edge of the FPGA controller clock F_{clk}. The right-hand-side counter counts digital coded switching time period v_R and at the rising edge of the high side gate signal u_H, it gets stored in the register as v_T. Thus the region covered by the blue dotted line represents the time to digital converter (TDC). The left-hand-side counter generates a fixed frequency clock F_s with the time period $T_s = (N_S + 1) \times t_{clk}$. The MUX takes this synchronous clock F_s and an asynchronous clock u_H as inputs and then generates the sampling clock F_{vs} for the A/D converter. During a large-signal transient, asynchronous sampling can cause the system to collapse, because the error voltage cannot be updated for few cycles, when u_H does not change its state. In the proposed scheme, if u_H remains same for a duration longer than $(N_M + 1) \times t_{clk}$, v_o is sampled using the fixed frequency clock F_s. Thus the proposed control scheme can successfully operate during large signal transients, and the high side gate signal is used as an asynchronous sampling clock during steady state operation to sample the output voltage.

C. Real-time hysteresis band adaptation

As v_o is sampled at the rising edge of the asynchronous sampling clock u_H which has the same frequency as v_o, peak current reference $i_{ref}[n]$ remains constant through out the periodic interval and thus inductor current ripple will be same as the hysteresis current band in the proposed scheme. So, the switching time period T_h is related to Δi_H as

$$T_h = \left[\frac{v_{in} L}{(v_{in} - v_o) v_o}\right] \times \Delta i_H. \tag{1}$$

Two hysteretic current bands Δi_{H1} and Δi_{H2} will generate two discrete time periods T_{h1} and T_{h2} respectively.

III. Power Spectral Density Analysis of the Proposed Scheme

It is cumbersome to analytically derive the Power Spectral Density of the input current or output voltage which give the measure of the conducted and radiated EMI respectively. To provide analytical insights of the PSD of input current, the Fourier series coefficients of the high side gate signal u_H are evaluated here.

Let us consider N consecutive cycles with hysteresis current band of Δi_{H1} and another N cycles with hysteresis band of Δi_{H2}, where $\Delta i_{H1} = \Delta i_H - \delta i_H$ and $\Delta i_{H2} = \Delta i_H + \delta i_H$. As the switching time period is proportional to the hysteresis current band, time periods are $T_{h1} = T_h - \Delta T_h$ for first N cycles and $T_{h2} = T_h + \Delta T_h$ for next N consecutive cycles (where $\delta i_H / \Delta i_H = \Delta T_h / T_h$). The modulating time period is

$$T_m = N(T_{h1} + T_{h2}) = 2NT_h = 1/f_m. \tag{2}$$

Fourier series coefficients of u_H can be expressed as

$$u_H(t) = \sum_{k=-\infty}^{\infty} c_k e^{j2\pi k f_m t}; \quad c_k = \frac{1}{T_m}\int_0^{T_m} u_H(t) e^{-j2\pi k f_m t} dt. \tag{3}$$

From (3) power spectrum $|c_k|^2$ using the Fourier coefficient can be derived, and it can be simplified as

$$|c_k|^2 = \left(\frac{1}{\pi k}\right)^2 \left[x_1 + (-1)^k x_2\right]^2; \quad \text{where, } x_r = \sin(\pi\, k\, f_m\, t_{hr}) \frac{\sin(\pi\, k\, f_m N\, T_{hr})}{\sin(\pi\, k\, f_m\, T_{hr})};\ r \in [1, 2]. \tag{4}$$

For k being an integral multiple of $2N$, the above expression of $|c_k|^2$ can be further simplified as

$$|c_{2Nk}|^2 = \left[\frac{\sin(\pi kNx) \times \sin(\pi kD) \times \cos(\pi kDx)}{\pi kN \times \sin(\pi kx)}\right]^2; \quad (5)$$
$$\text{where}, x = \delta i_{\rm H}/\Delta i_{\rm H} \text{ and } D = t_{\rm h1}/T_{\rm h1} = t_{\rm h2}/T_{\rm h2}.$$

In the case of fixed frequency DPWM, power spectrum becomes

$$|c_k|^2 = D^2 \text{sinc}^2(\pi kD). \quad (6)$$

From (5), and (6), the ratio of these two PSDs can be obtained as

$$\frac{|c_{2Nk}|^2_{\rm PSD-MBMSHCC}}{|c_k|^2_{\rm PSD-PWM}} = \left[\frac{\sin(\pi kNx) \times \cos(\pi kDx)}{N \times \sin(\pi kx)}\right]^2. \quad (7)$$

From (7), it is clear that the DC (at $k = 0$) PSD component of the gate signal $u_{\rm H}$ under the unmodulated PWM scheme remains unaffected by the proposed MBMSHCC schemes. Further, (7) also shows that PSD under the unmodulated PWM scheme will be attenuated at fundamental and harmonic components using the proposed scheme.

IV. Stability Analysis and Design of the Proposed Scheme

A. Stability analysis under peak current mode control

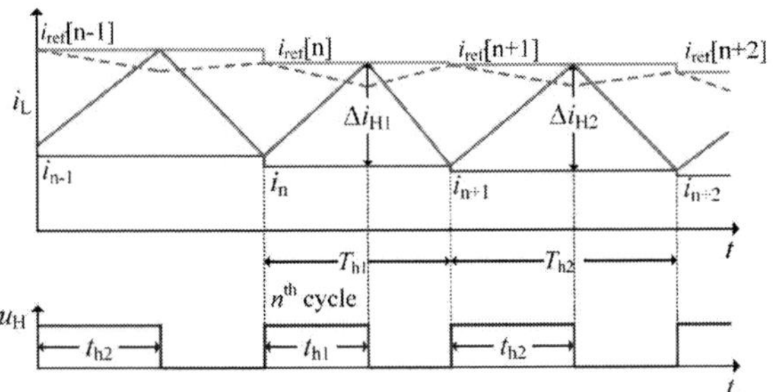

Fig. 3. Control waveforms of a buck converter under the proposed MBMSHCC scheme with peak current mode control

For an analytical study of the multi-band mixed-signal HCC, a simple proportional controller with an ESR dominated output voltage ripple is considered. Under this approximation, the output voltage ripple will be linear as well as in phase with the inductor current ripple. Thus the rising and falling slopes of $v_{\rm o}$ are $m_1 r_{\rm C}$ and $m_2 r_{\rm C}$, where m_1 and m_2 are the rising and falling slopes of $i_{\rm L}$, and $r_{\rm C}$ is the capacitor ESR. Small integral gain is added during steady state operation; however, discrete time integral gain is very less as compared to proportional gain, and hence its impact is not significant.

From Fig. 3, the inductor current $i_{\rm L}$ and output voltage $v_{\rm o}$ at the $n^{\rm th}$ rising edge of the high side gate signal $u_{\rm H}$ be $i_{\rm n}$ and $v_{\rm n}$, and those at the next rising edge of $u_{\rm H}$ be $i_{\rm n+1}$ and $v_{\rm n+1}$ respectively. Valley current $i_{\rm n}$ during the $n^{\rm th}$ cycle is

$$i_{\rm n} = i_{\rm ref}[n-1] - \Delta i_{\rm H2}. \quad (8)$$

Referring to Fig. 3, on-time $t_{\rm h1}$ during the $n^{\rm th}$ cycle is

$$\begin{aligned} t_{\rm h1} &= \frac{(i_{\rm ref}[{\rm n}] - i_{\rm n})}{m_1} = \frac{\Delta i_{\rm ref}[{\rm n}]}{m_1} + \frac{\Delta i_{\rm H2}}{m_1} \\ \text{where } &\Delta i_{\rm ref}[{\rm n}] = i_{\rm ref}[{\rm n}] - i_{\rm ref}[{\rm n}-1] = -k_{\rm p}\Delta v_{\rm n}; \\ &\Delta v_{\rm n} = v_{\rm n} - v_{\rm n-1}. \end{aligned} \quad (9)$$

Off-time $T_{\rm h1} - t_{\rm h1}$ during this period is

$$T_{\rm h1} - t_{\rm h1} = \frac{\Delta i_{\rm H1}}{m_2}. \quad (10)$$

The valley current during the $n + 1^{\rm th}$ cycle will be

$$i_{\rm n+1} = i_{\rm ref}[{\rm n}] - \Delta i_{\rm H1}. \quad (11)$$

The on-time $t_{\rm h2}$ and off-time $T_{\rm h2} - t_{\rm h2}$ during this period are

$$t_{\rm h2} = \frac{\Delta i_{\rm ref}[{\rm n}+1]}{m_1} + \frac{\Delta i_{\rm H1}}{m_1}; \; T_{\rm h2} - t_{\rm h2} = \frac{\Delta i_{\rm H2}}{m_2}. \quad (12)$$

Final current after the $n + 1^{\rm th}$ cycle will become

$$i_{\rm n+2} = i_{\rm ref}[{\rm n}+1] - \Delta i_{\rm H2}. \quad (13)$$

This shows that the current-loop is inherently stable irrespective of the input voltage conditions, provided that the outer voltage-loop remains stable. Thus the voltage-loop stability needs to be studied. Under the ESR-dominated linear ripple assumption, the final voltage $v_{\rm n+2}$ during the $n + 2^{\rm th}$ cycle will be

$$\begin{aligned} v_{\rm n+2} = v_{\rm n} &+ m_1 r_{\rm C} t_{\rm h1} - m_2 r_{\rm C}(T_{\rm h1} - t_{\rm h1}) \\ &+ m_1 r_{\rm C} t_{\rm h2} - m_2 r_{\rm C}(T_{\rm h2} - t_{\rm h2}). \end{aligned} \quad (14)$$

After simplification, it can derived that

$$v_{\rm n+2} - v_{\rm n} = -k_{\rm p} r_{\rm C}[v_{\rm n+1} - v_{\rm n-1}]. \quad (15)$$

From (15), we can conclude that the closed loop system using the proposed control scheme can be stabilized for

$$k_{\rm p} < 1/r_{\rm C}. \quad (16)$$

B. Analysing the effect due to comparator time delay

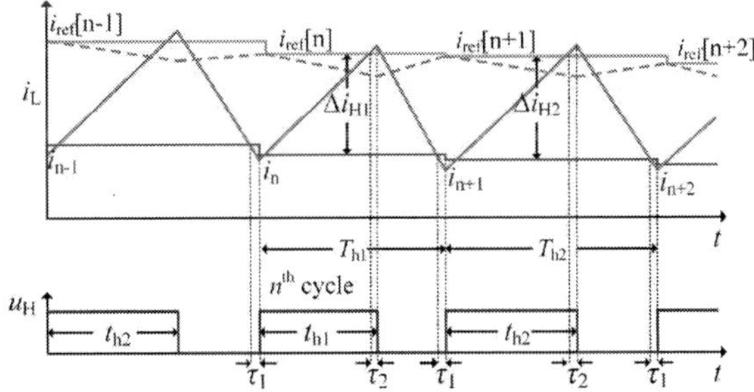

Fig. 4. Control waveforms with comparator turn-on and turn-off delay as τ_1 and τ_2 respectively

It is very important to investigate the effect of comparator time delay in the proposed multi-band MSHCC because closed loop operation completely depends on the current comparator.

Turn-on delay of the comparator is assumed to be τ_1 and turn-off delay is τ_2.

Referring to Fig. 4, valley current at the beginning of nth cycle is

$$i_{\text{n}} = i_{\text{ref}}[\text{n}-1] - \Delta i_{\text{H2}}. \tag{17}$$

On-time t_{h1} and off-time $T_{\text{h1}} - t_{\text{h1}}$ during the nth cycle can be derived as

$$\begin{aligned} t_{\text{h1}} &= \frac{i_{\text{ref}}[\text{n}] + m_1\tau_2 - i_{\text{n}}}{m_1} = \frac{\Delta i_{\text{ref}}[\text{n}]}{m_1} + \frac{\Delta i_{\text{H2}}}{m_1} \\ &\quad + \frac{m_1\tau_2 + m_2\tau_1}{m_1}; \\ T_{\text{h1}} - t_{\text{h1}} &= \frac{\Delta i_{\text{H1}}}{m_2} + \frac{m_1\tau_2 + m_2\tau_1}{m_2}. \end{aligned} \tag{18}$$

Similarly, on-time t_{h2} and off-time $T_{\text{h2}} - t_{\text{h2}}$ during the $n+1^{\text{th}}$ cycle can be derived as

$$\begin{aligned} t_{\text{h2}} &= \frac{\Delta i_{\text{ref}}[\text{n}+1]}{m_1} + \frac{\Delta i_{\text{H1}}}{m_1} + \frac{m_1\tau_2 + m_2\tau_1}{m_1}; \\ T_{\text{h2}} - t_{\text{h2}} &= \frac{\Delta i_{\text{H2}}}{m_2} + \frac{m_1\tau_2 + m_2\tau_1}{m_2}. \end{aligned} \tag{19}$$

Final inductor valley current $i_{\text{n}+2}$ after the $n+1^{\text{th}}$ cycle will be

$$i_{\text{n}+2} = i_{\text{ref}}[\text{n}+1] - \Delta i_{\text{H2}}. \tag{20}$$

Output voltage $v_{\text{n}+2}$ during the $n+2^{\text{th}}$ cycle can be derived in the same way described above, and can be simplified as

$$v_{\text{n}+2} - v_{\text{n}} = -k_{\text{p}} r_{\text{C}} [v_{\text{n}+1} - v_{\text{n}-1}]. \tag{21}$$

This shows that stability boundary remains same as (16), and hence it is unaffected due to time delay of the comparator under the proposed scheme.

Also for valley and average current mode control, closed loop stability condition can be derived in the same way and in all cases it remains as (16).

C. Stability analysis for a general case

From the above analysis we have observed that the stability condition remains unchanged for two consecutive variable hysteresis bands, of which one is Δi_{H1}, and another is Δi_{H2}. The above method can be extended for stability analysis in a periodic multi-band modulation of N_1 consecutive cycles with hysteresis band of Δi_{H1}, and N_2 cycles with hysteresis band of Δi_{H2}. The sampled output voltage $v_{\text{n}+\text{N}_1+\text{N}_2}$ at the end of $n + N_1 + N_2{}^{\text{th}}$ can be derived as

$$v_{\text{n}+\text{N}_1+\text{N}_2} - v_{\text{n}} = -k_{\text{p}} r_{\text{C}} [v_{\text{n}+\text{N}_1+\text{N}_2-1} - v_{\text{n}-1}]. \tag{22}$$

Thus the stability boundary is similar to that in (16), even under a periodic multi-band modulation.

D. Controller design

A simple PI controller is implemented in digital domain based on the small signal model of buck converter, operating under hysteretic current mode control. The controller implementation is as shown below.

$$\begin{aligned} v_{\text{c}}[n] &= k_{\text{p}}\left(v_{\text{ref}} - v_{\text{o}}[n]\right) + u_{\text{int}}[n] \\ u_{\text{int}}[n] &= u_{\text{int}}[n-1] + k_{\text{i}}\left(v_{\text{ref}} - v_{\text{o}}[n]\right) \end{aligned} \tag{23}$$

where k_{p} is he proportional gain, and k_{i} is the discrete time integral gain. The magnitude of the discrete time integral gain of a practical design is generally much smaller than that of the proportional gain; thus the latter is primarily responsible for the transient recovery. The above implementation provides an anti-windup arrangement by setting a saturation limit on u_{int}. It also provides the extra flexibility to completely disable the integral action, and also to store the value of $u_{\text{int}}[n-1]$ in the memory for future use. This can be used as the initial condition after re-initiating the integral action, which attempts to prevent an integrator wind-up. A voltage hysteresis band Δv_{th} can be used to detect a small-signal and/or a large-signal transient for the activation and/or the deactivation of the integral action. Also load current feed-forward is added to improve the transient response, and enables near load invariant regulation; thus a small integral gain is sufficient in the steady state.

E. Real time reconfigurability

The proposed control scheme can be reconfigured to a valley current mode control by generating the peak current reference as $i_{\text{ref}}[n]$+(Δi_{Hr}), and also can be extended to average current mode control with current references as $i_{\text{ref}}[n]$+($\Delta i_{\text{Hr}}/2$) and $i_{\text{ref}}[n]$-($\Delta i_{\text{Hr}}/2$). Hence it provides complete flexibility to set the desired current reference.

V. Hardware Implementation

A buck converter prototype has been made, and the proposed control technique is implemented using a (Virtex-5) FPGA device. The nominal parameters are $v_{\text{in}} = 8-12\,\text{V}$, $L = 10\,\mu\text{H}$, $C = 470\,\mu\text{F}$, nominal $\Delta i_{\text{H}} = 2\,\text{A}$, and capacitor ESR is $r_{\text{C}} \cong 30-50\,\text{m}\Omega$. The time period of the FPGA controller clock is $t_{\text{clk}} = 10\,\text{ns}$. In order to sample the output voltage a 10-bit differential pipeline A/D converter (AD9215) is used, and a differential amplifier (AD8138) reduces the common mode noise in ADC. The current reference is converted into an analog voltage v_{A} using a 12-bit D/A converter (AD9762) followed by a differential receiver amplifier (AD8130). The output of the DAC v_{A} is compared with the sensed inductor current using a high speed analog comparator (TLV3501). A current sense resistor of 10 mΩ is used to sense the inductor current and the sensed voltage is amplified using a high-side current sense amplifier (ADM4073) with a gain of 20.

A. Spectral shaping using the proposed MBMSHCC scheme

Figure 5 indicates steady-state operation under fixed frequency DPWM current mode control. Under the same test conditions, steady-state waveforms using the proposed multi-band mixed-signal HCC are shown in Figs. 6, 7 and 8 for peak, valley and average current mode control respectively. It can be clearly observed that the harmonic contents of the inductor current using fixed-frequency DPWM can be significantly attenuated using the proposed scheme, and this is true for all

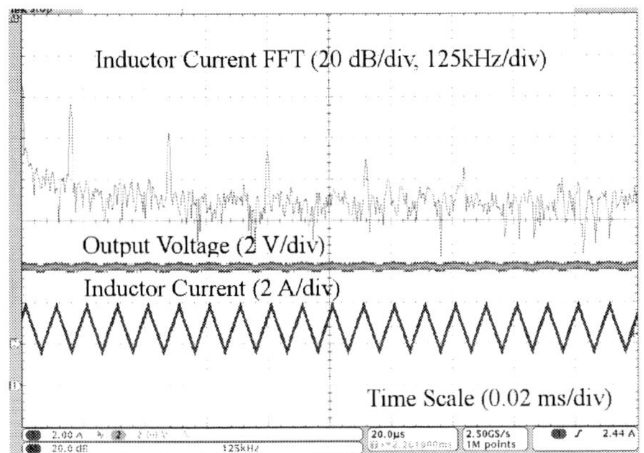

Fig. 5. Spectral behavior of a synchronous buck converter using the fixed frequency DPWM with $v_{\mathrm{ref}} = 4\,\mathrm{V}$

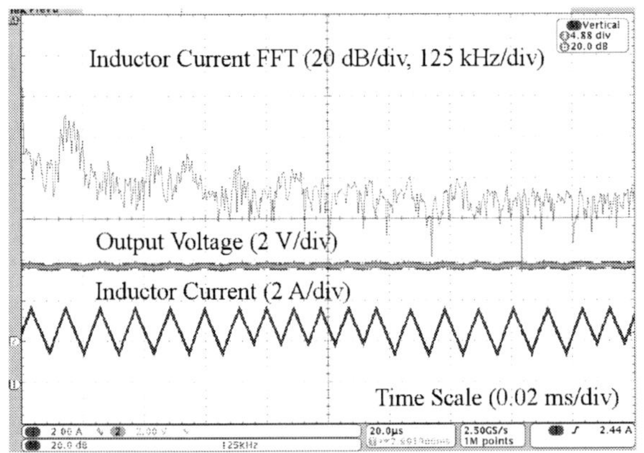

Fig. 6. Spectral behavior of a synchronous buck converter using the proposed MBMSHCC under peak current mode control with $N_1 = 5, N_2 = 5, \Delta i_{\mathrm{Hr}} = 2\,\mathrm{A} \pm 10\%$ and $\mathrm{v_{ref}} = 4\,\mathrm{V}$

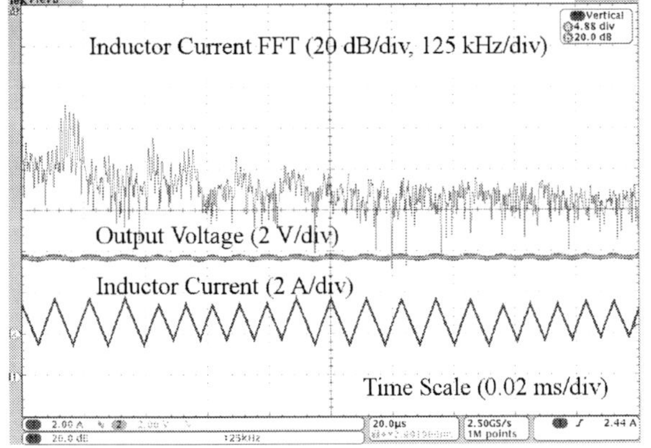

Fig. 7. Spectral behavior of a synchronous buck converter using the proposed MBMSHCC under average current mode control with $N_1 = 5, N_2 = 5, \Delta i_{\mathrm{Hr}} = 2\,\mathrm{A} \pm 10\%$ and $\mathrm{v_{ref}} = 4\,\mathrm{V}$

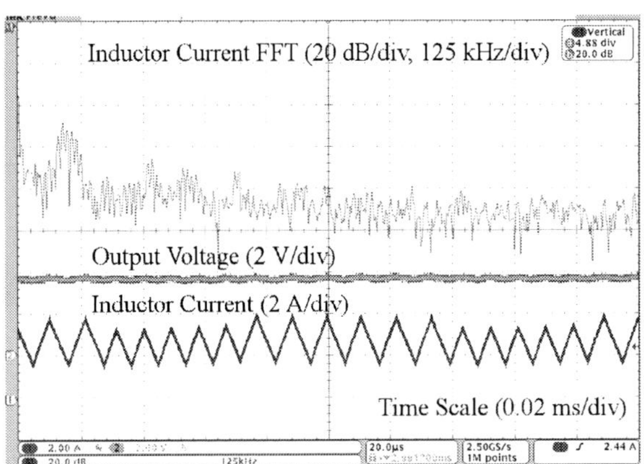

Fig. 8. Spectral behavior of a synchronous buck converter using the proposed MBMSHCC under valley current mode control with $N_1 = 5, N_2 = 5, \Delta i_{\mathrm{Hr}} = 2\,\mathrm{A} \pm 10\%$ and $\mathrm{v_{ref}} = 4\,\mathrm{V}$

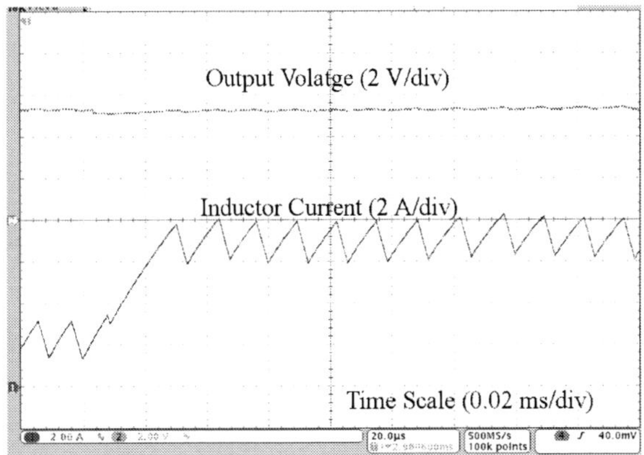

Fig. 9. Load (step) transient performance in a synchronous buck converter using the conventional MSHCC scheme for a load step of 2.2 A to 7 A with $v_{\mathrm{ref}} = 5\,\mathrm{V}$

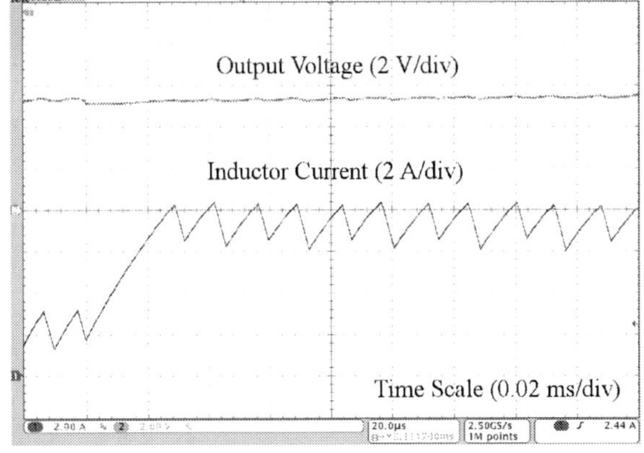

Fig. 10. Load (step) transient performance in a synchronous buck converter using the proposed MBMSHCC scheme for a load step of 2.2 A to 7 A with $N_1 = 1, N_2 = 1, \Delta i_{\mathrm{Hr}} = 2\,\mathrm{A} \pm 10\%$ and $\mathrm{v_{ref}} = 5\,\mathrm{V}$

three current control modes: peak, valley and average. Under the proposed MBMSHCC operation, the spectral peaks are reduced and spread over a wide frequency band. Also, in all the cases robust stability is guaranteed.

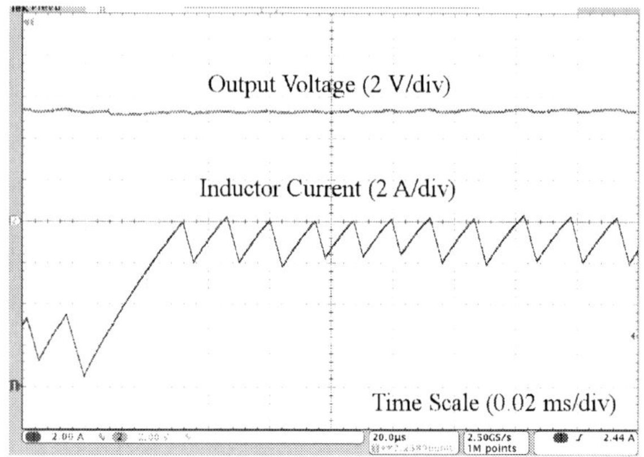

Fig. 11. Load (step) transient performance in a synchronous buck converter using the proposed MBMSHCC scheme for a load step of 2.2 A to 7 A with $N_1 = 3$, $N_2 = 3$, $\Delta i_{Hr} = 2\,A \pm 10\%$ and $v_{ref} = 5\,V$

B. Load transient performance

Figure 9 shows load (step-up) transient performance in a synchronous buck converter using the mixed signal hysteresis current control scheme for a step change in load current from 2.2 A to 7 A and $v_{ref} = 5\,V$. Settling time is approximately 35 μs, and voltage undershoot is around 200 mV. Moreover, a stable behaviour is obtained under steady state. Figures 10 and 11 show load (step-up) transient performance in a synchronous buck converter using the proposed MBMSHCC scheme under the same test conditions as above. Two discrete hysteresis current bands appear alternatively in the former case, while in the later case, each hysteresis current band occurs three consecutive times and repeats periodically. Figures show that the transient performance using the multi-band periodic modulation is almost unaffected compared to that using the unmodulated HCC scheme. Also, in all the cases a stable periodic behaviour is achieved using the proposed scheme.

VI. Conclusion

In this paper, a multi-band mixed-signal hysteresis current control scheme was proposed in a synchronous buck converter. Multi-band adaptation using the mixed-signal HCC scheme reduces the spectral peaks and completely controls the inductor current ripple with an insignificant performance impact. Also robust stability and parameter-insensitive current ripple are achieved using an asynchronous output voltage sampling. The proposed scheme reduces the hardware resources, and facilitates real time tuning of the controller to achieve fast transient recovery. These will be useful for high-frequency switching power converters with fast transient recovery and inherent current limiting. The proposed scheme is realized using an FPGA device and tested in a synchronous buck converter.

References

[1] K. Mainali and R. Oruganti, "Conducted EMI mitigation techniques for switch-mode power converters: a survey," *IEEE Trans. Power Electron.*, vol. 25, no. 9, pp. 2344–2356, Sept. 2010.

[2] F. Mihalic and D. Kos, "Reduced conductive EMI in switched-mode DCDC power converters without EMI filters: PWM versus randomized PWM," *IEEE Trans. Power Electron.*, vol. 21, no. 6, pp. 1783–1794, Nov. 2006.

[3] R. Mukherjee, A. Patra, and S. Banerjee, "Impact of a frequency modulated pulsewidth modulation (PWM) switching converter on the input power system quality," *IEEE Trans. Power Electron.*, vol. 25, no. 6, pp.1450–1459, Jun. 2010.

[4] F. Lin and D. Y. Chen, "Reduction of power supply EMI emission by switching frequency modulation," *IEEE Trans. Power Electron.*, vol. 9, no. 1, pp. 132–137, Jan. 1994.

[5] Y. F. Zhang, L. Yang, and C. Q. Lee, "EMI reduction of power supplies by Bi-Frequency modulation," *in Proc. IEEE APEC*, pp.601–607, Feb. 1994.

[6] S. Kapat, "Reconfigurable bi-frequency DPWM with custom spectral shaping in a synchronous buck converter," *in Proc. IEEE APEC*, pp.635–640, Mar. 2015.

[7] S. Kapat, "Reconfigurable Periodic Bi-frequency DPWM with Custom Harmonic Reduction in DC-DC Converters," *IEEE Trans. Power Electron.*, vol.PP, no.99, pp.1-1

[8] K. K. S. Leung and H. S. H. Chung, "Dynamic hysteresis band control of the buck converter with fast transient response," *IEEE Trans. Cir. Syst.-II*, vol. 52, no. 7, pp. 398–402, Jul. 2005.

[9] R. Redl and J. Sun, "Riple-based control of switching regulatorsan overview," *IEEE Trans. Power Electron.*, vol. 24, no. 12, pp. 2669–2680, Dec. 2009.

[10] J. C. Tsai, C. L. Chen, Y. H. Lee, H. Y. Yang, M. S. Hsu, and K. H. Chen, "Modified hysteretic current control (MHCC) for improving transient response of boost converter," *IEEE Trans. Cir. Syst.-I*, vol. 58, no. 8, pp. 1967–1979, Aug. 2011..

[11] S. Buso, S. Fasolo, L. Malesani, and P. Mattavelli, "A dead-beat adaptive hysteresis current control," *IEEE Trans. Ind. Appl.*, vol. 36, no. 4, pp. 1174–1180, Jul./Aug. 2000.

[12] W. Stefanutti and P. Mattavelli, "Fully digital hysteresis modulation with switching-time prediction," *IEEE Trans. Ind. Appl.*, vol. 42, no. 3, pp. 763–769, May/Jun. 2006.

[13] S. C. Huerta, P. Alou, J. A. Oliver, O. Garcia, J. Cobos, and A. A. Alfotouh, "Nonlinear control for DCDC converters based on hysteresis of the COUT current with a frequency loop to operate at constant frequency," *IEEE Trans. Ind. Electron.*, vol. 58, no. 3, pp. 1036–1043, Mar. 2011.

[14] S. C. Huerta, P. Alou, O. Garcia, J. A. Oliver, R. Prieto, and J. Cobos, "Hysteretic mixed-signal controller for high-frequency DC-DC converters operating at constant switching frequency," *IEEE Trans. Power Electron.*, vol. 27, no. 6, pp. 2690–2696, Jun. 2012.

[15] P. T. Krein and R. M. Bass, "Multiple limit cycle phenomena in switching power converters," *in Proc. IEEE APEC*, pp. 143–148, Mar. 1989.

[16] S. Saggini, M. Ghioni, and A. Geraci, "An Innovative Digital Control Architecture for Low-Voltage High-Current DC-DC converters with Tight Load Regulation," *IEEE Trans. Power Electron.*, vol. 19, no. 1, pp. 210–218, Jan. 2004.

[17] O. Trescases, A. Prodic, and W. T. Ng, "Digitally Controlled Current-Mode DC-DC Converter IC," *IEEE Trans. Cir. Syst.-I*, vol. 58, no. 1, pp. 219–231, Jan. 2011.

[18] M. Hallworth and S. Shirsavar, "Microcontroller Based Peak Current Mode Control Using Digital Slope Compensation," *IEEE Trans. Power Electron.*, vol. 27, no. 7, pp. 3340–3351, Jul. 2012.

[19] C. Tao and A. Fayed, "PWM Control Architecture With Constant Cycle Frequency Hopping and Phase Chopping for Spur-Free Operation in Buck Regulators," *IEEE Trans. Very Large Scale Integr. (VLSI) Syst.*, vol. 21, no. 9, pp. 1596–1607, Sept. 2013.

[20] M. Nashed and A. Fayed, "Variable switching noise mitigation in hysteretic power converters using spur-free control," *in Proc. IEEE APEC*, pp. 410–413, Mar. 2015.

A Parabolic Current Control Based Digital Current Control Strategy for High Switching Frequency Voltage Source Inverters

Lanhua Zhang[1], Rachael Born[1], Xiaonan Zhao[1], Jih-Sheng Lai[1], Hongbo Ma[2]

[1]Future Energy Electronics Center
Virginia Polytechnic Institute and State University
Blacksburg, VA, USA

[2]School of Electrical Engineering
Southwest Jiaotong University
Chengdu, Sichuan, China

Abstract—**Since parabolic current control has fast transient response and high control precision with constant switching frequency, it is an attractive current control strategy for voltage source inverters. The employment of parabolic current control delivers great performance on current waveform control or voltage waveform control. The implementation of parabolic current control usually requires high speed digital-to-analog converters (DACs) to generate a pair of parabolic carriers as the control band. With increases in switching frequency, this requirement becomes more critical because the speed of DAC limits the update rates of the parabolic carriers. This impacts the smoothness and punctuality of the parabolic control band. To solve this issue, a parabolic current control based digital current control strategy is proposed in this paper. The proposed method uses the sensed current of the output inductor to calculate the next duty-cycle. To improve the transient response speed, a duty-cycle adjustment is presented. The performance of the proposed digital control method is verified in simulation.**

I. INTRODUCTION

Current control strategy assists voltage source inverters (VSIs) in implementing the required system functions, such as minimizing the total-harmonic-distortion (THD) of the current or voltage waveforms. For example, standalone inverters such as uninterruptible power supplies (UPS) rely on the current control loop to maintain the required output voltage, especially when a non-linear load is in the system. An example of a grid-tied VSI is an active power filter (APF), which relies on a fast current response to compensate the harmonic current. To improve the performance of current control loops, numerous non-linear current control methods have been discussed and presented [1]-[20]. One well-known non-linear current control method is hysteresis current control, which boasts of a fast transient response and good control precision [11]-[15]. A well-known disadvantage of hysteresis control is the switching frequency variation problem [12]-[15]. Parabolic current control (PCC) solves the frequency variation problem by replacing the constant control band with a pair of parabolic carriers [1]-[6]. Gating signals are generated by comparing the current tracking error with the parabolic control band. So constant switching frequency and fast response are both obtained.

With the rapid development of power electronic devices, power density has become increasingly important [16]. One effective method to improve the power density is to increase the switching frequency of power converters. Thus, it is necessary to study how to implement PCC with high switching frequency.

Some implementation methods of parabolic current control are studied and discussed in [1]-[4]. In literature [1], digital-to-analog converters (DACs) generate both the current reference and parabolic carriers. Because PCC requires real time current tracking error to be compared with parabolic carriers, DAC chips with high speed and high resolution bits are required. In addition, the current tracking error is compared with parabolic carriers by the use of analog comparators, requiring the propagation delay of comparators to be minimized to ensure control precision. With higher switching frequency, this solution becomes less cost effective because the requirement on both DACs and comparator become higher. Furthermore, when compared with digital solutions of PCC, the analog comparison is more sensitive to noise. The operational amplifier based solution in [2] has a similar problem, requiring higher band width analog chips when switching frequency is higher. Compared with the solution in [1], the method in [2] is less flexible when adjusting the control parameters of PCC. Therefore the operational amplifier based solution is not popularly employed. A current sensorless PCC strategy is proposed in [3], solving the high cost problems caused by DACs, current sensor, and analog comparators. Because digital comparison is used, noise immunity performance is improved. However, because the current emulator operates at hundreds of times the switching frequency to simulate the inductor current information, this sensorless solution is not suitable for high switching frequency applications either.

In this paper, to solve the implementation problems of PCC for high frequency application, a new PCC based digital current control technology is proposed. The derivation process is conducted in detail. PSIM simulation results proved the effectiveness the proposed current control method.

978-1-4673-9551-9/16 $31.00 © 2016 IEEE

II. PCC OPERATION PRINCIPLES

The simplified diagram of a PCC controlled H-bridge inverter with bipolar modulation is shown in Fig. 1. δ represents current tracking error which is defined as [2]:

$$\delta = i_L - i_{ref}, \tag{1}$$

where i_L is the sensed inductor current and i_{ref} is the reference of the inductor current. Fig. 1 (b) shows the operation of the typical PCC. The positive parabolic carrier is [2]

$$\begin{cases} F_{PA}(t) = A_m[\frac{t}{T^*} - (\frac{t}{T^*})^2], & 0 \le t \le T^* \\ F_{PA}(\tau) = A_m[\frac{\tau}{T^*} - (\frac{\tau}{T^*})^2], & 0 \le \tau \le T^* \end{cases} \tag{2}$$

Where $A_m = \frac{T^*}{2L}(U_p - U_n)$, and T^* is the desired switching period. The period of the positive (or negative) parabolic carrier is also T^*. During T_p, S_p is on and the current tacking error increases until it intersects with the positive carrier. Then S_p is turned off and S_n is turned on. The current tracking error decreases until it reaches the negative parabolic carrier, then S_n is turned off and S_p is turned on again. As a result, the current tracking error is precisely controlled in the control band. Positive peak and negative peak of δ are defined as $\delta_{p,peak}$ and $\delta_{n,peak}$, respectively. $\delta_{p,peak}$ = - $\delta_{n,peak}$ holds true in steady state [2]. $\delta_{p,peak}$, and $\delta_{n,peak}$ are obtained by the following equation,

$$\delta_{p.peak} = -\delta_{n.peak} = F_{PA}(D^* T^*) = A_m(D^* - D^{*2}). \tag{3}$$

In (3), D^* is the converged duty-cycle in steady state.

III. PROPOSED DIGITAL CONTROL METHOD

As shown in Fig. 3, the inductor current is sensed at the sampling instant to determine the current tracking error δ. The converged current tracking error with PCC in steady state is indicated as δ^* and the converged turn-on time and turn-off time in steady state are T_p^* and T_n^*, respectively. The peak to peak current tracking ripple in steady state is defined as $\Delta\delta_{p-p}^*$. ΔT is the adjustment time of turn-on or turn-off time. The following equations hold true in Fig. 3:

$$\Delta\delta = \delta - \delta^* \tag{4}$$

$$\Delta\delta = \Delta\delta_1 + \Delta\delta_2 \tag{5}$$

$$\frac{\Delta\delta_1}{\Delta T} = \frac{\Delta\delta_{p-p}^*}{T_n^*}; \frac{\Delta\delta_2}{\Delta T} = \frac{\Delta\delta_{p-p}^*}{T_p^*}, \tag{6}$$

$$\frac{\Delta\delta}{\Delta T} = \delta_{p-p}^*(\frac{1}{T_p^*} + \frac{1}{T_n^*}) \tag{7}$$

$$\Delta T = \frac{\Delta\delta}{\delta_{p-p}^*} \bullet \frac{T_p^* T_n^*}{T^*} = \frac{\Delta\delta}{\delta_{p-p}^*}(D^*(1-D^*)T^*) \tag{8}$$

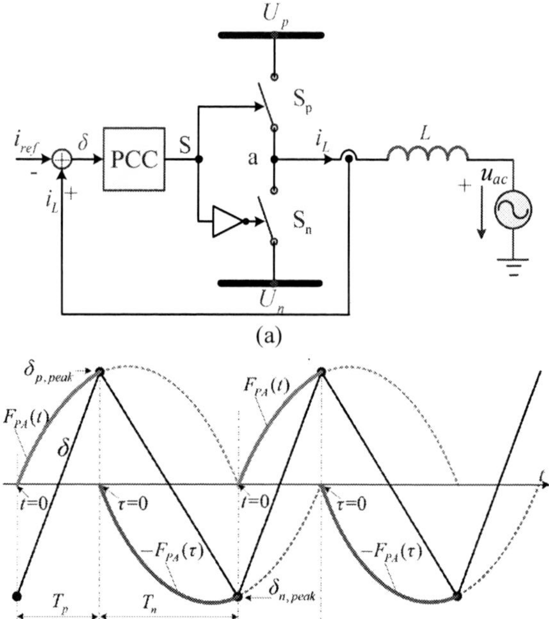

Fig. 1 Simpified diagram of a VSI with PCC

Combining (3), the adjustment time is

$$\Delta T = \frac{\Delta\delta L}{U_p - U_n}. \tag{9}$$

Then the turn-on and turn-off time are

$$\begin{cases} T_p = T_p^* - \Delta T \\ T_n = T_n^* + \Delta T \end{cases}. \tag{10}$$

By the use of (10), the duty-cycle adjustment is obtained. To increase the response speed, the duty-cycle adjustment is conducted twice in a switching period, indicated as time instant 1 and instant 2 in Fig. 2. Since the sampling instant is at the zero-crossing point of δ^*, the converged current tracking error equals the current tracking error.

$$\Delta\delta = \delta - \delta^* = \delta. \tag{11}$$

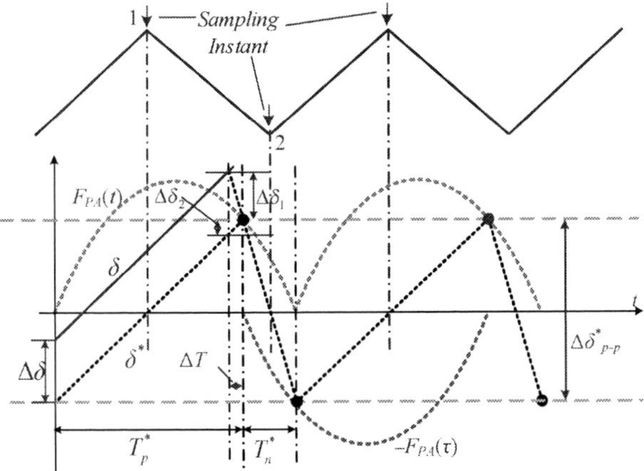

Fig. 2 Proposed digital current control strategy

IV. Simulation Analysis

A. Simulation schematic and parameters

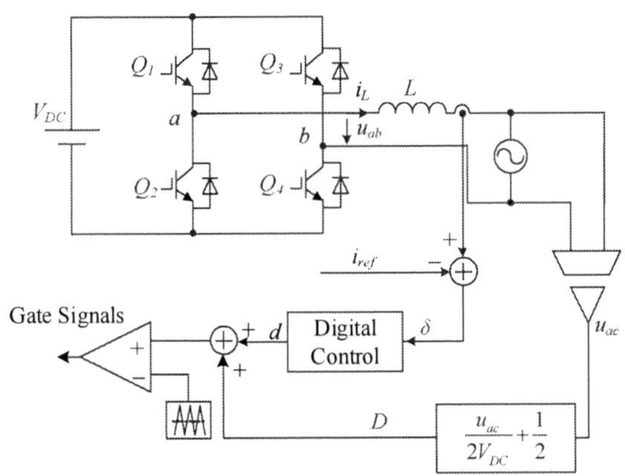

Fig. 3 Simulation diagram

To verify the performance of the proposed digital control method, the simulation is conducted in PSIM and the simulation diagram is shown in Fig. 3. D is the static duty-cycle and it can be calculated by the use of ac side voltage and dc side voltage. For bipolar modulation of a typical H-bridge inverter,

$$D = \frac{u_{ac}}{2V_{DC}} + \frac{1}{2}. \tag{12}$$

As shown in the figure, the output inductor current is sensed by an analog-to-digital converter (ADC), then δ and $\Delta\delta$ can both be obtained with (11). By combining (9) and (10), the adjustment of duty-cycle d can be calculated. Finally, by adding D and d together, the duty-cycle command is obtained, which is compared with the triangle carrier to generate the required gating signals. A simplified C-block function of PSIM can be used to implement a flexible digital control sequence and logic. Other simulation parameters are shown in Table 1.

Table 1 Simulation parameters

Description	Value
Output inductor (L)	0.33 mH
Dc link voltage (V_{DC})	400 V
Ac source voltage	240 V
Ac side line frequency	60 Hz
Switching frequency	100 kHz
Sampling frequncy	200 kHz

B. Steady state simulation results

The inductor current reference is set to be a sinusoidal waveform with 10A amplitude. The phase of the current reference is the same as the ac-side voltage source. The simulation results are shown in Fig. 4. Duty-cycle, current reference i_{ref}, inductor current i_L, and current tracking error δ are indicated in the figure. As can be seen from the simulation results, inductor current is well controlled and current tracking error matches the analysis in (3).

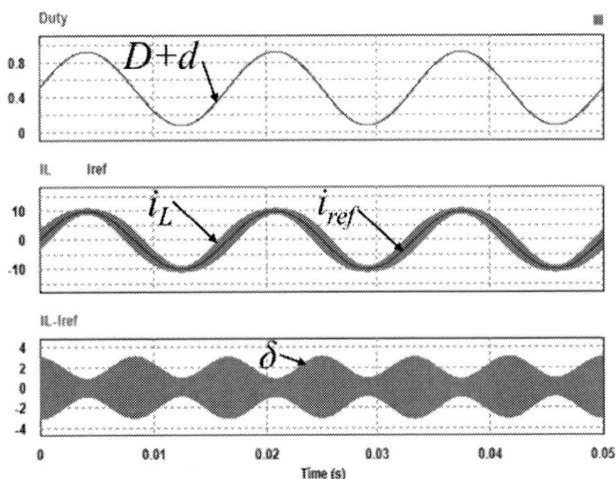

Fig. 4 Steady state waveforms

C. Improvement on transient response

To compare the transient response between the proposed digital current control strategy and traditional PCC, simulation is conducted in PSIM and simulated results are shown in Fig. 5 and Fig. 6. To simulate the transient response, the current reference changes from 10 A to 0 A at time instant 1. The response time is defined from the reference change to the intersection between inductor current and current reference. As Fig. 5 and Fig. 6 indicated, the response time of the proposed digital control method is 25 μs and the response time of traditional PCC is 4.7 μs. Two reasons are responsible for the slow response of the proposed digital control method. First, the digital method does not react to the current reference change until the next sampling instant, indicated by 2 in the figure, while traditional PCC can react immediately because of the analog comparison. Secondly, traditional PCC uses full duty-cycle or zero duty-cycle to follow the current reference, whereas the duty-cycle of digital method is not large enough during transient period. This also can be noticed in Fig. 5.

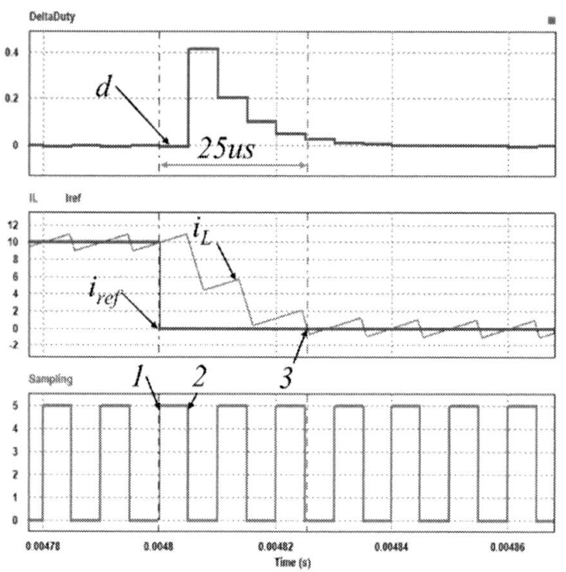

Fig. 5 Transient response of digital control method

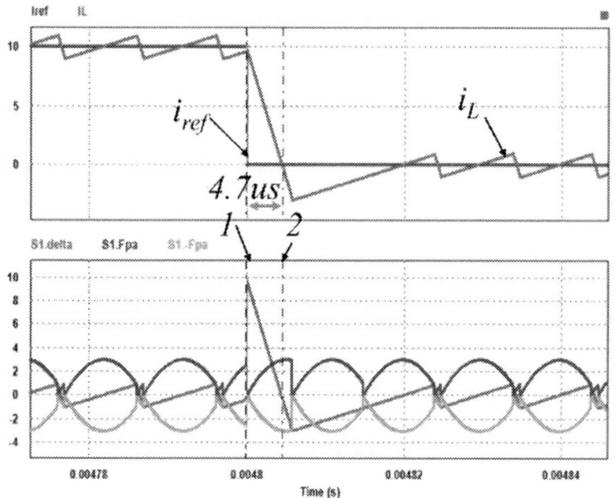

Fig. 6 Transient response of traditional PCC

To improve the transient response speed in the proposed current control strategy, the duty-cycle can be raised to full duty-cycle if $\Delta\delta < -\frac{A_m}{4}$ and be adjusted to zero duty-cycle if $\Delta\delta > \frac{A_m}{4}$. In other words, if the current tracking error is not in the range $(-\frac{A_m}{4}, \frac{A_m}{4})$, full duty-cycle or zero duty-cycle should be used to speed up the transition. This duty-cycle adjustment is easy to implement in the simplified C-block in PSIM. The simulation is conducted and the results are shown in Fig. 7. The response time is improved from 25 μs to 9.4 μs.

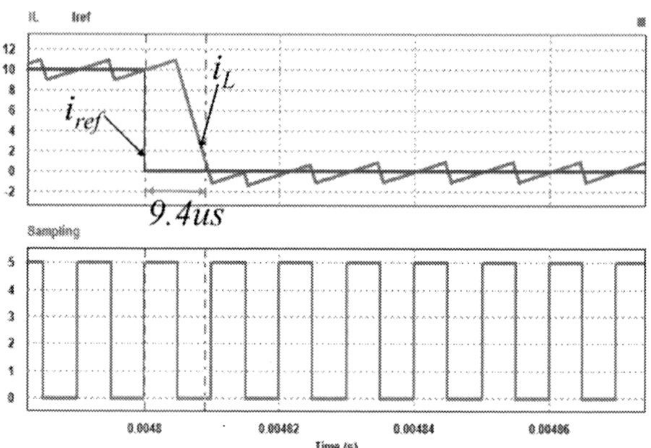

Fig. 7 Transient response with improved duty-cycle control

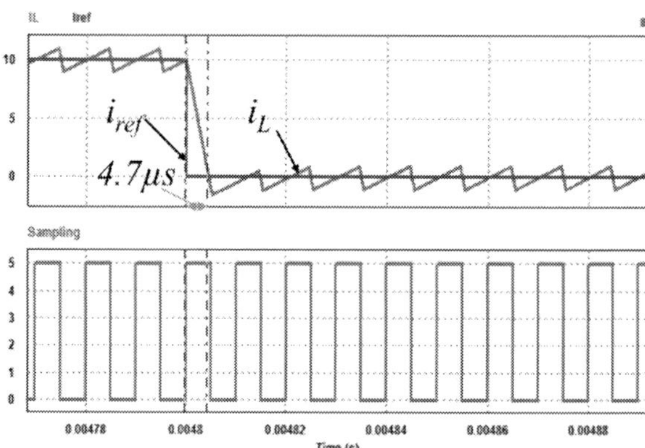

Fig. 8 Transient response with improved duty-cycle control and adjusted sampling sequence

In order to further increase the transient response speed, the duty-cycle adjustment should act immediately when the step change of current reference occurs. This can be implemented in the software by properly arranging the sequence between the update of current reference and digital current control. After the re-arrangement of the control sequence, the simulated results are shown in Fig. 8. The proposed digital control method has the same transient respond time as the traditional PCC. Also, with the specific dc side, ac side voltage, and output inductor, this is the fastest response for a step change. If the response speed needs further improvement, a smaller output filter inductance or higher dc side voltage are helpful.

V. CONCLUSIONS

In this paper, a parabolic current control based digital current control method is proposed, solving the high frequency application issue of parabolic current control and providing a digital fast current control method for voltage source inverters. The derivation of the proposed digital control method is discussed in detail. To verify the proposed control method, simulation is conducted in PSIM and comparison with conventional parabolic current control is discussed. By adjusting the control sequence and employing full duty-cycle or zero duty-cycle during transition, the response speed is same as the conventional analog comparison based parabolic voltage control. Due to the advantages of the proposed digital control method, it is an attractive current control method for high switching frequency voltage source inverters.

REFERENCES

[1] L. Zhang, B. Gu, J. Dominic, B. Chen, C. Zheng, and J. -S. Lai, "A dead-time compensation method for parabolic current control with improved current tracking and enhanced stability range," *IEEE Trans. Power Electron.*, vol. 30, no. 7, pp. 3892-3902, Jul. 2015

[2] G. Wang and Y. W. Li, "Parabolic PWM for Current Control of Voltage-Source Converters (VSCs)," *IEEE Trans. Ind. Electron.*, vol. 57, no. 10, pp. 3491-3496, 2010.

[3] L. Zhang, R. Born, B. Gu, B. Chen, C. Zheng, X. Zhao, and J.-S. Lai, "A Sensorless Implementation of the Parabolic Current Control for Single Phase Standalone Inverters", *IEEE Trans. Power Electron.*, vol. PP, no. 99, Aug. 2015

[4] L. Zhang, B. Gu, and J. -S. Lai, "The implementation of parabolic current control for dual-carrier PWM", in *Proc. of IEEE APEC 2015.*, Mar. 2015

[5] G. Wang, "A Novel PWM Current Control Technique Using Parabola Function," in *Proc. of the CSEE*, vol. 26, p. 6, 2006.

[6] G. Wang, "On the Stability of Parabolic PWM Current Control Technique," in *Proc. of the CSEE*, vol. 31, p. 6, Aug. 2011.

[7] K. M. Smedley and S. Cuk, "One-cycle control of switching converter," in *Proc. IEEE PESC*, 1991, pp. 886-896.

[8] K. M. Smedley, L. Zhou, and C. Qiao, "Unified constant-frequency integration control of active power filters-steady-state and dynamics," *IEEE Trans. Power Electron.*, vol. 16, no. 3, pp. 428-436, May 2001.

[9] D. Maksimovic, Y. Jang, and R. W. Erickson, "Nonlinear-carrier control for high-power-factor boost rectifiers," *IEEE Trans. Power Electron.*, vol. 11, pp. 578-584, July 1996

[10] R. Zane and D. Maksimovic, "Nonlinear-carrier control for high-power-factor rectifiers based on up-down switching converters," *IEEE Trans. Power Electron.*, vol. 13, pp. 213-221, Mar. 1998.

[11] Foreschle T. A. "Current controlled two-state modulation," U.S. patent 4456872. June, 1984

[12] L. Malesani, and P. Tenti, "A Novel hysteresis control method for current-controlled voltage-source PWM inverters with constant modulation frequency," *IEEE Trans. on Ind. Appl.*, Vol. 26, pp 88-92, Feb. 1990.

[13] R. Gupta, "Generalized Frequency Domain Formulation of the Switching Frequency for Hysteresis Current Controlled VSI Used for Load Compensation," in *IEEE Trans. Power Electron.*, vol.27, no.5, pp.2526-2535, May 2012

[14] Z. Yao and L. Xiao, "Two-Switch Dual-Buck Grid-Connected Inverter With Hysteresis Current Control," in *IEEE Trans. Power Electron.*, vol.27, no.7, pp.3310-3318, July 2012

[15] H. Mao, X. Yang, Z. Chen and Z. Wang, "A Hysteresis Current Controller for Single-Phase Three-Level Voltage Source Inverters," in *IEEE Trans. Power Electron.*, vol.27, no.7, pp.3330-3339, July 2012

[16] Shadmand, M.B.; Balog, R.S.; Abu-Rub, H., "Model Predictive Control of PV Sources in a Smart DC Distribution System: Maximum Power Point Tracking and Droop Control," in *Energy Conversion, IEEE Transactions on* , vol.29, no.4, pp.913-921, Dec. 2014

[17] Shadmand, M.; Balog, R.S.; Abu Rub, H., "Maximum Power Point Tracking using Model Predictive Control of a flyback converter for photovoltaic applications," in *Power and Energy Conference at Illinois (PECI), 2014* , pp.1-5, Feb. 28 2014-March 1 2014

[18] W. Cai, B. Liu, S. Duan and L. Jiang, "Power flow control and optimization of a three-port converter for photovoltaic-storage hybrid system," in *IEEE Energy Conversion Congress and Exposition (ECCE)*, pp. 4121-4128. Sep. 2012

[19] F. Yi, and W. Cai, "Repetitive control-based current ripple reduction method with a multi-port power converter for SRM drive", *IEEE Transportation Electrification Conference and Expo (ITEC2015)*, pp. 1-6, Jun. 2015

[20] Y. Zhao, Z. Zhang, C. Ma, W. Qiao, and L. Qu, "Sensorless control of surface-mounted permanent-magnet synchronous machines for low-speed operation based on high-frequency square-wave voltage injection," in *2013 IEEE Industry Applications Society Annual Meeting*, pp. 1-8, Oct. 2013.

[21] The little box challenge. Available online https://www.littleboxchallenge.com/.

Finite Control Set Model Predictive Control of Dual-Output Four-Leg Indirect Matrix Converter Under Unbalanced Load and Supply Conditions

Ozan Gulbudak, Enrico Santi
Department of Electrical Engineering
University of South Carolina
Columbia, USA
gulbudao@email.sc.edu, santi@engr.sc.edu

***Abstract*— Model Predictive Control (MPC) is an optimization based control method that solves a multi-objective optimization problem to determine the optimal control action that minimizes a given cost function. This control technique is very promising for switching converters because its implementation is straightforward and it has been shown to provide good performance. This paper presents predictive control scheme for dual-output four-leg Indirect Matrix Converter (IMC). This two-stage topology consists of a bidirectional three-phase rectifier cascaded by a four-leg dual-output inverter. The two three-phase output load currents can be controlled independently by solving a single multi-objective optimization problem. The proposed method aims to control the two ac load currents under unbalanced load condition and unbalanced supply condition. Simulation results show that the proposed method performs quite well under both steady-state and transient conditions.**

***Keywords*— Model Predictive Control, Dual-Output Converter, Indirect Matrix Converter, Optimal Control**

I. INTRODUCTION

Model Predictive Control is an attractive alternative to classical control methods, due to its fast dynamic response, simple concept, and ability to include nonlinearities. In particular, the Finite Control Set Model Predictive Control (FCS-MPC) is a well-known control technique in power converter systems and it has became an attractive control method for high-switch-count power electronics systems, such as Direct Matrix Converter (DMC), Modular Multilevel Converters (MMC) and Indirect Matrix Converter (IMC) topologies [1]-[3]. In the MPC approach the control objectives are specified as a multi-term cost function, which usually incorporates a linear or quadratic penalty on control variable errors. MPC uses a discrete-time model of the system to predict future behavior and calculates the cost function for all possible switching combinations. The switching combination that minimizes the cost function is chosen as the optimal control action. The MPC technique allows directly controlling more than one control objective simultaneously and this is achieved with a single control loop. Note that, in order to implement a multi-objective controller using linear control technique, a multi-loop controller is typically required with all the associated complications.

AC-AC power conversion systems have been widely used in industry. In conventional conversion systems, i.e., ac-dc-ac systems, a dc-link energy storage element, such as an electrolytic capacitor, is required for balancing the instantaneous difference between input source power and output load power. However, these energy storage elements are usually large in size and liable to cause reliability issues, thus reducing power converter expected lifetime. Moreover, passive components may have a significant impact on the overall converter losses. For these reasons, eliminating the dc-link storage elements is desirable to improve reliability and efficiency.

Four-leg converters have the capability of controlling the zero sequence current resulting in better converter operation under unbalanced load condition. In four-wire systems, output load currents (or voltages) can be regulated independently since the fourth wire provides a path for zero-sequence current to close. Four-leg converters are suitable for dealing with supply imbalance and can provide fault tolerant capability [5]-[7]. In general, four-wire systems are useful in applications that require a precise zero-sequence current control. In this work, a predictive control scheme is proposed for the dual-output four-leg indirect matrix converter shown in Fig.1,

978-1-4673-9551-9/16 $31.00 © 2016 IEEE

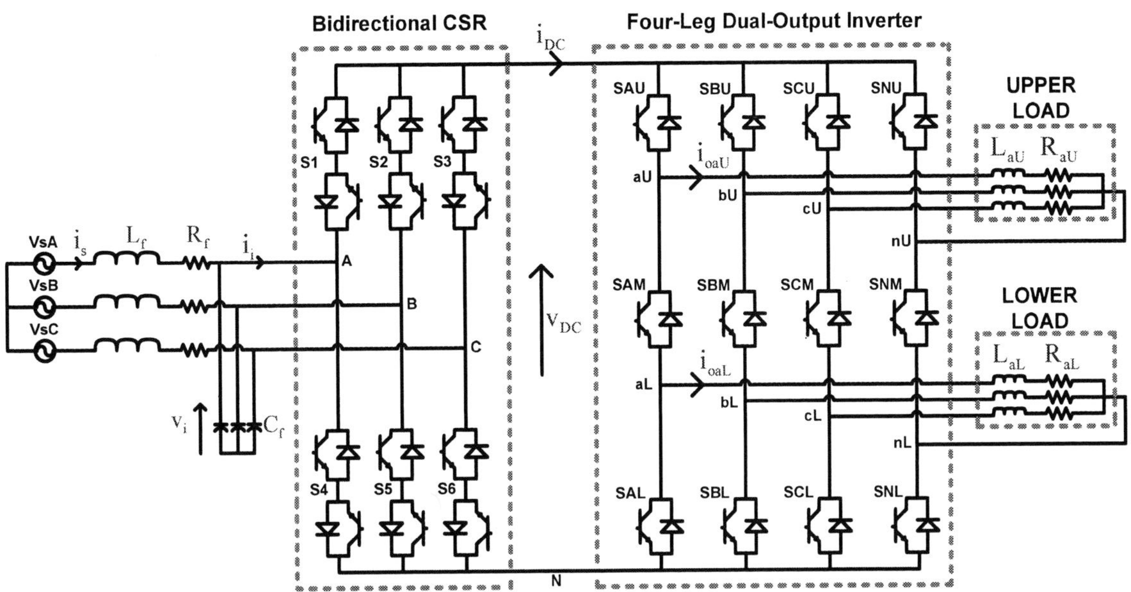

Figure 1. Dual-Output Four-Leg Indirect Matrix Converter (IMC) topology

and performance of the predictive controller investigated under unbalanced load condition and unbalanced supply condition.

II.SYSTEM MODEL

The dual-output four-leg IMC has two stages: Current Source Rectifier (CSR) stage and dual-output four-leg inverter stage. The converter topology shown in Fig. 1 has no dc-link energy storage element and the rectifier stage output is directly connected to the inverter stage. A filter is used at the input of the rectifier stage input to reduce input current harmonics. The models of the two stages are now described.

For the rectifier stage, producing positive dc-link voltage is critical for the proper operation of the inverter stage that it supplies. The interconnection matrix T_{CSR} is given by (1).

$$\mathbf{T}_{CSR} = [S_1 - S_4 \quad S_2 - S_5 \quad S_3 - S_6]^T \tag{1}$$

The input voltage vector $\mathbf{v}_i$ and the input current vector $\mathbf{i}_i$ are defined as (2) and (3), respectively.

$$\mathbf{v}_i = [v_{iA} \quad v_{iB} \quad v_{iC}]^T \tag{2}$$

$$\mathbf{i}_i = [i_{iA} \quad i_{iB} \quad i_{iC}]^T \tag{3}$$

The rectifier model is used to predict dc-link voltage v_{DC}. The dc-link voltage is given by (4).

$$\mathbf{v}_{DC} = \mathbf{T}_{CSR}^T \mathbf{v}_i \tag{4}$$

The relationship between the input current and dc-link current is given by (5).

$$\mathbf{i}_i = \mathbf{T}_{CSR} i_{DC} \tag{5}$$

For proper rectifier operation, the input phases of the rectifier cannot be short circuited. Thus, only nine switching combinations are valid for the rectifier stage. For the inverter stage, all switches on the same leg cannot be turned on at the same time to avoid dc bus short circuit. Another switching restriction is that at least two switches on the same leg must be on, so that floating of connected load is avoided. Considering these two switching restrictions, the inverter stage has 81 possible switching combinations, but since some of them are redundant, only 45 of these switching states are sufficient to control the two ac loads independently. To describe the inverter model, two interconnection matrices are defined as (6) and (7).

$$\mathbf{T}_U = [S_{AU} - S_{NU} \quad S_{BU} - S_{NU} \quad S_{CU} - S_{NU}]^T \tag{6}$$

$$\mathbf{T}_L = [S_{NL} - S_{AL} \quad S_{NL} - S_{BL} \quad S_{NL} - S_{CL}]^T \tag{7}$$

Output voltage vectors for the two three-phase ac loads are defined as in (8) and (9).

$$\mathbf{v}_{oU} = [v_{oaU} \quad v_{obU} \quad v_{ocU}]^T \tag{8}$$

$$\mathbf{v}_{oL} = [v_{oaL} \quad v_{obL} \quad v_{ocL}]^T \tag{9}$$

The relationship between load voltage and dc-link voltage is given by

$$\mathbf{v}_o = \begin{bmatrix} \mathbf{v}_{oU} \\ \mathbf{v}_{oL} \end{bmatrix} = v_{DC} \begin{bmatrix} \mathbf{T}_U \\ \mathbf{T}_L \end{bmatrix} \tag{10}$$

The continuous dynamic model of the load can be expressed as

$$v_o = Ri_o + L\frac{di_o}{dt} \quad (11)$$

In order to obtain a discrete-time model of the load, the classical Euler approximation is used

$$\frac{di_o}{dt} \approx \frac{i_o(k+1) - i_o(k)}{Ts} \quad (12)$$

An expression for the future value of the load current can be derived using (11) and (12).

$$i_{ojp}(k+1) = \frac{Ts}{L_{jp}} v_{ojp}(k) + i_{ojp}(k)\left[1 - \frac{R_{jp}Ts}{L_{jp}}\right] \quad (13)$$

where j=a, b, c and p= U, L.

III. MODEL PREDICTIVE CONTROL

Model Predictive Control (MPC) solves a multi-objective cost function by making an exhaustive search over the finite control set and determining the optimal control action [8]-[10]. The user-defined cost function, which is function of the errors with respect to the reference values, is solved for each switching state and the switching combination that minimizes the objective function is selected [11]-[13]. The choice of the objective function is critical since it determines the desired system behavior. In this work, the controller aims to minimize the load current errors of two ac loads. The cost function is defined as

$$g = \sum_{\substack{j=a,b,c \\ p=U,L}} \left| i^*_{ojp}(k+1) - i_{ojp}(k+1) \right|^2 \quad (14)$$

Superscript "*" refers to reference value. Three-phase load current errors are calculated in the abc frame. Producing a positive dc-link voltage is necessary for the inverter stage operation. The state elimination process is responsible for selecting rectifier switching combinations that provide positive dc-link voltage.

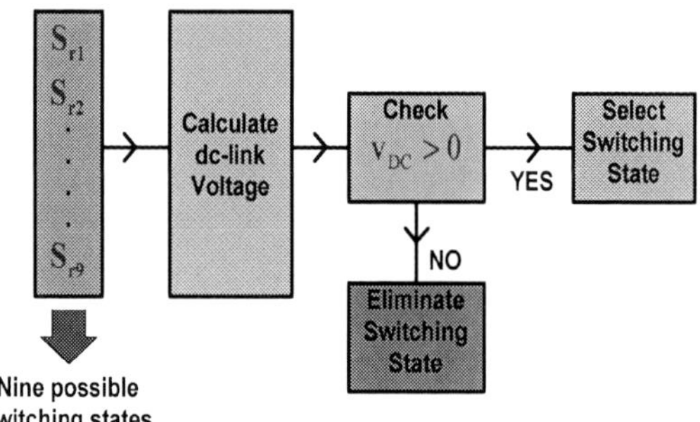

Figure 2. Switching state elimination process

Fig. 2 shows the flow diagram of the switching state elimination process that ensures that only rectifier switching states that provide positive voltage are used for calculating the future load current. Fig. 3 shows the predictive control scheme for the system. At instant k the input voltage vector and the output current vector are measured. The State Elimination block identifies valid rectifier switch combinations and the corresponding value of the DC link voltage is sent to the Future Current Calculation block that calculates future output current values. The cost function minimization block calculates the cost function for that switch combination. The process is repeated for all possible switch combinations and the optimal switch combination is found and applied to the converter.

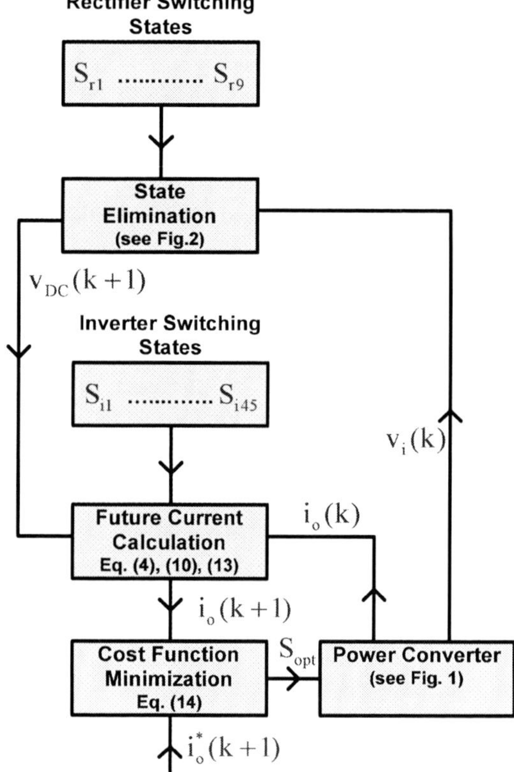

Figure 3. Predictive control scheme

I. SIMULATION RESULTS

A simulation study was performed to validate the proposed method. The predictive controller was tested under unbalanced load condition and unbalanced supply condition. Simulations are carried out using MATLAB/Simulink.

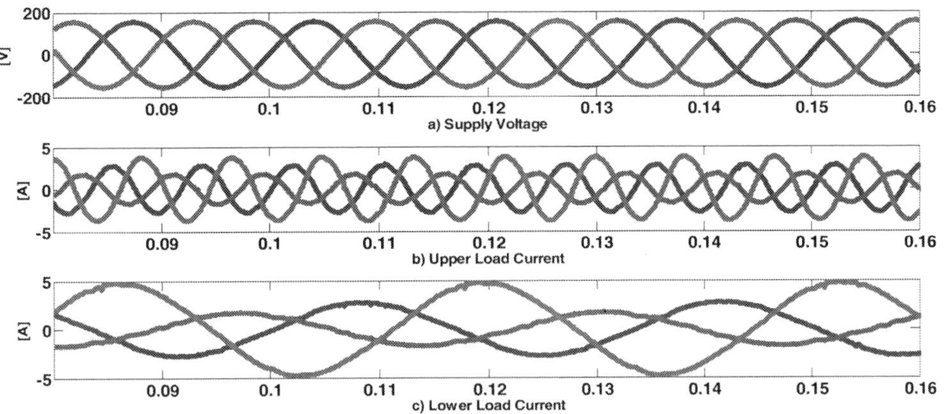

Figure 4. Upper load current and lower load current under unbalanced load condition

A. Unbalanced load condition

In this case, the predictive controller performance is investigated when two ac loads are unbalanced. Simulation parameters are listed in Table I. According to simulation results, the predictive controller can control the system in case of upper load and lower load both unbalanced. Fig. 4 shows that upper load and lower load are controlled independently and the predictive controller works well under steady-state condition.

Table I
Simulation Parameters

Supply	110 V rms/60 Hz
Filter	$R_f = 0.05\ \Omega$ $C_f = 28\ \mu F$ $L_f = 264$ mH
Sampling period	50 μs
Upper load reference	$i^*_{oaU} = 4\sin(2\pi*120t)$ $i^*_{obU} = 3\sin(2\pi*120t)$ $i^*_{ocU} = 2\sin(2\pi*120t)$
Lower load reference	$i^*_{oaL} = 5\sin(2\pi*30t)$ $i^*_{obL} = 3\sin(2\pi*30t)$ $i^*_{ocL} = 2\sin(2\pi*30t)$

Fig. 5 and Fig. 6 show the frequency spectrums of the load currents obtained with predictive control up to 2kHz. According to Fig. 5, the upper load current Total Harmonic Distortion (THD) is 2.77% and it is due mainly to the third harmonic component. Fig. 6 shows that lower load current quality is slightly better, with a total harmonic distortion of 2.36%, mostly due to the third and fifth harmonic. These values of THD are quite good. The lower THD of the lower load current can be attributed to the lower frequency (30Hz vs 120Hz for the upper load current). The predictive controller performance is tested under four different load conditions tabulated in Table II. In Fig. 7, load currents and zero sequence currents are shown for all four different cases according to the times shown in the last column of Table II.

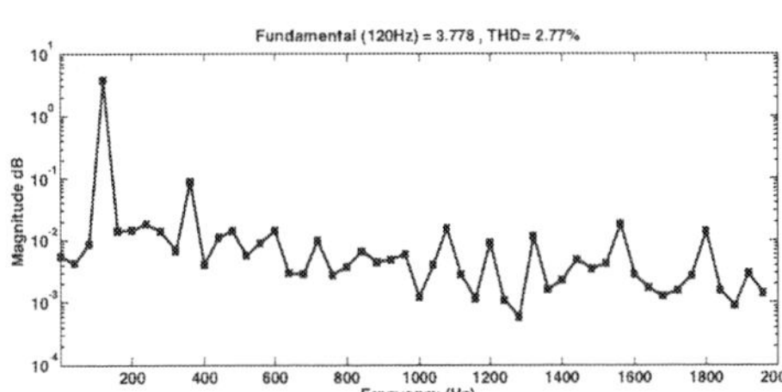

Figure 5. Frequency spectrum of upper load current for unbalanced load case

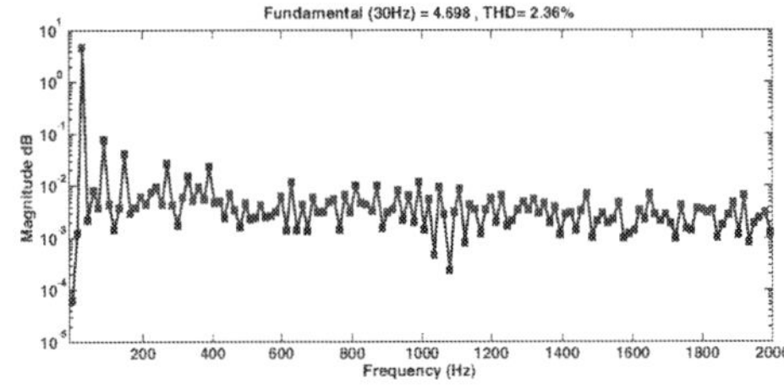

Figure 6. Frequency spectrum of lower load current for unbalanced load case

Table II
Load Conditions

CASE	UPPER LOAD	LOWER LOAD	Time
1	Balanced	Balanced	0s-0.02s
2	Balanced	Unbalanced	0.02s-0.043s
3	Unbalanced	Unbalanced	0.043s-0.06s
4	Unbalanced	Balanced	0.06s-0.08s

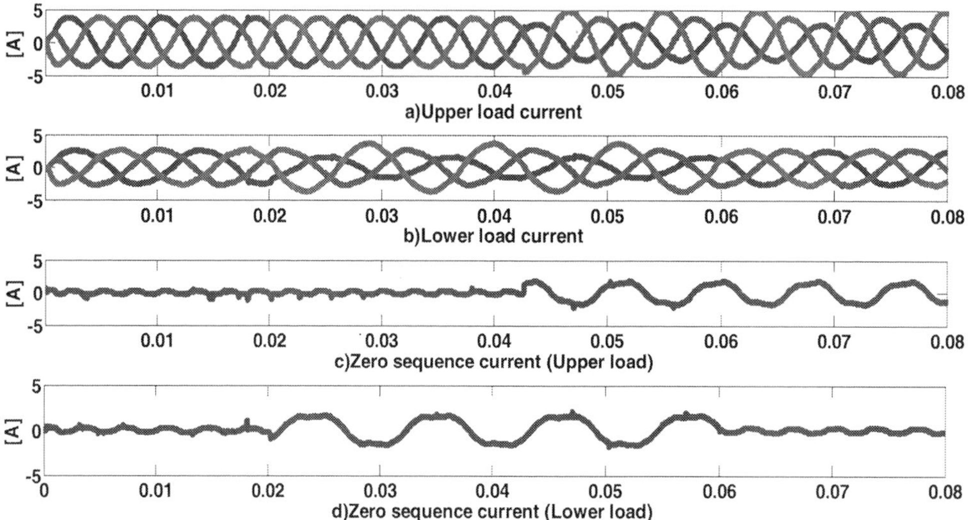

Figure 7. Simulation results for different load conditions

B. Unbalanced Supply Condition

Any source voltage disturbance, for example unbalances, can affect the load. In power electronics systems, it is important to provide clean power to the load even if supply is polluted or unbalanced. In this part, unbalanced supply condition for dual-output IMC is considered as a case study to evaluate the effectiveness of predictive control scheme. The IMC topology combined with predictive control scheme behaves as a buffer for supply voltage unbalances so that they do not reach the load. Fig. 10 shows that two ac loads are controlled independently under unbalanced supply condition. Switching elimination process works well, see Fig. 10(d), ensuring that the rectifier always provides positive dc-link voltage to the inverter stage. In order to evaluate the quality of load current, Fast Fourier Transform (FFT) analysis was carried out up to 4 kHz to calculate the spectral content. Upper load current spectrum is presented in Fig. 8 and magnitudes of frequency contents are plotted in log scale. In Fig. 9, lower load current frequency spectrum is shown. According to FFT results, upper load current THD is 3.62% and lower load current THD is 3.97%. The THD is mostly caused by the first few harmonics, even if high frequency harmonics do not drop off as fast as for the unbalanced load case of Figs. 5-6. In conclusion, dual-output IMC provides good load current quality and handles zero sequence current using predictive control scheme.

In order to test dynamic behavior of the proposed method, two reference steps are applied to the system. Fig. 11 shows the dynamic response of the predictive controller. Good tracking of the references is observed. The first reference step is applied to upper load at time 0.023s and the second reference step is applied to the lower load at time 0.042s. According to Fig. 11, fast dynamic response is obtained and the controller maintains control of each load current independently. In the second transient some interaction is noticeable, where the minimization of the cost function temporarily requires a non-zero error in both currents.

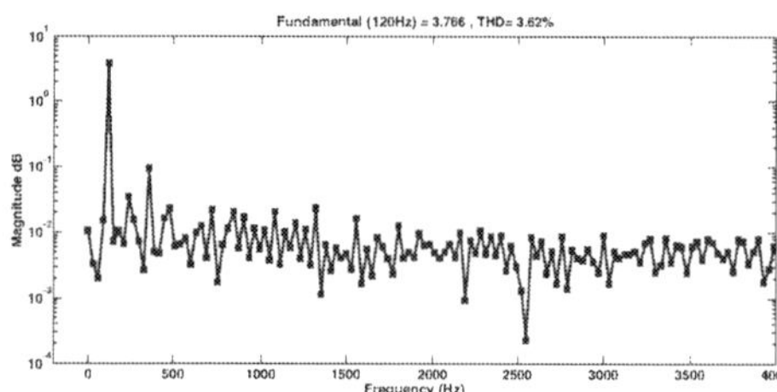

Figure 8. Frequency spectrum of upper load current for unbalanced supply case

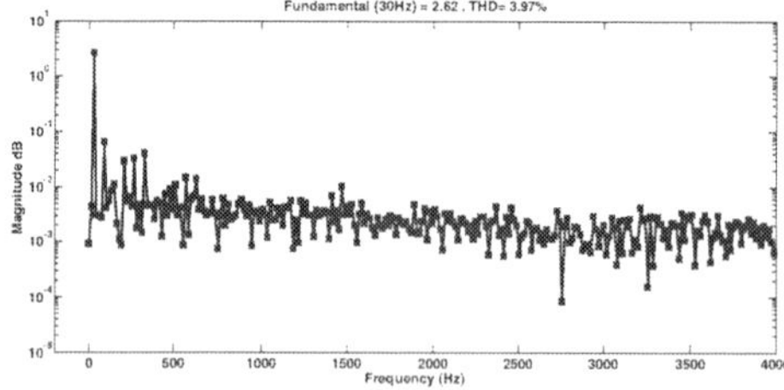

Figure 9. Frequency spectrum of lower load current for unbalanced supply case

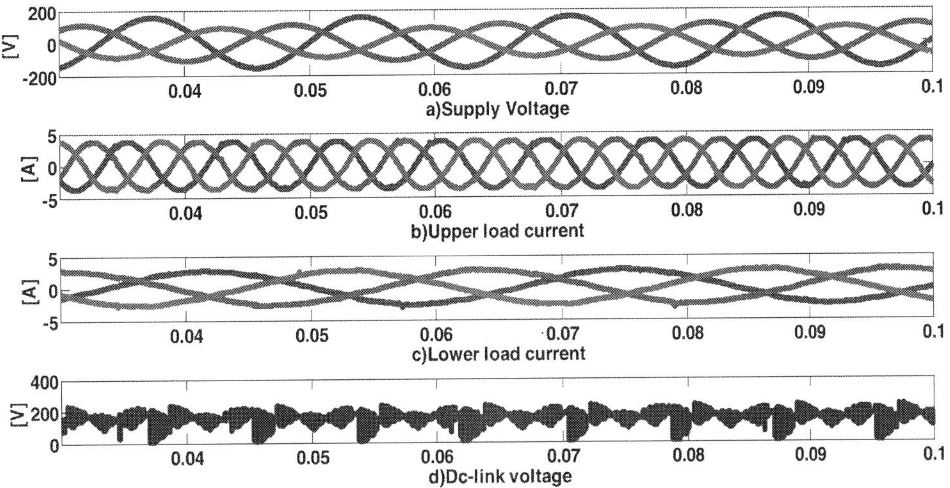

Figure 10. Simulation results under unbalanced supply condition

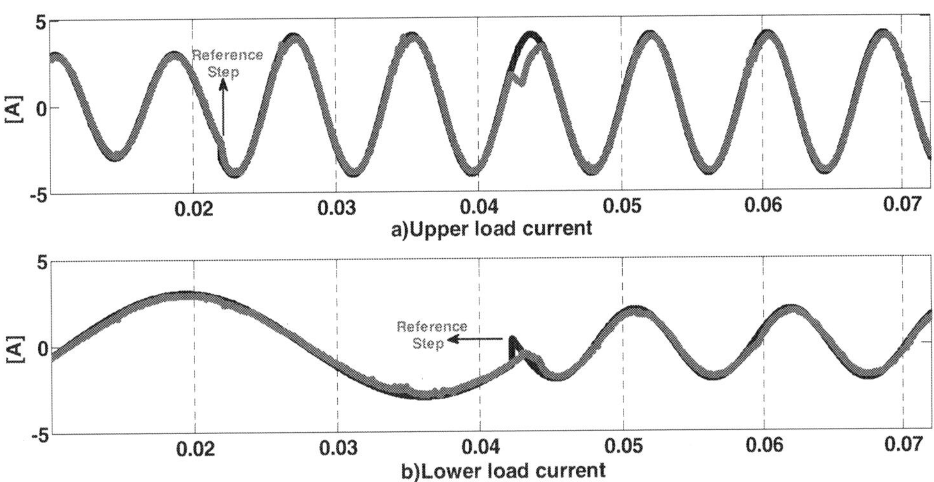

Figure 11. Dynamic behavior of predictive controller under unbalanced supply condition

II. CONCLUSION

In this paper, model predictive control scheme for dual-output four-leg indirect matrix converter is presented. The predictive control method does not require any modulator and it can effectively control two ac loads. This method has fast dynamic response and it is simpler compared to the classical methods. The control scheme uses the future values of load current to evaluate the best-suited switching combination considering the output load current errors.

Simulation results indicate that converter can be controlled under unbalanced load condition and unbalanced supply condition. It has been shown that predictive controller provides load current with low harmonic content and the zero sequence current is effectively controlled.

REFERENCES

[1] Karamanakos P., Pavlou K., Manias S., "An Enumeration-Based Model Predictive Control Strategy for the Cascaded H-Bridge Multilevel Rectifier", *IEEE Transactions on Inductrial Electronics*, Vol. 61, No. 7, July 2014

[2] Lopez M., Rodriguez J., Silva C., Rivera M.,"Predictive Torque Control of a Multidrive System Fed by a Dual Indirect Matrix Converter", *IEEE Transactions on Industrial Electronics"*, Vol. 61, No. 7, July 2014

[3] Gulbudak O., Marquart J., Santi E., "FPGA-based Model Predictive Current Controller for 3x3 Direct Matrix Converter", IEEE Energy Conversion Congress and Exposion (ECCE), Montreal, Canada, 2015, pp.4307-4314

[4] Friedli T., Kolar J., Rodriguez J., Wheeler P.,"Comparative Evaluation of Three-Phase AC-AC Matrix Converter and Voltage DC-Link Back-to-Back Converter Systems", IEEE Transactions on Industrial Electronics, Vol. 59, No. 12, December 2012, pp. 4487-4510

[5] Demirkutlu E., Hava A.,"A Scalar Resonant-Filter-Bank-Based Output-Voltage Control Method and a Scalar Minimum-Switching-Loss Discontinuous PWM Method for

the Four-Leg-Inverter-Based Three-Phase Four-Wire Power Supply", IEEE Transactions on Industrial Applications, Vol. 45, No. 3, MAY/JUNE 2009, pp. 982-991

[6] Cardenes R., Juri C., Pena R., Clare J., Wheeler P., "Analysis and Experimental Validation od Control Systems for Four-Leg MatrixConverter Applications", *IEEE Transactions on Industrial Electronics*, Vol. 59, No. 1, January 2012, pp. 141-143

[7] Liu Z., Liu J., Jin L.,"Modeling, Analysis, and Mitigation of Load Neutral Point Voltage for Three-Phase Four-Leg Inverter", *IEEE Transactions on Industrial Electronics*, Vol. 60, No. 5, May 2013, pp. 2010-2021

[8] Gulbudak O., Santi E., "Model Predictive Control of Dual-Output Nine-Switch Inverter with Output Filter", IEEE Energy Conversion Congress and Exposion (ECCE), Montreal, Canada, 2015, pp.1582-1589

[9] Gulbudak O, Santi E., Marquart J., "Finite State Model Predictive Control for 3x3 Matrix Converter based on Switching State Elimination", IEEE Energy Conversion Congress and Exposion (ECCE), Pittsburg, USA, 2014, pp.5805-5812

[10] Rivera M., Venkata Y., Llor A., Rodriguez J., Wu B., Fadel M., "Digital Predictive Current Controlof a Three-Phase Four-Leg Inverter", *IEEE Transactions on Industrial Electronics*, Vol. 60, No. 11, November 2013, pp. 4903-2013

[11] Gulbudak O., Santi E., "A Predictive Control Scheme for a Dual-Output Indirect Matrix Converter", IEEE Applied Power Electronics Conference and Exposition (APEC), Charlotte, USA, 2015, pp. 2828-2834

[12] Yaramasu V., Rivera M., Wu B., Rodriguez J., "Model Predictive Current Control of Two-Level Four-Leg Inverters-Part I: Concept, Algorithm, and Simulation Analysis", IEEE Transactions of Power Electronics, Vol. 28, No. 7, July 2013, pp. 3459-3468

[13] Rodriguez J., Kazmierkowski M., Espinoza J., Zanchetta P., Abu-Rub H., Young H., Rojas C., "State of the Art of Finite Control Set Model Predictive Control in Power Electronics", *IEEE Transactions on Industrial Informatics*, Vol. 9, No. 2, May 2013, pp. 1003-1016

A Silicon Carbide Integrated Circuit Implementing Nonlinear-Carrier Control for Boost Converter Applications

Richard Kyle Harris, Benjamin M. McCue, Benjamin D. Roehrs, Charles Roberts II, Benjamin J. Blalock, Daniel J. Costinett
University of Tennessee-Knoxville
Integrated Circuits and Systems Laboratory
Knoxville, TN

Kouros Sariri, George Megyei
Frequency Management International
Huntington Beach, CA

Cheng-Po Chen, Avinash Kashyap, Reza Ghandi
GE Global Research
Niskayuna, NY

***Abstract*—The properties of silicon carbide (SiC) integrated circuit (IC) processes are discussed and nonlinear-carrier control is proposed as a controller topology that can work within the design challenges presented by SiC. A boost converter with NLC controller is demonstrated using circuit blocks built with SiC IC models.**

***Keywords*— silicon carbide, PWM controller, nonlinear-carrier control, boost converter, high temperature**

I. Introduction

Due to the intrinsic material properties, SiC has been shown to have advantages over silicon based devices, including higher temperature capability and higher voltage operation [1]. SiC power devices provide an opportunity to create very high temperature power electronics circuits such as DC-DC converters. This high temperature capability allows for functionality in extreme environments such as space exploration, high temperature electric motor sensors and control, down-hole instrumentation, etc. NASA proposes a high temperature power processing unit that will use SiC power devices to create switch mode power supplies (SMPS) [9]. However, the high temperature functionality of these power electronics circuits can be limited by the high temperature capability of their support electronics, such as PWM controllers and gate drivers. SiC IC technology that can provide highly integrated systems but still maintain the high temperature capability inherent to SiC could be very impactful toward the advancement of highly integrated high temperature power electronics systems.

SMPS control options include voltage mode control and current-programmed control. Voltage mode control monitors the output voltage of a SMPS and directly manipulates the duty cycle in order to drive the output voltage to a match a reference. While controlling the output voltage of the SMPS is important, voltage mode control does not provide any current limiting in the SMPS, and currents that are too high could damage devices or impact long term reliability. Current-programmed controllers monitor inductor current and compare it to a reference current to control duty cycle while still driving the output voltage to match a voltage reference. There are also different ways to implement current-programmed control. Conventionally, inductor current is compared to a linear reference current. We will refer to this conventional approach as linear-carrier (LC) control. Another option is nonlinear-carrier (NLC) control which integrates inductor or switch current to compare to a nonlinear reference, such as an exponential carrier. This work aims to demonstrate a SiC IC nonlinear-carrier controller implemented for a SiC boost converter for high temperature applications.

This paper is organized as follows. Section II discusses the current generation of silicon carbide integrated circuit technology and the relative low performance of circuits designed in SiC IC processes. Section III describes how nonlinearity can impact the stability of a SMPS using a conventional current-programmed controller. Section IV presents nonlinear-carrier control as a potential alternative and discusses the advantages it has in light of the limitations of SiC IC technologies. Section V shows a schematic and simulation results of a boost converter with a nonlinear-carrier controller in which all controller components have been designed with SiC IC device models. Section VI concludes the paper and discusses future work in this research.

II. Characteristics of Current Generation Silicon Carbide Integrated Circuits

SiC IC design has properties that are both advantageous and disadvantageous in terms of designing high temperature integrated circuits for power electronics applications compared to silicon. GE reports that their SiC integrated analog and digital circuits can operate reliably in a 300°C environment [2]. Though high temperature operation is the primary advantage, GE goes on to describe the primary disadvantages of their SiC MOSFETs: low mobility, negative temperature coefficient of resistance, and no mature PMOS device [2]. Op-amps designed with SiC NMOS devices demonstrated just over 40 dB of gain at low frequencies at both room temperature and 300°C with a −3 dB frequency of about 1 kHz at room temperature and 10 kHz at 300°C [2]. GE demonstrates higher gain op-amps at

978-1-4673-9551-9/16 $31.00 © 2016 IEEE

high temperature, but they require supply voltages of 30 or 40 V [3].

Raytheon has demonstrated SiC CMOS circuits (including PMOS devices) up to 350°C [4] [5], but note that SiC MOS work has primarily focused on NMOS technology because bulk electron mobility is 8 to 9.5 times higher than hole mobility [4]. Aside from low hole mobility, another limitation of Raytheon's process is SiC MOSFET threshold instability, which they have mitigated with auto-zeroing switched capacitor approaches in their analog circuit designs [5] [6].

III. Effects of Nonlinearity On Current-Programmed Controllers

As mentioned in the introduction, conventional current-programmed controllers, or LC controllers, use linear waveforms. Large swing, linear waveforms are hard to produce in SiC IC processes for two reasons. Analog building blocks such as op-amps have very low gain and bandwidth. And process limitations mean that there is often no PMOS device available, or the available PMOS device is low performance even compared to the SiC NMOS. While a linear current-programmed controller might have been the conventional choice, this limitation of the analog building blocks made designing a system that depends on highly linear signals difficult.

Conventional linear-carrier current-programmed controller theory assumes comparing linear inductor current of the converter to a generated linear control signal called a compensating ramp that has the slope M_c. The slope of the control signal M_c contributes directly to the stability of the system [11]. Tan and Middlebrook define a LC current-programmed converter's stability parameter, Q_s, as

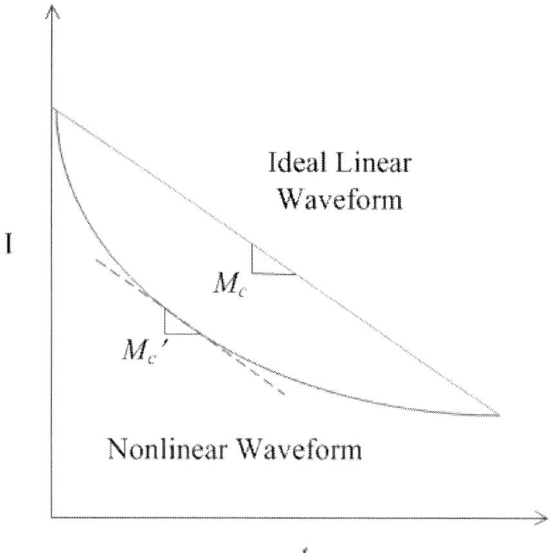

Fig. 1. Compensating ramp ideal linear waveform (blue) and nonlinear waveform (red) from low performance circuits

$$Q_s \equiv \frac{2}{\pi\left(\frac{D'}{D'_{min}} - 1\right)} \tag{1}$$

where

$$D'_{min} \equiv \frac{0.5}{\left(1 + \frac{M_C}{M_1}\right)} \tag{2}$$

and M_1 is the rising slope of the inductor current. As Q_s approaches infinity, the system becomes less stable and will eventually oscillate. It can be observed that as M_c decreases, D'_{min} increases, and as D'_{min} approaches D', the system becomes less stable, eventually oscillating if the condition $D' > D'_{min}$ is violated [11].

The stability of converters that use LC current-programmed controllers as presented in [11] depends on the assumption that a linear compensating ramp with a known slope is available. But with SiC op-amps, linearity cannot be assumed. If the compensating ramp is nonlinear, depending on where the inductor current waveform intersects it, it could have an effective slope, M_c', that is significantly different from M_c, impacting stability. Fig. 1 shows how M_c' could differ from M_c if the compensating ramp is assumed to be linear but is not.

IV. Nonlinear-Carrier Controller Advantages for Silicon Carbide

Rather than placing significant design effort into overcoming the nonlinearities of the SiC process, the controller designed in this work instead employs a nonlinear carrier. Conceptually, allowing for an exponential carrier waveform will inherently reduce design complexity. Whereas a highly linear triangular or sawtooth waveform is difficult to generate with neither PMOS devices nor linear amplifier circuits, the exponential carrier can be generated using only an NMOS switch, resistor, and capacitor.

Analysis of NLC control for boost converters was first presented in [12]. A generalized schematic for a NLC controlled power converter is given in Fig. 2.

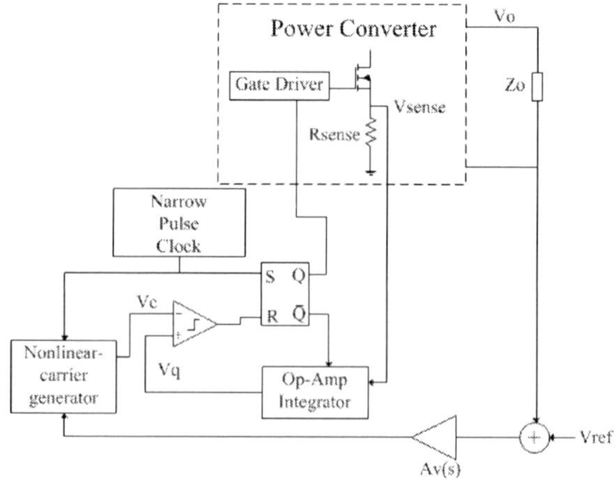

Fig. 2. Nonlinear-carrier controller block diagram

978-1-4673-9551-9/16 $31.00 © 2016 IEEE

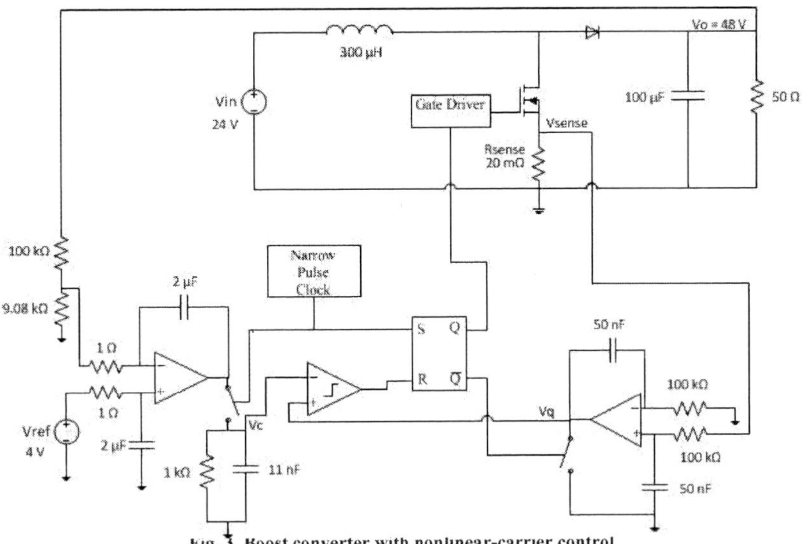

Fig. 3. Boost converter with nonlinear-carrier control

The controller in Fig. 2 is a generic NLC controller. The nonlinear-carrier generator is implemented as an exponential carrier that is realized with an RC network, which is a simpler solution than generating a parabolic carrier for which the method was derived in [7]. Additionally, instead of using a current transformer to sense and integrate switch current, a small sense resistor is used to generate a voltage signal proportional to the switch current that is integrated using a noninverting op-amp based integrator that is mathematically equivalent to integrating a current into a capacitor. Measuring the switch current with a sense resistor instead of a current transformer is much simpler and easier to implement in the boost converter compared with [7].

In addition to the properties of SiC MOSFETs described above, SiC MOSFETs also tend to have very large feature sizes. Large feature size coupled with low transconductance means that SiC transistor level layouts tend to be large compared to Si, so layout area is at a premium. Therefore, simplicity is important when designing an integrated SiC controller. A complete NLC controller, as seen in Fig. 2, is realized very simply: 1 op-amp is used as a noninverting integrator and 1 op-amp is used as a summer, $A_v(s)$, and nonlinear-carrier generator.

V. A Boost Converter with NLC Controller Using SiC Components

Fig. 3 shows a boost converter with R_{sense} to measure a voltage, V_{sense}, proportional to the switch current. V_{sense} is used to generate v_q with an op-amp integrator and switch. A second op-amp integrator compares the boost converter output and reference voltage and switches the op-amp output to an RC network to generate an exponential carrier, v_c, as demonstrated in [7]. The boost converter boosts a 24 V input to 48 V and produces about 46 W of output power with a switching frequency of 20 kHz. This boost converter uses a Cree CMF20120D SiC power MOSFET model for the power switch and a Microsemi MSiCSN10120CC SiC schottky diode model for the boost converter diode.

UT has designed all the necessary circuit blocks to build a

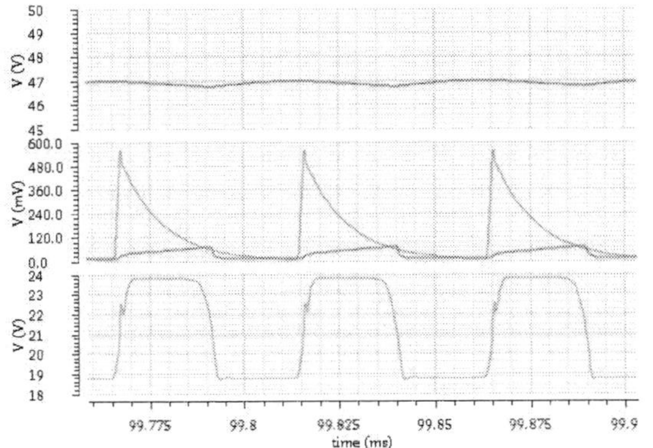

Fig. 4. Boost converter and NLC controller steady state waveforms: top – V_{out}, middle – v_q and v_c, bottom - Q.

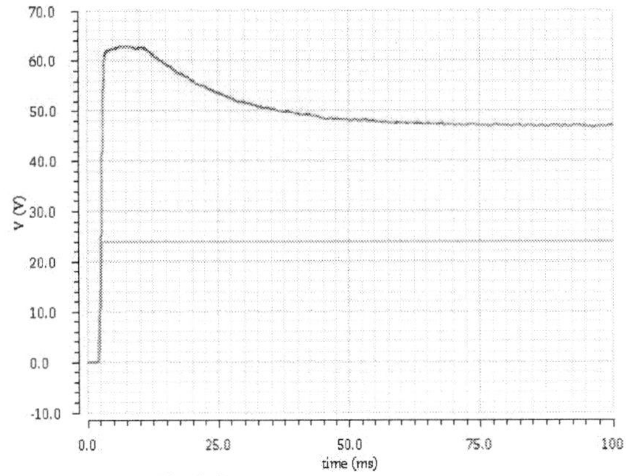

Fig. 5. Boost converter step response.

PWM controller in GE's NMOS-only SiC process. Included in the SiC IC circuits are op-amps, a comparator, an SR latch, clock signal generators (a multivibrator or crystal oscillator), a voltage reference, and NMOS switches with required switch drivers.

Simulations of the schematic seen in Fig. 3 using the SiC circuit blocks are seen in Fig. 4 and Fig. 5. In Fig. 4, the top waveform shows the steady state output voltage of the boost converter. The expected output voltage is 48 V, and the simulated result is 47 V, representing a 2% error. The middle waveform shows the carrier voltage, v_c, and integrated switch current, v_q, generated with op-amp integrator circuits. The bottom waveform shows the output voltage of the SR latch, which will drive the gate driver and control the boost converter's power switch. Fig. 5 shows the boost converter's output step response.

VI. Conclusions and Future Work

Fabricated SiC ICs with building blocks to construct a NLC controller are expected to be delivered in early 2016. At that time, hardware testing can be performed to verify the designs demonstrated with simulations here.

This research contributes to high temperature SiC power electronics systems such as [8] and [9], and if incorporated with a SiC gate driver [10], a high temperature power electronics system made entirely of SiC components can be realized.

Additionally, because NLC is shown to work with lower performance components, this technique could potentially benefit power electronics in other scenarios where analog circuits become low performance. For example, NLC control could be used to extend the operation of Si power electronics systems at high temperature or in radiation environments since high temperature and total ionizing dose degrade the performance of conventional Si electronics.

Acknowledgment

This work was supported by NASA, under grant NNX14CL20C. We are grateful for the support of Y. Chen of Langley research Center, M. Mojarradi and E. Carr of JPL.

References

[1] J. B. Casady and R. W. Johnson, "Status of silicon carbide (SiC) as a wide-bandgap semicondcutor for high-temperature applications: A review," *Solid-State Electron.*, vol. 39, no. 10, pp.1409–1422, Oct. 1996.

[2] A. S. Kashyap, *et al.*, "Silicon carbide integrated circuits for extreme environments," *2013 IEEE Workshop on Wide Bandgap Power Devices and Applications (WiPDA)*, pp. 60–63, Oct. 27–29, 2013

[3] A. Vert, *et al.*, "Silicon carbide high temperature operation amplifier." *Proc. Int. Conf. Exhibit. High Temperature Electron. (HiTEC)*, 2012.

[4] D. Clark, *et al.*, "High temperature silicon carbide CMOS integrated circuits." *Materials Science Forum*. Vol. 697, 2011.

[5] E. P. Ramsay, *et al.*, "Digital and Analogue Integrated Circuits in Silicon Carbide for High Tmeperature Operation." *IMAPS High Temperature Electron. Conf.*, Albuquerque, NM, USA, 2012.

[6] A. Lelis, R. Green, and D. Habersat, "Effect of Threshold-Voltage Instability on SiC Power MOSFET High-Temperature Reliability." *ECS Transactions*, 41.8 (2011): 203–214.

[7] R. Zane, D. Maksimovic, "Nonlinear-carrier control for high-power-factor rectifiers based on up-down switching converters," *IEEE Transactions on Power Electronics*, vol. 13, no. 2, pp. 213–221, Mar. 1998.

[8] Ruxi Wang, *et al.*, "Design of high-temperature SiC three-phase AC-DC converter for >100°C ambient temperature," *2010 IEEE Energy Conversion Congress and Exposition (ECCE)*, pp. 1283–1289, Sept. 12–16, 2010.

[9] G. A. Carr, *et al.*, "Extreme environment capable, modular and scalable power processing unit for solar electric propulsion," *2013 IEEE Aerospace Conference*, pp. 1–9, March 2–9, 2013.

[10] N. Ericson, *et al.*, "A 4H Silicon Carbide Gate Buffer for Integrated Power Systems," *IEEE Transacions on Power Electronics*, vol. 29, no. 2, pp. 539–542, Feb. 2014.

[11] F. Dong Tan and R. D. Middlebrook, "A Unified Model for Current-Programmed Converters," *IEEE Transactions on Power Electronics*, vol. 10, no. 4, July 1995.

[12] Maksimovic, D.; Yungtaek Jang; Erickson, R., "Nonlinear-carrier control for high power factor boost rectifiers," in *Applied Power Electronics Conference and Exposition*, 1995. APEC '95. Conference Proceedings 1995., Tenth Annual , vol., no.0, pp.635-641 vol.2, 5-9 Mar 1995.

A New Current Mode Constant On Time Control with Ultrafast Load Transient Response

Syed Bari, Qiang Li, and Fred C. Lee

Center for Power Electronics Systems (CPES)
The Bradley Department of Electrical and Computer Engineering
Virginia Polytechnic Institute and State University
Blacksburg, VA 24061 USA
Email: syedbari@vt.edu

***Abstract*— These days inductor current ripple based 'current mode constant on time (CMCOT)' control is widely used in voltage regulators (VRs) for its higher light load efficiency, higher bandwidth and simple compensation network. In today's VRs, high load current with very high slew rate is a very common requirement in microprocessor or memory applications. One issue of this ripple based CMCOT control is that, in the heavy load step up transient, inductor current increment becomes limited by on time (T_{on}) and minimum off time (T_{off_min}) ratio in each cycle, which can create large undershoot at the output. On the other hand, in the load step down case, if load change occurs at the beginning of the fixed Ton, a large overshoot can occur as well. For the multiphase operation case, its limited pulse overlapping capability is an issue at heavy load step up transient. States of art controllers add various nonlinear controls to the system to solve these issues. In this paper, a new charge based current mode control is proposed to improve the transient performance in CMCOT by naturally increasing or decreasing the on time in load step up or step down transient and by naturally overlapping these extended pulses in multiphase operation without adding any nonlinear control. The proposed concept is also extended to variable on-time variable off-time control to improve it transient performance. Simulation and test results of proposed control with CMCOT in standard VR platform are also presented in this paper.**

Keywords—constant on time control, transient response, multi-phase, pulse overlapping, on time extension.

I. Introduction

In today's world, constant on time control is widely used in the VR applications for its high light load efficiency and large bandwidth design capability, [1-2]. The benefit of constant on control has been explained in [3-4] by using a unified three terminal switch model. This model shows that double pole is located at high frequency and the quality factor, Q is constant; which allows to design the control bandwidth very high to achieve faster transient response. Besides, constant on time control also has smaller switching delay than peak current mode (PCM) control because of peak current mode's fixed frequency operation [5].

This work is supported by the Power Management Consortium of CPES Industry Partnership Program.

This work was conducted with the use of SIMPLIS donated in kind by Simplis Technologies of the CPES Industry Consortium Program.

In today's world, to support the latest high performance microprocessors and memory cards, their VRs often need to supply high load current with very high slew rate (up to 1000A/µs) [6], which, as a result, urges a constant transient performance improvement in today's controls. Though widely used, one issue in the conventional CMCOT control is that when a large and fast load step up is applied, regular feedback loop lost control and T_{on} starts occurring after predefined T_{off_min}. Hence, the I_L increment becomes limited by the fixed T_{on} and T_{off_min} ratio and will create a large undershoot at the output (V_o). On the other hand, while a load step down occurs, overshoot can be very large if load release is occurred at the beginning of the T_{on}, as inductor current will keep rising till fixed T_{on} expire, as shown in Fig. 3. Furthermore, a major problem for COT control is- its large frequency variation with wide duty cycle changes in many applications [7-8]. To solve this problem, constant on time control with adaptive on time (AOT) [9-11] is popularly used in the industry, where T_{on} changes with V_{in} and V_{out} at steady state condition, to keep the switching frequency almost constant. From transient response point of view, the undershoot problem in AOT control will become severe when T_{on} becomes very small at low duty cycle and becomes comparable with T_{off_min} of the system. On the other hand, overshoot will become worse if load step down occurs at the beginning of adaptive larger Ton at higher duty cycle. Several works have been done to address these issues by adding various nonlinear controls to the system. The concept presented in [12-13] is to overwrite the regular T_{on} pulses at load step up by a long pulse, equal to the time as long as V_o stays out of regulation. Another concept is - when undershoot crosses a preset threshold level at load step up, the regular T_{on} is replaced by number of predefined larger T_{on} to reduce undershoot [14-15], while one limitation of this concept that when that undershoot is small and just crosses threshold, a 'ring-back' problem can be created. In multiphase operation, nonlinear control is commonly applied to overlap the phases [16]. Author proposed 'Fast Adaptive On time (FastAOT)' control in [17] that increases or decreases the T_{on} in COT control immediately after load step up or down to reduce the undershoot or overshoot without any threshold detection and 'ring-back' issue; but still uses an additional dv/dt block.

In this paper, concept and implementation of a new constant on time current mode control is presented to improve the transient performance without any threshold detection and ring-back problem like [17]; but without adding any additional nonlinear circuit to the system. In this new control, instead of

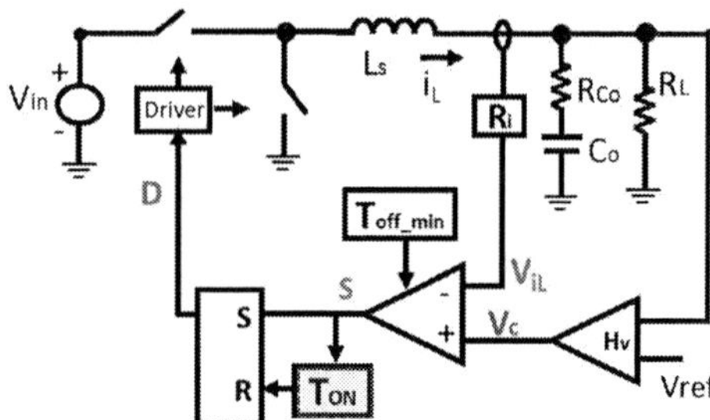

Fig. 1. Current mode constant on time (CMCOT) control structure

ripple based control, charge based control is used where T_{on} pulses can naturally merge together to create a larger pulse at heavy load step up or can truncate T_{on} at load step down to reduce undershoot or overshoot at the output and therefore able to reduce the significant amount of output capacitors.

The organization of the paper is as follows. Section II describes constant on time control and explains the limitations in transient response for both load step up and down in these controls in single and multi-phase operations. Section III presents the concept and implementations of the proposed charge control based constant on time current mode control while section IV shows the transient improvement by the proposed method in both load step up and down for single and multiphase operation. Section V presents the transient improvement in variable-on variable-off time control using proposed charge based control concept. Experimental test results to show the transient performance improvement using this proposed method are presented in section VI and finally conclusion is given in section VII.

II. Transient Response Issues in Constant On time Control

A. *Transient Issue at load step up*

A typical current mode constant on time (CMCOT) control architecture is shown in Fig. 1 while the operational waveforms in load step up case is shown in Fig. 2. At the steady state, while inductor current sense voltage $I_L{*}R_i$ touches compensator output voltage V_c, new pulse is triggered at comparator output S and then the Ton generator block produces a fixed on time (T_{on}) at gate driver D. At Ton time, as high side gate is turned on and low side gate is turned off, inductor current, I_L increases. After

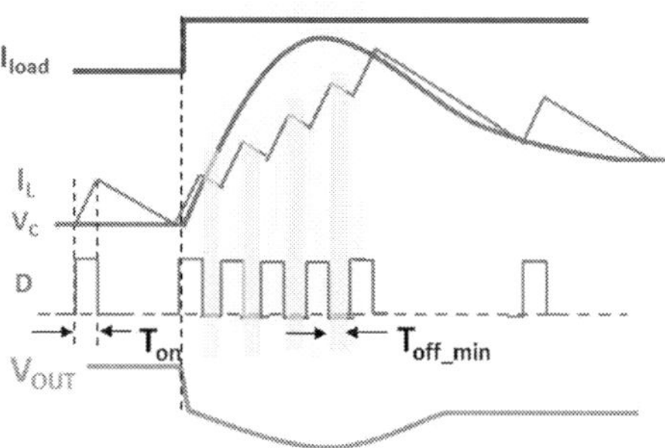

Fig. 2. Load step up transient issue in COT control

this fixed Ton, high side gate turned off and again I_L starts decreasing and sense voltage $I_L{*}R_i$ touches V_c again to start another new pulse.

However, CMCOT control has some limitations in load step up transient response. While a fast and large load step up is applied, shown in Fig. 2, first the Vout falls very quickly and following that compensator output V_c also goes up very quickly. If the inductor current I_L cannot increase quick enough to follow V_c, V_c goes above I_L and therefore, regular control law lost control (box area in Fig. 2). In this circumstance, new pulses are not initiated by the incidence of V_c and I_L intersection anymore as V_c stays higher than I_L. Rather pulses are generated by some non-linear control, which automatically generate Ton pulses with a preset fixed off time (T_{off}) between them without any feedback form control loop. From Fig. 2, it can clearly be understood that, inductor current increment is limited by the length of Ton and T_{off_min}. If Ton is small or T_{off_min} is large, a large undershoot can occur at the output.

B. *Transient Issue at load step down*

On the other hand, in the load step down case, for a given power stage, overshoot can be very large if load release is occurred at the beginning of the Ton which is shown in Fig. 3. In this Fig. 3, a load step down is occurred at the beginning of the second pulse and inductor current is expected to start decreasing right at that point. But because of fixed Ton time, I_L

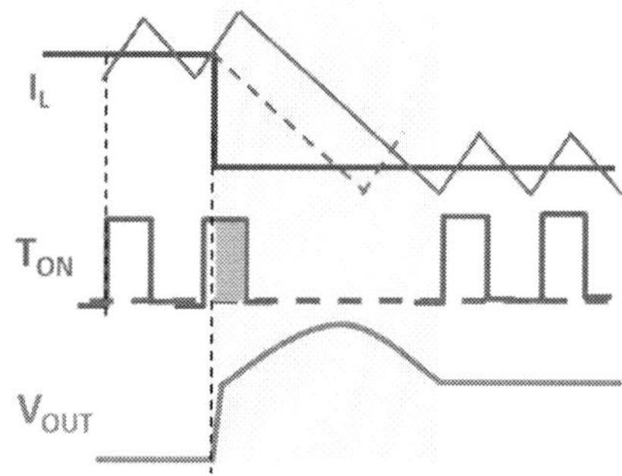

Fig. 3. Load step down transient issue in COT control

keeps increasing till Ton expires and this extra inductor energy will be dumped in output capacitor and hence, create an overshoot at the output. It is easily understood that, value of overshoot will be higher for larger Ton, provided that load step down happens at the beginning of the Ton.

C. *Transient Issue at multi-phase operation*

In constant on time control, for multiphase operation, pulse distribution method [10][11] is very popular because of its design simplicity. In Fig. 4, pulse distribution structure for 2-phase operation is shown where the basic control method is – when the summation of phase inductor currents (i_{Sum}) is equal to the compensator output Vc, a Ton pulse is initiated which is mentioned as D_{Sum} and then the phase manager block distribute this D_{Sum} in phases. Though this structure is widely used for its many advantages, it suffers from some limitations due to its aforementioned interleaving method. One issue is noise sensitivity under ripple cancellation effect of i_{Sum}. For example, in case of four phase operation at D=0.25, the current ripple of

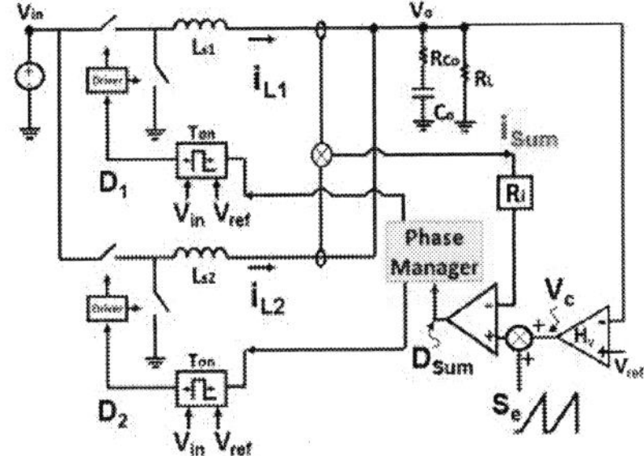

Fig. 4. Multiphase CMCOT control with external ramp

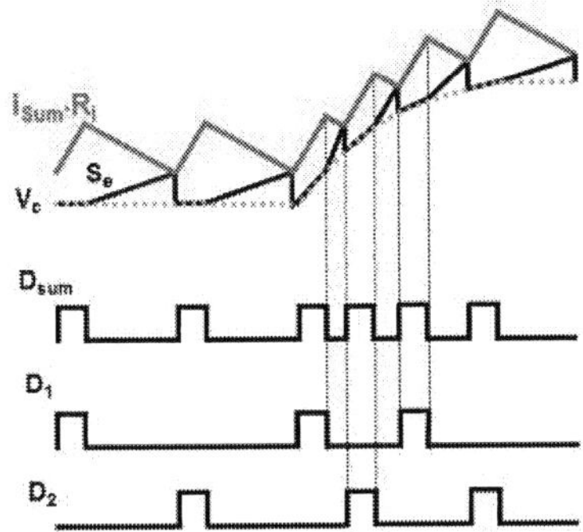

Fig. 5. Load step up transient waveforms in multiphase operation

i_{Sum} at the cancellation point is zero. Therefore control cannot produce any duty cycle signal as there is no ripple to generate PWM output. In common industry practice, an external ramp (S_e) is added in the control to generate duty cycle in that case [10], but this degrades the transient performance of CMCOT control by introducing output impedance peaking from the small signal point of view [18] and limit the phase overlapping in heavy load transient. In Fig. 5, the transient response waveforms in load step up are shown where pulses could not overlap because of pulse distribution method with external ramp and might create a large undershoot at load step up. As a result, more output capacitors are needed to meet the transient specification which will increase the board space and cost.

III. Concept of Proposed Charge Control Based Constant On-time Control

Basic concept of the proposed Charge based CMCOT control (Fig. 6) is – instead of comparing inductor current

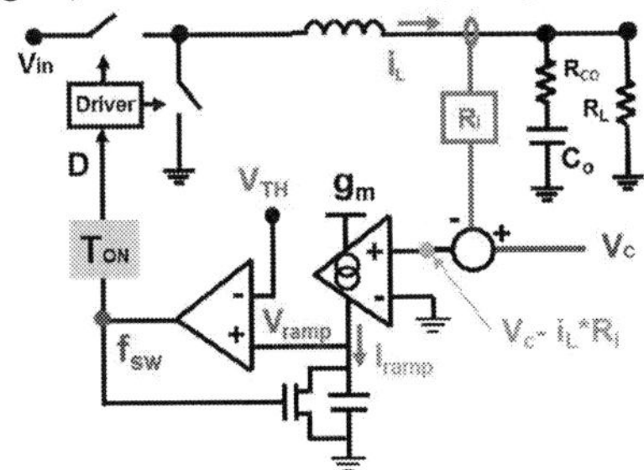

Fig.6. Basic structure of proposed CMCOT control

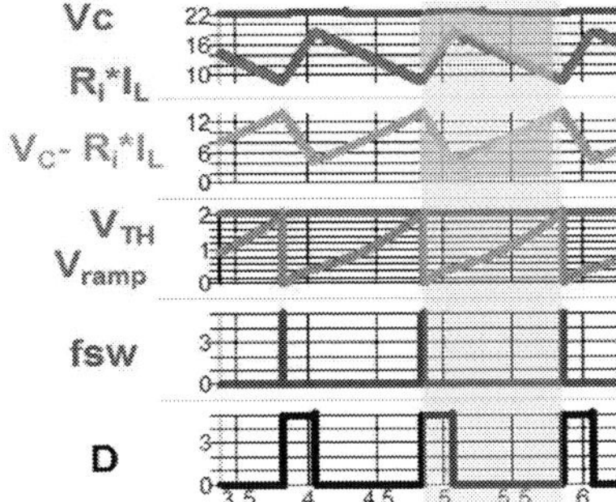

Fig. 7. Basic Waveforms of the proposed control

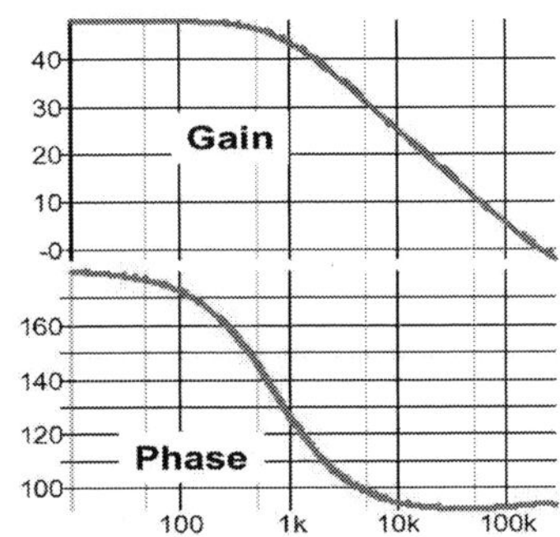

Fig. 8. Plant characteristics of the proposed control

(I_L*Ri) ripple with compensator output (V_c) like ripple based control, the difference between V_c and I_L*R_i is converted into current by using a Gm amplifier and this current is used to charge a capacitor. Then this capacitor voltage (V_{ramp}) is compared with a fixed threshold voltage (V_{TH}) in every cycle to create the pulse frequency f_{sw}, shown in the simulation waveforms in Fig. 7. The small signal property is presented in Fig. 8, which shows a single pole system in low frequency like conventional current mode COT control. The more detailed description of the proposed concept is given in [19] where author presented charge based current mode control to improve both the jittering (for ripple cancellation effect) and transient performance in multiphase operations.

IV. Proposed Control for Transient Response Improvement

A. Transient Improvement at Load Step Up

Unlike the load step up response of conventional CMCOT control (shown in Fig. 2), proposed control can merge duty pulses to achieve a long single pulse in heavy load step up. To achieve this feature of pulse merging, the state of art on time generator (Ton Gen) can be modified by adding an OR gate to control V_T signal shown in Fig. 9. In the heavy load step up (simulation results in Fig. 10), if inductor current I_L cannot cope up with V_c, (V_c-I_L*R_i) becomes very large, and f_{sw} pulse frequency becomes very high. As this pulse frequency is not dependent on ripple, consecutive pulse distance can be smaller than T_{on} time and can merge together, shown in Fig. 10. In that case, inductor current, I_L will be able to rise much faster and

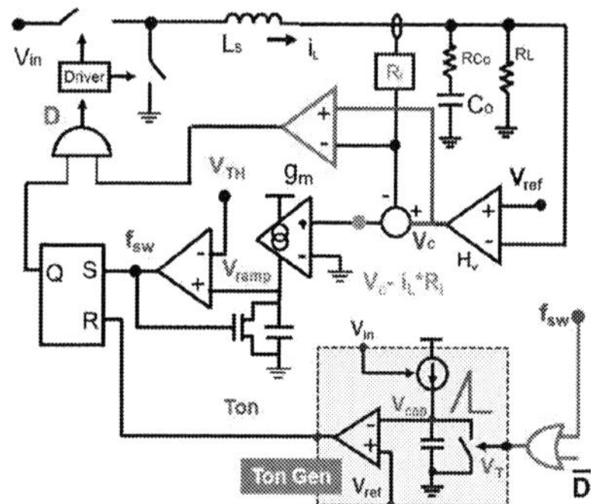

Fig. 9. Proposed control with modified Ton generator for pulse merging and comparator for Ton truncation

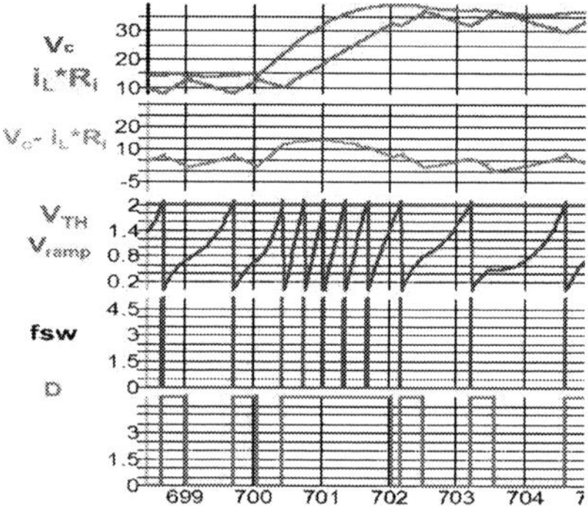

Fig. 10. Load step up waveforms with proposed control

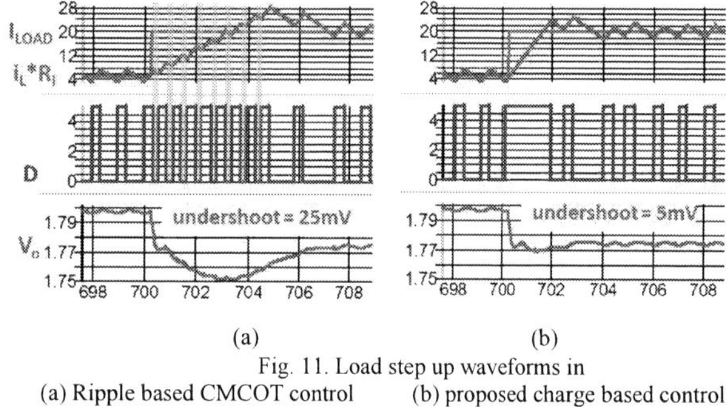

(a) (b)

Fig. 11. Load step up waveforms in
(a) Ripple based CMCOT control (b) proposed charge based control

significant undershoot reduction can be done at the output. Transient response simulation waveforms at load step up are compared between ripple based CMCOT (Fig.11a) and proposed control (Fig.11b) in a conventional VR setup, which shows that undershoot can be reduced 20mV by naturally extended Ton.

B. Transient Improvement at Load Step Down

The proposed structure can also improve the load step down transient response. When an overshoot is created at load step down transient, V_c goes low very quickly. If the inductor current cannot reduce very quickly, then V_c will cross the I_L* Ri immediately and make (V_c-I_L*R_i) negative (Fig. 12) which is always positive at regular steady state operation. Thus control

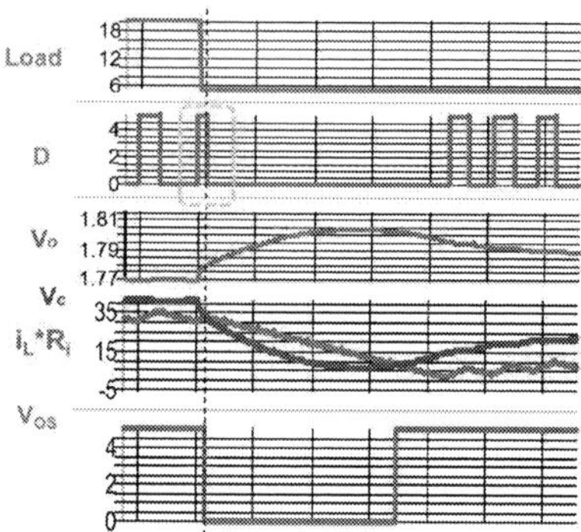

Fig. 12. Load step down waveforms with proposed control

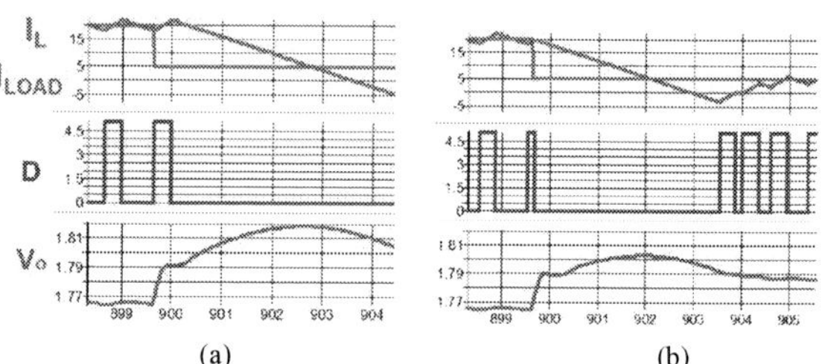

(a) (b)

Fig. 13. Load step down waveforms in
(a) Ripple based CMCOT control (b) proposed charge based control

can detect an overshoot at load step down without using any addition nonlinear circuit, unlike [20], that compares V_c and I_L difference with predefined threshold to create pulse truncation logic. If this logic is ANDed with duty cycle D, then predefined fixed Ton pulse can be truncated immediately and hence will reduce the overshoot at the output. Simulation comparison of load step down response is presented in Fig. 13 (a) and Fig. 13(b) where, it can be seen that around 20mV overshoot can be reduced by using the proposed control.

C. Transient Improvement at Multi-phase Operation

In multiphase operation, the proposed control can improve the performance by achieving natural and proportional pulse overlapping with naturally extended Ton pulses mentioned earlier. Fig. 14 shows a 2-phase structure of buck converter with

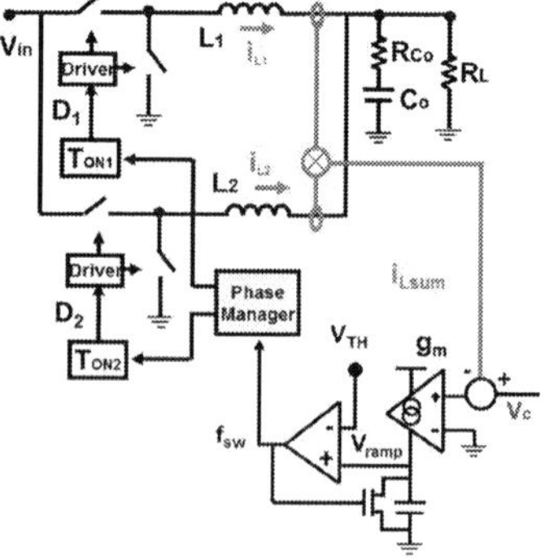

Fig. 14. 2-phase operation of proposed COTCM control

978-1-4673-9551-9/16 $31.00 © 2016 IEEE

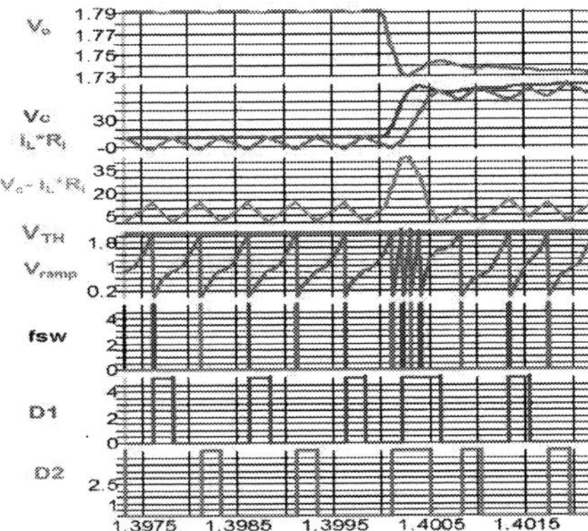

Fig. 15. Load step up transient with proposed control in Multi-phase operation

proposed control and Fig. 14 presents the transient performance simulation results at load step up where (V_c - I_{Lsum}*R_i) goes very high immediately after load step up and makes the f_{sw} frequency very high. As there is no limitation in f_{sw} frequency increment, when they are distributed in 2 phases, they can easily overlap with each other and if the distributed pulses of each phase are close enough, they can merge to create a longer pulse as seen in Fig. 15. In this way natural pulse overlapping with natural Ton extension can reduce the undershoot or reduce the number of output capacitors in the board significantly and as both features are natural and linearly proportional to output undershoot, any chance of ring-back will also be eliminated.

V. Extension of Proposed Charge Control for Multiphase Variable-on Variable-off time Control

In the variable-on variable-off time control both on time and off time are variable with duty cycle change. Among this type of control, Ramp Pulse Modulation (RPM) [21]-[23] is very popular now a days and therefore, RPM control, shown in Fig. 16, is taken as an example. In the Fig. 17, multiphase operation of RPM with proposed charge control is presented where summation of all phase current are summed and subtracted from V_c to generated V_{com} for multiphase operation. In can be seen that when V_{com} touches V_{th} pulses are generated to initiate the on time and then distributed in the phases by the phase manager block. In the load step up case, V_{com} goes above V_{th} and duty cycle becomes saturated and minimum distance between pulses becomes limited by minimum off time requirement. Therefore, in multiphase, RPM will suffer from the same limited overlapping problem like COT control.

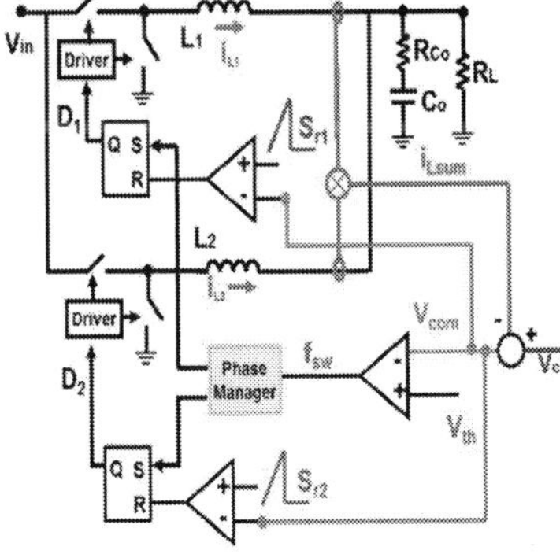

Fig. 16. Conventional Variable-on variable-off time control for multi-phase operation

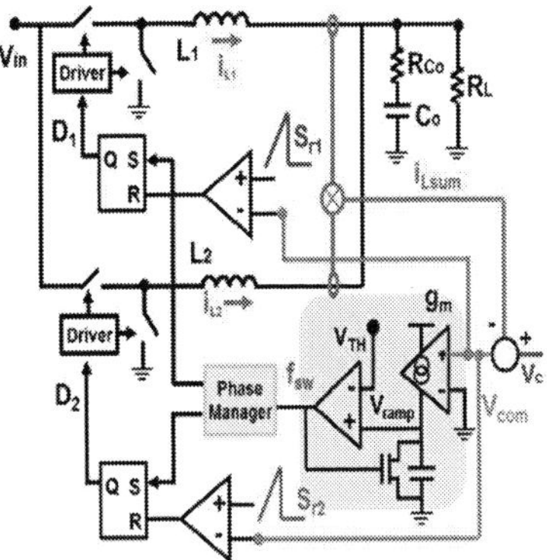

Fig. 17. Proposed structure for variable-on variable-off time control with charge based control concept for multi-phase operation

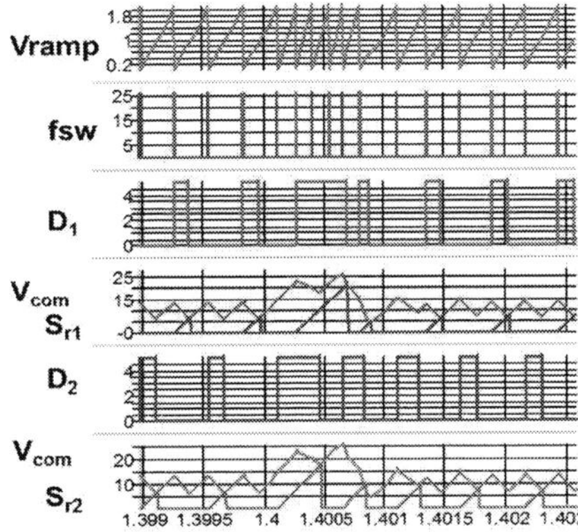

Fig. 18. Waveforms of proposed structure for variable-on variable-off time control with charge based control concept for multi-phase operation

Fig. 18 shows the waveforms of multi-phase steady state and transient response of the proposed structure. First advantage of this cgarge based RPM control is that the ripple cancellation effect is eliminated here like CMCOT and no external ramp is required for proper operation. From Fig. 18 it can also be clearly seen that at load transient, frequency of the f_{sw} increases very high and when they are distributed pulses overlap linearly in proportion to transient response. Moreover, as the Ton pulses can be extended, the overlapping of pulses becomes more effective.

VI. Experimental Verification

The proposed control method is implemented in both single and multi-phase operation. The implementation of proposed control is done with standard VR power stage and output filter model. The test condition is: Vin = 7V, Vo = 1.8V, switching

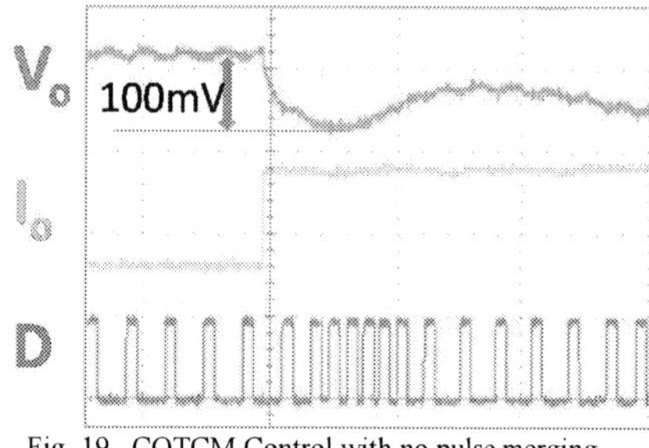

Fig. 19. COTCM Control with no pulse merging

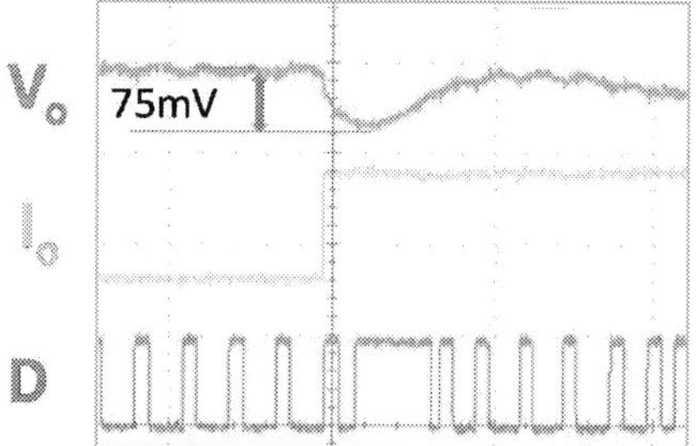

Fig. 20. COTCM Control with pulse merging by proposed charge based control method

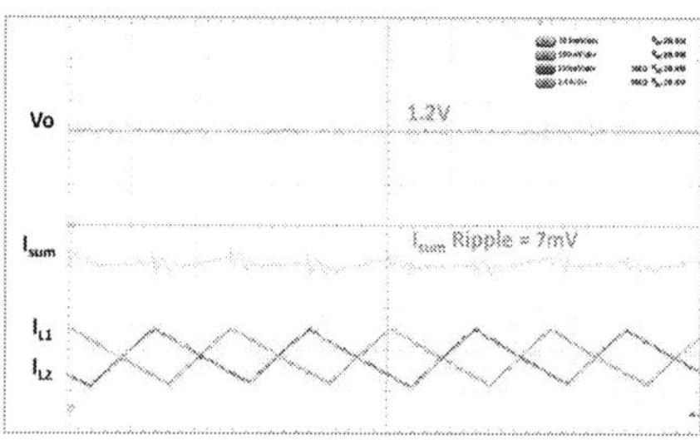

Fig. 21. Waveform at close to ripple cancellation point (D ≈50% for 2 phase operation)

frequency = 500kHz, L=360nH, Co_bulk = 2x470uF, Co_cer = 12x22uF, and the load step given ΔI_{LOAD}=20A. In Fig. 19, the waveforms are presented without pulse merging at load step up where 100mV undershoot is being seen. But in the load step up response with proposed control method, in Fig.20, shows the pulse merging feature with 75mV undershoot. Hence, it can also be said, for a given undershoot requirement, the proposed control can save lots of output capacitors also. A 2-phase operation in implemented in hardware to demonstrate natural and linear pulse overlapping with naturally merged T_{on}. In this operation the output impedance is set to zero (No AVP case). Fig. 21, the waveforms of the proposed charge based control are presented where we can see that the control can regulate the V_o even when the I_{sum} ripple is very small. The Fig. 22 shows that in a ΔIo=20A load step up, phases D1 and D2 can overlap naturally to supply the load demand. When the load step up is increased to ΔIo=30A, shown in Fig. 23, the Ton pulses in each phases can marge also to create larger Ton besides overlapping with each other to supply the load to the output. Although, with the increment of load step up amplitude, the output undershoot was supposed to be larger, but because of Ton pulse merging of each phase, the undershoot did not increase.

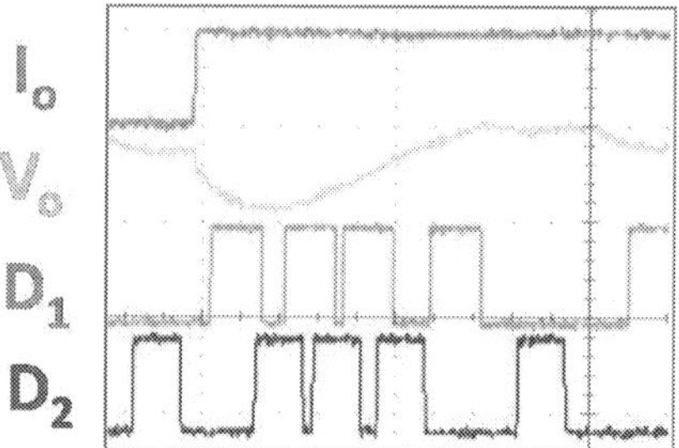

Fig. 22. Pulse overlapping in 2-phase with ΔIo=20A

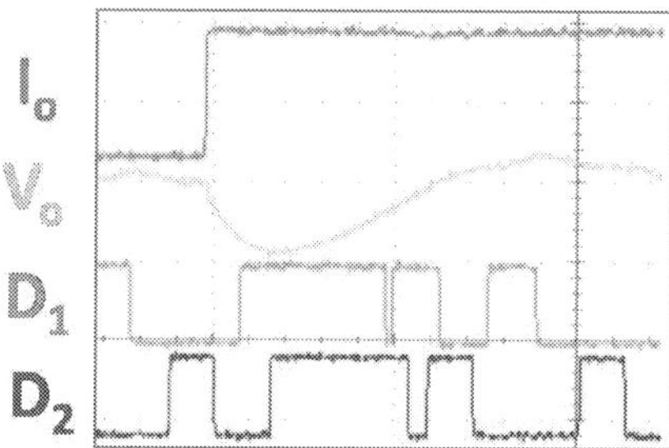

Fig. 23. Pulse overlapping with Ton extension in each phase in 2-phase with ΔIo=30A

VII. Conclusion

One issue of this ripple based current mode constant on time control is that, in the heavy load step up transient, inductor current increment becomes limited by on time and minimum off time ratio in each cycle, which can create large undershoot at the output. On the other hand, in the load step down case, if load change occurs at the beginning of fixed Ton, large overshoot can occur at the output as well. For the multiphase operation case, limited pulse overlapping capability of different phases becomes an issue at heavy load step up transient. In this paper, current mode control using charge control concept is proposed to improve the transient performance in CMCOT by naturally increasing or decreasing the on time in load step up or step down transient and also help the extended pulsed to overlap very easily in multiphase operation without adding any nonlinear control like present states of art solutions do. Simulations with proposed control are done in standard VR platform with standard output filter model and results are compared with state of art solutions. Hardware verifications in single and multi-phase operation are also presented in the paper.

References

[1] J. Sun, "Characterization and performance comparison of ripple based control for voltage regulator modules," IEEE Trans. Power Electron.,vol. 21, no. 2, pp. 347-353, Mar. 2006.

[2] X. Duan, A. Q. Huang, "Current-mode variable-frequency control architecture for high-current low-voltage DC-DC converters" in *Proc. IEEE Trans*, 2006, pp. 1133−1137.

[3] Y. Yan, F. C. Lee, and P. Mattavelli, "Comparison of Small Signal Characteristics in Current Mode Control Schemes for Point-of-Load Buck Converter Applications," in *Proc. IEEE Trans*, 2013, pp. 3405−3414

[4] Y. Yan, F. C. Lee, and P. Mattavelli, "Unified Three-Terminal Switch Model for Current Mode Controls," *in proc. IEEE Trans,* Power Electron. Sept. 2012. vol.27, no.9, pp. 4060-4070,

[5] Pit-Leong Wong and Fred C. Lee, "Switching Action Delays in Voltage Regulator Modules," in proc. IEEE APEC 2002, pp. 675 - 678 vol. 2.

[6] VR12/IMVP7 Pulse Width Modulation Specification, Intel Corporation, 2009

[7] X. Li, P. Li, and G. Jayakanthan, "Evaluation of Narrow Vdc-Based Power Delivery Architecture in Mobile Computing System," IEEE Trans. Industry Applications., vol.47, no.6, Nov. 2011

[8] P. Liu, "Advanced Control Schemes for High-Bandwidth Multiphase Voltage Regulators," Virginia Tech Dissertation, 2015.

[9] Hung-Chih Lin, Bou-Ching Fung, and Tsin-Yuan Chang "A Current Mode Adaptive On-Time Control Scheme for Fast Transient DC-DC Converters" in Proc. IEEE ISCAS'08, pp. 2602-2605.

[10] K. Cheng, F. Yu, Y. Yan, and F. C. Lee, "Analysis of Multi-phase Hybrid Ripple-Based Adaptive On-time Control for Voltage Regulator Modules," *in Proc. IEEE APEC'12*, pp.1088−1095.

[11] Texas Instruments, "TPS51650A Datasheet: Dual-Channel (3-Phase CPU/1-Phase GPU) SVID, D-CAP+TM Step-Down Controllers for IMVP-7 VCORE with Two Integrated Drivers," www.ti.com

[12] US 20040257056 A1: "Switching regulator with improved load transient efficiency and method thereof" by Jian-Rong Huang (Hsinchu City, Taiwan) Dec 23, 2004

[13] J. Quintero, M. Sanz, A. Lázaro, C. Fernández, A. Barrado, "Reduction of the Switching Frequency and the Number of Phases in multiphase VRM by applying Linear – Non – Linear Control," PESC06.

[14] US 20130314060 A1 : "Load Transient of a Cot Mode Power Supply" by Richtek Corp. Pub: Nov 2013

[15] Maxim Integrated Products, MAX17085B datasheet, 'Integrated Charger, Dual Main Step-Down Controllers, and Dual LDO regulators', http://www.maxim-ic.com

[16] Richtek Technology Corporation, RT8888C datasheet, '3-Phase Controller with Triple Integrated Driver for VR12.5 Mobile CPU Core Power Supply', http://www.richtek.com

[17] S. Bari, Q. Li, and F. C. Lee, "Fast adaptive on time control for transient performance improvement," in Proc. IEEE APEC 2015, pp. 397-403.

[18] S. Tian, F. C. Lee, Q. Li, J. Li, and P. Liu, "Equivalent Circuit Model of Constant On-time Current Mode Control with External Ramp Compensation," in Proc. IEEE ECCE'14, pp.3747-3754.

[19] S. Bari, Q. Li, and F. C. Lee, "A new current mode control for higher noise immunity and faster transient response in multi-phase operation" in Proc. IEEE ECCE 2015, pp. 2078-2083.

[20] Texas Instruments, "TPS53625 Datasheet: 2-Phase, D-CAP+TM Step-Down V-Core Controller for VR 12.0 Microserver," www.ti.com

[21] On Semiconductor, "ADP3211: 7-Bit, Programmable Single-Phase Synchronous Buck Controller," www.onsemi.com, Nov. 2009.

[22] K. Lee, and H. Zou, "Comparison between Ramp Pulse Modulation (RPM) and Constant Frequency Modulation for the Beat Frequency Oscillation in Voltage Regulators," in Proc. IEEE ECCE'10, pp. 3101-3106, Sept. 2010.

[23] G. Chen, T. F. Schiff, "Power Supply Control Circuits Including Enhanced Ramp Pulse Modulation," U.S Patent, no. US 8,148,966, Apr. 2012.

A Web-based tool for Compensation Design of Power Converters using Hybrid Optimization

Srikanth Pam[†], Yudhister Satija[†], Pradeep Chawda[*], Makram Mansour[*], Robert Hanrahan[*], Jeff Perry[*]
WEBENCH® Design Center, Texas Instruments,
[†]Bangalore, India; [*]Santa Clara, California, USA.
Email: {srikanth.pam, satija, pkc, makram.mansour, robert.hanrahan, jeff.perry}@ti.com

Abstract— **Power converter design and compensation design using numerical optimization techniques have been presented before in the literature. However, most of the work is limited to a particular topology and ideal component values. Also, prior works only focus on a proof of concept and do not demonstrate a full-fledged design tool aimed at the compensation design across multiple compensation types and converter topologies. This paper presents a web based, easy to use compensation design tool (Re-Comp), integrated into an industry leading online power design tool, WEBENCH Power Designer, used by thousands of power engineers worldwide. Re-Comp allows users to quickly design and optimize compensation networks to meet desired loop specifications. The proposed tool provides a robust design and ensures loop stability over the entire operating range of converter and perturbations in component parameters. Hybrid optimization algorithm used in the backend to design the compensation network combines the benefits of fast convergence of Conjugate Gradient method and robustness of Differential Evolution. Re-Comp also allows users to design with real components from a large database of passive components from all major vendors.**

Keywords—Compensation Designer; Power Converter Design; Control Design; Hybrid Optimization; WEBENCH Power Designer; Re-Comp;

I. INTRODUCTION

Typically, designing a switching power regulator consists of two stages, the power stage design and the compensation design. Power stage design mostly consists of selecting the appropriate Inductor and Capacitor for the output filter, power switches, the controller IC (some ICs may have integrated switches), input capacitor and the input EMI filter (if needed). The power stage design is typically based on requirements such as the input and output voltage, maximum output current, converter efficiency, total cost, foot-print size, etc. Compensation design usually consists of choosing the control topology and feedback compensation network to meet loop specifications such as phase margin (PM), gain cross-over frequency (Fco), low frequency gain (LFG) and gain margin (GM).

Compensation design for switching power converters is often viewed as difficult and time consuming process requiring a lot of manual iterations to meet all the desired loop specifications [1, 2]. Manual designs are not only time consuming but also prone to inaccuracies, as key design parameters like Op-Amp bandwidth and finite open loop gain or converter parasitics usually get ignored to simplify design equations. Also, the non-availability of the obtained compensation components values mean we need to choose between the closest available components leading to further deviation from desired specifications [3]. Manually ensuring that the compensation works well across the entire operating range and power stage component variations is not trivial. A poorly designed compensation can lead to unstable operation that can result in undesired oscillations in output voltage, audible noise from the inductor and capacitors, over heating of the switches, inductor and output capacitor, reduced life of the power supply, possible damage to the circuits powered by the converter [2]. Hence it is very important to properly design the compensation network to get stable operation across the entire operating range of the power supply and including variations of the component parameters. This paper focuses on designing the feedback compensation networks for voltage and current mode control topologies. It is assumed that the control mode is chosen based on the controller IC and also the inner current loop in case of current mode converters is also fixed. The key contributions of this paper are:

- Re-Comp is an easy to use and comprehensive web-based tool for compensation design of power converters, integrated into industry leading online power design tool WEBENCH® Power Designer.
- Re-Comp is tailored to many users with different levels of expertise:
  - For the non-power designer experts, the fix-it-for-me solution solves a lot of their compensation design problems with a single push-of-a-button.
  - For intermediate designers who want to focus on getting more performance out of the design, the auto-mode will allow them to tweak the specifications and trade-off between performance and robustness.
  - For advanced power designers who want full control, the manual mode with pole-zero analysis allows them to directly set the compensator values or pole-zero locations.

978-1-4673-9551-9/16 $31.00 © 2016 IEEE

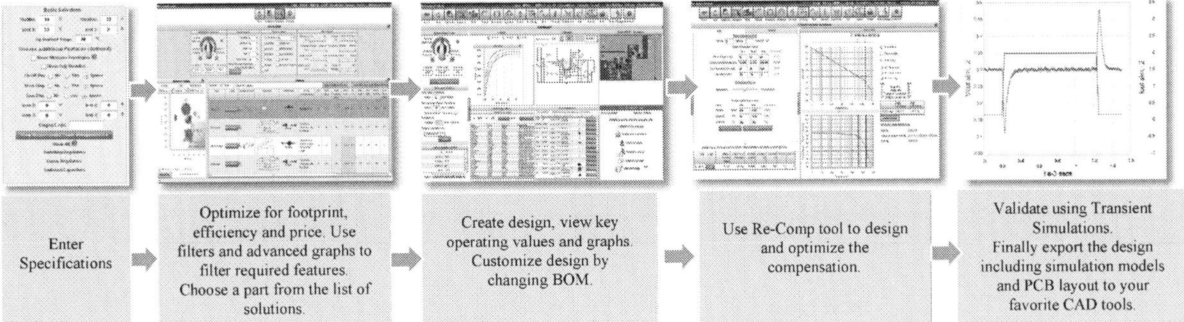

Fig. 1. Power Design flow using WEBENCH® Power Designer

- Re-Comp outputs real components from a database of over 40,000 passive components from all major vendors and thus the solution can be directly implemented on board.
- The loop models are derived including the power stage parasitics and feedback amplifier non-idealities. For this work, unified model by Tan and Middlebrook [5, 6] is used as this model is shown to be most accurate [7].
- Re-Comp makes sure that compensation is stable across the entire operating range of power converters and the variation of power stage component values.
- The Hybrid algorithm used in the backend for optimization combines the benefits of fast convergence of Conjugate Gradient method [8] and robustness of Differential Evolution [9].

II. Design Flow

The top level power design flow using WEBENCH® Power Designer with Re-Comp is shown in Fig. 1. During the initial solution finding stage, standard compensation design equations based on K-factor or simple pole-zero placement are used, as optimization method requires relatively longer design time and in this stage hundreds of of possible solutions are computed in the WEBENCH® Visualizer which can be sorted based on efficiency, cost or size of the design. Once a particular solution is chosen, the tool allows further manual customization of different BOM components, and here Re-Comp can be used to fine-tune the compensation network to meet the desired specifications. Re-Comp can be used in three different modes of operation based on user expertise as shown in Fig. 2:

- Fully automated one click method for beginners: In this mode the required loop specifications are auto-configured based on design conditions like switching frequency, system dominant pole, RHP zero if any, error amplifier gain / bandwidth limitations. Then the hybrid optimization is used to meet the specifications.
- Advanced Auto-mode: In this mode user can customize the specifications, change the range of compensation components.
- Manual-mode: In this mode user can directly change compensation components or edit locations of poles and zeros using text input or visual pole-zero placement method which also allows drag-and-drop poles and zeros on the bode plot for power design experts.

Re-Comp tool also allows user to go back and forth between the manual and auto modes and improve the designs. The tool allows user to compare multiple design using bode plot and choose the one which best suits the design requirement. The optimization of compensation design using combination of manual and auto modes is shown in Fig. 3.

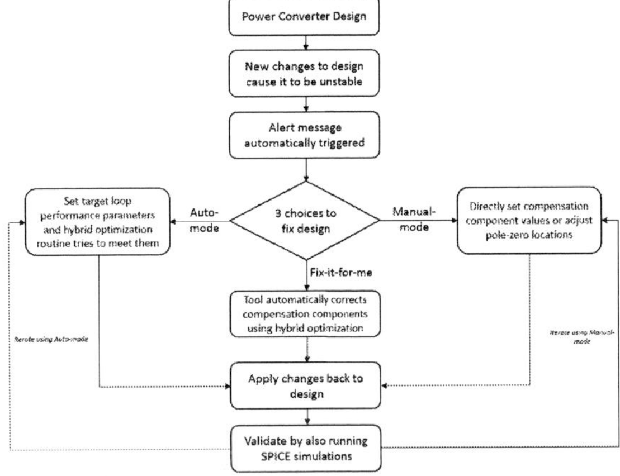

Fig. 2. Top Level flow showing use model for Re-Comp tool

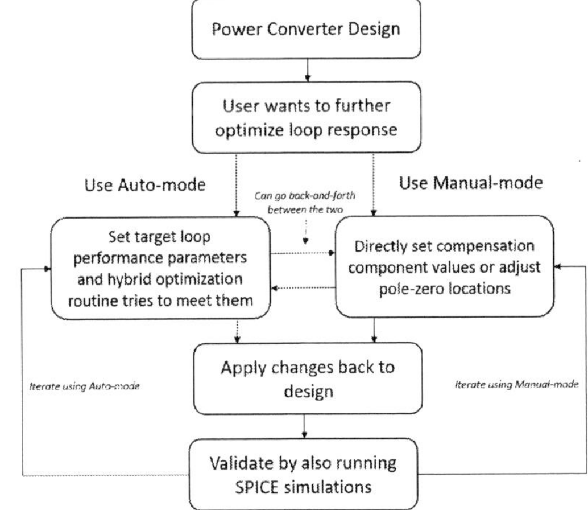

Fig. 3. Compensation Design using Re-Comp

III. Compensation Designer User Interface

The user interface is designed keeping ease of use as main goal and at the same time to allow freedom for expert designers to do advanced analysis. The auto-mode UI is showin in Fig. 4 and the manual-mode UI is shown in Fig. 5. In the auto-mode clicking on auto compensate runs hybrid optimization algorithm to design the compensation values. The obtained values correspond to real BOM components from the passive database. In the manual mode, pole-zero placement back computes the compensation components. Applying changes to design applies the current compensation design to the main design.

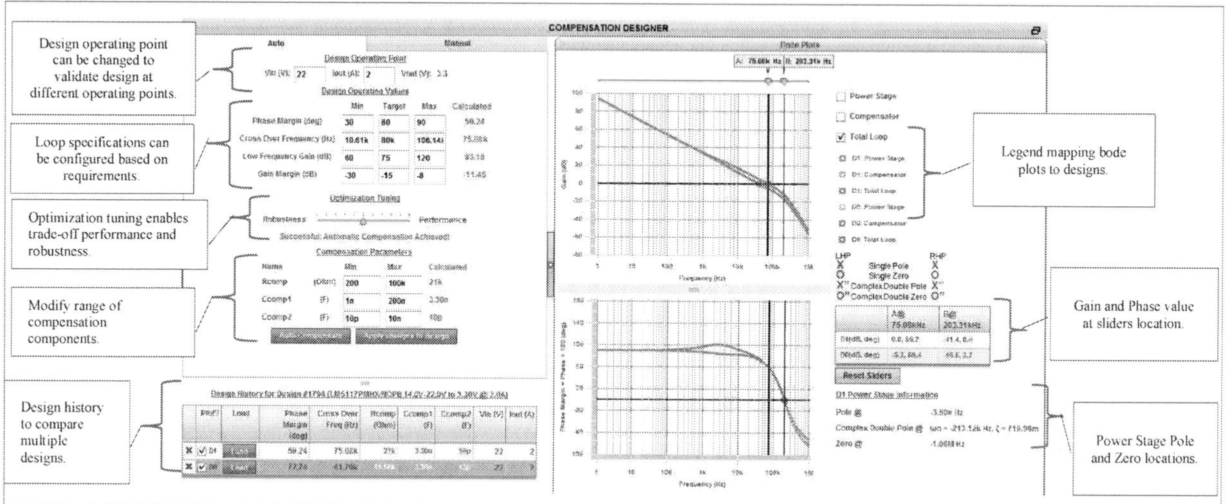

Fig. 4. Compensation Designer Auto-Mode UI

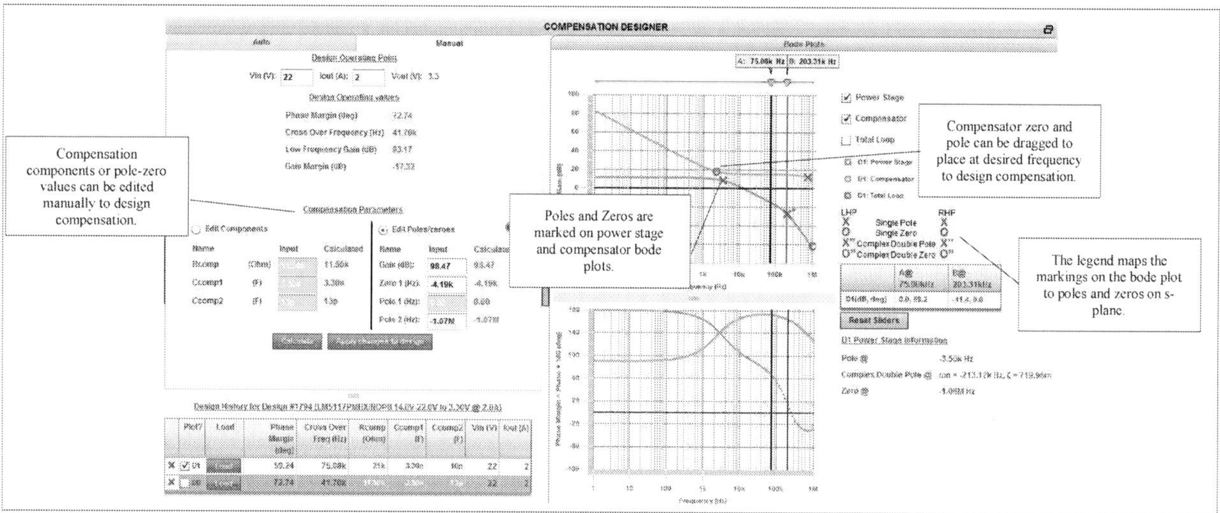

Fig. 5. Compensation Designer Manual-Mode UI and Pole-Zero Placement

IV. Objective Function Formulation

A. Compensation Design Requirements:

The important requirements of compensation design are to meet the following specifications:

- Load Regulation: Maximum variation in steady state output voltage for given load current variation. This determines the minimum low frequency gain.
- Line Regulation: Maximum variation in steady state output voltage for given change in input voltage. This determines the minimum low frequency gain.
- Overshoot: Maximum deviation of output voltage from steady state during load or line transient. This determines the minimum crossover frequency.
- Settling time: The time taken to come back to steady state during load or line transient. This determines the

maximum RC time constants of compensator components.

- Ringing: Steady or slowly decaying oscillations in output voltage during load / line transients. This determines the required Phase Margin.
- Noise rejection: The closed loop system needs to have sufficiently low gain at higher frequencies. This determines the maximum crossover frequency.

Based on the above requirements and the constraints of the power stage the target, min and max values of the key loop specifications like crossover frequency (Fco), Phase Margin (PM), Gain Margin (GM), Low Frequency Gain (LFG) are chosen. The objective function is formulated based on these specifications as given in the next section.

B. Objective function formulation

The key challenge for compensation design using optimization is formulation of a proper objective function which can be minimized and the constraints which need to be satisfied to arrive at the final design. A simple formulation which meets the desired loop specifications such as Fco, PM, GM and LFG may not be sufficient to guarantee a robust design. While the overshoot is minimized by increasing the Fco [11], the settling time depends on the compensation zero, where decreasing the zero location increases the settling time [11,12]. Thus minimizing the settling time should also be taken into account during the formulation. Also, meeting the loop specifications at a single operating point does not guarantee the stability or performance over a full operating range. In view of the above requirements the design problem is formulated as:

If $\mathbf{c} = [Rc_1, Cc_1, Rc_2, Cc_2 \ldots]$ are the compensation parameters, the compensation design problem can be formulated as the minimization of the objective function $O(\mathbf{c})$ given by

$$O(\mathbf{c}) = Se(\mathbf{c}) + T_{RC}(\mathbf{c}) + S_{PM}(\mathbf{c}) + P(\mathbf{c}), \quad (1)$$

where the weighted square error $Se(\mathbf{c})$ of the loop specifications is given by

$$\begin{aligned} Se(\mathbf{c}) &= \omega_{PM} \times \left(PM(\mathbf{c}) - PM_{spec}\right)^2 + \omega_{Fco} \times \left(Fco(\mathbf{c}) - Fco_{spec}\right)^2 \\ &+ \omega_{GM} \times \left(GM(\mathbf{c}) - GM_{spec}\right)^2 + \omega_{LFG} \times \left(LFG(\mathbf{c}) - LFG_{spec}\right)^2, \end{aligned} \quad (2)$$

and the weighted sum of RC time constants of the compensation network $T_{RC}(\mathbf{c})$ is given by

$$T_{RC}(\mathbf{c}) = \omega_1 \times Rc_1 \times Cc_1 + \omega_2 \times Rc_2 \times Cc_2. \quad (3)$$

$S_{PM}(\mathbf{c})$, the weighted slope of phase curve at the gain cross-over frequency is given by

$$S_{PM}(\mathbf{c}) = \omega_s \frac{dPM(\mathbf{c})}{df}\Big|_{f=Fco}, \quad (4)$$

and finally the penalty function $P(\mathbf{c})$ given in (5) includes an exponential penalty to the objective function when the specification goes lower than min limit or greater than max limit over the given load current and input voltage range. Minimum, nominal and maximum values of voltage and current are used for summation over vin and $Iload$.

$$\begin{aligned} P(c) &= \sum_{vin}\sum_{Iload} H\left(PM_{min} - PM(\mathbf{c})\right)\left(e^{\gamma_1(PM_{min} - PM(\mathbf{c}))} - 1\right) \\ &+ \sum_{vin}\sum_{Iload} H\left(Fco_{min} - Fco(\mathbf{c})\right)\left(e^{\gamma_2(Fco_{min} - Fco(\mathbf{c}))} - 1\right) \\ &+ \sum_{vin}\sum_{Iload} H\left(DCG_{min} - DCG(\mathbf{c})\right)\left(e^{\gamma_3(DCG_{min} - DCG(\mathbf{c}))} - 1\right) \\ &+ \sum_{vin}\sum_{Iload} H\left(GM_{min} - GM(\mathbf{c})\right)\left(e^{\gamma_4(GM_{min} - GM(\mathbf{c}))} - 1\right) \end{aligned} \quad (5)$$

where $H(x)$ is the unit step function such that $H(x) = 0$ if $x < 0$ and $H(x) = 1$ if $x \geq 0$.

$Se(\mathbf{c})$ is used to meet the target loop specifications at nominal operating point, $T_{RC}(\mathbf{c})$ is used to minimize the transient settling time for load and line transients, $S_{PM}(\mathbf{c})$ is used to improve the robustness of the solution w.r.t small perturbations in the power stage component parameters and $P(\mathbf{c})$ is used to make sure the design remain within desired specification limits over the entire operating range of the converter.

C. Optimization Tuning:

Optimization tuning allows user to trade-off performance and robustness. This is achieved by tuning the weights of the objective function. When the tuner is set to the left most position i.e. towards robustness, the weights of PM and GM functions as well as the weight for the slope of the Phase curve at Fco are set to higher value. Similarly if the tuner is set to right most position the weights of the cross over function and DC gain are set to higher value. Thus when it is not possible to meet all the desired specifications, in performance mode the design tries to achieve results closer to target Fco and DCG and in robustness mode the design tries to achieve results closer to target PM, GM.

V. HYBRID OPTIMIZATION ALGORITHM

The Hybrid optimization used in this work is a combination of Differential Evolution (DE) and Conjugate Gradient method (CG) as proposed in [10]. DE optimizes the problem by maintaining a population of candidate solutions and creating new candidate solutions by combining the existing ones using formulae as given in [9], the solutions with better scores are retained and iteratively new candidate solutions are formed. DE is robust and converges to global optima even in presence of local optimum points and noise in the function. Compared to gradient based methods DE is slow to converge. CG is fast converging but may give erroneous results in presence of noise and only converges to local minima. By combining DE with CG, where CG is applied for some of the candidate solutions to arrive at the new candidate solutions, the convergence can be significantly faster as demonstrated in [10]. The detailed algorithm is also given in [10].

VI. Results

Re-Comp is available online for use with hundreds of different controller ICs covering all major power topologies like Buck, Buck-Boost, Boost, different control modes like voltage mode, peak current mode, emulated peak current mode, and different compensation types like PI, type-II and type-III etc. Here, are two example cases comparing the compensation design using pole-zero cancelation to the Optimization based method. The comparative bode plots along with transient simulation results are included.

A. *Example 1:*

Voltage-Mode Buck Converter with type-III compensation using LM21212-2 controller IC is used with the specifications and power stage components as given in Table I. The power stage bode plot is shown in Fig. 6 and the power stage transfer function (control to output) is given by (6). Table-II shows two compensation designs one done manually using traditional pole-zero cancellation method and the other using hybrid optimization. The bode plots of loop gain transfer functions of both designs are given in Fig. 7. The load transient results for the two cases are shown in Fig. 8, which clearly shows the improved overshoot for the load transient compared to the original design.

TABLE I. Converter Power Stage Component Values and Operating Values

Operating Values		Power Stage Components		Feedback parameters	
Vin	3V-5V	Lout	330nH, 3.2mΩ	Rfbt	10kΩ
Vout	1V	Cout	100uF, 4.88mΩ	Rfbb	15kΩ
$Iout_{max}$	12A	R_{dshs}	7mΩ	Vref	0.6V
F_{sw}	1530kHz	R_{dsls}	4.3mΩ	Error Amp (DCG, BW)	95dB, 11MHz

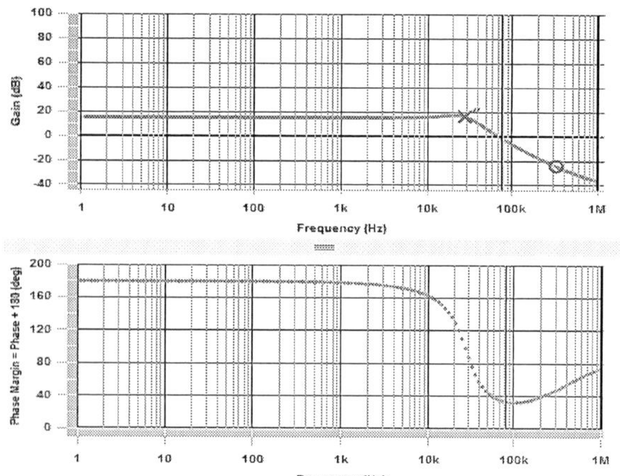

Fig. 6. Power Stage Bode Plot for Example 1.

$$Gps(s) = K * \frac{(1 + s / w_{zesr})}{(1 + s / Qw_{plc} + s^2 / w_{plc}^2)}, \quad (6)$$

where w_{zesr} = $2.05e6$ rad/s, w_{plc} = $178e3$ rad/s, Q = 1.14.

TABLE II. Compensation Components through Pole-Zero Cancellation and through the Re-Comp Tool

Pole-Zero cancellation Method (D0)			Re-Comp Tool (D1)		
Targeted Spec	Comp Values	Actual Spec	Targeted Spec	Comp Values	Actual Spec
Pole-Zero placement method.	Rc1 = 4.99kΩ	Fco = 80kHz	Fco_{spec} = 200kHz	Rc1=6.98k Ω	Fco = 200kHz
	Cc1= 1.2nF	PM=47 deg	PM_{spec} = 45 deg	Cc1=1nF	PM= 46deg
	Cc2= 220pF	GM = 45dB	GM_{spec} = 30dB	Cc2=15pF	GM = 45dB
	Cc3 = 560pF	LFG = 95dB	LFG_{spec} = 100dB	Cc3=1.2n F	LFG = 98dB
	Rc2 = 953 Ω			Rc2=787 Ω	

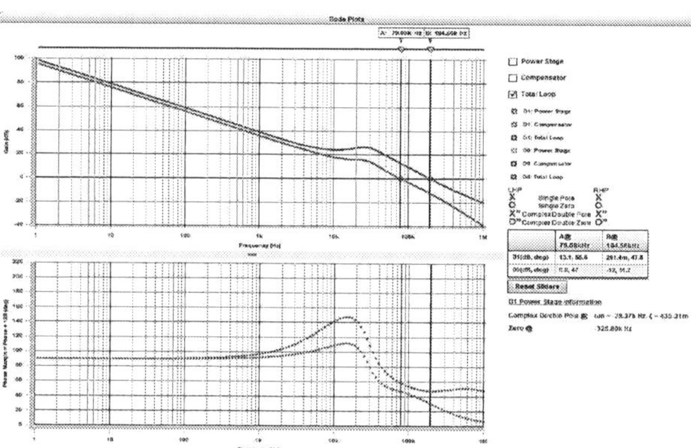

Fig. 7. Bode Plot of the loop gain and phase for D0 (pole-zero cancellation method) and D1 (Re-Comp).

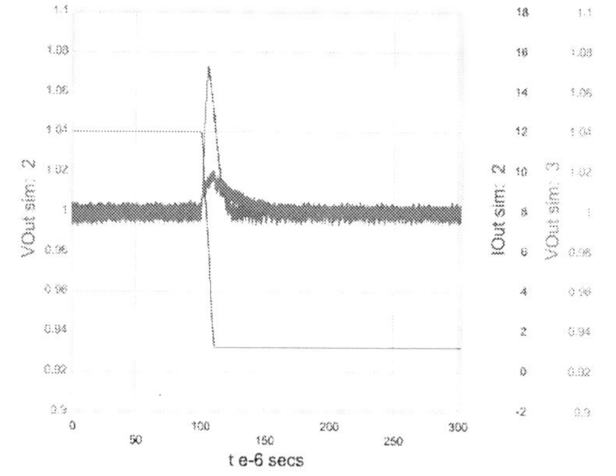

Fig. 8. Transient responses for a load step from 12A-1.2A for D0 (Vout sim:2) and D1 (Vout sim:3).

B. *Example 2: Current Mode Boost Converter TPS40210*

Current-Mode Boost Converter with type-II compensation using TPS40210 controller IC is used with the specifications and power stage components as given in Table III. Note that the output stage has mixed capacitors and hence cannot be lumped to single capacitor and ESR. The power stage bode plot is shown in Fig. 9 and the power stage transfer function (control to output) is given by (7). The mixed output capacitor gives rise to additional zero and pole in the transfer function. The presence of RHP zero limits the achievable bandwidth, and hence we cannot increase the BW significantly using Hybrid optimization algorithm. Table-IV shows two compensation designs one done manually using traditional pole-zero cancellation method and the other using Hybrid optimization. Fig. 10 gives the loop gain bode plots for both designs and Fig. 11 gives the transient results for the two cases, due to RHP zero we only see a marginal improvement using hybrid optimization method.

TABLE III. CONVERTER POWER STAGE COMPONENT VALUES AND OPERATING VALUES

Operating Values		Power Stage Components		Feedback parameters	
Vin	5V-6V	Lout	2.2uH, 4.1mΩ	Rfbt	53.6kΩ
Vout	12V	Cout	390uF, 14mΩ 88uF, 0.5mΩ	Rfbb	3.32kΩ
$Iout_{max}$	5A	V_D	0.510V	Vref	0.7V
F_{sw}	512kHz	R_{dsls}	7.5mΩ		

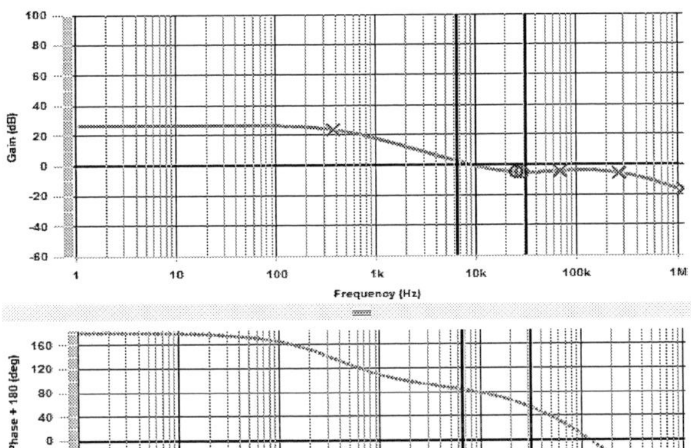

Fig. 9. Power Stage Bode Plot for Example 1.

$$Gps(s) = K * \frac{(1 - s/w_{zrhp})(1 + s/w_{zesr1})(1 + s/w_{zesr2})}{(1 + s/w_p)(1 + s/Qw_n + s^2/w_n^2)(1 + s/w_{pesr})}, \quad (7)$$

Where, $w_{zrhp} = 161e3$ rad/s, $w_p = 2.35e3$ rad/s, $w_{zesr1} = 182e3\frac{rad}{s}$, $w_{zesr2} = 45e6\frac{rad}{s}, w_n = 1.78e6\frac{rad}{s}, Q = 0.2311, w_{pesr} = 1.7e6\frac{rad}{s}$

TABLE IV. COMPENSATION COMPONENTS THROUGH POLE-ZERO CANCELLATION AND THROUGH THE RE-COMP TOOL

Pole-Zero cancellation Method (D0)			Re-Comp Tool (D1)		
Targeted Spec	Comp Values	Actual Spec	Targeted Spec	Comp Values	Actual Spec
Pole-Zero placement method.	Rc1 = 44.2k	Fco = 6.58 kHz	Fco_{spec} = 10kHz	Rc1=49.9k Ω	Fco = 8kHz
	Cc1= 2nF	PM=55 deg	PM_{spec} = 60 deg	Cc1=1.5nF	PM= 60deg
	Cc2= 150pF	GM = 12dB	GM_{spec} = 10dB	Cc2=56pF	GM = 8dB
		LFG = 88dB	LFG_{spec} = 100dB		LFG = 92dB

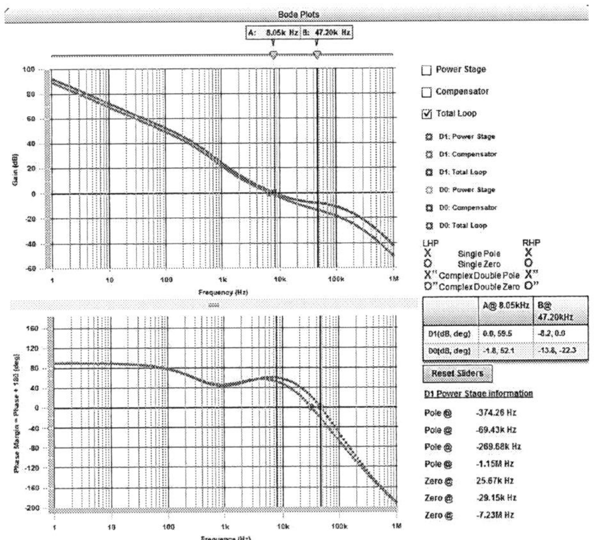

Fig. 10. Bode Plot of the loop gain and phase for D0 (pole-zero cancellation method) and D1 (Re-Comp).

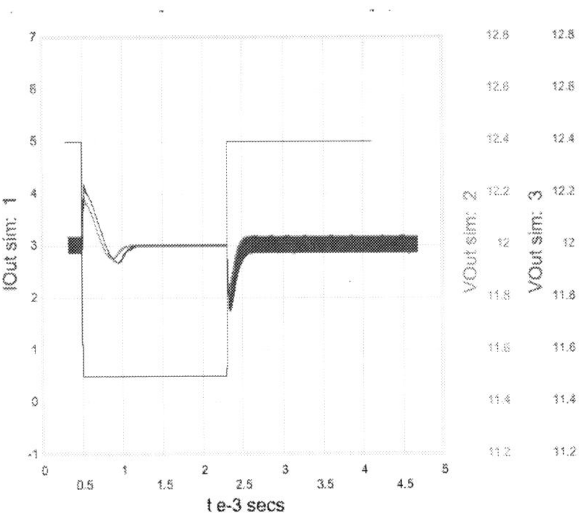

Fig. 11. Transient responses for a load step from 12A-1.2A for D0 (Vout sim:3) and D1 (Vout sim:2).

VII. Conclusion

A comprehensive web based tool for compensation design of power converters is presented, which includes a fully automated approach based on Hybrid optimization and a manual interactive pole-zero placement method. It is shown that the tool directly outputs real components from the database which can be directly implemented on board. The formulation used not only tries to meet the loop specifications at a given operating point, but also minimizes settling time and improves robustness over operating range and component perturbations. The design results for a voltage mode buck regulator show the improved transient performance of the design obtained through Re-Comp versus a design using pole-zero cancellation method, while the results for Boost converter show some marginal improvement over pole-zero cancelation method.

References

[1] Timothy Hegarty, "Voltage-mode control and compensation: Intricacies for buck regulators" Article, EDN Network June 20, 2008.

[2] Henry Zhang, "Modeling and Loop Compensation Design of Switching Mode Power Supplies, Part1 and Part2" Article, EDN Network, March 14, 2015.

[3] S. Samanta, S Pam, R Sheehan, S. Mukhopadhyay, "Constrained optimization based compensation design for power converters using method of feasible directions" APEC 2012.

[4] S. Ghosh, S. Samanta, "Fixed structure compensator design using a constrained hybrid evolutionary optimization approach" ISA Transactions, April 2014.

[5] Middlebrook R.D., "Modeling Current-Programmed Buck and Boost Regulators." IEEE, Transactions on Power Electronics, Vol. 4. No. 1, January 1989.

[6] Tan F.Dong and Middlebrook R.D., "A Unified Model for Current-Programmed Converters." IEEE, Transactions on Power Electronics, Vol. 10. No. 4, July 1995.

[7] S.S. Hong, B. Choi and H. Ahn., "The Unified Model For Current-Mode Control: An Alternative Derivation", Journal Of Circuits, Systems, and Computers, Vol. 13, No.4, 2004.

[8] Singiresu.S. Rao, "Engineering Optimization: Theory and Practice" Book, A Wiley-Interscience Publication 1996.

[9] Ken Price, Rainer Storn, and Jouni Lampinen, "Differential Evolution - A Practical Approach to Global Optimization" Book, Springer Publication, ISBN: 3-540-20950-6, 2005.

[10] Krzystof Bandurski, Wojciech Kwedlo, "A Lamarckian Hybrid of Differential Evolution and Conjugate Gradients for Neural Network Training", Springer Neural Processing Letters, Vol 32, Issue 1, August 2010.

[11] Kaiwei Yao, Yuancheng Ren and Fred C. Lee, "Critical Bandwidth for the Load Transient Response of Voltage Regulator Modules" IEEE Trans. Power Electron., vol.19, pp.1454-1461, Nov.2004.

[12] Byungcho Choi, "Step load response of a current-mode-controlled DC-to-DC converter" IEEE Transactions on Aerospace and Electronic Systems, Vol. 33, Issue 4, Oct 1997.

Second Order Sliding Mode Controlled Point of Load Power Supply

Prasanta K. Achanta, David C. Jones, Dragan Maksimovic
Colorado Power Electronics Center
University of Colorado Boulder
Boulder, USA

Serhii M. Zhak
Linear Technology Corporation
North Chelmsford, MA, USA
szhak@linear.com

Brett Miwa, Cory Arnold
Maxim Integrated
brett.miwa@maximintegrated.com
cory.arnold@maximintegrated.com

***Abstract*—This paper describes the implementation of a second order sliding mode (SOSM) controller for a practical point-of-load (POL) power supply with realistic values of output capacitor equivalent series resistance (ESR) and equivalent series inductance (ESL). The SOSM control presented in this paper results in fast transient responses, has a current limiting feature, supports light load operation, and achieves constant frequency steady state operation. Experimental results are given for a 2.5 V - 5.5 V input, 0.6 V – 1.5 V output, 3 A prototype.**

Keywords — digital control, sliding mode control, DC-DC converters, buck converter.

I. INTRODUCTION

Voltage-mode control techniques based on small signal modelling [1] restrict the maximum possible closed loop bandwidth of switched mode power converters. To improve the large-signal dynamics in point-of-load (POL) power supplies such as the buck converter shown in Fig.1, several nonlinear solutions, including charge balance principles [2], nonlinear state feedback [3-4], or more advanced nonlinear digital time optimal control techniques [5-10], have been reported. These approaches are either sensitive to power stage component parameters (e.g., output filter inductor and output capacitor) or face difficulty making transitions between controller modes in real time. Sliding mode control (SMC), which has long been of interest to power electronics engineers [11], is a powerful technique used to achieve superior transient performance while providing robustness against parameter uncertainties. SMC operates by splitting the state space of the system into distinct non overlapping regions separated by the sliding surface [12]. The control objective is to bring the system state variables onto the sliding surface and keep them on the sliding surface until the state variables reach a new equilibrium point. A second order sliding mode (SOSM) controller with near-optimal transient response was described in [13]. This controller undergoes performance degradation in the presence of realistic values of output capacitor ESR/ESL. The reason for this degradation and a modified SOSM controller were presented in [14]. Based on the approach presented in [14], this paper gives implementation details for SOSM controller, with added practical features of current limiting, enhanced light load operation, and constant steady state switching frequency.

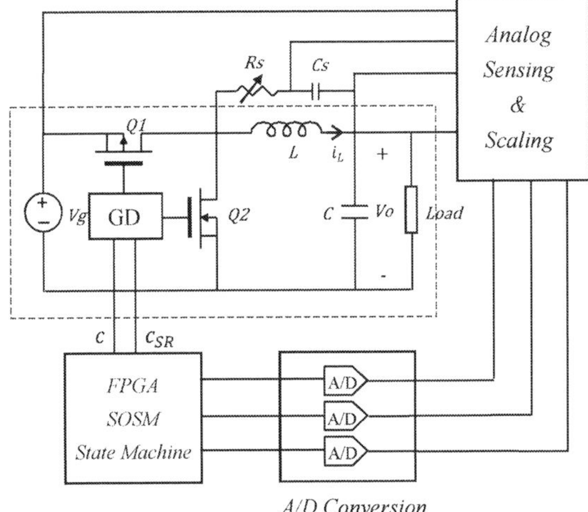

Fig. 1. Block diagram of SOSM controlled POL power supply.

This paper is organized as follows. The state machine used to implement SOSM control is introduced in Section II. An extension of the SOSM state machine to include constant frequency steady state operation is described in Section III. A prototype controller implementation on a 2.5 V – 5.5 V input, 0.6 V – 1.5 V output, 3 A buck converter and associated experimental results are presented in Section IV. Finally, Section V summarizes and concludes the paper.

II. SOSM STATE MACHINE

Unlike most first order SMC algorithms developed for power electronics, the SMC algorithms presented in [13-15] are of second order. Second order sliding mode control only requires voltage sensing to provide near optimal transient responses. The functionality of SOSM control can be best understood with the help of a state machine. The SOSM state machine shown in Fig. 2 has four toggle states ($T_{ON}^{+/-}, T_{OFF}^{+/-}$), four hold states ($H_{ON}^{+/-}, H_{OFF}^{+/-}$) and two discontinuous conduction mode states ($D_{OFF}^{+/-}$). These $D_{OFF}^{+/-}$ states proposed here, which improve light load performance, are an addition to the SOSM state machine proposed in [14]. The high-side switch of the buck converter is turned on (off) whenever the machine resides in a

978-1-4673-9551-9/16 $31.00 © 2016 IEEE

state with subscript ON (OFF). Both the high-side and the low-side switches are turned off when the machine resides in either of the discontinuous conduction mode (DCM) states.

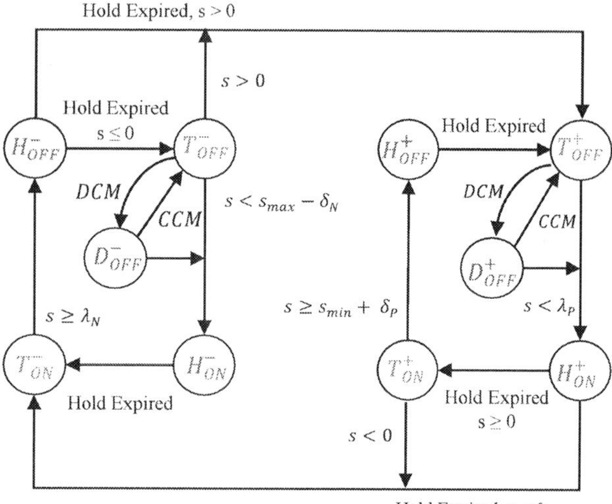

Fig. 2. SOSM controller state machine.

The voltage error $s = V_o - V_{ref}$ is the difference between the output voltage V_o and the reference voltage V_{ref}. s_{max} and s_{min} hold the most recent value of the error maximum and error minimum, respectively. The magnitude and sign of the error s, together with the voltage thresholds $\lambda_P, \delta_P, \lambda_N, \delta_N$ computed from Table I, trigger the transitions between states [14].

TABLE I. EQUATIONS AND CONSTANTS USED IN THE CALCULATION OF VOLTAGE THRESHOLDS IN THE STATE MACHINE OF FIG. 2.

$$\lambda_N = \alpha_{N1} + \alpha_{N2}\sigma_N - \alpha_{N3}\sigma_N V_{ref} + \alpha_{N4}\sigma_N^2$$

$$\sigma_N = -s_{min}\left[s_{min} + 2\left(V_{ref} - v_g\right)\right]/2v_g$$

$$\lambda_P = \alpha_{P1} + \alpha_{P2}\sigma_P + \alpha_{P3}\sigma_P\left(v_g - V_{ref}\right) + \alpha_{P4}\sigma_P^2$$

$$\sigma_P = s_{max}\left(s_{max} + 2V_{ref}\right)/2v_g$$

Parameter	Value	Units
δ_N	30	mV
α_{N1}	9.46e-3	V
α_{N2}	9.91e-1	
α_{N3}	3.60e-2	V^{-1}
α_{N4}	-2.95e-2	V^{-1}
δ_P	30	mV
α_{P1}	-1.34e-2	V
α_{P2}	9.97e-1	
α_{P3}	-3.36e-2	V^{-1}
α_{P4}	3.00e-2	V^{-1}

As an example, consider a startup transient shown in Fig. 3. When the system is powered on, the state machine sees a negative error, jumps to T^-_{ON} turning on the high-side switch Q1, of the buck converter shown in Fig. 1. When the state machine resides in T^-_{ON}, the output voltage begins to increase. When the condition $s \geq \lambda_N$ is met, the state machine enters H^-_{OFF} and turns the buck high-side switch off. The output capacitor's ESR and ESL cause trajectory discontinuities at these switching instants as shown in Fig. 4 [14]. The state machine is placed in hold states to avoid making decisions based on the indeterminate voltage samples that the state machine sees during these switching instants. The state machine stays in the hold state H^-_{OFF} for a fixed number of clock cycles (H) (e.g., 12 clock cycles), after which it transitions to T^-_{OFF} state and stays in this state until the condition $s < s_{max} - \delta_N$ is met. As soon as this condition is met, the state machine transitions to H^-_{ON} where it stays for a fixed duration of time (e.g., 12 clock cycles). This cycle $T^-_{ON} \rightarrow H^-_{OFF} \rightarrow T^-_{OFF} \rightarrow H^-_{ON} \rightarrow T^-_{ON}$ continues and the error converges to the neighborhood of zero. According to the simulations, the output voltage V_o reaches V_{ref} in less than two such cycles as shown in Fig. 3. Similar analysis can be applied

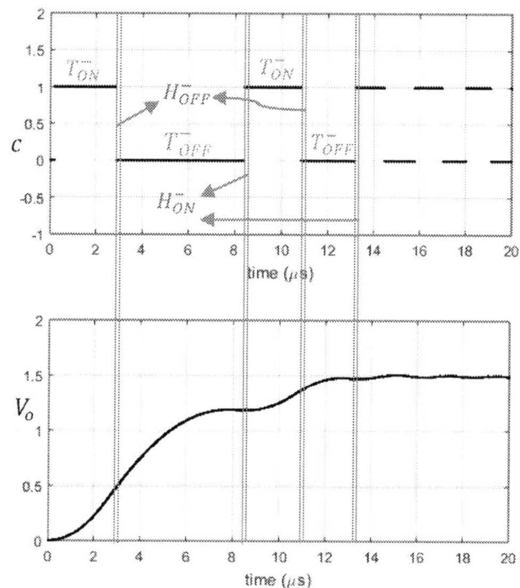

Fig. 3. Simulated startup for V_{ref} = 1.5 V, V_g = 3 V, load = 1 Ω. Top: high-side switch logic. Bottom: output voltage in Volts.

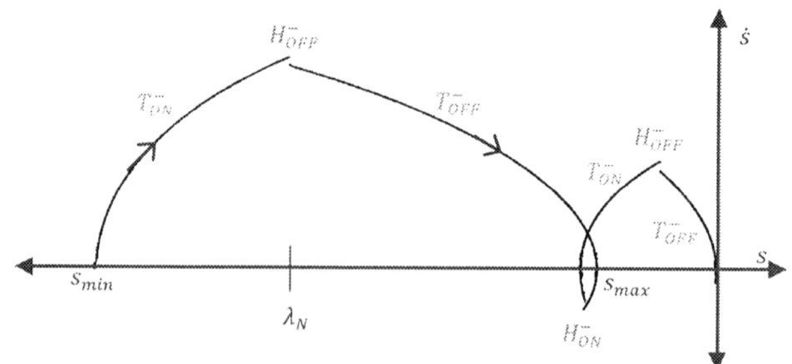

Fig. 4. State plane trajectory for the startup transient shown in Fig. 3. In this example, ESR = 5 mΩ, ESL = 2 nH.

in the case of a step down in V_{ref}, which engages the cycle $T^+_{OFF} \rightarrow H^+_{ON} \rightarrow T^+_{ON} \rightarrow H^+_{OFF} \rightarrow T^+_{OFF}$.

This state machine implementation only requires input voltage and output voltage sensing. However, if current limiting is required, the inductor current must also be sensed. In this case the sate machine transition equations of Table I are augmented to include terms involving the measured current. With these terms enabled, when the current reaches a threshold (e.g., 4.8 A) during the transient, the state machine accelerates the transitions between toggle and hold states, keeping the current within the limit (e.g., 5 A) while still maintaining a fast response.

To demonstrate the light load performance of SOSM control on the sychronous buck converter, consider the $1\ \Omega - 10\ \Omega$ load transient simulation shown in Fig. 5. When the load step occurs, the state machine trasitions to T^+_{OFF} as a result of the positive voltage error s. The high-side is turned off and the low-side is turned on in T^+_{OFF} state. The current ramps down and when a zero crossing of the current is sensed, the state machine transitions to D^+_{OFF} state, turning off both switches and preventing the inductor current from going negative. When the error s becomes less than λ_P, the state machine comes out of the DCM state and enters H^+_{ON} state turning the high-side switch on.

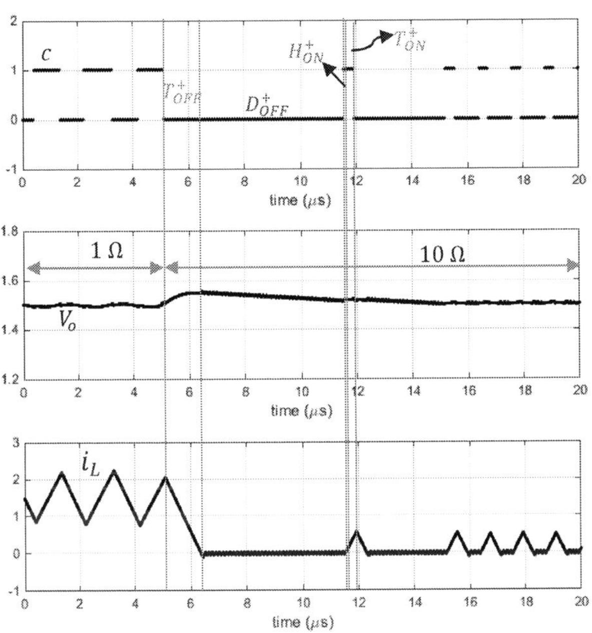

Fig. 5. Load transient ($1\ \Omega - 10\ \Omega$) simulation. Top: high-side switch logic. Middle: output voltage waveform in Volts. Bottom: inductor current waveform in Amperes.

III. Constant Frequency Steady State Operation

In any sliding mode control technique, chattering makes it difficult to achieve a constant frequency steady state operation. In the implementation described in this paper, constant frequency steady state operation is achieved by the addition of four transitions to the SOSM state machine shown in Fig. 2. Consider a simple digital pulse width modulator (DPWM) state machine shown in Fig. 6(a). It has two states, T_{ON} (high side switch is on) and T_{OFF} (high side switch is off). The state machine stays in T_{ON} (T_{OFF}) state until the D ($\overline{D}$) counter expires, where $\frac{D}{D+\overline{D}}$ corresponds to the duty cycle (d) of the converter. To make it easier for integration into the original state machine, two additional hold states are added to the DPWM state machine as shown in Fig. 6(b). For a DPWM clock resolution (t_{clk}) of 4 ns and a steady state switching period (T_s) of 1 μs, the timing diagram corresponding to the DPWM state machine is given in Fig. 6(c). The DPWM state machine can now be integrated into the SOSM state machine of Fig. 2. The extended SOSM state machine shown in Fig. 7, achieves constant frequency steady state operation. Once the SOSM resolves the transient and the error falls within a certain window ($\pm\Delta$), the control is transferred to a simple DPWM based feedback. This simple DPWM feedback increments (decrements) the D ($\overline{D}$) counter value if the error is negative and

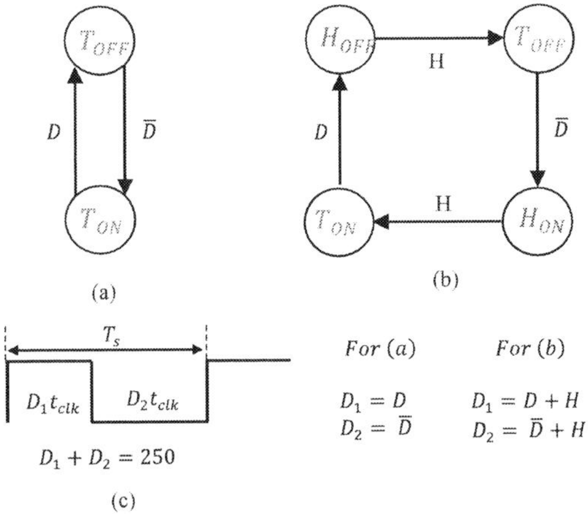

Fig. 6. DPWM state machine (a) two states, (b) four states, and (c) timing diagram.

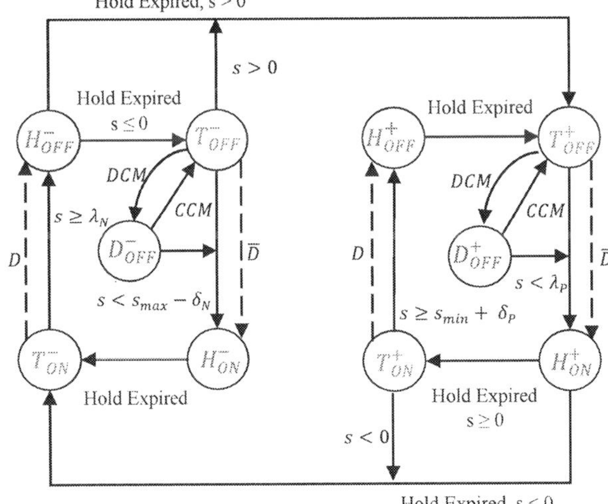

Fig. 7. Extended version of SOSM state machine capable of achieving constant frequency steady state operation. Additional transitions are shown as dotted lines.

decrements (increments) the D ($\overline{D}$) counter value if the error is positive similar to the approach described in [16].

To expedite the process of achieving constant frequency steady state operation after the transient is resolved by SOSM, D and $\overline{D}$ counter values are updated upon startup or V_{ref} transients in a feedforward manner as follows:

$$V_o = d\,V_g \tag{1}$$

$$V_o = (\,d\,\times 250\,)\frac{V_g}{250} \tag{2}$$

$$V_o \approx D_{ff}\,\frac{V_g}{256} = D_{ff}\,\frac{V_g}{2^8} \tag{3}$$

Put in words, $\frac{V_g}{2^8}$ is added D_{ff} times until the sensed value of the output voltage V_o is obtained. This process results in the required feedforward value D_{ff}, of the D counter.

IV. Experimental Results

The architecture of the prototype used to implement the extended SOSM controller is shown in Fig. 1. A tuned RC sense network is used for current sensing. For experimentation and debugging purposes, three 250-Msps, 10-bit A/D converters were used to sense the input voltage, output voltage and the inductor current, although inductor current sensing can be achieved by using simple comparators to achieve current limiting and light load operation. The state machine was configured to run at a clock rate of 62.5 MHz. The parameters of the synchronous buck converter prototype are listed in Table II. The results presented in this section are for the synchronous buck converter shown in Fig. 1, supplied with $V_g = 3$ V and V_{ref} set to 1.5 V unless otherwise stated.

A. Startup transients and light load operation

The SOSM state machine of Fig. 2 is sufficient to handle all the transients and achieve enhanced light load operation. A 0 V – 1.5 V startup transient for a load of 1 Ω with the current limiting feature turned off is shown in Fig. 8. With the current limiting feature turned on, the high-side switch is turned off as soon as the current threshold is detected as shown in Fig. 9. A

Table II. SOSM controlled POL power supply parameters

Ref.	Description	Value	Part no.	Manufacturer
C1	Input cap	10 µF	JMK107 series	Taiyo Yuden
C2	Output cap	22 µF	CL series	Samsung
L	Inductor	1 µH	MPI4040 series	Eaton
Cs	Sense cap	39 nF	C0805C series	Kemet
Rs	Sense resistor	1 KΩ ± 30%	PVZ2A series	Murata
GD	Gate driver		MAX19602	MAXIM
Q1	PMOS		CSD25310Q2	TI
Q2	NMOS		CSD13202Q2	TI
A/D	ADC		LTC2242-10	Linear
FPGA			Stratix IV series	Altera

0 V – 1.5 V startup transient for a load of 7 Ω is shown in Fig. 10. Note that the SOSM state machine allows the current to go negative to speed up the transient recovery. After the startup transient is resolved, whenever a zero crossing in the inductor current is detected, the state machine enters D_{OFF}^{-} or D_{OFF}^{+} depending on the polarity of the error. As shown in Fig. 11, in

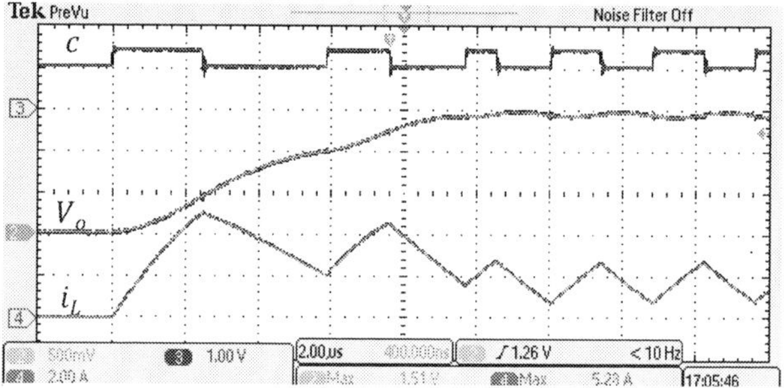

Fig.8. Startup transient (0 V – 1.5 V) for a load of 1 Ω without current limiting. Rise time = 10 µs, no overshoot.

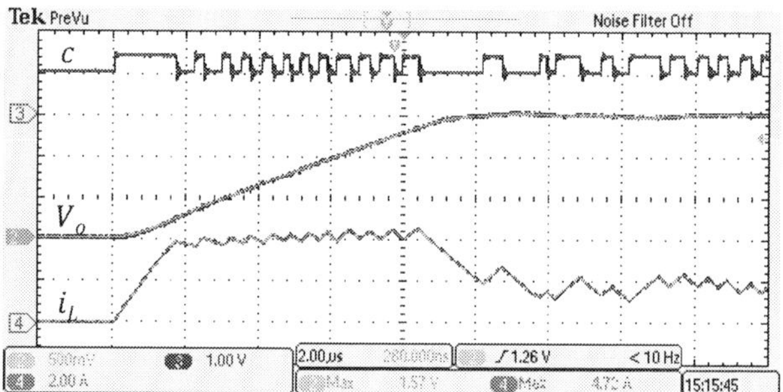

Fig.9. Startup transient (0 V – 1.5 V) for a load of 1 Ω with current limiting. Rise time = 10 µs.

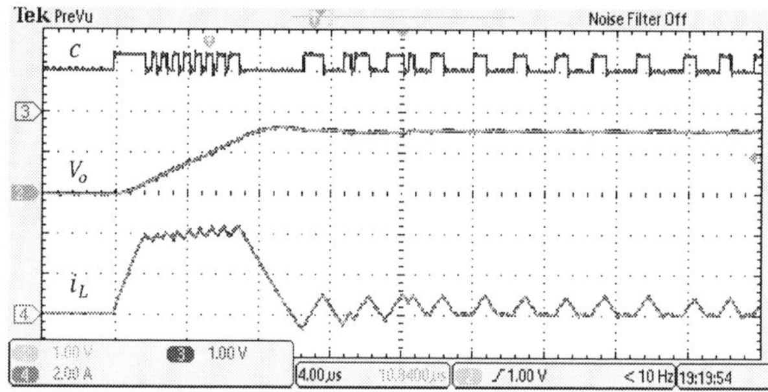

Fig. 10. Startup transient (0 V – 1.5 V) for a load of 7 Ω. Note that SOSM allows current to go negative while resolving transients.

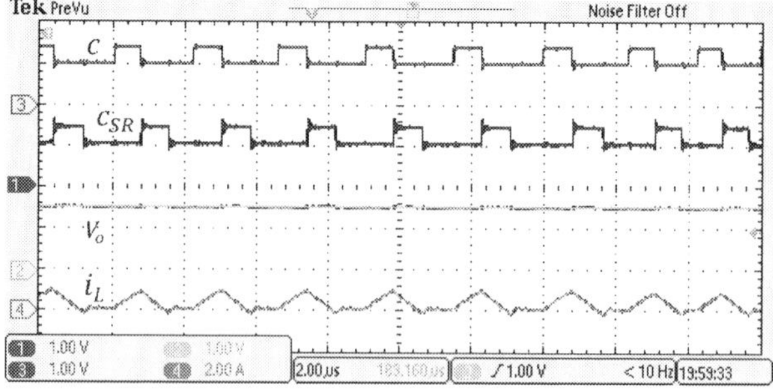

Fig. 11. Steady state light load operation. High-side and low-side switches are turned off at the zero crossing of inductor current.

either of these states, both the high-side and the low-side switches are turned off, which improves light load efficiency. Experimental waveforms showing the transition from continuous conduction mode (CCM) to discontinuous conduction mode (DCM) are shown in Fig. 12.

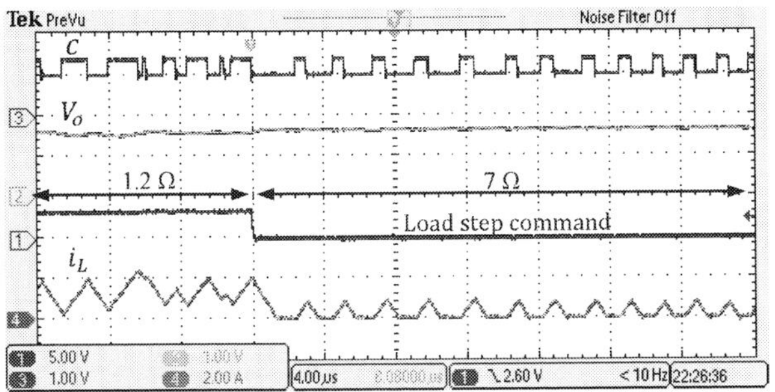

Fig.12. Load step transient (1.2 Ω – 7 Ω) showing the transition from CCM operation to DCM operation when inductor current zero crossing is detected.

B. V_{ref} step and load step responses with constant frequency steady state operation

To achieve a 1 MHz constant frequency steady state operation, the extended SOSM state machine of Fig. 7 is used. V_{ref} step down and step up transient response waveforms are given in Fig. 13 and Fig. 14 respectively. The transient responses corresponding to load steps are shown in Fig. 15 and Fig. 16. Note that there is a delay involved in detecting the load transient. After the load step command is given, the error takes a finite amount of time to come out of the error window, Δ. When the error becomes greater than the specified Δ, the control is handed over to SOSM which resolves the transient. The size of Δ and the load step size decide the amount of delay.

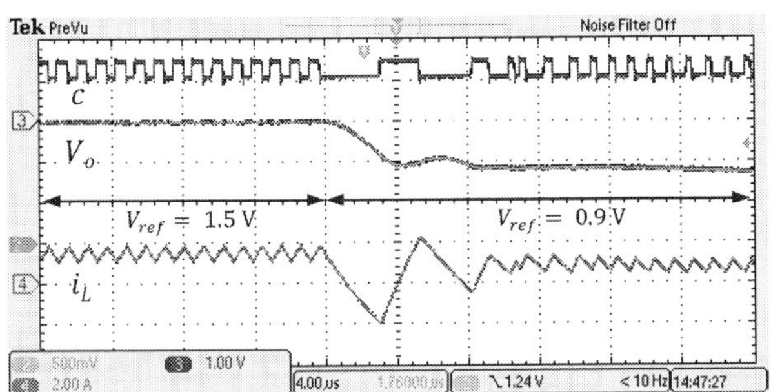

Fig.13. V_{ref} step down response (1.5 V - 0.9 V). Transient recovery time = 8 μs. SOSM resolves the transient after which the control is handed over to DPWM with the feedforward value $D = D_{ff}$.

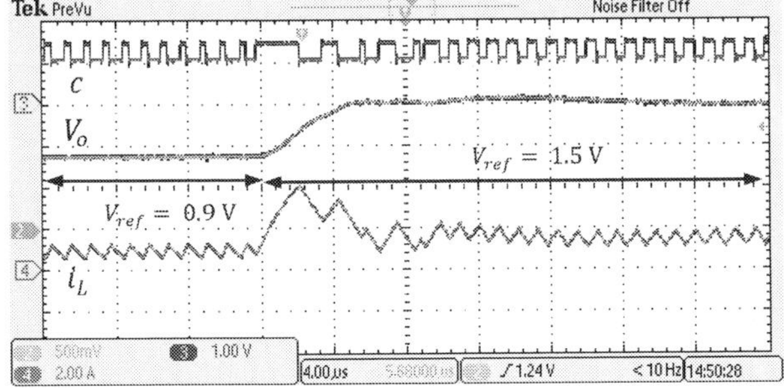

Fig.14. V_{ref} step up response (0.9 V - 1.5 V). Transient recovery time = 8 μs. SOSM resolves the transient after which the control is handed over to DPWM with the feedforward value of $D = D_{ff}$.

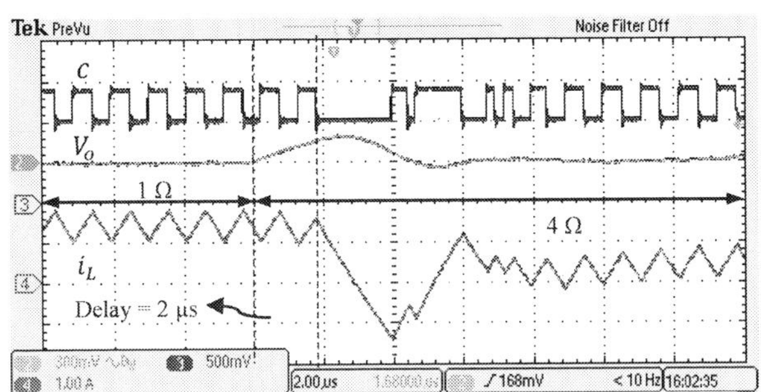

Fig.15. Load transient 1 Ω – 4 Ω. Transient recovery time = 6 μs.
Top: high side gate switching waveform.
Middle: AC coupled output voltage waveform.
Bottom: inductor current waveform.

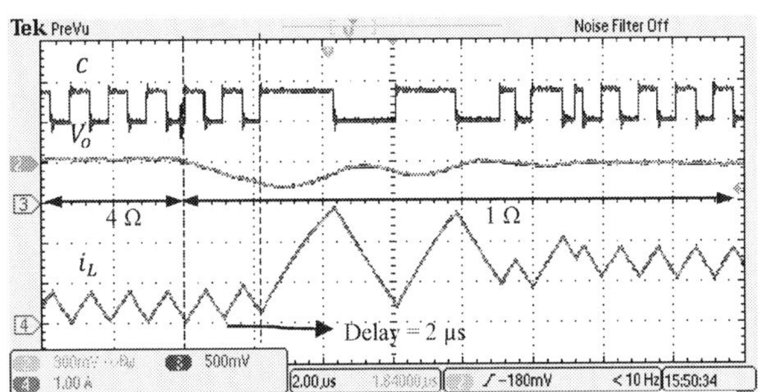

Fig.16. Load transient 4 Ω – 1 Ω. Transient recovery time = 8 μs.
Top: high side gate switching waveform.
Middle: AC coupled output voltage waveform.
Bottom: inductor current waveform.

C. Transition from transient to steady state operation

For the case where the calculated feedforward value of the D counter differs from the value that results in zero steady-state error, the DPWM feedback drives the D counter value to reach the steady state value. This process, however, involves transfer of control to and from SOSM and the DPWM feedback control either once or multiple times depending upon how far the calculated feedforward value is away from the steady state value of D. As an example, consider the 0 V – 1.5 V startup transient waveform shown in Fig. 17. The corresponding logic analyzer data is given in Fig. 18. At t = 0 μs, when the system is turned on, the controller sees a negative error of 1.5 V. The SOSM begins to resolve this transient. The feedforward value for the D counter is computed during this time and is ready for use at t = 0.42 μs. When the error falls within the error window Δ, the control is transferred to the DPWM feedback. The DPWM feedback sees a positive error and begins to decrement the D counter. This transfer of control between SOSM and the DPWM feedback continues until the steady state value of D counter is reached. Note that throughout this process, the error is maintained within the error window. In the experimental prototype, the error window is relatively large ($\pm \Delta = \pm$ 150 mV). This is mainly due to the limited resolution of the DPWM, which is based on a standard counter-based approach. This error window can be made much smaller using one of the high-resolution DPWM implementation techniques [17].

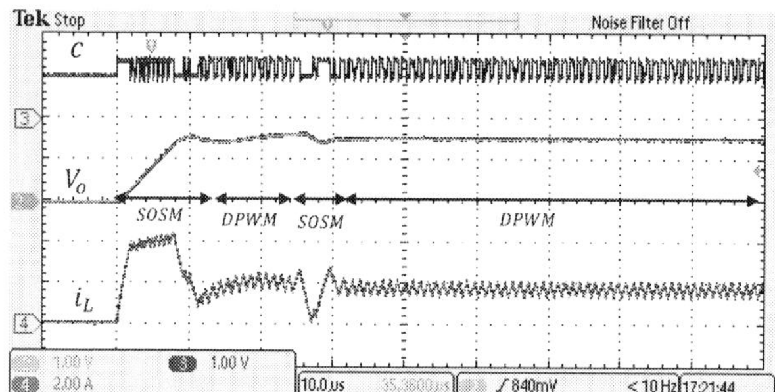

Fig. 17. Startup transient (0 V – 1.5 V) when the calculated feedforward value D_{ff}, is different from the steady state value of the D counter.

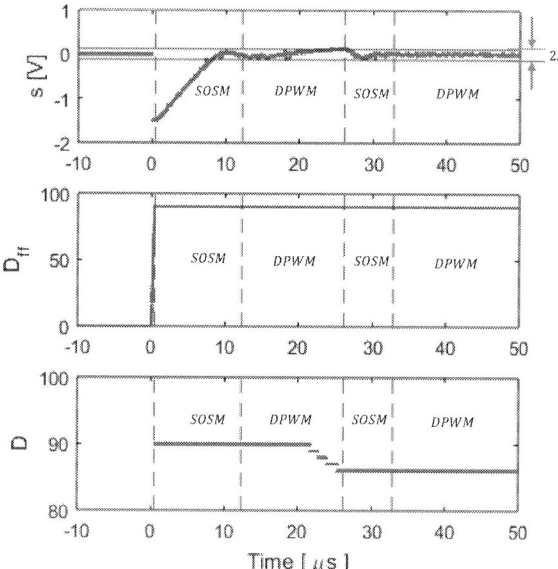

Fig. 18. Waveforms from the logic analyzer data corresponding to the experimental result shown in Fig. 17.
Top: output voltage error.
Middle: feedforward value D_{ff} of D counter.
Bottom: D counter value.

V. Conclusions

This paper demonstrates successful implementation of a second order sliding mode control capable of all the features desired of a digitally controlled point-of-load power supply, including fast transient responses, current limiting, and light load operation. The extended SOSM state machine can also achieve constant frequency steady state operation. All these features are implemented in a single relatively simple state machine.

References

[1] R. W. Erickson and D. Maksimovic, *Fundamentals of Power Electronics*, 2nd ed.: New York Springer-Verlag, 2001.

[2] E. Meyer and Y. Liu, "A practical minimum time control method for buck converters based on capacitor charge balance," in *Proc. IEEE Appl. Power Electron. Conf. Expo. (APEC 2008)*, pp. 10–16.

[3] A. Capel and D. O'Sullivan, "Very high frequency regulator for space applications," in *Proc. IEEE Power Electron. Spec. Conf. (PESC 1996)*, vol. 1, pp. 846–852.

[4] P. Gupta and A. Patra, "Super-stable energy based switching control scheme for DC-DC buck converter circuits," in *Proc. IEEE Int. Symp. Circuits and Systems (ISCAS* 2005), vol. 4, pp. 3063–3066.

[5] A. Stupar, Z. Lukic, A. Prodic, "Digitally-controlled steered-inductor buck converter for improving heavy-to-light load transient response", *Power Electron. Spec. Conf. (PESC 2008)*, pp. 3950 – 3954.

[6] V. Yousefzadeh, A. Babazadeh, B. Ramachandran, E. Alarcon, L. Pao, D. Maksimovic, "Proximate Time-Optimal Digital Control for Synchronous Buck DC–DC Converters," *IEEE Trans. Power Electron.*, vol.23, no.4, pp.2018,2026, July 2008.

[7] J. Cortes, V. Svikovic, P. Alou, J. A. Oliver, J. A. Cobos, " v^1 Concept: Designing a Voltage-Mode Control as Current Mode With Near Time-Optimal Response for Buck-Type Converters," *IEEE Trans. Power Electron.*, vol.30, no.10, pp.5829,5841, Oct. 2015.

[8] A. Babazadeh, L. Corradini and D. Maksimovic, "Near time-optimal transient response in DC-DC buck converters taking into account the inductor current limit", *Proc. IEEE Energy Convers. Congr. Expo. (ECCE 2009)*, pp.3328 -3335.

[9] L. Corradini, A. Costabeber, P. Mattavelli and S. Saggini, "Parameter-Independent time-optimal digital control for point-of-load converters", *IEEE Trans. Power Electron.*, vol. 24, no. 10, pp.2235 - 2248, 2009.

[10] S. Kapat and P. Krein, "Improved time optimal control of a buck converter based on capacitor current", *IEEE Trans. Power Electron.*, vol. 27, no. 3, pp.1444 -1454, 2012.

[11] R. Venkatamaran, A. Sabanovic, and S. Cuk, "Sliding mode control of dc-dc converters," in *Proc. IEEE IECON*, 1985, pp. 251-258.

[12] R. Venkataramanan, *Sliding Mode Control of Power Converters*, California Institute of Technology Ph. D. Thesis, May 1986.

[13] R. Ling, D. Maksimovic, and R. Leyva, "State-machine realization of second-order sliding-mode control for synchronous buck dc-dc converters," in *Proc. IEEE Energy Convers. Congr. and Expo. (ECCE 2013)*.

[14] D. C. Jones, D. Maksimovic, "Second order sliding mode control of a buck converter with output capacitor ESR and ESL", in *Proc. IEEE Workshop on Control and Modeling in Power Electron.* (COMPEL 2014).

[15] G. Bartolini, A. Pisano, and E. Usai, "An improved second-order sliding-mode control scheme robust against the measurement noise." *IEEE Trans. Automat. Contr.*, vol. 49, no. 10, pp. 1731-1737, Oct. 2004.

[16] A. Syed, E. Ahmed, D. Maksimovic, "Digital PWM controller with feed-forward compensation," in *Proc. IEEE Appl. Power Electron. Conf. Expo.* (APEC 2004).

[17] L. Corradini, D. Maksimovic, P. Mattavelli, R. Zane, *Digital Control of High-Frequency Switched-Mode Power Converters*. Piscataway, NJ: Wiley, 2015, ch.5, pp. 167-190.

Vibration and Torque Ripple Reduction of Switched Reluctance Motors Through Current Profile Optimization

Cong Ma and Liyan Qu
Department of Electrical and Computer Engineering
University of Nebraska–Lincoln
Lincoln, NE 68588-0511 USA
cma9@unl.edu; lqu2@unl.edu

Rakesh Mitra, Prerit Pramod, and Rakib Islam
Nexteer Automotive
Saginaw, MI 48601 USA
rakesh.mitra@nexteer.com; prerit.pramod@nexteer.com; rakib.islam@nexteer.com

Abstract—**This paper proposes a differential evolution (DE) optimization-based current profiling method for simultaneous reduction of the torque ripple and vibration of switched reluctance motors (SRMs). The mechanism of torque generation in SRMs produces radial forces in addition to the required tangential force. It has been shown that the radial forces acting on the stator are the main vibration source in SRMs and keeping the sum of the radial forces constant can reduce the magnitude of the significant harmonics of the sum of radial forces and further reduce vibration by avoiding the resonance caused by those harmonics. A simple method is proposed to model the torque and radial forces generated in the SRMs while considering the saturation effects. The resulting torque and radial force models are then used in the DE optimization process to generate the current profile of each phase in the form of Fourier series, where the Fourier coefficients of each phase current profile are determined to minimize the torque ripple and significant harmonics in the sum of the radial forces. The proposed method significantly reduces the computational cost of the finite element analysis (FEA)-based methods. The effectiveness of the proposed method is verified through both FEA simulation and experimental results.**

Keywords—Current profiling, switched reluctance motor (SRM), torque ripple, vibration.

I. Introduction

Switched reluctance motors (SRMs) have received substantial attention in the past few decades due to their high robustness, simple structure, and low manufacturing cost [1]-[3]. On the other hand, the drawbacks such as relatively high torque ripple and high vibration and acoustic noise impeded SRMs for high performance applications such as precision tooling, robotics or automotive steering. Therefore, significant studies have been reported in the literature on the reduction of torque ripple and vibration and acoustic noise in SRMs. Generally, the study of torque ripple has been separated and disconnected from the study of the vibration and acoustic noise. Therefore, most of the existing methods used to reduce torque ripple might result in large mechanical vibration and acoustic noise in the SRMs and vice versa.

Both machine design and control methods are responsible for the torque tipple and vibration and acoustic noise generated in SRMs. A comprehensive multiobjective design optimization framework was presented in [4] for torque ripple reduction of SRMs. While this paper proposes a current profile optimization method to simultaneously reduce the torque ripple and the vibration of SRMs from the control prospective.

Different control strategies for the reduction of torque ripple and vibration in SRMs can be found in the literature. Since the reluctance torque produced by an SRM is highly nonlinear depending on its electromagnetic characteristics, it is necessary to properly define the phase current command to minimize the torque ripple. Torque sharing function (TSF) is a technique to indirectly define current profiles. A TSF defines the distribution of the torque reference in each phase of an SRM with respect to the rotor position. Then, the machine characteristics can be used to calculate the corresponding current profile [5]-[9]. The waveforms of the TSF are based on simple mathematic functions [5]-[7] or the electromagnetic characteristics of the SRM [8]-[9]. However, it may be necessary to extend the TSF to the negative torque producing region in order to reduce the torque ripple during the commutation between phases in high speed conditions, which makes the TSF difficult to define. A current profile defines the current reference for each phase. In [10], a linear phase current profile stored in a look-up table is initially obtained based on the electromagnetic characteristics of an SRM and then the current profile is fine-tuned based on the error between reference torque and actual torque. In [11], the current waveform is represented as a set of harmonic coefficients instead of discrete points, and the four-phase SRM is converted into an equivalent two-phase SRM to obtain the desired current profiles. In [12], an optimization process using a small number of decision variables is performed to obtain the optimal current profile.

The vibration of an SRM is primarily caused by the radial force between the stator and rotor due to the electromagnetic excitation. A two-stage commutation method was proposed to cancel the vibration in the time domain by applying a voltage

978-1-4673-9551-9/16 $31.00 © 2016 IEEE

across a phase winding in two successive steps where the interval of each step equals a half mechanical resonant cycle, and the second vibration was utilized to cancel the first vibration [13]-[14]. In [15], the vibration reduction is achieved by analyzing the harmonic contents of the PWM voltage output and spreading the spectrum over the whole frequency range by pulse frequency modulation. These methods might not effectively reduce vibration because it is difficult to ensure that none of the exciting forces matches any resonant frequencies, particularly when the SRM is in the variable speed operation. More recently, vibration reduction techniques which keep the sum of the radial forces constant have been presented [16]-[17]. Those techniques can reduce the magnitude of the significant harmonics of the sum of radial forces and further reduce vibration by avoiding the resonance due to those harmonics [17]. The constant sum of the radial forces can be achieved by properly defining the current reference injected to the SRM. In [16], a force-sharing algorithm and a direct instantaneous force control method were proposed to achieve a constant sum of the radial forces. In [17]-[18], analytical current profiling methods were proposed to eliminate of the third harmonic in the sum of the radial forces. The methods first represent the relationship between current harmonics and radial force harmonics and then analytically calculate the magnitude and phase angle of the harmonics based on some assumptions. However, in [17]-[18], the methods do not consider the magnetic saturation in SRMs, and the current profile determined by the analytical calculation is not optimal. In addition, none of the aforementioned techniques considered simultaneous minimization of the torque ripple and vibration of SRMs.

This paper proposes a new current profiling technique to reduce vibration and torque ripple of an SRM. In the proposed current profiling method, the nonlinear characteristics of an SRM are analytically modeled and the current profile is defined as a sum of Fourier series components. The current profile is tuned by considering the high nonlinearity of an SRM and optimized based on a differential evolution (DE) optimization technique, which is capable of managing the tradeoff between vibration reduction and torque ripple reduction and is computationally efficient. The proposed current profiling technique does not require the costly FEA calculation during the optimization process. The effectiveness of the proposed current profiling method is verified by both simulation and experiment results.

II. TORQUE AND VIBRATION CHARATERISTICS OF AN SRM

The basic operating principle of an SRM is simple. A reluctance torque is generated by the tendency of the rotor to align with the excited stator pole due to the double salient structure of the SRM. In fact, the torque can be considered to be caused by the tangential component of the electromagnetic force between the stator and the rotor cores. While the radial component of the electromagnetic force, the radial force, is considered to be the main course of the vibration and acoustic noise.

A. Torque Ripple Characteristics of an SRM

Due to highly nonlinear magnetic characteristics of the double salient poles, the torque generated in an SRM is a function of the rotor position θ with respect to the energized phase. If the machine is unsaturated, the torque of an SRM with phase n excited can be expressed as:

$$T_n(\theta) = \frac{1}{2}\frac{dL(\theta)}{d\theta} i_n^2 \tag{1}$$

where $L(\theta)$ represents the self-inductance which is a function of the rotor position θ for a given current, and i_n is phase current. If the machine goes into saturation, the torque can be calculated based on the coenergy, $W_c(i_n, \theta)$, as

$$T_n(\theta, i_n) = \frac{\partial W_c(i_n, \theta)}{\partial \theta} \tag{2}$$

The coenergy is defined as

$$W_c(i_n, \theta) = \int \lambda(i_n, \theta)\, di_n \tag{3}$$

where the flux linkage $\lambda(i_n, \theta)$ is a function of both current and rotor position. Therefore, the torque cannot be simply express as a function of the square of phase current as (1).

The total torque T generated in an SRM is a sum of the torque produced by each phase if there is more than one phase excited in the machine. The torque ripple T_r is defined as

$$T_r = \frac{T_{max} - T_{min}}{T_{avg}} \tag{4}$$

where T_{max}, T_{min} and T_{avg} are the maximum, minimum and average values of the total torque, respectively. The torque ripple of an SRM stems from different sources, such as phase commutation, sensor errors, manufacturer tolerance, and current tracking errors. The excitation frequency (the frequency of the phase current) is expressed as

$$f_{ex} = \frac{N_r \omega_r}{60} \tag{5}$$

where ω_r is the rotor speed in rpm and N_r is the number of rotor poles. For an N_p-phase SRM, different sources lead to different frequency components of torque ripple as shown in Table I. k represents the kth harmonics.

TABLE I. FREQUENCY COMPONENTS OF THE TORQUE RIPPLE

Cause of torque ripple	Frequencies of the components
Rotor eccentiricity	kf_{ex}/N_r
Phase variation	kf_{ex}
Phase commutation	kN_pf_{ex}

B. Vibration Characteristics of an SRM

The vibration in an SRM is found to be directly correlated to the vibration of the stator yoke due to the radial magnetic force [19]. The radial force applied on a stator tooth is produced by the magnetic flux crossing the air gap, and it can be calculated using Maxwell stress tensor (MST) method [20]

$$F_{rn} = \int \frac{B_r^2 - B_t^2}{2\mu_0} dA \tag{6}$$

where B_r and B_t are radial and tangential magnetic flux density, respectively, μ_0 is the permeability of free space, and A is the individual pole surface. The sum of radial forces F_{rS} is expressed as

$$F_{rS} = \sum_{n=1}^{N_p} F_{rn} \tag{7}$$

where F_{rn} is the radial force applied on a single stator pole surface of the excited phase n. Due to the high nonlinearity of the magnetic characteristics in SRMs, the radial force is also a function of both current and rotor position. Radial forces lead to vibration on the stator, and may excite resonance with various vibration modes if any harmonics of the sum of radial forces match the natural frequencies of the vibration modes. Different vibration modes have different natural frequencies which depend on the machine geometry and material properties. An SRM is more prone to low-frequency vibration modes due to the resonance with radial force harmonics. Similarly, due to the phase commutation in SRMs, the kN_pf_{ex}-th harmonics exist in the sum of radial force and they are dominant. Reducing these harmonics in the sum of radial force will lead to the reduction of vibration in SRMs.

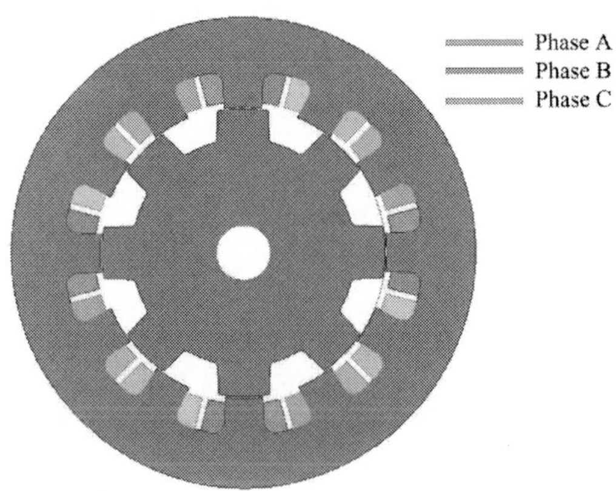

Fig. 1. The cross section of a 12/8 SRM.

III. Proposed Optimal Current Profiling Method

A 1-kW, 12-stator-pole/8-rotor-pole (12/8), three-phase SRM as shown in Fig. 1, was designed for high performance applications and is used as an example machine in this paper. The design parameters of the SRM are listed in Table II.

A. Modeling of the Torque and Radial Force in the SRM

The torque and the radial force in the SRM can be calculated based on (2) and (6) by the FEA. In order to reduce the computational cost by avoiding the FEA during the optimization process, a simple method is proposed to model the torque and radial forces generated in the SRMs while

Table II. Design Parameters of the 12/8 SRM

Design variables	Value	Unit
Outer diameter of stator D_{so}	85	mm
Outer diameter of rotor D_{ro}	52	mm
Inner diameter of rotor D_{ri}	10	mm
Length of stator yoke L_{sy}	9.5	mm
Length of rotor yoke L_{ry}	15.5	mm
Number of stator pole N_s	12	N/A
Number of rotor pole N_r	8	N/A
Angle of stator pole arc θ_s	15	deg
Angle of rotor pole arc θ_r	15.5	deg
Stack length L_{stk}	32	mm

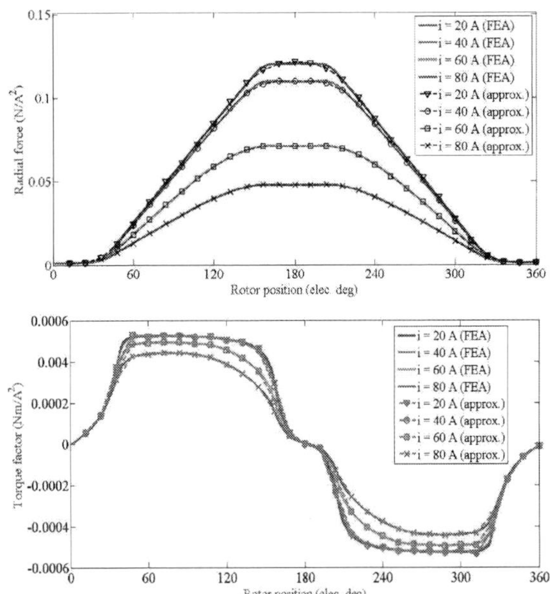

Fig. 1. Fig. 2. Radial force and torque factors of the 12/8 SRM.

considering the saturation effects. In [17], the torque and the radial force were assumed to be proportional to the square of the current and the torque and force factors introduced were functions of only rotor position without considering saturation. In this paper, the torque and the radial force are expressed as:

$$T_n(\theta, i_n) = k_{tn}(\theta, i_n) i_n^2 \tag{8}$$

$$F_{rn}(\theta, i_n) = k_{rn}(\theta, i_n) i_n^2 \tag{9}$$

where $k_{tn}(\theta, i_n)$ and $k_{rn}(\theta, i_n)$ are torque and force factors, respectively. These two factors are functions of both phase current and the rotor position. The plots of force and torque factors versus the rotor position for the 12/8 SRM with different excitation current obtained by the FEA are shown as the solid lines in Fig. 2. As shown in Fig. 2, $k_{tn}(\theta, i_n)$ and $k_{rn}(\theta, i_n)$ can be approximated by

$$k_{tn}\left(\theta,i_n\right)=\sum_{m=1}^{m_{kt\max}} a_{tm}\left(i_n\right)\cdot\sin\left(m\theta\right) \tag{10}$$

$$k_{rn}\left(\theta,i_n\right)=a_{r0}\left(i_n\right)+\sum_{m=1}^{m_{kr\max}} a_{rm}\left(i_n\right)\cdot\cos\left(m\theta\right) \tag{11}$$

where m_{ktmax} and m_{krmax} are the highest order considered in the equations. The coefficients $a_{r0}(i_n)$, $a_{rm}(i_n)$, and $a_{tm}(i_n)$ are functions of current. For a given current, the coefficients can be determined by using curve fitting techniques. The plots shown as the dashed lines in Fig. 2 are obtained based on (10) and (11) and do agree very well with the plots obtained by the FEA.

B. Optimization Model

In high performance applications, it is necessary to maintain both a low torque ripple and a low vibration. This can be achieved by injecting appropriate harmonic components in the phase current waveforms to generate the optimal current profiles. However, it is difficult to generate the optimal current profiles while considering the highly nonlinear characteristics of an SRM and tradeoff among different objectives. This section proposes a computationally efficient and reliable solution to obtain the optimal current profiles via optimization. The objective of the optimization is to obtain the optimal harmonic coefficients of the current profiles injected into the SRM to simultaneously minimize the torque ripple and vibration.

In this paper, for a given torque reference, the current profile of each phase is modeled as a Fourier series as follows

$$i_{ref}\left(\theta,T_{ref}\right)=c_0\left(T_{ref}\right)+\sum_{m=1}^{m_{\max}} c_m\left(T_{ref}\right)\cdot\cos\left(m\theta-\varphi_{cm}\left(T_{ref}\right)\right) \tag{12}$$

where m_{max} is the highest order of harmonics included in the model and can be determined according to the maximum bandwidth of the controller and base speed. The coefficients $c_0(T_{ref})$, $c_m(T_{ref})$ and the phase angle of the mth harmonic $\varphi_{cm}(T_{ref})$ are functions of the torque reference. Based on the current profile defined by (12), the torque and radial force applied on a single excited-phase pole surface can be determined using (8) and (9), respectively.

The design variables, objectives, and constraints of the optimization model are described in the reminder of this subsection. In this paper, $2m_{max}+1$ coefficients of the current profile for each torque reference are the design variables and represented by a vector $\boldsymbol{x}$.

The sum of the magnitude of the $kN_p f_{ex}$-th harmonic of the sum of radial forces is defined as:

$$H_{F_{rS}}=\sum_{k=1}^{3} F_{kN_p f_{ex}} \tag{13}$$

where F_{kNpfex} are the magnitude of $kN_p f_{ex}$-th harmonic of the sum of radial forces. The first three dominant harmonics due to the phase commutation in SRMs are considered during optimization in this paper. Actually, more harmonics (e.g., the harmonics corresponding to different vibration modes) of the sum of radial forces could be included. The optimization is subjected to the following constraints.

$$\left(1-\xi\right)T_{req}\leq T_{\text{avg}}\leq\left(1+\xi\right)T_{req} \tag{14}$$

$$\left(I_n\right)_{\max}\leq I_{\max} \tag{15}$$

$$\left(\left|\frac{d\lambda}{d\theta}\right|\right)_{\max}<k_\lambda \tag{16}$$

Eq. (14) defines the required average torque level T_{req} and ξ is the tolerance factor. Eq. (15) limits the maximum currents. Eq. (16) limits $d\lambda/d\theta$, which is the rate of change of flux linkage with respect to the rotor position [9] to guarantee the torque-speed performance of the SRM, where

$$k_\lambda=\frac{V_{dc}}{\omega_b} \tag{17}$$

ω_b is the basic speed and V_{dc} is the dc-link voltage.

The optimization model is constructed as follows:

$$\min_x\ y(\boldsymbol{x})=w_1T_r^{'}\left(\boldsymbol{x}\right)+w_2H_{F_{rS}}^{'}\left(\boldsymbol{x}\right) \tag{18}$$

subject to: (12)-(14)

$$\underline{x_i}\leq x_i\leq\overline{x_i},\ i=1, 2, ..., 2m_{max}+1 \tag{19}$$

where the objective is to minimize both T_r and $H_{F_{rS}}$; $T_r'(\boldsymbol{x})$ and $H_{F_{rS}}^{'}\left(\boldsymbol{x}\right)$ are the normalized values; $\boldsymbol{x}$ is the vector of the decision variables, which are the Fourier coefficients of the phase current profile in (10); w_1 and w_2 are weight factors satisfying $w_1+w_2=1$; $\underline{x_i}$ and $\overline{x_i}$ are the lower and upper bounds of each design variable, respectively. The optimization model (18)-(19) can be simplified as the following form using the Lagrangian relaxation technique [21]

$$\min_x\ y(\boldsymbol{x})=w_1T_r^{'}\left(\boldsymbol{x}\right)+w_2H_{F_{rS}}^{'}\left(\boldsymbol{x}\right)+p\left(\boldsymbol{x}\right) \tag{20}$$

subject to: (19)

where the constraints (14)-(16) are considered as the penalty function $p(\boldsymbol{x})$ in the objective function of (20). If any constraint is not satisfied, a positive value of $p(\boldsymbol{x})$ will be added to the objection function; otherwise, the value of $p(\boldsymbol{x})$ is zero. The optimization model (19)-(20) will be used to generate the optimal current profiles for the SRM.

C. Implementation of the DE Algorithm

The DE algorithm, which is a population-based stochastic optimization technique, is implemented to solve the

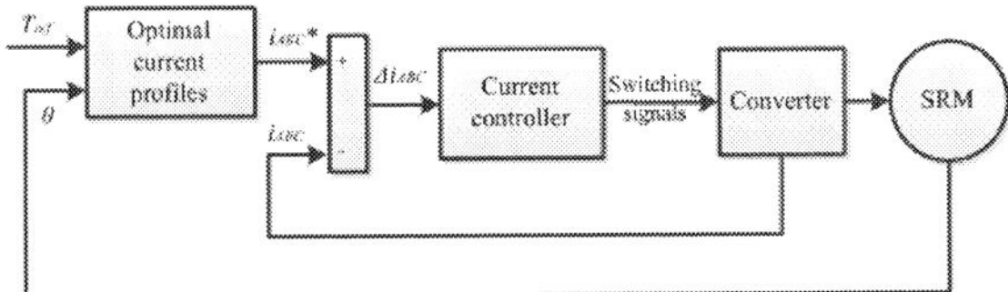

Fig. 3. block diagram of the SRM drive system.

optimization model (19)-(20). The DE is an evolutionary optimization algorithm which applies crossover, mutation, and selection operations on a population of the design variable vector to iteratively search for the optimal solution. Compared to the genetic algorithm, the DE is more robust and simpler to implement [22], [23].

In the initialization procedure of the DE algorithm, a random process is utilized to produce the first generation of the candidate design variable vector, and (18) is calculated for each candidate. Then, mutation is performed by arithmetic operations to obtain the mutants of the population of the decision variable vector. The mutants are then used in the crossover process to generate trial vectors based on the user-defined values of the crossover probabilities. The next-generation candidates are selected by comparing the performance calculated using (18) of the trial vectors and the target vectors in the current generation. These processes are repeated until the optimization is converged or the maximum iteration number is reached.

IV. VALIDATION

The effectiveness of the proposed method is verified through both FEA simulation and experimental results in this section. In the DE optimization algorithm, the maximum generation, number of individual designs, and number of design variables are 80, 40 and 13, respectively.

A. Simulation Results

In order to verify the effectiveness of the optimal current profiles generated by the proposed method, the total torque and the sum of radial forces in the SRM are calculated using the FEA. For a given torque reference of 1 Nm, a non-optimization-based current profile generated by the method presented in [24] and the optimal current profile obtained by the proposed method are used to perform the simulations. Figure 4 shows the simulated waveforms of the torque and the sum of radial forces with the non-optimization-based current profile. The simulated waveforms of the torque and the sum of radial forces with the optimal current profile obtained by the proposed method are shown in Fig. 5. Both current profiles have the same conduction angle of 240°. As shown in Fig. 4, the torque ripple is 8.11% and the variation in the sum of radial forces is 70.92%. Figure 5 shows that the torque ripple is 8.78% and the variation in the sum of radial forces is reduced to 45.66% with the optimal current profile. From Figs. 4 and 5, the variation in the sum of radial forces is reduced by using the optimal current profile while the torque ripple is almost kept at the same level. An optimal current profile with the conduction angle of 270° is generated by the proposed method for the same torque reference. Figure 6 shows that the torque ripple is increased to 11.96% and the variation in the sum of

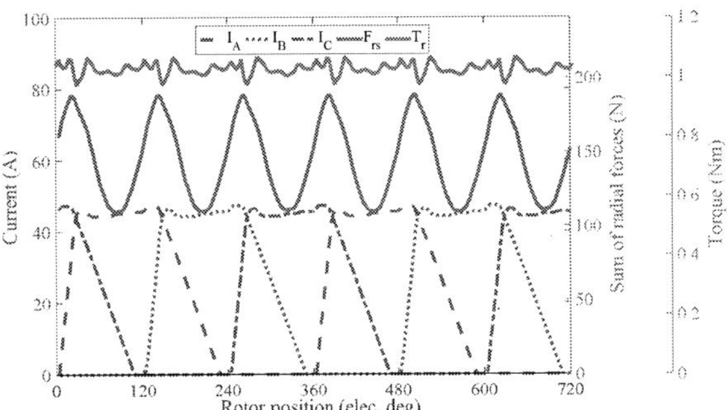

Fig. 4. The waveforms of the torque, the sum of radial forces, and the non-optimized phase currents (conduction angle = 240°).

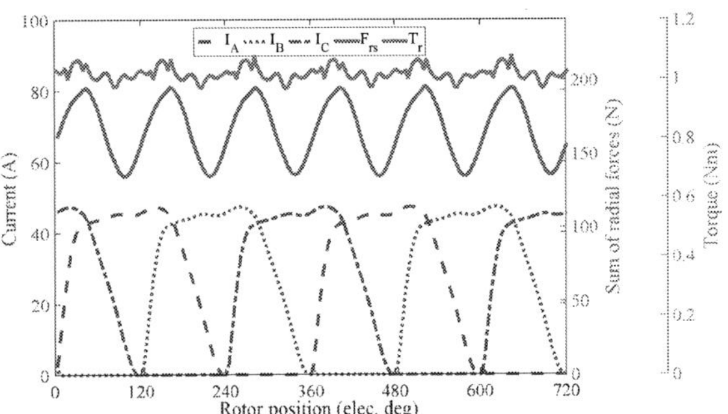

Fig. 5. The waveforms of the torque, the sum of radial forces, and the optimal phase currents (conduction angle = 240°).

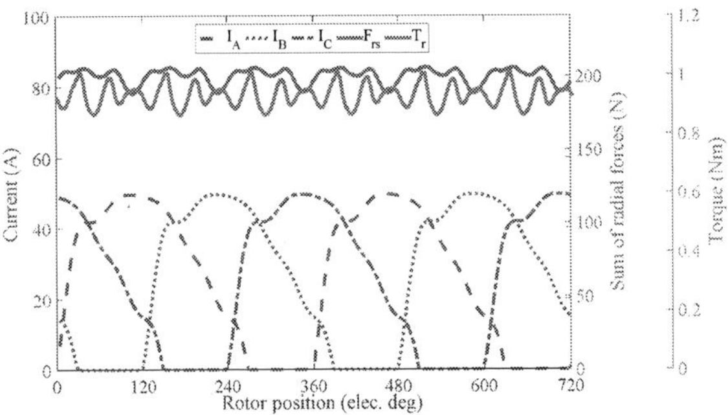

Fig. 6. The waveforms of the torque, the sum of radial forces, and the optimal phase currents (conduction angle = 270°).

radial forces is reduced significantly to 15.82% by increasing the conduction angle of the current profile. The first three significant harmonics of the sum of radial forces generated in the SRM with the three different current files are shown in Fig. 7. It can be seen that the 24th harmonic is significantly reduced by increasing the conduction angle. Due to the shift of the maximum values of $k_{tn}(\theta,i)$ and $k_{rn}(\theta,i)$ with respective to the rotor position, the extension of the conduction angle allows the proposed current profiling method to have more freedom to reduce the torque ripple and vibration simultaneously. This, however, is achieved at the expense of a relatively higher copper loss. The copper loss of the SRM with the 270° conduction angle is 14.51% higher than that with the 240° conduction angle. Results also show that a trade-off exists between the torque ripple and the variation of the sum of radial forces. The results have proved that the proposed current profiling method is capable of significantly reducing vibration while retaining a low torque ripple for an SRM.

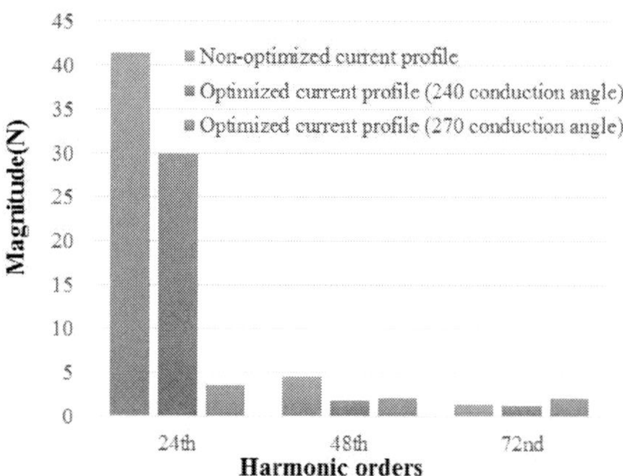

Fig. 7. Comparison of the the first three harmonics of the sum of radial forces generated in the SRM with the three different current profiles.

B. Experimental Results

In the experiments, the SRM is connected to an induction motor operating at 20 rpm in speed control mode, and the torque generated in the SRM is measured by a torque sensor. A block diagram of the SRM drive system is shown in Fig. 3. The current profiles are stored in a lookup table. A predictive current controller [24] is implemented to accurately track the current reference of each phase generated based on the lookup table. The two current profiles shown in Figs. 4 and 5 are used to conduct the experiments. The measured torque and the actual phase currents are plotted in Figs. 8 and 9. The torque ripples are 6.11% and 9.80%, respectively. The torque ripple associated with the optimal current profile is higher than that associated with the nonoptimized current profile as shown by the simulation results.

Vibration tests have been performed with these two current profiles. The frequency spectrum of the vibration tested at a constant speed of 400 rpm is shown in Fig. 10. As shown in Fig. 10, the magnitudes of the 24th, 48th, and 72nd harmonics of the vibration caused by the corresponding harmonics of the sum of radial forces are reduced by using the optimal current profile, which is consistent with the results shown in Fig. 7. As shown in Fig. 7, the magnitudes of these three harmonics of the sum of radial forces are reduced by using the optimal current profile. The frequency spectrum of the vibration shows that a dominant harmonic is near 1,200 Hz, which is considered to be a natural frequency of one vibration mode of the SRM. The vibration modes of the SRM will be investigated in the future work.

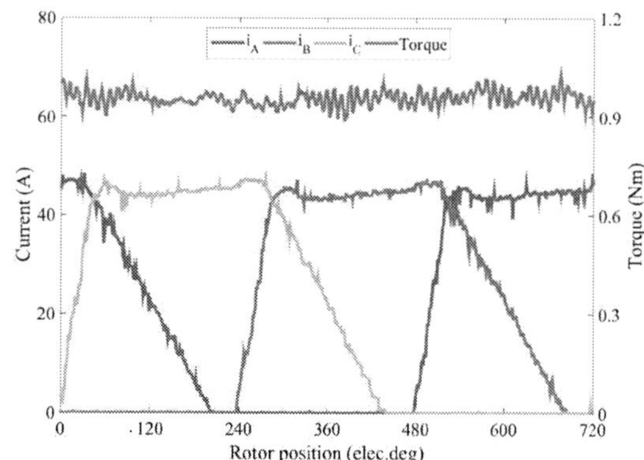

Fig. 8. The waveforms of the measured torque and the actual non-optimized phase currents (conduction angle = 240°).

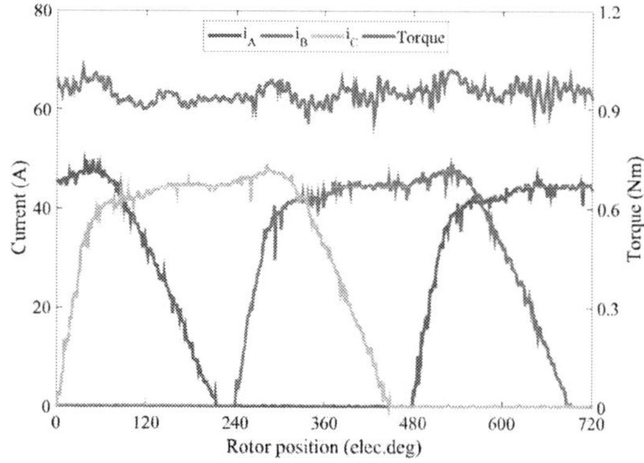

Fig. 9. The waveforms of the measured torque and the actual optimal phase currents (conduction angle = 240°).

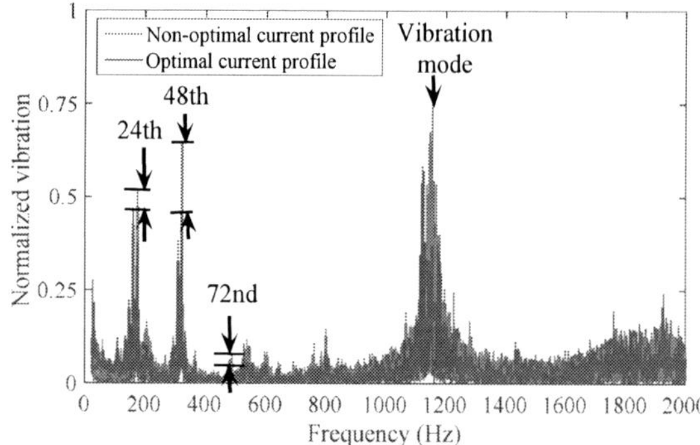

Fig. 10. The frequency spectrum of the vibration tested with a non-optimal current profile and an optimal current profile at 400 rpm.

V. Conclusions

This paper has proposed a simple and effective optimal current profiling algorithm to minimize the vibration and torque ripple of an SRM. Analytical models of the torque and radial force of the SRM have been proposed while considering the magnetic saturation of the SRM. A DE optimization algorithm has been implemented to search for the optimal Fourier coefficients of the current profile. The effectiveness of the proposed method has been verified by both FEA simulation and experimental results. The results have shown a significant reduction of the variation of the sum of the radial forces while retaining a low torque ripple by the proposed method compared to a nonoptimal current profile.

REFERENCES

[1] T. J. E Miller. *Switched reluctance motors and their control*, Magna Physics Publications/Oxford university press, 1993.

[2] C. Ma, L. Qu, and Z. Tang, "Torque ripple reduction for mutually coupled switched reluctance motor by bipolar excitations," in *Proc. IEEE International Electric Machines & Drives Conference*, Chicago, IL, USA, May 12-15, 2013, pp. 1211-1217.

[3] L. Zhang, R. Born, X. Zhao, J. -S. Lai, "A parabolic voltage control strategy for burst mode converters with constant burst frequency and eliminated audible noise", in *Proc. IEEE International Future Energy Electron. Conf.*, Nov. 2015.

[4] C. Ma and L. Qu, "Multiobjective optimization of switched reluctance motors based on design of experiments and particle swarm optimization," *IEEE Trans. Energy Conv.*, vol. 30, no. 3, pp. 1144-1153, Sept. 2015.

[5] X. D. Xue, K. W. E. Cheng, and S. L. Ho, "Optimization and evaluation of torque-sharing functions for torque ripple minimization in switched reluctance motor drives," *IEEE Trans. Power Electron.*, vol. 24, no. 9, pp. 2076-2090, Sep. 2009.

[6] D. H. Lee, J. Liang, Z. G. Lee, and J. W. Ahn, "A simple nonlinear logical torque sharing function for low-torque ripple SR drive," *IEEE Trans. Ind. Electron.*, vol. 56, no. 8, pp. 3021-3028, Aug. 2009.

[7] N. C. Sahoo, J. X Xu, and S. K. Panda, "Determination of current waveforms for torque ripple minimisation in switched reluctance motors using iterative learning: an investigation," *IEE Proc. Electr. Power Appl.*, vol. 146, no. 4, Jul. 1999, pp. 369-377.

[8] V. P. Vujicic, "Minimization of torque ripple and copper losses in switched reluctance drive," *IEEE Trans. Power Electron.*, vol. 27, no. 1, pp. 388-399, Jan. 2012.

[9] J. Ye, B. Bilgin, and A. Emadi, "An offline torque sharing function for torque ripple reduction in switched reluctance motor drives," *IEEE Trans. Energy Conv.*, vol. 30, no. 2, pp. 726-735, Jun. 2015.

[10] R. Mikail, I. Husain, Y. Sozer, M.S. Islam, and T. Sebastian, "Torque-ripple minimization of switched reluctance machines through current profiling," *IEEE Trans. Ind. Applic*, vol. 49, no. 3, pp. 1258-1267, May-Jun. 2013.

[11] P. L. Chapman and S. D. Sudhoff, "Design and precise realization of optimized current waveforms for an 8/6 switched reluctance drive," *IEEE Trans. Power Electron*, vol. 17, no. 1, pp. 76-83, Jan. 2002.

[12] N. T. Shaked and R. Rabinovici, "New procedures for minimizing the torque ripple in switched reluctance motors by optimizing the phasecurrent profile," *IEEE Trans. Magn.*, vol. 41, no. 3, pp. 1184-1192, Mar. 2005.

[13] C. Y. Wu and C. Pollock, "Time domain analysis of vibration and acoustic noise in the switched reluctance drive," in *Proc. Sixth International Conference on Electrical Machines and Drives*, Sept. 1993, pp. 558-563.

[14] Z. Q. Zhu, X. Liu, and Z. Pan, "Analytical model for predicting maximum reduction levels of acoustic noise and vibration in switched reluctance machine by active vibration cancellation," *IEEE Trans. Energy Conv.*, vol. 26, no. 1, pp. 36-45, Mar. 2011.

[15] F. Blaabjerg and J. K. Pedersen, "Digital implemented random modulation strategies for AC and switched reluctance drives," in *Proc. IEEE Annual Conf. Industr. Electron. Society*, Nov. 1993, pp. 676–682.

[16] A. Hofmann, A. Al-Dajani, M. Bosing, and R. W. De Doncker, "Direct instantaneous force control: A method to eliminate mode-0-borne noise in switched reluctance machines," in *Proc. International Electrical Machine & Drives Conference*, May 2013, pp. 1009-1016.

[17] M. Takiguchi, H. Sugimoto, N. Kurihara, and A. Chiba, "Acoustic noise and vibration reduction of SRM by elimination of third harmonic component in sum of radial forces," *IEEE Trans. Energy Conv.*, vol. 30, no. 3, pp. 883-891, Sept. 2015.

[18] N. Kurihara, J. Bayless, and A. Chiba, "Noise and vibration reduction of switched reluctance motor with novel simplified current waveform to reduce force sum variation," in *Proc. IEEE International Electric Machines and Drives Conference*, May 2015, pp. 1794-1800.

[19] D. E. Cameron, J. H. Lang, and S. D. Umans, "The origin and reduction of acoustic noise in doubly salient variable-reluctance motors," *IEEE Trans. Ind. Applicat.*, vol. 28, no. 6, pp. 1250-1255, Nov/Dec 1992.

[20] W.N. Fu, P. Zhou, D. Lin, S. Stanton, and Z.J. Cendes, "Magnetic force computation in permanent magnets using a local energy coordinate derivative method," *IEEE Trans. Magn.*, vol. 40, no. 2, pp. 683-686, Mar. 2004.

[21] A. E. Smith and D. W. Coit, *Handbook of evolutionary computation*, 1st ed. Oxford University Press and Instustrial Physics Publication, Sept. 1995.

[22] A. D. Lilla, M. A. Khan, and P.Barendse, "Comparison of differential evolution and genetic algorithm in the design of permanent magnet generators," in *Proc. IEEE International Conf. Industr. Tech.*, Feb. 2013, pp. 266-271.

[23] Y. Duan and D. M. Ionel,. "A review of recent developments in electrical machine design optimization methods with a permanent-magnet synchronous motor benchmark study," *IEEE Trans. Ind. Applic.*, vol. 49, no. 3, pp. 1268-1275.

[24] R. Mikail, I. Husain, Y. Sozer, M. S. Islam, and T. Sebastian, "A fixed switching frequency predictive current control method for switched reluctance machines," *IEEE Trans. Ind. Applic.*, vol. 50, no. 6, pp. 3717-3726, Nov.-Dec. 2014.

978-1-4673-9551-9/16 $31.00 © 2016 IEEE

Modified Predictive Current Control of Neutral-Point Clamped Converter with Reduced Switching Frequency

Dinto Mathew*, Anshuman Shukla† and Santanu Bandyopadhyay*

Email: dintomathew@iitb.ac.in, ashukla@ee.iitb.ac.in, santanub@iitb.ac.in

*Department of Energy Science and Engineering, †Department of Electrical Engineering

Indian Institute of Technology Bombay

Mumbai-400076, India

***Abstract*— This paper presents a modified predictive algorithm for the current control of Neutral-Point Clamped (NPC) converter with reduced switching frequency. The predictive controller uses discrete time model of the converter and load system for predicting the future states of the controlled variable. The reduction in the average switching frequency of the converter is achieved by modifying the finite control set (FCS) predictive algorithm to make use of the available redundant voltage vectors and thereby reducing the total number of switching transitions at each sampling instant to the maximum possible extent. This modified algorithm achieves reduced switching frequency without adding any additional terms in the cost function and provides excellent reference current tracking capabilities. Effectiveness of the proposed modified predictive algorithm in comparison with finite control set predictive algorithm has been validated through the MATLAB/Simulink simulation results.**

Keywords*— *Model Predictive Control (MPC), Power Converters, Switching Frequency, Redundant Voltage Vectors

I. INTRODUCTION

Nowadays, power electronic converters have become integral part of industrial drives, hybrid power vehicles, renewable energy sector and it leads to the wide exploration of new and advanced power converter topologies as well as control techniques. The main current control techniques applied to these power electronic converters are hysteresis control, linear control, sliding mode control, artificial intelligence based control (neural network, fuzzy control, neuro-fuzzy control) and predictive control. The predictive control technique is simple and the concepts are easy to understand. Predictive control approach can be applied to variety of systems incorporating various system constraints and nonlinearities [1]. Also, predictive control approach has the provision for controlling multiple variables simultaneously without making the controller much complex. The model predictive control (MPC) technique is based on the prediction of future states of controlled variable based on the plant model and selection of the best actuation which optimizes the performance index or cost function ensuring the constraints on input and output variables. The optimization of cost function is repeated at every sampling instant and the optimal actuation is applied [2]-[3]. The reduction in current ripple and better reference current tracking of the neutral-point clamped (NPC) converter using model predictive control require a higher sampling frequency which will result in increased switching frequency of the converter and it may force the power semiconductor switches to go beyond its thermal limits. Hence it is very important to use effective algorithms to reduce the switching frequency of the converter even at higher sampling frequencies. A predictive control algorithm with modified cost function considering a reduced set of possible voltage vectors to reduce switching frequency is described in [5]. Since the cost function is modified by including the switching frequency reduction term with weighting factor, the proper selection of weighting factor itself is a challenge when applied to complex systems. Some examples of reduced switching frequency predictive algorithms applied to different power electronic converters are discussed in [6]-[9].

The proposed modified predictive control algorithm in this paper ensures lower switching frequency of the converter without modifying the existing cost function defined in terms of current reference tracking. This avoids the use of weighting factors as in case of finite control set predictive control algorithms with additional terms in the cost function. The modified algorithm adopts the possibilities of using available redundant voltage vectors to reduce the average switching frequency.

II. SYSTEM DESCRIPTION

A. Neutral-point Clamped Converter

The neutral point clamped converters are commonly used for medium voltage power applications. The configuration of three phase NPC converter is shown in Fig. 1. Each leg of the NPC converter consists of four series connected controlled semiconductor power switches with antiparallel diodes and two clamping diodes. The voltage across upper and lower dc link capacitors (C_1 and C_2) is $V_{dc}/2$ volts each. The topology of NPC converter allows three possible voltage levels ($+V_{dc}/2$, 0, $-V_{dc}/2$) at the output of each phase with respect to the neutral point 'o' of the dc-link. The various output voltage vectors corresponding to different switching combinations of the four series connected controlled switches (H_{x1}, H_{x2}, H_{x3} and H_{x4}) per

978-1-4673-9551-9/16 $31.00 © 2016 IEEE

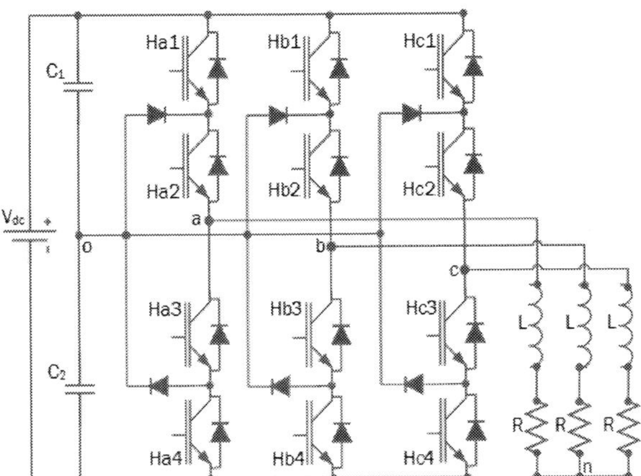

Fig. 1. NPC converter with Load System

Table 1. Output Voltage Vectors of NPC Converter

$v_0 = 0$	$v_1 = (1/6 + j\sqrt{3}/6)\mathrm{V}_{dc}$
$v_2 = (1/3 + j\,1/\sqrt{3})\mathrm{V}_{dc}$	$v_3 = (1/6 - j\sqrt{3}/6)\mathrm{V}_{dc}$
$v_4 = (1/3)\mathrm{V}_{dc}$	$v_5 = (1/2 + j\sqrt{3}/6)\mathrm{V}_{dc}$
$v_6 = (1/3 - j\,1/\sqrt{3})\mathrm{V}_{dc}$	$v_7 = (1/2 - j\sqrt{3}/6)\mathrm{V}_{dc}$
$v_8 = (2/3)\mathrm{V}_{dc}$	$v_9 = (-1/3)\mathrm{V}_{dc}$
$v_{10} = (-1/6 + j\sqrt{3}/6)\mathrm{V}_{dc}$	$v_{11} = (j\,1/\sqrt{3})\mathrm{V}_{dc}$
$v_{12} = (-1/6 - j\sqrt{3}/6)\mathrm{V}_{dc}$	$v_{13} = 0$
$v_{14} = (1/6 + j\sqrt{3}/6)\mathrm{V}_{dc}\mathrm{V}_{dc}$	$v_{15} = (-j\,1/\sqrt{3})\mathrm{V}_{dc}$
$v_{16} = (1/6 - j\sqrt{3}/6)\mathrm{V}_{dc}$	$v_{17} = (1/3)\mathrm{V}_{dc}$
$v_{18} = (-2/3)\mathrm{V}_{dc}$	$v_{19} = (-1/2 + j\sqrt{3}/6)\mathrm{V}_{dc}$
$v_{20} = (-1/3 + j\,1/\sqrt{3})\mathrm{V}_{dc}$	$v_{21} = (-1/2 - j\sqrt{3}/6)\mathrm{V}_{dc}$
$v_{22} = (-1/3)\mathrm{V}_{dc}$	$v_{23} = (-1/6 + j\sqrt{3}/6)\mathrm{V}_{dc}$
$v_{24} = (-1/3 - j\,1/\sqrt{3})\mathrm{V}_{dc}$	$v_{25} = (-1/6 - j\sqrt{3}/6)\mathrm{V}_{dc}$
$v_{26} = 0$	

phase are listed in Table 1. As compared to the common three phase voltage source converter, NPC has more number of possible switching combinations and corresponding output voltage vectors. Hence NPC converter has higher freedom over the control of output current quality, reduction in switching frequency etc. by using different possible switching combinations of the power switches. Among the total 27 voltage vectors, there are 19 distinct and 8 redundant vectors. The various families of redundant voltage vectors of the NPC converter are (v_0, v_{13}, v_{26}), (v_1, v_{14}), (v_3, v_{16}), (v_4, v_{17}), (v_9, v_{22}), (v_{10}, v_{23}) and (v_{12}, v_{25}). Each voltage vector in a particular redundant voltage vector family has same voltage magnitude and phase.

B. Load System

The output of the NPC converter is connected to RL load. Depending on the switching combinations selected, the output voltage of NPC converter differs. Using the space vector definition, the output load current dynamics and the voltage vector can be described by the differential equation (1).

$$v = {}^{2}/_{3}\left(v_{an} + \boldsymbol{a}v_{bn} + \boldsymbol{a}^2 v_{cn}\right)$$

$$= {}^{2}/_{3}L\left(\frac{di_a}{dt} + \boldsymbol{a}\frac{di_b}{dt} + \boldsymbol{a}^2\frac{di_c}{dt}\right) + {}^{2}/_{3}R\left(i_a + \boldsymbol{a}i_b + \boldsymbol{a}^2 i_c\right)$$

$$= L\frac{di}{dt} + Ri \qquad (1)$$

where the unit vector $\boldsymbol{a} = e^{j2\pi/3} = \left(-\frac{1}{2} + j\frac{\sqrt{3}}{2}\right)$, i_a, i_b & i_c are the phase A, phase B and phase C load currents respectively and the current vector $i = {}^{2}/_{3}\left(i_a + \boldsymbol{a}i_b + \boldsymbol{a}^2 i_c\right)$.

The discrete time model of the system is used to predict the future values of the controlled variable using the measured system parameters at the sampling instant. The discretization of the considered system can be done by using simple forward Euler method. The change in output load current between k^{th} and $(k+1)^{th}$ sampling instants (ie, during one sampling period, T) can be written as in (2) and can be used to predict the value of current at the $(k+1)^{th}$ sampling instant (3).

$$\frac{di}{dt} \approx \frac{i(k+1) - i(k)}{T} \qquad (2)$$

$$\Rightarrow i(k+1) = i(k)\left(1 - \frac{RT}{L}\right) + \frac{T}{L}v(k) \qquad (3)$$

where the current vector $i(k)$ is the measured vector at the kth sampling instant.

III. Model Predictive Control

The basic principles of model predictive control are the prediction of future states of the controlled variable over the prediction horizon using a discrete time model of the system and the selection of optimal actuation by minimizing the cost function defined in terms of desired performance of the system for all possible input actuations. In case of current controlled NPC converter, the output current of the converter must follow the reference current waveform. Hence the cost function is to be defined in terms of current reference tracking error as the optimal voltage actuation is selected based on the minimization of cost function. The sum of absolute errors between the reference and the predicted values of load current in alpha-beta rotating frame is defined as the cost function (C) for the current controlled NPC converter system (4).

$$C = \left|i_{\alpha ref}(k+1) - i_{\alpha p}(k+1)\right| + \left|i_{\beta ref}(k+1) - i_{\beta p}(k+1)\right| \quad (4)$$

where $i_{\alpha ref}(k+1)$ and $i_{\beta ref}(k+1)$ are the alpha-beta components of the reference current and $i_{\alpha p}(k+1)$ and $i_{\beta p}(k+1)$ are the alpha-beta components of the predicted load current for the future sampling instant (k+1).

In finite control set model predictive control (FCS-MPC) technique, the load current at the $(k+1)^{th}$ sampling instant corresponding to all possible voltage vectors is determined using the current prediction equation. The cost function values are calculated from the reference and predicted currents using (4) and the voltage vector which results in minimum value of cost function is selected and applied.

A. Modified Predictive Control Algorithm

The drawback of the existing finite control set predictive algorithm [4] is that it fails to effectively utilize the available eight redundant voltage vectors to reduce the average switching frequency without any compromise in the quality of reference current tracking. The redundant output voltage vectors have the same value and hence the cost function associated with each of

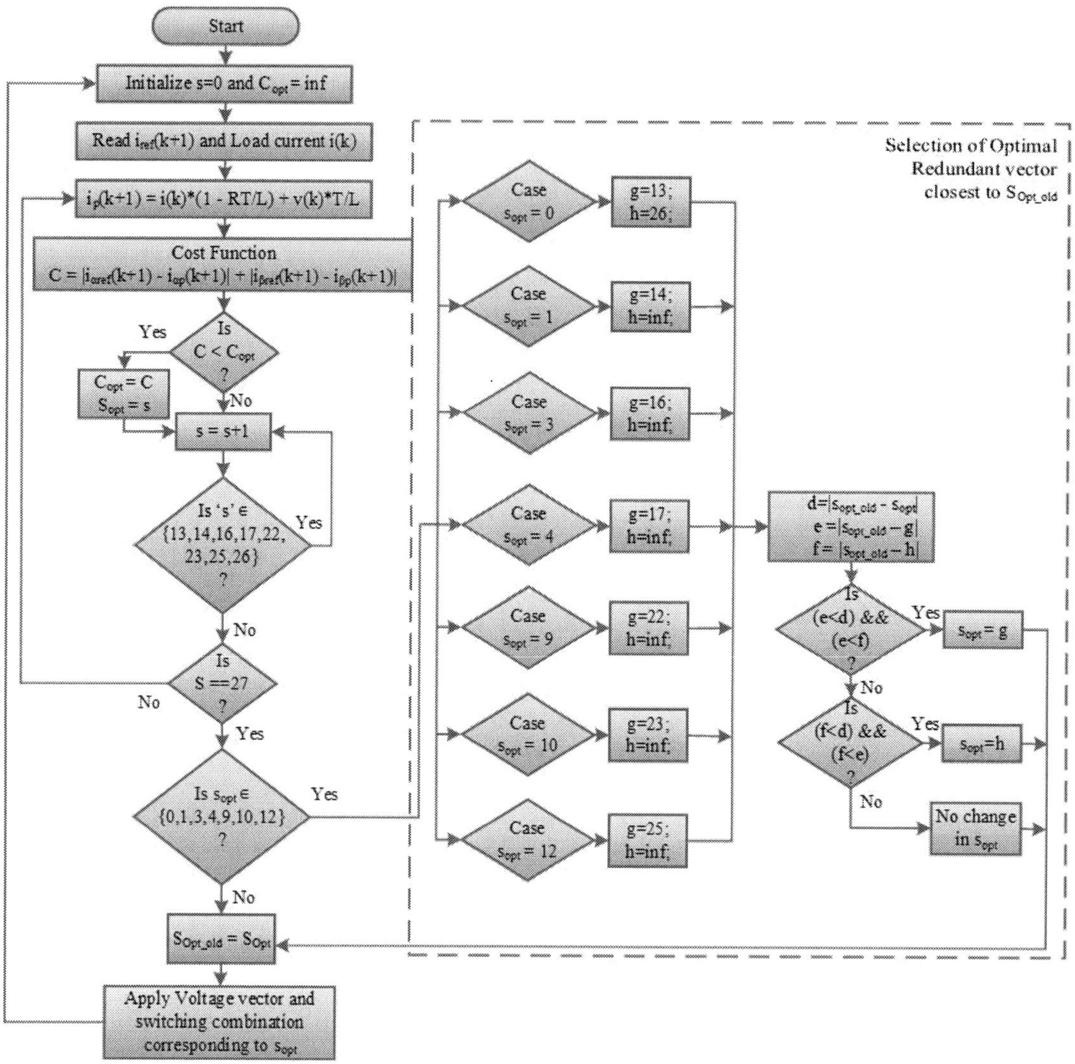

Fig. 2. Flowchart of Modified Predictive Control Algorithm

the redundant vector will be the same. Since C_{opt} is updated only if the present value of cost function is less than that of minimum cost function calculated so far, all redundant vectors are not taken into consideration. Always the redundant vector with lower index value is considered. The proposed modified predictive control algorithm considers each of the redundant voltage vectors separately and also the optimal voltage vector (switching combination) applied in the preceding sampling period for finding the optimal vector for the future sampling instant as shown in the flowchart (Fig. 2).

The basic principle for achieving reduction in switching frequency of the converter through modified predictive control algorithm is that the redundant voltage vectors having the same magnitude can be generated by different switching combinations. For example the redundant voltage vector v_0 can be generated by three different switching combinations corresponding to v_0, v_{13} and v_{26}. As shown in flowchart, if the optimal voltage vector index (s_{opt}) calculated after considering all the 27 possible vectors corresponds to any of the family of redundant vectors, then two additional checks are executed. Firstly, we check whether s_{opt} is same as the optimal vector index applied in the last sampling period (s_{opt_old}). In that case the s_{opt} is directly considered for the next sampling period and interestingly there is no switching transition between the two consecutive sampling periods and it directly results in reduced switching frequency. Secondly, if s_{opt} differs from the optimal vector index applied in the last sampling period (s_{opt_old}), then the redundant voltage vector index located nearer to s_{opt_old} is selected as the new optimal voltage vector. For example, consider the case where $s_{opt\_old} = 3$ and $s_{opt}=13$. Here, $s_{opt}=13$ corresponds to the family of redundant voltage vectors v_0, v_{13} and v_{26}. The required number of switching transitions increases if the preceding and succeeding vector states are located far away. Hence here the s_{opt} is updated as $s_{opt}=0$ as vector v_0 lies nearer to the preceding vector v_3. This makes sure the minimization of required number of switching transitions to apply the new optimal voltage vector for the future sampling period. These modifications in the finite control set predictive algorithm will definitely reduce the average switching frequency per controlled switch of the converter. The proposed modified predictive control algorithm is easy to implement and can achieve in reduction in average switching frequency without modifying the cost function.

IV. SIMULATION RESULTS

The effectiveness of modified predictive control algorithm is verified through MATLAB/Simulink simulations. The various simulation parameters are; V_{dc}=520V, R=10Ω, L=10mH. The performance of the proposed modified predictive control algorithm in achieving reduced switching frequency is compared with that of the conventional finite control set predictive algorithm in terms of the defined parameters [4] like average switching frequency per controlled switch (f_{sw}) and the absolute reference tracking error (e) considering different sampling times (T) in range of 5μs to 300μs. Two different current reference waveforms are considered in the simulation study. In case1, a sine wave with peak value of 25A is considered as the reference current wave, and in case2 a random wave with THD of 25% is taken. The output load current and voltage waveforms are shown in Fig. 3 and Fig. 4.

Fig.3(a) shows the load current for the case1 and excellent reference current tracking can be seen with a mean absolute reference tracking error of 0.1392A. The load current has good quality with a lower THD of 0.5873%. More interestingly, the average switching frequency (f_{sw}) is obtained as 3.797kHz showing a significant 6.4% reduction as compared to the finite control set predictive algorithm technique. The simulation results are summarized in Table 2. Moreover, similar performance is also observed in case2 with the different current reference wave. The proposed modified predictive control algorithm brings about 7.7% reduction in the average switching frequency with good reference tracking as shown in Fig. 3(b).

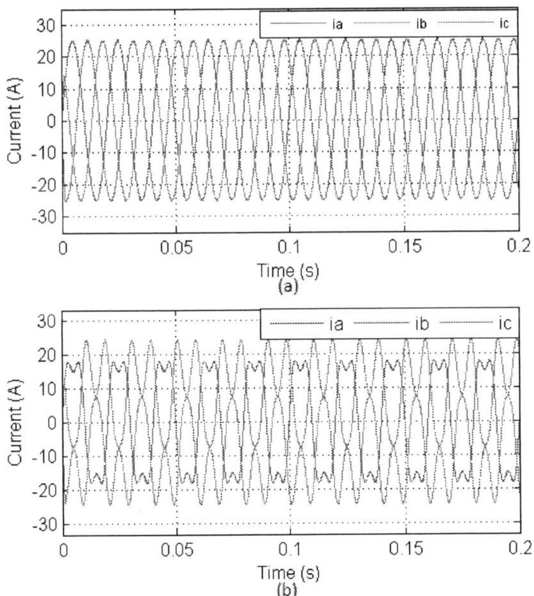

Fig. 3. Load current(T=25μs); (a) Modified MPC with Case1 (b) Modified MPC with Case2

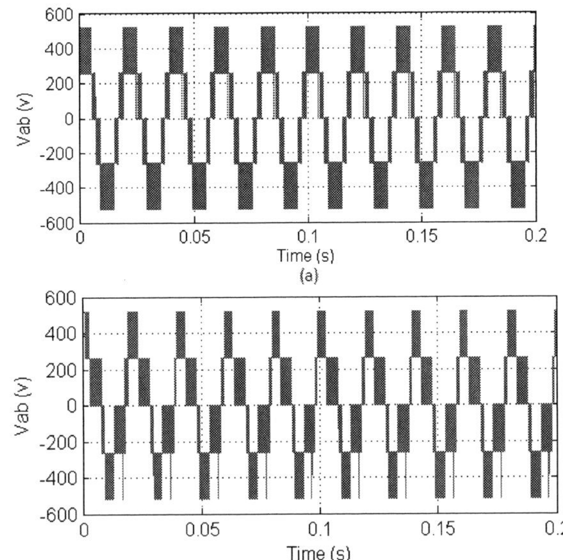

Fig. 4. Load voltage(T=25μs); (a) Modified MPC with Case1 (b) Modified MPC with Case2

The performance of the modified MPC algorithm is well analyzed over wide range of sampling periods and the results are summarized in Fig. 5, Fig. 6 and Fig. 7. The Fig. 5 shows the variation of absolute reference tracking error (e) Vs average switching frequency (f_{sw}) with conventional finite control set predictive scheme [4] and modified predictive control schemes. As the sampling time decreases, average switching frequency increases with better reference current tracking and it is to be noted that the absolute error remains in acceptable range even with lower average switching frequency of 0.5kHz.

The effectiveness of the modified MPC algorithm in reducing average switching frequency over a range of sampling time is shown in Fig. 6 and it is clear that, on an average, the modified MPC algorithm reduces 5-7% of f_{sw} as compared to conventional MPC algorithm. This considerable average switching frequency reduction is achieved without causing any additional cost to the system. Fig. 7 presents the variation of THD Vs sampling time with conventional FCS-MPC and

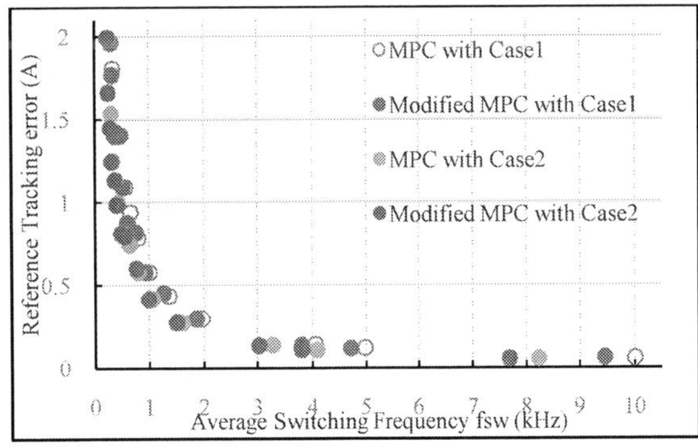

Fig. 5. Reference Tracking error(e) Vs Average switching Frequency(f_{sw})

Table 2. Summary of simulation results (T=25μs)

Reference wave	Case1 (Sine wave with peak of 25A)			Case2 (Random wave with THD of 25%)		
	f_{sw} (kHz)	e (A)	THD (%)	f_{sw} (kHz)	e (A)	THD (%)
Predictive Control (FCS-MPC)	4.058	0.1399	0.5837	3.277	0.1387	24.94
Modified Predictive Control	3.797	0.1392	0.5873	3.024	0.1361	24.95
	Reduction in f_{sw} = 6.4%			**Reduction in f_{sw} = 7.7%**		

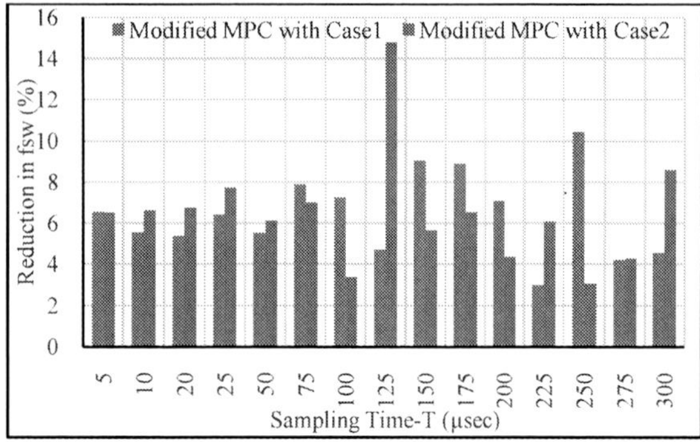

Fig. 6. %Reduction in f_{sw} Vs Sampling Time

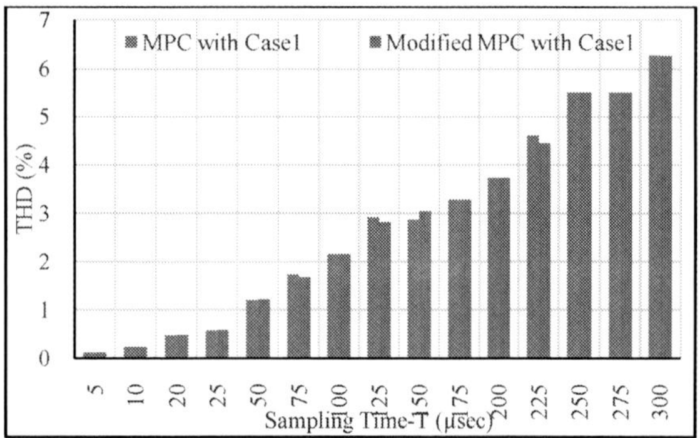

Fig. 7. THD Vs Sampling Time

modified MPC algorithm for the case1 and it verifies the efficient performance of modified MPC.

From the Fig. 7 we can conclude that the proposed modification in the conventional MPC algorithm does not deteriorate the quality of output current but provides similar/better performance over the wide range of sampling period. Also it is to be noted that increasing sampling time to higher values (>240μs) leads to unacceptable THD values though average switching frequency is reduced.

V. CONCLUSION

The model predictive control method uses a different approach from the conventional control techniques for the current control of power electronic converters. The conventional predictive control algorithm provides a good reference current tracking but with a higher average switching frequency. The proposed modified predictive control algorithm succeeds in achieving significant reduction in average switching frequency without any compromise in the reference current tracking. The new proposed algorithm eliminates the need of changing the cost function as done in other reduced switching frequency predictive algorithms. Also selection of proper weighting factor is not required since no additional terms are added in the cost function. Instead the reduction in switching frequency is achieved through smart utilization of available redundant voltage vectors and switching combinations. Moreover the proposed modified algorithm is simple and easy to implement which will considerably reduce the overall switching losses of the power electronic converters especially in active power filtering and high power applications.

ACKNOWLEDGMENT

The authors would like to thank IRCC-IIT Bombay and MHRD, Government of India for the financial support.

REFERENCES

[1] S. Kouro, P. Cortes, R. Vargas, U. Ammann, and J. Rodriguez, "Model Predictive Control-ASimple and Powerful Method to Control Power Converters," IEEE Trans. Ind. Electron., vol. 56, pp. 1826–1838, June 2009.

[2] J. Rodriguez, J. Pontt, and C. Silva, "Predictive current control of a voltage source inverter," IEEE Trans. Ind. Electron., vol. 54, pp. 495–503, Feb. 2007.

[3] P. Cortes, M. P. Kazmierkowski, R. M. Kennel, D. E. Quevedo, and J. Rodriguez, "Predictive Control in Power Electronics and Drives," IEEE Trans. Ind. Electron., vol. 55, no. 12, pp. 4312–4324, 2008.

[4] R. Vargas, P. Cort´es, U. Ammann, J. Rodr´ıguez, and J. Pontt, "Predictive control of a three-phase neutral-point-clamped inverter," IEEE Trans. Ind. Electron., vol. 54, no. 5, pp. 2697–2705, 2007.

[5] P. Cortes, L. Vattuone, and J. Rodriguez, "Predictive current control with reduction of switching frequency for three phase voltage source inverters," in EEE Int. Symp. Ind. Electron., pp. 1817–1822, 2011.

[6] P. Zanchetta, D. B. Gerry, V. G. Monopoli, J. C. Clare, and P. W. Wheeler, "Predictive current control for multilevel active rectifiers with reduced switching frequency," IEEE Trans. Ind. Electron., vol. 55, no. 1, pp. 163–172, 2008.

[7] R. Ramirez, J. R. Espinoza, F. Villarroel, E. Maurelia, and M. E. Reyes, "A novel hybrid finite control set model predictive control scheme with reduced switching," IEEE Trans. Ind. Electron., vol. 61, no. 11, pp. 5912–5920, 2013.

[8] S. Vazquez, J. I. Leon, L. G. Franquelo, J. M. Carrasco, O. Martinez, J. Rodriguez, P. Cortes, and S. Kouro, "Model Predictive Control with Constant Switching Frequency Using a Discrete Space Vector Modulation with Virtual State Vectors," in IEEE Int. Conf. Ind. Technol., pp. 1–6, 2009.

[9] M. Vatani, B. Bahrani, M. Saeedifard, and M. Hovd, "Indirect Finite Control Set Model Predictive Control of Modular Multilevel Converters," IEEE Trans. Smart Grid, vol. 6, no. 3, pp. 1520–1529, 2015.

Implicit Finite Control Set Model Predictive Current Control for Modular Multilevel Converter Based on IPA-SQP Algorithm

Hamed Nademi
Process Automation division
ABB AS
Oslo-0603, Norway

Lars Einar Norum
Department of Electrical Power Engineering
Norwegian University of Science and Technology
Trondheim-7491, Norway

***Abstract*—This paper investigates a Finite-Control-Set Model Predictive Control (FCS-MPC) for the precise control of (un)balanced load currents in Modular Multilevel Converter (MMC). The control objectives are circulating currents minimization inside the converter arms, achieve a capacitors voltage balance and load current control. To achieve the converter constrained optimization and facilitate the implementation on embedded systems, an integrated perturbation analysis and sequential quadratic programming (IPA-SQP) solver is also utilized. As a case study the proposed approach is applied to a grid-connected five-level MMC. The introduced FCS-MPC formulation reduces sensitivity of the converter output voltage to disturbances in grid side and measurement noise with reducing the computational burden. Simulation results reveal the effectiveness of the developed control scheme in cases when operational objectives, e.g., load current reference tracking and disturbance rejection are considered under system model uncertainties.**

Keywords—Model Predictive Control; Sequential Quadratic Programming; Modular Converter System; Current Control

I. INTRODUCTION

Model Predictive Control (MPC) has been received popularity in applications of multilevel power converter topologies due to its several merits [1-4]. In particular, Finite-Control-Set Model Predictive Control (FCS-MPC) seems promising does not need a modulator and provides reduced computational burden [4, 5].

The rising popularity of Modular Multilevel Converter (MMC) topology has recently evident at both industry and academia because of its modularity, low device rating, and fault tolerant capacity [6-8]. The usage of this topology in dc-ac power conversion e.g. for ac drives [9-11] and grid-connected renewable energy resources [8, 12] is currently being developed.

In general, the control objectives in case of a MMC in most applications are output voltage, floating capacitors voltage and I/O current [13]. Despite solutions introduced in available literature [5], [14-17], the aforementioned issues are still remaining along with high computational effort which in terns resulting a practical realization difficulties.

The main intention of this study is to formulate circulating current suppression as FCS-MPC problem and regulate the capacitor voltage ripples. Furthermore, the analysis is conducted to achieve either (un)balanced load current tracking and capacitor voltage balancing with reasonable performance. To derive the one-step-ahead prediction of the controlled variables, Midpoint Euler method is used as a discretization technique.

Multiple numerical optimization techniques have been introduced based on advanced algorithms [18-20]. Most of them employ single iteration of root finding over the sampling instant, thus lead to deterioration of the closed-loop performance and required accuracy. Due to the fact that in power converters significant variations may occur in the operating bounds, those optimization solutions become intractable.

This study introduces a novel numerical optimization strategy named as an integrated perturbation analysis and sequential quadratic programming (IPA-SQP) solver to reach lower sampling time easily run on FPGAs. In principal, IPA-SQP method comprises the computational merits of perturbation analysis and optimality of the SQP approach dealing with the optimization formulation at *k*th instant as a perturbed problem at the *k-1* instant [21]. As a result, IPA-SQP drastically enhances computational efficiency over the receding horizon. As shown in [21, 22], the required computation time to calculate one iteration of IPA-SQP approach is of order length of horizon, appears promising for a realization of MPC in power conversion systems.

The sensitivity of the developed controller relating to parameter mismatch, measurement noise as well as sudden changes in utility grid voltage are discussed in this paper. As a case study, the proposed method is applied to a 5-level MMC topology and some simulation results are reported to end this paper.

978-1-4673-9551-9/16 $31.00 © 2016 IEEE

II. MMC DESCRIPTION AND FCS-MPC FORMULATION

A. Physical Model of the MMCs

The circuit diagram of a three-phase (n+1)-level MMC is shown in Fig. 1. The load is modeled with inductor (L), resistor (R) and sinusoidal voltage source (e) representing a motor or utility grid. All the capacitors will be charged up to V_{dc}/N at the MMC start-up operation [23] and their voltage level should be kept within this reference value using the control scheme. Thus, the output pole voltage of MMC, V_{AO}, varies between $-V_{dc}/2$ and $+V_{dc}/2$ when all upper/lower arm cells are inserted or bypassed implying charging or discharging states, respectively. Neglecting the voltage drop across arm inductor and resistor, v_j (j= a, b and c) can be calculated by:

$$v_j = \frac{n_{lowj} - n_{upj}}{2N} V_{dc} \quad (1)$$

According to [13, 24] the upper and lower arm currents, i_p and i_n are given by:

$$i_{pj} = \frac{I_{dc}}{3} + \frac{i_j}{2} + i_{cir} \quad (2)$$

$$i_{nj} = \frac{I_{dc}}{3} - \frac{i_j}{2} + i_{cir} \quad (3)$$

where I_{dc} is the dc component of the dc line current, i_j is the output phase current and i_{cir} is the circulating current flowing through phase j.

The circulating current does not flow to the output load currents, while negatively influence on the converter switching losses, capacitor voltage fluctuation and components ratings [8-10]. From Fig.1 and employing KVL law, and some mathematical efforts, two main first order differential equations based on the controlled variables i.e., i_j, i_{cir} and capacitor voltages can be written as [24]:

$$\frac{di_j}{dt} = \frac{1}{(L_1 + 2L)}\left[v_{lowj} - v_{upj} - (R_1 + 2R) i_j - 2e_j\right]$$

$$\frac{di_{cirj}}{dt} = \frac{1}{2L_1}[V_{dc} - v_{upj} - v_{lowj} - 2R_1 i_{cirj} - \frac{2}{3} R_1 I_{dc}]$$

$$\frac{dv_{C_{ij}}}{dt} = \frac{1}{C} i_{mj} \qquad i = 1, \ldots 2N$$

(4)

Where v_{upj} and v_{lowj} are the number of upper and lower cells which are turned on and R_1 states the arm inductor and IGBTs resistivity. It is obvious that when the cell is turned off the current i_m is zero.

In the rest of this section, the formulation of the proposed FCS-MPC controller together with the optimization algorithm IPA-SQP will be elaborated.

B. Improved FCS-MPC Strategy with Numerical Optimization IPA-SQP

From the earliest conducted studies dealing with FCS-MPC control scheme, it is evident that it can be promising solution to be deployed in power conversion systems due mainly to the discrete feature of these systems. In this work, a perturbation analysis (PA) is used to get the knowledge of a change in the optimal solution in case of parameters variation, such as initial conditions.

However, the PA solution does not ensure complete optimality when this algorithm needs to update a nominal solution in an iterative manner. For this reason, an SQP update based on linearization and quadratic cost estimation can be utilized. As a result, a closed-form solution and computationally efficient optimization are provided.

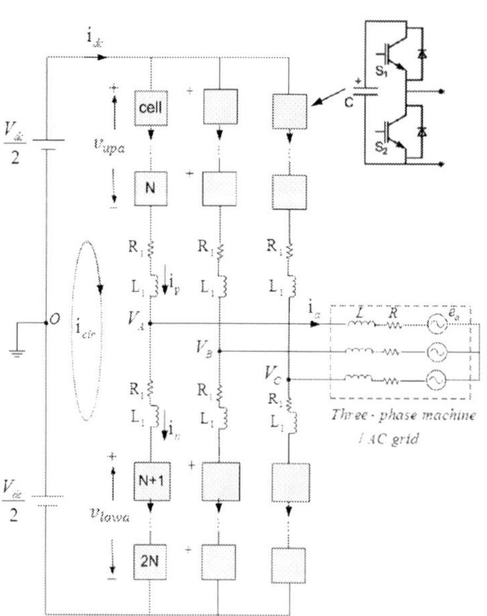

Fig. 1. Circuit configuration of three-phase grid-connected MMC acting as a dc-ac conversion.

In an *(n+1)*-level MMC, there are C_{2n}^{n} possible switching states. In order to predict the one prediction horizon of these controlled variables, then (1) should be discretized by Midpoint Euler method. For simplification purposes, e_j is assumed to be constant during a sampling period $\left(e_j(k+1) \approx e_j(k)\right)$. Therefore, the phase currents can be expressed

$$i_j(k+1) = i_j(k)\frac{2(L_1+2L)-T_s(R_1+2R)}{2(L_1+2L)+T_s(R_1+2R)} + \frac{T_s}{2(L_1+2L)+T_s(R_1+2R)}\begin{bmatrix} v_{lowj}(k+1)+ \\ v_{lowj}(k)- \\ \begin{bmatrix} v_{upj}(k+1) \\ +v_{upj}(k) \end{bmatrix} \\ -4e_j(k) \end{bmatrix} \tag{5}$$

where T_s is the sampling time.

Another controlled variable that must be predicted is the circulating currents that can be obtained by

$$i_{zj}(k+1) = i_{zj}(k)\frac{2L_1 - R_1T_s}{2L_1 + R_1T_s} + \frac{T_s}{4L_1+2R_1T_s}\begin{bmatrix} 2V_{dc} - \left(v_{upj}(k+1) + v_{upj}(k)\right) \\ -\left(v_{lowj}(k+1) + v_{lowj}(k)\right) \\ -\frac{4R_1}{3}\left(I_{dc}(k)\right) \end{bmatrix} \tag{6}$$

Backward Euler method gives a satisfying approximation to find $v_{Cij}(k+1)$,

$$v_{Cij}(k+1) = v_{Cij}(k) + \frac{T_s}{C}i_{mj}(k+1) \tag{7}$$

The following cost function *J* will be defined as [17]:

$$J = \left|i_{jref}(k+1) - i_j(k+1)\right| + \alpha_1 \cdot \sum_{i=1}^{2n}\left|v_{Cij}(k+1) - \frac{V_{dc}}{N}\right| + \alpha_2 \cdot \left|i_{cirj}(k+1)\right| \tag{8}$$

The block diagram of the proposed FCS-MPC controller for a MMC is depicted in Fig. 2.

This work proposes the integrated perturbation analysis and sequential quadratic programming (IPA-SQP) aiming to take into account the computational challenges for implicit MPC strategy. In doing so, the impact of nonlinear constraints optimal control problem are effectively reduced. As the name implies, the mentioned numerical optimization takes advantage of integrated PA and SQP algorithms, where at one sampling time *t*, the sequence of an optimal control for measured control variable *x(t)* is determined from the value of *x(t-1)*. With the help of this algorithm, the sequence of predicted states is optimally updated for each MPC iteration [21, 22]. The optimal control sequence *u(k)* is derived fulfilling the conditional expectation of *J(k)* to achieve reduced computational complexity.

It should be noticed that in the FCS-MPC approach, the switching frequency fluctuates depending on the converter working points. This feature can be assumed as a constraint in the cost function tuning process [1].

The summary of the procedures for IPA-SQP algorithm which is combined with the FCS-MPC problem formulation is shown in Fig. 3.

III. SIMULATION STUDY

A 5-level 6MW MMC is simulated using MATLAB/SIMULINK software to verify the proposed controller and theoretical findings. The detailed circuit specifications are given in Table I.

The computer simulations carried out with proposed MPC controller incorporating an optimization solution in the development of a three-phase MMC-based shunt active power filter.

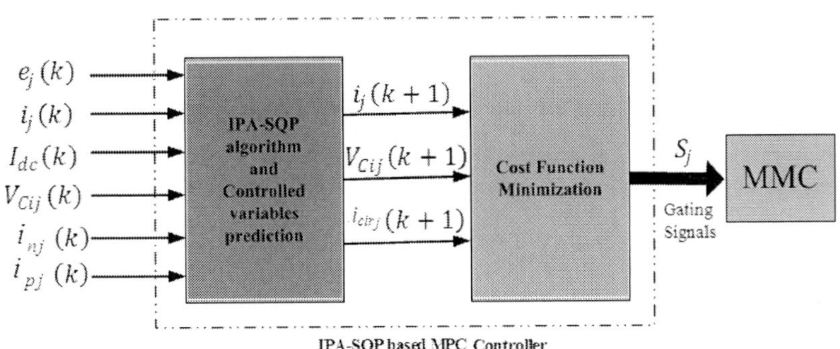

Fig. 2. Overall control structure of the proposed FCS-MPC scheme.

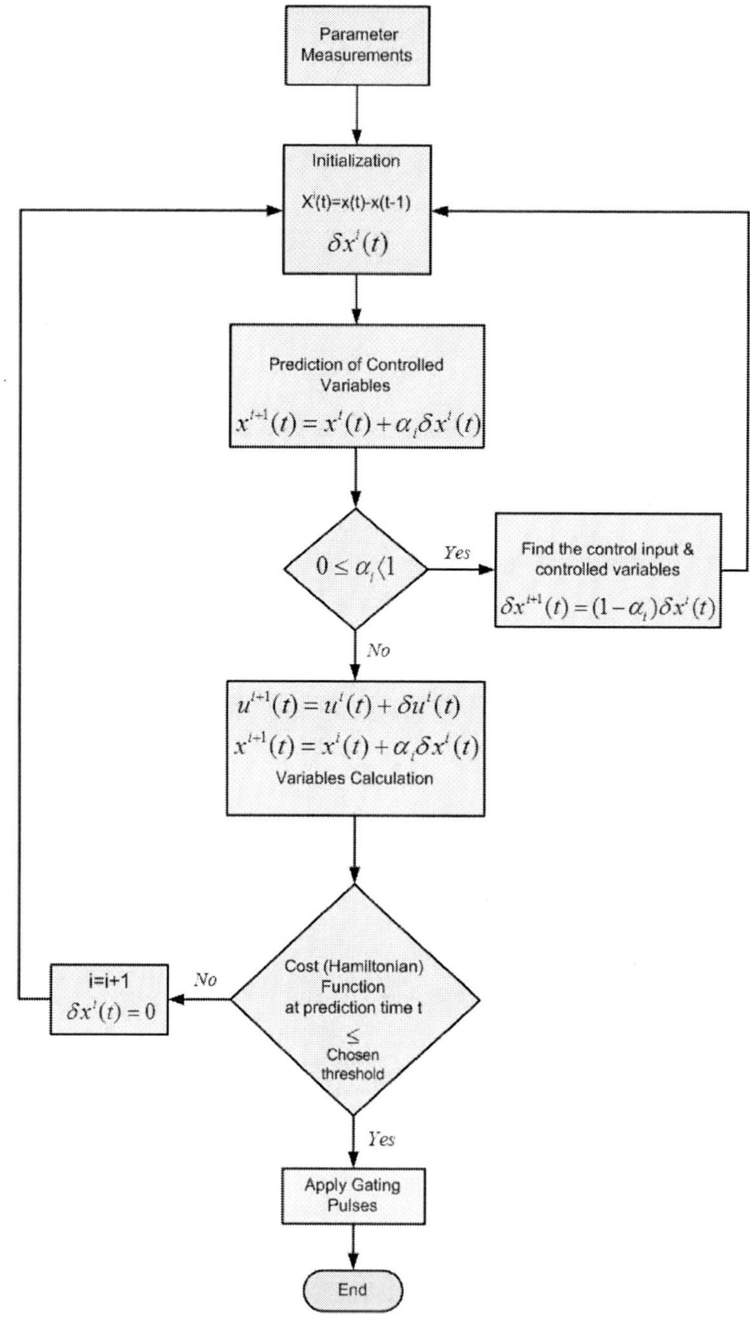

Fig. 3. IPA-SQP approach flowchart implemented in FCS-MPC formulation to include perturbation.

Fig. 4 illustrates actual and reference source current behavior in steady state for MMC in active power filter applications. It is observed that the current reference tracking is well performed. The current command is provided for FCS-MPC considering the required active power by the converter to regulate the dc-link voltage, and the reactive power to keep unity power factor, while delivering the sinusoidal current waveform to the grid side.

TABLE I. PARAMETERS OF THE SIMULATED MMC

Parameters	Value	Parameters	Value
Number of cells in each arm	4	DC link voltage	10 kV
Rated current	200 A	Sampling time	50 μs
Grid resistor	6 Ω	Arm inductance	1.2 mH
Grid inductor	10 mH	Arm resistance	0.04 Ω
Output frequency	50 Hz	Cell capacitance	4.7 mF

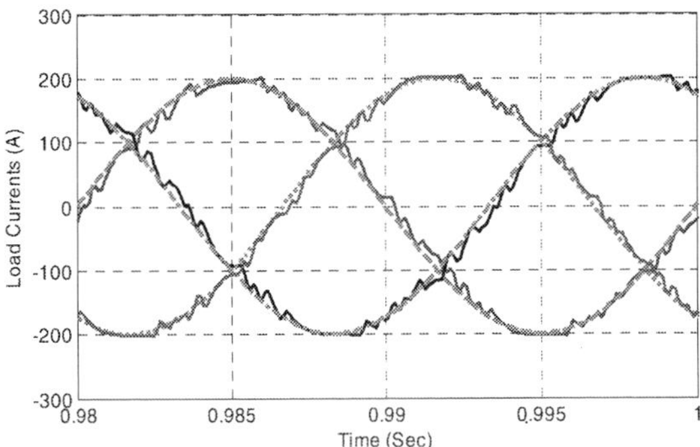

Fig. 4. Simulated three phase current references (dashed lines) with measured currents (solid lines) in proposed MPC controller.

To evaluate the robustness of discussed control structure, simulation is run in the presence of measurement errors in load parameters. Consequently, due to the parameters uncertainties, the load resistor and inductor values (R, L) are overestimated by 20% of their nominal values. In addition, the grid voltage disturbance steps down to 70% of its nominal value is imposed at t=1sec.

As shown in Fig. 5 the capacitor voltages (10kV/4) remain unaffected and developed scheme provides the new switching algorithm with fast response to balance capacitor voltages.

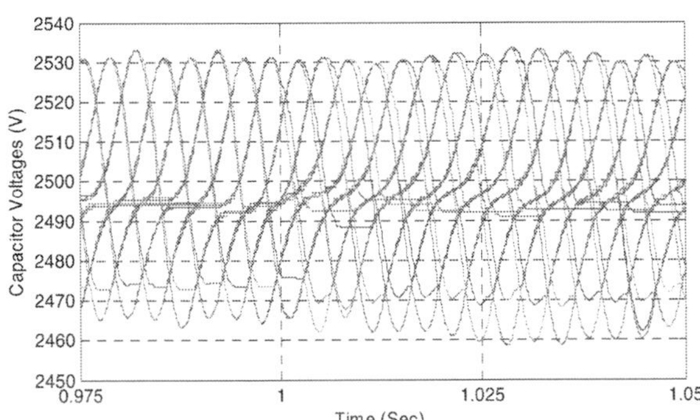

Fig. 5. Capacitor voltages in Phase A, when the grid voltage (load) drops by 30% at t=1 sec.

Moreover, the three-phase current reference tracking for aforementioned operating condition is shown in Fig. 6. As can be seen from the waveforms, the load current tracking objective is achieved without losing the dynamical performance under sudden changes and parameter sensitivity in the system.

The relevant load current spectrum for the same operating points is plotted in Fig. 7. The harmonic amplitudes are scaled associated with the rated load current. Obviously the harmonic contents are within an acceptable limits.

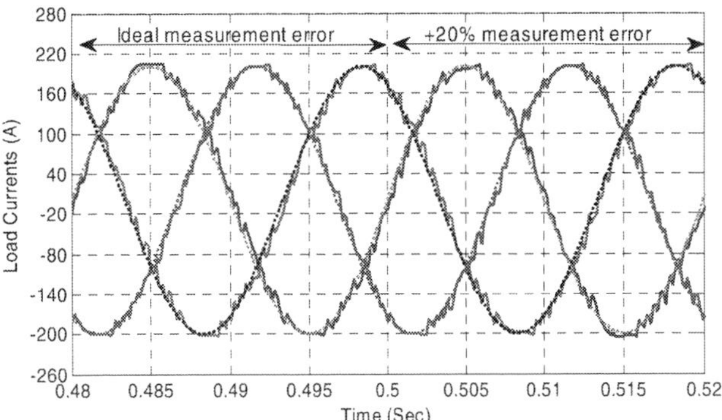

Fig. 6. Simulated three phase current references (dashed lines) with measured currents (solid lines) when the grid voltage drops by 30% at t=0.5 sec.

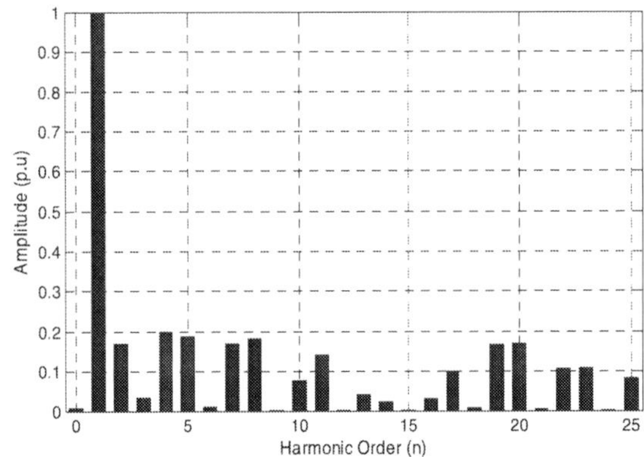

Fig. 7. Harmonic spectrum of grid current for phase A.

Finally, the feasibility of the IPA-SQP optimization algorithm in practical implementations has been assessed using real-time simulator Opal-RT. Fig. 8 presents the performance comparison of the closed-loop system with respect to the computational time for conventional FCS-MPC strategy and proposed controller.

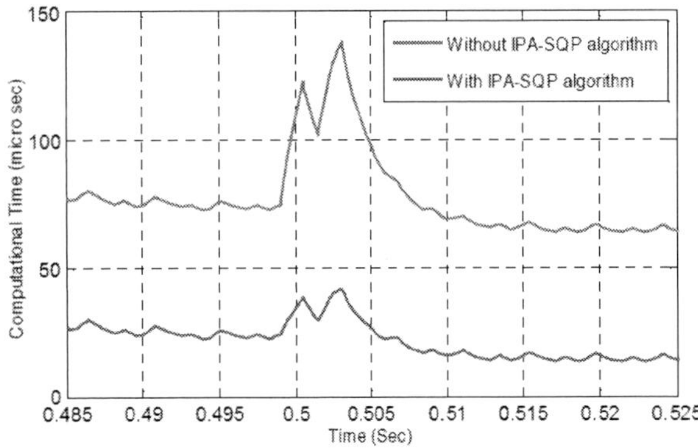

Fig. 8. Computational burden comparison of the proposed and conventional FCS-MPC schemes during step-down changes in grid voltage.

978-1-4673-9551-9/16 $31.00 © 2016 IEEE

It is found out that the developed FCS-MPC comprising IPA-SQP solver enhances both the steady-state and transient behavior with speed up the calculation of the optimal solution for a MMC.

IV. CONCLUSION

This paper has demonstrated a FCS-MPC scheme employing a new IPA-SQP optimization solution to speed up the computational efficiency for MMC topology. The proposed approach addresses either balanced or unbalanced reference current tracking including circulating current suppressing, balancing the capacitor voltages and load current control. Obtained results verify the superior steady-state and transient response, when system operates in the presence of parameter mismatch and load voltage dips. The proposed technique enables to overcome the high processing time associated with the MPC cost function, introducing an alternative for practical implementation.

REFERENCES

[1] J. Rodriguez and P. Cortes, "Predictive Control of Power Converters and Electrical Drives," First Edition., *IEEE press-Wiley*, Mar. 2012.

[2] M. A. Perez, J. Rodriguez, E. J. Fuentes and F. Kammerer, "Predictive Control of AC-AC Modular Multilevel Converters," *IEEE Trans. Ind. Electron.*, vol. 59, no. 7, pp. 2832-2839, July 2012.

[3] J. Rodriguez, M. P. Kazmierkowski, J. R. Espinoza, P. Zanchetta *et al.*, "State of the Art of Finite Control Set Model Predictive Control in Power Electronics," *IEEE Trans. Ind. Informat.*, vol. 9, no. 2, pp. 1003-1016, May 2013.

[4] R. P. Aguilera, P. Lezana and D. E. Quevedo, "Finite-Control-Set Model Predictive Control with Improved Steady-State Performance," *IEEE Trans. Ind. Informat.*, vol. 9, no. 2, pp. 658-667, May 2013.

[5] M. Vatani, B. Bahrani, M. Saeedifard and M. Hovd, "Indirect Finite Control Set Model Predictive Control of Modular Multilevel Converters," *IEEE Trans. Smart Grid*, vol. 6, no. 3, pp. 1520-1529, May 2015.

[6] M. Glinka and R. Marquardt, "A new AC/AC multilevel converter family," *IEEE Trans. Ind. Electron.*, vol. 52, no. 3, pp. 662-669, June 2005.

[7] H. Nademi, A. Das and L. Norum, "Modular Multilevel Converter with an Adaptive Observer of Capacitor Voltages," *IEEE Trans. Power Electron*, vol. 30, no. 1, pp. 235-248, Jan. 2015.

[8] M. A. Perez, S. Bernet, J. Rodriguez, S. Kouro, R. Lizana, "Circuit Topologies, Modelling, Control Schemes and Applications of Modular Multilevel Converters," *IEEE Trans. Power Electron*, vol. 30, no. 1, pp. 4-17, Jan. 2015.

[9] M. Hagiwara and H. Akagi, "Control and experiment of pulse-width modulated modular multilevel converters," *IEEE Trans. Power Electron*, vol. 24, no. 7, pp. 1737-1746, Jul. 2009.

[10] J. Kolb, F. Kammerer, M. Braun, "Dimensioning and Design of a Modular Multilevel Converter for Drive Applications," in *Proc. of IEEE EPE-PEMC* 2012, pp. 1-8.

[11] Antonopoulos, A., Angquist, L., Norrga, S., Ilves, K., Harnefors, L., Nee, H.-P.: 'Modular Multilevel Converter AC Motor Drives With Constant Torque From Zero to Nominal Speed', *IEEE Trans. Ind. Appl.*, vol. 50, no. 3, pp. 1982-1993, May/Jun. 2014.

[12] L. Liu, H. Li, Y. Xue *et al.*, "Decoupled Active and Reactive Power Control for Large-Scale Grid Connected Photovoltaic Systems Using Cascaded Modular Multilevel Converters," *IEEE Trans. Power Electron*, vol. 30, no. 1, pp. 176-187, Jan. 2015.

[13] H. Nademi and L. Norum "Analytical Circuit Oriented Modeling and Performance Assessment of Modular Multilevel Converter", *IET Power Electronics*, vol. 8, Issue 9, pp. 1625-1635, Sep. 2015.

[14] H. Akagi, "Classification, Terminology, and Application of the Modular Multilevel Cascade Converter (MMCC)," *IEEE Trans. Power Electron*, vol. 26, no. 11, pp. 3119-3130, Nov. 2011.

[15] M. A. Perez, J. Rodriguez, E. J. Fuentes and F. Kammerer, "Predictive Control of AC-AC Modular Multilevel Converters," *IEEE Trans. Ind. Electron.*, vol. 59, no. 7, pp. 2832-2839, July 2012.

[16] J. W. Moon, J. S. Gwon, J. W. Park, D. W. Kang, and J. M. Kim, "Model predictive control with a reduced number of considered states in a modular multilevel converter for HVDC system," *IEEE Trans. Power Del.*, vol. 30, no. 2, pp. 608-617, April 2015.

[17] R. N. Fard, H. Nademi and L. Norum, "Analysis of a Modular Multilevel Inverter Under the Predicted Current Control Based on Finite-Control-Set Strategy", in *Proc. of IEEE EPECS* 2013, Istanbul, Turkey, Oct. 2013, pp. 1-6.

[18] P. Karamanakos, T. Geyer, N. Oikonomou, F. D. Kieferndorf, and S. Manias, "Direct Model Predictive Control: A Review of Strategies That Achieve Long Prediction Intervals for Power Electronics," *IEEE Ind. Electron. Magaz.*, vol. 8, no. , pp. 32-43, Mar. 2014.

[19] H. J. Ferreau, H. G. Bock, and M. Diehl, "An online active set strategy to overcome the limitations of explicit MPC," *Int. J. Robust Nonlinear Control*, vol. 18, Issue 8, pp. 816-830, May 2008.

[20] V. M. Zavala and L. T. Biegler, "The advanced-step NMPC controller: Optimality, stability and robustness," *Automatica*, vol. 45, no. 1, pp. 86-93, Jan. 2009.

[21] Y. Xie, R. Ghaemi, J. Sun, and J. S. Freudenberg, "Model predictive control for a full bridge DC/DC converter," *IEEE Trans. Control Syst. Technol.*, vol. 20, no. 1, pp. 164-172, Jan. 2012.

[22] R. Ghaemi, J. Sun, and I. V. Kolmanovsky, "An integrated perturbation analysis and sequential quadratic programming approach for model predictive control," *Automatica*, vol. 45, no. 10, pp. 2412-2418, Sep. 2009.

[23] A. Das, H. Nademi and L. Norum, "A Method for Charging and Discharging Capacitors in Modular Multilevel Converter", in *Proc. of IEEE IECON* 2011, Melbourne, Australia, Nov. 2011, pp. 1058-1062.

[24] Q. Jiangchao, M. Saeedifard, "Predictive Control of a Modular Multilevel Converter for a Back-to-Back HVDC System," *IEEE Trans. Power Del.*, vol. 27, no. 3, pp. 1538-1547, July 2012.

Resolution Requirements to Avoid Limit Cycling in LLC Resonant Converter

Shadi Dashmiz, Behzad Mahdavikhah and Aleksandar Prodic
Laboratory for Power Management and Integrated SMPS, ECE Department
University of Toronto, 10 King's College Road
Toronto, ON, M5S 3G4, CANADA E-mail: prodic@ele.utoronto.ca

Brent McDonald
Texas Instruments
Dallas, Texas, USA

Abstract— **This paper introduces a general method for finding minimum resolution of digital pulse frequency modulators (DPFM) of LLC converters to avoid limit cycling oscillation (LCO) in the output voltage under worst operating conditions. This work is built on the prior art [1], [2] where detailed analysis of limit cycling has been investigated for different hard switching converters and for constant load current controlled soft switching converters. The analysis introduced here takes into account that both the switching frequency and the gain characteristic of the converter change with operating conditions and that the resolution of the DPFM can be selected from a constrained set of the combinations of those two values.**
Experimental results obtained from a 350 W, 400 V to 12 V LLC converter prove the effectiveness of the introduced analysis by varying DPFM resolution and comparison of LCO with the theoretically predicted results for different loads.

I. Introduction

LLC isolated resonant converters are broadly used in dc-dc applications where a high voltage needs to be stepped down to a much lower level. Some instances include front-end dc-dc stages decreasing a voltage acquired from a PFC rectifier to a bus voltage level and telecom applications. High efficiency, high power density and low electromagnetic interference has made LLC converter widely adopted [3], [4].

If LLC converter is controlled appropriately narrow range of switching frequencies, zero voltage switching (ZVS) of the primary side switches, and zero current switching (ZCS) of the secondary side switches can be achieved [5], [6]. In the past, various digital control schemes have been proposed for regulation of LLC converters, to satisfy its fairly complex control requirements, [7], [8]. Digital control offers features that can improve characteristics of LLC converters. These include flexibility, smaller size and lower counts of passive components [9], [10], [11] larger immunity to external disturbances [12], [13] and, reconfigurable design which is easily transferable from one implementation to another among many more.

This work of Laboratory for Power Management and Integrated SMPS is supported by Texas Instruments, Dallas, TX, USA.

One of the shortcomings of the digital control is a potential presence of limit cycling [14], [15], which is a direct result of improperly selected resolutions of digital components, including analog-to-digital converter (ADC) and digital pulse frequency/width modulator (DPF/WM). In the past it was shown [16] that if the minimum attainable output voltage increment due to the variation of DPF/WM is larger than the voltage quantization step of the ADC, limit cycling occurs.

The issue of limit cycling in hard-switching converters has been thoroughly analyzed in [17], [18] and solutions for its elimination is proposed. However, the presented analysis and solutions are not directly applicable to finding the minimum resolution of the DPFM unit of resonant converters. The reason is that the gain of an LLC converter alters with both changes of frequency and the load. A method proposed in [2] has examined the limit cycling condition for DPFM element of a digital controller for a variable current constant output load resonant converter. However, a drawback of that method is that often it cannot be directly applied to the constant output voltage controlled LLC systems where the load changes as well, since both the characteristic and frequency of the operation change in these cases.

This paper introduces a new general method for finding the resolution of DPFM of various resonant converters such that effect of both frequency and load are taken into account and no limit cycling conditions are satisfied over the entire range of converter operation. The case study is LLC converter operating in closed loop with voltage mode controller.

II. Theory of Finding Limit Cycling for Changing Loads

A digital controller can exhibit limit cycling due to the presence of nonlinear elements, such as ADC and PFM quantizers. As discussed in [19] and shown in Fig.1, limit cycling oscillations can be present in the case one LSB change of the control output $f[n]$, i.e. minimum frequency step, results in a larger than one quantization step, V_q, change at the input of the ADC. Therefore, the key to avoid limit cycling is to make

978-1-4673-9551-9/16 $31.00 © 2016 IEEE

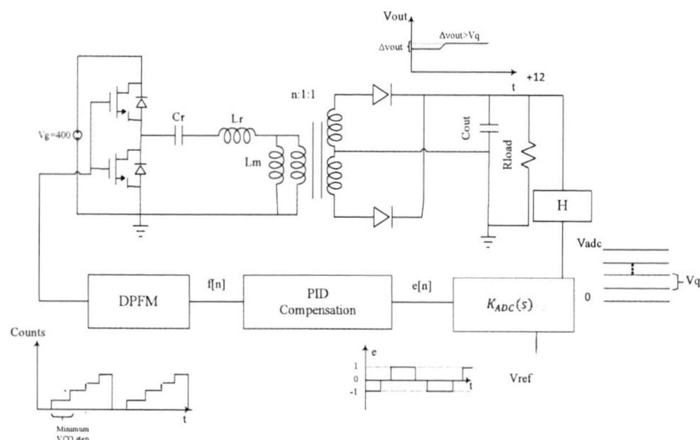

Fig. 1: Limit cycle oscillation-font size in the figure must be comparable to that of the sub-title.

sure that the quantization step is larger than the smallest step in the output voltage (ΔV_o) caused by the frequency step change satisfies the following equation:

$$V_q > H.\Delta V_o \quad (1)$$

Where H is the output voltage sensor gain , $\frac{V_{ref}}{V_o}$.

The condition to eliminate limit cycling from hard switching converters control loop is described in [16]. The principal of avoiding limit cycling is the same for the soft switching converters, meaning that the minimum output voltage generated from the minimum change in the frequency of voltage controlled oscillator (VCO) should be less than ΔV_o. Underlying factor to find the minimum frequency resolution is determining the load and frequency of operation through frequency-to-voltage gain function.

Frequency synthesizer or VCO can basically be observed as a counter with a time step ΔT that will reset when reaches a specified limit. The steady state switching frequency in that case can be defined as:

$$f_{Sw} = \frac{1}{2n\Delta T} \quad (2)$$

Where n (typically a large number) defines is the reset limit for half cycle of the VCO counter. The minimum possible frequency step produced by the counter is:

$$\Delta f_{min} = \frac{1}{2(n+1)\Delta T} - \frac{1}{2n\Delta T} = 2{f_{Sw}}^2\Delta T. \quad (3)$$

Equation (3) shows that the minimum possible change in the switching frequency is dependent on the steady state switching frequency, so the resolution is finer for higher switching frequencies and coarser for lower frequencies. Still, this is not the only criterion in deciding the limit cycling restraining condition. One other major criterion for finding needed resolution comes from the resonant tank gain ($M_g(f)$) which is the ratio of dc output to input voltage and is both frequency and load dependent. Therefore, both the load and frequency have impact on deciding the DPFM resolution. The LLC voltage gain can be calculated based on the first harmonic approximation (FHA) [19] as:

$$M_g(f) = \left|\frac{\omega^2 L_m.R_e.C_r}{j(\omega^3 L_r L_m - \omega.L_m) + (\omega^2(L_m.R_e.C_r + L_m.R_e) - R_e)}\right| \quad (4)$$

where L_m is transformer's magnetizing inductance, C_r and L_r are resonant tank capacitor and inductance, respectively, and R_e is the secondary side ac equivalent resistance, which can be calculated as: $R_e = \frac{8 \times n^2}{\pi^2} \times R_L$, and R_L is the load resistance of the LLC converter.

Fig. 2 shows the tank gain for different loads resistances and normalized frequencies ($\frac{f_{sw}}{f_o}$), $f_o = \frac{1}{\sqrt{L_r.C_r}}$ is the resonant frequency at short load condition.

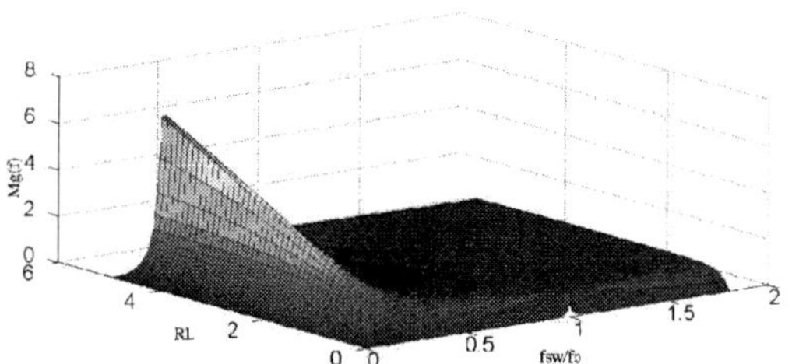

Fig. 2: LLC tank gain for continuously changing load and frequency. There are two peaks one at f_o and one at around $0.2f_o$ the latter being open circuit resonance.

Using Eqs. (1) and (4) the following inequality to prevent limit cycling should hold true:

$$|\Delta M_g(f).\Delta f| < \frac{V_q.2.n_{wid}}{H.V_g} \quad (5)$$

Where , V_g is dc input and n_{wid} is transformer ratio. Combining (5) and (3) minimum change in frequency step of frequency synthesizer can be calculated as:

$$|\Delta M_g(f).\Delta T.2{f_{Sw}}^2| < \frac{V_q.2.n_{wid}}{H.V_g} \quad (6)$$

The worst case condition for frequency step of counter is obtained when $|\Delta M_g(f).2{f_{Sw}}^2|$ is largest as shown in Fig.3.

However the worst case condition cannot simply be selected for the highest value on the graph. There is a constrained set of combinations of R_L and f_{Swt} that will lead to a specific combination of current and voltage for each load. So, the following equation should be solved in order to find pairs of possible load and switching frequency:

$$\frac{V_{oRL}}{V_{inRL}} = \frac{1}{2.n_{wid}}\left|\frac{\omega^2 L_m.R_e.C_r}{j(\omega^3 L_r L_m - \omega.L_m) + (\omega^2(L_m.R_e.C_r + L_m.R_e) - R_e)}\right| \quad (7)$$

Where V_{oRL} and V_{inRL} are values of output and input voltage for each specific load respectively. V_{oRL} and V_{inRL} are constant values to satisfy the regulated output voltage for each input voltage to the system.

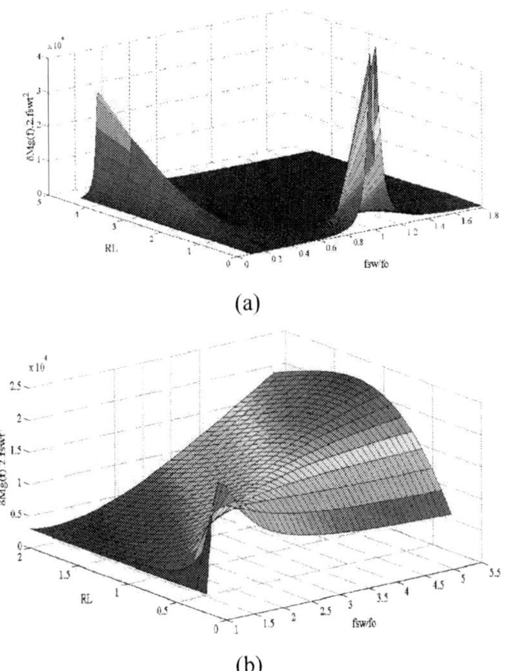

Fig. 3: (a) Gain characteristic $|\Delta M_g(f).2f_{sw}{}^2|$ is for continuously changing values of load and switching frequency. (b) Zoomed-in view of the region of practical operation.

III. DESIGN EXAMPLE, EXPERIMENTAL VERIFICATION AND EXTENSION OF THE MODEL

In this section accuracy of model is proven experimentally using a 400 V to 12 V, 300 W commercial LLC resonant converter prototype [20]. Also, an extension of the model to nonlinear loads is presented. The prototype operates in the frequency range above f_o (184.7 KHz). In the prototype an existing analog controller is replaced with an FPGA-based implementation, similar to the one presented in [11]. The parameters of the LLC filter are C_r=22.5nF, L_r=33µH and L_m=520 µH. Quality factor of series resonant circuit is $Q = \frac{\sqrt{L_r/C_r}}{R_e}$ and 8 bits ADC is used.

A. Accuracy of first harmonic approximation

The following subsection verifies accuracy of FHA to avoid limit cycling under all the operating conditions of examined LLC converter. The converter operation frequency and maximum time step required by the VCO to avoid LCO have been compared for two different cases. In the first scenario, those are obtained theoretically using the FHA as shown in Table 1. Table 2 shows the required converter switching frequency and maximum VCO time step obtained from a realistic simulation model. Results of comparison shows for the given operating points and range of load, FHA results in fairly accurate estimation of digital circuitry minimum required resolution. Table 1 and 2 show result of comparison between minimum resolutions acquired by two different methods.

Table 1: Possible switching frequencies and corresponding resolutions obtained for different load resistances to keep the output voltage at 12*v* using FHA.

Load *A* (*Q*)	Switching Frequency (Matlab)	Maximum VCO Time Step (Matlab)
12A(Q=0.18)	$1.435f_o$	78*ns*
3A(Q=0.046)	$1.394f_o$	183*ns*
1.2A(Q=0.0185)	$1.698f_o$	193*ns*

Table 2: Possible switching frequencies and corresponding resolutions obtained for different load resistances to keep the output voltage at 12*v* using non-ideal LLC simulation.

Load *A* (*Q*)	Switching Frequency (Non Ideal LLC)	Maximum VCO Time Step (Realistic LLC)
12A(Q=0.18)	$1.178f_o$	93*ns*
3A(Q=0.046)	$1.340f_o$	133*ns*
1.2A(Q=0.0185)	$1.447f_o$	146*ns*

B. Resolution for the worst case non-constant load operation

In this section of case study, the values are selected to correspond to the experimental system. For each load resistance two possible solutions exist, and solution above f_o is chosen to take benefits of ZVS of the primary side switches. Each different frequency requires different resolution of time step to prevent LCO. Table3 shows maximum time step of counter required to make sure each of these operating frequencies will be clear of LCO.

Table 3. Possible switching frequencies obtained for different load resistances to keep the output voltage at 12*v*

Load *A* (*Q*)	Switching Frequency	Maximum VCO Time Step
10A(Q=0.5)	$1.422f_o$	94*ns*
15A(Q=0.22)	$1.394f_o$	75*ns*

To verify the resolution requirement obtained in to table3, experiments have been carried. Resolution of the DPFM is selected to be 83 ns. In the experimental result shown in Fig.4, as an example, operation of the converter for 10 A and 15 A is shown. It can be seen that for 10 A (requiring resolution of 94 ns) no LCO will happen and for 15 A (requiring resolution of 75 ns) the oscillations will occur.

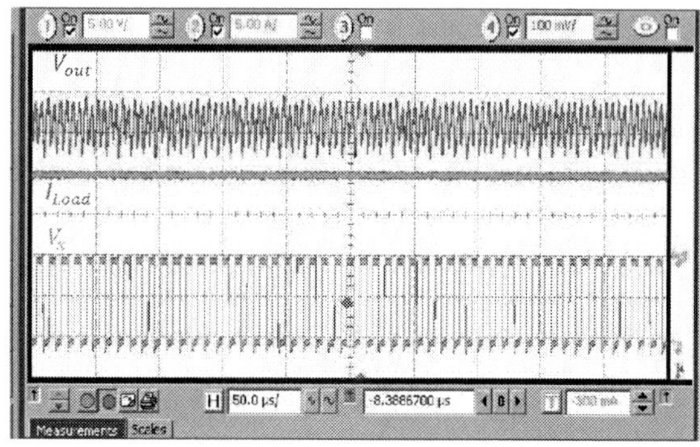

(a)

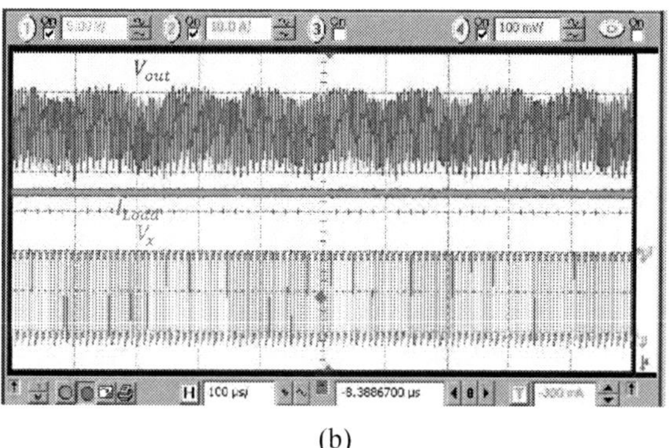

(b)

Fig 4. (a)10A showing no oscillation since maximum time step is larger than time step of system being tested(b) 15A showing oscillation since maximum time step is smaller than resolution of system being tested

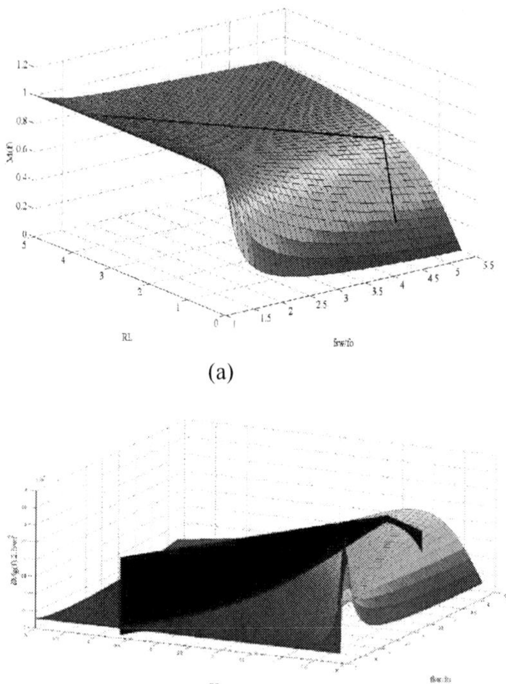

(a)

(b)

Fig 5. Mapping HID IV characteristics onto LLC gain (a) and derivation function (b) to find minimum frequency resolution

C. *model for non linear loads*

Results of previous sections are for linear loads resistances. The same arguments hold true if the load is changing nonlinearly and/or having negative incremental resistances, such as HID lamps [21]. Conventional HID lamps often have a long warm up time which can be as long as several minutes. During this time a high current is drawn to maintain constant power while the initial voltage is low. When the voltage increases to operating point current decreases leading to maintaining constant power resulting in a nonlinear load. Fig. 5 shows mapping of load of HID onto LLC derivation function. The mapping is done through finding corresponding point on the LLC gain and derivation function which results in the same load condition of HID lamp. The highest point on the derivation function is being used to find the finest resolution of DPFM.

According to Fig.5, which is obtained using the analysis from the previous section, change of load has a noticeable impact on the minimum requirement of VCO resolution. Here as an example load changes from 0.35 Ω to 0.546 Ω, 1.4 Ω and 4.11 Ω. For this case of study, the worst case frequency time step can be calculated as 7.3 ns, which ensures that for any combination of load and frequency limit cycling is be avoided.

IV. Conclusion

A general method for finding resolution of the digital pulse frequency modulators (DPFM) in digitally controlled voltage-mode LLC converters is demonstrated. The method takes into account that any change of load simultaneously results in the change of the converter gain and the frequency of operation. It has been shown that for a given input and output voltage conditions only a certain set of combinations of the gain and operating frequency could exist and, based on that set, the resolution of the DPFM can be determined. This method is applicable for both linear and non-linear loads and other types of resonant converters. Validity of the method has been experimentally verified.

References

[1] Prodic, Aleksandar, Dragan Maksimovic, and Robert W. Erickson. "Design and implementation of a digital PWM controller for a high-frequency switching DC-DC power converter." Industrial Electronics Society, 2001. IECON'01. The 27th Annual Conference of the IEEE. Vol. 2. IEEE, 2001.

[2] Peretz, M. M., & Ben-Yaakov, S. (2010). Digital control of resonant converters: resolution effects on limit cycles. Power Electronics, IEEE Transactions on, 25(6), 1652-1661.

[3] Bo, Yang, and F.C. Lee, "LLC resonant converter for front end DC/DC conversion, " in Seventeenth Annual IEEE Applied Power Electronics Conference and Exposition, APEC 2002, 2002. vol.2: pp. 1 108- 1 1 12.

[4] Mattavelli, P.; Rossetto, L.; Spiazzi, G.; Tenti, P., "General-purpose fuzzy controller for DC/DC converters," Applied Power Electronics Conference and Exposition, 1995. APEC '95. Conference Proceedings

[5] R. D. Middlebrook and S. Cuk, "A general unified approach to modeling switching-converter power stages," in Proc. IEEE Power Electron. Spec. Conf., 1976, pp. 36–57.

[6] Weiyi Feng; Lee, F.C.; Mattavelli, P., "Simplified Optimal Trajectory Control (SOTC) for LLC Resonant Converters," Power Electronics, IEEE Transactions on, vol.28, no.5, pp.2415,2426, May 2013.

[7] Fei, Chao, et al. "State-trajectory control of LLC converter implemented by microcontroller." Applied Power Electronics Conference and Exposition (APEC), 2014 Twenty-Ninth Annual IEEE. IEEE, 2014.

[8] Buccella, Concettina, et al. "Comparative transient response analysis of LLC resonant converter controlled by adaptive pid and fuzzy logic controllers." IECON 2012-38th Annual Conference on IEEE Industrial Electronics Society. IEEE, 2012

[9] Patella, B.J.; Prodic, A.; Zirger, A.; Maksimovic, D., "High-frequency digital PWM controller IC for DC-DC converters," Power Electronics, IEEE Transactions on , vol.18, no.1, pp.438,446, Jan 2003

[10] O'Malley, E.; Rinne, K., "A programmable digital pulse width modulator providing versatile pulse patterns and supporting switching frequencies beyond 15 MHz," Applied Power Electronics Conference and Exposition, 2004. APEC '04. Nineteenth Annual IEEE , vol.1, no., pp.53,59 Vol.1, 2004

[11] Amouzandeh, Maryam S.; Mahdavikhah, Behzad; Prodic, Aleksandar; McDonald, Brent, "Auto-tuned minimum-deviation digital controller for LLC resonant converters," inEnergy Conversion Congress and Exposition (ECCE), 2015 IEEE, vol., no., pp.1723-1728, 20-24 Sept. 2015 doi: 10.1109/ECCE.2015.7309903

[12] Xiao, J., Peterchev, A., Zhang, J., & Sanders, S. (2004). An ultra-low-power digitally-controlled buck converter IC for cellular phone applications. In Applied Power Electronics Conference and Exposition, 2004. APEC'04. Nineteenth Annual IEEE (Vol. 1, pp. 383-391). IEEE.

[13] Prodic, Aleksandar, Dragan Maksimovic, and Robert W. Erickson. "Digital controller chip set for isolated DC power supplies." Applied Power Electronics Conference and Exposition, 2003. APEC'03. Eighteenth Annual IEEE. Vol. 2. IEEE, 2003.

[14] Zhao, Zhenyu, and Aleksandar Prodic. "Limit-cycle oscillations based auto-tuning system for digitally controlled DC–DC power supplies." Power Electronics, IEEE Transactions on 22.6 (2007): 2211-2222.

[15] Zhao, Zhenyu, Huawei Li, A. Feizmohammadi, and A. Prodic. "Limit-cycle based auto-tuning system for digitally controlled low-power SMPS." In Applied Power Electronics Conference and Exposition, 2006. APEC'06. Twenty-First Annual IEEE, pp. 1143-1147. IEEE, 2006.

[16] Prodic, Aleksandar, Dragan Maksimovic, and Robert W. Erickson. "Design and implementation of a digital PWM controller for a high-frequency switching DC-DC power converter." Industrial Electronics Society, 2001. IECON'01. The 27th Annual Conference of the IEEE. Vol. 2. IEEE, 2001.

[17] Peterchev, Angel V., and Seth R. Sanders. "Quantization resolution and limit cycling in digitally controlled PWM converters." Power Electronics Specialists Conference, 2001. PESC. 2001 IEEE 32nd Annual. Vol. 2. IEEE, 2001.

[18] Peng, H., Prodić, A., Alarcón, E., & Maksimović, D. (2007). Modeling of quantization effects in digitally controlled dc–dc converters. Power Electronics, IEEE Transactions on, 22(1), 208-215.

[19] Erickson, Robert W., and Dragan Maksimović. Fundamentals of Power Electronics. Norwell, MA: Kluwer Academic, 2001. Print.

[20] UCD3138LLCEVM-028 User's Guide, Digitally Controlled LLC Resonant Half-Bridge DC-DC Converter, (Texas Instruments Literature Number SLUU979A), 2013.

[21] "12V 55W H13-3 Canbus Hid Xenon Kit , Slim Ballast Canbus Hid Fog Light Kit." - Automotivediagnostictool. N.p., n.d. Web. 24 July 2015. http://automotivediagnostictool.sell.curiousexpeditions.org/pz508f72a-12v-55w-h13-3-canbus-hid-xenon-kit-slim-ballast-canbus-hid-fog-light-kit.html

978-1-4673-9551-9/16 $31.00 © 2016 IEEE

Reduction of Storage Capacity in DC Microgrids Using PV-Embedded Series DC Electric Springs

Ming-Hao Wang* and Siew-Chong Tan*
*Department of Electrical and Electronic Engineering
The University of Hong Kong, Hong Kong
Email: mhwang@eee.hku.hk, sctan@eee.hku.hk.

Shu-Yuen (Ron) Hui*†
†Department of Electrical and Electronic Engineering
Imperial College London, U.K.
Email: r.hui@imperial.ac.uk

***Abstract*—In the integration of photovoltaics (PV) with the power grids, battery energy storage systems (BESS) are often adopted to buffer the intermittent solar power. For the PV-battery combination, the capacity of the BESS increases with the scale of PV integration, which increases the cost. In this paper, a PV-embedded series DC electric spring (PVES), which is a new combination of non-critical load, PV, and series DC electric spring (series ES), is proposed as a means for reducing the storage capacity as typically desired in microgrids. A comparison among the BESS, the series ES, and the PVES integrated DC microgrid is provided. Experimental works on a 48 V DC grid with constant and variable load have been performed to verify the storage reduction achievable with the PVES. The results show that the PVES can tackle the intermittency of the solar power with less energy storage capacity.**

***Index Terms*—Smart load, PV system, electric springs, storage reduction, DC grids.**

I. Introduction

The DC distribution grid is a promising grid system for user-end power dispatch due to its convenient incorporation with distributed renewable generations, plug-in hybrid electric vehicles (PHEVs), and electronic loads [1]. In a PV-integrated DC grid, the battery energy storage system (BESS) is usually installed alongside with the PV to form a PV-battery unit to buffer the intermittent solar power and to stabilize the voltage of critical loads. It can be operated to shave the load peak, reduce the electric bills [2], and improve the efficiency of the grid. However, this comes at the expense of the need of batteries, which ironically are also expensive, and are typically environmentally harmful [3]. The contradiction between reducing battery cost (i.e., using lead-acid battery) and improving environmental friendliness (i.e., using li-ion battery) can only be alleviated through the reduction of the BESS capacity.

Existing battery capacity reduction technologies can be categorised into three approaches, namely (i) optimal sizing and control of BESS, (ii) generation side management (GSM) and (iii) demand side management (DSM). In standalone PV system, achieving full autonomy is the priority. A battery sizing procedure for standalone PV system is provided in IEEE Standard 1013-2007 [4]. For grid-connected PV systems, the optimization of storage sizing with consideration of the time-of-use (TOU) tariff and electric bills reduction is reported in [5]. A cascaded coordinative control for storage capacity reduction is introduced in [6]. In [7], a shared energy storage scheme is proposed to maximize the usage of the storage capacity. These technologies are useful for minimizing the storage capacity and maximizing the cost-effectiveness of the BESS. However, they do not fundamentally reduce the mismatch of solar generation [8] and load demand profile [9]. Thus, the capacity of the BESS is still positively associated with the scale of PV integration [10].

With the GSM approach, the stable power source and the renewable energy source (RES) [11] are controlled to reduce the energy imbalance between the power supply and demand. Consequently, a smaller battery capacity can be used for energy buffering. A generation power scheduling method is reported in [12] to predict the power reference of the RES and utility grid interfaced rectifier to reduce the size and cost of the battery storage. In [13], [14], the implementation of peak solar power curtailment control for avoiding the over-production of solar generation for reducing the required BESS capacity is reported. However, with this method, a reduction of the solar peak generation will cause a reduction of the total solar power generation [13], which means that the utilization of the PV is suboptimal.

With the DSM approach, loads are adaptively manipulated to fit the generation profile so that the BESS can be reduced. Besides typical methods of adopting the TOU tariff and remote switching on/off of loads, the concept of controllable loads has also been reported [15] and heat pump air conditioner is programmed as an energy buffer to reduce the BESS capacity. In [16], it is demonstrated that loads can be programmed into bus regulating unit to compensate the power fluctuation in DC grids. These DSM technologies make use of the flexibility of loads to reduce the mismatch between the generation and loads without hindering the operation of RES. Therefore, they can be considered as cost-effective means of BESS capacity reduction.

The series DC electric springs (series ES) have been proposed in DC grids as a power-electronic-based DSM [17] technology. Each series ES is connected in series with a non-critical load (NCL) to form a smart load which can compensate the voltage variation in the DC bus. It distinguishes itself from other DSM technologies with its on-site voltage support, the independency of ICT, and the real-time manipulation of NCL. The series ES can reduce the mismatch between the power

978-1-4673-9551-9/16 $31.00 © 2016 IEEE

supply and demand profiles through the manipulation of NCL. However, it has to output a significant amount of energy when it is used to suppress the system voltage. This increases its storage capacity [18].

In this paper, a PV-embedded series ES (PVES) is proposed to reduce the storage capacity of PV-integrated DC grids. By shifting part of the PV panels from the grid to the series ES to form a PV-ES-NCL unit, the shifted part of the PV panels can help the series ES to output energy when suppressing the system voltage. Thus, the overall storage capacity can be significantly reduced. The energy exchange statuses of the PVES in (i) boosting-discharging (BD), (ii) boosting-charging (BC), and (iii) suppressing-discharging (SD) mode are analyzed. The results show that the new combination of PV-ES-NCL unit requires less storage capacity than both the original series ES (ES-NCL unit) and the BESS (PV-battery unit).

II. PV-INTEGRATED DC MICROGRID

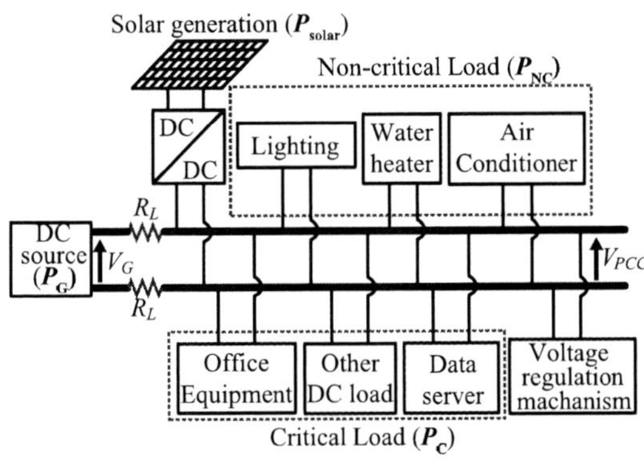

(a) Structure of a PV-integrated DC microgrid

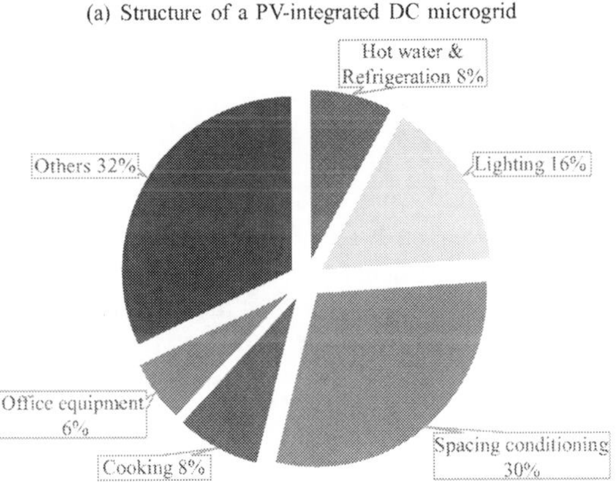

(b) Proportions of commercial energy end-use [19]

Fig. 1. A PV-integrated DC microgrid and commercial energy end-use information.

Fig. 1(a) shows an example of a PV-integrated DC microgrid. The DC source, which can be a bidirectional AC–DC converter or a DC generator, delivers a stable power P_G to the local DC bus through the distribution lines with a resistance of R_L. The DC–DC converter interfaces the PV panels with the DC bus and operates the PVs at the maximum power point (MPP) with maximum power point tracking (MPPT) control. All loads draw power from the DC bus at the point of common coupling (PCC). To buffer the intermittent solar generation and regulate the DC bus voltage, a voltage stabilization mechanism, which can be the BESS, the series ES or the PVES, is installed at the PCC.

A. PV Generation System and Solar Power Profile

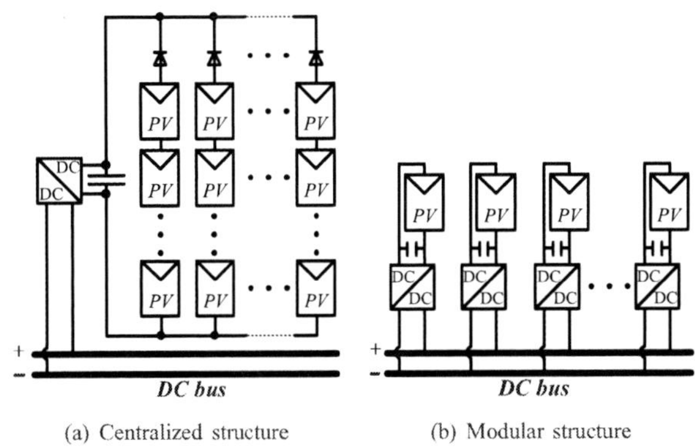

(a) Centralized structure (b) Modular structure

Fig. 2. Configurations of PV generation in DC grid.

The solar generation system can be categorized into centralized structure and modular structure as shown in Fig. 2 [20], [21]. The centralized structure shown in Fig. 2(a) comprises multiple PV strings forming an array. A single DC–DC converter interfaces the array with the DC bus. Although such a PV structure is easy to be implemented, its efficiency can be low when the output of the paralleled PV strings are not at a similar operating point. This also means that the scalability of this structure is low.

In the modular structure that is shown in Fig. 2(b), the PV array is disassembled into individual modules. The DC–DC converter in each module independently operates at its respective MPP. Thus, the overall conversion efficiency of the modular structure is relatively higher than that of the centralized structure. Also, the independency of each module ensures the system's high scalability and plug-and-play feature.

Conventionally, regardless of the kind of structure implemented, all solar power is injected into the DC bus and the PV panels are always operated at MPP. The maximum solar power P_m can be described by

$$P_m = \eta G A, \tag{1}$$

where G represents the solar irradiance in W/m^2 and A represents the surface area of the solar cell in m^2. Thus, the daily solar generation P_{solar} is proportional to the irradiance.

Fig. 3(a) depicts a typical set of solar irradiation patterns that is observable in January and July. Although their peak irradiation are different, the pattern of the profiles are similar and the solar power is mainly generated during 7:00 to 18:00.

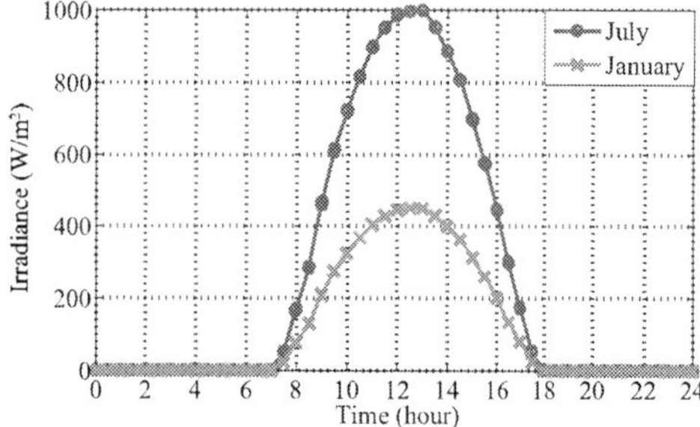

(a) Solar irradiation in July and January [22]

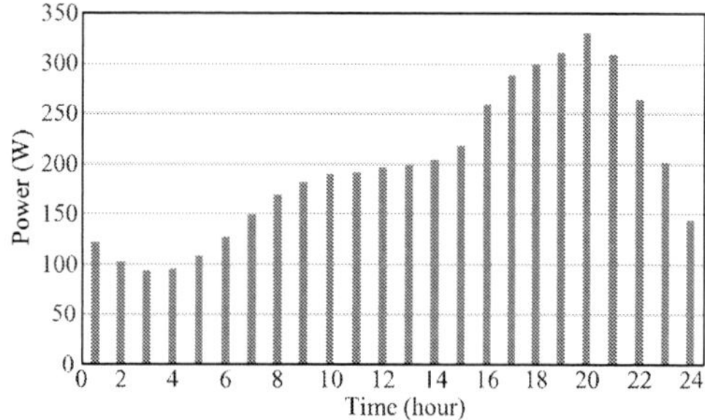

(b) Hourly-average load profile during a day [9]

Fig. 3. Solar irradiation profiles and hourly-average load profile.

B. Load Classification and Load Profile

As shown in Fig. 1(a), the loads can be classified into (i) critical load (CL), of which its applied voltage must be tightly regulated, (e.g., office equipment and server), and (ii) non-critical load (NCL), of which its applied voltage can be varied within a permissible range (e.g., air conditioner and water heater). An example of the proportions of commercial energy end-use is shown in Fig. 1(b). The NCL occupies around 50% of the total energy consumption. The load profiles are affected by human activities and they are regionally alike as shown in Fig. 3(b). The peak load always occurs during 18:00 to 22:00 when P_{solar} is zero.

III. Bus Voltage Stabilization Mechanisms

In Fig. 1(a), it is shown that the stable power supply has a regulated output voltage V_G. With the bus voltage V_{PCC} regulated to the reference V_{PCC_ref} by the voltage regulation mechanism, it delivers a constant power of $P_G = \frac{V_G - V_{PCC\_ref}}{R_L} \times V_G$. Neglecting the power loss on the distribution cable, the fluctuating part of total power can be written as

$$P_{OS}(t) = P_G + P_{solar}(t) - P_C(t) - P_{NC}(t) \ . \tag{2}$$

The mismatch between the solar generation P_{solar} and the load can cause a large variation of P_{OS}. It can be buffered by the BESS, the series ES, and the PVES with the equivalent circuits respectively shown in Fig. 4(a), 4(b), and 4(c).

A. DC Microgrid with the BESS

In Fig. 4(a), it is shown that the BESS is installed as a dedicated shunt auxiliary source. The instantaneous power and

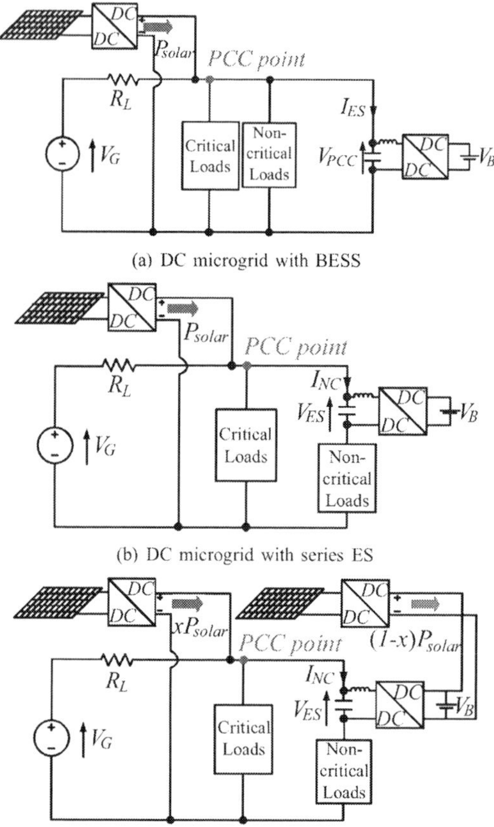

(a) DC microgrid with BESS

(b) DC microgrid with series ES

(c) DC microgrid with PV-embedded series ES (PVES)

Fig. 4. Equivalent circuits of PV-integrated DC microgrids. $1-x$ is the PV shifting coefficient, which quantifies the amount of the original grid-connected PVs that is shifted to the series ES, $0 \leq x \leq 1$.

battery capacity requirement of the BESS can be calculated using equation (3). The BESS stores (or outputs) one unit of power to compensate one unit of power fluctuation. The battery capacity of the BESS must be greater than the maximum energy fluctuation throughout a day. The operating curves of the BESS are illustrated in Fig. 5(a). The operating point of the BESS (Fig. 5(a)-i) moves according to the variation of P_{OS} (Fig. 5(a)-ii). The corresponding power curve and energy profile of the BESS are plotted in Fig. 5(a)-iii and 5(a)-iv. Throughout the day, the BESS has to store energy during the solar peak hour and deliver energy during the load peak hour, which results in the BESS capacity being positively correlated to the PV capacity, i.e., more PV installed leads to a larger BESS capacity.

$$\begin{cases} P_{BESS} = P_{OS} \\ E_{BESS} \geq max[|\int_0^t P_{OS}(t)dt|], \ t \in (0, 24) \end{cases} \tag{3}$$

B. DC Microgrid with the Series ES

As shown in Fig. 4(b), the series ES essentially acts as a controllable voltage source. It adjusts the applied voltage of the

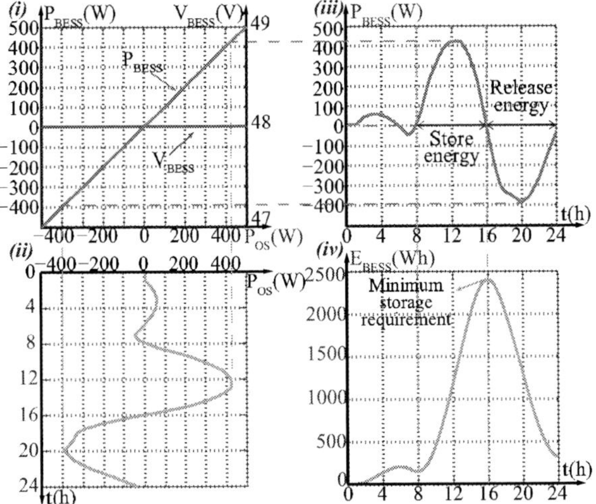

(a) Operating curves of the BESS

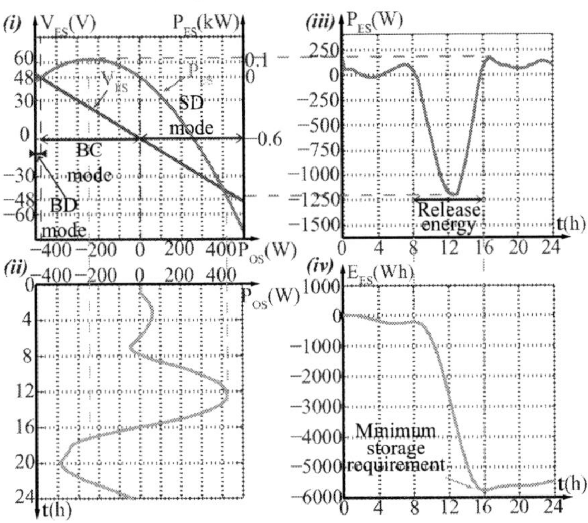

(b) Operating curves of the series ES

Fig. 5. Power and energy curves of the BESS and the series ES.

NCL to manipulate the NCL current I_{NC}. Thus, the series ES and the NCL form a smart load unit to automatically regulate V_{PCC} against the variation of P_{OS}. The series ES can reshape the NCL power profile P_{NC} accordingly to balance the load and the power generation. In this work, the NCL is considered to be purely resistive and passive (e.g., water heater and electric heater). When V_{PCC} is regulated to V_{PCC_ref}, the corresponding P_{ES} and V_{ES} can be derived using equation (4). P_{OS} and the nominal power of the NCL P_{NC_nom} will affect the profile of P_{ES}.

$$\begin{cases} P_{ES} = V_{ES} \times I_{NC} = (P_{OS} + P_{NC\_nom}) - \frac{(P_{OS}+P_{NC\_nom})^2}{P_{NC\_nom}} \\ V_{ES} = V_{PCC\_ref} - \frac{V_{PCC\_ref}(P_{OS}+P_{NC\_nom})}{P_{NC\_nom}} \end{cases} \quad (4)$$

Fig. 5(b)-i shows the operating curves of the series ES. Based on whether the series ES is boosting (when $P_{OS} < 0$) or suppressing (when $P_{OS} > 0$) V_{PCC} and whether its battery is charging or discharging power, the series ES will be operated in boosting-discharging (BD), boosting-charging (BC), and suppressing-discharging (SD) mode. The power exchange rate κ is defined as the ratio between the power of series ES P_{ES} and the change of smart load power ΔP_{SL} as shown in equation (5).

$$\begin{aligned} \kappa &= \left| \frac{P_{ES}}{\Delta P_{SL}} \right| \times 100\% \\ &= \left| \frac{V_{ES} \times I_{NC}}{V_{PCC\_ref}(I_{NC} - I_{NC\_nom})} \right| \times 100\% \\ &= \left| \frac{V_{PCC\_ref} - V_{ES}}{V_{PCC\_ref}} \right| \times 100\% \end{aligned} \quad (5)$$

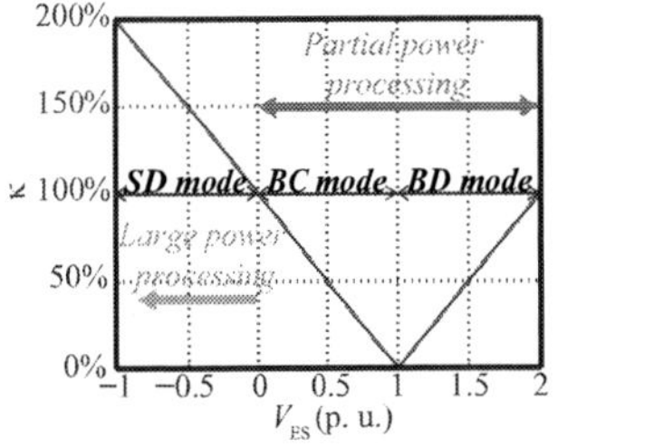

Fig. 6. Power exchange rate of series ES with respect to V_{ES}.

Fig. 6 shows the graph of κ versus V_{ES} in p. u. under the base of V_{PCC_ref}. When the series ES is boosting V_{PCC}, it processes partial power with the ability to shed the NCL. Thus, κ is below 100% and this saves its storage capacity. However, when the series ES is suppressing V_{PCC}, $\kappa > 100\%$. The series ES processes more power than the power change of the smart load, which results in a higher storage requirement for the voltage suppression operation.

As shown in Fig. 5(b), the operating point of the series ES (Fig. 5(b)-i) moves with respect to the variation of P_{OS} (Fig. 5(b)-ii). The corresponding power curve and energy profile of the series ES are shown in Fig. 5(b)-iii and 5(b)-iv. Comparing Fig. 5(a) and 5(b), with the same P_{OS} profile, the series ES processes less power than the BESS during the low power condition (i.e., voltage boosting condition). However, the large discharging power required in SD mode will significantly increase its storage capacity.

C. DC Microgrid with the PV-Embedded Series ES

According to the analysis on the series ES, it is the voltage suppression mode that increases its storage capacity requirement. However, the series ES discharges power under over-voltage condition, which is opposite to that of the BESS. This characteristic offers the possibility to a new solution that facilitates the absorption of excessive solar energy through smaller storage capacity, which is the solution offered through the proposed PV-embedded series ES.

Fig. 4(c) shows the PVES system with $(1-x)$ portion of the total PV panels being shifted to the ES. As discussed in Section II, the centralized structure is inapplicable for PV shifting. The PV modules can be conveniently shifted from the PCC to the ES with a decentralized modular structure. The voltage and power of the PVES can be derived using

$$\begin{cases} P_{PVES} = P_{ES} + P_{solar} = (P_G + P_{solar} - P_C) - \frac{(P_G + xP_{solar} - P_C)^2}{P_{NC\_nom}} \\ V_{PVES} = V_{PCC\_ref} - \frac{V_{PCC\_ref}[P_{OS} + P_{NC\_nom} - (1-x)P_{solar}]}{P_{NC\_nom}} \end{cases} \tag{6}$$

In (6), the variable x represents the proportion of the traditional grid-tied PVs' generation in Fig. 4(c). P_{PVES} is therefore a tunable variable. Due to the reduction of the solar power being injected into the PCC, the mean of P_{OS} is negative and P_{OS} tends to cause more voltage sag. Therefore, on one hand, the possibility that the ES is operated in the BC mode is increased, which makes the ES process partial power to reduce the storage requirement. On the other hand, the reduction of the peak solar power injection will alleviate the over-voltage condition and the ES need not to be operated deeply in the SD mode. Meanwhile, as shown in Table I, the shifted PV panels can supplement the ES with energy in SD and BD mode to reduce the battery power output. However, the ES battery must store energy from both the grid and the shifted PVs in the BC mode, which increases the storage capacity.

IV. Optimization of PV Shifting Coefficient

The minimum storage capacity and the corresponding optimal shifting coefficient $1-x$ can be derived by solving the optimization problem given by

$$\min\{\max[E(t)]\} = \min\left\{\max\left(\left|\int_0^t P_{PVES}dt\right|\right)\right\}. \tag{7}$$

Equation (7) is aimed at minimizing the peak value of the ES battery energy profile. The optimization procedure is shown in Fig. 7. $P_{solar}(t)$ and $P_C(t)$, which can be estimated through weather forecasting and energy end-use history, are input together with $P_G(t)$ into equation (6). Based on these inputs, the battery power curve $P^n_{PVES}(t)$ of the PVES corresponding to a shifting coefficient $1-x_n$ at n time's iteration can be derived using equation (6). In every iteration, $1-x_n$ will be increased by one step, which equals the proportion of the single PV module among the PV system. By comparing the calculated peak value of the energy profile $E^n_{PVES}(t)$, the minimum storage capacity C_{global} and the corresponding $1-x_{global}$ can be found.

V. Experimental Results and Discussions

The experimental circuit topologies of the BESS, series ES, and PVES integrated systems are respectively shown in Fig. 8(a), 8(b), and 8(c). A full-bridge converter TMD-SHV1PHINVKIT (Texas Instrument Inc.) is used for both the BESS and the ES applications. The solar generation is achieved by a grid-interfaced full-bridge converter programmed with a power profile which is proportional to the solar irradiation. The corresponding specifications are shown in Table II. The experiments are divided into two sets.

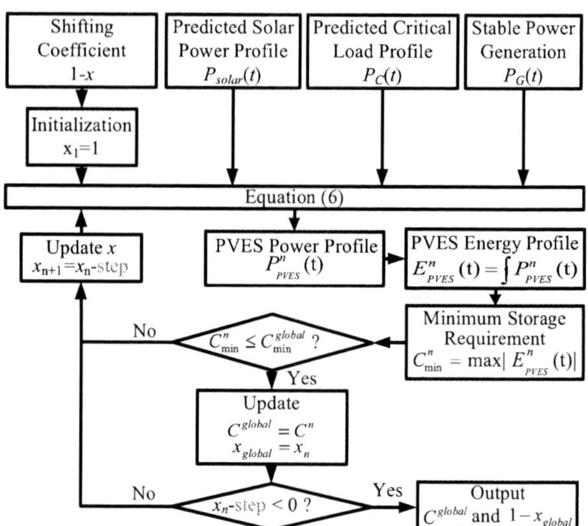

Fig. 7. Flow chart of the storage optimization procedure.

TABLE II
System Parameters in the Experiment of Variable Critical Load

Description	Description	Value
Stable power source	V_G	50.3 V
Nominal PCC voltage	V_{PCC_ref}	48 V
Distribution cable resistance	R_L	0.8 Ω
IGBT switches	$S_{1,2,3,4}$	IRG4PC30FDPBF
Critical load resistance	R_{CL}	10.6 Ω
Non-critical load	R_{NC}	25.4 Ω
Filter capacitor	C_{ES}	21 μF
Filter capacitor	C_{CL}, C_{PV}	1 μF
Filter capacitor	C_{DC}	470 μF
Filter inductor	L_f	3.3 mH
Batteries	V_{PV}	LC-P1228NA (5 × 12 V)
Batteries	V_{B1}	LC-P1228NA (8 × 12 V)
Batteries	V_{B2}	LC-P1228NA (10 × 12 V)
DSP controller		TMS320F28069
Maximum solar power	P_{solar_max}	189.33 W

A. PV System with Constant Critical Load

In this experiment, we use a timescale of 10 seconds to represent 1 hour. To simulate a day, the experiment is conducted for 240 s. As shown in Fig. 9(a), the critical load draws a constant power of 90.76 W. The solar generation follows the profile of daily solar irradiation and power is generated in the time period from 70 s to 180 s. The optimum shifting coefficient $1-x$ is calculated as 31%.

As shown in Fig. 9(b), when the solar power P_{solar} is low in the periods of 0 s to 70 s and 180 s to 240 s, the BESS generates a constant power of around 60 W into the system. Meanwhile, since there is no solar generation during these periods, both the series ES and PVES operate in the BC mode to store a constant power of 24.3 W from the grid system. As shown in Fig. 9(c), the NCL power P_{NC} with both the series ES and PVES are shed to 30 W in the periods from 0 s to 70 s and 180 s to 240 s, while the NCL with BESS dissipates a constant power of 90.7 W. Thus, with the ability to manipulate the NCL, the ES can process less power than the BESS to

TABLE I
ENERGY EXCHANGE STATUS OF A PV-EMBEDDED SERIES ES

Operating Modes	V_{ES}	P_{OS}	P_{solar}	Impact on Storage Capacity
Boosting–discharging	$> V_{PCC}$	$< -P_{NC\_nom}$	Very low	Reduce battery capacity insignificantly
Boosting–charging	$(0, V_{PCC})$	$(-P_{NC\_nom}, 0)$	Low	Increase battery capacity
Suppressing–discharging	< 0	> 0	High	Reduce battery capacity significantly

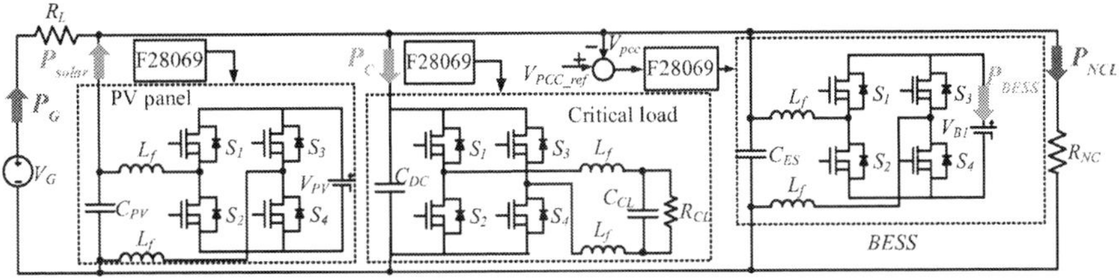

(a) BESS integrated DC system.

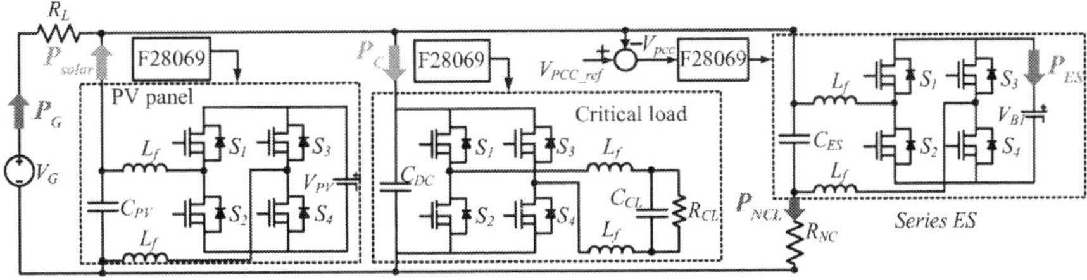

(b) Series ES integrated DC system.

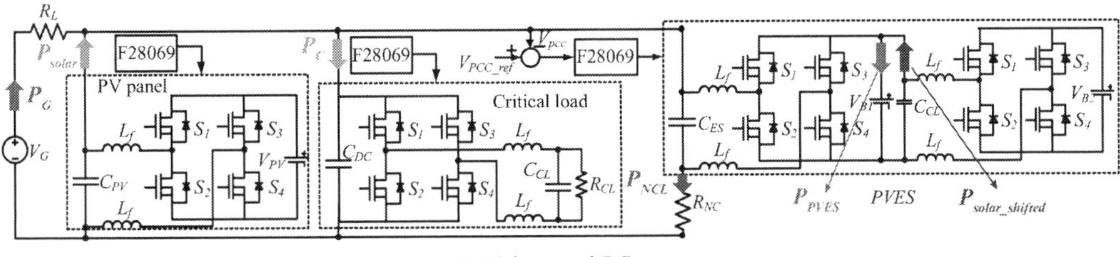

(c) PVES integrated DC system.

Fig. 8. Experimental setup of BESS, series ES and PVES integrated DC grid.

regulate the PCC voltage.

Comparing the power profiles of the series ES and PVES in Fig. 9(b), with the help of the embedded PV, the PVES delivers much less power than the series ES in the SD mode during the solar peak time from 80 s to 170 s. Correspondingly, as shown in Fig. 9(c), the NCL power with the series ES is boosted severely and its peak power is as large as 630.9 W, which is almost 7 times of that of the BESS (90.7 W). The NCL with PVES is not boosted as high as that of the series ES, which means the reduction of grid-connected PV alleviates the over-voltage condition. The NCL peak power with PVES is 305 W, which is almost half of that of the series ES.

Overall, the maximum value of the energy profile for the BESS, the series ES and the PVES are respectively 5.5233 kJ, −20.61 kJ, and 2.0457 kJ as shown in Fig. 9(d). With 31% PV being shifted to ES, not only is the net energy exchange of the PVES regulated to 0 J, but also the storage capacity has been significantly reduced (37.04% of the BESS and 9.93% of the series ES).

B. PV System with Variable Critical Load

In this experiment, the critical load is programmed to vary between 11.2 W and 177.4 W, of which the peak load occurs in the period of 160 s to 220 s, which emulates the daily load profile variation.

Fig. 10 shows a design example based on the data given in Fig. 11(a). It illustrates the relationship of the minimum storage capacity against the change of $1-x$. Clearly, the curve is convex and has a global minimum value of 1710.9 J when $1-x = 48.6\%$. When $1-x < 48.6\%$, the majority of P_{solar} is injected into PCC and it forces the ES to operate deeply in SD mode. Thus, the energy discharging status is dominant and increases the storage capacity. When $1-x > 48.6\%$, the solar power injected into PCC is significantly reduced to operate the ES more frequently in the BC mode and the shifted PVs is kept charging the ES battery. Thus, the energy charging status is dominant and a larger battery is needed to store the solar power.

The shifting coefficient is set at 48.6%. As shown in Fig.

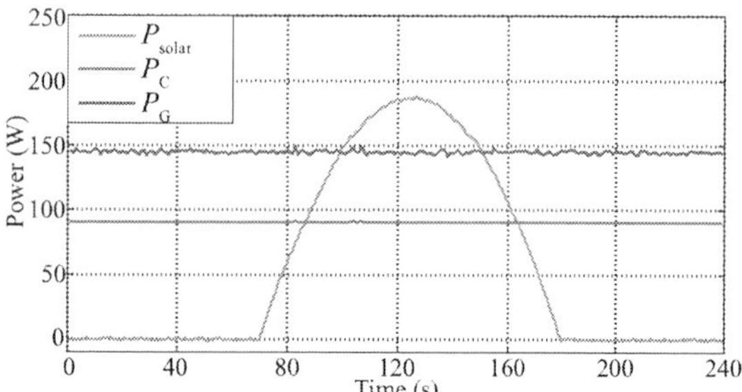

(a) Power profiles of stable power supply, PV panel and critical load

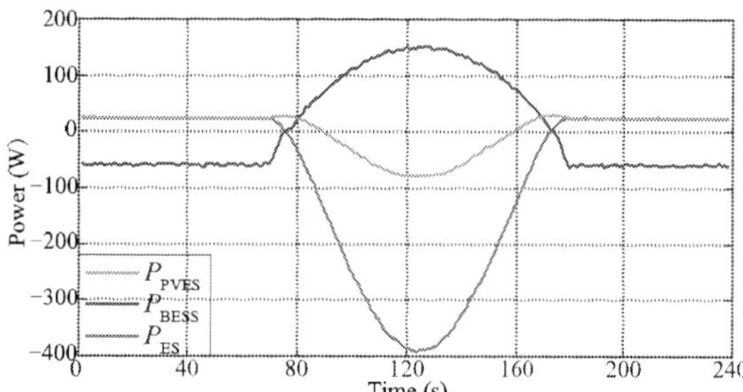

(b) Power profiles of BESS, series ES and PVES

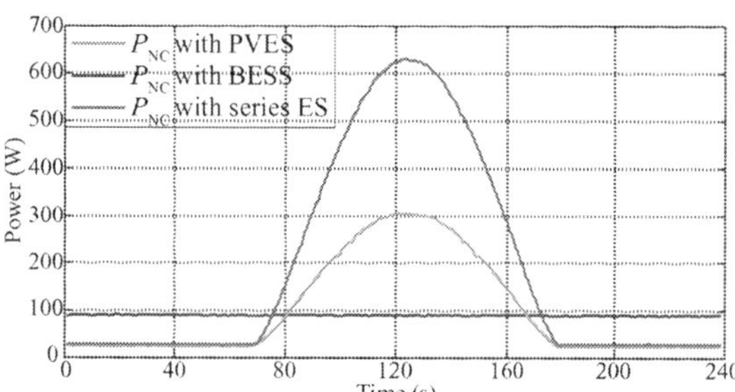

(c) Power profiles of NCL in BESS, series ES and PVES system

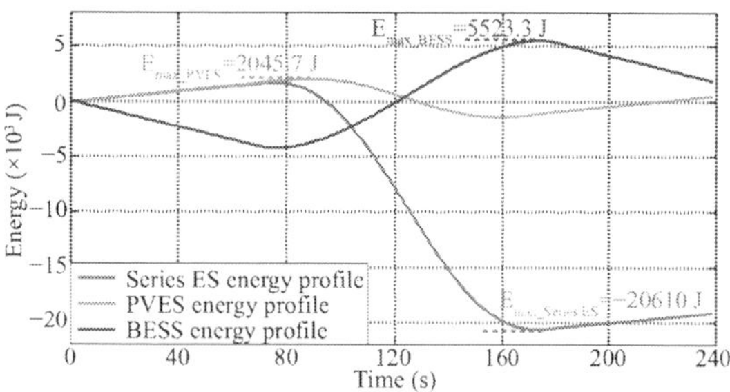

(d) Energy profiles of BESS, series ES and PVES

Fig. 9. Waveforms in the experiment with constant critical load.

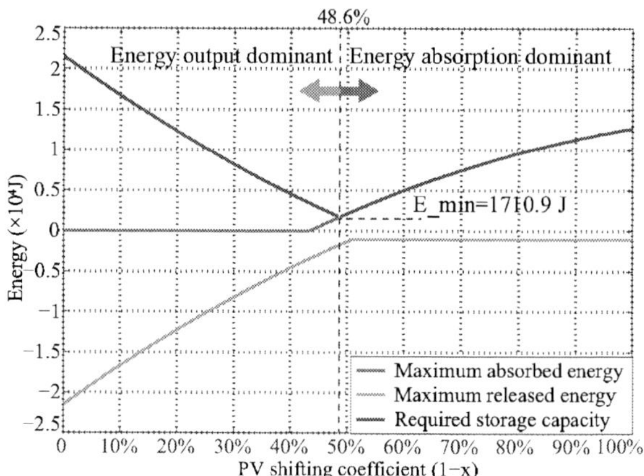

Fig. 10. Storage requirement with respect to PV shifting coefficient $1-x$.

11(b), without the ability to control the load profile, the BESS will passively store the energy during the solar peak time and delivers energy during the load peak time with a maximum charging power of 123.2 W and a maximum discharging power of 104.8 W. For the series ES, in order to boost the NCL to suppress the V_{PCC}, the series ES will operate deeply in the SD mode with a maximum discharging power of 345.3 W. Compared with the series ES, the PVES has the same power profile with the series ES for all time period except during the solar peak time from 70 s to 170 s, when the PVES operates in the suppressing-charging mode with the shifted PVs injecting power to the ES.

Fig. 11(c) shows the NCL power profiles with the BESS, the series ES, and the PVES operating to regulate V_{PCC}. Comparing the three curves, it is clear that both the series and the PVES can reshape the NCL and move the load peak to the solar peak time while the BESS cannot. The maximum NCL with the series ES is 580.6 W, which is larger than that of the PVES (231 W). Also, the accumulated NCL energy with the series ES is 40.23 kJ, which is 186.4% of that of the NCL in the BESS integrated system (21.587 kJ). For the PVES, the accumulative NCL energy is 22.424 kJ, which is 104% of that of the NCL in the BESS integrated system. Thus, the PVES can actually reschedule the power consumption of NCL rather than excessively powering it.

Overall, as shown in Fig. 11(d), the maximum values of the energy profile for the BESS, the series ES and the PVES are respectively 6.43 kJ, −18.341 kJ, and 1.825 kJ. Thus, the PVES can effectively reduce the storage capacity in a DC grid with a variable CL (28.38% of the BESS and 9.95% of the series ES).

VI. Conclusions

In this paper, a PV-embedded series ES (PVES) is proposed to reduce the energy storage capacity required in DC microgrids. A comparison among the BESS, series ES system and PVES system have been performed. The energy exchange status of the PVES is analyzed. The optimization of shifting coefficient is done to minimize the storage capacity. The experiment results prove that the PVES can automatically reschedule the non-critical loads and regulate the PCC voltage with less storage capacity under both the constant CL and the variable CL conditions. The PVES, which is a newly proposed combination of PV-ES-NCL unit, is a cost-effective measure for tackling the intermittency of solar power using a

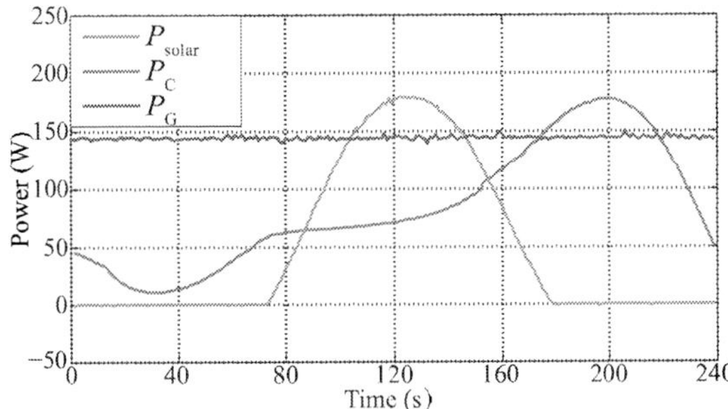

(a) Power profiles of stable power supply, PV panel and critical load

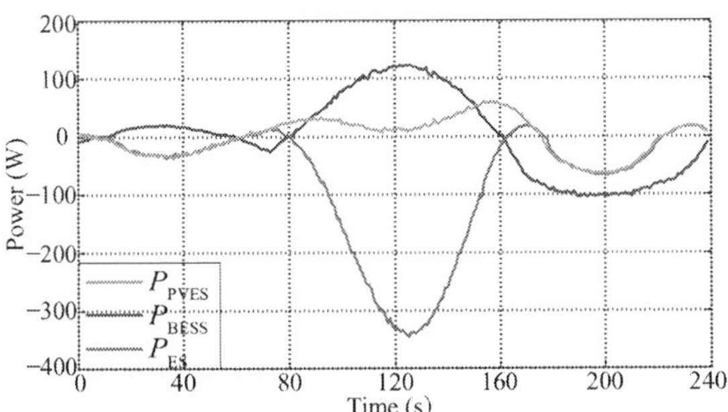

(b) Power profiles of BESS, series ES and PVES

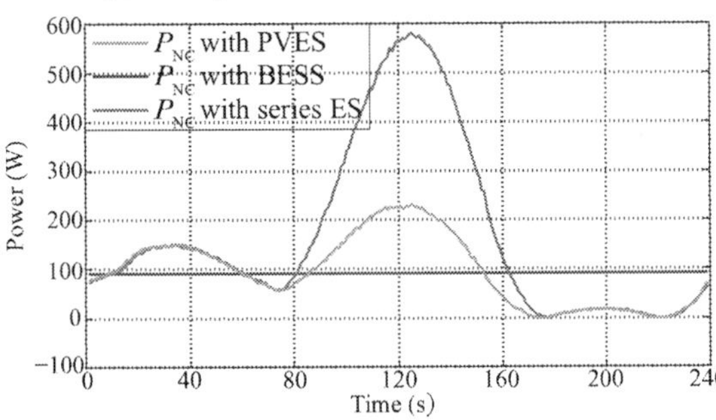

(c) Power profiles of NCL in BESS, series ES and PVES system

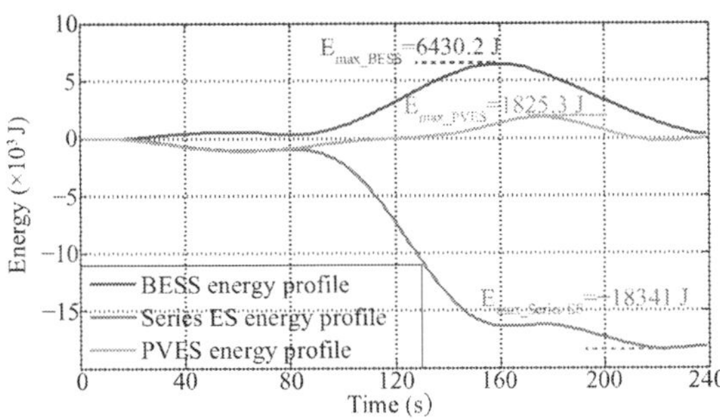

(d) Energy profiles of BESS, series ES and PVES

Fig. 11. Waveforms in the experiment with variable critical load.

significantly smaller energy storage capacity in DC grids.

VII. Acknowledgment

This work was supported by the Hong Kong Research Grant Council under the Theme-based Project T23-701/14-N.

References

[1] G. F. Reed, "DC technologies: solutions to electric power system advancements [guest editorial]," *IEEE Power Energy Mag.*, vol. 10, pp. 10–17, Nov. 2012.

[2] J. Rajasekharan and V. Koivunen, "Optimal energy consumption model for smart grid households with energy storage," *IEEE J. on Sel. Topics Signal Process.*, vol. 8, pp. 1154–1166, Dec. 2014.

[3] J. Matheys, J.-M. Timmermans, J. Van Mierlo, S. Meyer, and P. Van den Bossche, "Comparison of the environmental impact of five electric vehicle battery technologies using LCA," *Int. J. of Sustain. Manuf.*, vol. 1, no. 3, pp. 318–329, 2009.

[4] "IEEE recommended practice for sizing lead-acid batteries for stand-alone photovoltaic (PV) systems - redline," *IEEE Std 1013-2007 (Revision of IEEE Std 1013-2000) - Redline*, pp. 1–58, Jul. 2007.

[5] Y. Ru, J. Kleissl, and S. Martinez, "Storage size determination for grid-connected photovoltaic systems," *IEEE Trans. on Sustain. Energy*, vol. 4, pp. 68–81, Jan. 2013.

[6] S. Kawachi, J. Baba, T. Kikuchi, E. Shimoda, S. Numata, E. Masada, and T. Nitta, "Energy capacity reduction of energy storage system in microgrid stabilized by cascade control system," in *Euro. Conf.on Pow. Electron. and Applica.*, pp. 1–10, Sep. 2009.

[7] Z. Wang, C. Gu, F. Li, P. Bale, and H. Sun, "Active demand response using shared energy storage for household energy management," *IEEE Trans. on Smart Grid*, vol. 4, pp. 1888–1897, Dec. 2013.

[8] E. Kaplani and S. Kaplanis, *Solar Power*, ch. 1: Prediction of solar radiation intensity for cost-effective PV sizing and intelligent energy buildings. Feb. 2012.

[9] J. V. Paatero and P. D. Lund, "A model for generating household electricity load profiles," in *Int. J. Energy Res.*, vol. 30, pp. 273–290, Apr. 2006.

[10] H. Mahmood, D. Michaelson, and J. Jiang, "Decentralized power management of a PV/battery hybrid unit in a droop-controlled islanded microgrid," *IEEE Trans. on Pow. Electron.*, vol. 30, pp. 7215–7229, Dec. 2015.

[11] M. Datta, T. Senjyu, A. Yona, T. Funabashi, and C.-H. Kim, "A coordinated control method for leveling PV output power fluctuations of PV–diesel hybrid systems connected to isolated power utility," *IEEE Trans. on Energy Convers.*, vol. 24, pp. 153–162, Mar. 2009.

[12] A. Luna, N. Diaz, L. Meng, M. Graells, J. Vasquez, and J. Guerrero, "Generation-side power scheduling in a grid-connected DC microgrid," in *IEEE Int. Conf. on DC Microgrids (ICDCM)*, pp. 327–332, Jun. 2015.

[13] Y. Yang, F. Blaabjerg, and H. Wang, "Constant power generation of photovoltaic systems considering the distributed grid capacity," in *Appl. Pow. Electron. Conf. and Expo. (APEC)*, pp. 379–385, Mar. 2014.

[14] A. Ahmed, R. Li, S. Moon, and J.-H. Park, "A fast PV power tracking control algorithm with reduced power mode," *IEEE Trans. on Energy Convers.*, vol. 28, pp. 565–575, Sep. 2013.

[15] S. Kawachi, J. Baba, H. Hagiwara, E. Shimoda, S. Numata, E. Masada, and T. Nitta, "Energy capacity reduction of energy storage system in microgrid by use of heat pump: characteristic study by use of actual machine," in *Int. Pow. Electro. and Motion Control Conf.*, pp. T11–52–T11–58, Sep. 2010.

[16] Y. Gu, X. Xiang, W. Li, and X. He, "Mode-adaptive decentralized control for renewable DC microgrid with enhanced reliability and flexibility," *IEEE Trans. on Pow. Electron.*, vol. 29, pp. 5072–5080, Sep. 2014.

[17] K. T. Mok, M. H. Wang, S. C. Tan, and S. Y. R. Hui, "DC electric springs – an emerging technology for DC grids," in *IEEE Appl. Pow. Electron. Conf. and Expo. (APEC)*, pp. 684–690, Mar. 2015.

[18] M.-H. Wang, K.-T. Mok, S. C. Tan, and S. Y. R. Hui, "Series and shunt DC electric springs," in *Energy Convers. Cong. and Expo. (ECCE)*, pp. 6683–6690, Sep. 2015.

[19] E. M. S. D. of Hong Kong Special Administration Region, "Hong Kong energy end-use data 2015," pp. 43–44, Sep. 2015.

[20] L. Zhang, K. Sun, Y. Xing, L. Feng, and H. Ge, "A modular grid-connected photovoltaic generation system based on DC bus," *IEEE Trans. on Pow. Electron.*, vol. 26, pp. 523–531, Feb. 2011.

[21] K. Sun, L. Zhang, Y. Xing, and J. M. Guerrero, "A distributed control strategy based on DC bus signaling for modular photovoltaic generation systems with battery energy storage," *IEEE Trans. on Pow. Electron.*, vol. 26, pp. 3032–3045, Oct. 2011.

[22] S. Kaplanis and E. Kaplani, "A model to predict expected mean and stochastic hourly global solar radiation i (h; nj) values," *Renewable Energy*, vol. 32, no. 8, pp. 1414–1425, 2007.

A Vector Control Strategy of Grid-Connected Brushless Doubly Fed Induction Generator Based on the Vector Control of Doubly Fed Induction Generator

Sheng Hu
dept. electrical engineering
Wuhan University of Technology
Wuhan, China
haifengxiaochi@163.com

Guorong Zhu
dept. electrical engineering
Wuhan University of Technology
Wuhan, China
zhgr_55@whut.edu.cn

Abstract—A vector control strategy of grid-connected brushless doubly fed induction generator (BDFIG) is proposed in this paper. Compared to other control strategies, proposed control can realize completely decoupled control, doesn't need to evaluate rotor flux, rotor current or control windings flux and has smaller dependency on BDFIG model and parameters, has a very similar controller structure as mature doubly fed induction generator(DFIG) vector control, therefore, it is much more applicable. In order to validate the proposed control strategy, verifications for a 64kW BDFIG are carried out, the results show that the proposed control strategy can effectively realize variable-speed-constant-frequency operation, active and reactive power decoupled control and have good dynamic and stable performance.

Keywords—BDFIG; DFIG; vector control; decoupled control; variable-speed-constant frequency

I. Introduction

Nowadays, doubly fed induction generator (DFIG) is the most employed generator in wind power system due to the fact that it has advantages of variable-speed-constant-frequency-based operation and decoupled control of active and reactive power; moreover, rotor side converter is rated at around 25%–35% of the generator rating [1], which can achieve merits of small size, light weight, low cost, and small losses compared to direct-drive wind power systems with a full-scale power converter. Generally, the stator winding of DFIG is connected directly to the electrical grid and the rotor winding of which is connected to a bidirectional power converter (include machine side converter (MSC) and grid side converter (GSC)). The power rating of the rotor winding, i.e. the converter size, depends on the required speed range and the reactive-power requirements. This fact can be of particular interest in systems with limited speed ranges, such as variable-speed wind turbines. The configuration of grid connected DFIG system is shown in Fig.1 [1]. The main problem is that the slip rings and wound-rotor arrangement render the rotor of a slip-ring doubly fed machine more vulnerable to faults than a cage induction machine. Among other solutions, the use of the so-called brushless doubly fed induction generator (BDFIG) could overcome this problem [2]. Hence, DFIG will be most likely replaced with BDFIG in the future.

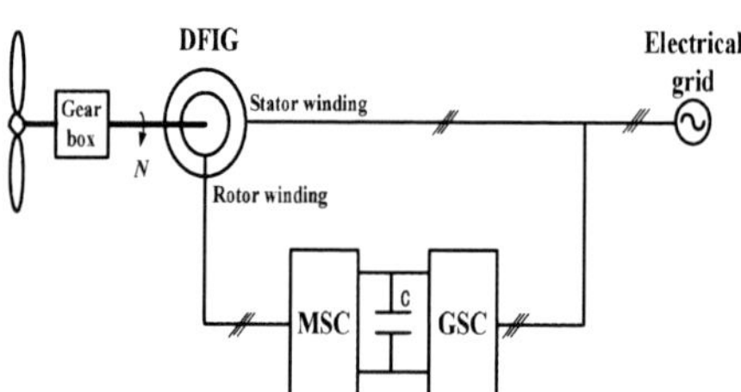

Fig.1. Grid connected DFIG system

The BDFIG is also known as a self-cascaded machine which is composed of two three-phase windings in the stator [called power winding (PW) and control winding (CW)] and a special rotor winding. The configuration of grid connected BDFIG wind power system is shown in Figure 2 [2-5], [7]. The existence of complex electro-magnetic relations between rotor and two stators, high order mathematical model and multiple reference frames related to the two-stator windings

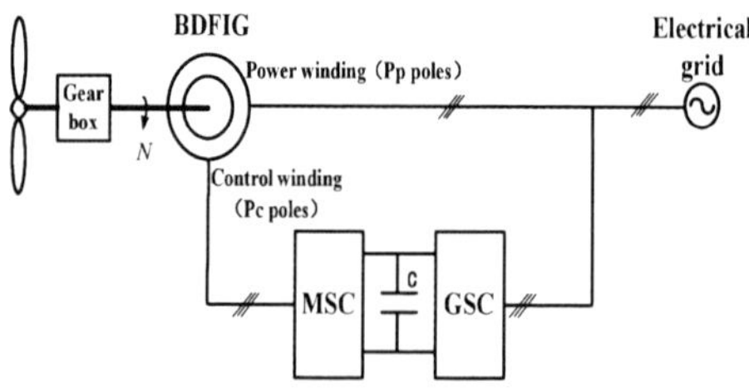

Fig. 2. Grid connected BDFIG system

and the rotor, makes it difficult to exploit the well-known standard induction machine control strategies.

Vector control of BDFIG mentioned in literature so far include vector control based on dual synchronous reference frame[6] and vector control based on unified power winding (PW) flux reference frame[2]-[5], [7]. Vector control based on dual synchronous reference frame [6] highly relies on machine model and increases the complexity and computational burden relative to conventional vector control schemes for standard ac machines. Vector controls based on unified PW flux reference frame [2]-[5], [7] have problems as follows: a) can't realize completely decoupled control of active power and reactive power of PW [2]-[5]; b) the controller structure is complex which make the control strategy has poor practicability [7]; c) rotor flux, rotor current or CW flux need to be estimated, thus, controllers have poor robustness to machine parameters [4]-[5].

A vector control strategy of grid connected BDFIG is proposed in this paper, as a difference with previous solutions, firstly, the structure of proposed control is very similar with the mature controller of DFIG, therefore, it has very good practicability; secondly, because forward control of all the cross perturbations between dq axis is used, completely decoupled control of active and reactive power can be realized; at last, rotor flux, rotor current or CW flux, don't need to be estimated, thus, it has good robustness to machine parameters and model.

In order to validate the effectiveness of proposed vector control, simulations are implemented by using 64kW BDFIG, the results demonstrate propose control can realize variable-speed-constant-frequency control, active and reactive power decoupled control and has good dynamic and stable control performance.

II. UNIFIED REFERENCE FRAME MODEL OF BDFIG AND EQUIVALENCE WITH DFIG

An unified dq model of the BDFIG in the PW flux synchronously rotating reference frame is obtained as [9]

$$\begin{cases} u_{dp} = r_p i_{dp} + p\psi_{dp} - \omega_p \psi_{qp} \\ u_{qp} = r_p i_{qp} + p\psi_{qp} + \omega_p \psi_{dp} \\ u_{dc} = r_c i_{dc} + p\psi_{dc} - \omega_c \psi_{qc} \\ u_{qc} = r_c i_{qc} + p\psi_{qc} + \omega_c \psi_{dc} \\ 0 = r_r i_{dr} + p\psi_{dr} - \omega_{pr} \psi_{qr} \\ 0 = r_r i_{qr} + p\psi_{qr} + \omega_{pr} \psi_{dr} \end{cases} \quad (1)$$

$$\begin{cases} \psi_{dp} = L_{sp} i_{dp} + M_{pr} i_{dr} \\ \psi_{qp} = L_{sp} i_{qp} + M_{pr} i_{qr} \\ \psi_{dc} = L_{sc} i_{dc} - M_{cr} i_{dr} \\ \psi_{qc} = L_{sc} i_{qc} - M_{cr} i_{qr} \\ \psi_{dr} = L_r i_{dr} + M_{pr} i_{dp} - M_{cr} i_{dc} \\ \psi_{qr} = L_r i_{qr} + M_{pr} i_{qp} - M_{cr} i_{qc} \end{cases} \quad (2)$$

However, the mechanical equation is:

$$\frac{d\omega_r}{dt} = \frac{1}{J}(T_m - T_e - K_d \omega_r) \quad (3)$$

The electromagnetic torque can be expressed as:

$$T_e = p_p M_{pr}(i_{qp} i_{dr} - i_{qp} i_{dr}) - p_c M_{cr}(i_{qc} i_{dr} + i_{dc} i_{dr}) \quad (4)$$

The active and reactive powers of the PW can be expressed as

$$\begin{cases} P = \frac{3}{2}(u_{qp} i_{qp} + u_{dp} i_{dp}) \\ Q = \frac{3}{2}(u_{qp} i_{dp} - u_{dp} i_{qp}) \end{cases} \quad (5)$$

Where p_p, r_p, L_{sp} and ω_p represent pole pair, resistance, self-inductance and angular frequency of PW, M_{pr} and M_{pr} represents mutual inductance between PW and rotor; p_c, r_c, L_{sc}, M_{cr}, ω_c represent pole pair, resistance, self-inductance and angular frequency of CW; r_r, L_r, ω_{pr} represent resistance, self-inductance and angular frequency of rotor; M_{pr} and M_{pr} represents mutual inductance between PW and rotor, u_{qp}, u_{dp}, ψ_{qp}, ψ_{dp}, i_{qp}, i_{dp}, u_{qc}, u_{dc}, ψ_{qc}, ψ_{dc}, i_{qc}, i_{dc}, ψ_{qr}, ψ_{dr}, i_{qr} and i_{dr} represent dq component of voltage, flux ,current of PW, CW and rotor, respectively; P and Q represent active power and reactive power of PW; Tm represents mechanical torque; T_e represents electromagnetic torque; ω_r represents mechanical angular speed of rotor; J represents inertia; K_d represents friction coefficient; subscripts "p" and "c" denote PW and CW, respectively; subscripts "s" and "r" denote stator and rotor; p represent differential operator.

The BDFIG model expressed from (1) ~ (5) in unified PW flux reference frame can be depicted in Fig. 3 [2-5], [7].

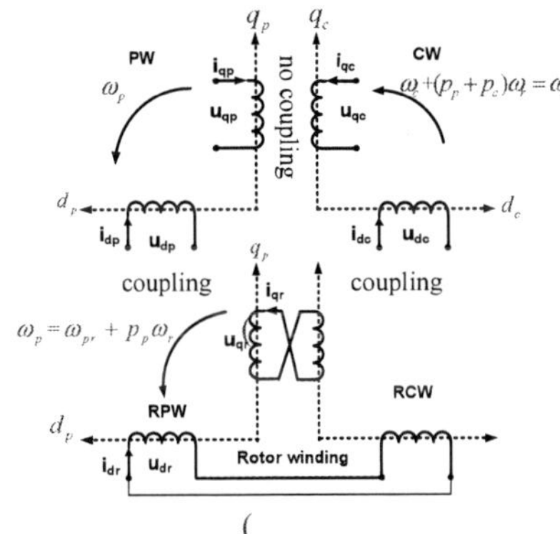

Fig.3. BDFIG model in unified PW flux reference frame

Compared Fig. 1 with Fig. 2, we know that the function of power winding (PW), control winding (CW) and rotating speed of BDFIG is in correspondence to the function of stator, rotor and rotating speed of DFIG, respectively. However, BDFIG is composed of three windings, namely, PW, CW and rotor windings. Besides, there is no any coupling between PW and CW, both PW and CW have coupling with rotor windings, separately [2-7]. DFIG just has two windings, i.e. stator winding and rotor winding and they are coupled [8]-[10]. In addition, the BDFIG model shown in Fig. 3 are much more complex than DFIG model shown in Fig. 4. Therefore, in order to make vector control strategy of BDFIG refer to the design of mature vector control of DFIG, the physical quantities of rotor windings in (1)~(5) should be eliminated by mathematical derivation, in this way, the equivalent coupling model between PW and CW in unified PW flux reference frame can be established. In order to simplify the derivation, the selected d axis is aligned with the PW-flux orientation, if r_p is neglected, according to (6) and (7), the steady-state component of the stator flux during grid faults can be expressed as

$$\begin{cases} \psi_{dp} = \psi_{sp} \\ \psi_{qp} = 0 \end{cases} \tag{6}$$

$$\begin{cases} u_{dp} \approx p\psi_{sp} = 0 \\ u_{qp} \approx \omega_p \psi_{sp} = u_{sp} \end{cases} \tag{7}$$

Where ψ_{sp} represents the amplitude of PW flux, u_{sp} represents the amplitude of PW flux. Hence, from (2), we can obtain:

$$\begin{cases} \psi_{sp} = L_{sp} i_{dp} + M_{pr} i_{dr} \\ 0 = L_{sp} i_{qp} + M_{pr} i_{qr} \\ \psi_{dc} = L_{sp} i_{dc} - M_{cr} i_{dr} \\ \psi_{qc} = L_{sp} i_{qc} - M_{pr} i_{qr} \end{cases} \tag{8}$$

According to (8), the dq component of CW flux can be derived as:

$$\begin{cases} \psi_{dc} = L_{sc} i_{dc} - \dfrac{M_{cr}}{M_{pr}} \psi_{sp} + \dfrac{M_{cr} L_{sp}}{M_{pr}} i_{dp} \\ \psi_{qc} = L_{sc} i_{qc} + \dfrac{M_{cr} L_{sp}}{M_{pr}} i_{qp} \end{cases} \tag{9}$$

By substituting (9) into (1), we can obtain

$$\begin{cases} u_{dc} = r_c i_{dc} + L_{sc} p i_{dc} + \dfrac{M_{cr} L_{sp}}{M_{pr}} p i_{dp} - \omega_c L_{sc} i_{qc} - \omega_c \dfrac{M_{cr} L_{sp}}{M_{pr}} i_{qp} \\ u_{qc} = r_c i_{qc} + L_{sc} p i_{qc} + \dfrac{M_{cr} L_{sp}}{M_{pr}} p i_{qp} + \omega_c L_{sc} i_{dc} - \omega_c \dfrac{M_{cr}}{M_{pr}} \psi_{sp} + \omega_c \dfrac{M_{cr} L_{sp}}{M_{pr}} i_{dp} \end{cases} \tag{10}$$

Above equation indicates the relation between CW voltage and CW current, therefore, CW current can be controlled by controlling CW voltage, in comparison with DFIG, the relation shown in (10) is very similar to the relation between rotor voltage and rotor current of DFIG [9].

In addition, according to (8) and (9), rotor flux can be derived, then, by substituting rotor flux into the fifth and sixth equation in (1), we can obtain:

$$\begin{cases} M_{cr} p i_{dc} = M_{cr} \omega_{pr} i_{qc} + L_M \omega_{pr} i_{qp} - (L_M p i_{dp} + \dfrac{r_r L_{sp}}{M_{pr}} i_{dp}) + \dfrac{r_r}{M_{pr}} \psi_{sp} \\ M_{cr} p i_{qc} = -M_{cr} \omega_{pr} i_{dc} - L_M \omega_{pr} i_{qp} - (L_M p i_{qp} + \dfrac{r_r L_{sp}}{M_{pr}} i_{qp}) + \dfrac{\omega_{pr} L_r}{M_{pr}} \psi_{sp} \end{cases} \tag{11}$$

Where $L_M = (L_r L_{sp} / M_{pr}) - M_{pr}$. By combing (11) and (10), equation (10) becomes

$$\begin{cases} u_{dc} = r_c i_{dc} + \sigma L_{sc} p i_{dc} + K_d \\ u_{qc} = r_c i_{qc} + \sigma L_{sc} p i_{qc} + K_q \end{cases} \tag{12}$$

In which

$$\sigma = 1 - \frac{M_{cr}^2 L_{sp}}{M_{pr} L_M L_{sc}} \tag{13}$$

$$K_d = -(\omega_c L_{sc} - \frac{\omega_{pr} M_{cr}^2 L_{sp}}{M_{pr} L_M}) i_{qc} - \frac{(\omega_c - \omega_{pr}) M_{cr} L_{sp}}{M_{pr}} i_{qp} - \frac{r_r L_{sp}^2 M_{cr}}{M_{pr}^2 L_M} i_{dp} + \frac{r_r M_{cr} L_{sp}}{M_{pr}^2 L_M} \psi_{dp} \tag{14}$$

$$K_q = (\omega_c L_{sc} - \frac{\omega_{pr} M_{cr}^2 L_{sp}}{M_{pr} L_M}) i_{dc} + \frac{(\omega_c - \omega_{pr}) M_{cr} L_{sp}}{M_{pr}} i_{dp} - \frac{r_r L_{sp}^2 M_{cr}}{M_{pr}^2 L_M} i_{qp} + (\frac{\omega_{pr} L_r M_{cr} L_{sp}}{M_{pr}^2 L_M} - \frac{\omega_c M_{cr}}{M_{pr}}) \psi_{dp} \tag{15}$$

Furthermore, in stable state, equation (11) becomes

$$\begin{cases} i_{qc} = -\dfrac{L_M}{M_{cr}} i_{qp} + \dfrac{r_r L_{sp}}{\omega_{pr} M_{pr} M_{cr}} i_{dp} - \dfrac{r_r}{\omega_{pr} M_{pr} M_{cr}} \psi_{sp} \\ i_{qc} = -\dfrac{L_M}{M_{cr}} i_{dp} + \dfrac{L_r}{M_{pr} M_{cr}} \psi_{sp} - \dfrac{r_r L_{sp}}{\omega_{pr} M_{pr} M_{cr}} i_{qp} \end{cases} \tag{16}$$

Generally, PW rotor slip ω_{pr} has large values along the rotor rotational velocity range and rr has small values. Evaluating the different factor values of (16), it can be concluded that terms containing ω_{pr} are negligible. Thus, the following approximate expressions can be defined:

$$\begin{cases} i_{qc} = -\dfrac{L_M}{M_{cr}} i_{qp} \\ i_{dc} = -\dfrac{L_M}{M_{cr}} i_{dp} + \dfrac{L_r}{M_{pr} M_{cr}} \psi_{sp} \end{cases} \tag{17}$$

Equation (17) indicates the relation between PW current and CW current, according to (17), *dq* current component of PW current can be controlled by controlling *dq* current component of CW current, respectively.

By positioning positive flux vector on d axis rotating at the speed of ω_p and according to (5), (6) and (7), active and reactive power of PW are derived as

$$\begin{cases} P = \dfrac{3}{2} u_{qp} i_{qp} \\ Q = \dfrac{3}{2} u_{qp} i_{dp} \end{cases} \tag{18}$$

Thus, according to (17) and (18), active and reactive power of PW can be controlled by controlling *q* and *d* component of CW current, separately. For grid connected BDFIG, PW voltage is determined by power grid, therefore, control target of BDFIG is active and reactive power exchanged with power grid. The reference of active power (or electromagnetic torque, rotating speed) can be given according to the requirement of power system or target based on MPPT(maximum power point track), the reference of reactive power can be given according to the requirement of reactive compensation of power system. Using (17) and (18), active power and reactive power can be expressed as

$$\begin{cases} P = -\dfrac{3}{2} \dfrac{M_{cr}}{L_M} u_{qp} i_{qc} \\ Q = -\dfrac{3}{2} \dfrac{M_{cr}}{L_M} u_{qp} (i_{dc} - \dfrac{L_r}{M_{pr} M_{cr}} \psi_{sp}) \end{cases} \tag{19}$$

Equation (19) indicates the relation between CW current and PW power, therefore, PW power can be controlled by controlling CW current, in comparison with DFIG, the relation shown in (19) is also very similar to the relation between rotor current and stator power of DFIG [9]. Hence, combining (12), (17), (18) and (19), the mathematical model of grid connected BDFIG can be shown in Fig. 5.

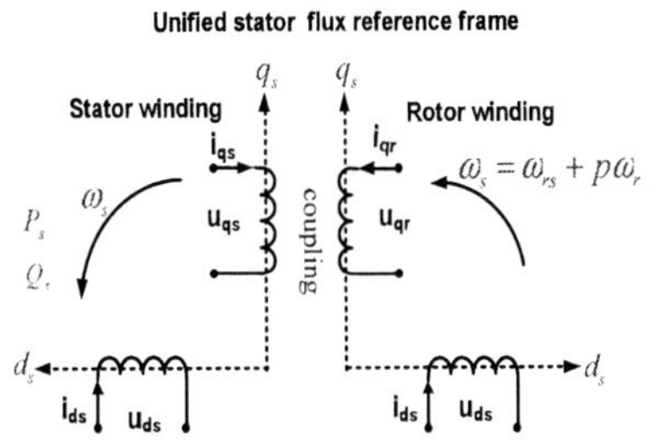

Unified stator flux reference frame

Fig. 4. DFIG model in stator flux

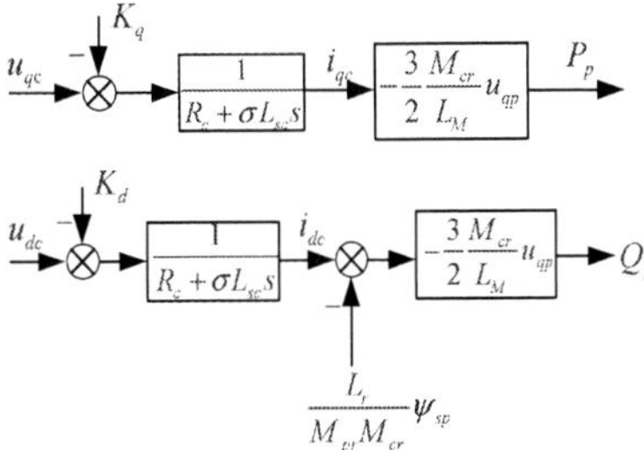

Fig. 5. Mathematical model of grid connected BDFIG

Unified PW flux reference frame

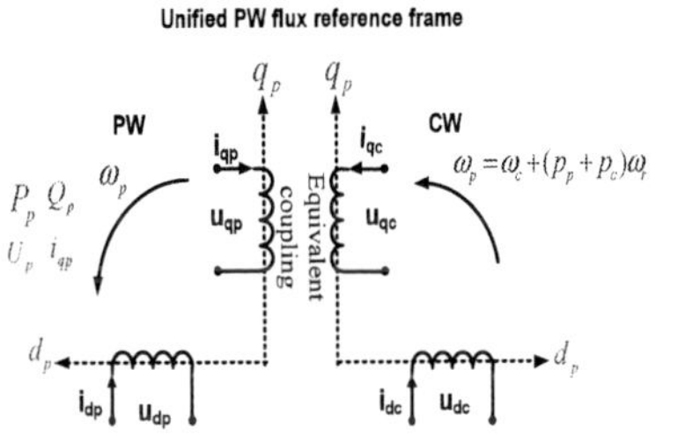

Fig. 6. Equivalent simplified model of BDFIG reference frame in PW flux reference frame

The BDFIG model shown in Fig. 5 in unified PW flux reference frame can be depicted in Fig. 6. Compared Fig. 6 with Fig. 4, it is concluded that the equivalent model of BDIFG in unified flux reference frame is very similar to the DFIG model in stator flux reference frame. Hence, vector control strategy of BDFIG can refer to the design of mature vector control of DFIG [8]-[10].

III. VECTOR CONTROL OF BDFIG BASED ON DFIG CONTROL

Based on Fig. 4~Fig.6 and made vector control of DFIG as a reference, the vector controller of BDFIG in PW flux synchronously frame are designed as Fig. 7. The controller includes two parts: a) PW active power P_p outer loop and CW current i_{qc} inner loop. b) PW reactive power Q_p outer loop and CW current i_{dc} inner loop.

In Fig. 7, K_q and K_d are crossing decoupled terms shown in (14) and (15) which is applied feed-forward compensation. In this manner, dq component of CW current and active and reactive power of PW can realize decoupled control. In

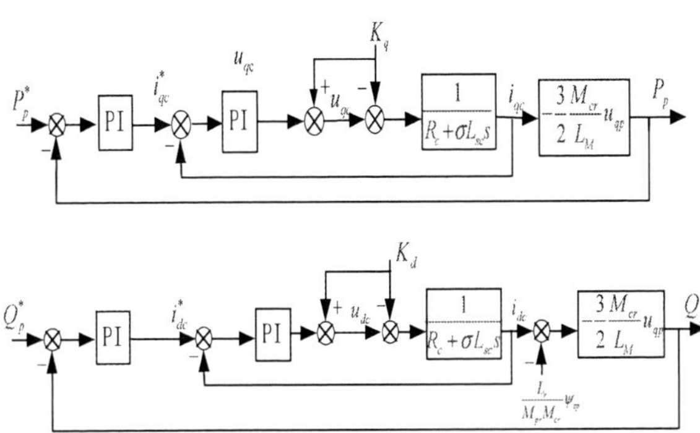

Fig.7. Block diagram of grid-connected BDFG controller in PW synchronously rotating frame

addition, dynamic control performance can be also improved.

To sum up, BDFIG and DFIG have following common points:

1) Variable speed constant frequency operation, the function of PW,CW and rotating speed of BDFIG agree with the function of stator, rotor and rotating speed of DFIG, respectively.

2) Control target for grid connected operation. Both of them control active power and reactive power which exchange with power grid.

3) Mathematical model of BDFIG is very similar to mathematical model of DFIG.

Considering above common points of BDFIG and DFIG and making reference to vector control design of DFIG, the complete vector control scheme is proposed, which is shown in Fig. 8. Three phase PW voltages, currents, and three-phase CW currents need to be measured.

An encoder is also required to provide rotor position information for the dq transformation. PW voltages and currents are used to estimate the angle of the forward sequence of the PW flux linkage through the PLL. Two dual PI controllers, namely, the active power and reactive controllers regulate active and reactive power of PW, respectively. After a 2/3 transformation, the three-phase CW reference voltage $u_{\alpha c}^*$, $u_{\beta c}^*$ is established as the input to the pulse width modulation (PWM) generator, which produces switching signals for the converter driving the CW of the BDFIG.

Considering above common points of BDFIG and DFIG and making reference to vector control design of DFIG [8]-[10], the complete vector control scheme is proposed, which is shown in Fig. 7 and has a very similar structure to the mature vector control of DFIG [8]-[10].

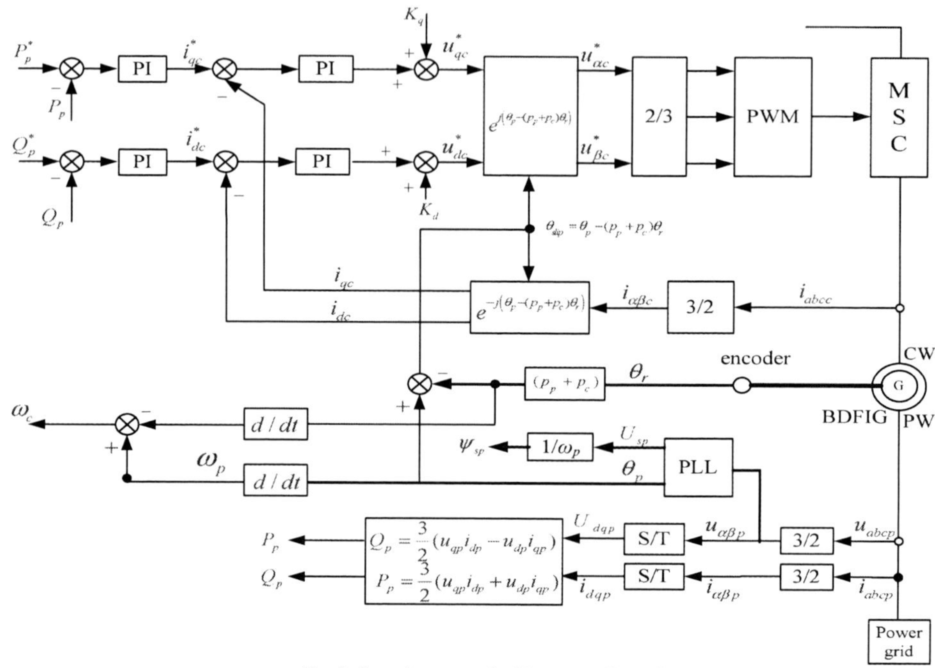

Fig. 8. Control strategy of grid-connected BDFG

IV. VECTOR CONTROL VERIFICATION

In order to evaluate the performance of the proposed scheme, simulations have been carried out by using a 64kW BDFIG, parameters of employed BDFIG are given as below:

Rated PW power is 64kW; rated PW voltage is 380V; p_p=2,r_p=0.07726Ω,L_{sp}=111.5mH,M_{pr}=143.2mH;p_c=4,r_c=0.1023 4Ω, L_{sc}=41.3mH, M_{cr}=44.3mH; r_r=0.174Ω, L_r=237 mH。

Parameters of CW current loop: K_{pi}= 48, K_{ii}= 816; Parameters of PW power loop: K_{pi}= 48, K_{ii}= 816; amplitude of grid voltage: 311V; frequency: 50Hz. This paper focuses on the control algorithm of the machine-side converter, assuming that the grid-side converter stabilizes the dc-link voltage well enough during grid connected operation.

Fig. 9~Fig. 18 show the control responses with a step of active power reference at 3.5s and reactive power reference at 4.4s. Simulation condition is as follows: rotating speed begins to vary from sub-synchronous state to super-synchronous state at 3s, rotor rotating speed is initially set to 450r/min and finally set to 600r/min. Active and reactive power of grid connected BDFIG is initially set to 20kW and 0kvar, respectively. Active power steps up to 40kW at 3.5 and reactive power steps up to 20kvar at 4.4s. Fig. 9 shows grid voltage u_{gabc}; Figure 10 shows PW current; Figure 11 shows phase a voltage and phase a current of PW; Fig. 12 shows *dq* current of PW; Fig. 13 shows three phase current of CW; Fig. 14 shows dq current of CW; Fig. 15 shows active power of PW; Fig. 16 shows reactive power of PW; Fig. 17 shows active power of CW; Figure 18 shows reactive power of CW; Fig. 19 shows three phase current of rotor i_{rabc}; Fig. 20 shows rotor rotating speed.

As can be seen from Fig. 11, Fig. 15 and Fig. 16, BDFIG only outputs active power before 3.5s. According to Fig.15, PW active power of BDFIG tracks with the reference well in stable state when active power steps up from 20kW to 40kW at 3.5s and the process of dynamic transition is very fast. From Fig.16, reactive power of BDFIG also tracks with the reference well in stable state when reactive power step up from 0kvar to 20kvar at 4.4s and the process of dynamic transition is also very fast, the value of reactive power is negative, which indicates BDFIG outputs inductive reactive power to electrical grid. Moreover, it is can be seen from Fig. 12, Fig. 15 and Fig. 16 that active and reactive power can be controlled separately, hence, decoupled control of active and reactive power is realized with proposed vector control scheme shown in Fig. 8. According to Fig. 11, PW phase voltage lagged behind PW phase current after 3.5s, this also indicates BDIFG output inductive reactive power. As can be seen from Fig. 13, the phase of CW current i_a, i_b and i_c is changed during process of sub-synchronous state to super-synchronous state, in addition, according to Fig. 14, the amplitude of CW current increases

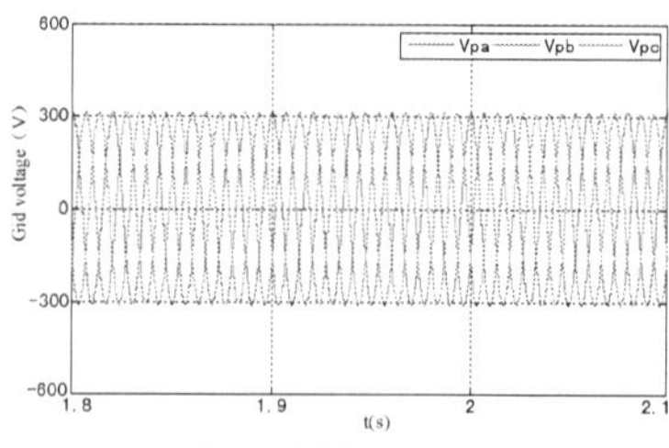

Fig. 9. Grid voltage

after a step of inductive power. It is can be seen from Fig. 17 and Fig. 18 that active and reactive power vary from positive to negative when BDFIG transits from the sub-synchronous

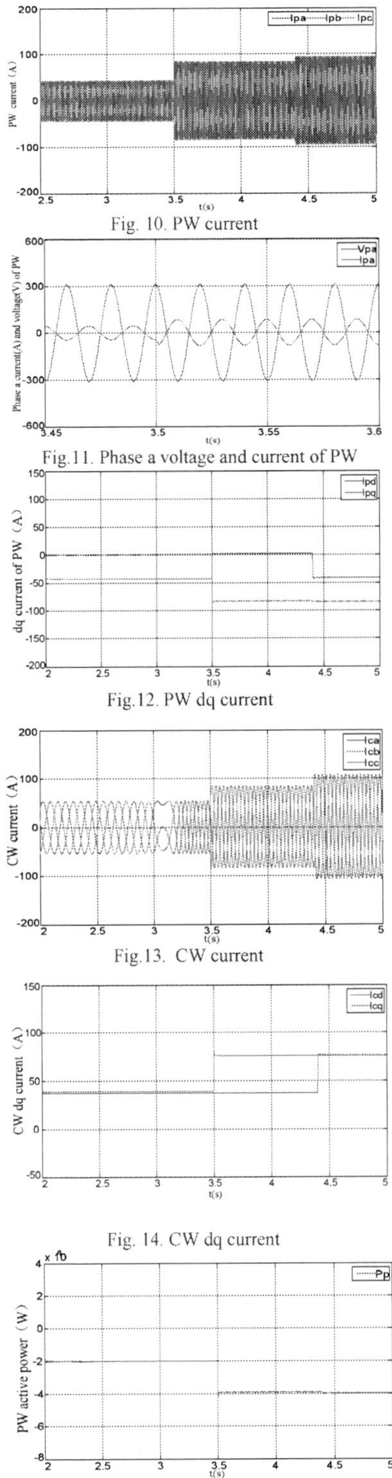

Fig. 10. PW current

Fig.11. Phase a voltage and current of PW

Fig.12. PW dq current

Fig.13. CW current

Fig. 14. CW dq current

Fig. 15. PW active power

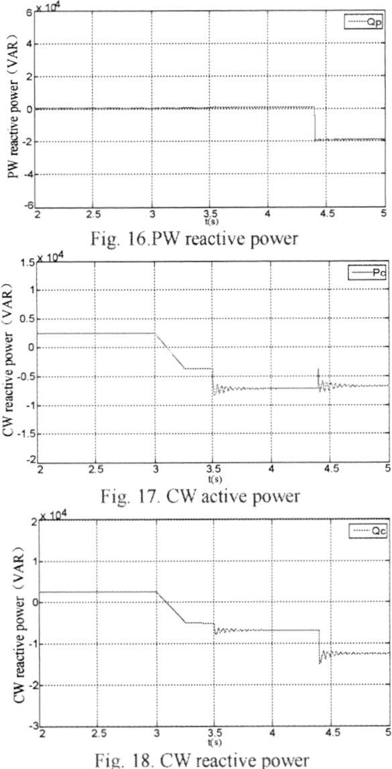

Fig. 16.PW reactive power

Fig. 17. CW active power

Fig. 18. CW reactive power

changes from absorbing power from grid side converter to sending out power to grid side converter. Furthermore, it is can also be seen from Fig. 18 that the reactive power increases when output inductive reactive power of PW increases, this reflects correlations between PW and CW. In conclusion, during the process of output power abrupt changes and transition from sub-synchronous state to super-synchronous state, both active power and reactive power track with the reference well and the control targets are realized, therefore, the effectiveness of proposed control scheme is validated.

V. CONCLUSIONS

The proposed control has following improvements: a) The structure of proposed control is very similar to the mature vector controller of DFIG, hence, proposed control has very good practicability. b) Proposed control can realize completely decoupled control of active and reactive power. c) It needs not to estimate rotor flux, rotor current or CW flux, thus, it has better robustness to machine parameters. In order to validate the effectiveness of proposed vector control, preliminary simulations for machine model and control is implemented by using 64kW BDFIG, the results demonstrate propose control can realize the control targets of grid connected BDFIG, keep active and reactive power tracking with the reference, realize active and reactive power decoupled control and has good dynamic and stable control performance.

ACKNOWLEDGMENT

This work is financially supported by the Fundamental Research Funds for the Central Universities and by the Natural

Postdoctoral Science Foundations of China under the Award Number 2013M540579

REFERENCES

[1] J. Hu, Y. He, L.Xu, B.W. Williams, "Improved Control of DFIG Systems During Network Unbalance Using PI–R Current Regulators," IEEE Transactions on Industrial Electronics, vol. 56, no. 2, pp.439–450, Feb. 2009.

[2] S. Shao, E.Abdi, F.Barati, R.McMahon, "Stator-Flux-Oriented Vector Control for Brushless Doubly Fed Induction Generator," IEEE Transactions on Industrial Electronics, vol. 56, no.10, pp. 1191–1197, Oct. 2009.

[3] S. Shao, T. Long. Dynamic Control of the Brushless Doubly Fed Induction Generator Under Unbalanced Operation. IEEE Transactions on Industrial Electronics, vol. 60, no. 6, pp. 2465–2476, Jun. 2013.

[4] K. rotsenko, D. Xu, "Modeling and Control of Brushless Doubly-Fed Induction Generators in Wind Energy Applications," IEEE Transactions on Power Electronics, vol. 23, no. 3, pp. 1191–1197, May. 2008.

[5] I. A. Gowaid, Ayman S. Abdel-Khalik, Ahmed M. Massoud, Shehab Ahmed, "Ride-Through Capability of Grid-Connected Brushless Cascade DFIG Wind Turbines in Faulty Grid Conditions—A Comparative Study," IEEE Transactions on Sustainable Energy, 2013, 4(4):1002–1015.

[6] D. Zhou, R. Sp'ee, G. Alexander, "Experimental Evaluation of a Rotor Flux Oriented Control Algorithm for Brushless Doubly-Fed Machines," IEEE Transactions on Power Electronics,1998, 12(1):72–78.

[7] J. Poza, E. Oyarbide, I. Sarasola, M. Rodriguez, "Vector control design and experimental evaluation for the brushless doubly fed machine," IET Elect. Power Appl., vol. 3, no. 4, pp. 247–256, Sep. 2009.

[8] Y. Tang, L. Xu, "A flexible active and reactive power control strategy for a variable speed constant frequency generating system," IEEE Transactions on Power Electronics, 1995, 10(4): 472-478

[9] A. Petersson, L. Harnefors, T. Thiringer, "Comparison between stator-flux and grid-flux-oriented rotor current control of doubly-fed induction generators," 35th IEEE Power Electronics Specialists Conference, 2004: 482-486

[10] A. Petersson, L. Harnefors, T. Thiringer, "Evaluation of current control methods for wind turbines using doubly-fed induction machines," IEEE Transactions On Energy Conversion, 2005, 20(1): 227-235

An Energy Router Based on Multi-Winding High-Frequency Transformer

Xianzhuo Liu, Zedong Zheng, Kui Wang, Yongdong Li
Department of Electrical Engineering
Tsinghua University
Beijing, China
liu-xz14@mails.tsinghua.edu.cn

Abstract—**The paper proposed a multi-port active bridge (MAB) converter applied to household energy router (ER), where a multi-winding high-frequency transformer is used for the isolation and energy redistribution. The ER can provide both DC and AC ports with adjustable voltages, thus can be used for various distributed renewable generations and energy storage access. This work also provides simplified equivalent model and control strategy of the multi-active bridge (MAB) using in the multi-winding high-frequency transformer. At the end of this paper, results of the simulation and part of the experiment have done are given.**

Keywords—Household energy router; MAB; multi-winding high-frequency transformer;

I. Introduction

In the recent decades, to face the contradiction between the shortage of traditional fossil energy and the growth of load, the renewable and distributed generation technologies have been developed rapidly. However, with a large amount of renewable and distributed generation devices accessing in, the traditional grid is going to face the problems of stability and reliability.

To solve these problems, the "Energy Internet" concept was proposed recently, and has drawn a wide attention of researchers over the world [1-4]. The Energy Internet aims at finding a way to dispatch the power more intelligent, also to utilize the renewable and distributed energy more efficient and flexible. As the core part of the Energy Internet, Energy Router's structure design and control strategy also become one of the research hotspots [5,6].

The present Energy Router proposed are mostly with two ports, which is not suitable for accessing variable sources and loads with different voltage. Some multi-port Energy Routers by paralleling devices or converters on the DC bus are also proposed, which are not flexible for the power flow control, and the devices voltages range are limited.

Reference [7] gives an AC-AC Energy Router based on MMC (modular multi-level converter) topology, which can reduce the volume and cost of traditional transformer. But MMC is very complex and it's still difficult to balance the voltages of flying capacities.

Reference [8] proposes a solid state transformer (SST) used in zonal DC microgrid. Distributed generation (DG) devices, DC and AC loads are all connected to the DC bus through converters. However all the devices or loads of this system are not electric isolated and the voltage range is limited by the DC bus' voltage.

Reference [9] presents a multi-winding high-frequency transformer applied in a multi-level converter. The main advantages of this multi-winding high-frequency transformer are:

- The generations and loads matching to the system are electric isolated;
- The transformer can access various voltage level by changing each winding's parameter;
- The transformer with single iron core would reduce the volume and weight of the system;
- Every two ports can match to each other via the transformer directly, thus the power flow will pass two converters at most, which can improve the efficiency of the system.

Reference [10] shows a quad-active-bridge (QAB) converter based multi-port SST integrating distributed generation and storage. However, the expressions and control methods related would become highly more complex with the increase of port amounts.

Reference [11] puts forward a more feasible power balancing control method of MAB, and we proposed a power and voltage combined control method basing on it.

In this paper, a MAB converter based Energy Router is proposed, where a multi-winding high-frequency transformer is used for isolation and power control. The simplified equivalent model and expressions of the high-frequency isolation unit is given. A power and voltage combined control method is put forward. Simulation and experiment are implemented to verify the analysis and proposed control method.

II. System Topopogy

The advantages of the topology with multi-winding transformer make the topology the optimum choice of a household Energy Router: various ports for DC or AC devices, high or low voltage; less volume and weight but more efficiency. Thus, a multi-port MAB converter for the Energy Router based on multi-winding high-frequency transformer is proposed as shown in Fig.1which can not only get power from the grid as tradition, but can also utilize the distributed and renewable sources more easily.

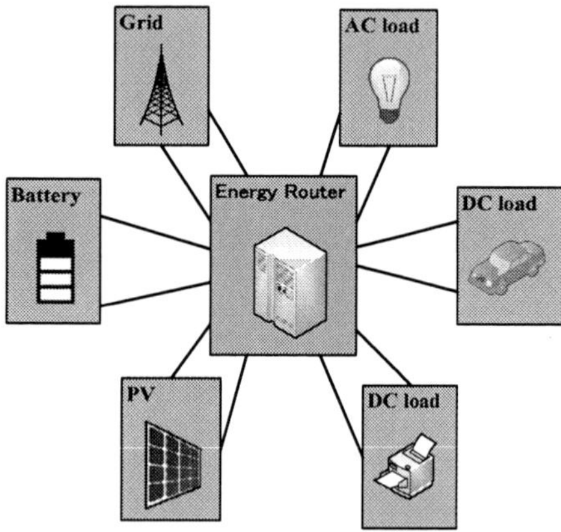

Fig.1 Household Energy Router.

The high-frequency isolation unit in the proposed Energy Router is shown as Fig.2, each winding of the transformer is connected to an H-bridge using simple phase modulation (SPM), meaning that the outputs of the H-bridges are all high-frequency 50% duty square wave.

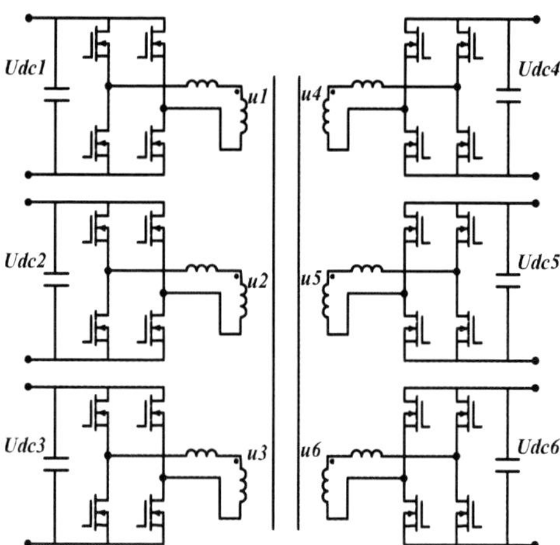

Fig.2 Multi-Winding High-frequency Transformer

The winding number and each winding's parameter can be adjusted according to the practical necessary. Here we set the parameters just for an experimental research. We use a transformer with six windings. Ports 1 to 6 links to six devices respectively: the rectifier connect to grid (220V single phase AC), battery (48V DC), PV (48V DC), the rectifier connect to load (220V AC), high voltage load (350V DC) and low voltage load (48V DC).

III. Operating Principle

In the proposed circuit, the voltage balance control of each port is very important and will affect the stability of the whole circuit. According to [11], the MAB converter can be simplified as the equation model shown in Fig3.

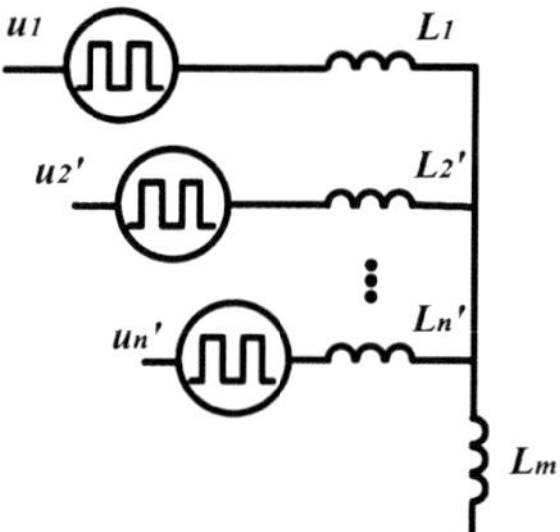

Fig.3 MAB's Simplified Model

Where the $u_2{}' \sim u_n{}'$ and $L_2{}' \sim L_6{}'$ are equivalent voltages and inductances converted from ports *2~n* to port *1*, and so the ports' equivalent amplitude voltages are $U_{dc2}{}' \sim U_{dc6}{}'$. And now the equations of these ports' voltages and currents can be given as (1).

$$\begin{cases} u_1 - L_1 \dfrac{di_1}{dt} = u_2{}' - L_2{}' \dfrac{di_2{}'}{dt} \\ u_1 - L_1 \dfrac{di_1}{dt} = u_3{}' - L_3{}' \dfrac{di_3{}'}{dt} \\ \cdots\cdots \\ u_1 - L_1 \dfrac{di_1}{dt} = u_n{}' - L_n{}' \dfrac{di_n{}'}{dt} \end{cases} \tag{1}$$

As i_j equals to the sum of all the branches' currents i_{ji}, so

the results of the equations is :

$$\sum_{i \neq j, i=1}^{6} L_j \frac{di_{ji}}{dt} = \sum_{i \neq j, i=1}^{6} \frac{L_p}{L_j L_i}(u_j - u_i) \tag{2}$$

Where $L_p = 1 / \sum_{k=1}^{6} L_k^{-1}$ is the parallel inductance value.

Thus the equal circuit between port i and j is seen in Fig.4 (a), where $L_{i,j}^{'} = L_p / L_i L_j$ and current waveforms of these two ports is shown in Fig.4 (b).

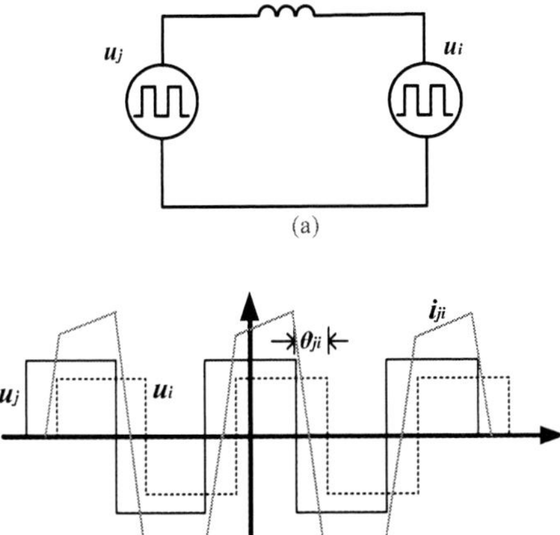

Fig.4 (a) Two Ports' Equal Circuit
(b) Voltage and Current Waveforms

Then it can be calculated out that the power's direction and magnitude flows between the two ports only depends on the phase difference as shown in (3).

$$P_{ji} \approx \frac{L_p}{2f} \sum_{i=1, i \neq j}^{n} \frac{U_{dci} U_{dcj}}{L_j L_i} (\theta_{ji} - \theta_{ji} | \theta_{ji} |) \tag{3}$$

As in the real condition, we can adjust the leak inductance according to (3) to make sure that the system's power flow range is enough with a small phase difference θ_{ji}. If the phase difference θ_{ji} is big, it will cause a large power reflow (from the load to the source) in each period, which is harmful to the system, and it makes the system analysis and control more difficult. When the phase differ is small enough, we can consider $\theta_{ji} | \theta_{ji} | \approx 0$.

For a further easier analysis and control strategy, we assume that all the ports' DC equivalent voltages and equivalent inductances are approximately equal. Then the equation (3) can be simplified to (4).

$$P_{ji} \approx \frac{U_{dc}^{2}}{12 f L_j} \sum_{i=1, i \neq j}^{6} \theta_{ji} \tag{4}$$

According to the analysis above, the relationship between the power flow P and phase difference θ_{ji} can be regarded linear approximately.

As for the DC bus's voltage control, a port's DC bus's capacity voltage is influenced by the power flow in and out as shown in Fig.5.

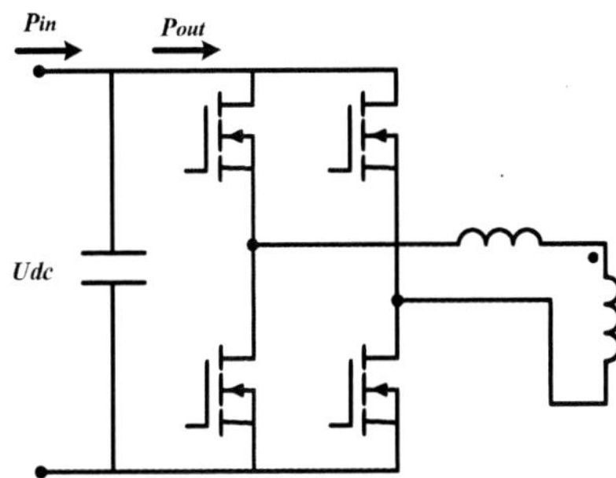

Fig.5 A Port's Voltage Control

When the DC voltage changes very little, the relation between power flow P and DC voltage change ΔU_{dc} is :

$$P_{in} - P_{out} \approx C \cdot U_{dc} \cdot \Delta U_{dc} \tag{5}$$

Thus the DC voltage control can be transformed to power flow control, and then the DC voltage can be controlled by the phase difference θ_{ji} too.

IV. Control Method

In the system, some ports need to keep the voltage steady (e.g. some load port with specific voltage demand), while others need to export or import given power (e.g. the battery port balancing the output and input power, the PV port running in MPPT mode). Thus, a power and voltage combined control method is needed.

When controlling the system, one port would be set as the referential port, which means its phase angle is 0. The referential port's voltage needs to be controlled to keep the system steady (e.g. the grid port in grid-connected condition or the PV port in the off-grid condition). Then, the other ports' voltage or power would be controlled by the phases according to the given commands.

When calculating the ports' phase angles, analytical method and iteration method would be more and more complex with the increasing of ports and it's hard to realize in the real system. Also errors cannot be avoided as the equations (4) and (5) are not accurate.

As analyzed above, the relationship between power flow and phase difference as well as the relationship between voltage and phase difference can be considered linear approximately in a reasonable range. Thus a power and voltage combined control method is put forward using PI closed loop as shown in Fig6.

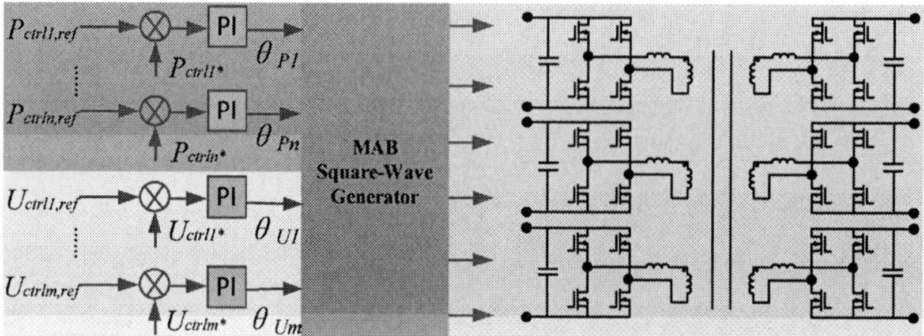

Fig.6 Power and Voltage Combined Control Method

V. SIMULATION AND EXPERIMENT

Simulation results on Matlab/Simulink are shown in Fig.7, there are six ports in the system: grid, PV, battery, high voltage DC load, low voltage DC load and AC load.

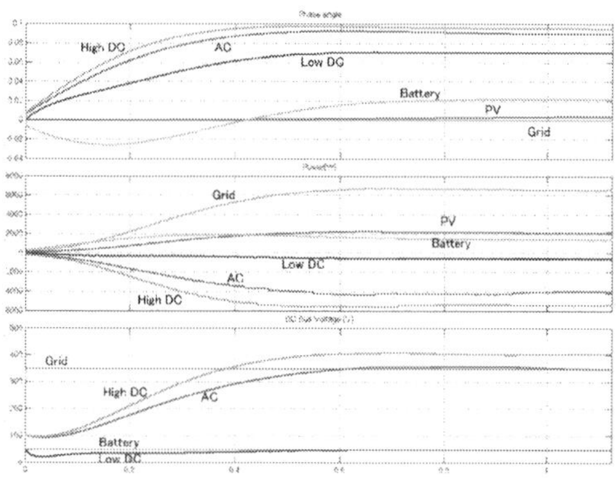

(a)

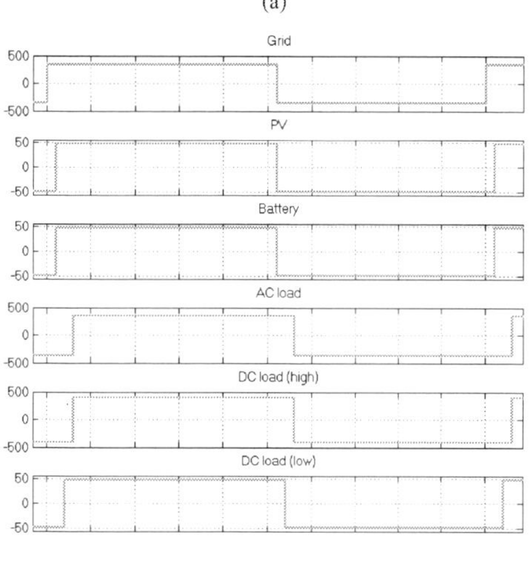

(b)

Fig.7 (a) Phases, Power and Voltages Simulation Results of Six Port

(b) Six ports' H-Bridges' AC output showing the phase differences

We set the grid port as the referential port. Then we control the battery and low DC load ports' voltages to 48V; AC load port's voltage to 350V; high DC load and PV ports' power to 4kW and 2kW respectively.

In (a), the system ports' phase angles, powers and DC bus voltages reach the expectant states within 1 sec. In (b), it shows that the sources' phases are advancer than the loads' phases. And the more power the source supplies, the advancer the port's phase is, which verify the theory about the relationship between power flow and phase difference.

We also build a 10 kVA experiment platform and part of the high-frequency isolation unit is shown in Fig.8 (each board is an H-bridge as the sample board upon the box).

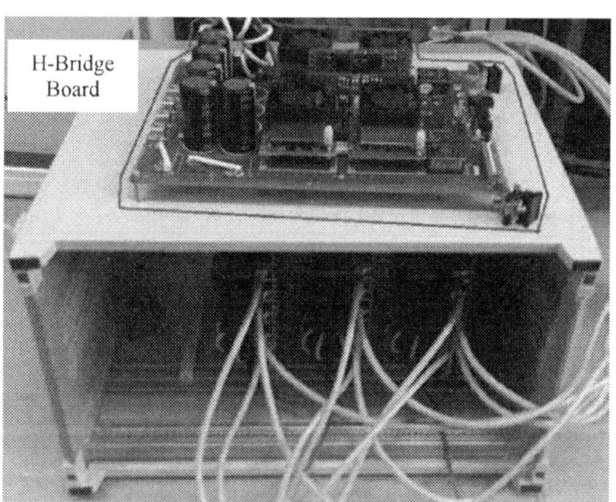

Fig.8 High-Frequency Isolation Unit with H-Bridge Boards

A two-port experiment is done, the ports' voltages and source port's current waveforms are shown in Fig.9, which meet with the expectation. The source port's DC voltage is 10V and we control the load port's DC voltage to 20V. According to the results, the voltage and current waveforms meet with the expectation.

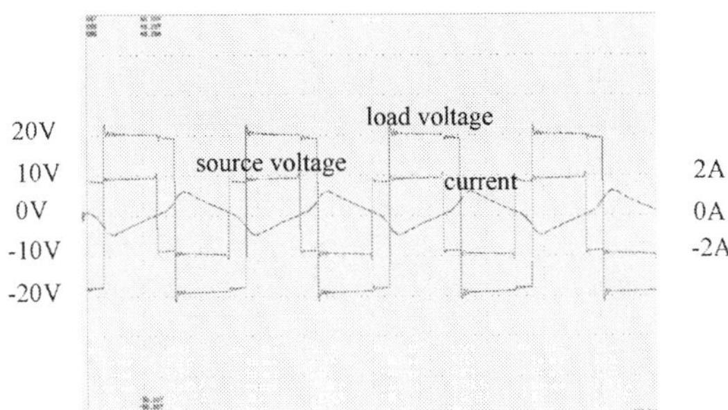

Fig.9 Two-port Experiment Voltages and Current Waveforms

Also a three-port (one source and two loads) testing experiment is done. The DC source's voltage is 15V, we first set the three ports' phases the same, then we give the two loads voltage commands of 20V and 25V respectively.

The three ports' DC bus voltage changes are shown in Fig.10, the two loads' voltage first turn to nearly 15V at the beginning with PWM open and no phase differences. Then, as the voltage commands are given, the load ports' phases begin to change, and the DC bus voltages turn to 20V and 25V as the commands within 0.5s.

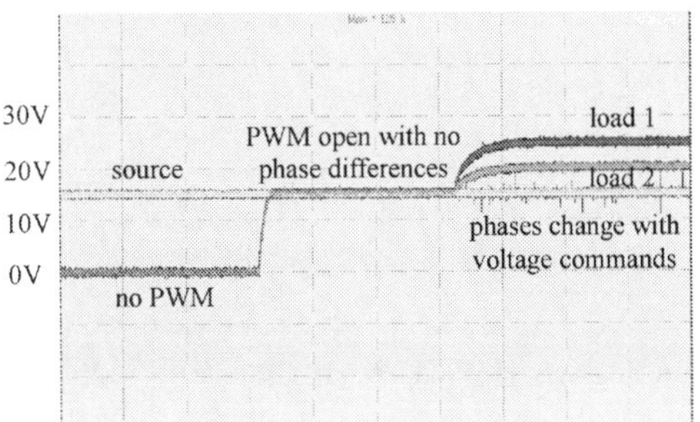

Fig.10 Three Ports' DC Bus Voltages

And the three H-bridges' AC output is shown in Fig.11. the loads' phases are behind the source's.

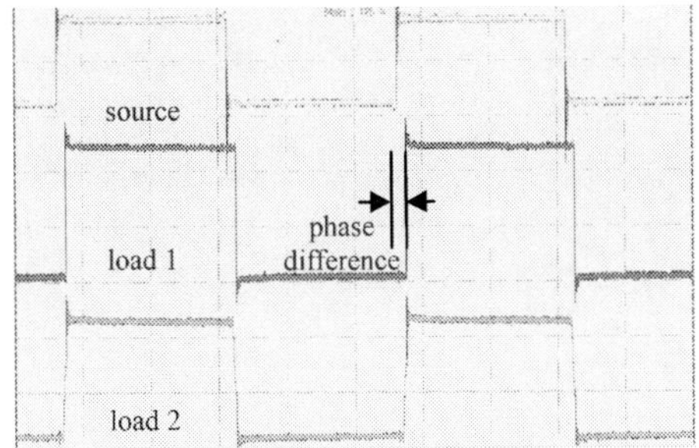

Fig.11 Three Ports' AC Voltages

VI. Conclusion

A novel multi-port household Energy Router with multi-winding high-frequency transformer is proposed which can realize the grid connection, distributed generation access and energy storage access. The equivalent circuit and mathematical model are analyzed, and an effective control strategy is proposed based on the analysis. Simulation and experimental results are given to verify the performance of the proposed circuit and the control strategy.

References

[1] Huang, A.Q., et al., The Future Renewable Electric Energy Delivery and Management (FREEDM) System: The Energy Internet. Proceedings of the IEEE, 2011. 99(1): p. 133-148.

[2] Su, W. and A.Q. Huang, Proposing a Electricity Market Framework for the Energy Internet. 2013 IEEE POWER AND ENERGY SOCIETY GENERAL MEETING (PES), 2013.

[3] Karady, G.G., A.Q. Huang and M. Baran, FREEDM System: An Electronic Smart Distribution Grid for the Future. 2012 IEEE PES TRANSMISSION AND DISTRIBUTION CONFERENCE AND EXPOSITION (T&D), 2012.

[4] Huang, A. FREEDM system - a vision for the future grid. in Power and Energy Society General Meeting, 2010 IEEE. 2010. Minneapolis, MN.

[5] Xu, Y., et al., Energy Router: Architectures and Functionalities toward Energy Internet. 2011 IEEE INTERNATIONAL CONFERENCE ON SMART GRID COMMUNICATIONS (SMARTGRIDCOMM), 2011.

[6] Jianhua, Z., W. Wenye and S. Bhattacharya, Architecture of solid state transformer-based energy router and models of energy traffic. 2012. p. 8 pp.-8 pp.

[7] Sizhao, L., et al., A new power circuit topology for energy router. 2014. p. 1921-5.

[8] She, X., et al., On Integration of Solid-State Transformer With Zonal DC Microgrid. IEEE Transactions on Smart Grid, 2012. 3(2): p. 975-985.

[9] Zheng, Z., et al., A New Multilevel Converter with Multi-Winding Medium-Frequency Transformer. PROCEEDINGS OF THE 2011-14TH EUROPEAN CONFERENCE ON POWER ELECTRONICS AND APPLICATIONS (EPE 2011), 2011.

[10] Falcones, S., R. Ayyanar and X. Mao, A DC−DC Multiport-Converter-Based Solid-State Transformer Integrating Distributed Generation and Storage. IEEE Transactions on Power Electronics, 2013. 28(5): p. 2192-2203.

[11] Chunyang, G., Z. Zedong and L. Yongdong, Control strategy of a multi-level converter with multi-winding MFT/HFT isolation. 2014. p. 3717-2

Noise Suppression of the DWT-based MRA on Mother Wavelet and Decomposition Level Optimization for a Robust Adaptive SOC Estimator in Multi-Cell Battery String

Jonghoon Kim
Department of Electrical Engineering
Chosun University
Gwangju, Republic of Korea
qwzxas@hanmail.net

Chang Yoon Chun
Department of Electrical and Computer Engineering
Seoul National University
Seoul, Republic of Korea
wobniw77@snu.ac.kr

Woonki Na
Electrical and Computer Engineering
California State University, Fresno
Fresno, CA, USA
wkna@csufresno.edu

Abstract— **This approach newly presents an innovative method that implements the noise suppression considering the discrete wavelet transform (DWT)-based multi-resolution analysis (MRA). Specifically, this approach elaborately carries out two comparative analyses of the noise suppression depending on information on mother wavelet and decomposition level in the MRA. The first is to select the decomposition level having a high performance of the noise suppression under the same condition of mother wavelet. On the contrary, with an identical condition of the decomposition level, the second is to compare the noise suppression of different mother wavelets under the same condition of the decomposition level. All comparative analyses are evaluated in terms of signal-to-noise ratio (SNR). This approach entirely makes a comparison between two noise suppression techniques considering hard- and soft-thresholding. From these results, our analytic results suggest the clear comparison by showing the SNR difference with various mother wavelets and decomposition levels. Consequently, it can be surely expected to optimize the mother wavelet and the decomposition level for noise suppression of the noise-riding EDCV signal in multi-cell battery string. Our experimental multi-cell battery string in series/parallel (2S3P) using 2.2Ah unit cells previously discriminated was used.**

I. Introduction

Nowadays, with great attention of multi-cell battery string for electric-powered transportation such as electric vehicle (EV) and hybrid electric vehicle (HEV), the necessity of the battery management system (BMS) for having a high confidence on operational performance in EV and HEV has been substantially increased together [1]. Among a number of researches for obtaining an enhanced BMS, researches on state-of-charge (SOC) estimation have been specifically remarkable. Through correct SOC information, it is completely possible to prevent from over-discharging and over-charging status in practical applications, thus it can be considerably expected to have major positive effects on BMS performance of multi-cell battery string [2]-[4]. For this goal, adaptive SOC estimators such as extended Kalman filter (EKF) [5][6] and sliding-mode observer (SMO) [7][8] generally take advantage of the equivalent-circuit model (ECM) that has outstanding strengths of being closed-loop and online in addition to the availability of the dynamic SOC estimation range. Physical parameterization of the ECM enables us to describe electrochemical characteristics of multi-cell battery string, as an estimated terminal voltage. Therefore, it stands to reason that the performance of adaptive SOC estimators is dependent on voltage difference between an experimental- and a terminal voltage. Little voltage difference is indispensable for high accuracy SOC estimation. Unfortunately, during discharging/charging process, there is a possibility of unexpected and instantaneous sensing of noisy, for example, noise-riding experimental discharging/charging voltage (EDCV) signal in the BMS. This phenomenon leads to the noise-riding EDCV signal that inevitably results in low BMS performance of an increased SOC estimation error and long run-time calculation. As a result, a new approach dealing with noise suppression of the EDCV signal should be preferentially required.

This approach gives insight to design and implementation of the discrete wavelet transform (DWT)-based multi-resolution analysis (MRA) [9]-[17] for noise suppression of the EDCV signal. Fortunately, with respect to the noise suppression of the EDCV in multi-cell battery string, there uniquely has been our comprehensive methodology [18]. In particular, this approach develops prior investigation one step further by showing two comparative analyses according to information on different mother wavelets [19][20] and decomposition levels in the MRA. The first analysis is to select the decomposition level that can provide a high performance of the noise suppression under the same condition of mother wavelet. On the other hand, with an identical condition of the decomposition level, the purpose of the second analysis is to compare the noise suppression applying different mother wavelets (Daubechies, Coiflet, Symlet, Meyer, etc.). All comparative analyses are evaluated with respect to the signal-

978-1-4673-9551-9/16 $31.00 © 2016 IEEE

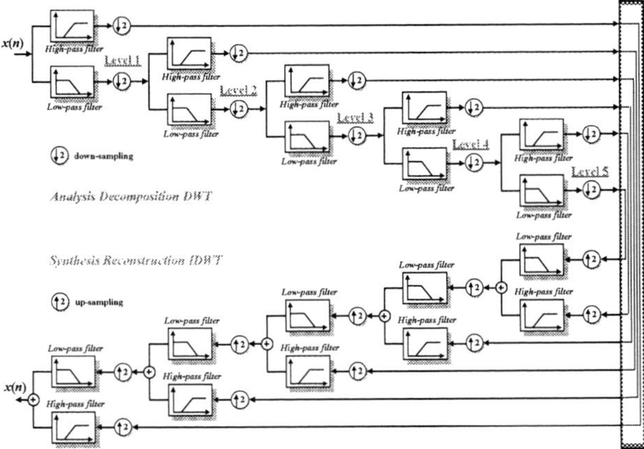

Figure 1. DWT-based decomposition and reconstruction processes.

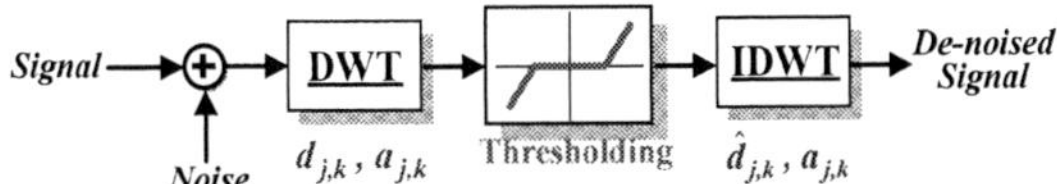

Figure 2. Three steps of the wavelet noise suppression technique.

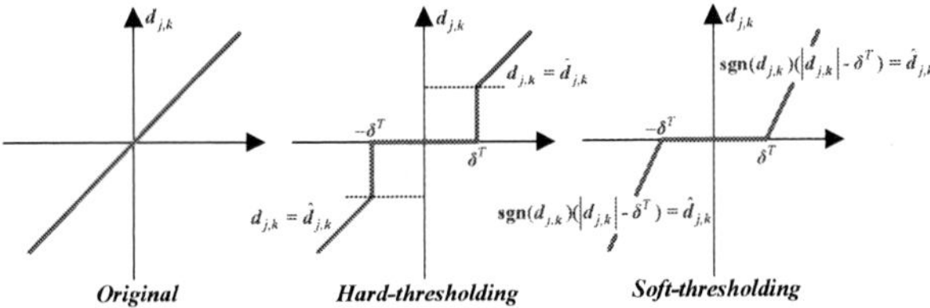

Figure 3. Original, hard-thresholding, and soft-thresholding functions for DWT-based noise suppression technique.

to-noise ratio (SNR) value. Our comparative analyses clearly show the SNR difference according to various mother wavelets and decomposition levels. From this overall perspective, it can be expected for optimization of the mother wavelet and decomposition level in the MRA for noise suppression of the noise-riding EDCV signal in multi-cell battery string. For reference, two noise suppression techniques such as hard- and soft-thresholding were considered and threshold value has been calculated by *VisuShrink* method. This approach has been extensively validated by experimental results of the multi-cell battery string (2S3P) consisting of six 2.2Ah Li-Ion cells produced by LG Chem [21]. These Li-Ion cells were previously discriminated by Ref. [22] in order to avoid the difference in electrochemical characteristics among Li-Ion cells in multi-cell battery string.

II. Fundamental Review of the Discrete Wavelet Transform (DWT) for Noise Suppression

A. *Brief introduction of the DWT*

Because of the requirement for analyzing the non-stationary signals, the DWT [9]-[17] has been extensively used as a new mathematical approach. The main feature of the DWT is to decompose a time domain signal into different frequency groups. Additionally, in both the time and frequency domain, the DWT involves a varied-time frequency window and can

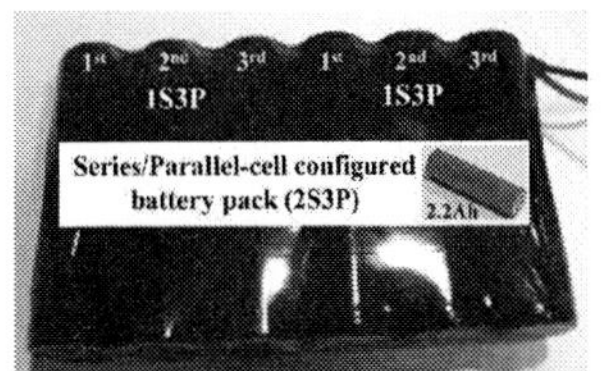

Figure 4. Experimental multi-cell battery string of 2S3P.

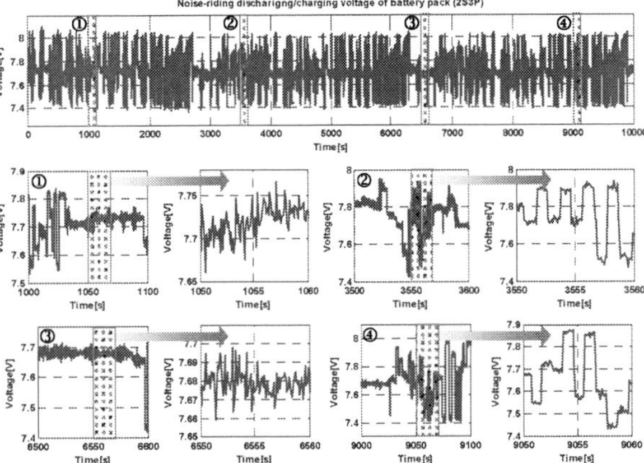

Figure 5. Four enlarged shapes of the noise-riding EDCV signal of multi-cell battery string.

provide localization property. Through the MRA, any non-stationary signal $x(t)$ can be fully decomposed related to approximation, provided by scaling function $\square_{j,k}(t)$ and the details, provided by the wavelet function $\psi_{j,k}(t)$. The decomposition and reconstruction processes in the DWT are displayed in Fig. 1. Additional concepts were introduced and detailed in Refs. [9]-[17].

B. *Procedure of noise suppreesion based on the DWT*

When unexpected and instantaneous noise is added to any non-stationary original signal $x(t)$, it is absolutely required to have the noise suppression technique based on the DWT [18]. Through elimination of the wavelet coefficients generated by the noise and reconstruction of the remaining wavelet coefficients, it is feasible to complete the noise suppression. Fig. 2 shows three steps of the wavelet noise suppression technique: (i) decomposition of the noisy signal into a set of wavelet coefficient, (ii) thresholding the resulting wavelet coefficients, (iii) reconstruction from the reduced set of wavelet coefficient. The performance of the noise suppression is determined by the threshold value δ^T. Regardless of the determination of the mother wavelet and MRA level, the noise suppression is intimately linked with the thresholding value and thresholding rule, such as hard-thresholding and soft-thresholding. As shown in Fig. 3 and expressed in Eq. (1), the difference between hard-thresholding and soft-thresholding is the magnitude of the coefficients decided by the threshold value δ^T. Ref [18] elaborately explained the concept of the noise suppression.

$$\hat{d}_{j,k(hard)} = \begin{cases} d_{j,k} \\ 0 \end{cases} \quad \hat{d}_{j,k(soft)} = \begin{cases} \mathrm{sgn}(d_{j,k})(|d_{j,k}| - \delta^T) & |d_{j,k}| \geq \text{threshold value } \delta^T \\ 0 & |d_{j,k}| < \text{threshold value } \delta^T \end{cases} \tag{1}$$

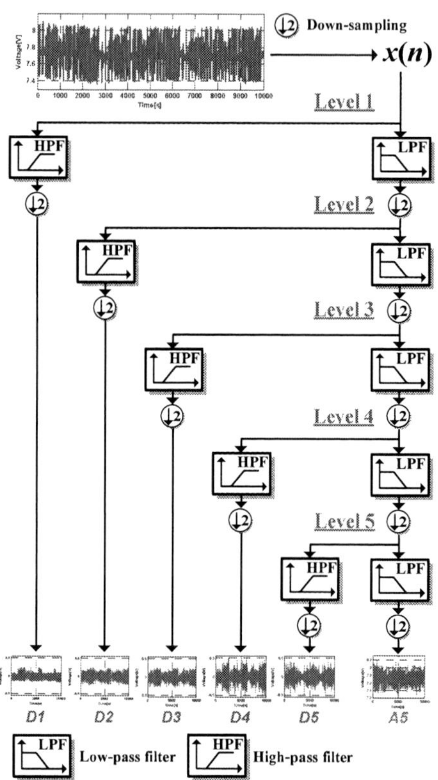

Figure 6. Decomposition process of the DWT-based five-level MRA(db3).

III. Proposed Approach

A. Decomposition of noise-riding EDCV signal for multi-cell battery string (2S3P)

Our experimental multi-cell battery string of 2S3P is shown in Fig. 4. The EDCV signal can be obtained from discharging/charging process, then for noise implementation, a filtering part in the BMS was arbitrary removed and a Gaussian noise is added to the EDCV (see Fig. 5). This noise-riding EDCV signal is decomposed by the DWT-based MRA. For example, Fig. 6 shows the decomposition process of the DWT-based five-level MRA using Daubechies wavelet (db). Unfortunately, after reconstruction process in the DWT, it stands to reason that it is impossible to avoid the noise influence on the EDCV signal without noise suppression. Therefore, Sections *3.B* and *3.C* will diversely do comparative analyses on the noise suppression performance with different mother wavelets and decomposition levels. For judgment on noise suppression performance, this approach considered the SNR value. For reference, for detailed comparison of the noise suppression performance, two techniques such as hard- and soft-thresholding were each implemented and *VisuShrink* method was applied for achievement of a threshold value δ^{T}.

B. Comparative analyses on noise suppression performance related to decomposition level

Considered in noise-riding EDCV signal in Fig. 5, this section implements the noise suppression with respect to the level-based hard- and soft-thresholding. In the DWT-based MRA, the order 3 Daubechies wavelet (db3) was properly used as an identical condition of mother wavelet [18]. Figs. 7 and 8 respectively present the noise suppression performance based on hard- and soft-thresholding at decomposition level 1 through 20. From obtained de-noised voltages in Figs. 7 and 8, each SNR values are compared in Figs. 9 and 10. These figures indicate that it can be defined as an optimum decomposition level 9 and 16 (hard/soft-thresholding) that have all maximum SNR values. The most interesting thing is that if the decomposition level is finally determined, the thresholding will effect on the main frequencies of the original EDCV signal, therefore the SNR has lower values for the level higher than determined decomposition level. Actually, as shown in Figs. 9 through and 11, it can be seen that the SNR values are decreasing if the decomposition levels are getting higher over 9 and 16. It can be additionally mentioned that an increase of decomposition level too high will decrease the SNR

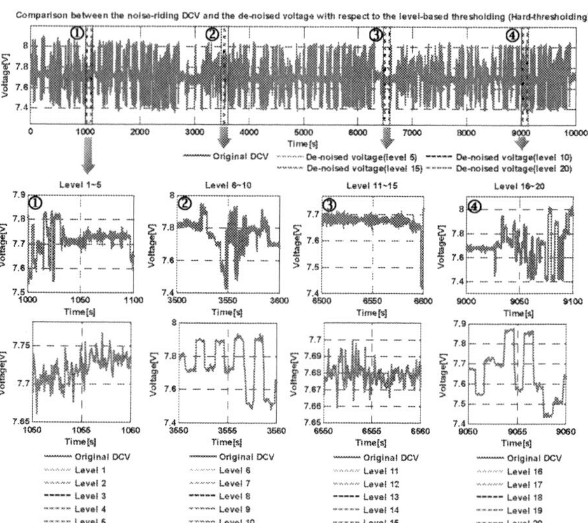

Figure 7. Comparison between noise-riding EDCV signal and de-noised voltages with respect to the level-based hard-thresholding.

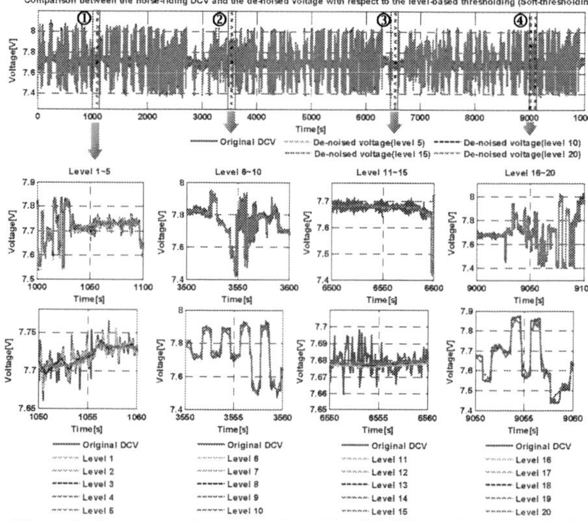

Figure 8. Comparison between noise-riding EDCV signal and de-noised voltages with respect to the level-based soft-thresholding.

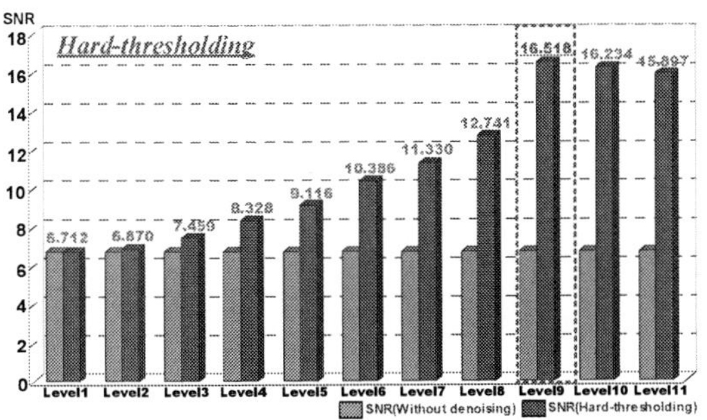

Figure 9. Each SNR values of de-noised voltages with respect to the level-based hard-thresholding (maximum value at level 9).

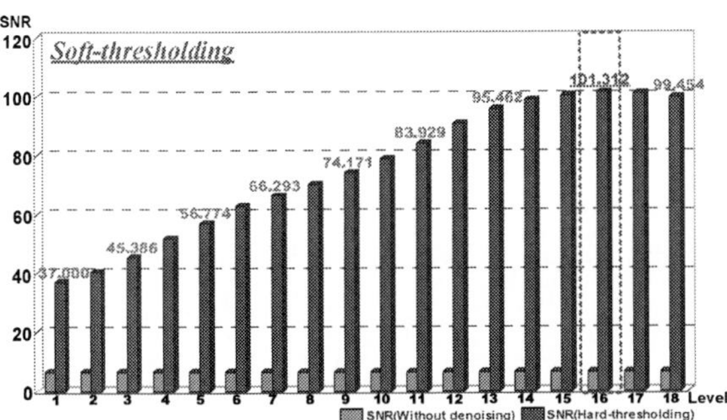

Figure 10. Each SNR values of de-noised voltages with respect to the level-based soft-thresholding (maximum value at level 16).

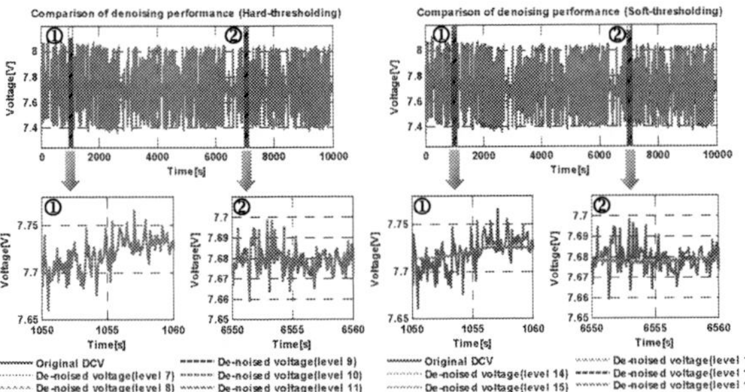

Figure 11. Comparison betwee noise-riding EDCV signal and de-noised voltages near maximum levels (hard : 9, soft : 16).

value after an optimum level and also increase the complexity of decomposition. Finally, the performance on noise suppression can be optimized at level 9 (hard-thresholding) and 16 (soft-thresholding) in order to provide the de-noised voltages for achievement of an improved SOC estimator. For reference, two threshold values for hard- and soft-thresholding were defined as δ^T=0.005783 and δ^T=0.041144, respectively.

C. Comparative analyses on noise performance related to mother wavelet type

By contrast to Section *3.B*, this section performs comparative analyses on noise suppression related to mother wavelet type under the same condition of decomposition level. All mother wavelets have each order range, therefore it is preferentially required to have an optimized order that have the maximum SNR value. For example, Fig. 12 shows each SNR values of the de-noised EDCV considering Daubechies mother wavelet at decomposition level 5. From this figure, it is surely known that the order of 9 has the maximum SNR value,

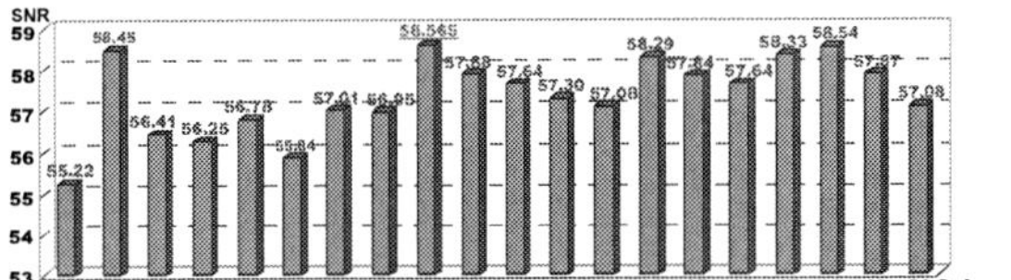

Figure 12. Each SNR values of the de-noised voltages with respect to the order (Daubechies mother wavelet, soft-thresholding).

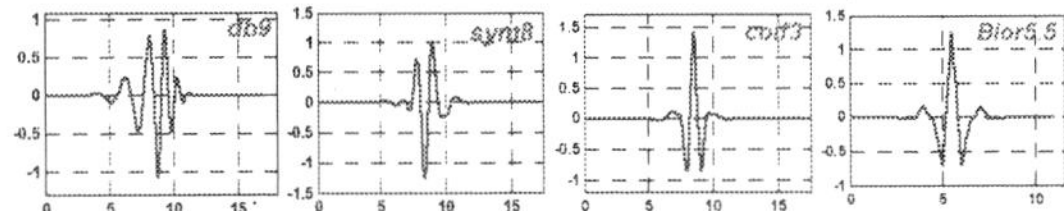

Figure 13. Four mother wavelet functions (db9, sym8, coif3, bior5.5).

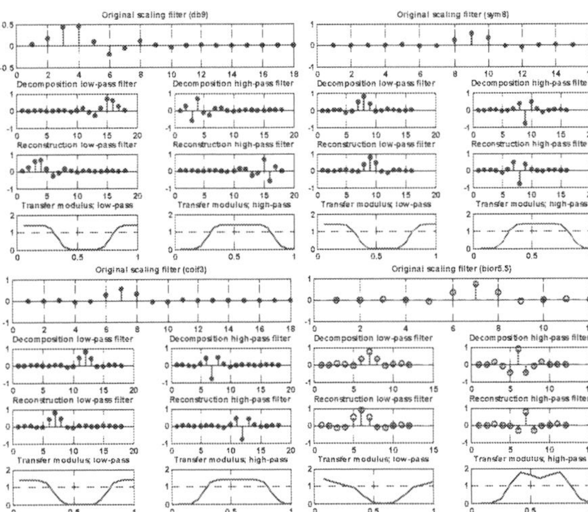

Figure 14. Original scaling filter coefficients, decomposition low-/high-pass filter coefficients, reconstruction low-/high-pass filter coefficients and transfer modulus of four mother wavelets (db9, sym8, coif3, bior5.5).

thus this value is set to a representative order of Daubechies mother wavelet (db9). In the same manner, an aforementioned rule can be applied to other mother wavelet. This approach considers a total of six mother wavelets including Symlet (sym), Coiflet (coif), Meyer (meyr), Biorthogonal (bior), and reverse Biorthogonal (rbio) wavelets, together with Daubechies wavelet (db). Considered in each mother wavelets at decomposition level 5, this section finally decided to use the following wavelet families: (i) Daubechies db9, (ii) Coiflet coif3, (iii) Biorthogonal bior55, (iv) reverse Biorthogonal rbio55, (v) Meyer meyr, and (vi) Symlet sym8. Among them, four mother wavelet functions of db9, sym8, coif3, and bior55 are shown in Fig. 13. In addition, original scaling filter coefficients, decomposition low-/high- pass filter coefficients, reconstruction low-/high-pass filter coefficients and transfer modulus low-/high-pass of four mother wavelets are displayed in Fig. 14. Two figures of 15 and 16 respectively showed the comparison between noise-riding EDCV signal and de-noised voltages with respect to mother wavelet db9 under the same decomposition level 5 considering hard- and soft-thresholding. This process was subsequently repeated in other mother wavelets. From this, Fig. 17 simultaneously shows the comparative analyses on noise suppression performance considering six mother wavelets between hard- and

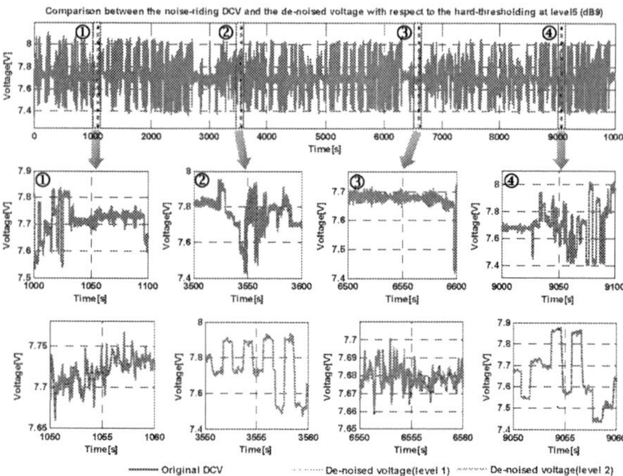

Figure 15. Comparison betweem noise-riding EDCV signal and de-noised voltages with respect to the mother wavelet db9 under the same decomposition level 5 (hard-thresholding).

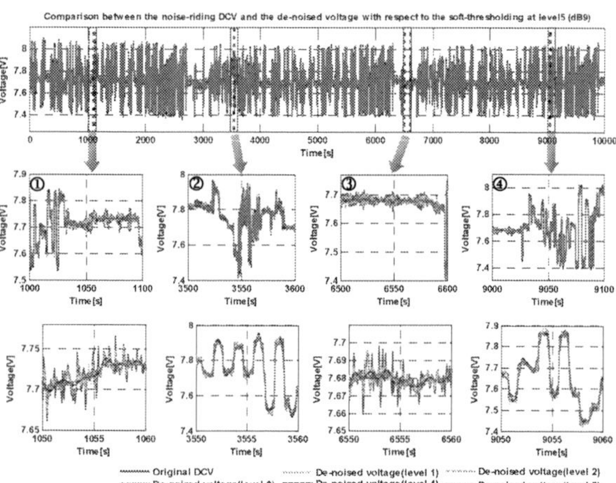

Figure 16. Comparison betweem noise-riding EDCV signal and de-noised voltages with respect to the mother wavelet db9 under the same decomposition level 5 (soft-thresholding).

soft-thresholdling. In addition, Fig. 18 totally displayed all detailed coefficients cD_5 of six mother wavelets. In case of hard-thresholding, it can be described as the process of setting the value of the detailed coefficient $d_{j,k}$ to zero, if it has absolute value is lower than the threshold value δ^T. In case of soft-thresholding, it can be considered as an extension of the hard-thresholding. It sets those detailed coefficients to zero if their absolute values are lower than the threshold value δ^T and shrinks the nonzero coefficients toward zero. As expected, this section considered δ^T as 0.005783 (hard) and 0.041144 (soft), respectively. Fig. 18 clearly displayed the detailed coefficient difference between hard- and soft-thresholding based on Fig. 3 and Eq. (1). Because of selection of the decomposition level 5, Fig. 18 showed the detailed coefficient cD_5. Final figure is introduced for showing each SNR values of noise-riding EDCV signal and de-noised voltage related to the mother wavelet with an identical decomposition level 5. From Fig. 19, two comparative analyses can be simply done. In case of hard-thresholding, an application of the mother wavelet of bior5.5 can surely guarantee the maximum SNR

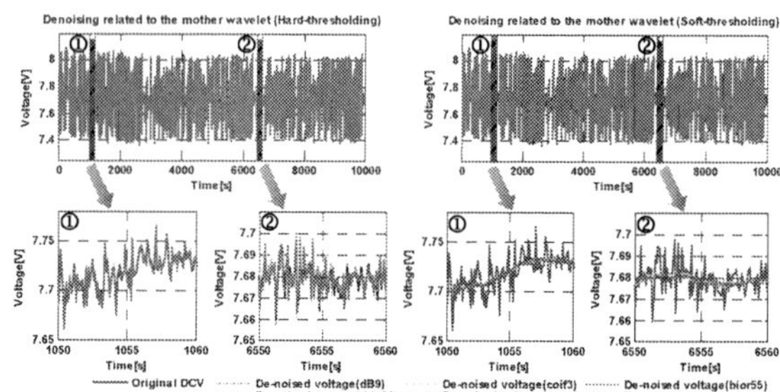

Figure 17. Comparison of noise suppression performance considering six mother wavelets (db9, coif3, bior55, rbio55, meyr and sym8) between hard- and soft-thresholding under the same decomposition level 5.

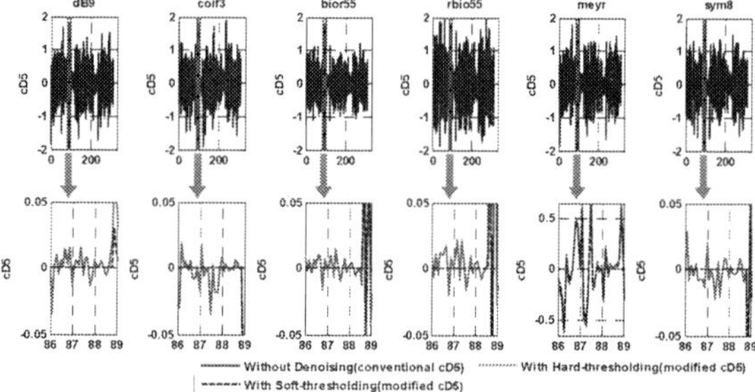

Figure 18. Comparison of detailed coefficients cD_5 considering six mother wavelets (db9, coif3, bior55, rbio55, meyr and sym8) between hard- and soft-thresholding under the same decomposition level 5.

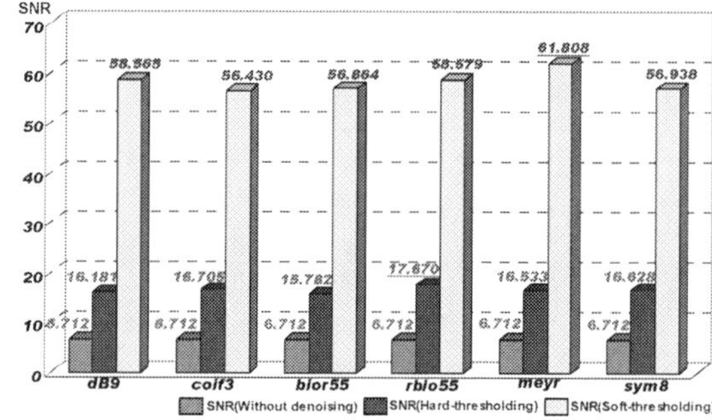

Figure 19. Each SNR values of noise-riding EDCV signal and de-noised voltages with respect to the mother wavelets (db9, coif3, bior55, rbio55, meyr, and sym8) under the same decomposition level 5.

values in comparison with other mother wavelets. It is expected to have the maximum SNR values when the mother wavelet of meyr is considered. Although the noise-riding EDCV signal is identically used as an original signal for noise suppression, there is a marked distinction of noise suppression performance according to selection of hard- or soft-thresholding. Therefore, it is certainly required to determine the mother wavelet that surely leads to high performance on noise suppression and by extension, an improved SOC estimator. For reference, under the same of mother wavelet, the performance on noise suppression of soft-thresholding is always far superior to that of hard-thresholding irrespective of mother wavelet type.

IV. CONCLUSION

This approach deals with an innovative approach for achievement of an improved SOC estimator through noise suppression of the noise-riding EDCV signal caused by unexpected sensing of noisy in the BMS. There are two major topics in this approach. The first is to select the decomposition level having a high performance of the noise suppression under the same condition of mother wavelet. On the contrary, with an identical condition of the decomposition level, the second is to compare the noise suppression of different mother wavelets under the same condition of the decomposition level. From these results, it can be surely expected to optimize the mother wavelet and the decomposition level for noise suppression of the noise-riding EDCV signal in multi-cell battery string.

ACKNOWLEDGMENT

This work was supported by the Human Resource Training Program for Regional Innovation and Creativity through the Ministry of Education and National Research Foundation of Korea(NRF-2014H1C1A1066713).

REFERENCES

[1] Y.-H. Liu and Y.-F. Luo, "Search for an optimal rapid-charging pattern for Li-ion batteries using the taguchi approach," *IEEE Trans. Ind. Electron.*, vol. 57, no. 12, pp. 3963–3971, Dec. 2010.

[2] L.Maharjan, S. Inoue, H. Akagi, and J. Asakura, "State-of-Charge (SOC)-balancing control of a battery energy storage system based on a cascade PWM converter," *IEEE Trans. Power Electron.*, vol. 24, no. 6, pp. 1628–1636, Jun. 2009.

[3] J. Kim, J. Shin, C. Chun, and B. H. Cho, "Stable Configuration of a Li-Ion Series Battery Pack Based on a Screening Process for Improved Voltage/SOC Balancing," *IEEE Trans. Power Electron.*, vol. 27, no. 1, pp. 411–424, Jan. 2012.

[4] H. He, R. Xiong, X. Zhang, F. Sun, and J. Fan, "State-of-Charge Estimation of the Lithium-Ion Battery Using an Adaptive Extended Kalman Filter Based on an Improved Thevenin Model," *IEEE Trans. Veh. Technol.*, vol. 60, no. 4, pp. 1461–1469, May. 2011.

[5] A. Vasebi, M. Partovibakhsh, and S. Bathaee, "A novel combined battery model for state-of-charge estimation in lead-acid batteries based on extended-Kalman filter for hybrid electric vehicle applications," *J. Power Sources*, vol. 174, no. 1, pp. 30–40, Nov. 2007.

[6] G. L. Plett, "Extended Kalman filtering for battery management systems of LiPB-based HEV battery packs-Part 3. State and parameter estimation," *J. Power Sources*, vol. 134, no. 2, pp. 277–292, Aug. 2004.

[7] I.-S. Kim, "A Technique for Estimating the State of Health of Lithium Batteries Through a Dual-Sliding-Mode Observer," *IEEE Trans. Power Electron.*, vol. 25, no. 4, pp. 1013–1022, Apr. 2010.

[8] F. A. Inthamoussou, R. J. Mantz, and H. D. Battista, "Flexible power control of fuel cells using sliding mode techniques," *J. Power Sources*, vol. 205, pp. 281–289, May. 2012.

[9] M. Uzunoglu and M. S. Alam, "Modeling and Analysis of an FC/UC Hybrid Vehicular Power System Using a Novel-Wavelet-Based Load Sharing Algorithm," *IEEE Trans. Energy Convers.*, vol. 23, no. 1, pp. 263–272, Mar. 2008.

[10] A. Y. Goharrizi and N. Sepehri, "A Wavelet-Based Approach to Internal Seal Damage Diagnosis in Hydraulic Actuators," *IEEE Trans. Ind. Electron.*, vol. 57, no. 5, pp. 1755–1763, May. 2010.

[11] M. R.-Guasp, J. A. A.-Daviu, M, P.-Sanchen, R. P.-Panadero, and J. P.-Cruz, "A General Approach for the Transient Detection of Slip-Dependent Fault Components Based on the Discrete Wavelet Transform," *IEEE Trans. Ind. Electron.*, vol. 55, no. 12, pp. 4167–4180, Dec. 2008.

[12] B. Chen, Z. Zhang, C. Sun, B. Li, Y. Zi, and Z. He, "Fault feature extraction of gearbox by using overcomplete rational dilation discrete wavelet transform on signals measured from vibration sensors," *Mech. Syst. Signal Pr.*, vol. 33, pp. 275-298, Nov. 2012.

[13] M. A. Farahani, M. T. V. Wylie, E, Castillo-Guerra, and B. G. Clopitts, "Reduction in the Number of Averages Required in BODTA Sensors Using Wavelet Denoising Technique," *J . Lightwave Technol.*, vol. 30, no. 8, pp. 1134–1142, Apr. 2012.

[14] C.-C. Liu, T.-Y. Sun, S.-J. Tsai, Y.-H. Yu, and S.-T. Hsieh, "Heuristic wavelet shrinkage for denoising," *Appl. Soft Comput.*, vol. 11, pp. 256–264, Jan. 2011.

[15] M. Beenamol, S. Prabavathy, and M. Mohanalin, "Wavelet based seismic signal de-noising using Shannon and Tsallis entropy," *Comput. Math. Appl.*, vol. 64, pp. 3580–3593, Dec. 2012.

[16] D. Giaouris, J. W. Finch, O. C. Ferreira, R. M. Kennel, and G. M,. El-Murr, "Wavelet Denoising for Electric Drives," *IEEE Trans. Ind. Electron.*, vol. 55, no. 2, pp. 543–550, Feb. 2008.

[17] Y. Kopsinis and S. McLaughlin, "Development of EMD-based Denoising Methods Inspired by Wavelet Thresholding," *IEEE Trans. Signal Proces.*, vol. 57, no. 4, pp. 1351–1362, Apr. 2009.

[1] [18] S. Lee and J. Kim, "Discrete wavelet transform-based denoising technique for advanced state-of-charge estimator of a lithium-ion battery in electric vehicles," *Energy*, vol. 83, pp. 462–473, Apr. 2015.

[19] J. Rafiee, P. W. Tse, A. Harifi, and M. H. Sadeghi, "A novel technique for selecting mother wavelet function using an intelligent fault diagnosis system," *Expert Syst. Appl.*, vol. 36, pp. 4862–4875, Apr. 2009.

[20] F. Adamo, G. Andria, F. Attivissimo, A. M. L. Lanzolla, and M. Spadavecchia, "A comparative study om mother wavelet selection in ultrasound image denoising," *Measurement*, vol. 46, pp. 2447–2456, Oct. 2013.

[21] http://www.lgchem.com/global/main

[22] J. Kim and B. H. Cho, "Screening process-based modeling of the multi-cell battery string in series and parallel connections for high accuracy state-of-charge estimation," *Energy*, vol. 57, pp. 581–599, Aug. 2013.

A Feedforward Control Based Power Decoupling Scheme for Voltage-Controlled Grid-Tied Inverters

Baojin Liu, Zeng Liu, Jinjun Liu, Teng Wu, Shike Wang
State Key Lab of Electrical Insulation and Power Equipment, School of Electrical Engineering
Xi'an Jiaotong University
Xi'an, China
liubaojin.pe@gmail.com

***Abstract*—Voltage-controlled inverters have been widely used in distributed generation due to their flexibility and smooth transition between grid-connected and islanded modes. However, in the grid-tied mode, the inherent coupling between active and reactive power will challenge the independent control of active power and reactive power, and therefore deteriorate system performance. To deal with this issue, this paper proposes a feedforward control based power decoupling scheme for voltage-controlled grid-tied inverters. A phase angle feedforward and a magnitude feedforward are introduced separately to the active power control loop and reactive power control loop, adjusting the output phase angle and magnitude to address the coupling problem. The feedforward phase angle and magnitude are regulated in every control cycle according to the power equation to cancel the coupling precisely. The effectiveness of the proposed decoupling scheme has been validated by simulation and experimental results.**

Keywords—Votage-controlled grid-tied inverter; power control; decoupling; feedforward control

I. Introduction

Grid-tied inverters play a critical role in utility application of power electronics [1]. According to the basic control mechanism adopted, the grid-tied inverter can mainly be classified into current-controlled one and voltage-controlled one [2]. Conventionally, current-controlled grid-tied inverter is widely applied in wind turbine and photovoltaic systems to easily achieve MPPT algorithm [3, 11]. However, voltage-controlled grid-tied inverter is more attractive in distributed generation as it can provide direct voltage and frequency support for islanding operation [4].

The coupling between active power and reactive power in the grid-tied inverter will challenge the independent control of active power and reactive power, and therefore deteriorate system performance [5]. For current-controlled grid-tied inverter, several power decoupling control methods have been proposed, including inductor current feed forward (ICFF) technique [6, 7], complex vector PI(VPI) controller [8, 9] and reference current feed-forward (RCFF) [10]. However, these power decoupling methods cannot be implemented directly into voltage-controlled grid-tied inverter due to the difference between their basic control mechanisms. Meanwhile the power decoupling control for voltage-controlled grid-tied inverter is seldom concerned in existing publications. In order to overcome aforementioned problem, this paper proposes a power decoupling scheme for voltage-controlled grid-tied inverter, which is based on the phase angle feedforward and voltage magnitude feedforward, and the coupling between active power and reactive power can be effectively suppressed. This work starts with description and coupling analysis of the studied system in section II, and the proposed decoupling control scheme is presented in section III. In section IV, simulation and experimental results are provided to verify its effectiveness.

This work was supported by the National Natural Science Foundation of China under Grant 51437007, and the Power Electronics Science and Education Development Program of Delta Environmental & Educational Foundation under Grant DREM2014002.

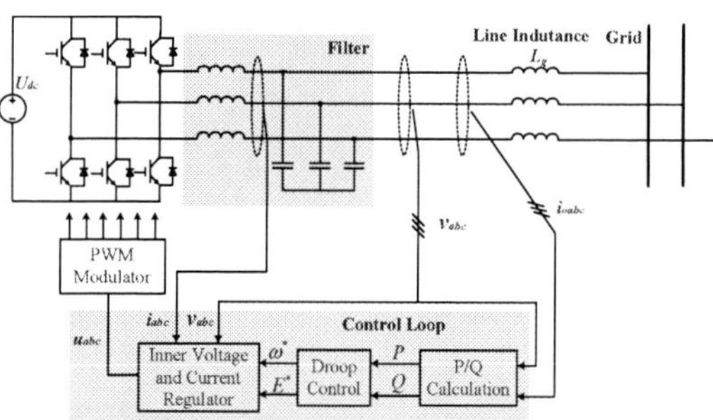

Fig.1 Configuration of grid-tied inverter

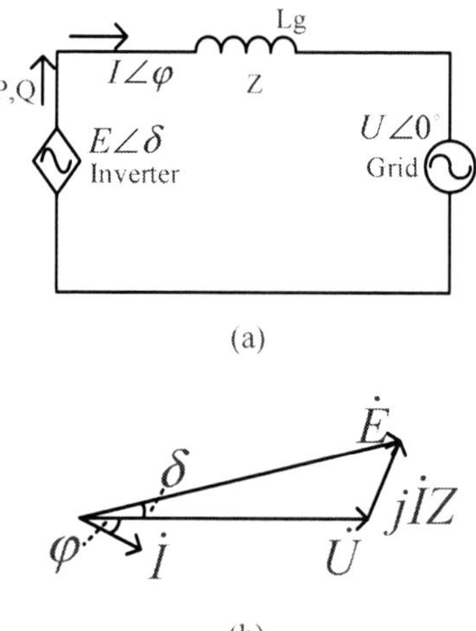

Fig.2 Simplified model of a grid-tied inverter. (a) Equivalent circuit. (b) Phasor diagram.

II. System Analysis

The configuration of a voltage-controlled grid-tied inverter is shown in Fig.1. The inverter is connected to the grid through a link inductance L_g. The output power injected to the grid is controlled by regulating the phase and magnitude of the capacitor voltage. Thus the system can be simplified as Fig.2 (a) shows. The inverter is regarded as a controllable voltage source and the grid is modeled as an ideal voltage source. The phasor diagram is illustrated in Fig.2 (b). The output active power (P) and reactive power (Q) can be calculated by (1).

$$\begin{cases} P = \dfrac{U}{Z} E \sin\delta \\ Q = \dfrac{1}{Z}\left(E^2 - UE\cos\delta\right) \end{cases} \tag{1}$$

Where $E\angle\delta$ is the output voltage of the inverter, $U\angle 0°$ is the voltage of the utility grid and the impedance of the transmission line is Z.

Usually the phase angle δ is very small, so it can be assumed that $\sin\delta \approx \delta$ and $\cos\delta \approx 1$. And the small signal linearization model can be expressed as:

$$\begin{cases} \hat{p} = \dfrac{UE}{Z}\hat{\delta} \\ \hat{q} = \dfrac{2E-U}{Z}\hat{e} \end{cases} \tag{2}$$

According (2), P is proportional to the angle δ, and Q is proportional to the magnitude E. P and Q are approximately decoupled in small signal model. In voltage-controlled inverter, the conventional P-f and Q-E droop control as Fig.3 shows is based on the small signal linearization model. However it is equation (1) that could be used to precisely depict the relationship between P, Q, δ and E. Therefore in large signal model (e.g. the steady state operating point is changed), P and Q are still coupled. For example, when the active power command changes to inject more active power into the grid and reactive power command remains constant, the output reactive will generate some fluctuation due to the coupling of power. This unwanted fluctuation may lead to grid voltage oscillation or some other stability issues. Therefore the decoupling mechanism is urgently required.

III. Proposed Power Decoupling Scheme

In order to overcome the coupling problem, a phase angle feedforward and a magnitude feedforward are respectively introduced to the active power control loop and reactive power control loop, adjusting the output phase angle and magnitude. The feedforward phase angle and magnitude are regulated in every control cycle according to (1). The calculation process is introduced below.

A. Decoupling control of reactive power

Assuming that P reference changes at t_0, the output reactive power at t_0 can be presented as:

$$Q(t_0) = \frac{1}{Z}\left(E_0^2 - UE_0\cos\delta_0\right) \tag{3}$$

After a very short time interval Δt ($t_1 = t_0 + \Delta t$), the voltage magnitude and phase angle becomes E_1 ($E_1 = E_0 + \Delta E$) and δ_1 ($\delta_1 = \delta_0 + \Delta\delta$), respectively. The output reactive power at t_1 is:

$$Q(t_1) = \frac{1}{Z}\left[(E_0 + \Delta E_1)^2 - U(E_0 + \Delta E_1)\cos(\delta_0 + \Delta\delta_1)\right] \tag{4}$$

Reactive power should not vary during this interval since the reference value of Q doesn't change. So

$$Q(t_1) = Q(t_0) \tag{5}$$

After ignoring the high order item ($\Delta E_1 \cdot \Delta\delta_1$), and assuming $\sin(\Delta\delta_1) \approx \Delta\delta_1$, $\cos(\Delta\delta_1) \approx \Delta\delta_1$, the following equation is obtained:

$$\Delta E_1 = \frac{UE_0 \sin\delta_0}{U\cos\delta_0 - 2E_0}\Delta\delta_1 \tag{6}$$

Similarly, after another short interval Δt ($t_2 = t_1 + \Delta t$), by making $Q(t_2) = Q(t_1)$, ΔE_2 can be calculated by (7).

$$\Delta E_2 = \frac{UE_1 \sin\delta_1}{U\cos\delta_1 - 2E_1}\Delta\delta_2 \tag{7}$$

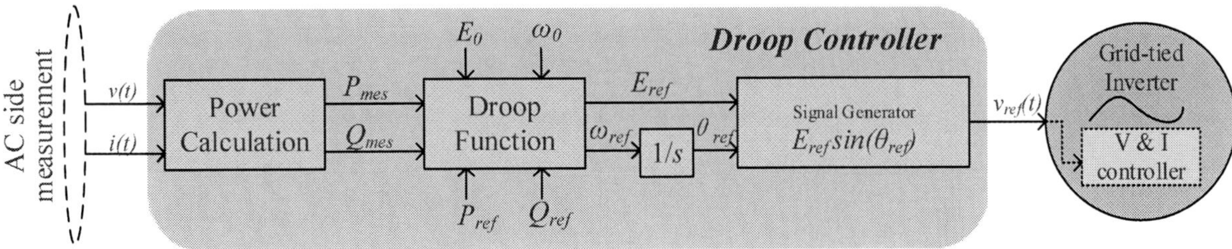

Fig.3 Conventional P-f and Q-E droop controller block diagram

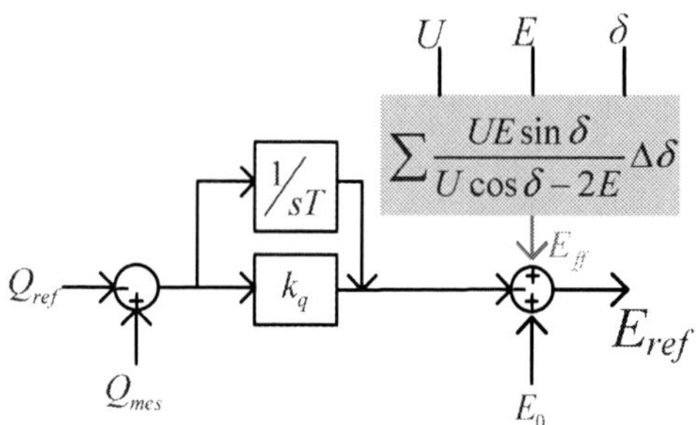

Fig.4 Control block diagram of reactive power with magnitude feedforward

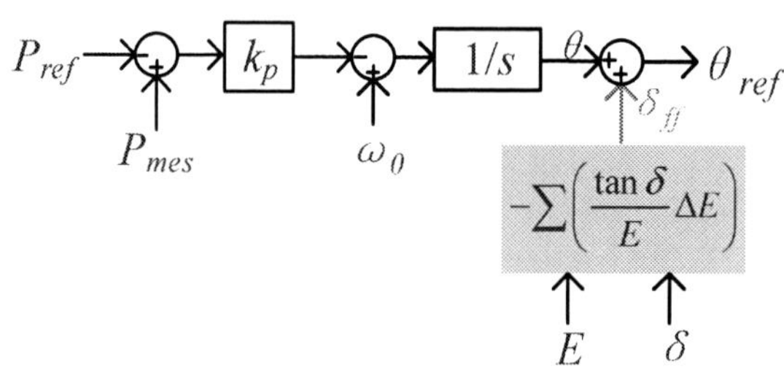

Fig.5 Control block diagram of active power with phase angel feedforward

Then, the voltage magnitude becomes:

$$E_2 = E_1 + \Delta E_2 = E_0 + \Delta E_1 + \Delta E_2 \tag{8}$$

So after n time steps, the magnitude reaches:

$$E_n = E_0 + \Delta E_1 + \Delta E_2 + \cdots + \Delta E_n = E_0 + \sum \Delta E \tag{9}$$

The aforementioned calculation indicates that the output reactive power can be kept as constant by adding a feedforward item E_{ff}, which can be calculated by (10).

$$E_{ff} = \sum \Delta E = \sum \frac{UE\sin\delta}{U\cos\delta - 2E}\Delta\delta \tag{10}$$

E_{ff} is introduced to the reactive power control loop as illustrated in Fig.4.

B. Decoupling control of active power

TABLE I. SIMULATION PARAMETERS

Parameters	Values	Parameters	Values
ω_{grid}	50.0Hz	U_{grid}	220V
R_g	1mΩ	L_g	10mH
k_p	0.0001Hz/W	k_q	0.0001V/var
ω_0	50Hz	E_0	310V

The feedforward item in active power control loop can be derived by following the same process as above. When Q reference changes while P reference doesn't vary, the output active power should be kept as constant.

$$P(t_0) = \frac{U}{Z}E_0\sin\delta_0 = P(t_1) = \frac{U}{Z}(E_0 + \Delta E)\sin(\delta_0 + \Delta\delta) \tag{11}$$

Ignoring the high order item and assuming that $\Delta\delta$ is small, (12) is obtained:

$$\Delta\delta = -\frac{\tan\delta}{E}\Delta E \tag{12}$$

Then the feedforward item δ_{ff} can be derived by:

$$\delta_{ff} = \sum \Delta\delta = -\sum \frac{\tan\delta}{E}\Delta E \tag{13}$$

δ_{ff} is introduced to the phase angle controller to regulate the output active power as Fig.5 shows.

The overall power control block diagram with the proposed feedforward control is shown in Fig.6. We use red color to emphasize the feedforward decoupling items.

IV. SIMULATION AND EXPERIMENTAL RESULTS

The proposed power decoupling method is verified through simulation in PSCAD and experiment on a 10kVA prototype.

A. Simulation Results

The system parameters used in the simulation are listed in Table I. Two simulation scenarios are designed as below. At first, the P reference is set to 10kW and the Q reference is set to 0kvar. At the moment t=1.5 sec, the P reference decreases to 5kW and then increases back to 10kW at t=2.5 sec. However the Q reference remains 0kvar during this

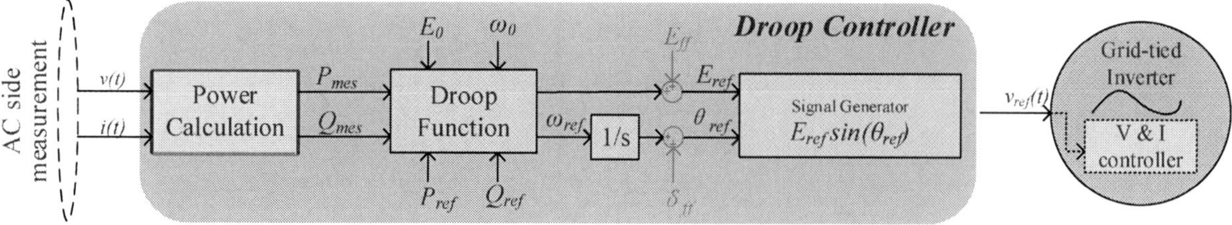

Fig.6 Overall power control block diagram with the proposed feedforward decoupling control scheme

period to investigate the decoupling effectiveness when P reference changes. The simulation result is shown in Fig.7. The second simulation is at t=3.5 sec, the Q reference increases to 6kvar and then change back to 0kvar at t=4.5 sec. he P Reference remains 10kW during this period to investigate the decoupling effectiveness when Q reference changes. The simulation result is shown in Fig.8.

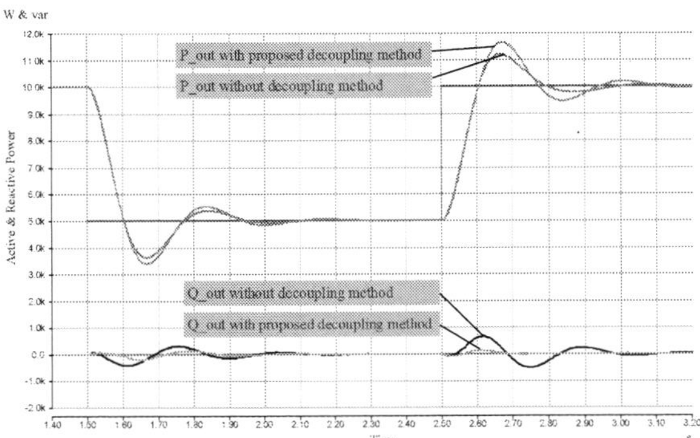

Fig.7 Simulation results of voltage-controlled grid-tied inverter when active power reference is changed

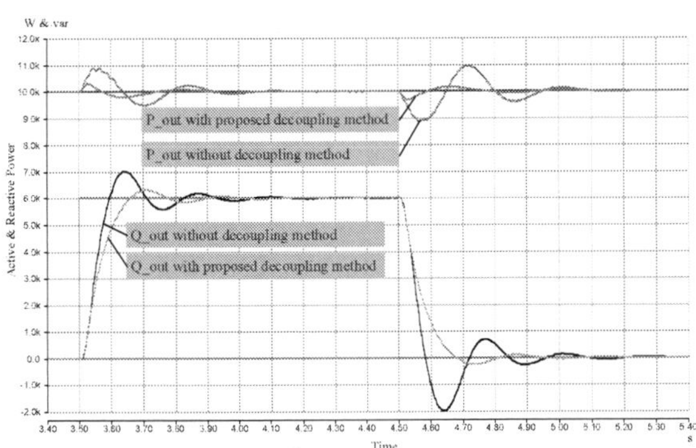

Fig.8 Simulation results of voltage-controlled grid-tied inverter when reactive power reference is changed

According to the result in Fig.7, the fluctuation in the output reactive power, which is caused by the change of P reference, can be effectively suppressed when the proposed decoupling method is adopted. In Fig.8, the fluctuation of active power in decoupled condition is much smaller than the one in coupled case, which means that the decoupled control scheme is effective.

TABLE II. EXPERIMENTAL PARAMERTERS

Parameters	Values	Parameters	Values
ω_{grid}	50.0Hz	U_{grid}	200V
R_g	1mΩ	L_g	2.6mH
k_p	10^{-4}Hz/W	k_q	10^{-3}V/var
ω_0	50Hz	E_0	240V

B. Experimental Results

Experiments are conducted to verify the proposed method based on a 10kVA prototype as shown in Fig.9. The system parameters are listed in Table II. The voltage and current waveform of the grid-tied inverter is shown in Fig.10. The phase and magnitude differences between inverter voltage and grid voltage can be easily noticed.

In the first experiment, the active power reference is stepped from 2kW to 3kW, and the reactive power reference is kept 0kvar as shown in Fig.11. The regulation of active power is similar with or without decoupling, but the fluctuation in reactive power under decoupled condition is much smaller than the case without decoupling.

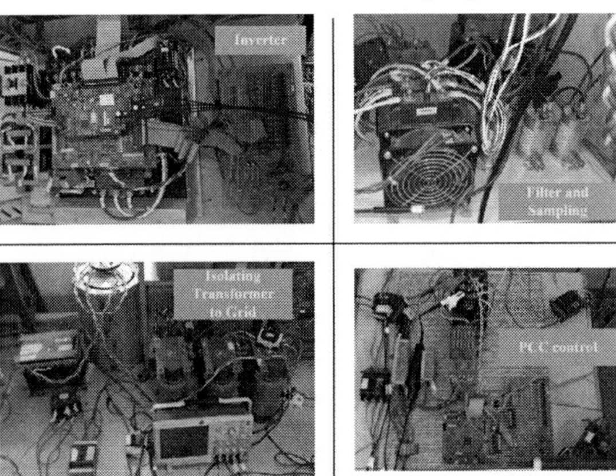

Fig.9 Experimental prototype

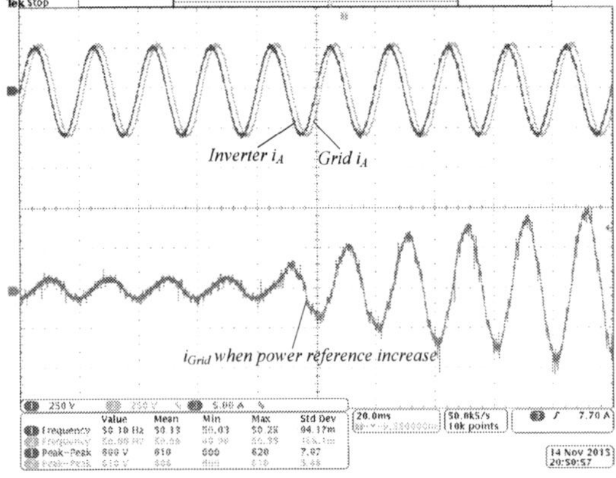

Fig.10 Voltage and current waveform of grid-tied inverter

The second experiment verifies the performance when the reactive power reference is changed to 1kvar and P reference is kept at 2kW. The experimental result is illustrated in Fig.12. We can see there is a valley in the curve of active power when there is no decoupling mechanism. But the active power can remain unchanged when the decoupling method is introduced to the control loop.

It is worth noticing that the deviation of active power from 2kW under decoupled condition in Fig.12 is caused by the gird frequency deviation from 50Hz. Since the grid voltage is not as stable as an ideal voltage source, this deviation is unavoidable.

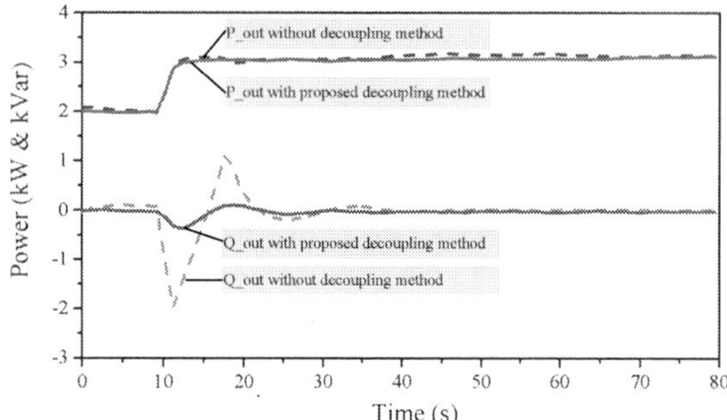

Fig.11 Experimental results of voltage-controlled grid-tied inverter when active power reference is changed from 2kW to 3kW

The experimental results show that the proposed method can effectively mitigate the coupling between active and reactive power.

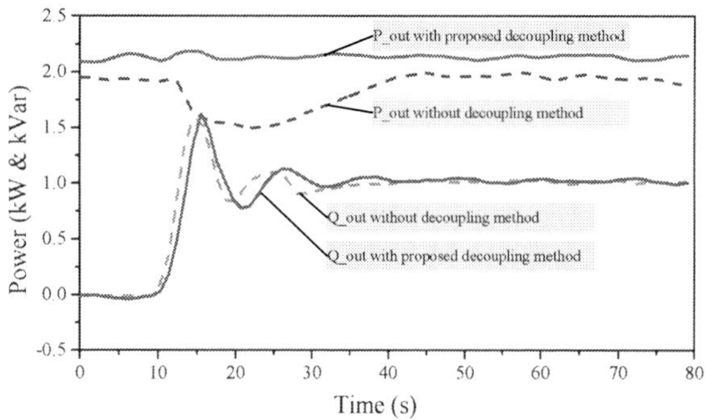

Fig.12 Experimental results of voltage-controlled grid-tied inverter when reactive power reference is changed from 0kvar to 1kvar

V. Conclusions

This paper proposes a novel power decoupling scheme for voltage-controlled grid-tied inverters. This control scheme can decouple the active and reactive power through adding a phase angle feedforward and a magnitude feedforward into active and reactive power control loops, respectively. The analysis and design methodology of the feedforward items are presented in detail. Simulation and experimental results have been obtained to validate the effectiveness of the proposed decoupling strategy.

References

[1] J. M. Carrasco, L. G. Franquelo, J. T. Bialasiewicz, E. Galvan, R. C. P. Guisado, M. A. M. Prats, *et al.*, "Power-Electronic Systems for the Grid Integration of Renewable Energy Sources: A Survey," *Industrial Electronics, IEEE Transactions on*, vol. 53, pp. 1002-1016, 2006.

[2] K. Sung-Hun, S. R. Lee, H. Dehbonei, and C. V. Nayar, "Application of voltage- and current-controlled voltage source inverters for distributed generation systems," *Energy Conversion, IEEE Transactions on*, vol. 21, pp. 782-792, 2006.

[3] J. Rocabert, A. Luna, F. Blaabjerg, Rodri, x, and P. guez, "Control of Power Converters in AC Microgrids," *Power Electronics, IEEE Transactions on*, vol. 27, pp. 4734-4749, 2012.

[4] Y. Guan, W. Wu, X. Guo, and W. Herong, "An improved droop controller for grid-connected voltage source inverter in microgrid," in *Power Electronics for Distributed Generation Systems (PEDG), 2010 2nd IEEE International Symposium on*, 2010, pp. 823-828.

[5] S. Xiaofeng, T. Yanjun, and C. Zhe, "Adaptive decoupled power control method for inverter connected DG," *Renewable Power Generation, IET*, vol. 8, pp. 171-182, 2014.

[6] J. G. Hwang, P. W. Lehn, and M. Winkelnkemper, "A Generalized Class of Stationary Frame-Current Controllers for Grid-Connected AC-DC Converters," *Power Delivery, IEEE Transactions on*, vol. 25, pp. 2742-2751, 2010.

[7] B. Bahrani, S. Kenzelmann, and A. Rufer, "Multivariable-PI-Based dq Current Control of Voltage Source Converters With Superior Axis Decoupling Capability," *Industrial Electronics, IEEE Transactions on*, vol. 58, pp. 3016-3026, 2011.

[8] F. Briz, M. W. Degner, and R. D. Lorenz, "Analysis and design of current regulators using complex vectors," in *Industry Applications Conference, 1997. Thirty-Second IAS Annual Meeting, IAS '97., Conference Record of the 1997 IEEE*, 1997, pp. 1504-1511 vol.2.

[9] J. Holtz, Q. Juntao, J. Pontt, J. Rodriguez, P. Newman, and H. Miranda, "Design of fast and robust current regulators for high-power drives based on complex state variables," *Industry Applications, IEEE Transactions on*, vol. 40, pp. 1388-1397, 2004.

[10] Z. Sizhan, L. Jinjun, Z. Linyuan, and S. Hongwei, "Cross-coupling and decoupling techniques in the current control of grid-connected voltage source converter," in *Applied Power Electronics Conference and Exposition (APEC), 2015 IEEE*, 2015, pp. 2821-2827.

[11] W. Bin, L. Jia, and Z. Fang, "The Micro-Grid fast simulation platform exploitation based on PSCAD," in *Applied Power Electronics Conference and Exposition (APEC), 2011 Twenty-Sixth Annual IEEE*, 2011, pp. 1737-1742.

Light Load Efficiency Improvement of Solar Farms Three-Phase Two-Stage Module Integrated Converter

Ahmadreza Amirahmadi, Utsav Somani, Mahmood Alharbi, Charlie Jourdan, Issa Batarseh

University of Central Florida
Department of Electrical and Computer Engineering

Abstract—**In this paper, a method of maintaining high power conversion efficiency across the entire load range of a three-phase micro-inverter is proposed. The proposed control method substantially increases the conversion efficiency at light loads by minimizing switching losses of semiconductor devices as well as core losses of magnetic components. This is accomplished by entering a phase skipping operating mode wherein two phases of an inverter are disabled and three inverters combined to form a new three phase system with minimal grid imbalance. The performance of the proposed control method was evaluated on a 400W, 208VAC three-phase micro-inverter.**

Keywords—Three-phase Two-Stage Module Integrated Converter, Solar Farm, Efficiency Improvement, Phase Skipping Control

I. Introduction

Micro-inverter converts and controls the DC power from each PV module into grid-compliant AC power. It has many advantages over the conventional centralized, string and multi-string inverters such as maximized energy harvest, improved safety, increased lifetime and reliability, enhanced performance monitoring, and simplified PV array design and installation [1]-[2].

Efficiency improvement of micro-inverters plays a significant role in gaining economical benefits, higher power density and reliability. Due to the intermittent feature of the PV generation, it is impossible for micro-inverter to always operate at its rated power. It usually operates at light or half load conditions. Therefore, European and CEC weighted-efficiency are mostly used to evaluate overall efficiency of these devices [3].

Soft switching techniques can be employed to maintain a high conversion efficiency and low EMI [4]. Considerable research has been conducted on the topologies and control methods of micro-inverters to achieve soft switching operation [2],[4]-[7]. ZVS BCM current control is an interesting soft switching candidate for micro-inverter applications due to its low cost, high power density and high efficiency [5]. The digital implementation of this soft switching approach known as Hybrid BCM current control has been verified in [2]. Different current modulation schemes of this control approach and their loss analysis have been presented in [6]. Although they can enhance the efficiency of the micro-inverter, they still suffer from poor light load efficiency.

This paper presents a method of maintaining high power-conversion efficiency across the entire load range of three-phase micro-inverter. The proposed method guarantees a flatter conversion efficiency curve at light load for the PV system.

II. Zero Voltage Switching Boundary Conduction Mode Current Control

Figure 1 shows the power circuit of the half bridge three-phase micro-inverter. In this figure the diode and capacitor in parallel with the switch are the body diode and the parasitic capacitance of the MOSFET respectively. In ZVS BCM control as shown in Fig. 2 the inductor current is intentionally operated in both directions within a switching cycle. It is controlled to pass through the body diode of the MOSFET after discharging its output capacitor to generate ZVS condition during commutations [5].

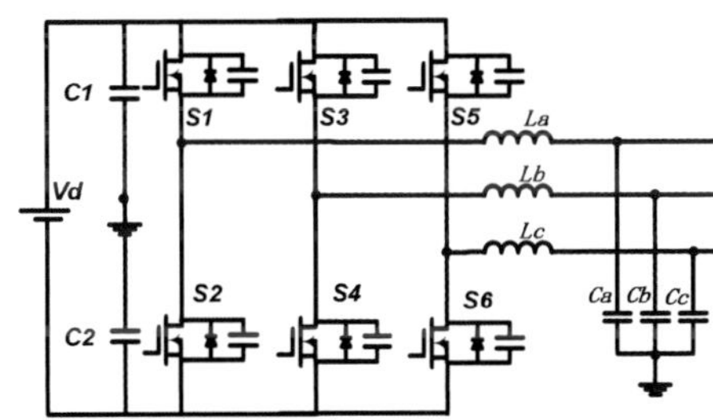

Fig. 1. Half Bridge Three-Phase Inverter

Implementation of the ZVS BCM control requires setting the upper and lower boundaries (limits) of the inductor current. The value of the reverse current during each half line period is determined by the amount of current required to charge the inductor such that its discharging current during the dead-time interval is large enough to discharge the parasitic capacitance and forward bias the body diode of the MOSFET about to turn

978-1-4673-9551-9/16 $31.00 © 2016 IEEE

ON, therefore ensuring ZVS operation for the switches. The other limit is set to ensure that the average current during each switching cycle is equal to the reference current [2]. The upper and lower boundaries of the inductor current with fixed reverse current modulation scheme are shown in Fig. 3 and expressed as equation (1).

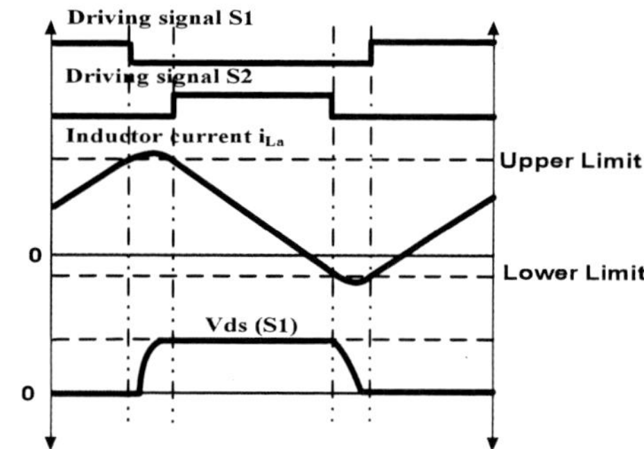

Fig. 2. Zero Voltage Transition with BCM ZVS Current Control

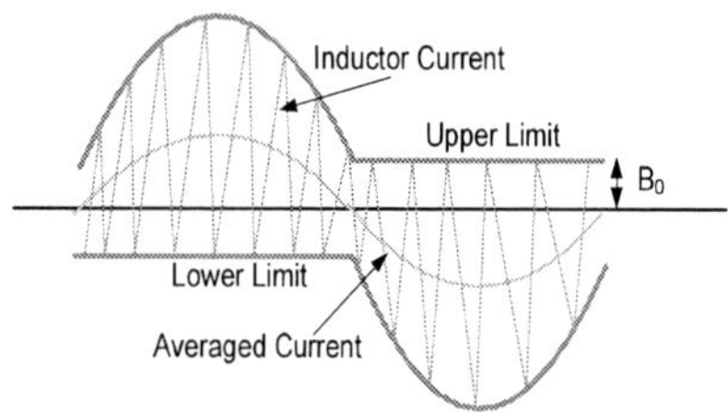

Fig. 3. Fixed Reverse Current Modulation Scheme

$$\begin{cases} i_{upper} = 2i_{ref}\sin(\omega t) + B_0 \\ i_{lower} = -B_0 \end{cases} \quad if\ i_{ref}\sin(wt) \geq 0$$
$$\begin{cases} i_{upper} = B_0 \\ i_{lower} = 2i_{ref}\sin(\omega t) - B_0 \end{cases} \quad if\ i_{ref}\sin(wt) < 0 \qquad (1)$$

Figure 4 shows the calculated loss distribution of a 400W three-phase micro-inverter with fixed reverse current modulation scheme. This figure illustrates the power losses under different load conditions. The power losses break-down includes the conduction and turn-off switching loss of the MOSFETs, core and copper loss of the inductors, and anti-parallel diode losses.

It can be clearly seen in Fig. 4 that turn-off switching losses and core losses of magnetic components are almost independent of the load. These fixed sources of power losses cause micro-inverter efficiency curve as a function of the load to exhibit a steep fall off at light loads. In this paper a new control method of reducing the variation of power conversion efficiency with fluctuations of the PV module's output power is presented.

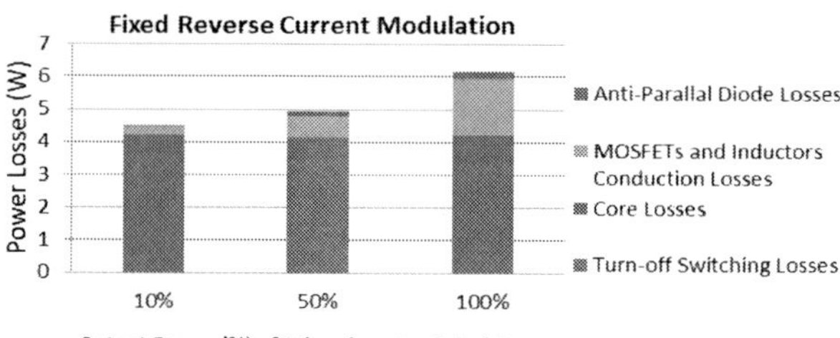

Fig. 4. Power Losses Break-down for ZVS BCM Current Control Three Phase Micro Inverter at 10%, 50% and 100% of Micro-inverter Rated Power (400W)

III. Light Load Efficiency Enhancement

The proposed method of light-load efficiency optimization is based on a simple observation that the maximum efficiency occurs when each phase of the three phase micro-inverter is operated close to full load.

Due to independent operation of each phase in the hybrid BCM current control micro-inverter, it is capable of disabling two phases at light loads thereby improving light load efficiency. Through a MPPT algorithm, available PV module's power can be determined. Once the available power falls below the preset light load point, two phases are turned off. Similarly, when available power increases above the light load point, the disabled phases will be enabled. Fig. 5 shows a two-stage micro-inverter under normal and phase skipping control operation. The DC/AC stage operation under phase skipping control is shown in Fig. 6.

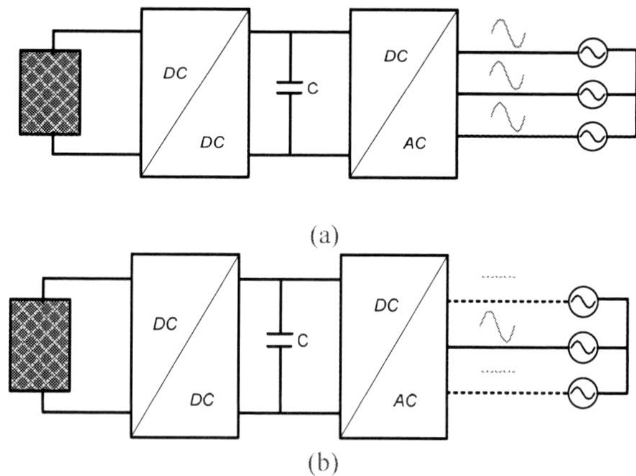

Fig. 5. Light Load Efficiency Enhancement for a Two-stage Three-Phase Micro-Inverter. (a) Normal Operation. (b) Phase Skipping Mode Operation

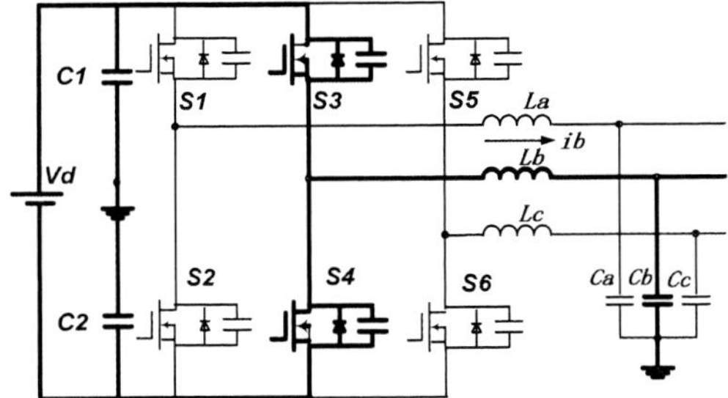

Fig. 6. DC/AC Stage of Three Phase Micro-inverter under Phase Skipping Mode

The idea here is to combine the advantages of three-phase systems and single-phase systems by employing phase skipping technology in order to optimize efficiency over a wide load range [8]-[10].

DC Link Capacitor

The transition from three-phase to single-phase happens in power levels less than 30% of the rated power of the micro-inverter. At these low power levels the voltage transient response and steady state ripples of the DC link capacitor does not cause any over or under voltage problems. For a single phase micro-inverter the size of the decoupling capacitor can be determined using equation (2) knowing that the power into the DC link is constant and that the power drawn from the DC link follows a sin2(ωt) waveform.

$$C_{DC} = \frac{P_{DC}}{2 \cdot \omega \cdot U_{DC} \cdot \hat{u}_{DC}} \quad (2)$$

Where P_{DC} is the average DC link power, ω is the grid frequency, U_{DC} is the average DC link voltage and $\hat{u}_{DC}$ is the amplitude of the ripple voltage. The grid current cannot be controlled if the DC link voltage is lower than the peak grid voltage plus the voltage drop across the semiconductors and output filter. Therefore the minimum DC link voltage at 10% over-voltage in the grid for the micro-inverter is calculated as 375V. Considering that UDC is 400V, DC link capacitor is calculated by equation (2) as 17μF at 130W.

Size calculation of the decoupling capacitor for the three-phase micro-inverter has been carried on in [7]. Considering grid voltage dips and surges, and disturbance response time it is determined as 35.3μF which meets the requirements of phase skipping operation mode.

IV. Solar Farm Architecture based on Three-phase Micro-inverter

The PV solar farm architecture based on three-phase micro-inverters is shown in Fig. 7. In this architecture the outputs of each micro-inverter are directly connected to a low voltage three-phase grid and then through a medium voltage transformer to the high voltage at power transmission line. Each micro-inverter operates independently regardless of failure of the other micro-inverters and carries out communication and MPPT functions.

Each micro-inverter comprises a node on a communication network and can exchange information with other micro-inverters on the network as well as a system level controller. This communication is an integral part of system maintenance and monitoring. ZigBee protocol enabled devices are often used in mesh network form to transmit data over longer distances, passing data through intermediate devices to reach more distant ones.

Each micro-inverter operates in normal control mode. Under light load conditions, it attempts to change its operating mode to phase skipping control but it is not allowed to until it finds two other micro-inverters in a similar state. After discovering each other, they build a three member cluster and then change their operating mode to phase skipping control simultaneously. Once they are in phase skipping mode, each one of them selects a unique phase. The idea here is to combine three different three-phase systems and create a new three-phase system out of three micro-inverters operating in single phase mode.

When one micro-inverter exits single phase operating mode, it commands the other two micro-inverters in its cluster to exit single phase mode regardless of their power output. It is important to remember that the default operating mode of each micro-inverter is three phase at all power levels and that phase skipping is strictly a method of improving the efficiency at low power.

It should be noted that each micro-inverter of a cluster operates below 30% of its rated power. At this low power any difference at their output power causes negligible imbalance in the grid.

V. Experimental Results

To verify the light load efficiency enhancement using the proposed phase skipping control method a 400W three-phase micro-inverter prototype was built. The specifications are as follows: input voltage: Vdc=400V; Grid voltage: Vac=208V; Grid frequency: F=60Hz; and the major components used in the power stage are as follows: inductor filter: La=270uH; capacitor filter: 1uF; MOSFETs: FCB20N60. The controller was implemented using a microchip DSP dsPIC33F.

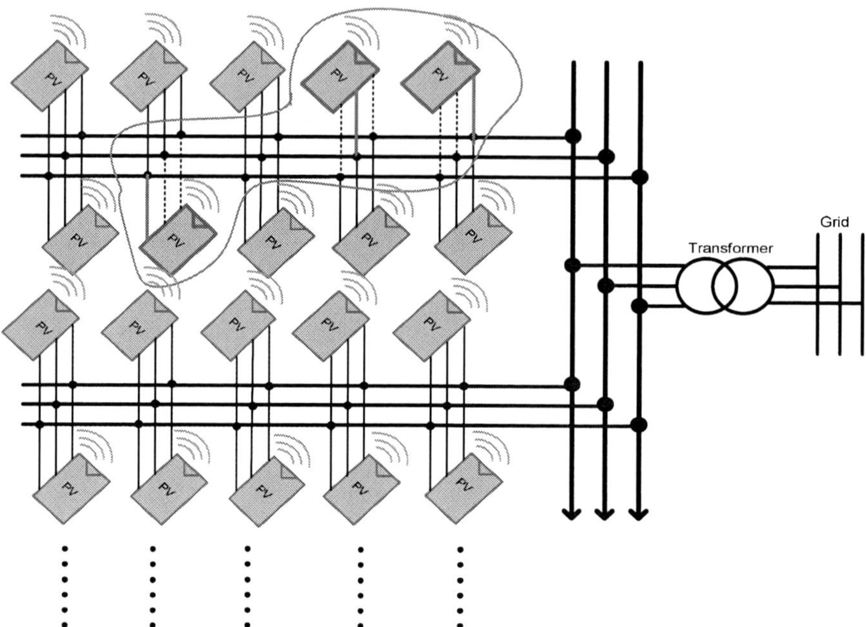

Fig. 7. Solar Farm Architecture based on Three Phase Micro-inverter, Three Micro-inverters are Operating in Phase Skipping Mode

Inductor current waveforms for ZVS BCM current control three-phase micro-inverter under the full-load condition is shown in Fig. 8(a). Fig. 8(b) illustrates the three phase output current. The output current THD is less than 2.5% and meets the industrial requirements.

Transition from normal operation mode to phase skipping mode is given in Fig. 9. Notice that the transition is very smooth and voltage ripple in the DC link capacitor meets the design requirements. THD of the output current remains less than 2.5% in the phase skipping mode.

Fig. 10 shows a comparison of the three-phase micro-inverter measured efficiency for original hybrid BCM current control and the one with phase skipping control. Owing to phase skipping control, higher efficiency can be achieved at light load without additional cost. All the efficiency measurements were performed using a Yokogawa PZ4000 power analyzer. The measured efficiency does not include auxiliary circuit power consumption.

VI. Conclusion

In this paper, a new control strategy for efficiency enhancement of solar farms based on three-phase micro-inverters was proposed. A three-phase micro-inverter employing the proposed control method has flat efficiency curve throughout its entire load range which improves its CEC weighted efficiency. In phase skipping control, switching losses of semiconductor devices, as well as core losses of magnetic components are minimized by disabling two of the three phases.

To reduce the effect of phase skipping control on the utility grid, three three-phase micro-inverters under light load conditions are combined to create a new three phase system with each micro-inverter feeding a different phase. The experimental results verified that with the phase skipping control method the light load efficiency is improved more than 2 percentage points.

References

[1] Z. Zhang, X. He, and Y. Liu, "An Optimal Control Method for Photovoltaic Grid-Tied-Interleaved Flyback Micro-inverters to Achieve High Efficiency in Wide Load Range" IEEE Transactions on Power Electronics, Vol. 28, No. 11, pp. 5074-5087, Nov. 2013

[2] A. Amirahmadi, H. Hu, A. Grishina, Q. Zhang, L. Chen, U. Somani, I. Batarseh "Hybrid ZVS BCM Current Controlled Three-phase Micro-inverter" IEEE Transactions on Power Electronics, Vol. 29, No. 4, pp. 2124-2134, April 2014

[3] L. Zhang, K. Sun, H. Hu, Y. Xing "A System-Level Control Strategy of Photovoltaic Grid-tied Generation Systems for European Efficiency Enhancement" Accepted on July 22, 2013 for publication in IEEE Transactions on Power Electronics

[4] D. Zhang, Q. Zhang, A. Grishina, A. Amirahmadi, H. Hu, J. shen, I. Batarseh, "A Comparison of Soft and Hard-switching Losses in Three phase Micro-inverters" in Proc. Energy conversion congress and exposition (ECCE) 2011, pp. 1076-1082

[5] Q. Zhang, H. Hu, D. Zhang, X. Fang, J. Shen, I. Bartarseh "A Controlled-type ZVS Technique without Auxiliary Components for the

Low Power DC/AC Inverter" IEEE Transactions on Power Electronics, Vol. 28, No. 7, pp. 3287-3296, 2013

[6] A. Amirahmadi, L. Chen, U. Somani, H. Hu, N. Kutkut, I. Batarseh "High Efficiency Dual-mode Current Modulation Method for Low Power DC/AC Inverters" Accepted on Oct 5, 2013 for publication in IEEE Transactions on Power Electronics

[7] L. Chen, A. Amirahmadi, Q. Zhang, N. Kutkut, I. Batarseh "Design and Implementation of Three-phase Two-stage Grid-connected Module Integrated Converter" Accepted on Dec 6, 2013 for publication in IEEE Transactions on Power Electronics

[8] U. Somani, A. Amirahmadi , C. Jourdan, I. Batarseh, Power Inverter Implementing Phase Skipping Control, Pending Patent, International Application No.: PCT/US14/37116, Application No.:14/271949

[9] U. somani, C. Jourdan, A. Amirahmadi, A. Grishina, H. Hu, I. Batarseh "Phase Skipping Control to Improve Light Load Efficiency of Three Phase Micro-Inverters" IEEE Applied Power Electronics Conference and Exposition 2014

[10] S. Milad Tayebi, Charles Jourdan, Issa Batarseh "Advanced phase-skipping control with improved efficiency of three-phase micro-inverters" IEEE Energy Conversion Congress and Exposition (ECCE), pp. 3802–3806, Sept. 2015.

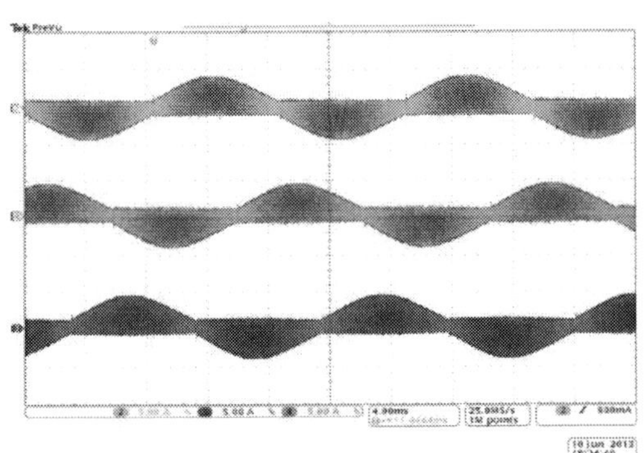

(a)

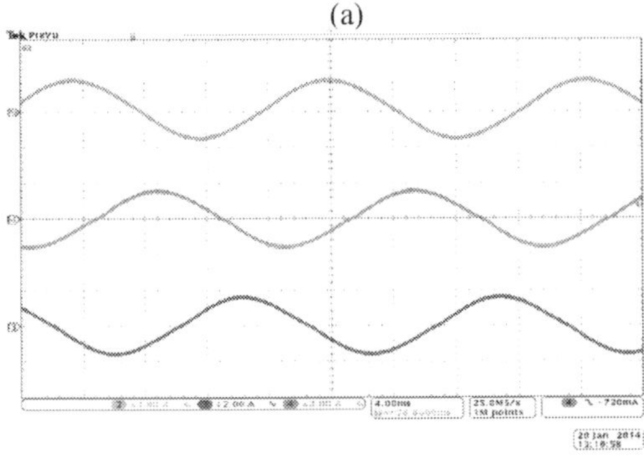

(b)

Fig. 8. Current Waveforms (a) Inductor Current Waveform of Hybrid BCM Current Control, (b) Micro-Inverter Three Phase Output Current

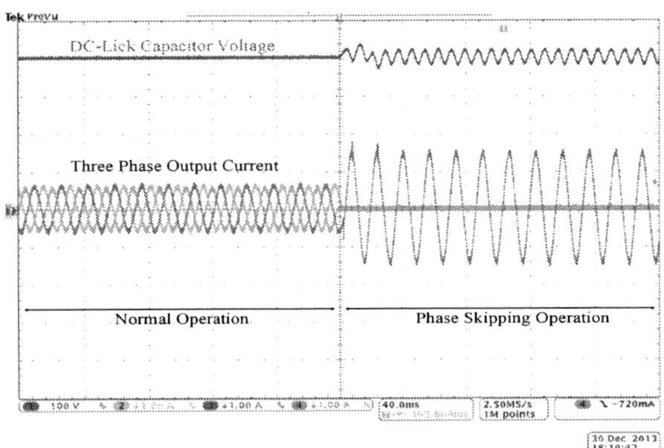

Fig. 9. Transition from Normal Operation Mode to Phase Skipping Mode

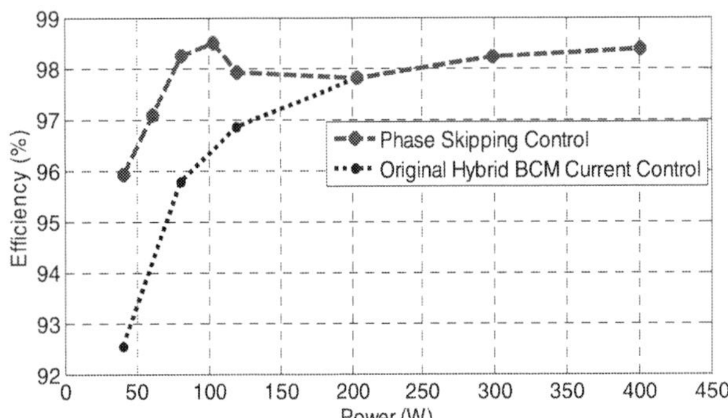

Fig. 10. Converter Efficiency versus Output Power

Switching system Stability Analysis of DC Microgrids with DBS Control

Na Zhi and Hui Zhang
Faculty of Automation and Information Engineering
Xi'an University of technology,
Xi'an, shannxi710048, China
E-mail: zhina@xaut.edu.cn, zhangh@xaut.edu.cn

Xi Xiao
State Key Lab of Control and Simulation of Power System and Generation Equipments
Tsinghua Universty, Beijing 100084, China
E-mail: xiao_xi@mail.tsinghua.edu.cn

***Abstract*—DC microgrids are becoming popular in distribution systems based on its higher reliability and system efficiency. DC bus signal (DBS) control as a common control strategy was widely used in dc microgrids, in which, control mode of each power electronic converter is triggered by monitoring the dc voltage variation to keep the power balance of the dc microgrid. Most of the dc microgrid stability analysis are linear and based on the impedance matching theory. But the dc microgrid will be a nonlinear system when the operation mode was automatically switched based on switching information. In this paper, the switched system theory was used to analyze the stability of dc microgrids under DBS control strategy. The stability conditions in each subsystem and switched system was present in this paper and simulation results verify the effectiveness of the proposed criterion.**

***Keywords*—DC microgrid; DBS control; switched system; stability analysis**

I. Introduction

With the increasing penetration of distributed renewable sources into the utility power grid, the concept of microgrid was proposed[1]. Compared with conventional ac microgrid, dc microgrid can achieve higher overall system efficiency by eliminating the ac/dc and dc/ac conversion stages, since many renewable distributed generations (DGs), such as PV, energy storage systems, and an increasing number of electric dc loads directly utilize dc power. Additionally, dc microgrid is easy to control because it does not suffer from frequency stability and reactive power compensation problems. Therefore, dc microgrid are more efficient for future residential buildings [2,3].

A typical configuration of dc microgrid is shown in Fig.1, it consists of four kinds of terminals [4], namely: grid connection converters (G-VSC), distributed generations (DGs), dc load with POL (point of load) converter and energy storage system(ESS). These terminals are decentralized connected to the common bus in parallel, the coordination control of various terminals is a key issue in the operation of dc microgrids. There are two control strategies have been used in dc microgrids: centralized[5] and decentralized control structure[6]. For the centralized strategy, all the microgrid terminals are regulated by energy management, it requires real-time feedback from each terminal to energy management, it relies heavily upon the communication line, the reliability and scalability of the system may be reduced. Compared to the centralized controller, the decentralized controller only needs the local variables to enable power sharing without communication. Droop control schemes are classical decentralized control method used on dc microgrids[7]. DBS is extended to control dc microgrid where the deviation of the DC bus voltage are predefined to distinguish different system operation modes and determine mode switching among different terminals[8]. Based on DBS, a mode-adaptive decentralized control scheme is proposed to realize flexible mode transition mechanism to react the changing condition[9].

This work was supported by National Natural Science Foundation of China (51277150/51307140); Key Science and Technology Program of Shaanxi Province (2013K07-05); Industrialization Cultivation-item of Shaanxi Province Educational Department (14JF020); Scientific research plan projects of Shaanxi Education Department(13JK0994);Provincial Foundation for the Construction of Key Disciplines of Shaanxi (105-5X1201).

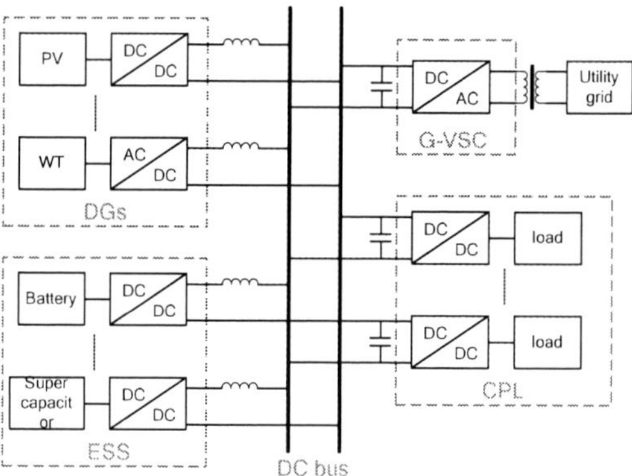

Fig. 1. A typical configuration of dc microgrid

The main issue during the design of a dc microgrid is its stability. Most of the dc microgrid stability analysis are linear and based on the Middlebrook criterion [10] and its various extensions, such as the Gain and Phase Margin Criterion [11], the Opposing Argument Criterion[12],the Energy Source Analysis Consortium(ESAC) Criterion[13], and so on. This linearization result in a model near to the voltage operating point, and that is suitable to local stability analysis [14]. The eigenvalue analysis was used in the stability analysis in dc microgrid with master-slave control structure [15] and droop control [16]. But in DBS decentralized structure of dc microgrid,

978-1-4673-9551-9/16 $31.00 © 2016 IEEE

every terminals have two operational status (droop control and max power control) which acknowledged via DC voltage variation[4], and the dc microgrid is a nonlinear system when the operation mode is switched even though the system can be considered as linear system when it work at every fixed operation mode. The influence of mode switching to dc microgrid stability was not mentioned in present papers.

To overcome this problem, linear stability operation conditions in every operation mode of DBS control and switching system stability analysis are present in this paper. Such approach models the POL converters and source terminals worked in max power control with a constant power characteristic [17]. The source terminals worked in droop control models as a voltage source series with virtual impedance. This paper is organized as follows: The principle of the mode-adaptive decentralized control strategy is explained in Section II. A general model for this control strategy adopted is described in Section III. The detailed stability analysis is presented in Sections IV. And Sections V present the simulation results based on the proposed DC MGs, respectively. The main contributions of this paper are summarized in the last section.

II. DC MICROGRID CONFIGURATION AND OPERATION MODE

In dc microgrid, dc bus voltage can be used as an effective indicator of power-flow status, real-time power balance at the dc bus is a complicated multiple objective control problem including maximizing DGs power generation, optimizing usages of BESS, and maintaining bus voltage within its limits under loads and resource variations[4,18,19]. DBS control strategy can realize the economical operation of dc microgrid and realize the adaptive mode switching. DGs (such as PV, wind generation and the other renewable energy) are controlled to supply power with higher priority. When the DGs power generation is not adequate to meet the demand of the local loads, the deficit can be compensated by importing from the utility or ESS when dc microgrid work in grid connect mode or island mode. The qualitative curves of the dc microgrid output V-I characteristics for all terminals in Fig. 1 are shown in the Fig. 2. The U_N is the normal dc voltage, k_{h1}, k_{h2}, and k_{l1}, k_{l2} are coefficients which determine the voltage band for each operation mode. The relationship among the threshold values is as follows:

$$k_{l2}U_N < k_{l1}U_N < U_N < k_{h1}U_N < k_{h2}U_N$$

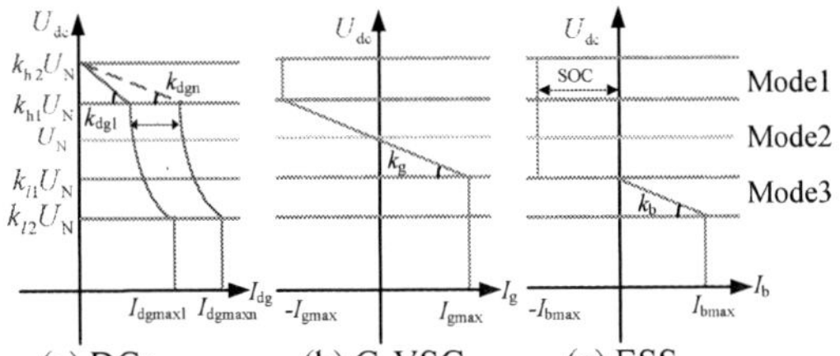

(a) DGs (b) G-VSC (c) ESS

Fig. 2. Qualitative static V-I characteristics of dc microgrid

The V-I characteristic of DGs is shown in Fig. 2(a). Since the DGs can only generate, the converter output current is always positive. For very light loads, when the voltage at the converter terminals is between $K_{h2}U_N$ V and $K_{h1}U_N$ V, this converter can regulate the bus by operating in the voltage droop mode with droop slope k_{dgn} and n is the number of DGs. Once it reaches the maximum available power, it continues to operate in the MPPT mode as a constant power source (CPS). The MPPT curve constantly changes depending on outside environment at the particular moment.

The V-I characteristic of bidirectional grid interface converter is shown in Fig. 2(b). In the first quadrant of the V-I plane, converter is taking energy from the grid, the current I_g shown is positive, while in the second quadrant it is sourcing energy back to the grid and I_g is negative. The droop coefficient (virtual dc output resistance) is k_g, it is a design parameter. If the converter output is overloaded, the converter limits the output current to I_{gmax} in Fig. 2(b). The converter changes its operation mode from voltage droop mode to the current regulation mode is at $k_{l1}U_N$ V.

The V-I characteristic of ESS interface converter is shown in Fig. 2(c). it can be anywhere according to the state of charge (SOC). Current I_b in Fig. 2(c) is positive present the battery is discharging and negative while the battery is charging. The vertical lines of the static characteristic in the first and second quadrant of the V-I plane are determined with the battery SOC, where I_{bmax} represent the maximum allowed charging or discharging currents, which could but do not have to be the same. The battery converter provides voltage droop with droop slope k_b to regulate the dc bus in the range in $k_{l1}U_N \sim k_{l2}U_N$ V.

If the demand increases even more, the dc microgrid could initiate a load shedding algorithm to prevent the dc bus voltage from dropping below $k_{l2}U_N$ V, which otherwise would cause the bus voltage collapse and system shutdown. The detailed characteristic of each operation mode and possible mode-switching patterns are summarized in Table I.

TABLE I. DC MICROGRID CONTROL MODE

Mode	DGs	G-VSC	ESS
1	Droop control (Bus regulation with k_{d1})	Constant power mode (CPL)	Constant power mode (CPL)
2	MPPT (CPS)	Droop control (Bus regulation with k_{d2})	Constant power mode (CPL)
3	MPPT (CPS)	Constant power mode (CPS)	Droop control (Bus regulation with k_{d3})

III. DC MICROGRID MODELING WITH EVERY OPERATION MODE

According to part II, the dc microgrid based on DBS control can be operated in three modes, and the mode designation and transition mechanisms can be implemented by detecting the change of dc bus voltage. As shown in Fig. 2, the voltage-current (V-I) characteristic for each terminal is divided into two sections. The drooped section corresponds to the bus regulating state, while the constant power section represents the maximum power state. In the next subsections, microsources in two operation mode and CPLs which connected to dc bus with POL converter are modeled to support the stability analysis.

A. Microsource(MS) modeling

The droop controller is a loop outer the current and voltage controller of the converter, and can be modeled as an ideal voltage source (U_N) in series with a resistance k_d, as illustrated in Fig. 3[].

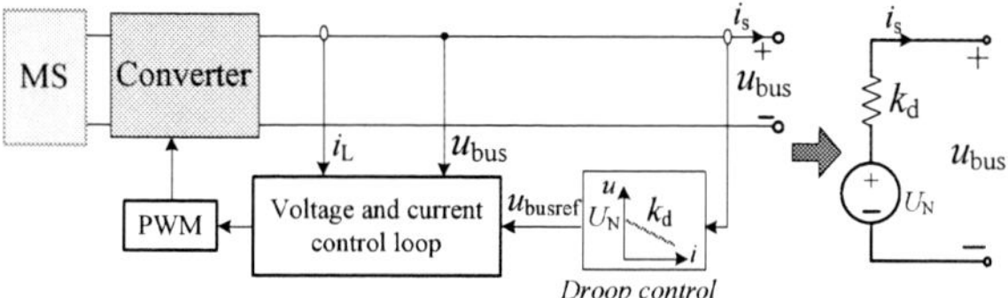

Fig. 3. Droop controlled converter models

The voltage and current droop characteristics can be mathematically expressed as

$$u_{\text{busref}} = U_N - k_d i_s \tag{1}$$

where U_N, i_s and u_{busref} are output voltage at no load, output current, and reference voltage of the MS converter. k_d is the virtual resistance, must be designed as follows:

$$k_d = \Delta u_{\text{bus}} / i_{\max} \tag{2}$$

The Δu_{bus} is the maximum allowed voltage deviation and $i_{\max}$ is the maximum output current, it depends on the maximum power of converter.

The dc microgrid, when the DGs, G-VSC and ESS operate at droop control mode, they have the role to work cooperatively to keep the bus voltage stable, so the current will equals to the sum of every terminals:

$$i_s = i_{s1} + i_{s2} + \ldots\ldots + i_{sn} \tag{3}$$

Assuming that all of the droop controlled sources have the same output reference voltage with no load and line losses between units are negligible (which is reasonable for a small isolated system):

$$U_{N1} = U_{N2} = \ldots = U_{Nn} = U_N$$

The n droop equations in parallel can be represented by a single equivalent equation:

$$U_{busref} = U_N - \frac{1}{\sum_{i=1}^{n} \frac{1}{k_{di}}} i_s = U_N - k_d i_s \tag{4}$$

it can be shown that the droop controlled converters can be approximated by an equivalent resistance k_d with an ideal voltage source U_N.

The terminals which work at constant power mode in dc microgrid can be divided into two types: CPSS (CPS_source) and CPSL(CPS_load). CPSS represent units inject constant power into dc bus (i.e. PV and WTG in MPPT mode or ESS and G-VSC work at first quadrant of the V-I plane with constant power control), while CPSL represent units extract constant power from the bus (i.e. batteries charging or G-VSC work at second quadrant of the V-I plane with constant power control). The equivalent circuit of CPSS and CPSL is illustrated in Fig. 4.

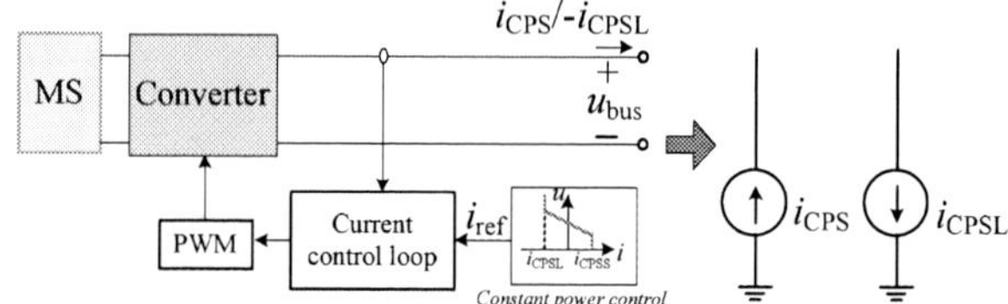

Fig. 4. Constant power controlled terminals models

In Fig.4 , where the P_{CPSS}、i_{CPSS} is the output power and output current of CPSS, P_{CPSL}、i_{CPSL} is the output power and output current of CPSL, respectively

$$i_{\text{CPSS}} = \frac{P_{\text{CPSS}}}{u_{\text{bus}}}, \quad i_{\text{CPSL}} = \frac{P_{\text{CPSL}}}{u_{\text{bus}}} \tag{5}$$

B. Constant power Load modeling

Each load in a dc microgrid needs a specific voltage level to work properly. This fact forces the system to have POL converters (always is buck converter) which are tightly controlled to keep a constant output voltage, as illustrated in Fig. 5. It is assumed that the output power of the POL converter is equal to the input power. Therefore, POL converters behave as CPL, because the control action reduces the input current if the input voltage increases and vice versa [20]. Ideal CPL can be mathematically expressed as a voltage controlled current source:

$$i_{CPL} = \frac{P_{CPL}}{u_{bus}} \tag{6}$$

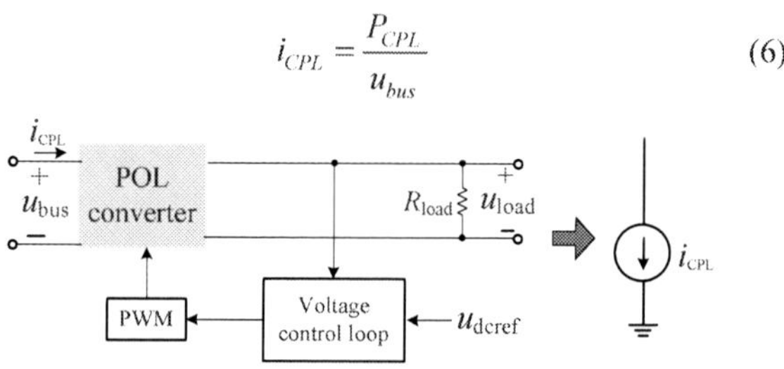

Fig. 5. Ideal CPLs models

C. Equivalent circuit model of dc microgrid

dc microgrid with DBS control, as illustrated in Fig.2, In every operating modes, some terminals are operating in droop mode and the others are operating in constant power mode, so the dc microgrid can be modeled by a single equivalent circuit, as shown in Fig. 6.

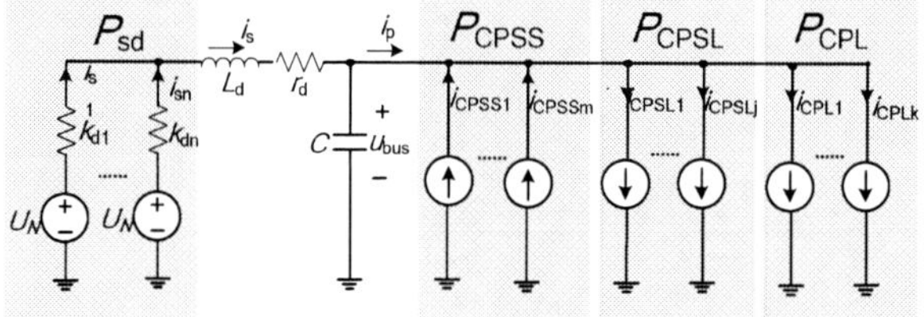

Fig. 6. Equivalent circuit model of dc microgrid

In Fig. 6, the total power of droop controlled terminals is represented by P_{sd}. Transmission line, all EMI filter and input capacitance of the POL converters, parameters on DC bus outside with a single equivalent LRC circuit, in which C , L_d and r_d correspond to the equivalent the lumped capacitance, equivalent inductive and resistance, respectively. The constant power controlled terminals behave similarly to a CPSS in the sense that injects the available power to the grid without caring for the system stability, the total power of the CPSS is P_{CPSS}. The total power of constant power controlled converters that consume power from DC micrgrid is P_{CPSL} and the total power of POL converters with powered devices is P_{CPL}. From the equivalent circuit model illustrated in Fig. 6, the dynamics of the DC MG with an ideal CPL can be represented by

$$\begin{cases} L_d \dfrac{di_s}{dt} = U_N - u_{bus} - k_d i_s - r_d i_s \\ C \dfrac{du_{bus}}{dt} = i_s - \dfrac{P}{u_{bus}} \end{cases} \tag{7}$$

Where

$$\begin{cases} i_s = i_{s1} + i_{s2} + \ldots\ldots + i_{sn} \\ k_d = \dfrac{1}{\sum_{i=1}^{n} \dfrac{1}{k_{di}}} \\ P = P_{CPL} + P_{CPSL} - P_{CPSS} \\ P_{CPSS} = (i_{CPSS1} + i_{CPSS2} + \ldots\ldots + i_{CPSSm}) \cdot u_{bus} \\ P_{CPSL} = (i_{CPSL1} + i_{CPSL2} + \ldots\ldots + i_{CPSLj}) \cdot u_{bus} \\ P_{CPL} = (i_{CPL1} + i_{CPL2} + \ldots\ldots + i_{CPLk}) \cdot u_{bus} \end{cases}$$

IV. SYSTEM STABILITY ANALYSIS

A switched system is a hybrid system that is composed of many continuous-time subsystems and a rule scheduling the switching between the subsystem [21]. At any given time, a particular subsystem may be chosen by switching rules, it may be a function of time or state, and we say that the system is autonomous switched. The stability of switching system is a common problem, which is affected by two factors: the stability of its subsystems and the switching rule. The dc microgrid studied in this paper is considered as an *autonomous switched system*, and its stability analysis is decomposed into two parts: analysis of the dynamics of the linear subsystems (mode1, mode2 and mode3), analysis of the stability of the switching system. As mentioned in the previous section, the variation of the dc bus voltage range causes the switch of the subsystems as illustrated in Fig. 7 according to Fig. 2.

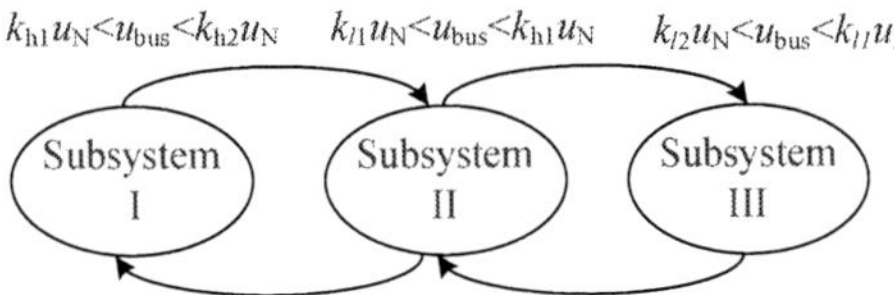

Fig. 7. Switching subsystem constructure

It is shown in Fig. 7, the dc microgrid is a switching system , and it can be mathematically represented by:

$$\dot{x} = A_{q(u_{bus})} \cdot x + B \cdot U_N ,$$

$$\text{and } q(u_{bus}) = \begin{cases} 1, \text{ if } k_{h1}u_N < u_{bus} < k_{h2}u_N \\ 2, \text{ if } k_{l1}u_N < u_{bus} < k_{h1}u_N \\ 3, \text{ if } k_{l2}u_N < u_{bus} < k_{l1}u_N \end{cases} \tag{9}$$

Where A is coefficient matrix of dc microgrid and it is different in different subsystem, $q(u_{bus})$ is the switching rule.

A. Subsystem stability

According to (7) we can get the system equilibrium points when the dc microgrid operate at stable state, the equilibrium are abtained by making (6) equals to zero. The equilibrium points are:

$$x_e = [I_{se}, U_{buse}] = [\frac{U_N - U_{bus}}{k_d}, \frac{U_N \pm \sqrt{U_N^2 - 4P(k_d + r_d)}}{2}] \tag{10}$$

Hence the necessary condition to the existence of equilibrium points is

$$P < \frac{U_N^2}{4(k_d + r_d)} \tag{11}$$

For our application, in order to use the Lyapunov method for subsystem(7), we suppose that input variable U_{ref} in (7) are constant, and move the equilibrium points to the coordinate origin:

$$x_1 = i_s - I_{se},\ x_2 = u_{bus} - U_{buse} \tag{12}$$

Then the Jacobian matrix A of the subsystems can be defined as:

$$\begin{bmatrix} \dot{x}_1 \\ \dot{x}_2 \end{bmatrix} = \begin{bmatrix} -\dfrac{k_{di} + r_d}{L_d} & -\dfrac{1}{L_d} \\ \dfrac{1}{C_i} & \dfrac{P_i}{C(x_2 + U_{buse})^2} \end{bmatrix} \begin{bmatrix} x_1 \\ x_2 \end{bmatrix} = A_i \cdot \begin{bmatrix} x_1 \\ x_2 \end{bmatrix} \quad i=1,2,3 \tag{13}$$

The subsystem stability is said to be asymptotically stable near equilibrium point x_e if it is Lyapunov stable and there exists a positive definite matrix T and I, and it satisfied

$$A_i^T T + T A_i = -I \tag{14}$$

When the $\|x\| \to \infty$, then Lyapunov function

$$V(x) = x^T T x \to \infty \tag{15}$$

the subsystem(13) is global asymptotic stability. When I accept as identity matrix, the matrix T can be get:

$$T = \frac{-0.5 \cdot (I + \det(A_i)(A_i A_i^T)^{-1})}{Tr(A_i)} \tag{16}$$

Where, $Tr(A_i) = -\dfrac{k_{di} + r_d}{L_d} + \dfrac{P_i}{C_i (U_{buse})^2}$

The subsystem is stable when the determinant of the T greater than zero(Δ>0) . Solving these inequalities results in two cases that guarantees the stability of this system:

1. $\Delta 1>0$, the system is stable when

$$-\frac{k_{di}+r_d}{L_d}+\frac{P_i}{C_i U_{buse}^2}>0 \tag{17}$$

2. $\Delta 2>0$, the system is stable when

$$-\frac{k_{di}+r_d}{L_d}\cdot\frac{P_i}{C_i U_{buse}^2}+\frac{1}{C_i L_d}>0 \tag{18}$$

In order to achieve the stable equilibrium point the following two conditions must be simultaneously met according to (17)(18):

$$P_i<\frac{U_{buse}^2}{k_{di}+r_d} \tag{19}$$

$$C_i>\frac{L_d\cdot P_i}{(k_{di}+r_d)\cdot U_{buse}^2} \tag{20}$$

(19), (20) suggests that bigger values of C_i and smaller P_i facilitate achieving a stable equilibrium point in every subsystem. Different subsystem has different voltage range in DBS control, the value of P_i based on (19) maybe bigger than the practical value because of the droop coefficient and maximum power correlation. In practice, according to the actual system to choose P_i and it must meet (19) and (11). So we can get the stable conditions of each subsystem in Table II.

TABLE II. STABLE CONDITIONS OF SUBSYSTEM

subsystem	P_{max}	C_{min}
I	$\min(\frac{k_{h1}^2 U_N^2}{k_{d1}+r_d},\frac{U_N^2}{4(k_{d1}+r_d)})$	$\frac{P_{max1}}{k_{h1}^2 U_N^2}\cdot\frac{L_d}{k_{d1}+r_d}$
II	$\min(\frac{k_{l1}^2 U_N^2}{k_{d2}+r_d},\frac{U_N^2}{4(k_{d2}+r_d)})$	$\frac{P_{max2}}{k_{l1}^2 U_N^2}\cdot\frac{L_d}{k_{d1}+r_d}$
III	$\min(\frac{k_{l2}^2 U_N^2}{k_{d3}+r_d},\frac{U_N^2}{4(k_{d3}+r_d)})$	$\frac{P_{max3}}{k_{l2}^2 U_N^2}\cdot\frac{L_d}{k_{d1}+r_d}$

B. Switching system stability

The switched system could result in instability even though each subsystem is stable in standalone operation. Switched system is a complicated nonlinear system, but the dc microgrid is a switched system which all the subsystems are linear time-varying systems. These system are termed as *switched linear systems* and are mathematically represented by

$$\dot{x}=A_{q(t)i}\cdot x+B\cdot u \quad (i=1,2\ldots m) \tag{15}$$

Where $x\in R^n$ is the state variables, $u\in R$ is the input variables. $A_i\in R^{n\times n}$, are constant matrices. m is the number of subsystem.

$q(t)$is the switching rule. In this subsection, we analyze the stability properties of the switched system expressed by (9).

According to the switching system stability theorem[23], under the switching sequence (9), when the system

$$\dot{x}=(A_1+A_2+\ldots..+A_m)\cdot x \tag{16}$$

is asymptotically stable , then the switched linear system(9) is asymptotically stable.

For our application, in order to use the Lyapunov method for switched system (9), we suppose that input variable U_{ref} in (5) are constant. It should be noted that this assumption does not imply a CPL. Indeed, U_{bus} and P may be highly variable because of the changed CPL load. dc microgrid has three operation mode in this paper as illustrated in Fig. 4. The coefficient matrix A of switching system can be defined as:

$$A=\sum_{i=1}^{3}A_i=\begin{bmatrix}-\frac{\sum_{i=1}^{3}k_{di}+3r_d}{L_d} & -\frac{3}{L_d}\\ \frac{3}{C} & \frac{P}{C(U_{busei})^2}\end{bmatrix} \tag{17}$$

The switching system is stable when the determinant of the T greater than zero ($\Delta>0$). Solving these inequalities results the conditions that guarantees the stability of this system:

1. $\Delta 1>0$, the system is stable when

$$\frac{\sum_{i=1}^{3}k_{di}+3r_d}{L_d}-\frac{P}{C(U_{busei})^2}>0 \tag{18}$$

2. $\Delta 2>0$, the system is stable when

$$\frac{\sum_{i=1}^{3}k_{di}+3r_d}{L_d}\cdot\frac{P}{C(U_{busei})^2}<\frac{9}{CL_d} \tag{19}$$

In order to achieve the stable equilibrium point the following two conditions must be simultaneously met:

$$P<\frac{9(U_{busei})^2}{\sum_{i=1}^{3}k_{di}+3r_d} \tag{20}$$

$$C>\frac{L_d P}{\sum_{i=1}^{3}k_{di}+3r_d}\cdot\frac{1}{(U_{busei})^2} \tag{21}$$

The selection range of dc bus voltage based on droop control is $k_{l2}U_N$-$k_{h2}U_N$ according to the change of P_i. If the switched system is global asymptotic stability in the voltage range from $k_{l2}U_N$ to $k_{h2}U_N$, then the switched system is stable at every operating point(x_e).

V. SWITCHED SYSTEM STABILITY ANALYSIS

In this section, the system stability analysis under load variations was validated based on simulation results. Two PV generations in parallel interfaced by dc-dc boost converters, ESS interfaced by bidirectional dc-dc buck-boost converter and an interfacing dc-ac converter connected to the utility grid are connected through transmission lines to a dc bus with resistive and CPL loads, as illustrated in Fig. 14. In order to simplify the analysis of the transmission line and all EMI filter, parameters on DC bus outside with a single equivalent RLC circuit. The parameters of this simple dc microgrid are given by Table III.

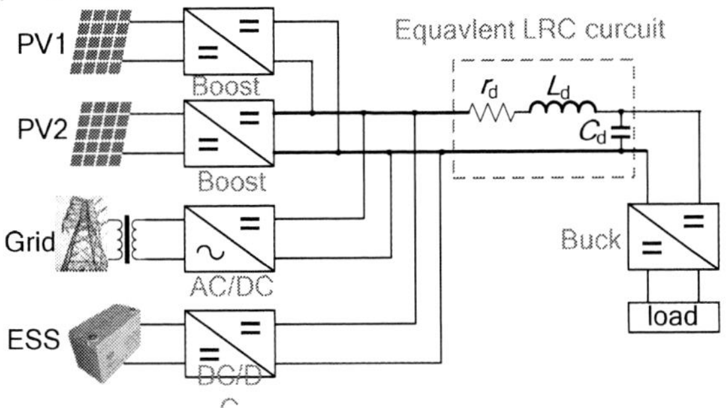

Fig. 8. Proposed dc microgrid

TABLE III. PRAMETER OF THE TEST DC MICROGRID

Modulation name and modulation parameters	*Droop coeffient*	*Max power*/kW	U_N /V
PV1	0.72	5	420
PV2	0.38	10	420
Grid	0.78	5	400
ESS	0.78	5	390

Therefore, voltage variations in the specified range can be acceptable for most of the dc loads. Based on the aforementioned references, in this paper, The bus voltage nominal value U_N is set at 400 V; k_{h1} and k_{h2} are configured at 2.5% and 5% above U_N, respectively, while k_{l1} and k_{l2} are 2.5% and 5% below U_N, respectively. So the operating voltage range of the bus is chosen to be between 380 V and 420 V to allow for power sharing and voltage regulation using the droop control. All converters are programmed to turn off after the voltage at the converter terminals stays below 380 V for some predetermined time, and to turn off instantaneously if the voltage exceeds 420 V by some margin. A POL buck converter driving a variable resistance R_L is modeled as a CPL load.

A. *Subsystem Parameter design*

This setup is designed to be a subsystem I with k_{d1} =0.25, L_d=100μH, r_d=20mΩ and the dc bus voltage range from 410 to 420V. According to TABLE II, the maximum P_{CPL} is 132.7kW, but in given system, the P_{CPL} must small than the maximum output power of converter. In subsystem I, the range of P_{CPL} is 0-5kW, using these values in (21), it is easy to conclude that the system is stable for C >40μF as illustrated in Fig. 9(a). Fig. 9(b) is the subsystem II with k_{d2} =0.38, L_d=100μH, r_d=20mΩ and the dc bus voltage range from 390 to 410V. The subsystem II is stable for P_{CPL} < 15kW and C >50μF. Fig. 9(c) is the subsystem III with k_{d3} =0.38, L_d=100μH, r_d=20mΩ and the dc bus voltage range from 380 to 390V. The subsystem III is stable for P_{CPL} < 30kW and C >102μF.

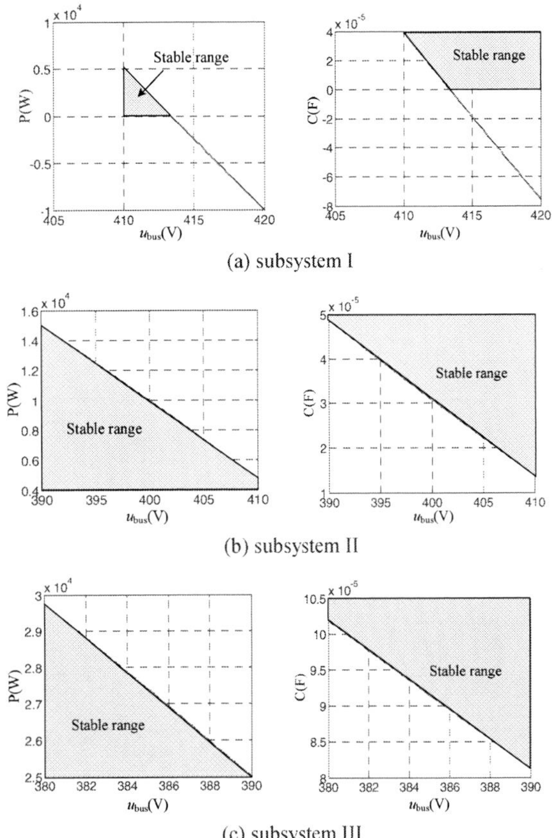

Fig. 9. Stable range of parameter P_{CPL} and C in subsystem

B. *Switching stability analysis*

According to (20), the limitaition of P is P_{CPL} < 1214kW , in this system, the total maximum power of terminals is 35kW, so the P_{CPL} must smaller than 35kW. According to (21), the limitation of C is C >786μF.

When the P_{CPL} is 2.5kW, 10kW and 27.5kW, the DC mcirogrid will worked at subsystem I ,II and III respectively. the state locus and state components curve of subsystems and switching system for C=800uF and C=80uF and the system initial state is $(x_1, x_2)^T = (5,2)^T$ was simulated in MATLAB.

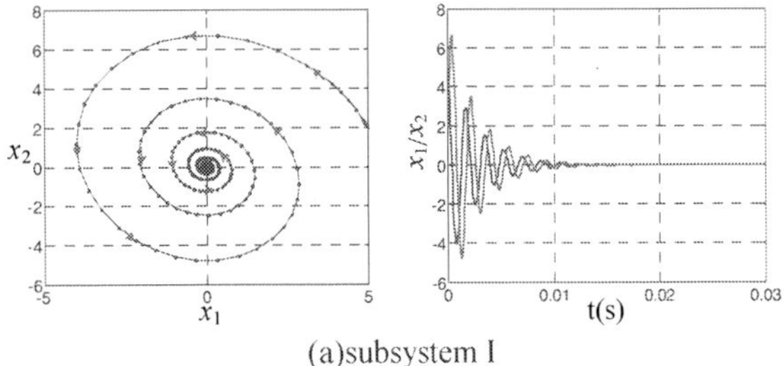

(a)subsystem I

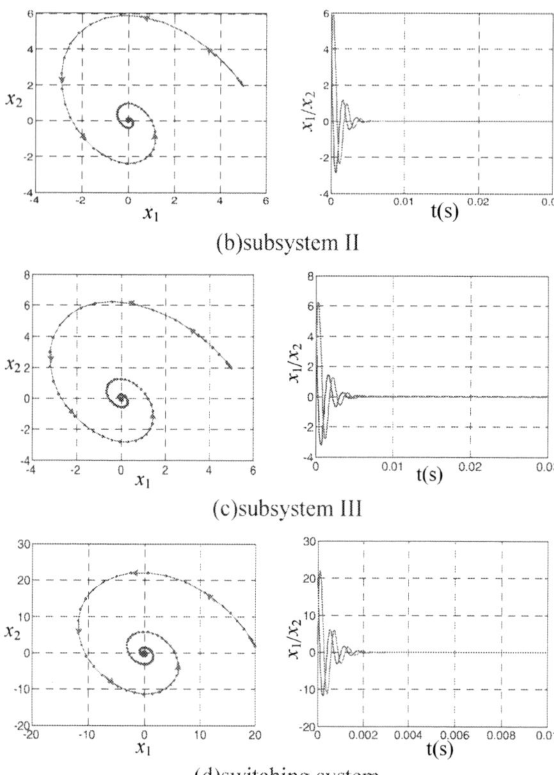

(b)subsystem II

(c)subsystem III

(d)switching system

Fig. 10. State movement track of subsystem and switching system with C=800uF

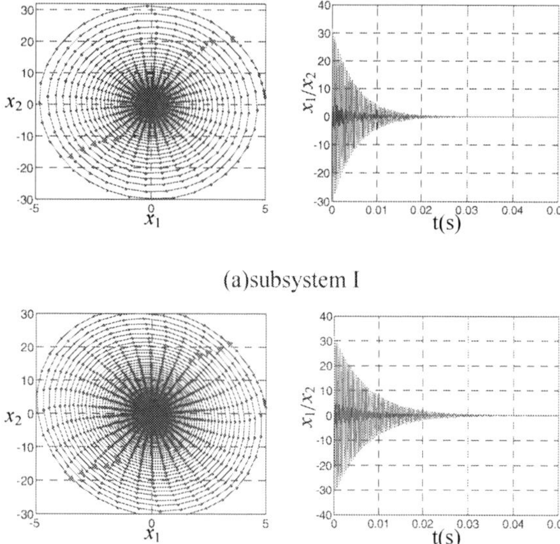

(a)subsystem I

(b)subsystem II

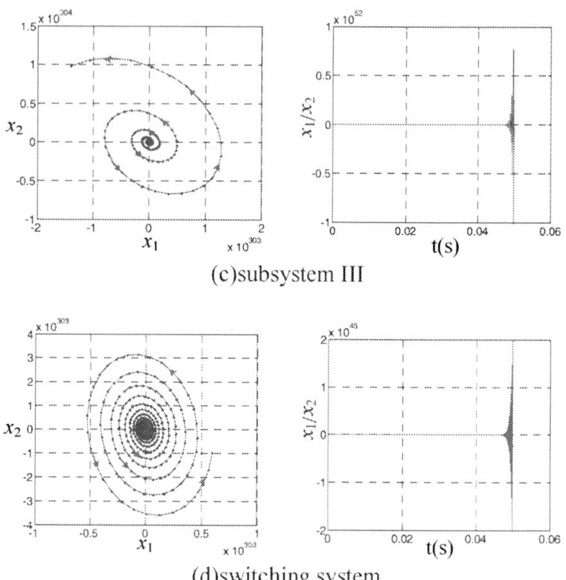

(c)subsystem III

(d)switching system

Fig. 11. State movement track of subsystem and switching system with C=80uF

Fig. 10 shows trajectories of states ultimately entering into some neighbor- hood of the coordinate origin, and the convergence of states is evident since C=800uF satisfied the stable condition of each subsystem and switched system. But when C=80uF does not meet the stability conditions of subsystem III and switched system, so the state components will far from equilibrium, resulting in a oscillation, as illustrated in Fig. 11. Although the subsystems I and II are stable, the convergence speed is slow, the overshoot and regulating time are large, the C is small, the dynamic characteristic of the system is worse.

VI. CONCLUSION

The DBS decentralized dc microgrid control strategy and its stability analysis presented in this paper. A simplified model of dc microgrid under droop control was addressed. And based on this model, linearize stability analysis for subsystems and Lyapunov method for switching system give us some relations among the DC MGs parameters, which allow as to determine the qualitative behavior of DC MGs. This design guideline can build reliable DC MGs under DBS control based on safe operation regions.

REFERENCES

[1] Lasseter R H, Paigi P. Microgrid: A Conceptual Solution[C]// In Proc Power Electronics Specialists Conference. 2004:4285 - 4290 Vol.6.

[2] Kwasinski A. Quantitative Evaluation of DC Microgrids Availability: Effects of System Architecture and Converter Topology Design Choices[J]. Power Electronics IEEE Transactions on, 2011, 26(3):835-851.

[3] Chang Y C, Liaw C M. Establishment of a Switched-Reluctance Generator-Based Common DC Microgrid System[J]. IEEE Transactions on Power Electronics, 2011, 26(9):2512-2527.

[4] Chen D, Xu L, Yao L. DC Voltage Variation Based Autonomous Control of DC Microgrids[J]. IEEE Transactions on Power Delivery, 2013, 28(2):637 - 648.

[5] Wang B, Sechilariu M, Locment F. Intelligent DC Microgrid With Smart Grid Communications: Control Strategy Consideration and Design[J]. Smart Grid IEEE Transactions on, 2012, 3(4):2148-2156.

[6] Karlsson P. DC bus voltage control for a distributed power system[J]. IEEE Transactions on Power Electronics, 2003, 18(6):1405 - 1412.

[7] Guerrero J M, Vasquez J C, Matas J, et al. Hierarchical Control Of Droop-Controlled Ac And Dc Microgrids—A General Approach Toward Standardization[J]. Industrial Electronics IEEE Transactions on, 2011, 58(1):158 - 172.

[8] Gu Y, Li W, He X. Frequency-Coordinating Virtual Impedance for Autonomous Power Management of DC Microgrid[J]. Power Electronics IEEE Transactions on, 2015, 30(4):2328-2337.

[9] Gu Y, Xiang X, Li W, et al. Mode-Adaptive Decentralized Control for Renewable DC Microgrid With Enhanced Reliability and Flexibility[J]. Power Electronics IEEE Transactions on, 2014, 29(9):5072-5080.

[10] R. Middlebrook and S. Cuk. "Input filter considerations in design and application of switching regulators". IEEE Industrial Applications Society Annual Meeting, Chicago, USA, 1976:366-382

[11] Jinjun Liu,Xiaogang Feng, Fred C.Lee."Stability Margin Monitoring for DC Distributed Power Systems via Perturbation Approaches". IEEE Trans on Power Electronics, 2003, 18(6): 1254-1261.

[12] Feng X, Ye Z, Xing K, et al. "Individual Load Impedance Specification for a Stable DC Distributed Power System". IEEE Annual Applied Power Electronics Conference and Exposition (APEC), 1999:923-929.

[13] Sudhoff S D, Glover S F, Lamm P T, et al. "Admittance Space Stability Analysis of Power Electronic Systems". IEEE Transaction on Aerospace and Electronic Systems, 2000, 36(3):965-973.

[14] Sudhoff S D, Crider J M. "Advancements in Generalized Immittance Based Distribution Systems". Electric Ship Technologies Symposium (ESTS), 2011:207-212.

[15] Li X, Guo L, Wang C. Stability Analysis in a Master-Slave Control Based Microgrid[J]. Transactions of China Electrotechnical Society, 2014, 29(2):24-34.

[16] Majumder R, Chaudhuri B, Ghosh A, et al. Improvement of Stability and Load Sharing in an Autonomous Microgrid Using Supplementary Droop Control Loop[J]. IEEE Transactions on Power Systems, 2010, 25(2):796-808.

[17] Kwasinski A, Onwuchekwa C N. Dynamic Behavior and Stabilization of DC Microgrids With Instantaneous Constant-Power Loads[J]. Power Electronics IEEE Transactions on, 2011, 26(3):822-834.

[18] Sun K, Zhang L, Xing Y, et al. A Distributed Control Strategy Based on DC Bus Signaling for Modular Photovoltaic Generation Systems With Battery Energy Storage[J]. IEEE Transactions on Power Electronics, 2011, 26(10):3032-3045.

[19] Boroyevich D, Dong D, Burgos R, et al. Future electronic power distribution systems a contemplative view[C]// Optimization of Electrical and Electronic Equipment (OPTIM), 2010 12th International Conference on. IEEE, 2010:1369-1380.

[20] Emadi A, Khaligh A, Rivetta C H, et al. Constant power loads and negative impedance instability in automotive systems: Definition, modeling, stability, and control of power electronic converters and motor drives[J]. IEEE Transactions on Vehicular Technology, 2006, 55(4):1112-1125.

[21] Zhai GS, Hu B, Yasuda K, Michel AN. Qualitative analysis of disvrete-time switched systems. In: Proceedings of the American Control Conference, Anchorage, AK May8-10,2002,pp.1880-1885

A GRID-CONNECTED WECS WITH POWER LIMITING CONTROL

Jéssica Santos Guimarães[1], Demercil de S. Oliveira Jr.[2], Juliano de Oliveira Pacheco[3], Paulo P. Peixoto[4]
Federal University of Ceará, Department of Electrical Engineering, Group of Energy Processing and Control
PO Box 6001, P. Code 60455-760, Tel.: +55 (85) 3366-9586 - Fortaleza, Ceará, Brazil
e-mail: jessica@dee.ufc.br, demercil@dee.ufc.br, julianop@dee.ufc.br, paulopp@dee.ufc.br

***Abstract*—This paper presents a grid-connected Wind Energy Conversion System (WECS) using a permanent magnet synchronous generator (PMSG) operating with variable speed. A maximum power point tracking (MPPT) algorithm is used to determinate the power to be extracted from the generator for each rotational speed while wind power remains lower than the PSMG rated power. Otherwise, the MPPT algorithm is disabled and the proposed external mechanical power control loop maintains the rated power condition. Both control techniques determine the proper current ratings through the PMSG, which is driven by a three-phase bridgeless boost rectifier using digitally-implemented self-control strategy. The rectifier supplies a three-phase full-bridge inverter that provides an ac output with the desired requirements. Simulations results on a 6-kW WECS are shown in order to validate the proposal.**

Keywords—MPPT; PMSG; self-control; three-phase bridgeless boost rectifier; WECS

I. INTRODUCTION

The increase in electricity consumption and global concerns about emissions of greenhouse gases resulted in the expansion of distributed generation associated to the use of renewable energy sources. Consequently, new conversion techniques and energy management strategies are necessary in order to ensure increasing power quality and grid reliability.

Overall statistics report the growing of WECS [1] - [3], where a wind turbine is used to convert kinetic energy into mechanical energy. A generator is then responsible for converting it into electric energy, as a power electronic converter is required in order to process the generated energy enabling power grid connection as seen in Fig. 1.

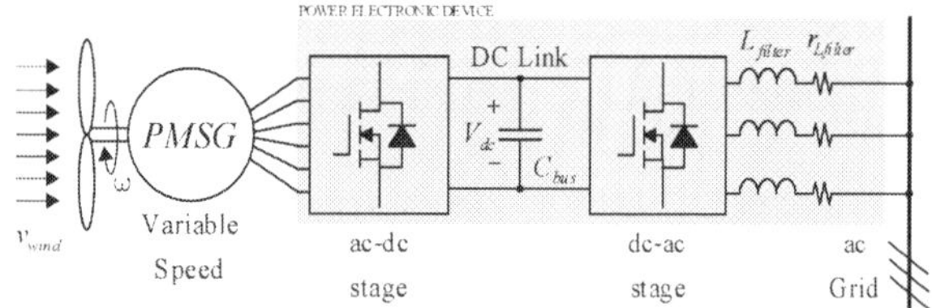

Fig. 1. WECS structure.

Among the possible technologies for wind power generation, the variable speed is a must because it is possible to extract the maximum power for a wide range of wind speeds [4]-[5].

The PMSG has become a trend in small WECSs since it avoids the use of brushes and can be implemented with high number of poles in order to eliminate the need for gearboxes. Thus, reduced weight and volume as well as less maintenance needs are intrinsic advantages [6]-[9]. Operation with variable speed and frequency is necessary considering that the amplitude of waveforms change continuously. It is worth to mention that such prominent advantages can be addressed to the proposed WECS.

II. PROPOSED TOPOLOGY

This WECS presents a power control strategy following the behavior in Fig. 2 in which four different operating regions can be addressed: the region I whose wind speed is below a threshold value and the system doesn't impose resistant torque enabling the acceleration of the turbine; the maximum power tracking region II; the region III above the rated wind speed in which a mechanical power limitation strategy is used; and the region IV whose wind speed is higher than that supported by the blades and the power generation is interrupted.

$$P_{mec} = \begin{cases} 0, & \text{if } 0 \le v_{wind} < 4 \\ P_{mec,max}, & \text{if } 4 \le v_{wind} < 10 \\ P_{mec(rated)}, & \text{if } 10 \le v_{wind} < 25 \\ 0, & \text{if } v_{wind} \ge 25 \end{cases} \quad (1)$$

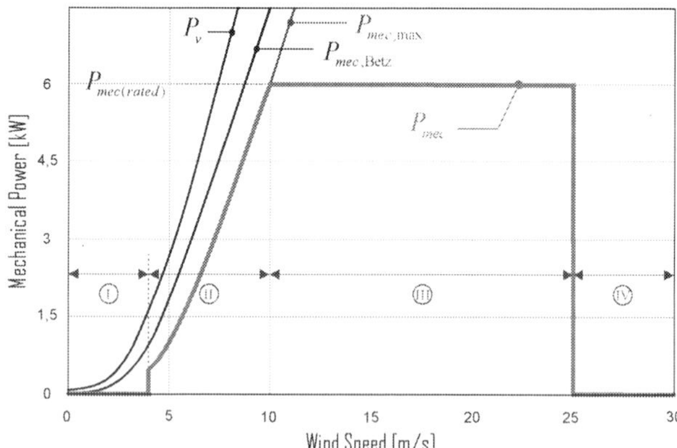

Fig. 2. Power extraction curve of the proposed WECS.

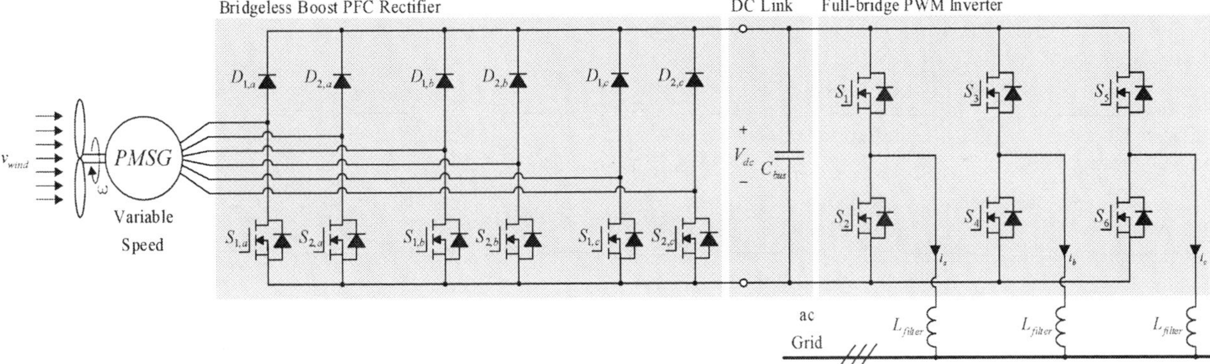

Fig. 3. Proposed grid-connected WECSs.

The mechanical power P_{mec} (2) represents only a part of the power available in the wind, P_v and depends on the turbine radius R $[m]$, the air mass density ρ $[kg/m^3]$, the wind speed v_{wind} $[m/s]$ and the power coefficient $C_p(\lambda,\beta)$. The power coefficient depends on the blade pitch angle β and the Tip-Speed-Ratio (TSR) λ given by $\lambda = R\cdot\omega/v_{wind}$ where ω $[rad/s]$ represents the shaft speed.

$$P_{mec} = P_v \cdot C_p(\lambda,\beta) = 0.5\cdot\rho\cdot\pi\cdot R^2\cdot v_{wind}^3\cdot C_p(\lambda,\beta) \quad (2)$$

The mechanical power is converted into electric power, $P_{3\Phi(generator)}$, by the PMSG and processed by the power electronic device which is composed into two stages (Fig. 3). The ac-dc stage is represented by a three-phase bridgeless boost rectifier [10]-[13] with self-control strategy [14]-[17] and the dc-ac stage consists of a three-phase bridge PWM inverter.

A. Bridgeless Boost PFC Rectifier with Self-Control Strategy

This indirect current control technique doesn't demand the generation of a current reference and the use of current compensators. The current signal feedback is multiplied by a gain k_i in order to generate the PWM (Pulse Width Modulation) signals.

Thus, k_i determines the resistant torque imposed by the generator to the wind turbine. As it can be seen in Fig. 4, the value of k_i is determined by the MPPT ($k_{i(mppt)}$) or by the mechanical power limiting control loop ($k_{i(power\ control)}$).

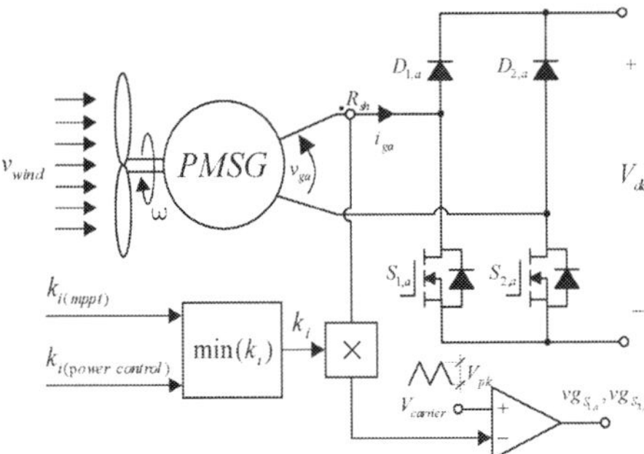

Fig. 4. Implementation of self-control strategy.

B. MPPT Technique

The literature provides several ways to implement the MPPT algorithm [18]. The optimal torque technique is simple, efficient and doesn't require the use of anemometers in order to measure the wind speed.

If the turbine is operating in region II with maximum power coefficient ($C_p=C_{p,max}$) and optimum TSR ($\lambda=\lambda_{optimum}$), the system is supposed to extract maximum power (3) and the torque imposed by the wind turbine is optimum (4).

$$P_{mec} = P_{mec,\max} = \frac{1}{2}\cdot\rho\cdot\pi\cdot R^5\cdot\frac{\omega^3}{\lambda_{optimum}^3}\cdot C_{p,\max} = K_{optimum}\cdot\omega^3 \quad (3)$$

$$T_{turbine} = T_{turbine_{optimum}} = \frac{P_{mec,\max}}{\omega} = K_{optimum}\cdot\omega^2 \quad (4)$$

This maximum mechanical power $P_{mec,max}$ (3) must be converted into electric power $P_{3\Phi(generator)}$ (5), which can be written in terms of the shaft speed and k_i. From (3) and (5), the equation that defines the operation of the MPPT is found in (6). In summary, the MPPT algorithm monitors the shaft speed ω and calculates $k_{i,mppt}$ (so current $I_{gi}(t)$ which causes the adequate resistant torque $T_{generator}$) so that the system will keep operating in optimal points.

$$P_{3\phi(generator)} = 3\cdot\frac{E_i}{\sqrt{2}}\cdot\frac{I_{gi}}{\sqrt{2}} = \frac{3}{2}\cdot\frac{K_\omega^2}{R_{sh}\cdot\left(1/V_{pk}\right)\cdot V_{dc}}\cdot\frac{\omega^2}{k_i} \quad (5)$$

$$k_{i(mppt)} = \frac{3\cdot K_\omega^2}{2\cdot R_{sh}\cdot\left(1/V_{pk}\right)\cdot V_{dc}\cdot K_{optimum}}\cdot\frac{1}{\omega} = K_1\cdot\frac{1}{\omega} \quad (6)$$

C. Limiting the Mechanical Power

When the wind speed reaches values greater than the rated conditions, power limiting mechanisms are required. The active pitch control, through hydraulic or electrical mechanisms, rotates the blades and positively increases the pitch angle so that the power remains at a nominal value. In the passive stall control, when the wind speed increases, the attack angle also increases (even though the pitch angle is kept constant) resulting in the loss of the lift forces and consequent

978-1-4673-9551-9/16 $31.00 © 2016 IEEE

decrease of the extracted mechanical power. In systems that operate with active stall control, there is a mechanical mechanism (usually springs) which causes the blades to rotate toward the plane of rotation with increasing wind speed. This phenomenon causes the pitch angle increases negatively and there is a decrease in the lift coefficient as well as in the extracted power [19].

In Fig. 5, the behavior differences between the aforementioned techniques are highlighted. The pitch control is the most accurate and robust technique as well presents greater energy efficiency. However, due to the high cost (because the actuators present in each blade), and robustness issues, it is rarely used in small WECS. If the extracted power is less than the nominal value, the blades are built with fixed nonzero pitch angle on systems with passive stall. Consequently, energy losses are inherent (region 1 in Fig. 5). The active stall control allows improved power conversion but causes overshoot (region 2) that can overload both the generator and the converter.

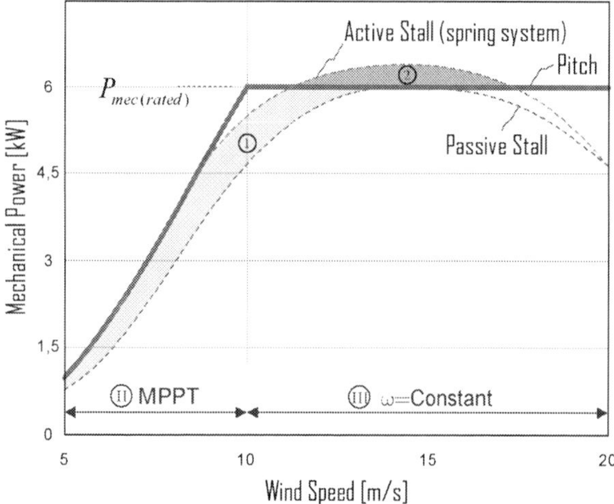

Fig. 5. Comparison of power limitation techniques.

This work proposes an electronic control that ensures mechanical power extraction curve with near the same behavior of the curve presented in systems with pitch control. The main advantage is the energy gain for wind speeds between 4 and 10 m/s where most of the energy is extracted. In addition, it eliminates the power overshoot and reduces the accuracy needed in the active stall mechanism.

This is performed by an external control loop which limits the mechanical power with a fixed reference $P_{mec(rated)}$ using the internal shaft speed control loop as it can be seen in Fig. 6. The transfer functions are obtained from (7), which describes the mechanical system composed by the wind turbine and generator. Parameters $T_{turbine}$ and $T_{generator}$ are the torques imposed by the wind turbine and the generator, respectively. The transfer function given in (8) is obtained considering $T_{turbine}$ as an external disturbance. Considering the electrical power $P_{3\phi(generator)}$ as an external disturbance and linearizing the equation around a quiescent operating point ($\omega = \omega_{operation\ point}$), the small signal model that relates mechanical power and speed is obtained, as (9).

$$T_{turbine} - T_{generator} = J \cdot \frac{d\omega}{dt} \tag{7}$$

$$\frac{\omega(s)}{I_{gi}(s)} = \frac{3 \cdot K_{\omega}}{2 \cdot J \cdot s} \tag{8}$$

$$P_{mec} - P_{3\phi(generator)} = J \cdot \frac{d\omega}{dt} \cdot \omega \Rightarrow \frac{P_{mec}(s)}{\omega(s)} = \left(J \cdot \omega_{\text{operation point}}\right) \cdot s \tag{9}$$

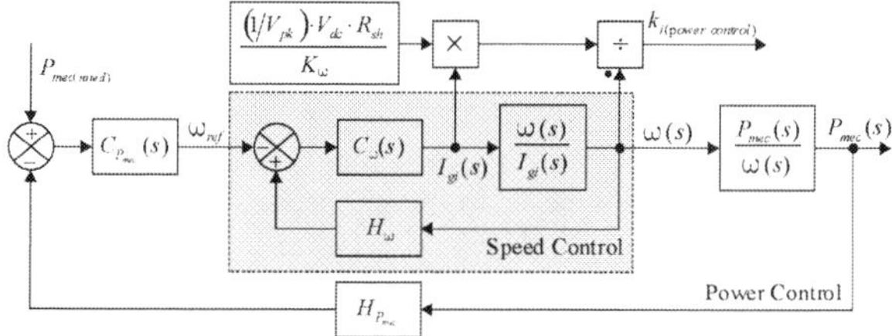

Fig. 6. Control strategy for the mechanical power.

D. *Three-phase Full-bridge inverter and Control Strategy*

The three-phase inverter is responsible for: control of dc-link voltage; grid synchronization; control of the power transferred from the dc-link to the grid; as well as the quality of this injected power [20]-[21].

This converter was modeled using the Park transformation as shown in [22]. The output currents (ia, ib, and ic) are sampled and a dq transform block is used to obtain id (d-axis current) and iq (q-axis current). At this point, a quadrature Phase-Locked Loop (q-PLL) [23]-[25] scheme is designed to ensure synchronization between the currents injected in the power grid and the ac mains voltage.

As it can be seen in Fig. 7, two control loops are required to control both currents. A current loop with zero reference value for the d-axis current component ensures that only active power is transferred to the grid, whereas the reference of the q-axis current loop is determined by an external voltage control loop that regulates the dc-link voltage.

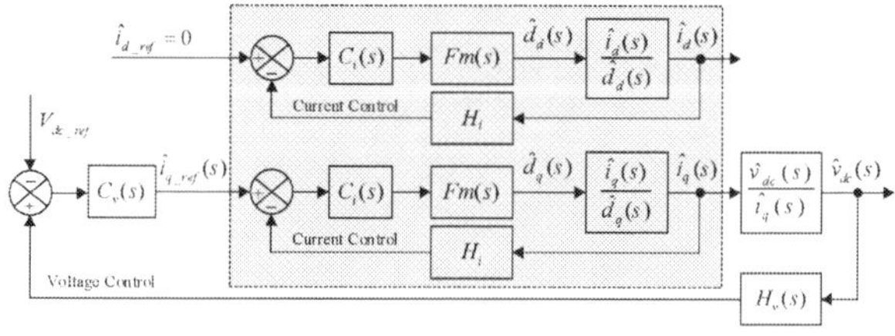

Fig. 7. Control scheme used in the grid-side converter.

III. SIMULATIONS RESULTS

Simulation results provided by PSIM software are presented considering the parameters shown in Table I and Table II.

TABLE I. POWER CONVERTER SPECIFICATIONS AND PARAMETERS

Maximum Output Power	6 kW
Dc link voltage	700 V
Grid three-phase rms voltages	380 V
Grid three-phase frequency	60 Hz
Switching frequency	21 kHz

TABLE II. WIND TURBINE SPECIFICATIONS AND PARAMETERS

Generator topology	PMSG Axial Flux
Number of poles	30
Rotor diameter	5.55 m
Type of control	Active Stall
Steering mechanism	By a rudder
Number of blades	3 twisty blades (5 airfoils along the profile)

Fig. 8 and Fig. 9 show the performance of the MPPT algorithm for various wind speed steps. When the wind speed increases, the torque imposed by the turbine becomes higher than the resistant torque causing the speed to increase. Thus, the value of $k_{i(mppt)}$ decreases implying higher currents through the generator i.e. higher resistant torque. A new operating point is obtained so that the power coefficient is maintained at the maximum value of 0.41. The current through the generator has a THD (total harmonic distortion) about 2.5% and the obtained power factor is equal to 0.997.

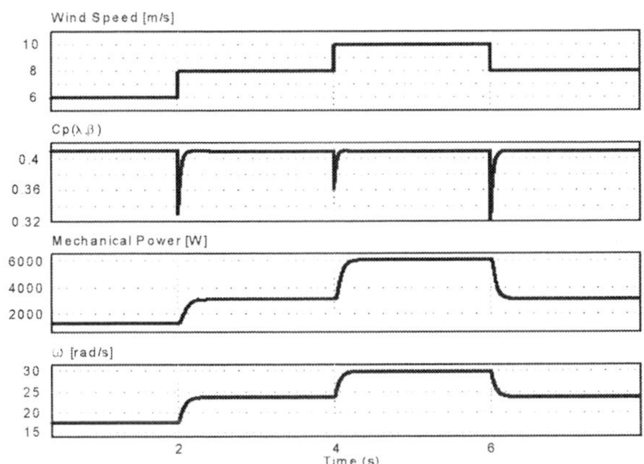

Fig. 8. MPPT response.

Fig. 10 validates the operation of the speed control loop with constant reference signal of 29 rad/s for a step in wind speed from 10 m/s to 12 m/s.

Although high wind speed is imposed in Fig. 11, the control loop is able to maintain the mechanical power constant and close to 7 kW. Thus, considering the losses in the generator windings, the converter processes a maximum power of 6 kW i.e. the rated power. Due to the kinetic energy, overshoot exists during transients but can be redirected to a crowbar circuit in the dc link.

A wind profile with wider range of values is applied in Fig. 12 so that it is possible to observe which value of k_i ($k_{i(mppt)}$ or $k_{i(power\ control)}$) defines the system operation. When the wind speed increases, the maximum extracted power can be higher than the rated one, but the mechanical power control generates a very high current reference (for very small values of k_i) so that the resistant torque does not allow the speed to be increased. In this case, $k_{i(power\ control)}$ is the minimum value. For wind speeds lower than about 11 m/s, $k_{i(mppt)}$ is the minimum value and the MPPT scheme operates properly.

Fig. 13 shows the steps responses of the dc link voltage, which is regulated at about 700 V with an overshoot of less than 0.6%.

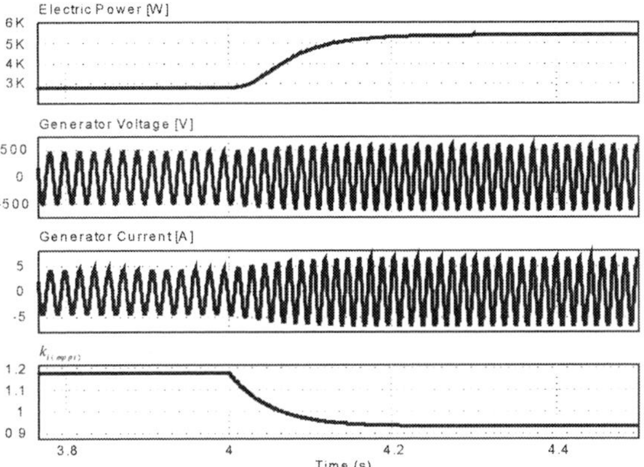

Fig. 9. PMSG waveforms for step in wind speed from 8 m/s to 10 m/s.

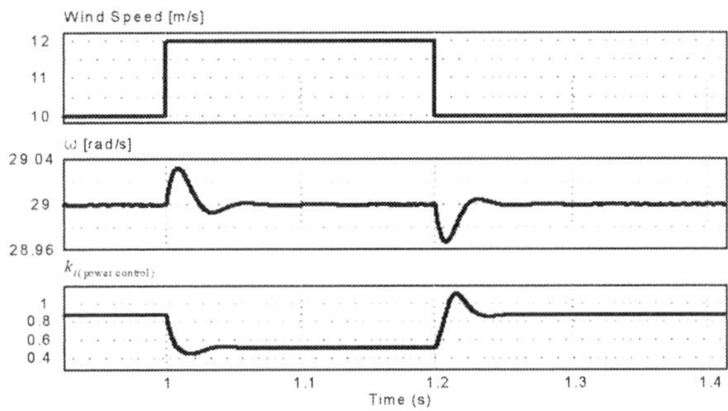

Fig. 10. Actuation of the shaft speed control loop.

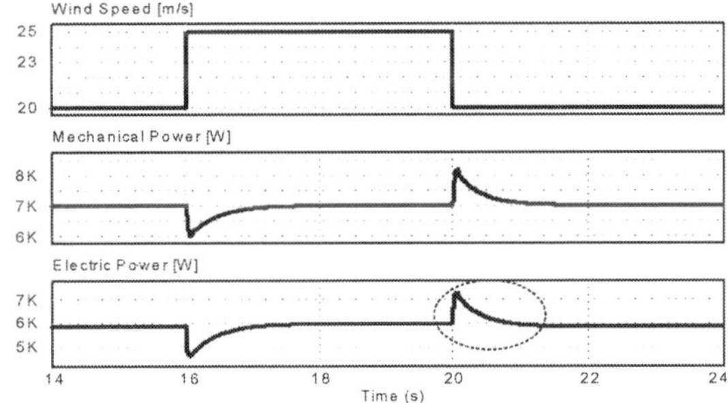

Fig. 11. Control of mechanical power for a step in wind speed from 20 m/s to 25 m/s.

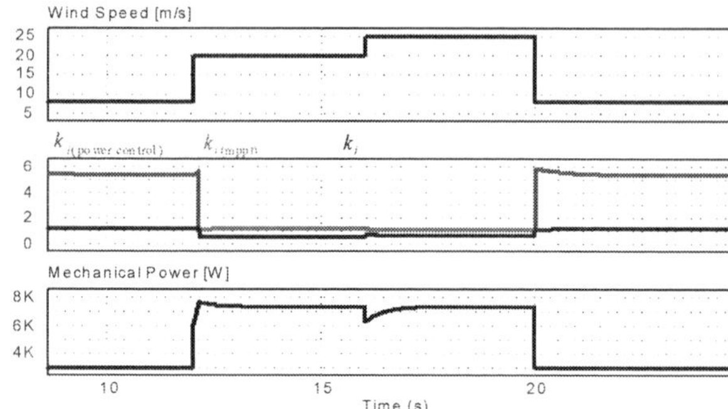

Fig. 12. Control of mechanical power for a variable wind step profile.

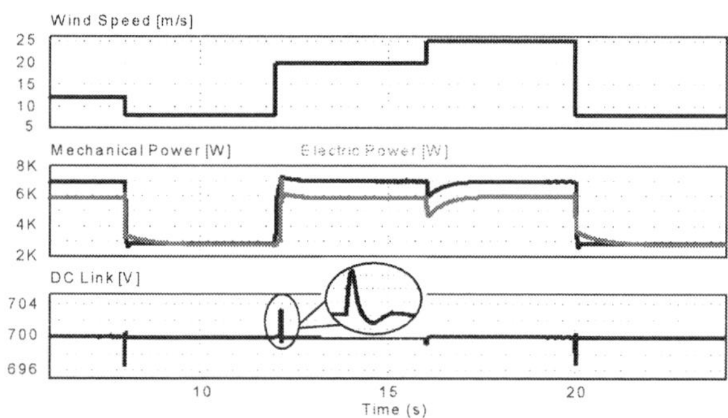

Fig. 13. Actuation of the dc-link voltage control loop.

IV. Experimental Prototype

A wind turbine model VERNE555 manufactured by Enersud was installed at the Federal University of Ceará with the characteristics presented in Table II.

Fig. 14 shows the details of the passive spring system responsible for the active stall control.

Fig. 14. Active stall spring system of the turbine model VERNE555.

The aforementioned converter is currently under test, while the prototype layout is shown in Fig. 15.

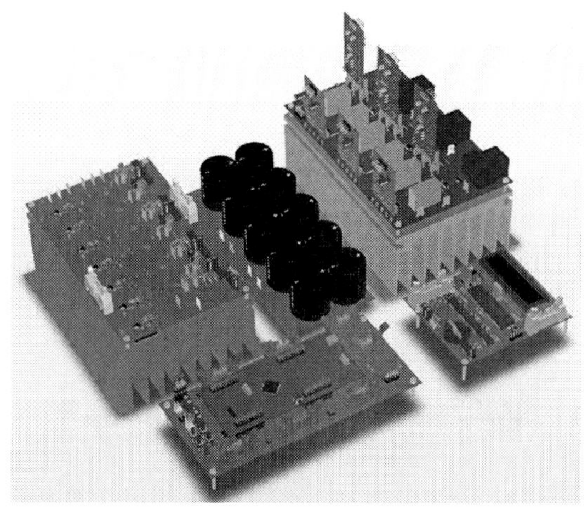

Fig. 15. 3-D view of the proposed power electronic device.

A DSP (digital signal processor) TMS320F28377S manufactured by Texas Instruments and a programmable microcontroller manufactured by Microchip model dsPIC30F4011 were used for control the total system.

IGBT module model GA35XCP12-247 (with Sic freewheeling diodes) manufactured by GeneSiC was used to built the inverter and the rectifier was assembled with the following semiconductors: IGBTs IRG4PF50WD by IR; CREE SiC Diodes C4D05120A.

In the following figures, can be seen some partial experimental results of the grid-side converter.

Fig. 16 shows the angular frequency detected from q-PLL according to the grid phase voltage. A frequency step of 45 Hz to 85 Hz is applied at t = 46 ms and the system tracking the phase angle.

In Fig. 17, waveforms of current in continuous operation at rated conditions (9.09 A rms) are present. The current has a THD of about 1.34%.

The power factor reaches 0.994. As desired by the loop current ($i_{d\_ref} = 0$), the current is in phase with the grid phase voltage, Fig. 18.

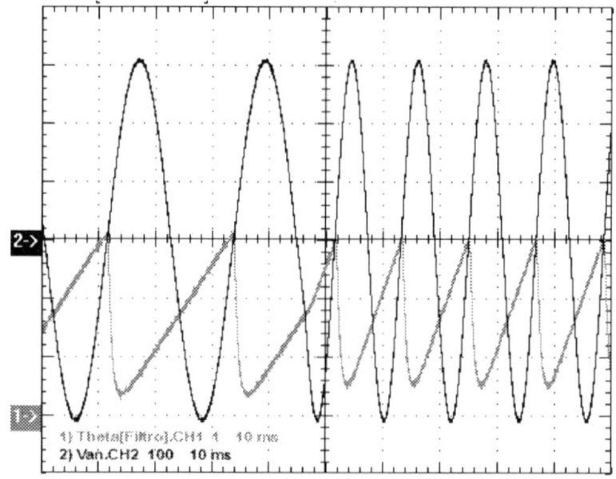

Fig. 16. PLL response for a step in frequency from 45 Hz to 85 Hz.

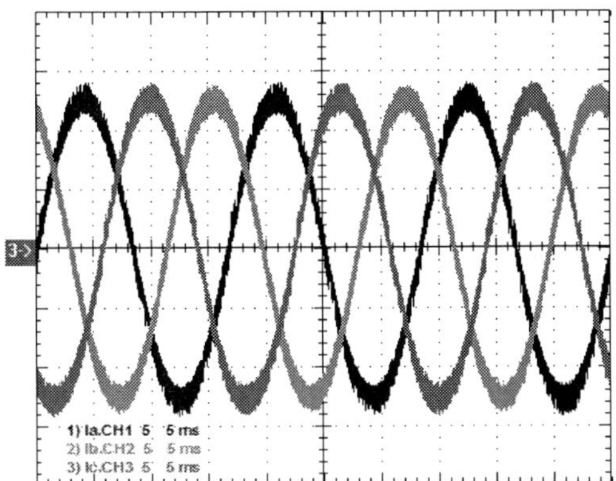

Fig. 17. Actuation of the current control with $i_{d\_ref} = 0$ and $i_{q\_ref} = 12.728$ A.

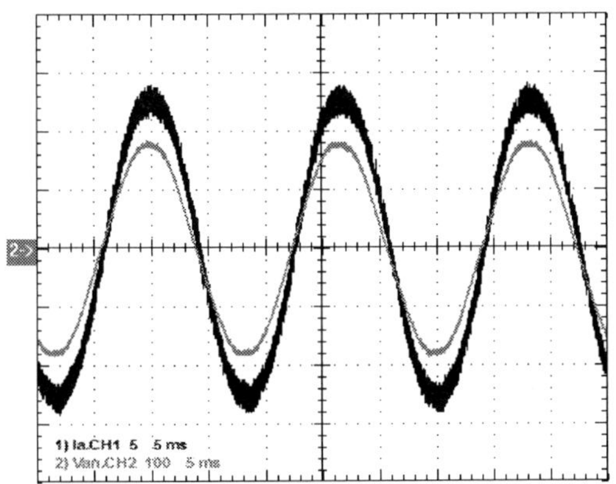

Fig. 18. Grid phase voltage V_{an} in phase with current I_a.

The Fig. 19 displays the dc link controlled by the loop voltage with reference $V_{dc\_ref} = 400$ V. A load step from 2.4 kW to 1.2 kW is applied at t = 180 ms and a load step from 1.2 kW to 2.4 kW at t = 620 ms. An acceptable overshoot/undershoot of 40 V (10%) and a response time of approximately 130 ms can be checked on loop voltage.

The inverter has efficiency of 94.6% to 3.4 kW (power achieved in the partial results).

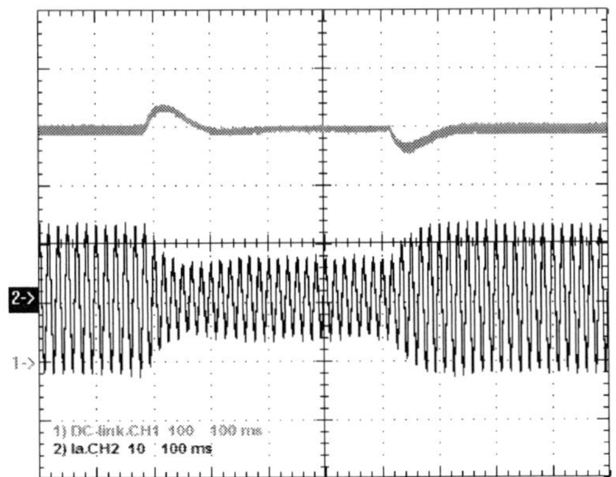

Fig. 19. Dynamic response of the dc-link control loop.

V. Conclusions

This paper has described the operation principle of a WECS, as well as simulation results in order to validate its respective operation. Considering the aforementioned advantages, reduction of cost associated to the proposed WECS due to the absence of anemometers (since the MPPT algorithm does not require the wind speed to be sensed), speed sensors (for the case of a synchronous generator, whose speed can be calculated from the current or voltage signals), interface inductances between the generator and the rectifier (since the synchronous inductance of the generator is high enough), and gearboxes (the PMSG can be built with high number of poles).

The MPPT algorithm or the electronic power limitation control that operate with the active stall strategy in wind turbine ensure maximum utilization of power for a wide range of wind speeds.

Acknowledgment

The authors acknowledge the Energy Processing and Control Group (GPEC) and also governmental support agencies CNPq and CAPES.

References

[1] Global Wind Energy Council. (2015, Mar.). *Global Wind Report: Annual Market Update*. Global Wind Energy Council. Brussels. [Online]. Available: http://www.gwec.net/wp-content/uploads/2015/03/GWEC_Global_Wind_2014_Report_LR.pdf

[2] U.S Energy Information Administration. (2015, Apr.). *Annual Energy Outlook 2015 with projections to 2040*. U.S Energy Information Administration. Washington. [Online]. Available: http://www.eia.gov/forecasts/aeo/pdf/0383(2015).pdf

[3] Empresa de Pesquisa Energética. (2015, June). *PDE 2024: Plano Decenal de Expansão de Energia 2024*. Empresa de Pesquisa Energética. Brasília. [Online]. Available: http://www.epe.gov.br/PDEE/PDE%202024.pdf

[4] J. M. Carrasco, L. G. Franquelo, J. T. Bialasiewicz, E. Galvan, R. C. P. Guisado, M. A. M. Prats, *et al.*, "Power-Electronic Systems for the Grid Integration of Renewable Energy Sources: A Survey", *Industrial Electronics, IEEE Transactions on*, vol. 53, pp. 1002-1016, 2006.

[5] C. Elmano, "Estudo e Desenvolvimento Experimental de um Sistema Eólico Interligado à Rede", Doctoral thesis, GPEC, UFC, Fortaleza, 2012.

[6] C. Tze-Fun,L. Loi Lei, "Permanent-Magnet Machines for Distributed Power Generation: A Review", in Power Engineering Society General Meeting, pp. 1-6, 2007.

[7] H. Kai-Wei,L. Chang-Ming, "Development of a Wind Interior Permanent-Magnet Synchronous Generator-Based Microgrid and Its Operation Control", Power Electronics, IEEE Transactions on, vol. 30, pp. 4973-4985, 2015.

[8] G. Tibola, "Sistema eólico de pequeno porte para geração de energia elétrica com rastreamento de máxima potência", master dissertation, INEP, UFSC, Florianópolis, 2009.

[9] I. R. Machado, D. S. Oliveira Jr., L. H. S. C. Barreto, H. M. Oliveira F., S. R. O. Souza, M. M. Reis, "Sistema Eólico de Pequeno Porte para Carregamento de Baterias," *Eletrônica de Potência*, vol. 12, no. 2, pp. 97-104, July 2007.M. Young, The Technical Writer's Handbook. Mill Valley, CA: University Science, 1989.

[10] P. N. Enjeti,R. Martinez, "A high performance single phase AC to DC rectifier with input power factor correction", in Applied Power Electronics Conference and Exposition, 1993. APEC '93. Conference Proceedings 1993., Eighth Annual, pp. 190-195, 1993.

[11] R. Martinez,P. N. Enjeti, "A high-performance single-phase rectifier with input power factor correction", *Power Electronics, IEEE Transactions on,* vol. 11, pp. 311-317, 1996.

[12] A. F. de Souza,I. Barbi, "High power factor rectifier with reduced conduction and commutation losses", in *Telecommunication Energy Conference, 1999. INTELEC '99. The 21st International,* p. 5 pp, 1999.

[13] L. Huber, J. Yungtaek,M. M. Jovanovic, "Performance Evaluation of Bridgeless PFC Boost Rectifiers", *Power Electronics, IEEE Transactions on,* vol. 23, pp. 1381-1390, 2008.

[14] B. R. de Almeida,D. S. Oliveira, "Power converter for vertical wind energy conversion system", in *Power Electronics Conference (COBEP), 2013 Brazilian,* pp. 468-473, 2013.

[15] D. Borgonovo, J. P. Remor, A. J. Perin,I. Barbi, "A Self-Controlled Power Factor Correction Single-Phase Boost Pre-Regulator", in *Power Electronics Specialists Conference, 2005. PESC '05. IEEE 36th,* pp. 2351-2357, 2005.

[16] A. L. P. Alcalde, D. Borgonovo,S. A. Mussa, "An FPGA control application: Self-control of current and linear control of DC link of PFC", in *Industrial Electronics, IECON. 34th Annual Conference of IEEE,* 2008, pp. 2387-2392, *2008.*

[17] C. E. A. Silva, D. S. Oliveira Jr., H. M. Oliveira Filho, L. H. S. C. Barreto, F. L. M. Antunes, "A three-phase Rectifier for WECS with Indirect Current Control," *Eletrônica de Potência*, vol. 16, no. 1, pp. 28-36, Feb. 2011.

[18] M. A. Abdullah, A. H. M. Yatim,T. Chee Wei, "A study of maximum power point tracking algorithms for wind energy system", in *Clean Energy and Technology (CET), 2011 IEEE First Conference on,* pp. 321-326, 2011.

[19] T. Burton, N. Jenkins, D. Sharpe,E. Bossanyi, *Wind Energy Handbook.* Chichester: Wiley, 2011.

[20] F. Blaabjerg, R. Teodorescu, M. Liserre,A. V. Timbus, "Overview of Control and Grid Synchronization for Distributed Power Generation Systems", *Industrial Electronics, IEEE Transactions on,* vol. 53, pp. 1398-1409, 2006.

[21] S. T. Sager, M. A. Khan,P. S. Barendse, "Laboratory setup of a grid-tied PM WECS for experimental investigation", in *Energy Conference and Exhibition (EnergyCon), 2010 IEEE International*, pp. 524-529, 2010.

[22] D. Borgonovo, "Modelagem e Controle de Retificadores PWM Trifásicos Empregando a Transformação de Park" , master dissertation, INEP, UFSC, Florianópolis, 2001.

[23] M. Karimi-Ghartemani, H. Karimi,M. R. Iravani, "A magnitude/phase-locked loop system based on estimation of frequency and in-phase/quadrature-phase amplitudes", *Industrial Electronics, IEEE Transactions on,* vol. 51, pp. 511-517, 2004.

[24] Sasso, E. M., Sotelo, G., Ferreira, A., Watanabe, E. H., Aredes, M., Barbosa, P. G. "Investigação dos Modelos de Circuitos de Sincronismo Trifásicos Baseados na Teoria das Potências Real e Imaginária Instantâneas (p-PLL e q-PLL)", in Proceedings of the 14th Brazilian Automatic Control Conference, pp. 480-485 (in Portuguese), 2002.

[25] A. Nagliero, R. A. Mastromauro, M. Liserre,A. Dell'Aquila, "Synchronization techniques for grid connected wind turbines", in *Industrial Electronics, IECON '09. 35th Annual Conference of IEEE,* pp. 4606-4613, 2009.

Overshoot Control of the Electromagnetic Torque During Fault Recovery for an SCIG with a STATCOM

Zahra Mahmoodzadeh, *Graduate Student Member, IEEE,* Mehrdad Yazdanian, *Graduate Student Member, IEEE,* Hooman Ghaffarzadeh, *Graduate Student Member, IEEE,* and Ali Mehrizi-Sani, *Senior Member, IEEE*
School of Electrical Engineering and Computer Science
Washington State University
Pullman, WA, 99164

Abstract—**Squirrel cage induction generators (SCIG) are one of the widely used generators in the wind turbines. The popularity of the SCIG is because of their relatively low price. However, speed and torque control in SCIGs is not as well studied as doubly fed induction generators (DFIG). This leads to more stress in the mechanical parts and more fatigue in the gearbox. In this paper a control method is proposed for limiting the electromagnetic torque overshoot during the recovery process subsequent to a fault. The idea behind the proposed method is to utilize the set point automatic adjustment with correction enabled (SPAACE) method to modify the voltage control loop of the static synchronous compensator (STATCOM) installed at the terminal of the SCIG to limit the torque overshoot.**

Index Terms—**Voltage control, reactive power compensation, wind energy.**

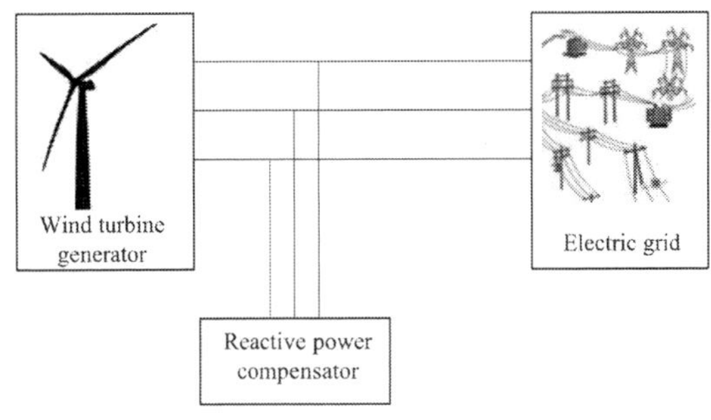

Fig. 1. Schematic configuration of the grid, wind turbine, and the STATCOM.

I. Introduction

Electromagnetic torque control by means of power electronic converters in variable speed wind generators is not a new subject [1]–[5]. However, there are few papers that study fixed speed wind generators electromagnetic torque control. Because many installed or planned wind generators are fixed speed generators [6], [7], it is beneficial to study the possibility of the torque control to reduce the mechanical stress in the drivetrain of the wind turbine.

Fixed speed wind generators are mostly squirrel cage induction generators (SCIG) connected to the grid directly without any converter. Since these generators are locked to the grid voltage and frequency, they are vulnerable to disturbances and can be easily affected by faults. Moreover, it is almost impossible to fully control the electromagnetic torque in an SCIG because torque is a function of the rotor current, and the rotor current is proportional to the terminal voltage. Therefore, controlling the electromagnetic torque is equivalent to controlling the terminal voltage.

On the other hand, sensitive parts of the drivetrain such as the gearbox suffer from the mechanical stress caused by the electromagnetic torque transients [8], [9]. The torque overload is harmful for the gearbox teeth and causes gearbox fatigue. As the torque increases beyond the rated value, it starts damaging the gearbox. The depth of the damage depends on how much the torque exceeds the nominal value and how long the overload persists. Reducing controllable torque overshoots results in a smoother accumulated duration of torque levels [10], which in turn significantly decreases the fatigue of the gearbox and therefore extends its lifetime.

What causes the torque overshoot during the voltage recovery after a fault clearance is the reactive power compensation device usually installed at the SCIG terminal. Grid codes are strict about the low voltage ride through (LVRT) capability of the wind generator. Therefore, it is common to use a STATCOM or other reactive power compensators near a wind generator. Fig. 1 shows the schematic configuration of a wind generator that has a reactive power compensator at its terminal. References [11]–[13] show that reactive power compensators can improve the LVRT capability of a wind generator. Specifically, [14] discusses and compares the LVRT capability of an SCIG with different reactive compensation devices including a STATCOM, a VSC, and a fixed capacitor. This comparison confirms that the STATCOM has a better performance for LVRT capability and stability.

The common practice for the STATCOM installed near an SCIG is to inject the maximum reactive current during the fault. Although it results in a higher maximum torque and consequently increases the transient stability margin and the LVRT capability, at the same time it causes overshoot of the torque after a fault. In this paper, we propose to add one additional control to the STATCOM that is originally installed at the SCIG terminal. The goal is to eliminate the overshoot

978-1-4673-9551-9/16 $31.00 © 2016 IEEE

of the torque after the fault clearance during the recovery.

There are not many papers that address this issue in the literature. Reference [15] suggests the indirect torque control (ITC) method to limit the maximum torque during the recovery process. The idea of ITC is to indirectly control the electromagnetic torque by changing the reference value of the voltage control loop of the STATCOM. In this method, whenever the torque value increases beyond the maximum torque limit set by the operator (e.g., 1.1 pu), the reference of the voltage control loop in the STATCOM switches to the value calculated by the ITC to maintain the torque within the limits. The reference value calculated by the ITC is a function of the rotor speed. Therefore, the rotor speed must be known to find the reference value. The voltage reference value is derived from KVL and KCL equations in the quasi-stationary equivalent circuit of the induction machine.

The ITC method can successfully eliminate torque overshoot after the fault clearance. Moreover, [10] shows that the method effectively alleviates harmful effects of the torque transients after the fault on gearbox. However, the ITC method has some drawbacks, which make it hard to implement in real wind generator. ITC is a model-based method and is highly sensitive to the system parameters. Therefore, any slight mismatch in the model parameters degrades the controller performance. Furthermore, the speed of the generator is used in the control loop and must be measured. Again any inaccuracy in the speed measurement has an adverse effect on the torque control. In addition, the main disadvantage of the ITC method is that it is necessary to detect the fault to be able to implement the ITC.

This paper proposes a control method for smoothing the electromagnetic torque overshoot after a fault clearance. The method utilizes set point automatic adjustment with correction enabled (SPAACE) method to achieve this goal. The contributions of the proposed method are as follows:

- The proposed method is robust against system parameters uncertainties;
- The proposed method does not require fault detection. Therefore, it is easier and cheaper to implement; and
- The proposed method uses only the electomagnetic torque measurements.

This paper is organized as follows. Section II introduces the SPAACE method. In Section III, the SPAACE unit is applied to the voltage control loop of the STATCOM to eliminate the electromagnetic torque overshoot after a fault clearance. Section IV presents the simulation results to confirm the performance of the proposed method. Finally, the paper is concluded in Section V.

II. SPAACE

SPAACE monitors the variable of interest and temporarily switches the set point to a scaled down value based on its real time (or predicted) value [16]–[18]. In this section, an explanation of the SPAACE method is provided. Then, SPAACE with prediction is demonstrated.

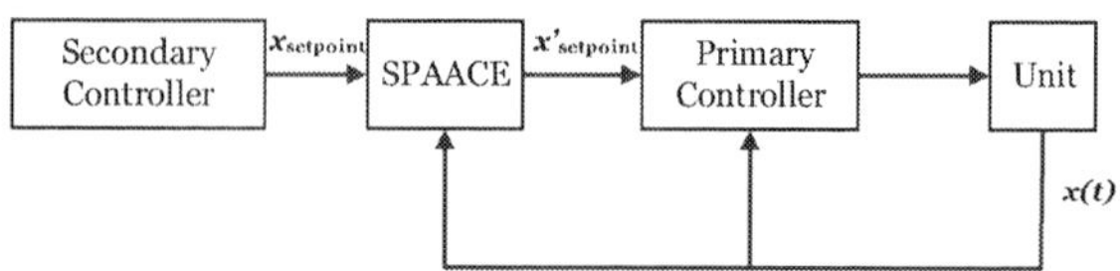

Fig. 2. Configuration of SPAACE [18].

A. SPAACE Based on Real Time Value

The main objective of the SPAACE method is to track the variable of interest and shape its trajectory to mitigate transients occurring in the system in real time. To achieve this goal, SPAACE compares the measured signal with the predetermined operational limits of the system in each time step and changes the set point according to the real time value of the response. The set point remains at this value while the violation continues and returns to the original value when the response gets back to the acceptable region.

Fig. 2 shows the configuration of the SPAACE among the existing controllers of a system. SPAACE acts as an independent unit and monitors the output signal of the secondary controller and the output signal of the system. According to the variation of the system output, SPAACE decides whether to switch the set point temporally between the original value and the scaled value.

For each signal, a minimum and maximum limit can be defined. When the set point increases from x_1 to x_2, $x(t)$ starts to increase in response to this variation in the input signal. Whenever $x(t)$ violates the maximum allowable limit, SPAACE imposes a temporary correction in the set point and decreases $x(t)$. The following equation shows the mathematical explanation of the SPAACE method:

$$x_2' = \begin{cases} (1+m)x_2, & x(t) < x_{\min} \\ (1-m)x_2, & x(t) > x_{\max} \\ x_2, & \text{otherwise} \end{cases}, \qquad (1)$$

where m is a heuristically determined scaling factor, and x_2' is the scaled set point imposed by SPAACE. This condition continues until the SPAACE detects that $x(t)$ does not violate the allowable limits anymore. At this time, set point changes from $(1-m)x_2$ to x_2. Similarly, if $x_2 < x_1$, the SPAACE temporarily scales up the set point from x_2 to $(1+m)x_2$ and when the trend of $x(t)$ indicates that it is within the acceptable region, SPAACE releases the set point [18].

B. SPAACE Based on Linear Prediction

The general SPAACE method only uses the real time value of $x(t)$ and decides whether to change the set point of the system based on this value. However, reference [18] uses a prediction approach to speed up the reaction of the SPAACE to the changes occurring within the system. In prediction-enabled SPAACE method, the real time value of $x(t)$ as well as the trend of $x(t)$ are monitored. SPAACE performs a prediction algorithm to calculate the future values of $x(t)$ and its trend.

The linear prediction is used by linear extrapolation on the output of the system. By assuming the current time is t_0, values of $x(t_0)$ and $x(t_0 - T_{pred})$ are used to calculate the average rate of change. The average rate of change of $x(t)$ is calculated as follows:

$$r = \frac{x(t_0) - x(t_0 - T_{pred})}{T_{pred}}. \tag{2}$$

Where T_{pred} and r are the prediction horizon and the average rate of change, respectively.

In order to limit the required storage space, only the last n sampled values of the output signal $x(t)$ are stored, where $n = T_{pred}/T_s$ and T_s is the sampling rate. Finally, the predicted value $x(t_0 + T_{pred})$ is calculated as follows:

$$x(t_0 + T_{pred}) = x(t_0) + rT_{pred}. \tag{3}$$

III. Implementation of the SPAACE Method in Voltage Control Loop

This section provides a SPAACE-based control method for limiting the torque during the recovery after a fault. The idea of the control method is to change the reference value of the voltage to keep the torque in the limits by means of the SPAACE method. Fig. 3 shows the torque during and after a fault at the terminal of an SCIG that has a STATCOM without a torque overshoot control. The torque has high-frequency transients exactly at the beginning and at the end of the fault. These transients are not within the scope of this paper. During the recovery process, there is another transient which is slower compare to the fault transients. The goal is to eliminate the overshoot of the electromagnetic torque during the slower transient. The configuration of the the SPAACE-based torque control is shown in Fig. 4. The structure of the STATCOM controllers and the inner current control loop are available in [19].

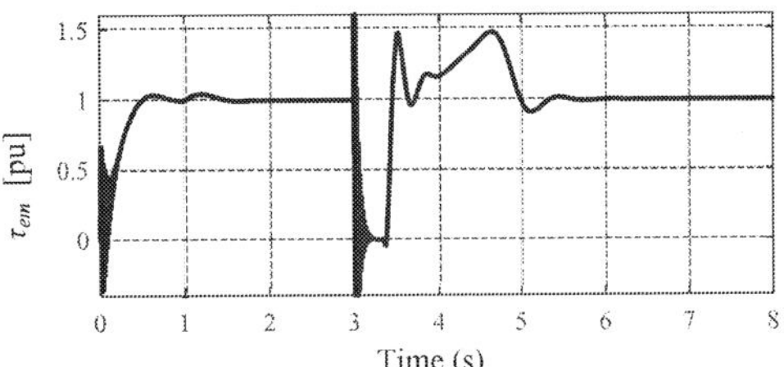

Fig. 3. The electromagnetic torque during and after a fault at the terminal of an SCIG with a STATCOM without torque control.

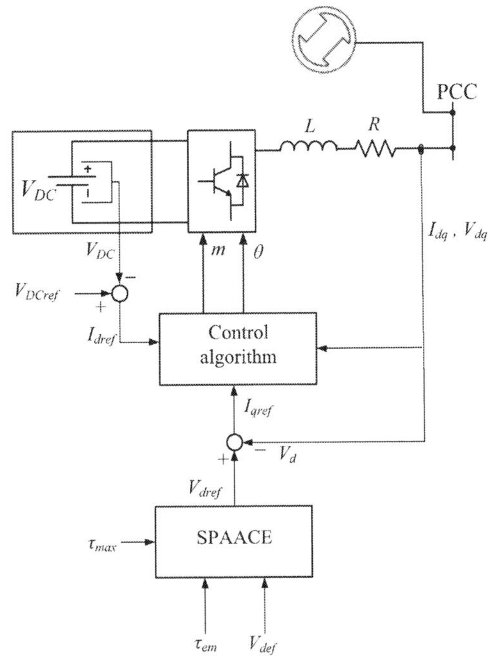

Fig. 4. STATCOM control system with a SPAACE-based torque control loop.

Fig. 4 shows that the inputs of the SPAACE unit are the real time value of the electromagnetic torque τ_{em} and the default set point of the voltage V_{def}, which is 1 pu in this study. The SPAACE unit predicts the torque value for the next prediction horizon and decides whether to change the voltage set point or use the default value based on the predicted value.

When a fault happens, the terminal voltage of the SCIG goes down. Therefore, the electromagnetic torque goes down as well. During the fault, the mechanical torque is much higher than the electromagnetic torque. Mechanical torque is the driving torque, and the electromagnetic torque is the resistive torque. Therefore, the turbine speeds up. If the speed does not increase beyond the stability limit during the fault, the generator is able to recover and brings back the terminal voltage to the nominal value. In this case, the turbine starts to slow down due to the presence of the resistive torque. As the turbine slows down, the electromagnetic torque increases due to the nature of the SCIG speed-torque curve and causes the overshoot discussed earlier. Once the SPAACE unit observes that the electromagnetic torque is going above the maximum limit, it scales down the voltage reference value to 80% of the command value to maintain the torque within the limit. SPAACE continues predicting the torque and keeps commanding the scaled down voltage reference as far as the predicted value of τ_{em} is out of the limit. Whenever the predicted value stops violating the maximum limit, SPAACE releases the voltage reference to the command value.

IV. Simulation Results

To verify the effectiveness of the proposed method, a simulation case study is performed in this section. Table I defines the parameters of the SCIG, STATCOM, and the grid Thevenin equivalent circuit. The parameters are extracted from [15]. The simulation is performed in PSCAD/EMTDC environment. The simulation results are shown for 12 s in Fig. 5. The first 0.5 s is related to the machine start up. At $t = 1$ s, the machine reaches the steady state, and the voltage control loop starts working.

TABLE I
PARAMETERS OF THE SCIG, STATCOM, AND THE GRID THEVENIN EQUIVALENT CIRCUIT

Parameters of the SCIG	
Nominal power	2.26 MVA
Nominal terminal voltage	0.69 kV mΩ
Generator inertia constant	6.5 s
Self damping	0.008 pu
Stator resistance	0.01 pu
Stator unsaturated leakage reactance	0.179 pu
Unsaturated magnetizing reactance	4.376 pu
Rotor resistance	0.008 pu
Rotor unsaturated leakage reactance	0.074 pu
Parameters of the STATCOM	
Filter inductance	0.2 mH
Filter resistance	0.15 Ω
Maximum inductive current	4 pu
Maximum capacitive current	4 pu
Parameters of the grid	
Rated voltage	.69 kV
Rated frequency	60 Hz
Equivalent resistance	0.1126 pu
Equivalent reactance	0.1126 pu

At $t = 3$ s, the system is subjected to a 350 ms three-phase fault at the terminal of the machine.

Fig. 5(a) shows the rotor speed (ω_r). It shows that the wind turbine is able to recover the speed to the nominal value in 5.65 s. Fig. 5(b) shows the electromagnetic torque (τ_{em}). Before the fault, τ_{em} is at the nominal value. During the fault, it decreases to zero since the terminal voltage of the generator is zero. After the fault clearance τ_{em} has some fast transients then it reaches 1.1 pu. τ_{em} remains 1.1 pu up to the end of the recovery process, when ω_r returns to the nominal value. Therefore, the method can successfully smooth the torque overshoot during the recovery process, and the electromagnetic torque does not exceed the maximum value of 1.1 pu. However, this overshoot elimination is with the cost of extending the recovery time. This recovery time extension is unavoidable in any torque control technique for SCIGs because the shaft needs more time to get back to the rated speed when the electromagnetic torque is kept close to the mechanical torque. Fig. 5(c) shows the terminal voltage of the SCIG (V_t). The SCIG terminal experiences a voltage sag with the minimum of 80% for 5.5 s. Fig. 5(d) shows the reactive current injected by the STATCOM (I_{qs}). The reactive current is saturated during the fault and at some time intervals during the recovery process.

There are many factors which determine the length of the recovery. Among those, the maximum allowable value for the torque and the maximum reactive current that the STATCOM can provide have the most influential effects. Table II shows the recovery time versus the maximum torque limit. This table indicates that the length of the recovery process is inversely proportional to the maximum allowable value of the torque. In other words, the stricter overshoot limit results in the longer recovery time.

Table III shows the length of the recovery time versus the maximum reactive current of the STATCOM. This table indicates that as the STATCOM gets larger, the recovery process gets faster.

TABLE II
RECOVERY TIME VERSUS MAXIMUM TORQUE LIMIT

Torque Limit (pu)	Recovery Duration (s)
1.06	8.15
1.1	5.65
1.13	5.03
1.16	4.22
1.18	4
1.2	3.68

TABLE III
RECOVERY TIME VERSUS MAXIMUM REACTIVE CURRENT OF THE STATCOM

STATCOM Max Reactive Current (pu)	Recovery Duration (s)
3.5	5.65
4	3.44
4.5	3.33
5	3.18
5.5	3.14

V. CONCLUSION

This paper proposes a SPAACE-based method to limit the torque overshoot during the recovery process after a fault for an SCIG with a STATCOM installed at its terminal. In the proposed method, the set point of the voltage is temporarily switched to a lower value based on the real time and predicted torque values. The advantages of the proposed method is that it is not sensitive to the model parameters uncertainties, it uses only the electromagnetic torque measurement to control the overshoot, and more importantly, there is no need for fault detection in the proposed method. The simulation results show that the method can successfully eliminate the torque overshoot during the recovery process.

REFERENCES

[1] F. Kanellos, S. Papathanassiou, and N. Hatziargyriou, "Dynamic analysis of a variable speed wind turbine equipped with a voltage source AC/DC/AC converter interface and a reactive current control loop," in *10th Mediterranean Electrotechnical Conf.*, vol. 3, May 2000, pp. 986–989.

[2] S. Papathanassiou and M. Papadopoulos, "Dynamic behavior of variable speed wind turbines under stochastic wind," *IEEE Trans. Energy Convers.*, vol. 14, no. 4, pp. 1617–1623, Dec. 1999.

[3] J. Carrasco, L. Franquelo, J. Bialasiewicz, E. Galvan, R. Guisado, M. Prats, J. Leon, and N. Moreno-Alfonso, "Power-electronic systems for the grid integration of renewable energy sources: A survey," *IEEE Trans. Ind. Electron.*, vol. 53, no. 4, pp. 1002–1016, Jun. 2006.

[4] F. Blaabjerg, R. Teodorescu, M. Liserre, and A. Timbus, "Overview of control and grid synchronization for distributed power generation systems," *IEEE Trans. Ind. Electron.*, vol. 53, no. 5, pp. 1398–1409, Oct. 2006.

[5] J. Baroudi, V. Dinavahi, and A. Knight, "A review of power converter topologies for wind generators," in *EEE Int. Conf. on Electric Mach. and Drives*, May 2005, pp. 458–465.

[6] M. Molinas, J. Suul, and T. Undeland, "Extending the life of gear box in wind generators by smoothing transient torque with STATCOM," *IEEE Trans. Ind. Electron.*, vol. 57, no. 2, pp. 476–484, Feb. 2010.

[7] L. H. Hansen, L. Helle, F. Blaabjerg, E. Ritchie, S. Munk-Nielsen, H. Bindner, P. Srensen, and B. Bak-Jensen, "Conceptual survey of generators and power electronics for wind turbines," *Tech. report for Riso National Laboratory, Roskilde, Denmark*, Jan. 2002.

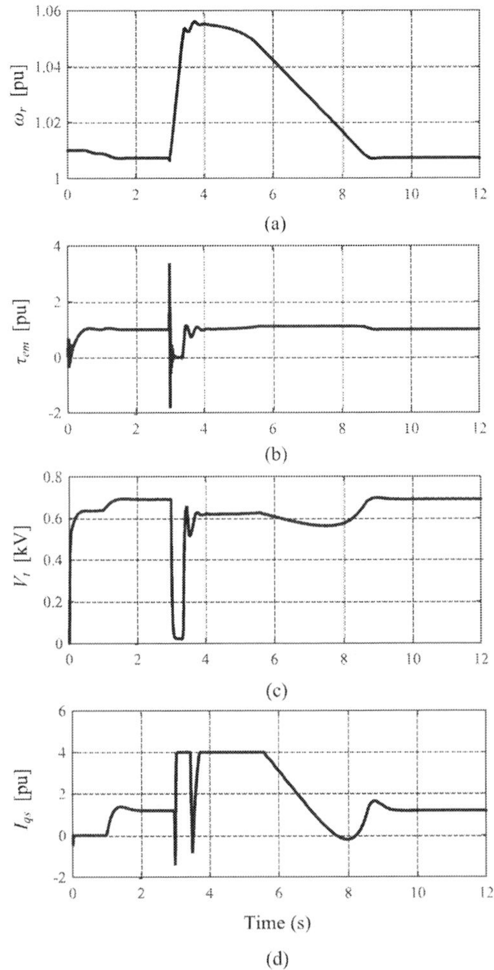

Fig. 5. Simulation results for the STATCOM with SPAACE-based voltage control loop. a) rotor speed, b) electromagnetic torque, c) reactive current injected by the STATCOM, and d) terminal voltage.

[8] J. Faiz, M. Ghaneei, and A. Keyhani, "Performance analysis of fast reclosing transients in induction motors," *IEEE Trans. Energy Convers.*, vol. 14, no. 1, pp. 101–107, Mar. 1999.

[9] S. Papathanassiou and M. Papadopoulos, "Mechanical stresses in fixed-speed wind turbines due to network disturbances," *IEEE Trans. Energy Convers.*, vol. 16, no. 4, pp. 361–367, Dec. 2001.

[10] M. Molinas, J. Suul, and T. Undeland, "Torque transient alleviation in fixed speed wind generators by indirect torque control with STATCOM," in *13th conf. on Power Electron. and Motion Control*, Sep. 2008, pp. 2318–2324.

[11] ——, "Low voltage ride through of wind farms with cage generators: STATCOM versus SVC," *IEEE Trans. Power Electron*, vol. 23, no. 3, pp. 1104–1117, May 2008.

[12] S. Salman and A. Teo, "Improvement of fault clearing time of wind farm using reactive power compensation," in *Power Tech Proceedings, Porto*, vol. 2, Sep. 2001, p. 6.

[13] ——, "Investigation into the estimation of the critical clearing time of a grid connected wind power based embedded generator," in *IEEE/PES Asia Pacific Transmission and Distribution Conf. and Exhibition 2002*, vol. 2, Oct. 2002, pp. 975–980.

[14] M. Molinas, J. Suul, and T. Undeland, "Low voltage ride through of wind farms with cage generators: STATCOM versus SVC," *IEEE Trans. Power Electron*, vol. 23, no. 3, pp. 1104–1117, May 2008.

[15] J. Suul, M. Molinas, and T. Undeland, "STATCOM-based indirect torque control of induction machines during voltage recovery after grid faults," *IEEE Trans. Power Electron.*, vol. 25, no. 5, pp. 1240–1250, May 2010.

[16] A. Mehrizi-Sani and R. Iravani, "Online set point adjustment for trajectory shaping in microgrid applications," *IEEE Trans. Power Syst.*, vol. 27, no. 1, pp. 216–223, Feb. 2012.

[17] A. Mehrizi-Sani, "A strategy to improve reference tracking of distributed energy resources," in *IEEE PES Innovative Smart Grid Technologies (ISGT)*, Feb. 2013, pp. 1–6.

[18] A. Mehrizi-Sani and R. Iravani, "Online set point modulation to enhance microgrid dynamic response: theoretical foundation," *IEEE Trans. Power Syst.*, vol. 27, no. 4, pp. 2167–2174, Nov. 2012.

[19] M. Yazdanian and A. Mehrizi-Sani, "Internal model-based current control of the RL filter-based voltage-sourced converter," *IEEE Trans. Energy Convers.*, vol. 29, no. 4, pp. 873–881, Dec. 2014.

A self-adaptive power balance control strategy for PV inverters in islanded microgrids

Zhenxiong Wang
zhxwang2014@163.com

Hao Yi
yi_hao@ xjtu.edu.cn

Fang Zhuo
zffz@mail.xjtu.edu.cn

Zhigang Zhang
y18719990@163.com

State Key Laboratory of Electrical Insulation and Power Equipment,
Xi'an Jiaotong University,
Xi'an, China.

Abstract—**For photovoltaic inverters in islanded microgrids, droop control is a preferable strategy due to its attractive ability in bus voltage regulation and load sharing without additional communication system. However, in islanded microgrids without redundant storages, further realization of power balance between droop controlled PV sources and dynamic loads, which is beneficial to enhance PV sources penetration and system reliability, remains a serious problem. This study focuses on this issue and improves conventional droop control method with self-adaptive power balance strategy. The proposed droop method can promise stable operation when total load consumption is under maximum capacity of all PV sources, which means PV arrays quit maximum power point and shift in accordance with load demands to avoid large storage component at dc-side of inverters while keeping voltage regulation ability from droop control.**

Keywords—photovoltaic inverter; islanded microgrid; power balance; voltage control mode

I. INTRODUCTION

Photovoltaic (PV) generations have attracted widely concerns in last decades due to its exploitation of sustainable and environment-friendly solar energy [1]. Besides, they provides practical benefits in powering remote regions [2] [3], such as mountainous areas and small islands. In these conditions, long distance power transmission can be avoided and solar energy highly participate in serving local loads, where the islanded microgrids are formed.

Although solar energy provides great profits, the stability problem still impedes its development. Therefore, large storage components and generations with fossil fuels are indispensable constituents in PV microgrids, which highly promotes investment. Normally, PV inverters in such kind of islanded microgrids can operate in current control mode (CCM) or voltage control mode (VCM) [4]-[6]. In the previous one, PV inverter works in master/slave mode where it is controlled as a current source to track the MPP of PV arrays and grid voltage is maintained by a master inverter or generators with fossil fuels[7]-[9]. It is a kind of "source driven" mode, where inverter output power is determined by the solar energy conditions and power difference between sources and loads should be buffered by storage systems or more reliable generators connected in microgrid AC bus. Storage system is almost completely responsible for the whole power management strategy. On the other hand, in VCM, PV inverters are controlled by droop strategy, where bus voltage is supported by all the PV inverters and their output power is driven by the load demands [10]-[12]. Hence VCM is a better opinion for voltage regulation, where high penetration of PV inverters operated in CCM will lead to instability in microgrids. However, large storage components should be implemented at each inverter dc-side to buffer the power difference between sources and loads in VCM. The inverter is highly dependent on its internal storage components to realize power balance.

Both the source-driven and load-driven methods are biased in power grid due to negligence of power balance and hence large storage components are always crucial in aforementioned strategies and predominate the stability and reliability of the microgrid. Storages must take responsibility for compensation of power difference, which dramatically restrains the utilization of solar energy in islanded microgrids [4] [13].

To reduce the participation of storages and improve performance of PV arrays and inverters in maintaining stability and reliability of microgrid, a novel self-adaptive power balance control strategy is proposed for PV inverters in islanded operation mode. The proposed strategy combines droop control at inverter side with power control in PV arrays where additional storage components and complex communication systems are no longer necessary. PV inverters with this strategy maintain capability of bus voltage regulation, thus improving stability of microgrids. Meanwhile, power balance control at PV arrays eliminates superfluous power from sources, which actually quit MPP of PV arrays and automatically move the operating point along the P-V curve according to real-time load demands. Consequently, islanded microgrids can stably operate as long as load demands are less than generation of PV sources. The detailed principle and realization are analyzed in following sections.

II. SYSTEM DESCRIPTION

The general structure of a PV generation is described in Fig. 1, which includes PV arrays, boost converter and inverter. In this figure, V_{PV}, I_{PV} and P_{PV} are output voltage, current and power of PV arrays; ω is voltage frequency of inverter, while ω^* is nominal frequency; V_{MPP} is the voltage of MPP, V_{dc} is dc-link capacitor voltage and P_{out} is output power of inverter.

This work was supported by the National High Technology Research and Development Program of China (863 Program) 2015AA050606; grants from the Power Electronics Science and Education Development Program of Delta Environmental & Educational Foundation DREG2015018.

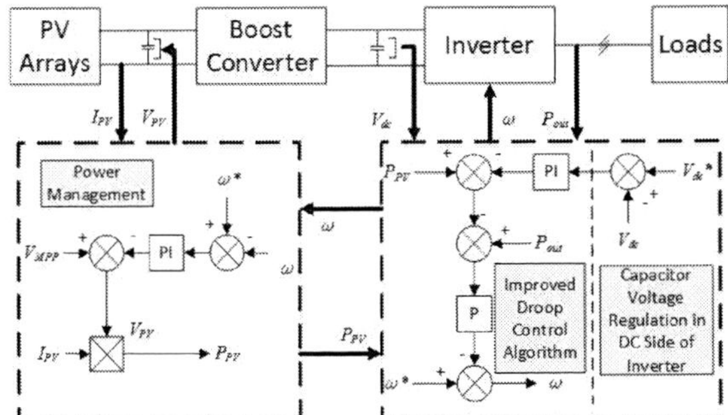

Fig.1. Structure of an integral System.

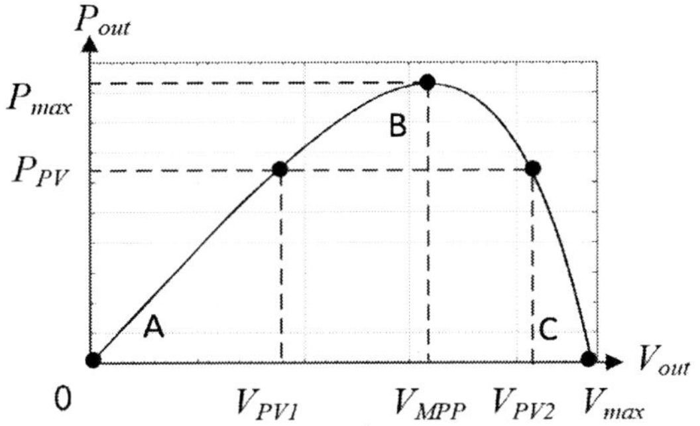

Fig.2. Characteristic power curve of PV array.

The boost converter is responsible for PV array voltage regulation to manage power flow. And the power balance strategy is realized through the control of boost converter. Therefore, large storage component is not employed at inverter DC side to assume the extreme condition that storage systems fail in participation of power management. PV generations applied with this strategy actively balance load demands instead of operating in source-driven or load-driven mode. After the PV arrays generate desired power, the inverter transports desired power to ac-bus through an improved droop control and dc-link capacitor voltage control strategy, which ensures PV generator's capability in ac-bus voltage regulation and hence improves stability, reliability and robustness of islanded microgrids. The detailed principle is introduced as following.

III. Power Management of Photovoltaic Arrays

In conventional MPPT algorithm, photovoltaic generation ignores the actual load demands and track its maximum power point even when the energy is superfluous, which ignores the fact that power balance is a critical problem in islanded microgrids. When PV sources cannot generate adequate amount of power corresponding to load demands in islanded mode, removing loads to guarantee critical users or resorting to back-up sources are advisable choices. On the other hand, when PV energy is superfluous, quitting MPP and searching for a proper operating point are advisable and reasonable for PV array power management.

A typical power curve of PV arrays is displayed in Fig. 2. The voltage V_{MPP} is obtained through a MPPT algorithm which disturbs PV array voltage and determines its operating voltage to output maximum power, thus achieving the best exploitation of solar energy [14]. When power is superfluous, PV arrays quit MPP and change the terminal voltage, therefore decreasing generated power. The regulation strategy of PV array voltage is showed in Fig. 1, a PV terminal voltage controller is implemented in boost converter, which is described as:

$$V_{PV} = V_{MPP} - \left(k_{pv} + \frac{k_{iv}}{s}\right)\left(\omega^* - \omega\right) \tag{1}$$

Where k_{pv} and k_{iv} are PI regulator parameters, ω is the output voltage frequency of inverter, ω^* is the idling frequency. Due to predesigned droop characters, mismatch between sources and loads will lead to grid frequency's deviation from nominal value. The deviation contains information of load demands and therefore can be used to manage output power of PV arrays. A PI regulator in (5) is applied to adjust terminal voltage and output power in the meantime. It is observed in Fig. 2 that the same power point P_{PV} is linked to a pair of voltage points V_{PV1} and V_{PV2} that situated in line AB and line BC respectively. Due to excessive terminal voltage and rapid variation, operating points in line BC are discarded with an upper voltage limit.

Moreover, it should be noted that PV generations with the proposed power management strategy can consequently compensate the frequency deviation introduced by droop control. The detailed problem will be analyzed in the next section. When grid frequency deviates from its nominal value, PV arrays change their output power and recover output frequency of inverter. With the participation of PV sources in frequency management, the steady-state error is accordingly eliminated.

With the aforementioned algorithm, the output power of PV sources can change self-adaptively in line with load demands. PV generation transports desired power to dc-side of inverter. The proposed strategy is effective even when large storage system is not assumed to compensate power difference between PV arrays and loads, as long as load demands are less than maximum power generation. After this, another control strategy is required to transfer assigned power from boost converter to islanded microgrid bus with good performance.

IV. Improved Droop Control Strategy

Inverters operated in CCM are highly dependent on other supporting generators and communication systems, which causes the gird extremely vulnerable to disturbance and failure. However, droop control in islanded mode enables a PV inverter to operate in VCM and behave like a voltage source, which is advantageous for good voltage regulation and load demands satisfaction performance. Therefore, droop control is a more reasonable and profitable strategy.

A. Improved Droop Control Algorithm

In conventional droop control method, paralleled inverters self-regulate their output power by predesigned droop curve. In the case of inductive line impedance, the p-f droop curve can be described as:

$$\omega = \omega^{*} - mP_{out} \quad (2).$$

Where m is the droop coefficient and P_{out} is output power of inverter. Power flow in microgrids with this control strategy is driven by the loads and should not be influenced by sources. Output power of each inverter is determined by the ratio of droop coefficients among different inverters. The overall power can exactly satisfy load demands, which cause it a self-adaptive strategy in ideal condition [15]. Therefore, large storage component is assumed at dc-side of inverter to buffer the power difference at any time. Unfortunately, the condition is particularly strict and does not always exist in practical field. A different control strategy is required to extend droop control's application in non-ideal conditions.

To solve the problem mentioned above, an improved droop control algorithm displayed in Fig. 1 is proposed. The droop controller can be described as:

$$\omega = \omega^{*} - k_{pm}\left(P_{out} - P_{ref}\right) \quad (3).$$

Where P_{ref} is the reference output power of the inverter, k_{pm} is a proportional regulator in droop control. In conventional droop control algorithm, since output power is driven by load demands, power reference is meaningless in inverter's control loop. However, when storage system fails to participate in power management due to the capacity limitation, system faults or some other problems in some cases, a reference power is necessary so that inverter's output power is controllable and independent of its predesigned power that determined by droop coefficients and load demands. In this paper, reference power is equal to output power of PV arrays to accomplish power balance of dc and ac-link. Inverters with the improved control method can track reference power without change of nominal frequency. The detailed principle is illustrated in Fig. 3. Divide (2) into a couple of new equations:

$$\omega = \underbrace{\left(\omega^{*} + k_{pm}P_{ref}\right)}_{\omega^{*}_{ref}} - k_{pm}P_{out} \quad (4).$$

Where ω^{*}_{ref} is the equivalent idling frequency. From the modified equation, mismatch between reference and output power leads to variation of idling frequency with introduce of reference power P_{ref}. An example is showed in Fig. 3. Idling frequency ω^{*} changes from ω^{*}_{refA} to ω^{*}_{refB} in accordance with reference power's deviation from P_{refA} to P_{refB}. Then the droop curve moves from line A to line B in the ω-P coordinates, resulting in increasing output power which finally reaches reference power P_{refB}. From the aforementioned analysis, inverters adopted with this improved droop control algorithm can track reference power by automatically modifying droop curve while maintaining the benefits of VCM. Moreover, the

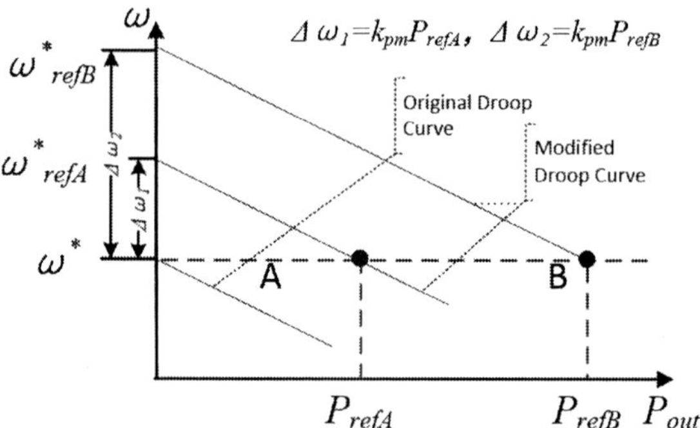

Fig.3. Illustration of improved droop control algorithm.

improved strategy doesn't change nominal frequency in steady state, which partly compensate for the inherent drawback of droop control. However, as loads are variable and unpredictable in islanded microgrids, the difference between reference and output power is accordingly inevitable even with adjustable droop curves which results in annoying steady-state error. Furthermore, disturbance such as computational and sampling error also causes a steady state deviation. The power management strategy introduced in previous section can resolve the problem.

B. DC Voltage Control in Inverter

Large voltage ripples in dc-link capacitor is disadvantageous to system stability. In PV inverter, capacitor voltage fluctuations are provoked in two aspects. On the one hand, the time constant difference between boost converter and inverter will lead to transient power imbalance inside a PV generation. On the other hand, load variation will also cause large transient power imbalance. Consequently, capacitor at dc-side of inverter may deeply charge and recharge to balance the transient power, which causes unexpected dc-link voltage fluctuations. Therefore, a control loop is designed to suppress such voltage ripples and the details are depicted in Fig. 1. With the DC voltage control loop, reference value of inverter output power is then modified as:

$$P^{*} = P_{ref} - \left(k_{p} + \frac{k_{i}}{s}\right)\left(V_{dc}^{*} - V_{dc}\right) \quad (5)$$

The algorithm regulates voltage at dc-side by slightly modifying inverter's transient output power. A PI regulator is implemented here to promise fast dynamic response and steady-state error elimination. As another control algorithm is applied to balance power flow, which will be introduced later, capacitor is adequate in balancing transient power fluctuation where additional storage component is not required in dc-link.

Then the inverter can operate in VCM and regulate bus voltage. However, an indispensable premise of such strategy is the balanced power flow between PV sources and microgrid's loads, which is realized in previous section. With the combination of droop control and power management strategy,

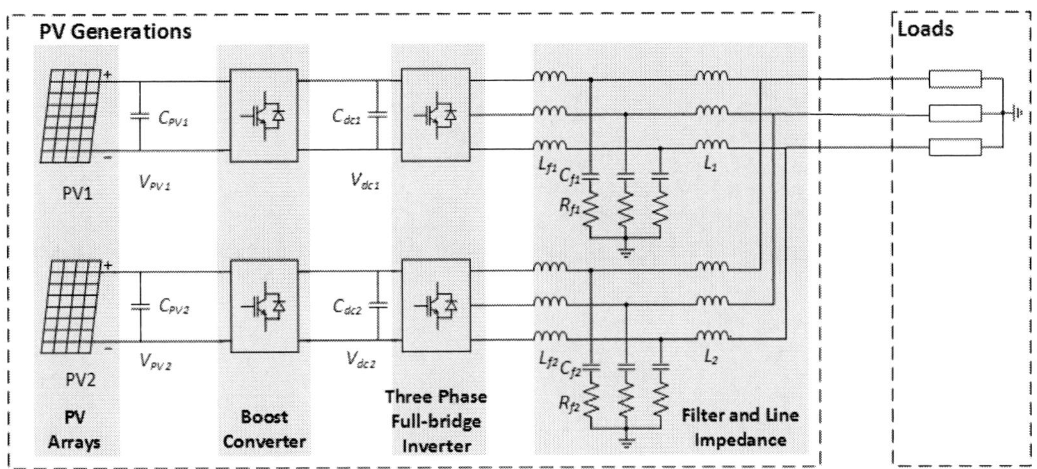

Fig.4. The main circuit applied in simulation.

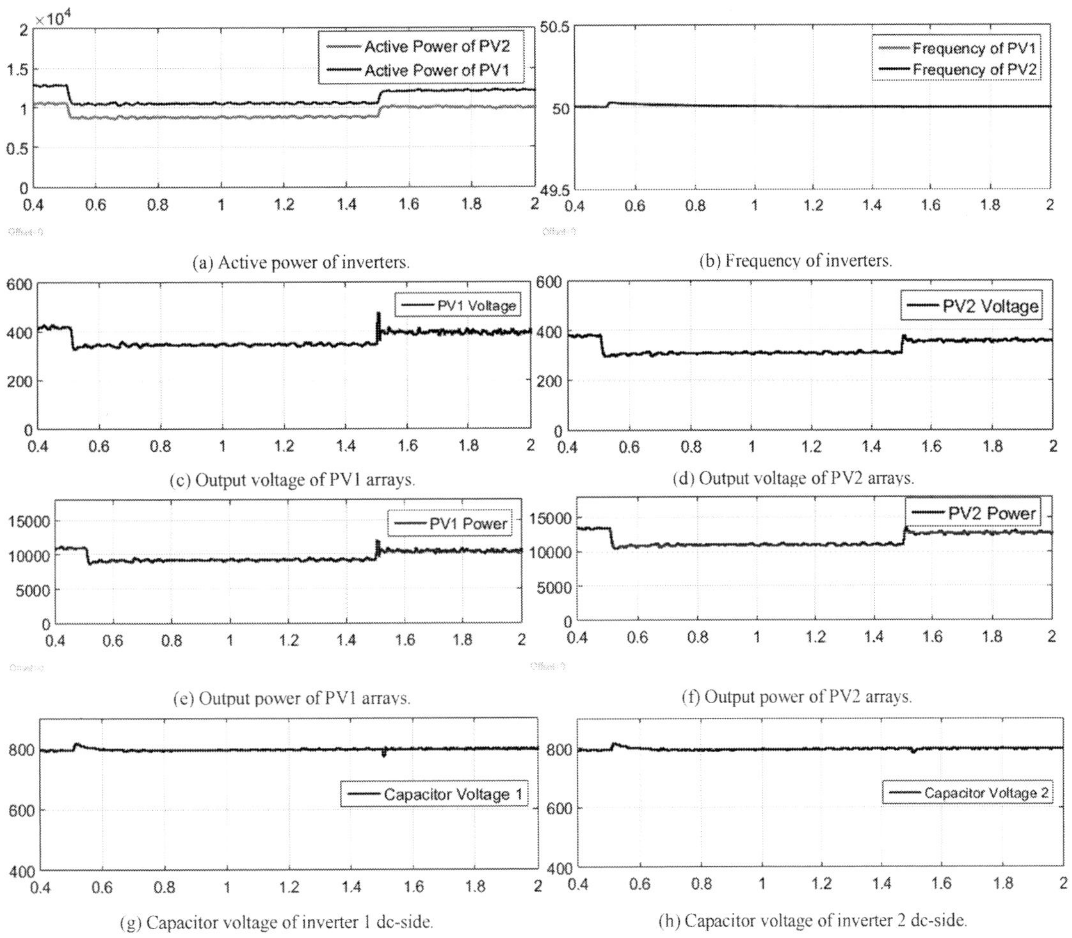

(a) Active power of inverters.

(b) Frequency of inverters.

(c) Output voltage of PV1 arrays.

(d) Output voltage of PV2 arrays.

(e) Output power of PV1 arrays.

(f) Output power of PV2 arrays.

(g) Capacitor voltage of inverter 1 dc-side.

(h) Capacitor voltage of inverter 2 dc-side.

Fig.5. Simulation Results.

power of PV generations is self-adaptively controlled in islanded microgrids.

V. Simulation

The proposed strategy is verified in MATLAB. The simplified schematic diagram is depicted in Fig. 4. Microgrid comprises two paralleled PV generations connected to the common loads, where each generation includes PV arrays, a boost converter and a three-phase full-bridge inverter. Microgrid operates at islanded mode and large storage component is not employed at dc-bus or ac-bus. Power management and improved droop control strategies are employed in both PV generations. Parameters of the main circuit are listed in Table I. PV array models are simulated with SUNTECH STP305W PV modules. The rating of the PV modules are U_{oc}=44.7V, I_{sc}=8.89A, U_{mpp}=36.2V, I_{mpp}=8.43A and P_{max}=305W on an insolation level of 1000 W/m^2 and 25°C temperature. PV1 arrays comprise 13 cells in series and 3 in parallel. PV2 arrays comprise 12 cells in series and 4 in parallel.

TABLE I. Main circuit Parameters

Symbol	Value
C_{PV1}	400μF
C_{PV2}	400μF
C_{dc1}	1000μF
C_{dc2}	1000μF
L_{f1}	0.4mH
L_{f2}	0.4mH
C_{f1}	50μF
C_{f2}	50μF
R_{f1}	0.5Ω
R_{f2}	0.5Ω
L_1	0.15mH
L_2	0.15mH

Simulation results are showed in Fig. 5 (a) to (h). At the beginning, inverters work in the vicinity of MPP, where load demands are 24kW and set slightly lower than maximum capacity of PV sources (11.89kW and 14.64kW respectively). Power flow in microgrid is balanced. Capacitor voltage in inverter dc-side is controlled to 800V. The inverters are operated stably with the proposed strategy.

The load is controlled to decrease suddenly from 24kw to 19kw in 0.5s. Output power of both inverters decreases according to load demands, which is depicted in Fig.5 (a). As a result of instant power imbalance, frequency varies but soon returns to its nominal value with the power management strategy, as showed in Fig. 5 (b). In Fig. 5 (c) to (f), PV array voltage and power decrease with load variation and they are controlled at a new operating point. Capacitor voltage instantly increases but finally returns to its reference value. The peak value of voltage is acceptable and voltage ripples are small, as showed in Fig. 5 (g) (h).

In 1.5s, the load is controlled to increase from 19kw to 22kw. Output power of both inverters increases corresponding to load demands. PV generations finally work at a new operating point with no frequency deviation. Voltage and power of PV arrays increase with a larger load demands. Capacitor voltage instantly increases due to instant power imbalance. However, with the dc-side capacitor voltage regulation strategy, the voltage varies at an acceptable level.

Simulation results suggest that the power can be balanced with the proposed power management strategy, and frequency does not deviate from nominal frequency in steady state, which highly improves the power quality of the islanded microgrid.

VI. Conclusions

In this paper, a self-adaptive power balance strategy is proposed for PV generations operated in islanded microgrids without redundant storage and communication systems. With the proposed strategy, PV generations keep voltage regulation ability with improved droop control strategy and control output power of PV arrays according to load demands with proposed power management strategy. Hence PV arrays and inverters participate in power management and bus voltage regulation, which highly improves stability, reliability and robustness of islanded microgrids. The proposed strategy is verified in simulation.

References

[1] F. Blaabjerg, R. Teodorescu, M. Liserre, and A. V. Timbus, "Overview of Control and Grid Synchronization for Distributed Power Generation Systems," Industrial Electronics, IEEE Transactions on, vol. 53, pp. 1398-1409, 2006.

[2] R. W. Wies, R. A. Johnson, A. N. Agrawal, and T. J. Chubb, "Simulink model for economic analysis and environmental impacts of a PV with diesel-battery system for remote villages," Power Systems, IEEE Transactions on, vol. 20, pp. 692-700, 2005.

[3] R. H. Lasseter, "MicroGrids," in Power Engineering Society Winter Meeting, 2002. IEEE, 2002, pp. 305-308 vol.1.

[4] J. A. Peas Lopes, C. L. Moreira, and A. G. Madureira, "Defining control strategies for MicroGrids islanded operation," Power Systems, IEEE Transactions on, vol. 21, pp. 916-924, 2006.

[5] H. Jinwei and L. Yun Wei, "Hybrid Voltage and Current Control Approach for DG-Grid Interfacing Converters With LCL filters," Industrial Electronics, IEEE Transactions on, vol. 60, pp. 1797-1809, 2013.

[6] H. Jinwei, L. Yun Wei, and M. S. Munir, "A Flexible Harmonic Control Approach Through Voltage-Controlled DG–Grid Interfacing Converters," Industrial Electronics, IEEE Transactions on, vol. 59, pp. 444-455, 2012.

[7] A. Elrayyah, Y. Sozer, and M. Elbuluk, "Microgrid-Connected PV-Based Sources: A Novel Autonomous Control Method for Maintaining Maximum Power," Industry Applications Magazine, IEEE, vol. 21, pp. 19-29, 2015.

[8] D. B. Bizarro, R. B. Godoy, P. E. M. J. Ribeiro, L. A. Carniato, J. O. Soares, G. Luigi, et al., "MPP tracking for grid connected inverters controlled by drooping curves," in Power Electronics Conference (COBEP), 2013 Brazilian, 2013, pp. 658-665.

[9] G. Dehnavi and H. L. Ginn, "Load sharing among converters in an autonomous microgrid in presence of wind and PV sources," in Innovative Smart Grid Technologies (ISGT), 2013 IEEE PES, 2013, pp. 1-6.

[10] M. B. Shadmand, R. S. Balog, and H. Abu-Rub, "Model Predictive Control of PV Sources in a Smart DC Distribution System: Maximum Power Point Tracking and Droop Control," Energy Conversion, IEEE Transactions on, vol. 29, pp. 913-921, 2014.N.

[11] Haichuan, J. Meng, Z. Daming, and J. Fletcher, "Autonomous micro-grid operation by employing weak droop control and PQ control," in Power Engineering Conference (AUPEC), 2014 Australasian Universities, 2014, pp. 1-5.

[12] K. De Brabandere, B. Bolsens, J. Van den Keybus, A. Woyte, J. Driesen, and R. Belmans, "A Voltage and Frequency Droop Control

Method for Parallel Inverters," Power Electronics, IEEE Transactions on, vol. 22, pp. 1107-1115, 2007.

[13] L. Baoquan, Z. Fang, Z. Yixin, and Y. Hao, "System Operation and Energy Management of a Renewable Energy-Based DC Micro-Grid for High Penetration Depth Application," *Smart Grid, IEEE Transactions on*, vol. 6, pp. 1147-1155, 2015.

[14] W. Bin and S. Keyue, "A current control MPPT method in high power solar energy conversion system," in Applied Power Electronics Conference and Exposition (APEC), 2014 Twenty-Ninth Annual IEEE, 2014, pp. 3021-3025.

[15] S. Hongtao, Z. Fang, Y. Hao, W. Feng, Z. Dong, and G. Zhiqing, "A Novel Real-Time Voltage and Frequency Compensation Strategy for Photovoltaic-Based Microgrid," Industrial Electronics, IEEE Transactions on, vol. 62, pp. 3545-3556, 2015.

High Performance ZVT with Bus Clamping Modulation Technique for Single Phase Full Bridge Inverters

Yinglai Xia, *Member, IEEE*, Raja Ayyanar, *Senior Member, IEEE*
School of Electrical, Computer and Energy Engineering
Arizona State University, Tempe, United States

***Abstract*—This paper proposes a topology based on bus clamping modulation and zero-voltage-transition (ZVT) technique to realize zero-voltage-switching (ZVS) for all the main switches of the full bridge inverters, and inherent ZVS and/or ZCS for the auxiliary switches. The advantages of the strategy include significant reduction in the turn-on loss of the ZVT auxiliary switches which typically account for a major part of the total loss in other ZVT circuits, and reduction in the voltage ratings of auxiliary switches. The modulation scheme and the commutation stages are analyzed in detail. Finally, a 1kW, 500 kHz switching frequency inverter of the proposed topology using SiC MOSFETs has been built to validate the theoretical analysis. The ZVT with bus clamping modulation technique of fixed timing and adaptive timing schemes are implemented in DSP TMS320F28335 resulting in full ZVS for the main switches in the full bridge inverter. The proposed scheme can save up to 33 % of the switching loss compared with no ZVT case.**

Keywords—ZVT, bus clamping, full bridge, switching loss saving

I. Introduction

The single phase full bridge inverter is widely used for power conversion in numerous applications [1]. High power density has become a very important metric in many of these applications in recent years. Increasing the switching frequency to significantly higher levels can lead to high power density, but it is at the expense of correspondingly higher switching losses and therefore lower efficiency. Accordingly, soft-switching full-bridge inverters have been pursued [2-5]. The application of soft-switching techniques to inverters can not only realize reduced switching loss, miniaturization, and lightweight, but also reduce the electromagnetic interference (EMI) and switching noise.

Among these soft-switching techniques, the zero voltage transition (ZVT) technique is particularly advantageous, especially when the active switches are implemented with majority carrier devices such as power MOSFETs [6]. While there are different approaches to achieving ZVT, a commonly used method involves an auxiliary active circuit to divert the current from the main circuit to discharge the switch parasitic capacitance, allowing its turn on under null voltage [7]. Many ZVT inverters have been proposed. Among them, the representative circuits are triangular or stellate resonant snubber inverters, coupled inductor inverters, transformer-assisted inverters, and resonant ac-link inverters [8-15].

However, one major concern in these ZVT converters is the turn-on loss of the auxiliary switch which can be a dominant part of the total loss in the auxiliary circuit especially in high switching frequency applications [16-17]. Also the voltage ratings of the auxiliary switches are usually comparable to the voltage ratings of the main switches which increases the volume and cost of the auxiliary circuit. In addition, these topologies need either two bulk capacitors or complex coupled inductors, or transformers with corresponding magnetic flux reset circuits [18]. And for the full bridge inverters using unipolar PWM, two separate auxiliary circuits - one for each leg of the inverter, are needed.

Though the mechanism by which soft switching is achieved is similar for the various ZVT converters, the exact configuration in terms of where the auxiliary circuit is connected to the main circuit, how the driving voltage for the auxiliary circuit is obtained and the switching strategy for the auxiliary circuit have a major impact on the performance of overall converter system.

This paper proposes a ZVT circuit implementation that significantly decreases the voltage stress and the switching loss in the auxiliary switches. A synchronous buck converter operating under inherent ZVS is used to obtain the driving voltage for the auxiliary circuit. In addition, bus clamping modulation is employed in this work to avoid the need for two separate bi-directional auxiliary circuits. In bus clamping modulation, also known as hybrid PWM modulation, two of the four switches are driven by high switching frequency PWM signals for high quality sinusoidal output and the other two are commutated at the low, fundamental frequency [19-24]. Bus clamping modulation technique will be used in combination with the proposed ZVT circuit to achieve reduced voltage rating of auxiliary switches. This paper designs the modulation scheme for the realization of ZVT of main switches, while achieving low turn-on voltage and low switching loss of the auxiliary switches.

978-1-4673-9551-9/16 $31.00 © 2016 IEEE

II. PROPOSED TOPOLOGY AND ANALYSIS OF OPERATION

Fig.1 shows the proposed topology employing the bus clamping modulation with low-loss ZVT technique for single phase full bridge inverters. The buck stage in the front provides a controllable, stable and relatively low voltage (designed to be around 20% of the input voltage in the experiment) for the auxiliary ZVT circuit. The buck stage needs to process only the power needed to realize ZVT for the main full bridge switches, and hence, its rating and the losses are quite small. Also the average value of the buck stage inductor current (i_b) is very low (almost 0), and the inductor is purposely chosen to be small enough to allow a relatively large ripple. The ripple current has the correct positive or negative direction at each switching interval, and hence, leads to natural ZVS for both the switches in the buck stage.

The two auxiliary switches S_1 and S_2 realize ZVT for Q_1 and Q_2 respectively. The voltage rating of the two auxiliary switches S_1 and S_2 are equal to the buck output voltage (V_b) which is small compared with previous ZVT topologies as discussed in [6] and [8] which is $V_{dc}/2$ and V_{dc} respectively. The turn-on loss of the auxiliary switches S_1 and S_2 which is a major part of the ZVT circuit can be decreased significantly. Also switches can be chosen with lower voltage rating which can save volume, cost and possibly conduction losses.

As the ZVT circuit can only achieve ZVS for the upper main switches Q_1 and Q_3, traditional unipolar modulation of the full bridge inverters cannot be used to achieve full ZVT of all 4 main switches of the full bridge inverter. Instead the bus clamping modulation is used in combination with the proposed ZVT auxiliary circuit.

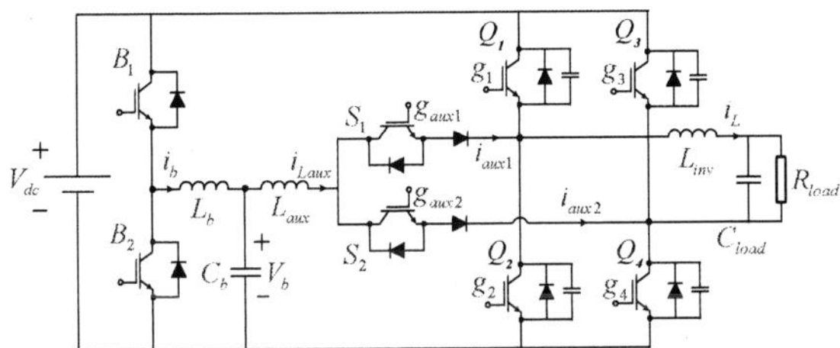

Fig. 1 Proposed topology of bus clamping modulation & ZVT technique for full bridge inverters

A. *Bus clamping modulation with ZVT realization*

The modulation scheme is shown in Table I where d is the duty cycle as same in the unipolar modulation (ranging from -1 to 1) and i_L is the output inductor current as shown in Fig. 1. The green waveforms labelled D_1 are the duty cycle given to Q_1 while the red waveforms labelled D_2 are the duty cycle given to Q_3. Q_2 and Q_4 are complementary to Q_1 and Q_3 respectively. d equals to $D_1 - D_2$.

This strategy can realize ZVS of all switches at any power factor. Take the condition of $d > 0$ and $i_L > 0$ for example, Q_2 and Q_3 can achieve natural ZVS because i_L is positive [4]. As $d > 0$ the positive state and zero state are needed which means Q_1 and Q_4 are needed. The strategy is that Q_4 is always ON which means the second bridge is bus clamped to the negative bus. Then Q_1 is ON when positive state is needed, Q_2 is ON when zero state is needed. As mentioned above, Q_2 can achieve natural ZVS by i_L while Q_1 achieves ZVT by S_1.

The scheme is capable of supporting non-unity power factors, but if the power factor is 1 the modulation strategy is relatively easier to implement compared to other power factors.

TABLE I. BUS CLAMPING MODULATION SCHEME WITH ZVT

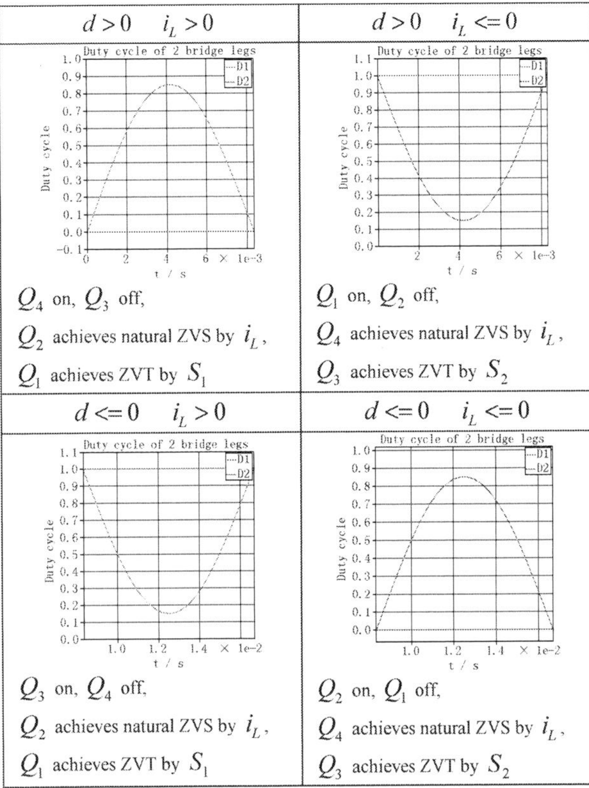

$d > 0 \quad i_L > 0$	$d > 0 \quad i_L <= 0$
Q_4 on, Q_3 off, Q_2 achieves natural ZVS by i_L, Q_1 achieves ZVT by S_1	Q_1 on, Q_2 off, Q_4 achieves natural ZVS by i_L, Q_3 achieves ZVT by S_2
$d <= 0 \quad i_L > 0$	$d <= 0 \quad i_L <= 0$
Q_3 on, Q_4 off, Q_2 achieves natural ZVS by i_L, Q_1 achieves ZVT by S_1	Q_2 on, Q_1 off, Q_4 achieves natural ZVS by i_L, Q_3 achieves ZVT by S_2

B. *Detailed analysis of commutation process*

The commutation process of the ZVT will be analyzed in this section. Take the condition of $d > 0$ and $i_L > 0$ for example where Q_4 is always ON and Q_3 is always OFF and Q_1 and Q_2 are switching to generate positive and zero states. Also Q_2 can achieve natural ZVS by i_L while Q_1 achieves ZVT by S_1.

Fig. 2 shows the timing of the proposed bus clamping modulation with ZVT scheme. Fig. 3 shows commutation stages corresponding to the timing in Fig. 2.

(a) Initial stage [before t_1]. Q_2 is ON, the main inductor current i_L flows through Q_2

(b) Pre-charging stage [t_1~t_2]. S_1 is turned-on at t_1, the buck output voltage V_b is applied to the resonant inductor L_{aux}, i_{aux1} increases linearly, current in Q_2 decreases to zero at t_2.

(c) Boost-charging stage [t_2~t_3]. I_{aux1} is linearly charged to a current well above the inductor current i_L. This will make sure $V_{ds,Q1}$ (the drain to source voltage of the Q_1 switch) resonates to zero in resonant stage t_3~t_4.

978-1-4673-9551-9/16 $31.00 © 2016 IEEE

(d) Resonant stage [t_3~t_4]. Q_2 is turned-off at t_3, L_{aux} starts to resonant with C_{ds1} and C_{ds2} until $V_{ds,Q1}$ resonates to zero.

(e) Clamping stage [t_4~t_5]. When $V_{ds,Q1}$ resonates to zero, D_1 (the body diode of the switch Q_1) naturally turns-on at t_4. The voltage of $-(V_{dc}-V_b)$ is applied to L_{aux}, i_{aux1} decreases. During this stage, $V_{ds,Q1}$ is clamped to 0 until t_5, Q_1 can be ZVS turned-on in this stage.

(f) Discharging stage [t_5~t_6]. I_{aux1} is discharged to zero, current in Q_1 linearly increases to i_L. S_1 can be ZCS turned-off before t_7.

(g) Steady-state stage [t_6~t_7]. Q_1 fully conducts the load current i_L.

(h) Snubber turn-off stage [t_7~t_8]. Q_1 is turned-off at t_7, i_L linearly charges C_{ds1} and discharges C_{ds2} until $V_{ds,Q1}$ reaches V_{dc}, dv/dt is limited by C_{ds1} and C_{ds2}, due to the extra C_{ds} added to the main switches, turn-off loss of Q_1 is reduced.

(i) Steady-state stage [after t_8]. When $V_{ds,Q2}$ reaches to zero, D_2 is naturally turned-on, and then Q_2 can be ZVS turned-on after t_8 resulting in natural ZVS of Q_2.

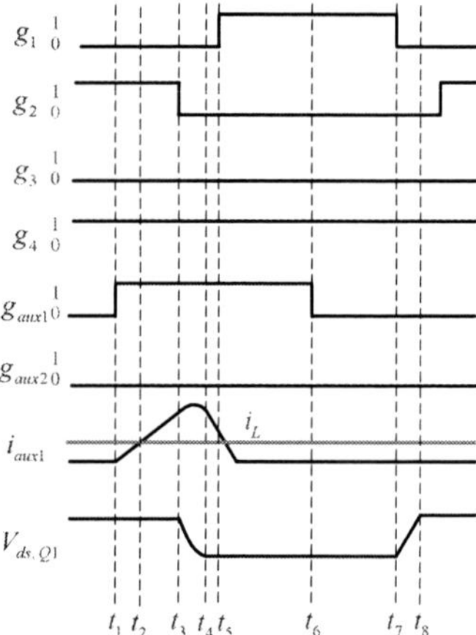

Fig. 2 Timing diagram of proposed ZVT with bus clamping modulation

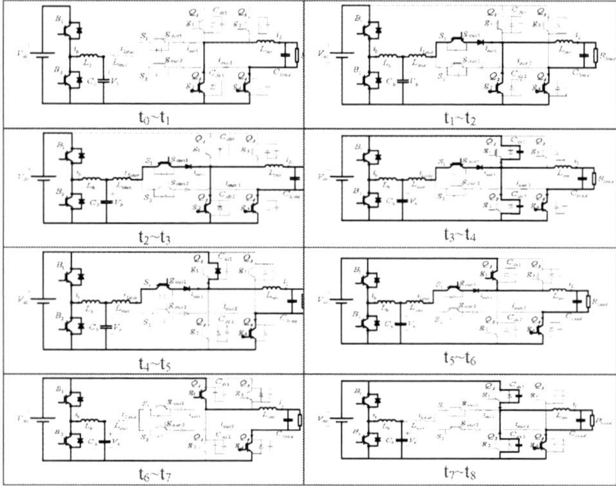

Fig. 3 Commutation stages of the proposed inverter

III. HARDWARE IMPLEMENTATION AND RESULTS

A 1kW full bridge inverter with LC filter and resistive load operating at 500 kHz switching frequency has been built to verify the proposed ZVT with bus clamping modulation technique. The power factor is 1 and the modulation ratio is 0.85 which represents a typical PV standalone application. The specifications of the inverter, the ZVT auxiliary circuit and details of switches are shown in Table II, Table III and Table IV respectively.

The blanking time which is the deadband between the gate signals of the upper switch and the lower switch (which equals to $t_5 - t_3$ in Fig. 2) is set to 90 ns. The ahead time, which is the time between the auxiliary switch gate signal and the lower switch gate signal (which is $t_3 - t_1$ in Fig. 2), is set to 210 ns in the fixed timing scheme. In the variable timing scheme, the ahead time will change with the inductor current value to further decrease the loss in the auxiliary circuit. External capacitor of 220 pF is put in parallel with C_{ds} of the main switches of the full bridge to decrease the turn-off loss of the main switches while maintaining the ZVS turn-on. The timing and auxiliary inductance, the buck output voltage are designed based on loss calculation and optimization.

SiC MOSFETS from CREE are used in full bridge main circuit, buck stage and auxiliary ZVT circuits. The driver ICs are chosen as Avago ACPL 337J. The positive gate voltage of the MOSFET (V_{gs}) is 20 V and the negative is -5V. In order to reduce the diode recovery loss, 4 SiC schottky diodes are put in parallel to the 4 main switches. LeCroy 6200A oscilloscope is used to capture waveforms. Power analyzer YOKOGAWA WT3000 is used to measure the input power at dc terminal and output power at load terminal to calculate efficiency. Ferrite core with Litz wire are used for the main output inductor.

TABLE II. MAIN CIRCUIT SPECIFICATIONS

Power	1 kW
Dc input voltage	230 V
Load voltage	140 V
Load current	8.3 A
Resistive load	17.2Ω
Power factor	1
Switching frequency	500 kHz

TABLE III. AUXILIARY AND BUCK CIRCUIT SPECIFICATIONS

Auxiliary inductance	270 nH
Blanking time	90 ns
Ahead time	210 ns
Buck switching frequency	50 kHz
Buck inductance	260 uH
Buck output capacitance	800 nF
Buck output voltage	50 V

TABLE IV. SWITCH DETAILS

Full bridge and buck	Type	SiC MOSFET
	Manufacturer	CREE
	Part no.	C2M0080120D
	C_{ds}	73 pF
Auxiliary circuit	Type	SiC MOSFET
	Manufacturer	CREE
	Part no.	C2M0160120D
	C_{ds}	43 pF

Fig. 4 shows the full bridge power board, the buck and ZVT auxiliary board and ezDSP28335 control board.

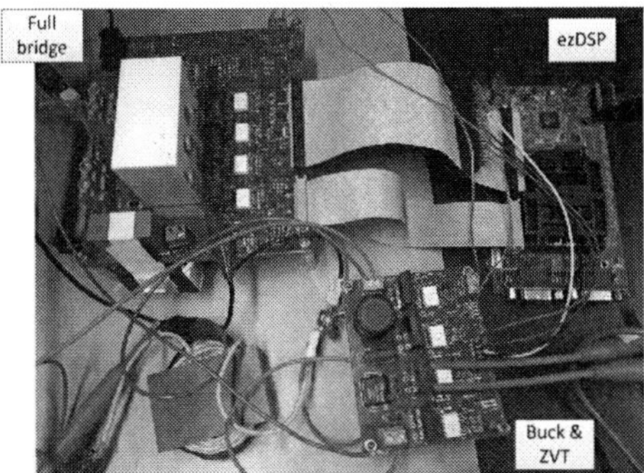

Fig. 4 H bridge, buck and ZVT board and eZDSP

Fig. 5 shows the bus clamping modulation scheme and ZVT gate signals along with the inductor current for reference. Fig. 6 shows the zoomed gate signals of the full bridge switches and the ZVT auxiliary switch. As the power factor is 1 which means the polarity of the duty ratio is almost equal to the polarity of the inductor current, the modulation strategy is easier to implement compared to other power factors though the scheme is capable of supporting other power factors as well.

It can be seen from Fig. 5 and Fig. 6 that when the inductor current is positive, for the 4 switches in the full bridge inverter, only Q_1 and Q_2 are switching while Q_4 is always ON and Q_3 is always OFF which means the midpoint of the second leg is bus clamped to the negative bus. In order to achieve ZVT for the Q_1, the auxiliary switch S_1 is switched each time before Q_1 is turned-on. While in the negative part of the inductor current, Q_2 is always ON and Q_1 is always OFF, only Q_3 and Q_4 are switching. Fig. 6 also shows that the switching frequency is 500 kHz.

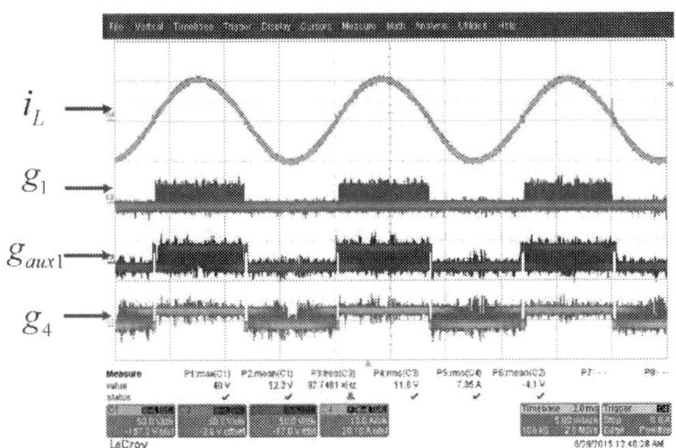

Fig. 5 ZVT with bus clamping modulation (time scale 5 ms/div, i_L scale 10 A/div, g_1, g_{aux1}, g_4 scale 50 V/div)

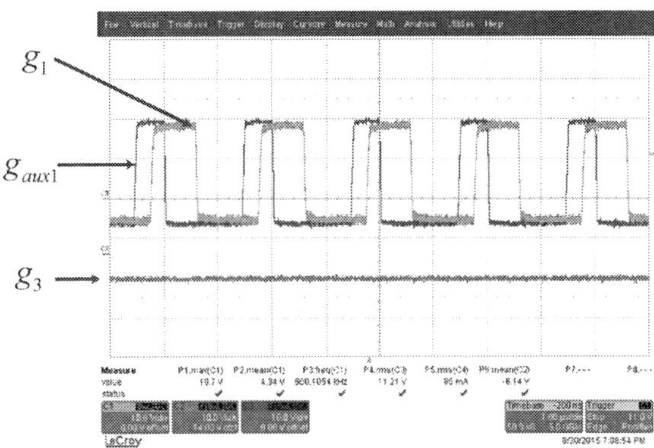

Fig. 6 Zoomed gate signal of ZVT with bus clamping modulation when inductor current is positive (time scale 1 us/div, g_1, g_{aux1}, g_3 scale 10 V/div)

Fig. 7 shows the fixed timing scheme of the ZVT where the blanking time is 90 ns and the ahead time is 210 ns. The adaptive timing method [25] is also implemented which means the ahead time changes with the inductor current adaptively to further decrease the loss in the auxiliary ZVT circuit.

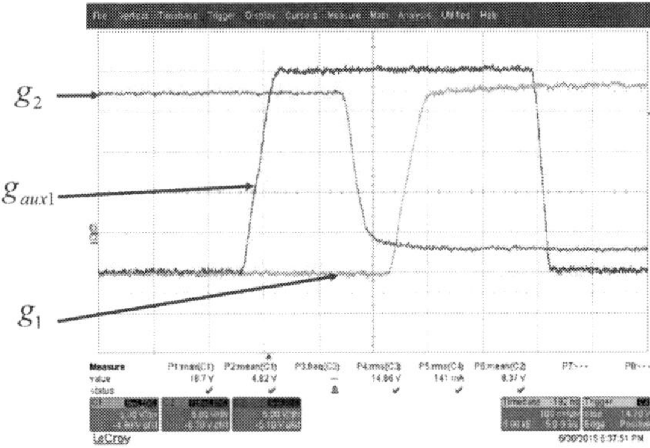

Fig. 7 ZVT timing with blanking time of 90 ns and ahead time of 210 ns (time scale 100 ns/div, g_2, g_{aux1}, g_1 scale 5 V/div)

Fig. 8 shows the waveforms of the ZVT process of the main switch Q_1 which include the drain to source voltage of the Q_1 switch ($V_{ds,Q1}$), the gate to source voltage of the Q_1 switch (g_1), the inductor current (i_L) and the auxiliary inductor current (i_{aux1}). It can be seen that the auxiliary inductor current is larger than the value of the inductor current for some time which means it can discharge the drain to source capacitance of the Q_1 switch (C_{ds1}) as analyzed in the Section II commutation stages part to achieve ZVT of Q_1. The drain to source voltage of the Q_1 switch has already fallen to 0 before the gate signal is applied proving that Q_1 realizes soft turn-on.

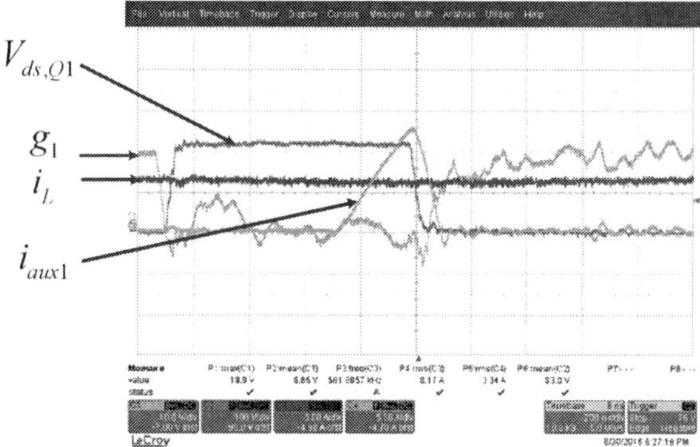

Fig. 8 ZVS of main switch (time scale 200 ns/div, $V_{ds,Q1}$ scale 100 V/div, g_1 scale 10 V/div, i_L scale 5 A/div, i_{aux1} scale 5 A/div)

Fig. 9 shows the measured loss using power analyzer YOKOGAWA WT3000 at different power levels for three different cases – base case (no ZVT), proposed ZVT with fixed timing and proposed ZVT with adaptive timing. It can be seen that the proposed ZVT can improve the system overall efficiency significantly.

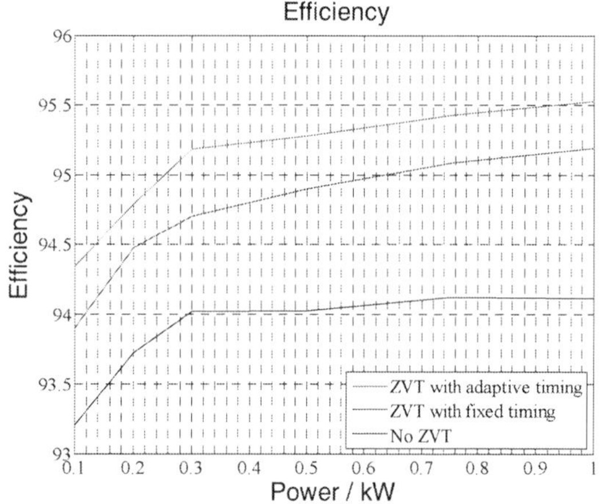

Fig. 9 Loss of full bridge without ZVT, with fixed timing ZVT and with adaptive timing ZVT proposed

The independent loss that cannot be changed by ZVT includes the switch conduction loss and inductor loss. The conduction loss of the main 4 switches which is same in all the above three schemes are calculated using $P_{con} = 2I^2R_{ds,on}{}^2 = 11.3$ W. The inductor loss is estimated to be 3 W. The core loss and copper loss are small because ferrite core with Litz wire are used and the switch frequency is high enough which results in the max current ripple of 0.2 A pk-zero.

Table V shows the calculation of the loss saving in the experimental prototype at 1 kW and 500 kHz switching frequency in which the conduction loss of the switches and inductor loss are subtracted from the total loss to obtain the rest loss for each scheme. The full bridge switching loss without ZVT is 48.3 W. The full bridge switching loss and ZVT auxiliary circuit and buck stage total loss is 36.3 W with the fixed timing ZVT and 32.6 W with the adaptive timing ZVT. After subtracting the independent loss, the fixed timing ZVT saves up to 25 % while the adaptive timing ZVT saves up to 33 % of the original switching loss of the full bridge inverter.

TABLE V. LOSS ANALYSIS IN HARDWARE PROTOTYPE

	No ZVT	Fixed timing ZVT	Adaptive timing ZVT
Total measured loss	62.6 W	50.6 W	46.9 W
Estimated conduction loss and inductor loss	11.3 W + 3 W = 14.3 W		
Switching losses and loss in ZVT circuits	48.3 W (Full bridge switching loss)	36.3 W (Full bridge switching loss & ZVT circuit & buck loss)	32.6 W (Full bridge switching loss & ZVT circuit & buck loss)
Reduction in the original switching loss	/	25 %	33 %

IV. CONCLUSION

This paper proposes a topology based on bus clamping modulation and zero voltage transition (ZVT) method to achieve zero voltage switching (ZVS) for all the main switches of the full bridge inverter. The turn-on loss of the ZVT auxiliary switches which typically is a major part of the total loss in other ZVT circuits is reduced by this strategy. The voltage rating can also be reduced which can decrease volume and cost of the auxiliary circuit. The modulation scheme under different power factor conditions and the commutation stages of the ZVT realization are analyzed in detail. A 1 kW, 500 kHz switching frequency inverter using the proposed topology and SiC MOSFETs has been built to validate the theoretical analysis. The bus clamping modulation and ZVT fixed timing and adaptive timing schemes are implemented resulting in full ZVS for the main switches in the full bridge inverter. The proposed scheme is shown in experiments to save up to 33 % of the switching loss compared with no ZVT case.

ACKNOWLEDGMENT

The information, data, or work presented herein was funded in part by the Office of Energy Efficiency and Renewable Energy (EERE), U.S. Department of Energy, under Award Number DE-EE0006521 with North Carolina State University, PowerAmerica Institute.

DISCLAIMER

The information, data, or work presented herein was funded in part by an agency of the United States Government. Neither the United States Government nor any agency thereof, nor any of their employees, makes any warranty, express or implied, or assumes any legal liability or responsibility for the accuracy, completeness, or usefulness of any information, apparatus, product, or process disclosed, or represents that its use would not infringe privately owned rights. Reference herein to any specific commercial product, process, or service by trade name, trademark, manufacturer, or otherwise does not necessarily constitute or imply its endorsement, recommendation, or favoring by the United States Government or any agency thereof. The views and opinions of authors expressed herein do not necessarily state or reflect those of the United States Government or any agency thereof.

REFERENCES

[1] H. Zhu, J. Lai, A.R. Hefner, Y. Tang, and C. Chen, "Modeling-based examination of conducted EMI emissions from hard and soft-switching PWM inverters," IEEE Transactions on Industry Applications, vol. 37, n. 5, pp. 1383-1393, 2001.

[2] F. J. Seixas and D. C. Martins, "The ZVS-PWM commutation cell applied to the DC-AC converter," IEEE Trans. Power Electron., vol. 12, no. 4, pp. 726–733, Jul. 1997.

[3] Y.-C. Chuang, "High-efficiency ZCS buck converter for rechargeable batteries," IEEE Trans. Ind. Electron., vol. 57, no. 7, pp. 2463–2472, Jul. 2010.

[4] K. M. Smith and K. M. Smedley, "A comparison of voltage-mode soft switching methods for PWM converters," IEEE Trans. Power Electron., vol. 12, no. 2, pp. 376–386, Mar. 1997.

[5] F. C. Lee, R. L. Lin, and Y. Zhao, "Improved soft-switching ZVT converters with active snubber," in Proc. IEEE APEC'98, vol. 2, 1998, pp. 1063–1069.

[6] R.W. De Doncker and J.P. Lyons, "The auxiliary resonant commutated pole converter," IEEE Industry Applications Society Annual Meeting, vol 2, pp. 1228-1235, 1990.

[7] Jih-Sheng Lai, "Fundamentals of A New Family of Auxiliary Resonant Snubber Inverters", 23rd International Conference on Industrial Electronics, Control and Instrumentation, Page(s): 645 -650, V01.2, 1997.

[8] Jih-Sheng Lai, Junhong Zhang, Huijie Yu and Kouns, H. Source and Load Adaptive Design for a High-Power Soft-Switching Inverter. IEEE Transactions on Power Electronics, Volume 21, Issue 6, Nov. 2006 Page(s):1667 – 1675

[9] M. L. Martins, J. L. Russi, and H. L. Hey, "Novel design methodology and comparative analysis for ZVT PWM converters with resonant auxiliary circuit," IEEE Trans. Ind. Appl., vol. 42, no. 3, pp. 779–796, May/Jun. 2006.

[10] C.-M. Wang, "Novel zero-voltage-transition PWM dc-dc converters," IEEE Trans. Ind. Electron., vol. 53, no. 1, pp. 254–262, Feb. 2006.

[11] J. Lai, S. R. W. Young, J. G. W. Ott, J. W. McKeever, and Z. Fang, "Adelta-configured auxiliary resonant snubber inverter," *IEEE Trans. Ind. Appl.*, vol. 32, no. 3, pp. 518–525, Jun. 1996.

[12] J. Lai, "Practical design methodology of auxiliary resonant snubber inverters," in *Proc. 27th Annu. IEEE Power Electron. Spec. Conf.*, Baveno, Italy, Jun. 1996, pp. 432–437.

[13] J. L. Russi, M. L. D. S. Martins, and H. L. Hey, "Coupled-filter-inductor soft-switching techniques: Principles and topologies," *IEEE Trans. Ind. Electron.*, vol. 55, no. 9, pp. 3361–3373, Sep. 2008.

[14] W. Yu, J. Lai, and S. Park, "An improved zero-voltage switching inverter using two coupled magnetics in one resonant pole," *IEEE Trans. Power Electron.*, vol. 25, no. 4, pp. 952–961, Apr. 2010.

[15] X. Yuan and I. Barbi, "Analysis, designing, and experimentation of a transformer-assisted PWM zero-voltage switching pole inverter," *IEEE Trans. Power Electron.*, vol. 15, no. 1, pp. 72–82, Jan. 2000.

[16] C. Y. Inaba, T. Yamazaki, M. Yoshida, E. Hiraki, Y. Konishi, and M. Nakaoka, "Three phase soft switching inverter with pulse current transformer-assisted resonant snubbers," in *Proc. IEEE Int. Conf. Ind. Electron.*, 2001, pp. 1106–1111.

[17] H. Iyomori, M. Yoshida, S. Chandhaket, E. Hiraki, and M. Nakaoka, "An active auxiliary resonant AC link snubber with a single active power switch for voltage source type soft switching sinewave PWM inverter," in *Proc. IEEE Int. Conf. Commun., Circuits Syst. West Sino Expo.*, 2002, pp. 1702–1706.

[18] M. Yoshida, Y. Fujii, E. Hiraki, and M. Nakaoka, "Electromagnetic noises and efficiency evaluations of three-phase ZVS-PWM inverter with new resonant AC link snubbers," in *Proc. 29th Annu. IEEE Ind. Electron. Soc. Conf.*, Nov. 2003, pp. 2911–2916.

[19] R.S. Lai and K. D. T. Ngo, "A PWM method for reduction of switching loss in a Full-bridge inverter," IEEE Trans. Power Electronics., vol. 10, pp. 326-332, May 1995.

[20] Panda, R.; Tripathi, R.K.; , "A symmetrical hybrid sine PWM switching technique for full bridge inverters," Power Electronics, 2006. IICPE 2006. India International Conference on , vol., no., pp.345-348, 19-21 Dec. 2006.

[21] Tsu-Hua Ai, Jiann-Fuh Chen, Tsorng-Juu Liang;"A Random Switching Method for HPWM Full-Bridge Inverter"; IEEE transactions on industrial electronics, vol. 49, no. 3, june 2002; pp.595-597.

[22] Yousefzadeh, V.; Maksimovic, D.; Qiong Li; , "A zero voltage switching single-phase inverter using hybrid pulse-width modulation technique," Power Electronics Specialists Conference, 2004. PESC 04. 2004 IEEE 35th Annual , vol.2, no., pp. 1274- 1279 Vol.2, 20-25 June 2004.

[23] G. Narayanan, H. K. Krishnamurthy, D. Zhao, R. Ayyanar, "Advanced bus-clamping PWM techniques based on space vector approach" IEEE Transactions on Power Electronics, vol. 21, No. 4, pp. 974-984, July. 2006.

[24] G. Narayanan, J. S. Siva Prasad, H. K. Krishnamurthy, R. Ayyanar, "Reduction in Torque ripple in induction motor drives using an advanced hybrid PWM technique" IEEE Transactions on Industrial Electronics, vol. 57, No. 6, pp. 2085-2091, June. 2010.

[25] Chan, C.C.; Chau, K.T., Chan, D.T.W., Jianming Yao, Jin-Sheng Lai, Yong Li, Switching characteristics and efficiency improvement with auxiliary resonant snubber based soft-switching inverters. PESC '98 Volume 1, 17-22 May 1998 Page(s):429 - 435 vol.1

Small AC Signal Droop Based Secondary Control for Microgrids

Teng Wu, Zeng Liu, Jinjun Liu, Baojin Liu, and Shike Wang

State Key Lab of Electrical Insulation and Power Equipment, School of Electrical Engineering,
Xi'an Jiaotong University
Xi'an, China
wuteng008@yeah.net

***Abstract*—The well-known active power-frequency and reactive power-voltage amplitude droop control is widely used in the coordinative control of parallel inverters in Microgrids. However, this conventional droop method may cause deviation on the frequency and voltage amplitude, which will affect the accuracy of power supply. This paper proposes a novel secondary control method to compensate this deviation. This method is based on the control of a small AC voltage signal, which serves as a communication link among the parallel inverters. The output power of the inverter can be automatically regulated by establishing a droop relationship between the frequency of the small AC signal and the bias of the inverter's droop characteristics. Through this method, frequency and voltage amplitude deviation can be eliminated and power sharing can be realized among the parallel inverters. Simulation and experimental results are provided to prove the effectiveness of the proposed control method.**

***Keywords*—droop control; small AC signal; voltage and frequency deviation; secondary control; microgrid**

I. Introduction

Droop control method [1]–[3] is widely applied in microgrids in consideration of the distributed geographical locations of the distributed energy resources (DERs). It can effectively realize power sharing among inverters without intercommunication. However, in conventional droop control, the drawback of output voltage frequency and amplitude deviation is a significant issue worth further study [4]. In order to eliminate or compensate the voltage frequency and amplitude deviation, a number of improved control methods were proposed [4]–[8]. Although these improving methods compensate to some degrees this drawback of conventional droop method, they also bring about other new flaws. In [4], an altered droop method based on the relationship between the virtual flux and the output powers has been proposed. However, this strategy cannot well compensate the voltage amplitude deviation. The strategy in [5] introduces a secondary level control but it cannot realize complete restoration of frequency or voltage amplitude in steady state with proportional controllers. The method in [6] is a secondary control method based on load power demand estimation. This method can compensate the voltage and frequency deviation effectively and at the same time share the load power demand among DGs. However, the dependence on dead zones in this method can result in a time consuming regulation process. In [7], the theory of hierarchical control is applied, from which the secondary control is improved for the compensation of voltage and frequency deviation. However, the dependence on the central controller in this method reduces the system reliability since the fail down of the central controller would lead to the collapse of the whole system. The method proposed in [8] has improved the system reliability by using distributed controllers. However, parameter inconsistency among controllers can result in different steady-state operating points for each inverter, which will affect the accuracy of power sharing. Moreover, the dependence on communication lines brings about the limitation in physical location, raises the cost and increases the sensibility for noises [10].

To overcome the limitations of distributed secondary control [8], this paper proposes a novel secondary control method without intercommunication among parallel inverters. This secondary control method is based on the control of a small AC voltage signal, which is inspired by the idea in [9] where harmonic current sharing is focused. A droop relationship between the frequency of the small AC signal and the bias of the inverter's droop characteristics is innovatively established to achieve the synchronization of the parallel inverters during the secondary control. Meanwhile, the frequency and the voltage amplitude are both regulated by PI controllers. Therefore, the proposed method can not only restore the frequency and voltage amplitude accurately, but can also rapidly realize equal power sharing among the parallel inverters without the help of communication lines. The effectiveness of the proposed method is verified by both simulation and experimental results.

II. Conventional Droop Control Method

For an AC system in HV or MV microgrids where the distribution line impedance is typically inductive, the real and the reactive power flows between two voltages $E\angle\phi$ and $U_L\angle 0$ shown in Fig. 1 obey the following relationship [1], [2], [6]:

$$P \approx \frac{EU_L\phi}{Z} \tag{1}$$

This work was supported by the National Natural Science Foundation of China under Grant 51437007, and the Power Electronics Science and Education Development Program of Delta Environmental & Educational Foundation under Grant DREM2014002.

$$Q \approx \frac{E(E-U_L)}{Z} \quad (2)$$

where E and U_L are respectively the DG's output terminal voltage and the PCC voltage. ϕ is the phase angle difference between E and U_L. Z is the value of the distribution line impedance.

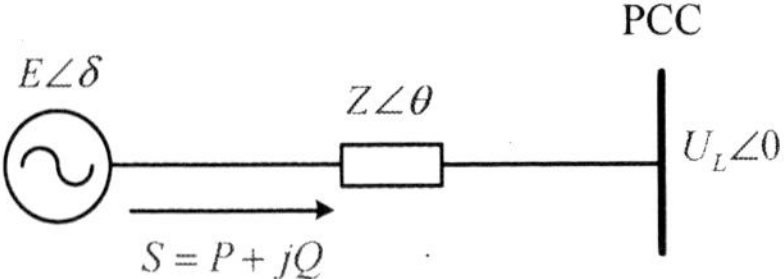

Fig. 1. Equivalent circuit of an inverter connected to the AC bus of the microgrid.

Such relationships in (1) and (2) lay the foundation of the well-known real power-frequency and reactive power-voltage amplitude droop control method [1], [2], [6]. The droop characteristics are mathematically expressed in (3) and (4):

$$\omega^* = \omega_0 - m \cdot (P - P_0) \quad (3)$$

$$E^* = E_0 - n \cdot (Q - Q_0) \quad (4)$$

where ω^* and E^* are the generated reference of frequency and voltage amplitude. P and Q are respectively the measured real power and reactive power. P_0 and Q_0 are respectively the bias of the P-ω and the Q-E droop characteristics. m and n (defined as positive) are the droop gains.

The droop control method contributes to share the load power demand among parallel inverters without the help of communication lines. However, there is an inherent tradeoff within the conventional droop method between the power sharing accuracy and voltage regulation rate. Since two droop characteristics are impossible to be designed strictly the same due to the parameter inconsistency or hardware differences, there will be a small gap d between the two droop curves as illustrated in Fig. 2. When the droop gains are small, the power cannot be equally shared between the two inverters, as illustrated in Fig. 2(a). On the contrary, if the droop gains are designed higher, larger frequency and voltage deviation will be produced when the load power demand changes, as illustrated in Fig. 2(b).

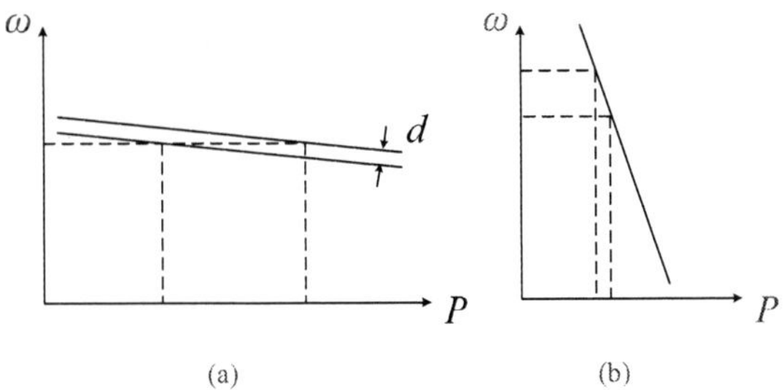

Fig. 2. Issues under different droop gains: (a) droop gains are small. (b) droop gains are large.

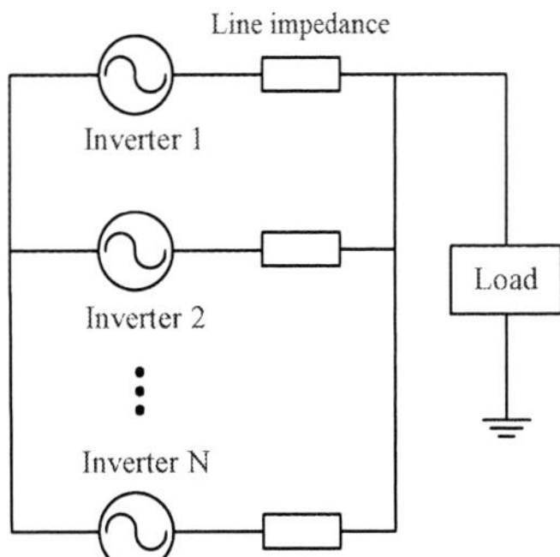

Fig. 3. Simplified microgrid system consisting of N parallel inverters

In order to achieve a high power sharing accuracy, the droop gains cannot be designed too small, which brings about the necessity for the compensation of the frequency and voltage deviation.

III. Proposed Secondary Control Method

The microgrid system consisting of N parallel inverters is illustrated in Fig. 3. The parallel inverters are droop controlled as voltage sources on the primary control level. The control equations are expressed in (3) and (4).

In the proposed secondary control method, both frequency and voltage amplitude of the point of common coupling (PCC) are regulated by PI controllers. Meanwhile, a small AC voltage signal is injected into the system as a control signal. The amplitude of the small AC signal is set to be 1% of the fundamental voltage amplitude. Droop relationships are established between the frequency of the small AC signal and the bias of droop characteristics for fundamental signal, which are expressed in (5) and (6).

$$\hat{\omega}_p^* = \hat{\omega}_{p0} - \hat{m} \cdot P_0 \quad (5)$$

$$\hat{\omega}_q^* = \hat{\omega}_{q0} - \hat{n} \cdot Q_0 \quad (6)$$

where $\hat{\omega}_p^*$ and $\hat{\omega}_q^*$ are the generated frequency for small AC signals. $\hat{\omega}_{p0}$ and $\hat{\omega}_{q0}$ are the base values for $\hat{\omega}_p$ and $\hat{\omega}_q$. $\hat{m}$ and $\hat{n}$ (defined as positive) are the droop gains for small AC signals. Note that $\hat{\omega}_{p0}$ and $\hat{\omega}_{q0}$ cannot be selected the same value when the frequency and voltage amplitude are expected to be restored simultaneously. However, in the simulations and experiments in this paper, $\hat{\omega}_{p0}$ and $\hat{\omega}_{q0}$ are identical because the frequency and voltage amplitude are restored in separate simulations and experiments.

The real power generated by the small AC signal is measured and is fed back to the output of PI controllers to generate the droop bias P_0 and Q_0, as expressed in (7) and (8).

$$P_0 = \left(k_{p\omega} + \frac{1}{k_{i\omega} \cdot s}\right) \cdot (\omega_r - \omega) + G_p \cdot \hat{P}_p \quad (7)$$

$$Q_0 = \left(k_{pE} + \frac{1}{k_{iE} \cdot s}\right) \cdot (E_r - E) + G_q \cdot \hat{P}_q \quad (8)$$

where ω_r and ω are respectively the rated and the measured frequency. E_r and E are respectively the rated and the measured voltage amplitude. $k_{p\omega}$, $k_{i\omega}$ and k_{pE}, k_{iE} are the parameters of PI controllers for the regulation of frequency and voltage amplitude. G_p and G_q are the gains from the real power $\hat{P}_p$ and $\hat{P}_q$ generated by the small AC signal to the droop bias P_0 and Q_0. $\hat{P}_p$ and $\hat{P}_q$ are calculated based on the injected voltages and the currents due to the injected signals, which could be extracted by a phase locking loop (PLL).

The control block diagram is illustrated in Fig. 4, where the red parts present the small AC signal droop mechanism. The droop control dashed block contains the droop equations (3)–(6). The composite voltage reference is given by (9).

$$v_{ref} = v_{fund} + \hat{v}_p + \hat{v}_q \quad (9)$$

where $v_{fund} = E\cos\omega t$, $\hat{v}_p = \hat{E}\cos\hat{\omega}_p t$ and $\hat{v}_q = \hat{E}\cos\hat{\omega}_q t$. $\hat{E}$ is the amplitude of the small AC signal, which is set to be 2V, 1% of the fundamental voltage amplitude.

In this method, the power lines are used as the control interconnection among the parallel inverters. The small AC signal functions as a communication link and helps to achieve the synchronization of parallel inverters during the secondary control. It can be seen from Fig. 4 that this method also establishes a droop relationship from the power due to the small AC signal to the frequency of the small AC signal. Therefore, at steady state, all the inverters are producing the same frequency for the small AC signal, which further ensures the droop bias of the parallel inverters being identical according to the droop relationships in (5) and (6). Through this mechanism, accurate power sharing can be realized. Moreover, both the fundamental frequency and voltage amplitude can be regulated to the rated values by the PI controllers.

IV. Simulation Results

PSCAD simulations are conducted to verify the efficiency of the proposed method. The essential control parameters are listed in Table I. In order to investigate the power sharing capability of the proposed method, the PI controller parameters for the two parallel inverters differ slightly, as depicted by the parameters in Table I. Moreover, the system adopting the distributed secondary control method in [8] is also simulated to show that the small difference of PI parameters could lead to poor power sharing performance.

TABLE I. Simulation Parameters

Parameters	Values	Parameters	Values
ω_0	50Hz	E_0	200V
ω_r	50Hz	E_r	200V
$\hat{\omega}_{p0}$	100Hz	$\hat{\omega}_{q0}$	100Hz
m	$1\cdot10^{-4}$Hz/W	n	$1\cdot10^{-3}$V/Var
$\hat{m}$	$1\cdot10^{-4}$Hz/W	$\hat{n}$	$1\cdot10^{-4}$V/Var
$k_{p\omega1}$	1000	$k_{i\omega1}$	$5\cdot10^{-6}$
$k_{p\omega2}$	1000	$k_{i\omega2}$	$1.5\cdot10^{-5}$
k_{pE1}	1000	k_{iE1}	$9\cdot10^{-5}$
k_{pE2}	1000	k_{iE2}	$1.1\cdot10^{-4}$
G_p	8000	G_q	30000
ΔP	8kW	ΔQ	4kVar

The first simulation is to investigate the frequency restoration performance of the novel secondary control. At the moment t=2s, the load real power demand increases by 8kW.

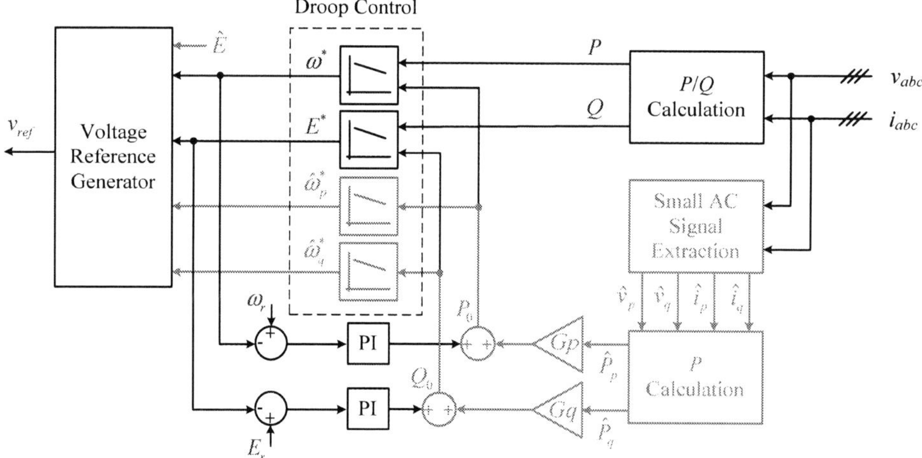

Fig. 4. Control block diagram of the system

Consequently, the frequency drops from 49.96Hz to 49.58Hz. The secondary control is triggered manually at the moment t=5s. (The secondary control should be continuous. It is triggered manually in the simulation just for clearer illustration.) The performance of the distributed secondary control in [8] is shown in Fig. 5. It can be seen that even though the frequency is drawn back to the rated value 50Hz, the real power is not equally shared among the inverters. On the contrary, the proposed method can realize both the frequency restoration and a good power sharing accuracy. The performance of the proposed secondary control method is

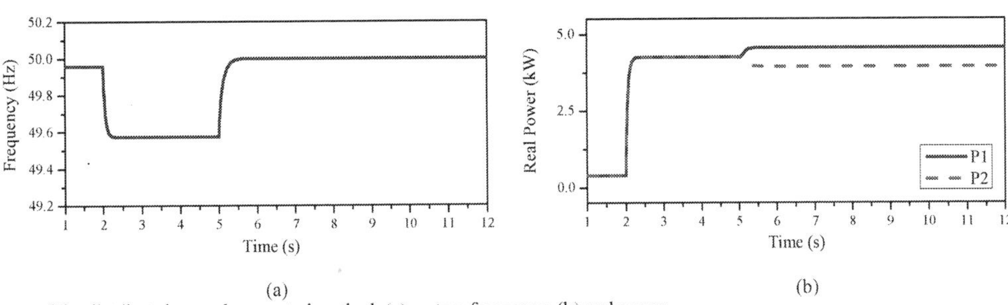

Fig. 5. Performance of the distributed secondary control method: (a) system frequency. (b) real power.

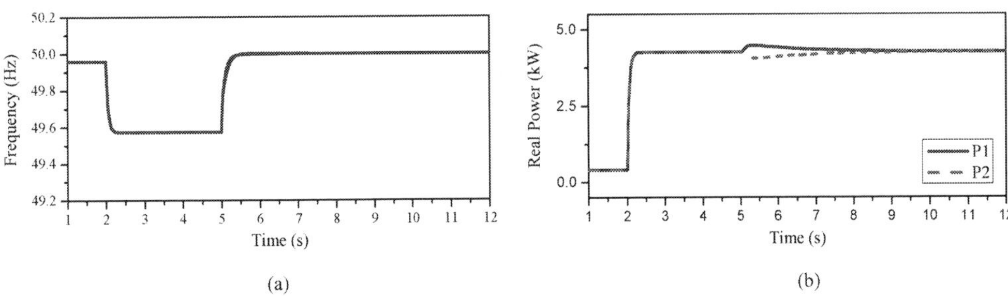

Fig. 6. Performance of the proposed secondary control method: (a) system frequency. (b) real power.

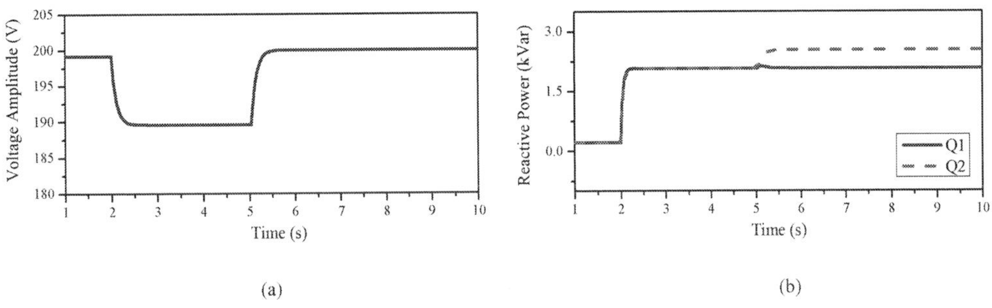

Fig. 7. Performance of the distributed secondary control method: (a) voltage amplitude. (b) reactive power.

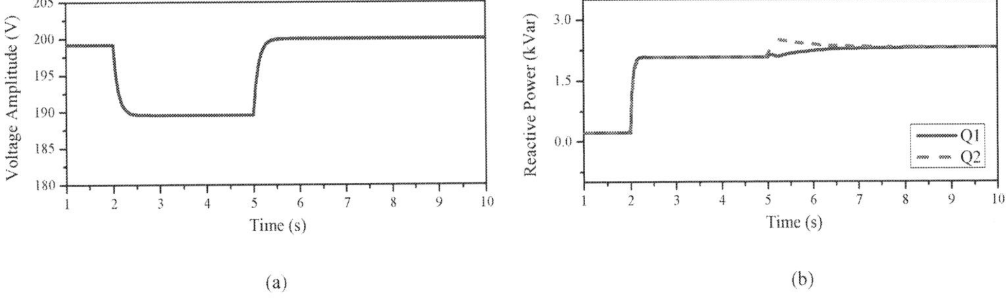

Fig. 8. Performance of the proposed secondary control method: (a) voltage amplitude. (b) reactive power.

shown in Fig. 6.

The second simulation is to investigate the voltage amplitude restoration performance of the novel secondary control. At the moment t=2s, the load reactive power demand increases by 4kVar. Consequently, the voltage amplitude drops from 199.2V to 189.6V. The secondary control is triggered manually at the moment t=5s. The performance of the distributed secondary control in [8] is shown in Fig. 7. It can be seen that even though the voltage amplitude is drawn back to the rated value 200V, the reactive power is not equally shared among the inverters. On the contrary, the proposed method can realize both the voltage amplitude restoration and a good power sharing accuracy. The performance of the proposed secondary control method is shown in Fig. 8.

V. EXPERIMENTAL VERIFICATION

The hardware experiments are conducted in a platform consisting of two similar parallel three-phase inverters. The essential control parameters are listed in Table II. The waveforms of the output power, the frequency and the voltage amplitude are drawn in Origin based on the recorded data by the HIOKI Power Analyzer 3390.

The first experiment is aimed at investigating the frequency restoration capability of the proposed secondary control method. Fig. 9 shows the waveforms of system frequency and real power during the whole process. At the beginning, both inverters are droop controlled. At the moment t=17s, the load real power demand is increased by 1.2kW. As a result, the system frequency drops from 49.76Hz to 49.47Hz. At the moment t=33s, the proposed secondary control is triggered and consequently, the system frequency is restored to 49.80Hz while the real power is equally shared between the two inverters. The steady-state voltage and current waveforms of the inverters with the proposed secondary control are shown in Fig. 10.

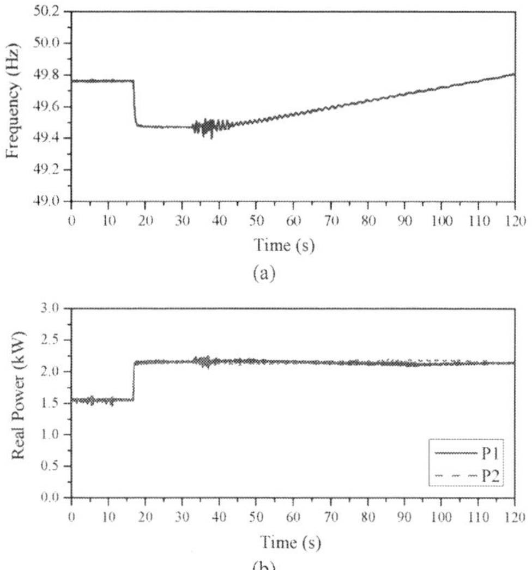

Fig. 9. Experimental waveforms of the system frequency and the output real power: (a) system frequency. (b) real power.

TABLE II. EXPERIMENT PARAMETERS

Parameters	Values	Parameters	Values
ω_0	50Hz	E_0	200V
ω_r	50Hz	E_r	200V
$\hat{\omega}_{p0}$	100Hz	$\hat{\omega}_{q0}$	100Hz
m	$5\cdot10^{-4}$Hz/W	n	$5\cdot10^{-3}$V/Var
$\hat{m}$	$1\cdot10^{-4}$Hz/W	$\hat{n}$	$1\cdot10^{-4}$V/Var
$k_{p\omega}$	1000	$k_{i\omega}$	10
k_{pE}	100	k_{iE}	10
G_p	2000	G_q	2000
ΔP	1.2kW	ΔQ	1.6kVar

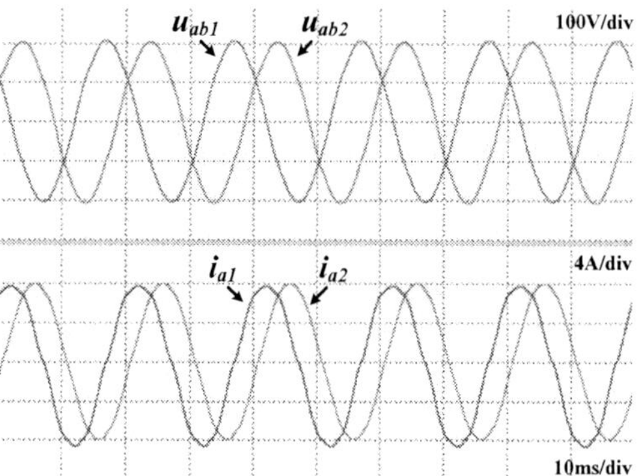

Fig. 10. Steady-state voltage and current waveforms of the inverters with the proposed frequency secondary control.

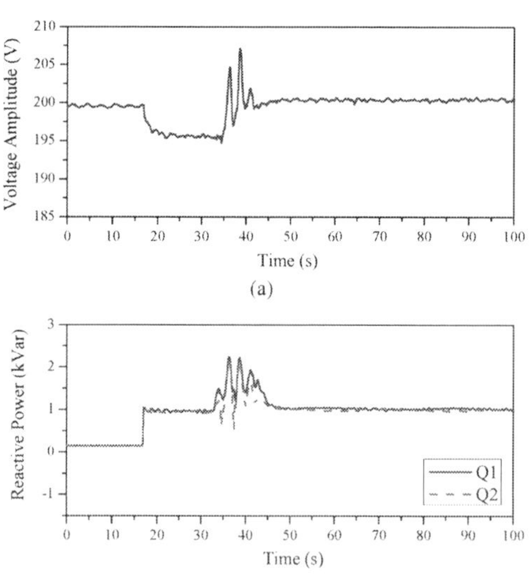

Fig. 11. Experimental waveforms of the voltage amplitude and the output reactive power: (a) PCC voltage amplitude. (b) reactive power.

The second experiment is aimed at investigating the voltage amplitude restoration capability of the proposed secondary control method. Fig. 11 shows the waveforms of voltage amplitude and reactive power during the whole process. At the beginning, both inverters are droop controlled. At the moment t=17s, the load reactive power demand is increased by 1.6kVar. As a result, the voltage amplitude drops from 199.4V to 195.5V. At the moment t=35s, the proposed secondary control is triggered and consequently, the voltage amplitude is restored to 200.5V while the reactive power is equally shared between the two inverters. The steady-state voltage and current waveforms of the inverters with the proposed secondary control are shown in Fig. 12.

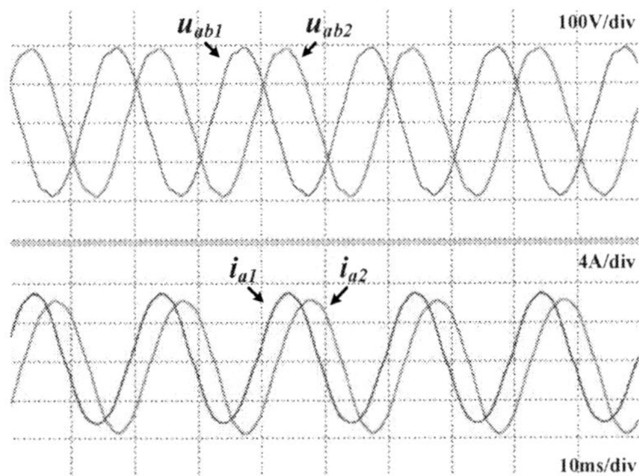

Fig. 12. Steady-state voltage and current waveforms of the inverters with the proposed voltage amplitude secondary control.

VI. CONCLUSION

This paper proposes a small AC signal droop based secondary control method for microgrids. The droop bias of the fundamental output voltage is adjusted online by controlling the frequency of the small AC voltage signal. This method can not only restore the frequency and voltage amplitude accurately, but can also rapidly realize equal power sharing among the parallel inverters without the help of intercommunications. The effectiveness of the proposed method is verified by both simulations and experiments.

REFERENCES

[1] J. Rocabert, A. Luna, F. Blaabjerg, and P. Rodriguez, "Control of power converters in AC microgrids," *IEEE Trans. Industrial Electronics,* vol. 27, no. 11, pp. 4734–4749, Nov. 2012.

[2] J. M. Guerrero, L. Hang, and J. Uceda, "Control of distributed uninterruptible power supply systems," *IEEE Trans. Industrial Electronics,* vol. 55, no. 8, pp. 2845–2859, Aug. 2008.

[3] J. M. Guerrero, J. C. Vasquez, J. Matas, M. Castilla, and L. G. de Vicuna, "Control strategy for flexible microgrid based on parallel line-interactive UPS systems," *IEEE Trans. Industrial Electronics,* vol. 56, no. 3, pp. 726–736, Mar. 2009.

[4] J. Hu, J. Zhu, D. J. Dorrell, and J. M. Guerrero, "Virtual flux droop method—a new control strategy of inverters in microgrids," *IEEE Trans. Power Electronics*, vol. 29, no. 9, pp. 4704–4711, Sep. 2014.

[5] M. Hua, H. Hu, Y. Xing, and J. M. Guerrero, "Multilayer control for inverters in parallel operation without intercommunications," *IEEE Trans. Power Electronics*, vol. 27, no. 8, pp. 3651–3663, Aug. 2012.

[6] T. Wu, Z. Liu, J. Liu, S. Wang, and B. Liu, "Load power estimation based secondary control for microgrids," in *Proc. ICPE-ECCE Asia,* Seoul, Korea, 2015, pp. 722–727.

[7] J. M. Guerrero, J. C. Vasquez, J. Matas, L. G. de Vicuna, and M. Castilla, "Hierarchical control of droop-controlled AC and DC microgrids—a general approach toward standardization," *IEEE Trans. Industrial Electronics*, vol. 58, no. 1, pp. 158–172, Jan. 2011.

[8] Q. Shafiee, J. M. Guerrero, and J. C. Vasquez, "Distributed secondary control for islanded microgrids–a novel approach," *IEEE Trans. Power Electronics*, vol. 29, no. 2, pp. 1018–1031, Feb. 2014.

[9] A. Tuladhar, H. Jin, T. Unger, and K. Mauch, "Control of parallel inverters in distributed AC power systems with consideration of line impedance effect," *IEEE Trans. Industrial Applications*, vol. 36, no. 1, pp. 131–138, Jan./Feb. 2000.

[10] B. Wu, and S. Keyue, "A current control MPPT method in high power solar energy conversion system," in *Proc. APEC,* 2014, no. 1, pp. 3021–3025.

Mode Transition Control Strategy for Multiple Inverter Based Distributed Generators Operating in Grid-Connected and Stand-Alone Mode

Onkar Vitthal Kulkarni
Department of
Energy Science and Engineering,
Indian Institute of Technology, Bombay
Email: onkarvkulkarni@gmail.com

Suryanarayana Doolla
Department of
Energy Science and Engineering,
Indian Institute of Technology, Bombay
Email: suryad@iitb.ac.in

B. G. Fernandes
Department of
Electrical Engineering,
Indian Institute of Technology, Bombay
Email: bgf@ee.iitb.ac.in

***Abstract*—This paper proposes a novel automatic mode transition control strategy for multiple inverters to operate in grid-connected and islanded modes without communication. When grid is available, all the inverters operate in grid-tied current control mode and transfer available power to the grid. On grid failure, they automatically shift to conventional droop control mode and shift back to grid-tied current control mode when the grid becomes available. The control signals for mode transition are generated by state machines specific to each inverter. The state machine has appropriate delays to facilitate smooth mode transition. Feasibility of the proposed control strategy is substantiated using MATLAB/SIMULINK simulation results.**

***Index Terms*—Multiple inverters, Automatic mode transition, PLL, State machine**

I. Introduction

Renewable energy sources (RESs) such as solar, wind and tidal are environmental friendly and sustainable. Generally, these sources are used to form small distributed generators (DGs) to generate power at the point of consumption. The concept of connecting these DGs to the public ac grids at low voltage (LV) level via power inverters is popular. Additionally, on grid failure they are expected to feed power available with them to selected loads. Hence, these inverter interfaced DGs must be able to operate in both the grid-connected (GC) and stand-alone (SA) control modes [1]–[3]. A lot of literature is available mainly for the seamless mode transition of single inverter systems [1]–[4]. In [1], [3] seamless mode transfer of single inverter system from GC to SA and vice-versa is proposed. In [1], static transfer switch (STS) is used for seamless transfer and hence SA to GC transition is executed only after phase-locked loop (PLL) convergence. In [2], [3], STS is eliminated at the expense of possibility of overcurrent operation while switching from SA to GC mode and hence requires inverter with current handling capability much greater than the rated value. Indirect grid current control based mode transition technique for single master inverter with modified connection topology is also discussed in the literature [5].

For low cost and/or high power applications, a number of DGs should be operated in parallel. Hence, mode transition of multiple parallel operating DGs (also known as a microgrid (MG)) with or without using communication is addressed in the literature. In [6] and [11], communication based multi-layer control structure is used for mode transition of multiple inverters.

In [7], all DGs in MG under consideration operate in droop controlled (DC) mode in both GC and SA conditions except for the DG which has control over STS acting as a current controlled dispatch unit. This strategy alongwith one proposed in [4] is valid only for DGs with integrated storage (perpetual DGs) and cannot be applied to RES based intermittent DGs. Also, strategy in [7] is valid only in developed countries where the grid frequency deviation is very small (of the order of ± 0.02 Hz). Because, to achieve frequency vs. active power droop control of DGs in GC mode in the countries where the frequency deviation is of the order of ± 0.3 Hz implies large droop coefficients leading to reduced small signal stability margin in SA mode [8]. Generally, in the presence of grid, both the intermittent and perpetual DGs operate in maximum power point tracking (MPPT) mode. During GC mode, it is undesirable to spend stored energy which is preserved for grid failure conditions [9]. In [5], [10], seamless mode transition of a microgrid (MG) consisting of multiple inverter interfaced DGs is proposed wherein only the master inverter has two modes of operation whereas all the slaves always operate in current control (CC) mode. In SA mode, there should not be any mismatch between generated and absorbed power. Hence, this system requires a large storage to absorb excess power under light load conditions in SA mode and thus is not applicable to intermittent DGs.

In this paper, a control scheme for multiple inverters operating in GC and SA modes is proposed. In GC mode, inverters inject maximum power available with them into the grid. In SA mode, inverters operate in virtual inductance based conventional droop control mode to achieve proportional power sharing. GC to SA mode transition of all the inverters is achieved using PLL. Similarly, SA to GC mode transition is achieved by disturbing the phase angle of droop control operation. The control signals for mode transition are generated by state machines specific to each inverter. The state machine has appropriate delays to facilitate smooth mode transition. This helps in avoiding overcurrent phenomenon as reported in [2] and [3].

The paper is organized as follows. In Section II, description of the system and details of control structures for both GC and SA modes are provided. In Section III, the details of state machines used for the inverters and aggregate pulse controller for controlling inverter switches are provided. In Section IV, simulation results confirming the validity of the proposed strategy are provided. Finally, the conclusion is provided in Section V.

II. System description

The system under consideration consists of two three-phase inverters with LCL filters which interface renewable energy sources with the ac grid, linear loads, a tie line connecting both the inverters and static transfer switch (STS) as shown in Fig. 1. The grid voltage v_{g1} is measured and is fed to inverter-1 controller. Based on this signal, inverter-1 controller decides the status of STS. In other words, whenever the grid is available and is in healthy condition, STS is turned ON and upon grid failure it is turned OFF. From control perspective, inverters in the system can be classified as

1) Principal inverter: is the inverter which is close to the point of common coupling (PCC). This inverter controls the STS and has direct access to grid voltage information. This inverter initiates the SA to GC mode transition of all the other inverters in the system by disturbing the phase angle reference of droop controller.
2) Auxiliary inverter: is the inverter without direct access to grid voltage information under SA control mode. Hence, this inverter has only one voltage signal v_{o2} available for its control algorithm and is dependent on principal inverter for SA to GC mode transition. For symmetry in describing the control structure, a signal v_{g2} is defined as shown in Fig. 1. From this figure, it is obvious that $v_{o2} = v_{g2}$ irrespective of the status of STS.

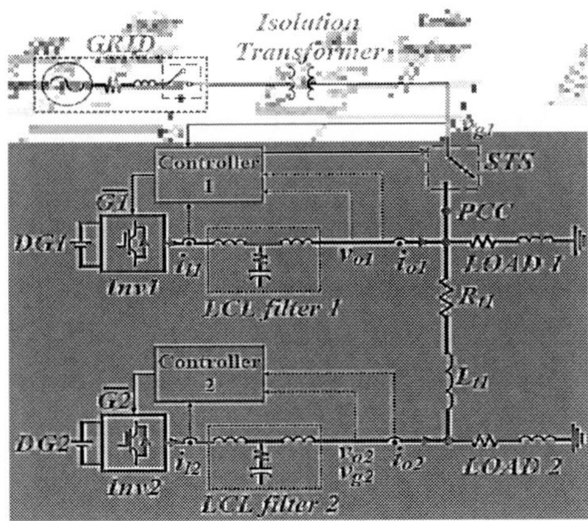

Fig. 1: System configuration

In GC mode, inverters inject available power into the grid. In SA mode, inverters operate in virtual inductance based conventional droop control mode to achieve proportional power sharing. Considering presence of intermittent DGs, the nominal voltage for SA operation is set at a predefined value (0.85 pu in this case) to ensure power supply to maximum loads. The control structures for these modes are provided in the following subsections.

A. Control structure for GC mode

This consists of two main stages: (a) PLL and (b) Inner current control loop as follows.

1) PLL: The synchronism with the grid is obtained using PLL. The principle of PLL used is in this paper is given in Fig. 2 [2]. In this figure, v_{xi} and v_{yi} are the stationary $\alpha\beta$ -

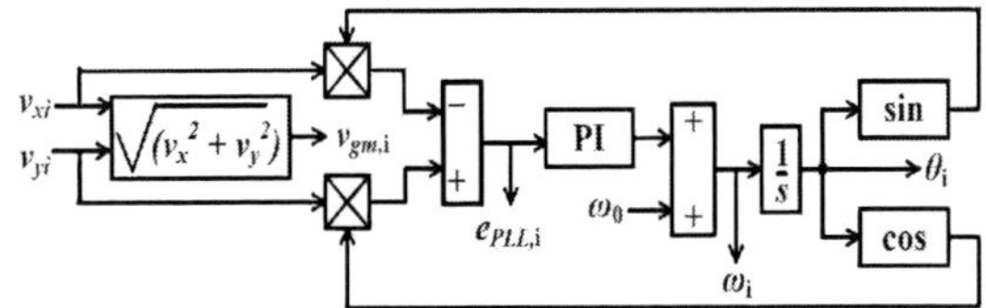

Fig. 2: Phase locked loop

reference frame components of the grid voltage corresponding to ith inverter measured in per unit. $v_{gm,i}$ is the grid voltage magnitude (in p.u.). $e_{PLL,i}$ is the error of the PI-controller and forms the basis of SA to GC mode transition of auxiliary inverters as described in section III. ω_0 and ω_i are the reference and actual grid frequency. θ_i is the desired inverter phase angle. PLL outputs $v_{gm,i}$, $e_{PLL,i}$ and ω_i are also used for mode transition control as described in section III.

2) Inner current control loop: Power controller for GC mode is depicted in Fig. 3. It is mainly based on the PI-controller based inverter side filter inductance current control. The reference inductor currents (i_{ld}^* and i_{lq}^*) are calculated

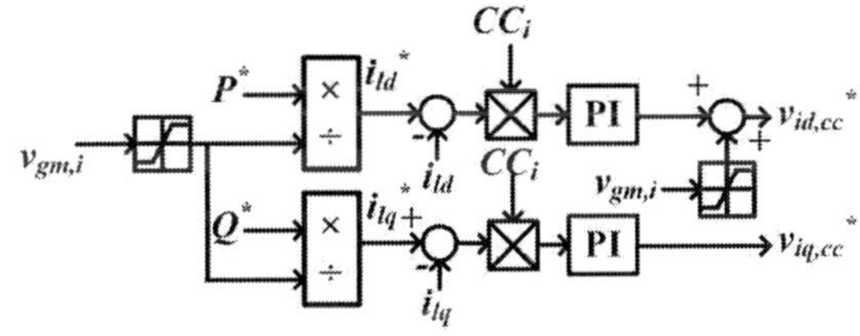

Fig. 3: Inner current control loop

based on P^* (which generally comes from MPPT algorithm), Q^* (is generally zero for grid-tie inverters operating at unity power factor) and grid voltage magnitude ($v_{gm,i}$). Feedforward control is provided to improve transient response of the system. Control signal CC_i (status of grid-tied current controller) is generated by state machine of ith inverter and handles the activation ($CC_i = 1$) and deactivation ($CC_i = 0$) of the controller. Thus, whenever current controller becomes inactive, inputs of both the PI regulators are reset to zero. Similarly, $v_{id,cc}^*$ and $v_{iq,cc}^*$ are also reset to zero.

978-1-4673-9551-9/16 $31.00 © 2016 IEEE

B. Control structure for SA mode

This consists of three main stages: (a) power controller, (b) virtual inductance loop and (c) outer voltage and inner current loop as described in the following subsections.

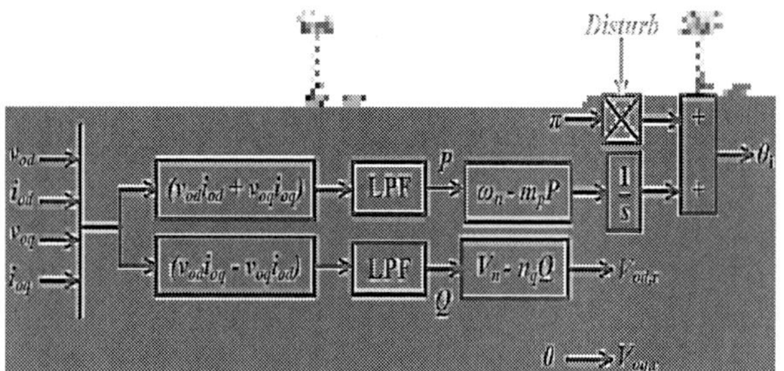

Fig. 4: Power controller for SA mode

1) Power controller: In SA mode, multiple decentralized inverters operating in parallel generate their own reference voltage using power controller as shown in Fig. 4. Once grid is restored, the $Disturb$ signal is generated by state machine of principal inverter to introduce phase angle disturbance of π radian among synchronized inverters.

2) Virtual inductance loop: In low voltage (LV) microgrids, parallel inverters share significant amount of reactive power due to the presence of predominantly resistive lines. Hence, the accuracy in proportional reactive power sharing among inverters is lost. This problem can be overcome by using virtual inductance loop shown in Fig. 5 as described in [12].

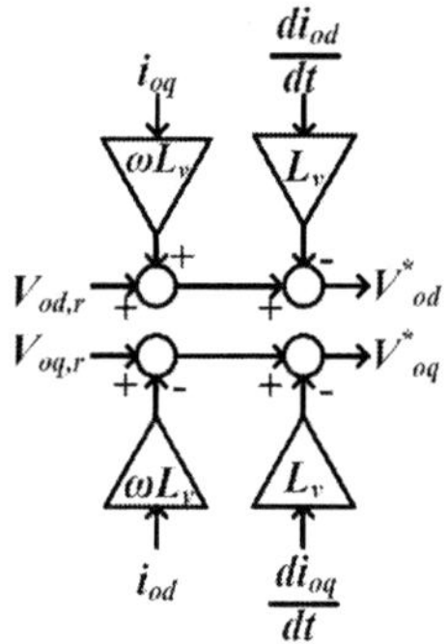

Fig. 5: Virtual inductance loop

3) Outer voltage and inner current control loop: PI-controller based multiple feedback control loops used in droop control are shown in Fig. 6. Control signal DC_i (status of droop controller) is generated by state machine of ith inverter and handles the activation (DC_i = 1) and deactivation (DC_i = 0) of the controller. Thus, whenever droop controller becomes inactive, inputs of all the four PI regulators are reset to zero. Similarly, $v^*_{id,dc}$ and $v^*_{iq,dc}$ are also reset to zero.

III. Mode Transition Control

As mentioned in section II, PLL outputs are also used to achieve stateflow based mode transition control for both the

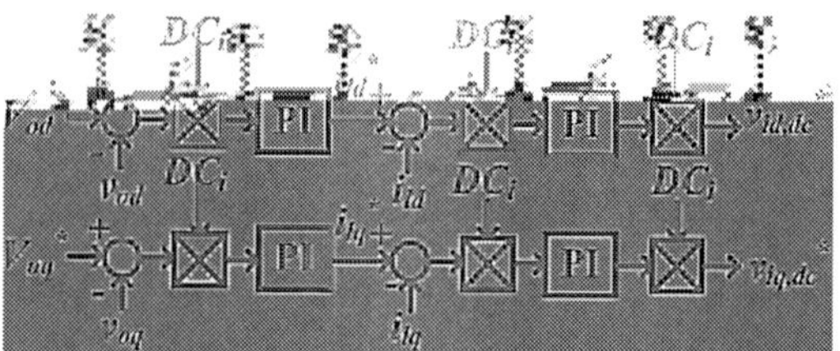

Fig. 6: Outer voltage and inner current controller for SA mode

principal as well as auxiliary inverters. The signals $v_{gm,i}$ and ω_i are used to check for grid availability. A grid available (GA_i) signal is defined for ith inverter using equations (1) and (2) such that if both the conditions given by (1) and (2) are satisfied simultaneously then $GA_i = 1$ otherwise it is 0.

$$V_{gm,min} < V_{gm,i} < V_{gm,max} \tag{1}$$

$$\omega_{min} < \omega_i < \omega_{max} \tag{2}$$

Principal inverter uses GA_i signal for both the GC to SA and SA to GC mode transitions. On the other hand, auxiliary inverters use GA_i signal for GC to SA mode transition and control signal $ePLL_i$ for SA to GC mode transition. During SA mode of operation, if PLL output $e_{PLL,i}$ crosses a threshold value e_{TH} then control signal $ePLL_i$ becomes one and SA to GC mode transition of auxiliary inverters is activated. The following subsections describe the state machine for principal inverter, state machine for auxiliary inverters and aggregate pulse controller.

A. State machine for principal inverter

This state machine consists of eight different states used for generating control signals CC, DC, STS and $Disturb$. As described earlier, CC controls GC mode of operation of inverter, DC controls SA mode of operation of inverter, STS controls the static transfer switch and $Disturb$ controls the phase angle disturbance of power controller of principal inverter in SA mode. The most prominent states are colored grey as shown in Fig. 7. The changes in control signal are marked in red color. The GC mode of operation is considered to be default operating mode. While operating in this mode (state GC3), if grid fails ($GA = 0$), inverter state SA1 is activated to ultimately reach state SA4 (GC3-SA1-SA2-SA3-SA4). Similarly, while operating in SA mode (state SA4), if grid arrives ($GA = 1$), inverter state SA5 is activated to ultimately reach state GC3 (SA4-SA5-GC1-GC2-GC3). Appropriate delays are inserted between some states in order to achieve smooth mode transition.

1) D2: delay to facilitate quick discharge of filter capacitance through grid-connected loads.
2) D3: delay to ensure that all the inverters start operating in droop control mode at almost the same instant.
3) D4: delay to ensure SA mode of operation for minimum duration D4.
4) D6: delay for which inverter disturbs the phase angle of synchronization during SA mode. This disturbance

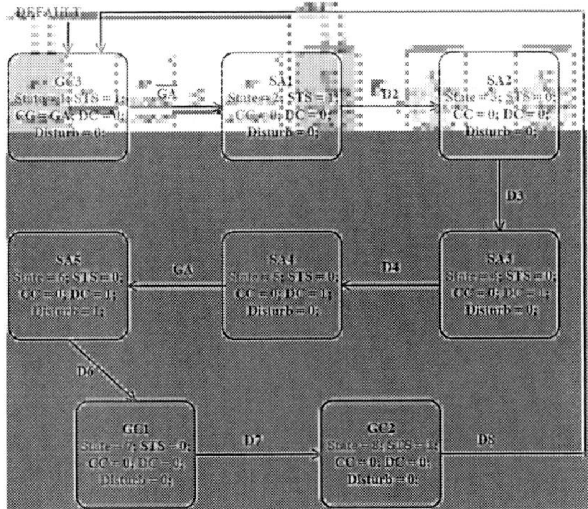

Fig. 7: State machine for principal inverter

causes the $e_{PLL,i}$ of auxiliary inverters to cross their threshold values and activates control signal $ePLL_i$ causing activation of their next state as described in following subsection.

5) D7: delay to facilitate discharge of filter capacitance through loads in microgrid before closing STS in order to avoid possibility of high rate of change of voltage across capacitor.
6) D8: delay to start GC mode of operation

B. State machine for auxiliary inverter

This state machine consists of five different states used for generating control signals CC and DC. The most prominent

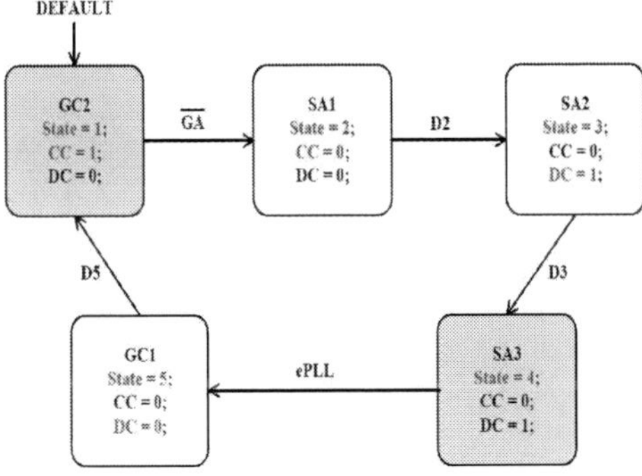

Fig. 8: State machine for auxiliary inverter

states are colored grey as shown in Fig. 8. The changes in control signal are marked in red color. The GC mode of operation is considered to be default operating mode. While operating in this mode (state GC2), if grid fails ($GA = 0$), inverter state SA1 is activated to ultimately reach state SA3 (GC2-SA1-SA2-SA3). Similarly, while operating in SA mode (state SA3), if $e_{PLL,i}$ crosses its threshold value (which is indication of grid arrival for auxiliary inverter), inverter state GC1 is activated to ultimately reach state GC2 (SA3-GC1-GC2). Appropriate delays are inserted between some states in order to achieve smooth mode transition.

1) D2: delay to ensure that all the inverters start operating in droop control mode at almost the same instant.
2) D3: delay to avoid transient duration of PLL so that inverter does not stop operating in droop control mode due to undesirable activation of $ePLL_i$ as elaborated in section IV and also to ensure SA mode of operation for minimum duration D3.
3) D5: delay to start GC mode of operation.

C. Aggregate pulse controller

The pulses for inverters are obtained using sinusoidal pulse width modulation technique (SPWM). However, as the mod-

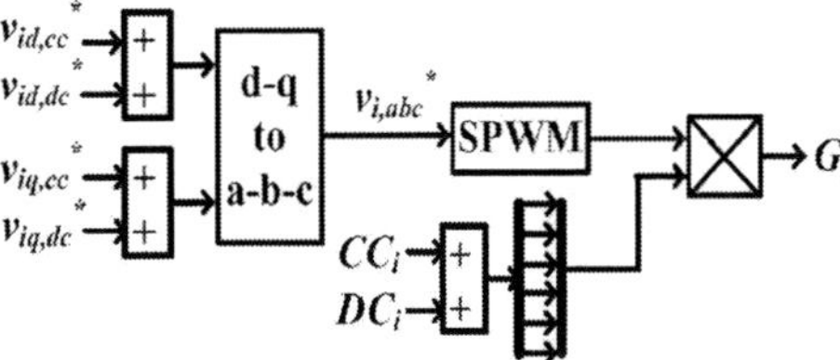

Fig. 9: Pulse controller

ulating signals are generated by two different controllers, the management and selection of the appropriate modulating signal is done as shown in Fig. 9. When $CC = 1$ and $DC = 0$, $v^*_{i,abc}$ is generated using $v^*_{id,cc}$ and $v^*_{iq,cc}$ as per Figs. 3, 6 and 9. Similarly, When $CC = 0$ and $DC = 1$, $v^*_{i,abc}$ is generated using $v^*_{id,dc}$ and $v^*_{iq,dc}$. When both CC and DC are zero, inverter pulses are blocked.

IV. Simulation Results and Discussions

In simulation, for the control purpose, all the measurements are done in per unit system. For this three phase system, the peak base values are given by $V_{base} = 325V$, $I_{base} = 41A$ and $P_{base} = 20kW$. The important parameters used for simulation and their values are listed in Table I.

A. Mode transition control of inverters

The simulation begins with both the DGs operating in GC mode and injecting maximum power available with them into the grid at unity power factor. At 0.2 sec, grid fails and both the inverters shift to droop control mode. In droop control mode, both the inverters operate at a set nominal voltage level and share the connected load in proportion to their power ratings. As can be seen from Table I, droop coefficients of inverter 1 are double of that of inverter 2. In other words, droop coefficients are selected such that rating of inverter 2 is double that of inverter 1. Similarly, local loads are selected such that there is almost no power flow in the tie line at steady state in SA mode. This forms worst case scenario for SA to GC mode transition for the proposed control strategy due to the presence of floating inverter-load combination.

Table I: Parameters for simulation study

Parameter	Symbol	Value	Units
Min. grid voltage	V_{gmin}	115	Vrms
Nominal grid voltage	V_g	230	Vrms
Nominal island voltage	V_n	195.5	Vrms
Max. grid voltage	V_{gmax}	276	Vrms
Min. grid frequency	ω_{gmin}	300	rad/s
Nominal grid frequency	ω_g	314.15	rad/s
Nominal island frequency	ω_n	314.15	rad/s
Max. grid frequency	ω_{gmax}	316	rad/s
Tie line resistance	R_{t1}	0.1	Ω
Tie line inductance	L_{t1}	20	μH
Frequency droop coefficient	m_{p1}	2e-5	W/rad
Frequency droop coefficient	m_{p2}	1e-5	W/rad
Voltage droop coefficient	n_{q1}	0.0013	Var/V
Voltage droop coefficient	n_{q2}	6.5e-4	Var/V
Virtual inductance	L_{v1}	0.75	mH
Virtual inductance	L_{v2}	1.5	mH
Load 1 at 0.85 pu	P_{11}	7000	W
Load 1 at 0.85 pu	Q_{11}	5500	Var
Load 2 at 0.85 pu	P_{21}	3500	W
Load 2 at 0.85 pu	Q_{21}	2750	Var
Load 1 at 1 pu	P_{12}	9800	W
Load 1 at 1 pu	Q_{12}	7600	Var
Load 2 at 1 pu	P_{22}	4900	W
Load 2 at 1 pu	Q_{22}	3800	Var
DC link voltage 1	V_{dc1}	700	V
DC link voltage 2	V_{dc2}	700	V

At 0.28 sec, grid is restored and both the inverters start injecting power after predefined delay. For considering worst case scenario grid angle is set at 180° phase shift to inverter output voltage phase angle in SA mode as can be observed from simulation results.

1) Principal inverter: The control signals and three phase output voltages alongwith output currents for the principal inverter are shown in Fig. 10 and 12, respectively. At 0.2 s, on grid failure, state machine sequentially moves from state GC3 to SA4 through states SA1-SA2-SA3 as indicated in Fig. 7. Thus, first current control mode is turned off ($CC = 0$, state SA1), then after delay D2 (2.5 ms), STS is opened ($STS = 0$, state SA2) and then after delay D3 (1 ms), operation in DC mode begins ($DC = 1$, state SA3). After delay D4 (60 ms), inverter becomes ready to shift back to GC mode (state SA4). Then on grid restoration, inverter first introduces an offset of π rad in the phase angle of its droop controller as depicted in Fig. 4 for the interval D6 (0.625 ms)($Disturb = 1$, state SA5). This causes $e_{PLL,i}$ of auxiliary inverters to cross threshold value. Then inverter stops its SA mode ($DC = 0$, state GC1) and after delay D7 (1 ms), closes STS ($STS = 1$, state GC2)and is ready to shift to GC mode. And after delay D8 (60 ms), inverter starts operating in GC mode ($CC = 1$,state GC3). In this way, on grid arrival inverter sequentially moves from state SA4 to GC3 through states SA5-GC1-GC2 as indicated in Fig. 7.

2) Auxiliary inverter: The control signals and three phase output voltages alongwith output currents for the auxiliary inverter are shown in Fig. 11 and 13. At 0.2 s, on grid failure, state machine sequentially moves from state GC2 to SA3 through states SA1 and SA2 as depicted in Fig. 8. Thus, first current control mode is turned off ($CC = 0$, state SA1), then after delay D2 (3 ms), droop control mode is activated ($DC = 1$, state SA2). After delay D3 (60 ms) which is minimum duration of operation in SA mode, state machine starts monitoring $ePLL$. As can be observed from Fig. 13, inverter simply continues to operate in droop control mode irrespective of large transients in e_{PLL} during GC to SA mode transition. When the principal inverter moves to SA5, e_{PLL} becomes greater than e_{TH} (0.05 p.u.) and hence control signal $ePLL$ becomes 1 and inverter stops its droop control mode (DC = 0, state GC1). Then after predefined delay D5 (60 ms), inverter starts operating in GC mode ($CC = 1$, state GC2).

B. Operation in SA mode

In this subsection, simulation results focusing on various scenarios in SA mode are presented.

1) Tripping of a large load: As microgrid is small system as compared to the main grid, generally only small load variations are allowed ($\pm$ 20 percent of its rated load) in it. However, there may arise a situation in which a large load may get disconnected. In such a situation, mode transition control should not get activated and inverters should continue to operate in SA mode upto the instant when grid arrives. This scenario is verified using the following simulation case study. In this scenario, inverters initially operating in GC mode, shift to SA mode at 0.2 s and at 0.28 s, a large load ($3500 + j2750$ VA at 0.85 p.u.) near inverter 2 is disconnected. It can be observed from Fig. 14 and 15 that inverters (especially auxiliary inverter) continue to operate in SA mode. This is because, e_{PLL} of auxiliary inverter does not cross its threshold value e_{TH} due to large load disconnection.

2) Tripping of principal inverter under light load conditions: While operating in SA mode, there may arise a situation wherein principal inverter trips due to some fault. In such a scenario, especially when the microgrid is operating under light load condition, auxiliary inverter must continue to supply power to these loads. This scenario is verified using the following simulation case study.

In this case, inverters initially operating in GC mode, shift to SA mode at 0.2 s. The total load connected to the system is approximately $5600+j3400$ VA at 0.85 pu. At 0.28 s, principal inverter stops working. Inverter output voltages and currents are depicted in Fig. 16 and 17. It can be observed from Fig. 17 that auxiliary inverter continues to supply power. This is because, e_{PLL} of auxiliary inverter does not cross its threshold value e_{TH} due to tripping of principal inverter under light load condition.

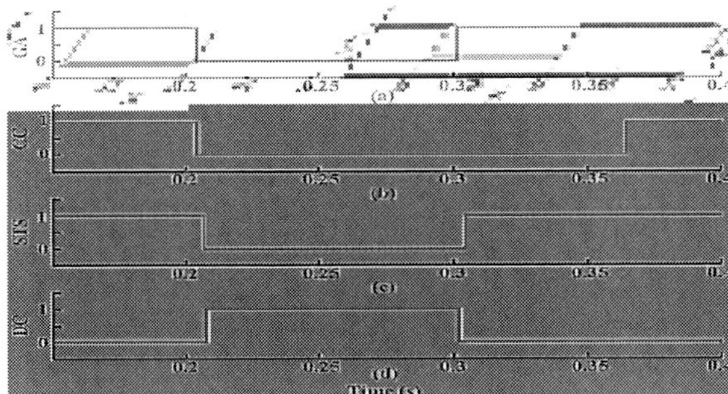

Fig. 10: Control signals for principal inverter (a) GA (b) CC (c) STS and (d) DC

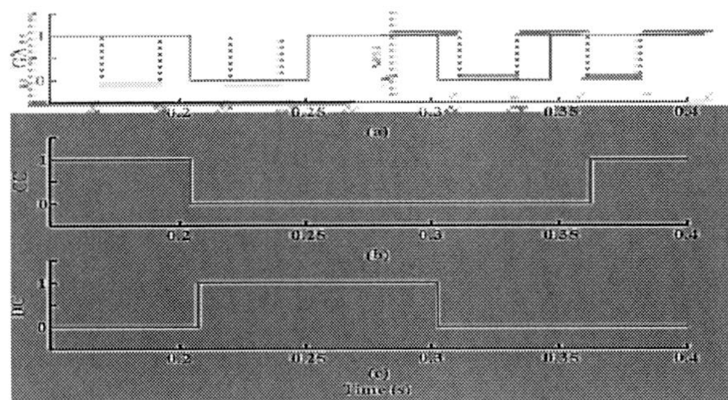

Fig. 11: Control signals for auxiliary inverter (a) GA (b) CC and (c) DC

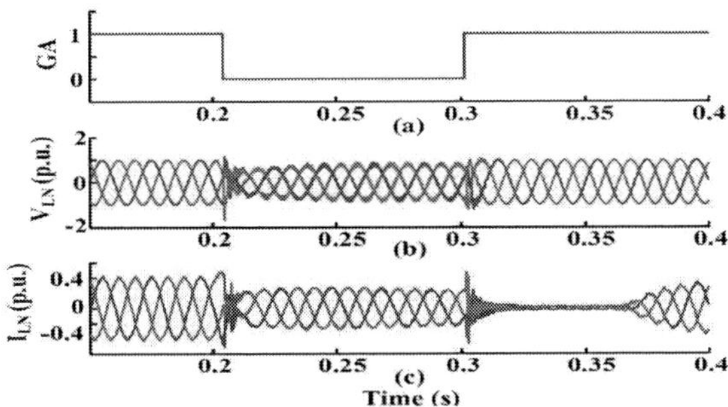

Fig. 12: Response of principal inverter during mode transition (a) GA signal (b) three phase output voltages and (c) three phase output currents

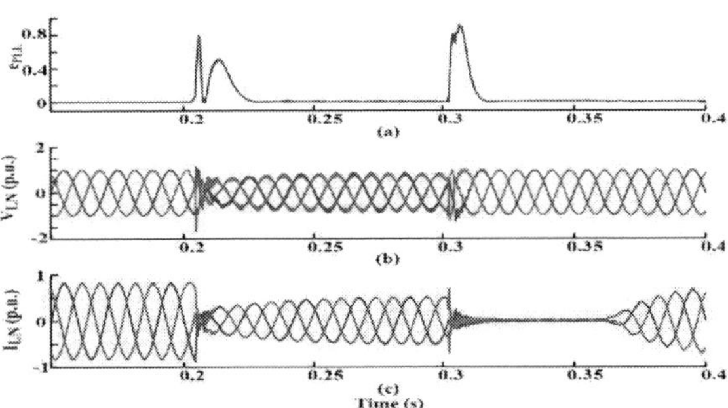

Fig. 13: Response of auxiliary inverter during mode transition (a) e_{PLL} (b) three phase output voltages and (c) three phase output currents

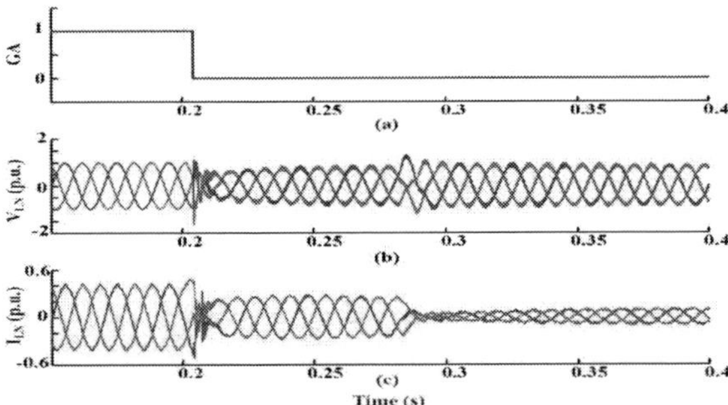

Fig. 14: Principal inverter response in case of load change (a) GA signal (b) three phase output voltages and (c) three phase output currents

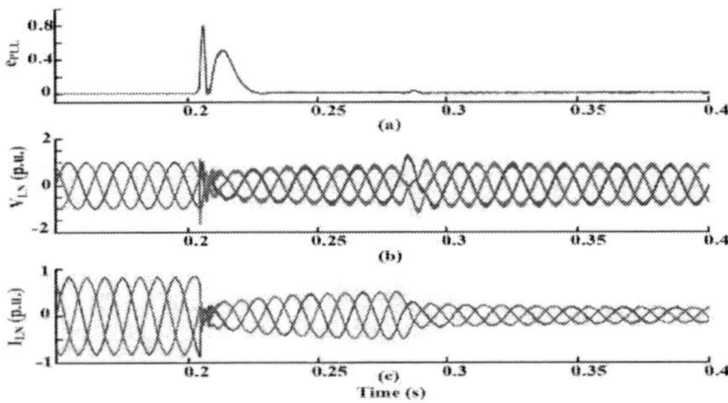

Fig. 15: Response of auxiliary inverter in case of load change (a) e_{PLL} (b) three phase output voltages and (c) three phase output currents

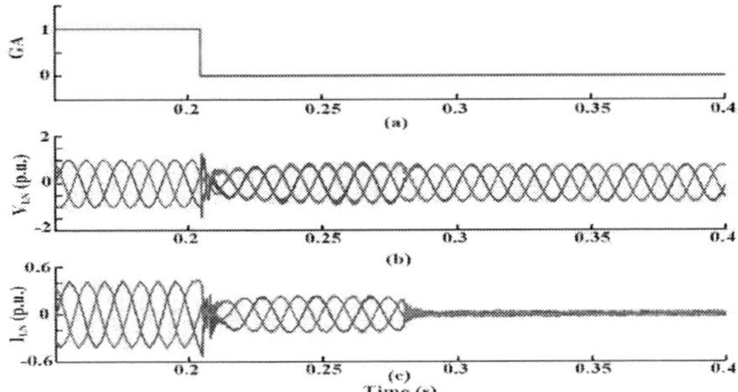

Fig. 16: Simulation results for the case of principal inverter tripping (a) GA signal (b) three phase output voltages and (c) three phase output currents

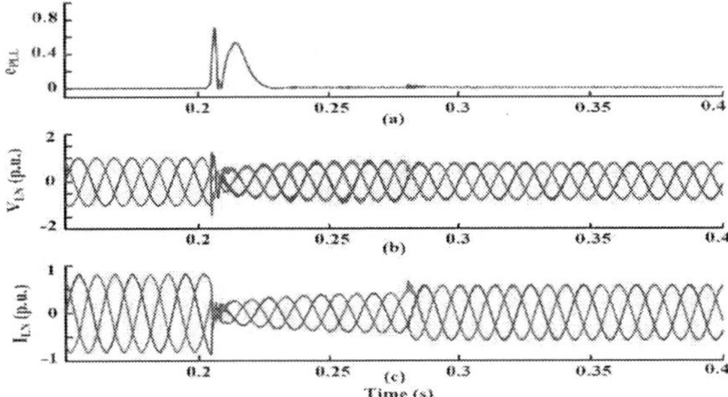

Fig. 17: Response of auxiliary inverter in case of principal inverter tripping (a) e_{PLL} (b) three phase output voltages and (c) three phase output currents

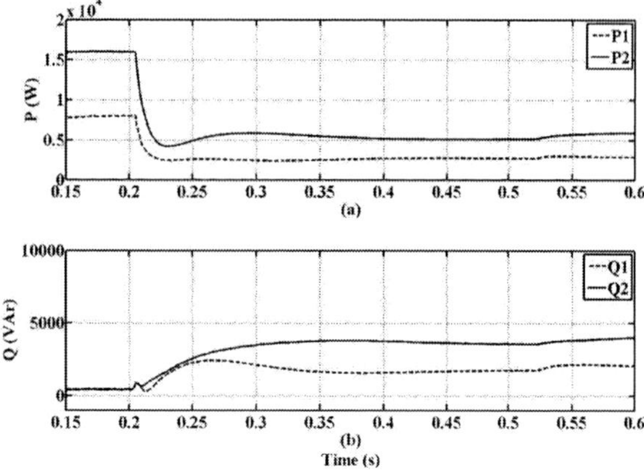

Fig. 18: Power sharing in SA mode (a) active power (b) reactive power

3) Proportional power sharing in SA mode: In this scenario, inverters initially operating in GC mode, shift to SA mode at 0.2 s and small increase in load ($1400 + j1100$ VA at 0.85 p.u) takes place at 0.52 s. It can be observed from Fig. 18 that even in the presence of predominantly resistive lines substantially accurate proportional active and reactive power sharing is achieved due to the presence of virtual inductance loop. Additionally, it can be observed that power sharing settles in around 0.25 s.

V. Conclusion

In this paper, a novel mode transition control strategy for multiple DGs able to operate in, grid-connected and islanding modes is presented. In the first case, the inverters operate in grid-tied current controlled mode, whereas in the later case they operate in virtual inductance based conventional droop control mode to achieve proportional active and reactive power sharing. The mode transition control structures of both the invertes consist of PLL and state machine. Appropriate delays are introduced between various states in order to facilitate smooth mode transition and elimination of overcurrent phenomenon for SA to GC transition. The proposed control strategy is verified using MATLAB/Simulink for various scenarios consisting of normal operation, large load changes during islanded operation and inverter tripping during SA operation. The simulation results show satisfactory performance. Hence, the proposed mode transition control strategy can be effectively used for multiple DGs operating in flexible microgrids.

VI. Acknowledgments

The authors would like to thank the support of NCPRE, IIT Bombay a project funded by MNRE, India.

References

[1] R. Tirumala, N. Mohan, and C. Henze, "Seamless transfer of grid-connected PWM inverters between utility-interactive and stand-alone modes," in *Proc. IEEE Appl. Power Electron. Conf.*, vol. 2, Mar. 2002, pp. 1081–1086.

[2] R. Teodorescu and F. Blaabjerg, "Flexible control of small wind turbines with grid failure detection operating in stand-alone and grid-connected mode," *IEEE Trans. Power Electron.*, vol. 19, no. 5, pp. 1323–1332, Sept. 2004.

[3] D. Ochs, B. Mirafzal, and P. Sotoodeh, "A method of seamless transitions between grid-tied and stand-alone modes of operation for utility-interactive three-phase inverters," *IEEE Trans. Ind. Appl.*, vol. 50, no. 3, pp. 1934–1941, May 2014.

[4] J. C. Vasquez, J. M. Guerrero, A. Luna, P. Rodriguez, and R. Teodorescu, "Adaptive droop control applied to voltage-source inverters operating in grid-connected and islanded modes," *IEEE Trans. Ind. Electron.*, vol. 56, no. 10, pp. 4088–4096, Oct. 2009.

[5] Z. Liu and J. Liu, "Indirect current control based seamless transfer of three-phase inverter in distributed generation," *IEEE Trans. Power Electron.*, vol. 29, no. 7, pp. 3368–3383, July 2014.

[6] Y. Xu, H. Li, and L. M. Tolbert, "Inverter-based microgrid control and stable islanding transition," in *Proc. IEEE Energy Convers.Congr. Expo.*, Sept 2012, pp. 2374–2380.

[7] M. N. Arafat, A. Elrayyah, and Y. Sozer, "An effective smooth transition control strategy using droop based synchronization for parallel inverters," *IEEE Trans. Ind. Appl.*, vol. PP, no. 99, pp. 1–1, 2014.

[8] N. Pogaku, M. Prodanovic, and T. C. Green, "Modeling, analysis and testing of autonomous operation of an inverter-based microgrid," *IEEE Trans. Power Electron.*, vol. 22, no. 2, pp. 613–625, Mar. 2007.

[9] S. Jiang, W. Wang, H. Jin, and D. Xu, "Power management strategy for microgrid with energy storage system," in *IEEE IECON Ind. Elec. Soc.*, Nov. 2011, pp. 1524–1529.

[10] W. C. Shan, L. X. Lin, G. Li, and L. Y. Wei, "A seamless operation mode transition control strategy for a microgrid based on master-slave control," in *Proc. CCC*, July 2012, pp. 6768–6775.

[11] J. Wang, N. C. P. Chang, X. Feng, and A. Monti, "Design of a generalized control algorithm for parallel inverters for smooth microgrid transition operation," *IEEE Trans. Ind. Electron.*, vol. 62, no. 8, pp. 4900–4914, Aug. 2015.

[12] J. M. Guerrero, J. Matas, L. de Vicua, M. Castilla, and J. Miret, "Wireless-control strategy for parallel operation of distributed-generation inverters," *IEEE Trans. Ind. Electron.*, vol. 53, no. 5, pp. 1461–1470, Oct. 2006.

An Autonomous Power Management Strategy Based on DC Bus Signaling for Solid-State Transformer Interfaced PMSG Wind Energy Conversion System

Rui Gao, Iqbal Husain, and Alex Q. Huang
FREEDM systems center
North Carolina State University
Raleigh, NC, US
rgao@ncsu.edu

Abstract—**The solid-state transformer (SST) enabled DC/AC Microgrid provides an effective solution for distributed renewable energy resources (DRER) integration with conventional utility grid. This paper investigates a DC network system consisting of wind turbines, SST, and DC loads. Without any energy storage devices, an autonomous power management strategy based on improved DC bus signaling (DBS) is proposed to achieve system stable operation and power balance under various scenarios, specifically system grid-connected mode, islanding mode, and the mode transition. The extreme conditions were emphasized and analyzed as a testament to verify the feasibility of proposed control. DC bus voltage level and its gradient information have been employed as the only indication for distinguishing different modes and control implementation. System power management competence has been simulated and verified with MATLAB/Simulink.**

Keywords—DC bus Signaling; solid-state transformer; PMSG; wind energy conversion

I. Introduction

In the past decade, significant research has been conducted on advanced configurations, power electronic topologies, energy storage devices and control solutions to facilitate the wind DRER integration with utility grid or remote islanding operation [1-3]. One interesting track is the control and operation of the power electronics enabled Microgrid system that formed by small-scale wind turbines and photovoltaic generation, energy storage devices and loads. The multiple converter paralleled structure greatly enhances the flexibility and redundancy of such system, while brings in the tradeoff of complexity as well as stability issues [4-6]. Meanwhile, medium-frequency transformer link based converter topology is proposed for wind generator and battery storage device in [12]. Transformer-less concept for wind turbine system has also been reported to eliminate the bulky line frequency transformers [7], [8]. Researchers in [9] and [10] have proposed and verified the SST functions and application for wind generation system integration. Meanwhile, the DC Microgrid system which enabled by SST is discussed in [11]; however, the dynamics and controls of wind renewables are not covered in such system. Normally, most AC/DC Microgrid systems that consider wind DRER have to rely on energy storage devices such as batteries to maintain system power balance or islanding mode operation [6], [13], while few papers report on controls and transient dynamics at the extreme condition like islanding or grid and islanding transitions without energy storage devices. In this paper, we consider a DC network consisting of PMSG wind turbines, SST, and critical and noncritical DC loads, of which the noncritical loads might be cut out depending on current system operation status. The motivation and contribution of this paper is to investigate how the wind turbines function during the transient conditions and how such system could manage power balance under various scenarios without any energy storage devices. Assuming that the wind power is beyond the critical loads demand, the foremost objective of such system is to ensure that the critical loads are prioritized and uninterrupted supplied at given conditions. In addition, the utilization of wind renewable resources are maximized under given condition. Moreover, a stable and balanced system operation is the prerequisite for any cases. This paper is organized as follows: Section II illustrates the system configuration, and Section III describes the proposed control methods. Case studies and simulation results under various scenarios are presented in Section IV for validation, and finally, Section V gives the conclusion.

II. System Configuration

Fig. 1 shows the overall system configuration. The investigated wind turbines might be equipped with a gear box or a direct drivetrain depending on the condition, which is marked with dash line for explanation. The three-phase voltage source rectifier (VSR) is used to connect the wind generator to the DC network. The SST which consists of dual active bridges (DAB) and voltage source inverter (VSI) with the former functioning as the isolated step-up transformer and the latter serving as the interface with the medium voltage at the point of interconnection. Both stages feature bidirectional power flow. Anticipated DC loads are prioritized into critical and noncritical parts. Depending on the actual system configuration and real application, multiple PMSG wind turbines might be added to this DC network shown as in Fig. 1. Extra generations could increase the system flexibility as well as the redundancy. However, the operation and function principles will largely be the same. Therefore, the study here only considers a paradigmatic wind turbine with all the required functionalities interfacing the SST.

This work was supported by the National Science Foundation under Award Number EEC-0812121.

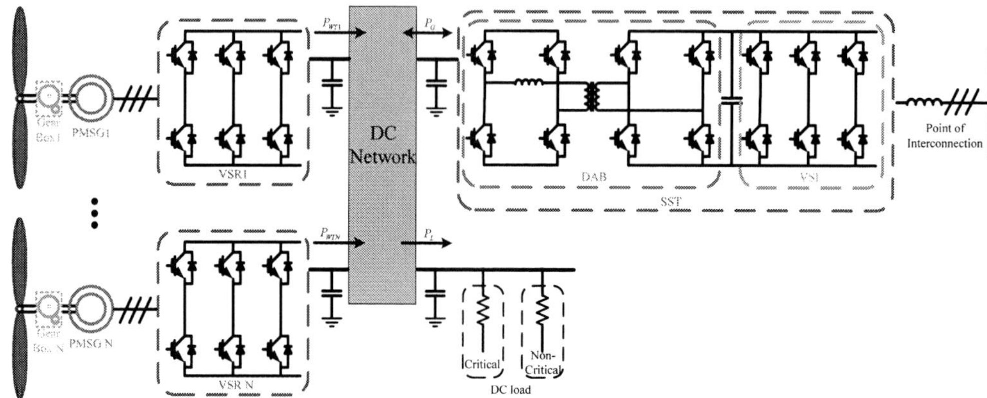

Fig. 1. Configuration of SST interfaced PMSG wind energy conversion system.

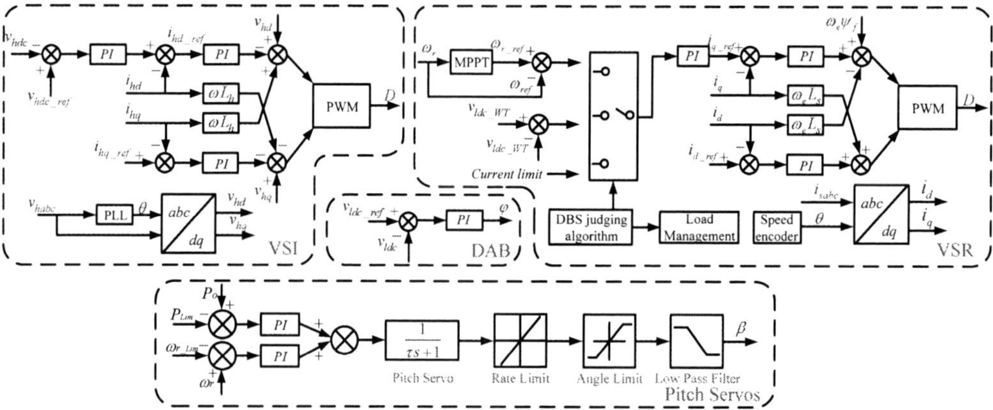

Fig. 2. System control block diagram.

III. System Operation and Control Principle

The system overall control design is complicated with this architecture for two reasons; first, unlike normal DC Microgrid system, maintaining a balanced power flow in islanding mode without energy storage devices is challenging, and second, it is inherently difficult to achieve a seamless transition between the two modes without knowing current load condition including full loads and critical loads without any communication link. To solve these two issues, an autonomous power management strategy based on improved DC bus signaling (DBS) is proposed to maintain the power balance and stable system operation under grid-connected mode, islanding mode as well as during the mode transition.

The system control block diagram is shown in Fig. 2. The VSI stage is used to regulate the high voltage DC bus and to provide the reactive power support under ac grid-connected mode. The DAB functions as isolated step-up transformer and regulates the DC network bus during grid-connected. In the VSR control, three options might apply to the I_{q_ref} depending on the system needs, and the control signal is generated by the DBS rule base. The basic idea of DBS is scheduling various distributed resources according to different voltage levels. This control strategy enjoys the inherent reliability by directly using DC bus itself as the communication link [14], [15]. In this paper, depending on the voltage threshold scheduled for different operation mode, the reference applied to I_{q_ref} varies. Specifically, under grid-connected mode, rotational speed ω_r is the control variable to implement maximum power point tracking (MPPT) since any power surplus and deficit within the DC network is automatically balanced by the SST. In scenario when the SST is disconnected from the ac grid and the dc network transits into islanding, the control objective will switch from MPPT to DC bus maintenance. This is especially critical for a multiple wind turbine system where a master VSR is used to maintain the DC bus while the other wind turbines might either work under MPPT or power curtail condition depending on current load demand. Current limit might trigger when the system encounters overload situation even with only critical load, under which the DC bus has to collapse. This condition has been assumed not to occur in this research as mentioned in Section I, and thus will not be detailed addressed further. On the other hand, the I_{d_ref} is set to 0 to achieve high efficiency operation considering that a surface-mounted PM generator is used. Meanwhile, a pitch actuator is also considered for the studied wind turbine system, which can activate to curtail extra wind power. Turbine pitch angle is the other key factor that influences wind power extraction besides the rotor speed, and it is also critical to utilize the pitch actuators to protect wind turbine system from severe weather conditions and unbalanced power flow [9]. As shown in Fig. 2 pitch servo control part, both input power and rotor velocity limitation have been taken into consideration in this paper, and specifically, the power limit is 1.1 p.u. and speed limit is 1.2 p.u.. The base values are given in Table I.

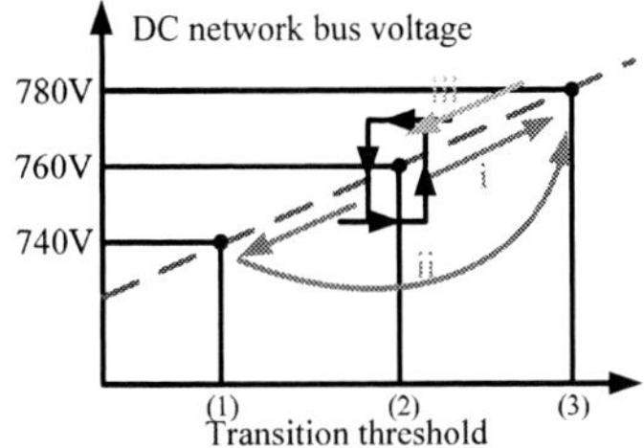

Fig. 3. DBS algorithm illustration.

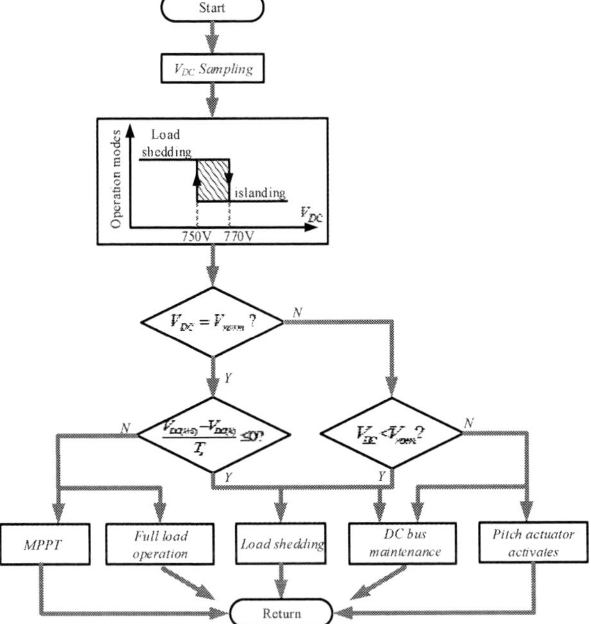

Fig. 4. DBS and load management flowchart.

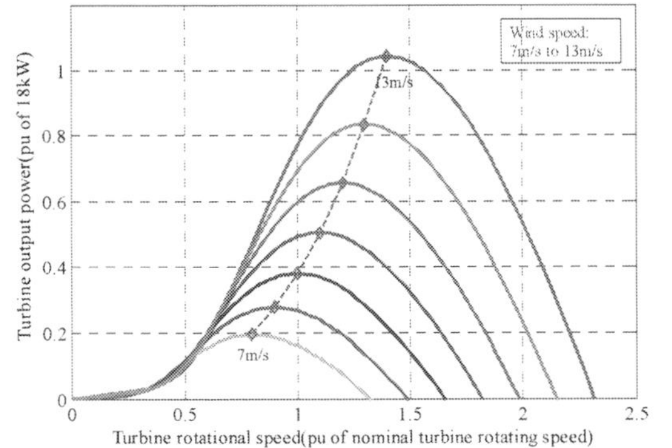

Fig. 5. Turbine power-rotational speed curve.

Fig. 3 illustrates the improved DBS rule based algorithm under various process, while Fig. 4 shows the flowchart for control implementation. The nominal DC bus voltage is set at V_{norm} =760V and a 10V hysteresis band is adopted for noise and ripple window. 20V difference is used to distinguish between different operation thresholds; specifically, 780V indicates the system is in islanding mode and VSR is regulating the DC bus voltage; 740V is the tripping indictor of a overload unbalanced situation. This algorithm adopts V_{norm} as the grid-connected indication, while 780V is the stable operation point for islanding mode. In Fig. 3, process (i) corresponds to grid-connected to islanding transition. When the system samples the DC bus voltage level has reached the threshold of 780V, the control reference will switch from MPPT to DC bus maintenance. If the bus voltage continues to increase, the pitch control will be activated to curtail input wind power, so that a new stable operation point is reached. The difference between process (i) and (ii) is judged from the current load situation; specifically, a heavier unbalanced load occurs and causes the DC bus level to decrease till it reaches the 740V threshold during process (ii) which will trigger the noncritical load cutting out. Then the DC bus level tends to settle down at 780V since the VSR has already taken over the control of the bus voltage. It is worthwhile pointing out that as the DC bus level passes V_{norm}, the system might assumedly switch into grid-connected mode, which will mistakenly lead to cutting in the noncritical loads. This could cause a severe power oscillation if inappropriately managed. To solve this problem, this paper proposes an improved DBS method which takes the DC bus voltage slope information into account. Besides the voltage level scheduling, the signal of its changing slope has to be utilized to judge whether the system recovers to grid-connected mode or it is still in the curve to settle down at islanding mode. The positive gradient is intentionally neglected in the algorithm, and only the negative gradient caused by SST re-connection will enable a full load online to maximize the energy utilization. This method can efficiently avoid the power oscillation and easy to implement. Process (iii) describes the system transitionings from islanding mode back to grid-connected mode. When the V_{norm} has been reliably measured, SST would take over the regulation of DC bus voltage and wind turbine switches into MPPT mode.

IV. CASE STUDIES

Proposed autonomous control under various scenarios are simulated and verified using MALAB/Simulink. Table I lists the system parameters of the simulated wind turbine, PMSG, and converters. 7.2 kW critical load and 4 kW noncritical load is used. Table II lists the system operations under different cases. Only the mode or function changes are recorded and shown in the table for simplicity, and the blank spaces indicate the same operation following above status.

1) Case I: This case study corresponds to AC grid connection via the SST; full load operation is allowed when there is enough power available from either wind or grid side. The wind turbine works at MPPT, which means the rotor velocity ω_r will be the control variable. With initial wind speed 10 m/s, the PMSG operates at 168.9 rad/s, which is 1.11 p.u. At 2 s, V_W starts to increase, which causes captured wind power

to go beyond the setting limitation point 1.1 p.u at 2.5 s, and the pitch actuator is thus activated to curtail undesired wind power as shown in Fig. 6 (c). At 4 s, the noncritical load is switched in, and VSI controlled at unity power factor, switches from regeneration to generation. This is verified from Fig. 6 (f). Figs. 6 (e) and (g) prove the robustness of regulated DC bus voltage under load variation dynamics. At 4.5 s, wind speed ramps down and settles at 7 m/s with the operating point at 118.26 rad/s, which is 0.77 p.u.. The MPPT operation is verified with the studied wind turbine characteristics shown in Fig. 5. MPPT control and details are not illustrated since this is not the focus of this paper.

2) Case II: This case describes the system mode transition from grid-connected to islanding mode. At 2 s, SST is disabled. The unbalanced power flow causes the DC link voltage level to keep increasing, and the VSR automatically switches to regulate the DC bus voltage when scheduled voltage threshold has been reached. During this transient, the rotor velocity will naturally ramp up to absorb certain amount of extra power. It is worthwhile to note that for a large wind turbine, this inertial stored energy can be beneficial for balancing power flow or supporting AC grid for frequency regulation. But an appropriate speed limitation is needed. In this paper, once it reaches the preset speed limit of 1.2 p.u., the pitch actuators would start functioning to cut off extra power by increasing pitch angle. Fig. 7 (a) shows the mimicked wind profile where some noise component has been intentionally inserted. Figs. 7 (c) and (d) demonstrate the wind power curtailment caused by the pitch actuator and the shifting from the optimum operation point. Figs. 7 (e) and (f) show the DC voltage performance under such transient. This mechanism can guarantee a secure and smooth mode transition without knowing load situation and energy storage devices.

3) Case III: This case verifies system islanding to grid-connection transition. The condition is set at 10m/s wind speed under full load operation. The simulation results are shown in Fig. 8. At 2 s, the SST is disabled. Captured wind power is insufficient for the current load demand, which is indicated by the DC bus voltage drop. The VSR would switch into DC bus regulation automatically after the judgement. Meanwhile, the noncritical load is tripped off and pitch servo is activated. After this transient, system could quickly recover to normal islanding operation status, as seen from the DC bus voltage performance. At 3 s, the SST comes online and after the transient, wind turbine will transit to MPPT mode from DC bus maintenance and deactivate the power curtailment function. Meanwhile, the load managing algorithm enables the noncritical load. Figs. 8 (d), (e) and (f) demonstrate the dc bus response under such situation.

TABLE I. SYSTEM PARAMETERS

Wind turbine		**PMSG**		**Converters**	
Air density	1.225 kg/m^3	No. of poles	5	VSI f_{sw}	1.2 kHz
Blade radius	3.5m	Rated speed	152.8 rad/s	DAB f_{sw}	3.6 kHz
V_w range	3-15 m/s	Rated power	10 kW	DAB transformer ratio	15:1
Gear ratio	1:7.3	Stator resistance	0.425 ohm	VSR f_{sw}	10 kHz
Rated power	18 kW	Flux linkage	0.3 Wb	V_{hac}	12 kV
V_{dc} sample rate	20 kHz	Stator inductance	8.5 mH	I_{hac} base	2A

TABLE II. OPERATION EVENTS FOR VARIOUS CASES

Case	Events	Time (s)	SST	VSR	Pitch actuator	Critical load	Noncritical load
I	V_w=10m/s	0	Enabled	ω_r control	Deactivate	Online	Offline
	V_w increases from 10 m/s to 11 m/s	2					
		2.5			Activate		
	Noncritical load switched in	4					Online
	V_w decreases from 11 m/s to 7 m/s	4.5			Deactivate		
II	V_w =10m/s with noise	0	Enabled	ω_r control	deactivate	Online	Offline
		2	Disabled	DC bus regulation	Activate		
III	V_w =10m/s without noise	0	Enabled	ω_r control	Deactivate	Online	Online
		2	Disabled	DC bus regulation	Activate		Offline
		3	Enabled	ω_r control	Deactivate		Online

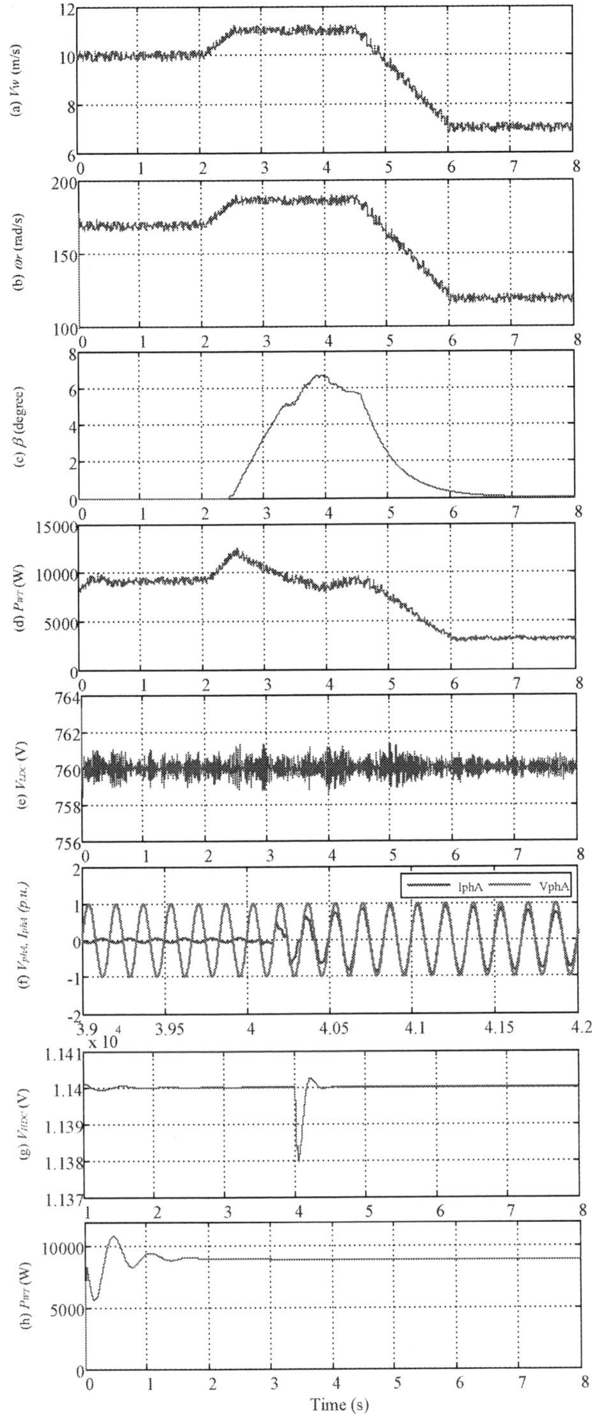

Fig. 6. Case I system operation. (a) Wind Speed profile. (b) Generator speed. (c) Pitch angle. (d) Turbine output power. (e) DC bus voltage. (f) Grid voltage and current. (g) High voltage dc bus. (h) 10m/s stable wind turbine output power.

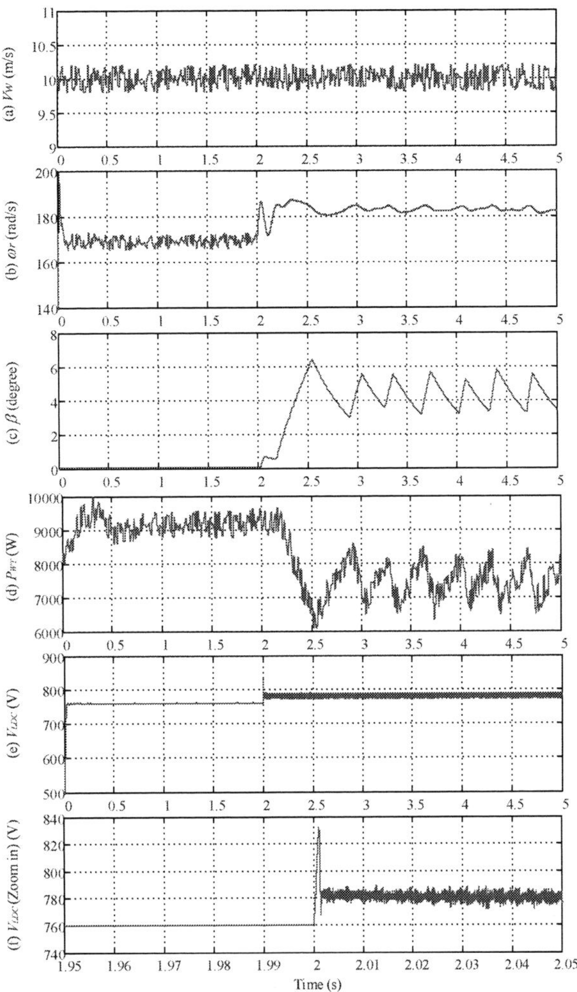

Fig. 7. Case II system operation. (a) Wind Speed profile. (b) Generator speed. (c) Pitch angle. (d) Turbine output power. (e) DC bus voltage. (f) Zoom in DC bus voltage.

V. Conclusion

Energy storage devices have been widely accepted and adopted in the DC/AC Microgrid system with the tradeoff of additional cost, control complexity as well as maintenance issues.Targeted at the SST interfaced wind generation system, an autonomous power management strategy has been proposed and verified. Without knowing the current load situation and any energy storage devices, such system has been justified of being able to achieve the power balance and stable operation under grid-connected mode and islanding mode. The challenging mode transition from both sides have been analyzed and verified with proposed autonomous control scheme. Simulation justification has been made to demonstrate the control feasibility for such wind DRER network system.

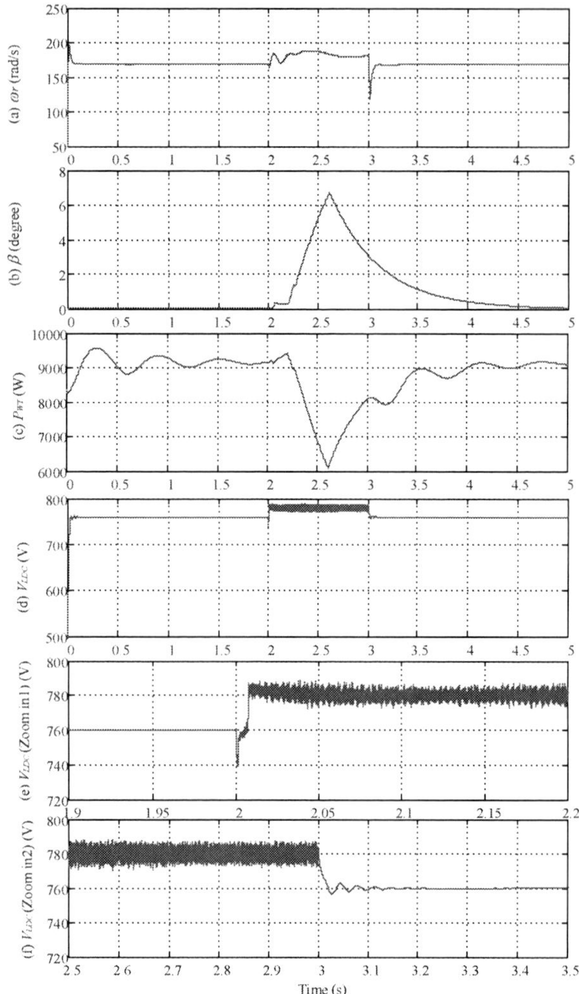

Fig. 8. Case III system operation. (a) Generator speed. (b) Pitch angle. (c) Turbine output power. (d) DC bus voltage. (e) Zoom in DC bus voltage (1). (f) Zoom in DC bus voltage (2).

REFERENCES

[1] F. Blaabjerg, K. Ma, "Future on Power Electronics for Wind Turbine Systems," IEEE Journal of Emerging and Selected Topics in Power Electronics, vol. 1, no. 3, pp. 139-152, Sep. 2013.

[2] Z. Chen, J.M. Guerrero, F. Blaabjerg, "A Review of the State of the Art of Power Electronics for Wind Turbines," IEEE Transactions on Power Electronics, vol.24, no.8, pp.1859-1875, Aug. 2009.

[3] M. Liserre, R. Cardenas, M. Molinas, and J. Rodriguez, "Overview of Multi-MW Wind Turbines and Wind Parks," IEEE Transactions on Industrial Electronics, vol. 58, no.4, pp. 1081-1095, Apr. 2011.

[4] F. Blaabjerg, Z. Chen, and S. B. Kjaer, "Power electronics as efficient interface in dispersed power generation system," IEEE Transactions on Power Electronics, vol. 19, no. 5, pp. 1184–1194, Sep. 2004.

[5] F. Blaabjerg, R. Teodorescu, M. Liserre, and A. V. Timbus, "Overview of control and grid synchronization for distributed power generation systems," IEEE Transactions on Industrial Electronics., vol. 53, no. 5, pp.1398–1409, Oct. 2006.

[6] L. Xu and D. Chen, "Control and operation of a dc microgrid with variable generation and energy storage," IEEE Transactions on Power Delivery, vol. 26, no. 4, pp. 2513–2522, Oct. 2011

[7] X. Yuan, "a set of multilevel modular medium-voltage high power converters for 10-MM wind turbines," IEEE Transactions on Sustainable Energy, vol. 5, no. 2, pp. 524-534, Apr. 2014.

[8] X. Yuan, J Chai, and Y. Li, "A Transformer-Less High-Power Converter for Large Permanent Magnet Wind Generator Systems," IEEE Transactions on Sustainable Energy, vol. 3, no. 3, pp. 318-329, Jul. 2012.

[9] R. Gao, I. Husain, F. Wang, and A.Q. Huang, "Solid-State Transformer Interfaced PMSG Wind Energy Conversion System," in IEEE Applied Power Electronics Conference, 2015, pp. 1310-1317.

[10] X. She, A.Q. Huang, F. Wang, and R. Burgos, "Wind Energy System with Integrated Functions of Active Power Transfer, Reactive Power Compensation, and Voltage Conversion," IEEE Transactions on Industrial Electronics, vol. 60, no. 10, pp. 4512-4524, Oct. 2013.

[11] X. Yu, X. She, X. Ni, and A.Q. Huang, "System Integration and Hierarchical Power Management Strategy for a Solid-State Transformer Interfaced Microgrid System," IEEE Transactions on Power Electronics, vol. 29, no. 8, pp. 4414-4425, Aug. 2014.

[12] H.S. Krishnamoorthy, D. Rana, P.Garg, P.N. Enjeti, and I.J. Pitel, "Wind Turbine Generator–Battery Energy Storage Utility Interface Converter Topology With Medium-Frequency Transformer Link," IEEE Transactions on Power Electronics, vol. 29, no. 8, pp. 4146-4155, Aug. 2014.

[13] S. Lu, L. Wang, T. Lo, and A. Prokhorov, "Integration of Wind Power and Wave Power Generation Systems Using a DC Microgrid," IEEE Transactions on Industry Applications, vol. 51, no. 4, pp. 2753-2761, Jul./Aug. 2015.

[14] J. Schonberger, R. Duke, and S. D. Round, "DC bus signaling: A distributed control strategy for a hybrid renewable nanogrid," IEEE Transactions on Industrial Electronics, vol. 53, no. 5, pp. 1453–1460, Oct. 2006.

[15] J. Bryan, R. Duke, and S. Round, "Decentralized generator scheduling in a nanogrid using DC bus signaling," in Proc. IEEE Power Engineering Society Summer Meeting, Jun. 2004, vol. 1, pp. 977–982.

An Isolated Buck-Boost Type High-frequency Link Photovoltaic Microinverter

Shiladri Chakraborty, *Student Member, IEEE* and Souvik Chattopadhyay, *Member, IEEE*
Department of Electrical Engineering
Indian Institute of Technology Kharagpur
Kharagpur - 721302, India
Email: shiladri@ee.iitkgp.ernet.in

Abstract—**This paper introduces a new topology and the associated control method of an isolated, high-frequency ac link inverter for photovoltaic applications. The circuit, which uses a buck-boost interface on the dc side and a half-bridge cycloconverter on the ac side, works on the principle of controlling the amount of charge delivered to the output over a switching period. The proposed circuit has advantages of low device count, soft-switching operation of switches and absence of reactive power flow. Circuit operation and details of the charge control algorithm are discussed followed by some relevant simulation and experimental results, that validate the approach.**

I. Introduction

In recent years, microinverters have become an increasingly attractive choice for small-scale residential and commercial photovoltaic (PV) installations. Compared to their central or string inverter counterparts, microinverters offer the advantages of modular operation, higher reliability, higher harvested energy (on account of module level maximum power extraction) and reduced safety hazard (due to absence of high voltage dc cabling) [1].

An extensive review of recently proposed single-phase PV microinverter architectures is presented in [2]- [3]. Broadly speaking, isolated single-phase topologies can be classified as two-stage or single-stage. Two-stage topologies comprise a front-end isolated dc-dc converter, responsible for maximum power point tracking (MPPT) and voltage boosting, followed by an inverter stage for synthesizing the line-frequency output. Such topologies have disadvantages of higher device count and lower efficiency (owing to cascaded power processing).

Single-stage topologies, on the other hand, realize the tasks of MPPT, voltage step-up and output shaping through a single power stage. Flyback-based solutions [4], which are a common form of such topologies, have advantages of low device count and simple control but cannot achieve high performance over a wide operating range. High-frequency ac link based solutions, such as those reported in [5]- [7], represent another category of single-stage topologies. Their basic architecture consists of a high-frequency resonant inverter, a high-frequency transformer, and a cycloconverter (with four-quadrant switches). Operationally, these circuits function like the series resonant dc-ac dual active bridge (DAB) converter [8]. Thus despite the advantage of inherent zero-voltage-switching (ZVS) operation, they suffer from the well-known problem of high reactive power present in any DAB-based architecture [9]. Moreover, the full-bridge version of these converters require twelve devices, which is the same as that needed in a typical two-stage solution.

In this paper, a new circuit topology of a non-DAB based, isolated, high-frequency ac link PV microinverter is proposed. The proposed topology uses only eight devices, only half of which switch at high frequency. Moreover, all devices are capacitively clamped, which obviates the need of auxiliary voltage clamping circuits. The remainder of this paper is organized as follows. Section II introduces the topology of the proposed converter and describes its different modes of operation. Details of the overall control structure, including the charge-mode control strategy adopted for synthesizing the desired sinusoidal output profile, are discussed in section III. Finally numerical simulation and experimental results are presented in sections IV and V respectively to validate the analysis.

II. Proposed Topology

The proposed topology, shown in Fig. 1, consists of a high-frequency transformer which is connected on its primary side to an integrated boost half-bridge converter [10], preceded by a half-bridge leg. Thus, the circuit topology on this side resembles a cascaded buck-boost architecture. On the secondary side, the transformer is connected to a half-bridge cycloconverter. The leakage inductor L_{lk}, though not a bulk-energy storage element as in a DAB converter, is necessary to prevent shorting of the primary and secondary side capacitors and is also needed for achieving ZVS of S_3 and S_4. The transformer's inherent leakage inductance is utilised to realize L_{lk}, instead of using an external inductor (as is often the case with DAB converters).

One of the key ideas behind the circuit operation is to keep the current (i_L) in the inductor L constant by adjusting the duty-ratio of the front-end buck converter. The other core principle is to use this current to control the amount of charge delivered to the output (over a switching period T_s) and vary it sinusoidally over the ac line cycle. The following modes of high-frequency circuit operation (considered for the positive ac line cycle), illustrated in Fig. 2, help to explain how this can be achieved. Key waveforms corresponding to these modes are depicted in Fig. 3. For the entire positive ac line cycle, S_6 and S_8 are continuously on, while over the negative line cycle, S_5 and S_7 are kept on. For simplicity of analysis, it is assumed that on account of the current control exercised by $S_1 - S_2$ and a large enough value of the the boost inductor L, i_L can be approximated as a dc current source.

*1) **Mode I** $(0 - t_1)$:* At the beginning of the switching period, S_3 is turned on so that C_{p1} is switched across the transformer primary making $v_p = v_{Cp1}$ and increasing the

978-1-4673-9551-9/16 $31.00 © 2016 IEEE

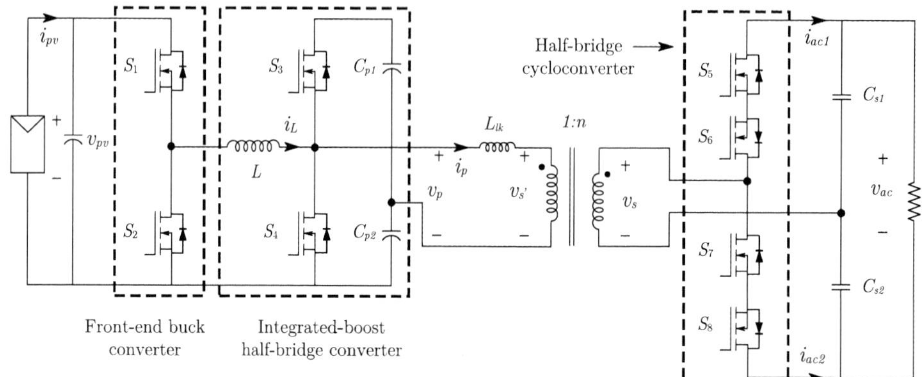

Fig. 1: Circuit topology of the proposed high-frequency link microinverter.

primary current i_p (from zero). The positive value of i_p means that on the secondary side, current flows through (channel of) S_6 and the body-diode of S_5, while no current flows through the bottom devices (S_7,S_8) since S_7 is off. Thus, $i_{ac1} = i_p/n$, $i_{ac2} = 0$ and the reflected voltage v_s' on the primary is approximately equal to $v_{ac}/2n$. At the end of this mode, S_3 is turned off. Subject to the necessary condition that $i_p(t_1) > i_L$, the net switch-node current $i_p(t_1) - i_L$ starts discharging and charging the output capacitors of S_4 and S_3 respectively.

2) ***Mode II*** *(t_1 − t_2):* Once the output capacitor of S_4 is fully discharged and body-diode conduction commences, it can be turned on under ZVS. As S_4 turns on and C_{p2} switches across the primary, a negative voltage ($-v_{Cp2} - v_{ac}/2n$) appears across the leakage inductor and i_p quickly comes down to zero. Thus, this mode essentially represents the commutation period due to the leakage inductance. It is evident, that the duration of this period and hence the commutation effect is more pronounced near the zero crossing of the ac line cycle, since the rate of fall of i_p is smallest there.

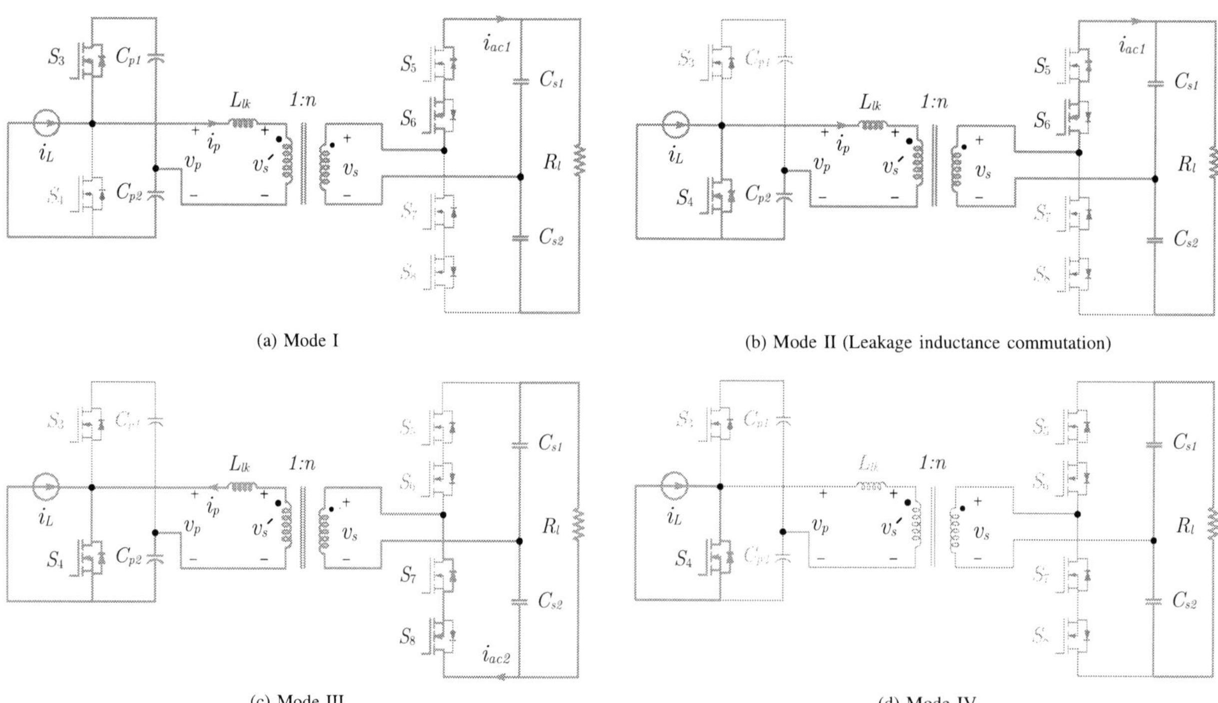

(a) Mode I

(b) Mode II (Leakage inductance commutation)

(c) Mode III

(d) Mode IV

Fig. 2: Modes of operation over a high-frequency switching time period for the positive ac line cycle. (The boost inductor L can be modeled as a constant current source.)

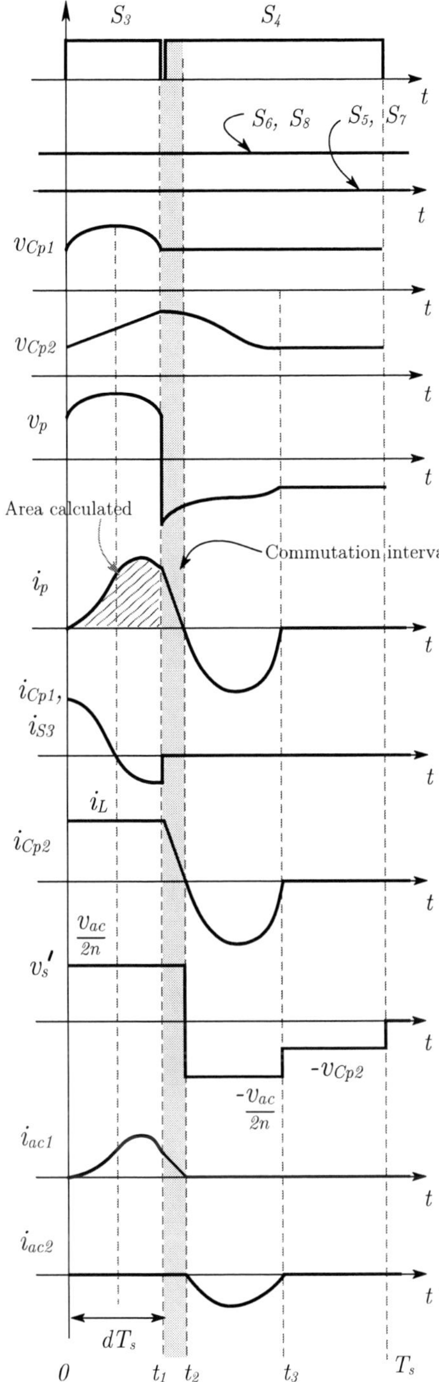

Fig. 3: Key waveforms over a high-frequency switching time period for the positive ac line cycle.

*3) **Mode III** (t_2-t_3):* As i_p becomes negative, (channel of) S_8 and body-diode of S_7 conduct. Thus $i_{ac1} = 0$, $i_{ac2} = i_p/n$ and $v_s' = v_{ac}/2n$. This mode ends when i_p reaches zero.

*4) **Mode IV** (t_3-T_s):* After reaching zero, i_p cannot become positive, since this would mean that S_6 and diode of S_5 would conduct and the reflected voltage v_s' would become $v_{ac}/2n$, which would again bring down i_p. Thus i_p stays at zero. The boost inductor current i_L discharges the output capacitor of S_3, which can be turned on with ZVS at the end of this mode.

III. Control Algorithm

A. Basic control philosophy

The overall control strategy is summarised in the flowchart of Fig. 4. The basic control objective of rendering a line-frequency sinusoidal profile to the output side current (i_{ac1}, i_{ac2}) is equivalent to delivering a controlled amount of charge (to be varied sinusoidally) at the output side over a fixed switching period T_s.

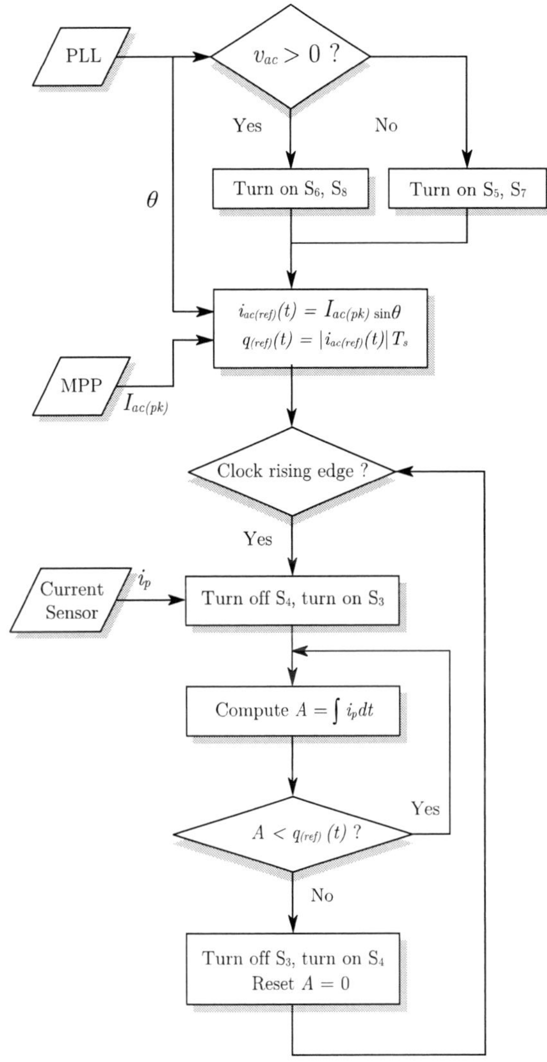

Fig. 4: Flowchart explaining the control algorithm.

Now, as discussed in the previous section, for the positive ac line cycle,

$$i_{ac1} = \begin{cases} i_p/n, & S_3 = 1 \\ 0, & S_3 = 0 \,. \end{cases}$$

Thus, it is possible to control the amount of charge delivered at the upper ac rail during the positive line cycle by controlling the amount of charge delivered through the transformer when S_3 is on. By a similar logic, it can be concluded that during the negative line cycle, the charge delivered at the lower ac rail can be controlled (for this S_5, S_7 are to be kept on) by controlling S_3. Note that, during both the half-line cycles, positive charge is to be delivered through the transformer when S_3 is on. Thus the objective of sinusoidal shaping of the output current over the entire line cycle ultimately translates to the following condition

$$\int_0^{dT_s} i_p \, dt = \left| i_{ac(ref)(t)} \right| T_s = I_{ac(pk)} \left| \sin \omega t \right| T_s \,. \qquad (1)$$

The above equation is used to determine the duty-ratio of the switch S_3 under closed-loop condition. S_3 is turned on at the beginning of each switching period and the current i_p is continuously integrated (reset integration) till the computed value equals the rectified sinusoidal profile reference value. When this happens, S_3 is turned off and S_4 is turned on.

The detailed control block diagram of the system is shown in Fig. 5. Apart from the cycloconverter control and charge-mode control blocks responsible for output voltage/current shaping, the peak current-mode control structure (incorporating the usual stabilising ramp [11]) of the front-end half-bridge is also depicted.

B. Compensation of leakage inductance commutation effect

As discussed in the previous section, the transformer current takes a finite amount of time to come down to zero after S_3 is turned off. This finite leakage inductance commutation duration implies that in any switching period, more than the desired charge reference will be delivered to the output ac rails. Hence, in practice it is not possible to synthesize a perfect sinusoidal output profile, without compensating for this error. In the absence of any compensating mechanism, the output voltage/current will have distortions, particularly near the zero crossings, where the effect of leakage inductance commutation is most pronounced.

Two ways in which such compensation may be implemented are shown in Fig. 6. The first technique, shown in Fig. 6a involves modifying the charge reference of the $(k+1)^{th}$ switching cycle, by subtracting from it the additional charge $q_{err}[k]$ delivered in the kth cycle. The second method, shown in Fig. 6b uses an outer voltage/current (integral) controller to decide the cycle by cycle charge reference.

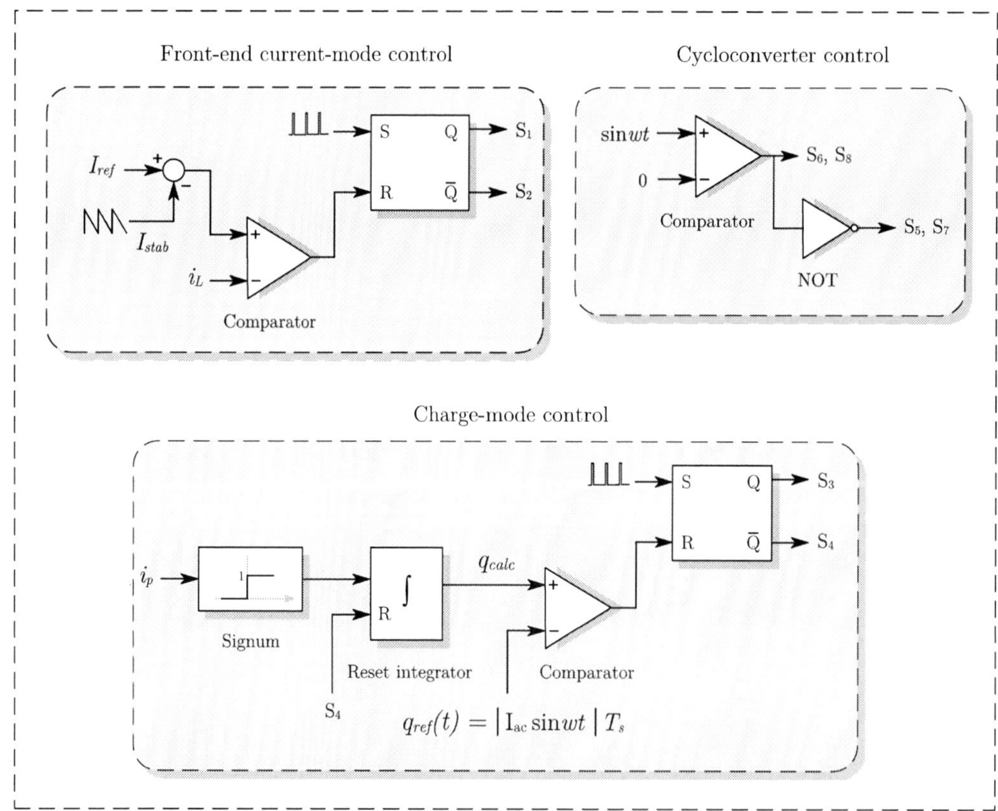

Fig. 5: Block diagram of the complete control architecture.

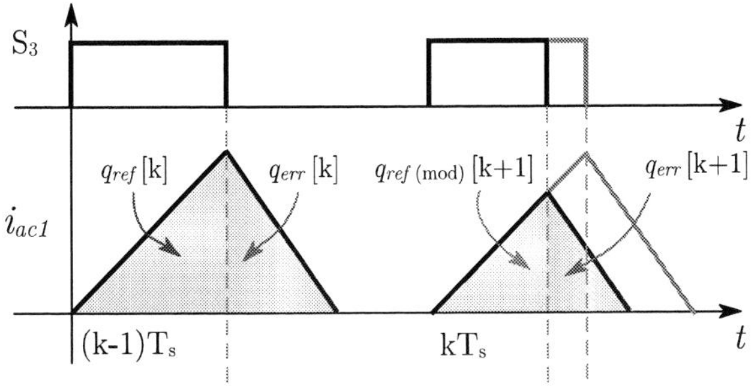

(a) Using cycle by cycle compensation.
$q_{ref}(\text{mod})[k+1] = q_{ref}[k] - q_{err}[k]$

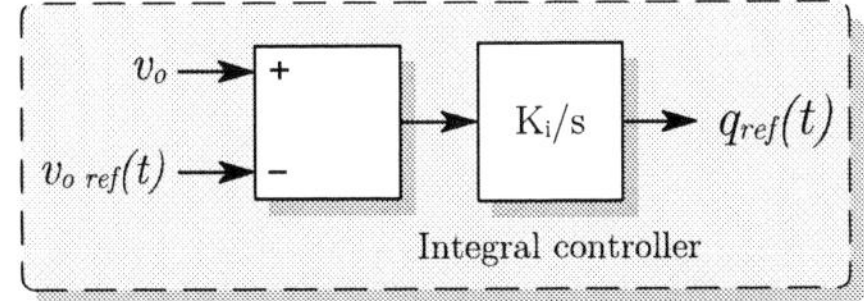

(b) Compensation using outer voltage/current controller.

Fig. 6: Two ways of compensating the charge reference tracking error arising due to finite leakage inductance commutation time.

This controller suitably adjusts the charge reference (which no longer has the open-loop rectified sinusoidal profile) to meet the control goal of tracking the sinusoidal reference.

IV. Simulation Results and Discussions

A model of the proposed converter working in standalone mode has been developed in the PLECS simulation platform. Relevant simulation waveforms for the parameter values listed in Tab. I are shown in Fig. 7-11.

TABLE I: Parameter values for the simulation model

V_g	f_s	f_{ac}	v_{ac}	L
48 V	50 KHz	50 Hz	230 V(rms)	200 μH

	C_{p1}, C_{p2}	C_{s1}, C_{s2}	$L_{lk}(pri)$	Turns-ratio (n)
	20 μF	2 μF	1 μH	1:4

Fig. 7a and 7a show the transformer voltage and current waveforms near the peak and zero crossing respectively of the ac line cycle. It can be seen that, as expected, the effect of leakage inductance commutation is much more pronounced near the zero crossing of the line cycle. Moreover, it can be observed that the transformer primary side voltage and current are in phase, indicating absence of reactive power flow in the circuit unlike a DAB-based solution. Gate pulses of the integrated half-bridge boost converter, the transformer current and the currents delivered at the output ac rails for the positive ac line cycle are shown in Fig. 8. The line-frequency gate pulses of the cycloconverter switches and the synthesized ac voltage waveform are shown in Fig. 9. Also, shown is the inductor current on which double-line-frequency oscillations can be observed. This low-frequency oscillations arise due to the effect of the stabilising ramp of the peak current-mode controller. A less steeper stabilising ramp signal would result in reduced magnitude of low-frequency ripple. From the current waveforms of the primary side switches, shown in Fig. 10 it can be seen that the current through S_2, S_3 and S_4 go in the negative direction as the switches begin conduction, implying that their respective parallel capacitors get discharged resulting in ZVS turn-on. Fig.11 illustrates how a modified switching pattern of the cycloconverter devices may be used to greatly minimize losses on the secondary-side.

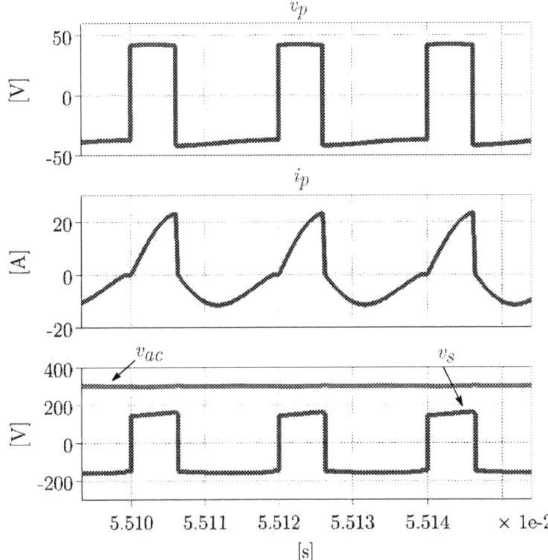

(a) Negligible leakage inductance commutation duration (near peak of line cycle)

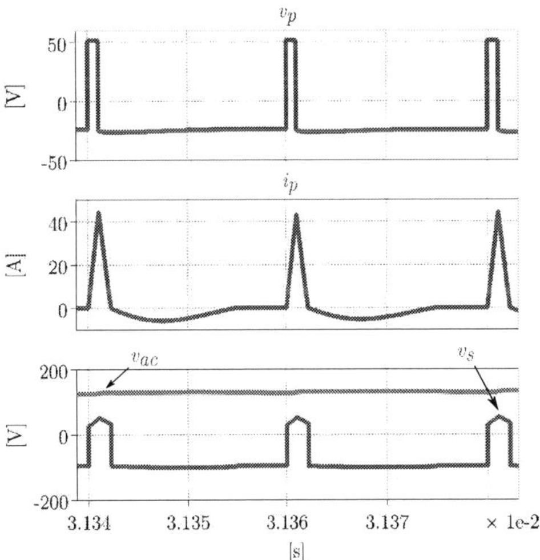

(b) Significant leakage inductance commutation duration (near zero of line cycle)

Fig. 7: Transformer primary voltage (v_p), primary current (i_p) and secondary voltage (v_s).

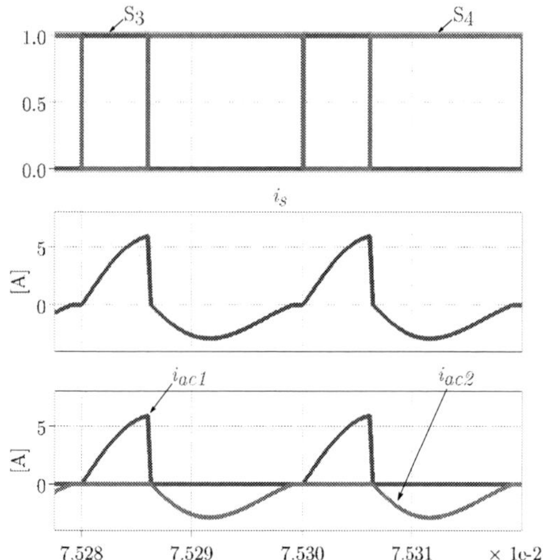

Fig. 8: Gate pulses of the integrated boost section, transformer secondary current (i_s) and the currents at the upper (i_{ac1}) and lower (i_{ac2}) ac rails.

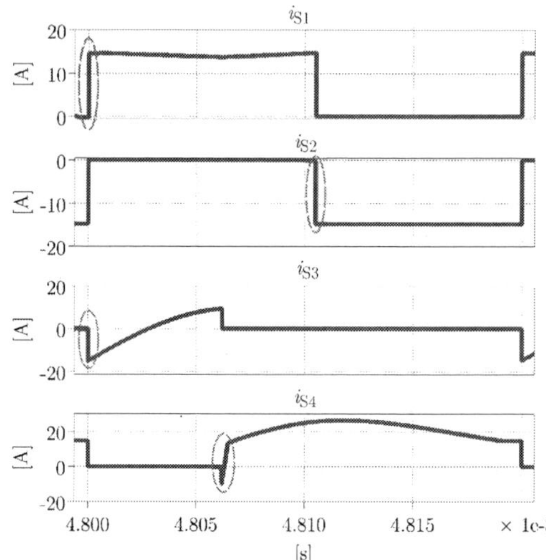

Fig. 10: Investigating switching transitions of the primary-side devices. Transitions encircled with green denote ZVS turn-on.

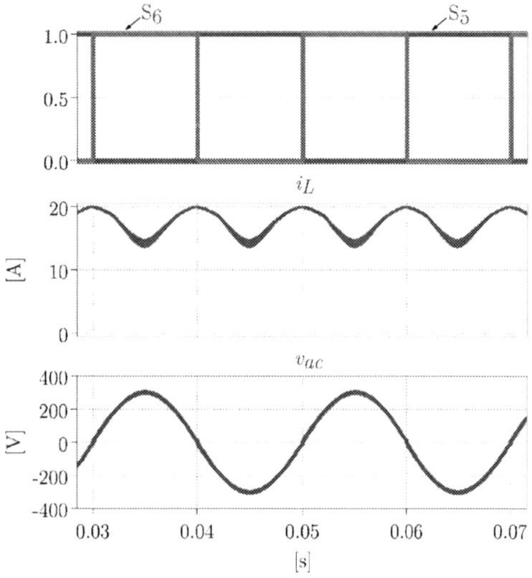

Fig. 9: Cycloconverter gate pulses, inductor current (i_L) and the output ac voltage (v_{ac}).

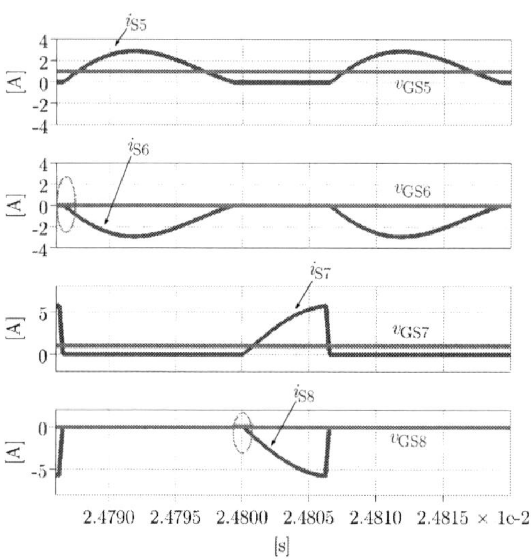

Fig. 11: Illustrating possibility of ZVS turn-on and synchronous rectification of the cycloconverter devices.

Since S_5 and S_7 are on for the entire half line-cycle, they entail no switching loss during this time. Also, since current naturally flows the body-diodes of S_6 and S_8 during this period, if high-frequency gate pulses are applied to these devices, they will turn-on with ZVS. Such lossless switching will also serve to reduce conduction losses due to diode drop.

V. Experimental Results

A 50 KHz, 30 V input, 110 V (rms), 2 A (rms) output proof-of-concept prototype of the converter has been built and tested in the standalone mode. Specifications of the components used in the prototype are listed in Tab. II. All power components and associated control circuitry including gate drivers and sensors are populated on a single dual-sided printed circuit board. Gate pulses are given to the converter using a Xilinx XC3S200 FPGA-based control card. Some waveforms of interest are shown in Fig. 12-16.

Fig. 12-13 depicts the low frequency waveforms including output voltage, current and cycloconverter gate pulses. From these, the double-line-frequency oscillations on the inductor current and the dc bus voltage (v_{Cp1}+v_{Cp2}) can also be observed. Low-frequency switching waveforms of the cycloconverter devices are shown in Fig. 14.

TABLE II: Components used in the hardware prototype

Component	Value	Specification/Part no.
S_1, S_2	100 V, 2.3 mΩ	CSD19536KCS
S_3, S_4	200 V, 8 mΩ	IRFP4668
L	190 μH, 20 A(rms)	-
C_{p1}, C_{p2}	20 μF, 500 V	MKP1848C62050JP4
L_{lk} (external)	1 μH	-
Turns-ratio	1 : 3	-
$S_5 - S_8$	500 V, 260 mΩ	FDA20N50F
C_{s1}, C_{s2}	2.2 μF, 850 V	B32656S8225K561

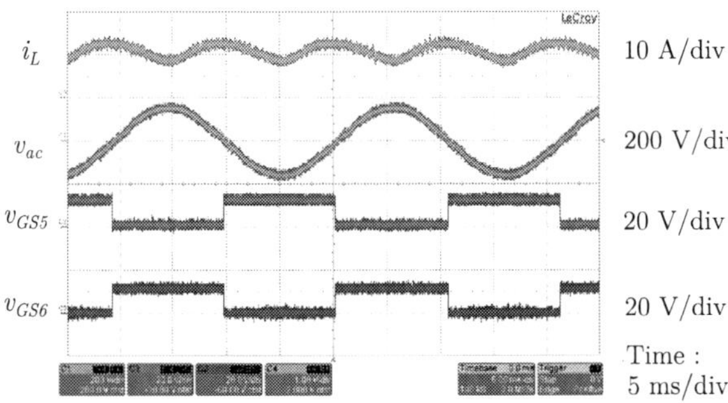

Fig. 12: Inductor current (i_L), output voltage (v_{ac}) and cycloconverter gate pulses.

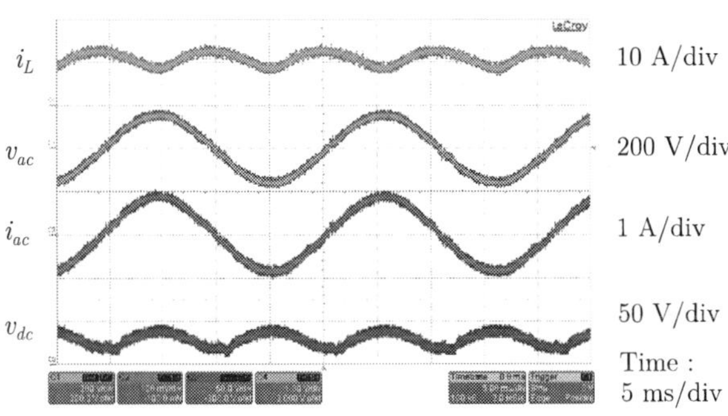

Fig. 13: Inductor current (i_L), output voltage, current(v_{ac}, i_{ac}) and primary-side dc bus voltage (v_{dc}).

From the high-frequency transformer waveforms shown in Fig. 15, the pronounced effect of leakage inductance commutation near the ac line cycle can be clearly noticed. Finally, ZVS operation of the $S_3 - S_4$ leg can be verified from Fig. 16. As can be observed from Fig. 16a, just prior to the turn-on of S_4, i_p being greater than i_L, the difference current $(ip - i_L)$ discharges its output capacitor resulting in ZVS. Similarly, S_3 always undergoes ZVS because prior to its turn-on, i_p becomes zero and i_L discharges its output capacitor. Though switching waveforms of $S_1 - S_2$ are not shown, it is evident that S_1 will be always hard-switched, whereas S_2 will always have ZVS turn-on, just as in a continuous conduction mode synchronous buck converter.

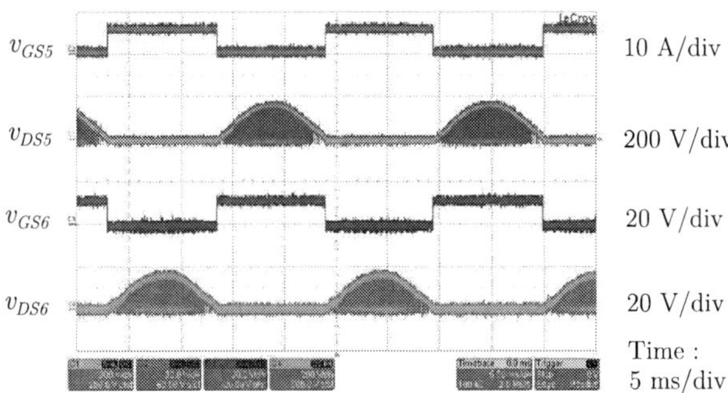

Fig. 14: Switching waveforms of the cycloconverter devices.

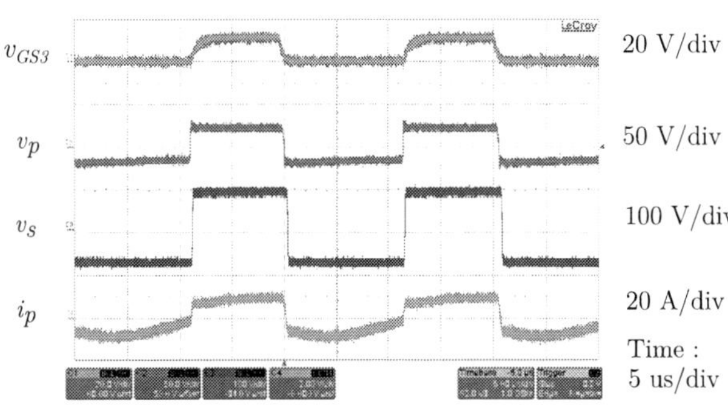

(a) Negligible leakage inductance commutation duration (near peak of line cycle)

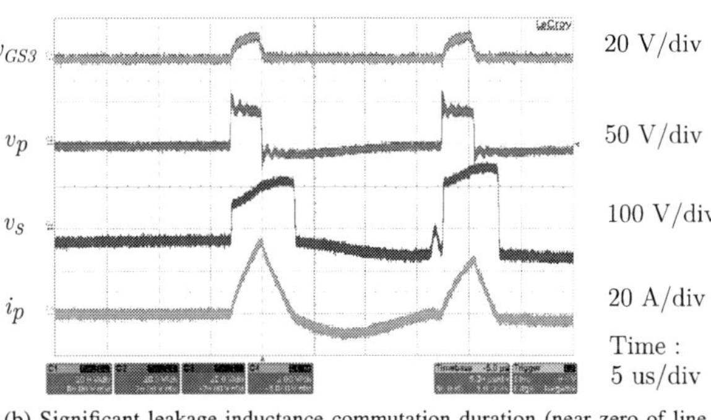

(b) Significant leakage inductance commutation duration (near zero of line cycle)

Fig. 15: Transformer voltage and current waveforms.

VI. Conclusion

In this paper, a new topology of a non-DAB-type, isolated, high-frequency ac link-based PV microinverter has been introduced. Details of a charge-mode control algorithm, used to incrementally calculate and control the amount of charge delivered to the ac output side over a switching period, are discussed. The great advantage of the proposed solution is the lack of circulating or reactive power flow as in classical DAB-based solutions, which reduces conduction losses on the primary side. Three of the four primary-side devices undergo ZVS operation, while by adopting a synchronous rectification based switching strategy, reduced conduction losses and ZVS operation of the secondary-side cycloconverter devices is possible.

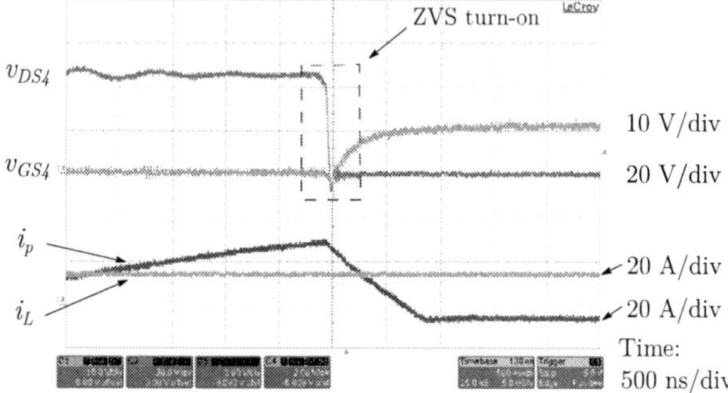

(a) ZVS of S_4. Note that i_p is greater than i_L at the turn-on instant.

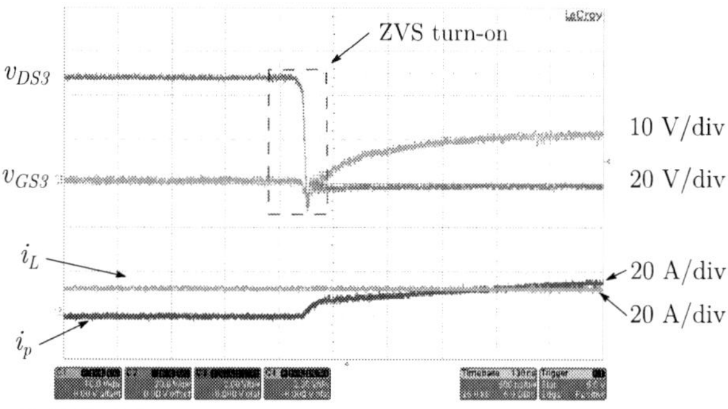

(b) ZVS of S_3. Note that i_L is greater than i_p (which is 0) at the turn-on instant.

Fig. 16: Illustrating ZVS turn-on of two of the primary side devices.

REFERENCES

[1] S. Harb, M. Kedia, H. Zhang and R. S. Balog, "Microinverter and string inverter grid-connected photovoltaic system A comprehensive study," *Photovoltaic Specialists Conference (PVSC) 2013*, pp. 2885-2890, June 2013.

[2] Soeren Baekhoej Kjaer, John K. Pedersen and Frede Blaabjerg, "A Review of Single-Phase Grid-Connected Inverters for Photovoltaic Modules," *IEEE Trans. on Ind. Applications*, vol 41(5), pp. 1292-1306, September-October 2005.

[3] Q. Li and P. Wolfs, "A review of the single phase photovoltaic module integrated converter topologies with three different DC link configurations,"*IEEE Trans. on Power Electron.*, vol 23(3), pp. 1320-1333, May 2008.

[4] H. Hu, S. Harb, N. H. Kutkut, Z. J. Shen and I. Batarseh, "A Single-Stage Microinverter Without Using Eletrolytic Capacitors," *IEEE Trans. Power Electron.*, 28(6), 2677-2687, June. 2013.

[5] A. Trubitsyn, J. Pierquet, A. K. Hayman, G. E. Gamache, C. R. Sullivan and D. J. Perreault , "High-Efficiency Inverter for Photovoltaic Applications," *IEEE Energy Conversion and Congress Exposition (ECCE) 2010*, pp. 2803-2810, Sep. 2010.

[6] H. Krishnaswami, "Photovoltaic Microinverter using Single-stage Isolated High-frequency link Series Resonant Topology," *IEEE Energy Conversion and Congress Exposition (ECCE) 2011*, pp. 495-500, Sep. 2011.

[7] M. Chen, K. K. Afridi and D. J. Perreault , "A Multilevel Energy Buffer and Voltage Modulator for Grid-Interfaced Micro-Inverters," *IEEE Energy Conversion and Congress Exposition (ECCE) 2012*, pp. 3070-3080, Sep. 2012.

[8] M. H. Kheraluwala, R.W. Gascoigne, D.M. Divan and E. D. Baumann, "Performance characterization of a high-power dual active bridge dc-to-dc converter," *IEEE Trans. Ind. Applications*, 28(6), pp. 1294-1301, Nov./Dec. 1992.

[9] Hua Bai and Chris Mi, "Eliminate Reactive Power and Increase System Efficiency of Isolated Bidirectional Dual-Active-Bridge DC-DC Converters Using Novel Dual-Phase-Shift Control," *IEEE Trans. Power Electron.*, 23(6), pp. 2905-2914, Nov. 2008.

[10] H. Li, F. Z. Peng and J. S. Lawler, "A Natural ZVS Medium-Power Bidirectional DC-DC Converter With Minimum Number of Devices," *IEEE Trans. Ind. Applications.*, 39(2), pp. 525-535, Mar. -Apr. 2003.

[11] R. W. Erickson and D. Maksimovic, " Fundamentals of Power Electronics," (2nd ed) Springer International Edition.

Energy management and stabilization of a hybrid DC microgrid for transportation applications

Mehdi Karbalaye Zadeh[1], Louis-Marie Saublet[3], Roghayeh Gavagsaz-Ghoachani[3],
Babak Nahid-Mobarakeh[3], Serge Pierfederici[3], and Marta Molinas[2]
[1]Department of Electric Power Engineering, Norwegian University of Science and Technology, Trondheim, Norway
[2]Department of Engineering Cybernetics, Norwegian University of Science and Technology, Trondheim, Norway
[3]GREEN Laboratory, University of Lorraine, Vandœuvre-lès-Nancy, France
Email: mehdi.zadeh@ntnu.no

***Abstract*— Power electronics-based AC/DC microgrids are attractive for clean energy applications, and are the enabling technology for hybrid energy systems in the more electric transportation. In such hybrid energy systems, storage units like battery and supercapacitor (SC) are used to store the surplus energy and to regulate the DC bus voltage. In on-board electric power systems, the converter controlled loads act as constant power loads (CPLs), imposing fast dynamics to the system. In this paper, a simple power control approach is proposed to provide the energy management and to ensure the stable operation of the system in the presence of CPLs. Furthermore, a laboratory prototype is built in order to validate the performance of the controller and the energy management strategy. The control system is implemented on dSPACE real-time cards. The simulated control scenarios are then experimented with the hardware prototype.**

Keywords—DC microgrid; power electronics; hybrid systems; energy management; transportation.

I. Introduction

DC microgrids have become attractive, since they facilitate integrating energy storage devices and modern electric loads [1-4]. Such distribution systems are introduced as optimized solutions, particularly for the next generation transportation systems like electric ships, more electric aircraft (MEA) and advanced automotive systems [5-10]. In this context, the power electronics converters play a key role in providing reliable, low emissions, and high-efficiency energy conversion for on-board power systems [11-13]. As an example of more electrification, mechanical actuators in aircrafts are replaced by converter-controlled electric drives, which are supplied by local DC distribution systems [14].

In such on-board energy systems, the main energy source is a mechanic-driven generator, as a unidirectional source. Hence, energy storage systems are used to store the surplus energy and to maintain the DC bus voltage [15-17]. The load converters transfer the DC bus voltage to tightly regulated power for the AC and DC loads. Different storage systems are used in the hybrid systems, which are mainly DC. Consequently, DC-DC converters are used to adapt the voltage level, and to control the power supplied by/stored in the storage unit.

Hybrid power converters are also merged with proton exchange membrane fuel cell (PEMFC) [18, 19]. In this topology, the dynamic of the electrical source is limited by a controlled converter to avoid an undesired phenomenon named oxygen starvation. It also cannot accept bidirectional operation [20]. More energy units like a battery or ultracapacitor are studied to regulate the load in such systems.

Batteries have a limited life cycles, slow transient response, and slow charging pattern. By contrast, supercapacitor (SC) banks are attractive for emerging hybrid energy systems, mainly because of the higher power density and lower time response in transient states, compared to the battery [21]. However, SC does not produce the electrical energy, and consequently, the energy density of SC is low, compared to the battery. Therefore, the SC-based energy management cannot deliver power in the extended transient loads. Energy-based controllers are studied to manage the transient load power, while regulating the DC bus voltage.

In this paper, a simple energy-based control strategy is developed for the hybrid electric transportation systems with SC bank as the main energy storage unit. The hybrid system is emulated using a laboratory hardware prototype comprised of a main energy source, a SC bank, and load converters. The main energy source is a balanced AC electric source and is controlled with a voltage source converter (VSC). The SC bank is controlled by a bidirectional DC-DC converter, and the load converters supply the loads with regulated power. A complete simulation model of the system in MATLAB is used to perform the control strategy. The proposed control strategy is then validated experimentally.

The rest of the paper is organized as follows: In Section II the studied system is presented; the energy-based control strategy and the inner-loop control of the converters are also presented. The simulation results of the system model are presented in Section III; while Section IV presents the laboratory test-bed and the experimental results.

978-1-4673-9551-9/16 $31.00 © 2016 IEEE

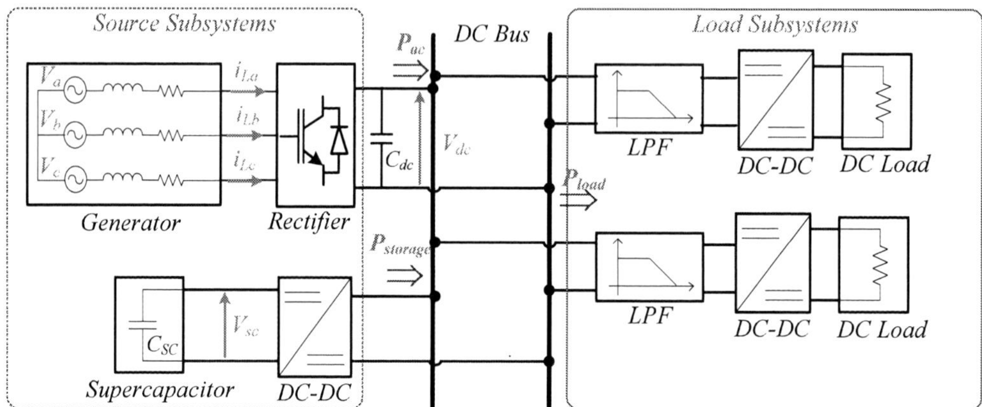

Fig. 1. Hybrid on-board power system with DC distribution architecture.

II. System Representation and Control

Fig. 1 indicates the scheme of the studied hybrid power system, comprising of: the main source, the storage system, and the CPLs. In this model, the AC source is provided by a three phase voltage source inverter. A boost rectifier converts the AC power to the DC voltage on the main distribution bus. A SC bank is used as the energy storage unit, and is controlled by a bidirectional boost DC-DC converter, as the SC converter. The storage system is controlled to regulate the DC bus voltage, and to provide the load power under transient loads and overloads. The load sub-system, comprised of DC-DC converters, absorbs controlled power. The load converters then regulate the load currents, as proportional to the variations of the filter voltages in a constant power P_{load}, behaving as constant power load (CPL).

A. Energy-based Control

The electrostatic energies stored in the DC bus capacitor (y_{dc}) and the SC (y_{sc}) are used as the control parameters. By using these control variables the reference values ($y_{dc_{ref}}$ and $y_{sc_{ref}}$) are defined by:

$$y_{dc} = \frac{1}{2} C_{dc} V_{dc}^2 \; ; \; y_{dc_{ref}} = \frac{1}{2} C_{dc} V_{dc_{ref}}^2 \; ; \qquad (1)$$
$$y_{sc} = \frac{1}{2} C_{sc} V_{sc}^2 \; ; \; y_{sc_{ref}} = \frac{1}{2} C_{sc} V_{sc_{ref}}^2$$

Then, using PI compensators, the control law for the DC bus and the SC can be written in the form:

$$\left(\dot{y}_{dc} - \dot{y}_{dcref}\right) + K_{p_{dc}}\left(y_{dc} - y_{dcref}\right) + K_{i_{dc}} \int \left(y_{dc} - y_{dcref}\right) = 0 \qquad (2)$$

$$\left(\dot{y}_{sc} - \dot{y}_{scref}\right) + K_{p_{sc}}\left(y_{sc} - y_{scref}\right) + K_{i_{sc}} \int \left(y_{sc} - y_{scref}\right) = 0 \qquad (3)$$

Using (2) and (3), the power control components for the SC converter $P_{storage_{ref}}$ and the rectifier $P_{ac_{ref}}$ are calculated in (4) and (5), respectively.

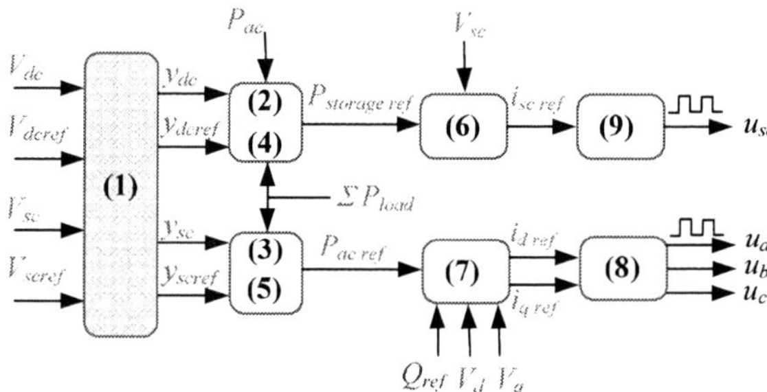

Fig. 2. Structure of the energy-based control strategy.

The energy based control strategy is summarized in the block diagram of Fig. 2.

$$P_{storage_{ref}} = \sum P_{load,k} - P_{ac} + \dot{y}_{dc_{ref}} - K_{p_{dc}}\left(y_{dc} - y_{dc_{ref}}\right) - K_{i_{dc}} \int \left(y_{dc} - y_{dc_{ref}}\right) \qquad (4)$$

$$P_{ac_{ref}} = \sum P_{load,k} + \dot{y}_{scref} + K_{p_{sc}}\left(y_{sc_{ref}} - y_{sc}\right) + K_{i_{sc}} \int \left(y_{sc_{ref}} - y_{sc}\right) \qquad (5)$$

B. Inner-loop Control

Using the reference powers, generated by the energy management unit, the reference current for the SC $i_{sc_{ref}}$ is calculated by the reference power $P_{storage_{ref}}$, as defined in (6).

$$i_{sc_{ref}} = \frac{2P_{max}}{V_{sc}}\left(1 - \sqrt{1 - \frac{P_{storage_{ref}}}{P_{max}}}\right) \qquad (6)$$

where: $P_{max} = \dfrac{V_{sc}^2}{4r_{sc}}$

The SC current i_{sc} is then regulated by an indirect sliding mode controller [22]. To this end, an active stabilization system is included in the current controller of SC converter, in order to damp the unstable oscillations, which are arising from the constant power dynamic of the load [23]. The duty cycle of the SC converter is generated as given in (9).

The reference currents of the rectifier result from the reference power $P_{ac_{ref}}$, and the current components are calculated in dq frame as follows:

$$i_{d_{ref}} = \frac{P_{ac_{ref}} * V_d + Q_{ref} * V_q}{{V_d}^2 + {V_q}^2} \qquad (7)$$
$$i_{q_{ref}} = \frac{P_{ac_{ref}} * V_q - Q_{ref} * V_d}{{V_d}^2 + {V_q}^2}$$

The current controller of the VSC is then implemented based on the decoupling control approach in dq frame, and uses a proportional integral (PI) compensator to regulate the current components i_d and i_q. The rectifier voltages in dq are calculated as in (8).

$$V_d^* = V_d + 2\xi\omega L(i_d - i_{dref}) + \omega^2 L \int i_d - i_{dref} + \omega_e L i_q \ ;$$
$$V_q^* = V_q + 2\xi\omega L(i_q - i_{qref}) + \omega^2 L \int i_q - i_{qref} - \omega_e L i_d \ ; \qquad (8)$$

The resulting voltage components generate the output voltage of the rectifier using a pulse width modulation in *ABC* frame.

III. Stability Analysis

The proposed control strategy is evaluated using a simulation model in MATLAB, which is constructed with the differential equations of the system model. Some parameters of the studied system, used in simulation and experiment, are given in Table I.

TABLE I. Parameters of The Studied Hybrid System

Parameter	Symbol	Value
Capacitance of the SC Bank	C_{sc}	82 F
Voltage of SC Bank	V_{sc}	50 V
DC Bus Capacitor Value	C_{dc}	1100 μF
DC Bus Voltage	V_{dc}	120 V
Generator RMS voltage	V_{rms}	40 V
AC side frequency	f_{AC}	50 Hz

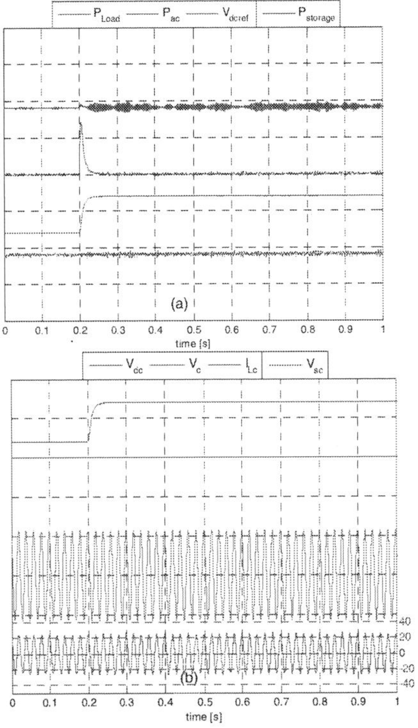

Fig. 3. System charactristics under the voltage change ($V_{dc_{ref}} = 120\ V \rightarrow 170V$): (a) power components; (b) voltage and currents.

First, a stepwise change is applied to the reference voltage of the DC bus: $V_{dc} = 120\ V \rightarrow 170$V, in a constant power: $P_{load} = 400W$. Fig. 3 shows the simulation result of the studied case: Fig. 3 (a) indicates the power components and Fig. 3 (b) shows the voltage and current waveforms. In Fig. 3 (b), V_c and is the voltage of phase C in AC side, and i_{Lc} is the line current of phase C. The simulation results show the successful performance of the controller to regulate the DC bus voltage.

In the next test, the DC bus voltage is fixed at $V_{dc} = 120\ V$, and the load power is changed stepwise: $P_{load} = 400\ W \rightarrow 600W$. Fig. 4 shows the simulation results: the load power P_{load}, the power of the storage system $P_{storage}$, and the power absorbed from the AC source P_{AC} are shown in Fig. 4 (a); DC and AC voltages and currents are shown in Fig. 8 (b). The results show the effectiveness of the controller to manage the power flow under the transient loads. During the transients, the load power is supplied by the storage system as seen from the blue curve in Fig. 4 (a). However, the voltage of the storage unit V_{sc} is constant as shown in Fig. 4 (b). Indeed, the controller can maintain the SC voltage during the transient loads, while the voltage of the SC will be dropped during prolonged overloads.

$$D = \frac{\left(L_{sc}\, K_v \left(\frac{i_{filter} - i_{sc}}{C_{dc}} - \omega_{filter}(V_{filter} - V_s)\right) + r_{sc}\, i_{sc} - V_e + V_s + L\,(K_i + \lambda)(i_{scref} - i_{sc}) - L\, K_i\, \lambda \int i_{sc} - i_{scref}\right)}{V_s - K_v\, \frac{L_{sc}}{C_{dc}}\, i_{sc}} \qquad (9)$$

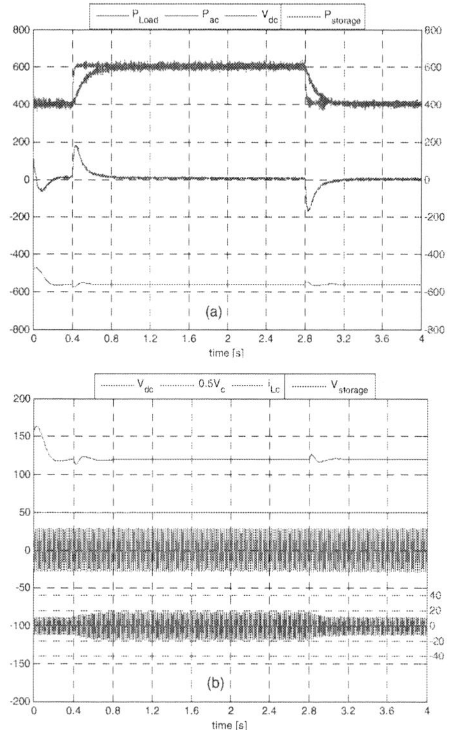

Fig. 4. System charactristics with the power change ($P_{load} = 400\ W \rightarrow 600W$): (a) power components; (b) voltage and currents.

To this end, the performance of the active stabilizer is examined under a large step change in the load power of CPLs. First the stabilizer is deactivated and the CPL power is increased stepwise: $P_{load} = 2000\ W \rightarrow 2500W$. The system characteristics such as the power and the voltages are shown in Fig. 5. Obviously, the system loses stability, and the system characteristics get unstable oscillations. The stabilizer is then activated and the same simulation scenario is repeated. In this case, the load power and the voltages are shown in Fig. 6: the system's characteristics remain stable, with tolerable oscillations. This validates the performance of the active stabilizer to damp the unstable oscillations caused by the constant power behavior of the load.

IV. Experimental Results

The energy-based control strategy is implemented in a laboratory hardware prototype as indicated in Fig. 7. In this setup, a SC bank is used as the storage unit with rated voltage of 96V, and the total capacitance of $C_{sc} = 82F$. The parameters of the laboratory test-bed are the same as those of the simulation model. The DC converters and the rectifier are built with the IGBT switches, and are controlled with dSPACE real-time controllers. The load converters are connected to the DC bus trough two LC filters, corresponding to the system model presented in Fig. 1. A series of experimental tests is performed, including DC voltage change and load power change.

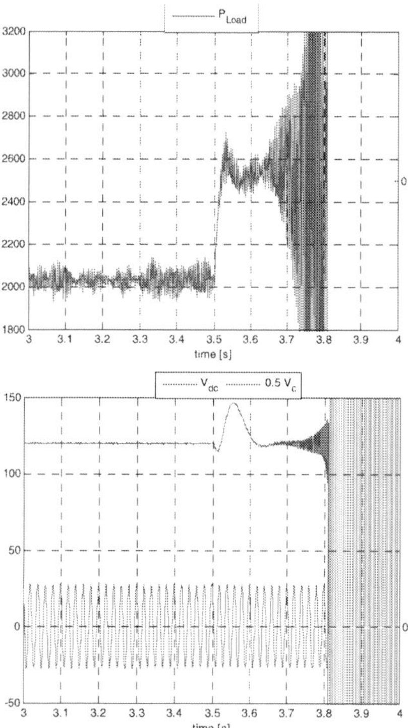

Fig. 5. Control performanc with the power change $P_{load} = 2000\ W \rightarrow 2500W$, without stabilizer: load power (upper); DC and AC voltages (lower).

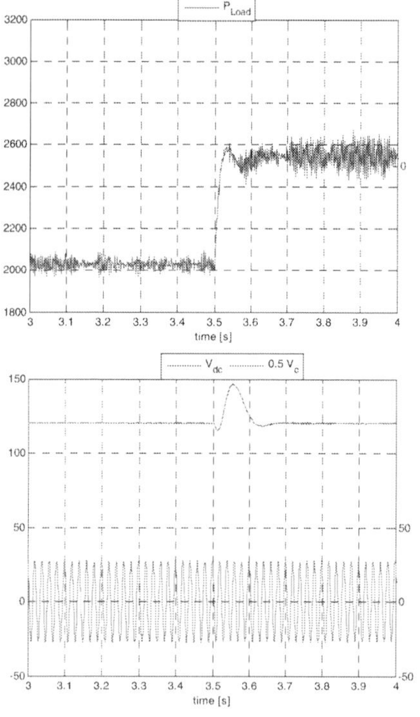

Fig. 6. Control performanc with the power change $P_{load} = 2000\ W \rightarrow 2500W$, with stabilizer: load power (upper); DC and AC voltages (lower).

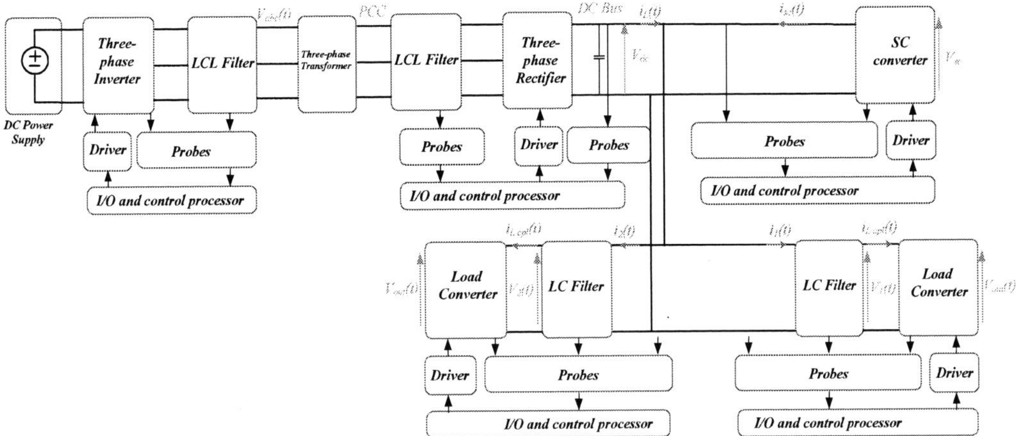

Fig. 7. Schematic of the laboratory test-bed.

First, a stepwise change is applied to the reference voltage of the DC bus: $V_{dc_{ref}} = 120\ V \rightarrow 170V$, in a constant power: $P_{load} = 400W$. Fig. 8 presents the experimental results: the total load power, the SC power, the AC power, and the reference voltage of the DC bus are indicated in Fig. 8 (a); AC and DC voltages and AC currents are shown in Fig. 8 (b). In this case, the experimental results are consistent with the simulations in Fig. 3.

Secondly, the experimental tests are repeated when the DC bus voltage is fixed in $V_{dc} = 120\ V$, and the load power is changed from $P_{load} = 400\ W \rightarrow 600W$. The corresponding experimental results are shown in Fig. 9. In this case, the load power is changed sharply, while the AC power is increased smoothly. The load power in the fast changing is compensated by the SC bank, as observed from Fig. 9 (a). Again, the experimental results, including the power profile of the system and the voltage and current waveforms are consistent with the simulations in Fig. 4. Clearly, the controller is able to compensate the transient load power and energy, without losing the control on the DC bus voltage and the SC voltage.

V. CONCLUSION

This paper has presented has presented an energy-based controller for hybrid DC microgrids with supercapacitor storages. The controller is implemented with simple PI compensators. The objective of the proposed controller is to ensure simultaneously the voltage stability, power management, storage availability, and stabilization of the system under increased CPL power.

The simulation results shown the effectiveness of the control system to manage the power flows between the SC, the rectifier and the load. Unstable oscillations caused by the CPLs are also damped with the active stabilizer embedded in the controller of SC converter. A laboratory hardware prototype is also constructed, in which the CPLs are emulated by DC-DC converters. The performance of the controller is then validated experimentally. The experimental results are consistent with the simulations.

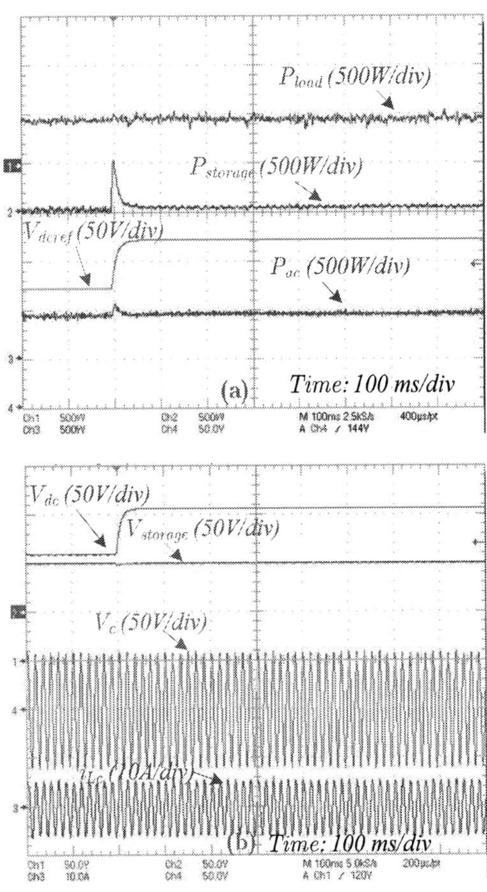

Fig. 8. Experimental results of the voltage change ($V_{dc_{ref}} = 120\ V \rightarrow 170V$): (a) power components; (b) voltage and currents.

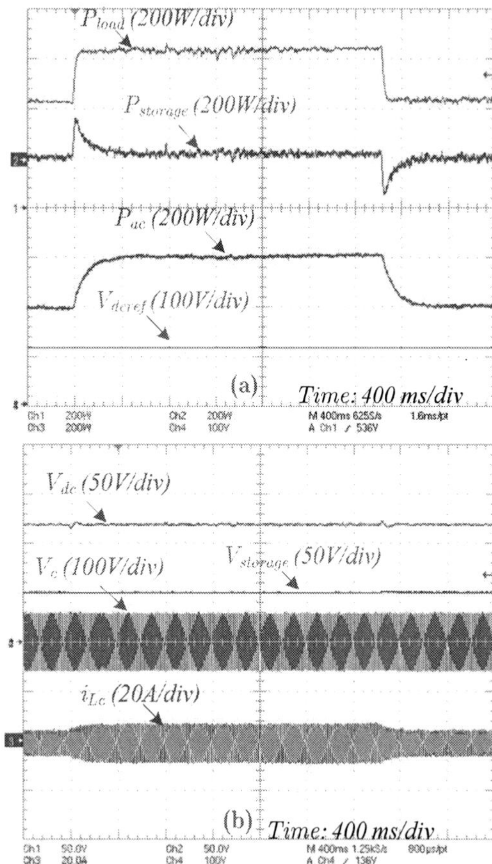

Fig. 9. Experimental results of the power change ($P_{load} = 400\ W \rightarrow 600W$): (a) power components; (b) voltage and currents.

REFERENCES

[1] P. C. Loh, D. Li, Y. K. Chai, and F. Blaabjerg, "Autonomous operation of hybrid microgrid with AC and DC subgrids," *IEEE Transactions on Power Electronics*, vol. 28, pp. 2214-2223, 2013.

[2] R. S. Balog and P. T. Krein, "Bus Selection in Multibus DC Microgrids," *IEEE Transactions on Power Electronics*, , vol. 26, pp. 860-867, 2011.

[3] A. Kwasinski and C. N. Onwuchekwa, "Dynamic behavior and stabilization of DC microgrids with instantaneous constant-power loads," *IEEE Transactions on Power Electronics*, vol. 26, pp. 822-834, 2011.

[4] M. K. Zadeh, R. Gavagsaz, S. Pierfederici, B. Nahid-Mobarakeh, and M. Molinas, "Stability analysis and dynamic performance evaluation of a power electronics-based DC distribution system with active stabilizer," *IEEE J. Emerg. Sel. Topics Power Electron.*, 2015.

[5] M. K. Zadeh, B. Zahedi, M. Molinas, and L. E. Norum, "Centralized stabilizer for marine DC microgrid," in *Proc. 39th Annual Conference of the IEEE Industrial Electronics Society, IECON 2013*, 2013, pp. 3359-3363.

[6] P. Magne, B. Nahid-Mobarakeh, and S. Pierfederici, "Active Stabilization of DC Microgrids Without Remote Sensors for More Electric Aircraft," *IEEE Transactions on Industry Applications*, vol. 49, pp. 2352-2360, 2013.

[7] A. Khaligh, "Realization of parasitics in stability of DC-DC converters loaded by constant power loads in advanced multiconverter automotive systems," *IEEE Transactions on Industrial Electronics*, vol. 55, pp. 2295-2305, 2008.

[8] J. Cao and A. Emadi, "A new battery/ultracapacitor hybrid energy storage system for electric, hybrid, and plug-in hybrid electric vehicles," *IEEE Transactions on Power Electronics*, , vol. 27, pp. 122-132, 2012.

[9] M. K. Zadeh, G.-G. Roghayeh, S. Pierfederici, N.-M. Babak, and M. Molinas, "A discrete-time tool to analyze the stability of weakly filtered active front-end PWM converters," in *Proc. IEEE Transportation Electrification Conf. & Expo. (ITEC)*, 2014.

[10] A. Ferreira, J. A. Pomilio, G. Spiazzi, and L. de Araujo Silva, "Energy management fuzzy logic supervisory for electric vehicle power supplies system," *IEEE Transactions on Power Electronics*, vol. 23, pp. 107-115, 2008.

[11] B. Sarlioglu and C. Morris, "More Electric Aircraft; Review, Challenges and Opportunities for Commercial Transport Aircraft," in *Proc. IEEE Transactions on Transportation Electrification*, vol. PP, pp. 1-1, 2015.

[12] R. Burgos, G. Chen, F. Wang, D. Boroyevich, W. G. Odendaal, and J. D. Van Wyk, "Reliability-oriented design of three-phase power converters for aircraft applications," *IEEE Transactions on Aerospace and Electronic Systems*, vol. 48, pp. 1249-1263, 2012.

[13] G. Gong, M. L. Heldwein, U. Drofenik, J. Miniböck, K. Mino, and J. W. Kolar, "Comparative evaluation of three-phase high-power-factor AC-DC converter concepts for application in future More Electric Aircraft," *IEEE Transactions on Industrial Electronics*, vol. 52, pp. 727-737, 2005.

[14] M. K. Zadeh, R. Gavagsaz-Ghoachani, J. P. Martin, S. Pierfederici, B. Nahid-Mobarakeh, and M. Molinas, "Discrete-time modelling, stability analysis, and active stabilization of dc distribution systems with constant power loads," in *Proc. 2015 IEEE Applied Power Electronics Conference and Exposition (APEC)*, 2015, pp. 323-329.

[15] H. Zhang, F. Mollet, C. Saudemont, and B. Robyns, "Experimental validation of energy storage system management strategies for a local dc distribution system of more electric aircraft," *IEEE Transactions on Industrial Electronics*, vol. 57, pp. 3905-3916, 2010.

[16] S. N. Motapon, L. Dessaint, and K. Al-Haddad, "A comparative study of energy management schemes for a fuel-cell hybrid emergency power system of more-electric aircraft," *IEEE Transactions on Industrial Electronics*, , vol. 61, pp. 1320-1334, 2014.

[17] M. B. Camara, B. Dakyo, and H. Gualous, "Polynomial control method of DC/DC converters for DC-bus voltage and currents management—Battery and supercapacitors," *IEEE Transactions on Power Electronics*, vol. 27, pp. 1455-1467, 2012.

[18] J. Ke and R. Xinbo, "Hybrid Full-Bridge Three-Level LLC Resonant Converter—A Novel DC–DC Converter Suitable for Fuel-Cell Power System," *IEEE Transactions on Industrial Electronics*, vol. 53, pp. 1492-1503, 2006.

[19] J.-S. Lai and D. J. Nelson, "Energy management power converters in hybrid electric and fuel cell vehicles," *Proceedings of the IEEE*, vol. 95, pp. 766-777, 2007.

[20] M. Phattanasak, R. Gavagsaz-Ghoachani, J. P. Martin, B. Nahid-Mobarakeh, S. Pierfederici, and B. Davat, "Control of a Hybrid Energy Source Comprising a Fuel Cell and Two Storage Devices Using Isolated Three-Port Bidirectional DC-DC Converters," *IEEE Transactions on Industry Applications*, vol. 51, pp. 491-497, 2015.

[21] M. Zandi, A. Payman, J.-P. Martin, S. Pierfederici, B. Davat, and F. Meibody-Tabar, "Energy management of a fuel cell/supercapacitor/battery power source for electric vehicular applications," *IEEE Trans. Veh. Tech.*, vol. 60, pp. 433-443, 2011.

[22] L. M. Saublet, R. Gavagsaz-Ghoachani, J. P. Martin, S. Pierfederici, B. Nahid-Mobarakeh, and J. Da Silva, "Stability analysis of a tightly controlled load supplied by a DC-DC boost converter with a modified sliding mode controller," in *Proc. 2014 IEEE Transportation Electrification Conference and Expo (ITEC)*, 2014, pp. 1-6.

[23] M. K. Zadeh, G.-G. Roghayeh, S. Pierfederici, N.-M. Babak, and M. Molinas, "A new discrete-time modelling of PWM converters for stability analysis of DC microgrid," in *Proc. 11th International Conference on Modeling and Simulation of Electric Machines, Converters and Systems (Electrimacs'14)*, 2014, pp. 336-341.

A Low-cost Solar Micro-inverter with Soft-switching Capability Utilizing Circulating Current

Xiaohu Liu, Mohammed Agamy, Dong Dong, Maja Harfman-Todorovic, Luis Garces
General Electric Global Research Center, Niskayuna, NY, US
Xiaohu.L@ge.com

***Abstract*— This paper presents a low-cost high-efficiency photovoltaic (PV) micro-inverter with soft-switching capability. The system is based on a partial power processing resonant front end dc-dc stage, followed by an interleaved inverter stage. The proposed new soft-switching modulation scheme enables significant cost and volume reduction of the magnetics in the inverter stage. In order to shrink the magnetics, the interleaved inverter modulation scheme adopts a real-time varying phase-shift delay control in a way such that a smaller inter-phase transformer (IPT) can be utilized due to reduced voltage-second and the additional conduction loss caused by the circulating current can be minimized. In addition, the zero-voltage-switching (ZVS) operation in the full operation range can be maintained. As a result, the power MOSFETs body-diode reverse-recovery loss can be eliminated and the power stage switching frequency can be further increased. The operation principle of new modulation scheme and ZVS operation are analyzed. Key simulation result and baseline experimental results are presented to validate the analysis.**

Keywords— solar micro-inverter, magnetics, ZVS soft-switching

I. INTRODUCTION

In the last few years various modes of renewable energy have experienced one of the largest growths (over 30% per year) compared with the much slower growth of coal and fossil fuels energy. Solar power penetration is relatively low compared with the wind systems but it is estimated by EPIA [1-3] that its market share could be as high as 12% in 2020.

Solar electric systems based on the integration of micro-inverters and photovoltaic (PV) modules now represent ~8% of the U.S. residential market and offer many advantages related to safety, performance, and simplified installation. Deficiencies in the performance and mechanical configuration of existing micro-inverters along with functional redundancies present barriers to achieving an unsubsidized residential system cost ($3.00/W). Moreover, it presents barriers to the corresponding levelized cost of electricity (LCOE=$0.13/kWh) competitive with the average retail price of electricity in the U.S.

A high efficiency PV micro-inverter with grid support functions was presented in [4]. Based on the previous research, the magnetics cost is around 15% of total system cost in the proposed solar micro-inverter. The magnetics mainly includes the inter-phase transformer (IPT) and the output LCL filter. This paper proposes a new soft-switching modulation scheme that can enable inverter stage utilizing a much lower cost IPT without sacrificing the power stage efficiency. Moreover, the proposed method can achieve the inverter stage zero-voltage-switching (ZVS) operation in the full operation range. As a result, the power MOSFETs body-diode reverse-recovery loss can be eliminated and the power stage switching frequency can be further increased. Therefore, the size and cost of output LCL filter can also be reduced. The trade-off design of the proposed method is to optimize the IPT size and phase shift angle in order to minimize the extra conduction loss. The new soft-switching modulation analysis and simulation validation is presented in the paper. The baseline experimental results are also given to validate the presented circuit analysis.

II. PROPOSED TWO-PHASE INTERLEAVED SOFT-SWITCHING INVERTER

A. System Description and Cost/Efficiency Breakdown Analysis

A solar micro-inverter topology is presented that is well suited to operate with modules of higher operating voltage such as multi-crystalline silicon mc-Si panels of 72-cell or more or to operate with multiple high voltages, low power thin-film modules. The proposed topology is composed of an input partial power processing resonant dc-dc converter, followed by an interleaved inverter stage. A block diagram of the proposed topology and the picture of prototype are shown in Fig. 1.

As highlighted in the Fig. 1, the IPT size is relatively large and its core is made of nano-crystalline silicon material, which is relatively expensive. Fig. 2 shows the interleaved inverter cost and power loss breakdown analysis. As shown, the magnetic cost is 14.6% of total system cost and the IPT cost is about 80% of total magnetic cost. The goal of proposed method is to shrink the IPT and LCL filter and reduce the magnetic cost by 90%. The proposed method can achieve ZVS operation of all MOSFETs. Therefore, the reverse recovery loss of the body diode of the MOSFET

This work was supported by the U.S. Department of Energy under Award Number DE-EE0005344.

978-1-4673-9551-9/16 $31.00 © 2016 IEEE

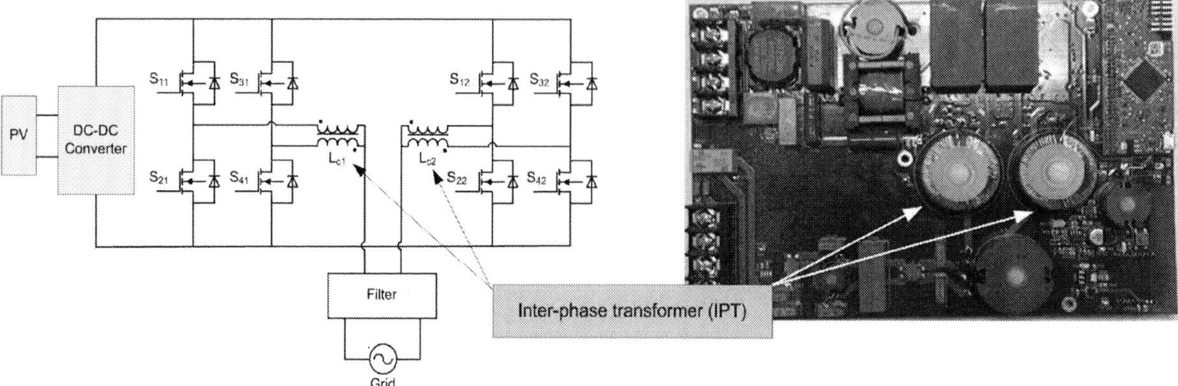

Fig. 1. The proposed solar micro-inverter hardware prototype and the circuit diagram of interleaved inverter power stage.

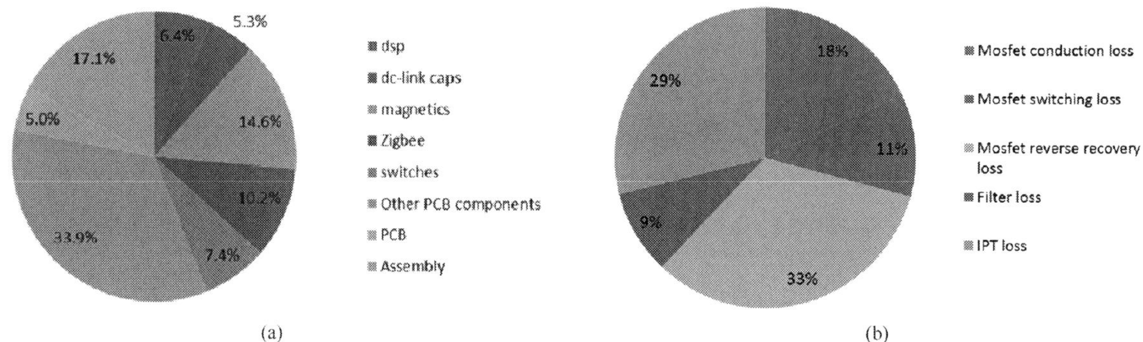

Fig. 2. Interleaved inverter profile analysis: (a) cost breakdown; (b) power loss breakdown

could be eliminated which help to balance the extra conduction loss introduced by the proposed method.

B. Proposed Soft-switching Modulation Analysis

Before introducing the proposed method, the interleaved inverter operation principle will be explained first. It consists of a parallel connection of two H-bridges whose outputs are joined to generate the desired voltage, as shown in Fig. 3. This connection adds the output currents of the two bridges but, if they are not switched at the same time in both legs, generates an undesired current circulating between the two circuits without going to the output. To limit this current, IPTs are connected at each of the bridge output terminals. The size of these IPTs is determined by the acceptable circulating current and by the maximum voltage difference at the H-bridge outputs. The H-bridge with the bipolar PWM already has a two level output. A parallel connection of two bridges allows for doubling the frequency of the effective ripple voltage at the output terminals. The minimum ripple for a given PWM carrier is obtained when both are switched at the same frequency, with the carriers phase shifted by 180 degrees. This ripple reduction is obtained at the cost of maximizing the volt-seconds across the inductors used to limit the circulating current. The following traces in Fig. 4 show the blue bridge output voltages at the points indicated in the circuit above. Shown are the leg voltages with respect to the negative DC-bus, the output voltages at the bridges and the voltage across one of the circulating current limiter inductors. In Fig. 4, m is the desired modulation index.

In order to reduce the IPT cost, the idea proposed in the paper is to use the much smaller IPT at the cost of relatively larger ciculating current compared to the orignal design. The phase shift angle is not fixed 180 degrees. In contrast, the phase shift angle will be given in real-time in a way such that the ZVS operation will be maintained and the additional conduction loss caused by the circulating current is minimized. The fundamental principle of ZVS operation is to gate the MOSFET while its antiparallel diode is conducting [5-7]. Therefore, the circulating current has to be large enough to discharge the device capacitor charge and turn on the MOSFET body-diode before enabling the device.

Based on the efficiency breakdown analysis shown in Fig. 2 (b), the reverser-recovery loss is 33% of total loss. Therefore, due to the elimination the body-diode reverse-recovery loss because of ZVS operation, the total power stage efficiency may still be maintained at higher level despite the increased conduction loss.

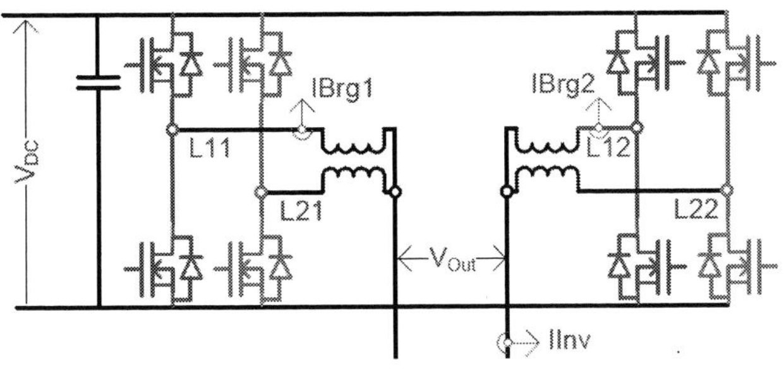

Fig. 3. Interleaved inverter parallel connection of two H-bridges.

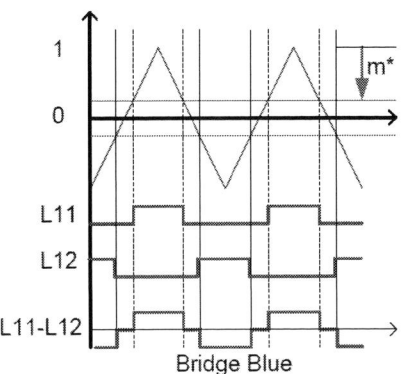

Fig. 4. Gate driver command signal generation.

Another trade-off of the proposed method is that the total output ripple frequency is not twice the switching freqeuncy since the phase shift angle is not 180 degrees.

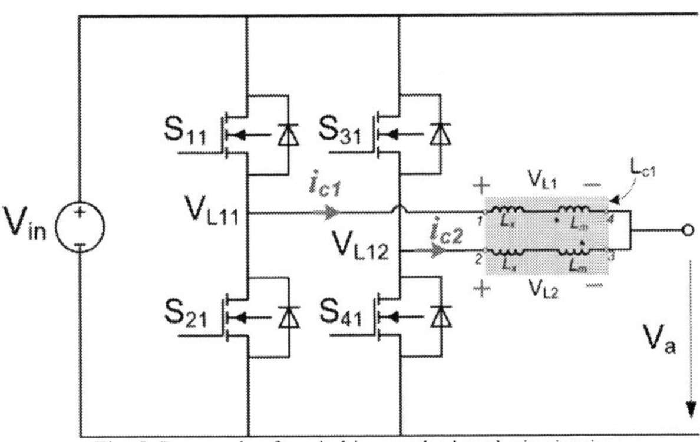

Fig. 5. Proposed soft-switching method analysis circuit.

However, the switching frequency can be pushed higher since the reverse-recovery loss is eliminated and switching loss is reduced. The total output ripple frequency could actually be higher than the original design. As a result, the IPT and LCL filter size and cost can also be further reduced. In order to illustrate the detailes of ZVS operatoin scheme in the proposed method, Fig. 5 shows one H-bridge of the circuit with the key variables. In Fig. 5, L_{c1} is the IPT, V_{L11} and V_{L12} are the midponit to ground voltage, V_{L1} and V_{L2} are the coupled two inductor votlges, V_a is the H-bridge total output to groud votlage, i_{c1} and i_{c2} are the circulating current.

In the original design, the phase shift angle φ is fixed at 180 degrees. The IPT is designed such that the circulating current between two legs is within 1%. Therefore, i_{c1} and i_{c2} would share the load current evenly. The downside of original design is that all the devices are hard-switching in most of the operation conditions. Fig. 6 shows the ideal key soft-switching operation waveforms in the proposed method for two cases with D=0.2 and D=0.8, respectively. D is the control output duty cycle. As illustrated in Fig. 6, the relatively smaller φ is selected to ensure the key corner points (I_1 and I_2) of i_{c1} and i_{c2} meet the ZVS operation conditions: I_1 is smaller than zero and I_2 is bigger than zero. S_{11} is tuned on at the corner point I_1. The body diode of S_{11} would be on if i_{c1} is smaller than zero referring to Fig. 5. Therefore, S_{11} can be ZVS turned on. The same soft-switching scheme can be applied to all the other devices. Moreover, the ZVS operation can be maintained in the whole operation range of duty cycle.

III. SIMULATION VERIFICATION

The simulation verification is conducted in the PLECS to verify the aforementioned analysis and the proposed soft-switching modulation scheme. The simulation conditions are that the DC-bus voltage is 400V ideal source. The interleaved inverter switches at 50 kHz. Fig. 7 shows the key waveforms to validate the proposed soft-switching modulation scheme with duty cycle D =0.2 and 0.8, respectively. As shown, the circulating current i_{c1} and i_{c2} waveform proves the analysis shown in Fig. 6. The highlighted corner points (I_1, I_2) has small delay with respects to the device turn-on timing. This is because the 400ns dead time is added in the simulation. During the dead-time interval, the bode-diode is freewheeling and the bridge mid-point is clamped to negative or positive rail of dc-link depending on the current direction.

Fig. 8 shows the performance comparison between the baseline method with a fixed 180 degrees phase shift angle and proposed method with varied phase shift angle. The IPT total circulating-loop inductance is 40mH (20mH per leg) and 40μH (20μH per leg) for the baseline method and proposed method, respectively. The root cause for the significant reduction on IPT is that the volt-second can be minimized with the proposed method. As shown in Fig. 8(a), the circulating current cannot meet the soft-switching conditions for most cases in the baseline method. In contrast, the proposed method can ensure the soft-switching conditions satisfied in all cases as validated in Fig. 8(b). The "glitch" shown on the i_{c1} in Fig. 8(b) is caused by the dynamic change of phase shift angle. The downside of the proposed method is that the circulation current is bigger which leads to larger conduction loss. The results shown in Fig. 8 validate the proposed soft-switching modulation scheme of the studied inverter stage.

IV. Experimental Test Bed Development

A 300 W micro-inverter circuit prototype has already been shown in Fig. 1. The input dc-dc stage operated at the resonant frequency of 120 kHz, and the interleaved inverter operated at the 15 kHz switching frequency. Experiments were conducted with the grid emulator set to 240V rms (PCR2000L AC programmable power supply from Kikusui). Input was connected to TerraSAS PV emulator from AMETEK.

The switching frequency was limited to 15 kHz because the modulation control adopts the fixed 180 degree phase shift delay that leads to large switching loss, especially the MOSFET body-diode loss. The baseline test has been done to validate the previous circuit analysis. Fig. 9 shows the experimental results with original control method. The

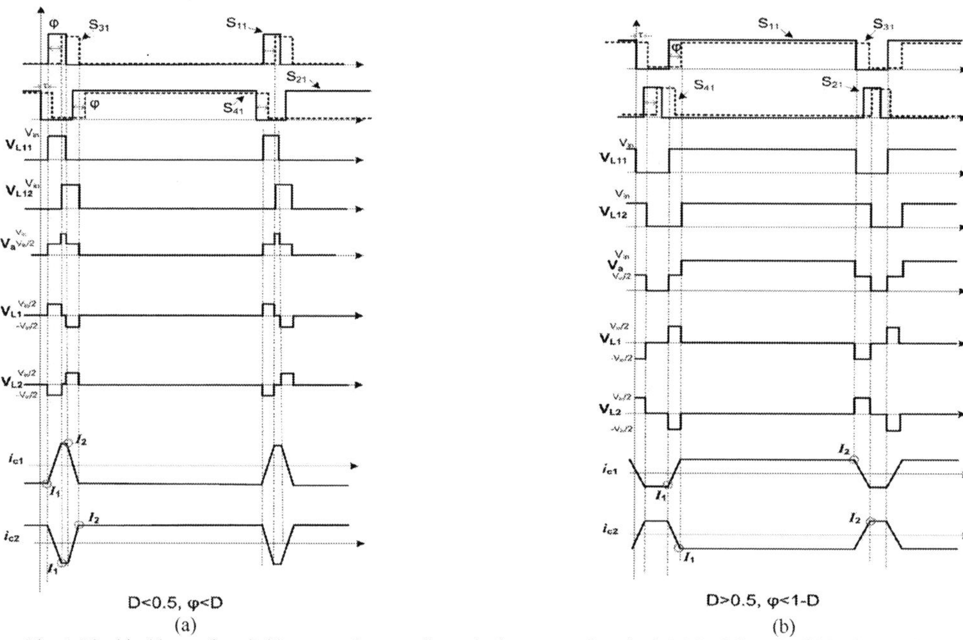

Fig. 6. The ideal key soft-switching operation waveforms in the proposed method: (a) $D<0.5$, $\varphi<D$; (b) $D>0.5$, $\varphi<1-D$.

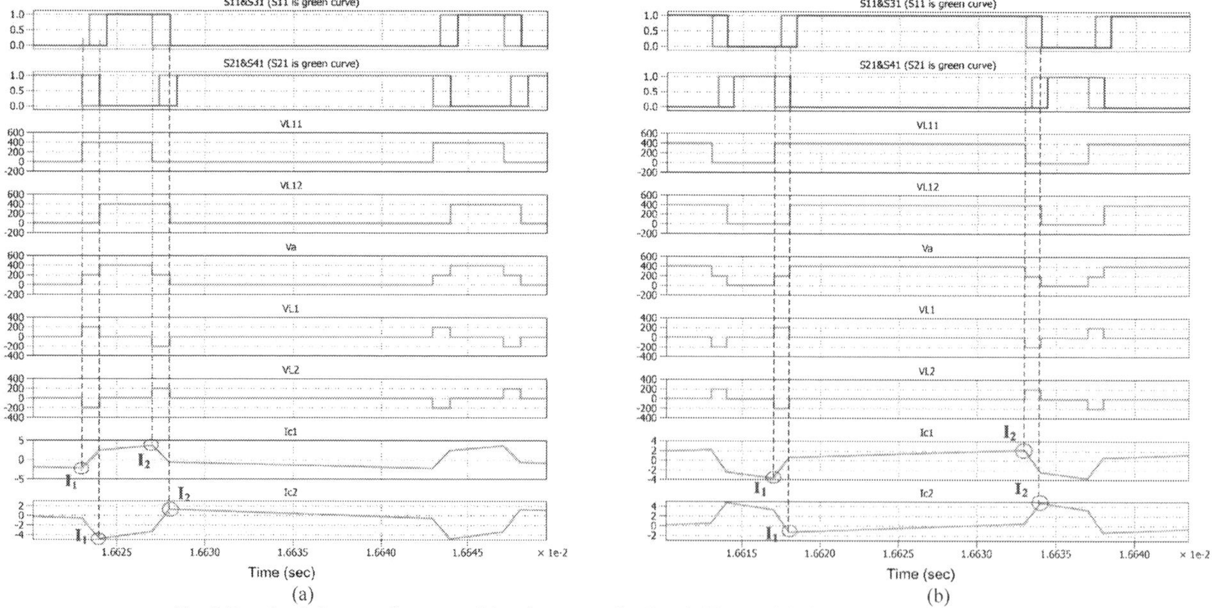

Fig. 7.Key simulation waveforms to validate the proposed soft-switching modulation scheme. (a) D=0.2; (b) D=0.8

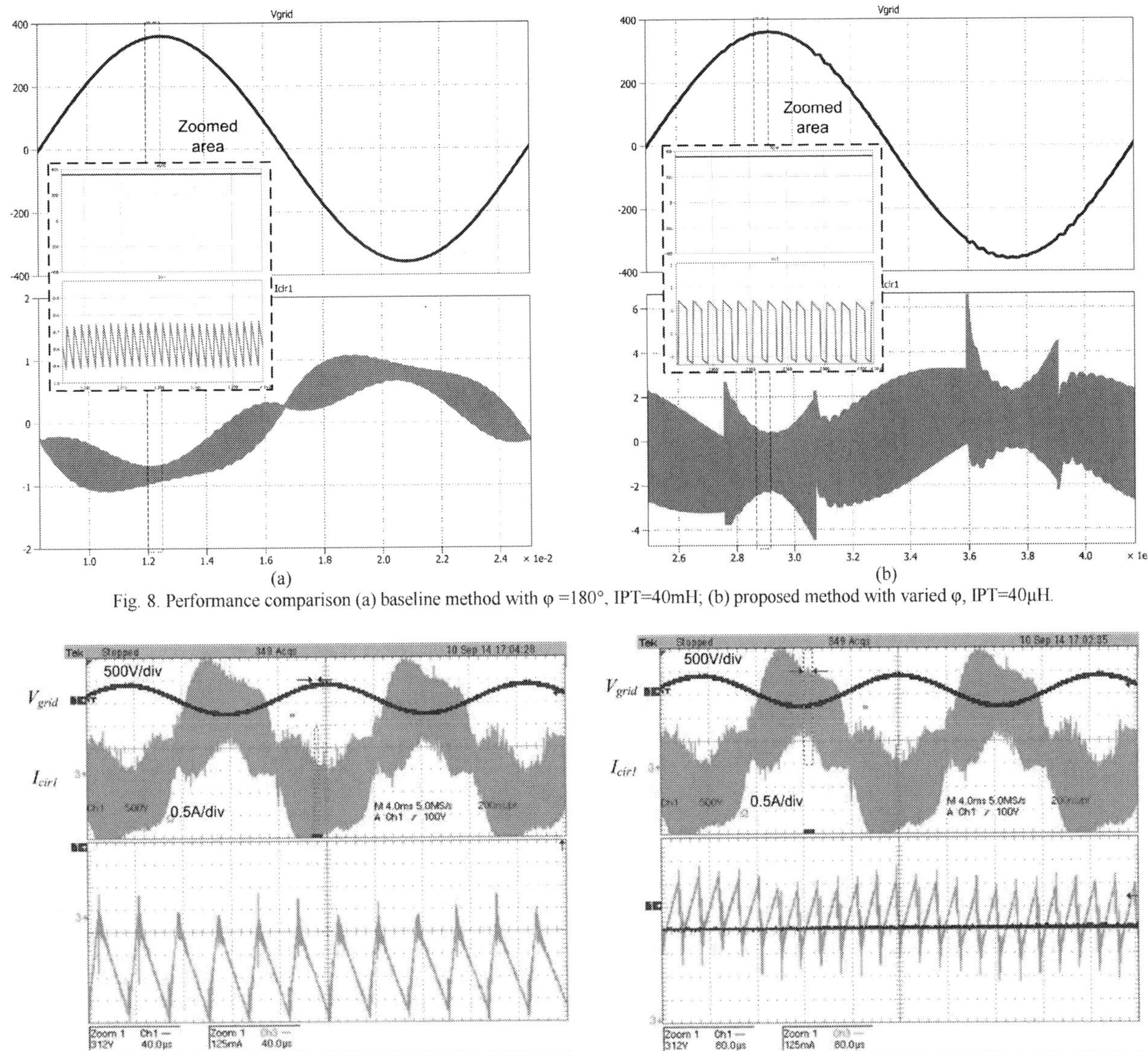

Fig. 8. Performance comparison (a) baseline method with φ =180°, IPT=40mH; (b) proposed method with varied φ, IPT=40μH.

Fig. 9. Baseline experimental results with fixed 180 degree phase-shift delay control. (a) bottom zoomed area of I_{cir1}, (b) top zoomed area of I_{cir1}.

circulating current i_{c1} is zoomed in both bottom and top area. By comparing Fig. 9 to Fig. 8(a), those two results match well. The difference mainly caused by the switching frequency. In Fig. 8(a), the switching frequency is 50 kHz instead of 15 kHz.

V. Conclusion

A new soft-switching modulation scheme for the interleaved inverter is proposed, which is capable of shrinking the magnetics. The proposed method adopts the real-time varied phase-shift delay control in a way such that the smaller inter-phase transformer (IPT) can be utilized and the additional conduction loss caused by the circulating current can be minimized. The detailed circuit analysis of proposed method soft-switching scheme has been described in the paper. The key simulation result and baseline experimental results have been presented to validate the analysis.

References

[1] A. Elasser, M. Agamy, J. Sabate, R. Steigerwald, R. Fisher & M. Harfman-Todorovic, "A Comparative Study of Central and Distributed MPPT Architectures for Megawatt Utility and Large Scale Commercial Photovoltaic Plants," Proceedings of Industrial Electronics Conference (IECON) 2010, pp. 2753-2758.

[2] Q. Li and P. Wolfs, "A review of the single phase photovoltaic module integrated converter topologies with three different DC link

configurations", IEEE Trans. on power electronics, vol. 23, May 2008, pp. 1320-1333

[3] S. B. Kjaer and F. Blaabjerg, "Design Optimization of a single phase inverter for photovoltaic applications", in proc. IEEE PESC'03, vol. 3, 2003, pp. 1183-1190

[4] Harfman-Todorovic, M.; Tao, F.; Agamy, M.; Dong, D.; Liu, X.; Garces, L.; Zhou, R.; Delgado, E.; Marabell, D.; Stephens, C.; Steigerwald, R., "A high efficiency PV micro-inverter with grid support functions," Energy Conversion Congress and Exposition (ECCE), 2014 IEEE , vol., no., pp.4244,4250, 14-18 Sept. 2014

[5] Xiaohu Liu; Hui Li; Zhan Wang, "A Fuel Cell Power Conditioning System With Low-Frequency Ripple-Free Input Current Using a Control-Oriented Power Pulsation Decoupling Strategy," Power Electronics, IEEE Transactions on , vol.29, no.1, pp.159,169, Jan. 2014

[6] Xiaohu Liu; Hui Li, "An Electrolytic-Capacitor-Free Single-Phase High-Power Fuel Cell Converter With Direct Double-Frequency Ripple Current Control," Industry Applications, IEEE Transactions on , vol.51, no.1, pp.297,308, Jan.-Feb. 2015

[7] Xiaohu Liu; Hui Li; Zhan Wang, "A Start-Up Scheme for a Three-Stage Solid-State Transformer With Minimized Transformer Current Response," Power Electronics, IEEE Transactions on , vol.27, no.12, pp.4832,4836, Dec. 2012

Disclaimer: This report was prepared as an account of work sponsored by an agency of the United States Government. Neither the United States Government nor any agency thereof, nor any of their employees, makes any warranty, express or implied, or assumes any legal liability or responsibility for the accuracy, completeness, or usefulness of any information, apparatus, product, or process disclosed, or represents that its use would not infringe privately owned rights. Reference herein to any specific commercial product, process, or service by trade name, trademark, manufacturer or otherwise does not necessarily constitute or imply its endorsement, recommendation, or favoring by the United States Government or any agency thereof. The views and opinions of authors expressed herein do not necessarily state or reflect those of the United States government or any agency thereof.

Design and Stability Analysis for an Autonomous DC Microgrid with Constant Power Load

Qianwen Xu[12] , Xiaolei Hu[2], Peng Wang[1], Jianfang Xiao[1], Leonardy Setyawan[1], Changyun Wen[1], Lee Meng Yeong[3]
1-School of Electric & Electronic Engineering, Nanang Technological University, 50 Nanyang Avenue, 639798, Singapore
2-Rolls Royce @ NTU Co-Lab, 65 Nanyang Drive, 637460, Singapore
3-Advanced Technology Centre, Rolls-Royce Singapore Pte. Ltd., 6 Seletar Aerospace Rise, 797575, Singapore
qxu007@e.ntu.edu.sg

***Abstract*— Fuel cell is a promising source in autonomous dc microgrid. Hybridization of fuel cell with battery is commonly implemented to overcome slow dynamic of fuel cell. Then battery compensates high frequency fluctuation and fuel cell provides consistent power at steady state. To achieve this objective, most control strategies require a centralized controller, which encounter reliability and scalability issues. This paper proposes a virtual capacitor droop controller to achieve autonomous dynamic power sharing at distributed level. An autonomous dc microgrid is designed to verify the proposed method. Considering the high penetration of constant power loads (CPLs) in dc microgrid and the destabilizing effect of CPLs, system stability is investigated with the implementation of the proposed controller. Both simulation and hardware experiment are conducted to validate the effectiveness of the proposed control method and analytical results.**

Keywords—Autonomous dc microgrid; virtual capacitor droop; transient power sharing; constant power load; stability analysis

I. Introduction

DC microgrid (MG) is gaining increasing attention recently due to its improved efficiency, controllability and reliability compared with its AC counterpart [1]–[3]. Fuel cell (FC) is a promising energy source which has less noise and near zero green-gas emissions compared to conventional generators [4]. It has wide applications in autonomous DC MG, such as renewable energy systems [5], electric vehicles [6] and more electric aircrafts [7]. It can provide consistent power when renewable energy source cannot match the load. However, fuel cell has issues working independently due to its low dynamic response to load demand [5], [6]. As battery and supercapacitor have fast dynamics, hybridization of FC with battery/supercapacitor is commonly implemented to provide a fast dynamic response to load variation and maintain bus voltage. In such hybrid system, battery or super-capacitor provides power during transient while FC provides consistent power to maintain a stable DC bus. This kind of dynamic power management is normally realized by centralized or hierarchical control [4], [6]–[9]. However, these methods require a centralized controller and communication lines. Due to the distributed nature of DC MG, this may lead to increased cost as well as reduced reliability and scalability [2], [3]. A communication-free DC MG that can manage power sharing at transient is therefore preferred [10]. As a classical decentralized method, droop control is usually applied for autonomous power sharing among energy sources and several improved droop control methods have been developed to improve system performance [11]–[14]. However, these droop control methods only achieve power sharing at steady state. As droop controller can be represented as a virtual resistor from its V-I characteristics, a virtual impedance droop control is proposed in [15] to realize dynamic power sharing autonomously. In this method, output current reference of energy source with fast/slow dynamics is obtained as voltage deviation with high-pass/low-pass filter. However, this method is in fact a proportional control in voltage control loop to generate load current reference with a high-pass/low-pass filter, which is not a kind of droop control by nature. As proportional controller introduces voltage deviation and increases overshoot [16], this method cannot stabilize bus voltage very well. Moreover, parameter design is not illustrated and stability analysis, which should be ensured at design phase, is not investigated in the literature.

In DC MG, with the high penetration of power electronic converters, stability issue is becoming significantly important [17]. Power electronic converter loads, when tightly controlled, behave as constant power loads (CPLs) [18]. CPLs have negative impedance characteristics, which may impact power quality and destabilize the MG. This impact becomes more significant when MG operates in standalone mode [19]. Therefore, to ensure MG stable operation under a developed control strategy, it is essential to investigate system stability with the penetration of CPLs.

In this paper, a virtual capacitor droop (VCD) controller is proposed for an autonomous DC MG, which consists of FC, battery, PV, resistive load and constant power load. Output voltage characteristics of the FC converter and battery converter are regulated by a virtual resistor and a virtual capacitor respectively. The proposed droop controllers inherently act as a low-pass filter for FC and a high-pass filter battery to load response. The proposed method achieves dynamic current sharing autonomously as well as stabilizes bus voltage. Then an autonomous DC MG is designed based on the proposed control method. During system transits, battery will compensate high frequency power fluctuation to maintain DC bus voltage, whereas in steady state FC will provide power to loads and battery current will reduce to zero eventually. By proper parameter design, system can be extended without impact of dynamics. To ensure system stable operation under the proposed controller, system stability is investigated with the penetration of CPLs. Simulation and experiment are conducted

This study was conducted within the Rolls-Royce@NTU Corporate Lab with support from the National Research Foundation (NRF) Singapore under the Corp Lab@University Scheme.

to verify the effectiveness of the VCD control method and the analytical results.

This paper is organized as follows. An autonomous DC MG with virtual impedance droop control is introduced in section II. Virtual capacitor droop control is proposed and the autonomous DC MG is designed based on this method in section III. In section IV, system stability under the proposed controller is investigated and stability boundary is analyzed with the penetration of CPLs. Simulations and experiments are presented in section V. Conclusions are drawn in section VI.

II. AUTONOMOUS DC MICROGRID

Fig. 1 presents an autonomous DC MG which consists of FC, battery, PV and various loads. FC is connected to dc bus by a boost converter and battery is tied to the bus by a bidirectional dc/dc converter. PV panels are connected to the dc bus through a boost converter and work at maximum power point tracking mode. DC load is connected to the bus. The hybrid fuel cell and battery system compensates power imbalance between PV and load to maintain bus voltage. However, there exists a challenge in load sharing between FC and battery that battery should provide power during transient to compensate slow dynamics of the FC and FC should provide power at steady state.

Droop control is commonly implemented for autonomous power sharing between fuel cell and battery. Its V-I relationship can be represented by a virtual resistor, which is expressed as

$$V = V_{ref} - R_d \cdot I \tag{1}$$

where V and I are output voltage and current of the converter, V_{ref} is the reference bus voltage, and R_d is the droop coefficient.

As droop control only allocates power at steady state, it can be extended to virtual impedance droop control to achieve transient power sharing. Virtual resistor R_d in (1) is extended to $Z_d(s)$ and VI relationship of virtual impedance droop control is expressed as

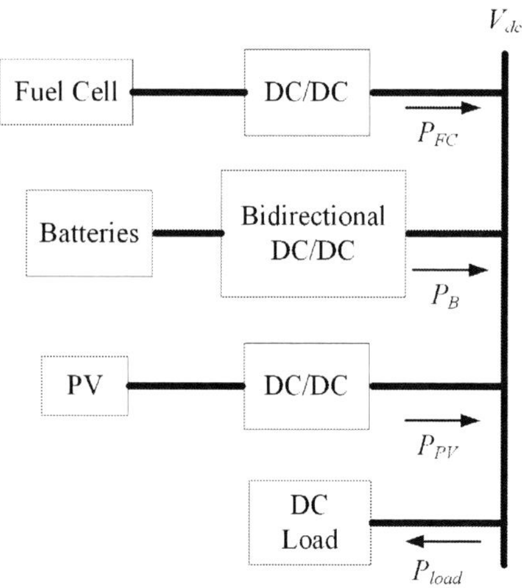

Fig. 1 Proposed DC microgrid integrating with fuel cell and battery.

$$v(s) = V_{ref} - Z_d(s) \cdot i(s) \tag{2}$$

where $v(s)$, $i(s)$ represent transient voltage and current in Laplace space; virtual impedances $Z_d(s)$ for FC and battery can be obtained by inserting low-pass and high-pass filters in series with virtual resistor R_d for FC and battery respectively, as expressed in (3).

$$\begin{cases} Z_{FC} = R_d \cdot \dfrac{\omega_c + s}{\omega_c} \\ Z_B = R_d \cdot \dfrac{\omega_c + s}{s} \end{cases} \tag{3}$$

where ω_c is the crossover frequency of the filters.

The virtual impedance droop control can achieve autonomous load sharing during transient as well as at steady state. But the implementation of low pass filter and high pass filter makes the parameter design and system analysis more complicated. To simplify parameter design and system analysis without any tradeoff in effectiveness of the autonomous power management and voltage stabilization, a virtual capacitor droop (VCD) controller is proposed. Output voltage characteristic of the FC is controlled by the conventional virtual resistor droop controller and the proposed VCD is implemented for battery. The proposed droop controllers inherently act as a low-pass filter and high-pass filter for FC and battery respectively. Details of the proposed VCD controller will be presented in the next section.

III. PROPOSED VCD CONTROL AND SYSTEM DESIGN

Fig. 2 shows the equivalent circuit of hybrid FC and battery system under the proposed VCD control. C_1 is the virtual capacitor implemented in battery control loop and R_2 is the virtual resistor implemented in FC control loop. Because of the frequency characteristics of virtual capacitor, battery responds to load variation immediately during transient and at steady state, output current of battery reduces to zero.

The output voltage of battery and fuel cell can be calculated as:

$$\begin{cases} V_1 = V_{ref} - \dfrac{1}{sC_1} \cdot i_1 \\ V_2 = V_{ref} - R_2 \cdot i_2 \end{cases} \tag{4}$$

The load current and PV current can be modeled as total load current of FC and battery:

$$i_o = i_{load} - i_{PV}, \tag{5}$$

and

$$i_1 + i_2 = i_o. \tag{6}$$

when $i_o > 0$, i_1 is always greater than zero and:

$$\begin{cases} i_1 = G_1(s) \cdot i_o = \dfrac{sC_1R_2}{sC_1R_2 + 1} \cdot i_o \\ i_2 = G_2(s) \cdot i_o = \dfrac{1}{sC_1R_2 + 1} \cdot i_o \end{cases} \tag{7}$$

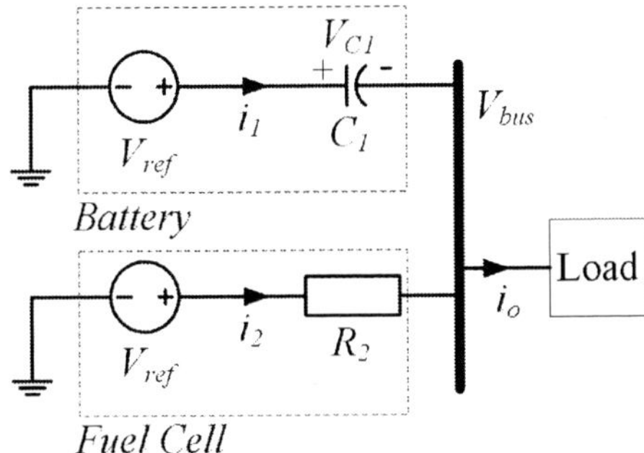

Fig. 2 Equivalent circuit of microgrid integrating with fuel cell and battery when VCD control is applied.

In (7), $G_1(s)$ acts like a high pass filter while $G_2(s)$ acts like a low pass filter. Thus, battery current (i_1) equals to the high frequency part of the load current while the FC current (i_2) equals to the low frequency part. Fig. 3 shows the dynamics of battery current (i_1) and FC current (i_2) when i_o is a step disturbance. It can be observed that battery current responds immediately and gradually reduces to zero. Meanwhile FC current increases gradually and finally reaches i_o. System dynamic is determined by the product of virtual resistance R_2 and capacitance C_1, which is expressed as

$$\tau = R_2 C_1 \tag{8}$$

where τ is the time constant which determines system dynamics.

The virtual resistor R_2 is calculated by

$$R_2 = \frac{V_{\max} - V_{\min}}{I_{\max}}, \tag{9}$$

where $V_{\max}$ $V_{\min}$ are the maximum and minimum output voltage and $I_{\max}$ is the maximum current of the FC converter.

Given the required system dynamics, virtual capacitor can be calculated by (8)(9). Therefore, a hybrid system under the proposed VCD control can be designed to achieve dynamic power sharing autonomously. This process is very similar to hybrid systems using a centralized controller and low pass filter to get the current reference of FC current, however, the proposed method doesn't require a centralized controller or a communication line to let power electronic converters (PECs) share load. It reduces the system complexity and cost.

The proposed control method can also be used for system with multiple FCs and batteries. Fig. 4 shows the equivalent circuit when multiple FCs and batteries are connected.

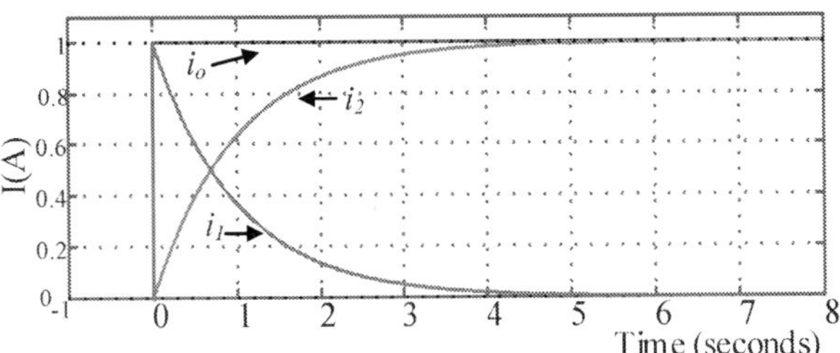

Fig. 3 Step responses of battery current (i_1) and FC current (i_2) when C_1R_2 = 1.

Fig. 4 can be simplified to Fig. 2 by taking into account that:

$$\frac{1}{R_2} = \sum_{m=1}^{M} \frac{1}{R_m}, C_1 = \sum_{n=1}^{N} C_n \tag{10}$$

When the capacity of the system is to be expanded, the virtual resistance (R_{M+1}) of the $(M+1)^{th}$ FC can be calculated by (9). The virtual capacitance of the new battery branch (C_{N+1}) is given by:

$$C_{N+1} = \frac{R_2}{R_{M+1}} \cdot C_1 \tag{11}$$

Substitution of (11) into (10) yields to

$$R_2' C_1' = \left(\frac{R_2 \cdot R_{M+1}}{R_2 + R_{M+1}}\right) \cdot \left(1 + \frac{R_2}{R_{M+1}}\right) \cdot C_1 = R_2 C_1 \tag{12}$$

It can be observed from (12) that the system dynamic won't change when the virtual resistance of the new battery branch is given by (11). Thus, the system capacity can be easily expanded without affecting the existing system dynamics.

IV. System Stability Analysis

Power electronic converters (PECs) are widely used in distributed electrical network for power conversion and motor drive. Loads associated with PECs can be treated as constant power loads. Constant power load (CPL) can be modeled as a current source paralleled with a negative resistor [20]. Constant power source (CPS) can be modeled the same way as CPL with the sign of current source and resistor reversed [14]. PV panels associated with PEC can be modeled as a CPS when it works in maximum power point tracking (MPPT)

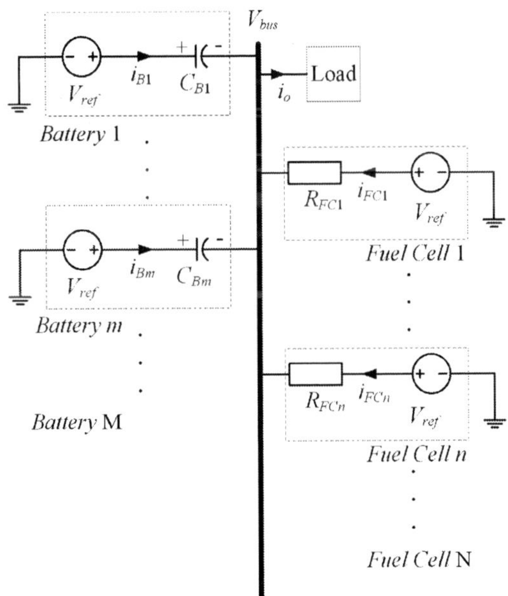

Fig. 4 Equivalent circuit of microgrid integrating with M FCs and N batteries when VCD control is applied.

mode. The electrical model of CPL and CPS are combined and is shown in Fig.5. System stability is to be investigated with the implementation of the proposed control method.

At steady state, V_{dc} can be expressed as

$$V_{dc}=\frac{V_{ref}\cdot\left(\frac{1}{R_2}+sC_1\right)-I_{CP}}{\frac{1}{R_2}+sC_1+\frac{1}{R_{CP}}+\frac{1}{R_{load}}} \tag{13}$$

where R_{CP} and I_{CP} are given by (14) according to [20].

$$\begin{aligned}R_{CP}&=-V_{dc}^{\ 2}/P_{CP}\\ I_{CP}&=2P_{CP}/V_{dc}\end{aligned} \tag{14}$$

When the inequality of (15) is valid, (13) has a right half panel (RHP) pole and the dc network will be unstable.

$$\left(1/R_2+1/R_{CP}+1/R_{load}\right)<0 \tag{15}$$

However, when R_{CP} is greater than zero, which means CPS rather than CPL is present, (15) is always invalid. Thus, the system is always stable and only the case of connecting CPL is needed for instability investigation.

When CPL is present, R_{CP} is less than zero and the inequality of (16) must be valid to ensure a stable system.

$$\frac{P_{CP}}{V_{dc}^{\ 2}}<\frac{1}{R_2}+\frac{1}{R_{load}} \tag{16}$$

Furthermore, in steady state i_2 is equal to i_o and the output power of the source versus the bus dc voltage is:

$$P_{out}=\frac{1}{R_2}\cdot\left(-V_{dc}^{\ 2}+V_{ref}V_{dc}\right) \tag{17}$$

It is easy to get the maximum output power of the source according to (17):

$$P_{out}(\max)=P_{\max}=\frac{V_{ref}^{\ 2}}{4R_2} \tag{18}$$

where $V_{dc}=0.5V_{ref}$.

Fig. 6 shows the output power (P_{out}) and current (i_o) versus dc bus voltage (V_{dc}). The maximum current (I_{max}) is V_{ref}/R_2 when V_{dc} is 0.

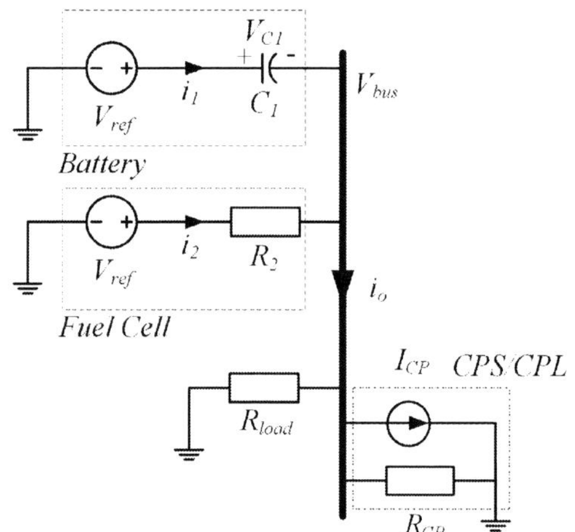

Fig. 5 Constant Power load / Constant Power Source.

When R_{load} approaches infinite, $P_{out}=P_{CP}$. DC bus voltage may have two solutions but only one is viable and system stability is explained in [12]. As shown in Fig. 6, system operates at point A and $P_{CP}=P_{old}$. When P_{old} jumps to P_{new}, dc bus voltage drops and i_o increases and system will work at point B. when system works at point C, when P_{CP} tries to increase to P_{new} and voltage drops while i_o increases. However, in this case, source power drops and dc voltage will decrease to zero. Thus system is unstable and only point A is a viable dc bus voltage. It can be concluded that when dc bus voltage is greater than $0.5V_{ref}$, the system is stable and this constraint can be used for protecting. When the PEC of the CPL detect the dc bus voltage is close to $0.5V_{ref}$, it may decrease the power demand to prevent crashing the overall system.

For more general cases, R_{load} is not infinite and (19) can be obtained by:

$$\left(\frac{1}{R_2}+\frac{1}{R_{load}}\right)\cdot V_{dc}^{\ 2}-\frac{V_{ref}}{R_2}\cdot V_{dc}+P_{CP}=0 \tag{19}$$

V_{dc} can be solved from (20).

$$V_{dc1,2}=\frac{\frac{V_{ref}}{R_2}\pm\sqrt{\left(\frac{V_{ref}}{R_2}\right)^2-4P_{CP}\left(\frac{1}{R_2}+\frac{1}{R_{load}}\right)}}{2(1/R_2+1/R_{load})} \tag{20}$$

When P_{CP} is less than zero, which means CPS is present, (20) has two real roots but only one is greater than zero. When CPL is present, P_{CP} is greater than zero and inequality of (21) must be valid to let (20) have real roots and stabilize the dc system.

$$P_{CP}\le\frac{1}{4}\left(\frac{V_{ref}}{R_2}\right)^2\cdot\left(\frac{R_2R_{load}}{R_2+R_{load}}\right)=P_{\max}\cdot\frac{1}{R_2/R_{load}+1} \tag{21}$$

It can be observed from (21) that increasing resistive load will decrease maximum possible CPL. The maximum possible power of CPL is P_{max}, which is defined by (18), when R_{load} is connected. The per-unit P_{CP} can be expressed by (22).

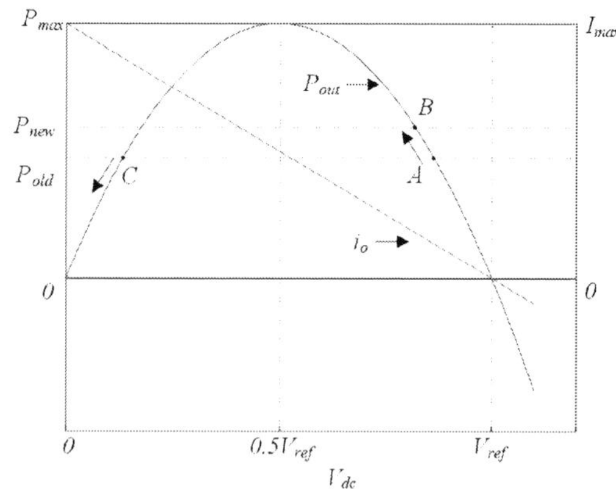

Fig. 6 Output power and current of the source verus dc bus voltage.

$$\frac{P_{CP}}{P_{\max}} = 4\left[-\left(1+\frac{R_2}{R_{load}}\right)\cdot\left(\frac{V_{dc}}{V_{ref}}\right)^2+\left(\frac{V_{dc}}{V_{ref}}\right)\right] \quad (22)$$

The 3D plot of P_{CP}/P_{max} versus R_2/R_{load} and V_{dc}/V_{ref} is shown in Fig. 7.

It can be observed from Fig. 7 and (22) that, when resistive load increase, the maximum possible power of CPL decrease and the lowest stable dc bus voltage decreases as well. System can operate under lower dc bus voltage than $0.5V_{ref}$. Fig. 8 shows the plot of P_{CP}/P_{max} versus V_{dc}/V_{ref} when R_{load} equals to R_2 and this figure can be obtained from Fig. 7 by setting $R_2/R_{load} = 1$.

Fig. 8 shows that, the maximum possible power of the CPL is $0.5P_{max}$ when $V_{dc} = 0.25V_{ref}$. However, when P_{CP} is 0, $V_{dc} = 0.5V_{ref}$, which is the voltage that will trigger protection. Even though the stable dc bus voltage can be extended by decreasing R_{load}, it is not practical to lower the dc voltage to trigger protection, since R_{load} may be disconnected any time and the lowest dc bus voltage, which may cause instability issue, will return to $0.5V_{ref}$. Thus,

$$R_{load} > R_2 \quad (23)$$

must be valid to ensure that the dc bus voltage is higher than $0.5V_{ref}$. The 3D plot of possible power of CPL versus V_{dc} and R_{load} is shown in Fig. 9 under the aforementioned constrains.

As a conclusion, (16), (21) and (23) ensure the stability of the proposed dc system and $0.5V_{ref}$ can be used as a criterion for protection.

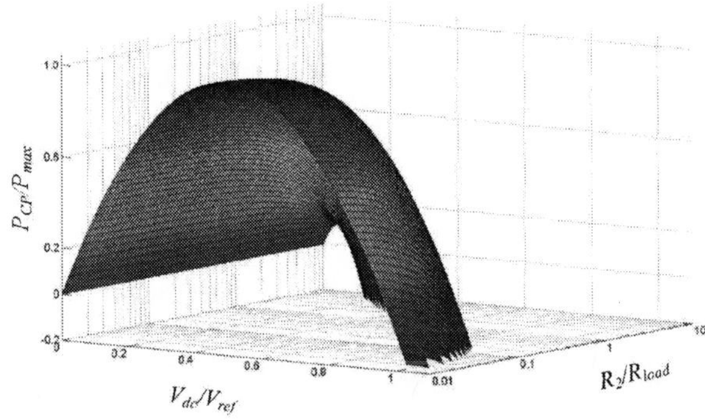

Fig.7 3D plot of P_{CP}/P_{max} versus R_2/R_{load} and V_{dc}/V_{ref}.

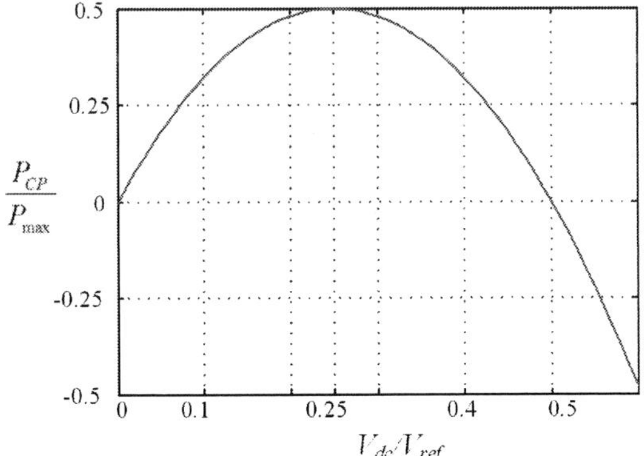

Fig. 8 Plot of P_{CP}/P_{max} versus R_2/R_{load} and V_{dc}/V_{ref}.

V. Simulation and Experimental Results

The proposed VCD control method is verified by PLECS. Fig. 10 shows the system diagram of the proposed dc microgrid for simulation and experiment. Table I shows the summary of the parameters for simulation and experiment.

TABLE I. SUMMARY OF PARAMETERS FOR SIMULATION

Symbol	Description	Value
V_{ref}	Normal DC bus voltage	270 V
C_1	Virtual Capacitance	$1/0.2\pi$ F
R_2	Virtual Resistance	1 Ω
V_B	Battery voltage	100V
V_{FC}	FC voltage	50V
L_{FC}, L_B	Inductors	0.001H
C_{FC},C_B	Capacitors	220uF

A. Simulation Results

1) Normal operation

Fig. 11 shows the simulation results of the proposed system under a step constant power load perturbation. I_{FC} is the current of FC and I_B is current of battery. At 5 s, the CPL jumps from 0 kW to 1 kW, I_B increases rapidly while I_{FC} increase slowly. After several seconds, output power of FC gradually reaches the total load power and I_B decreases to zero. At 15 s, constant power load changes from 1 kW to 0, I_B increases rapidly in the negative direction to absorb the excess current from the FC. I_{FC} decreases gradually with I_B and both of them reach zero eventually. This shows the effectiveness of the proposed controller.

2) Unstable situation

Fig. 12 shows simulation result of unstable situation. According to (18), the maximum CPL is calculated as 18.225kW and dc voltage should be maintained above 135V. When CPL is 18 kW, system operates normally. At 20s, CPL is increased to 19 kW and bus voltage will gradually drop below 135V to 100V, which is the battery voltage. This indicates that battery converter doesn't work anymore and the boost diode is always conducting. At steady state, battery supplies part of load power, which is unexpected. In real case,

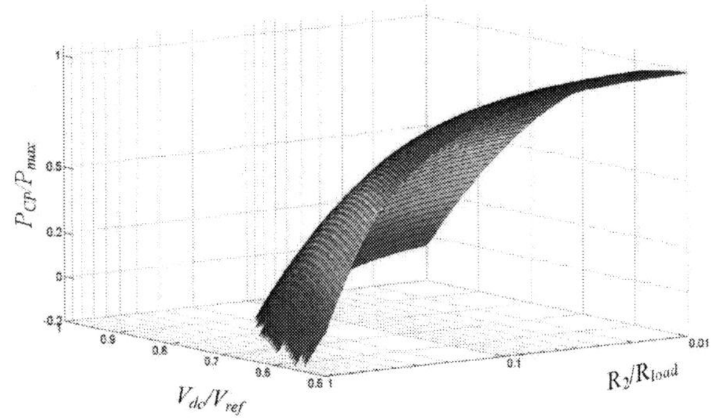

Fig.9 3D plot of P_{CP}/P_{max} versus R_2/R_{load} and V_{dc}/V_{ref} when $R_{load} > R_2$ and $V_{dc} > V_{ref}$.

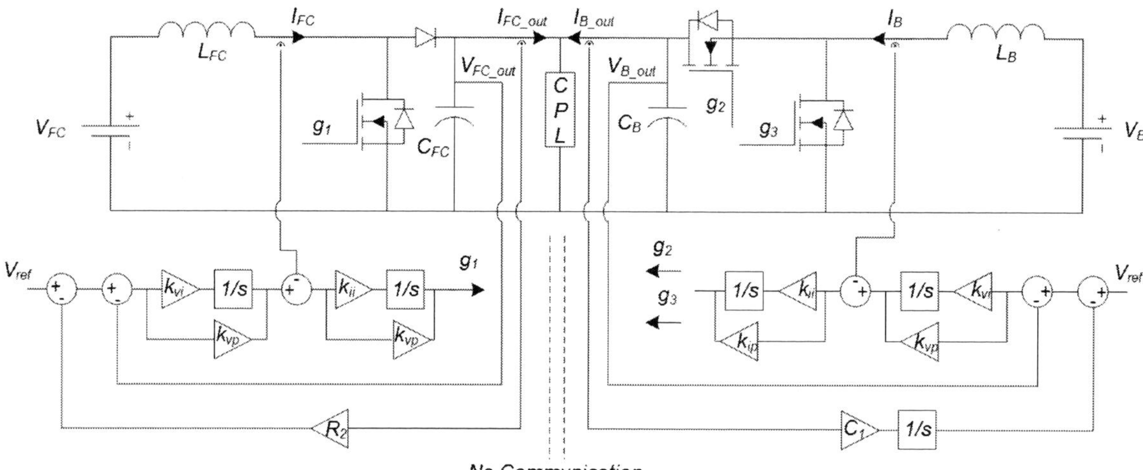

Fig. 10 DC microgrid for simulation and hardware experiment.

because of constraints such as converter capacity, output power won't exceed maximum CPL and bus voltage will not drop below $0.5V_{ref}$. Therefore, stable operation can always be ensured.

B. Hardware Experimental Results

A laboratory test platform is built to verify the effectiveness of proposed VCD control. A 95V Li-ion battery bank which has fast dynamic response and high power density is to compensate transient fluctuations and represents battery in the paper. A 195V lead-acid battery bank which has slow dynamic response is to provide consistent power at steady and represents FC. Experimental results under normal operation are shown in Fig. 13. Fig. 13(a) shows dynamic response of FC voltage V_{FC}, dc bus voltage V_{dc}, output current of FC converter I_{FC_out}, FC current I_{FC}. Fig. 13(b) shows dynamic response of battery voltage V_B, dc bus voltage V_{dc}, output current of battery converter I_{B_out}, battery current I_B. When CPL increases 300W, FC current I_{FC} increases gradually and becomes constant at steady state steady state; battery current I_B increases rapidly at transient and decreases to zero during steady state. When CPL decreases 300W, FC current I_{FC} decreases gradually and becomes constant at steady state steady state; battery current I_{B_L} increases rapidly in the negative direction at transient and decreases to zero during steady state. Therefore, battery provides power at transient and FC provides power at steady state. This validates the effectiveness of the proposed VCD control method.

It can also be observed that in experiment result, FC current dynamic response is not as ideal as simulation result. It is because current response is also impacted by voltage control, which will be illustrated in the future work.

VI. Conclusion

This paper proposes a virtual capacitor droop (VCD) control method for transient and steady state power sharing between fuel cell and battery. Output voltage characteristics

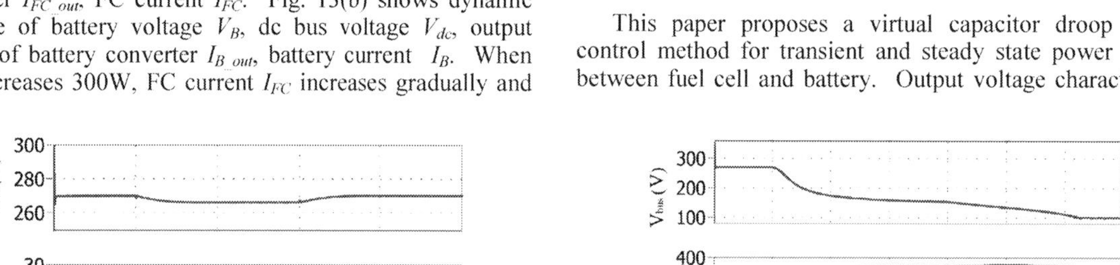

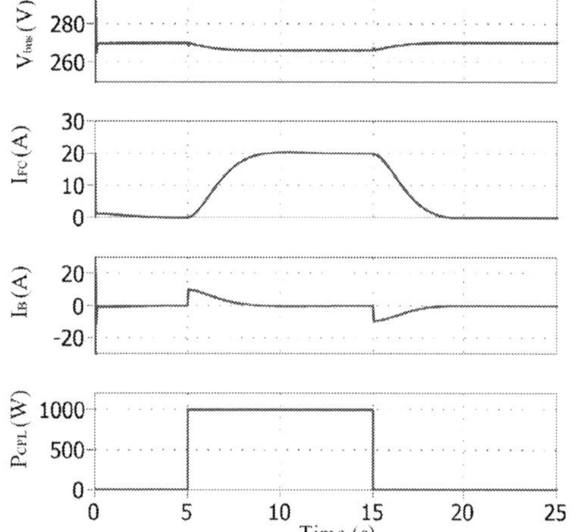

Fig. 11 Simulation results of normal situation.

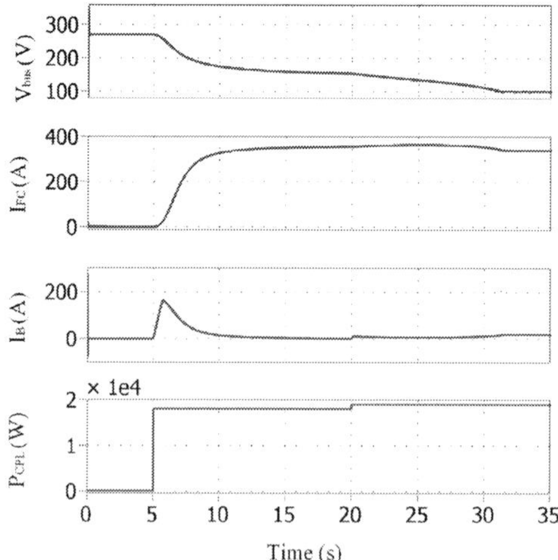

Fig. 12 Simulation results of unstable situation.

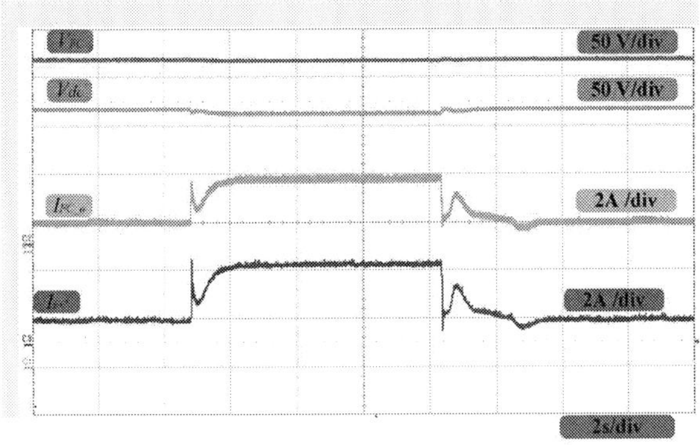

(a)

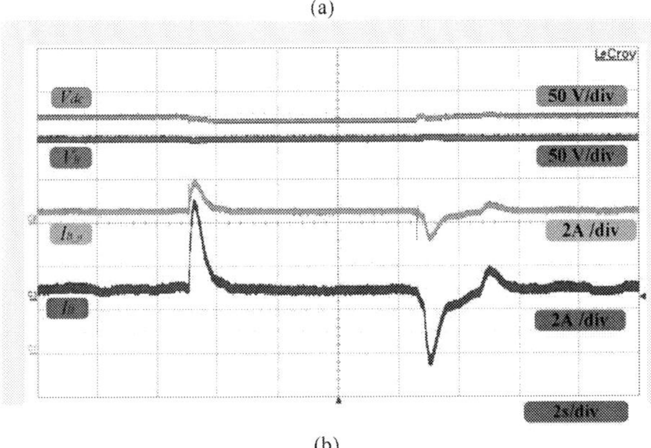

(b)

Fig.13 Experiment results. (a) Dynamic response of FC; (b) Dynamic response of battery.

of the FC and battery are regulated by a virtual resistor and a virtual capacitor respectively. This method achieves autonomous power sharing during system transits and steady state. Battery will compensate transit power to maintain DC bus voltage and FC will provide power during steady state. The proposed droop controllers inherently act as a low-pass filter and high-pass filter for FC and battery to load response respectively. Compared with virtual impedance control method which adds low-pass and high-pass filters, the proposed method simplifies the controller design and makes system analysis easily without any tradeoff in the effectiveness of power management strategy and stabilization of bus voltage. An autonomous DC MG is designed based on this method. The proposed VCD control method for DC MG is also extended to integrate any number of FCs and batteries without impacting system dynamics. System stability with the penetration of CPL is analyzed and system constraints are discussed. Simulations and experiments are conducted to validate the proposed VCD control method and the analytical results.

REFERENCES

[1] A. T. Ghareeb, A. a. Mohamed, and O. a. Mohammed, "DC microgrids and distribution systems: An overview," *IEEE Power Energy Soc. Gen. Meet.*, 2013.

[2] V. Nasirian, S. Moayedi, A. Davoudi, and F. Lewis, "Distributed Cooperative Control of DC Microgrids," *IEEE Trans. Power Electron.*, vol. 8993, no. c, pp. 1–1, 2014.

[3] Q. Shafiee, T. Dragicevic, J. C. Vasquez, and J. M. Guerrero, "Hierarchical control for multiple DC-microgrids clusters," *IEEE Trans. Energy Convers.*, vol. 29, no. 4, pp. 922–933, 2014.

[4] J. Torreglosa, P. Garcia, L. Fernandez, and F. Jurado, "Predictive Control for the Energy Management of a Fuel Cell-Battery-Supercapacitor Tramway," *IEEE Trans. Ind. Informatics*, vol. 10, no. 1, pp. 1–1, 2013.

[5] A. M. O. Haruni, M. Negnevitsky, M. E. Haque, and A. Gargoom, "A Novel Operation and Control Strategy for a Standalone Hybrid Renewable Power System," *IEEE Trans. Sustain. Energy*, vol. 4, no. 2, pp. 402–413, Apr. 2013.

[6] M. Zandi, A. Payman, J. Martin, S. Pierfederici, B. Davat, and F. Meibody-Tabar, "Energy Management of a Fuel Cell / Supercapacitor / Battery Power Source for Electric Vehicular Applications," *IEEE Trans. Veh. Technol.*, vol. 60, no. 2, pp. 433–443, 2011.

[7] S. Njoya Motapon, L. Dessaint, and K. Al-Haddad, "A Comparative Study of Energy Management Schemes for a Fuel-Cell Hybrid Emergency Power System of More-Electric Aircraft," *IEEE Trans. Ind. Electron.*, vol. 61, no. 3, pp. 1320–1334, Mar. 2014.

[8] J. Shen and A. Khaligh, "A Supervisory Energy Management Control Strategy for a Battery/ Ultracapacitor Hybrid Energy Storage System," *IEEE Trans. Transp. Electrif.*, vol. 7782, no. 1, pp. 1–1, 2015.

[9] J. Li, Y. Liu, S. Wang, J. Shi, L. Ren, K. Gong, and Y. Tang, "Design and advanced control strategies of a hybrid energy storage system for the grid integration of wind power generations," *IET Renew. Power Gener.*, vol. 9, no. 2, pp. 89–98, 2015.

[10] Y. Gu, X. Xiang, W. Li, and X. He, "Mode-Adaptive Decentralized Control for Renewable DC Microgrid With Enhanced Reliability and Flexibility," *IEEE Trans. Power Electron.*, vol. 29, no. 9, pp. 5072–5080, 2014.

[11] P. Huang, P. Liu, S. S. Member, W. Xiao, and S. S. Member, "A Novel Droop-Based Average Voltage Sharing Control Strategy for DC Microgrids," *IEEE Trans. Smart Grid*, vol. 6, no. 3, pp. 1096–1106, 2015.

[12] T. Dragicevic, J. M. Guerrero, J. C. Vasquez, and D. Skrlec, "Supervisory control of an adaptive-droop regulated DC microgrid with battery management capability," *IEEE Trans. Power Electron.*, vol. 29, no. 2, pp. 695–706, 2014.

[13] V. Nasirian, A. Davoudi, F. L. Lewis, and J. M. Guerrero, "Distributed Adaptive Droop Control for DC Distribution Systems," *IEEE Trans. Energy Convers.*, vol. 29, no. 4, pp. 944–956, Dec. 2014.

[14] S. Anand, B. G. Fernandes, and J. Guerrero, "Distributed Control to Ensure Proportional Load Sharing and Improve Voltage Regulation in Low-Voltage DC Microgrids," *IEEE Trans. Power Electron.*, vol. 28, no. 4, pp. 1900–1913, Apr. 2013.

[15] Y. Gu, W. Li, and X. He, "Frequency-Coordinating Virtual Impedance for Autonomous Power Management of DC Microgrid," *IEEE Trans. Power Electron.*, vol. 30, no. 4, pp. 2328–2337, 2015.

[16] Z. Bubnicki, *Modern control theory*. Springer Science & Business Media, 2005.

[17] X. Lu, K. Sun, L. Huang, J. M. Guerrero, J. C. Vasquez, and Y. Xing, "Virtual impedance based stability improvement for DC microgrids with constant power loads," *Energy Convers. Congr. Expo. (ECCE), 2014 IEEE*, no. 51177083, pp. 2670–2675, 2014.

[18] A. Emadi, A. Khaligh, C. H. Rivetta, and G. A. Williamson, "Constant power loads and negative impedance instability in automotive systems: definition, modeling, stability, and control of power electronic converters and motor drives," *IEEE Trans. Veh. Technol.*, vol. 55, no. 4, pp. 1112–1125, 2006.

[19] J. Zeng, Z. Zhang, and W. Qiao, "An interconnection and damping assignment passivity-based controller for a DC-DC boost converter with a constant power load," *IEEE Trans. Ind. Appl.*, vol. 50, no. 4, pp. 2314–2322, 2014.

[20] L. Guo, Y. Feng, X. Li, C. Wang, and Y. Li, "Stability Analysis of a DC Microgrid with Master- Slave Control Structure," in *Energy Conversion Congress and Exposition (ECCE), 2014 IEEE*, 2014, pp. 5682–5689.

MPC-SVM Method for Vienna Rectifier with PMSG used in Wind Turbine Systems

June-Seok Lee
Korea Railroad Research Institute
Uiwang, South Korea
ljs@krri.re.kr

Yeongsu Bak and Kyo-Beum Lee
Department of Electrical and Computer Engineering
Ajou University
Suwon, South Korea
wov2@ajou.ac.kr, kyl@ajou.ac.kr

Frede Blaabjerg
Department of Energy Technology
Aalborg University
Aalborg, Denmark
fbl@et.aau.dk

***Abstract*—Using a Vienna rectifier as the machine-side rectifier of back-to-back converter is advantageous in terms of size and cost compared to three-level topologies and for this reason, the Vienna rectifier has been used in Wind Turbine Systems (WTS). This paper proposes a Model Predictive Control (MPC) method for the Vienna rectifier used in WTS with a Permanent Magnet Synchronous Generator (PMSG). The proposed MPC method considers the feasible eight-voltage vectors of the Vienna rectifier. In addition, the voltage vectors, which are the center voltage vectors of two feasible adjacent voltage vectors, are taken into consideration to improve the performance of the MPC method. The optimized voltage vector for the ripple minimization of PMSG currents is determined by cost function. Then, the neutral-point voltage unbalancing problem is considered for selecting the final switching set, which is generated by the space modulation method. The effectiveness and performance of the proposed MPC method is verified by simulations.**

I. Introduction

Applications using Vienna rectifier have been expanded from grid connected systems such as the telecommunication systems [1] to Wind Turbine Systems (WTS) [2]-[3]. Various Vienna rectifiers, which are realized from many topologies [4]-[5], have the same performances as three-level topologies (neutral-point clamped, T-type); nevertheless, the Vienna rectifier can reduce the topology components [1]-[3]. This gives some advantages in terms of size and cost. However, the Vienna rectifier is boost-type unidirectional rectifier; therefore, it has some limitations in the variation of the power factor [6].

Many control methods for the Vienna rectifier have been proposed [1]-[3],[6]-[8]. In the Vienna rectifier, one of the main control aims is to make the input currents to have sinusoidal waveforms. The hysteresis current control method is used in [1] and it is a classic control method. The switching set is determined by the rule defined from the hysteresis current control. The classic linear controllers such as the proportional-integral controller are used in [3],[6]-[8] and the switching set is generated by comparing the reference voltages with the carrier voltages [6]-[7] or by the space vector modulation method [8]. The aforementioned control methods have been widely used in other topologies such as two-level converter and three-level converter. The Direct Torque Control (DTC) method and the Model Predictive Control (MPC) method with cost function have been proposed for the

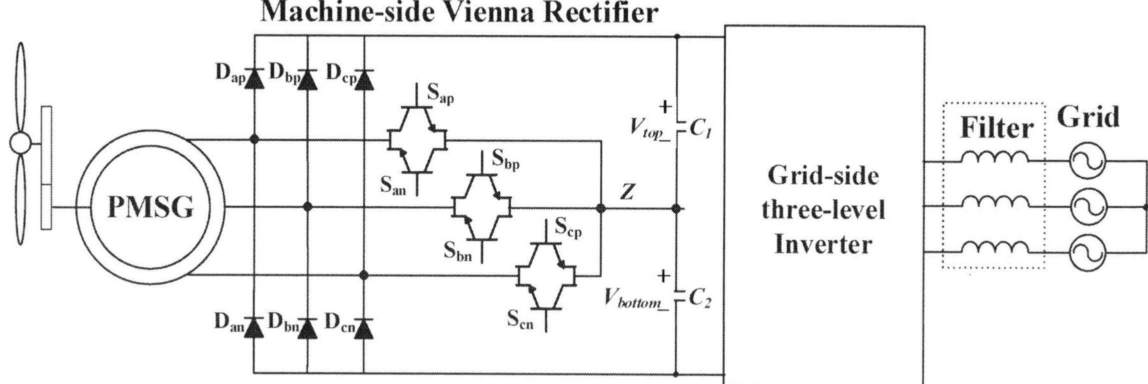

Fig. 1. Back-to-back converter using a Vienna rectifier in a wind turbine system.

applications using a motor or generator [9]-[10]. Research on the DTC and MPC methods at two-level converter, three-level converter, and matrix converter are being carried out consistently. Recently, DTC method for the Vienna rectifier was proposed in [2].

This paper proposes the MPC method for the Vienna rectifier used in WTS with a Permanent Magnet Synchronous Generator (PMSG). The proposed MPC method considers the feasible eight-voltage vectors of the Vienna rectifier; moreover, the additional voltage vectors, which are the center voltage vectors of two feasible adjacent voltage vectors, are taken into consideration to improve the performance of the MPC method. In the proposed MPC method, the errors in d/q-axis current are predicted based on model of the PMSG, and then, the optimized voltage vector is selected by cost function for the current ripple minimization. Additionally, from the optimized voltage vector, N-type and P-type voltage vectors, which give the different effect on the neutral-point voltage, are considered in selecting the final switching set to solve the neutral-point voltage unbalancing problem. The final switching set is generated by the Space Vector Modulation (SVM) method. Simulations are conducted to show the performance of the proposed MPC method.

II. PMSG Modle

The PMSG voltage equations in the d/q-axis synchronous reference frame can be expressed as

$$V_{sd}(t)=\omega_e L_{sq} i_{sq}(t)-R_s i_{sd}(t)-L_d\frac{di_{sd}(t)}{dt}$$
$$V_{sq}(t)=\omega_e\lambda_r+\omega_e L_{sd} i_{sd}(t)-R_s i_{sq}(t)-L_q\frac{di_{sd}(t)}{dt}, \quad (1)$$

where V_{sd} and V_{sq} are the d-axis and q-axis voltages generated by rectifier, I_{ds} and I_{qs} are d-axis and q-axis currents of the PMSG, R_s is the stator resistance of the PMSG, L_{sd} and L_{sq} are the d-axis and q-axis inductances of the PMSG stator, ω_e is the angular frequency of the PMSG, and λ_r is the flux of the PMSG rotor magnet.

Equation (1) can be represented as new set equations in discrete terms and these are expressed as

$$\frac{i_{sd}(t+1)-i_{sd}(t)}{T_s}=\frac{1}{L_d}(\omega_e L_{sq} i_{sq}(t)-R_s i_{sd}(t)-V_{sd}(t))$$
$$\frac{i_{sq}(t+1)-i_{sq}(t)}{T_s}=\frac{1}{L_q}(\omega_e\lambda_r+\omega_e L_{sd} i_{sd}(t)-R_s i_{sq}(t)-V_{sq}(t)) \quad ; (2)$$

where T_s is the sampling time.

Depending on $V_{sd}(t)$ and $V_{sq}(t)$, the predicted currents $i_{sd}(t+1)$ and $i_{dq}(t+1)$ can be represented as

$$i_{sd}(t+1)=$$
$$=\frac{T_s}{L_{sd}}(\omega_e L_{sq} i_{sq}(t)-V_{sd}(t))+(1-\frac{T_s R_s}{L_{sd}})i_{sd}(t)$$
$$i_{sq}(t+1)$$
$$=\frac{T_s}{L_{sq}}(\omega_e\lambda_r+\omega_e L_{sd} i_{sd}(t)-V_{sq}(t))+(1-\frac{T_s R_s}{L_{sq}})i_{sq}(t) \quad . (3)$$

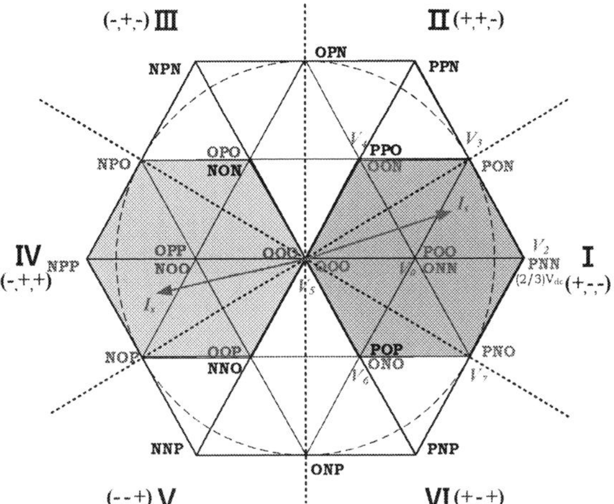

Fig. 2. Space vector diagram of the Vienna rectifier.

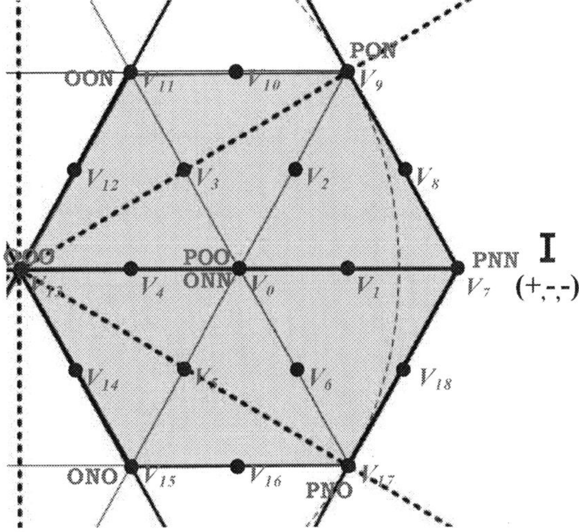

Fig. 3. Feasible voltage vectors in the proposed MPC method.

III. Model Predictive Control Using Vienna Rectifier for PMSG

In the Vienna rectifier, the switching set should be selected by considering the polarity of the currents. Therefore, the feasible switching sets are limited compared with the three-level converter. The space vector diagram of the Vienna rectifier can be expressed as shown Fig. 2 and this is similar to that of the three-level converter.

The space vector diagram is divided into six Sectors depending on the polarity of currents. The OOO, OON, POO, ONN, ONO, PON, PNO, and PNN switching sets can be selected if the current vector I_s is in Sector I. This complies with the special operation requirement of a Vienna rectifier [5]: the polarity of a input voltage of a Vienna rectifier should be same as the polarity of a corresponding phase current. Therefore, in contrast to the MPC method using all switching sets (27-switching sets) in three-level converter, the MPC

TABLE I. FEASIBLE VOLTAGE VECTORS IN SECTOR I.

Vector	V_{ds}	V_{qs}	Vector	V_{ds}	V_{qs}
V_0	$2V_{dc}/3 \times 2/4$	0	V_{10}	$2V_{dc}/3 \times 4/8$	$\sqrt{3}\,V_{dc}/3 \times 1/2$
V_1	$2V_{dc}/3 \times 3/4$	0	V_{11}	$2V_{dc}/3 \times 2/8$	$\sqrt{3}\,V_{dc}/3 \times 1/2$
V_2	$2V_{dc}/3 \times 5/8$	$\sqrt{3}\,V_{dc}/3 \times 1/4$	V_{12}	$2V_{dc}/3 \times 1/8$	$\sqrt{3}\,V_{dc}/3 \times 1/4$
V_3	$2V_{dc}/3 \times 3/8$	$\sqrt{3}\,V_{dc}/3 \times 1/4$	V_{13}	0	0
V_4	$2V_{dc}/3 \times 1/4$	0	V_{14}	$2V_{dc}/3 \times 1/8$	$-\sqrt{3}\,V_{dc}/3 \times 1/4$
V_5	$2V_{dc}/3 \times 3/8$	$-\sqrt{3}\,V_{dc}/3 \times 1/4$	V_{15}	$2V_{dc}/3 \times 2/8$	$-\sqrt{3}\,V_{dc}/3 \times 1/2$
V_6	$2V_{dc}/3 \times 5/8$	$-\sqrt{3}\,V_{dc}/3 \times 1/4$	V_{16}	$2V_{dc}/3 \times 4/8$	$-\sqrt{3}\,V_{dc}/3 \times 1/2$
V_7	$2V_{dc}/3$	0	V_{17}	$2V_{dc}/3 \times 6/8$	$-\sqrt{3}\,V_{dc}/3 \times 1/2$
V_8	$2V_{dc}/3 \times 7/8$	$\sqrt{3}\,V_{dc}/3 \times 1/4$	V_{18}	$2V_{dc}/3 \times 7/8$	$-\sqrt{3}\,V_{dc}/3 \times 1/4$
V_9	$2V_{dc}/3 \times 3/4$	$\sqrt{3}\,V_{dc}/3 \times 1/2$			

TABLE II. SWITCHING SETS USED IN THE PROPOSED MPC METHOD.

Sector I			Sector II		
Vector	Switching set		Vector	Switching set	
	$V_{top} > V_{Bottom}$	$V_{top} < V_{Bottom}$		$V_{top} > V_{Bottom}$	$V_{top} > V_{Bottom}$
V_0	[ONN]	[POO]	V_0	[OON]	[PPO]
V_1	[ONN]-[PNN]	[POO]-[PNN]	V_1	[OON]-[PPN]	[PPO]-[PNN]
V_2	[ONN]-[PON]	[POO]-[PON]	V_2	[OON]-[OPN]	[PPO]-[OPN]
V_3	[OON]-[ONN]	Cannot be selected	V_3	Cannot be selected	[PPO]-[PPO]
V_4	[ONN]-[OOO]	[POO]-[OOO]	V_4	[OON]-[OOO]	[PPO]-[OOO]
V_5	[NON]-[ONN]	Cannot be selected	V_5	Cannot be selected	[PPO] -[PPO]
V_6	[ONN]-[PNO]	[POO]-[PNO]	V_6	[OON]-[PON]	[PPO]-[PON]
V_7	[PNN]		V_7	[PPN]	
V_8	[PON]-[PNN]		V_8	[OPN]-[PPN]	
V_9	[PON]		V_9	[OPN]	
V_{10}	[OON]-[PON]	Cannot be selected	V_{10}	Cannot be selected	[OPN]-[OPO]
V_{11}	[ONN]	Cannot be selected	V_{11}	Cannot be selected	[OPO]
V_{12}	[OON]-[OOO]	Cannot be selected	V_{12}	Cannot be selected	[OPO]-[OOO]
V_{13}	[OOO]		V_{13}	[OOO]	
V_{14}	[POP]-[OOO]	Cannot be selected	V_{14}	Cannot be selected	[POO]-[OOO]
V_{15}	[ONO]	Cannot be selected	V_{15}	Cannot be selected	[OOP]
V_{16}	[PNO]	Cannot be selected	V_{16}	Cannot be selected	[POO]-[PON]
V_{17}	[PNO]		V_{17}	[PON]	
V_{18}	[PNO]-[PNN]		V_{18}	[PON]-[PPN]	

method for the Vienna rectifier can only consider the eight switching sets according to Sector. The feasible switching sets can be represented as the d/q-axis voltage vectors. Then, only the seven voltage vectors can be used to calculate the current prediction in (3).

In this paper, ten voltage vectors are additionally considered to improve the performance of the MPC method as shown in Fig. 3. The d/q-axis components of the total voltage vectors are summarized in Table I.

By considering the nineteen voltage vectors, (3) can be represented as

$$
\begin{aligned}
& i_{sd}^{k}(t+1) = F_d(V_{sd}^{k}(t)) \\
& = \frac{T_s}{L_{sd}}(\omega_e L_{sq} i_{sq}(t)) + (1 - \frac{T_s R_s}{L_{sd}}) i_{sd}(t) - \frac{T_s}{L_{sd}} V_{sd}^{k}(t) \\
& i_{sq}^{k}(t+1) = F_q(V_{sq}^{k}(t)) \\
& = \frac{T_s}{L_{sq}}(\omega_e \lambda_r + \omega_e L_{sd} i_{sd}(t)) + (1 - \frac{T_s R_s}{L_{sq}}) i_{sq}(t) - \frac{T_s}{L_{sq}} V_{sq}^{k}(t)
\end{aligned}
, \quad (4)
$$

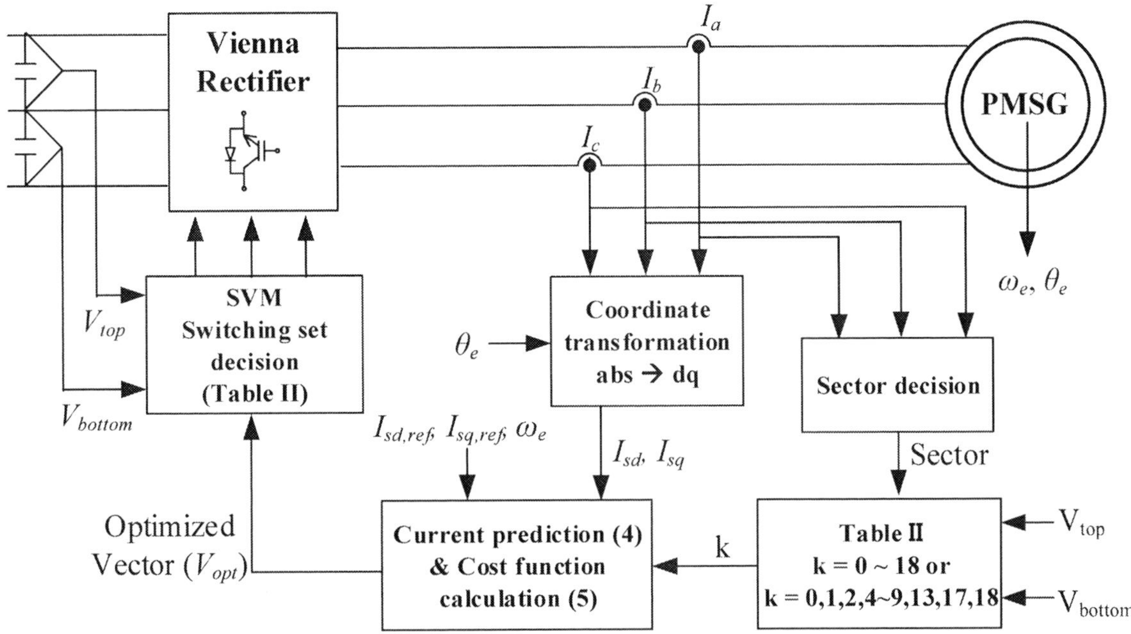

Fig. 4. Block diagram of machine-side Vienna rectifier control using the proposed MPC method.

where k is the vector number from 0 to 18.

The predicted currents of (4) are used in the cost function (CF). The CF has two error terms of the d/q-axis current and it is expressed as

$$CF = \left| i_{sd,ref} - i_{sd}^{k}(t+1) \right| + \left| i_{sq,ref} - i_{sq}^{k}(t+1) \right|, \tag{5}$$

where the $I_{sd,ref}$ and $I_{sq,ref}$ are the d-axis reference current and q-axis reference current in the d/q-axis synchronous reference frame, respectively. The voltage vector leading to the minimum CF is determined as the optimized voltage vector (V_{opt}) for the switching set of the Vienna rectifier.

The optimized voltage vector does not guarantee the neutral-point voltage balancing which is an important issue in topologies using two dc-link capacitors such as the Vienna rectifier and three-level converter. Each voltage vector causes different effect on the neutral-point voltage which is the difference between the top dc-link capacitor voltage (V_{top}) and the bottom dc-link capacitor voltage (V_{bottom}). Especially, the voltage vectors represented by the small vector influence the variation of the neutral-point voltage largely. In Fig. 3, V_0 can be realized by only POO switching set (the P-type small vector), only ONN switching set (the N-type small), or the sum of POO and ONN switching sets. V_1, V_2, V_4, and V_6 have the same characteristic with V_0. However, V_3, V_5, V_{10}, V_{11}, V_{12}, V_{14}, V_{15}, and V_{16} in Sector I should be realized by the switching sets represented by the N-type small vector because only N-type small vectors satisfy the operation requirement of the Vienna rectifier. On the other hand, these vectors in Sector II should be realized by the switching sets represented by the P-type small vector. Other vectors, V_7, V_8, V_9, V_{13}, V_{17}, and V_{18} do not relate with the small vectors; therefore, these do not have the effect on the neutral-point voltage. In conclusion, the principle of the proposed MPC method for the neutral-point voltage balancing is summarized as:

- The unidirectional neutral-point voltage control: V_3, V_5, V_{10}, V_{11}, V_{12}, V_{14}, V_{15} and V_{16} should be considered before the current prediction (4) because these voltage vectors can cause one-sided variation in the neutral-point voltage according to Sector.
- The bidirectional neutral-point voltage control: V_0, V_1, V_2, V_4 and V_6 should be considered after the current prediction (4) because these voltage vectors can control the neutral-point voltage by adjusting the ratio of P-type and N-type small vectors.
- No neutral-point voltage control capacity: V_7, V_8, V_9, V_{13}, V_{17} and V_{18} are considered in the current prediction (4) regardless of the neutral-point voltage because these voltage vectors have the minor influence on the variation of the neutral-point voltage.

The SVM method is used to implement the final switching set by following the principle for the neutral-point voltage balancing. The switching sets of each voltage vector used in the proposed MPC method are represented in Table II. In odd Sectors (I, III, and V), V_3, V_5, V_{10}, V_{11}, V_{12}, V_{14}, V_{15} and V_{16} can be selected in the case when V_{top} is larger than V_{bottom} and vice versa in even Sectors (II, IV, and VI). The block diagram of the proposed MPC method is shown in Fig. 4.

IV. Simulation Results

The simulations are performed to identify the performance of the proposed MPC method. It is implemented by using

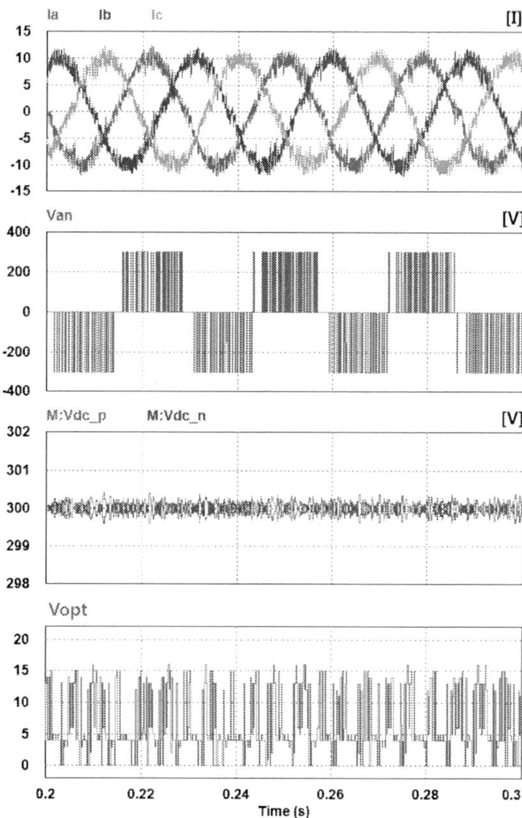

Fig. 5. Simulation result of the proposed MPC method when the PMSG speed is 700 rpm.

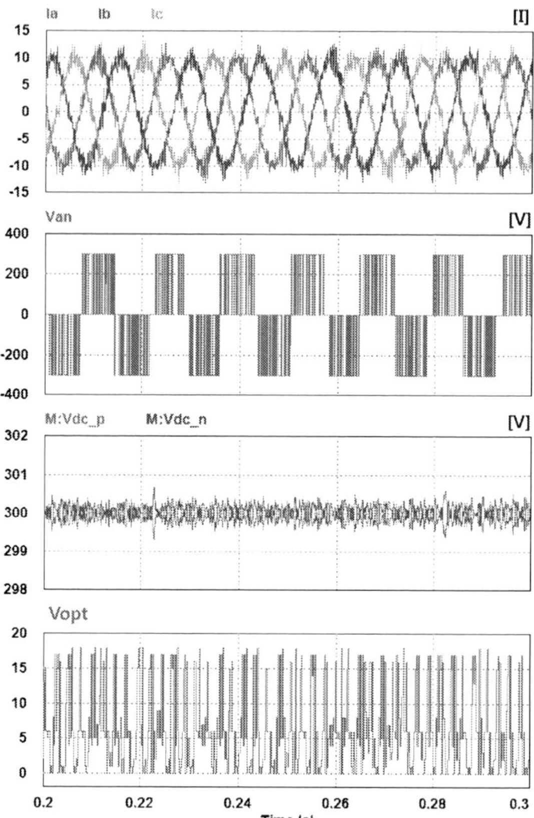

Fig. 6. Simulation result of the proposed MPC method when the PMSG speed is 1400 rpm.

TABLE III. PMSG PARAMETERS.

Parameter	Value
Rated power	11 kW
Number of pole	6
Rated voltage (line-to-line)	380 V_{rms}
Rated current	19.9 A_{rms}
Rated speed	1750 rpm
Resistance	0.349 Ω
q-inductance	15.6 mH
d-inductance	13.16 mH

PSIM tool. The simulation circuit is the same as shown Fig. 1. The PMSG parameters are given in Table III and simulation parameters are as follows: the sampling period is 200 us, the dc-link voltage is 600 V, and two dc-link capacitors are 2200 uF respectively.

Fig. 5 and Fig. 6 show the performance of the proposed MPC method at different speeds. The d-axis reference current is 0 A and the q-axis reference current ($I_{sq,ref}$) is 10 A. The polarity of the a-phase current (I_a) is the same as the polarity of the a-phase pole voltage (V_{an}) in both figures. Through these results, it can be seen that the operation requirement of the Vienna rectifier is satisfied and the Vienna rectifier operates properly. In addition, two dc-link voltages (V_{top}, V_{bottom}) is controlled as the same value by the proposed MPC method with the sinusoidal three phase currents. In addition, the selecting tendency of the optimized voltage vector is shown in both figures. It can be seen that the selecting tendency at 700 rpm is not the same as that of 1400 rpm. It is due to the optimized voltage vector selected from the results of (5) is related to the PMSG speed.

In Fig. 7, the q-axis reference current ($I_{sq,ref}$) is changed from 5 A to 10 A at 0.2 s and from 10 A to 5 A at 0.22 s. Three-phase currents are controlled according to the reference currents and the neutral-point voltage balancing is also achieved. In addition, it is known that the operation requirement of the Vienna rectifier is satisfied through the a-phase pole voltage (V_{an}) and a-phase current (I_a) constantly.

Fig. 8 shows the simulation results when a resistor (300 Ω) is connected to bottom-side dc-link capacitor (V_{bottom}) at 0.4 s to demonstrated the control capabilities. It can be seen that the power unbalance of two dc-link capacitors caused by the resistor is existing through the currents (I_{top}, I_{bottom}) flowing into two capacitors. Nevertheless, the neutral-point voltage balancing is maintained constant.

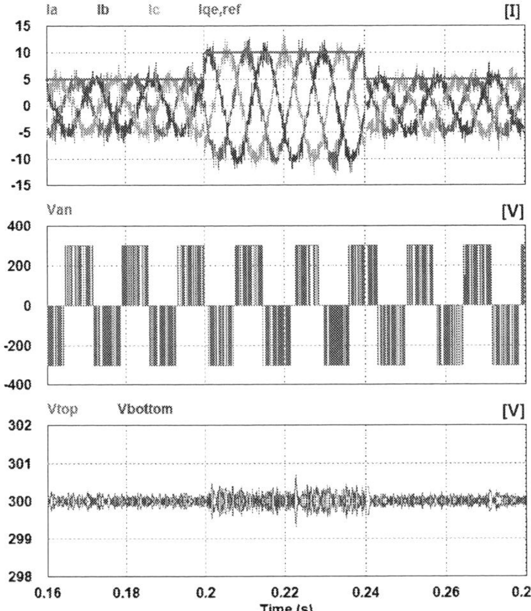

Fig. 7. Simulation result of the proposed MPC method when the q-axis reference current is changed at 1400 rpm.

V. CONCLUSION

This paper proposes a MPC method for a Vienna rectifier used in wind turbine system with PMSG. The additional voltage vectors, which are the center voltage vectors of two feasible adjacent voltage vectors, are taken into consideration to improve the performance of the MPC method. The operation requirement for the Vienna rectifier is considered in selecting the candidates of the voltage vectors to calculate cost function. In addition, the effect of each voltage vector on the neutral-point voltage is reflected for selecting the candidates of voltage vectors. The final switching set is generated by the SVM method by considering the neutral-point voltage balancing. The simulations are conducted to identify the performances of the proposed MPC method for the Vienna rectifier based PMSG.

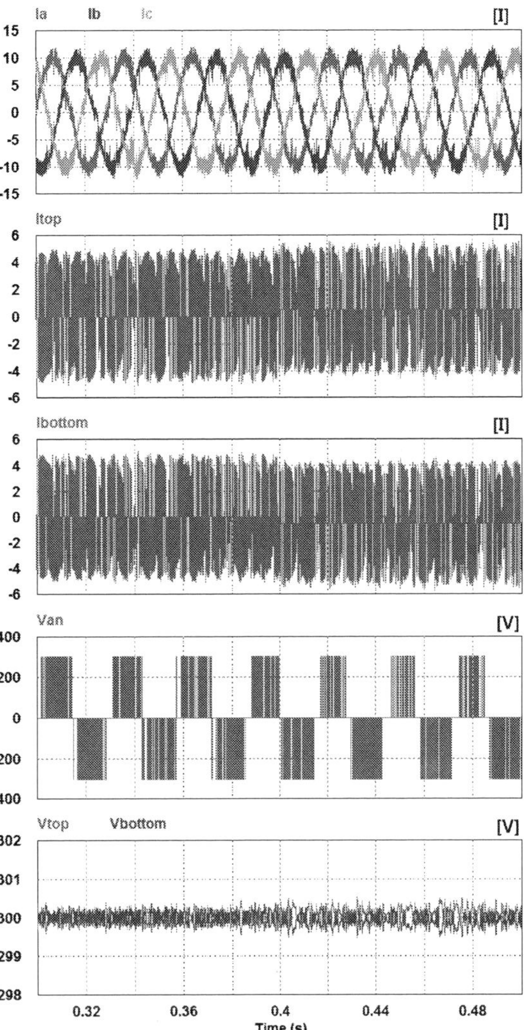

Fig. 8. Simulation results of the proposed MPC method when an additional registor (300 Ω) is connected to bottom dc-link capacitor at 0.4 s.

REFERENCES

[1] J. W. Kolar, and F. C. Zach, "A Novel Three-Phase Utility Interface Minimizing Line Current Harmonics of High-Power Telecommunications Rectifier Modules," IEEE Trans. Ind. Electron., vol. 44, no. 4, pp. 456-466, Aug. 1997.

[2] A. Rajaei, M. Mohamadian, A. Y. Varjani, "Vienna-Rectifier-Based Direct Torque Control of PMSG for Wind Energy Application," IEEE Trans. Ind. Electron., vol. 60, no. 7, pp. 2919-2929, Jul. 2013.

[3] H. Chen, and D. C. Aliprantis, "Analysis of Squirrel-Cage Induction Generator with Vienna Rectifier for Wind Energy Conversion System," IEEE Trans Energy Conv., vol. 26, no. 3, pp. 967-975, Sep. 2011.

[4] T. B. Soeiro, and J. W. Kolar, "Analysis of High-Efficiency Three-Phase Two- and Three-Level Unidirectional Hybrid Rectifiers," IEEE Trans. Ind. Electron., vol. 60, no. 9, pp. 3589-3601, Sep. 2013.

[5] M. S.a Ortmann, S. A. Mussa, and M. L. Heldwein, "Three-Phase Multilevel PFC Rectifier Based on Multistate Switching Cells," IEEE Trans. Power Electron., vol. 30, no. 4, pp. 1843-1854, Apr. 2015.

[6] J. S. Lee and K. B. Lee, "Carrier-Based Discontinuous PWM Method for Vienna Rectifiers," IEEE Trans. Power Electron., vol. 30, no. 6, pp. 2896-2900, Jun. 2015.

[7] L. Hang, B. Li, M. Zhang, Y. Wang, and L. M. Tolbert, "Equivalence of SVM and Carrier-Based PWM in Three-Phase/Wire/Level Vienna Rectifier and Capability of Unbalanced-Load Control," IEEE Trans. Ind. Electron., vol. 61, no. 1, pp. 20-28, Jan. 2014.

[8] L. Dalessandro, S. D. Round, U. Drofenik, and J. W. Kolar, "Discontinuous Space-Vector Modulation for Three-Level PWM Rectifiers," IEEE Trans. Power Electron., vol. 23, no. 2, pp. 530-542, Mar. 2008.

[9] S. Kouro, P. Cortés, R. Vargas, U. Ammann, and J. Rodríguez, "Model Predictive Control—A Simple and Powerful Method to Control Power Converters," IEEE Trans. Ind. Electron., vol. 56, no. 6, pp. 1826-1838, Jun. 2009.

[10] V. Yaramasu, and B. Wu, "Predictive Control of a Three-Level Boost Converter and an NPC Inverter for High-Power PMSG-Based Medium Voltage Wind Energy Conversion Systems," IEEE Trans. Power Electron., vol. 29, no. 10, pp. 5308-5322, Oct. 2014.

An Equivalent Circuit Model for State of Energy Estimation of Lithium-ion Battery

Kaiyuan Li
School of Electrical and Electronic Engineering,
Nanyang Technological University
Singapore
kli1@e.ntu.edu.sg

King Jet Tseng
School of Electrical and Electronic Engineering,
Nanyang Technological University
Singapore
k.j.tseng@pmail.ntu.edu.sg

***Abstract*—For a Li-ion battery based energy storage system, the state of energy is a key index for the energy management and operation optimization. In the actual Li-ion batteries based energy storage system in building environment, the varying load current will lead to the estimated battery's residual energy deviated from the real value. To solve this problem, an equivalent circuit model is presented for the Li-ion battery's state of energy estimation during and after discharging process. The state of energy is introduced in the proposed approach to replace the traditional index SOC, to describe the electrical behaviors and the residual energy of the Li-ion battery under dynamic load current conditions, so that the battery's internal energy loss on its internal resistance and during the electrochemical reaction processes, as well as the effects of the decrease of the OCV during discharging are taken into consideration. Experimental data of the previous study regarding energy efficiency of Li-ion batteries used as energy storage devices in building is utilized in the present study. Finally, the simulation results on 20 Ah LTO batteries indicate that the proposed equivalent circuit is competent to predict the state of energy accurately. What's more, state of energy appears an improvement over SOC in terms of accuracy in capturing battery's energy changes. Thus, this study provides a viable solution for estimating Li-ion battery's SOE and capturing the battery's energy behaviors under dynamic loads.**

Keywords—Li-ion battery; LTO battery; energy storage device; SOE; SOC; dynamic load currents; equivalent circuit model

I. Introduction

Li-ion batteries are playing an increasingly important role in our daily life. Since this decade, Li-ion batteries have been widely used in energy storage systems, such as urban building nano-grid energy storage systems [1]. The state of energy (SOE) [2] of Li-ion battery, which provides the significant information of energy management and operation optimization of the battery energy storage systems, is a critical index. Additionally, to all Li-ion battery operated devices and system requiring continuous operations, estimation of the battery's SOE is significant since it provides consumers a clear clue on the remaining usable time their battery can further supply power. Uncertainties in SOE estimation can cause the consumers to be highly conservative, however, frequent re-charging can shorten battery's life-span [3]. On the other hand, if the re-charging is done too late, an over-discharging may occur which will also shorten battery's life.

Traditionally, the state of charge (SOC) is utilized to represent Li-ion battery's residual energy. In recent decades, to simulate Li-ion battery's electrical behavior while operating, diverse battery models with varying degrees of complexity can be found in the literature. Battery models for SOC estimation can be classified by the technology used, as the electrochemical models, the mathematical models and the electrical models [4, 5, and 6]. Thanagasundram et al. summarized the three kinds of battery model in [6] as: electrochemical models are derived from porous electrode theory and concentrated solution theory proposed by Newman and Tiedemann [7] and Doyle et al. [8]. These models have the advantages to accurately describe the electrical behaviors of Li-ion batteries. However a detailed understanding of the complex electrochemical processes is required, which often includes many algebraic equations and state variables. The main demerit of electrochemical models is that they are very complex and very computationally time-consuming, thus are not suitable in the energy storage application due to the feature of the real-time demanding.

Mathematical models are developed based on the Shepherd model and the further modifications to the Shepherd relation [9, 10]. A basic instance of the mathematical model is the battery model in the SimPower Systems Toolbox within the Simulink/Matlab environment [11]. These models are easy to understand and to set up but they are application-dependent and sometimes are not precise enough to simulate the varying characteristics of the battery behaviors under dynamic currents.

As to the electrical models, equivalent circuit battery models are established to simulate the battery's electrical behaviors under different working conditions. Being the most commonly used method within IEEE, the equivalent circuit models are quite straight-forward and competent since the complex electrochemical reactions inside the battery are replaced by the electrical circuit elements in order to describe the battery's electrical behaviors under dynamic current loads [4, 12 and 13]. Electrical battery models use basic circuit components like voltage sources, capacitors and resistors to capture battery's electrical behaviors under different current conditions. Various electrical models [14] have been presented based upon the battery electrical performances.

Nevertheless, with the more widely applications of Li-ion battery used as energy storage device, the functional demand of the battery system requires a more precise capture of battery's

978-1-4673-9551-9/16 $31.00 © 2016 IEEE

energy behaviors. An example is that, during the Li-ion battery discharging process in real applications, the lower the battery's SOC, the shorter the time taken for the SOC to reduce for the same interval, especially when the battery's SOC is below 40%. Take for example, the time taken for the Li-ion battery to discharge from 90% to 80% is always longer than that for the battery to discharge from 30% to 20%, indicating that the battery seems to be "weaker" when the SOC is low. In general, the discharged energy that the battery is able to output is less for the same SOC interval along with the battery's SOC is decreasing during discharging. Therefore, the disadvantages become more prominent when battery's SOC is taken into use for representing the battery's residual available energy and to describe the battery's electrical behaviors. Liu et al. [15] proposed to use SOE to estimate battery's residual energy and developed a method for SOE estimation using the mathematical neural network. The simulation results of the SOE estimation of [15] is quite acceptable while the calculating procedures are very complex and very algorithm dependent. Liu et al. concludes the advantages of SOE over SOC for capturing and describing the battery's electrical behaviors under dynamic current loads in [15]. Firstly, SOE is different from SOC: SOC defines the residual capacity ratio of Li-ion batteries, namely, SOC describes the battery's capacity status other than its energy status, on which the practical battery energy storage system is dependent. In order to get a more comprehensive understanding of the Li-ion battery used as energy storage devices, the battery's energy efficiency and its residual energy are required; however, SOC estimation can provide no more energy information about the battery. Some works took into consideration on the battery's residual capacity instead of SOC to describe battery's available residual energy [16, 17]. Secondly, although a positive relationship can be found between SOC and SOE, there is no explicit quantitative relationship between each other. Battery's SOC decreases almost linearly during discharging, however the battery's energy value is the results of the capacity times the voltage. There is a difference between SOC and SOE because the fact that the battery's energy loss on its internal resistance and during its internal electrochemical reaction processes, and the effects of the decrease of battery's open circuit voltage (OCV) during discharging are not taken into consideration in the SOC estimation [3-8, 15, 18, and 19]. Last but not least, in the practical battery energy storage systems, where the discharging current usually dramatically changed, the battery acted poorly [20-22]. At the same SOC, SOE values may be different because that the discharging efficiency depends on the discharging current rates. Therefore, it is desirable to set a more comprehensive investigation on the estimation of the Li-ion batteries' residual energy using the SOE, instead of using the traditional parameter SOC.

In the present study, SOE is introduced to describe the Li-ion battery's residual energy, in order to take into consideration of battery's energy loss on its internal resistance and during the internal electrochemical reaction processes, and the effect of the OCV decreasing during discharging. And an equivalent circuit battery model is proposed to realize the estimation of the SOE under dynamic operating current conditions. In Section II, a clear definition of SOE is interpreted for Li-ion batteries and the discharged energy of the Li-ion batteries is calculated. Experimental data of the previous studies regarding energy efficiency of the same Li-ion batteries as energy storage devices is utilized. In Section III and IV, an equivalent circuit model is proposed and established. Parameters of the model of 20 Ah lithium titanate (LTO) batteries are extracted in Section V. The energy loss of the same batteries from the previous study are compared with the simulation results based on the proposed equivalent circuit to verify the reliability and accuracy of the battery model, and the simulation results of SOE estimation is presented and compared to the SOC estimation results in Section VI. Finally, Section VII gives the conclusion.

II. State of Energy

A. Definition of SOE

SOE represents the available residual energy ratio of the Li-ion battery [23, 24], thus being a critical index of the energy management and operation optimization for the battery energy storage systems. Herein, SOE is defined as:

$$\mathrm{SOE}(t) = \frac{E_t - E_d(t)}{E_t} \tag{1}$$

where *SOE(t)* is the ratio of the battery's remaining energy at time t, E_t is battery's total available energy while $E_d(t)$ is the consumed energy during discharging until time t. Basically, SOE reaches the maximum value of 100% after the battery is fully charged, and reaches 0% when the battery is discharged to its lower cut-off voltage threshold.

Previous studies in [25] indicate that battery's discharged energy consists of the output part for powering loads, the lost energy on the battery's internal resistance and during the internal electrochemical reactions. The output electrical energy is used for powering loads, and is indicated by the variation of battery's SOC in the arts. The battery's energy loss on its internal resistance and during the electrochemical reaction processes is in the form of heat thus will heat up the battery during discharging.

In the present study, decrease of the battery's SOE during discharging process will be investigated, with the establishment of an equivalent circuit battery model. Under high discharging current rates, batteries will demonstrate fast empty behavior, which means that in that circumstances, the battery is discharged to its lower cut-off voltage threshold under large discharging current, however, the battery is still able to provide energy at a lower discharging current rate. This fast empty characteristic makes it difficult to estimate the battery's SOE accurately.

B. OCV test and discharged energy

After the battery has relaxed for 2 hours, its terminal voltage will achieve the instantaneous OCV value. Results from the previous study [25] is utilized to acquire the OCV value at different SOC, and the SOC-OCV curve is plotted in Fig. 1.

According to Fig. 1, although following an approximately linear trend, OCV value of the battery decreases rapidly with the discharging process, especially at the beginning and the end. For SOC changes from 100% to 0%, OCV changes from 2.63 V to 2.04 V specifically. Since energy equals to the results

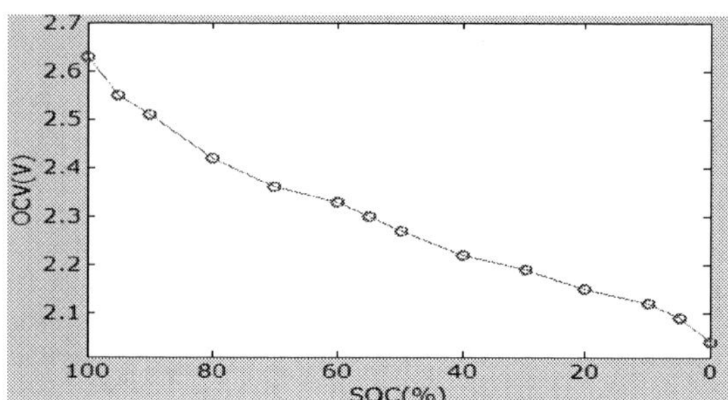

Fig. 1 OCV-SOC characteristic of the 20Ah battery.

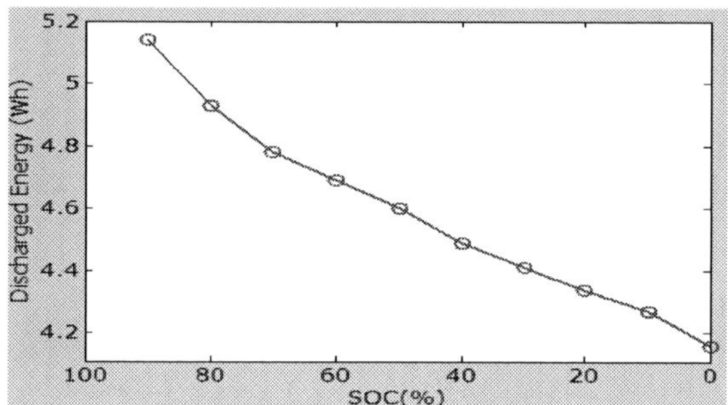

Fig. 2 Discharged energy for each 10% SOC interval at different SOC value.

of OCV times capacity, for uniform SOC interval, the discharged energy from the battery is not constant caused by the OCV change. Additionally, the battery's discharged energy for each 10% SOC interval from 100% to 0% is calculated using Eq. (2) to show the deviation of battery's discharged energy from the discharged capacity:

$$\text{Energy} = (20 \text{ Ah} \times 10\%) \times \frac{OCV_f + OCV_l}{2} \tag{2}$$

where (20 Ah×10%) means battery's changing capacity is 2 Ah for each 10% SOC interval, and OCV_f represents battery's OCV value at the former SOC before discharged, while OCV_l stands for the latter OCV after discharged for a 10% SOC interval. Fig. 2 illustrates the different values of the discharged energy when SOC changes, where the *Y*-axis describes the discharged energy of the battery. Take as an example, in Fig. 2, the discharged energy at 80% SOC is 4.93 Wh represents that, the discharged energy is 4.93 Wh for SOC decreasing from 90% by a 10% interval down to 80%. As can be observed from Fig. 2, the battery's discharged energy for each 10% SOC interval decreases while the SOC is decreasing. For example, according to Eq. (2), when SOC changes from 100% to 90%, 5.14 Wh energy is discharged; on the other hand, when SOC changes from 10% to 0%, the discharged energy for this 10% SOC interval is only 4.16 Wh. Thus we can conclude that the discharged energy for the same SOC interval is not uniform at different SOC. Furthermore, in Li-ion batteries based energy storage applications, typical battery energy storage systems always consist of a large amounts of battery cells through electrical connections, resulting the differences in Fig. 2 to be larger and more significant in actual operations. Thus, SOC is not able to accurately capture the energy changes of the battery during discharging.

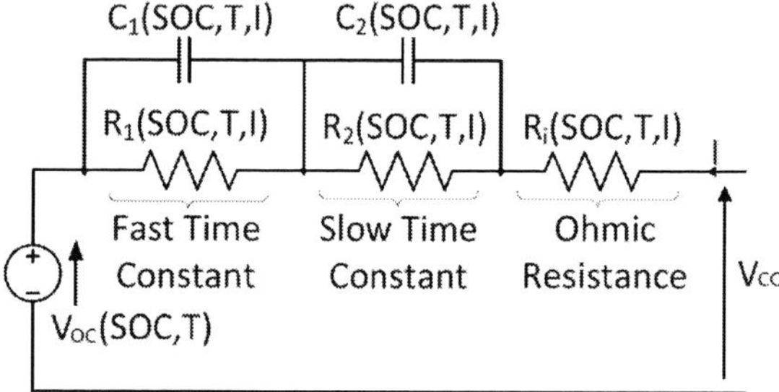

Fig. 3 Second Order Equivalent Circuit of a Li-ion battery cell.

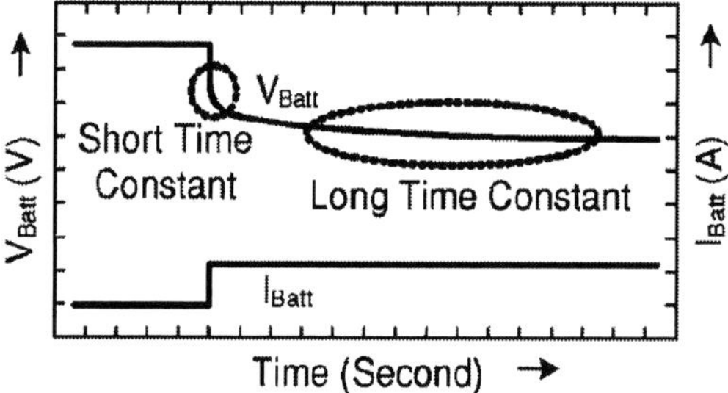

Fig. 4 Typical Li-ion battery characteristic curve of transient response under a pulse current event.

III. Proposed Model

The fact that battery's SOE depends on the load conditions due to the above mentioned fast empty characteristic makes it more meaningful to accurately investigate an equivalent circuit model for the battery in order to estimate the SOE. Electrical battery model is selected in the present study. In the arts, electrical battery models can be further classified into two categories, and Chen summarized this classification in [26] as: the Thevenin-based and the impedance-based electrical battery model. For the Thevenin-based model, its corresponding parameters are identified by the curve fitting technique on the battery's experimental voltage and current characteristics at various operating conditions, whereas the impedance-based models are developed by using the electrochemical impedance spectrum (EIS) technique [26]; nonetheless, in practical applications, performing the EIS measurements is impractical in a cell and it is time consuming and often subject to human intervention. Thus, in the present study, the Thevenin-based model is selected.

A. The structure of the equivalent circuit

Plenty of Thevenin-based battery models are proposed in recent decades, among which the second order equivalent circuit model is adopted in the present study. Fig. 3 shows the simple schematic of the second order equivalent circuit mode. In summary, the model has 7 components which are: V_{OC}, V_{CC}, R_i, R_1, R_2, C_1 and C_2. Each of the components represents a different aspects inside the battery. Specifically, V_{OC} is OCV and it is one of the most critical battery parameters, while V_{CC} is the close circuit voltage (CCV), which is the easiest and most straight-forward parameter to measure and monitor. R_i is the internal or over-potential resistance, and R_1, C_1, R_2 and C_2 represents for the two parallel polarization RC network

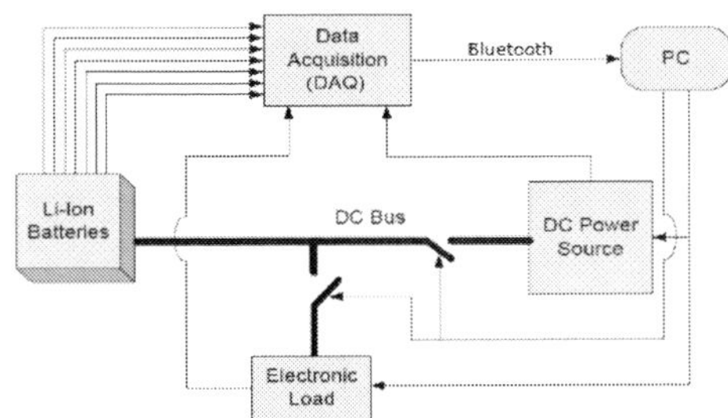

Fig. 5 Schematic of the experimental setup.

combination, which are responsible for the battery's transient response characteristics [26-28].

B. Open circuit voltage

The OCV $f(V_{OC})$ is dependent on the changing SOC, as illustrated in Fig. 1. The non-linear relationship between the OCV and SOC is important and it is essential to be included in the model in order to accurately describe the battery's electrical behaviors. Thus, V_{OC} is represented by a voltage-controlled voltage source in the schematic diagram of the battery model.

C. Transient response characteristic

In pulse discharging tests, the response of the battery's terminal voltage has two parts depending the response timing when the discharge is cut-off, i.e., the fast response and the slow response. As can been observed from Fig. 4, the response curve of the battery's terminal voltage usually presents an instantaneous together with a curve-dependent voltage drops. For this reason, the battery's transient responses are characterized by the two RC networks in Fig. 3 [6], the slow and the fast transient response, respectively.

Series resistor R_i represents the battery's internal resistance, and since the purely resistor follows the Ohm's Law, it leads to the immediate terminal voltage rise and drop when the charging and discharging test is started. Circled by the dotted lines in Fig. 4, R_1 and C_1 represent the above mentioned fast transient response of the battery depicting surface effects on the electrodes and reaction kinetics [26]. R_1 stands for the charge transfer resistance and C_1 is the double layer capacitance. Meanwhile, R_2 and C_2 describe the slow transient response in the battery, the duration of which is in the order of hour. The slow dynamics represent more of the diffusion processes in the electrolyte and active material [26]. It is indicated in [29] that using two RC time constants, rather than one or three, can achieve the best tradeoff between complexity and accuracy because two RC time constants is able to keep the error within 1 mV for all the curve fittings results [6, 29 and 30], the detailed comparison results can be found in [29, 30] with numerous experimental curves.

Theoretically, all of the model parameters are multi-variable functions of SOC, temperature (T), current (I) and cycle number. However, some parameters can be simplified to be independent or linear functions of some variables for the specific batteries within certain error tolerance, indicated by Chen et al. in [26]. For instance, the influences caused by temperature can be ignored when a large-capacity battery is operated in a constant ambient temperature such as in a laboratory, and the 5% self-discharging rate can be ignored without suffering significant errors when a battery is used frequently [26]. Thus, temperature and self-discharge effects will be not be considered through the present study.

IV. Test System and Procedure

A. Test bench

In order to extract all the parameters of the proposed model, a test bench is built. The schematic of the experimental setup is demonstrated in Fig. 5. HIOKI measurement station LR8410-20 works as data logger in the experiments. Ambient temperature, battery's surface temperature, terminal voltage and operating current are monitored and recorded to prevent the battery from overheating and to collect the experimental data. The data acquiring time is set as 1 second to obtain sufficient number of data points used in the curve fitting process. The experimental data is stored in the database of HIOKI measurement station and is copied to PC via Bluetooth. Being able to keep the current through it at a constant value, Chroma programmable AC/DC electronic load Model 63803 3.6KW/36A/350V is connected in series with the battery to maintain the charging/discharging current at a constant rate. As to the battery, lithium titanate (LTO) type battery produced by Toshiba CO. Ltd. is used in the present study, and the rated capacity of the LTO battery is 20.0 Ah. The battery's parameters are given in Table 1 [31].

Table 1
Parameters of the 20 Ah LTO battery cell.

Battery	***Characteristics***
Chemical system	Lithium Titanate (LTO) rechargeable battery
Nominal voltage	2.3 V
Capacity	20.0 Ah Typical
Charging condition	CVCC 2.7V 20.0 A, 1.0A cut-off 25°C
Discharging condition	Constant current 6.67 A, 1.5 V cut-off 25 °C
Low voltage cutoff	1.5 V
Upper voltage limit	2.7V
Operating temperature	-30 °C – 55 °C

B. Experimental procedure

The test procedure is similar to that provided in [6] and [32]. The LTO battery is loaded in the test bench shown in Fig. 5, and pulse discharging tests are taken. The OCV value of each cells will be measured before loaded into the test bench. All the parameters such as the pulse discharge current rates, battery's terminal voltage, surface temperature and ambient temperature are collected by the data logger and finally send to PC. All the pulse discharging tests are cut-off when the battery's terminal voltage reaches the low cut-off voltage threshold of 1.5 V. After the pulse discharging tests, the discharged battery is relaxed with no load connected for 2 hours before its OCV is measured for the after-discharge SOC determination. The parameters of the battery during discharging is recorded in the PC and the experimental curves are drawn accordingly. Major objectives of the tests is to obtain the voltage curves similar to that in Fig. 4 using pulse discharging, thereby extracting the values of all the model parameters. Fig. 4 shows the voltage curves obtained by pulse discharging tests, where the discharging currents are set as 0.1-

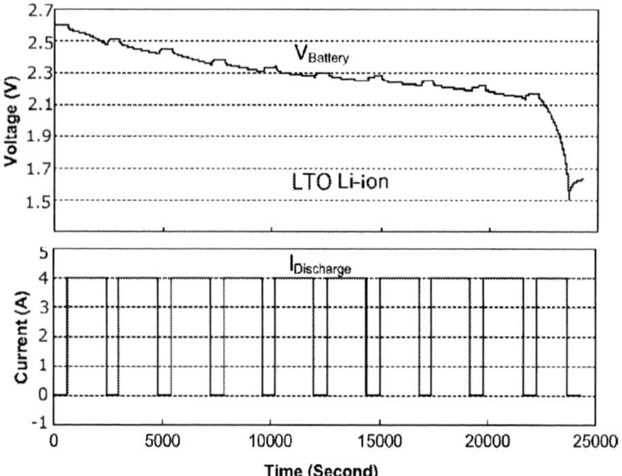

Fig. 6 The voltage curve when battery is under pulse discharging.

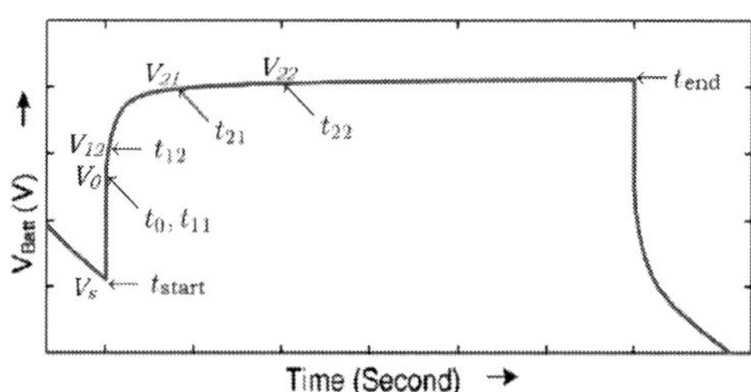

Fig. 7 Voltage relaxation characteristic during a pulse discharge test.

1 C [9] (herein, 'C' represents the discharging current that will discharge a battery from 100% to 0% SOC in 1 h time). The extraction and calculation procedures are detailed in Section V.

V. Model Extraction

A second order equivalent circuit battery model is established in order to estimate the battery's SOE, and the model parameters are extracted and identified experimentally in this section. 20 Ah LTO battery is used in the present study. As aforementioned, all the model parameters are multivariable functions of battery's SOC, temperature (T), current (I) and cycle numbers. And under the current circumstances, some sub-ordinate variables are herein simplified or ignored not only because it eases the validating process but also because these variables will influence the battery negligibly, such as self-discharging rate (2%-8% per month), indicated by Chen et al. in [26]. Additionally, the dependence on temperature of the parameters is ignored in the present study because the battery based energy storage systems are usually mounted in places where the ambient temperature is almost constant. In the tests, the LTO batteries are mounted in a laboratory where the ambient temperature is kept constant at 23 °C.

A. Model extraction

Ten pieces of fresh 20 Ah Toshiba LTO Li-ion battery cells are tested with suitably spaced pulse discharge currents (4, 8, 12, 16 A in the present study) at lab temperature. As specified by the instruction manual [31], these batteries are first fully charged with constant current of 20 A, 2.7 V constant voltage

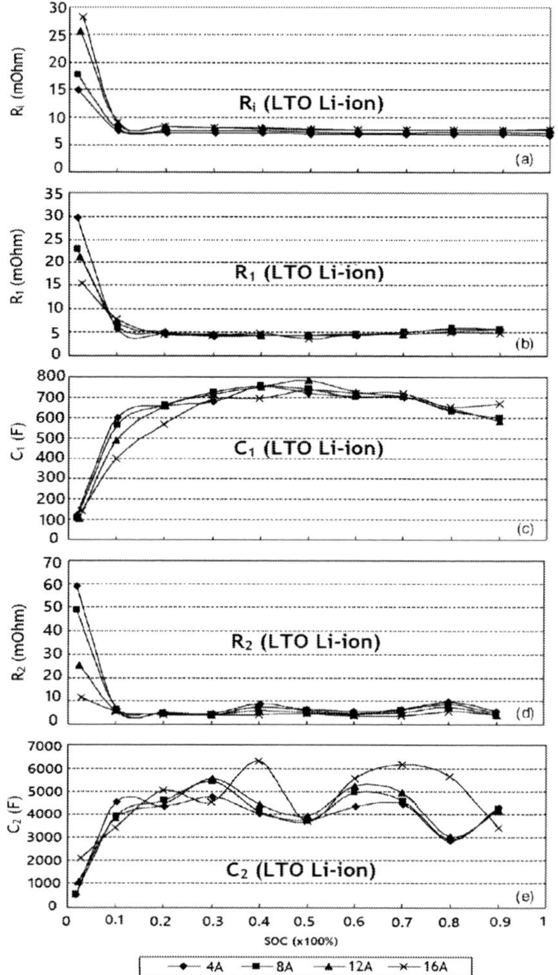

Fig. 8 Experimentally identified parameters of the LTO Li-ion battery in the present study.

condition, and 1.0 A completing current, and then discharged under pulse currents aforementioned with 1.5 V completing voltage after 2 hours' relaxation.

Experimental data from previous study [25] indicates that these LTO Li-ion batteries from Toshiba have great performance against cell unbalancing when operated in series connection, and their discharging curves during normal discharging and pulse discharging under the same current rates and the same temperature conditions are very close to each other and with negligible deviations for SOC range 10%-100%.

One piece of LTO battery cell is chosen for the extraction of the model parameters. The LTO battery is first fully charged and rested for 2 hours, before each pulse discharge tests.

Fig. 6 shows the discharging curve under 4 A (0.2 C) pulse discharging current on the piece of LTO battery. The pulse width herein is selected in order to ensure the enough number of "humps" to guarantee sufficient plotting points, and the cut-off time is chosen as 10 mins in this case so that the terminal

Table 2
RMSE between the extracted parameters and previous experimentally data

Discharging current rates (C)	0.33	0.5	0.75	1.0	1.25	1.5	1.75
Root-mean-square error (%)	1.66	1.85	0.88	1.90	0.98	1.83	1.88

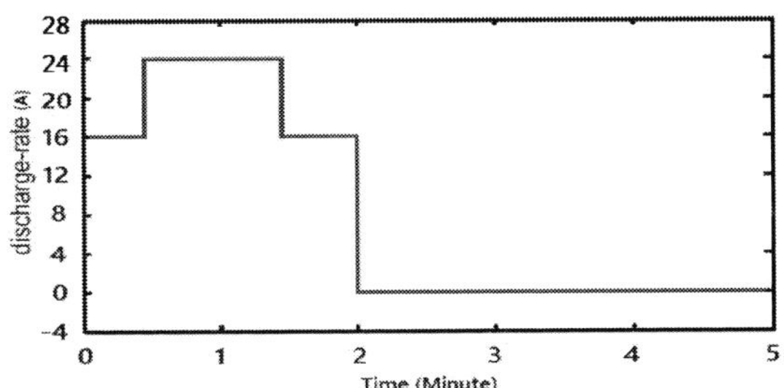

Fig. 9 Diagram of the current during the simulation operation conditions.

voltage can reach to the corresponding steady state conditions gradually.

Take as an example, Fig. 7 illustrates the zoomed voltage relaxation timing and voltage characteristic of the battery when the discharging current is just off during a pulse discharge test, and the battery's terminal voltage tends to gradually recover to the OCV value at that SOC. The fast and slow dynamics can be observed in Fig. 7.

Hentunen et al. discussed and provided the definition of the timing and voltage elements in Fig. 7 in [32], as:

1) t_{start} is the instant when the pulse discharging completes and currents starts to reduce. V_s is the voltage at t_{start}.

2) t_0 is the instant when the current reduced to zero, i.e., the starting of the relaxation process. This time instant will be considered as $t = 0$ s in the following. V_0 is battery's terminal voltage at t_0.

3) t_{11} is the beginning instant of the first relaxation time window, i.e., the fast dynamic, and typically, $t_{11} = t_0$.

4) t_{12} is the completing instant of the fast dynamic window. V_{12} is battery's terminal voltage at t_{12}.

5) t_{21} is the beginning instant of the second relaxation, i.e., the slow dynamic. V_{21} is battery's terminal voltage at t_{21}.

6) t_{22} is the completing instant of the slow dynamic window. V_{22} is battery's terminal voltage at t_{22}.

7) t_{end} is the completing instant of the whole relax period.

The natural beginning instant for the fast dynamic window is t_0, namely, $t = 0$ s. The beginning instant for the slow dynamic window is supposed to be no earlier than three times of the fast dynamic time constant, which ensures that the voltage of the fast or short time-constant process has already reduced to below 5% of its initial value, and thus it will have a negligible influence on the parameter extraction results of the slow or long time-constant process. During the parameter extraction processes, the value of R_i, R_1, C_1, R_2 and C_2 are identified using the equations derived in [6] by Thanagasundram et al. and in [33] by Abu-Sharkh et al. These parameters values are then substituted back into Eq. (3) to establish the equivalent circuit battery model and to predict battery's SOE under dynamic currents.

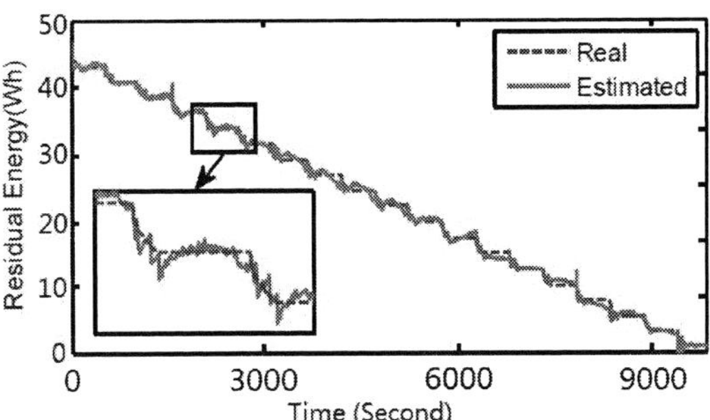

Fig. 10 SOE estimation of the LTO battery under dynamic load current conditions, comparing the estimated SOE and the real SOE.

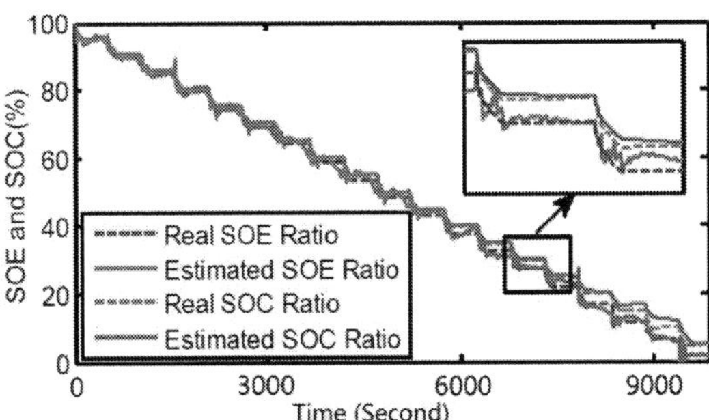

Fig. 11 SOE estimation of the LTO battery under dynamic load current conditions, comparing the SOE and the SOC.

$$V_{CC} = V_{OC} + I(t)R_i + I(t)R_1\left(1 - e^{-\frac{t}{R_1C_1}}\right) + I(t)R_2\left(1 - e^{-\frac{t}{R_2C_2}}\right) \quad (3)$$

Fig. 8 shows the extracted nonlinear series resistance (R_i) and RC network (R_1, C_1, R_2 and C_2) as functions of the battery's SOC and the discharging current rates. It can be concluded from the figure that values of all the RC parameters are of approximately constant value when the SOC is between 20-100% and change exponentially when the SOC is falls within 0-20%, which phenomena is due to the battery's internal electrochemical reactions. A little parameter difference between the curves at different discharging current rates indicate that these extracted parameters are roughly independent of the operating current rates, which helps to ease the equations expression of these parameters. Thus, SOC based functions with only one variable are utilized to quantify these extracted curves, as shown by Eq. (4)-(8). V_{OC} is also quantified as single-variable functions of SOC in Eq. (9), the plotting diagram of which is shown in Fig. 1.

$$R_i = 15.62 \cdot e^{-24.37 \cdot SOC} + 7.75 \quad (4)$$
$$R_1 = 18.08 \cdot e^{-25.14 \cdot SOC} + 4.67 \quad (5)$$
$$C_1 = -596 \cdot e^{-11.01 \cdot SOC} + 703.6 \quad (6)$$
$$R_2 = 31.1 \cdot e^{-155.2 \cdot SOC} + 4.98 \quad (7)$$
$$C_2 = -3956 \cdot e^{-27.12 \cdot SOC} + 4475 \quad (8)$$

$$V_{OC} = -0.018e^{-83 \cdot SOC} + 2.0633 + 0.5673 \cdot SOC$$
$$-0.642 \cdot SOC^2 + 0.642 \cdot SOC^3 \quad (9)$$

B. Limitations

As this is the first attempt to relate the SOE estimation of Li-ion batteries to the equivalent circuit model, this work is limited to the following situation:

- The temperature of the battery during discharging and charging is assumed to be constant at the lab temperature of 23°C. Temperature effects of the parameter extraction will be considered in the future work.
- Self-discharge behavior of the battery is not considered. This can be considered only if the electrochemistry process of the self-discharging is well understood.
- No high discharging current (i.e. < 30A) so that Peukert effect is insignificant.

VI. Simulation and Discussion

In this section, the proposed model is validated experimentally. 20 Ah LTO batteries are utilized for the validation test to verify the performance of the equivalent circuit model, which is to accurately estimate the battery's SOE under various load currents, as well as the improvement of SOE over SOC in describing Li-ion electrical behaviors under dynamic discharging current conditions at lab temperature.

A. Model validation based on battery's energy efficiency and energy loss

Previous study in [25] investigated the energy efficiency during charging/discharging process of the same Li-ion batteries. Battery's energy loss during pulse discharging is considered to be totally consumed on its internal resistance and during the electrochemical reaction processes, i.e., on R_i, R_1, and R_2 of the equivalent circuit model in the form of heat. Therefore, battery's energy loss during pulse discharging can be used as an indicator of the summation of R_i, R_1, and R_2. Since all the energy loss is on R_i, R_1, and R_2 in the end, regardless of the hysteresis phenomenon and the fast and slow dynamics, Eq. (10) is introduced to verify the summation of the experimentally extracted value of R_i, R_1, and R_2 in the proposed model. In Eq. (10), $I(t)$ is the current rate of the pulse discharge, and t is the total time being when the discharging current is "on" and the battery is being discharged.

$$Energy\ loss = (I(t))^2 \times (R_i + R_1 + R_2) \times t \quad (10)$$

The comparison of the summation of R_i, R_1, and R_2 from the experimental data in [25] using Eq. (10) and the extracted parameters in the present study is showed in Table 2. The root-mean-square error (RMSE) between the extracted summation of R_i, R_1, and R_2 and the calculated summation using Eq. (10) from the battery's energy efficiency and discharging energy loss perspective is taken at each discharging current rates. The results show that the error is no more than 1.9% for all the current rates. This close agreement verifies that the extracted parameter values of the proposed model is very accurate.

B. Comparison of the simulated and measured SOE and SOC

As analyzed in Section IV, an SOE estimation of the Li-ion battery approach based on an equivalent circuit model is proposed in the present study. In order to verify the proposed model, the extracted parameters are applied to a model in the Matlab/Simulink environment to simulate the battery's behaviors under dynamic operating currents. Since the load current in energy storage applications in micro-grid is not as variable as that in EVs, the testing battery is loaded repetitively at a relative stable load current conditions as shown in Fig. 9, until the terminal voltage of the battery reached to the low cut-off voltage threshold 1.5 V. The discharging load current waveform for real time simulation is shown in Fig. 9, where currents for discharging is positive value while current for charging is negative.

Fig. 10 shows the comparison of the estimated SOE with the real SOE curves. The estimated SOE is obtained by loading the equivalent circuit model repetitively under the current curve shown in Fig. 9. The real SOE value is calculated by the SOC-based method introduced in Section II B, and the real SOC value is calculated from the Coulomb counting method provided in [5]. The real SOE is indicated in blue dotted line, while the estimated SOE is indicated in red line. The RMSE between the estimated and real SOE is 1.15, therefore the equivalent circuit model gives acceptable SOE estimation results.

Furthermore, so as to verify the improvement of SOE compared to SOC in capturing battery's electrical behaviors under dynamic load current conditions, Fig. 11 provides the comparison results of the estimated SOE, the real SOE, the estimated SOC and the real SOC. The real SOC is obtained by Coulomb counting method and the estimated SOC is calculated using the unscented Kalman filter [3]. In Fig. 11, the real SOC is indicated by the brown dotted line and the estimated SOC is indicated by the purple line. It can be observed that battery's SOC values show a deviation from the SOE values, particularly at the latter parts when the discharging is about to finish. Generally, the SOC values are higher than the SOE values, which can be explained by Section II B and Fig. 2 as: during the battery discharging process, for the same decreased SOC interval during discharging process, battery's discharged energy is not uniform at different SOC level; specifically, the lower the SOC, the lower the discharged energy per the same decreased SOC interval. Thus, it can be concluded that the SOC values show a deviation from the true SOE values, and the SOC values are always larger than the SOE values, especially at the latter parts of discharging when the SOC is low. For real applications of the Li-ion battery as energy storage devices in buildings, constant power discharging is always required, thus, using SOE to indicate the battery's residual energy is more accurate than using SOC, because the determination of SOE takes into account the fact that the OCV value of the battery is decreasing with the discharging process.

Based on the above experimental results and analysis, SOE has an obvious advantage over SOC in terms of the accuracy in capturing the battery's energy behaviors under dynamic load currents. The major reason can be explained as that, the battery's energy loss on its internal resistance and during its internal electrochemical reaction processes, and the effect of the decrease of the OCV during discharging are taken into consideration in the estimation of battery's SOE. In real applications, a typical battery system as energy storage devices is always composed of a large numbers of battery cells connected in series and parallel, thus the deviation of SOC from SOE shown in Fig. 11 will become more significant in real applications.

Based on the above verification results and discussion, it can be concluded that the proposed equivalent circuit battery model can achieve an accurate SOE estimation for Li-ion battery under dynamic operating currents, and SOE has an obvious improvement over SOC, in terms of accuracy in capturing the energy behaviors of the Li-ion batteries under dynamic load currents.

VII. Conclusion

With the fast developing of the battery technology, the Li-ion battery based energy storage systems in building environment are drawing more and more public attention. In the practical applications, the variable discharging current of the battery due to dynamic loads raises a big challenge for the energy management and the operation optimization of the battery systems. The concept of SOE is utilized in the present study for accurately describing the battery's electrical behaviors and the residual available energy, instead of the traditional parameter SOC. In the determination process of SOE, the battery's energy loss on its internal resistance and during its internal electrochemical reaction processes, and the effect of the decrease of battery's OCV during discharging are taken into consideration more comprehensively. A second order equivalent circuit model is established for the SOE estimation, the corresponding model parameters of which are extracted and identified experimentally. In the end, the proposed battery model is verified to be competent to capture the battery's electrical behaviors under dynamic load current conditions and to estimate the battery's residual energy and SOE. Based on the experimental results and comparison, SOE appears an obvious improvement over SOC in terms of accuracy in capturing the energy behaviors of the Li-ion batteries under dynamic load currents. Therefore, a viable solution is provided in the present study for the Li-ion battery's SOE estimation when used as energy storage devices. Limitations in Section V need to be further addressed in future studies.

Acknowledgment

This research is funded by the Republic of Singapore's National Research Foundation through a grant to the Berkeley Education Alliance for Research in Singapore (BEARS) for the Singapore-Berkeley Building Efficiency and Sustainability in the Tropics (SinBerBEST) Program. BEARS has been established by the University of California, Berkeley as a center for intellectual excellence in research and education in Singapore.

References

[1] M. Armand, J.M. Tarascon, "Building better batteries", *Nature* 451 (2008) 652-657.

[2] K. Mamadou, A. Delaille, E. Lemaire-Potteau, Prof. Y. Bultel, "New method for the energetic performances evaluation of electrochemical storage devices", *ECS Trans.* 2010 volume 25, issue 35,105-112.

[3] V. Pop, H.J. Bergveld, D. Danilov, P.H.L. Notten, P.P.L. Regtien, Battery Management Systems: Accurate State-of-charge Indication for Battery-powered Applications, *Springer Science*, B.V, 2008.

[4] S. Piller, M. Perrin, A. Jossen, "Methods for state-of-charge determination and their applications", *J. Power Sources* 96 (2001) 113-120.

[5] M. Coleman, C. K. Lee, C. Zhu, W.G. Hurley, "State-of-charge determination from EMF voltage estimation: using impedance, terminal voltage, and current for lead-acid and lithium-ion batteries", *IEEE Trans. Ind. Electron.* 54 (2007) 2550-2557

[6] S. Thanagasundram, R. Arunachala, K. Makimejad, T. Teutsch, A. Jossen, "A cell level model for battery simulation", *European Electric Vehicle Congress Brussels, Belgium*, 20th-22nd November 2012.

[7] J. Newman, W. Tiedemann, "Porous-Elevtrode theory ith battery applications", *AIChE Journal*, 21 (1975) 25-41.

[8] M. Doyle, T.Fuller, J.Newman, "Modelling of galvanostatic charge and discharge of the lithium/polymer/insertion cell", *Journal of the Electrochemical society*, 140 (1993) 1526-1533.

[9] C.M. Shepherd, "Design of primary and secondary cells", *Journal of the Electrochemical Society*, 112 (1965), 657-664.

[10] O. Tremblay, L.A. Dessaint, A.I. Dekkiche, "A generic battery model for the dynamic simulation of hybrid electric vehicles", *IEEE Vehicles Power and Propulsion Conference*, 2007.

[11] Mathworks Inc. MATLAB SimPowerSystems User's Guide, Version 5.4 (R2013a).

[12] V.H. Johnson, "Battery performance models in ADVISOR", *J. Power Sources*, 110 (2002) 321-329.

[13] S. Jiang, "A parameter identification method for a battery equivalent circuit model", *SAE Technical Papre*, 2011, 2011-01-1367.

[14] X. Hu, S. Li, H. Peng, "A comparative study of equivalent circuit models for Li-ion batteries", *J. Power Sources* 198 (2012) 359-367.

[15] X. Liu, W. Ji, C. Zhang, Z. Chen, "A method for state of energy estimation of lithium-ion batteries at dynamic currents and temperatures", *J. Power Sources* 270 (2014) 151-157.

[16] W. Wagg, D.U. Sauer, "Adaptive estimation of the electromotive force of the lithium-ion battery after current interruption for an accurate state-of–charge and capacity", *Appl. Energy* 111 (2013) 416-427.

[17] A. Hausmann, C. Depcik, "Expanding the Peukert equation for battery capacity modeling through inclusion of a temperature dependency", *J. Power Sources* 235 (2013) 148-158.

[18] B. Wu, V.Yufit, M. Marinescu, G.J. Offer, R.F. Martinez-Botas, N.P. Brandon, "Coupled thermal–electrochemical modelling of uneven heat generation in lithium-ion battery packs", *J. Power Sources* 243 (2013) 544-554.

[19] L.H. Saw, K. Somasundaram, Y. Ye, A.A.O. Tay, "Electro-thermal analysis of Lithium Iron Phosphate battery for electric vehicles", *J. Power Sources* 249 (2014) 231-238.

[20] X. Liu, Z. Chen, C. Zhang, J. Wu, "A novel temperature-compensated model for power Li-ion batteries with dual-particle-filter state of charge estimation", *Appl. Energy* 123 (2014) 263-272.

[21] J. Yi, U.S. Kim, C.B. Shin, T. Han, S. Park, "Modeling the temperature dependence of the discharge behavior of a lithium-ion battery in low environmental temperature", *J. Power Sources* 244 (2013) 143-148.

[22] X.H. Rui, Y. Jin, X.Y. Feng, L.C. Zhang, C.H. Chen, "A comparative study on the low-temperature performance of $LiFePO_4$/C and $Li_3V_2(PO_4)_3$/C cathodes for lithium-ion batteries", *J. Power Sources* 196 (2011) 2109-2114.

[23] K. Mamadou, E. Lemaire, A. Delaille, D. Riu, S.E. Hing, Y. Bultel, "Definition of a State-of-Energy Indicator (SoE) for Electrochemical Storage Devices: Application for Energetic Availability Forecasting", *J. Electrochem. Soc.* 159 (2012) A1298-A1307.

[24] T. Dong, M. Montaru, A. Kirchev, M. Perrin, F. Lambert, Y. Bultel, "Modeling of Lithium Iron Phosphate Batteries by an Equivalent Electrical Circuit: Part II - Model Parameterization as Function of Power and State of Energy (SOE)", *ECS Trans.* 35 (2011) 229-237.

[25] K. Li, K.J. Tseng, "Energy efficiency of lithium-ion batteries used as energy storage devices in micro-grid", *41st IEEE IECON*, Yokohama, Japan, 2015.

[26] M. Chen, G.A. Rincon-Mora, "Accurate electrical battery model capable of predicting runtime and I-V performance", *IEEE Transactions on Energy Conversion*, 21 (2006) 504-511.

[27] S. Lee, J. Kim, J. Lee, B.H. Cho, "State-of-charge and capacity estimtion of lithium-ion battery using a new open-circuit voltage versus state-of-charge", *J. Power Sources*, 185 (2008) 1367-1373.

[28] M. Dubarry, B. Y. Liaw, "Development of a universal modeling tool for rechargeable lithium batteries", *J. Power Sources* 174 (2007) 856-860.

[29] B. Schweighofer, K.M. Raab, G. Brasseur, "Modeling of high power automotive batteries by the use of an automated test system", *IEEE Trans. Instrum. Meas.*, vol. 52 (2003) 1087-1091.

[30] Farag K. Abo-Elyousr, F. N. Abd-Elbar, H. A. Abo-Zaid, G. H. Rim, "Accurate Modeling of Prismatic Type High Current Lithium-Iron-Phosophate (LiFePO4) Battery for Automotive Applications", *Energy and Power Engineering* 4 (2012) 465-481.

[31] "Toshiba SCiB lithium-ion rechargeable battery cell specification, NP2211F10FHA", ed. 2014.

[32] A. Hentunen, T. Lehmuspelto, J. Suomela, "Time-Domain Parameter Extraction Method for Thévenin-Equivalent Circuit Battery Models", *IEEE Transactions on Energy Conversion*, 29 (2014) 558-566.

[33] S. Abu-Sharkh, D. Doerffel, "Rapid test and non-linear model characterisation of solid-state lithium-ion batteries", *J. Power Sources* 130 (2004) 266-274.

978-1-4673-9551-9/16 $31.00 © 2016 IEEE

Distributed Optimal Control of Reactive Power and Voltage in Islanded Microgrids

Yanbo Wang [*†], Xiongfei Wang [*], Zhe Chen [*], Frede Blaabjerg [*]

Department of Energy Technology, Aalborg University, Aalborg, Denmark [*],
Electrical Engineering School, Southwest Jiaotong University, Chengdu, China [†],
ywa@et.aau.dk, xwa@et.aau.dk, zch@et.aau.dk, fbl@et.aau.dk

Abstract—**The paper proposes a distributed optimal control strategy for islanded microgrids, which allows to perform reactive power sharing and voltage regulation without communication system. To realize the twofold objective, an improved small signal model is developed first to reconstruct the system input-output relationship, where the relationship is evaluated with sensitivity analysis. A state estimator is then constructed based on the new input-output relationship in order to observe reactive power distribution and system voltages by local measurement. An optimal regulator is developed to perform both reactive power sharing and system voltage restoration. And the dynamic performance of the optimal controller is analyzed, from which the guideline for choosing controller parameters is formulated. The results obtained from sensitivity analysis, simulations and experiments show that the proposed approach provides the expected reliability and flexibility for optimizing the reactive power sharing and system voltages restoration,**

*Keywords—**Distributed optimal control; dynamic performance; voltage regulation; islanded microgrid; state estimator***

I. INTRODUCTION

The deployments of renewable energy sources such as photovoltaics, wind turbine and fuel cell in the form of Distributed Generation (DG) systems have been increasing. Microgrids have been emerged to systematically integrate DG units, local loads and energy storage devices [1]. A microgrid may be allowed to operate either in a grid-connected mode or in an islanded mode [2]. During the islanded operation, DG units are generally controlled by the active power-frequency (*P-f*) droop and reactive power-voltage (*Q-V*) droop relationships, where the terminal frequency and voltage of DG unit is changed to track the output power with the load changes. The absence of critical communication links in the droop control method enhances the reliability of microgrid power management [2]. However, the accuracy of reactive power sharing is usually degraded by the unequal line impedance or local reactive power loads [3].

To improve the accuracy of reactive power sharing, a number of improved droop control methods [4]-[5] have been reported. An accurate power sharing strategy has been developed in [3], where the controller is able to regulate the load voltage with the reduced voltage deviation. An enhanced reactive power control strategy was proposed in [4]. Those two methods are, however, achieved at the expense of converter terminal voltage drop.

Another promising approach to mitigating the reactive power sharing error is the virtual impedance-based control [6]-[8]. An enhanced virtual impedance control approach for a distributed generation unit was discussed in [4], which is able to compensate for the impacts of mismatched physical feeder impedances. Although the virtual impedance-based droop control methods can improve the performance of reactive power sharing, the voltage droop caused by the droop controller and virtual impedance degrades the terminal voltages quality [8]. In addition, virtual impedance-based control fails to deal with the effect of local reactive power loads.

To alleviate the tradeoff between reactive power sharing and bus voltage deviations, a number of voltage control methods, either centralized or distributed, have been introduced to regulate the system voltage [9]-[10]. The basic principle underlying these methods is to introduce an additional voltage offset for compensating the voltage deviation. Yet, these voltage control strategies often rely on communication systems, where the time delay or data loss may cause voltage stability problem [3]. In contrast, a distributed voltage control on the basis of state estimation is recently introduced in [11]-[12], which is capable of dealing with system voltage restoration without communication facilities, but the performance of reactive power sharing has not yet been addressed.

This paper therefore presents a distributed optimal control strategy, which is composed by a Kalman Filter-based estimator and optimal regulator as shown in Fig. 1. The contributions of this work can be summarized as follows: (1) a new input-output relationship between the reactive power sharing and the terminal voltage output is constructed. (2) The state estimator is proposed to observe reactive power error and voltage amplitude. Meanwhile, the optimal controller is developed to eliminate the reactive power error and perform voltage restoration.

II. OPERATION PRINCIPLE OF OPTIMAL CONTROL STRATEGY

Fig. 2 shows the operation principle of the optimal controller, which consists of the state estimator and optimal regulator. The state estimator is adopted to observe the reactive power error and network voltage by the local voltage and current. And the optimal regulator is responsible for computing the optimal control command to perform reactive power sharing and voltage restoration.

978-1-4673-9551-9/16 $31.00 © 2016 IEEE

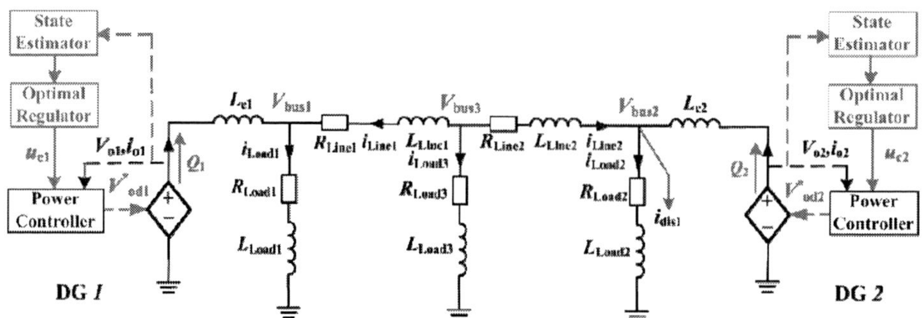

Fig. 1. The decentralized control-based islanded microgrid configuration.

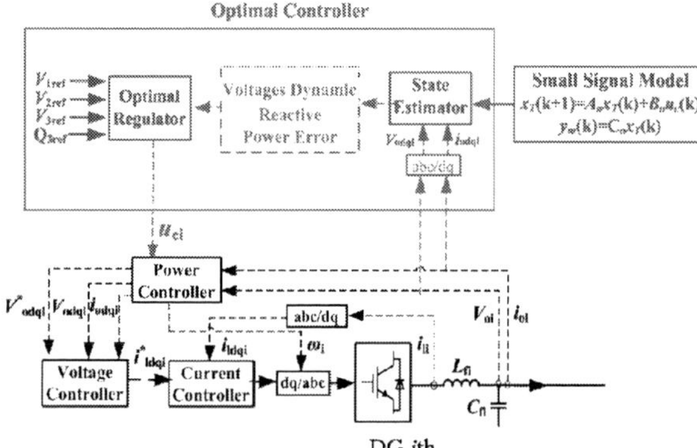

Fig. 2. Operation principle of the proposed optimal controller using state estimation.

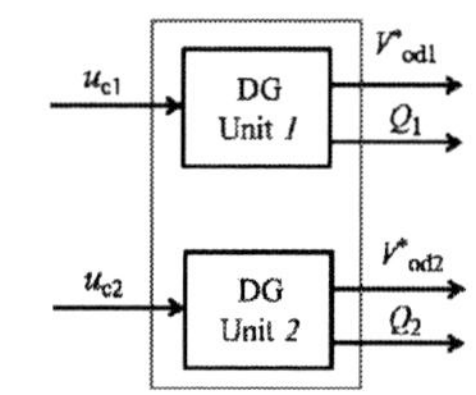

(a). Diagram of the original input-output relationship.

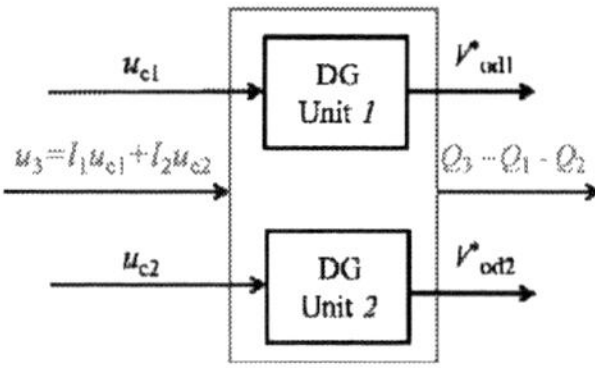

(b). Diagram of the proposed input-output relationship.

Fig. 3. Diagram of the system configuration.

A. Proposed System Input-Output Relationship

A droop controlled islanded microgrid is shown in Fig. 1, which is composed of two DG units and three loads. The diagram of input-output relationship for the whole system can be illustrated as Fig. 3(a). It can be observed that only two control inputs (u_{c1} and u_{c2}) are developed to regulate 4 output variables (Q_1, Q_2, V^*_{od1} and V^*_{od2}). In fact, in terms of a MIMO (Multiple Input Multiple Output) system, it is impossible to perform an asymptotic reference tracking and disturbance rejection, if the dimension of control inputs is less than that of the output variables. Hence, for the islanded microgrid, the control degree of freedom is not sufficient for performing both reactive power control and voltage regulation simultaneously with conventional droop control methods.

In this work, an improved small signal model is presented to reconstruct the system input-output relationship as shown in Fig. 3(b). Original small signal models have been intensively proposed in [2],[11]-[12]. Fig. 4 shows the block diagram of the droop-based power controller [2], including power calculation, low-pass filter and droop controller.

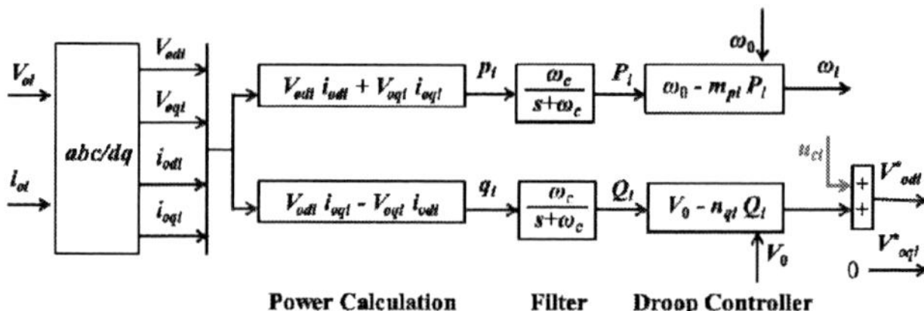

Fig. 4. Power controller of i th DG unit

The reactive power-voltage (Q-V) droop control method considering the voltage restoration can be written as (1).

$$V^*_{odi} = V_0 - n_{qi}Q_i + u_{ci} \tag{1}$$

Then the small signal voltage dynamic may be represented as (2) by combining a low pass filter and linearizing (1)

$$\Delta\dot{V}^*_{odi} = -n_{qi}\omega_c\Delta q_i - \omega_c V^*_{odi} + \omega_c\Delta u_{ci} \tag{2}$$

The details of the network and load models can be found in [11]-[12]. On the basis of this, the small signal dynamics of power controller, loads and network can be represented in a common D-Q frame as

$$\Delta\dot{x} = A\Delta x + B\Delta i_{dis} + B_c\Delta u_c \tag{3}$$

$$\Delta y_i = C\Delta x + D\Delta i_{dis} \tag{4}$$

where Δx is the overall state of the microgrid, Δi_{dis} are the unmeasured disturbances. Δu_c are the optimal control inputs of DG units. $\Delta u_c = [\Delta u_{c1}, \Delta u_{c2}]^T$ is the voltage control input vector. A, B, B_c, C, D can be found in [12]. The sign 'Δ' is omitted in the following for the sake of simplicity.

In (3) and (4), DG output reactive powers (Q_1 and Q_2) are modelled independently. To reconstruct the system input-

978-1-4673-9551-9/16 $31.00 © 2016 IEEE

output relationship, an additional control output Q_3 is defined to indicate the reactive power error as (5)

$$Q_3 = Q_1 - Q_2 \tag{5}$$

where Q_3 indicates the reactive power sharing error under the droop control method. Besides, control input u_3 is introduced by a linear combination of the known control inputs u_{c1} and u_{c2}.

As shown in Fig. 3(b), the dimension of the control inputs is equal to the number of output variables in the improved small signal model, which thus allows to achieve the twofold objective.

An improved discrete state-space model can be obtained by combining and discretizing (3)-(5). Moreover, to improve the robustness under load disturbances, disturbance models are augmented to the improved small signal model according to internal model principle. The measurement noise is also added to measured output channels for imitating measurement environment. The details of discrete small signal model, disturbance model and measurement noise model can be seen in [12]. Consequently, the overall model combining the system states, disturbances and measurement noise model can be represented as given in (6) and (7)

$$x_T(k+1) = A_o x_T(k) + B_o u_c(k) + W_o w_T \tag{6}$$

$$y_m(k) = C_o x_T(k) + D_o w_T \tag{7}$$

$x_T = [x \quad x_{dis} \quad x_m]^T$. x_{dis} is the disturbance state, x_m is the noise state. The improved small signal model will be utilized to present state estimator later.

B. Observability, Controllability and Sensitivity Analysis of the Improved System Model

Observability and controllability are critical aspects for estimator design and system synthesis, which are thus analyzed in the following in terms of mathematic criterion and circuit topology.

A system is observable if, for any possible sequence of state and control vectors, current states can be determined in finite time using only the outputs [13]. Observability of the system may be validated easily by PBH criterion in linear control theory, which means that system is fully observable when the augmented matrix $[C, \lambda I\text{-}A]^T$ is full rank. On the other hand, if the system is observable in part, it can be decomposed as observable subsystem and unobservable subsystem as follow [13].

$$\bar{A} = \begin{bmatrix} A_{no} & A_{12} \\ 0 & A_o \end{bmatrix}, \ \bar{C} = [0 \quad C_o] \tag{8}$$

where A_{no} is state matrix of unobservable subsystem, A_o is state matrix of observable subsystem. State estimator can be implemented if real part of eigenvalues in unobservable subsystem (A_{no}) is negative [13]. For the improved system model, eigenvalues of the unobservable subsystem are computed and listed as Table I. It can be seen that real part of all the eigenvalues in unobservable subsystem (A_{no}) is negative. State estimator is thus applicable.

Controllability means that state equation (3) is controllable if there exists an input that is able to move initial state to final state [13]. Controllability of a system can be validated by PBH criterion in linear control theory, which means that system is controllable if augmented matrix $[\lambda I\text{-}A, B]$ is full rank. However, if the system is controllable in part, it can be decomposed as controllable subsystem and uncontrollable subsystem [13] as (9).

$$\bar{A} = \begin{bmatrix} A_{uc} & 0 \\ A_{21} & A_c \end{bmatrix}, \ \bar{B} = \begin{bmatrix} 0 \\ B_c \end{bmatrix} \tag{9}$$

where A_{uc} is state matrix of uncontrollable subsystem, A_c is state matrix of controllable subsystem. Controller synthesis can be implemented if real part of eigenvalues in uncontrollable subsystem (A_{uc}) is negative [13]. Eigenvalues of the uncontrollable subsystem are listed as Table II.

Table I
EIGENVALUES OF UNOBSERVABLE SUBSYSTEM

Eigenvalues of unobservable subsystem	
- 410.26 + 314.08*i	- 408.62 - 314.04*i
- 410.26 - 314.08*i	- 292.48 + 335.81*i
- 416.13 + 314.08*i	- 292.48 - 335.81*i
- 416.13 - 314.08*i	- 329.55 + 146.02*i
- 326.53 + 314.08*i	- 329.55 - 146.02*i
- 326.53 - 314.08*i	-31.4
- 408.62 + 314.04*i	-35.44

Table II
EIGENVALUES OF UNCONTROLLABLE SUBSYSTEM

Eigenvalues of uncontrollable subsystem	
- 346.44 - 317.0*i	-0.0027686
- 410.26 + 314.07*i	-6.5185
- 410.26 - 314.07*i	-31.392
- 409.51 + 314.88*i	-24.871
- 409.51 - 314.88*i	- 326.53 + 314.08*i
- 416.13 + 314.08*i	- 326.53 - 314.08*i
- 416.13 - 314.08*i	- 346.44 + 317.0*i

For the improved model, optimal control strategy can be performed since real part of eigenvalues for uncontrollable subsystem is negative [13].

In addition, sensitivity analysis may provide clear insights into relationship among system states, inputs and outputs. So it can be adopted to explain system controllability and observability in terms of microgrid topology. In this paper, sensitivity analysis is developed to investigate reactive power sharing and voltages response under optimal control actions.

In general, a dynamic system can be represented as the following set of differential algebraic equation [14], as described in (10) and (11).

$$\dot{x} = f(x, y, u) \quad x(t_0) = x_0 \tag{10}$$

$$0 = g(x, y, u) \quad y(t_0) = y_0 \tag{11}$$

where x, u, y are vector state variables, control inputs and system outputs, x_0, u_0 and y_0 are the initial values. The sensitivity trajectory is normally based on the first-order linearization of the system, which can be obtained by a Talor expansion of (10) and (11) [14]. Thus, the trajectory of the system can be represented by (12) and (13)

$$x(t, x_0, u) = x(t, x_0, u_0) + x_u(t)(u - u_0) \tag{12}$$

$$y(t, x_0, u) = y(t, x_0, u_0) + y_u(t)(u - u_0) \tag{13}$$

where $x_u(t)$ and $y_u(t)$ are the sensitivity of $x(t)$ and $y(t)$ for control inputs. Eq. (14) represents the first-order approximation which changes with respect to the control inputs along the system trajectory. Likewise, the output change y is given by (15).

$$\dot{x}_u = \frac{\partial f}{\partial x} x_u(t) + \frac{\partial f}{\partial y} y_u(t) + \frac{\partial f}{\partial u} \tag{14}$$

$$0=\frac{\partial g}{\partial x}x_u(t)+\frac{\partial g}{\partial y}y_u(t)+\frac{\partial g}{\partial u} \tag{15}$$

The theoretical basis for dynamic sensitivity analysis is detailed in [14], where it accounts for the relationship between the system states and control inputs.

The sensitivity of the system voltages and reactive power sharing error Q_3 with respect to each control input are shown in Fig. 5. Fig. 5(a) depicts the voltage sensitivity at bus2, which indicates that the bus 2 voltage will boost 0.68 V and 0.3 V when 1 V control increment of DG1 (u_{c1}) and DG2 (u_{c2}) are performed, respectively. Also, Fig. 5(b) illustrates the voltage sensitivity at bus3, where the bus3 voltage will increase about 0.5 V for 1 V voltage increment of either DG1 (u_{c1}) or DG2 (u_{c2}). Similarly, the reactive power sharing sensitivity is shown in Fig. 5(c). It can be seen that a +300Var sharing error occurs when 1 V voltage increment of DG1 (u_{c1}) is performed, whereas a -300Var reactive error happens under 1 V voltage increment of DG2 (u_{c2}).

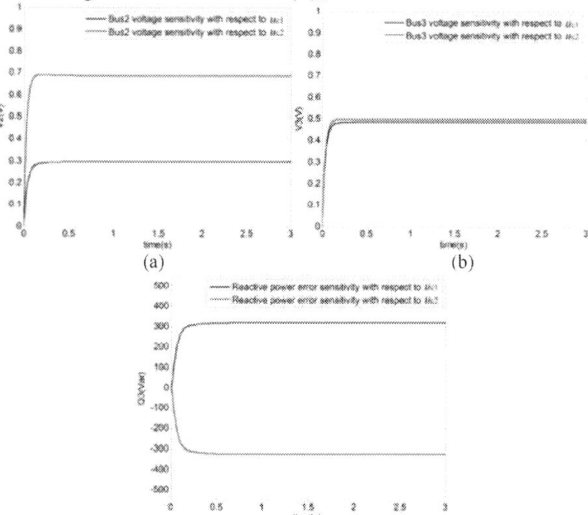
(a) (b)

(c) Reactive power sharing sensitivity for control inputs

Fig. 5. Sensitivity analysis for control inputs. (a) Bus2 voltage sensitivity for control inputs. (b) Bus3 voltage sensitivity for control inputs. (c) Reactive power sharing sensitivity for control inputs

The sensitivity results listed in Table III show the mathematical relationship of control inputs on reactive power sharing and voltages, which are formulated as (16)-(18)

$$k_1u_{c1}+k_2u_{c2}=\Delta V_{bus2} \tag{16}$$

$$k_3u_{c1}+k_4u_{c2}=\Delta V_{bus3} \tag{17}$$

$$k_5u_{c1}+k_6u_{c2}=Q_3 \tag{18}$$

ΔV_{bus2}, ΔV_{bus3} are voltage deviation at bus2 and bus3, Q_3 is the reactive power sharing error.

Table III
SENSITIVITY ANALYSIS FOR CONTROL INPUTS

Sensitivity \ Input	DG1 controller (u_{c1})	DG2 controller (u_{c2})
Bus2 Voltage	k_1= 0.68 V/V	k_2= 0.3 V/V
Bus3 Voltage	k_3= 0.49 V/V	k_4=0.5 V/V
Reactive power sharing error	k_5= +310 Var/V	k_6= -310 Var/V

The sensitivity results from Fig. 5 show that the system outputs are able to reflect effectively estimated states associated with the bus voltages and reactive power flows, thus the system is observable for these states. Also, optimal inputs (u_{c1} and u_{c2}) have different effects on system states associated with network voltage and reactive power, which implies that these states are controllable.

III. OPERATION PRINCIPLE OF OPTIMAL CONTROL STRATEGY

A. State Estimator Principle

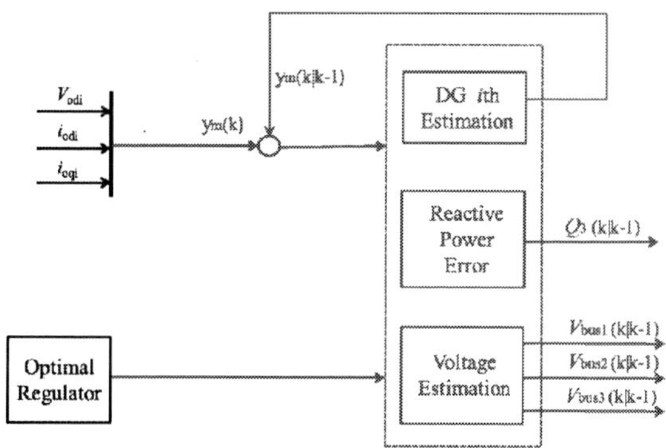

Fig. 6. Operation principle of the proposed state estimator.

A Kalman filter-based state estimator is proposed to observe the reactive power error and network voltages. Fig. 5 shows the operation principle of the proposed state estimator. It can be seen that the state estimator performs the estimation of reactive power and bus voltages based on the local voltage, current and optimal control commands. The state estimation equation can be given [15] by

$$x_T(k|k)=x_T(k|k-1)+K_T(y_m(k)-y_m(k|k-1)) \tag{19}$$

where $K_T=[K_1\ \ K_2\ \ K_3]^T$ is the Kalman gain, which is the solution of the Ricatti matrix equation [13]. Then, the state update equation and estimated output equation are given as (20) and (21) from small signal model as shown in Fig. 1.

$$x_T(k+1|k)=A_ox_T(k|k)+B_ou_c(k) \tag{20}$$

$$y_m(k|k-1)=C_ox_T(k|k-1) \tag{21}$$

These estimated states are updated and compensated by the measured voltage, current as well as the optimal control commands. Then, the state estimation equation can be obtained as follows:

$$x_T(k+1|k)=A_Tx_T(k|k-1)+B_Ty_m(k)+B_ou_c(k) \tag{22}$$

where $A_T=A_o-A_o*K_T*C_o$, $B_T=A_o*K_T$. As shown in Fig. 6, the estimator output consists of two parts: $y_m(k|k-1)=[V_{odi}(k|k-1),i_{odi}(k|k-1),i_{oqi}(k|k-1)]^T$ is the vector of measured local voltage and current state, which is viewed as feedback signal to compensate the estimation error, while $y_{um}(k|k-1)=[Q_3(k|k-1)\ \ V_{bus1}(k|k-1)\ \ V_{bus2}(k|k-1)\ \ V_{bus3}(k|k-1)]^T$ is the vector of the reactive power error and network voltages.

B. Optimal Regulator Principle

Optimal regulators are proposed to compute control commands according to the estimated information, which is an optimization cost function. The desired control action is obtained by computing the cost function that minimizes reactive power sharing error and voltage error, which can be formulated as (23):

$$\min_{\Delta u(k|k),\cdots,\Delta u(m-1+k|k)} J=\left\{\sum_{i=0}^{p-1}\left(\sum_{j=0}^{n_b}(w_b(V_{busj}(k+i+1|k)-V_{jref}))^2\right)\right. \quad (23)$$
$$\left.+w_Q(Q_3(k+i+1|k)-Q_{3ref})^2+\sum_{j=1}^{n_c}(w_c\Delta u_{cj}(k+i+1|k)^2\right\}$$

In the cost function, w_b, w_Q, w_c are weights for network voltages, reactive power error and control increments, respectively. p is the prediction horizon. It has been chosen that p=10; n_b=3 is the number of bus; n_c=2 is the number of optimal control inputs; $V_{busj}(k+i+1|k)$ denotes the voltages information predicted for time $k+i+1$ based on the measured information available at time k. Q_{3ref} and V_{jref}(j=1,2,3) are set to 0 (initial state defined in origin). Besides, the control horizon is also a critical parameter related to the control action. In the implementation of simulations and experiments, the control horizon has been chosen to be m=2.

C. *Dynamic Performance Analysis of the Optimal Controller*

To analyze the relationship between the dynamic performance and the parameters in the optimal controller, the closed-loop system poles are evaluated by varying the weight coefficients in the cost function.

In Fig. 7, the closed-loop poles are plotted when varying the bus3 voltage error weight of the DG1 controller. It can be seen that as the increase of the weight, one real pole and a conjugate pole pair move toward the origin (improving voltage control dynamics), and another real pole moves toward one (slowing down the dynamic response).

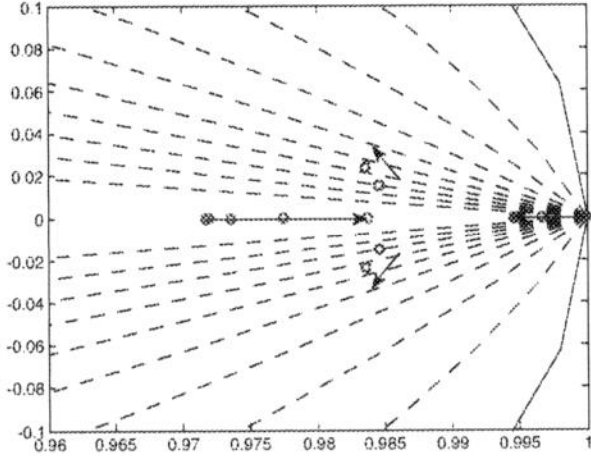

Fig. 7. The poles map for bus3 voltage error weight variation of DG1 controller. w_b = 10, 50, 100, 200, 600, 1000, 2000.

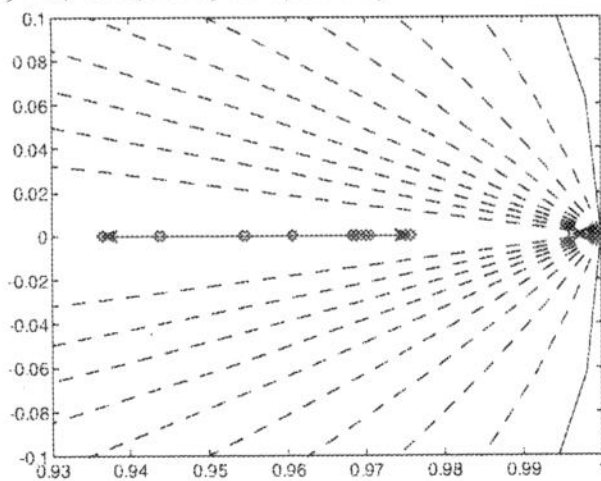

Fig. 8. The poles map for reactive power error weight variation of DG1 controller. w_Q = 1, 5, 10, 20, 50, 100.

The dynamic performance for the reactive power error weight variation is shown in Fig. 8. The two real poles move towards origin and a real pole moves toward one as the reactive power error weight increases.

Similarly, the closed-loop poles are plotted when increasing the bus3 voltage error weight of the DG2 controller as shown in Fig. 9, where a real pole moves towards one (causing the system to be unstable) and another has a slight movement towards zero (improving the voltage control dynamics). In Fig. 10, the dynamic performance is investigated when increasing the reactive power error weight. It can be seen that a real pole moves towards origin (improving dynamic response), while a couple of conjugate poles and a real pole move towards one (causing the controller to be unstable).

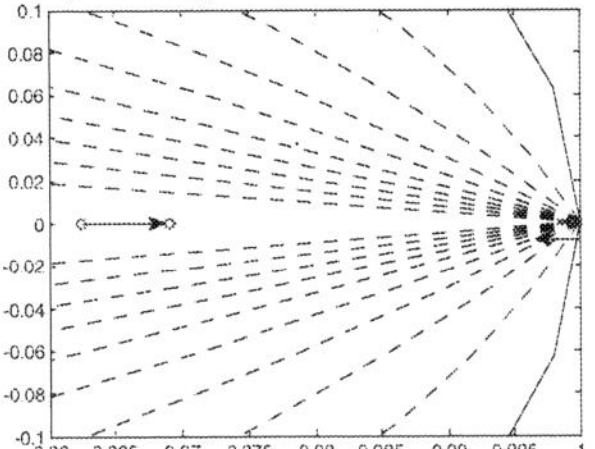

Fig. 9. The poles map for bus3 voltage error weight variation of DG2 controller. w_b = 10, 50, 100, 200, 600, 1000, 2000.

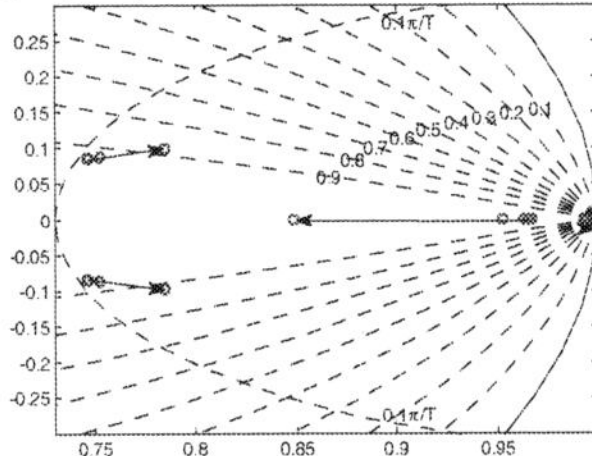

Fig. 10. The poles map for reactive power error weight variation of DG2 controller. w_Q = 1, 5, 10, 20, 50, 100.

IV. Simulation and Experimental Verification

To verify effectiveness of the proposed control approach, simulations and experiments have been carried out in this section. As shown in Fig. 11, the microgrid comprises two DG units in parallel and three linear RL loads. Also, the experimental setup is shown in Fig. 12, which is controlled by DS1006 dSPACE system. The circuit parameters of the configuration are given in Table IV.

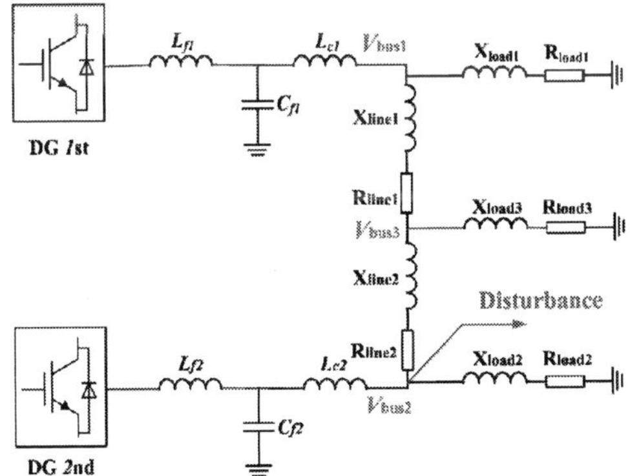

Fig. 11. The microgrid configuration.

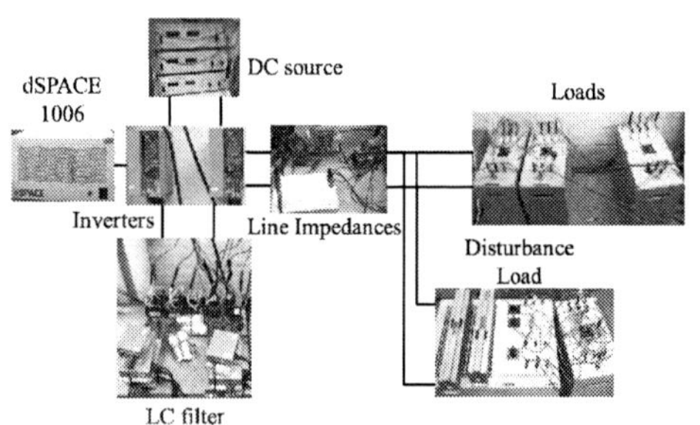

Fig. 12 The experimental setup.

TABLE IV
PARAMETERS FOR SIMULATIONS AND EXPERIMENTS

Parameters	Value	Parameters	Value
Inverter Rating	10 kVA	L_{c1}/L_{c2}	1.8mH/1.8mH
f_s	10 kHz	$R_{Line1}+jX_{Line1}$	0.2+j0.628
L_{f1}/L_{f2}	1.5mH/1.5mH	$R_{Line2}+jX_{Line2}$	0.2+j0.628
C_{f1}/C_{f2}	25μF/25μF	$R_{Load1}+jX_{Load1}$	64.5+j48.67
m_{p1}/m_{p2}	2.5e-5/1e-4	$R_{Load2}+jX_{Load2}$	64+j48.98
n_{q1}/n_{q2}	1e-3/1e-3	$R_{Load3}+jX_{Load3}$	80+j76.93

A. Case I: Reactive Power Sharing and Bus3 Voltage Control

In case I, the objective of the optimal control strategy is to perform reactive power sharing proportionally and hold bus3 voltage at its original state in the presence of load disturbance. The voltage error weight w_b, reactive power error weight w_Q and input rate weight w_c of DG1 are set to 600, 5 and 30. The voltage error weight w_b, reactive power error weight w_Q and input rate weight w_c of DG2 are set to 600, 1 and 20.

(1) Simulation Verification

To validate the optimal control strategy, a disturbance load (L_{d1} = 70 mH, R_{d1} = 10 Ω) is exerted at bus2. Fig. 13(a) shows the reactive power sharing performance under the conventional droop control method. It can be seen that an obvious reactive power sharing error (about 2400 Var) happens. Also, voltage deviations is depicted in Fig. 14 (blue curves), where a rough 6 V voltage deviation occur at bus3.

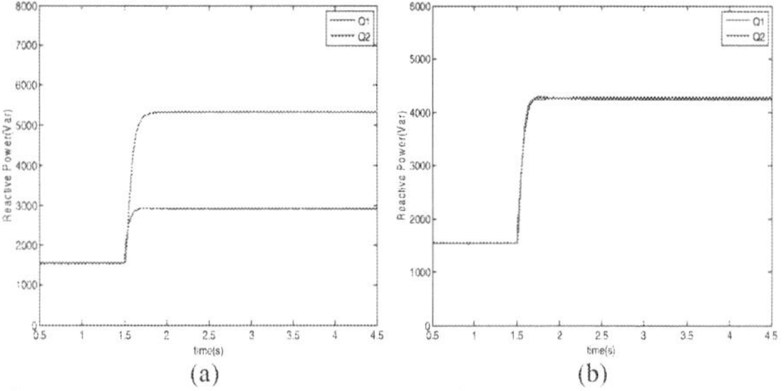
Fig. 13. The simulation results. (a) Reactive power sharing performance without the proposed control scheme. (b) Reactive power sharing performance with the proposed control scheme.

Furthermore, simulations are performed to validate the optimal control strategy. It can be observed that the reactive power sharing error is compensated to almost zero once the control strategy is activated as shown in Fig. 13(b). Meanwhile, Fig. 14(b) shows that the bus3 voltage is also restored to nearly its original state once the control strategy is exerted (green curves). In addition, the optimal control actions of DG 1st and DG 2nd can be seen in Fig. 15.

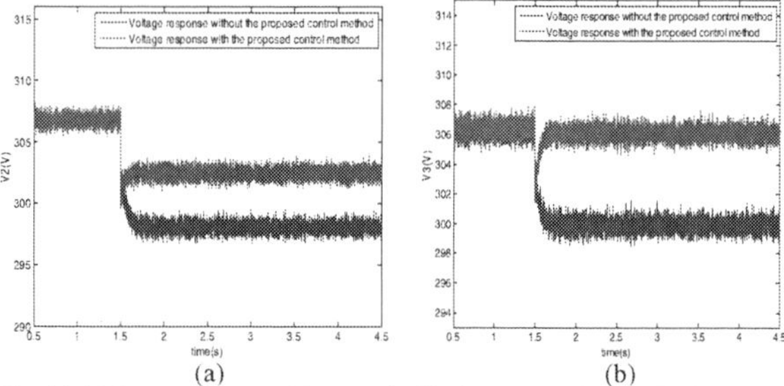
Fig. 14. Voltage response without and with the proposed control scheme. (a) Voltage response at bus 2. (b) Voltage response at bus 3.

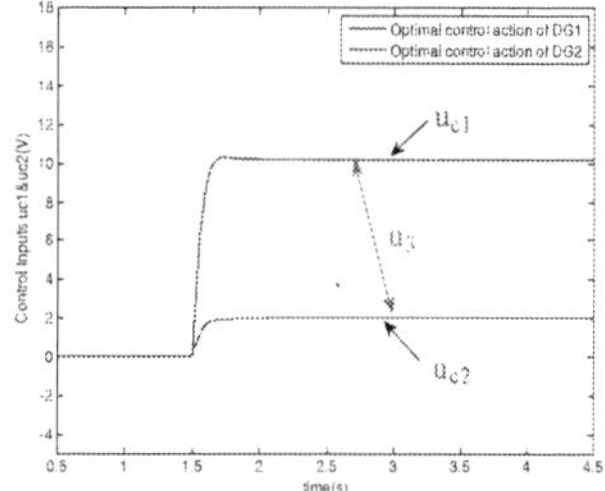
Fig. 15. The optimal control action of DG units

The load effect is also an important issue. To further investigate the proposed optimal control strategy under heavy loading condition, a serious disturbance (L_{d1} = 30mH, R_{d1} = 3 Ω) is exerted at bus2. It can be seen from Fig. 16(a) and Fig. 17(b) (blue curve) that the disturbance results in a 5400 Var reactive power error and 15 V voltage drop at bus3.

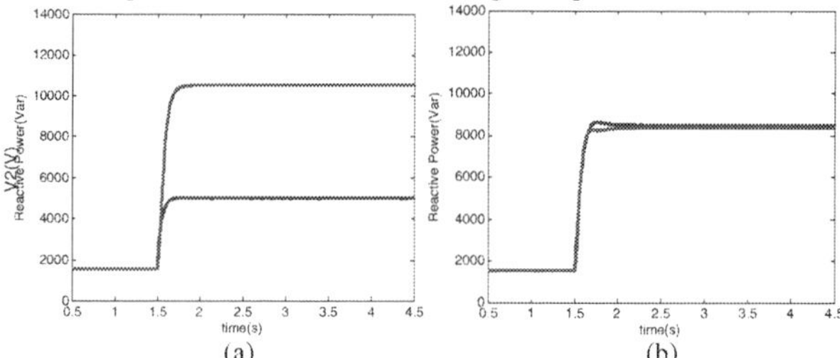
Fig. 16. Simulation results. (a) Reactive power sharing performance without the proposed control scheme. (b) Reactive power sharing performance with the proposed control scheme.

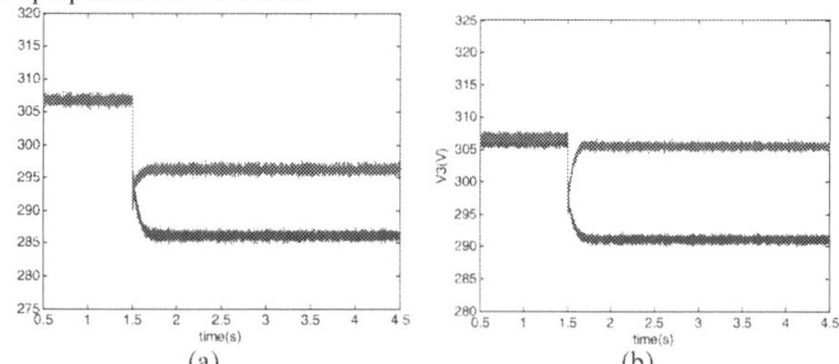
Fig. 17. Voltage response without and with the proposed control scheme. (a) Voltage response at bus2. (b) Voltage response at bus3.

As shown in Fig. 16(b), once the proposed optimal controller is activated, the reactive power error will be eliminated, and the bus3 voltage drop is restored to the original state as shown in Fig. 17(b) (green curve).

Hence, the results from Fig. 16- Fig. 17 show that the proposed control scheme is still able to handle serious load disturbance.

(2) Experimental Verification

To further investigate the proposed optimal control strategy, an accompanying experiment is carried out, where disturbance load (L_{dl} = 70 mH ± 5 mH, R_{dl}=10 Ω ± 0.5 Ω) is exerted at bus2.

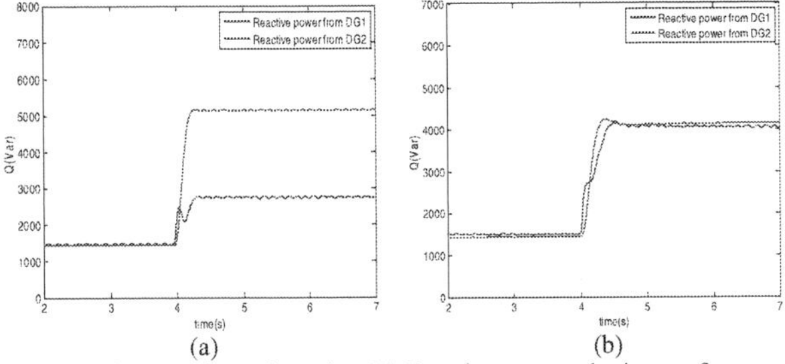

(a) (b)

Fig. 18. The experimental results. (a) Reactive power sharing performance without the proposed control scheme. (b) Reactive power sharing performance with the proposed control scheme.

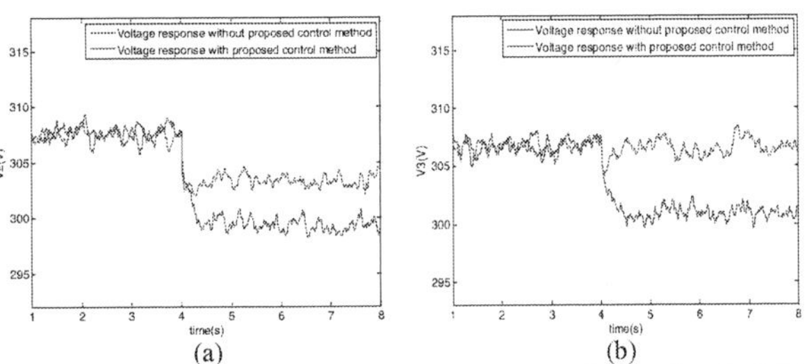

(a) (b)

Fig. 19. Voltages response without and with the proposed optimal method. (a) Voltage response at bus2. (b) Voltage response at bus3.

Fig. 18(a) and Fig. 19 demonstrate the reactive power sharing performance and voltage response (blue curves) under conventional droop control methods. On the other hand, the experimental results depict that the reactive power sharing error is eliminated once the proposed optimal control strategy is exerted as shown in Fig. 18(b). Fig. 19(b) shows that the proposed controller brings bus3 voltage back to the original state.

Comparing Fig. 13 to Fig. 19, the good agreements between the simulation and experimental results can be observed. The effectiveness of the proposed optimal control strategy is thus confirmed.

B. Case II: Reactive Power Sharing and Bus2 Voltage Control

To further investigate effectiveness of the proposed control strategy, a case2 is conducted, where the control objective is to perform reactive power sharing proportionally and hold bus2 voltage at its nominal value.

(1) Simulation Verification.

Disturbance load (L_{dl} = 70 mH, R_{dl} = 10 Ω) is exerted at bus2. Fig. 20(a) shows the reactive power sharing error (Q_3= 2400 Var) under the conventional droop control method. Also, Fig. 21 depicts the voltage response at different buses (blue curves). It can be seen that bus2 voltage has a dramatic decline (ΔV_{bus2} = 10V). Sensitivity analysis is first adopted to account for the optimal control actions by (16) and (18). The solution of (16) and (18) show that the optimal control objective can be performed when the optimal control actions of DG1 (u_{c1}) and DG2 (u_{c2}) are 14 and 6, respectively.

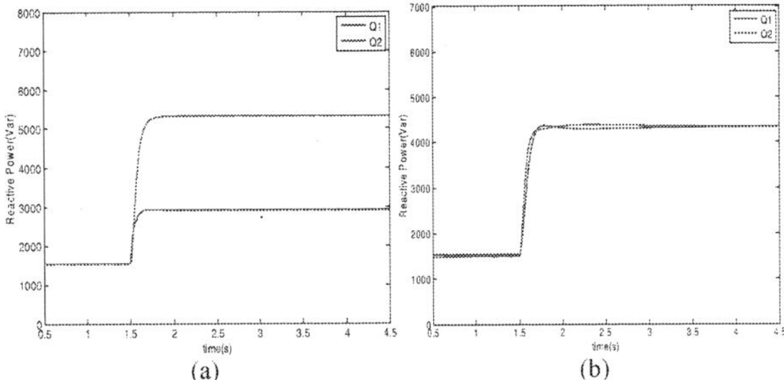

(a) (b)

Fig. 20. The simulation results. (a) Reactive power sharing performance without the proposed control scheme. (b) Reactive power sharing performance with the proposed control scheme.

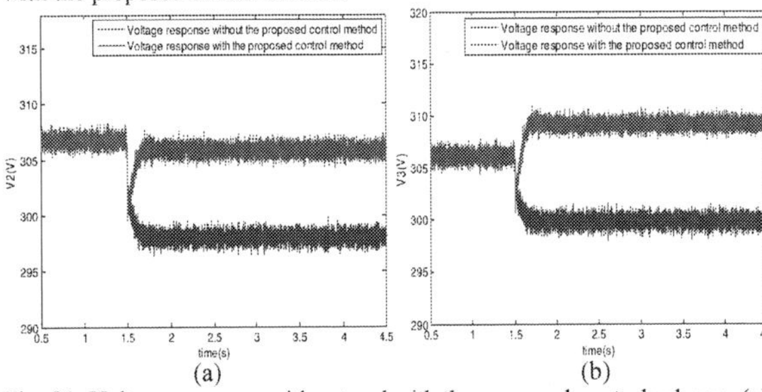

(a) (b)

Fig. 21. Voltage response without and with the proposed control scheme. (a) Voltage response at bus2. (b) Voltage response at bus3.

Simulation analysis is conducted to validate the proposed optimal control strategy. When the proposed optimal control strategy is activated, the reactive power is shared proportionally as shown in Fig. 20(b). At the same time, Fig.21(a) shows that the bus2 voltage is also restored to nearly the original state after the control scheme is exerted (green curves). In addition, the optimal control actions of DG 1 and DG 2 can be seen in Fig. 22.

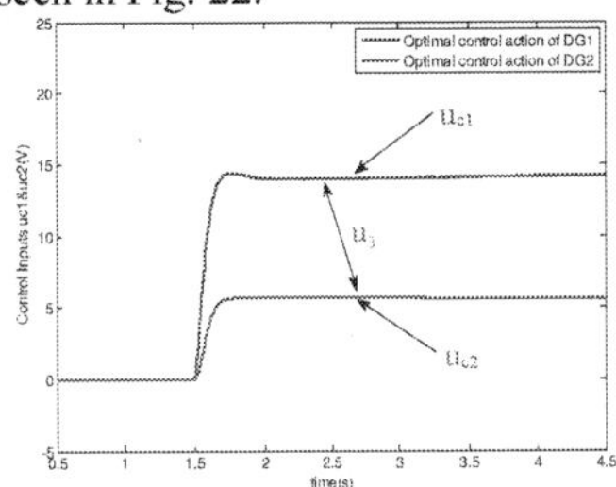

Fig. 22. Optimal control action of DG units

(2) Experimental Verification.

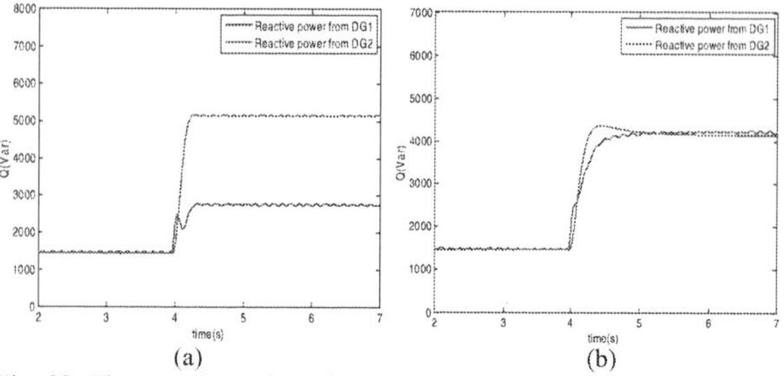

Fig. 23. The experimental results. (a) Reactive power sharing performance without the proposed control scheme. (b) Reactive power sharing performance with the proposed control scheme.

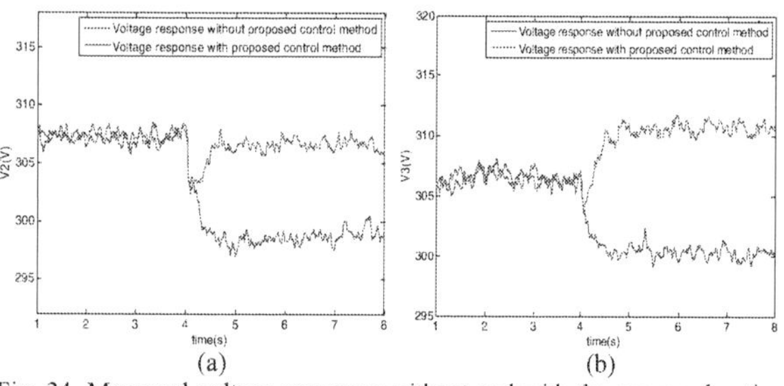

Fig. 24. Measured voltage responses without and with the proposed optimal method. (a) Voltage response at bus2. (b) Voltage response at bus3.

In the accompanying experiment, a disturbance load (L_{dl} = 70mH ± 5mH, R_{dl} = 10Ω ± 0.5Ω) is exerted at bus2. The experimental results shown in Fig. 23(b) demonstrate that the proposed optimal control strategy is able to eliminate the reactive power sharing error. At the same time, Fig. 24(a) shows that the proposed optimal controller brings bus2 voltage to the original state.

Hence, the sensitivity analysis, simulations as well as experimental results confirm effectiveness of the proposed optimal control strategy.

V. CONCLUSION

In this paper, a distributed optimal control strategy of reactive power and voltage has been proposed for islanded microgrids. First, an improved small signal model is developed to reconstruct the system input-output relationship. Then, the sensitivity analysis is employed to calculate the reactive power sharing and voltages response. Second, an optimal controller with state estimator is employed to perform reactive power sharing and voltage restoration. Then, dynamic performance of the optimal controller is analyzed. The results of the dynamic performance analysis show that the reactive power error weight coefficient and voltage weight coefficient have a critical effect on the controller dynamic performance. The results from the sensitivity analysis, simulation and experimental results show that the optimal control strategy is able to perform reactive power sharing and voltage restoration for different buses. The distributed optimal control strategy, hence, provides the much better reliability and flexibility for reactive power sharing and voltages control due to communication-less operation.

ACKNOWLEDGMENT

The authors would like to thank the Danish Council for Strategic Research for providing the financial support for the project "Development of a Secure, Economic and Environmentally-friendly Modern Power Systems" (DSF 09-067255)

REFERENCES

[1] R. H. Lasseter, "Smart distribution: Coupled microgrids," Proc. IEEE, vol. 99, no. 6, pp. 1074-1082, Jun. 2011.

[2] N. Pogaku, M. Prodanovic, and T. C. Green, "Modeling, analysis and testing of autonomous operation of an inverter-based microgrid," IEEE Trans. Power Electron., vol. 22, no. 2, pp. 613-625, Mar. 2007.

[3] Q. C. Zhong, "Robust droop controller for accurate proportional load sharing among inverters operated in parallel," IEEE Trans. Ind. Electron., vol. 60, no. 4, pp. 1281-1290, Apr. 2013.

[4] J. W. He, and Y. W. Li, "An enhanced microgrid load demand sharing strategy," IEEE Trans. Power Electron., vol. 27, no. 9, pp. 3984-3995, Sep. 2012.

[5] C. T. Lee, C. C. Chu, and P. T. Cheng, "A new droop control method for the autonomous operation of distributed energy resource interface converters," IEEE Trans. Power Electron., vol. 28, no. 4, pp. 1980-1993, Apr. 2013.

[6] J. Matas, M. Castilla, L. G. de Vicuna, J. Miret, and J. C. Vasquez, "Virtual impedance loop for droop-controlled single-phase parallel inverters using a second-order general-integrator scheme," IEEE Trans. Power Electron., vol. 25, no. 12, pp. 2993-3002, Dec. 2010.

[7] X. Wang, Y. Li, F. Blaabjerg, and P. C. Loh, "Virtual-impedance-based control for voltage-source and current-source converters," IEEE Trans. Power Electron, 2015. (Early access)

[8] M. Hua, H. Hu, Y. Xing, and J. M. Guerrero, "Multilayer control for inverters in parallel operation without intercommunications," IEEE Trans. Power Electron., vol. 27, no. 8, pp. 3651-3663, Aug. 2012.

[9] J. Rocabert, A. Luna, F. Blaabjerg, and P. Rodriguez, "Control of power converters in AC microgrids," IEEE Trans. Power Electron., vol. 27, no. 11, pp. 4734-4749, Nov, 2012.

[10] X. Wang, F. Blaabjerg, Z. Chen, and J. M. Guerrero, "A centralized control architecture for harmonic voltage suppression in islanded microgrids," Proc. of IECON, pp. 3070–3075, 2011.

[11] Y. B. Wang, X. F. Wang, Z. Chen, Y. J. Tian, and Y. D. Tan, "A communication-less distributed voltage control strategy for a multi-bus islanded AC microgrid," The international power electronics conference-ECCE Asia- Japan, May. 18-21. 2014.

[12] Y. B. Wang, Z. Chen, X. F. Wang, Y. J. Tian, Y. D. Tan, and C. Yang, "An estimator-based distributed voltage predictive control strategy for AC islanded microgrids," IEEE Trans. Power Electron., vol. 30, no. 7, pp. 3934-3951, Jul. 2015.

[13] C. T. Chen, "Linear system theory and design," 3rd ed. Oxford University Press, 1999.

[14] H. K. Khalil, "Nonlinear System," 3rd ed. Englewood Cliffs, NJ: Prentice-Hall, 2002.

[15] G. Welch, and G. Bishop, "An introduction to the kalman filter," University of North Carolina at Chapel Hill, Chapel Hill, NC, 1995.

New Start-up Scheme for HF Transformer Link Photovoltaic Inverter

Abhijit Kulkarni and Vinod John
Department of Electrical Engineering,
Indian Institute of Science, Bangalore - 560012, India
Email: abhijitk@ee.iisc.ernet.in, vjohn@ee.iisc.ernet.in

Abstract—**Safe start-up procedure is critical for reliable operation of power converters. The challenges in start-up schemes of high-frequency (HF) transformer link inverters involve pre-charge of the capacitors without significant inrush currents and over-voltages. The pre-charge is normally followed by enabling of the closed-loop control. The over-voltage or inrush current problems in conventional start-up schemes available in literature are identified in this paper. A new start-up scheme is proposed which achieves the objectives of limiting inrush current and start-up over-voltages. This method does not require any additional dedicated pre-charge circuitry. The proposed scheme is validated using simulation and experimental results which agree with the theoretical prediction.**

Index Terms—**Photovoltaic inverter, high-frequency transformer, start-up scheme, dc-ac converter.**

I. Introduction

Photovoltaic (PV) inverters of few kilowatt power ratings normally have low input voltage. This is specially true when PV panels are connected in predominantly parallel configurations to achieve better performance for rapidly changing insolation and partial shading problems [1]. Transformers are used in such cases to provide the required voltage conversion ratio. High-frequency (HF) transformers are popular over the line-frequency transformers due to their compactness and reduced cost. Many power circuit topologies are proposed with HF transformer link inverters for various applications including PV inverters [2]–[10].

It is observed that the work done on the modulation and closed-loop control of different HF transformer link inverters [2]–[10] is considerably higher than the work done on the start-up schemes. In [11], the strategy of slowly increasing the duty ratio of the HF inverter devices from 0 to 0.5 is described for dual active bridge (DAB) converters. The rectification is done using the body diodes of the HF rectifier during start-up. Though this method will not have inrush current problem, it is sensitive to the stray elements in the circuit that can cause over-voltage problem due to resonance. This problem of over-voltage is detailed in this paper. Another method discussed in [12] directly applies a duty ratio of $D = 0.5$ in the HF inverter devices. This method is affected by the inrush current problem. If this method uses a passive rectifier, it will be affected by both inrush current and over-voltage problems. The gradual increase of duty ratio in both primary and secondary side bridges discussed in [12], [13] has the disadvantage of the requirement of increased number of gate drive control channels during start-up. In [14], increasing the switching frequency during pre-charge is suggested. The inrush current peak, which is normally tens of orders of magnitude higher than nominal current, can linearly decrease with the increase in switching frequency. However, switching frequency cannot be increased to very high value as it is limited by the gate-drive circuitry and the device switching speed. There can still be over-voltage problem when diode rectification is used. The works reported in [15]–[17] discuss the start-up method for full-bridge converters. However, these methods require additional circuitry and modification to the power circuit which add to system cost and complexity.

The start-up schemes investigated in this paper are applicable to full-bridge HF transformer link topologies. A HF transformer AC link topology with lossless snubber fed by PV modules is shown in Fig. 1. This is similar to a DAB topology [5] except that the stiff dc link in the secondary side of DAB topology is replaced by a lossless snubber circuit.

Before closing the output switch $SW2$ in Fig. 1, there are two dc voltages that must be pre-charged. These are marked as V_{dc} and V_{link} in Fig. 1. The objectives of a good start-up scheme are as follows.

1) The dc voltages must be pre-charged without very high inrush currents.
2) The pre-charge must not cause very high over-voltages due to circuit parasitics and unwanted resonances.
3) It is desirable to have no additional control or power circuitry for start-up .

All these aspects are investigated in this paper for the HF transformer link inverter. A new start-up scheme that satisfies the above objectives is proposed and validated using simulations and experimental results. This start-up scheme is applicable to any HF transformer link topology that contains a full bridge HF rectifier section and is fed by PV modules.

This paper is organized as follows. Section II discusses the pre-charge of the input side dc voltage V_{dc} using PV modules. Conventional methods for secondary link voltage pre-charge, and their limitations are detailed in Section III. Proposed start-up method is discussed in Section IV. Conclusion is presented in Section V.

II. Pre-charge of Input DC Bus from PV Modules

Input dc bus voltage is marked as V_{dc} in Fig. 1. The input side capacitor bank has to be pre-charged to the rated voltage

978-1-4673-9551-9/16 $31.00 © 2016 IEEE

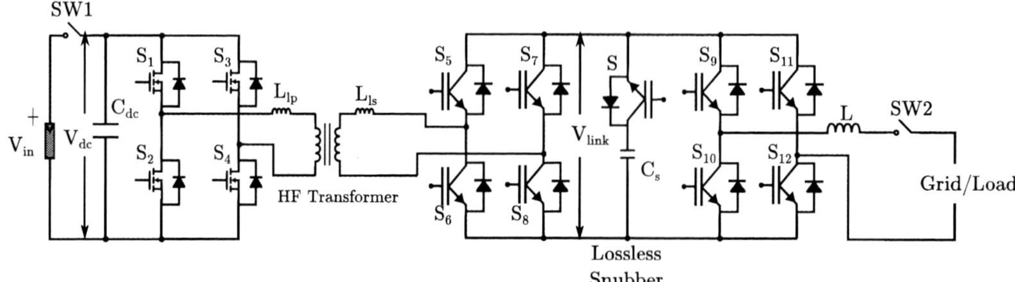

Fig. 1. HF transformer ac Link inverter with lossless snubber.

from the PV source V_{in}. Note that all the semiconductor switches are OFF.

Consider that the switch $SW1$ is closed directly to pre-charge C_{dc} from the PV modules. As C_{dc} is initially uncharged, the PV modules would see practically short-circuit condition. Hence, the PV modules would act as current sources and charge-up the input dc bus. As the voltage builds closer to the open circuit voltage, the PV current reduces and eventually goes to zero when C_{dc} is charged upto the open circuit voltage (V_{oc}). Hence, direct charging up of C_{dc} is possible as the PV modules are current limited.

Practically, there will be ringing observed in the initial current from the PV modules. This can exceed the short circuit current. The reason for this ringing is the resonance between the PV module capacitance and the effective wire inductance from the PV modules to the power converter [18]. Fig. 2(a) shows the pre-charge of C_{dc} directly from the PV modules. The pulsed shape of the pre-charging current indicates that the PV modules were charging the C_{dc} as a current source till the final voltage of V_{oc} is reached. The initial ringing in the pre-charge current due to resonance effect is shown in Fig. 2(b).

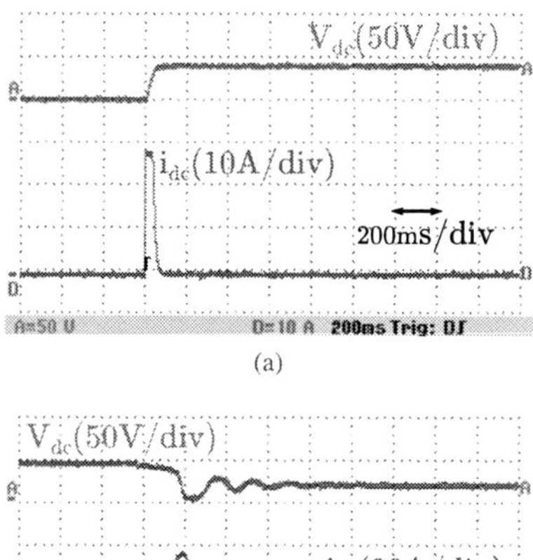

(a)

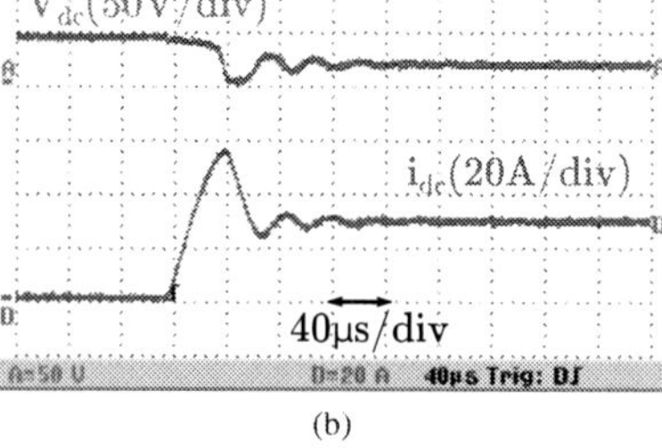

(b)

Fig. 2. Input dc bus pre-charged directly with PV modules as source. (a) Higher time scale indicating full pre-charge, (b) Smaller time scale showing the resonant peaks in the PV current.

It is possible to eliminate the ringing and pre-charge C_{dc} at a known maximum current. This can be achieved by including a pre-charge resistor in series with the PV modules. This resistor must be shorted once the pre-charge is complete. Including pre-charge resistor slows down the process. Fig. 3 shows the pre-charge when a resistor of 22Ω is used. It can be seen that the pre-charge current is very small and consequently the time taken is very high to complete the pre-charge.

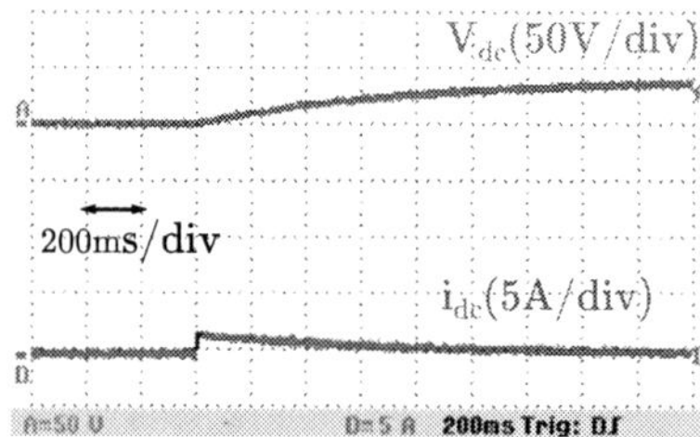

Fig. 3. Input dc bus pre-charged using a 22Ω pre-charge resistor in series with the PV modules.

III. Conventional Pre-charge of the Secondary Side Link Voltage

It is assumed that the input dc bus V_{dc} is pre-charged. The snubber switch S in Fig. 1 is assumed to be ON so that the snubber capacitor can be pre-charged to a dc voltage. Conventional approaches reported in literature for pre-charging the link voltage [11]–[13] are analyzed in the following subsections and the related issues are detailed. The start-up methods in literature having additional start-up circuitry [15]–[17] are not evaluated. These methods increase the system cost and complexity. It must be noted that the start-up method proposed in this paper (discussed in Section IV) does not require any additional circuit.

A. Square-Wave Modulation with Diode Rectifier

In this method, the HF inverter devices are switched in square-wave mode. In other words, the transformer primary voltage is a square wave. The HF inverter switches have a duty ratio $D = 0.5$. The HF rectifier IGBTs are all OFF. Hence, the rectification is done by the diodes only. This is a simplest start-up method that can be considered for HF transformer link inverters. However, this method has significant limitations which are explained below.

1) Over-voltage problem: The start-up equivalent circuit for this case is shown in Fig. 4. The primary voltage is replaced by

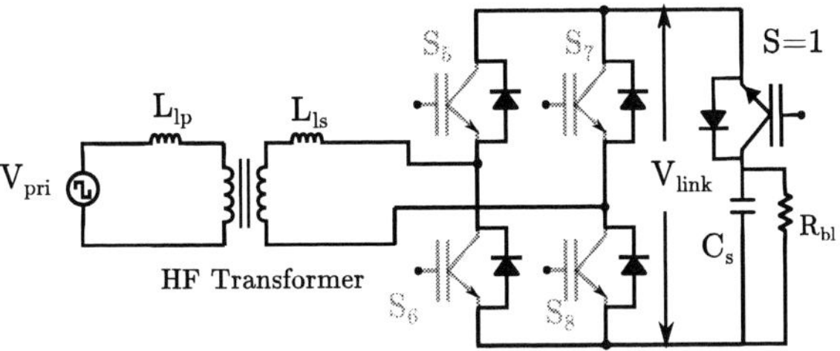

Fig. 4. Equivalent start-up circuit for $D = 0.5$ for HF inverter devices. The rectifier is passive with only diodes taking part in rectification. Bleeder resistor R_{bl} across the capacitor C_s is shown.

a square wave voltage source. The output inverter is inactive as all the switches are OFF and hence it is not shown in the equivalent circuit. The rectifier IGBTs are switched OFF as only the diodes are used for rectification. Hence the IGBTs are shown in light-shade.

With an ideal transformer, the rectified voltage will be purely dc with amplitude of the primary voltage. However, practically the HF transformer has a stray capacitance appearing in its equivalent circuit shown in Fig. 5. This leads to a resonance when the input voltage has step-change. The parameters of the HF transformer equivalent circuit model are determined using a network analyzer. The HF transformer is connected in open circuit and short circuit configurations to the network analyzer to determine the parameters of the equivalent circuit. These parameters are listed in Table I.

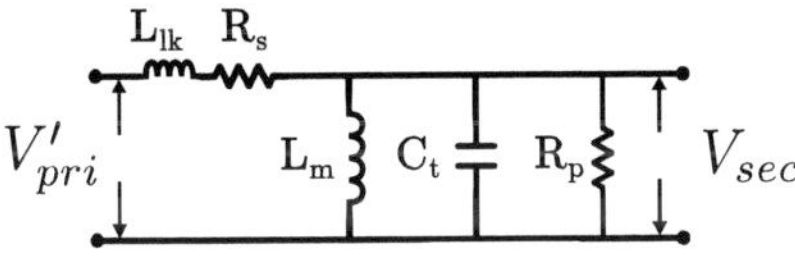

Fig. 5. Practical equivalent circuit of HF transformer from secondary side showing stray capacitance.

TABLE I
EQUIVALENT CIRCUIT PARAMETERS OF THE HF TRANSFORMER REFERRED TO THE SECONDARY SIDE.

Parameter	Notation	Value
Turns ratio	$1 : N_t$	$1 : 10$
Leakage inductance	L_{lk}	$66\mu H$
Winding resistance	R_s	0.68Ω
Magnetizing inductance	L_m	$10.8mH$
Effective capacitance	C_t	$0.266nF$
Core loss resistance	R_p	$150k\Omega$

Due to the resonance, the reflected secondary voltage goes beyond the primary voltage and reverse biases the diodes in HF rectifier. This results in the link capacitance charging to a higher voltage than the amplitude of primary voltage. As PV inverter topologies have to operate between open-circuit voltage to maximum power point (MPP) voltage, the start-up over-voltage can exceed the voltage rating or can considerably reduce the safe operating margin of semiconductor devices.

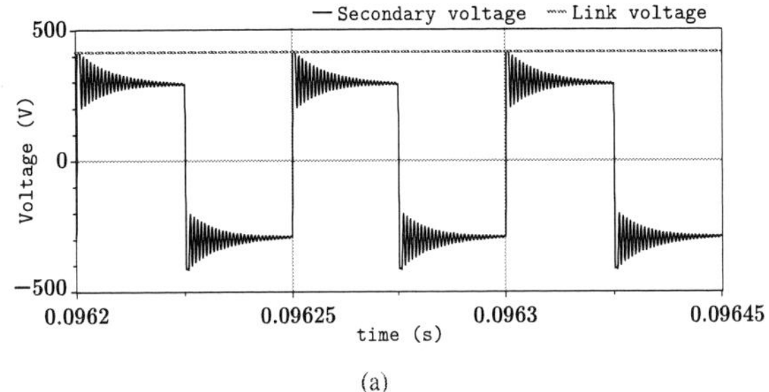

(a)

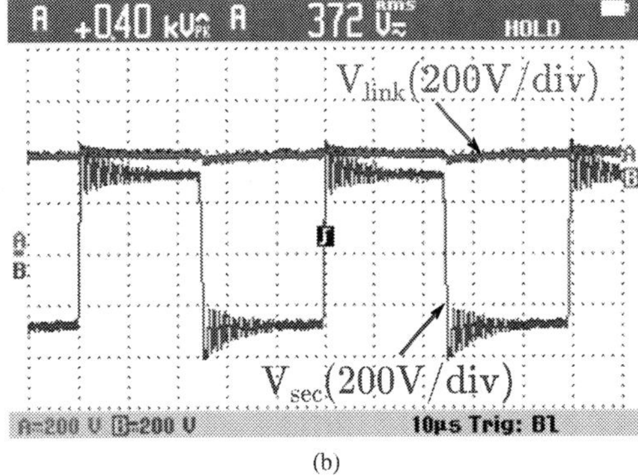

(b)

Fig. 6. Pre-charge of the link voltage V_{link} with $D = 0.5$ and passive rectifier. $V_{in} = 30V$. (a) Simulation result (b) Experimental result (time scale: $10\mu s/div$).

Fig. 6(a) shows the simulation result for this method. The resonance in the secondary voltage and its effect on increasing the link voltage beyond the expected value due to reverse bias of the rectifier diodes can be observed. The input voltage is $30V$. After the resonance is damped, it can be observed that the secondary voltage approaches $300V$ as expected due to the $1 : 10$ transformer. However, due to resonance, the link voltage reaches a value of $400V$ which is considerably higher.

The experimental result corresponding to this case is shown in Fig. 6(b). In the experimental result also the input dc voltage is $30V$. As it can be observed from Fig. 6(b), the link voltage goes upto $400V$. Consider the 72-cell PV modules with open circuit voltage of about $45V$ [19]. With this pre-charge method, the link voltage can go upto $45 \times 400/30 = 600V$. In single-phase application, the secondary side devices are normally rated for $600V$. Hence with this method, the link voltage can reach the device withstand limit. This would necessitate oversizing of the devices which would increase the cost and affect system efficiency. Note that the results shown in Fig. 6 correspond to the case when R_{bl} or the bleeder-resistance across secondary link is $40k\Omega$. For higher values of R_{bl}, the peaks in V_{link} will be even higher.

2) Inrush current problem: The pre-charge of the secondary side link voltage can cause significant inrush currents when the input dc bus capacitance C_{dc} is already pre-charged [12]. This is another problem is in addition to the over-voltage problem described in the previous subsection. The inrush current peak can be estimated theoretically using an equivalent circuit model. Consider that the input dc bus is pre-charged and the devices S_1 and S_4 are turned ON. Depending on the duty ratio D, the HF inverter devices S_1 and S_4 will be ON for a duration DT_s. Assuming the corresponding secondary side diodes are conducting, the equivalent circuit for this duration is shown in Fig. 7. For the inrush current estimation, the transformer model is simplified to leakage inductance and winding resistance.

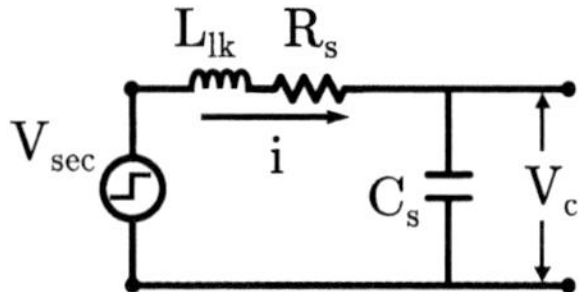

Fig. 7. Simplified equivalent circuit model for estimating the inrush current.

The equivalent circuit describes a linear second-order system with zero initial conditions ($i(0) = 0$ and $V_c(0) = 0$). Its response can be derived by solving the governing differential equation. The final expression for the current $i(t)$ is as follows.

$$i(t) = V_{sec}\sqrt{\frac{C_s}{L_{lk}}}e^{-\frac{R_s t}{2L_{lk}}}\sin(\omega_r t) \tag{1}$$

The damped resonance frequency ω_r is given by,

$$\omega_r = \frac{1}{\sqrt{L_{lk}C_s}}\sqrt{1 - \frac{R_s^2 C_s}{4L_{lk}}} \tag{2}$$

Using the parameters from Table I for Eqns. (1) and (2), the maximum inrush current can be calculated. The maximum current occurs just before the HF inverter devices are switched to apply zero-state or negative voltage in the primary side. Hence, at $t = DT_s$ the inrush current peak can be determined. That is,

$$\begin{aligned} i_{max} &= V_{sec}\sqrt{\frac{C_s}{L_{lk}}}e^{-\frac{R_s DT_s}{2L_{lk}}}\sin(\omega_r DT_s) \\ &= N_t V_{in}\sqrt{\frac{C_s}{L_{lk}}}e^{-\frac{R_s DT_s}{2L_{lk}}}\sin(\omega_r DT_s) \end{aligned} \tag{3}$$

From (3), it can be observed that the maximum inrush current is proportional to the input voltage (scaled as V_{sec}) and D. Consider the case when input voltage is $V_{in} = 10V$. Expected secondary voltage will be $100V$ for the $1:10$ transformer. For $D = 0.5$, and snubber capacitance value $C_s = 7.5\mu F$, the maximum inrush current can be calculated using (3) as

$$i_{max} = 26.6A \tag{4}$$

Note that this is the secondary current. Hence, the primary current peak would be $267A$ which is very high and it will invariably trip the protection system of the power converter.

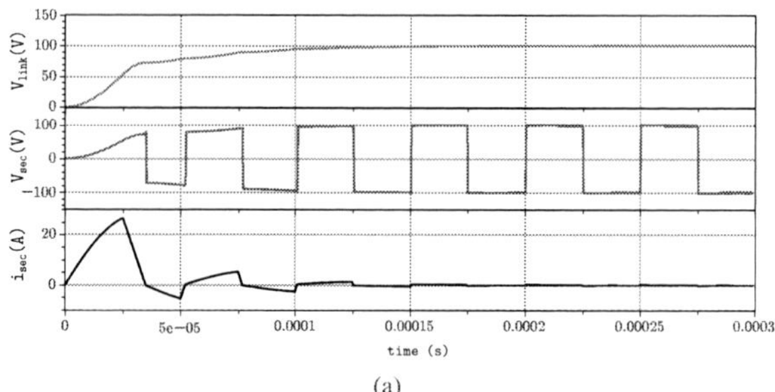

(a)

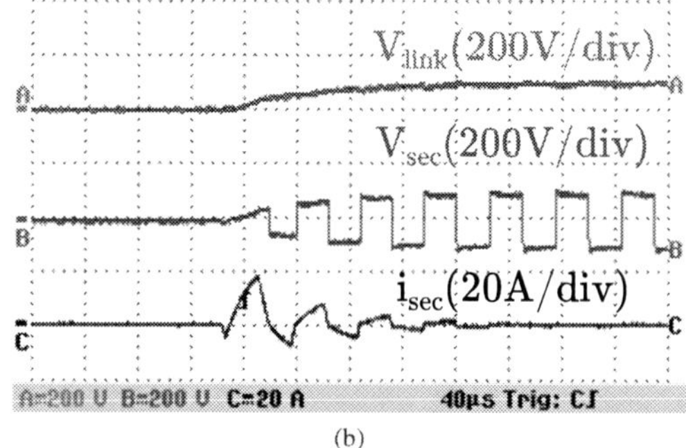

(b)

Fig. 8. Inrush secondary current and building up of secondary/link voltage for $D = 0.5$ and input voltage $V_{in} = 10V$. (a) Simulation result (b) Experimental result (time scale: $40\mu s/div$).

Fig. 8(a) shows the simulation result when input voltage is $10V$. The circuit simulation is performed using PSIM. The build-up of the secondary and link voltage is very fast and occurs in about $150\mu s$. The secondary current peak can be observed to be $26.5A$ which agrees with the analytical value in Eqn. (4). Corresponding experimental result is shown in Fig. 8(b). It can be observed that it is in agreement with the simulation result. The peak secondary current is measured as $17.8A$ (primary current = $178A$) which is lesser than the theoretical estimate. The deviation is observed because the resistances of the devices and the respective voltage drops are ignored in theoretical estimation. Even with the additional damping obtained practically, the observed peak current in primary is as high as $178A$ which would trip the protection system. The inverter protection was disabled while measuring the inrush current experimentally. It must be noted that for an input voltage of $10V$, the theoretical inrush current is $266A$. Hence, for the maximum input of about $45V$, the inrush current would be as high as $1.2kA$. Thus, it is important to address the issue of both inrush current and over-voltage.

B. Quasi Square Wave Modulation with Diode Rectifier

In this method, the HF inverter will apply a quasi square wave voltage across the transformer primary. The pulse width of the quasi square wave is determined by the duty ratio D. From (3), it can be observed that the maximum current depends on D. Hence, by selecting a very small D, the inrush current can be limited. The pre-charge duration will be larger than the case with square wave modulation but there will not be any inrush currents. Another approach is to slowly increment the duty ratio D from 0 to the value of 0.5 [11]. This method will also not have inrush current but it is affected by the over-voltage problem discussed in Section III-A(1) when diode bridge rectifier is used.

Fig. 9(a) shows the simulated secondary voltage and the link voltage for quasi square wave input with $D = 0.08$. The resonance effect in causing the over-voltage in the link voltage can be observed from the simulation result. Corresponding experimental result is shown in Fig. 9(b).

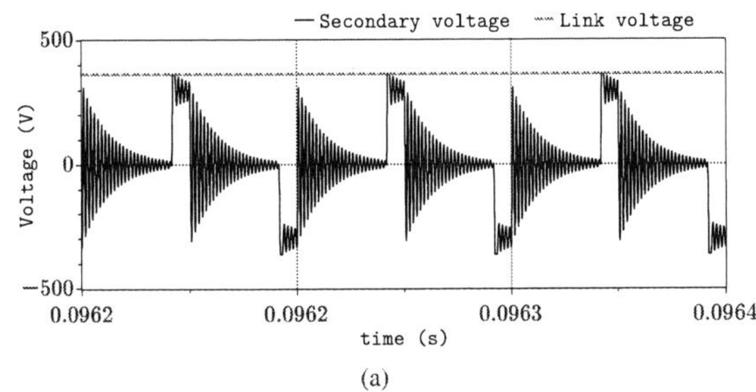

(a)

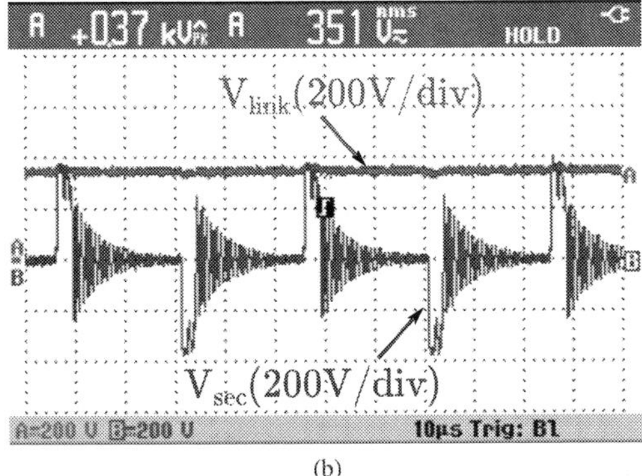

(b)

Fig. 9. Pre-charge of the link voltage V_{link} with $D = 0.08$ and passive rectifier. $V_{in} = 30V$. (a) Simulation result (b) Experimental result (time scale: $10\mu s$/div).

For an input voltage of $30V$, the link voltage peak has a value of $370V$ experimentally. Hence, for the rated open circuit voltage of $45V$, the link voltage can go upto $555V$ with very less margin from the rated breakdown voltage of $600V$.

The quasi-square wave method has another disadvantage in addition to the over-voltage problem. As the normal operation of the HF inverter is with $D = 0.5$ [10], only two unique signals are needed from the gate-drive control card or also known as, protection and delay (PD) card of the inverter hardware. Two unique signals correspond to the gating pulses for S_1 and S_2. Ideally they are complements of each other. However, practically they are realized to have a finite dead-time and hence, they are unique. The switching pulses for S_4 and S_3 will be derived from those of S_1 and S_2 respectively due to $D = 0.5$ operation. However, to realize quasi square wave modulation, four unique signals will be required from the gate-drive control card to switch the HF inverter devices during start-up. This is because, the switches S_1, S_4 and S_2, S_3 no longer have identical switching states and the dead-time is still required between the complementary pairs. The methods reported in [12], [13] for DAB converters use quasi square wave modulation for both HF inverter and rectifier during start-up. Hence for this case, eight unique gate drive control channels will be required during start-up. This can increase the overall cost of the system.

IV. Proposed Start-Up Method

By operating the HF rectifier as an active rectifier synchronized with the HF inverter with $D = 0.5$, the problem of over-voltage can be solved. In this method, the HF inverter and rectifier devices will have synchronized switching. That is, S_1, S_4 of the HF inverter will be switched simultaneously with S_5, S_6 of the HF rectifier. Similarly the other two pairs are switched together. Thus, the IGBTs of the secondary side are also involved in the rectification. When the HF rectifier is switched as an active rectifier, there is a path for the link capacitance to transfer the energy due to resonance back to source. This is absent when only diode rectifier is used resulting in over-voltage problem. This has been validated using circuit simulation and experimental results.

Fig. 10(a) shows the steady state simulation result with the proposed pre-charge method. The input voltage is $30V$. It can be observed that the secondary voltage and link voltage have a value of $300V$ which is the ideally expected value.

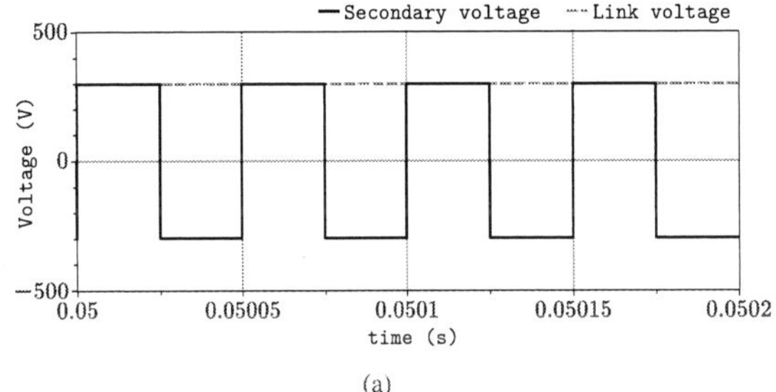

(a)

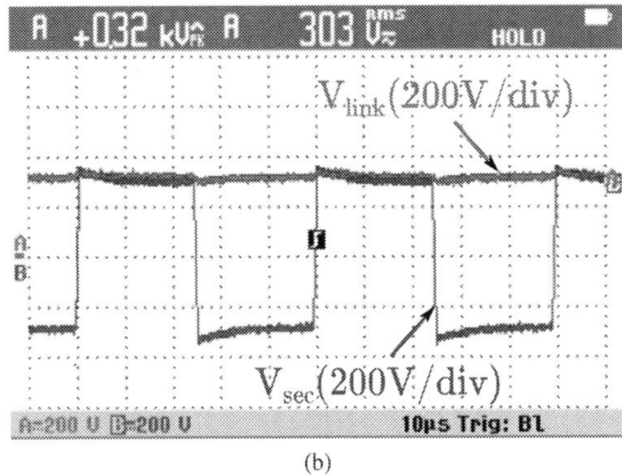

(b)

Fig. 10. Performance with the proposed pre-charge method of the link voltage V_{link} with $D = 0.5$ and active rectifier. $V_{in} = 30V$. (a) Simulation result (b) Experimental result (time scale: $10\mu s$/div).

TABLE II
COMPARISON OF PROPOSED START-UP METHOD WITH THE CONVENTIONAL METHODS*

S.No.	Start-up method	$V_{peak}(V)$	$i_{inrush}(A)$	Additional gate-drive control channels for start-up modulation
1	Square-wave primary voltage + diode bridge rectifier	600	1197	0
2	Quasi square wave primary voltage + diode bridge rectifier [11]	555	0	2
3	Quasi square wave primary voltage + active rectifier [12], [13]	X†	0	4
4	Proposed method	480	0	0

* The methods in [15]–[17] modify the power circuit to ensure smooth start-up. The methods compared above do not change the power circuit, hence do not increase the cost in power circuit.
† The over-voltage value is not reported in the references. It can be assumed that the value will be comparable to the proposed method as active rectification is used.

The corresponding experimental result is shown in Fig. 10(b). It can be observed that the link voltage peak is $320V$ for an input voltage of $30V$. Hence for the maximum open circuit voltage of $45V$, the link voltage peak goes upto $480V$. As a result, there will be a significant safety margin of $120V$ for the $600V$ rated IGBTs.

The application of $D = 0.5$, when the input dc bus is already pre-charged by the PV modules, will cause inrush current problem as explained in Section III-A. This can be solved by using the following sequence of events for pre-charging both the input dc bus and the link voltage.

1) SW1 and SW2 in Fig. 1 are open. Release the pulses of $S_1 - S_8$ with synchronized inversion and rectification with $D = 0.5$. The snubber switch S is given fully ON command. This means that the pulses are released but the PV modules are not yet connected to the inverter.
2) Close SW1. The PV module current is limited to be around short circuit current i_{sc}. The slow time constant due to the heavy input capacitance C_{dc} causes the voltage and currents to rise very slowly. The start-up dynamics will be determined by this time constant.
3) When the pre-charge is complete, SW2 can be closed and control algorithm can be enabled in the digital controller.

The validity of this approach in eliminating the inrush currents is explained as follows. Consider equation (3). It can be observed that, if V_{in} is small and increased at a slow rate starting from zero, there will not be any inrush current. The slow time constant due to the input dc capacitance C_{dc} ensures that this condition is met and hence there will not be any inrush current problem. Fig. 11(a) shows the simulated pre-charge of link voltage, secondary voltage and primary current with the proposed start-up method. It can be observed that the primary current increases slowly without any inrush. Fig. 11(b) shows the experimental result of the proposed pre-charge method. It can be observed that there is no inrush current. The primary current peak is observed to be $2.3A$. The overall start-up time takes about $120ms$. Hence, the proposed start-up method mitigates the problems of over-voltage and inrush currents without the need of additional start-up circuitry. As the start-up method does not involve any quasi square wave voltage application, extra gate-drive control channels are also not necessary.

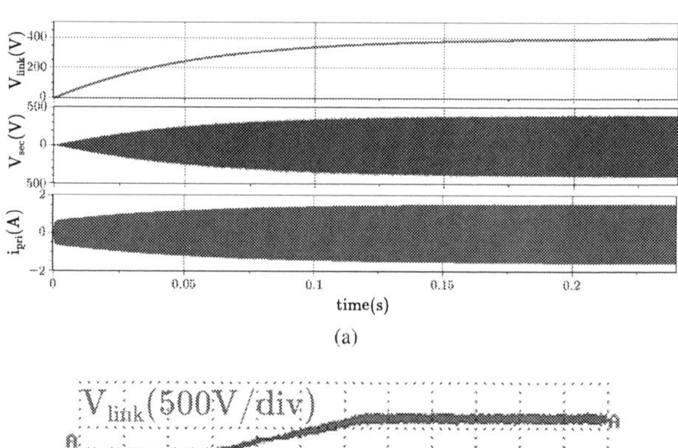

(a)

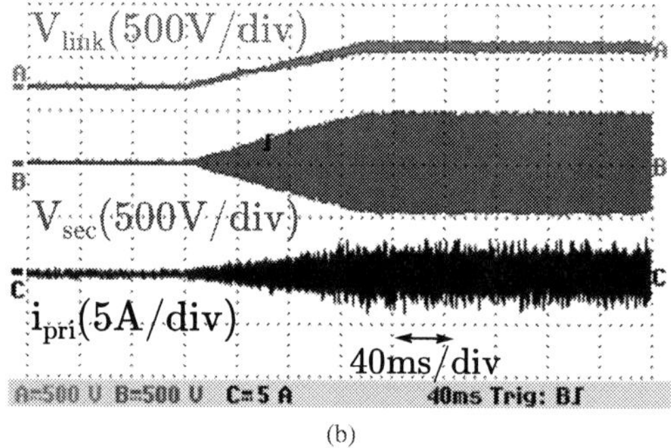

(b)

Fig. 11. Smooth pre-charge with the proposed method without any inrush current. $V_{in} = 40V$. (a) Simulation result (b) Experimental result (time scale: $40ms$/div).

The summary of the advantages of the proposed start-up scheme compared to other schemes available in literature is given in Table II. The input voltage for the HF transformer link inverter is considered as $V_{in} = 45V$ during start-up which is the rated maximum voltage equal to the open circuit voltage of PV modules. Ideally, the link voltage peak should be $450V$ as $1:10$ transformer is used. The actual peak link voltage V_{peak} is indicated in Table II for the different start-up techniques.

Thus, the percentage over-voltage for the proposed method is only 6.7% while it is 33.3% and 23.3% for the first two methods. The third method can have an over-voltage in the order of the proposed method as active rectification is used, but it uses extra channels for the gate-drive control card during start-up. The inrush current for these start-up methods is also tabulated in Table II. The first method with direct application of $D = 0.5$ has a high inrush current. For a HF transformer link inverter with a power rating of $1kW$ and maximum power point (MPP) input voltage of $37V$, the input rated current is $27A$. Hence, the inrush current is as high as 44 times with this method. The remaining three methods do not have the problem of inrush current.

Overall, it can be observed from Table II that the proposed method has the best performance while keeping the hardware complexity to the minimum.

V. Conclusion

In this paper, start-up schemes for high frequency transformer link inverter for PV application are analyzed. The input dc bus can be pre-charged directly from the PV modules as they are current limited. The pre-charge of secondary side link voltage poses challenges as both over-voltage and inrush current problems can be observed. The approach of using diode rectifier, reported in literature for HF transformer link inverters, causes severe over-voltages due to resonance. The quasi square wave voltage application method with diode rectifier does not have inrush current problem but suffers from the over-voltage problem, significantly reducing the operating safe voltage margin for the semiconductor devices. This method, when used along with active rectification, does not have the over-voltage and inrush current problems but requires additional channels in generating the required gate pulses used only during the start-up. There are some other methods reported which modify the power circuit for smooth start-up. These methods increase the system cost and operational complexity.

In this paper, a new start-up method is proposed which uses active rectification synchronized with the HF inverter switching in square wave mode. This method eliminates the over-voltage problem. The inrush current problem is mitigated by closing the input voltage contactor only after releasing the HF inverter and rectifier pulses. The performance of the proposed start-up is validated and compared with conventional start-up schemes using simulation and experimental results. The proposed method does not require any additional power or control circuitry for start-up and hence does not increase the converter cost and complexity.

References

[1] L. Gao, R. Dougal, S. Liu, and A. Iotova, "Parallel-connected solar PV system to address partial and rapidly fluctuating shadow conditions," *IEEE Transactions on Industrial Electronics*, vol. 56, pp. 1548–1556, May 2009.

[2] K. Wang, F. Lee, and W. Dong, "A new soft-switched quasi-single-stage (qss) bi-directional inverter/charger," in *Thirty-Fourth IEEE IAS Annual Meeting, Industry Applications Conference*, vol. 3, pp. 2031–2038, 1999.

[3] P. Krein, R. Balog, and X. Geng, "High-frequency link inverter for fuel cells based on multiple-carrier PWM," *IEEE Transactions on Power Electronics*, vol. 19, pp. 1279–1288, Sept 2004.

[4] S. Mazumder and A. Rathore, "Performance evaluation of a new hybrid-modulation scheme for high-frequency-ac-link inverter: Applications for PV, wind, fuel-cell, and DER/storage applications," in *IEEE Energy Conversion Congress and Exposition (ECCE)*, pp. 2529–2534, Sept 2010.

[5] H. Qin and J. Kimball, "Closed-loop control of DC-DC dual-active-bridge converters driving single-phase inverters," *IEEE Transactions on Power Electronics*, vol. 29, pp. 1006–1017, Feb 2014.

[6] K. Basu and N. Mohan, "A high-frequency link single-stage PWM inverter with common-mode voltage suppression and source-based commutation of leakage energy," *IEEE Transactions on Power Electronics*, vol. 29, pp. 3907–3918, Aug 2014.

[7] H. Keyhani and H. Toliyat, "Single-stage multistring PV inverter with an isolated high-frequency link and soft-switching operation," *IEEE Transactions on Power Electronics*, vol. 29, pp. 3919–3929, Aug 2014.

[8] E. Koutroulis, J. Chatzakis, K. Kalaitzakis, and N. Voulgaris, "A bidirectional, sinusoidal, high-frequency inverter design," *IEE Proceedings - Electric Power Applications*, vol. 148, pp. 315–321, Jul 2001.

[9] A. Kulkarni and V. John, "HF transformer based grid-connected inverter topology for photovoltaic systems," *IETE Technical Review*, 2015.

[10] A. Kulkarni and V. John, "High performance modulation of high-frequency ac link inverter," in *IEEE International Transport Electrification Conference India (ITEC India)*, pp. 1–6, Aug 2015.

[11] H. Tao, A. Kotsopoulos, J. Duarte, and M. Hendrix, "Design of a soft-switched three-port converter with dsp control for power flow management in hybrid fuel cell systems," in *Power Electronics and Applications, 2005 European Conference on*, pp. 10 pp.–P.10, Sept 2005.

[12] H. Bai and C. Mi, "Eliminate reactive power and increase system efficiency of isolated bidirectional dual-active-bridge DC-DC converters using novel dual-phase-shift control," *IEEE Transactions on Power Electronics*, vol. 23, pp. 2905–2914, Nov 2008.

[13] H. Zhou, T. Duong, S. T. Sing, and A. Khambadkone, "Interleaved bi-directional dual active bridge dc-dc converter for interfacing ultracapacitor in micro-grid application," in *IEEE International Symposium on Industrial Electronics (ISIE)*, pp. 2229–2234, July 2010.

[14] G. D. Demetriades, *On small-signal analysis and control of the single- and the dual-active bridge topologies*. PhD thesis, KTH Sweden, 2005.

[15] K. Wang, F. Lee, and J. Lai, "Operation principles of bi-directional full-bridge DC/DC converter with unified soft-switching scheme and soft-starting capability," in *Fifteenth Annual IEEE Applied Power Electronics Conference and Exposition, (APEC)*, vol. 1, pp. 111–118, 2000.

[16] L. Zhu, K. Wang, F. Lee, and J.-S. Lai, "New start-up schemes for isolated full-bridge boost converters," *IEEE Transactions on Power Electronics*, vol. 18, pp. 946–951, July 2003.

[17] F. Mihalic and A. Hren, "Safe start-up procedures of isolated bi-directional DC-DC converter," in *14th International Power Electronics and Motion Control Conference (EPE/PEMC)*, pp. T2–67–T2–173, Sept 2010.

[18] F. Spertino, J. Sumaili, H. Andrei, and G. Chicco, "PV module parameter characterization from the transient charge of an external capacitor," *IEEE Journal of Photovoltaics*, vol. 3, pp. 1325–1333, Oct 2013.

[19] EMMVEE Photovoltaics, *Photovoltaic module - ES 300 M72 S*. Datasheet.

Analysis and Improvement of Harmonic Quasi Resonant Control for *LCL*-filtered Grid-Connected Inverters in Weak Grid

Qiang Qian, Jinming Xu, Shaojun Xie and Lin Ji
Department of Automation Engineering
Nanjing University of Aeronautics and Astronautics
Nanjing, Jiangsu 211106, China
Email: nuaaqianqiang@nuaa.edu.cn

***Abstract*—Harmonic Quasi Resonant (HQR) control is widely adopted to suppress the harmonic currents for LCL-filtered inverters in stiff grid. However, inverters with HQR control may observe instability when the point of the common coupling (PCC) behaves uncertain impedance or the HQR control frequency is out of the bandwidth of the system control loop. Besides, it could not effectively eliminate the harmonics which is beyond the control frequency. Therefore, to improve the stability performance and enhance the harmonic suppression for inverters, this paper first discusses the merits and limitations of the conventional HQR control based on the impedance model and then introduces a phase compensation (PC) method for conventional HQR control. The proposed method can improve the stability and suppress a wide frequency range of harmonics for inverter when the weak grid connected. The correctness of the mathematical model for the inverter output impedance and the adverse effect of the conventional HQR control due to the existence of the grid impedance is verified by simulations and experimental results respectively. Applicability of inverters with the proposed method to the wide grid impedance variation is also validated on a single-phase PV inverter.**

Keywords—harmonic quasi resonant; phase compensation; grid impedance; stability

I. Introduction

Conventional HQR control can effectively suppress the harmonic current caused by the grid voltage background distortion in stiff grid [1]. However, with the increasing penetration level of the Distribution Generation Power System (DGPS), a large set of grid impedance appears due to the long distance power cables, which challenges the stable operation of electric equipment connected to the weak grid [2]. Therefore, how to maintain the stability and harmonics suppression in weak grid for grid-connected inverters which acts as a significant interface between the grid and the renewable energy resources draws a large popularity recently [3-4].

Generally, the control model with the grid impedance included is adopted to analyze the performance of the inverter with HQR controller adopted in weak grid. Reference [5] indicates that the grid impedance significantly decreases the bandwidth of the system control loop and the phase margin. Therefore, only few HQR controllers can be plugged in to guarantee the stability of the inverter. Otherwise, the selected harmonic frequency may locate outside the system bandwidth, which leads to the system instability [6]. So it is unfavorable that only several low order harmonics could be eliminated. In other words, the conventional HQR can't guarantee both the system stability and high quality for the injected current.

To avoid this instability phenomenon caused by the conventional HQR control, researches usually parallel only few HQR controllers [5] or remove the HQR controller [7-8]. Reference [7] proposed inverter output impedance shaping method to extend the stable operation region of the grid-connected inverter instead of using the HQR controller. However, this method is to some extent a trade-off between the stability and the harmonic suppression. Second order derivative terms are also included in the feed-forward link, which impedes the accomplishment in practice. Reference [8] proposes the self-adaptive method to enable the grid-connected inverter to operate over a wide range of grid impedance. It is based on the on-line grid impedance detection, then the potential positive feed-forward link is removed to maintain the inverter's stability in weak grid. This method is mainly limited to the performance of the grid impedance detection as the grid is time variant.

Effective methods to suppress harmonics and guarantee system stability simultaneously haven't been seen in literatures before. As the conventional HQR is an effective way to improve the magnitude of output impedance at the selected harmonic frequencies, the harmonic suppression for the injected current can be guaranteed. Basically, the reason why inverters with conventional HQR controllers are unsatisfied is the phase margin is not enough. Therefore, this paper tries to focus on the enhancement of the conventional HQR control.

Refers to the harmonic resonant (HR) controller with phase compensation (PC) which is used to track the high frequency harmonic currents in active power filter (APF) field [9-10]. This method adds a zero to the numerator of the HR controller to improve the phase margin. However, it could not adapt to the grid voltage frequency variation, and it needs on-line trigonometric function calculation. Therefore, this paper adds a

This work was supported by "China National Natural Science Foundation" under grants 51477077.

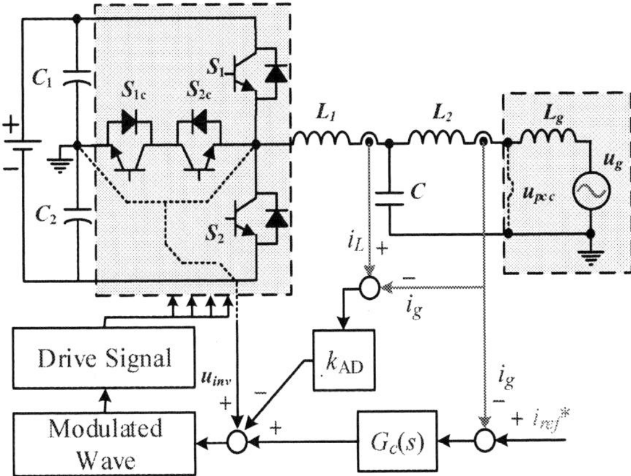

Fig.1. T-NPC based LCL-filtered grid-connected inverter

damping factor to the denominator of HR+PC, and HQR+PC is proposed. As the phase can be compensated due to PC and the high magnitude of output impedance is inherited by HQR, more HQR+PC controllers can be paralleled together to expand the harmonic frequency range with high magnitude for the inverter output impedance. Accordingly, the gird-connected inverter can adapt to weak grid with the wide grid impedance variation. Meanwhile, the high quality for the injected current is guaranteed.

This paper aims to seek for a method to improve the output impedance and the stability of grid-connected inverter in weak grid. And the rest is organized as follows. In Section II, the LCL-filtered inverter based on the impedance model is depicted. The merits and limitations of the conventional HQR control in Section III. In Section IV, the performance of the proposed method HQR+PC is thoroughly analyzed. The parameter design procedure is scheduled in Section V. In Section VI, the experiments are provided to verify the effectiveness of the proposed method. Finally, Section VII concludes this paper.

II. Impedance Model of the LCL-Filtered Grid-Conncected Inverter

Fig.1 shows the general T-NPC based LCL-filtered grid-connected inverter. C_1 and C_2 are the dc-side capacitor. The inverter-side inductor L_1, the grid-side inductor L_2 and the filter capacitor C_f constitute the LCL filter. The voltage of the Point of Common Coupling (PCC) u_{pcc} is sensed for the purpose of generating the synchronous current reference i_{ref}*. The phase-lock-loop (PLL) is implemented on the second-order-generalized-integrator (SOGI) [11]. Moreover, the bandwidth for PLL is chosen narrow to avoid the instability phenomenon caused by the effect of the grid synchronization method [11]. Grid-side current i_g and inverter-side current i_L are measured to make a subtraction to facilitate the active damping based on the capacitor current [12]. Besides, u_{inv} represents the middle point voltage of bridge leg. For simplification, the grid voltage u_g cascaded with inductance L_g is considered here as the inverter is connected to the weak grid. The error of i_g and i_{ref}* is sent to the current controller $G_c(s)$. The product of the active damping factor k_{AD} and the capacitor current is subtracted from the output of $G_c(s)$ to generate the modulated wave.

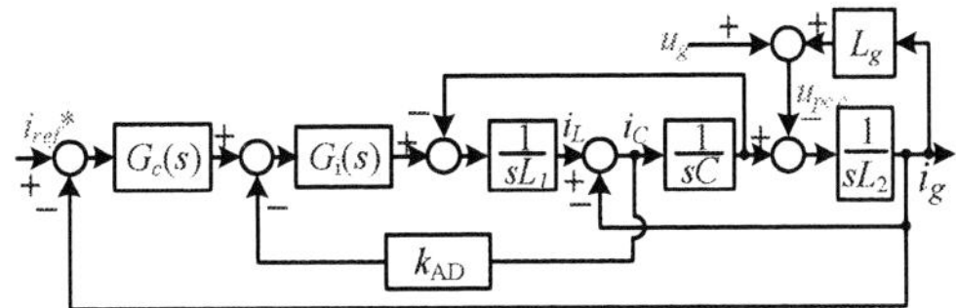

Fig.2. Control block diagram of LCL-filtered grid-connected inverter.

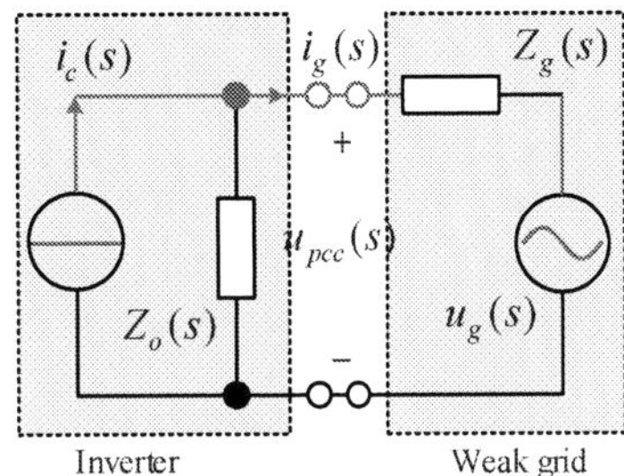

Fig.3. The cascaded system impedance model.

Generally, the sampling-computational delay, which is the time between the sampling instant and the PWM updating instant, and a PWM delay are concerned in the digital control system [13]. Sampling-computational delay is half of the sampling period T_s when the symmetrical regular sampling method is adopted. Meanwhile, the sampler [7] which is the effect of the analog-digital-conversion (ADC) in digital signal processor (DSP) can be modelled as $1/T_s$. Besides, the PWM delay is usually modeled as the zero-order-hold (ZOH) for the reason that the PWM keeps constant after it has been updated. Therefore, the delay transfer function $G_i(s)$ can be expressed as

$$G_i(s) = e^{-0.5T_s s} \cdot \frac{1}{T_s} \cdot G_{zoh} \tag{1}$$

Accordingly, the mathematical control block diagram of the LCL-filtered inverter can be obtained in Fig.2.

According to Fig.2, the injected current i_g can be derived as (2) which has a relation with the current reference i_{ref}* and the voltage of PCC u_{pcc}.

$$i_g(s) = G_{i_{ref}*}^{i_g}(s) I_{ref}^*(s) + G_{u_{pcc}}^{i_g}(s) u_{pcc}(s) \tag{2}$$

where $G_{i_{ref}*}^{i_g}(s)$ is the closed-loop transfer function from the current reference i_{ref}* to the injected current i_g and $G_{u_{pcc}}^{i_g}(s)$ is the closed-loop transfer function from u_{pcc} to i_g. According to (2), $G_{i_{ref}*}^{i_g}(s)$, $G_{u_{pcc}}^{i_g}(s)$ can be respectively derived as

$$G_{i_{ref}*}^{i_g}(s) = \frac{G_c(s)G_i(s)}{s^3 L_1 L_2 C + sL_2 + sL_1 + s^2 L_2 C \cdot k_{AD} G_i(s) + G_c(s) G_i(s)} \tag{3}$$

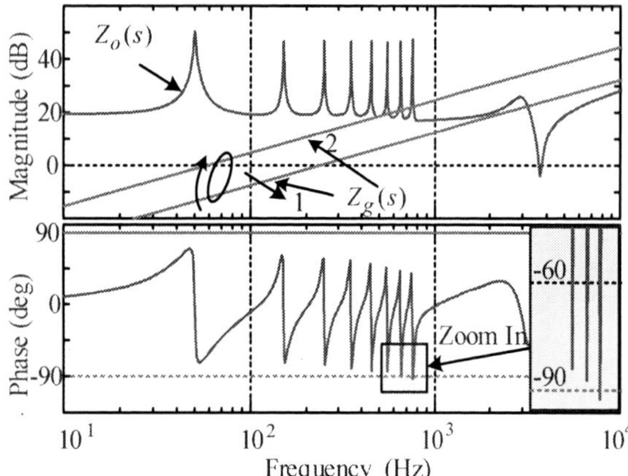

Fig.4. Bode of $Z_g(s)$ and $Z_o(s)$ with up to 15th HQR controller.

$$G_{u_{pcc}}^{i_g}(s) = \frac{-s^2 L_1 C - s k_{AD} C \cdot G_i(s) - 1}{s^3 L_1 L_2 C + sL_2 + sL_1 + s^2 L_2 C \cdot k_{AD} G_i(s) + G_c(s) G_i(s)} \quad (4)$$

According to [14], the impedance model is simple and intuitive as it can decouple the grid from the inverter system. So the following analysis is done with this powerful tool. And the cascaded system impedance model based on the inverter output impedance $Z_o(s)$ and the grid impedance $Z_g(s)$ is shown in Fig.3, where $i_c(s)$ is the equivalent controlled current source. Therefore, $i_c(s)$, $Z_o(s)$ and $i_g(s)$ can be derived as

$$i_c(s) = G_{i_{ref}^*}^{i_g}(s) i_{ref}^*(s) \quad (5)$$

$$Z_o(s) = -\frac{1}{G_{u_{pcc}}^{i_g}(s)} = \frac{s^3 L_1 L_2 C + sL_2 + sL_1 + s^2 L_2 C \cdot k_{AD} G_i(s) + G_c(s) G_i(s)}{s^2 L_1 C + s k_{AD} C \cdot G_i(s) + 1} \quad (6)$$

$$i_g(s) = \frac{Z_o(s)}{Z_o(s) + Z_g(s)} \cdot i_c(s) + \frac{1}{Z_o(s) + Z_g(s)} \cdot u_g(s) \quad (7)$$

If the cascaded system is stable, three following requirements need to be satisfied [15].

1) The controlled current source $i_c(s)$ is stable when a stiff grid is connected, which is usually guaranteed during the procedure of the control loop design.

2) The equivalent system $Z_g(s)/Z_o(s)$ satisfies the Nyquist stability criterion. This criterion is challenged by the magnitude of the grid impedance $Z_g(s)$.

3) $Z_g(s)$ is stable. It is assured when $Z_g(s)$ is modelled as a pure inductor for simplification.

III. Merits and Limitations of The Conventional HQR Control

In light of (7), the injected current i_g is easily distorted by u_g which contains abundant harmonics in weak grid. Enlarging the magnitude of $Z_o(s)$ can be an effective way to impair the adverse effect in respect of u_g. Conventionally, HQR control can provide high resonant gain at the selected harmonic frequencies, which in turn enlarges the magnitude of $Z_o(s)$.

Therefore, the conventional PQR+HQR shown as (8) is adopted as the regulator $G_c(s)$.

$$G_c(s) = k_p + \sum_{n=1,3,5,...} \frac{k_r s}{s^2 + \alpha s + (n\omega_f)^2} \quad (8)$$

where k_p, k_r represents the proportional gain and the resonance gain, respectively. ω_f means the fundamental angular frequency, n is the harmonic order, and α is the damping factor to adapt to the grid frequency variation.

Fig.4 depicts the bode diagram of $Z_g(s)$ and $Z_o(s)$ with up to 15th HQR controller plugged-in. The characteristics of the conventional HQR control can be summarized as follows:

1) The magnitude of $Z_o(s)$ experiences significant increase at the desired harmonic frequency, which facilitates the suppression of harmonics caused by grid harmonics. The magnitude 40dB means that 1V harmonic voltage will only cause maximum 10mA injected current.

2) The cascaded system may enter the instable region as the phase curve of $Z_o(s)$ intersects the $-\pi/2$ line at 750Hz [7]. This instability happens when the magnitude of $Z_g(s)$ equals $Z_o(s)$ around the cross frequency (see curve 2 in Fig.4). It can be mitigated by decreasing the resonance gain k_r, then the magnitude dips significantly at the selected harmonic frequency. Therefore, system stability and the good quality for the injected current i_g can't be guaranteed with the conventional HQR controller simultaneously.

Basically, the limitations of the conventional HQR control is that the phase margin is not sufficient to stabilize the cascaded system over a wide range of grid impedance $Z_g(s)$. Therefore, the optimized orientation for improving the performance of the conventional HQR control is that the high magnitude should be inherited and the phase margin should be improved. Then the inverter with the conventional HQR controller can stably operate over a wide range of grid impedance.

IV. HQR with Phase Comepensation to Ehance Stability and Harmonic Suppression

The Harmonic Resonant (HR) with Phase Compensation (PC) controller which is proposed in [9] is adopted to cope with the tracking of high frequency current reference in Active Power Filter (APF). This controller $G_{HR\_PC}(s)$ is provided as

$$G_{HR\_PC}(s) = k_{rh} \cdot \frac{s\cos(\varphi_c) - n\omega_f \sin(\varphi_c)}{s^2 + (n\omega_f)^2} \quad (9)$$

where k_{rh} represents the resonance gain at the harmonic frequency. The compensated angle at the harmonic frequency can be adjusted by the parameter φ_c.

Although $G_{HR\_PC}(s)$ could provide adjustable phase correction due to the existence of the zero and the infinite gain at the harmonic frequency, it could not effectively compensate the phase curve around the harmonic frequency with the consideration of the grid frequency fluctuation. Therefore, a

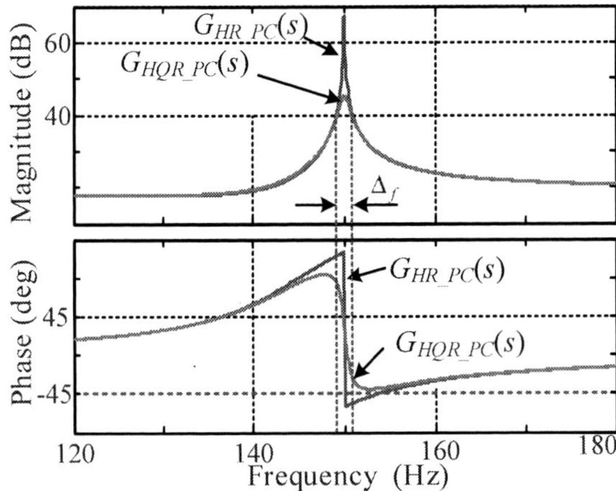

Fig.5. Bode comparison of $G_{HQR\_PC}(s)$ and $G_{HR\_PC}(s)$ at 150Hz.

damping factor α_h is added to the denominator of $G_{HR\_PC}(s)$ to impair the effect of the conjugated pole located at the imaginary axis. Then the Harmonic Quasi Resonant with Phase Compensation (HQR+PC) controller is obtained as

$$G_{HQR\_PC}(s) = k_{rh} \cdot \frac{s\cos(\varphi_c) - n\omega_f \sin(\varphi_c)}{s^2 + \alpha_h s + (n\omega_f)^2} \tag{10}$$

where $G_{HQR\_PC}(s)$ is the transfer function of the proposed controller. And k_{rh} represents the resonance gain at the harmonic frequency.

Combined with PQR at the fundamental frequency, then PQR+HQR+PC is adopted as $G_c(s)$.

$$\begin{aligned} G_c(s) = k_p &+ \frac{k_{rb} s}{s^2 + \alpha_b s + (n\omega_f)^2} \\ &+ \sum_{n=3,5,...} k_{rh} \cdot \frac{s\cos(\varphi_c) - n\omega_f \sin(\varphi_c)}{s^2 + \alpha_h s + (n\omega_f)^2} \end{aligned} \tag{11}$$

where k_p dominates the crossover frequency.

To intuitively discuss the advantages of $G_{HQR\_PC}(s)$ compared with $G_{HR\_PC}(s)$, Fig.5 plots the bode diagram for both $G_{HR\_PC}(s)$ and $G_{HQR\_PC}(s)$ with k_p=9， k_{rh}=1120， α_h=6.3， n=3， φ_c=30°.

From Fig.5, it can be inferred that the performance of $G_{HQR\_PC}(s)$ is much better than $G_{HR\_PC}(s)$. First, enough magnitude (larger than 40dB) around the selected harmonic frequency is provided. Then good quality can be guaranteed as the harmonics are eliminated. Second, the phase curve is mitigated with the effect of the damping factor α_h in the region of Δf. Moreover, it is lifted above -45 degree around the selected harmonic frequency. Therefore, $G_{HQR\_PC}(s)$ controller could provide high magnitude and adequate phase margin when the grid frequency fluctuates.

It is notable that the compensated angle is not only determined by the parameter φ_c due to the damping factor α_h, which is not as the same as $G_{HR\_PC}(s)$. Then the accurate compensated angle θ^* can be derived from (11), and it satisfies

TABLE I. PARAMETERS OF INVETER

parameter	value
DC input U_{in}/V	800
Grid voltage U_g/V	220
Output Power P/kW	5
Sample frequency f_s/kHz	15
converter-side inductor L_1/mH	0.75
filter capacitor C/μF	6.8
grid-side inductor L_2/mH	0.45

$$\tan\theta^* = \frac{\sin(\varphi_c)}{\frac{\alpha_h \cdot k_p}{k_{rh}} + \cos(\varphi_c)} \tag{12}$$

Solving (12), the parameter φ_c can be designed to accurately compensate the angle θ^* at the harmonic frequency with (13).

$$\varphi_c = \arcsin \frac{\frac{\alpha_h \cdot k_p}{k_{rh} \cdot \tan\theta^*} + \sqrt{1 - \left(\frac{\alpha_h \cdot k_p}{k_{rh}}\right)^2 + \left(\frac{1}{\tan\theta^*}\right)^2}}{1 + \left(\frac{1}{\tan\theta^*}\right)^2} \tag{13}$$

With (13), the phase of the plant can be accurately compensated to provide satisfied phase margin at the selected harmonic frequency. And it should be noticed that the compensated angle θ^* is related to the harmonic frequency where the phase lag is different for a given control plant. Besides, the magnitude of $Z_o(s)$ at the desired frequency is assured. Therefore, system stability and favorable harmonic suppression under varied grid frequency is expected.

Thus, with the consideration of the software consumption and the practical grid voltage harmonics, this paper adopts PQR+HQR (up to 23th) with phase compensation as the controller. It could guarantee both the system stability and good quality of injected current.

V. PARAMETER DESIGN METHOD AND PERFORMANCE VERIFICATION

The parameter design procedure of the current controller PQR+HQR(3rd~23th)+PC (see (11)) is discussed in this section.

It should be bear in mind that the principle of the design mainly has two parts.

1) The magnitude of $Z_o(s)$ should be large enough (40dB) to suppress the harmonics caused by grid voltage distortion.

2) The phase of $Z_o(s)$ is above -45 degree so that the equivalent system $Z_g(s)/Z_o(s)$ has sufficient phase margin to stabilize the cascaded system.

A. Parameter Design Method

The parameters of the inverter is listed in Table I.

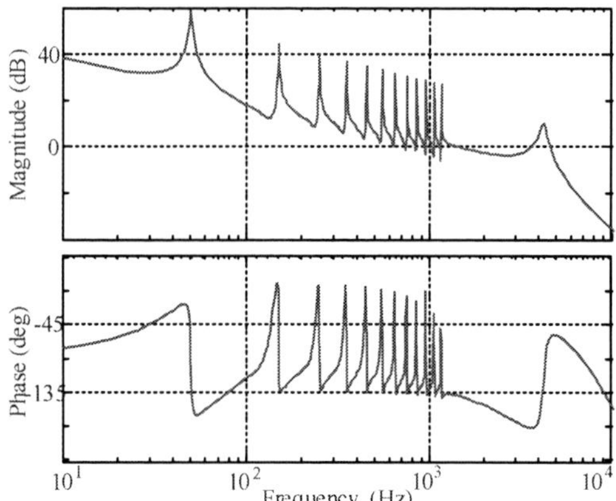

Fig.6. Bode diagram of open-loop transfer function with the proposed control method..

TABLE II. CONTROL PARAMETERS OF INVETER

parameter	value
proportional gain k_p	9
damping factor α_b, α_h	6.3
resonance gain k_{rb} at ω_f	2000
resonance gain k_{rh} at $n\omega_f$	1120

According to [3], the proportional gain k_p, the resonance gain k_{rb}, the damping factor α_b is designed under stiff grid to satisfy the requirement 1) in section II. Then the logarithm quotient of the resonance gain k_{rh} divided by the damping factor α_h should be larger than 40dB to improve the magnitude of $Z_o(s)$. Therefore, k_p, k_{rb}, k_{rh}, α_b, α_h can be derived as shown in Table II.

From (6), the phase lag of $Z_o(s)$ consists of three dominant components.

1) θ_1: The phase lag caused by the adverse effect of the digital control needs to be well compensated.

2) θ_2: Although a damping factor is added to mitigate the phase curve at the selected harmonic frequency, the phase still needs to be compensated to boost the phase around the selected harmonic frequency (see Fig.4). Here $\pi/6$ is chosen.

3) θ_3: The denominator of $Z_o(s)$ induces another phase lag caused by the LCL-filtered grid-connected inverter.

Therefore, the compensated angle reference θ^* at the selected harmonic frequency can be calculated as

$$\begin{aligned}\theta^* &= \theta_1+\theta_2+\theta_3 \\ &= \frac{2\pi n\omega_f}{\omega_s}+\frac{\pi}{6}+\arctan\frac{\omega k_{AD}C\cdot\cos(\omega T_s)}{1-\omega^2 L_1 C+\omega k_{AD}C\cdot\sin(\omega T_s)}\end{aligned} \tag{14}$$

where ω_s is the sampling angular frequency.

B. Performance Verification

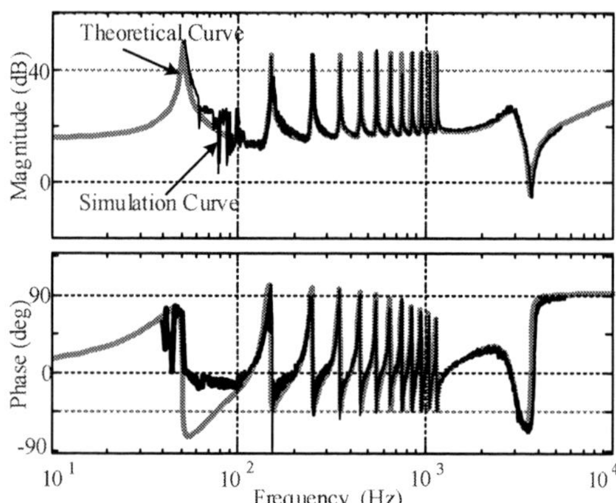

Fig.7. $Z_o(s)$ Comparison of simulation results and theoretical analysis.

To evaluate the performance of inverter output impedance $Z_o(s)$ with PQR+HQR+PC control, the bode diagram of open-loop transfer function is first plotted in Fig.6.

From Fig.6, the phase curve lays above the -135 degree so that the proposed controller has sufficient phase margin to stabilize the grid-connected inverter when stiff grid connected. Therefore, the requirement 1) in Section II is satisfied.

Fig.7 plots the inverter output impedance $Z_o(s)$ (see Theoretical Curve). It can be seen that the magnitude experiences a magnitude of 40dB at all the selected harmonic frequency. Moreover, the phase curve is above the -45 degree, which means that the inverter output impedance $Z_o(s)$ can guarantee the stable operation over a wide range of grid impedance. Therefore, the parameter design method in Section V-A is feasible by the theoretical curve in Fig.7.

VI. SIMULATION AND EXPERIMENTAL RESULTS

A. Saber Simulation

TDSA module in simulation software Saber is an effective way to accomplish the verification of the mathematical model $Z_o(s)$. This method injects a wide frequency voltage at PCC, then the injected current i_g is measured. The simulation results is shown in Fig.7 in Section V-B.

The simulation curve matches the theoretical curve well in Fig.7, which means the inverter output impedance model is reliable and the parameter design method in Section V-A is feasible. Therefore, the inverter output impedance can be reshaped to be more satisfied. Then inverter is more applicable to wide grid impedance based on the proposed method.

B. Experimental Results

To experimentally validate the feasibility of the proposed control method, a 5kW single-phase T-type NPC inverter is implemented with insulated gate bipolar transistor modules (F3L75R12W1H3-Infineon Technology). The parameter of the prototype is listed in Table I. The type of DSP is TMS320F28335. Chroma 62150H-1000S is adopted as the DC voltage source. Power spectrum analyzer YOKOGAWA

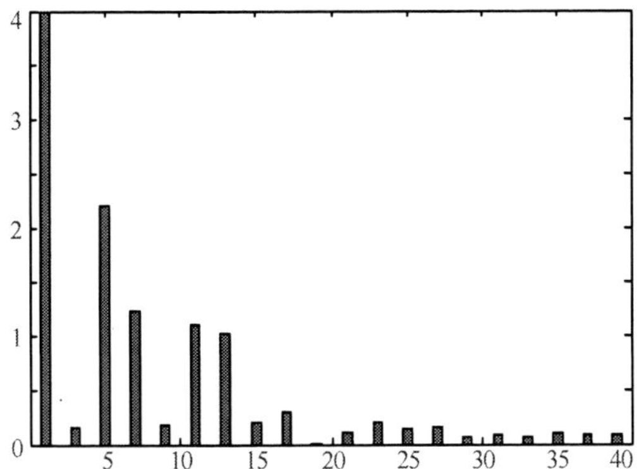

Fig.8. Spectrum of grid voltage.

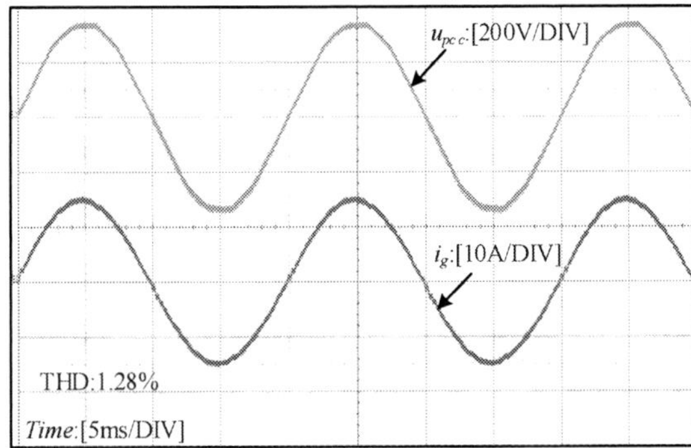

Fig.9. Conventional control PQR+HQR(3rd~15th) with $Z_g(s)$=0.

WT1800 is used to analyze the harmonics of the injected current. Grid series with the pure inductor is emulated as the weak grid.

Fig.8 shows the harmonic distribution for grid voltage. It can be seen that the grid voltage mainly contains 5th, 7th, 11th, 13th and 17th. The inverter output impedance should be large enough at these selected harmonic frequencies to eliminate the injected current harmonics.

Fig.9 is the experimental result of the conventional control PQR+HQR (3th~15th) in stiff grid. And the measured current THD is 1.28%. It shows that the conventional control can stabilize the inverter and acquire good quality for injected current in stiff grid.

Fig.10 shows the half-load experimental results of the conventional control with an inductor 4.46mH (0.145pu) in series with the grid. The system is unstable due to the unsatisfied output impedance (see Fig.4). From the magnified curve in Fig.10, it can be seen that the dominant harmonic is 15th which is in accordance with the intersected frequency where the phase curve cross $-\pi/2$ in Fig.4.

Therefore, according to Fig.9 and Fig.10, the inverter with the conventional control couldn't operate stably with a wide range of grid impedance, which verifies the analysis based on inverter output impedance.

Fig.11 and Fig.12 show the experimental results with the

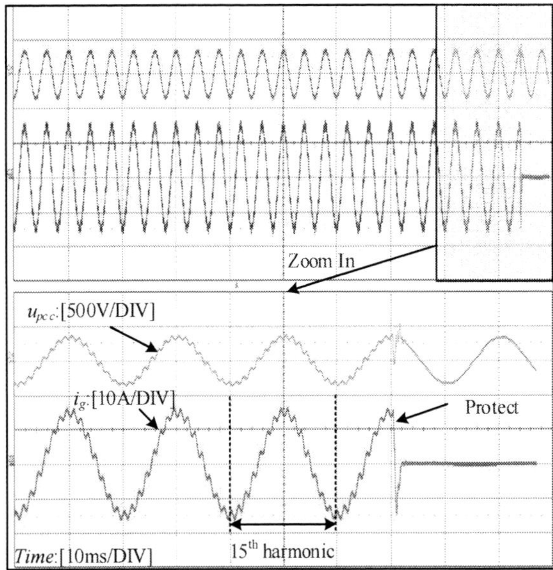

Fig.10. Conventional control PQR+HQR(3rd~15th) with $Z_g(s)$=4.46mH.

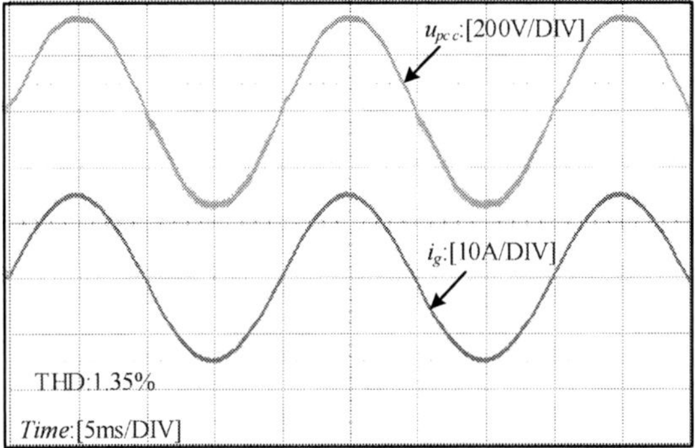

Fig.11. Proposed control PQR+HQR(3rd~23th) with $Z_g(s)$=4.46mH.

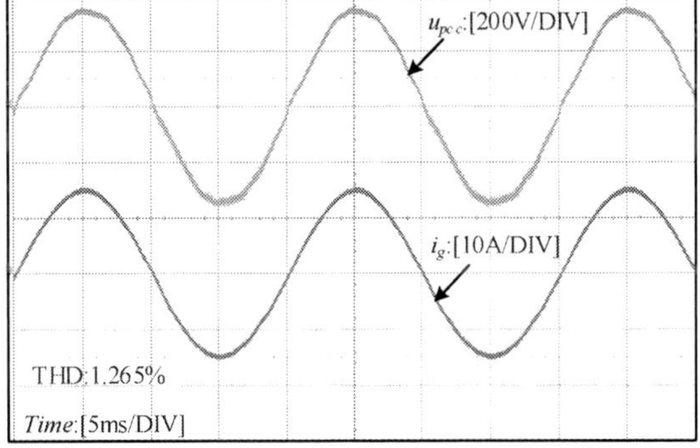

Fig.12. Proposed control PQR+HQR(3rd~23th) with $Z_g(s)$=10.24mH.

proposed control PQR+HQR (3th~23th)+PC over a wide range of grid impedance. Moreover, the THD for grid impedance 4.46mH (0.145pu), 10.24mH (0.333pu) is 1.35%, 1.265%, respectively. Therefore, the inverter operates stably, and high quality current is injected into the weak grid. The proposed method is effective for inverter when weak grid is connected.

VII. Conclusion

This paper focuses on the stability enhancement of inverters with conventional HQR controller in weak grid based on the cascaded system criterion and the mathematical model of output impedance. The research indicates that the conventional HQR control faces difficulties in stability with the increasing grid impedance. The model of output impedance which is verified by simulation provides an effective method to analyze similar control on inverter. The inverter with the proposed strategy could reshape the inverter output impedance, which can not only guarantee stability with varied grid impedance but also suppress the low order harmonics for the injected current.

Acknowledgment

This research has been supported by "National Natural Science Foundation of China (51477077)".

References

[1] J. Xu, S. Xie, T. Tang, "Research on Low-Order Current Harmonics Rejections for Grid-Connected LCL-filtered Inverters," *IET Power Electronics*, vol. 6, no. 2, pp. 227-234, Jun. 2013.

[2] L. Macro, T. Remus, B. Frede, "Stability of Photovoltaic and Wind Turbine Grid-Connected Inverters for a Large Set of Grid Impedance Values," *IEEE Trans on Power Electronics*, vol.21, no. 1, pp. 263-272, Nov. 2006.

[3] J. Xu, S. Xie, T. Tang, "Evaluations of current control in weak grid case for grid-connected LCL-filtered inverter," *IET Power Electronics*, vol. 6, no. 2, pp. 227-234, Feb. 2013.

[4] W. Jing, S. Yulun, A. Monti, "A Study of Feedforward Control on Stability of Grid-parallel Inverter With Various Grid Impedance," in *Power Electronics for Distributed Generation System (PEDG)*, 2014, pp. 24-27.

[5] M. Castilla, J. Miret, J. Matas, "Control design guidesline for single-phase grid-connected photovoltaic inverters with damped resonant harmonic compensators," *IEEE Trans on Industrial Electronics*, vol.56, no. 11, pp. 4492-4501, Nov. 2009.

[6] C. Miguel, M. Jaume, C. Antonio, "Reduction of Current Harmonic Distortion in Three-Phase Grid-Connected Photovoltaic Inverters via Resonant Current Control," *IEEE Trans on Industrial Electronics*, vol.60, no. 4, pp. 1464-1472, Apr. 2013.

[7] D. Yang, X Ruan, H. Wu, "Impedance Shaping of the Grid-Connected Inverter with LCL Filter to improve Its Adaptability to Weak Grid Condition," *IEEE Transactions on Power Electronics*, vol. 29, no. 11, pp. 5795-5804, Nov. 2014.

[8] J. Xu, S. Xie, T. Tang, "Improved control strategy with grid-voltage feedforward for LCL-filter-based inverter connected to weak grid," *IET Power Electronics*, vol. 7, no. 10, pp. 2660-2671, Oct. 2014.

[9] A.G. Yepes, F.D. Freijedo, O. Lopez, J. Doval-Gandoy, "High-Performance Digital Resonant Controllers Implemented With Two Integrators," *IEEE Transactions on Power Electronics*, vol. 26, no. 2, pp. 563-575, Fen. 2011.

[10] B.Radu, R. Leonardo, P. Francesco, "Analysis of current controllers for active power filters using selective harmonic compensation schemes," *IEEJ Transactions on Electrical and Electronic Engineering*, vol. 10, no. 4, pp. 139-157, Apr. 2009.

[11] C. Zhang, X. Wang, F. Blaabjerg, "Analysis of phase-locked loop influence on the stability of single-phase grid-connected inverter," in *Power Electronics for Distributed Generation System (PEDG)* 2015, pp.1-8.

[12] D. Pan, X. Ruan, C. Bao, "Capacitor-Current-Feedback Active Damping With Reduced Computation Delay for Improving Robustness of *LCL*-Type Grid-Connected Inverter," *IEEE Transactions on Power Electronics*, vol. 29, no. 7, pp. 3414-3418, Jul. 2014.

[13] D. Heng, O Ramesh, S. Dipti, "PWM Methods to Handle Time Delay in Digital Control of a UPS Inverter," *IEEE Transactions on Power Electronics*, vol. 3, no. 1, pp. 1-6, Jan. 2005.

[14] J. Sun, "Impedance-Based Stability Criterion for Grid-Connected Inverters," *IEEE Transactions on Power Electronics*, vol. 26, no. 11, pp. 3075-3078, Nov. 2011.

[15] C. Mauricio, J. Sun, "Impedance Modeling and Analysis of Grid-Connected Voltage-Source Converters," *IEEE Transactions on Power Electronics*, vol. 29, no. 3, pp. 1254-1261, Mar. 2014.

Model Predictive Control Method to Reduce Common-Mode Voltage and Balance the Neutral-point Voltage in Three-Level T-type Inverter

Xiangyang Xing, Alian Chen, Zicheng Zhang, Jie Chen, Chenghui Zhang
School of Control Science and Engineering
Shandong University
Jinan 250061, P. R. China

***Abstract*—The three-level T-type inverter has outstanding performances and better in the switching device selection than a two-level inverter. However, the problem of common-mode voltages and neutral-point potential (NP) unbalance is generated. This paper proposes a model predictive control method for reducing the common mode voltage and balancing the neutral-point potential (NP). Based on the relationship between the switching states and common-mode voltage (CMV), the 19 voltage switching vectors are adopted to reduce the CMV. In addition, the P-type and N-type small vectors are selected to control the NP balance. Without utilizing the redundancy zero and small vectors, the common-mode voltage controlled by the proposed MPC algorithms can be restricted within $\pm V_{dc}/6$. Furthermore, the proposed technique can balance the DC-link voltage with reduced CMV. The proposed method has been verified by the lab experiment.**

Keywords—model predictive current control, three-level T-type inverters, common-mode voltage, neutral-point potential (NP) balance.

I. INTRODUCTION

THREE-PHASE AC/DC inverters have been extensively in renewable energy application[1]-[15]. For such a high power application, the three-level converter is most commonly used[2]-[13]. Recently, the three-level T-type inverter ($3LT^2I$) topology has been widely employed in low-voltage power applications, as shown in Fig. 1. The $3LT^2I$ introduces an active bidirectional switch to the DC-link midpoint compared with the conventional two-level inverter.

The $3LT^2I$ combines the advantages of two-level inverter such as low conduction loss, less part count and simple operation principles and the neutral point clamped (NPC) such as low switching loss and superior output voltage quality. These advantages make $3LT^2I$ as a good alternative for distributed generation in low-voltage power system [6].

However, common-mode voltage (CMV) generated by fast switching operation has been known to introduce overvoltage stress, which will affect the system stability [10]-[12]. In addition, in order to generate different voltage levels, the $3LT^2I$ comprises two DC-link capacitors connected in series. The topology will lead to neutral-point potential (NP) unbalance, which will increase the THD of output currents [16]. The PWM algorithms for balancing the NP voltage and CMV can be incorporated with the PI controller to control the reference current [6].

Recently, a finite control set model predictive control (FCS-MPC) has been proposed as an effective current tracking method owing to its simplicity without using any PWM blocks as well as its control flexibility [16]-[19]. Consequently, an improved FCS-MPC algorithm is proposed to reduce CMV, track the currents and balance NP of the $3LT^2I$, as shown in Fig. 1. The control block diagram of grid-connected $3LT^2I$ using the improved FCS-MPC is shown in Fig. 2.

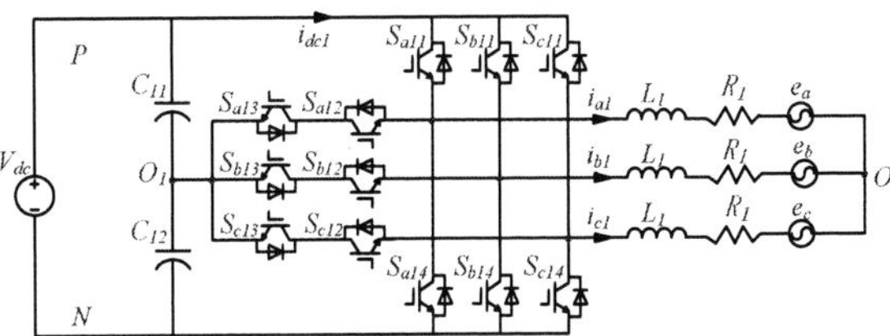

Fig. 1. The topology structure of three-phase $3LT^2I$.

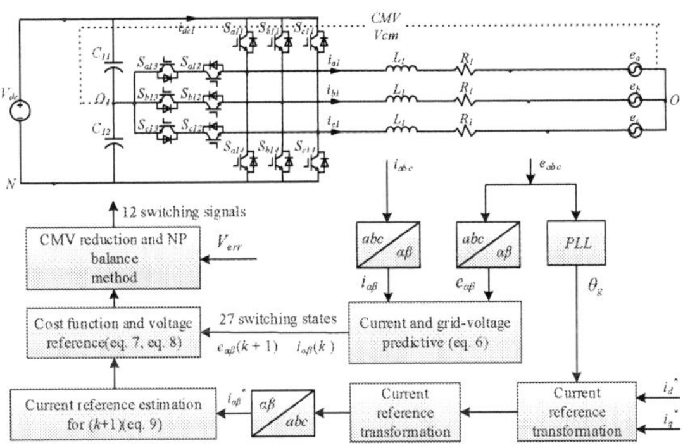

Fig. 2. The predictive control block diagram of grid-connected $3LT^2I$.

II. THE CONVENTIONAL FCS-MPC METHOD FOR $3LT^2I$

The middle pole of DC-link voltage is selected as reference point. And the average model of $3LT^2I$ corresponding to three-phase stationary reference frame is as follows:

$$\frac{d}{dt}\begin{bmatrix} i_{a1} \\ i_{b1} \\ i_{c1} \end{bmatrix} = -\frac{R_1}{L_1}\begin{bmatrix} i_{a1} \\ i_{b1} \\ i_{c1} \end{bmatrix} - \frac{1}{L_1}\begin{bmatrix} e_a \\ e_b \\ e_c \end{bmatrix} + \frac{1}{L_1}\begin{bmatrix} u_{a1} \\ u_{b1} \\ u_{c1} \end{bmatrix} + \frac{1}{L_1}\begin{bmatrix} u_{N1} \\ u_{N1} \\ u_{N1} \end{bmatrix} - \frac{1}{L_1}\begin{bmatrix} u_{ON} \\ u_{ON} \\ u_{ON} \end{bmatrix} \quad (1)$$

where i_{a1}, i_{b1} and i_{c1} are output currents of $3LT^2I$; u_{ai}, u_{bi} and u_{ci} are the output voltages; e_a, e_b and e_c are the grid voltages;

u_{ON} is the voltage of point O and N; u_{N1} are the output voltage of point O_1 and N.

The mathematical model of three-phase 3LT[2]I in the two-phase synchronous stationary coordinate can be derived from the $abc/\alpha\beta$ transformation of (1) as:

$$\frac{d}{dt}\begin{bmatrix} i_{\alpha 1} \\ i_{\beta 1} \end{bmatrix} = -\frac{R_1}{L_1}\begin{bmatrix} i_{\alpha 1} \\ i_{\beta 1} \end{bmatrix} + \frac{1}{L_1}\begin{bmatrix} u_{\alpha 1} - e_{\alpha} \\ u_{\beta 1} - e_{\beta} \end{bmatrix} \quad (2)$$

Considering V_{dc} is a constant DC-link voltage, the output voltage of the 3LT[2]I is expressed as:

$$u_{xi} = S_x V_{dc} / 2 \quad (3)$$

where x=a,b,c.

The switching function S_x is defined as the switching signals for each leg, which can be expressed as:

$$S_x = \begin{cases} 1, S_{x1} = 1, S_{x2} = 1, S_{x3} = 0, S_{x4} = 0 \\ 0, S_{x1} = 0, S_{x2} = 1, S_{x3} = 1, S_{x4} = 0 \\ -1, S_{x1} = 0, S_{x2} = 0, S_{x3} = 1, S_{x4} = 1 \end{cases} \quad (4)$$

From (2), the discrete form of the average model is expressed as:

$$\begin{bmatrix} i_{\alpha}(\mathrm{k}) \\ i_{\beta}(\mathrm{k}) \end{bmatrix} = \frac{L}{RT_s + L}\begin{bmatrix} i_{\alpha}(\mathrm{k}-1) \\ i_{\beta}(\mathrm{k}-1) \end{bmatrix} + \frac{T_s}{RT_s + L}\begin{bmatrix} u_{\alpha}(\mathrm{k}) - e_{\alpha}(\mathrm{k}) \\ u_{\beta}(\mathrm{k}) - e_{\beta}(\mathrm{k}) \end{bmatrix} \quad (5)$$

where $\frac{di}{dt} = \frac{i(k) - i(k-1)}{T_s}$.

The practical implementation of FCS-MPC requires considering the time delay due to the time needed for the measurements through algorithm calculations and analog-to-digital. Therefore, the discrete time equation of the model is shifted one step forward to eliminate the time delay. The formula (5) can be revised as:

$$\begin{bmatrix} i_{\alpha}(\mathrm{k}+1) \\ i_{\beta}(\mathrm{k}+1) \end{bmatrix} = \frac{L}{RT_s + L}\begin{bmatrix} i_{\alpha}(\mathrm{k}) \\ i_{\beta}(\mathrm{k}) \end{bmatrix} + \frac{T_s}{RT_s + L}\begin{bmatrix} u_{\alpha}(\mathrm{k}+1) - e_{\alpha}(\mathrm{k}+1) \\ u_{\beta}(\mathrm{k}+1) - e_{\beta}(\mathrm{k}+1) \end{bmatrix} \quad (6)$$

In order to reduce CMV, track the currents and balance NP of the 3LT[2]I, the proposed cost function is shown as the following equation.

$$g = |u_{\alpha}(k+1) - u_{\alpha}| + |u_{\beta}(k+1) - u_{\beta}| \quad (7)$$

The $u_{\alpha}(k+1)$, $u_{\beta}(k+1)$ can be obtained from (6) as:

$$\begin{bmatrix} u_{\alpha}(\mathrm{k}+1) \\ u_{\beta}(\mathrm{k}+1) \end{bmatrix} = \frac{RT_s + L}{T_s}\begin{bmatrix} i^*_{\alpha}(\mathrm{k}+1) \\ i^*_{\beta}(\mathrm{k}+1) \end{bmatrix} - \frac{L}{T_s}\begin{bmatrix} i_{\alpha}(\mathrm{k}) \\ i_{\beta}(\mathrm{k}) \end{bmatrix} + \begin{bmatrix} e_{\alpha}(\mathrm{k}+1) \\ e_{\beta}(\mathrm{k}+1) \end{bmatrix} \quad (8)$$

The $e_{\alpha}(k+1)$, $e_{\beta}(k+1)$, $i_{\alpha}*(k+1)$, $i_{\beta}*(k+1)$ are the next sampling time values can be obtained from the second-degree Lagrange interpolation, i.e. The above currents and voltages can be estimated from:

$$\begin{aligned} & e(\mathrm{k}+1) = 3e(\mathrm{k}) - 3e(\mathrm{k}-1) + e(\mathrm{k}-2) \\ & i_{\alpha}*(\mathrm{k}+1) = 3i_{\alpha}*(\mathrm{k}) - 3i_{\alpha}*(\mathrm{k}-1) + i_{\alpha}*(\mathrm{k}-2) \\ & i_{\beta}*(\mathrm{k}+1) = 3i_{\beta}*(\mathrm{k}) - 3i_{\beta}*(\mathrm{k}-1) + i_{\beta}*(\mathrm{k}-2) \end{aligned} \quad (9)$$

The needed voltage vector can be obtained in equation (8). The u_{α} and u_{β} are the nearest voltage vector corresponding to the vectors of 3LT[2]I.

The flow diagram of improved Predictive control algorithm is shown in Fig. 3.

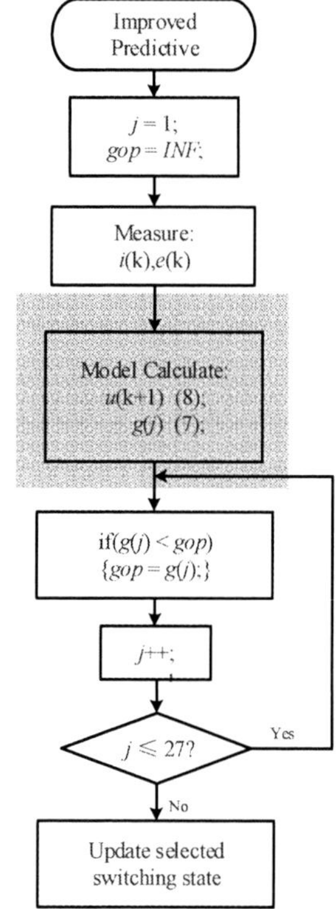

Fig. 3. Improved predictive control algorithm flow diagram in the inverter.

III. THE IMPROVED FCS-MPC METHOD FOR CMV REDUCTION

The CMV is defined as one-third of the summation of the three phase output voltage $V_{CM} = (u_{a1} + u_{b1} + u_{c1})/3$. Different types of vectors result in different CMV shown in TABLE I.

Conventional PWM for the three-level inverter uses all vectors to control its output voltage. Therefore, V_{CM} can be from 0 V to V_{dc} as shown in Table. I.

This paper proposes a method using the vectors generate $\pm 1/6V_{dc}$ and 0, which can totally eliminate the CMV as shown in Table II.

IV. THE BALANCE OF NP VOLTAGE CONTROL

The 3LT[2]I has zero, small, medium and large vectors based on the magnitude. There are total 27 switching status including six large vectors, six medium vectors, six small vectors and one zero vectors which is shown in Table III. Both the zero and small vectors have redundant switching status. The relationship between the switching states and space voltage vectors is shown in Fig. 4.

TABLE I

THE RELATIONSHIP BETWEEN SWITCHING STATES AND CMV

Vsmall(N-type)	V_{CM}	Vmedium	V_{CM}
V_1(ONN)	-2Vdc/6	V_7(PON)	0
V_2(OON)	-Vdc/6	V_8(OPN)	0
V_3(NON)	-2Vdc/6	V_9(NPO)	0
V_4(NOO)	-Vdc/6	V_{10}(NOP)	0
V_5(NNO)	-2Vdc/6	V_{11}(NOP)	0
V_6(ONO)	-Vdc/6	V_{12}(PON)	0
Vsmall (p-type)		Vlarge	V_{CM}
V_1(POO)	Vdc/6	V_{13}(PNN)	-Vdc/6
V_2(PPO)	2Vdc/6	V_{14}(PPN)	Vdc/6
V_3(OPO)	Vdc/6	V_{15}(NPN)	-Vdc/6
V_4(OPP)	2Vdc/6	V_{16}(NPP)	Vdc/6
V_5(OOP)	Vdc/6	V_{17}(NNP)	-Vdc/6
V_6(POP)	2Vdc/6	V_{18}(PNP)	Vdc/6
Vzero	V_{CM}		
V_0(OOO)	0		
V_0(PPP)	3Vdc/6		
V_0(NNN)	-3Vdc/6		

TABLE II

THE RELATIONSHIP BETWEEN SWITCHING STATES AND CMV BASED ON THE PROPOSED METHOD

Vsmall(N-type)	V_{CM}	V_{10}(NOP)	0
V_2(OON)	-Vdc/6	V_{11}(NOP)	0
V_4(NOO)	-Vdc/6	V_{12}(PON)	0
V_6(ONO)	-Vdc/6	Vlarge	V_{CM}
Vsmall (p-type)		V_{13}(PNN)	-Vdc/6
V_1(POO)	Vdc/6	V_{14}(PPN)	Vdc/6
V_3(OPO)	Vdc/6	V_{15}(NPN)	-Vdc/6
V_5(OOP)	Vdc/6	V_{16}(NPP)	Vdc/6
Vzero	V_{CM}	V_{17}(NNP)	-Vdc/6
V_0(OOO)	0	V_{18}(PNP)	Vdc/6
Vmedium	V_{CM}		
V_7(PON)	0		
V_8(OPN)	0		
V_9(NPO)	0		

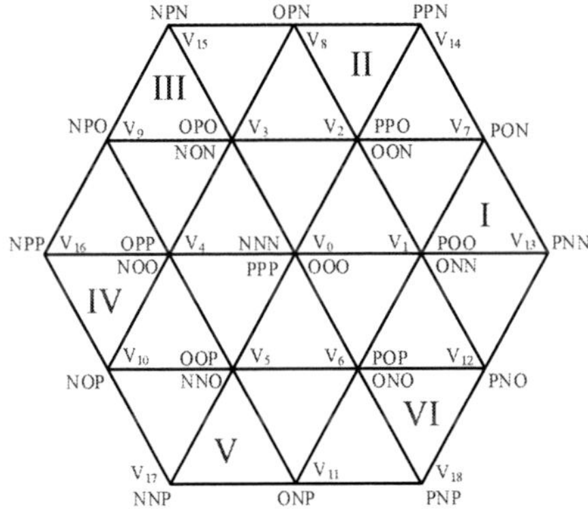

Fig. 4. Space–vector diagram of a 3LT²I

The reference vector V^* is located sector I shown in Fig. 5. Large voltage vector [PNN] does not affect the NP voltage because the NP O unconnected with right as shown in Fig. 6(a). Although the zero voltage vector [OOO] is connected to NP O, it does not affect the NP voltage because the sum of three phase output currents is equal to zero. The zero voltage vector [OOO] is shown in Fig. 6(b). If the 3LT²I is operated with the small vector [POO] shown in Fig. 6(c), the neutral current flows from the positive DC-link point P to neutral-point O, causing the upper capacitor voltage to decrease. Similarly, the other P-type small vectors are shown in Table III which will decrease the upper capacitor voltage. Adversely, the switching state [ONN] of small vectors decreases the lower capacitor voltage since the neutral current flows from the neutral-point O to negative DC-link point N as shown in Fig. 6(d). Similarly, the other N-type switching states of small vectors are shown in Table III causing the lower capacitor voltage to decrease. The switching state [PON] of medium voltage vector is shown in Fig. 6(e). This is difficult to control the NP voltage because the direction of neutral current is uncertain.

TABLE III

SWITCHING STATES AND VOLTAGE VECTORS

Vector	Magnitude	Switching State	
Zero vector	0	[PPP] [OOO] [NNN]	
Small vector	$\frac{1}{3}V_{dc}$	P-type [POO] [PPO] [OPO] [OPP] [POP] [OOP]	N-type [ONN] [OON] [NON] [NOO] [ONO] [NNO]
Medium vector	$\frac{\sqrt{3}}{3}V_{dc}$	[PON] [OPN] [NPO] [NOP] [ONP] [PNO]	
Large vector	$\frac{2}{3}V_{dc}$	[PNN] [PPN] [NPN] [NPP] [NNP] [PNP]	

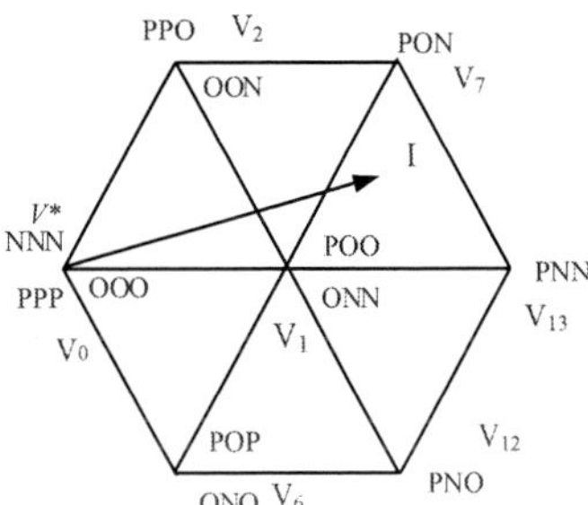

Fig. 5. Space-vector diagram of sector I for a 3LT²I.

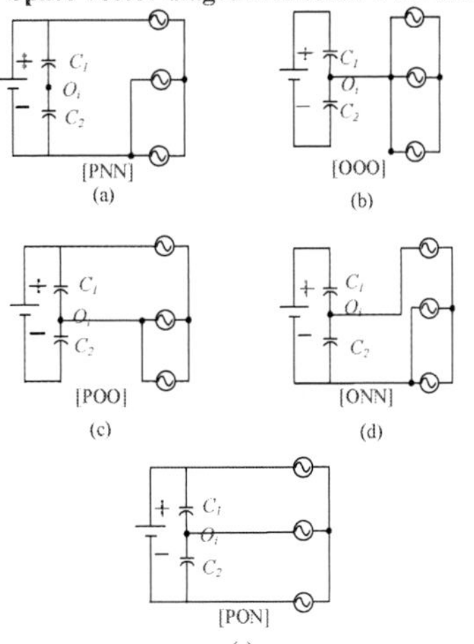

Fig. 6. Effect of the switching state on the NP voltage in the 3LT²I. (a) Large voltage vector [PNN],(b) zero voltage vector [OOO],(c) small voltage vector [POO], (d) small voltage vector [ONN], (e)The medium voltage vector [PON].

Therefore, the small vectors are employed to neutral-point voltage balance control. Based on the relationship between the switching states and common-mode voltage (CMV), the 19 voltage switching vectors are adopted to reduce the CMV as shown in Table II. If the bottom side capacitor voltage is larger than the top side capacitor voltage, the N-type vector V_2, V_4, V_6 are selected to control the NP balance. On the contrary, if the top side capacitor voltage is larger than the bottom side capacitor voltage, the P-type vector V_1, V_3, V_5 are selected to control the NP balance. The neutral-point voltage balancing control is shown in Fig. 7.

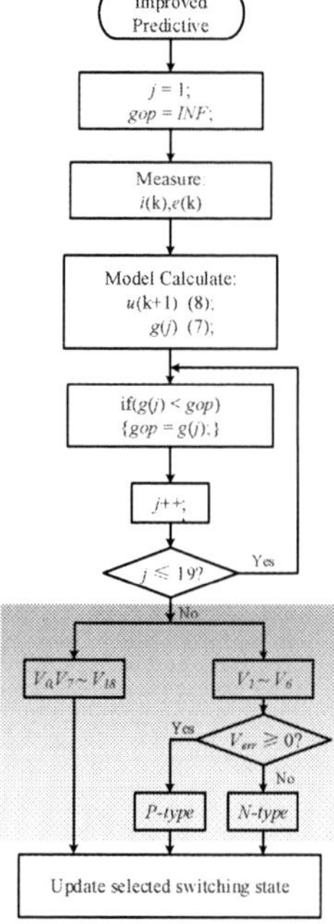

Fig. 7. Improved predictive control algorithm for NP balance control

V. EXPERMENTAL RESULTS

To verify the proposed approach, the experiments are performed in the two parallel 3LT²Is as shown in Fig. 8. The parameters for experimental are shown in Table VI.

TABLE VI

THE PARAMETERS FOR EXPERIMENTAL

Grid voltage e_{ab}	40Vac
DC-link voltage	100Vdc
DC-link capacitor	3300μF
Rated current	6A
L-filter	4mH
Switching frequency	10kHz
Period time T_s	100 μ s
Resistor	0.5Ω

The system of three-level inverter contains several parts, including main circuit, AC contactor, L filter and dSPACE 1005. The communication of main circuit is interfaced with the controller dSPACE 1005 by an interface board to provide gate signal, enable signal and protective signals. The main circuit of three-level inverter includes a sensor board, gate signal which is also connected to the controller dSPACE 1005 as feedback for the NP balance, CMV reduction and currents tracking control. The 3LT²Is are connected to 40-V line-to-line ac grid via the L filter and contactor. And the dead time is set to 2.5 μs.

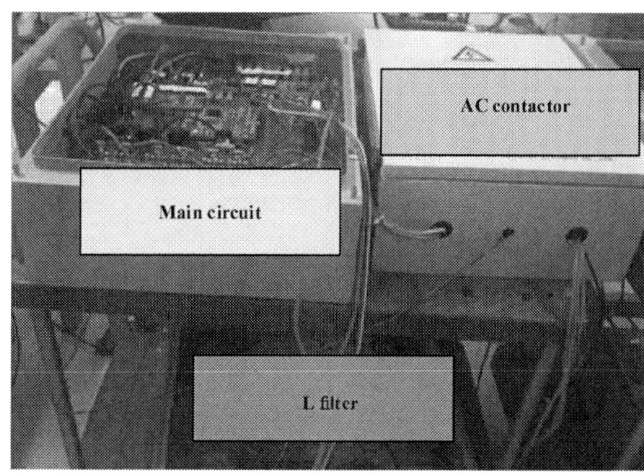

Fig. 8. Experimental platform of a 3LT²I.

The proposed FCS-MPC method with current tracking, NP balance and CMV reduction is verified on an experimental prototype as shown in Fig. 8. For the NP balance result, the line-to-line output voltage, dc-link capacitor voltage and three-phase grid current for the proposed FCS-MPC is shown in Figs. 9 and 10. Fig. 11 shows the current generated by the FCS-MPC methods accurately track their references. In addition, it is observed that the CMV of the proposed method is limited to $\pm V_{dc}/6$, whereas the conventional method generates a CMV oscillating between $-V_{dc}/2$ and $+V_{dc}/2$ owing to the utilization of all small vectors and zero vectors in Table I.

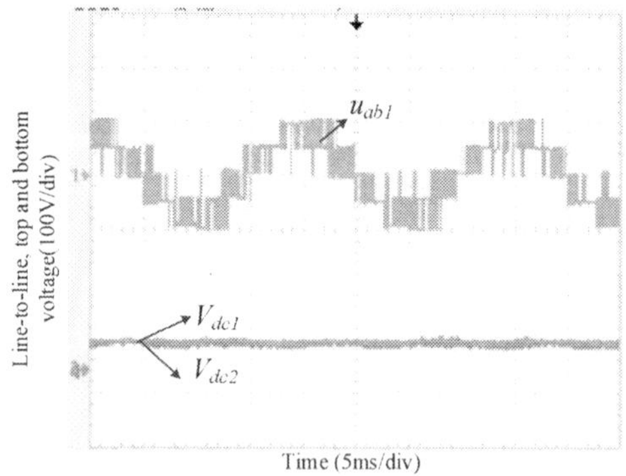

Fig. 9 Line-to-line voltage dc-link voltage for FCS-MPC

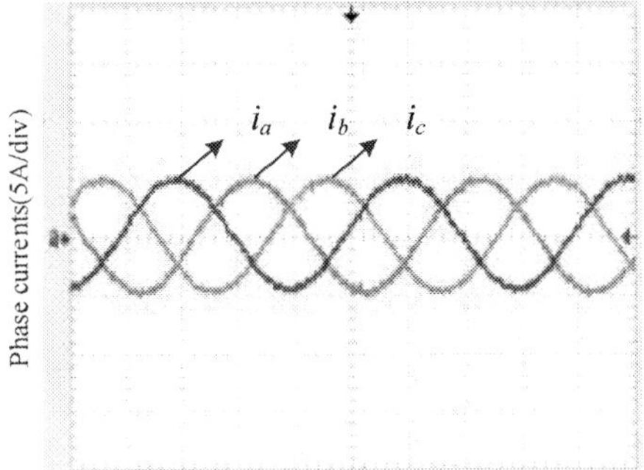

Fig. 10 three-phase current for FCS-MPC

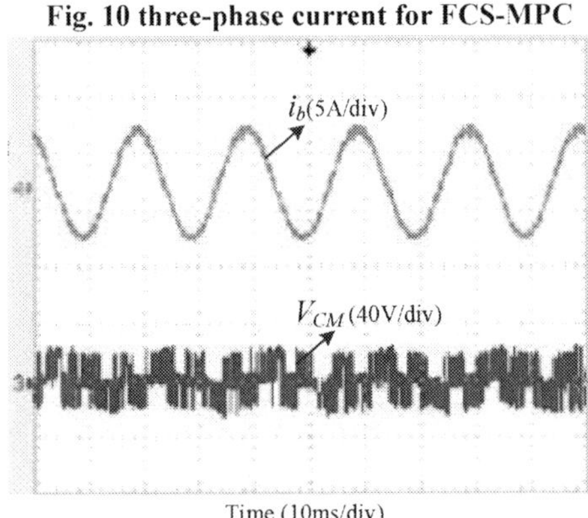

(a)

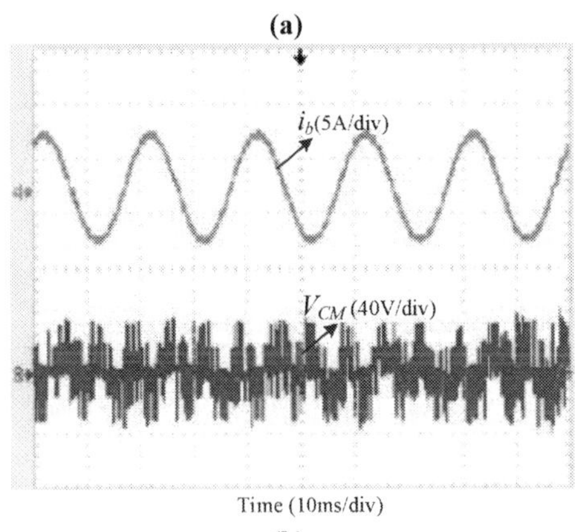

(b)

Fig. 11 Experimental waveforms with T_s = 100 µs, (a) the proposed method; (b) conventional method.

VI. CONCLUSIONS

In a three-level inverter system, the CMV and the NP voltage balancing control are considered for the normal inverter operation. This paper proposes FCS-MPC method to reduce the CMV and balance the NP voltage with fast dynamics of the 3LT^{2}I. In the proposed FCS-MPC method, only $\pm 1/6V_{dc}$ and 0 vectors are selected to reduce the CMV by avoiding the zero vectors and small vectors that generate the higher CMV. In addition, the NP voltage balance is also achieved by means of the improved model current predictive control. The paper proposes an improved model current predictive current control to balance the NP voltage, reduce the common-mode voltage (CMV) and track the reference current, simultaneously.

ACKNOWLEDGMENT

This work was supported by the National Natural Science Foundation of China under Grant 61320106011, Grant 51377101, and Grant 61527809 and by the Program for New Century Excellent Talents in University (NCET-13-0339).

REFERENCES

[1] B. Chen, B. Gu, L. Zhang, Z. U. Zahid, J. -S. Lai, Z. Liao, and R. Hao, "A high efficiency MOSFET transformerless inverter for non-isolated micro-inverter applications," IEEE Trans. Power Electron., vol. 30, pp. 3610-3622, Jul. 2015

[2] X. Xing, A. Chen, W. Wang, C. Zhang, Y. Li, C. Du, "Space-vector-modulated for Z-source three-level T-type converter with neutral voltage balancing," IEEE Conf. Appl. Power Electron. Conf., Mar, 2015, pp. 833 – 840

[3] M. Schweizer and J. W. Kolar, Design and implementation of a highly efficient 3-level T-type inverter for low-voltage application. IEEE Trans.Power Electron., vol. 28, no. 2, pp. 899–907, Feb. 2013.

[4] Park Y, Sul S K, Lim C H, et al. Asymmetric control of DC-link voltages for separate MPPTs in three-level inverters. IEEE Trans. Power Electron., vol. 28, no. 6, pp. 2760-2769, Feb. 2013.

[5] U. M. Choi, H. H. Lee, and K. B. Lee, "Simple Neutral-Point Voltage Control for Three-Level Inverters Using a Discontinuous Pulse Width Modulation," IEEE Trans. Energy Convers., vol. 28, no. 2, pp. 434-443, Jun. 2013.

[6] J. S. Lee and K. B. Lee, "New modulation techniques for a leakage current reduction and a neutral-point voltage balance in transformerless photovoltaic systems using a three-level inverter," IEEE Trans. Power Electron., vol. 29, no. 4, pp. 1720–1732, Apr. 2014.

[7] Park Y, Sul S K, Lim C H, et al. "Reliability Improvement of a T-Type Three-Level Inverter With Fault-Tolerant Control Strategy". Trans. Power Electron., vol. 30, no. 5, pp. 2660–2673, May. 2015.

[8] U. M. Choi and K. B. Lee, "Space vector modulation strategy for neutral-point voltage balancing in three-level inverter systems, IET Power Elec-tron., vol. 6, no. 7, pp. 1390–1398, Aug. 2013.

[9] Ui-Min Choi, June-Seok Lee ,and Kyo-Beum Lee. New Modulation Strategy to Balance the Neutral-Point Voltage for Three-Level Neutral-Clamped Inverter Systems, IEEE Trans. Energy Conversion, vol. 29, no. 1, pp. 91–100, Mar. 2014

[10] J. D. Barros, J. F. A. Silva and É. G. A. Jesus, "Fast-predictive optimal control of NPC multilevel converters," IEEE Trans. Ind. Electron., vol. 60, no. 2, pp. 619–627, Feb. 2013.

[11] J. Rodriguez, J. Pontt, C. A. Silva, P. Correa, P. Lezana, P. Cortés and U. Ammann, "Switching strategy based on model predictive control of VSI to obtain high efficiency and balanced loss distribution," IEEE Trans. Power Electron., vol. 29, no. 9, Sep. 2014.

[12] S. Kwak and J. C. Park, "Predictive control method with future zero-sequence voltage to reduce switching losses in Three-Phase voltage source inverters," IEEE Trans. Power Electron., vol. 30, no. 3, Mar. 2015.

[13] B. Chen, B. Gu, L. Zhang, J. -S. Lai, "A Novel Pulse Width Modulation Method for Reactive Power Generation on a CoolMOS and SiC Diode based Transformerless Inverter", IEEE Trans. Indus. Electron. accepted.

[14] L. Zhang, R. Born, X. Zhao, J.-S. Lai, "A High Efficiency Inverter Design for Google Little Box Challenge", in 2015 IEEE Wide Bandgap Power Semiconductor Workshop. Blacksburg. Nov. 2015

[15] B. Chen, B. Gu, L. Zhang, Z. U. Zahid, J. -S. Lai, Z. Liao, and R. Hao, "A high efficiency MOSFET transformerless inverter for non-isolated micro-inverter applications," IEEE Trans. Power Electron., vol. 30, pp. 3610-3622, Jul. 2015.

[16] M. A. Perez, J. Rodriguez, E. J. Fuentes and F. Kammerer, "Predictive control of ac–ac modular multilevel converters," IEEE Trans. Ind Electron., vol. 59, no. 7, pp. 2832–2839, Jul 2012.

[17] M. Duran, J. A. Riveros, F. Barrero, H. Guzman, and J. Prieto, "Reduction of common-mode voltage in five-phase induction motor drives using predictive control techniques," IEEE Trans. Ind. Appl., vol. 48, no. 6,pp. 2059–2067, Nov. 2012.

[18] D. Zhang, F. (Fred)Wang,R.Burgos, and D.Boroyevich, "Common-mode circulating current control of paralleled interleaved three-phase two-level voltage-source converters with discontinuous space-vector modulation,"IEEE Trans. Power Electron., vol. 26, no. 12, pp. 3925–3935, Dec. 2011.

[19] C.-C. Hou, C.-C. Shih, P.-T. Cheng, and A. M. Hava, "Common-mode voltage reduction pulsewidth modulation techniques for three-phase grid-connected converters," IEEE Trans. Power Electron., vol. 28, no. 4,pp. 1971–1979, Apr. 2013.

Convergence Analysis of Distributed Control for Operation Cost Minimization of Droop Controlled DC Microgrid Based on Multiagent

Chendan Li, Juan C. Vasquez, and Josep M. Guerrero
Department of Energy Technology, Aalborg University
Aalborg, Denmark
{che, juq , joz}@et.aau.dk
(http://www.et.aau.dk/research-programmes/microgrids)

Abstract—**In this paper we present a distributed control method for minimizing the operation cost in DC microgrid based on multiagent system. Each agent is autonomous and controls the local converter in a hierarchical way through droop control, voltage scheduling and collective decision making. The collective decision for the whole system is made by proposed incremental cost consensus, and only nearest-neighbor communication is needed. The convergence characteristics of the consensus algorithm are analyzed considering different communication topologies and control parameters. Case studies verified the proposed method by comparing it without traditional methods. The robustness of system is tested under different communication latency and plug and play operation.**

Keywords— ***economic dispatch problem; operation cost minimization; droop control; DC microgrid; voltage scheduling; incremental cost consensus; multiagent.***

I. Introduction

Microgrids are the future of the distribution system in smart grid. The cluster of the loads, distributed generators and energy storage system within a defined boundary, through coordinated control, can provide higher reliability, flexibility and efficiency to both the end-users within the grid and the upper grid through ancillary services. Although most of the existed microgrids are based on AC delivery scheme since this scheme has been the paradigm dominated in the traditional power system, DC microgrid is continuously demonstrating the advantages in terms of higher flexibility, power quality and lower power loss compared with AC microgrid [1]-[3]. In fact, most of end-user loads need embedded rectifiers to convert the power to DC from the original AC, such as computer, LED lights, and so on. Moreover, on the generation side, solar generation system and energy storage system are all DC sources. A DC distribution system therefore can reduce the capital cost as well as provide added valued mentioned above.

Besides providing increased reliability, power quality, the potential of a microgrid to offer increased efficiency is gaining more interests with the technology becoming more and more mature [4]-[11]. As the generation resources of microgrid can be quite heterogeneous, one of the optimization objects is to reduce the operation cost of the system through optimal dispatch according to the different operation cost of each generator. To achieve this goal, control strategies can be either centralized or decentralized. Authors in [4] used a centralized controller to reduce the operation cost through reduced gradient method. Authors in [5] provided optimized droop parameter through optimal power flow for the DC microgrid, which needs a centralized controller to run the power flow and optimization. Although applications realized in centralized way with a single centralized controller can achieve the operation cost minimization of system, they suffer from the single point of failure [6].

Instead, reference [7] and [8] provided the cost-based droop to optimize the operation cost in the primary control level in a distributed way. However, although it will reduce the total cost, the optimum cannot be guaranteed since it makes the linearization of the cost function by approximation. In [9], author adopted incremental cost consensus to optimize the system in a smart grid context, but the details of power regulation realization are not given. In [10] and [11], authors implemented a distributed optimization method based on incremental cost for the system, but it was applied to AC microgrid.

In this paper, a multiagent system is designed aiming at minimizing the operation cost of the DC microgrid. Each local controller for each converter is taken as an agent. The control strategy is detailed to the primary control level based on droop, and the power regulation is realized through voltage scheduling. The optimal power command is generated in a distributed way. Practical issues regarding the interoperation of the low band communication are investigated.

The rest of the paper is organized as follows. Section II introduces the general structure of a DC microgrid. The overall control system is described in Section III, which elaborate the problem formulation, incremental consensus algorithm and power regulation using voltage scheduling method. Case studies of the proposed design follow in Section IV. In the end, Section V concludes the paper.

978-1-4673-9551-9/16 $31.00 © 2016 IEEE

II. DC MICROGRID CONFIGURATION

Fig. 1 shows a typical DC microgrid with distributed control system studied in this work. The DC microgrid contains renewable resource and load, and other dispatchable distributed generators acting as the backup for the intermittent renewables. Each dispatchable generator is interfaced with a DC/DC converter between the primary source and the DC bus. This DC microgrid is a cyber-physical system which has the controllers and communication system being the computational elements upon the physical components in the fleet.

Conventionally, there is an extra centralized controller which computes for the upper layer functions above the primary control, such as economic dispatch and voltage regulation. In this work, optimization is applied to the dispatchable distributed generators. Each their local controller itself will work as an autonomous controller to collaboratively accomplish the optimization of the system.

III. PROPOSED MULTIAGENT SYSTEM

A. Optimization Problem Formulation

The generation costs of different DGs (fuel cells, batteries, diesel generators, etc.) include many factors, which are surely not the same, but they might have a similar pattern which can be generalized as quadratic cost function [6]-[11].

During operation, assume the cost only incurred due to the fuel cost or the power loss incurred by charging/discharging efficiency of the energy storage system. The generation cost of each dispatchable generator can be generalized into:

$$C_i(P_{G,i}) = \alpha_i P_{G,i}^2 + \beta_i P_{G,i} + \gamma_i \tag{1}$$

where α_i, β_i and γ_i are the coefficient of cost function of DG unit i,

Assuming there are n dispatchapble generators in the microgrid. The total cost of operation of a microgrid can be expressed as

$$C_{total} = \sum_{i=1}^{n} C_i(P_{G,i}) = \sum_{i=1}^{n} \alpha_i P_{G,i}^2 + \beta_i P_{G,i} + \gamma_i \tag{2}$$

The optimization problem should be constrained by the power balance equation and power generation limitation, and therefore the objective function for minimizing the operation cost can be written as

$$\begin{aligned} &Min \sum_{i=1}^{n} \alpha_i P_{G,i}^2 + \beta_i P_{G,i} + \gamma_i \\ &s.t. \ \sum_{i=1}^{n} P_{G,i} = P_D \\ &P_{G,i}^{\min} \le P_{G,i} \le P_{G,i}^{\max} \end{aligned} \tag{3}$$

where $P_{G,i}$ denotes the output power of unit i, and P_D denotes the total power demand of the system.

This optimization above can be solved in a centralized control system using Lagrange multiplier method [9]. In this work, a distributed control method based on multiagent is adopted using incremental cost consensus, which is elaborated in the next part.

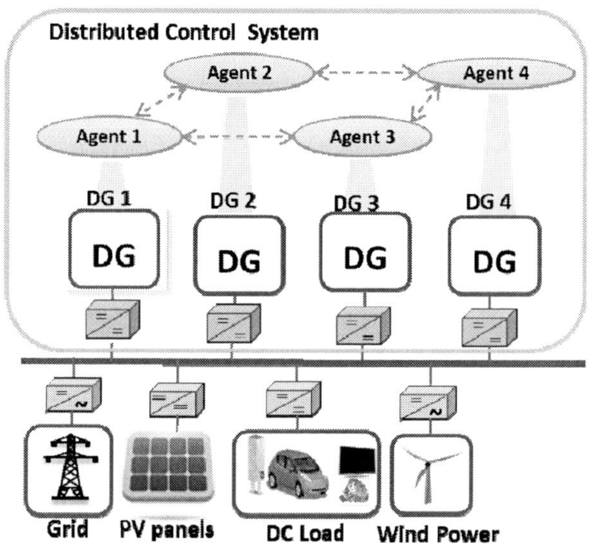

Fig. 1 Example DC microgrid controlled by multiagent system

B. Incremental Cost Consensus

1) Incremental Cost

Same as conventional economic dispatch method, the incremental cost of each DG is defined as

$$r_i = \frac{\partial C_i(P_{G,i})}{\partial P_{G,i}} = 2\alpha_i P_{G,i} + \beta_i \tag{4}$$

Without generation capacity constraints, when the incremental cost reaches equality, it is the solution to (3). The common optimal r* can be expressed as

$$r^* = [\sum_{i=1}^{n} \frac{\beta_i}{2\alpha_i} + P_D] / (\sum_{i=1}^{n} \frac{1}{2\alpha_i}) \tag{5}$$

2) Update Rules of Consensus Algorithm

In this distributed control strategy, the update rule of proposed incremental consensus algorithm is as follows.

$$r_i[t+1] = \sum_{i \in N_i} d_{ij} r_j + \varepsilon P_{D,i}[t] \tag{6}$$

$$P_{G,i}[t+1] = \frac{r_i[t+1] - \beta_i}{2\alpha_i} \tag{7}$$

$$P'_{D,i}[t+1] = P_{D,i}[t] - (P_{G,i}[t+1] - P_{G,i}[t]) \tag{8}$$

$$P_{D,i}[t+1] = \sum_{i \in N_i} d_{ij} P'_{D,i}[t+1] \tag{9}$$

where $r_i[t]$ is the incremental cost of agent i at iteration t, ε is the feedback coefficients which controls the convergence of the consensus, $P_{D,i}[t]$ is the estimation of the global supply-demand mismatch, and d_{ij} is defined as

$$d_{ij} = \begin{cases} 2/(n_i + n_j + 1) & j \in N_i \\ 1 - \sum_{j \in N_i} 2/(n_i + n_j + 1) & i = j \\ 0 & otherwise \end{cases} \tag{10}$$

3) Convergence characteristics of consensus algorithm

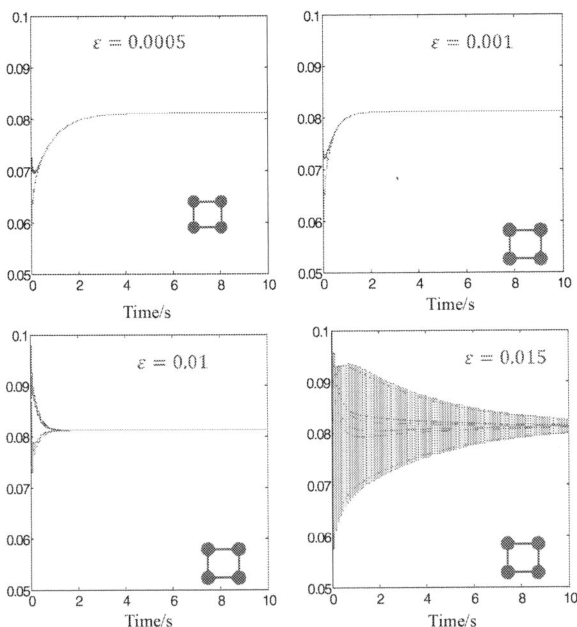

Fig. 2 Convergence analysis with different ε

To analyze the dynamics of the proposed consensus algorithm, several factors that influence the system are investigated in this part in order to guide the system design.

Firstly, the impact of the control parameter in the consensus algorithm is analyzed. Fig. 2 shows the convergence characteristic with different ε. It can be seen that, as ε becomes smaller the convergence will become slower. However, it cannot be unlimitedly large. When parameter ε reaches 0.015, the consensus algorithm cannot converge. Thus, the convergence speed and stability of the consensus algorithm is a trade-off when choosing the parameter.

Secondly, the impact of different topologies to the consensus is also studied. Fig. 3 shows that the convergence characteristic of the proposed algorithm under different topologies. As it can be seen, the more connected, the faster the system converges. When designing the communication system, the latency tolerance and the cost should be weighed.

Since the latency in the communication only influence how fast in each update during the convergence, it will not affect the convergence of the system. However, in the whole control system, the effect of it should not be overlooked and it will be addressed in the case studies of the whole system.

C. Power regulation base on voltage scheduling

The basic droop control for DC microgrid uses a virtual impedance to regulate the output voltage so as to regulate output power of DG units, which can be expressed as [12]-[14]

$$V_{o\_i} = V_{ref,i} - R_{d,i} i_{o,i} \tag{11}$$

where V_{o_i} is the voltage command given to the voltage loop of the converter i, $V_{ref,i}$ is the voltage reference for the droop controller and $R_{d,i}$ is the virtual impedance, and $i_{o,i}$ is the output current.

Instead of changing $R_{d,i}$, $V_{ref,i}$ is modified directly based on the power command in this work. This control strategy actually does not only maintains the benefit of traditional droop control to avoid power circulation, but also realizes accurate power sharing if power command is correctly given. The voltage reference is modified as (12), and the control diagram is given in Fig. 4.

$$V_{ref,i} = V^* + Kp(P_{G,i}^* - P_{G,i}) + Ki\int(P_{G,i}^* - P_{G,i}) \tag{12}$$

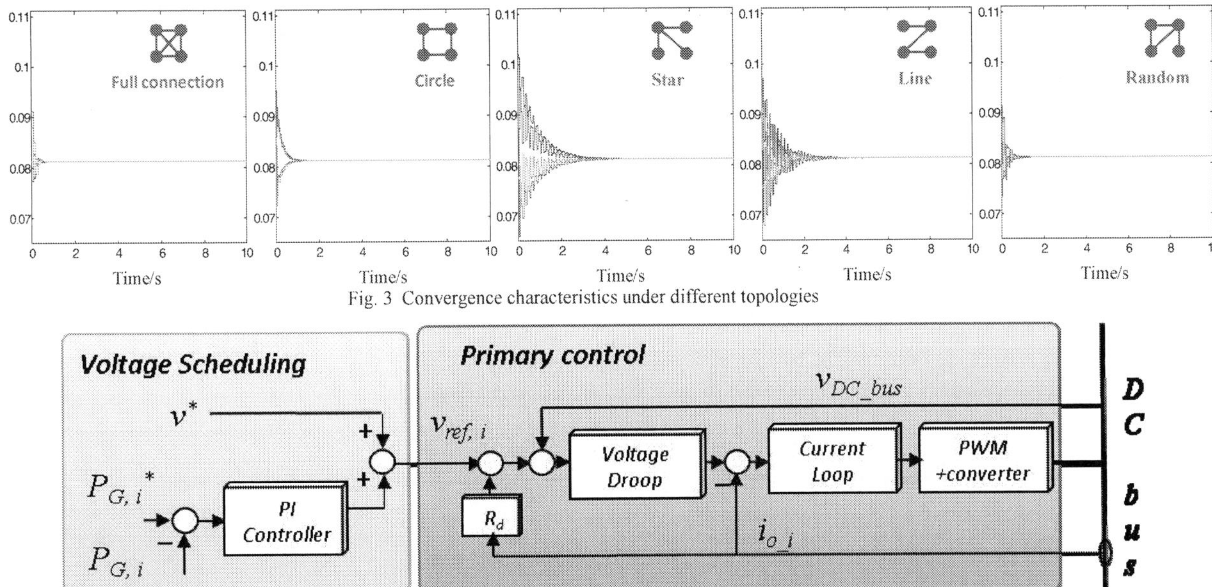

Fig. 3 Convergence characteristics under different topologies

Fig. 4 Control diagram of power regulation

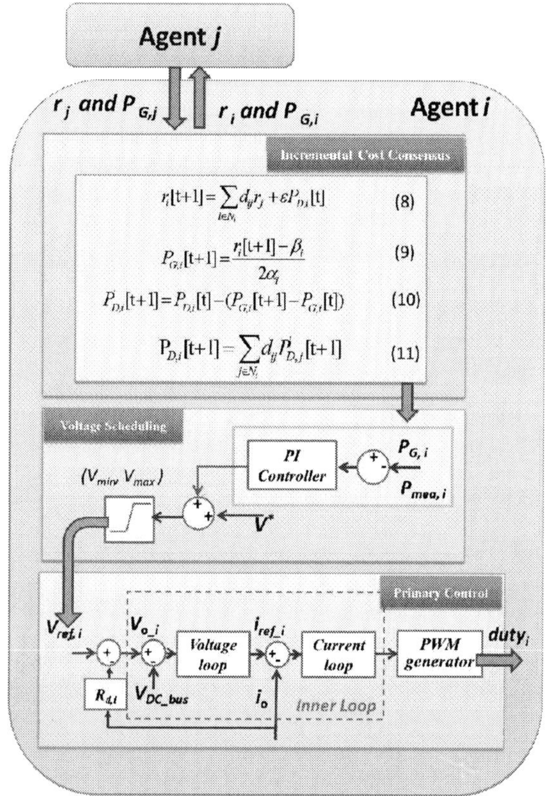

$$r_i[t+1] = \sum_{i \in N_i} d_{ij} r_j + \varepsilon P_{D,i}[t] \quad (8)$$

$$P_{G,i}[t+1] = \frac{r_i[t+1] - \beta_i}{2\alpha_i} \quad (9)$$

$$P_{D,i}[t+1] = P_{D,i}[t] - (P_{G,i}[t+1] - P_{G,i}[t]) \quad (10)$$

$$P_{D,i}[t+1] = \sum_{j \in N_i} d_{ij} P'_{D,j}[t+1] \quad (11)$$

Fig. 5 Proposed multiagent system

TABLE I CONTROL PARAMETER

Item	Symbol	Value
Nominal bus voltage	V^*	400V
Minimum bus voltage	$Vmin$	420V
Maximum bus voltage	$Vmax$	380V
Virtual impedance for DG1,3	$R_{d,1}, R_{d,3}$	0.2Ω
Virtual impedance for DG2,4	$R_{d,2}, R_{d,4}$	0.5Ω
Voltage scheduling proportional term	Kp	0.0009
Voltage scheduling integral term	Ki	0.001s-1
Consensus convergence coefficient	ε	0.001

TABLE II COEFFICIENTS OF THE OPERATION COST FUNCTION

Unit	α_i	β_i	γ_i
1	7.15e-3	0.77	0.002
2	4.75e-3	0.78	0.005
3	3.75e-3	0.55	0.001
4	3.45e-3	0.51	0.001

D. Overall distributed control system

The control diagram of the proposed strategy is illustrated in Fig. 5. Here we illustrate the multiagent system using two arbitrary agents which have communication connection between each other.

Each agent is controlling their local converter through hierarchical control structure, and only its own incremental cost and power command generated in the iteration of the consensus algorithm are exchanging between them, which are needed for the convergence of the consensus. Bellow consensus layer using communication, there is the power regulation loop which uses voltage scheduling to track the optimal power command. In the primary control, each converter is using virtual impedance to sharing the power.

IV. CASE STUDIES

A. Performance comparison with traditional method

To verify the effectiveness of the control strategy, the proposed method is simulated in a tested DC microgrid with four different dispatchable generators. The control parameters of the DC microgrid are listed in Table I. The cost coefficients of the generators are given in Table II.

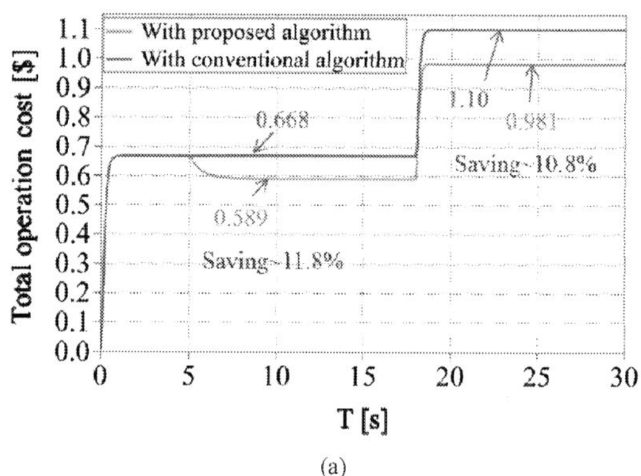

(a)

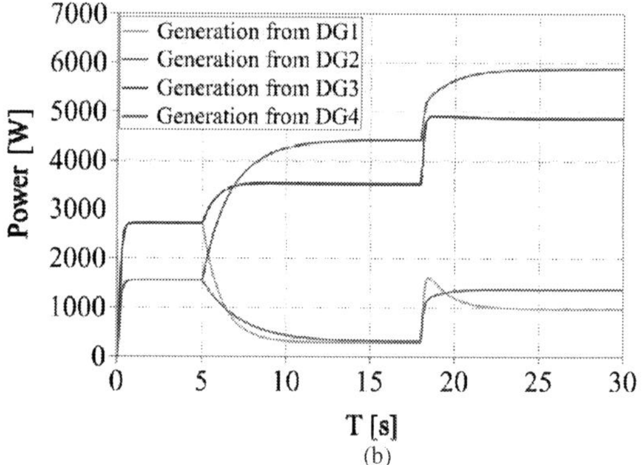

(b)

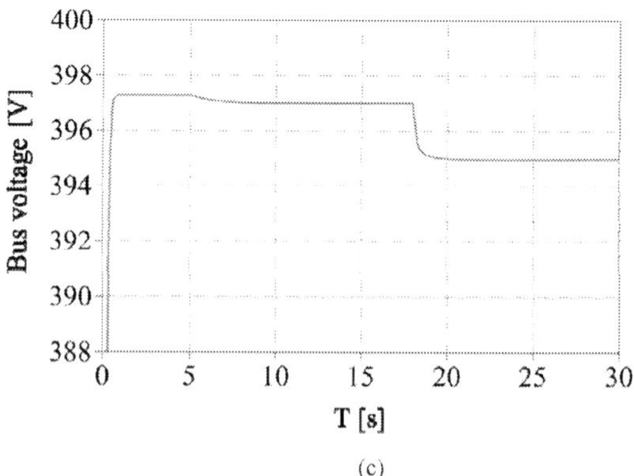

(c)

Fig. 6 Controller performance (a) Total operation cost, (b) Generation power of each DG unit and (c) DC bus voltage

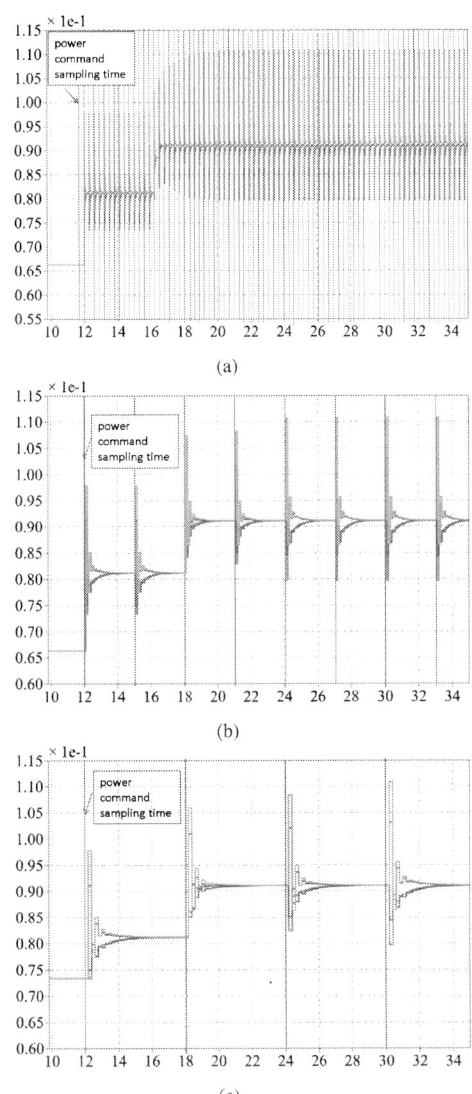

Fig. 7 Power command with (a) 10ms, (b) 100ms and (c) 200ms latency

Firstly, only traditional droop is adopted at the beginning. The proposed operation cost minimization method is activated at t = 5s. To test the system during the load change, at t = 18s, the total load of the system is changed from 8.5kW to 13kW. Fig. 6 (a) to (c) show the total operation cost of the system, generation of each DG unit and DC bus voltage. The total cost is reduced up to 11.8% compared with that using only the droop control in each converter with virtual impedance configured as in Table I. Fig. 6 (b) shows power shared among converters respectively, which are not identical because of the optimization.

B. *Impact of Communication Latency*

In this cyber-physical system, although the communication load is light, the condition of communication would influence the performances of the system. In this part, the effect of communication latency is investigated. Same as last part, the load is changed from 8.5kW to 13kW at 16s. The system is tested under the communication with 10ms, 100ms and 200ms latency, respectively. The power command generated from consensus under different scenarios is shown in Fig. 7, which is passed to the lower level controller synchronously after consensus gets converged. It can be seen from Fig. 8, although the responses with longer latency get slower, the output power will reach the same value in the steady state.

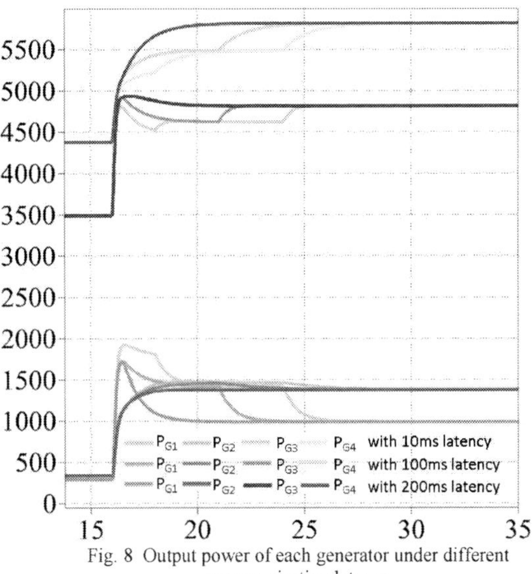

Fig. 8 Output power of each generator under different communication latency

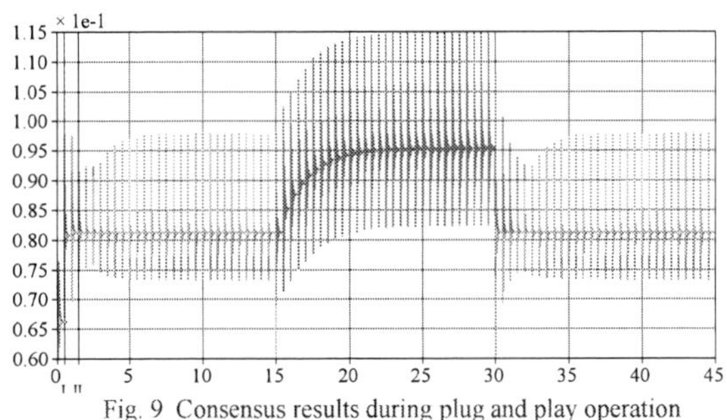
Fig. 9 Consensus results during plug and play operation

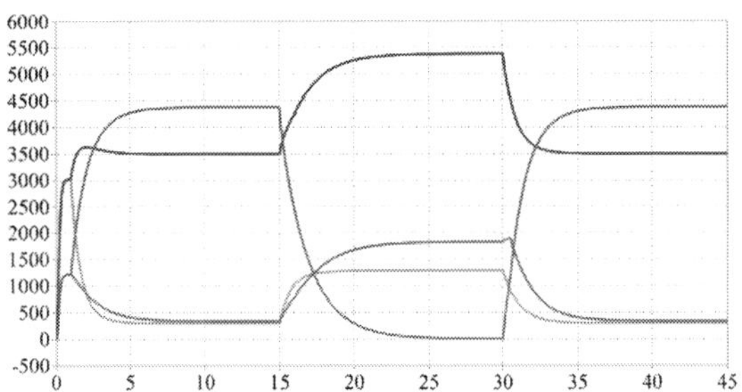
Fig. 10 Output power results during plug and play operation

C. Plug and play operation

To test the plug and play functionality of the multiagent system, DG unit 4 is cut in and out at t=15s and t=30s. The optimization is activated at t=1s. Fig. 9 shows the consensus results, which give the optimal power command. Fig. 10 shows the output power of the four converters. It can be seen from Fig. 10, during plug and play, the transients of the out power from one state to another are smooth.

V. Conclusion

In this paper, a multiagent system is proposed aiming at minimizing the operation cost for DC microgrids. Each local controller for each converter is taken as an agent, which optimizes the local converter autonomously in a hierarchical way with only communication with nearest neighbour. Compared with method without optimization, the operation cost is reduced effectively under different load conditions. The impact of communication issues on the multiagent system convergence is investigated to shed light on the system design. Experimental results are expected in the future work.

References

[1] Guerrero, J.M.; Vasquez, J.C.; Matas, J.; de Vicuña, L.G.; Castilla, M., "Hierarchical Control of Droop-Controlled AC and DC Microgrids—A General Approach Toward Standardization," *Industrial Electronics, IEEE Transactions on* , vol.58, no.1, pp.158,172, Jan. 2011; doi: 10.1109/TIE.2010.2066534

[2] D. Salomonsson, Modeling , Control and Protection of Low-Voltage DC Microgrids. 2008, p. 66.

[3] Yunjie Gu; Xin Xiang; Wuhua Li; Xiangning He, "Mode-Adaptive Decentralized Control for Renewable DC Microgrid With Enhanced Reliability and Flexibility," *Power Electronics, IEEE Transactions on* , vol.29, no.9, pp.5072,5080, Sept. 201R. Nicole, "Title of paper with only first word capitalized," J. Name Stand. Abbrev., in press.

[4] N. Augustine, S. Suresh, P. Moghe, and K. Sheikh, "Economic dispatch for a microgrid considering renewable energy cost functions," *in 2012 IEEE PES Innovative Smart Grid Technologies (ISGT)*, 2012, pp. 1–7.

[5] Chendan Li; de Bosio, F.; Chaudhary, S.K.; Graells, M.; Vasquez, J.C.; Guerrero, J.M., "Operation Cost Minimization of Droop-Controlled DC Microgrids Based on Real-Time Pricing and Optimal Power Flow," in Industrial Electronics Society, IECON 2015- 40th Annual *Conference of the IEEE* , in press.

[6] Meng, L.; Dragicevic, T.; Roldan-Perez, J.; Vasquez, J.C.; Guerrero, J.M., "Modeling and Sensitivity Study of Consensus Algorithm-Based Distributed Hierarchical Control for DC Microgrids," *in Smart Grid, IEEE Transactions on* , vol.PP, no.99, pp.1-1

[7] Nutkani, I.U.; Poh Chiang Loh; Blaabjerg, F., "Droop Scheme With Consideration of Operating Costs," *Power Electronics, IEEE Transactions on* , vol.29, no.3, pp.1047,1052, March 2014

[8] Nutkani, I.U.; Wang Peng; Poh Chiang Loh; Blaabjerg, F., "Cost-based droop scheme for DC microgrid," *in Energy Conversion Congress and Exposition (ECCE), 2014 IEEE* , vol., no., pp.765-769, 14-18 Sept. 2014

[9] Ziang Zhang; Mo-Yuen Chow, "Convergence Analysis of the Incremental Cost Consensus Algorithm Under Different Communication Network Topologies in a Smart Grid," *Power Systems, IEEE Transactions on* , vol.27, no.4, pp.1761,1768, Nov. 2012

[10] S. Yang, S. Tan, and J.-X. Xu, "Consensus Based Approach for Economic Dispatch Problem in a Smart Grid," *IEEE Trans. Power Syst.*, vol. 28, pp. 4416–4426, 2013.

[11] Chendan Li; Savaghebi, M.; Vasquez, J.C.; Guerrero, J.M., "Multiagent based distributed control for operation cost minimization of droop controlled AC microgrid using incremental cost consensus," *in Power Electronics and Applications (EPE'15 ECCE-Europe)*, 2015 17th European Conference on , vol., no., pp.1-9, 8-10 Sept. 2015

[12] Shafiee, Q.; Dragicevic, T.; Vasquez, J.C.; Guerrero, J.M., "Hierarchical Control for Multiple DC-Microgrids Clusters," in *Energy Conversion, IEEE Transactions on* , vol.29, no.4, pp.922-933, Dec. 2014

[13] Dong Chen; Lie Xu; Liangzhong Yao, "DC network stability and dynamic analysis using virtual impedance method," *IECON 2012 - 38th Annual Conference on IEEE Industrial Electronics Society , vol.*, no., pp.5625,5630, 25-28 Oct. 2012

[14] Dragicevic, T.; Lu, X.; Vasquez, J.; Guerrero, J., "DC Microgrids–Part I: A Review of Control Strategies and Stabilization Techniques," in *Power Electronics, IEEE Transactions on* , vol.PP, no.99, pp.1-1

A Novel Model Predictive Control Algorithm to Suppress the Zero-Sequence Circulating Currents for Parallel Three-Phase Voltage Source Inverters

Zicheng Zhang, Alian Chen, Xiangyang Xing, Chenghui Zhang
School of Control Science and Engineering
Shandong University
Jinan 250061, P. R. China

***Abstract*—The topology of parallel three-phase voltage source inverters (VSIs) has been widely utilized to raise system power rating, but zero-sequence circulating currents (ZSCCs) are generated by control effect and hardware parameter differences. ZSCCs could lead to current distortion and impact the system stability. The model predictive control (MPC) method has been applied to the inverters to get high robustness, fast dynamic response and low switching frequency. However, the MPC method is rarely used in parallel inverters because of the ZSCCs problem. This paper proposes an improved MPC algorithm for parallel system to track the reference currents as well as suppress the ZSCCs. The contribution of each space vector to ZSCCs is analyzed and the cost function is redesigned in the new method. The cost function will pick out the optimal vectors to guarantee the control requirements. Experimental results verified that the improved algorithm is effective and performs well in both current tracking and ZSCCs suppression.**

Keywords*—parallel three-phase voltage source inverters (VSIs), model predictive control (MPC), current tracking, zero-sequence circulating currents (ZSCCs) suppression

I. INTRODUCTION

In power electronic applications, parallel voltage source inverters (VSIs) have lots of advantages such as high power capacity, high efficiency and low harmonics [1]. However, the zero-sequence circulating currents (ZSCCs) will be generated between different inverters when their ac and dc sides are connected directly. ZSCCs will lead to current distortion, additional system loss and efficiency reduction. Consequently, the ZSCCs suppression has become a focus in the parallel inverters [1]–[15].

The ZSCCs can be eliminated by using separate dc power supplies or an isolating transformer at ac side for the parallel inverters [2]-[4], but it will make the system bulky and expensive. The medium and high frequency circulating currents were suppressed in [5], while the low-frequency circulating currents were not efficiently suppressed.

The discontinuous space vector modulation (DSVM) based interleaved PWM method was proposed to suppress the ZSCCs, but it would result in high output current ripple [6]-[7]. Nonlinear control methods were also presented to suppress ZSCCs, but the algorithms were too complicated to implement

This work is supported by the National Natural Science Foundation of China under Grant 61320106011, Grant 51377101, and Grant 61527809 and by the Program for New Century Excellent Talents in University (NCET-13-0339).

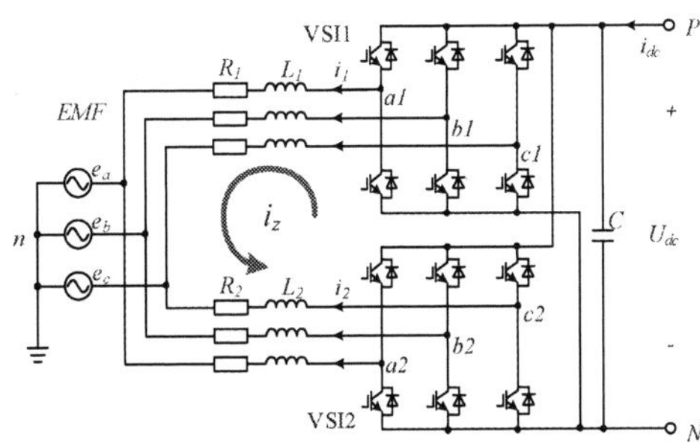

Fig. 1 Parallel VSIs

in the project [8]-[10]. The method of ZSCCs suppression in harmonic elimination PWM (HEPWM) was proposed by Chen [11], but this method relies on zero vectors and would not perform well when zero vectors were not applied.

In addition, the ZSCCs control method for SVM technique was discussed in [12]-[14]. A control variable was introduced to adjust the distribution of zero vectors in each PWM cycle with a proportional integral (PI) or a deadbeat controller. But these methods need a large amount of calculation when multiple inverters are paralleled. The ZSCCs and the common mode voltage were suppressed by the multicarrier PWM scheme which can mitigate the ZSCCs by abolishing the uses of zero vectors [15]. However, the modulation of this method is complicated.

Predictive control has been widely utilized in power converters and electrical drives [16]-[23]. Among various predictive control methods, finite control set model predictive control (FCS-MPC) is most commonly used because of the easy implementation, fast dynamic response, good current tracking accuracy and low switching frequency [16]. Based on the calculation of system dynamic model and minimization of the cost function which could be designed according to different requirements, the optimal vector is applied to the inverter to get required performances [17].

The FCS-MPC method was applied to various conditions such as rotational speed control in motor drives [17], direct power control in power converters [20] and low-voltage ride-through control in neutral-point-clamped inverters [21]-[22]. Nevertheless, there are few papers discussing the FCS-MPC based ZSCCs suppression methods. Therefore, this paper proposes a novel FCS-MPC algorithm by redesigning the cost function to retain the advantages of predictive control as well as suppress the circulating currents. And the term MPC in latter description refers to FSC-MPC.

This paper is organized as follow. Section II gives the average model of the inverter as well as the conventional MPC algorithm. In Sections III, the contributions of voltage vectors on ZSCCs are analyzed and a novel MPC algorithm is proposed to suppress the ZSCCs. In Section IV, corresponding experiments are designed to validate the proposed control strategy. Finally, conclusions are given in Section V.

II. CONVENTIONAL MPC ALGORITHM

A. Average Model of the VSI

The topology of parallel voltage source inverters is shown in Fig.1, while the block for the ZSCCs suppression and currents tracking control is shown in the latter parts. The inverters are connected with inductance, resistance, and ac back electromotive force (EMF), which represent most application fields. The two VSIs share the same dc bus and EMF, causing the ZSCC i_z. This paper selects the negative pole of dc-link as the reference point. The average model of the parallel inverters in the three-phase stationary reference frame is represented as:

$$\frac{d}{dt}\begin{bmatrix} i_{ax} \\ i_{bx} \\ i_{cx} \end{bmatrix} = -\frac{R_x}{L_x}\begin{bmatrix} i_{ax} \\ i_{bx} \\ i_{cx} \end{bmatrix} + \frac{1}{L_x}\begin{bmatrix} u_{axN} - e_a + u_{Nn} \\ u_{bxN} - e_b + u_{Nn} \\ u_{cxN} - e_c + u_{Nn} \end{bmatrix} \quad (1)$$

where $x=\{1, 2\}$; e_a, e_b and e_c are the EMF voltages; i_{ax}, i_{bx} and i_{cx} are the AC side currents; u_{Nn} is the voltage between the point N and n; L_x are the filter inductors in the ac side; R_x are the circuit resistors in the ac side; u_{axN}, u_{bxN} and u_{cxN} are the output voltages of inverters.

The aforementioned model is based on the three-phase stationary reference frame. It is complex to design the MPC because the common mode voltage u_{Nn} is difficult to get. Therefore, the model should be transformed to two-phase stationary coordinate system.

For the paralleling inverters, the zero-axis component can be independent controlled. So 2-D stationary coordinate transform is used here, the α-axis and β-axis coordinate transform matrix can be defined as:

$$C_{3s/2s} = \frac{2}{3}\begin{bmatrix} 1 & -1/2 & -1/2 \\ 0 & \sqrt{3}/2 & -\sqrt{3}/2 \end{bmatrix} \quad (2)$$

The mathematical model should also be discretized for predictive control and digital implementation, in which the current derivative form can be expressed from backward Euler discretization as:

$$\frac{di}{dt} = \frac{i(\mathrm{k}) - i(\mathrm{k}-1)}{T_s} \quad (3)$$

where T_s is the sampling interval.

Then the mathematical model in the two-phase stationary coordinate system can be derived from the abc/αβ transformation and backward Euler discretization as:

$$\begin{bmatrix} i_{\alpha x}(\mathrm{k}+1) \\ i_{\beta x}(\mathrm{k}+1) \end{bmatrix} = \frac{L_x}{R_x T_s + L_x}\begin{bmatrix} i_{\alpha x}(\mathrm{k}) \\ i_{\beta x}(\mathrm{k}) \end{bmatrix} + \frac{T_s}{R_x T_s + L_x}\begin{bmatrix} u_{\alpha x} - e_\alpha(\mathrm{k}+1) \\ u_{\beta x} - e_\beta(\mathrm{k}+1) \end{bmatrix} \quad (4)$$

The equation has been move one step forward for delay compensation using a second-order extrapolation:

$$e(\mathrm{k}+1) = 3e(\mathrm{k}) - 3e(\mathrm{k}-1) + e(\mathrm{k}-2) \quad (5)$$

In equation (4), $u_{\alpha x}$ and $u_{\beta x}$ are α-axis and β-axis components of the output voltages corresponding to the two inverters. The output voltage can be expressed by voltage vector which defined as:

$$\boldsymbol{u} = \frac{2}{3}(u_{aN} + r \times u_{bN} + r^2 \times u_{cN}) \quad (6)$$

where the operator $r=e^{2\pi j/3}$.

As for three-phase two level inverters shown in Fig.1, there are totally eight vectors and include two zero vectors.

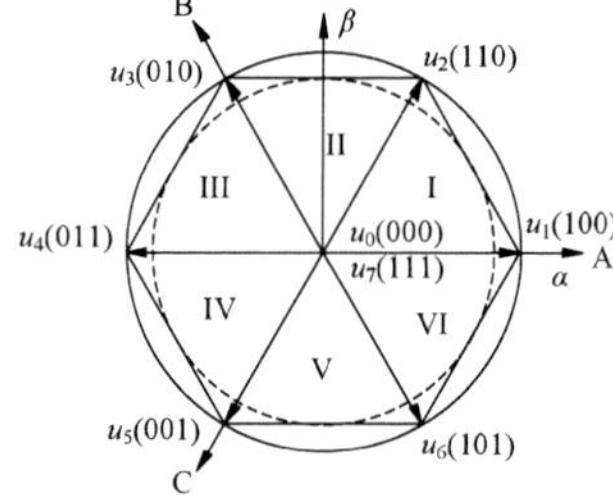

Fig. 2 Space vector diagram of a two level inverter

Then the α-axis and β-axis components $u_{\alpha x}$ and $u_{\beta x}$ of eight vectors can be derived from Fig. 2 and are listed in Table I.

TABLE I. COMPONENTS OF EIGHT VECTORS

Vector	u_0	u_1	u_2	u_3	u_4	u_5	u_6	u_7
$u_{\alpha x}/U_{dc}$	0	2/3	1/3	-1/3	-2/3	-1/3	1/3	0
$u_{\beta x}/U_{dc}$	0	0	$\sqrt{3}/3$	$\sqrt{3}/3$	0	$-\sqrt{3}/3$	$-\sqrt{3}/3$	0

B. Conventional MPC Algorithm Description

The future value of the currents $i_{\alpha x}$(k+1), $i_{\beta x}$(k+1) are predicted for the 8 switching states generated by the parallel inverters(see Fig. 4(a)), by using (4). In order to predict the above currents, it is necessary to measure the present current value $i_{\alpha x}$(k), $i_{\beta x}$(k) and grid voltages $e_{\alpha x}$(k), $e_{\beta x}$(k) through the sensors. The estimated values $e_{\alpha x}$(k+1), $e_{\beta x}$(k+1) at the instant k+1 are calculated by using (5) according to the measurements at the instant k, k−1 and k−2. Therefore, eight pairs of estimated future currents could be given corresponding to different vectors. But here are actually seven pairs of values since $\boldsymbol{u}_0$ and $\boldsymbol{u}_7$ have the same output voltages.

A cost function is evaluated for all the switching states when the predictive values are calculated in the parallel inverters. Therefore, the optimal switching state is selected to be applied to the parallel inverters at the beginning of the next

978-1-4673-9551-9/16 $31.00 © 2016 IEEE

sampling period which minimizes the cost function. The control scheme is described in Fig. 3.

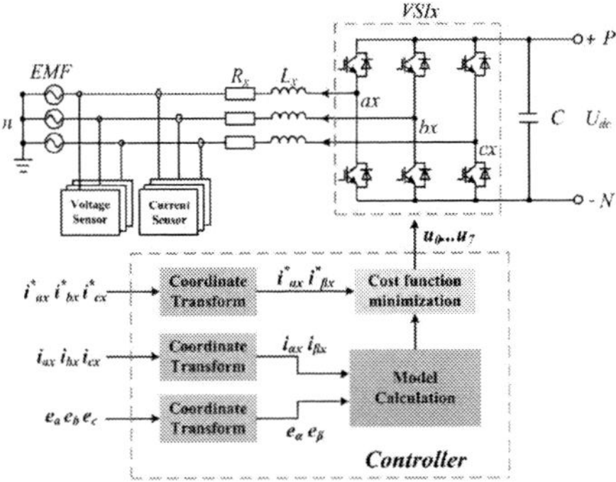

Fig. 3 Control scheme of conventional MPC algorithm

The cost function for each inverter in the conventional predictive control is expressed as:

$$g_x = \left|i^*_{\alpha x}(k+1) - i_{\alpha x}(k+1)\right| + \left|i^*_{\beta x}(k+1) - i_{\beta x}(k+1)\right| \tag{7}$$

where x={1, 2}; $i^*_{\alpha x}$(k+1), $i^*_{\beta x}$(k+1) are the α-axis and β-axis components of the reference currents at the instant k+1 and are also estimated by using (5).

The conventional predictive controllers choose the vector to minimize the cost function g_x of the variables to the reference values. Therefore, the phase currents could track the references.

III. Improved MPC Algorithm

A. Analysis of ZSCCs in Parallel Inverters

Referring to Fig. 1, the parallel inverters are linked by the same dc bus and the ac side directly. For the single inverter, no ZSCCs exist in the inverter. However, in two parallel inverters, the ZSCCs will be generated by the paths of the loop circuit. The ZSCCs of the two inverters have the same magnitude but opposite direction and i_z is defined as:

$$i_z = i_{z1} = -i_{z2} = i_{a1} + i_{b1} + i_{c1} \tag{8}$$

From [14] and considering circuit resistors, the relationship between ZSCCs and circuit parameters is derived as:

$$i_z(\text{k}+1) = \frac{L - RT_s}{L} i_z(\text{k}) + \frac{U_{dc}T_s}{L}\left(\sum_{j=a,b,c} S_{j1} - \sum_{j=a,b,c} S_{j2}\right) \tag{9}$$

where i_z(k+1), i_z(k) are the circulating currents at the instant k+1 and k; $L=L_1+L_2$; $R=R_1+R_2$; U_{dc} is the DC link voltage; S_{j1}, S_{j2} are the switching states of each bridge in parallel inverters which are expressed as:

$$S_{jx} = \begin{cases} 1 & u_{jxN} = U_{dc} \\ 0 & u_{jxN} = 0 \end{cases} \tag{10}$$

It can be seen from the mathematic model of ZSCCs in parallel three-phase PWM inverters that different vector combinations have different contributions on ZSCCs. For simplicity, the contributions of all vector combinations between the two inverters are listed in Table II.

TABLE II. The Effects Of Different Vector Combinations On Circulating Currents

		VSI2							
		G2				G1			
		u_0 (000)	u_1 (100)	u_3 (010)	u_5 (001)	u_2 (110)	u_4 (011)	u_6 (101)	u_7 (111)
VSI1		u_0 (000)	0	−1	−1	−1	−2	−2	−2	−3
	G2	u_1 (100)	+1	0	0	0	−1	−1	−1	−2
		u_3 (010)	+1	0	0	0	−1	−1	−1	−2
		u_5 (001)	+1	0	0	0	−1	−1	−1	−2
	G1	u_2 (110)	+2	+1	+1	+1	0	0	0	−1
		u_4 (011)	+2	+1	+1	+1	0	0	0	−1
		u_6 (101)	+2	+1	+1	+1	0	0	0	−1
		u_7 (111)	+3	+2	+2	+2	+1	+1	+1	0

As given in Table II, a plus sign means that the combination of two vectors has positive contribution on ZSCCs while a minus sign has the opposite meaning. The amplitude of the value in Table II represents the degree of contribution on ZSCCs. Higher amplitude means greater contribution. For further analysis, the eight vectors are divided into two groups respectively named G1and G2. Vector $\boldsymbol{u}_2$, $\boldsymbol{u}_4$, $\boldsymbol{u}_6$, $\boldsymbol{u}_7$ are included in G1 and the others are included in G2.

As shown in Table II, when vectors from G1 are applied to VSI1 and vectors from both G1 and G2 are applied to VSI2, the contributions are almost positive. When vectors from G2 are applied to VSI1, the contributions are almost negative. It can be concluded that for VSI1, the average contribution of vectors from G1 is positive while it is negative when from G2. As for VSI2, since i_{z2} is opposite to i_z, the average contribution on i_{z2} can also be summarized.

The previous analysis shows that for one inverter, when vectors from G1 are applied, the average contribution on its own ZSCC is positive while it is negative when vectors from G2 are applied. That means it is possible to control the ZSCCs by changing the weights of different vectors. Besides, the design of each inverter is the same which will reduce the difficulty of programming.

B. Improved MPC Algorithm for ZSCCs Suppression

In conventional MPC algorithm, the optimal vector is selected by minimizing the cost function g_x, but it is not easy to add the weights to different vectors. Therefore, the cost function needs to be redesigned.

In [17], the optimal control signal has been solved using the least square minimization method according to (4) and (7):

$$\begin{bmatrix} u^{opt}_{\alpha x} \\ u^{opt}_{\beta x} \end{bmatrix} = \frac{R_x T_s + L_x}{T_s} \begin{bmatrix} i^*_{\alpha x}(\text{k}+1) \\ i^*_{\beta x}(\text{k}+1) \end{bmatrix} - \frac{L_x}{T_s} \begin{bmatrix} i_{\alpha x}(\text{k}) \\ i_{\beta x}(\text{k}) \end{bmatrix} + \begin{bmatrix} e_\alpha(\text{k}+1) \\ e_\beta(\text{k}+1) \end{bmatrix} \tag{11}$$

Then the cost function could be designed as:

$$g_{ux} = \left|u^{opt}_{\alpha x} - u_{\alpha x}\right| + \left|u^{opt}_{\beta x} - u_{\beta x}\right| \tag{12}$$

In this design, the optimal vector is selected minimizing the cost function g_{ux}, thus it is nearest to the optimal control signal among the eight vectors. Then the different weighting factors could be added.

Another advantage using the redesigned cost function is that calculation burden of digital processor is reduced. The former method computes the mathematical model eight times in one sampling interval for each inverter while the latter one needs to compute just once which reduces the calculation significantly.

To suppress the ZSCCs, different weights should be added to the vectors. According to the analysis before, the improved cost function should be designed as:

$$g_{zux} = \begin{cases} \lambda_1\left(\left|u_{\alpha x}^{opt} - u_{\alpha x}\right| + \left|u_{\beta x}^{opt} - u_{\beta x}\right|\right) & u \in G1 \\ \lambda_2\left(\left|u_{\alpha x}^{opt} - u_{\alpha x}\right| + \left|u_{\beta x}^{opt} - u_{\beta x}\right|\right) & u \in G2 \end{cases} \quad (13)$$

where λ_1, λ_2 are the weighting factors of G1 vectors and G2 vectors, and $\lambda_1 + \lambda_2 = 1$.

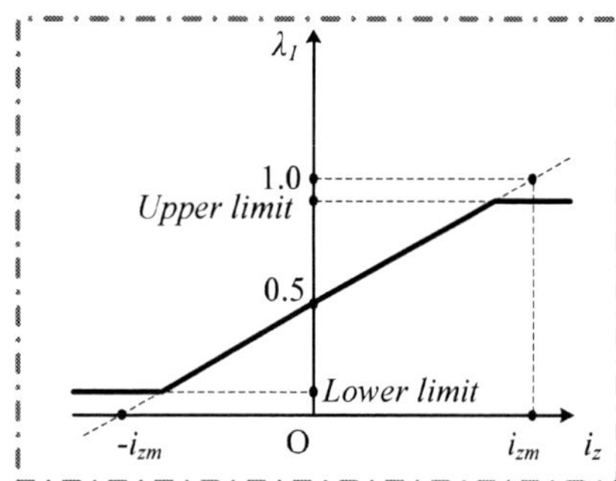

Fig. 4 Calculation of λ_1.

As the design of VSI2 is the same as VSI1, take VSI1 for example. Fig. 3 shows the calculating method of λ_1. The relationship function between λ_1 and i_z can be derived as:

$$\lambda_1 = \frac{1}{2i_{zm}} i_z + 0.5 \quad (14)$$

where i_{zm} is the maximum tolerable value of the circulating current i_z.

At the beginning of each sampling interval, the digital processor calculates the VSI model and gets the optimal control signal. Then there are eight weighted cost values corresponding to eight vectors. Finally the optimal vector which minimizes the cost function is picked and applied to the inverter. The other inverter works the same.

The ZSCCs are suppressed by adding different weights to two groups of vectors. When the circulating current i_z is near to zero, the weighting factors λ_1 and λ_2 are near to 0.5, and the vector selection is just like the conventional MPC algorithm. But when the circulating current i_z is near to i_{zm}, λ_1 is near to 1 and λ_2 is near to 0, then optimal vectors are mainly selected from G2. Accordingly, the condition could be analyzed in the same way when i_z is near to $-i_{zm}$.

But if the weighting factors λ_1 and λ_2 are 0, the optimal vectors could not be selected from eight vectors which will decrease the accuracy of current tracking. In order to avoid this, upper and lower saturation are introduced to limit the value of λ_1 and λ_2. Generally, λ_1 and λ_2 change from 0.35 to 0.65 to guarantee both current tracking and ZSCCs suppression. The controller of proposed algorithm is shown in Fig. 5.

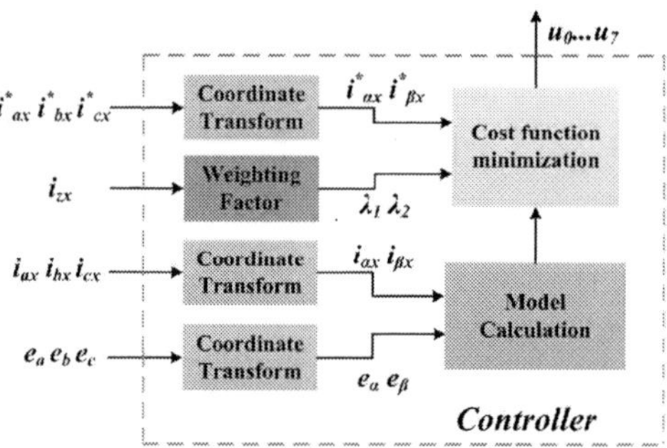

Fig. 5 Control scheme of improved MPC algorithm

IV. Experimental Results

To verify the proposed algorithm, experiments are performed on a parallel VSI platform. The platform uses dSPACE1005 as the controller which is connected by an interface board to the inverters. The dSPACE provides gate signal, enable signal and protective signals and the dead time is set to 2.5 μs. The parameters for the experiments are shown in Table III where E_m, I_{1m}, I_{2m} refer to peak values.

TABLE III. Experimental Parameters

C	3300μF	E_m	150V
L_1	4mH	U_{dc}	300V
L_2	4mH/6mH	I_{1m}	10A
R_1	0.4Ω	I_{2m}	10A/8A
R_2	0.4Ω	T_s	50μs
Upper lim	0.99	*Lower lim*	0.01

It can be seen from Table III that different conditions such as various filter inductors and reference currents have been considered to make the results more persuasive. The sampling interval T_s is set to 50μs to ensure the control accuracy.

The experimental results are shown as follow. Fig. 6 shows the experimental waveforms on conventional MPC algorithm and improved algorithm under the condition that the parameters are $L_1 = L_2 = 4$mH and $I_{1m}=I_{2m}=10$A. As we can see in the Fig. 6, the ZSCC problem is serious though the reference currents and filter inductors are the same. The little difference can generate the high ZSCCs between the parallel inverters which distort the waveforms of phase currents. It is shown that the ZSCCs become smaller and the currents have good performance when improved MPC algorithm for ZSCCs suppression is applied.

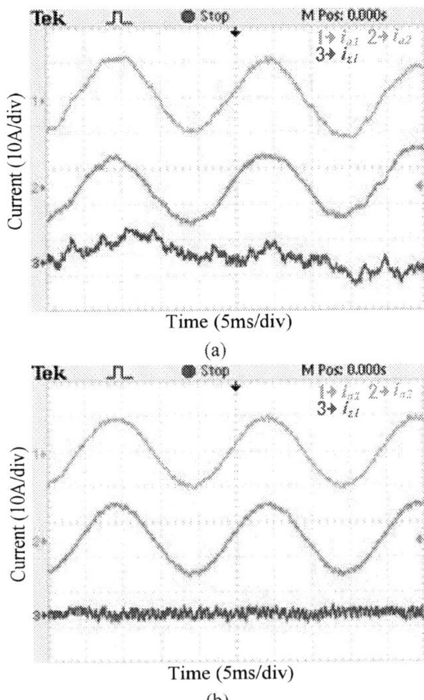

Fig. 6 Experiment results with L_1=L_2=4mH, I_{1m}=I_{2m}=10A
(a) Waveforms on conventional MPC algorithm.
(b) Waveforms on improved algorithm.

Fig. 7 shows the experimental waveforms with equal reference currents and different filter inductors. The parameters of two parallel inverters are L_1=4mH, L_2=6mH and I_{1m}=I_{2m}=10A. Fig. 7 shows the ZSCCs and current tracking before and after the proposed method is applied. It is seen that the phase currents are distorted without controlling ZSCCs. After using the improved algorithm with ZSCCs control, the qualities of phase currents are superior. Besides, the ZSCCs are suppressed effectively.

Fig. 8 shows the experimental waveforms with different reference currents and filter inductors. The parameters of two parallel inverters are L_1=4mH, L_2=6mH and I_{1m}=10A, I_{2m}=8A. It is more difficult to control the ZSCCs due to the differences between both reference currents and filter inductors. But by applying the proposed algorithm, both the current tracking accuracy and ZSCCs suppression could be guaranteed.

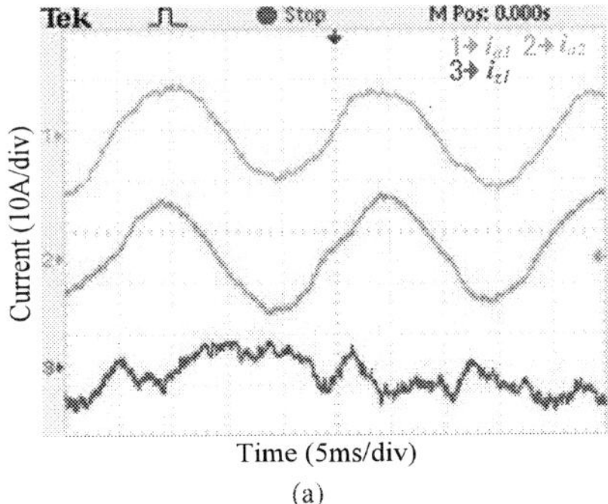

(a)

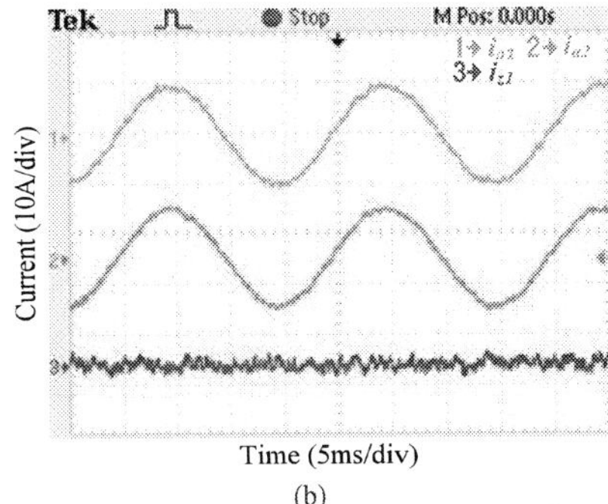

(b)

Fig. 7 Experiment results with L_1=4mH, L_2=6mH,I_{1m}=I_{2m}=10A
(a) Waveforms on conventional MPC algorithm.
(b) Waveforms on improved algorithm.

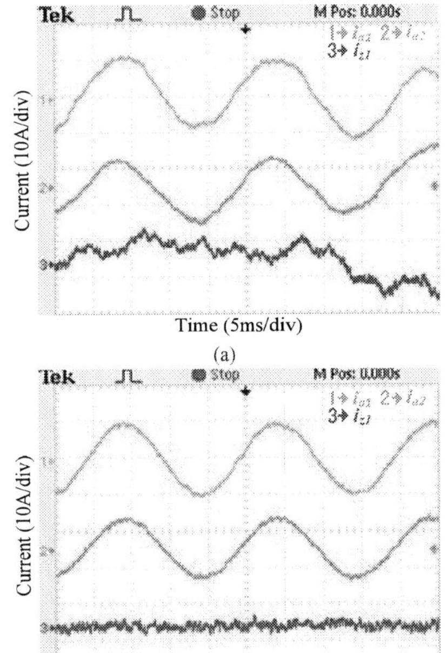

Fig. 8 Experiment results with L_1=4mH,L_2=6mH,I_{1m}=10A I_{2m}=8A
(a) Waveforms on conventional MPC algorithm.
(b) Waveforms on improved algorithm.

V. Conclusion

This paper proposes a novel predictive MPC algorithm for ZSCCs suppression in the parallel VSIs. The eight vectors are divided into two groups based on the average contribution on ZSCCs. Half of the vectors have negative effect to ZSCCs while the others have positive effect. Then a weighted cost function algorithm is designed which is able to control ZSCCs. Moreover, the new method requires only one time calculation of system model rather than eight times in the conventional method. The experiments are conducted in various conditions of same parameters, different current references and different filter inductors. The experimental results prove that the improved algorithm is effective and performs well in both current tracking and ZSCCs suppression. The quality of current waveforms is superior after implementing the improved algorithm for all the three cases.

REFERENCES

[1] Jiang, Yan, Shan Xiong, Shou Dao Huang, Ke Yuan Huang, Lei Xiao, "Control of circulating current in parallel three-phase inverter in MW wind power system," Electrical Machines and Systems (ICEMS), 2010 International Conference on. IEEE, 2010.

[2] Y. Zhang, Y. Kang and J. Chen, "The zero-sequence circulating currents between parallel three-phase inverters with three-pole transformers and reactors," in Proc. IEEE Conf. Appl. Power Electron. Conf., 2006,pp. 1709–1715.

[3] J.-K. Ji and S.-K. Sul, "Operation analysis and new current control of parallel connected dual converter system without interphase reactors," in Proc. 25th Annu. Conf. Ind. Electron. Soc., San Jose, America, 1999, pp. 235–240.

[4] R. Inzunza, T. Sumiya, Y. Fujii and E. Ikawa, "Parallel connection of grid-connected LCL inverters for MW-scaled photovoltaic systems," in Proc.Int. Conf. Power Electron, Sapporo, Japan, Jun 2010, pp. 1988–1993.

[5] Y. Sato and T. Kataoka, "Simplified control strategy to improve AC-input-current waveform of parallel-connected current-type PWM rectifiers," in Proc. Inst. Electr. Eng. Electr. Power Appl., Jul 1995, vol. 142, no. 4,pp. 246–254.

[6] Z. Ye, K. Xing, S. Mazumder, D. Borojevic and F. Lee, "Modeling and control of parallel three-phase PWM boost rectifiers in PEBB-based DC distributed power systems," in Proc. Appl. Power Electron. Conf. Expo.,Feb 1998, vol. 2, pp. 1126–1132.

[7] D. Zhang, F. Wang, R. Burgos and D. Boroyevich, "Common-mode circulating current control of paralleled interleaved three-phase two-level voltage-source converters with discontinuous space-vector modulation," IEEE Trans. Power Electron., vol. 26, no. 12, pp. 3925–3935, Dec 2011.

[8] S. K. Mazumder, "A novel discrete control strategy for independent stabilization of parallel three-phase boost converters by combining space-vector modulation with variable-structure control," IEEE Trans. Power Electron. , vol. 18, no. 4, pp. 1070–1083, Jul 2003.

[9] L. Zhang, B. Gu, J. Dominic, B. Chen, C. Zheng, and J. -S. Lai, "A dead-time compensation method for parabolic current control with improved current tracking and enhanced stability range," IEEE Trans. Power Electron., vol. 30, no. 7, pp. 3892-3902, Jul. 2015

[10] S. K. Mazumder, K. Acharya and M. Tahir, "Joint optimization of control performance and network resource utilization in homogeneous power networks," IEEE Trans. Ind. Electron., vol. 56, no. 5, pp. 1736–1745,May 2009.

[11] T. P. Chen, "Zero-sequence circulating current reduction method for parallel HEPWM inverters between AC bus and DC bus," IEEE Trans. Ind. Electron., vol. 59, no. 1, pp. 290–300, Jan. 2012.

[12] Z. Ye, D. Boroyevich, J.Y. Choi and F. C. Lee, "Control of circulating current in two parallel three-phase boost rectifiers," IEEE Trans. Power Electron., vol. 17, no. 5, pp. 609–615, Sep 2002.

[13] X. Zhang, J. Chen, Y. Ma, Y. Wang and D. Xu, "Bandwidth expansion method for circulating current control in parallel three-phase PWM converter connection system," IEEE Trans. Power Electron.,vol. 29, no. 12, Dec 2014.

[14] X. Zhang, W. Zhang, J. Chen and D. Xu, "Deadbeat control strategy of circulating currents in parallel connection system of three-phase PWM converter," IEEE Trans. Energy Conversion, vol. 29, no. 2, Jun 2014.

[15] C. C. Hou, "A multicarrier PWM for parallel Three-Phase active front-end converters," *IEEE Trans. Power Electron.*, vol. 28, no. 6, Jun 2013.

[16] Rodriguez, Jose, and Patricio Cortes. "Predictive control of power converters and electrical drives." Vol. 40. John Wiley & Sons, 2012.

[17] Wang, L., Chai, S., Yoo, D., Gan, L., "PID and Predictive Control of Electrical Drives and Power Converters Using MATLAB/Simulink." John Wiley & Sons, 2014.

[18] S. S. Davari, D. Khaburi, and R. Kennel, "An improved FCS–MPC algorithm for an induction motor with an imposed optimized weighting factor," IEEE Trans. Power Electron., vol. 27, no. 3, pp. 1540–1551, Mar.2012.

[19] S. Kouro, P. Cortes, R. Vargas, U. Ammann and J. Rodriguez, "Model predictive control—A simple and powerful method to control power converters," IEEE Trans. Ind. Electron., vol. 56, no. 6, pp. 1826–1838, Jun 2009.

[20] J. R. Fischer, S. A. Gonzalez, I. Carugati, M. A. Herran, M. G. Judewicz and D. O. Carrica, "Robust predictive control of grid-tied converters based on direct power control," IEEE Trans. Power. Electron., vol. 29, no. 10, pp. 5634-5643, Oct 2014.

[21] A. Calle-Prado, S. Alepuz, J. Bordonau, J. Nicolas-Apruzzese, P. Cortés and J. Rodriguez, "Model predictive current control of grid-connected neutral-point-clamped converters to meet low-voltage ride-through requirements," IEEE Trans. Ind. Electron., vol. 62, no. 3, pp. 1503-1514, Mar 2015.

[22] V. Yaramas, B. Wu, S. Alepuz and S. Kouro, "Predictive control for low-voltage ride-through enhancement of Three-Level-Boost and NPC-Converter-Based PMSG wind turbine," IEEE Trans. Ind. Electron., vol. 61, no. 12, pp. 6832-6843, Dec 2014.

[23] L. Zhang, R. Born, B. Gu, B. Chen, C. Zheng, X. Zhao, and J.-S. Lai, "A Sensorless Implementation of the Parabolic Current Control for Single Phase Standalone Inverters", IEEE Trans. Power Electron., accepted.

978-1-4673-9551-9/16 $31.00 © 2016 IEEE

Design of Dynamic Voltage Restorer and Active Power Filter for Wind Power Systems Subject to Unbalanced and Harmonic Distorted Grid

Woei-Luen Chen and Meng-Jie Wang
Department of Electrical Engineering
School of Electrical and Computer Engineering, College of Engineering, Chang Gung University
Tao-Yuan, Taiwan, R.O.C.

Abstract—**Nowadays unbalanced voltage and harmonic distortion are in the majority of the power quality problems in wind energy conversion systems (WECS). To alleviate the impact registered by unbalanced and harmonic distortion, a new circuit topology comprising a dynamic voltage restorer (DVR) and an active power filter (APF) is presented. A frequency shifting technique based on coordinate transformation is employed to unify the positive and negative sequence harmonics into a resonant current controller. To improve the accuracy for harmonic detection, a second-order generalized integrator (SOGI), characterized with large bandwidth at specific frequency, is capable of separating harmonics from feeder current. The fundamental and harmonic current controllers can be individually realized by the resonant current controller and combined to form voltage command for a voltage-sourced inverter (VSI) based on superposition theorem. Not only the computing time but also the harmonic currents in the feeder can be effectively reduced along with the proposed approach. To share the dc-circuit with the DVR, the APF and DVR are in back-to-back connection in favor of unbalanced and harmonic compensation for specific grid bus. A 3kW squirrel cage induction generator (SCIG) driven by a permanent synchronous motor (PMSM) is performed for wind-turbine generator emulation. The experimental results validate that the grid-tied SCIG power quality can be refined when grid bus is subjected to unbalanced and harmonic distortion.**

Keywords—active power filter; dynamic voltage restorer; squirrel cage induction generator

I. Introduction

Wind power penetration in the load center vicinity is beneficial in reducing power transmission loss and alleviating the amount of electrical power fed by fossil fuel based power plants. However, the negative impacts resulted from wind power generation including vibration and irritating acoustic noise under a distorted distribution system are still in the stage of exploration. The switch–mode voltage-sourced inverter (VSI) is capable of changing the voltage format for adapting the power sources of various types to electrical appliance, which is indispensable to many industrial applications such as electric vehicle, UPS, and renewable energy conversion. By modulating the magnitude and the phase of the VSI ac voltage, the desired VSI ac current is achievable and consequently enables bidirectional power flow control. In additional to the power flow management, the VSI can be also used to improve the power quality for the grid utility. The dynamic voltage restorer (DVR) [1-5] and the active power filter (APF) [6-12] are the typical applications for voltage magnitude and current harmonic compensations, respectively.

Two sinusoidal internal models in Laplace transform of a cosine function and a sine function were proposed as the resonant controller [13-14]. It is found that using the sine function would highly underdamp the feedback control system due to 90° short in phase margin compared with the cosine function. To account for the transient and steady-state tracking performance, a pole assignment technique [15] applied to an undamped PR controller was proposed to improve the closed-loop damping. The resonant frequency of the PR controller can also be assigned at a specific frequency for compensating harmonic current [16-17] caused by the nonlinear loads.

This paper presents a combination of the DVR and APF for refining the SCIG terminal power quality. The positive sequence component at grid frequency is first extracted from the distorted SCIG terminal voltage for VSI synchronization. The unbalanced components can also be found by subtracting the positive sequence component from the distorted voltage. Once the positive and negative sequence components are determined, the defect of unbalance in voltage as well as the positive sequence component can both be compensated by the DVR. To avert the SCIG stator from specific harmonic currents distortion, an APF control design that focuses on the low-order harmonic currents was presented. The SCIG torque pulsation incurred by the distortions such as the harmonic current and unbalanced voltage can then be alleviated with the compensation of the DVR and the APF.

II. Controller Design

A. APF Controller Design

Fig. 1 shows an ac circuit of a VSI. The voltage across the coupled inductor can be altered by modulating the inverter output voltage (e_f). Therefore, the current fed from the inverter is controllable with appropriate inverter output voltage control. The relation among voltages and currents in Fig. 1 can be represented in the stationary reference frame ($\alpha\beta$-axis) in time domain

This work was supported in part by the Ministry of Science and Technology under Contracts MOST 104-3113-E-002-007 and MOST 104-2221-E-182-036 and in part by Green Technology Research Center of Chang Gung University.

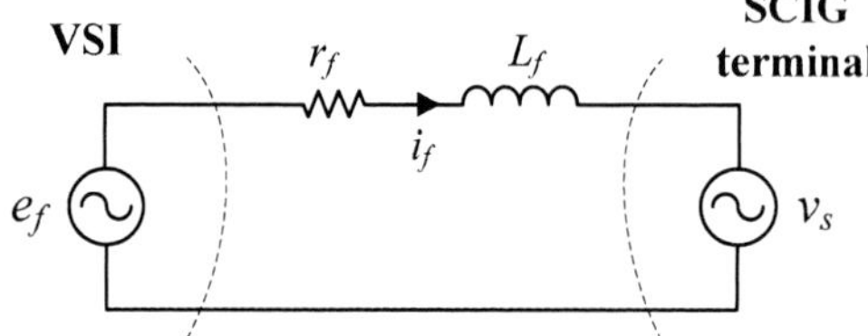

Fig. 1. One-line diagram of a two-bus circuit.

$$\mathbf{e}_f^{\alpha\beta} - \mathbf{v}_s^{\alpha\beta} = r_f \mathbf{i}_f^{\alpha\beta} + L_f \frac{d\mathbf{i}_f^{\alpha\beta}}{dt} \tag{1}$$

or in s-domain

$$\mathbf{e}_f^{\alpha\beta} - \mathbf{v}_s^{\alpha\beta} = r_f \mathbf{i}_f^{\alpha\beta} + sL_f \mathbf{i}_f^{\alpha\beta} \tag{2}$$

By using the frequency-shifting theorem, (2) can be transformed to the synchronous reference frame (*dq*-axis)

$$\mathbf{e}_f^{dq} - \mathbf{v}_s^{dq} = r_f \mathbf{i}_f^{dq} + (s + j\omega_0) L_f \mathbf{i}_f^{dq} \tag{3}$$

To enable the VSI absorbing the specific current harmonics presented at SCIG terminal, a controller simplification [8] that transforms the current harmonic pair with rotating speed of $(6n\pm1)\omega_0$ to the *dq*-axis and causing the current vector with respect to the *dq*-axis with the same rotating speed $(6n\omega_0)$. (As shown in Fig. 2, both the relative rotating speeds of the fifth- and seventh-order harmonics corresponding to the *dq*-axis are equal to $6\omega_0$). To achieve sinusoidal current tracking for the model expressed in the *dq*-axis, a harmonic compensator in the fashion of the PI controller can be modified in the form as

$$G_c = K_p + \frac{(K_i + \Delta)}{s} = \frac{sK_p + (K_i + \Delta)}{s} \tag{4}$$

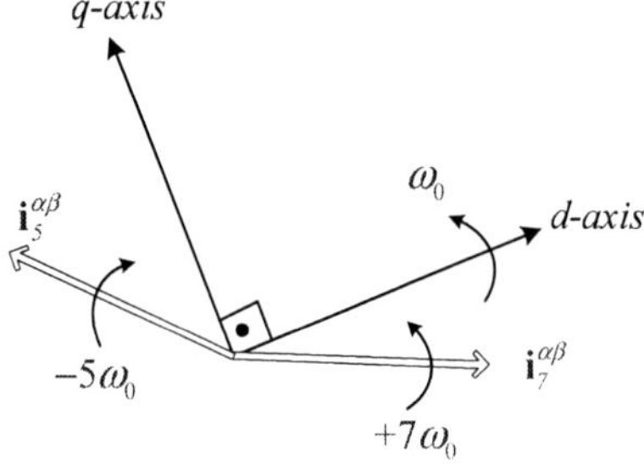

Fig. 2. Harmonic vectors in dq-axis.

Applying the frequency-shifting theorem, (4) can be tuned capable of compensating the harmonics vectors with the rotating speeds of $+6n\omega_0$ and □$6n\omega_0$, respectively, as

$$G_c^+ = \frac{(s - j6n\omega_0)K_p + (K_i + \Delta)}{(s - j6n\omega_0)} = \frac{K_p s + K_i}{s - j6n\omega_0} \tag{5}$$

and

$$G_c^- = \frac{K_p s + K_i}{s + j6\omega_0} \tag{6}$$

where $\Delta=\pm j6n\omega_0$ for the purpose of simplification. Summation of (5) and (6) gives a harmonic compensator for enabling the VSI to absorb one current harmonic pair $(6n\pm1)\omega_0$ at the SCIG terminal:

$$G_c = G_c^+ + G_c^- = 2\frac{K_p s^2 + K_i s}{s^2 + (6n\omega_0)^2} \tag{7}$$

B. Vector Representation of Symmetrical Components [5]

Eq. (8) shows the vector representation for each phase voltage of a three-phase system.

$$\begin{aligned} \mathbf{v}_a(t) &= v_{\alpha a}(t) + jv_{\beta a}(t) \\ \mathbf{v}_b(t) &= v_{\alpha b}(t) + jv_{\beta b}(t) \\ \mathbf{v}_c(t) &= v_{\alpha c}(t) + jv_{\beta c}(t) \end{aligned} \tag{8}$$

where the real- and imaginary- parts (or components in the $\alpha\beta$-axis) can be extracted by the a second order generalized integrator (SOGI) that is used to convert the three-phase voltages (v_a, v_b, v_c) to three voltage vectors ($v_a^{\alpha\beta}$, $v_b^{\alpha\beta}$, $v_c^{\alpha\beta}$) where the components in αβ-axis are the same magnitude and orthogonal to each other. The SOGI also behaves similar to a band-pass filter which have large gain at grid frequency and is beneficial to reject the harmonic distortion and typically circuit disturbance.

Because the vector representation in (8) involves the information of angular frequency, the associated symmetrical components are time dependent. Expansion of (8) gives the zero sequence vector, positive sequence vector, and negative sequence vector as

$$\begin{aligned} \mathbf{v}_a^0(t) &= v_{\alpha a}^0 + jv_{\beta a}^0 \\ &= \frac{1}{3}(v_{\alpha a} + v_{\alpha b} + v_{\alpha c}) \\ &\quad + j\frac{1}{3}(v_{\beta a} + v_{\beta b} + v_{\beta c}) \end{aligned} \tag{9}$$

$$\begin{aligned} \mathbf{v}_a^+(t) &= v_{\alpha a}^+ + jv_{\beta a}^+ \\ &= \frac{1}{3}\left(v_{\alpha a} - \tfrac{1}{2}v_{\alpha b} - \tfrac{\sqrt{3}}{2}v_{\beta b} - \tfrac{1}{2}v_{\alpha c} + \tfrac{\sqrt{3}}{2}v_{\beta c}\right) \\ &\quad + j\frac{1}{3}\left(v_{\beta a} + \tfrac{\sqrt{3}}{2}v_{\alpha b} - \tfrac{1}{2}v_{\beta b} - \tfrac{\sqrt{3}}{2}v_{\alpha c} - \tfrac{1}{2}v_{\beta c}\right) \end{aligned} \tag{10}$$

$$\begin{aligned} \mathbf{v}_a^-(t) &= v_{\alpha a}^- + jv_{\beta a}^- \\ &= \frac{1}{3}\left(v_{\alpha a} - \tfrac{1}{2}v_{\alpha b} + \tfrac{\sqrt{3}}{2}v_{\beta b} - \tfrac{1}{2}v_{\alpha c} - \tfrac{\sqrt{3}}{2}v_{\beta c}\right) \\ &\quad + j\frac{1}{3}\left(v_{\beta a} - \tfrac{\sqrt{3}}{2}v_{\alpha b} - \tfrac{1}{2}v_{\beta b} + \tfrac{\sqrt{3}}{2}v_{\alpha c} - \tfrac{1}{2}v_{\beta c}\right) \end{aligned} \tag{11}$$

The positive sequence component derived from (10) is the key technique to resolve the dynamic symmetrical components from the instantaneous voltage vector. With the voltage vectors in hand, the phase of the positive sequence voltage can be estimated by a software phase locked loop (SPLL) and regarded as the phase command for the DVR.

III. Experimental Setup

The experimental configuration and setup for an APF and a DVR coupled SCIG system are shown in Fig. 3. The SCIG is driven by a torque-controlled servomotor which emulates a wind turbine. The torque command of the servomotor drive is calculated based on the rotor speed and the mechanical power which represents the wind power at specific wind speed. The DVR has a common dc-link with the APF and is capable of providing three independent voltages to compensate the unbalanced voltages. As shown in Fig. 3, the current harmonic orders: 5th, 7th, 11th, and 13th are the specific harmonics scheduled to be removed from the SCIG stator circuit. In addition to absorb the specific current harmonics at the SCIG terminals, the APF is designated to regulate the dc-link voltage for the DVR and APF itself in favor of stable sinusoidal PWM modulation. The positive sequence voltage phase angle (θ_{v+}) is estimated by a dynamic symmetrical components analysis technique [5]. A PR voltage controller with double feedforward loops is introduced to restore the SCIG stator voltage in the fast way under diverse operating conditions.

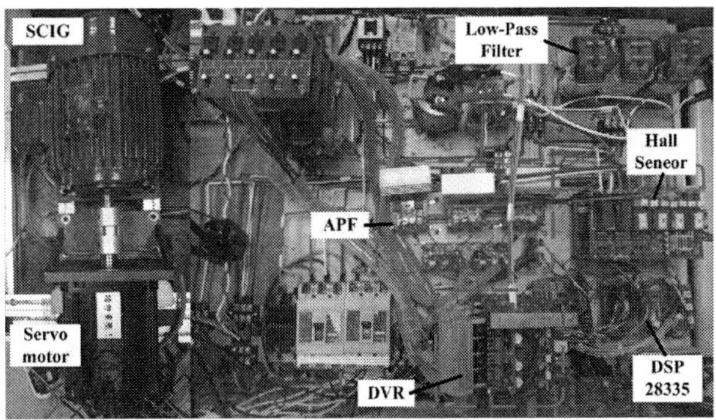

Fig. 4. Test rig for SCIG power quality compensation.

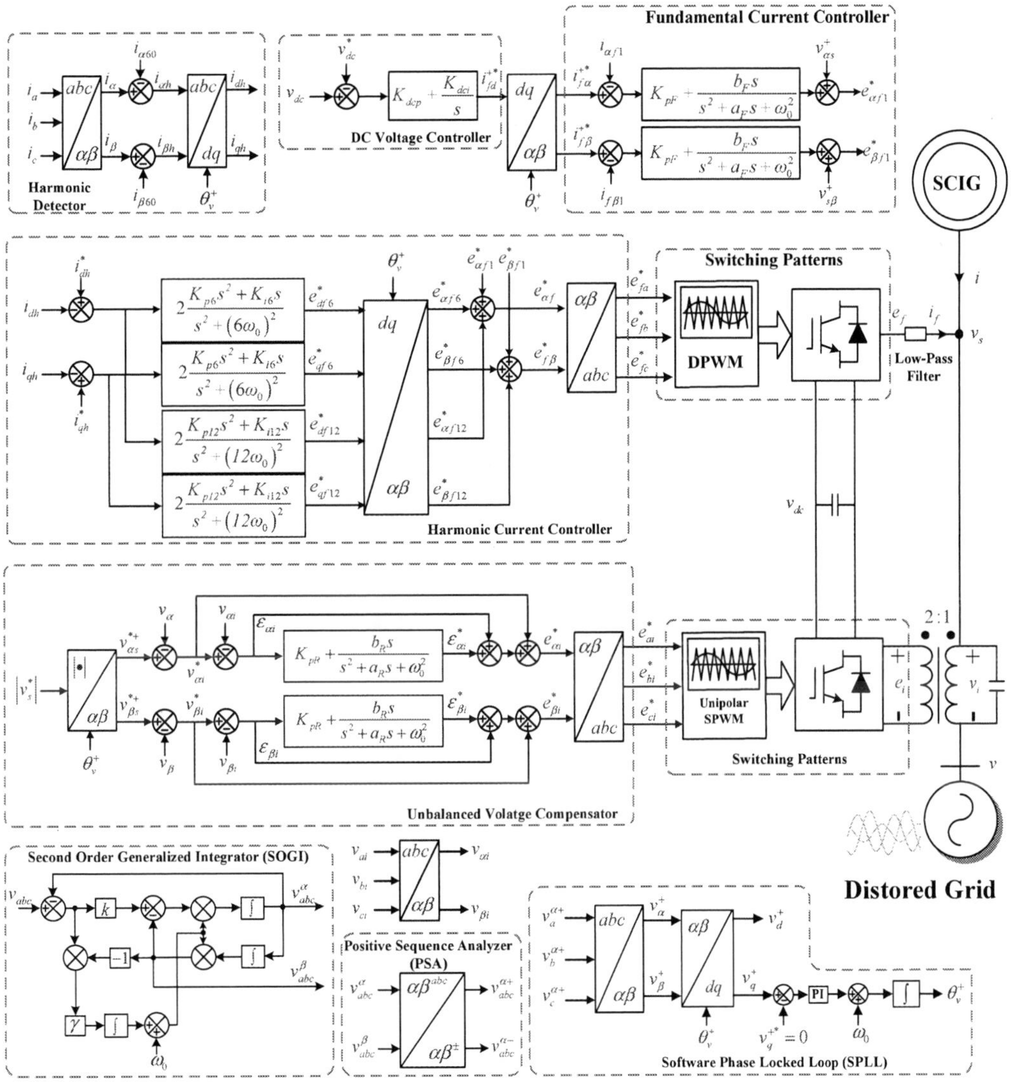

Fig. 3. Block diagram of the DVR and APF controllers for a grid-tied SCIG system.

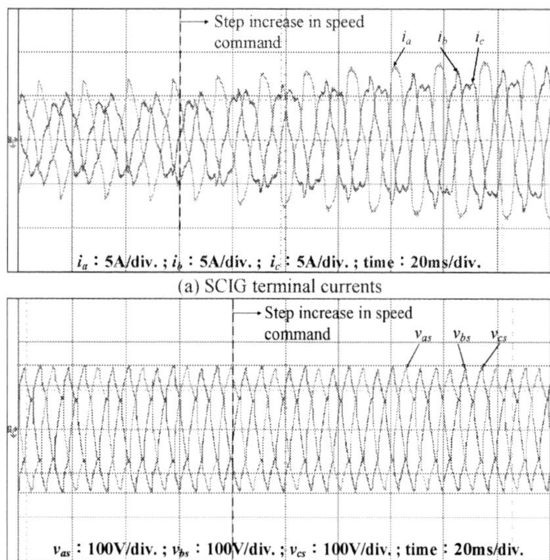

(a) SCIG terminal currents

(b) SCIG terminal voltages

Fig. 5. System response of the uncompensated SCIG.

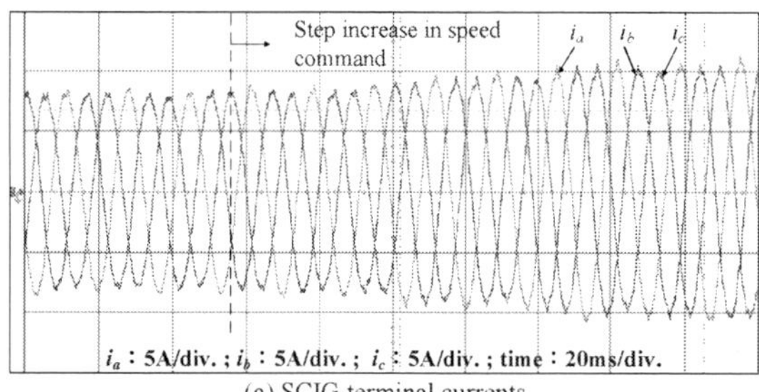

(a) SCIG terminal currents

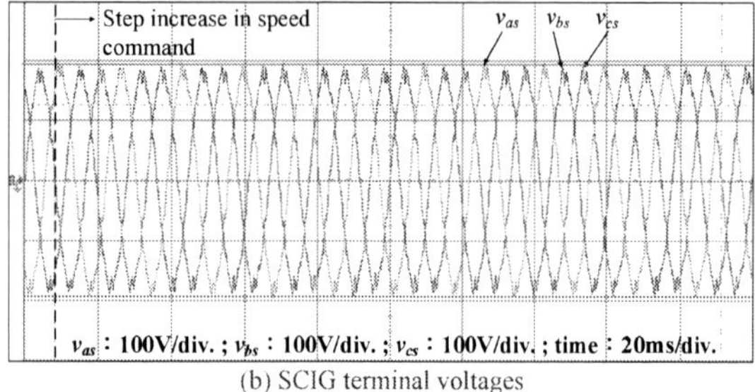

(b) SCIG terminal voltages

Fig. 6. System response of the compensated SCIG.

IV. EXPERIMENTAL RESULTS

To validate that the DVR can reduce the percent voltage and current unbalance for the SCIG under diverse operating conditions, a ramp change in SCIG rotor speed was performed to emulate the wind turbine subjected to random wind speed. Fig. 5 shows the SCIG terminal current and voltage responses without using the DVR and the APF. It is clear that the SCIG current would experience severe current unbalance even a slight voltage unbalance presents in the SCIG stator terminal. Fig. 6 shows that the current and voltage qualities of the SCIG with DVR and APF compensation are superior to those of the SCIG alone. The unbalanced components and harmonics in current and voltage could be greatly improved even in the case of the variable speed condition. In addition, the SCIG terminal voltage could also be promoted to its nominal value and would not decline due to the line impedance. Fig. 6 shows that higher stator voltage level and better power quality at SCIG terminal are beneficial to penetrate higher power to the grid and prolong the SCIG durability.

V. CONCLUSION

The SCIG system tapped to the distorted power grid would suffer from the torque pulsation and which would damage the rotor bearings and shorten the SCIG durability. The difficulty to cope with the power quality problem lies in how to identify the distortion component from the distorted voltage or current signal. The unwanted distortions can be divided into two categories that is the voltage unbalance and the current harmonic. Because the definition of the voltage unbalance is independent to the harmonic distortion, the fundamental component should to be resolved from original voltage signal with the first priority. In this paper, the three-phase fundamental voltages are first resolved and vectorized by the SOGI and the positive sequence component is then extracted by (10). The fundamental positive sequence voltage provides with not only the grid voltage phase for the VSI to parallel with the power grid but also the degree of the voltage unbalance for the DVR to fast compensate the unbalanced component. The application of the frequency-shifting theorem and the coordinate transformation to the APF controller can simplify the harmonic compensator design and reduce the computing time. The experimental results show that the combination of the DVR and the APF can effectively improve the SCIG terminal power quality and prolong the SCIG durability. The vibration measurement is the next work for verifying the necessity of the DVR and the APF for the wind-driven SCIG.

REFERENCES

[1] D. N. Zmood, D. G. Holmes, and G. H. Bode, "Frequency-domain analysis of three-phase linear current regulators," *IEEE Trans. on Industry Applications*, vol. 37, no. 2, pp. 601-610, 2001.

[2] D. N. Zmood and D. G. Holmes, "Stationary frame current regulation of PWM inverters with zero steady-state error," *IEEE Trans. on Power Electronics*, vol. 18, no. 3, pp. 814-822, 2003.

[3] Y. Sato, T. Ishizuka, K. Nezu, and T. Kataoka, "A new control strategy for voltage-type PWM rectifiers to realize zero steady-state control error in input current," *IEEE Trans. on Industry Applications*, vol. 34, no. 3, pp. 480-486, 1998.

[4] C. Lascu, L. Asiminoaei, I. Boldea, and F. Blaabjerg, "Frequency response analysis of current controllers for selective harmonic compensation in active power filters," *IEEE Trans. on Industrial Electronics*, vol. 56, no. 2, pp. 337-347, 2009.

[5] W.-L. Chen, B.-S. Lin and M.-J. Wang, "Dynamic symmetrical components analysis and compensation for unbalanced distribution systems," *The 9th International Conference on Power Electronics-ECCE Asia*, pp. 2221 - 2226, 2015.

[6] C. Meyer, R. W. De Doncker, Y. W. Li, and F. Blaabjerg, "Optimized control strategy for a medium-voltage DVR-theoretical investigations and experimental results," *IEEE Trans. on Power Electronics*, vol. 23, no. 6, pp. 2746-2754, 2008.

[7] S. R. Naidu and D. A. Fernandes, "Dynamic voltage restorer based on a four-leg voltage source converter," *IET Generation Transmission Distribution*, vol. 3, no. 5, pp. 437-447, 2009.

[8] C. Lascu, L. Asiminoaei, I. Boldea, and F. Blaabjerg, "High performance current controller for selective harmonic compensation in active power filters," *IEEE Trans. on Power Electronics*, vol. 22, no. 5, pp. 1826-1835, 2007.

[9] H. K. Al-Hadidi, A. M. Gole, and D. A. Jacobson, "Minimum power operation of cascade inverter-based dynamic voltage restorer," *IEEE Trans. on Power Delivery*, vol. 23, no. 2, pp. 889-898, 2008.

[10] Y. W. Li, P. C. Loh, F. Blaabjerg, and D. M. Vilathgamuwa, "Investigation and improvement of transient response of DVR at medium voltage level," *IEEE Trans. on Industry Application*, vol. 43, no. 5, pp. 1309-1319, 2007.

[11] Y. W. Li, D. Mahinda Vilathgamuwa, F. Blaabjerg, and P. C. Loh, "A robust control scheme for medium-voltage-level DVR implementation," *IEEE Trans. on Industrial Electronics*, vol. 54, no. 4, pp. 2249-2261, 2007.

[12] P. Cheng and H. Nian, "An improved control strategy for DFIG system and dynamic voltage restorer under grid voltage dip," *in Proc. IEEE International Symposium on Industrial Electronics*, pp. 1868-1873, 2012.

[13] S. Fukuda and T. Yoda, "A novel current-tracking method for active filters based on a sinusoidal internal model," *IEEE Trans. on Industry Applications*, vol. 37, no. 3, pp. 888-895, 2001.

[14] S. Fukuda and R. Imamura, "Application of a sinusoidal internal model to current control of three-phase utility-interface converters," *IEEE Trans. on Industrial Electronics*, vol. 52, no. 2, pp. 420-426, 2005.

[15] W. L. Chen and J. S. Lin, "One-dimensional optimization for proportional-resonant controllers design against the change in source impedance and solar irradiation of PV systems," *IEEE Trans. on Industrial Electronics*, vol. 61, no. 4, pp. 1845-1854, 2014.

[16] R. Teodorescu, F. Blaabjerg, M. Liserre, and P.C. Loh, "Proportional-resonant controllers and filters for grid-connected voltage-source converters," *IEE Proc. Electric Power Applications*, vol. 153, no. 5, pp. 750-762, 2006.

[17] L. Herman, I. Papic, and B. Blazic, "A proportional-resonant current controller for selective harmonic compensation in a hybrid active power filter," *IEEE Trans. on Power Delivery*, vol. 29, no. 5, pp. 2055-2065, 2014.

Dynamic variable coupling analysis and modeling of proton exchange membrane fuel cells for water and thermal management

Daming Zhou, Elena Breaz, Alexandre Ravey, Fei Gao, Abdellatif Miraoui
Institut de recherche sur les transports l'énergie et la société (IRTES) - EA 7274
Université de Technologie de Belfort-Montbéliard
Belfort 90000, France
(daming.zhou, elena.breaz, alexandre.ravey, fei.gao, abdellatif.miraoui)@utbm.fr

Ke Zhang
School of Astronautics
Northwestern Polytechnical University
Xi' an 710072, PR China
zhangke@nwpu.edu.cn

***Abstract*—Dynamic variable coupling analysis is an important tool to properly design a control structure for complex multivariable and multi-physique systems, such as fuel cells. Fuel cell is an electrochemical energy conversion device which contains different inter-coupled dynamic phenomena in electrochemical, fluidic and thermal domains. Among those dynamic phenomena, the management of water and thermal dynamics play an important role to achieve the optimized fuel cell performance. In this paper, a control variable coupling analysis of fuel cell dynamic behaviors are presented and discussed based on a dynamic proton exchange membrane (PEM) fuel cell model, which considers in particular the transient behaviors of membrane water content and cell temperature. Through the analysis based on the relative gain array (RGA) method, the effects of nonlinear coupling among different fuel cell control variables are shown. The analysis results can be used to optimize the controller design for fuel cell system.**

***Keywords*—dynamic variable; coupling analysis; management of water and thermal; transient behaviors; membrane water content**

I. INTRODUCTION

Fuel cell is considered as a potential candidate for clean energy production in the near future [1]. In the last decades, the research in the field of fuel cells is increasing in continuous. Before its commercial deployment, one of the major problems in fuel cell research is the development of optimal control methods for fuel cell system. The control of the fuel cell system (air compressor, cooling circuit, power converter, etc.) is a complicate work because it incorporates different control variables in different physical domains [2]. During fuel cell operation, different dynamic phenomena with different time constant ranges, such as voltage transient due to double layer capacitance, gas pressure variation due to the volume of gas manifold, water content variation due to the water absorption in membrane and temperature variation due to the cell thermal capacity, can be clearly observed during load transient [3]-[5]. These dynamic phenomena in different physical domains are indeed inter-coupled between each other and the variation of one would influence another. This inter-coupling effect is especially important between the dynamic phenomena which have similar transient time constants. Thus, the variable coupling analysis should be performed for fuel cell system in order to develop an optimized control algorithm (decoupling control for example).

In order to perform such analysis, a dynamic, 1D, multi-physique model of a proton exchange membrane fuel cell is developed at first. The presented model is based on a previously developed multi-physical PEMFC model [6], with a major modeling improvement by considering in addition the dynamic behavior of water content in the membrane (rarely addressed in the literature) [7] [8]. The explicit expression of time constant of membrane water content variation is then deduced in order to analyze the interactions between dynamics of membrane water content and temperature. Based on the results, the dynamic coupling analyses for different fuel cell operational parameters which have the similar dynamic time constants is further discussed using Relative Gain Array [9] (RGA) method to properly design a control structure for optimal water and thermal management for fuel cell systems.

II. IMPROVED DYNAMIC PEM FUEL CELL MODEL

A. Multi-Physique PEM Fuel Cell Model in Electrochemical, Fluidic and Thermal Domains

The previously developed model (without dynamic of membrane water content) can be found in [6].

1) Electrical Domain Modeling

The cell output voltage is given from the thermodynamic voltage E_{thermo} (V) and the different over-potentials during fuel cell operation: the cell activation loss η_{act} (V) in the catalyst layers and the ohmic losses η_{ohmic} (V) in the membrane layer:

$$V_{cell} = E_{thermo} - \eta_{act} - \eta_{ohmic} \quad (1)$$

The thermodynamic voltage E_{thermo} is derived from the Nernst equation:

$$E_{thermo} = 1.229 - 0.85 \cdot 10^{-3}(T_{cata} - 298.15) + \frac{R \cdot T_{cata}}{2F} \ln\left(\sqrt{P_{O_2,cata}} \cdot P_{H_2,cata}\right) \quad (2)$$

where T_{cata} is the temperature (K) of the catalyst layer, $R = 8.314$ is the perfect gas constant, $F = 96485$ is the Faraday constant. $P_{O_2,cata}$ is the oxygen pressure (Pa) at the interface of the cathode catalyst layer, $P_{H_2,cata}$ is the hydrogen pressure (Pa) at the interface of the anode catalyst layer.

The cell electrochemical activation loss η_{act} can be modeled using implicit Butler-Volmer equation [6]:

$$i = i_0 A_{cata}\left[e^{\left(\frac{\alpha_c nF}{RT_{cata}}\eta_{act}\right)} - e^{\left(-\frac{(1-\alpha_c)nF}{RT_{cata}}\eta_{act}\right)}\right] \quad (3)$$

where i is the stack current (A), i_0 is the exchange current density (A/m^2), A_{cata} is the section surface of catalyst layer (m^2), α_c is the charge transfer coefficient, and n is the number of electrons involved in the electrochemical reaction.

The resistance of the membrane R_{mem} (Ω) is calculated by:

$$R_{mem} = \frac{1}{\int_0^{\delta_{mem}} \sigma_{mem} dz \cdot A_{mem}} \quad (4)$$

where σ_{mem} is the membrane conductivity (S/m), which is given by [5]. A_{mem} is the membrane section surface (m^2), δ_{mem} is the thickness of the membrane (m). The Ohmic losses η_{ohmic} can thus be obtained from Ohm's law:

$$\eta_{ohmic} = i \cdot R_{mem} = \frac{i\delta_{mem}}{\sigma_{mem}} \quad (5)$$

Since the conductivity σ_{mem} is highly dependent on water content λ_w in polymer membrane, a more detailed knowledge of transient behavior of the λ_m would give more accurate value of Ohmic losses η_{ohmic}. Moreover, the dynamic phenomena of λ_w plays an important role on the dynamic performance of PEM fuel cell due to its relatively long transient time.

2) Fluidic Domain Modeling

The gas pressure dynamic response in the fuel cell is generally due to the gas supply channels volume. It can be given by the mass balance equation [10]:

$$\frac{M_{gas}V_{ch}}{RT_{ch}}\left(\frac{d}{dt}P_{ch}\right) = \sum_{in/out}^{ch} q_{fluid} \quad (6)$$

where M_{gas} is the gas molar mass (kg/mol), V_{ch} is the volume of the channels (m^3), T_{ch} is the gas supply channel temperature (K), P_{ch} is the gas pressure in the channels (Pa) and q_{fluid} is the fluid mass flow rate (kg/s) entering or leaving the channels.

To obtain the reactant gas pressure at interface of the catalyst layer, the diffusion phenomenon in the gas diffusion layer can be described by Fick law:

$$\frac{dp_{GDL,x}}{dz} = \frac{\Delta p_{GDL,x}}{\delta_{GDL}} = \frac{-RT_{GDL}\dot{N}_{GDL,x}}{A_{GDL}\cdot D_{GDL,x-y}^{eff}} \quad (7)$$

where $p_{GDL,x}$ is pressure of specie x in gas diffusion layer (Pa), δ_{GDL} is thickness of gas diffusion layer (m), T_{GDL} is the gas diffusion layer temperature (K), $\dot{N}_{GDL,x}$ is the gas molar flow rate of specie x (mol/s), A_{GDL} is the gas diffusion layer area (m^2), $D_{GDL,x-y}^{eff}$ is the gas diffusion coefficient (m^2/s) between the species x and y can be calculated from [5]:

$$D_{GDL,x-y}^{eff} = \frac{101325}{P_{tot}}\cdot a\cdot\left(\frac{T_{GDL}}{\sqrt{T_{crit,i}T_{crit,j}}}\right)^b\cdot\left(\frac{P_{crit,i}P_{crit,j}}{101325}\right)^{\frac{1}{3}} \cdot (T_{crit,i}T_{crit,j})^{\frac{12}{5}}\cdot(\frac{10^{-3}}{M_x}+\frac{10^{-3}}{M_y})^{\frac{1}{2}} \quad (8)$$

where P_{tot} is the total pressure of species (Pa), T_{crit} is the critical temperature (K), P_{crit} is the critical pressure (Pa), and M is the molar mass of species (kg/mol). Coefficients a and b are determined on the gases are polar gases or not, which are given in [5].

3) Thermal Domain Modeling

The fuel cell temperature transient behavior in the thermal domain is due to the heat generation and thermal conduction and convection phenomena. Like dynamic behavior of membrane water content, dynamics of fuel cell temperature is an important phenomenon due to its relatively long transient time. This dynamic behavior can be generally described as follows [10]:

$$\rho V C_p \frac{dT_{CV}}{dt} = \dot{Q}_{cond} + \dot{Q}_{mass} + \dot{Q}_{forced\_conv} + \dot{Q}_{source} \quad (9)$$

where ρ is the mean layer volume density of stack (kg/m3), V is the layer volume of stack (m3), C_p is the layer thermal capacity (J/kg.K), T_{CV} is the temperature (K) of each control volume and $\dot{Q}$ stands for the different types of heat (J) flows entering or leaving the layer respectively: conduction, convective flow, forced convection and internal heat sources.

The heat flows due to conduction can be expressed according to Fourier's law

$$\dot{Q}_{cond} = \frac{2\lambda_{CV}S_{section,CV}}{\delta_{CV}}(T_{boundary} - T_{CV}) \quad (10)$$

where λ_{CV} is the control volume thermal conductivity (W/m.K), $S_{section,CV}$ is the section of the control volume in heat transfer direction (m2), and δ_{CV} is the control volume thickness (m).

The convective heat flow due to the mass transfer entering or leaving the control volume can be calculated by

$$\dot{Q}_{mass} = \left[\sum_{specie}(q_{specie}C_{p,specie})\right](T_{boundary} - T_{CV}) \quad (11)$$

where q_{specie} is the mass flow rate (kg/s).

The heat transfers by forced convection $\dot{Q}_{forced\_conv}$ can be written according to Newton's cooling law:

$$\dot{Q}_{forced\_conv} = h_{forced}S_{forced}(T_{cool} - T_{cell}) \quad (12)$$

where h_{forced} is the forced convection heat transfer coefficient (W/m2.K), T_{cool} is the coolant temperature and S_{forced} is the contact area (m2).

At last, the linear expression of the heat sources $\dot{Q}_{source}$ as a function of the cell temperature can be obtained [5]:

$$\dot{Q}_{source} = k_{source}T_{cell} + i\eta_{act} + 4.2i^2 k_{mem} \quad (13)$$

B. Dynamic Modeling of Membrane Water Content

The dynamics of the membrane water content is generally described by two effects, the electro-osmotic driving forces due to an electrochemical potential difference between the anode and the cathode, and the water diffusion caused by the concentration gradient at each side boundary. Then the dynamics of the water content λ_w in the membrane can be obtained considering the water mass flows at two boundaries of at each side (cathode and anode) and conservation of water mass in the membrane, as shown in equation (1):

$$\frac{A_{mem}\delta_{mem}\rho_{dry,mem}}{M_{mem}}\frac{d\lambda_w}{dt} = W_{dra,an} - W_{dra,ca} + W_{diff,ca-\mathrm{mem}} - W_{diff,mem-ca} \quad (14)$$

where A_{mem} is area of membrane, δ_{mem} is the thickness of membrane, $\rho_{dry,mem}$ is the membrane dry density, M_{mem} is the molecular mass of membrane, and W is the different water mass flow entering or leaving the membrane due to electro-osmotic drag and back-diffusion, as shown in Fig. 1.

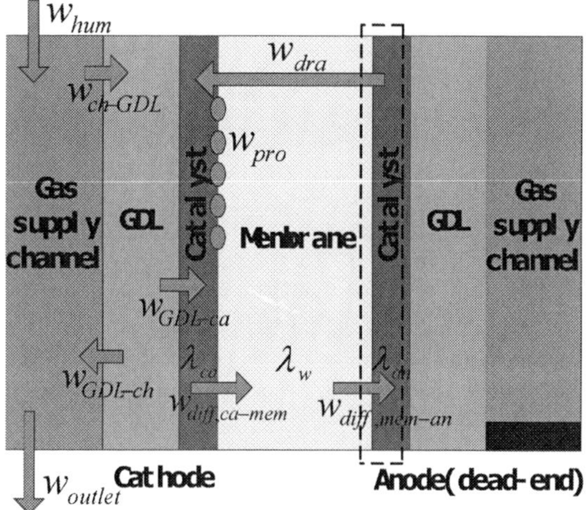

Fig. 1. Dynamic water flow behaviors in membrane.

The water mass entering or leaving the membrane due to electro-osmotic drag W_{dra} can be described by:

$$W_{dra} = \frac{\lambda_w i}{22F} \cdot n_{drag}^{SAT} \quad (15)$$

where the $n_{drag}^{SAT} \approx 2.5$ is the coefficient of electro-osmotic drag for maximum hydration conditions. The water mass entering (from cathode to membrane) or leaving (from membrane to anode) the membrane $W_{diff,i}$ due to back diffusion flow can be described by:

$$W_{diff,i} = \frac{A_{mem}D_{water}\rho_{dry,mem}|\lambda_i - \lambda_w|}{M_{mem}\delta_{mem}} \quad (16)$$

where the membrane water diffusion coefficient D_{water} (m^2/s) can be calculated from the empirical equations [5]. $\rho_{dry,mem}$ is the dry density of the membrane (kg/m^3), M_{mem} is the equivalent mass of the membrane (kg/mol). The boundary water content λ_i at anode and cathode side can be expressed as a function of water activity α_i which can be deduced from the water vapor partial pressure [5].

As shown in Fig. 1, in the case of non-humidified hydrogen supply and anode dead-end mode operation, the anode side water accumulation is only caused by the water diffusion from the membrane to the anode $W_{diff,mem-an}$, and the cathode side water accumulation depends on three factors: W_{GDL-ca} comes from humidified air supply at cathode W_{hum}, the produced water at cathode side W_{pro} during electrochemical reaction and the electro-osmotic drag flow from the anode to the cathode W_{dra}.

Under the dead-end mode operation (no water accumulation at anode side), the water mass flow entering into the membrane from anode due to electro-osmotic drag $W_{dra,an}$ is equal to the water mass leaving out the membrane to anode due to back-diffusion $W_{diff,mem-\mathrm{an}}$ (dashed portion as shown in Fig. 1). Thus, the dynamics of the membrane water content λ_w can thus be simplified by:

$$\frac{A_{mem}\delta_{mem}\rho_{dry,mem}}{M_{mem}}\frac{d\lambda_w}{dt} = \frac{A_{mem}D_{water}\rho_{dry,mem}(\lambda_{ca} - \lambda_w)}{M_{mem}\delta_{mem}} - \frac{\lambda_w i}{22F} \cdot n_{drag}^{SAT} \quad (17)$$

with anode side water content λ_{an}:

$$\lambda_{an} = \lambda_w - \frac{\lambda_w i \cdot n_{drag}^{SAT}\ M_{mem}\delta_{mem}}{22F \cdot A_{mem}D_{water}\rho_{dry,mem}} \quad (18)$$

C. Derivation of Time Constant Value For Dynamic of Water Content in the Membrane

Using similar analysis method discussed in [10], the expression of time constant of water content τ_{water} in membrane can be expressed by equation:

$$\tau_{water} = \frac{A_{mem}\rho_{dry,mem}\delta_{mem}^2 F}{A_{mem}D_{water}\rho_{dry,mem}F + \frac{n_{drag}^{SAT}}{22} iM_{mem}\delta_{mem}} \quad (19)$$

Based on the fuel cell stack properties data in [6], this time constant value for water content dynamic in the membrane τ_{water} can be estimated to around 110 s.

The explicit expression of thermal time constant can be found in [10]:

$$\tau_{thermal} = \frac{\rho_{cell}V_{cell}C_p}{h_{forced}S_{forced} + h_{nat_{radia}}S_{ext} - k_{source}} \quad (20)$$

Thus, the time constant of cell temperature dynamic $\tau_{thermal}$ can be estimated between 37.15s and 124.3s using the same fuel cell properties data in [6].

III. Experimental Validation

The presented dynamic fuel cell model has been validated through simulation and experiments with a commercial 1.2 kW Ballard NEXA 47 cells stack. The quick current step variation profiles are presented in Fig. 2.

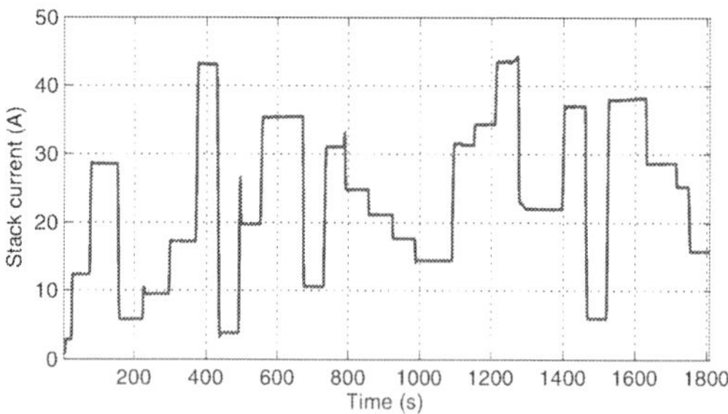

Fig. 2. Stack current profile.

A. Experimental Validation of Improved Dynamic Model

The measured real stack voltage and model predicted are compared in Fig. 3. It can be seen from Fig. 3 that, the predicted voltage values from the model show a good agreement with the experimentation.

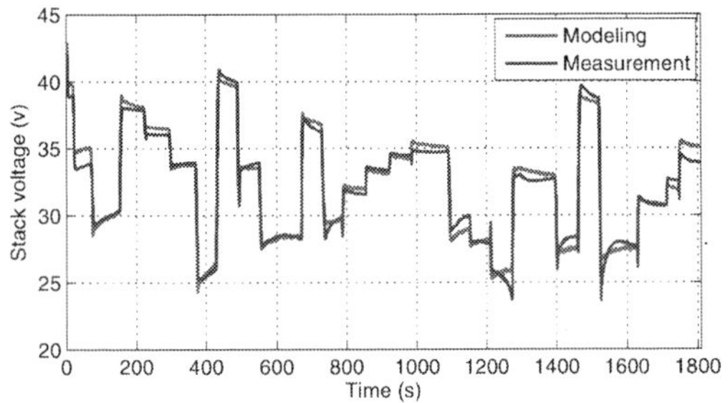

Fig. 3. Experimental validation of stack voltage

B. Dynamic Membrane Water Content Results and Discussion

Fig. 4 shows the dynamic response of membrane water content at both cathode and anode side following fuel cell current step change in one cell.

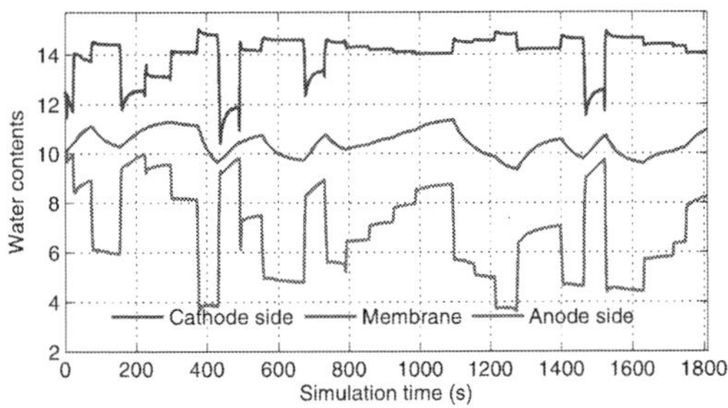

Fig. 4. Dynamic variation of water content in the membrane at different current step

It can be seen from Fig. 4 that, the water content of anode side λ_{an} (red line) decreases sharply because of the main factor electro-osmotic drag caused by a high current step. The water content of anode side λ_{ca} (blue line) decreases due to the main factor water diffusion at low current step. It can also be observed that, the dynamics of membrane water content λ_{w} (green line) are directly related to the stack current variations (shown in Fig. 2).

IV. Dynamic Variable Coupling Analysis

The dynamic behaviors in different physical domains are inter-coupled between each other, especially the dynamic phenomena which have the similar dynamic time constants. In this section, analysis of step responses are presented first, then the coupling analysis is discussed based on RGA method.

A. Analyses of Step Responses

In order to understand the effects of coupling characteristic for fuel cell system, the analysis of step responses are performed in this part. The coolant inlet mass flow rate $q_{cool,inlet}$ (controlled by coolant circulating pump) and the gas supply channel inlet water vapor pressure $P_{H_2o,inlet}$ (controlled by gas humidifier) are considered as two input variables to form a 2×2 inputs-outputs models multivariable system, and controlled output variables are MEA temperature $T_{membrane}$ and the membrane water content $\lambda_{membrane}$, since this two control input variables $q_{cool,inlet}$ and $P_{H_2o,inlet}$ can directly affect the two fuel cell state variables $T_{membrane}$ and $\lambda_{membrane}$. The state space equations of this MIMO system can then be expressed as follows:

$$\begin{cases} \dot{T}_{membrane} = f_2(T_{membrane}, q_{cool,inlet}) \\ \dot{\lambda}_{membrane} = f_1(\lambda_{membrane}, P_{H_2o,inlet}, q_{cool,inlet}) \\ y = \begin{pmatrix} T_{membrane} \\ \lambda_{membrane} \end{pmatrix} \end{cases} \quad (21)$$

As the MIMO system state space Eq. (21) shown, the vector form representations of the control input variables and state variables are $U = [q_{cool,inlet}, P_{H_2o,inlet}]^T$ and $X = [T_{membrane}, \lambda_{membrane}]^T$, respectively. f_1 and f_2 can be derived using the Eq. (9) and Eq. (14).

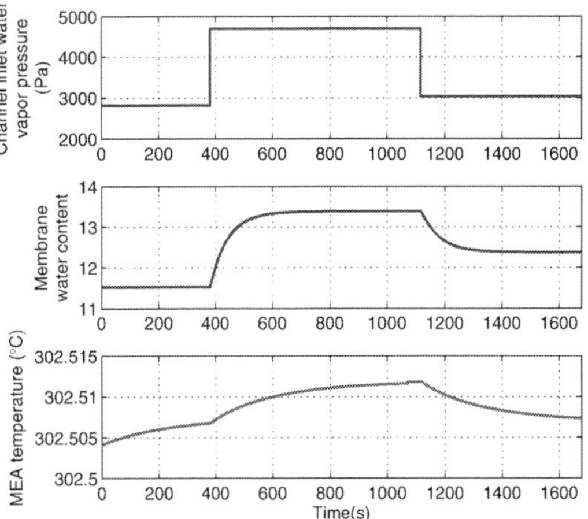

Fig. 5. The fluidic and thermal domain system outputs response after step change of the gas supply channel inlet water vapor pressure.

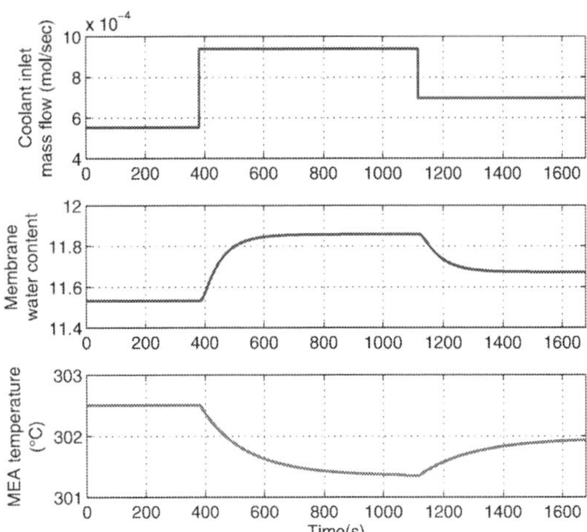

Fig. 6. The fluidic and thermal domain system outputs response after step change of the coolant inlet mass flow rate.

Fig. 5 shows the dynamic response after step changes of the gas supply channel inlet water vapor pressure for the system (21). It can be seen from the Fig. 5 that, membrane water content increases due to an increase of water vapor inlet pressure at 381 s. Similarly, when the water vapor inlet pressure decreases at 1100 s, it results in a decrease in the membrane water content. On the other hand, a step change of the gas supply channel inlet water vapor pressure has no significant effect on the MEA temperature.

Fig. 6 shows the fluidic and thermal domain system outputs response after step changes of the coolant inlet mass flow rate. The cathode catalyst water content increases at 381 s because the water activity α_i increases based on its expression [5]. Therefore, the membrane water content increases as a result. The MEA temperature decreases due to an increase of the coolant air inlet mass flow rate at 381 s, which results in a decrease of the saturation vapor pressure.

B. Coupling Analysis

The Relative Gain Array (RGA) method [9] is used for the coupling analysis in this section (RGA is a valuable tool which can point out the suitable pairings amongst input and output variables for use in the control of a multi-physique modeling system). For the two-input $U = [q_{cool,inlet}, P_{H_2o,inlet}]^T$ and two-output $Y = [\lambda_{membrane}, T_{membrane}]^T$ inter-coupled system (21), around an operating point, the RGA matrix is shown in Table 1.

TABLE I. RELATIVE GAIN ARRAY OF TWO INPUTS TWO OUTPUTS SYSTEM BETWEEN FLUIDIC AND THERMAL DOMAIN

	$q_{cool,inlet}$	$P_{H_2o,inlet}$
$T_{membrane}$	1.2273	-0.2273
$\lambda_{membrane}$	-0.2273	1.2273

The RGA matrix shows that there exist an intermediate coupling effect between the control variable pairs, due to the difference of diagonal element compared with unity. It means that, a change of coolant mass flow rate affects both MEA temperature and membrane water content, a decoupling method thus need to be used to achieve optimized water and thermal management in fuel cell.

V. CONCLUSION

Although the coupling of dynamic phenomena in a fuel cell has an important influence on the control design of fuel cell system, very little information has yet been published in the literature on the analysis of interaction between dynamic variables. This paper investigates in particular the coupling effect between the dynamic behavior of membrane water content and cell temperature, based on an improved dynamic multi-physique proton exchange membrane fuel cell model. Compared to the previously published model, the dynamic behaviors of membrane water content have been modeled in addition to the commonly considered thermal dynamic. Furthermore, to help engineers design and optimize fuel cell water and thermal control strategies, the dynamic coupling analyses for different fuel cell operational parameters which have the similar dynamic time constants are presented and discussed.

REFERENCES

[1] S. Njoya Motapon, L. Dessaint, and K. Al-Haddad, "A Comparative Study of Energy Management Schemes for a Fuel-Cell Hybrid Emergency Power System of More-Electric Aircraft," IEEE transactions on industrial electronics, vol. 61, no. 3, pp. 1320-1334, 2014

[2] X. Liu, H. Li and Z. Wang, "A Fuel Cell Power Conditioning System With Low-Frequency Ripple-Free Input Current Using a Control-Oriented Power Pulsation Decoupling Strategy," IEEE transactions on industrial electronics, vol. 29, no. 1, pp. 159-169, 2014.

[3] K. Vijayaraghavana, J. DeVaalb and M. Narimani, "Dynamic model of oxygen starved proton exchange membrane fuel-cell using hybrid analytical-numerical method," Journal of Power Sources, vol. 285, pp. 291-302, 2015

[4] J. Jia, Q. Li, Y. Wang, Y. T. Cham, and M. Han, "Modeling and Dynamic Characteristic Simulation of a Proton Exchange Membrane Fuel Cell," IEEE Transactions on energy conversion, vol. 24, no. 1, pp. 283-291, 2009.

[5] F. Gao, B. Blunier, A. Miraoui, and A. E. moudni, "Cell layer level generalized dynamic modeling of a PEMFC stack using VHDL-AMS language," International Journal of Hydrogen Energy, vol. 34, no. 13, pp. 5498–5521, 2009.

[6] F. Gao, B. Blunier, A.Miraoui, and A. E. moudni, "Amultiphysic dynamic 1D model of a proton exchange membrane fuel cell stack for real time simulation," IEEE transactions on industrial electronics, vol. 57, no. 6, pp. 1853–1864, 2010.

[7] T. Colinart, A. Chenu, S. Didierjean, O. Lottin and S. Besse, "Experimental study on water transport coefficient in Proton Exchange Membrane Fuel Cell, " Journal of Power Sources, vol. 190, pp. 230–240, 2009

[8] B. Cheng, O. Minggao and Y. Baolian, "Analysis of Water Management in Proton Exchange Membrane Fuel Cells, " Tsinghua Science and Technology, vol. 11, no. 1, pp. 54 - 64, 2006

[9] M.T.Tham, "Multivariable control: an introduction to decoupling control, "An introduction to Decoupling control/MTT/July, 1999

[10] F. Gao, B. Blunier, A. Miraoui, and A. E. moudni, "Proton exchange membrane fuel cell multi-physical dynamics and stack spatial non-homogeneity analyses, " Journal of Power Sources,vol. 195, pp. 7609–7626, 2010

Voltage and Frequency Control of Electric Spring Based Smart Loads

Yun Yang[*], Siew-Chong Tan[*]
[*]Department of Electrical and Electronic Engineering
The University of Hong Kong, Hong Kong
Email: sctan@eee.hku.hk

Shu-Yuen (Ron) Hui[*†]
[†]Department of Electrical and Electronic Engineering
Imperial College London, U.K.

***Abstract*—With the increasing adoption of renewable energy sources into power grids, the issue of voltage and frequency fluctuation of the network becomes a more critical problem that needs to be resolved urgently. The so-called demand-side management (DSM) of power system provides the opportunity for the prospective grids to intelligently balance the power supply and consumption. As one of the crucial roles of DSM, smart loads are naturally committed to manage the power flow of the network. Other than previous communication-based smart loads, the recently proposed electric springs (ES) provide an opportunity to turn some conventional loads into smart loads without the need for communication. In this paper, a simple voltage and frequency (v&f) control method is applied to such a smart load for tackling the voltage and frequency fluctuations of the power grids. Simulation results reveal that with the proposed v&f control, the smart load can achieve better voltage and frequency regulation performance than that with the conventional control and smart load. Experimental results validating the feasibility of the proposed v&f controller on regulating the voltage amplitude and frequency of the common-coupling point of a power grid are provided.**

***Keywords*—Smart load, electric springs, non-communication-based, voltage and frequency (v&f) control, power grid.**

I. Introduction

Renewable energy sources like wind energy and photovoltaic energy are an important part of future power grid systems. The intermittency and uncertainty of these renewable energy sources give rise to difficulty in achieving instantaneous balance in power supply and demand. To tackle this problem, energy storage and demand-side management (DSM) methods have been proposed [1]–[8]. DSM is a portfolio of measures to improve the energy system on the side of utility consumption. Ranging from improving energy efficiency to sophisticated real-time control of distributed energy resources, DSM attempts to bring the electricity demand and supply closer to a perceived optimum [7]. Thereinto, smart loads play irreplaceable roles for DSM. Traditional concept of smart loads is to add affordable communication infrastructure on conventional loads. Hence, the communication-based smart loads will introduce issue of network complexity and counter-measures against cyber-attack, which will increase the cost of smart grids [8]. In respond to this, the recently proposed concept of electric springs (ES) provide the opportunity to turn some conventional loads into smart loads without the need for communication [6]. ES have shown to be highly effective in providing reactive power compensation [9]–[11] and along with the incorporation of batteries, in providing active power compensation [12].

Several previous works have demonstrated the use of various control strategies on ES to achieve grid voltage regulation [6], [9]–[16], power factor correction [14], [17], power balancing [18], neutral current minimization [19], and frequency stabilization [15]. However, in some isolated grid systems with very low rotor inertia and intermittent renewable power source, e.g. in microgrids that are powered by small diesel generators and wind turbines (without or with limited energy storage), the frequency variation is highly sensitive to supply-demand imbalance, and the requirement for frequency stabilization is higher than that for voltage amplitude stabilization. While ES control methods have been reported previously in [15], [20], fast control of ES for frequency stability of microgrid with very small inertia has not been specifically addressed.

In this paper, a v&f control strategy that is catered to distinctly prioritize the frequency regulation over the voltage regulation of the grid, is proposed for the ES-based smart load. For any power imbalance, the frequency of the system is always controlled through active power compensation before the voltage amplitude of the common-coupling point is regulated through reactive power compensation. The underlying assumption for the success of this approach is that the adopted smart loads possess sufficiently an overall capacity that meets all power generation surplus and shortage. The proposed v&f control has been verified by both simulation and experimental results.

II. Proposed v&f Control for the ES-Based Smart Load

Fig. 1 shows a simplified diagram of a microgrid incorporated with an ES-based smart load implemented using a voltage source single-phase half-bridge inverter. The renewable source can be wind energy, solar energy, and tidal energy, etc. The non-critical load Z_n is an electric load (such as a refrigerator, a water heating system, or a lighting system), which can be of resistive, inductive-resistive, or capacitive-resistive nature. The analysis in this work adopts the inductive-resistive load (as most of the common loads are of this type). "Non-critical loads" are referred to those appliances

978-1-4673-9551-9/16 $31.00 © 2016 IEEE

that can tolerate a wider range of voltage and power fluctuation. By adopting the ES, the power consumption of Z_n follows the variation of the renewable power generation, therefore achieving instantaneous power balance in the microgrid. Z_c is a critical load (voltage and frequency sensitive equipment), which requires a well-regulated mains voltage (the voltage of the common-coupling point in Fig. 1).

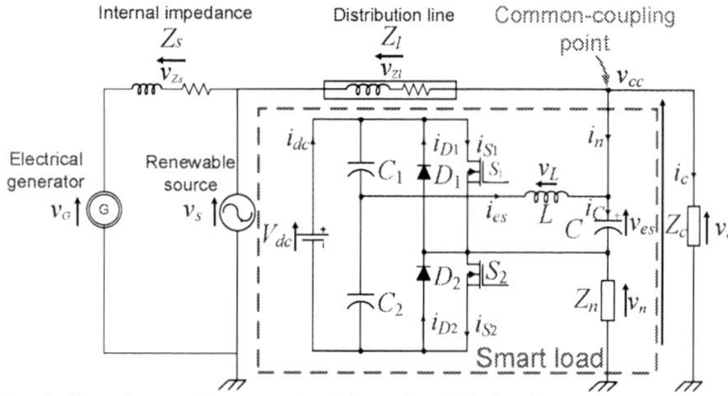

Fig. 1. Topology of a smart load in a simplified microgrid.

Without the ES, the load is passive as v_{cc} follows the fluctuation of the renewable source, i.e.,

$$\vec{v}_{cc} = \vec{v}_n. \tag{1}$$

If the variable power source delivers electric power higher than the rated power of the existing loads, v_{cc} goes above the reference v_{ccref}. Conversely, if the power delivered is less than the rated power, v_{cc} goes below v_{ccref}, as shown in Fig. 2(a). In Fig. 2, the voltage and current are plotted as vectors (in RMS values) in the *d-q* coordinates. The unit circle represents the reference v_{ccref} for all phase angles.

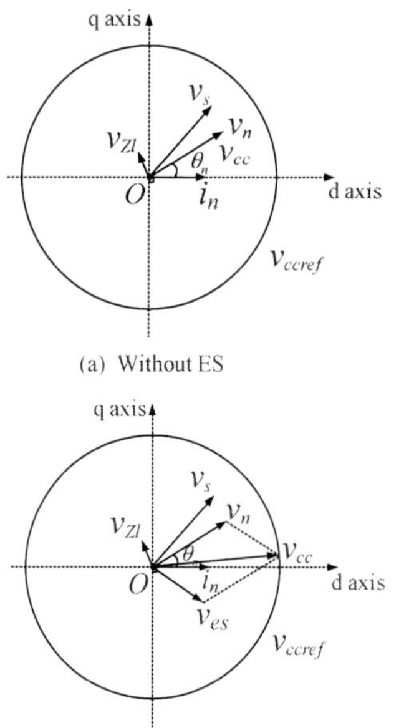

(b) With ES

Fig. 2. Phase diagram of the smart-load voltage and current in *d-q* coordinates.

With the utilization of the ES, the load becomes a smart load such that

$$\vec{v}_{cc} = \vec{v}_{es} + \vec{v}_n. \tag{2}$$

An ES with batteries can be controlled to achieve a v_{es} of any phase angle. Thus, many designs of different amplitudes and phase angles of v_{es} are possible. Fig. 2(b) presents one possible design of ES for regulating v_{cc} in reaching v_{ccref}. In practice, an ES can be flexibly controlled to have smart loads operating in different modes, which correspond to various situations. In Fig. 3, several operating modes of ES-based smart loads are presented. Fig. 3(a) and (b) show the smart load operating in unity power factor (UPF) mode when v_{cc} exceeds and deceeds v_{ccref}, respectively. Fig. 3(c) and (d) show the smart load operating as active loads (AL mode) when v_{cc} exceeds and deceeds v_{ccref}, respectively. Fig. 3(e) and (f) show the smart load operating in reactive power compensating (RPC) mode when v_{cc} exceeds and deceeds v_{ccref}, respectively. Hence, for the case in Fig. 3(e), the smart load is not controllable if v_{cc} significantly exceeds v_{ccref}. Besides, in RPC mode, the smart load has two operation modes. Fig. 3(g) and (h) show the smart load operating in minimum voltage of battery (MV) mode when v_{cc} exceeds and deceeds v_{ccref}, respectively. In MV mode, the battery can operate with a minimum voltage level.

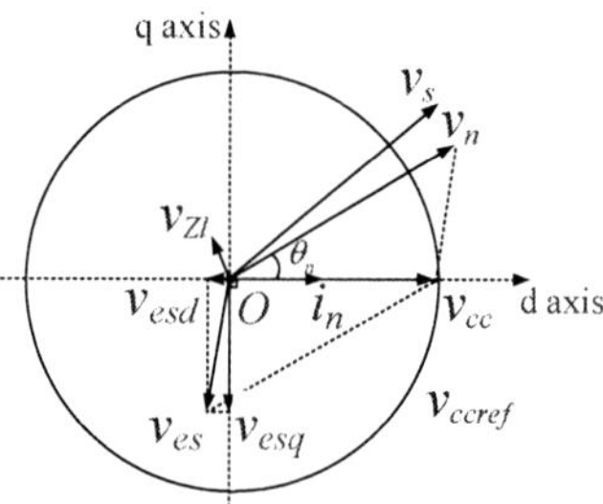

(a) UPF mode (v_{cc} exceeds v_{ccref})

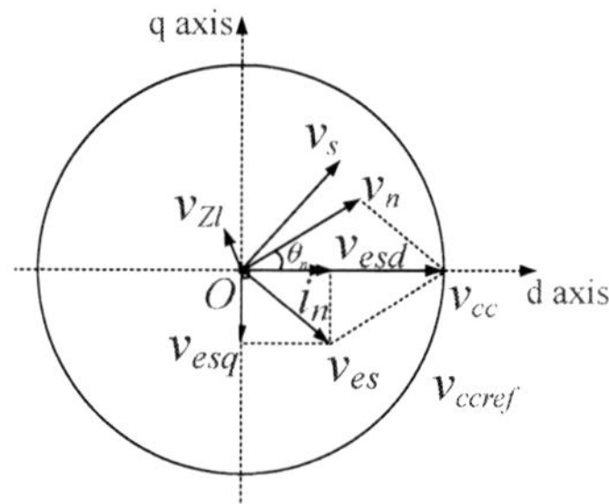

(b) UPF mode (v_{cc} deceeds v_{ccref})

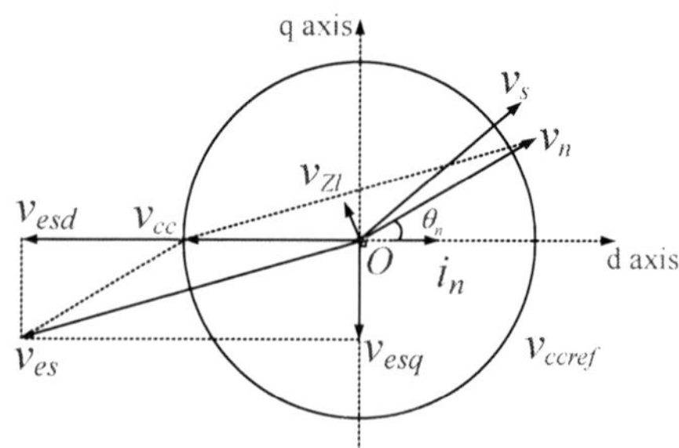

(c) AL mode (v_{cc} exceeds v_{ccref})

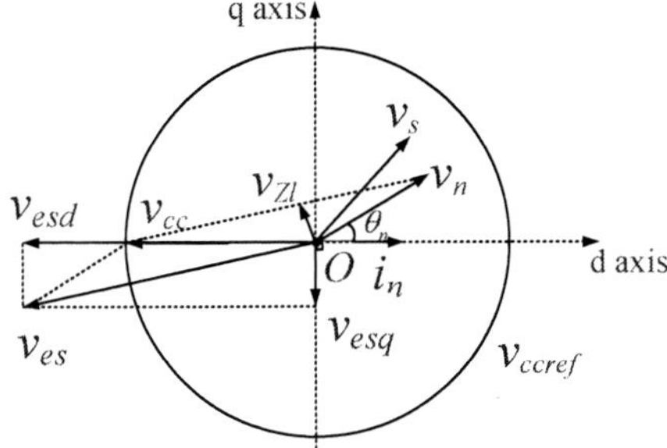

(d) AL mode (v_{cc} deceeds v_{ccref})

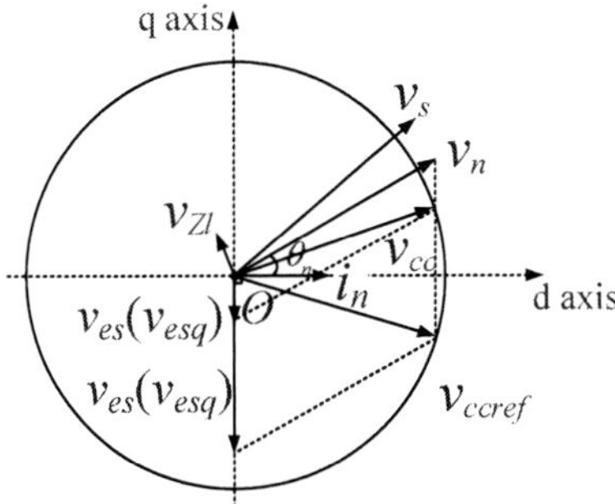

(e) RPC mode (v_{cc} exceeds v_{ccref})

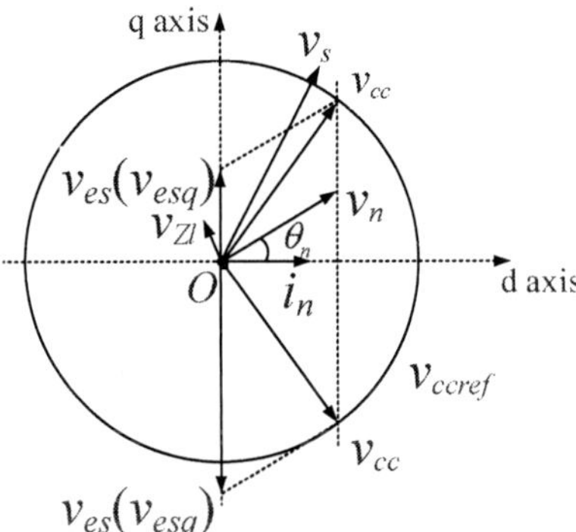

(f) RPC mode (v_{cc} deceeds v_{ccref})

q axis v_s v_n v_{cc} v_{Zl} v_{esa} θ_n i_n d axis v_{es} v_{esq} v_{ccref}

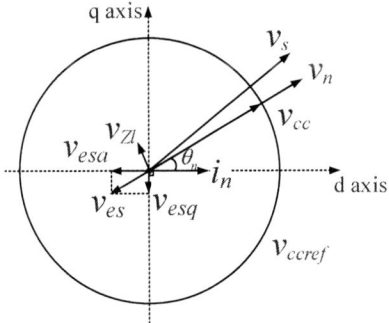

(g) MV mode (v_{cc} exceeds v_{ccref})

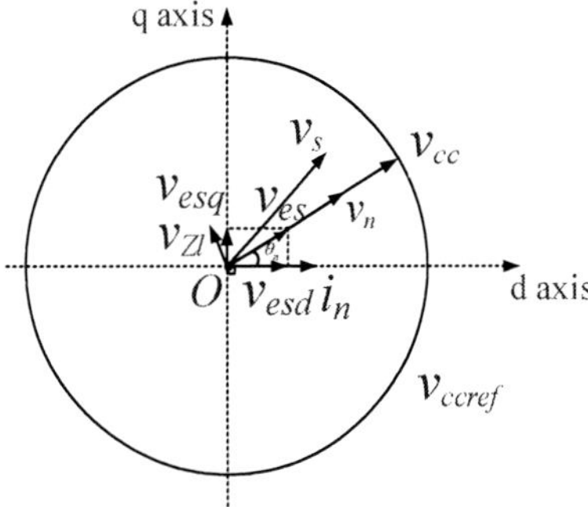

(h) MV mode (v_{cc} deceeds v_{ccref})

Fig. 3. Various operation modes of ES-based smart loads.

Apparently, in all diagrams in Fig. 3, v_{es} can be decoupled into v_{esd} and v_{esq}, where v_{esd} is designed for active power compensation and v_{esq} for reactive power compensation.

The relationship of power and frequency of a variable power source based on the electrical generator is governed by the swing equation [21].

$$J\omega_r \frac{d\omega_r}{dt} = P_m - P_e, \tag{3}$$

where J is the rotational inertia of the rotor; ω_r is the angular frequency of the generator; P_m is the mechanical turbine power; and P_e is the electrical power. For microgrid comprising small generators, the priority of keeping the frequency stability is more important than regulating the mains voltage if there is a conflict between them. Since the frequency of the network is

$$f = \frac{\omega_r}{2\pi}, \tag{4}$$

and based on (3), the frequency is directly related to the active power P_e. Therefore, the frequency control of the smart load can be achieved solely by regulating v_{esd}. However, as the change of v_{esd} will affect the amplitude of v_{cc}, it is necessary that the voltage amplitude of the common-coupling point is controlled only after the frequency is well-regulated. Hence, the proposed v&f control is a two-step approach with frequency control being the priority.

Fig. 4 shows that v_{esd} is initially designed to compensate the change of frequency and amplitude of v_{cc}. Based on Cartesian Geometry,

$$\theta_n = \arctan\left(\frac{\omega L_n}{R_n}\right), \tag{5}$$

where ω is the nominal angular frequency, L_n is the inductance, and R_n is the resistance of the non-critical load. Then, V_n (RMS value of v_n) can be measured. The coordinate of V_n is $V_n(V_n \cos\theta_n, V_n \sin\theta_n)$. Obviously, the coordinate of point A in Fig. 3 is $A(V_n \cos\theta_n - V_{esd}, V_n \sin\theta_n)$. The coordinate of V_{cc} is set as $V_{cc}(V_{cc} \cos\theta_{cc}, V_{cc} \sin\theta_{cc})$. Then,

$$V_n \cos\theta_n - V_{esd} = V_{cc} \cos\theta_{cc} \tag{6}$$

can be acquired and phase angle between V_{cc} and V_n can be further obtained as

$$\theta = \left| \arccos\left(\frac{V_n}{V_{cc}} \cos\theta_n - \frac{V_{esd}}{V_{cc}} \right) - \theta_n \right|. \qquad (7)$$

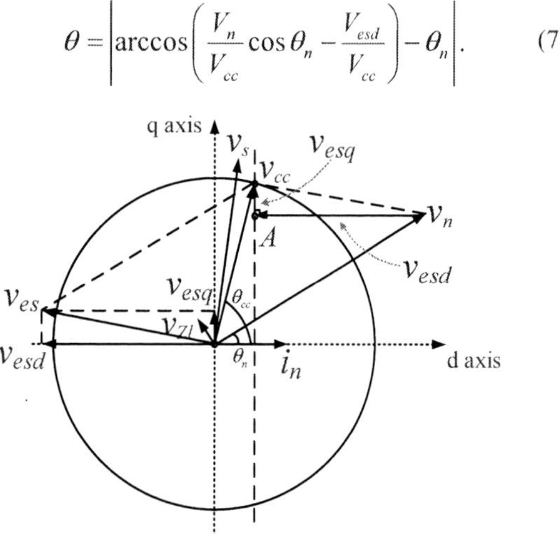

Fig. 4. v&f controller design with both voltage amplitude and frequency deviation in phase diagram.

The proposed v&f control diagram for the smart load is shown in Fig. 5. f_{ref} denotes the reference of the grid frequency. V_{ccref} denotes the reference of the RMS value of v_{cc}. PLL is short for phase-locked loop.

III. SIMULATION RESULTS

Without the loss of generality, the simulation is conducted via MATLAB/SIMULINK on a microgrid of 220 V (RMS), which is operated in isolated mode from the IEEE 13-node test feeder, as shown in Fig. 6. The reference of the frequency is set at 50 Hz. The specifications of main components used in the simulation are listed in Table I. Fig. 7(a) gives a comparison of the simulated waveforms of v_{cc} and v_{cc}(RMS) of the grid with the conventional load, the smart load with conventional control, and the smart load with the proposed control, when v_s(RMS) is changed from 230 V to 250 V, while frequency is kept constant at 50 Hz. Apparently, with the smart load being controlled by the conventional method or by the proposed method, the reference of v_{cc} is always tracked. But this is not the case for the conventional load. Fig. 7(b) shows that when the frequency is shifted (by increasing power) to 50.5 Hz while v_{cc}(RMS) is kept at 220 V, the smart load with the proposed control is able to regulate the frequency at 50 Hz while keeping v_{cc}(RMS) at 220 V. However, the conventional load and the smart load with the conventional control are incapable of enforcing the frequency of the system to follow the variation of the frequency of the renewable generation. Fig. 7(c) shows that with both variation of the voltage amplitude (V_s(RMS) changed from 230 V to 250 V) and frequency (f shifted to 50.5 Hz), the smart load with the proposed v&f control can effectively regulate both the voltage(RMS) and frequency simultaneously. However, this is not the case for the conventional load (without ES). On the other hand, the smart load with the conventional control possesses good steady-state tracking ability for the voltage(RMS), but not a good frequency tracking ability. . Besides, by using v&f control, the transient performance of v_{cc}(RMS) of the smart load is much better than the performance of smart load being controlled by the conventional method, which is presented in Fig. 7(c).

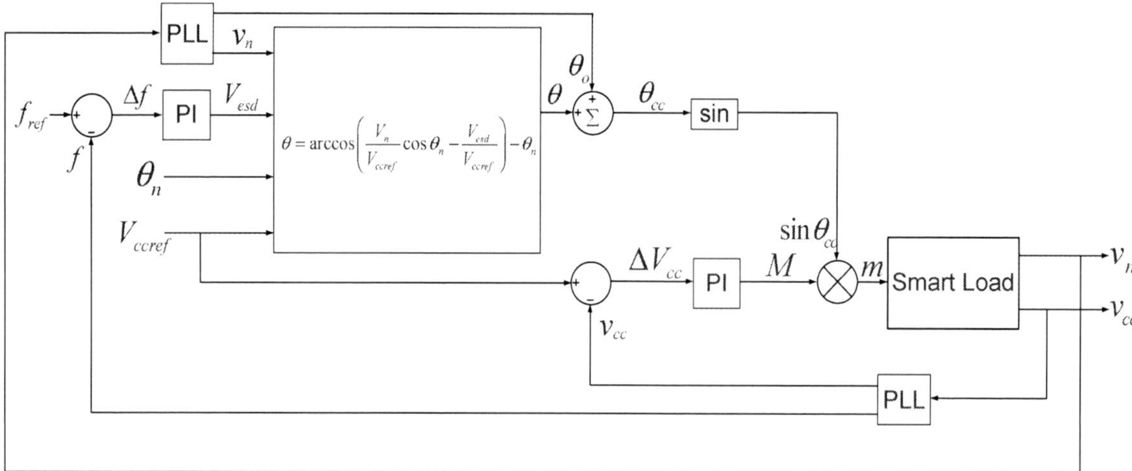

Fig. 5. Comprehensive v&f control diagram for smart loads.

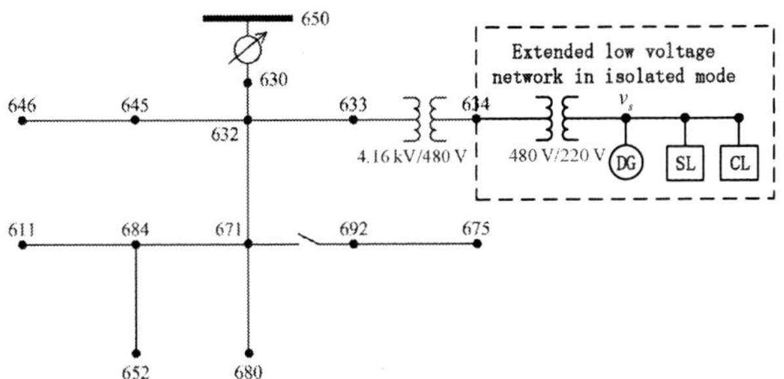

Fig. 6. IEEE 13-node test feeder with extended low voltage network in isolated mode.

TABLE I. SPECIFICATIONS OF SMART LOAD AND RENEWABLE SOURCE FOR SIMULATION

Parameters	Values	Parameters	Values
V_{dc}	400 V	C_1	3000 μF
C_2	3000 μF	L	500 μH
C	13.2 μF	Z_l	0.1+0.38j Ω
v_s(RMS)	230 V	J	3.5 kg×m^2
v_{ccref}(RMS)	220 V	f_{ref}	50 Hz
Z_n	17.17+9.42j Ω	Z_c	53 Ω
f_s	20 kHz		

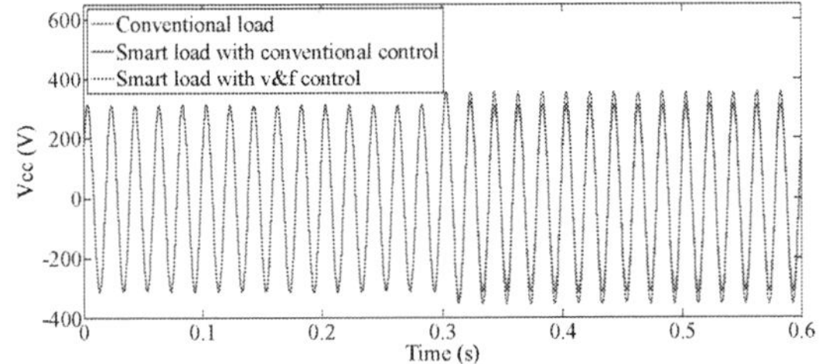

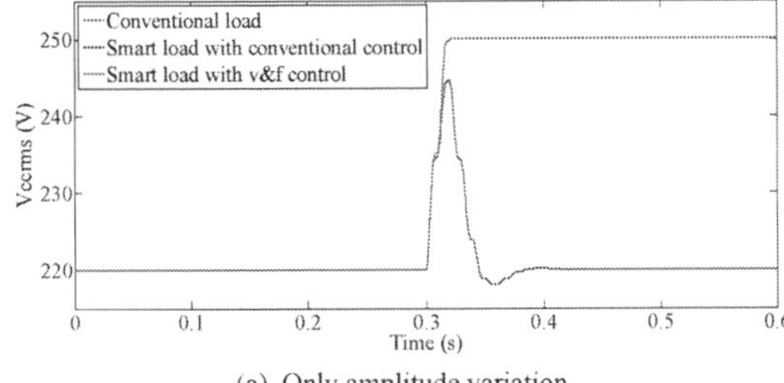

(a) Only amplitude variation

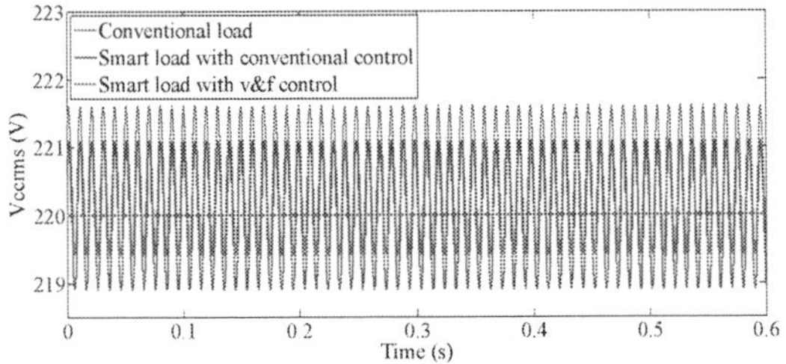

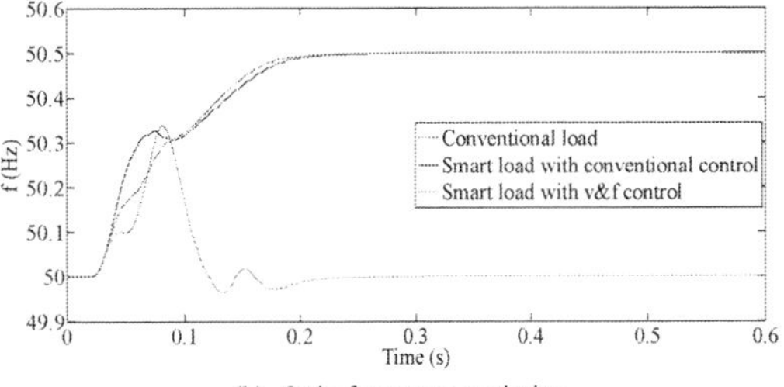

(b) Only frequency variation

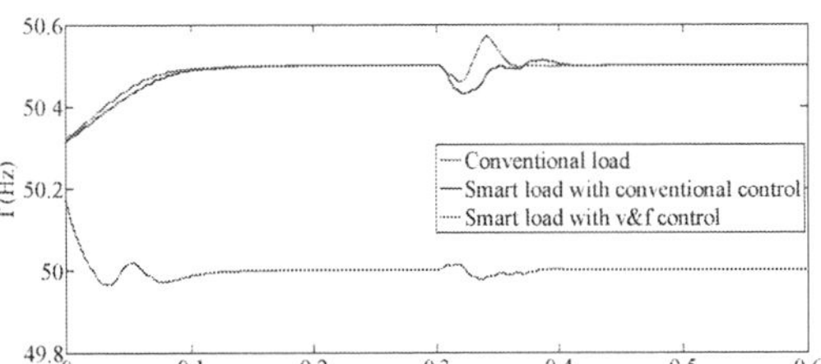

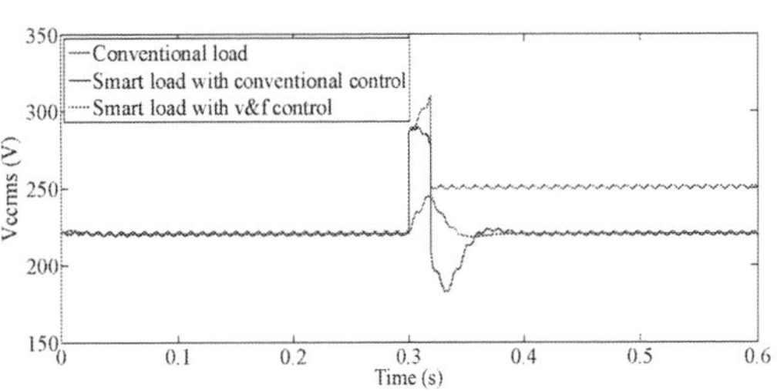

(c) Both frequency and amplitude variation

Fig. 7. Simulation results of the comparisons among the conventional load, the smart load with conventional control, and the smart load with v&f control.

IV. EXPERIMENTAL VERIFICATION

Experiments are conducted with the non-critical load being a nominal 60 W bulb and a nominal 200 W bulb in parallel. The critical load is a pure 53 Ω resistor. The renewable energy source is emulated by a California Instruments CSW5550 AC/DC power supply with MATLAB/SIMULINK. A modified T&K's voltage source inverter (VSI) is used as an ES. The v&f control is implemented using TI's TMS320F28069 digital signal processor (DSP). The main parameters of the experimental setup are presented in Table II.

TABLE II. SPECIFICATIONS OF SMART LOAD AND RENEWABLE SOURCE FOR EXPERIMENTS

Parameters	Values	Parameters	Values
V_{dc}	48 V	C_1	1640 μF
C_2	1640 μF	L	500 μH
C	13.2 μF	Z_l	0.1+0.38j Ω
v_s(RMS)	31.32 V	J	3.5 kg×m^2
v_{ccref}(RMS)	30 V	f_{ref}	50 Hz
Z_n	17.17+9.42j Ω	Z_c	53 Ω
f_s	20 kHz		

Fig. 8 shows the obvious change of v_{cc}(RMS) from 26 V to 30 V when the conventional load is converted into a smart load through the activation of the ES. In Fig. 8, the renewable energy source is operated at about 27.2 V of v_s(RMS) and 50 Hz of the frequency. The smart load is controlled by the proposed V&f control method.

Then, the frequency deviation is also considered, such that the frequency of the renewable energy source is shifted from 50 Hz to 50.5 Hz. In Fig. 9, the voltage over the conventional load is about 26.01 V of v_{cc}(RMS) and 50.558 Hz of the frequency. However, by updating the conventional load into the smart load with V&f control, v_{cc}(RMS) can be regulated at 29.99 V and the frequency of the system at 50.092 Hz.

V. Conclusions

The recently proposed electric springs (ES) provide an opportunity for converting conventional non-critical loads into non-communication based smart loads that will intelligently manage the power balance and quality of future smart grids comprising a high penetration of renewable sources. So far, most studies on the smart loads with ES are focused on the case that only the voltage amplitude of the common-coupling point fluctuates. However, in grid systems where significant portion of the power is provided by smaller electrical generators, a variation of the power generation or loads will also result in a significant fluctuation of the frequency. Therefore, in this paper, a simple two-step approach of v&f control is proposed. Both simulation and experimental results have verified that the system performance of the proposed v&f controller is better than the conventional controller in terms of mitigating voltage and frequency fluctuations at the common-coupling point of a small microgrid with small generator inertia.

Acknowledgment

This project is supported in part by the Hong Kong Research Grant Council under the Theme-Based Research Scheme T23-701/14-N.

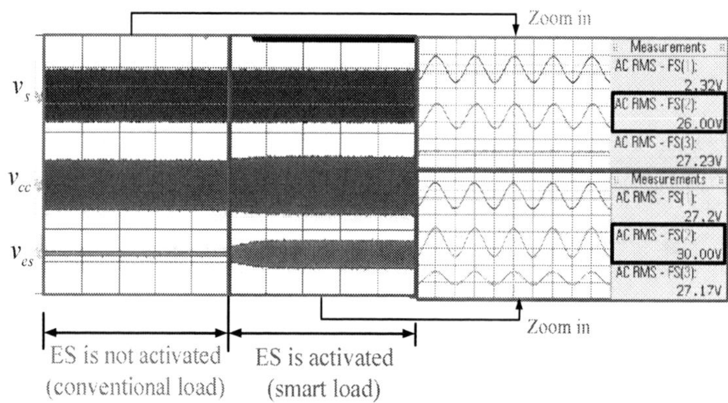

Fig. 8. Waveforms of v_s, v_{cc}, and v_{es} before and after the activation of the smart load with v&f control (only voltage deviation).

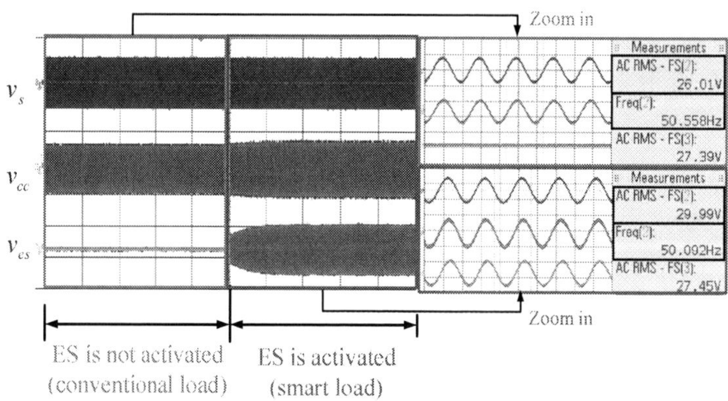

Fig. 9. Waveforms of v_s, v_{cc}, and v_{es} before and after the activation of the smart load with v&f control (both voltage and frequency deviation).

References

[1] M. Parvania and M. Fotuhi-Firuzabad, "Demand response scheduling by stochastic SCUC," *IEEE Trans. Smart Grid*, vol. 1, no. 1, pp. 89-98, Jun. 2010.

[2] M. Pedrasa, T. D. Spooner, and I. F. MacGill, "Scheduling of demand side resources using binary particle swarm optimization," *IEEE Trans. Power Syst.*, vol. 24, no. 3, pp. 1173-1181, Aug. 2009.

[3] A. Mohsenian-Rad, V. W. S. Wong, J. Jatskevich, R. Schober, and A. Leon-Garcia, "Autonomous demand-side management based on game-

978-1-4673-9551-9/16 $31.00 © 2016 IEEE

theoretic energy consumption scheduling for the future smart grid," *IEEE Trans. Smart Grid*, vol. 1, no. 3, pp. 320-331, Dec. 2010.

[4] A. Brooks, E. Lu, D. Reicher, C. Spirakis, and B. Weihl, "Demand dispatch," *IEEE Power Energy. Mag.*, vol. 8, no. 3, pp. 20-29, May 2010.

[5] S. C. Lee, S. J. Kim, and S. H. Kim, "Demand side management with air conditioner loads based on the queuing system model," *IEEE Trans. Power Syst.*, vol. 26, no. 2, pp. 661-668, Sept. 2010.

[6] S. Y. R. Hui, C. K. Lee, and F. F. Wu, "Electric springs-A new smart grid technology," *IEEE Trans. Smart Grid*, vol. 3, no. 3, pp. 1552-1561, Sept. 2012.

[7] P. Palensky and D. Dietrich, "Demand side management: demand response, intellgent energy systems, and smart loads," *IEEE Trans. Ind. Informat.*, vol. 7, no. 3, pp. 381-388, Jun. 2011.

[8] E. Hossain, Z. Han, and H. Poor, *Smart Gird Communicatons and Networkings*, Cambridge, UK; New York : Cambridge University Press, 2012.

[9] C. K. Lee, N. R. Chaudhuri, B. Chaudhuri, and S. Y. R. Hui, "Droop control of distributed electric springs for stabilizing future power grid," *IEEE Trans. Smart Grid*, vol. 4, no. 3, pp. 1558-1566, Sept. 2013.

[10] N. R. Chaudhuri, C. K. Lee, B. Chaudhuri, and S. Y. R. Hui, "Dynamic modeling of electric springs," *IEEE Trans. Smart Grid*, vol. 5, no. 5, pp. 2450-2458, Sept. 2013.

[11] C. K. Lee and S. Y. R. Hui, "Reduction of energy storage requirements in future smart grid using electric springs," *IEEE Trans. Smart Grid*, vol. 4, no. 3, pp. 1282-1288, Sept. 2013.

[12] S. C. Tan, C. K. Lee, and S. Y. R. Hui, "General steady-state analysis and control principle of electric springs with active and reactive compensations," *IEEE Trans. Power Electron.*, vol. 28, no. 8, pp. 3958-3969, Aug. 2013.

[13] X. Luo, Z. Akhtar, C. K. Lee, B. Chaudhuri, S. C. Tan, and S. Y. R. Hui, "Distributed voltage control with electric springs: Comparisons with STATCOM," *IEEE Trans. Smart Grid*, vol. 6, no. 1, pp. 209-219, Jan. 2015.

[14] K. T. Mok, S. C. Tan, and S. Y. R. Hui, "Decoupled power angle and voltage control of electric springs," *IEEE Trans. Power Electron.*, vol. PP, no. 99, Apr. 2015.

[15] X. Chen, Y. Hou, S. C. Tan, C. K. Lee, and S. Y. R. Hui, "Mitigating voltage and frequency fluctuation in microgrids using electric springs," *IEEE Trans. Smart Grid*, vol. 6, no. 2, pp. 508-515, Mar. 2015.

[16] J. Soni, K. R. Krishnanand, and S. K. Panda, "Load-side demand management in buildings using controlled electric springs," in *Ind. Electron. Society, IECON 2014*, Oct. 2014, pp. 5376–5381.

[17] Q. Wang, M. Cheng, Z. Chen, and Z. Wang, "Steady-state analysis of electric springs with a novel δ control," *IEEE Trans. Power Electron.*, vol. PP, no. 99, Jan. 2015.

[18] S. Yan, S. C. Tan, C. K. Lee, B. Chaudhuri, and S. Y. Hui, "Electric spring for reducing power imbalance in three-phase power systems," *IEEE Trans. Power Electron.*, vol. 30, no. 7, pp. 3601-3609, Aug. 2014.

[19] K. R. Krishnanand, S. M. F. Hasani, J. Soni, and S. K. Panda, "Neutral current mitigation using controlled electric springs connected to microgrids within build environment," in *Proc. IEEE Energy Convers. Congr. Expo.*, Sep. 2014, pp. 2947–2951.

[20] K. T. Mok, S. C. Tan, and S. Y. R. Hui, "Power controller and power control method," Nov. 13 2014, PCT Application No.: PCT/CN2014/090992.

[21] P. M. Anderson and A. A. Fouad, *Power System Control and Stability*, Ames, IA : Iowa State Univ. Press, 1977.

Second Harmonic Current Compensator With Improved One-Cycle-Control

Li Zhang, Xinbo Ruan, Xiaoyong Ren
College of Automation Engineering
Nanjing University of Aeronautics and Astronautics
zhangli_frank@nuaa.edu.cn

Abstract—**The instantaneous output power of the two-stage single inverter pulsates at twice the output voltage frequency, generating second harmonic current (SHC) in the input dc voltage source and the front-end dc-dc converter. To eliminate the SHC, a second harmonic current compensator (SHCC), whose input current is controlled by the improved one-cycle-control (OCC) proposed in this paper, is applied to compensate the SHC in the input of the downstream inverter. For the purpose of maximizing the direct reactive power transfer between the dc-link and the storage capacitor in the SHCC, peak voltage control is proposed to regulate the storage capacitor voltage such that power losses of the SHCC could be reduced at light load. Finally, simulation results are presented to verify the effectiveness of the proposed improved OCC on input current control and the peak voltage control on storage capacitor voltage regulation.**

I. Introduction

Two-stage inverters, composed by front-end dc-dc converters and downstream dc-ac inverters, never fail to fascinate researchers or engineers in the application where the input dc voltage source experiences a wide variation or mismatches with the required output ac voltage.

As commonly known, the instantaneous output power of a two-stage single-phase inverter pulsates at twice the output voltage frequency, giving rise to the SHC in the input of the downstream dc-ac inverter. This SHC will propagate back into the input dc voltage source and the front-end dc-dc converter, exerting a negative effect on their operational performance. The SHC in the input dc voltage sources like fuel cells and batteries will shorten their lifetime [1], while the SHC in the front-end dc-dc converter will produce extra power losses [2]. Hence, it is of necessity to eliminate the SHC in the input dc voltage source and the front-end dc-dc converter.

The approaches to eliminating the SHC can be preliminarily categorized into three types. The simplest approach is to increase the intermediate dc bus capacitor [3]. But relatively large capacitance is required to limit the SHC within certain extent. Normally, electrolytic capacitors, featured with high energy storage density, are considered as the best candidate for this large intermediate bus capacitor. Unfortunately, the lifespan of electrolytic capacitors is fairly short compared with that of the input voltage source like fuel cells or PV modules and declines by half with the rise of environmental temperature of every 10 °C [4]. Such trait makes electrolytic capacitors become the component that limits the lifetime of the inverter. Thus, it is recommended to be removed for the sake of extending the inverter's lifespan.

The second approach is to prevent the SHC in the downstream inverter from propagating back into the front-end dc-dc converter by applying elaborately designed control schemes to the front-end dc-dc converter. The most representative work is the current mode control proposed by C. Liu and J. Lai in [5]. This control scheme has very good SHC reduction performance. Owning to relatively low outer voltage loop bandwidth, however, the front-end dc-dc converter would suffer from poor dynamic respond during load transient. To make up this defect, a variety of control schemes which could give considerations to both of the SHC reduction and dynamic performance are put forward in [2], [6-8]. Unfortunately, all these control schemes enjoy superior SHC reduction performance on the basis of a smooth dc bus, which implies that large intermediate bus capacitor is still required and electrolytic capacitors are inevitable to be adopted as the intermediate bus capacitor.

The third approach is to counteract the SHC by active power decoupling method. With this approach, the pulsating power is dealt with a SHCC which involves a long lifetime auxiliary storage element instead of the large intermediate dc bus capacitor constituted by electrolytic capacitors. In [9], a ripple-port terminated with a film capacitor is embedded into the inverter. To realize SHC compensation, the voltage across the film capacitor is controlled to be sinusoidal with a leading phase of $\pi/4$ with respect to the inverter's output voltage. Nevertheless, the voltage reference is a function of the capacitance of the film capacitor. Thus, the capacitance variation would affect the SHC compensating performance. A bidirectional buck/boost converter with feed-forward digital control is proposed in [10] to compensate the SHC. This solution features with high power density. But the converter has to operate in DCM, which results in high current stress and large conduction losses in power switches. Moreover, the proposed feed-forward digital control does not work before

978-1-4673-9551-9/16 $31.00 © 2016 IEEE

the information of the inductance of the auxiliary inductor is fed back to the controller. Thus, the inductance variation caused by the load variation and environmental temperature variation would negatively affect the performance of SHC compensation. With the same control scheme, a buck-boost converter is suggested as the ripple eliminator to smooth dc-bus voltage [11]. This solution would face the same challenge as that proposed in [10]. Since the reactive power is indirectly transferred between the dc-link and the storage capacitor in the buck-boost converter, this solution may suffer from poor conversion efficiency.

In this paper, a SHCC with improved OCC is proposed to compensate the SHC in the downstream inverter. The paper is organized as follows. Section II gives the basic operation principle of the SHCC. After that, the improved OCC is proposed in Section III. To further reduce the power losses produced by the SHCC at light load, Section IV puts forward a peak voltage control to regulate the storage capacitor voltage. The simulation results are presented in Section V, followed by the conclusions in Section VI.

II. Operation Principle of the Second Harmonic Current Compensator

Fig. 1 presents the schematic diagram of the two-stage inverter with SHCC in this paper, in which power switch Q_b, rectifier diode D_b, filter inductor L_b and filter capacitor C_{bus} which also serves as the intermediate dc bus capacitor form the front-end boost converter; power switches Q_1~Q_4, filter inductor L_f, filter capacitor C_f and Load R_L constitute the downstream full-bridge inverter; power switches Q_{a1}~Q_{a2}, filter inductor L_s and storage capacitor C_s make up the SHCC and the input of the SHCC is equipped in parallel with the intermediate dc bus capacitor C_{bus}.

Fig. 2 gives the key operation waveforms of the investigated two-stage inverter with SHCC, in which v_o and i_o are the output voltage and current of the two-stage inverter respectively; P_{d2d} and p_o are the output power of the front-end dc-dc converter and the downstream dc-ac inverter respectively; i_{inv} and i_{d2d} are the input current of the downstream inverter (with harmonic current at switching frequency neglected) and the output current of the front-end dc-dc converter respectively; i_s is the current provided by the SHCC; v_{Cs} is the voltage across the storage capacitor and Δv_{Cs} is the ripple of v_{Cs}.

As seen from Fig.2, when $p_o > P_{d2d}$, the insufficient power will be provided by the SHCC. In this case, the SHCC will source current, i.e., i_s >0 and the storage capacitor is discharged. We call such operation as discharging mode. On the contrary, when $p_o < P_{d2d}$, the redundant power will be stored by the SHCC. In this scenario, the SHCC will sink current, i.e., i_s < 0 and the storage capacitor is charged. We call such operation as charging mode. Evidently, if the i_s equals to the SHC in the input of the downstream inverter, the front-end dc-dc converter would only needs to output dc component in i_{inv}. In the meantime, the bus capacitor will be only responsible for filtering the current harmonics at switching frequency which contributes to capacitance reduction to allow for the usage of film capacitor.

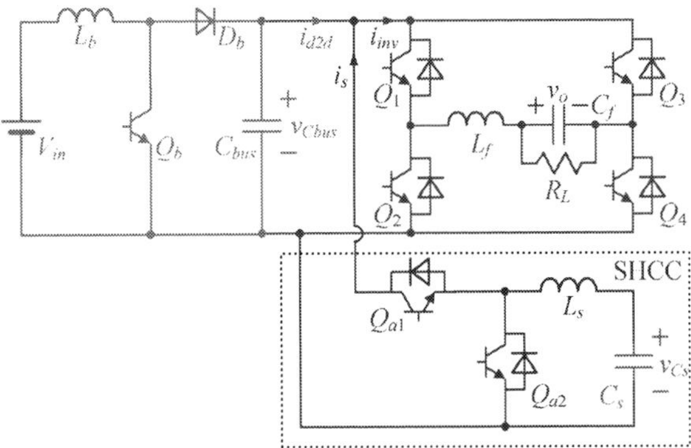

Fig. 1 Schematic diagram of the two-stage inverter with SHCC.

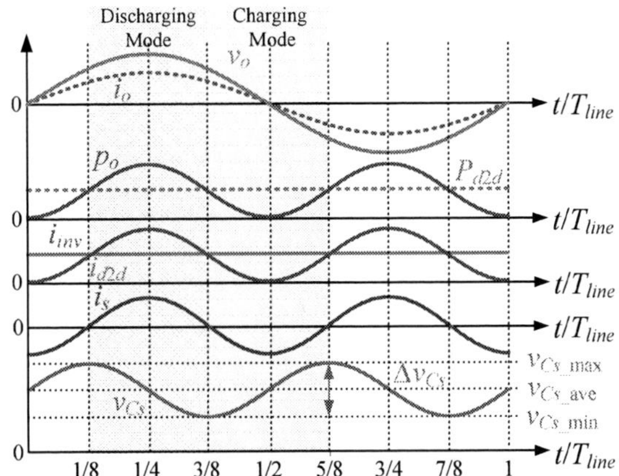

Fig. 2 Key operation waveforms.

III. Improved One-Cycle-Control Scheme

To make the SHCC provide the required SHC in i_{inv}, the average of the current in Q_{a1}, i.e., i_{Qa1} (or i_s), should equal to the SHC in i_{inv}. Based on such consideration, an improved OCC is proposed to control the input current of the SHCC.

Fig. 4 (a) and (b) give the key operation waveforms (Solid line) when the SHCC operates with the proposed improved OCC in charging mode and discharging mode respectively, in which i_{ref} is current reference; i_{sbias} and i_{rbias} are sample bias and reference bias added into the current sample of i_{Qa1}, i.e., i_{Qa1_s} and current reference respectively; i_{int} is the integral of the sum of i_{Qa1_s} and i_{sbias}. It should be emphasized that i_{sbias} is elaborated designed such that the sum of i_{Qa1_s} and i_{sbias} is always positive regardless of the load variation. As seen from Fig. 3, Q_{a1} and Q_{a2} complementarily conduct. At the beginning of each switching cycle, Q_{a1} is turned on. Q_{a1} is not turned off until i_{int} reaches the sum of i_{ref} and i_{rbias}. When i_{int} equals to the sum of i_{ref} and i_{rbias}, we have

$$
\begin{aligned}
i_{int} &= \frac{1}{T_s}\int_0^{d_y T_s}\left(H_i i_{Qa1} + i_{sbais}\right)dt \\
&= i_{ref} + i_{rbais} = H_i i_{SHC} + i_{rbais}
\end{aligned} \tag{1}
$$

where i_{SHC} is the SHC in i_{inv} and H_i is sampling factor of i_{Qa1}.

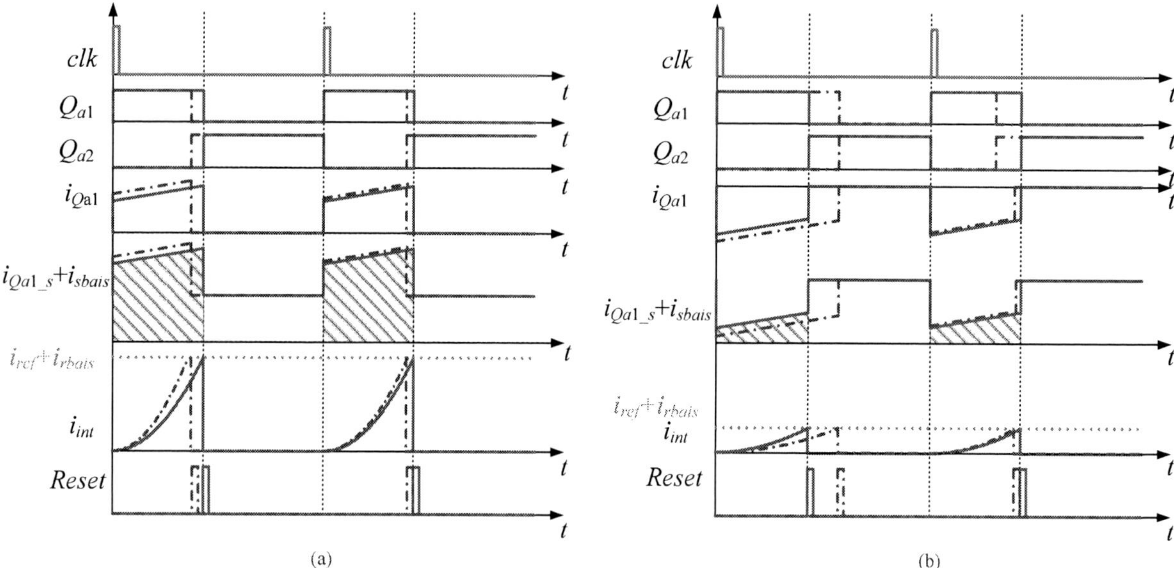

Fig. 3. Key operation waveforms of the SHCC with improved OCC. (a) Charging mode. (b) Discharging mode.

On the basis of (2), we could have

$$i_{rbais} = d_y i_{sbais} = \frac{v_{Cs}}{v_{Cbus}} i_{sbais} \tag{2}$$

IV. Regulation of the Voltage Across the Storage Capacitor

To possibly reduce the capacitance of the storage capacitor to allow for the usage of film capacitor, the voltage ripple across the storage capacitor is intentionally increased. Nevertheless, it should be noted that the voltage ripple is proportional to the inverter output power [12-14] and that the peak of the storage capacitor voltage is required to be within the dc bus voltage to maintain the normal operation of the SHCC.

If the average of the storage capacitor voltage is regulated as presented in Fig. 4, this average voltage should be designed on the basis of the full-load operation condition to keep the peak of the storage capacitor voltage within the dc bus voltage. On the other hand, however, the voltage ripple across the storage capacitor is expected to be relatively small with the designed average voltage under the light-load operation condition. Hence, the average voltage can be intentionally increased under such circumstance without affection of the normal operation of the SHCC.

Based on such consideration, a peak voltage control is proposed to regulate the voltage across the storage capacitor. With the peak voltage control, the average of the storage capacitor voltage could be adaptively increased with heavier inverter output power as shown in Fig. 5. Such practice could maximize the direct reactive power transfer between the dc-link and the storage capacitor [15], which could help save power losses produced by the SHCC under light-load operation condition.

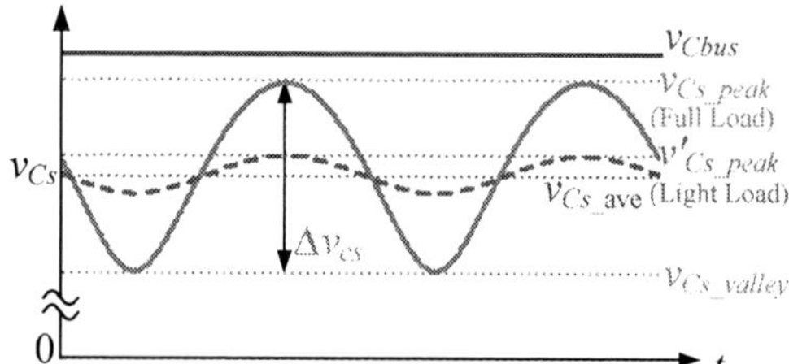

Fig. 4. Average voltage control.

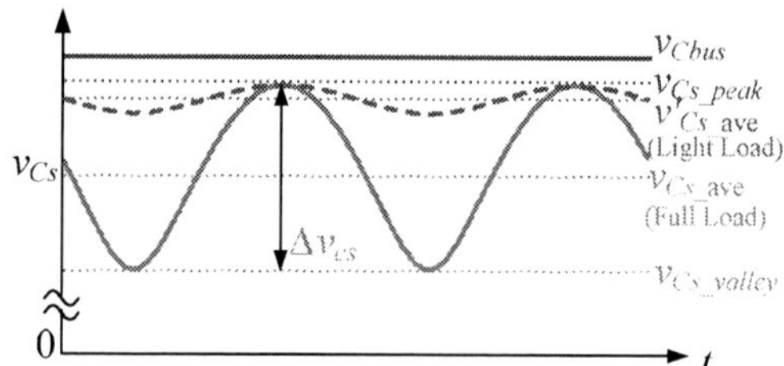

Fig.5. Peak voltage control.

V. Simulation Results

In this section, simulation results are presented to verify the effectiveness of the proposed improved OCC and the peak voltage control.

The main parameters in the the Saber simulation model are listed as follows.

- Input voltage: V_{in} = 110V±10%
- Intermediate bus voltage: V_{Cbus} = 400V
- Ouput voltage: V_o = 220V/50Hz
- Average of v_{Cs}: v_{Cs_ave} = 330V (Full-load)
- Peak of v_{Cs}: v_{Cs_peak} = 380V (Full-load)
- Output power: P_o = 1kVA
- Filter inductor in the boost converter: L_b = 360μH
- Intermediate bus capacitor: C_{bus} = 16μF

- Filter inductor in the SHCC: $L_s = 2\text{mH}$
- Storage capacitor: $C_s = 100\mu\text{F}$
- Sampled bias: $i_{sbias} = 5\text{V}$

Fig.6 gives the simulation waveforms of the SHCC with improved OCC and average voltage control, in which i_{SHC} is the extracted SHC in i_{inv}; i_{s_SHC} is the SHC provided by the SHCC; i_{err} is the error between i_{SHC} and i_{s_SHC}; v_{Cs} is the storage capacitor voltage; i_{Qa1} is the current in Q_{a1}; and i_{int} is the integral of the sum of i_{Qa1} and i_{sbias}. As seen from Fig. 6 (a) and (b), the SHCC could stably operate in both of the charging mode and discharging mode. In the meantime, it can be found from Fig.6 (c) and (d) that the SHCC could source and sink the SHC in i_{inv} with an error within 8% and the average of v_{Cs} is regulated at 330V regardless of load variation. Such results indicate that the proposed improved OCC enjoy good performance on controlling the input current of SHCC.

Fig.7 gives the simulation waveforms of the SHCC with improved OCC and peak voltage control. It can be found from Fig.7 (a) and (b) that the SHCC could also stably operate when the peak of v_{Cs} is regulated. Seen from Fig.7 (c) and (d), we can find that the peak of v_{Cs} maintains at 380V regardless of load variation and the average of v_{Cs} adaptively increase as inverter output power becomes heavier. This result could validate the feasibility of the proposed peak voltage control.

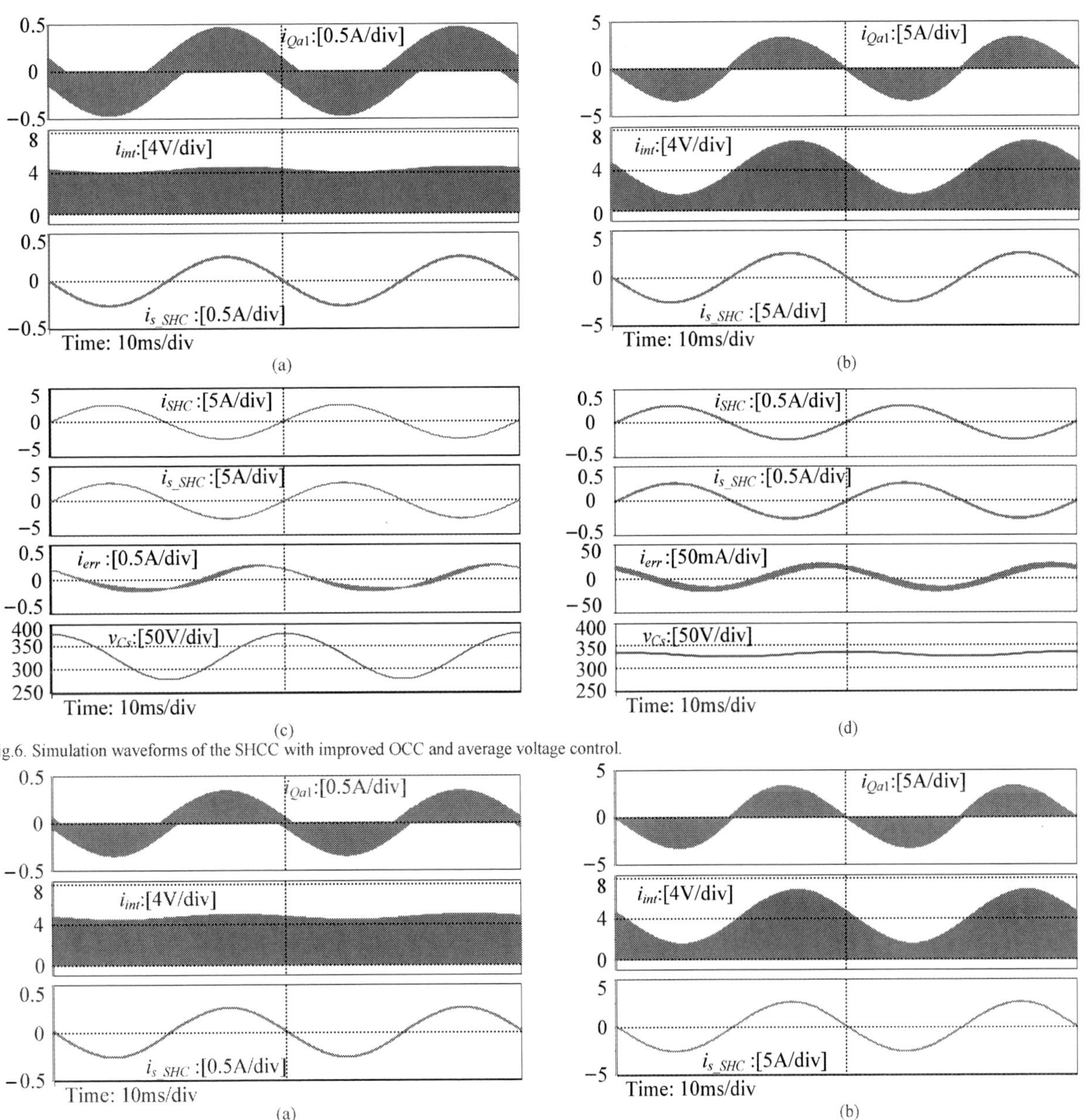

Fig.6. Simulation waveforms of the SHCC with improved OCC and average voltage control.

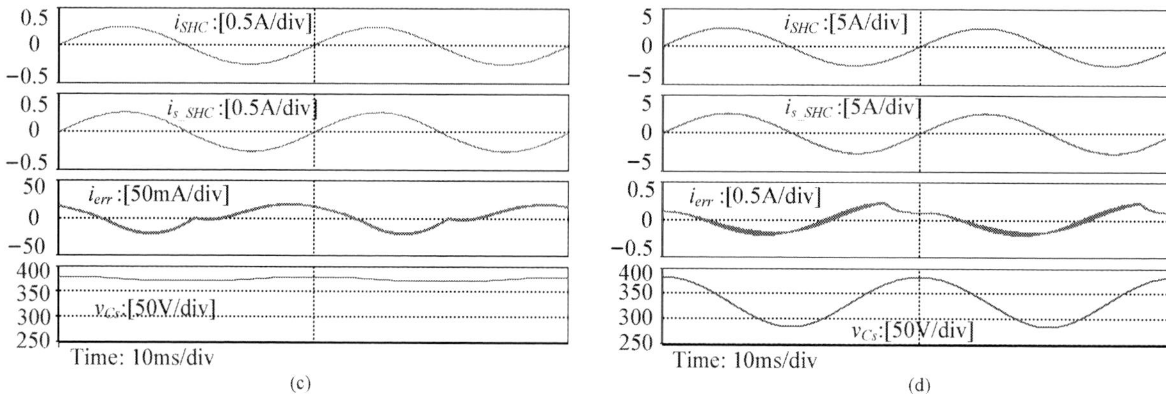

Fig.7. Simulation waveforms of the SHCC with improved OCC and peak voltage control.

VI. CONCLUSION

In this paper, a SHCC with improved OCC is proposed to help eliminate the SHC in the front-end dc-dc converter and the input dc voltage source. The improved OCC has good performance on the SHCC's input current control and promises the SHCC to operate in CCM to save conduction losses. To maximize the direct reactive power transfer between the dc-link and the storage capacitor, a peak voltage control is proposed to regulate the storage capacitor voltage. Finally, simulation results verify the feasibility of the improved OCC and peak voltage control.

REFERENCES

[1] G. Fontes, C. Turpin, S. A. Astier, et al, "Interactions between fuel cells and power converters: Influence of current harmonics on a fuel cell stack," *IEEE Trans. Power Electron.*, 2007, 22(2): 670–678.

[2] L. Zhang, X. Ren and X. Ruan, " A band-pass filter incorporated into the inductor current feedback path for improving dynamic performance of the front-end dc-dc converter in two-stage inverter," *IEEE Trans. Ind. Electron.*, 2014,61(5): 2316–2325.

[3] M. E. Schenck, J. Lai, K. Stanton, "Fuel cell and power conditioning system interactions ,"*Proc. IEEE Applied Power Electron. Conf. Expo.*, 2005: 114–120.

[4] Electrolytic Capacitors Application Guide, Espoo, Finland: Evox Rifa, 2001.

[5] C. Liu, J. Lai, "Low frequency current ripple reduction technique with active control in a fuel cell power system with inverter load,"*IEEE Trans. Power Electron.*, 2007, 22(4): 1429–1436.

[6] L. Zhang, X. Ruan and X. Ren, " Second-harmonic current reduction and dynamic performance improvement in the two-stage inverters: an output impedance perspective" *IEEE Trans. Ind. Electron.*, 2015,62(1): 394–404.

[7] G. Zhu, X. Ruan, L. Zhang and X. Wang, " On the reduction of second harmonic current and improvement of dynamic response for two-stage single-phase inverter," *IEEE Trans. Power Electron.*, 2015, 30(2): 1028–1041.

[8] L. Zhang, X. Ruan and X. Ren, " A generalized control scheme derivation approach for frond-end dc-dc converters in two-stage inverters," *Proc. IEEE Energy Conv. Congress Expo.*, 2015: 1563–1570.

[9] P. Krein, R. Balog and M. Mirjafari, " Minimum energy and capacitance requirements for single-phase inverters and rectifiers using a ripple port," *IEEE Trans. Power Electron.*, 2012, 27(11): 4690–4698.

[10] R. Wang, F. Wang, D. Boroyevich, et al, " A high power density single-phase PWM rectifier with active ripple energy storage ," *IEEE Trans. Power Electron.*, 2011, 26(5): 1430–1443.

[11] X. Cao, Q. Zhong and W. Ming, "Ripple eliminator to smooth dc-bus voltage and reduce the total capacitance required," *IEEE Trans. Ind. Electron.*, 2015, 62(4): 2224–2235.

[12] L. Gu, X. Ruan, M. Xu, and K. Yao, "Means of eliminating electrolytic capacitor in ac/dc power supplies for LED lighting," *IEEE Trans. Power Electron.*, 2009, 24(5): 1399–1408.

[13] S. Wang, X. Ruan, K. Yao, S-C. Tan, Y. Yang, and Z. Ye, "A flicker-free electrolytic capacitor-less AC-DC LED driver," *IEEE Trans. Power Electron.*, 2013, 27(11): 4540–4548.

[14] Y. Yang, X. Ruan, L. Zhang, J. Xiu and Z. Ye, " A feed-forward scheme for an electrolytic capacitor-less AC-DC LED driver to reduce output current ripple," *IEEE Trans. Power Electron.*, 2014, 29 (10): 5508–5517.

[15] X. Ren, X. Ruan , H. Qian, M. Li and Q. Chen, " Three-mode dual-frequency two-edge modulation scheme for four-switch buck-boost converter," *IEEE Trans. Power Electron.*, 2009, 24 (2): 499–509.

Frequency Adaptive Control of a Smart Transformer-Fed Distribution Grid

Zhi-Xiang Zou, Giovanni De Carne, Giampaolo Buticchi, Marco Liserre
Chair of Power Electronics
Christian-Albrechts University of Kiel
Kiel, Germany
zz, gdc, gibu, ml@tf.uni-kiel.de

***Abstract*—An advanced service provided by the Smart Transformer (ST) is the decoupling between the Medium Voltage (MV) and Low Voltage (LV) grids. In the LV side, the ST can modify the waveform's frequency in order to interact with the droop controllers of the local generators to control the power demand among the sources without affecting the MV grid. However, most of the existing controllers for power converters cannot guarantee good performances under variable frequency conditions. To address this issue, a frequency adaptive control scheme based on the Fractional-Order Repetitive Control (FORC) as well as Frequency Locked Loop (FLL) is proposed in this paper. This proposed scheme provides fast online parameter tuning capability in order to be highly adaptive to variable frequencies, and it can be implemented either in a ST converter or in distributed generators. In this work simulation and experimental results are provided to demonstrate the effectiveness and advantages of the proposed scheme.**

Keywords—Solid state transformer; distribution grid; frequency adaptive; frequency-locked loop; stability

I. Introduction

The increasing integration of Distributed Energy Resources (DERs) challenges the hosting capacity of distribution grids. As pointed out in recent literature, these intermittent distributed appliances will cause severe problems on the existing distribution grids in terms of voltage and current capacity limit violations, unintentional islanding, increased harmonic pollution, and reliability issue [1-3]. Moreover, these can propagate in other voltage level grids, like the MV ones. It is the case of the current imbalance in the LV grids, due to the presence of single-phase loads and generators, or voltage sags in the MV grids, that can affect sensitive loads in LV grids (*e.g.*, data center or high-precision electronic devices).

The ST, being a three-stage power electronics-based transformer, can be the solution to the aforementioned issues. In addition to transforming the voltage levels, it also provides ancillary services to the distribution grid, both in MV and LV side: improvement of the current and voltage profile [4], balancing services [5-6], and reliability improvement in the whole system [7]. In particular, the ST is able to decouple completely the LV grid from the MV grid thanks to the presence of the DC Link. While in the LV grid the ST controls the voltage waveform, in the MV grid it requests the active power demanded by the loads and regulates the injection of reactive power, (*i.e.* for voltage control). Of particular interest is the possibility to adapt the frequency to interact with the distributed generation [8] in order to change the power sharing between ST and distributed generation [9-10]. It is worth noticing that this solution has very limited impact on the main grid thanks to the electrical decoupling between the MV grid and LV grid.

Currently, most of the existing controllers for power electronics converters are designed to regulate signals with fixed fundamental frequency (*e.g.* 50/60 Hz signals) and offer accurate voltage/current control solutions [11]. However, in ST-fed distribution grid applications with variable frequency, these conventional control schemes incur in performance degradation [12-13]. To deal with this problem, the Fractional-Order Repetitive Control (FORC) plus Frequency Locked Loop (FLL)-based frequency adaptive control is proposed and implemented in both ST and DERs. Equipped with this proposed frequency adaptive control, the ST-fed distribution is able to modify online the power production and demand by changing the grid frequency while maintaining good performance.

This paper is organized as follows: the detailed frequency control of ST-fed distribution grid is given in Section II. In Section III, the stability of the control system is investigated considering the FLL. The proposed controller has been verified in Section IV with simulations performed in PLECS and experimental results have been performed in a Real Time Dynamic Simulator (RTDS) by means of the Control-Hardware-In-Loop (CHIL) method. Finally, conclusions are drawn in Section V.

II. Frequency Adaptive ST-Fed Distribution Grid

A. Frequency Control in Distribution Grid

Different from traditional distribution grid, the LV side voltage frequency of ST-fed distribution grid can be flexibly modified within a certain range according to the local grid codes (for example, ± 0.5Hz in Germany). Therefore, the power production and consumption of frequency-dependent appliances/loads can be adjusted as well following the frequency changes. Fig. 1 shows the power/frequency characteristics of DER and typical load in a distribution grid.

978-1-4673-9551-9/16 $31.00 © 2016 IEEE

To mimic the power system inertia, the grid-interfaced power electronic converters usually adopt droop controllers reproducing the *P-f* droop behavior of synchronous generators used in the conventional power system [14-15]. The *P-f* droop curve with inductive output impedance is shown as the red curve in Fig. 1. Meanwhile, it is well known that the load magnitude (especially the motor load) is also strongly dependent on the system frequency as the blue curves. Within this context, the main reason to employ a frequency control is to reestablish the balance between power generation and load consumption. In the case of a load increment ΔP (for example, from load curve 1 to curve 2), if the distribution grid decreases its frequency from ω_1 to ω_2, the contribution of DER will increase while the load consumption will decrease. The frequency will continuously decrease until a new power balance between power production and consumption is established, and vice versa. Two main advantages are provided by the frequency control: 1) the current flow through the transformer can be limited to avoid overloads; 2) the power rating of controllable load or storage systems installed in a distribution grid can be partly reduced according to the droop coefficients of DERs and loads.

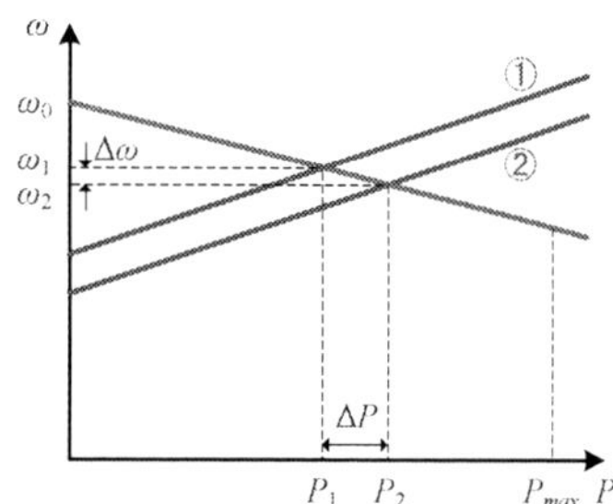

Fig. 1. Power/frequency characteristics of DER (red curve) and load (blue curves)

The simplified system configuration of the ST-fed distribution grid with the frequency control scheme is presented in Fig. 2. Droop control is employed both in ST and DER to provide frequency and power reference, respectively. In normal operation scenario, both the frequency and voltage amplitude of the LV side of the ST are kept constant at their nominal values. When the load increases, the frequency of LV side will be regulated following the droop curve as soon as the RMS current of ST exceeds the security limit. On the other hand, the variable frequency will be detected by the FLL of DER, whose dynamic response is much faster than the maximum rate of grid frequency [16]. According to the *P-f* droop curve, the DER is solicited to output desired real power/current under variable frequency condition.

B. Frequency Aaptive Control Strategy

To achieve good control performance, a frequency-adaptive voltage/current control scheme is proposed for both the ST LV side inverter and the DER grid-interfaced inverter. The overall control system is presented in Fig. 3(a), in which a FORC controller $G_{forc}(z^{-1})$ is plugged into the feed-forward channel of

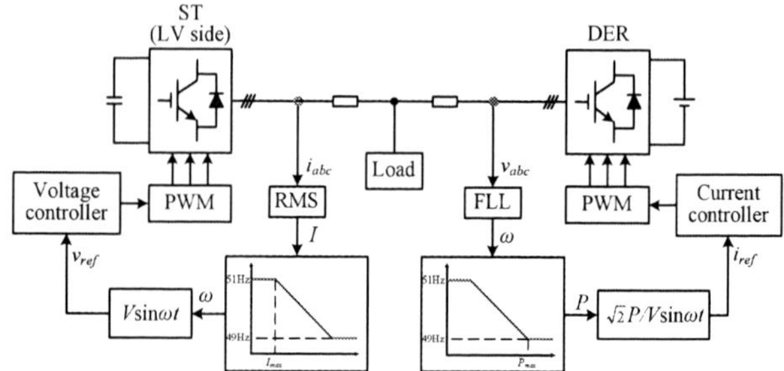

Fig. 2. Frequency control of a ST-fed distribution grid

the system. $G_c(z^{-1})$ and $G_p(z^{-1})$ are the transfer functions of conventional voltage/current controller, and control plant (DC/AC inverter); $R(z^{-1})$, $Y(z^{-1})$, and $D(z^{-1})$ are the reference, output, and disturbance of the control system; $E(z^{-1})$ and $U_r(z^{-1})$ are the tracking error and the FORC output. Study cases on the FORC-based system have been done in literature for several power applications, such as the grid-interfaced power converters [13], programmable AC power supply [12], active power filters [17], and active noise cancelations. The main idea of FORC controller is derived from Conventional Repetitive Control (CRC), whose transfer function is

$$G_r(z^{-1}) = k_r \frac{z^{-N}Q(z^{-1})}{1 - z^{-N}Q(z^{-1})} \tag{1}$$

where k_r is the control gain; $Q(z^{-1})$ is the transfer function of a low-pass filter that improves the system stability; N is the order of CRC and can be expressed as f_s / f. Here, f_s and f are the sampling frequency and fundamental frequency, respectively. When f is fixed, a proper choice of f_s allows to make sure that N is an integer value. However, when the fundamental frequency is variable, the N would often be fractional in the case of fixed sampling rate. A fractional order N which indicates a Fractional Delay (FD) cannot be directly implemented in the practical digital controller. To address this issue, the FORC with a fixed sampling rate is proposed and the schematic diagram of the proposed control system is shown in Fig. 3(b), where $G_f(z^{-1})$ is the transfer functions of a phase lead compensation filter, N_i and F are the integral and fraction orders of FORC ($N = N_i + F$).

As mentioned before, the main problem is to properly deal with the potential FD in a FORC system caused by the time-varying fundamental frequency. According to the FD filters design method [18-19], the FD z^{-F} of Fig. 3(b) can be approximated by a Lagrange Interpolation polynomial Finite-Impulse-Response (FIR) filter as following

$$z^{-F} \approx \sum_{k=0}^{n} A_k z^{-k} \tag{2}$$

where $k \in \mathbf{N}$, and the Lagrange coefficient A_k of (2) can be calculated as

$$A_k = \prod_{\substack{i=0 \\ i \neq k}}^{n} \frac{F-i}{k-i}, \qquad k, i = 0, 1, \ldots, n. \tag{3}$$

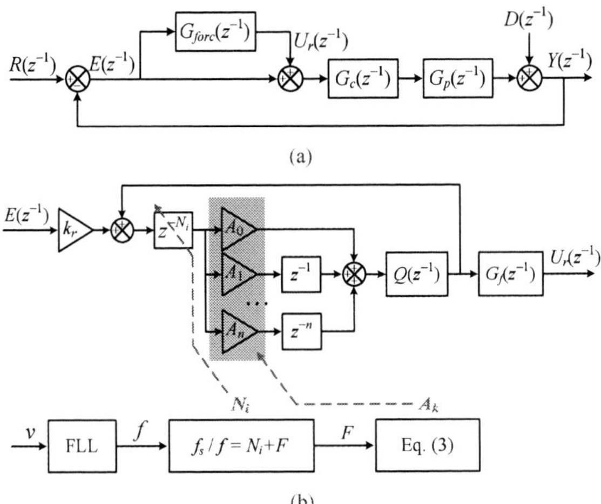

Fig. 3. Schematic diagram of frequency adaptive control system: (a) overall control system, (b) FORC

Substituting (2) and (3) into (1), the transfer function of FORC will be obtained as following

$$G_{forc}(z^{-1}) = \frac{U_r(z^{-1})}{E(z^{-1})} = k_r \frac{z^{-N_i}\sum_{k=0}^{n} A_k z^{-k} Q(z^{-1})}{1 - z^{-N_i}\sum_{k=0}^{n} A_k z^{-k} Q(z^{-1})} G_f(z^{-1}) \quad (4)$$

According to the properties of the Lagrange Interpolation polynomial, there is always an approximation remainder between the theoretical FD and the polynomial. The remainder term of FD can be written as follows

$$R_n = z^{-F} - \sum_{k=0}^{n} A_k z^{-k} = \frac{\xi^{-F-n}\prod_{i=0}^{n-1}(-F-i)}{(n+1)!}\prod_{i=0}^{n}(F-i) \quad (5)$$

where $\xi \in [T_k, T_{k+1}]$ with T_k and T_{k+1} being the kth and $k+1$th sampling intervals, respectively. Fig. 4(a) compares the magnitude responses of Lagrange Interpolation-based FD with different polynomial degrees. With the increase of the polynomial degree n, better approximation accuracy can be achieved. It can be seen that the approximated FIR filter with the degree $n \geq 2$ can achieve a good approximation towards the FD z^{-F} within the Nyquist frequency. For example, the magnitudes are close to the nominal value (abs = 1) within the bandwidth of 75% Nyquist frequency when n = 3, which indicates the proposed FORC can exactly track fractional period signals within the frequency range. To evaluate the tracking ability, the approximation remainder norms of FORC with different sampling frequencies are analyzed as shown in Fig. 4(b) and also compared with those of CRC. When the fundamental frequency varies in a certain range, the remainder norms of CRC change drastically and some of them are relatively high, indicating the tracking error could not be eliminated under variable frequency condition. However, the norms of FORC always remain in a lower level at the neighborhood of fundamental frequency. In addition, the averaging norm at the frequency neighborhood will decrease when the sampling frequency increases. It can be concluded that a higher sampling frequency should be employed to reduce the RMS value of tracking error of FORC or CRC.

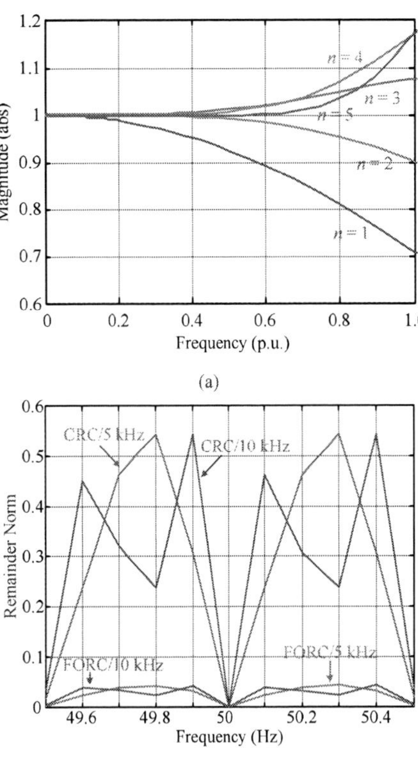

Fig. 4. Performance evaluation: (a) magnitude response of Lagrange Interpolation-based FD with different degrees, and (b) comparisons of remainder norm between CRC and FORC under different sampling frequencies

The FORC scheme provides a general approach to track or eliminate of any periodic signal with an arbitrary fundamental frequency and fixed sampling frequency. It is evident that the transfer function (4) becomes a conventional repetitive control as (1) in the case of an integral order ($F = 0$). From Fig. 3(b), it is shown that two different time scales have been utilized in the control system: 1) the changing rate of delay orders (N_i and F) is highly dependent on the dynamic response of frequency detector, *e.g.* FLL in this case; 2) the update rate of A_k depends on the sampling frequency. In the distribution grid control applications, both N_i and F change relatively slowly compared with the sampling rate of the FORC system. Moreover, the Lagrange Interpolation-based FD only needs a small number of sums and multiplications for a fast online update of A_k so that it can be easily implemented in the real-time control system along with high approximation accuracy.

III. Stability Analysis

A. System Stability without FLL

The closed-loop transfer function of the frequency adaptive control system can be derived from Fig. 3(a). Before the FORC plugs in, the transfer function of the former closed-loop system is

$$H(z^{-1}) = \frac{Y(z^{-1})}{R(z^{-1})} = \frac{G_c(z^{-1})G_p(z^{-1})}{1+G_c(z^{-1})G_p(z^{-1})} \tag{6}$$

When the FORC is employed, the error transfer function of the overall system can be derived as following

$$\begin{aligned} G_e(z) &= \frac{E(z^{-1})}{R(z^{-1}) - D(z^{-1})} \\ &= \frac{(1-z^{-N_i}\sum_{k=0}^{n} A_k z^{-k})\cdot(1+G_c(z^{-1})G_p(z^{-1}))^{-1}}{1-z^{-N_i}\sum_{k=0}^{n} A_k z^{-k} Q(z^{-1})(1-k_r G_f(z^{-1})H(z^{-1}))} \end{aligned} \tag{7}$$

From (7), the FORC-based frequency adaptive control system is asymptotically stable if the following two conditions hold: 1) the roots of $1 + G_c(z^{-1})G_p(z^{-1}) = 0$ are inside the unity circle; 2) the roots of $1-z^{-N_i}\sum_{k=0}^{n} A_k z^{-k} Q(z^{-1})(1-k_r G_f(z^{-1}) \cdot H(z^{-1}))$ are inside the unity circle.

The first stability condition requests the former closed-loop system to be stable before the FORC plugs in. The second stability condition can be rewritten as

$$\begin{aligned} \left\|1-k_r G_f(z^{-1})H(z^{-1})\right\| &< \left\|Q(z^{-1})\cdot\sum_{k=0}^{n} A_k z^{-k}\right\|^{-1} \\ &\le \left\|Q(z^{-1})\right\|^{-1}\cdot\left\|\sum_{k=0}^{n} A_k z^{-k}\right\|^{-1} \end{aligned} \tag{8}$$

As seen from Fig. 4(b), the magnitudes of FIR filters are always closed to 1 within the band-pass, so that

$$\left\|\sum_{k=0}^{n} A_k z^{-k}\right\|^{-1} \approx \left\|\sum_{k=0}^{n} A_k z^{-k}\right\| \to 1 \tag{9}$$

Substituting (9) into (8), it can be seen that the stability condition of FORC-based system turns to be the same as that of CRC systems. Moreover, considering that the characteristics of the proposed FD filter is way similar to that of $Q(z^{-1})$, the low-pass filter can be omitted in some special cases. For example, when the designed bandwidth of FD is close to the cut-off frequency of $Q(z^{-1})$, $\left\|Q(z^{-1})\right\|\cdot\left\|\sum_{k=0}^{n} A_k z^{-k}\right\| \approx \left\|\sum_{k=0}^{n} A_k z^{-k}\right\|$ within the cut-off frequency/bandwidth. In this regard, the stability condition of FORC-based system is still fulfilled without the need of $Q(z^{-1})$. In the normal conditions, since the cut-off frequency of $Q(z^{-1})$ is usually lower than the bandwidth of the FD, the characteristics of $Q(z^{-1})$ is more decisive to the system stability. With this consideration, most of design criteria of the well-known CRC-based systems can be also used for the FORC analysis and design.

B. System Stability with FLL

As discussed in Section III-B, a FLL is employed in DER to provide the real-time measurement of frequency for the online update of the Lagrange coefficients. With properly designed parameters, FLL can provide faster dynamic response than other phase locked loops under variable frequency conditions [20]. For example, the gains of FLL $k = \sqrt{2}$ and $\gamma = 100$ are used in our system and the detailed characteristics are shown in TABLE I.

TABLE I. CHARACTERISTICS OF FLL

Parameter	Value
Nominal frequency	50 Hz
Settling time	45 ms
Overshoot (to nominal value)	3.1 %
Frequency range	47~52 Hz

According to the FLL characteristics, the settling time is around 50ms which is much faster than the maximum rate of grid frequency variation (*e. g.* 1 Hz/s) and much slower than the sampling frequency of the current loop of DER system (typically from 5 kHz to 20 kHz). Considering a 5 kHz sampling frequency system as an example, in case of 1 Hz frequency variation, the FLL takes 500 ms to increase or decrease the order of integer delay N_i by 1, and at the same time the order of fractional delay would change only 4×10^{-4} over one sampling interval. Substituting this value of 4×10^{-4} into (3), the variations of A_k are less than 1×10^{-4}. In other words, the change of the Lagrange coefficients in one sampling interval is negligible so that has very limited effects on the FORC output As a result, the overall control system can be treated as a Linear Time-Invariant (LTI) system in the stability analysis. With these considerations, a schematic diagram of the overall system is shown in Fig. 5, where an outer-loop incorporating the FLL is added, and the input signal R is the amplitude of the reference, v_{pcc} is the PCC voltage.

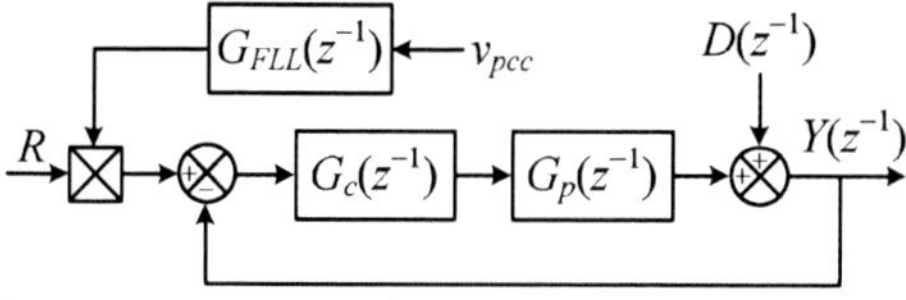

Fig. 5. Block diagram of DER control system with FLL

It is worth noting that the stability conditions in Section III-A are derived from the generic closed-loop system so that they are still applicable to the FLL-considered system. However, the closed-loop transfer function of (6) needs to be revised as following:

$$Y(z^{-1}) = H(z^{-1})G_{FLL}(z^{-1})\cdot R\cdot v_{pcc} \tag{10}$$

where $G_{FLL}(z^{-1})$ is the transfer function of FLL and its *s*-domain representation is given in (11). From [16], it can be seen that the FLL behaves as a nonlinear system. To obtain the transfer function, linearization procedure must be carried out and the transfer function can be derived as follows

$$G_{FLL}(s) = \frac{k^2\omega_0 s}{s^3 + k\omega_0 s^2 + \omega_0^2 s + k\gamma\omega_0} \quad (11)$$

where k and γ are the gains of FLL, which are $\sqrt{2}$ and 100, respectively, for the 47-52Hz applications, ω_0 is the fundamental frequency with 50Hz. Discretizing (11) and substituting that into (10), a pole-map can be drawn based on the transfer function of (10) and shown in Fig. 6. Compared with the pole-map of the control system without FLL, three additional poles (red poles) are introduced and are on the left half plane which fulfills the first stability condition. It can be seen that the system with a FLL has lower bandwidth but better rejection of grid disturbances.

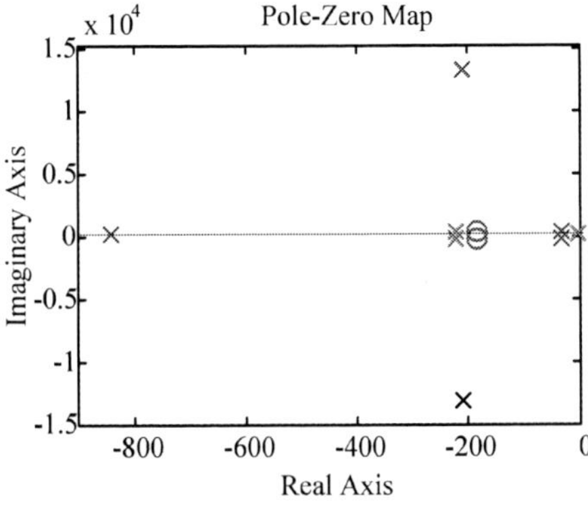

Fig. 6. Pole-zero map of the overall control system with FLL: closed-loop system poles without FLL (blue crosses), FLL additional poles (red crosses)

IV. Simulation and Experimental Verifications

This section provides the simulation results obtained implementing the CRC and FORC control in MATLAB and the verification of the proposed method with the Control-Hardwar-In-Loop technique by means of a Real Time Digital Simulator.

A. Simulation Results

To validate the proposed strategy, the ST-fed distribution grid of Fig. 2 has been simulated. The nominal voltage of ST LV side is 230V (RMS), 50Hz, and the security limit of the ST LV side current is set as 14.2A (RMS) in this study case. As first approximation, it has been assumed that both the voltage and current of ST LV side are sinusoidal, thus the RMS values are adopted for the thresholds of frequency control. The power generation of DER is 3.5kW (Power Factor (PF) = 1) at the nominal frequency, and it is connected to the Point of Common Coupling (PCC) via a feeder, whose line impedance is Z_g = 0.03 + j0.06 Ω. The total load demand is 7kW (PF = 0.85), while the linear load and nonlinear load demands are both 3.5kW. At t = 0.5s, the total load demand increases from 7kW to 8.75kW. As soon as the ST LV side current exceeds the security limit, the frequency control is activated. As shown in Fig. 7(a), the current of the ST LV side decreases till below the security limit and reaches its new steady-state by means of frequency control. At the same time, the voltage amplitude of ST reverts to its normal state in less than one cycle and remains the same during all the frequency control stage. Fig. 7(b) shows frequency and power production details during the frequency control stage. When the load increases, the frequency reference of ST LV side voltage starts to decrease (with the rate of 0.28Hz/sec) instantaneously in order to reestablish the power balance within the distribution grid. After the frequency drop, the FLL can fast and accurately respond to the frequency variation and provides good frequency performance to the DER. Following the frequency, the power production of DER proportionally increases according to the given P-f droop curve. At t = 1.2s, the power balance of the distribution grid is established again and the final frequency holds the line of 49.8Hz.

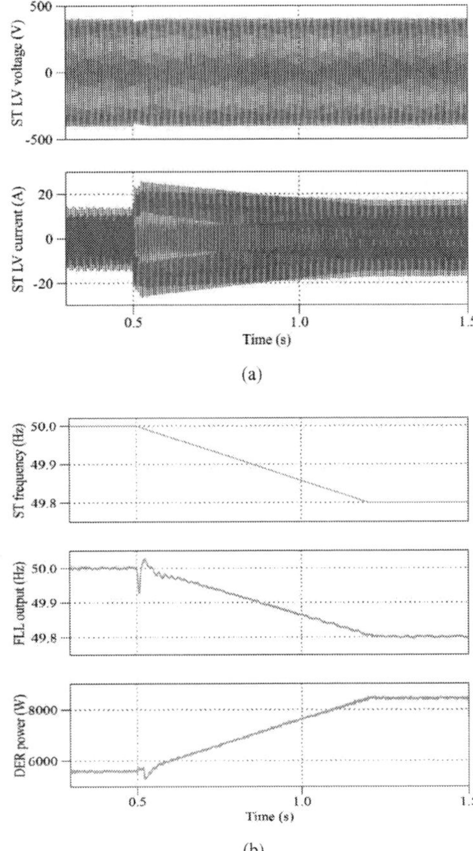

Fig. 7. The ST-fed distribution grid with frequency control: (a) ST LV side voltage and current, (b) reference frequency of ST LV voltage, freuqnecy detected by FLL of DER, and DER output power (from top to bottom)

Fig. 8 shows the performance comparisons of the ST-fed distribution grid at the neighboring frequency around 49.8Hz when different control strategies are used. In the first case, conventional repetitive controllers are employed in both the ST voltage controller and the DER current controller. From Fig. 8(a) and (b), it is clear that the ST voltage and the DER current contain considerable harmonics under variable frequency condition. In the second case, the proposed FORC controllers are implemented in the ST voltage controller and the DER current controller. Although a considerable amount of nonlinear loads are connected to the grid, the FORC enables the ST and the DER to provide much better quality sinusoidal voltage and current and offers very good voltage/current harmonic mitigation capacity. Generally speaking, Fig. 8(c) and (d) demonstrate that the FORC is adaptive to frequency variation, whereas conventional controllers are not.

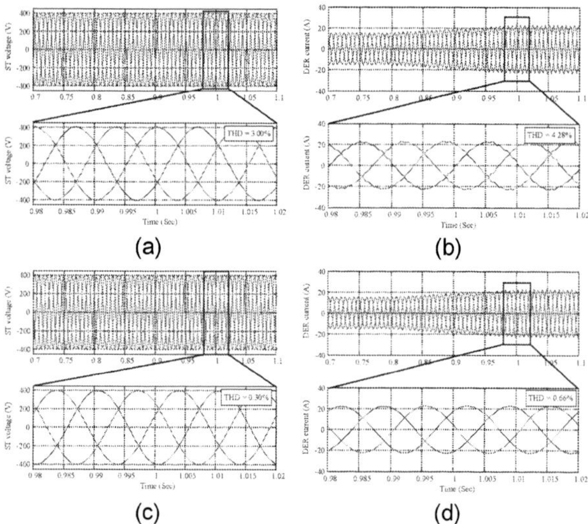

Fig. 8. Voltage/Current waveforms of a ST-fed distribution grid during the frequency control: (a) CRC-controlled ST LV side voltage, (b) CRC-controlled DER output current, (c) FORC-controlled ST LV side voltage, and (d) FORC-controlled DER output current

B. Experimental Results

The effectiveness of the FORC controller has been validated experimentally by means the Control Hardware In-Loop method using RTDS. This method allows simulating the converter behavior using a real controller implemented in an external device. The lab facility that performs this experiment is shown in Fig. 9. The voltage measurement at the ST bus are taken from the RTDS and sent to the dSPACE, where the control is implemented. The dSPACE system reads the new voltage measurements and generates the PWM signals for RTDS by the designed controller. Here the PWM signals are directly imposed to the ST LV side simulated in a small-time step subsystem. This enables a more realistic simulation of the power semiconductors, due to the reduced time steps (1-2 μs) with respect to the main grid (50 μs). The LV grid implemented in RTDS is a modified European LV distribution network benchmark [21], shown in Fig. 9 as well. With respect to the original grid the loads are balanced and the wind turbine has been neglected in order to reduce the computational effort in RTDS. The controls of local PV and battery systems have been simplified and the frequency-adaptive control scheme and FLL have not been utilized in this grid. The effectiveness of the proposed method has been proved in presence of disturbances, as a three-phase 5th harmonic current source of 10 A, simulating a non-linear load, installed at the bus 14.

The experimental results have been carried out comparing the performances of the CRC and the FORC in nominal conditions (50 Hz) and with lower grid frequency (49.8 Hz).

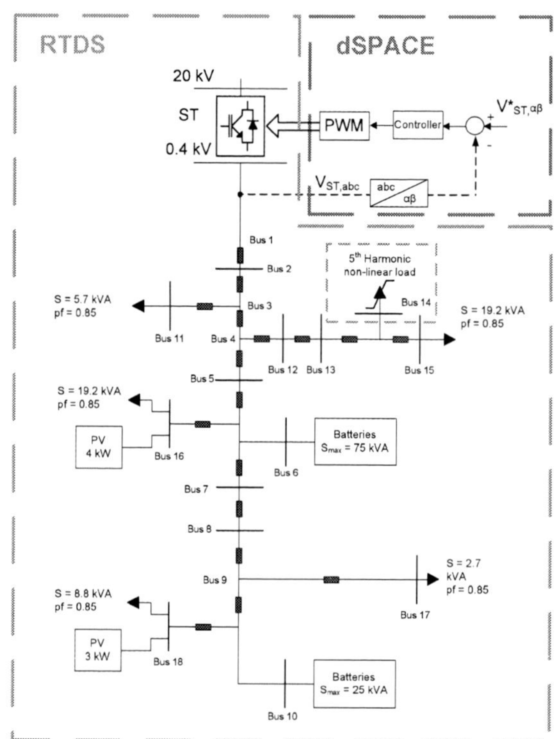

Fig. 9. Modified CIGRE European LV distribution network benchmark implemented in RTDS (blue square), LV side ST control implemented in dSPACE (red square), 5th harmonic non-linear load (green square)

As can be noticed from Fig. 10(a) and (c), where the voltage waveforms over a time window of 40 ms have been plotted, the CRC and the FORC have comparable performances at the nominal frequency. When the frequency changes, the performance of the two methods differs substantially, as shown in Fig 10(b) and (d). The CRC causes the voltage waveform to decrease its amplitude *e.g.* 290. When the FORC is implemented in the ST control the amplitude is kept to the nominal value of 325V peak and no harmonic content is present in the voltage waveform. Similar consideration can be drawn from the current waveform analysis. Although the performance is similar at the nominal frequency, Fig. 11(a) and (c), the amplitude of the current is heavily affected when the CRC is applied.

Taken in account the constant impedance nature of the loads installed in the simulated grid, a lower voltage value leads to a lower current request from the loads, affecting the power quality. When the FORC is employed as controller, the current has the same amplitude of the 50 Hz case, guaranteeing the same power quality standard like the 50 Hz case.

V. Conclusions

This paper proposed a frequency-adaptive ST-fed distribution grid. In this scenario, the ST is able to flexibly modify the LV side frequency with the purpose of avoiding overload and reducing the power rating of controllable load or storage system. The proposed FORC technology ensures better performance of the ST and the power electronics interfaced-DERs in the distribution grid under variable frequency condition. Comprehensive control design and stability analysis considering FLL are presented in this paper. A study case of the proposed frequency-adaptive ST-fed distribution grid is performed in both the simulation and experiment platform. The results prove the power sharing ability of frequency-adaptive ST-fed distribution grid and the effectiveness of the proposed control strategy under variable frequency conditions.

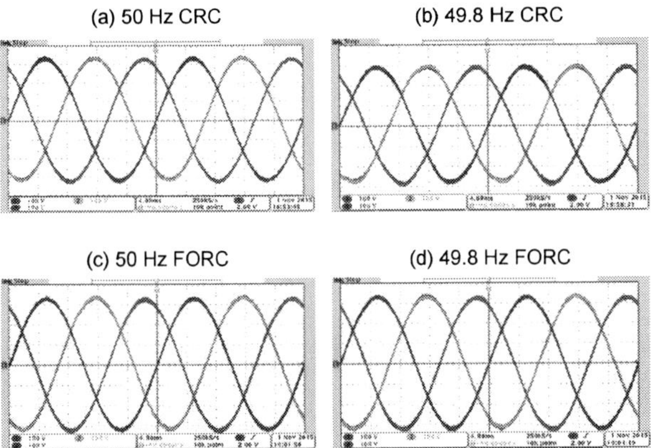

Fig. 10. Voltage waveforms of a ST-fed distribution grid during the frequency control: (a) CRC 50Hz, (b) CRC 49.8 Hz, (c) FORC 50Hz, and (d) FORC 49.8 Hz

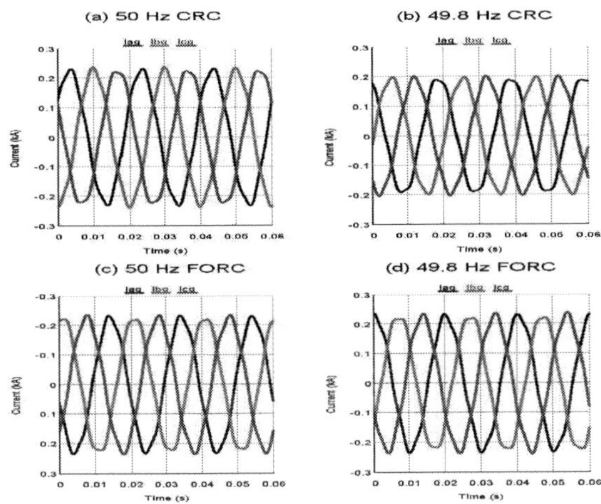

Fig. 11. Current waveforms of a ST-fed distribution grid during the frequency control: (a) Classical Control 50Hz, (b) Classical Control 49.8 Hz, (c) FORC 50Hz, and (d) FORC 49.8 Hz

References

[1] A. Bhowmik, A. Maitra, S.M. Halpin, and J.E. Schatz, "Determination of allowable penetration levels of distributed generation resources based on harmonic limit considerations," *IEEE Transactions on Power Delivery*, vol. 18, no. 2, pp. 619-624, Apr. 2003.

[2] V.H.M. Quezada, J.R. Abbad, and T.G.S. Román, "Assessment of energy distribution losses for increasing penetration of distributed generation," *IEEE Transactions on Power Systems*, vol. 21, no. 2, pp. 533-540, May 2006.

[3] E.J. Coster, J.M.A. Myrzik, B. Kruimer, and W.L. Kling, "Integration issues of distributed generation in distribution grids," *Proceedings of the IEEE*, vol. 99, no. 1, pp. 28-39, Jan. 2011.

[4] G. De Carne, M. Liserre, K. Christakou, and M. Paolone, "Integrated voltage control and line congestion management in active distribution networks by means of smart transformers," in *IEEE 2014 International Symposium on Industrial Electronics*, 2014, pp. 2613-2619.

[5] G. De Carne, G. Buticchi, M. Liserre, C. Yoon, and F. Blaabjerg, "Voltage and current balancing in low and medium voltage grid by means of smart transformer," in *IEEE 2015 Power & Energy Society General Meeting*, 2015, pp. 1-5.

[6] C. Kumar and M. Liserre, "Operation and control of smart transformer for improving performance of medium voltage power distribution system," in *IEEE 6th International Symposium on Power Electronics for Distributed Generation Systems*, 2015, pp. 1-6.

[7] A.Q. Huang, M.L. Crow, G.T. Heydt, J.P. Zheng, and S.J. Dale, "The future renewable electric energy delivery and management (FREEDM) system: The energy internet," *Proceedings of the IEEE*, vol. 99, no. 1, pp. 133-148, Jan. 2011.

[8] G. Buticchi, M. Liserre, D. Barater, C. Concari, A. Soldati, and G. Franceschini, "Frequency-based control of a micro-grid with multiple renewable energy sources," in *IEEE 2014 Energy Conversion Congress and Exposition*, 2014, pp. 5273-5280.

[9] G. De Carne, G. Buticchi, M. Liserre, and C. Vournas, "Frequency-based overload control of smart transformers," in *IEEE 2015 PowerTech*, 2015, pp. 1-5.

[10] G. De Carne, G. Buticchi, M. Liserre, P. Marinakis, and C. Vournas, "Coordinated frequency and voltage overload control of smart transformers," in *IEEE 2015 PowerTech*, 2015, 1-5.

[11] J. Rocabert, A. Luna, F. Blaabjerg, and P. Rodríguez, "Control of power converters in AC microgrids," *IEEE Transactions on Power Electronics*, vol. 27, no. 11, pp. 4734-4749, Nov. 2012.

[12] Z. Zou, K. Zhou, Z. Wang, and M. Cheng, "Fractional-order repetitive control of programmable AC power sources," *IET Power Electronics*, vol. 7, no. 2, pp. 431-438, Mar. 2015.

[13] Y. Yang, K. Zhou, H. Wang, F. Blaabjerg, D. Wang, and B. Zhang, "Frequency adaptive selective harmonic control for grid-connected inverters," *IEEE Transactions on Power Electronics*, vol. 30, no. 7, pp. 3912-3924, Jul. 2015.

[14] R. Brandt, "Theoretical approach to speed and tie line control," *Transactions of the American Institute of Electrical Engineers*, vol. 66, no. 1, pp. 24-30, Jan. 1947.

[15] N. Cohn, "Recollections of the evolution of realtime control applications to power systems," *Automatica*, vol. 20, no. 2, pp. 145-162, Mar. 1984.

[16] P. Rodríguez, A. Luna, I. Candela, R. Mujal, R. Teodorescu, and F. Blaabjerg, "Multiresonant frequency-locked loop for grid synchronization of power converters under distorted grid conditions," *IEEE Transactions on Industrial Electronics*, vol. 58, no. 1, pp. 127-138, Jan. 2011.

[17] Z. Zou, K. Zhou, Z. Wang, and M. Cheng, "Frequency-adaptive fractional-order repetitive control of shunt active power filters," *IEEE Transactions on Industrial Electronics*, vol. 62, no. 3, pp. 1659-1668, Mar. 2015.

[18] T. I. Laakso, V. Valimaki, M. Karjalainen, and U. K. Laine, "Splitting the unit delay," *IEEE Signal Processing Mag.*, vol. 13, no. 1, pp. 30–60, Jan. 1996.

[19] Y. Wang, D. Wang, B. Zhang, and K. Zhou, "Fractional delay based repetitive control with application to PWM DC/AC converters," in *16th IEEE International Conference on Control Applications*, 2007, pp.928-933.

[20] R. Teodorescu, M. Liserre, and P. Rodriguez, *Grid converters for photovoltaic and wind power systems*, Wiley, Chichester, 2011.

[21] "*Benchmark system for network integration of renewable and distributed energy resources c06.04.02*," CIGRE, Tech. Rep., 2014.

A Synchronization Scheme for Single-Phase Grid-Tied Inverters under Harmonic Distortion and Grid Disturbances

Lenos Hadjidemetriou, Elias Kyriakides
Department of Electrical and Computer Engineering
KIOS Research Center, University of Cyprus
Nicosia, Cyprus
hadjidemetriou.lenos@ucy.ac.cy, kyriakides@ieee.org

Yongheng Yang, Frede Blaabjerg
Department of Energy Technology
Aalborg University
Aalborg, Denmark
yoy@et.aau.dk, fbl@et.aau.dk

***Abstract*—Synchronization is a crucial aspect in grid-tied systems, including single-phase photovoltaic inverters, and it can affect the overall performance of the system. Among prior-art synchronization schemes, the Multi Harmonic Decoupling Cell Phase-Locked Loop (MHDC-PLL) presents a fast response under grid disturbances and high accuracy under harmonic distortions. However, major drawbacks of the MHDC-PLL include increased complexity and inaccurate response under non-nominal frequencies, which may occur in practical applications. Thus, this paper proposes strategies to address these issues. At first, a novel re-formulation of an equivalent decoupling cell is proposed for reducing the implementation complexity, and then a frequency adaptive quadrature signal generator for the MHDC-PLL is proposed to enable an accurate response even under non-nominal frequencies. Simulation and experimental results are provided, which show that the proposed synchronization (i.e., the frequency adaptive MHDC-PLL) can achieve a fast and accurate response under any grid disturbance and/or severe harmonic conditions.**

***Keywords*—grid tied inverter; harmonic distortion; phase locked loop; photovoltaics; synchonization.**

I. INTRODUCTION

The integration of roof-top photovoltaic (PV) systems has rapidly increased in recent years, where a single-phase configuration is preferable. Such single-phase PV systems employ grid-side power electronic inverters to properly inject the produced energy into the power grid [1], as it is exemplified in Fig. 1. The inverter controller ensures a proper operation of the grid-tied PV systems [1]-[5]. The inverter controller is based on a synchronization unit, an active (P) and reactive (Q) power controller, a current controller and a Maximum Power Point Tracker (MPPT). As shown in Fig. 1, the synchronization unit is a crucial part of the inverter controller, since it can affect the inverter controllers and as a result the entire system operation.

The synchronization unit is usually performed by a Phase-Locked Loop (PLL) algorithm and it is responsible for a fast and accurate estimation of the grid voltage and phase angle at the Point of Common Coupling (PCC) under any grid condition (e.g., voltage sags). A fast synchronization enables a proper dynamic performance of the inverter, which is essential for providing Fault Ride Through (FRT) support under low-voltage grid faults that is required by the grid regulations [6], [7]. It should be noted that Japan [8] and Italy [9] have already issued FRT regulations even for single-phase inverters. Moreover, the accuracy of the synchronization against voltage harmonic distortions can benefit the power quality of the PV system as discussed in [10]. Thus, it is vital to propose an advanced PLL scheme in terms of fast and good dynamic response under grid faults as well as high robustness against harmonics in these applications.

In the prior-art work, several advanced PLL-based synchronization methods can be found. An Inverse Park Transformation (IPT) based PLL [11], [12] and synchronization based on adaptive filtering and generalized integrators, such as the Enhanced PLL (EPLL) [2], [12] and the Second Order Generalized Integrator (SOGI-PLL) [12], [13] respectively, present fast dynamic responses. However, the performance of these PLL schemes is poor under low-order voltage harmonic distortions. It is more often to observe such performance degradation in the EPLL. In order to improve the robustness against low-order harmonic distortions, synchronization techniques based on adaptive or notch filtering techniques [14], [15], solutions using repetitive and multi-resonant controllers [16], and schemes based on moving average filters [17], [18] have been proposed. Nevertheless, the harmonic robustness of these methods is achieved at the cost of the dynamic response of the corresponding synchronization.

An interesting PLL scheme is proposed in [19], [20] based on a novel Multi Harmonic Decoupling Cell (MHDC) PLL,

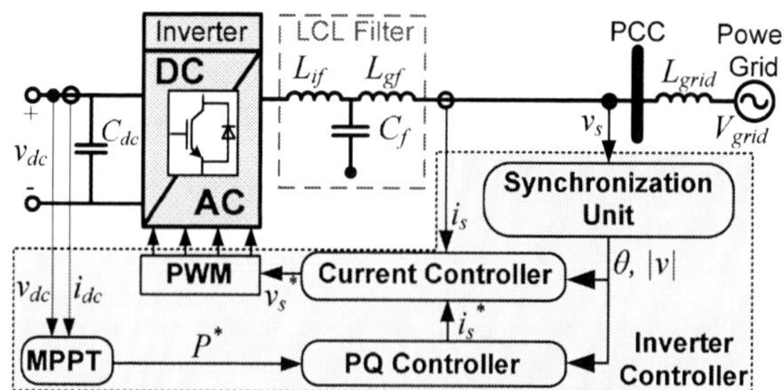

Fig. 1. Structure of a grid-tied inverter and its controller with a synchronization unit.

This work was supported by the Research Promotion Foundation (RPF, Cyprus, Project KOINA/SOLAR-ERA.NET/0114/02), by Energinet.dk (ForskEL, Denmark) and the SOLAR-ERA.NET (European Union's Seventh Framework Programme).

which can achieve an accurate performance under harmonic distortions without affecting its dynamic responses. Major disadvantages of the MHDC-PLL are, however, the increased complexity in terms of heavy computation burden and the inaccurate response under non-nominal frequencies (e.g., a frequency jump).

This paper aims to address these major issues and thus to improve the performance of the MHDC-PLL by proposing a frequency adaptive MHDC-PLL. In Section II.A, a frequency adaptive Quadrature Signal Generator (QSG) is developed to enable an accurate response under non-nominal frequencies. In Section II.B, the MHDC is reformulated in such way to achieve an equivalent fast and accurate performance under harmonic distortion and grid disturbances, but with a decrease in complexity and required processing time. Finally, in Section II.B, the development of the frequency adaptive MHDC-PLL is proposed. A performance analysis and a complexity assessment of the proposed synchronization scheme is performed in Section III, while Section IV demonstrates simulation and experimental results of the frequency adaptive MHDC-PLL. Finally, conclusions are drawn in Section V.

II. FREQUENCY ADAPTIVE MHDC-PLL

The MHDC-PLL proposed in [19], [20] employs a Quadrature Signal Generator (QSG) unit to generate the in-quadrature voltage vector ($\mathbf{v}_{\alpha\beta}$) that is free of high-order harmonics, the MHDC to dynamically decouple the effect of low-order harmonics, and the *dq*-PLL algorithm to estimate the phase angle of the fundamental voltage component ($\mathbf{v}^{+1}$). The MHDC-PLL of [19], [20] suffers from the increased implementation complexity of the MHDC and the non-ideal or unsatisfactory responses of the QSG under frequency changes as mentioned in [13]. Hence, the following propose schemes to address these issues.

A. Frequency Adaptive Quadrature Signal Generator

The QSG of the MHDC-PLL in [19], [20] is based on a combination of an Inverse Park Transformation (IPT) method for cancelling the high-order harmonics and on a $T/4$ delay unit, where T is the period of the grid voltage, to generate the in-quadrature voltage vector. The IPT method is based on a forward Park's transformation, on a low-pass filter $\omega_{f1}/(s+\omega_{f1})$, and on a backward Park's transformation as described in [19]. It should be noted that the IPT method is not used for the in-quadrature vector $v_{\alpha\beta}$ generation, since the filtering effect of IPT on v_α and v_β is different. Thus, a more complicated design of the MHDC is required for cancelling out the low-order harmonics. Here, the IPT is only used for filtering the high-order harmonics of the grid voltage v_s. Therefore, the v_α is free of high-order harmonics and then a T/4 delay unit is used for generating the in-quadrature voltage vector $\mathbf{v}_{\alpha\beta}$.

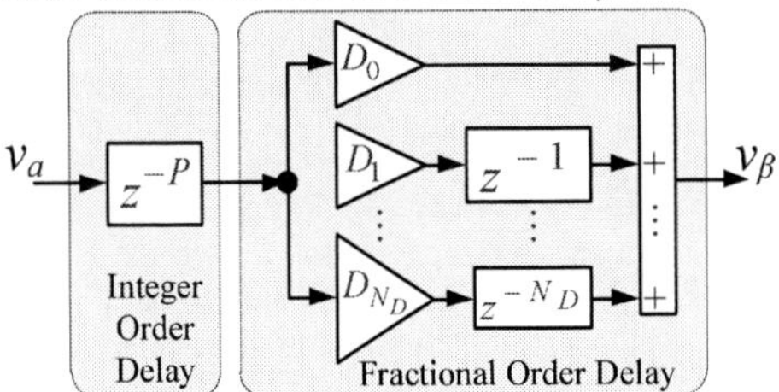

Fig. 2. Structure of the proposed frequency adaptive MHDC-PLL.

The $T/4$ delay unit of [19], [20] can accurately be performed in a digital controller, only when the ratio between the sampling rate (f_s) and the grid frequency (f_{grid}) is an integer. Thus, in the case where the ratio ($f_s/4 f_{grid}$) is an integer, the in-quadrature voltage vector $\mathbf{v}_{\alpha\beta}$ can equivalently be expressed in the continuous and in the discrete-time domains as,

$$\begin{aligned}&\text{Continuous-time} \rightarrow \mathbf{v}_{\alpha\beta}(t) = \begin{bmatrix} v_\alpha(t) \\ v_\beta(t) \end{bmatrix} = \begin{bmatrix} v_\alpha(t) \\ v_\alpha(t - T/4) \end{bmatrix} \\ &\text{Discrete-time} \rightarrow \mathbf{v}_{\alpha\beta}(k) = \begin{bmatrix} v_\alpha(k) \\ v_\beta(k) \end{bmatrix} = \begin{bmatrix} v_\alpha(k) \\ v_\alpha(k - f_s/4\, f_{grid}) \end{bmatrix}\end{aligned} \tag{1}$$

Unfortunately, in the case where the ratio ($f_s/4 f_{grid}$) is not an integer, the $T/4$ delay unit of [19], [20] can only approach the closest sample ($k_{T/4}$) to the $T/4$ delayed signal, where $k_{T/4}$ is the rounded $f_s/(4 f_{grid})$. Hence, when the grid frequency varies, the generated discrete-time $v_\beta(k)$ can present a phase shift error $\Delta\varphi$ from its desired continuous-time signal $v_\beta(t)$ as explained by,

$$v_\beta(k) = v_\alpha(k - k_{T/4}) \Leftrightarrow v_\beta(t) = v_\alpha(t - T/4 \pm \Delta\varphi) \tag{2}$$

In the worst case scenario, the phase error $\Delta\varphi$ is equal to $180 f_s/f_{grid}$. For instance, when f_s=10 kHz, the phase error can reach 0.89 degrees, when f_{grid}=49.505 Hz. Such a phase shift error on the in-quadrature voltage vector $\mathbf{v}_{\alpha\beta}$ will cause significant oscillations on the voltage signals expressed in the synchronous reference frame. As a result, undesired inaccuracies on the synchronization signals will be raised. A straight forward way to overcome this issue is to use a variable sampling rate as discussed in [21] to ensure that the $f_s/(4 f_{grid})$ is always an integer under any grid frequency. However, in such grid-tied inverter applications, the variable sampling rate is usually not an option, due to several restrictions on the controller design.

Hence, a novel frequency adaptive implementation of the $T/4$ delay unit is introduced in the following to enable an accurate operation of the MHDC-PLL under any grid frequency. The proposed frequency adaptive $T/4$ delay unit is developed in a digital controller by splitting the $T/4$ delay unit (i.e., $z^{-(fs/fgrid)/4}$) into an integer order delay (z^{-P}) and a fractional order delay (z^{-F}) as shown in Fig. 2. Then, the fractional order delay (z^{-F}) is approximated using the *Langrange* interpolation polynomial finite-impulse-response filter [22] as,

$$\begin{aligned}&z^{-(f_s/f_{grid})/4} = z^{-(P+F)} \\ &\text{where:} \quad z^{-P} = \text{Integer Order Delay}, \; z^{-F} \approx \sum_{l=0}^{N_D} D_l \cdot z^{-l}; \\ &D_l = \prod_{\substack{i=0 \\ i \neq l}}^{N_D} \frac{F-i}{l-i} \quad \text{and } l = 0,1,2\ldots,N_D\end{aligned} \tag{3}$$

with N_D being the *Langrange* interpolation polynomial order, and N_D=3 being selected in this paper.

Such an adaptive $T/4$ delay unit according to (3) and Fig. 2 is then employed in the digital controller in order to generate the in-quadrature voltage vector $\mathbf{v}_{\alpha\beta}$. The discrete-time $v_\beta(k)$

can accurately estimate the continuous-time $v_\beta(t)=v_\alpha(t-T/4)$ due to the *Langrange* interpolation and thus, the use of the adaptive $T/4$ delay on the structure of the proposed frequency adaptive MHDC-PLL (as shown in Fig. 3) can enable an accurate synchronization at any grid frequency.

B. Reformulation of the MHDC for Complexity Reduction

The decoupling network (MHDC) of [19], [20] can achieve a dynamic cancellation of the low-order voltage harmonics and hence a fast and accurate synchronization can be ensured under any grid conditions (i.e., grid faults, high harmonic distortion). Another major disadvantage of the MHDC of [19], [20] is the increased complexity (processing burden) of the decoupling network, which may be sufficiently high in such real-time applications. Hence, a re-formulation of the decoupling network is proposed in the following in order to minimize the required processing time of the algorithm and still to achieve an exact equivalent response with the conventional MHDC [19], [20]. The complexity minimization is achieved by re-designing the decoupling network in the stationary reference frame ($\alpha\beta$-frame) instead of performing this in each synchronous reference frame (dq^n-frame) as it has initially been proposed in [19], [20].

The re-design of the decoupling network for dynamically cancelling out the low-order harmonics requires the analysis of the grid voltage under high harmonic distortion. As already proved in [19], the QSG is filtering out only the high-order harmonics (due to the IPT) and then it generates the v_β component by delaying the v_α for a period of $T/4$. Thus, under harmonic distorted conditions, the vector $\mathbf{v}_{\alpha\beta}$ can be expressed as a summation of the fundamental component (n=1) and of the low-order harmonics (n=3, 5, 7, 9, 11, 13, ...), as given by,

$$\mathbf{v}_{\alpha\beta}=\mathbf{v}^1_{\alpha\beta}+\mathbf{v}^3_{\alpha\beta}+\mathbf{v}^5_{\alpha\beta}+\ldots=V^1\begin{bmatrix}\cos(\omega t+\theta_1)\\ \cos(\omega(t-\frac{T}{4})+\theta_1)\end{bmatrix}+\sum_{n=3,5,7,\ldots}V^n\begin{bmatrix}\cos(n\omega t+\theta_n)\\ \cos\left(n\omega(t-\frac{T}{4})+\theta_n\right)\end{bmatrix},\tag{4}$$

where V^n and θ_n represent the amplitude and the initial phase angle of each voltage component respectively and ω is the angular grid frequency. Based on the QSG of the MHDC-PLL [19], [20] and by using basic trigonometric identities, (4) can be analyzed into,

$$\mathbf{v}_{\alpha\beta}=\sum_{n=1,3,5,7,\ldots}\mathbf{v}^n_{\alpha\beta}=\sum_{n=1,3,5,7,\ldots}V^n\begin{bmatrix}\cos(\mathrm{sgn}(n)\cdot n\omega t+\theta_n)\\ \sin\left(\mathrm{sgn}(n)\cdot n\omega t+\theta_n\right)\end{bmatrix}\tag{5}$$

$$\text{where }\ \mathrm{sgn}(n)=\sin\left(\frac{n\pi}{2}\right)=\begin{cases}+1 & \text{for } n=1,5,9,13,\ldots\\ -1 & \text{for } n=3,7,11,\ldots\end{cases}$$

where $\mathrm{sgn}(n)$ defines the speed direction of each component. It worth noticing that the $\mathrm{sgn}(n)$ is inserted in this analysis due to the QSG-based $\mathbf{v}_{\alpha\beta}$.

According to (5) the direct calculation of each voltage component $\mathbf{v}^n_{\alpha\beta}$ is not possible due to the coupling effect among the existing components. However, an accurate estimation especially of the fundamental voltage component $\mathbf{v}^1_{\alpha\beta}$ is necessary for the grid synchronization of a grid-tied inverter and thus, a novel decoupling network is proposed hereafter. The development of the decoupling network requires to express (5) in terms of any voltage component $\mathbf{v}^n_{\alpha\beta}$ as,

$$\mathbf{v}^n_{\alpha\beta}=\mathbf{v}_{\alpha\beta}-\sum_{m\neq n}\mathbf{v}^m_{\alpha\beta}\Leftrightarrow\begin{bmatrix}v^n_\alpha\\ v^n_\beta\end{bmatrix}=\begin{bmatrix}v_\alpha\\ v_\beta\end{bmatrix}-\sum_{m\neq n}\begin{bmatrix}v^m_\alpha\\ v^m_\beta\end{bmatrix}\Leftrightarrow\tag{6}$$

Now, since the vectors $\mathbf{v}^m_{\alpha\beta}$ are unknown, the estimation of the voltage component $\mathbf{v}^n_{\alpha\beta}$ by $\mathbf{v}^{*n}_{\alpha\beta}$ is enabled in (7) by replacing the unknown vectors $\mathbf{v}^m_{\alpha\beta}$ of (6) with the corresponding filtered estimation vectors $\bar{\mathbf{v}}^{*m}_{\alpha\beta}$.

$$\mathbf{v}^{*n}_{\alpha\beta}=\mathbf{v}_{\alpha\beta}-\sum_{m\neq n}\bar{\mathbf{v}}^{*m}_{\alpha\beta}\tag{7}$$

The filtered estimation vectors $\bar{\mathbf{v}}^{*m}_{\alpha\beta}$ are produced by filtering out the corresponding estimation vector $\mathbf{v}^{*m}_{\alpha\beta}$ in order to eliminate any remaining oscillations after subtracting the coupling effect caused by the existence of other harmonic components. Since each estimation vector $\mathbf{v}^{*m}_{\alpha\beta}$ is rotated with a different $\mathrm{sgn}(m)\,m\,\omega$, an equivalent filtering cannot be directly achieved for each component. Therefore, for an equivalent filtering of all the voltage components, it is first necessary to express each estimated vector $\mathbf{v}^{*m}_{\alpha\beta}$ into a corresponding synchronous reference frame rotating with a $\mathrm{sgn}(m)\,m\,\omega$ speed ($dq^{\mathrm{sgn}(m)m}$-frame) by using Park's transformation theory. Thus, the m component of the estimation voltage vector expressed in the corresponding $dq^{\mathrm{sgn}(m)m}$-frame ($\mathbf{v}^{*m}_{dq^{\mathrm{sgn}(m)m}}$) is given by,

$$\mathbf{v}^{*m}_{dq^{\mathrm{sgn}(m)m}}=\begin{bmatrix}v_{d^{\mathrm{sgn}(m)\cdot m}}\\ v_{q^{\mathrm{sgn}(m)\cdot m}}\end{bmatrix}=\left[T_{dq^{\mathrm{sgn}(m)\cdot m}}\right]\mathbf{v}^{*m}_{\alpha\beta}\tag{8}$$

$$\text{where}\left[T_{dq^{\mathrm{sgn}(m)\cdot m}}\right]=\begin{bmatrix}\cos(\mathrm{sgn}(m)m\omega t) & \sin(\mathrm{sgn}(m)m\omega t)\\ -\sin(\mathrm{sgn}(m)m\omega t) & \cos(\mathrm{sgn}(m)m\omega t)\end{bmatrix}$$

It should be noted that each vector $\mathbf{v}^{*m}_{dq^{\mathrm{sgn}(m)m}}$ is actually described as a DC/non-rotating vector, since both the $dq^{\mathrm{sgn}(m)m}$-frame and the m voltage component are rotated with the same $\mathrm{sgn}(m)\,m\,\omega$ speed. Therefore, an equivalent filtering can now be achieved since all the voltage components can be expressed as non-rotating vectors according to (8). Therefore, the filtered estimation vector $\bar{\mathbf{v}}^{*m}_{dq^{\mathrm{sgn}(m)m}}$ expressed in a corresponding $dq^{\mathrm{sgn}(m)m}$-frame is generated by filtering the corresponding estimation vector component $\mathbf{v}^{*m}_{dq^{\mathrm{sgn}(m)m}}$ according to,

$$\bar{\mathbf{v}}^{*m}_{dq^{\mathrm{sgn}(m)m}}=\begin{bmatrix}\bar{v}^{*m}_{d^{\mathrm{sgn}(m)m}}\\ \bar{v}^{*m}_{q^{\mathrm{sgn}(m)m}}\end{bmatrix}=\left[F(s)\right]\mathbf{v}^{*m}_{dq^{\mathrm{sgn}(m)m}}\tag{9}$$

$$\text{where}\left[F(s)\right]=\frac{\omega_{f2}}{s+\omega_{f2}}\begin{bmatrix}1 & 0\\ 0 & 1\end{bmatrix}$$

and ω_{f2} is the cut-off frequency of the low-pass filter $[F(s)]$ and should be set to $2\pi 50/3$ rad/s according to the theoretical analysis in [19]. Finally, the filtered estimation vector $\bar{\mathbf{v}}^{*m}_{\alpha\beta}$,

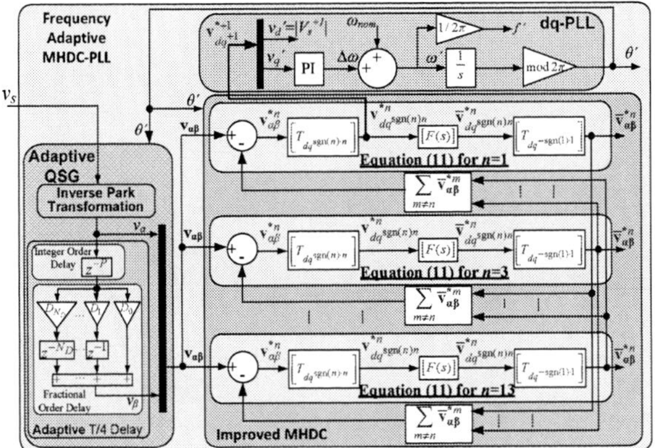

Fig. 3. Structure of the proposed frequency adaptive MHDC-PLL.

TABLE I. DESIGN PARAMETERS OF FREQUENCY ADAPTIVE MHDC-PLL

IPT unit	Filtering parameter → $\omega_{f1}=2\pi 50\sqrt{2}$ rad/s
Adaptive T/4 delay unit	*Langrange* interpolation order → $N_D=3$
Improved MHDC	Multiple use of (11) for n=1, 3, 5, 7 , 9, 11, 13 Filtering parameter → $\omega_{f2}=2\pi 50/3$ rad/s
dq-PLL	Tuning parameters → k_P=92, T_I=0.000235

required by (7) to develop the decoupling network, can be calculated based on Park's transformations and is given by,

$$\bar{\mathbf{v}}_{\alpha\beta}^{*m} = \begin{bmatrix} \bar{v}_{\alpha}^{*m} \\ \bar{v}_{\beta}^{*m} \end{bmatrix} = \begin{bmatrix} T_{dq^{-\mathrm{sgn}(m)\cdot m}} \end{bmatrix} \bar{\mathbf{v}}_{dq^{\mathrm{sgn}(m)m}}^{*m} \tag{10}$$

When submitting (8) to (9) and then (9) to (10), (7) can be re-written as,

$$\mathbf{v}_{\alpha\beta}^{*n} = \mathbf{v}_{\alpha\beta} - \sum_{m\neq n} \begin{bmatrix} T_{dq^{-\mathrm{sgn}(m)m}} \end{bmatrix} [F(s)] \begin{bmatrix} T_{dq^{\mathrm{sgn}(m)m}} \end{bmatrix} \mathbf{v}_{\alpha\beta}^{*m} \tag{11}$$

Finally, (11) is the cornerstone of the proposed decoupling network (improved MHDC) as shown in Fig. 3. Hence, the multiple use of (11) for n=1, 3, 5, 7, ... in a cross feed-back decoupling network enables a dynamic estimation of each voltage component $\mathbf{v}_{\alpha\beta}^{*n}$. It should be pointed out that the cross feed-back network is required to dynamically eliminate the coupling effect between the fundamental voltage component and all the significant low-order harmonic components. It is important to mention that the improved MHDC can achieve an exact equivalent performance with the MHDC in [19], [20], but the complexity of the improved MHDC is significantly decreased, as it will be proved in Section III.

C. Development of the Frequency Adaptive MHDC-PLL

The proposed frequency adaptive MHDC-PLL can now be developed based on the adaptive QSG (as described in Section II.A), the improved MHDC (as described in Section II.B) and on the *dq*-PLL algorithm [23] for estimating the phase angle of the grid voltage as shown in Fig. 3. The new frequency adaptive MHDC-PLL can achieve a fast and accurate synchronization under any grid disturbance and under highly harmonic distortion. The complexity of the synchronization has been significantly decreased compared to the MHDC-PLL of [19], [20], as it will be proved in Section III.

The adaptive QSG is based on the IPT method in order to filter out the high-order harmonics and then the proposed frequency adaptive T/4 delay unit is used to generate the voltage vector $\mathbf{v}_{\alpha\beta}$. The proposed frequency adaptive method overcomes the inaccuracies under non-nominal frequency caused by the initial QSG used in [19], [20] and thus, an accurate operation can be achieved under any grid frequency.

Then, the $\mathbf{v}_{\alpha\beta}$ is fed on the proposed improved MHDC as shown in Fig. 3, in order to dynamically cancel out the effect of the low-order harmonics with minimized computation burden. For a proper design of the improved MHDC, it is first necessary to define the number of harmonic-orders that are considered and eliminated by the decoupling network. An investigation is performed on the accuracy of the frequency adaptive MHDC-PLL under the worst-case voltage harmonic distortion according to EN50160 standards (see HC3 on Table II). The investigation shows that a very accurate synchronization (error in phase angle estimation $\theta_{error}=\theta_{grid}-\theta'$ less than 0.00035 rad) can be achieved, when the improved MHDC is designed for N=7, considering the fundamental (n=1) and the effect of the six most significant harmonic components (n=3, 5, 7, 9, 11, 13) in single-phase systems. The effect of the higher order harmonics is minimized due to the second-order band-pass filtering characteristics of IPT as mentioned in [19]. Therefore, the decoupling network of the frequency adaptive MHDC-PLL is designed for N=7 as shown in Fig. 3 in order to achieve an accurate synchronization under any harmonic distorted grid conditions. Thus, the fast and accurate estimation of the fundamental voltage component $\mathbf{v}_{dq^{+1}}^{*1}$ expressed in the dq^{+1}-frame is enabled by the improved MHDC. This estimation voltage vector is free of any low- and high-order harmonics and therefore, the conventional dq-PLL algorithm can be used to accurately estimate the phase angle and the amplitude of the grid voltage, as demonstrated in Fig. 3.

The conventional dq-PLL system [23] is a closed-loop synchronization system, which aims to force the per unit $v_q' = v_{q^{+1}}^{*1}$ to track zero. The synchronization algorithm is based on a Proportional-Integral (PI) controller, whose transfer function is given by k_p+1/($T_i s$), with k_p and T_i being the controller parameters. Therefore, based on a linearized small signal analysis, the transfer function of the PLL is given by the second order transfer function of (12), where it is obvious that the tuning parameters (k_p and T_i) can affect the response of the synchronization unit.

$$\frac{\theta'}{\theta} = \frac{k_p \cdot s + 1/T_i}{s^2 + k_p \cdot s + 1/T_i} \tag{12}$$

where: $k_p = \dfrac{9.2}{ST}$ and $T_i = 0.047 \cdot \zeta^2 \cdot ST^2$

For an optimally damped response of the PLL, the damping coefficient ζ should be set to $1/\sqrt{2}$; the Settling Time (ST) has been set to 100 ms for the purposes of this work. Hence, the tuning parameters k_p and T_i of the frequency adaptive MHDC-PLL have been set to 92 and $2.35 \cdot 10^{-4}$, respectively.

The proposed frequency adaptive MHDC-PLL can achieve a fast and accurate response under any harmonic distortion and under any grid faults and the required processing time of the proposed synchronization algorithm is significantly less compared to the original MHDC of [19], [20], as it will be demonstrated in Section III and IV. All the design parameters for developing the proposed frequency adaptive MHDC-PLL are summarized in Table I.

III. Performance Analysis - Complexity Evaluation

This section aims to prove that the proposed frequency adaptive MHDC-PLL (based on the improved MHDC) and the MHDC-PLL of [19], [20] can achieve an equivalent performance under any harmonic distorted grid voltage and that the proposed improved MHDC (Section II.B) requires significantly less processing time in each control loop.

The proposed improved MHDC, as shown in Fig. 3, is based on the multiple use of (11) for accurately estimating each voltage component. By analyzing (11), it can be proved that the improved MHDC can achieve an exact equivalent response with the original MHDC proposed in [19], [20]. Thus, by multiplying both sides of (11) with the transformation matrix $\left[T_{dq^{sgn(n)n}}\right]$ and applying Park's transformation theory, (11) can be equivalently re-written as (13).

$$\mathbf{v}^{*n}_{dq^{\mathrm{sgn}(n)n}} = \left[T_{dq^{\mathrm{sgn}(n)n}}\right]\mathbf{v}_{\alpha\beta} - \sum_{m\neq n}\left[T_{dq^{\mathrm{sgn}(n)n}}\right]\left[T_{dq^{-\mathrm{sgn}(m)m}}\right]\left[F(s)\right]\mathbf{v}^{*m}_{dq^{\mathrm{sgn}(m)m}} \quad (13)$$

Then, by filtering both side of (13) with the filtering matrix $[F(s)]$, by using (9), and by merging the transformation matrices, (13) can be re-written as (14), which is the cornerstone equation of MHDC according to [19], [20].

$$\bar{\mathbf{v}}^{*n}_{dq^{\mathrm{sgn}(n)n}} = \left[F(s)\right]\left\{\mathbf{v}_{dq^{\mathrm{sgn}(n)n}} - \sum_{m\neq n}\left[T_{dq^{\mathrm{sgn}(n)n-\mathrm{sgn}(m)m}}\right]\bar{\mathbf{v}}^{*m}_{dq^{\mathrm{sgn}(m)m}}\right\} \quad (14)$$

Since (11) can be re-written in the form of (14), it is proven that the MHDC of [19], [20] and the improved MHDC proposed in this paper can achieve an exact equivalent response in terms of estimating the voltage vector of the fundamental voltage component. Furthermore, simulation results of Fig. 4, verify that under several Harmonic Conditions (HC), where HC1, HC2 and HC3 are defined in Table II, both the MHDC of [19], [20] and the improved MHDC present an exact equivalent response on the accurate estimation of the $\mathbf{v}^{*1}_{dq^{+1}}$. Although the two decoupling networks present an equivalent response, the improved MHDC proposed in this paper requires significantly less processing time compared to the MHDC of [19], [20], as it is analyzed below.

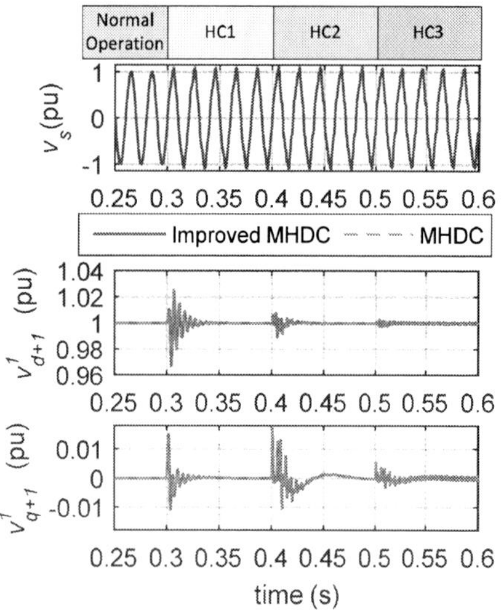

Fig. 4. Comparison of the performance of the proposed improved MHDC and of the MHDC of [19] and [20].

TABLE II. Definition of Several Harmonic Conditions (HC)

Normal Operation	Pure sinusoidal grid voltage
HC1	$\lvert V_3\rvert$=5% and $\lvert V_5\rvert$=6%
HC2	$\lvert V_3\rvert$=5%, $\lvert V_5\rvert$=6%, $\lvert V_7\rvert$=5%, $\lvert V_9\rvert$=1.5%, $\lvert V_{11}\rvert$=3.5%
HC3	Worst case harmonic distortion according to EN50160: $\lvert V_3\rvert$=5%, $\lvert V_5\rvert$=6%, $\lvert V_7\rvert$=5%, $\lvert V_9\rvert$=1.5%, $\lvert V_{11}\rvert$=3.5%, $\lvert V_{13}\rvert$=3%, $\lvert V_{15}\rvert$=0.5%, $\lvert V_{17}\rvert$=2%, $\lvert V_{19}\rvert$=1.5%, $\lvert V_{h>20}\rvert$=0.3%

For a fair complexity analysis, both decoupling networks need to be designed for N=7, to consider the fundamental component (n=1) and the effect of the six most significant harmonic components (n=3, 5, 7, 9, 11, 13). Therefore, the process of the MHDC of [19], [20] for N=7 requires the repeated processing of (14) for seven times. As a consequence, in each control step the decoupling network should process N multi-subtractions $\left(\mathbf{v}_{dq^{\mathrm{sgn}(n)n}} - \sum_{m\neq n}\bar{\mathbf{v}}^{*m}_{dq^{\mathrm{sgn}(m)m}}\right)$, N^2 transformation matrices $\left[T_{dq^{\mathrm{sgn}(n)n}}\right]$, and N low-pass filtering matrices $[F(s)]$. In the case of the improved MHDC, for N=7, (11) must be repeated for seven times as well. At each control step the improved MHDC should process N multi-subtractions $\left(\mathbf{v}_{\alpha\beta} - \sum_{m\neq n}\bar{\mathbf{v}}^{*m}_{\alpha\beta}\right)$, $2N$ transformation matrices, and N low-pass filtering matrices $[F(s)]$. It is obvious that the proposed improved MHDC requires the processing of $2N$=14 transformation matrices instead N^2=49 matrices of the MHDC of [19], [20]. The complexity comparison is summarized in Table III. A further complexity analysis is also presented in Table III, where the two decoupling networks have been analyzed in terms of the required additions, subtractions and multiplications at each control step. It is obvious that the re-formulation of the decoupling cell according to the proposed improved MHDC (Section II.B) can significantly decrease the required processing time of the synchronization algorithm. The decreased complexity of the new improved MHDC is particularly useful, since the real-time operation of the inverter is enabled and a higher sampling rate can be adapted by the

TABLE III. Complexity Comparison of the Decoupling Networks

Decoupling Network	Complexity analysis in each control loop			
	Multi-subtractions	$\left[T_{dq^{\mathrm{sgn}(n)n}}\right]$	$[F(s)]$	Total mathematical operations
MHDC of [19] (N=7)	N	N^2	N	322 Multiplications 63 Additions 133 Subtractions
Improved MHDC (N=7)	N	N^2	N	112 Multiplications 28 Additions 98 Subtractions

TABLE IV. PERFORMANCE COMPARISON OF TWO PLLS

PLL algorithm	Required Processing Time (%)	Dynamic Response under grid faults	Accurate Resposne under		
			Voltage sag/phase shift	Frequency jump ($f \neq 50$ Hz)	Harmonic distortion
MHDC-PLL	100%	Fast	+	-	+
Freq. Adapt. MHDC-PLL	39%	Fast	+	+	+

inverter controller for an improved performance.

Further, an in-depth experimental complexity analysis has been performed, based on a widely used microcontroller, such as the Texas Instrument TMS320F28335 digital signal processor. The investigation demonstrates a significant improvement with regards to the algorithm complexity, since the process of the frequency adaptive MHDC-PLL at each control loop requires 54.6 μs instead of 140 μs in the case of the initial MHDC-PLL of [19], [20]. Therefore, the proposed synchronization method requires 61% less processing time compared to the initial MHDC-PLL, as summarized in Table IV. It is important to mention that an inverter controller based on the frequency adaptive MHDC-PLL can achieve a sampling rate of 8 kHz on TMS320F28335 (54.6 μs for the synchronization and 63 μs for the rest units of the inverter controller). On the other hand, if the inverter controller is based on the initial MHDC-PLL of [19], [20] the sampling rate should be decreased to 4 kHz on TMS320F28335 (140 μs for the synchronization and 63 μs for the rest of the units of the inverter controller). It is to be noted that the reduction on sampling rate of the controller can negatively affect the accuracy and the performance of the inverter.

IV. SIMULATION AND EXPERIMENTAL RESULTS

The performance of the proposed frequency adaptive MHDC-PLL has been tested through simulation and experimental results in order to demonstrate the outstanding response of the new synchronization method. The investigation based simulations have been performed in MATLAB/Simulink. The experimental setup is based on a TMS320F28335 microcontroller (where the inverter controller has been applied), a Semikron SEMITeach (B6CI) inverter, a California Instrument 2253IX AC power source for emulating the grid, and an Elektro-Automatik EA-PSI 9750-20 DC power supply for emulating the DC bus of a PV system. It is worth mentioning that in both simulation and experimental studies, an 8 kHz sampling rate has been used for the inverter controller.

Simulations performed in MATLAB/Simulink show the synchronization response of the proposed frequency adaptive MHDC-PLL, as demonstrated in Fig. 5. The results of Fig. 5 show the synchronization response of the frequency adaptive MHDC-PLL and the MHDC-PLL of [19], [20] under the worst-case harmonic distortion (HC3 of Table II) at 0.3 s, a 10° phase jump at 0.4 s, a 25% voltage sag at 0.5 s and -1.5 Hz frequency change at 0.6 s. The performance comparison of Fig. 5 between the two PLLs, proves that the frequency adaptive implementation of the QSG (Section II.A) enables the accurate synchronization under non-nominal frequencies ($0.6<t<0.75$ s). Therefore, the frequency adaptive MHDC-PLL not only requires a significantly less processing time as shown in Section III, but additionally, it can also achieve a superior performance under any grid conditions, as summarized in Table IV. From Fig. 5, it is obvious that the new frequency adaptive MHDC-PLL presents a very accurate response (even under the worst-case harmonic distortion for $t>0.3$ s) and a

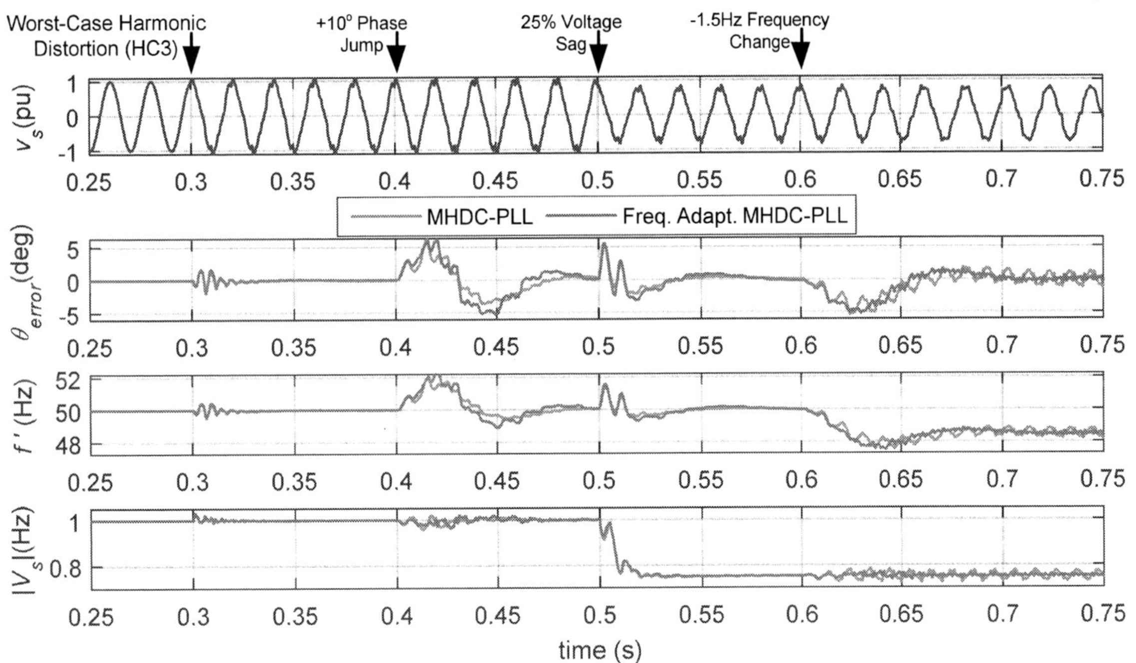

Fig. 5. Simulation results for the response of the frequency adaptive MHDC-PLL (new) and the MHDC-PLL under harmonic distortion, phase jump, voltage sag and frequency change events.

very fast synchronization under any grid disturbance (e.g., phase jump, voltage sag, frequency change).

The superior synchronization performance of the proposed frequency adaptive MHDC-PLL has also been experimentally verified as shown in Fig. 6. The experimental results of Fig. 6(a) demonstrate the response of the new synchronization method when the worst-case harmonic distortion (HC3 of Table II) occurs at the grid voltage. The experiments show that the proposed PLL can decouple the effect of harmonics within 10 ms. The initial voltage in Fig. 6(b)-(d) has harmonic distortion with the worst-case harmonics (HC3) and then a grid disturbance is applied. According to Fig. 6(b), a fast and accurate synchronization is achieved under a 25% voltage sag, which can enable a proper FRT operation of the grid tied inverter. Fig. 6(c) demonstrates the operation of the new PLL under a 1 Hz frequency change. It is worth mentioning that the frequency adaptive MHDC-PLL presents a very accurate response under non-nominal frequencies due to the adaptive QSG. Finally, the fast and accurate synchronization response under a 10° phase jump is demonstrated in Fig. 6(d). According to the experiments of Fig. 6, the proposed frequency adaptive MHDC-PLL can achieve a fast and accurate synchronization under any grid conditions. Such an advanced PLL based synchronization method can be an ideal solution for the synchronization of grid tied inverters.

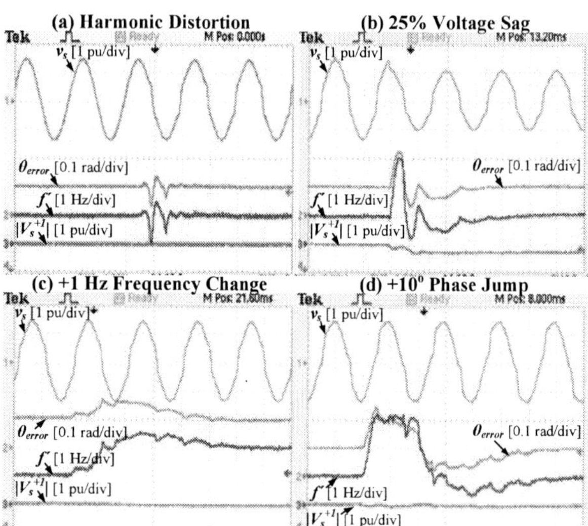

Fig. 6. Experimental results for the synchronization response of the frequency adaptive MHDC-PLL under: (a) a harmonic distorted voltage, (b) a 25% voltage sag, (c) an 1 Hz frequency change, and (d) a 10° phase jump. The time division of the results is 10 ms/div.

The synchronization accuracy under harmonic distortion can be very beneficial for the power quality of the grid-tied inverter as discussed in [10] and [20], while the dynamic response of the synchronization, especially under voltage sags, can enable a proper FRT operation of the inverter in order to enhance the stability of the power system under grid faults. A simulation based investigation has been performed here to demonstrate the beneficial effect of a fast and accurate synchronization on the operation of the grid-tied inverter as shown in Fig. 7. For this investigation, an open loop PQ controller designed in the dq-frame [24] is used to generate the reference currents $\mathbf{I}_{dq}^{*} = [I_d^{*}\ I_q^{*}]^{*}$. Then, a current controller based on a PI controller and also designed in the dq-frame [24], [25] is used to regulate the inverter current. It is to be noted that the current controller is enhanced with a harmonic compensation module [25] in order to minimize the distortion of the injected current. Further, the PQ controller is enhanced with FRT capabilities [26] in order to provide a proper voltage support under voltage sags, in terms of proper reactive current injection I_Q, according to $I_Q=k(v_N-v_s)$, where k is set to 2 for the purposes of this investigation and v_N is the nominal grid voltage [26].

Therefore, the simulation results of Fig. 7 present the inverter operation when a) a SOGI-PLL (t<0.5 s) and b) when the proposed Frequency Adaptive MHDC-PLL (t>0.5s) are used for the grid synchronization of the inverter. It should be noted that the SOGI-PLL does not present immunity against harmonic distortion in contrast with the new PLL which presents a great robustness against harmonics. The inverter operation is demonstrated under highly harmonic distorted grid voltage (HC2 of Table II) and under a 25% voltage sag. It is obvious that the accuracy of the new synchronization against harmonics enables the accurate generation of reference currents ($\mathbf{I}_{dq}^{*}$) while the inaccuracies of SOGI-PLL cause undesired oscillations on the reference currents. The oscillations on the reference currents cause an increased total harmonic distortion of the current (THD_i) in the case of the SOGI-PLL, with a THD_i=2.15% when there is no voltage sag and THD_i=4.5% when a voltage sag occurs. On the other hand, the accuracy of the proposed frequency adaptive MHDC-PLL enables an oscillation-free generation of the reference currents and thus, a high quality current injection is achieved, with a THD_i=1.5% under normal operation and voltage sag. Further, the dynamic response of the frequency adaptive MHDC-PLL enables a fast and a proper FRT operation of the inverter, where a reactive power support is properly injected into the grid within 15 ms, for enhancing the stability of the power system. During the FRT operation the injected active power is decreased to maintain the injected current within the inverter limits. Hence, the results of Fig. 7 demonstrate that the accurate and fast response of the frequency adaptive MHDC-PLL is particularly beneficial for the operation of the grid tied inverter, in terms of increasing the power quality and of enabling an appropriate dynamic and FRT operation of the inverter.

V. CONCLUSION

In this paper, two new methods have been proposed: one for decreasing the complexity of the MHDC-PLL and one for enhancing its performance against non-nominal grid frequencies. Hence, the proposed frequency adaptive MHDC-PLL requires a significantly less processing time and presents a superior performance in contrast to the original one. Thus, the proposed frequency adaptive MHDC-PLL can achieve a fast and accurate response under any grid disturbances and under highly harmonic distorted voltage. As a consequence, the proposed synchronization method can be beneficial for the operation of a grid-tied inverter in terms of power quality and also of dynamic performance.

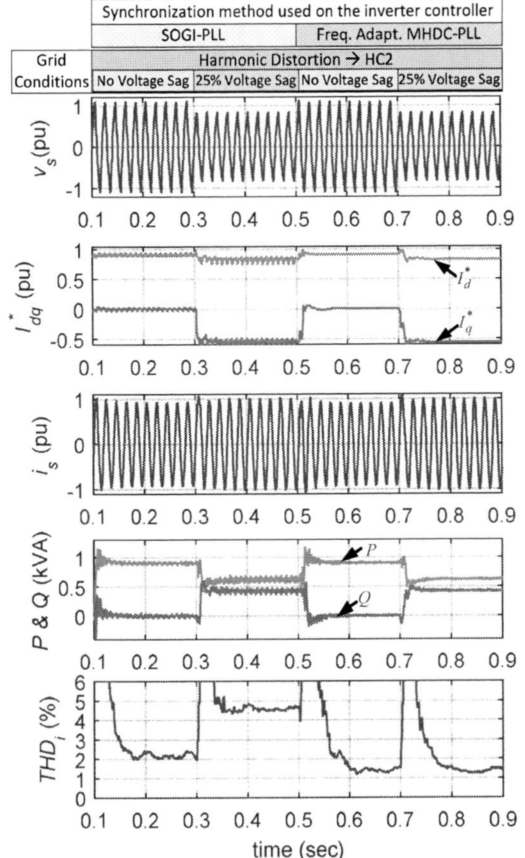

Fig. 7. Simulation results for the inverter operation when (a) a SOGI-PLL and (b) a frequency adaptive MHDC-PLL is used for the synchronziation of the inverter under harmonic distorted voltage and a voltage sag event.

References

[1] L. Zhang, K. Sun, H. Hu and Y. Xing, "A system-level control strategy of photovoltaic grid-tied generation systems for european efficiency enhancement," *IEEE Trans. Power Electronics*, vol. 29, no. 7, pp. 3445-53, Jul. 2014.

[2] S.A. Khajehoddin, M. Karimi-Ghartemani, A. Bakhshai and P. Jain, "A power control method with simple structure and fast dynamic response for single-phase grid-connected DG systems," *IEEE Trans. Power Electronics*, vol. 28, no. 1, pp. 221-233, Jan. 2013.

[3] B. Bahrani, A. Rufer, S. Kenzelmann and L.A.C. Lopes, "Vector control of single-phase voltage-source converters based on fictive-axis emulation," *IEEE Trans. Industry Applications*, vol. 47, no. 2, pp. 831-840, Apr. 2011.

[4] H. Hu, S. Harb, N. Kutkut, I. Batarseh and Z.J. Shen, "A review of power decoupling techniques for microinverters with three different decoupling capacitor locations in PV systems," *IEEE Trans. Power Electronics*, vol. 28, no. 6, pp. 2711-2726, June 2013.

[5] R.A. Mastromauro, M. Liserre and A. Dell'Aquila, "Control issues in single-stage photovoltaic systems: MPPT, current and voltage control," *IEEE Trans. Industrial Informatics*, vol. 8, no. 2, pp. 241-254, May 2012.

[6] R. Teodorescu, M. Liserre and P. Rodriguez, *Grid Converters for Photovoltaic and Wind Power Systems*, John Wiley & Sons, 2011.

[7] B. Craciun, T. Kerekes, D. Sera and R. Teodorescu, "Overview of recent grid codes for PV power integration," in *Proc. OPTIM*, Brasov, 2012, pp. 959-965.

[8] H. Kobayashi, "Fault ride through requirements and measures of distributed PV systems in Japan," in *Proc. IEEE-PES General Meeting*, San Diego (CA), 2012, pp. 1-6.

[9] CEI, "CEI 0-21: Reference Technical Rules for Connecting Users to the Active and Passive LV Distribution Companies of Electricity," Comitato Elettrotecnico Italiano 2012.

[10] L. Hadjidemetriou, E. Kyriakides and F. Blaabjerg, "A robust synchronization to enhance the power quality of renewable energy systems," *IEEE Trans. Industrial Electronics*, vol. 62, no. 8, pp. 4858-4868, Aug. 2015.

[11] L.N. Arruda, S.M. Silva and B.J.C. Filho, "PLL structures for utility connected systems," in *Proc. IEEE IAS Annual Meeting*, 2001, vol.4, pp. 2655-2660.

[12] Y. Yang and F. Blaabjerg, "Synchronization in single-phase grid-connected photovoltaic systems under grid faults," in *Proc. IEEE PEDG*, 2012, pp. 476-482.

[13] Y. Yang, L. Hadjidemetriou, F. Blaabjerg and E. Kyriakides, "Benchmarking of phase locked loop based synchronization techniques for grid-connected inverter systems," in *Proc. ICPE ECCE Asia*, 2015, pp. 2517-2524.

[14] F. Gonzalez-Espin, G. Garcera, I. Patrao and E. Figueres, "An adaptive control system for three-phase photovoltaic inverters working in a polluted and variable frequency electric grid," *IEEE Trans. Power Electronics*, vol. 27, no. 10, pp. 4248-4261, Oct. 2012.

[15] K. Lee, J. Lee, D. Shin, D. Yoo and H. Kim, "A novel grid synchronization PLL method based on adaptive low-pass notch filter for grid-connected PCS," *IEEE Trans. Industrial Electronics*, vol. 61, no. 1, pp. 292-301, Jan. 2014.

[16] B. Zhang, K. Zhou and D. Wang, "Multirate repetitive control for PWM DC/AC converters," *IEEE Trans. Industrial Electronics*, vol. 61, no. 6, pp. 2883-2890, June 2014.

[17] Y. Han, M. Luo, X. Zhao, J. Guerrero and L. Xu, "Comparative performance evaluation of orthogonal-signal-generators based single-phase PLL algorithms-A survey," *IEEE Trans. Power Electronics*, pp. 1-12 2015 (in press).

[18] S. Golestan, M. Ramezani, J.M. Guerrero, F.D. Freijedo and M. Monfared, "Moving average filter based phase-locked loops: Performance analysis and design guidelines," *IEEE Trans. Power Electronics*, vol. 29, no. 6, pp. 2750-2763, June 2014.

[19] L. Hadjidemetriou, E. Kyriakides, Y. Yang and F. Blaabjerg, "A synchronization method for single-phase grid-tied inverters," *IEEE Trans. Power Electronics*, vol. 31, no. 3, pp. 2139-2149, Mar. 2016.

[20] L. Hadjidemetriou, E. Kyriakides, Y. Yang and F. Blaabjerg, "Power quality improvement of single-phase photovoltaic systems through a robust synchronization method," in *Proc. IEEE ECCE*, Pittsburgh (PA), 2014, pp. 2625-2632.

[21] I. Carugati, S. Maestri, P.G. Donato, D. Carrica and M. Benedetti, "Variable sampling period filter PLL for distorted three-phase systems," *IEEE Trans. Power Electronics*, vol. 27, no. 1, pp. 321-330, Jan. 2012.

[22] Y. Yang, K. Zhou, H. Wang, F. Blaabjerg, D. Wang and B. Zhang, "Frequency adaptive selective harmonic control for grid-connected inverters," *IEEE Trans. Power Electronics*, vol. 30, no. 7, pp. 3912-3924, July 2015.

[23] V. Kaura and V. Blasko, "Operation of a phase locked loop system under distorted utility conditions," *IEEE Trans. Industry Applications*, vol. 33, no. 1, pp. 58-63, Jan./Feb. 1997.

[24] Y. Yang, F. Blaabjerg and Z. Zou, "Benchmarking of grid fault modes in single-phase grid-connected photovoltaic systems," *IEEE Trans. Industry Applications*, vol. 49, no. 5, pp. 2167-2176, Sep. 2013.

[25] L. Hadjidemetriou, E. Kyriakides and F. Blaabjerg, "A grid side converter current controller for accurate current injection under normal and fault ride through operation," in *Proc. IEEE IECON*, Vienna, 2013, pp. 1454-1459.

[26] L. Hadjidemetriou, P. Demetriou and E. Kyriakides, "Investigation of different fault ride through strategies for renewable energy sources," in *Proc. IEEE POWERTECH*, Eindhoven, 2015, pp. 1-6.

Series-Parallel Connection of Low-Voltage Sources for Integration of Galvanically Isolated Energy Storage Systems

Ramy Georgious, Jorge Garcia, Angel Navarro, Sarah Saeed, Pablo Garcia
LEMUR Research Team, Dept. of Electrical Engineering,
University of Oviedo
Gijon, 33204, Spain
Email: georgiousramy@uniovi.es

***Abstract*—This work explores the Series-Parallel Connection of a Low Voltage Supercapacitor Module to obtain a Hybrid Energy Storage System for grid support applications. The Hybrid System is formed by the Supercapacitor Module itself, intended to ensure fast performance upon peak power requirements, together with a battery that provides the energy requirements. In the full system, the front end converter and the load interfacing converter share a common DC link. The battery is connected to the DC link by means of a Full-Bridge Current-Source bidirectional DC-DC converter. The Supercapacitor Module is connected to the system using a Series-Parallel Configuration, which overcomes the main problems that arise with the most common topologies found in the literature. The full operation of the system has been demonstrated theoretically and by simulations. A demonstration of such connection is shown experimentally, in a converter operating at reduced power levels, in order to validate the feasibility of the system. Conclusions show how this scheme can be used in Hybrid Storage Systems.**

Keywords— Energy Storage Systems, Battery, Supercapacitor, Hybrid Storage Systems, Power Electronic Converters

I. Introduction

The increasing penetration of distributed generation and Energy Storage Systems (ESS) into the distribution grid is boosting the growth of Microgrids and Smartgrids [1]-[4]. The system under consideration is shown in Fig. 1, in which a Front-End Converter (FEC) connects a given load (which can be generalized as a Microgrid) to the three-phase distribution line. The main ESS, formed by the battery and a Main Storage Converter, is intended to support the DC link voltage in case of load variations or line fluctuations. In this work, the Full-Bridge Current-Source (FBCS) converter, shown in Fig. 2, has been selected as a suitable topology. This main ESS and its associated converter might have limitations in the power rating and bandwidth (due reliability, expected operating life or efficiency constraints). These limitations might affect the transient behavior of the storage system. Upon a power demand from the load, the power will initially be given by the DC link capacitor, which in turn will be discharged. It will take some time, depending on the design, for the ESS to be able to provide the required power to the DC link. This voltage drift might cause problems in the performance of the full system.

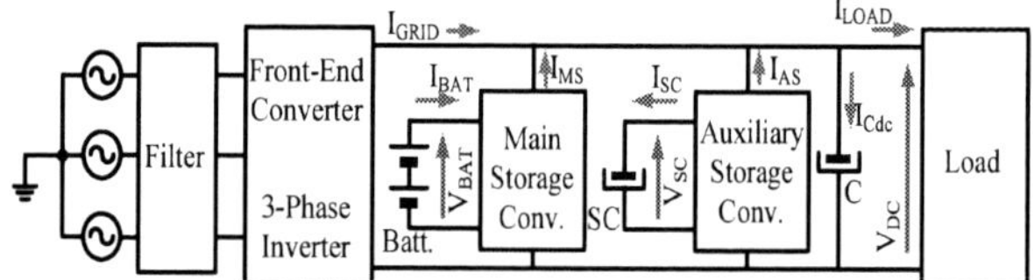

Fig. 1. Block Diagram of the system under consideration (Front-End Converter with ESS).

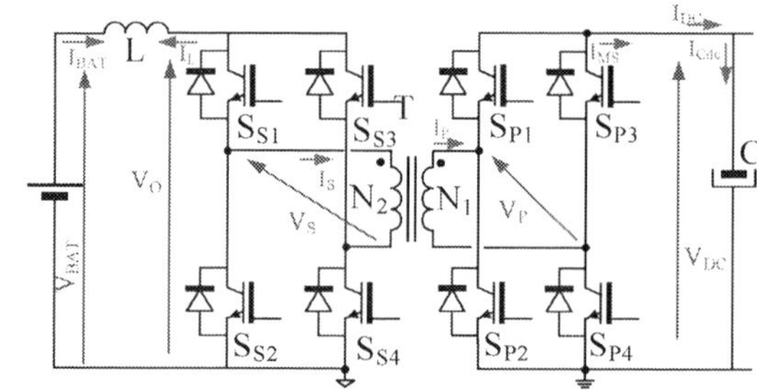

Fig. 2. Full Bridge Current Source (FBCS) converter for interfacing of a DC-link and a battery.

The most current solution is to add in parallel an additional auxiliary storage system, in order to form a Hybrid Energy Storage System (HESS), able to supplement the needed transient power requirements with a very fast dynamics [5]-[11]. This paper is focused on proposing an alternative for this application, based on a Supercapacitor (SC) module, connected to the DC link and the main ESS through a Series Parallel Connection (SPC), shown in Fig. 3 (shaded).

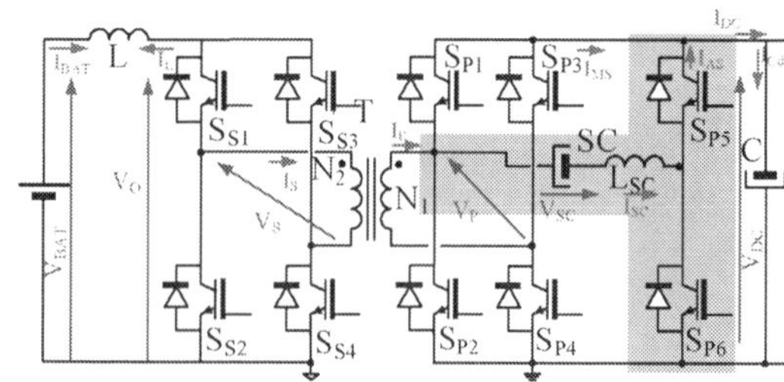

Fig. 3. Full Bridge Current Source (FBCS) converter with SPC connection of Supercapacitor Bank.

II. THE MAIN ENERGY STORAGE SYSTEM

The Main Energy Storage System is connected to the DC link by means of a DC to DC bidirectional Converter. Among the most used isolated topologies, the Double Active Bridge (DAB) converter outstands as a mature, well-known and versatile solution. It provides possibility of soft commutations within some operation margins, and high power density, plus overall good performance. The main flaws of this converter come from the loss of performance at low power levels, and the high current harmonics in the input and output current waveforms [refs required]. In order to avoid such harmonics to flow though the output port (in this case the ESS), in this work, the Current Source version of the DAB has been selected.

The FBCS converter depicted in Fig. 2 is formed by a Full-Bridge structure (switches SP1 to SP4), connected to the primary side of a transformer [12]-[13]. At the secondary side, another Full-Bridge (switches SS1 to SS4) interfaces with a filter inductor, L in series with the battery itself. Unlike in a standard Voltage-Source Double Active Bridge (VSDAB), this inductor operates with high DC current values and relatively small ripples, yielding to a current source behavior in the battery side. This also yields to a more compact design of the inductor, as the AC magnetic field is smaller. This inductor is therefore filtering the current to the battery, thus increasing the system reliability. In addition, the current control through this inductor is also the battery current control, which simplifies the control design.

The switching scheme is depicted in Fig. 4.a, along with the main waveforms of the FBCS converter in steady state. The only control parameter is the duty ratio D that defines the modulation pattern. Fig. 4.b shows simulations of the FBCS operation. Fig. 5.a outlines the control scheme used for the full system. In order to compare the results, the performance of the FEC dynamics has been made equal in both cases (with and without battery). The current reference for the battery is calculated from the actual load step, as shown in Fig. 5.b.

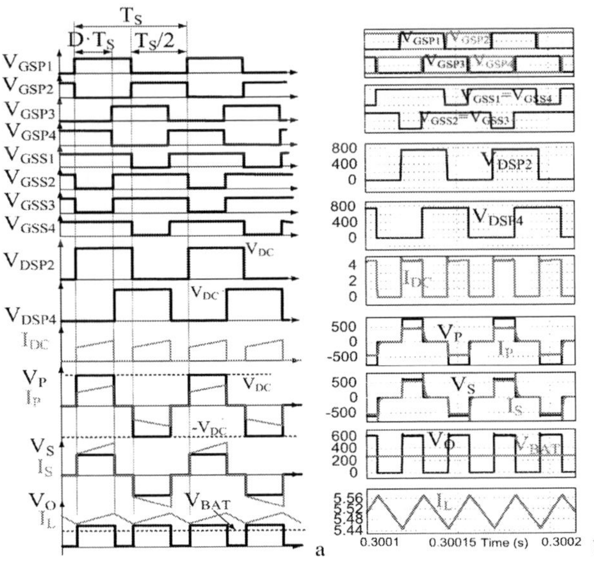

Fig. 4. Main theoretical (a) and simulated (b) waveforms of FBCS operation in steady state

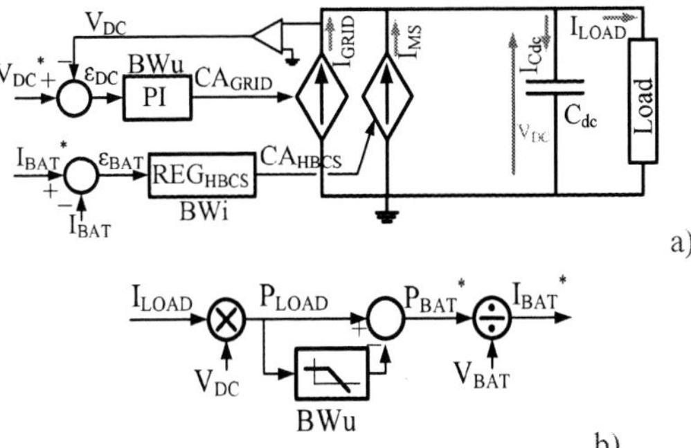

Fig. 5.a) Control scheme for main ESS alone. b) Generation of reference for battery current

Fig. 6 and Fig. 7 depicts the performance of this initial system, without and with the main ESS, for a load step of 1 kW, in a 700 VDC link voltage system. It can be seen how the DC voltage drift is substantially smaller in the system with the main ESS. The rest of the operating parameters are shown in Table I. A closer look to Fig. 7 shows a high di/dt in the initial response after the load step. This might yield to battery operating life shortening, etc. In order to solve this issue, a SC module can be placed in the system, to cope with such transient efforts.

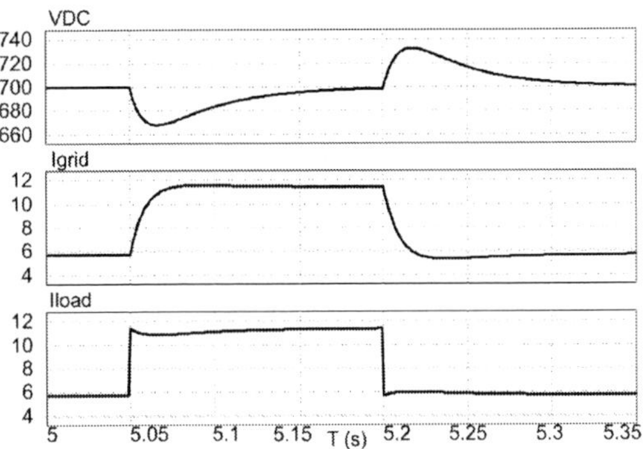

Fig. 6. Simulated waveforms without the main ESS, upon 1kW load step up and down.

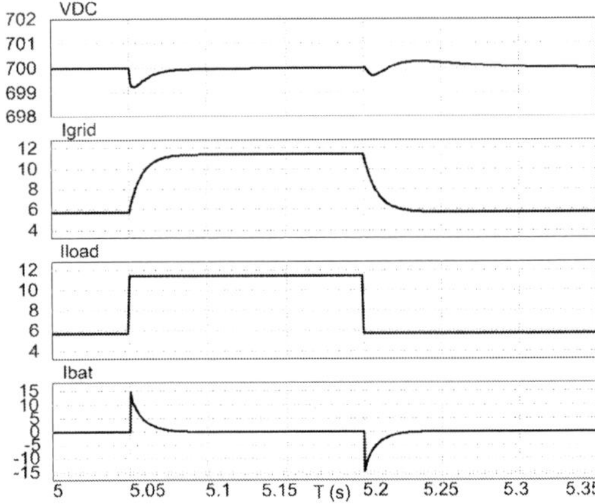

Fig. 7. Simulated waveforms with the main ESS, upon 1kW load step up and down.

TABLE I. PARAMETERS OF THE SIMULATED SYSTEM.

Parameter	Value
DC link ratings	700 VDC
Bandwidth of DC link Control	20 Hz.
Nominal Power (FEC)	10 kW
Battery Nominal Voltage	300 V
Bandwidth of Battery Current Loop	1 kHz.
Load Step	1 kW

III. THE PROPOSED INTEGRATION OF SUPERCAPACITOR

One option for this integration would be the use of a Solid State Transformer (SST), with an additional port for the supercapacitor bank [14]-[15]. This is especially suitable for this application, given the high mismatch voltage ratings among the three ports, which can be solved by selecting the proper turns-ratio of the transformer. However, this would provide high stresses in the transformer, as it must therefore be defined for high currents to avoid saturation during the transients. This work explores the integration of the SC module as an auxiliary leg in the primary side of the converter through SPC, different from the series connection presented in [16], as depicted in Fig. 3.

Ahead is a summary of the principle of operation of the SPC. The simplest connection of the HESS to manage the power flow from and into the storage devices is the Parallel Connection, shown in Fig. 8, based on a dedicated bidirectional boost converter connected to a DC link [17]-[24].

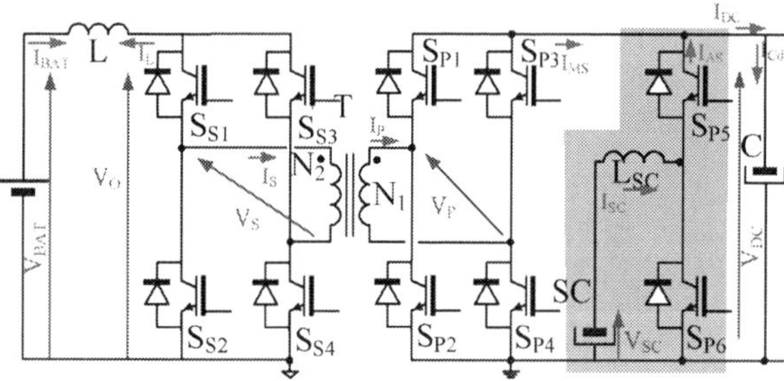

Fig. 8. Full Bridge Current Source (FBCS) converter with direct parallel connection of Supercapacitor Bank.

The main limitations of this connection come from the high voltage mismatch between the supercapacitor voltage ratings and the fixed DC bus voltage. Considering a battery nominal voltage of around half the DC link, then the duty ratio of the battery leg will be around 50%, which optimizes the converter performance, in terms of stresses, design complexity and control capability. However, provided that in some applications the supercapacitor nominal voltage ratings are ranged from 20V to 40V, then for DC bus voltage ratings ranged from 500V to 900V the required gain for the converter might reach values of 40:1 and higher. Such requirements prevent the use of such topology. To overcome this drawback, the Series-Parallel Connection of the supercapacitor modules (Fig. 3) can be carried out. The gain required for the supercapacitor module leg of the converter results in a much more reasonable value than before. Indeed, the drawback of the voltage mismatch at the bidirectional boost configuration is solved, as the average voltage at the midpoint of the supercapacitor module leg results in the addition of the supercapacitors plus the battery voltages. Assuming that the supercapacitor ratings are much smaller than the battery ones, this leg operates with a duty ratio similar to the battery leg. In addition, if the switching pulses are adequately synchronized, the current ripples in the converter might decrease.

Considering steady state, and given that the average voltage at the midpoint of the leg formed by switches S_{P1} and S_{P2} at the primary side equals $V_{DC}/2$, then the average voltage value at the midpoint of the SPC leg (Switches S_{P5} and S_{P6}) will be equal to $(V_{DC}/2)+V_{SC}$, as the average value of the SC inductor voltage is zero. This means that the duty ratio of S_{P5}, D_{SC}, can be calculated as follows:

$$D_{SC} = \frac{V_{SC}+\frac{V_{DC}}{2}}{V_{DC}} \quad (1)$$

Yielding to a value close to 50%, which allows for the operation of this leg with a high gain between the SC module and the DC link, but within the most adequate values for the duty ratio (current and voltage stresses matched, high control margin).

IV. RESULTS AND VALIDATION OF THE PROPOSED SCHEME

Fig. 9 shows theoretical and simulated waveforms of the FBCS converter with the SPC scheme. As it can be seen, the DC current is now the addition of the current given by the main ESS (I_{MS}, from the battery) plus the current given by the Auxiliary ESS (I_{AS}, from the SC module). Fig. 10 shows the control scheme implemented for the control of the converter. As it can be seen, the error of the battery current loop is the parameter that will provide the reference for the supercapacitor current. Fig. 11 shows the performance of the HESS. It can be seen how the main operating parameters are pretty close to the simple battery ones depicted in Fig. 7. However, now the current that the battery is providing has a much slower di/dt, given the peak power is supplied initially by the SC.

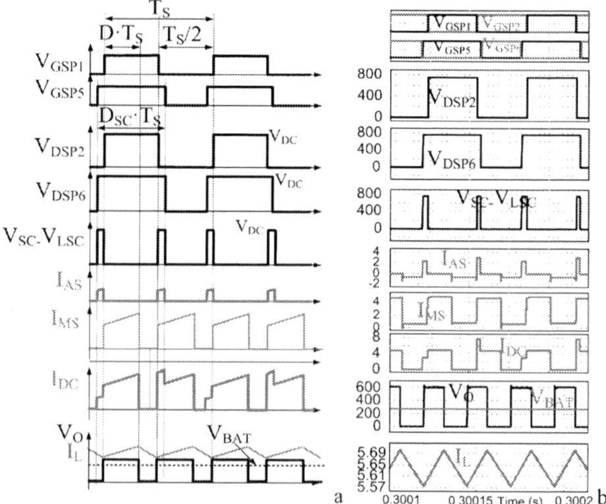

Fig. 9. Main theoretical (a) and simulated (b) waveforms of FBCS operation in steady state with SPC of Supercapacitors.

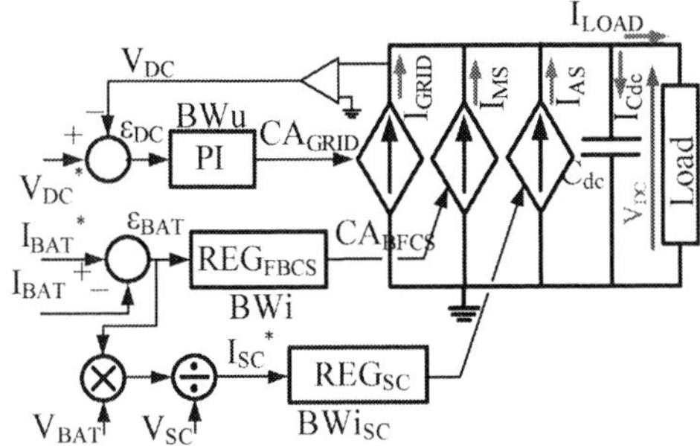

Fig. 10. Control scheme for main and auxiliary ESSs.

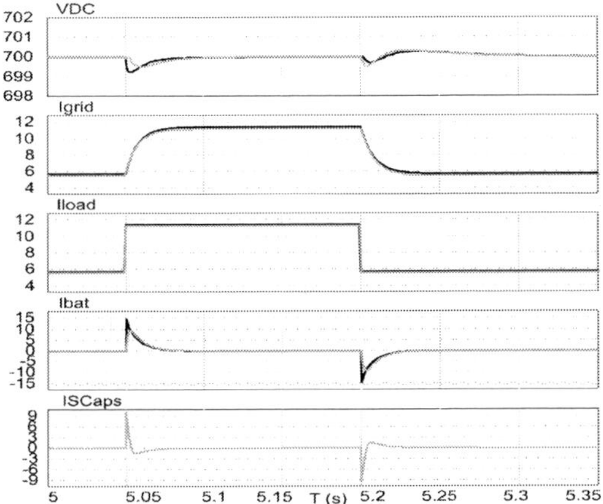

Fig. 11. Simulated waveforms of main ESS alone (black) vs SPC of Supercapacitor, SC, (green), upon 1kW load step up and down.

In order to demonstrate the feasibility of the system, several experimental tests have been carried out in two demonstrator. The first demonstrator consists of a 1kW, $600V_{DC}$ system, with a supercapacitor bank voltage of $30V_{DC}$. Fig. 12 shows the captured data of the SPC scheme operating in steady state. As it can be seen, the leg of the SC (CH7) operates with a duty ratio close to the battery leg (CH6), for 30 V_{DC} and 300 V_{DC} operating voltages, respectively, for a DC link of around 600V.

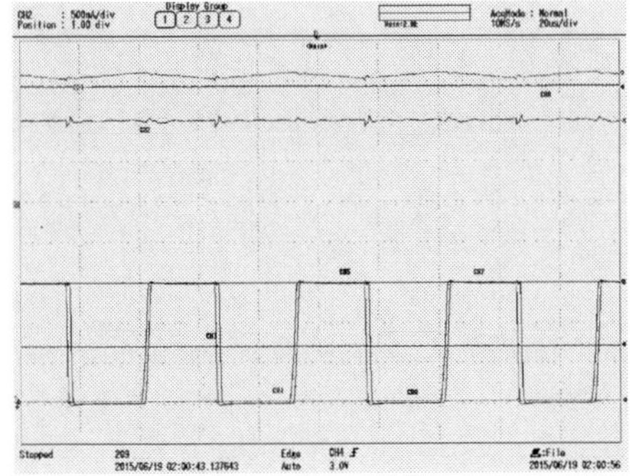

Fig 12. SPC operation of a demonstrator converter. CH1 (yellow): I_{BAT}, 1A/div. CH2 (green): I_{SCaps}, 0.5A/div. CH3 (magenta): V_{BAT}, 200V/div. CH4(cyan): V_{SC}, 50V/div. CH5 (red): V_{DC} , 200V/div, CH6(orange): u_{DSP1}, 200V/div. CH7 (blue): u_{DSP5}, 200V/div. CH8(violet): I_{LOAD},1AV/div. Time: 10us/div

The control system performance has been validated in a scaled prototype, with $200/300V_{DC}$ in the DC link, around 200V in the battery, 20 kHz switching frequency, and 30 V_{DC} in the supercapacitor bank. The main parameters of the scaled demonstrator are depicted in Table II

TABLE II. PARAMETERS OF THE EXPERIMENTAL SYSTEM.

Parameter	Value
DC link ratings	200/300 V_{DC}
Bandwidth of DC link Control	10 Hz.
Nominal Power (FEC)	1 kW
Battery Nominal Voltage	200 V
Supercaps	20 V
Bandwidth of Supercaps Current Loop	1 kHz.
Load Step	400 W

Fig. 13. shows the performance of a load step in the 300 V_{DC} DC link, and the 10 Hz dynamics of the grid converter. The first trace shows the DC link variation due this load current step (fourth trace), and a voltage ripple of around 10 V can be appreciated.

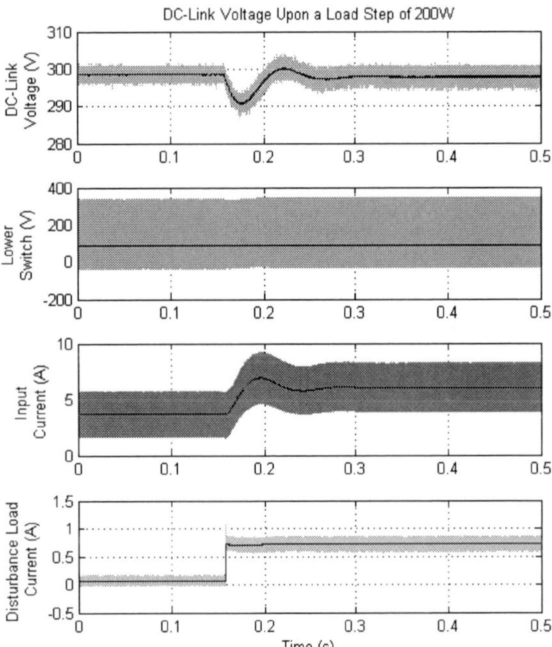

Fig 13. Load step in the DC link, without the supercapacitor module.

Fig. 14. shows the steady state operation of the battery charge, for a 1A charging current reference. The approximately 90° degrees phase shift in the primary side legs of the converter can be seen. Also, the battery current and the battery voltage itself are depicted. The V_{DC} link is in this case equal to 200 V_{DC}.

Fig. 15 shows the operation of the supercapacitor bank under SPC connection, again for 200 V_{DC} link voltage. A current step reference is provided, and the implemented control loop with a bandwidth of 1kHz is shown. As it can be seen, the system is able to track the reference within the desired 1 kHz bandwidth.

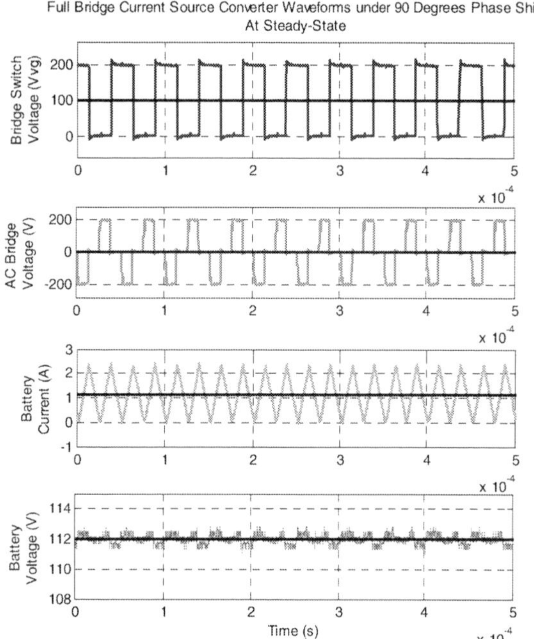

Fig 14. FBCS Converter. Traces (from upper to lower): Voltage in one of the bridge legs (blue); voltage in the primary of the transformer (green); battery current (blue). Battery voltage (red).

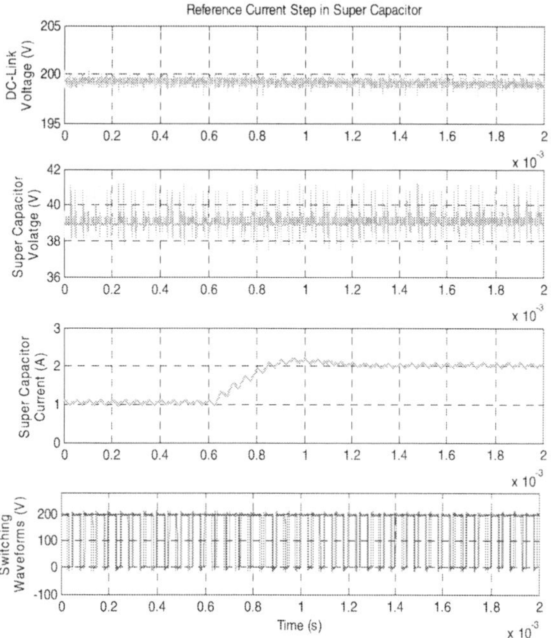

Fig 15. Supercapacitor reference current step. Traces (from upper to lower): DC link voltage (red), Supercaps voltage (green), supercaps current (blue), voltage at both legs of the inverter (lower plot).

V. CONCLUSION

This work has demonstrated theoretically and through simulations the feasibility of the SPC of a Low Voltage SC Module for interfacing with an ESS formed by a DC link and a Battery connected through the FBCS converter. The main features of this Hybrid System include operation of the system with a high voltage mismatch from the supercaps module to the DC link. It also shows the easy control scheme, design and tuning (fully decoupled control parameters for the main and the auxiliary ESSs). Theoretically this scheme provides higher efficiency than other isolated connections (as the SC energy does not need to flow though the transformer). These characteristics allow this solution to be considered for Hybrid Systems of medium voltage and power ratings.

ACKNOWLEDGMENT

This work has been partially supported by the Spanish Government, Innovation Development and Research Office (MEC), under research grant ENE2013-44245-R, Project "Microholo", and by the European Union through ERFD Structural Funds (FEDER). This work has been partially supported by the government of Principality of Asturias, Foundation for the Promotion in Asturias of Applied Scientific Research and Technology (FICYT), under Severo Ochoa research grant, PA-13-PF-BP13138.

REFERENCES

[1] P. Thounthong, and S. Rael, "The benefits of hybridization," Industrial Electronics Magazine, IEEE , vol. 3, no. 3, pp. 25-37, Sept. 2009.

[2] A. Hintz, U.R. Prasanna, and K. Rajashekara., "Novel modular Multiple-Input Bidirectional DC-DC Power Converter (MIPC)," In International Power Electronics Conference (IPEC-Hiroshima 2014 - ECCE-ASIA), pp. 2343–50, 18-21 May 2014.

[3] S.D.G. Jayasinghe, and D.M. Vilathgamuwa. "Flying supercapacitors as power smoothing elements in wind generation," IEEE Transactions on Industrial Electronics, vol. 60, no. 7, pp. 2909–18, July 2013.

[4] J. Leuchter, "Review of energy storage for small photovoltaic power source," In Renewable Power Generation Conference (RPG), 3rd, pp. 1–8, 24-25 Sept. 2014.

[5] M. S. Kahn, and M. R. Iravani, "Hybrid control of a grid-interactive wind energy conversion system," IEEE Transactions on Energy Conversion, vol. 23, no. 3, pp. 895-902, Sept. 2008.

[6] I. Vechiu, A. Etxeberria, H. Camblong, and J. M. Vinassa, "Three-level neutral point clamped inverter interface for flow battery/supercapacitor energy storage system used for microgrids," 2nd IEEE PES International Conference and Exhibition on Innovative Smart Grid Technologies (ISGT Europe), pp. 1-6, 5-7 Dec. 2011.

[7] B. Wang, B. Zhang, and Z. Hao, "Control of composite energy storage system in wind and PV hybrid microgrid," IEEE Region 10 Conference (TENCON), pp. 1-5, 22-25 Oct. 2013.

[8] L. Guo, Y. Zhang, and C. S. Wang, "A new battery energy storage system control method based on SOC and variable filter time constant," IEEE PES Innovative Smart Grid Technologies (ISGT), pp. 1-7, 16-20 Jan. 2012.

[9] T. Asao, R. Takahashi, T. Murata, J. Tamura, M. Kubo, A. Kuwayama, and T. Matsumoto, "Smoothing control of wind power generator output by superconducting magnetic energy storage system," International Conference on Electrical Machines and Systems (ICEMS), pp. 302-307, 8-11 Oct. 2007.

[10] X. Han, F. Chen, X. Cui, Y. Li, and X. Li, "A power smoothing control strategy and optimized allocation of battery capacity based on hybrid storage energy technology," Energies, vol. 5, pp. 1593–1612, May 2012.

[11] K. Yoshimoto, T. Nanahara, G. Koshimizu, and Y. Uchida, "New control method for regulating state-of-charge of a battery in hybrid wind power/battery energy storage system," IEEE PES Power Systems Conference and Exposition (PSCE), pp. 1244-1251, 29 Oct. 2006.

[12] D. Maiti, N. Mondal, and S. K. Biswas, "Optimization of a bi-directional hybrid current-fed-voltage-fed converter link," Joint International Conference on Power Electronics, Drives and Energy Systems (PEDES) & 2010 Power India, pp. 1-9, 20-23 Dec. 2010.

[13] P. Xuewei, and A. K. Rathore, "Comparison of bi-directional voltage-fed and current-fed dual active bridge isolated DC/DC converters low voltage high current applications," IEEE 23rd International Symposium on Industrial Electronics (ISIE), pp. 2566–2571, 1-4 June 2014.

[14] Hongfei Wu, Peng Xu, Haibing Hu, Zihu Zhou, and Yan Xing, "Multiport converters based on integration of full-bridge and bidirectional DC-DC topologies for renewable generation systems," IEEE Transactions on Industrial Electronics, vol. 61, no. 2, pp. 856-869, Feb. 2014.

[15] H. Tao, A. Kotsopoulos, J. L. Duarte, and M. A. M. Hendrix, "Family of multiport bidirectional DC-DC converters," IEEE Proceedings Electric Power Applications, vol. 153, no. 3, pp. 451-458, 1 May 2006.

[16] H. Zhou, T. Duong, S. T. Sing, and A. M. Khambadkone, "Interleaved bi-directional dual active bridge DC-DC converter for interfacing ultracapacitor in micro-grid application," IEEE International Symposium on Industrial Electronics (ISIE), pp. 2229–2234, 4-7 July 2010.

[17] W. Li, and G. Joos, "A power electronic interface for a battery supercapacitor hybrid energy storage system for wind applications," IEEE Power Electronics Specialists Conference (PESC), pp. 1762 - 1768, 15-19 June 2008.

[18] H. Yoo, S.-K. Sul, Y. Park, and J. Jeong, "System integration and power-flow management for a series hybrid electric vehicle using supercapacitors and batteries," IEEE Transactions on Industry Applications, vol.44, no.1, pp. 108-114, Jan.-Feb. 2008.

[19] S. D. G. Jayasinghe, D. M. Vilathgamuwa, and U. K. Madawala, "A direct integration scheme for battery-supercapacitor hybrid energy storage systems with the use of grid side inverter," Twenty-Sixth Annual IEEE Applied Power Electronics Conference and Exposition (APEC), pp. 1388-1393, 6-11 March 2011.

[20] Li. Wei, G. Joos, and J. Belanger, "Real-time simulation of a wind turbine generator coupled with a battery supercapacitor energy storage system," IEEE Transactions on Industrial Electronics, vol. 57, no. 4, pp. 1137-1145, April 2010.

[21] N. R. Tummuru, M. K. Mishra, and S. Srinivas, "Dynamic energy management of hybrid energy storage system with high-gain PV converter," IEEE Transactions on Energy Conversion, vol. 30, no. 1, pp. 150-160, March 2015.

[22] S. K. Kollimalla, M. K. Mishra, and L.Narasamma N., "Design and analysis of novel control strategy for battery and supercapacitor storage system," IEEE Transactions on Sustainable Energy, vol. 5, no. 4, pp. 1137-1144, Oct. 2014.

[23] S. K. Kollimalla, M. K. Mishra, and L. Narasamma N., "Coordinated control and energy management of hybrid energy storage system in PV system," International Conference on Computation of Power, Energy, Information and Communication (ICCPEIC), pp. 363-368, 16-17 April 2014.

[24] K. Yamamoto, A. Imakiire, R. Lin, and K. Iimori, "Comparison of configurations of voltage boosters in PWM inverter with voltage boosters with regenerating circuit augmented by electric double-layer capacitor," International Conference on Electrical Machines and Systems (ICEMS), pp. 1-6, 15-18 Nov. 2009.

Saturation Controller-Based Direct Power Control for Doubly-Fed Induction Generator

Chun Wei[1], Zhe Zhang[2], Wei Qiao[1], and Liyan Qu[1]
cwei@huskers.unl.edu; zhe.zhang@nexteer.com; wqiao3@unl.edu; lqu2@unl.edu
[1]Power and Energy Systems Laboratory, Department of Electrical and Computer Engineering, University of Nebraska-Lincoln, Lincoln, NE, 68588 USA
[2]Nexteer Automotive, Saginaw, MI, 48601 USA

***Abstract*—This paper proposes a novel direct power control (DPC) scheme for doubly-fed induction generator (DFIG). Unlike the conventional DPC using hysteresis comparators, the proposed DPC utilizes nonlinear saturation controllers for the voltage vector selection and duty ratio modulation. This makes the calculation of the duty ratios of the voltage vectors much simpler and independent of DFIG parameters. The effect of DFIG rotor speed variations on the power ripples is considered in the proposed DPC by designing an adaptive bandwidth for the saturation controller. The proposed DPC scheme can significantly reduce the power ripples using a fixed switching frequency while preserving most of the intrinsic merits of the conventional DPC, such as fast dynamics, low parameter dependence, and easy implementation. The effectiveness of the proposed DPC scheme is confirmed by simulation results for a 2-MW DFIG-based WECS in MATLAB/Simulink.**

Keywords*—*Direct power control (DPC); doubly-fed induction generator (DFIG); saturation controller

I. INTRODUCTION

The doubly-fed induction generator (DFIG) is one of the most widely used generators in variable-speed wind energy conversion systems (WECSs) owing to their merits of a decoupled control of the active and reactive powers and a low power converter rating [1]-[2]. The DFIG in a WECS is usually controlled by a stator flux/voltage-oriented vector control method, which, however, is subject to the problems of slow dynamic response, high sensitivity to system parameter variations, and tedious parameter tuning for the proportional-integral (PI) controllers [3]-[4]. A more efficient and robust power control algorithm is therefore desired for the DFIG-based WECS.

The direct power control (DPC) is a promising technique for the control of DFIG-based WECSs owing to its attractive features, such as faster dynamic response because of no slow-response PI controllers, lower complexity by removing the need for current decoupling computation, and lower parameter dependence when compared to the conventional vector control [5]. However, large ripples are found in active and reactive power waveforms when using the conventional DPC.

Recently, much research effort has gone into reducing active and reactive power ripples in DPC systems. A popular solution is to use the space vector modulation (SVM), where the desired rotor voltage vectors for the SVM were mainly obtained by using the voltage-power equations based on the mathematical model of the DFIG [6]-[7], the model-based predictive method [8], or the sliding-mode control approach [9]. A new model predictive DPC method was proposed in [10], where the desired voltage vector was selected according to an optimal cost function. However, these methods need complicated voltage vector computations or accurate machine parameters, thus eroding several basic features of the DPC, e.g., simplicity.

Three-vector-based duty cycle control methods were used in the DPC to reduce the power ripples of a DFIG and maintain the simplicity and robustness of the DPC [11]-[12]. In these methods, the calculation of the duration of each voltage vector in one sampling cycle is relatively complicated and depends on machine parameters. Moreover, there were steady-state errors in the active power when using the duty cycle control and the power ripples were influenced by the machine rotor speed value. These issues were not considered in the existing DPC schemes for DFIGs.

This paper proposes a novel saturation controller-based DPC scheme for DFIGs. By replacing the traditional hysteresis comparators with the proposed saturation controllers in the DPC, the selection of the three voltage vectors and their durations becomes flexible and adjustable [13]. The calculation of the duration of each voltage vector in one sampling cycle becomes simple and independent of DFIG parameters. An adaptive bandwidth is used in the saturation controllers to minimize the steady-state error and the effect of the DFIG rotor speed on the power ripples. The proposed DPC scheme greatly reduces the power ripples with a fixed switching frequency while preserving most of the intrinsic properties of the conventional DPC, such as fast dynamic response, no PI controller, high robustness to machine parameter variations, and easy implementation. The effectiveness of the proposed DPC scheme is confirmed by simulation results for a 2-MW DFIG WECS in MATLAB/Simulink.

II. DYNAMIC MODEL OF A DFIG

The dynamic equation of a three-phase DFIG is written by space vectors in the stationary reference frame as [1],

$$u_s^s = R_s i_s^s + \frac{d\lambda_s^s}{dt} \tag{1}$$

$$u_r^s = R_r i_r^s + \frac{d\lambda_r^s}{dt} - j\omega_r \lambda_r^s \tag{2}$$

This work was supported in part by the U.S. National Science Foundation under CAREER Award ECCS-0954938.

where u_s^s, i_s^s, u_r^s, and i_r^s are the stator voltage vector, stator current vector, rotor voltage vector, and rotor current vector, respectively; ω_r is the electrical rotor speed; the stator flux linkage vector λ_s^s and rotor flux linkage vector λ_r^s are given by

$$\lambda_s^s = L_s i_s^s + L_m i_r^s \tag{3}$$

$$\lambda_r^s = L_m i_s^s + L_r i_r^s \tag{4}$$

where $L_s = L_{ls} + L_m$; $L_r = L_{lr} + L_m$; L_{ls}, L_{lr} and L_m are the stator leakage, rotor leakage and mutual inductances, respectively.

According to (3) and (4), the rotor flux vector in the stationary reference frame can be expressed using the stator flux and current vectors as

$$\lambda_r^s = \frac{L_r}{L_m}\lambda_s^s - \frac{1}{\gamma L_m} i_s^s \tag{5}$$

where $\gamma = 1/(L_s L_r - L_m^2)$. Then the rotor flux in the rotor reference frame can be obtained to estimate the rotor flux angle θ_λ in the rotor reference frame and is given by

$$\lambda_r^r = |\lambda_r^r| \angle\theta_\lambda = \lambda_r^s \cdot e^{-j\theta_r} \tag{6}$$

where θ_r is the rotor position angle. Neglecting the power losses associated with the stator resistances, the active and reactive stator powers are given by [3]

$$P_s = -\frac{3}{2}\frac{L_m}{\sigma L_s L_r}\omega_s |\lambda_s^r||\lambda_r^r|\sin\delta \tag{7}$$

$$Q_s = \frac{3}{2}\frac{\omega_s}{\sigma L_s}|\lambda_s^r|\left[\frac{L_m}{L_r}|\lambda_r^r|\cos\delta - |\lambda_s^r|\right] \tag{8}$$

where $\sigma = (L_s L_r - L_m^2)/L_s L_r$ and δ is the relative angle between the stator and rotor flux space vectors. It can be seen from (7) and (8) that the stator active and reactive powers can be controlled by adjusting the relative angle between and the amplitudes of the stator and rotor flux space vectors.

The differentiating forms of (7) and (8) are given by

$$\frac{dP_s}{dt} = -\frac{3}{2}\frac{L_m}{\sigma L_s L_r}\omega_s |\lambda_s^r| \frac{d(|\lambda_r^r|\sin\delta)}{dt} \tag{9}$$

$$\frac{dQ_s}{dt} = \frac{3}{2}\frac{L_m}{\sigma L_s L_r}\omega_s |\lambda_s^r| \frac{d(|\lambda_r^r|\cos\delta)}{dt} \tag{10}$$

The differentiating terms on the right hand side of (9) and (10) can be expressed by

$$\frac{d(|\lambda_r^r|\sin\delta)}{dt} = \frac{d(|\lambda_r^r|)}{dt}\sin\delta + |\lambda_r^r|(\omega_s - \omega_r)\cos\delta \tag{11}$$

$$\frac{d(|\lambda_r^r|\cos\delta)}{dt} = \frac{d(|\lambda_r^r|)}{dt}\cos\delta - |\lambda_r^r|(\omega_s - \omega_r)\sin\delta \tag{12}$$

From (9) to (12), it can be seen that besides the relative angle between the stator and rotor flux space vectors and the rotor flux amplitude, the rotor speed also has an impact on the change of the stator active and reactive powers, which makes the power ripples dependent on the operating point of the DFIG.

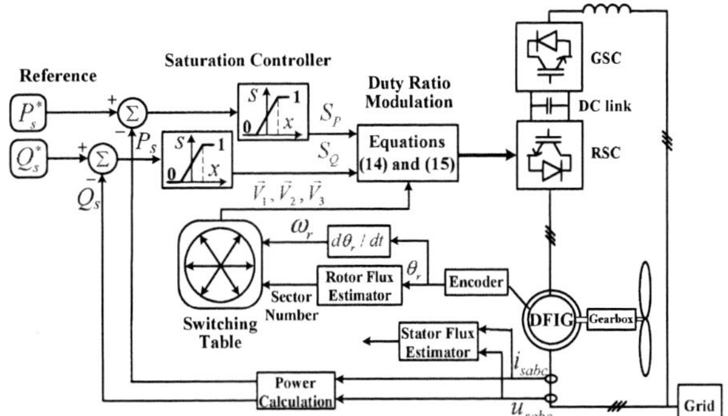

Fig. 1. The schematic of the proposed saturation controller-based DPC scheme for the DFIG.

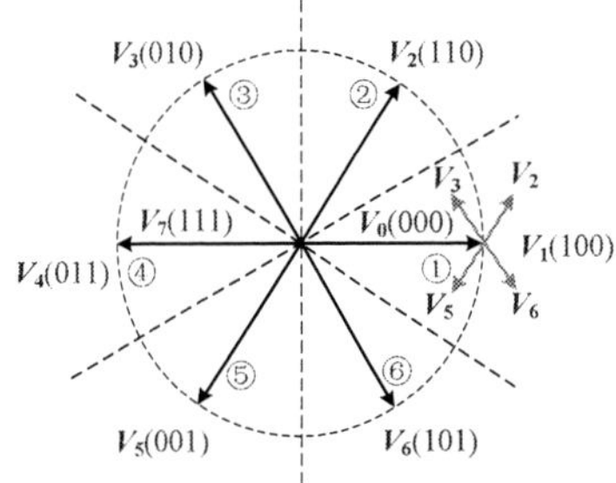

Fig. 2. Basic voltage vectors and sector divisions for the DPC of the DFIG.

III. Proposed Saturation Controller-Based DPC

The schematic of the proposed DPC is shown in Fig. 1, where two saturation controllers and a duty ratio modulation algorithm are used to replace the hysteresis comparators and the switching table in the conventional DPC. Compared with the hysteresis comparators, the saturation controllers consider not only the signs of the errors but also their magnitudes, which can avoid overusing the nonzero voltage vectors in one switching period. The sector number is determined according to θ_λ in (6).

A. Saturation Controller

A normalized saturation controller defined as follows is used in this paper.

$$sat(x, B_w) = \begin{cases} 0.5\times[\mathrm{sgn}(x)+1], & |x| > B_w \\ 0.5\times(x/B_w + 1), & |x| \le B_w \end{cases} \tag{13}$$

where x is the input, which is the error between the reference power and the actual power; B_w is the bandwidth; and sgn(x) is the sign function. When the absolute value of x exceeds the bandwidth B_w, the saturation function will only output a discrete state 1 or 0, which is the same as a hysteresis comparator. However, when x is within the bandwidth, the input will be scaled linearly to the range of [0, 1]. Compared with the hysteresis comparator which typically only outputs 1 or 0 (or −1), the output of the saturation controller has a

TABLE I
SWITCHING TABLE FOR THE SATURATION CONTROLLER-BASED DPC

Sector	Effect	S1	S2	S3	S4	S5	S6
Super-synchronous speed	↑	V5 (001)	V6 (101)	V1 (100)	V2 (110)	V3 (010)	V4 (011)
	↑	V6 (101)	V1 (100)	V2 (110)	V3 (010)	V4 (011)	V5 (001)
	↓	V0 (000)	V0 (000)	V0 (000)	V0 (000)	V0 (000)	V0 (000)
Sub-synchronous speed	↓	V3 (010)	V4 (011)	V5 (001)	V6 (101)	V1 (100)	V2 (110)
	↓	V2 (110)	V3 (010)	V4 (011)	V5 (001)	V6 (101)	V1 (100)
	↑	V0 (000)	V0 (000)	V0 (000)	V0 (000)	V0 (000)	V0 (000)

continuous region between the upper and lower limits. This helps achieve a continuously adjustable duty ratio modulation in the range of [0, 1] for the PWM control of the rotor-side converter (RSC) of the DFIG.

It is known that the pulse width of a voltage vector (see Fig. 2) within one sampling cycle is fixed in the conventional DPC. The duty ratio of a switch is either 1 or 0 if neglecting the dead-time effect. This mechanism guarantees a fast dynamic response of the DFIG-power converter system but needs a relatively high sampling frequency to achieve a satisfactory steady-state performance. However, if the saturation controller is adopted in the DPC scheme, the duration of a voltage vector in each sampling cycle can be adjusted continuously and smoothly. It is possible to combine multiple voltage vectors within one sampling cycle to improve the system's steady-state performance with a relatively low sampling frequency. Meanwhile, when a large change in the stator active or reactive power is needed, the saturation controller will function as a hysteresis comparator to make full use of the physical limitation of the RSC, thus possessing the advantage of fast dynamic response as the conventional DPC. The saturation controller allows the DPC to take the magnitudes of the stator active and reactive power errors into consideration, thereby achieving good system performance in both transient and steady states.

B. Duty Ratio Modulation Algorithm

The rotor speed information and a modified switching table are used to determine the voltage vectors in the proposed duty ratio modulation algorithm. The rotor speed information is used to judge whether the DFIG is operating at the sub-synchronous or super-synchronous speed. Based on this signal, proper voltage vectors are chosen to make the rotor flux rotate in the correct direction. To maximally preserve the merits of the conventional DPC, a switching table shown in Table I modified based on the conventional switching table in [5] is used. To explain how the voltage vectors are selected, first assume that the DFIG is operating in the sub-synchronous speed and the rotor flux is rotating anticlockwise and lying in the first sector. Three voltage vectors are chosen in one switching cycle. The first two voltage vectors will lead to a decrease of the stator active power and a zero vector will serve as the third voltage vector to prevent large stator active power ripples. It is noticed that both V_2 and V_3 can lead to a decrease of the stator active power, but their effects on the stator reactive power are opposite. Therefore, both V_2 and V_3 are candidate voltage vectors to be selected when the stator active power needs to be decreased. Their durations in one switching cycle are assigned according to the output of the reactive power saturation controller, S_Q, to obtain a smooth and circular rotor flux trajectory. However, it is not necessary to decrease the stator active power in the whole switching cycle and the stator active power can be increased for a certain period during the switching cycle to reduce its ripple. There are two ways to increase the stator active power by applying either a zero vector V_0 or V_7 or a nonzero vector V_5 or V_6. However, since V_5 and V_6 will force the rotor flux to rotate clockwise, the stator active power will increase much more than that when V_0 or V_7 is applied. To prevent too much increase of the stator active power to cause large power ripples, a zero voltage vector instead of V_5 or V_6 is used. The execution time for V_0 or V_7 is determined by the output of the active power saturation controller S_P.

In this way, the duty ratios for the three legs of the RSC are determined as follows

$$\vec{d} = (1 - S_P)\times(S_Q \times \vec{V}_1 + (1 - S_Q)\times\vec{V}_2) + S_P \times \vec{V}_3 \tag{14}$$

$$\vec{d} = S_P \times(S_Q \times \vec{V}_1 + (1 - S_Q)\times\vec{V}_2) + (1 - S_P)\times\vec{V}_3 \tag{15}$$

where S_P and S_Q are the outputs of the active and reactive power saturation controllers, respectively; $\vec{V}_1$, $\vec{V}_2$ and $\vec{V}_3$ are the three voltage vectors in the same order as shown in Table I. Equation (14) or (15) is applied when the DFIG operates in the sub-synchronous or super-synchronous speed, respectively.

C. Adaptive Bandwidth for Saturation Controllers

The second terms in the right hand side of (11) and (12) show that the rotor speed also has an impact on the changes of the active and reactive powers besides the voltage vectors. A larger slip speed leads to larger power ripples when a voltage vector is applied. When the DFIG operates at the synchronous speed with a zero slip speed, the rotor speed has no impact on the power changes and, therefore, the power ripples are relatively small. In addition, the bandwidths of the saturation controllers have significant influences on the power ripples and the steady-state errors of the stator active power. A large bandwidth will make full use of the error information between the reference and actual power values, thus making the duration of a voltage vector continuous and smooth. This decreases the power ripples but causes a deviation of the actual power from its reference. A small bandwidth will make the saturation controller more like a hysteresis comparator, which causes larger power ripples. However, this decreases the steady-state error of the output power. To reduce both the ripple and the steady-state error of the power, an adaptive bandwidth is designed for the saturation controllers. A larger bandwidth is used for an operating point with a larger slip speed; while a small bandwidth is used when the rotor speed approaches the synchronous speed. In this paper, the bandwidth is designed to change linearly with the slip speed as follows to make the DPC scheme simple and effective.

$$G(\omega_r) = \begin{cases} -\dfrac{a(\omega_r - \omega_s)}{\omega_s} + b & \omega_r \le \omega_s \\ \dfrac{a(\omega_r - \omega_s)}{\omega_s} + b & \omega_r > \omega_s \end{cases} \quad (16)$$

where a and b are control parameters. In this paper, $a = 0.9$ and $b = 0.05$.

IV. Simulation Results

Simulation studies are carried out in MATLAB/Simulink to validate the proposed saturation controller-based DPC scheme for a 2-MW DFIG-based WECS connected to a power grid shown in Fig. 3. The parameters of the DFIG system are listed in Table II. A resistor-capacitor (RC) filter is connected in shunt with the stator of the DFIG to absorb the switching harmonics generated by the RSC and the grid-side converter (GSC). The RSC uses the proposed DPC scheme to control the stator active and reactive powers. The GSC is used mainly to maintain a constant dc-link voltage. In this paper, the GSC is controlled by the conventional grid voltage-oriented vector control. The sampling frequency of the saturation controller-based DPC and the conventional DPC is 10 kHz and the simulation step is 5 μs.

A. Steady-State and Dynamic Performance of Proposed DPC

To validate the proposed DPC in both sub-synchronous and super-synchronous speed modes, the rotor speed of the DFIG is initially 150 rad/s and then increases linearly from 150 rad/s at 0.15 s to 220 rad/s at 0.45 s, which are approximately −20% and 20% around the synchronous speed of 188 rad/s, respectively, as shown in Fig. 4. The stator reactive power reference changes from zero to −0.5 MVar at t = 0.25 s and then changes to 0.5 MVar at t = 0.45 s; the stator active power reference changes from −0.5 MW to −1.8 MW at t = 0.2 s and changes to −1.5 MW at t = 0.4 s. All of the changes are step changes. A negative or positive value indicates that the DFIG generate or absorb power, respectively. Fig. 4 shows the dynamic responses of the DFIG controlled by the proposed DPC, including the stator reactive power Q_s and active power P_s, stator phase currents I_{sabc}, and rotor phase currents I_{rabc}. Both the stator active and reactive powers follow their reference values well during the steady states and step changes. There are no abrupt changes on the power curves during the transient step changes. The stator and rotor currents also vary with the active and reactive powers smoothly and there is no overshoot in the current waveforms during the power step changes. As a consequence, the DFIG operates with good steady-state and dynamic performance in both the sub-synchronous and the super-synchronous speed modes using the proposed DPC scheme.

The saturation controllers are designed to obtain an adjustable duty ratio mainly to minimize the active power ripples. As a result, the amplitudes of the stator active power ripples are less than 2.5%, which are negligible; while the stator reactive power oscillates around the reference value with 5% ripples.

Comparing to the dynamic responses of the DFIG controlled by the conventional DPC scheme in Fig. 5, the active and reactive power ripples in Fig. 4 are largely reduced by applying the proposed DPC scheme. Specifically, the amplitudes of the active and reactive power ripples are reduced from 0.347 MW and 0.245 MVar when using the conventional DPC to 0.042 MW and 0.105 MVar when using the proposed DPC, respectively. The transition from the sub-synchronous speed mode to the super-synchronous speed mode is smooth by using the proposed DPC.

Figs. 6 and 7 compare the frequency spectra of the stator and rotor phase currents of the DFIG controlled by the proposed and conventional DPC schemes, respectively. The results clearly show that there are less harmonic contents in the stator and rotor phase currents when using the proposed DPC scheme. Specifically, the total harmonic distortion (THD) of the stator phase current is reduced from 8.67% when using the conventional DPC to 4.48% when using the proposed DPC; and the THD of the rotor phase current is reduced from 12.18% to 4.95%.

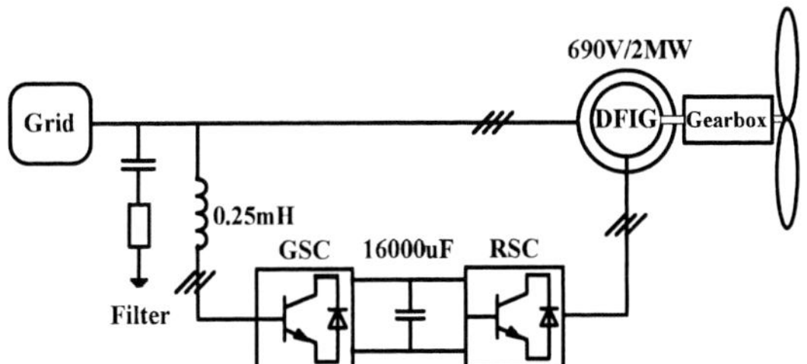

Fig. 3. Schematic of a grid-connected DFIG-based WECS.

TABLE II
THE PARAMETERS OF THE DFIG-BASED WECS

Parameter	Value
Rated power	2 MW
Stator voltage	690 V
Stator frequency	60 Hz
Stator-to-rotor ratio	0.3
Stator resistance	0.0108 pu
Rotor resistance	0.0121 pu
Mutual inductance	3.362 pu
Stator leakage inductance	0.102 pu
Rotor leakage inductance	0.11 pu
Lumped inertia constant	32.4 kg·m^2
Number of pole pairs	2
DC bus voltage	1200 V
DC capacitor	16000 uF
Grid-side inductance	0.25 mH

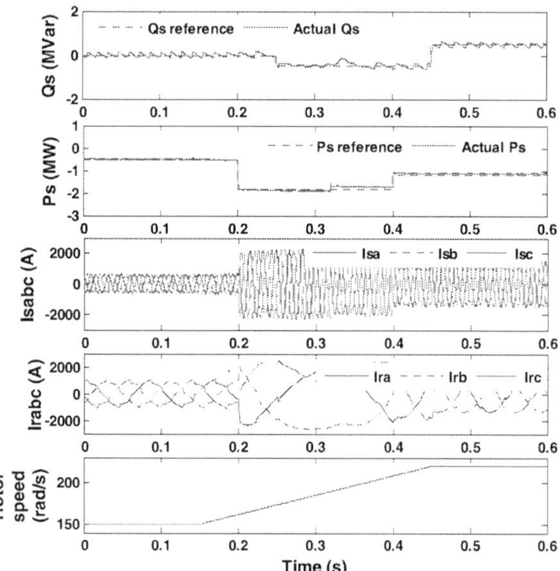

Fig. 4. Dynamic responses of the DFIG controlled by the proposed DPC scheme.

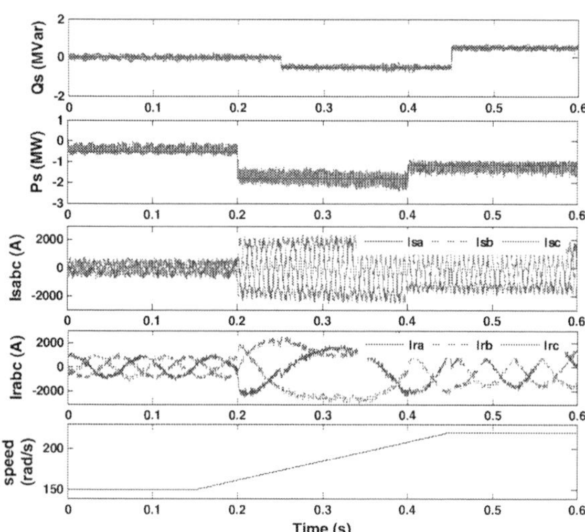

Fig. 5. Dynamic responses of the DFIG controlled by the conventional DPC scheme.

The harmonics in the rotor phase currents also have an impact on the rotor flux. Fig. 8 shows that the trajectories of the rotor fluxes are circular when using both the proposed and the conventional DPC schemes. However, the rotor flux trajectory when using the proposed DPC scheme is much smoother than that when using the conventional DPC scheme. The multiple circles in the trajectories are caused by the step changes of the active and reactive power references.

The influence of the rotor speed on the power ripples is also considered in this paper and an adaptive bandwidth is used in the saturation controller to handle this issue. Fig. 9

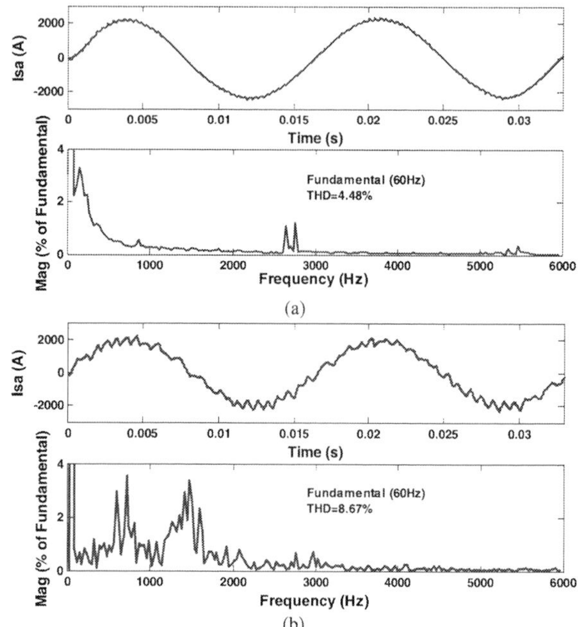

Fig. 6. Stator phase current and its frequency spectrum: (a) the proposed DPC scheme and (b) the conventional DPC scheme.

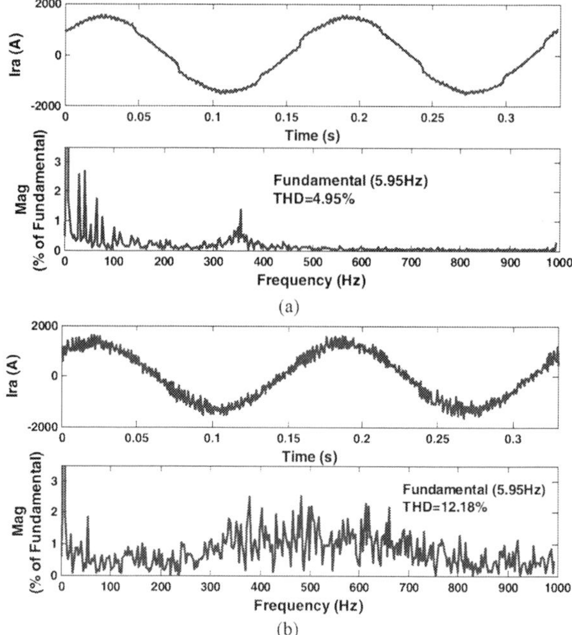

Fig. 7. Rotor phase current and its frequency spectrum: (a) the proposed DPC scheme and (b) the conventional DPC scheme.

shows the stator active power ripples of the DFIG in varying rotor speed conditions using constant and adaptive bandwidths for the saturation controllers. The rotor speed changes linearly from 150 rad/s (80% of the synchronous speed) to the

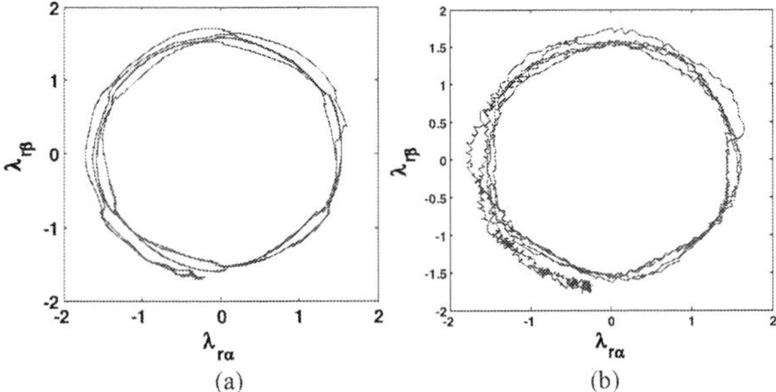

Fig. 8. Rotor flux trajectory: (a) the proposed DPC scheme and (b) the conventional DPC scheme.

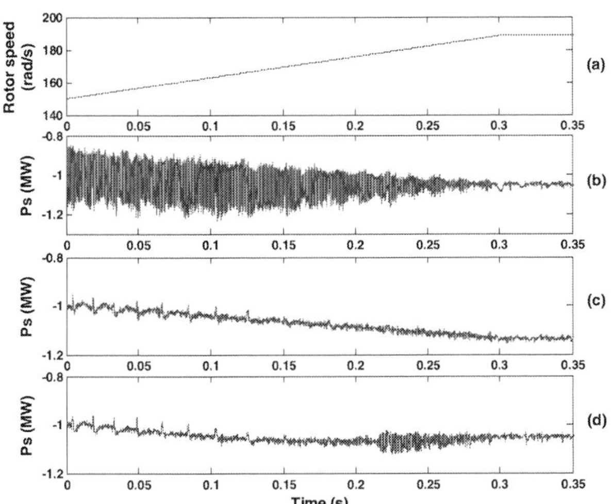

Fig. 9. Stator active power ripples of the DFIG with varying rotor speed using different bandwidths in the saturation controllers: (a) rotor speed, (b) B_w=0.05, (c) B_w=0.14, and (d) $B_w=0.9(\omega_r-\omega_s)/\omega_s+0.05$.

synchronous speed from 0 s to 0.3 s and the stator active power reference remains constant at −1 MW during this process. As shown in Fig. 9(b), when a small bandwidth of 0.05 is used, the active power ripple is relatively large (about 0.3 MW) and decreases to a small value as the slip speed decreases to zero. When a larger bandwidth of 0.14 is used in Fig. 9(c), the active power ripple is small (about 0.02 MW). However, a large steady-state error can be seen in Fig. 9(c), which is about 0.175 MW. By using the proposed adaptive bandwidth for the saturation controllers, the active power ripple is small (about 0.02 MW) and the steady-state error is reduced to about 0.04 MW as well during the change of the rotor speed, as shown in Fig. 9(d). The adaptive bandwidth makes a good tradeoff to control the ripple and the steady-state error of the power.

B. MPPT Performance

In the wind energy application, the DFIG is usually operated in the MPPT control mode to extract the maximum energy from wind. To test the performance of the DFIG when using the proposed DPC scheme for the MPPT control, a measured 9-second wind speed profile is used, in which the wind speed changes randomly between 9 m/s and 11 m/s. During the operation of the DFIG WECS, the reference of the stator reactive power is set to zero, and the reference of the stator active power is determined by a MPPT algorithm proposed in [14]. The performance of the DFIG during this MPPT process is shown in Fig. 10. The actual power coefficient C_p of the wind turbine tracks the optimal value of 0.438 well during the wind speed variations. The stator reactive power maintains at zero and the stator active power tracks its reference value well. The actual total output electrical power also tracks the maximum power well, where the small high-frequency deviations are caused by the large system inertia. These results clearly show that the proposed DPC scheme works well for the MPPT control of the DFIG.

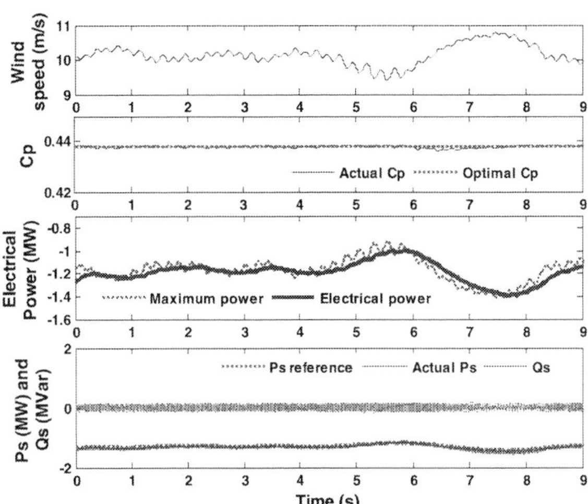

Fig. 10. MPPT performance of the DFIG WECS controlled by the proposed sensorless DPC scheme with random wind speed variations.

V. Conclusion

This paper has proposed a novel DPC scheme for the DFIG-based WECSs in which a saturation controller has been proposed to replace the hysteresis comparator in the conventional DPC. This made the calculation of the duty ratios of the voltage vectors much simpler and independent of DFIG parameters. The influence of the rotor speed variations on the power ripples has been considered. An adaptive bandwidth has been designed for the saturation controllers to make a good tradeoff to control the ripples and the steady-state errors of the stator active power. The proposed DPC scheme has significantly reduced the power ripples using a fixed switching frequency while preserving most of the intrinsic properties of the conventional DPC, such as fast dynamics, low parameter dependence, no PI controller, and easy to implement.

References

[1] C. Wei, Z. Zhang, J. Zeng, and W. Qiao, "Stator current-based sliding-mode observer for sensorless vector contorl of doubly-fed induction

geneartors," *in Proc. IEEE Energy Conversion Congress and Exposition*, Sept. 2015, pp. 4165-4171.

[2] C. Wei, Z. Zhang, W. Qiao, and L. Qu, "Reinforcement learning-based intelligent maximum power point tracking control for wind energy conversion systems," *IEEE Trans. Industrial Electronics*, vol. 62, no. 10, pp. 6360-6370, Oct. 2015.

[3] L. Xu and P. Cartwright, "Direct active and reactive power control of DFIG for wind energy generation," *IEEE Trans. Energy Convers.*, vol. 21, no. 3, pp. 750-758, Sep. 2006.

[4] C. Wei, W. Qiao, and Y. Zhao, "Sliding-mode observer-based sensorless direct power control of DFIG for wind power applications," *in Proc. IEEE PES general meeting*, July 2015, pp. 1-5.

[5] R. Datta and V.T. Ranganathan, "Direct power control of grid-connected wound rotor induction machine without rotor position sensors," *IEEE Trans. Power. Electron.*, vol. 16, no. 3, pp. 390-399, May 2001.

[6] D. Zhi, and L. Xu, "Direct power control of DFIG with constant switching frequency and improved transient performance," *IEEE Trans. Energy Convers.*, vol. 22, no. 1, pp. 110-118, Mar. 2007.

[7] H. Nian, Y. Song, and Y. He, "Improved direct power control of a wind turbine driven doubly fed induction generator during transient grid voltage unbalance," *IEEE Trans. Energy Convers.*, vol. 26, no. 3, pp. 976-986, Sep. 2011.

[8] D. Zhi, L. Xu, and B. W. Williams, "Model-based predictive direct power control of doubly fed induction generators," *IEEE Trans. Power Electron.*, vol. 25, no. 2, pp. 341-351, Feb. 2010.

[9] J. Hu, H. Nian, B. Hu, Y. He, and Z. Q. Zhu, "Direct active and reactive power regulation of DFIG using sliding-mode control approach," *IEEE Trans. Energy Convers.*, vol. 25, no. 4, pp. 1028-1039, Dec. 2010.

[10] J. Hu, J. Zhu, and D. G. Dorrell, "Predictive direct power control of doubly fed induction generators under unbalanced grid voltage conditions for power quality improvement," *IEEE Trans. Sustainable Energy*, vol. 6, no. 3, pp. 943-950, Jul. 2015.

[11] G. Abad, M.Rodriguez, and J. Poza, "Two-level VSC-based predictive direct power control of the doubly fed induction machine with reduced power ripple at low constant switching frequency," *IEEE Trans. Energy Convers.*, vol. 23, no. 2, pp. 570-580, Jun. 2008.

[12] Y. Zhang, J. Hu, and J. Zhu, "Three-vectors-based predictive direct power control of the doubly fed induction generator for wind energy applications," *IEEE Trans. Power Electron.*, vol. 29, no.7, pp. 3485-3500, Jul. 2014.

[13] Z. Zhang, W. Qiao, and L. Qu, "Saturation controller-based ripple reduction for direct torque controlled permanent-magnet synchronous machines," in *Proc. IEEE Applied Power Electronics Conference and Exposition*, Mar. 2014, pp. 799-805.

[14] W. Qiao, X. Gong, and L. Qu, "Output maximization control for DFIG wind turbines without using wind and shaft speed measurements," in *Proc. IEEE Energy Convers. Congr. Exposit.*, Sep. 2009, pp. 404-410.

Inductance-Simulating Control for DFIG-Based Wind Turbine to Ride-through Grid Faults

Donghai Zhu, *Student Member, IEEE,* Xudong Zou, and Yong Kang
State Key Laboratory of Advanced Electromagnetic Engineering and Technology
Huazhong University of Science and Technology
Wuhan, China
zhudh@hust.edu.cn

Lu Deng
Wuhan NARI Limited Company of State Grid Electric Power Research Institute
Wuhan, China

Qingjun Huang
State Key Laboratory of Disaster Prevention & Reduction for Power Grid Transmission and Distribution Equipment
Changsha, China

***Abstract*—For doubly fed induction generator (DFIG)-based wind turbine (WT), large electromotive force will be induced in the rotor circuit under grid faults, and the rotor side of DFIG is prone to suffering from overcurrent and overvoltage. To mitigate this problem, an inductance-simulating control strategy is proposed to enhance its low-voltage ride through (LVRT) capability. In this method, the rotor side converter (RSC) is controlled to simulating as an inductance. Furthermore, under proper inductance value, both the required rotor voltage and postfault rotor current can be reduced within the permissible range. In addition, the electromagnetic torque ripple can be effectively eliminated. Finally, the simulation and experimental results clearly demonstrate the effectiveness of the proposed method.**

Keywords—doubly fed induction generator (DFIG); inductance-simulating control; low-voltage ride through (LVRT); wind turbine (WT)

I. Introduction

With the increasing penetration of wind energy into grid, the grid codes require wind energy conversion systems (WECSs) to have low-voltage ride through (LVRT) capability [1]. Namely, the WECS must remain connected to the power system and even need to provide reactive power within certain period during grid faults.

Among several types of WECSs, doubly fed induction generator (DFIG)-based wind turbine (WT) is the most employed generator [2]. However, it is extremely sensitive to grid voltage disturbances, due to the stator of DFIG is directly connected to the grid. When grid faults occur, the stator flux will contain transient and negative-sequence components which can induce large electromotive force (EMF) in rotor circuit. Without proper protection scheme, the rotor-side converter (RSC) is prone to suffering from overcurrent, which may lead to destruction of the converter [3].

The most popular LVRT solution is to employ crowbar circuit to bypass RSC due to the advantages of simple structure and low cost [4]. However, once the crowbar is activated, the DFIG works as squirrel-cage induction generator which absorbs reactive power from the grid. To solve this problem, the method of series grid-side converter (SGSC) [5], dynamic voltage restorer (DVR) [6], and dynamic resistor [7] are proposed to fulfill LVRT requirement. Nevertheless, control complexity and high cost discourage their utilization.

The other solutions are to achieve LVRT requirement by improving control strategies for RSC. In [8], a feedforward transient compensation (FFTC) control is proposed to suppress EMF disturbance. Nevertheless, it requires RSC outputting enough voltage to offset the EMF, which will increase the required rotor voltage. In [9], the required rotor voltage can be reduced by controlling rotor current to contain both transient and negative-sequence components which relate to the stator flux. However, the demagnetization current is very large. In [10], a virtual inductance self-demagnetization control is proposed to suppress the transient rotor voltage surge, but the same problem as the method in [9] cannot be avoided. In [11], the postfault rotor current can be effectively suppressed by controlling rotor current to tracking stator current. However, the required rotor voltage is also very large. Unfortunately, these methods are only focus on how to reduce one of the required rotor voltage or the postfault rotor current.

In this paper, an inductance-simulating control strategy is proposed to coordinate the constraints of both rotor voltage and rotor current. Both required rotor voltage and postfault rotor current can be reduced within the permissible range by means of dynamic adjustment of the inductance value. Moreover, electromagnetic torque ripple is effectively eliminated.

II. DFIG Performances Under Grid Faults

A. DFIG Model

For simplicity, all parameters are referred to the stator. Motor convention is used for both the stator and rotor. In the stationary reference frame, the voltage and flux equations can be expressed as

This work is financially supported by the National Natural Science Foundations of China (NSFC) under Award Number 51477064, Lite-On Power Electronics Technology Research Fund under Award Number 2013-04, and the National Basic Research Program of China under Award 2012CB215100.

$$\begin{cases} \overrightarrow{u_s} = R_s \vec{i_s} + d\overrightarrow{\psi_s} / dt \\ \overrightarrow{u_r} = R_r \vec{i_r} + d\overrightarrow{\psi_r} / dt - j\omega_r \overrightarrow{\psi_r} \end{cases} \tag{1}$$

$$\begin{cases} \overrightarrow{\psi_s} = L_s \vec{i_s} + L_m \vec{i_r} \\ \overrightarrow{\psi_r} = L_r \vec{i_r} + L_m \vec{i_s} \end{cases} \tag{2}$$

where u, i and ψ are the voltage, current and flux, respectively, R and L are resistance and inductance, respectively. Subscripts "s" and "r" distinguish the stator and rotor variables. L_m is the mutual inductance. ω_r is the rotor electrical frequency.

According to (1) and (2), the rotor voltage can also be expressed as

$$\overrightarrow{u_r} = \underbrace{\frac{L_m}{L_s}\frac{d}{dt}\overrightarrow{\psi_s}}_{\overrightarrow{e_r}} + \underbrace{(R_r + \sigma L_r \frac{d}{dt})\vec{i_r}}_{u_{RL}} \tag{3}$$

where σ is the leakage factor and $\sigma = 1 - L_m^2/(L_s L_r)$.

From (3), it can be obtained that the rotor voltage consists of two parts: the EMF e_r inducted by stator flux, and the voltage drop u_{RL} on both rotor resistance R_r and transient inductance σL_r. Moreover, the equivalent circuit of DFIG viewed from rotor side can be achieved, as shown in Fig. 1 [3].

B. Transient EMF Characteristics Under Grid Faults

The EMF during normal condition can be expressed as [12]

$$\overrightarrow{e_r} = \frac{L_m}{L_s} j\omega_r \overrightarrow{\psi_s} = \frac{L_m}{L_s} sU_{sN} e^{j\omega_s t} \tag{4}$$

where U_{sN} is the rated stator voltage.

According to (4), and because the slip s is generally in range of -0.3 to 0.3, the amplitude of EMF e_r cannot exceed 30% of the rated stator voltage U_{sN}.

When a symmetrical grid faults occur, the stator flux will contain some transient components, and can be expressed as [12]

$$\overrightarrow{\psi_s} = \frac{(1-p)U_{sN}}{j\omega_s} e^{j\omega_s t} + \frac{pU_{sN}}{j\omega_s} e^{-t/\tau_s} \tag{5}$$

where p is the depth of the voltage dip, and τ_s is the time constant of the stator flux. ω_s is the grid angular frequency.

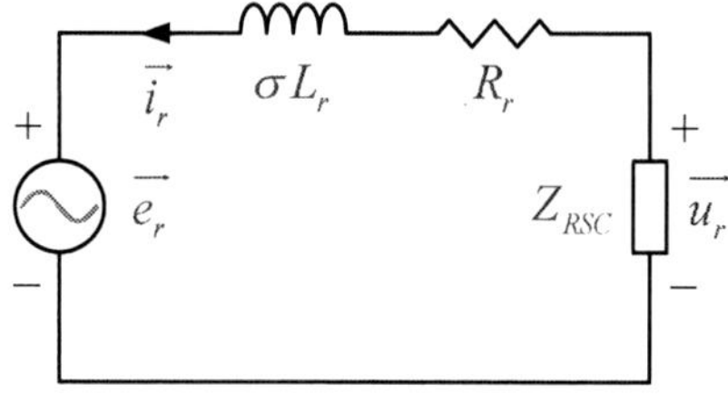

Fig. 1. Equivalent circuit of DFIG viewed from rotor side.

According to (3) and (5), neglecting $1/\tau_s$, the EMF under symmetrical grid faults can be expressed as [12]

$$\overrightarrow{e_r} = \frac{L_m}{L_s}[s(1-p)U_{sN}e^{j\omega_s t} - (1-s)pU_{sN}e^{-t/\tau_s}] \tag{6}$$

The second item is the transient components of EMF, which inducted by the transient components of stator flux.

As (6) indicated, the initial amplitude of EMF is usually relatively large. For instant, under a full voltage dip with s=-0.3, the initial amplitude of EMF is 4.3 times of its normal value.

III. The Proposed Inductance-Simulating Control Stratgy

A. Basic Principle

The equivalent impedance of RSC can be expressed as

$$Z_{RSC} = -\frac{\overrightarrow{u_r}}{\vec{i_r}} \tag{7}$$

By substituting (7) into (3), it results in

$$\overrightarrow{e_r} = -(R_r + jX_{\sigma L_r} + Z_{RSC})\vec{i_r} \tag{8}$$

From (8), it can be obtained that different Z_{RSC} indirectly corresponding to the different angle θ between rotor current vector i_r and EMF vector e_r. Supposing the constant rotor current and neglecting the tiny R_r, the influence of different Z_{RSC} on rotor voltage is shown in Fig. 2.

According to Fig. 2, it indicates that if $\theta=\pi/2$, namely Z_{RSC} is pure inductance, the u_{RL} will be in the opposite direction with e_r. Meanwhile, the required rotor voltage u_r is minimal. Similarly, if Z_{RSC} is pure inductance with same rotor voltage, the postfault rotor current is also minimal. Therefore, it can be concluded that, if the equivalent impedance of RSC is controlled to be simulated as inductance, both the required rotor voltage and postfault rotor current will be reduced.

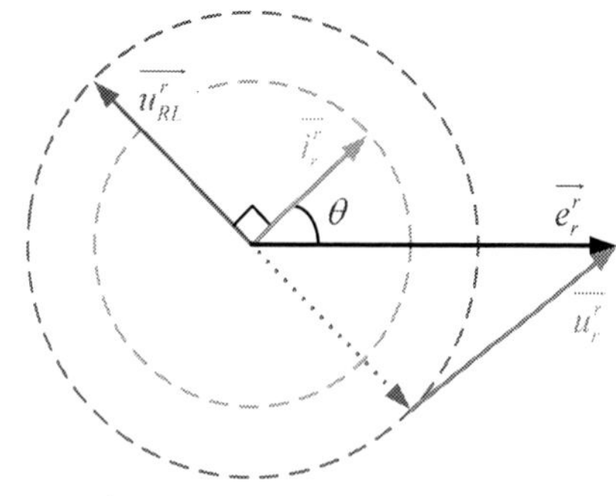

Fig. 2. Rotor voltage with different equivalent impedance of RSC.

B. Control System Design

For simplicity, neglecting the tiny R_r and merging the equivalent inductance L_{RSC} of RSC with σL_r into equivalent inductance L_{eq} of rotor side. The rotor current reference under the proposed method can be given as

$$\vec{i}_r^* = -\frac{L_m}{L_s L_{eq}} \vec{\psi}_s \tag{9}$$

By substituting (8) into (3), the rotor voltage can be expressed as

$$\vec{u}_r = (1 - \frac{\sigma L_r}{L_{eq}})\vec{e}_r \tag{10}$$

According to (9) and (10), it can be concluded that, the larger L_{eq} is, the lower postfault rotor current is. However, a higher rotor voltage is required. On the contrary, the required rotor voltage can be reduced with a smaller L_{eq}, but the postfault rotor current will be increased. Therefore, the proper selection of L_{eq} is important for the control performance. Then, considering the constraint of both rotor voltage and rotor current, the maximal and minimal value of L_{eq} can be expressed as

$$L_{eq\max} = \frac{\sigma L_r |\vec{e}_r|}{|\vec{e}_r| - U_{r\max}} \tag{11}$$

$$L_{eq\min} = \frac{L_m}{L_s I_{r\max}} |\vec{\psi}_s| \tag{12}$$

where U_{rmax} is RSC's maximal output voltage, I_{rmax} is the allowed maximal rotor current.

In order to coordinate the constraints of both rotor voltage and rotor current, the value of L_{eq} can be solved as

$$L_{eq} = 0.5(L_{eq\min} + L_{eq\max}) \tag{13}$$

The overall control diagram of inductance-simulating control strategy for DFIG is shown in Fig. 3.

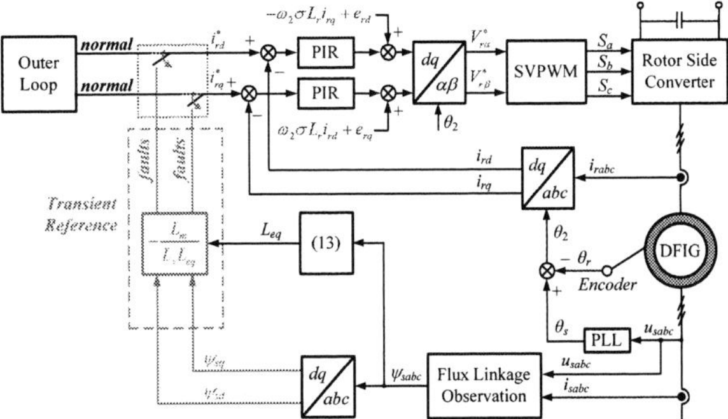

Fig. 3. Control diagram of inductance-simulating control.

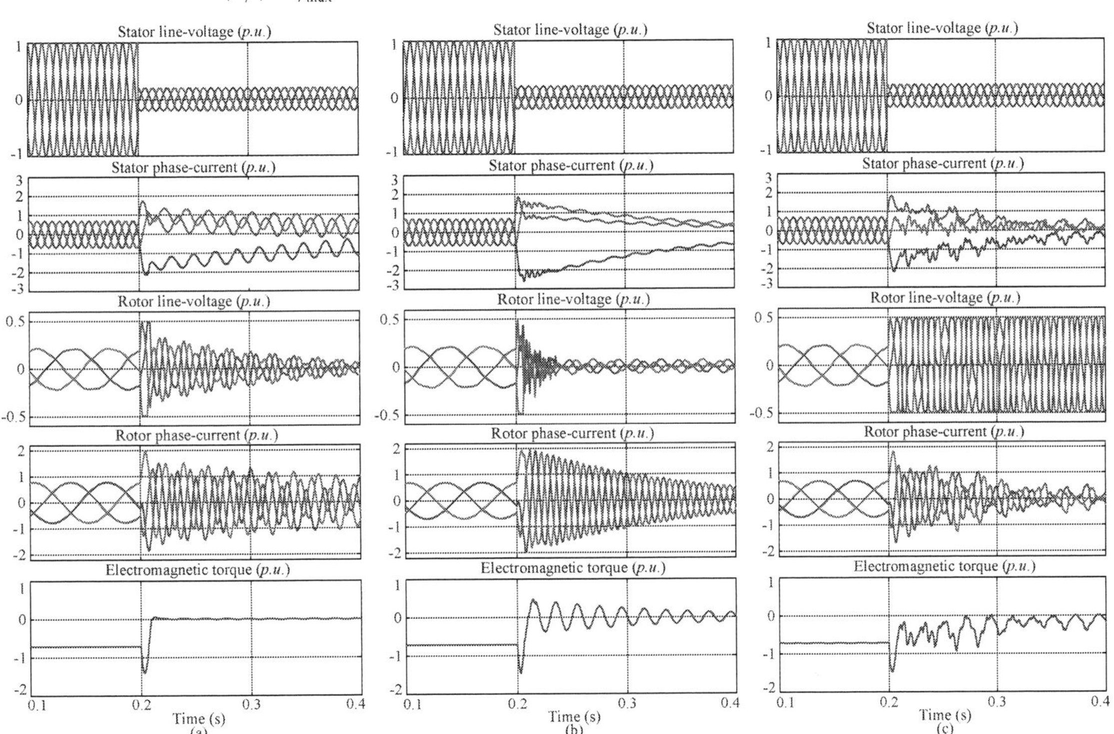

Fig. 4. Simulation results under 80% symmetrical faults. (a) Proposed method. (b) Demagnetization control [9]. (c) Method in [11].

IV. Simulation and Experiment Verification

A. Simulated Results

A 1.5-MW DFIG-based wind energy conversion system is built in MATLAB/Simulink to verify the effectiveness of the proposed method, and the parameters of simulation system are listed in Table I.

Initially, DFIG working at the rated voltage condition, the output active power P_s is 0.7 *p.u.* and output reactive power Q_s is 0 *p.u.* The slip of DFIG is -0.2. At t=0.2 s, a symmetrical fault with an 80% depth of voltage dip occurs, and the LVRT control is enabled as soon as detecting the grid fault. Moreover, the performance of three different LVRT control method are compared.

Fig. 4 shows the simulation results under an 80% symmetrical fault, and the value of equivalent inductance is shown in Fig. 5. It can be seen that all methods can effectively suppress the postfault rotor current within 2 *p.u.* From Fig. 4(b), it can be obtained that the advantage of the method in [9] is that a lower rotor voltage is required, but a large rotor current is needed to accelerate flux attenuation. Although the postfault rotor current can be reduced under method in [11], the required rotor voltage is also very large and almost arrives to the saturation values, as shown in Fig. 4(c). From Fig. 4(a) and Fig. 5, it can be obtained that the value of equivalent inductance is also large to reduce the rotor transient current under initial condition, due to larger the initial amplitude of EMF. With the decay of the stator flux, the value of equivalent inductance is decreasing to ease the demand pressure of required rotor voltage. Therefore, both required rotor voltage and postfault rotor current can be reduced within the permissible range, due to adjusting the value of the inductance dynamically. Moreover, the electromagnetic torque ripple can be effectively suppressed.

B. Experimental Results

The 2.5-kW experimental setup is carried out to verify the effectiveness of proposed inductance-simulating control strategy. The parameters of the system are listed in Table II and the schematic diagram of the experimental setup is shown in Fig. 6. The DFIG is driven by a squirrel cage induction motor (SCIM). The fault generator is based on transformer and switching devices [2]. RSC and GSC are independently controlled by two DSPs (TMS320F28335). The switching and sampling frequencies are 2.5 kHz and 5.0 kHz, respectively.

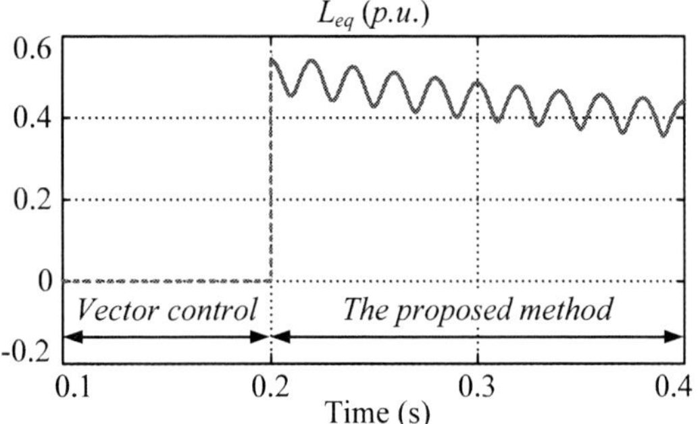

Fig. 5. The value of equivalent inductance under 80% symmetrical faults.

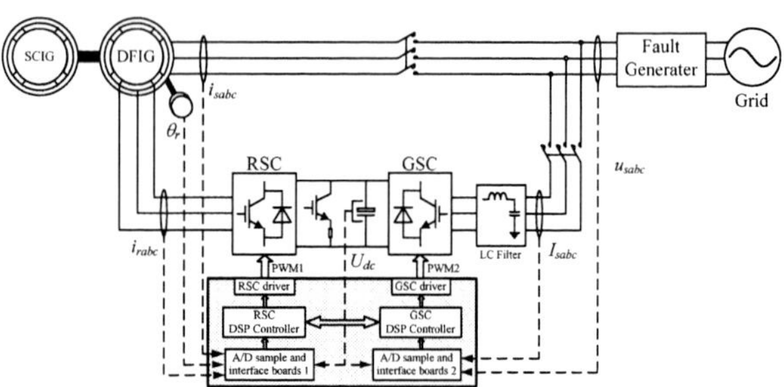

Fig. 6. Schematic diagram of the experimental setup.

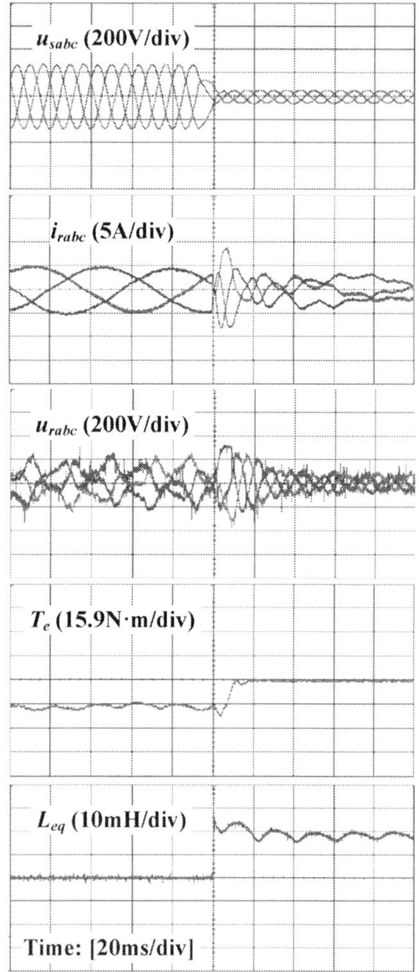

Fig. 7. Experiment results under 80% symmetrical faults.

Initially, DFIG is under vector control, the output active power P_s and reactive power Q_s are 1.0 *p.u.* and 0 *p.u.*, respectively. The slip of DFIG is -0.2. Once the grid faults is detected, the proposed control method is activated.

The experiment results for an 80% symmetrical fault are shown in Fig. 7. It can be obtained that, 1) the postfault rotor current is limited within 2 *p.u.* and it protect the converter safety effectively. 2) Because the initial amplitude of EMF is larger under the early stage of grid faults, the value of equivalent inductance is also large to reduce the postfault rotor current. Moreover, the value of equivalent inductance is decreasing with the decay of the transient compensations of EMF, which will reduce the demand pressure of required rotor voltage. 3) The electromagnetic torque can be effectively eliminated.

V. CONCLUSION

In this paper, an inductance-simulating control strategy is proposed for DFIG to enhance its LVRT capability with considering the constraints of both rotor voltage and rotor current. By controlling RSC to simulating as an inductance and dynamically adjusting the value of the inductance, both required rotor voltage and postfault rotor current can be suppressed within the permissible range. Moreover, it helps to suppress the oscillations of torque.

APEENDIX

TABLE I. THE SIMULATION PARAMETERS OF 1.5-MW DFIG SYSTEM

Symbol	Parameters	Values
P_N	Rated power	1.5 MW
U_N	Rated stator voltage	690 V
f_n	Rated stator frequency	50 Hz
n_p	Pairs of poles	2
N_s/N_r	Turn ratio	0.4
R_s	Stator resistance	0.007 *p.u.*
L_{ls}	Stator leakage inductance	0.171 *p.u.*
L_m	Mutual inductance	2.902 *p.u.*
R_r	Rotor resistance	0.005 *p.u.*
L_{lr}	Rotor leakage inductance	0.156 *p.u.*
U_{dc}	Rated DC-bus voltage	1200 V

TABLE II. THE EXPERIMENTAL PARAMETERS OF 2.5-KW DFIG SYSTEM

Symbol	Parameters	Values
P_N	Rated power	2.5 kW
U_N	Rated stator voltage	190 V
f_n	Rated stator frequency	50 Hz
n_p	Pairs of poles	2
N_s/N_r	Turn ratio	0.4
R_s	Stator resistance	0.006 *p.u.*
L_{ls}	Stator leakage inductance	0.072 *p.u.*
L_m	Mutual inductance	2.619 *p.u.*
R_r	Rotor resistance	0.010 *p.u.*
L_{ls}	Rotor leakage inductance	0.072 *p.u.*
U_{dc}	Rated DC-bus voltage	325 V

REFERENCES

[1] M. Tsili and S. Papathanassiou, "A review of grid code technical requirements for wind farms," *IET Renew. Power Generat.*, vol. 3, no. 3, pp. 308-332, Sep. 2009.

[2] S. Hu, X. Lin, Y. Kang, and X. Zou, "An improved low-voltage ride-through control strategy of doubly fed induction generator during grid faults," *IEEE Trans. Power Electron.*, vol. 26, no. 12, pp. 3653-3665, Dec. 2011.

[3] S. Xiao, G. Yang, H. Zhou, and H. Geng, "An LVRT control strategy based on flux linkage tracking for DFIG-Based WECS, " *IEEE Trans. Ind. Electron.*, vol. 60, no. 7, pp. 2820-2832, Jul. 2013.

[4] J. Morren and S. W. H. de Haan, "Ridethrough of wind turbines with doubly-Fed induction generator during a voltage dip," *IEEE Trans. Energy Convers.*, vol. 20, no. 2, pp. 435-441, Jun. 2005.

[5] P. S. Flannery and G. Venkataramanan, "Unbalanced voltage sag ride-through of a doubly fed induction generator wind turbine with series grid-side converter," *IEEE Trans. Ind. Appl.*, vol. 45, no. 5, pp. 1879-1887, Sep./Oct. 2009.

[6] C. Wessels, F. Gebhardt, and F. W. Fuchs, "Fault ride-through of a DFIG wind turbine using a dynamic voltage restorer during symmetrical and asymmetrical grid faults," *IEEE Trans. Power Electron.*, vol. 26, no. 3, pp. 807-815, Mar. 2011.

[7] X. Yan, G. Venkataramanan, Y. Wang, Q. Dong, and B. Zhang, "Grid-fault tolerant operation of a DFIG wind turbine generator using a passive resistance network," *IEEE Trans. Power Electron.*, vol. 26, no. 10, pp. 2896-2905, Oct. 2011.

[8] J. Liang, D. Howard, and J. Restrepo, "Feedforward transient compensation control for DFIG wind turbines during both balanced and unbalanced grid disturbances," *IEEE Trans. Ind. Appl.*, vol. 49, no. 3, pp. 1452-1462, May/Jun. 2013.

[9] D. Xiang, L. Ran, J. Peter, and S. Yang, "Control of a doubly fed induction generator in a wind turbine during grid fault ride-through," *IEEE Trans. Energy Convers.*, vol. 21, no. 3, pp. 652-662, Sep. 2006.

[10] S. Yang, Y. Chen, L. Chang, Z. Xie, and X. Zhang, "Virtual inductance self-demagnetization based LVRT control strategy for doubly fed WTs," in *proc. 30th IEEE Appl. Power Electron. Conf. Expo.*, Charlottee, America, Mar. 15-19, 2015, pp. 2988-2992.

[11] F. K. A. Lima, A. Luna, P. Rodriguez, E. H. Watanabe, and F. Blaabjerg, "Rotor voltage dynamics in the doubly fed induction generator during grid faults, " *IEEE Trans. Power Electron.*, vol. 25, no. 1, pp. 118-130, Jan. 2010.

[12] J. Lopez, P. Sanchis, X. Roboam, and L. Marroyo, "Dynamic behavior of the doubly fed induction generator during three-phase voltage dips," *IEEE Trans. Energy Convers*, vol. 22, no. 3, pp. 709-717, Sep. 2007.

Misalignment Effect on Efficiency of Wireless Power Transfer for Electric Vehicles

Yabiao Gao[1], Antonio Ginart[1,2], Kathleen Blair Farley[3], Zion Tsz Ho Tse[1]

[1]College of Engineering, The University of Georgia, Athens, GA 30602, USA

[2]Sonnenbatterie GmbH, Norcross, GA 30093, USA

[3]Southern Company Services, Inc. Birmingham, AL, 35291, USA

ygao@uga.edu, ziontse@uga.edu

***Abstract*—In order to consistently provide efficient EV wireless charging, changes in efficiency caused by coil misalignment should be investigated. This body of work uses a vector network analyzer to record the power transmission coefficients then calculate the efficiency and coupling factors. An automated 3-axis platform was built for coil motion control. The test result shows that the efficiency can remain at 90% as long as the misalignment ratio is within 0.5 and the air gap ratio is from 0 to 0.25.**

Keywords—Wireless power transfer; electric vehicles; misalignment

I. Introduction

Driven by consumer demand and government incentives to further the development of environmentally friendly, highly efficient transportation, Electric Vehicle (EV) markets are growing for consumers and businesses alike. Convenient EV charging utilizing wireless power transfer (WPT) technologies could enhance user experience and reduce associated infrastructure cost [1-4]. Compared with traditional plug-in charging, wireless charging (WC) can be implemented sub-surface with no external street equipment, which helps eliminate hassle and unwieldy charging cables [5, 6]. WPT has the potential to increase EV range for future owners by increasing opportunities to charge EVs while away from home. Currently, some research institutes and universities are developing EVWC technologies, including University of Auckland, ORNL, Clemson University, University of Michigan at Dearborn, KAIST, etc. [1, 7-9].

While WC holds the key to solving battery problems of EVs, coil misalignment can cause charging efficiency to drop when parking a vehicle over the WC equipment or in dynamic EV charging [1, 10]. An automated 3-axis platform was built to investigate the effects of misalignment on efficiency and power transfer capabilities under diverse configurations of misalignment and air gap. The motion curve and spatial distribution of two coils can be programmed in the computer interface of the platform. A network analyzer was used to measure the transmission and reflection coefficients, from which the efficiency and coupling factors are calculated.

II. Gapping and Misalignment Effects on Efficiency

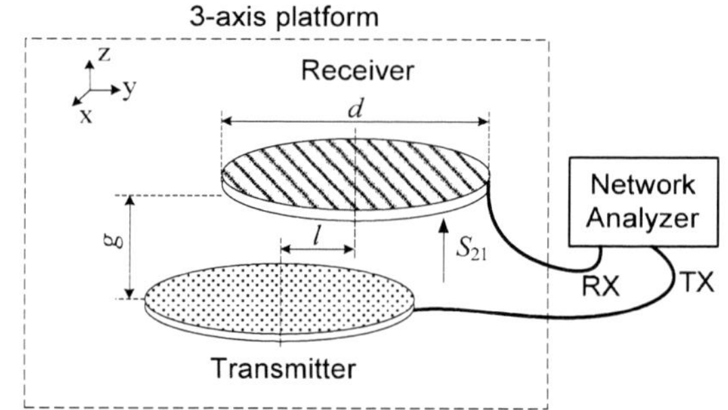

(a)

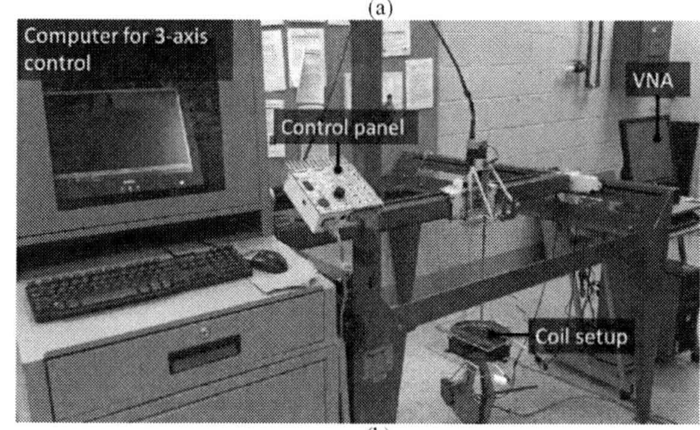

(b)

Fig. 1 (a) Principle of the testing platform; (b) Experimental setup for misalignment study

Fig. 1 shows a 3-axis motorized platform that tests efficiency changes of wireless power transfer due to coil alignment. In Figure 1, a vector network analyzer (VNA) was applied to analyze the power transfer efficiency, coupling coefficient, and resonant frequency. The power transmission efficiency η_{21} and reflection ratio η_{11} were calculated from equations (1) and (2). The transmission S_{21} and reflection wave ratio S_{11} in equations (1) and (2) were outputs of the VNA. The coupling coefficient k can be calculated from equation (3), where f_e and f_m ($f_e > f_m$) are two resonant frequencies of the coil setup. The two resonant frequencies were acquired from the VNA. For a detailed theoretical analysis of these equations, refer to Imura et al [11].

978-1-4673-9551-9/16 $31.00 © 2016 IEEE

$$\eta_{21} = S_{21}^2 \quad (1)$$

$$\eta_{11} = S_{11}^2 \quad (2)$$

$$k = \frac{f_e^2 - f_m^2}{f_e^2 + f_m^2} \quad (3)$$

The 3-axis platform shown in Fig. 1(b) is based on a CNC machine (Model DHC, PlasmaCAM, Inc., Colorado City, CO, U.S.). It has a width of 1.75m, a height of 1.65m, and a depth of 1.65m. The maximum speeds are 25m/min in the horizontal direction and 2m/min in the vertical direction. The moveable ranges are 1.2m×1.2m in the horizontal plane and 0.6m in the vertical direction (Fig. 1(b)). The speed and motion trail can be programmed in a computer interface. The misalignment test was only required to be performed in a single direction in the horizontal plane since the coil shape is circular.

Relative air gap ratio m and misalignment ratio t are defined to normalize our testing results. Air gap ratio m equals gap g divided by coil diameter d, while t is lateral offset l divided by d (Fig. 1(a)). During the experiment, the air gap ratio was set at 0, 0.25, 0.5, and 0.75, and the misalignment ratio was set from -1.5 to 1.5 in the platform's computer interface. The coupling coefficient k is defined as

$$k = L_{12} / \sqrt{L_1 L_2} \quad (4)$$

where L_{12} is the mutual inductance between the transmitter and receiver in wireless power transfer, and L_1 and L_2 are the self-inductance of the transmitter and receiver.

III. Results and Discussions

A. Air gaps

Fig. 2(a) shows the maximum efficiency value at different normalized air gaps. Fig. 2(b) is the relationship between coil coupling and air gaps, where the coupling becomes weaker with the increase in gap ratio.

The system impedance is adjusted by the spatial distribution of the two coils, which affects the resonance and efficiency. Combining Fig. 2(a) and 2(b), the maximum efficiency stays at 90% even if the coupling factor is as low as 0.1. The efficiency drops to 80% when the coupling factor is around 0.01, indicating that the efficiency of wireless power transfer can still maintain a high value at a weak coupling. This feature of the magnetic resonance coupling allows for wireless power transfer over a long distance at a reasonable efficiency.

Fig. 3 presents the power transmission and reflection ratios, η_{21} and η_{11} respectively, versus changes in frequency with air gap ratios of m = 0, 0.25, 0.5 and 0.75. There are two resonant frequencies when m ranges from 0 to 0.5. Fig. 3 shows that the two resonant frequencies occur when the air gap distances are small and the associated coupling factors are large. In Fig. 3, the first, smaller resonant frequency has a higher efficiency than the second one. Based on Fig. 3, as the air gap increases, the two resonant frequencies move closer and eventually merge into one, followed by a dramatic decrease in the corresponding efficiency.

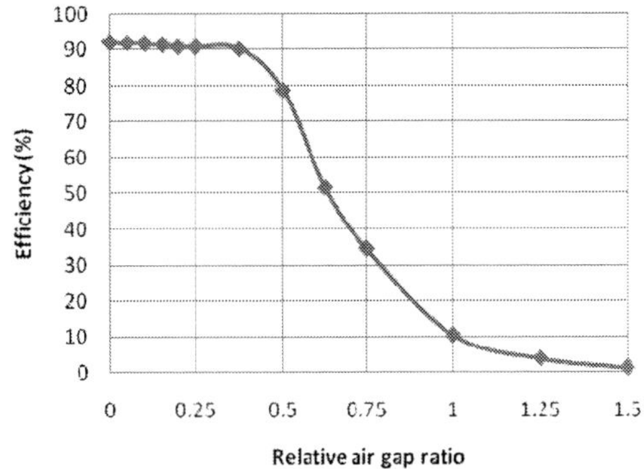

(a)

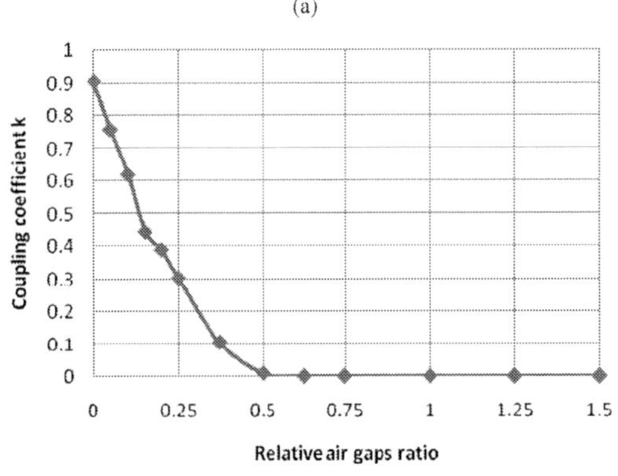

(b)

Fig. 2 (a) Power transfer efficiency vs. normalized air gaps; (b) Coupling coefficient vs. normalized air gaps.

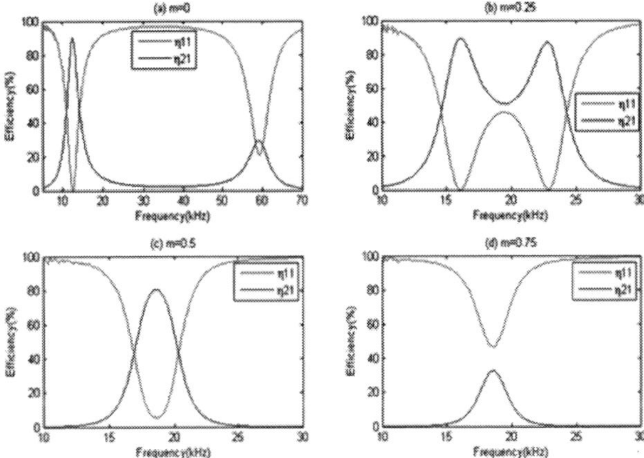

Fig. 3 Relationship between efficiency and frequency at varying air gap ratios of m = 0, 0.25, 0.5, and 0.75.

B. Misalignment

Fig. 4(a) shows the relationship between efficiency and misalignment ratio at four different air gaps. When the air gap ratio is set at 0 and 0.25, the efficiency remains at 90% as long as the misalignment ratio is within 0.5, which shows that the magnetic resonance coupling allows a relatively large misalignment. Fig. 4(b) shows the magnetic coupling curve for different air gap ratios. Combining Fig. 4(a) and 4(b), the efficiency still maintains a high value even if the coupling factor is small, which is similar to the efficiency-gap ratio curve (Fig. 2).

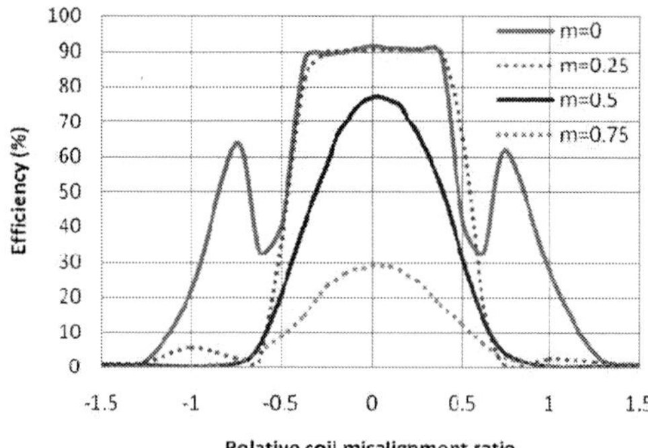

(a)

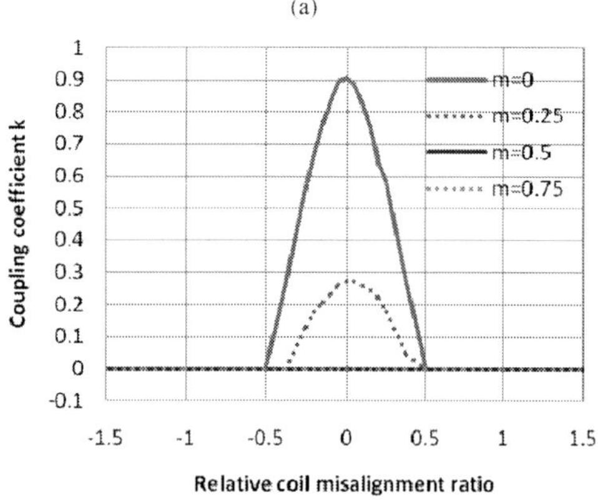

(b)

Fig. 4 (a) Efficiency vs. misalignment ratio at relative air gap ratios of 0, 0.25, 0.5, and 0.75; (b) Coupling coefficient vs. misalignment ratio at relative air gap ratios of 0, 0.25, 0.5, and 0.75.

IV. Conclusions and Future Work

A motorized 3-axis platform and network analyzer were applied to study the effects on efficiency caused by misalignment. The results indicate that although magnetic resonance allows a high efficiency for WPT, coil misalignment could have a significant influence on the power transfer efficiency. In order to eliminate misalignment during EV parking, an automated vehicle alignment system should be in place to navigate parking accurately.

References

[1] G. A. Covic and J. T. Boys, "Modern trends in inductive power transfer for transportation applications," *Emerging and Selected Topics in Power Electronics, IEEE Journal of,* vol. 1, pp. 28-41, 2013.

[2] C. Duan, C. Jiang, A. Taylor, and K. H. Bai, "Design of a zero-voltage-switching large-air-gap wireless charger with low electric stress for electric vehicles," *IET Power Electronics,* vol. 6, pp. 1742-1750, 2013.

[3] A. P. Sample, D. A. Meyer, and J. R. Smith, "Analysis, experimental results, and range adaptation of magnetically coupled resonators for wireless power transfer," *Industrial Electronics, IEEE Transactions on,* vol. 58, pp. 544-554, 2011.

[4] Y. Gao, K. B. Farley, and Z. T. H. Tse, "A Uniform Voltage Gain Control for Alignment Robustness in Wireless EV Charging," *Energies,* vol. 8, pp. 8355-8370, 2015.

[5] H. H. Wu, A. Gilchrist, K. D. Sealy, and D. Bronson, "A high efficiency 5 kW inductive charger for EVs using dual side control," *Industrial Informatics, IEEE Transactions on,* vol. 8, pp. 585-595, 2012.

[6] T. M. Fisher, K. B. Farley, Y. Gao, H. Bai, and Z. T. H. Tse, "Electric vehicle wireless charging technology: a state-of-the-art review of magnetic coupling systems," *Wireless Power Transfer,* vol. 1, pp. 87-96, 2014.

[7] J. Miller, O. Onar, C. White, S. Campbell, C. Coomer, L. Seiber, *et al.*, "Demonstrating Dynamic Wireless Charging of an Electric Vehicle: The Benefit of Electrochemical Capacitor Smoothing," *Power Electronics Magazine, IEEE,* vol. 1, pp. 12-24, 2014.

[8] T.-D. Nguyen, S. Li, W. Li, and C. C. Mi, "Feasibility study on bipolar pads for efficient wireless power chargers," in *Applied Power Electronics Conference and Exposition (APEC), 2014 Twenty-Ninth Annual IEEE,* 2014, pp. 1676-1682.

[9] A. Gil and J. Taiber, "A Literature Review in Dynamic Wireless Power Transfer for Electric Vehicles: Technology and Infrastructure Integration Challenges," in *Sustainable Automotive Technologies 2013,* ed: Springer, 2014, pp. 289-298.

[10] Y. Gao, K. B. Farley, A. Ginart, and Z. T. H. Tse, "Safety and efficiency of the wireless charging of electric vehicles," *Proceedings of the Institution of Mechanical Engineers, Part D: Journal of Automobile Engineering,* p. 0954407015603863, 2015.

[11] T. Imura, H. Okabe, and Y. Hori, "Basic experimental study on helical antennas of wireless power transfer for electric vehicles by using magnetic resonant couplings," in *Vehicle Power and Propulsion Conference, 2009. VPPC'09. IEEE,* 2009, pp. 936-940.

Genetic Algorithm Design of a 3D Printed Heat Sink

Tong Wu[1], Burak Ozpineci[1,2], Curtis Ayers[2]
wut1@ornl.gov, burak@ornl.gov, ayerscw@ornl.gov

[1]Bredesen Center
The University of Tennessee
Knoxville, TN 37996 USA

[2]Power Electronics and Electric Machinery Research Center
Oak Ridge National Laboratory
Knoxville, TN 37932 USA

Abstract— **In this paper, a genetic algorithm- (GA-) based approach is discussed for designing heat sinks based on total heat generation and dissipation for a pre-specified size and shape. This approach combines random iteration processes and genetic algorithms with finite element analysis (FEA) to design the optimized heat sink. With an approach that prefers "survival of the fittest", a more powerful heat sink can be designed which can cool power electronics more efficiently. Some of the resulting designs can only be 3D printed due to their complexity. In addition to describing the methodology, this paper also includes comparisons of different cases to evaluate the performance of the newly designed heat sink compared to commercially available heat sinks.**

Keywords*— *Thermal management, Genetic Algorithm, 3D Printing, Liquid Cooled Heat sink, Cold Plate

I. Introduction

Electric vehicle (EV) sales have been increasing exponentially [1] and so is the importance of higher power density, more efficient electric drive systems for EVs [2]. The developments in the power semiconductor technology have recently shown significant improvements in the power density especially with the use of silicon carbide power devices. Further improvements can be achieved using better optimized more compact heat sinks that allow intimate cooling of power devices. The advances in 3D printing of aluminum allow newer approaches for heat sink designs. Even though this paper focuses on heat sink for electric vehicles, the proposed approach can be used for designing heat exchangers for any application both air and liquid-cooled systems, including dual phase systems.

The conventional manufacturing processes, such as milling, drilling, and casting have limitations for constructing one-piece heat sinks with internal complex shapes and geometries and multi-piece heat sinks are prone to leaks. Using 3D printing technology, it is possible to manufacture one-piece heat sinks with complex internal structures with no additional cost for complexity. Instead of drilling and milling a whole block where material is subtracted, 3D printing adds layers of material to build parts. 3D metal printing is available by melting metal powder with laser or fine e-beam to build parts layer by layer following a computer generated design [3]. Similar technology has already been used by Oak Ridge National Laboratory (ORNL) to build the first 50% 3D printed inverter with a 3D printed heat sink and power module package resulting in a high efficiency inverter with a compact design [4].

Based on the total heat generation, heat distribution pre-specified size, shape, and liquid flow amount, a unique and better performance heat sink can be designed automatically using genetic algorithms (GA). Previous GA applications in power electronics literature have focused on switching time optimization [5-7] and in control of power converters [8].

II. Conventional Heat Sink Designs

Over the past several decades, plenty of work has been done in designing and optimizing heat sinks for multiple purposes. Normally, a mathematic model is built to simplify and calculate the heat transfer ability. The model depends on the structure and material being used and can be complicated to derive. Take an air-cooling heat sink optimization from [9] as an example:

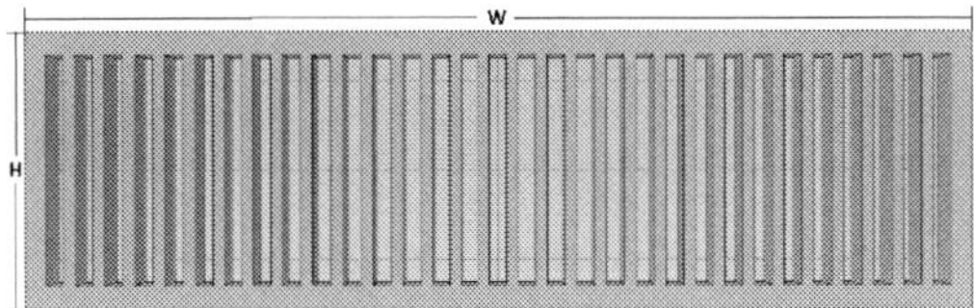

Fig. 1. A finned heat sink for optimization

This manuscript has been authored by UT-Battelle, LLC under Contract No. DE-AC05-00OR22725 with the U.S. Department of Energy. The United States Government retains and the publisher, by accepting the article for publication, acknowledges that the United States Government retains a non-exclusive, paid-up, irrevocable, world-wide license to publish or reproduce the published form of this manuscript, or allow others to do so, for United States Government purposes. The Department of Energy will provide public access to these results of federally sponsored research in accordance with the DOE Public Access Plan (http://energy.gov/downloads/doe-public-access-plan).

a) First, the basic structure is pre-specified, in this case, a finned heat sink shown in Fig. 1. For a given size and operating conditions, the optimization goal is to maximize the total heat dissipation rate.

b) Then, make an initial guess for the plate-to-plate (fin-to-fin) distance to determine the channel spacing (S) and the Rayleigh number (Ra_s) by

$$S = p - th_c \tag{1}$$

$$Ra_s = \frac{gS^3\beta(T_s-T_\infty)}{\upsilon\alpha} \tag{2}$$

where p is the pressure drop, th_c is the fin thickness, T_s is maximum junction temperature, T_∞ is ambient temperature, g is gravity constant, k, υ, α, β are properties of the coolant.

c) The average Nusselt number ($\overline{Nu_s}$) and the average heat transfer coefficient($\bar{h}$) are calculated based on (1) and (2)

$$\overline{Nu_s} = Ra_s S(1 - e^{(\frac{-35L}{Ra_s S})^{0.75}/_{24L}}) \tag{3}$$

$$\bar{h} = \frac{\overline{Nu_s}}{S} \tag{4}$$

Calculating the heat transfer coefficient for an open channel flow

$$\dot{q_{plate}} = 2\bar{h}LW(T_s - T_\infty) \tag{5}$$

for a single plate results in the total heat transfer rate

$$\dot{q_{total}} = N\dot{q_{plate}} \tag{6}$$

d) Based on the calculations above, iterations are used to come up with a design with the maximum total heat transfer and best performance.

The steps described above can be summarized in a flow chart shown as Fig. 2.

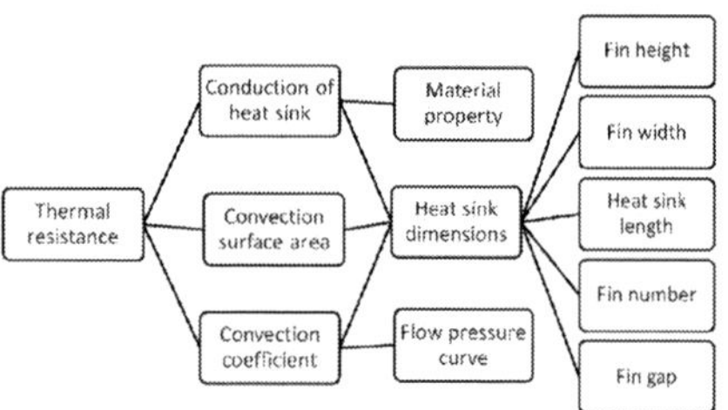

Fig. 2. Design process flow chart of air-cooled heat sink

The approach is easy to implement, and the primary goal of the design is to decrease the thermal resistance in the equivalent circuit; however, several draw backs should be noticed here. First, the overall design freedom is limited, because of the type of heat sink is predefined. Moreover, the optimization accuracy is not high enough, due to the approximations in mathematic analytical model. Finally, the heat sink is designed based on the thermal resistance only. This is a general approach, without considering unique thermal distribution for different power conversion systems scarifying the efficiency of the heat sink.

Several other ways of designing heat sinks exist, for example, the Entropy Generation Minimization Method [10] uses the effects of thermal resistance in combination with pressure drop to evaluate the entropy generation and comes up with a mathematic model for optimization. However, the same limitations apply here as the previously mentioned approach.

Overall, conventional heat sink designing processes mentioned above are powerful for simple heat sink shapes where the thermal modeling equations are typically known or are simple to derive. The heat sink structures are pre-selected by designers and the optimization calculations are completed on these structures without considering the actual heat distribution.

Automotive manufacturers have developed complex heat sinks to cool their electric drive technologies spending a lot of time and money to come up with analytical models and them optimizing them.

In this paper, a new process will be shown to design liquid-cooled heat sinks without empirical principles and analytical models but focusing on FEA modeling of power devices and heat sinks. The developed algorithm can be applied to any application or any condition as long as the maximum allowed heat sink size, device and system loss characteristics, and the coolant information in supplied. The resulting heat sinks are automatically generated and are typically complex structures. Overall, the new design process is expected to dramatically reduce the heat sink design cycle time and improve the thermal performance of the power electronics systems.

III. Thermal Equivalent Circuit

A. Heat Transfer Path

A typical heat transfer path for a liquid-cooled system is shown as Fig. 3. The heat generated by the devices transfer through the thermal grease to the heat sink body, heat up the contact surface and transfer heat to the cooling media. Then the heat is moved away by the liquid to be dissipated.

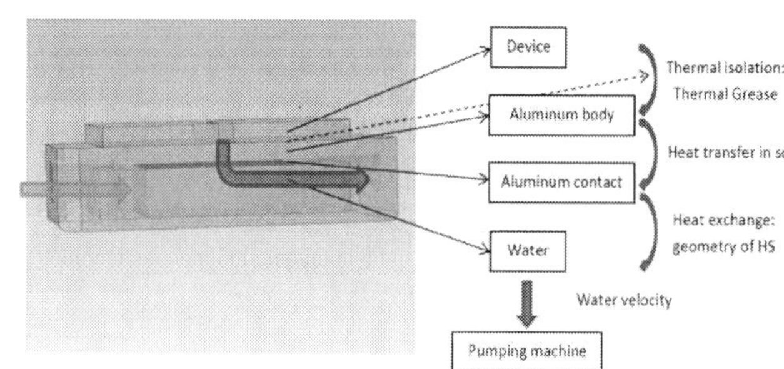

Fig. 3. Heat transfer path of liquid-cooled heat sink

B. Steady State Thermal Equivalent Circuit

Based on the analysis of heat transfer path, a thermal equivalent circuit is shown as Fig. 4.

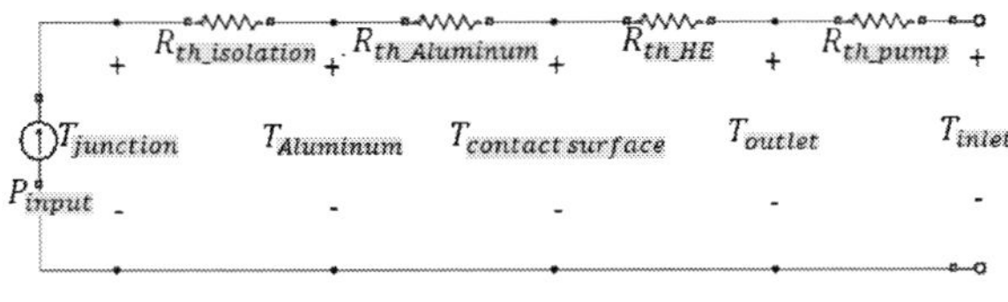

Fig. 4. Steady state thermal equivalent circuit

At steady state, the equivalent voltage on each node represents the corresponding steady state temperature.

Based on this circuit, several conclusions can be made using the energy conversion and the junction temperature equations below:

$$T_{outlet} = T_{inlet} + R_{th\_pump} \times P_{input} \tag{13}$$

$$T_{junction} = T_{inlet} + (R_{th_{isolation}} + R_{th_{Aluminum}} + R_{th\_HE} + R_{th\_pump}) \times P_{input} \tag{14}$$

As shown above, the junction temperature could be reduced for the same operating conditions by:

- Reducing $R_{th\_isolation}$ or eliminating the thermal grease
- Reducing $R_{th\_Aluminum}$, by improving thermal conductivity. $R = \frac{1}{\sigma} \cdot \frac{l}{A}$, where $\frac{l}{A}$ is fixed for the given size and the σ represents the thermal conductivity. This paper will consider bulk Aluminum for design optimization. 3D printed metal technology can eventually result in higher thermal conductivity low-cost heat sinks but this is not within the scope of this paper.
- Reducing R_{th_pump} by increasing the liquid inlet speed. However, this will increase power consumption and levels of acoustic noise and pressure drop.
- Decreasing T_{inlet} which will again introduce higher cooling costs.

Besides the approaches above, the only viable option that will not increase the cost of the system is reducing the R_{th_HE} , which can be achieved by designing a better performing heat sink.

To evaluate and show the improvement of the heat sink designed in this paper, a commercially available comparison model is introduced in this section. The corresponding FEA simulation model and the working environment are also stated.

IV. Fundamental Model

A. Comparison heat sink

A commercial heat sink, Lytron's standard tubed CP15 is selected for comparison purposes. It is an aluminum-based heat sink, with copper tubes shown in Fig. 5 and is a typical heat sink used in the Power Electronics and Electric Machinery Research Center labs for applications such as the SiC MOSFET based inverter for wireless charging built by ORNL PEEM group [11]. The inlet and outlet tubes are connected to the pump which allows a liquid flow loop with the fixed inlet temperature and inlet velocity.

Fig. 5. Lytron heat sink for the SiC MOSFET based inverter

B. Finite element analysis model

Based on the actually dimension measurements of the heat sink in Fig. 5, a FEA simulation model is built as shown in Fig. 6. The body is made of Aluminum 6061, the tube is made by copper, and the liquid inside is assumed to be water. All materials are defined by the material library built in COMSOL. The active area is in the middle of the flat surface. The dimensions of the heat sink are $86mm \times 64mm \times 8mm$. Based on the measurement of the

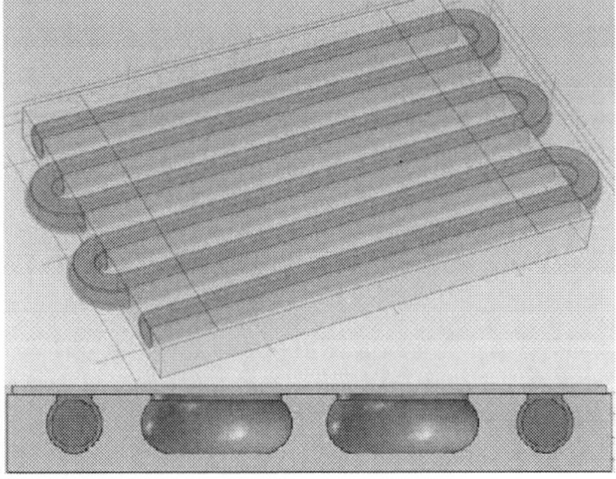

Fig. 6. COMSOL commercial heat sink model and the cross section of the copper tube

dimensions, the copper weight is estimated to be $56.1\ cm^3 \times cm\ {}^{g}/_{cm^3} = 502g$ and the aluminum weight is: $292.5\ cm^3 \times c7\ {}^{g}/_{cm^3} = 790g$ with a total weight of 1292g.

C. *Parameter, environment and assumption*

The models are simulated in COMSOL for 3D stationary conjugate heat transfer ignoring the radiation to the environment whose contribution is negligible compared with the conjugate heat transfer.

Case one simulates a large heat source on the heat sink assuming a SiC MOSFET H-bridge based inverter module with a baseplate. The heat source is defined to be $64mm \times 64mm \times 1mm$ and the total power dissipation is assumed to be 2 kW. The ambient temperature is the room temperature 20°C and the inlet water is set to be laminar flow with a velocity of 0.036L/s and temperature of 20°C.

Case two simulates an H-Bridge inverter with four separate die instead of using a baseplate. This case simulates four individual devices on a heat sink rather than on one module. To dissipate the heat generated by four power electronics components, instead of using the base plate, thermal interface material is placed as an isolation layer between chips and heat sink. This topology is the most common design for the commercial electric vehicle drive system [12]. Four chips are sparsely placed on the surface of the heat sink and each of them generates power 250W that is a total 1000W power loss. The flow parameters are the same as the first case.

V. Genetic Algorithm Approach

Genetic algorithm (GA) is known as a solution space searching algorithm that imitates the process of natural evolution following the rule of "the survival of the fittest". After iterations, the optimized object evolves automatically based on the selection rules and the fitness function. Finally, a desired solution is obtained by exploring a large space of random solutions to find the optimum solution efficiently instead searching for every possibility [13].

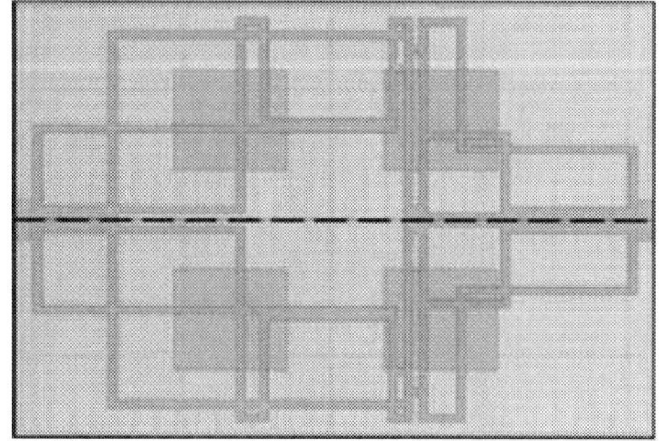

Fig. 8. Symmetric geometry approach

The proposed approach uses an algorithm that combines a newly developed random walking process and genetic algorithms to design the heat sink. First, to save computation time, it is assumed that the system is horizontally symmetrical, so the algorithm only focuses on half of the heat sink, shown in Fig. 8.

The overall process is a self-evolution process containing two stages. The first stage is a fundamental genetic algorithm iteration based on the fitness function to select the targets for the second stage. The second stage then, acts a perturbation function on targets to optimize for better individuals. The proposed process then replaces the targets by the better ones and repeats the same steps until the given number of iterations or the goal is reached.

A. *First stage genetic algorithm approach*

First stage GA has four steps that are iterated: Initialization, Evaluation and Selection, Crossover and Mutation, and Reproduction.

1) Initialize Population: Generated by the random walking process which randomly digs flow channels inside the aluminum block creating a heat sink automatically. Random walking process, which is implemented in steps with random direction and distance, is used for creating increased variety of different geometries within the heat sink. The process works like mining and digging tunnels underground. Before digging, each model contains a given size of aluminum heat sink block with inlet and outlet locations predefined and a heat source attached. Then, starting from the inlet, two to ten random water channels are formed, randomly with the walking process and are merged at the outlet of the heat sink. Each channel is represented as an n×2 matrix, as shown in Fig. 9 with first row being the starting point or the inlet which is designated to start at a location between (1, 1) and (1, 10) . n is the number of the turns (changes in direction or elbows) the channel has and the matrix presents where these turns occur. For example, in the matrix in Fig. 9, 15 turns are generated starting from location (1, 5), going horizontal first to (1, 10); continuing from (1, 10) and going vertical (down) two points to (3, 10). The same process is repeated until the channel reaches the end point (outlet) which is at (80, 5). The rows in Fig. 10

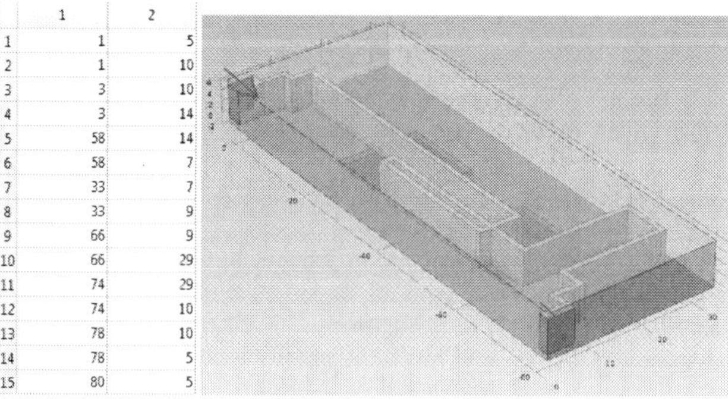

	1	2
1	1	5
2	1	10
3	3	10
4	3	14
5	58	14
6	58	7
7	33	7
8	33	9
9	66	9
10	66	29
11	74	29
12	74	10
13	78	10
14	78	5
15	80	5

Fig. 9. Channel matrix and the generated channel

represent each heat sink solution, where the columns represent starting points for each channels. "[]" means there are no channels starting from that point. The numbers in the cells represent the n×2 matrices representing the channels starting from that point. 15 rows in Fig. 10 form an initial solution population of 15.

'Start from{1,1}'	'Start from{1,2}'	'Start from{1,3}'	'Start from{1,4}'	'Start from{1,5}'	'Start from{1,6}'	'Start from{1,7}'	'Start from{1,8}'	'Start from{1,9}'	'Start from{1,10}'
24x2 double	[1,2;1,1;31,1;31...	7x2 double	[]	27x2 double	27x2 double	7x2 double	8x2 double	27x2 double	12x2 double
[]	[]	[]	6x2 double	17x2 double	[]	[]	[]	[]	[]
12x2 double	[]	14x2 double	7x2 double	[]	[]	18x2 double	[1,8;1,1,78,1;78...	[]	7x2 double
[]	[]	[]	[]	14x2 double	[]	[]	[]	[1,9;79,9;80,9]	[]
[1,1;1,29;77,29;...	[]	7x2 double	[]	[]	[]	[]	8x2 double	23x2 double	[]
9x2 double	7x2 double	7x2 double	13x2 double	8x2 double	60x2 double	10x2 double	7x2 double	9x2 double	11x2 double
[]	[]	[]	[]	[]	[]	21x2 double	9x2 double	23x2 double	12x2 double
[]	[]	[]	[]	[1,5;77,5;80,5]	[]	21x2 double	[]	[]	11x2 double
[]	33x2 double	[]	[]	27x2 double	[]	15x2 double	12x2 double	[]	[1,10;1,30;21,30...
29x2 double	7x2 double	16x2 double	7x2 double	9x2 double	[1,6;1,1;45,1,45...	[]	[]	[]	14x2 double
33x2 double	10x2 double	37x2 double	9x2 double	[1,5;1,9;39,9;39...	7x2 double	36x2 double	7x2 double	37x2 double	[1,10;1,27;78,27...
12x2 double	[1,2;79,2;80,2]	10x2 double	26x2 double	11x2 double	11x2 double	6x2 double	12x2 double	[1,9;1,28;25,28;...	11x2 double
37x2 double	14x2 double	[]	[]	33x2 double	[1,6;1,3;71,3;71...	7x2 double	7x2 double	[1,9;1,7;63,7;63...	9x2 double
[]	20x2 double	[1,3;1,27;6,27;5...	[]	[]	[]	10x2 double	[]	15x2 double	[]
[1,1;79,1;80,1]	35x2 double	[]	[1,4;1,23;19,23;...	[]	[]	10x2 double	[]	[]	31x2 double

Fig. 10. Initial generation matrix cells

2) Evaluation and Selection: Each individual in the population is evaluated under the same boundary conditions. The maximum heat sink (shown as junction temperature in Fig. 4) temperature and the inlet force (defining the pressure drop) are calculated using COMSOL finite element simulation. A fitness score (cost function) is set as a function of the heat sink temperature rise along with the inlet force

$$Fitness = \Delta T + F_{inlet}/2^{\gamma} \tag{15}$$

where γ is a weight factor between the temperature and the force causing pressure drop (defined as force instead of pressure because of the variation of the inlet opening for different solutions) and can be varied depending on the application. If the importance of the inlet force is higher than the temperature rise, a relatively small γ can be used. In the opposite condition a relatively large γ can be used. In this paper, γ is selected as 0, so that the importance of the force index is around 3% of the temperature.

The individual solutions are then ranked from high to low in a decreasing order of the fitness scores. The individual (i)'s survival possibility equals to the value of$(Rank(i)/Sum(rank))^{\alpha}$, where α is the control factor for the convergence speed. At the beginning, α is small so that the convergence speed is slow to generate a larger variety of solution; at the end, α is higher to ensure the presence of the "good" individuals. The higher possibility means a higher possibility to be selected for the next round. Generally, the individual with lower temperature rise and an acceptable inlet force will be more likely to survive, that is, it will have a higher survival possibility. This selection process follows the "survival of the fittest" rule. For our example, for a population of 21, 6 of them will be eliminated and 15 survivors will go to the next step.

3) Crossover and Mutation: 15 survivors (surviving solutions) are evenly split into 5 groups. Each group contains 3 survivors and will go through crossover and produce three individuals to replace them as the next generation. Crossover is implemented by exchanging the channel matrix cells (chromosome) between pairs of group members. After crossover, if some of the newborns contain less than 2 channels, they will be replaced by a newly generated solution. This process ensures that all the surviving individuals are qualified and at the same time, better distributes the chromosomes. Also, during the crossover, mutations may happen, which means the parts of channels are randomly changed. The possibility of a mutation is a function of the convergence level. A high convergence level will bring a larger possibility of mutation to prevent the premature convergence and being stuck in a local minimum instead of a global one.

4) Reproduction: The new population is formed by 15 new-born individuals inheriting genetic information from the ancestors' generations, 5 new randomly generated individuals, and one of the best individuals from the previous generation. This will be used for the next iteration of GA. The algorithm will run until the target temperature is achieved or the maximum iteration number is reached.

The overall process of GA is shown as Fig. 11 and the convergence plot is shown in Fig. 12.

The steps shown in Fig. 12 demonstrate the evolution

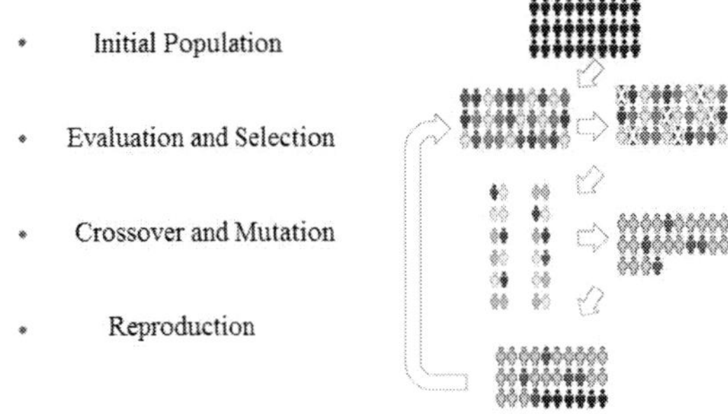

Fig. 11. The overall GA approach

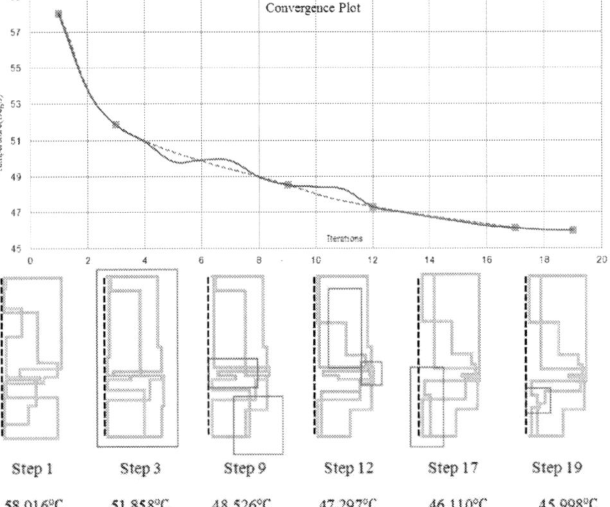

Fig. 12. Convergence plot and the evolved individuals

process. Between steps 1 and 3, a total new geometry is generated by the algorithm and the system temperature drops a significant seven degrees. The transitions from Step 3 to 12 are mostly carried out by crossover. After Step 12, the possibility of the mutation increases and some small changes can be observed from Step 17 to Step 19 due to the mutation. The overall convergence plot trend is shown by the purple dashed line which is an exponentially decrease. For this example, in 19 iterations, the minimum temperature of the heat sink is decreased by 12 degrees for the same load and the same size heat sink.

B. Second stage genetic algorithm approach

After the iterations in the first stage, several "pretty good" geometries of heat sink survive in the final population. Based on the results from the first stage, second stage optimization is applied. The second stage is a perturbation function of the optimized geometries and it includes Translation, Connection, Creation, and Deletion steps shown as Fig. 13 below for a three channel heat sink design:

Each type of perturbation can be applied up to 3 times for each individual (solution):

The Translation perturbation is performed as in Fig. 13(a) by randomly selecting parts of a channel and translating (moving) it to another position and still keeping the connections to other parts. In this case, three parts (two horizontal and one vertical), one per channel have been translated.

The Creation perturbation is performed as Fig. 13(b) by selecting points beyond the original channels and connecting them to the nearest channels.

The Connection perturbation randomly picks a possible short cut two parallel channels and connects them directly as shown in Fig. 13(c).

The Deletion perturbation picks a random point on the existing channel and deletes the potentially blocked or circulating channels around this point.

After the first iteration of the second stage optimization, the worst performing individuals are replaced with the better ones and the process is repeated until a stable group of individuals and obtained.

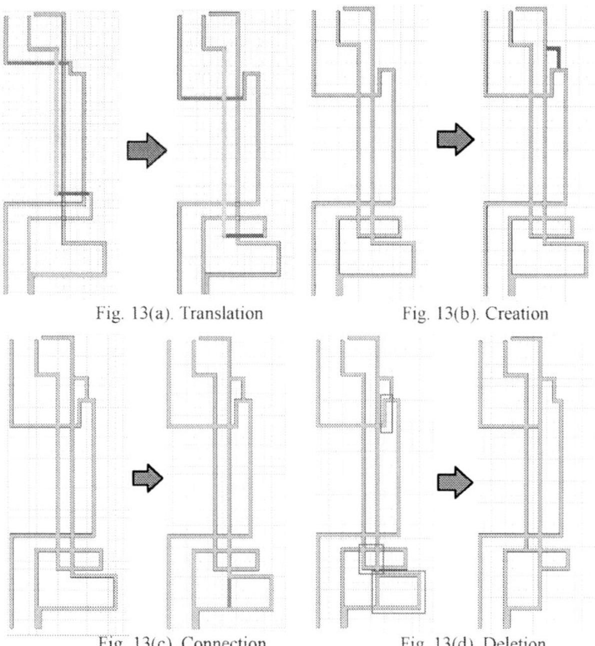

Fig. 13(a). Translation
Fig. 13(b). Creation
Fig. 13(c). Connection
Fig. 13(d). Deletion

Fig. 13. Perturbations

VI. Simulation Results

A. Case one, large plate heat source:

1) Commercial Simulation

The commercial heat sink is simulated in COMSOL as shown in Fig. 14 which also shows the resulting temperature distribution. The inlet water was set to have a velocity of 0.036L/s and a temperature of 20°C. For a total power of 2kW applied uniformly to the large plate heat source, the max junction temperature on the heat sink is 53.78°C, which is a temperature rise of 33.78°C. The correesponding inlet force, which is the integral of the pressure drop of the inlet, is 0.56N. Applying the fitness function (15), the fitness value is 33.78 + 0.56 = 34.34.

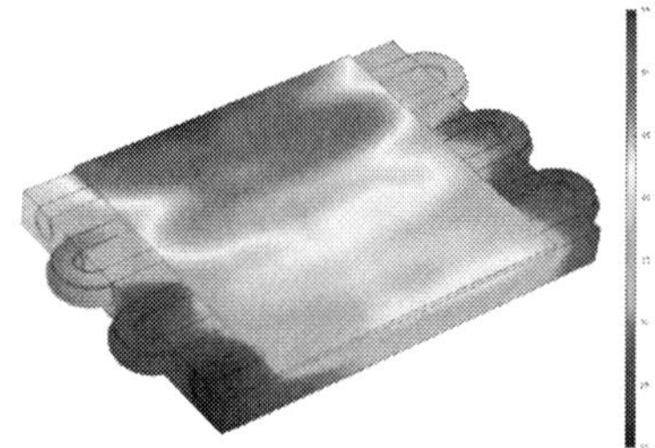

Fig. 14. Commercial large heat source heat sink performance

2) 1st stage optimization

After the first stage of the algorithm is applied, the solution in Fig. 15 is obtained with the same temperature scale as in Fig 14 using the same load and flow parameters as in the commercial heat sink simulation. The resulting max junction temperature is 49.21°C which is a 29.21°C rise

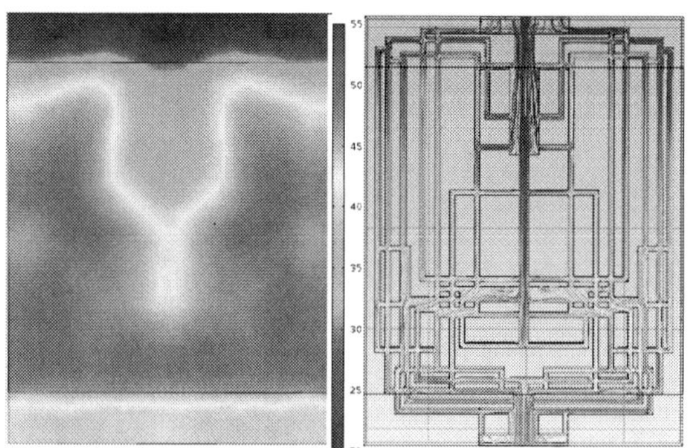

Fig. 15. 1st stage optimized heat sink

in temperature with an inlet force of 0.58N. Applying the fitness function (15), the fitness value is $29.21 + 0.58 = 29.79$ which is lower than the commercial case since this solution has a lower temperature and lower inlet force performance.

3) 2nd stage optimization

Applying the second stage optimization algorithm, the resulting solution in Fig. 16 is obtained. In this case, the max temperature is 47.92 °C, with a temperature rise of 27.92°C and a inlet force of 0.55N corresponding to a fitness value is $27.92 + 0.55 = 28.47$. Both the temperature and the force are lower than the commercial unit. The heat sink is composed only by aluminum and the total weight is $315.5\ cm^3 \times 2.7g/cm^3 = 852g$.

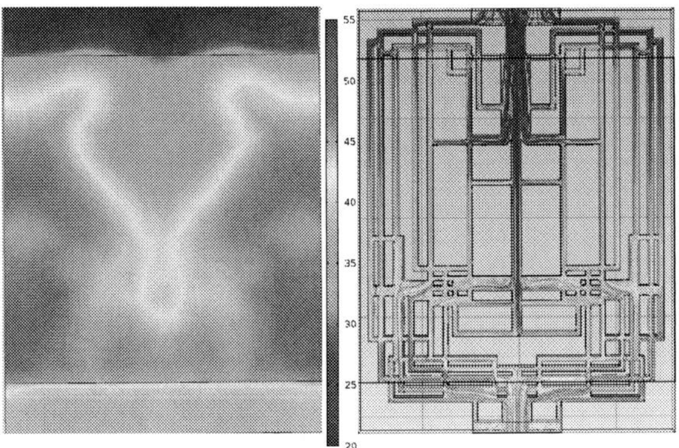

Fig. 16. 2nd stage optimized heat sink

4) Analysis

The fitness values for the commercial heat sink and the GA designed unit after the first and second stage optimization are 34.34, 29.79, and 28.47 showing 13.3% and 17.1% improvements compared with the commercial heat sink. Moreover, the weight is about 2/3 of the commercial one and better performance is obtained even with no copper. Based on the heat distribution plots, it is clear to see that the optimized heat sink evenly distributes the heat avoiding possible heat spots leading to lower heat sink temperatures.

It is interesting to note that the algorithm decided to design the channels to bring the cooler water to the bottom half (closer to the outlet) of the heat sink preventing the water to circulate in the upper half and warming up. On the bottom half (closer to the bottom half), however, it decided to insert more channels to take full advantage of any cooler water coming directly from the inlet. In addition, it can be noted that the vertical channels are more or less uniformly distributed. After the second optimization, several channels are deleted and others are created in the upper half. This strategy which is already built into the algorithm, distributes colder water to the lower half, bringing a more uniform temperature distribution.

B. Case two, distributed heat sources:

1) Commercial Simulation

The same commercial heat sink is used in this case with four distributed heat sources with each generating 250W loss for a total of 1 kW. The devices are evenly placed on the surface of the heat sink. The simulation results are shown in Fig. 17. The max junction temperature is 53.02°C, with a temperature rise of 33.02°C and an inlet force of 0.56N. Applying the fitness function (15), the fitness value is $33.02 + 0.56 = 33.58$.

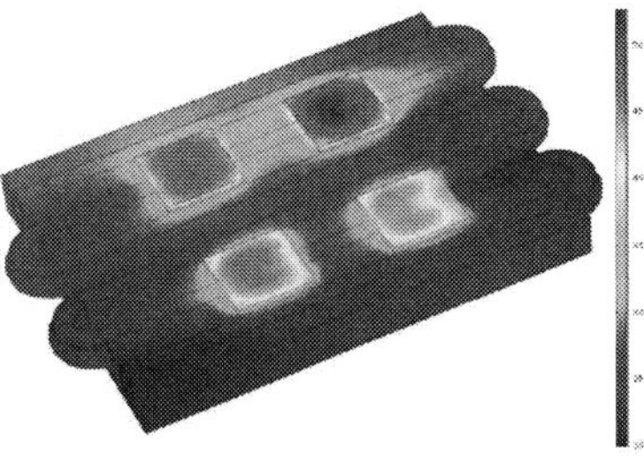

Fig. 17. Commercial separated heat source heat sink performance

2) 1st stage optimization

After the first stage of the algorithm developed in this paper, the solution obtained is shown in Fig. 18 with the same temperature scale as in Fig 17 using the same load and flow parameters as in the commercial heat sink simulation. The resulting max junction temperature is 46.1°C, which is a 26.1°C rise in temperature with an inlet force of 1.63N and a fitness score of $26.1 + 1.63 = 27.73$.

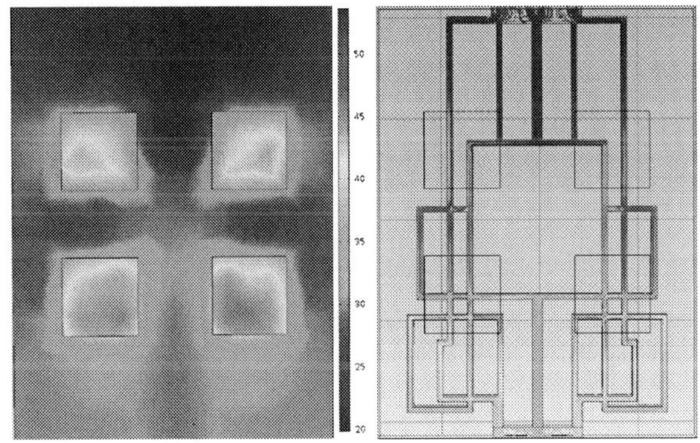

Fig. 18. 1st stage optimized heat sink

3) 2nd stage optimization

Applying the second stage optimization, the solution in Fig. 19 is obtained. It is interesting to note that the algorithm designed bypass channels to the right of the upper right heat source and the left of the upper left heat source to bypass these and send cooler water to the lower heat sources. The result is a heat sink temperature of 44.85 °C with a temperature rise of 24.85°C and an inlet force of 1.3N. The fitness value then becomes $24.85 + 1.3 = 26.15$ and the weight of the heat sink is calculated to be $383.3\text{cm}^3 \times 2.7\text{g}/\text{cm}^3 = 1035\text{g}$.

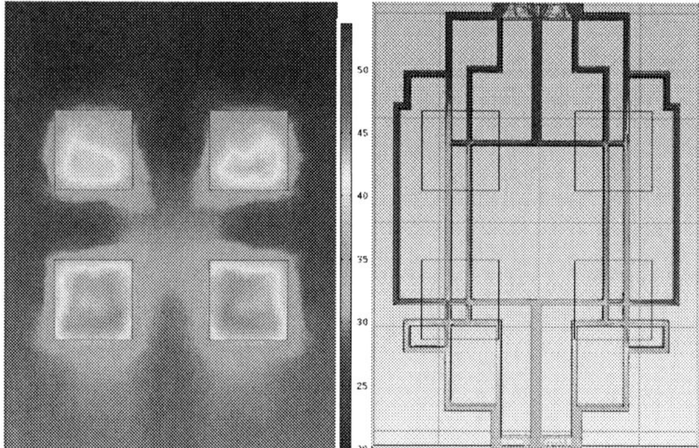

Fig. 19. 2nd stage optimized heat sink

4) *Analysis*

The fitness values for the commercial heat sink and the GA designed unit after the first and second stage optimization are 33.85, 27.73, and 26.15 showing 18.1% and 22.8% improvements compared with the commercial heat sink. The overall strategy of the algorithm is balancing the temperatures of the upper and lower halves of the heat sink. A relatively fast flow bypasses the upper heat sources and brings cooler water (as indicated by a darker streamline in Fig. 19) to the lower half. Concentrating the cooler liquid in "#" shaped channels under the lower half heat sources result in a slower speed flow and larger heat exchange surface. This increases the heat transfer rate at the lower half and ensures that the local temperature stays balanced. However, at the same time, the inlet force is sacrificed. The designer has to pick the relative importance of junction temperature and the inlet force (which is proportional to the pressure drop) through the use of γ. For a smaller γ, with a more importance on pressure drop, a different solution would be obtained.

Evaluating the results for both cases, it can be observed that the algorithm is trying to evenly distribute the heat throughout the heat sink. To avoid, water being heated up when flowing under the upper half heat sources, two different strategies are implemented. A uniform distribution is developed in the first case and more simplified and concentrated channels are used in the second case. Overall, the strategies ensure the even temperature distribution. Actually, the maximum junction temperature difference between upper and lower parts is about 1°C compared with the commercial case, which is about 10°C.

Also, comparing the weight of the optimized heat sink with the commercial one, a 16% to 30% reduction is observed for both cases.

VII. Conclusion

A genetic algorithm optimization process is developed in this paper. Several simulation comparisons are shown and more than 15% improvement in thermal performance is observed. The algorithm approach provides an automatic way for a unique and better heat sink design without the need for complicated analytical models. Comparing with the common optimization methods in the literature, the proposed algorithm allows for a higher degree of freedom.

In this paper, the volume of the heat sink to be optimized was pre-specified. Instead, if a target heat sink temperature is selected, the algorithm can be designed to optimize the size of the heat sink, increasing the power density of power conversion systems. This will be a part of the future work

VIII. References

[1] Z. Wang and R. Paranjape, "An evaluation of electric vehicle penetration under demand response in a multi-agent based simulation," *Electrical Power and Energy Conference (EPEC), 12-14 Nov. 2014* .pp. 220 - 225

[2] D.S. Gautam; F. Musavi; D. Wager; M. Edington, "A comparison of thermal vs. patterns used for thermal management in power converter," *Energy Conversion Congress and Exposition (ECCE)*, 15-19 Sept. 2013, pp. 2214 – 2218

[3] T. A. Burress, "Evaluation of 2004 Toyota Prius hybrid electric drive system," Oak Ridge National Laboratory, Oak Ridge, TN, USA, Tech. Rep. ORNL/TM-2006/423, May 2006

[4] Michael Molitch-Hou (2014) Oak Ridge National Lab controls 3D printed metal at the microscale, available at: http://3dprintingindustry.com/2014/10/16/oak-ridge-national-lab-controls-3d-printed-metal-microscale/

[5] M. Chinthavali, C. Ayers, S. Campbell, R. Wiles, B. Ozpineci, "A 10-kW SiC inverter with a novel printed metal power module with integrated cooling using additive manufacturing," *Wide Bandgap Power Devices and Applications (WiPDA)*, 13-15 Oct. 2014, pp. 48 – 54

[6] B. Ozpineci, J. O. P. Pinto, and L. M. Tolbert, "Pulse width optimization in a pulse density modulated high frequency ac-ac converter using genetic algorithms," 2001 IEEE International Conference, Arizona, USA, Oct. 2001, pp. 1924-1929

[7] B. Ozpineci; L.M. Tolbert; J.N. Chiasson, "Harmonic optimization of multilevel converters using genetic algorithms," *Power Electronics Specialists Conference, IEEE 35th Annual*, 20-25 June 2004, pp. 3911 – 3916.

[8] M. J. Schutten, D. A. Torrey, "Genetic algorithms for control of power converters," *Conference Proceedings of IEEE Power Electronics Specialists Conference*, 1995, pp. 1321-1326

[9] W. R. Hamburgen "Optimal finned heat sinks". WRL research report 86/4. Available at: http://www.hpl.hp.com/techreports/Compaq-DEC/WRL-86-4.pdf

[10] W. A. Khan, M. M. Yovanovich, and J. R. Culham, "Optimization of microchannel heat sinks using entropy generation minimization method," *Semiconductor Thermal Measurement and Management Symposium, 14-16 March 2006*, pp. 78 - 86

[11] O. Onar, M. Chinthavali, S. Campbell, N. Puqi, C.P. White, J. M. Miller, "A SiC MOSFET based inverter for wireless power transfer applications," *Applied Power Electronics Conference and Exposition (APEC)*, 16-20 March 2014, pp. 1690 - 1696

[12] T. A. Burress, "Evaluation of the 2010 Toyota Prius hybrid synergy drive system," U.S. Department of Energy, Oak Ridge National Laboratory, Oak Ridge, TN, USA

[13] W. Banzhaf, P. Nordin, R. Keller, F. Francone, "Genetic Programming – An Introduction". San Francisco, CA: Morgan Kaufmann. ISBN 978-1558605107

Evaluation of Power Flow Control for an All-Electric Warship Power System with Pulsed Load Applications

J. Neely, *Member IEEE*, L. Rashkin, *Member, IEEE*, M. Cook, D. Wilson, *Member, IEEE*, S. Glover, *Senior Member, IEEE*

Sandia National Laboratories
Albuquerque, NM, USA
jneely@sandia.gov

***Abstract*— Future U.S. Navy ships will require power systems that meet more stringent agility, efficiency, scalability, controllability and resiliency requirements. Modularity and the ability to interconnect power systems having their own energy storage, generation, and loads is an enabling capability. To aid in the design of power system controls, much of what has been learned from advances in the control of networked microgrids is being applied. Developing alternative methods for controlling and analyzing these systems will provide insight into tradeoffs that can be made during the design phase. This paper considers the problem of electric ship power disturbances in response to pulsed loads, in particular, to electromagnetic launch systems. Recent literature has indicated that there exists a trade-off in information and power flow and that intelligent, coordinated control of power flow in a microgrid system (i.e. such as an electric ship) can modify energy storage hardware requirements. The control presented herein was developed to provide the necessary flexibility with little computational burden. It is described analytically and then demonstrated in simulation and hardware.**

Keywords—Navy all-electric ship, pulsed loads, power system stability, EMALS

I. Introduction

The all-electric warship provides the potential for unprecedented flexibility and system capability [1]. However, pulsed power launch, weapon and radar systems present design challenges [2]-[15]. Pulsed power subsystem integration is a technology challenge today and is a good motivating example for the development of a rational control approach for ship microgrids with information and power sharing. Numerous pulsed power studies have been conducted. Many focus on the bus voltage effects and methods for mitigation [5]. Some studies have postulated harmonizing pulsed power operation with other power intensive ship operations - typically the power plant of an electric ship [14]. Harmonizing all the ship's power assets through closed loop control has not been solved, but offers tremendous promise for enabling not only pulsed power technology, but also future technology with nonstandard power requirements.

Achieving regulation and power balance in a system with highly variable loads is a key capability addressed by this research. Specifically, a hierarchical networked microgrid control concept is being applied to the U.S. Navy's Electric Ship program. The controls approach is developed using a Hamiltonian-based power flow control methodology developed in [16]-[19] and expanded for use in electric ship power systems in [20]. Hamiltonian-based control allows for the kinetic and potential energy stored in the system state to be accounted for explicitly. In particular, since the control is applied herein to a DC-based system, the generators are not synchronized, and the power output of the generator is controlled through power electronics, allowing for some energy to be borrowed from the generator inertia and made available through high-bandwidth control. This can allow a reduction in the size of energy storage necessary but results in state disturbances, ie. greater generator speed deviations.

The control scheme has three layers including a guidance control, a Hamiltonian-based control and a servo control. The guidance control receives measurement information and computes changes to converter reference commands; these can be quasi-static (approx. 1 Hz) or rapid updates (approx.50-100 Hz). The Hamiltonian control is implemented herein as a proportional plus integral (PI) control. The servo control refers to the inner most control loop and pulse width modulation (PWM) scheme residing on each power converter; it operates on the fastest timescale to implement commands from the Hamiltonian control layer.

In this work, a notional electric ship reference model was generated using typical values from literature, the reference model was scaled to be represented using the Secure Scalable Microgrid testbed [21], and experiments were conducted to evaluate the use of hierarchical control. In particular, this work considers the implementation of a simple Guidance control scheme that adjusts power flow through filtering when pulsed loads are used. This scheme is an indirect way of affecting power flow but has a reduced computational burden compared to dynamic optimization planning schemes such as the one presented in [20]. The trade-off between energy storage and generator response is quantified in simulation and hardware.

The next section describes the microgrid testbed used for experimentation. Descriptions of the electric ship reference model and Electromagnetic Launch System (EMALS) pulsed load follow. Section IV reviews the Hamiltonian power flow control scheme. Section V describes the Guidance Control approach used in this study. Section VI provides simulation results for the notional system. Section VII provides

978-1-4673-9551-9/16 $31.00 © 2016 IEEE

experimental results using the microgrid testbed. Finally, conclusions and discussion are provided in Section VIII.

II. Microgrid Testbed Description

The Secure Scalable MicroGrid Test Bed (SSMTB) was designed to conduct experiments on networked microgrids that share information flow and power flow [21]-[23]. The testbed includes three microgrid systems, a central bus cabinet for interconnecting the components, and computers used for control, data acquisition, and situational awareness. In total, the system components include: a reconfigurable bus cabinet, five permanent magnet generators, nine energy storage emulators capable of sourcing or sinking 5kW of power, seven 600V commercial power supplies, mechanical source emulators based on commercial motor drives, a DC/AC converter, a three-phase resistive load, three high-power digital resistors rated to 6.7 kW at 400V bus voltage, and a master control console that scripts the experiments with designated source and load profiles. Some key components are shown in Fig. 1. An example screenshot of the master control software is shown in Fig. 2 highlighting the coordination of several components and monitoring of several quantities for each scripted experiment. Additional information may be found in [20]-[23].

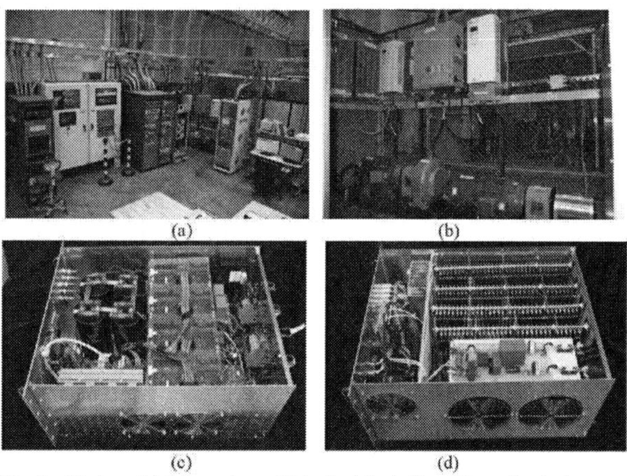

Fig. 1. Photos of (a) the microgrid testbed including (b) mechanical source emulators, (c) energy storage emulator, and (d) high power digital resistor

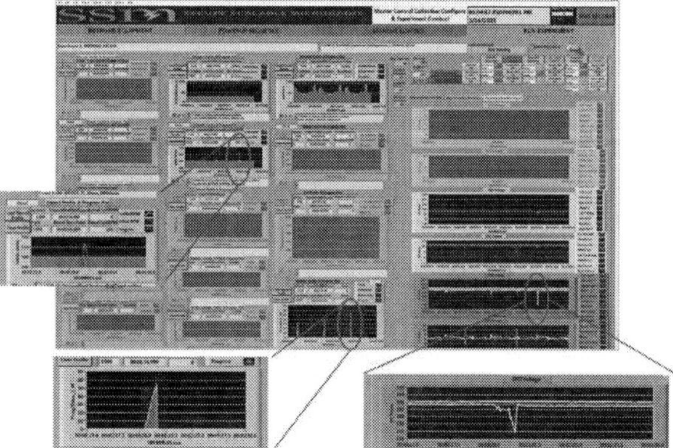

Fig. 2. Shows screen capture for the Master control computer summarizing experiment profile and outcomes with (top center) bus 3 load profile (left) generator 2 speed transients, (bottom left) pulsed load current, (bottom right) dc bus voltage

III. Electric Ship Reference Model

A notional electric ship was developed based on typical power levels and architectures identified in open literature [1]-[15]. The notional system is depicted in Fig. 3a. This notional system uses a high-voltage DC bus, relies heavily on power electronic converters to provide flexible power flow control, and includes a network to enable communication. The system was scaled down and emulated using components in the SSMTB, shown in Fig. 3b, thus providing a laboratory scale model of a 3-zone electric ship networked power system. To evaluate the response of bus voltage, generator speed transients and energy storage effort in response to an EMALS launch, the EMALs system was modeled and an emulator was built and integrated into the testbed.

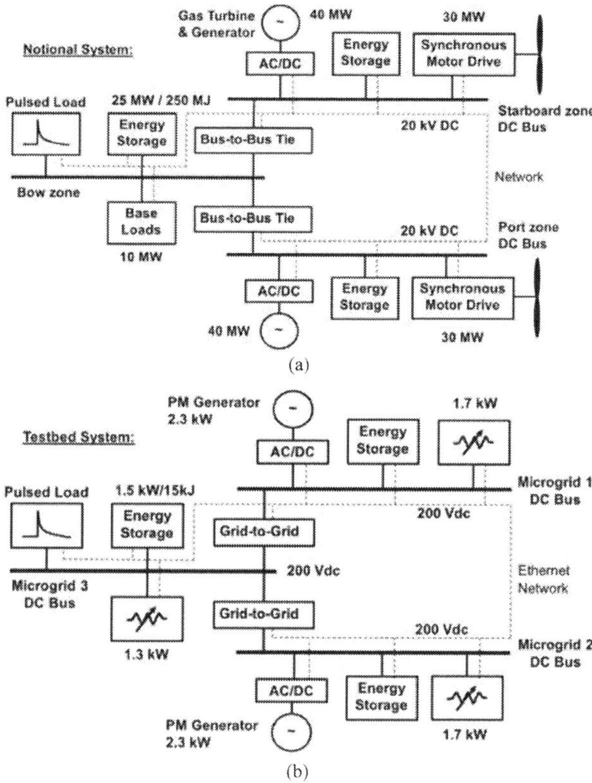

Fig. 3. Illustrates the development of a candidate (a) notional electric ship architecture and (b) possible hardware implementation using SSMTB components

A. *Electromagnetic Launch System (EMALS)*

The EMALS is a major component of the electric ship that will replace more conventional steam powered launch systems. The system must be capable of accelerating an aircraft to a launch speed of up to 103 m/s over the course of a ship's runway, approximately 73 m, [24] no matter the aircraft or how heavily loaded it is. Fig. 4 shows the expected application of force required of an EMALS system reported in [24] and the desired speed profile, which roughly scales with the square root of the position on the runway.

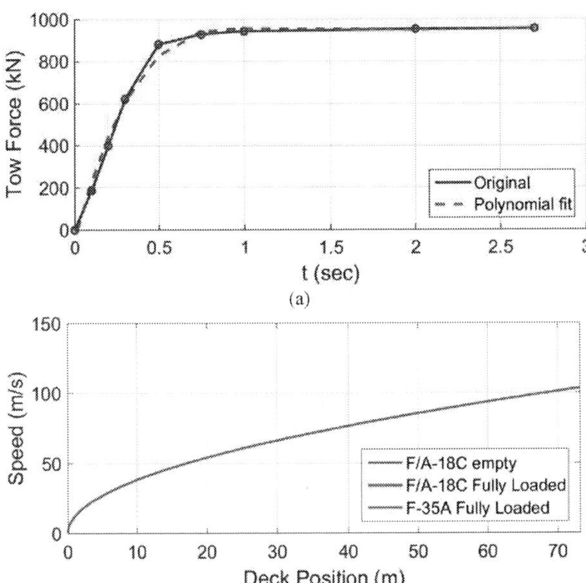

Fig. 4. Shows (top) EMALS force profile (from [24]) and (bottom) desired aircraft speed as a function of deck position

To accomplish this, a feedback controller was implemented to track the speed reference

$$v_{ref}(p) = v_{\max}\sqrt{\frac{p}{L}} \tag{1}$$

where p is the aircraft position on the runway, $v_{\max}$ is the final takeoff speed, and L is the effective length of the runway. The model also accounted for drag, but this was a small component of the force. The model was run for three different aircraft configurations and an end speed of 103 m/s: an unloaded F/A-18C with a mass of 10,400 kg, a fully loaded F/A-18C with a mass of 16,770 kg, and a fully loaded F-35A with a mass of 31,800 kg [25]-[26].

The resulting power expended on each aircraft configuration during a launch is shown in Fig. 5. This graph can be used to derive currents and voltages for modelling the behavior of the EMALS in the electrical system.

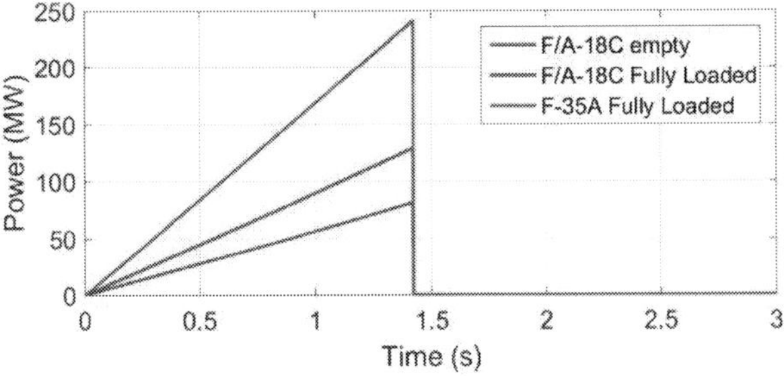

Fig. 5. Power output of EMALS system as a function of time for three different aircraft configurations assuming a take-off speed of 103 m/s

Assuming a relatively steady voltage, the power profile was emulated on the SSMTB using a custom pulsed load assembly that included several resistors connected in parallel through controlled semiconductor switches.

In addition to the EMALS load, the deck of the ship is a dynamic environment; the preparation, loading and movement of aircraft in advance of a launch requires electrical power and is assumed for the purposes of this study to have a similar profile surrounding each launch. Thus a baseload was also defined with a repeated profile.

IV. Power Flow Control Control Strategy

A nonlinear control design architecture based on *Hamiltonian Surface Shaping and Power Flow Control* (HSSPFC) has been employed [16],[27]. The HSSPFC scheme uses a power flow control approach that balances generation and dissipation subject to energy storage (kinetic and potential energies) which define the Hamiltonian for the system. Both static and dynamic stability conditions are determined. The model consists of three microgrid circuit models which are developed in references [20],[27]. Briefly, the model can be defined in matrix form as

$$M\dot{x} = Rx + v + u \tag{2}$$

where the M matrix consists of the passive energy storage elements (inductance, capacitance) of the circuit and the R matrix consists of the resistive elements of the circuit, and x is the system state. The v-vector consists of the general source inputs to the network and the u-vector contains the controller inputs. The u inputs are intended to be actuated by energy storage systems. The R matrix is decomposed further as the sum of a diagonal and skew-symmetric matrix components.

$$M\dot{x} = \left[\overline{R}(R_{load}) + \widetilde{R}(\lambda)\right]x + v + u \tag{3}$$

The vector x is composed of the system state (inductor currents, capacitor voltages). The error state along with the reference control are defined as

$$\Delta x = x_{ref} - x \tag{4}$$

It is assumed that the reference state vector is constant with $\dot{x}_{ref} = 0$, e.g., operating at some desired steady-state condition, and the reference control signal becomes

$$u_{ref} = -\left[\overline{R} + \widetilde{R}\right]x_{ref} - v \quad . \tag{5}$$

This steady-state reference relationship forms the basis from which the feedforward or guidance control stategy is discussed in the next section. Next, based on the error-state the Hamiltonian or energy surface is defined as

$$H = \frac{1}{2}\Delta x^T M \Delta x + \frac{1}{2}\left(\int \Delta x(t)dt\right)^T K_I \left(\int \Delta x(t)dt\right) \quad \forall \Delta x = 0 \tag{6}$$

where the controller integral term provides a control potential energy to help design or shape the energy surface to meet the static stability condition. Note the integral controller diagonal gain matrix K_I is positive definite.

The Hamiltonian time derivative (or power flow) becomes:

$$\dot{H} = \Delta x^T \overline{R} \Delta x + \Delta x^T \Delta u + \Delta x^T K_I \int \Delta x(t) dt \quad (7)$$

where the skew-symmetric portion of the R matrix is zero and the feedback controller is determined by

$$\Delta u = u_{ref} - u \,. \quad (8)$$

In the next step, a Proportional-Integral (PI) controller is proposed as

$$\Delta u = -K_P \Delta x - K_I \int \Delta x(t) dt \,. \quad (9)$$

Substituting and simplifying the Hamiltonian time derivative (power flow) yields

$$\dot{H} = -\Delta x^T \left[K_P - \overline{R}\right] \Delta x < 0 \quad (10)$$

which is the dynamic stability condition. Performance is determined by the selection of the proportional controller diagonal gain matrix K_P, defined as positive definite. The PI feedback controller designs have been integrated into the energy storage systems and the controller gains selected to provide the desired transient performances. The state model formulation in this work is similar to that provided in [20].

V. Guidance Control Strategy

The guidance controller provides regular updates to the power converters in the system. The generators and energy storage communicate current and voltage measurements as well as state-of-charge (SOC) estimates. The guidance controller sends voltage references to the energy storage units which then regulate bus voltage and their own SOC. The guidance controller also sends power commands to the converters that connect generators to the bus. Power requirements are computed using a realtime load estimator, voltage setpoints and energy storage SOC values.

Since the load is assumed to be purely resistive and all bus voltages are set to the same reference value, the load estimator estimates instantaneous load admittance based on bus currents and voltages transmitted to the guidance controller as

$$\hat{G}_{load} \cong N \frac{\sum_{i=1}^{N} \hat{i}_{b,i}}{\sum_{i=1}^{N} \hat{v}_{b,i}} \quad (11)$$

where $i_{b,i}$ is output current into the bus and $v_{b,i}$ is the output voltage at the bus connection of the ith component, N is the number of components, and the hatted quantities denote measured values. The estimated power required by the resistive load is thus

$$\hat{P}_{load} \cong \left(V_b^*\right)^2 \hat{G}'_{load} \quad (12)$$

where V_b^* is the bus voltage setpoint and $\hat{G}'_{load}$ is the filtered load admittance estimate. The total commanded power, P_{tot}^*, is computed by summing the load power requirement, $\hat{P}_{load}$, and the desired charge/discharge power, P_{ES}^*, needed by the various energy storage resources, expressed as

$$P_{tot}^* = \hat{P}_{load} + P_{ES}^* \,. \quad (13)$$

The duty-cycles and commanded currents are computed for the DC-DC boost converters using

$$\lambda_i^* = \frac{1}{2V_b^*}\left(\hat{v}_{si} + \sqrt{\hat{v}_{si}^2 - 4r_{Li}\alpha_i\left(P_{tot}^*\right)}\right) \quad (14)$$

$$i_{Li}^* = \frac{\alpha_i\left(P_{tot}^*\right)}{\lambda_i^* V_b^*} \quad (15)$$

where λ_i^* is the "right-hand side" duty cycle of the boost converter (the duty cycle of the *top* switch), $\hat{v}_{si}$ is the measured boost converter input voltage, r_{Li} is the boost converter input resistance, and i_{Li}^* is the commanded boost converter input current. The converter then uses the servo control to regulate $i_{Li} \rightarrow i_{Li}^*$.

The Guidance Control is depicted in Fig. 6. For the sake of brevity and clarity, we assumed $P_{ES}^* = 0$ in this work, and the SOC management is not applied in the simulation and experimental results herein; however, the signal is still included among the feedback signals shown in Fig. 6 and is expected to be part of future work.

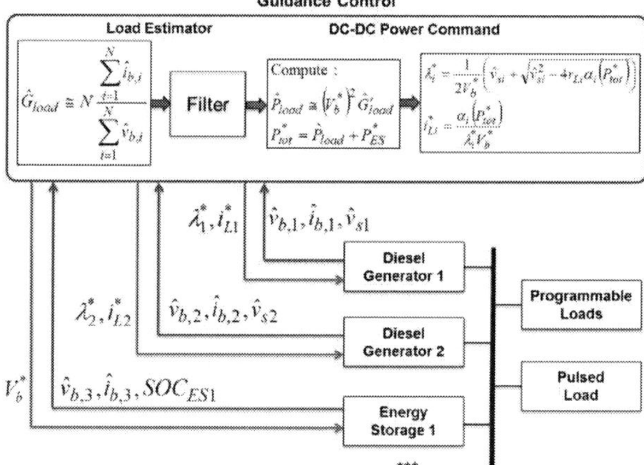

Fig. 6. Illustration of the guidance control strategy

VI. Simulation Results

Simulink simulation was performed on the scaled system in Fig. 3b. The simulations used a first-order continuous-time filter to regulate response of the AC/DC converters connecting diesel generators to the bus. It is noted that although gas turbine models would be preferable in this application, these were not available at the time; so, diesel generator models were used [21]. The system was tested in five simulation experiments with filter time constants of (1) 0.1356, (2) 0.5299, (3) 2.005, (4) 7.512, and (5) 28.56 seconds.

Each simulation experiment ran for 180 seconds. The SSMTB model buses 1 and 2 each had a constant resistance load and a diesel engine with governor set to 950 RPM, the energy storage units regulated bus voltage using Hamiltonian-based control, and power flow through the bus-to-bus converters was controlled based on the AC/DC converter command and relative bus voltage errors. The pulsed load on Microgrid 3 (Bow zone) was varied to represent three simulated EMALS launches; the pulsed load was engaged at t = 20 sec for the F/A-18C empty case, then at t=80 sec for the F/A-18C fully loaded case, and finally at t=140 sec for the F-35A fully loaded case. An additional base load on Microgrid 3 was varied to simulate ancillary electrical systems that support each launch and thus repeated three times. See the bus 3 power profile in Fig. 7.

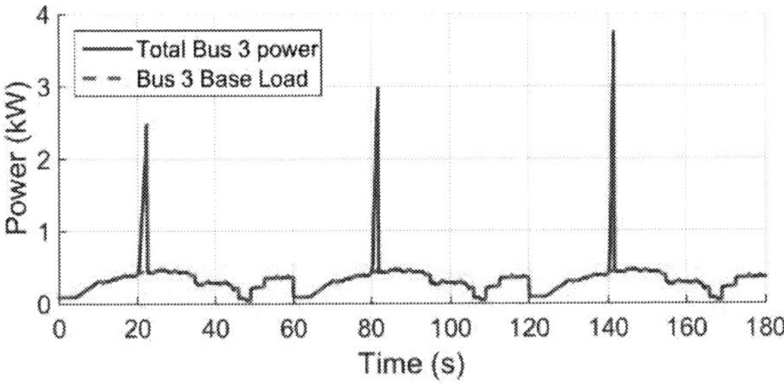

Fig. 7. The bow (Microgrid 3) load profile including the variable bus load and pulsed loads from simulated EMALS launches, scaled down for the SSMTB

In general, when the diesel generator converter output was bandwidth limited, the energy storage units that shared a bus with generation provided more control energy to mitigate high-frequency voltage disturbances which also resulted in a mitigation of generator speed disturbances. Fig. 8 shows the simulated bus currents (positive into the bus) from all the generator converters, energy storage converters, and the pulsed load. The figure shows that the filter does effect the response from the diesel generator converters and the port and starboard energy storage converters as expected. With a smaller time constant, the diesel generators are able to respond more quickly to the pulsed load. With a larger time constant the diesel generator responds to changes more slowly, and the pulsed load must be supplied by the energy storage units that respond to bus voltage error. It is noted that the energy storage responses are greatest in the case with the longest time constant: bow energy storage has a peak value of 5.65 A, and port/starboard energy storages have a peak of 8.11 A. In the case of the shortest time constant, bow energy storage has a peak value of 4.24 A and port/starboard units have a peak value of 1.71 A. The trade-off between generator and energy storage control energies is greatest when they share a bus or are electrically close since the energy storage is responding to bus voltage deviations.

Fig. 9 shows the generator speeds, relative to their 950 RPM reference, for the largest and smallest time constants. For the shortest time constant the generator speeds dip to 917, 913 and 909 RPM. For the 28.56 second time constant, there is far less variation with dips to about 948 RPM for each. This shows that the filter time constant has an appreciable effect on generator speed response.

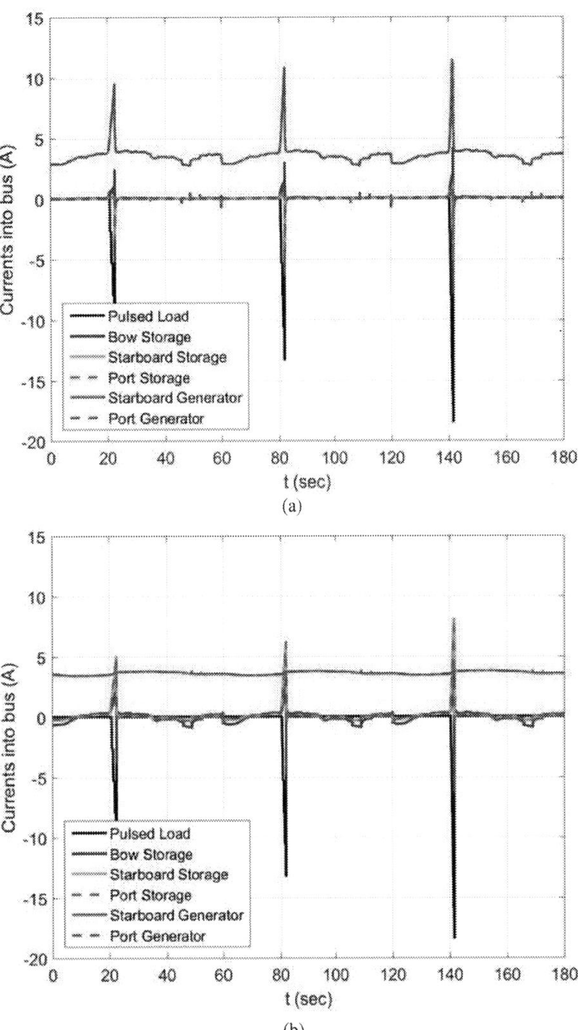

Fig. 8. Shows bus currents from generator-sourced converters and the energy storage emulator for (a) a time constant of 0.1356 seconds and (b) a time constant of 28.56 seconds.

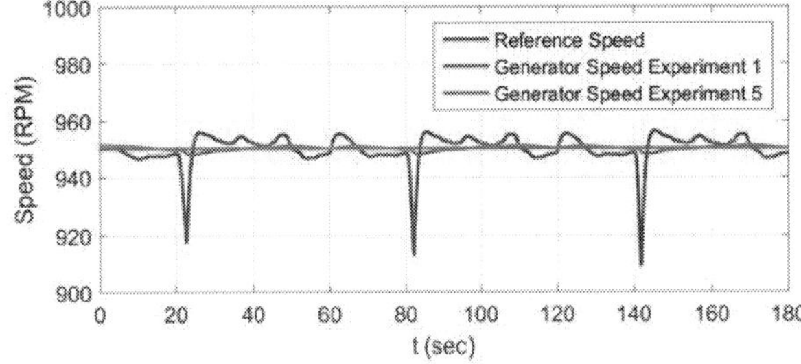

Fig. 9. Shows the speed of the starboard generator as compared to the reference speed for Experiments 1 and 5.

A. Design Trade-off

To help quantify the affect of the filter time constant on performance, a Pareto frontier was identified to investigate the tradeoff between energy storage and generator control effort; the two performance quantities J_1 and J_2 are defined as

$$J_1 = \int_{t_0}^{t_f} \left(\sum_{i}^{N_{Gens}} \left(i_{bi}(\tau) - \hat{i}_{bi} \right)^2 \right) d\tau \quad (16)$$

$$J_2 = \int_{t_0}^{t_f} \left(\sum_{i}^{N_{ES}} \left(i_{ESi}(\tau) - \hat{i}_{ESi} \right)^2 \right) d\tau \quad (17)$$

where $i_{bi}(t)$ are the currents delivered to the respective busses by the starboard and port generator converters as a function of time, N_{Gens} is the number of generators, $i_{ESi}(t)$ are the bus currents from the N_{ES} energy storage systems, and the hatted quantities represent average currents over the interval $t \in [t_0, t_f]$ and are constant in (16) and (17). Fig. 10 shows the Pareto frontier between generator control effort and energy storage control effort for the simulated system with two generators and three energy storage systems. There is a clear trade-off between the two values with a nearly linear relationship.

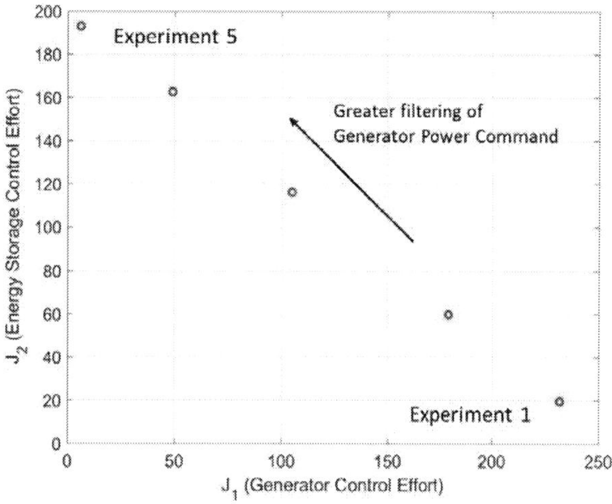

Fig. 10. Simulated trade space (Pareto Frontier) illustrating trade-off between generator control effort and energy storage control effort

VII. Experimental Results

As in simulation, the control was evaluated in a series of five hardware experiments that utilized select components from the testbed system in Fig. 3b at a reduced voltage of 150 V and generator speed of 650 RPM. The system configuration used for the hardware experiments is illustrated in Fig. 11. This system had fewer components than what was modeled in the simulation, but the results are consistent.

A. Experiment Description

As with the simulation, each experiment ran for 180 seconds with both generators engaged, the propulsion load was represented as a constant resistance, the variable load on Microgrid 3 (Bow zone) was varied according to a predesignated profile to represent three simulated EMALS launches and a varying base load was implemented on the SSMTB bus 3. The power profile shown in Fig. 7 was used to program the pulsed load and programmable load (scaled down at lower voltage). In each experiment, the generator converter power command was filtered by a digital low-pass filter with different equivalent time constants, corresponding to those used in simulation, before being transmitted to the generators by the Guidance controller. In Experiment 1, the fast response of the power electronics effectively supplied the pulse energy by extracting that energy from the generator inertia, and the energy storage control effort was minimized. Experiment 5 had the largest time constant; in this case, the generators provide a more averaged power, and the pulsed load is supplied primarily by the energy storage system. In this way, the control effort priority could be selected by the user and regulated by the Guidance controller through simple separation of time scales.

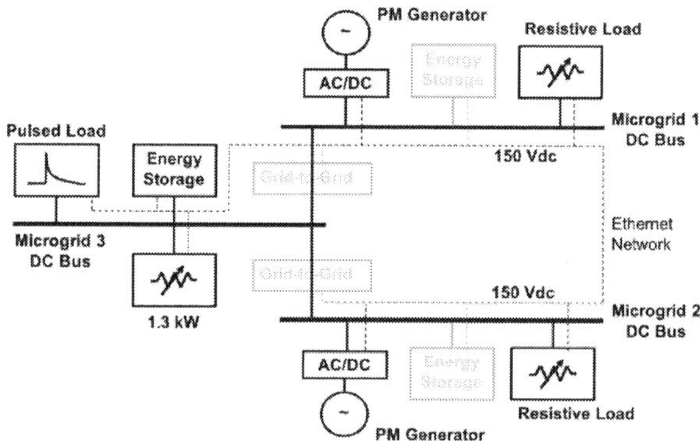

Fig. 11. Illustration of the dispatch control strategy used controlling the diesel generators. The faded components were not connected.

B. Experimental Results

Fig. 12 illustrates the component currents and bus voltages. Fig 12a shows the two generator currents, the energy storage current and the pulsed load current for Experiment 1. Therein, it is noted that the energy storage current is minimal, providing some current for the base load and when the pulsed load is active. The majority of the baseload power and pulsed load are tracked by the generators, which effectively comes from energy stored in the generator inertia. Fig 12b illustrates the Experiment 5 results; therein, the generator currents vary slowly, and the baseload and pulsed power demands are tracked by the energy storage. Fig. 12c shows the bus voltage, indicating that the system achieves very good voltage regulation despite the highly variable load. This is due to the fact that the energy storage directly regulates bus voltage; this is achievable because the guidance control system coordinates the energy resources rather than relying on droop control to balance the contributions from different generators. Both voltages have a mean of 148 V. The worst-case voltage deviation occurs for the 0.1356 sec time constant; the voltage spikes to 156.3 V (5.6% error) and immediately returns to steady state within 125 msec.

Fig. 13 shows generator speed transients (blue traces) relative to the speed reference (green traces) for Experiments 1 and 5 respectively. Therein, it is noted that the Experiment 1 case results in greater deviation in generator speed during a transient, including nadirs of 631 RPM, 620 RPM and 618 RPM, or 2.9%, 4.6% and 4.9% deviation. In addition, some variation is seen in the generator speed due to the variable baseload. As the load reference used to command generator power is filtered, the energy storage provides a greater share of the pulsed power, resulting in less generator deviation. Fig. 13b shows the generator speed nadirs at 642, 642, and 637 RPM respectively, with less variation seen due to base load as well.

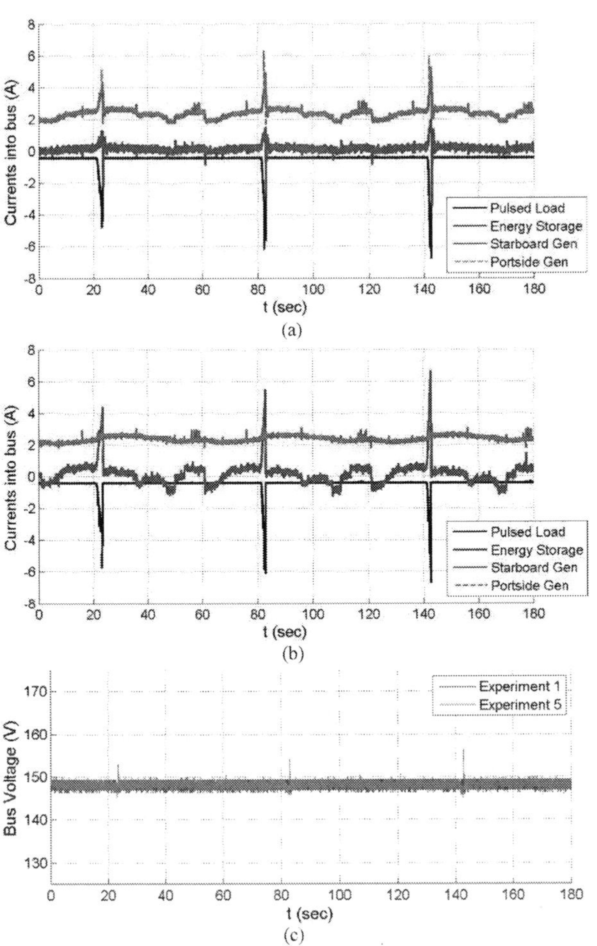

Fig. 12. Measured bus currents for (a) Experiment 1 and (b) Experiment 5 and (c) bus voltages for each

The resulting Pareto frontier obtained from the hardware experiment, comparing control effort cost from the two generators and one energy storage emulator, is shown in Fig. 14. The hardware results are intuitive and consistent with simulation; the Pareto frontier for the two cost measures shows a clear trade-off with an almost linear frontier.

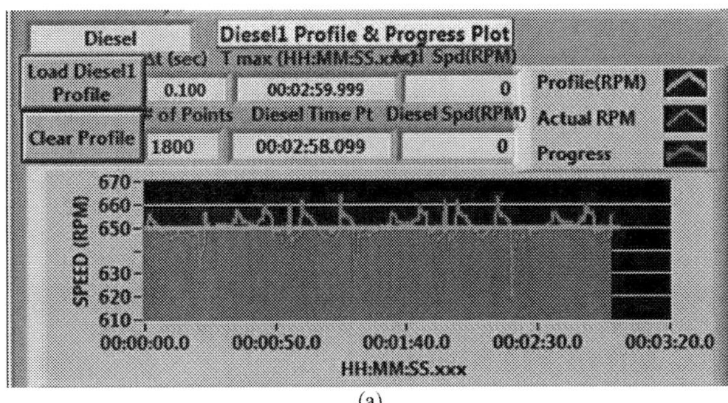

(a)

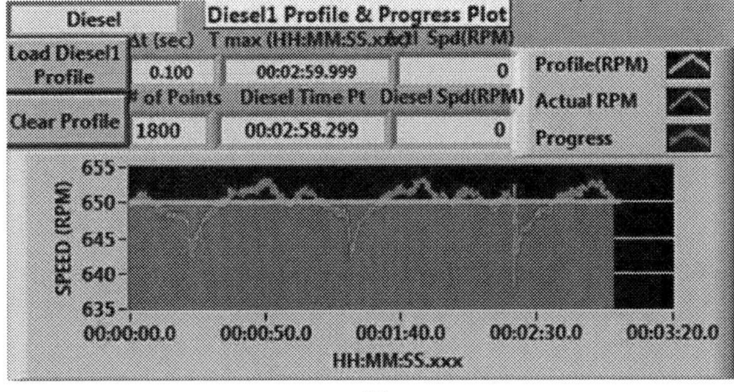

(b)

Fig. 13. Shows master control console close-up of the generator 1 speed deviations relative to the reference for (a) Experiment 1 and (b) Experiment 5.

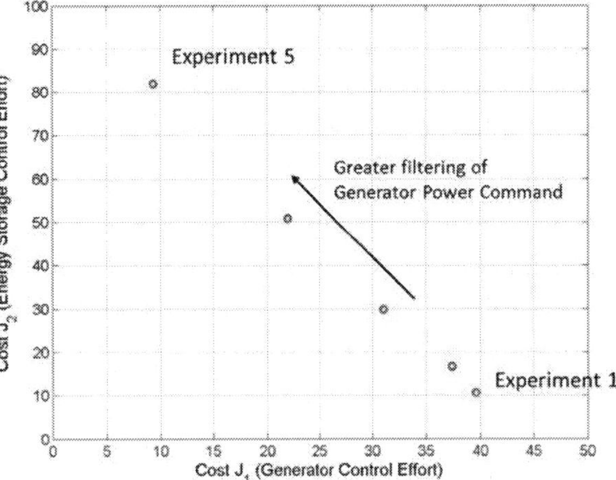

Fig. 14. Trade space (Pareto Frontier) illustrating trade-off between generator control effort and energy storage control effort

VIII. Conclusions

In this paper, a hierarchical control, developed for use in networked microgrids, was applied to the problem of coordinating energy sources and energy storage onboard an all-electric ship. The problem of pulsed load management, in particular the electromagnetic launch system (EMALS), was addressed. A notional electric ship model was emulated at laboratory scale by the SSMTB, including an EMALS emulator. A method of control presented in previous works [16]-[20] was modified to include a simple guidance control function for power flow control and evaluated in simulation and in hardware to mitigate disturbances caused by EMALS power consumption during launch. This guidance control function is capable of designating where the pulsed power is supplied (energy storage units or generator inertia) through manipulation of a filter time constant. The trade-off in control effort is easily visualized and intuitive. The results suggest that the Pareto frontier may be navigated very easily by the Guidance control by simply manipulating a filter time constant rather than computationally-intensive time-domain dynamic optimization planning [20]. For systems with pulsed loads, this allows for fast on-the-fly changes to power flow control in response to dynamic time-changing priorities.

ACKNOWLEDGMENTS

This work was supported by NAVSEA for a project entitled *Nonlinear Power Flow Control Design for NGIP Energy Storage Requirements*, PR# 1400354102.

The authors wish to thank Forest White, Jazmin Pedroza, Michael Horry, Joseph Rudys, Peter Foster, and John Brown for their contributions to the hardware testbed.

Sandia National Laboratories is a multi-program laboratory managed and operated by Sandia Corporation, a wholly owned subsidiary of Lockheed Martin Corporation, for the U.S. Department of Energy's National Nuclear Security Administration under contract DE-AC04-94AL85000.

REFERENCES

[1] D. Schneider; "The Electric Warship"; IEEE Spectrum; July 2013. URL: http://spectrum.ieee.org/aerospace/military/the-electric-warship

[2] Kulkarni, S.; Santoso, S., "Impact of pulse loads on electric ship power system: With and without flywheel energy storage systems," *Electric Ship Technologies Symposium, 2009. ESTS 2009. IEEE*, pp.568-573, 20-22 April 2009.

[3] Domaschk, L.N.; Ouroua, A.; Hebner, R.E.; Bowlin, O.E.; Colson, W.B., "Coordination of Large Pulsed Loads on Future Electric Ships," *Magnetics, IEEE Transactions on*, vol.43, no.1, pp.450-455, Jan. 2007.

[4] Webb, T.W.; Kiehne, T.M.; Haag, S.T., "System-Level Thermal Management of Pulsed Loads on an All-Electric Ship," *Magnetics, IEEE Transactions on*, vol.43, no.1, pp.469-473, Jan. 2007

[5] Crider, J.M.; Sudhoff, S.D., "Reducing Impact of Pulsed Power Loads on Microgrid Power Systems," *Smart Grid, IEEE Transactions on*, vol.1, no.3, pp.270-277, Dec. 2010.

[6] Mitra, P. ; Venayagamoorthy, G.K., "An Adaptive Control Strategy for DSTATCOM Applications in an Electric Ship Power System," *Power Electronics, IEEE Transactions on*, vol.25, no.1, pp.95-104, Jan.2010

[7] Woodruff, S.L.; Li Qi; Sloderbeck, M.J., "Hardware-in-the-Loop Experiments on the Use of Propulsion Motors to Reduce Pulse-Load System Disturbances," *Electric Ship Technologies Symposium, 2007. ESTS '07. IEEE*, 21-23 May 2007.

[8] Salehi, V.; Mirafzal, B.; Mohammed, O., "Pulse-load effects on ship power system stability," *IECON 2010 - 36th Annual Conference on IEEE Industrial Electronics Society*, pp.3353-3358, 7-10 Nov. 2010.

[9] Doyle, M.R.; Samuel, D.J.; Conway, T.; Klimowski, R.R., "Electromagnetic aircraft launch system-EMALS," *Magnetics, IEEE Transactions on*, vol.31, no.1, pp.528-533, Jan. 1995.

[10] Steurer, M.; Andrus, M.; Langston, J.; Qi, L.; Suryanarayanan, S.; Woodruff, S.; Ribeiro, P.F., "Investigating the Impact of Pulsed Power Charging Demands on Shipboard Power Quality," *Electric Ship Technologies Symposium, 2007. ESTS '07. IEEE*, pp.315-321, 21-23 May 2007

[11] Langston, J.; Suryanarayanan, S.; Steurer, M.; Andrus, M.; Woodruff, S.; Ribeiro, P.F., "Experiences with the simulation of a notional all-electric ship integrated power system on a large-scale high-speed electromagnetic transient simulator," *Power Engineering Society General Meeting, 2006.*

[12] Balathandayuthapani, S.; Edrington, C.S.; Henry, S., "Study on converter topologies for capacitive pulse forming network and energy storage units in electric ship," *Electric Ship Technologies Symposium (ESTS), 2011 IEEE*, pp.459-462, 10-13 April 2011

[13] Jizhu Liu; Shuanghui Hao; Zili Tang; Minghui Hao, "Autonomous control of electromagnetic aircraft launch system based on decentralized architecture," *Robotics and Biomimetics, 2008. ROBIO 2008. IEEE International Conference on*, pp.2089-2092, 22-25 Feb. 2009

[14] Domaschk, L.N.; Ouroua, A.; Hebner, R.E.; Bowlin, O.E.; Colson, W.B., "Coordination of Large Pulsed Loads on Future Electric Ships," *Magnetics, IEEE Transactions on*, vol.43, no.1, pp.450-455, Jan. 2007

[15] Mitra, P.; Venayagamoorthy, G.K., "A DSTATCOM controller tuned by Particle Swarm Optimization for an Electric Ship Power System," *Power and Energy Society General Meeting - Conversion and Delivery of Electrical Energy in the 21st Century, 2008 IEEE*, pp.1-6, 20-24 July 2008

[16] Robinett, R.D.; Wilson, D.G.; Nonlinear Power Flow Control Design: Utilizing Exergy, Entropy, Static and Dynamic Stability, and Lyapunov Analysis; Springer-Verlag, London, 2011.

[17] Robinett, R.D.; Wilson, D.G.; "Nonlinear power flow control design for combined conventional and variable generation systems: Part I-theory," *Control Applications (CCA), 2011 IEEE International Conference on*, pp.61-64, 28-30 Sept. 2011.

[18] Wilson, D.G.; Robinett, R.D.; "Transient stability and performance based on nonlinear power flow control design of renewable energy systems," *Control Applications (CCA), 2011 IEEE International Conference on*, pp.881-886, 28-30 Sept. 2011.

[19] Wilson D. G., Robinett III R. D., Goldsmith S. Y. "Renewable energy microgrid control with energy storage integration," *International Symposium on Power Electronics, Electrical Drives, Automation and Motion (SPEEDAM)*, June 20th-22nd, 2012, Sorrento, Italy.

[20] Wilson, D.G.; Neely, J.C.; Cook, M.A.; Glover, S.F.; Young, J.; Robinett, R.D., "Hamiltonian control design for DC microgrids with stochastic sources and loads with applications," *Power Electronics, Electrical Drives, Automation and Motion (SPEEDAM), 2014 International Symposium on*, pp.1264-1271, 18-20 June 2014.

[21] S. Glover, J. Neely, A. Lentine, J. Finn, F. White, P. Foster, O. Wasynczuk, S. Pekarek, B. Loop, "Secure Scalable Microgrid Test Bed at Sandia National Laboratories," IEEE Cyber 2012 Conference, May 27th-31st, 2012, Bangkok, Thailand.

[22] J. Neely, S. Pekarek, S. Glover, J. Finn, O. Wasynczuk, and B. Loop, "An economical diesel engine emulator for micro-grid research," *International Symposium on Power Electronics, Electrical Drives, Automation and Motion (SPEEDAM)*, June 20th-22nd, 2012, Sorrento, Italy.

[23] J. Neely, S. Glover, O. Wasynczuk, B. Loop, "Wind turbine emulation for intelligent microgrid development," IEEE Cyber 2012 Conference, May 27th-31st, 2012, Bangkok, Thailand.

[24] Doyle, Michael R.; Samuel, Douglas J.; Conway, Thomas; Klimowski, Robert R; "Electromagnetic Aircraft Launch System – EMALS," *IEEE Transactions on Magnetics*, Vol. 31, No. 1, January, 1995.

[25] URL: https://en.wikipedia.org/wiki/McDonnell_Douglas_F/A-18_Hornet

[26] URL: https://en.wikipedia.org/wiki/Lockheed_Martin_F-35_Lightning_II

[27] Weaver, W.W., Robinett III, R.D., Parker, G.G., and Wilson, D.G., Distributed Control and Energy Storage Requirements of Networked DC Microgrids, Control Engineering Practice 44 (2015), pg.10-19,June 2015.

Reduced Active Switch AC to DC Rectifier with High Frequency Isolation for Electric Vehicle Chargers

José Juan Sandoval
Student Member, IEEE

Taeyong Kang
Student Member, IEEE

Prasad Enjeti
Fellow, IEEE

Department of Electrical & Computer Engineering, Texas A&M University, College Station, USA
email: enjeti@tamu.edu

***Abstract*— This paper presents a reduced active switch count push-pull based DC fast charger for electric vehicles. The proposed system takes a 3-phase utility line voltage and produces a high frequency (HF) AC link employing only two active switches. A 5-limb HF transformer, in zig-zag connection, is employed for isolation. The primary side of the transformer is connected to a clamp circuit and two active semiconductor devices which provide a 3-phase HF AC link. The secondary side of the transformer is connected to a twelve pulse rectifier to achieve AC to DC conversion. The output of the rectifier is then interfaced to an electric vehicle's battery bank. The advantages of the proposed system include low active switch count, simple control strategy, and high input current quality. Furthermore, employing zig-zag connection in the secondary side helps to balance transformer leakage inductance. A design example rated at 50 kW charging a 500V battery is presented. Mathematical analysis and simulation results demonstrate the performance of the proposed system. Experimental results, with a medium frequency zig-zag transformer, at 1.6 kW validate the operation of the system.**

Keywords—DC fast charger; zig zag; high frequency; electric vehicle;

I. INTRODUCTION

Recently, the automobile transportation sector has seen an increased production and sales of plug-in electric (PEV) and electric vehicles (EV) [1]. In fact, one goal of the United States is to have one million EVs on the roads by the end of 2015 [2]. The success of this goal is dependent upon improvements in battery technology, electric drivetrain systems and battery charging infrastructure. Furthermore, mass adoption of this technology is dependent upon alleviating public concerns such as range anxiety and charging times [3, 4]. A strategy to alleviate such concerns is the deployment of a public charging infrastructure with Level 3 DC battery fast chargers. Such DC fast chargers, which add 60 to 80 miles of driving range with a 20 minute charge [5], must include the following features: high efficiency, high input current quality, high power density, and cost-effectiveness [6].

A conventional unidirectional topology for EV DC Fast Charging employs a 3-phase diode rectifier at the input followed by an isolated DC/DC converter as shown in Fig. 1. In this topology, the line input current has 5th and 7th harmonics resulting in poor power factor operation. A DC fast charging topology which employs parallel power processing is proposed in [7]. However, this topology still suffers from a low input current quality.

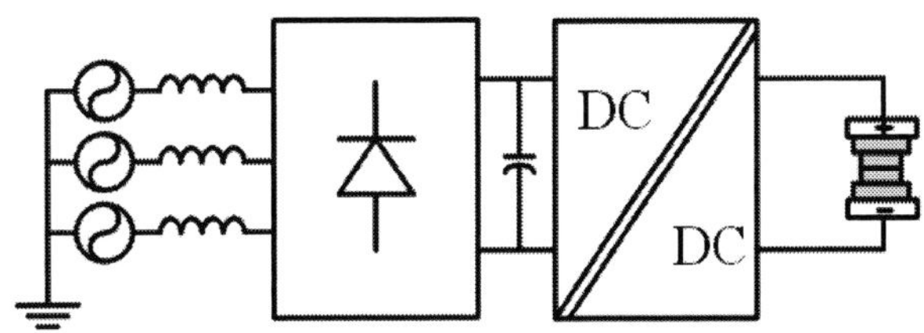

Fig. 1: Conventional topology for EV DC Fast Charging.

Various topologies with improved power factor are presented in [8-10] but often employ a large active switch count and require complex control and modulation strategies. Matrix converter based topologies are proposed for bidirectional battery fast chargers in [11, 12]. Although these topologies have good input current quality and have the capability of vehicle-to-grid (V2G) operation, they also use a relatively high number of active semiconductor devices and require complex control strategies.

The DC battery fast charger proposed in this paper, as shown in Fig. 2, addresses the existing drawbacks by introducing a push-pull based topology with a reduced active switch count and simple control strategy. The proposed system has the following advantages:

- Only two active switches are employed to perform 3-phase AC-DC rectification.
- Using a high frequency (HF) transformer increases the power density of the system while providing galvanic isolation [13, 14].
- Zig-zag winding connection helps to balance leakage inductance in the transformer's secondary side.
- Simple square wave modulation (variable duty cycle) control strategy.
- The topology can be extended to use higher pulse rectifiers on the secondary side.

978-1-4673-9551-9/16 $31.00 © 2016 IEEE

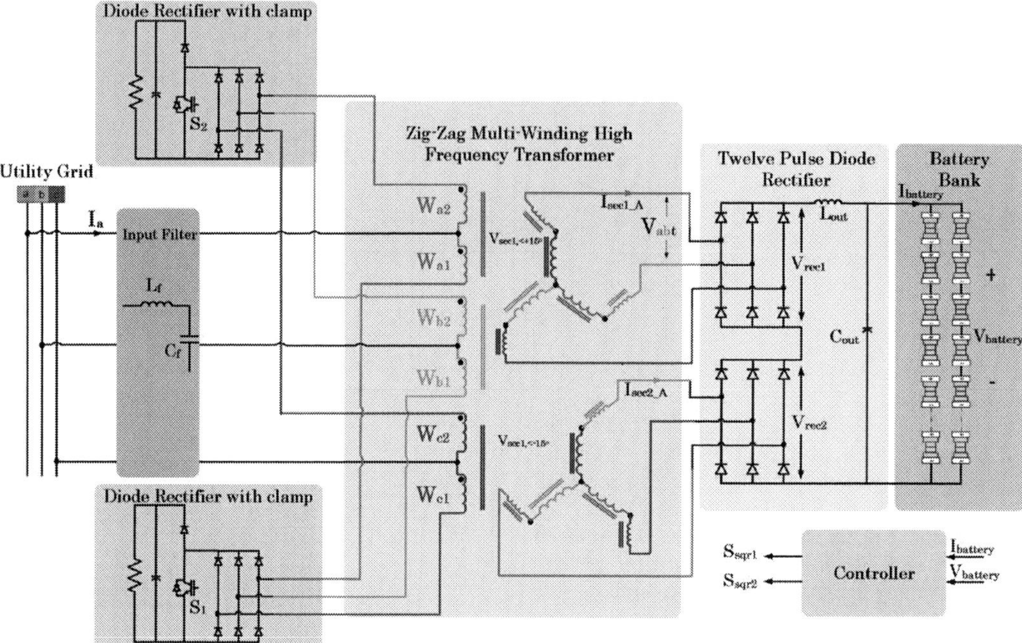

Fig. 2: Proposed push-pull based topology with high frequency isolation for EV DC fast charging. Active semiconductor devices S_1/S_2 create a 3-phase AC link across the transformer windings. The transformer has 3 primary windings (with center tap) and twelve secondary windings.

II. Proposed Electric Vehicle charger with High Frequency Isolation

The proposed topology for EV DC fast charging is shown in Fig. 2. The operation of the system can be divided into the following subsections: (A) diode rectifiers with clamp circuit, (B) high frequency zig-zag transformer, (C) twelve pulse diode rectifier and (D) modulation and control scheme.

A. Diode Rectifier with Clamp Circuit

The creation of the 3-phase AC link across the transformer windings is achieved by switching S_1 and S_2 [15, 16]. The primary windings of the zig-zag transformer can be divided into two sets that are 180° phase shifted in magnetic coupling, namely windings (W_{a1}, W_{b1}, W_{c1}) and windings (W_{a2}, W_{b2}, W_{c2}). As shown in Fig. 2, the center tap of each primary winding is connected to the utility grid through an input filter. In addition, the switching terminals of windings (W_{a1}, W_{b1}, W_{c1}) are connected to a diode rectifier whose output is in turn connected to S_1. Similarly, the switching terminals of windings (W_{a2}, W_{b2}, W_{c2}) are connected to a diode rectifier whose output is in turn connected to S_2.

The active devices S_1 and S_2 operate with simple square wave modulation. The switching function of S_2 has a 180° phase shift compared to the switching function of S_1. When S_1 is gated ON, the switching terminals of windings (W_{a1}, W_{b1}, W_{c1}) are shorted through the diode rectifier. In essence, the switching terminals are shorted to the neutral point as shown in Fig. 3. At this instant, the line-to-neutral voltages V_{an}, V_{bn}, and V_{cn} appear across windings W_{a1}, W_{b1}, and W_{c1} respectively. The switching terminals of the other set of windings (W_{a2}, W_{b2}, W_{c2}) are open and the energy stored in their leakage inductance is dissipated through the capacitor and bleeding resistor. The voltages across windings (W_{a2}, W_{b2}, W_{c2}) have the opposite polarity compared to the voltages across windings (W_{a1}, W_{b1}, W_{c1}) because they are 180° in magnetic coupling. Meanwhile, the induced voltages on the secondary side have the same polarity as the voltages across windings (W_{a1}, W_{b1}, W_{c1}).

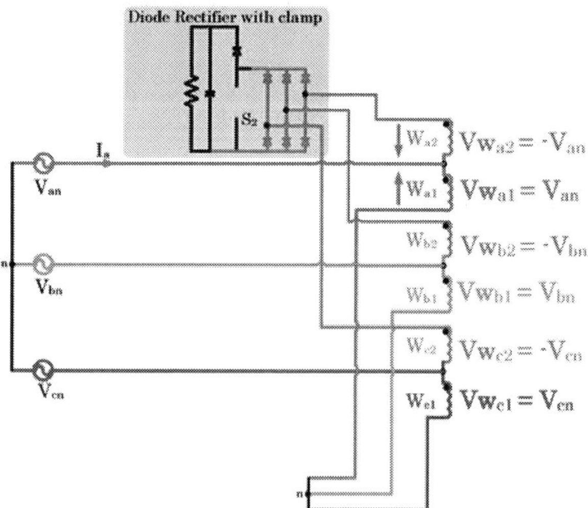

Fig. 3: Operation when S_1 is gated ON and S_2 is gated OFF. The switching terminals of W_{a1}, W_{b1}, W_{c1} are shorted while the switching terminals of W_{a2}, W_{b2}, W_{c2} are open.

When S_2 is gated ON, the switching terminals of windings (W_{a2}, W_{b2}, W_{c2}) are shorted and connected to the neutral point as shown in Fig. 4. At this instant, the line-to-neutral voltages V_{an}, V_{bn}, and V_{cn} appear across windings W_{a2}, W_{b2}, and W_{c2} respectively. The switching terminals of the other set of windings (W_{a1}, W_{b1}, W_{c1}) are open and the energy stored in their leakage inductance is dissipated through the capacitor and bleeding resistor. Now the voltages across windings (W_{a1}, W_{b1}, W_{c1}) and the induced voltages on the secondary side have opposite polarity compared to the utility grid line-to-neutral voltages. It can be noted that the voltage polarity across each winding changes as S_1 and S_2 are switched. By switching S_1 and S_2 at HF a 3-phase AC link is created.

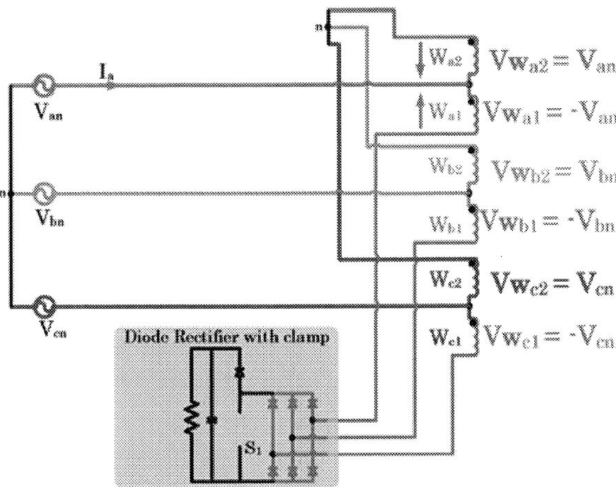

Fig. 4: Operation when S_1 is gated OFF and S_2 is gated ON. The switching terminals of W_{a2}, W_{b2}, W_{c2} are shorted while the switching terminals of W_{a1}, W_{b1}, W_{c1} are open.

Mathematically speaking, the 3-phase AC link is simply the multiplication of the line-to-neutral voltages with a square wave switching function. Fig. 5(b) shows the resulting HF AC link across winding W_{a1} when the system operates at 50% duty cycle. With a line-to-neutral voltage as in (1) and a square wave switching function described by (2), the resulting voltage across winding W_{a1} can be expressed as in (3). It can be noted that the frequency of the square wave switching function determines the fundamental frequency of the AC link created across the two sets of windings. The voltages across windings W_{b1} and W_{c1} have a similar expression as in (3) but are 120° and 240° phase shifted respectively. The expression at (3) is valid only for 50% duty cycle operation.

$$V_{an} = \sqrt{\frac{2}{3}} V_{LL} \sin(\omega_s t) \tag{1}$$

$$S_{sqr} = \frac{4}{\pi} \sum_{n=1,3,5,\dots}^{\infty} \frac{1}{n} \sin(n\omega_{sqr} t) \tag{2}$$

$$V_{Wa1} = \sqrt{\frac{2}{3}} V_{LL} \sum_{n=1,3,5,7,\dots}^{\infty} \frac{2}{n\pi} \sin(\{n\omega_{sqr} \pm \omega_s\} \cdot t) \tag{3}$$

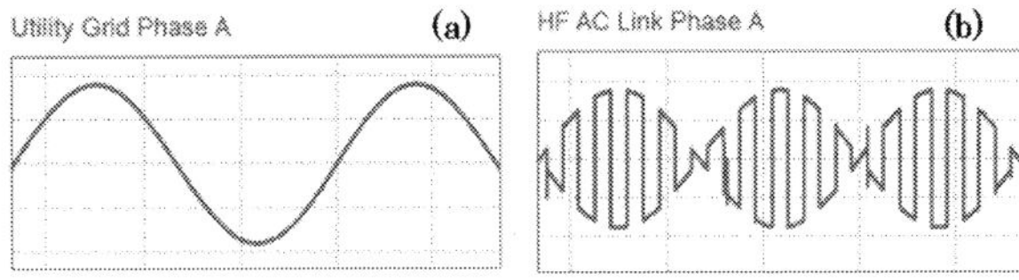

Fig. 5: (a) Utility line-to-neutral input voltage V_{an}; (b) HF AC link created by switching S_1 and S_2 at 50% duty cycle operation.

B. High Frequency Zig-Zag Transformer

The operation of a medium frequency zig-zag transformer is described in [16]. For this EV charging application, a HF transformer zig-zag arrangement is considered. As in [16], the switching frequency of S_1 and S_2 determines the transformer's frequency of operation. Operating the transformer at HF reduces its core size [13], thus increasing the system's power density. Due to its relative low losses at the kHz switching frequency range [17], a ferrite core material is envisioned for the HF transformer. The zig-zag multi-winding transformer can be built using three 1-phase multi-winding transformers or it can be built using a single 3-phase multi-winding transformer. To achieve a more compact design, a single 3-phase, 5-limb, transformer is proposed for this system as shown in Fig. 6. The primary and secondary windings are wound around the interior three limbs of the transformer. The outer limbs provide a magnetic path for any residual flux in the transformer which helps to avoid magnetic core saturation.

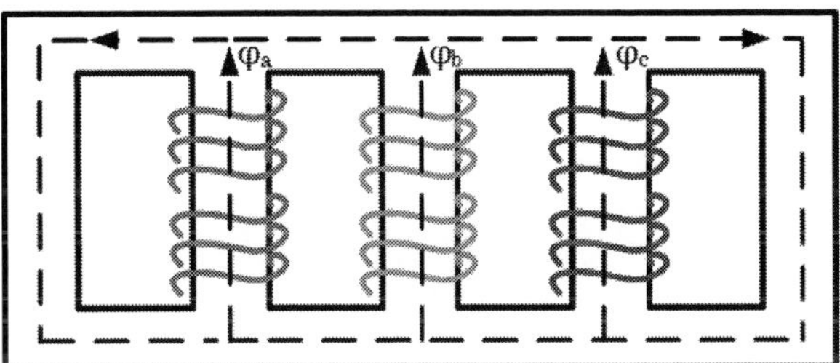

Fig. 6: Multi-winding 3-phase, 5-limb, transformer

In conventional 12-pulse operation, multi-winding transformers in star-delta connection are often employed to create a 30° phase shift. In order for the two output diode rectifiers to process the same amount of power, the delta connected windings must have a higher number of turns compared to the star connected windings. This difference in turns ratio creates different leakage inductances on the secondary side. To mitigate this problem, the secondary side of the HF transformer in the proposed system is connected in zig-zag. With zig-zag arrangement the leakage inductances on the secondary side are balanced. The secondary side windings are connected in such a way that two sets of 3-phase voltages with a net 30° phase difference are fed to the 12-pulse diode rectifier. One set of 3-phase voltages is displaced by +15° with respect to the primary windings, while the second set of 3-phase voltages is displaced by -15° with respect to the primary windings. A vector diagram of the line-to-line voltages is shown in Fig. 7.

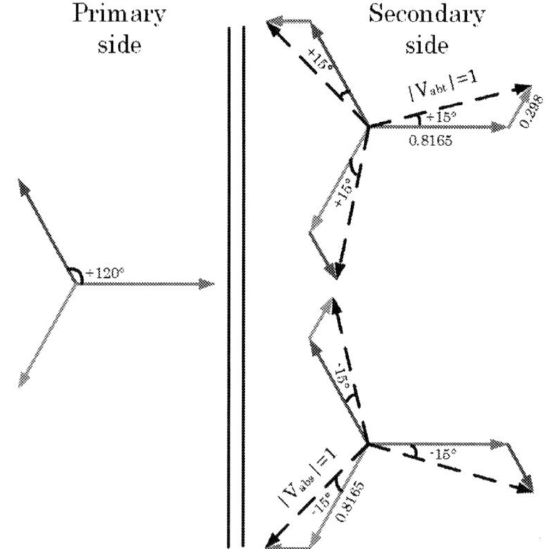

Fig. 7: Vector diagram to obtain two sets of 3-phase voltages, in the secondary side, with a net 30° phase shift.

The magnitudes of the vectors on the secondary side correspond to the turns ratio of the secondary windings. In order to achieve line-to-line vectors which have a unity magnitude, relative to the utility grid line-to-line voltage, and with a ±15° phase difference the transformer' turns ratio must be set as in (4). The turns ratio of the secondary windings can be found using phasor operations as described in [16]. For each phase, there is a primary winding with center-tap and four secondary windings. This accounts for a total of three primary windings with center-tap and 12 secondary windings.

$$N_{P1}:N_{P2}:N_{S1}:N_{S2}:N_{S3}:N_{S4}= \\ 1:1:0.298:0.8165:0.298:0.8165 \quad (4)$$

C. Twelve Pulse Diode Rectifier

The voltages across the zig-zag secondary windings are fed to the output 12-pulse diode rectifier. As shown in Fig.2, the set of voltages (V_{abt}, V_{bct}, V_{cat}) with +15° phase difference with respect to the primary windings are fed to one of the output 3-phase diode rectifiers. The other set of voltages (V_{abs}, V_{bcs}, V_{cas}), which are -15° in phase shift with respect to the primary windings, are fed to the second 3-phase diode rectifier. The output voltages of the diode rectifiers, V_{rec1} and V_{rec2}, are displaced by a net 30° with respect to each other producing a smooth 12-pulse DC output voltage. The output DC voltage is then interfaced to the EV battery bank. The operation of the 12-pulse diode rectifier is similar to the conventional line frequency 12-pulse rectifier. However, the diodes in this system must be able to operate at HF to avoid reverse recovery switching loses. Silicon carbide diodes, which are fast switching and have no reverse recovery current, can be used for this application [18].

The output DC battery voltage of the twelve pulse diode rectifier can be expressed as in (5), where the line-to-line voltage $V_{abt,rms}$ depends on the duty cycle of active devices S_1 and S_2. Therefore the battery voltage can be regulated by varying the duty cycle of S_1 and S_2.

$$V_{battery} \approx \frac{6\sqrt{2}}{\pi} V_{abt,rms} \quad (5)$$

D. Modulation and Control Scheme

As stated in the previous sub-section, the output DC voltage, which is fed to the battery bank, is controlled by varying the duty cycle of S_1 and S_2. The duty cycle, D, can be varied theoretically from 0-50%. Active switch S_1 operates at duty cycle D and at frequency f_{sqr}, its switching function S_{sqr1} can be observed in Fig. 8(a). Similarly, active switch S_2 also operates at the same duty cycle and frequency but its switching function S_{sqr2} is phase shifted by 180° as shown in Fig. 8(b). The overall switching function of the proposed system is the subtraction of S_{sqr1} and S_{sqr2} and can be appreciated in Fig. 8(c). Operating at duty cycle less than 50% introduces zero states in the HF AC link as shown in Fig. 8(e), decreasing the overall output DC voltage. Therefore, maximum DC output voltage is obtained at 50% duty cycle operation. If the duty cycle is increased beyond 50%, catastrophic short circuits occur across the transformer windings. Consequently, the modulation becomes simple (variable duty cycle operation) and robust; in contrast to other systems that employ complicated modulation strategies.

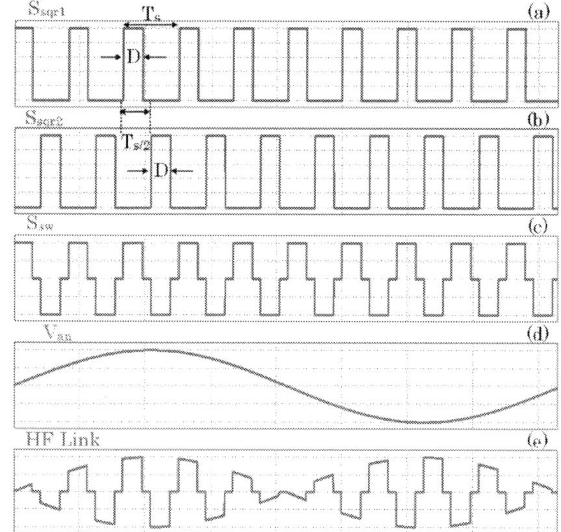

Fig. 8: (a) S_{sqr1}, switching function for S_1; (b) S_{sqr2}, switching function for S_2, it is 180° phase shifted compared to S_{sqr1}; (c) S_{sw}, overall system switching function; (d) V_{an}, input line-to-neutral voltage; (e) HF AC Link, multiplication of S_{sw} and V_{an}.

The switching function S_{sqr1} can be expressed as in (6); the expression for S_{sqr2} is similar to (6) but is phase shifted by 180°. The overall system's switching function is given by (7). When the line-to-neutral input voltage described by (1) is multiplied to S_{sw}, the resulting voltage across winding W_{a1} is given by (8). From (8), it can be noted that the rms voltage across the transformer windings depends on the duty cycle D.

$$S_{sqr1} = D + \frac{2}{\pi}\sum_{n=1}^{\infty}\frac{1}{n}\sin(n\pi D)\cdot\sin(n\omega_{sqr}t) \tag{6}$$

$$S_{sw} = \frac{4}{\pi}\sum_{n=1,3,5,7,...}^{\infty}\frac{1}{n}\sin(n\pi D)\cdot\sin(n\omega_{sqr}t) \tag{7}$$

$$V_{Wa1} = \sqrt{\frac{2}{3}}V_{LL}\sum_{n=1,3,5,7,..}^{\infty}\frac{2}{n\pi}\sin(n\pi D)\sin(\{n\omega_{sqr}\pm\omega_s\}\cdot t) \tag{8}$$

The battery voltage and battery current can be regulated using a closed loop control scheme. Fig. 9 shows a block diagram of the control scheme. There are two controllers; a voltage controller and a current controller. The proportional plus integral (PI) current controller determines the duty cycle of active switches S_1 and S_2. The switching controller then creates the 180° phase delay between the switching functions S_{sqr1} and S_{sqr2} as shown in Fig. 8.

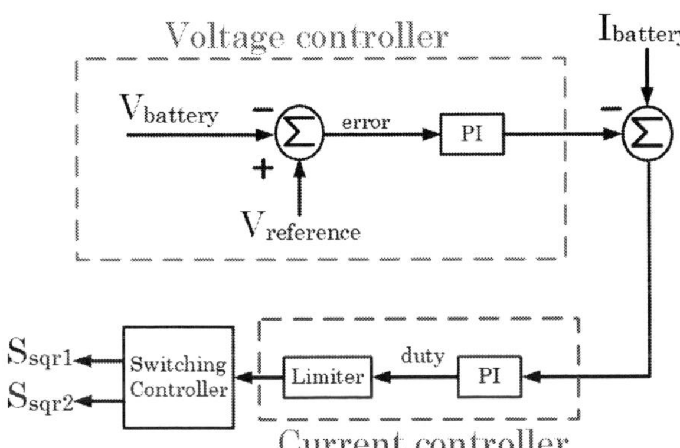

Fig. 9: Closed loop control scheme for proposed topology.

III. Design Example and Simulation Results

A 50 kW/ 500V design example was simulated using PSIM. The parameters in Table 1 were used for simulation. Fig. 10(a) shows the battery voltage and Fig. 10(b) shows the battery current. It can be observed that both the battery voltage and current have very small ripple. The secondary side HF AC link voltages V_{abs}, V_{bcs}, V_{cas} are shown in Fig. 11(a). The voltages are displaced by 120° as in 3-phase operation. The FFT of the secondary side voltage V_{abs} is given in Fig. 11(b); the fundamental components of V_{abs} appear at 12 kHz ± 60 Hz enabling HF operation. Twelve pulse operation can be appreciated from Fig. 12. The voltages V_{abs} and V_{abt}, in Fig. 12(a), have a 30° phase shift compared to each other. Similarly, the input currents to the 12-pulse diode rectifier, shown in Fig. 12(b), have a 30° phase shift due to 12-pulse operation.

High power factor operation can be observed from Fig. 13. The utility line-to-neutral voltages are in phase with the line input currents. The simulated power factor is 0.98 and simulated line current THD is <5%. The input filter was designed to remove the switching frequency components from the input currents and also to remove the 11th and 13th harmonics inherent in 12-pulse rectifier systems.

A step response of the battery voltage, $V_{battery}$, is obtained to verify the performance of the control scheme. The battery voltage reference is changed from 450V to 500V. It is anticipated for the duty cycle to increase during the step change in order to increase the output battery voltage to 500V. The battery voltage and current during the step transient are shown in Fig. 14(a) and Fig. 14(b) respectively. Fig. 14(c) shows how the duty cycle of active switches S_1 and S_2 increases to adjust the output battery voltage. Furthermore, Fig. 14(d) shows the line input currents increasing in amplitude during the step response. The simulated power factor remains close to unity during the transient.

Table 1: Specification & Operating Conditions used for the system in Fig. 2

Grid voltage (line-to-line rms)	480 V
Rectifier output voltage	500 V_{dc}
Rated power	50 kW
Switching frequency (f_{sqr})	12 kHz
Output Inductor (L_{out})	2 mH
Output Capacitor (C_{out})	200 μF

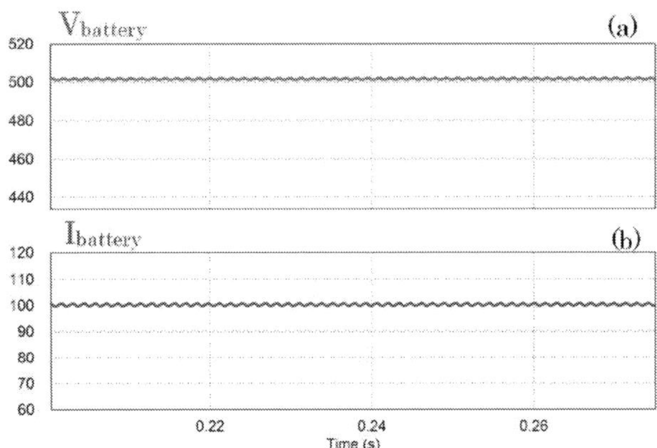

Fig. 10: (a) Battery voltage constant at 500V; (b) Battery current is 100A to accomplish 50 kW charging operation. Very small ripple can be appreciated on both the battery voltage and current.

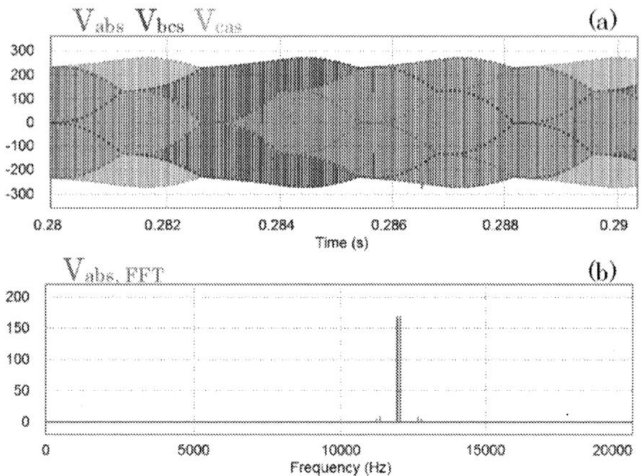

Fig. 11: (a) Secondary side HF AC link voltages V_{abs}, V_{bcs}, V_{cas}, (185 V_{rms}); (b) FFT of voltage V_{abs}, fundamental components appear at 12 kHz ±60 Hz.

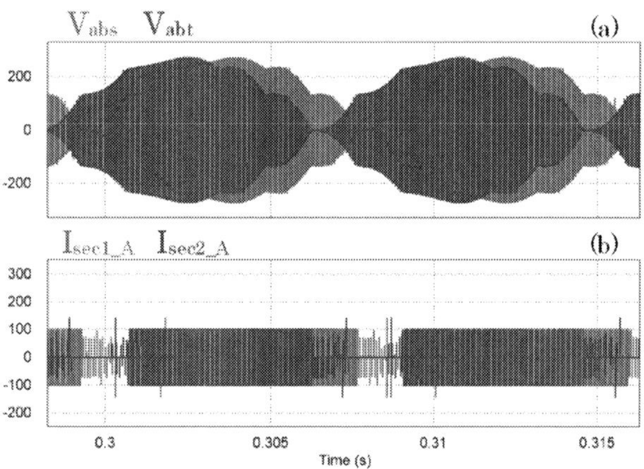

Fig. 12: (a) Secondary side voltages V_{abs} and V_{abt} (185 V_{rms}), they are displaced by 30° as in 12-pulse operation; (b) secondary side currents I_{sec1_A} and I_{sec2_A} are also displaced by 30°.

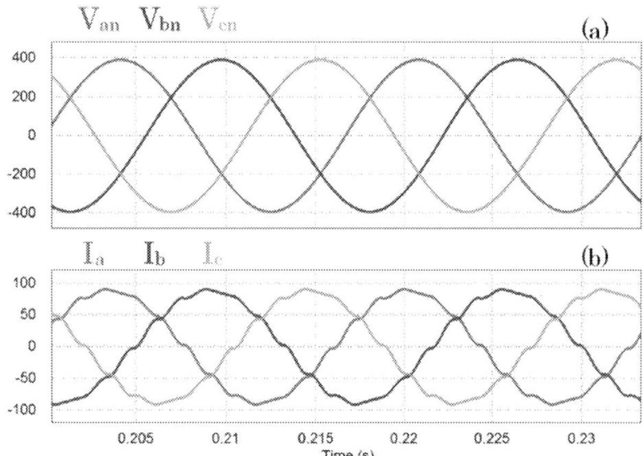

Fig. 13: (a) Line-to-neutral voltages (277V_{rms}); (b) Line input currents (60A_{rms}), simulated power factor is 0.98 and THD is <5%.

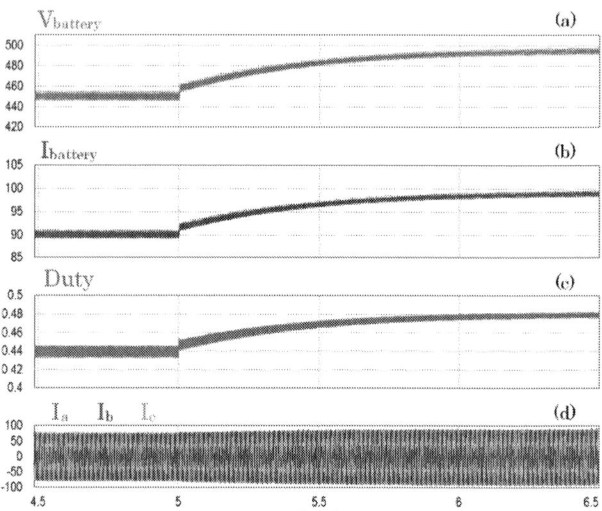

Fig. 14: (a) Step change in the reference battery voltage from 450V to 500V; (b) Battery current $I_{battery}$ increases from 90A to 100A during the step change; (c) Duty cycle of active switches S_1 and S_2 increases to adjust the output battery voltage. (d) Line input currents increase in amplitude from 48A_{rms} to 60A_{rms} during the transient

IV. Experimental Results

A scaled down 1.6 kW laboratory prototype of the push pull based AC-DC rectifier system for EV charging was built and tested in order to validate the proposed topology. The input 3-phase line-to-line voltages are provided by a California Instruments power conditioner. The line frequency of the input voltages is set to 50 Hz. The switching frequency of the active devices S_1 and S_2 is set to 600 Hz. The duty cycle of operation is set to 50%. The gate drive signals for the active devices were generated using a Texas Instruments TMS320F28335 microcontroller.

The 3-phase zig-zag connected medium frequency transformer is designed to operate at 600 Hz and is built using silicon steel material with magnetic flux density of 0.8T. The transformer parameters are given in Table 2. Fig. 15 shows an image of the transformer employed for the experiments. As shown in the image, the transformer consists of 5-limbs with the windings wound around the interior three limbs.

The obtained experimental results are similar to simulation results. Fig. 16 shows the 3-phase voltages feeding one of the 12-pulse diode rectifiers, namely V_{abs}, V_{bcs}, and V_{cas}. It is evident that the set of 3-phase voltages are displaced by 120° from each other. Operation at medium frequency is confirmed from Fig. 17. The FFT of voltage V_{abs} shows fundamental components at 550 Hz and 650 Hz enabling the transformer to operate at medium frequency. From Fig. 18, 12-pulse operation can be observed; the voltages V_{abs} and V_{abt} show a net 30° phase shift with respect to each other as expected. This figure also shows a smooth DC link output voltage as in 12-pulse operation. Furthermore, 12-pulse operation is confirmed by the line input current shown in Fig. 19. The FFT of the unfiltered line input current shows the dominant harmonics to be 11th (550 Hz) and 13th (650 Hz).

Table 2: Medium Frequency Zig-Zag Transformer Characteristics

Core Type	Silicon Steel
Flux Density, B	0.8 T
Frequency	600 Hz
Dimensions	39.5 cm x 11 cm x 28 cm
VA rating	7 kVA
Power Density	580 W/L

Fig. 15: Medium frequency transformer used for laboratory prototype. The HF transformer is currently under construction.

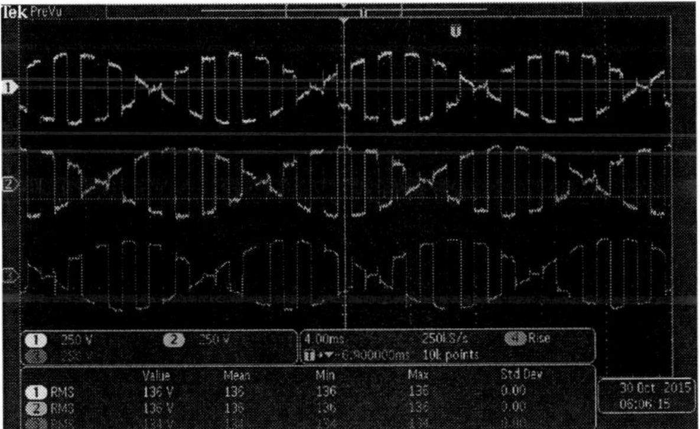

Fig. 16: 3-phase AC link. Voltages V_{abs}, V_{bcs}, V_{cas} fed to one of the output diode rectifiers. Voltages are displaced by 120° with respect to each other as in 3-phase operation.

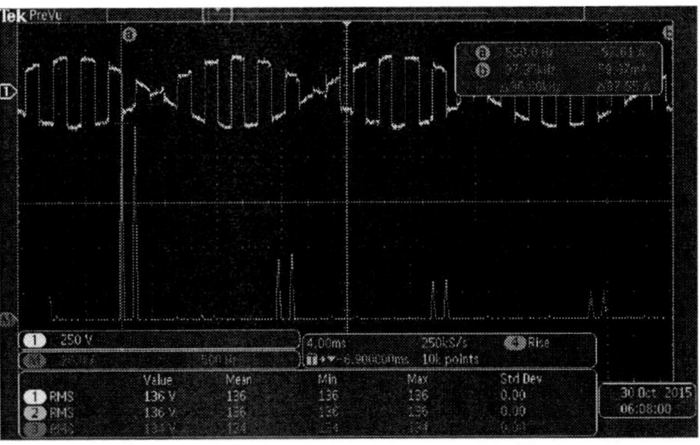

Fig. 17: FFT of V_{abs} shows fundamental components at 550 Hz and 650 Hz enabling MF operation.

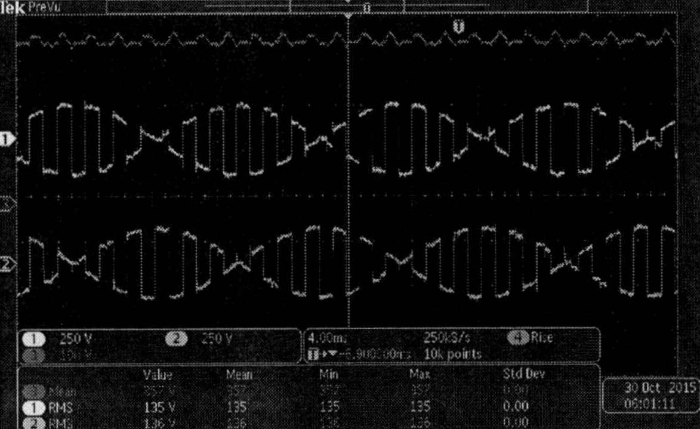

Fig. 18: 600 Hz operation waveforms- (Ch. 1) Line-to-line voltage V_{abs}, (Ch. 2) line-to-line voltage V_{abt}, (Ch. 3) DC link output voltage.

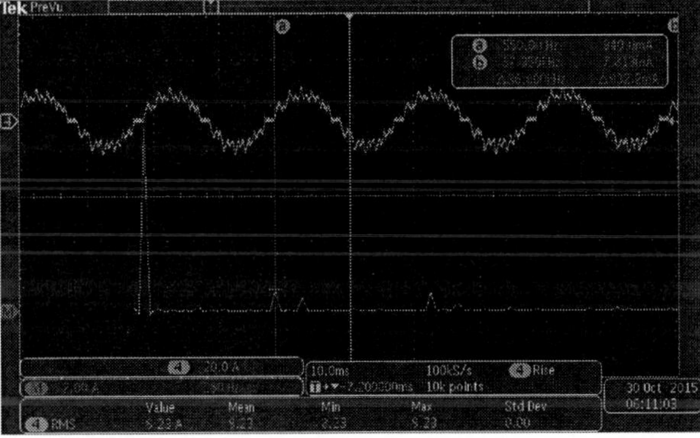

Fig. 19: Line input current I_a (8.23Arms). FFT of I_a shows dominant harmonics at 550 Hz and 650 Hz, which correspond to the 11th and 13th harmonic respectively, as in 12-pulse operation.

V. Conclusion

A reduced active switch push-pull based AC to DC rectifier system for high power electric vehicle battery charging was introduced in this paper. The system is able to provide a 3-phase high frequency AC link using only two active high voltage switches. Galvanic isolation is achieved through a high frequency zig-zag transformer. The operation of the proposed topology was described through detailed analysis. A simple control scheme for the proposed system was also introduced. Simulation results of a 50kW/500V design example demonstrate the topology's feasibility. In addition, 1.6 kW experimental results with a medium frequency prototype illustrate the functionality and practicality of the system. Overall the advantages of the proposed system include a reduced active switch count, high power density, and simple modulation and control scheme. The high frequency zig-zag transformer is currently under development.

References

[1] United States Department of Energy, "Vehicle Technologies Office: Plug-in Electric Vehicles and Batteries". Available: http://energy.gov/eere/vehicles/vehicle-technologies-office-plug-electric-vehicles-and-batteries

[2] United States Department of Energy, "One Million Electric Vehicles by 2015". Available: https://www1.eere.energy.gov/vehiclesandfuels/pdfs/1_million_electric_vehicles_rpt.pdf

[3] Wang Shuo, R. Crosier, and Chu Yongbin, "Investigating the power architectures and circuit topologies for megawatt superfast electric vehicle charging stations with enhanced grid support functionality," in *IEEE International Electric Vehicle Conference (IEVC)*, 2012, pp. 1-8.

[4] A. G. Boulanger, A. C. Chu, S. Maxx, and D. L. Waltz, "Vehicle Electrification: Status and Issues," *Proceedings of the IEEE*, vol. 99, pp. 1116-1138, 2011.

[5] United States Department of Energy, "Plug-in Electric Vehicle Handbook for Public Charging Station Hosts". Available: http://www.afdc.energy.gov/pdfs/51227.pdf

[6] Yilmaz, M.; Krein, P.T., "Review of Battery Charger Topologies, Charging Power Levels, and Infrastructure for Plug-In Electric and Hybrid Vehicles," *IEEE Transactions on Power Electronics*, vol.28, no.5, pp.2151,2169, May 2013.

[7] A. S. Yilmaz, M. Badawi, Y. Sozer, and I. Husain, "A fast battery charger topology for charging of electric vehicles," in *IEEE International Electric Vehicle Conference (IEVC)*, 2012, pp. 1-5.

[8] Zhang Zuzhi, Xu Haiping, Shi Lei, Li Dongxu, and Han Yuchen, "A unit power factor DC fast charger for electric vehicle charging station," in *7th International Power Electronics and Motion Control Conference (IPEMC)*, 2012, pp. 411-415.

[9] T. Soeiro, T. Friedli, and J. W. Kolar, "Three-phase high power factor mains interface concepts for Electric Vehicle battery charging systems," in *Twenty-Seventh Annual IEEE Applied Power Electronics Conference and Exposition (APEC)*, 2012, pp. 2603-2610.

[10] Suk-Ho Ahn; Ji-woong Gong; Hong-Je Ryoo; Sung-Roc Jang, "Implementation of 60-kW fast charging system for electric vehicle," in *39th Annual Conference of the IEEE Industrial Electronics Society, IECON 2013* - , vol., no., pp.7256-7261, 10-13 Nov. 2013

[11] H. S. Krishnamoorthy, P. Garg, and P. N. Enjeti, "A matrix converter-based topology for high power electric vehicle battery charging and V2G application," in *38th Annual Conference on IEEE Industrial Electronics Society*, 2012, pp. 2866-2871.

[12] Sandoval, J.J.; Essakiappan, S.; Enjeti, P., "A bidirectional series resonant matrix converter topology for electric vehicle DC fast charging," *2015 IEEE Applied Power Electronics Conference and Exposition (APEC)*, vol., no., pp.3109,3116, 15-19 March 2015.

[13] Krishnamoorthy, H.S.; Enjeti, P.N.; Garg, P., "Simplified medium/high frequency transformer isolation approach for multi-pulse diode rectifier front-end adjustable speed drives," *2015 IEEE Applied Power Electronics Conference and Exposition (APEC)*, vol., no., pp.527,534, 15-19 March 2015.

[14] Taeyong Kang; Essakiappan, S.; Enjeti, P.; Sewan Choi, "Towards a smart distribution transformer for smart grid," in *2015 9th International Conference on Power Electronics and ECCE Asia (ICPE-ECCE Asia)*, vol., no., pp.1997-2003, 1-5 June 2015.

[15] Gupta, R.K.; Mohapatra, K.K.; Mohan, N., "A novel three-phase switched multi-winding power electronic transformer," *IEEE 2009 Energy Conversion Congress and Exposition, 2009.* vol., no., pp.2696,2703, 20-24 Sept. 2009.

[16] Sandoval, Jose Juan; Krishnamoorthy, Harish; Enjeti, Prasad; Choi, Sewan, "High power density adjustable speed drive topology with medium frequency transformer isolation," in *2015 IEEE Energy Conversion Congress and Exposition (ECCE)*, vol., no., pp.6986-6992, 20-24 Sept. 2015.

[17] She Xu, A. Q. Huang, and R. Burgos, "Review of Solid-State Transformer Technologies and Their Application in Power Distribution Systems," *IEEE Journal of Emerging and Selected Topics in Power Electronics*, vol. 1, pp. 186-198, 2013.

[18] Cree Inc, "Silicon Carbide Schottky Diode," C4D40120D datasheet, 2014. Rev. E.

978-1-4673-9551-9/16 $31.00 © 2016 IEEE

A Wide Bandgap Device Based Multilevel Switched-Capacitor Converter

Diogo Cesar Santos de Moura, Boris Curuvija, and Dong Cao
Department of Electrical and Computer Engineering
North Dakota State University
Fargo, USA
dong.cao@ndsu.edu

Abstract— **This paper presents a new bidirectional multilevel modular switched-capacitor dc-dc converter for high voltage and high conversion ratio application with low capacitor voltage rating. For the high voltage and high conversion ratio application, high voltage capacitor with good performance will not be available either, like the switching devices. For the traditional multilevel modular switched-capacitor dc-dc converter (MMSCC), the switching devices maximum voltage rating is low, but the capacitor voltage rating increases linearly as the conversion ratio increases. Compared with the linear increasing capacitor voltage stress like MMSCC, the proposed circuit maximum capacitor voltage stress is only twice of the low input side voltage. Besides, the proposed circuit also inherits all the good features from the traditional MMSCC, like modular structure, simple control strategy, low device conduction loss, and soft switching. Therefore, it is possible to build a high voltage and high conversion ratio switched-capacitor dc-dc converter with low voltage rating switching device and capacitors to achieve high power density, high efficiency.**

Keywords—GaN; wide bandgap; multilevel; switched-capacitor

I. Introduction

Many industry applications requires small size, light weight, and high efficiency dc-dc converters operating at high temperature, such as aerospace, automotive, airplane, oil and gas. With the development of SiC and GaN semiconductor switching devices and ceramic capacitors, high temperature operation becomes possible. In order to significantly increase the converter power density, high switching frequency is normally used to reduce the passive component size. With the superior switching performance of new SiC and GaN devices, 10x faster switching frequency becomes possible for switching devices. The magnetic core suffers high power loss at high frequency and has relatively low operating temperature. It becomes the bottleneck that limits the further increase of switching frequency of the dc-dc converter and environmental operating temperature. Therefore, a switched-capacitor dc-dc converter without using a magnetic core becomes important for future wide bandgap device applications.

Switched-capacitor dc-dc converter without using magnetic core has been used for low power on-chip applications since 1970s [1]–[3]. Since 2003, several new types of magnetic-less multilevel dc-dc converter derived from high power multilevel inverter structure has been proposed for high power automotive applications [4], [5]. By properly design the multilevel dc-dc converter, several very high power high current dc-dc converter prototypes have been built and tested, i.e. 10 kW and 55 kW, the reported converter efficiency is up to 99% [6], [7]. But these multilevel dc-dc converters suffer high device conduction loss and high device current stress issues, especially with high voltage conversion ratio. A multilevel modular capacitor clamped dc-dc converter based on Dickson charge pump has been proposed to overcome the issue for high power and high voltage conversion ratio application [8]. In order to achieve high efficiency, a bulky capacitor bank with high capacitance has to be used [9], [10]. And these above mentioned multilevel dc-dc converter all work at hard switching mode, when the device turn off current is high, a huge voltage overshoot may appear across the device, which limits the converter application to even higher power level and higher switching frequency. By inserting a small inductor with magnetic-core to the switched-capacitor dc-dc converter, the converter can achieve resonant operation, the zero current switching of the switching devices can also be reached [11]. By properly design the inserted inductor, the converter rating can be pushed to MWs level [12]. But inserting a magnetic-core may limit the benefits of the switched-capacitor dc-dc converter with high temperature and high frequency based on wide-bandgap devices [13], [14]. Recently a new breed of resonant multilevel dc-dc converters utilizing the properly designed distributed stray inductance or air core inductor has been proposed [15]–[17]. These new dc-dc converters are based on multilevel or Dickson charge-pump structure, all the switching devices can achieve zero current switching. Different from traditional multilevel dc-dc converter with a huge capacitor bank to achieve high efficiency, these new proposed converters can utilize low capacitance ceramic capacitors to achieve high efficiency and high power density design due to the resonant operation. However, the structure of these multilevel dc-dc converter is still quite complicated with large device number and high capacitor voltage stress. Many efforts have been done in order to simplify the structure to reduce the switching device number without losing other advantages [18]–[21]. Similar efforts have also been done to reduce the capacitor voltage rating, but the peak voltage rating of the capacitor is still half of the maximum output voltage [18], [22]. Therefore, these converters are still not very suitable for the high power applications with high output voltage.

This paper presents a new type of multilevel switched-capacitor dc-dc converter with combined advantages of above mentioned multilevel dc-dc converter and resonant switched-

978-1-4673-9551-9/16 $31.00 © 2016 IEEE

capacitor dc-dc converters. The proposed multilevel switched-capacitor dc-dc converter inherent all the good features of multilevel modular capacitor clamped dc-dc converter with low device current and current stress, simple control, and modular structure. It can achieve zero current or zero voltage switching like the resonant switched-capacitor dc-dc converter. Compared with resonant switched-capacitor dc-dc converter, the distributed stray inductance or air-core inductor can be used to achieve the resonant, thus no magnetic core will be used. Compared with the multilevel dc-dc converter, no huge capacitor bank is required for the design, the low capacitance ceramic capacitor can also be utilized to achieve high temperature and high power density design. At the same time, all the recent advance technique of reducing switching device number, reducing capacitor voltage stress are all utilized. The total device number is about three times less than the traditional multilevel modular capacitor clamped dc-dc converter. Most importantly, the capacitor voltage rating of the proposed circuit will not increase linearly with the voltage conversion ratio, as the multilevel dc-dc converters mentioned in the literature [8], [18], [19]. The maximum capacitor voltage stress is the same as the switching device voltage stress, which is only twice of the low side voltage. Therefore, the proposed circuit combines the advantages of previously mentioned circuits in the literature with minimum device number, capacitor and device voltage stress, low device current stress, soft-switching capability, low capacitance requirement. The proposed circuit is especially for high power, high output voltage, and high voltage conversion ratio application using wide-bandgap device and ceramic capacitor with high power density and high efficiency. Simulation results will be provided, although the proposed converter is more suitable for very high voltage application where very high voltage rating high performance capacitor is not available, due to the lab limitation a scale down 1 kW prototype based on GaN from EPC is being built for the experiment results.

II. PROPOSED CIRCUITS AND OPERATING PRINCIPLE

Fig. 1 shows two proposed converter generalized structure with conversion ratio equals to N. Fig. 1(a) shows the case that the input and output don't share the same ground. Fig. 1(b) shows the case that the input and output share the same ground. Both structures are able to achieve N times conversion ratio, with N capacitors, and N+4 active switches. The proposed converter structure has bidirectional power flow features which is able to achieve either boost or buck functions with input on the low voltage or high voltage side. Fig. 1 shows the case that the input voltage source is on the low voltage side, and the converter is able to achieve N times boost function. Although the proposed converter is able to use the simplified control strategy with two complementary control signals, in order to achieve resonant operation and zero current switching easily, a detailed control strategy is proposed in the following section for the detailed design consideration of resonant operation. The maximum capacitor voltage rating and maximum active device voltage stress are both 2 times of the low side voltage, which is $2V_{in}$ as shown in the case of Fig. 1. All the active switches current stress are as low as the high voltage side load current assuming the input voltage is on the low voltage side except the 4 switches S_{T1}, S_{T2}, S_{B1}, S_{B2} directly connected, as shown in Fig. 1. The low voltage side devices have higher current stress, which is not an issue since the low voltage side devices will normally have a better performance.

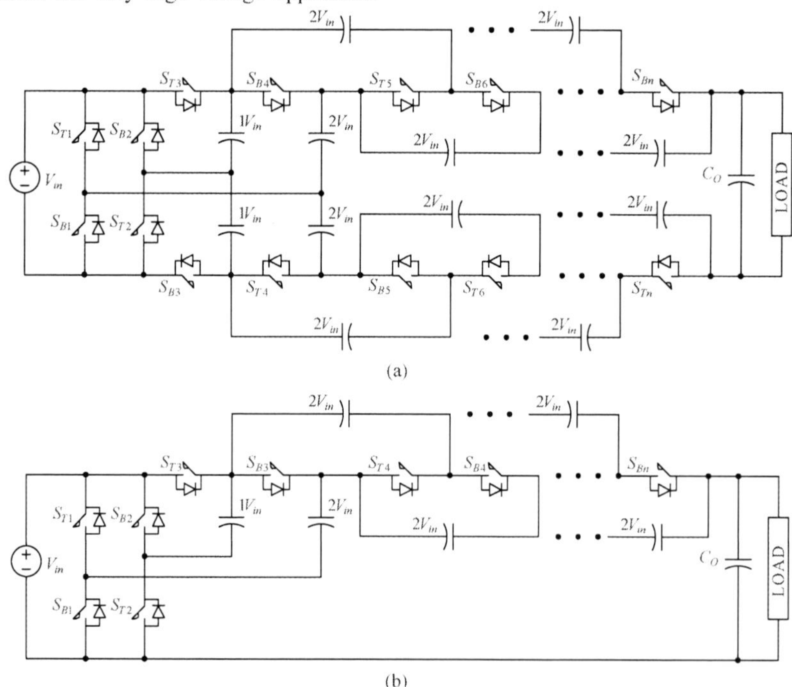

Fig. 1 Proposed New Multilevel Modular Switched-Capacitor DC-DC Converter with the Low Capacitor Voltage Stress. (a)Double wing structure without the same ground. (b) Single wing structure with the same ground.

978-1-4673-9551-9/16 $31.00 © 2016 IEEE

III. Design Consideration and Resonant Operation

By properly inserting two air core inductors in the circuit, and select the capacitance and switching frequency properly, the resonant operation of the proposed circuit could be achieved, zero current switching for all the switches could be achieved too. Fig. 2 shows an example of the proposed circuit in Fig. 1(a) with 8 times conversion ratio, and two inserted air core inductors. Due to the low inductance requirement of the inserted inductors, air core inductor without magnetic core could be used. As shown in Fig. 2, all the capacitor voltages are $2V_{in}$ besides C_2 and C_3. And the switches $S_{T1\sim3}$, $S_{B1\sim3}$, have the voltage stress equal to V_{in}, while all the other switches $S_{T4\sim6}$, $S_{B4\sim6}$ have the voltage stress that is equal to $2V_{in}$. Fig. 3 shows one possible gate signals operation arrangement with eight different operation modes.

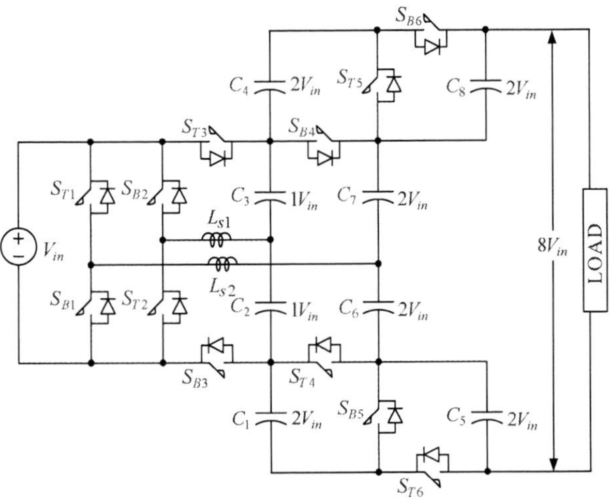

Fig. 2 Proposed multilevel switched-capacitor dc-dc converter with 8 times conversion ratio of the case in Fig. 1(a).

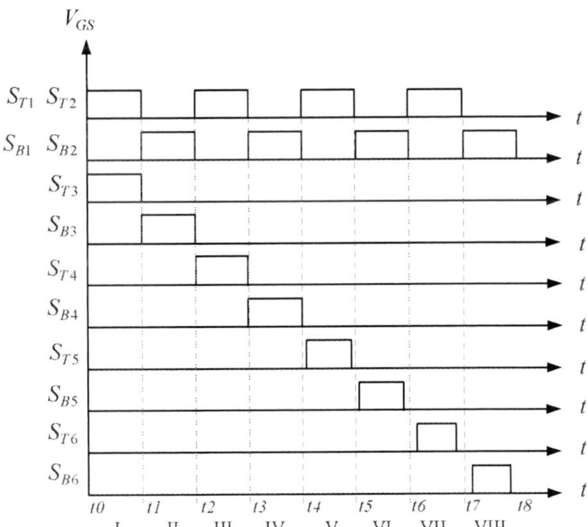

Fig. 3 One possible gate control signal of the proposed dc-dc converter in Fig. 2.

The Project is sponsored by ND-EPSCoR and ND NASA EPSCoR.

Much simpler gate signals arrangement is also possible if no resonant operation is required. If assume all the capacitor have the same capacitance, some of the resonant loops will have less capacitance while many capacitors are connected in series during the operation. Therefore, the resonant frequency will be higher than the switching frequency, and the turn on time of the corresponding switches should be reduced. The proposed control signal arrangement has the capability to meet the control signal needs while no resonant frequency is the same with the switching frequency in the extreme case, and the turn on time of each resonant loop could be fine tuned by controlling the turn on time of the switches $S_{T3\sim6}$ $S_{B3\sim6}$. The detailed resonant frequency calculation and detailed operation analysis of each resonant loop will be provided in the next section.

IV. Steady State Resonant Operating Principle

In every switching state, the simplified equivalent circuits of the proposed converter can be considered as a series RLC circuit with a voltage source. Switch conduction loss is modeled using R_{on} and the capacitor power loss is modeled using R_{ESR}. L can be considered as the sum of the stray inductance, in one switching loop $L = \sum L_{SN}$. C can be considered as the value of all the capacitance in series in one switching loop. R can be considered as the sum of all the turn ON resistance and the capacitor ESR in one switching loop, $R = \sum R_{on} + \sum R_{ESR}$. In every switching state, the circuit behavior can be considered as the step response of this series RLC circuit. The initial value of capacitor voltage is V_C, and the initial value of inductor current is zero. When the value of R, L, and C meets the underdamped circuit operation, the circuit is able to operate in the resonant case with zero current switching for the switching devices. The natural frequency is $\omega_0 = \frac{1}{\sqrt{LC}}$. For the underdamped case, the resonant frequency of each loop is $\omega_d = \sqrt{\omega_0{}^2 - \alpha^2}$, where $\alpha = \frac{R}{2L}$ is the neper frequency. If gate signal switching algorithm in Fig. 2 is used, the proposed multilevel switched-capacitor dc-dc converter can be separated in eight different loops and can be analyzed separated.

A. State I ($t_0 - t_1$):

During state I, the switches S_{T2} and S_{T3} are conducting and capacitor C_3 is charged to $1\,V_{in}$. Moreover, as mentioned previously to model the conduction loss and capacitor power loss, assuming the turn on resistance of the switches S_{T2} and S_{T3} is Rt_{on2} and Rt_{on3} respectively, and the Equivalent Series Resistance (ESR) of the capacitor C_3 is R_{ESR3}. The required resonant frequency for the soft switching can be calculated using the equations mentioned above. The equivalent inductance of the loop is L_{S1}. The resonant frequency is shown in (1) ~ (3).

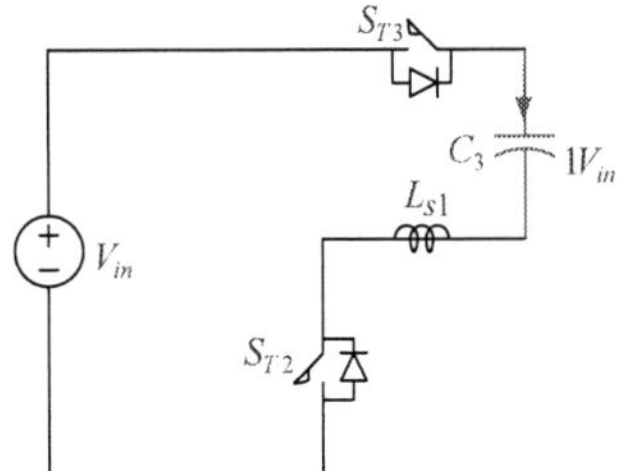

Fig. 4 Simplified equivalent circuit during state I.

$$\omega_{01} = \frac{1}{\sqrt{LC}} = \frac{1}{\sqrt{(L_{S1})(C_3)}} \quad (1)$$

$$\alpha_{01} = \frac{(Rt_{on2}+Rt_{on3}+R_{ESR3})}{2L_{S1}} \quad (2)$$

$$\omega_{d01} = \sqrt{\omega_{01}^2 - \alpha_{01}^2} \quad (3)$$

B. State II ($t_1 - t_2$):

During state II, S_{B2} and S_{B3} are turned on and capacitor C_2 is being charged by the voltage source to IV_{in}. Similarly as in previous state, conduction loss model of the switches S_{B2} and S_{B3} is Rb_{on2} and Rb_{on3} and the ESR of the capacitor C_2 is R_{ESR2}. The equations for calculating the resonant frequency are listed below as shown in (4) ~ (6):

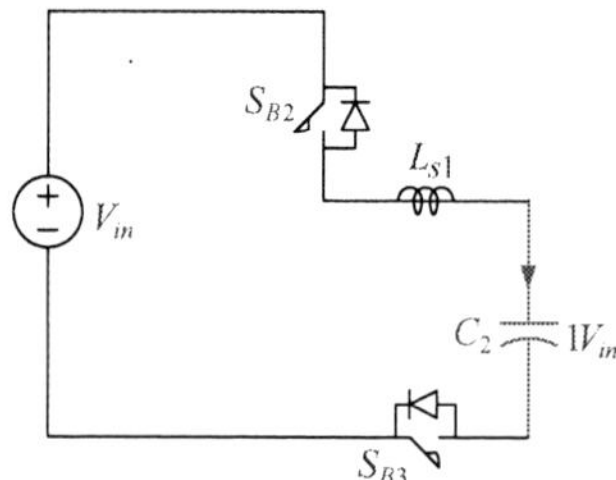

Fig. 5 Simplified equivalent circuit during state II.

$$\omega_{02} = \frac{1}{\sqrt{LC}} = \frac{1}{\sqrt{(L_{S1})(C_2)}} \quad (4)$$

$$\alpha_{02} = \frac{(Rb_{on2}+Rb_{on3}+R_{ESR2})}{2L_{S1}} \quad (5)$$

$$\omega_{d02} = \sqrt{\omega_{02}^2 - \alpha_{02}^2} \quad (6)$$

C. State III ($t_2 - t_3$):

During the state III, the switching devices S_{T1}, S_{T2} and S_{T4} are turned on. Capacitor C_2 and the voltage source are connected in series to charge the capacitor C_6 to $2V_{in}$. Turn on resistance of the switches S_{T1}, S_{T2} and S_{T4} are Rt_{on1}, Rt_{on2}, and Rt_{on4}, and the ESR of the capacitors C_2 and C_6 are R_{ESR2} and R_{ESR6} respectively. Similarly, the equations for calculating the resonant frequency are listed below in (7)~(9). Where equivalent inductance for the loop is the sum of L_{S1} and L_{S2}, equivalent capacitance is C_2 in series with C_3, and equivalent resistance is the sum of the Rt_{on1}, Rt_{on2}, Rt_{on4}, R_{ESR2} and R_{ESR6}.

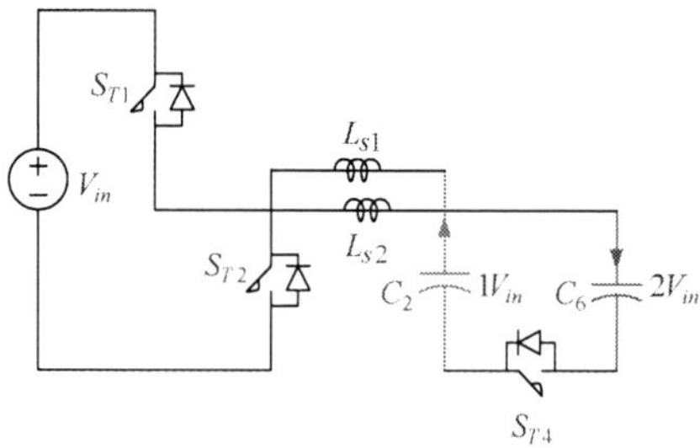

Fig. 6 Simplified equivalent circuit during state III.

$$\omega_{03} = \frac{1}{\sqrt{LC}} = \frac{1}{\sqrt{(L_{S1}+L_{S2})\frac{1}{\frac{1}{C_2}+\frac{1}{C_3}}}} \quad (7)$$

$$\alpha_{03} = \frac{(Rt_{on1}+Rt_{on2}+Rt_{on4}+R_{ESR2}+R_{ESR6})}{2(L_{S1}+L_{S2})} \quad (8)$$

$$\omega_{d03} = \sqrt{\omega_{03}^2 - \alpha_{03}^2} \quad (9)$$

D. State IV ($t_3 - t_4$):

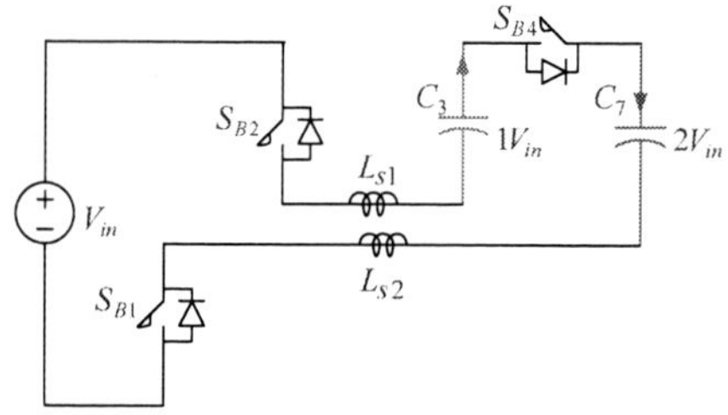

Fig. 7 Simplified equivalent circuit during state IV.

During state IV, the switches S_{B1}, S_{B2} and S_{B4} are conducting. Capacitor C_3 and the voltage source are connected in series to charge the capacitor C_7 to $2V_{in}$. Conduction loss model of the switches S_{B1}, S_{B2} and S_{B4} are Rb_{on1}, Rb_{on2}, and Rb_{on4}, and the ESR of the capacitors C_3 and C_7 are R_{ESR3} and R_{ESR7} respectively. The equations for calculating the resonant frequency are showed below. Where equivalent inductance for the loop is the same as in state three. Equivalent capacitance is C_3 in series with C_7, and equivalent resistance is the sum of Rb_{on1}, Rb_{on2}, Rb_{on4}, R_{ESR3} and R_{ESR7}.

$$\omega_{04} = \frac{1}{\sqrt{LC}} = \frac{1}{\sqrt{(L_{S1}+L_{S2})\frac{1}{\frac{1}{C_3}+\frac{1}{C_7}}}} \quad (10)$$

$$\alpha_{04} = \frac{(Rb_{on1}+Rb_{on2}+Rb_{on4}+R_{ESR3}+R_{ESR7})}{2(L_{S1}+L_{S2})} \quad (11)$$

$$\omega_{d04} = \sqrt{\omega_{04}^2 - \alpha_{04}^2} \quad (12)$$

E. State V ($t_4 - t_5$):

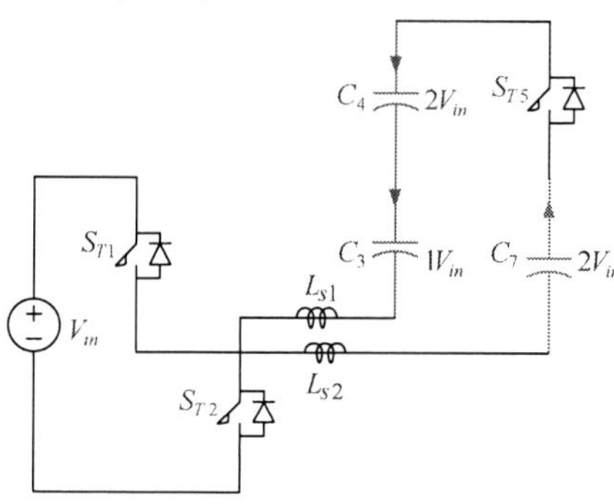

Fig. 8 Simplified equivalent circuit during state V.

During state V, the switches S_{T1}, S_{T2} and S_{T5} are conducting. Voltage source and capacitor C_7 are connected in series to charge the capacitors C_3 and C_4. Since C_3 is previously charged to IV_{in}, C_4 is being charged to $2V_{in}$. The equivalent inductance is the same as in state three and the equivalent capacitance is C_3, C_4, and C_7 in series with each other. Equivalent resistance is the sum of Rt_{on1}, Rt_{on2}, Rt_{on5}, R_{ESR3}, R_{ESR4} and R_{ESR7}. Assuming turn on resistance of the switches S_{T1}, S_{T2} and S_{T5} are Rt_{on1}, Rt_{on2}, and Rt_{on5}, and the ESR of the capacitors C_3, C_4 and C_7 are R_{ESR3}, R_{ESR4} *and* R_{ESR7} respectively. Since more capacitor are connected in series in this case, the resonant frequency is higher than all the previous cases without considering the turn on resistant influences.

$$\omega_{05} = \frac{1}{\sqrt{LC}} = \frac{1}{\sqrt{(L_{S1}+L_{S2})\frac{1}{\frac{1}{C_3}+\frac{1}{C_4}+\frac{1}{C_7}}}} \quad (13)$$

$$\alpha_{05} = \frac{(Rt_{on1}+Rt_{on2}+Rt_{on5}+R_{ESR3}+R_{ESR4}+R_{ESR7})}{2(L_{S1}+L_{S2})} \quad (14)$$

$$\omega_{d05} = \sqrt{\omega_{05}{}^2 - \alpha_{05}{}^2} \quad (15)$$

F. State VI ($t_5 - t_6$):

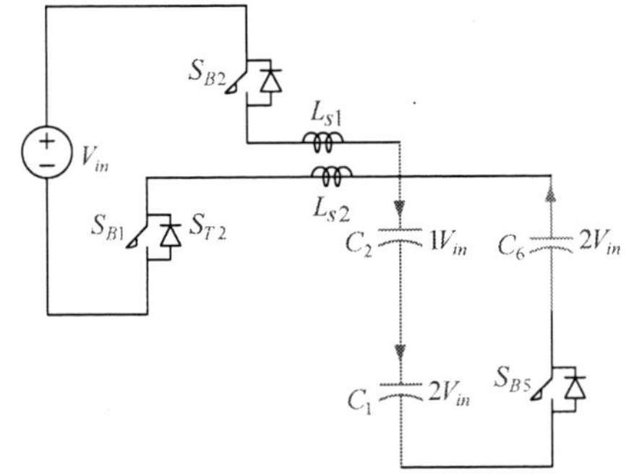

Fig. 9 Simplified equivalent circuit during state VI.

During state VI, the switching devices S_{B1}, S_{B2} and S_{B5} are conducting. Voltage source and capacitor C_6 are connected in series to charge the capacitors C_1 and C_2. Since C_2 is previously charged to $1V_{in}$, C_1 is being charged to $2V_{in}$. Similarly, the equivalent inductance is the same as in state III. Assuming conduction loss model of the switches S_{B1}, S_{B2} and S_{B5} are Rb_{on1}, Rb_{on2}, and Rb_{on5}, and the ESR of the capacitors C_1, C_2 and C_6 are R_{ESR1}, R_{ESR2} and R_{ESR6} respectively. Equivalent capacitance is C_1, C_2, and C_6 in series with each other and equivalent resistance is the sum of Rb_{on1}, Rb_{on2}, Rb_{on5}, R_{ESR1}, R_{ESR2} and R_{ESR6}. The resonant frequency equations are listed below.

$$\omega_{06} = \frac{1}{\sqrt{LC}} = \frac{1}{\sqrt{(L_{S1}+L_{S2})\frac{1}{\frac{1}{C_1}+\frac{1}{C_2}+\frac{1}{C_6}}}} \quad (16)$$

$$\alpha_{06} = \frac{(Rb_{on1}+Rb_{on2}+Rb_{on5}+R_{ESR1}+R_{ESR2}+R_{ESR6})}{2(L_{S1}+L_{S2})} \quad (17)$$

$$\omega_{d06} = \sqrt{\omega_{06}{}^2 - \alpha_{06}{}^2} \quad (18)$$

G. State VII ($t_6 - t_7$):

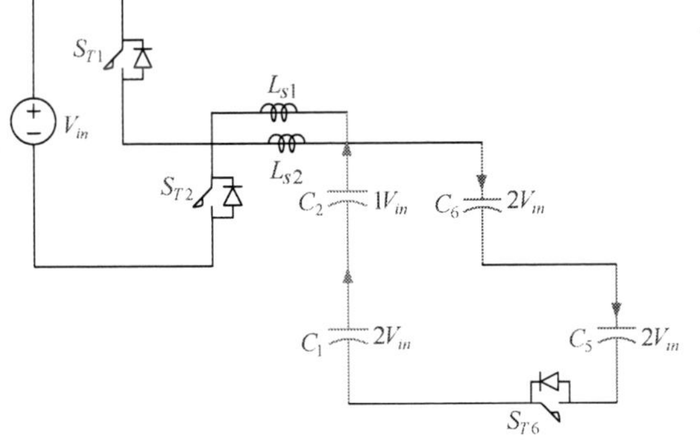

Fig. 10 Simplified equivalent circuit during state VII.

During state VII, the switches S_{T1}, S_{T2} and S_{T6} are conducting. During this state, C_5 and C_6 are being charged while the C_1, C_2 and the input voltage sources are connected in series to discharge. The equivalent inductance is the same as in the previous state and the equivalent capacitance is C_1, C_2, C_5 and C_6 in series with each other. Equivalent resistance is the sum of the Rt_{on1}, Rt_{on2}, Rt_{on6}, R_{ESR1}, R_{ESR2}, R_{ESR5} and R_{ESR6}. Assuming conduction loss model of the switches S_{T1}, S_{T2} and S_{T6} are Rt_{on1}, Rt_{on2} and Rt_{on6}, and the ESR of the capacitors C_1, C_2, C_5 and C_6 are R_{ESR1}, R_{ESR2} R_{ESR5} and R_{ESR6} respectively. The resonant frequency equations are showed below. For this case, more capacitors are connected in series to resonant, the resonant frequency is even higher than all the previous cases, but since the turn on time of the switch S_{T6} could be controlled independently, it would not be an issue.

$$\omega_{07} = \frac{1}{\sqrt{LC}} = \frac{1}{\sqrt{(L_{S1}+L_{S2})\frac{1}{\frac{1}{C_1}+\frac{1}{C_2}+\frac{1}{C_5}+\frac{1}{C_6}}}} \quad (19)$$

$$\alpha_{07} = \frac{(Rt_{on1}+Rt_{on2}+Rt_{on6}+R_{ESR1}+R_{ESR2}+R_{ESR5}+R_{ESR6})}{2(L_{S1}+L_{S2})} \quad (20)$$

$$\omega_{d07} = \sqrt{\omega_{07}{}^2 - \alpha_{07}{}^2} \quad (21)$$

H. State VIII ($t_7 - t_8$):

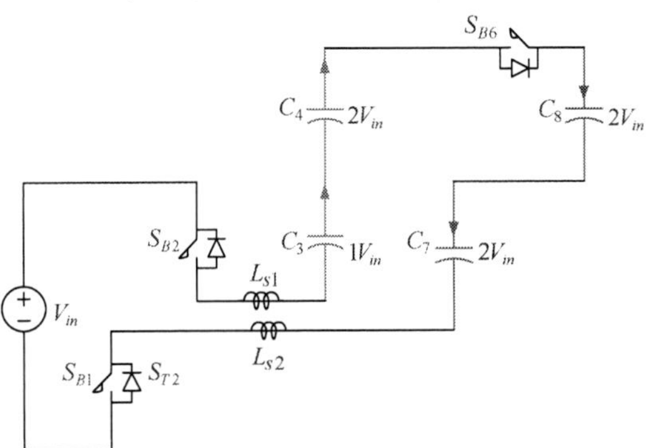

Fig. 11 Simplified equivalent circuit during state VIII.

During state VIII, the switches S_{B1}, S_{B2} and S_{B6} are conducting. During this state, the capacitors C_7 and C_8 are being charged while the C_3, C_4 and the input voltage source are connected in series to discharge. The equivalent inductance is the same as in state three and the equivalent capacitance is C_3, C_4, C_7 and C_8 in series with each other. Assuming turn on resistance of the switches S_{B1}, S_{B2} and S_{B6} are Rb_{on1}, Rb_{on2} and Rb_{on6}, and the ESR of the capacitors C_3, C_4, C_7, and C_8 are R_{ESR3}, R_{ESR4}, R_{ESR7}, and R_{ESR8} respectively. Equivalent resistance is the sum of the Rb_{on1}, Rb_{on2}, Rb_{on6}, R_{ESR3}, R_{ESR4}, R_{ESR7}, and R_{ESR8}. The resonant frequency of this case should be the same as the state VII.

$$\omega_{08} = \frac{1}{\sqrt{LC}} = \frac{1}{\sqrt{(L_{S1}+L_{S2})\frac{1}{\frac{1}{C_3}+\frac{1}{C_4}+\frac{1}{C_7}+\frac{1}{C_8}}}} \quad (22)$$

$$\alpha_{08} = \frac{(Rb_{on1}+Rb_{on2}+Rb_{on6}+R_{ESR3}+R_{ESR4}+R_{ESR7}+R_{ESR8})}{2(L_{S1}+L_{S2})} \quad (23)$$

$$\omega_{d08} = \sqrt{\omega_{08}{}^2 - \alpha_{08}{}^2} \quad (24)$$

Table I shows a summary of all the switches conducting and the capacitor charging and discharging path.

Table I Summary table of the capacitor charging path

↑ = Capacitor discharging, ↓ = Capacitor charging.

Capacitor charging path	On-state switch	Capacitor charging path	On-state switch
State I		State II	
C_3 ↓	S_{T2},S_{T3}	C_2 ↓	S_{B2},S_{B3}
State III		State IV	
C_6 ↓, C_2 ↑	S_{T1},S_{T2},S_{T4}	C_7 ↓ C_3 ↑	S_{B1},S_{B2},S_{B4}
State V		State VI	
C_3,C_4, ↓ C_7 ↑	S_{T1},S_{T2},S_{T5}	C_1,C_2 ↓ C_6 ↑	S_{B1},S_{B2},S_{B5}
State VII		State VIII	
C_5,C_6 ↓ C_1,C_2 ↑	S_{T1},S_{T2},S_{T6}	C_7,C_8 ↓ C_3,C_4 ↑	S_{B1},S_{B2},S_{B6}

V. SIMULATION RESULTS

A simulation model of the proposed converter as shown in Fig. 2, with resonant operation and zero current switching has been built. The power rating of the designed converter is 1 kW. The input voltage is assumed to be 15 V, and the expected output voltage is 120 V. The capacitance of all the capacitors $C_1 \sim C_8$ are assumed to be 47 uF, and the two resonant air core inductors L_{S1}, L_{S2} are 216 nH. The switching frequency of the proposed converter is 50 kHz. The turn on resistance of the low voltage switching devices are assumed to be 0.5 Ohm, and the turn on resistance of the high voltage device $S_{T4,5,6}$, $S_{B4,5,6}$. Fig. 12 shows the simulation result of this converter. T1 ~ 6 and B1 ~ 6 are the gate signals of the proposed converter. Vo is the output voltage, Vin is the input voltage, I_ST2 ~ 6 are the current through the switches $S_{T2\sim6}$, S_{B2}. The output voltage ripple is relatively large for the simulated case without an output capacitor. However, for the practical application, there will be large capacitor banks on both input and output side. While considering the large capacitor bank influence, the resonant frequency will be slightly different from the previous calculation, but it could be done in a similar manner. Fig. 13 shows the simulation results where an additional 470 uF on the high voltage output side is added to simulate the real case with a large output dc capacitor bank. The additional output capacitor bank has very little impact on the output current resonant as shown from the simulation results. And the output voltage ripple is reduced about 1 V in this case compared with almost 50V output capacitor ripple while no output capacitor bank is considered. A 1 kW prototype based on GaN FET from EPC with 15V and 120V output has been being built. Experimental results will be provided in the future papers.

VI. CONCLUSION

A new multilevel modular dc-dc converter without magnetic core has been proposed for wide bandgap devices enabled applications. The voltage stress of all the capacitors are as high as twice of the low voltage side voltages with N times conversion ratio. And the device voltage stress of the proposed converter is the same as the traditional multilevel modular dc-dc converter. Resonant operation can also be achieved for the proposed converter using the air core inductor concept. By properly design the switching frequency, capacitance, and air core inductance, a very high efficiency and high power density design can be achieved.

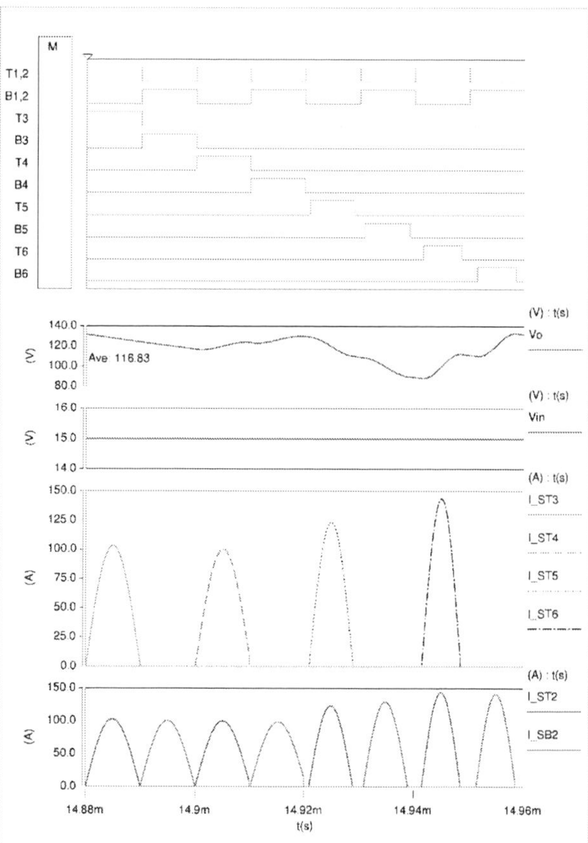

Fig. 12 Simulation results of the proposed 1 kW dc-dc converter with 8 times conversion ratio.

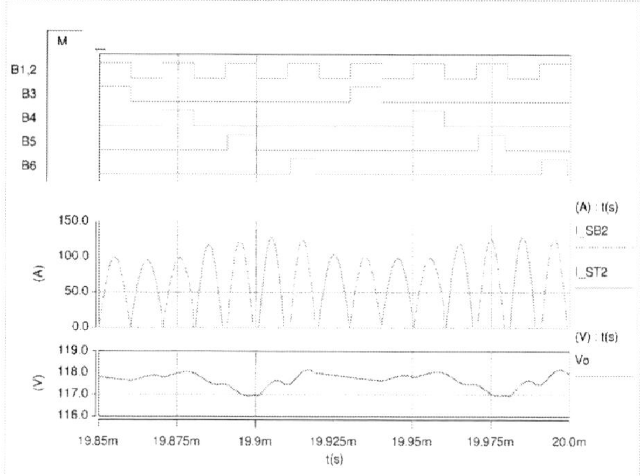

Fig. 13. Simulation results of the proposed 1 kW dc-dc converter with 8 times conversion ratio with a 470 uF output capacitor added.

REFERENCES

[1] J. F. Dickson, "On-chip high-voltage generation in MNOS integrated circuits using an improved voltage multiplier technique," *Solid-State Circuits, IEEE J.*, vol. 11, no. 3, pp. 374–378, 1976.

[2] D. Maksimovic and S. Dhar, "Switched-capacitor DC-DC converters for low-power on-chip applications," in *Power Electronics Specialists Conference, 1999. PESC 99. 30th Annual IEEE*, 1999, vol. 1, pp. 54–59 vol.1.

[3] A. Ioinovici, "Switched-capacitor power electronics circuits," *Circuits Syst. Mag. IEEE*, vol. 1, no. 3, pp. 37–42, 2001.

[4] F. Z. Peng, F. Zhang, and Z. Qian, "A magnetic-less DC-DC converter for dual-voltage automotive systems," *IEEE Trans. Ind. Appl.*, vol. 39, no. 2, pp. 511–518, 2003.

[5] F. Z. Peng, F. Zhang, and Z. Qian, "A novel compact DC-DC converter for 42 V systems," in *Power Electronics Specialist Conference, 2003. PESC '03. 2003 IEEE 34th Annual*, 2003, vol. 1, pp. 33–38 vol.1.

[6] M. Shen, F. Z. Peng, and L. M. Tolbert, "Multilevel DC–DC Power Conversion System With Multiple DC Sources," *IEEE Trans. Power Electron.*, vol. 23, no. 1, pp. 420–426, Jan. 2008.

[7] W. Qian, H. Cha, F. Z. Peng, and L. M. Tolbert, "55-kW Variable 3X DC-DC Converter for Plug-in Hybrid Electric Vehicles," *IEEE Trans. Power Electron.*, vol. 27, no. 4, pp. 1668–1678, Apr. 2012.

[8] F. H. Khan and L. M. Tolbert, "A Multilevel Modular Capacitor-Clamped DC – DC Converter," *IEEE Trans. Ind. Appl.*, vol. 43, no. 6, pp. 1628–1638, 2007.

[9] F. Zhang, L. Du, F. Z. Peng, and Z. Qian, "A New Design Method for High Efficiency DC-DC Converters with Flying Capacitor Technology," pp. 92–96, 2006.

[10] D. Cao, S. Jiang, and F. Z. Peng, "Optimal design of a multilevel modular capacitor-clamped DC-DC converter," *IEEE Trans. Power Electron.*, vol. 28, no. 8, pp. 3816–3826, 2013.

[11] K. W. E. Cheng, "Zero-current-switching switched-capacitor converters," *Electr. Power Appl. IEE Proc. -*, vol. 148, no. 5, pp. 403–409, 2001.

[12] O. Keiser, P. K. Steimer, and J. W. Kolar, "High power resonant Switched-Capacitor step-down converter," in *Power Electronics Specialists Conference, 2008. PESC 2008. IEEE*, 2008, pp. 2772–2777.

[13] A. Ioinovici, H. S. H. Chung, M. S. Makowski, and C. K. Tse, "Comments on 'Unified analysis of switched-capacitor resonant converters,'" *IEEE Trans. Ind. Electron.*, vol. 54, no. 1, pp. 684–685, 2007.

[14] A. Ioinovici, C. K. Tse, and H. S. H. Chung, "Comments on 'Design and analysis of switched-capacitor-based step-up resonant Converters,'" *Circuits Syst. I Regul. Pap. IEEE Trans.*, vol. 53, no. 6, p. 1403, 2006.

[15] D. Cao and F. Z. Peng, "Zero-Current-Switching Multilevel Modular Switched-Capacitor DC-DC Converter," *Ind. Appl. IEEE Trans.*, vol. 46, no. 6, pp. 2536–2544, 2010.

[16] D. Cao and F. Z. Peng, "Multiphase Multilevel Modular DC-DC Converter for High-Current High-Gain TEG Application," *Ind. Appl. IEEE Trans.*, vol. 47, no. 3, pp. 1400–1408, 2011.

[17] D. Cao and F. Z. Peng, "A Family of Z-source and Quasi-Z-source DC-DC Converters," in *Applied Power Electronics Conference and Exposition, 2009. APEC 2009. Twenty-Fourth Annual IEEE*, 2009, pp. 1097–1101.

[18] W. Qian, D. Cao, J. G. Cintron-Rivera, M. Gebben, D. Wey, and F. Z. Peng, "A Switched-Capacitor DC–DC Converter With High Voltage Gain and Reduced Component Rating and Count," IEEE Trans. Ind. Appl., vol. 48, no. 4, pp. 1397–1406, Jul. 2012.

[19] L. He, "A novel quasi-resonant bridge modular switched-capacitor converter with enhanced efficiency and reduced output voltage ripple," *IEEE Trans. Power Electron.*, vol. 29, no. 4, pp. 1881–1893, 2014.

[20] L. He, "High-Performance Bridge Modular Switched-Capacitor Converter With Small Component Requirement Based on Output Impedance Analysis for Low Loss," vol. 28, no. 10, pp. 4668–4680, 2013.

[21] D. Cao, X. Lyu, and Y. Li, "Multilevel Modular Converter with Reduced Device Count for Hybrid and Electric Vehicle," in *2015 IEEE Transportation Electrification Conference and Expo (ITEC)*, 2015.

[22] D. Cao, X. Yu, X. Lu, and F. Z. Peng, "A Double-Wing Multilevel Modular Capacitor-Clamped DC-DC Converter with Reduced Capacitor Voltage Stress," in *Energy Conversion Congress and Exposition, 2011. ECCE 2011. IEEE*, 2011.

Novel Circulating Current Suppression Strategy for MMC Based on Quasi-PR Controller

Shengbao Geng, Yiliang Gan, Yungui Li, Lijun Hang, Guojie Li
Department of Electrical Engineering, Shanghai Jiao Tong University,
Shanghai 200240, China

***Abstract*—According to the advantages and disadvantages of the traditional control strategy used in modular multilevel converter (MMC), an improved control method is discussed in this paper. In the proposed method, the outer loop of traditional control strategy is still in use to balance submodule capacitor voltage. However, the dc component extraction and quasi-proportional resonant controller (quasi-PR) controller are added to modify the inner loop to eliminate second-order circulating harmonic current. Besides, design principles of parameters related to the improved part are given in discrete domain and the stability analysis are also discussed. Finally, experiment results under traditional and improved control strategy are analyzed and compared, which indicate that the improved control method has the merit of capacitor voltage balance and performs well in suppressing circulating harmonic current.**

Keywords—Modular multilevel converter; capacitor voltage balance; circulating harmonic current suppression

I. INTRODUCTION

Modular multilevel converter (MMC) was first introduced in 2001 by Germany scholar [1]. Due to its modularization design, low switching frequency and low output harmonics, ect. [2-4], the converter gradually gets focus from the industrial and academe. MMC can flexibly change the output voltage levels by changing the number of submodules [5-6]. Through constant development, MMC has been widely used in medium/high power applications, especially in high-voltage direct current transmission (HVDC) [7-9].

However, MMC also has its own problems. Because the on or off states of IGBT frequently switches, submodule capacitors are charged or discharged, unavoidably resulting in the voltage fluctuation [10]. The variation of capacitor voltages leads to inequality between the three phase legs of MMC. This introduces large harmonic content into circulating current, which increases system losses and reduces device lifetime, even affects system stability [11-14]. Thus, it is necessary to balance submodule capacitor voltage and eliminate/minimize the circulating harmonic current. According to the directions of arm currents, reference [15], [16] proposed improved sort algorithms to decide the conduction state of submodule capacitors. Reference [17] described the design principles of arm inductors to reduce the amplitude of circulating current. Although increasing the arm inductance can reduce the circulating current, it will deteriorate system performance caused by large voltage drop and slowing frequency response l. Reference [18] analyzed the harmonic components in circulating current and applied double line-frequency, negative-sequence rotational frame to suppress the second-order circulating harmonic current. However, it is infeasible in single-phase or multi-phase system. By regulating the total capacitor energy within the same leg and the unbalance energy between the upper and the lower arms, reference [19] compensated the voltage difference between the dc-link input and phase leg output to control circulating current. But this method needs precise angle of arm equivalent impedance, so the performance in practical engineering will be slightly affected. Reference [20] proposed an average capacitor voltage control model, which consists of dual closed loop. The outer controller is responsible for balancing submodule capacitor voltage. The inner loop forces the actual circulating current to track its reference command generated by the outer loop. The combination of capacitor voltage balance and circulating current suppression makes this model widely used. However the PI controller adopted by the inner loop cannot suppress the twice of fundamental frequency component of the circulating current.

This paper proposes an improved circulating current control method based on the model proposed in reference [20]. First, the mathematical model of MMC is built to show its outer and inner operation characters. Then, the relationship between the average voltage of submodule capacitor and circulating current is discussed to indicate that the output of outer controller in average capacitor voltage control model in [20] contains the information of dc component of circulating current. Afterward, by extracting the dc component of circulating current and based on quasi-PR controller, the second-order circulating harmonic current is eliminated. As to the improved parts, related parameter design principles are given in discrete domain. Finally, experimental results are provided to validate the feasibility and good performance of the proposed control method.

II. BASIC STRUCTURE AND MATHEMATICAL MODEL OF THE MMC

The equivalent circuit of the MMC is shown in Fig.1. We can see that a three phase MMC has six arms, each of which is constituted by N series-connected submodules (SMs) and one arm inductor L_0. R_0 is the equivalent arm resistance. u_{kp} and u_{kn} (k=a, b, c. upper and lower arms are marked as p and n, respectively) are arm voltages generated by N series-connected SMs. i_{kp} and i_{kn} are the arm currents of phase k and i_{diffk} is the corresponding circulating current, which

National Key Technology R&D Program of China (2015BAA01B02), Natural Science Foundation of Shanghai Science and Technology Commission (14ZR1422200), Power Electronics Science and Education Development Program of Delta Environmental & Education Foundation (DREG201505).

flows through both the upper and lower arms.

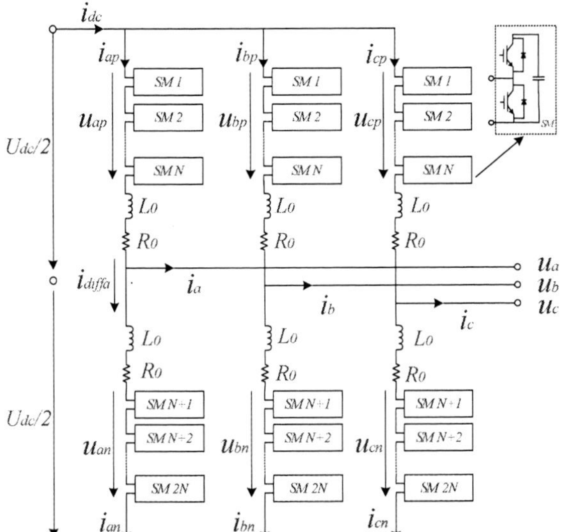

Fig. 1. Diagram of the MMC.

From reference [21], the relationship between arm currents and circulating current is expressed as

$$i_{diffk} = \frac{i_{kp} + i_{kn}}{2} \tag{1}$$

According to Kirchhoff's voltage law, the math model of MMC can be derived by the two following equations

$$u_k = \frac{1}{2}(u_{kn} - u_{kp}) - \frac{1}{2}(L_0 \frac{di_k}{dt} + R_0 i_k) \tag{2}$$

$$2(L_0 \frac{di_{diffk}}{dt} + R_0 i_{diffk}) = U_{dc} - (u_{kn} + u_{kp}) \tag{3}$$

Define e_k as the inner electromotive force (emf). From (2), it can be written as

$$e_k = \frac{u_{kn} - u_{kp}}{2} \tag{4}$$

According to (2), the regulation of ac network voltage u_k can be realized by controlling e_k. Equation (2), which resembles the two-level converter, describes the external behavior of the MMC.

From (3) we can see that the sum of upper and lower arm voltages mismatch the dc-link voltage, which generates circulating current i_{diffk}. Redefine the left of (3) as the equation (5), in which u_{diffk} is regard as the inner unbalance voltage.

$$u_{diffk} = L_0 \frac{di_{diffk}}{dt} + R_0 i_{diffk} \tag{5}$$

Apparently, i_{diffk} can be controlled by regulating u_{diffk} and equation (5) expresses the inner character of the MMC.

III. Control Strategy of the MMC under CPS-PWM

Define M as the voltage modulation ratio, then

$$M = \frac{E_k}{U_{dc}/2} \tag{6}$$

where $0 \le M \le 1$ and E_k is the magnitude of the inner emf e_k. The modulation function of upper and lower arms can be derived as

$$\begin{cases} m_p = \frac{1}{2}[1 - M\cos(\omega_0 t + \theta)] \\ m_n = \frac{1}{2}[1 + M\cos(\omega_0 t + \theta)] \end{cases} \tag{7}$$

where ω_0 is the fundamental frequency and θ is the initial phase angle.

The sum of N capacitor voltages within the upper arm in the phase k is marked as u_{Ckp}. Similarly, u_{Ckn} expresses the sum of N capacitor voltages within the lower arm in the phase k. And they satisfy the following equation

$$\begin{cases} C_{SM}\frac{du_{Ckp}}{dt} = C_{SM}\frac{d(\sum u_{Cki})}{dt} = i_{kp} \cdot N \cdot m_p \\ C_{SM}\frac{du_{Ckn}}{dt} = C_{SM}\frac{d(\sum u_{Ckj})}{dt} = i_{kn} \cdot N \cdot m_n \end{cases} \tag{8}$$

where u_{Cki} is the ith (i=1...N) capacitor voltage in phase k, and u_{Ckj} is the jth (j=N+1...2N) capacitor voltage in phase k.

In steady state, the sum of all the $2N$ submodule capacitor voltages within the same phase leg should be equal to $2U_{dc}$, as shown as the following equation

$$u_{Cks} = u_{Ckp} + u_{Ckn} = 2U_{dc} \tag{9}$$

Based on (8), (9), the following equation can be obtained.

$$\frac{du_{Cks}}{dt} = \frac{1}{C_{arm}} i_{diffk} - \frac{M}{2C_{arm}} i_k \cdot \cos(\omega_0 t + \theta) \tag{10}$$

Define C_{SM} as the capacitance of submodule capacitor. Then, the equivalent capacitance of the phase leg k can derived as $C_{arm} = C_{SM}/N$.

On the basis of (10), the average capacitor voltage of phase leg k can be derived as

$$\bar{u}_{Cks} = \frac{1}{C_{arm}} I_{diffk} - \frac{M}{2C_{arm}T}\int_0^T i_k \cdot \cos(\omega_0 t + \theta)dt \tag{11}$$

where T is the fundamental period and I_{diffk} is the dc component of circulating current i_{diffk}. It can be seen that the balance of submodule capacitor voltage within the same phase leg is depended on the dc component of circulating current.

IV. TRADITIONAL CONTROL STRATEGY OF THE MMC

Reference [20] proposed an average capacitor voltage control model which is widely used in MMC under CPS-PWM.

The basic structure of this model is shown as Fig.2, in which we can see the outer loop controller makes average capacitor voltage $\overline{u}_{Ck}$ track its reference u_C^* and generates circulating current command i_{diffk}^*. The inner unbalance voltage u_{diffk} can be obtained by regulating actual circulating current i_{diffk} to follow i_{diffk}^*.

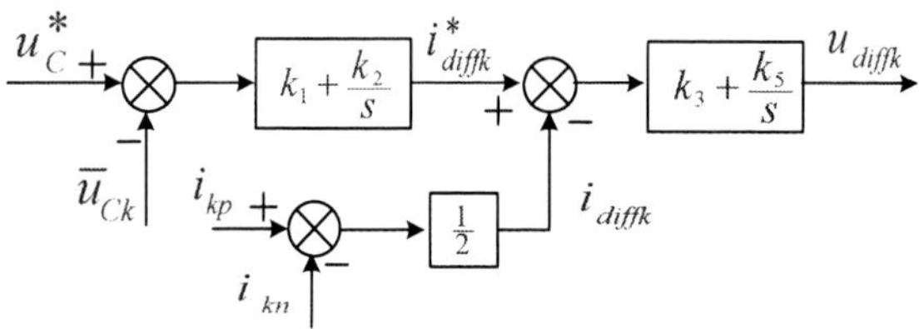

Fig. 2. Diagram of average submodule capacitor voltage control model.

$$\overline{u}_{Ck} = \frac{1}{2N}\left(\sum_{i=1}^{N} u_{Cki} + \sum_{j=N+1}^{2N} u_{Ckj}\right) \tag{12}$$

Since circulating current is mainly composed of dc component and second-order, negative-sequence harmonic component, the difference between i_{diffk}^* and i_{diffk} has a great deal of second-order harmonic component. However, the proportional integral controller adopted by inner loop is inability to control ac component. Therefore, this model performs well in balancing submodule capacitor voltage but cannot effectively suppress circulating current.

A. Improved control strategy of the MMC

First, use low-pass filter to extract dc component from the outer loop output, as shown as Fig.3.

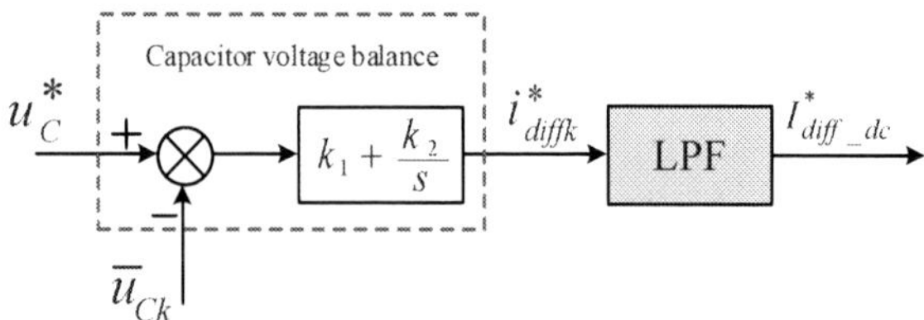

Fig. 3. Diagram of dc component extraction

The low-pass filter is in the form of first order and its transfer function is given by

$$F(s) = \frac{1}{s \cdot T_s + 1} \tag{13}$$

where T_s is the time constant, and the cut-off frequency of low-pass filter is equivalent to $1/(2\pi \cdot T_s)$.

Then, eliminate circulating harmonic current by applying quasi-PR controller, as shown as Fig.4. The transfer function of quasi-PR can be written as

$$F_{PR}(s) = K_p + \frac{2K_r\omega_c s}{s^2 + 2\omega_c s + (2\omega_0)^2} \tag{14}$$

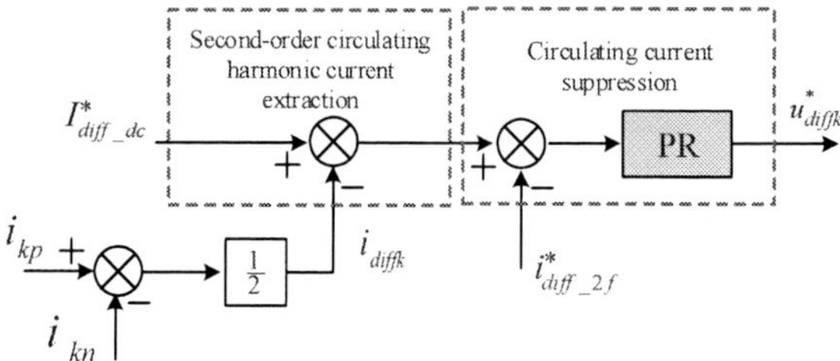

Fig. 4. Diagram of circulating current suppression.

where K_p and K_r are the proportional coefficient and the gain of second-order resonance, respectively. ω_c is used to decrease the sensitivity of resonant controller when frequency shift happens. ω_0 is fundamental angular frequency.

From the above, the arm voltage references of phase k are given as

$$\begin{cases} u_{kp}^* = \frac{1}{2}U_{dc} - e_k - u_{diffk} \\ u_{kn}^* = \frac{1}{2}U_{dc} + e_k - u_{diffk} \end{cases} \tag{15}$$

V. PRINCIPLES OF PARAMETER DESIGN

Principles of parameter design are given in discrete domain for the convenience of engineering application.

As previously mentioned in Section II, circulating current i_{diffk} can be regulated by inner unbalance voltage u_{diffk}. Based on the controllers described in Fig.3 and Fig.4, the circulating current suppression model can be modified as Fig.6, in which u_{kh} is the equivalent harmonic voltage source in phase k (refer to [21] for more analysis).

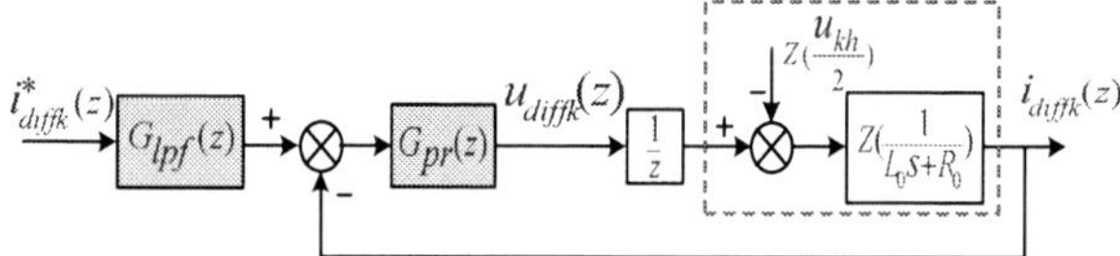

Fig. 6. Diagram of circulating current suppression model.

Define $G_s(z) = Z(1/(L_0 s + R))$, then the expression of circulating current and its error can be written as

$$i_{diffk}(z) = \frac{G_{pr}(z)G_s(z)z^{-1}}{1 + G_{pr}(z)G_s(z)z^{-1}} \cdot (G_{lpf}(z)i_{diffk}^*(z))$$

$$- \frac{G_s(z)}{1 + G_{pr}(z)G_s(z)z^{-1}} \cdot \frac{u_{kh}(z)}{2} \tag{16}$$

$$e(z)=\frac{G_{lpf}(z)i^{*}_{difk}(z)+G_{s}\frac{u_{kh}(z)}{2}}{1+G_{pr}G_{s}z^{-1}} \tag{17}$$

For better extracting the dc component and filtering the second-order harmonic component, LPF's cutoff frequency is set to be 20Hz. By applying Tustin transformation, its transfer function can be modified as

$$G_{lpf}(z)=\frac{0.0141(z+1)}{z-0.9718} \tag{18}$$

If the grid frequency ripple is (50±0.5) Hz, then ($2\omega_0$) in (14) equals to 200Hz and (ω_c/π)=2Hz. Employing Tustin transformation, the z-domian form of quasi-PR can be seen as

$$G_{pr}(z)=K_p+\frac{A(z^2-1)}{z^2-1.982z+0.9974} \tag{19}$$

Taking T_s as the sample period, the expression of A is described as

$$A=\frac{K_r\omega_c T_s}{1+\omega_c T_s+(\omega_0 T_s)^2} \tag{20}$$

In experiment, arm inductance L_0 equals to 5mH, and its equivalent resistance is set to be 0.5Ω. By applying ZOH transmission, the impedance transfer function can be derived as

$$G_s(z)=\frac{0.0396}{z-0.982} \tag{21}$$

Only considering the action of disturbance, system's response should tend to be zero and the error equation is given by (23).

$$E_h(z)=0-i_{difk}(z)=\frac{G_s(z)}{1+G_{pr}(z)G_s(z)z^{-1}}\cdot\frac{u_{kh}}{2}(z) \tag{22}$$

Then, under unit step disturbance, the steady state error can be written as

$$\lim_{n\to\infty}e_h(nT_s)=\lim_{z\to 1}(1-z^{-1})E_h(z)=\frac{2}{1+2K_p} \tag{23}$$

If the relative error is required to less than 0.1, then $k_p>9.5$. Based on (19), (20), (22), the open-loop transfer function can be written as

$$G_o(z)=0.0396\frac{(K_p+A)z^2-1.982z+(0.9974K_p-A)}{z(z-0.9802)(z^2-1.982z+0.9974)} \tag{24}$$

When $K_p=10$, the bode diagram of open-loop system with the variation of A can be seen Fig.7.

From Fig.7, it can be seen that A only affect the magnitude around double line-frequency, and the magnitude increases with A. However, with the increase of A, the phase lag around resonant frequency also increases. Comprehensively considering magnitude and the corresponding phase lag, set A=0.13.

When $A=0.13$, we can see that the open-loop gain increases with K_p, and there is no significant difference at the point of resonant frequency, as shown as Fig.8. The larger K_p is beneficial to improve system stability precision, but it's not favorable to stability caused by the decrease of phase margin. Considering the requirement that the phase margin is usually 45°, the controller parameter K_p can set to be 12. When $A=0.13$, $K_p=12$, the phase margin and magnitude margin are 48° and 6.3 dB, respectively, which indicates that the system is stable.

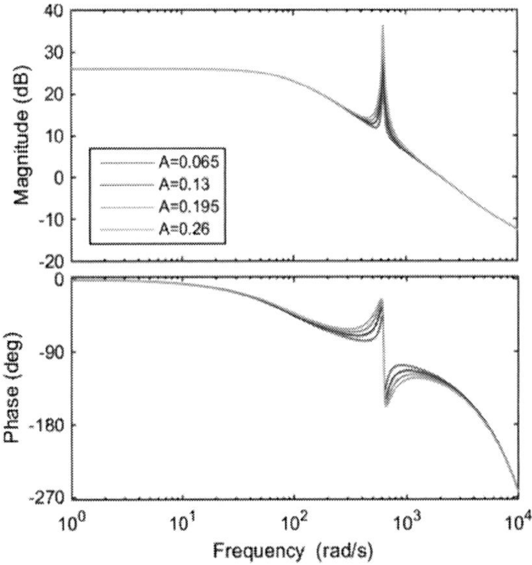

Fig. 7. Bode diagram of open-loop system ($K_p=10$)

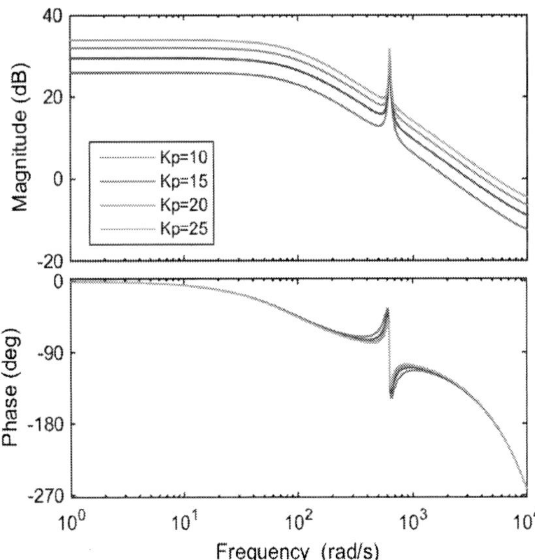

Fig. 8. Bode diagram of open-loop system (A=0.13)

VI. EXPERIMENTAL RESULTS

To validate the effectiveness of the improved method, a three phase MMC with passive load was built. The main circuit parameters and operating conditions are listed in Table I.

TABLE I. PROTOTYPE PARAMETERS

Items	Symbol	Value
Number SMs per arm	N	4
Arm inductance	L_0	5mH
Submodule capacitance	C	1.36mF
DC supply voltage	U_{dc}	400V
Submodular capacitor voltage	U_{Cki}	100V
Output RMS voltage	u_{ab}	240V
Output RMS current	i_a	5.1A
Output inductance	L_f	4mH
Output frequency	f_o	50Hz
Carrier frequency	f_s	5000Hz
Rated load	R_0	27 Ω

A. Apply traditional control strategy

By applying traditional control strategy, the experimental results are shown as Fig.10. Fig.10(a) gives the phase voltage between the passive load end. From Fig.10(b), the peak-peak value of submodule capacitor voltage u_{aC1}, u_{aC5} are all about 6V. Upper and lower arm currents of phase a are shown in Fig.10(c). Fig.10(d) gives the circulating current i_{diffa} and its peak-peak value is about 1.6A.

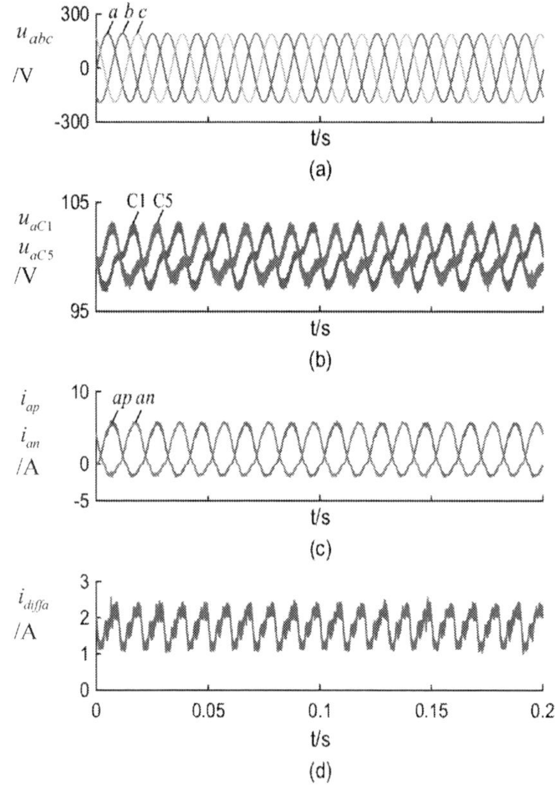

Fig. 9. Experimental results of MMC under the traditional control strategy.

B. Apply improved control strategy

By applying the improved control strategy, the experimental results are shown as Fig.11. Comparing Fig.11(a) with Fig.10(a), it can be seen that there is no obvious differences between the two ones. The peak-peak value of capacitor voltage in Fig.11(b) is about 7V, 1V larger than the voltage in Fig.10(b). However, from Fig.11(d) and Fig.10(d), it can be seen that the peak-peak value of circulating current is about 1.6A under the traditional control strategy and is reduced to 0.7A after the improved strategy being adopted. It is clear that the improved strategy has the ability to suppress circulating current and balance submodule capacitor voltage.

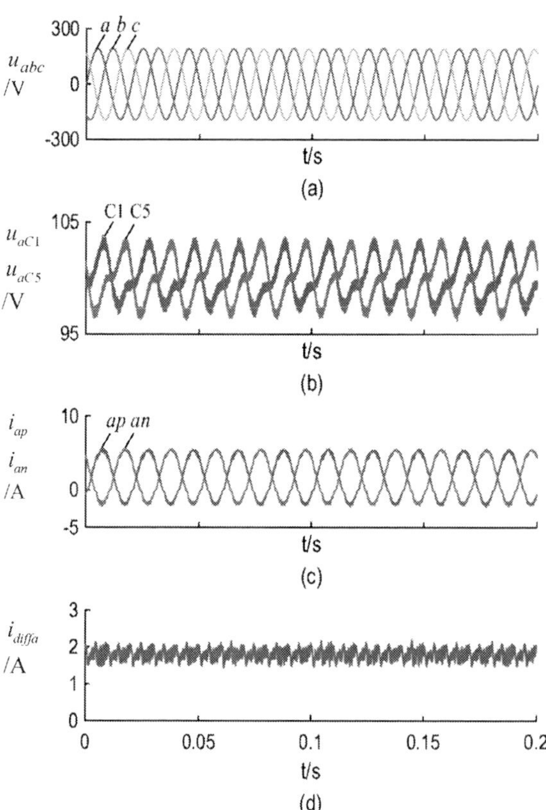

Fig. 10. Experimental results of MMC under the improved control

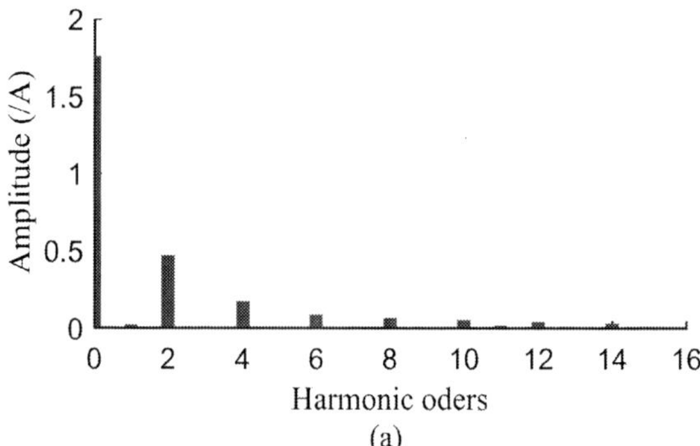

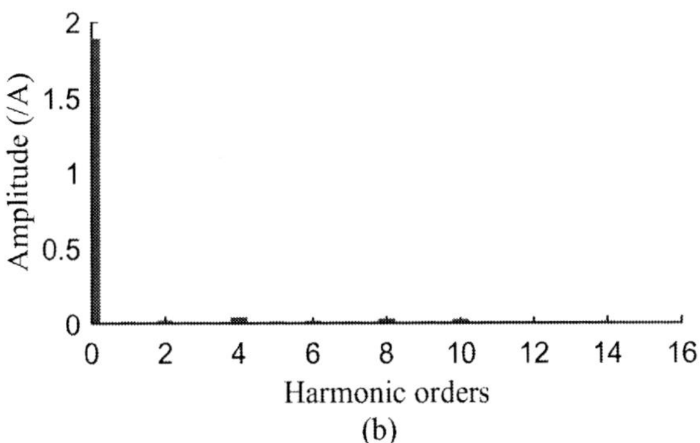

Fig.11. Spectrums of circulating current.

Fig.11.(a) represents the spectrum of circulating current under traditional control strategy and the one in Figure.11.(b) is the spectrum of circulating current under improved control strategy. It can be seen that the amplitude of second-order circulating harmonic current is about 0.47A under the traditional control strategy and is reduced to 0.02A after the improved strategy being applied.

VII. CONCLUSION

In this paper, the outer loop of the traditional control strategy is inherited by the improved control strategy for balancing submodule capacitor voltage. To suppress circulating current, the inner loop is modified by adding dc component extraction unit and quasi-PR controller. In addition, parameter design principles are introduced in discrete-domain for obtaining better effectiveness, and the stability of system are also analyzed. According to the comparison of experimental results before and after improvement, the feasibility and effectiveness of the improved control method are verified.

REFERENCES

[1]J. Rodriguez, J.-S Lai, Z. Fang, "Multilevel inverters: a survey of topologies, controls, and applications," *IEEE Trans. Ind. Electron.*, vol. 49, no. 4, pp. 724-738, Aug. 2002.

[2]S. Debnath, J. Qin, B. Bahrani, M. Saeedifard. "Operation, control, and applications of the modular multilevel converter: A Review," IEEE Trans. Power Electro, 2015, vol. 30, no. 1, pp. 37-53, Jan. 2015.

[3] L. Harnefors, A. Antonopoulos, S. Norrga, L. Angquist, "Dynamic analysis of modular multilevel converters," IEEE Trans. Ind. Electron., vol. 60, no. 7, pp. 2526-2537, Jul. 2013.

[4]D. Siemaszko, A. Antonopoulos, K. V, M. Vasiladiotis, L. Angquist, H.-P. Lennart, "Evaluation of control and modulation methods for modular multilevel converters," in Proc. Int. Power Electro. Conf. Jun. 2010, pp. 746-753

[5] A. Legnica, R.Marquardt, "An innovative modular multilevel converter topology suitable for a wide power range," Power Tech. Conf. Pro., Jun. 2003, pp. 23-26

[6] D. Siemaszko, A. Antonopoulos, K. Ilves, M. Vasiladiotis, L. Angquist, H.-P. Nee, "Evaluation of control and modulation methods for modular multilevel converters," Int. Power Elec. Conf., Jun. 2010, pp. 21-24.

[7] H. Liu, L. Poh, F. Blaabjerg, "Generalized modular multilevel converter and modulation," Int. Power Elec. Conf., May. 2014, pp. 2034-2038.

[8] I.A. Gowaid, G.P. Adam, S. Ahmed, D. Holliday, B.W. Williams, "Analysis and design of a modular multilevel converter with trapezoidal modulation for medium andhigh voltage DC-DC transformers," IEEE Trans. Power Electron., vol. 20, no. 10, pp. 5439-5457, May. 2015.

[9] A. Nami, J. Liang; F. Dijkhuizen, G.D. Demetriades, "Modular multilevel converters for HVDC applications: Review on converter cells and functionalities," IEEE Trans. Power Electron., vol. 30, no. 1, pp. 18-36, Jan. 2015.

[10] R. Feldman, M. Tomasini, E. Amankwah, J.C. Clare, P.W. Wheeler, D.R.

[11] Y. Guan, Z. Xu, "Modeling and Control of a modular multilevel converter-based HVDC system under unbalanced grid conditions," IEEE Trans. Power Electron., vol. 27, no. 12, pp. 4858-4867, Dec. 2012.

[12] P.M. Meshram, V.B. Borghate, "A simplifiednearest level control (NLC) voltage balancing method for modular multilevel converter (MMC)," IEEE Trans. Power Electron., vol. 30, no. 1, pp. 450-462, Jan. 2015.

[13] C. Gao; X. Jiang, Y. Li, Z. Chen, J. Liu, "A dc-link voltage self-balance method for a diode-clamped modular multilevel converter with minimum number of voltage sensors," IEEE Trans. Power Electron., vol. 28, no. 5, pp. 2125-2139, May. 2013.

[14] A. Dekka, B. Wu, N.R. Zargari, "A novel modulation scheme and voltage balancing algorithm for modular multilevel converter," IEEE Appl. Power Electro. Conf. Exp., pp.1227-1233, Mar. 2015.

[15] M. Moranchel, E.J. Bueno, F.J. Rodriguez, I. Sanz, "Novel capacitor voltage balancing algorithm for modular multilevel converter," Ann. Conf. Ind. Electron. Society, Oct. 2014, pp. 4697 – 4701.

[16] S. Rohner, S. Bernet, M. Hiller, R. Sommer, "Modulation, losses and semiconductor requirement of modular multilevel converters," IEEE Trans. Ind. Electron., vol. 57, no. 8, pp. 2633-2642, Aug. 2010.

[17] Q. TU, Z. Xu, H. Huang, J. Zhang, "Parameters design principles of the arm inductor in modular multilevel converter based HVDC", Proc. Int. Conf. Power Syst. Technol., 2010, pp. 1-6.

[18] Q. TU, Z. Xu, L. Xu, "Reduced switching-frequency modulation and circulating current suppression for modular multilevel converters," IEEE Trans. Power Del., vol. 26, no. 3, pp. 2009-2016, July. 2011

[19] A. Antonopoulos, L. Angquist, H.-P. Nee, "On dynamics and voltage control of the Modular Multilevel Converter," Eur. Conf. Power Electron. Appl., Sept. 2009, pp. 1-10.

[20] M. Hagiwara, H. Akagi. "Control and experiment of pulsewidth-modulated modular multilevel converters," IEEE Trans. Power Electron., vol. 24, no. 7, pp. 1737-1746, Jul. 2009.

[21] M. Zhang, L. Huang, W. Yao, Z. Lu. "Circulating harmonic current elimination of a CPS-PWM based modular multilevel converter with a Plug-In repetitive controller," IEEE Trans. Power Electron., vol. 29, no. 9, pp. 2083-2097, Apr. 2009.

Assymmetric Duty-Cycle Phase-Shift Modulation for Power Management in Double Half-Bridge Inverter with Partly Coupled Inductive Loads

C. Carretero
Dept. Fisica Aplicada
Universidad de Zaragoza
Zaragoza (Spain)
e-mail: ccar@unizar.es

H. Sarnago, O. Lucia, J. Acero, J.M. Burdio
Dept. Ingenieria Electronica y Comunicaciones
Universidad de Zaragoza
Zaragoza (Spain)

***Abstract*—Concentric coils are used in domestic induction heating system to improve the range of cookware sizes suitable to be heated up because the coils are switched on or switched off depending on the degree of their covering by the pot. The power is supplied by power converters, typically single-switch or half-bridge inverters, feeding the coils with a medium frequency current. However, the switching frequencies of the inverters connected to the coils has to be equal due to the coupling between the coils implies intermodulation noise which must be avoided in order to achieve a user-friendly product. The preceding switching frequency constraint limits the flexibility in the selection of the power delivered to each load. Phase-Shift Modulation has been proposed to overcome in part this drawback, but, in this work, the complimentary Asymmetric Duty Cycle Control has been also included in the modulation in order to enhance the power sharing between the loads. The proposed modulation scheme is analyzed in this paper, and its benefits and feasibility are tested experimentally using domestic induction heating inverter prototype.**

***Keywords*—induction heating, home appliances, power converters**

I. INTRODUCTION

Multiple coil induction heating system have been used with the purpose to improve the power distribution in the induction heating load where usually uniform surface power densities are required to achieve high-quality results [1]. Different coil arrangements have been proposed: small-size coils [2], aligned solenoid coils [3], and concentric circular coils [4]. The coils of the preceding are partly coupled due to the shared magnetic flux between coils [5, 6]. Each coil system can be connected to a single inverter; in that case, the power delivered to the load can be controlled by means of different parameters: switching frequency, phase-shift between the inverters, and, duty cycle of the inverters. With the sake of simplicity, the inverters works at the same switching frequency and the phase-shift between are modified in order to achieve the desired surface power density in the load, [7-9]. Synchronous switching frequency of the Phase-Shift Modulation (PSM), [10, 11], implies some advantages: the unsafe switching conditions are easy to be detected and the acoustic noise is reduced because the intermodulation products are avoided. However, this implementation implies the use of redundant elements, leading to an increased cost, and the output power control is severely limited by the switching condition and frequency constraints. In this paper, the previous works devoted to PSM applied to a double inverter feeding partly coupled induction heating loads, [12, 13], are revisited in order to include the duty cycle variation as a control parameter, [14], to achieve flexible power sharing between the loads. The Asymmetric Duty-Cycle Phase-Shift Modulation (ADPSM) [15-17] is described for the proposed system.

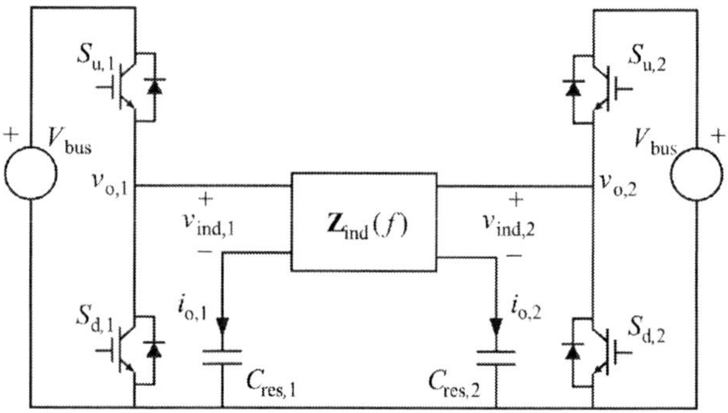

Fig. 1. Schematic of the double half-bridge inverter.

The remainder of this paper is organized as follows. Section II describes the dual resonant inverter topology under study and its partly-coupled induction heating load. Section III summarizes the main experimental results that prove the feasibility of this proposal. Finally, Section IV summarizes the main conclusions of this paper.

This work was partly supported by the Spanish MINECO under Project TEC2013-42937-R, Project CSD2009-00046, and Project RTC-2014-1847-6, by the DGA-FSE, by the University of Zaragoza under Project JIUZ-2014-TEC-08, and by the BSH Home Appliances Group.

978-1-4673-9551-9/16 $31.00 © 2016 IEEE

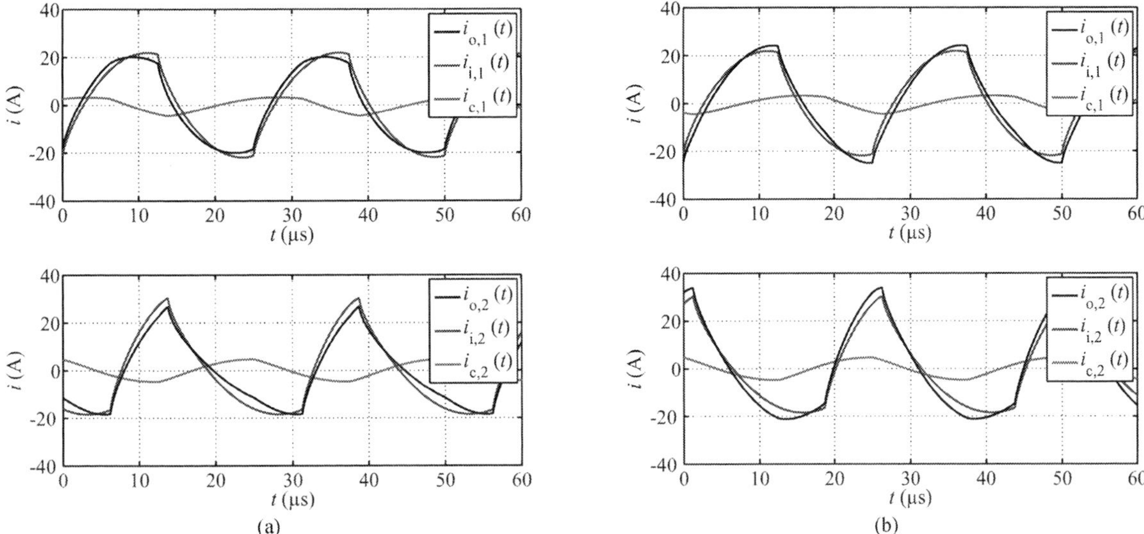

Fig. 2. Current waveforms at the switching frequency, f_{sw}, of 40 kHz (top: $i_{o,1}(t)$, bottom: $i_{o,2}(t)$. 10 μs/div, 20 A/div) with $D_1 = 0.5$ and $D_2 = 0.3$: (a) phase-shift, ϕ=90°, (b) phase-shift, ϕ=270°.

II. Double Half-Bridge Inverter with Partly-Coupled Inductive Loads

The analyzed induction heating system, shown in Fig. 1, is composed of a double half-bridge inverter feeding a coupled inductive load. Compared with previous analysis, this system features a partly coupled induction load which determines the final converter operation. Since the proposed system uses series resonance, the equivalent load will determine the resonant converter operating conditions and, thus, its output power, efficiency and reliable operation.

The equivalent matrix impedance $\mathbf{Z}_{ind}(f)$ of the system previously described is given by:

$$\mathbf{Z}_{ind}(f)=\begin{bmatrix} Z_{ind,1}(f) & Z_c(f) \\ Z_c(f) & Z_{ind,2}(f) \end{bmatrix} \tag{1}$$

where $Z_{ind,1}(f)$ and $Z_{ind,2}(f)$ are the independent self-impedance of the inductive loads (composed of resistive and inductive parts), and $Z_c(f)$ is the coupling term between both coils. Each load is series connected to resonant capacitor $C_{res,1}$ and $C_{res,2}$, respectively. At this point, it is important to remark that these parameters will be considered to be constant in the operating range. As a result, the equivalent impedance matrix of the resonant tank is:

$$\mathbf{Z}_{tank}(\omega)=\mathbf{Z}_{ind}(\omega)+\mathbf{Z}_{C_{res}}(\omega) = \begin{bmatrix} Z_{ind,1}(\omega)+\dfrac{1}{j2\pi f C_{res,1}} & Z_c(\omega) \\ Z_c(\omega) & Z_{ind,2}(\omega)+\dfrac{1}{j2\pi f C_{res,2}} \end{bmatrix} \tag{2}$$

The half-bridge inverters are voltage sources [18, 19], $v_{o,1}(t)$ and $v_{o,2}(t)$, respectively which are pulse train waveforms modulated by the switching frequency, f_{sw}, the same in both inverters for simplicity, and the duty cycle, D_1 and D_2 respectively. Moreover, both inverters are delayed a time t_d between, or, equivalently, they are phase-shifted $\phi=2\pi f_s \omega_d$. The voltages, $v_{o,1}(t)$ and $v_{o,2}(t-t_d)$, can be decomposed by the Fourier Transform. Consequently, since thee load impedance is known, the harmonics of the current driven by the load are provided by the following expression:

$$\begin{bmatrix} I_{o,1}(f) \\ I_{o,2}(f) \end{bmatrix} = \mathbf{Y}_{tank}(f)\cdot\mathbf{V}_o(f) = \begin{bmatrix} Y_{tank,1}(f) & Y_{tank,c}(f) \\ Y_{tank,c}(f) & Y_{tank,2}(f) \end{bmatrix}\cdot\begin{bmatrix} V_{o,1}(f) \\ V_{o,2}(f) \end{bmatrix} \tag{3}$$

where tank admittance matrix $\mathbf{Y}_{tank}(f)$ is the inverse of $\mathbf{Z}_{tank}(f)$. Applying the Inverse Fourier Transform to the current harmonics, we have:

$$i_{o,1}(t)=i_{i,1}(t)+i_{c,1}(t-t_d) \tag{4}$$

and

$$i_{o,2}(t)=i_{i,2}(t)+i_{c,2}(t+t_d) \tag{5}$$

where $i_{i,1}(t)$ and $i_{i,2}(t)$ are the independent currents carried by the loads when the opposite is switched off, whereas $i_{c,1}(t)$ and $i_{c,2}(t)$ are the zero-delay coupling currents between the loads. Note that the switching frequency and duty cycle modifies the waveforms of the currents, but, the phase-shift just delays the coupling currents with respect to the independent currents. Thus, the phase shift affects only to one current component and its analysis and influence on the output power can be simplified.

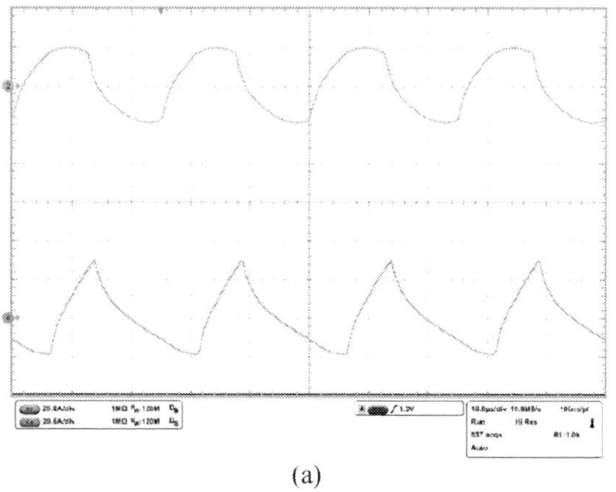

(a)

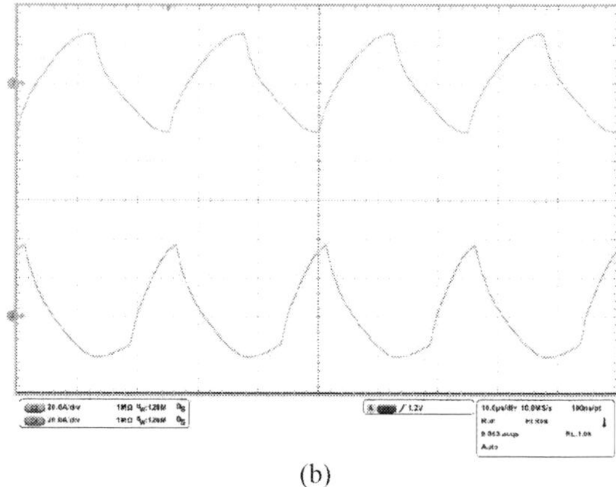

(b)

Fig. 3. Experimental waveforms at the switching frequency, f_{sw}, of 40 kHz (top: $i_{o,1}(t)$, bottom: $i_{o,2}(t)$. 10 μs/div, 20 A/div) with $D_1 = 0.5$ and $D_2 = 0.3$: (a) phase-shift, ϕ=90°, (b) phase-shift, ϕ=270°.

The currents carried by the two inductive loads are represented in Fig. 2. The switching frequency is 40 kHz, and the duty cycles are D_1=0.5 and D_2=0.3, respectively. In Fig. 2(a) are depicted the different current contribution at ϕ=90° and in Fig. 2(b) shows the currents at ϕ=270°. The waveforms components $i_{i,1}$(t), $i_{i,2}(t)$, $i_{c,1}(t)$ and $i_{c,2}(t)$ are independent to the phase-shift except for the relative delay between them. Both output current waveforms depends on the phase-shift because the addition of the independent current and the coupling current are different due to the relative delay between them.

The power provided by each power converter j is calculated as follows:

$$\begin{aligned} P_{o,j}(t) &= \langle v_{o,j}(t) i_{o,j}(t) \rangle \\ &= \langle v_{o,i}(t) i_{i,i}(t) \rangle + \langle v_{o,i}(t) i_{c,i}(t \mp t_d) \rangle \\ &= P_{i,j} + P_{c,j}(\mp t_d) \end{aligned} \quad (6)$$

where $\langle\ \rangle$ is the mean operator in a switching period, $P_{i,j}$ is the independent power which depends on f_{sw} and D_j, and $P_{c,j}(t_d)$ is the coupling power which also depends on the preceding parameters as well as the phase-shift between the inverters. As it has been established in the current analysis, it is important to remark that the phase-shift contribution to the output power can be analyzed independently. The adequate switching conditions basically are determined by the output switching current which varies with the phase-shift due to the coupling current effects.

III. Experimental measurements

In order to experimentally verify the analysis of the proposed control scheme, an experimental set-up composed of two synchronized half-bridge inverters has been built. The inverters are composed of Fairchild HGTG20N60 insulated gate bipolar transistors (IGBTs) with antiparallel diode. Both converters are controlled using a digital programmable devices which ensures frequency synchronization while allows generating the required modulations for the converters.

The inductive loads are a double concentric-coil induction heating systems which heats up a ferromagnetic pan. The measurements have been performed with a DPO-7354 Tektronix digital oscilloscope by means of two voltage differential probes, and two PEMUK ultra mini Rogowski current probes.

The experimental waveforms of the current driven by each load are shown in Fig. 3. The switching frequency has been selected to 40 kHz, and the duty cycles are D_1 = 0.5 and D_2 = 0.3, respectively. Two different phase-shifts have been applied: ϕ=90° shown in Fig. 3(a) and ϕ=270° depicted in Fig. 3(b). As it can be seen in both figures, the phase-shift modifies the current waveforms which also imply a different power delivered to each load.

Fig. 4 shows the power distribution between the loads at the switching frequency of 40 kHz when the variable phase-shift is applied. Fig. 4(a) shows the power sharing for duty cycles of D_1=0.5 and D_2=0.3, and Fig. 4(b) shows the power sharing for duty cycles of D_1=0.35 and D_2=0.45. The agreement between experimental data and the numerical results perfectly matches, verifying the proposed analysis model of the system. Moreover, as it can be seen in the figures, the different duty cycles constitutes a control parameter of the power delivered to the loads. Besides, the synchronous switching frequency implies a smooth modification of the working conditions with respect to the control variables, thus, the ease of system control is achieved. For this reason, this modulation scheme is proposed to effectively control output power in coupled induction heating systems.

IV. Conclusions and Future Work

Power management in a partly coupled double load system becomes a complex task with the converters associated with each loads must work at the same switching frequency due to noise reduction requirements. This is a common issue in domestic induction heating systems, where compact and flexible implementations lead to significant load coupling. The working conditions has to guarantee safe operation of the electronics given in soft-switching conditions, but, the coupling

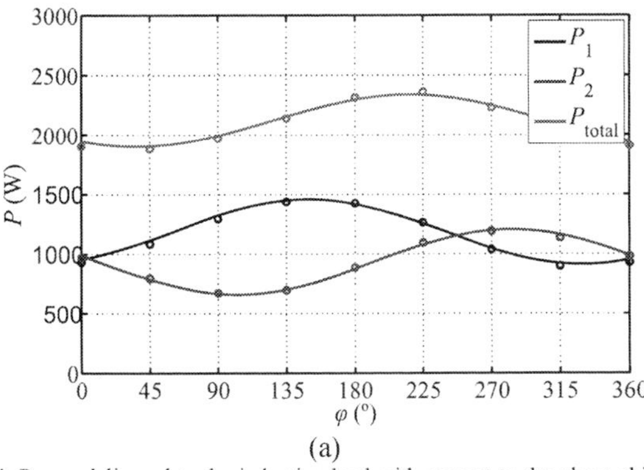

(a)

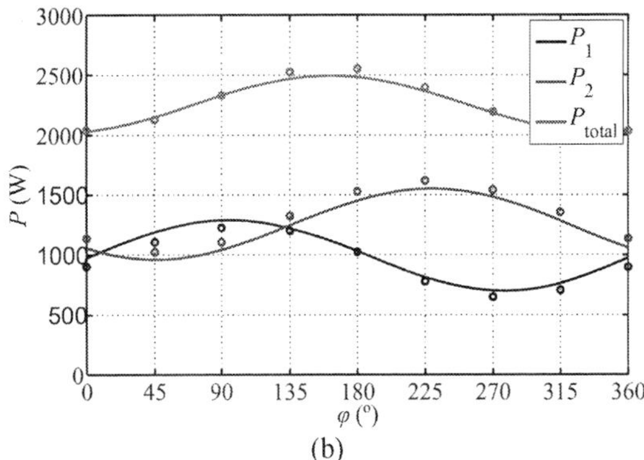

(b)

Fig. 4. Power delivered to the inductive load with respect to the phase-shift, ϕ, at the switching frequency, f_{sw}, of 40 kHz (circles: experimental measurements, continuous line: simulation values): (a) $D_1 = 0.5$ and $D_2 = 0.3$, (b) $D_1 = 0.35$ and $D_2 = 0.45$.

between the loads can imply inadequate switching conditions with the result of failure in the converter. In this paper, the ADPSM is proposed to achieve a high degree of flexibility in the share of the power delivered to the load under controlled operational conditions. The proposed analysis has extracted the main expressions, with special emphasis on the independent contribution of the phase shift to the waveforms and output power. Besides, the proposed description has been also validated experimentally.

The ADPSM provided in this work permits an in-deep study of soft-switching conditions at different frequencies, duty cycles and phase-shift suitable to be employed by the double half-bridge inverter in order to deliver the desired power to the partly coupled inductive loads.

References

[1] O. Lucía,J. Acero,C. Carretero, andJ.M. Burdío, "Induction heating appliances: Towards more flexible cooking surfaces," *IEEE Industrial Electronics Magazine,* vol. 7, pp. 35-47, September 2013.

[2] O. Lucia,J. Acero,C. Carretero, andJ.M. Burdio, "Induction heating appliances: towards more flexible cooking surfaces," *Industrial Electronics Magazine,* vol. 7, pp. 35 - 47, Sep. 2013.

[3] H. Pham,H. Fujita,K. Ozaki, andN. Uchida, "Estimating method of heat distribution using three-dimensional resistance matrix for Zone-Control Induction Heating systems," *IEEE Transactions on Power Electronics,* vol. 27, pp. 3374-3382 Jul. 2012.

[4] J. Egalon,S. Caux,P. Maussion,M. Souley, andO. Pateau, "Multiphase system for metal disc induction heating: Modeling and RMS current control," *IEEE Transactions on Industry Applications,* vol. 48, pp. 1692-1699, September/October 2012.

[5] J. Acero,C. Carretero,I. Millán,O. Lucía,R. Alonso, andJ. M. Burdío, "Analysis and modeling of planar concentric windings forming adaptable-diameter burners for induction heating appliances," *IEEE Transactions on Power Electronics,* vol. 26, pp. 1546-1558, 2011.

[6] W. G. Hurley, andM. C. Duffy, "Calculation of self and mutual impedances in planar magnetic structures," *IEEE Transactions on Magnetics,* vol. 31, pp. 2416-2422, Jul. 1995.

[7] Ha Ngoc Pham,H. Fujita,K. Ozaki, andN. Uchida, "Dynamic analysis and control for resonant currents in a Zone-Control Induction Heating system," *IEEE Transactions on Power Electronics,* vol. 28, pp. 1297-1307, Mar. 2013.

[8] M. Souley,S. Caux,J. Egalon,O. Pateau,Y. Lefevre, andP. Maussion, "Optimization of the settings of multiphase induction heating system," *IEEE Transactions on Industry Applications,* vol. 49, pp. 2444-2450, 2013.

[9] H. Fujita,N. Uchida, andK. Ozaki, "A new zone-control induction heating system using multiple inverter units applicable under mutual magnetic coupling conditions," *IEEE Transactions on Power Electronics,* vol. 26, pp. 2009-2017, Jul. 2011.

[10] H. Kifune,Y. Hatanaka, andM. Nakaoka, "Cost effective phase shifted pulse modulation soft switching high frequency inverter for induction heating applications," *IEE Proceedings - Electric Power Applications,* vol. 151, pp. 19-25, Jan. 2004.

[11] T. Mishima, andM. Nakaoka, "A load-power adaptive dual pulse modulated current phasor-controlled ZVS high-frequency resonant inverter for induction heating applications," *IEEE Transactions on Power Electronics,* vol. 29, pp. 3864-3888, 2014.

[12] C. Carretero,O. Lucia,J. Acero, andJ. M. Burdio, "Phase-shift control of dual half-bridge inverter feeding coupled loads for induction heating purposes," *Electronics Letters,* vol. 47, pp. 670-671, May 2011.

[13] C. Carretero,O. Lucía,J. Acero, andJ.M. Burdío, "Computational modeling of two partly-coupled coils supplied by a double half-bridge resonant inverter for induction heating appliances," *IEEE Transactions on Industrial Electronics,* vol. 60, pp. 3092-3105, August 2013.

[14] J. M. Burdio,L. A. Barragan,F. Monterde,D. Navarro, andJ. Acero, "Asymmetrical voltage-cancellation control for full-bridge series resonant inverters," *IEEE Transactions on Power Electronics,* vol. 19, pp. 461-469, Feb. 2004.

[15] E. Asa,K. Colak,M. Bojarski, andD. Czarkowski, "Asymmetrical Duty-Cycle and Phase-Shift Control of a novel multi-port CLL resonant converter," *IEEE Journal of Emerging and Selected Topics in Power Electronics,* vol. 3, pp. 1122 - 1131 Dec. 2015.

[16] Hu Jingang,Bi Chuang,Jia Kelin, andXiang Yong, "Power Control of Asymmetrical Frequency Modulation in a Full-Bridge Series Resonant Inverter," *Power Electronics, IEEE Transactions on,* vol. 30, pp. 7051-7059, 2015.

[17] J. Hu,c bi,K. Jia, andY. Xiang, "Power Control of Asymmetric Frequency Modulation in Full-Bridge Series Resonant Inverter," *Power Electronics, IEEE Transactions on,* vol. PP, pp. 1-1, 2015.

[18] H. Sarnago,O. Lucía,A. Mediano, andJ.M. Burdío, "Analytical model of the half-bridge series resonant inverter for improved power conversion efficiency and performance," *IEEE Transactions on Power Electronics,* vol. 30, pp. 4128-4143, August 2015.

[19] V. Esteve,J. Jordán,E. Sanchis-Kilders,E. J. Dede,E. Maset,J. B. Ejea, andA. Ferreres, "Comparative Study of a Single Inverter Bridge for Dual-Frequency Induction Heating Using Si and SiC MOSFETs," *IEEE Transactions on Industrial Electronics,* vol. 62, pp. 1440-1450, 2015.

Control implementation for a wide voltage range high efficiency power supply utilizing low voltage MOSFETs

Werner Konrad
Electric Drives and
Machines Institute
Technical University of Graz
Graz, Austria

Gerald Deboy
Infineon Technologies AG
Villach, Austria

Annette Muetze
Electric Drives and
Machines Institute
Technical University of Graz
Graz, Austria

Abstract—**The ever-increasing number of data centers requires huge amounts of electrical energy and the amount of renewable energy, of which only a fraction is renewable energy, requiring a considerable amount of additional combustion or nuclear power plants. To minimize the waste of energy and further reduce the energy costs of a modern data center, high conversion efficiencies are sought. This paper proposes a new concept for a power supply including details on the control strategy for a Switched Mode Power Supply (SMPS), achieving independent of high or low line input voltage the currently strictest efficiency classification of titanium, which is defined for high line by the 80 PLUS® organization. The key idea behind the proposed solution is the utilization of low voltage MOSFETs by splitting the input voltage, if required with a capacitive voltage divider, which makes it possible to utilize low voltage MOSFETs with a much better Figure of Merits (FOM) compared to high voltage MOSFETs. The paper presents details about the implementation of the control along with measurement results from a prototype in order to prove this new concept.**

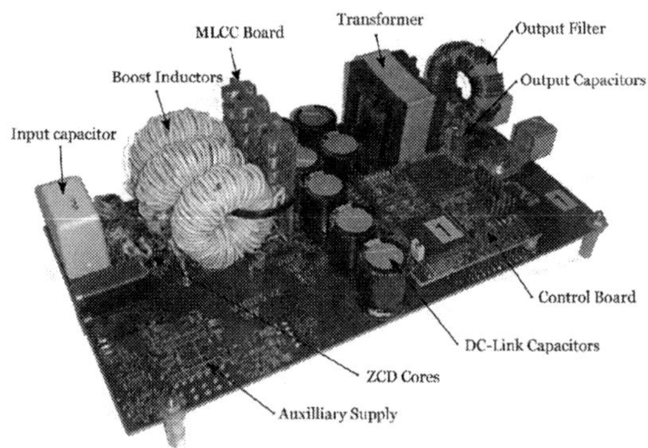

Fig. 1: Image of the prototype for one module.

I. Introduction

Over the last century different standards have evolved concerning the attributes of the mains, such as the amplitude and frequency. Nevertheless, a modern SMPS must be compatible with all possible grid configurations. One way of characterizing SMPS is to set efficiency and power factor requirements over the load range, which is done by the 80 PLUS® organization [1]. They set up different requirements and publish test results, making the choice easier for customers. The mains voltage can range anywhere from 110 to 240 volts, for low and high line mains respectively, depending on the country. Due to this international inconstancy, separate categories of classifications have been put in place by the 80 PLUS® organization. These requirements for high line are usually stricter than those for low line, since supplying the same amount of power requires only half the current, compared to a low line input, thus lowering the conduction loss at high line and thereby increasing the efficiency.

A new method was investigated that satisfies the high line titanium requirements for the 80 PLUS® organization by taking advantage of the better FOM of low voltage MOSFETs. In order to experimentally verify this concept a prototype was developed which is shown for one module in Fig. 1. With this new concept the stricter high line titanium criteria was met at both low and high line, where such a power supply is to date to the best of the authors knowledge not yet commercially available. The best listed power supply from the 80 PLUS® organization for titanium low line achieves, for example, an efficiency of 95.24% at 50% of its rated load. This is from Superflower®, in which 96% would be required in order to fulfill the titanium high line requirements. FOM offers a way to describe the performance of a power MOSFET, chosen here by the product of the total gate charge (Q_g) and on-resistance (R_{on}) which is commonly used in soft switched topologies. The proposed power supply consists of two modules [2], see Fig. 2 and Fig. 3 for the schematic, in which the modules can be connected in series, to divide the input voltage capacitively, or in parallel.

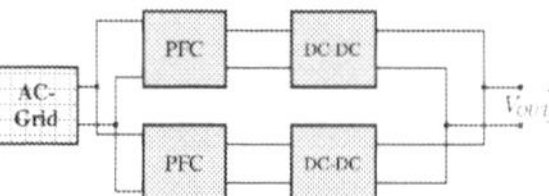

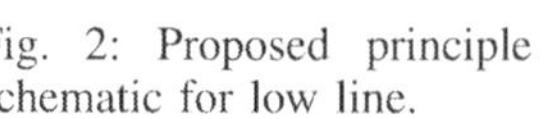

Fig. 2: Proposed principle schematic for low line.

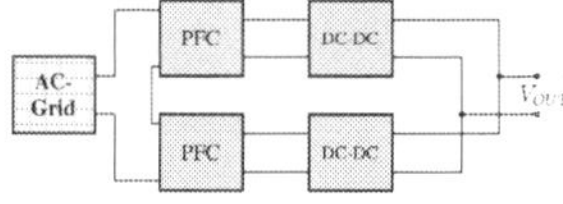

Fig. 3: Proposed principle schematic for high line.

To switch between the arrangements of the modules, semiconductor assisted relays can be used, which results in near zero conduction losses. One module consists of a Triangular

978-1-4673-9551-9/16 $31.00 © 2016 IEEE

Current Mode (TCM) Power Factor Correction (PFC) stage [3]–[7], followed by a traditional Phase Shifted Zero Voltage Switched (PS-ZVS) bridge for isolated DC/DC conversion, a control unit and an auxiliary power supply. With the utilization of a PS-ZVS bridge, the required efficiencies can be achieved with comparatively simple control without the need of more complicated DC/DC converters, such as an LLC converter.

The goal in the component placement was to stack two modules into each other for maximum power density. The total efficiency of the whole power supply is equal to the efficiency of a single module, since the modules are switched in series (high line) or in parallel (low line), depending on the line voltage. Therefore each module always sees the same input voltage. In the case of high line input and serial connection of the modules, an additional control was superimposed to keep input voltages balanced across the small input capacitors. The input capacitors which divide the grid input voltage are relatively small ($4.7\,\mu$F) in order to minimize the phase-shift between the line voltage and the input current, as well as save on space and cost.

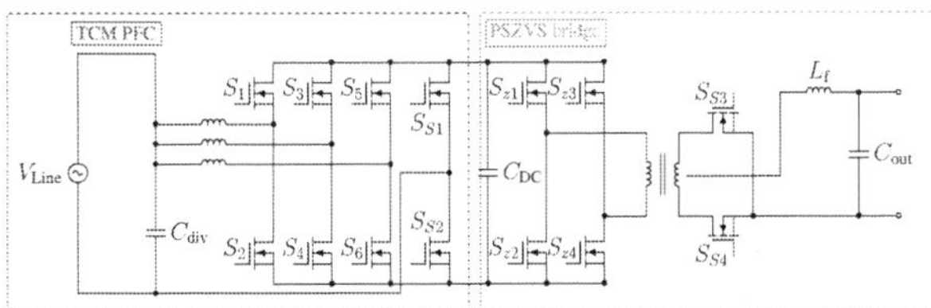

Fig. 4: Basic schematic for a single module consisting of the resonant triangular current mode PFC and a PSZVS bridge with center tap rectification on the secondary side.

II. Control Strategy

Each module is controlled by a combination of a microcontroller and a Field Programmable Gate Array (FPGA), in which all calculations for the control are performed on the microcontroller and the FPGA is used for generating the actual switching signals by receiving the switch timings from the microcontroller. The microcontroller used is a XMC 4500 from Infineon Technologies and the FPGA is an iCE40HX4k from Lattice. The control algorithm on the microcontroller has a sampling time of $50\,\mu$s which is much faster than the grid frequency. Therefore, for the calculation of the switching times, the grid voltage and the DC-link voltages can be viewed as DC-sources.

A. PFC

For the chosen control of the PFC, it is required to measure the line voltage, the DC-link voltage and the zero crossing events through each boost inductor. A simplified schematic, outlining the control is shown in Fig. 5, including a control for the DC-link voltage, a phase-shift control for setting a desired phase-shift between the phases to lower the Electromagnetic Interference (EMI) filter requirements, a voltage balancing control to balance the voltage across the input capacitors of the modules, and a control which ensures that the current drawn from the grid is sinusoidal.

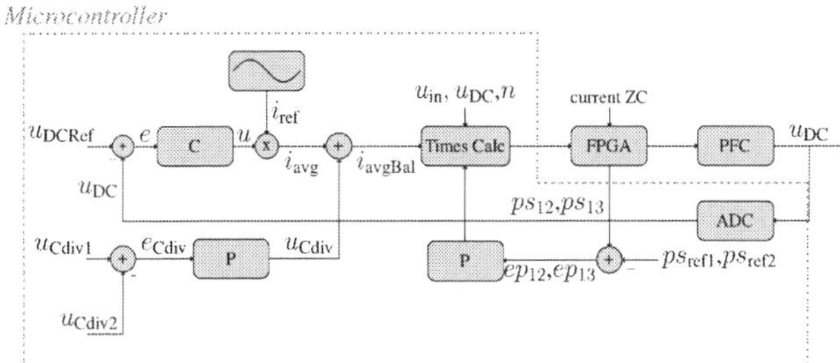

Fig. 5: Simplified block diagram of the PFC control.

1) Control strategy: In Fig. 5 the reference phase-shifts ps_{ref1} and ps_{ref2} are subtracted from the phase-shifts between the phases ps_{12} and ps_{13}, which are the calculated phase-shifts from the FPGA in reference to the master phase which is chosen as the first phase. The phase-shift error signals ep_{12} and ep_{13} are the input of proportional controllers which have an influence on the calculation of the switching times. The options for reducing the phase-shift errors are slowing down a phase, which is possible by increasing the switching period for a phase by simultaneously increasing the on and off time or speeding up a phase, which is only possible to a certain limit. In the case of slowing down a phase, both times must be increased in a manner that allows the average current for the switching period to stay the same. The case of a negative phase-shift error indicates that that phase needs to increase in frequency, which is limited by the average current i_{avg} requirement, since the required power can no longer be delivered. In this case the master phase is slowed down, providing the same effect. The timings are then calculated depending on the input voltage u_{in}, the DC-link voltage u_{DC}, the amount of active phases n and the required average current modified by the balancing controller i_{avgBal}.

The average current i_{avg} is derived by subtracting the current DC-link voltage u_{DC} from the desired DC-link voltage u_{DCref} which is the input of a compensator. The compensator can be either a PID or a state space compensator. The compensator output u is then multiplied by the sinusoidal reference current i_{ref}, which is in phase with the grid voltage.

The proposed power supply always operates with both modules active, depending on whether it operates at high line, in which case both modules are in series, or at low line, with both modules in parallel, forming either an Input Serial Output Parallel (ISOP) or Input Parallel Output Parallel (IPOP) system, respectively. For serial operation at high line input, the input voltages across each module must be balanced with a superimposed control, requiring that the input voltages for both modules are known. It is possible to implement this control on both modules or on only one module. For balancing the input voltage of one module, u_{Cdiv1} is subtracted from the input voltage of the other module u_{Cdiv2} and the error e_{Cdiv} is the input of a P-type controller. The output of the P-controller u_{Cdiv} is added to the required average current i_{ref} equalizing the module voltages. Fig. 6 shows a measurement of both modules running at high line, indicating that the voltage balancing control is working. In order to share the output current between the modules, one of both modules takes the role of a master and the other one is the slave. This configuration is important for both high and low line input cases. [8]

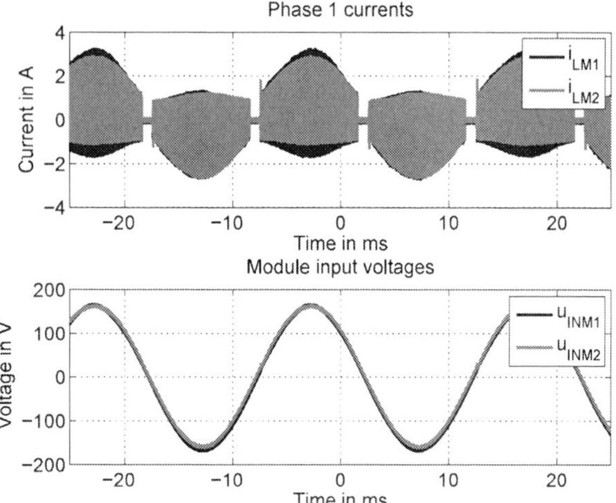

Fig. 6: Measurement of both modules running at 230 V in series connection; (top) currents for phase 1 of module 1 and 2; (bottom) module input voltages.

In addition to achieving high efficiencies at low loads, phases are removed or added, depending on the load which enables the PFC to remain in a high efficiency region. The phase shedding control consists of a state-machine and if i_{avg} exceeds or falls below a specified threshold, a phase is added or removed. The amount of active phases is only allowed to change during the zero crossings of the grid voltage. Furthermore, at each zero crossing only one phase can be added or removed at a time.

2) Zero current detection: For the Zero Current Detection (ZCD) a solution with a saturable transformer core was chosen, [9], [10]. The circuit is shown in Fig. 7 with the response in Fig. 8. For the core of the transformer, a high permeability material, namely Vitroperm 500f from Vacuumschmelze, was chosen. A more accurate response for the zero current events is achieved if the core saturates already at small currents. The point of saturation is adjusted with the core size and the amount of primary turns. The core thickness of the smallest available toroid-shaped core out of Vitroperm 500f was 1.5 mm. In order to lower the saturation current and also the core losses, the plastic mantle of the core was removed in order to peel the inner tape off until the thickness was reduced to only 0.3 mm.

3) Voltage measurements: The input voltages for both modules as well as for the DC-link voltage are all located on different grounds compared to the microcontroller. Therefore these voltages cannot be measured directly. Hence, these voltages are measured via an isolation amplifier and in order to keep the costs low, the ADUM 3190 from Analog devices was chosen. To compensate for the non-linearity towards the upper and lower end of its output range, a polynomial fit was used.

4) FPGA implementation: The FPGA receives the switch timings from the microcontroller via Quad Serial Peripheral Interface (qSPI) running at 10 MHz. For every phase of the PFC, a state machine, as shown in Fig. 9, is implemented

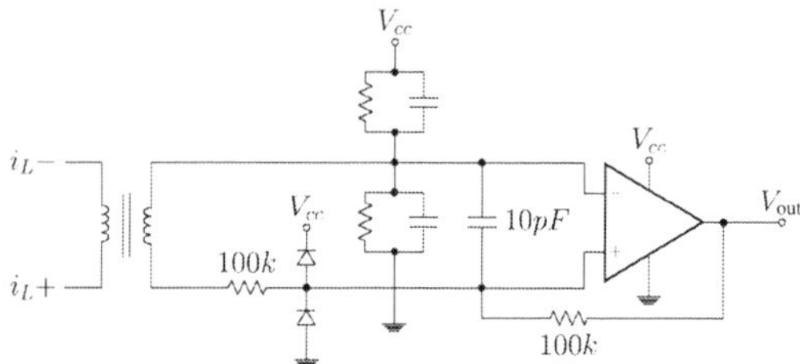

Fig. 7: Inductor current zero crossing detection circuit with a saturable transformer and a fast comparator with positive feedback.

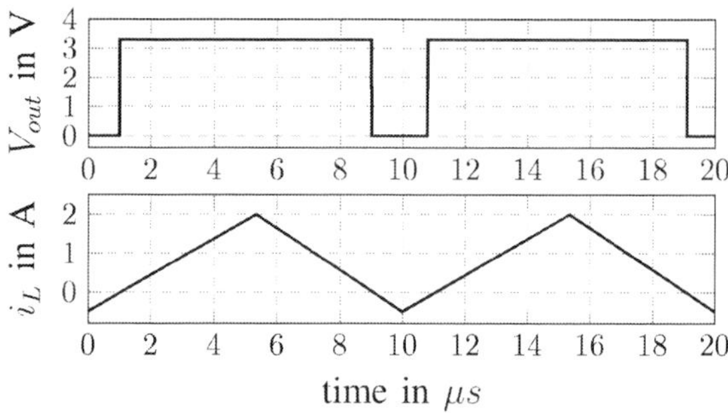

Fig. 8: Simulated output signal of the ZCD circuit; (above) output voltage of the ZCD circuit; (below) current through the transformer primary.

which transitions from energizing the inductor, to a resonant state, to de-energizing the inductor and again to a resonant state. During the resonant states, all switches for one leg are off (upper and lower switch). It is possible to calculate the resonant times T_{res1} and T_{res2} exactly. However, in order to save calculation power, they were implemented as constant times with enough time to allow for resonant switching in the worst case. Additionally, with the ZCD events, the phase-shifts between the phases of the PFC are measured and transmitted to the microcontroller.

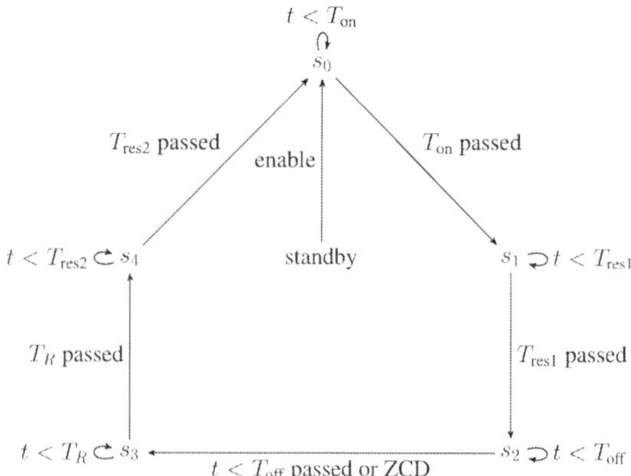

Fig. 9: State machine for each phase inside the FPGA.

B. PSZVS bridge

To convert the DC-link voltage from 200 V down to 12 V, a PS-ZVS bridge with center tap rectification on the secondary was implemented for the isolated DC/DC conversion. In order to allow resonant transitions down to low load conditions, an inductor in series to the primary of the transformer is required. The disadvantage to using the inductor is that it increases the duty-cycle loss, which also decreases the hold-up time, and thus requires a compromise, since the PS-ZVS bridge had a design constraint in order to achieve a hold-up time of 10 ms. Achieving zero voltage switching for the primary switches requires enough energy to be stored in the resonant inductor and the leakage inductance of the transformer which may be a problem for low load conditions. Nevertheless, since only 250 V MOSFETs were utilized and the DC-link voltage is regulated to 200 V, not turning on at zero, or near zero voltage will destroy the primary switches due to the voltage overshoot and the small voltage headroom of only 50 V. Therefore, it is imperative that the devices always turn on at zero or near zero voltage for all load conditions. Solutions to this end will be explained in the following section.

1) Control strategy: The PS-ZVS bridge is well documented [11] and known in industry and therefore various integrated solutions already exist on the market. For the presented prototype the control was implemented on the PFC required microcontroller and the FPGA. The measurement of the output voltage and adjustment of the duty cycle via a PI controller is handled by the microcontroller. The duty-cycle is then transmitted to the FPGA, where the PWM pattern for the switches is then generated.

2) Resonant inductor: The resonant inductor must be able to store enough energy at light load to discharge the resonance capacitor C_R, which is the sum of the parasitic MOSFETs' output capacitance and the transformer capacitance, see eq. 1, in which C_{MOS} is the parasitic output capacitance of the MOSFET and C_{Tr} is the parasitic capacitance of the transformer. [12]

$$E = \frac{1}{2}L_\sigma I_P^2 > \frac{4}{3}C_{MOS}U_{DC}^2 + \frac{1}{2}C_{Tr}U_{DC}^2, \qquad (1)$$

The stored energy in the resonance inductance must be greater than the required energy to charge/discharge C_{MOS} and C_{Tr} during low load. The formula shows that both the current and the inductance can be increased to store more energy. The insertion of an additional resonant inductor increases size and cost of the whole circuit; therefore integrating the resonant inductor into the transformer is preferred. By redirecting the flux away from the secondary, the leakage inductance of the transformer can be increased. This is done by inserting a flux conducting material between the primary and the secondary. Further, with the material parameters, the cross sectional area and the airgap, the inductance and the point of saturation can be adjusted. Adjusting the saturation point to a desired level allows a further minimization of the duty cycle loss along with the hold-up time loss. Fig. 10 shows how the material Vitroperm 500f was wrapped around the primary of the transformer. The secondary consists of copper foils, 2x200 μm thick, wrapped around the primary with the Vitroperm mounted on top. The transformer leakage inductance was thus experimentally adjusted to 10 μH.

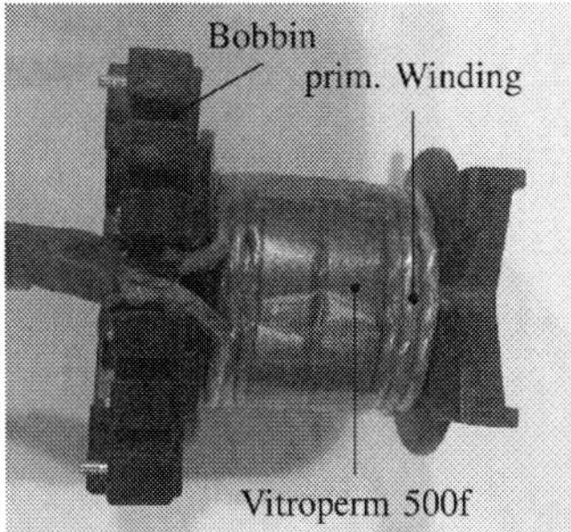

Fig. 10: Increasing the leakage inductance by reducing the magnetic resistance between the primary and the secondary with Vitroperm 500f.

3) Modified switching pattern: Increasing the primary current before the resonant transition occurs also increases the stored energy in resonance inductance. The primary current can be increased by turning both synchronous rectifier switches on during the power delivery phase (S_{z1} and S_{z4} or S_{z2} and S_{z3} are on simultaneously). Since the output of the transformer is then shortened, the primary current rises rapidly with the resonance inductance as the only limitation for the di/dt. During low load, which is indicated to the FPGA by only one phase of the PFC being active, the typical switching pattern for the PS-ZVS is modified, see Fig. 11. This figure shows the gate signals of the primary switches as well as from the secondary switches the primary voltage across the transformer U_P and the primary current through the transformer I_P. During the power delivery phase normally only one synchronous rectification switch is turned on at a time. In this proposed modified switching pattern the second synchronous rectification switch is turned on as well, shortly before the power delivery phase ends, shortening the secondary of the transformer which leads to a rapid increase of the current in the transformer. This is illustrated in Fig. 11, in which the primary current I_P for normal operation, shown as a black solid line, and for the modified operation, shown as a dashed line, are compared. This short circuit happens only for a very short time and was 20 ns for the presented prototype, which is also the clock period for the FPGA. This proposed switching pattern showed an improvement in the possible achievable zero voltage switching (ZVS) range as well as an improvement of efficiency for low loads.

Fig. 12 shows the experimental verification for the modified switching pattern for 10% load.

4) Burst-mode operation: With the modified switching pattern at low loads, the PS-ZVS bridge can be operated at low loads while still achieving zero voltage turn-ons. For small loads of 10% and below, a burst-mode was implemented to improve the efficiency. During burst-mode only a few cycles are switched and then for a time all switches cease their switching operation. The burst-mode offers the advantage of decreasing the switching and transformer core losses at the cost of a bit more conduction loss. If the PS-ZVS bridge does not switch during the pause in the burst-mode, the voltages

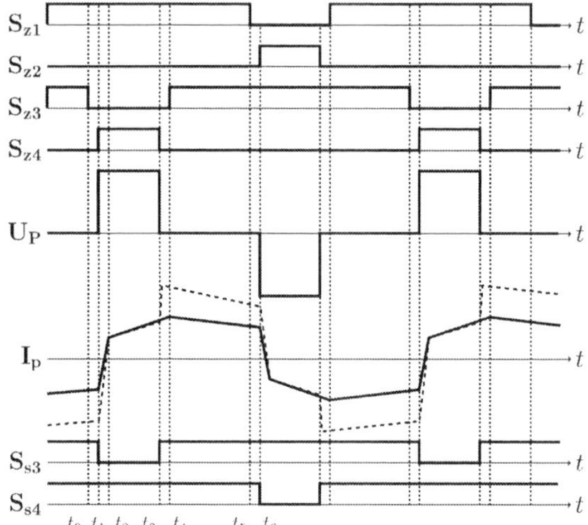

Fig. 11: Extending the zero-voltage turn-on range by slightly changing the synchronous rectifier switching pattern to increase the current in the leakage inductance (dashed gray line for the primary current); (top) primary transformer current; (middle) voltages across S_1 and S_2; (bottom) digital signals for the switches from S_1 (top most) to S_6 (lowest).

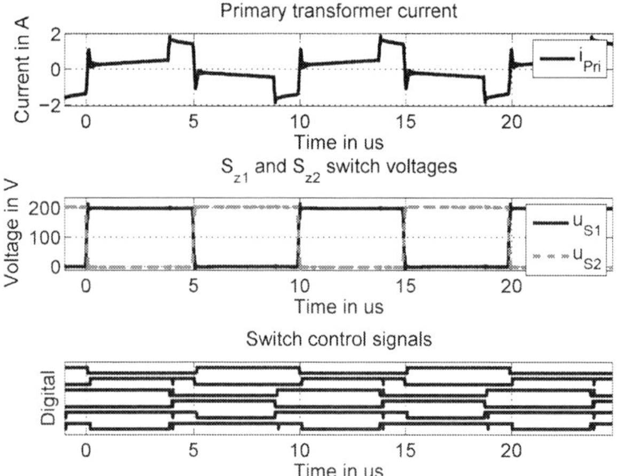

Fig. 12: Measurement for normal operation with extended zero voltage switching at 10% load, (top) primary transformer current; (middle) voltages across S_{z1} and S_{z2}; (bottom) digital signals for the switches S_{z1} (top most) to S_{z4} and further S_{s3} and S_{s4} (lowest).

across the switches become half the DC-link voltage due to the capacitive voltage divider which is formed with the parasitic output capacitors of the MOSFETs. In a traditional PS-ZVS application with a DC-link voltage of around 400 V and devices which have a breakdown voltage of 600 or 650 V, the voltage across one MOSFET during the pause in burst-mode is at half the DC-link voltage at around 200 V. In this scenario, it is possible to just turn on the MOSFET without reaching the breakdown voltage of the device, since there is enough headroom available. However, in this proposed solution the headroom for reaching the breakdown MOSFETs is much smaller and a turn-on at half the DC-link voltages would hit the breakdown voltage of the device and would in turn cause device failure or unwanted additional losses. In burst mode, when no switch is operating during the burst pause, the voltages across the switches settle at half the DC-link voltage and a turn-on of one MOSFET would destroy the other MOSFET connected in series due to an overvoltage spike because of the stray inductance and the high di/dt. Therefore some modification to the burst-mode switching pattern is needed to reduce the voltage across the switches at the first turn-on after the pause in burst-mode. After the inactive state of the PS-ZVS bridge, meaning no switch has switched for a longer time and all switches are in their off-state, one of the synchronous rectifier switches is turned on before any other switching operation happens. In case there is already an established output voltage, energy is transferred back to the primary. It is then possible to almost completely discharge the output capacitors of the primary switches. This is verified with measurement results in Fig. 13. It is shown that, for example, if the synchronous rectifier S_5 is switched on, the voltage across S_2 decreases and the voltage across S_1 increases, this enables S_2 to be turned on at nearly zero voltage, reducing the overvoltage spike at the first turn-on after the burst break dramatically. [13]

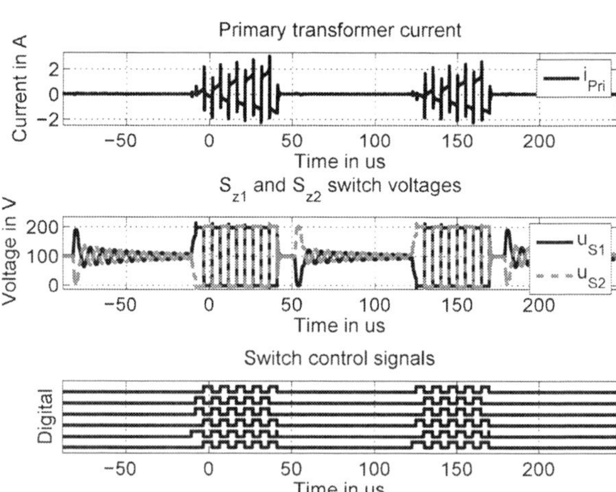

Fig. 13: Measurement of the burst-mode with the modified burst-mode pattern to reduce the voltage across the primary switches after the break; (top) primary transformer current; (middle) voltages across S_{z1} and S_{z2}; (bottom) digital signals for the switches S_{z1} (top most) to S_{z4} and further S_{s3} and S_{s4} (lowest).

Fig. 13 further shows that the primary current increases with each switching cycle. This is because the current in the output filter drops to zero during the pause in burst-mode and it is necessary to build up the current again. Due to this, the peak currents in burst-mode are higher than in the normal operation mode, thus causing higher resistive losses. This, in turn, implies that for higher loads, burst-mode is no longer a viable solution since the resistive losses preponderate the switching and transformer loss. Another disadvantage of burst-mode at higher loads is the ripple of the output voltage which might reach unacceptable values.

C. Complete module

Both modules need to operate together and certain parameters are shared between the modules, depending on whether they are operating in series at high line or in parallel at low line. At low line the whole system becomes an IPOP system in which the output current is shared and at high line becomes an ISOP system, in which the output current as well as the input voltage are shared. For stable operation it is required that these parameters are always shared equally, [14]; whereby one possibility is by measuring and controlling these parameters, [15]–[17]. In [14] it is analytically and experimentally shown that by balancing the input voltages of the modules, output current sharing for ISOP or output voltage sharing for Input Serial Output Serial (ISOS) modules is achieved without any current measurement, if the efficiencies of the modules are equal. For the IPOP case at low line, another possibility is to use a master/slave concept in which one module is the master with active control and the slave receives the required controller output values from the master module. In this case the slave receives the required average current i_{avg} for the PFC from the master. Therefore, the DC-link voltage control is disabled on the slave. In addition, the duty cycle for the PS-ZVS bridge is as well received from the master, meaning both output voltage controllers are disabled on the slave. Experiments further showed a sharing of the DC-link voltage between the modules. This was tested by disabling the PFC of one module but still operating both of the PS-ZVS bridges. For the slave module in which the PFC was disabled, the DC-link voltage was nearly equal to the DC-link voltage of the master module (only slightly lower). Since the PS-ZVS of the slave was still operating, the power flow in the PS-ZVS bridge from the slave was reversed, indicating a strong natural balancing effect for the DC-link voltage.

III. Measurement Results

In order to prove the concept, a prototype was constructed with a maximum output power of 1 kW at 12 V output in which each module delivers 500 W with a power density of 1 kW/dm^3.

1) Efficiency measurement: The efficiency for the whole module was measured with a Norma 5000 power analyzer and is shown in Fig. 14. It can be investigated that at 10% load burst-mode is required to meet the titanium requirements. For higher loads the more strict titanium criteria is achieved in normal operation.

2) Thermal measurements: Fig. 15 shows thermal images of one module in operation at 50% load and at 100% load. Further, it shows that the transformer of the PS-ZVS bridge

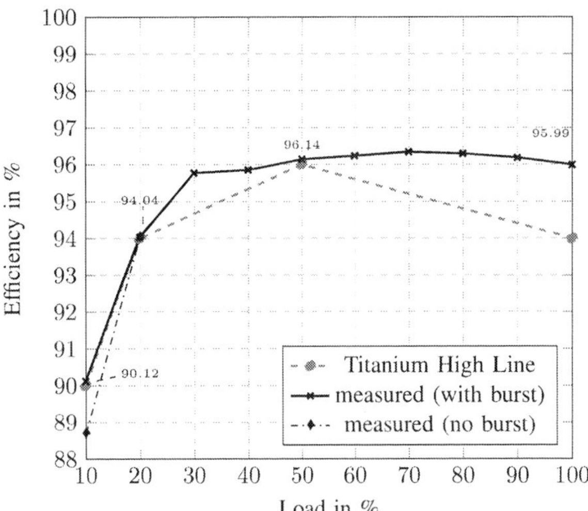

Fig. 14: Measured efficiency for the complete module.

is always the hotspot. This suggests that an improvement to the transformer design, for example, with a professional transformer construction, compared to the handmade manufactured one could reduce the transformer losses due to an increased copper fill factor.

IV. Conclusion

This paper presented a new concept for a server power supply meeting the currently strictest efficiency requirement of titanium for high line independent of the line voltage by utilizing the better FOM in low voltage MOSFETs. The concept is further experimentally demonstrated with a prototype backed up with measurements, such as the total efficiency, waveforms and thermal images. Furthermore, it demonstrates that for achieving such efficiency levels next generation semiconductors like Gallium Nitride (GaN) or Silicone Carbide (SiC) are not necessary and that the DC/DC conversion can be done with a robust and well-known PS-ZVS bridge. This leaves room for future research, for example, by replacing the DC/DC converter with an LLC converter which should then achieve even higher efficiencies. Another possibility is an increase in the amount of modules to use even lower voltage MOSFETs, which is currently another active research topic for telecom power supplies [18]–[20]. In the future, wide bandgap devices and new multi-cell topologies are expected to contribute in pushing the limits of power electronics further, by continuing the trend established in 1970 of doubling the power density of converters every 10 years, [21].

References

[1] 80plus, "Plugloadsolutions." http://www.plugloadsolutions.com. accessed on 06-10-2014.

[2] W. Konrad, C. Rainer, G. Deboy, and A. Muetze, "A power supply reaching titanium level efficiency for a wide range of input voltages," in *Energy Conversion Congress and Exposition (ECCE), 2013 IEEE*, pp. 3398–3405, Sept 2013.

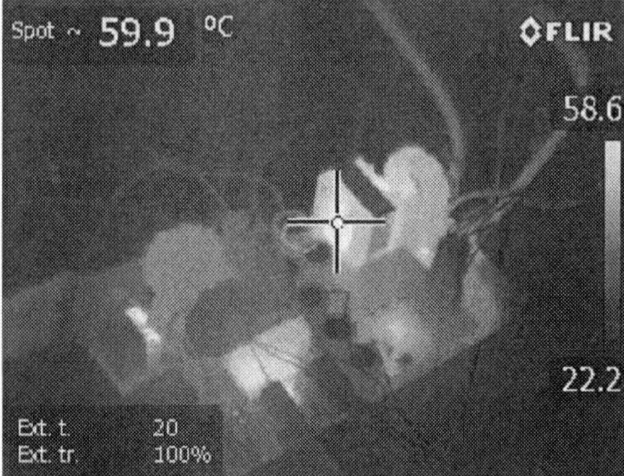

(a) 50% load

(b) 100% load

Fig. 15: Thermal images of one module in operation; (above) at 50% load; (below) at 100% load.

[3] C. Marxgut, J. Biela, and J. Kolar, "Interleaved triangular current mode (tcm) resonant transition, single phase pfc rectifier with high efficiency and high power density," in *Power Electronics Conference (IPEC), 2010 International*, pp. 1725–1732, June 2010.

[4] C. Marxgut, J. Biela, and J. Kolar", "Design of a multi-cell, DCM PFC rectifier for a 1mm thick, 200W off-line power supply; the power sheet," in *Integrated Power Electronics Systems (CIPS), 2010 6th International Conference on*, pp. 1–6, March 2010.

[5] G. Deboy and H. Kapels, "System and method for controlling a converter." Patent US 8,026,704.

[6] J. Biela, J. Kolar, and G. Deboy", "Optimal design of a compact 99.3% efficient single-phase PFC rectifier," in *Applied Power Electronics Conference and Exposition (APEC), 2010 Twenty-Fifth Annual IEEE*, pp. 1397–1404, Feb 2010.

[7] A. Stupar, T. Friedli, J. Miniboeck, M. Schweizer, and J. Kolar", "Towards a 99% efficient three-phase buck-type PFC rectifier for 400 V DC distribution systems," in *Applied Power Electronics Conference and Exposition (APEC), 2011 Twenty-Sixth Annual IEEE*, pp. 505–512, March 2011.

[8] S. Luo, Z. Ye, R.-L. Lin, and F. Lee, "A classification and evaluation of paralleling methods for power supply modules," in *Power Electronics Specialists Conference, 1999. PESC 99. 30th Annual IEEE*, vol. 2, pp. 901–908 vol.2, 1999.

[9] B. Wrzecionko, L. Steinmann, and J. Kolar, "Fast high-temperature (250°C / 500°F) isolated DC and AC current measurement: Bidirectionally saturated current transformer," in *Power Electronics and Motion Control Conference (IPEMC), 2012 7th International*, vol. 1, pp. 480–487, June 2012.

[10] J. Biela, D. Hassler, J. Miniboek, and J. Kolar, "Optimal design of a 5kw/dm3 / 98.3% efficient TCM resonant transition single-phase PFC rectifier," in *Power Electronics Conference (IPEC), 2010 International*, pp. 1709–1716, June 2010.

[11] M. Jovanovic, W. Tabisz, and F. Lee, "Zero-voltage-switching technique in high-frequency off-line converters," in *Applied Power Electronics Conference and Exposition, 1988. APEC '88. Conference Proceedings 1988., Third Annual IEEE*, pp. 23–32, Feb 1988.

[12] J. Sabate, V. Vlatkovic, R. Ridley, F. Lee, and B. Cho, "Design considerations for high-voltage high-power full-bridge zero-voltage-switched pwm converter," in *Applied Power Electronics Conference and Exposition, 1990. APEC '90, Conference Proceedings 1990., Fifth Annual*, pp. 275–284, March 1990.

[13] W. Konrad, *A high efficiency modular wide input voltage range power supply*. Dissertation, Technische Universitt Graz, 2015.

[14] W. Chen, X. Ruan, H. Yan, and C. Tse, "Dc/dc conversion systems consisting of multiple converter modules: Stability, control, and experimental verifications," *Power Electronics, IEEE Transactions on*, vol. 24, pp. 1463–1474, June 2009.

[15] R. Ayyanar, R. Giri, and N. Mohan, "Active input-voltage and load-current sharing in input-series and output-parallel connected modular dc-dc converters using dynamic input-voltage reference scheme," *Power Electronics, IEEE Transactions on*, vol. 19, pp. 1462–1473, Nov 2004.

[16] A. Bhinge, N. Mohan, R. Giri, and R. Ayyanar, "Series-parallel connection of dc-dc converter modules with active sharing of input voltage and load current," in *Applied Power Electronics Conference and Exposition, 2002. APEC 2002. Seventeenth Annual IEEE*, vol. 2, pp. 648–653 vol.2, 2002.

[17] R. Giri, R. Ayyanar, and N. Mohan, "Common duty ratio control of input series connected modular dc-dc converters with active input voltage and load current sharing," in *Applied Power Electronics Conference and Exposition, 2003. APEC '03. Eighteenth Annual IEEE*, vol. 1, pp. 322–326 vol.1, Feb 2003.

[18] M. Kasper, C. Chen, D. Bortis, J. Kolar, and G. Deboy, "Hardware verification of a hyper-efficient (98%) and super-compact (2.2kw/dm3) isolated ac/dc telecom power supply module based on multi-cell converter approach," in *Proceedings of the 30th Applied Power Electronics Conference and Exposition (APEC 2015), Charlotte, NC, USA, March 15-19, 2015*, March 2015.

[19] J. Huber and J. Kolar, "Optimum number of cascaded cells for high-power medium-voltage multilevel converters," in *Energy Conversion Congress and Exposition (ECCE), 2013 IEEE*, pp. 359–366, Sept 2013.

[20] M. Kasper, D. Bortis, and J. Kolar, "Scaling and balancing of multi-cell converters," in *Power Electronics Conference (IPEC-Hiroshima 2014 - ECCE-ASIA), 2014 International*, pp. 2079–2086, May 2014.

[21] J. Kolar, U. Drofenik, J. Biela, M. Heldwein, H. Ertl, T. Friedli, and S. Round, "PWM converter power density barriers," in *Power Conversion Conference - Nagoya, 2007. PCC '07*, pp. P-9–P-29, April 2007.

A Single-Phase Dual Frequency Inverter Based on Multi-Frequency Selective Harmonic Elimination

Chongwen Zhao, Daniel Costinett, Brad Trento
EECS department, University of Tennessee
Knoxville, TN, USA

Daniel Friedrichs
Medtronic
Boulder, CO, USA

Abstract: **Simultaneous generation of two AC outputs at different frequencies from a single-phase inverter offers practical benefits and control flexibility for many industrial applications. In this paper, a dual-frequency selective harmonic elimination (DFSHE) modulation method is proposed to generate and control two individual frequencies in an H-bridge, while diminishing undesired harmonics. Two AC elements are synthesized independently in the modulation scheme, and thus flexible individual power regulation is achieved via the proposed method. In addition, both unipolar and bipolar DFSHE cases are investigated. Characteristics of the two methods are compared and alternatives are provided for different applications. The generation algorithms of the DFSHE are also studied in this paper, and can be applied to a variety of DC/AC topologies without adding extra switching devices. Finally, the experimental results from a 100W dual-load single-phase inverter verify the effectiveness of proposed method, where 50 kHz and 450 kHz AC outputs are generated and individually regulated.**

Keywords—dual frequency; inverter; SHE; GaN.

I. INTRODUCTION

In conventional single-phase DC/AC inverters, an AC waveform at a single-frequency is generated to feed loads using PWM-based modulation schemes. Generating a dual-frequency AC output, compared to a single-frequency AC output, has advantages in many industrial applications. Dual frequency induction heating (DFIH), for example, is desirable for hardening ununiformed part surfaces [1][2]. Electrosurgical power supplies [3], multi-load inverters [4][5], and multi-pickup wireless power transfer systems [6] also stand to benefit from simultaneous dual-frequency conversion.

In the electrosurgical power supply application in particular, clinical applications require precise control over the frequency and power of each output. A single phase inverter is designed for electrosurgical application in this paper, where a 50 kHz ultrasonic (US) output for dissection and sealing, and a 500 kHz RF output to cut/coagulate tissue. Also, low order harmonics at RF output is dangerous and zero harmonics are required for surgical performance, since those harmonics below 100 kHz will cause muscle contraction.

To provide a waveform with two significant AC frequency components, a simple solution is to use two resonant inverters with series or parallel connected outputs [2][7]. The dual-frequency, dual-inverter configuration fundamentally requires two independent inverters and filters, which increases the numbers of switching devices and passive components and overall cost compared to a single-inverter counterpart.

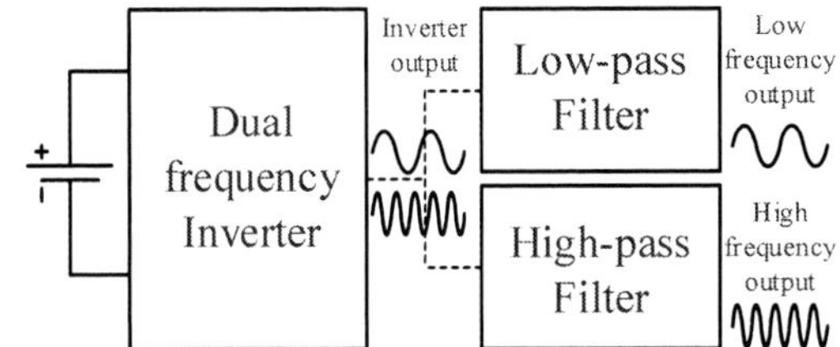

Figure. 1. An ideal dual-frequency inverter for electrosurgical application.

A variety of dual-frequency non-simultaneous AC output techniques, based on a single-inverter configuration, have been reported for varying applications [1,8-9]. However, these dual or multi-frequency generation methods are based on time division multiplexing, which make them unsuitable for simultaneous dual-frequency AC output applications. Simultaneous dual-frequency inversion has been demonstrated for induction heating [2, 4, 10] and wireless power transfer [6, 11], but these works have largely done so either through additional hardware, or through harmonic injection with a spread spectrum, resulting in significant low-order harmonics and coupling between the two frequencies.

In this work, it is desired to generate two distinct frequency outputs from a single inverter, as it shown in Fig.1, where two different frequencies are generated with individual regulation, and two filters are employed to separate them and feed different loads. Minimizing power is achieved at all other frequencies. To decouple and control the amplitude of simultaneous dual-frequency outputs, a dual-frequency selective harmonic elimination (DFSHE) modulation scheme is proposed. In this modulation method, the fundamental and certain harmonics are independently controlled, allowing individual power regulation. Also, undesired harmonics can be easily attenuated, which reduces extra losses and EMI issues. Moreover, the proposed DFSHE modulation schemes are applicable to a variety of DC/AC topologies, such as half-bridge, full bridge, multilevel inverter, and resonant type inverters, and facilitate power regulation for multi-load system.

The paper is organized as follows. The principle of the proposed DFSHE modulation method is presented in Section II. The characteristic investigation of two DFSHE cases are presented in Section III. The experimental prototype and experimental results are presented in Section IV. Conclusions are stated in Section V.

978-1-4673-9551-9/16 $31.00 © 2016 IEEE

II. DFSHE PRINCIPLE

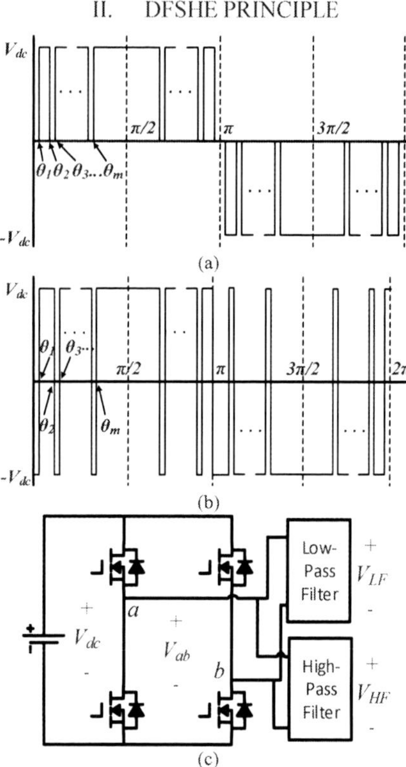

Figure. 2. Three level unipolar (a) and two-level bipolar (b) DFSHE ¼ symmetric waveforms, and the full bridge inverter(c).

Compared to traditional PWM modulation, which controls only the fundamental frequency of the output waveform through modulation of duty cycle at constant frequency, selective harmonic elimination (SHE) [12] modulation varies both switching frequency and duty cycle per switching period in order to generate an output in which a large range of the output spectrum is directly controlled. The SHE method uses Fourier analysis, based on the desired output spectrum, to synthesize a pulse train, consisting of a number of discrete switching instances. Each switching instance is defined by switching angles relative to the fundamental period. Traditional single-frequency SHE modulation was developed to control a fundamental, often line-frequency, element while eliminating a number of low order harmonics, limited by the number of switching angles.

For example, if the modulation consists of m switching angles is per quarter-period, a finite number of harmonics of the fundamental $n=f(m)$ can be controlled, where n is a function of the number of switching angles and modulation scheme (unipolar, bipolar, multi-level, etc.). Harmonics at higher frequency than this are unregulated. Note that $f(m)$ is a monotonic function, indicating that more harmonics may be suppressed by increasing the number of switching angles, m, which increases the equivalent switching frequency of the converter.

The equivalent switching frequency f_{eq} is defined as:

$$f_{eq} = (2m-1)\cdot f_{LF} \qquad (1)$$

The f_{LF} in (1) refers to the fundamental frequency in DFSHE modulation scheme.

In this work, the SHE method is extended to a dual-frequency scheme. Rather than generating a single fundamental frequency and cancelling n harmonics, the DFSHE approach regulates the amplitude of the fundamental and a kth harmonic to controlled values, while cancelling all harmonics in between and, possibly, a number of harmonics above the kth.

A standard full bridge DC/AC inverter is selected to demonstrate proposed DFSHE modulation methods. The PWM output waveform of the converter under unipolar modulation is shown in Fig. 1(a). The switching angles θ_1, θ_2, θ_3 ... θ_m of the quarter symmetric unipolar waveform are calculated using proposed DFSHE algorithms. For this case, it is possible to control $n=f(m)=2m$-1 harmonics. The Fourier expansion of this quarter symmetric unipolar waveform is

$$v(\omega t) = \sum_{n=1,3,5,...}^{\infty} \frac{4V_{dc}}{n\pi}[\cos(n\theta_1) - \cos(n\theta_2) + \cos(n\theta_3) - ... + \cos(n\theta_m)]\cdot \sin(n\omega t) \qquad (2)$$

In order to apply DFSHE, the two frequency components to be synthesized are assumed to be a fundamental element and its kth harmonic. Under these conditions, the Fourier expansion of (1) can be rearranged to form a system of equations

$$\begin{cases} \frac{4V_{dc}}{\pi}(\cos\theta_1 - \cos\theta_2 \cdots + \cos\theta_m) = V_{LF} \\ \frac{4V_{dc}}{\pi}(\cos 3\theta_1 - \cos 3\theta_2 \cdots + \cos 3\theta_m) = 0 \\ \frac{4V_{dc}}{\pi}(\cos 5\theta_1 - \cos 5\theta_2 \cdots + \cos 5\theta_m) = 0 \\ \ldots\ldots \\ \frac{4V_{dc}}{k\pi}(\cos k\theta_1 - \cos k\theta_2 \cdots + \cos k\theta_m) = V_{HF} \\ \ldots\ldots \\ \frac{4V_{dc}}{\pi}(\cos n\theta_1 - \cos n\theta_2 \cdots + \cos n\theta_m) = 0 \end{cases} \qquad (3)$$

where each equation in the system represents a condition necessary to regulate a specific harmonic. V_{LF} and V_{HF} are the desired amplitudes of the fundamental and k^{th} harmonic, respectively, and all other harmonics are set to zero. When solved, (3) will yield the switching angles necessary to synthesize the desired dual-frequency output. A modified iterative solution based on Newton-Raphson method [13]-[15], in which two solver loops are employed, is used to solve the transcendental equations of (3).

Since two distinct frequencies are individually modulated, the modulation index of the low frequency element is defined as $M_{i(LF)}$, whose modulation range is from 0 to 1. The modulation index of the high frequency output is given as $M_{i(HF)}$, whose range is influenced by the switching angles, waveforms and $M_{i(LF)}$.

Using the Newton-Raphson method, solution convergence is dependent upon an initial guess being sufficiently close to the exact solution. Qualitatively, the algorithm used to solve (3) which yields good, though not guaranteed, convergence includes following steps:

1. Predefine initial values of switching angles in (3) based on the number of switching angles and modulation scheme (e.g. unipolar, ¼ symmetric), and assuming $V_{HF} = 0$ in (3).
2. Determine the modulation index of the low frequency element, $M_{i(LF)}$, and use numeric iteration method to calculate the initial switching angles $\theta_{i1}...\theta_{im}$ when the modulation index of the high frequency $M_{i(HF)} = 0$, and $M_{i(LF)} = V_{LF}/V_{DC}$.
3. Determine the modulation index of the high frequency element, $M_{i(HF)} = V_{HF}/V_{DC}$. Use calculated initial switching angles, $\theta_{i1}...\theta_{im}$, in step 1 as initial values for the second iteration loop, and then perform numeric iteration method to find final switching angles, $\theta_1...\theta_m$, of desired DFSHE waveforms.
4. Repeat (1)–(3) if multiple solution sets exit.

The Fourier expansion of the bipolar DFSHE waveforms, shown in Fig. 1(b), is

$$v(\omega t) = \sum_{n=1,3,5...}^{\infty} \frac{4V_{dc}}{n\pi}[1 - 2\cos(n\theta_1) + 2\cos(n\theta_2) - 2\cos(n\theta_3) + ... + 2\cos(n\theta_m)]\cdot\sin(n\omega t) \quad (4)$$

The transcendental equations for bipolar modulation are

$$\begin{cases} \frac{4V_{dc}}{\pi}(1 - 2\cos\theta_1 + 2\cos\theta_2 \cdots + 2\cos\theta_m) = V_{LF} \\ \frac{4V_{dc}}{\pi}(1 - 2\cos 3\theta_1 + 2\cos 3\theta_2 \cdots + 2\cos 3\theta_m) = 0 \\ \frac{4V_{dc}}{\pi}(1 - 2\cos 5\theta_1 + 2\cos 5\theta_2 \cdots + 2\cos 5\theta_m) = 0 \\ ...... \\ \frac{4V_{dc}}{k\pi}(1 - 2\cos k\theta_1 + 2\cos k\theta_2 \cdots + 2\cos k\theta_m) = V_{HF} \\ ...... \\ \frac{4V_{dc}}{\pi}(1 - 2\cos n\theta_1 + \cos n\theta_2 \cdots + 2\cos n\theta_m) = 0 \end{cases} \quad (5)$$

The previous algorithm for solving (3) is applicable to the bipolar case in (5).

Examining (3) and (5) it is clear that in order to control the fundamental and k^{th} harmonic, at least $m = (k+1)/2$ switching angles are necessary. However, selecting $m > (k+1)/2$ allows harmonics above the high frequency carrier to be regulated. This selection can be advantageous, as it allows for more simple filter design to attenuate harmonics beyond the controlled range.

Characteristics of the unipolar and the bipolar DFSHE are investigated in the following sections with focus on the achievable range of $M_{i(HF)}$ and the distribution of unregulated high-order harmonics.

III. DFSHE ANALYSIS

A. Unipolar SHE

To compare the impacts that switching angles have on output characteristics, 5 and 11 switching angles of a unipolar SHE modulation scheme are selected and illustrated in time domain and frequency domain respectively, as shown in Fig. 4 and Fig. 5. In both cases, the fundamental and 9th harmonic are chosen as low-frequency (LF) and high-frequency (HF) AC outputs, respectively, and the analysis is performed in normalized fashion, independent of the value of the fundamental frequency.

From Fig. 4 and Fig. 5, it can be observed that as the number of switching angles increases, the characteristics of the output waveform are altered in three ways

1. Additional harmonics beyond the HF output can be eliminated
2. $M_{i(HF)}$ has narrower range, reducing the achievable amplitude of the HF output
3. The equivalent switching frequency of the inverter increases.

In Table I, the maximum limit of $M_{i(HF)}$ at different switching angles is calculated as $M_{i(LF)}$ has 0.1 increment from 0 to 1. It is also found that the HF modulation range is strictly limited when LF modulation index is also low in the unipolar case.

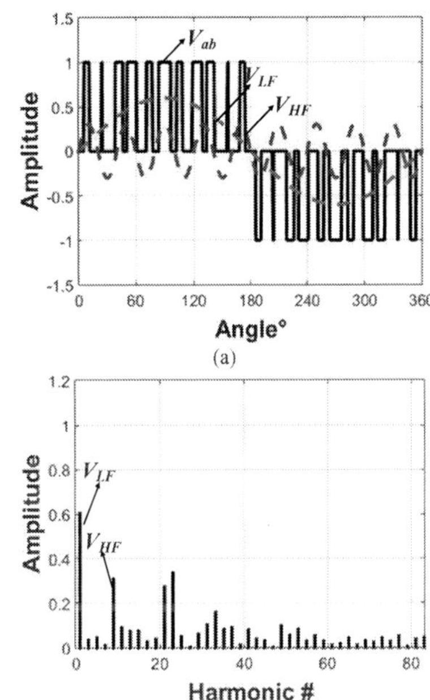

Figure. 3. PWM based dual frequency modulation with 11 equivalent switching angles.

TABLE I. RELATIONSHIP BETWEEN $M_{i(LF)}$ AND $M_{i(HF)}$ FOR UNIPOLAR DFSHE.

	Number of Switching Angles	$M_{i(LF)}$ 0.1	0.2	0.3	0.4	0.5	0.6	0.7	0.8	0.9	1
$M_{i(HF)}$ Limit	5	0.15	0.3	0.4	0.5	0.55	0.59	0.6	0.55	0.47	0.22
	7	0.12	0.24	0.35	0.44	0.5	0.55	0.56	0.45	0.28	0.05
	11	0.07	0.15	0.22	0.29	0.36	0.42	0.43	0.31	0.17	0.01

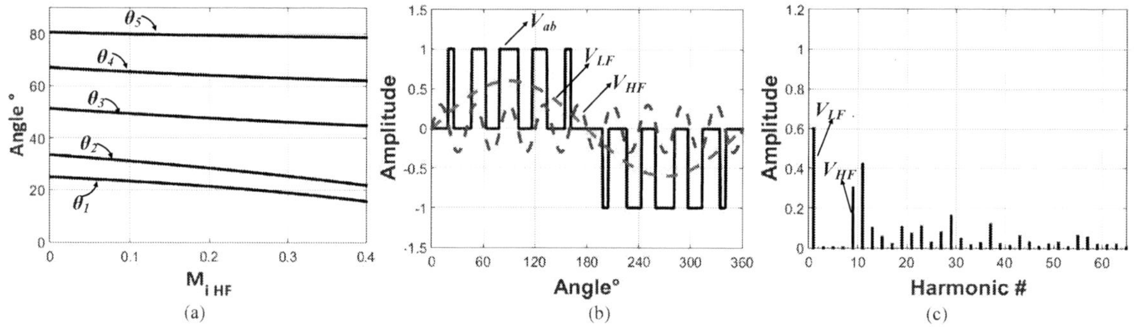

Figure. 4. Unipolar DFSHE 5-switching-angle case. (a) Modulation range of HF element when $M_{i(LF)} = 0.6$; (b) Time domain waveforms (c) Frequency domain spectrum.

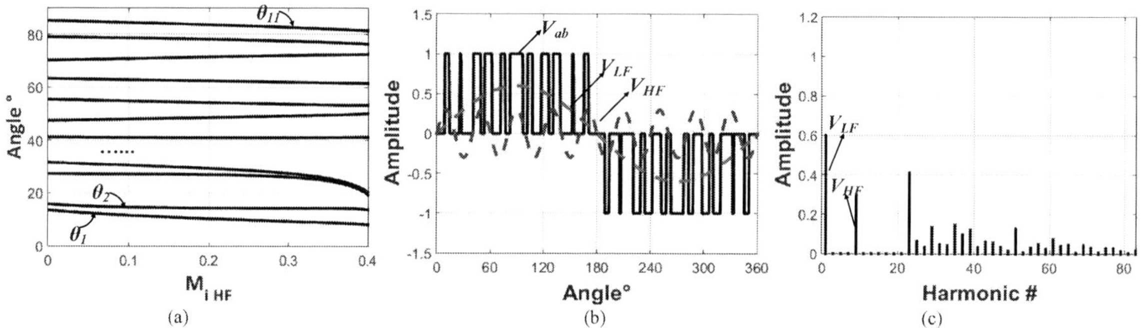

Figure. 5. Unipolar DFSHE 11-switching-angle case. (a) Modulation range of HF element when $M_{i(LF)} = 0.6$; (b) Time domain waveforms (c) Frequency domain spectrum.

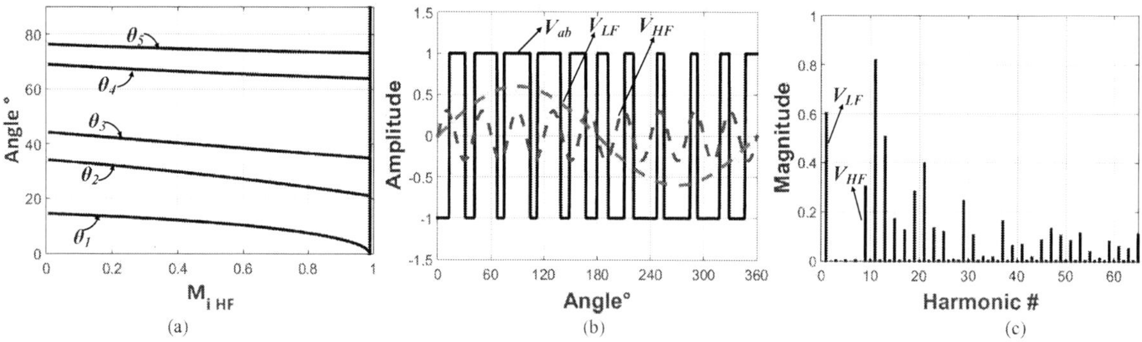

Figure. 6. Bipolar DFSHE 5-switching-angle case. (a) Modulation range of HF element when $M_{i(LF)} = 0.3$; (b) Time domain waveforms (c) Frequency domain spectrum.

TABLE II. RELATIONSHIP BETWEEN $M_{i(LF)}$ AND $M_{i(HF)}$ FOR BIPOLAR DFSHE.

	Number of Switching Angles	$M_{i(LF)}$ 0.1	0.2	0.3	0.4	0.5	0.6	0.7	0.8	0.9	1
$M_{i(HF)}$ Limit	5	1	1	1	1	1	0.99	0.88	0.75	0.58	0.28
	7	1	1	1	1	0.94	0.8	0.66	0.48	0.29	N/A

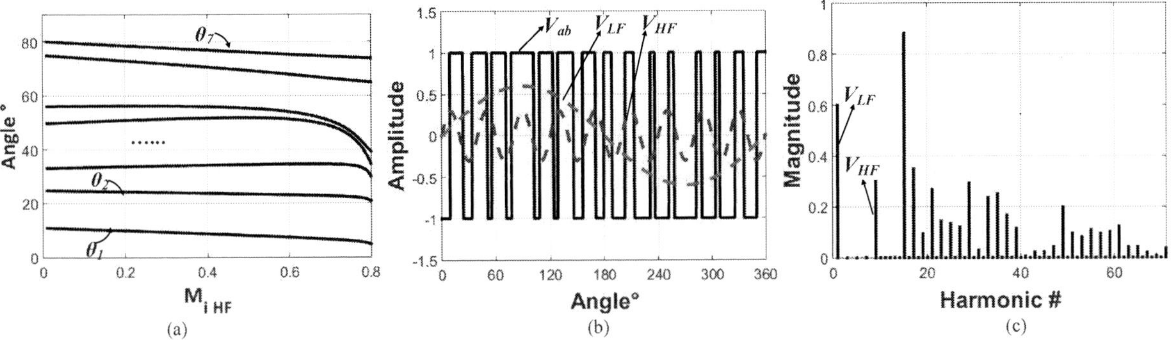

Figure. 7. Bipolar DFSHE 7-switching-angle case. (a) Modulation range of HF element when $M_{i(LF)}$ = 0.3; (b) Time domain waveforms (c) Frequency domain spectrum.

The THD of DFSHE waveform indicates the level of unregulated harmonics content above HF output, and those harmonics may cause circulating power flow and increase power loss if not filtered. The better THD is, the easier it is to design the output filter and minimizing the loss on it. Following equation is adopted as two different frequencies are regarded as "effective" outputs and other harmonics are "noise":

$$THD_{DF} = \sqrt{\frac{\sum_{n_{LF}+1}^{n_{HF}-1} V^2 + \sum_{n_{HF}+1}^{H} V^2}{V_{LF}^2 + V_{HF}^2}} \quad (6)$$

The low order harmonics between LF and HF, and unregulated high order harmonics up to a certain H^{th} are considered in (6). Ideally, the low order harmonics are 0 for DFSHE, and thus (6) can be simplified as

$$THD_{DF} = \sqrt{\frac{\sum_{n_{HF}+1}^{H} V^2}{V_{LF}^2 + V_{HF}^2}} \quad (7)$$

Assume H = 30, (7) is utilized to calculate the THD for DFSHE and the results are given in Table III. In the PWM based method shown in Fig.3, the pulse train is generated via sinusoidal waveforms and sawtooth waveform comparison, and its equivalent switching angles are 11. Because of the unregulated side-band harmonics, (6) is adopted and the THD_{DF} = 0.681 in PWM based method.

TABLE III. THD_{DF} FOR DIFFERENT SWITCHING ANGLES DFSHE WHEN $M_{i(LF)}$ = 0.6 AND $M_{i(HF)}$=0.3.

Switching angles	THD_{DF}
5	0.743
7	0.736
11	0.649

From Table III, the more switching angles there are, the better THD is achieved, which indicates that high order harmonics are easier to be filtered. At the same switching angles, the unipolar DFSHE has better THD than that in PWM. Moreover, the PWM based method is not applicable to the electrosurgical inverter due to those hazardous low order harmonics.

B. Bipolar case

In order to extend the modulation range of the HF output, expanding achievable operation points, bipolar DFSHE is one potential alternative. As in the unipolar analysis previously, the fundamental and the 9th harmonic are selected as LF and HF outputs, respectively; the harmonics between these two frequencies are suppressed to zero as in unipolar case. Compared to the unipolar counterpart, bipolar DFSHE has a wider HF modulation range when the LF modulation index is the same, which is illustrated in Fig. 6.(a) and Fig.7.(a), and Table II.

On the other hand, the unregulated harmonics close to the desired HF element, are notably higher than that in unipolar case, which is shown in Fig.6.(c) and Fig.7.(c). Equation (7) is also employed to calculate the THD of bipolar DFSHE, the sample cases shown in Fig.6 and Fig.7 are used as input. For 5 switching angle case, the THD_{DF} = 1.75 and THD_{DF} = 0.839 for 7 switching angle case, which implies that the filter design for bipolar DFSHE is more difficult than it for unipolar case to achieve the same filtering effect.

C. Discussion

In a standard full bridge configuration, less switching loss is expected for unipolar DFSHE than in the bipolar case, due to fewer numbers of switching actions per device. For the dual-frequency application that does not require extensive HF output range, the unipolar DFSHE offers better THD and reduced switching losses, while achieving similar regulation performance, as it shown in the Fig. 4 and Fig. 6.

The number of switching angles is critical to the performance of DFSHE. The benefits of less switching angles is that lower switching loss can be achieved since the equivalent switching frequency drops. On the other hand, less switching angles result in higher THD for both unipolar and bipolar case, and unregulated harmonics close to the HF output are hard to be filtered due to the small difference in frequency. As a result, compromises need to be made for different applications.

In this work, a prototype dual frequency inverter, which integrates an US output and an RF output into one device, for surgical applications [16] is the focus of the design. Combining US and RF surgical power sources into a single inverter requires flexible dual frequency outputs and no harmonics between two output frequencies to avoid adverse clinical effects. The unregulated high order harmonics are not major concern in this case, as they are easily filtered, and the bipolar DFSHE is employed for this application.

IV. EXPERIMENTAL RESULTS

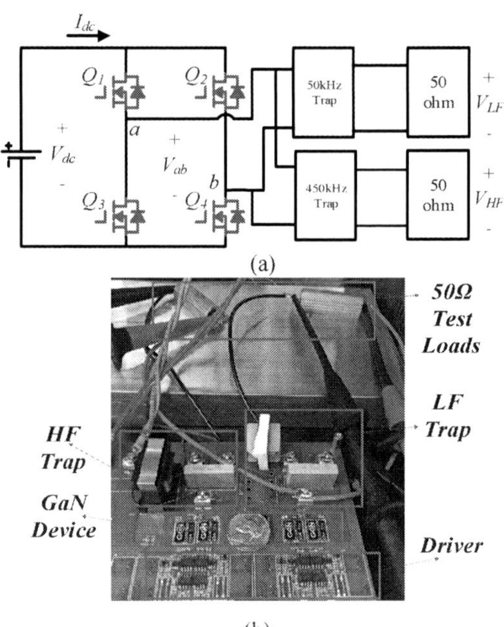

(b)

Figure. 8. (a) Schematic circuits of two-load inverter using DFSHE; (b) Hardware platform.

To verify the effectiveness of the proposed DFSHE modulation schemes, a 100 W full bridge inverter based on Gallium-Nitride (GaN) devices has been constructed and verified for the high frequency operation capability of GaN [17]. This prototype uses 50 kHz LF and 450 kHz HF frequency components to power two individual loads simultaneously. To extract the LF and HF elements from dual-frequency inverter output, two shunt LC filters connected to the inverter output, and tuned at their individual resonant frequencies. The resulting 50 kHz trap and 450 kHz trap are shown in Fig. 8(a). The schematic circuit and test platform are shown in Fig. 8(b). The parameters of the dual-frequency inverter are provided in Table IV.

TABLE IV. SPECIFICATION OF PROPOSED DUAL-FREQUENCY INVERTER

Parameter	Value/model
Rated Power (P)	100W
Input Voltage (V_{dc})	100V
50 kHz trap inductance	500μH
50 kHz trap capacitance	20.2nF
450 kHz trap inductance	50μH
450 kHz trap capacitance	2.5nF
MOSFET (Q_1-Q_4)	GS66508P

The experimental time domain waveforms under DFSHE modulation and their corresponding specturm by FFT analysis are shown in Fig. 9 and Fig. 10. In Fig. 9(a) and Fig. 10(a), the waveforms of mid-point voltage of the full bridge, V_{ab}, the voltage of the LF output, V_{LF}, and the HF output, V_{HF}, are demonstrated, and their spectrums are illustrated in Fig. 9(b) and Fig. 10(b). Sampled experimental waveforms are recorde with a MSO5104B ocilloscope; the FFT analysis is performed using Matlab.

In Fig.9(b) and Fig.10(b), there is no low order harmonics at the HF output, which diminshing the dangers of muscel contraction of clinical surgery. The amplitude of LF and HF can be altered by the proposed DFSHE. In addition, the aliasing is noticed in FFT analysis, and the amplitude of two frequencies are influenced with different sampling rate, as it shown in Fig.11(b) and (c).

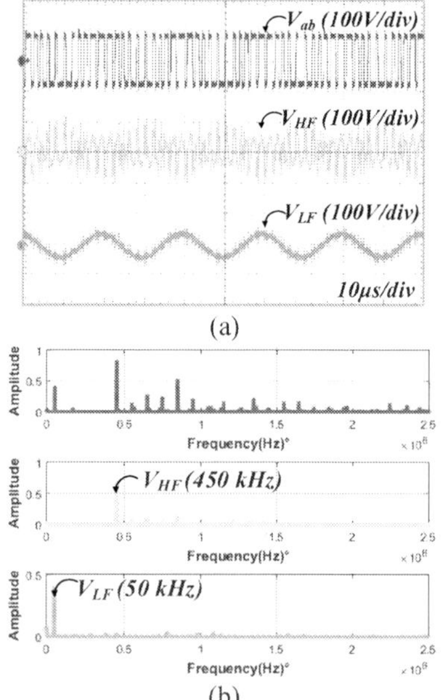

(b)

Figure. 9. (a) Waveforms of inverter output, LF and HF outputs when $M_{i(LF)}$ =0.5 and $M_{i(HF)}$ =0.9; (b) Spectrum (sampling rate = 5M/s).

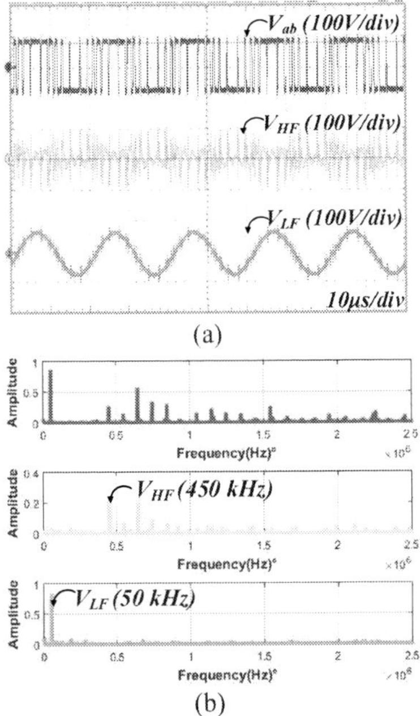

Figure. 10. (a) Waveforms of inverter output, LF and HF outputs when $M_{i(LF)} = 0.9$ and $M_{i(HF)} = 0.3$; (b) Spectrum (sampling rate = 5M/s).

The LF and HF elements generated via the proposed DFSHE methods agree with theoretical prediction and simulation results, as it shown in Fig.11 and Fig.12. After the LF and HF trap, two different AC outputs feed their individual load with suppressed low and high order harmonics. Further, the experimental prototype has verified that the platform can generate and independently regulate the two frequency outputs with controllable power flow to each by regulating the voltage of LF and HF.

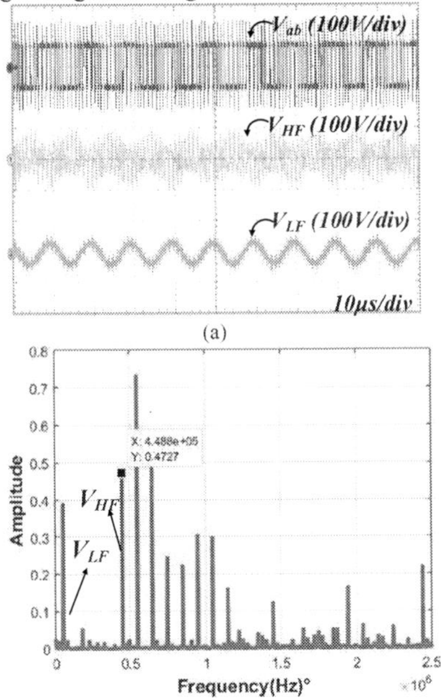

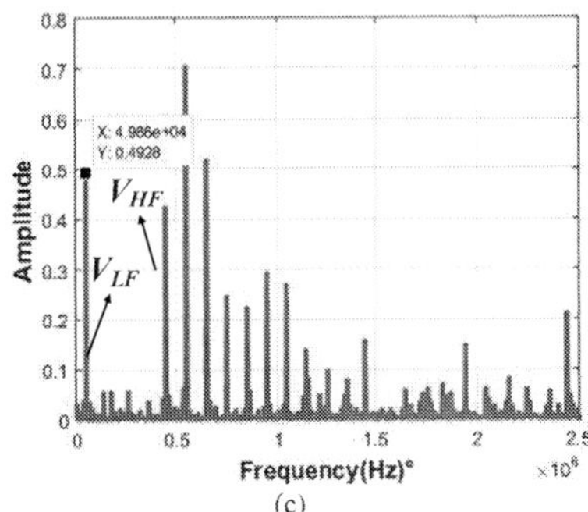

Figure. 11. FFT analysis when $M_{i(LF)} = 0.5$ and $M_{i(HF)} = 0.5$. (a) Time domain waveforms, (b) FFT result under sampling rate = 10M/s, (c) FFT result under sampling rate = 5M/s.

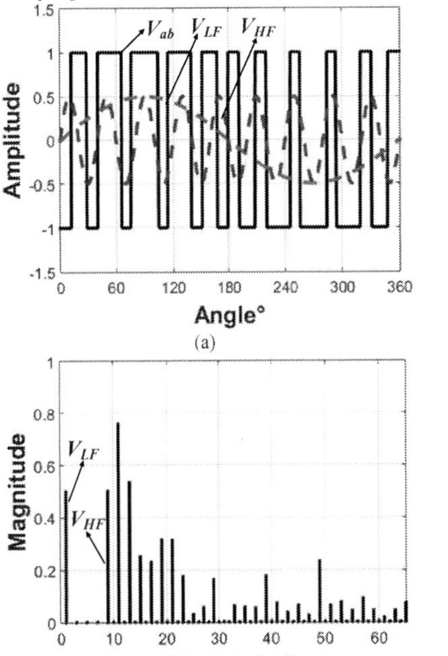

Figure. 12. Simulation result when when $M_{i(LF)} = 0.5$ and $M_{i(HF)} = 0.5$. (a) Time domain waveforms, (b) spectrum.

V. CONCLUSION

In this paper, a simultaneous dual-frequency modulation scheme for single-phase inverters is proposed, in which output power at different frequencies are controlled independently. One valuable application of this technology is combining US and RF surgical power supplies to enable improved performance and simultaneous usage. Two modulation schemes, unipolar and bipolar DFSHE, are investigated and compared in HF modulating range, expected losses, and harmonics content. The experimental results from a 100W inverter confirm the effectiveness of the proposed modulation methods, enabling the inverter to generate two simultaneous high frequency AC outputs with flexible power control. Since the outputs of the proposed inverter are high frequency AC and frequency-sensitive trap filters are adopted, it is necessary to design and implement such trap inductors to enable less proximity effect, less core loss and precise tuning to selected frequency.

In traditional PWM or SHE modulation schemes, only the fundamental element is generated and controlled, while unregulated harmonics are filtered. In the proposed DFSHE modulation, however, both fundamental and the kth harmonics are simultaneously generated and accurately modulated, and undesired harmonics are eliminated. Moreover, unlike harmonic utilization techniques in [2][4][6], the fundamental and the kth harmonics employed in this method are decoupled in modulation, which is able to accomplish better individual power regulations. Those features of the DFSHE modulation promotes the dual-frequency applications such as electrosurgical power supplies in particularly. Another advantages of the proposed method is that no additional switching devices are required, which significantly reduces overall costs while achieving equivalent regulation performance.

ACKNOWLEDGMENT

This work made use of the Engineering Research Center Shared Facilities supported by the Engineering Research Center Program of the National Science Foundation and DOE under NSF Award Number EEC-1041877 and the CURENT Industry Partnership Program.

REFERENCES

[1] Okudaira, S.; Matsuse, K., "Adjustable Frequency Quasi-Resonant Inverter Circuits Having Short-Circuit SwitchAcross Resonant Capacitor," *Power Electronics, IEEE Transactions on*, vol.23, no.4, pp.1830,1838, July 2008

[2] Esteve, V.; Jordán, J.; Sanchis-Kilders, E.; Dede, E.J.; Maset, E.; Ejea, J.B.; Ferreres, A., "Comparative Study of a Single Inverter Bridge for Dual-Frequency Induction Heating Using Si and SiC MOSFETs," *Industrial Electronics, IEEE Transactions on*, vol.62, no.3, pp.1440,1450, March 2015

[3] Friedrichs, D.A.; Erickson, R.W.; Gilbert, J., "A New Dual Current-Mode Controller Improves Power Regulation in Electrosurgical Generators," *Biomedical Circuits and Systems, IEEE Transactions on*, vol.6, no.1, pp.39,44, Feb. 2012

[4] Papani, S.K.; Neti, V.; Murthy, B.K., "Dual frequency inverter configuration for multiple-load induction cooking application," *Power Electronics, IET*, vol.8, no.4, pp.591,601, 4 2015

[5] Hirokawa, T.; Hiraki, E.; Tanaka, T.; Imai, M.; Yasui, K.; Sumiyoshi, S., "Dual-frequency multiple-output resonant soft-switching inverter for induction heating cooking appliances," *Industrial Electronics Society, IECON 2013 - 39th Annual Conference of the IEEE*, vol., no., pp.5028,5033, 10-13 Nov. 2013

[6] Pantic, Z.; Lee, K.; Lukic, S.M., "Receivers for Multifrequency Wireless Power Transfer: Design for Minimum Interference," *Emerging and Selected Topics in Power Electronics, IEEE Journal of*, vol.3, no.1, pp.234,241, March 2015

[7] Fujita, H.; Uchida, N.; Ozaki, K., "A New Zone-Control Induction Heating System Using Multiple Inverter Units Applicable Under Mutual Magnetic Coupling Conditions," *Power Electronics, IEEE Transactions on*, vol.26, no.7, pp.2009,2017, July 2011

[8] Yuequan Hu; Jovanovic, M.M., "High-Intensity-Discharge Lamp Ballast With Igniter Driven by Dual-Frequency Inverter," *Applied Power Electronics Conference, APEC 2007 - Twenty Second Annual IEEE*, vol., no., pp.268,273, Feb. 25 2007-March 1 2007

[9] Hua Cai; Liming Shi; Yaohua Li, "Harmonic-Based Phase-Shifted Control of Inductively Coupled Power Transfer," *Power Electronics, IEEE Transactions on*, vol.29, no.2, pp.594,602, Feb. 2014

[10] Diong, B.; Corzine, K.; Basireddy, S.; Shuai Lu, "Multilevel inverter-based dual-frequency power supply," *Power Electronics Letters, IEEE*, vol.1, no.4, pp.115,119, Dec. 2003

[11] Pantic, Z.; Lee, K.; Lukic, S.M., "Multifrequency Inductive Power Transfer," *Power Electronics, IEEE Transactions on*, vol.29, no.11, pp.5995,6005, Nov. 2014

[12] Enjeti, P.N.; Ziogas, P.D.; Lindsay, J.F., "Programmed PWM techniques to eliminate harmonics: a critical evaluation," in *Industry Applications, IEEE Transactions on*, vol.26, no.2, pp.302-316, Mar/Apr 1990

[13] Chiasson, J.N.; Tolbert, L.M.; McKenzie, K.J.; Zhong Du, "A complete solution to the harmonic elimination problem," *Power Electronics, IEEE Transactions on*, vol.19, no.2, pp.491,499, March 2004

[14] Agelidis, V.G.; Balouktsis, A.; Balouktsis, I.; Cossar, C., "Multiple sets of solutions for harmonic elimination PWM bipolar waveforms: analysis and experimental verification," *Power Electronics, IEEE Transactions on*, vol.21, no.2, pp.415,421, March 2006

[15] Ahmadi, D.; Ke Zou; Cong Li; Yi Huang; Jin Wang, "A Universal Selective Harmonic Elimination Method for High-Power Inverters," *Power Electronics, IEEE Transactions on*, vol.26, no.10, pp.2743,2752, Oct. 2011

[16] Jensen, S.; Maksimovic, D.; Friedrichs, D.; Gilbert, J., "Fast tracking electrosurgical generator using GaN switches," in *Applied Power Electronics Conference and Exposition (APEC), 2015 IEEE*, vol., no., pp.1404-1408, 15-19 March 2015

[17] W. Zhang, Y. Cui, F. Wang, L. M. Tolbert, B. J. Blalock, D. J. Costinett, "Investigation of Gallium Nitride devices benefits on LLC resonant DC-DC converter," in Proc. *IEEE Applied Power Electronics Conference and Exposition (APEC)*, 2015, pp.146-153, 15-19 March 2015

GRID CONNECTED DC DISTRIBUTION NETWORK DEPLOYING HIGH POWER DENSITY RECTIFIER FOR DC VOLTAGE STABILIZATION

Danillo B. Rodrigues[1], Paulo R. Silva[2], Gustavo B. Lima[1], Ernane A. A. Coelho[2], Luiz C. G. Freitas[2]
[1] Universidade Federal do Triângulo Mineiro (UFTM), Uberaba – MG, Brazil
[2] Núcleo de Pesquisa em Eletrônica de Potência (NUPEP) - Faculdade Engenharia Elétrica (FEELT)
Universidade Federal de Uberlândia (UFU), Uberlândia – MG, Brazil
e-mail: danillorodrigues@yahoo.com.br, gustavo.brito.28@gmail.com, lcgfreitas@yahoo.com.br

Abstract – **This paper proposes the use of a high power density ac-dc converter to provide higher efficiency and a tight dc bus for connection of electronic loads as well as renewable energy sources. As a new contribution, the authors present a novel hybrid rectifier structure with a series dc-link voltage compensation technique. The proposed hybrid solution is characterized by the fact that a three-phase Boost Converter in parallel input connection and an isolated dc-dc converter in series output connection with an ordinary non-controlled rectifier provides power factor correction and dc bus voltage stabilization. This structure is also capable of imposing line currents with low total harmonic distortion and voltage sag ride-through capability.**

Keywords – ***microgrids; series DC-Link voltage compensation; high power density.***

I. INTRODUCTION

Grid connected dc distribution systems has aroused as a great choice for modern electrical power delivery systems at low and medium voltage levels due to the facility of integration of different renewable sources and higher efficiency provided [1]-[10]. In order to achieve these accomplishments, the dc distribution system requires accurate dc bus voltage control to balance the power flow among renewable sources, dc loads, and ac grid. Therefore, a dc distribution system capable of dealing with severe step load changes and the natural instability of renewable sources such as wind and photovoltaic can be implemented. Island operation of the system requires energy storage element such Supercapacitors (SC) and Fuel Cells (FC) [2].

In this context, this work focuses on the study of a feasible and high efficiency solution for the control of the dc bus voltage in a grid connected dc microgrid such as portrayed in Fig. 1. It is based on the use of a high power density rectifier to provide not only power factor correction and low current harmonic distortion, but also dc voltage stabilization. Moreover, high-quality power can be supplied continuously to the dc loads even during the occurrence of voltage sags in the utility grid. If the energy provided by the renewable sources is higher than the load requirements, the excess of energy must be injected into the ac grid through a bi-directional inverter, however, it is not the target of the proposed paper. On the contrary, the lack of renewable energy is compensated by the ac grid through the proposed hybrid ac-dc converter.

The main feature of the proposed solution is the use of a single and high power density ac-dc conversion stage to provide higher efficiency and a tight dc bus for connection of electronic loads as well as renewable energy sources. As a new contribution, the authors present a novel hybrid rectifier structure with a Series DC-Link Voltage Compensation (SDCVC) technique. The main topological arrangement is portrayed in Fig. 2.

In this scenario, the design project is currently focused on keeping the power sharing between the rectifiers groups so that 70% of the rated power is provided by the non-controlled rectifier and 30% is provided by the switched converters (Boost + Isolated Full-Bridge). Therefore, the voltage and current ratings of the switched converters that are responsible for sinusoidal input current imposition and dc bus voltage control are drastically reduced when compared to other solutions [2], [4].

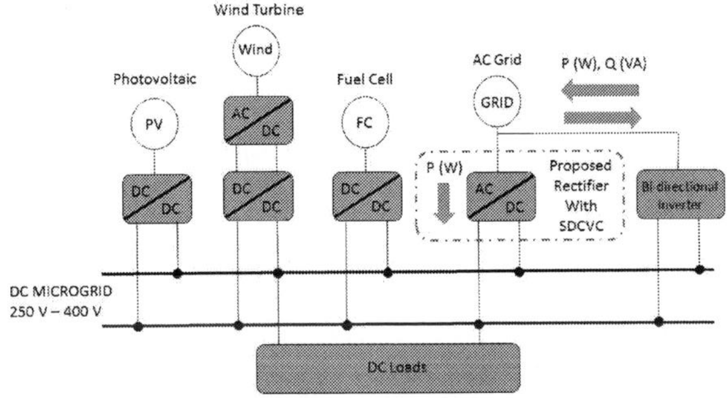

Fig 1. Example of a Grid-connected DC Distribution System.

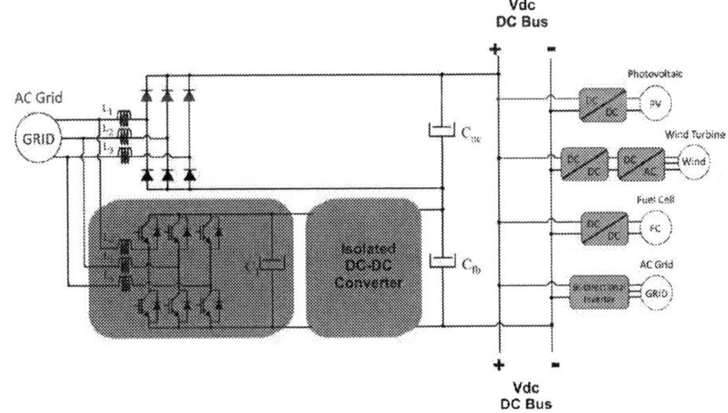

Fig. 2. Schematic diagram of a dc microgrid with the proposed rectifier using three-phase Boost converter in parallel input connection and an isolated DC-DC converter in series output connection.

978-1-4673-9551-9/16 $31.00 © 2016 IEEE

II. OPERATING PRINCIPLE

Under normal operating conditions, the proposed hybrid rectifier structure operate in power factor correction mode by imposing sinusoidal input line currents and controlling the dc bus voltage. To achieve this goal, the uncontrolled rectifier process 70% of the total output power while the switched converter process the remainder amount of energy demanded by the dc load and not supplied by the renewable energy sources. The DC bus voltage is fully controlled, therefore, even during the occurrence of a voltage sag on the ac grid, or during severe step load changes the isolated dc-dc converter provides the dc bus voltage stabilization since the DC-link voltage is a result of the series composition of V_{Cnc} and V_{Cfb}, as illustrated in Fig. 2.

Fig. 3 illustrates a graph obtained from computer simulation results, showing the effect of increasing the voltage imposed on the DC link on the processing power division between the grinding structures Rect-1 and Rect-2.

Note that as the DC bus voltage is increased from 300 V to 600 V, there is an increase in processing power by the Rect-2 structure, whereas for a 600 V voltage, the processing division is approximately 50% for each rectifier structure, demonstrating that the processing power by the switched converters is small in terms of the wide range of voltage that the three-phase hybrid rectifier can control in the DC bus. For the application shown in this paper, it was considered as operating point the imposition of a 400 V voltage in the DC bus, which implies in a power sharing of the total power delivered to the load by about 73% to Rect-1 and 27 % to Rect-2.

Therefore, a tight dc bus voltage can be achieved through a very simple and low cost compensation system, which is called in this paper as Series DC Voltage Compensation (SDCVC). Comparing to ordinary solutions with the same purpose, one can conclude that a very simple and effective control technique of the dc bus voltage is provided deploying a robust and high efficient hybrid rectifier structure, since the switched converters must be rated at a small fraction of the total output power. Thus, with the percentage of 70% of the power processing by the not controlled rectifier, besides being sufficient to ensure that the input line currents drained of the grid presents a level of harmonic distortion in accordance with standards IEC 61000-3-2 [11] and IEC 61000-3-4 [12], also determines that the overall structure of the hybrid rectifier provide high operational reliability and robustness, which makes this topological structure very attractive for applications in high power levels. In other words, this means that the three-phase hybrid rectifier is a structure that combines robustness, simplicity and reliability of the three-phase six-switch rectifier with high frequency operation of switched converters, reducing the volume, weight and size of the proposed structure.

In order to prove the effectiveness of the proposed SDCVC technique, a prototype was built and has been analyzed in laboratory and the proposed control technique was implemented as illustrated in Fig. 4.

For instance, one can note that aiming at imposing sinusoidal input line current and in synchronism with the input voltage, a rectified reference signal of current (I_{ref}) is compared to a sample of the total input line current (i_{in}), being I_{ref} obtained through the multiplication of i_{sin} (digitally generated) by the control signal K_v, which is provided by the PI controller of voltage. Thus, the magnitude of I_{ref} is determined by the error of the intermediate DC link voltage ($V_{o(Boost)}$). In order to avoid over currents, the reference current is limited in accordance with the maximum power contribution allowed for three-phase Boost converter.

Regarding the control strategy developed for the isolated dc-dc converter responsible for providing the control of the main dc bus voltage (V_o), the gate-drive signals are obtained through comparison between the output signal provided by the PI voltage controller and a saw tooth signal with fixed frequency, generating the PWM gate-drive signals. The higher value of series dc voltage compensation, the higher will be the power contribution of the isolated dc-dc converter during the disturbance.

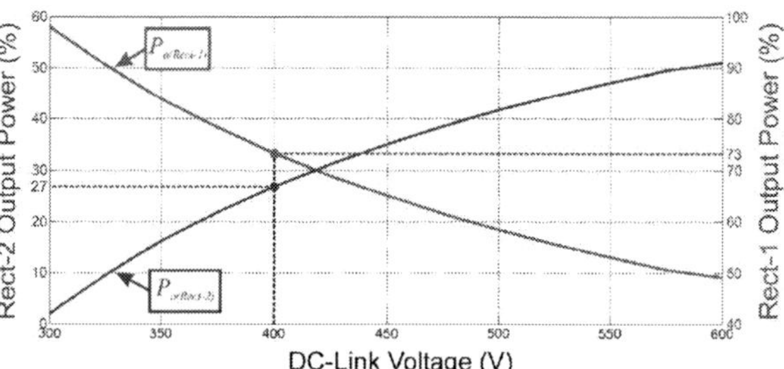

Fig. 3. Processing power division between Rect-1 and Rect-2 with increased DC bus voltage.

III. SIMULATION AND EXPERIMENTAL RESULTS

A summary of simulation results is presented herein. In Fig. 5 the input line currents before and after the occurrence of a severe step load change are presented. One can observe that sinusoidal input currents are obtained and the dip of voltage on the main dc bus voltage is minimized due to the action of the SDCVC technique.

As illustrated in Fig. 6, during the occurrence of a dip voltage sag on the mains the main dc bus voltage (V_o) is fully controlled assuring that the electronics loads connected to the dc microgrid can still operate during the disturbance. The value of the SDCVC depends on the dc microgrid requirements and defines the isolated Full-Bridge converter power contribution, hence, the final cost of the overall structure. Concerning the volume, weight, and size of the proposed rectifier and, hence, the final power density, a precise design procedure must be followed in order to minimize the size of passive elements of the isolated dc-dc converter.

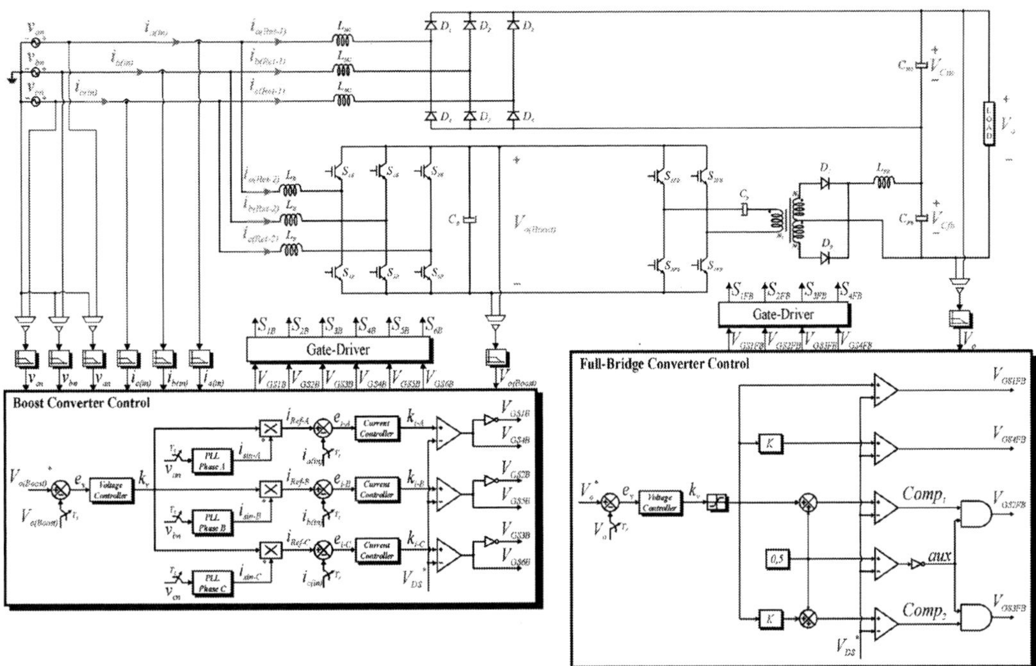
Fig. 4. Control diagram of Boost converter and Full-Bridge Converter.

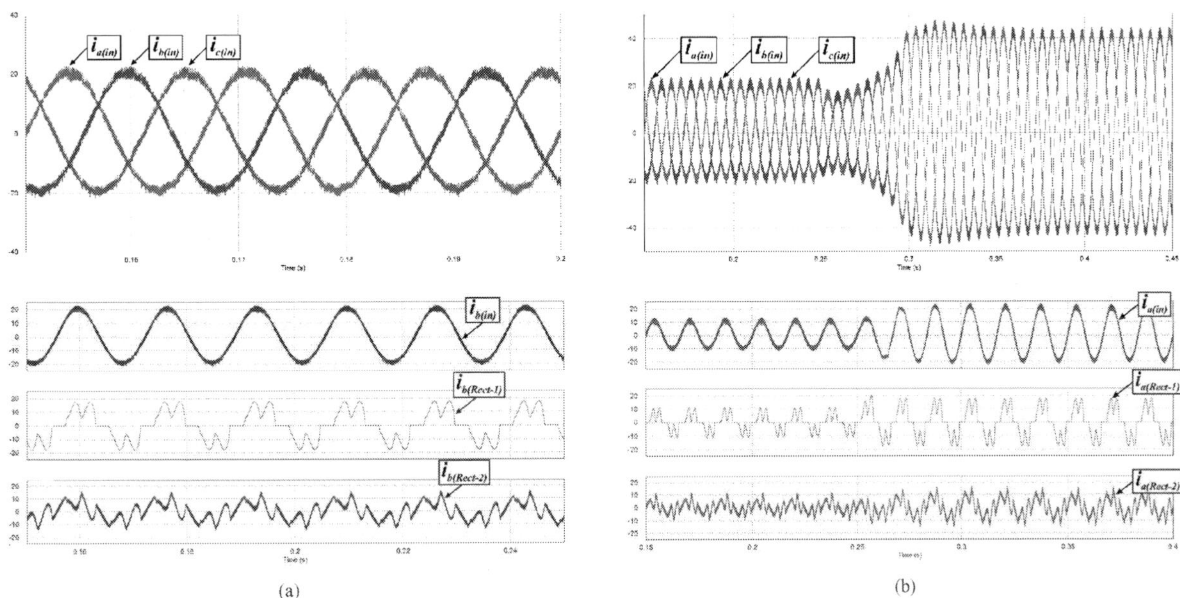
Fig. 5. Input line currents and input line current composition (a) before and (b) after the occurrence of a step load change.

In Fig. 7 one can observe the behavior of the proposed system when the renewable energy is not enough to supply the dc loads and the necessary energy is delivered by the ac grid through the hybrid rectifier without any significant disturbance on the dc bus voltage.

Fig. 8 shows the effect of increase in power generated by a photovoltaic system connected to microgrid simulating a condition in which the total power generated from the alternative energy sources is greater than the power required by the load.

It can be observed that in these conditions the voltage in the DC bus voltage is kept constant, controlled at 400 V, a parcel of the power generated by the photovoltaic system is used to supply the load and the energy excess in the DC bus is transferred to the AC power supply (P_{CA} grid) through the inverters connected to the microgrid. In this situation, the hybrid rectifier out of operation and three-phase current injected into the AC grid through inverters connected to the DC bus, as illustrative reference phase A ($i_{a(AC)}$), are out of phase 180° relative to the Line-Neutral voltages, demonstrating the reverse power flow, flowing from DC bus to the AC mains supply.

It is worth mentioning that future works to be developed by the authors of this article aim at the possibility of promoting bi-directionality in the power flow through its own hybrid rectifier structure with characteristic SDCVC. This would be possible by changing the topology of the Full-Bridge Converter, where the diode structure would be replaced with power switches, thus having a further elaboration in control of switched converters in order to promote the injection of energy excess on DC bus in AC power supply. In Fig. 9 the experimental setup is presented and in Figs. 10 and 11 the authors present preliminary experimental results. In Fig. 10 the obtained input line currents in normal condition of operation are presented.

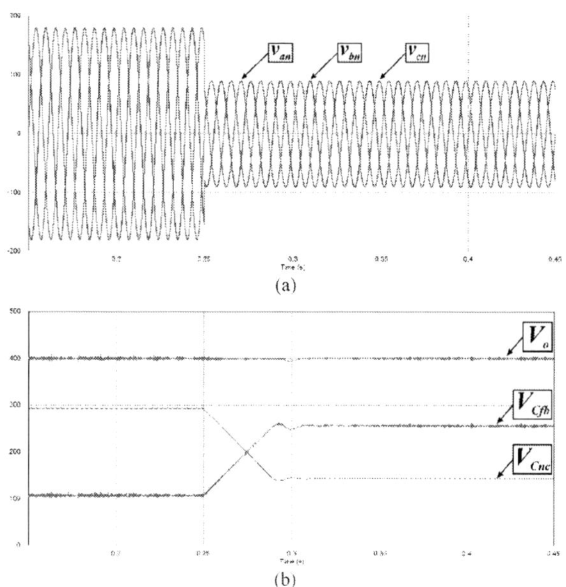

Fig. 6. Sag condition in the ac grid (a) input line-to-neutral voltages (b) dc bus voltage.

In Fig. 11 the authors present the results obtained when there is a voltage sag in the ac grid. One can observe that the main dc bus voltage ($V_o = V_{Cnc} + V_{Cfb}$) is stabilized as desired. Importantly, the preliminary test of the hybrid rectifier structure proposed were performed with voltage values lower than those shown in the results of computer simulation, since the objective was to validate the ideas of achievement imposing input line current sinusoidal and compensation voltage range in DC bus for a 50% sag in power three-phase voltages, while ensuring safer structure operators. Future works will contemplate experimental results with nominal voltage values.

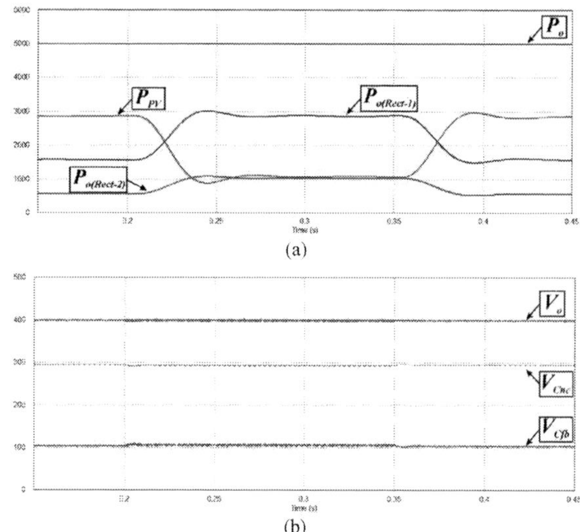

Fig. 7. Lack of necessary renewable energy (a) power contribution of each rectifier group and the photovoltaic system (b) dc bus voltages.

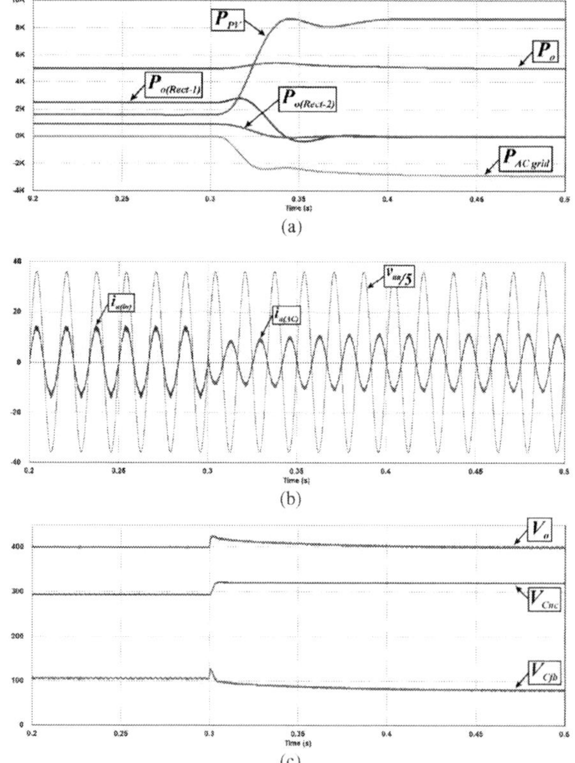

Fig. 8 (a) Energy surplus in the DC bus and power injected into the AC mains, (b) Current drained by the hybrid rectifier and current injected into the AC grid for phase A, (c) DC bus voltage during the surplus energy.

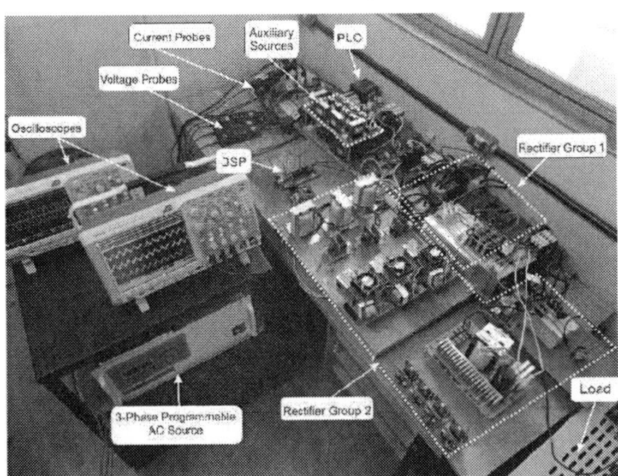

Fig. 9. Experimental set-up - 5 kW hybrid three-phase rectifier.

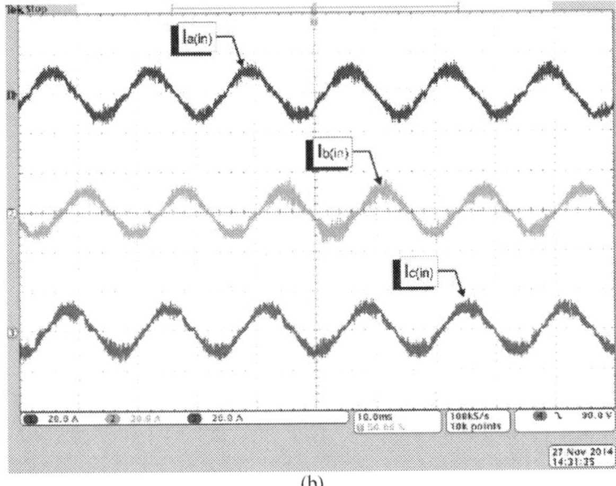

(b)

Fig. 10. Waveforms of the input line currents.

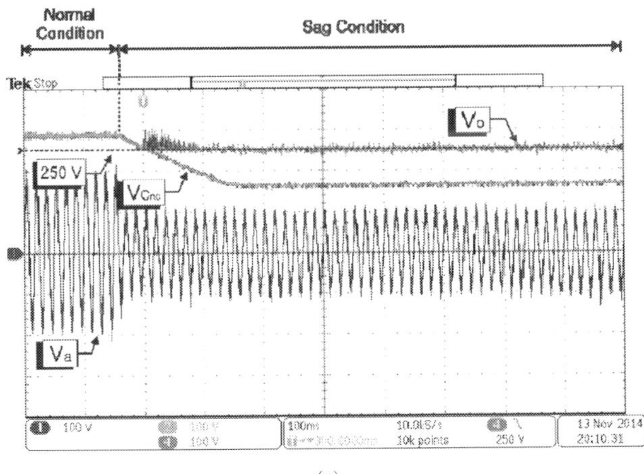

(a)

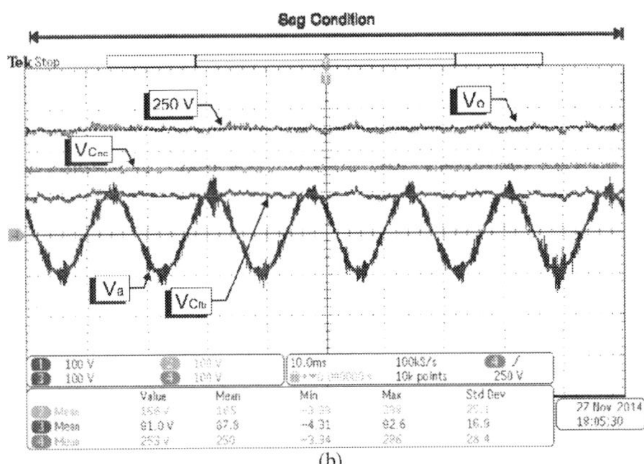

(b)

Fig. 11. waveforms of the input voltage, main dc bus voltage (V_o), and the voltage across capacitors V_{Cnc} (b) input line voltage and current after the occurrence of the disturbance and the dc bus voltages V_{Cnc}, V_{Cfb}, and V_o.

IV. CONCLUSIONS

This work presented, in general, the main features of a new topological structure of three-phase hybrid rectifier for imposing sinusoidal input line current and voltage range compensation in the DC bus with voltage sag ride-through capability to special loads connected to the DC bus.

The computer simulation results were presented, as well as the preliminary experimental results to show the operation of the hybrid rectifier in both operating conditions, i.e. normal mains supply and under temporary ac voltage sags.

One could observe the effectiveness of the rectifier structure in terms of wide dc voltage regulation range which can be very attractive for dc microgrids, since it is also able to ensure a tight dc bus voltage even under the occurrence of severe power oscillations due to the inherent intermittent characteristic of renewable energy sources such as wind and solar.

Currently new studies are been carried out in order to evaluate the performance of the proposed solution in a experimental dc microgrid with grid-tied photovoltaic systems, taking into account the unidirectional power flow of solar inverters commonly found at the market.

ACKNOWLEDGEMENTS

The authors thank FAPEMIG under processes APQ-00023-12 and TEC-PPM00031-14 and CNPq under processes 474055/2013-2 and 304307/2013-0 for the financial support.

REFERENCES

[1] D. Salomonsson, L. Söder, and A. Sannino, "Protection of Low-Voltage DC Microgrids", in *IEEE Transactions On Power Delivery*, Vol. 24, No. 3, July 2009

[2] H. Kakigano, Y. Miura, and T. Ise, "Low-Voltage Bipolar-Type DC Microgrid for Super High Quality

Distribution", in *IEEE Transactions On Power Electronics*, Vol. 25, No. 12, December 2010;

[3] P. C. Loh, D. Li, Y. K. Chai, and F. Blaabjerg, "Hybrid AC–DC Microgrids With Energy Storages and Progressive Energy Flow Tuning", *IEEE Transactions On Power Electronics*, Vol. 28, No. 4, April 2013

[4] M. Elshaer, A. Mohamed, and O. Mohammed, "Grid Connected DC Distribution System for Efficient Integration of Sustainable Energy Sources", in Proc. Of *IEEE Power Systems Conference and Exposition* 2011, pp. 1-9;

[5] D. M. Vilathgamuwa, P. C. Loh, and Y. Li, "Protection of Microgrids During Utility Voltage Sags", *IEEE Transactions On Industrial Electronics*, Vol. 53, No. 5, October 2006.

[6] Salomonsson, D., and Sannino, A., "Low-Voltage DC Distribution System for Commercial Power Systems With Sensitive Electronic Loads," Power Delivery, IEEE Transactions on , vol.22, no.3, pp.1620-1627, July 2007.

[7] Imecs, M., Szabo, C., and Incze, I .I. "Modelling and simulation of controlled bi-directional power electronic converters in a DC energy distribution line with AC grid- and motor-side active filtering" European Conference on Power Electronics and Applications, pp.1-10, 2-5 Sept. 2007.

[8] Steimer, P. K. "Enabled by high power electronics - Energy efficiency, renewables and smart grids" Power Electronics Conference (IPEC), 2010 International , pp.11-15, 21-24 June 2010.

[9] Wu, T., Chang, C., Lin, L., Yu, G. and Chang, Y. "DC-Bus Voltage Control with a Three-Phase Bi-directional Inverter for DC Distribution Systems" in IEEE Transactions on Power Electronics, vol. PP, n.99, pp.1.

[10] Zhao, B., Qingguang, Y., and Weixin, S. "Extended-Phase-Shift Control of Isolated Bidirectional DC–DC Converter for Power Distribution in Microgrid" IEEE Transactions on Power Electronics, vol.27, no.11, pp.4667-4680, Nov. 2012.

[11] IEC 61000-3-2, "Part 3-2: Limits for harmonic current emissions (equipment input current lower than 16A per phase)", International Electrotechnical Commission, second edition, 2000-08.

[12] IEC 61000-3-4, "Part 3-4: Limits for harmonic current emissions (equipment input current greather than 16A per phase)", International Electrotechnical Commission, first edition, 1998-10.

Even-Harmonic Repetitive Control for Circulating Current Suppression in Modular Multilevel Converters

Shunfeng Yang[1,2], Peng Wang[1], Yi Tang[1], Michael Zagrodnik[3], Xiaolei Hu[2] and King Jet Tseng[1]
Email: syang012@e.ntu.edu.sg epwang@ntu.edu.sg yitang@ntu.edu.sg
[1] School of Electrical and Electronic Engineering, Nanyang Technological University, Singapore
[2] Rolls-Royce@NTU Corporate Lab, Nanyang Technological University, Singapore
[3] Rolls-Royce Singapore Pte Ltd, Singapore

***Abstract*—The differential current in Modular Multilevel Converters (MMCs) is inherently subjected to even order harmonics, which are mainly caused by the voltage mismatch between phase legs and dc bus. This issue may exist in both single-phase and multi-phase MMC systems and can be eased by repetitive control based active harmonic suppression methods, which regulate the circulating current through legitimate adjustment of the voltage applied on the arm inductance. Nevertheless, conventional repetitive controllers have relatively slow dynamic response because of the presence of one fundamental period delay, and all the sampled errors in the past one cycle have to be stored for closed-loop control. This paper presents an improved even-harmonic repetitive control scheme that requires less data memory with faster dynamic response. The steady state performance can also be enhanced if system disturbances e.g. frequency deviation are considered. In order to verify its practicality and effectiveness, PLECS simulation and experimental results are presented to show the excellent dynamic response and disturbance rejection of the proposed even-harmonic elimination scheme.**

Keywords*—*Modular Multilevel Converter (MMC); Repetitive Control; Even-harmonic

I. Introduction

Modular Multilevel Converter (MMC) is one of the most attractive topologies in recent years for medium or high voltage industrial applications [1, 2], such as high voltage dc transmission (HVDC) [3], medium voltage variable speed motor drives [4] and static synchronous compensators [5]. The wide adoption of MMCs in industry is mainly due to its flexible expandability, transformer-less configuration, common dc bus, high reliability from redundancy, and so on. A well-known problem within an MMC system is that the differential current in phase legs may be distorted by low order harmonics [6], and this is because of the inherent mismatch between the inserted voltage of each phase leg and the dc bus voltage. Such low order harmonics may introduce more power losses and in turn disturbances on the sub-module capacitor voltage, and consequently deteriorate the performance of the MMC system [6-9].

In most applications, the harmonics in the differential current of an MMC are undesirable from efficiency and controllability points of view. Nevertheless, it is difficult to find a practical passive method to mitigate these low-order harmonics in the differential current. Alternatively, some active harmonic suppression methods were proposed in recent literatures such as open-loop control strategy that compensates the inserted voltage in each phase leg [10], feedforward control that injects adequate harmonics to reduce the second order harmonics in the differential current [8]. However, these methods highly rely on accurate MMC models and are sensitive to the disturbance and model parameter variation. Feedback control in the d-q or rotating frames were proposed in [11, 12] for second order harmonic suppression of a three phase MMC system. Proportional resonant (PR) controllers can also be adopted to deal with harmonics in differential current as discussed in [13]. The limitation of those methods is that they can only cope with specific order harmonics.

Recently, plug in repetitive controllers are reported in [14, 15] to eliminate multiple harmonics in the differential current. In these papers, the conventional repetitive controllers with one fundamental period delay are employed to handle both odd and even order harmonics. However, according to the steady-state analysis in [6], the harmonics in the differential current of an MMC are dominated by only even order components. Therefore, a repetitive controller being able to suppress even harmonics is sufficient in such applications [16, 17].

In this paper, an improved repetitive controller with half fundamental period delay is presented to eliminate the even harmonics in the inner differential current of an MMC system. The proposed repetitive controller has the same even-harmonic elimination performance and system stability comparing to conventional ones, with additional benefits of less memory occupation, doubled low frequency gain, and faster dynamic response. Moreover, it features higher frequency disturbance rejection because of the wider bandwidth at specific frequencies. The control strategy of the differential current in an MMC and the even-harmonic repetitive control scheme with its frequency response analysis are detailed in this paper. A

This work was conducted with support from the National Research Foundation (NRF) Singapore under the Corp Lab@University Scheme.

978-1-4673-9551-9/16 $31.00 © 2016 IEEE

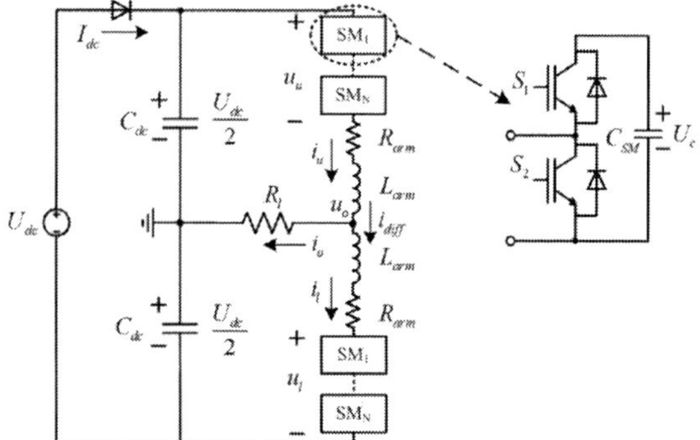

Fig. 1 Structure of a single phase MMC based inverter.

single-phase MMC prototype was built in the laboratory to verify the proposed repetitive control scheme. Both simulation and experimental results are finally presented to prove the effectiveness of this concept and theoretic analysis, as well as the excellent performance of the even-harmonic repetitive controller in dynamic response and disturbance rejection.

II. MMC Differential Current Control Strategy

The basic structure and operation of an MMC has been extensively explained in literatures [2, 7, 18] and will not be discussed in this paper. A single-phase MMC based inverter shown in Fig. 1 is adopted to demonstrate the proposed even-harmonic repetitive control scheme. There are N sub-modules connected in series in each arm, equipped with an arm inductor L_{arm}. An equivalent resistor R_{arm} is employed in each arm to represent the losses in each arm. The load resistor is connected between the middle points of the two arms and dc bus capacitors. u_u and u_l are voltages inserted into the upper and lower arm respectively, and U_c is the voltage across the sub-module capacitor.

A. Inner Differential Current in an MMC

Assuming the two arms are identical, the currents flowing through the upper and lower arms (i_u and i_l) are described as in (1), where i_{diff} is the inner differential current in the phase leg [6] and i_o is the output current of the MMC.

$$\begin{cases} i_u = i_{diff} + \dfrac{i_o}{2} \\ i_l = i_{diff} - \dfrac{i_o}{2} \end{cases} \tag{1}$$

Applying Kirchhoff's voltage law (KVL) to the dc loop of the MMC in Fig. 1, the equation in (2) can be obtained. It is clear in (2) that the voltage mismatch between the dc bus voltage U_{dc} and the inserted voltage in the phase leg will be applied onto the arm inductors and resistors, and consequently introduces the inner differential current in the phase leg. In steady state, i_{diff} normally consists of two parts as in (3), i.e. a dc current I_{dc} which ensures the power balance between the dc input and ac output [4], and a circulating current which is dominated by the 2nd order harmonic current [6, 9]. In most cases, only the dc component I_{dc} is preferred to maintain the stable operation of the MMC and minimize the losses. Therefore, besides controlling the output power of the MMC, special efforts are also devoted onto the inner differential current control to remove i_{cir}. According to (2) and (3), i_{diff} can be controlled by legitimately adjusting the voltage applied on L_{arm} and R_{arm} [11].

$$L_{arm}\frac{di_{diff}}{dt} + R_{arm}i_{diff} = \frac{U_{dc}}{2} - \frac{u_u + u_l}{2} \tag{2}$$

$$i_{diff} = I_{dc} + i_{cir} \tag{3}$$

B. Control Strategy for Inner Differential Current

The differential current reference i_{diff}^* is comprised by three components, i.e. a dc current reference $i_{diff\_dc}^*$ for power balance, a current reference $i_{diff\_va}^*$ for voltage averaging control [7] to maintain the sub-module capacitor voltages at the same level, and a fundamental frequency current reference $i_{diff\_vd}^*$ for the differential voltage control [19] between the upper and lower arms. Assuming the output voltage u_o and current i_o are well controlled as in (4), the dc current reference in (5) can be obtained, where U_o and I_o are the magnitudes of the output voltage and current respectively, ω_o is the fundamental angular frequency and ϕ_o is the phase displacement between the output voltage and current. The block diagram of generating the differential current reference is shown inside the red dashed block in Fig. 2. u_c^* is the reference of the sub-module capacitor voltage described in (6). $\overline{u}_{cu}$ and $\overline{u}_{cl}$ are the average capacitor

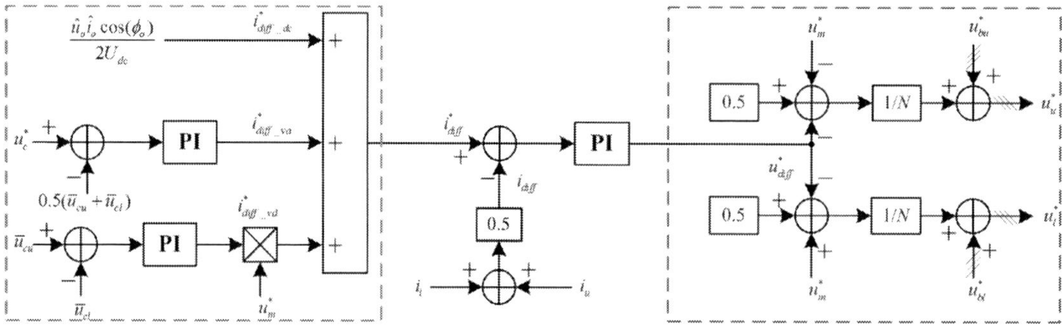

Fig. 2 Block diagram of difference current control loop of the MMC.

voltages in upper and lower arms respectively, as in (7). u_{cuk} and u_{clk} are the voltages across the kth sub-module capacitor in corresponding arms.

$$\begin{cases} u_o = U_o \sin(\omega_o t) \\ i_o = I_o \sin(\omega_o t + \phi_o) \end{cases} \tag{4}$$

$$i_{diff\_dc}^* = \frac{\hat{u}_o \hat{i}_o \cos(\phi_o)}{2U_{dc}} \tag{5}$$

$$u_c^* = \frac{U_{dc}}{N} \tag{6}$$

$$\begin{cases} \bar{u}_{cu} = \frac{1}{N}\sum_{k=1}^{N} u_{cuk} \\ \bar{u}_{cl} = \frac{1}{N}\sum_{k=1}^{N} u_{clk} \end{cases} \tag{7}$$

The blue dashed block in Fig. 2 shows the process of generating the reference signals for individual sub-module as in (8), where u_{diff}^* is the reference voltage that will be applied to the arm inductor to control the differential current, and u_m^* is the modulation signal of the output voltage in (9). The relationship between u_{diff} and i_{diff} and the s-domain transfer function of the plant, according to (2), can be expressed as in (10) and (11) respectively. The transfer function of the PI controller can be found in (12). Moreover, individual capacitor voltage balancing control [7] shown in Fig. 3 is also applied to ensure that the voltage across each sub-module capacitor is balanced.

$$u_{uk/lk}^* = \frac{1}{2}(1 \mp u_m^*) - u_{diff}^* + u_{buk/blk}^* \tag{8}$$

$$u_m^* = \frac{2u_o^*}{U_{dc}} \tag{9}$$

$$u_{diff} = R_{arm} i_{diff} + L_{arm}\frac{di_{diff}}{dt} \tag{10}$$

$$G(s) = \frac{1}{L_{arm}s + R_{arm}} \tag{11}$$

$$G_{PI}(s) = K_P + \frac{K_i}{s} \tag{12}$$

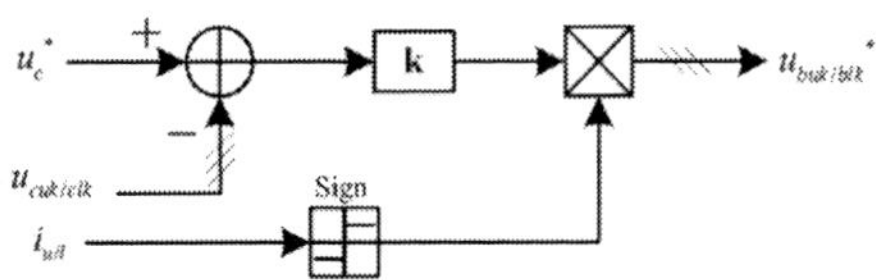

Fig. 3 Block diagram of individual voltage balancing control.

III. EVEN-HARMONIC REPETITIVE CONTROLLER

A. Design of Even-harmonic Repetitive Controller

The circulating current i_{cir} cannot be completely removed if only a conventional PI controller is employed as in Fig. 2. In order to regulate the differential current, a repetitive controller is adopted in the inner differential current control scheme instead of the single PI controller as shown in Fig. 4, where $PI(z)$ and $G(z)$ are the z-domain transfer functions of the PI controller and the plant respectively. Since only even order harmonics exist in the circulating current, the repetitive controller is designed to cope with even order harmonics by setting the number of error samples N_s to be half of that in one fundamental period. In this paper, the number of sampling periods $N_s = f_s/(2f_o)$ is adopted rather than $N_s = f_s/f_o$ as in the conventional repetitive controllers, where f_o is the fundamental frequency of the MMC output. By doing this, an even-harmonic repetitive controller is achieved. The memory cells required to store these samples are halved, and the performance of the repetitive controller is also improved, which will be discussed later.

The detailed parameters of the MMC system are listed in

TABLE I. PARAMETERS OF THE MMC SYSTEM

Parameters	Values
DC bus voltage: U_{dc}	200 V
Load Resistor: R_l	10 Ω
Rated output frequency: f_o	50 Hz
Arm inductance: L_{arm}	10 mH
Arm resistor: R_{arm}	0.025 Ω
No. of SMs in each arm: N	2
SM capacitor: C_{SM}	470 uF
Carrier frequency: f_c	2 kHz
PI parameters of the difference current control	$K_p = 5$, $K_i = 10$
Repetitive controller gain: K_r	0.8

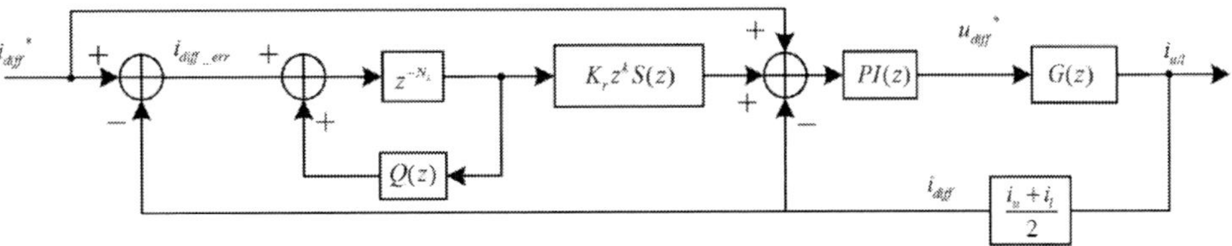

Fig. 4 Block diagram of the repetitive control scheme.

Table I. As the equivalent switching frequency of the MMC with phase-shifted PWM modulation scheme is $4f_c = 8$ kHz [7], the sampling frequency is designed to be $f_s = 8$ kHz and the control system is synchronized with it, having a period of $T_s = 1/f_s$. N_s is chosen to be 80 to perform the even-harmonic repetitive controller. In Fig. 4, $Q(z)$ is a low pass filter with unit gain at low frequencies and no phase delay, whose transfer function is expressed in (13) with frequency response as $Q(e^{j\omega T_s}) = 0.25*[2\cos^2(\omega T_s)+\cos(\omega T_s)+1]$. The cutoff frequency of $Q(z)$ is designed to be around 940 Hz.

$$Q(z) = \frac{z^2 + z + 4 + z^{-1} + z^{-2}}{8} \tag{13}$$

The phase lead filter, consisting of a second order low pass filter $S(z)$ and a phase delay compensation unit z^k [20], is elaborately designed such that the natural frequency of $S(z)$ is f_n=800 Hz and k=8. The system dynamic response and stability will be mainly determined by the repetitive gain K_r, which should be carefully designed. Larger K_r provides faster dynamic response of the repetitive controller at the expense of less stability margin.

B. Frequency Response Analysis of the Repetitive Controller

In order to analyze the differential current control scheme more conveniently and effectively, the overall differential current control loop is eventually simplified as in Fig. 5. The closed-loop transfer function from the reference to the differential current can be derived from Fig. 5 as in (14), where $P(z)=PI(z)G(z)/(1+PI(z)G(z))$. Since one sufficient condition for the stability of the repetitive control system is $|Q(z)-K_r z^k S(z)P(z)|<1$ [21], the stability of this control system will not be influenced by choosing different values of N_s, as N_s does not appear in this inequality. The Nyquist plot in Fig. 6 indicates that the control system is stable based on the parameters given in this subsection and Table I.

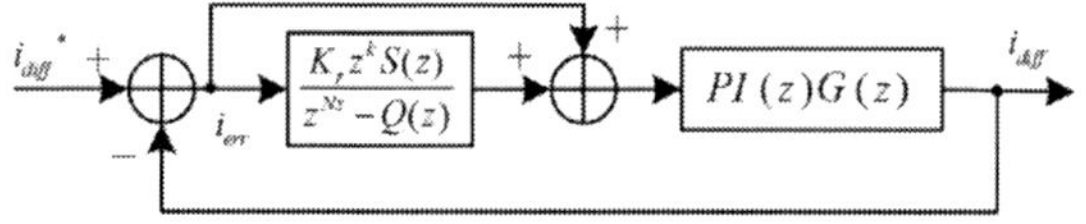

Fig. 5 Simplified block diagram of the control system.

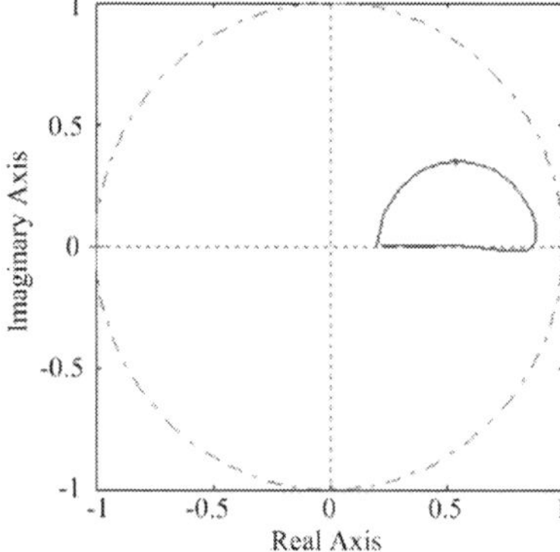

Fig. 6 Nyquist plot of z^{-N_s}-$Q(z)$+$K_r z^k S(z)P(z)$.

$$i_{diff} = \frac{P(z)\left[z^{N_s} - Q(z) + K_r z^k S(z)\right]}{z^{N_s} - Q(z) + K_r z^k S(z)P(z)} i_{diff}^* \tag{14}$$

As shown in Fig. 5, the overall control system can be simply divided into two parallel forward paths, i.e. a simple PI controller applied to the plant model, and a repetitive controller cascaded with the PI controller and the plant model. In this paper, only the path including the repetitive controller is analyzed and discussed. The transfer function of the repetitive controller path and the controller itself are described as in (15) and (16) respectively.

$$G_{RP}(z) = G_{rc}(z)PI(z)G(z) \tag{15}$$

$$G_{rc}(z) = \frac{K_r z^k S(z)}{z^{N_s} - Q(z)} \tag{16}$$

It is clear in (15) and (16) that N_s will affect the open loop gain of the repetitive control path. Since the open loop gain of the repetitive control path is proportional to that of $G_{rc}(z)$, only $G_{rc}(z)$ is discussed here. The frequency response of $G_{rc}(z)$ can be expressed as in (17).

$$G_{rc}(e^{j\omega T_s}) = \frac{K_r e^{jk\omega T_s} S(e^{j\omega T_s})}{e^{j\omega N_s T_s} - Q(e^{j\omega T_s})} \tag{17}$$

As the high frequency gain of a repetitive controller is normally designed as small as possible to prevent oscillation, only the low frequency gain is investigated. The time advance unit z^k will not affect the gain of the repetitive controller and the low pass filter $S(z)$ is also dedicatedly designed with almost unity gain at low frequencies. Therefore, the gain of $G_{rc}(z)$ can be rewritten as in (18) when $\omega \ll 2\pi*800$ rad/s.

$$|G_{rc}(e^{j\omega T_s})| \approx \frac{K_r}{|e^{j\omega N_s T_s} - Q(e^{j\omega T_s})|} = \frac{K_r}{|e^{j\omega N_s T_s} - 0.25(2\cos^2(\omega T_s) + \cos(\omega T_s) + 1)|} \tag{18}$$

Note that when ω is low, the discrete frequency ωT_s is very close to zero, so that the open loop gain of this repetitive controller can be further simplified as in (19). Equation (20) shows that the crossover frequency of the repetitive controller f_{cross} is inversely proportional to N_s.

$$|G_{rc}(e^{j\omega T_s})| \approx \frac{K_r}{|e^{j\omega N_s T_s} - 1|} = \frac{K_r}{|\cos(\omega N_s T_s) + j\sin(\omega N_s T_s) - 1|} = \frac{K_r}{2} \frac{1}{|\sin(\omega N_s T_s/2)|} \tag{19}$$

$$f_{cross} = \left.\frac{\arcsin(K_r/2)}{\pi N_s T_s}\right|_{0 \le K_r \le 2,\ \omega T_s \approx 0} \tag{20}$$

Substituting $N_s = f_s/f_o$ and $N_s = f_s/2f_o$ into (19) respectively and noting that $T_s f_s = 1$, it is obvious that the gain of the even-harmonic repetitive controller is almost twice of that of the conventional repetitive controller when $f \ll f_s$. Furthermore, the phase-frequency response of the repetitive control can be

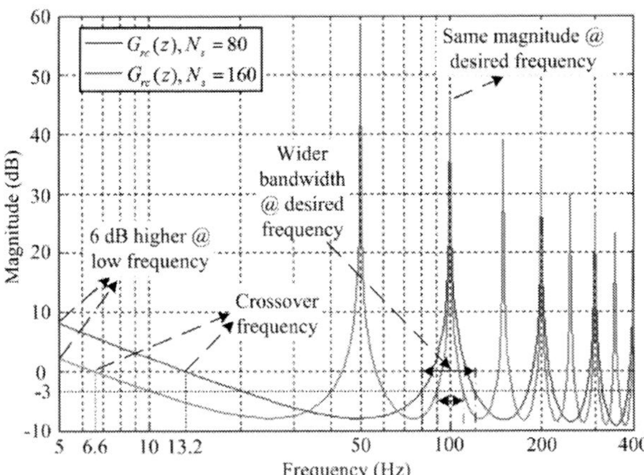

Fig. 7 The magnitude of the repetitive controller with different Ns.

derived as in (21) based on (17). The phase margins of the proposed and conventional repetitive controllers are almost the same at frequencies nf_o with n being an even integer. It means that the dynamic response of the proposed repetitive control is better while its closed-loop system stability is basically not affected. Meanwhile, it could be verified by (18) that the gains of $G_{rc}(z)$ at desired frequencies (e.g. 100 Hz, 200 Hz...) are also not affected no matter the value of N_s is selected to be f_s/f_o or $f_s/2f_o$.

$$\begin{aligned}\angle G_{rc}(e^{j\omega T_s}) &= k\omega T_s + \angle S(e^{j\omega T_s}) - \angle(e^{j\omega N_s T_s} - Q(e^{j\omega T_s})) \\ &\approx k\omega T_s + \angle S(e^{j\omega T_s}) - \angle(e^{j\omega N_s T_s} - 1)\end{aligned} \tag{21}$$

Furthermore, if there is a frequency disturbance f_d in the output frequency of the MMC which makes the output frequency to be $f = f_o \pm f_d$, the magnitude of $G_{rc}(e^{j\omega T_s})$ at relatively low frequencies can be found by (22), where n is an even integer and $n\pi f_d << f_o$. Substituting $N_s = f_s/f_o$ and $N_s = f_s/2f_o$ into (22), the results indicate that the gains of the even-harmonic repetitive controller at desired frequencies are twice of those of the conventional controller. In other words, the proposed repetitive controller has wider bandwidth at critical even-order frequencies and consequently better compensating performance when there is a system frequency deviation.

$$|G_{rc}(e^{j\omega T_s})| \approx \frac{K_r}{2}\frac{1}{|\sin(n\pi(f_o \pm f_d)N_s T_s)|} \approx \frac{K_r}{2n\pi f_d N_s T_s} \tag{22}$$

The amplitude-frequency responses of the repetitive controller $G_{rc}(z)$ with N_s=160 and N_s=80 can be found in Fig. 7, which shows that the frequency responses of the proposed repetitive controller are in accordance with the mathematical analysis.

IV. Simulation and Experimental Results

Simulations with PLECS as well as experimental validation of the proposed even-harmonic repetitive control scheme were conducted based on the single-phase MMC configuration shown in Fig. 1, and the system parameters are listed in Table I. The main objectives were to show the effectiveness of the proposed repetitive control scheme in even-harmonic suppression, and its improved dynamic response and frequency disturbance rejection.

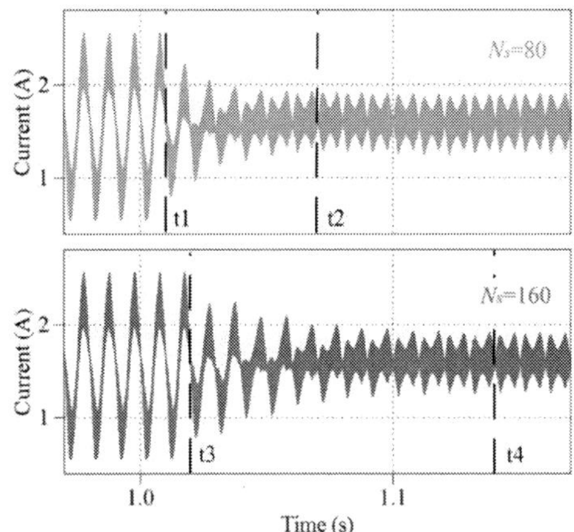

Fig. 8 Dynamic response of the repetitive controller with Ns=80 and Ns=160.

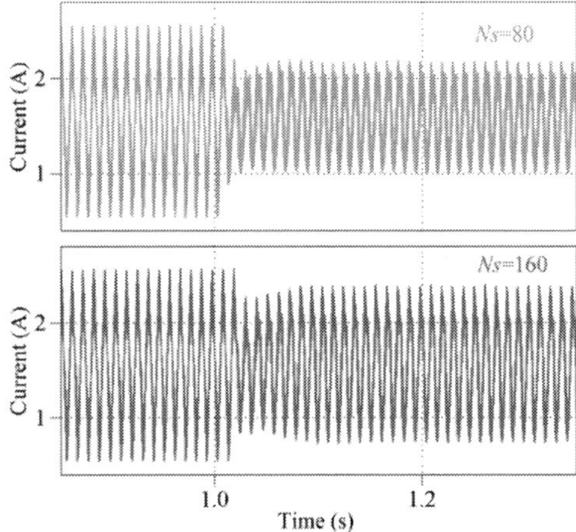

Fig. 9 Performance of the repetitive control system @ fo = 47.5 Hz in simulation.

A. Simulation Results

The dynamic response of the proposed and conventional repetitive controllers are illustrated in Fig. 8, where the repetitive controllers are activated at t1 and t3 respectively. K_r is intentionally set to be 0.5 to demonstrate the transient process after the repetitive controller was enabled. It is obvious that the proposed repetitive controller only requires half period to achieve steady state after being activated as compared to that of the conventional ones. The halved N_s not only reduces the memory cells required to store these error samplings, but also speeds up the dynamic response of the system. The performances of the both repetitive controllers under the condition of 5% frequency variation in the MMC output are shown in Fig. 9. The total harmonic distortion (THD), relative to the dc component in the differential current, of the differential current is reduced by 33.94% if the proposed repetitive controller is used instead of the conventional one.

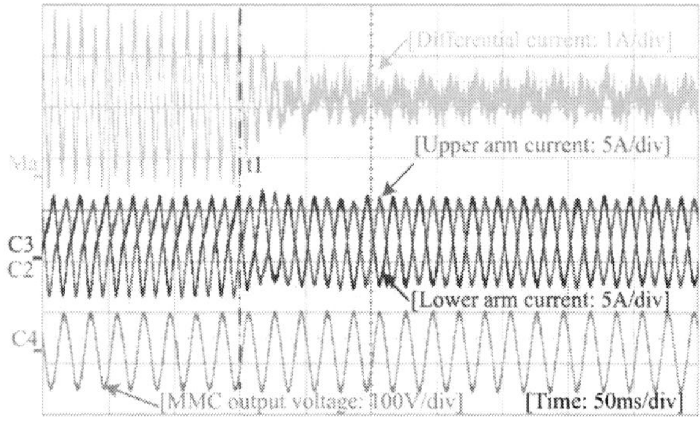

Fig. 10 Experimental results of the repetitive control system with Ns = 80.

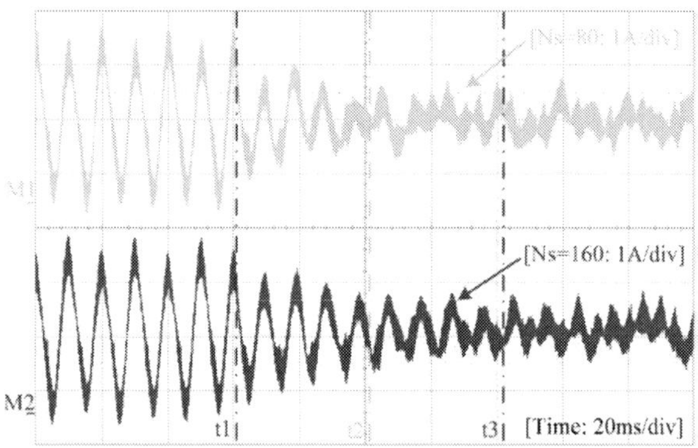

Fig. 11 Dynamic response of the repetitive control system.

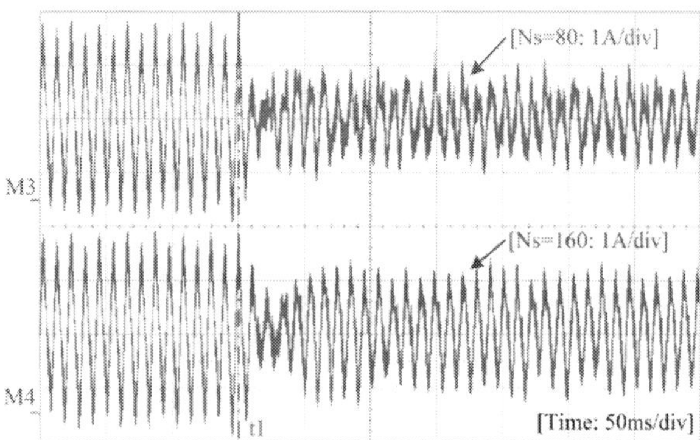

Fig. 12 Performance of the repetitive control system @ fo = 47.5 Hz.

B. Experimental Results

The modulation index of the output voltage was set to be 0.8 in all experiments. The experimental results are shown in Fig. 10 to Fig. 12, where the repetitive controller was activated at t1. Fig. 10 presents the effectiveness of the even-harmonic repetitive control scheme for eliminating the harmonics in the differential current. It is obvious that the low frequency harmonics in the circulating current are almost completely removed when the repetitive control is applied, leaving only high frequency ones mainly introduced by the switches. The peak-to-peak value of the circulating current is reduced from 3.5 A to less than 1 A. The distortion in the arm current is also removed, and only the dc and fundamental components are retained as seen in Fig. 10.

The dynamic response of the repetitive control system is illustrated in Fig. 11. The differential current is regarded as settled down if no distinct low frequency component can be observed. After enabling the repetitive controller at t = t1, the differential current regulated by the even-harmonic repetitive controller is converged smoothly by t = t2 (40ms), while the harmonic contents are removed by the conventional controller by t = t3 (80ms). The result shows that the dynamic response of the proposed repetitive control system is doubled.

In Fig. 12, a frequency deviation of f_d = -2.5 Hz (-5%) is intentionally applied to the MMC output voltage reference while the repetitive controller is still designed for f_s = 50 Hz. The experimental result suggests that the even-harmonic repetitive controller has better frequency disturbance rejection ability than that of the conventional one. The harmonic contents in the differential current can be reduced by 31% if the conventional repetitive controller is replaced by the proposed one when f_o = 47.5 Hz.

V. Conclusion

This paper has presented an even-harmonic repetitive control scheme that incorporates half delayed control cycle to suppress the harmonics in the differential current of an MMC. The proposed control scheme not only removes the harmonics from the differential current, but also exhibits faster dynamic response and better ability to withstand frequency variation. The analysis suggests that the repetitive control system is improved in terms of higher low frequency gain, higher crossover frequency, and wider control bandwidth at desired frequencies. The mathematical analysis, PLECS simulations, and experimental results agree with each other very well. The results show that the harmonics in the differential current are well suppressed, the dynamic response of the differential current control system is doubled, and the performance under frequency variation is almost improved by one third.

References

[1] S. Kouro, M. Malinowski, K. Gopakumar, J. Pou, L. G. Franquelo, W. Bin, *et al.*, "Recent Advances and Industrial Applications of Multilevel Converters," *IEEE Trans. Ind. Electron.*, vol. 57, pp. 2553-2580, 2010.

[2] S. Debnath, J. Qin, B. Bahrani, M. Saeedifard, and P. Barbosa, "Operation, Control, and Applications of the Modular Multilevel Converter: A Review," *IEEE Trans. Power Electron.*, vol. 30, pp. 37-53, 2015.

[3] G. Bergna, E. Berne, P. Egrot, P. Lefranc, A. Arzande, J. C. Vannier, *et al.*, "An Energy-Based Controller for HVDC Modular Multilevel Converter in Decoupled Double Synchronous Reference Frame for Voltage Oscillation Reduction," *IEEE Trans. Ind. Electron.*, vol. 60, pp. 2360-2371, 2013.

[4] M. Hagiwara, K. Nishimura, and H. Akagi, "A Medium-Voltage Motor Drive With a Modular Multilevel PWM Inverter," *IEEE Trans. Power Electron.*, vol. 25, pp. 1786-1799, 2010.

[5] H. P. Mohammadi and M. T. Bina, "A Transformerless Medium-Voltage STATCOM Topology Based on Extended Modular Multilevel Converters," *IEEE Trans. Power Electron.*, vol. 26, pp. 1534-1545, 2011.

[6] K. Ilves, A. Antonopoulos, S. Norrga, and H. P. Nee, "Steady-State Analysis of Interaction Between Harmonic Components of Arm and

Line Quantities of Modular Multilevel Converters," *IEEE Trans. Power Electron.*, vol. 27, Jan 2012.

[7] M. Hagiwara and H. Akagi, "Control and Experiment of Pulsewidth-Modulated Modular Multilevel Converters," *IEEE Trans. Power Electron.*, vol. 24, pp. 1737-1746, Jul 2009.

[8] C. Wang, Q. R. Hao, and B. T. Ooi, "Reduction of low-frequency harmonics in modular multilevel converters (MMCs) by harmonic function analysis," *Iet Generation Transmission & Distribution*, vol. 8, pp. 328-338, 2014.

[9] Q. Song, W. H. Liu, X. Q. Li, H. Rao, S. K. Xu, and L. C. Li, "A Steady-State Analysis Method for a Modular Multilevel Converter," *IEEE Trans. Power Electron.*, vol. 28, pp. 3702-3713, Aug 2013.

[10] L. Angquist, A. Antonopoulos, D. Siemaszko, K. Ilves, M. Vasiladiotis, and H. P. Nee, "Open-Loop Control of Modular Multilevel Converters Using Estimation of Stored Energy," *IEEE Trans. Ind. Appl.*, vol. 47, pp. 2516-2524, 2011.

[11] Q. Tu, Z. Xu, and L. Xu, "Reduced Switching-Frequency Modulation and Circulating Current Suppression for Modular Multilevel Converters," *IEEE Trans. Power Del.*, vol. 26, pp. 2009-2017, 2011.

[12] B. Bahrani, S. Debnath, and M. Saeedifard, "Circulating Current Suppression of the Modular Multilevel Converter in a Double-Frequency Rotating Reference Frame," *IEEE Trans. Power Electron.*, vol. PP, pp. 1-1, 2015.

[13] Z. Li, P. Wang, Z. Chu, H. Zhu, Y. Luo, and Y. Li, "An Inner Current Suppressing Method for Modular Multilevel Converters," *IEEE Trans. Power Electron.*, vol. 28, pp. 4873-4879, 2013.

[14] M. Zhang, L. Huang, W. Yao, and Z. Lu, "Circulating Harmonic Current Elimination of a CPS-PWM-Based Modular Multilevel Converter With a Plug-In Repetitive Controller," *IEEE Trans. Power Electron.*, vol. 29, pp. 2083-2097, 2014.

[15] L. He, K. Zhang, J. Xiong, and S. Fan, "A Repetitive Control Scheme for Harmonic Suppression of Circulating Current in Modular Multilevel Converters," *IEEE Trans. Power Electron.*, vol. 30, pp. 471-481, 2015.

[16] K. Zhou, K. S. Low, D. Wang, F. Luo, B. Zhang, and Y. Wang, "Zero-phase odd-harmonic repetitive controller for a single-phase PWM inverter," *IEEE Trans. Power Electron.*, vol. 21, pp. 193-201, 2006.

[17] P. C. Loh, Y. Tang, F. Blaabjerg, and P. Wang, "Mixed-frame and stationary-frame repetitive control schemes for compensating typical load and grid harmonics," *Power Electronics, IET*, vol. 4, pp. 218-226, 2011.

[18] A. Lesnicar and R. Marquardt, "An innovative modular multilevel converter topology suitable for a wide power range," in *Power Tech Conference Proceedings, 2003 IEEE Bologna*, 2003, p. 6 pp. Vol.3.

[19] L. Harnefors, A. Antonopoulos, S. Norrga, L. Angquist, and H. P. Nee, "Dynamic Analysis of Modular Multilevel Converters," *IEEE Trans. Ind. Electron.*, vol. 60, pp. 2526-2537, Jul 2013.

[20] K. Zhang, Y. Kang, J. Xiong, and J. Chen, "Direct repetitive control of SPWM inverter for UPS purpose," *IEEE Trans. Power Electron.*, vol. 18, pp. 784-792, 2003.

[21] G. F. Franklin, J. D. Powell, and A. Emami-Naeini, *Feedback Control of Dynamic Systems*. MA: Addison-Wesley, 1991.

978-1-4673-9551-9/16 $31.00 © 2016 IEEE

A new DSC-PLL using Recursive Discrete Fourier Transform for Robustness to Frequency Variation

Jaedo Lee[1] and Hanju Cha[2]

[1] Dept. of Instrument control & Electrical System, Korea Institute of Nuclear Safety, Daejeon, Korea

[2] Dept. of Electrical Engineering, Chungnam National University, Daejeon, Korea

Abstract—**In this paper, a new DSC-PLL(Delayed Signal Cancellation Phase Locked Loop) with FLL(Frequency Locked Loop) based on RDFT(recursive discrete fourier transform) method is proposed for coping with frequency variation. This method shows significant performance improvement for detection of fundamental positive sequence component voltage when the grid voltage is polluted by grid harmonics and frequency variation. The frequency detection technique of DSC-PLL tracks frequency drift by the RDFT. These compensation algorithms can correct for discrepancies of changing the frequency within maximum 28ms and improve traditional DSC-PLL. To verify feasibility of the proposed DSC-PLL, Matlab/Simulink results show good agreement with the analysis.**

I. INTRODUCTION

Grid code requirements for the grid connection of distributed generation systems become demanding, and system operators are concerned about the Low Voltage Ride Through (LVRT) requirements. The demand of these technologies in the electrical network has reinforced about their influence in the grid stability. The grid connection standards are becoming more restrictive for distribution generation systems [1]. The distribution generation systems Grid codes demand control requisites similar to that demanded in conventional power plants, which are basically focused on achieving a stable and reliable operation of the electricity network. Control strategy for the power converters under voltage sag have been proposed [2]. In this case, fast fundamental component phase angle detection under the fault conditions contributes to improve grid converter control systems. Thus, the Phase Locked Loop algorithms are crucial. In three-phase systems, PLL based on a synchronization reference frame (SRF-PLL) have become a conventional synchronization technique [3]. Although these systems have shown to be fast and accurate under balanced conditions, its response is deficient when the utility voltage is distorted. Advanced synchronization systems, able to deal with this grid conditions, have been proposed as the alternative methods such as the decoupled double synchronous reference frame PLL (DDSRF-PLL) [4], the three phases enhanced PLL (3PH-EPLL) [5], the dual second order generalized integrator resting on a frequency locked loop (DSOGI-FLL) [6], Delayed Signal Cancellation PLL (DSC-PLL) [7] and multiple second order generalized integrators with frequency-locked loop(MSOGI-FLL)[8]. A DSC PLL represents a three phase selective harmonic detection method based on a delayed signal cancellation phase locked loop [9]. It uses DSC operation to extract harmonics of interest and the undesired harmonics can be completely eliminated, resulting in zero steady-state detection error. Despite successful phase locked loop methods, the existing selective harmonic detection methods have a few drawbacks [10]. DSC PLL is sensitive to frequency variation and need to be designed to handle considerable frequency shift with frequency locked loop. Due to the frequency variation, DSC PLL system causes the difficulty of removing particular harmonics. This paper proposes a FLL method by using a recursive discrete fourier transform(RDFT) [11]. This paper is organized as follows. The proposed DSC-PLL with RDFT is presented and analyzed in section II. Performance of the proposed DSC-PLL is evaluated by experiment and simulation in section III. Conclusion is given in section IV.

II. DELAYED SIGNAL CANCELLATIONS(DSC) PLL USING RDFT

A. Configuration of DSC-PLL

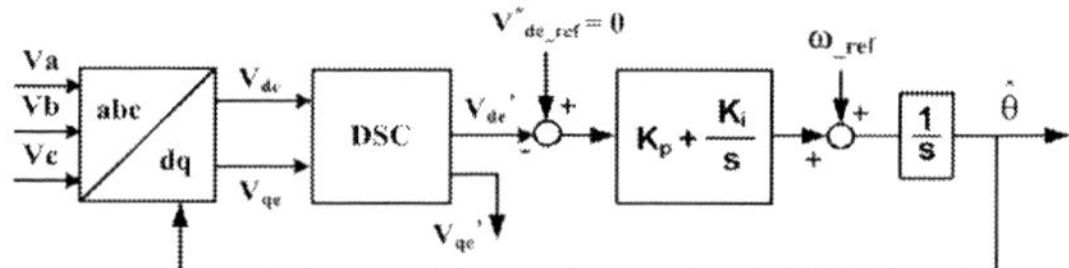

Fig.1. Configuration of DSC-PLL

Fig. 1 shows configuration of DSC-PLL that is combined SRF-PLL and DSC. Three-phase voltage are converted to dq voltage components by Park's transformation as shown in (1), (2), (3), where θ is the actual grid phase and $\hat{\theta}$ is the estimated phase. As $\hat{\theta}$ approaches to θ, V_{de} goes to zero and V_{qe} goes to $-V_m$. Therefore, dq voltage components are to dc term when grid voltage is balanced conditions and It means that PLL detects grid's phase information accurately. If three-phase voltages are distorted, dq voltage components contains oscillatory terms due to the influence of harmonics. Under these conditions, DSC performs important role that eliminates oscillatory terms. Therefore, DSC-PLL can rapidly obtain phase and frequency of the fundamental positive-sequence voltage.

$$\begin{bmatrix} v_{de} \\ v_{qe} \end{bmatrix} = \frac{2}{3}\begin{bmatrix} \cos\hat{\theta} & \cos(\hat{\theta}-\frac{2}{3}\pi) & \cos(\hat{\theta}+\frac{2}{3}\pi) \\ -\sin\hat{\theta} & -\sin(\hat{\theta}-\frac{2}{3}\pi) & -\sin(\hat{\theta}-\frac{2}{3}\pi) \\ \frac{1}{2} & \frac{1}{2} & \frac{1}{2} \end{bmatrix}\begin{bmatrix} V_m \sin\theta \\ V_m \sin(\theta-\frac{2}{3}\pi) \\ V_m \sin(\theta+\frac{2}{3}\pi) \end{bmatrix} \tag{1}$$

$$V_{de} = V_m \sin(\theta - \hat{\theta}) \tag{2}$$

$$V_{qe} = -V_m \cos(\theta - \hat{\theta}) \tag{3}$$

B. Small signal model of PLL

Fig. 2 shows small signal model of PLL. Using small signal model, PI controller parameter of PLL is derived. Open-loop transfer function and closed-loop transfer function of PLL are described in (4), (5). K_p and T_i are the proportional gain and time constant of PI controller, respectively.

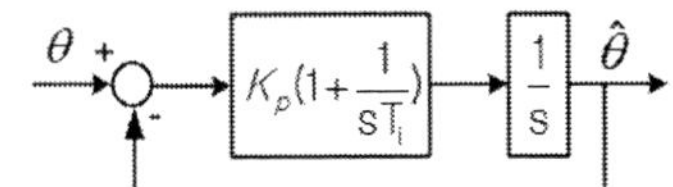

Fig.2. Small Signal model of PLL

$$T_{OL}(s) = \frac{K_p s + \frac{K_p}{T_i}}{s^2} \tag{4}$$

$$T_{CL}(s) = \frac{K_p s + \frac{K_p}{T_i}}{s^2 + K_p s + \frac{K_p}{T_i}} \tag{5}$$

Fig. 3 shows frequency response characteristics of open loop and closed loop transfer function for DSC-PLL, respectively. From bode plot of open loop transfer function, Designed PLL system is stable system as it has sufficient phase margin of 65°.

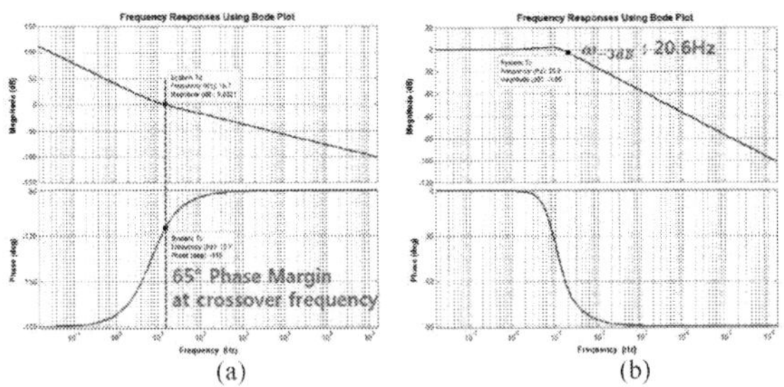

Fig. 3. (a) Open Loop transfer fucntion (b) Closed Loop transfer function

Table I shows PLL PI controller parameters. Critical damping factor of 0.707 and natural frequency of 10Hz is used and bandwidth is 20.6Hz.

TABLE I. DSC-PLL PI CONTORLLER PARAMETER

Parameter		Value
Damping ratio	ζ	0.707
Natural Frequency	ω_n	10Hz
Proportional gain	K_p	88.4
Time constant	T_i	0.025
Integrator gain	K_i	3497
Crossover frequency	f_c	15.7H
Phase margin	ϕ	65°
Bandwidth	ω_{-3dB}	20.6Hz

C. Characteristic of DSC method

Fig. 4 shows configuration of DSC and the even and triple harmonics in utility voltage are assumed to be zero in general. On the other hand, unbalance three-phase voltage and the odd harmonic components existing in utility voltage are respectively transformed into 2nd harmonic and even harmonic components in synchronous reference frame by park's transformation. In other words, Negative fundamental sequence appears as 2nd harmonic in dq axis voltage. In addition, 5th and 7th harmonics in three-phase grid voltage bring 6th harmonic term. Similarly, 11th and 13th harmonics appear as 12th harmonic term. Therefore, if utility voltage is polluted by unbalance or harmonics, the PLL phase has a certain amount of ripple. In this case, DSC method can rapidly remove harmonic terms in synchronous reference frame. Table II shows delay time for elimination of each harmonic component [16].

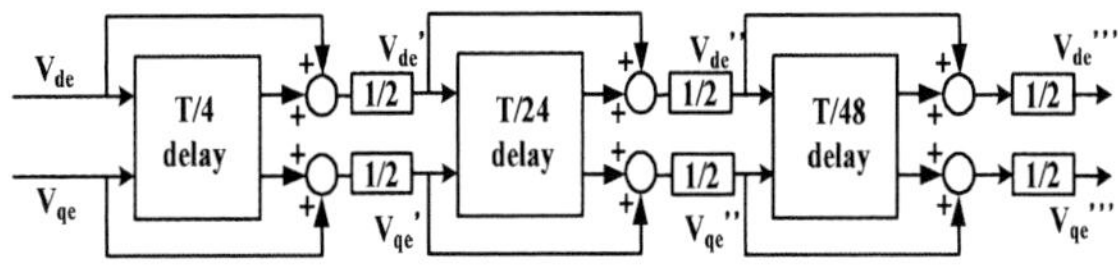

Fig.4. Configuration of DSC

TABLE II. DELAY TIME FOR HARMONICS SEQUENCE ELIMINATION

three-phase sequence	dq harmonic	delay time
Negative fundamental	2nd harmonic	T / 4
5th harmonic (Neg)	6th harmonic	
7th harmonic (Pos)		
11th harmonic (Neg)	12th harmonic	T / 24
13th harmonic (Pos)		
17th harmonic (Neg)	18th harmonic	T / 4
19th harmonic (Pos)		
23th harmonic (Neg)	24th harmonic	T / 48
25th harmonic (Pos)		

Using d axis voltage in synchronous reference frame, each output voltage of DSC block is represented by the following. Under distorted voltage condition, V_{de} consist of the dc term and the harmonics terms up to n^{th} order in synchronous reference frame as shown (6), where k is the harmonic's order transformed by park's transformation. Output of the first delay block is represented as (7). 2nd and 6th harmonics in dq axis voltage components are eliminated. Second and third delay blocks eliminate higher harmonics as shown in (8), (9) [16].

$$V_{de}(t) = V_{dc} + \sum_{k=2,6,12\cdots}^{n} V_k \sin(k\omega t + \phi_k) \tag{6}$$

$$V_{de}'(t) = \frac{1}{2}\left[V_{de}(t) + V_{de}(t - \frac{\pi}{2\omega})\right] \quad \because \frac{T}{4} = \frac{\pi}{2\omega} \tag{7}$$

$$V_{de}''(t) = \frac{1}{2}\left[V_{de}'(t) + V_{de}'(t - \frac{\pi}{12\omega})\right] \quad \because \frac{T}{24} = \frac{\pi}{12\omega} \tag{8}$$

$$V_{de}'''(t) = \frac{1}{2}\left[V_{de}''(t) + V_{de}''(t - \frac{\pi}{24\omega})\right] \quad \because \frac{T}{48} = \frac{\pi}{24\omega} \tag{9}$$

In the case of a constant frequency of grid fundamental voltage, DSC PLL becomes to use accurately constant delay times for DSC delay blocks to eliminate perfectly unbalance and harmonic components. Thus, DSC PLL can be performed and calculated rapidly and precisely phase angle value other than PLL methods.

III. PROPOSED FLL USING RDFT

A. FLL using RDFT

Consider a time signal v(t) that is sampled at the rate N/T_s to produce the time sequence{v(n)}. where N and T_s are data points in one period, sampling time period. The DFT producing the kth harmonic of {v(n)} at the step n-1 can be written as following

$$V_k(n-1) = \sum_{i=n-N}^{n-1} v(i)\exp(-j2\pi k(i-1)/N) \tag{10}$$

One time step later, a new sample of v(t) is taken and the DFT at the time step n becomes

$$V_k(n) = \sum_{i=n-N+1}^{n} v(i)\exp(-j2\pi k(i-1)/N) \tag{11}$$

Subtracting (10) from (11) yields

$$V_k(n) = V_k(n-1) + [v(n) - v(n-N)]\exp(-j2\pi k(n-1)/N) \tag{12}$$

Equation (12) is called the recursive DFT and can be used to estimate the spectral content present in v(t). recursive DFT equation is based on the difference between the nth and $(n-N)^{th}$ values in the time sequence. All that is needed to use the transform as a filter is the inverse transform of the harmonic of interest. For the kth harmonic at the frequency $k \cdot f$ this inverse RDFT is given by

$$v_k(n) = \Gamma(k)V_k(n)\exp(j2\pi k(n-1)/N) \tag{13}$$

$$\Gamma(k) = \begin{cases} N^{-1}, & k = 0, N/2 \\ 2N^{-1}, & otherwise \end{cases}$$

$$v_1(n) = \frac{2}{N}V_1(n)\exp(j2\pi(n-1)/N) \tag{14}$$

For system line filtering, the harmonic of interest is $k = 1$. The frequency response of the RDFT can be obtained from the z-transform of (12) and (13). Therefore, the z-transform of the transfer function $\{v_1(n)/v(n), k = 1\}$ can be expressed by[11]

$$H(z) = \frac{1}{N}\left[\frac{1 - z^{-N}}{1 - e^{j2\pi/N}z^{-1}} + \frac{1 - z^{-N}}{1 - e^{-j2\pi/N}z^{-1}}\right] \tag{15}$$

The Bode plot of (15) is shown in fig. 5. The output of the presented $\{v_1(n)\}$ at the nth sampling instant can be expressed by

$$v_1(n) = A_1(n)\sin\{\omega(n)nT_s + \theta_1\} \tag{16}$$

The fundmental voltage amplitude $A_1(n)$ estimated from $\mathrm{Re}\{V_1(n)\}$ and $\mathrm{Im}\{V_1(n)\}$. Therefore, the amplitude normalized instantaneous fundamental voltage component can be obtained by [15]

$$v_1^u(n) = \frac{v_1(n)}{\frac{2}{N}\sqrt{[\mathrm{Re}\{V_1(n)\}]^2 + [\mathrm{Im}\{V_1(n)\}]^2}} = \sin\{\omega(n)nT_s + \theta_1\} \tag{17}$$

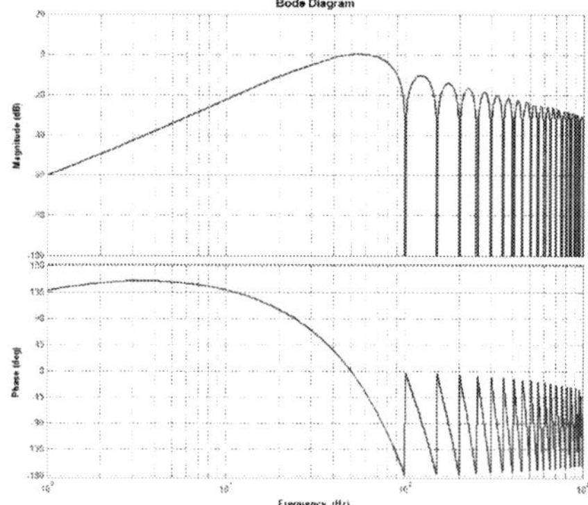

Fig. 5. Bode plot of the transfer function H(z)

The amplitued accumulation error is rejected due to the use of the amplitude normalized instantaneous fundamental voltage waveform. Fig.6 shows recursive DFT filter based on this concpet, with three outputs representing the real and imaginary components of the DFT as well as the discrete filtered output waveform [11].

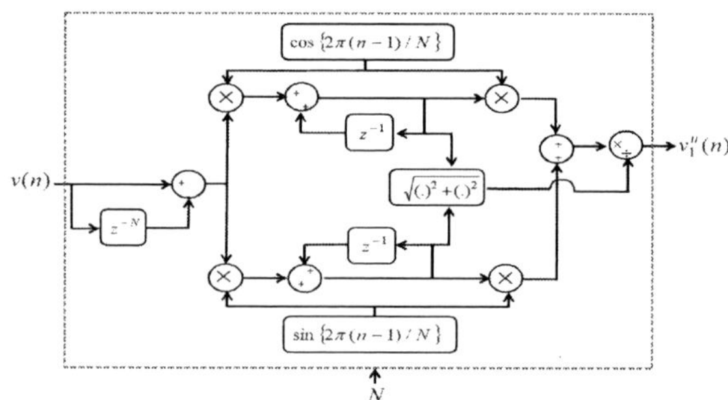

Fig. 6. Block diagram for the recursive DFT filter

B. Fundamental Frequency Estimation

Based on the constant fundamental parameters within three consecutive samples of $\{v_1(n)\}$, the relation can be expressed by Teager Energy Operator concept [13]~[15].

$$[\sin\{\omega(n)T_s\}]^2 = \{v_1^u(n-1)\}^2 - v_1^u(n)v_1^u(n-2) \quad (18)$$

Therefore, the fundamental frquency can be estimated from (18).

$$\hat{f}(n) = \frac{\sin^{-1}\left[\left\{\sqrt{\{v_1^u(n-1)\}^2 - v_1^u(n)v_1^u(n-2)}\right\}\right]}{2\pi T_s} \quad (19)$$

The proposed algorithm, as shown in Fig. 7, is used to track the fundamental frequency of the grid voltage waveform given by (6). The adaptive digital BPF based on the recursive DFT and inverse recursive DFT is used to extract the amplitude normalized instantaneous grid voltage fundamental component. The actual fundamental frequency information required by the adaptive BPF is also updated by the estimated frequency. It can be noticed that the number of samples in one fundamental time period is calculated as $N = f_s / \hat{f}$.

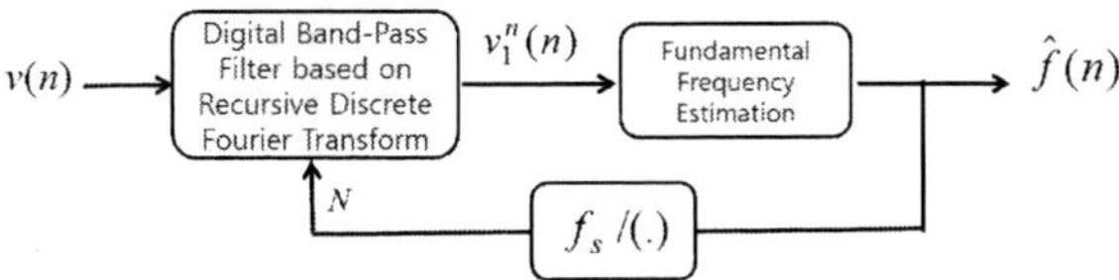

Fig.7. Block diagram of the grid voltage fundamental frequency estimation

In the case of a considerable frequency variation, period of grid fundamental voltage is varied from T and it is no longer accurate to use a constant T to calculate the delay times for DSC delay blocks. Therefore, it needs to adjust T to fit the grid frequency variation. For this purpose, FLL(Frequency Locked Loop) proposed in Fig. 8, and a method of tracking the fundamental frequency of the grid voltage is presented, based on phase predictions calculated T is obtained from the output frequency based on the Recursive Discrete Fourier Transform(RDFT) algorithm. ,

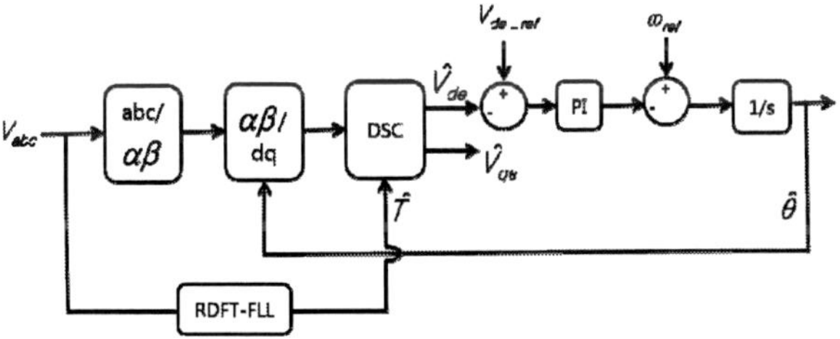

Fig.8. Proposed DSC-PLL block diagram with RDFT-FLL

IV. SIMULATION RESULTS

The proposed DSC PLL with RDFT FLL is evaluated by Matlab/Simulink simulations considering balanced, unbalanced and harmonics conditions such as frequency deviations in a from 50Hz to 47Hz. Matlab/Simulink results are provided by several cases as follows :

i. Frequency deviation with balanced condition

ii. Frequency deviation with unbalanced condition

iii. Frequency deviation with unbalanced and harmonics

Unbalanced and harmonic grid voltage used in the simulation is shown in Table III .

TABLE III. THREE-PHASE CONDITION FOR DSC-PLL VERIFICATION

Element	Magnitude	Frequency
Positive fundamental	180 V	50 Hz
Negative fundamental	50 V	50 Hz
5th harmonic (Neg)	18 V	250 Hz
7th harmonic (Pos)	9 V	350 Hz
11th harmonic (Neg)	3.6 V	550 Hz
13th harmonic (Pos)	3.6 V	650 Hz

Fig. 9 shows the Matlab/Simulink block diagram DSC PLL with RDFT FLL. Block diagram constitutes subsystems for DSC, fundamental extraction, etc.

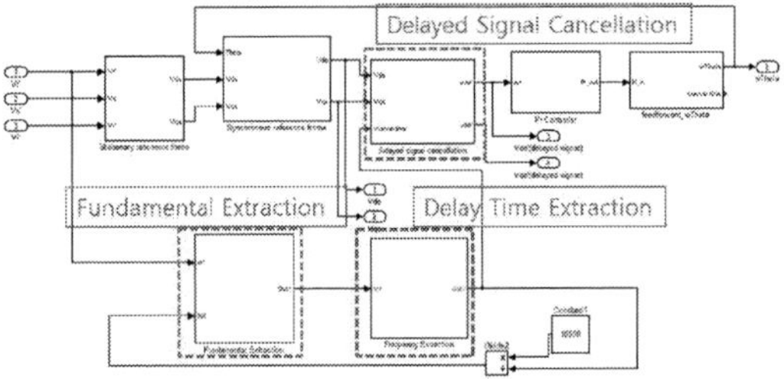

Fig.9. Matlab/Simulink DSC PLL with RDFT FLL

A. *Frequency deviation with balanced condition*

Fig. 10 shows a response of DSC PLL to frequency deviation from 50Hz to 47Hz in the balanced three phase voltage balance at 0.2 second. It takes about 25ms to reach to zero steady state error.

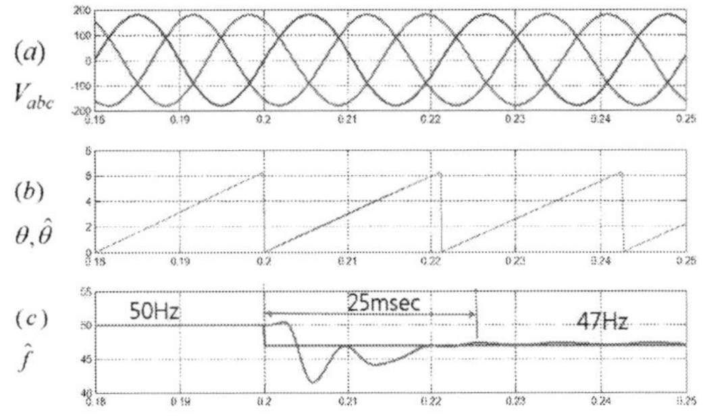

Fig.10. 3Hz frequency deviation with balanced condition at t=0.2 sec. (a) grid voltage waveform. (b) phase reference and estimation. (c) tracking frequency reference.

B. Frequency deviation with unbalanced condition

Fig. 11 shows a response of DSC PLL to frequency deviation form 50Hz to 47Hz in the unbalanced three phase voltage at 0.2 second. It takes about 25ms to reach to zero steady state error.

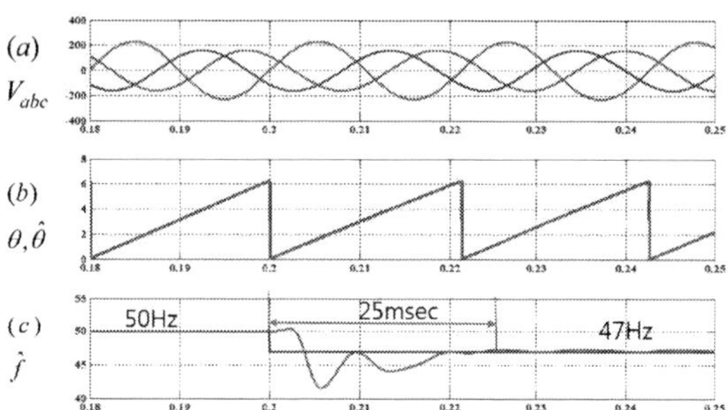

Fig.11. 3Hz frequency deviation with unbalanced condition at 0.2 sec. (a) grid voltage waveform. (b) phase reference and estimation. (c) tracking frequency reference.

C. Frequency deviation with unbalanced and harmonics.

Fig. 12 shows a response of DSC PLL to frequency deviation form 50Hz to 47Hz in the unbalanced and 5th, 7th, 11th, 13th harmonics three phase voltage at 0.2 second. It takes about 28ms to reach to zero steady state error.

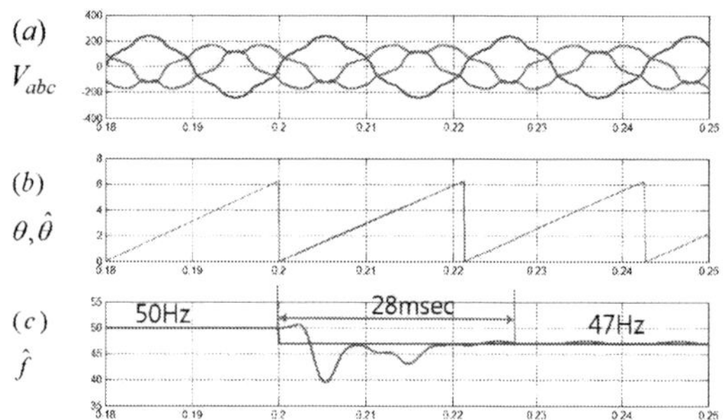

Fig.12. 3Hz frequency deviation with unbalanced and harmonic condition at 0.2 sec. (a) grid voltage waveform. (b) phase reference and estimation. (c) tracking frequency reference.

V. CONCULSION

This paper presented a new DSC-PLL to compensate frequency variation by using recursive discrete fourier transform method for grid synchronization. In spite of harmonics and unbalances, the proposed PLL has responded fast to sharp changes of grid voltage and shows precise steady-state output. In specially, if it sharply changes the frequency deviation, the sample period calculation regulates the value of a new sample frequency through RDFT frequency locked loop. The harmonic elimination and compensation with frequency variation technique of DSC has been analyzed and DSC-PLL with RDFT FLL method has been simulated by Matlab /Simulink under balanced, unbalanced and distorted conditions respectively. The results showed the DSC PLL with RDFT FLL has perfectly eliminated any harmonics and compensated frequency variation within maximum 28ms and has detected the fundamental positive sequence successfully under distorted three-phase voltage. The proposed DSC-PLL could manage to be steady frequency deviation as conventional DSC-PLL.

REFERENCES

[1] IEEE1547, "IEEE standard for interconnecting distributed resources with electric power systems". 2003.

[2] Lima, K., Luna, A., Rodriguez, P., Watanabe, E., Teodorescu, R. and Blaabjerg, F., "Doubly-Fed Induction Generator Control Under Voltage Sags," *Energy 2030 Conference, 2008. ENERGY 2008. IEEE*, vol., no 1., Nov, pp. 1-6.

[3] Blaabjerg, F., Teodorescu, F., Liserre, M. and Timbus, A.V., "Overview of Control and Grid Synchronization for Distributed Power Generation Systems," *IEEE Trans. on Industrial Electronics*, vol. 53, no. 5, pp. 1398-1409, Oct. 2006.

[4] Rodriguez, J. Pou, J. Bergas, J. I. Candela, R. P. Burgo, and D. Boroyevinh., "Decoupled Double Synchronous Reference Frame PLL for Power Converters Control," *IEEE Trans. Power Electron*, vol.22, pp.584-592, May. 2007.

[5] Karimi-Ghartemani, M. and Iravani, M. "A nonlinear adaptive filter for online signal analysis in power systems: applications," *Power Delivery, IEEE Transactions* on, vol. 17, no. 2, Apr, 2002, pp. 617-622.

[6] Blaabjerg, F., Teodorescu, F., Candela, I., Timbus, A., Liserre, M. and Blaabjerg, F., "New positive sequence voltage detector for grid synchronization of power converters under faulty grid conditions," in *Power Electronics Specialists, 2006 IEEE 37th Conference on*, 2006, pp. 1492-1498.

[7] J. S vensson, M. Bongiorno, A. Sannino, "Practical implementation of delayed signal cancellation method for phase-sequence separation," *IEEE Trans. Power Del.*, vol. 22, no.1, pp.18-26, Jan. 2007.

[8] P. Rodriguez, A. Luna, I. Etxeberria, J. R. Hermoso and R. Teodorescu, "Multiple second order generalized integrators for harmonic sysnchronization of power converters," *IEEE Energy Conversion Congress and Exposition*, pp. 2239-2246, 2009.

[9] Liang Wang, Oirong Jiang, Lucheng Hong, "A Novel Three-Phase Software Phase-Locked Loop Based on Frequency-Locked Loop and Initial Phase Angle Detection Phase-Locked Loop", *IECON 2012-38th Annual Conference on IEEE Industrial Electronics Society*, 2012, pp 150-155.

[10] Yi Fei Wang, Yun Wei Li, "Three-Phase Cascaded Delayed Signal Cancellation PLL for Fast Selective Harmonic Detection", *IEEE Transactions on Industrial Electronics*, Vol. 60, No. 4, April 2013, pp 1452-1463.

[11] McGrath, B.P., Holmes, D. G. and Galloway, J., "Improved Power Converter Line Synchronisation Using an Adaptive Discrete Fourier Transform(DFT)". In *Power Electronics Specialists Conference*, 2002. pesc. 02. 2002 IEEE 33rd Annual, Vol. 2, 2002, pp.821-826.

[12] Remus Teodorescu, Marco Liserre, Pedro Rodriguez, "Grid Converters for Photovoltaic and Wind Power Systems", A John Wiley and Sons, Ltd, Publication, 2011.

[13] I.Kamwa,A.K.Pradhan, and G,Joos, "Robust detetion and analysis of power system oscillations using the Teager-Kaiser energy operator", *IEEE Trans. Power Sys.*, vol. 26, no. 1, pp. 323-333, Feb. 2011.

[14] A. Subasi, A. S. Yilmaz, and K. Tufan, "Detection of generated and measured transient power quality event using Teager energy operator", *Elsevier Energy Conv. Amd Management*, Vol. 52, pp. 1959-1967, 2011.

[15] Md. Shamim Reza, Mihai Ciobotaru, Vassilios G. Agelidis, "A Recursive DFT Based Technique for Accurate Estimation of Grid Voltage Frequency", *Industrial Electronics Society, IECON 2013-39th Annual Conference of the IEEE*, pp 6420-6425, 2013.

[16] Jongmin Jo, Byung-Moon Han and Hanju Cha, "FPGA based DSC-PLL for grid harmonics and voltage unbalance effect elimination", *2015 IEEE Applied Power Electronics Conference and Exposition(APEC)*, pp 2212-2216, 2015.

A Four-Quadrant Modulation Technique for Cascaded Multilevel Inverters to Extend Solution Range for Selective Harmonic Elimination / Compensation

Hui Zhao and Shuo Wang

Power Electronic and Electrical Power Research Laboratory (PEEPRL)

University of Florida

Gainesville, FL 32601, United States

Email: zhaohui@ufl.edu, shuo.wang@ece.ufl.edu

Abstract—**One critical drawback of Selective Harmonic Elimination (SHE) and Compensation (SHC) in Cascaded Multilevel Inverter (CMI) topology is that the solution range is very limited and in an irregular shape. As a result, the compensation capacity and system parameters are hard to determine. This paper proposed a voltage phasor diagram to explain the reason of the limited solution range, proposed a four-quadrant modulation method to achieve an enlarged and regular solution range, and discussed guidelines for system parameters design. Simulation and experimental results verified the method's effectiveness.**

Keywords—Selective Harmonic Compensation; Modulation; Four-Quadrant; Solution Range; Cascaded Inverter;

I. INTRODUCTION

The Selective Elimination (SHE) method can eliminate specific order harmonics by controlling the switching transitions with line switching frequency. The Selective Harmonic Compensation / Injection [1-5] (SHC or SHI) is extended from the SHE method, but instead of eliminating the specific order harmonics, SHC method can control the magnitude and phase of the selected order harmonics so that it can draw / generate active power from / to the grid, compensate reactive power [1, 5], compensate high frequency harmonics [3] with line switching frequency. One of the critical problems to apply SHE / SHC is that the solution range is very limited and in an irregular shape [6, 7]. Conventional solutions including only adapting the SHE / SHC method in the regions that effective solution exists [1], or sacrificing the number of harmonics to be controlled where effective solution does not exist [5, 8]. Some authors proved that non-symmetry modulation [9] and unequal DC-link voltage [10] can extend the solution range, but these method would increase increasing number of equations that need to be solved, and increase the system complexity. Moreover, the extended solution range is still limited. Another problem precluding SHC's industrial application is that the parameter design guideline for SHC still follows the steps of designing PWM inverters[11] without considering its features caused by the low switching frequency, especially for the compensation capacity.

This paper is organized as follows. Section I analyzed the Cascaded Multilevel Inverter's (CMI) output voltage by using phasor diagrams, explained the reasons for limited solutions, proposed a four-quadrant modulation method for solution range extension and discussed the during implementing the proposed modulation method. Section II discussed the relationship between the compensation capacity and the parameters and provided the guidelines for parameter design.

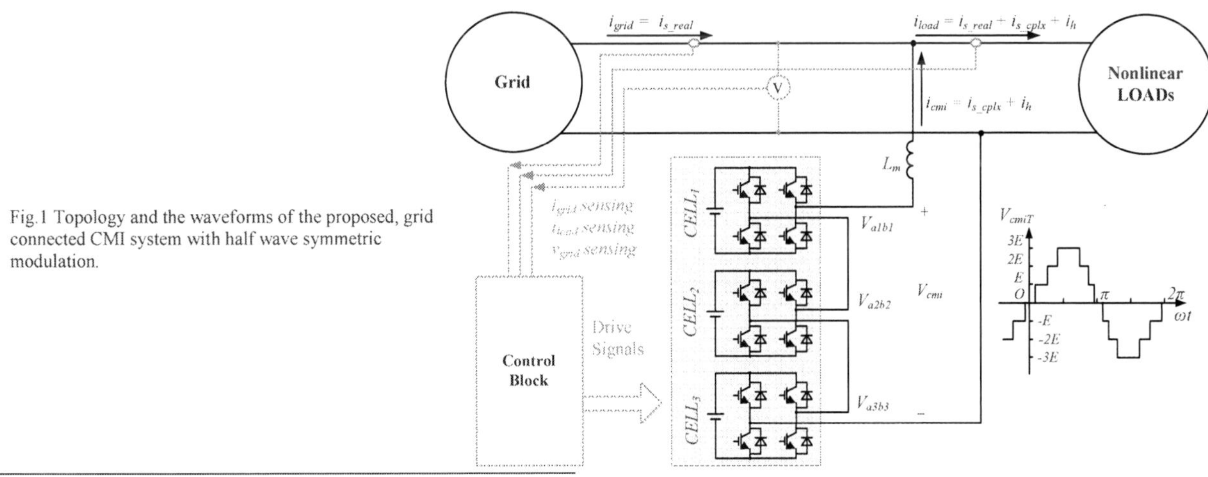

Fig.1 Topology and the waveforms of the proposed, grid connected CMI system with half wave symmetric modulation.

This work was supported by National Science Foundation under award number ECCS-1540118.

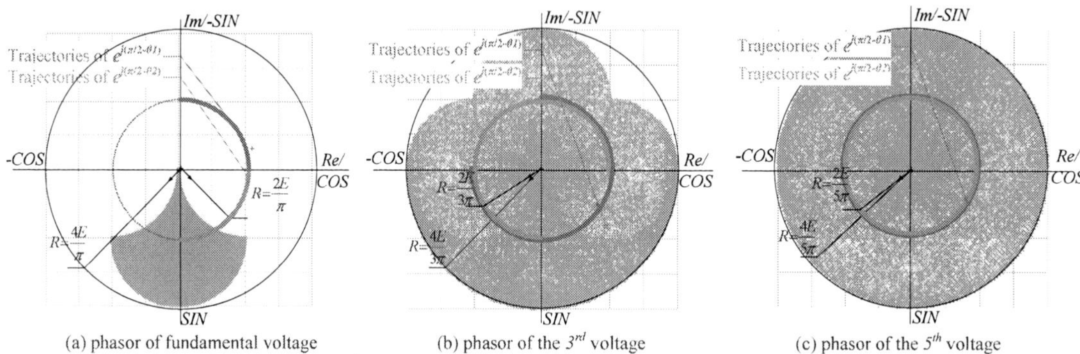

(a) phasor of fundamental voltage (b) phasor of the 3rd voltage (c) phasor of the 5th voltage

Fig. 2 voltage phasor diagram (per unit) of one H bridge with $\theta_1 \in [0, \pi/2]$, $\theta_2 \in [\pi/2, \pi]$

Section III validated the effectiveness of the proposed four-quadrant modulation by the simulation and experimental results. Because SHE can be viewed as a specified case of SHC whose references of the reactive power and selective high order harmonics are set as zero, all the properties of SHC can be directly used in SHE. Only SHC methods are discussed if not explicitly mentioned.

II. Four-Quadrant Modulation Method

A. Voltage Phasor Diagram and Four Quadrant Modulation

Half-wave symmetry modulation is adopted to compensate the odd harmonics in the power system with the minimum number of equations [1, 5]. The topology and waveform is as fig. 1. For single H Bridge, the output voltage waveform is,

$$v_{cmi}(t) = \begin{cases} E, & 2n\pi + \theta_1 < \omega_g t < 2n\pi + \theta_2 \\ -E, & (2n+1)\pi + \theta_1 < \omega_g t < (2n+1) + \theta_2 \\ 0, & otherwise \end{cases} \quad (1)$$

where E is the DC bus voltage for H bridge;

θ_1 and θ_2 are the two switching transitions generated by H bridge;

ω_g is the grid angular frequency, which is $2\pi(60)$ *rad/s* in this paper;

n can be any integer.

The Fourier Series for $v_{cmi}(t)$ can be expressed as:

$$v_{cmi}(t) = \sum_{k=-\infty}^{\infty} c_k e^{jk\omega_0 t},$$

$$c_k = \begin{cases} \frac{E}{\pi k}\left(e^{j\left(\frac{\pi}{2}-k\theta_2\right)} - e^{j\left(\frac{\pi}{2}-k\theta_1\right)}\right), & k \text{ is odd} \\ 0, & k \text{ is even} \end{cases} \quad (2)$$

where

k can be any integer, representing the harmonic order;

$e^{jk\omega_0 t}$ is a unit complex number, the phase is $k\omega_g t$, representing the angular velocity for k^{th} order harmonic is $k\omega_g$;

c_k is a complex number, the magnitude and phase are some function of θ_1 and θ_2.

For any waveform, c_{-k} and c_k are conjugate pair, hence $v_{cmi}(t)$ can be expressed as,

$$\begin{aligned} v_{cmi}(t) &= \sum_{k=-\infty}^{\infty} c_k e^{jk\omega_0 t} \\ &= \begin{cases} 0, & k \text{ is even} \\ \sum_{k=1}^{\infty}\begin{pmatrix} (\mathrm{Re}(c_k) + j\,\mathrm{Im}(c_k))(\cos(k\omega_0 t) + j\sin(k\omega_0 t)e^{jk\omega_0 t}) \\ +(\mathrm{Re}(c_k) - j\,\mathrm{Im}(c_k))(\cos(k\omega_0 t) - j\sin(k\omega_0 t)e^{jk\omega_0 t}) \end{pmatrix}, & k \text{ is odd} \end{cases} \\ &= \begin{cases} 0, & k \text{ is even} \\ 2\,\mathrm{Re}(c_k)\cos(k\omega_0 t) - 2\,\mathrm{Im}(c_k)\sin(k\omega_0 t), & k = 1, 3, 5, \ldots \end{cases} \end{aligned} \quad (3)$$

The above equation shows that even for the complex Fourier Transform, $v_{cmi}(t)$ is still a real value. The magnitude of c_k is half of the magnitude of the kth order harmonic, the phase of c_k represents the initial phase of the kth order harmonics.

Voltage phasor diagram can be derived from (2) shown in fig. 2. Conventional limitation $\theta_1 \in [0, \pi/2]$, $\theta_2 \in [\pi/2, \pi]$ [6, 7] is adopted. Fig. 2(a) shows the phasors distribution of fundamental voltage: trajectories of $e^{j(\pi/2-\theta 1)}$ / $e^{j(\pi/2-\theta 2)}$ are blue / red quarter arcs, green shadowed region is the range of the overall voltage phasor synthetized by the two arcs. This phasor diagram is similar as the vector diagram in the Space Vector PWM modulation (SVPWM) method and is intuitive because two linear-independent segment synthetize a region in two-dimension plane. Similarly, fig. 2(b) shows the phasors distribution of the 3rd voltage. The trajectories of $e^{j(\pi/2-3*\theta 1)}$ and $e^{j(\pi/2-3*\theta 2)}$ are the 3 quarters green / red arcs, and the area of the synthetic phasor is also increased but still irregular. Fig. 2(c) shows the phasor of the 5th voltage phasor. Because both the $5\theta_1/5\theta_2$ exceed 4 quart arcs, the overall output phasor can cover the whole circle. For all other voltage phasors with frequency higher than 5th order would cover whole circle and are hence not given in the paper. It should be noted that fig. 1 is a per unit diagram, the actual amplitude of the voltage phasor with different frequency are the not the same and are labeled in the figures. Fig. 2 can explain the reason why solution for SHI is limited and irregular. As shown in fig. 2, the area of the overall voltage phasors generated by one H-Bridge is limited, hence the synthetic phasor of the cascaded H bridges are limited, which would limit the range for SHI's solution. For example, the fundamental phasor cannot generate negative sinusoidal voltage phasors nor phasors with large *COS* component. Because the coverage of the voltage phasor is related to the solution range. Extending the range of switching angles can enlarge the range of the overall synthetic phasor. Fig.3 shows the fundamental voltage phasors with extending θ_1, θ_2 to $[-\pi, \pi]$. Either the trajectories of $e^{j(\pi/2-\theta 1)}$ or $e^{j(\pi/2-\theta 2)}$ covers 4 arcs, and the synthetic overall voltage phasor can cover the whole circle. Because the extended range for both fundamental component and the high order harmonic component can cover the whole circle, an enlarged and regular solution range can be reached.

To verify the effectiveness of the four-quart modulation, this method is applied in a 7-level cascaded inverter as shown in fig. 1. MATLAB's Genetic Algorithm (GA) toolbox is adopted to find the solution [1, 12]. The Objective Function (OF) adopted is $OF = \frac{\sqrt{E_{1st}^2 + E_{3rd}^2 + E_{5th}^2}}{V_{1\max}}$, where E_{1st}, E_{3rd} and E_{5th} are the voltage errors between the references and actual CMI's voltage's spectrum, V_{1max} is the maximum fundamental voltage. Because the solution data is huge [4], only part of the solution is given in Fig. 4. The references of 3rd and 5th harmonics are set to zero and the phase and magnitude of the reference of the fundamental component are changed. The X-axis is M_{1_COS}, representing the modulation index of fundamental's *COS* component; Y-axis is M_{1_SIN}, representing the modulation index of the fundamental *SIN* component. For the full range ($M_{1\_\sin}^2 + M_{1\_\cos}^2 \le 3^2$), the generated *OF*'s value is below 4 percent as shown in fig. 4. For conventional

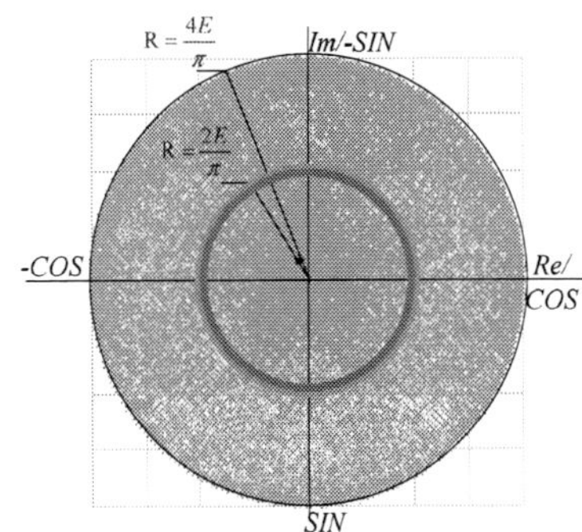

Fig. 3 fundamental voltage phasor of the extended modulation method

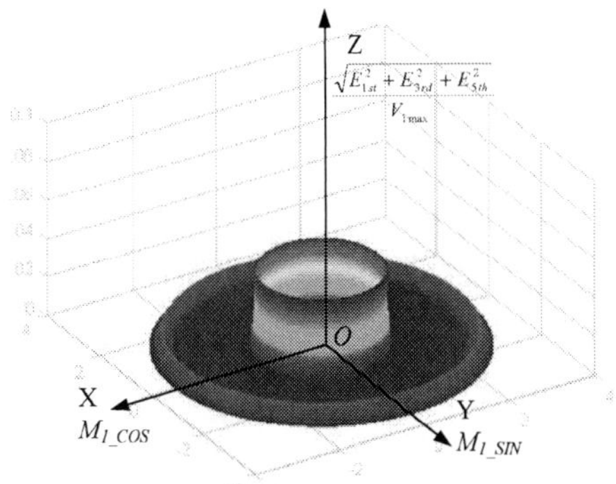

Fig. 4 extended solution range illustration

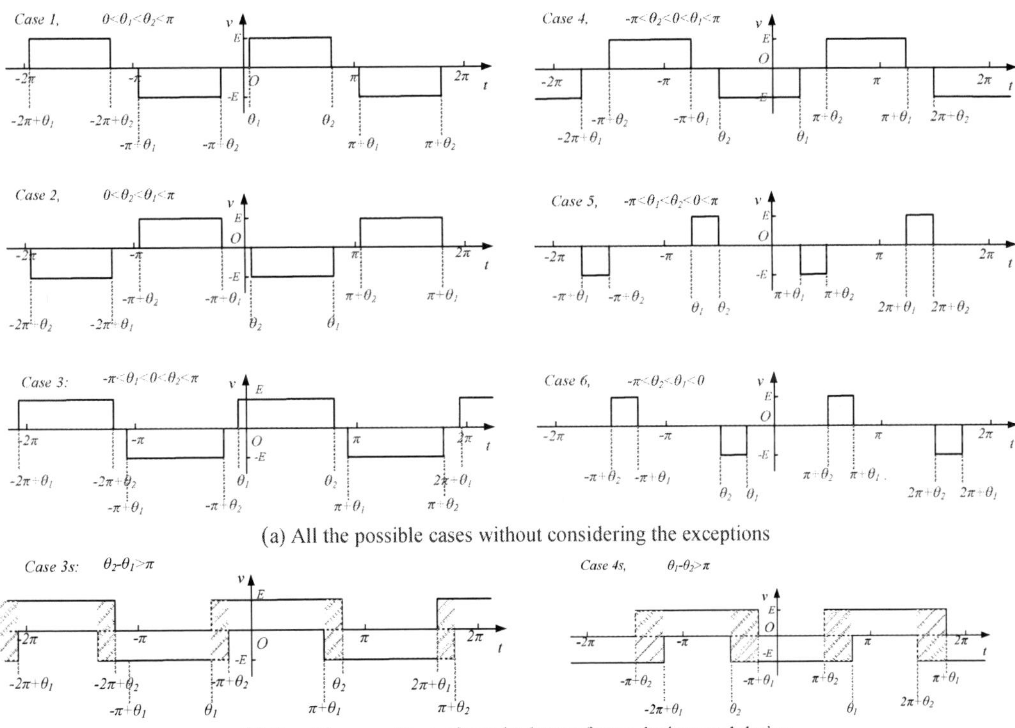

(a) All the possible cases without considering the exceptions

(b) Possible exceptions of required waveforms during modulation

Fig. 5 all the possible cases and exceptions when implementing the proposed 4 quart modulation

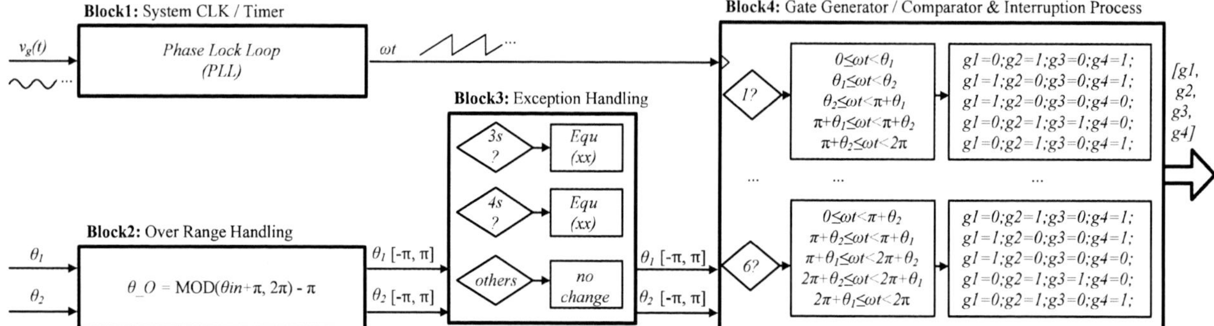

Fig. 6 Digital Implementation of the 4-Qudrant Modulation for single H-Bridge with Exception Process

modulation method, the valid solution only can be found in $1.65 < M_{1\_SIN} < 2.0$, and $M_{1\_COS} = 0$. This enlargement on solution range is significant.

B. Exceptions and Implementation

There are 6 switching states in total considering all the permutations of θ_1, θ_2 and 0's relative positions, and are listed in fig. 5 (a). It should be noted that for H-Bridge, the waveform with $|\theta_2 - \theta_1| > \pi$ cannot be generated, which can only exist in case 3 and case 4. This paper names the two abnormal conditions as exceptions and listed them as case 3s and case 4s separately as shown in fig. 5(b). To prevent the possible errors, one possible solution is considering the two exceptions as constrains in all the design processes including angles calculation, and the modulation blocks do not need to dealt with the two exceptions. This method would increase the system complexity. Considering that the objective of the modulation is to generate the required synthetic voltage phasor rather than the time-domain waveforms, this paper proposed a method to replace the impractical waveform to practical ones with same voltage phasor for all the frequency. For exception 3, a new switching angle pair θ_1' *and* θ_2' are generated as

$$\begin{cases}\theta_1' = -\pi + \theta_2 \\ \theta_2' = \pi + \theta_1\end{cases} \tag{4}$$

Since $\theta_1 \in [-\pi, 0]$ *and* $\theta_2 \in [0, \pi]$, the new angles are $\theta_1' \in [-\pi, 0]$ *and* $\theta_2' \in [0, \pi]$. Hence the new generated waveform still belongs to case 3 which is as shown in fig. 5(a). The difference between the new generated switching angles is $(\theta_2' - \theta_1') = -(\theta_1 - \theta_2) + 2\pi$. Because $(\theta_2 - \theta_1) \in [\pi, 2\pi]$, $(\theta_2' - \theta_1') \in [-\pi, \pi]$, which means that the waveform is practical.

The new synthetic phasor for k^{th} (when k is odd number) order harmonic is unchanged as shown in (5).

$$\begin{aligned} c_k' &= \frac{E}{\pi k}\left(e^{j\left(\frac{\pi}{2}-k\theta_2'\right)} - e^{j\left(\frac{\pi}{2}-k\theta_1'\right)}\right) = \frac{E}{\pi k}\left(e^{j\left(\frac{\pi}{2}-k(\pi+\theta_1)\right)} - e^{j\left(\frac{\pi}{2}-k(-\pi+\theta_2)\right)}\right) \\ &= \frac{E}{\pi k}\left(e^{-jk\pi}e^{j\left(\frac{\pi}{2}-k\theta_1\right)} - e^{jk\pi}e^{j\left(\frac{\pi}{2}-k\theta_2\right)}\right) \\ &= \frac{E}{\pi k}\left(e^{j\left(\frac{\pi}{2}-k\theta_2\right)} - e^{j\left(\frac{\pi}{2}-k\theta_1\right)}\right), \quad Since\ e^{-jk\pi} = e^{jk\pi} = -1,\ when\ k\ is\ odd \\ &= c_k \end{aligned} \tag{5}$$

Similar methods can be adapted for case 4s, generating θ_1' and θ_2' as,

$$\begin{cases}\theta_1' = \pi + \theta_2 \\ \theta_2' = -\pi + \theta_1\end{cases} \tag{6}$$

Because $\theta_1 \in [0, \pi]$ *and* $\theta_2 \in [-\pi, 0]$, the range of θ_1' and θ_2' is $\theta_1' \in [0, \pi]$ *and* $\theta_2' \in [-\pi, 0]$. The replaced waveform still belongs to case 4 as shown in fig. 5(a). Because $(\theta_1 - \theta_2) \in [\pi, 2\pi]$, the difference of the new switching angles is $(\theta_1' - \theta_2') = -(\theta_1 - \theta_2) + 2\pi \in [-\pi, \pi]$. The generated waveform is practical. The proof that the modified voltage phasor remains the same is as shown in (7).

$$\begin{aligned} c_k' &= \frac{E}{\pi k}\left(e^{j\left(\frac{\pi}{2}-k\theta_2'\right)} - e^{j\left(\frac{\pi}{2}-k\theta_1'\right)}\right) = \frac{E}{\pi k}\left(e^{j\left(\frac{\pi}{2}-k(-\pi+\theta_1)\right)} - e^{j\left(\frac{\pi}{2}-k(\pi+\theta_2)\right)}\right) \\ &= \frac{E}{\pi k}\left(e^{jk\pi}e^{j\left(\frac{\pi}{2}-k\theta_1\right)} - e^{-jk\pi}e^{j\left(\frac{\pi}{2}-k\theta_2\right)}\right) \\ &= \frac{E}{\pi k}\left(e^{j\left(\frac{\pi}{2}-k\theta_2\right)} - e^{j\left(\frac{\pi}{2}-k\theta_1\right)}\right), \quad Since\ e^{-jk\pi} = e^{jk\pi} = -1,\ when\ k\ is\ odd \\ &= c_k \end{aligned} \tag{7}$$

The digital implementation diagram is as shown in fig. 6. The overall inputs are v_g(t), which is measured in circuit; and the switching angles, θ_1 and θ_2, which can be pre-stored in Look-up Table or generated in Real Time mode[2]. The outputs are the 4 driver signals for single H-bridge. The Block 1 is the PLL block, which obtains the phase information of v_g(t, which is usually implemented using the DSP's Timer Block. Block 2 is to transfer the switching angles out of the range $[-\pi, \pi]$ into this range and keep the voltage phasor the same. Block 3 is the exceptions handling process to detect the case 3s and case 4s and transfer them to normal case 3 and case 4 using (4) and (6). Block 4 is to generate the time domain waveform with phase signals and switching angles using DSP's comparator and Interruptions. This implementation process is simple and clear.

III. Parameter Design

From (2), the overall output voltage phasors generated by N cell's CMI is as below.

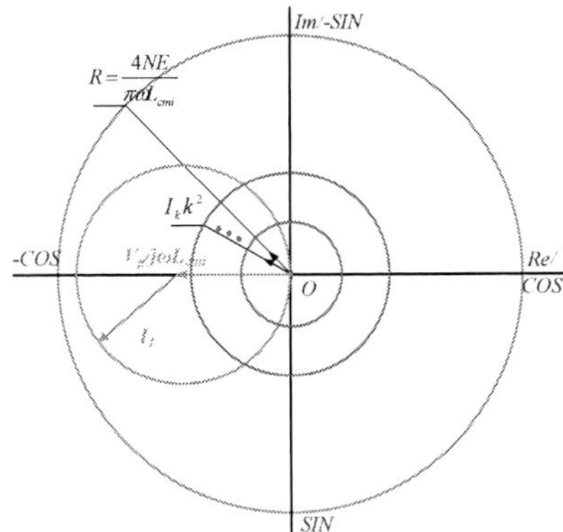

Fig. 7 Compensation Capacity vs requirement for general SHC purpose

$$V_{cmiT\_k} \in cir\left(\frac{4NE}{k\pi}, 0+0\mathrm{j}\right)$$
(8)

where V_{cmiT_k} is the total voltage phasor generated;
cir(x,y) is a function to define a region enclosed by a circle with radius equal to x and circle is y.

All the fundamental and high order harmonics are fully controlled if and only if the equation below can be meet:

$$V_{cmiT\_k} = V_{g\_k} + I_{cmi\_k} jk\omega_g L_{cmi} \tag{9}$$

where,
V_{g_k} is the required k^{th} order harmonic current;
I_{cmi_k} is the required k^{th} order harmonic current.

Noting that V_{g_k} = 0 for $k \neq 1$, the fundamental part and high order frequency part should be considered separately as,

$$\begin{cases} V_{cmiT\_1} = V_g + I_{cmi\_1} j\omega_g L_{cmi} \\ V_{cmiT\_k'} = I_{cmi\_k'} jk'\omega_g L_{cmi}, \quad k' = 3, 5, 7, ... \end{cases} \tag{10}$$

where
V_{cmi_1} and $V_{cmi\_k'}$ are the required total CMI output voltage phasor for *first* and k'^{th} harmonic;
V_g is the fundamental grid voltage phasor;
I_{cmi_1} and $I_{cmi\_k'}$ are the required output current phasor for the CMI topology.

The limitation equation, (11) and (12) can be derived from (2) and (10). The left hand side (LHS) of the equations represent the required phasors need to be compensated, and the right hand side (RHS) of the equations represent compensation capacity.

$$\frac{V_g}{j\omega_g L_{cmi}} + I_{cmi\_1} = \frac{V_{cmiT\_1}}{j\omega_g L_{cmi}} \in cir\left(\frac{4NE}{\pi\omega_g L_{cmi}}, 0+0\mathrm{j}\right) \tag{11}$$

$$I_{cmi\_k'} k'^2 = \frac{k' V_{cmiT\_k'}}{j\omega L_{cmi}} \in cir\left(\frac{4NE}{\pi\omega_g L_{cmi}}, 0+0j\right), \quad k' = 3,5,... \tag{12}$$

The corresponding phasor diagram is as shown in fig. 8. The blue circle is the phasor of $\frac{V_g}{j\omega L_{cmi}} + I_1$, assuming that V_g is in phase with *SIN*; the red circles are the phasors of $I_k k'^2$, representing the compensation requirement of high order harmonics; Green circle is $cir\left(\frac{4NE}{\pi\omega_g L_{cmi}}, 0+0\mathrm{j}\right)$, representing

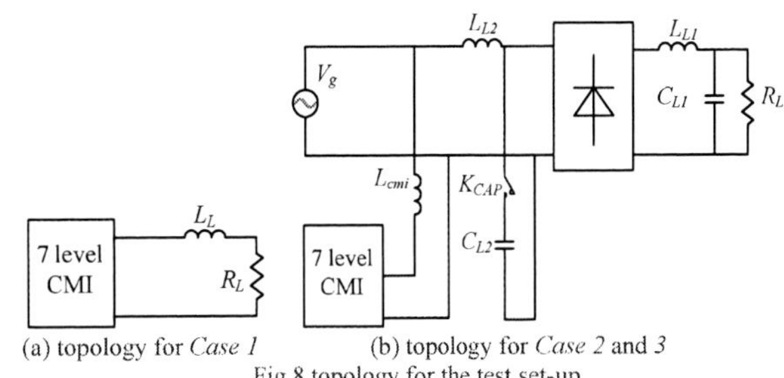

(a) topology for *Case 1* (b) topology for *Case 2* and *3*
Fig.8 topology for the test set-up

TABLE I. PARAMETERS FOR SIMULATION AND EXPERIMENT

Grid Parameters	Grid Voltage, V_g	110 V / 60 Hz
	Rated Power, P	1650 VA
	Required Power Factor, PF	0.9
CMI Parameters	DC bus voltage, *E*	50 V
	Number of cells, *N*	3
	CMI Inductance (in SHC), L_{cmi}	5 mH
	Compensation Capacity, *S*	1500 VA
Case 1	Inductance L_L	15 mH
	Resistance R_L	1 ohm
Case 2	Inductance L_{L1}	5 mH
	Inductance L_{L2}	2.5 mH
	Capacitance C_{L1}	1000 uF
	Resistance R_{L1}	10 ohm
Case 3	Capacitance C_{L2}	200 uF

the compensation capacity of both the fundamental and high order harmonics. If all the red circles and the blue circle is within the green one, the system has the ability to compensate it.

Because the harmonics SHC can control is limited, considering the uncontrolled order harmonics, $I_{cmi\_k'}$ is not necessary to compensate the load harmonic current, but might act as current source and increase the grid harmonic current. The worst case happens when $I_{cmi\_k'}$ is in the opposite direction as $I_{l\_k'}$ and the magnitude reach its maximum value, which can be derived from (12):

$$I_{g\_k'} = I_{l\_k'} - I_{cmi\_k'} \le \left|I_{l\_k'}\right| + \left|I_{cmi\_k'}\right| \le \left|I_{l\_k'}\right| + \frac{4NE}{k'^2 \pi\omega_g L_{cmi}} \tag{13}$$

where the $\frac{4NE}{k'^2 \pi\omega_g L_{cmi}}$ is the extra current source.

Small L_{CMI} means large compensation capacity and small volume, but might increase the extra harmonics, whose order are not specified. A compromise between the compensation capacity and unintended harmonic current source effect need be considered, and the compensation capacity should not exceed the harmonic too much. Moreover, small L_{cmi} would increase the offset of the fundamental's requirement circle as shown in fig. 7, so that L_{cmi} cannot be too small.

IV. SIMULATION AND EXPERIMENT VERIFICATION

As stated in the introduction, the modulation method can work for general applications for both SHE and SHC method; for SHC method, the bidirectional active power flow and both the inductive and the capacitive reactive power compensation need to be considered. Four scale-down test set-ups are built for a 7 level CMI topology with 50V DC voltage. Simulations were performed using power system module in MATLAB simulink. In experiments, the time domain waveforums were recorded with Rigol MSO4054 digital oscilloscope. FFT was conducted by MATLAB.

Case 1: topology is as shown in fig. 8(a); works in stand-alone mode with 100 W RL load, using SHE method to eliminate 3rd and 5th harmonics;

Case 2: topology is as shown in fig. 8(b) with switch K_{CAP} off; CMI works in SHC mode, need to generate 1000 W active power to the grid, compensate inductive reactive power and also the 3rd and 5th harmonics;

Case 3: topology is as shown in fig. 8(b) with switch K_{CAP} on; CMI works in SHC mode need to draw 500 W active power from the grid, compensate the capacitive power and also the 3rd and 5th harmonics;

A. Parameter verification

For *case 1*, system works in SHE mode, no inductance need to be designed.

for *case 2*:

the CMI compensation capacity is $\frac{4NE}{\pi\omega_g L_{cmi}} = 101.3$;

the nonlinear load's parameter is, $I_{1L} = 11.9\angle{-38.1}°$, $I_{3L} = 3.6\angle 52.3°$, $I_{5L} = 0.81\angle 25.9°$;

the compensation requirement is: $I_{1CMI}{}^* = 1000/110{*}2 + j{*}\text{imag}(I_{1L}) = 20.60\angle{-21.0}°$, $I_{3CMI}{}^* = 3.6\angle 52.3°$, $I_{5CMI}{}^* = 0.81\angle 25.9°$.

substituting all the variables to (11) and (12), the inequality equations are satisfied;

For *case 3*:

the compensation requirement remains 101.3;

the nonlinear load's parameter is $I_{1L} = 13.46\angle 34.0°$, $I_{3L} = 4.96\angle 49.3°$, $I_{5L} = 0.87\angle{-81.8}°$;

the compensation requirement is: $I_{1CMI}{}^* = -500/110{*}2 + j{*}\text{imag}(I_{1L}) = 12.21\angle 141.94°$, $I_{3CMI}{}^* = 4.96\angle 49.3°$, $I_{5CMI}{}^* = 0.\ 0.87\angle{-81.8}°$;

substituting all the above variables to (11) and (12), the inequality equations are satisfied.

B. Simulaiton

For *case 1*, because the low voltage reference, conventional method can not obtain the solution[7, 12]. A compromised method is to always keep two cell in off (bypass) state, and only control the fundamental component. The voltage and current waveform is as shown in fig. 10.

The proposed four quarter modulation can obtain the solutions as shown in fig. 10. Fig. 9(a) shows the separate cell's output voltages, the sum of them would synthize the overall voltage as shown in fig. 9(b). Instead of the conventional step waveform, the overall voltage is- similar as 3-level voltages with 6 switching transitions in one fundamental period. When the required fundamental voltage increases, the overall voltage can appeare as 5 level or 7 level. Fig. 11 shows the current harmonic comparison of the two method. The 3rd and 5th harmonic is decreased from 6.01% and 5.09% to almost zeros. Moreover, the THD is also improved from 8.21% to 4.45%. It should be noted that for the un-specific harmonics, the magnitude is not necessary decreased, which is as predicted in section III.

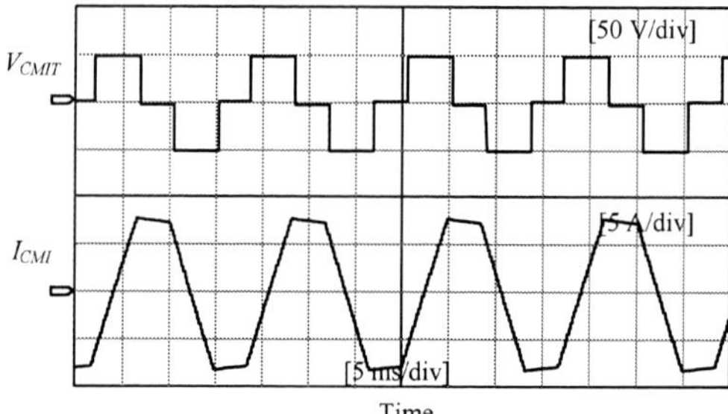

Fig. 9 Conventional modulation waveform in case 1

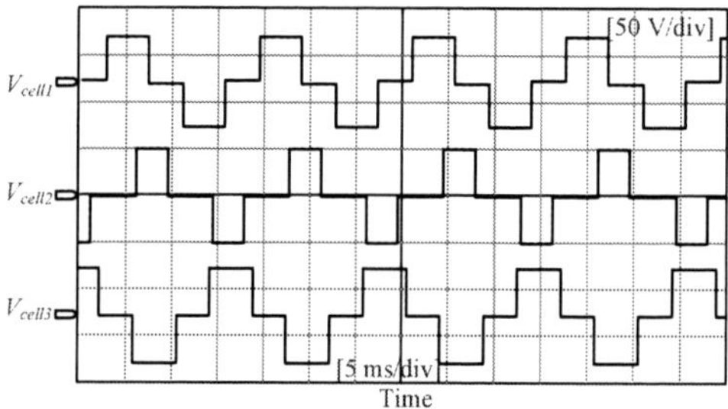

(a) separate CMI cell's output voltage

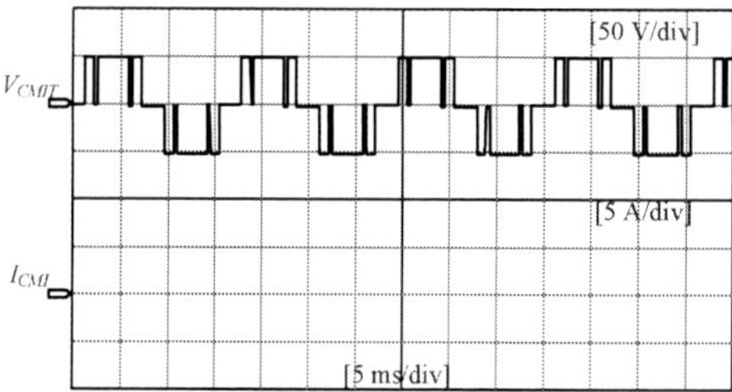

Fig. 10 Proposed modulation waveform in case 1

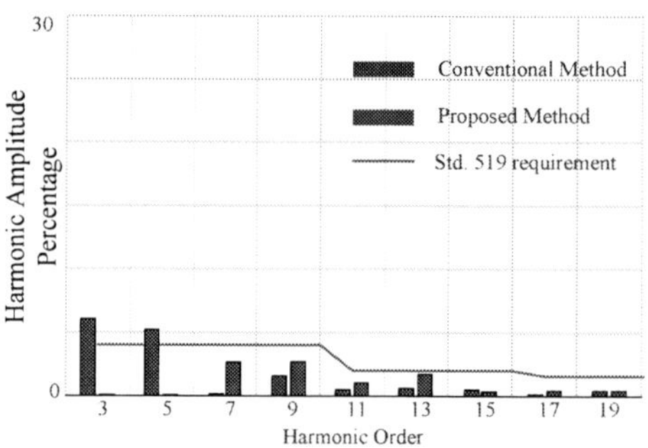

Fig. 11 CMI current harmonic comparison between proposed method and conventional method in *case 1*

SHC's compensation result in case 2 is as shown in fig. 12. Unlike SHE method, the conventional theta's limitation would decrease the range of fundamental, 3rd and 5th overall voltage phasors. For case 2, even only the fundamental voltage phasor is considered while 3rd and 5th harmonic are not as case 1 did, the solution can not be obtained with conventional limitation. With proposed method, the waveform is as shown in fig. 12, and spectrum result is as fig. 13. The 3rd and 5th harmonic are compensated. Because the CMI topology provide active power, the current after compensation is in the opposite phase as the grid voltage. Power factor increased from 0.81 to -1; TDD also decreased from 17.72 to 6.12. It should be noted that some un-sepecific order harmonics increases. 7th and 13th harmonic are larger than the standard limitation, which indicates that the current source effect should be further considered in the future work.

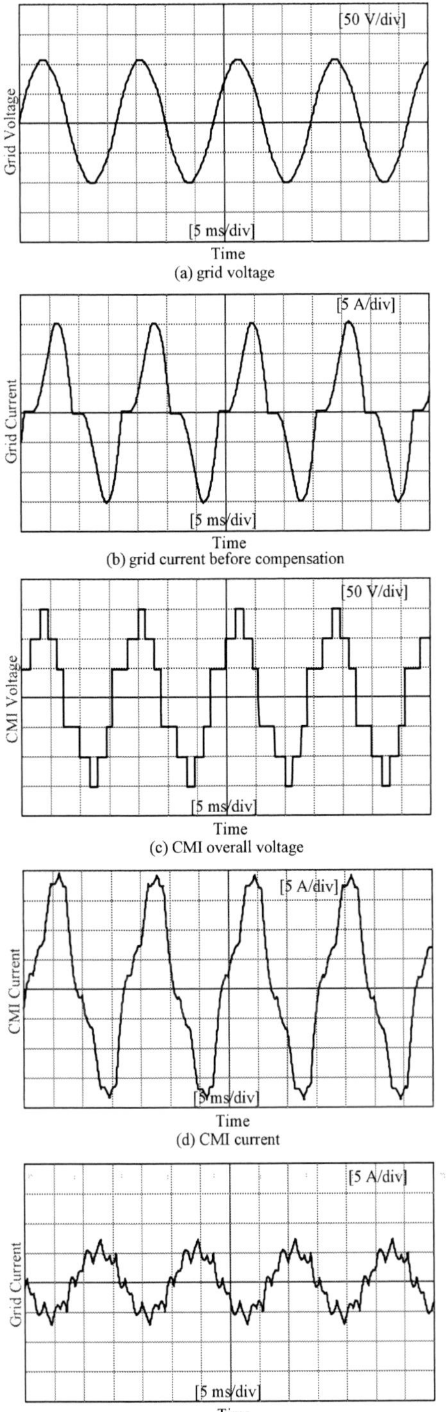

(a) grid voltage

(b) grid current before compensation

(c) CMI overall voltage

(d) CMI current

(e) grid current after compensation

Fig. 12 voltage and current waveforms in case 2

For case 3, the conventional method can achieve a solution, and the proposed method's solution is exactly the same as conventional one. Time domain waveform is as fig. 14, spectrum result is as fig. 15. With compensation, PF increased

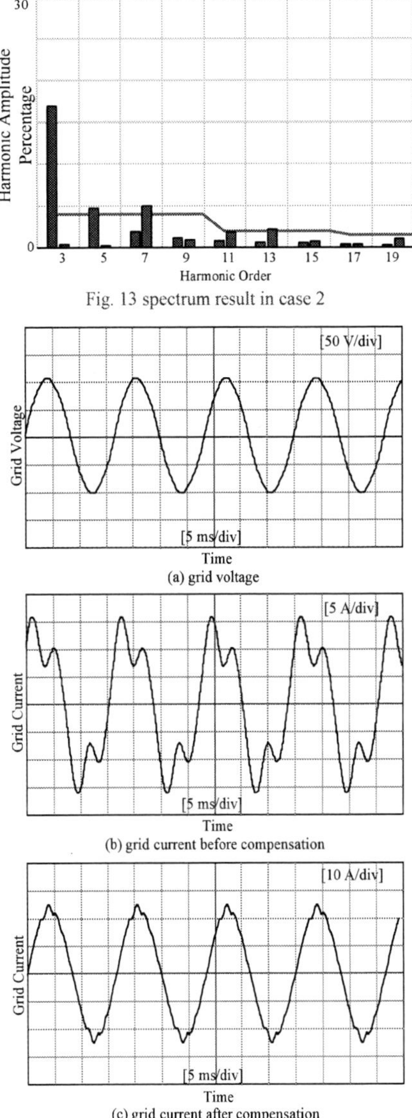

Fig. 13 spectrum result in case 2

(a) grid voltage

(b) grid current before compensation

(c) grid current after compensation

Fig. 14 voltage and current waveform in case 3

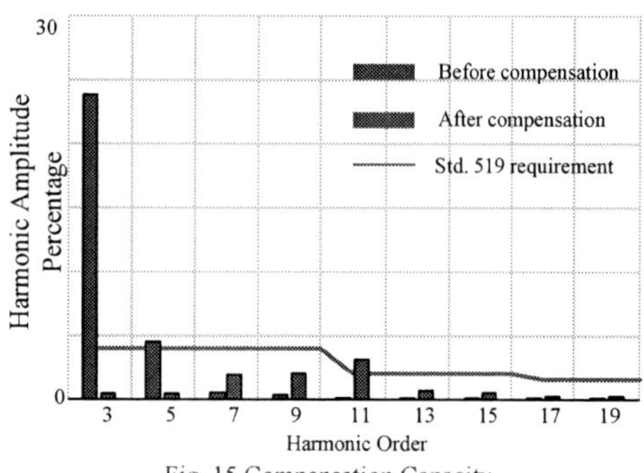

Fig. 15 Compensation Capacity

from 0.79 to 1, TDD decreased from 24.28% to 4.49%. Because CMI's voltage and current is similar as fig. 14 and not shown in the paper.

C. Experiment

Experimental results are as shown in fig. 16 and fig. 17, which match with the simulation results. Experimental results of case 3 is not provided because it is not special and can be achieved by conventional methods.

V. CONCLUSION & FUTURE WORK

To expand the solution range of SHE and SHC method in the cascaded multilevel inverter, a four quadrant modulation method is proposed and analyzed. The CMI's inductance design based on the compensation capacity is derived. The simulation and experimental results verified the proposed method in both SHE and SHC cases.

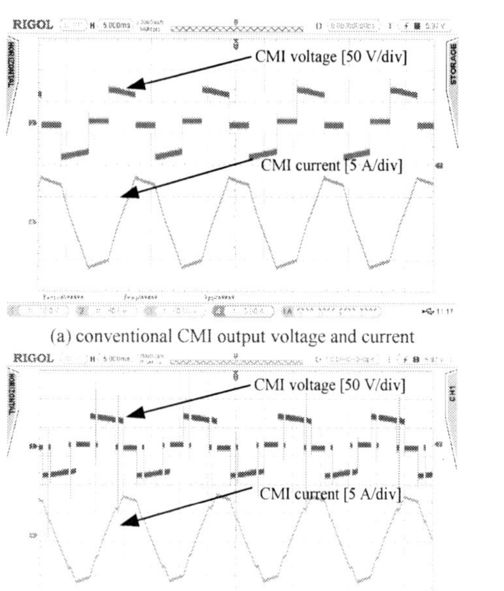

(a) conventional CMI output voltage and current

(b) proposed CMI output voltage and current
Fig. 16 Experimental result in case 1

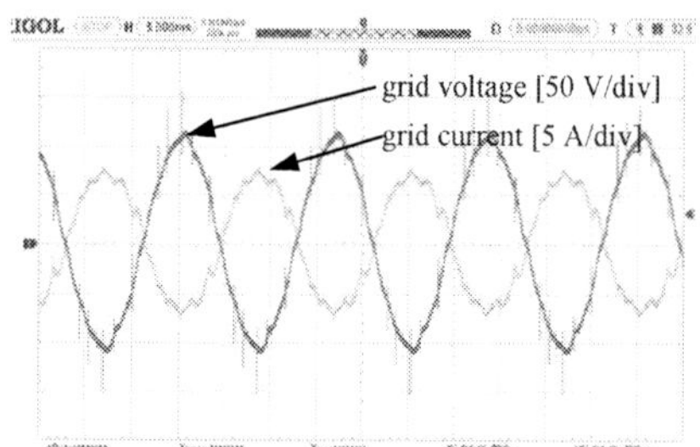

Fig. 17 Experimental results after compensation in case 2

REFERENCES

[1] Y. Chu, S. Wang, and R. Crosier, "Grid active power filters using cascaded multilevel inverters with direct asymmetric switching angle control for grid support functions," in *Applied Power Electronics Conference and Exposition (APEC), 2013 Twenty-Eighth Annual IEEE*, 2013, pp. 1332-1338.

[2] Z. Hui and W. Shuo, "A Real-Time Selective Harmonic Compensation (SHC) based on Asymmetric Switching Angle Modulation and Current Feedback Control for Cascaded Modular Multilevel Inverters," presented at the Applied Power Electronics Conference and Exposition (APEC), 2015 Thirty Annual IEEE, 2015.

[3] D. Ahmadi and W. Jin, "Online Selective Harmonic Compensation and Power Generation With Distributed Energy Resources," *Power Electronics, IEEE Transactions on*, vol. 29, pp. 3738-3747, 2014.

[4] Z. Hua, L. Yun Wei, N. R. Zargari, C. Zhongyaun, N. Ruoshui, and Z. Ye, "Selective Harmonic Compensation (SHC) PWM for Grid-Interfacing High-Power Converters," *Power Electronics, IEEE Transactions on*, vol. 29, pp. 1118-1127, 2014.

[5] Z. Hui, J. Tian, W. Shuo, and S. Liang, "A Real-Time Selected Harmonic Elimination Based on Transient-free Inner Closed-Loop for Cascaded Modular Multilevel Inverter," *Power Electronics, IEEE Transactions on*, 2015.

[6] M. S. A. Dahidah, G. Konstantinou, and V. G. Agelidis, "A Review of Multilevel Selective Harmonic Elimination PWM: Formulations, Solving Algorithms, Implementation and Applications," *Power Electronics, IEEE Transactions on*, vol. 30, pp. 4091-4106, 2015.

[7] D. Hong, S. Bai, and S. M. Lukic, "Closed-Form Expressions for Minimizing Total Harmonic Distortion in Three-Phase Multilevel Converters," *Power Electronics, IEEE Transactions on*, vol. 29, pp. 5229-5241, 2014.

[8] W. Jin and D. Ahmadi, "A Precise and Practical Harmonic Elimination Method for Multilevel Inverters," *Industry Applications, IEEE Transactions on*, vol. 46, pp. 857-865, 2010.

[9] G. Konstantinou and V. G. Agelidis, "On re-examining symmetry of two-level selective harmonic elimination PWM: Novel formulations, solutions and performance evaluation," *Electric Power Systems Research*, vol. 108, pp. 185-197, 3// 2014.

[10] A. Moeini, H. Iman-Eini, and M. Bakhshizadeh, "Selective harmonic mitigation-pulse-width modulation technique with variable DC-link voltages in single and three-phase cascaded H-bridge inverters," *Power Electronics, IET*, vol. 7, pp. 924-932, 2014.

[11] M. Jaksic, S. Zhiyu, I. Cvetkovic, D. Boroyevich, R. Burgos, and P. Mattavelli, "Multi-level single-phase shunt current injection converter used in small-signal dq impedance identification," in *Applied Power Electronics Conference and Exposition (APEC), 2014 Twenty-Ninth Annual IEEE*, 2014, pp. 2775-2782.

[12] B. Ozpineci, L. M. Tolbert, and J. N. Chiasson, "Harmonic optimization of multilevel converters using genetic algorithms," *Power Electronics Letters, IEEE*, vol. 3, pp. 92-95, 2005.

Online Battery Impedance Spectrum Measurement Method

Jaber A. Abu Qahouq, *Senior Member, IEEE*

The University of Alabama
Department of Electrical and Computer Engineering
Tuscaloosa, Alabama 35487, USA

***Abstract*— The electrochemical impedance spectroscopy (EIS) measurement is a method used for offline battery impedance spectrum measurement, which usually requires a specialized and costly equipment. Recent work presented an online method to measure the AC impedance of batteries by utilizing the DC-DC power converter that usually follows the battery for regulation purposes. In this method, the duty cycle of the power converter is sinusoidally perturbed, each time at a given frequency, in order to generate sinusoidally varying voltage and current in the battery to be used for AC impedance measurements/calculations at a given frequency. This paper presents a development of a method which measures the impedance spectrum of a battery for different frequency values at the same time rather than at one single frequency each time. This provides the advantages of being able to measure a spectrum of frequencies online under the same battery conditions such as State-Of-Charge (SOC) and temperature and with faster speed and shorter disturbance time duration. The advantages of the original method are also maintained such as there is no need for additional AC signal injection circuits.**

Index Terms* — *Impedance, Battery, DC-DC, State of Charge, State of Health, Control, Electrochemical Impedance, Impedance Spectroscopy, Power, Converter, Electronic System, Power Management, Fourier Transform.

I. Introduction

Battery systems performance and health is important for many applications such as plug-in hybrid electric vehicles (PHEVs), electric vehicles (EVs), smart and distributed grid, and consumer and portable electronics, among others [1-5].

The electrochemical impedance spectroscopy (EIS) measurement is a method which can be used to characterize batteries [4-8]. Conventionally, EIS measurements is performed offline on batteries by using specialized and relatively costly equipment. However, Battery systems would benefit from low cost online EIS measurements, for examples to evaluate their performance and health.

In [4], an online AC impedance measurement method for batteries is presented. The method is based on sinusoidal perturbation of the duty cycle of DC-DC power converter at a given frequency. The method does not require signal injection or complicated circuities and is suitable for practical applications. However, in [4], the measurement of impedance spectrum which includes different frequencies needs to be done sequentially.

In this paper the method in [4] is taken a step further with a method that allows AC impedance measurement for a spectrum of frequencies at the same time rather than sequentially. This provides advantages that includes faster online measurements, shorter system disturbance time, more consistent results because the impedances at several frequencies are measure under the same battery conditions, no external injection signal circuits are needed, and impedance spectrum measurement under real operating conditions.

II. Principle of AC Impedance Spectrum Online Measurement Method Using DC-DC Power Converter

Fig. 1 shows an illustration diagram bidirectional DC-DC boost/buck power converter with online impedance spectrum measurement.

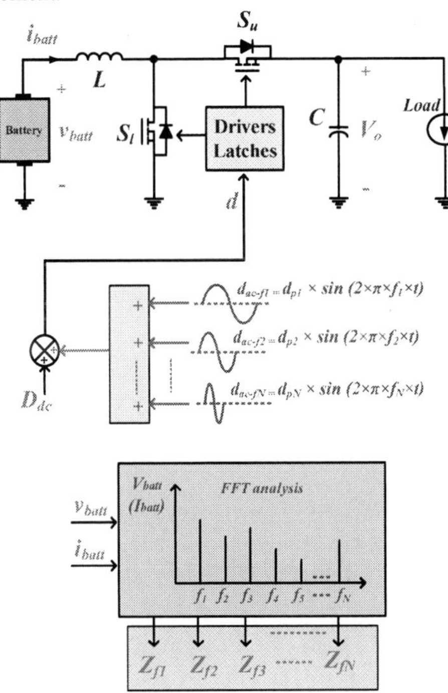

Fig. 1. Illustration diagram bidirectional DC-DC boost/buck power converter with online impedance spectrum measurement.

In Fig. 1, several sinusoid perturbation functions for the duty cycle, each with different frequency, are added to form the summation as given by Equation (1) and as illustrated in Fig. 2 for an example with three frequencies of f1 = 1,000 Hz, f2 = 5,000 Hz, and f3 = 10,000 Hz. The peak values of the duty cycle sinusoid functions are dp1 through dpN and in Fig. 2 dp1 = dp2 = dp3 = dp = 0.02. Ddc is the steady-state duty cycle needed for a desired output voltage and in Fig. 2 it is set to 0.5.

$$d = D_{dc} + d_{ac-f1} + d_{ac-f2} + \cdots + d_{ac-fN}$$

$$\begin{aligned} d_{ac-f1} &= d_{p1} \times \sin(2 \times \pi \times f_1 \times t) \\ d_{ac-f2} &= d_{p2} \times \sin(2 \times \pi \times f_2 \times t) \\ &\cdot \\ &\cdot \\ &\cdot \\ d_{ac-fN} &= d_{pN} \times \sin(2 \times \pi \times f_N \times t) \end{aligned} \tag{1}$$

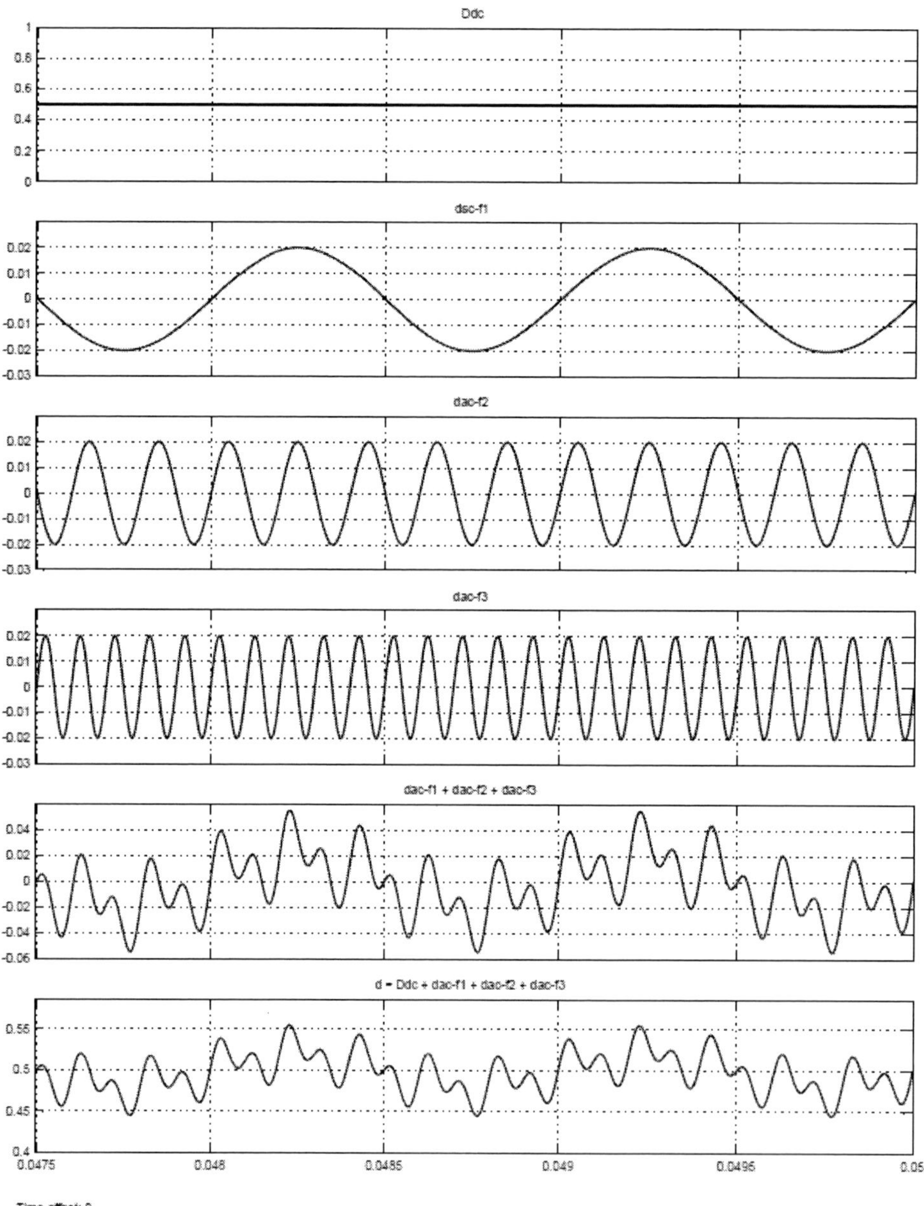

Fig. 2. There sinusoid functions with three frequencies are added to generate the duty cycle function used for battery AC impedance spectrum measurement.

The duty cycle function (d) which controls the DC-DC power converter switches results in multi-frequency changes/ripples in the battery's voltage and current in addition to the output voltage of the power converter.

Fast Fourier Transform is then obtained for the voltage and for the current of the battery in order to obtain the characteristics of each sinusoid frequency. This information at each frequency for the voltage and current is then used to calculate the impedance at each frequency as described in [4] and given by Equation (2). This yields to an impedance spectrum of the battery.

$$Z(f) = \frac{v_{batt-p}(f)}{i_{batt-p}(f)}\ e^{j\,\theta z(f)} = |Z(f)|\ e^{j\,\theta z(f)} = R + j\,X \quad (2)$$

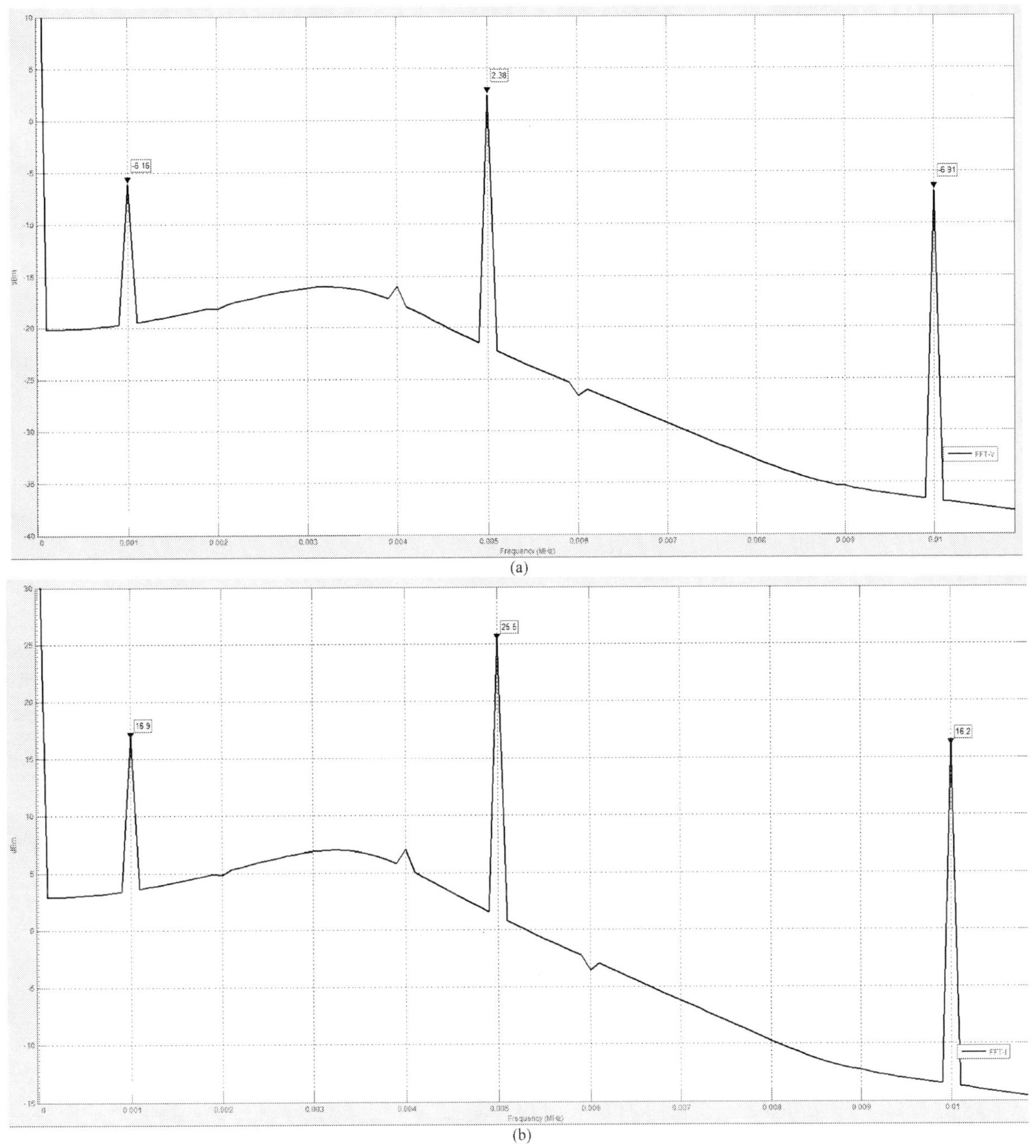

Fig. 3. Matlab/Simulink Spectrum Analyzer output of FFT for voltage and current of battery under duty cycle function of Fig. 2 using the Simulink/Matlab model that only has resistance: (a) Voltage and (b) Current.

Where $Z(f)$ is the impedance at a given frequency, $V_{batt-p}(f)$ is the peak value of the battery voltage at a given frequency component, $i_{batt-p}(f)$ is the peak value of the battery current at a given frequency component, $\theta z(f) = \theta v(f) - \theta i(f)$ is the angle/phase shift between the voltage and current of the battery at a given frequency, $|Z(f)|$ is the magnitude of the complex impedance at a given frequency, R is the real part of the complex impedance and X is the imaginary part of the complex impedance

III. Preliminary Results

The duty cycle function shown in Fig. 2 is used for a DC-DC boost power converter which receive an input from a Lithium-Ion (Li-Ion) battery model of Simulink/Matlab software package. The battery voltage is 4V and the power converter output voltage is ~8V with a load current of 2A. The used battery model in Simulink/Matlab allows for only a resistive/DC impedance which initially was set to 70 mΩ. This implies that it is expected that the model will provide same impedance at all frequencies.

Fig. 3 shows the output of the spectrum analyzer of Simulink/Matlab for both voltage and current of the battery at the three frequencies of interest (1 kHz, 5k Hz, and 10 kHz). At each of these three frequencies, the impedance can be calculated by using Equation (2) and is found to well match the value used in the simulation of 70 mΩ.

The minimum duration needed to obtain FFT for impedance calculations is equal to the time duration of the smallest frequency, which is 1 kHz in the example of this paper.

In order to simulate the performance of the presented concept with a battery model includes capacitance (given that the available Simulink/Matlab battery model has only resistive component of a model), a 2 μF capacitor (values selected just for demonstration purposes which does not imply an actual battery parameter values) in parallel with a 35 mΩ resistor are connected in series with the Simulink/Matlab battery model and the Simulink/Matlab battery model resistance is set to 35 mΩ, which is effectively in series with the added external capacitor in parallel with resistor. This combination results in impedance magnitude values of 65 mΩ, 43 mΩ, and 38 mΩ for the three frequencies of 1 kHz, 5 kHz, and 10 kHz, respectively.

Fig. 4 shows the output of the spectrum analyzer of Simulink/Matlab for both voltage and current of the battery at the three frequencies of interest (1 kHz, 5k Hz, and 10 kHz). At each of these three frequencies, the impedance can be calculated by using Equation (2) and is found to well match the expected values.

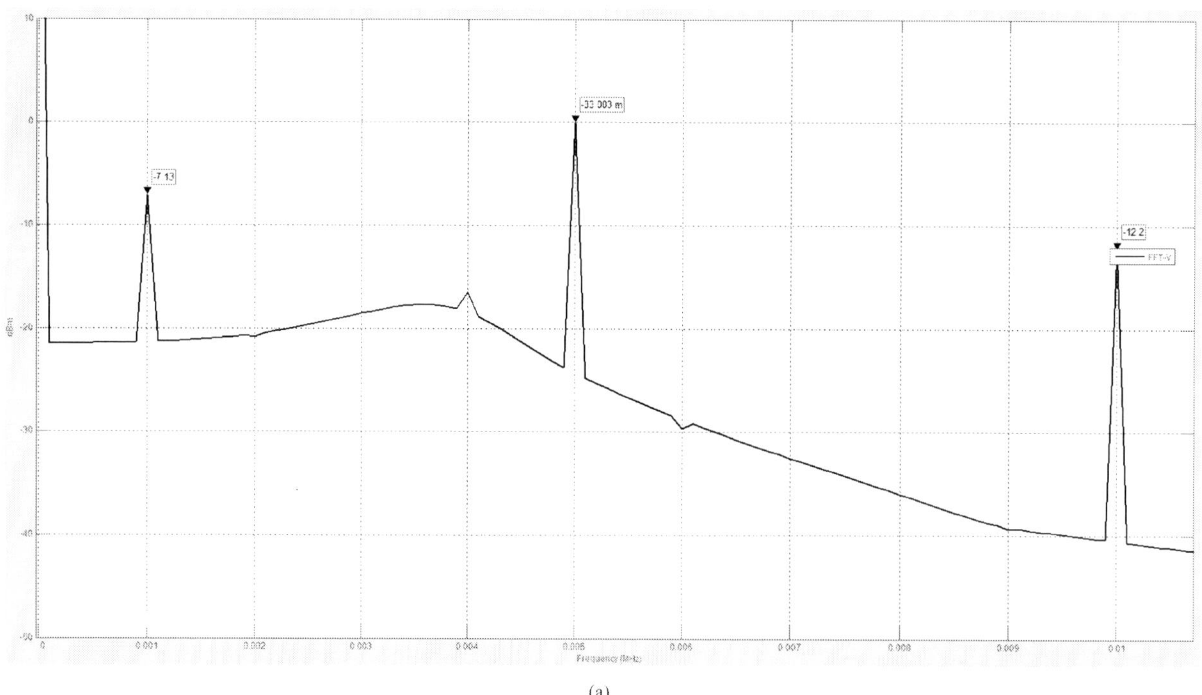

(a)

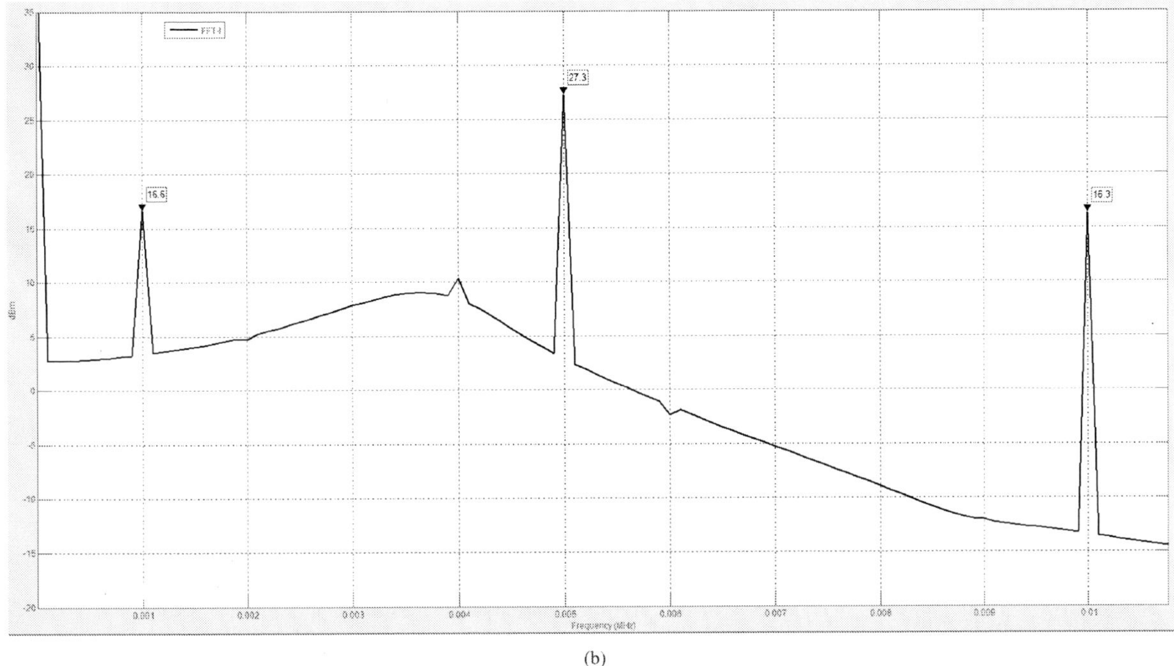

(b)

Fig. 4. Matlab/Simulink Spectrum Analyzer output of FFT for voltage and current of battery under duty cycle function of Fig. 2 by adding external capacitor and resistor to the Simulink/Matlab model that only has resistance: (a) Voltage and (b) Current.

IV. Conclusion

This paper presented an online method to measure an impedance spectrum of battery for range of frequency at the same time rather than one frequency at a time. The method is based on perturbing the duty cycle of a power converter with a sum of several sinusoid function with different frequencies. In this manner, the impedance values of a battery can be measured at the same time for several frequencies. This yields to advantages such as faster online measurements, shorter system disturbance time, more consistent results because the impedances at several frequencies are measure under the same battery conditions, no external injection signal circuits are needed, and impedance spectrum measurement under real operating conditions. The paper discussed the basics of the methods and presented preliminary results. Future work includes but is not limited to completing experimental hardware development and using the results for State-Of-Health and State-Of-Charge estimations, among others.

Acknowledgement

This work is supported in part by the National Science Foundation (NSF) of USA under Grant No. 1509824. Any opinions, findings and conclusions or recommendations expressed in this material are those of the author(s) and do not necessarily reflect the views of the National Science Foundation.

References

[1] M. Yilmaz and P. T. Krein. "Review of Battery Charger Topologies, Charging Power Levels, and Infrastructure for Plug-In Electric and Hybrid Vehicles," IEEE Transactions on Power Electronics, vol. 28, no. 5, pp. 2151 - 2169, August 2012.

[2] David Velasco de la Fuente, Cesar L. Trujillo Rodriguez, and Gabriel Garcera, "Photovoltaic Power System with Battery Backup with Grid-Connection and Islanded Operation Capabilities," IEEE Transactions on Industrial Electronics, vol. 60, no. 4, pp. 1571-1581, April 2013.

[3] W. Huang and J. A. Abu Qahouq. "Energy Sharing Control Scheme for State-of-Charge Balancing of Distributed Battery Energy Storage System," IEEE Transactions on Industrial Electronics, vol. 62, no. 5, pp. 2764-2776, May 2015.

[4] W. Huang and J. A. Abu Qahouq. "An Online Battery Impedance Measurement Method Using DC–DC Power Converter Control," IEEE Trans. on Ind. Electronics, vol. 61, no. 11, pp. 5987-5995, Jun. 2014.

[5] F. Huet, "A Review of Impedance Measurements for Determination of the State-of-Charge or State-of-Health of Secondary Batteries," J. Power Sources, vol. 70, pp. 59-69, 1998.

[6] Peter A. Lindahl, Matthew A. Cornachione, and Steven R. Shaw, "A Time-Domain Least Squares Approach to Electrochemical Impedance Spectroscopy," IEEE Tran. Instrum. and Meas., vol. 61, no. 12, pp. 3303-3311, Dec. 2012.

[7] Jon P. Christophersen, David F. Glenn, Chester G. Moltoch, Randy B. Wright, Chinh D. Ho, and Vincent S. Battaglia, "Electrochemical impedance spectroscopy testing on the advanced technology development program lithium-ion cells," in Proc. IEEE Veh. Technol. Conf., pp. 1851-1855, 2002.

[8] J. L. Jespersen, A. E. Tonnese, K. Norregaard, L. Overgaard, and F. Elefsen, "Capacity Measurements of Li-ion Batteries Using AC Impedance Spectroscopy," World Electric Vehicle J. , vol. 3, pp. 1-7, May 2009.

Analysis and Control of a Reduced Switch Converter for Active Magnetic Bearings

Dong Jiang
Huazhong University of Science and Technology
Wuhan, China, 430074
jiangd@hust.edu.cn

Parag Kshirsagar
United Technologies Research Center
411 Silver Ln, E. Hartford, CT-06108
kshirsp@utrc.utc.com , parag@vt.edu

***Abstract*—A reduced switch converter for active magnetic bearing system is analyzed and its performance is experimentally validated at 24,000 rpm on a 4-axis magnetic bearing test-rig. Based on the coupled coil arrangement and unidirectional current operation, the reduced switch converter is derived and its modes of operation are elaborated. In comparison to conventional topologies, this converter reduces the requirement on number of switches, gate drives and pulse width modulation signals stemming from the controller. The analysis and experimental results successfully validates the operation of reduced switch converter to meet the magnetic bearing operation requirements.**

I. INTRODUCTION

Recent progress in power electronics has pushed the motor drives to a wide speed range for many applications including high speed turbo compressors which are widely used in heating ventilation and cooling (HVAC), oil and gas, water filtration and several other applications[1]. However, bigger friction in the bearing also appears with higher rotational speeds, which impairs the lifetime and reliability of the bearings. With this consideration, Active Magnetic Bearing (AMB) has been developed in the recent times as an alternative for the conventional rolling element and foil bearings. AMB is based on the concept of using electromagnetic force to levitate the rotor shaft in suspension without contacting the stator [1], [3].

Fig 1 (a) shows the concept of a 4-axis magnetic bearing electromechanical system. The currents in the coils of the left and the right bearings are controlled by a buck converter. The magnitude of the current controlled in each coil is determined by a proportional, integral and derivative (PID) controller. The PID controller regulates the position of the rotor shaft based on the feedback signals from the position sensors for the given axis. For the illustration in Fig 1(a), each axis is controlled independently to maintain the rotor position within certain air-gap clearance at high rotor velocities.

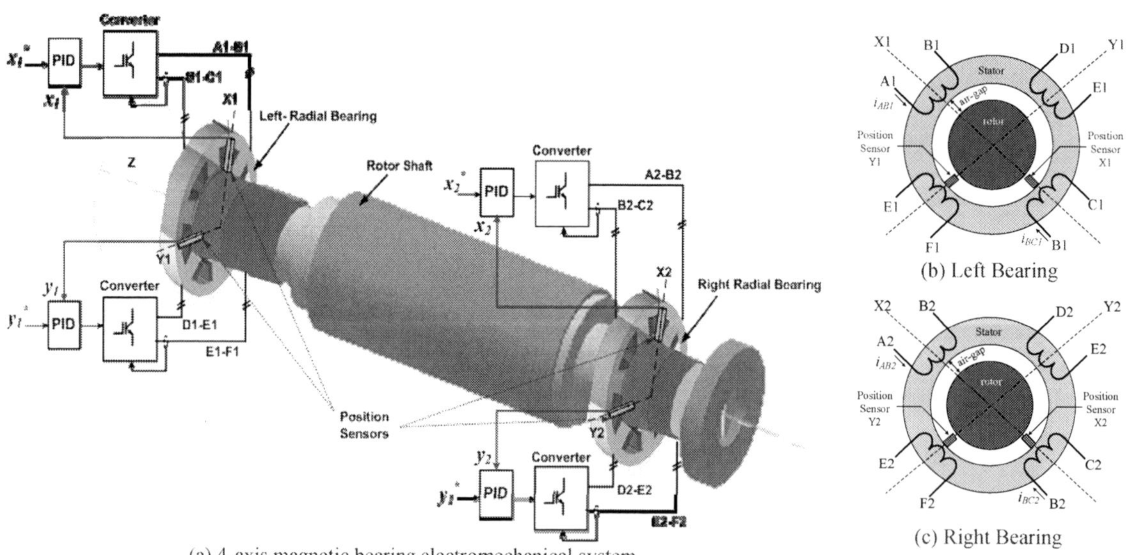

(a) 4-axis magnetic bearing electromechanical system.

Fig.1. (a) 4-axis magnetic bearing electromechanical system. (b) and (c) Coupled coil arrangement and location of position sensors in each axis.

978-1-4673-9551-9/16 $31.00 © 2016 IEEE

Fig.1 (b) and (c) shows that each bearing has two set of orthogonally displaced windings along the X and Y axes. The currents in these two set of windings results in an electromagnetic force that attracts the rotor towards the stator along that axis. By actively balancing the forces acting on the rotor through control of the current in the windings, a contactless levitation can be realized.

The current in the coil of magnetic bearings is typically controlled separately. In this study, the coils forming opposite pole pairs along the X and Y axes are coupled and configured to have a common return current path. Illustration of the same is shown in Fig .1(b) where the coil A1-B1 and the coil B1-C1 along the X1 axis are coupled and have a common point B1. The same can be seen for the coils along Y1, X2 and Y2 axes. The coupling of the coil is possible because the electrical time constant is much smaller than the mechanical time constant of the system. Hence any transient on the mechanical system does not significantly affect the current control performance of the active magnetic bearing system. With this approach, different converter topologies can be used for the magnetic bearing system. Accordingly, this research is focused on analysis and control of a reduced switch active magnetic bearing converter topology which was proposed in [2]. The single axis configuration of the topology in [2] is shown in Fig 2 (d). Such architecture reduces the requirement on the number of switches, their gate drives and the number of pulse width modulation signals stemming from the control board. The topology is well known for switched reluctance motor drives [4], however detailed analysis and validation of such a converter for magnetic bearing system has yet to appear in the current literature and hence considered in this study. Since the current in the coil of the AMB is unidirectional and DC, it can be controlled by a buck type of DC-DC converter. Accordingly, Fig 2 (a) and (b) shows the most commonly used full H-bridge and half bridge topologies which can control the current in the coil A1-B1 [1], [3]. In [5], an interleaved H-bridge converter topology was evaluated for a bearingless slice motor. Due to the AC current requirement for the bearingless two-phase motor, the switching devices are arranged for bi-directional current as shown in Fig 2(c). AMBs with heteropolar design require unidirectional DC current. Hence, opportunity for further reduction in the number of switching devices exists. The comparison between the topologies in Fig 2(a-d) with respect to the number of switching devices and diodes for a five axis heteropolar magnetic bearing system is given in Table 1. The table shows that the reduced switch converter has lowest switch count amongst the topologies compared. This feature significantly reduces the cost of the system. With this background it is essential to evaluate the control and performance of the reduced switch converter. Accordingly, the research in this paper is organized as follows.

- A step by step operation of the reduced switch converter topology is explained section II.A.
- Various pulse width modulation methods, and cross coupling effect of the coils are discussed section II.B.
- Operation of the reduced switch converter in a magnetic bearing test-rig is successfully demonstrated in section III.
- Contributions and conclusions are drawn in section IV.

Table.1 Comparison of components with different topologies for a five-axis magnetic bearing system

Topology	Switches	Diodes
Full H-bridge	40	40
Unidirectional H-bridge	20	40
Full three phase-legs	30	30
Reduced switch converter	15	30

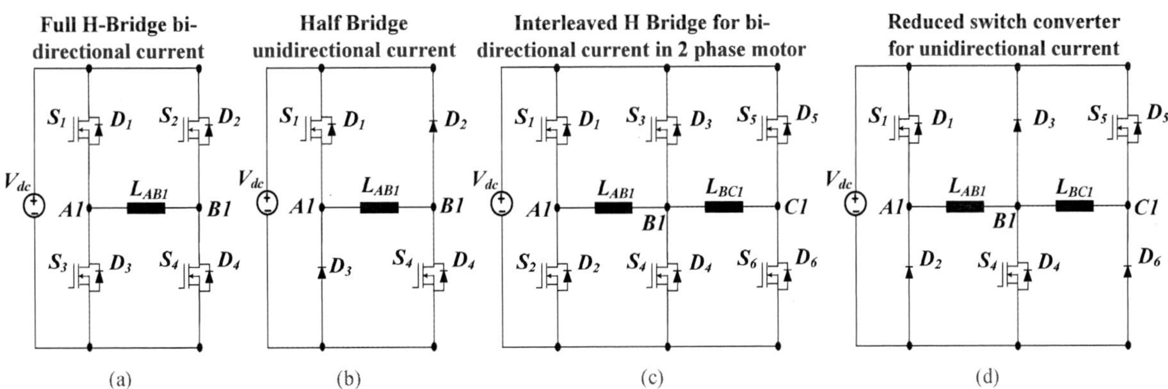

Fig 2. Active magnetic bearing topologies: (a) Full H-bridge [3], (b) Half-bridge [3], (c) Bidirectional AC current interleaved H-bridge [5] and (d) reduced switch unidirectional current [2].

II. Reduced Switch Converter For Magnetic Bearing System

Fig 2 (d) shows the topology of a reduced switch converter for control of X1 axis of the AMB. The phase leg of the converter has a diode D2 in series with the switch S1. This arrangement prevents possibility of shoot through in the phase leg thus making the topology very reliable. Since the current in the central leg is the sum of the two sides' phase current in the three-phase-leg based topologies, the conduction and switching losses are similar to the losses in the H-bridge based converter.

There are various modes of operation of the converter with coupled coil approach for considering single axis control and they are described in the following section.

A. Modes of operation

The modes of operation of the reduced switch converter are shown in Fig. 3. The modes are explained using the circuit across the coil A1-B1. Considering unidirectional current i_{AB1}, there are four modes of operation that can be used to control the current in the coil A1-B1.

1. During mode I, the DC link voltage is applied across the coil by turning the switches S_1 and S_4 on. This will increase the current and also develop a positive DC-link voltage across the coil.
2. Modes II (a) and (b) are freewheeling modes where the residual energy in the coil is dissipated through the switch and diode combination. Accordingly, in mode II (a), the current is directed through the diode D_2 and the switch S_4, while in mode II(b), it passes through the diode D_3 and switch S_1. The modes II (a) and (b) are essential to maintain the average magnitude of the current in the coil.
3. During mode III, the switches are turned off and the residual energy in the coil is freewheeled through diodes D_2 and D_3 into the DC bus. In this mode, a negative DC bus voltage appears across the coils resulting in rapid reduction of the current in it.

The operating modes are similar with respect to coil B1-C1 for the unidirectional current i_{BC1} except that the switch S_5, and diodes D_3 and D_6 are in effect.

B. Modulation methods

Due to the interleaved architecture of the reduced switch converter, there are two methods of modulating the active switches and they are (a) out of phase modulation and (b) in-phase modulation. Given the coil ends A1, B1 and C1, the corresponding legs of the converter can be defined as phase A, phase B and phase C respectively. For the analysis, the modulation of phase B through switch S_4 is kept constant and at 50 percent duty cycle. Then the amplitude of the current in coils A1-B1 or C1-B1 can be varied by modulating the switches S_1 and S_5 respectively. Accordingly, Fig 4 (a) and (b) shows the out of phase and in phase modulation methods respectively for the given converter. For the case with out-of-phase modulation, the current is predominantly controlled through modes I and III where in large voltage vectors are applied. Hence the corresponding current ripple amplitude is high.

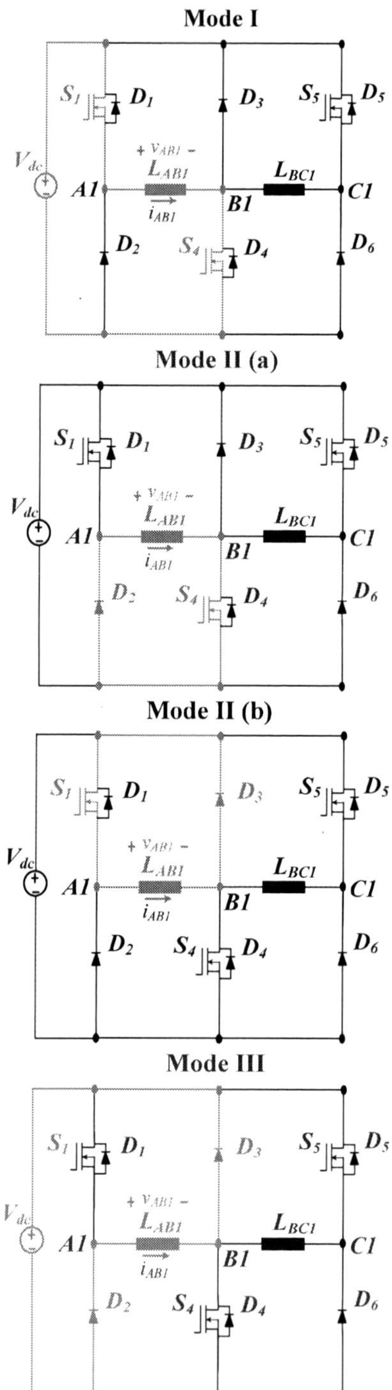

Fig 3. Modes of operation of reduced switch converter with coupled coil.

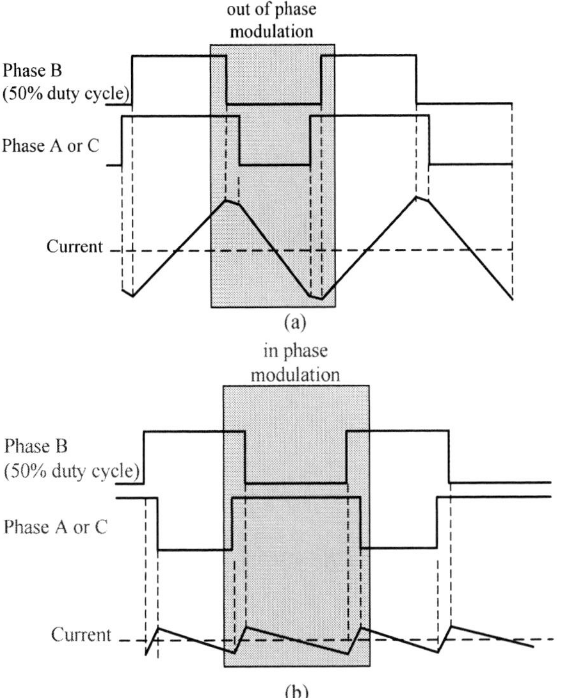

Fig 4.(a) Out of phase and (b) in phase modulation methods in reduced switch converter.

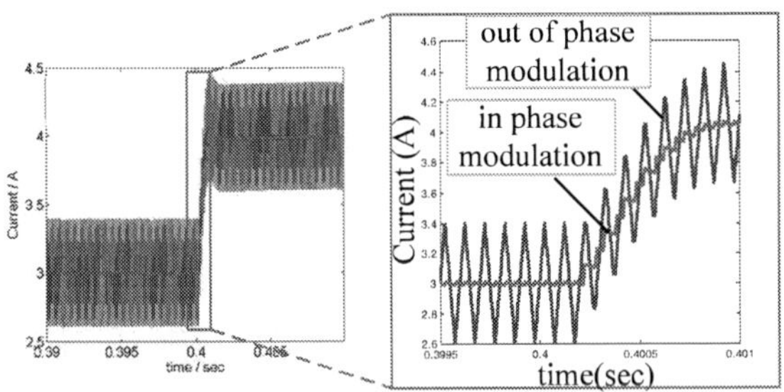

Fig.5. Simulation results: current dynamic response comparison between in phase and out of phase modulation.

For the case with in-phase modulation, the modes II (a) and (b) are exercised more often in combination with modes I and III thereby reducing the current ripple significantly. The amplitude of the current ripple can be deduced as follows. Given the switching frequency of PWM as fs and switching period as Ts, with coil inductance as L, then the peak-to-peak current ripple for out of phase modulation is approximately given as Vdc Ts/2L. If the voltage drop in the freewheeling loop (switch, diode and resistance) is ΔV, then the peak-to-peak current ripple for in-phase modulation is ΔV·Ts/2L. Since the freewheeling voltage drop ΔV is much smaller than Vdc, the current ripple due to out of phase modulation is much smaller.

In order to verify the same, simulation of the reduced switch converter is made with DC link voltage of 160V, coil inductance of 10mH, resistance of 0.1Ω and converter switching frequency of 10 kHz. Results of the simulation are shown in Fig 5. It is seen that for a step change in current amplitude, the in-phase modulation can achieve the similar dynamic performance as that of the out-phase modulation, but with much smaller current ripple validating the above analysis.

C. Center leg coupled and de-coupled control

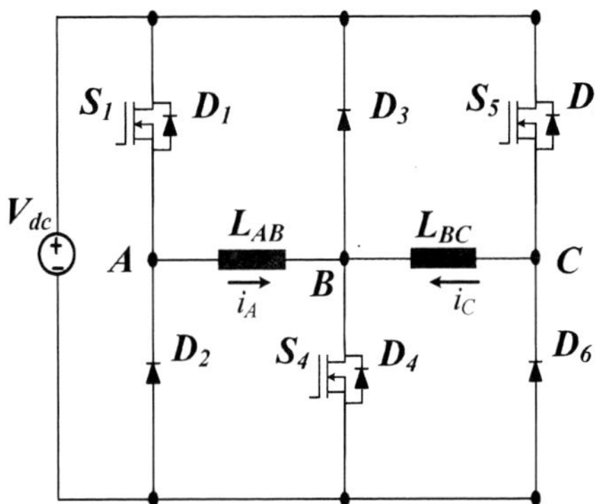

Fig. 6. Reduced switch converter for single axis control

The control method in Fig.5 is illustrated considering fixed duty cycle of 50% in the central phase-leg (B). With reference to Fig 6, the duty cycles of phases A and C are varied to control the current in the coil. The 50 % of fixed duty cycle in phase-B provides equal utilization of the DC-link voltage for dynamic control to increase and decrease of the current in either coil. Assuming d_a and d_c are the duty cycles of the two switches in phases A and C and d_b = 0.5, then the voltages across the two coils are given as,

$$\begin{cases} V_{AB} = (2d_a - 1)\dfrac{V_{dc}}{2} \\ V_{CB} = (2d_c - 1)\dfrac{V_{dc}}{2} \end{cases} \quad (1)$$

In this case, the maximum and minimum voltage on each coil is $\pm V_{dc}/2$. An alternate PWM method is modulating the central phase-leg duty cycle as d_b=-(d_a+d_c), which implies coupling of the A and C phase legs. Then the voltage on the two windings is given as,

$$\begin{cases} V_{AB} = (4d_a + 2d_c - 2)\dfrac{V_{dc}}{2} \\ V_{CB} = (4d_c + 2d_a - 2)\dfrac{V_{dc}}{2} \end{cases} \quad (2)$$

For each winding, the modulation of d_a or d_c can generate a voltage of $\pm V_{dc}$ to change the inductor current, which means doubling of the control bandwidth. Fig 7 shows the open loop current control bandwidth response of two control methods based on modulation of the middle leg switch S_4. In first case, the switch S_4 has fixed duty cycle of 50 % (α=0) while in the second case, it has a variable duty cycle (α=1) given by d_b = -(d_a+d_c), where d_a, d_b and d_c are the average duty cycles for the phase legs corresponding to the coils A1, B1 and C1 respectively. From Fig 7, it is seen that the variable duty cycle method can result in almost twice the current control bandwidth than the constant duty cycle method.

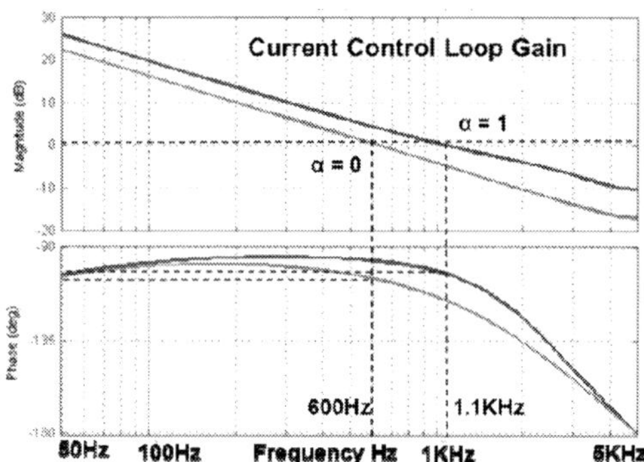

Fig. 7. Open loop current control bandwidth with fixed(α=0) and variable (α=1) duty cycle control

Considering control of X1 axis of the radial magnetic bearing, the coils A1-B1 and B1-C1 are coupled at node B1. The switch S4 commonly controls the current in these two coils. Therefore variation of current in coil A1-B1 affects the control of current in coil B1-C1. Such a case is shown in Fig 6. In the illustration, current in coil A1-B1 is varied from 0 to 10 A and back to zero every 2.5 msec. The current in coil B1-C1 is maintained at zero. It is seen that during the transient events, variation of current in coil A1-B1 causes a 2 A transient in coil B1-C1 for a period of 250 μsec. Due to the small time duration of the transient, it does not affect the mechanical performance thus making the coupled coil approach feasible. In case of a half bridge independent coil control such cross coupling does not exist and is to be noted in the figure.

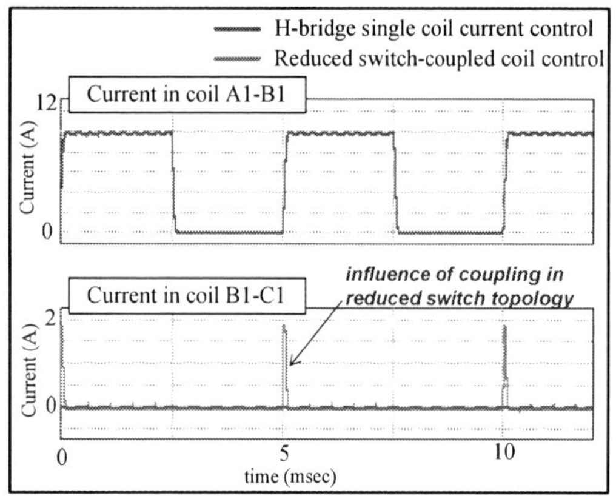

Fig.8. Simulation results: (a) Influence of coupled coil implementation in the reduced switch topology

D. Experimental results: converter modes of operation

For proof of concept of the topology, a converter prototype is made. It comprises of four intelligent power modules (IPMs) with only 3 gate drives activated for each IPM. The four IPMs work as four three-half-phase-legs converters for 4 pairs of windings in a four-axis magnetic bearing drive. DC-link voltage is 160V. 8 Hall-effect current sensors and 4 eddy current based position sensors are used for current and position feedback control. The control of the system is implemented on a commercial motion control microcontroller.

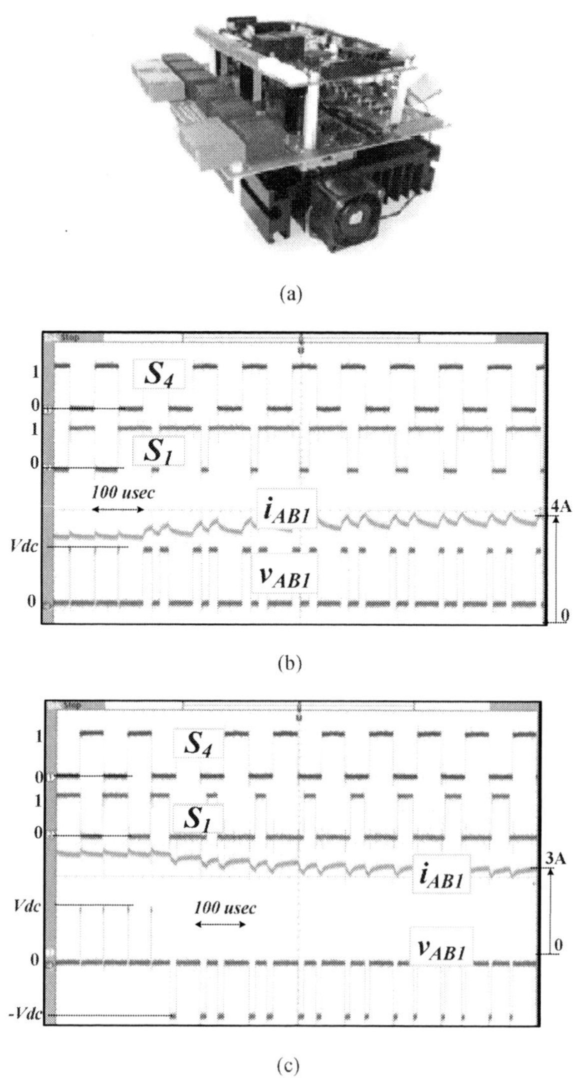

Fig 9.Experimental validation of the modes of operation of reduced switch converter (a) Converter prototype (b) in-phase modulation showing increase in current. (c) in-phase modulation showing decrease in current (note mode III where negative DC bus voltage is across the coil).

To validate the operating modes in section II.A, the switch S4 of central leg has fixed duty cycle of 50 percent. To increase the current, the duty cycle of switch S1 is increased as shown in Fig 9 (b). The corresponding voltage across the coil A1-B1 is increases and the current in the coil increases from 3A to 4A. To decrease the current in the coil, the duty cycle of switch S1 is reduced below 50 percent during the transient condition. This results in negative DC voltage across the coil and reduction in the current. Once the steady state has been reached, the duty cycle of switch S1 is close to 50 percent. Accordingly, Fig 9(c) shows the dynamic response where the current in the coil reduces from 4 A to 3 A. The experimental results show the capability of current dynamic control for the reduced switch converter for magnetic bearing. Based on the 50% duty cycle control of central leg, the closed loop current control bandwidth is measured and is shown in Fig 10. At 10 kHz switching and sampling frequency, the current closed loop bandwidth achieved is 250 Hz.

This completes the converter analysis and validation of current control considering the coupled coil approach used in this work. The next section will discuss on high speed position control of the shaft using the reduced switch converter topology

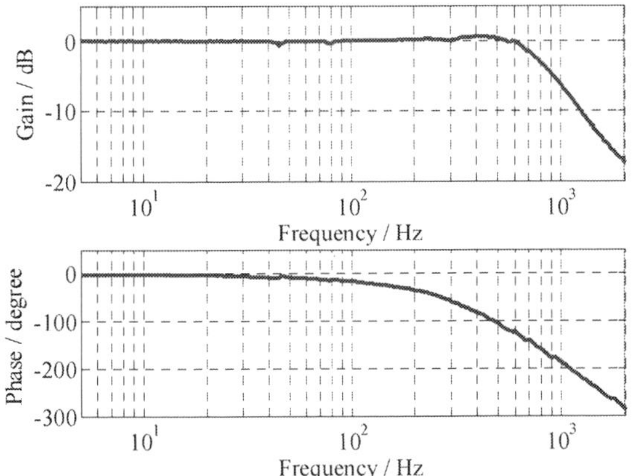

Fig.10. Experimental results: close loop Bode plot of the inner current loop with the novel converter

III. HIGH SPEED DEMONSTRATION OF AMB SYSTEM USING REDUCED SWITCH CONVERTER

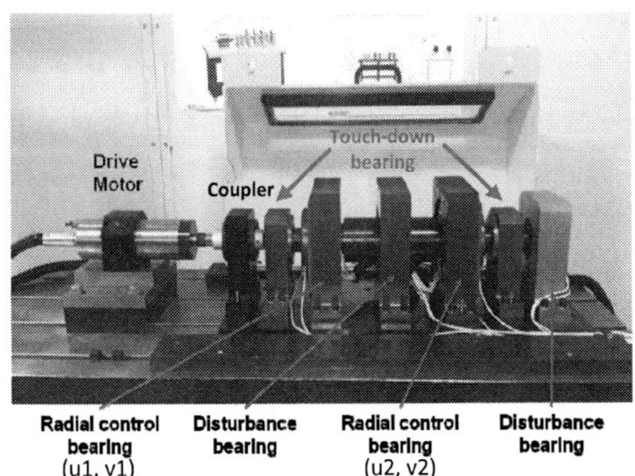

Fig.11. Magnetic bearing test-rig for validation of reduced switch converter

In order to verify the performance of the magnetic bearing and its drive, a test-rig is setup and is shown in Fig.11. It comprises of a 4 kW induction machine which drives a 20 kg shaft through a flex coupling. Two sets of radial magnetic bearings are used to control the shaft position with 250μm magnetic air gap. In order to protect the magnetic bearing and the shaft in case of the control failure, two touch-down bearings with 150μm clearance are used. The touchdown bearings limit the position error tolerance to ±150μm.

Fig.12 shows the control diagram for the magnetic bearing system with the reduced switch converter. The inner loop is the current controller and the outer loop is the position controller. The reference control current i_x is generated from position controller. The control current i_x is added/subtracted from the bias current (I_bias) to generate the reference winding current (i_a*, i_c*) for each pair of the winding. The modulation and the current control methods are based on the description in sections II. B and C respectively.

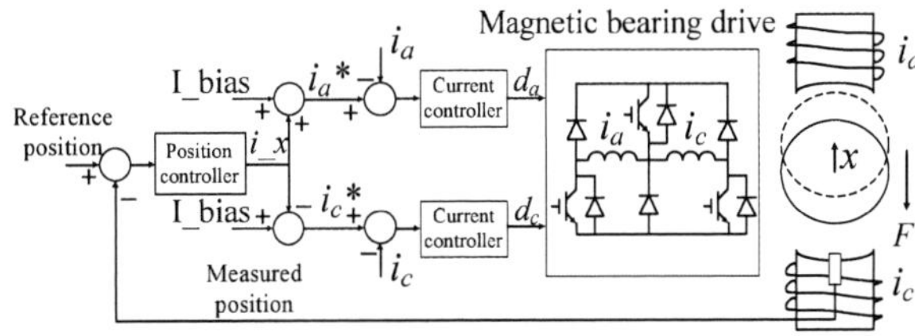

Fig.12. System control diagram for the magnetic bearing with the reduced switch converter

Table.2. Test-rig parameters

Parameter	Value
Shaft Mass	20 kg
Magnetic bearing air gap	250μm (total 500 μm)
Touch-down bearing air gap	150μm (total 300 μm)
Stiffness K_i	85 N/A
Stiffness K_x	500kN/m
Rotating inertia	0.0128kg.m^2
Motor pole pairs	2
Maximum speed	30,000rpm
Motor power rating	4 kW

For the magnetic bearing system, the position control dynamic performance is assessed based on ISO 14839-3 standard [6]. The performance parameter is sensitivity which is essentially the frequency response of position error over reference position signal (*err/ref*). It represents the ability to reject rotational disturbances. Higher sensitivity means that the disturbance at the corresponding frequency cannot be attenuated adequately. Hence with smaller sensitivity, the peak position error is smaller. In ISO 14839-3, sensitivity peak is required to be less than 9.5 dB for newly commissioned machines. Accordingly, Fig.13 shows the measured position sensitivity for the four axes (x1, y1, x2, y2) magnetic bearing system with the reduced switch converter. Since all the four sensitivity plots have peak amplitude below 9.5 dB, the results show compliance with the ISO standards.

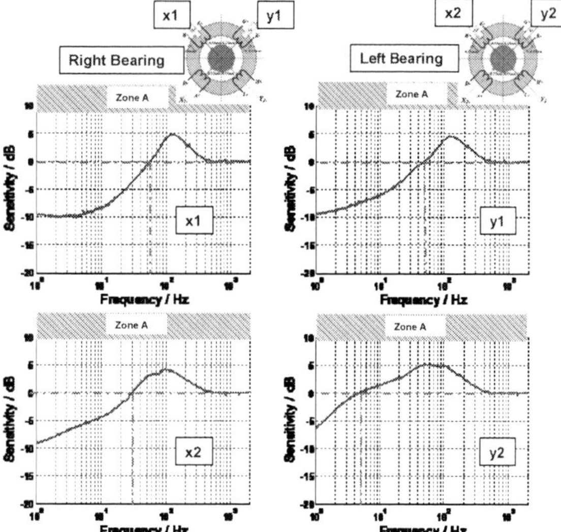

Fig.13.Sensitivity plot (*err/ref*) of position control loop of the four axes magnetic bearing system.

After the sensitivity tests, the rotor is ramped up to 24,000 rpm in 120s. The position errors in the ramping-up experiment are shown in Fig.14. During the ramp-up, the peak position errors are less than 30μm for all the four axes, which is 20% of the clearance for the touch-down bearing. The results validate successful operation of the magnetic bearing system incorporating the reduced switch converter presented and analyzed in this research.

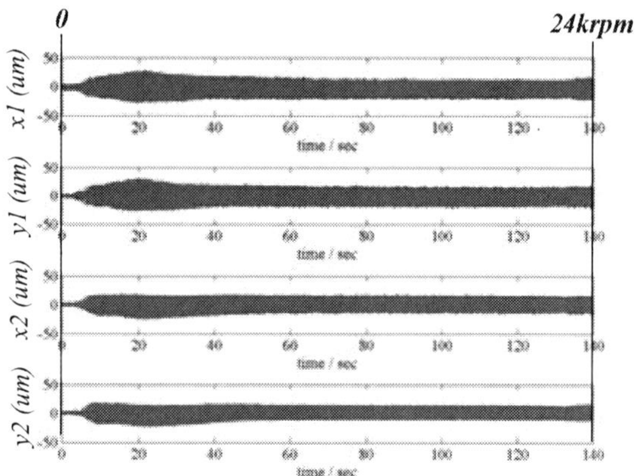

Fig.14. Position sensor response showing ramp-up operation from 0 to 24000 rpm. The total air-gap clearance is 150 um, while the peak position signal error is below 30 um.

IV. CONCLUSIONS

In this paper a reduced switch converter for active magnetic bearing system proposed in [2] is analyzed and its performance is experimentally validated at 24,000 rpm on a 4-axis magnetic bearing test-rig. Based on the coupled coil arrangement and unidirectional current operation, the reduced switch converter is elaborated. In comparison to conventional topologies, this converter reduces the requirement on number of switches, gate drives and pulse width modulation signals stemming from the controller and hence the cost of the system. The simulation and experimental results successfully validate the operation of reduced switch converter to meet the magnetic bearing operation requirements. Following are the salient contributions of the work:

1. Detailed analysis of the reduced switch converter is made with respect to the operating modes and the modulation methods. The analysis and results show that the in-phase modulation results in smaller current ripple than out of phase modulation.
2. The coupled and de-coupled central leg control methods show that are the current control bandwidth with coupled control is twice that of the de-coupled (50% duty cycle) control method. The higher bandwidth makes the reduced switch topology competitive with the H-bridge topology.
3. The influence of cross coupling due to the coupled coil approach on the electromechanical system is not significant making the proposed converter a viable option.

REFERENCES

[1] Gerhard Schweitzer, Eric H. Maslen, H. Bleuler and M. Cole, "*Magnetic Bearings: Theory, Design, and Application to Rotating Machinery*", Springer Press, 2010

[2] Kshirsagar Parag, Yan Peng, "Multiple-Axis Magnetic Bearing and Control of the Magnetic Bearing with Active Switch Topologies", WO2013US35052A

[3] Akira Chiba, Tadashi Fukao, Osamu Ichikawa and Masahide Oshima, "*Magnetic Bearings and Bearingless Drives*", Elsevier Press, 2005

[4] R. Krishnan, "*Switched Reluctance Motor Drives: Modeling, Simulation, Analysis, Design, and Applications*," USA, CRC Press LLC, June 2001

[5] Bartholet. T. M, Nussbaumer, T, Kolar, J , " Comparison of Voltage-Source Inverter Topologies for Two-Phase Bearingless Slice Motor" , *IEEE Transactions on Industrial Electronics*, Vol 58, No.5, May 2011

[6] ISO 14839-3 "Mechanical vibration—Vibration of rotating machinery equipped with active magnetic bearings" ISO 2006

A Novel Balanced Winding Topology to Mitigate EMI without the Need for a Y-capacitor

Yongjiang Bai[1], Xu Yang[1], Xinlei Li[2], Dan Zhang[2], Wenjie Chen[1]
Email: baiyj@mail.xjtu.edu.cn yangxu@mail.xjtu.edu.cn cwj@mail.xjtu.edu.cn
[1]Institute of Electrical Engineering, Xi'an JiaoTong University, Xi'an, China
[2]Silergy Corp. A206, Ascends Innovation Hub, XHTZ, Xi'an, China

***Abstract*—Due to the interaction between human and the power supply equipment, more and more Switch-mode power supplies (SMPS) require lower leakage current. In the AC-DC SMPS, the leakage current is mainly caused by the Y-capacitors. It is better to remove the Y-caps in this situation. But connecting Y-caps between the primary return line terminal and the secondary return line terminal is a very common method in EMI design. Certain works should be done to acquire good EMI design without the help of Y-cap. In this paper, a novel balanced winding topology is proposed to mitigate EMI noise without the need for Y-cap. To have a better understanding of the operational principle, the EMI noise transmission path with and without Y-cap are analyzed. The design procedure of the proposed topology is discussed in detail. The experimental results demonstrated the prototype has excellent EMI performance without the help of Y-cap. It also can simplify the transformer design and be easy to implement mass production.**

***Keywords*—EMI; Transformer; balance winding; Flyback**

I. INTRODUCTION

As we know, reducing the original CM noise emission of power converters can help to simplify the filter design and reduce the total filter size. Significant efforts, such as additional large CM chokes and shielded EMI filters, are made in order to filter the CM noise in today's power converter.[10] However, the attenuation of common mode (CM) current in 2-wires AC input SMPS is more difficult than those 3-wires SMPS [1]. For these 2-wires isolated SMPS, the Y-capacitor is an essential component to reduce the CM interference.

On the other hand, most of the AD/DC SMPS realize the separation between high voltage AC input and low voltage DC output to meet the safety standards, such as UL1950, which specified the dielectric strength and the maximum leakage current at the same time. Standards in leakage current ensure human safety, and prevent the user from becoming part of a path for substantial current to ground when touching the output or the enclosure of a power supply [3]. For example, person who uses a charging cell phone with the metal case will have the risk of getting an electric shock owing to the existence of Y capacitor. And the leakage current will affect the performance of the touch screen too. So for the cell phone charger, the leakage current is mainly caused by the Y-cap. From this point of view, it is required to get rid of Y capacitor in these applications. However, it will generate big challenge in EMI design for these equipments.

Up to now, few studies have been done to reduce CM noise without the help of Y-cap [4-8]. In [1-3], the Y-cap is used to bypass the CM noise from the propagation path. But the Y-cap is unacceptable in those applications. In [1], [4], [5], [8], an additional balanced winding is added to cancel the CM current goes through parasitic capacitance between transformer primary side and secondary side. In [1], [6], a shield winding is added in series of a compensation capacitor to cancel the CM current goes through inter-winding capacitance. However, these methods can only reduce the CM noise that goes through one propagation path. Moreover, it requires accurately control of parasitic capacitance, which is not only very difficult but also time consuming in mass production. In [8-9], new topologies are proposed to reduce the CM noise at the expense of cost and complexity. Although these methods are effective in reducing EMI, there is a lack of manuscript that investigating the No Y-cap EMI design method from topology optimization viewpoint.

Considering the above mentioned problems, this paper pr oposed a novel balance winding topology as well as the design rules to reduce the EMI noise without the help of tradition Y-cap. The impact of Y-cap on EMI noise and system stabilization is analyzed in detail. A more precise transformer model is also proposed. Based upon that, the impacts of topologies with dedicated transformer on the CM noise are evaluated. The effectiveness and feasibility of the proposed topology is verified by the experimental results in 5W cell phone charger.

II. THE IMPACT OF Y-CAP ON EMI NOISE AND STABILIZATION

It is well known that common-mode (CM) conducted EMI is caused by the common mode current flowing through the parasitic capacitance of transistors, diodes, and transformers to earth in the power circuit. As shown in Fig.1, the CM noise is accounted by the current flowing through the LISN resistor.

To find solution to reduce the CM current, the secondary side CM noise propagation path is also illustrated in Fig.1.When a Y-cap is added between the primary return terminal GND and secondary return terminal GNDS, the CM noise current will change original path from C_{se} to Y2. By

This work was supported by the National Natural Science Foundation of China under Project 51277145.

means of this method, the CM noise flowing to the secondary side will be shunted and better EMI performance will be obtained. On the contrary, if the Y-cap is removed for the leakage current standard requirements, all the CM noise has only one path by the C_{se} and it must cause worse EMI results.

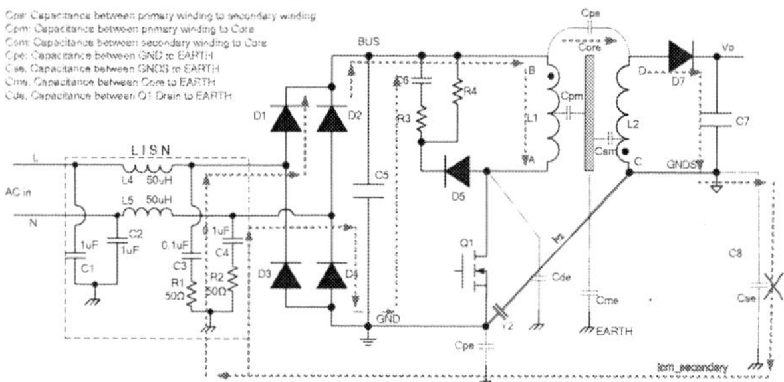

Fig. 1: Shunting effects of Y2 for Common mode noise when Q1 is being turned off

In order to remove the Y-cap without additional EMI filter, this paper proposed a novel balanced winding topology. Comparing with other cancellation techniques, it is much simpler and requires no additional transistors and gate-drive circuitry.

III. PROPOSED TOPOLOGY WITH BALANCED WINDING

EMC improvement methods in the power section usually do not change the topology and are implemented on the sideline. In this paper, a better EMC behavior along with other desirable converter parameters is considered in the topology design. Since the flyback converter is widely used as an isolated power supply, this paper takes it as example to introduce a balanced winding power converter as shown Fig.2 (a). It just operates as a flyback with primary winding divided into two branches and places the switch and control logic in the middle. Based on a dedicated transformer structure, the EMC CM current in the two branches has cancellation. So it will have excellent EMI performance and the Y-cap can be removed easily.

First of all, the operational principle of the balanced winding flyback will be introduced. When the switch is turning on, the magnetic current will increase linearly from zero. The energy will store in the primary winding just as inductor. The voltage of winding P1 and P2 is proportional to V_{in}. After a certain time, the switch turns off and then the energy stored in the primary winding will transfer to the secondary by output rectifier diode. At then, the switch voltage will be $V_{in}+(N_1+N_2)\cdot V_O$ (N_1 and N_2 is the transformer ratio). After the secondary winding current goes to zero, the primary switch voltage will be clamped to V_{in}.The operational principle of the proposed topology is shown in Fig.2 (b), (c) and (d). The real time waveforms are shown in Fig.2 (e).

In order to compare the EMI performance, a traditional flyback converter without the auxiliary winding shown in Fig.1 is analyzed firstly. The primary winding is evenly divided into four layers winding. The secondary winding is distributed symmetrically in a single layer. There is no auxiliary winding due to the self-powered flyback. Fig. 3(a) shows the half window of transformer structure implemented with EE core. The terminal B, E and C are corresponding terminals. The winding to core capacitors can be neglected because its capacitance is too small and the primary winding inner layers capacitors can also be neglected because they do not contribute to CM noise [8]. And it is assumed that all the winding layers cover the whole window width and there is 2 layer taps between different layers. Due to the different distance between the different primary winding layers to the secondary winding layer, the parasitic capacitance is different and there is a k ratio between them. According to current formula of capacitor $I_C = C\frac{dV}{dt}$, the displacement current between primary side and secondary side only depends on the voltage jump slope and the parasitic capacitance.

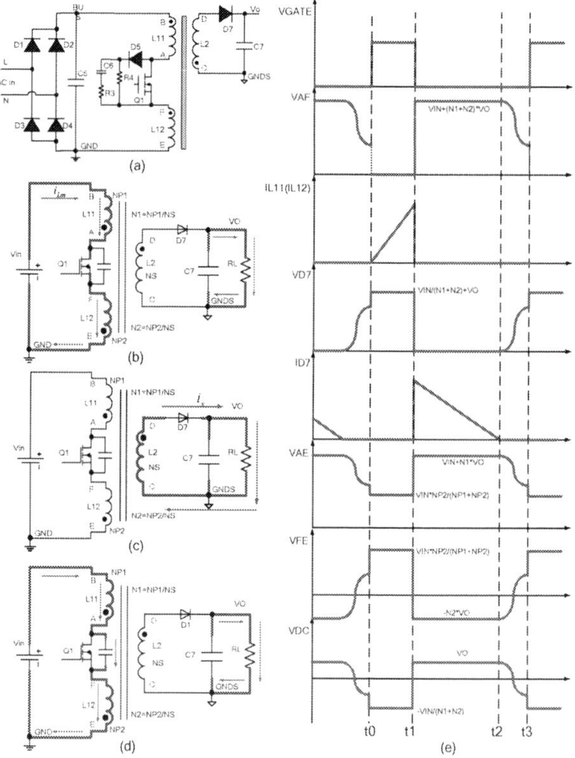

Fig. 2 The proposed balanced winding topology

In a traditional flyback converter, when the Q1 is in switching mode, the point A can be considered as a pulse voltage source which amplitude is V_A from turn on state to turn off state. And it can be expressed as following:

$$V_A = V_{in} + \frac{N_P}{N_S} * V_O \quad (1)$$

Similarly, the point D can be considered as a pulse voltage source which amplitude is V_D from turn on state to turn off state. And it can be expressed as following:

$$V_D = V_O + \frac{N_S}{N_P} V_{in} = \frac{N_S}{N_P} V_A \qquad (2)$$

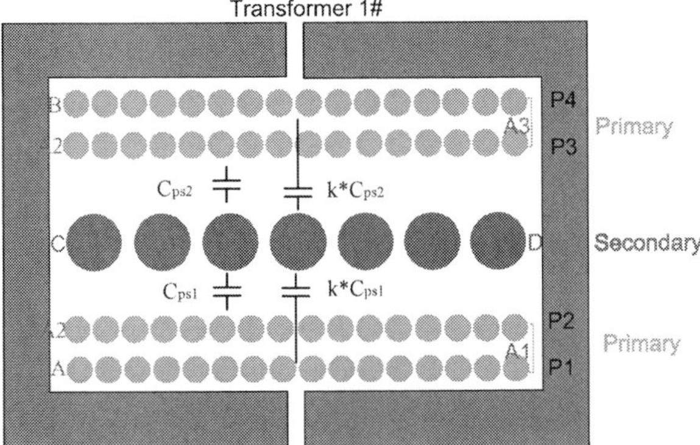

Fig. 3(a) Inter-winding of transformer

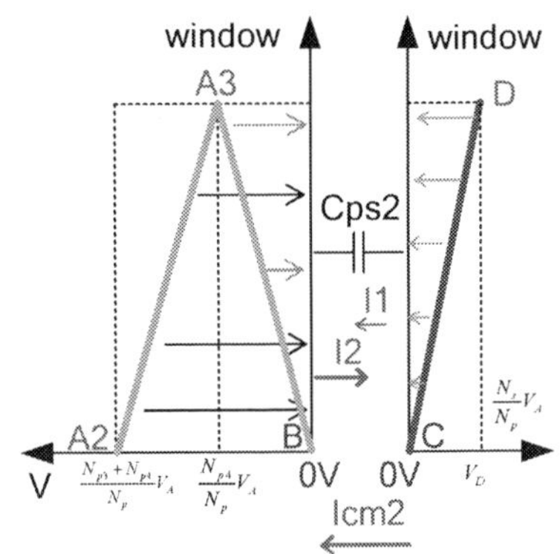

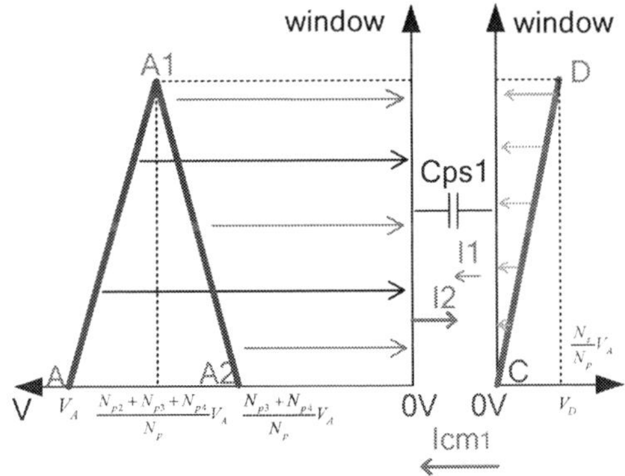

Fig. 3 (b) Voltage and current distribution of winding layers

The voltage of the four layers winding at different point is shown in Fig.3 (b) assuming the voltage is evenly distributed. When the Q1 is turned off, the voltage of point A1 and D is rising up relative to point B and point C respectively. By the means of the above equations, the displacement current is shown in Fig.4 (b). I1 and I2 have different direction assuming that the current orient to primary side is positive. The displacement current between inner primary winding and the secondary winding is expressed as:

$$\begin{aligned} I_{CM1} &= C_{ps1} \frac{N_{p2} + 2*(N_{p3} + N_{p4}) - N_S}{2*N_P*\Delta t} V_A \\ &+ k*C_{ps1} \frac{N_P + N_{p2} + N_{p3} + N_{p4} - N_S}{2*N_P*\Delta t} V_A \end{aligned} \qquad (3)$$

Similarly, the displacement current between outer primary winding and the secondary winding is given as:

$$\begin{aligned} I_{CM2} &= C_{ps2} \frac{N_{p3} + 2*N_{p4} - N_S}{2*N_P*\Delta t} V_A \\ &+ k*C_{ps2} \frac{N_{p4} - N_S}{2*N_P*\Delta t} V_A \end{aligned} \qquad (4)$$

From formula (1) and (2), supposing: $N_{p1} = N_{p2} = N_{p3} = N_{p4} = 0.25*N_p$ and $C_{ps1} = C_{ps2} = C_{ps}$ the total displacement current from primary and secondary can be expressed by:

$$I_{CM} = I_{CM1} + I_{CM2} = C_{ps}(1 - \frac{N_S}{N_P})(k+1)\frac{V_A}{\Delta t} \qquad (5)$$

Therefore, based on equation (5), N_S is much smaller than N_P in a traditional flyback converter, the CM displacement current is very large and the EMI turn to be worse if the Y-cap is removed. In a similar way, the balanced winding topology is analyzed as follow. The transformer structure is just the same as the above traditional flyback which is shown in Fig.4 (a).

In the balanced winding topology, when the Q1 is in switching mode, the point A can be considered as a pulse voltage source which amplitude is V_A from turn on state to turn off state. And it can be expressed as following:

$$V_A = V_{in} * \frac{N_{P1}}{N_{P1} + N_{P2}} + N_1 * V_O \qquad (6)$$

Similarly, the point F can be considered as a pulse voltage source which amplitude is V_F from turn on state to turn off state. And it can be expressed as following:

$$V_F = -(V_{in} * \frac{N_{P2}}{N_{P1} + N_{P2}} + N_2 * V_O) \qquad (7)$$

At the same, the point D can be considered as a pulse voltage source which amplitude is V_D from turn on state to turn off state. And it can be expressed as following:

$$V_D = V_O + \frac{V_{in}}{N_1 + N_2} = \frac{V_A}{N_1} = -\frac{V_F}{N_2} \qquad (8)$$

The voltage of the four layers winding at different point is shown in Fig.4 (b). When the Q1 is turned off, the voltage of point F and D is rising up relative to point E and point C separately. However, the voltage of point B is decreasing down relative to point A. The displacement current is shown in Fig.4 (b). The displacement current between inner primary winding and the secondary winding is given as:

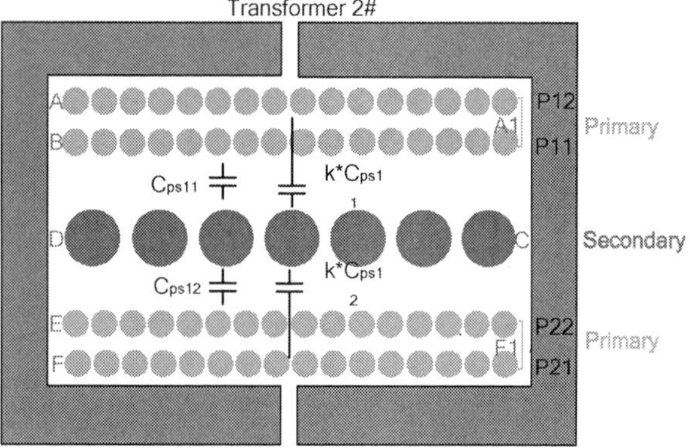

Fig. 4(a) Inter-winding of transformer

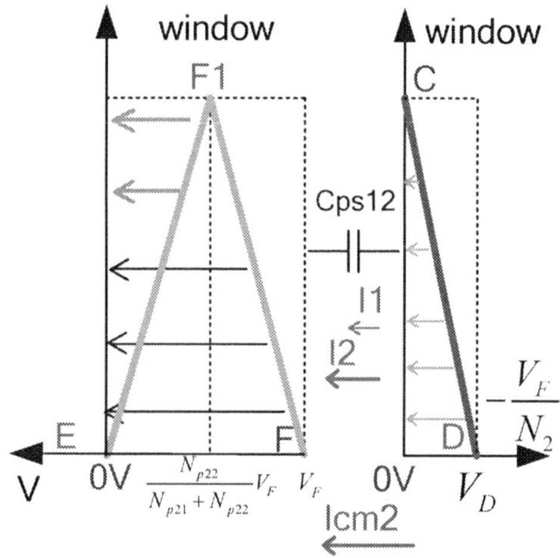

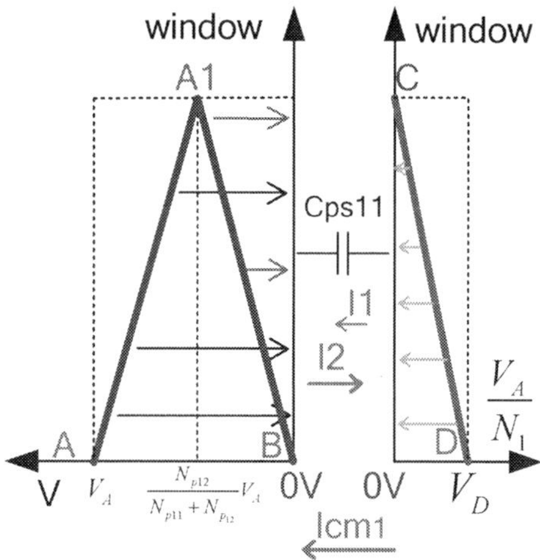

Fig. 4 (b) Voltage and current distribution of winding layers

$$I_{CM1} = C_{ps11}\frac{N_{p12} - N_S}{2*N_{P1}*\Delta t}V_A + k*C_{ps11}\frac{N_{P1} + N_{p12} - N_S}{2*N_{P1}*\Delta t}V_A \tag{9}$$

Similarly, the displacement current between outer primary winding and the secondary winding is given as:

$$I_{CM2} = C_{ps12}\frac{N_{p22} + N_S}{2*N_{P2}*\Delta t}V_F + k*C_{ps12}\frac{N_{P2} + N_{p22} + N_S}{2*N_{P2}*\Delta t}V_F \tag{10}$$

From formula (6) and (7), supposing: $C_{ps11} = C_{ps12} = C_{ps}$, $N_{p11} = N_{p12} = 0.5*N_{P1}$ and $N_{p21} = N_{p22} = 0.5*N_{P2}$ total displacement current from primary and secondary can be expressed by:

$$I_{CM} = I_{CM1} + I_{CM2} = \frac{C_{ps}}{4*\Delta t}* \left[(3k+1)(N_1 - N_2) - 4(k+1)\right]\left(\frac{V_{in}}{N_1 + N_2} + V_O\right) \tag{11}$$

The CM noise from the primary to secondary can be made to zero by a proper design of N1 and N2. Therefore, there is no difference whether the Y-cap is removed or not. At the same time, due V_A and V_F has the different voltage pulsating, so the CM noise from the point A and Point F will be partial cancelled.

IV. IV. EXPERIMENTAL RESULTS

In order to verify the proposed CM noise reduction technique to get rid of the Y-cap, a flyback converter prototype is built. The basic specification of the prototype is shown in Tab.1.

Tab. 1 Prototype Specification

Input	L and N 2-wire, 85V~264Vac 47Hz~63Hz,
Output	5V, 1A ,V_{ripple}<100mV and I_{OCP}<1.3A with USB port
Efficiency	Energy star Level 6 and No load loss <70mW
EMC and leakage current	EN555022 Class B limits; Leakage current <0.1mA
Size	W*L*H=36mm*36mm*18mm
Switching frequency	Quasi-Resonant flyback with 70kHz at full load in 115Vac
Transformer	EE13: N_{P1}=98Ts, N_{P2}=62Ts, N_S=10Ts

The prototype is shown in Fig.5 (a). The flyack converter operation waveforms are shown in Fig.5 (b). The voltage of pulsating points V_F, V_A and V_D and the V_{OUT} are shown in CH1, CH2, CH3 and CH4 respectively. Fig.5 (c) shows the test environment for the conducted EMI of home information technology device, as governed by European Union. Normally we test quasi-peak level limits as well as average level limits. All the experiments are test based on this platform.

Fig. 6 (a) shows EMI CE spectrum at 220Vac@50Hz by a traditional sandwich transformer with 470pF Y-cap connected between the primary and the secondary side. With the help of Y-cap, the EMI can meet the EN55022 EMC standard easily. When the Y-cap is removed for lower leakage current, the test EMI spectrum is shown in figure6 (b). It turns to be worse and can't meet the EN55022 EMC standard. It is over the limitation line over 10dB. Therefore the Y-cap can't be removed easily and it will cause the EMI worse.

Based on analysis in section III, in order to reduce the CM noise, a novel balanced winding topology is proposed to solve this issue.

Fig. 5 (a) 5V, 1A prototype

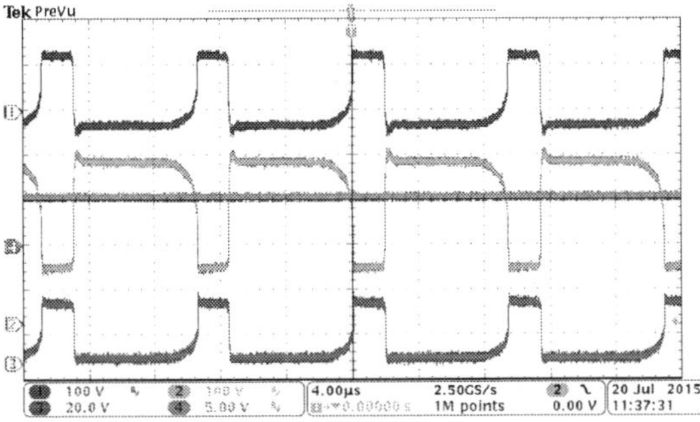

Fig. 5 (b) CH1: VFE, CH2: VAE, CH3: VDC, CH4: VOUT

Fig. 5 (c) EMC test platform

As expected, a proper transformer is designed for the proposed topology. The CM noise is significantly reduced by this balanced winding. The EMI test results in Fig.7 are much better and it can pass the EN55022 standard easily without the help of Y-cap.

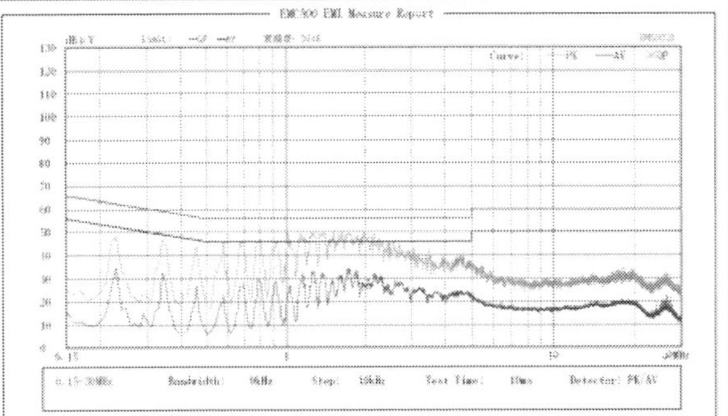

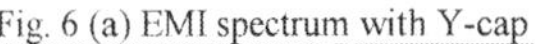

Fig. 6 (a) EMI spectrum with Y-cap

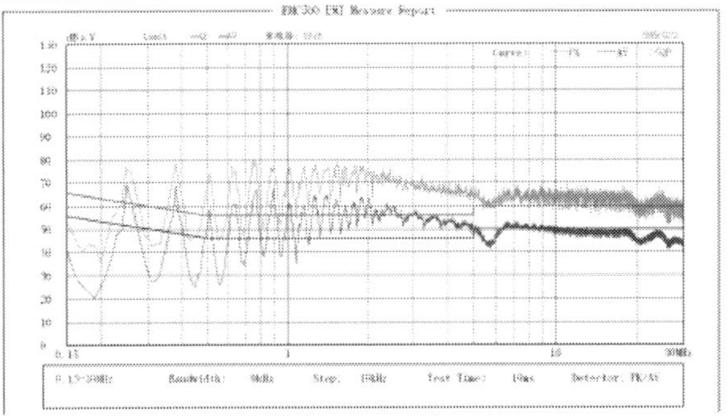

Fig. 6 (b) EMI spectrum without Y-cap

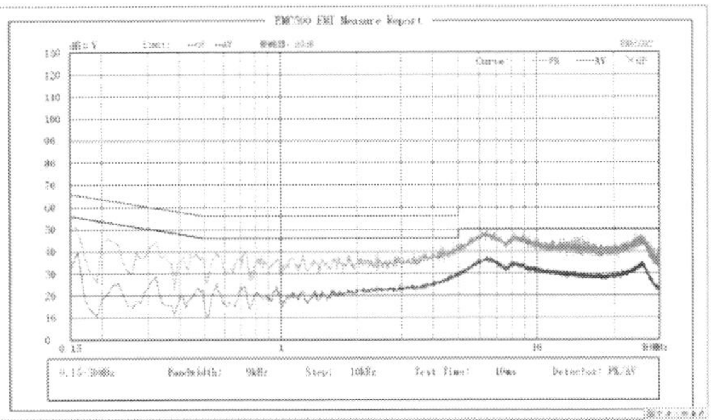

Fig. 7 (a) Proposed topology EMI spectrums without Y-cap in 115Vac

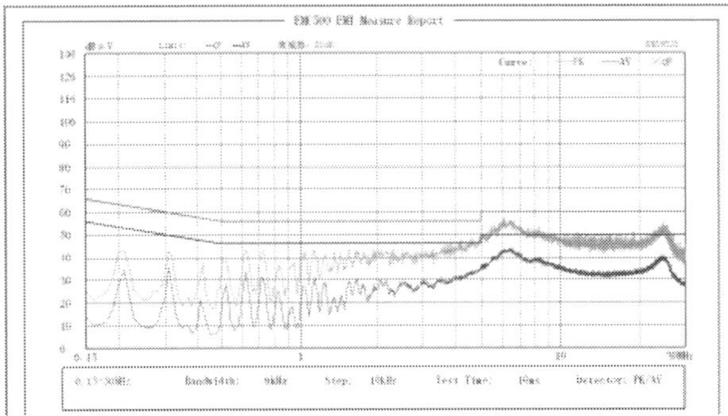

Fig. 7 (b) Proposed topology EMI spectrums without Y-cap in 220Vac

V. V. CONCLUSION

In this paper, the effects of Y-cap for EMI noise of isolated power converter are analyzed. Characteristics of inter-winding capacitances are first summarized as well as key rules for an isolated power converter without the need of Y-cap. The effects of transformer winding structure are discussed and CM noise reduction by novel balanced winding structure is proposed. Analysis of the novel balanced winding topology for EMI cancellation effect has been carried out. Experimental results validate the above analysis and the proposed topology EMI performance without Y-cap.

REFERENCES

[1] Jin-ho Choi, Majid Madafshar, Kevin Parmenter, "Designing common-mode (CM) EMI noise cancellation without Y-Capacitor" in IEEE APEC 2007, pp. 936 - 940

[2] Pingping Chen, Honghao Zhong, Zhaoming Qian, Zhengyu Lu, "The Passive EMI Cancellation Effects of Y Capacitor and CM Model of Transformers Used in Switching Mode Power Supplies (SMPS) " in PESC 04. 2004 IEEE 35th Annual pp: 1076-1079

[3] Milind M. Jha, Kunj Behari Naik and Shyama P. Das, "Types of Electro Magnetic Interferences in SMPS and Using Y-Capacitor for Mitigation of Mixed Mode Noise" in ICPCES,2010 International Conference on,pp1-6

[4] Hung-I Hsieh, Sheng-Fang Shih, "Effects of Transformer Structures on the Noise Balancing and Cancellation Mechanisms of Switching Power Converters" in IPEC-Hiroshima 2014 - ECCE-ASIA, pp: 2380-2384

[5] Yuchen Yang, Daocheng Huang, Fred C. Lee and Qiang Li, "Transformer shielding technique for common mode noise reduction in isolated converters" in IEEE 2013 ECCE, pp. 4149 - 4153

[6] Zengyi Lu, Wei Chen, "Common mode EMI noise reduction technique by noise path configuration of high frequency power transformer," in Proc. IEEE Conf. on Power Electronics and Motion Control, 2009, pp. 954-956

[7] Pengju Kong and Fred C. Lee, "Transformer Structure and Its Effects on Common Mode EMI Noise in Isolated Power Converters" in IEEE APEC 2010, VOL. 2, pp: 1424-1429

[8] Yick Po Chan, Bryan Man Hay Pong, Ngai Kit Poon, and Joe Chui Pong Liu, "Common-Mode noise cancellation in Switching-Mode Power Supplies using an equipotential transformer modeling technique" in IEEE Transactions on Electronmagnetic Compatibility Vol.54,No.3,June 2012

[9] Mohammad Rouhollah Yazdani, Hosein Farzanefard, and Jawad Faiz, "EMI Analysis and Evaluation of an Improved ZCT Flyback Converter" in Power Electronics, IEEE Transactions on Year

[10] Wenjie Chen, Xu Yang, Jing Xue and Fred Wang."A novel filter topology with active motor CM impedance regulator in PWM ASD system",IEEE Transactions on Industrial Electronics.vol.61, no.12, Dec.2014, pp.6938-3946.

Topology and Control Strategy for Accelerated Lifetime Test Setup of DC-Link Capacitor of Wind Turbine Converter

Youngjong Ko, Holger Jedtberg, Giampaolo Buticchi, and Marco Liserre
Chair of Power Electronics
Christian-Albrechts-Universität zu Kiel
Kaiserstr. 2, 24143 Kiel, Germany
yoko@tf.uni-kiel.de, hje@tf.uni-kiel.de, gibu@tf.uni-kiel.de, ml@tf.uni-kiel.de

***Abstract*—Back-to-back converters for wind turbine systems (WTS) feature capacitors in the DC-link to maintain a stable DC-link voltage and to decouple the generator from the grid. Electrolytic and film capacitors can be chosen to this purpose. Long-term field experience and recorded failure data reveal that capacitor failures are one of the main reasons for the downtime of WTS. The ripple current accelerates the wear-out of the capacitors, which is strongly dependent on the mission profile of the system. Therefore, it becomes important to estimate the useful lifetime of a capacitor as a function of the ripple current. In this paper, aging characteristics of electrolytic capacitors are described, and a power converter topology and its control strategy are designed to perform accelerated lifetime tests. The voltage and ripple current of the capacitor under test (CUT) can be controlled with low THD to correlate wear-out with ripple.**

Keywords*—Reliabilty, Accelerated lifetime test, Electrolytic capacitor, Wind turbine converter, Controller design

I. Introduction

Many efforts have been devoted to the research of WTS in terms of system integration, optimization of the employed power converter topologies (such as bidirectional, multilevel, and parallel configurations), as well as the application of novel control strategies in order to maximize the efficiency of the energy conversion [1]-[6]. However, challenging operating conditions due to variable wind power and increasing grid code requirements lead to a high stress variability for the electrical system of the WTS. Therefore, unexpected failures occur that come along with high costs due to downtime and maintenance of the WTS.

As shown in Fig. 1, the electrical system is one of the main reasons for failures of WTS. More precisely, electrolytic capacitors, which are typically used as DC-link in back-to-back converters of WTS since they offer the advantages of higher rated voltage, energy density, and capacitance, respectively, at lower costs compared to other capacitor types, exhibit the highest failure rate inside the electrical system [7], [8]. Research papers on capacitor reliability have been mainly focused on degradation diagnosis methods by analysis of internal parameters, such as the equivalent series resistance (ESR) and the capacitance, including complex algorithms and models [9]-[11]. The lifetime of electrolytic capacitors is influenced by the ambient temperature, the ripple current flowing through the capacitor, and the applied voltage, respectively [12]. To date, capacitors' lifetime models take into account the operating temperature and the current RMS value. The operating temperature is then dependent on the power dissipated by the capacitor. However, a clear correlation between the ripple current frequency and the lifetime of the capacitors is missing. The analytic lifetime model based on thermos-mechanical characteristics can more precisely consider the correlation between the lifetime and the operating conditions [13], [14]. For this model the number of cycle to failure is required and the test is accomplished under over-rated condition, which is called the accelerated lifetime test.

This paper proposes a power converter setup and its control strategy to subject the capacitors to arbitrary current/voltage profiles. By applying a stress greater than the nominal conditions, accelerated lifetime tests can be performed. The main challenge is to generate current waveforms with a high fundamental frequency, great amplitude and a low THD, in order to obtain a better correlation between the single frequency and the capacitor damage.

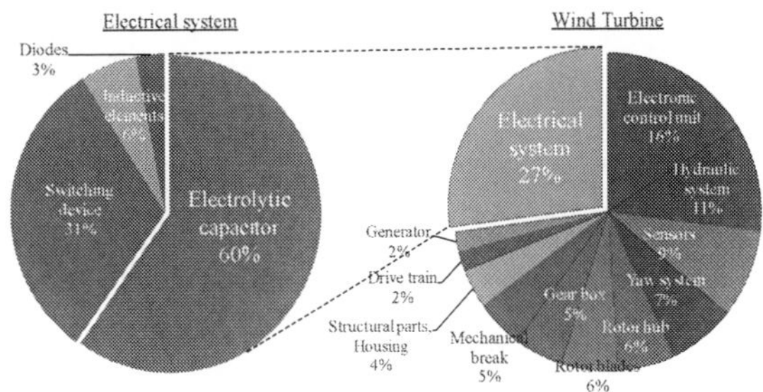

Fig. 1. Components failure rate in wind turbine [7], [8].

II. Electrolytic Capacitor

A. Electrolytic Capacitor Model

A simplified model of an electrolytic capacitor is depicted in Fig. 2, where C is the terminal capacitance, C_D and R_D account for the losses caused by the dielectric, R_0 is the combined constant resistance of terminals, tabs and foils, respectively, and R_t is the resistance of the electrolyte [10]. It should be noted that the capacitor model is also containing an inductive part in series, that is dominant, when the capacitor is operated above its resonance frequency. For the following analysis this inductance will be neglected, following the assumption that the capacitor is operated well below its resonance frequency. The equivalent series resistor (ESR) represents the real part of the capacitor's impedance. Therefore, it can be expressed as

$$R_{ESR} = R_0 + R_f + R_t \tag{1}$$

where R_f represents the frequency dependent resistance of the dielectric layer, which is composed by C_D and R_D, and R_t is the temperature dependent part due to the characteristics of the employed electrolyte [10], [15]. Since with increasing temperature the viscosity of the electrolyte changes, its conductivity increases, which in turn leads to a reduction in ESR due to R_t. This effect can be modeled by applying (2), where $R_{t,b}$ is R_t at base temperature T_b, T_s is the surface temperature of the capacitor, and SF denotes a capacitor dependent temperature sensitivity factor [10].

$$R_t = R_{t,b}(T_b) \cdot e^{\frac{T_b - T_s}{SF}} \tag{2}$$

In Fig. 3 the ESR is plotted against frequency for different temperatures in accordance to [10]. As discussed before, it can be seen that the ESR decreases as temperature increases. Moreover, it can be noted that increasing frequencies also lead to a decrease in ESR due to R_f.

B. Impact of Ripple Current on Lifetime

The lifetime of an electrolytic capacitor is strongly influenced by its operating conditions or mission profiles, respectively (such as temperature, humidity, pressure, vibration, voltage, and current). Apart from ambient temperature, which accelerates electrochemical reactions, and the applied voltage level, the ripple current flowing through the capacitor is one of most critical stressors as discussed in [12]. The latter is of crucial importance, since it causes additional internal heating due to the power dissipation caused by ESR. The surface temperature of the capacitor can be estimated by applying (3), where T_a denotes the ambient temperature, ΔT is the additional heating caused by the power

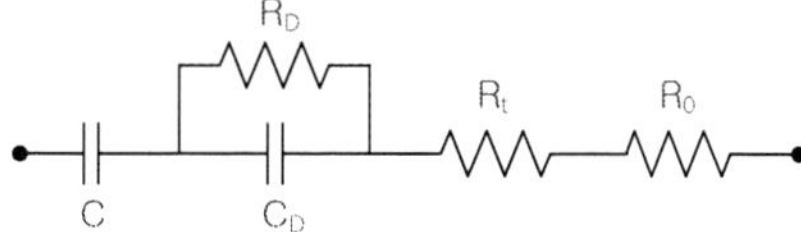

Fig. 2. Equivalent model of electrolytic capacitor.

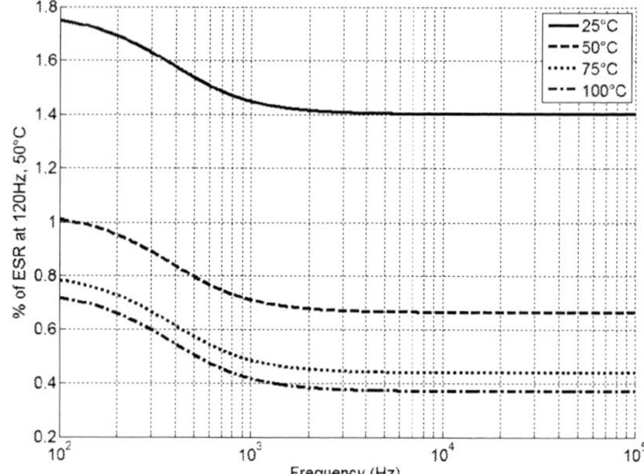

Fig. 3. Variation of ESR with respect to ripple current frequency and temperature [10].

dissipation P_d and the thermal resistance of the capacitor R_{th}.

$$T_s = T_a + \Delta T = T_a + P_d \cdot R_{th} \tag{3}$$

Since the ESR exhibits a frequency dependent behavior, for the correct calculation of P_d the knowledge of the RMS capacitor currents at the corresponding frequencies is needed [16], [17]. Therefore, a spectral analysis of the ripple current has to be carried out in order to derive its harmonic content. By knowing the current harmonics, P_d can be calculated by

$$P_d = \sum_{i=1}^{n} I_{R,i}^2 \cdot R_{ESR,i} \tag{4}$$

where $I_{R,i}$ and $R_{ESR,i}$ denote the RMS ripple current and the ESR at frequency harmonic i, respectively, with $i = 1, 2, \ldots, n$.

$$L = L_0 \cdot 2^{\frac{T_{\max} - T_a}{10\text{K}}} \cdot 2^{-\left(\frac{I_a}{I_0}\right)^2 \frac{\Delta T_0}{A}} \cdot \left(\frac{V_a}{V_0}\right)^{-m} \tag{5}$$

The lifetime model of an electrolytic capacitor is shown in (5), where L_0 denotes the lifetime at nominal values as given in the manufacturer's datasheets, T_{max} is the maximum permissible temperature, I_a is the applied ripple current, I_0 is the rated ripple current, ΔT_0 is the temperature increase when I_0 is applied, A is a temperature coefficient, V_a is the applied voltage, V_0 is the nominal voltage, and m accounts for the manufacturer dependent voltage factor as discussed in [18]. The model consists of three parts, where each part is considering one of the three major stressors. The impact of the ambient temperature follows the well-known rule of Arrhenius, which constitutes a doubling in lifetime for each 10 K temperature decrease. Moreover, the applied voltage is taken into account in this model, since an increasing voltage level causes degradation due to electrolyte evaporation effects, which in turn affects the lifetime of the capacitor. As mentioned above, the ripple current is influencing the capacitor's lifetime by acting on the temperature rise ΔT, which is also taken into account in (5). The factor A accounts for the higher impact of ΔT on the lifetime, since it is evaluated experimentally by some manufacturers, that the lifetime is bisected by an increase of, e.g., 7…10 K in ΔT [19].

Degradation of the capacitor is not included in (5). Since the heat rise due to the ripple current causes electrolyte evaporation, it leads to a degradation of the capacitor's parameters, being the ESR and the capacitance C, respectively. As discussed in [12] the ESR will increase with time, while C might decrease. An increasing ESR will have an impact on the power dissipation as shown in (4) and, thus, affects the lifetime of the capacitor as well. For this reason, more research efforts have to be devoted to better understand the correlation between parameter degradation and lifetime of capacitors.

DC-link capacitors of power converters are subject to very different ripple current profiles. The ripple current flowing through the DC-link capacitors in a back-to-back converter of a wind turbine system, for example, is dependent on the operating conditions at both ends of the converter. As presented in [16] the frequency spectrum of the capacitor current is depending on factors such as the modulation scheme, the employed carrier-frequency, the power factor, amplitude, and frequency of the output currents of the converter. Based on the above formulas, from Fig. 3 it becomes evident that ripple currents of lower frequencies have a bigger impact on the capacitor's lifetime than currents of higher frequencies, since the ESR is higher in the low frequency region. This effect is also discussed in [18].

In order to achieve a better understanding of degradation methods due to mission profiles, it is necessary to investigate the impacts of ripple current harmonics (and their combinations) on the capacitor's parameters and lifetime. Therefore, a careful analysis of different operating conditions has to be carried out in order to enable the application-oriented verification of both, the thermal model and the lifetime model of the capacitor, which can be of assistance to create a tool that can help to reach a more reliable and cost effective system. The proposed system is useful for detailed analysis of ripple current stresses, which are derived from mission profiles of wind turbine converters, since it provides the application of combined stresses as will be discussed in the following sections.

III. Proposed Accelerated Lifetime Test Setup

A. Topology Description

The goal of the test setup is to apply accelerated stress tests to electrolytic capacitors by means of ripple currents of both, high RMS amplitude and frequency, in order to investigate thermal effects and aging characteristics based on mission profile data. In order to investigate the impact of single frequency harmonic ripple currents, a low THD and therefore proper filtering of the switching frequency components is required. Moreover, since the focus is put on application-oriented stress tests for DC-link capacitors in WTS, the setup needs to be able to provide a high DC voltage to the capacitor under test (CUT). Therefore, the test setup has to be carefully designed in order to meet the required specifications and to ensure that no components other than the CUT fail during the tests. Apart from the test specifications, for the reasons of cost effectiveness and lower system complexity, standard switching devices (IGBTs) are to be implemented into the test setup. Based on this, two possible solutions are discussed in this section, which can be used to achieve high frequency ripple currents in combination with a DC voltage at the output side of the converter.

The first possible configuration, the cascaded H-bridge (CHB), is shown in Fig. 4 (a). It consists of three H-bridge converter cells connected in series, allowing the output voltage to be three times the DC source voltage of each cell. The modulation is accomplished with the phase shifted PWM as shown in [20]. With this topology, the effective inverter switching frequency is six times higher than the device switching frequency of each cell. This offers the advantage of using standard switching devices in the test setup. Moreover, with a higher switching frequency ripple on the output side of the converter a smaller filter can be realized, since the filter inductance is inversely proportional to the switching

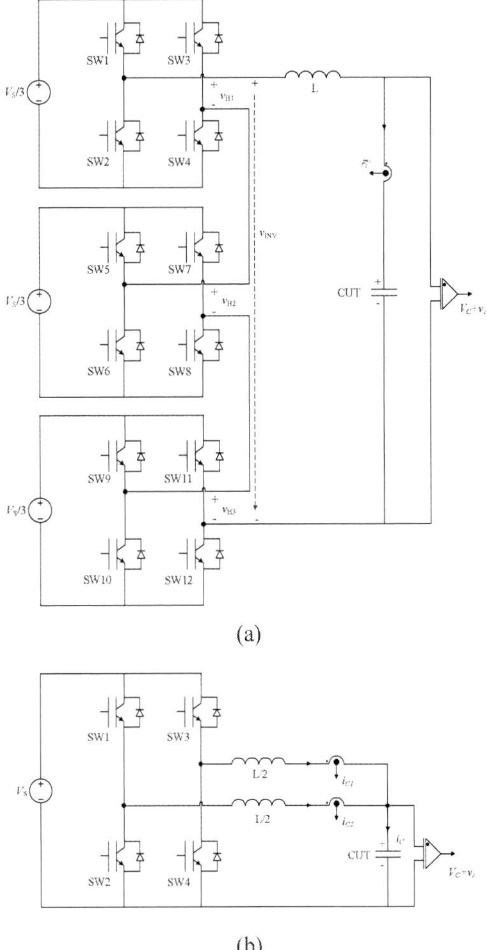

Fig. 4. Possible topologies for test setup. (a) cascade half-bride (CHB), (b) interleaved bidirectional (IB).

frequency (see below).

From the system complexity's point of view, a much simpler solution could be used. In Fig. 4 (b) the configuration of an interleaved bidirectional DC/DC converter (IB) is shown [21]. It is based on an H-bridge and can therefore be realized with less switching devices than the CHB, which reduces the overall complexity of the system by means of control and hardware efforts. Nevertheless, in order to fulfill the above mentioned test requirements the CHB topology is chosen, since the IB configuration has certain limitations as discussed in the following paragraph.

Equations (6) and (7) describe the relation between the switching frequency and the filter inductor in case of IB and CHB, respectively [22], [23].

$$\begin{cases} f_{sw} = \dfrac{V_S}{\Delta I_L \cdot L} \dfrac{(1-2D)D}{1-D} & for\ 0.0 < D < 0.5 \\ f_{sw} = \dfrac{V_S}{\Delta I_L \cdot L} \dfrac{(2D-1)(1-D)}{D} & for\ 0.5 < D < 1.0 \end{cases} \quad (6)$$

$$f_{sw} = \frac{V_S}{18 \cdot \Delta I_L \cdot L} D(1-D) \quad for\ 0.0 < D < 1.0 \quad (7)$$

where f_{sw} is the switching frequency, V_S is the DC source, L is the filter inductor, ΔI_L is the switching ripple component which determines the THD of the output ripple current, and D is the duty ratio. The DC voltage and the frequency of the output current, both of them required for the determination of the required filter inductor size, are given by the test requirements and can therefore be considered as constants for the following analysis. Resulting from this the switching frequency is a function of the inductance and the duty ratio. Moreover, V_S is composed of the DC voltage on the CUT, V_C, and the voltage drop on the filter inductor, $2\pi \cdot f \cdot L \cdot I_C$, as

$$V_S = V_C + 2\pi \cdot f \cdot L \cdot I_C \quad (8)$$

Considering (8), Fig. 5 shows the switching frequency against the inductance, when V_S is 1.5 kV, V_C is 1 kV, ΔI_L is 1 A (5% of 20 A), the frequency of the ripple current, f, is 3 kHz, and D is 0.75 for the IB and 0.5 for the CHB where each topology reaches its maximum THD at this duty ratio. The values are arbitrarily chosen as an example, but a DC-link voltage within the range of 1 kV is typical for WTS applications. As can be seen, in order to achieve the same THD and, thus, the same quality of the ripple current applied to the CUT, either a bigger filter inductance needs to be chosen or the switching frequency needs to be increased in case of the IB. Since the target of this paper is to propose a setup which makes use of standard IGBT devices, a very high switching frequency is not applicable. Due to this fact, for the operation of the IB at the same THD a bigger filter would be required which, in turn, would lead to a higher voltage drop across the filter and, thus, to a lower effective DC voltage for the CUT. Since one requirement for the setup is the ability to provide high a DC voltage to the CUT, a bigger DC source would be required in order to compensate for the voltage drop across the filter. Consequently, this would lead to the fact that

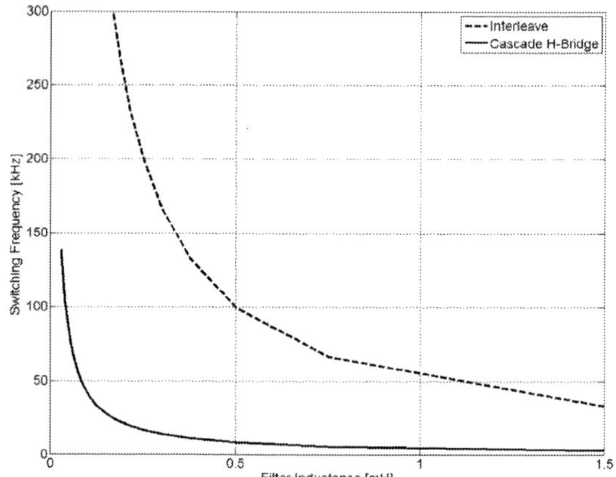

Fig. 5. Switching frequency against filter inductance to satisfy the required THD.

the blocking voltage of the IGBT devices would limit the capability of the setup. For this reason, the choice of the CHB topology is reasonable, since the output voltage can be shared equally by the three converter cells, thus, requiring smaller DC sources. On the one hand, the choice of the CHB leads to the burden of a higher system complexity but, on the other hand, it makes it possible to realize the system with standard devices.

B. Control Strategy and Analysis

The proposed controller of the accelerated lifetime test setup is shown in Fig. 6. As can be seen, it consists of a voltage control loop, which is employed to control the DC voltage of the CUT, and a current control loop for the application of the ripple current stress.

The current control loop consists of a PI current controller, which regulates the RMS amplitude of the applied ripple current, a frequency control block, that is needed to achieve the desired ripple current frequency, and the RMS feed-back. The measured capacitor ripple current i_C is fed back to the current controller via the RMS feed-back block that filters the RMS amplitude of the ripple current from the measured signal. Since the measured ripple current, as expressed as (9), can be converted by help of square operation and triangular functions into (10).

$$i_C = I_C \cdot \sin(\omega t + \alpha) \quad (9)$$

$$i_C^2 = I_C^2 \sin^2(\omega t + \alpha) = \frac{1}{2} I_C^2 \left[1 - \cos\left(\frac{\omega t + \alpha}{2}\right)\right] \quad (10)$$

where i_C denotes the measured ripple current, I_C is the amplitude, ω is the angular frequency of the ripple current, and α denotes the phase displacement angle of the current with respect to the output voltage of the CHB. It is evident that the DC quantity of the current can be found by the application of a simple low-pass filter (LPF). Thus, the RMS value can be found as the square root of the LPF output value as

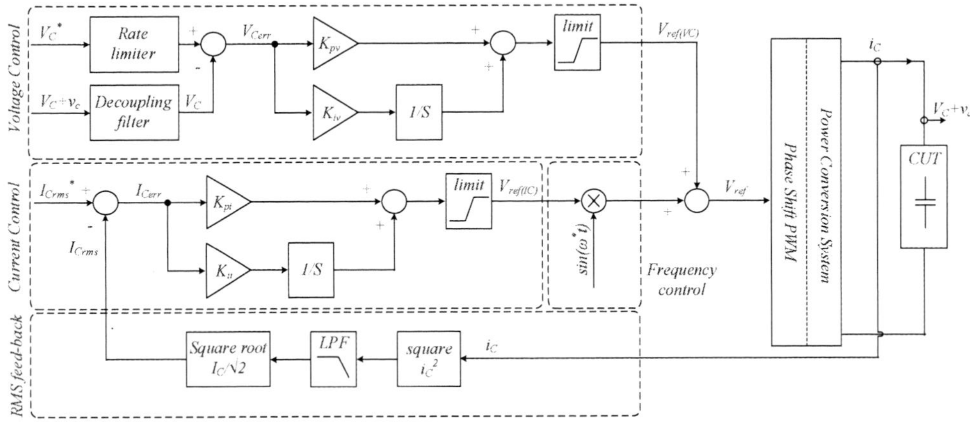

Fig. 6. Control block diagram of accelerated lifetime test setup.

$$\frac{1}{2} I_C^2 \Rightarrow \frac{1}{\sqrt{2}} I_C \tag{11}$$

Due to the fact that the PI controller is only acting on the RMS amplitude of the ripple current, its output value is multiplied with a unity sinusoidal signal that accounts for the reference ripple current frequency. Therefore, the frequency of the ripple current can be controlled by adjusting the frequency of the sinusoidal signal.

For the control of the DC voltage, which is applied to the CUT, a voltage control loop is required. In order to prevent the possibility of surge currents, that might occur because step changes in the output voltage across the CUT, the voltage reference is processed by a rate limiter. The ripple current through the CUT is causing a ripple component v_C that is superimposing the DC voltage V_C. Therefore, a decoupling filter is adopted to filter the ripple voltage components. Thus, the DC quantity V_C is fed into the voltage control, so that the voltage control can be considered decoupled from the current control. In addition, the regulation of the DC voltage can now be done by means of a PI voltage controller.

Due to its simplicity, the proposed control scheme offers certain advantages compared to other methods. Instead of the employed RMS feed-back block a conventional phase-locked loop (PLL), such as the SRF-PLL method presented in [24], could be used in order to obtain the RMS value of the ripple current. However, this method would require an additional feed-back control loop, which would lead to an unnecessary increase in the complexity of the system. Moreover, the control of both, the amplitude and the frequency, of the ripple current could be realized by implementing a proportional resonant (PR) controller. However, a PR controller is tuned to a certain frequency, which would mean for the application at hand that a number of PR controllers would be needed corresponding to the required frequency range of the controlled ripple current [25].

Before designing the controllers, the output load composed of the filter inductor and the capacitor is described as

$$T_S(s) = \frac{1}{L} \cdot \frac{s}{s^2 + 1/LC} \tag{12}$$

From the above transfer function, it becomes evident that the design of filter inductance is dependent on the capacitance of the CUT as well as on the required frequency range of the stress ripple current. In this application, the filter inductance is chosen such that the test setup is operated within the inductive region and to achieve the required damping respect to the CUT.

The simplified linear models of the voltage controller and the current controller are shown in Fig. 7 (a) and (b), respectively. To simplify the model, the rate limiter and the decoupling filter are not considered in the voltage controller model, and the RMS feed-back block is not involved in the current controller model. The open loop transfer function and the closed loop transfer function for the voltage controller are represented as (13) and (14), respectively.

$$G_v^o(s) = G_{cv}(s) \cdot G_{pv}(s) = (K_{pv} + \frac{K_{iv}}{s}) \cdot \frac{1}{sC} \tag{13}$$

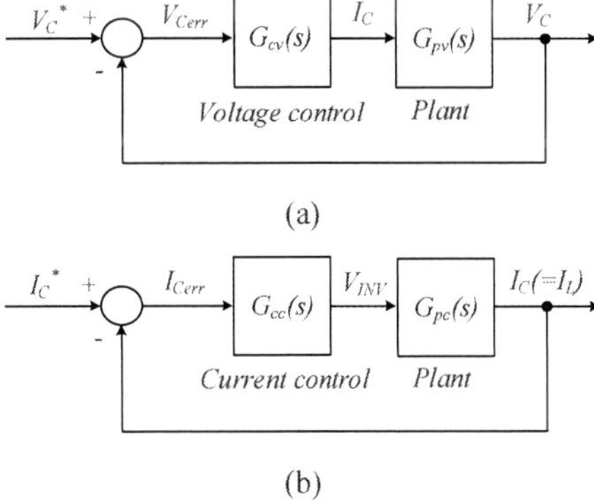

Fig. 7. Simplified linear model of (a) voltage controller and (b) current controller.

$$G_v^c(s) = \frac{G_v^o(s)}{1+G_v^o(s)} = \frac{\frac{K_{pv}}{C}s + \frac{K_{iv}}{C}}{s^2 + \frac{K_{pv}}{C}s + \frac{K_{iv}}{C}} \quad (14)$$

where G_{cv} is the transfer function of voltage controller, G_{pv} is the transfer function of load, K_{pv} and K_{iv} are the proportional and the integral gain of the voltage controller, respectively, and C is the capacitance of the CUT. The load is considered as only capacitive load in G_{pv} since the DC quantity is controlled by means of the decoupling filter and the inductor works as short circuit at DC. The gains can be found by comparing the coefficients of (14) with the typical 2nd order transfer function in (15).

$$T(s) = \frac{2\varsigma\omega_n s + \omega_n^2}{s^2 + 2\varsigma\omega_n s + \omega_n^2} \quad (15)$$

where ω_n is the cut-off frequency and ς is the damping ratio. Finally, the gains for the voltage control are found as

$$K_{pv} = 2 \cdot \varsigma \cdot \omega_n \cdot C$$
$$K_{iv} = \omega_n^2 \cdot C \quad (16)$$

The open loop transfer function and the closed loop transfer function of the current controller are expressed in (17) and (18), respectively.

$$G_c^o(s) = G_{cc}(s) \cdot G_{pc}(s) = \left(K_{pc} + \frac{K_{ic}}{s}\right) \cdot \frac{1}{sL^*} \quad (17)$$

$$G_c^c(s) = \frac{G_c^o(s)}{1+G_c^o(s)} = \frac{\frac{K_{pc}}{L^*}s + \frac{K_{ic}}{L^*}}{s^2 + \frac{K_{pc}}{L^*}s + \frac{K_{ic}}{L^*}} \quad (18)$$

where G_{cc} is the transfer function of current control, G_{pc} is the transfer function of the load, K_{pc} is the proportional gain, K_{ic} is the integral gain of the current controller, and L^* is the equivalent inductance at the ripple current frequency. For the current controller design, the load is considered as inductive load because the inductance is designed such that the setup is operated in the inductive region with respect to capacitance of the CUT as mentioned before. The equivalent inductive load at ripple current frequency can be calculated as

$$L^* = L - \frac{1}{C \cdot \omega^{*2}} \quad (19)$$

The gains can be obtained as (20) with the same procedure as for the voltage control design.

$$K_{pc} = 2 \cdot \varsigma \cdot \omega_n \cdot L^*$$
$$K_{ic} = \omega_n^2 \cdot L^* \quad (20)$$

To define the control ability of the DC voltage on the CUT, the control strategy is mathematically analyzed in Table I, where C and I_{Cmax} are the required values and L and V_S are the designed values for satisfying the requirements mentioned in this section.

TABLE I. ANALYSIS OF CONTROL STRATEGY

Impedance	$Z = \omega L - \frac{1}{\omega C}$
Maximum duty ratio of the maximum ripple current	$D_{\max,I_C} = \frac{I_{C\max} \cdot Z}{V_S}$
Maximum DC voltage on CUT	$V_C \le V_S(1 - D_{\max,I_C})$

IV. EXPERIMENTAL RESULTS

In order to verify the analytical findings experimentally, a first prototype of the proposed accelerated lifetime test setup has been built. Thereby, the CHB topology is evaluated with respect to the proposed control strategy. Moreover, the ability to control the ripple current frequency as well as the controller design are verified. Table II describes the parameters of the test setup, the specifications of the CUT, the test requirements, and the designed setup parameters and controllers, respectively.

With the determined specification of the CUT and the test requirements, the DC source and the inductance are chosen as explained in the previous section. The cut-off frequency of the decoupling filter is designed at 25 Hz in order to obtain only the DC quantity from the measured voltage. Subsequently, the bandwidth of the voltage controller is set to 2 Hz which is lower than 10 times of the cut-off frequency of the decoupling filter for the stable feed-back control. The cut-off frequency of the RMS feed-back is configured at 50 Hz for eliminating the signal which has 0.5ω in (10). For the same reasons as for the voltage controller, the bandwidth of the current controller is chosen as 5 Hz.

TABLE II. SYSTEM PARAMETERS

Parameter		Value
Power conversion system	Device Switching frequency (f_{sw})	20kHz
	DC source (V_S)	300V (100V*3EA)
Capacitor under test	Capacitance	500uF
	Rated voltage	250V
Test requirement	Ripple current (I_{Crms})	I_{Crms}: 10~15A$_{rms}$ ω: 2~3kHz
Design	Inductance (L)	0.3mH
	Maximum DC voltage ($V_{C,max}$)	224.1V @2kHz 183.4V @3kHz
	Voltage controller	ω_{nv}: 2Hz / ζ: 0.707 decoupling filter: 25Hz rate limiter: 1000V/s
	Current controller	ω_{ni}: 5Hz / ζ: 0.707 RMS feed-back: 50Hz

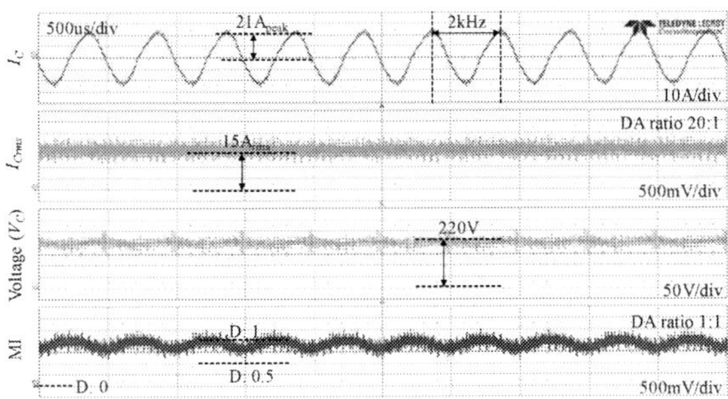

Fig. 8 Verification of DC voltage control boundary.

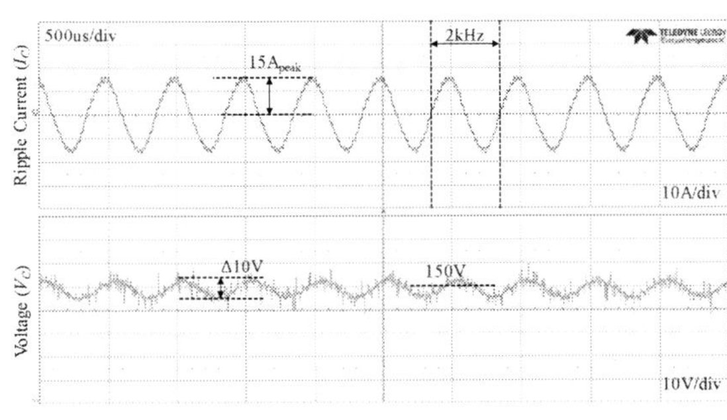

(a)

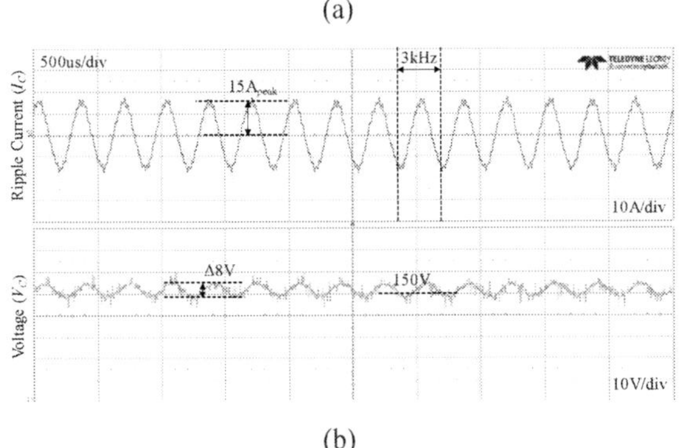

(b)

Fig. 9. Verification of frequency control and steady state (a) at 2kHz, (b) at 3kHz.

A. Verification of Control Boundary

As shown in Fig. 8, the analyzed control boundary is verified by applying a 2 kHz ripple current in order to ensure whether the designed setup can cover the stress requirements. The duty ratio reaches near its maximum value when the controlled stress is 224 V/15 A_{rms} as described in Table II. The theoretical boundary and the experimental boundary are identical.

B. Verification of Frequncy Control and Steady State

The performance of the frequency feed-forward control is presented in Fig. 9. The ripple currents of 2 kHz and 3 kHz are stably regulated at 10 A_{rms} as shown in Fig. 9 (a) and (b), respectively. For both cases the V_C is set to 15 V, which lies well below the control ability limits (see Table II).

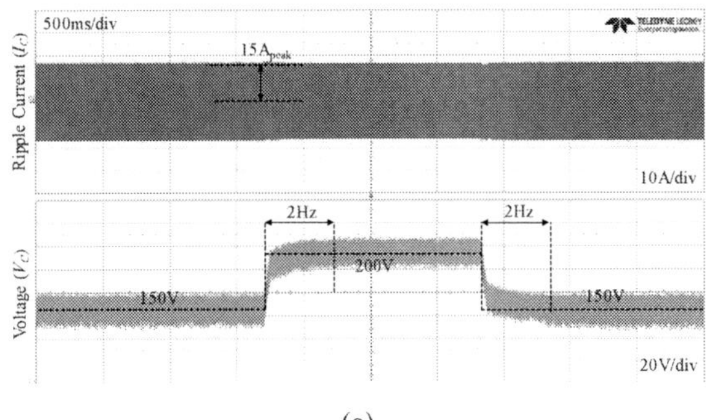

(a)

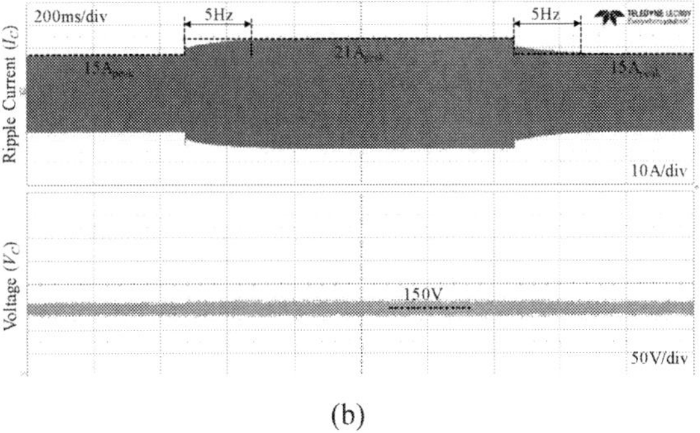

(b)

Fig. 10. Verification of controller design. (a) voltage controller, (b) current controller.

C. Verification of Controller Design

Fig. 10 (a) indicates the dynamic response of the voltage controller. The voltage is varied from 150 V to 200 V and vice versa while the current is regulated at 10 A_{rms}. In Fig. 10 (b), the current is changed from 10 A_{rms} to 15 A_{rms} and vice versa while the voltage is controlled at 150 V. The results are suitable for the designed dynamic response requirements (bandwidth for voltage control is 2 Hz and for current control is 5 Hz).

V. Conclusion

This paper presented the design of a test setup, developed to apply accelerated stress tests to capacitors based on mission profile data of WTS. Therefore, aging characteristics of electrolytic capacitors and effective factors which affect its lifetime were discussed. It was concluded that the ripple current stress of DC-link capacitors in WTS is the most important factor with respect to lifetime. Based on this, the objective of this paper was to design a test setup based on standard switching devices with which the correlation of both, ripple current frequency and amplitude, in combination with a high DC-voltage offset with lifetime can be achieved. Hence, it was demonstrated that the CHB combined with the proposed control scheme is a feasible solution. Even though it offers an increased complexity compared to other possible topologies, a high DC-voltage offset as well as low THD in the output current can be achieved with low rated standard switching devices. Finally, experimental results verified the applicability of the proposed test setup for application-oriented capacitor ripple current stress tests.

REFERENCES

[1] I. A. Gowaid, A. S. Abdel-Khalik, A. M. Massoud, and S. Ahmed, "Ride-Through Capability of Grid-Connected Brushless Cascade DFIG Wind Turbine in Faulty Grid Conditions-A Comparative Study," IEEE Transactions on, Sustainable Energy , Vol. 4, No. 4, pp, 1002-1015, 2013.

[2] Q. Jiang, M. Xue, and G. Geng, "Energy Management of Microgrid in Grid-Connected and Stand-Alone Modes," IEEE Transactions on, Power Systems, Vol. 28, No. 3, pp. 3380-3389, 2013.

[3] L. H. Kocewiak, J. Hjerrild, and C. L. Bak, "Wind Turbine Converter Control Interaction with Complex Wind Farm Systems," IET, Renewable Power Generation, Vol. 7, No. 4, pp. 380-389, 2013.

[4] A.S. Makinen and H. Tuusa, "Analysis, Comparison and Performance Evaluation of Wind Turbine Grid Synchronizing Methods," IEEE Conference, EUROCON, pp. 1108-1115, 2013.

[5] S. Debnath and M. Saeedifard, "A New Hybrid Modular Multilevel Converter for Grid Connected of Large Wind Turbines," IEEE Transaction on, Sustainable Energy, Vol. 4, No, 4, pp. 1051-1064, 2013.

[6] M. Tsai, C. Tseng, and Y. Hung, " A Novel MPPT Control Design for Wind-Turbine Generation System using Neural Network Compensator," IEEE Conference, IECON, pp. 3521-3526, 2012.

[7] Adam M. Regheb and Magdi Ragheb, "Wind Turbine Gearbox Technologies, Fundamental and Advanced Topics in Wind Power," ISBN: 978-953-307-508-2, In Tech, 2011, Available from: http://www.intechopen.com/books/fundamental-and-advanced-topics-in-wind-power/wind-turbine-gearbox-technologies.

[8] P. Venet, F. Perisse, M.H. EL-Husseini, and G. Rojat, "Realization of a Smart Electrolytic Capacitor Circuit," IEEE Industrial Application Magazine, pp. 16-20, Jan. and Feb., 2002.

[9] A. M. Imam, T. G. Habetler, R. G. Harley, and D. M. Divan, "LMS based Condition Monitoring of Electrolytic Capacitor," IEEE Conference, IECON, pp. 848-853, 2005.

[10] M. L. Gasperi, "Life Prediction Modeling of Bus Capacitors in AC Variable-Frequency Drives," Pulp and Paper Industry Technical Conference, pp. 141-146, 2005.

[11] E. C. Aeloiya, J. H. Kim, R. Ruminot, and P. N. Enjeti, "A Real Time Method to Estimate Electrolytic Capacitor Condition in PWM Adjustable Speed Drives and Uninterruptible Power Supplies," IEEE Conference, PESC, pp. 2867-2872, 2005.

[12] Arne Albertsen, "Electrolytic Capacitor Lifetime Estimation," Available from: http://jianghai-europe.com/wp-content/uploads/1-Jianghai-Europe-E-Cap-Lifetime-Estimation-long-AAL-2012-10-30.pdf.

[13] H. Wang, M. Liserre, and F. Blaabjerg, "Toward Reliable Power Electronics: Challenges, Design Tools, and Opportunities," IEEE Industrial Electronics Magazine, Vol. 7, No. 2, pp. 17-26, 2013.

[14] H. Wang, M. Liserre, F. Blaabjerg, P. de Place Rimmen, J. B. Jacobsen, T. Kvisgaard, and J. Landkildehus, "Transitioning to Physics of Failure as a Reliability Driver in Power Electronics," IEEE Journal of Emerging and Selected Topics in Power Electronics, Vol. 2, No. 1, pp. 97-114, 2014.

[15] H. Wang and F. Blaabjerg, "Reliability of Capacitors for DC-Link Applications in Power Electronic Converters-An Overview," IEEE Transactions on Industry Applications, Vol. 50, No.5, pp. 3569-3578, 2014.

[16] Z. Qin, H. Wang, F. Blaabjerg, and L.C. Poh, "Investigation into the Control Methods to Reduce the DC-Link Capacitor Ripple Current in a Back-to-Back Converter," IEEE Conference, ECCE, pp. 203-210, 2014.

[17] J.W. Kolar and S.D. Round, "Analytical Calculation of the RMS Current Stress on the DC-Link Capacitor of Voltage-PWM Converter Systems," IEE Proceedings – Electric Power Applications, Vol. 153, No. 4, pp. 535-543, 2006.

[18] Sam G. Parler and P.E. Jr., "Deriving Life Multipliers for Electrolytic Capacitors," IEEE Power Electronics Society Newsletter, Vol. 16, No. 1, pp. 11-12, 2004.

[19] Rubycon corporation, "Technical Notes for Electrolytic Capacitor," Available from: http://www.rubycon.co.jp/de/products/alumi/pdf/Life.pdf

[20] P. Omer, J. Kumar, and B. S. Surjan, "Comparison of Multicarrier PWM Techniques for Cascaded H-bridge Inverter," IEEE Students' Conference, Electrical, Electronics and Computer Science (SCEECS), pp. 1-6, 2014.

[21] J. Elmes, R. Kersten, I. Batarseh, M. Pepper, and K. Mansfield, "Modular Bidirectional DC-DC Converter for Hybrid/Electric Vehicles with Variable-frequency Interleaved Soft-switching," Vehicle Power and Propulsion Conference (VPPC), pp.448-454, 2009.

[22] R. Buerger, A. Peres, R. Hausmann, R. A. Reiter, and A. L. Stankiewicz, "Ripple Analyze and Design Considerations for an Interleaved Boost Converter (IBC) for a PV Source," International Conference, ICREPQ, 2014.

[23] Michael Salcone and Joe Bond, "Selecting Film Bus Link Capacitors for High Performance Inverter Applications," IEEE International Conference, IEMDC, 1692-1699, 2009.

[24] Y. J. Ko, K. W. Park, K. B. Lee, and F. Blaabjerg, "A new PLL system using full order observer and PLL system modeling in a single phase grid-connected inverter," IEEE International Conference, ICPE&ECCE, pp. 803-808, 2011.

[25] B. Li, M. Zhang, L. Huang, L. Hang, and Leon M. Tolbert, "A Robust Multi-Resonant PR Regulator for Three-Phase Grid-Connected VSI using Direct Pole Placement Design Strategy," Twenty-Eighth Annual IEEE Conference, Applied Power Electronics Conference and Exposition (APEC), pp. 960-966, 2013.

Voltage Droop Compensation Based on Resonant Circuit for Generalized High Voltage Solid-State Marx Modulator

Hiren Canacsinh[1,2], Luís M. Redondo[2]
[1]INESC-ID, Lisboa
[2]Instituto Superior de Engenharia de Lisboa, ISEL, GIAAPP
Lisboa, Portugal

J. Fernando Silva[1,3], Beatriz Borges[2,3]
[1]INESC-ID, Lisboa
[2]Intituto de Telecomunicações, IT, Lisboa
[3]Instituto Superior Técnico, Universidade de Lisboa
Lisboa, Portugal

***Abstract*— The purpose of this paper is to present and discuss the voltage droop compensation associated with long pulses generated by solid–state bipolar or unipolar high-voltage Marx modulators. In particular a novel design scheme for voltage droop compensation based on resonant circuit in solid-state bipolar (positive and/or negative pulses) high-voltage Marx generator, using off-the-shelf components, design and control, is described. The compensation consists in adding one auxiliary resonant stage to the existing Marx stages, without changing the modularity and topology of the circuit. The auxiliary compensation voltage is added to the output voltage to compensate the pulse voltage droop. Experimental results are presented for five stages Marx circuit, 10% voltage droop, using 1 kV pulse amplitude, 100 μs pulse width, 9.5 ms relaxation time and with 50 Hz pulse repetition rate.**

I. Introduction

The Marx modulator concept is a widely used technique to generate either unipolar or bipolar transient high-voltage pulses from relatively low d-c voltage sources. Recent technological upgrade in the mature Marx modulator concept lead to the intensive use of semiconductor switches, replacing the passive elements and contributing to the compactness of the Marx modulators, capability to generate square pulse shapes, possibility to control the number of pulses, pulse-width, frequency, and the ability to deal with different load conditions [1-4]. In effect, growing variety of environmental, biological, medical, food and industrial applications using repetitive unipolar and/or bipolar high-voltage pulses, for enhancing the characteristics of a final product or method, impose various conditions to the Marx modulators.

One of the most complex conditions to manage is the pulse voltage droop when long pulses are used (i.e. higher then dozens of microseconds), where the output voltage plateau will droop significantly as the Marx capacitors are discharged, unless large size high valued capacitances are used. Instead, several authors have come up with different solutions in unipolar Marx type circuits [5,6], using auxiliary independent cells that are charged during the charging mode of the main capacitors and turned on sequentially to generate corrections to the droop in the main Marx cells. This technique produces discontinuities in the flat top voltage waveform as well as a very complex control system. Other authors [7,8], created an auxiliary circuit called "bouncer", which employs an additional capacitor and inductor to provide an oscillatory, in which part of the resonant voltage is added to the main pulse voltage to compensate the discharge voltage droop. Under this scheme, besides the necessity of an auxiliary power supply to charge the bouncer capacitor this auxiliary circuit has generally a significant design component and cost implication. However, the principle of resonant type voltage droop compensation has not yet been studied in bipolar type generators where sequences of positive and negative pulses are generated.

This work was supported by national funds through Fundação para a Ciência e a Tecnologia (FCT) with reference UID/CEC/50021/2013.

This paper presents the first results of a novel design scheme of resonant type voltage droop compensation for solid-state bipolar Marx, S^{2}BM, modulator shown in Fig. 1 and described in detail elsewhere [3, 9].

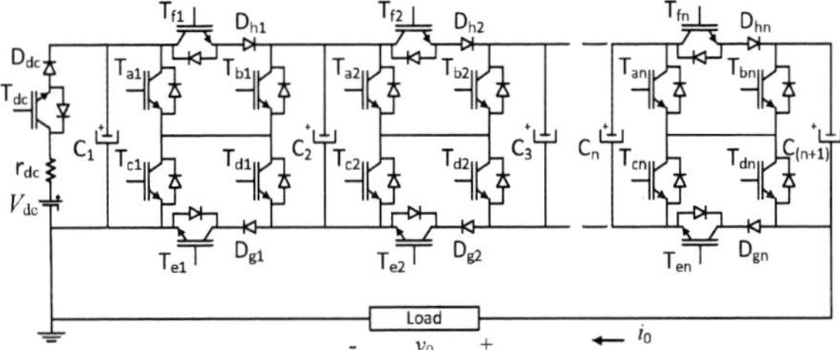

Fig. 1. S^{2}BM modulator with n stages for bipolar or unipolar pulse generation.

To maintain the well known modular characteristic of this topology, one independent auxiliary modulator stage was added, which includes an inductor, as shown in Fig. 2 (particular case of circuit of Fig. 1, without semiconductors D_{hi} and T_{fi}), to perform as a resonant circuit whose voltage is added to the positive or negative pulses to compensate the

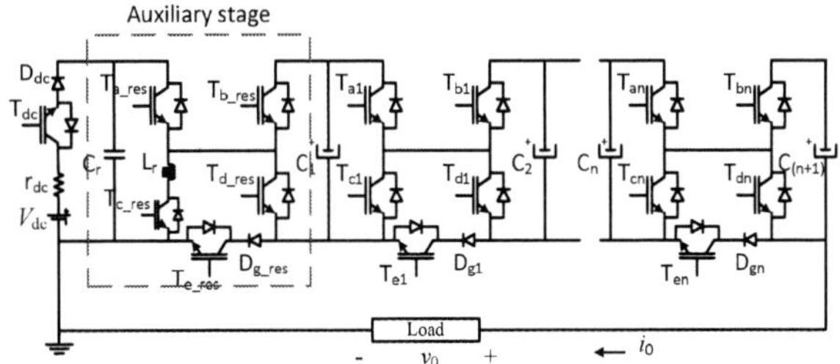

Fig. 2. Tested HV topology of the n stages S^{2}BM modulator, with auxiliary resonant based voltage droop compensation stage for bipolar or unipolar pulse generation, particular case of circuit of Fig. 1.

discharge voltage droop.

The concept of resonant voltage droop compensation consists in taking advantage of the almost linear part of the sine waveform of the capacitor C_r voltage, where (1) holds.

$$sin(\omega t) \approx \omega t \quad \Rightarrow \quad -\pi/6 < \omega t < \pi/6 \qquad (1)$$

In order to use only part of the almost linear area of the sine waveform, it was considered that pulses starts at zero resonant voltage, as shown in Fig. 3.

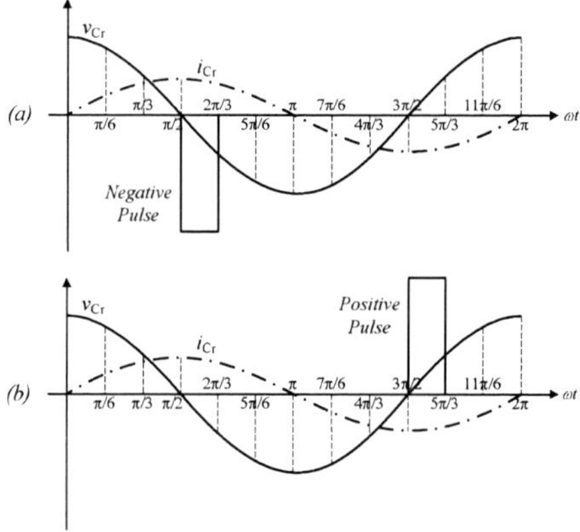

Fig. 3. Resonant circuit voltage and current waveforms and (a) negative and (b) positive pulse starting at zero resonant voltage under almost linear part of the sine waveform.

II. Circuit Operation

Fig. 2 shows the proposed circuit topology to deliver repetitive unipolar or bipolar high-voltage output pulses with compensation, comprising *n* stages and one voltage droop compensation stage, with IGBTs as on-off switches. The circuit of Fig. 2 presents several operating modes. The first one consists in charging the capacitors C_i and C_r through the path consisting by switches T_{dc}, D_{dc}, T_{a_res}, T_{ai}, D_{b_res}, D_{bi}, D_{g_res}, D_{gi}, T_{e_res} and T_{ei} on and all others switches off, as shown in Fig. 4. The next operating mode depends on the required output characteristic imposed by the load. Thus, before applying negative or positive pulse with compensation into a load, it is necessary to trigger on switches, T_{a_res} and T_{c_res} to connect the capacitor C_r in series with inductor L_r to start the resonance through its single cycle sine wave, as shown in Fig. 3. To apply negative pulse with compensation accordingly with Fig. 3a), switches T_{bi} and T_{ci} are turned on at $\omega t=\pi/2$ (zero resonant voltage). After negative pulse, turning off switches T_{a_res} and T_{c_res} must obey the condition in (2), assuring the continuity of the current and recovery of the energy in inductor L_r back to capacitor C_r, as can be seen in Fig. 3a) and Fig. 4.

$$\pi < \omega\tau < 2\pi \qquad (2)$$

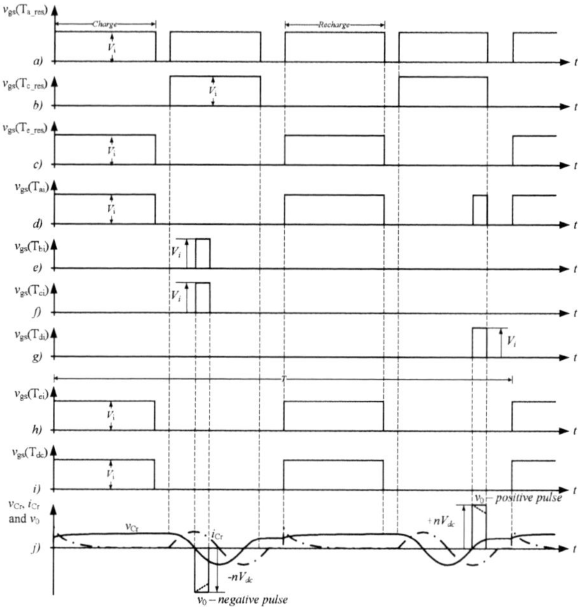

Fig. 4. Theoretical wave forms for the operation of theS[2]BM of Fig. 2, considering a resistive load. Drive signal of semiconductors: a) T_{a_res}; b) T_{c_res}; c) T_{ai}; d) T_{bi}; e) T_{ci}; f) T_{di}; g) T_{ei}; h) T_{dc}; i) v_{Cr}, i_{Cr} and output v_0 voltage.

The positive pulse with compensation can be obtained (assuming that capacitor C_r is previously connected in series with inductor L_r, see Fig. 3b)) by switching on T_{ai} and T_{di} at $\omega t=3\pi/2$ after starting the resonance. After the positive pulse, switches T_{a_res} and T_{c_res} can be turned off safely because the freewheeling diodes of these switches will assure the continuity of the current and recovery the energy in inductor L_r back to capacitor C_r, as shown in Fig. 3 b) and Fig. 4.

III. Resonant parameters

In the resonant voltage droop compensation stage the capacitor C_r has same capacitance of Marx stages capacitors C_i. The inductance L_r was calculated taking into account the RLC series equivalent of the resonant circuit. Thus, solving the generic solution of RLC series circuit, (3) is obtained considering that the resistive part (the *on* resistance of the semiconductors) is negligible and the initial conditions of the capacitor C_r voltage and inductor L_r current are V_{dc} and zero respectively.

$$i_{Lr} = V_{dc} sin(\omega_0 t)/(L_r \omega_0) \qquad (3)$$

The inductance L_r was obtained, (4), considering that the negative and positive pulses lengths are restrained into intervals given by (5) and (6), for negative and positive pulse respectively, (Fig. 3).

$$L_r > 36 t_{on}^2/(\pi^2 C_r) \qquad (4)$$

978-1-4673-9551-9/16 $31.00 © 2016 IEEE

$$\pi/2 < \omega t < 2\pi/3 \tag{5}$$

$$3\pi/2 < \omega t < 5\pi/3 \tag{6}$$

IV. Results

A laboratory prototype with five stages of the circuit presented in Fig. 2 was assembled using 1200 V IGBTs (SKW15N120) and diodes (IDB09E120), 4.5 μF for main capacitors and capacitor Cr, 8 mH for inductor L_r, operating with 50 Hz pulse repetition rate, giving 1 kV bipolar pulses, with 100 μs pulse width, and 9.5 ms relaxation time (i.e. time between negative and positive pulses) into resistive load.

Fig. 5 a) and b) shows the experimental results for the output voltage v_0, into the load, without compensation and 10% voltage droop at the end of the negative and positive pulses.

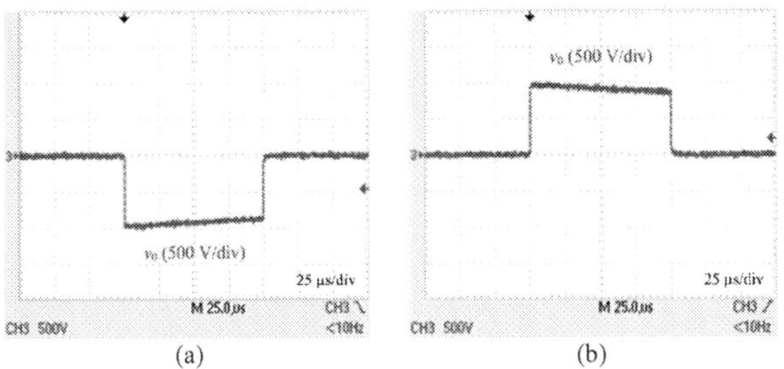

(a) (b)

Fig. 5. Experimental results for the output voltage of the Fig. 2 circuit, into the load, without compensation and 10% voltage droop at the end of the negative (a) and positive (b) pulse. The scales are 25 μs/div (horizontal) and 500 V/div (vertical).

Fig. 6 a) and b) shows the experimental results for the capacitor C_r voltage, current in inductor L_r and the output v_0 voltage with compensation.

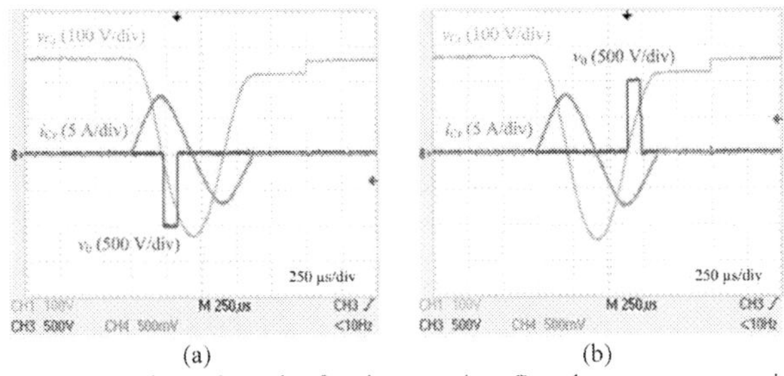

(a) (b)

Fig. 6. Experimental results for the capacitor C_r voltage, v_{Cr}, current in inductor L_r i_{Lr}, and the output v_0 voltage with compensation of the circuit of Fig. 2, for negative (a) and positive (b) pulse. The scales are 250 μs/div (horizontal) and 100 V/div (vertical).

In Fig. 7 a) and b) it is shown the experimental results of the output negative and positive pulse voltage with voltage droop compensation with zoomed horizontal scale.

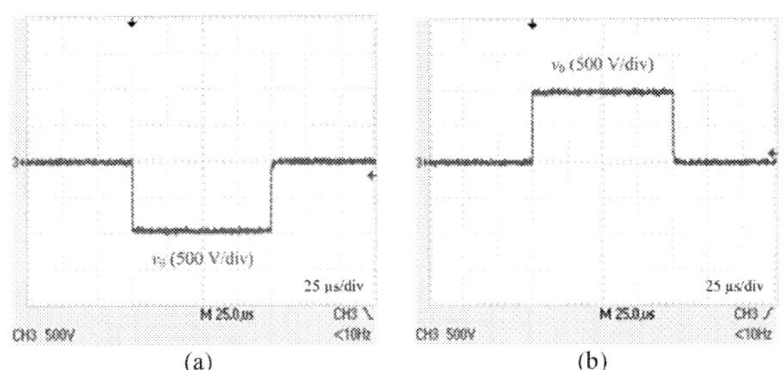

(a) (b)

Fig. 7. Experimental results of the obtained output (a) negative and (b) positive pulse voltage with resonant compensation with zoomed horizontal scale, 25 μs/div (horizontal) and 500V/div (vertical).

V. Conclusions

A new design scheme for voltage droop compensation based on resonant circuit in generalized solid-state bipolar Marx modulator of Fig. 2 was proposed for high-voltage repetitive pulsed power applications.

Keeping the topology of the Marx circuit, an auxiliary stage with inductor was added to the existing Marx stages, which compensates both the positive and negative output pulses. One characteristic of this scheme of compensation is the compromise between pulse repetition rate and the resonant frequency. Other aspect of the circuit of Fig. 2 is that the semiconductor T_{dc} has to hold-off two times the power supply voltage amplitude, V_{dc}. However, the presented topology avoids the need for auxiliary power supply to charge the capacitor of the voltage droop compensation stage, for both the positive and negative voltage pulses. The experimental results obtained from a laboratory prototype, for bipolar output voltage v_0 into a resistive load, shows that the resonant based voltage droop compensation stage is able to compensate a 10% voltage droop both on the negative and positive pulses.

References

[1] D. Tastekin, F. Blank, A. Lunk, and J. Roth-Stielow, "Power supply with bipolar pulsed output voltage and high repetition rate based on a solid state Marx topology", in Pulsed Power Conference, June 2011, pp. 1377 – 1381.

[2] T. Sakamoto, A. Nami, M. Akiyama and H. Akiyama, "A Repetitive Solid State Marx-Type Pulsed Power Generator Using Multistage Switch-Capacitor Cells", IEEE Transactions on Plasma Science, Vol. 40, No. 10, pp. 2316 – 2321, October 2012

[3] H. Canacsinh, L. M. Redondo and J. Fernando Silva, "New Solid-State Marx Topology for Bipolar Repetitive High-Voltage Pulses", in 39th IEEE Power Electronics Specialists Conference, Greece, June 2008.

[4] R. L. Cassel, "A Solid State High Voltage Pulse Modulator which is Compact and without oil or pulse transformer", in Power Modulator Conference, San Francisco, CA, USA, pp. 72-74, 2004.

[5] T. Tang, C. Burkhart and N. Nguyen, "A vernier regulator for ILC Marx droop compensation", in IEEE Pulsed Power Conference, June 2009, pp. 1402-1405.

[6] C. Burkhart, T. Beukers, M. Kemp, R. Larsen, K. Macken; M. Nguyen, J. Olsen and T. Tang, "ILC Marx Modulator Development Program Status", in IEEE Pulsed Power Conference, July 2009, pp. 807 – 810.

[7] H. Pfeffer, L. Bartelson, K. Bourkland, C. Jensen, Q. Kerns, P. Prieto, G. Saewert, and D. Wolff, "A Long Pulse Modulator For Reduced Size And Cost", in Twenty-First International Power Modulator Symposium, 1994, pp. 48 – 51.

[8] R. L. Cassel, "Pulsed Voltage Droop Compensation for Solid State Marx Modulator", in Proceedings of the IEEE International Power Modulators and High Voltage Conference, May 2008, pp. 117 – 119.

[9] H. Canacsinh, L. M. Redondo and J. Fernando Silva, "Marx type solid-state bipolar modulator topologies: performance comparison", in IEEE Transaction on Plasma Science, Vol. 40, Issue: 10, October 2012, pp. 2603-2610.

Four H-Bridge Based Shunt Active Power Filter for Three-Phase Four Wire System

Edgard L. L. Fabricio[1], Cursino B. Jacobina[2], Gregory A. A. Carlos[2,3], and Maurício B. R. Correa[2]
[1]Academic Unity of Control and Industrial Processes
Federal Institute of Paraíba (IFPB) – 58.015-430, João Pessoa - PB - Brazil
[2]Electrical Engineering Department (DEE)
Federal University of Campina Grande (UFCG) – 58429-900, Campina Grande - PB - Brazil
[3]Post-Graduate Program in Electrical Engineering - PPgEE - COPELE
e-mails: edgard.fabricio@ifpb.edu.br, [jacobina, mbrcorrea]@dee.ufcg.edu.br, gregory.ska@gmail.com

***Abstract*—This paper proposes a shunt active power filter (SAPF) composed of four H-bridge converters. Each of them is connected by means of isolation transformers. Compared to conventional topologies, the proposed converter permits reducing: dc-link voltage rating, grid current harmonic distortion and/or switching frequency (switching losses). A suitable control strategy to adjust the filter currents as well as dc-link voltage are presented. Two PWM strategies are designed to command the converter according to the reference voltage provided by the controllers. Simulation and experimental results are presented in order to validate theoretical approaches.**

I. Introduction

Nowadays power quality is a concern between suppliers and consumers of energy due to widespread usage of non-linear loads and their effects in utility networks [1]. In this context, the use of shunt active power filter (SAPF) has been stimulated. Several topologies have been investigated in the literature. Among them, four-leg (C-4L) [2]–[4] and three H-bridge (C-3HB) [4]–[6] are the most common topologies at four wire systems. They are denominated here as conventional ones and are shown in Fig. 1. The C-4L (Fig. 1(a)) is suitable for low and medium power applications while C-3HB (Fig. 1(b)) is suitable for a wider range of application, i.e. until high power. Other topology of four-wire SAPF is presented in [7], [8], this one does not use transformers nevertheless it is not capable to compensate zero sequence. In four wire systems, it is worth noting that excessive current in four-wire conductor may cause damage in distribution transformers (overheating), losses, malfunction of sensible loads, and degrading the voltage quality [9], [10].

Connections of H-bridge standard converter have been used in a wide range of applications. These connections can be done either in Y or Δ types, presenting advantages such as simpler manufacturing, easy maintenance and extension to more levels [11]. Inspired in this statement, this paper presents a SAPF for three-phase four wire systems, as shown in Fig. 2. This SAPF is denominated here as P-4HB and is composed of four H-bridge converters, four isolation transformers, and one dc-link capacitor. This structure is similar to C-3HB, however the use of an additional H-bridge (converter 4) in the proposed topology permits increasing the number of steps in the converter output voltages, reducing the harmonic content of grid currents compared to conventional topologies. These voltages can be changed by means of the turns ratio of transformer connected to converter 4. Other similar structure has been presented in [12], which is composed by four converters and four dc-link capacitors. However, this topology requires an additional converter to control the dc-link voltage of fourth converter.

The P-4HB as well as C-3HB is suitable to be applied in high power applications and both can operate with an independent phase control, i.e. each converter operates and is controlled independently.

The following topics concerning the proposed system that will be addressed in this paper are: i) dynamic system modelling; ii) PWM strategy, using Single-carrier PWM (SCPWM) or Level-shifted PWM (LSPWM); iii) minimum dc-link voltage analysis; iv) overall system control; v) harmonic distortion analysis; and vi) semiconductor losses evaluation.

II. System Model

The proposed system in Fig. 2 is constituted of four H-bridge converters associated with four transformers. For this topology, transformer leakage inductance it is taken into consideration, as well as, external interfacing shunt inductor, represented by L_f (inductance l_f and load dependent loss of transformer denoted by r_f). The number of turns on the secondary coil of transformer connected to converter 4 is given by N. For the other three transformers, the number of turns on secondary is the same of the primary. Two simplified circuits are derived from Fig. 2. They are illustrated in Fig. 3. One of them considers the transformers, [see Fig. 3(a)] while in the another one, circuit parameters, as well as, electrical variables are referred to the grid side [see Fig. 3(b)].

The system model equations are obtained from Fig. 3(b) as follow

$$v_{gk} = -z_f i_{ak} + v_{sk} \tag{1}$$

$$v_{sk} = v_{a_k0} - v_{b_k0} + v_{4n} \tag{2}$$

$$v_{4n} = (-z_f i_{a4} + v_{a_40} - v_{b_40})/N \tag{3}$$

with $z_f = r_f + pl_f$ ($p = \frac{d}{dt}$), along the text $k = 1, 2, 3$, v_{sk} are the converter voltages.

Assuming that $\sum_{k=1}^{3} v_{gk} = 0$ and

$$\sum_{k=1}^{3} i_{ak} = N i_{a4} \tag{4}$$

978-1-4673-9551-9/16 $31.00 © 2016 IEEE

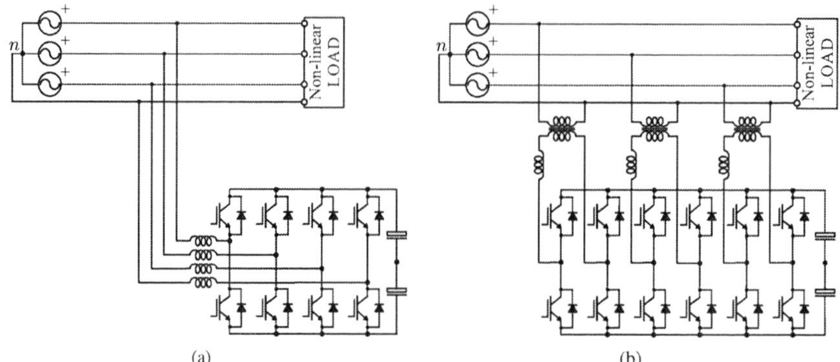

Fig. 1. Conventional topologies. (a) Four-leg topology (C-4L). (b) Three H-bridge topology (C-3HB).

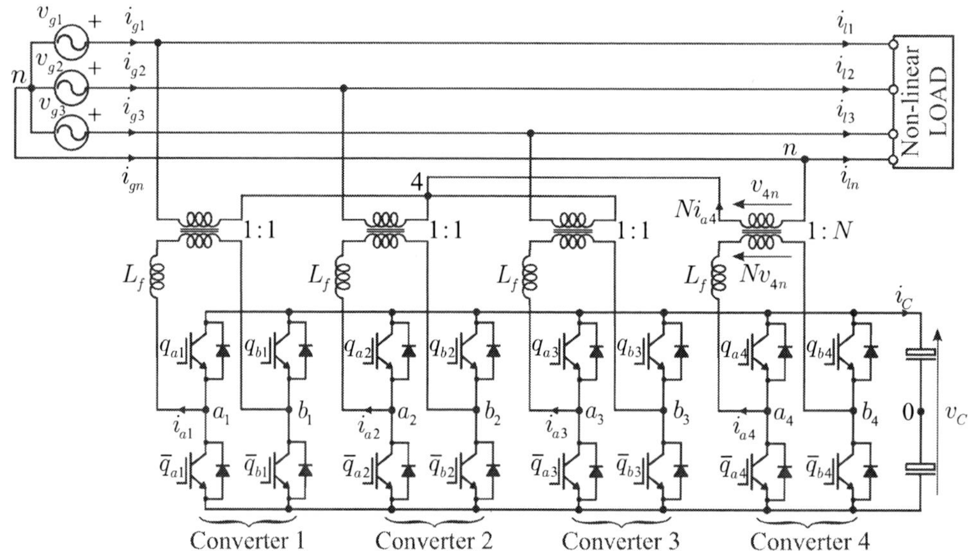

Fig. 2. Proposed four H-bridge (P-4HB) based shunt active power filter.

from (1)-(4) is obtained that

$$\begin{aligned} v_{4n} &= \frac{v_{b_1 0} - v_{a_1 0} + v_{b_2 0} - v_{a_2 0} + v_{b_3 0} - v_{a_3 0}}{N^2+3} + \\ &+ \frac{(v_{a_4 0} - v_{b_4 0})N}{N^2+3} \end{aligned} \quad (5)$$

Due to v_{4n} voltage, the proposed topology is able to obtain more than three level steps at output converter voltages (v_{sk}). Three level steps are obtained by C-3HB topology, that is equivalent to make $v_{4n} = 0$. It is worth noting that the converter 4 is disabled when $v_{4n} = 0$. This can be helpful in case of failure in the converter 4.

III. PWM Strategy

Converter power switches are commanded by using an adequate pulse-width modulation strategy (PWM). This strategy defines state (s_{a_k}, s_{b_k}, s_{a_4}, and s_{b_4}) of the power switches. In this way, the pole voltages ($v_{a_k 0}$, $v_{b_k 0}$, $v_{a_4 0}$, and $v_{b_4 0}$) are obtained, which are given by,

$$v_{a_k 0} = (2s_{a_k} - 1)\frac{v_C}{2} \quad (6)$$

$$v_{b_k 0} = (2s_{b_k} - 1)\frac{v_C}{2} \quad (7)$$

$$v_{a_4 0} = (2s_{a_4} - 1)\frac{v_C}{2} \quad (8)$$

$$v_{b_4 0} = (2s_{b_4} - 1)\frac{v_C}{2} \quad (9)$$

where v_C is dc-link voltage.

The converter controller outputs are v_{s1}^*, v_{s2}^*, and v_{s3}^*. It is worth to note that these voltages are synthesized by the eight converter pole voltages, which means that, besides the three given references phase voltages, five auxiliary voltages are necessary to compute the reference pole voltage. In this paper they will be named: v_{xk}^* (v_{x1}^*, v_{x2}^* and v_{x3}^*), v_{x4}^*, and v_{4n}^*. These voltages are given by

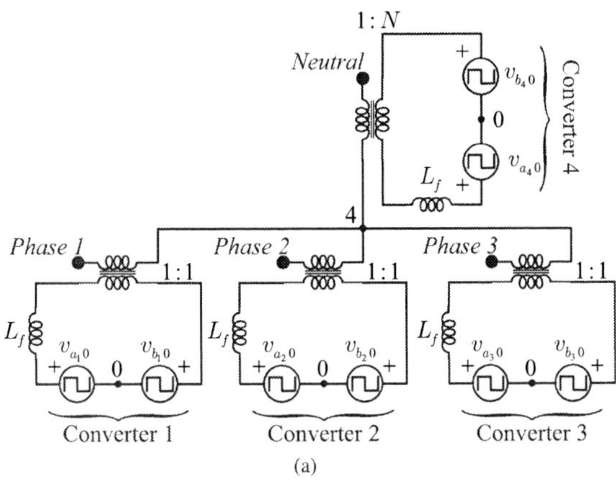

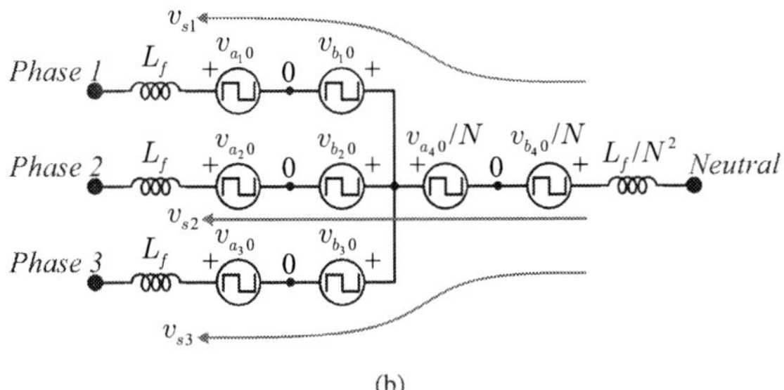

Fig. 3. Equivalent circuit for proposed four H-bridge (P-4HB). (a) Equivalent circuit with transformers. (b) Equivalent primary circuit including reflected impedances.

$$v_{xk}^{*} = v_{a_k0}^{*} \tag{10}$$
$$v_{x4}^{*} = v_{a_40}^{*} \tag{11}$$
$$v_{4n}^{*} = \frac{v_{b_10}^{*} - v_{a_10}^{*} + v_{b_20}^{*} - v_{a_20}^{*} + v_{b_30}^{*} - v_{a_30}^{*}}{N^2+3} + \frac{(v_{a_40}^{*} - v_{b_40}^{*})N}{N^2+3} \tag{12}$$

The pole voltage references are obtained from system model and auxiliary voltages as below:

$$v_{a_k0}^{*} = v_{xk}^{*} \tag{13}$$
$$v_{b_k0}^{*} = -v_{sk}^{*} + v_{4n}^{*} + v_{xk}^{*} \tag{14}$$
$$v_{a_40}^{*} = v_{x4}^{*} \tag{15}$$
$$v_{b_40}^{*} = -\frac{v_{s1}^{*} + v_{s2}^{*} + v_{s3}^{*}}{N} - Nv_{4n}^{*} + v_{x4}^{*} \tag{16}$$

The auxiliary voltages must be chosen respecting their maximum and minimum limits. Those limits must be calculated by taking into account the maximum ($v_C^*/2$) and minimum ($-v_C^*/2$) values of reference pole voltages. Then $v_{4n\max}^*$, $v_{4n\min}^*$, $v_{xk\max}^*$, $v_{xk\min}^*$, $v_{x4\max}^*$ and $v_{x4\min}^*$ are given by:

$$v_{4n\max}^{*} = \min(v_C^* - \max(\vartheta), \frac{v_C^*}{N} - \frac{v_{s1}^{*} + v_{s2}^{*} + v_{s3}^{*}}{N^2}) \tag{17}$$
$$v_{4n\min}^{*} = \max(-v_C^* - \min(\vartheta), -\frac{v_C^*}{N} - \frac{v_{s1}^{*} + v_{s2}^{*} + v_{s3}^{*}}{N^2}) \tag{18}$$

$$v_{xk\max}^{*} = v_C^* - \max(\vartheta_k) \tag{19}$$
$$v_{xk\min}^{*} = -v_C^* - \min(\vartheta_k) \tag{20}$$
$$v_{x4\max}^{*} = v_C^* - \max(\vartheta_4) \tag{21}$$
$$v_{x4\min}^{*} = -v_C^* - \min(\vartheta_4) \tag{22}$$

where

$$\vartheta = \{-v_{s1}^*, -v_{s2}^*, -v_{s3}^*\} \tag{23}$$
$$\vartheta_k = \{0, -v_{sk}^* + v_{4n}^*\} \tag{24}$$
$$\vartheta_4 = \{0, -\frac{v_{s1}^{*} + v_{s2}^{*} + v_{s3}^{*}}{N} - Nv_{4n}^*\} \tag{25}$$

Variation of v_{4n}^*, v_{xk}^* and v_{x4}^* can be normalized introducing distribution parameters μ_{4n}^*, μ_{xk}^* and μ_{x4}^*, respectively. Then, v_{4n}^*, v_{xk}^* and v_{x4}^* can be calculated by using

$$v_{4n}^{*} = \mu_{4n}^* v_{4n\max}^* + (1 - \mu_{4n}^*) v_{4n\min}^* \tag{26}$$
$$v_{xk}^{*} = \mu_{xk}^* v_{xk\max}^* + (1 - \mu_{xk}^*) v_{xk\min}^* \tag{27}$$
$$v_{x4}^{*} = \mu_{x4}^* v_{x4\max}^* + (1 - \mu_{x4}^*) v_{x4\min}^* \tag{28}$$

The parameters μ allow to change pulse's position in the pole voltages generated by the converter, modifying also their high-frequency distortion components.

The state of switches are defined by comparing the reference signal with high frequency carrier. In this paper, two techniques are used: SCPWM and LSPWM. In SCPWM, one high frequency carrier is compared to a reference signal. In LSPWM, two high frequency carriers are compared to reference signals, as shown in Fig. 4. In this case, some combinations of switch states can be used generating the same output voltage, one of them allows one leg per converter to operate with low frequency, the other combinations of state switches produce a average switching frequency equal per leg (equal to half of carrier frequency).

IV. DC-Link Voltage

An important setting of a power converter is the dc-link voltage value. Such an incorrect choice of this value affect the converter controllability and peak inverse voltage (PIV) at each power switch. In four-wire systems, the minimum required dc-link value can be chosen dynamically, due to the load changes that occur in these systems. For the proposed topology the minimum dc-link voltage value is defined from the circuit shown in Fig. 5. This circuit shows a specific operation state, i.e. a state which provide maximum voltage of converter. Then, the minimum dc-link voltage value is limited by one of the two branch circuits. The first one (most common) is defined by a unbalanced phase (e.g. *Phase k+1*), 4 node and a secondary phase (*Phase k*). Other branch circuit which can define the dc-link is composed of the most unbalanced phase (*Phase k+1*)), 4 node and neutral node (*Neutral*). Take into account only fundamental frequency can be obtained the following expression

$$v_{C\min}^* = \max\{\frac{|\mathbf{V}_{gk+1} + \mathbf{L}_f\mathbf{I}_{fk+1} - \mathbf{V}_{gk} - \mathbf{L}_f\mathbf{I}_{fk}|}{2}, N\frac{|\mathbf{V}_{gk+1} + \mathbf{L}_f\mathbf{I}_{fk+1} - \mathbf{L}_f\mathbf{I}_{fn}/N^2|}{N+1}\} \tag{29}$$

It is worth knowing that the definition of dc-link depends on harmonic content then it is necessary to give a incrementing

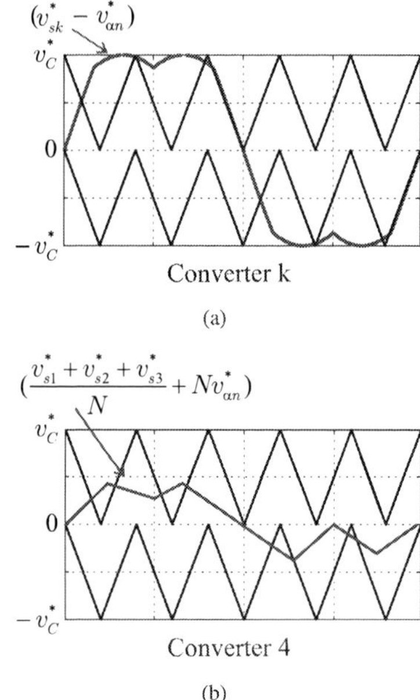

Fig. 4. Modulation waveform and carriers used in LSPWM technique.(a) Comparison used by converter k. (b) Comparison used by converter 4.

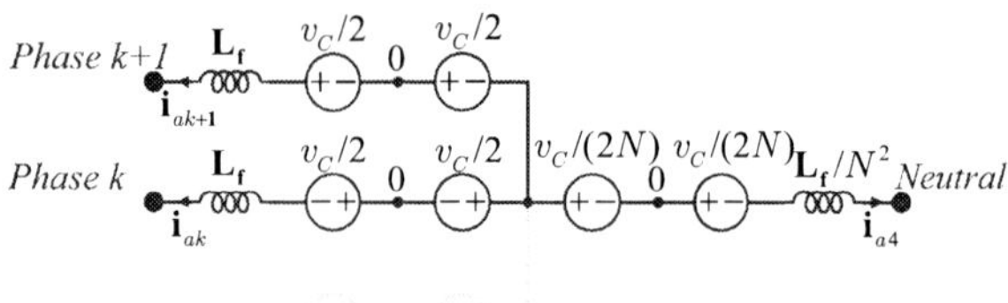

Fig. 5. Ideal equivalent circuit with maximum dc-link voltage applicable .

(e.g. 10%) in calculated value (29). Other important aspect is that, the number of ratios (N) affects the calculation of dc-link voltage value, mainly in unbalanced situations. For some values of N the dc-link voltage have to be increased. Then, to find the most adequate value for a four wire system is a tough task. However, comparisons ca be done, and compared with conventional topologies (C-4L and C-3HB), the P-4HB reduces dc-link voltage in 50 % and 15 %, respectively.

V. Control Strategy

The control block diagram of the proposed topology is shown in Fig. 6. The dc-link capacitor voltage (v_C) is controlled by means of a PI controller (R_C block), whose output is the reference amplitude of active current parcel I_{at}^*, which is synchronized with grid voltages (block S_{in}). The reactive and harmonic load current are detected from RH detec. block by mean of algorithm used in [13], it can be used other techniques, as presented in [14]. The reference filter current is determined as shown in the figure. These currents is controlled by mean of controllers (R_{i123} block) composed of some resonant PI, ensuring null error at different frequencies. R_{i123} defines the reference voltages v_{s1}^*, v_{s2}^*, and v_{s3}^*. From these voltages and, with μ_{x1}^*, μ_{x2}^*, μ_{x3}^*, μ_{x4}^*, and μ_{4n}^*, the PWM strategy defines the state of switches (q_{aj} and q_{bj}).

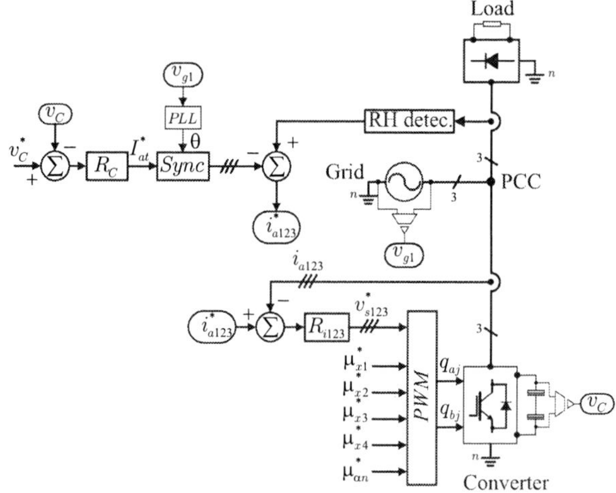

Fig. 6. Control block diagram of proposed system.

VI. Harmonic Distortion

In this paper, THD (Total Harmonic Distortion) is used to evaluate grid current distortions. For this, THD is calculated by

$$THD(p)\% = \frac{100}{Y_1}\sqrt{\sum_{n=2}^{p}(Y_n)^2} \tag{30}$$

where Y_1 is the amplitude of the fundamental component, Y_n is the amplitude of nth component harmonic and p is the number of harmonics taken in consideration.

The goal of this analyze is compare the impact of converters in grid currents. Therefore it is considered all disturb of current (harmonics, reactive, and zero sequence) are compensated by the APF in open-loop simulation. The THD of grid current for conventional and proposed topologies can be observed in Table I. For this analyze is considered the same operation conditions, i.e. same modulation index, and same average switching frequency. From Table I can be noticed that proposed topology presents lower THD value than conventional ones, for different N values and PWM techniques. The best (minimum) THD value is achieved with proposed topology with LSPWM and $N = 1/3$. In this case, P-4HB achieves a reduction of 80.15 % for C-4L and 67.7 % for C-3HB.

TABLE I. Grid current THD analysis

Topology	$THD(\%)$
C-4L, see Fig. 1(a)	2.62
C-3HB, see Fig. 1(b)	1.61
P-4HB with SCPWM and $N = 1$	1.47
P-4HB with LSPWM and $N = 1$	0.65
P-4HB with LSPWM and $N = 1/3$	0.52
P-4HB with LSPWM and $N = 2$	0.91

The converter voltages (v_{sk}) is responsible to generate the filter currents, see (1), and consequently, the grid currents too. Then to justify the THD obtained we can look for v_{sk} voltages.

Aspects as number of steps and distribution of the pulses affect directly the THD of currents. The converter voltage obtained using the SCPWM technique are shown in Fig. 7(a), (b), and (c) for C-4L, C-3HB and P-4HB (N=1) topologies, respectively. Notice that the better pattern of pulses, among them, is obtained by P-4HB (Fig. 4(c)). However, the better pattern voltages of P-4HB is obtained when is used LSPWM. These voltages are shown in Fig. 4(d), (e) and (f) for $N = 1$, $N = 1/3$ and $N = 1/5$, respectively.

VII. Semiconductor Losses

The loss estimation is taken to evaluate comparatively the topology efficiency. For this, it is used a regression model based on experimental tests, as presented in [15], [16]. The switch loss model includes: i) IGBT and diode conduction losses; ii) IGBT turn-ON and turn-OFF losses; and iii) diode turn-OFF energy.

Figure 8 shows conduction (P_{cd}), switching (P_{sw}), and total ($P_t = P_{cd} + P_{sw}$) losses of proposed P-4HB topology (denoted here as P), including different N values, and conventional ones (C-4L and C-3HB). The losses are shown as a percentage of C-4L topology losses (51.8 W) compensating a load of 1 kW. Power switching losses analyze was performed by assuming two scenarios, the first one is imposed same average switching frequency in all conventional and proposed topologies. This case is shown in Fig. 8(a), where we can be noticed the proposed topology presents a small reduction total losses, except in P-N=1/3 due to the variation in index modulation and current at converter 4 that increase its losses. However, in this case P-4HB presents THD advantages, as shown in previously section. The second scenario, shown in Fig. 8(b), imposes the same THD in analyzed topologies. For this is reduced switching frequency in all configurations to equal the THD of C-4L topology. The best losses result was obtained for the proposed topology with N=1, which reaches a reduction of 60 %. These results were obtained for a specific situation of unbalanced load, however for a balanced load another value of N can be more favorable.

VIII. Simulation and Experimental Results

Simulation results are presented in Fig. 9. These are obtained from PSIM $v9.0$ software, they show the system performance when a non-linear is compensated and power load is incremented of 50 % in $t = 1.2$ s. With this, it can be verified dynamic response of the system. Furthermore, from these results can be noticed correction of power factor and harmonics (Fig. 9(a) and (b)), besides mitigation of neutral current (Fig. 9(c)) and controlled dc-link voltage (Fig.9(d)).

Experimental results were obtained from a experimental development platform (shown in Fig. 10(a)) controlled by a digital signal processor, TMS320F28335 using carrier frequency of 10 kHz. These results are presented in Fig. 10, which presents open-loop (Fig. 10(b) and (c)) and closed-loop results (Fig. 10(d), (e), (f) and (g)). The open-loop results show the convert voltage output (v_{s1}) and pole voltages of P-4HB with N=1 using LSPWM. In this case the converter is operating as a inverter providing sinusoidal voltage for a resistive load. High resolution waveforms are shown in Fig. 10(c). Note that in this case a leg is working with low frequence (v_{a_10}).

On the other hand, the system supplies two single-phase non-linear load, and a linear load (RL), all of them are compensated by proposed converter in closed-loop. The proposed converter provides harmonic and reactive power, as well as zero sequence component. This closed-loop results show grid currents (Fig. 10(c)), shunt filter (i_{a1234}) currents (Fig. 10(d)), load (i_{labcn}) currents (Fig. 10(e)), and dc-link voltage (v_C) (Fig. 10(f)). Despite the load current being very distorted, the grid current is controlled adequately.

IX. Conclusions

A four bridge based four-wire shunt compensator has been proposed in this paper. System modelling, PWM strategies, and control have been presented. Simulation and experimental results support the theoretical approaches. Comparison between proposed and conventional configurations has been carried out. Proposed topology permits to reduce dc-link voltage value. As consequence, the blocking-reverse voltage at switches is reduced too and additionally the grid current THD is reduced due to the increased number of steps at the output converter voltages. Such an advantage can be converted to switching losses reduction. However the proposed configuration had a higher number of components, which can be justified by the afore mentioned advantages. This way the proposed topology is a suitable option mainly considering its applicability in a wide power range (since low power until high power systems). Furthermore, the dc-link voltage reduction can save cost of the converter if buying lower rated reverse blocking voltage switches is a priority for the SAPF design.

Acknowledgment

The authors would like to thank IFPB campus João Pessoa, PPgEE/UFCG, CAPES, CNPq for the financial support and research grants.

References

[1] J. Stones and A. Collinson, "Power quality," *Power Engineering Journal*, vol. 15, pp. 58–64, Apr. 2001.

[2] C. Quinn and N. Mohan, "Active filtering of harmonic currents in three-phase, four-wire systems with three-phase and single-phase nonlinear loads," *Applied Power Electronics Conference and Exposition, 1992. APEC '92. Conference Proceedings 1992., Seventh Annual*, pp. 829 – 836, 1992.

[3] J. Pinto, P. Neves, D. Goncalves, and J. Afonso, "Field results on developed three-phase four-wire shunt active power filters," *Industrial Electronics, 2009. IECON '09. 35th Annual Conference of IEEE*, pp. 480 – 485, 2009.

[4] B. Singh, K. Al-Haddad, and A. Chandra, "A review of active filters for power quality improvement," *Industrial Electronics, IEEE Transactions on*, vol. 46, pp. 960 – 971, 1999.

[5] L. Garcia Campanhol, S. Oliveira da Silva, and A. Goedtel, "Application of shunt active power filter for harmonic reduction and reactive power compensation in three-phase four-wire systems," *Power Electronics, IET*, vol. 7, no. 11, pp. 2825–2836, 2014.

[6] V. Khadkikar and A. Chandra, "An independent control approach for three-phase four-wire shunt active filter based on three h-bridge topology under unbalanced load conditions," *Power Electronics Specialists Conference, 2008. PESC 2008. IEEE*, pp. 4643 – 4649, 2008.

[7] C.-C. Hou and Y.-F. Huang, "Design of single-phase shunt active filter for three-phase four-wire distribution systems," *Energy Conversion Congress and Exposition (ECCE), 2010 IEEE*, pp. 1525 – 1528, 2010.

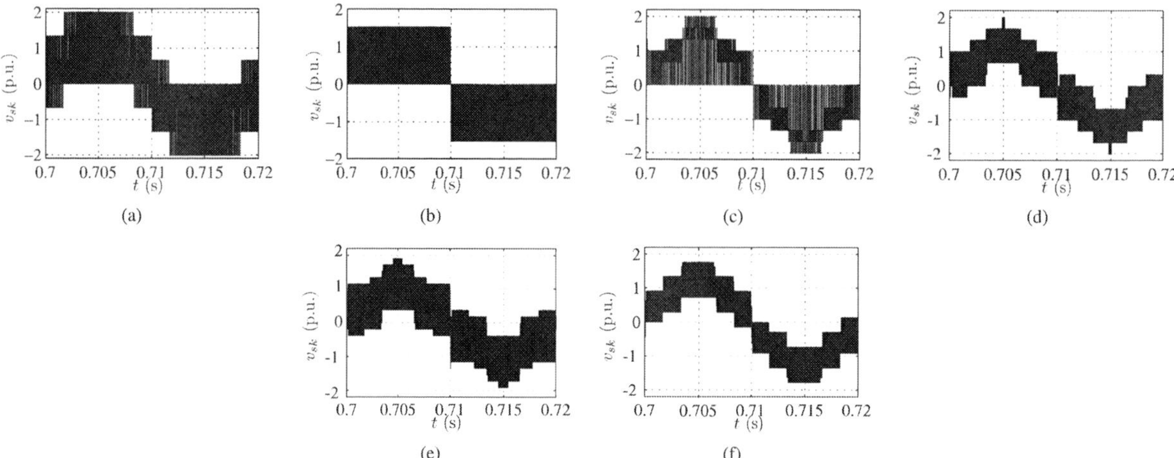

Fig. 7. v_{sk} voltage obtained by each one topology. (a) C-4L topology. (b) C-3HB topology. (c) P-4HB using SCPWM technique with $N = 1$. (d) P-4HB using LSPWM with $N = 1$. (e) P-4HB using LSPWM with $N = 1/3$. (e) P-4HB using LSPWM with $N = 1/5$.

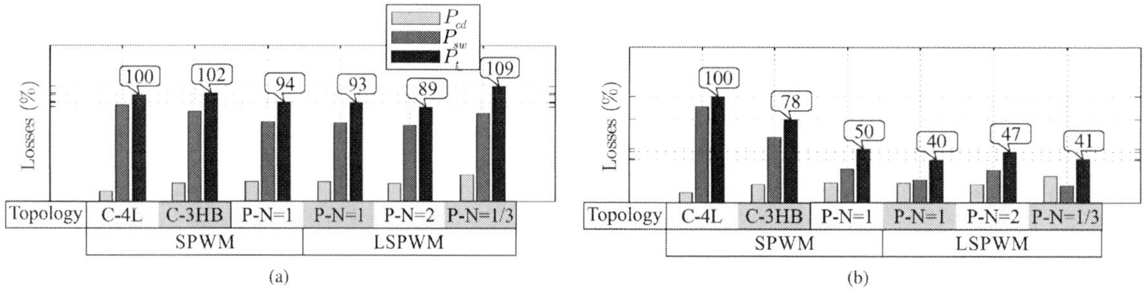

Fig. 8. Losses evaluation. (a) Same switching frequency scenarios. (b) Same THD scenarios.

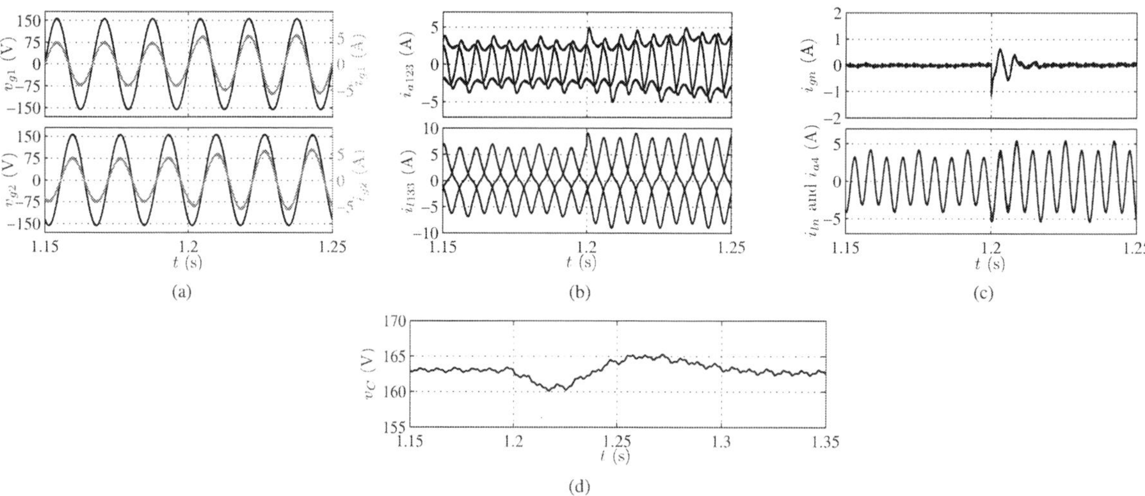

Fig. 9. Simulation results including transient response. (a) Grid voltages and currents. (b) Load and filter currents. (c) Neutral currents of grid side, injected by filter, and load side. (d) Dc-link voltage.

[8] A. Dell'Aquila and A. Lecci, "A shunt active filter control for one-phase and three-phase four-wire systems," *Industrial Electronics, 2002. ISIE 2002. Proceedings of the 2002 IEEE International Symposium on*, vol. 4, pp. 1301 – 1305, 2002.

[9] E. Pashajavid, K. Kanzi, and M. Bina, "Practical issues concerned with zero sequence component and harmonic compensation in four-wire systems," in *Power Electronics and Drive Systems, 2007. PEDS '07. 7th International Conference on*, Nov 2007, pp. 234–238.

(a)

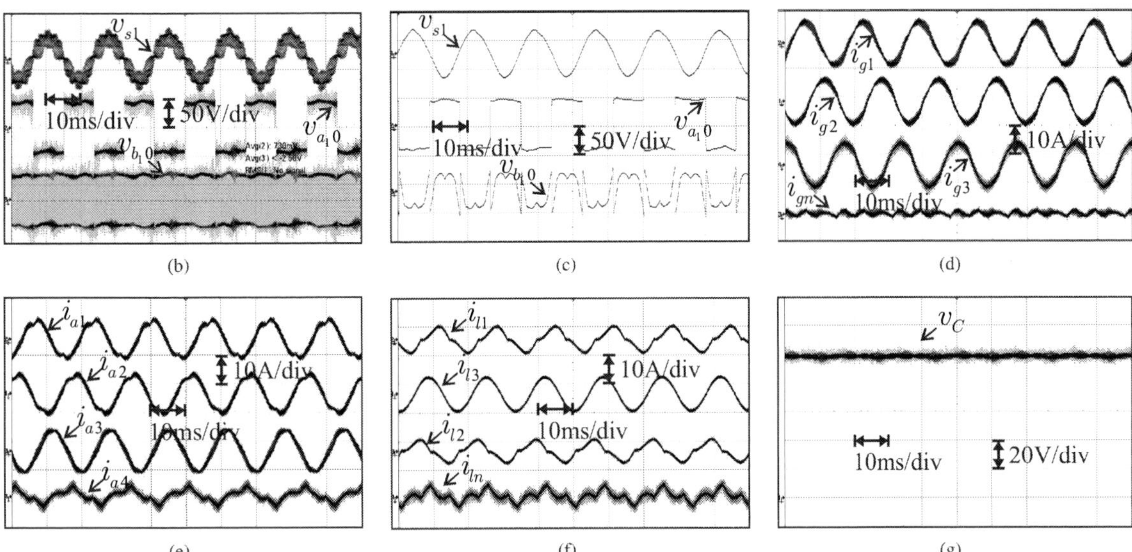

Fig. 10. Experimental results. (a) Development platform. Open-loop setup: (b) Voltage converter (v_{s1}) and pole converter voltage. (c) High resolution waveforms. Closed-loop setup, for $N = 1$: (d) Grid currents. (e) Shunt filter currents (i_{a1234}). (f) Load currents. (g) Dc-link voltage.

[10] J. Desmet, I. Sweertvaegher, G. Vanalme, K. Stockman, and R. Belmans, "Analysis of the neutral conductor current in a three-phase supplied network with nonlinear single-phase loads," *Industry Applications, IEEE Transactions on*, vol. 39, no. 3, pp. 587–593, May 2003.

[11] J. Wen and K. M. Smedley, "Synthesis of multilevel converters based on single and/or three-phase converter building blocks," *IEEE Trans. Power Electron.*, vol. 23, no. 3, pp. 1247–1256, May 2008.

[12] E. Fabricio, C. Jacobina, and V. Nobrega, "Four-wire shunt compensator based on h-bridge y-connected converters," in *Industrial Electronics Society, IECON 2014 - 40th Annual Conference of the IEEE*, Oct 2014, pp. 5184–5190.

[13] V. Soares, P. Verdelho, and G. Marques, "Active power filter control circuit based on the instantaneous active and reactive current id-iq method," in *Power Electronics Specialists Conference, 1997. PESC '97 Record., 28th Annual IEEE*, vol. 2, Jun 1997, pp. 1096–1101 vol.2.

[14] L. Asiminoaei, F. Blaabjerg, and S. Hansen, "Detection is key - harmonic detection methods for active power filter applications," *Industry Applications Magazine, IEEE*, vol. 13, no. 4, pp. 22–33, July 2007.

[15] J. Dias, E. dos Santos, C. Jacobina, and E. da Silva, "Application of single-phase to three-phase converter motor drive systems with igbt dual module losses reduction," *Power Electronics Conference, 2009. COBEP '09. Brazilian*, pp. 1155 – 1162, Sept./Oct. 2009.

[16] M. Cavalcanti, E. da Silva, D. Boroyevich, W. Dong, and C. Jacobina, "A feasible loss model for igbt in soft-switching inverters," *Power Electronics Specialist Conference, 2003. PESC '03. 2003 IEEE 34th Annual*, vol. 4, pp. 1845 – 1850, Jun. 2003.

High-frequency AC Distributed Power Delivery System

Mengqi Wang
Department of Electrical and Computer Engineering
University of Michigan-Dearborn
Dearborn, USA
mengqiw@umich.edu

Qingyun Huang, Wensong Yu and Alex Q. Huang
Department of Electrical and Computer Engineering
North Carolina State University
Raleigh, USA
aqhuang@ncsu.edu

***Abstract*—A novel high-frequency-AC (HFAC) distributed power delivery system is presented in this paper. The target applications of the proposed system are the aerial and ground transportation, and renewable energy systems. Unlike traditional line-frequency-AC (LFAC) or DC distributed power delivery system, the bus frequency of the proposed system can be set at tens or hundreds of kilo-Hertz, so as the power transmission frequency. As a result, 50 or 60 Hz transformers will be eliminated from the system. DC arcing flash will not exist as well with the utilization of HFAC bus. In this paper, single-stage isolated bi-directional power electronics interfaces are developed which enables distributed AC and DC sources or loads to access the main HFAC bus. Soft-switching is also realized for all the semiconductor devices in the power electronics interface. Distributed control architecture is developed so that each source or load can operate independently. Simulation results are provided to validate the distributed control architecture. A scaled-down hardware demonstration, with the bus specification of 54 kHz and 400 VAC, is presented to validate the proposed architecture of the HFAC distributed power delivery system.**

Keywords—high-frequency-AC, power delivery, distributed control, high-frequency AC parallel operation, single-stage power electronics interface, soft-switching

I. INTRODUCTION

For traditional LFAC power delivery system, multiple power conversion stages are required for distributed sources or loads to access the bus. Fig. 1 (a) and (b) illustrates two typical power conversion options for an AC source or load to access the bus. In Fig. 1 (a), the LFAC voltage is stepped-down through an LF transformer. The power then needs to be converted to DC power through an AC/DC converter. A second-stage DC/AC converter will then invert the DC power to LFAC power. For the DC/LFAC inverter, a two-stage flyback plus an unfolding bridge is a common choice. When not using an LF transformer, more stages of conversion are required in order to realize galvanic isolation, as shown in Fig. 1 (b). For the case of a DC source or load accessing the LFAC bus, similar power conversion stages are required, as indicated in Fig. 2 (a) and (b). Power conversion efficiency, power density, and system stability are impacted by the multiple power conversion stages.

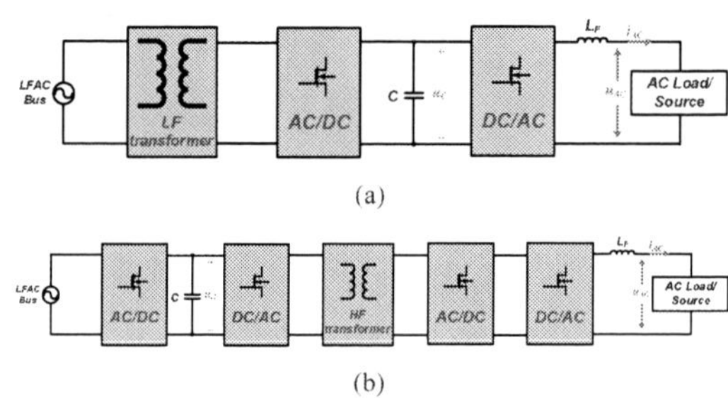

Figure 1: Power conversion stages between LFAC bus and AC source /load (a) with LF transformer; (b) without LF transformer

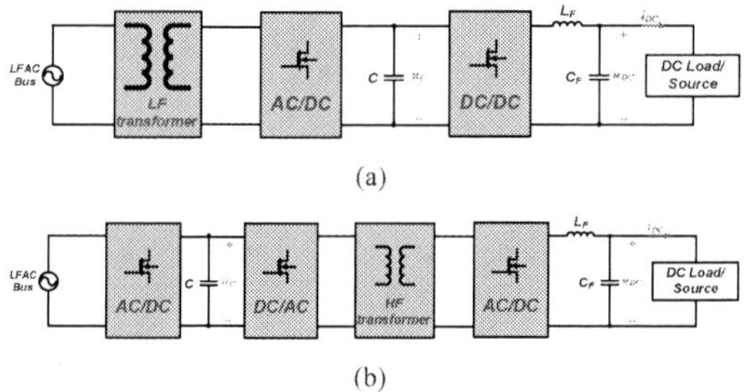

Figure 2: Power conversion stages between LFAC bus and DC source /load (a) with LF transformer; (b) without LF transformer

NASA has first proposed the concept of high-frequency AC (HFAC) power distribution system for astronautics applications, with the objectives of reduce size and weight, minimize both system complexity and the total power conversion steps [1], [2]. The conceptual system diagram is shown in Figure 3. The limitation for this approach is that there can be only one source to power the distributed loads through the HFAC bus. The major challenge for multi-source operating in AC parallel is that voltage amplitude, frequency and phase must be identical to avoid circulating current. [3], [4] have proposed an HFAC power distribution architecture, shown in Figure 4, for two inverter modules operating in parallel. 3 stages are needed for the DC source to access the HFAC bus: DC to DC, DC to AC and a resonant stage. The major issue for this approach is that multiple stages (3 stages for DC source, 4 stages for AC source) are required for the power electronics interface between the source and HFAC bus. Also the resonant stage will increase the system sensitivity. The scalability of the two-inverter system is yet to verify. In addition, bi-directional power flow is not achieved in either of the approaches mentioned above. This paper proposes a novel HFAC distributed power delivery system which enable various and multiple sources or loads to access the main HFAC power bus through the advanced power electronics interfaces. Cyclo and cyclo-oriented topologies are employed for AC and DC interfaces respectively, both of which features in: single-stage conversion, bi-directional power flow, soft-switching and galvanic isolation.

The major challenge for its application in the proposed HFAC power delivery system is that there are two current sources at both ends of the cycloconverter: the transformer leakage inductor and output filter inductor, which makes it a "current breaker". This will endure high voltage spikes due to the cut-off of the leakage inductor current. A unipolar-SPWM-oriented modulation technique proposed by the author in [5] is employed to achieve zero current switching (ZCS), which greatly mitigates the voltage spikes on switches. Most common devices used in present applications are Silicon-controlled rectifier (SCR) and Insulated-gate bipolar transistor (IGBT), which are only suitable for medium frequency, i.e., 5-10 kHz, because of high switching losses and conduction losses [6]. Si MOSFETs is suitable for high frequency applications. But for high voltage rating devices, the reverse recovery loss is still large thus the switching frequency is quite limited [7]. We proposed to use SiC MOSFETs because the on-resistance of the SiC device is low and the switching performance is superior. Especially the body diode reverse-recovery speed is ultra-fast [8]-[11]. All these features enable the SiC MOSFETs to work at a much higher frequency for high-voltage high-power applications while achieving high power-density and high efficiency as well.

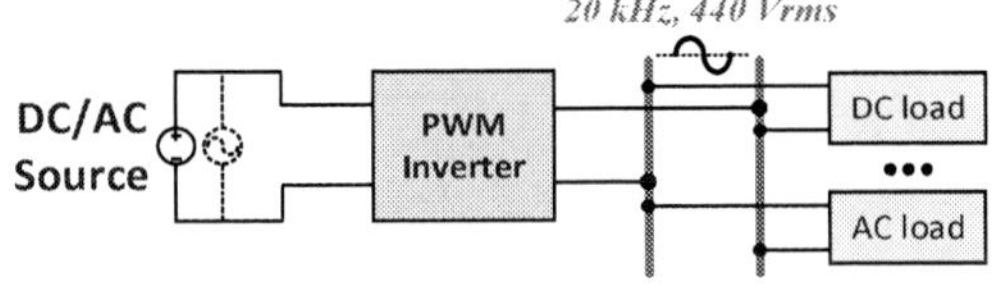

Figure 3: Conceptual system diagram for the HFAC distribution system proposed by NASA.

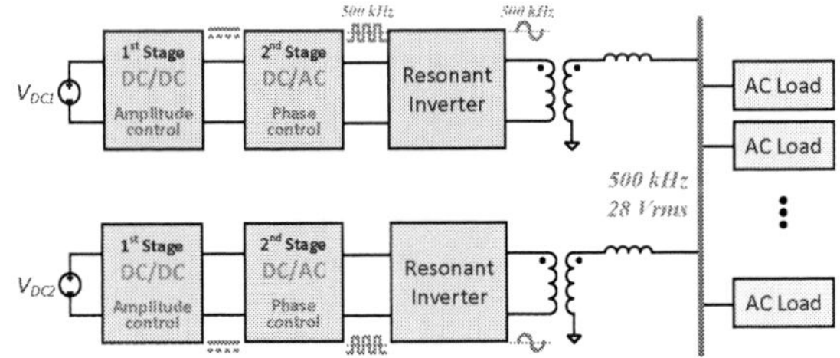

Figure 4: Conceptual system diagram for two-source HFAC distribution system in [3], [4].

II. Proposed HFAC Distributed Power Delivery System

In this paper, we propose a novel HFAC power delivery system with the bus frequency ranging from tens of kHz to hundreds of kHz. The conceptual system configuration of the proposed HFAC distributed power delivery system is shown in Fig. 5. Various distributed energy or power sources and controllable loads are connected to the HFAC bus through the innovative power electronics interfaces with plug-and-play functionality. High-efficiency and high-power-density LFAC/HFAC converters are proposed to serve as the power electronics interface for traditional AC loads or sources, and DC/HFAC converters for the DC loads or sources. All power electronics interfaces are able to operate in parallel and controlled by an independent local controller. Thus the architecture is scalable to form a high power system quickly. To ensure high power density and high power efficiency, the innovative converters have single-stage structure, galvanic isolation, and bi-directional power flow capability and are soft-switched. The overall driving force behind is to make integration of distributed power and energy resources significantly simplified, and to make the power adaptors for various loads and appliances ultra-compact. For example, a commercial 2 kW solar inverter for LFAC power delivery system weights 50 lb and the volume is 28 Liter. While in the proposed HFAC power delivery architecture, when the bus frequency is above 50 kHz, both the weight and volume of a new 2 kW solar inverter are expected to be reduced by 95%.

Compared to traditional LFAC power delivery system, the proposed HFAC architecture has the following advantages:

(1) The size and weight of transformers can be greatly reduced. The volume and weight reduction of a 50 kHz transformer is around 99% comparing with that of a 60 Hz transformer [12].
(2) Passive components, such as filter inductors and capacitors, have smaller value and size; thus, the interface converters/inverters can be made very compact.
(3) Acoustic noise is greatly reduced, in that the link frequency is above 20 kHz.
(4) Soft-switching can be achieved to reduce the voltage stresses on the switches, minimize the switching losses, and improve the converter performance and efficiency.
(5) It further provides the possibility of wireless connection to the HFAC bus by the distributed energy sources when utilizing contactor-less transformers.

Compared to DC power delivery system, the proposed HFAC architecture has the following advantages:

(1) Simpler form of power electronics circuits.
(2) Fewer component counts.
(3) Less complex control scheme.
(4) Higher reliability because of simpler structure, fewer component counts.
(5) Elimination of large electrolytic capacitors, which helps extend the system life span.
(6) Plug-and-play functionality can be easily realized when utilizing transformer connector-less contact.

The proposed HFAC power delivery system has a wide range of power/voltage/frequency selection depends on different applications, i.e., ground transportation, aerial transportation, naval and astronautics. In the case study presented in this paper, the specifications are set as follows: the bus frequency is pushed to 54 kHz as well as the power transmission frequency; the bus voltage waveform is quasi-square (trapezoidal) and the peak value is 400 V. All the power electronics interfaces are realized by single-stage, bi-directional and soft-switched DC/DC or DC/AC converters. A shunt bus compensator is equipped to the major power source in the system to help build up the UFHAC bus and also to stabilize the HFAC bus voltage. It is composed of a full bridge converter with a high frequency capacitor. The bus compensator switches at the bus frequency, which is 54 kHz in this paper.

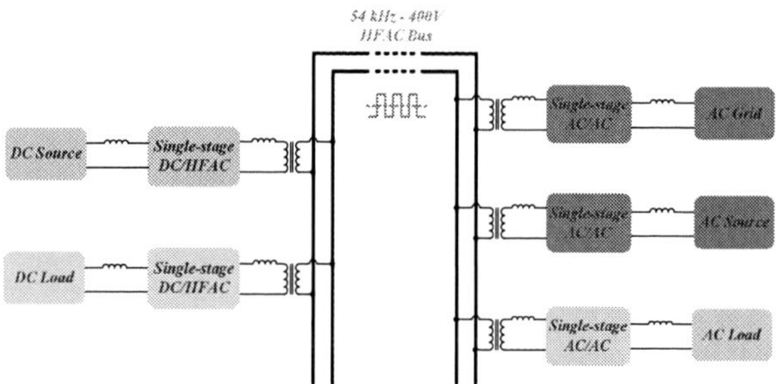

Figure 5: Conceptual system configuration of the proposed HFAC power delivery system.

III. Advanced Power Electronics Interfaces for the Proposed System

The topology of the power electronics interface for AC load or source is shown in Fig. 6 (a). We proposed to use SiC MOSFETs for all the switches because of the superior switching characteristics. The DC power electronics interface is proposed in this paper as shown in Fig. 6 (b). It is a bi-direction single-stage DC/AC converter derived from the AC/AC cycloconverter. PWM control is utilized for the DC power electronics interface. For the utility grid interface, a bus compensator is connected in parallel with the cycloconverter, as shown in Fig. 6 (c). The most important functionalities for the utility grid interface converter are to start up and stabilize the HFAC bus, and to serve as a master for the synchronization as well. The shunt bus compensator is included in the utility grid interface in order to stabilize the HFAC bus square-wave voltage.

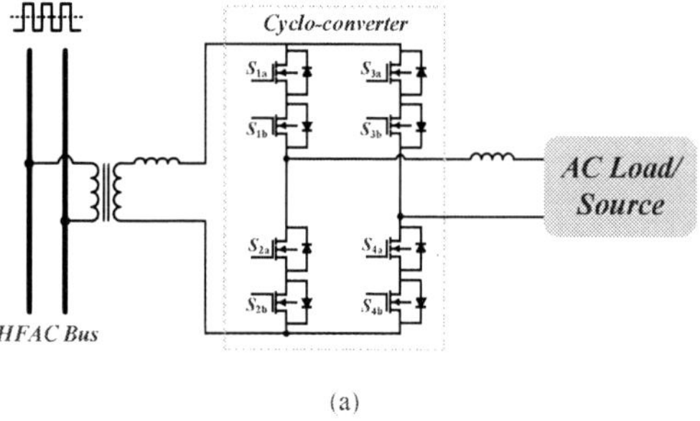

(a)

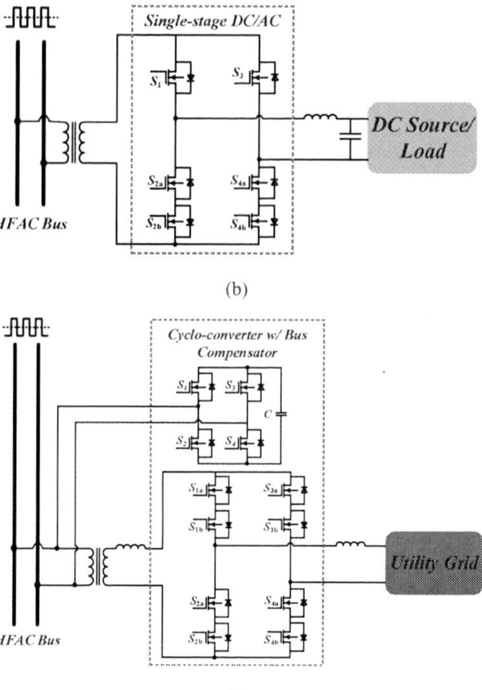

(c)

Figure 6: Power electronics interfaces in the proposed system. (a) for AC load/source; (b) for DC load/source; and (c) for utility grid.

IV. Distributed Control Architecture

For the proposed HFAC power delivery system in this paper, we have developed a distributed control architecture as shown in Fig. 7. All the power electronics interfaces can link the HFAC bus with any types of loads or sources through a local controller. For simplification purposes, we have included a synchronization bus in the system instead of a zero-crossing or slope detection module. Fig. 8 (a) to (d) demonstrates the controller structure for each type of source or load. As can be seen from the Fig. 8, each load/source has its own local controller, which enables the independent start-up, operation and shut down. Furthermore, the system is capable of plug-n-play as well with the distributed control architecture. Dual-loop-compensator has been developed for the AC utility grid-interface with the bus compensator. The outer voltage loop (G_{cv}) regulates the capacitor voltage in the bus compensator. The output of the voltage controller serves as the current reference of the inner grid current control loop (G_{ci}). A feed-forward block is added before the inner control loop, which is the grid current controller, to improve the dynamic of current tracking. Another feed-forward block is added to the gating signal generation block, to improve the dynamic of duty-cycle control. The controller for the DC source interface, AC or DC load interface are all single voltage/current loop depends on the application.

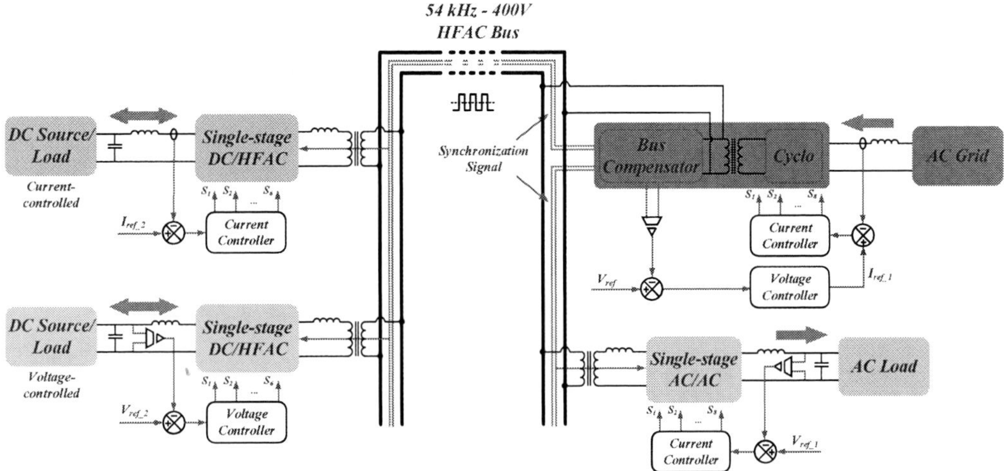

Figure 7: Proposed distributed system control architecture

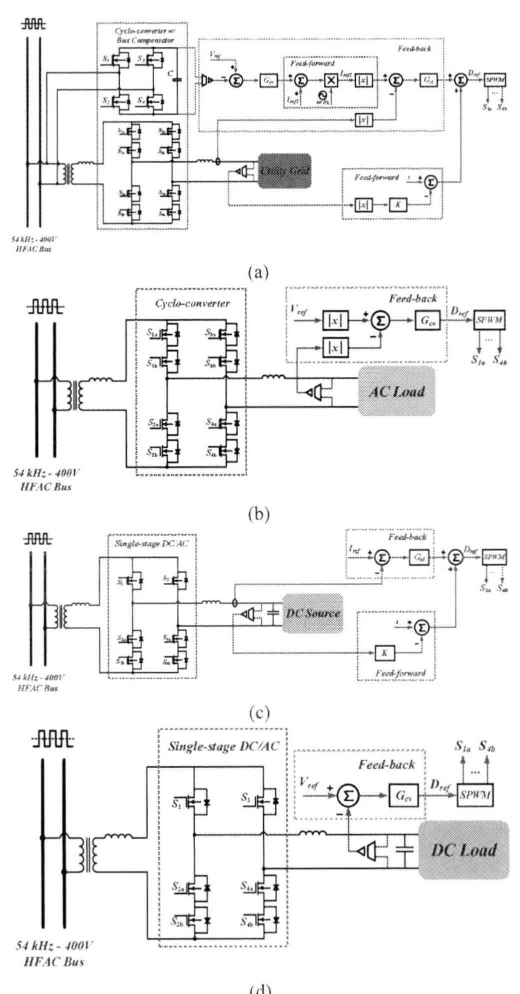

Figure 8: Local control structure for (a) utility grid interface; (b) AC load interface; (c) DC source interface; (d) DC load interface.

V. SIMULATION AND EXPERIMENTAL VERIFICATIONS

The system simulation model is developed in PSIM as shown in Fig. 9. 4 major components are included to demonstrate the functionality of the proposed HFAC power delivery system: Utility AC grid, DC source, LF AC load and DC load. The interface converter topologies all follow the design proposed in Section III as shown in Fig. 6 (a) to (d): for the LFAC source/load interface, cycloconverter is utilized and for the DC source/load interface, cyclo-oriented single-stage DC/AC converter is used. Close-loop controllers for all the source/load interfaces are developed to form a functional system. Dual-loop-compensators have been developed for the AC utility grid-interface with the bus compensator. As the output of the bus compensator voltage controller serves as the current reference of the AC utility grid-interface. The controller for the DC source interface, AC or DC load interface are all single voltage/current loop depends on the application.

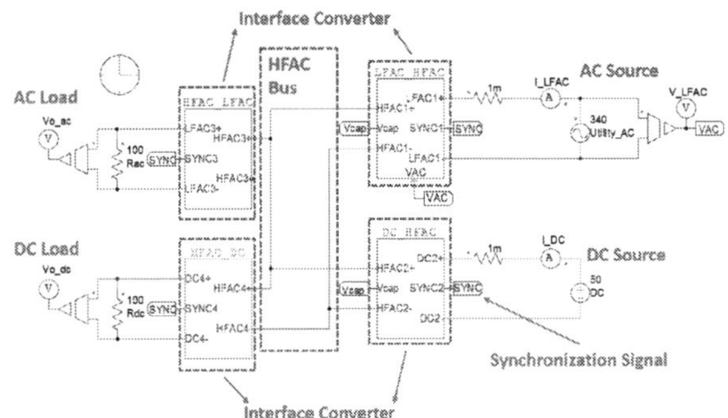

Figure 9: Simulation model for the proposed HFAC power delivery system

Fig. 10 shows the steady-state waveforms of the simulation model. From the top to bottom, the waveforms in the graph are: 1) HFAC bus voltage and the capacitor voltage in the bus compensator; 2) AC current reference and actual AC current output from the Utility AC grid; 3) DC current reference and actual DC current output from the DC source; 4) AC load voltage reference and actual AC load voltage; 5) DC load voltage reference and actual DC output voltage.

As can be observed from the simulation result, the controlled current or voltage can follow the reference very well. The HFAC bus voltage is also regulated for the desirable shape, frequency and amplitude. The functionality and effectiveness of the proposed HFAC power delivery system is well validated.

The Start-up sequence and transient response simulation results are illustrated in Fig. 11. From the top to bottom, the waveforms in the graph are: 1) HFAC bus voltage and the capacitor voltage in the bus compensator; 2) AC current reference and actual AC current output from the Utility AC grid; 3) DC current reference and actual DC current output from the DC source; 4) AC load current; 5) AC load voltage reference and actual AC load voltage; 6) DC load current; 7) DC load voltage reference and actual DC output voltage.

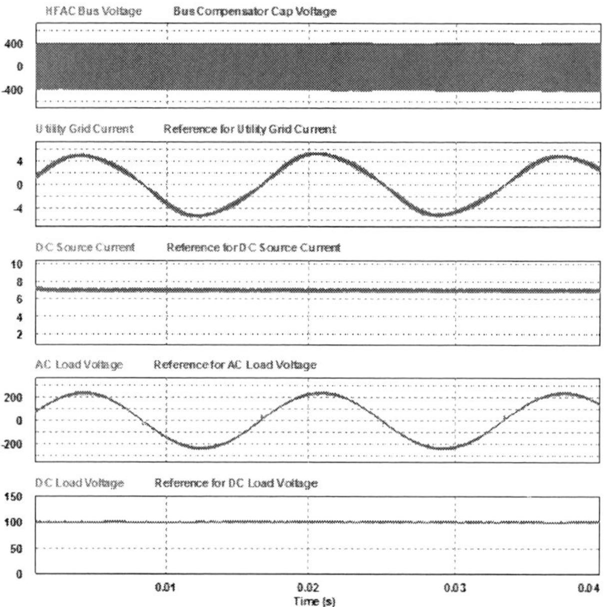

Figure 10: Simulation result of the HFAC power delivery system in steady state

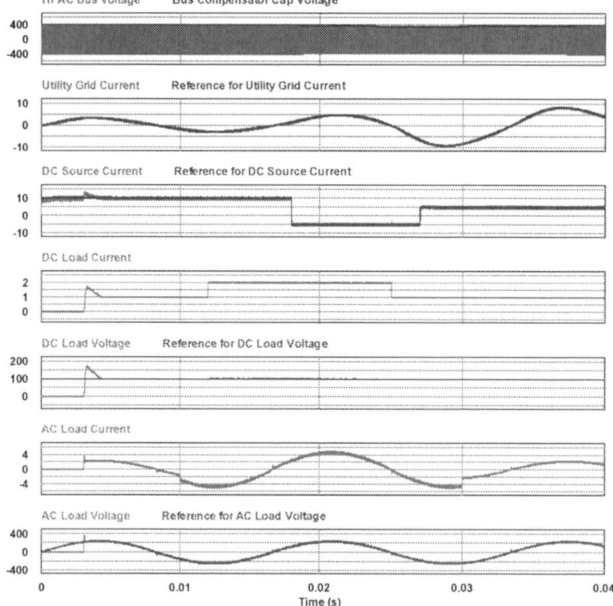

Figure 11: Simulation result of the HFAC power delivery system in start-up and transient state

For the system operation, the interface converter between the utility grid and HFAC bus starts first. At 0.003 s, the rest source/load interface converters start. The source/load changes in the simulation are made as follows:

- At 0.01 s, AC load is increased from half load to full load;
- At 0.012 s, DC load is increased from half load to full load;
- At 0.018 s, reference for DC source current is changed from 10 A to -5 A, which means the power flow is changed from the feeding the HFAC bus to being supplied from the HFAC bus;
- At 0.025 s, DC load is changed from full to half load;
- At 0.027 s, reference for DC source current is changed from -5 A to 5 A, which means the power flow is changed from being supplied from the HFAC bus to feeding the HFAC bus;
- At 0.03 s, AC load is changed from full to half load.

As can be seen from the simulation waveform, all the interface converters can follow the references and the dynamic response is fairly well. Further optimizing the controller parameters will improve the dynamic response in terms of lowering the voltage/current overshoot and reaching steady-state faster.

In order to carry out the experiment, we need a full bridge DC/HFAC converter to generate the HFAC power from a DC power supply, since there is no readily available HFAC power source. In addition, the full bridge will also serve as a shunt bus compensator. The first stage of the experiment is to test the power electronics interface for AC loads or sources, which is the cycloconverter. The hardware prototype of the interface converter is shown in Fig. 12. Fig. 13 is the experimental waveforms of the cycloconverter gating, input voltage of the cycloconverter, output voltage of the cycloconverter before the filter, and output filter inductor current. The output current is a sinusoidal waveform with amplitude of 7.5 A. The functionality of the hardware setup has been well validated.

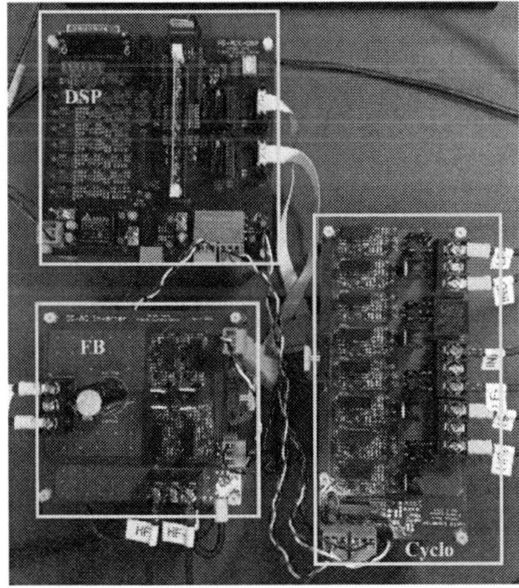

Figure 12: Hardware prototype of the power electronics interface between HFAC bus and LFAC source/load

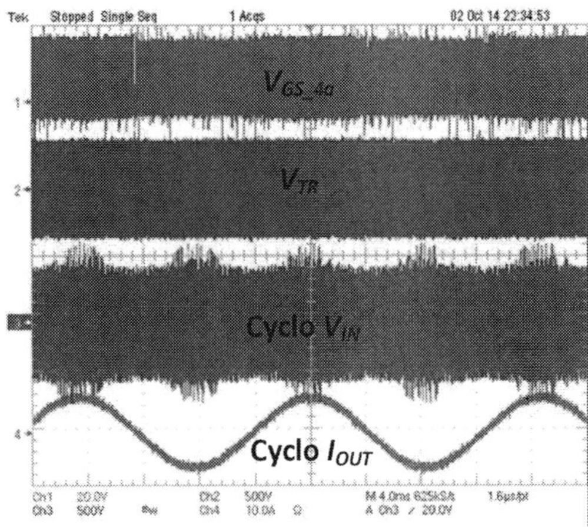

Figure 13: AC interface (cycloconverter) experiment waveforms

With the developed interface prototype, we have set up a hardware platform, as shown in Fig. 14, to demonstrate the scenario of a DC source powering one LFAC load and one DC load. Each interface is equipped with a DSP controller, shown stacked in Fig. 14. For current demonstration, the synchronization of the gating signals for each interface is realized by the master-slave architecture of the DSP controller through a synchronization bus. The rising edge of the master gating signal serves as the trigger for the gating of the rest of the interfaces. For the preliminary prototype, the HFAC bus voltage is a square waveform with 400 V amplitude, the bus frequency is 54 kHz, and the power source for the system is 400 V DC. The experimental result is shown in Fig. 15. As can be observed from the figure, the DC load voltage is stabilized at 100 V and the AC load current is in sinusoidal waveform with a peak amplitude of 2 A. The functionality of the proposed HFAC power delivery architecture is partially validated. Future work will include more comprehensive testing with more AC or DC sources and loads under various scenarios.

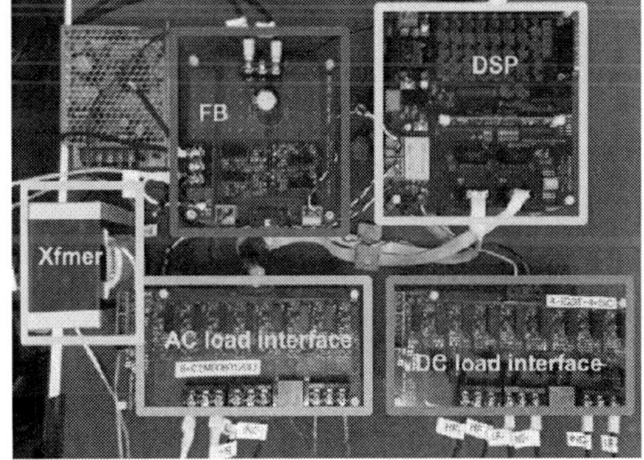

Figure 14: Platform setup for the scenario of DC power supplying AC and DC loads;

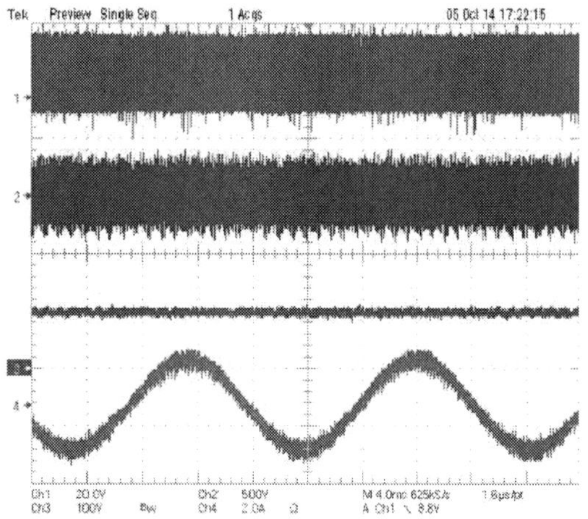

Figure 15: Experiment waveforms of DC source powering DC and AC loads.

VI. Conclusion

A novel high-frequency AC (HFAC) distributed power delivery system is proposed in this paper. The contributions are listed as follows: 1). Proposes a system structure that enables multiple and various sources (AC or DC) to operate in parallel under HFAC through a single-stage converter; 2). Proposes a cycloconverter-oriented DC/HFAC converter which serves as the power electronics interface for DC load or source; 3). Proposes a distributed control architecture for the HFAC power delivery system; 4). Simulation verification is provided to validate the control scheme; 5). Experimental results are presented to verify the functionality of the power electronics interface.

References

[1] Hansen, Irving. G. "Status of 20kHz space station power distribution technology." *NASA Technical Memorandum* 100781, 1988.

[2] Tsai, F-S., and Fred C. Lee. "High-frequency AC power distribution in space station," *Aerospace and Electronics Systems, IEEE Transactions on*, vol. 26, no. 2, pp. 239-253, 1990.

[3] Ye, Zhongming, Praveen K. Jain, and Paresh C. Sen. "A two-stage resonant inverter with control of the phase angle and magnitude of the output voltage," *Industrial Electronics, IEEE Transactions on*, vol. 54, no.5, pp. 2797-2812, 2007.

[4] Ye, Zhongming, Praveen K. Jain, and Paresh C. Sen. "Circulating current minimization in high-frequency AC power distribution architecture with multiple inverter modules operated in parallel," *Industrial Electronics, IEEE Transactions on, vol.* 54, no. 5, pp. 2673-2687, 2007.

[5] Wang, Mengqi, Qingyun Huang, Wensong Yu and Alex Q. Huang, "An isolated bi0directional soft-switched DC-AC converter using wide-band-gap devices with novel carrier-based unipolar modulation technique under synchronous rectification," In *Applied Power Electronics Conference and Exposition (APEC), 2015 Thirtieth Annual IEEE*, pp. 2317-2324. IEEE, 2015.

[6] Masserant, Brian J., Beans, E. William, and Stuart, Thomas A., " A study of volume versus frequency for soft switching IGBT converters," *Aerospace and Electronics Systems, IEEE Transactions on*, vol. 31, no. 1, pp. 280-287, 1995.

[7] Bhatnagar, Mohit, and B. Jayant Baliga. "Comparison of 6H-SiC, 3C-SiC, and Si for power devices," *Electron Devices, IEEE Transactions on*, vol. 40, no. 3, pp 645-655, 1993.

[8] Ning, Puqi, Di Zhang, Rixin Lai, Dong Jiang, Fred Wang, Dushan Boroyevich, Rolando Burgos, Kamiar Karimi, Vikram D. Immanuel, and Eugene V. Solodovnik. "High-Temperature Hardware: Development of a 10-kW High-Temperature, High-Power-Density Three-Phase ac-dc-ac SiC Converter." *Industrial Electronics Magazine, IEEE*, vol. 7, no. 1, pp. 6-17, 2013.

[9] Zhang, Hui, Leon M. Tolbert, Burak Ozpineci, and Madhu S. Chinthavali. "A SiC-based converter as a utility interface for a battery system," In *Conference Record, 41st Industry Applications Conference IAS Annual Meeting*, vol. 1, pp. 346-350, 2006.

[10] L. M. Tolbert, B. Ozpineci, S. K. Islam, M. Chinthavali, "Wide bandgap semiconductors for utility applications," *IASTED International Conference on Power and Energy Systems (PES 2003)*, Palm Springs, California, pp. 317-321, February 24-26, 2003.

[11] T. Ericsen, "Future navy application of wide bandgap power semiconductor devices," *Proceedings of the IEEE*, vol. 90, Issue 6, pp. 1077-1082, June 2002.

[12] Wen-Jian Gu and Rui Liu, "A study of volume and weight vs. frequency for high-frequency transformers", *IEEE Power Electronics Specialists Conference (PESC)*, pp. 1123-1129, 1993.

Effect of the capacitance distribution on the output impedance of the half-wave Cockcroft-Walton voltage multiplier

Liran Katzir and Doron Shmilovitz
School of Electrical Engineering
Tel-Aviv University
Tel Aviv, Israel

***Abstract*—High voltages are typically generated by voltage multipliers that consist of several multiplying stages, where each stage is composed of a few capacitors and diodes. The Half-Wave Cockcroft-Walton voltage multiplier is probably the most common multiplier. We compare different methods of capacitance distribution in the Half-Wave Cockcroft-Walton voltage multiplier and show that the output impedance can be decreased by more than three times compared to the case of the most commonly used method, i.e., the equal capacitance throughout all stages, and a higher output voltage is consequently gained (while yet using an equal total capacitance). For example, our experiment displayed a voltage increase from 5.8 kV to 9.7 kV across the load when the capacitance distributions were changed and the voltage multiplier was split into two sections.**

Keywords*— *Half-Wave Cockcroft-Walton; Voltage multiplier; High Voltage

I. INTRODUCTION

High voltages are required in many industry applications such as scanning electron microscopy, medical x-ray imaging, particle filtering systems, etc. [1]-[4]. The use of a notably high-turn-ratio transformer to achieve the high voltage is limited by the large leakage inductance and parasitic capacitance of the transformer. Therefore, the voltage multiplier circuits are commonly used to attain high voltages [5]. The Half-Wave Cockcroft-Walton voltage multiplier (HWCW) is probably the most commonly used voltage multiplier because it boosts high voltages while maintaining a low voltage stress on the diodes and capacitors, and it is cost efficient [6]. The major shortcomings of the HWCW voltage multiplier are the voltage drop ΔVo and peak-to-peak voltage ripple δVo, which appears on top of the DC output voltage. The voltage drop and ripple occur because of the ac impedance of the capacitors [7], [8] and become more prominent under high loads. Some previous works on HWCW circuits assumed a very low load current and consequently neglected the voltage drop and ripple [9]-[11] effects. Other works did not apply this assumption and referred to the mathematical details of the voltage drop and ripple [2], [12]. Recent efforts to improve the output voltage of the HWCW are reported in [13]-[15]. The capacitance in the voltage multiplier was rearranged to reduce the ripple and voltage drop in [13]. In [14], the authors proposed to feed the voltage multiplier by two frequencies in order to decrease the output ripple, whereas in [15], the authors proposed to split the voltage multiplier into several sections.

In this paper, we compare several methods of capacitance distribution in the HWCW voltage multiplier and observe the effect of the capacitance distribution on the output impedance and output voltage under load. Additional improvement is accomplished by splitting the voltage multiplier into several sections and connecting the outputs of these sections in series, where the output voltages and power of the sections accumulate across the load. Other examples of source splitting and power accumulation over a single load have referred to photovoltaic applications [15], [16].

The simulation and experimental results show that a decrease by more than a factor of three is possible while incorporating the identical total capacitance, which is distributed differently in the voltage multiplier.

II. HALF WAVE COCKCROFT-WALTON VOLTAGE MULTIPLIER

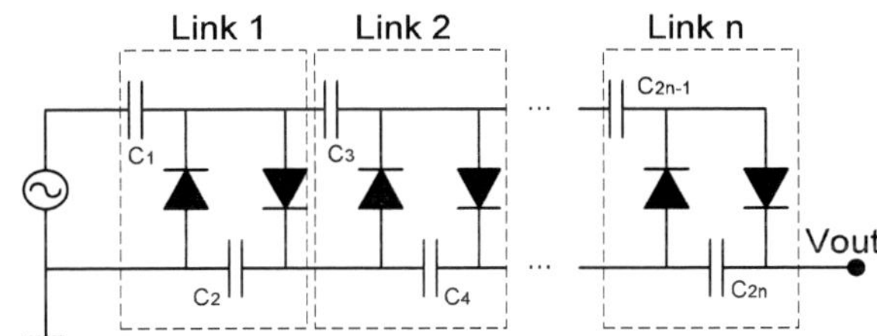

Fig. 1 - A half wave Cockcroft-Walton voltage multiplier with n links

In an unloaded HWCW voltage multiplier, the output voltage is given by:

$$V_{out} = 2nV_{amp} \tag{1}$$

where V_{amp} is the peak value of the input voltage to the multiplier, and n represents the total number of existing links in the multiplier. However, when the circuit is loaded and draws a current, the multiplier suffers from a voltage drop ΔVo and a voltage ripple δVo.

A diagram of the HWCW is shown in Fig. 1, whereas the voltage drop is expressed by (2) [7]:

$$\Delta V = I_{Load} \cdot \left\{ \frac{1}{f_{sw}} \left(\sum_{i=1}^{n} \frac{(n+1-i)^2}{C_{2i-1}} + \sum_{i=1}^{n-1} \frac{(n+1-i)(n-i)}{C_{2i}} \right) \right\} \quad (2)$$

Assuming the most common practice, with equal capacitors through all links ($C_1 = C_2 = C_3 = \cdots = C_n = C$), the voltage drop is:

$$\Delta V_O = \frac{I_{Load}}{f_s C} \left(\frac{2n^3}{3} + \frac{n^2}{2} - \frac{n}{6} \right) \quad (3)$$

where I_{Load} is the output current, C_i is the capacitance in each link, and f_s is the switching frequency of the input power supply. Equations (1) and (3) yield the output voltage of the voltage multiplier circuit as a function of the amplitude of the input voltage V_{amp}:

$$V_{out} = 2nV_{amp} - \Delta V_0 = = 2nV_{amp} - I_{Load} \cdot \underbrace{\frac{1}{f_s C} \left(\frac{2n^3}{3} + \frac{n^2}{2} - \frac{n}{6} \right)}_{Ro} \quad (4)$$

where I_{Load} is the output voltage divided by the output load used.

Thus, the output impedance for identical capacitances in all stages is:

$$R_o = \frac{1}{f_s C} \left(\frac{2n^3}{3} + \frac{n^2}{2} - \frac{n}{6} \right), \quad (5)$$

And the total capacitance is $2nC$

III. Effect of the Capacitance Distribution on Output Impedance and Voltage

Although the most common capacitance distribution in the HWCW voltage multiplier is to use equal capacitance in all stages [7], [8], it is also possible to distribute the capacitance in other configurations [13]. An example of a different capacitance distribution is the double first capacitor method, where the first capacitor is twice the size of all other capacitors: $C1 = 2C$ and $C2i = C2i-1 = C \mid i \neq 1$. In this case, the voltage drop is given by (6).

$$\Delta V_O = \frac{I_{Load}(4n^3 - n)}{6 f_s C} \quad (6)$$

The output impedance is:

$$Z_{out} = \frac{4n^3 - n}{6 f_s C} \quad (7)$$

And the output voltage in this case is:

$$V_{out} = V_{amp} \frac{12nf_s CR_{Load}}{6 f_s CR_{Load} + 4n^3 - n} \quad (8)$$

The total capacitance of the voltage multiplier is $(2n+1)C$

Another example is the increased low-stage capacitance method, where the capacitances of two capacitors of each stage are equal but lower than the capacitance of the previous stage: $C2i = C2i-1 = (n + 1 - i)C$. As before, n is the number of stages, and C is the capacitance of the last stage. The voltage drop in this case is as follows:

$$\Delta V_O = \frac{I_{Load} n^2}{f_s C} \quad (9)$$

The output impedance is:

$$Z_{out} = \frac{n^2}{f_s C} \quad (10)$$

And the output voltage in this case is:

$$V_{out} = V_{amp} \frac{2nf_s CR_{Load}}{f_s CR_{Load} + n^2} \quad (11)$$

The total capacitance of the voltage multiplier is $n(n+1)C$.

IV. Split-Source Voltage Multiplier Output Impedance

A recently proposed topology suggested splitting the input voltage of the voltage multiplier and connecting the outputs in series as shown in Fig. 2 [15]. In this method, the output impedance can be further reduced.

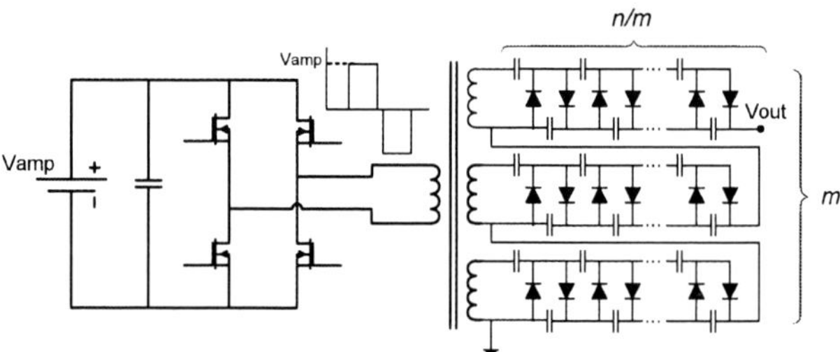

Figure 2– A split-source HWCW volatge multiplier with m sections

In the simple case when the capacitance is identical throughout all stages, the voltage drop in a split source multiplier is given by:

$$\Delta V_O = \frac{I_{Load}}{6 f_{sw} C} \left(4\frac{n^3}{m^2} + 3\frac{n^2}{m} - n \right) \quad (11)$$

The output impedance is:

$$Z_O = \frac{I_{Load}}{6f_{sw}C}\left(4\frac{n^3}{m^2}+3\frac{n^2}{m}-n\right) = \\ = \frac{4n^3+3n^2m-nm^2}{6fCm^2} \qquad (11)$$

The output voltage is:

$$V_{out} = V_{amp}\left(\frac{12nm^2 f_s CR_{Load}}{6m^2 f_s CR_{Load}+4n^3+3mn^2-m^2n}\right) \qquad (11)$$

The total capacitance is $2nC$, where C is the capacitance of each capacitor.

The output impedance and output voltage for the four capacitance distribution methods are compared in Fig. 3.

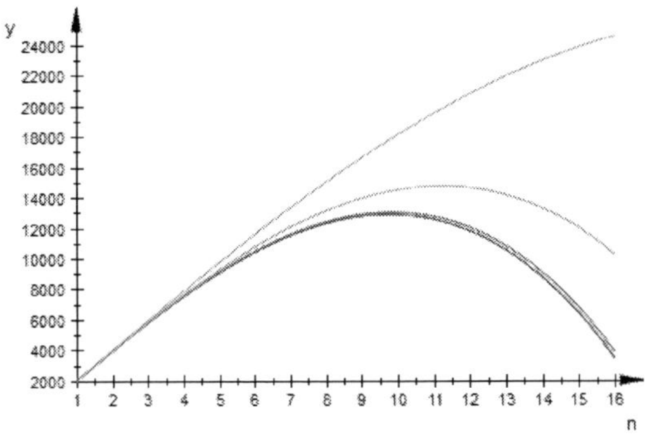

(a)

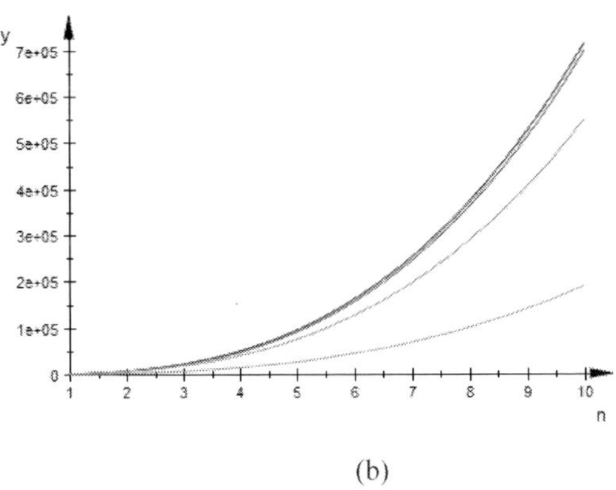

(b)

Figure 3 – Simulation based comparison of the (a) output impedance and (b) output voltage of four capacitance distribution methods. Blue: equal capacitance in all stages; red: double first capacitor method; green: increased low-stage capacitance method; yellow: equal capacitance with a split source.

The simulation was performed with an n-stage voltage multiplier, a voltage amplitude of 1000 V, a frequency of 100 kHz, and a total capacitance of 5 nF per cell (i.e., 50 nF for the 10-cell case and 80 nF for the 16-cell case). The split-source multiplier is divided into two sections.

V. Experimental Results

An experimental setup with a 12-link voltage multiplier was constructed and tested. The load was a high-voltage-resistor of 5 MΩ and was connected to the output of stages from n=3 to n=12 during the test. The input stage consisted of a full-bridge, which fed a high-frequency isolation transformer (Fig. 4), see details in table 1.

TABLE 1: Main experiment details

Full-Bridge power stage	
Frequency	100 khz
Duty	50%
Dead time	200 nS
Power Transformer	
Core	3F3
Ratio	1:30
Voltage Multiplier	
Diode	Vishay GP02-30
Total Cap	60nF
High Voltage Probe	
Tektronix	P6015A

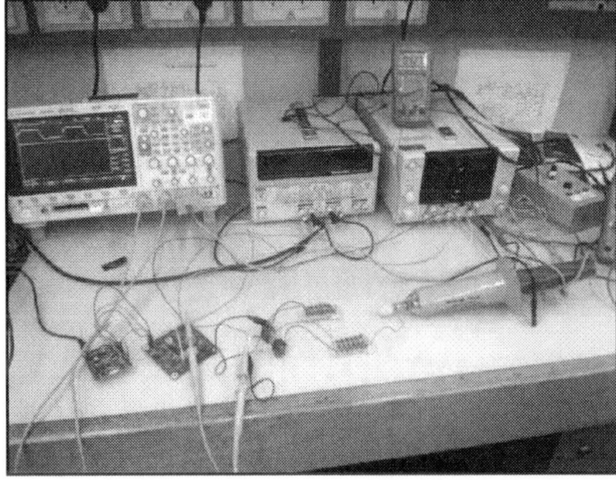

Figure 4 –Experiment setup with a split-source HWCW volatge multiplier with two sections

The comparative results are shown in Fig. 5. The comparison was based on the same total capacitance in the capacitance configuration methods. The total capacitance was based on 5 nF per stage. Thus, if n=12, the total capacitance in all methods was approximately 60 nF: in the equal-capacitance method, each capacitor was 2.55 nF (two 5.1nF in series); in the double first capacitor method, the base capacitance was 2.4 nF, and the capacitance of the first capacitor was 4.8 nF; in the increased low-stage capacitance method, the last capacitor was

0.4 nF, and the first capacitor was 4.7 nF; in the split source, each capacitance was 2.55 nF.

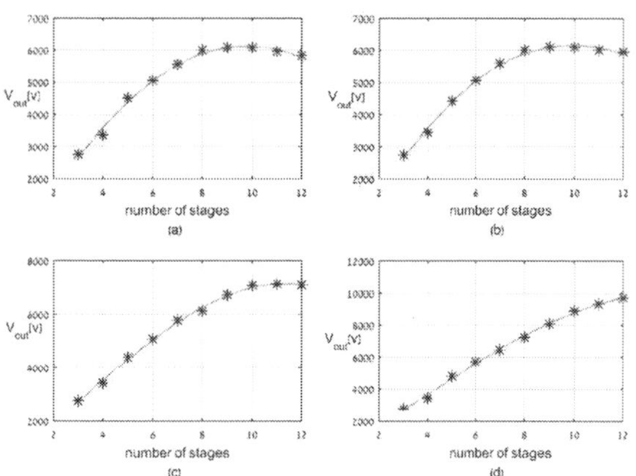

Figure 5 – Output voltage experimental results with identical total capacitances in all methods: (a) equal-capacitance method, (b) double first capacitor method, (c) increased low-stage capacitor method and (d) split-source method.

VI. Conclusions and Future Work

Several capacitance distribution methods for the HWCW voltage multiplier were compared. The most common method of equal capacitance in all stages was found to be non-optimal. Other methods are more efficient in obtaining high voltages. The theoretical analysis was verified using simulation and experiments. Particularly encouraging results were displayed under high output loads. Specifically, the proposed method of splitting the source significantly reduces the output impedance by more than a factor of three and increases the maximum output voltage of the multiplier compared with the conventional multipliers with equal total capacitance and diode count. However, the multi-section topology requires isolated transformers.

References

[1] B. P. Israelsen, J. R. Martin, C. R. Reeve, and V. S. Scown, "A 2.5 kV High-Reliability TWT Power Supply : Design Techniques for High Efficiency and Low Ripple," *Proceedings of IEEE PESC*, 1977, pp. 212-222.

[2] S. D. Johnson, A. F. Witulski, and R. W. Erickson, "Comparison of Resonant Topologies in High-Voltage DC Applications," *IEEE Trans. on Aerospace and Electronic System*, pp.263-274, 1988.

[3] H. Hino, T. Hatakeyama, and M. Nakaoka, "Resonant PWM Inverter Linked DC-DC Converter using Parasitic Impedance of High-Voltage Transformer and its Applications to X-ray Generator," *Proceedings of IEEE PESC*, 1988, pp. 1212-1219.

[4] V. Garcia, M. Rico, J. Sebastian, M. M. Hernando, and J. Uceda, "An Optimized DC-to-DC Converter Topology for High-Voltage Pulse-Load Applications," *Proceedings of IEEE PESC*, 1994, pp. 1413-1421.

[5] S. D. Johnson, A. F. Witulski, and R. W. Erickson, "Comparison of resonant topologies in high voltage DC applications," *IEEE Trans. on Aero. And Elect. Systems*, vol.24, No.3, May 1988, pp.263 – 274.

[6] J. Sun, H. Konishi, Y. Ogino, and M. Nakaoka, "Series resonant high voltage ZCS-PFM, DC-DC converter for medical power electronics," in *Proc. IEEE 31st Annu. Power Electron. Spec. Conf.*, Galway, Ireland, vol. 3, Jun. 18–23, 2000, pp. 1247–1252.

[7] M. Khalifa, "High-voltage engineering, theory and practice," in *Electrical Engineering and Electronics, A Series of Reference Books and Textbooks*, vol. 63. New York: Marcel Decker, Mar. 1990, ch. 16.

[8] E. Kuffel and W. S. Zaengl, *High Voltage Engineering Fundamentals*. New York: Pergamon International Library, 1984, ch. 2.

[9] L. C. Franko, L. L. Pfitscher, and R. Gules, "A new high static gain nonisolated DC-DC converter," in *Proc. IEEE 34th Annu. Power Electron. Spec. Conf.*, Jun. 15–19,, 2003, vol. 3, pp. 1367–1372.

[10] D. F. Spencer, R. Aryaeinejad, and E. L. Rebel, "Using the Cockcroft–Walton voltage multiplier with small photomultipliers," *IEEE Trans. Nuclear Sci.*, vol. 49, no. 3, pp. 1152–1155, Jun. 2002.

[11] M. Prudente, L. L. Pfitscher, G. Emmendoerfer, E. F. Romaneli, and R. Gules, "Voltage multiplier cells applied to non-isolated DC-DC converters," *IEEE Trans. Power Electronics.*, vol. 23, no. 2, pp. 871–887, Mar. 2008.

[12] K. Ogura, E. Chu, M. Ishitobi, M. Nakamura, and M. Nakaoka, "Inductor snubber-assisted series resonant ZCS-PFM high frequency inverter link DC-DC converter with voltage multiplier," in *Proc. Power Convers. Conf.*, Osaka, Japan, vol. 1, Apr. 2–5, 2002, pp. 110–114.

[13] Kobougias, I.C.; Tatakis, E.C., "Optimal Design of a Half-Wave Cockcroft–Walton Voltage Multiplier With Minimum Total Capacitance," *IEEE Trans. Power Electronics*, vol.25, no.9, pp.2460,2468, Sept. 2010.

[14] Young, C.; Chen, H.; Chen, M., "A Cockcroft–Walton Voltage Multiplier Fed by a Three-Phase-to-Single-Phase Matrix Converter With PFC," *IEEE Trans. Industry Applications*, vol.50, no.3, pp.1994,2004, May-June 2014.

[15] L. Katzir, D. Shmilovitz, "A Matrix-like Topology for High Voltage Generation," *Plasma Science, IEEE Transactions on*, vol.43, no.10, pp.3681-3687, Oct. 2015.

[16] Martinez-Salamero, L.; Leyva, R.; Calvente, J.; Giral, R., "Design of photovoltaic-based current sources for maximum power transfer by means of power gyrators," *Power Electronics, IET*, vol.4, no.6, pp.674-682, July 2011.

978-1-4673-9551-9/16 $31.00 © 2016 IEEE

A Cost Effective High Performance LED Driver Powered by Electronic Ballasts

Jianwen Shao
STMicroelectronics
Schaumburg, IL, USA
jianwen.shao@st.com

Thomas Stamm
STMicroelectronics
Schaumburg, IL, USA
thomas.stamm@st.com

***Abstract*— Many T8 LED tubes are currently available in the market; however, the majority of commercial products are passive converters, so the lamp current is not well controlled. This paper presents a low voltage, current-fed Buck converter for a T8 LED tube application. A very simple control method with zero-crossing synchronization is proposed; the operation of the circuit is described, and the performance of prototype proves the concept. The current-fed Buck converter is compatible with most commercial electronic ballasts.**

Keywords—T8 LED tube; Current-fed Buck; Electronic Ballasts

I. INTRODUCTION

With the rapid development of high brightness LEDs, SSL (Solid State Lighting) is no longer a niche market. In commercial buildings and department stores, LED fixtures are chosen over fluorescents because of efficiency, fast startup, easy dimming, low maintenance, and even style. However, the cost of replacement of a fluorescent light is not only the cost of the LED fixture, but also the cost of labor. It takes a long time to remove the fluorescent ballast, so the cost of labor and down time is a significant portion of the total cost. Even though there are some apparent drawbacks like the unknown life time of the existing ballast and lower efficiency than a direct AC power LED lamp, the plug-and-play LED tube concept is gaining some attraction. There are a few commercial products available on the market [1] [2].

Since the output of electronic ballast is constant current, it is possible to connect an LED load directly to the electronic ballast. The simplest way to power an LED from electronic ballast is to use a full bridge rectifier and capacitor, called a passive converter [3]. A more complicated passive circuit with impedance transformation [4] was also reported. In passive converters, the LED current is determined by the ballast output. Since there are different types of ballast, from different manufacturers, the current output will vary from ballast to ballast. An active controlled converter is necessary to provide stable and consistent light output from LED tubes. Several papers have been published discussing how to improve LED current control and maintain the compatibility [5] [6] [7] [8]. In [5], a linear regulator to limit LED current was used. Hence the efficiency for the system was quite low. In [6], a switched capacitor was controlled to modulate the resonant frequency of electronic ballast to regulate output current. In [7], a phase shift resonant converter was proposed to control the LED current. In [8], Choi et al proposed a current-fed Boost to drive the LED. But they all required complex control and even digital controllers to do the job. The cost of the driver is high as well. In this paper, a simple current-fed Buck is proposed as the driver. The paper is arranged as follows: in section II, the operation of electronic ballast and the passive LED converter are reviewed; in section III, the principle of operation of the current-fed Buck is introduced; experimental results are shown in section IV.

II. OPERATION OF ELECTRONIC BALLAST AND PASSIVE CONVERTER

A. Electronic Ballast

There are three types of electronic ballast: instant-start, rapid-start, and programmed-start. An instant-start ballast does not preheat the electrodes; instead, it uses a relatively high voltage (~600 V) to initiate the discharge arc. A rapid-start ballast applies voltage and heats the cathodes simultaneously. A programmed-start ballast is a more advanced version of rapid-start. This ballast applies power to the filaments first, it allows the cathodes to preheat, and then it applies voltage to the lamps to strike an arc. Typically, the electronic ballast consists of two stages of converters, a PFC stage followed by a resonant inverter. The main topologies are current-fed half bridge [9] and voltage-fed half bridge [10], both utilizing parallel resonant network. The parallel resonant inverter behaves like a current source, which can be used to stabilize fluorescent lamp current. Therefore, no matter what topology is used for the ballast, the simplest model of ballast can be looked at as a high frequency current source, shown in Fig.1, with a typical frequency of 40~50KHz.

B. Passive Converter to be compatible with Electronic Ballast

Since the ballast has current source characteristics, it can directly drive an LED. A simple bridge rectifier circuit will allow operation of an LED string from the ballast output, shown in Fig.2, if the LED current and voltage are reasonably close to the fluorescent tube's characteristics.

If the output current of the ballast is $I_{ballast}$, which typically is rated in RMS value, we can calculate the LED current.

978-1-4673-9551-9/16 $31.00 © 2016 IEEE

$$I_{LED} = \frac{2*\sqrt{2}}{\pi} * I_{ballast} \quad (1)$$

The output voltage of the ballast is automatically clamped by the LED voltage. So, all the component voltage ratings just need to be chosen according to the LED voltage. If the voltage of the LED is significantly different from that of the T8 fluorescent lamp, an autotransformer can be used to match the voltage. This passive converter is very simple and low cost.

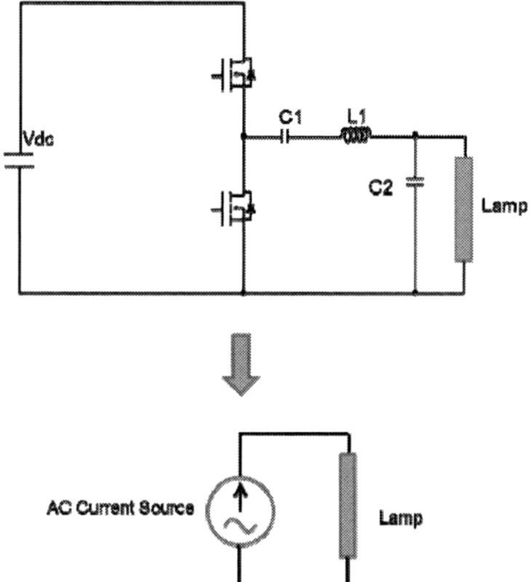

Fig.1. Simplified equivalent ballast model: AC current source

The problem for this passive circuit is that the LED current, and resulting brightness, varies from ballast to ballast since the LED current is not controlled.

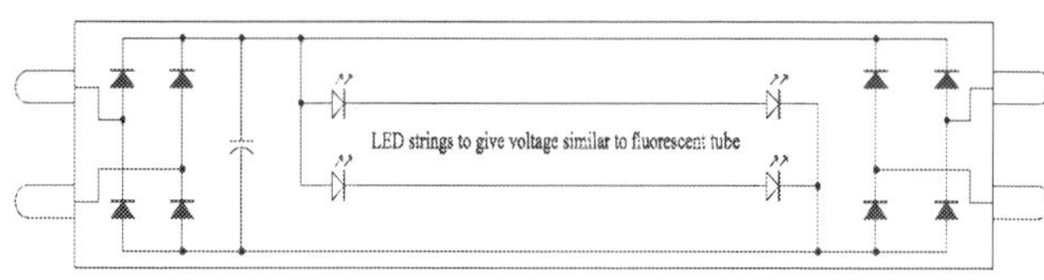

Fig.2. Passive converter to rectify ballast AC current to DC current

III. A Current-fed Buck Converter Compatible with Electronic Ballast

The typical power source like an AC main is a voltage source. Common power converters use an AC main as an input. Those converters are referred to as voltage-fed converters. Since the output of electronic ballast is a current source, commonly used voltage-fed converters are not suitable for this application, but a current-fed converter is. Current-fed converters can be derived from voltage-fed counterparts [8] [11]. Reference [8] gives an excellent overview for the possible topologies to control LED current from ballast. Two possible topologies are current-fed Buck or current-fed Boost, shown in Fig.3.

For current-fed Buck, the operation is as simple as this: when the switch is on, the current source is shorted; when the switch is off, the current source feeds the load. If the duty cycle is D, the load current is

$$I_{load} = (1-D) * I_s \quad (2)$$

Where I_{load} is load current, and I_s is the current from current source.

For current-fed Boost, the operation is as follows: when the switch is off, the current source will charge input capacitor to store the energy; when the switch is on, the capacitor will be discharged to load. Based on the charge balance of the capacitor, it is easy to derive the equation:

$$I_{load} = \frac{1}{D} * I_s \quad (3)$$

The authors of reference [8] have worked on current-fed Boost topology based on the theory that switching loss of current-fed Boost may be lower than current-fed Buck. However, current-fed Boost has to endure the high striking voltage during the electronic startup phase since the switch is off at that time. Hence, all the components have to have a high voltage rating.

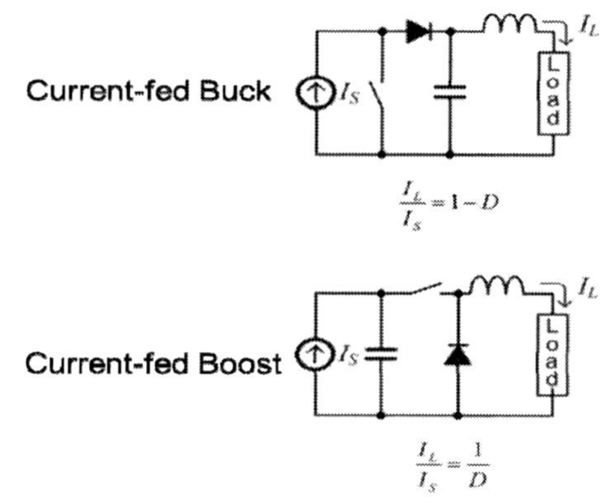

Fig.3. Current-fed Buck and current-fed Boost topology

On the other hand, in current-fed Buck, when the switch is off, the voltage is clamped by the LED voltage. Therefore, the max voltage, even during the ballast startup phase, is limited by the LED string. Another benefit of current-fed Buck is that the LED will be lighted if the switch is failed open.

For the above reasons, we have decided to work with the current-fed Buck converter. The example LED load is 500mA/40V. The circuit is shown in Fig4. The max voltage is clamped by LED voltage; therefore, all components have a

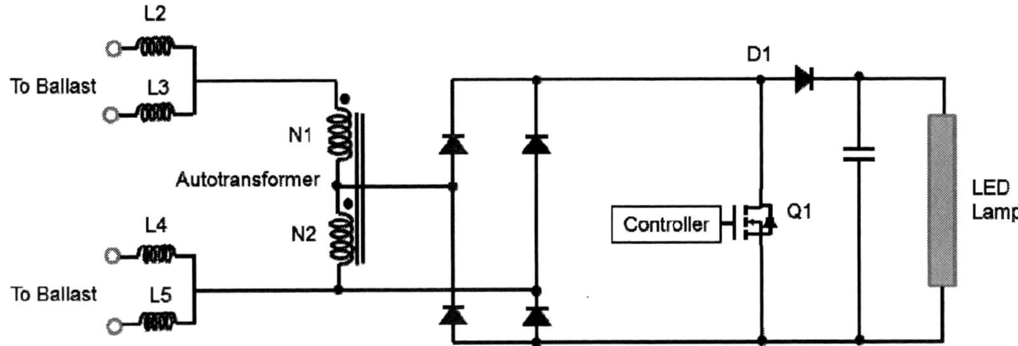

Fig.4. Current-fed Buck circuit power stage

low voltage rating, i.e. 60V. In order to match the LED voltage to the typical T8 fluorescent lamp voltage, a high frequency auto-transformer with a turn's ratio of N1:N2 is used. The small inductors, L2, L3, L4 and L5, are used to emulate filaments for programmed-start ballasts.

The control block diagram is shown in Fig.5 and the main control timing in Fig.6.

The switching action of Q1 is synchronized with the zero crossing of current from the ballast output. The switch is turned on at zero crossing, minimizing the switching loss. Another benefit of the zero crossing synchronization is that the disturbance to the resonant circuit of the ballast is reduced to a minimum. Typically, the switching frequency of electronic ballast is around 40~50KHz. The switching frequency for the current-fed Buck will be doubled. Therefore, the output filter can be small for 80~100 KHz.

The synchronization is implemented through zero crossing detection of input current. The zero crossing detection circuit will generate a pulse to start the control ramp. At the beginning of the ramp, the switch is turned on. When the switch Q1 is on, the output of the ballast is shorted through the switch. There is no power transferred to LED during the "on" time. The current in LED will decrease. When the ramp reaches the value from the error amplifier, the switch will be turned off, and the ramp will be reset. The circuit will wait until the next zero crossing pulse.

When the voltage on the control ramp reaches the output of the error amplifier, Q1 is turned off; the output current from the ballast will flow through the LED load. The LED current will increase during "off" time. So the duration of "on time", duty cycle D, will determine the average LED current, according to Equation (2).

The detail circuit and operation are described in [12].

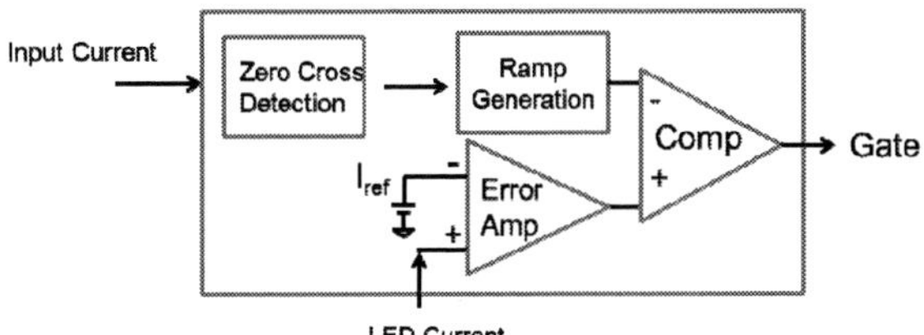

Fig.5. Control block diagram

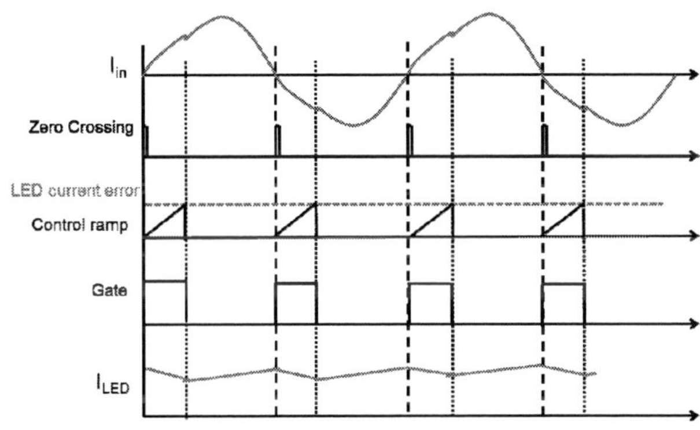

Fig.6. Current-fed Buck control timing

IV. Experiment Results

A prototype circuit was built to prove the concept. The control circuit was implemented by inexpensive analog logic devices. The power devices (MOSFET and diodes) have a low voltage rating, 60v. The key waveforms of the converter are shown in Fig.7.

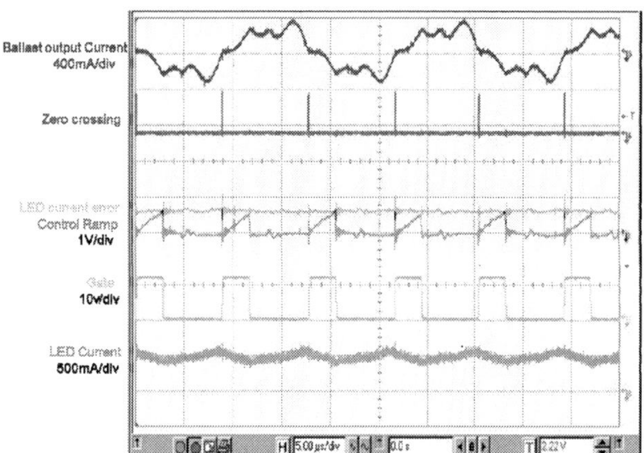

Fig.7. Experimental results

The top trace is the output current from electronic ballast which is also the input current into the current-fed Buck converter . The second trace is the zero crossing detection pulse of the input current. The third trace is the LED current error amplifier output, compared with the forth trace, which is the control ramp signal. The fifth trace is the gate signal to the switch, and the last trace is the LED current. The logic in the current-fed converter is complementary to the voltage-fed converter. In the current-fed converter, increasing the duty cycle will result in lower output current.

It is observed that the output current of the ballast has significant high frequency content. L2~L5 and the parasitic capacitance of the diode bridge introduce this higher order of resonant frequency to the output current of the ballast. Resistors can be used to replace L2~L5 to damp the ringing, but the efficiency will be compromised.

A few 2-lamp ballasts from different manufacturers have been tested in the lab with the prototype current-fed Buck converter. One ballast drives two current-fed Buck converters, each with 14 LEDs as the load. The current-fed Buck converter was able to operate from all the ballasts. The efficiency data also were recorded in Table 1. Overall efficiency is around 80%, which is lower than the typical converter that is directly powered from the AC main. However, since an LED lamp still consumes less power than a fluorescent lamp, the total power consumption is reduced nevertheless.

V. Conclusion

In this paper, a current-fed Buck converter is proposed to drive a T8 LED lamp compatible with electronic ballasts. The current-fed Buck is very easy to control. The zero crossing synchronization reduces the switching loss and stabilizes the operation of the LED driver and ballast operation. The LED current is well controlled and the cost of the low voltage system is lower than other active high voltage converters, e.g. current-fed Boost converters. The prototype circuit proves the concept. The efficiency of the system is lower than an AC main powered LED driver, as expected.

References

[1] T8 Series 4' Linear LED Lamp, www.cree.com.

[2] Philips InstantFit T8 LED, www.Philips.com

[3] T. Stamm, and J. Shao, " Drive LEDs with Electronic Ballasts (part1), " EDN, Dec. 2013.

[4] B. Lee, H. Kim, and C. Rim, " Robust Passive LED Driver Compatible With Conventional Rapid-Start Ballast, " IEEE Trans on Power Electronics, VOl. 26, No.12, Dec. 2011.

[5] T. Liang, W.Tseng, W. Chen, and J. Chen, " Design and Implementation of Retrofit LED Lamp for Fluorescent Lamp Driven by Eletronic, Electromagnetic Ballast and AC Mains," IEEE IFEEC, Nov. 2013, PP. 585-589.

[6] E. Lee, B. Choi, D. Nguyen, G. Jang, C. Rim, " Versatile LED Drivers for Various Eletronic Ballasts by Variable Switched Capacitor," IEEE Trans on Power Electronics, Vol.31, No.2, Feb. 2016.

[7] N.Chen, and H Chung, "A Universal Driving Technology for Retrofit LED Lamp for Fluorescent Lighting Fixtures," APEC 2012, Page 980-986.

[8] J.Choi, H. Han, and K. Lee, "A Current-Sourced LED Driver Compatible with Fluorescent Lamp Ballast," IEEE Trans. Power Electronics, VOl.30, No.8, Aug.2015

[9] T.Wu, T.Yu and H. Huang, "Complete analysis and performance-characteristic compromise for self-excited half-bridge parallel resonant electronic ballasts," PESC'94, page 124-131 vol.1.

[10] M. Cosby, and R. Nelms, " A Resonant Inverter for Eletronic Ballast Applications," IEEE Trans. IE, VOl. 41, No.4, Aug. 1994.

[11] S. Cuk, " General topologies properties of switching structures," IEEE PESC 1979, page 109-130.

[12] US patent application, US20140320007 A1.

Ballast Manufacture	Input voltage(V)	Pin(W)	Vout1(V)	Iout1(A)	Vout2(V)	Iout2(A)	Efficiency
Philips	120.00	53.70	45.70	0.49	45.00	0.48	81.8%
Sylvania	120.00	56.20	45.70	0.49	45.10	0.50	79.6%
Keystone	120.00	58.20	44.80	0.49	44.40	0.49	74.9%
LOA	120.00	55.00	45.60	0.49	45.30	0.49	80.5%

Table 1 Efficiency data with different ballasts

Model Predictive Control of Z-Source Four-Leg Inverter for Standalone Photovoltaic System with Unbalanced Load

Sertac Bayhan[1,2], *Member, IEEE,* Mohamed Trabelsi[2], *Member, IEEE,* and Haitham Abu-Rub[2,3], *Senior Member, IEEE*

[1] Department of Electronics and Automation, Gazi University, Ankara, Turkey

[2]Department of Electrical and Computer Engineering, Texas A&M University at Qatar, Doha, Qatar

[3]Qatar Environment and Energy Research Institute, Hamad Bin Khalifa University, Qatar Foundation, Doha, Qatar

e-mails: sbayhan@gazi.edu.tr, mohamed.trabelsi@qatar.tamu.edu, haitham.abu-rub@qatar.tamu.edu

***Abstract*—In this paper, Z-Source (ZS) four-leg inverter equipped with a Model Predictive Control (MPC) strategy for standalone Photovoltaic (PV) system under various load conditions is presented. The inverter is aimed to supply simultaneously three-phase and single-phase AC loads with balanced voltage and constant frequency. In this topology, a simple LC (Z) network is used instead of the boost converter to obtain single stage power conversion system. The MPC scheme is developed to control output currents of the inverter and ZS network current and voltage. Hardware-In-The-Loop (HIL) tests have been performed to verify steady-state and transient performances of the proposed system under balance/unbalanced load and/or reference current conditions.**

***Index Terms*—Model predictive control, four-leg inverter, Z-Source Inverter (ZSI), Hardware-In-the-Loop (HIL).**

I. Introduction

The performance of the PV based standalone system entirely depends on the power converter topologies and their control structure [1]. Recently, number of power converter topologies have been proposed to extract maximum power from PV modules [2], [3]. Each of these converters present particular advantages and disadvantages [4]. The Z-source inverter (ZSI) is appeared as an emerging power converter topology for PV systems. The ZSI is a new single-stage power converter, and it can fulfill the two-stage converter functions [5]–[7]. With proper control, the ZSI can boost voltage to a desired magnitude, which might be greater than the available dc-link voltage. The main advantage of this topology is that the number of switches is lower than conventional two stages (DC/DC+DC/AC) power conversion system.

Three-phase three-leg ZSI is designed for balanced three-phase loads. However, unbalanced load conditions are common in renewable energy based standalone power generation systems. Unbalanced load condition also creates unbalanced current circulating on the neutral wire, causing overheat of the neutral wire, and harmonic distortion on the output voltage [8]. One way to get rid of these drawbacks is to use four-leg topology instead of three-leg [9]. In this study, the Z-source (ZS) network is employed in four-leg voltage source inverter (VSI) to obtain reliable single-stage power converter topology for standalone PV systems.

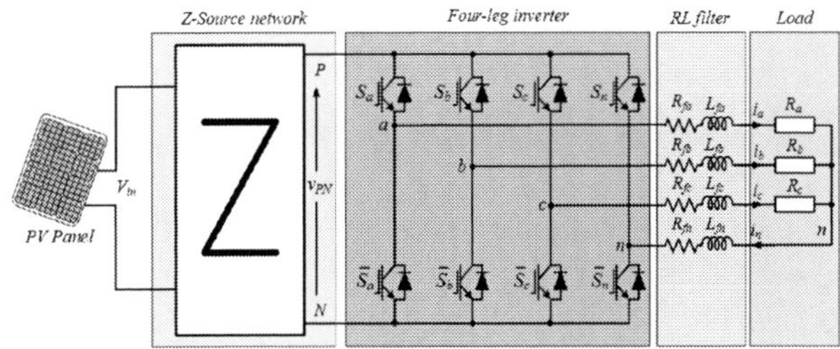

Fig. 1. Z-source three-phase four-leg inverter topology.

The control strategy of the power converter plays a crucial role to ensure reliable and efficient operation of the renewable energy based power generation systems. To control ZSI and four-leg VSI in such systems a number of control techniques have been proposed [10]–[13]. One of interesting control techniques is model predictive control (MPC) that has appeared as an attractive solution for the control of power converters due to its fast response and increased control accuracy [14]. The major advantage of MPC lies in the direct application of the control action to the converter, without requiring a complex modulation stage. A number of studies have been presented under the name of MPC for current control of VSI [15], multilevel inverters [16], [17], qZSI [18], and several electrical machine drives [19], [20].

This paper presents MPC technique for ZS four-leg inverter to improve control performances of the load currents and ZS network capacitor voltage and inductor current. The main contribution of this study is to improve reliability of the standalone PV system under unbalanced load conditions with single-stage power converter topology.

II. Z-Source Four Leg Inverter Model

A. Topology

The ZS four-leg inverter topology with $R-L$ output filter and loads is shown in Fig. 1. This topology can be investigated in two stages:the Z-source network and four-leg inverter with $R-L$ filter and loads.

978-1-4673-9551-9/16 $31.00 © 2016 IEEE

B. Mathematical Model of the Z-Source Network

The equivalent circuit of the Z-source network with its non-shoot-through state and shoot-through states are given in Figs. 2 (a)-(c), when assuming that the inductors (L and L) and capacitors (C and C) have the same inductance and capacitance, respectively. Thus, the ZS network becomes symmetrical. From the equivalent circuits, we have

$$\left.\begin{array}{lll} V_C & V_C & V_C, \\ v_L & v_L & v_L. \end{array}\right\} \quad (1)$$

During the non-shoot-through state (T) the inverter model is represented by a constant current source; it can be seen from Fig. 2 (b). From the equivalent circuit, Fig. 2 (b), the inductor voltage (v_L), dc-link voltage (v_{PN}), and diode voltage (v_d) are

$$\left.\begin{array}{ll} v_L & V - V_C, \\ v_d & V_{in}, \\ v_{PN} & V_C - v_L \quad V_C - V_{in}. \end{array}\right\} \quad (2)$$

where V_{in} is the input dc voltage.

During the shoot-through state (T), the inverter model is represented by a short-circuit that can be seen from Fig. 2 (c). By applying Kirchhoff's voltage law to Fig. 2 (c), inductor voltage, dc link voltage, and diode voltage are

$$\left.\begin{array}{ll} v_L & V_C, \\ v_d & V_C, \\ v_{PN} & . \end{array}\right\} \quad (3)$$

The average voltage of the inductors over one switching period (T T T) should be zero in steady-state, from (2) and (3), following is achieved

$$V_C \quad \frac{T}{T - T} V_{in} \quad (4)$$

Similarly, the average dc-link voltage across the inverter bridge can be found as follows

$$v_{PN} \quad \frac{T}{T - T} V_{in} \quad V_c \quad (5)$$

The peak dc-link voltage across the inverter bridge is expressed in (2) and can be rewritten as

$$\hat{v}_{PN} \quad V_C - v_L \quad V_C - V_{in} \quad \frac{T}{T - T} V_{in} \quad BV_{in} \quad (6)$$

where B is the boost factor.

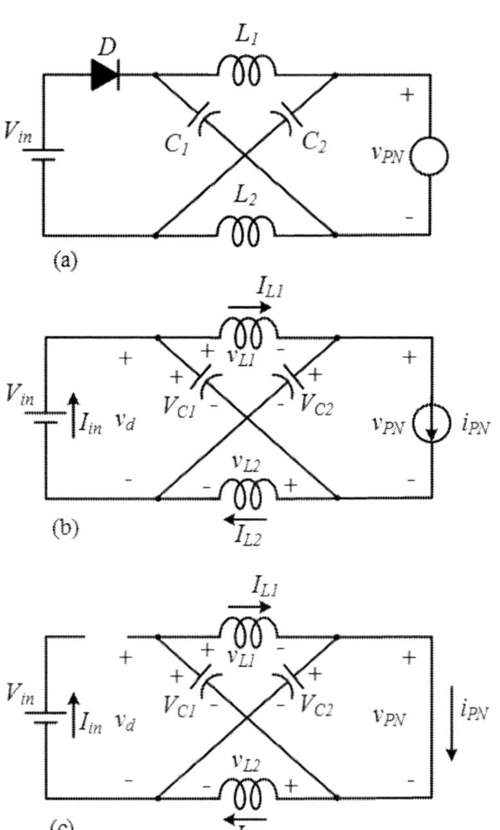

Fig. 2. (a) Equivalent circuit of the Z-source network. (b) in the non-shoot-through state, (c) in the shoot-through state.

C. Mathematical Model of the Four-Leg Inverter

The equivalent circuit of the four-leg inverter with the output $R-L$ filter and load is shown in Fig. 3, where the L_{fj} is filter inductance, R_{fj} is filter resistance, and R_j is load resistance (j a, b, c).

The voltage in each leg of the four-leg inverter can be expressed as

$$\left.\begin{array}{ll} v_{aN} & S_a v_{dc}, \\ v_{bN} & S_b v_{dc}, \\ v_{cN} & S_c v_{dc}, \\ v_{nN} & S_n v_{dc} \end{array}\right\} \quad (7)$$

where S_a, S_b, S_c, and S_n are the switching states, v_{dc} and v_{nN} are dc-link and load neutral voltages, respectively.

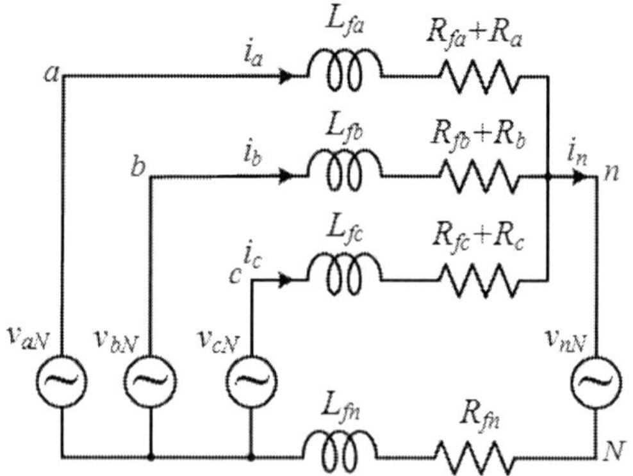

Fig. 3. Equivalent circuit of the three-phase four-leg inverter.

The output voltage of this inverter can be written in terms of the previous inverter voltages

$$\left.\begin{aligned} v_{an} &\quad S_a - S_n \; v_{dc}, \\ v_{bn} &\quad S_b - S_n \; v_{dc}, \\ v_{cn} &\quad S_c - S_n \; v_{dc}. \end{aligned}\right\} \tag{8}$$

By applying Kirchhoffs voltage law to Fig. 3, the inverter voltages can be expressed in terms of load-neutral voltages and load currents as, in the following:

$$\left.\begin{aligned} v_{aN} &\quad R_{fa} \quad R_a \; i_a \quad L_{fa}\tfrac{di_a}{dt} \quad v_{nN}, \\ v_{bN} &\quad R_{fb} \quad R_b \; i_b \quad L_{fb}\tfrac{di_b}{dt} \quad v_{nN}, \\ v_{cN} &\quad R_{fc} \quad R_c \; i_c \quad L_{fc}\tfrac{di_c}{dt} \quad v_{nN}. \end{aligned}\right\} \tag{9}$$

From (8) and (9), the output (load) voltages can be expressed as

$$\left.\begin{aligned} v_{an} &\quad R_{fa} \quad R_a \; i_a \quad L_{fa}\tfrac{di_a}{dt}, \\ v_{bn} &\quad R_{fb} \quad R_b \; i_b \quad L_{fb}\tfrac{di_b}{dt}, \\ v_{cn} &\quad R_{fc} \quad R_c \; i_c \quad L_{fc}\tfrac{di_c}{dt}. \end{aligned}\right\} \tag{10}$$

which can be simplified to

$$v_j \quad R_{fj} \quad R_j \; i_j \quad L_{fj}\frac{di_j}{dt}, \quad j \quad a,b,c. \tag{11}$$

and neutral current i_n can be written as

$$i_n \quad i_a \quad i_b \quad i_c \tag{12}$$

The output (load) current can be derived from (11) as

$$\frac{di_j}{dt} \quad \frac{\ }{L_f} \; v_j - R_{fj} \quad R_j \; i_j \;, \quad j \quad a,b,c. \tag{13}$$

III. The Proposed MPC Scheme

The proposed MPC scheme is shown in Fig. 4. It has two main layers, consisting of predictive model, and cost function optimization.

A. Predictive models of the system

To control ZS four-leg inverter output currents (i_a, i_b, i_c), input current (i_L), and capacitor voltage (V_C), three discrete-time models must be defined from the continuous-time equations. To do that, the general structure of the forward-difference Euler equation [18] can be used to compute the differential equations.

1) Predictive Model I: This model is used to predict future behavior of the output currents. The continuous-time expression for the output currents is given in (13). By using forward-difference Euler equation, the discrete-time model of the i_a, i_b, i_c can be obtained as

$$\begin{aligned} i_j \; k \quad & \frac{T_s}{L_f \quad R \quad R_f \; T_s} v_j \; k \\ & \frac{L_f}{L_f \quad R \quad R_f \; T_s} i_j \; k \;, \quad j \quad a,b,c. \end{aligned} \tag{14}$$

where $i_j \; k$ is the predicted output current vector at the next sampling time.

2) Predictive model II: This model is used to predict future behavior of the inductor current (I_L). The continuous-time model of the inductor voltage can be expressed as

$$v_L \quad L\frac{dI_L}{dt} \tag{15}$$

where L is the inductance of the inductor. Based on (15), the inductor current is derived as

$$\frac{dI_L}{dt} \quad \frac{\ }{L} v_L \tag{16}$$

By using forward-difference Euler equation, the discrete-time model of the I_L can be obtained as

$$I_L \; k \qquad I_L \; k \quad \frac{T_s}{L} v_L \; k \tag{17}$$

where $I_L \; k$ is the predicted inductor current at the next sampling time, and $v_L \; k$ is inductor voltage that depends on the states of the ZSI topology. For non-shoot-through and shoot-through states, inductor voltage can be defined as follows:

i. During non-shoot-through state:

$$v_L \quad V_{in} - V_C \tag{18}$$

ii. During shoot-through state:

$$v_L \quad V_{in} \quad V_C \tag{19}$$

3) Predictive model III: This model is used to predict future behavior of the capacitor voltage (V_C). The continuous-time model of the capacitor current can be expressed as

$$i_C \quad C\frac{dV_C}{dt} \tag{20}$$

where C is the capacitance of the capacitor. Based on (20), the capacitor voltage is derived as,

$$\frac{dV_C}{dt} \quad \frac{\ }{C} i_C \tag{21}$$

By using forward-difference Euler equation, the V_C is obtained as

$$V_C \; k \qquad V_C \; k \quad \frac{T_s}{C} i_C \; k \tag{22}$$

where $V_C \; k$ is the predicted capacitor voltage at the next sampling time, and $i_C \; k$ is the capacitor current that depends on the states of the ZSI topology. For non-shoot-through and shoot-through states, capacitor current can be defined as follows:

i. During non-shoot-through state:

$$i_C \quad I_L - S_a i_a \quad S_b i_b \quad S_c i_c \tag{23}$$

ii. During shoot-through state:

$$i_C \quad -I_L \tag{24}$$

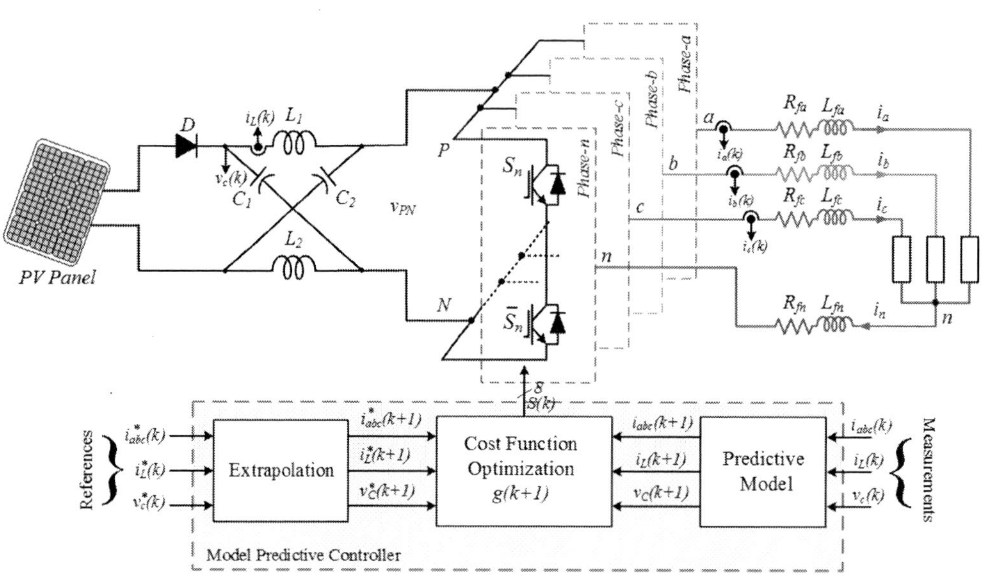

Fig. 4. Block diagram of the proposed model predictive control scheme.

B. Cost Function Optimization

The selection of the cost function is a key element of the MPC technique. The proposed MPC scheme has three cost functions that are used to minimize output current, inductor current, and capacitor voltage errors in the next sampling time.

The cost functions of output currents, inductor current and capacitor voltage are

$$g_{i_o} \quad i_a^* \; k \quad - i_a \; k \qquad i_b^* \; k \quad - i_b \; k \qquad i_c^* \; k \quad - i_c \; k \quad . \tag{25}$$

$$g_{I_L} \quad \lambda_i \, |I_L^* \; k \quad - I_L \; k \quad | \tag{26}$$

$$g_{v_C} \quad \lambda_v \, |v_C^* \; k \quad - v_C \; k \quad | \tag{27}$$

where λ_i and λ_v are the weighting factors. The total cost function is

$$g \; k \qquad g_{i_o} \; k \qquad g_{I_L} \; k \qquad g_{v_C} \; k \tag{28}$$

IV. Hardware-In-The-Loop Test Results

In order to verify the proposed MPC technique for ZS four-leg inverter, Hardware-In-the-Loop (HIL) studies carried out for different reference and load conditions. The HIL test-bench setup of the system is shown in Fig. 5. The proposed control structure is implemented using dSPACE DS1006 microprocessor board with DS2004 high-speed A/D board, and DS4003 digital I/O board. In the implementation of the controller a minimum sampling time of T_s=20 μs is achieved without any overrun. However, further optimization in the programming is possible, but it is not the objective of this paper. The HIL test model of ZS four-leg inverter is built based on Matlab&Simulink. Then, this model is compiled and downloaded into the OPAL-RT simulator, which executed the HIL tests. The parameters used in the tests are summarized in Table I.

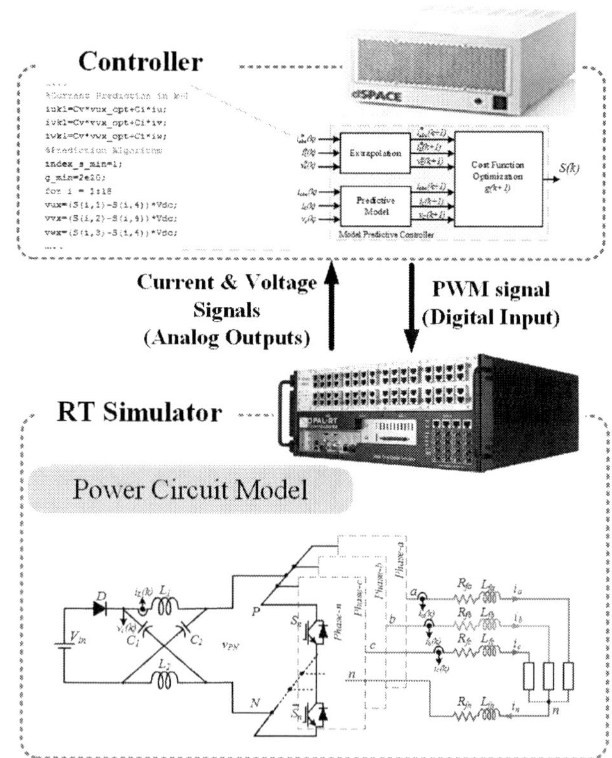

Fig. 5. HIL setup of ZS four-leg inverter.

Table I
SYSTEM PARAMETERS

Parameter	Value
Input DC voltage (V_{in})	150-250 V
Reference Capacitor voltage (V_C^*)	300 V
Z-source inductances (L_1, L_2)	1.5 mH
Z-source network capacitors (C_1, C_2)	470 μF
Load resistance (R)	5-10 Ω
Filter inductance, L_f	10 mH
Filter resistance R_f	0.05 Ω
Nominal frequency (f_o)	50 Hz
Sampling time (T_s)	20 μs

The proposed MPC scheme is tested to verify the robustness of the proposed technique against unbalanced reference currents (i_a^*=15 A, i_b^*=5 A, and i_c^*=15 A) and unbalanced loads (R_a=5 Ω, R_b=R_c=10 Ω). Steady-state results under unbalanced reference output currents and unbalanced loads are shown in Fig. 6 (a). It can be observed from results that the proposed system can control each phase current independently while the capacitor and dc-link voltages are kept constant. Moreover, the neutral current flows through the fourth leg because of the unbalance reference currents.

The transient performance of the proposed MPC scheme for the reference output currents steps from 10 to 15 A are shown in Fig. 6 (b). For this test, reference currents and loads (R_a=R_b=R_c=10 Ω) are balanced. It can be observed from these results that the transient time is very short with no overshoot. The neutral current is zero because of balanced reference output currents. In Fig. 6 (c), the results are presented with an unbalanced reference current step change and unbalanced loads (R_a=5 Ω, R_b=R_c=10 Ω). For this test, all reference output currents are set to 10 A at the beginning. Then, reference currents are set to i_a^*=15 A, i_b^*=10 A, and i_c^*=5 A. Results of this test show that the output currents, inductor current, and the capacitor voltage track their references with high accuracy. However, the double-line frequency ($*f_o$=100 Hz) ripple exist on the ZS network due to unbalanced currents.

V. CONCLUSION

This paper presents model predictive control (MPC) scheme for Z-source (ZS) three-phase four-leg inverter. The main aim of this paper is to achieve single-stage power converter topology for renewable energy based power generation systems under balanced and unbalanced conditions with high control capability. To do that, ZS three-phase four-leg inverter topology was proposed in this study. To improve control capability of the controller, the MPC scheme was employed in the controller stage. Hardware-In-the-Loop (HIL) studies were performed to verify the performance of the proposed inverter topology and its control strategy. The results show that the proposed technique has an excellent steady-state and transient performances.

ACKNOWLEDGMENT

This work was supported by NPRP grant No. NPRP-EP X-033-2-007 from the Qatar National Research Fund (a member of Qatar Foundation). The statements made herein are solely the responsibility of the authors.

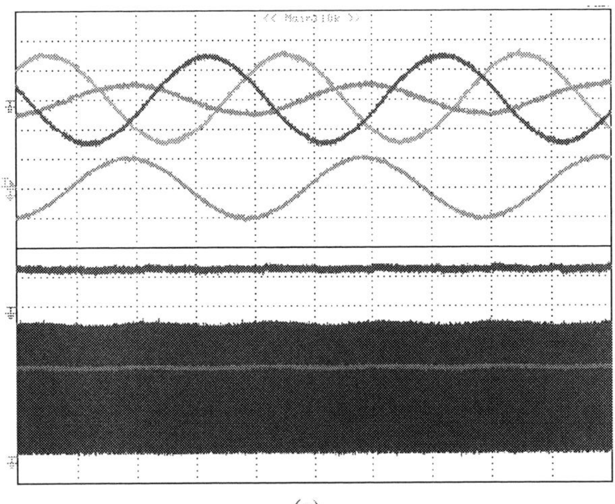

(a)

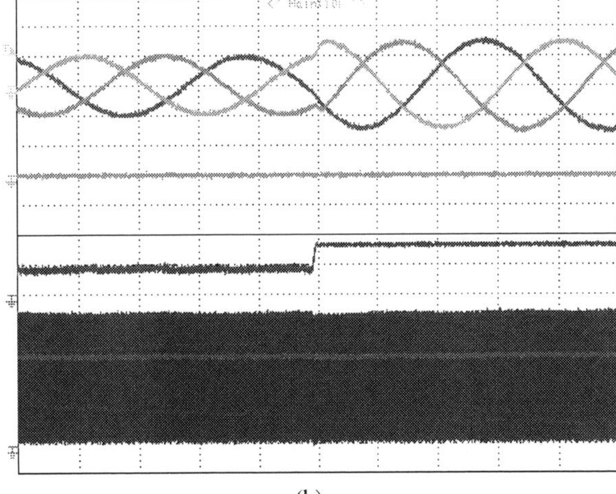

(b)

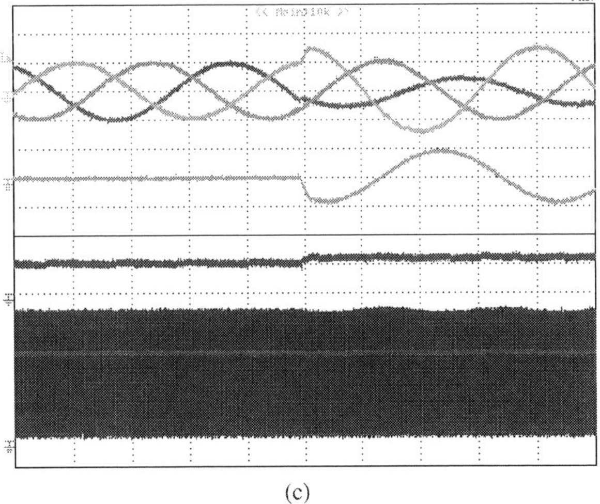

(c)

Fig. 6. HIL test results (a) steady-state analysis with unbalanced reference currents and unbalanced loads; (b) transient response with balanced reference currents and balanced loads; (c) transient response with unbalanced reference currents and unbalanced loads.

REFERENCES

[1] H. Abu-Rub, M. Malinowski, and K. Al-Haddad, *Power Electronics for Renewable Energy Systems, Transportation and Industrial Applications*. A John Wiley&Sons Ltd., 2014.

[2] S. Bayhan, "A labview-based remote laboratory experiments for multi-mode single-leg converter," *Journal of Power Electronics*, vol. 14, no. 5, pp. 1069–1078, Sept 2014.

[3] S. Ozdemir, S. Bayhan, I. Sefa, and N. Altin, "Three-phase multilevel grid interactive inverter for pv systems with reactive power support capability," in *Smart Grid and Renewable Energy (SGRE), 2015 First Workshop on*, 2015, pp. 1–6.

[4] S. Kouro, J. Leon, D. Vinnikov, and L. Franquelo, "Grid-connected photovoltaic systems: An overview of recent research and emerging pv converter technology," *IEEE Industrial Electronics Magazine*, vol. 9, no. 1, pp. 47–61, March 2015.

[5] H. Abu-Rub, A. Iqbal, S. Moin Ahmed, F. Peng, Y. Li, and G. Baoming, "Quasi-z-source inverter-based photovoltaic generation system with maximum power tracking control using anfis," *IEEE Transactions on Sustainable Energy*, vol. 4, no. 1, pp. 11–20, Jan 2013.

[6] Y. Liu, B. Ge, and H. Abu-Rub, "Theoretical and experimental evaluation of four spacevector modulations applied to quasi-z-source inverters," *IET Power Electronics*, vol. 6, no. 7, pp. 1257–1269, August 2013.

[7] O. Ellabban, H. Abu-Rub, and S. Bayhan, "Z-source matrix converter: An overview," *IEEE Transactions on Power Electronics*, vol. PP, no. 99, pp. 1–1, 2015.

[8] I. Vechiu, O. Curea, and H. Camblong, "Transient operation of a four-leg inverter for autonomous applications with unbalanced load," *IEEE Transactions on Power Electronics*, vol. 25, no. 2, pp. 399–407, Feb 2010.

[9] E. Demirkutlu and A. Hava, "A scalar resonant-filter-bank-based output-voltage control method and a scalar minimum-switching-loss discontinuous pwm method for the four-leg-inverter-based three-phase four-wire power supply," *IEEE Transactions on Industry Applications*, vol. 45, no. 3, pp. 982–991, May 2009.

[10] S. Bifaretti, A. Lidozzi, L. Solero, and F. Crescimbini, "Comparison of modulation techniques for active split dc-bus three-phase four-leg inverters," in *2014 IEEE Energy Conversion Congress and Exposition (ECCE)*, Sept 2014, pp. 5631–5638.

[11] L. Yushan, G. Baoming, H. Abu-Rub, and Z. P. Fang, "Modelling and controller design of quasi-z-source inverter with battery-based photovoltaic power system," *IET Power Electronics*, vol. 7, no. 7, pp. 1665–1674, July 2014.

[12] Y. Liu, B. Ge, H. Abu-Rub, and F. Z. Peng, "An effective control method for three-phase quasi-z-source cascaded multilevel inverter based grid-tie photovoltaic power system," *IEEE Transactions on Industrial Electronics*, vol. 61, no. 12, pp. 6794–6802, Dec 2014.

[13] D. Patel, R. Sawant, and M. Chandorkar, "Three-dimensional flux vector modulation of four-leg sine-wave output inverters," *IEEE Transactions on Industrial Electronics*, vol. 57, no. 4, pp. 1261–1269, April 2010.

[14] J. Rodriguez, M. Kazmierkowski, J. Espinoza, P. Zanchetta, H. Abu-Rub, H. Young, and C. Rojas, "State of the art of finite control set model predictive control in power electronics," *IEEE Transactions on Industrial Informatics*, vol. 9, no. 2, pp. 1003–1016, May 2013.

[15] P. Cortes, G. Ortiz, J. Yuz, J. Rodriguez, S. Vazquez, and L. Franquelo, "Model predictive control of an inverter with output lc filter for ups applications," *IEEE Transactions on Industrial Electronics*, vol. 56, no. 6, pp. 1875–1883, June 2009.

[16] P. Cortes, A. Wilson, S. Kouro, J. Rodriguez, and H. Abu-Rub, "Model predictive control of multilevel cascaded h-bridge inverters," *IEEE Transactions on Industrial Electronics*, vol. 57, no. 8, pp. 2691–2699, Aug 2010.

[17] V. Yaramasu and B. Wu, "Model predictive decoupled active and reactive power control for high-power grid-connected four-level diode-clamped inverters," *IEEE Transactions on Industrial Electronics*, vol. 61, no. 7, pp. 3407–3416, July 2014.

[18] S. Bayhan and H. Abu-Rub, "Model predictive control of quasi-z source three-phase four-leg inverter," in *41th Annual Conference of the IEEE Industrial Electronics Society, IECON 2015*, Nov 2015, p. to appear.

[19] J. Guzinski and H. Abu-Rub, "Speed sensorless induction motor drive with predictive current controller," *IEEE Transactions on Industrial Electronics*, vol. 60, no. 2, pp. 699–709, Feb 2013.

[20] S. Bayhan and H. Abu-Rub, "Model predictive sensorless control of standalone doubly fed induction generator," in *40th Annual Conference of the IEEE Industrial Electronics Society, IECON 2014*, Oct 2014, pp. 2166–2172.

Efficiency Optimization of an Integrated Wireless Power Transfer System by a Genetic Algorithm

Rosario Pagano, Siamak Abedinpour

Analog and Power Division,
Integrated Device Technology,
1140 W. Warner Rd, Tempe, AZ, 85224, USA

Angelo Raciti, Salvatore Musumeci

Department of Electrical, Electronics and Computer Engineering, University of Catania,
Viale A. Doria, 6 – 95125 – Catania, Italy

Abstract – **This paper presents a multi-objective genetic algorithm (GA) for a wireless power transfer system composed of a full-bridge series-resonant inverter (FBSRI), a synchronous full-bridge rectifier (SFBR) and a buck converter. The proposed algorithm maximizes the total system efficiency by optimizing the switching frequency of the FBSRI and the SFBR output voltage, subject to their respective design constraints. Two application-specific integrated circuits (ASICs) respectively integrating the transmitter and receiver power-stages along with their controllers were optimized by the new algorithm. A discrete-time small-signal model of the system has been developed to analyze the stability of the wireless charger. Experimental results validating the above analysis are reported along with simulation data. The wireless power transfer system analyzed in this paper conforms to Wireless Power Consortium (WPC) specifications.**

Index terms - *Wireless Power Charger, Integrated Circuits, Magnetic Coupling, Genetic Algorithm, Full-Bridge Series-Resonant Inverter, Synchronous Full-Bridge Rectifier, MOSFET.*

I. INTRODUCTION

The adoption of wireless charging is massively expanding from modern mobile platforms to tablets, phablets and wearable technology, as a result of its capability to support different consumer electronics applications. *Magnetic induction charging* is nowadays a popular technology which is being heavily employed due to earlier adoption of its specifications compared to *magnetic resonance technology* [1]-[5]. Even so, the former charging technique demands the transmitter and receiver coils to be set few millimeters apart and optimally aligned. This disadvantage did not harm the mass production of integrated wireless charging systems based on magnetic induction, and only in recent years integrated circuits based on magnetic resonance have started to emerge in the wireless power ecosystem. A significant number of features must be provided in order to enable and control power transfer from the primary to the secondary coil, such as back-channel communication, over-voltage, over-current and thermal runaway protection, foreign object detection, etc. Consequently, the research and development effort has been focused on system integration to increase power density and reduce bill of material cost [6].

The design and optimization of isolated power systems, including wireless power transfer systems (WPTSs), has been extensively performed, paying attention to discrete-component solutions [7]. However, the need for a highly integrated power system in wireless charging applications has motivated the research of a new method of optimization that also considers the parasitics inherent to application-specific integrated circuits (ASICs) [8]-[9]. Owing to the stray resistance and capacitance associated with the layout of the power stages, a new optimization procedure must be sorted out aiming at maximizing system efficiency. As a wireless charging system is usually composed of a DC-AC stage (half-bridge or full-bridge series-resonant inverter), two coupled coils, a full bridge rectifier and a DC-DC converter, *i.e.* buck regulator, the impact of the power stages parasitics on the total power dissipation cannot be disregarded.

In the above discussion, we need to mention the tradeoff concerning the input voltage of the buck converter, as this influences the efficiencies of both the DC-DC regulator and rectifier in opposing directions. In detail, the buck efficiency decreases at increasing input voltage because of switching losses. So, it could be preferable to operate the buck converter at low input voltage (assuming that switching losses are dominant over conduction losses). However, this will cause even more conduction losses in the rectifier stage since the buck input current must increase to provide the same amount of required power. Accordingly, the required optimization algorithm must observe the afore-mentioned compromise in the search for maximum efficiency.

Non-linear optimization methods are widely adopted to solve complex problems in power electronics circuits and systems. In [10], the *decoupled optimization technique* is improved by introducing two parallel co-adapted evolutionary training processes. An *ant colony optimization* algorithm which dynamically updates the database of components manufactured in discrete and continuous values is presented in [11]. *Weighted gradient direction search* based on chaos optimization has been applied to non-linear constraint optimization problems [12]. In [13], a *fuzzy particle swarm optimization* algorithm based on computational intelligence has been reported. Finally, *Pareto optimization* has been applied to the thermal design of resonant converters [14], and an *electro-thermal PSpice model* of power modules was proposed in [15].

With reference to the optimization problem addressed in this paper, a genetic algorithm (GA) based procedure which maximizes system efficiency was applied to WPTSs. The proposed algorithm is based on a state-space model representing the WPTS and coupled with the model of the power MOSFETs integrated on chip. Layout parasitics and thermal effects have also been included. A discrete-time

978-1-4673-9551-9/16 $31.00 © 2016 IEEE

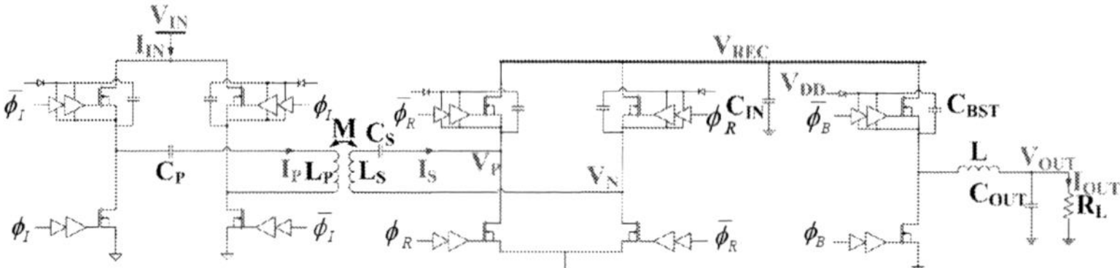

Fig. 1. Schematic of the integrated wireless power transfer system.

small-signal model of the WPTS has subsequently been derived to assess the stability of the optimized system.

II. SYSTEM MODELING

According to the voltage-gain characteristic of a WPTS, power delivered to the load is controlled by adopting variable-frequency control of the TX full-bridge inverter when the frequency is lower than 205kHz, and fixed-frequency variable duty-cycle control when the frequency exceeds 205kHz [6]. Three power conversion stages can be discerned when the system in Fig. 1 is observed:

1) *Full-bridge series-resonant inverter* – this stage performs DC-AC conversion and drives an LC tank, composed of L_P and C_P, which is in turn coupled with a secondary LC tank formed by L_S and C_S. Power switches are turned on diagonally in a complimentary fashion under zero-voltage switching. NMOS devices are employed for all four switches to reduce the on-resistance and die size. As a consequence, two bootstrap circuits are required to supply the drivers of the high-side switches.
2) *Synchronous full-bridge rectifier* – AC-DC conversion is performed by adopting a four-switch topology wherein NMOS devices are diagonally turned on and off based on V_P and V_N. Zero-voltage and zero-current detection assist respectively the turn on and turn off of diagonal switches pairs to reduce the conduction time of the body diodes built-in in the four NMOSs. Once again, bootstrapped drivers are used for the high-side FETs.
3) *Buck converter* – this last stage of the WPTS down-converts V_{REC} to 5V, delivering an output power of 5-7.5W. The DC-DC converter operates at 3MHz and its input voltage ranges from 6V to 20V. Since a minimum off-time must be provided to charge up the bootstrap capacitor C_{BST} when the low-side NMOS is on, the duty-cycle of the buck regulator is limited to nearly 90%.

The switching waveforms of the power system can be derived by considering the simplified electrical schematic in Fig. 2. Input voltage V_{IN} is 5V and FBSRI's switches are modeled by their on-state resistance $r_{ds,I}$. r_P and r_S represent the stray resistance of primary and secondary coil windings, respectively. At secondary side, $r_{ds,R}$ models the on-resistance of the SFBR switches, while r_{OFF} is the off-resistance. The buck regulator is represented by its input

Stage 1: [t₁, t₂]

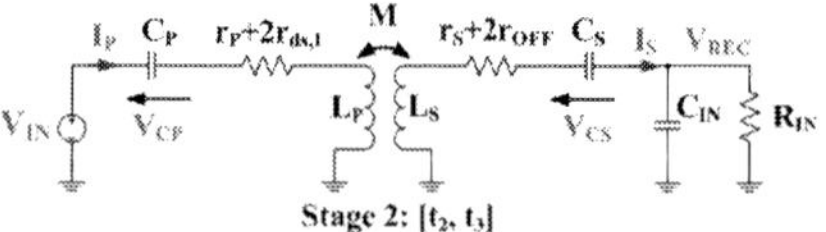

Stage 3: [t₃, t₄]

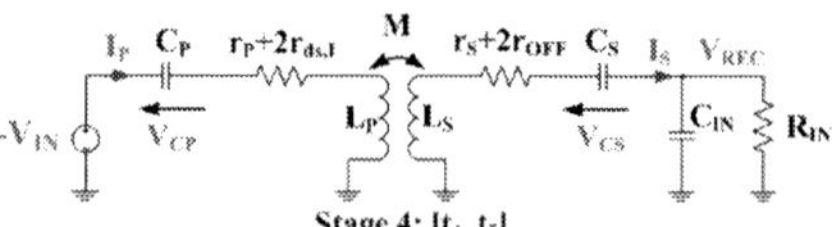

Fig. 2. Equivalent electrical circuit of the wireless power transfer system shown in Fig. 1.

DC resistance, namely R_{IN}. As $V_{REC} = R_{IN}|I_S|_{avg}$ and $\eta_B V_{REC}|I_S|_{avg} = V_{OUT}I_{OUT}$, we can express the input resistance of the buck converter as in (1):

$$R_{IN} = \frac{\eta_B V_{REC}^2}{V_{OUT}I_{OUT}} \tag{1}$$

The above parameter reflects the amount of power required by the buck load and takes into consideration the efficiency of the DC-DC regulator. In the following analysis, a state-space modeling approach is applied to derive the operating waveforms of the WPTS. By defining $\alpha = (M^2 - L_P L_S)^{-1}$, the state matrix and input vector relative to stage 1 are:

$$\boldsymbol{A}_{1,2} = \begin{bmatrix} \alpha L_S R_P & \alpha M R_S & \alpha L_S & \alpha M & \alpha M \\ \alpha M R_P & \alpha L_P R_S & \alpha M & \alpha L_P & \alpha L_P \\ C_P^{-1} & 0 & 0 & 0 & 0 \\ 0 & C_S^{-1} & 0 & 0 & 0 \\ 0 & C_{IN}^{-1} & 0 & 0 & -(R_{IN}C_{IN})^{-1} \end{bmatrix} \tag{2}$$

$$\boldsymbol{B}_{1,2} = \begin{bmatrix} -\alpha L_S V_S \\ -\alpha M V_S \\ 0 \\ 0 \\ 0 \end{bmatrix} \tag{3}$$

where $R_P = r_P + 2r_{DS,I}$ and $R_S = r_S + 2r_{DS,R}$. In this phase, the primary side is coupled with the secondary side, and power transfer occurs. Eqs. (2) and (3) also hold during stage 2, where now $R_S = r_S + 2r_{OFF}$, since the SFBR is fully off. Accordingly, the primary coil sees a large impedance when looking at the secondary side, and the buck input capacitor gets discharged while providing power to the load. When stage 3 is considered, we define the following state and input matrices:

$$A_{3,4} = \begin{bmatrix} \alpha L_S R_P & \alpha M R_S & \alpha L_S & \alpha M & -\alpha M \\ \alpha M R_P & \alpha L_P R_S & \alpha M & \alpha L_P & -\alpha L_P \\ C_P^{-1} & 0 & 0 & 0 & 0 \\ 0 & C_S^{-1} & 0 & 0 & 0 \\ 0 & -C_{IN}^{-1} & 0 & 0 & -(R_{IN}C_{IN})^{-1} \end{bmatrix} \quad (4)$$

$$B_{3,4} = \begin{bmatrix} \alpha L_S V_S \\ \alpha M V_S \\ 0 \\ 0 \\ 0 \end{bmatrix} \quad (5)$$

with $R_S = r_S + 2r_{DS,R}$. Eqs. (4) and (5) also apply during stage 4, where $R_S = r_S + 2r_{OFF}$ (fully off SFBR). The solution of a linear time-invariant system during stage $[t_{j+1}, t_j]$ is (see [8]):

$$X(t) = e^{A_j(t-t_j)}X(t_j) + A_j^{-1}(e^{A_j(t-t_j)} - I)B_j \quad (6)$$

Switching period T_S represents a variable of the system under analysis, while t_3 and t_5 are consequently determined as $t_3 = DT_S$ and $t_5 = T_S$. By defining the state-space vector X as:

$$X = \begin{bmatrix} I_P \\ I_S \\ V_{CP} \\ V_{CS} \\ V_{REC} \end{bmatrix} \quad (7)$$

we can determine the initial solution by equating $X(t_1)$ and $X(t_5)$, as we are searching for a periodic solution, yielding:

$$X(t_1) = \frac{\sum_{j=1}^{4} \prod_{k=j+1}^{4} e^{A_k(t_{k+1}-t_k)} A_j^{-1}(e^{A_j(t_{j+1}-t_j)} - I)B_j}{I - \prod_{j=1}^{4} e^{A_j(t_{j+1}-t_j)}} \quad (8)$$

With the aid of Fig. 3, we notice that at $t=t_2$ and $t=t_4$, secondary current I_S approaches 0A (zero-current detection). If T_S is known, we can derive the previous time instants using a Newton-Raphson algorithm. Setting $k_2 = [0 \;\; 1 \;\; 0 \;\; 0 \;\; 0]$, we need to find the roots of the following system:

$$F = \begin{bmatrix} k_2 X(t_2) \\ k_2 X(t_4) \end{bmatrix} \quad (9)$$

whose Jacobian is:

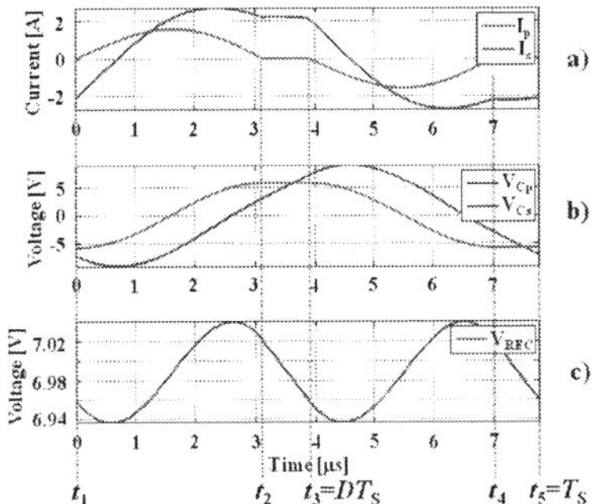

Fig. 3. Simulation results obtained by the equivalent electrical circuit in Fig. 2. a): I_P and I_S; b): V_{CP} and V_{CS}; c): V_{REC}.

$$J = \begin{bmatrix} k_2 \dot{X}(t_2) & 0 \\ 0 & k_2 \dot{X}(t_4) \end{bmatrix} \quad (10)$$

By iteration, the new solution pair can be found from the old one as:

$$\begin{bmatrix} t_{2,new} \\ t_{4,new} \end{bmatrix} = \begin{bmatrix} t_{2,old} \\ t_{4,old} \end{bmatrix} - J^{-1}F \quad (11)$$

The algorithm stops when the norm of the error vector is contained within threshold ε:

$$\| t_{2,new} - t_{2,old}, t_{4,new} - t_{4,old} \| \leq \varepsilon \quad (12)$$

Using the previously determined values of t_2 and t_4, along with $t_5=T_S$ and D=0.5, the initial solution in (8) can be determined. As a result, the state-space vector can be fully generated within the switching period. It should be noticed that the above sub-routine takes T_S as an input, which is solved by the genetic algorithm presented in Section III. Accordingly, system (1)-(12) is solved self-consistently with the GA later discussed.

To complete the modeling of the WPTS, we need to determine the average input current of the system:

$$I_{IN,avg} = \frac{1}{T_S}\left\{\int_{t_1}^{DT_S} I_P dt - \int_{DT_S}^{T_S} I_P dt\right\} \quad (13)$$

which is used to compute the total system efficiency η_T as:

$$\eta_T = \frac{V_{OUT} I_{OUT}}{V_{IN} I_{IN,avg}} \quad (14)$$

III. OPTIMIZATION ALGORITHM

In this section, the GA is described using the model equations developed in Section II. The purpose of the algorithm is to find an optimal pair of V_{REC} and F_S in order to maximize system efficiency. Based on the information provided by the algorithm, a feedback-control loop is designed to set V_{REC} to its optimal value. It is worth to be noticed that V_{REC} is regulated by controlling F_S as it will be explained when the communication protocol of the WPTS is discussed in detail. Junction temperatures T_j^T and T_j^R of the transmitter and receiver, respectively, are also re-evaluated

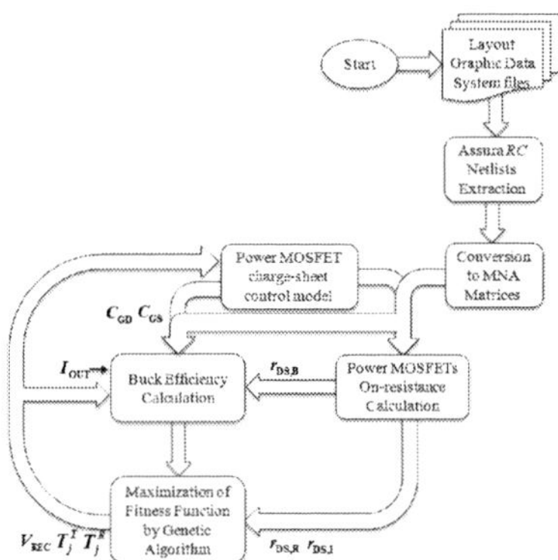

Fig. 4. Flow-diagram illustrating the steps involved in the development of the proposed optimization algorithm.

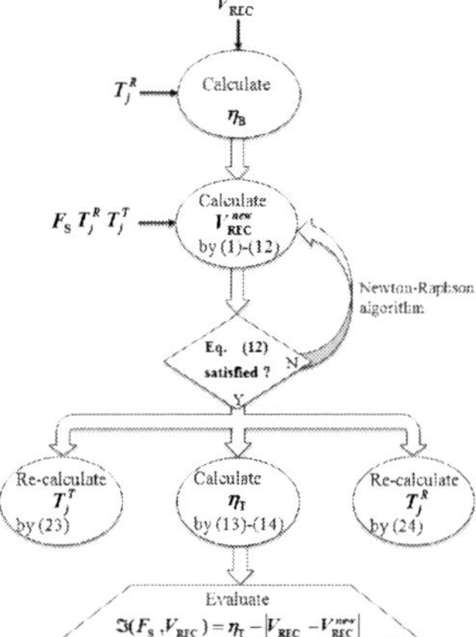

Fig. 5. Fitness function invoked by the genetic algorithm.

every-time V_{REC} is updated to compute buck efficiency and power dissipations in the ASICs.

A) Genetic algorithm

The optimization algorithm proposed in this paper was developed in Matlab and includes the following steps as described in Fig. 4:

- The graphical data system files relative to the power MOSFETs in the WPTS are converted into *RC*-extracted SPICE netlists.
- The generated SPICE netlists are subsequently converted into trans-conductance matrices using MNA (Modified Node Analysis), and integrated with the power MOSFET model for on-resistance and capacitance calculation.
- Buck efficiency for a given load current is computed using an initial guess solution for V_{REC}.
- The fitness function is then maximized by the genetic algorithm, while the difference between V_{REC} and its update from an inner Newton-Raphson algorithm tends to a small value.

The genetic algorithm is articulated as depicted in Fig. 5. It is noticed that maximizing the fitness function means

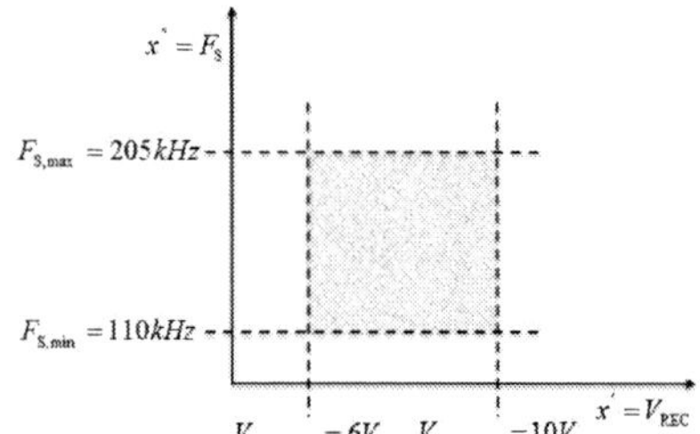

Fig. 6. Boundary intervals of the two variables V_{REC} and F_S.

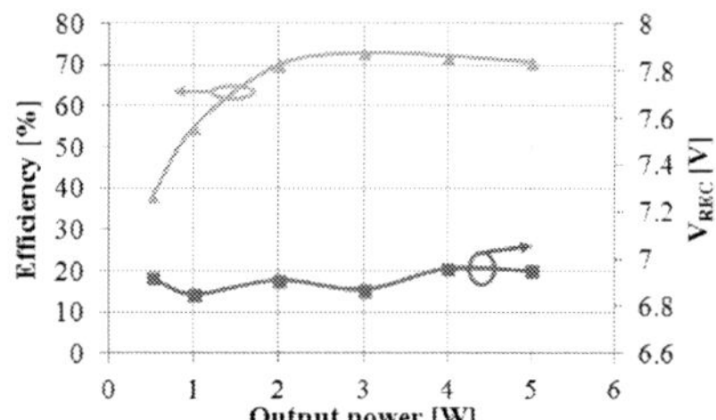

Fig. 7. Model results relative to the proposed GA: total system efficiency and V_{REC}.

maximizing η_T, and minimizing the difference between V_{REC} and its update from the inner Newton-Raphson loop. So, the minimization and maximization tasks are embedded together in the proposed GA. As shown in Fig. 6, variables V_{REC} and F_S, encoded into 8-bit chromosomes, are bounded within their respective intervals. The interval relative to F_S is defined by Wireless Power Consortium (WPC) specifications, while the minimum value of V_{REC} arises from the need to guarantee a minimum off-time in the buck converter's PWM logic. A maximum value has also been specified as buck efficiency drops to 85% at 12V. In Fig. 7, the model results relative to the proposed GA showing total system efficiency and V_{REC} are plotted. Notice that the optimal value of V_{REC} providing maximum efficiency is around 6.9V. F_S varied negligibly from 135 to 145kHz across the output power range specified in the figure.

B) ASICs Communication

Based on the above analysis, the WPTS control has been designed in such a way to provide the buck converter with an input voltage of 6.9V. Fig. 8 illustrates the implementation of the communication strategy in our system. V_{REC} is sensed along with the buck output current and converted by an ADC, which sends in turn the binary-code representation of the voltage and current to a microprocessor unit. Next, a logic circuit drives two MOSFETs to perform capacitive modulation onto V_P and V_N. The modulation signal is then sent back to the transmitter, and separated from the 110-205kHz switching waveform using a high-pass filer. The high-frequency signal is subsequently demodulated and combined in the TX microprocessor unit with the sensed input current of the FBSRI. Eventually, the switching frequency of the DC-AC converter is updated until V_{REC} sets to 6.9V. The PID algorithm embedded into the TX microprocessor unit is

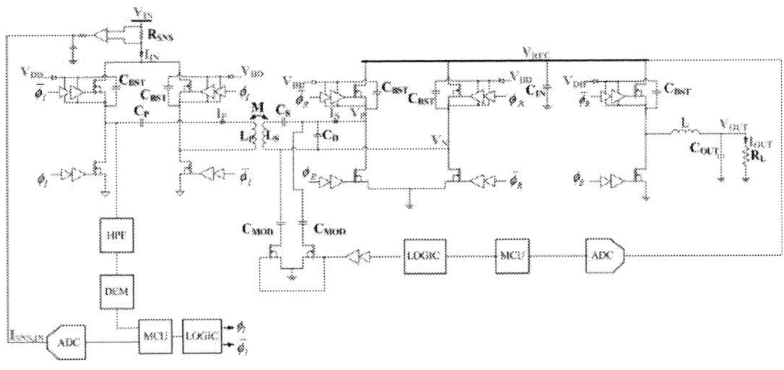

Fig. 8. The receiver integrates the SFBR and buck converter with their controllers and driving stages. The ADC, MCU and AC-modulation NMOSs are also integrated on chip. C_{MOD}, C_{BST}, L, C_{OUT} and R_{SNS} are external components. The transmitter integrates the FBSRI, the controller and driving stage. The high-pass filter (HPF), demodulator, ADC, MCU and logic are also integrated on chip.

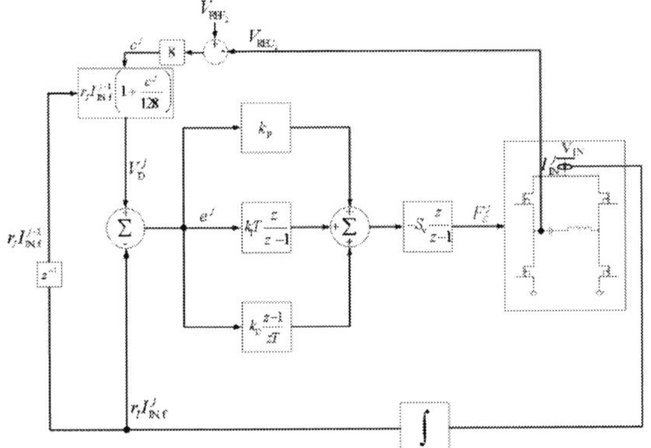

Fig. 9. PID circuit at the transmitter side implementing power transfer control protocol. T represents the time to execute a single loop iteration, and is constrained between 1ms and 5ms.

shown in Fig 9. The j-th control error packet referred to as c^j represents the high-frequency content of the inverter switching waveform generated by the receiver. Using this information along with the previously sampled average primary cell current I_{IN}^{j-1}, V_D^j is derived. The latter is subsequently compared against $r_I I_{IN}^j$, denoting the current average primary-cell current, to yield error voltage e^j. The output of the PID block is then converted into control variable F_S^j using scaling factor S_V, and the new primary-cell current is eventually obtained.

IV. STABILITY ANALYSIS

Aiming at analyzing the stability of the WPTS, a small-signal model of the system in Fig. 8 is derived by resorting to sampled-data modeling technique. Taking advantage of the discrete-time modeling focus on the switching actions, the effects of the variable switching frequency on the system dynamics are fully captured. According to WPC standards, the duty-cycle is constant and equal to 50% when the operating frequency of the FBSRI is below 205kHz. As shown in Fig. 9, a digital integrator implementing a pole in the origin was employed to generate the clock frequency. Consequently, in the following model derivation, only

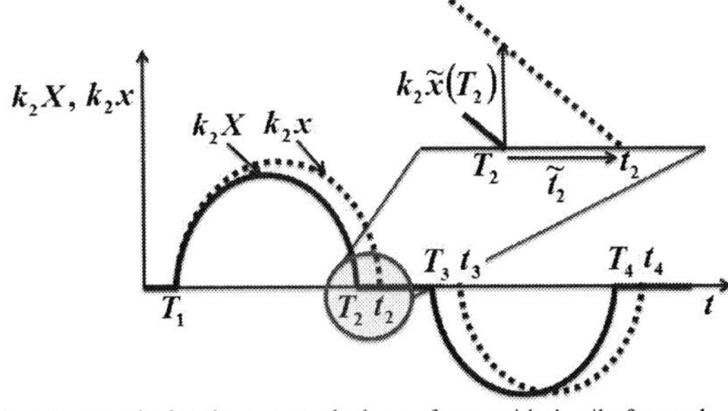

Fig. 10. Perturbed and non-perturbed waveforms with detail of perturbation interval $[T_2, t_2]$.

variable switching frequency is considered. By perturbing (6) over interval $[t_{j+1}, t_j]$, the state vector perturbation at time instant T_n can be expressed as in (15) (see [8]):

$$\tilde{x}(T_n) = \Phi_{n,1}\tilde{x}(T_1) + \textstyle\sum_{j=1}^{n-1} \Gamma_{j,n}\, \tilde{t}_{j+1} \tag{15}$$

where the following identities are used:

$$\Phi_{n,1} = \textstyle\prod_{j=n-1}^{1} \phi_{j+1,j} \tag{16}$$

$$\Gamma_{j,n} = \textstyle\prod_{k=j+1}^{n-1} \phi_{k+1,k}\, \gamma_{j+1} \tag{17}$$

$$\phi_{j+1,j} = e^{A_j(T_{j+1}-T_j)} \tag{18}$$

$$\gamma_{j+1} = (A_j - A_{j+1})X(T_{j+1}) + B_j - B_{j+1} \tag{19}$$

In (15)-(17), n=5. The perturbations superimposed on time instants T_2 and T_4 can be determined by considering that in Fig. 10 $i_S = k_2 x$ is zero. We can thus write:

$$i_S(t_2) = k_2 x(t_2) = 0 \tag{20}$$

Defining $k_5 = [0 \quad 0 \quad 0 \quad 0 \quad 1]$, the plant transfer function is eventually obtained as:

$$P(z) = \frac{\tilde{v}_{REC}}{\tilde{f}_S} = -\frac{k_5}{F_S^2}(zI - \Phi_T)^{-1}\Lambda_5 \tag{21}$$

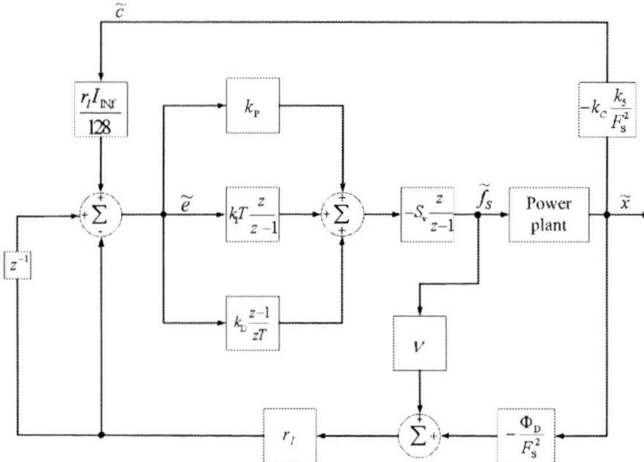

Fig. 11. Small-signal block diagram of the WPTS.

where the identity $\tilde{t}_S = -\frac{1}{F_S^2}\tilde{f}_S$ has been applied The derivation of parameters $\mathbf{\Phi}_T$ and $\mathbf{\Lambda}_5$ in (21) are not reported.

After having determined the small-signal model of the plant, we now proceed with the derivation of the control transfer function. In this respect, in Fig. 8, we notice that V_{REC} is scaled by a resistive divider having a gain of 1/8 and is subsequently input to an 8-bit ADC. Considering that the ADC input voltage range is [0V, 2.5V], the decimal representation of the binary output of the ADC is:

$$V_{\text{REC}_2} = \frac{2^8}{8}\frac{V_{\text{REC}}}{2.5} = \frac{k_C}{8}V_{\text{REC}} \tag{22}$$

At the transmitter side, the control packet referred to as c_j in Fig. 9 is calculated as:

$$c_j = 8\left(V_{\text{REC}_2} - V_{\text{REF}_2}\right) \tag{23}$$

where

$$V_{\text{REF}_2} = \frac{2^8}{8}\frac{6.9}{2.5} \cong 88 \tag{24}$$

In (23), the multiplication factor 8 accounts for the resistive divider gain. Accordingly, the conversion gain from V_{REC} to the input of the transmitter controller is k_C=102.5 LSB/V. As shown in Fig. 9, an inner current loop and outer voltage loop are implemented to control F_S. The current control loop senses and then averages the input current of the FBSRI. Parameter r_I represents the product of R_{SNS} and the gain of the operational amplifier in Fig. 8 (transmitter side). The error voltage perturbation at the input of the PID controller is then expressed as:

$$\tilde{e} = r_I(z^{-1} - 1)\tilde{\imath}_{\text{IN,f}} + \frac{r_I I_{\text{IN,f}}}{128}\tilde{c} \tag{25}$$

From (22) and (23), we obtain $\tilde{c} = k_C\tilde{v}_{\text{REC}}$. Using (21), we can thus write:

$$\frac{\tilde{c}}{\tilde{f}_S} = -\frac{k_C}{F_S^2}\mathbf{k}_5(z\mathbf{I} - \mathbf{\Phi}_{\text{T}})^{-1}\mathbf{\Lambda}_5 \tag{26}$$

The derivation of $\tilde{\imath}_{\text{IN,f}}$ (perturbation on average input current) in (25) involves more mathematical steps, and starts

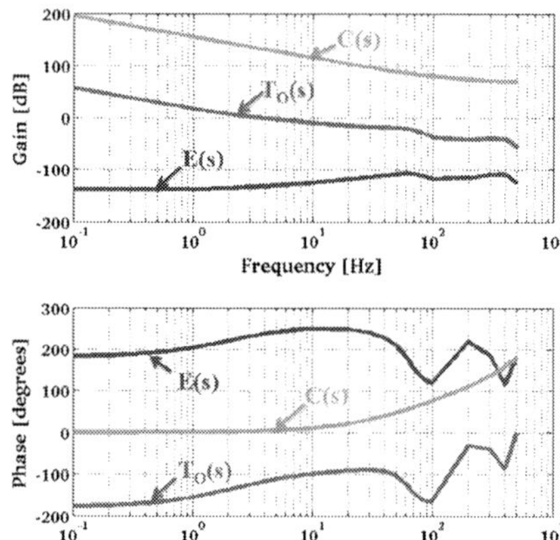

Fig. 12. Bode plots of the frequency to error-voltage transfer function, control transfer function and open loop transfer function.

by considering that the integral of the state vector in $[t_{j+1}, t_j]$ is:

$$\int_{t_j}^{t_{j+1}} \boldsymbol{x}(t)dt = \boldsymbol{A}_j^{-1}\left[\boldsymbol{x}(t_{j+1}) - \boldsymbol{x}(t_j) - \boldsymbol{B}_j(t_{j+1} - t_j)\right] \tag{27}$$

Defining $\boldsymbol{k}_1 = [1 \quad 0 \quad 0 \quad 0 \quad 0]$ and using (13) along with (15), the average input current can be expressed as:

$$i_{\text{IN,f}} = \frac{\boldsymbol{k}_1}{t_5 - t_1}\sum_{j=1}^{4}\alpha_j \boldsymbol{A}_j^{-1}\left[\boldsymbol{x}(t_{j+1}) - \boldsymbol{x}(t_j) - \boldsymbol{B}_j(t_{j+1} - t_j)\right] \tag{28}$$

where $\alpha_j = \begin{cases} 1, \ j = 1,2 \\ -1, \ j = 3,4 \end{cases}$. By perturbing (28), we eventually come up with the following equation:

$$\frac{\tilde{\imath}_{\text{IN,f}}}{\tilde{f}_S} = -\frac{\mathbf{\Phi}_{\text{D}}}{F_S^2}(z\mathbf{I} - \mathbf{\Phi}_{\text{T}})^{-1}\mathbf{\Lambda}_5 + \boldsymbol{\upsilon} \tag{29}$$

where $\mathbf{\Phi}_{\text{D}}$ and $\boldsymbol{\upsilon}$ deriving from the perturbation procedure are also not shown in this paper. It should be finally noticed that replacing (26) and (29) in (25) yields the frequency to error-voltage transfer function, namely $E(z)$. The control transfer function is the product of the PID transfer function and the one associated with the voltage-to-frequency conversion, that is:

$$C(z) = \frac{\tilde{f}_S}{\tilde{e}} = -S_V\frac{z}{z-1}\left(k_P + k_D\frac{z-1}{zT} + k_I T\frac{z}{z-1}\right) \tag{30}$$

In Fig. 11, the complete small-signal model of the WPTS is reported, while the loop gain plots are shown in Fig. 12, with details of the frequency to error-voltage transfer function and control transfer function. By visual inspection, the following observations can be made:

- The magnitude of $E(s)$ is below -100dB as a result of the control variable being the switching frequency. As a matter of fact, in (26) and (29) and consequently in (25), the term F_S^2 appears at denominator. The phase of

Fig. 13. Transmitter (top) and receiver (bottom) ASICs with their respective coils.

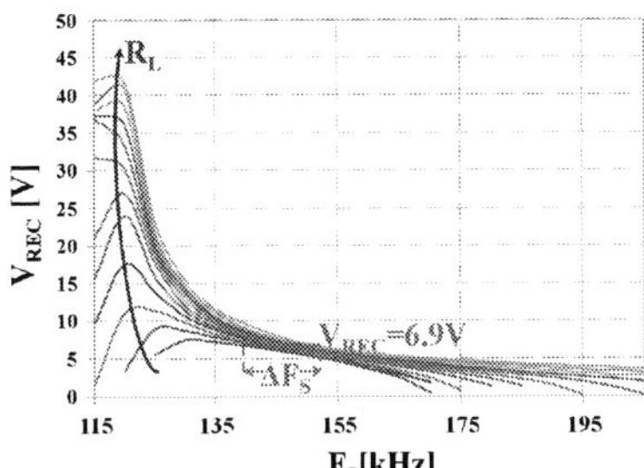

Fig. 14. Experimental voltage-frequency characteristics of the WPTS at several load steps. I_{OUT} = 0.1A to 1.2A at 0.1A step.

Table I. Control parameters.

Parameter	Value
k_P	$1mV^{-1}$
k_I	$0.05mV^{-1}ms^{-1}$
k_D	$0mV^{-1}ms$
S_V	1.5Hz
T	1ms

$E(s)$ starts from 180 degrees because of the minus sign in (26) and (29).

- $C(s)$ exhibits two poles in the origin and a zero around 40Hz. Notice that the phase of $C(s)$ starts from zero degrees due to the inversion sign in (30).
- Based on the compensator design, the open loop transfer function, referred to as $T_O(s)$ in Fig. 12, has a gain cross-over frequency of 2.5Hz and a phase margin of 60 degrees.

In Table I, the parameters of the PID controller are reported. As $E(s)$ already contains a zero near the cross-over frequency (see term $(z^{-1} - 1)$ in (25)), the derivative gain of the PID controller has been set to zero. In accordance with the above analysis, the MCU inside the transmitter chip has been designed to have the control parameters collected in Table I.

V. Experimental Results

In Fig. 13, TX and RX ASICs are shown with their respective development boards. The primary ASIC adopts a quad-flat no-leads package with a size of 36mm^2, while the secondary ASIC uses a wafer-level chip scale package having a size of 16 mm^2. The buck regulator operates at 3MHz and uses a voltage-mode control loop complemented with a feed-forward control loop. Break-before-make technique is adopted to ensure no cross-conduction occurs between the power MOSFETs. The same consideration applies to the power switches of the SFBR and FBSRI. It should be mentioned that a standard start-up routine implemented by programming an internal ROM is automatically loaded when a 5V power supply is enabled.

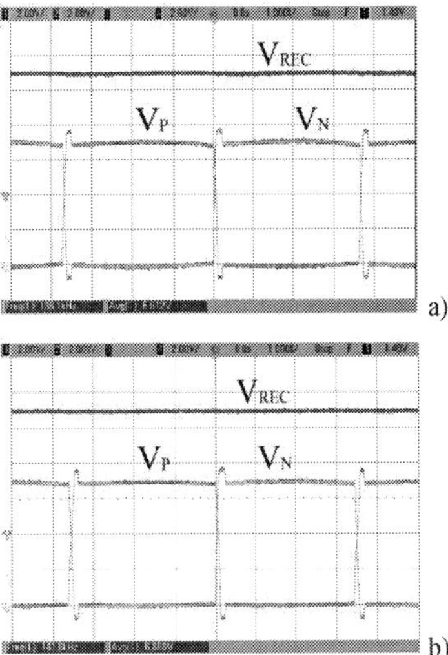

Fig. 15. Experimental waveforms pertaining to the ASIC wireless power system: a) I_{OUT}=1A, b) I_{OUT}=0.5A. V_P=V_N=V_{REC}=2V/div.

The voltage frequency characteristics of the WPTS are show in Fig. 14 at several steps of the load current. As highlighted, the frequency range corresponding to V_{REC} = 6.9V is [140kHz, 150kHz] across the load current range under interest. Such a narrow variation of the switching frequency allows avoiding pole-splitting phenomena affecting system stability. According to previous discussions, capacitive modulation on the rectifier inputs is performed to send communication packets to the transmitter and regulate the rectified voltage to the optimal value of 6.9V. In Fig. 15, V_{REC}, V_P and V_N are shown, respectively at I_{OUT}=1A and I_{OUT}=0.5A. In both cases, the rectified voltage is a about 6.9V in accordance with the model results shown in Fig. 7. In Fig. 16, a load-step transient at the buck output is shown. The load current is stepped up from 0 to 1A, and both V_{REC} and V_{OUT} are observed. Before the step is applied, the load current is 0A and V_{REC} is 8V. The reason is that at no load (<50mA), the control loop sets the output voltage to 8V in such a way to guarantee a minimum voltage of 6V at the input of the buck regulator, when the positive load step occurs. As a matter of fact, when the load increases up to 1A, V_{REC} suddenly drops down to 6V and then slowly rises. The high-frequency ripple superimposed on V_{REC} for communication can be noticed lasting 30ms. During this time interval, control error packets are communicated to the TX side using capacitive modulation on the rectifier inputs.

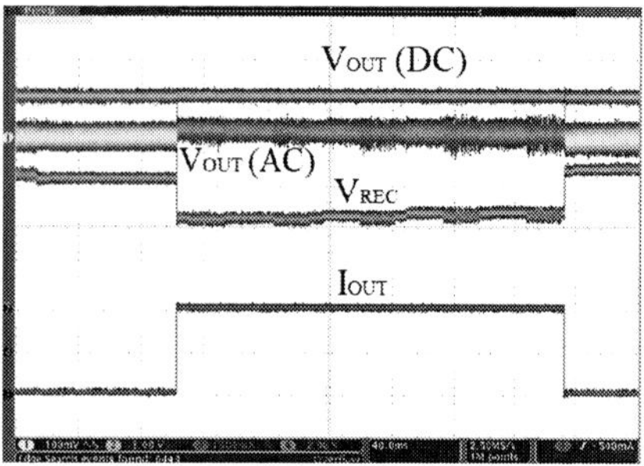

Fig. 16. 0A-to-1A load step at buck output. V_{OUT} (DC) = 1V/div, V_{OUT} (AC) = 100mV/div, V_{REC} = 2V/div, I_{OUT} = 0.5A/div, t = 40ms/div.

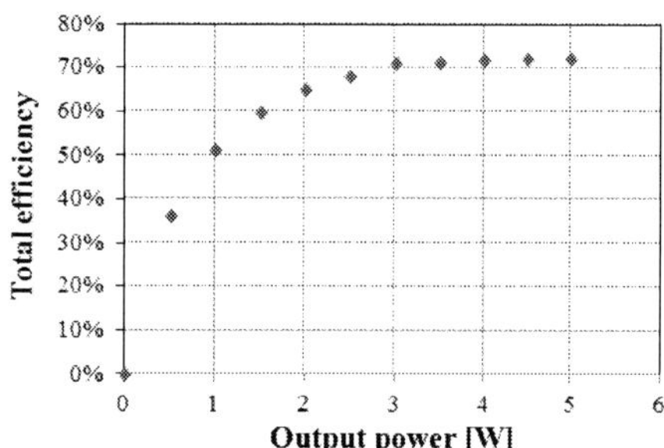

Fig. 17. Total system efficiency versus output power.

While the rectified voltage rises, the load is completely removed, leading V_{REC} to step up to 8V again.

The experimental total system efficiency versus output power is shown in Fig. 17. The efficiency is above 70% in the 3-5W output power range. As the output power decreases below 3W, the efficiency reduces since it is strongly dominated by the buck efficiency. By comparing Fig. 17 with Fig. 7, it emerges that a good correlation between measurement and model results exists, thus conferring validity on the proposed approach.

VI. CONCLUSIONS

In this paper a multi-objective genetic algorithm for a wireless power transfer system composed of a full-bridge series-resonant inverter (FBSRI), a synchronous full-bridge rectifier (SFBR) and a buck converter, has been presented. The proposed algorithm maximizes the total system efficiency by optimizing the switching frequency of the FBSRI and the SFBR output voltage, subject to their respective design constraints. Two ASICs respectively integrating the transmitter and receiver power-stages along with their relative controllers have been used to test the new algorithm. A stability analysis of the WPTS has been also carried out to assess system stability. Experimental results have been provided to validate the proposed approach. The wireless power transfer system analyzed in this work has an efficiency higher than 70% at full load, and adopts a communication protocol between the power transmitter and receiver.

REFERENCES

[1] Choi J.H., Yeo S.K., Park S., Lee J. S., Cho G. H, "Resonant Regulating Rectifiers (3R) Operating for 6.78 MHz Resonant Wireless Power Transfer (RWPT)," in *IEEE J. Solid-State Circuits*, vol. 48, no. 12, pp. 2989–3001, Dec. 2013.

[2] Yin J., Lin D., Lee C.K., Hui S.Y, "A Systematic Approach for Load Monitoring and Power Control in Wireless Power Transfer Systems Without Any Direct Output Measurement," in *IEEE Trans. Power Electron.*, vol. 30, no. 3, Mar, 2015.

[3] Sample A. P., David A. Meyer D. A, Smith J. R, "Analysis, Experimental Results, and Range Adaptation of Magnetically Coupled Resonators for Wireless Power Transfer," in *IEEE Trans. Ind. Electron.*, vol. 58, no. 2, pp. 544-554, Feb. 2011.

[4] Galizzi M., Caldara M., Re V., Vitali A., "A Novel Qi-standard Compliant Full-bridge Wireless Power Charger for Low Power Devices," in *IEEE Wireless Power Transfer (WPT)*, pp. 44-47, Perugia, Italy, May 15-16, 2013.

[5] Li X., Tsui C.Y., Ki W.H., "A 13.56 MHz Wireless Power Transfer System With Reconfigurable Resonant Regulating Rectifier and Wireless Power Control for Implantable Medical Devices," in *IEEE J. Solid-State Circuits*, vol. 50, no. 4, pp. 978–989, 2015.

[6] System Description Wireless Power Transfer. Vol I: Low Power, Part 1: Interface Definition Version 1.1.2 June 2013. *Design specifications*. Available online.

[7] de Groot H., Janssen E., Pagano R., Schetters K., "Design of a 1 MHz LLC resonant converter based on a DSP-driven SOI half-bridge power MOS module," in *IEEE Trans. Power Electron.*, vol. 22, no. 6, pp. 2307–2320, Nov. 2007.

[8] Pagano R., "Sampled-data Modeling of Hysteretic Converters Accounting for Intra-Cycle Waveform Propagation," in *IEEE Trans. Circuits Syst. I, Reg. Papers*, vol. 58, no. 3, pp. 619–632, Mar. 2011.

[9] Degen P., de Groot H., Pagano R., Pansier F., Schetters K., "A Power Management Unit With Integrated Stand-by Supply for Computer Motherboards: Control and VLSI Design," in *IEEE Trans. Power Electron.*, vol. 23, no. 4, pp. 2093-2105, Jul. 2008.

[10] Zhang J., Chung H.S.H, Lo W.L., "Pseudocoevolutionary genetic algorithms for power electronic circuits optimization," in *IEEE Trans. System, Man, Cybern. B, Cybern, vol. 36, no. 4, pp. 590-598, Jul. 2006.*

[11] Zhang J., Chung H.S.H, Lo A.W.L, Huang T., "Extended ant colony optimization algorithm for power electronic circuit design," in *IEEE Trans. Power Electron.*, vol. 24, no. 1, pp. 147-162, Jan. 2009.

[12] Liu S., Hou Z., "Weighted gradient direction based chaos optimization algorithm for nonlinear programming problem," *World Congress on Intell. Control and Autom.*, pp. 1779-1783, Shangai, China, Jun. 10-14, 2002.

[13] Wang A., Huang Z., Lei C., Chao Z., Wu E., "Parameters optimization study and analysis of PID controller in buck converter based on fuzzy particle swarm optimization algorithm," *Int. Conf. Inf. Science, Electron., Electrical Eng.*, pp. 1562-1566, Sapporo, Japan, Apr. 26-28, 2014.

[14] Lucía Ó., Sarnago H., Betés F., Burdío J. M., "Thermal design optimization of a high-efficiency resonant converter based on multi MOSFET cells using the Pareto analysis," *Appl. Power Electron. Conf. Exp.*, pp. 3418-3422, Fort Worth, TX, USA, Mar. 16-20, 2014.

[15] Raciti A., Cristaldi D., Greco G., Vinci G., Bazzano G., "Electro-Thermal PSpice Modeling and Simulation of Power Modules," in *IEEE Trans. Ind. Electron.*, vol. 62, no. 10, pp. 6260-6271, Oct. 2015.

Loss Analysis of a High Efficiency GaN and Si Device Mixed Isolated Bidirectional DC-DC Converter

Fei Xue, Ruiyang Yu, Alex Q. Huang
NSF FREEDM Systems Center
North Carolina State University
Raleigh, NC 27606, USA
fxue@ncsu.edu

Abstract—**High conversion efficiency is always desired in energy storage device (ESD). In this work a high efficiency GaN and Si device mixed isolated bidirectional dc-dc converter is proposed in the distributed ESD application. To optimize the efficiency of the bidirectional half-bridge push-pull active clamp converter over a wide input/output voltage and load range, it is necessary to accurately predict the dissipated power for each power component so as to identify and properly design the heavily loaded components. This paper describes a universal method to predict the power losses of the converter. Loss models are provided to calculate total component losses using the current and voltage information derived from the steady state inductor current calculator. Details of loss breakdown are given. With the presented converter prototype, a top efficiency of 98.3% and an output power of 1 kW in a wide input/output voltage range is achieved. The loss analysis provides valuable information for designing an efficiency optimized converters in the application.**

I. INTRODUCTION

The ever-increasing demand for fossil fuels combined with a worldwide increasing consumption of electric energy, and the ongoing discussion about global warming and emissions reduction of CO_2 present strong incentives for current power grid to further improve the penetration of renewable energies such as solar and wind, to gradually displace conventional power plants, so as to lower the dependent on fossil fuels. DC microgrids that integrate various renewable energy resources is one of the future trends that draw the attention of researchers and engineers. In the DC microgrid, DC ESD is one key enabling technology that compensates the intermittent nature of renewable energy resources and stabilize the whole DC microgrid.

Bidirectional power conversion is always required for ESD connected to DC microgrids. In these applications, high power conversion efficiency and high power density are strongly required. A 98.7% efficiency, 750V 100kW 20kHz bidirectional isolated dual-active-bridge (DAB) dc–dc converter using four 1.2kV 400A SiC MOSFET/SBD dual modules were designed and tested in [1]. Two sets of loss break down results were given under two different assumptions $P_{sw} \propto I^2$ or $P_{sw} \propto I$. An unknown loss was used to represent the temperature dependent conduction loss component related to the SiC modules and other distributed losses. References [2] and [3] combined conventional loss analysis with measured efficiency values, and showed that conventional converter analysis predicts substantially inaccurate efficiency for the given converter. The main reasons why the conventional method fails were described and different steps required to predict the power losses more accurately were documented. An efficiency of more than 92% was achieved at an output power of 2 kW in a wide input/output voltage range. Although improved loss models are given, the efficiency can be further increased and an universal optimization design is needed.

Our ESD design utilizes 12V $LiFePO_4$ battery pack, and the 1C charge/discharge current is as high as 100A. Accordingly, a 1 kW converter is required in order to fully use the battery pack to interface with a high voltage DC microgrid such as the 380Vdc microgrid. In order to design a high-efficiency converter with a low-voltage high-current port, a power loss model is necessary to identify and quantify the critical converter parts with respect to power dissipation. The possibility to accurately estimate the total power losses enables the prediction of efficiency for different operating conditions; the precise calculation of the distribution of the power losses allows for future optimization of converter components. Up to now, no precise and easy-to-apply method for the power loss calculation or optimization has been presented in the literature for this type of application.

One of the conventional methods used to predict magnetic components is the fundamental harmonic approximation method. However, this method only considers fundamental frequency component and produces errors when the inductor current is abundant in high frequency harmonics. With the development of numerical computational techniques, the present research work utilizes nonlinear programming techniques to solve converter steady-state equations. An improved magnetic loss model is proposed in this paper. The key equations are solved by numerical method.

This paper presents a loss analysis of a GaN and Si device mixed novel isolated bidirectional dc-dc converter based on half-bridge center-tap active clamp power stage. The operation principle of main power stage is introduced in section II. A detailed loss model is investigated in Section III. Section IV presents the loss distribution and optimization results. Section V presents experimental results for a 400V-to-12V, 1kW bidirectional dc-dc converter for ESD. Section VI concludes the paper.

978-1-4673-9551-9/16 $31.00 © 2016 IEEE

II. Half-Bridge Push-Pull Active Clamp Converter

Fig. 1 shows a bidirectional isolated half-bridge push-pull active clamp converter suitable for high conversion ratio and low-voltage high-current applications. Its operation principle is similar to that of a DAB dc-dc converter when single phase shift control is applied [5]. Actually, the two bridge converters produce two square wave v_{ab} and v_{cd} (or v_{ec}), with 0.5 duty ratio. There is a phase shift difference of φ and different amplitudes. The inductor between the two bridges is applied with a voltage difference between v_{ab} and v_{cd}. The inductor current can therefore be obtained by

$$i_l(t)=\frac{1}{4\pi f_s L_s}[-\frac{\pi V_{HV}}{2}+n(\pi-2\varphi)V_{LV}]+\frac{V_{HV}/2+nV_{LV}}{L_s}(t-t_1) \tag{1}$$

where $t_1 \leq t \leq t_3$, and

$$i_l(t)=\frac{1}{4\pi f_s L_s}[V_{HV}(\varphi-\frac{\pi}{2})+n\pi V_{LV}]+\frac{V_{HV}/2-nV_{LV}}{L_s}(t-t_3) \tag{2}$$

where $t_3 \leq t \leq t_4$.

The operation principles of the proposed converter can be explained referring to the timing diagram and idealized waveform shown in Fig. 2.

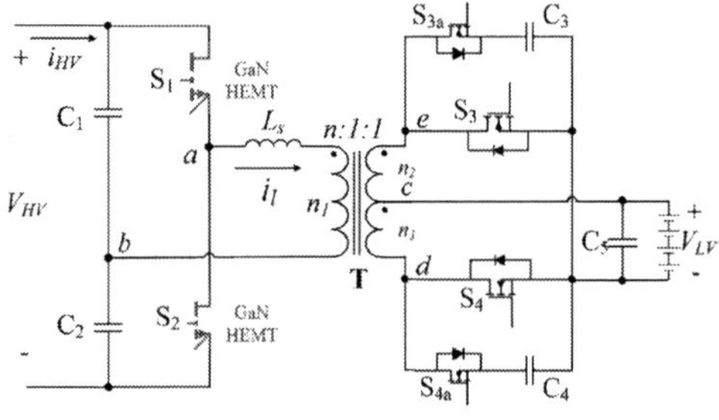

Fig. 1 Topology of proposed converter based on GaN transistors.

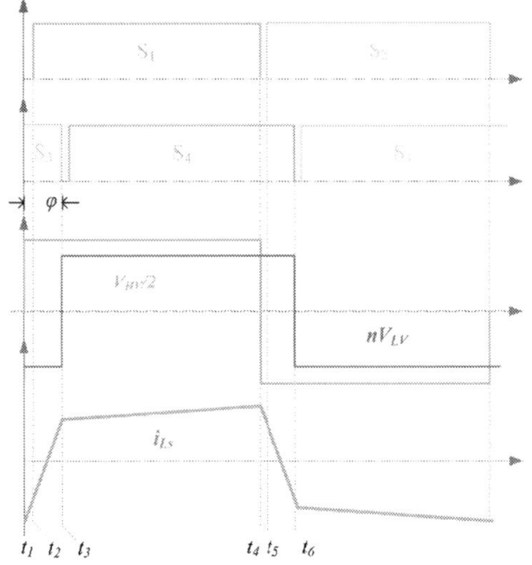

Fig. 2 Steady-state waveform of the proposed converter in charging mode

III. Loss Models

The current waveform of half-bridge push-pull converter are determined by the phase shift angle and calculated by equation (1) and (2). Current harmonics are calculated to predict losses. A numerical method is used to sample a switching cycle with n sample points. The current harmonics are calculated by fast Fourier transform, as shown in Fig. 3. Fig. 3(a) is in light load and 3(b) is in full load situation.

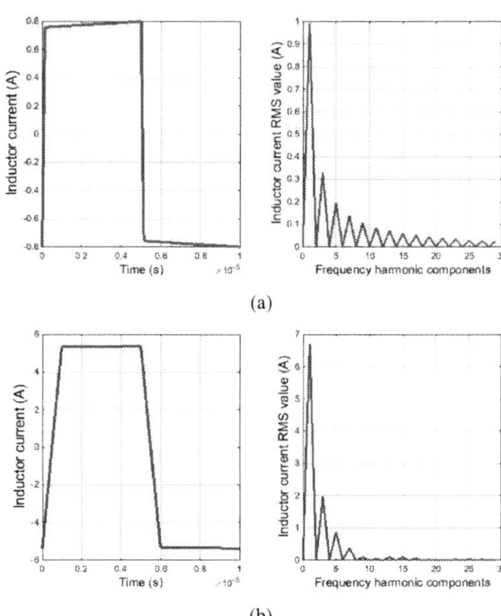

Fig. 3. Inductor current waveform and harmonic components.

A. Primary Side GaN HEMT Loss

GaN HEMT GS66508P-E03 from GaN Systems was selected as the primary side switch. Although GaN HEMT does not have an intrinsic body diode, the device is naturally capable of reverse conduction and exhibits different characteristics depending on the gate voltage.

A simple and effective GaN HEMT switching loss is proposed for the prediction of turn-off switching loss at different turn-off currents, as shown in Fig. 4. This model utilizes a curve fitting method to record SPICE simulation results of turn-off switching loss $E_{off}(I_{off})$. The input voltage is fixed at 360V and 400V in SPICE simulation.

$$E_{off}(I_d)=a\cdot I_d^2+b\cdot I_d+c \tag{3}$$

$$P_{off\_pri}=E_{off}\cdot f_s \tag{4}$$

Since every switch conducts current during half the period in steady state, the RMS switch current is easily obtained from the inductor current I_{rms_Ls}

$$I_{rms\_pri}=\frac{\sqrt{2}}{2}I_{rms\_L_s} \tag{5}$$

$$P_{cond\_pri}=I_{rms\_pri}^2\cdot R_{ds(on)\_pri} \tag{6}$$

$$P_{dr\_pri} = Q_{g\_pri} \cdot V_{g\_pri} \cdot f_s \tag{7}$$

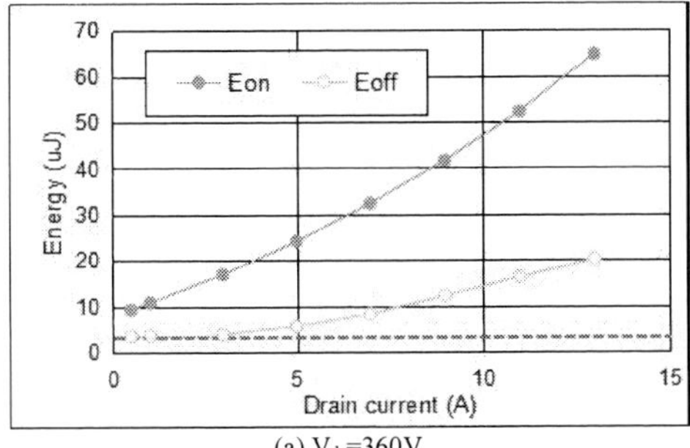

(a) V_{ds}=360V

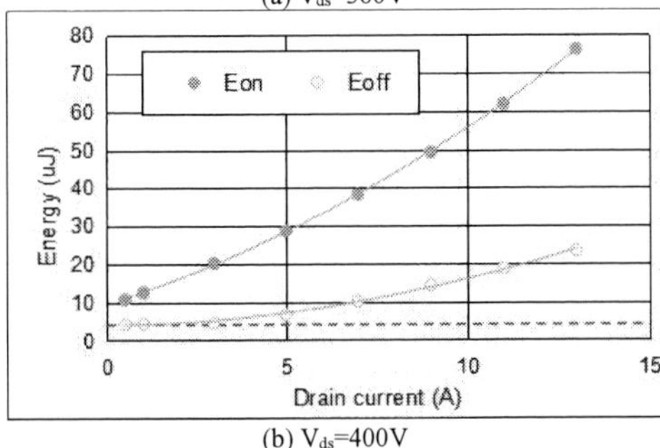

(b) V_{ds}=400V

Fig. 4. PSPICE-simulated primary GaN HEMT switching loss.

It should be noticed that the turn-off switching loss is overestimated if we directly use the data. According to [4], the typical tested E_{off} result includes the energy stored in the junction capacitors. In the meanwhile, the turn-on switching loss is underestimated since the energy stored in the junction capacitor is dissipated by the channel, which cannot be measured. Fig. 4 clearly shows that turn-on switching energy increases exponentially with current, while the turn-off switching energy remains nearly constant and negligible over the tested current range. These important switching characteristics suggest that a nearly zero power switching can be achieved if the GaN HEMT operates in ZVS mode.

In the reverse conduction state, the reverse characteristic is different from that of silicon MOSFET. The device then exhibits a channel resistance similar to $R_{ds(on)}$, thus it can be modeled as a "body diode" with zero reverse recovery charge but higher forward voltage. Thus the turn-on signal on synchronous rectifying mode should be carefully designed and optimized.

B. *Isolation Transformer*

Transformer design of half-bridge push-pull converter is an important task toward achieving high efficiency. Here, litz wire and copper foil winding are implemented in order to reduce the AC and DC resistance of the transformer. The push-pull (or center tap) configuration is applied at secondary side with copper foils for high-current low-voltage applications. Magnetizing inductance is integrated in the isolation transformer with distributed air gaps.

The planar transformer structure is shown in Fig. 5(a). The primary and secondary DC resistance R_{XF_pri} and R_{XF_sec} can be directly calculated by the winding geometry. The skin depth of the *nth* harmonics frequency is given by

$$\delta(n) = \sqrt{\frac{\rho_{cu}}{n\pi f_s \mu_{cu}}} \tag{8}$$

The AC-to-DC resistance ratio F_R at nth harmonic frequency is calculated by Dowell's equation, given by [7]

$$F_R(X) = X\frac{\sinh(2X)+2\sin(2X)}{\cosh(2X)-2\cos(2X)} + 2X\frac{p^2-1}{3}\frac{\sinh(X)-2\sin(X)}{\cosh(X)+2\cos(X)} \tag{9}$$

where $F_{R\_pri}=F_R(X)$ is for primary round litz wire with p equals to the layer of primary winding. The number of portions and the layer per portion p is shown in Fig. 5(b). It is shown that the sandwich winding structure has a low p value and hence lower F_r value.

$$X = \frac{\sqrt{\pi}d_{pri}}{2\delta(n)} \tag{10}$$

$F_{R\_sec}(n) = F_R(X)$ is for secondary side copper foils with $p = n_s/2$ and $X = h_{foil}/\delta(n)$

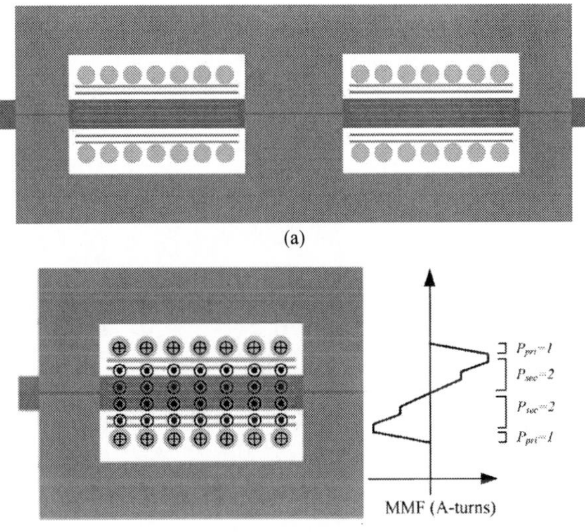

Fig. 5. Isolation transformer. (a) Winding structure of planar transformer. (b) MMF and p value of the planar transformer.

The AC copper loss at each harmonic frequency is calculated by summing the losses from dc (0th harmonic) to 20th harmonics. The primary side and secondary copper losses of the transformer are given by

$$P_{cu\_xf\_pri} = R_{XF\_pri}\sum_{n=0}^{20} F_{R\_pri}(n) I_{n\_pri}^2 \tag{11}$$

$$P_{cu\_xf\_sec} = R_{XF\_sec} \sum_{n=0}^{20} F_{R\_sec}(n) I_{n\_sec}^2 \quad (12)$$

where I_{n_pri} and I_{n_sec} denote the nth order harmonic current at the primary side and the secondary side of the transformer.

Flux swing of the half-bridge push-pull converter is bidirectional. The peak-to-peak flux density is given by

$$B_{m\_XF} = \frac{V_{LV}}{2 f_s A_{e\_XF}} \quad (13)$$

The empirical Steinmetz equation is applied to calculate the core loss of the transformer, given by

$$P_{core\_XF} = V_{e\_XF} k_{core1} f_s^{\alpha 1} \Delta B_{m\_XF}^{\beta 1} \quad (14)$$

where $\Delta B_{m\_XF} = 1/2 B_{m\_XF}$ is the flux swing, and k_{core1}, α_1 and β_1 are the Steinmetz coefficients provided by the manufacturer [6].

C. Interface Inductor

A separate interface inductor is applied in the half-bridge push-pull converter. Integrated transformer may lead to totally different loss models, designs, and optimization procedures. The losses in the interface inductor are copper loss and core loss. The DC resistance of interface inductor is calculated according to its geometry. Dowell's equation (9) is also applied to calculate AC resistance. $F_{R\_Ls}(n) = F_R(X)$ is for interface inductor, with p equals to the layer number of the inductor.

$$X = \frac{\sqrt{\pi} d_{L_s}}{2\delta(n)} \quad (15)$$

The copper loss of interface inductor is given by

$$P_{cu\_L_s} = R_{L_s} \sum_{n=0}^{20} F_{R\_L_s}(n) I_{n\_pri}^2 \quad (16)$$

Core loss of the inductor is given by

$$P_{core\_L_s} = V_{e\_L_s} k_{core2} f_s^{\alpha 2} \Delta B_{m\_L_s}^{\beta 2} \quad (17)$$

Where

$$\Delta B_{m\_L_s} = \frac{L_s I_{L_s\_max}}{n_{L_s} A_{e\_L_s}} \quad (18)$$

is the flux swing of the inductor. k_{core2}, α_2 and β_2 are the Steinmetz coefficients.

D. Secondary Si MOSFET Loss

We assume that the secondary Si MOSFETs work under a timing scheme that current does not flow through synchronous rectifier body diode. The major losses in rectifying mode are the conduction loss, turn-off switching loss, and the gate-drive loss. Turn-off switching loss is the energy stored in the stray inductance and dissipated by the circuit. The simplified model for the turn-off loss and the gate-driving loss are denoted as P_{sw_sec} and P_{dr_sec}, respectively. The conduction loss of secondary MOSFETs is denoted as P_{cond_sec}.

$$P_{off\_sec} = E_{off}(I_{d\_sec}) \cdot f_s = (a' \cdot I_d^2 + b' \cdot I_d + c') \cdot f_s \quad (19)$$

$$P_{cond\_sec} = I_{rms\_sec}^2 \cdot R_{ds(on)\_sec} \quad (20)$$

$$P_{dr\_sec} = Q_{g\_sec} \cdot V_{g\_sec} \cdot f_s \quad (21)$$

The secondary side switching losses are measured again by using the PSPICE double pulse test simulation. The result is shown in Fig. 6. For each of the MOSFETs, the drain currents and the drain to source voltages are measured simultaneously in order to calculate the respective switching losses. Fig. 6 presents the simulated turn-off losses under 30V drain-source voltage. In the converter, the secondary side is equipped with two parallel MOSFETs per switch.

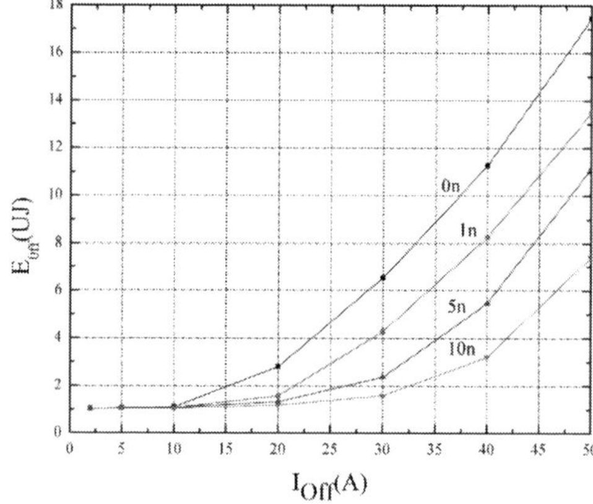

Fig. 6. PSPICE-simulated secondary Si MOSFET turn-off loss.

E. Active Clamp MOSFET

Each active clamp MOSFETs only conducts in two short intervals as shown in Fig. 7. Take S_{3a} for example. At t_2, the main secondary S_3 turns off, the current stored in the parasitic loop inductance and transformer leakage inductance will continue flow into the body diode of S_{3a}. If we turn S_{3a} at the beginning of this interval, the current will flow through the trench of S_{3a} instead of the body diode. The resonant circuit is designed that the current will resonant to zero at the end of the dead-time. Thus ZVS turn on and ZCS turn off can be achieved during the first half period and the loss is only conduction loss.

In the second half period , S_{3a} is actively turned on to release the energy stored in the active-clamp capacitor. Again, the current will resonant from zero between the inductor and capacitor. The MOSFET is ZCS turn-on hence the loss is conduction loss and turn-off loss.

$$P_{sw\_ac} = (V_{LV} + V_{HV} / n) \cdot Q_{oss\_ac} \cdot f_s \quad (22)$$

$$P_{cond\_ac} = I_{rms\_ac}^2 \cdot R_{ds(on)\_ac} \quad (23)$$

$$P_{dr\_ac} = Q_{g\_ac} \cdot V_{g\_ac} \cdot f_s \quad (24)$$

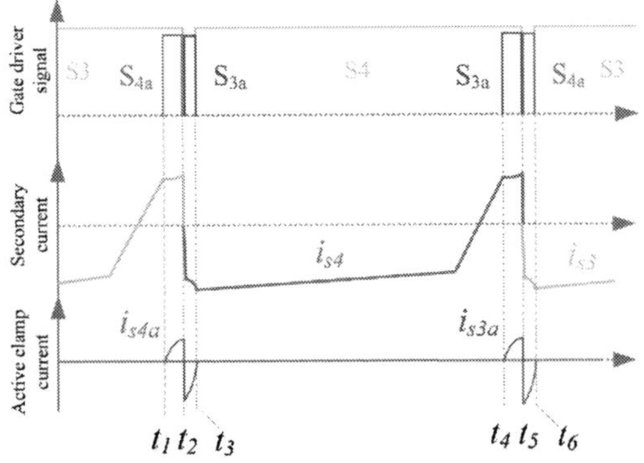

Fig. 7. Waveform of active clamp circuit.

F. Other Losses

The losses generated from high current flowing through EMI filters, PCB traces, input/output capacitors are also taken into account.

IV. Discussion and Efficiency Optimization

The loss analysis of the converter operating in soft switching are conducted in section III. Since GaN HEMT has very low output capacitance, the turn-off loss is close to zero and can be neglected within the current range of our converter. Therefore the loss of the primary side switches are conduction loss, which is determined by the output power and primary side voltage. On the secondary side, however, the Si devices will have both conduction loss and switching loss (mainly turn-off) loss. The turn-off loss has a relationship with the transformer turns ratio. The turns ratio is a parameter of the transform that directly determines the ZVS range of the converter, therefore there's a trade-off between the secondary side switching loss and the ZVS range of the converter.

Fig. 8 shows the result of the power-loss breakdown among different components. In the figure, the losses for 1kW output power operation are presented for the active clamp circuit, the primary high voltage side GaN HEMTs, secondary low voltage side Si MOSFETs, the transformer and the interface inductor. The largest loss of 19W occur for the transformer, 14.2W of which are the winding loss at the secondary side. However, we didn't see a very high temperature rise on the transformer windings. The reasons are we are using sandwich structure to reduce the magneto-motive force (MMF) and multi-layer copper foils to reduce the winding's dc resistance. There are 8 layer of copper in the secondary side, 4 of which are printed on the PCB and the other 4 are additional copper foils soldered onto the PCB. The structure and MMF are as shown in Fig. 5.

Another information that Fig. 8 has shown is that the critical parts of the converter are the resistive losses on the LV side as well as switching losses. Accordingly, the losses on the secondary side are significantly higher than on the primary side (about 6 times higher), even though the secondary side contains 8 switches in total (2 switches in parallel and 2 sets of secondary windings in parallel).

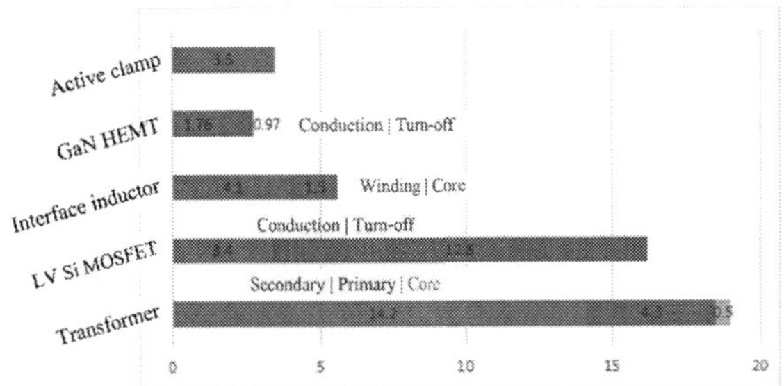

Fig. 8. Power-loss breakdown profile at 380V-12V 1kW.

A universal optimization procedure is proposed to improve the efficiency. Since the secondary side turn-off loss and secondary winding loss occupies half of the total loss, it is reasonable that the optimization focuses on the secondary side. While the resistance of transformer winding is not easy to reduce, however the turn-off loss is related to the turn-off current. We noticed that the turn-off current is affected by the transformer turns ratio. Fig. 9 shows the relationship between the total transistor loss, including primary GaN HEMTs and secondary MOSFETs. Two types of assumption are made. One is assuming that the turn-off loss is proportional to turn-off current. The aim of Fig. 9(b) is to look into how the switching loss is affected by turn-off current. An assumption that the switching loss is proportional to the switching frequency and is proportional to a square of the turn-off current was made in [1].

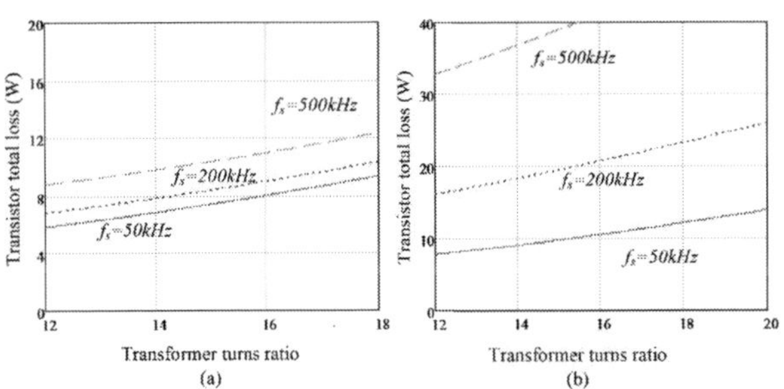

Fig. 9 Transistor total power loss versus transformer turns ratio and switching frequency. (a) $P_{sw} \propto I^2$; (b) $P_{sw} \propto I$.

Fig. 9 indicates that smaller turns ratio can reduce the total transistor loss. The reduced turn-off loss is the major contributor. It should be noted that reducing the turns ratio will not change the loss breakdown profile significantly. First, the high voltage and low voltage only varies between 360~420V and 11.2~14.4V respectively, the turns ratio can be between 14 to16 to guarantee a proper ZVS range. Second, the major loss contributors are taken into considered in the optimization, the other parts such as active clamp circuit and transformer primary winding loss will also have a positive effect on loss reduction. The transformer core loss and inductor loss is only a small percentage.

V. Experimental Results

Two prototype of the proposed isolated bidirectional half-bridge push-pull dc-dc converter is shown in Fig. 10(a). The thermal infrared image of the prototype converter operating at full load 12 V 100 A (for 0.5 hour) is also shown in Fig. 10(b). Three modules of distributed ESDs are shown in Fig. 10(c). The three modules are connected in parallel in a dc microgrid. The current of each module is automatically equally shared by droop control. The circuit parameters are presented in Table I.

(a)

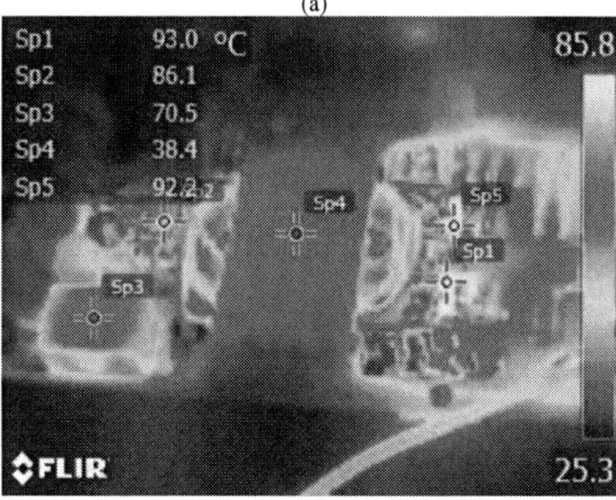

(b)

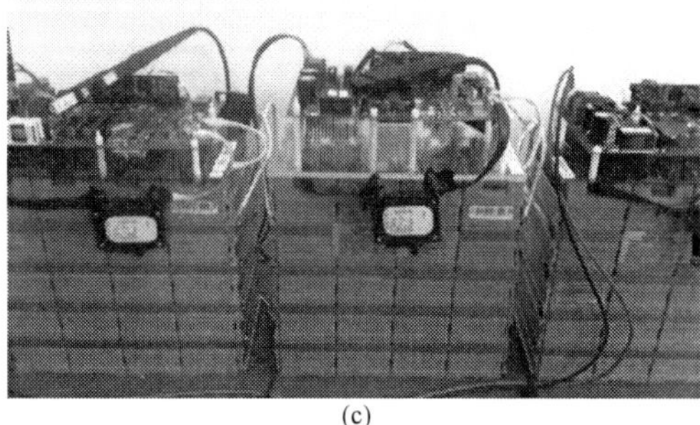
(c)

Fig. 10 Prototype converter and thermal images. (a) Prototype converter. (b) Thermal image of prototype converter. (c) Three 12V 100A energy storage device prototypes.

Fig. 11 shows ZVS turn on and turn off of GaN device S_2 in 100W charging mode. It shows that ZVS can be achieved in light load situation.

Fig. 12 shows efficiency test results under different load conditions. The controller power is excluded, which is about 2.4W. The measured efficiency is compared to the calculated efficiency for different operating points in this figure. Very good agreement between the results obtained from the theoretical model and the experimental values is observed.

Table I

Component List

Optimized half-bridge push-pull converter with active clamp V_{HV}=380V, V_{LV}=12V, I_{out}=100A		
Primary GaN HEMT	GS66508P-E03	$R_{ds(on)}$=52mΩ
Secondary MOSFET	BSC014N06NS	$R_{ds(on)}$=1.45mΩ
Active clamp MOSFET	CSD18537NQ5A	$R_{ds(on)}$=11mΩ
Isolation transformer	Turn ratio 14:1:1 Customized E64/10/50 3C90	Primary: 1050/#44 litz wire Secondary: 0.25mm foils
Interface inductor	26μH EQ30/8/20 3F35	750/#44 litz wire N_{Ls}=12
Active clamp capacitor	100nF 100V	101X18W104MV4E
Input capacitor	47μF*2 500V	
Output capacitor	220μF*2 25V	ST203C227MAJ10

The efficiency decreases faster as load becomes heavier. The main reasons are the transformer secondary winding loss and secondary side switching losses increase significantly under heavy load. Therefore, it would be beneficial to parallel more MOSFETs on the secondary side in the future work.

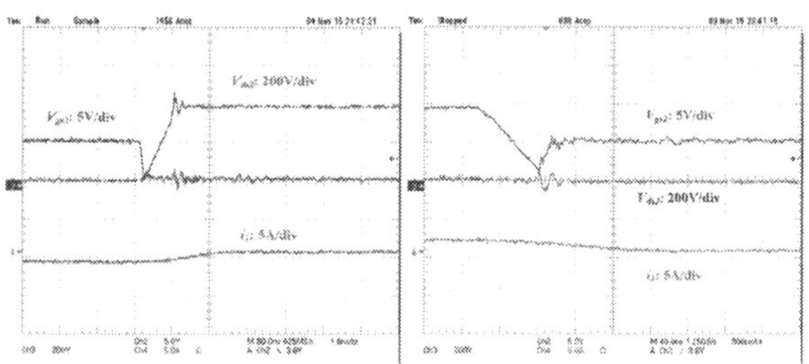

Fig. 11 Waveform of GaN HEMT S_2 turn off (left) and turn on (right). Ch2: G-S voltage; Ch3: D-S voltage; Ch4: Inductor current i_L.

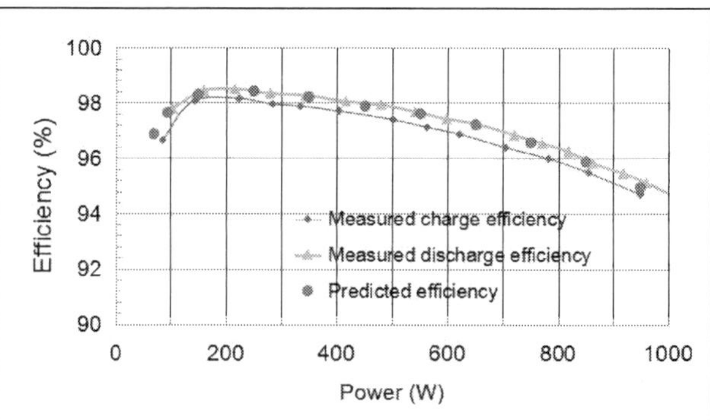

Fig. 12 Predicted and measured efficiency under different load (control power excluded).

VI. CONCLUSION

This paper describes a computer aided method to predict the power losses of the proposed half-bridge push-pull converter with active clamp. Loss models are provided to calculate total component losses using the current and voltage information derived from the steady state inductor current. Details of loss distributions are given. With the presented converter prototype, a top efficiency of 98.3% and an output power of 1 kW in a wide input/output voltage range are achieved. The loss analysis provides valuable information for designing an efficiency optimized converters in the application. Both the analysis and experiment show that the available GaN and Si device mixed converter are capable of achieving high efficiency and high power density in the considered converter design.

REFERENCES

[1] Akagi, H.; Yamagishi, T.; Tan, N.M.L.; Kinouchi, S.-I.; Miyazaki, Y.; Koyama, M., "Power-Loss Breakdown of a 750-V 100-kW 20-kHz Bidirectional Isolated DC–DC Converter Using SiC-MOSFET/SBD Dual Modules," Industry Applications, IEEE Transactions on , vol.51, no.1, pp.420,428, Jan.-Feb. 2015

[2] Krismer, F.; Kolar, J.W., "Accurate Power Loss Model Derivation of a High-Current Dual Active Bridge Converter for an Automotive Application," in Industrial Electronics, IEEE Transactions on , vol.57, no.3, pp.881-891, March 2010

[3] Yu, Ruiyang; Ho, Godwin Kwun Yuan; Pong, Bryan Man Hay; Ling, B.W.-K.; Lam, J., "Computer-Aided Design and Optimization of High-Efficiency LLC Series Resonant Converter," in Power Electronics, IEEE Transactions on , vol.27, no.7, pp.3243-3256, July 2012

[4] Suxuan Guo, Liqi Zhang, Yang Lei, Xuan Li, Fei Xue, Wensong Yu, Alex Q. Huang, "3.38 Mhz Operation of 1.2kV SiC MOSFET With Integrated Ultra-Fast Gate Drive," In in Wide Bandgap Power Devices and Applications (WiPDA), 2015 IEEE Workshop on , vol., no., pp. Nov. 2015

[5] Fei Xue, Ruiyang Yu, Wensong Yu, Alex Huang, "A Novel Bi-Directional DC-DC Converter for Distributed Energy Storage Device," In Applied Power Electronics Conference and Exposition (APEC), 2015 Thirtieth Annual IEEE (pp. 1126-1130). IEEE.

[6] Ferroxcube application note, "Design of planar power transformers," [Online]. Available: www.ferroxcube.com.

[7] Suxuan Guo, Xijun Ni, Kai Tan, Huang, A.Q., "Operation principles of bidirectional isolated AC/DC converter with natural clamping soft switching scheme," Industrial Electronics Society, IECON 2014 - 40th Annual Conference of the IEEE , vol., no., pp.4866,4872, Oct. 29 2014-Nov. 1 2014

[8] Hurley, W.G.; Gath, E.; Breslin, J.G., "Optimizing the AC resistance of multilayer transformer windings with arbitrary current waveforms," in Power Electronics, IEEE Transactions on , vol.15, no.2, pp.369-376, Mar 2000

Dynamic Efficiency Tracking Controller for Reconfigurable Four-coil Wireless Power Transfer System

Yuan Cao, *Student Member, IEEE,* Zhigang Dang, *Student Member, IEEE,*
Jaber A. Abu Qahouq, *Senior Member, IEEE* and Evan Phillips

The University of Alabama
Department of Electrical and Computer Engineering
Tuscaloosa, Alabama 35487, USA

Abstract— **This paper presents a dynamic efficiency tracking control strategy for reconfigurable four-coil wireless power transfer (R-WPT) system. The proposed R-WPT system includes an array of drive loops, a transmitter (Tx coil), a receiver (Rx coil), an array of load loops, two relay switch arrays, a rectifier and a control unit. By sensing voltage and/or current from the power source and the load, Tx and Rx side switches are adaptively controlled such that one drive loop out of several drive loops with different diameters is connected to the power source and one load loop out of several load loops with different diameters is connected to the load. Since drive loops and load loops have different sizes, the optimum coupling factor with corresponding drive loop and load loop is adaptively varied and selected, such that the maximum efficiency is achieved at a given distance/misalignment condition. Proof of concept experimental results demonstrate and verify the operation of the system achieving high efficiency.**

Keywords—Adaptive control; Closed-loop; Charging; Consumer Electronics; Electric Vehicles; Perturbation and Observation; Reconfigure; Wireless Charging; Wireless Power transfer

I. INTRODUCTION

There are variety of applications that can use either wired or wireless power or battery charging [1-16]. Wireless power transfer (WPT) technology increasingly attracts attention for its potential applications such as in electric vehicles (EVs), battery charging and consumer electronics including cell phones and laptop, among others [1-3, 5-8, 10-16]. However, due to the limited power received by Rx coil, transmission efficiency is a main factor restraining the development of WPT technology. In order to improve transmission efficiency and extend transmission distance, different tuning methods have been extensively studied [17-25]. Increasing coupling factor and/or quality factor are two major ways to realize higher efficiency WPT systems. Frequency tuning method, capacitor tuning method and impedance matching method, among others, are some typical methods to improve transmission efficiency and extend transmission distance [2, 17-25].

Reference [17] uses frequency tuning method to improve efficiency under different distances (DIS) or misalignment conditions (MIS). Similarly, [19] adopts frequency tuning in charging electrical vehicles (EVs). In [20], impedance matching method is presented as being effective in improving efficiency by automatically controlling the impedance seen by the power source.

For four-coil WPT system, since extra drive loop and load loop are introduced, besides the tuning methods above, optimizing coupling factor between drive loop and Tx coil and between load loop and Rx coil is another candidate method to achieve higher efficiency [25-27]. However, because of the mechanical movement of drive loop and load loop in these methods, extra volume is needed.

In order to overcome the disadvantage of mechanical movement, reference [2] proposes a reconfigurable WPT system to improve the transmission efficiency by choosing the optimum drive loop and load loop. However, in [2] the reconfigurable tuning method is realized manually, which is not practical in real application.

In this paper, a dynamic efficiency tracking control strategy for a reconfigurable four-coil WPT (R-WPT) system presented in [2] is proposed in order to adaptively improve transmission efficiency under different distances (DIS) and/or misalignment conditions (MIS). By adaptively tuning different sizes of drive loops and load loops, optimal transmission efficiency can be obtained. The proposed R-WPT controller and system do not require mechanical movement of coils and loops.

Next section reviews the R-WPT concept and describes the R-WPT design parameters used in this paper. Section III presents the dynamic efficiency tracking controller of this paper and its algorithm. Section IV presents results of a

978-1-4673-9551-9/16 $31.00 © 2016 IEEE

proof of concept experimental prototype and Section V gives the paper conclusion.

II. Overview Of Reconfigrable Four-coil WPT System

Fig. 1a illustrates the physical model of the four-coil R-WPT system [2], which includes an array of drive loops, a Tx coil, a Rx coil, an array of load loops, two switch arrays, a rectifier and the proposed controller. One switch array is used to switch between drive loops and another switch array is used to switch between load loops. The proposed controller is used to obtain voltage and/or current of power source/input and load/output to calculate the transmission efficiency, which is defined as the power received by load loop divided by the power injected to drive loop. In the R-WPT system, config_*ab* implies that a^{th} drive loop (DLoop_*a*, *a*=1, 2..., *n*) at the primary/Tx side and b^{th} load loop (LLoop_*b*, *b*=1, 2..., *n*) at the secondary/Rx side are turned on. Since *a* is not necessarily equal to *b*, more configurations can be provided compared to symmetrical system where *a*=*b*. When adaptively tuning the switches, there is no mechanical movement of coils and loops.

Fig. 1b illustrates the circuit model of the R-WPT system. Drive loop is modeled as an LR (Inductive-Resistive) loop which is connected to a power source at the left side and inductively coupled to Tx coil with coupling factor K_{DT} defined in Equation (1), where M_{xy} is mutual inductance. Tx and Rx coils are modeled as a capacitor in series with an inductor with a parasitic resistor. Tx and Rx are linked with a coupling factor K_{TR}. Similar to drive loop array, load loop is inductively coupled to Rx coil with a coupling factor K_{RL} at one side and is connected to load at the other side. Equation (2) gives the self-resonance frequency of Rx and Tx coils, where *L* and *C* are coil inductance and capacitance, respectively. A full-bridge rectifier circuits is connected to load loop on one side and load on the other side.

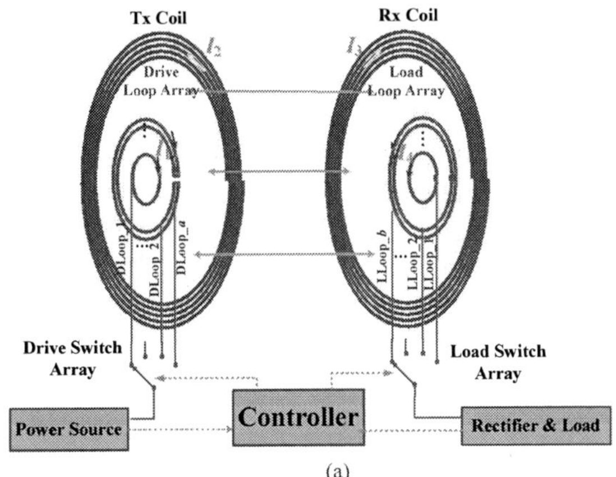

(a)

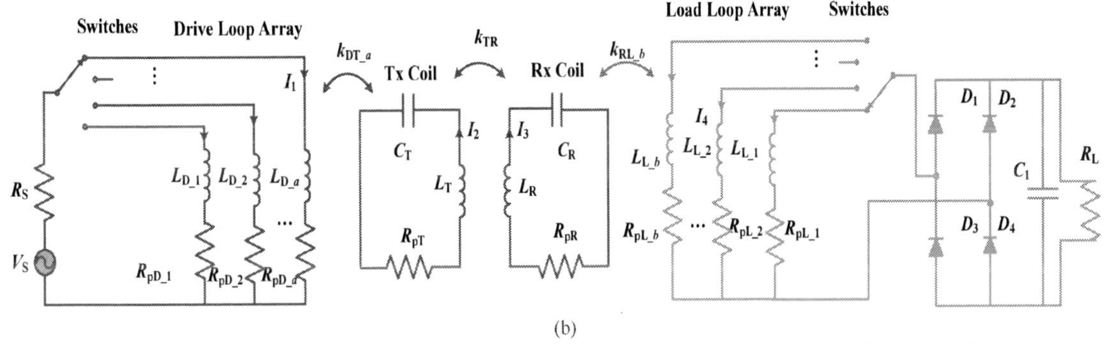

(b)

Fig. 1. (a) Physical model of reconfigurable four-coil WPT system, (b) Circuit model of reconfigurable four-coil WPT system

$$k_{xy} = \frac{M_{xy}}{\sqrt{L_x L_y}} \quad (1)$$

$$f_o = \frac{1}{2\pi\sqrt{LC}} \quad (2)$$

$$S_{21} = 2\frac{V_L}{V_S}\sqrt{\frac{R_S}{R_L}} \quad (4)$$

$$\eta = |S_{21}|^2 \times 100\% \quad (5)$$

$$\frac{V_L}{V_S} = \frac{j\omega^3 k_{DT}k_{TR}k_{RL}L_T L_R\sqrt{L_{D\_a}L_{L\_b}}R_L}{(k_{DT}^2 k_{RL}^2 L_{D\_a}L_T L_R L_{L\_b}\omega^4 + Z_{D\_a}Z_T Z_R Z_{L\_b} + \omega^2(k_{DT}^2 L_{D\_a}L_T Z_R Z_{L\_b} + k_{TR}^2 L_T L_R Z_{D\_a}Z_{L\_b} + k_{RL}^2 L_R L_{L\_b}Z_{D\_a}Z_T))} \quad (3)$$

By solving the four KVL equations of drive loop, Tx coil, Rx coil and load loop, the ratio of V_L and V_S can be derived as in Equation (3) [2], which can be further utilized to calculate S_{21} parameter as in Equation (4) and transmission efficiency η as in Equation (5) when $R_S=R_L$ [2, 8]. With the experimental parameters provided in Table I, a 3-D view of S_{21} is plotted as shown in Fig. 2. It can be observed that within a short distance between Tx and Rx, S_{21} can remain almost constant in the over coupled region. As K_{TR} continues to decrease, S_{21} drops significantly beyond the critical coupling point.

By controlling the switches attached to drive loops and load loops, the optimum coupling factor k_{DT_a} and k_{RL_b} can be obtained to achieve/track maximum transmission efficiency under various distance and misalignment conditions.

Table I. Design simulation parameters

Parameters	Values	Parameters	Values
R_S	50 Ω	R_L	50 Ω
L_T	11.6 μH	L_R	11.8 μH
C_T	240 pF	C_R	246 pF
R_{pT}	0.42 Ω	R_{pR}	0.41 Ω
L_D	0.6 μH	L_L	0.6 μH
R_{pD}	0.1 Ω	R_{pD}	0.1 Ω
K_{DT}	0.1	K_{RL}	0.1
K_{TR}	0.001-0.6	Frequency range	$0 \leq f \leq 6$ MHz
Natural frequency	3 MHz	-	-

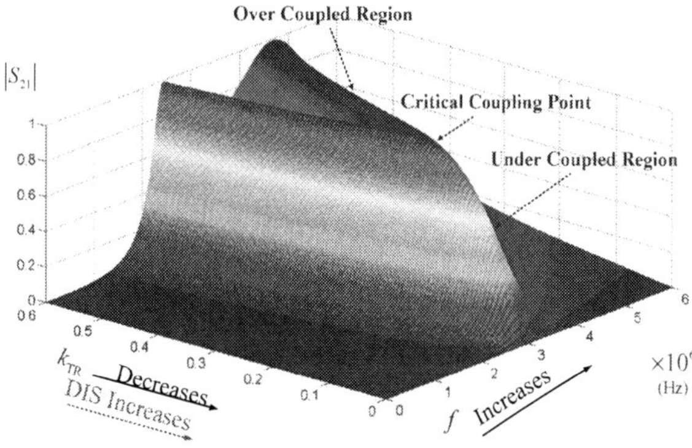

Fig. 2. S_{21} as a function of K_{TR} and frequency

III. Dynamic Efficiency Tracking Control Strategy For The R-WPT System

This section describes how the control strategy is utilized to dynamically track maximum efficiency point. If the numbers of drive loops and load loops are a and b respectively, then the total number of configurations in R-WPT system is $a \times b$. To obtain highest transmission efficiency, one simple way is the exhaust method which tries (sweeps through) every configuration option. However, this method results in large convergence time and high complexity. In order to overcome these disadvantages, a dynamic efficiency tracking control strategy is used by the controller of the R-WPT system.

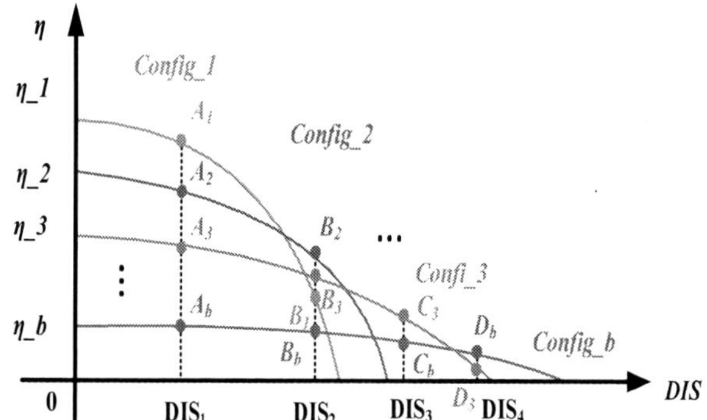

Fig. 3. Efficiency curves illustration for different configurations (Drive/load loop sizes)

The objective of the proposed control strategy is to find the optimum operation point (where optimum sizes of drive loop and load loop are chosen in order to obtain maximum efficiency) under various distances and misalignment conditions. When the ath drive loop is fixed and turned on, the transmission efficiency as a function of distance for different configurations (sizes of load loops are different and $r_1>r_2>r_3>r_4$) is illustrated in Fig. 3. It can be observed that at various distances (DIS_1 through DIS_4), optimum configurations and operation points are different. The proposed control strategy utilizes perturb and observe (P&O) method to adaptively tune different size of load loops to dynamically track maximum efficiency. For example in Fig. 3, initial configuration is selected as Config_3, then at DIS_1 and DIS_2, the size of load loop is tuned to a larger size in order to obtain higher efficiency. At DIS_4, the size of load loop is tuned to a smaller size since transmission efficiency decreases with larger coil size configurations. In this way, maximized transmission efficiency can be obtained with the corresponding operation point (A_1, B_2, C_3 and D_b shown in Fig. 3) at certain distance.

In the proposed controller for the R-WPT system, both Tx and Rx sides use reconfiguration mechanisms as illustrated in Fig. 1. The detailed control flowchart for R-WPT system is illustrated in Fig.4. There are two algorithm loops for the control strategy: The inter algorithm loop which utilizes P&O method to tune load loops when drive loop is fixed, and the outer algorithm loop which sweeps all of drive loops while repeating inner algorithm loop. Therefore, after the best size of load loop is found in one algorithm inner loop cycle for a given fixed drive loop, the algorithm outer algorithm loop cycle repeats the process for each drive loop.

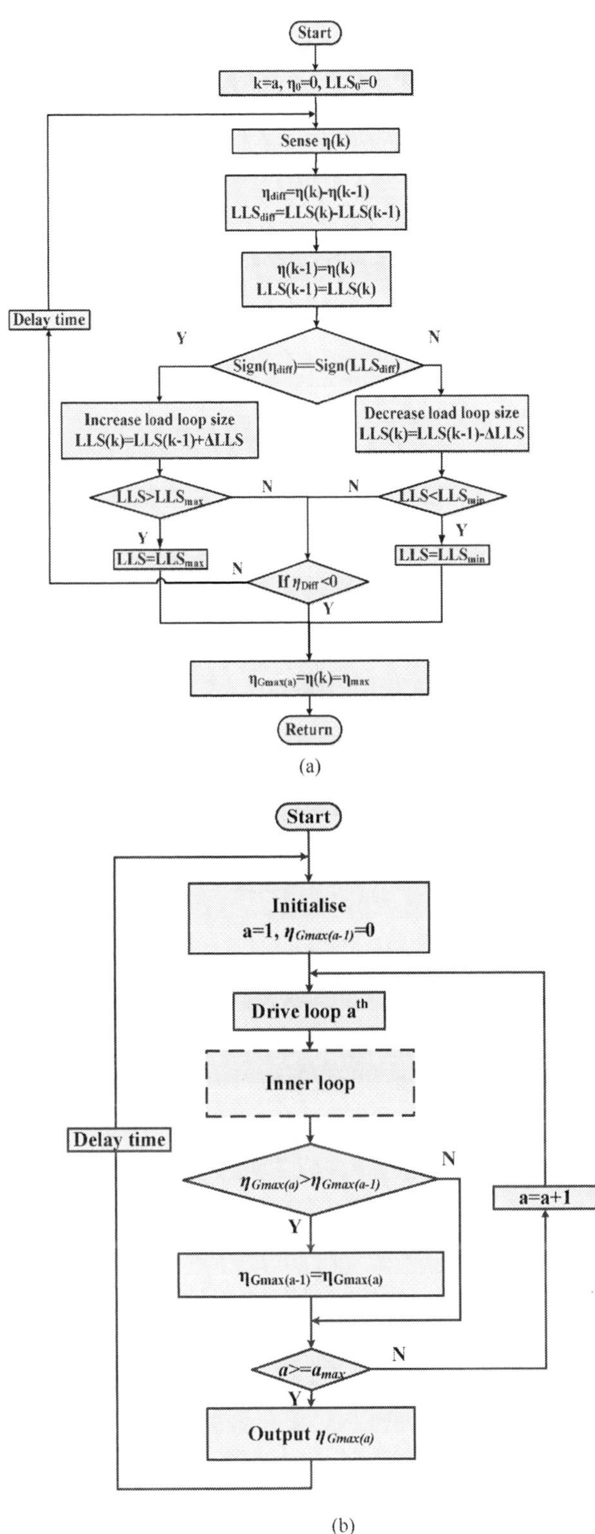

Fig. 4. Control flowchart for R-WPT system: (a) Inner algorithm loop and (b) Outer algorithm loop

In the inner algorithm loop, the efficiency value η(k) is compared with its previous value η(k-1) to yield the change in efficiency η_{diff}=η(k)-η(k-1). Similarly, the change of load loop size LLS_{diff} is achieved as LLS_{diff}=LLS(k)-LLS(k-1). If the signs of η_{diff} and LLS_{diff} are the same, then the load loop size is increased by ΔLLS (change of load loop size by one step). Otherwise, if the signs of η_{diff} and LLS_{diff} are the different, the load loop size is decreased by ΔLLS. This increment or decrement of load loop size (inner algorithm cycle) ends either when maximum efficiency is reached or when the size reaches its maximum or minimum value. Once the maximum efficiency value of inter loop $\eta_{\max}$ is reached, the global maximum value η_{Gmax} is updated to be $\eta_{\text{Gmax}} = \eta_{\max}$.

When one cycle of the inner algorithm loop is completed, the controller operation returns to the outer algorithm loop where the drive loop is reconfigured (replaced by or switched) to the next load loop and the inner loop process is repeated. During this process, η_{Gmax} variable is always updated by the latest global maximum efficiency value.

IV. Proof of Concept Experimental Results

A proof of concept experimental R-WPT system with the presented controller is developed with the main design parameters as shown in Table II. There are three drive loops and three load loops, respectively. One drive loop and one load loop are connected/selected by the controller at any given time. The voltage and current of power source/input and load/output are sensed and sampled to calculate efficiency under different configurations in order for the microcontroller to control switch arrays to different sizes of drive loop and load loop using the algorithm discussed in Section III (Fig. 4).

In the experiment, an AC source signal is generated by a power source to supply power to a drive loop, ADC module of TI microcontroller TMS320F28335 is used to sample the output voltage/current, and TE 1462050-1 relays are used to switch between different drive loops and load loops.

Table II.
Parameter specifications for the experimental R-WPT system

Parameter	Values	Parameter	Values
Coil structure	Spiral	Coil material	copper
Coil outer diameter	35cm	Coil inner diameter	15cm
r_{coil}	2mm	Coil turns (N_{coil})	7
Dloop_1	35cm	*Lloop_1*	35cm
Dloop_2	30cm	*Lloop_2*	30cm
Dloop_3	25cm	*Lloop_3*	25cm
f_o	3MHz	Rs, R_L	50 Ω

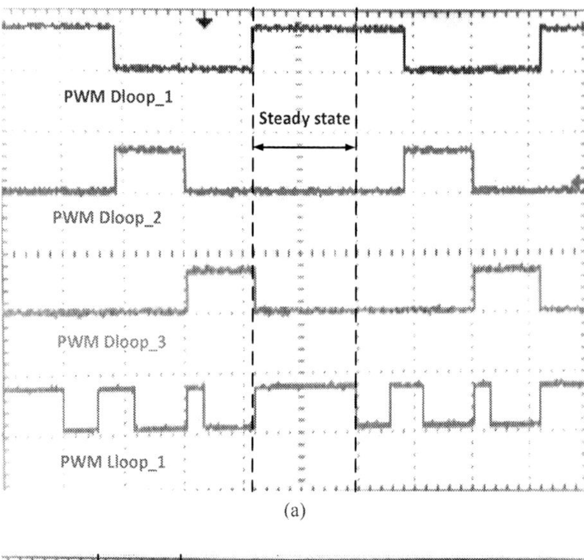

(a)

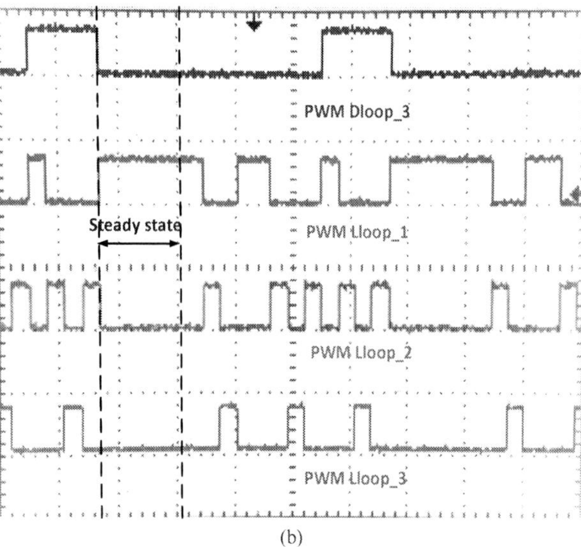

(b)

Fig. 5. PWM control signal for three drive loops and three load loops.

Fig. 5 shows the PWM signals for three dive loops and the three load loops, respectively. Dive loops at Tx side are reconfigured by the outer control loop and load loops at Rx side are reconfigured by the inner control loop as shown from Fig. 4. After the algorithm completes its operation to reconfigure the system to the drive loop and load loop which will give highest efficiency under a given condition of operation (e.g. distance between Tx and Rx), the system enters a steady-state period as indicated on Fig. 5, before the process is repeated either after a delay time or once a change is detected.

It can be observed from Fig. 5 that at any time only one drive loop and one load loop are turned on. In order to minimize the disturbance during switch changing period, there is a small overlap between two PWM control signals of two adjacent loops such that a loop is connected before another is disconnected.

Fig. 6 shows the transmission efficiency results for the R-WPT system when the dynamic efficiency tracking is used to select the drive loop and load loop that will result in maximized transmission efficiency. Transmission efficiency is measured under perfectly aligned conditions (0-80cm) and laterally misaligned conditions (0-50cm). Furthermore, the larger numbers of configurations are used in the proposed R-WPT system, the higher transmission efficiency can be achieved under different DIS and MIS conditions.

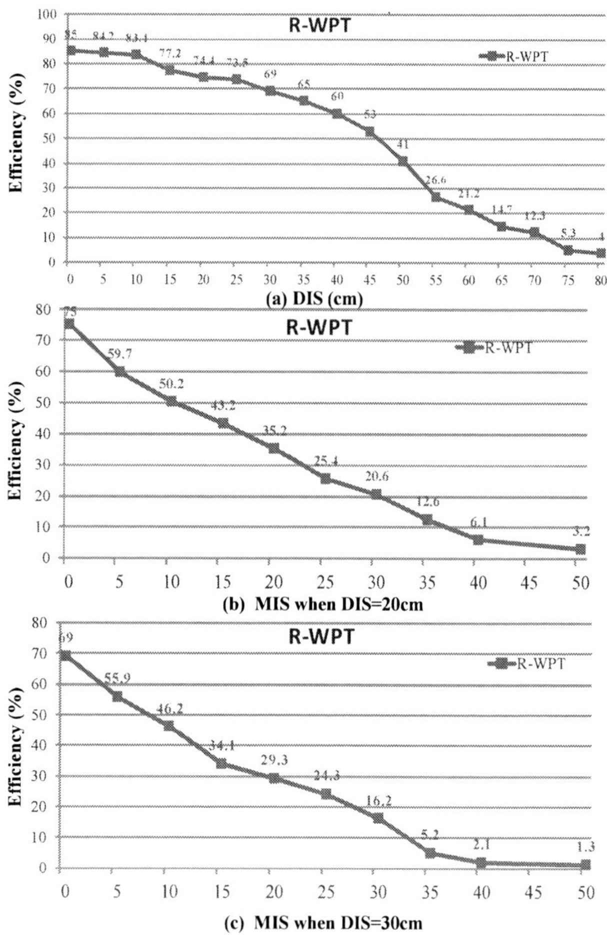

Fig. 6. Transmission efficiency of R-WPT system: (a) DIS when MIS=0cm, (b) MIS when DIS=20cm and (c) MIS when DIS=30cm

V. Conclusion

This paper presented a dynamic efficiency tracking control for reconfigurable four-coil wireless power transfer system in order to obtain maximized transmission efficiency. By using different sizes of drive loops and load loops, higher efficiency can be obtained under varying DIS conditions, and/or MIS conditions by dynamically tracking/finding the drive loop and load loop for highest transmission efficiency. The concept experimental results presented in the paper verify the proposed control strategy for the R-WPT system.

REFERENCES

[1] J. Kim, H. Son, D. Kim and Y. Park, "Optimal design of a wireless power transfer system with multiple self-resonators for an LED TV," IEEE Trans. Consumer Electronics, vol. 58, no. 3, pp. 775-780, Aug. 2012.

[2] Z. Dang, Y. Cao and J. A. Abu Qahouq, "Reconfigurable Magnetic Resonance-Coupled Wireless Power Transfer System," IEEE Transactions on Power Electronics, vol. 30, no. 11, pp. 6057-6069, Nov. 2015.

[3] Z. Dang, J and A. Abu Qahouq, "Modelling and Simulation of Magnetic Resonance Coupled Wireless Power Transfer Systems," International Review of Modelling and Simulation (I.RE.MO.S), vol. 6, no. 5, pp. 1607-1617, Oct. 2013.

[4] W. Huang and J. A. Abu Qahouq, "Energy Sharing Control Scheme for State-of-Charge Balancing of Distributed Battery Energy Storage System." Industrial Electronics, IEEE Transactions on, vol. 62, no. 5, pp. 2764-2776, May 2015.

[5] Z. Dang and J. A. Abu Qahouq, "Range and Misalignment Tolerance Comparisons between Two-coil and Four-coil Wireless Power Transfer Systems," The 2015 IEEE Applied Power Electronics Conference and Exposition, APEC'2015, pp. 1234-1240, Mar. 2015.

[6] Z. Dang and J. A. Abu Qahouq, "Elimination Method for the Transmission Efficiency Valley of Death in Laterally Misaligned Wireless Power Transfer Systems," 2015 IEEE Applied Power Electronics Conference and Exposition, APEC'2015, pp. 1644-1649, Mar. 2015.

[7] J. A. Abu Qahouq and V. Arikatla. "Online Closed-Loop Autotuning Digital Controller for Switching Power Converters." Industrial Electronics, IEEE Transactions on, vol. 60, no. 5, pp. 1747-1758, May 2013.

[8] Z. Dang and J. A. Abu Qahouq, "Modeling and Investigation of Magnetic Resonance Coupled Wireless Power Transfer System with Lateral Misalignment," The 2014 IEEE Applied Power Electronics Conference and Exposition, APEC'2014, pp 1317-1322, Mar. 2014.

[9] W. Huang and J. A. Abu Qahouq, "An Online Battery Impedance Measurement Method Using DC-DC Power Converter Control." Industrial Electronics, IEEE Transactions on, vol. 61, no. 11, pp. 5987-5995, Jun. 2014.

[10] Z. Dang and J. A. Abu Qahouq, "Evaluation of High-Current Toroid Power Inductor With NdFeB Magnet for DC-DC Power Converters." Industrial Electronics, IEEE Transactions on, vol. 62, no. 11, pp. 6868-6876, Nov. 2015.

[11] R. Wu, W. Li, H. Luo, J.K.O. Sin and C.C. Yue, "Design and Characterization of Wireless Power Links for Brain-Machine Interface Applications," IEEE Trans. Power Electronics, vol.29, no.10, pp. 5462-5471, Oct. 2014.

[12] W. Chwei-Sen, O. H. Stielau, and G. A. Covic, "Design considerations for a contactless electric vehicle battery charger," IEEE Trans. Ind. Electron., vol. 52, no. 5, pp. 1308-1314, Oct. 2005.

[13] G. A. Landis, "Reevaluating Satellite Solar Power Systems for Earth," IEEE 4th World Conference on Photovoltaic Energy Conversion, pp.1939-1942, May 7-12, 2006.

[14] A. Sahai and D. Graham, "Optical wireless power transmission at long wavelengths," International Conference on Space Optical Systems and Applications (ICSOS), pp. 164-170, 11-13, May 2011.

[15] U. K. Madawala and D. J. Thrimawithana, "A bidirectional inductive power interface for electric vehicles in V2G systems," IEEE Trans. Ind. Electron., vol. 58, no. 10, pp. 4789-4796, Oct. 2011.

[16] M. G. Eagen, D. L. O'Sullivan, J. G. Hayes, M. J. Willers, and C. P. Henze, "Power factor corrected single stage inductive charger for electric vehicle batteries," IEEE Trans. Ind. Electron., vol. 54, no. 2, pp. 1217-1226, Apr. 2007.

[17] A. P. Sample, D. A. Meyer and J. R. Smith, "Analysis, experimental results, and range adaptation of magnetically coupled resonators for wireless power transfer," IEEE Trans. Ind. Electron., vol. 58, no. 2, pp. 544–554, Feb. 2011.

[18] L.H. Chen, S. Liu, Y.C. Zhou and T.J. Cui, "An Optimizable Circuit Structure for High-Efficiency Wireless Power Transfer," IEEE Trans. Ind. Electron.,vol. 60, no. 1, pp. 339-349, Jan. 2013.

[19] J. M. Miller, O. C. Onar and M. Chinthavali, "Primary-Side Power Flow Control of Wireless Power Transfer for Electric Vehicle Charging." Emerging and Selected Topics in Power Electronics, IEEE Journal of 3, no. 1, pp 147-162, Jan. 2015.

[20] L. Yongseok, T. Hoyoung, L. Seungok and P. Jongsun, "An adaptive impedance-matching network based on a novel capacitor matrix for wireless power transfer," Power Electronics, IEEE Transactions on, vol. 29, no. 8, pp. 4403-4413, Aug. 2014.

[21] Y. Zhang, Z. Zhao and K. Chen, "Frequency Decrease Analysis of Resonant Wireless Power Transfer," IEEE Transaction on Power Electronics, Vol. 29, No. 3, Mar. 2014.

[22] D. Ahn and S. Hong, "A Study on Magnetic Field Repeater in Wireless Power Transfer," IEEE Trans. Ind. Electron., Vol. 60, No. 1, pp. 360-371, Jan. 2013.

[23] W. Zhong, C. Lee and S. Y. R. Hui, "General Analysis on the Use of Tesla's Resonators in Domino Forms for Wireless Power Transfer," IEEE Trans. Ind. Electron., Vol. 60, No. 1, pp. 261-270, Jan. 2013.

[24] S. Hui, W. Zhong and C. Lee, "A Critical Review of Recent Progress in Mid-Range Wireless Power Transfer," IEEE Trans. Power Electron. no. 99, 2013.

[25] T. P. Duong and J. W. Lee, "Experimental results of high-efficiency resonant coupling wireless power transfer using a variable coupling method," IEEE Microw. Wireless Components Lett., vol. 21, no. 8, pp. 442-444, Aug. 2011.

[26] R. Huang and B. Zhang. "Frequency, Impedance Characteristics and HF Converters of Two-Coil and Four-Coil Wireless Power Transfer." Emerging and Selected Topics in Power Electronics, IEEE Journal, vol. 3, no. 1, pp. 177-183, Mar. 2015.

[27] Y. Zhang, Z. Zhao, T. Lu. Zhang, Yiming, Zhengming Zhao and Ting Lu,+ "Quantitative Analysis of System Efficiency and Output Power of Four-Coil Resonant Wireless Power Transfer."Emerging and Selected Topics in Power Electronics, IEEE Journal, vol. 3, no. 1, pp. 184-190, Mar. 2015.

Wireless Power and Data Transfer System for Smart Bridge Sensors

Yujin Jang∗ , Jung Kyu Han†, Shin Young Cho†, Gun-Woo Moon∗†, Ji-Min Kim‡ and Hoon Sohn‡
Email: yujinjang@kaist.ac.kr, Hanjk715@angel.kaist.ac.kr, martin@angel.kaist.ac.kr, gwmoon@kaist.ac.kr, jimin.kim@kaist.ac.kr, hoonsohn@kaist.ac.kr
∗ Department of Future Vehicle, KAIST, Daejeon, Republic of Korea.
†School of Electrical Engineering, KAIST, Daejeon, Republic of Korea.
‡Department of Civil and Environmental Engineering, KAIST, Daejeon, Republic of Korea.

***Abstract*— This paper proposes a novel system for smart bridge sensors that enables power and information data to be transferred simultaneously. For power transfer, the proposed system uses a half-bridge converter including a series-series compensator, and it provides power to the sensors through the concrete structure. For data transmission, the proposed system detects physical information of the smart bridge by sensing the voltage variation of the resonant capacitor in the transmitter. The proposed system thereby can reduce management cost of the smart bridge by using a wireless power transfer system for rechargeable batteries and eliminating the communication module for data transmission. A prototype system is validated by experiments. The efficiency is 43.6% with a distance of 300 mm and data transmission is properly conducted with a 1500bps data rate.**

Keywords—wireless power and data transfer, load variation, concrete structure, digital packet transmission.

I. Introduction

Bridges for highways and railways deteriorate over time. This deterioration of bridges worsens due to continuous stress in accordance with vehicle use or a harsh external environment. If this situation is not addressed, damage or collapse of the bridge may occur. Therefore, structural health monitoring (SHM) for the detection of the bridge status is a very important issue for preventing these problems. SHM entails continuous monitoring of the concrete structure by using accelerometers, strain sensors, and piezoelectric sensors. These sensors are installed on the bridge, and the control center receives physical information such as concrete cracks, vibration, and temperature of the bridge from the sensors. Abnormal status can then be detected from the data, which facilitates management of the bridge. Bridges that incorporate a maintenance system by using sensors for condition measurement and data transmission are known as smart bridges [1]-[3].

The conventional smart bridge system with SHM consists of sensors for obtaining physical information of the bridge, wired communication modules for transmitting the data, and cables for supplying power to these devices. Hundreds of sensors are installed at the bottom or internal site of the bridge. The wired communication modules transmit the information obtained by the sensors to a data acquisition system. Theses cables are linked over long distances to the bridge [4].

However, these systems accompany some problems. The cables for communication and power supply occupy additional space. In addition, wired data transmission and the power supply system are vulnerable to external noise due to continuous connection to cables. A key disadvantage of this is low reliability. As a result, the cables for this system increase the cost of the bridge construction and maintenance due to the long length of the bridge. In recent years, wireless sensors for SHM have been studies as a means to solve this problem. By using these, it is possible to transmit data obtained by the sensors wirelessly. However, the wireless sensors mainly depend on battery power. Use of disposable batteries as a power source has been considered in order to eliminate the cables. This could more readily allow numerous sensors to be applied on the bridge as compared to the case of using cables. However, battery power is not guaranteed for the lifespan of the wireless sensors. Furthermore, it is difficult to replace the battery when it expires because the battery is located at the bottom of the bridge and should be replaced manually.

To solve this problem, energy harvesting for supplying power to the wireless sensor has been studied [5]-[6]. Energy harvesting is a means of obtaining electric energy by using wind energy or vibration. This affords the advantage that the battery does not have to be exchanged after installing wireless sensors. However, the electric power obtained by energy harvesting is in a range from μW to mW. This is not sufficient to operate the wireless sensors. Furthermore, installation of the energy harvesting facility is costly. Thus, wireless power transfer (WPT) technology has been widely studied for SHM application. Battery charging is simple and convenient by using WPT technology.

Numerous wireless communication modules in order to transmit the data related to the status of the bridge are also expensive. Generally, a WPT system requires a communication module for feedback control and load condition monitoring, which should be stably operated for SHM applications with WPT. For realizing these, a radio frequency (RF) system can perform data communication between a transmitter and a receiver. Note that the communication module must be installed to the sensors for SHM individually. However, data transmission by using this system requires high maintenance costs.

978-1-4673-9551-9/16 $31.00 © 2016 IEEE

Therefore, research on a new wireless power and data transfer (WPDT) system for smart bridge sensors that simultaneously charges rechargeable batteries and transmits physical information without communication modules is urgently needed. However, very few academic studies have been conducted on this.

Several papers have proposed multi-resonator systems for WPDT systems [7]-[8]. These systems are separated by each resonator for power transfer and data transmission. However, this method results in magnetic interference between the two resonators. In addition, the construction cost of the overall system is high. To address these problems, recent research has proposed a common inductive link for WPDT [9]. For realizing this, a coupled inductor and an additional circuit are added instead of using a multi inductive link. However, this method has the drawback that the coupled inductor and additional circuit increase the volume of the WPDT system.

The present paper proposes a novel WPDT system that not only supplies power to sensors through the concrete structure, but also transmits sensed data on the status of the bridge to the control center simultaneously. The proposed data transmission method can reduce the cost by eliminating an additional communication module.

II. Operational Principle

A. System configuration

The structure of the power and data transfer system for smart bridge sensors can be seen in Fig. 1. The power and data transfer start when an inspection vehicle moves to the point where a sensor is installed. This vehicle comprises a resonator for WPT and a MCU for processing data of a wireless sensor node (WSN). WPT is performed through the resonators between the vehicle and the sensor for the smart bridge. When the battery is charged more than threshold voltage that is able to operate MCU and WSN, the WSN transmits start signal. At the same time, the physical data of the bridge that are detected by the sensor for a period of time are transmitted to the inspection vehicle. This signal means that the WSN can transmit physical data of the bridge to the transmitter. Then this signal is modulated to a packetized digital signal, and transmitted to the transmitter through the resonators for WPT. When this signal is input to the MCU of the WSN, transmission of tendon force data commences. The MCU of the inspection vehicle translates the packetized digital data into tendon force. After data transmission is completed, the inspection vehicle consistently performs battery charging for the WSN. When the battery charging is finished, the inspection vehicle moves to another WSN. The timing diagram shown in fig.2 illustrates the above scenario.

B. System Specifications

These wireless power and data transfer systems must meet several criteria. (1) The efficiency and power of the WPT should be more than 30% and 10W. (2) The diameter of the resonator is limited to 300mm due to the limitation of space in the inspection vehicle. (3) The distance between the resonators is determined as being up to 300mm based on the construction specifications of the bridge. (4) The power and data should be transferred through the concrete structure. (5) The sensors for the smart bridge convert the physical data of the bridge such as concrete cracks, vibration, and temperature into an eight

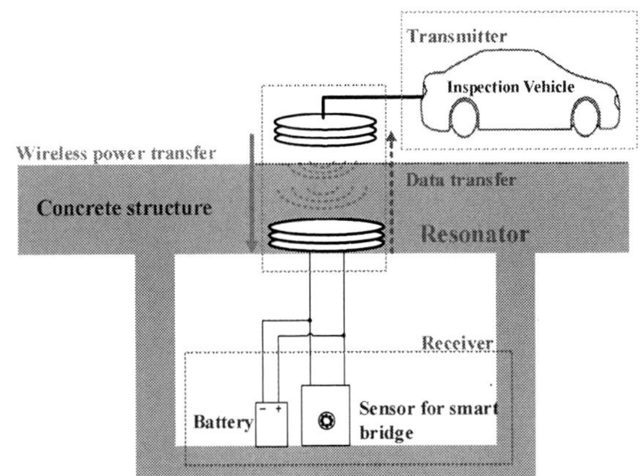

Fig. 1. Concept diagram of a wireless power and data transfer system using the inspection vehicle for SHM sensors

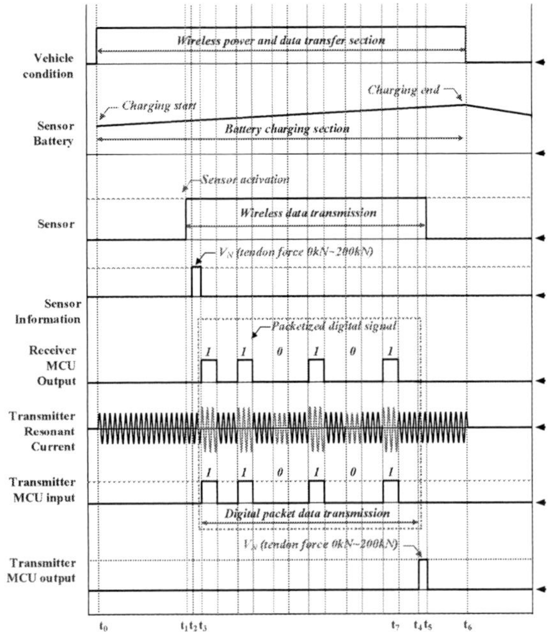

Fig. 2. Timing diagram of WPDT system

TABLE I. Specifications of Proposed WPDT System

	Concrete structure width	*Diameter of resonators*	*Efficiency*	*Power*
Power transfer	300mm	300mm	> 30%	> 10W
Data transfer	*Data steps*			
	8			

step analog voltage. Therefore, as shown in TABLE I, these five criteria should be considered in the proposed system for the smart bridge.

III. DESIGN OF WPDT SYSTEM

A. The design of resonator for high efficiency WPT

Power transfer using the coupling inductor with an air gap usually has low efficiency due to the high circulating current caused by the low coupling coefficient k. To achieve high efficiency by reducing the circulating current, the imaginary part of the input impedance should be designed to be as small as possible. Thus, as shown in Fig.3, an L-C resonance circuit is required in the WPT system. Meanwhile, the efficiency η of a WPT system employing L-C resonance can be expressed as follows.

$$\eta = \frac{1}{1+\frac{1}{k^2 Q_1 Q_2}} \cdot \frac{R_L}{R_2 + R_L} \quad (1)$$

$$Q_1 = \frac{\omega L_1}{R_1}, \quad Q_2 = \frac{\omega L_2}{R_2 + R_L} \quad (2)$$

where L_1 and L_2 are the self-inductances, R_1 and R_2 are the resistances of the transmitter and receiver resonators, and R_L is the output load, and Q_1 and Q_2 are the quality factor of the primary and secondary wingding, respectively. From (1), η of a resonant-coupling-based WPT system depends on two factors: k factor and Q factor. The coupling coefficient k can be expressed as follows [10].

$$k = \frac{M}{\sqrt{L_1 L_2}} = \frac{r_R^2 \cdot r_T^2 \cdot d}{\sqrt{r_R r_T}\left(\sqrt{d^2 + r_T^2}\right)^3} \quad (3)$$

where M is mutual inductance, r_R and r_T are the outer radii of the transmitter and receiver resonators, and *d* is the distance between the resonators. From (3), k is related to the radius of the resonator and the distance between the resonators. As mentioned in the system configuration section, since the radius of the resonator is determined as 300mm, the k factor is inversely proportional to the distance. In the case of Q, from (1), it is proportional to the self-inductance, which can be expressed as follows [11].

$$L_{self} = \frac{C_1 \mu_o N^2 r_{avg}}{2}\left[\ln\left(\frac{C_2}{\varphi}\right) + C_3\varphi + C_4\varphi^2\right] \quad (4)$$

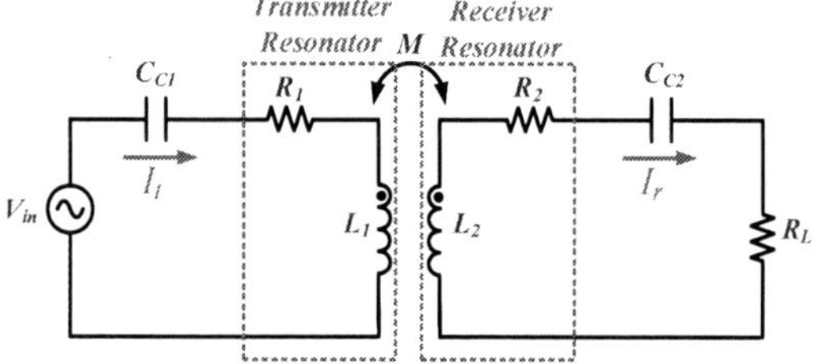

Fig. 3. Schematic of the inductive link system for wireless power transfer

$$r_{avg} = 0.5 \times (r_{out} + r_{in}) \quad (5)$$

$$\varphi = \frac{(r_{out} - r_{in})}{(r_{out} + r_{in})} \quad (6)$$

where L_{self} is L_1 and L_2, r_{out} and r_{in} are the outer radius and the inner radius of the resonators, N is the number of turns, and C_1, C_2, C_3,and C_4 are 1.0, 2.46, 0,and 0.2 respectively. From (4), L_{self} is proportional to the number of turns of the resonators. Therefore, from (2) and (4), the Q value has a trade-off relationship between the number of turns and resistance of the resonator due to the limited diameter of the resonators. For example, if a wire with a large cross-section is used in the resonator, the number of turns is decreased and the resistance of the resonator is also decreased. Therefore, by considering these constraints, the optimal design of the resonator can be achieved for a high efficiency WPT system.

B. The design of wireless data transmission

In a WPT system, there is a notable characteristic that a no load condition of the receiver causes large circulating current of the transmitter, which also results in a large fluctuation of the resonant capacitor voltage. Therefore, provided that the physical information of the smart bridge can be presented as the ON/OFF status of the load, the transmitter can obtain the information of the sensor by detecting the resonant capacitor voltage of the transmitter. For realizing data transmission, Fig.4 shows the proposed wireless power and data transfer system. This system is composed of transmitter with AC voltage sensing block, resonator, receiver with load ON/OFF control switch, MCU, and sensor. As shown in Fig. 5, the physical information is detected by the sensor of the receiver within a range of 0~200kN. This information is then divided by 20kN and converted to an eight step DC voltage value. For example, V_5 signifies the physical information of 100kN~125kN. Each analog DC voltage is converted to eight kinds of packetized digital signal by the ADC block in the receiver. As shown in Fig. 6, packetized digital data consist of six bits. The first bit and final bit mean start and end of data, respectively. The remaining four bits of data are physical information of the bridge. Then, to convert the digital signal into the load ON/OFF status, a load ON/OFF control switch (S_L) is added to the proposed system, as shown in Fig.4. S_L is controlled by the packetized digital signal. Because the digital signal is not controlled directly, a gate driver is added between the ADC and S_L. The digital signal is input to the gate driver, and converted to a gate driving signal.

According to turn ON/OFF of S_L, the circulating current of the transmitter is varied. As circulating current is varied by load variation of the receiver, the voltage of the compensation capacitor of the transmitter is also varied. For obtaining the voltage difference of the compensation capacitor, the impedance of the receiver has to be converted to impedance of the transmitter. When Kirchhoff's circuit laws are applied to the transmitter and the receiver in Fig.4, the relationship of the voltage and the current is as follows.

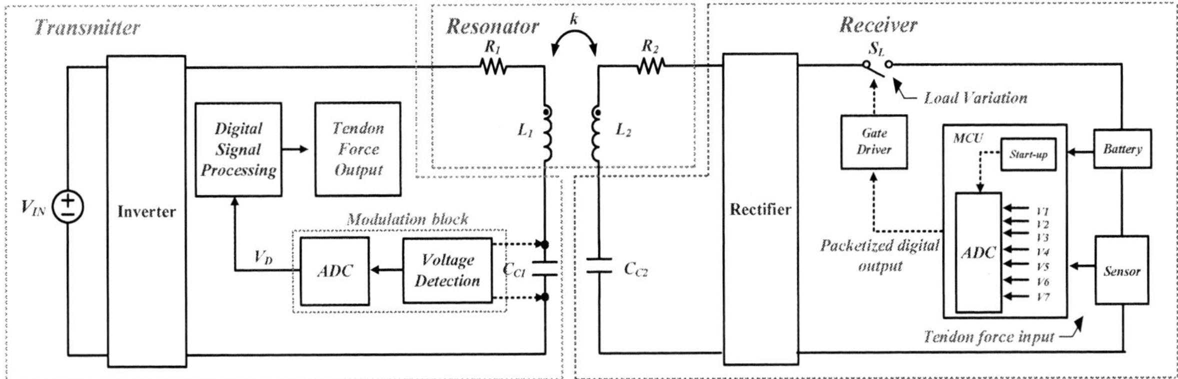

Fig. 4. Proposed wireless power and data transfer system

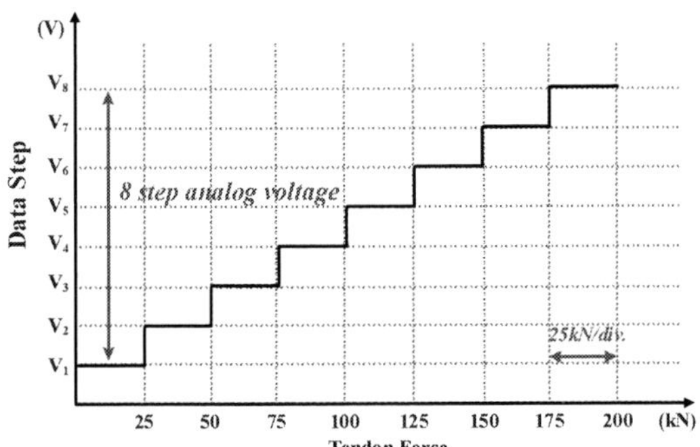

Fig. 5. Relationship between tendon force and data

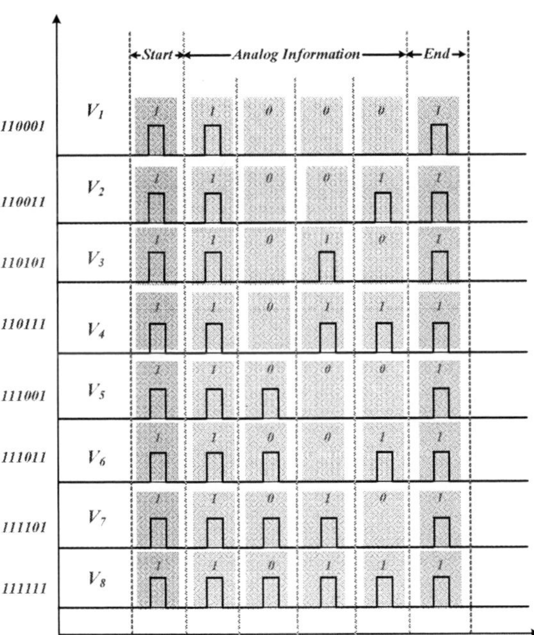

Fig. 6. Relationship between analog voltage and packetized digital data

$$\begin{bmatrix} V_{in} \\ 0 \end{bmatrix} = \begin{bmatrix} Z_t & sM \\ sM & Z_r \end{bmatrix} \begin{bmatrix} I_t \\ I_r \end{bmatrix} \tag{7}$$

$$Z_t = sL_1 + R_1 + \frac{1}{sC_{C1}} \tag{8}$$

$$Z_r = sL_2 + R_2 + \frac{1}{sC_{C2}} + R_L \tag{9}$$

C_{C1} and C_{C2} are capacitors for L-C resonance. I_t and I_r are transmitter and receiver current. Z_t and Z_r are transmitter and receiver impedance. By using (7), the relationship of I_t and I_r is as follow.

$$I_r = -\frac{sM}{Z_r} \cdot I_t \tag{10}$$

The reflected impedance (Z_{ref}) to the transmitter is then obtained by using (7), (10).

$$Z_{ref} = -\frac{s^2 M^2}{Z_r} \tag{11}$$

The real part (Re Z_{ref}) and the imaginary part (Im Z_{ref}) of the reflected impedance can be expressed by (7), (9) [12].

$$\operatorname{Re} Z_{ref} = \frac{\omega^4 C_{C2}^2 M^2 (R_2 + R_L)}{(\omega^2 C_{C2} L_2 - 1)^2 + \omega^2 C_{C2}^2 (R_2 + R_L)^2} \tag{12}$$

$$\operatorname{Im} Z_{ref} = \frac{-\omega^3 C_{C2} M^2 (\omega^2 C_{C2} L_2 - 1)}{(\omega^2 C_{C2} L_2 - 1)^2 + \omega^2 C_{C2}^2 (R_2 + R_L)^2} \tag{13}$$

The proposed WPDT system should be operated at L-C resonant frequency for high WPT efficiency. Therefore, the above formulas (12), (13) are simplified as follows at the L-C resonant frequency of the receiver.

$$\operatorname{Re} Z_{ref} = \frac{\omega^2 M^2}{R_L}, \ \operatorname{Im} Z_{ref} = 0 \tag{14}$$

As shown in Fig.7, the reflection of the receiver impedance (equation 11) is set by ON/OFF control of S_L for load variation detection. That is, when the S_L is ON, the impedance of the receiver is reflected to the transmitter. And, when the S_L is OFF, the impedance of the receiver is not reflected due to open of the receiver circuit. Then, the compensation capacitor voltage by S_L control is calculated in the case of S_L ON and S_L OFF, respectively. The compensation capacitor voltage (V_{C1F}) voltage at S_L ON is as follows.

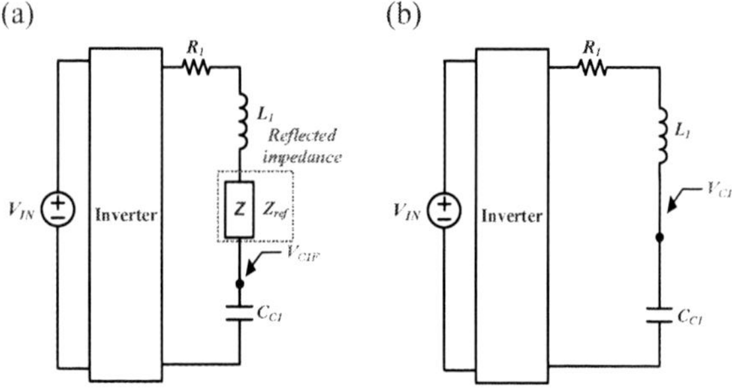

Fig. 7. Transmitter equivalent circuit (a) with reflected impedance (S_L ON) (b) without reflected impedance (S_L OFF)

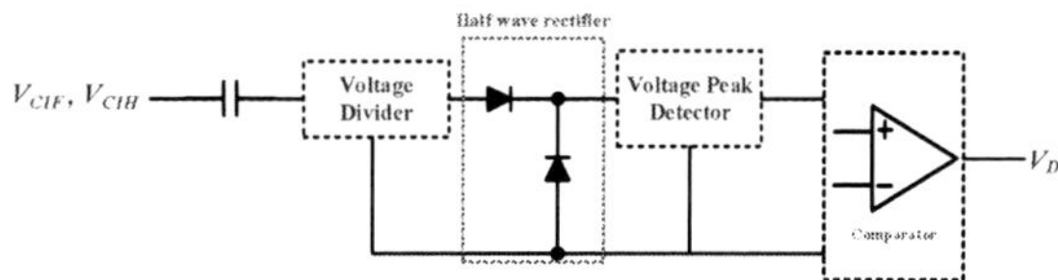

Fig. 8. Schematic of the modulation circuit for data transfer

$$V_{C1F} = \frac{1}{s^2 L_1 C_{C1} + s(C_{C1}R_1 + C_{C1}Z_{ref}) + 1} V_{IN} \tag{15}$$

The compensation capacitor voltage (V_{C1N}) at S_L OFF is as follows.

$$V_{C1N} = \frac{1}{s^2 L_1 C_{C1} + sC_{C1}R_1 + 1} V_{IN} \tag{16}$$

By using (15) and (16), voltage variation (V_{C1D}) of the compensation capacitor voltage is calculated as follow.

$$\begin{aligned} V_{C1D}(s) &= V_{C1F}(s) - V_{C1N}(s) \\ &= \left(\frac{1}{s^2 L_1 C_{C1} + sC_{C1}R_1 + 1} - \frac{1}{s^2 L_1 C_{C1} + s(C_{C1}R_1 + C_{C1}(\mathrm{Re}\, Z_{ref})) + 1} \right) V_{IN} \\ &= \left(\frac{sC_{C1}(\mathrm{Re}\, Z_{ref})}{\left| \begin{array}{l} s^4 L_1^2 + s^3(2L_1 C_{C1} R_1 + L_1 C_{C1}(\mathrm{Re}\, Z_{ref})) + s^2(C_{C1}^2 R_1^2 + 2L_1 \\ + C_{C1}^2(\mathrm{Re}\, Z_{ref})R_1) + s(2C_{C1}R_1 + C_{C1}) + 1 \end{array} \right|} \right) \end{aligned} \tag{17}$$

For $s = j\omega$, equation (15) can be rewritten as follow.

$$\frac{1}{A(\omega) + j\left(\dfrac{(\omega^2 L_1 C_{C1} - 1)^2 + B(\omega)}{L_1^2 C_{C1} / R_1} \right)} \tag{18}$$

where,
$$A(\omega) = \frac{-\omega^4 L_1 M^4 - 2\omega^2 L_1 C_{C1} R_1 R_L + R_L(2R_1 + 1)}{\omega^2 M^2} \tag{19}$$

$$\begin{aligned} B(\omega) &= \omega^2 \left(\frac{2L_1 C_{C1} R_1 + L_1^2 C_{C1}^2 R_1^2 R_L - L_1^4 R_L}{R_1 M^2} \right) \\ &+ \omega \frac{R_L L_1^2}{R_1 M^2}(2L_1 - 1) - 1 \end{aligned} \tag{20}$$

In (20), the difference of the compensation capacitor voltage is maximized by the L-C resonance frequency of the transmitter. That is, when the WPDT system is operated at resonance frequency of the transmitter and the receiver, the AC voltage variation of the resonant capacitor of the transmitter is easily detected.

As shown in Fig.8, AC voltage variation detected at the resonant capacitor is divided by the resistor. It is converted into DC voltage by a half-wave rectifying circuit using diodes and capacitors. This is applied to the input of the comparator, and it is modulated into eight kinds of digital signal. This method to detect the data from the receiver is easily implemented. In addition, since the proposed method uses a voltage detection circuit, it has no influence on the efficiency of the WPT system. The data transmission method using voltage variation of the resonant capacitor has higher accuracy as the output load increases. The reason is that the voltage variation of the resonant capacitor becomes larger. In addition, since the data transmission time is shorter than battery charging time, it has little influence on the efficiency of WPT.

IV. EXPERIMENTAL RESULTS

WPDT system including resonators and data transmission circuit is experimented to concrete structure. The concrete structure setup for the WPDT system is as shown in Fig. 9. It is a reinforced concrete structure with steel rebar. As shown in Fig.9, steel rebar is located at a 50mm distance from the bottom of the reinforced concrete structure. The distance between steel rebar is 150mm. These measurements correspond with real specifications of a bridge construction. Finally, the concrete is deposited into a mold. The size of the concrete structure is 600 × 600 × 300mm^3 (width × length × height) with steel rebar. The overall experimental setup is shown in Fig.10. The resonators are located between the concrete structures. They are designed with 10 turns and a diameter of 300mm. The values of the self-inductance and the resistance are 42μH and 0.2Ω, respectively. k is 0.035 at a distance of 300mm. The transmitter circuit for WPT is a half-bridge topology with series-series capacitor compensation. The receiver circuit consists of the resonator and full-bridge rectifier and connects a load to the output. Also, the load ON/OFF control switch (S_L) for data transmission is connected to the output. When the WPT circuit is operated by the switching frequency corresponding to the resonant frequency, it is possible to transfer maximum power. The geometric information of the concrete structure and the resonator is shown in Table II.

A. *WPT efficiency*

The effectiveness of the proposed WPDT system is verified by an experiment. The resonators are located between the top and bottom of the concrete structure. For convenience of the experiment, an electric load is connected to the output of the receiver, and the power source and controller are connected to the transmitter. The efficiency of the WPT system with an air-gap (type1), the efficiency of the WPT system with a concrete structure (type2), and the efficiency of the WPDT system with a concrete structure (type3) are compared.

TABLE II. EOMETRIC INFORMATION OF CONCRETE STRUCTURES AND RESONATORS

	Geometric information	*Value*
Concrete structures	Concrete size (W × L × H)	600 × 600 × 300mm^3
	Diameter of steel rebar	15mm
	distance between steel rebar	150mm
resonators	Outer diameter	300mm
	Diameter of wire	3.1mm
	Turns	10

Fig. 9. Rebar spacing of concrete structure, Depositing concrete structure

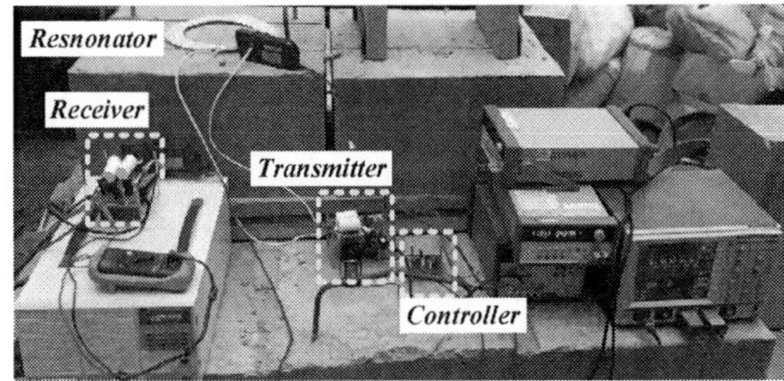

Fig. 10. Experimental setup for the WPDT system

TABLE III. MEASURED RESULTS OF WPT EFFICIENCY

Measured at 100kHz	**WPT efficiency**		
Power transfer	*Air (type1)*	*Concrete (type2)*	*Concrete with Data transfer circuit (type3)*
100mm	66.7%	63.6%	63.2%
200mm	60.1%	57.7%	57.2%
300mm	46.3%	44.1%	43.6%

The measured results with type1, type2, and type3 are listed in Table III. WPT efficiency of 43.6% at 300mm distance is obtained for type3. The efficiency of type3 is similar to that of type2. That is, the data transmission circuit has little influence on the efficiency of WPT. This means that there is almost no loss caused by the data transmission circuit to the transmitter. In addition, this also demonstrates higher efficiency than conventional works on WPT systems in concrete structures by designing optimal resonators [13]-[14].

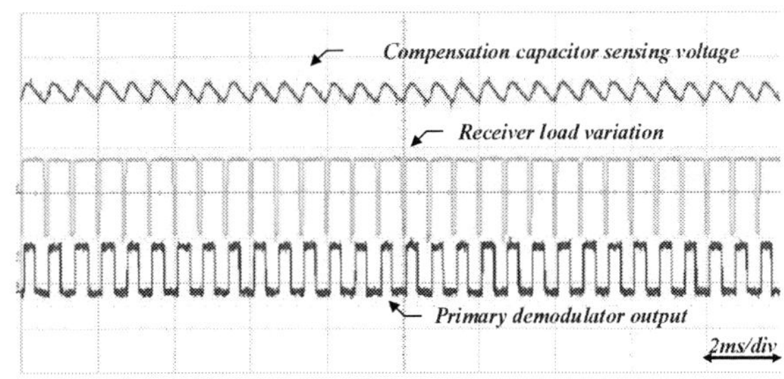

Fig. 11. 1500bps data transmission through the resonator

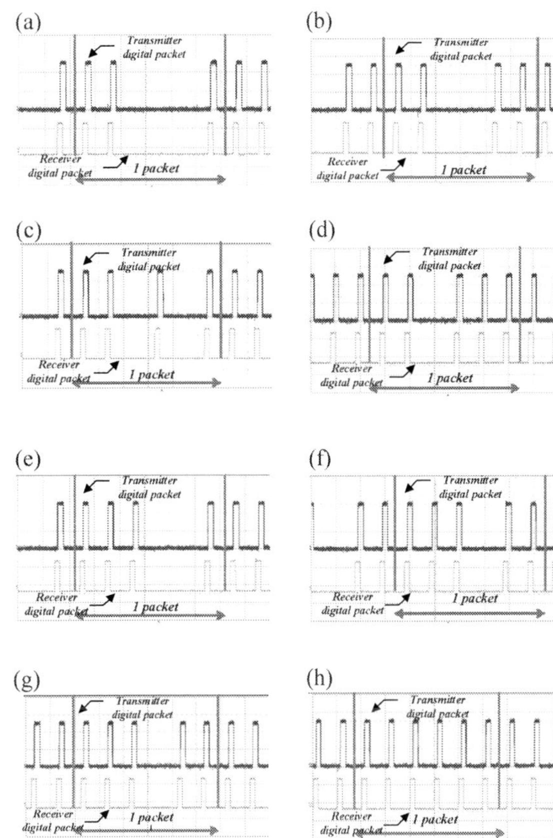

Fig. 12. Digital signal transmission (a) 110001 (b) 110011 (c) 110101 (d) 110111 (e) 111001 (f) 111011 (g) 111101 (f) 111111

B. Data transmission

The data transmission experiment is performed in a concrete structure. The function generator in the receiver makes a signal of 1.5 kHz frequency for this experiment. The generated signal is input to the load variation switch. This varies the load output voltage. ON/OFF control information of the load is displayed as voltage variation of the compensation capacitor. Then, the synchronous signal of the compensation capacitor in the transmitter is measured. Fig.11 shows that it is possible to transmit from the receiver to the transmitter with a data rate of 1500bps. It is confirmed that data are transmitted from the receiver to the transmitter without distortion. Next, a digital packet transmission experiment is performed. The packetized data are generated by the MCU in the receiver and input to the load variation switch. It is confirmed that 8 kinds of packetized digital signal (110001, 110011, 110101, 110111, 111001, 111011, 111101, 111111) are transmitted from the receiver to the transmitter without any distortion, as shown in Fig.12. The data rate of the transmitted digital signal is set to 1500bps. This data rate is fast enough to transmit 6-bit data. As a result, the signal data generated by the receiver are converted into a digital signal and transmitted through the resonator of the transmitter with WPT. The obtained results indicate that the proposed data transmission method by using load variation is effective to deliver physical information of a concrete structure. Also, the proposed method has little effect on the WPT efficiency.

V. CONCLUSIONS

This paper presents a novel WPT system incorporating a digital data transmission circuit for smart bridge sensors. The proposed system performs WPT from the transmitter to the receiver. It achieves high efficiency of WPT by using resonators with an optimized number of turns and resistance based on the design of the resonator section. At the same time, it can transmit eight kinds of packetized digital signal with a data rate of 1500bps. The data transmission circuit has no influence on the WPT efficiency. In addition, a WPDT system transmits the data from the receiver to the transmitter without any distortion. However, this approach has the limitation that variable data of the smart bridge sensors cannot be transmitted simultaneously. Further work to transmit data of more than two sensors should be carried out. Nevertheless, the proposed method has notable advantages. The principle of data transmission is simple and it is possible to eliminate the additional communication modules. Therefore, the proposed WPT system including simultaneous data transmission is suitable for low-cost bridge sensor systems.

ACKNOWLEDGMENT

This work was supported by the National Research Foundation of Korea (NRF) grant funded by the Korea government (MSIP) (No.2010-0028680).

REFERENCES

[1] Chandrashekhara. K,Watkins. S.E., Nanni. A., and Prakash Kumar,. "Design and technologies for a smart composite bridge" in IEEE 2004 Intelligent Transportation Systems, 2004, pp. 954-959

[2] http://www.economist.com/node/17647603

[3] Setijadi, E.; Suwadi; Slamet, B.P.; Muntaqo, A.A.; In'am; Nur, A.E.; Suprobo, P.; Faimun; Febry, F.A. "Design of large scale structural health monitoring system for long-span bridges based on wireless sensor network" in IEEE 2013 Awareness Science and Technology and Ubi-Media Computing (iCAST-UMEDIA), 2013, pp. 169-174

[4] Lynch, J. P., Law, K. H., Kiremidjian, A. S., Kenny, T., & Carryer, E.. "A wireless modular monitoring system for civil structures". In Proceedings of the International Modal Analysis Conference–IMAC , 2002, February

[5] H. J. Jung, I. H. Kim, and S. J. Jang, "An energy harvesting system using the wind-induced vibration of a stay cable for powering a wireless sensor node," Smart Materials and Structures, vol. 20, no. 7, article no. 075001, 2011.

[6] H. Kulah and K. Najafi, "Energy scavenging from lowfrequency vibrations by using frequency up-conversion for wireless sensor applications," IEEE Sensors Journal, vol. 8, no. 3, pp. 261-268, 2008.

[7] G. Simard, M. Sawan and D. Massicotte, "High-speed OQPSK and efficient power transfer through inductive link for biomedical implants," IEEE Trans. Biomed. Circuits Syst., vol. 4, no.3, pp. 192-200, Jun. 2010.dd

[8] G. Wang, P. Wang, Y. Tang, and W. Liu, "Analysis of dual band power and data telemetry for biomedical implants," IEEE Trans. Biomed. Circuits Syst., vol. 6, no.3, pp. 208-215, Jun. 2012.

[9] Wu, J., Zhao, C., Du, J., Lin, Z., Hu, Y., & He, X. "Wireless Power and Data Transfer via a Common Inductive Link using Frequency Division Multiplexing." IEEE Transactions on Industrial Electronics, vol. 62, no. 12 pp. 7810 - 7820, 2015

[10] Saad Mutashar, Mahammad A. Hannan, Salina A. Samad, and Aini Hussain, "Analysis and Optimization of Spiral Circular Inductive Coupling Link for Bio-Implanted Applications on Air and within Human Tissue", Sensors, vol. 14, no. 7, pp. 11522-11541, June, 2014

[11] S. S. Mohan, M. M. Hershenson, S. P. Boyd, and T.H. Lee, "Simple accurate expressions for spiral inductances," IEEE Journal of Solid-state Circuits, Vol. 34, No. 10, pp.1419-1424, 1999.

[12] X. Liu, W. Ng, C. Lee, and S. Hui, "Optimal operation of contactless transformers with resonance in secondary circuits," in Proc IEEE Appl. Power Electron. Conf. Expo., 2008, pp. 645–650

[13] Jonah. O., Georgakopoulos. S.V. "Efficient wireless powering of sensors embedded in concrete via magnetic resonance" in IEEE 2011 Antennas and Propagation (APSURSI), 2011, pp. 1425 – 1428

[14] Ishida. H,Furukawa.H,"Wireless Power Transmission Through Concrete Using Circuits Resonating at Utility Frequency of 60 Hz"IEEE Transactions on Power Electronics, vol. 30, no. 3 pp.1220 - 1229, May. 2015

Inrush Transient Current Analysis and Suppression of Photovoltaic Grid-Connected Inverters During Voltage Sag

Zhongyu Li, Rende Zhao
Electrical Engineering Department
China University of Petroleum (Eastern China)
Qingdao China
lzyupc619@gmail.com, zhaorende@126.com

Zhen Xin, Josep M.Guerrero, Mehdi Savaghebi
Energy Department
Aalborg University
Aalborg Denmark
zxi@et.aau.dk, joz@et.aau.dk, mes@et.aau.dk

Peide Li
Shandong Jinan Power Equipment Factory Co., LTD
Jinan China
peidezhilin@163.com

Abstract—**The Inrush Transient Current (ITC) in the output of the photovoltaic grid-connected inverters is usually generated when grid voltage sag occurs, which can trigger the protection of the grid-connected inverters, and even destroy the semiconductor switches. Then, the grid-connected inverters will thus fail to ride through the voltage sag and even further cause more serious grid faults. This paper analyzes the generation principle of ITC and explores its influence factors, upon which, the suppression approaches are presented. Simulation and experimental results validate the theoretical analysis and the effectiveness of the proposed method.**

Keywords—the inrush transient current (ITC); photovoltaic grid-connected inverters; voltage sag; influence factors; suppression approaches

I. INTRODUCTION

The recent interest in solar energy is increasing tremendously. And the installation of large photovoltaic (PV) power generation systems, which are interconnected with the utility grid, is accelerating [1], [2]. However, the PV power generation systems may lead to instability when grid disturbances occur. The mainly reason is that the photovoltaic grid-connected inverters (PGI) is sensitive to grid disturbances, especially, the voltage sags [3], [4]. Therefore, to maintain the stability of the power system, one important grid code issued in many countries is that the PGI should success riding through voltage sags [5-8]. In order to meet the requirement of the grid code, firstly, PGI shouldn't disconnect from the grid by the reason of the fault current.

The fault current studies of grid-connected inverters have been widely reported in many literatures [2], [9-13]. In [9], a fault current limiter is used to reduce overcurrent in the converter, but it has the drawback of absorbing active power. Also, in the field of overcurrent reduction, there are other methods relying on the control scheme, which can avoid any increase of devices cost in the system. A study reported in [10] proposes a proportional-resonant (PR) current controller for the current limiter to make sure the output current without overcurrent, this current limiter is acquired indirectly by using an active power limiter. In [11] and [12], a proper reference current is selected to ride through grid faults and achieve requirements of different power quality. Among these methods, although low currents can be acquired, the control scheme does not assure its minimum value, but [13] overcomes this problem. On the other hand, for unbalanced voltage sags, there is also a control algorithm [2] to limit the peak current.

Obviously, the above mentioned technical literatures are all having concerns about the fault current limitation when the digital controller can take control of the output voltage of PGI. However, when a sudden decrease of grid voltage occurs, the output voltage of the PGI cannot change instantaneously. This is because of the time delay in its digital control system. Therefore, a large voltage drop will be generated on the output L-filters or LCL-filters of the inverter, which will further lead to an inrush transient current (ITC). This ITC may trigger the overcurrent protection of the inverter and in the worst case may even destroy the semiconductor switches, causing the disconnection of the PGI from the grid. The disconnection of too many PGI from the grid system will aggravate the voltage sags and lead to more severe grid faults.

Therefore, it is of great importance to study the ITC suppression solutions for the PGI. Obviously, a good understanding of the ITC influence factors will contribute a lot to this work. Actually, although the generation principle of ITC is not complex, the ITC influence factors have not been fully investigated so far. Therefore, in this paper, a thorough study of the ITC influence factors is carried out. Based on these findings, the appropriate ITC suppression approaches are then discussed. Finally, some experimental results are given to validate the theoretical analysis and the effectiveness of the proposed ITC suppression method.

II. PRINCIPLE AND CAUSE OF INRUSH TRANSIENT CURRENT

Fig.1 shows the basic control structure of the three-phase PGI. To explain the generation principle of ITC, the grid-connected inverter with L filter is analyzed for example. u, v, i, L, R represent the grid phase voltage, inverter output voltage, phase current, grid-connected inductance and its resistance,

This work was supported by Shandong Provincial Natural Science Foundation, China (ZR2014EEM012).

978-1-4673-9551-9/16 $31.00 © 2016 IEEE

respectively. In this paper, a proportional-integral (PI) current controller is used, which is denoted by $G(z)$ in Fig.1.

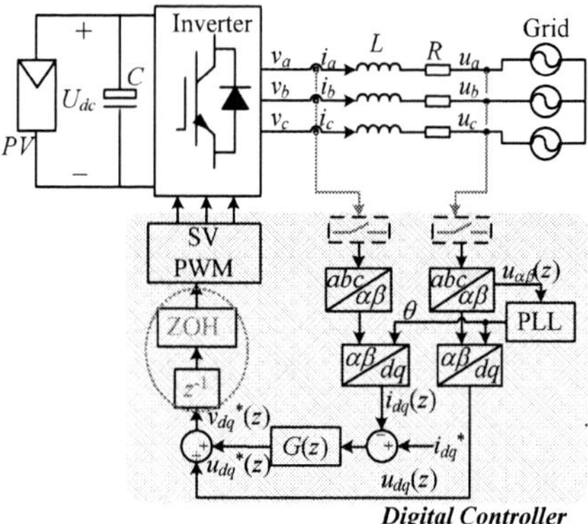

Fig. 1 The basic control structure of the three-phase PGI

As shown in Fig.1, the inverter output voltage reference v^*_{dq} is influenced by two components, which are the output of current controller u^*_{dq} and the direct feed-forward gird voltage u_{dq}. Due to the time inertial of the current PI controller, the direct feed-forward gird voltage is the effective component to control the inductor current during the process of the grid voltage sag.

Traditionally, the single-sampling approach is widely used in PGI digital control system, then the state variables (i.e., inductor currents and grid voltages) needed in the control algorithm are sampled once per switching period [14]. Because of this sampling approach, it is hard to guarantee that the instantaneous voltage can be sampled instantly, and with considering the zero-order hold (ZOH) effect of the PWM [14], [15], the valid output voltage, which can limit the ITC, need at least one and half switching period to be produced by DSP. Due to this delay, the voltage difference on the grid-connected inductor is large and will result in the rapidly rising of the inductor current in this short time. Thus, ITC is generated.

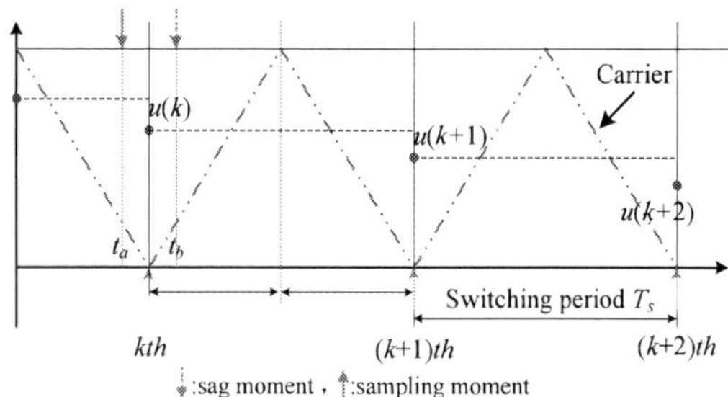

Fig. 2 Generation mechanism of the digital control delay

The generation mechanism of the digital control delay is shown in Fig.2. T_s is the switching period which is equal to the sampling period by using the single-sampling approach. In Fig.2, suppose that the voltage sags at t_a just before the kth sampling, the controller could catch the voltage. After one sampling period, the digital regulator could update the command signal to $u(k+1)$ based on the newly sampled state variables of the fault grid voltage, together with the ZOH effect of PWM, the digital control delay is at least $1.5T_s$ when the inverter could control the PGI current. However, as the sag occurs at t_b which lags behind the kth sampling, it's impossible to capture the fault voltage at this period. Only during the $(k+1)th$ sampling period, the effective voltage command $u(k+2)$ could be produced by the controller. Therefore, the delay expands to $2.5T_s$. During this $1.5\sim2.5T_s$ time delay, the DSP is not able to trace the variation of voltage, hence the PGI current rapidly increases to form the ITC. In the event the ITC was large enough and beyond the device component limit, a destructive damage of semiconductors could occur. Thus, the analysis the value of ITC is imperative in the next part.

III. Influence Factors Of Inrush Transient Current

According to Fig. 1, the mathematical model of the PGI in the three phase stationary reference frame is shown as following

$$\begin{cases} S_a U_{dc} - u_a - Ri_a - L\dfrac{di_a}{dt} \\ = S_b U_{dc} - u_b - Ri_b - L\dfrac{di_b}{dt} \\ = S_c U_{dc} - u_c - Ri_c - L\dfrac{di_c}{dt} \end{cases} \tag{1}$$

where S_a, S_b, S_c are switch function. In the three-phase three-wire system, the sum of the current is $i_a+i_b+i_c=0$, so (1) becomes

$$\begin{cases} L\dfrac{di_a}{dt} = v_a - u_a - Ri_a + \dfrac{1}{3}u_x \\ L\dfrac{di_b}{dt} = v_b - u_b - Ri_b + \dfrac{1}{3}u_x \\ L\dfrac{di_c}{dt} = v_c - u_c - Ri_c + \dfrac{1}{3}u_x \end{cases} \tag{2}$$

where v_a, v_b, v_c are the inverter output voltage. And their expression can be given by

$$\begin{cases} v_a = \left[S_a - \dfrac{S_a + S_b + S_c}{3} \right] U_{dc} \\ v_b = \left[S_b - \dfrac{S_a + S_b + S_c}{3} \right] U_{dc} \\ v_c = \left[S_c - \dfrac{S_a + S_b + S_c}{3} \right] U_{dc} \end{cases} \tag{3}$$

u_x is the sum of the three phase voltage and is given by

$$u_x = u_a + u_b + u_c \tag{4}$$

According to (2), any single phase equivalent circuit diagram can be seen in Fig. 3 when considering the balance sag.

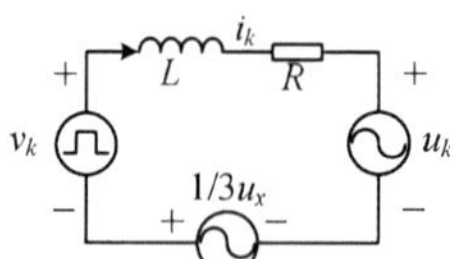

Fig. 3 Single phase equivalent circuit diagram

where k is the symbol of a, b, c phase. Before the voltage sags, according to Fig.3, the single phase voltage expression for this system can be given by

$$L\frac{di_k}{dt} = v_k - u_k - Ri_k + \frac{1}{3}u_x \quad (5)$$

When the voltage sag occurs, (5) changes to

$$L\frac{di_{k\_f}}{dt} = v_{k\_f} - u_{k\_f} - Ri_{k\_f} + \frac{1}{3}u_{x\_f} \quad (6)$$

where v_{k_f}, u_{k_f} and u_{x_f} are the output voltage of PGI, grid phase voltage and the sum of the three phase voltage, respectively.

In (6), the resistance of the grid-connected inductor is so small that its influence to the ITC can be neglected in the following analysis.

When a sudden decreasing of the grid voltage occurs, DSP may not sample the voltage instantly. Hence, the PGI output voltage remains unchanged, with $v_{k\_f}=v_k$. However, the grid voltage could be significantly different and (5) changes to:

$$L\frac{di_{k\_f}}{dt} = v_k + \frac{1}{3}u_{x\_f} - u_{k\_f} \quad (7)$$

In (7), for different sag type, the value of u_x is different. According to [16], the value of u_x under the four typical sag conductions are shown in Table I, where p_u is the sag depth and $p_u \in (0, 1]$.

TABLE I u_x VALUE UNDER DIFFERENT SAG TYPE

Sag type	u_x	Sag type	u_x
(A)1Φn	$(p_u-1)U_m$	(C)2Φn	$(1-p_u)U_m$
(B)2Φ	0	(D)3Φ	0

where U_m is the normal phase voltage. Assuming that the three voltage symmetrical sag occurs in the grid, then $u_{x\_f}=0$ and $u_{k\_f}=p_u u_k$. At the sag moment, the current in the inductor can be given by

$$i_{k\_f} = \frac{1}{L}\int_{t_1}^{t_2}(v_k - p_u u_k)dt + i_0 \quad (8)$$

Thus, the ITC Δi_k is given by

$$\Delta i_k = \frac{1}{L}\int_{t_1}^{t_2}(v_k - p_u u_k)dt = \frac{1}{L}\Delta u \Delta t \quad (9)$$

where $\Delta t=t_2-t_1$ represents the digital control delay. $\Delta u=v_k-p_u u_k$ represents the voltage difference of the inductor. It shows obviously that ITC is influenced by the digital control delay Δt, the voltage difference Δu and the inductor L. v_k is the PGI output voltage and kept unchanged during Δt, while the grid voltage is changed to $p_u u_k$.

Under normal grid-connected condition, assuming that the power factor is unit, hence the difference between u_k and v_k is very small compared with the fault condition, approximately, $u_k \approx v_k$. So (9) can be rewritten by

$$\Delta i_k \approx \frac{1}{L}\int_{t_1}^{t_2}(u_k - p_u u_k)dt = \frac{1}{L}(1-p_u)u_k \Delta t \quad (10)$$

According to (10), Figs.4-6 show how the ITC are influenced by these factors.

1) The influence of grid-connected inductor L and the digital control delay Δt to the ITC: The influencing factors were set by following conditions (see Fig. 4):

a) CASE A:

u_k=311V p_u=50% L=2mH Δt=2.5T_s=75μs

b) CASE B:

u_k=311V p_u=50% L=4mH Δt=2.5T_s=125μs

c) CASE C:

u_k=311V p_u=50% L=2mH Δt=1.5T_s=125μs

By the comparison between *CASE B* and *CASE C* of Fig. 4, given that the grid-connected inductor is 2mH, the ITC is larger. As shown in *CASE A* and *CASE C*, the only difference is the sampling time which leads to the digital control delay are 75μs and 125μs, respectively. Clearly, the longer the digital control delay is, the larger the ITC is.

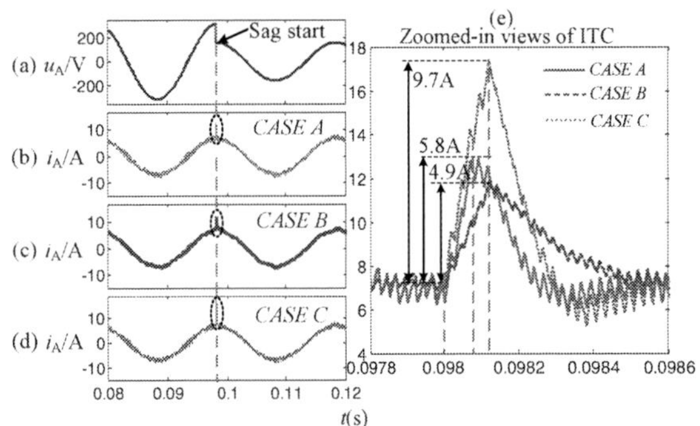

Fig. 4 The influence of L and the digital control delay Δt to the ITC: (a) Phase A grid voltage, (b)-(d) phase A grid-connected current, and (e) zoomed-in views of ITC.

2) The influence of sag depth p_u to the ITC: The influence factors were set by the following conditions (see Fig. 5):

a) CASE A:

u_k=311V p_u=70% L=2mH Δt=125μs

b) CASE B:

u_k=311V p_u=50% L=2mH Δt=125μs

c) CASE C:

u_k=311V p_u=30% L=2mH Δt=125μs

As shown in Fig. 5, it's clearly to see that the ITC is 13.6A when the sag depth is 70%, which is larger than the other two cases (9.7A and 5.8A).

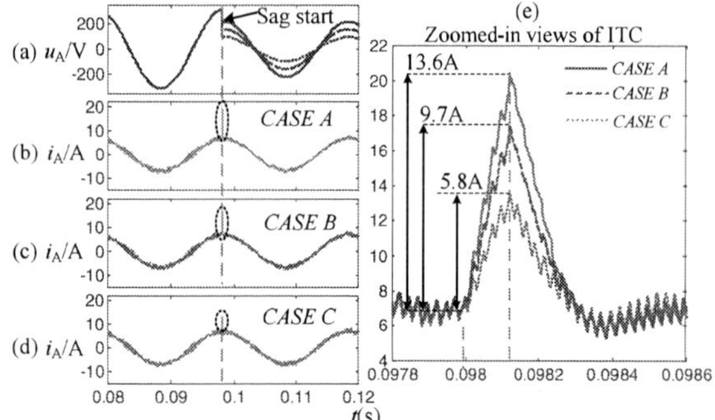

Fig. 5 The influence of sag depth p_u to the ITC: (a) Phase A grid voltage, (b)-(d) phase A grid-connected current, and (e) zoomed-in views of ITC.

3) The influence of sag time to the ITC: The influence factors were set by the following conditions (see Fig. 6):

a) CASE A:

u_k=200V p_u=70% L=2mH Δt=125μs

b) CASE B:

u_k=100V p_u=70% L=2mH Δt=125μs

c) CASE C:

u_k=0V p_u=70% L=2mH Δt=125μs

Fig. 6 shows the ITC at different sag time. The ITC is 8.8A, 4.4A, 0A, respectively. It's easy to find out that the ITC is much larger as the voltage reaches the peak value.

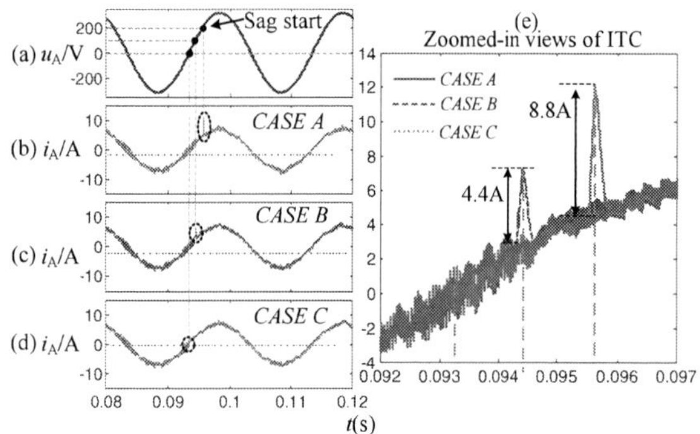

Fig. 6 The influence of sag time to the ITC: (a) Phase A grid voltage, (b)-(d) phase A grid-connected current, and (e) zoomed-in views of ITC.

The validity of the (10) is demonstrated by the simulation results above.

In the high power PGI, the switching frequency can be hundreds hertz per second [17]-[18]. Moreover, considering the worst voltage sag condition, the ITC could reach the destructive value of the semiconductor.

IV. Current Suppression Method

From the above analysis, obviously, the ITC can be suppressed by regulating the influence factors. The ITC can be substantially reduced by using a larger grid-connected inductance, adopting the multi-sampling scheme approach, etc. However, if the system is not able to change these influence factors, the methods mentioned above will be invalid. Due to the time delay of ITC can be dozens of μs, to suppress this inrush current, this paper proposed a hardware suppression method to limit the ITC under the safe range.

As shown in Fig.7, the hardware current limiter in PGI is positioned between the DSP controller and the IGBT driver, when a phase current is sampled with a value exceeding the limiting value, the limiter can thus set a low level of the PWM signals to the IGBT driver. By shutting down the IGBT about dozens of μs, the ITC is suppressed and then the PWM signals resume normal again. Thus, the PGI overcurrent protection will not be triggered and the system will continuously work in the grid-connected condition. This is biggest different between the hardware current limiter and the overcurrent protection circuit. Meanwhile, the limiter does not influence the normal condition. The block diagram of hardware current limiter is shown in Fig.8.

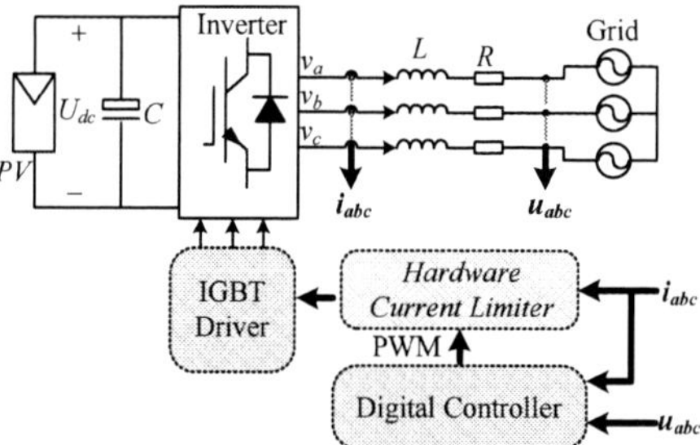

Fig. 7 The location of the hardware current limiter

As shown in Fig. 8, the first part of hardware current limiter is the current sampling circuit whose function is removing the higher harmonics from the sampled current and properly scaling the sampled current. Both the reference current circuit output I_{ref} and the processed current signal i_{sample} are sent to a current comparing circuit. The kernel of the comparing circuit is the hysteresis comparator, which has two thresholds (I_{T1}, I_{T2} and $I_{T1}> I_{T2}$). This hysteresis comparator is realized by a differential comparator, like LM339. When i_{sample} is larger than I_{T1}, the comparator output Out_C goes low. When i_{sample} is lower than the value of I_{T2}, Out_C will go high. Suppose that Out_C is low, when the PWM signals pass through the PWM blocking circuit which is realized by an *AND* gate, the PWM signals would be set to low level and until Out_C is changed to a high level, the PWM signals then become normal. According to this mechanism, the ITC can be suppressed. This is quite different from the overcurrent protection circuit which will shut down the PGI. Although the above discussion is with one phase case, it also applies to three-phase system.

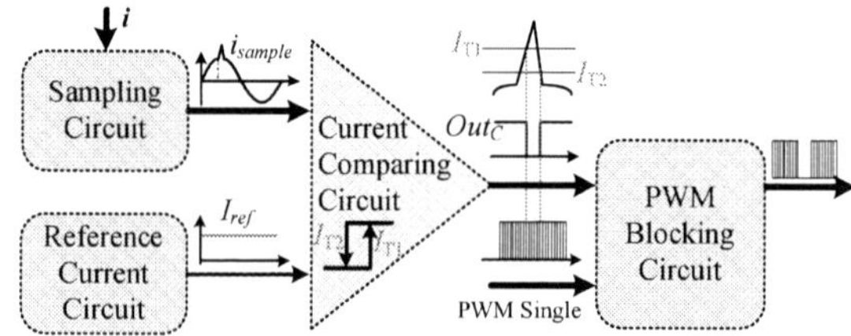

Fig. 8 The block diagram of hardware current limiter

The hardware of the current limiter is shown in Fig. 10. Because the devices used in the hardware limiter is inexpensive, the additional cost of the limiter is quite little.

The simulation results of ITC without and with the hardware current limiter are shown in Fig. 9, where the influence factors were set by the following condition

$$u_k=311\text{V} \quad p_u=70\% \quad L=2\text{mH} \quad \Delta t=125\mu s$$

In Fig. 9, *CASE A* and *B* are the ITC waveform without and with the hardware current limiter. In *CASE B*, the current is suppressed under 11.2A due to the hysteresis of comparator. When the current goes beyond the first threshold (11.2A), the PWM signals are blocked, and this blocking status remains until the current goes down below the second threshold (8.6A). Then the PWM signals resume normal.

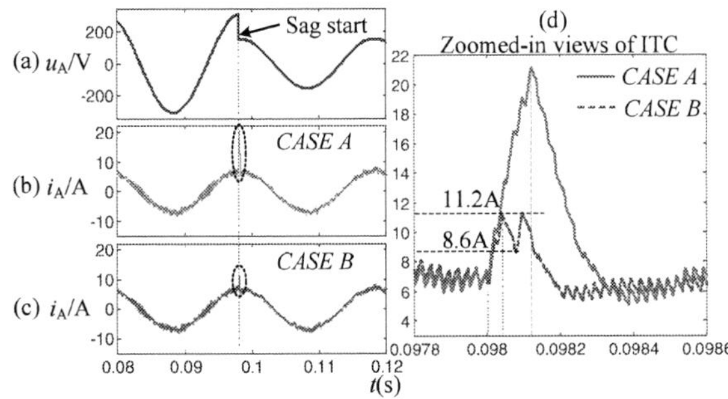

Fig. 9 ITC without and with the hardware current limiter: (a) Phase A grid voltage, (b)-(c) phase A grid-connected current, and (d) zoomed-in views of ITC.

V. Experimental Result

To examine the theoretical analysis and the current limiter proposed in this paper, the experiments with a PGI system are performed shown in Fig.10. In the experimental setup, a voltage sag generator (VSG) is used to simulate the voltage sags. And the inverter is controlled by DSP TMS320F28335. The photovoltaic power source is implemented by a programmed Chroma DC source. The parameters of the actual system are given in Table II.

Fig. 10 Experimental system

TABLE II Experimental Parameters

Parameter	Value
Grid phase voltage amplitude	311 V
Grid frequency	50 Hz
Grid inductance	2.0 mH
DC filter capacitance	1660 μF
Switching frequency	20 kHz
Voltage sag type	Three phase voltage sags
Sag duration time	0.5 s

In the actual system, the grid voltage and current are sampled by the hardware circuit. Due to the existence of the low-pass filter circuit, a hardware delay would be inevitably introduced in digital control. In this PGI system, the hardware delay is $t_d=50\mu s$.

In Fig. 11, the influence factors of ITC were set by the following condition:

$$u_k=311\text{V} \quad p_u=50\% \quad L=2\text{mH} \quad \Delta t=1.5T_s+t_d=125\mu s$$

As shown in Fig. 11, under the normal condition, the grid-connected current is 7.0A, yet the ITC is 9.7A as the voltage sags. Substituting the influence factors (L=2.0mH, p_u=50%, u_{ks}=311V, Δt=125μs) into (10), the theoretical result of ITC is 9.69A, which is consistent with the experimental result.

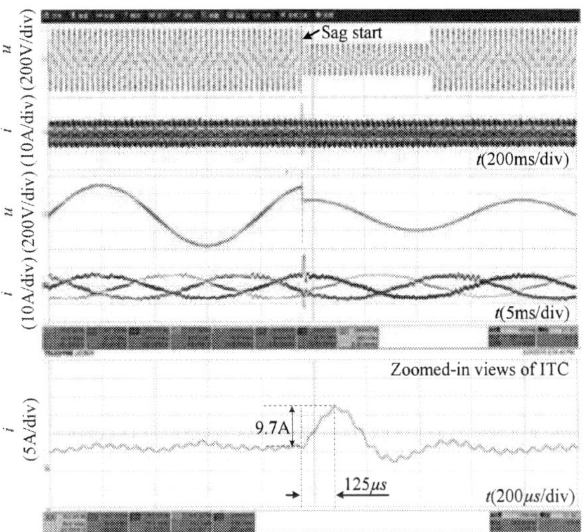

Fig. 11 The experiment waveform when voltage sags at peak value

In Fig. 12, the influence factors of ITC were set as below

$$u_k=190\text{V} \quad p_u=50\% \quad L=2\text{mH} \quad \Delta t=1.5T_s+t_d=125\mu s$$

Under this conduction, according to (10), the calculated theoretical result of ITC is 6.0A, which is similar to the experimental result (6.2A). By comparing Fig.11 and Fig.12, the analysis of Fig. 6 is verified.

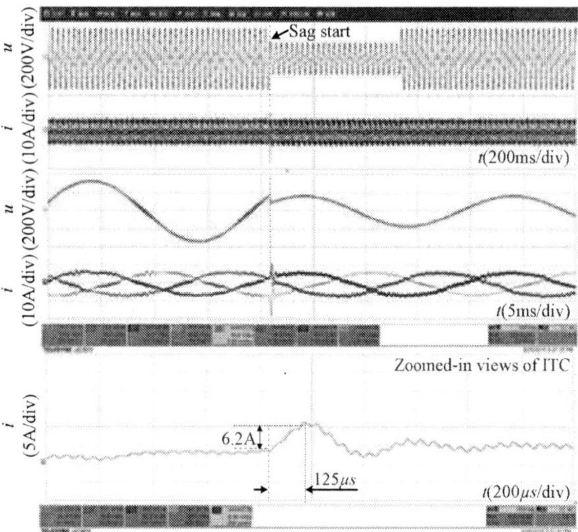

Fig. 12 The experiment waveform when voltage sags at 190V

In Fig.13 and Fig.14, the influence factors of ITC were set by the following conditions:

u_k=311V p_u=70% L=2mH Δt=2.5T_s+ t_d=175μs

u_k=311V p_u=50% L=2mH Δt=2.5T_s+ t_d=175μs

As shown in Fig. 13, when the sag depth is 70%, the ITC is 19.0A which is larger than the ITC in Fig. 14. The only difference between two Figures is that sag depth is changed to 175μs in Fig. 14 from 125μs in Fig. 11, consequently the ITC increases to 13.6A from 9.7A. By substituting the above influence factors into (10), the ITC are 19.05A and 13.56A, respectively. By observing the results given in Fig. 11, Fig. 13 and Fig. 14, obviously, the deeper of the voltage sag and the longer of the digital control delay, the greater destructive of ITC.

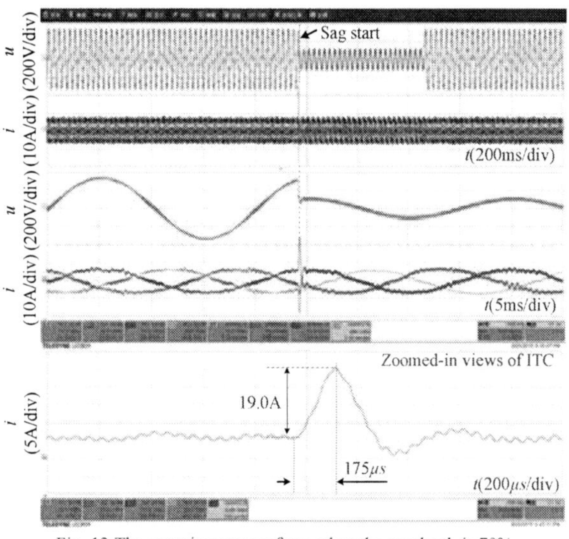

Fig. 13 The experiment waveform when the sag depth is 70%

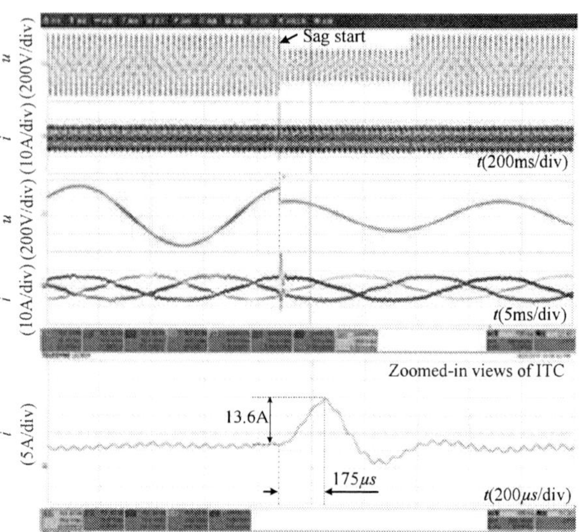

Fig. 14 The experiment waveform when the digital control delay is 175us

Fig. 15 shows the effect of the hardware current limiter, the influence factors of ITC were the same as that of Fig. 13. The current limiter suppresses the current under 13.1A, which should be 20.56A in Fig. 13. Meanwhile, because of the hysteresis comparator, as the current is less than 8A, the current limiter releases the PWM signals and the current increases again. In this case, obviously, the PGI does not disconnect from grid. Therefore, Fig. 15 demonstrates the effectiveness of the hardware current limiter.

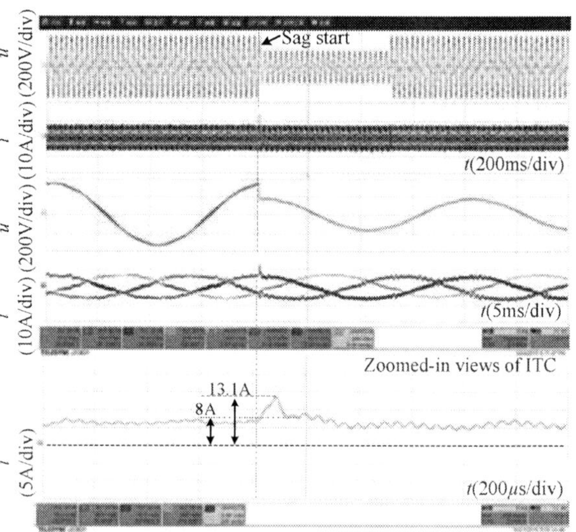

Fig. 15 The experiment waveform with the hardware current limiter

VI. Conclusion

In this paper, the influence factors of the Inrush Transient Current (ITC) and its suppression method are presented. By using a hysteresis comparator, the proposed hardware current limiter has fast and excellent suppression ability to suppress the ITC. Thus, the PGI can uninterruptedly work in the grid-connected condition without the influence of the ITC. Both simulation and experimental results validate the theoretical analysis and the proposed suppression method. Moreover, the proposed methods in this paper can also be used in other applications, where PWM inverters are used.

REFERENCES

[1] G. M. S. Islam, A. Al-Durra, S. M. Muyeen, and J. Tamura, "Low voltage ride through capability enhancement of grid connected large scale photovoltaic system," in *IECON 2011 - 37th Annual Conference on IEEE Industrial Electronics Society*, 2011, pp. 884-889.

[2] J. Miret, M. Castilla, A. Camacho, L. G. Vicuna, and J. Matas, "Control scheme for photovoltaic three-phase inverters to minimize peak currents during unbalanced grid-voltage sags," *IEEE Trans. Power Electron.*, vol. 27, no. 10, pp. 4262–4271, Oct. 2012.

[3] I. J. Gabe, F'. F. K. Palha and H. Pinheiro, "Grid Connected Voltage Source Inverter Control During Voltage Dips," in *IECON 2013 - 35th Annual Conference on IEEE Industrial Electronics Society*, 2009, pp. 4571-4576.

[4] G. Saccomando and J. Svenssan, "Transient Operation of Grid Connected Voltage Source Converter Under Unbalanced Voltage Conditions," in *IEEE Industry Applications Society Annual Meeting*, 2001, vol. 4. pp. 2419 -2424.

[5] F. K. A. Lima, A. Luna, P. Rodriguez, E. H. Watanabe, and F. Blaabjerg, "Rotor voltage dynamics in the doubly fed induction generator during grid faults," *IEEE Trans. on Power Electron.*, vol. 25, no. 1, 2010, pp. 118--130.

[6] M. Tsili and S. Papathanassiou, "A review of grid code technical requirements for wind farms," *IET Renew. Power Gener.*, vol. 3, no. 3, 2009, pp. 308--332.

[7] M. Molinas, J. A. Suul, and T. Undeland, "Low voltage ride through of wind farms with cage generators: STATCOM versus SVC," *IEEE Trans. on Power Electron.*, vol. 23, no. 3, 2008, pp. 1104--1117.

[8] "Grid Code: High and Extra High Voltage," *E.ON Netz GmbH*, Bayreuth, Germany, Apr. 2006.

[9] M. Taghizadeh, J. Sadeh, and E. Kamyab, "Protection of grid connected photovoltaic system during voltage sag," in *2011 Inter. Conference on Advanced Power System Automation and Protection.*, vol. 3, pp. 2030-2035.

[10] G. M. S. Azevedo, G. Vazquez, A. Luna, D. Aguilar and A. Rolan, "Photovoltaic Inverters with Fault Ride-Through Capability," in *IEEE International Symposium on Industrial Electronics*, 2009, pp. 549-553.

[11] P. Rodriguez, AV. Timbus, R. Teodorescu, M. Liserre, and F. Blaabjerg, "Flexible active power control of distributed power generation systems during grid faults," *IEEE Trans. Ind. Electron.*, vol. 54, no. 5, pp. 2583-2592, Oct. 2007.

[12] M. Castilla, J. Miret, J. L. Sosa, J. Matas, and L. García de Vicuña, "Grid-fault control scheme for three-phase photovoltaic inverters with adjustable power quality characteristics," *IEEE Trans. Power Electron.*, vol. 25, no. 12, pp. 2930–2940, Dec. 2010.

[13] A. Camacho, M. Castilla, J. Miret, A. Borrell, and L. G. de. Vicuña, "Active and Reactive Power Strategies With Peak Current Limitation for Distributed Generation Inverters During Unbalanced Grid Faults," *IEEE Trans. Ind. Appl.*, vol. 62, no. 3, pp. 1515-1525, Dec. 2009. Mar. 2015.

[14] S. Buso, and P. Mattavelli. "Digital Control in Power Electronics,"Synthesis Lectures on Power Electronics.2006.

[15] L. Corradini, W. Stefanutti, and P. Mattavelli, "Analysis of multisampled current control for active filters," *IEEE Trans. Ind. Appl.*, vol. 44, no. 6, pp. 1785–1794, Nov./Dec. 2008.

[16] R. Zeng, H. Nian, and P. Zhou, "A three-phase programmable voltage sag generator for low voltage ride-through capability test of wind turbines," *Energy Conversion Congress and Exposition (ECCE)*, 2010, pp. 305-311.

[17] A. Edpuganti, and A. K. Rathore. "A Survey of Low Switching Frequency Modulation Techniques for Medium-Voltage Multilevel Converters," *IEEE Trans. Ind. Appl.*, vol. 51, no. 5, pp. 4212-4228, Sep./Oct. 2015.

[18] A. Khambadkone, and J. Holtz. "Low switching frequency and high dynamic pulsewidth modulation based on field-orientation for high-power inverter drive," *IEEE Trans. on Power Electron.*, vol. 7, no. 4, 1992,pp.627-632.

A Highly Reliable Single-Stage Converter for Electric Vehicle Applications

S.A.KH. Mozaffari Niapour
Department of Electrical and Computer Engineering
Northeastern University
Boston, MA
Email: s.a.kh.mozaffari.niapour@ieee.org

Mahshid Amirabadi
Department of Electrical and Computer Engineering
Northeastern University
Boston, MA
Email: m.amirabadi@neu.edu

***Abstract*—The parallel ac link universal converter is an extension of a buck-boost converter in which the current of the inductor is alternating with zero dc component. Charging an inductor in both positive and negative directions allows better utilizing it. The input and output in parallel ac-link universal converter can have any number of phases with dc or ac voltages, and the switches benefit from the soft switching. The main limitation of the parallel ac-link universal power converter is that it requires large number of switches, i.e. twice the number of switches required for the same converter with dc link, to be able to charge and discharge the inductor in both positive and negative directions. Large number of switches decreases the reliability of the converter and increases its cost. Several modified configurations that require fewer switches than the original configuration and offer the same advantages have been proposed. In this paper, the application of one of these topologies, called "extremely sparse parallel ac-link universal converter", in Electric Vehicle (EV) is studied. This single-stage converter fulfills the task of the dc-dc converter, inverter, and battery charger in an EV. In contrast to the conventional configuration in which a bulky electrolytic capacitor is used, the proposed approach uses a small inductor and a small ac capacitor at the link that increases the reliability of the system. In this paper, principles of operation, analysis and design procedure of this converter are presented, and its performance in EV application is evaluated.**

Keywords—universal converter; soft switching; electric vehicle; high-power-density converters

I. Introduction

Rapidly increasing energy consumption and depletion of fossil fuels are main motivations for utilizing Electric Vehicles (EVs). The interest in developing EVs as a potential and appealing solution for transportation has been continually increased over the last decade. In pursuit of this goal, extensive studies are carried out on increasing power density, reliability, and lifetime of power converters used in EVs.

Parallel ac-link universal power converters are a class of power converters that extends the principles of the operation of the dc-dc buck-boost converters into dc-ac, ac-dc and ac-ac configurations. In this family of converters, in order to increase the utilization of the inductor, the link inductor is charged and discharged in both positive and negative directions leading to a high frequency alternating link current with no dc component. The ac-link current in parallel ac-link universal power converter is achieved at the expense of doubling number of switches. This converter was proposed in [1]. The main disadvantage of this family of converters is large number of switches, resulting in lower reliability and more complexity in control scheme. Several solutions have been proposed to overcome this shortcoming of parallel ac-link universal converters. In [2], a modified configuration called sparse parallel ac-link universal converter using two intermediate cross-over switching circuits, one circuit for the input side and another circuit for the output side, was proposed. This converter has all the advantages of the parallel ac-link universal power converter, but it requires fewer switches, leading to higher reliability and compactness. In [3], the ultra-sparse parallel ac-link universal power converter, which further decreases the number of switches, was proposed for systems that do not require bidirectional flow of power. This converter imposes some limitations on the load power factor. In [4], a topology similar to the sparse configuration was introduced for PV applications by combining input and output intermediate cross-over switching circuits. Unlike the original configuration, in this converter transformer cannot be added to the link. In [5], the converter proposed in [4] was modified to allow bidirectional flow of power.

In this paper, the converter proposed in [5], called extremely sparse parallel ac-link universal converter, is adapted to EV application, and its performance is evaluated to show its superiority over the state of the art converters. Although this converter uses unidirectional switches, it is capable of transferring power in both directions. The detailed principles of operation, analysis, simulation and experimental results of extremely sparse parallel ac-link universal converter for EV application are presented in this paper.

II. Dynamic Modeling of Electric Vehicle

An EV is mainly composed of a battery, a power electronic converter, and an electric motor/generator. Tractive effort, which is the force propelling the vehicle forward, is key factor for dynamic modelling of an EV. The tractive effort needs to overcome the rolling resistance, aerodynamic drag, the component of the vehicle's weight acting down the slope, and

This work was supported by UIC Chancellor's Innovation Fund in partnership with IllinoisVENTURES.

also to accelerate the EV, as shown in Fig. 1. These forces are explained as follows [6]:

Rolling resistance force: The rolling resistance is mainly because of the friction of the EV tires, bearings and the gearing system. This force, which is approximately constant and proportional to vehicle weight, can be determined as follows:

$$F_{rr} = \mu_{rr} Mg \tag{1}$$

where μ_{rr}, M, and g are coefficient of rolling resistance, mass of the vehicle, and gravitational acceleration, respectively. The coefficient depends on type of tire and the tire pressure, and the typical coefficient range between 0.005 and 0.015.

Aerodynamic drag: This force is caused by the friction of the EV body moving through the air, and it generally depends on the whole shape of the EV. The force can be expressed by:

$$F_{ad} = 0.5 \rho A C_d v^2 \tag{2}$$

where ρ, A, *and* v *are* density of the air, frontal area, and velocity of the vehicle, respectively. C_d is a constant value called drag coefficient.

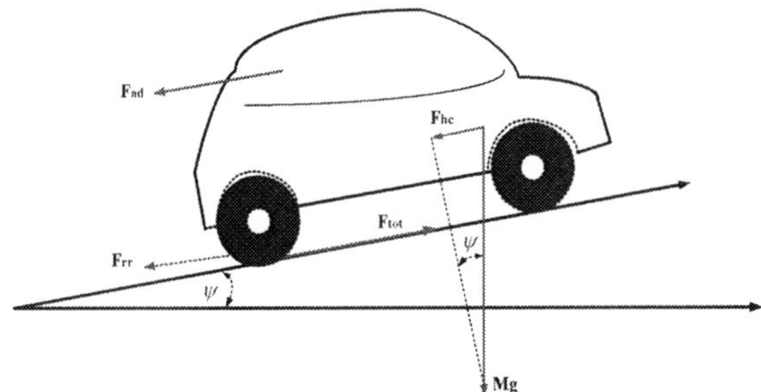

Fig. 1. The forces involving on an EV traveling along a slope

Hill climbing force: This force is caused by EV weight for driving it up on a slope with slope angle of ψ, and it is calculated as follows:

$$F_{hc} = Mg\sin(\psi) \tag{3}$$

Acceleration force: In order to change EV's speed, acceleration force needs to be applied to the EV. This force which is derived from Newton's second law creates linear acceleration (a) as follows:

$$F_{la} = Ma \tag{4}$$

The total tractive effort is equal to summation of the above mentioned forces and can be expressed by:

$$F_{tot} = F_{rr} + F_{ad} + F_{hc} + F_{la} \tag{5}$$

F_{tot} needs to be provided by electric motor torque. This force is converted to wheel torque, as shown in Fig. 2.

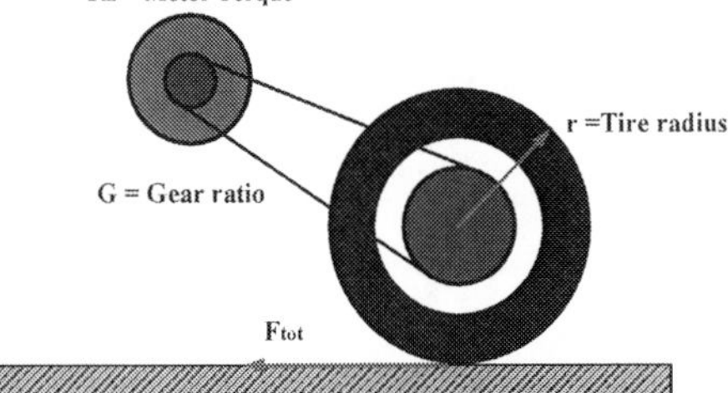

Fig. 2. Arrangement of connecting a motor to a drive wheel

Supposing that the wheel radius is r, then the wheel torque can be calculated as follows:

$$T_{wheel} = F_{tot} r \tag{6}$$

If G is the transmission gearbox ratio, the motor torque (T_m) can be obtained as follows:

$$T_m = \frac{F_{tot} r}{G} = \frac{T_{wheel}}{G} \tag{7}$$

Figs. 3 and 4 illustrate speed and torque of the motor using typical parameters for an EV which are tabulated in Table I. By utilizing urban moving cycle including acceleration, constant speed, regenerative braking, constant speed, acceleration, constant speed, acceleration, constant speed, regenerative braking, constant speed, and barking, as shown in Fig. 3, the EV wheel applies appropriate torque to the shaft of the motor as depicted in Fig. 4.

TABLE I
PARAMETERS OF A TYPICAL EV [7]

Parameter	Value
Weight	1020 Kg
Coefficient of rolling resistance	0.008
Density of the air	0.3
Frontal area	1.965 m^2
Tire radius	0.305 m
Maximum speed	48 m/s
Gear ratio	2

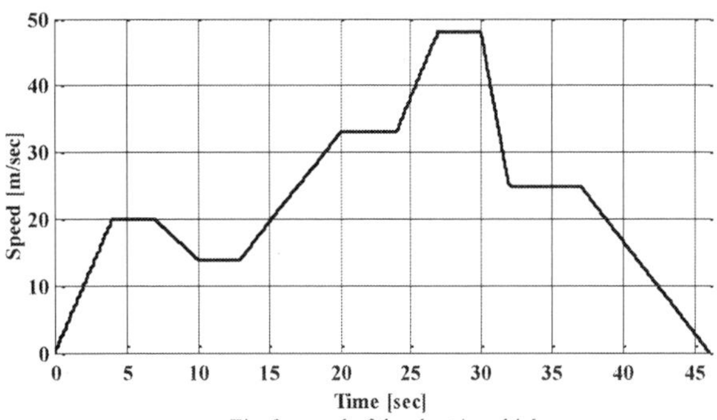

Fig. 3. speed of the electric vehicle

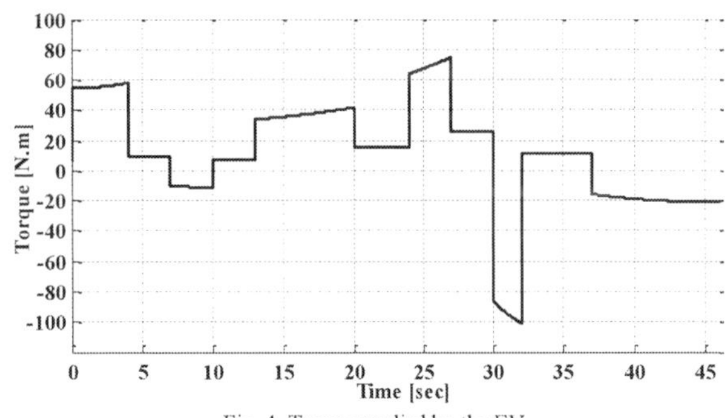

Fig. 4. Torque applied by the EV

III. PRINCIPLES OF OPERATION

Fig. 5 represents schematic of the extremely sparse parallel ac-link universal power converter used in EV application. This converter transfers power entirely through the link inductor.

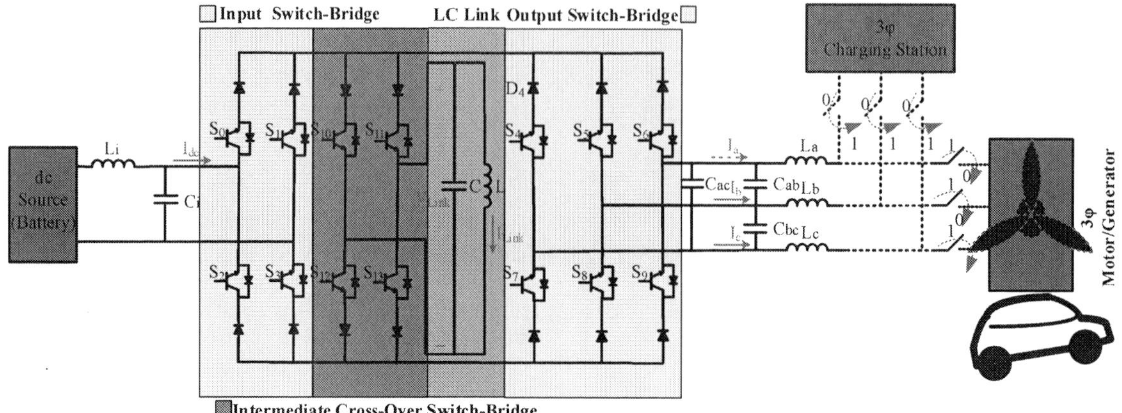

Fig. 5. The proposed system

Link capacitor is very small and is added to allow the switches to have zero voltage turn-on and soft turn-off. The proposed system is composed of an input dc source (battery), input switch-bridge (switches S0-S3), LC link, intermediate cross-over switch-bridge (switches S10-S13), output switch-bridge (switches S4-S9), three-phase motor/generator, and three-phase ac source (charging station). In this configuration, the input and output switch bridges contain forward conducting bidirectional blocking switches.

The link inductor is first charged by connecting to the battery through the input switch-bridge, and the intermediate cross-over switch-bridge and then discharges into the three-phase motor through output switch-bridge, and the intermediate cross-over switch-bridge. Similarly, in the regenerative braking or charging state, the link is first charged by connecting to three-phase generator or three-phase ac source from charging station through output and intermediate cross-over switch-bridges and then discharged into the battery through input and intermediate cross-over switch-bridges.

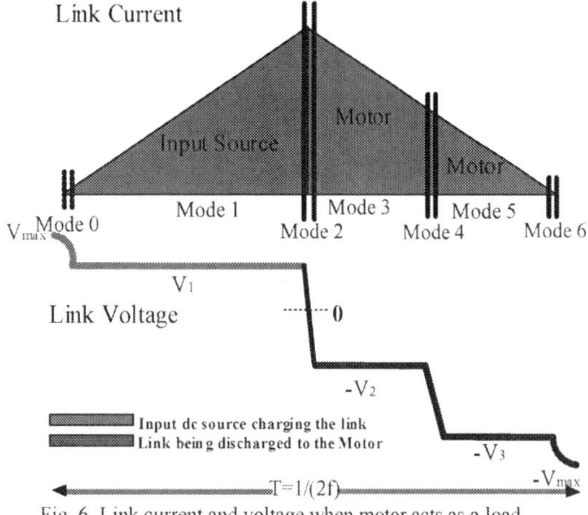

Fig. 6. Link current and voltage when motor acts as a load

Charging and discharging take place alternately in such a way that between each charging and discharging mode, there is a resonating mode during which none of the switches conduct, and the LC link resonates to facilitate zero voltage turn-on and soft turn-off. The link frequency is typically much higher than the output frequency. The intermediate cross-over switch-bridge allows charging and discharging of the link in both positive and negative directions, leading to a high frequency alternating link current with zero dc component.

Figs. 6 and 7 illustrate half cycle of the link current and voltage for the proposed configuration when three-phase motor acts as a load and when three-phase generator or charging station act as a source, respectively.

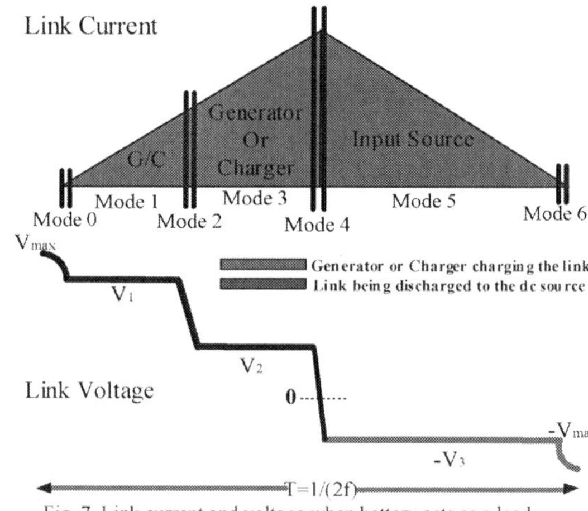

Fig. 7. Link current and voltage when battery acts as a load

When the battery supplies the three-phase motor, called motor state, the proper switches at the input switch bridge, namely, switches S0, S3 when link current direction is positive, and switches S1, S2 when link current direction is negative, connect the battery to the link inductor during mode 1. Furthermore, during modes that link is carrying current in the positive direction switches S11, S12 should be turned on, and similarly when link current is negative S10, S13 needs to be turned on. Fig. 8 illustrates the behavior of the converter when source is connecting to the link with the positive direction. All these switches are turned on before mode 1 starts (mode 0); however, they are reverse-biased before mode 1, because the voltage across the link, which is resonating, is higher than input voltage.

As shown in the figure, the link inductor is charged in a positive direction through two switches at the input switch bridge (S0 and S3), which connect the link to the input dc source, and proper switches at the intermediate cross-over switch-bridge (S11 and S12), which allow the link to be charged in the positive direction. The link is charged until the dc input source current averaged over a cycle meets its reference value [5] and [8].

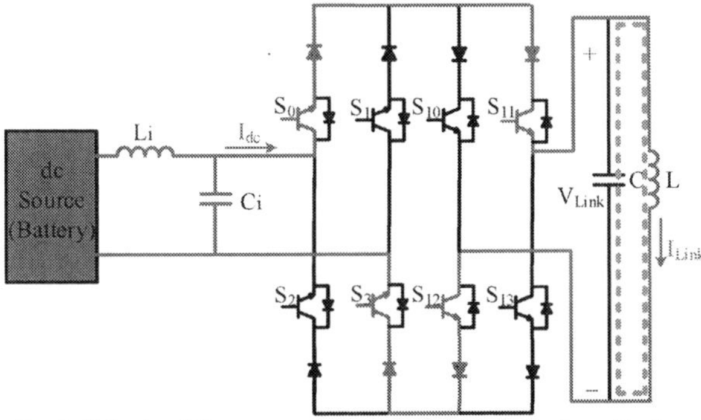

Fig. 8. Behavior of the converter when the link inductor is charged in the positive direction (red line) and when the link resonates (blue line)

Turning off switches S0 and S3 initiates mode 2, which is a resonating mode. Current flow during this mode is shown in Fig. 8 by dashed line in blue. The other switches that were conducting during mode 1 become reverse-biased during mode 2, and the link voltage starts to decrease due to the resonance. When the polarity of link voltage changes proper switches at the output switch-bridge are turned on. However, they cannot conduct immediately as they are reverse-biased. In order to better control the currents and increase their quality, the link discharging mode is split into two modes. Although there are three phase-pairs in a three-phase system, considering the polarity of the current in each phase, only two of these phase-pairs can provide a path for the current when connected to the link. If output phase a has the highest absolute value of reference current among the output reference currents, then phase pairs ab and ac are chosen to be charged from the link during modes 3 and 5. If I_a is negative, switches S5, S9, and S4 are turned on. Supposing that $|V_{ac}|$ is higher than $|V_{ab}|$, once the link voltage becomes equal to the voltage across the output phase pair ab (V_{ab}) switches S5, S9, S11, and S12 become forward biased initiating mode 3. During mode 3, the voltage across the link is negative and the link current decreases linearly. When the current of phase b (I_b) meets its reference, switch S5 is turned off to allow the link to resonate during mode 4. When the link voltage becomes equal to V_{ac}, switches S4, S9, S11, and S12 become forward biased and they start to conduct, initiating mode 5. The link continues to be discharged during mode 5 until the energy remained in the link is sufficient for the link voltage to swing to a minimum voltage which is lower than any input or output line to line voltages ($-V_{max}$). At the end of mode 5, all the switches are turned off and the link resonates during mode 6. Proper switches that are supposed to conduct during modes 3 and 5 are demonstrated in Fig. 9. During modes 7-12, the operation is similar to that of modes 1-6; however, the polarities of the link current and voltage are opposite. Therefore, switches S0, S3, S5, S9, S4, S10, and S13 conduct during modes 7-12 for the same situation explained.

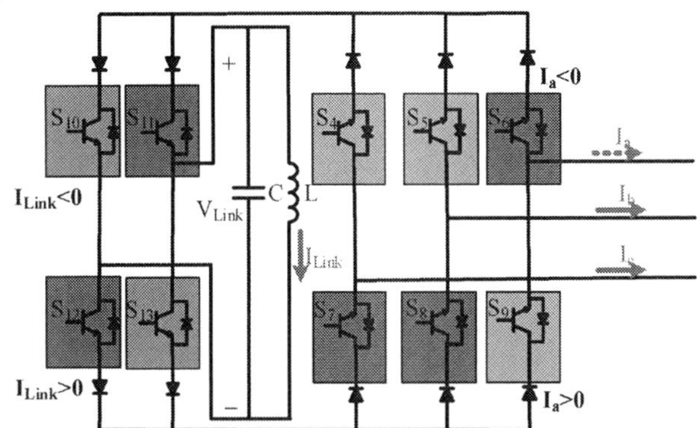

Fig. 9: Proper switch selection of the three-phase motor when I_a has the maximum current

During regenerative braking state or battery charging sate, three-phase generator or the three-phase ac source at charging station act as three-phase input source, and the battery acts as load. In this case there are 12 modes, as well. Before the beginning of mode 1 all possible input switches needs to be turned on. Supposing that reference of I_a has maximum current with positive polarity, and the link requires to be charged in the positive direction, switches S6, S7, S8, S11, and S12 need to be turned on. If $|V_{ac}|$ is higher than $|V_{ab}|$, during mode 1 switches S6, S7, S11, and S12 conduct, and when the current of phase c meets its reference, switch S7 is turned off. This initiates mode 2 which is a resonating mode. The other switches which were conducting during mode 1, become reverse-biased during mode 2, and the voltage across the link decreases due to link resonance.

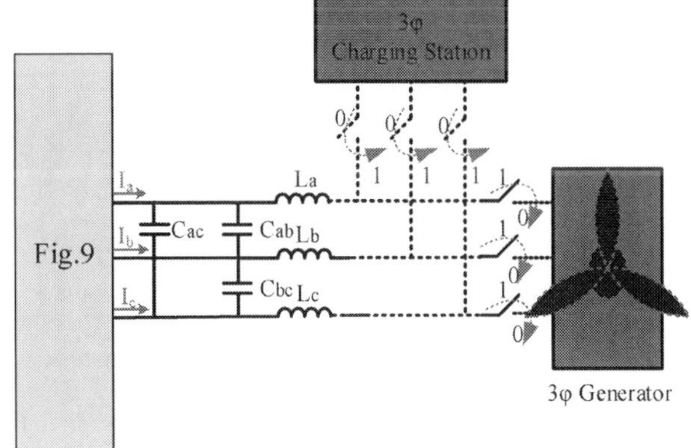

Fig. 10: Behavior of the converter when battery is being charged and I_a has the maximum current

When then link voltage becomes equal to $|V_{ab}|$ switches S6, S8, S11, and S12 become forward biased and they start to conduct, initiating mode 3. During mode 3, the link continues to be charged in the positive direction until the current of phase a meets its reference. S6 and S8 are then turned off, and the link starts resonating, initiating mode 4. Fig. 10 illustrates operation of the converter in the mentioned modes. It should be mentioned that the switch selection in this figure is the same as Fig.9.

During mode 4, the link voltage decreases until its polarity changes, and this is the time to turn on proper battery switches, i.e. S1, S2, as depicted in Fig. 11. Mode 5 starts when the link voltage becomes equal to the battery voltage. During this mode, the link discharges. At the end of this mode all the switches are turned off and the link resonates during mode 6. During modes 7-12, the scenario is the same except that polarities of the link current and voltage are opposite to those of modes 1-6.

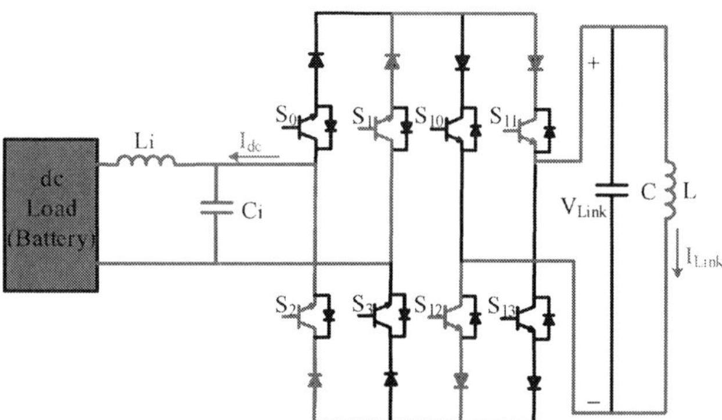

Fig. 11: when the battery is discharging the link in the positive direction

Design procedure of the extremely sparse parallel ac-link universal converter is similar to that of parallel ac-link universal converter. Using the equations derived in [9], the link peak current ($I_{Link,peak}$), inductance (L) and capacitance (C) of the link can be determined as follows:

$$I_{Link,peak} = 2(I_{dc} + \frac{3I_{o,peak}}{\pi}) \tag{8}$$

$$L = \frac{P_{dc}}{I^2_{Link,peak} f} \tag{9}$$

$$C \langle\langle \frac{1}{4\pi^2 f^2 L} \tag{10}$$

where I_{dc}, $I_{o,peak}$, P_{dc}, and f are dc input current, filtered output peak phase current, extracted power of the dc source, and link frequency, respectively.

Since during the resonating modes no power is transferred, it is preferred to keep these modes as short as possible. Thus, link capacitance in (10) is chosen such that the resonating periods are retained within a small percentage of the link cycle. Maximum voltage of the link (V_{max}) is another factor for controlling duration of mode 6. In this converter when the difference between the energy remained in the link becomes equal to the energy required for the link voltage to swing to V_{max}, mode 6 starts. Choosing a higher value for V_{max} results in starting mode 6 earlier and a longer resonating time. The difference between the energy remained in the link and the required energy is as follows:

$$\Delta E = 0.5\left(LI^2_{Link} + C(V^2_{Link} - V^2_{max})\right) \tag{11}$$

To minimize duration of mode 6, V_{max} should be considered only slightly higher than the maximum of dc voltage and peak of line to line ac voltages.

IV. Simulation and Experimental Results

A. *Simulation Results*

The proposed EV system, as shown in Fig. 1, has been simulated in PSIM software for the maximum motor supplying power of 80 kW, and the results are presented in this section. Parameters of the simulated system are tabulated in Table II.

Table II

Simulation Parameters of the Proposed EV System for 80 kW

Parameter	Value
Battery Voltage	375 V
Link inductance	10 μH
Link capacitance	20 nF
Battery side filter inductance	50 μH
Battery side filter capacitance	1000 μF
Motor side filter inductance	50 μH
Motor side filter capacitance	30 μF
Number of poles	20
Rated speed	2600 r/min
Phase resistance	0.008 Ω
Phase inductance (Ld/Lq)	(75/80) μH
Load torque	185 Nm
Peak load torque	250 Nm
Torque constant	90.68 Vpk/krpm
Moment of inertia	0.31 kgm^2

Figs. 12-18 demonstrate the simulation results when the motor is operating at maximum power of 80 kW. Fig. 12 illustrates rotor speed of the motor for a moving cycle including constant speed, acceleration, and constant speed when the motor is supplied by the battery with 375 V. In the motor state, in the first constant speed state, the applied load torque to the shaft is 185 Nm, and in the second acceleration stage, motor experiences a maximum load torque of 250 Nm with the reference speed of 3200 r/min. Battery and motor current (phase "a") are demonstrated in Figs. 13, and 14, respectively.

As shown in Figs. 13 and 14, the current control strategy is capable of following reference currents even in high torque situation. Figs. 15 and 16 show current of ac source (phase "a") and battery when the battery is being charged (6.6 kW) with the three-phase ac source. Fig. 17 shows link current and voltage for the proposed system when the applied torque to the shaft is equal to 185 Nm. Voltage across S4 and D4 (see Fig. 5), and current passing through the switch as the indicator of the soft switching technique are illustrated in Fig. 18.

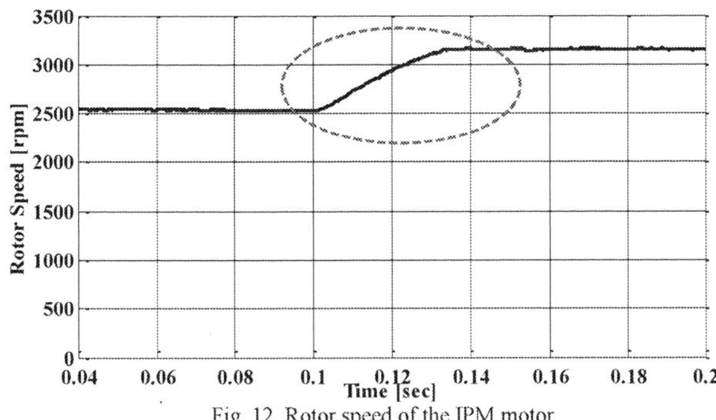

Fig. 12. Rotor speed of the IPM motor

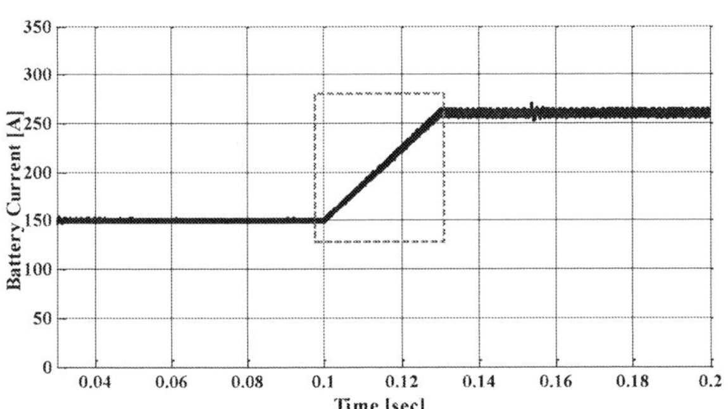

Fig. 13. Battery current of the proposed system

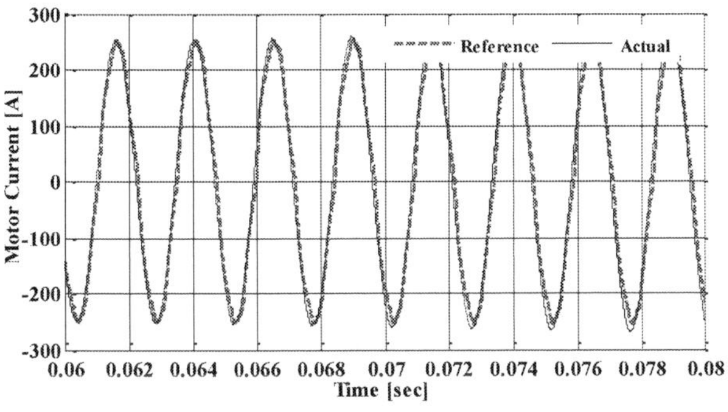

Fig. 14. Actual and reference IPM motor current (phase a) of the proposed system

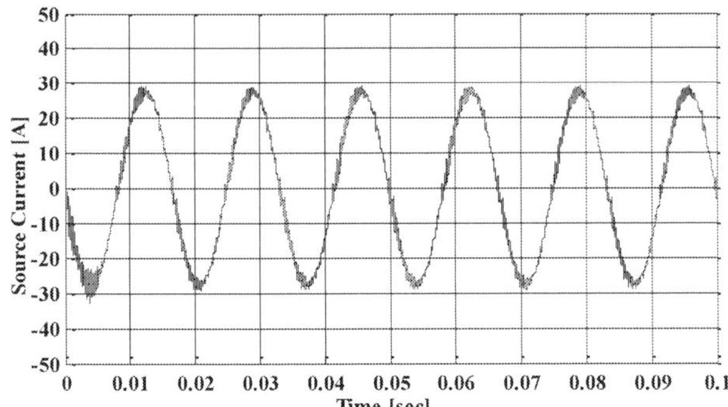

Fig. 15. Current (phase a) of theree-phase ac source (charging station) while charging the battery

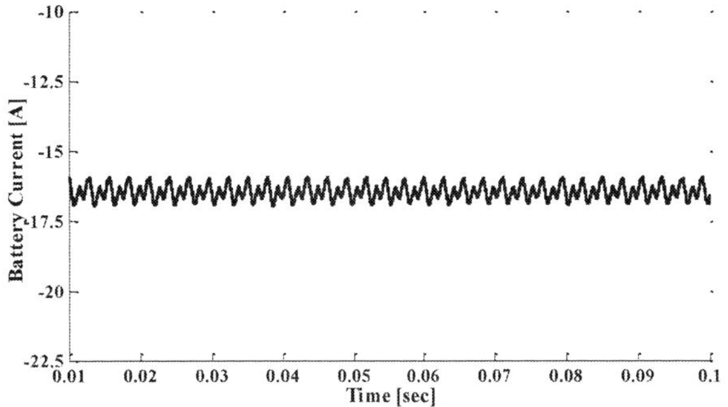

Fig. 16. Battery current while being charged with three-phase ac source (charging station)

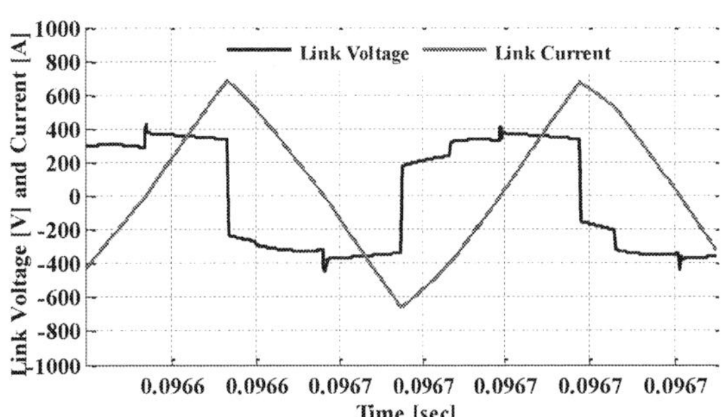

Fig. 17. Link voltage and current of the proposed system

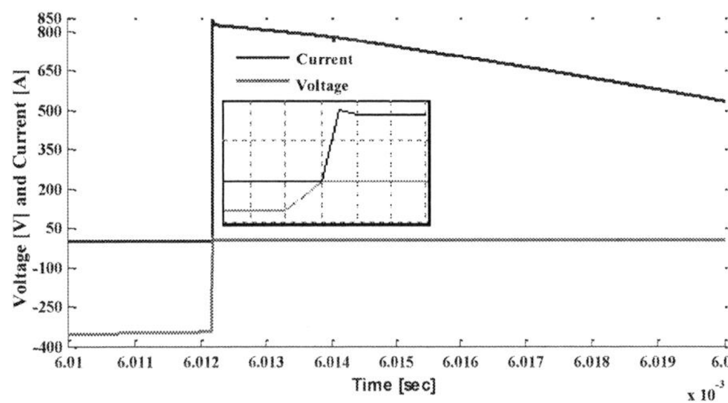

Fig. 18. Voltage acrosss the S4 and D4, and current passing through the S4 of the proposed system

B. Preliminary Experimental Results

In order to evaluate the performance of the extremely-sparse converter, a low power laboratory prototype was built. The parameters of the converter prototype are listed in Table III. A resistive load has been considered in this test. The input dc voltage, and the input power have been considered to be about Vdc=60 V, and Pin=150W. In the prototype, one isolated voltage sensor along with five isolated current sensors are employed to feedback the converter link voltage, link current, and unfiltered input and three-phase ac currents.

The current drawn by the dc input source has been set at 2.5 A, while the converter is operating at 150W, as shown in Fig. 19. Figs. 20 and 21 represent the link current and voltage, and the current (phase "a") of the there-phase ac load, respectively.

The soft switching technique by showing the voltage across S4, and D4, and current passing through S4 is depicted in Fig. 22.

TABLE III
PROTOTYPE PARAMETERS OF THE PROPOSED SYSTEM FOR 150 W

Parameter	Value
Battery Voltage	60 V
Link inductance	845 μH
Link capacitance	700 nF
Dc source side filter inductance	95 μH
Dc source filter capacitance	1000 μF
Three-phase ac side filter inductance	563 μH
Three-phase ac side filter capacitance	15 μF

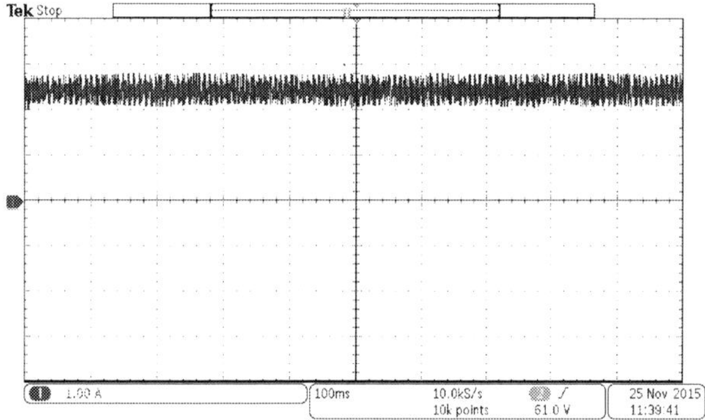

Fig. 19. The input dc current (1 A/div) for the designed experiment with the time scale of 100 ms/div

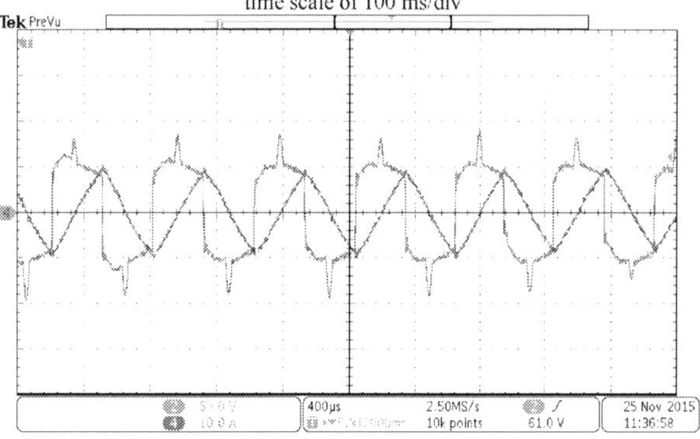

Fig. 20. The link voltage (50 V/div) and current (10 A/div) for the designed experiment with the time scale of 400 μs/div

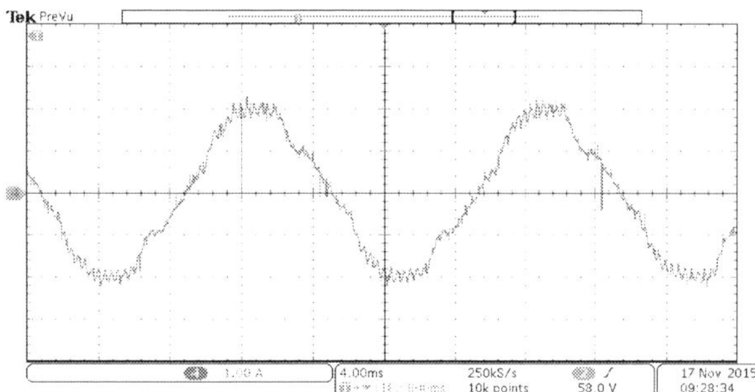

Fig. 21. The current (phase a) of there-phase ac load (1 A/div) for the designed experiment with the time scale of 4 ms/div

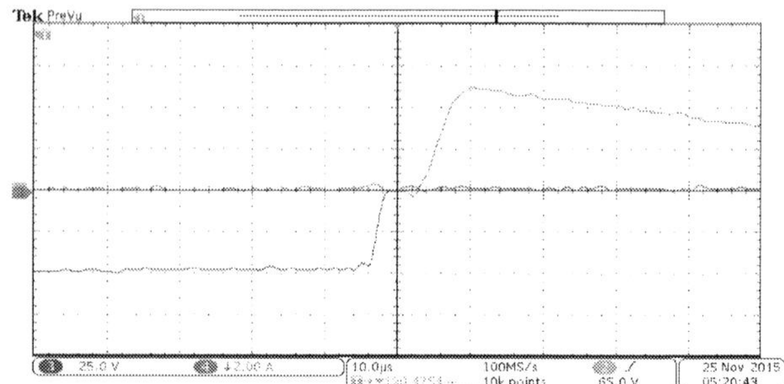

Fig. 22. The voltage (25 V/div) acrosss the S4 and D4, and current (2 A/div) passing through the S4 of the designed experiment with the time scale of 10 μs/div

V. CONCLUSION

The application of a new single-stage converter in EV was studied in this paper. The proposed system uses a single-stage converter to fulfil the task of dc-dc converter, inverter, and the rectifier that are usually used in an EV. Furthermore, the proposed configuration eliminates the need for electrolytic capacitors which are the most unreliable components in a power electronic converter and are responsible for most of the failures, particularly at high temperatures. In this converter all switches are turned on at zero voltage and their turn off is capacitance buffered, resulting in negligible switching losses and minimized voltage stress over switches. Additional sources, e.g., PV or FC, can be added to the system through only a switch. In this paper, the performance of the proposed system was validated by analysis and simulation, and also preliminary experimental results were presented.

ACKNOWLEDGMENT

The financial support of this work by UIC Chancellor's Innovation Fund in partnership with IllinoisVENTURES is gratefully acknowledged. The authors would also like to thank the office of technology management at University of Illinois at Chicago, especially Mr. Krivchenia.

REFERENCES

[1] W. Alexander, "Universal Power Converter," U.S. Patent 11/758,970, February 2009.

[2] M. Amirabadi, J. Baek, and H. A. Toliyat, "Sparse ac-link buck-boost inverter," *IEEE Trans. Power Electron.*, vol. 29, pp. 3942-3953, 2014.

[3] J M. Amirabadi, H. A. Toliyat, and J. Baek, "Ultra-sparse ac-link converters," *IEEE Trans. Ind. Appl.*, vol.51, no.1, pp.448-458, 2015.

[4] I. Pecelj, S. W. H. de Haan, and J. A. Ferreira, "Soft-switched, transformerless inverter suitable for photovoltaic panels," in *Proc. IEEE EPE'13*, Lille, France, 2-6 Sep., 2013, pp. 1-9.

[5] M. Amirabadi, "Extremely sparse parallel ac-link universal power converters," in *Proc. IEEE ECCE'14*, Pittsburgh, PA, USA, 14-18 Sep., 2014, pp. 1534-1541.

[6] J. Larminie, J. Lowry, "Electric Vehicle Technology Explained," *John Wiley and Sons Ltd*, pp. 303, 2003.

[7] S. Sadeghi, J. Milimonfared, M. Mirsalim, M. Jalalifar, " Dynamic modeling and simulation of a switched reluctance motor in electric vehicles," in *Proc. IEEE IEA'06*, Singapore, 24-26 May, 2006, pp. 1-6.

[8] M. Amirabadi, J. Baek, H. Toliyat, and W. Alexander, "Soft-switching ac-link three-phase ac–ac buck–boost converter," *IEEE Trans. Ind. Electron.*, vol. 62, no.1 pp. 3–14, Jan. 2015.

[9] M. Amirabadi, A. Balakrishnan, H. Toliyat, and W. Alexander, "High frequency AC-link PV inverter," *IEEE Trans. Ind. Electron.*, vol. 61, pp. 281-291, 2014.

Simple and Efficient Low Power Photovoltaic Emulator for Evaluation of Power Conditioning Systems

Jesus Gonzalez-Llorente, Andres Rambal-Vecino
Escuela de Ciencias Exactas e Ingeniería.
Universidad Sergio Arboleda
Bogotá D.C., Colombia
jesus.gonzalez@usa.edu.co

Luciano A. Garcia-Rodriguez, Juan C. Balda
Department of Electrical Engineering
University of Arkansas
Fayetteville, AR, USA
{lgarciar, jbalda}@uark.edu

Eduardo I. Ortiz-Rivera
Department of Electrical and Computer Engineering
University of Puerto Rico at Mayaguez
Mayaguez, Puerto Rico
eduardo.ortiz7@upr.edu

***Abstract*—Low power conditioning systems are required for photovoltaic (PV) applications in portable devices, small satellites, solar vehicles, etc. The research and development of the power conditioning unit for this application such as MPPT requires of controllable environment conditions, which are difficult to obtain due to weather variability. Thus, a photovoltaic emulator is developed to obtain a power source that presents the same current-voltage (I-V) relationship of PV cells according to the connected-load, irradiance and temperature. This paper presents the design of a PV system based on a synchronous buck converter, current and voltage sensors, and a microcontroller to control the PWM duty cycle to adjust the output voltage and current according to the I-V characteristic. The results show a good agreement between the output current of the emulator and the expected characteristic curve of the PV module; the achieved efficiency was between 80 % and 95 % over the entire operating range.**

Keywords—photovoltaic, emulator, DC-DC converter, synchronous buck converter.

I. INTRODUCTION

In addition to the large PV plants for grid-connected systems and house applications, solar energy is used in low power applications such as small satellites, portable devices, solar vehicles, etc. Research and development in these solar systems involve the design of new techniques to obtain the maximum energy from the photovoltaic modules and to reduce losses in the power conditioning stage [1].

Among these techniques are maximum power point tracking algorithms and voltage and power regulation; the evaluation of these systems requires constant conditions for the irradiance and temperature in order to obtain a reference to compare the performance of the proposed techniques. Thus, it is necessary a power source that can mimic the electric behavior of the *PV* modules, and thus testing the power conditioning system for whatever conditions of irradiance and temperature.

The current-voltage relationship of *PV* modules (Fig. 1) is described by the equation

$$I(V) = \frac{I_{sc}}{1-e^{-\frac{1}{b}}}\left[1 - e^{\frac{V}{bV_{oc}}-\frac{1}{b}}\right], \tag{1}$$

where b is characteristic constant of the PV module, I_{sc} and V_{oc} are the short-circuit current and the open-circuit voltage [2], which mainly depend on the irradiance and temperature conditions according to following equations:

$$I_{sc} = I_{sc,STC}\frac{G}{G_{STC}} + K_i(T - T_{STC}), \tag{2}$$

$$V_{oc} = V_{oc,STC} + K_v(T - T_{STC}) + a_m V_t \ln\frac{G}{G_{STC}}, \tag{3}$$

where G is the irradiance, T the PV module temperature, K_i and K_v the temperature coefficients for current and voltage respectively, and V_t the thermal voltage [3]. STC stands for the values at standard test conditions.

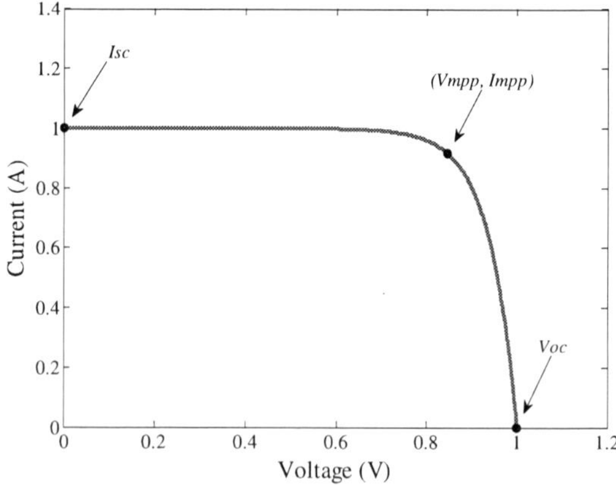

Fig 1. Current – voltage curve of photovoltaic module.

A low power emulator have been presented in [4]; however this was an analog implementation for a 12 W thin-film PV module based on Op-Amps which limits the operation. A 10 W PV emulator with digital implementation is presented in [5]; nevertheless, the characteristic curve is not exactly followed because a liner approach was implemented and the maximum efficiency was just 80 %.

This paper presents a digital implementation based on a buck converter that emulates a 5 W PV module whose measured efficiency is between 80 % and 95 % for all its operating range. Other designs in literature use more complex topologies than the proposed one such as full-bridge, push-pull, forward and ac-dc cascaded with interleaved dc-dc converters [6-8]; however, these topologies were implemented for higher power levels.

II. Architecture of the Photovoltaic Emulator

The photovoltaic emulator consists of a microcontroller and a synchronous buck converter including gate driver and sensors of its output voltage and output current; the architecture is shown in Fig. 2. The output port of the buck converter is controlled in such a way that its voltage and current agree with the characteristic curve of the emulated photovoltaic module; therefore, current and voltage sensors are required as the feedback signals to determine if these variables coincide with the modeled PV current-voltage characteristics.

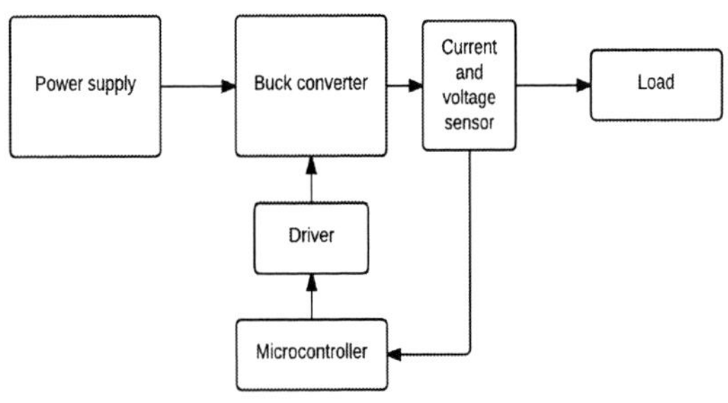

Fig 2. PV emulator architecture.

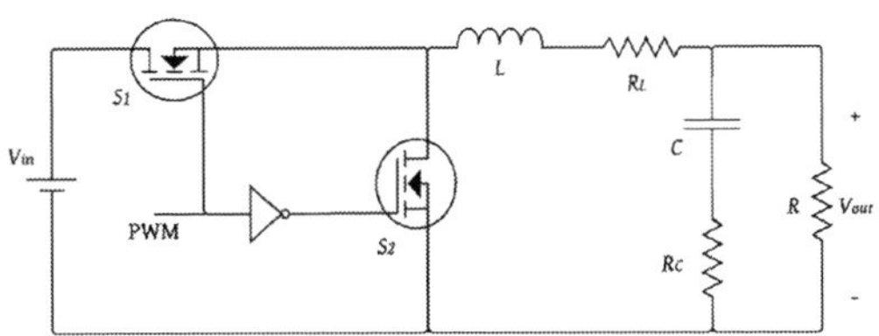

Fig 3. Synchronous buck converter including parasitic elements for the inductor and capacitor

A. Buck converter

The synchronous buck converter shown in Fig. 3 generates an output voltage (V_{out}) lower than the input voltage (V_{in}) according to the duty ratio,

$$V_{out} = DV_{in}, \tag{4}$$

when it operates in continuous conduction mode. In addition, the dynamic behavior of the buck converter is modeled by using state-space equations with the inductor current (I_L) and the capacitor voltage (V_C) as state variables. The model is given by

$$\frac{dV_C}{dt} = i_L\left(\frac{1}{C} - \frac{R_C}{R+R_C}\right) - V_C\left(\frac{1}{C(R+R_C)}\right), \tag{5}$$

$$\frac{di_L}{dt} = \frac{V_{in}D}{L} + i_L\left(-\frac{RR_C}{L(R+R_C)} - \frac{R_L}{L}\right) - V_C\left(\frac{R}{L(R+R_C)}\right), \tag{6}$$

$$V_{out} = i_L\left(\frac{RR_C}{R+R_C}\right) + V_C\left(\frac{R}{R+R_C}\right), \tag{7}$$

where L is the inductance, C the capacitance, R is the load resistance, R_C the capacitor series resistance and R_L the inductor series resistance. The synchronous buck converter is built using the CSD87331Q3D NexFET™ half bridge power MOSFET block to realize the converter switches. The TPS28225 synchronous MOSFET high-speed driver is used to drive both buck converters switches.

B. Current and voltage sensors

As output current sensor, the high precision op-amp INA225 amplifies the voltage drop across a 0.01 Ω shunt resistor connected at the output terminal. The Op-Amp circuit is designed to have a gain up to 50 V/V; thus, the maximum current that can be measured is 3 A to obtain 3 V at the input of the analog to digital converter of the microcontroller. As output voltage sensor, a voltage divider is adjusted to obtain 3 V when the output voltage is 20 V.

C. Microcontroller

The microcontroller, an ARM Cortex M4, executes the algorithm that adjusts the duty cycle to vary the operating point of the buck converter until this point coincides with the calculation of the exponential function that describes the PV characteristics. This algorithm presents the functional description of the photovoltaic emulator and is explained in the following section.

III. Functional Description Of The Photovoltaic Emulator

The PV emulator functionality is based on a simple searching algorithm that adjusts the duty cycle until the buck converter operating point agrees with the current-voltage characteristic of the PV module as shown in Fig. 4.

During initialization, the microcontroller sends an initial duty cycle and reads a set of values of the output current and the output voltage. These measurements correspond to the output of the PV emulator (V_{PV}, I_{PV}) and are filtered to obtain a stable measurement due to the present noise.

With the measured output voltage (V_{PV}), the microcontroller calculates the current that corresponds to the actual PV module for that voltage according to the current-voltage relationship, shown in (1). The purpose of this calculation is to determine if the output of the emulator agrees with the expected value of the PV module, given by the mathematical model of the PV module. Thus, if the calculated current value (I) is lower than the measured value (I_{PV}), then the duty cycle is decreased; conversely, when the calculated current (I) is greater than the measured value, the duty cycle in increased.

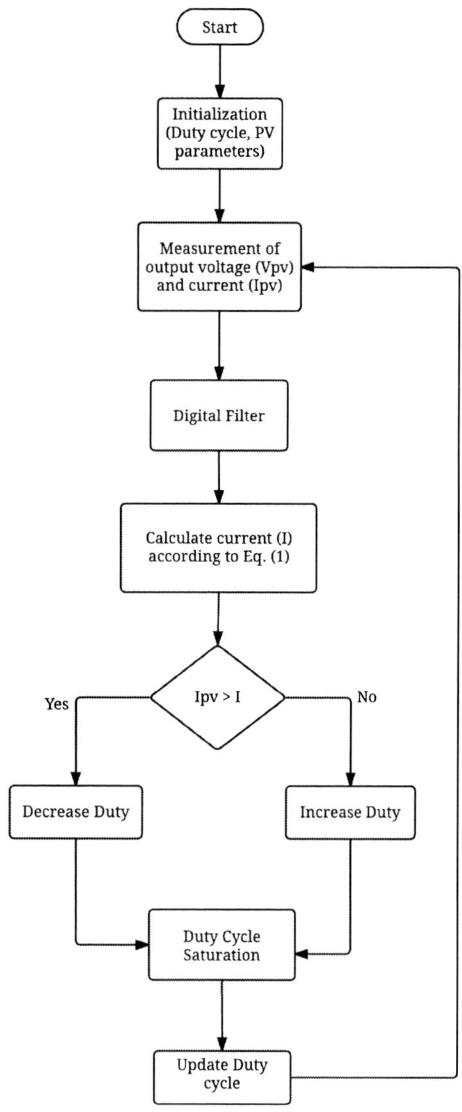

Fig 4. PV emulator control algorithm.

In addition, the duty cycle must be limited between 0 and 1, therefore a saturation limit is included to achieve this condition. Once the duty cycle is updated, the process continues with the measurement stage and all the process is repeated as described by the flow chart in Fig. 4. After several iterations the desired values of output voltage and output current are achieved.

IV. Results

The functionality of the photovoltaic emulator in steady-state conditions was validated using a rheostat as a load; thus, the emulator was evaluated for different resistance values, covering the full current-voltage range for a constant irradiance and temperature. The experimental setup is show in Fig. 5 and the characteristic of the emulated PV module are shown in Table I.

The current and voltage measurements of the emulator agree with the expected theoretical behavior as shown in Fig. 6. Therefore, the power curve corresponds to the characteristic of the photovoltaic module where a maximum power point exists according with the manufacturer specification (Fig. 7).

The efficiency of the PV emulator was also calculated, considering the ratio of the output power and the input power when the load was varied by using the rheostat. This efficiency correspond to the efficiency of the buck converter that implements the power stage; even though this is a low power stage and it has a wide operating range, the efficiency is between 80 % and 96 % when the load is increased (Fig. 8).

The dynamic behavior was evaluated by observing the buck output voltage and current for two conditions: load increasing and load decreasing. In both cases, the emulator correctly followed the theoretical voltage and current with a settling time of about 700 ms. The load decreasing and the load increasing conditions were evaluated changing the resistance value from 2 Ω to 10 Ω (Fig. 9) and from 30 Ω to 13 Ω (Fig. 10), respectively. The current sensor output with a gain equal to 0.5 V/A is shown in red and the PV emulator output voltage in blue.

TABLE I. Parameter of the Emulated Solar Cells

Parameter	Symbol	Value
Open-circuit voltage	V_{oc}	5.334 V
Short-circuit current	I_{sc}	0.506 A
Voltage at maximum power	V_{mpp}	4.742 V
Current at maximum power	I_{mpp}	0.487 A
Characteristic constant	b	0.0338

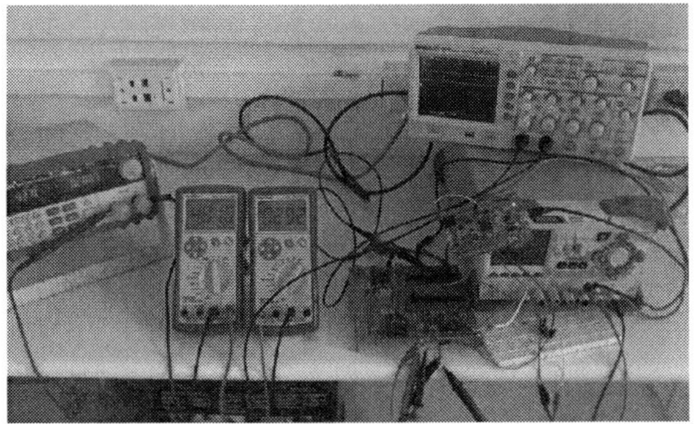

Fig 5. PV emulator experimental setup.

V. Conclusions and Future Work

The low power PV emulator was developed using a buck converter whose output voltage and output current are controlled to mimic the current-voltage relationship of PV modules. The control algorithm is based on a searching algorithm that adjusts the duty cycle until the operating condition agrees with the behavior of photovoltaic modules.

The photovoltaic emulator obtained a good agreement with the photovoltaic behavior and high efficiency over a wide operating range. It is expected to improve the transient response with an adaptive step size of the duty cycle and the testing using an electronic load will be included. This technique could also be used for emulations of other energy sources such as fuel cell, thermoelectric generators, etc.

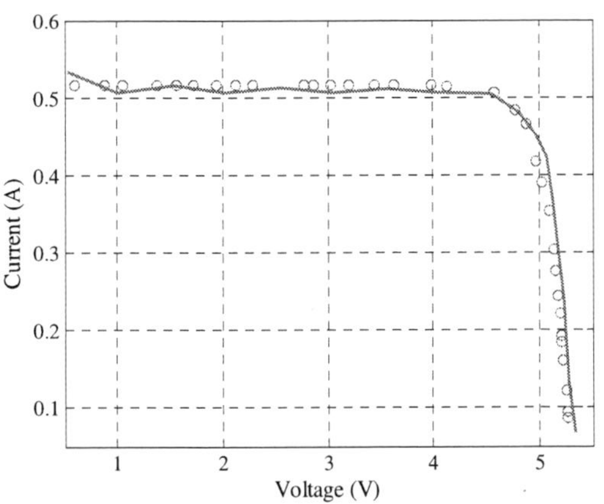

Fig 6. Ideal (continue) vs emulated (circle) current voltage.

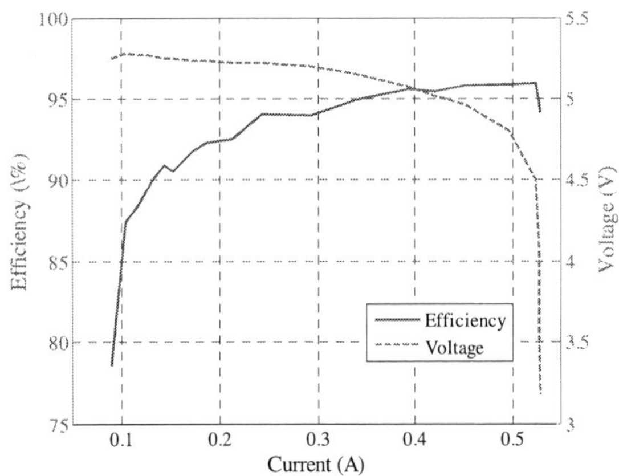

Fig 8. PV emulator efficiency of a 5W PV module.

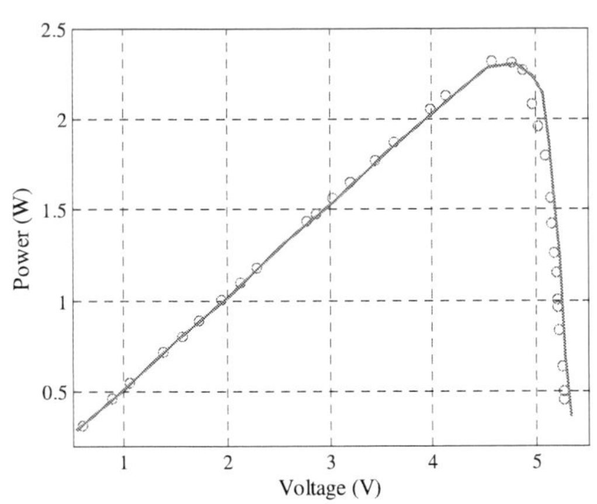

Fig 7. Ideal (continue) vs emulated (circle) power curve.

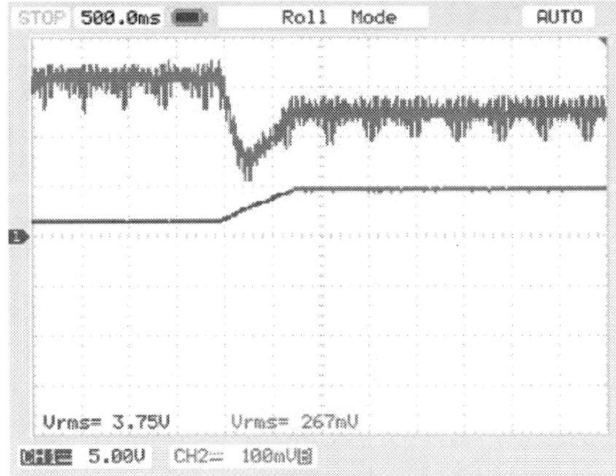

Fig 9. PV emulator output voltage and measured current with gain equal to 0.5 V/A under a decreasing load step (from 2 Ω to 10 Ω).

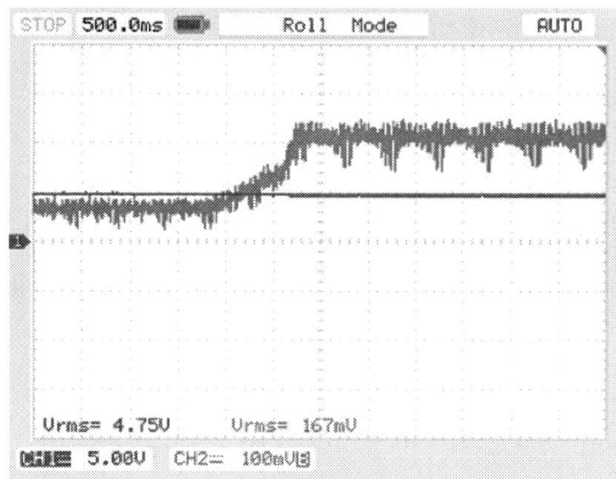

Fig 10. PV emulator output voltage and measured current with gain equal to 0.5 V/A under an increasing load step (from 30 Ω to 13 Ω).

ACKNOWLEDGMENT

The authors are grateful for financial support from the COLCIENCIAS under research grant No. 0208-2013 with resources of El Patrimonio Autónomo Fondo Nacional de Financiamiento para la Ciencia, la Tecnología y la Innovación, Francisco José de Caldas and from the Universidad Sergio Arboleda through Fondo de Investigación e Innovación (FIIUSA) and Escuela de Ciencias Exactas e Ingeniería.

REFERENCES

[1] J. Gonzalez-Llorente, D. Rodriguez-Duarte, S. Sanchez-Sanjuan, and A. Rambal-Vecino, "Improving the efficiency of 3U CubeSat EPS by selecting operating conditions for power converters," in Aerospace Conference, 2015 IEEE, 2015, pp. 1–7.

[2] E. I. Ortiz-Rivera and F. Z. Peng, "Analytical Model for a Photovoltaic Module using the Electrical Characteristics provided by the Manufacturer Data Sheet," in IEEE 36th Conference on Power Electronics Specialists, 2005., 2005, pp. 2087–2091.

[3] V. J. Chin, Z. Salam, and K. Ishaque, "Cell modelling and model parameters estimation techniques for photovoltaic simulator application: A review," Appl. Energy, vol. 154, pp. 500–519, Sep. 2015.

[4] D. M. K. Schofield, M. P. Foster, and D. A. Stone, "Low-cost solar emulator for evaluation of maximum power point tracking methods," Electron. Lett., vol. 47, no. 3, pp. 208–209, 2011.

[5] D. D. C. Lu and Q. N. Nguyen, "A photovoltaic panel emulator using a buck-boost DC/DC converter and a low cost micro-controller," Sol. Energy, vol. 86, no. 5, pp. 1477–1484, 2012.

[6] C. Liu, J. Feng, Z. Liu, W. Tong, and Y. Ji, "Development of a digital solar simulator based on full-bridge converter," in Proc. SPIE 9142, Selected Papers from Conferences of the Photoelectronic Technology Committee of the Chinese Society of Astronautics: Optical Imaging, Remote Sensing, and Laser-Matter Interaction, 2014, vol. 9142, pp. 914222–914226.

[7] J. Zhang, S. Wang, Z. Wang, and L. Tian, "Design and Realization of a Digital PV Simulator with a Push-Pull Forward Circuit," J. Power Electron., vol. 14, no. 3, pp. 444–457, May 2014.

[8] A. Koran and T. LaBella, "High Efficiency Photovoltaic Source Simulator with Fast Response Time for Solar Power Conditioning Systems Evaluation," IEEE Trans. Power Electron., vol. 29, no. 3, pp. 1285–1297, Mar. 2014.

Data Transmission Method without Additional Circuits in Bidirectional Wireless Power Transfer System

Yeongrack Son
Dept. of Electrical and Computer Engineering
Seoul National University
Seoul, Korea
syrident@snu.ac.kr

Jung-Ik Ha
Dept. of Electrical and Computer Engineering
Seoul National University
Seoul, Korea
jungikha@snu.ac.kr

***Abstract*— In the bidirectional wireless power transfer (WPT) system, each power circuit of the transmitter and the receiver part is controlled by different controllers. Therefore, to control and maintain the entire system, the output voltage phase synchronization and data transmission functions between the controllers should be implemented. The communication circuit is normally used to realize them, but this additional circuit increases overall cost of the system. Therefore, this paper presents the data transmission method without communication circuits for eliminating the dedicated circuit in the bidirectional WPT system. Proposed data transmission is implemented by perturbing peak of the fundamental output voltage at the transmitter and monitoring the signal fluctuation due to the perturbation at the receiver. The operation of the proposed method is verified by the simulation result.**

Keywords*— *Bidirectional converter, Cost reduction, Data transmission, Vehicle-to-grid system, Wireless Power Transfer (WPT).

I. Introduction

Most power supply systems use contacted wires for transferring the power. These cables can cause some problems: the users might suffer the electric shock when plug on or off the device, and the contamination of the wire due to water or dust can reduce the efficiency and the reliability of the system. The power transfer through the coupled inductive link needs no contact, so it has some advantages, such as convenience, reliability, and safety, compared to the power transfer through contacted wires [1]-[3]. Due to these advantages, the wireless power transfer (WPT) systems have been researched for many applications, including electric vehicle charger, portable device charger, industrial automation, and electrical transmission system [4]-[6].

The WPT system is classified into the unidirectional WPT system and the bidirectional WPT system. The bidirectional WPT system is composed of the transmitter and the receiver units all of which has active switches and can manipulate the output voltage. This system is suitable to the applications which need two-way power transfer, such as EV charger in vehicle-to-grid systems and bidirectional wireless charging applications between portable devices [1]-[11].

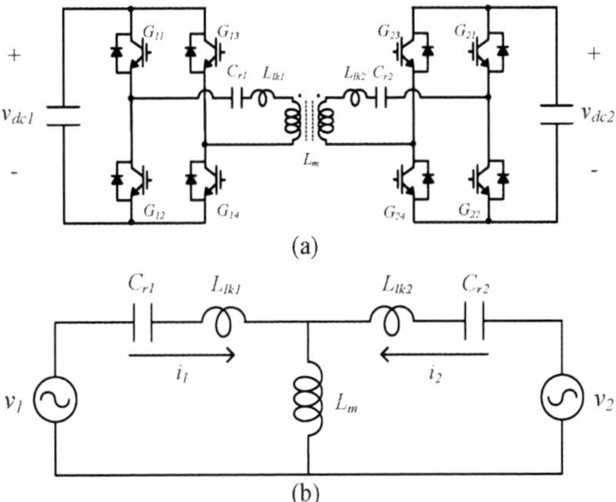

Fig. 1: The proposed bidirectional WPT system.
(a) The circuit diagram, (b) the equivalent circuit diagram.

For the control of the bidirectional WPT system, the frequencies and the phases of the output voltages of the transmitter and the receiver are required to be synchronized during the power transfer operation. Since each transmitter and receiver unit is controlled by different controller in this system, bidirectional communication between controllers is also needed for control and maintenance of the bidirectional WPT system. These two functions are widely implemented by using the commercial communication modules [7, 8]. It is simple, but using dedicated communication circuit makes additional cost.

This paper proposes the data transmission method without communication modules for the cost reduction of the bidirectional WPT system. The conventional research about the output voltage synchronization without dedicated circuit [7] calculates the output voltage phase of the transmitter from the induced current and the output voltage at the receiver. The peak of the signal for synchronization is proportional to the fundamental output voltage peak of the transmitter, and data transmission can be implemented by perturbing the output voltage peak at the transmitter and monitoring the signal fluctuation at the receiver. This paper is organized as the following. Steady-state operation of the proposed WPT system is briefly introduced in section II, and the proposed data transmission method is explained in section III. Finally, the

978-1-4673-9551-9/16 $31.00 © 2016 IEEE

simulation result for verifying the operation of the proposed method are presented in section IV.

II. Steady-State Operation of the Proposed System

The circuit diagram and the equivalent circuit of the proposed WPT system are shown in Fig. 1(a) and 1(b). Here, the system is classified into the inductive WPT system and the resonant WPT system, based on the presence of the series-connected capacitor C_{r1} and C_{r2}. In the resonant system, the capacitances of C_{r1} and C_{r2} are determined so that resonant frequency between the inductive link and the resonant capacitor is tuned to the switching frequency. The relationship among the voltages and currents of the system is derived as in (1), where A_{11}, A_{12}, A_{21}, and A_{22} are the admittances of the system. Fig. x shows bode plots of the admittances A_{21} and A_{22} in inductive case and resonant case.

The output voltage waveforms generated by the proposed WPT system are shown in Fig. 3. The output voltage equations are given by

$$v_1 = v_{dc1}\frac{4}{\pi}\sum_{n=1,3,\ldots}^{\infty}\frac{1}{n}\cos(n\omega_s t)\sin\left(\frac{nd_1}{2}\right), \quad (2)$$

$$v_2 = v_{dc2}\frac{4}{\pi}\sum_{n=1,3,\ldots}^{\infty}\frac{1}{n}\cos(n\omega_s t + nd_2)\sin\left(\frac{nd_1}{2}\right), \quad (3)$$

where v_{dc1} and v_{dc2} are the DC link voltages of the system. From (1) to (3), the active power and the reactive power transferred in this system calculated from as in (4) and (5). Here, $\angle A_{21}$ and $\angle A_{22}$ is $3\pi/2$ and 0 at the switching frequency

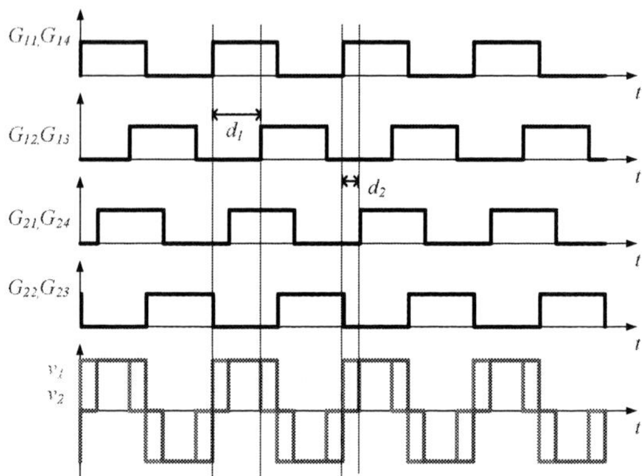

Fig. 3: The output voltage waveforms of the proposed WPT system.

in resonant case. Therefore, the active power can be delivered with minimum reactive power at $d_2 = \pm\,\pi/2$ points in $v_{dc1} = v_{dc2}$ case.

$$\overline{P}_2 = |v_2||A_{21}||v_1|\cos(d_2 - \angle A_{21}) + |v_2||A_{22}||v_2|\cos(-\angle A_{22}). \quad (4)$$

$$\overline{Q}_2 = |v_2||A_{21}||v_1|\sin(d_2 - \angle A_{21}) + |v_2||A_{22}||v_2|\sin(-\angle A_{22}). \quad (5)$$

The output voltage synchronization is implementable by sensing the coil current and the output voltage at the receiver and calculating them [7]. The calculated term with the phase of the transmitter voltage can be derived as in (6).

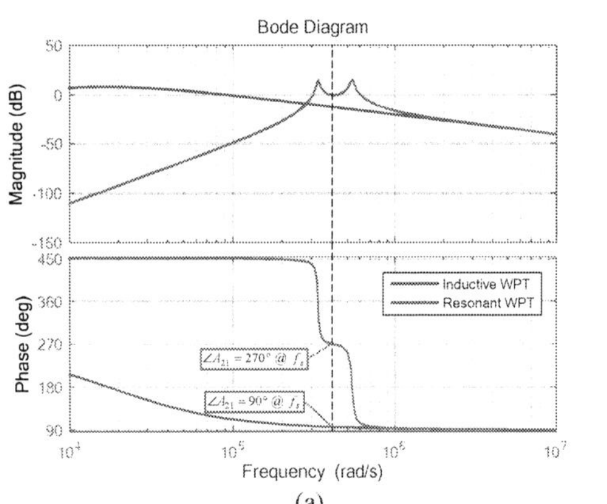

(a)

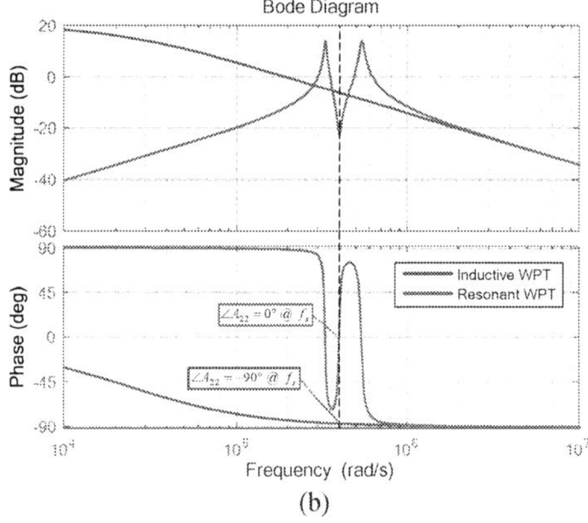

(b)

Fig. 2: Bode plots of (a) A_{21}, and (b) A_{22} of the proposed WPT system.

$$\begin{bmatrix} i_1 \\ i_2 \end{bmatrix} = \begin{bmatrix} s(L_m + L_{lk1}) + 1/sC_{r1} & sL_m \\ sL_m & s(L_m + L_{lk2}) + 1/sC_{r2} \end{bmatrix}^{-1} \begin{bmatrix} v_1 \\ v_2 \end{bmatrix} = \begin{bmatrix} A_{11} & A_{12} \\ A_{21} & A_{22} \end{bmatrix} \begin{bmatrix} v_1 \\ v_2 \end{bmatrix}. \quad (1)$$

$$\begin{aligned} i_2 - |A_{22}|v_2 &= |A_{21}||v_1|\cos(\omega_s t + \angle A_{21}) + |A_{22}||v_2|\cos(\omega_s t + d_2 + \angle A_{22}) - |A_{22}||v_2|\cos(\omega_s t + d_2) \\ &= |A_{21}||v_1|\sin(\omega_s t) = v_{dc1}\frac{4}{\pi}|A_{21}|\sin\left(\frac{d_1}{2}\right)\sin(\omega_s t) \end{aligned} \quad (6)$$

The band-pass filter with the cutoff frequency of f_s is used to extract the fundamental component of the secondary output voltage, and the magnitude of the admittance A_{22} is calculated from the known parameter. The carrier signal of the receiver can be synchronized to the output voltage of the transmitter by resetting the carrier at the zero-crossing point of (6).

III. PROPOSED DATA TRANSMISSION METHOD

The amplitude of the signal for synchronization, (6), is proportional to the peak of the fundamental voltage of the transmitter. Using this feature, simple data transmission is implementable through the power transfer circuit. Fig. 4 shows the waveforms of the transmitter and the receiver during the data transmission. d_1 for the proposed data transmission at transmitter is shown as below.

$$d_{1,tr} = d_{1,pow} - (data\,bit)\cdot \Delta d_1, \tag{7}$$

where $d_{1,pow}$ and Δd_1 are d_1 for power transfer and Δd_1 is perturbed value with respect to the data bit, respectively, and data bit is set to 0 or 1. This perturbation fluctuates the peak of (6), $|A_{21}||v_1|$, which can be monitored at the receiver. The fluctuated $|A_{21}||v_1|$ is shown as in (8). The receiver obtains the data from the transmitter by calculating (8) from $|A_{21}||v_1|$ and decoding it. Data transmission in the opposite direction, from the receiver to the transmitter, is also possible by using the same method.

The perturbation of d_1 at the transmitter causes the ripple of the transferred power which can affect the power transfer operation. The power ripple due to Δd_1 can be derived as in (9). Maximum $\Delta P / P$ is at $d_1 = d_2 = \pi/2$, and Δd_1 is selected so that maximum $\Delta P / P$ does not exceed certain limit.

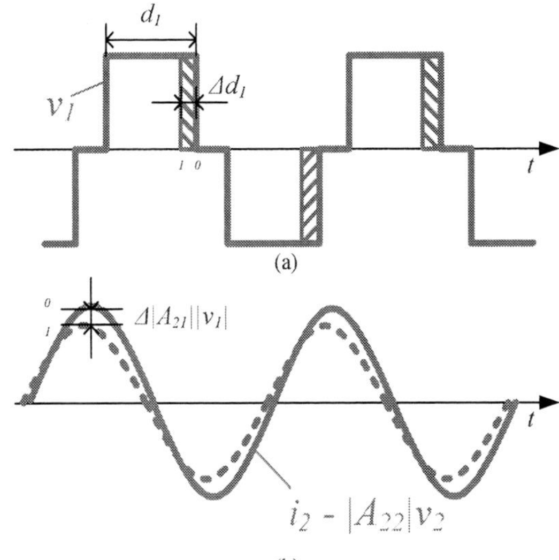

Fig.4. The diagram of the proposed data transmission.
(a) d_1 perturbation for data transmission,
(b) The fluctuation of (6) with respect to d_1 perturbation.

Fig. 5 shows the overall control block diagram of the proposed WPT system including the synchronization and the data transmission. Here, Data12 and Data21 are the data from the transmitter to the receiver, and the data from the receiver to the transmitter, respectively.

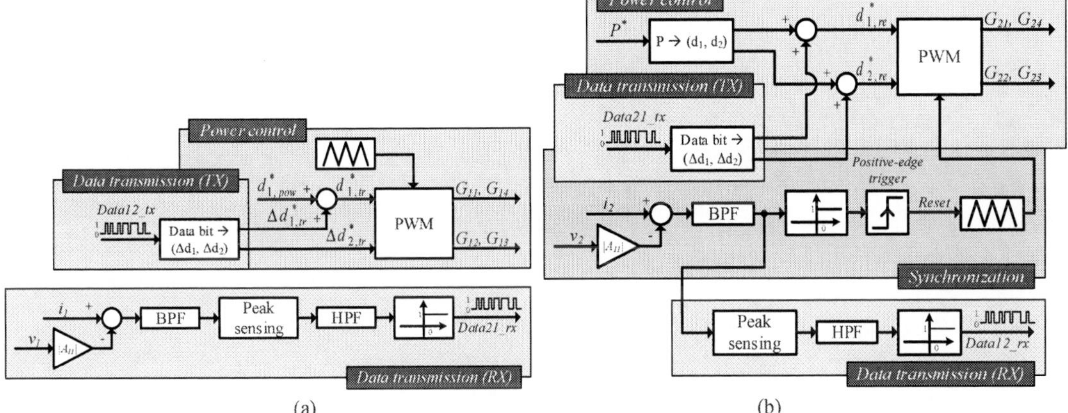

Fig. 5: Overall control block diagram of the proposed WPT system.
(a) The power transmitter side, (b) the power receiver side.

$$\Delta\left[|A_{21}||v_1|\right] = v_{dc1}\frac{4}{\pi}|A_{21}|\left[\sin\left(\frac{d_{1,tr}}{2}\right) - \sin\left(\frac{d_{1,power}}{2}\right)\right] \cong v_{dc1}\frac{4}{\pi}|A_{21}|\cos\left(\frac{d_{1,power}}{2}\right)\sin\left(\frac{\Delta d_1}{2}\right) \cong v_{dc1}\frac{2}{\pi}|A_{21}|\cos\left(\frac{d_{1,power}}{2}\right)\Delta d_1. \tag{8}$$

$$\Delta P = \frac{1}{2}|P_0|\left[\cos(d_1)\sin(d_2)\{1-\cos(\Delta d_1)\} + \sin(d_1)\sin(d_2)\sin(\Delta d_1)\right] + \frac{1}{2}|P_1|\left[\cos(d_1)\{\cos(\Delta d_1) - 1\}\right]. \tag{9}$$

978-1-4673-9551-9/16 $31.00 © 2016 IEEE

TABLE I: PARAMETERS OF THE PROPOSED WPT SYSTEM.

PARAMETER	VALUE
L_1 / L_2	6.6 [μH]
k	0.45
C_{r1} / C_{r2}	0.97 [μF]
v_{dc1} / v_{dc2}	10 [V]
f_s	63 [kHz]

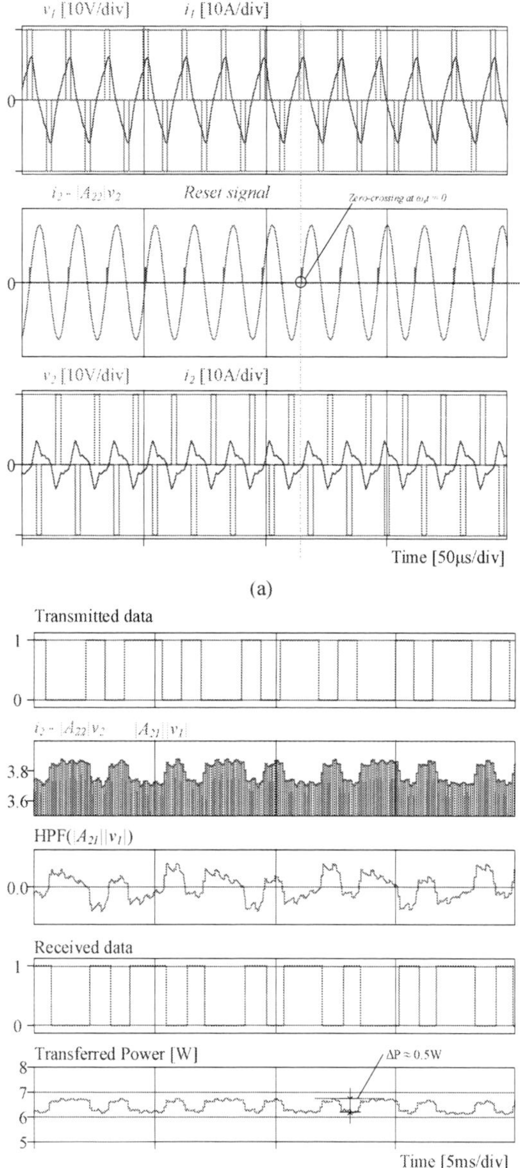

Fig. 6: Simulation waveforms of proposed WPT system. (a) Output voltage synchronization, (b) Proposed data transmission.

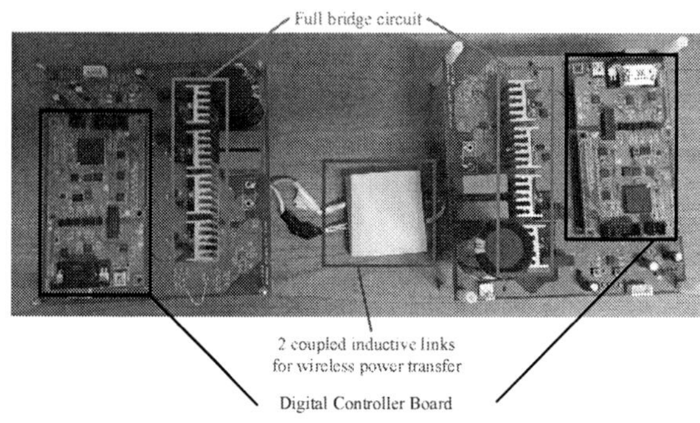

Fig. 7. The hardware configuration for the experiment.

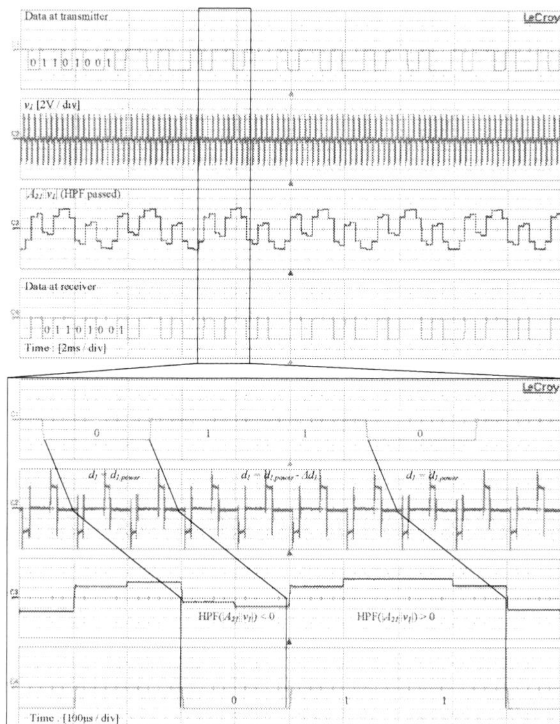

Fig. 8: Experimental result of the proposed data transmission method.

IV. SIMULATION RESULT

The proposed data transmission method is verified by the simulation result. TABLE I shows the parameters of the simulated WPT system. Fig. 6 shows the waveforms about the output voltage synchronization and the proposed data transmission. As shown in the figure, the zero-crossing points of (6) are synchronized to the output voltage phase of the transmitter. The amount of perturbation Δd_1 is set to 0.01, and the data rate is set to 6.3 Kbits/s, sending the data "00101101" repeatedly. Power ripple due to the perturbation is about 0.5W, which is about 7.7% of the transferred power.

V. Experimental Result

The experiment is also performed to verify the proposed data transmission method. Fig. 7 shows the experimental setup. The resonant capacitor C_{r1} and C_{r2} are removed and the switching frequency is reduced to 10 kHz in the experiment to secure sufficient sampling points for the digital filters. Fig. 8 shows the experimental result of the proposed data transmission method. As shown in this figure, the data can be transmitted to the receiver by perturbing d_1 of the output voltage at the transmitter and filtering the fluctuated peak of (6) and decoding at the receiver.

VI. Conclusions

The bidirectional data transmission method without communication circuits in bidirectional WPT system is proposed in this paper. Proposed method use the current component induced by the other side to calculate the voltage magnitude and phase. By applying this method, the bidirectional data transmission is realizable without dedicated communication circuits. Therefore, the cost and volume of the WPT system are expected to be reduced with the proposed method by removing the communication circuits.

References

[1] J. Hirai, T. W. Kim, and A. Kawamura, "Study on Intelligent Battery Charging using Inductive Transmission of Power and Information," *IEEE Trans. Power Electron.*, vol. 15, no. 2, pp. 335–345, Mar. 2000.

[2] W. P. Choi, W. C. Ho, X. Liu, and S. Y. Hui, "Comparative Study on Power Conversion Methods for Wireless Battery Charging Platform," *in Proc. 14th Int. Power Electron. Motion Control Conf. (EPE/PEMC)*, 2010, pp. S15-9–S15-16.

[3] K. K. Shyu, K. W. Jwo, Z. Y. Chen, and C. H. Lo, "Inductive Power Supply System with Fast Full-Duplex Information Rate Device," *in Proc. Int. Conf. on EUROCON 2007*, Sept. 2007, pp. 1382-1386.

[4] A. Zaheer, H. Hao, G. A. Covic, and D. Kacprzak, "Investigation of Multiple Decoupled Coil Primary Pad Topologies in Lumped IPT Systems for Interoperable Electric Vehicle Charging," *IEEE Trans. Power Electron.*, vol. 30, no. 4, pp. 1937–1955, Jun. 2014.

[5] N. B. Carvalho, *et al.*, "Wireless Power Transmission: R&D Activities within Europe", *IEEE Trans. Microwave Theory and Techniques*, vol. 62, no. 4, pp. 1031-1045, Apr. 2014.

[6] S. Y. Hui, "Planar Wireless Charging Technology for Portable Electronic Products and Qi ", *Proc. IEEE*, vol. 101, no. 6, pp. 1290–1301, Jun. 2013.

[7] D. J. Thrimawithana, U. K. Madawala, and M. Neath, "A Synchronization Technique for Bidirectional IPT Systems," *IEEE Trans. Ind. Electron.*, vol. 60, no. 1, pp. 301–309, Jan. 2013

[8] U. K. Madawala and D. J. Thrimawithana, "A Bidirectional Inductive Power Interface for Electric Vehicles in V2G Systems," *IEEE Trans. Ind. Electron.*, vol. 58, no. 10, pp. 4789–4796, Oct. 2011.

[9] G.Simard, M.Sawan, and D.Massicotte, "High-speed OQPSK and Efficient Power Transfer through Inductive Link for Biomedical Implants," *IEEE Trans. Biomed. Circuits Syst.*, vol. 4, no. 3, pp. 192–200, Jun. 2010.

[10] M. E. -Kantarci, J. H. Sarker, and H. T. Mouftah, "Quality of Service in Plug-in Electric Vehicle Charging Infrastructure," *in Proc. IEEE Int. Electric Vehicle Conf. (IEVC)*, 2012, pp. 1–5.

[11] S. Miura, K. Nishijima, and T. Nabeshima, "Bi-directional Wireless Charging between Portable Devices," *in Proc. Int. Conf. on Renewable Energy Research and Applications (ICRERA)*, 2013, pp. 775-778.

Improved Impedance Source Inverter for Hybrid/Electric Vehicle Application with Continuous Conduction Operation

Thilak Senanayake, Ryuji Iijima, Takanori Isobe, Hiroshi Tadano

Faculty of Pure and Applied Sciences, University of Tsukuba, Tsukuba, Japan

***Abstract*—This paper proposes an improved impedance (Z) source inverter topology. Compared to the traditional impedance-source inverter, it can reduce Z-source inverter device voltage stress significantly to perform the same voltage boost, and provides the controllable and stable output power in all load condition. The control strategy of the proposed Z-source inverter is exactly the same as the traditional one. An input filter is also proposed to suppress the power source stress due to pulse input current. The operation principle of the improved topology is analyzed in detail. Experimental results are given to demonstrate the new features of improved topology. Compared to conventional topology 32% reduction of dc-link voltage stress and 75% of input current ripple reduction is achieved.**

Keywords—CCM, DCM, Z source, Inverter

I. Introduction

The Z-source inverter shown in Fig.1 has been reached actively especially on application in the fuel-cell system and photovoltaic system [1,2,9,10]. The 3-phase Z-source inverter has advantages that it can boost or buck input DC voltage and it enhances the reliability of the inverter for avoiding influence of the shoot-through by EMI. However, at light load operation or with small inductances, the Z-source inverter works in discontinuous conduction mode (DCM), the DC bus voltage is increasing infinitely, the output voltage will be uncontrollable and system is unstable. There are number of controllers proposed for the Z-source inverter [3-7,11,12], these controllers control the capacitor voltage and output voltage using separate control loops, which theoretically difficult to guarantee the system stability. Also, Z-source inverter has some significant drawbacks due to discontinuous input current in the boost mode and that the capacitors must sustain a high voltage. This discontinuous input current further increases voltage stress across the input diode due to parasitic inductance between DC power source and the input diode of Z-source inverter. This discontinuous (pulse) input current limited the life-time of input power source. Recent research on Quasi Z source topology features continuous input current [13]. However, it is difficult to reduce the input ripple current because both inductors in the Quasi Z-source network has to increase to obtain desired input ripple current.

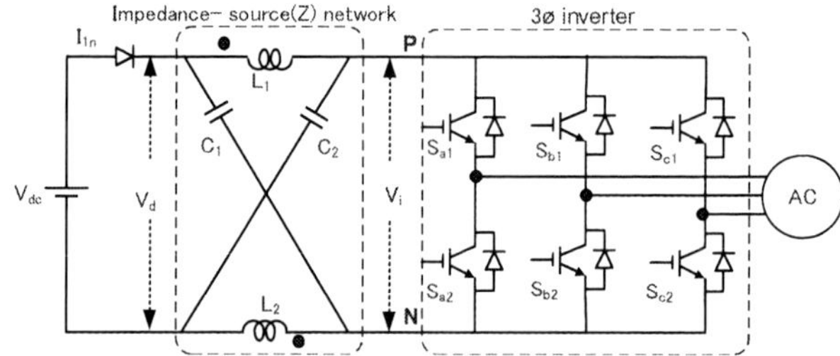

Fig. 1 Z source inverter

Furthermore, input-diode exits in between Z-source network and DC power source limited the energy transfer only to the one direction for DC to AC. The bi-directional energy transfer is important requirement in Hybrid / electric vehicle applications and traditional Z-source inverter has to improve. To overcome the above difficulties, improved. Z-source inverter topology is proposed. It is possible to operate in continuous conduction mode with small inductor or light load condition with regenerative power flow and, low stress of DC power source with input LC filter.

II. Uncontrollable and Unstable Operation of Z-Source Inverter

Z-source inverter shown Fig.1 utilizes the shoot-through zero states to boost voltage in addition to the traditional 6 active-states and 2 zero-states. Compared to the conventional VSI, only difference is dead-time is converted into the shoot-through state. As illustrated in Fig.2, the three switching modes of "shoot-through", "active" and "zero" states are available and different types of switching transitions can be occurred within a switching cycle. All these switching state transitions can be simplified into two possible state sequences as in Table I.

TABLE. I Simplified inverter-switching state sequence of Z-source inverter

Description	State sequence
State sequence 1	Shoot-through -> Active -> Zero
State sequence 2	Shoot-through -> Zero -> Active

978-1-4673-9551-9/16 $31.00 © 2016 IEEE

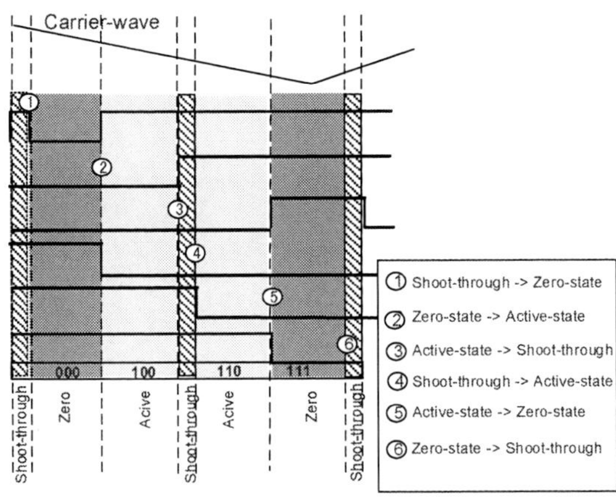

Fig. 2 Possible switching state transitions of 3-phase Z-source inverter

The inductor current behavior is analyzed for each state sequence by considering the possible simplified circuits shown in Fig.4. The inverter bridge is assumed as an AC current source and six operation modes (M1- M6) are available.

As seen from the Fig.3, inductor current is in DCM condition at mode M5 (Fig.4(e)) for both switching state-sequences; "Shoot-through -> Zero -> Active state" and "Shoot-through -> Active -> Zero state". It can be explained by the attempt at light load or small inductances, inductor current become 0A and remains until meet the next switching cycle Therefore, conventional Z-source inverter DC-link voltage (Vi) is uncontrollable and output voltage become unstable.

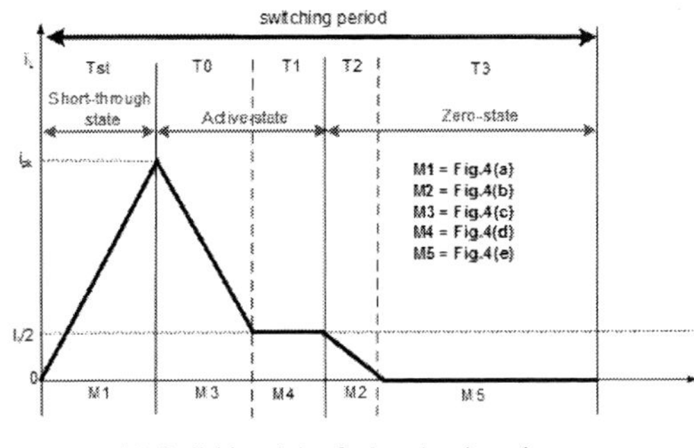

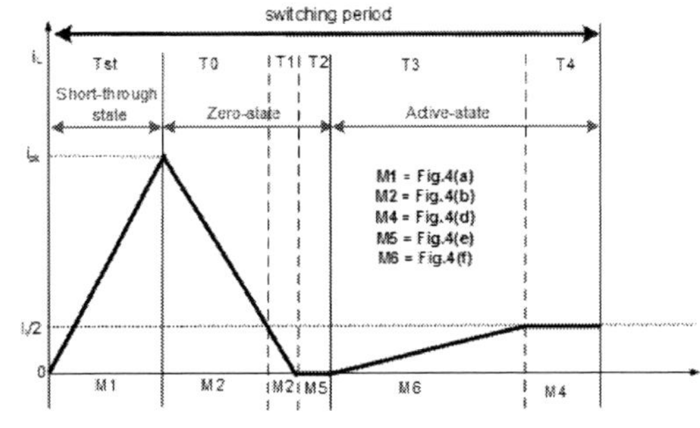

Fig.3 Inductor current of conventional Z source inverter in DCM condition

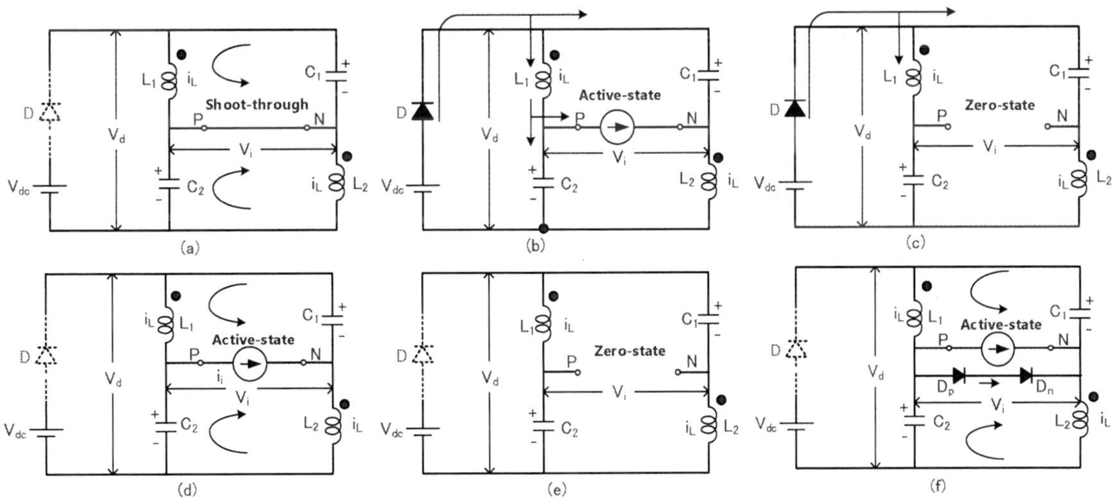

Fig.4 Simplified operation modes of conventional Z-source inverter

III. Improved Impedance Source Inverter

Improved impedance source inverter shown in Fig.5 overcome the topological barriers in conventional Z-source inverter. The proposed inverter is comprised of input LC filter, a semiconductor switch and the impedance-source network. LC filter provides the continuous input current and reduces the stress on power source

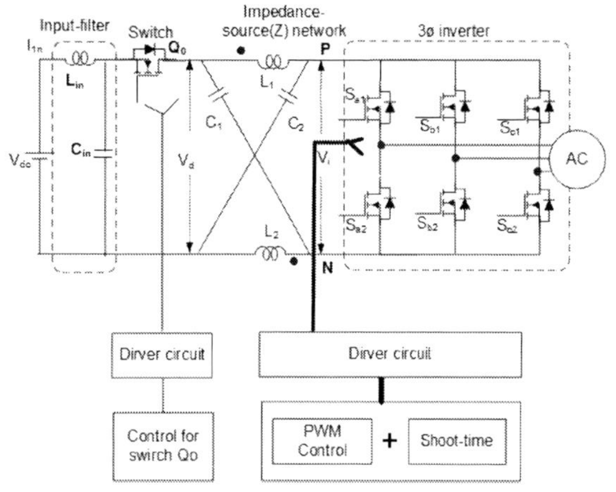

Fig. 5 Improved impedance source inverter

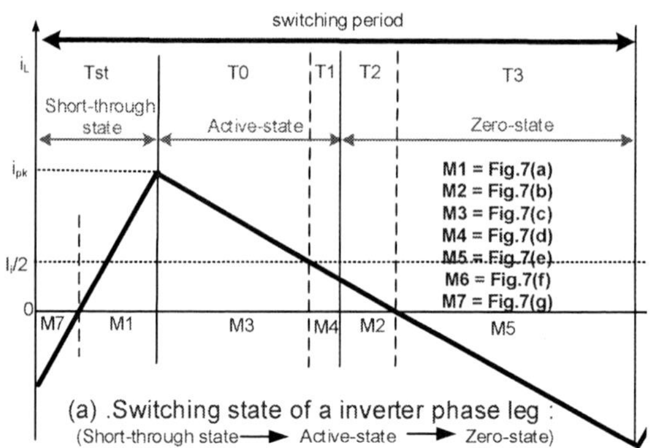

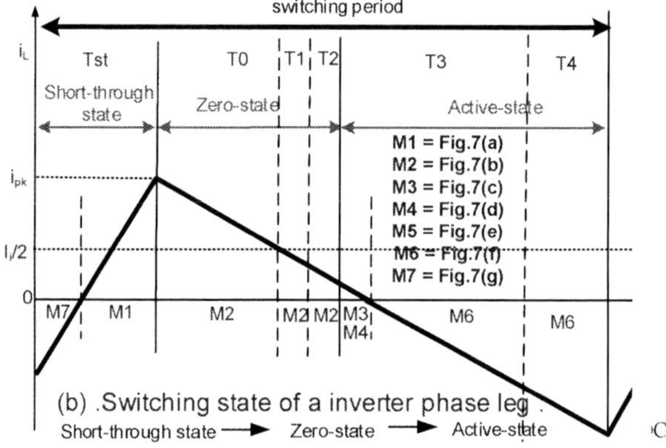

Fig. 6 Inductor current variation of improved impedance source inverter (inductor current in CCM condition)

A. Controlable and Stable Operation

By turning on of the added semiconductor switch (Q_0), discontinuous conduction mode (DCM) in inductor current can be eliminated. As a result, DC link voltage (Vi) become controllable with stable output voltage. Also features the bi-directional power flow which is necessary requirement for hybrid/electric vehicular applications.

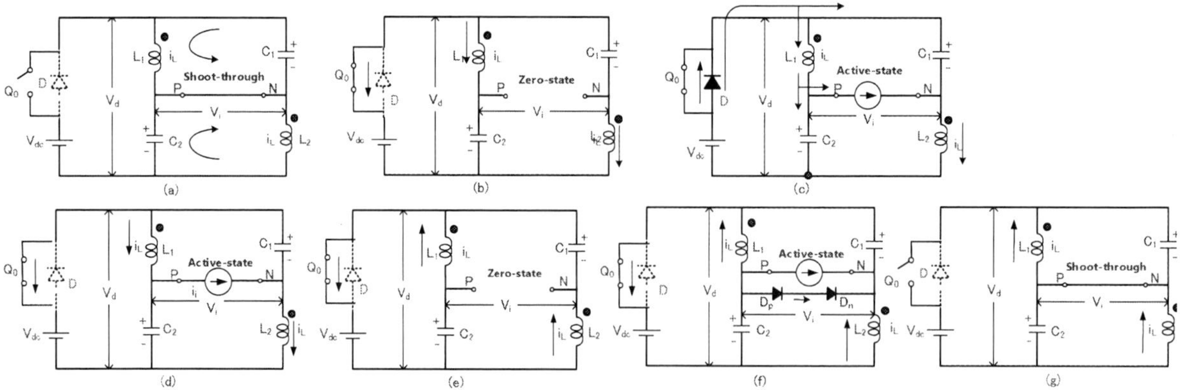

Fig.7 Simplified operation modes of improved impedance source inverter

The elimination of DCM condition in inductor current is analyzed for all possible state-sequences of switching transitions, "shoot-through -> Active -> Zero-state" and "shoot-through -> Zero -> Active state". The inductor current variation in continuous conduction mode (CCM) for both switching state-sequences of an inverter phase-leg in a switching cycle is shown in Fig. 6. As shown in Fig.7, there are seven operational modes (M1- M7) for proposed impedance source inverter. In all these operational modes, inductor current behaves only charge and discharge mode with in a switching cycle. Therefore with activation of switch Q_0, all operation modes can be simplified into two equivalent modes such as

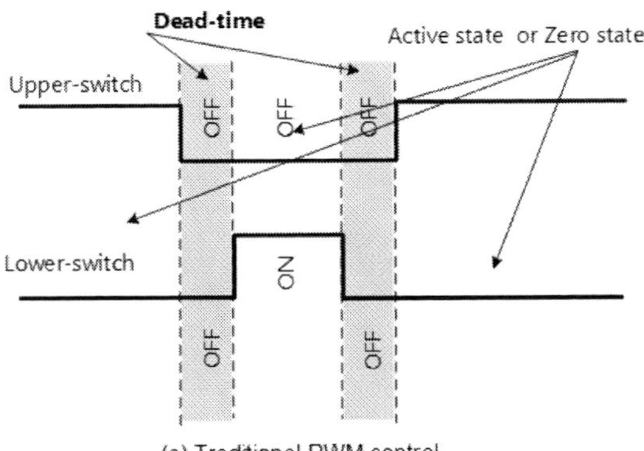

(a) Traditional PWM control

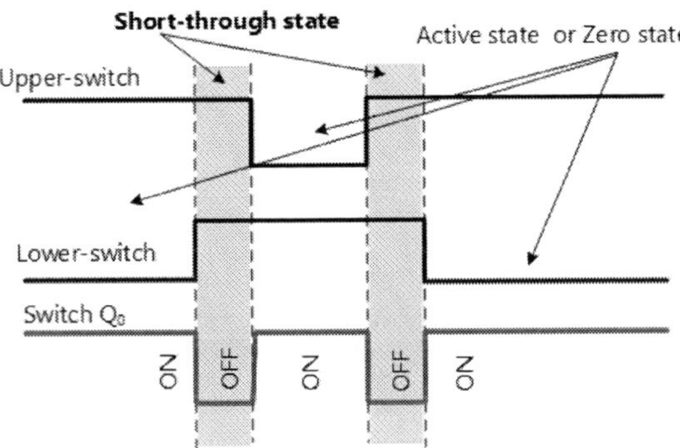

(b) Modified PWM with additional switch control

Fig.8 Control strategy of an inverter phase-leg for traditional PWM and modified PWM with additional switch (Q_0).

"shoot-through" and "non-shoot-through". It is clear that the DCM condition can be eliminated by proposed impedance source inverter.

Control strategy of the proposed impedance source inverter is very simple. The existing PWM control method can be used with small modification [6-8]. The switching waveform of an inverter phase leg is shown in Fig.8. The dead-time is converted into the shoot-through time and the additional switch (Q_0) is turns OFF in shoot-through time interval. All other switching are similar to traditional PWM control.

B. Continous input current with LC filter

Generally, input filter capacitor is connected parallel to the switch network in traditional inverter. This will smoothing the input current and this capacitor may remove or small values can be used. However Z-source inverter, input current become discontinuous in the boost mode due to the shoot-through switching. High frequency current harmonics of substantial amplitude are injected back into the power source. These harmonics can interfere with operation of nearby equipment and limited the life time of the input power source. Further increases the voltage stress across the input diode due to parasitic inductance between DC power source and input diode of Z source inverter. Therefore, input LC filter is necessary requirement for Z-source inverter to overcome the above difficulties. The LC filter connected in proposed inverter lower the stress of DC power source.by providing continuous input current.

IV. EXPERIMENTAL VERIFICATION

As shown in Fig.9, small scale experimental circuit is built according to the available facilities using the specification in Table II. Fig.10 shows the inductor current, DC link voltage (V_i), input current for both conventional and proposed topologies. As seen from Fig. 10(a) inductor current is in DCM condition and DC bus voltage is unstable. For the proposed topology, Fig.10 (b) shows the inductor current is in CCM condition and DC-link voltage is controllable. As shown

TABLE. II Specification for Experimental setup of improved impedance source inverter

Description	Symbol	Value
Inductor	L1,L2	400uH
Capacitor	C1,C2	100uF
Input voltage	Vin	50V
Boost Voltage	Vi	150V
Switching frequency	fs	10kHz
Load resistance	Light/Heavy	66.6Ω / 16.5Ω

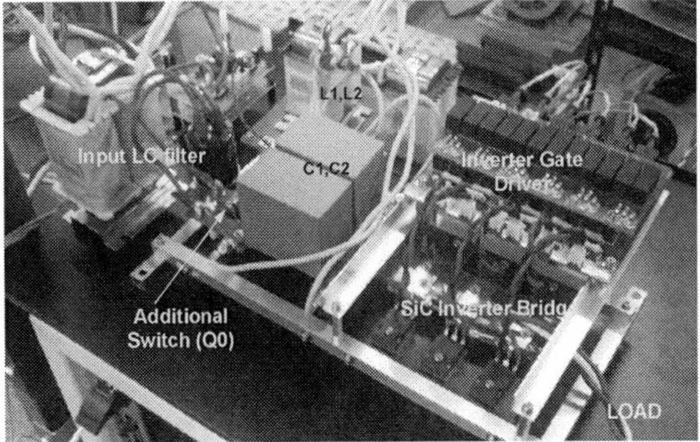

Fig.9 Experimental setup of modified impedance source inverter.

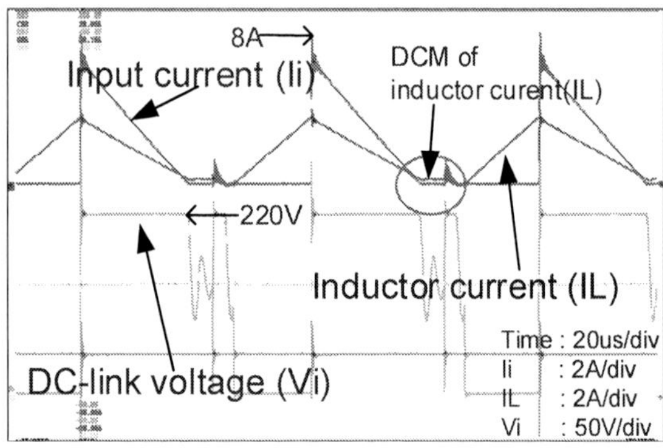

(a).

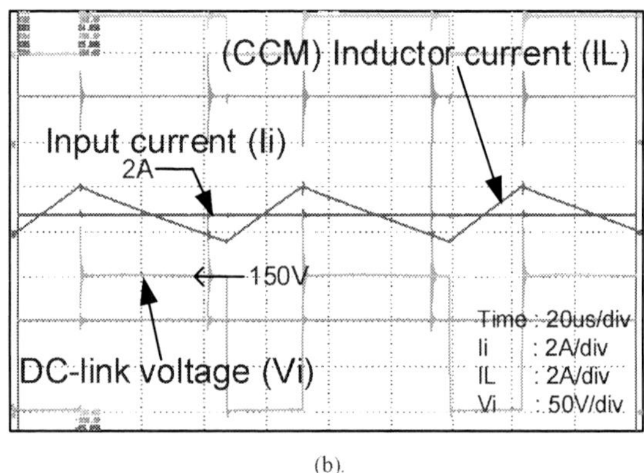

(b).

Fig.10 Experimental results: Inductor current, DC-link voltage and input current (a). Conventional Z-source inverter (b). Modified impedance source inverter.

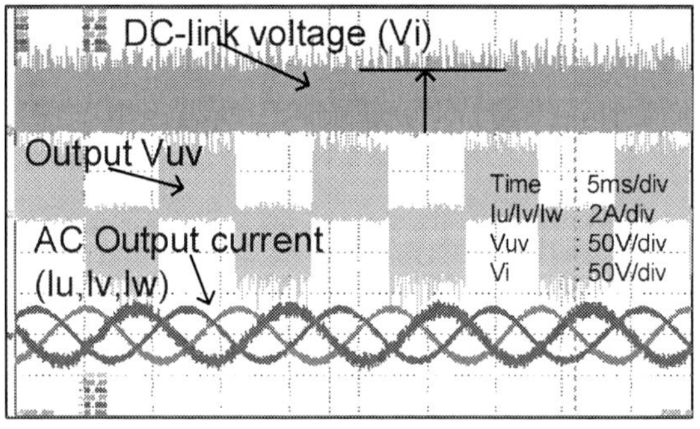

Fig.11 Experimental results: AC output waveform of modified impedance source inverter

in Fig.11, it is verified that inverter output is stable with controllable DC-link voltage. Fig.10 (a), (b) confirm that input current is continuous with LC input filter, leading to the low stress on power source.

V. Conclusion

An improved topology of Z source inverter is proposed. It is possible to operate in continuous conduction mode (CCM) with small inductor or light load condition. DC-link voltage is controllable and overcome the uncontrollable and unstable limitations of the conventional Z source inverter. Voltage stress of the DC-link is reduced by 32% with improved impedance source inverter. The LC filter is necessary requirement for impedance source inverter and 75% reduction of input-ripple current is achieved. Furthermore, the proposed topology provides the circuit with bi-directional power flow which is important requirement for Hybrid/Electric vehicles. This proposed topology is good candidate for traction drive application and suitable for use in the power control unit of the hybrid/electric vehicles.

Acknowledgment

This work was supported by Council for Science, Technology and Innovation (CSTI), Cross-ministerial Strategic Innovation Promotion Program (SIP), "Next-generation power electronics" (funding agency: NEDO).

References

[1] F. Z. Peng; "Z-source inverter" IEEE Transactions on Industry Applications Volume 39, Issue 2, Mar.2003, pp.504 – 510.

[2] Peng, Fang Z, "Z-source inverter for motor drives," 35th Annual IEEE Power Electronics Specialists Conference, pp. 249-254, June 2004.

[3] C. Gajanayake, D. Vilathgamuwa, and P. Loh, "Small-signal and signal-flow-graph modeling of switched Z source impedance network" IEEE Power Electronics Letters, Volume 3, Issue 3, pp.111-116, Sept. 2005.

[4] J. Liu, J. Hu, and L. Xu;" A modified space vector PWM for Z-source inverter - modeling and design" Proc. of International Conference on Electrical Machines and Systems, Vol. 2, pp.1242 – 1247, 27-29 Sept. 2005.

[5] T. Vinh, T. Chun, J. Ahn, and H. Lee, "Algorithms for controlling both the DC boost and AC output voltage of the Z-source inverter" in Proc of IEEE Industrial Electronics Society Conference, 2005, pp. 970-974, 22-25 Nov. 2005.

[6]. P.C.Loh,D.M.Vilathgamuwa,Y.S.Lai,G.T.Chua,Y.Li, "Pulse width modulation of Z-source inverters," IEEE Transactions on Power Electronics, vol. 20, no. 6, pp. 1346–1355,2005

[7]. P.C.Loh,D.M.Vilathgamuwa, C.J.Gajanayake,Y.R.Lim,C.WTeo, "Transient modeling and analysis of pulse width modulated Z-source inverter ," IEEE Transactions on Power Electronics, vol. 22, no. 2, pp. 498–507,2007

[8]. Rivera E.I.O,Rodriguez L.A, "The Z-source converter as an introduction to power electronics and undergraduate research," Frontiers in Education Conference, pp T2C-5-10.,2007

[9] Peng, Fang Z,M.Shen,K.Holland, "Application of Z-source inverter for traction drive of fuel cell-battery hybrid electric vehicles," IEEE Transaction of Power Electronics,vol.22,no.3, pp. 1054-1061, J2007.

[10] M. Shen, A.Joseph,J.Wang, F. Z. Peng, "Comparison of traditional inverters and Z-source inverters for fuel cell vehicles," IEEE Transaction of Power Electronics,vol.22,no.4, pp. 1453-1463, July 2007.

[11].Kun Yu,Fang L.Luo,Miao Zhu,"Study of improved z-source inverter: small signal analysis",5th IEEE Conference of Industrial Electronics and Applications, vol. 2, no. 4, pp. 2169-2174,2010.

[12].D.Amudhavalli,L.Narendran "Improved Z source inverter for speed control of an induction motor," International Journal of Advanced Research in Electrical Electronics and Instrumentation Engineering, vol. 2, no. 4, pp. 1240–1247,2013

[13].Y. Li, S. Jiang, J. Cintron-Rivera, and F. Z. Peng, "Modeling and control of quasi-z-source inverter for distributed generation applications," IEEE Transactions on Industrial Electronics, vol. 60, no. 4, pp. 1532–1541,2013.

AUTHOR INDEX

A

B

C

D

E

F

G

H

I

J

K

L

M

N

O

P

Q

R

S

T

U

V

W

X

Y

Z

IEEE
445 Hoes Lane
Piscataway, NJ 08854-4141